T0389123

TRANSPORT INFRASTRUCTURE AND SYSTEMS

PROCEEDINGS OF THE AIIT INTERNATIONAL CONGRESS ON TRANSPORT INFRASTRUCTURE AND SYSTEMS (TIS 2017), ROME, ITALY, 10–12 APRIL 2017

Transport Infrastructure and Systems

Editors

Gianluca Dell'Acqua
University of Napoli Federico II, Napoli, Italy

Fred Wegman
Delft University of Technology, Delft, The Netherlands

CRC Press
Taylor & Francis Group
Boca Raton London New York Leiden

CRC Press is an imprint of the
Taylor & Francis Group, an **informa** business

A BALKEMA BOOK

On the book cover: *Concrete relief entitled "Fari, pistoni e semafori" (1959) located in the building of the ACI directorate.*

Author: Imre Tóth (Fehérvárcsurgó, Austria-Hungary 1909; Rome, Italy 1984).

CRC Press/Balkema is an imprint of the Taylor & Francis Group, an informa business

© 2017 Taylor & Francis Group, London, UK

Typeset by V Publishing Solutions Pvt Ltd., Chennai, India
Printed and bound in Great Britain by CPI Group (UK) Ltd, Croydon, CR0 4YY

Published by: CRC Press/Balkema
       P.O. Box 11320, 2301 EH Leiden, The Netherlands
       e-mail: Pub.NL@taylorandfrancis.com
       www.crcpress.com – www.taylorandfrancis.com

ISBN: 978-1-138-03009-1 (Hbk + USB-card)
ISBN: 978-1-315-28189-6 (eBook)

Table of contents

Transport systems

Transport Infrastructure and Systems – Dell'Acqua & Wegman (Eds)
© 2017 Taylor & Francis Group, London, ISBN 978-1-138-03009-1

Preface

The Transport Infrastructure and Systems TIS International Congress aims to promote and discuss efficient planning, design, construction and maintenance of transport infrastructure and systems by addressing important issues related to roads, railways, airports, maritime and intermodal systems.

The 1st TIS International Congress organized by the Italian Association for Traffic and Transport Engineering AIIT will be held April 10–12, 2017 in Rome. The TIS 2017 International Congress is focused on emerging technologies to enable smarter, greener and more efficient movement of people and goods around the world.

The congress aims to present and discuss current knowledge of the ever-changing challenges to scientists, engineers, managers, and professionals from around the world who are involved in sustainable development and maintenance of transport infrastructure and systems.

In many countries, transport infrastructure is currently strained to a state of limited functionality. Increased ridership and depleting funds have caused many Agencies to become concerned with the future of facilities. Much maintenance is required on infrastructure, and their capacity is in desperate need of expansion.

Along with the declining state of transport infrastructure is a growing concern for the need to recognize and mitigate mankind's impact on the natural environment.

A more sustainable solution to transport infrastructure and systems problems must be assessed. Infrastructure can have a large negative impact on surrounding ecosystems and overall environmental quality. The next step in infrastructure's advancement needs to include practices that reduce their effect on the natural environment, increase capacity, and benefit society beyond the ability of current infrastructure. This can be achieved by thinking to transport infrastructure that mitigate the negative impact on the environment, include more sustainable practices than modern construction techniques, and consist of maximizing the lifetime of an infrastructure. Sustainable construction techniques include the use of recycled materials, ecosystem management, energy reduction, increasing the water quality of storm water runoff, and maximizing overall societal benefits. Dynamic and liveable cities rely on efficient mobility systems, and road safety plays a large part in this.

To meet the transport infrastructure and systems needs of future generations, new goals must be set. Research will have to be given more attention, and a new generation of infrastructure designs will have to be developed. The TIS 2017 will provide such a forum for new concepts and innovative solutions. The meeting program will cover all transportation modes, with more than 150 presentations in over 20 sessions and workshops, addressing topics of interest to policy makers, administrators, practitioners, researchers, and representatives of government, industry, and academic institutions.

This proceedings book includes submissions to the congress in the areas of Asset management in transport infrastructure, financial viability of transport engineering projects/Life cycle Cost Analysis, Life-Cycle Assessment and Sustainability Assessment of transport infrastructure/Infrastructures financing and pricing with equity appraisal, operation optimization and energy management/Low-Volume roads: planning, maintenance, operations, environmental and social issues/Public-Private Partnership (PPP) experience in transport infrastructure in different countries and economic conditions/Airport Pavement Management Systems, runway design and maintenance/Port maintenance and development issues, technology relating to cargo handling, landside access, cruise operations/Infrastructure Building Information Modelling (I-BIM)/Pavement design and innovative bituminous materials/Recycling and re-use in road pavements, environmentally sustainable technologies/Stone pavements, ancient roads and historic railways/Cementitious stabilization of materials used in the rehabilitation of transportation infrastructure/Sustainable transport and the environment protection including green vehicles/Urban transport, land use development, spatial and transport planning/Bicycling, bike, bike-sharing systems, cycling mobility/Human factor in transport systems/Intelligent Mobility: emerging technologies to enable the smarter movement of people and goods/Airport landside: access roads, parking facilities, terminal facilities, aircraft apron and the azdjacent taxiway/Transportation policy, planning and design, modelling and decision making/Transport economics, finance and pricing issues, optimization problems, equity appraisal/Road safety impact assessments, road safety audits, the management of road network safety and safety inspections/Tunnels and underground structures: preventing incidents-accidents mitigating their effects for both people and goods/Traffic flow characteristics, traffic control devices, work zone traffic control, highway capacity and quality of service/Track-vehicle interactions in railway systems, capacity analysis of railway networks/ Risk assessment and safety in air and railway transport, reliability aspects/Maritime transport and inland waterways transport research/Intermodal freight transport: terminals and logistics.

At least two, but often three reviewers, including members of the Scientific Committee, subjected all submitted contributions to an exhaustive refereed peer review procedure. Based on the reviewers' recommendations, those contributions which best suited the conference goals and objectives were chosen for inclusion in the proceedings.

The Editors would like to thank the Scientific Committee members and individual reviewers for their dedication and contributions of their time and efforts to ensure the high technical quality of the accepted papers. In addition, sincere thanks are extended to Salvatore Antonio Biancardo for ensuring that final manuscripts were in accordance with the publication format requirements.

The guidance and continuing input from the AIIT Committee members were essential in planning of this congress, and highly appreciated. Finally, we would like to gratefully acknowledge the Organizing Committee members for their help, suggestions and contributions to the management of the Congress affairs.

April 10, 2017
Rome

Gianluca Dell'Acqua and Fred Wegman
Chairs of the TIS International Congress 2017

Committees

SCIENTIFIC COMMITTEE

Gianluca Dell'Acqua, *Chairman*
Fred Wegman, *co-Chairman*

(by alphabetical order)

Andrus Aavic, *Tallinn University of Technology, Estonia*
Vladimír Adamec, *Brno University of Technology, Czech Republic*
Tomas Apeltauer, *Brno University of Technology, Czech Republic*
John Andrews, *The University of Nottingham, UK*
Orazio Baglieri, *Polytechnic of Torino, Italy*
Marco Bassani, *Polytechnic of Torino, Italy*
Ana Maria César Bastos Silva, *University of Coimbra, Portugal*
Shlomo Bekhor, *Technion—Haifa, Israel*
Salvatore Antonio Biancardo, *Federico II University of Napoli, Italy*
Werner Brilon, *Ruhr-University Bochum, Germany*
Paolo Budetta, *Federico II University of Napoli, Italy*
Francesco Canestrari, *Polytechnic University of Marche, Italy*
Giuseppe Cantisani, *Sapienza University of Roma, Italy*
Armando Cartenì, *Federtico II University of Napoli, Italy*
Maria Castro Malpica, *Technical University of Madrid, Spain*
Christine Chaloupka-Risser, *FACTUM, Austria*
Vicent de Esteban Chapapría, *University of Valencia, Spain*
Jack C.P. Cheng, *The Hong Kong University of Science and Technology, Hong Kong*
Pasquale Colonna, *Polytechnic of Bari, Italy*
Olja Čokorilo, *University of Belgrade, Serbia*
Mauro Cozzolino, *University of Salerno, Italy*
Kevin Cullinane, *University of Gothenburg, Sweden*
Antonio D'Andrea, *Sapienza University of Roma, Italy*
Essam Dabbour, *Abu Dhabi University, United Arab Emirates*
Mario De Luca, *Federico II University of Napoli, Italy*
Gianluca Dell'Acqua, *Federico II University of Napoli, Italy*
Miguel Angel del Val Melus, *Tech. University of Madrid, Spain*
Paola Di Mascio, *Sapienza University of Roma, Italy*
Coenraad Esveld, *ECS BV, The Netherlands*
Nicolau Dionísio Fares Gualda, *University of São Paulo, Brasil*
Bruna Festa, *Federico II University of Napoli, Italy*
Demetrio Carmine Festa, *Calabria University, Italy*
Gaetano Fusco, *Sapienza University of Roma, Italy*
Fabio Galatioto, *Transport Systems Catapult, UK*
Felice Giuliani, *University of Parma, Italy*
Hercules Haralambides, *Erasmus University Rotterdam, The Netherlands*
Imine Hocine, *Inst. of Science and Tech. for Transport, France*
Felix Huber, *Wuppertal University, Germany*
Matteo Ignaccolo, *University of Catania, Italy*

Renato Lamberti, *Federico II University of Napoli, Italy*
Davide Lo Presti, *The University of Nottingham, UK*
Giuseppe Loprencipe, *Sapienza University of Roma, Italy*
Roberto Maja, *Polytechnic of Milano, Italy*
Gabriele Malavasi, *Sapienza University of Roma, Italy*
Giulio Maternini, *University of Brescia, Italy*
Raffaele Mauro, *University of Trento, Italy*
Kazuaki Miyamoto, *Tokyo City University, Japan*
Goran Mladenovic, *University of Belgrade, Serbia*
Charles Musselwhite, *Swansea University, UK*
G.P. Jayaprakash, *Transportation Research Board, USA*
David Jones, *University of California Davis, USA*
Maja Krcum, *University of Split, Croatia*
Raimundas Junevicius, *Vilnius Gediminas Tech. University, Lithuania*
Robert Parsons, *University of Kansas, USA*
Hrvoje Pilko, *University of Zagreb, Croatia*
Ioannis Politis, *Aristotle University of Thessaloniki, Greece*
Carlo Prato, *The University of Queensland, Australia*
Thanos Pallis, *University of the Aegean, Greece*
Marco Pasetto, *University of Padova, Italy*
Luca Persia, *Sapienza University of Roma, Italy*
Anand J. Puppala, *The University of Texas at Arlington, USA*
Cesar Queiroz, *former World Bank Highways Adviser, USA*
Thomas Kjær Rasmussen, *Technical University of Denmark, Denmark*
Danijel Rebolj, *University of Maribor, Slovenia*
Stefano Ricci, *Sapienza University of Roma, Italy*
Ralf Risser, *FACTUM, Austria*
Guo Runhua, *Tsinghua University, China*
Francesca Russo, *Federico II University of Napoli, Italy*
Giuseppe Salvo, *University of Palermo, Italy*
Ezio Santagata, *Polytechnic of Torino, Italy*
Piotr Sawicki, *Poznan University of Technology, Poland*
Kungrabat Sharipov, *TT Polytechnic University, Uzbekistan*
Valentin Silyanov, *State Technical University, Russia*
Andrea Simone, *University of Bologna, Italy*
Wayne K. Talley, *Old Dominion University, USA*
Bagdat Teltayev, *Highway Research Institute, Kazakhstan*
Susan L. Tighe, *University of Waterloo, Canada*
Tomaz Tollazzi, *University of Maribor, Slovenia*
Elen Twrdy, *University of Ljubljana, Slovenia*
András Várhelyi, *Lund University, Sweden*
Ashish Verma, *Indian Institute of Science, India*
Alex Visser, *University of Pretoria, South Africa*
Zhong Wang, *Dalian University of Technology, China*
Fred Wegman, *Delft University of Technology, The Netherlands*
Alfred Weninger-Vycudil, *Viagroup, Switzerland*
Jean François Wounba, *Université Libre de Bruxelles, Belgium*
Thomas J. Yager, *NASA Langley Research Center, USA*
Stefano Zampino, *Road and Transport Engineer, Italy*
Daiva Zilioniene, *Vilnius Gediminas Tech. University, Lithuania*

ORGANIZING COMMITTEE

(by alphabetical order)

Stefania Balestrieri
Sonia Briglia
Lauragrazia Daidone
Gianluca Dell'Acqua
Mario Magnanelli
Edoardo Mazzia
Alessandro Ruperto
Rocco Sorropago

ORGANIZER

Associazione Italiana
per l'Ingegneria del Traffico e dei Trasporti

HOSTING INSTITUTION

Automobile Club d'Italia

UNDER THE AUSPICES OF

Transportation Research Board
Università degli Studi di Napoli Federico II
Universidade Federal do Rio Grande do Sul
Società Italiana Infrastrutture Viarie
Nottingham Transportation Engineering Centre
Transportation Research Group of India
ANAS, SITEB, CIFI, FASTIGI, FINCO, WISTA Italia
Sapienza Università di Roma
University of Pretoria

REVIEWERS

Andrus Aavic
Francesco Abbondati
Vladimír Adamec
Maja Ahac
Gordon Airey
Francesco Alberti
Rami M. Alfaqawi
Sharid Amiri
Scott Anderson
Boris Antic
Tomas Apeltauer
Daniela Aresu
Federico Autelitano
Orazio Baglieri
Nicola Baldo
Edita Baltrenaite
Marco Bassani
Ana Bastos

Milan Batista
Gianfranco Battiato
Michelle Beiler
Shlomo Bekhor
Francesco Bella
Dario Bellini
Andrea Benedetto
Luca Bernardini
Šime Bezina
Salvatore Antonio Biancardo
Arianna Bichicchi
Luigi Biggiero
Edoardo Bocci
Guido Bonin
Gerardo Botas
Marilisa Botte
Davor Brcic
Sonia Briglia

Werner Brilon
Lelio Brito
Massimiliano Bruner
James Bryce
Luciano Bucharles
Paolo Budetta
Mariarosaria Busiello
Daniel Jorge Caetano
Salvatore Cafiso
Francesco Canestrari
Giuseppe Cantisani
Francesco Saverio Capaldo
Agostino Cappelli
Angela Carboni
Fabrizio Cardone
Armando Carteni
Maria Castro
Alberto Castro Fernandez

Selene Cattani
Clara Celauro
Gianluca Cerni
Eleonora Cesolini
Christine Chaloupka-Risse
T. Donna Chen
Tao Chen
Jack Cheng
Laura Chiacchiari
Giuseppe Chiappinelli
Sandro Colagrande
Pasquale Colonna
Andrea Conca
Giovanni Coraggio
Maria Vittoria Corazza
Alessandro Corradini
Marcello Corsi
Daniele Cortis
Marco Costa
Kevin Cullinane
Luca D'Acierno
Pierpaolo D'Agostino
Essam Dabbour
Andrew Dawson
Shirley Ddamba
Vicent de Esteban
    Chapapria
Francesco De Florio
Maria Luisa De Guglielmo
Mario De Luca
Stefano de Luca
Gianluca Dell'Acqua
Camilla De Micheli
Vittorio De Riso di
    Carpinone
César De Santos-Berbel
Ilaria Del Ponte
Miguel Angel Del Val Melus
Roberto Devoto
Paola Di Mascio
Roberta Di Pace
Marco Diana
Nuria Diaz Maroto
    Llorente
Natalia Distefano
Marilyn Dodson
Lorenzo Domenichini
Laura Eboli
Manuela Esposito
Coenraad Esveld
Massimiliano Fabbricino
Aldo Fabregas
Asif Faiz
Gianfranco Fancello
Nicolau Dionísio Fares
    Gualda
Javed Farhan
Enrico Fattorini

Luís Fernandes
Jose Leomar Fernandes Jr.
Silvia Ferrini
Gilda Ferrotti
Paola Firmi
Gerardo Flintsch
Alessandro Focaracci
Anna Formisano
Giovanni Forte
Mateus Fratoni Souza
Sanja Fric
Francesca Frigio
Stanisław Gaca
Fabio Galatioto
Stipe Galić
Vincenzo Gallelli
Mariano Gallo
Alfredo García García
Lilli Gargiulo
Andrea Giaccherini
Giovanni Giacomello
Nadia Giuffrida
Orazio Giuffrè
Tullio Giuffrè
Felice Giuliani
Marinella Giunta
Christos Gkartzonikas
Carlotta Godenzoni
Marcelo Gonzalez
Hernan Dario Gonzalez
    Zapata
Anna Granà
Andrea Graziani
James Grenfell
Andrea Grilli
Marco Guerrieri
Inbal Haas
Michael Hendry
Theuns.F.P Henning
Imine Hocine
Felix Huber
Luis Iglesias
Matteo Ignaccolo
Paolo Intini
Giuseppe Inturri
Alibay Iskakbayev
G.P. Jayaprakash
Marlova Johnston
David Jones
Raimundas Junevicius
Jwan Kamla
Marianne Karlsson
Nariman Khalil
Paraskevi Kladaki
Katerina Konsta
Maja Krcum
Srećko Krile
Przemysław Kupka

Rosa Anna La Rocca
Claudio Lantieri
Michela Le Pira
Matheus Lemos
    Nogueira
Giovanni Leonardi
Salvatore Leonardi
Michele Leone
Riccardo Licciardello
Jose Manuel Lizarraga
Aleksander Ljubic
Davide Lo Presti
Simone Lopes
Adelino Jorge Lopes
    Ferreira
Giuseppe Loprencipe
Lucia Losasso
Jurate Kumpieneuniversity
    Lulea
Renato Macciotta Pulisci
Igor Majstorović
Gabriele Malavasi
Francesca Maltinti
Laura Mancini
Sadko Mandžuka
Mohammad Manghrour
Giovanni Mantovani
Edoardo Marcucci
Andrea Marella
Alessandro Marradi
Ana Marušić
Vittorio Marzano
Giulio Maternini
Raffaele Mauro
Federico Mazzetta
Giorgia Mazzoni
Gabriella Mazzulla
Silvio Memoli
Rodrigo Mesa-Arango
Nenad Milutinovic
Kazuaki Miyamoto
Goran Mladenovic
Douglas Mocelin
Laura Moretti
Miso Mudric
Steve Muench
Charles Musselwhite
Lorenzo Mussone
Dmitry Nemchinov
Dusan Nicolic
Rubina Normahomed
Manabu Nose
Nozomi Nose
Nurit Oliker
Sara Guerra-Oliveira
Lucas Ortiz
Masafumi Ota
Rudra P. Pradhan

Francesca Pagliara
Enrico Pagliari
Athanasios Pallis
Roberto Palmarini
Aimilia A. Papachristou
Panagiotis Papaioannou
Andrea Papola
Luigi Pariota
Tony Parry
Robert Parsons
Stefania Pascale
Marco Pasetto
Emiliano Pasquini
Ujwalkumar Patil
Aravind Pedarla
Orazio Pellegrino
Jorge Augusto Pereira
   Ceratti
Federico Perrotta
Dalibor Pesic
Marco Petrelli
Massimiliano Petri
Cristiana Piccioni
Hrvoje Pilko
Francesco Pirozzi
Antonio Placido
Maria Jose Poblaciones
Ioannis Politis
Silvia Portas
Annalisa Pranno
Filippo Praticò
Carlo Prato
Pavel Pribyl
Christos Pyrgidis
Cesar Queiroz
Vittorio Ranieri
Thomas Kjær Rasmussen
Danijel Rebolj
Francesco Rebora
Dv Reddy

Marko Rencelj
Oskar Rexfelt
Stefano Ricci
Ralf Risser
Keith Robinson
Jorge Luis Rodriguez
   Navarro
Matteo Rossi
Francesca Russo
Emanuele Sacchi
Nicola Sacco
Frank Saccomanno
Giuseppe Salvo
Jose San Martin
Xiomara Sanchez-Castillo
Meg Sangimino
Cesare Sangiorgi
Ezio Santagata
João Santos
Piotr Sawicki
Francesco Sdao
Ramoel Serafini
Antonino Sferlazza
Manuel Silvestri
Valentin Silyanov
Andrea Simone
Zuzanna Simonova
Giuseppe Sollazzo
Everton Cesar Souza
Andrea Spinosa
Tatjana Stanivuk
Arianna Stimilli
Roger Surdahl
Yusak Susilo
Luca Tefa
Bagdat Teltayev
Stefano Terribile
Andrej Tibaut
Tomaz Tollazzi
Mateusz Tomczak

Ivano Toni
Alex Torday
Adam Torok
Enza Torrisi
Riccardo Troisi
Filip Trpcevski
Laura Trupia
Lucia Tsantillis
Marialuisa Tummiello
Marjan Tušar
Elen Twrdy
George K. Vaggelas
Rosolino Vaiana
Andrius Vaitkus
Luís Vasconcelos
Flavio Vaz
Loretta Venturini
Ashish Verma
Carlos Videla
Valeria Vignali
Tejo Vikash Bheemasetti
Carolina Villa
Francesco Viola
Alex Visser
Alessandro Vitale
Antonino Vitetta
Francesco Viti
Martin Vsetecka
András Várhelyi
Ting Wang
Zhong Wang
Fred Wegman
Hongbo Wu
Hai Yan
Stefano Zampino
Marina Zanne
Nataša Zavrtanik
Olja Čokorilo

Transport infrastructure

Transport Infrastructure and Systems – Dell'Acqua & Wegman (Eds)
© 2017 Taylor & Francis Group, London, ISBN 978-1-138-03009-1

Deformation and strength of asphalt concrete under static and step loadings

A.I. Iskakbayev & B.B. Teltayev
Kazakhstan Highway Research Institute, Almaty, Kazakhstan

C. Oliviero Rossi
University of Calabria, Rende, Italy

ABSTRACT: Deformation and strength of fine-grained asphalt concrete were investigated under static and step loadings experimentally in the paper. Temperature varied within the range of 20 ± 2°C. The stress varied within the range of 0.055 MPa and 0.311 MPa during creep test. The results of experiments showed that creep curves have three sites within specified ranges of stress—site of unstabilized creep, site of stabilized creep and site of accelerating creep. The long-term strength curve is described successfully by exponential function. The stress affects greatly the long-term strength of asphalt concrete: the increase of stress for single order reduces failure time for four orders. The sequence of impacts also affects greatly the long-term strength of asphalt concrete.

1 INTRODUCTION

Asphalt concrete is one of the main materials for highway pavements. Mechanical properties of an asphalt concrete are highly depending on temperature, value, duration, and rate of loading (MS-4 2008, Papagiannakis & Masad 2008, Yoder & Witczak 1975). In real road conditions temperature in points of asphalt concrete layers of pavement structures due to variations of ambient temperature, track wheels load values, their action duration and rate varies within wide limits. Therefore, determination of mechanical behavior of an asphalt concrete taking into account the variation of the above mentioned factors has important practical value.

It is known that the basic methods for evaluation of mechanical behavior of viscoelastic materials are tests on creep and relaxation (Cristensen 1971, Ferry 1980, Tschoegl 1989). Technically, realization of creep test is easier. It is possible to construct creep curves and long-term strength with using its results. Relaxation curves can be obtained from the creep curves by using known methods (Tschoegl 1989, Hopkins & Hamming 1957). The long-term strength curves enable to determine service life of a road asphalt concrete pavement.

In this paper test results of hot fine-grained asphalt concrete samples on creep are presented. Creep tests were carried out by the direct tensile scheme until complete fracture of the asphalt concrete samples. Test temperature was 20 ± 2°C. The applied stress was changed from 0.055 to 0.311 MPa. Creep curves under different loads and long-term strength curve of the asphalt concrete have been constructed. Three characteristic sites of creep curves—the unstabilized, stabilized and accelerating creep sites are shown.

2 MODELS FOR PREDICTION OF ASPHALT CONCRETE DAMAGE

The vehicles, running along the highways, have different number of axles, the loads on which varies within the wide range. It is natural that the axles with different values of loads cause different damage level for pavement. One of the prediction models for materials and structures damage, considering load impact of different values, is well-known Miner's law (Miner 1945). Miner's law is also known as the principle of linear summation of damages (Talreja & Sing 2012) and found wide application in engineering calculations. For example, in the USA it is used for prediction of fatigue life for asphalt concrete pavement of highways (ARA, ERES 2004). In this regard fatigue damage is calculated under the following equation:

$$D = \sum_{i=1}^{T} \frac{n_i}{N_i} \qquad (1)$$

where D = damage; T = total number of periods; n_i = actual traffic period i; and N_i = traffic allowed under conditions prevailing in i.

Vehicles run along the highways with different speed, i.e. they affect the pavement structure with different duration. As under real road conditions the speed of the vehicles varies within the wide ranges, axle load duration differs greatly. Unfortunately, equation (1), based on Miner's law, does not consider load duration, which can be a source of systematic large inaccuracies during determination of fatigue life for asphalt concrete pavement of the highway.

At present, the so-called Bailey's criterion is well-known in science and engineering practice (Bailey 1939), which can be described in the following form:

$$\int_0^{t_p} \frac{dt}{\tau[\sigma(t)]} = 1 \qquad (2)$$

where t_p = failure time; $\sigma(t)$ = stress, varied in time; and $\tau[\sigma(t)]$ = dependence of material failure time on stress.

Contrary to Miner's law (1), Bailey's criterion (2) considers load duration, i.e. the speed of the vehicle along the highway. Dependence $\tau[\sigma(t)]$ of Bailey's criterion represents by itself the analytical equation for the so-called curve of long-term strength, which is constructed based on test results of the material according to the creep scheme (Kachanov 1986, Rabotnov 1987).

3 MATERIALS

3.1 Bitumen

In this paper bitumen of grade 100–130 has been used which meets the requirements of the Kazakhstan standard (ST RK 1373 2013). The bitumen grade on Superpave is PG 64–40 (Superpave series No.1 2003). Basic standard indicators of the bitumen are shown in Table 1. Bitumen has been produced by Pavlodar processing plant from crude oil of Western Siberia (Russia) by the direct oxidation method.

3.2 Asphalt concrete

Hot dense asphalt concrete of type B that meets the requirements of the Kazakhstan Standard (ST RK 1225 2013) was prepared with use of aggregate fractions of 5–10 mm (20%), 10–15 mm (13%), 15–20 mm (10%) from Novo-Alekseevsk rock pit (Almaty region), sand of fraction 0–5 mm (50%) from the plant "Asphaltconcrete-1" (Almaty city) and activated mineral powder (7%) from Kordai rock pit (Zhambyl region).

Bitumen content of grade 100–130 in the asphalt concrete is 4,8% by weight of dry mineral material. Basic standard indicators of the aggregate and the asphalt concrete are shown in Tables 2 and 3 respectively. Granulometric composition curve for mineral part of the asphalt concrete is shown in Figure 1.

Table 1. Basic standard indicators of the bitumen.

Indicator	Measurement unit	Requirements of ST RK 1373	Value
Penetration, 25°C, 100 gr, 5 s	0.1 mm	101–130	104
Penetration Index PI	–	–1.0... +1.0	–0.34
Tensility at temperature:	cm		
25°C		≥ 90	140
0°C		≥ 4,0	5.7
Softening point	°C	≥ 43	46.0
Fraas point	°C	≤ –22	–25.9
Dynamic viscosity, 60°C	Pa·s	≥ 120	175.0
Kinematic viscosity	mm²/s	≥ 180	398.0

Table 2. Basic standard indicators of the crushed stone.

Indicator	Measurement unit	Requirements of ST RK 1284	Value fraction 5–10 mm	fraction 10–20 mm
Average density	g/cm³	–	2.55	2.62
Elongated particle content	%	≤ 25	13	9
Clay particle content	%	≤ 1,0	0.3	0.2
Bitumen adhesion	–	–	satisf.	satisf.
Water absorption	%	–	1.93	0.90

4

Table 3. Basic standard indicators of the asphalt concrete.

Indicator	Measurement unit	Requirements of ST RK 1225	Value
Average density	g/cm³	–	2.39
Water saturation	%	1.5–4.0	2.3
Voids in mineral aggregate	%	≤ 19	14
Air void content in asphalt concrete	%	2.5–5.0	3.8
Compression strength at temperature:	MPa		
0°C		≤ 13.0	7.0
20°C			3.4
50°C		≥ 1.3	1.4
Water stability	–	≥ 0.85	0.92
Shear stability	MPa	≥ 0.38	0.39
Crack stability	MPa	4.0–6.5	4.1

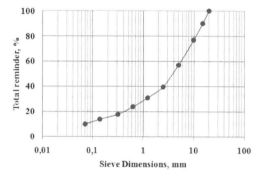

Figure 1. Granulometric curve of mineral part of the asphalt concrete.

4 TEST METHODS

4.1 Sample preparation

Samples of the hot asphalt concrete in form of a rectangular prism with length of 150 mm, width of 50 mm and height of 50 mm were manufactured in the following way. Firstly the asphalt concrete samples were prepared in form of a square slab by means of the Cooper compactor (UK, model CRT-RC2S) according to the standard (EN 12697-33 2003). Then the samples were cut from the asphalt concrete slabs in form of a prism. Deviations in sizes of the beams didn't exceed 2 mm.

4.2 Test

Tests of hot asphalt concrete samples in a form of rectangular prism on creep were carried out according to the direct tensile scheme until a complete failure. The test temperature was equal to 20 ± 2 °C, stress was changing from 0.055 to 0.311 MPa·s. The tests were carried out in a special assembled installation. The sample strain was measured by

means of two clock typed indicators while data was recorded in a video camera.

5 EXPERIMENT

5.1 Creep curves

Our previous works (Iskakbayev et al. 2016a, b, Teltayev et al. 2016) showed that an asphalt concrete creep curve as the most of viscoelastic materials has three characteristic sites: the site I of unstabilized creep with decreasing rate, the site II of stabilized creep with constant (minimum) rate and the site III of accelerating creep with increasing rate which precedes failure. The above works show test results of asphalt concrete for creep with relatively narrow ranges of stress variation. This work includes test results for seven values of stress from 0.055 MPa to 0.311 MPa. Practically five samples of asphalt concrete were tested for each value of stress.

Figures 2 and 3 show creep curves for asphalt concrete with stresses 0.055 MPa and 0.260 MPa respectively. It is clearly seen that at small stress (0.055 MPa), as well as at high stress (0.260 MPa) creep curves have all three sites. It is important to remember that since the beginning of loading to the failure moment (t_f) asphalt concrete passes three sites of deformation.

Figure 4 shows creep curves for five samples of asphalt concrete with stress 0.188 MPa. This figure shows well the impact of structural non-homogeneity of asphalt concrete on its deformability and durability: values of long-term strength and also deformation characteristics have static dispersion.

5.2 Long-term strength

The dependence of failure time on stress is called long-term strength of a material (Kachanov 1986, Rabotnov 1987). The long-term strength curve of

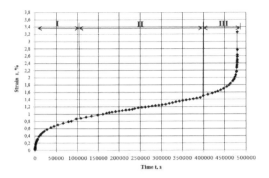

Figure 2. The asphalt concrete creep curve at stress 0,055 MPa.

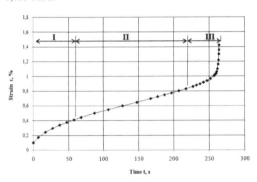

Figure 3. The asphalt concrete creep curve at stress 0,260 MPa.

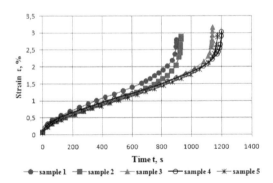

Figure 4. The asphalt concrete creep curves at stress 0,188 MPa.

the asphalt concrete constructed by results of the test on creep is shown in Figure 5. As it can be seen, the long-term strength curve is successfully described by exponential function:

$$t_f = 1,4072 \ \sigma^{-4,063} \tag{3}$$

T_f – failure time, s; and σ = stress, MPa.

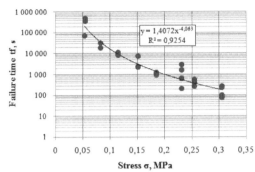

Figure 5. The long-term strength curve of the asphalt concrete.

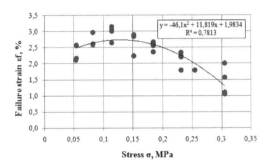

Figure 6. Dependence of failure strain on stress.

Equation shows that the stress impacts greatly on long-term strength of asphalt concrete: stress increase for single-order reduces long-term strength for four orders.

5.3 Failure strain

It is known that one of the first criteria for strength in mechanics of materials is connected with limit value of strain. Due to the above it is important to evaluate the values of failure strain at different stresses. Figure 6 shows the dependence of failure strain for asphalt concrete on stress, where you can see that failure strain value is not constant, i.e. it cannot be the criterion of strength. In addition, this dependence is of complicated character and described by quadric polynomial: failure strain is increased monotonously up to the stress of 0.12–0.13 MPa and reduces with further stress increase.

5.4 Step loading

As it was said previously, vehicles with different axle load run along the highways. Even loads on certain axles of one vehicle differ significantly. The sequence of impacts for these axle loads on

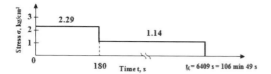

Figure 7. The first scheme of loading for the sample of the asphalt concrete.

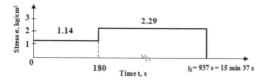

Figure 8. The second scheme of loading for the sample of the asphalt concrete.

pavement, differing in value, is of changeable character. Therefore, to evaluate the sequence of impacts on asphalt concrete strength, the following tests were carried out under the scheme of step loading.

The first five samples were loaded by stress equal to 2.29 kg/cm² which had not changed for 180 seconds. Then the stress was reduced till 1.14 kg/cm² which was constant till the sample failure. The samples average failure time was 6409seconds (Figure 7).

The next five samples were loaded firstly by stress equal to 1.14 kg/cm² which was constant during 180 seconds. Then the stress was increased till 2.29 kg/cm² which was constant till the sample failed. The samples average failure time was 937 seconds (Figure 8).

There was established that changes in small and big tension stresses sequences eventually influence on the asphalt concrete failure time. In the case when the sample was firstly loaded by big stresses and then by small ones the failure time increases almost for seven times than the sample which was loaded firstly by small stresses and then by big ones. This fact can be explained by the fact that in the first case the duration of acting of big stress is essentially less than duration of acting of small stresses.

As the sequence of load impact affects greatly the failure time of asphalt concrete, it is required to develop a new strength criterion, which differs from Miner's law and Bailey's criterion, mentioned above.

6 CONCLUSION

The results of experimental investigation for strain and strength of fine-grained asphalt concrete at static and step loading at temperature 20 ± 2°C and at stresses from 0.055 MPa to 0.311 MPa showed that:

– creep curves have three sites—unstabilized creep, stabilized creep and accelerating creep. It is important to know that since the beginning of loading to failure point the deformation of asphalt concrete increases with different rate, which is characteristic for certain segments of the creep curve;
– the long-term strength curve for asphalt concrete is successfully described by exponential function. The stress impacts greatly on long-term strength of asphalt concrete: stress increase for single-order reduces failure time for four orders;
– the sequence of impacts affects greatly the strength of asphalt concrete. It requires the development of a new strength criterion for asphalt concretes;
– due to structural non-homogeneity of asphalt concrete its long-term strength and charac-teristics of deformability have essential static dispersion.

REFERENCES

ARA ERES. 2004. *Guide for mechanistic-empirical design of new and rehabilitated pavement structures. Final report. Part 3. Design analysis. Chapter 3. Design of new and reconstructed flexible pavements.* Washington: Transportation Research Board.
Bailey, J. 1939. An attempt to correlate some tensile strength measurements on glass: III. *Glass Industry* 20: 95–99.
Cristensen, R.M. 1971. *Theory of viscoelasticity: An introduction.* New York: Academic Press.
EN 12697-33. 2003. *Bituminous Mixtures. Test Methods for Hot Mix Asphalt. Part 33: Specimen prepared by roller compactor.* Brussels: European Committee for Standardization.
Ferry, J.D. 1980. *Viscoelastic Properties of Polymers.* New York: John Wiley & Sons, Inc.
Hopkins, I.L. & Hamming, R.W. 1957. On creep and relaxation. *Journal of Applied Physics* 28: 906–909.
Iskakbayev, A., Teltayev, B. & Alexandrov, S. 2016a. Deter-mination of the creep parameters of linear viscoelastic materials. *Journal of Applied Mathematics* 2016: 1–6.
Iskakbayev, A., Teltayev, B., Andriadi, F., Estayev, K., Suppes, E. & Iskakbayeva, A., 2016b. Experimental research of creep, recovery and fracture processes of asphalt concrete under tension. *Proc. 24th intern. con-gress on theoretical and applied mechanics, Montreal, 21–26 August 2016.*
Kachanov, L. 1986. *Introduction to continuum damage mechanics.* Dordrecht: Martinus Nijhoff Publishers.
Miner, M.A.1945. Cumulative damage in fatigue. *Journal of Applied Mechanics.* 12: A159-A164.
MS-4. 2008. *The Asphalt Handbook. 7th Edition.* Lexing-ton: Asphalt Institute. Papagiannakis, A.T. & Masad, E.A. 2008. *Pavement Design and Materials.* New Jer-sey: John Wiley & Sons, Inc.

Rabotnov, Yu.N. 1987. *Introduction to fracture mechanics*. Moscow: Nauka.

ST RK 1225. 2013. *Hot mix asphalt for roads and airfields. Technical specifications*. Astana.

ST RK 1284. 2004. *Crushed stone and gravel of dense rock for construction works. Technical specifications*. Astana.

ST RK 1373. 2013. *Bitumens and bitumen binders. Oil road viscous bitumens. Technical specifications*. Astana.

Superpave series No. 1. 2003. *Performance graded asphalt binder specification and testing*. Lexington: Asphalt Institute.

Talreja, R. & Sing, C.V. 2012. *Damage and failure of composite materials*. Cambridge: Cambridge University Press.

Teltayev, B.B., Iskakbayev, A.I. & Rossi, C.O. 2016. Regularities of creep and long-term strength of hot asphalt concrete under tensile. In Sandra et al. (eds.), *Functional pavement design; Proc. 4th Chinese-European workshop, Delft, 29 June-1 July 2016*. London: Taylor & Francis Group.

Tschoegl, N.W. 1989. *The Phenomenological Theory of Linear Viscoelastic Behavior. An Introduction*. Berlin: Springer-Verlag.

Yoder, E.J. & Witczak, M.W. 1975. *Principles of Pavement Design*. New Jersey: John Wiley & Sons, Inc.

Transport Infrastructure and Systems – Dell'Acqua & Wegman (Eds)
© 2017 Taylor & Francis Group, London, ISBN 978-1-138-03009-1

Roundabout design guidelines: Case study of Croatia

H. Pilko

Department of Road Transport, Faculty of Transport and Traffic Science, University of Zagreb, Zagreb, Croatia

ABSTRACT: The popularity of roundabout application around the world is evident. Due to the inexperience of construction companies and the lack of proper national guidelines, distinctiveness in design is noticeable. In some intersections this led to reduction of Traffic (operational) Efficiency (TE). The purpose of this paper is to analyze: 1) the current state of roundabouts in Croatia; (2) known approaches to using geometry elements of roundabouts to predict TE; (3) overview and comparison of selected design guidelines; and (4) to present and comment the latest Croatian *Roundabout Design Guidelines on State Roads 2014* and show examples of good practice. Research results will serve to disseminate the knowledge for proper application and implementation of national roundabouts in order to compare it with international design practice and standards.

1 INTRODUCTION

The popularity of roundabouts around the world has driven substantial efforts to optimize their planning and modeling. Analyses of how geometric elements, traffic flow movements, driver behavior and other factors influence roundabout Traffic (operational) Efficiency (TE) have led to the development of numerous simulation-based computational models. In addition, they use primarily empirical and gap-acceptance mathematical models to determine TE (mainly focused on capacity and delay).

Croatia has slightly more than 200 roundabouts, of which more than 60% lie within or on the edge of urban areas, and many of them deviate substantially from international standards for roundabout planning, design and modeling, which compromises their TE and traffic safety (Pilko 2014).

These deviations are due primarily to the inexperience of construction companies and the use of outdated design guidelines.

The country lacks a system for monitoring and analyzing TE, and other parameters, though the government has called for the building and reconstruction of roundabouts as part of its National Traffic Safety Plan 2011–2020 (Pilko 2014).

The present research aims to describe and analyze (1) the current state of roundabouts in Croatia; (2) known approaches to using geometry elements of roundabouts to predict TE; (3) overview and comparison of selected design guidelines; and (4) to present and comment the latest Croatian *Roundabout Design Guidelines on State Roads 2014* and show examples of good practice.

The reminder of the paper is organized as follows. Section 2 contains the literature review on roundabout geometry and TE.

It also gives a brief overview of the literature on roundabout design and TE in Croatia. Section 3 gives a brief overview and comparison of European, American and Croatian guidelines.

In Section 4 recent Croatian roundabout geometry design practice is shown. Section 5 reports the conclusions.

2 BACKGROUND

2.1 *Roundabout geometry and efficiency*

Numerous models for determining roundabout capacity under mixed-traffic conditions suggest that it is strongly affected by geometric elements (Dahl & Lee 2012). Studies of roundabouts in various countries, particularly of single-lane roundabouts in urban areas, have shown that proper design and modeling can significantly improve Traffic (operational) Efficiency (TE) (Vasconcelos et al. 2013; Mauro & Cattani 2012). The most important geometric elements influencing TE (i.e. entry capacity) are entry width, entry radius, flare length, entry angle, Inscribed Circle Diameter (ICD) and number of entry lanes (Kimber 1980; Al-Omari et al. 2004; Dahl & Lee 2012; Yap et al. 2013; Barić et al. 2016).

Geometry considerations have also proven useful for analyzing driver behavior in roundabout situations and implications for traffic safety (Muffert et al., 2013; Wang et al., 2002). Methods for predicting traffic accidents have been described based on geometric elements (Maycock & Hall 1984), sight distance (Turner et al., 2009; Zirkel et al., 2013), as well as traffic dynamics and driver behavior when passing through the „potential conflict" zone of the intersection (Mauro & Cattani, 2004).

Studies of roundabout TE, conducted primarily in Western Europe and Australia, have led to several computational mathematical models that have been integrated into various roundabout software engineering simulation tools (e.g. ARCADY and PTV VISSIM). These mathematical models can be classified as (1) empirical, (2) gap acceptance, and (3) microsimulation. Each category has its disadvantages (Yap et al. 2013). Generally countries with updated roundabout design guidelines apply Highway Capacity Manual (HCM) models for analyzing roundabout capacity (Chodur 2005). These models take into account empirical and/or gap-acceptance models, but they do not address the Level of Service (LOS). The capacity formulation for urban single-lane roundabouts developed in HCM is based on the mathematical formulation of geometry parameters developed by Kimber (1980) and formulation of queue length developed by Ning Wu (Wu 2001). Differences among these models in how data are collected and analyzed, as well as deviations between predicted and actual driver behavior, make it difficult to identify the most suitable ones for given conditions (Mauro 2010). Planners and designers should be aware of the specific limitations of these models and the selected model(s) must be calibrated against field data or other validated models to ensure accuracy. Unfortunately, this calibration step is often neglected. In situations where the model is not even developed, it may be advisable to analyze roundabout TE using various approaches, and HCM models may be the most appropriate for such work (Mauro, 2010).

2.2 Roundabouts in Croatia

The *in situ* work by Legac et al. (2008) on 30 roundabouts in Zagreb examined relationships between main geometric elements and the numbers and types of traffic accidents, traffic flow demand, and numerous other determinants of capacity. That study confirmed that unsignalized roundabouts in Croatia are safer than classical intersections and it showed that geometric elements strongly influence safety.

Other authors have focused on applying roundabout models from outside Croatia to the Croatian situation. Their results suggest that imported models can work well, as long as they are calibrated for local conditions (Otković Ištoka 2008; Ištoka Otković & Dadić 2009; Šubić et al. 2012). For example, Ištoka Otković et al. (2013) used neural networks to calibrate a traffic microsimulation model for two urban single-lane roundabouts in Osijek. Šurdonja et al. (2013) optimized geometric elements such as inscribed circle radii, entry/exit radii, entry/exit approach width, and vehicle path trajectory, while Pilko et al. (2014) examined the relationship between vehicle trajectory design speeds through the roundabout and observed vehicle speeds.

3 OVERVIEW AND COMPARISON OF DESIGN GUIDELINES

3.1 *International design guidelines*

Worldwide positive experiences with modern single-lane roundabouts contributed to further research, development and implementation of other types of roundabouts. Substantial practical experiences initiated the formation of roundabout design guidelines. The recent achievements in this field for European countries like Austria, Germany, Switzerland, Poland, UK and the American are presented here in a brief overview.

In general, types of roundabout are defined by spatial limitation, location and traffic capacity. Mini roundabouts are characterized by a small external diameter and traversable central island for large vehicles. They are commonly used in urban environments with average operating speeds of 40 km/h or less. Single-lane roundabouts represent a standard solution and they are characterized by single entry lane, exist lane and circulatory lane. There is a non-traversable central island and they are used both in urban and rural environments. Multi-lane roundabouts have two or more entry/exit and circulatory lanes. Due to a possibility of path ovelap at the entry and the exit as well as higher speeds these types of roundabouts are less safe in comparison with mini and single-lane roundabouts.

According to *Austrian guidelines* (Osterreichische Forschungsgemeinschaft Strasse und Verkehr 2010) the geometry elements in mini roundabouts are determined on the basis of curve of the course of design vehicles. It is recommended that single-lane roundabouts have the external diameter ≥ 26 m (35–40 m) and that multi-lane roundabouts have the external diameter ≥ 40 m (50 to 60 m).

German guidelines (Forschungsgesellschaft fur Strassen und Verkherswesen 2006) have determined that mini roundabouts have a diameter of 13 to 22 m and capacity of 18000 veh/day. A central island is traversable and has truck apron raised by 4 to 5 cm. The width of circulatory lane is beween 4 and 6 m with transversal inclination of 2.5% outwards. The entry/exit radii is from 8 to 10 m. Small single-lane roundabouts have the capacity of 25000 veh/day. The external diameter is from 26 to 45 m, the entry radius is from 10 to 16 m and the exit radius is from 12 to 18 m. These dimensions provide great traffic safety. Circulatory lane being 6.5 to 9 m wide consists of driving and traversable part used by large vehicles. Small double-lane roundabouts have one or two lane entry with the entry radii of 12 to16 m, depending on the traffic load and one lane exit with the exit radii of 12 to 18 m. In the circulating area there are two traffic lanes with the total width of 8 to 10 m, but they are not marked with the horizontal signalization. Diameter varies from 40 to 60 m with the maximum capacity of 32000 veh/day. Their diameter is more than 60 m and they have two or more lanes in entry, exit and circulatory lanes (Forschungsgesellschaft fur Strassen und Verkherswesen 2006). According to the *German guideliness* the designing of small double-lane roundabouts is not allowed due to the reduced safety. The solution of the problem of increased through traffic is found in the use of roundabouts with the spiral traffic course, the turbo roundabouts.

Swiss guidelines (Vereinigung Schweizerischer Strassenfachleute (VSS) 1999) have determined that mini roundabouts and single-lane roundabouts are designed with the entry radii of 10 to 12 m, while the approach radius is five times larger. In a properly designed entry the entry angle α has to be as large as possible. The exit radius is from 12 to 14 m. Mini roundabouts with circulatory lane width of 7 to 8 m are characterized by external diameter of 14 to 16 m. Single-lane roundabouts have external diameter of 26 to 40 m.

USA guidelines (NCHRP-National Cooperative Highway Research Program 2010) have determined that splitter islands can be raised, traversable or only marked and designed according to design vehicle. The width of circulatory lane in single-lane roundabouts varies from 4.8 to 6 m. Circular shape of a central island is recommended, but oval, irregular or raindrop shapes can also be used. The entry radius is from 15 to 30 m and the exit radius is from 15 to 60 m. The traversable portion of a central island is 50 to 75 mm raised. Multi-lane roundabouts have at least one entry with two or more lanes which requires a wider roadway in circulating part of the intersection so that at least two vehicles can travel side by side. The width of circulating double lane is from 8.5 to 9.8 m and of circulating triple lane is from 12.8 to 14.6 m. Firstly an entry is designed with a smaller radius of 20 to 35 m, and then with a radius of 45 m and more. The entry lane can be moved to the left in order to obtain increased deflection which reduces it at exit. Radius of the fastest path is between 53 and 84 m which results in design speed of 40 to 50 km/h.

According to *Polish guidelines* (Generalna Dyrekcja Dróg Krajowych i Autostrad 2004) the geometry elements have determined that mini roundabouts have diameter of 14 to 22 m (exceptionally 25 m), fully traversable central island, entry/exit radii of 6 to 12 m and capacity of 15000–17000 veh/day. Urban single-lane roundabouts are designed with the external diameter ranging from 14 to 45 m and from 30 to 50 m for rural areas.

Urban multiple (two)-lane roundabouts are supposed to have external diameter from 37.5 m to over 55 m and from 40 m to 65 m for rural areas. Single-lane roundabouts are the solutions which, in Polish conditions, ensure a very high level of safety to the users and high capacity. Furthermore, two-lane roundabouts are not entirely optimal solutions since substantial number of drivers use mainly the right lane, both at the approach entry and on the circulatory roadway (Macioszek 2013).

The *UK guidelines* (UK Goverment 2007) not only define geometry elements for mini, single and multi-lane roundabouts, but they also give instructions to: traffic safety, road users specific requirements, the assessment procedure, and conspicuity. The main geometry parameters for mini, single and multi-lane roundabouts are shown in Table 2.

3.2 Croatian guidelines

The procedures of planning and designing the roundabouts in the Republic of Croatia are based on the current national (Institut prometa i veza 2002) and applied foreign guidelines, especially German (Forschungsgesellschaft fur Strassen und Verkherswesen 2006), Austrian (Osterreichische Forschungsgemeinschaft Strasse und Verkehr 2010) and Switzerland (Vereinigung Schweizerischer Strassenfachleute 1999), positive examples of world practice and empirical practice of designers (especially from the Netherlands). One of the first steps in creating national regulations for roundabout designing were the *Guidelines for the Design of Circular Intersection* (*Guidelines 2002*) from 2002 (Institut prometa i veza 2002). The main objective of these guidelines was the standardization of design and implementation of roundabouts on public roads in the country. Then, in the absence of local regulations, designers applied the aforementioned foreign regulations and practices, which resulted in a large number of non-compliant roundabout design. Also, insufficient attention has been made to the criteria of TE.

Roundabout Guidelines for State Roads (*Guidelines 2014*) from 2014 (Deluka-Tibljaš et al. 2014) represent a logical step on the path of creating national regulations for planning and designing of all roundabout types, especially in urban areas. They also represent a significant upgrade of *Guidelines 2002* in terms of geometry design elements, TE criteria, defining, planning and implementation of turbo roundabouts, and the importance and necessity for trajectory checking of the relevant design vehicle in the design phase. However, the guidelines do not indicate what models or simulation software should be used for analyzing TE. The *Guidelines2014* only briefly presents possible methods and their pros and cons of

Austrian (Osterreichische Forschungsgemeinschaft Strasse und Verkehr 2010), British linear regression, Australian (by National Association of Road authorities) and German (Wu 2001) method that can be used for analyzing TE parameters.

As the process of roundabout designing is very complex and specific, it is not always possible to apply the optimal geometry design elements and meet all requirements. However, planning and designing in accordance with *Guidelines2014* enables unification at the national level, which derives complete and optimal solutions. By making the final legislative framework for designing roundabouts, it is necessary to gradually adopt and implement certified foreign methodologies and practices, and to introduce them to state regulation, with obligatory consideration of all national features.

3.3 *Guideline comparison*

From the brief overview (Table 1.a.b and Table 2) we can conclude that all European guidelines are more less similar in terms of main geometry design elements. Only the UK geometry parameters are slightly different than the other European. This is also evident when we compare the concepts and dimension with the latest Croatian guidelines. The most evident difference is approximate capacity. The USA guidelines have bigger values of geometry design when compared to European and Croatian guidelines. This is because of longer design vehicle and dimensions related to specific national conditions and driver behavior.

Table 1.a.　Guideline comparison of roundabout type and main geometry design elements.

Design guidelines/Design Environment	Roundabout type									
	Mini					Single-lane				
	A[1]	D[2]	CH[3]	USA	CRO[4]	A[1]	D[2]	CH[3]	USA	CRO[4]
			urban					urban or rural		
External diameter	<26	13–22	14–16	13–27	13–25	≥26	26–45	26–40	27–55	30–40
Entry lane (m)			1					1		
Exit lane (m)			1					1		
Circulatory lane			1					1		
Central island			traversable					non-traversable		
Entry radii (m)	–	8–10	10–12	15–20	10–12		10–16	10–12	15–30	10–12
Exit radii (m)	–	8–10	12–14	≤30	12–14	12–25	12–18	12–14	30–60	12–14
Circulatory lane width (m)	–	4–6	7–8	4–10	4.5–5.0		6.5–9.0		4.8–6.1	5.5–7.0
Maximum recommended entry design speed (km/h)			25–30					30–40		
Capacity (veh/day)	10000		18000		≤15000		25000			20000

Table 1.b.　Guideline comparison of roundabout type and main geometry design elements.

Design guidelines/Design Environment	Roundabout type				
	Multi-lane				
	A[1]	D[2]	CH[3]	USA	CRO[4]
			urban or rural		
External diameter	≥40	40–60	≥60	46–91	50–90
Entry lane (m)			≥1		
Exit lane (m)			≥1		
Circulatory lane			≥2		
Central island			non-traversable		
Entry radii (m)	10–16	12–16		20	10–14
Exit radii (m)	12–25	12–18		≥20	12–16
Circulatory lane width (m)	8–10	8–10		4.3–14.6	5.5–7.7
Maximum recommended entry design speed (km/h)		50–60		40–50	
Capacity (veh/day)	30000	32000		≤45000	≥250000

1 – Austrian, 2 – German, 3 – Swiss, 4 – Croatian.

Table 2. Guideline comparison of roundabout type and main geometry design elements.

	Roundabout type					
	Mini		Single-lane		Multi-lane	
Design guidelines/Design	PL[1]	UK[2]	PL[1]	UK[2]	PL[1]	UK[2]
Environment	urban				urban or rural	
External diameter			22–45*		≥ 37.5–55*	
	14–22(25)	<28	30–50**	<28	≥ 40–65**	<100
Entry lane (m)	1		1		≥ 1	
Exit lane (m)	1		1		> 1	
Circulatory lane	1		1		≥ 2	
Central island	traversable				non-traversable	
Entry radii (m)	6–12	6–20	10–15	10(20)	12–15	12–16
Exit radii (m)	6–12	8–10	12–15	15–20	14–17.5	20–100
Circulatory lane width (m)	5.0	6.0	5.5–8.75	<15	9.1–10	8–10
Maximum recommended entry design speed (km/h)	25–30		30–40		40–50	>40
Capacity (veh/day)	15000–17000	18000	20000–25000		40000	

1 – Polish, 2 – United Kingdom, *urban areas, **rural areas.

4 CROATIAN EXAMPLES

Given the above in section 2.2 and 3.2, the following will give a brief overview of recent implemented and planned forms of urban roundabouts. The presented examples of design practice are designed in accordance with the *Guidelines2002*, *Guidelines2014*, and according to aforementioned foreign regulations.

Mounted roundabouts are also presented in the latest guidelines, and the first examples of its usage can be seen in the City of Zagreb. Figure 1 shows the latest implementation of mounted roundabout in urban area.

Reconstruction of State Road S517 on a road section through the city of Belišće is performed using two roundabouts (Figure 2). The reconstruction eliminates existing dangerous small turn radius and adverse horizontal elements.

External diameter of Roundabout 1. (to the right in Figure 3) is 38.00 m, and is the busiest and most important intersection in the city. Roundabout 2. (to the left in Figure 3) has external diameter of 36.00 m. The circulatory roadway width is 6.6 meters, with an additional truck apron width of 1.8 m. The entry/exit radii were designed using vehicle design trajectories for a truck with a trailer length of 18.00 m (Pilko et al. 2015).

The first turbo roundabout in Croatia was carried out in the city of Osijek in 2014. (Figure 2). Intersection is designed in accordance with the *Guidelines2002* as well as the empirical experiences with turbo roundabouts from Netherland. The most important design elements are: external diameter is 5.15 m (outer space between shifted centers

Figure 1. Roundabout Vončinina—Voćarska, City of Zagreb, Croatia—in traffic 2016.

Figure 2. Turbo roundabout in the City of Osijek, Croatia—in traffic 2014.

Figure 3. Roundabouts at State road S517 in city of Belišće, Croatia.

of circular segments); inscribed central diameter is 4.95 m (inner space between shifted centers of circular segments); R1 = 15 m (radius of the inner edge of the road surface); R2 = 20 m (radius of the outer edge of the outer lane); R3 = 20.3 m (radius of the inner edge of the outer lane); R4 = 25.2 m (radius of the outer edge of the road surface). Cycling lanes and tram lines are also designed making this intersection more complex for all road users.

5 CONCLUSION

There is an obvious increase in the application and design of all types of roundabouts in urban areas worldwide. From the brief overview of the selected West European and USA roundabout guidelines we can conclude that West European guidelines are more less similar in terms of main geometry design elements. This is also evident when we compare the concepts and dimension with the latest Croatian guidelines. The most evident difference is approximate capacity. The USA guidelines have bigger values of geometry design when compared to West European and Croatian guidelines. This is because of longer design vehicle and dimensions related to specific national conditions and driver behavior.

Due to the lack of Croatian regulations, some studies and designers warn of the distinctiveness of performance, which is at some roundabouts resulted in reduction of TE. *Guidelines2014* enables better preparation of national legislation for planning and design of all types of roundabouts. However, the guidelines do not indicate what models or simulation software should be used for analyzing TE. Future studies should make comparative comparison of capacity formulations (TE—Measure of Effectiveness (MOE) of presented guidelines by incorporating real traffic data collected from roundabouts in Croatia. This would enable possible

implementation of the existing or development of a new method and its application and validation for Croatia. However, planning and designing in accordance with *Guidelines2014* enables unification at the national level, which derives complete and optimal solutions. By developing the final legislative framework for designing roundabouts, it is necessary to conduct a comprehensive and systematic field-analytical research on existing roundabouts. In addition, current national studies may serve as a certain database, as well as a source of empirical knowledge. It is necessary to involve all stakeholders and international experts from this field. Results of this paper can be used for dissemination of the knowledge for proper application and implementation of national roundabouts in order to compare it with international design practice and standards.

REFERENCES

Al-Omari, B., Al-Masaeid, H. & Al-Shawabkah, Y., 2004. Development of a Delay Model for Roundabouts in Jordan. *Journal of Transportation Engineering*, 130(1): 76–82.

Barić, D., Pilko, H. & Strujić, J., 2016. An Analytic Hierarchy Process Model to Evaluate Road Section Design. *Transport*, 31(3): 312–321.

Chodur, J., 2005. Capacity Models and Parameters for Unsignalized Urban Intersections in Poland. *Journal of Transportation Engineering*, 131(12): 924–930.

Dahl, J. & Lee, C., 2012. Empirical Estimation of Capacity for Roundabouts Using Adjusted Gap-Acceptance Parameters for Trucks. *Transportation Research Record*, 2312: 34–45.

Deluka-Tibljaš, A. et al., 2014. *Smjernice za projektiranje kružnih raskrižja na državnim cestama*, University of Rijeka, Faculty of Civil Engineering, Rijeka (in Croatian).

Forschungsgesellschaft fur Strassen und Verkherswesen, 2006. *Merkblatt fur die Anlage von Kreisverkehren*, Koln (in German).

Generalna Dyrekcja Dróg Krajowych i Autostrad, 2004. *Wytyczne projektowania skrzyżowań drogowych*, Warszawa.

Institut prometa i veza, 2002. *Smjernice za projektiranje i opremanje raskrižja kružnog oblika—rotora*, Institute of Transport and Communications, Zagreb (in Croatian).

Ištoka Otković, I. & Dadić, I., 2009. Comparison of Delays at Signal-controlled Intersection and Roundabout. *PROMET—Traffic & Transportation*, 21(3): 157–165.

Ištoka Otković, I., Tollazzi, T. & Šraml, M., 2013. Calibration of Microsimulation Traffic Model Using Neural Network Approach. *Expert Systems with Applications*, 40(15): 5965–5974.

Kimber, R., 1980. *The traffic capacity of roundabouts*, Transportation and Road Research Laboratory, Crowthorne.

Legac, I. et al., 2008. *Korelacija oblikovnosti i sigurnosti u raskrižjima s kružnim tokom prometa (MZOŠ br. 135-0000000-3313)*, Ministry of Science, Education and Sports, Zagreb (in Croatian).

Macioszek, E., 2013. Analysis of the Effect of Congestion in the Lanes at the Inlet to the Two-Lane Roundabout on Traffic Capacity of the Inlet. *Activities of Transport Telematics*: 97–104.

Mauro, R., 2010. *Calculation of Roundabouts—Capacity, Waiting Phenomena and Reliability*, Berlin, Heidelberg: Springer Science & Business Media.

Mauro, R. & Cattani, M., 2012. Functional and Economic Evaluations for Choosing Road Intersection Layout. *PROMET—Traffic & Transportation*, 24(5): 441–448.

Mauro, R. & Cattani, M., 2004. Model to Evaluate Potential Accident Rate at Roundabouts. *Journal of Transportation Engineering*, 130(5): 602–609.

Maycock, G. & Hall, R.D., 1984. *Accidents at 4-arm roundabouts*, Transportation and Road Research Laboratory, Crowthorne, Berkshire.

Muffert, M., Pfeiffer, D. & Franke, U., 2013. A Stereovision Based Object Tracking Approach at Roundabouts. *IEEE Intelligent Transportation Systems Magazine*, 5(2): 22–32.

NCHRP-National Cooperative Highway Research Program, 2010. *Roundabouts: An Informational Guide*, Washington DC.

Osterreichische Forschungsgemeinschaft Strasse und Verkehr, 2010. *RVS 03.05.14 Strassenplanung, Plangleiche Knoten—Kreisverkehr*, Wien (in German).

Otković Ištoka, I., 2008. Capacity Modelling of Roundabouts in Osijek. *Tehnički vjesnik-Technical Gazzette*, 15(3): 41–47.

Pilko, H., 2014. *Optimization of Roundabout Design and Safety Component* (doctoral thesis), University of Zagreb, Faculty of Transport and Traffic Sciences, Zagreb (in Croatian).

Pilko, H., Barišić, I. & Bošnjak, H., 2015. Urban Roundabouts. In *VI. Croatian Congress on Roads*. Opatija: Croatian Road Association Vi-vita.

Pilko, H., Brčić, D. & Šubić, N., 2014. Study of Vehicle Speed in the Design of Roundabouts. *GRAĐEVINAR*, 66(5): 407–416.

Šubić, N., Legac, I. & Pilko, H., 2012. Analysis of Capacity of Roundabouts in the City of Zagreb according to HCMC–2006 and Ning Wu Methods. *Tehnički vjesnik—Technical Gazette*, 19(2): 451–457.

Šurdonja, S., Deluka-Tibljaš, A. & Babić, S., 2013. Optimization of Roundabout Design Elements. *Tehnički vjesnik-Technical Gazzette*, 20(3): 533–539.

Turner, S.A., Roozenburg, A.P. & Smith A W, 2009. *Roundabout crash prediction models*, Wellington.

UK Goverment, 2007. *Design manual for roads and bridges, Volume 6: Road Geometry, Section 2: Junctions, Part 2 TD 16/07; Geometric Design of Roundabouts*, London.

Vasconcelos, A., Seco, A. & Silva, A., 2013. Comparison of Procedures to Estimate Critical Headways at Roundabouts. *PROMET—Traffic & Transportation*, 25(1): 43–53.

Vereinigung Schweizerischer Strassenfachleute (VSS), 1999. *Schwizer Norm 640263 - Knoten mit Kreisverkehren*, Zurich (in German).

Wang, B., Hensher, D.A. & Ton, T., 2002. Safety in the Road Environment: A Driver Behavioural Response Perspective. *Transportation*, 29(3): 253–270.

Wu, N., 2001. A Universal Procedure for Capacity Determination at Unsignalized (priority-controlled) Intersections. *Transportation Research Part B: Methodological*, 35(6): 593–623.

Yap, Y.H., Gibson, H.M. & Waterson, B.J., 2013. An International Review of Roundabout Capacity Modelling. *Transport Reviews*, 33(5): 593–616.

Zirkel, B. et al., 2013. Analysis of Sight Distance, Crash Rate and Operating Speed Relationships for Low-Volume Single Lane Roundabouts in the United States. *Journal of Transportation Engineering*, 139(6): 565–573.

Transport Infrastructure and Systems – Dell'Acqua & Wegman (Eds)
© 2017 Taylor & Francis Group, London, ISBN 978-1-138-03009-1

High modulus asphalt concrete for Ljubljana airport apron

A. Ljubič
IGMAT d.d. Building Materials Institute, Ljubljana, Slovenia

ABSTRACT: Central part of the main apron at the Ljubljana airport from 1978 was scheduled for reconstruction in 2014 because of the increased traffic loads and subsequent distress of the existing asphalt pavement. Because of the traffic situation and faster completion time the asphalt pavement was chosen for the reconstruction and for the planned high traffic loading and severe climatic conditions a suitable asphalt mix had to be selected. A high stiffness modulus asphalt mix was chosen for base layers and the article describes the mix design, its testing and finally completed reconstruction of the airport apron area.

1 INTRODUCTION

1.1 *Apron area*

Central part of the Main Apron at the Ljubljana airport from 1978 was scheduled for reconstruction in 2014 because of the increased traffic loads and subsequent distress of the existing asphalt pavement. It comprises about 19.000 m^2 of asphalt pavements, built in year 1978 during last major reconstruction of the airport.

It consisted of two asphalt layers—base layer in 10 cm thickness and surface layer 5 cm thick—mainly built over the concrete 26 cm thick apron from the year 1962 and in lesser part over asphalt pavement enlargement of the apron area from 1975.

In 1993 an antiskid surface dressing was laid over the entire area of asphalt surface serving as protection against oil derivatives and at the same time improving surface friction characteristics.

Very heavy loaded parts of Main Apron at airplane parking positions PSN 3, 4, 5 and 6 were processed from asphalt pavement to concrete pavement in 2006, the same was done for parking postions PSN 7, 33, 34 and 35 in year 2011, because of the high permanent deformations of asphalt layers.

Area planned for reconstruction in year 2014 comprised:

– Longitudinal driving axis of airplanes (TWY E2, partly E1 and E3) during their scheduling for parking in the first and second parking rows,
– Drive-ins and drive-outs to single 'push back' parking positions in the first row parking (mainly parking positions with avio-bridges),
– Drive-ins and drive-outs to single 'push back' parking positions in the second row parking,

1.2 *Condition before the reconstruction*

Condition of asphalt surface on the Apron area before the reconstruction was as follows:

– Extensive surface cracks mostly filled with hot applied sealants and joint filler (Figure 2),

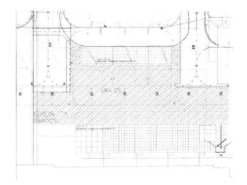

Figure 1. Area of Main Apron TWY E reconstruction.

Figure 2. Typical asphalt surface cracks.

Figure 3. Permanent deformations in form of wheel tracks.

– Crushing of the surface layer and loss of particles on some places,
– Permanent deformations in the shape of wheel tracks, especially in airplane trajectories along the TWY E to the first row of parking positions (Figure 3).

2 RECONSTRUCTION PROJECT

2.1 First area

At the first area (over the completely asphalt pavement) the design of new asphalt pavement was as follows:

- AC 22 base PmB PG 64-16 with high stiffness modulus in 8 cm thickness
- AC 22 bin PmB PG 64-16 with high stiffness modulus in 8 cm thickness
- SMA 11 PmB PG 70-22 in 4 cm thickness (or AC 16 surf PmB PG 70-22 in 6 cm thickness on the area of 8362 m² where only substitution of surface layer was scheduled).

2.2 Second area

At the second area (over the concrete pavement) the design of new asphalt pavement was as follows:

- AC 22 base PmB PG 64-16 with high stiffness modulus in 12 cm thickness
- SMA 11 PmB PG 70-22 in 4 cm thickness

The drive-ins to parking positions PSN 3, 4, 5, 6 and 7 were designed as concrete pavement:

2.3 Additional specifications

Additional specifications for High Stiffness Modulus Asphalt Concrete (or HMAC) besides already existing specifications in standards EN 13108-1

and SIST 1038-1 Bituminous Concrete were the following:

– resistance to permanent deformations according to EN 12697-22, method B in air at the temperature of 60°C—maximum allowed wheel tracking rate 0,10 mm/10^3 passes ($WTS_{AIR\,0,10}$) and maximum allowed proportional rut depth of 3% ($PRD_{AIR\,3,0}$)
– stiffness, on prismatic specimens according to EN 12697-26, 4PB-PR at the temperature of 10°C and frequency of 10 Hz should be minimum of 14000 Mpa ($S_{min14000}$)
– resistance to fatigue on prismatic specimens according to EN 12697-24, 4PB-PR at the temperature of 10°C and frequency of 10 Hz the minimum strain corresponding to 10^6 loading cycles should be $\varepsilon = 130\ \mu m/m$ ($\varepsilon_{6\,130}$).

3 HIGH MODULUS ASPHALT CONCRETE

3.1 Definitions and specifications

Asphalt concrete with high stiffness modulus (HMAC) is intended for base or binder asphalt layers for heavy traffic roads, airports and other trafficked surfaces where we want a higher resistance to permanent deformations and to fatigue compared to bituminous concrete, for example on slow traffic lanes, road crossings, bus lanes etc.

Higher resistance to permanent deformations and to fatigue is being reached by different mix design and functional specifications. Characteristic feature of HMAC is a relatively fine grading and higher bitumen content which contribute to higher fatigue resistance, the use of harder type of bitumen and/or polymer modified bitumen at the same time enables higher resistance to permanent deformations. Higher stiffness modulus of asphalt layers decreases tension stresses in lower layers of pavement.

This technology was developed in France under the name of EME (Enrobé à Module Elevé) already in the nineties of former century and characteristic for it was the use of very hard bitumen in high content (till 6%).

There is a high interest for the use of HMAC all over the world, a lot of countries already made use of this technology, some of the countries also meaningfully adapted French specifications to their needs and boundary conditions. There is however always a high content of relatively hard bitumen and Richness factor K greater than 3,4.

In chapter A7 of Requirements for HMAC mix design and pavement design for Slovenia, as a result of SPENS Project, the use of five types of bitumen is provided (20/30, 15/25, 1020, PmB 10/40–60 and PmB 25/55-65) with a minimal content 4,8% by mass, nominal aggregate size of 11 mm and 16 mm, air voids content from 2 to 4%, maximum proportional rut depth at 60°C 3%, stiffness modulus (at 4PB, 10°C and 10 Hz) minimum 14000 MPa and resistance to fatigue (at 4PB, 10°C in 10 Hz) minimum ε_{6-130}.

3.2 HMAC design

Additional specifications for high stiffness modulus asphalt concrete (resistance to permanent deformations, high stiffness and resistance to fatigue) together with demands for high resistance to climatic conditions at Ljubljana airport area, where continental climate with low winter temperatures is predominant, dictated very careful selection of constituent materials, especially of the bitumen.

After the market survey of potential suppliers and their products we decided together with the asphalt manufacturer TAČ (Ljubljana) for PmB Starfalt 10/40-65 from Austria with a declared performance grade index according to AASHTO of PG 82-28. With this bitumen and stone aggregate from the quarry Črnotiče and filler Stahovica in the IGMAT asphalt laboratory we made a job-mix formula for HMAC 22 PmB with optimum binder content of 5,0% by mass and analysed all the required parameters from additional HMAC specifications.

The testing results of additional characteristics were as following:

- resistance to permanent deformations according to EN 12697-22, method B in air at the temperature of 60°C—wheel tracking rate 0,04 mm/10^3 passes ($WTS_{AIR\ 0,04}$) and proportional rut depth of 1,8% ($PRD_{AIR\ 1,8}$)
- stiffness, on prismatic specimens according to EN 12697-26, 4PB-PR at the temperature of 10°C and frequency of 10 Hz of 14196 Mpa ($S_{min14196}$)
- resistance to fatigue on prismatic specimens according to EN 12697-24, 4PB-PR at the temperature of 10°C and frequency of 10 Hz- the strain corresponding to 10^6 loading cycles $\varepsilon = 183,1\ \mu m/m$ (ε_{6-183}).

All determined additional characteristics therefore satisfied the project criteria and also all specifications for asphalt concrete with high stiffness modulus, some characteristics quite clearly exceeded the requirements.

3.3 HMAC production and laydown

Start of the HMAC 22 base/bin PmB 10/40-65 production was on 21.10.2014 in TAČ asphalt plant and test section for laydown on Zaloška street in the city of Ljubljana.

On the test section we determined the optimum speed of the paver, number and type of rollers and compaction pattern. It turned out that minimum temperature of the mix on site should be 170°C and also heavier than usual rollers were needed, with mass over 10 ton. We also tried laying the HMAC in different lifts, thickness range was from 60 mm till 120 mm, with good results—compaction rate over 98%—therefore satisfying the requirements.

After the satisfactory results from the HMAC test production and laydown the Phase 1 of reconstruction works on the airport Apron began on 30.10.2014.

Works have been carried out in late autumn, with short daylight time, every morning we usually had to wait for the fog to lift up and the air temperature to rise before starting the asphalt works. Despite not the optimal environmental conditions in the first phase of the project the works were done technically correct and the required minimum compaction rate of 98% has been exceeded on all of the first layer, on average it was at 100,3%, and on the second layer of HMAC reached on average 101,0% of reference asphalt mix bulk density in the laboratory.

Average thickness of the first HMAC layer was at 93 mm, of the second layer 87 mm, together therefore on average 180 mm.

After completion of Phase 1with surface course SMA 11 PmB 45/80-65 FR and AC 16 PmB 45/80-65 FR in year 2014, we continued Phase 2 of airport Main Apron reconstruction in April next year and finished the project in the end of May 2015.

All the quality control result for Phase 2 were also satisfactory, thickness of the built-in HMAC layer was at 130 mm, with average compaction rate of 102,1%, and air voids content of 2,8%.

3.4 HMAC quality

We checked the produced and built-in HMAC for additional criteria of the project in terms of resistance to permanent deformations, stiffness and resistance to fatigue. All the tests were done at IGMAT Building Materials Institute laboratories, on the most up-to-date testing equipment for dynamic testing of asphalt. Testing was done on the cut prismatic specimens from laboratory prepared slabs with roller compactor according to EN 12697-33. Asphalt layer specimens were than tested according to EN 12697-26 and EN 12697-24 using the four point bending test on prismatic specimens (4PB-PR) method.

Test results on additional characteristics of built-in HMAC on airport Apron were as following:

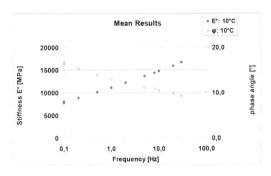

Figure 4. Stiffness E* and phase angle dependent on test frequency graph.

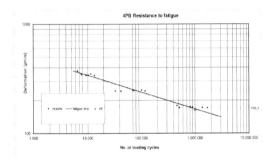

Figure 5. Resistance to fatigue graph (deformation dependent on load cycles number).

– resistance to permanent deformations according to EN 12697-22, method B in air at the temperature of 60°C—wheel tracking rate 0,04 mm/10^3 passes (WTS$_{AIR\,0,04}$) and proportional rut depth of 1,7% (PRD$_{AIR\,1,7}$).
– stiffness, on prismatic specimens according to EN 12697-26, 4PB-PR at the temperature of 10°C and frequency of 10 Hz of 14913 Mpa (S$_{min14913}$).
– resistance to fatigue on prismatic specimens according to EN 12697-24, 4PB-PR at the temperature of 10°C and frequency of 10 Hz- the strain corresponding to 10^6 loading cycles ε = 166,8 µm/m ($\varepsilon_{6\,167}$).

All the above mentioned results were inside the required specifications of the project and therefore proving the quality of HMAC asphalt mixes and layers on airport Main Apron.

We also analysed used bitumen Starfalt PmB 10/40-65 in company Ramtech laboratory for satisfying the AASHTO M320 criteria for performance

grade index PG 64-16 class. PG class could not be determined because of the too high initial kinematic viscosity, however the determined critical high temperature was at 88°C and critical lowest temperature was at –20,9°C, that exceeds the requirements for the Ljubljana airport.

4 CONCLUSIONS

During airport surfaces reconstruction we are always facing special conditions and requirements for asphalt pavement, be it greater loads and traffic situation or special working conditions. This case was the first use of high stiffness modulus asphalt concrete in Slovenia and went down successfully. This proves that with a lot of knowledge and hard work, some courage and a little bit of luck, if not elsewhere then with the weather, it is possible to get the base course suitable for the high traffic loading and severe climatic conditions on airport Apron and not necessarily always implementing concrete technology for that.

REFERENCES

AP-T283-14 Technical Report, 2014 High Modulus High Fatigue Resistance Asphalt (EME2) Technology Transfer, Austroads.
Bankowski W. et al., 2009 Laboratory and field implementation of high modulus asphalt concrete. Requirements for HMAC mix design and pavement design, SPENS.
Horak, E., High Modulus Asphalt (HiMA) French development, Total.
IGMAT d.d., 2014 Elaborat k projektu za izvedbo—obnova dela glavne letališke ploščadi (TWY E).
IGMAT d.d., 275-POA-14 Odpornost proti kolesnicam.
IGMAT d.d., 281-POA-14 PSAZ AC 22 base PmB 10/40-65.
IGMAT d.d., 302-POA-14 Togost.
IGMAT d.d., 303-POA-14 Odpornost proti utrujanju.
IGMAT d.d., 321-POA-14 Odpornost proti kolesnicam.
IGMAT d.d., 359-POA-14 Delno poročilo o kakovosti asfalta AC 22 base PmB 10/40-65.
IGMAT d.d., 8-POA-15 Togost.
IGMAT d.d., 9-POA-15 Odpornost proti utrujanju
IGMAT d.d., 78-POA-15 Poročilo o kakovosti asfalta AC 22 base PmB 10/40-65.
Le Bouteiller, E. 2012 High Modulus Asphalt The French experience, Colas, AAPA Melbourne.
Ramtech d.o.o., I-2015-77 PG-10-40-65 Test Report on Performance Grade.
RD-JN-2014-G2-A/3/AP 2014 Obnova dela glavne letališke ploščadi in izgradnja 2. faze ceste okoli letališkega perimetra na letališču Jožeta Pučnika Ljubljana, razpisna dokumentacija.

Transport Infrastructure and Systems – Dell'Acqua & Wegman (Eds)
© 2017 Taylor & Francis Group, London, ISBN 978-1-138-03009-1

Assessment of technologies for roadway energy harvesting

S. Colagrande & M. Patermo
University of L'Aquila, L'Aquila, Italy

ABSTRACT: This paper deals with the electric energy production from roads, through an overview of the existing and experimental technologies that could guarantee, in a not so distant future, a substantial source of power. The study analyzes the performances of several possible technologies for energy generation from road infrastructures as follows: (i) piezoelectric device able to generate electricity from the surface movement of the road platform caused by automobile traffic; (ii) photovoltaic systems achieved through the installation of photovoltaic panels both on noise barriers placed by the roadside and on shelters of parking service stations; (iii) photovoltaic panel devices distributed on the road surface that allow to convert solar energy into electricity and sustain high loads and stresses generated by vehicular traffic. Finally, a practical application of these technologies on a A24 highway section located in central Italy has been evaluated and the results are compared in terms of energy production.

1 INTRODUCTION

Modern Italian highway network is the result of a design and development model conceived in the sixties. This implies that, from the environmental sustainability point of view, we are actually dealing with the technologies and the road conception developed 55 years ago. Nowadays, on the contrary, the highway should not be intended as a pollution source because, when properly designed and requalified through the integration of specific systems, it can become a valuable energetic resource for the entire civil community (Cappelletti 2011).

It is well known that, since 1966, Italian transport (both commercial and non-commercial) mainly has been taking place by road and has been noticeably increasing year by year. In particular, the middle-range and long-range transport is practiced on the highway network.

Since the energetic problem closely regards all the people and Italy, as well as European Union states, has important goals to pursue within 2020 in terms of energy conservation and reduction of CO_2 emissions in the atmosphere (ENEA 2014), why the possibility to transform such a mechanic energy into electricity should not be considered? Again, why not increasing the utilization of the road spaces for installing photovoltaic plants? The answer to these questions brought to explore the topic of the sustainable road infrastructure (Diamantini et al. 2011).

2 TECHNOLOGIES FOR ENERGY PRODUCTION FROM THE ROADS

2.1 Piezoelectric production

An interesting answer for the production of piezoelectric energy from roadways comes directly from an Israeli company, Innowatech, with its research offices located in the Institution of Technion, Haifa (Israel). The company has created a piezoelectric device suitable for producing electricity from the traffic-related movement of the road platform (Ronchi 2014). Using the principle of piezoelectricity, for which some materials with crystalline structure (named piezoelectric crystals) generate electric voltage from elastic deformations, researchers developed a product to be installed under the asphalt pavement (or inside of the railway sleepers), able to generate electricity from the transit of vehicles (or trains) and accumulate it in a battery system. According to the company, this technology is easy and fast to be implemented both in the case of construction of new roads and during the maintenance operations of the existing ones.

In particular, the vertical load by vehicle tires produces a compressive stress that proportionally decreases with depth (Vigo Majello 2013). For this reason, generators designed by Innowattech are placed in the road pavement to a depth of 5 centimeters, where stresses produced by vehicles are more intense (Fig. 1).

This solution, when applied to a highway, produces electricity as a function of the number of

Figure 1. Piezoelectric energy generator for road applications.

vehicles, of their weight and speed. Consequently, the greater is the traffic and the more convenient will be this solution.

The operational steps for the installation of piezoelectric generators are the following:

- Cutting of the pavement surface,
- Laydown of a quick-setting concrete,
- Positioning of the piezoelectric generators (30 × 30 cm) and drowning in quick-setting concrete,
- Connecting the cables,
- Overlaying the generators with an asphalt layer (to provide better adhesion between concrete and asphalt layer),
- Laydown of final asphalt wearing layer.

Once completed the installation, generators collect the mechanical energy produced by vehicles and convert it into electricity, which is then transferred in storage systems installed along the roadway, every 500 meters.

This system provides many advantages:

- It does not require the occupation of areas adjacent to the roadway,
- It is suitable to operate in any weather conditions,
- It requires little maintenance (replacement or repair after about 30 years),
- It allows the power supply for street lighting and/or vertical road signs.

Its application is suitable for multiple urban and suburban applications: local streets, roads characterized by heavy traffic but especially highways, railways and subways.

First tests by Innowattech dealt with the installation of piezoelectric nano-generators along a stretch of 10 meters in a road asphalt pavement. In this case the generators could potentially produce about 2 kWh, which means that one lane could generate about 200 kWh per km. This trial allowed to experimentally verify that the system works better when traffic is at least 600 vehicles/hour with an average speed of about 72 km/h.

The life cycle of a piezoelectric system is about thirty years, but actually the system is under testing and it is characterized by high implementation costs that could be reduced if a production in mass will be promoted (Edery-Azulay 2010, Kurzweilai 2011).

2.2 Photovoltaic systems

Nowadays, photovoltaic systems are commonly known; they can be distinguished in rooftop or ground-mounted systems and are able to convert solar energy into electricity. Among the multiple uses, there are very interesting photovoltaic panel applications on roadways. In particular, these panels can be integrated on noise barriers, shelters of service stations parking or even the road surface according to the project by the American engineer Scott Brusaw.

2.2.1 Photovoltaic noise barriers

Photovoltaic Noise Barriers (PVNB) can be considered not only as a noise isolation device, but also as a street furniture element; they are able to improve the visual impact (strictly linked to noise barriers installation), protect the surrounding area from the noise of the highway or train routes and concurrently produce clean and renewable energy. (Fig. 2).

The energy output of each linear meter of this type of barrier is on average 169 kWh per year. This value refers to an annual yield according to Enea Data for the province of Pisa, considering southern exposure of the panels (Redazione Ingegneri 2009).

In a traditional photovoltaic system costs are strongly related to the initial investment for its implementation; in particular, the construction of supporting structures, site purchase and preparation determine 10% of the above-mentioned costs. In this case, costs related to the use of the area and supporting structures are attributed to the noise barriers.

A considerable potential saving income of total costs is obtained with the standardization of the production phase. From a more general point of

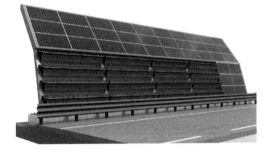

Figure 2. Photovoltaic Noise Barriers (PVNB).

view, a considerable saving for the electricity distribution network derives from the possibility to install energy production facilities near to the points where the energy is actually used, particularly if the daily production is close to the market demand. Moreover, PVNB should be placed close to urban area also for a proper acoustic protection performing.

In Germany and Switzerland numerous PVNB structures have been installed across different highways and railway tracks; the longest PVNB in the world is actually in Italy, along the Brenner highway (A22), and it consists of 3944 panels for an exposition surface of 5035 m² on a length of 1070 m, which generate 690 MWh of energy per year (Area Tecnica Autobrennero 2015).

2.2.2 Photovoltaic shelters

Photovoltaic shelters are structures covered with photovoltaic panels (Fig. 3), designed to cover car parks, bus service or storage areas (Massa 2015). In particular, photovoltaic shelters are used to protect vehicles against the weather and concurrently produce green electricity which makes them environmentally responsible. Thus, these sites, which are generally dedicated exclusively to parking, become solar electricity generating areas and their energy output could be used for equipment power supply (mobile phone charging, Wi-Fi access points, billboards lighting or even ventilation systems for the creation of refreshment areas).

A photovoltaic shelter, which has a surface of 30 m² for 3 kW of capacity and shades a parking place for 2 cars, annually produces 3610 kWh of electricity (Far Systems 2014). In general, this solution is easy to be installed because it does not require any structure construction and is adaptable to any soil. However, a project for panel installation is necessary in order to determine the proper orientation and dimensioning and achieve a better efficiency.

2.2.3 Photovoltaic panels as a road surface

As previously mentioned, the photovoltaic panel placed on the road surface is a pioneering idea by the American engineer Scott Brusaw who, supported by his working team, realized the "Solar Panel Road". This device is a photovoltaic panel able to convert solar energy into electricity and concurrently bear load and stresses caused by road traffic (Fig. 4).

The panel, designed to substitute the asphalt wearing course, is composed of three layers:

- Surface layer made of a rough glass, anti-abrasive, self-cleaning and highly resistant, which contains photovoltaic cells and led diodes; this layer has the main function to resist weathering and protect the electronic apparatus located underneath;
- Intermediate electronic layer, which contains a microprocessor for controlling and monitoring loadings and lighting;
- Bottom layer, which carries the energy collected by the intermediate layer to various storage systems connected to the roadway.

The produced energy is carried to storage systems located near the road surface every 12 m and can be directed to a primary network for satisfying various energy requirements (e.g. homes, street lighting and road signs, service stations).

The proceeds would be significant in terms of energy amount: it has been estimated that, for an average daily solar irradiation of 4 hours, each photovoltaic panel should be able to produce around 7.6 kWh per day. Considering a road section 1.6 km long and 10 m large, it could be possible to satisfy the energy requirement of about 500 families (Verdi Ambiente e Società 2014). However, maintenance procedures for dust accumulation, the duration of photovoltaic cells and the high costs still make the photovoltaic panel for road surfaces in need of improvements.

The Solar Roadways is currently being tested in a section of a highway (70 km long) located between Coeur D'Alene and Sandpoint in Idaho. Also in France, a photovoltaic pavement has been realized and named with the explicit term Wattway. The latter has been constructed by the National Institute of Solar Energy (INES) in cooperation with Colas

Figure 3. Photovoltaic shelters.

Figure 4. Photovoltaic panels as a road surface.

(company specialized in transport infrastructure). This construction represents an example of a unified concept of photovoltaic road surface: the Wattway slabs include photovoltaic cells made of polycrystalline silicon incorporated in a substrate few millimeters thick. At the side, the system is connected to a case that contains electronic security components. The slabs are antiskid, resistant, adaptable to every surface and have been designed to bear the traffic of all types of vehicles, including heavy trucks. The installation of these slabs is very easy: they can be directly glued on the existing pavement surface without any further constructions (Energie Rinnovabili 2015).

3 PRACTICAL APPLICATION

Bearing in mind what described so far, an example of application of such technologies has been hypothesized for the Italian highway A24 (section Roma-L'Aquila-Teramo, located in central Italy) and the potential energy performance has been evaluated (Fig. 5).

3.1 Piezoelectric system

As previously described, on the basis of tests carried out by the University of Israel in cooperation with Innowattech company, generators have the maximum productivity when they work on road sections with a traffic higher than 600 vehicles/h characterized by a speed of 72 km/h. With these conditions, the energy production is estimated around 200 kWh in 1 km per lane. Considering the traffic data provided by "Strada dei Parchi SpA", managing company of A24 Highway, a

theoretical estimation of the energy production per single primary segment (i.e. road segment between two tollbooths) of both lanes has been performed.

First, the evaluation of the average annual traffic per hour in each primary segment (reference year 2014) allowed the identification of all road segments with more than 600 vehicles/h (considered for the energy production estimation). Only the first three out of fourteen primary segments resulted suitable for the installation in the pavement structure of the piezoelectric device. They are comprised between the following tollbooths: Roma East—Connection A1/A24 (segment A); Connection A1/A24 – Tivoli (segment B); Tivoli—Castel Madama (segment C). As a consequence, only these three segments have been considered for the estimation of the energy production as illustrated in Table 1.

In particular, for calculating the energy produced by the piezoelectric device, the energy production has been considered proportional to the traffic volume. As reported in the table, the energy production value is higher than 200 kWh for 1 km of each segment. Finally, by multiplying the so calculated energy values by the length of each corresponding segment, the potential amount of piezoelectric energy produced in the A24 highway results equal to 7.7 MWh.

3.2 Anti-noise photovoltaic barrier

In this paragraph, an ideal anti-noise photovoltaic barrier plant is evaluated. Based on the orography of the territory where the highway is situated, only viaducts nearby existing towns were considered. As already described, in central Italy each meter of those barriers facing south can produce on average 169 kWh per year. Hence, considering that the total length of the barriers is 6390 m, the valuation of the production is given in Table 2. The total production results about 1081 MWh/year.

3.3 Photovoltaic shelters

This section evaluates the hypothesis of an installation of photovoltaic shelters in the parking lots situated along the A24 Highway. The calculation was based on the squared meters of parking lots that could be covered with shelters, equal to 13583 m^2 (Table 3). Considering that 30 m^2 of photovoltaic shelters can produce 3610 kWh per year, it is possible to estimate about 1634 MWh/year.

3.4 Photovoltaic panel for road pavements

The installation of photovoltaic panel for road pavements was hypothesized in the highway segment between L'Aquila West and L'Aquila East characterized by a total length of 4.1 km (without considering the length of the three tunnels). Such a

Figure 5.   A24 Highway.

Table 1. Energy produced by the piezoelectric device in the A24 Highway.

Sect.	Traffic vol. [vehicles/h]		Energy per km [kWh]		Length [km]	Tot. Energy [MWh]	
	Right lane	Left lane	Right lane	Left lane		Right lane	Left lane
A	1066	944	355	315	3.2	1.1	1.0
B	881	850	294	283	1.5	0,5	0.4
C	630	624	210	208	11.2	2.4	2.3
			SUM			4.0	3.7
			TOTAL			7.7	

Table 2. Energy produced by the anti-noise photo-voltaic barriers in the A24 Highway.

Viaduct name	Length [m]	Energy produced [MWh/year]
Licenza II	180	30
Roviano	175	30
Colle Alto	350	59
Peschieto 2	245	41
Peschieto 1	85	14
Valle Mura	270	46
Fiume Aterno	105	18
SS17	80	14
Fosse Vetoio	200	34
Pettino	485	82
San Sisto	1755	297
San Giacomo	390	66
Cerchiara	1280	216
Villa Ilii	790	134
TOTAL	6390	1081

Table 3. Summary of parking lots and service stations and energy produced by the photovoltaic shelters in the A24 Highway.

Type	Name	Surface [m²]	Tot. Energy [MWh/year]
Service station	Colle Tasso North	450	54.15
Service station	Colle Tasso South	1042	125.33
Parking lot	Roviano East	1675	201.56
Parking lot	Roviano West	3406	409.81
Service station	Civita North	798	96.03
Service station	Civita South	1436	173
Service station	Valle Aterno West	498	59.93
Service station	Valle Aterno East	582	70.03
Parking lot	Gran Sasso	3696	444.75
TOTAL		13583	1634.59

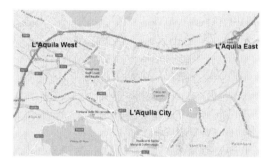

Figure 6. L'Aquila West—L'Aquila East Highway section (source: Google Maps).

project will be justified by the presence of the near city of L'Aquila that will directly use the energy produced (Fig. 6).

It was considered that the two separate carriageways (one for each direction) would be covered by photovoltaic panels; each carriageway width is 10 m (two lanes with the width of 3.5 m and one shoulder with the width of 3.0 m). Considering that a single panel with an area of 1 m² should be able to produce 7.6 KWh/day (based on an average daily irradiation of 4 hours), the selected segment (4100 m of total length and 20 m of total width) covered by photovoltaic panel should produce a total of 623.20 MWh/day.

4 CONCLUSION

From the result analysis it can be deduced that the use of photovoltaic systems can determine the production of significant amounts of energy. It is clear that the piezoelectric technology is also very promising, even if the benefits/costs ratio must be considered.

This study aimed to demonstrate the potentiality of a highway to become an energy generator for the territory that it crosses. The proposed hypotheses are purely theoretical and they have not been deepened on a proper scale to verify their applicability. However, the paper describes new technological systems for the next-generation design of the binomial infrastructure/territory, with the objective to provide a different approach and modify the timeworn concept of the highway as a pollution source.

Lastly the energy produced will be used as much as possible during its production, otherwise must be stored in appropriate batteries.

REFERENCES

Area Tecnica Autobrennero 2015. A22 Viaggia verso il sole verso il futuro—barriera antirumore fotovoltaica. URL autobrennero.it.

Cappelletti, G. 2011. L'autostrada come condotto energetico: scenari per la produzione di energia rinnovabile lungo la pedemontana lombarda. Graduation Thesis, Faculty of Architecture, Politecnico di Milano.

Diamantini, C., Scaglione, G., Rizzi, C., Staniscia, S., Ricci, M. & Cribari, V. 2011 Reinventing A22 Eco-boulevard. Verso infrastrutture osmotiche. XIV SIU Conference – 24/26 March.

Edery-Azulay, L. 2010. Innowattech: Harvesting Energy and Data. First International Symposium The Highway to Innovation, Israel national roads company Ltd., 1–3 November, Tel Aviv, Israel.

Enea 2014. Documento di predisposizione del Piano d'Azione italiano per l'Efficienza Energetica (PAEE2014).

Energie Rinnovabili 2015. Wattway—anche in Francia arriva la strada fotovoltaica. URL rinnovabili.it.

Far Systems SpA 2014. Pensilina Fotovoltaica "Boston". URL farsystems.it.

Kurzweilai 2011. Innowattech attach harvests mechanical energy from roadways. URL kurzweilai.net.

Massa, M.F. 2015. Pensiline fotovoltaiche: come alimentare auto elettriche, cellulari, pannelli pubblicitari con una pensilina fotovoltaica. URL fotovoltaicosulweb.it.

Redazione Ingegneri 2009. Barriere Antirumore Fotovoltaiche. URL ingegneri.cc.

Ronchi, F. 2014. Da Israele l'Asfalto che ricarica le batterie—Energia gratuita da strade e ferrovie. Scienza e Tecnologia. URL lacritica.org.

Verdi Ambiente e Società 2014. In USA pannelli solari al posto delle strade asfaltate. URL vasonlus.it

Vigo Majello, M.C. 2013. Sistemi innovativi per l'implementazione energetica del patrimonio edificato: dispositivi piezoelettrici applicati alle pavimentazioni. PhD Thesis. Università degli studi di Napoli "Federico II".

Transport Infrastructure and Systems – Dell'Acqua & Wegman (Eds)
© 2017 Taylor & Francis Group, London, ISBN 978-1-138-03009-1

Loss of life risk due to impacts of boulders on vehicles traveling along a very busy road

P. Budetta, G. Forte & M. Nappi
Department of Civil, Architectural and Environmental Engineering, University of Naples "Federico II", Naples, Italy

ABSTRACT: The paper is aimed to describe the used approach for calculating the risk along a road stretch belonging to a very busy coastal road in Southern Italy. During the time span 1969–2013, 22 rockfalls affecting this road were inventoried. On 18th February 2014 a new rockfall happened and several boulders reached the northern lane of the road. On the basis of collected data concerning the landslide hazard and road vulnerability, a procedure for the probability evaluation of a fatal accident—for a road user—is presented, discussed in details and compared. The analysis is meant to allow the design of appropriate protection devices along the cliffs overhanging the road.

1 INTRODUCTION

Rockfalls on transportation corridors can cause casualties mainly due to direct impact on vehicles or by impact of vehicles with deposed material, as well as a large amount of economic consequences due to traffic interruptions. The rockfall risk evaluation affecting roads is a highly complex operation requiring the assessment of the hazard (triggering mechanisms and the run out parameters) and the vulnerability of vehicles on the roads along the foothills. Vulnerability mainly depends on the vehicle speed and length, the available decision sight distance, the traffic volume, the length of the rockfall risk section of the route, the number of occupants in a vehicle, and the type of vehicle. In quantitative terms, the annual probability of loss of life to an individual is given by multiplying the annual probability of occurrence of a rockfall event (of a given magnitude) by the probability of a falling rock hitting the moving vehicle, and by the vulnerability of the person given a block of size m impacting the vehicle. In order to apply this approach, the number of boulders that may hit the road must be obtained through trajectory simulations by calculating the percentage of all trajectories that could fall on the road or that are not interfering with it.

In the literature several methods concerning the Quantitative Risk Analysis (QRA) affecting roads have been proposed (Bunce et al. 1997; Hungr et al. 1999; Budetta 2002; Peila & Guardini 2008; Ferlisi et al. 2012; Mignelli et al. 2012; Budetta et al., in press). Here, we show results obtained by means of application of the ROckfall risk MAnagement (RO.MA.) method by Peila & Guardini (2008), concerning a very busy road affected by recurrent rockfalls.

Figure 1 presents a schematic operational flowchart of the RO.MA. approach.

This method develops through five steps, including (Figure 1): identification of unstable areas and the number of rocks per year that may hit the road (N_r); road vulnerability assessment; event tree analysis; risk assessment; evaluation of the efficiency of rockfall protective measures and calculation of the residual risk (Peila & Guardini 2008; Mignelli et al. 2012). The number of boulders that may hit the road or that may stop upstream (N_s), is obtained from field data or, alternatively, through trajectory simulations (Jaboyedoff et al. 2005). For the vulnerability evaluation it is need to have data concerning the length of the hazardous road stretch (L_r), the average (or limit) speed of the vehicles (V_v), the average vehicle length (L_v), and the number of vehicles travelling on the road per hour (N_v). The complete sequence of events which may result from a rockfall up to the killing of a road user (fatal accident) is evaluated by means of the event tree analysis. This analysis develops along twelve different paths (Figure 1A) and the probability of occurrence of each of them can be calculated from the product of each single event that constitutes the path itself. The final value of each path is given by the sum of probabilities concerning paths with the same final result (i.e. fatal accident, non-fatal accident, and no accident).

We partially modify the approach given by Peila & Guardini (2008), to calculate the spatio-temporal probability (P_{ST}) for a vehicle to be in the path of the

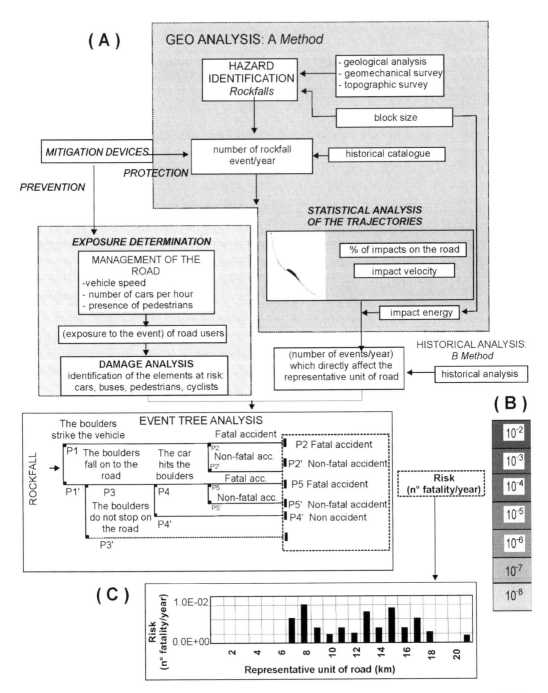

Figure 1. Procedural flow chart for ROckfall risk MAnagement assessment (A); Abacus defining the threshold values of rockfall risk. Acceptable values are lower than 10^{-5}, the ALARP band is 10^{-5}, unacceptable risk values are upper than 10^{-5} (B); histogram representing risk values versus the representative road sections (C). After Peila & Guardini (2008) and Mignelli et al. (2012) modified.

falling boulder, when the mass falls, using both the dimension of the falling block (*We*) and the length of the vehicle (*Lv*) (Nicolet et al. 2015) as follows:

$$P_{ST} = \frac{N_V x (W_e + L_v)}{V_v}$$

(1)

where P_{ST} = spatio-temporal probability; N_v = number of vehicles travelling on the road per hour; We = dimension of the falling block; Lv = length of the vehicle; V_v = average speed of the vehicles.

The probability of a fatal accident due to a direct impact of a boulder on a moving vehicle was calculated according to Bunce et al. (1997) whereas considering the impact of a block on the road surface, the probability that a travelling vehicle has an accident due to the damaged road surface was evaluated as a function of the deposited volume because generally volumes more than 0.1 m³ may cause deep potholes (Budetta et al., in press).

Another reason for which we preferred the suggested approach by Peila & Guardini (2008) is given by the possibility to evaluate the ability of installed protection measures along the cliffs (e.g. rockfall barriers, reinforced wire rope nets, mesh drapes, etc.) to stop the falling rocks, reducing the number of those that may hit the road. The number of retained blocks (N'_r) can be calculated on the basis of the catching capacities of these devices, by means of the percentage of rocks (C) that can be stopped by the protection device. N'_r is given by:

$$N'_r = (1 - C) N_r \qquad (2)$$

In this way, we can compare calculated risk values (without protection measures) with those obtained taking into account the efficiency of these protections, or with risk criteria defined in the international literature. In this respect, Mignelli et al. (2012) suggested the use of an abacus defining the threshold values of acceptable and unacceptable rockfall risks (Figure 1B).

The above-mentioned approach, which has been partially modified with respect to the original one by Peila & Guardini (2008), was applied to a 350 m-long road stretch belonging to a coastal road that links the Vietri sul Mare resort with Salerno.

2 GEOMORPHOLOGICAL AND GEOLOGICAL SETTING

The studied road crosses a coastal area characterized by high reliefs lying near the harbor of Salerno (Fig. 2). Given the near vertical cliffs and very steep slopes, in a short distance from the coast, the relief goes from the sea level to heights greater than 700 m ASL. On slopes flanking the road, heavily fractured dolomites rocks outcrop. Some low-angle normal faults and several pervasive mutually intersecting joints affect the rock mass. These joints identify wedges whose lines of intersection are almost normal to the cliff face. Other randomly oriented deep tension cracks are caused by tensile stresses along cliffs. The intersections of these

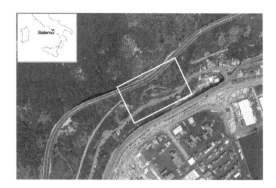

Figure 2. The harbor of Salerno and the surrounding hills. The white rectangle shows the study area.

Figure 3. The rockfall occurred on 18th February 2014.

joints with the cross-bedded dolomitic layers result in isolated, potentially unstable boulders. Almost vertical slopes contribute to the free fall of boulders on the road, whereas in the remaining cases irregular rock faces, due to the presence of ridges or benches with lower slopes, cause launching and rebounding phenomena.

Along the entire road stretch linking Vietri sul Mare with Salerno, between 1969 and 2013, 22 rockfalls were inventoried based on the information in the Italian Landslide Inventories (AVI and IFFI Catalogues), historical documents, and newspapers. Some rockfalls of about several cubic metres caused prolonged traffic interruptions. About 13 rockfalls occurred during the period 1993–1997, whereas only five landslide events are reported in the subsequent period up to 2013. Likely, this is due to passive devices (wire nets) which, during that time, were installed along cliffs above the road. Despite the considerable number of recorded rockfalls, there is incomplete information regarding the magnitudes as well trajectories and stopping points. On 18th February 2014, a new rockfall of about 40 m³ affected the road at the kilometer 51+300 (Figure 3).

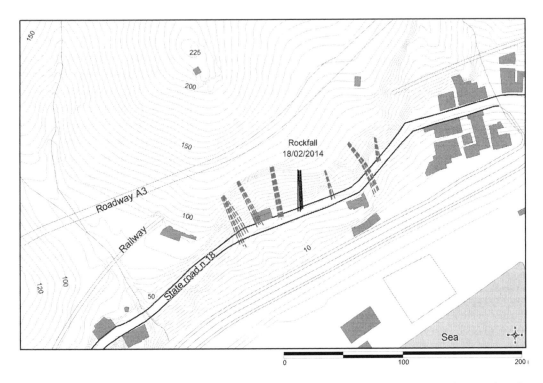

Figure 4. The trajectories that could stop on the road or that are going beyond it. With the black lines, we show the real trajectories traveled by the boulders of the rockfall of 2014. With the dotted gray lines, we show trajectory simulations obtained using kinematic parameters inferred by means of the back-analysis of the rockfall. Only trajectories that may interfere with the road are shown.

Several boulders falling from the overhanging cliff reached the northern lane of the road (in the direction towards Vietri sul Mare) threatening an adjacent petrol station and causing a new prolonged traffic interruption. The average volume of the deposed boulders on the road was evaluated about 65 dm^3. It is worth to observe that the collapsed cliff was already protected by wire nets which, evidently, were unfit to arrest the blocks.

3 TRAJECTORY SIMULATIONS

On the basis of deposed block volumes as well run-out distances and cliff geometries, several trajectory simulations were performed using a 3D approach in order to obtain more suitable energy restitution (normal, R_N and tangential, R_T) and rolling (friction angle, ϕ) coefficients as well kinetic energies, velocities, and bounce-height along trajectories travelled by boulders. In such a way, a detailed back-analysis of the rockfall of 2014 was performed. Trajectory simulations were performed using a three-dimensional code (AZTECROCK 10.0, by Aztec Informatica Inc. 2012) designed to analyze topographic spatial models by means of deterministic and/or probabilistic (by a Monte Carlo sampling technique that uses a normal distribution) approaches. In this code the topographic surface is generated using the Delaunay triangulation method. As the program uses a lumped-mass method, possible boulder fragmentation during the fall is not considered.

Afterward, using kinematic parameters obtained from the back-analysis, a hazard scenario concerning probable trajectories arising from other potentially unstable rockfall sources, has been prepared for falling blocks with volumes equal to 65 dm^3 (Figure 4).

The rockfall sources have been carefully chosen on the basis of the morphology of the area and the starting point for boulders were conventionally located at the highest point of each source (in most cases corresponding with the cliff top). Each simulation consisted of releasing 10,000 blocks from any potential rockfall source. In order to facilitate the reading of the map, in Figure 4 only boulder trajectories that could stop on the road or that are going beyond it have been shown. In such a way, along the total length of the road potentially prone to rockfalls (about 350 m) it is possible to measure the lengths of sections exposed to boulder impacts and more precisely calculate the probability that a moving vehicle is located in the section covered by the event when it happen (Figure 4). Excluding the effective length of the road section affected by blocks coming from the rockfall of 2014 (about 5.0 m), other calculated potential lengths range between 9.0 and 3.6 metres. It is worth noting that, at this stage, the effect of installed protection devices (wire nets) flanking the road that are able to stop and/or alter boulder trajectories, is not taken into account.

4 FATAL ACCIDENT RISK CALCULATION

In order to calculate the risk, expressed as the annual probability of a fatal accident for a road user, vulnerability concerning this stretch of the road firstly was evaluated. On the basis of traffic data furnished by the Municipality of Salerno, a mean hourly traffic density (number of vehicles travelling on the road per hour – N_v) of 4500 was established. The mean velocity (V_v) and length (L_v) of vehicles were 50 km h^{-1} and 4.5 m, respectively. The mean block diameter (W_e) inferred from field surveys was 0.5 m. Regarding other hazard data needed for risk calculation, on the basis of trajectory simulations the number of rocks hitting the road per year (N_r) was assumed equal to 1.15. According to eq. 1, the probability that a vehicle travelling on the road is hit by a falling rock (P_{ST}) is 1.87×10^{-2}. For simplicity, we assume that throughout the year the traffic is constant.

Associated with the 12 paths of the event tree (Figure 1A), the fatal accident probability value was calculated for the above-mentioned hazard scenario, using the approach by Peila & Guardini (2008). Unlike the original method, we used the calculated P_{ST} value because neglecting the dimension of the block or the dimension of the vehicle might lead to inexact results (Nicolet et al. 2015).

According to Bunce et al. (1997), probability values of a fatal accident due to moving vehicle/ falling rock and moving vehicle/fallen rock interactions (Figure 1A) of 0.2 and 0.1 were assumed, respectively. Furthermore, the probability of a fatal accident due to an impact of a falling rock on a stationary vehicle was assumed equal to 0.125 (Bunce et al. 1997). Finally, the probability of serious damages on the road surface and consequent fatal accident, due to the damaged road surface, was 0.7. This value assesses the likelihood that a pothole produced by a falling and rebounding block on the asphalt is more than the wheel diameter. In other words, 70% of rebounding blocks,

with volumes of 65 dm^3, result in dangerous potholes that may cause serious damage to vehicles (Budetta et al., in press).

In order to perform repeated and complex calculations, an Excel spreadsheet was prepared that uses the above-mentioned input data. In such a way, it was possible to obtain probability values concerning the 12 paths of the event tree. Finally, by the sum of values of identical outcomes, probability values of a fatal accident, non-fatal accident, and no accident were calculated.

As the rockfall of 2014 already proved that the existing wire nets were unable to retain block volumes of 65 dm^3, it was not considered necessary to recalculate N_r on the basis of the percentage of rocks (C) that can be stopped by this protection device.

Finally, for the entire road stretch potentially prone to rockfalls, the annual risk of fatal accident for a road user is 4.52×10^{-3}. According to the abacus defining values of acceptable and unacceptable rockfall risks (Figure 1C) suggested by Mignelli et al. (2012), the individual risk is not acceptable, and some actions are required in order to lower it.

5 CONCLUSIONS

It is useful to compare the fatal accident risk value calculated by means of the modified RO.MA. method with those concerning all car accidents resulting in loss of life in Campania (the region where the studied area is located). During the time span 1996–2008, available data from the Italian Institute for Statistics (ISTAT 2014) shows a mean value of about 330 fatalities/year or 3.41×10^{-2} fatalities year^{-1} km^{-1}: the length of the entire Campania road network (motorways, national and provincial roads) being about 9652 kilometres. Another comparison can be done with the fatal risk caused by rockfalls on the Amalfitana road since its construction, in the second half of the 19th century (1.65×10^{-4}) (Budetta et al., in press). This state road crosses an area characterized by a geomorphological and geological layout similar to the studied one. The rockfall risk value calculated for the study area is also above the acceptability limit defined for "involuntary" risk such as rockfalls, as proposed by Geotechnical Engineering Office Hong Kong (1998). This means that the individual risk is not acceptable, and more effective countermeasures are required such as rockfall barriers or reinforced wire rope nets.

The rockfall hazard evaluation and the assessment of the risk performed by means of the above-mentioned approach are affected by uncertainties and limitations that we must bear in mind in order to use the results for risk mitigation and planning purposes. The quality of the rockfall hazard scenario depends on the preciseness of trajectory simulations and the accurate identification of rockfall sources.

With reference to the risk analysis, major difficulties concern the assessment of more exact probability values, which must be assigned to interactions between rocks and vehicles. Sometimes the definition of these values was heuristic, and to some extent arbitrary.

Even though this quantitative risk analysis must be improved, the applied methodology demonstrates that it is possible to perform a good assessment of the expected individual loss of life, since it is based on the essential elements concerning the rockfall hazard evaluation and road vulnerability.

REFERENCES

Aztec Informatica Inc.: AZTECROCK v. 10.0 2010. Caduta massi. Casole Bruzio, Cosenza, Italy.

Budetta, P. 2002. Risk assessment from debris flows in pyroclastic deposits along a motorway, Italy. *Bull Eng Geol Env* 61:293–301.

Budetta, P., De Luca, C. & Nappi, M. In press. Quantitative rockfall risk assessment for an important road by means of the rockfall risk management (RO.MA.) method. *Bull Eng Geol Environ*. DOI 10.1007/s10064-015-0798-6.

Bunce, C.M., Cruden, D.M. & Morgenstern, N.R. 1997. Assessment of the hazard from rockfall on a highway. *Can Geotech J* 34:344–356.

Ferlisi, S., Cascini, L., Corominas. J. & Matano, F. 2012. Rockfall risk assessment to persons travelling in vehicles along a road: the case study of the Amalfi coastal road (southern Italy). *Nat Hazards* 62:691–721.

Geotechnical Engineering Office 1998. Landslides and boulder falls from natural terrain: interim risk guidelines. GEO report no. 75. Geotechnical Engineering Office, The Government of the Hong Kong Special Administrative Region: pp 183.

Hungr, O., Evans, S.G. & Hazzard, J. 1999. Magnitude and frequency of rock falls and rock slides along the main transportation corridors of south-western British Columbia. *Can Geotech J* 36:224–238.

Italian Institute for Statistics ISTAT 2014. Anno 2013 incidenti stradali in Campania. Statistiche focus. http://www.istat.it. Accessed 18 June 2015 (in Italian).

Jaboyedoff, M., Dudt, J.P. & Labiouse, V. 2005. An attempt to refine rockfall hazard zoning based on the kinetic energy, frequency and fragmentation degree. *Nat Hazards Earth Syst Sci* 5:621–632.

Mignelli, C., Lo Russo, S. & Peila, D. 2012. Rockfall risk management assessment: the RO.MA. approach. *Nat Hazards* 62:1109–1123.

Nicolet, P., Jaboyedoff, M., Cloutier, C., Crosta, G.B. & Lévy, S. 2015. Brief communication: on direct impact probability of landslides on vehicles. *Nat. hazards Earth Syst Sci Discuss*. DOI 10.5194/nhessd-3-7311-2015.

Peila. D. & Guardini, C. 2008. Use of the event tree to assess the risk reduction obtained from rockfall protection devices. *Nat Hazards Earth Syst Sci* 8:1441–1450.

Transport Infrastructure and Systems – Dell'Acqua & Wegman (Eds)
© 2017 Taylor & Francis Group, London, ISBN 978-1-138-03009-1

A big data approach to assess the influence of road pavement condition on truck fleet fuel consumption

F. Perrotta, T. Parry & L. Neves
Faculty of Engineering, Nottingham Transportation Engineering Centre, University of Nottingham, Nottingham, UK

ABSTRACT: In Europe, the road network is the most extensive and valuable infrastructure asset. In England, for example, its value has been estimated at around £344 billion and every year the government spends approximately £4 billion on highway maintenance (House of Commons, 2011).

Fuel efficiency depends on a wide range of factors, including vehicle characteristics, road geometry, driving pattern and pavement condition. The latter has been addressed, in the past, by many studies showing that a smoother pavement improves vehicle fuel efficiency. A recent study estimated that road roughness affects around 5% of fuel consumption (Zaabar & Chatti, 2010). However, previous studies were based on experiments using few instrumented vehicles, tested under controlled conditions (e.g. steady speed, no gradient etc.) on selected test sections. For this reason, the impact of pavement condition on vehicle fleet fuel economy, under real driving conditions, at network level still remains to be verified.

A 2% improvement in fuel efficiency would mean that up to about 720 million litres of fuel (~£1 billion) could be saved every year in the UK. It means that maintaining roads in better condition could lead to cost savings and reduction of greenhouse gas emissions.

Modern trucks use many sensors, installed as standard, to measure data on a wide range of parameters including fuel consumption. This data is mostly used to inform fleet managers about maintenance and driver training requirements. In the present work, a 'Big Data' approach is used to estimate the impact of road surface conditions on truck fleet fuel economy for many trucks along a motorway in England. Assessing the impact of pavement conditions on fuel consumption at truck fleet and road network level would be useful for road authorities, helping them prioritise maintenance and design decisions.

1 INTRODUCTION

One of the most important challenges that the world is facing nowadays is the reduction of Greenhouse Gas (GHG) emissions. This problem has been highlighted in the road transport industry and by road agencies asking themselves how they can play a role in this. Agency decisions influence design, construction and maintenance of roads, but since European road networks are fairly complete, one major impact of agencies is in reducing the emissions of GHGs through adequate maintenance policies.

In England—as in other countries across Europe—the road network is the most extensive and valuable infrastructure asset. Recent estimates assess that in England its value is about £344 billion and every year the government spends approximately £4 billion for maintaining highways (House of Commons, 2011).

Although the transport industry is one of the major contributors to the economic growth of a country, recent estimates assessed that a quarter of the total energy consumption in Europe is used in transportation, with more than 80% of it consumed by road vehicles (Haider et al. 2011). Due to continuous reliance on fuel oil derivatives, road transport is an important cause of energy consumption and GHG emissions.

Fuel economy is an essential part of the Life-Cycle Assessment (LCA) analysis of road pavements but frequently not taken into account due to inadequacy of existing models (Zaabar and Chatti, 2010) and lack of a standard and widely accepted methodology (Santero, Harvey and Horvath 2011). Therefore, it is difficult to conduct a complete and reliable assessment.

Fuel consumption is influenced by vehicle technology, road geometry, weather conditions, vehicle speed, and pavement condition among other factors (Beuving et al., 2004; Zaabar and Chatti, 2010; and Haider et al. 2011). Sandberg (1990) and Laganier and Lucas (1990), among others, have concluded that rougher roads lead to higher fuel consumption. In particular the impact of road roughness on fuel consumption has been assessed at approximately 5% (Zaabar and Chatti 2010) of the whole vehicle fuel consumption. This means that well designed and well maintained roads can influence road vehicle fuel economy, reducing their impact on the environment.

Looking at the latest available statistics it is possible to see that about 36 billion litres of fuel are used every year by road vehicles circulating in England (Department for Transport, 2015), so road condition accounts for around £1 billion every year (depending on the price of fuel). This is equivalent to 25% of the total annual investment in road maintenance on highways in England. Maintaining some roads in a better condition could therefore lead to environmental mitigation and overall cost savings.

However, conclusions of previous studies of the impact of road condition on fuel consumption cannot be considered completely exhaustive. In fact previous studies consider just a few vehicles driven on a few selected road segments, in specific road conditions, with specific road geometry. Because of this, the impact of road conditions on vehicle fuel consumption at network level under real driving conditions still remains unclear.

In the era of 'Big Data', large amounts of data are collected every day and this includes for road vehicles. Modern trucks are equipped with a wide range of sensors defined in SAE J1939 (SAE International 2002). These sensors collect data about vehicle performance (fuel consumption, vehicle speed and other factors) and they are usually analysed to help truck fleet managers in their truck maintenance and driver training decision making processes. At the same time, every year, road agencies collect data on road condition (roughness, skid resistance, texture, and stiffness among other parameters).

The objective of this work is to compare truck fuel consumption and road condition data using a Big Data approach. This will lead to the definition of more accurate models for the LCA of pavements. Using this approach, almost real time data can be used and any generated model can be progressively updated with time, following improvements in vehicle and road technology. This paper reports some initial results exploring the feasibility of this approach.

2 AIM AND OBJECTIVES

The main aim of this study is to investigate the feasibility of using a Big Data approach to assess the impact of road pavement conditions on truck fleet fuel consumption at road network level under real driving conditions.

3 DATA

A limitation of previous studies of vehicle fuel consumption and road conditions is that measurements have been taken under carefully controlled conditions, on flat roads, testing only a few vehicles driving on a few selected road segments. For this study data from 43 trucks driving for one year over various segments of the M18 motorway in England are considered. 1970 data points for these articulated trucks equipped with 12,419cc, euro 5 engine, driving on M18 in 2015 at 85 (±2.5) km/h using gear 12 are available. These data were chosen for this initial study from hundreds of thousands of records, to remove the influence of truck type, speed and gear from the analysis.

This and other sensor data are collected daily and analysed to assist truck fleet managers in decision making about vehicle maintenance and driver training.

Each record contains data recorded at an 'event' during the journey of each truck (license number and tracker ID define the specific truck). At any brake, stop or anomaly in vehicle performance etc., or each 2 minutes (120 s) or 2 miles (~3,219 m), an event is generated and a data record taken.

Each vehicle's performance database contains hundreds of gigabytes of data but not every record is stored or is useful for the data analysis. The data accessed during this study are:

- the vehicle profile, identifying the vehicle,
- the tracker ID, reference for the system of sensors installed on the vehicle,
- the geographical position of the truck, given in the British National Grid (latitude and longitude) coordinate system (5 metres GPS precision),
- the distance travelled by the vehicle since the previous event (metres),
- the time spent by the vehicle to travel to the current position from the previous event (seconds),
- the total fuel consumed until the current event is recorded (0.001 litres precision, rounded to 0.1 litres for the purpose of reducing database size),
- the air temperature (0.1°C precision),
- the current gear,
- the current engine torque percentage,
- the engine revolutions (revs/min).

HAPMS (the Highway Asset Performance Management System) is a database owned by Highways England (the agency) which contains all data collected during annual surveys of the condition of the road network. The database contains historical information about each road of the network, including:

- a road identifier code,
- a direction code,
- the year of construction,
- the latest date of significant maintenance,
- the construction materials,
- roughness measurements,

- texture measurement,
- skid resistance measurement,
- deflection measurement (not considered in this study).

Each record refers to a 10 metre segment of road. Data are linked to their geographical position by coordinates (in WGS84 coordinate system). Because of the different coordinate systems used in data collection for vehicle performance and road pavement condition, the coordinates of vehicle performance are transformed to the WGS84 coordinate reference system in order to compare the data.

In this initial study only information about road surface condition is taken from HAPMS. This includes measurements of roughness (Longitudinal Profile Variance, LPV, at 3 and 10 metres in mm^2), texture (sensor-measured texture depth in mm), skid resistance (SCRIM Coefficient), and road gradient (0.1% resolution). Although previous studies used measurements like IRI (International Roughness Index) and MPD (Mean Profile Depth) as roughness and texture measurements they are similar to the LPV and texture measurements used in this study. Road surface roughness and texture have been shown to have a significant effect on fuel consumption in previous experimental studies (Sandberg, 1990, Laganier and Lucas, 1990, du Plessis, Visser and Curtayne, 1990, Beuving et al., 2004, Zaabar and Chatti, 2010, Haider et al., 2011, Zaabar and Chatti, 2012).

4 METHOD

A Big Data approach has been taken in this study. 910,591 events have been recorded and are available in total from trucks driven on M18 in 2015.

In order to limit the variables at this initial stage, the data were filtered for truck model, vehicle speed, engine type, road geometry, etc.

In the analysis reported here the following subset of data was used where the records show:

- records for which the trigger event was the default time or distance (i.e. no other driving event (e.g. harsh braking or cornering) triggered the record),
- an average speed of 85 km/h
- the initial and final speed were similar to the average speed (calculated as the travelled distance divided by the travelled time) and do not differ by more than 2.5 km/h (this assumption makes sure that the vehicle speed is steady between the two considered events),
- no gear change.

No data about the truck payload is currently available. Applying these filters, all the data at

steady speed, for a certain type of vehicle can be separately considered and analysed.

In this study only data for 3-axle tractor with 3-axle trailer articulated trucks at 85 km/h using gear 12, equipped with 12,419cc euro 5 engines are considered. This has been done in order to avoid the influence of speed, engine performance and vehicle type, and results in 1970 records.

5 RESULTS

Comparing the vehicle fuel consumption (as litres per 100 kilometres) and the road condition, a predictive model has been generated. Data about the engine torque percentage (as a substitute for the vehicle payload), the road gradient, LPV at 3 and 10 metres wavelength and texture measurements are considered. First each variable has been related to fuel consumption separately in order to estimate the correlation between each single variable and fuel economy.

$$FC = 25.46 + 0.12T\% \tag{1}$$

$$FC = 26.15 + 7.18g\% \tag{2}$$

$$FC = 28.36 + 3.14LPV03 \tag{3}$$

$$FC = 28.46 + 0.53LPV10m \tag{4}$$

$$FC = 26.17 + 1.80t \tag{5}$$

where, FC = predicted fuel consumption [l/100 km]; $T\%$ = engine torque percentage on maximum available [%]; $g\%$ = road gradient [%]; $LPV03$ = Longitudinal Profile Variance at 3 m wavelength [mm^2]; $LPV10$ = Longitudinal Profile Variance at 10 m wavelength [mm^2]; t = texture depth [mm].

Based on the literature review the payload and the gradient are two of the most influential variables (Beuving et al., 2004). However, as already mentioned, no data about the payload is currently available for this study. Therefore the torque generated by the engine for moving the vehicle (moving a certain payload) is used instead.

The relationship between the fuel consumption and the road gradient is shown in Figure 1.

Including only two variables, $T\%$ and $g\%$ in the model we obtain:

$$FC = 24.34 + 0.068T\% + 6.65g\% \tag{6}$$

Then, performing a backward analysis based on the Aikake Information Criterion (AIC, Aikake, 1973) the profile variance and texture data have been included in the model. Among all the models generated the one that shows the lowest AIC coefficient is:

Scatter plot for Road Gradient and Fuel Consumption

referred to 1970 travels at 85 km/h avg speed, for trucks with 3 AXLE + 3 AXLE ARTIC and engine 12419 euro 5 using gear 12

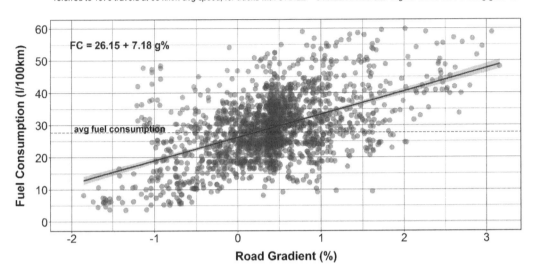

Figure 1. Plot of the direct impact of road gradient [%] on truck fleet fuel consumption [l/100 km] for articulated trucks driven at 85 km/h equipped with 12,419cc euro 5 engine.

Scatter plot for Predicted Fuel Consumption and Real Measurements

referred to 1970 travels at 85 km/h avg speed, for trucks with 3 AXLE + 3 AXLE ARTIC and engine 12419 euro 5 using gear 12

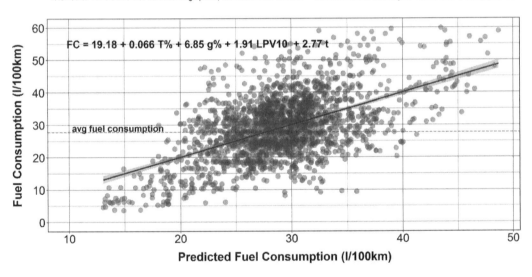

Figure 2. Plot of the predicted value of fuel consumption [l/100 km] with the real measurements for articulated trucks driven at 85 km/h equipped with 12,419cc euro 5 engine.

$$FC = 19.18 + 0.066T\% + 6.85g\% + 1.91LPV10 + 2.77t \tag{7}$$

The correlation coefficient (r) between the predicted and the measured fuel consumption is 0.54. The correlation between the predicted

and the measured fuel consumption is shown in Figure 2.

6 DISCUSSION

The generated model considers a limited number of road condition variables (avoiding overfitting) including those ones that previous studies considered to be the most influential on fuel economy. It assesses the impact of road roughness on fuel consumption to be up to 4.1% and up to 1.25% for the texture. Zaabar and Chatti (2010 and 2012) estimated the impact of roughness on fuel consumption at 5% and our first estimate from this initial work lies slightly below that.

Beuving (2004) reported that for a truck driving at 80 km/h the impact of other variables such as the vehicle mass and the gradient of the road is about 40% of the entire fuel consumption. The model estimates that the impact of torque (payload) and gradient measured for this data set is about 41.6%. Compared to experimental studies this gives confidence in this Big Data approach.

Although the results match what was found previously by other authors in experimental studies, they cannot be considered valid as yet. In fact, in this initial study only specific conditions were selected: only one vehicle model was considered, driven at steady speed, 85 km/h, etc. However, previous studies include many assumptions which may not be valid at network level under real driving conditions. Previous studies in fact considered only few vehicles tested at selected steady speeds on about ten flat road sections. They base their conclusions on a significantly lower amount of data without considering the impact of the considered variables in real driving conditions.

The correlation coefficient between the measured fuel consumption and the predicted value calculated using the generated model is 0.54. This is a low level of correlation compared to the more controlled experimental studies in the literature but is the result of analysing 1970 data records for 2 mile long sections over the majority of the 42.6 km long motorway. Models derived in this way represent real driving conditions (albeit at constant speed and gear in this initial study) for many trucks and over a route length. For these reasons they may be considered more representative of real world fuel consumption than the limited experimental studies. This demonstrates the strength and weakness of the Big Data approach. Although techniques for reducing the noise included in the data exist (and will be implemented soon) the lack of precision in the model estimate is an inevitable consequence of using this approach. Further research using this approach is underway. A better estimate of payload is being developed using a mechanical model based on the engine performance using further sensor data. Also, by considering single trips performed by individual constantly loaded trucks, it might be possible to remove the effect of payload on the fuel consumption estimates. Another potential solution is to identify from position and time data, when and where trucks discharge their loads

and consider the fuel consumption of the unladen vehicles for the rest of their journey. It may also be possible to collect data records more frequently and to record the fuel consumption more precisely for a limited number of trucks over a limited time period but this approach will be limited by the size of the generated data set and hence cost. Furthermore, it may be possible to add meteorological data (e.g. wind speed and direction) but because this data is routinely reported only over extended time periods (e.g. average per hour), it may be of limited use.

It is anticipated that including more reliable data using these approaches will improve the generated model, which can then be extended to further types of truck and speeds and a wider road network.

These models of the influence of road condition on fuel consumption can be used to validate those from experimental studies and extend them to truck fleet and road network level. This will allow more confidence in the use of the results in road maintenance decision making, for instance to include vehicle fuel cost and GHG emissions in determining maintenance strategies.

ACKNOWLEDGMENTS

The authors would like to thank Alex Tam (Highways England) for giving permission to use data from the HAPMS database, Mohammad Mesgarpour and Ian Dickinson (Microlise Ltd) for allowing us to use an anonymized part of their truck telematics database and Emma Benbow, David Peeling and Helen Viner (TRL Ltd) for their help in the interpretation of results and for their support in this initial part of the research.

 This project has received funding from the European Union Horizon 2020 research and innovation programme under the Marie Sklodowska-Curie grant agreement No. 642453 and it is part of the Training in Reducing Uncertainty in Structural Safety project (TRUSS Innovative Training Network, www.trussitn.eu).

REFERENCES

Akaike, H. 1973. Information theory and an extension of the maximum likelihood principle. In Second International Symposium on Information Theory, ed. B. N. Petrov and F. Csaki, 267–281. Budapest: Akailseoniai–Kiudo.

Beuving, E., De Jonghe, T., Goos, D., Lindhal, T., and Stawiarski, A., 2004. Environmental Impacts and Fuel Efficiency of Road Pavements. Industry report. Eurobitume & EAPA Brussels.

Chatti, K. & Zaabar, I., 2012. Estimating the Effects of Pavement Condition on Vehicle Operating Costs, National Cooperative Highway Research Program, Report 720. Washington, DC.

Du Plessis, H.W., Visser, A.T., and Curtayne P.C. 1990. Fuel Consumption of Vehicles as Affected by Road-Surface Characteristics. ASTM STP 1031: 480–496.

Haider, M., Conter, M., and Glaeser, K.P. 2011. Discussion paper what are rolling resistance and other influencing parameters on energy consumption in road transport, Models for Rolling Resistance in Road Infrastructure Asset Management Systems (MIRIAM), AIT, Austria.

House of Commons, 2011. House of Commons Committee of Public Accounts Departmental Business Planning.

Laganier R. & Lucas J. 1990. The Influence of Pavement Evenness and Macrotexture on Fuel Consumption. ASTM STP 1031 pp. 454–459, USA 1990. SAE International, 2002. *Surface Vehicle Recommended Practice*, 4970: 724–776.

SAE International 2002. Vehicle Application Layer—J1939-71—Surface Vehicle Recommended Practice Rev. Aug. 2002.

Sandberg, Ulf S. I. 1990. Road Macro-and Megatexture Influence on Fuel Consumption. ASTM STP 1031: 460–479.

Santero, N.J., Harvey, J. & Horvath, A., 2011. Environmental policy for long-life pavements. *Transportation Research Part D: Transport and Environment*, 16(2): 129–136.

Zaabar, I. & Chatti, K., 2010. Calibration of HDM-4 models for estimating the effect of pavement roughness on fuel consumption for U.S. conditions. *Transportation Research Record*, 2155: 105–116.

Zaniewski, J.P., 1989. Effect of Pavement Surface Type on Fuel Consumption. Portland Cement Association—Research & Development Information. Special report. FHWA.

Transparency and good governance as success factors in public private partnerships

C. Queiroz
Claret Consulting LLC, Washington, DC, USA

P. Reddel
Brisbane, Australia

ABSTRACT: Many governments do not have all the financial resources required to meet their country's infrastructure needs. Facing limited resources and searching for more efficient and effective delivery of their infrastructure, many countries have involved the private sector under long term Public-Private Partnership (PPP) contractual arrangements. Typically, PPPs involve a private sector company or consortium (the 'concessionaire') contracting with government to finance, build, operate, and then transfer the asset back to the public sector at the end of a concession. On the premise that good governance can lead to more successful PPPs, this paper reviews some key requirements for good governance in such projects, including competitive selection of the concessionaire, public disclosure of relevant information, and regulatory oversight of the concession contract. Because PPPs in infrastructure tend to have monopolistic features, good governance in managing them is essential to ensure that the private sector's involvement yields the optimum benefit for society.

1   INTRODUCTION

Many governments do not have all the financial resources required to expand, maintain, and operate their country's transport networks and other forms of infrastructure. Because of such limited resources and a search for more investment efficiency, in a number of countries the private sector has been involved in delivering infrastructure through concessions under PPP programs.

While there is no widely agreed, single definition or model for PPPs, the PPIAF Public-Private Partnerships Reference Guide: Version 2.0 (PPIAF 2014) provides a useful, broad description: "A long-term contract between a private party and a government entity, for providing a public asset or service, in which the private party bears significant risk and management responsibility, and remuneration is linked to performance."

PPPs have some key features around the different roles for the government and private sector, the length of the contracts, how they are financed and the risk transfers involved. They are long-term associations between the public and private sectors that usually involve the private sector undertaking investment projects that traditionally have been executed (or at least financed) and owned by the public sector. A central feature is that the private sector's investment and long-term operation of the asset are funded through the revenue it receives for providing a flow of services from that asset. This revenue comes either directly from the government or through the concessionaire imposing a charge/toll/tariff on users. This differs from the conventional public sector model where the government directly funds the building of a physical asset and then employs the personnel for (or contracts out) its long-term maintenance and operation.

The difference in the PPP delivery model can be illustrated in Figure 1 showing the varying roles for government and the private sector (the concessionaire, usually through a Special Purpose Vehicle or SPV) over the planning and preparation for a road project, its long-term operation by the private sector over the concession period, and the considerations for government in the post-concession period.

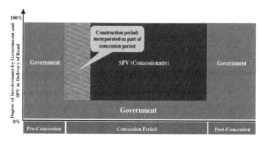

Figure 1.   Government and concessionaire roles.

For the Pre-Concession Period:

- Government funds all expenditure on construction of road and operation and maintenance
- Government starts of planning and preparation for possible PPP: pre-feasibility and feasibility studies
- PPP transaction go-ahead if viable, with selection of successful bid

For the Concession Period:

- A Special Purpose Vehicle (SPV, i.e. the Concessionaire) is competitively selected for delivery of road project
- SPV responsible for financing and delivery of road project and ongoing maintenance and operations over the concession period
- Ownership of assets remains with government
- SPV responsible for keeping road in acceptable condition, meeting Key Performance Indicators (KPIs) specified in the Concession Agreement
- Revenue to the SPV would come from payments by government (through availability payments or shadow tolls), from users (i.e. tolls), or some hybrid from both government and users
- Government role through the concession period to monitor performance of Concessionaire against KPIs
- Towards end of concession period, assess performance of Concessionaire and decide if to extend concession or plan for handback

For the Post-Concession Period:

- Following handback, government would be responsible for funding of all expenditure on road, any further upgrading and operations and maintenance
- If government decides to extend concession, terms of agreement would be reset:

  – Toll or availability payment
  – Concession period
  – KPIs

Rather than directly controlling the financing, building, operation and maintenance of a road, government takes responsibility for its delivery and the overall enabling environment for PPPs while having the private sector contribute to reducing the overall cost of delivering infrastructure services through increased efficiency and better management of some risks (such as construction). In successful PPP projects, the private sector's higher cost of financing and the need for a return on its investments are offset by the benefits provided by the private participation.

There are two general types of PPPs (Utz 2013). The first is where the main revenue stream or source of funding that repays the private sector finance used to build the asset takes the form of a service or availability payment from government. These are sometimes termed social infrastructure PPPs because this model is typically used for schools, hospitals, prisons and other 'social' (i.e., non-income producing) infrastructure. The second is where the primary source of funding takes the form of charges paid by the users of the infrastructure, such as tolls paid by the users of a toll road. These economic infrastructure PPPs are typically used for roads, railways and other 'economic' (i.e., income producing) infrastructure.

The archetypical PPP is a Build-Operate-Transfer (BOT) project where the Concessionaire finances, builds, operates and transfers the asset back to the public sector at the end of the concession life. The Concessionaire sells the final service to the public sector or to the public (i.e., the users) under a government concession. Figure 2 shows how such a road PPP concession would typically be structured where tolls were imposed on road users.

Key features of such a PPP include:

- SPV would be competitively selected for the road project.
- SPV would be responsible for constructing the road, ongoing maintenance and operations including the collection of tolls, and would retain that revenue.
- SPV would be responsible for keeping the road in acceptable condition, meeting the KPIs specified in the Concession Agreement.
- The government/Road Agency could make an upfront payment to the SPV and, if triggered, any payments from a Minimum Revenue Guarantee (MRG), if specified in the Concession Agreement.
- SPV would manage the road over the concession period while the ownership of assets remains with government/Road Agency.
- SPV would borrow from banks, as needed. The government/Road Agency could consider providing guarantees such as an MRG, if agreed and needed.

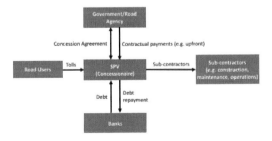

Figure 2. Structure of a toll road concession.

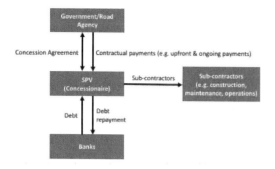

Figure 3. Road concession structure without tolling.

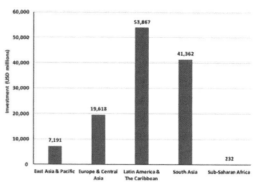

Figure 4. PPI investment in road projects by region: 2011–2015.

Alternatively, a road PPP concession could be structured so that there was no tolling of road users, as shown in Figure 3.

Key features of this type of road PPP concession would include:

- An SPV would be competitively selected for the road project.
- SPV would be responsible for upgrading the road and ongoing maintenance and operations over the concession period.
- SPV would be responsible for keeping the road in acceptable condition, meeting the KPIs specified in the Concession Agreement.
- The government/Road Agency would make ongoing payments to the SPV over the concession period, based on meeting the KPIs. The government/Road Agency could also make an upfront payment to the SPV to reduce the level of ongoing payments.
- SPV would manage the road assets transferred to it over the concession period while the ownership of assets remains with the government/Road Agency.
- SPV would borrow from banks, as needed. The government/Road Agency could consider providing guarantees, if agreed and needed.

The impact of PPP projects has been mixed. As stated by McCarthy (2013), at their best, PPPs can provide rapid injections of cash from private financiers, timely and within-budget delivery of quality services, and overall cost-effectiveness the public sector can't achieve on its own. But at their worst, PPPs can also drive up costs, under-deliver services, harm the public interest, and introduce new opportunities for fraud, collusion, and corruption.

The World Bank's Private Participation in Infrastructure (PPI) Database (PPI 2016) provides some useful insights into the extent of recent road investment in developing countries. Figure 4 shows the regional breakdown of total PPI roads investment of $122 billion over the last 5 years (2011–2015). The largest regional investment share has been in Latin America and The Caribbean with 44%. Investment has grown strongly over the last 10 years, with a trend in more recent years of fewer but larger deals.

KPMG in its June 2015 review of emerging global trends in PPPs notes that "A particular challenge in developing markets, such as Asia and the Middle East, is uncertainty around the ease of business and the transparency of governance arrangements."

In the past, governance has often referred to hierarchical, top-down systems of control by governments and institutions (Brinkerhoff & Brinkerhoff 2011). Increasingly under the emerging environment for PPPs, there is also a less hierarchical model where governance is without government; or at least without a hierarchical control-based role for government (Stoker 1998): "Governance is ultimately concerned with creating the conditions for ordered rule and collective action. The outputs of governance are not therefore different from those of government. It is rather a matter of difference in process."

On the premise that good governance can lead to more successful PPP projects, this paper reviews some key requirements for good governance in such projects. Because PPPs in infrastructure tend to have monopolistic features, good governance in managing them is essential to ensure that the private sector's involvement yields the maximum benefit for the society.

Crucial requirements for good governance in PPPs include (Queiroz & Izaguirre 2008): (i) competitively selecting the strategic private investor, (ii) properly disclosing relevant information to the public, and (iii) having a regulatory entity appropriately oversee the contractual agreements over the life of the concession. Additionally, an appropriate legal framework helps reduce the need for public sector guarantees, thus facilitating the transfer of risks to the private sector, which is a key feature of PPPs (Queiroz & Lopez-Martinez 2013).

2 COMPETITIVE SELECTION OF CONCESSIONAIRES

Open and transparent competitive bidding is usually perceived as a prerequisite to ensuring the efficient allocation and use of scarce public resources. The World Bank Procurement Guidelines recommend the use of International Competitive Bidding (ICB) to select the concessionaire or entrepreneur under BOO (Build, Own, Operate), BOT (Build, Operate, Transfer), BOOT (Build, Own, Operate, Transfer) or similar types of concessions for projects such as toll roads, tunnels, harbors, bridges (World Bank 2011). The Guidelines state that the ICB procedures may include several stages to arrive at the optimal combination of evaluation criteria, such as the cost and magnitude of the financing offered, the performance specifications of the facilities offered, the cost charged to the user, other income generated by the facility, and the period of the facility's depreciation. Competition can help to reduce prices and expand access, and should be used to the maximum extent possible (Harris 2003). To facilitate bid evaluation, it is highly recommended that a simple criterion be applied to compare the price proposals. Examples of such criterion include minimum toll rate to be charged to the users, minimum availability payments to be paid by the government to the concessionaire, or maximum present value of the payments to be made by the concessionaire to the government during the concession life.

The European Union encourages the use of the 'competitive dialogue procedure' for major projects (Freshfields 2005). Such competitive dialogue is somewhat like the World Bank ICB procedure that involves two stages (World Bank 2011).

Steps in the selection process are likely to include (Queiroz & Lopez-Martinez 2013):

a. *Advertising.* A notice requesting expressions of interest to prequalify should be published in at least one international newspaper and one of national circulation and should include the scheduled date for availability of prequalification documents.
b. *Investor Feedback.* Meeting with selected potential investors/concessionaires to solicit feedback on the options being analyzed as well as on the key parameters and assumptions underpinning the conclusions of financial feasibility.
c. *Public Information.* Implement an appropriate program to disseminate information to the public on the financing and construction of the proposed facility (or project).
d. *Prequalification of Concessionaires.* Develop operational and financial criteria to be used in judging the suitability of prospective bidders, and conduct a transparent pre-qualification process. Pre-qualification ensures that invitations to bid are extended only to those who have adequate capabilities and resources. Prequalification shall be based entirely upon the capability and resources of prospective bidders to perform the particular concession contract satisfactorily, taking into account their (i) experience and past performance on similar contracts, (ii) capabilities with respect to personnel, equipment, and construction, and (iii) financial position. All bidders that pass this stage are by definition qualified and should be considered for the next phase.
e. *Inviting pre-qualified firms/consortiums to submit bids.* Define the procedures for the pre-qualified bidders to carry out their own due diligence of the proposed project. In addition, a Data Room prepared by the Client should be made available to potential investors, to enable them to fully assess the investment opportunity.
f. *Bidders' review and comments.* In order to minimize opportunities for post-bid negotiations on substantive issues with the winning bidder, major transaction documents (such as concession contract, shareholders' agreement) should be circulated to the bidders for review and comments before bids are submitted. The clear understanding to bidders should be that the period designated for review and providing comments will be their only opportunity to influence the terms of the bidding process.
g. *Competitive Bidding Process.* Once the structure of the transaction has been approved, organize a competitive bidding process to award the concession contract to a strategic investor. Steps in the bidding process include preparation of the tender documents, administering the offering period for bidders' due diligence, and preparing the bid procedures and selection criteria.
h. *Bid Evaluation.* Evaluate the bids received based on the agreed, transparent selection criteria, and recommend award to the best evaluated bidder. As an example, in the case of road projects, typically the final selection criterion is based on: (i) the minimum toll rate proposed by the bidders; or (ii) the minimum public contribution, or subsidies, to the construction cost of the project, which is required by the bidders; or (iii) the minimum availability fee or annuity proposed by the bidders.
i. *Transaction Closure.* The principal parties complete and execute the concession contract, shareholders' agreement and other documents necessary for the satisfactory closing of the transaction.
j. *Public disclosure of the concession agreement.* By providing a further check on corruption, this can enhance the legitimacy of private sector involvement.

The selection process described above follows general international good practice. However, in cases where competition is perceived to be relatively limited and the country environment is prone to collusion of prospective bidders, some innovative approach in the procurement process may lead to better results. A good example is provided in the selection of concessionaires under the second phase of the Brazilian federal road concession program. The first phase, in the 1990s, had followed a more traditional approach, which led to relatively high toll rates. In the second phase, around 2008, the selection was carried out by the Sao Paulo Stock Exchange (BOVESPA) through an auction (Amorelli 2009). The process consisted of the following main steps:

a. Public consultation on the draft bidding documents.
b. Preparation of the final bidding documents.
c. Adopting as the selection criterion the lowest toll rate (as in the first concession phase).
d. BOVESPA held a simultaneous auction for seven road concessions, without prequalification, defining the bidder offering the lowest toll rate for each project road, including foreign firms.
e. The bid evaluation committee reviewed compliance of the technical offer and bid guarantee of the best-ranked candidate for each project, while other technical offers were not opened.
f. Award of each concession contract to the respective lowest bidder, without negotiations.

Compared to the first phase of road concessions, the innovative approach used in the second phase showed considerable advantages, by (i) reducing the time between invitation to bid and contract signing, and (ii) leading to substantially lower toll rates, with an average of about US$0.02/car-km, compared to about US$0.04/car-km in the first phase. Additionally, the sequencing of bidding steps (financial offer first, technical second) limited the impact of judicial recourses—a critical factor in traditional tendering processes in Brazil (Véron & Cellier 2010). A toll booth showing toll rates for a concession bid under the first phase is given in Figure 5.

While this innovative approach worked well for these relatively simple projects, it is likely that the traditional approach, including prequalification, will be the most appropriate for more complex projects.

Nevertheless, the above example illustrates that even in a country with a well-established legal framework it is important to keep some flexibility for applying innovative solutions to specific problems (such as collusion of potential concessionaires).

Figure 5. Toll booth, Rio de Janeiro to Juiz de Fora expressway, Brazil. Good governance helps assure that the minimum possible toll rates are charged to road users. *Source: Queiroz and Kerali 2010.*

3 UNSOLICITED PROPOSALS

Unsolicited proposals, which seem attractive to some governments in their wish to accelerate the implementation of infrastructure projects in the country, tend to be so controversial (usually involving allegations of corruption), that in fact they usually take longer to award than an open, competitive tender procedure. In theory, unsolicited proposals could generate beneficial ideas; in practice, there have been several unfavorable experiences, mostly as a result of exclusive negotiations behind closed doors. There have been cases where a contract signed between a government and a private company included a clause that prohibits any leakage of the signed contract.

Several countries have adopted specific legislation to deal with such proposals, while some governments have simply forbidden unsolicited proposals to reduce public sector corruption and opportunistic behavior by private sector companies. The general experience with unsolicited proposals is often negative, reflecting the fact that projects of this type have usually represented poor value for money, and were frequently incompatible with the actual development needs of the countries, and their ability to pay. They also often lead to allegations of corruption. Corruption has been shown to be associated with the lack of adequate transport infrastructure in a country, as well as a low degree of economic development (Queiroz & Visser 2001). It is essential to eliminate or minimize the perception, as well as the reality, of corruption in PPP programs so that such programs can best contribute to a country's economic development.

Some governments have adopted procedures to transform unsolicited proposals for private infrastructure projects into competitively tendered projects. Such countries include Chile, the Republic of Korea, the Philippines, and South Africa (Hodges

2003a and 2003b). How to respond to unsolicited bids so as to protect transparency in the procurement process and recognize the initiative of the proponent, is typically difficult. The World Bank considers that unsolicited proposals should be dealt with extreme caution and does not permit the use of unsolicited proposals in World Bank-funded projects.

A number of approaches on managing unsolicited proposals are available at the *World Bank PPP in Infrastructure Resource Center for Contracts, Laws and Regulation* (World Bank 2013). Such approaches include:

a. UNCITRAL Approach. UNCITRAL sets out suggested legislative language in provisions 20 to 23 of its Model Legislative Provisions (UNCITRAL 2004). Whenever a host authority receives an unsolicited bid, UNCITRAL recommends that the authority first consider whether the proposal is potentially in the public interest. If so, the authority then requests further information from the proponent in order to make a full evaluation. If the authority decides to go ahead with the project, it determines whether the project involves intellectual property, trade secrets or other exclusive rights of the proponent. For projects that do not involve these rights, a full selection procedure is followed, with the proponent being invited to take part in the selection. If it does involve the proponent's intellectual property, a full selection procedure does not need to be followed. In this case, the contracting authority may request the submission of other proposals, subject to any incentive or other benefit that may be given to the person who submitted the unsolicited proposal.
b. Chilean law approach. Chile has adopted an approach whereby the project proponent is required to take part in a fully competitive tender process, but is given bonus points in relation to the evaluation (Chile 1997).

In view of the risks involved with unsolicited proposals, it seems essential that the legal framework in each country include a clear provision to deal with this type of offers. The absence of competitive pressure on bidders also leads to concerns of less commercial outcomes from such proposals.

4 DISCLOSURE OF CONCESSION AGREEMENTS

Full disclosure of concession agreements, an indication of good governance, helps ensure that the users know what to expect from the facility under concession, thus increasing transparency in the role of the regulator. Inherent PPP risks can be reduced if the terms, costs, and benefits of the PPP project are made more understandable and accessible to governments, private parties, and consumers (McCarthy 2013).

Nevertheless, not all concession contracts are open to public scrutiny. Excuses range from a claimed need for confidentiality to the cost of photocopying (The Economist 2007).

In one country in Central and Eastern Europe, the main text of a concession agreement was published but key annexes including financial and technical obligations of the concessionaire were not open to the public. In a Latin American country, the full final draft of the concession agreements is published, but the signed version is kept confidential. As a result, potential last minute negotiations conducted behind closed doors between the successful bidder (i.e., the concessionaire) and the agency responsible for the project, if inserted in the contract, are not made available to the public or to the other contenders in the competitive bidding process (Queiroz 2009). Full disclosure, in every case, increases accountability of both the concessionaire and the regulator.

Based on a review and an assessment of practices conducted by the World Bank Institute (WBI 2013), some key elements of proactive disclosure should be:

- The disclosure of the current PPP contract (identifying any changes made since the contract was originally signed) and relevant side agreements including government guarantees, with minimal redactions which reflect commercially confidential information
- The disclosure of future stream of payments and government commitments under PPP contracts
- The publication of a summary which provides in plain language the most important elements of the contract and project and key information on the rationalization of the project, selection as a PPP, and procurement
- Information on a regular basis on the performance of the project
- A process by which information is authenticated/validated

Bloomgarden and Neumann (2016) stress the benefits of open data and increased disclosure in leading to better PPPs through improved user feedback, improved decisions around contracts and performance (particularly on contract renegotiations and adjustments), and increasing competition from bidders having improved information about PPPs at all stages from planning onwards.

5 LONG TERM MONITORING OF CONCESSION AGREEMENTS

More than two centuries ago Adam Smith (1776) wrote that "a high road, though entirely neglected, does not become altogether impassable. The proprietors of the tolls upon a high road, therefore, might neglect altogether the repair of the road, and yet continue to levy very nearly the same tolls."

To avoid such situations, which might occur even today, many countries have established regulatory agencies that monitor the performance of infrastructure facilities under concession. For example, in 2001 Brazil established the National Agency for Land Transport, which, inter alia, monitors federal road concessions (Brazil 2001).

Roads and other infrastructure concession contracts typically include required standards for construction, operation, maintenance, and toll (or fee) collection. For monitoring the quality of the facility during the life of the concession, several indicators of condition are usual. In the case of roads, such indicators include roughness, skid resistance, luminescence of pavement markings, and the presence and condition of signs, lighting, and other safety features.

Monitoring indicators that fall outside the boundaries of acceptability may lead to penalties for the concessionaire. Enforcing such standards helps the government and the users to reap maximum benefits of PPP projects. The long-term monitoring of PPP contracts by government was identified in Figure 1 above as part of the changing role for government under such arrangements.

Sensible negotiation of disputes that may occur during the life of a PPP project should assure the continuation of services and prevent the collapse of projects and consequent public waste (UNECE 2008).

6 SUMMARY AND CONCLUSIONS

This paper presented a review of some key requirements for good governance in PPP projects. Because such infrastructure projects tend to have monopolistic features, transparency and good governance in managing them is essential to ensure that uncertainty is reduced, encouraging the private sector's involvement to yield the maximum benefit for the public. Good governance in this case requires, *inter alia*: (i) competitively selecting the strategic private investor, including an adequate treatment of unsolicited proposals; (ii) properly disclosing relevant information to the public; and (iii) having a regulatory entity oversee the contractual agreements over the life of the concession.

It seems fair to conclude that governments willing to implement successful PPP programs should consider properly implementing the three key factors discussed in the paper.

REFERENCES

Amorelli, L. 2009. Brazilian Federal Road Concessions: New challenges to the regulatory framework. *Institute of Brazilian Business and Management Issues, Minerva Program*, George Washington University, Washington, D.C., USA. http://www.gwu.edu/~ibi/minerva/Spring2009/Lara.pdf

Bloomgarden, D. & Neumann, G. 2016. Open data + increased disclosure = better PPPs. *Handshake*, Issue 17 https://handshake.pppknowledgelab.org/features/open-data-increased-disclosure-better-ppps/

Brazil. 2001. National Agency for Land Transport. Brazil. http://www.antt.gov.br

Brinkerhoff, D. & Brinkerhoff, J. 2011. Public–Private Partnerships: Perspectives on Purposes, Publicness, and Good Governance. *Public Administration and Development*, Volume 31, Issue 1, February 2011, pp 2–14 http://onlinelibrary.wiley.com/doi/10.1002/pad.584/full

Chile. 1997. *Chile Concession Regulation No.956*. Ministerio de Obras Publicas. Santiago, Chile. http://siteresources.worldbank.org/INTINFANDLAW/Resources/Chileanconcessionregulations.pdf

Clayton Utz. 2013. *Improving the Outcomes of Public Private Partnerships*. https://www.claytonutz.com/ArticleDocuments/178/Clayton-Utz-Improving-The-Outcomes-Of-Public-Private-Partnerships–2013.pdf.aspx?Embed = Y

Economist, The. 2007. *Who Benefits from the Minerals?* Vol 384, Number 8547, September 22, page 62. London, UK.

Freshfields Bruckhaus Deringer. 2005. *Competitive dialogue: the EU's new procurement procedure*. http://www.freshfields.com/publications/pdfs/2005/13494.pdf

Harris, C. 2003. Private Participation in Infrastructure in Developing Countries: Trends, Impacts, and Policy Lessons. *World Bank Working Paper No. 5*. The World Bank. Washington, D.C., USA. Page 25. http://rru.worldbank.org/Documents/PapersLinks/1481.pdf

Hodges, J. 2003a. Unsolicited Proposals—Competitive Solutions for Private Infrastructure. *Public Policy for the Private Sector, Note No 258*. The World Bank. Washington, D.C., USA. http://rru.worldbank.org/Documents/PublicPolicyJournal/258Hodge-031103.pdf

Hodges, J. 2003b. Unsolicited Proposals—The Issues for Private Infrastructure Projects. *Public Policy for the Private Sector*, Note No 257. The World Bank. Washington, D.C., USA. http://rru.worldbank.org/Documents/PublicPolicyJournal/257Hodge-031103.pdf

KPMG. 2015. *Public Private Partnerships: Emerging Global Trends and the implications for future infrastructure development in Australia*. https://home.kpmg.com/au/en/home/insights/2015/06/public-private-partnerships-global-trends.html

McCarthy, L. 2013. Fixing Fraud in Public-Private Projects. *Voices: Perspectives on development*. The World Bank, Washington, D.C., USA. http://blogs.worldbank.org/voices/fixing-fraud-in-public-private-projects

Public-Private Infrastructure Advisory Facility-PPIAF. 2014. *Public-Private Partnerships Reference Guide: Version 2.0.* Washington, D.C., USA. https://ppp.worldbank.org/public-private-partnership/library/public-private-partnerships-reference-guide-version-20

Queiroz, C. & Izaguirre. A.K. 2008. Worldwide Trends in Private Participation in Roads: Growing Activity, Growing Government Support. *Gridlines series*, no. 37, Public-Private Infrastructure Advisory Facility-PPIAF, Washington, D.C. USA. https://www.ppiaf.org/sites/ppiaf.org/files/publication/Gridlines-37-Worldwide%20Trends%20in%20Private%20-%20CQueiroz%20AIzaguirre.pdf

Queiroz, C. & Kerali, H. 2010. A Review of Institutional Arrangements for Road Asset Management: Lessons for the Developing World. *World Bank Transport Paper No. TP-32*. Washington, D.C., USA. http://go.worldbank.org/6HDCYBMRT0

Queiroz, C. & Lopez Martinez, A. 2013. Legal Framework for Successful Public-Private Partnerships. In *"The Routledge Companion to Public-Private Partnerships,"* edited by de Vries, P. & Yehoue, E.B. pp 75–94. http://www.routledge.com/books/details/9780415781992/

Queiroz, C. & Visser, A. 2001. Corruption, Transport Infrastructure Stock and Economic Development. Infrastructure and Poverty *Briefing for the World Bank Infrastructure Forum*, CD-ROM. World Markets Research Centre Ltd. The World Bank. Washington, D.C., USA. http://www2.udec.cl/~provial/expo/WB%20Infrast%20Forum%20May%202001%20TranspInfra&Corrup%20CQ%20AV.pdf

Queiroz, C. 2009. Financing of Road Infrastructure. *Proceedings of the 5th Symposium on Strait Crossings*. June 21–24, 2009. ISBN 978-82-92506-69-1. pp 45–57. Trondheim, Norway. http://www.straitcrossings.com

Smith, A. 1776. *An Inquiry into the Nature and Causes of the Wealth of Nations*. Pennsylvania State University, Electronic Classics Series, Jim Manis, Faculty Editor, Hazleton, PA USA, 2005, page 593. http://www2.hn.psu.edu/faculty/jmanis/adam-smith/Wealth-Nations.pdf

Stoker, G. 1998. Governance as theory: five propositions. *International Social Science Journal*, Volume 50, Issue 155, March 1998, pp 17–28 https://www.researchgate.net/profile/Gerry_Stoker/publication/227980445_Governance_as_theory_five_propositions/links/53f5bc4f0cf2888a7491d638.pdf

UNCITRAL. 2004. *Model Legislative Provisions on Privately Financed Infrastructure Projects*. United Nations, NY, USA. http://www.uncitral.org/pdf/english/texts/procurem/pfip/model/03-90621_Ebook.pdf

UNECE. 2008. *Guidebook on Promoting Good Governance in Public-Private Partnerships*. United Nations Economic Commission for Europe, United Nations, New York and Geneva. http://www.unece.org/fileadmin/DAM/ceci/publications/ppp.pdf

Véron, A, & Cellier, J. 2010. Private Participation in the Road Sector in Brazil: Recent Evolution and Next Steps. *World Bank Transport Paper No. TP-30*. Washington, D.C., USA. http://go.worldbank.org/6HDCYBMRT0

World Bank Institute. 2013. *Disclosure of Project and Contract Information in Public-Private Partnerships*. World Bank, Washington, D.C., USA. http://wbi.worldbank.org/wbi/document/disclosure-project-and-contract-information-public-private-partnerships

World Bank. 2004. *Reducing the 'Economic Distance' to Market: A Framework for the Development of the Transport System in South East Europe*. Washington, D.C., USA. http://wbln0018.worldbank.org/ECA/Transport.nsf/ECADocByLink/BEF3FC761FF49D0785256FB200508860?Opendocument

World Bank. 2011. *Guidelines: Procurement of Goods, Works, and Non-Consulting Services under IBRD Loans and IDA Credits & Grants*. The World Bank. Washington, D.C., USA. http://go.worldbank.org/1KKD1KNT40

World Bank. 2013. PPP in Infrastructure Resource Center for Contracts, Laws and Regulation (PPPIRC). Washington, D.C., USA. http://www.worldbank.org/pppiresource

World Bank. 2016. "Private Participation in Infrastructure Database" Washington, D.C., USA. http://ppi.worldbank.org/

Transport Infrastructure and Systems – Dell'Acqua & Wegman (Eds)
© 2017 Taylor & Francis Group, London, ISBN 978-1-138-03009-1

SUP&R ITN: An international training network on sustainable pavements and railways

D. Lo Presti & G. Airey
University of Nottingham, Nottingham, UK

M.C. Rubio
Universidad de Granada, Granada, Spain

P. Marsac
IFSTTAR, Nantes, France

ABSTRACT: The Sustainable Pavements & Railways Initial Training Network (www.superitn.eu) is a training-through-research programme that is empowering Europe by forming a new generation of multi-disciplinary professionals capable of conceiving, planning and executing sustainable road and railway infrastructures. The SUP&R ITN, started at the beginning of October 2013, it's a 4 million € effort entirely funded by the Euroepan Commission through its Marie Curie Actions 2012 of the FP7 programme 2006–2013. The University of Nottingham is leading this effort which is the first of its kind and involves 29 partners between universities, research centres and companies/industries, from five EU countries (UK, Italy, France, Ireland and Spain) and USA. This paper aims to be a reference for those who plan to structure similar programmes and will present the overall idea, the research framework, the training platform and a summary of the main scientific results obtained so far.

1 INTRODUCTION

The design, construction, maintenance, use and end-of-life management of road pavements and railways is associated with a number of important impacts on the environment; namely the consequences of energy consumption, unsustainable use of materials/resources, waste generation and release of hazardous substances into the environment. Whilst environmental sustainability legislation requires road and rail infrastructure to deliver more sustainable solutions, progress to date has been limited. This is the result of a lack of well-founded scientific tools to measure sustainable solutions, the lack of suitably trained personnel to design and use such solutions, and the fragmented approach generally taken by researchers and industrial stakeholders across Europe. Industry and the private sector are well aware of this problem and have been highly supportive of a coherent research and training approach that can address and rectify this situation.

SUP&R (Sustainable Pavement & Railways) ITN, through a coherent research and training approach involving close collaboration between research institutions and industrial stakeholders across Europe, will allow this step change in the sustainability of road and rail infrastructure to be addressed by targeting the following overall aim:

To setup a multidisciplinary and multi-sectorial network in order to form a new generation of engineers versed in sustainable technologies and to provide, to both academia and industry, design procedures and sustainability assessment methodologies to certify the sustainability of the studied technologies to the benefit of the European community.

To overcome these barriers, the Nottingham Transportation Engineering Centre (NTEC) shaped the programme and it's in charge of coordinating a consortium of 29 partners between universities, research centres and companies/industries, from five EU countries (UK, Italy, France, Ireland and Spain) and USA. The SUP&R ITN started at the beginning of October 2013 and will empower Europe by forming a new generation of multi-disciplinary researchers and professionals capable of conceiving, planning and executing sustainable road and railway infrastructures, and deliver long-term benefits in terms of:

1. Eco-designed road and rail infrastructure that maximizes the recycling of waste materials and ensures best performance characteristics to suit the diverse set of European environments;
2. Reduced installation, maintenance and operating costs as well as long term sustainable solutions;

3. A bespoke sustainability assessment tool, tailored to needs of product development in the road pavement and rail infrastructure sector.

This paper will present the overall programme, the planned training and dissemination activities, together with the details of each of the main scientific Work Packages as well as a summary of the main results. More details in the website: www.superitn.eu

2 SUP&R ITN PROJECT

2.1 Project consortium and formation strategy

The SUP&R International Training Network (ITN) will offer a training-through-research programme to selected young researchers across Europe. The project includes a high level four year training programme which will be international, multi-disciplinary and multi-sectors.

The network of SUP&R ITN is comprised of partners (Tables 1 and 2) who, in addition to having outstanding expertise and state-of the-art facilities, complement each other to bring together the necessary scientific and professional power needed to provide high quality training and cutting edge research in designing sustainable transport infrastructures.

The industry partners cover the whole stakeholders' chain of road pavements and railways from transport infrastructure managers (IRAIL, ADIF, NR), engineering designers (URS), contractors (SACYR, ETP) and manufacturers (REPSOL, NYNAS, AI, PI). This consortium is complemented by partners ranging from research and training bodies (TUD, UW, VTTI), Professional European Associations (CEDR, EAPA, ECOPN, EIM), experts in sustainability (GR, CEEQUAL), online training of sustainability (ILL) and communication in transport engineering field (FEHRL). The choice of collaborators from academic sector was driven by excellence and based on a selection of specific expertise of each partners and successful past collaborations.

Table 1. SUP&R ITN full partners.

University of Nottingham NTEC— (Coordinator)	UNOTT
Universita' degli Studi di Palermo DICAM	UNIPA
University College Dublin CSEE	UCD
IFSTTAR	IFSTTAR
Universidad de Granada LABIC	UGR
Universidad de Huelva CFEL	UHU
AECOM—URS	AECOM
IrishRail	IRAIL
Sacyr Vallehermoso	SACYR
REPSOL	REPSOL

Table 2. SUP&R ITN associate partners.

Eiffage Travaux Public	ETP	Ecopneus	ECOPN
Network Rail	NR	Euro Rail Infrastructure Managers	EIM
Nynas	NYNAS	Forum of European National Highway Research Laboratories	FEHRL
Universidad politécnica de Madrid	UPM	Ceequal	CEEQUAL
Slovenian National Building and Civil Engineering Institute	ZAG	GreenRoads	GR
Aggregate Industries	AI	University of Illinois Urbana-Champaign	ILL
Phoenix Industries	PI	University of Washington	UW
Administrador de Infrae-structuras Ferroviarias	ADIF	Virginia Tech Transportation Institute	VTTI
Euro Asphalt Pavement Association	EAPA		

New partners have been introduced to add new complementary skills and expertise to the existing structure. Instead, the choice of the private sector full partners has been strategically planned in order to cover the whole stakeholder chain of road pavements and railways within transport infrastructure. Table 3 summaries the exploitation of synergies and complementary expertise within the whole SUP&R ITN consortium.

2.2 Project structure

To achieve the overall research and training objectives, the SUP&R ITN has been implemented with 6 Work Packages (WPs). Training-through-research projects will be delivered within three scientific research work packages (WP1-WP3), while three additional WPs will handle respectively; Network-wide Training (WP4), Management (WP5) and the external and internal Communication (WP6).

The Coordination at The University of Nottingham is in charge of the management of the SUP&R ITN (WP5) as well as shaping the international, multi-disciplinary and multi-sectoral training platform

Table 3. SUP&R ITN expertise matrix.

Expertise, Role in the ITN /Partner	Full partners										Associate partners																	
	UNOTT	UN/IPA	UCD	IFFSTAR	UHU	UGR	URS	IRAIL	SACYR	REPSOL	ETP	NYNAS	NR	UPM	ZAG	AI	AKZO	PI	CEDR	EAPA	ECOPX	EIM	FEHRL	CEEQUAL	GR	ILL	UW	VTTI
Material chemistry and manufacturing	○	○		○	○					○	○							○										
Material mechanical characterisation of	○	○	○	○	○	○				○	○						○											
Material behaviour modelling	○	○			○										○													
Infrastructure design	○		○			○									○													
Geotechnics															○													
Infrastructure implementation	○		○	○		○	○			○								○										
Infrastructure management						○	○				○											○						
Sustainability assessment	○	○		○			○																		○	○		
Materials' supplier								○			○				○	○	○				○							
Data supplier for lifecycle inventories						○	○	○	○				○	○	○	○	○			○	○				○	○	○	○
Visiting/online Scient.																									○	○	○	
Training at Workshops	○	○	○	○	○	○	○	○	○	○	○	○	○	○	○				○	○	○	○	○	○	○	○	○	○
Outreach/ Dissemination	○																		○	○	○	○						

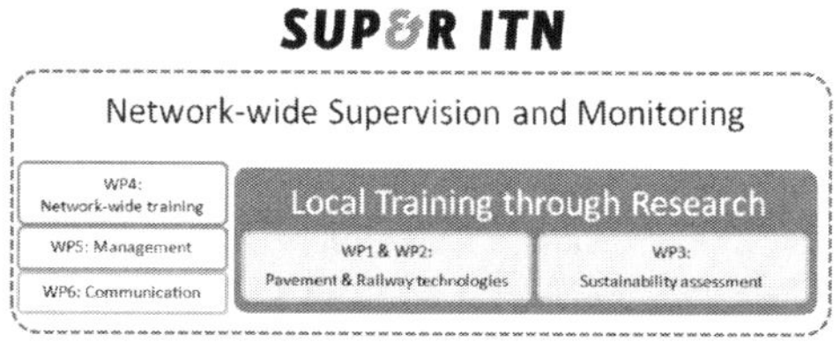

SUP&R ITN

Network-wide Supervision and Monitoring

WP4: Network-wide training	Local Training through Research	
WP5: Management	WP1 & WP2: Pavement & Railway technologies	WP3: Sustainability assessment
WP6: Communication		

Figure 1. SUP&R ITN project structure.

which consists in: a local and network-wide training activities (WP4) and scientific dissemination and public outreach activities of each fellow and the consortium. (WP6). The next sections will provide more details on the contents and quality of the research framework and the training platform, as well as a summary of the main results obtained so far.

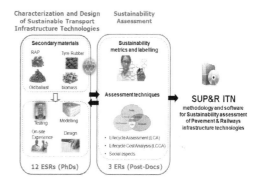

Figure 2. SUP&R ITN research framework.

3 RESEARCH FRAMEWORK

SUP&R ITN will drive the researchers on their multidisciplinary training-through-research pathway by closely investigating promising pavements and railway technologies with practical methods, models and tools for estimating their sustainability. This will be carried out through a systematic integration of sustainability aspects at a very early stage in the technology design. The SUP&R ITN programme will achieve these results by developing inter-linked research projects aiming at:

• Developing new sustainable technologies and materials for use in road pavements (highways)

and railway transport infrastructure (standard railway infrastructure or high speed railways).

• Providing advanced characterisation of recycled and reused materials generated from road and railway infrastructure and/or other production processes for use in road and railwaya.

• Developing detailed material modelling and design approaches for immediate uptake by industrial stakeholders working in the area of transport infrastructure.

• Developing and refining estimation tools so that industry can assess where it currently stands in key sustainability indicators and determine how much it can improve using low-energy and more recycled materials. Such estimation tools will allow industry, road and railway agencies to recognize the sustainability of different technologies and methodologies already at the design stage.

3.1 Characterization and design of transport infrastructures technologies (WP1 and WP2)

WP1 and WP2 will be carried out predominantly by the ESRs through the development of 12 research projects with the common goal of closely investigating promising sustainable technologies, respectively for road pavements and railways. The selected road technologies will be investigated for specific application in highways, while the railway technologies will be designed to be applied either on standard railway infrastructure or high-speed tracks. Performance prediction will be validated through simulative tests in laboratories and, for some research projects, through possible validation results from field trials. Thanks to the joint contribution of the universities and industrial partners involved in each project, the ESRs will characterise and design the proposed technologies and at the same time will also contribute to building a Lifecycle Inventory to be used in WP3. This data, indeed, will represent the most critical step to accurate building the LifeCycle Inventories and for the validation of the sustainability assessment methodology developed in WP3.

Sustainable Pavements (WP1). This series of projects will focus on investigating several sustainable technologies for road pavements. The academic leader is IFSTTAR (P. Marsac) with the Industrial deputy leader being REPSOL (E. Moreno). WP1 includes the following projects that will be all carried out within the timeframe 2014–2017:

- Project ESR1: Pavement design for cold in-situ recycled materials
- Project ESR2: Design and characterization of bituminous mixes manufactured with biomass
- Project ESR3: Long term performance of low-temperature asphalt mixes containing reclaimed asphalt
- Project ESR4: Rubberised binder and asphalt mixes for wearing course
- Project ESR5: Binder design for low-temperature asphalt mixes
- Project ESR6: Half Warm Mixes Asphalt Recycling Asphalt in urban roads

Sustainable Railways (WP2). This series of projects will focus on investigating several sustainable technologies for railway infrastructure with academic leadership by UGR (MC. Rubio) and Industrial leadership by IRAIL (C. Bowe). WP2 will include the following projects that will be all carried out within the timeframe 2014–2017:

- Project ESR7: Optimization of trackbed design and maintenance
- Project ESR8: Characterisation of rubberised asphalt for railways sub-ballasts
- Project ESR9: Modelling and Design of rubberised asphalt for railways sub-ballast

- Project ESR10: Settlement monitoring and prediction in railway tracks
- Project ESR11: Optimisation techniques for geophysical assessment of rail support structures
- Project ESR12: The use of waste materials in railways

3.2 SUP&R tool: Development of the sustainability assessment methodology (WP3)

In WP3, results arising from both previous phases will allow researchers and partners involved in the development of the projects, to understand the possibility of widespread use of the proposed sustainable technologies. Once the methodology is developed and tested, the assessment system will be incorporated in a methodology, to be included in software tool, for a wider group of users within the industry. This will assist product developers in industry to assess and compare alternative products and approaches. While similar tools are being developed for the operational phase, this tool will be focused on the product development stage. Unlike existing infrastructure sustainability assessment tools, it will be limited to the road and rail product sector and this will allow it to be tailored specifically to the assessment needs of these products, and include well defined methodologies which will assist in quantification and comparison of results. Its use will assist in early assessment of emerging products and practices and in this way help in prioritizing the most promising technologies at an early stage of development. WP3 has UNOTT (D. Lo Presti & T. Parry) as the academic leader with IFSTTAR (A. Julien) as the deputy leader and will include the following tasks:

- Project ER1: Definition of sustainability assessment factors and current state-of-the-art in sustainable practices—2014–2016
- Project ER2: Sustainability assessment of Railways and SUP&R ITN technologies—2016–2017
- Project ER3: Life cycle impact assessment of Road and Railway Practices—methodology and tools—2016–2017

4 TRAINING PLATFORM

The overarching aim of SUP&R ITN is to provide high quality interdisciplinary, inter-sectors and international training to young researchers, thereby contributing to the development of a new generation of multidisciplinary engineers, fully able to work within the field of sustainable road and rail technologies. These objectives will be achieved through the following a structured programme of training and dissemination activities.

4.1 Training activities

Supervision arrangements and Personal Career Development Plans: Once the fellows have officially started the outcome of their initial evaluation and their career options and own ideas will lead to the design of a Personal Training and Career Development Plan (PTCDP) detailing the knowledge gaps and the need for both local and network-wide training. The team leaders then guide the fellows locally, but they also receive a network-wide supervision directly from the Training Responsible at the coordination. However, The overall responsibility for the planning and implementation of the training programme lies with the network through the Supervisory Board (SB). SB is composed by the representative of the partners and shall make sure that there is an adequate balance in the training activities through personalised research and complementary training such that useful skills for career development and contribution to society are achieved at the end of the project.

Local training activities: The SUP&R ITN fellows will conduct their research primarily at their host institutions where the local training features a number of important elements, including:

- Each ESR/ER has individual scientific and on-the-job training provided by academic and industry partners and tailored to his/her specific research project.
- The ESR/ER are encouraged to interact with the local research group at their host institution. This will include research forums, national symposiums and meetings. This approach will ensure that the ESR/ER achieves a true perspective of the research being conducted at the local level. Similarly, perspectives acquired during this secondment can be brought back to the host institution and shared with fellow researchers.
- Most of the partners are located at universities with strong post-graduate programmes with modules on various topics related to managing transport infrastructure. The ESRs have access to these modules, which include scientific lectures, but also generic research methods, project management, etc.
- At the very initial stage of the programme, if needed all the fellows are provided language courses from the host institutions.

Network-wide training activities: This training will be offered network-wide and will focus on the skills that fellows will need to develop during and after the development of the their PhDs. The training programme started with a kick-off meeting to introduce the fellows to SUP&R ITN, to build up team spirit and to deliver the basic scientific training. Since then an average of two network-wide events between workshops, spring/summer schools and outreach activities are organised every year. This training offers the opportunity to fellow to team up and to provide them with a great exposure to the industrial partners. A specific training programme is shaped for "complentary skills" and focuses on the skills that fellows will need to develop at different stages in the development of the network and their PhDs and will involve the consultancy and industrial partners raising an awareness of the demands of organizations outside academia for the fellows. Furthermore, to complement the expertise of the consortium world-side experts are invited as visiting lecturer is some of the training weeks; amongst those: Dr. A. Scarpas of Delft University; Prof. Steve Muench, from UW and GR; Prof. G. Flinch, VTTI, USA and Dr. J. Tomkin, ILL, USA.

Network-wide training structured activities are coordinated by the University of Nottingham with the collaborations of the various local partner. In fact, the structured training takes place at the different institutions of the consortium. Here the calendar and details of the on-going training programme which is constantly updated and generally opened also to external participants:

- 2014 Sep—Nottingham (UK)—"Scientific Introduction on Sustainable roads and railways"
- 2015 Mar—Dublin (IE)—"Railway engineering"
- 2015 Jan and Mar—Nottingham (UK)—"Outreach and Complementary Skills"
- 2015 Jun—Delft (NL)—Advanced Scientific training with Dr. Scarpas of TU Delft
- 2015 Sep—Palermo (IT)—"Sustainability Assessment of Transport Infrastrctures"
- 2016 Feb—Granada (ES)—"From research to Sustainable Pavement and Railway to implementation in Practical engineering Projects"
- 2016 Oct—Madrid/Huelva (ES)—"Bituminous Products for Sustainable Asphalt Technologies"
- 2017 Jan/Feb—Nantes (FR)—Still to be planned

4.2 Dissemination and outreach activities

SUP&R ITN involves a series of outreach activities for public audiences around Europe. These activities, are carried out by the SUP&R ITN fellows and supported by coordination at UNOTT.

Outreach programme: coordinates the communication between the scientific community and the general public to convey the importance of and increase the awareness of sustainability in road and rail infrastructure. Here is a list of the outreach activities (updated regularly):

- *Outreach Training:* To prepare for outreach activities, all fellows attended the course Public Understanding of Science—Communication Skills for Researchers

- *Workshop Day:* The fellows showed their project directly at schools to kids with an age of 15–17
- *SUP&R ITN Open days:* During training weeks for the fellows, the public and other entities (like industry) are invited to an open day. These have been held already in Italy and Spain
- *Multimedia Releases:* SUP&R ITN fellows have produced videos to introduce their research and created a Youtube channel to distribute them.
- *SUP&R ITN movie.* The project will end with a final event will be an occasion to share the achievements of the entire consortium to the scientific community and to the general public. For that event, the consortium will premier al illustrated film that will present the sustainability aspects of the studied technologies for a wide audience. This movie will summaries the learned principles of sustainable pavements and railways and will represent an educational resource to everyone from teachers to business to community leaders.

Scientific Dissemination: It is also important that SUP&R ITN fellows disseminate their research to the scientific community. This is done as follows:

- *ESR specific dissemination:* the coordination provides support with advices on interesting scientific conferences and symposiums as well as other industry-related meetings, so that each fellow could have the chance to participate.
- *Annual dissemination event*: the programme has an annual dissemination event which is a group talk where the whole project is presented by the fellows in several venues where they can increase their exposure. The final outreach event will be the SUP&R ITN Symposium in 2017.

5 SUMMARY OF RESULTS UP-TO DATE

5.1 *WP1 sustainable pavements*

The WP1 is composed by 6 complementary individual projects attempting at improving sustainability of road pavements by means of actions that can be grouped in 4 broad categories: materials, manufacturing process, structure design and durability improvement. The aim of ESR1 is to integrate materials cold recycled with foamed asphalt and cement in pavement design methods. As few empirical data are available for this type of material, a mechanistic approach with a layered elastic model was chosen to relate stress, strain and deflection field within the pavement to load conditions. The following relevant input material parameters were identified: stiffness, fatigue and permanent deformation. Adapted tests and procedures were evaluated and notably an Indirect Tensile Fatigue Test associated with a dissipated energy approach. The experimental data obtained show that further study is needed for fatigue criteria as failure mode for the cold recycled material is distinct from usual hot materials.

The objective of ESR2 is to promote the use of binders obtained from renewable resources (bio-binders) through undertaking their mechanical characterization and design optimization for mixture with Reclaimed Asphalt (RA). Bio-binders are characterized and their interactions with RA binder and their rejuvenating effect are studied and compared to standard and polymer modified binders for 50% RA mixtures. Rheological characterization reveals that all the bio-binders studied produced a rejuvenating effect. Fatigue ($G^*\sin\varphi$) and rutting ($G^*/\sin\varphi$) parameters suggests that bio-binders can enhance fatigue resistance without compromising rutting behavior.

The evaluation of the long term performance of low-temperature asphalt containing reclaimed asphalt is the objective of ESR3. An experimental plan combining different recycling rate and ageing step on a reference hot mix asphalt and warm mixes (additives process and foaming process) was completed. Binders' ageing was characterized through infrared spectroscopy. For each combination the rheological properties of the binder were evaluated and the modulus and fatigue resistance were measured on the mixes. Compared to the hot process the warm processes induce a lower ageing but this tend to be compensated by a stronger long-term ageing. For all processes, modulus and fatigue performances tend to improve with RA addition and ageing. Furthermore, a new method to estimate the apparent molecular weight distribution of a binder was validated to relate mixes and binders properties (Perez-Martinez et al. 2015).

The main objective of the ESR4 research is to investigate the design of a hybrid system consisting of crumb rubber and polymers to form a new binder. Extensive laboratory tests were carried out with the same crumb rubber content but for different bituminous base, polymer type and polymer concentration in order to select the blending according to storage stability parameters. The physico-chemical, thermo-mechanical and rheological properties of these selected blending were characterized. The selected hybrid system will be then used to manufacture three type of hot asphalt mixes for wearing courses and results will be compared with at least one control for each type of mix.

The task of ESR5 is to design new binders for low-temperature asphalt mixes incorporating high amount of RA (Yuliestyan et al. 2015). As the ratio binder to aggregate need to stay almost constant in an asphalt mixture, high recycling rate implies low amount of new binder added. To insure a decent coating, a new emulsion process (bitumen in the form of micro droplets) was selected. A Kraft lignin bio-polymer (an abundant by-product of the paper pulp industry) was modified by amination in order to obtain a suitable emulsifier to be used with tailored soft bitumen. Taking advantage of these innovative constituents, a pioneering emulsion was designed and successfully used to produce 100% cold recycled asphalt mixes. The firsts mechanical tests conducted on samples of the produced mixes show very promising results.

Half-warm mix can be considered as an intermediate step between cold and warm mix potentially associating the reduced manufacturing energy of the first and the mechanical properties of the second. To rule on uncertainties that tend to prevent their widespread use, ESR6 investigate their actual performances in laboratory and in the field when they are combined with high RA content. Mixtures with 70% RA and 100% RA and emulsion were compared to a reference hot mix. Results show discrepancies between tests on laboratory and in the field, notably as regard to the stiffness modulus, probably linked to different laboratory compaction methods (gyratory and Marshal). The three mixes tested showed equivalent in field performances. Research is carried out to further evaluate the long-term performances in field.

5.2 WP2 sustainable railways

Sustainable Railway Projects aimed to optimize materials, design and manufacturing techniques to improve problems associated with railway tracks such as settlement, bearing capacity reduction, vibrations and noise among others; from a sustainable approach. For this purpose innovative methods and tests have been carried out, including field measurements in real conditions.

Four of the six projects that compose WP2 deal with new materials to be used in the granular layers. Some of these materials are: bitumen emulsions, low temperature bitumen sub-ballast, and crumb rubber to design bitumen sub-ballast. As part of the project: "ESR7: *Optimization of trackbed design and maintenance*" the study of bitumen emulsion to be used as a ballast stabilisation method for newly constructed or exiting track has been carried out (D'Angelo et al. 2016). For this purpose, different emulsions were tried and tests at different scales reproducing rail loads where performed. Results obtained from both half and full-scale box tests highlighted the potential for this solution to reduce ballast problems associated with its unbound nature such as settlement and particle deterioration, limiting therefore the need for maintenance and increasing trackbed durability. Results have shown that the application of bitumen emulsion during track initial construction can lead to a clear decrease in ballast permanent deformation (−25%) and deformation rate (−30%), indicating its effectiveness for both short and long-term behavior.

Considering that bituminous subballast has shown to be an alternative to granular subballast,

three WP2 projects aimed to optimise the bituminous subballast design by introducing sustainable techniques which include the use of waste such as crumb rubber (wet and dry processes), and the reduction of the asphalt mixes manufacturing temperature (WMA). These have being carried out through the projects: "*ESR8 Characterisation of rubberised asphalt for railways sub-ballasts*", "*ESR9 Modelling and Design of rubberised asphalt for railways sub-ballast*", "*ESR12 The use of waste materials in railways*".

In this sense a WMA was designed to be used as bituminous subballast, and tested under different railway loads and temperature conditions. Results have proved an adequate performance to be used in the track. In the same way, two types of bituminous subballast (wet and dry processes) containing crumb rubber have been designed and are being tested. For this porpoise the adaptation of Superpave asphalt mix design was needed, once Superpave considers traffic loads while railway loads were the ones to consider in this case. Also, a model bases on the fractional calculus is being developed to predict the mechanical behaviour of the crumb rubber asphalt subballast.

On the other hand, WP2 includes projects focused on improving prediction techniques to obtain more accurate information about the deterioration of the track. The project "*ESR10 Settlement monitoring and prediction in railway tracks*" develops and tests methods whereby sensors in a train can be used to detect tract settlement and bridge damage; numerical models of the dynamic interaction of a train with track, sleepers, ballast and bridge are used to analyse vehicle response to detect damages. This is the first time optimisation techniques have been used to process vehicle measurements to identify track stiffness, track profile and bridge damage.

Finally, the project "*ESR11 Optimisation techniques for geophysical assessment of rail support structures*" seeks to employ non-destructive, vibration based monitoring of railway substructure. One of the challenges consists in detecting weak spots in the substructure in transversal and longitudinal direction of railway track. By now, numerical studies and field work have been carried out. Results obtained shows that impact hammer testing could be used to assess the substructure condition.

5.3 WP3 SUP&R sustainability assessment methodology and tool

SUP&R ITN sustainability assessment tool is being built by three complementary projects: ER1 project began with a state-of-the-art review of sustainability assessment and rating tools. This review identified key criteria and subject areas that are found in sustainability assessment tools for civil engineering infrastructure and critically reviewed

Figure 3. SUP&R Multi-Criteria Decision Analysis tool.

the information across the tools. The results helped to identify limitations and areas of contributions for the project, and in combination with a review of the literature, they supported the development of the conceptual framework for the SUP&R ITN tool. The state-of-the-art review was also the first step in developing a set of indicators for the SUP&R ITN tool (Brice et al 2016). By focusing specifically on pavement-related indicators used in existing tools, we have developed a shortlist of indicators for the SUP&R ITN tool. It is important to note that there are no assessment tools specific to railway pavement or construction. As a result, Project 3.2 (ER2) is developing a comprehensive library of best practices in sustainable construction for railways and this will assist the identification of indicators specific to railways. Both railway and roadway indicators are organized into higher level domains, which the conceptual framework terms "means objectives," or objectives critical for comprehensive sustainability. The means objectives will be weighted in the tool and indicators classified under them will receive that weighting. This is being developed within project 3.3 (ER3) that is also programming and coding the tool to calculate weightings using two processes, the Analytical Hierarchy Process and the PROMETHEE method. The tool will also have a default weighting that will be established based on a survey of industry professionals.

Finally, through Multi-Criteria Decision Analysis (MCDA), the SUP&R ITN WP3 will produce a tool that will be freely available and will support decision making for sustainable pavements at least at the design stage. At last, once the tool is defined, WP3 will then use it to perform the sustainability assessment of the investigated technologies in WP1 and WP2.

6 CONCLUSIONS

SUP&R ITN programme is the first international training programme at PhD level on Sustainable Transport Infrastructures and one of the firsts ITNs dedicated to transport infrastructure engineering. The programme is entering its last year and as a result of this experience the following points play in favour of this kind of educational programme:

- Fellows are clearly experiencing something different from conventional PhD students. They conceive research already as a collaborative, multi-disciplinary and international effort.
- Secondments are a key part of this type of training and allow strengthening relation between research institutions as well as exploiting collaborations between academic and industrial partners. Furthermore, they provide fellows with great exposure to several sectors and in some cases their research products are already being up taken from industrial partners.
- Network-wide activities are another fundamental part that allow personal and professional development of all the participants as well as fostering collegiality amongst fellows, advisors and research groups as a whole.
- SUP&R ITN is highlighting the importance of the introduction of lifecycle approach and sustainability concepts into transport infrastructure engineering. This is significantly changing the research and engineering practices of the partners involved. As example, it has inspired a new master course on Sustainable Highways at University of Nottingham and it has also created a solid group whom core as already ensured the funding for another training programme of this type: SMARTI ETN 2017–2021

ACKNOWLEDGEMENTS

This project has received funding from the European Union's Seventh Framework Programme for research, technological development and demonstration under grant agreement n. 607524.

REFERENCES

Bryce J., Brodie S., Parry T., Lo Presti D. A Systematic Assessment of Road Pavement Sustainability through a Review of Rating Tools, Resources, Conservation and Recycling, Elsevier, (accepted 2016).

Colinas N., Di Paola M., Pinnola F.P., Fractional characteristic times and dissipated energy in fractional linear viscoelasticity, Communications in Nonlinear Science and Numerical Simulation, Vol. 37, 2016. DOI: 10.1016/j.cnsns.2016.01.003

D'Angelo, G., N. Thom, D. Lo Presti, Bitumen stabilised ballast: A potential solution for railway track-bed, Construction and Building Materials 124(2016):118–126. July 2016. doi: 10.1016/j.conbuildmat.2016.07.067

Perez-Martinez, M., P. Marsac, T. Gabet, M. Lopes, S. Pouget, F. Hammoum (2015). Durability Analysis Of Different Warm Mix Asphalt Containing Reclaimed Asphalt Pavement. Proceedings of the XXVth Road World Congress, Seoul.

Yuliestyan, A., Cuadri, A. A., García-Morales, M., & Partal, P. (2016). Influence of polymer melting point and Melt Flow Index on the performance of ethylene-vinyl-acetate modified bitumen for reduced-temperature application. Materials & Design, 96, 180–188.

Climate resilient slope stabilization for transport infrastructures

A. Faiz
Consultant, World Bank, Washington DC, USA

B.H. Shah
Consultant, National Highway Authority and Punjab Forest Department, Islamabad, Pakistan

A. Faiz
Analyst, Faiz and Associates, LLC, Arlington VA, USA

ABSTRACT: In regions subject to climate stress from increased storm intensity and frequency, soil bioengineering coupled with appropriate water management techniques can help prevent slope failures in transport infrastructure facilities. Soil bioengineering uses locally available plant and vegetative materials to deter slope failure and treat slope instability. In soil bioengineering systems, grasses and plants, especially deep-rooted species, are an important structural component to reduce the risk of soil erosion and to stabilize slopes. Where technically feasible, use of soil bioengineering alternatives produces equal or better economic and environmental results than the sole application of traditional geotechnical solutions. The paper includes a case study to show soil bioengineering and biotechnical applications in remediating failed slopes on an expressway in Pakistan. The underlying bioengineering principles and techniques for erosion control and slope stabilization are equally applicable to railways, airports, ports and other physical infrastructure.

1 INTRODUCTION

Soil bioengineering combined with appropriate water management can be the key to preventing slope failures in transport infrastructure facilities, especially in regions subject to climate stress from increased storm intensity and frequency. Soil bioengineering is a technique that uses plants and vegetative material alone, whereas biotechnical techniques use plants in conjunction with more traditional engineering measures and structures to stabilize slopes and control erosion of stream banks. Moreover, soil bioengineering and biotechnical techniques contribute to sustainable development practices by reducing the environmental impacts of highway construction, operations and maintenance (Shah 2008, Fay et al. 2012) Vegetative or inert structures can be used alone or jointly to stabilize engineered or natural slopes and reduce the risk of landslides. Stabilization measures utilizing soil bioengineering are designed to aid or enhance the reestablishment of vegetation.

Planting trees and vegetative growth alone can play an important role in stabilizing slopes by intercepting and absorbing water, retaining soil below ground with roots and above ground with stems, controlling runoff by providing a break in the path of the water and increasing surface roughness, and improving water infiltration rates, soil porosity, and permeability (Fay et al. 2012).

Different types of vegetation serve different functions. Grasses and other herbaceous cover protect sloped surfaces from rain and wind erosion. Shrubs, trees, and other vegetation with deeper roots are effective at preventing shallow soil failures, as they provide mechanical reinforcement with the roots and stems and modify the slope hydrology by root uptake and foliage interception (Schor & Gray 2007). The main function of the structural elements is to help vegetation to become established and take over the role of slope stabilization; these inert structural components eventually deteriorate or disintegrate. Appropriately designed and installed vegetative systems can be self-repairing, with only minor maintenance to maintain healthy and vigorous vegetation (USDA 1992).

As with all engineering works, it is most important that the techniques selected are correct for the site to be treated, and that the work is carried out with all due care and attention (Howell 1999a).

In general, combined slope protection systems have proven to be more cost effective than the use of vegetative treatments or structural solutions alone. The overall cost of the soil bioengineering or biotechnical components in slope stabilization projects is often little more than one percent of the total project budget (Fay et al. 2012). Where technically feasible, use of soil bioengineering alternatives tends to produce more sustainable outcomes than the sole application of traditional geotechnical solutions (Faiz et al. 2015).

2 CLIMATE RESILIENT SLOPE STABILIZATION

2.1 Slope stabilization

Slope stabilization is the science and art of managing slopes (both natural and manmade) that pose a hazard to the built environment including roads, railways and other physical infrastructure.

Slope instability is generally a mass-wasting process, while erosion is typically a surficial process. Stabilization solutions address both processes and can range from allowing grass to reestablish on a disturbed slope to building an engineered retaining structure. The treatment depends on the affected area, technical feasibility, and cost. The 'one-size-fits-all' approach does not work as site conditions and constraints vary greatly.

2.2 Causes and characteristics of slope instability

Most landslides and slope stability problems involve water-related factors–presence of natural springs, seepage from water channels and septic tanks, improper disposal of black water, intense rainfall (cloudbursts), rapid snowmelt, floods, rapid changes in water table and soil pore pressures, as well as stream erosion. Other isolated causes include earthquakes, volcanic eruptions, but above all human activity, including improper road construction techniques, e.g. uncontrolled clearing and grubbing, blockage of natural drainage paths, side-casting of excavated earth and rock, poor placement of fill material, uncontrolled blasting, and vibrations from passing vehicles.

The shape of slope is often a defining factor in slope stability. Natural slopes are generally concave (most stable and least susceptible to erosion). Man-made slopes, on the other hand, are linear (in many cases, these slopes erode until they assume a concave shape). Slope linearity is often broken by benching and terracing, but poorly designed benches can cause severe gully erosion at water outfalls (Schor & Gray 2007, Shah 2008).

Other typical features of unstable slopes include: (i) excessive slope angle or height; (ii) increased slope loading often due to slope toe cutting by streams and rivers and accumulated material/man-made loadings at crest; (ii) unstable geological strata and presence of joints, faults, disconformities, creep, and weak foundations; (iii) poor drainage, seepage, natural springs, high soil moisture and soil lique-

faction; and (iv) eroded stony surfaces, stripped of vegetation and trees, resulting from uncontrolled logging, wildfires, animal grazing, foraging for fuelwood, solid waste disposal and construction activities (Shah 2008).

2.3 Ensuring climate-resilient slopes

A basic premise for obtaining stable and climate resilient slopes in transport infrastructure is to address slope protection and drainage (both surface runoff and groundwater) issues during location, design and construction—and not as an afterthought. Slope instability has its genesis for the most part in shoddy engineering or cost cutting design and construction shortcuts. Climate resilience is enhanced by using a combination of drainage, slope protection and stabilization measures, including both inert (engineered) and vegetative components. Appropriate technologies include retaining structures and earthworks (e.g. benched slopes), mechanical stabilization (e.g. mechanically stabilized earth walls), vegetative erosion control (e.g. hydroseeding), soil bioengineering, and biotechnical stabilization.

3 SOIL BIOENGINEERING

Soil bioengineering uses locally available plants and plant material and a minimum of equipment to inhibit slope failure and treat unstable critical slopes. It attempts to mimic nature (Lewis et al. 2001). In bioengineering applications, it is necessary to distinguish between slope protection and slope stabilization, the former is aimed at controlling slope erosion and shallow slope failure, while the latter is aimed at remediating failures associated with deeper slope movement. Bioengineering applications combine elements of protection and stabilization to prevent long-term, permanent slope deterioration. They are most effective in combating erosion (both rill and gully), soil flowage (up to a depth of 150 mm) and localized translational shear failure of soil slopes (TRL/ODA 1997). Deep rooted grasses (e.g. Vetiver) and hydrophilic trees (such as poplar and acacias) have shown to be effective in soil fixing and protecting slopes up to a depth of about 2–3 meters.

3.1 Vegetative stabilization mechanisms

Soil bioengineering utilizes six stabilization mechanisms, conceptually similar to how a soil mass is fixed by a deep-rooted tree as shown in Figure 1. These six functions serve: (i) to catch eroded materials with physical barriers (e.g. vegetated walls, plants, trees); (ii) to armor the slope from erosion

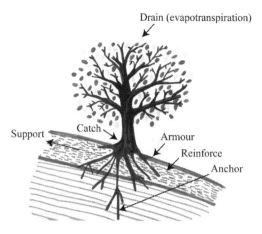

Figure 1. Stabilization mechanisms involved in soil mass fixation by deep-rooted vegetation.

Table 1. Comparison of geotechnical engineering and soil bioengineering systems relative to stabilization function.

Function	Geotechnical engineering system	Soil bioengineering system
Catch	Gabion wire bolster Check dams Catch walls	Contour lines (grass lines; brush layers) Live check dam Shrubs and bamboo (many stems)
Armor	Revetment walls Stone pitching	Mixed plant stories, giving complete cover Grass carpet (dense, fibrous roots)
Reinforce	Reinforced earth Soil nailing	Densely-rooting grasses, shrubs, and trees Most vegetation structures
Anchor	Rock anchors by bolting	Deeply-rooting trees and shrubs (long string roots)
Support	Retaining walls Breast walls Prop walls	Large trees and bamboo (deep and dense root system)
Drain	Surface drains French drains	Down slope and diagonal vegetation lines Angle fascines, pole drains

Source: Shah 2008, Howell 1999.

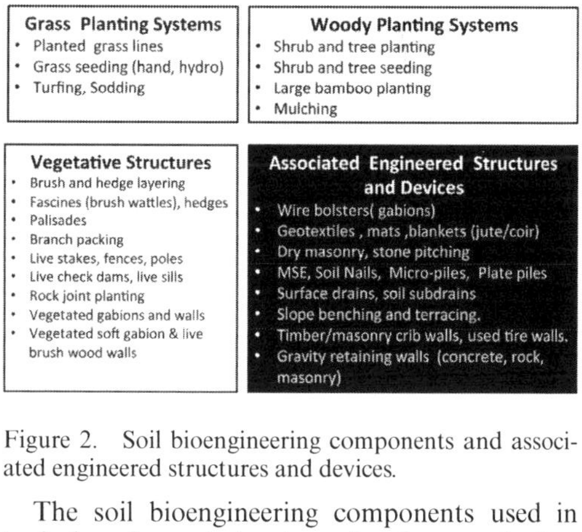

caused by runoff or rain splash using vegetative cover, partial armoring using lines of vegetation; (iii) to reinforce soil physically with plant roots; (iv) to anchor surface material to deeper layers using large vegetation with deep roots; (v) to support soil by buttressing with vegetated walls or large vegetation; and (vi) to drain excess water from the slope through evapotranspiration and the use of drains (Howell 1999a, Schor & Gray 2007).

A comparison of geotechnical and soil bioengineering techniques and their combined use in relation to the six stabilization functions is summarized in Table 1. The choice of stabilization techniques (both geotechnical and bioengineering) depends on an identification of the functions needed to stabilize and protect the slope. A detailed description of these techniques and their application is provided by Howell 1999b.

3.2 Soil bioengineering components and structures

There is a wide range of techniques associated with bioengineering systems and their management as reported by Coppin & Richards 1990, Gray & Sotir 1996, Schiechtl & Stem 1996, Howell 1999b, Shah 2008, and Fay et al. 2012. Grasses, shrubs and trees are the building blocks of soil bioengineering. As shown in Figure 2, they can be used directly for slope stabilization or transformed into vegetative structural components with hedge and brush layering, branch packing, stakes, palisades, fascines, and so on. More advanced components include vegetated walls and drains, and live fences, sills, check dams. The effectiveness of soil bioengineering systems is significantly enhanced when they are used in conjunction with engineered structures and devices.

Grass Planting Systems	Woody Planting Systems
• Planted grass lines • Grass seeding (hand, hydro) • Turfing, Sodding	• Shrub and tree planting • Shrub and tree seeding • Large bamboo planting • Mulching

Vegetative Structures	Associated Engineered Structures and Devices
• Brush and hedge layering • Fascines (brush wattles), hedges • Palisades • Branch packing • Live stakes, fences, poles • Live check dams, live sills • Rock joint planting • Vegetated gabions and walls • Vegetated soft gabion & live brush wood walls	• Wire bolsters(gabions) • Geotextiles , mats ,blankets (jute/coir) • Dry masonry, stone pitching • MSE, Soil Nails, Micro-piles, Plate piles • Surface drains, soil subdrains • Slope benching and terracing. • Timber/masonry crib walls, used tire walls. • Gravity retaining walls (concrete, rock, masonry)

Figure 2. Soil bioengineering components and associated engineered structures and devices.

The soil bioengineering components used in typical engineered structures for slope protection and stabilization are summarized in Table 2.

The selection of a soil bioengineering technique for protecting or stabilizing a particular slope depends on many site-related factors. In order to select the most relevant and appropriate technique, it is necessary to consider these factors in as

Table 2. Soil bioengineering applications in slope stabilization structures.

Structure	Bioengineering component
Retaining walls	Vegetated soft gabions, timber crib walls and used tire walls; Live brushwood
Diversion channels	Sodding treatment and stone pitching
Surface drains	Sodding treatment and stone pitching with live sills
Check dams	Vegetated poles, live brush wood, palisades, vegetated brush wood
Barriers for fixing loose debris	Brush wattles, brush layering, brush fences, semi-dead fences with live hedges, seeding and turfing
Retaining walls	Vegetated soft gabions, timber crib walls and used tire walls; Live brushwood

Figure 3. Soil bioengineering treatment of a landslide in Northern Pakistan.
A) Treatment plan for landslide remediation. B) Actual application of remediation measures. C) Before bioengineering treatment. D) After bioengineering treatment.

much detail as time and skill permit. Assessments rely mainly on experience as it not practicable to quantify the many variables involved. Although vegetation cannot be designed or built to engineering specification in the conventional sense, it can be selected and arranged on the slope to perform a specific engineering function. This function should be identified as part of the process of bioengineering design (TRL/ODA 1997). Figure 3 illustrates the treatment plan and application of soil bioengineering techniques for landslide remediation in northern Pakistan, as well as 'before' and 'after' condition of the treated site.

3.3 Advantages and disadvantages of soil bioengineering

Soil bioengineering provides a sustainable means of protecting and stabilizing slopes for reasons of availability, relatively low cost, appropriateness of installation techniques and compatibility with rural environment. It is particularly appropriate in situations where large areas of slope are affected, a common situation on road cuttings and over unstable mountain slopes. The enhancement of roadside vegetation also has a positive aesthetic effect both visually and in terms of plant diversity (TRL/ODA 1997). Soil bioengineering applications are labor intensive and a source of job creation, especially for rural women, a distinct advantage in labor abundant economies. As native plant material is used, it can be sourced locally without the risk of introducing invasive plant species. Technically it is simple and does not require sophisticated engineering designs. The bioengineered structures are long lasting and get stronger with passage of time. Once mature, they become a productive asset providing timber, fuelwood and fodder, and may also provide a habitat for wildlife. Soil bioengineering is environmentally and ecologically friendly and could have a significant role in reducing GHG emissions.

Soil bioengineering, however, has some disadvantages. It is a season bound activity and cannot be carried year round. Moreover, it is affected by high and low temperatures as well as dry and wet spells. The bioengineered structures can be adversely affected by droughts and intense storms (damage from wind, rain, hail and snow) especially during the first 2–3 years. The timing of implementation also becomes a critical factor, with planting mostly confined to the dormant season (late fall or early spring in temperate zones, and just before the rainy season in tropical zones). Its implementation requires on-the-job training and skilled oversight, as well and support and vigilance of local communities to protect newly established vegetation and structures. Plants need protective fencing from foraging and grazing animals for 4–5 years, and from humans seeking fodder, firewood, and forage. The design and implementation of bioengineering programs requires interdisciplinary skills and interagency coordination (e.g. between forestry and road departments) and may be difficult to obtain in institutionally weak settings.

4 CASE STUDY: ISLAMABAD-MURREE EXPRESSWAY (PAKISTAN)

The 43 km Islamabad—Murree Expressway (E75) connects Islamabad, the country's capital, with Lower Topa (Murree District) in Pakistan. This four-lane divided expressway was constructed by the National Highway Authority (NHA) over the period 2004–2012. The E75 traverses the Murree Formation, a geological zone of young, steep and fragile mountains (Bossart & Ottiger 1989), consisting primarily of cross bedded sandstone and claystone/shale strata, with extensive springs and seepage zones. The rock structure varies from fine to coarse grain, with parallel laminations, amalgamation, and calcitic veins (Khan & Niazi 2010). Unstable geology, a wet climate (torrential monsoon rains in summer), and human activity (deforestation, unstable and poorly drained terraces for cultivation and housing, and improper disposal of sewage) are some of the causal factors responsible for the landslide hazard along the route. The geographical location of E75 is shown in Figure 4.

The E75 has an average roadway formation width of about 25 m, which required extensive clearing of natural vegetative cover and deep linear cuts (with side casting of waste material) in steep (30–40% gradients) natural slopes, making them unstable and triggering huge landslides even during the construction period. Other landslides resulted from bank cutting by mountain streams due to inadequate drainage and lack of slope protection measures such as provision of cut-off drains and check dams. Based only on rudimentary geotechnical investigation and design, resistance structures (massive reinforced concrete retaining walls) were constructed at points where the landslides were triggered during construction. The height, length, and width of these walls were dimensioned as a function of the severity and extent of the landslide (Figure 5).

Control structures such as diversion channels, and surface and subsurface drains were not used despite their critical role in landslide management and control. Measures to control rill and gully erosion of exposed slopes were also absent during expressway construction. As a result landslides remained active along the expressway during and after construction and many reinforced concrete walls failed with increased water content and pore pressures in the backfill during the wet season.

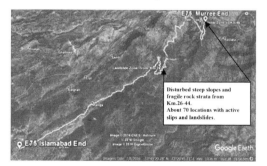

Figure 4. Physical location of Islamabad—Murree Expressway, E75, with concentration of unstable slopes between Km 26 and 44.

Figure 5. Massive reinforced concrete walls used as resistance structures to contain landslides.

The monsoon seasons since 2010 have been characterized by more intense and frequent rainfall, with severe weather events, resulting in landslides at some 70 locations along an 18 km section of the expressway. The roadway has subsided at several sites resulting in cracking and collapse of the massive retaining walls and closure of one carriageway in a few sections. Remedial treatment utilizing a combination of geotechnical and biotechnical measures has helped to control the landslides and stabilize the disturbed slopes at most of the affected locations. NHA, however, continues to struggle with the E75 slope instability problem. It owes its genesis to poor route location, design and construction practices, and will now require continuous monitoring and intensive routine and emergency maintenance.

4.1 Landslide stabilization project

In January 2014, NHA in collaboration with the provincial Punjab Forest Department initiated a project for the stabilization and follow-up maintenance of active landsides with biotechnical, soil bioengineering and biological methods. In total, about 70 active landslides between Km 26 and Km 44 (18 km long section) were treated during 2014, 2015 and spring of 2016. In addition gabion and vegetated gabion check dams were installed on mountain streams endangering the expressway's right of way through bank cutting. The project was allocated a budget of PKR 42.5 million (USD 453,000), at an average cost of about $75,000 per landslide treatment location.

4.2 Measures used for slope stabilization

The slope stabilization measures utilized by the project included engineering, biotechnical, soil bioengineering and biological (vegetative) applications. The primary focus was on soil bioengineering applications.

4.2.1 Engineering measures
These consisted of *gabion check dams* on 15 mountain streams to reduce bank cutting and common *gabion retaining walls* to further strengthen the Reinforced Concrete (RCC) retaining structures. Each stream

was treated with 3–5 check dams, with the height ranging from 2–3 m. The main purpose of the gabion walls, placed in one or two layers, was to prevent landslide debris from overtopping the RCC walls.

4.2.2 Biotechnical measures

When living plants are incorporated in engineering structures they are called biotechnical structures. Such biotechnical structures were used in the project for retaining walls, check dams, diversion channels and surface drains.

Retaining walls. For constructing a *vegetated dry masonry wall* brush and hedge layering was applied at every 0.5 m height of the dry masonry wall. Brushwood of poplar, willow, mulberry and seedlings of <u>*Robinea pseudoacacia*</u> were used in the brush layering treatment. Although used vehicle tires have been used for constructing retaining walls in different parts of the world, the use of a <u>*vegetated used tire wall*</u> was pioneered under this project for the first time in 2015. Used tires were placed in staggered layers with stakes and seedlings planted in the earth-filled tire. In addition to the plantings, brush layering treatment was also applied to each layer of tires (Figure 6).

Vegetated check dams. Both gabions and dry masonry were utilized for construction of check dams. V*egetated gabion check dams* were constructed on streams running laterally to the dual carriageway formation. These check dams, one meter in height, consisted of three layers of gabion wire boxes. Each gabion layer was covered by soil and brushwood of poplar willow *Ipomea* with the basal ends of brushwood extending beyond the check dam wall. The check dams were back filled with soil for rooting of the brush wood (Figure 7). *Vegetated dry masonry check dams* were constructed in the gullies within the landslide-affected slopes and proved effective in stabilizing them.

Drains. Biotechnical treatment was applied to *diversion channels* to divert runoff water from entering the slide area and to *surface drains* to remove rainwater from the slide area. The diversion

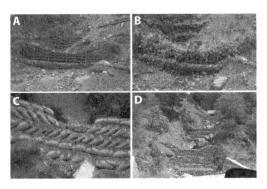

Figure 7. Vegetated check dams. A) Pole construction before sprouting and B) after sprouting. C) Soft gabion. D) Dry masonry.

channels and surface drains both were treated with stone pitching and supported with willow stakes to help retain the stones in place. The stakes after sprouting help stabilize the channels and the drains.

4.2.3 Soil bioengineering measures

A variety of soil bioengineering structures were put to use as retaining walls, drainage structures, and mechanical barriers to fix the landslide debris before planting.

Retaining Walls. The structural elements for these walls consisted of soft gabions made from used empty cement bags filled with landslide debris or timber crib walls. The *vegetated soft gabion walls* were used on most of the landslides because they are an economical alternative to standard wire gabions filled with stone, especially where stones are not available on site or it is cumbersome to haul stones up steep eroded slopes. The bags were filled with debris at the toe of the landslides and used as the foundation for constructing the retaining wall with brush and hedge layering. After sprouting, the brushwood provides a strong and durable retaining structure. The bags eventually rot and plant roots, trunks and branches serve as a vegetated wall (Figure 8).

Although *vegetated timber crib walls* are not economical in Pakistan they were used at two landslide locations in this project as there was no other suitable technical alternative. The timber cribs were constructed using brushwood of native poplar and willow species (Figure 9).

Drainage structures. Soil bioengineering was used for construction of diversion channels, surface and pole drains and check dams. Extensive *diversion channels* were excavated at ten project sites in the undisturbed area above the active landslide. Some excavated channels were lined with stone pitching but others were treated with sodding and supported with vegetated soft gabion walls on downhill side. Sodding is quite effective as the grass cover protects the channel from scoring

Figure 6. Vegetated used tire wall at time of construction and after sprouting of vegetation.

A - Newly constructed; Winter

B - Sprouted vegetation; Spring

C - Fully vegetated; Late summer

Figure 8. Vegetated soft gabion wall (one seasonal cycle).

Figure 9. Timber crib walls used in combination with gabions.

and helps stabilize the channel quickly. Excavated *surface drains* were sodded and supported by soft gabion walls at five sites to drain runoff produced on the landslide. On steep slopes surface drains are not stable and *pole drains* were utilized for draining surface seepage water. Pole drains were installed at four sites by excavating a trench and placing bundles of live brushwood in the trench and covering it with soil, with one fifth portion of the brush wood bundle left uncovered for sprouting. Four types of soil bioengineering check dams were used for stabilizing gullies on the bigger landslides, namely, *palisades, vegetated soft gabion, vegetated pole, and vegetated brushwood check dams.*

Mechanical barriers (contour walls). It is necessary to fix the landslide mass debris before planting; otherwise seedlings planted in the upper portion of the landslide are uprooted while plantings on lower side are buried under the debris. The normative engineering solution for fixation of debris is the construction of stone *contour bunds* (contour walls). The soil bioengineering techniques for creating mechanical barriers are *brush layering, bush hedge layering, semi-dead fences with live hedges and sodding.* The use of these live mechanical barriers in the project also helped to inhibit rill and gully formation.

4.2.4 *Biological (vegetative) measures*

Purely biological applications for erosion control and slope stabilization used in the project included planting of bare rooted seedlings and tube plants, staking with live willow and poplar branches, and sowing seeds of indigenous plant species. A mixture

of plant types was used to provide a range of rooting depths. As far as possible, local species rather than imported material were used because native plants are better adapted to grow in the hostile conditions found on bare sites and are resistant to local diseases. Fast growing broad leave tree species such as *Robinea pseudoacacia, Ailanthus altisimma,* Poplar (*Populus detoides*) and Willow (*Salix tetrasperma*) were planted on landslide debris mass to provide quick vegetation cover. It was also necessary to plant local conifer tree species, Chir pine (*Pinus roxburghi*) and at higher altitude Deodar (*Cedrus deodara*), in combination with broad leave tree species to encourage ultimate colonization by the indigenous varieties. Stakes were prepared by cutting fresh branches of willow and poplar (*Populus deltoides* and *Populus ciliata*) to supplement the planting of seedlings. To supplement these plantings, the seeds of indigenous trees and bushy plants were also sown, namely Chir pine, *Acacia modesta, Wendlandia exserta* (Ukan), *Leucania leucocephalla* (Ipel ipel) and *Dodenia viscosa* (Sanatha).

4.3 *Results of slope stabilization project*

The E75 Slope Stabilization Project treated 70 landslides ranging from earth slumps to rock and debris slides to earth subsidence: 51 in 2014, 17 in 2015 and 2 new ones in 2016. The remedial work was carried out in spring to permit the vegetative elements to root before the monsoons. The typical treatment (other than diversional channels, drains and check dams as needed) consisted of vegetated dry stone masonry and soft gabion walls, brush layering and planting, staking, and sowing of indigenous tree species (Figure 10). This basic soil bioengineering treatment helped stabilize all the treated landslides except at three locations, despite unusually heavy rains both during the two ensuing monsoons as well as in the winter of 2015 and spring of 2016 (which triggered several new slides).

Vegetation recovery was impressive at all the three failed locations. However, the presence of

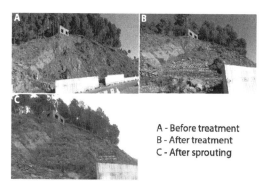

A - Before treatment

B - After treatment

C - After sprouting

Figure 10. Treatment of landslide at kilometer 29 + 250 on Route E75 below a damaged house.

a spring in the middle of the landslide and seepage from the septic tanks of two houses above the landslide caused the soil bioengineering treatment to fail in one location. The second failure was due to the landowner not permitting the proper installation of surface drains on his property. And the third failure was due to water seepage into the slide area from the drainage runoff from roofs of houses located above the landslide; the application of a pole drain, however, proved effective in draining the seepage water and the landslide was eventually stabilized with vegetation. At six locations, cattle grazing could not be controlled and vegetation recovery was not satisfactory.

5 CONCLUSIONS AND RECOMMENDATIONS

With increasing rainfall intensity and shorter return periods of climate-induced extreme weather events, unstable slopes and landslides pose a serious risk to transport infrastructure assets, especially in mountainous areas. There is an urgent need to deliver cost-effective and sustainable solutions to mitigate the landslide risk.

As shown by the case study included in this paper, barring catastrophic events like earthquakes or floods, most landslide problems affecting transportation facilities have their origin in faulty route location, design that is not sensitive to environmental and climate change considerations or construction and maintenance practices that worsen the vulnerability to the landslide hazard, especially in mountainous terrain. The study shows that a majority of slope instability problems can be rectified by the application of fairly simple and environment friendly slope stabilization measures such as vegetative erosion control, soil bioengineering, and biotechnical stabilization. Climate resilience is significantly enhanced by using a combination of drainage, slope protection and stabilization measures, including both inert (engineered retaining structures and earthworks and mechanical stabilization) and vegetative components.

Most landslides and slope stability problems involve water-related factors. Soil bioengineering combined with appropriate water management can be the key to preventing slope failures in transport infrastructure facilities, especially in regions subject to climate stress from increased storm intensity and frequency. As recommended by Howell (1999a), soil bio-engineering techniques for stabilizing slopes should be used on:

i. all areas of bare soil on cut face and embankment slopes;
ii. wherever there is a risk of gullying;
iii. all slopes where there is a risk of shallow slumps or planar slips of less than 500 mm depth;

iv. any slope segment in which civil engineering structures are planned or have been built, and the surface remains bare;
v. any area that has failed and needs to be restored, other than rock slopes; and
vi. any area, such as tipping and quarry sites, or camp compounds, that requires rehabilitation.

ACKNOWLEDGEMENT

Maryam Faiz prepared all the art work for the paper. Her technical support is gratefully acknowledged.

REFERENCES

Bossart, P. & Ottiger, R. 1989. Rocks of the Murree Formation in Northern Pakistan: Indicators of a descending foreland basin of late Paleocene to middle Eocene age. *Eclogae Geologicae Helvetiae 82(1):133–165.*
Coppin, N.J. & Richards, I.G. 1990. *Use of vegetation in civil engineering.* London: Butterworths.
Faiz, A. et al. 2015. Prevention is better than cure: bioengineering applications for climate resilient slope stabilization of transport infrastructure assets. *Proc. of First International Conference on Surface Transportation System Resilience to Climate Change and Extreme Weather Events. 16–18 September, 2015.* Washington, DC. Transportation Research Board.
Fay, L. et al. 2012. Cost-Effective and Sustainable Road Slope Stabilization and Erosion Control. *NCHRP Synthesis 430.* Washington, DC. Transportation Research Board.
Gray, D.H. & Sotir, R.B.1996. *Biotechnical and soil bioengineering slope stabilisation. A practical guide for erosion control.* New York: Wiley & Sons.
Howell, J. 1999a. Bioengineering in Bhutan: Interim manual for the road sector on bioengineering techniques for slope protection and stabilization. Thimpu. Department of Roads.
Howell, J. 1999b. *Roadside Bio-engineering: Site Handbook*, Kathmandu. His Majesty's Government of Nepal (Department of Roads).
Khan, I. & Niazi, M.J. 2010. *Geological Report of Hazara Division.* Abbottabad, Pakistan. COMSATS Institute of Information Technology.
Shah, B.H. 2008. *Field Manual on Slope Stabilization,* Islamabad, United Nations Development Program.
Schiechtl, H.M. & Stern, R. 1996. *Ground bioengineering techniques for slope protection and erosion control.* Oxford: Blackwell Science.
Schor, B. & Gray, D.H. 2007. *Landforming: An Environmental Approach to Hillside Development, Mine Reclamation and Watershed Restoration.* Hoboken, NJ. John Wiley & Sons. J.
TRL/ODA 1997. Principles of low cost road engineering in mountainous regions. *Overseas Road Note 16.* Crowthorne, Berkshire. Transportation Research Laboratory.
USDA.1992, Soil Bioengineering for Upland Slope Protection and Erosion Reduction, *National Engineering Handbook, Part 650, Engineering Field Handbook, Chapter 18.* Washington DC. Natural Resources Conservation Service, US Department of Agriculture.

Transport Infrastructure and Systems – Dell'Acqua & Wegman (Eds)
© 2017 Taylor & Francis Group, London, ISBN 978-1-138-03009-1

Monitoring of railway track with light high efficiency systems

S. Cafiso, B. Capace, C. D'Agostino, E. Delfino & A. Di Graziano
Department of Civil Engineering and Architecture, via Santa Sofia, University of Catania, Catania, Italy

ABSTRACT: Rail Track performance is highly dependent upon the magnitude and variation of differential track geometry. Consequently, there is an increasing interest in obtaining a consistent measure of trackbed stiffness and track geometry in the physical and time restrictions applicable on live track for assessing potential maintenance requirement and subsequent design of remedial measures. Furthermore, a lack of a systematic monitoring brings to the impossibility to produce an effective long term track management system, by allocating budget where emergencies come. In this framework, in the case of local railway track with reduced gauge, the use of the traditional high speed track monitoring systems is not feasible. This paper details a site investigation comprising trial Ground Penetrating Radar (GPR), Light Falling Weight (LWD) and Laser Measurement System (LCMS) testing. These systems are currently used in road pavement maintenance where have shown their reliability and effectiveness. Application of such Non Destructive Tests in railway maintenance are promising but in the early stage of investigation. In the paper literature review and trial site testing are used to identify Strengths, Weaknesses, Opportunities and Threats (SWOT analysis) of the application of GPR, LWD and LCMS for the assessment of trackbed stiffness and geometry.

1 INTRODUCTION

Railway lines are investments with very long life. Today many tracks are over 100 years old. Of course components and mainly rolling stock have been exchanged during the years, but parts of the track might remain the same—especially the substructure. Typical lifetimes of for rails are 30–60 years and turnouts 20–30 years (H. Sundquist, Byggande, 2000). However, to ensure this long life a large amount of maintenance is necessary. There are several, quite obvious reasons for maintenance, for example:

- Safety—Probability for accidents needs to be low.
- Comfort—Comfort is important, both for passengers and freight as well as for the environment in terms of noise and vibration.
- Serviceability—With lots of failures and speed restrictions due to safety etc. the serviceability of the track will be low.
- Economy—A track with low quality is cost driving, since the deterioration of both track and trains will be higher.

At the same time maintenance is expensive. Optimization and Life Cycle Cost (LCC) planning is needed. Since many railway tracks are quite old, the demands put on the tracks today are different from the ones when the tracks were built. There is a clear trend towards higher speeds and higher capacity (more trains on the tracks and heavier trains). Nowadays more trains occupy the track and the competition with other means of transportation becomes harder. To face the new circumstances, more effort has to be put on track maintenance to ensure the issues of safety, comfort, serviceability and economy. Moreover, there is also a trend towards decreased time for maintenance and decreased funds for maintenance.

Theoretically, the number one solution for optimal maintenance is to do the right measure at the right time to fulfil the requirements of safety, comfort and serviceability in terms of LCC. This task is practically impossible, since it requires complete knowledge about the current condition of the track and what effect different kinds of maintenance, or no maintenance, will have on the track. Instead, the goal of condition based maintenance is to come as close to this optimum as possible. This could be done with the help of measurements of important parameters which are analyzed to give knowledge about the condition of the track. After that, regulations, budget constraints and knowledge about the deterioration (from models and/or skilled engineers) are used to make decisions about maintenance. Unfortunately, again this kind of approach require a control of the network, or of those parameters able to represent the real condition of the track and substructure.

2 CONTROL TEST IN RAILWAY

Track failures is one of the main problems that railroads have faced since the earliest days is the prevention of service. The North American railroads have been inspecting their costliest infrastructure asset, the rail, since the late 1920's; and, with increased traffic and more frequent failures on railroads, rail inspection is more important today than it has ever been (C. NDT, 2013). Although the focus of the inspection seems like a fairly well-defined piece of steel, the testing variables present are significant and make the inspection process challenging. To keep railroads safe and prevent any high maintenance costs caused by failures on the railroads, scheduled inspections must be performed on rail tracks, soil, and bridges. There are several types of track inspections such as soil inspection, railroad bridge inspection, and railroad inspection, and each track inspection type has their subcategories shown in the following Figure 1.

Soil inspection investigates the Ballast, Subgrade, and Roadway (BSR) component, which includes all earthen materials on the track structure, tracks, and embankments (Uzarski et al., 1993). It focuses on the thickness of the ballast, subsoil material and geotechnical properties of subgrade. In addition, the inspection is executed mainly by digging trenches at evenly spaced intervals and in locations of special interest (Hugenschmid, 1999). Soil inspection also investigates the plants on the railroads. For instance, Eriksen et al. (2004) tried to improve the reliability of soil inspection by adding the investigation of plant growth in the inspection process. Rail inspection investigates the rail heads, switch blades, bolt holes, foot of the rail, rail gauge, thermite welds, etc. (H. NDT, 2014). Latest technology can improve the rail inspection with higher cost of the equipment (Cerniglia et al., 2006). This technology is mainly developed directly from the railway Agency to monitor their network. The great advantage is related to the speed and precision of the measurement, avoiding to interfere with the normal use of the infrastructure.

Generally local railways have not the most advanced equipment to survey the network as well as the economic possibility for a systematic global survey. The result is that the maintenance is carried out to avoid track failures with visual and manual inspection of the track based on the experience of the technicians.

Local railway tracks, usually with reduced gauge have not the potentiality to develop their own high efficiency monitoring systems. Unfortunately, the lack of a systematic control and the economic difficulties, especially in local railway track, lead to the impossibility of producing a long-term effective track management system. Specifically, in presence of local tracks with reduced gauge, the use of traditional monitoring systems track at high speed is not feasible. In this framework the preventive monitoring of the track with alternative Non-Destructive Techniques (NDT), such as GPR, LWD and LCMS are promising for soil and rail inspection and to plan a preventive maintenance from the early stage of investigations. In recent decades, these devices have already proven their effectiveness in the field of road pavement engineering and prospects are the same in the rail sector which is increasingly growing (Cafiso et al. 2016a, Cafiso et al. 2016b).

3 SWOT ANALYSIS

Because of the interest in the introduction of new NDT for the investigation of railway track, a SWOT analysis will be performed basing on the literature review and trial tests of GPR, LWD and LCMS.

Originated by Albert S. Humphrey in the 1960s, SWOT analysis is a basic, straightforward model that assesses what an organization can and cannot do as well as its potential opportunities and threats. It is an extremely useful tool for understanding and decision-making for all sorts of situations and disciplines in project planning, management and business for investigating problems from a strategic perspective.

SWOT is an acronym for Strengths, Weaknesses, Opportunities and Threats that, specifically, takes information from an environmental analysis and

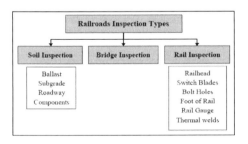

Figure 1. Types of track inspections in a railroad.

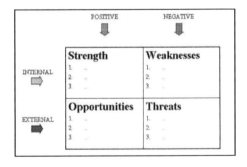

Figure 2. SWOT analysis general schema.

separates it into internal strengths and weaknesses, as well as its external opportunities and threats.

SWOT Analyses are often arranged as a 2 by 2 matrices with the lists of strength and weaknesses in the first two boxes in the first row and the lists of opportunities and threats in the second row. By arranging the analysis this fashion, the lists are separated into internal factors that can affect a project on the first row and external factors on the second row. In addition, the first column consists of the positive factors (strengths and opportunities) and the second column consists of negative actors (weaknesses and threats.). This method provides a simple framework to keep lists organized and conceptualize how the lists are related. SWOT analysis is presented in the present research work, for the application of GPR, LWD and LCMS to survey a railway track.

The analysis was conducted for each single equipment after a trial on a real railway environment and post processing of the results.

SWOT analysis resulted particularly interesting to evaluate the feasibility of application of those equipment in a different context than their usual test environment.

4 GROUND PENETRATING RADAR (GPR)

The ground penetrating radar is a geophysical radar system with two antennas and receivers used to perform non-destructive investigations of underground characteristics with high resolution and in depth (up to 3.2 m from the surface).

GPR operation principle is based on electromagnetic theory, its functioning consists in sending short electromagnetic pulses into a medium and when pulses achieve an interface they are reflected back partially and collected by the receiving antenna (Figure 3). The reflected energy is displayed in waveforms and the greatest amplitudes represent the interfaces between layers with distinct dielectric characteristics (Daniels, 2004;). Therefore, GPR measures the travel time between the transmission of the energy pulses and its reception. Transmission and reception of radar pulses are performed from one or more antennas that are moved on the investigated medium (Sussmann et al., 2003). The collected data are processed and saved on a control unit, that is also used to generate the necessary pulses for the operation of the antennas.

As mentioned GPR can provide a fast, nondestructive measurement technique for evaluating railway track substructure condition (Olhoeft & Selig, 2002; Selig et al. 2003).

Main application of GPR in railway track are:

– monitoring the condition of railway ballast, and detect zones of clay fouling leading to track instability

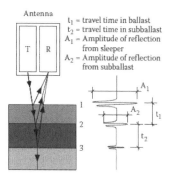

Figure 3.   GPR principles (Saarenketo, 2006).

– Mapping soil, rock or fill layers in geological and geotechnical investigations, or for foundation design.

The track substructure, consisting of the ballast, sub ballast, and subgrade layers, has a profound influence on track performance. The substructure performance is significantly affected by moisture accumulation and thickness of the roadbed layers (Selig & Waters, 1994).

Accurate knowledge of the substructure condition is important in effectively assessing the potential for service interruptions and the need for slow orders. A significant part of a railroad's track maintenance budget is allocated to correct track roughness that is caused by movements in the substructure under repeated train loading. In this field, the GPR method seems to be a good alternative to traditional core inspection techniques. Methods of applying GPR to railways are being developed to provide a continuous evaluation of the track substructure conditions relative to subsurface layering, material type, moisture content and density.

GPR application to railways, during construction and monitoring phase, is relatively recent.

In Germany, Saarenketo (2006) referred that Göbel et al. (1994) performed GPR tests for determining: ballast thickness, layers interfaces, ballast pockets and mudholes location.

Jack and Jackson (1999) studied ballast layer along a track using two antennas with different frequencies, they found: clearer ballast interfaces indicating a clean ballast and thickness variations, affirmed GPR as a useful tool for identifying track sections with urgent necessity of rehabilitation.

Gallagher et al. (1999) researches found positive results for survey of ballast/subgrade interface, such as anomalies detection. Hugenschmidt (2000) reports a study developed on different alignments, evaluating the ballast thickness, fouled zones and ballast/subgrade interface depth. Their conclusion was that radar survey is useful combined with traditional inspection methods. Recently, to determine

Figure 4.   Ground Penetrating Radar.

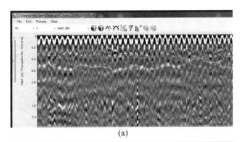

(a)

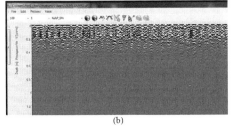

(b)

Figure 5.   Radar scans of the trial test with 600 kHz (a) and 2 GHz (b).

the correlation between water content or fouling of a railroad track and GPR signals, a full-scale railway track model was designed and constructed at the University of Massachusetts Amherst (Hamed et al. 2016). Different models were tested with moisture content conditions of dry, saturated and two points between these extremes. 450 MHz and 2 GHz frequency antennas were used to evaluate the different conditions. The results shown that the dielectric permittivity and frequency spectrum can be used as an indicator of fouling percentage and moisture content in a track. In addition, a linear correlation is observed between the fouling percentage and moisture content under saturation conditions.

4.1   GPR in field trials

The system is composed by two antennas of 600 MHz and 2 GHz frequency. The frame was adapted for the test by modifying the wheel system and the Distance Measurement Instrument (DMI) configuration. Wheels are built with cone shape and made with resin for running on the railway track and avoiding electrical interferences (Figure 4).

Figure 5 shows the radar cans obtained from the two antennas. As expected 2 GHz gives better resolution for the upper layers' interface (e.g. 0.3 m) complemented by the 600 kHz for layers' interface at lower depth (e.g. 1.0 m).

Basing on the literature review and the practical in field trial, the SWOT analysis for GPR was developed and it is reported in Table 1.

5   LIGHT WEIGHT DEFLECTOMETER

The Light Weight Deflectometer (LWD) (Figure 6), is a hand portable device that was developed in Germany to measure the soil in situ dynamic

Table 1.   SWOT analysis for GPR System.

STRENGHT (internal, positive factor)	WEAKNESSES (internal, negative factor)
• Reliable and established technology • NDT testing • Speed of execution • Reduced number of operators • More issues investigated • Potential data on the whole structure	• Radar analysis subject to interpretation of the operator • Still provisional models • Need of technologically and advanced equipment • Need of calibration tests for data deduction • Interference of magnetic materials
OPPORTUNITIES (external, positive factor)	THREATS (external, negative factor)
• Development of new models for extrapolation of results. • Approach to a field of engineering technique that is not "saturated" • Possibility of agreements with public authorities or private companies that need of an advanced know-how for the development of its own procedures	• Results distorted by unexpected and/or unforeseen conditions • Difficulties in the interpretation of the data obtained • Ignorance and mistrust of the Public Administrations in the use of these devices for NDT on railways

modulus. The LWD consists of a circular plate (150, 200, 300 mm diameter) loaded by a falling mass (10–15 – 20 kg).

The load resulting from the drop is dampen by a series of buffers, producing a transient force, dynamic in nature, simulating the effect of the vehicles moving on the road surface. At the same time the resulting deflection is measured directly under the plate. The LWD, used for the trial test, had one geophone positioned in the centre of the plate and 2 additional geophones that can be used for specific measures outside the plate.

The LWD is ideal for Quality Assurance/Quality Control on subgrade, subbase and thin flexible pavement constructions to verify that specifications are met. It can also be used to identify weaknesses, leading to further tests using FWDs and other material analysis techniques.

When applied directly on the subgrade, data processing from LWD test is carried out according to Boussinesq theory for the estimation of the Elastic modulus E by the formula:

$$E = \frac{f \cdot (1 - v^2) \sigma_0 \cdot a}{d_0}$$

where (f) is the plate rigidity factor, (a) is the load plate radius, (d_0) is the measured deflection under the center of the plate, v is the Poisson's ratio.

The influence of soil depth of LWD tests is considered to be 1÷1.5 diameters of the plate.

Neupane et al. (2016) from the university of Kansas estimated with LWD the resilient modulus of the substructure considering different combination of fouled ballast (10–40% by weight) and moisture content (1–10%). The test was conducted by reproducing in laboratory the real condition. Considering the various combination, they concluded that moisture content has the highest influence in reducing the bearing capacity.

Horníček et al. (2014) used the LWD for the long-term evaluation of the condition of trial sections with the application of under-ballast geocomposites useful to avoid long-term problems of the track geometric position caused by the pushing of fine-grained soil from the subballast into the ballast bed (so-called pumping effect) and by the missing base layer between the ballast and the subgrade.

The exploitation potential of the Lightweight Falling Deflectometer is still increasing, and the results are aptly combined with other analytical and mathematical methods (Fernandes, et al., 2012; Burrow et al, 2007). In some cases, the impact device is used for a specific assessment of spots with problems resulting from the unstable geometric track position (Sharpe & Govan, 2014).

5.1 In field trials

LWD was used to test the response in terms of deflection in railway track with different characteristics of ballast and sleepers and with the loading plate on the ballast or on the sleeper. Particularly the trial was conducted on four different sites with the same LWD configuration (plate 150 mm, weight 10 kg, drop height of about 68 cm–27 inches):

- Site 1: concrete sleepers on ballast with partial renovation and hilling.
- Site 2: concrete and wood sleepers on ballast in bad conditions.
- Site 3: concrete sleepers on just renewed ballast;

Results, reported in Figure 7, show a clear variability in the measured deflection among the sites with different ballast conditions.

Site with ballast of good quality (site 3) reports lower deflections and higher uniformity.

The SWOT analysis for LWD is reported in Table 2.

6 LASER CRACK MEASUREMENT SYSTEM (LCMS)

The LCMS system employs high speed cameras, custom optics, and laser line projectors to acquire 2D images and high resolution transversal profiles of road surfaces 4 meters wide. LCMS was originally for automatic detection of cracks, rut depth, macrotexture and other road pavement distress. The LCMS can be operated at speeds of up to 100 km/h.

Figure 6. Light Weight Deflectometer.

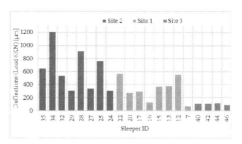

Figure 7. Light Weight Deflectometer.

Table 2. SWOT analysis for LWD System.

STRENGHT *(internal, positive factor)*	WEAKNESSES *(internal, negative factor)*
• Reliable technology • NDT testing • Reduced number of operators • Possibility to investigate the various components of the infrastructure • Application of mathematical and analytical models	• Techniques of analysis often subject to comparison • Variability of the results • Loads insufficient for the optimal evaluation of the ballast bearing capacity • Need of calibration tests for the data acquisition • Speed of execution
OPPORTUNITIES *(external, positive factor)*	THREATS *(external, negative factor)*
• Development of new models for the interpretation of the results • Approach to a field of engineering technique that is not "saturated" • Possibility of agreements with public authorities or private companies that need an advanced know-how for development of its own procedures	• Results distorted by unexpected and/or unforeseen conditions and on-board effects cannot be assessed • Returning of untruthful data about the bearing capacity of the superstructure • Ignorance and mistrust of the Public Administrations in the use of these devices for NDT on railways

The only experimental trials of LCMS on railway tracks were carried out by Pavemetrics®. The surveys were conducted by using a multifunction vehicle adapted for running on a railway track with standard gauge (Figure 8).

That configuration minimize the needs for sensors placement representing the optimum in terms of sensor distances from the ground DMI synchronization.

There are not yet other experiences reported in literature. Most probably it is related to two key factors, (i) the technology of LCMS was developed only in the last decade and (ii) the high cost of equipment does not allow a widespread diffusion in a such short time.

6.1 In field trials

LCMS are actually installed on the Automatic Road Analyzer of the Department of Civil Engineering & Architecture of University of Catania. The test was conducted putting ARAN on ad hoc prepared railway wagon (Figure 9). This configuration is not the optimum can be achieved because of the increased distance of the laser from the track, but, at this stage of the research, that was the only configuration available to survey the reduced railway gauge which does not allow ARAN to travel with its own wheel on the railroad track.

Another issue to be solved, was the coupling of LCMS with the DMI which control the frequency of acquisition.

A medium resolution DMI (2048 pulse/round) was installed and connected with the acquisition unit in ARAN. Before the trial the system was re-

Figure 8. Multifunctional vehicles used by Pavemetrics® to test the LCMS performance on surveying railway tracks.

Figure 9. LCMS configuration and DMI installed on the rail wagon wheel.

calibrated for taking into account the new configuration (LCMS height and DMI resolution).

As mentioned above, basic LCMS output are images and cross section profiles. High resolution images can be used to replace a visual rail inspection (e.g. sleepers, fast, ballast). Cross section profile allows to detect gauge values and track geometry. A sample output of is reported in Figure 10.

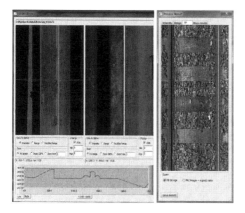

Figure 10.   Output from LCMS survey.

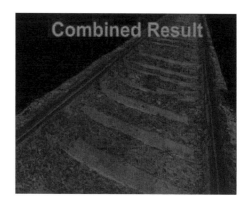

Figure 11.   3D rendering of railroad track (courtesy of Pavemetrics®).

Table 3.   SWOT analysis for LCMS system.

STRENGHT (internal, positive factor)	WEAKNESSES (internal, negative factor)
• Established technology with increasing diffusion • NDT testing • High performance in surveys and post-analysis • Reduced number of operators • Chance to investigate more features of the infrastructure • Data or information about the most important track issues • Higher precision of data	• Technologically advanced equipment (expensive and complex) • Require high operator training for survey and post elaboration • difficulty adapting configurations used in road surveys to railway tracks • Not suitable for adverse climatic conditions
OPPORTUNITIES (external, positive factor)	THREATS (external, negative factor)
• Development of new models for the extrapolation of indirect index from surveys data • Approach to a field of engineering technique that is not "saturated" • Possibility of agreements with public authorities or private companies that need of an advanced know-how for the development of its own procedures	• Ignorance and mistrust of the Public Administrations in the use of these devices for NDT on railways • Partial lack of interest of railway companies to use new technologies

Due to the laser height and DMI resolution the resolution of output was of 3 mm in transversal profile with a longitudinal acquisition step of 5 mm.

Coupling the system with an Inertial Measurement Unit (IMU) it will be possible to correct the undesired wagon motion (roll, pitch, yaw) joining the successive section to obtain a complete 3D scan of the track.

This data can be combined with the digital images with surprising results in terms of resolution and details (Figure 11).

The SWOT analysis for LCMS is reported in Table 3.

7   CONCLUSIONS

The main railway lines are by now monitored with high efficiency equipment based on the most advanced technologies in the field. In the case of local railway, which are not part of the main network, monitoring is still challenging in terms of costs and maintenance needs. Furthermore, a lack of a systematic monitoring brings to the impossibility to produce an effective long term track management system, by allocating budget where emergencies come. The present paper has explored new and promising solutions to overcome this limitation by using the equipment well known for road pavement monitoring, but not yet diffused in the railway management. Particularly the paper focused on the adaptation needed to use GPR, LWD and LCMS on a railway track to test the bearing capacity and quality of ballast and track geometry. The results are presented in terms of a SWOT analysis based literature review and on site trials.

More specifically, the field tests were carried out on a local railway with reduced gauge of 900 mm. That condition gave the opportunity to test the system in an environment open to the introduction of such system, but posed also specific limitations for the use of the systems.

GPR and LWD are not new in such application and confirmed the suitability of the equipment for testing the quality of ballast. More specifically, LWD, despite of the limited load (6 kN in the trial test, while an optimum configuration required heavier loads) applied directly on the sleeper, was able to detect defects in the bearing capacity of the Sleeper/Ballast system.

More challenging is the introduction of LCMS that showed high potentialities, but also equipment costs and installation issues that need further investigations.

ACKNOWLEDGEMENT

The authors wish to express their gratitude to Pavemetrics for the support in the development of the trial test with LCMS. ARAN, LWD and GPR were acquired by the University of Catania in the framework of the Regional RESET project, funded by the Sicily Region Authority. The authors wish to express their gratitude also to "Ferrovia Circumetnea" and Ventura Company for them in field assistance during the surveys and to make available the railway sites.

REFERENCES

Burrow, M.P.N., Chan, A.H.C. & Shein, A. 2007. Deflectometer based analysis of ballasted railway tracks, In *Proceedings of The Institution of Civil Engineers geotechnical Engineering*, 160(3): 169–177.

Cafiso, S., Capace, B., D'Agostino, C., Delfino, M., & Di Graziano A. 2016a. Application of NDT to railway track inspection. *International Conference on Traffic and Transport Engineering*, 24th – 25th November 2016, Belgrade (Serbia).

Cafiso, S., D'Agostino, C., Capace, B., Motta, E., & Capilleri, P. 2016b. Comparison of in situ devices for the assessment of pavement subgrade stiffness. *1st IMEKO TC4 International Workshop on Metrology for Geotechnics,* March 17–18, 2016, Benevento, Italy.

Cerniglia, D., Garcia G., Kalay, S. & Prior, F. 2006. Application of Laser Induced Ultrasound for Rail Inspection. *Railway Research Center.*

Daniels, D.J. 2004. Ground Penetrating Radar, 2nd Edition. *IET.*

Eriksen, A., Gascoyne, J., & Al-Nuaimy, W. 2004. Improved Productivity & Reliability of Ballast Inspection using Road-Rail Multi-Channel GPR. *Proceedings of Railway Engineering.* London, UK: IEEE.

Fernandes, J., Paixão, S. Fontul, & Fortunato. E. 2012. The Falling Weight Deflectometer: Application to Railway Substructure Evaluation, In *Proceedings of the First International Conference on Railway Technology: Research, Development and Maintenance*, Paper 130, Civil-Comp Press, Stirlingshire, UK.

Gallagher, G.P., Leiper, Q., Williamson, R., Clark, M.R., & Forde, M.C., 1999. The application of time domain ground penetrating radar to evaluate railway track ballast. *NDT E Int.* 32: 463–468.

Göbel, C., Hellmann, R., & Petzold, H., 1994. Georadar-model and in-situ investigations for inspection of railway tracks, In *Fifth International Conferention on Ground Penetrating Radar.*

Sundquist, H. Byggande. 2000. Drift och Underhåll av Järnvägsbanor, *TRITA-BKN, Rapport 57, KTH*, Compendium in Swedish, Stockholm.

Hamed F. Kashani, Carlton L. Ho, William P. Clement, & Charles Oden, 2016. Evaluating The Correlation Between Geotechnical Index and Electromagnetic Properties of Fouled Ballasted Track By Full Scale Laboratory Model, *TRB 2016 Annual Meeting.*

Horníček, L. & Břešťovský, P. 2014. Using the Lightweight Falling Deflectometer for Monitoring Trial Railway Sections with Under-Ballast Geocomposites, In *Railway Condition Monitoring (RCM 2014), 6th IET Conference on.*

Hugenschmid, J. 1999. Railway track inspection using GPR. *Journal of Applied Geophysics*: 147–155.

Hugenschmidt, J., 2000. Railway track inspection using GPR. *J. Appl. Geophys.* 43:147–155.

Humphrey, A. December 2005. SWOT Analysis for Management Consulting. *SRI Alumni Newsletter. SRI International*: 7–8.

Jack, R. & Jackson, P. 1999. Imaging attributes of railway track formation and ballast using ground probing radar. *NDT&E International*, No. 32:457–462.

NDT, C. 2013. Rail Inspection, Retrieved from *NDT Resource Center*: http://www.ndted.org/AboutNDT/SelectedApplications/RailInspection/RailInspection.htm

NDT, H. 2014. A document outlining the emergence of the eddy current NDT inspection method as an important part of rail maintenance and safety.

Neupane, M., Parsons, R.L. & Han, J. 2016. Rapid estimation of fouled ballast material properties. *Transportation Research Board of the National Accademy*, Washington DC, Annual Meeting 2016.

Olhoeft, G. R., & Selig, E. T., 2002. Ground penetrating radar evaluation of railroad track substructure conditions. Proc. Of the *9th Int'l Conf. on Ground Penetrating Radar*, Santa Barbara, CA, April, S. K. Koppenjan and H. Lee, eds., Proc. of SPIE, vol. 4758: 48–53.

Saaranketo, T. 2006. Electrical properties of road materials and subgrade soils and the use of ground penetrating radar in traffic infrastructure surveys (PhD Thesis). University of Oulu.

Selig, E. T. & Waters, J. M. 1994. Track Geotechnology and Substructure Management. Thomas Telford Services Ltd., London.

Selig, E. T., Hyslip, J. P., Olhoeft, G. R., & Smith, Stan, 2003. Ground penetrating radar for track substructure condition assessment. Proc. Of *Implementation of Heavy Haul Technology for Network Efficiency*, Dallas, TX, May, pp. 6.27–6.33.

Sharpe, P.C. & Govan, R. 2014. The use of Falling Weight Deflectometer to Assess the Suitability of Routes for Upgrading", In *Proceedings of the Second International Conference on Railway Technology: Research, Development and Maintenance*, Civil-Comp Press, Paper 134, Stirlingshire, UK.

Sussmann, T.R., Selig, E.T. & Hyslip, J.P. 2003. Railway track condition indicators from ground penetrating radar. *NDT E Int.* 36:157–167.

Uzarski, D.R., Brown, D.G., Harris, R.W. & Plotkin, D.E. 1993. Maintenance Management of U.S. Army Railroad Networks-the RAILER System: Detailed Track Inspection Manual. *Champaign, IL: US Army Corps of Engineers*; Construction Engineering Research Laboratories.

Reference trajectories of vehicles for road alignment design

G. Cantisani & G. Loprencipe
Dipartimento di Ingegneria Civile, Edile e Ambientale, Sapienza, University of Rome, Rome, Italy

ABSTRACT: The geometric design principles traditionally hypothesize that trajectories of vehicles can be assumed corresponding to the road longitudinal axle of the road or, more correctly, to the median line of each allowed lane. In real conditions, instead, vehicles always travel along trajectories each other different; the variability is due to dynamic actions affecting the motion and, in a great measure, because the control of vehicle trajectories, performed by users, is not perfect. In order to consider if theoretical models can be effective for safety and comfort verifications in design process, it is important to evaluate how a reference trajectory can statistically represent the whole population of road users. In fact, the difference between a real trajectory of a generic vehicle and the theoretical one can emphasize the safety problems related to geometric characteristics of roads. To deal with these problems, it appears interesting to analyse the dispersion of trajectories in various road sections; in this way, in fact, the "reference trajectory" along a road alignment can be recognized by means of a statistical approach. Starting from surveys on real road elements, the paper presents a method aimed to obtain trajectories that have formal geometric expression and that can correctly represent the scattering of vehicles' position, because reference lines are defined after a statistical analysis of collected data.

1 INTRODUCTION

Vehicle trajectories, in real conditions, are generally scattered about the geometric reference alignments of the given road section; in order to use design models that can better represent vehicle trajectories of real vehicles in motion, it is important to evaluate how reference trajectories can statistically represent the whole population of road users.

Traditionally, in fact, the design theories assume that the median line of each allowed lane represents the reference curve for an "ideal" trajectory. As a consequence, considering that the lanes are usually parallel, the centre of the carriageway—that is equicentric to the median lines of all allowed lanes—can be assumed as the design reference line or, more briefly, as the road axis. In this assumption, it is implicitly established that the geometry of design line has to be consistent with vehicle dynamics (in particular speed and speed rate) and various possible manoeuvres along road section.

Most of road design guidelines in the world (AASHTO 2001; VSS 1991; IT-DM 1991; TAC 1999), are based on this design hypothesis.

In addition, theoretical studies on road geometric design generally correspond to the above presented basic principles. For example, the works focused on the Design Speed, Operating Speed and Design Criteria (Leisch & Leisch, 1997; Lamm et al, 1988; Cantisani & Di Vito, 2012) provide speed values and speed profiles as the result of the analysis of road alignments. In these models the dynamic equilibrium of vehicles refers to circular curves corresponding to the reference road axis.

In other research papers the same basic concepts are used for calculating and evaluating speed profiles (Messer, 1980; Fitzpatrick et al, 2000, Hassan, 2004, Cafiso et al, 2005), in order to ensure consistency and homogeneity of road alignments (Gibreel et al, 1999).

With regard to safety issues, vehicle stability along theoretical trajectories is noted as a critical problem, because conventional models are sometimes deemed too approximate. In particular, the effect of a vertical curve or grade has not been considered (Furtado et al, 2002). Other considerations are referred to the analysis of horizontal curves (Reinfurt et al, 1991) that can increase speed reduction and edgeline encroachments on the inside lane.

Some studies underline that the width of lanes cannot be neglected with regard to the influence on speed and vehicle lateral placement (Neuhardt et al, 1971; McLean, 1974); comparisons between real road conditions and simulator environment confirm these outcomes (Blana & Golias, 2002). Moreover, the effects of lane widths on road safety are not so easy to recognize: Hauer (2005) affirms that lane width plays a different role in single and multi-lane roads, because for single-lane roads it has a bigger influence on driver behaviour, in terms of trajectory and selected lateral position.

Considering various phenomena that can occur where vehicles travel along a road, real trajectories are always different to the "ideal" one (Glennon & Weaver, 1972); in fact, the dynamic actions on vehicle body, tires and mechanical parts (Bonneson et al, 2007), in addition to the intrinsic variability due to imperfect user control of steering and speed (Rosey & Auberlet, 2012; Cantisani et al., 2013), significantly affect the vehicle's motion or the user's positioning of their vehicle in the available cross-section or both. As a result, designers need to be cautious when using current design theories to improve safety and comfort of road sections.

In particular, a probabilistic approach could be incorporated in the road design process to account for vehicle positioning and trajectory, which in turn impacts vehicle speed and safety. According to (Hirsh et al, 1986), designers should note that «*current practice is based on a deterministic approach whereas the factors involved in the geometric design process (e.g, speed, friction, reaction time) are stochastic in nature and vary among road users*». In this way it would be possible to evaluate how a reference trajectory can statistically represent the road user population; in other studies (Blana & Golias, 2002), it is recommended that designers explore a more "rational style", rather than a "pragmatic" one, for road safety management, including a review of the criteria aimed to obtain a correct design and assessment of road geometric characteristics.

To fulfil the above recommendations, the knowledge of users' behaviour and the study of their vehicle trajectories are essential in order to determine the driver's likely behaviour on the actual project being analysed.

2 STUDY, METHODS AND OBJECTIVES

To deal with the above presented problems, it is interesting to analyse the dispersion of trajectories at various road sections. The "reference trajectories", in fact, can be assumed as a statistical representations of the actual dispersed ones, but it is necessary to know the limits and the accuracy of this assumption.

In other terms, many important design verifications, directly related to the safety of the infrastructure, like the dynamic equilibrium in a curve and the sight distance assessment, are developed starting from the assumption that the position of any vehicle along a road section agrees to an ideal trajectory. Therefore, it is important to know how the real position of vehicles in a traffic stream differs from the ideal one, because in this way a statistical evaluation regarding the percentage of vehicles that have a risk exposure greater that the theoretical one can be obtained. At the same time, considering that the transversal dimension of road elements and the distances between vehicles and lateral obstacles can influence some important functional parameters, like the average of vehicular speeds and the capacity of the road sections, the statistical distribution of the lateral displacement is significant in order to consider the functional performances of a road section.

In particular, the scattering of vehicles' positions in each cross section, belonging to a road alignment, has to be investigated with the aim of defining a reference line as the result of statistical analyses on collected data. For these purposes, various researches focused on the collection of experimental data about operating conditions in traffic flows and/or single vehicles motion (Yu et al, 2001; Harlow & Peng, 2001; Coifman et al, 1998).

Important tools for performing surveys over vehicular traffic are now available. It is possible to obtain measurements and observations by means of automatic detection sensors like microwave radar and infrared detectors (Cantisani et al, 2012), or video cameras, and to store and process the data by a number of electronic resources. In particular, the AVIP—Automatic Video Image Processing (Semertzidis, 2010; Song & Tai, 2007; Laparmonpinyo & Chitsobhuk, 2010)—are integrated systems that allow to continuously detect the traffic scenes on a section of road and obtain traffic parameters and relevant lane positioning information.

The data provided by traffic monitoring systems can be presented as a set of points (like scattered seeds) that represents the sequence of positions of a representative part of each vehicle that is following its particular path. Linking these points (Figure 1), after a coordinates transformation process (Wei et al, 2005; García & Romero, 2009), it is possible to obtain curves that give a continuous representation of trajectory: in this way, a geometric approximation of the position of the vehicle, between two points recorded by the monitoring system, is produced, but the error results not significant for the usual frequency sampling of the recording devices.

It is also possible to obtain the distribution of lateral vehicle displacements (Figure 2), at each point along the road element, as the intersection of each vehicle's trajectory with cross section lines. The distribution can be analysed by means of statistical techniques, so obtaining a representation of the way in which the road user (driver) population travels along the road infrastructure.

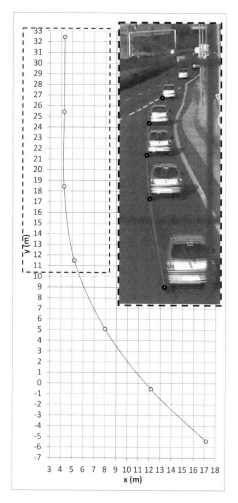

Figure 1. Example of vehicle trajectory, obtained as a continuous line that interpolates a points' seeding corresponding to the left rear tyre of vehicle.

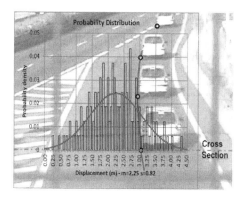

Figure 2. Distribution of vehicle displacement obtained by crossing the observed trajectories with section lines.

3 DATA ANALYSIS

Statistical analyses were performed on real vehicle trajectory data; the data came from video camera surveys on various road elements, collected at critical points (curves, ramps, intersections, …) along a road, and processed with an original method. In particular, in various cross sections along the examined roads, the distribution of lateral displacement was extracted as previously described.

The methodology for successive treatments was previously developed and presented by the same Authors (Cantisani & Loprencipe, 2013); it is interesting to consider, here, what is the theoretical and practical interpretation of the obtained results.

In particular, the statistical meaning of data distribution can be expressed as following: if the lateral displacements present a normal distribution, it is possible to individuate the mean (μ) as the value that corresponds to a point, in the road cross section, where the probability a vehicle exactly passes there reaches the maximum. Other statistical parameters (like standard deviation σ, etc.) can allow to establish where are the points related to some characteristic percentiles, with reference to the probability that vehicles, passing through the analysed sections, fall into the space included between these points. For example, it is possible to consider the characteristic points defining the intervals where a certain percentile of users is included: this is the so-called 68-95-99.7 empirical rule (see Figure 3): for a sample of data normally distributed, about 68% of values fall into the interval [μ-σ, μ+σ], about 95% into the interval [μ-2σ, μ+2σ] and about 99.7% into the interval [μ-3σ, μ+3σ]. The statistical parameters can be used also for safety and functional design evaluations (Figure 4).

To obtain trajectories having a formal geometric expression (but also representing the scattering of vehicles' positions) the means of statistical distributions in selected sections can be linked by

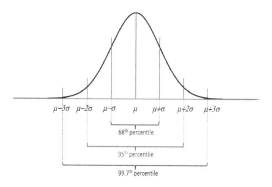

Figure 3. Percentiles of population falling into the characteristic intervals for the 68-95-99.7 rule.

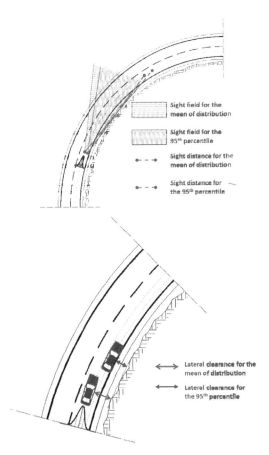

Figure 4. Examples of design evaluation of sight conditions and lateral clearance for the mean and for the 95th percentile of users (Cantisani & Loprencipe, 2013).

a curve that interpolates these points. The curve should be derivable and should maintain a continuity of order of at least equal to 2, in order to satisfy the real variation of the kinematic properties related to the vehicle motion. Furthermore, for a generic couple of points in sequence, it is required that no sign change has to occur, in the value of the curvature $\rho = 1/R$.

In this research the best results, as regard to the geometric representation, were obtained by interpolating the points by means of polynomial curves "in pieces" (Hermite Parametric Cubic Spline, see Figure 5).

The meaning of the described treatments is the following: the resulting lines join the points having the maximum probability of vehicle passages, so, in a statistical sense, they are the most representative lines that one can assume as "reference trajectories" (Figure 6).

In the Figure 6 is possible to note that the normal distributions are not the same for all the considered

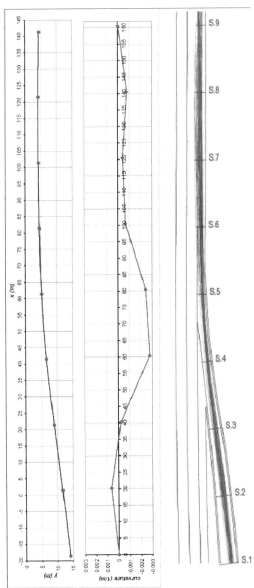

Figure 5. The diagram shows, in the sequence: the cubic spline interpolation of the maximum probability points ("reference trajectory"), the curvature diagram and the dispersion of real trajectories.

road sections. For example, the vehicle trajectories in the section no.6 are more scattered than in the section no.3 where the vehicles adopt a trajectory cutting the curve in order to minimize the centripetal acceleration forces and to maximize the speed. In addition, the large scattering of trajectories in the section no.6 probably is due to the presence of a ramp entrance where two traffic flows are in

74

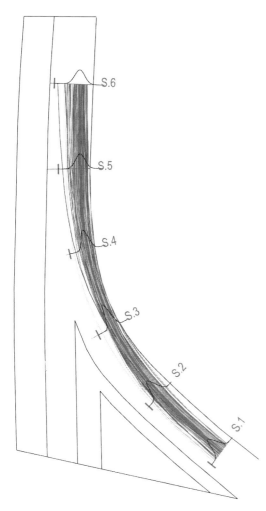

Figure 6. Example of reference trajectory for the case of a ramp terminal in a grade-separated intersection.

Table 1. Average values and the standard deviations of the six normal distributions of vehicle trajectories.

Section	Average of vehicle positions μ (m)	Standard deviation of vehicle positions σ (m)
1	2.032	0.450
2	1.968	0.448
3	2.068	0.341
4	2.188	0.427
5	2.640	0.573
6	3.003	0.597

conflict and there are different opportunities to occupy the main lane for merging vehicles.

In the Table 1 the average values and the standard deviations of the six normal distributions

are reported: they are calculated starting by the trajectories of about 200 vehicles recorded on a road ramp. The average values are calculated with reference to different points in the cross sections because the dimension of lane is different along the road. In fact, the zero reference point used for calculating the displacement in each section is the edge of the internal line delimiting the ramp.

4 DISCUSSION OF RESULTS

The above presented results, obtained by statistical treatments on collected experimental data, deal with some remarks about the actual behaviour of drivers and the way they interact with the road infrastructure. In particular, two main observations can be presented, that result in good agreement with other studies presented in the literature review (see previous chapter 1).

First of all, we can see that if we link the maximum probability points, in each cross section along a road alignment, the obtained line can significantly differ from the theoretical one. This circumstance frequently occurs and in particular, referring to the presented examples, it can be noted in the case of the ramp terminal belonging to an intersection.

In this road section, in fact, the radius (R2) of the osculating circle obtained by linking the point of maximum probability in the cross sections no.2, no.3 and no.4, is larger than the one (R1) considered for the geometric design (or theoretical assessment) of the road element (see Figure 7, R2 >> R1). This circumstance entails that the actual speed vehicles can reach by travelling along the ramp terminal, even if they maintain a safe condition with regard to the dynamic equilibrium, can result significantly higher than the one resulting by theoretical analyses. The underestimation of speed, in this particular road section (ramp terminal in a grade-separated intersection), can also produce a not negligible influence in the merging process between primary and secondary traffic flow. Thus, in the design analyses, the problem should be taken in account in order to prevent possible conflicts in the traffic manoeuvres. As a suggestion, we can deduce that in similar cases the design could consider a set of possible input data (referring to geometrical and cinematic parameters of vehicle's motion), ensuring that also the most critical ones will be compatible with road infrastructure and traffic conditions.

In addition, another remark emerges by considering the variability of standard deviation values in various sections. In the sections no.1 and no.2, in fact, the standard deviation is higher than in the sections in the middle of the curve (no.4 and,

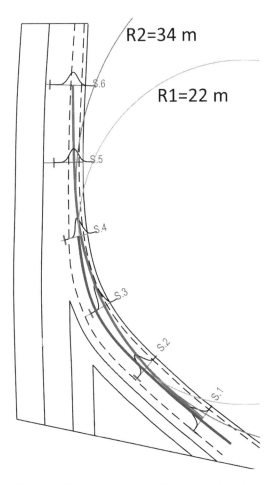

R2=34 m

R1=22 m

S.6

S.5

S.4

S.3

S.2

S.1

Figure 7. Comparison between the reference trajectory (in red) and the theoretical one (in blue) in the case of a ramp terminal in a grade-separated intersection.

especially, no.3). This depends essentially by the more effective trajectory control that a sharper curvature radius can determine in comparison to a larger one. Moreover, also the lane width represents an essential factor, because the displacement typically presents a larger variance when the lane is larger. Consequently, we can assume that the lane width should be correctly assessed in the geometric design process, also considering the special characteristics or the function of the road element (a ramp terminal, in this case), in order to ensure a more effective control of trajectories.

Different considerations have to be reserved to the standard deviation in the sections no.5 and no.6, where the high values are primarily due to the entrance manoeuvre accomplishment; therefore, they seem quite independent respect to road geometry.

5 CONCLUSIONS

The vehicular trajectories, along a road section, are important because their configuration can influence the development of design safety and functional verifications (dynamic equilibrium, sight distance, road capacity or average speed of a traffic stream). However, it is important to consider that a theoretical trajectory can only have a statistical meaning, because in real conditions the vehicles can move through the road section also passing in a different position (in comparison to the "ideal" one).

Therefore the analyses of vehicle paths, along road sections, can be used to identify the "reference trajectories" for the design of road alignments, based on statistical analysis of experimental data. In fact, it is possible to observe the distribution of lateral displacement in each cross section along a road element. After a geometric treatment of the data, the reference trajectories can be obtained as the lines that continuously interpolate the points where the probability to find a vehicle, when it passes through each cross section line, reaches the maximum. In the same examined points, other statistical properties can be observed, like the shape of distribution and its parameters (standard deviation, etc.).

The possible uses of these reference lines deal with the safety or functional problems, related to road design procedures. In fact, a reference trajectory can be assumed as the median line of each allowed lane, but often it would be also necessary to examine other conditions, for the design verifications accomplishment, so considering the scattering of vehicles' positions in the real trajectories.

As a practical deployment, it is possible to observe in the Figure 4 how the design verifications, regarding sight conditions and lateral clearance respectively for the mean and for the 95th percentile of users, are different; so, the knowledge of trajectories distribution allows to consider and evaluate the risk exposure conditions for the whole population of users.

In addition, based on the observed data, some remarks regarding main implications for the design and the checks for ramps are presented in the paper, in agreement with other literature studies.

More general, the application of presented principles allows improving the knowledge of real road operations; this opportunity is showed in the paper, considering various case studies, especially referred to some critical points (curves, ramps, intersections …) along examined road elements.

REFERENCES

AASHTO (2001). Policy on Geometric Design of Highways and Streets. Washington, DC: American Association of State Highway and Transportation Officials, 1: 990, 2001.

Blana, E., & Golias, J. (2002). Differences Between Vehicle Lateral Displacement on the Road and in a Fixed-Base Simulator. In Human Factors: *The Journal of the Human Factors and Ergonomics Society*, 44(2), pp 303–313.

Bonneson, J. A., Pratt, M., Miles, J., & Carlson, P. (2007). *Development of Guidelines for Establishing Effective Curve Advisory Speeds*. Texas Transportation Institute, Texas A&M University System.

Cafiso, S., Di Graziano, A., & La Cava, G. (2005). Actual Driving Data Analysis for Design Consistency Evaluation. In Transportation Research Record: *Journal of the Transportation Research Board*, No. 1912, pp 19–30.

Cantisani, G., & Di Vito, M. (2012). CCV: A New Model for S85 Prediction. In *Procedia-Social and Behavioral Sciences*, No 53, pp 765–776.

Cantisani, G., Di Vito, M., & Luteri, P. (2012). VPL Project'09: An Integrated Station for Vehicles' Operating Conditions Survey. In *Procedia-Social and Behavioral Sciences*, 53, pp 777–788.

Cantisani, G., & Loprencipe, G. (2013). A statistics based approach for defining reference trajectories on road sections. In *Modern Applied Science* 7 (9), pp. 32–46.

Cantisani, G., Loprencipe, G., & Primieri, F. (2011). The integrated design of urban road intersections: A case study. In *The International Conference on Sustainable Design and Construction* (pp. 722–728).

Coifman, B., Beymer, D., McLauchlan, P., & Malik, J. (1998). A Real-time Computer Vision System for Vehicle Tracking and Traffic Surveillance. In *Transportation Research Part C: Emerging Technologies*, 6(4), pp 271–288.

Fitzpatrick, K., Wooldridge, M.D., Tsimhoni, O., Collins, J.M., Green, P., Bauer, K. Parma, K.D., Koppa, R., Harwood, D.W., Anderson, I., Krammes, R.A., & Poggioli, B. (2000). *Alternative Design Consistency Rating Methods for Two-lane Rural Highways*. In FHWA-RD-99-172, pp 1–154.

Furtado, G.A., Easa, S.M., & Abd El Halim, A.O. (2002). A Vehicle Stability on Combined Horizontal and Vertical Alignments. Carleton University. *Annual Conference of the Canadian Society for Civil Engineering. Montreal, Quebec, Canada, June 5–8.*

García, A., & Romero, M. (2009). Discussion of 'Video-Capture-Based Approach to Extract Multiple Vehicular Trajectory Data for Traffic Modeling' by Heng Wei, Chuen Feng, Eric Meyer, and Joe Lee. In *Journal of Transportation Engineering*, 135(3), pp 149–150.

Gibreel, G.M., Easa, S.M., Hassan, Y., & El-Dimeery, I.A. (1999). State of the Art of Highway Geometric Design Consistency. In *Journal of Transportation Engineering*, 125(4), pp 305–313.

Glennon, J.C., & Weaver, G.D. (1972). *Highway Curve Design for Safe Vehicle Operations*. In Highway Research Record, No. 371.

Harlow, C., & Peng, S. (2001). Automatic Vehicle Classification System with Range Sensors. In *Transportation Research Part C: Emerging Technologies*, 9(4), pp 231–247.

Hassan, Y. (2004). Highway Design Consistency: Refining the State of Knowledge and Practice. In *Transportation Research Record: Journal of the Transportation Research Board* No. 1881.1, pp 63–71.

Hauer, E. (2005). The Road Ahead. In *Journal of Transportation Engineering*, 131(5), pp 333–339.

Hirsh, M., & Prashker, J.N., and Ben-Akiva, M. (1986). New Approach to Geometric Design of Highways. In *Transportation Research Record: Journal of the Transportation Research Board*, No. 1100, pp 50–57.

IT, D.M. (2001). Norme Funzionali e Geometriche per la Costruzione delle Strade. (*Italian Standard for Geometric and Functional Road Design*)—D.M. 05/11/2001. Rome: Ministry of Infrastructures and Transportation.

Lamm, R., Choueiri, E.M., Hayward, J.C., & Paluri, A. (1988). Possible Design Procedure to Promote Design Consistency in Highway Geometric Design on Two-lane Rural Roads. In *Transportation Research Record: Journal of the Transportation Research Board*, No. 1195, pp 111–122.

Laparmonpinyo, P., & Chitsobhuk, O. (2010). A Video-based Traffic Monitoring System Based on the Novel Gradient-edge and Detection Window Techniques. *The 2nd International Conference. In Computer and Automation Engineering (ICCAE)*, 3, pp 30–34.

Leisch, J.E., & Leisch, J.P. (1977) New Concepts in Design-Speed Application. In *Transportation Research Record: Journal of the Transportation Research Board*, No. 631, pp. 4–14.

McLean, J.R. (1974) Driver Behaviour on Curves—A Review. In *Australian Road Research Board (ARRB), Conference, Adelaide*. Vol. 7.

Messer, C.J. (1980) Methodology for Evaluating Geometric Design Consistency. In *Transportation Research Record: Journal of the Transportation Research Board*, No. 757, pp. 7–14.

Neuhardt, J.B., Herrin, G.D., & Rockwell, T.H. (1971). *Demonstration of a test-driver technique to assess the effects of roadway geometrics and development on speed selection*. Federal Highway Administration, Study No Hpr-1(4).

Reinfurt, D.W., Zegeer, C.V., Shelton, B.J., & Neuman, T.R. (1991). Analysis of Vehicle Operations on Horizontal Curves. In *Transportation Research Record: Journal of the Transportation Research Board*, No. 1318, pp 43–50.

Rosey, F. & Auberlet, J.-M. (2012) Trajectory variability: Road geometry difficulty indicator. In *Safety Science*, 50(9), pp. 1818–1828.

Semertzidis, T., Dimitropoulos, K., Koutsia, A., & Grammalidis, N. (2010). Video Sensor Network for Real-time Traffic Monitoring and Surveillance. In *Intelligent Transport Systems*, IET Vol. 4 Issue 2, pp 103–112.

Song, K.T., & Tai, J.C. (2007). Image-based Traffic Monitoring with Shadow Suppression. In *Proceedings of the IEEE*, 95(2), pp. 413–426.

TAC. *Geometric Design Guide for Canadian Roads*. Ottawa, Canada: Transportation Association of Canada. 1999.

VSS. (1991) Schweizer Norm SN 640-080-b, *Projektierung Grundlagen—Geschwindigkeit Als Projektierungselement. Vereinigung Schweizerrischer Strassenfachleute*, Zurich, Switzerland.

Wei, H., Feng, C., Meyer, E., & Lee, J. (2005) Video-capture-based Approach to Extract Multiple Vehicular Trajectory Data for Traffic Modeling. In *Journal of Transportation Engineering*, 131(7), pp 496–505.

Yu, X., Sulijoadikusumo, G., Li, H., & Prevedouros, P. (2011). Reliability of Automatic Traffic Monitoring with Non-Intrusive Sensors. In *Proceedings of the 11th International Conference of Chinese Transportation Professionals. Nanjing, China, August 14–17.*

Transport Infrastructure and Systems – Dell'Acqua & Wegman (Eds)
© 2017 Taylor & Francis Group, London, ISBN 978-1-138-03009-1

Decoupling of wheel-rail lateral contact forces from wayside measurements

D. Cortis, M. Bruner & G. Malavasi
Department of Civil, Building and Environmental Engineering, Sapienza University of Rome, Rome, Italy

ABSTRACT: Running safety and stability of railway vehicles are provided by respect of track geometry and by observance of the limits of wheel-rail contact forces. The ratio between lateral and vertical forces has an effect on the guiding forces provided by the rail, which prevent the wheel flange climbing and the derailment. This highlights the importance of the development of measurement methods of wheel-rail contact forces. In this paper, we present an experimental method to estimate the lateral contact force starting from the measurement of the strains on the rail foot surface. A suitable combination of the recorded strains allows to reproduce the same continuous signal of the applied lateral force, decoupling the effects of the vertical one. These studies, based on finite element simulations and laboratory tests, show how to find a constant ratio between the applied lateral load and the recorded strains on the rail foot surface.

1 INTRODUCTION

Running safety and stability of railway vehicles are provided by respect of track geometry and by observance of the limits of wheel-rail contact forces. The ratio between lateral and vertical forces has an effect on the guiding forces provided by the rail, which prevent the wheel flange climbing and the derailment. The safety against derailment depends moreover by the wheel-rail contact condition as the contact area, the shape of the wheels and rail's profile, the wear of the contact surfaces, the friction factor and the influence of the environmental conditions. All these physical quantities change the distribution and the magnitude of the contact forces. The evaluation of the wheel-rail contact forces plays also a key role on the maintenance of the track and rolling stock. The knowledge of the history of the contact forces allows the development of mathematical models to predict the progressive degradation of the track and to improve the maintenance planning. The measurement of the wheel-rail contact forces is still required for the homologation and the acceptance of the new vehicles. The design of rolling stocks in fact needs the knowledge of the loads exchanged by the wheels and the rail during all types of the running conditions (Malavasi, 2014). The European Standard EN14363 regulates the experimental tests for the homologation.

These considerations highlight the importance of the development of measurement methods of wheel-rail contact forces. In the last years, the wayside monitoring systems of track loads have indeed attracted the interest of the railways infrastructure managers. The identification in real-time of the magnitude of the forces exchanged between the wheel and the rail allows the monitoring of the rail traffic and the reporting of any non-compliance to the respective railways companies. The main issue of this type of continuous measurements is the decoupling of the effects of the wheel-rail contact forces. The simultaneous presence of bending moments and torques produce on the rail surfaces a complex strain state that makes difficult identifying the effects produced by the corresponding forces.

One of the strategies to decouple the wheel-rail contact forces and evaluate their magnitude is to discover, on the rail surface, specific areas in which to measure the strains. These specific areas could be identified through the analysis and the study of the structural properties of the rail. Regarding the measurement of the vertical force (Q), there are several well-developed technical solutions. For example, the Italian Infrastructure Manager (RFI, Rete Ferroviaria Italiana) is experiencing and testing the SMCV system (Accattatis et al. 2014) to monitor the vertical load. The evaluation of the lateral force (Y) is, instead, a less developed field and there are less known solutions. In the last decades, some authors have deal with these subjects, as Ahlbeck & Harrison (1977, 1980, 1981) and Moreau (1987). Recently, there have been some studies concern the evaluation of lateral forces through different experimental approaches. Milkovic et al. (2013) have tried to decouple vertical and lateral forces using a method based on independent component analysis of recorded strain signals on the rail surfaces. Delprete et al. (2009), Bracciali et al. (2001, 2004) and Di Benedetto et al. (2010) have studied a sensor to measure at the same time vertical and the lateral contact forces applying transducers inside the rail web close to the barycentre axes. These and other techniques use strain gauges placed on the rail surface, in order to measure shear or bending effects.

In this paper, we present an experimental method to estimate the lateral contact force starting from the measure of the strains on the rail foot surface. The experimental measurement system (SMCT) is composed by four shear strain gauges symmetrically arranged on the rail foot and connected together through a full Wheatstone bridge. The full bridge configuration, with an opportune combination of the recorded strains, is able to reproduce the same continuous signal of the applied lateral force decoupling the effects of the vertical one. The present research draws on the study of Yifan et al. (2001) and the investigations made by the research group in railway engineering of the Department of Civil, Building and Environmental Engineering, Sapienza University of Rome (Bruner et al. 2015). These studies show a methodology to find a constant ratio between the applied lateral load and the recorded strains on the rail foot (Bruner et al. 2016). They also highlight the robustness of this experimental procedure, representing with finite element simulations the not influence of the boundary conditions (thermal gradient, inclination of the rail, status of the rail profile, stiffness of fastenings and ballast) on the measurements.

2 MEASUREMENT SYSTEM

The monitoring system for lateral loads (SMCT) consists of four V-shaped strain gauges, with two measuring grids arranged at an angle of about 45°, placed on the rail foot surface. Typical applications for these sensors include measurements of shear stresses. The strain gauges are symmetrically disposed on the rail foot and they are connected together by a Wheatstone circuit.

Considering one-meter long rail segment, the sensors are symmetrically located respect the y-axis of the rail at 120 mm from each other and at 20 mm from the outer edge of the rail (Fig. 1). An opportune combination of the measuring grids inside a full Wheatstone bridge circuit generates as output a strain having the same signal of the lateral force (Y) decoupling the effects of the vertical one (Q). In fact, the symmetrical position of the measuring grids removes the contribution of the vertical force on the recorded strains (Fig. 2).

Figure 3 represents the full Wheatstone bridge circuit configuration: V_A is the supply voltage, V_U is

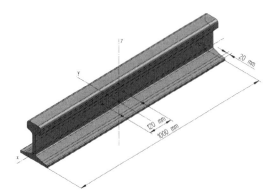

Figure 1. V-shaped strain gauges.

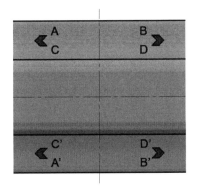

Figure 2. V-shaped strain gauges: measuring grids.

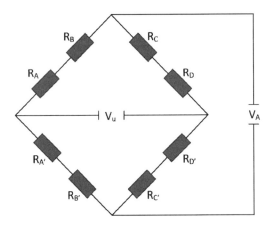

Figure 3. Wheatstone bridge circuit configuration.

the output voltage, R_A, R_B, R_C, R_D, $R_{A'}$, $R_{B'}$, $R_{C'}$ and $R_{D'}$ are the measuring grids of the four V-shaped strain gauges.

The following expressions show the relation of the full Wheatstone bridge for this specific application, where f is the gauge factor of the sensor and ε the recorded strain.

$$\frac{V_U}{V_A} = \frac{1}{4} \cdot \left(\frac{\Delta R_{AB}}{R_{AB}} - \frac{\Delta R_{A'B'}}{R_{A'B'}} + \frac{\Delta R_{C'D'}}{R_{C'D'}} - \frac{\Delta R_{CD}}{R_{CD}} \right) \quad (1)$$

$$\frac{\Delta R}{R} = f \varepsilon \quad (2)$$

For a single V-shaped strain gauge, the imbalance is:

$$\frac{\Delta R_{AB}}{R_{AB}} = \left(\frac{\Delta R_A + \Delta R_B}{R_{AB}} \right) = \left(\frac{R_A f \varepsilon_A + R_B f \varepsilon_B}{R_A + R_B} \right)$$
$$= \frac{Rf(\varepsilon_A + \varepsilon_B)}{2R} = \frac{f}{2}(\varepsilon_A + \varepsilon_B) \quad (3)$$

Considering the other strain gauges:

$$\frac{\Delta R_{A'B'}}{R_{A'B'}} = \frac{f}{2}(\varepsilon_{A'} + \varepsilon_{B'}) \quad (4)$$

$$\frac{\Delta R_{C'D'}}{R_{C'D'}} = \frac{f}{2}(\varepsilon_{C'} + \varepsilon_{D'}) \quad (5)$$

$$\frac{\Delta R_{CD}}{R_{CD}} = \frac{f}{2}(\varepsilon_C + \varepsilon_D) \quad (6)$$

Replacing the previous expressions into the equation (1):

$$\frac{V_U}{V_A} = \frac{f}{8}(\varepsilon_A + \varepsilon_B - \varepsilon_{A'} - \varepsilon_{B'} + \varepsilon_{C'} + \varepsilon_{D'} - \varepsilon_C - \varepsilon_D)$$
$$= \frac{f}{8} \varepsilon_{tot} \quad (7)$$

Because of the SMCT measurement system produces as output a total strain (ε_{tot}) that has the same signal of the applied lateral force (Y), it is possible find a constant K, define as follow:

$$K = \frac{Y}{\varepsilon_{tot}} \quad (8)$$

This constant allows to evaluate the lateral force (Y) applied by the wheel on the rail in a generic situation, starting from the measurement of the rail foot strains (9).

$$Y = K \cdot \varepsilon_{tot} \quad (9)$$

3 FINITE ELEMENT SIMULATION

3.1 Finite element model

The finite element simulations were performed to check and validate the results of the SMCT meas-

urement system. One-meter long rail segment (type UIC60/60E1) was reproduced with a 3D model. The model was meshed using hexahedral elements that changes their size along the rail longitudinal axis.

The smaller size of the volume mesh is 5 mm, corresponding to the location of the four V-shaped strain gauges. The mesh was refined in this area in order to improve the numerical solutions. Moreover, in this rail section the hexahedral elements were covered and glued with 2D shell elements (Fig. 4). This finite element modelling technique was used to compare the rail foot strains with the experimental results of the four V-shaped strain gauges.

Concerning structural constrain conditions, the model was considered fixed between two sleepers (spaced 600 mm from each other). The bottom side of the rail foot was connected to the ground with spring-damper elements (mesh size 1 mm). These elements generate a vertical displacement of the rail proportionally to the stiffness of elastic fastenings and rail pad (300 kN/mm). Moreover, the rail foot surface was constrained for the cross displacement. Finally, in order to simulate a wheel-rail contact in curving, the vertical (Q) and lateral (Y) forces were applied on the railhead in the mid-section of the model, 20 mm from the z-axis and 14 mm under the rolling plane (Fig. 5). The simulations were performed with a static analysis using ANSYS Code. The static analysis has been chosen because of, at the beginning, the SMCT measurement system is design to work in areas where rolling stocks transit at very reduced speed (below 30 km/h), like stations, depots and marshalling yards. The finite element model of the rail segment has 67,349 elements and 75,669.

3.2 Results of simulations

Simulations results, which consider a distance (d) between the strain gauges of 120 mm (Fig. 4), are

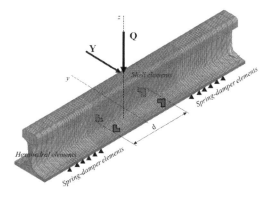

Figure 4.   Finite element model.

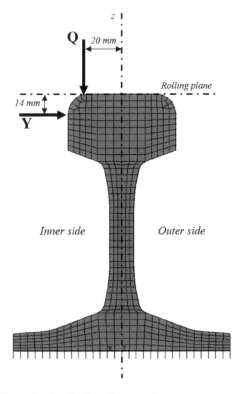

Figure 5. Application of contact forces.

Table 1. Results of simulations (d = 120 mm).

Vertical Force Q kN	Lateral Force Y kN	Ratio Y/Q –	Total strain ε_{tot} μɛ	Constant $K = Y/\varepsilon_{tot}$ kN/μɛ
100	10	0.1	40.5	0.247
100	20	0.2	88.0	0.227
100	30	0.3	135.5	0.221
100	40	0.4	183.0	0.219
100	50	0.5	230.6	0.218
100	60	0.6	278.1	0.216
100	70	0.7	325.6	0.215
100	80	0.8	373.1	0.214
100	90	0.9	420.7	0.214
100	100	1.0	468.2	0.214
100	110	1.1	515.7	0.213
100	120	1.2	563.2	0.213
100	130	1.3	610.7	0.213
100	140	1.4	658.3	0.213

reported in Table 1. The ratio between the lateral force (Y) and the recorded strains (ε_{tot}) is almost constant changing the load conditions. The average value is 0.218 kN/μɛ. Test runs started with a

Y/Q ratio of 0.1 and ended with 1.4. The highest value allowed by the European Standard EN 14363 is 1.2.

Other tests are carried out by changing the distance (d) between the V-shaped strain gauges. The value of constant K changes modifying this distance (Fig. 6). The results prove that K depends by the distance of strain gauges, remaining almost constant. Increasing the distance (d), the absolute value of the rail foot strains raises up and the constant K decreases proportionally.

In order to evaluate the influence of the vertical force on the strain measures, Figure 7 reports the results of simulations considering applied on the railhead only the vertical force (Q). Two different schemes were considered: the first with the force applied directly on the z-axis of the rail (Fig. 7, solid line) and the second with the force applied at a distance of 20 mm from the z-axis (Fig. 7, dotted line).

The first condition shows the SMCT method removing the effects of the vertical force from the recorded strains combination. The recorded strain (ε_{tot}) is about zero with any distance (d) of the strain gauges. The second condition, typical scenario for the vertical force in a generic wheel-rail contact,

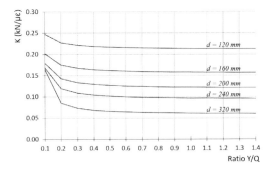

Figure 6. Constant K with a different distance of strain gauges.

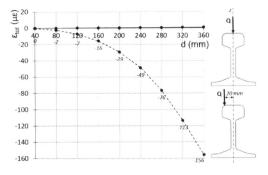

Figure 7. Influence of the vertical force on the strain measures.

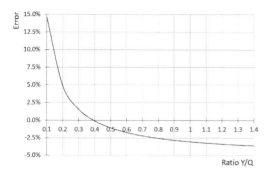

Figure 8. Error curve for the evaluation of the lateral force.

Figure 9. Wheelset and test bench.

displays that when the distance (d) between the strain gauges increases, the recorded strain (ε_{tot}) increases too. This means that by increasing the distance between strain gauges proportionally increases the measurement error. Therefore, the distance (d) should be chosen minimizing the measurement error and maximizing the accuracy of strain gauges measurement location.

Taking into account the above considerations, we fixed the distance (d) equal to 120 mm and assessed the percentage error (Fig. 8) for the evaluation of the lateral force (Y). The diagram shows that the percentage error grows exponentially for low values of Y/Q, whereas for the others (≥ 0.3) remains below 4%.

This trend depends by the low influence that the lateral force (Y) has on rail foot strains for low values of Y/Q. In this case, the twist strain on the rail foot, produced by the vertical force, is greater than the one generated by the lateral force and the SMCT measurement system could not properly decouple lateral and vertical forces.

4 EXPERIMENTAL TESTS

4.1 Test bench configuration

Experimental results were reached on the test bench (Fig. 9). As the finite element simulations, one-meter long rail segment (type UIC60/60E1) was used, fixed between two sleepers spaced of 600 mm from each other.

The rail was connected to the support by four indirect elastic fastenings (type Vossloh), two for each sleeper. The distance (d) between the V-shaped strain gauges was set to 120 mm.

The technical scheme of the test bench (Fig. 10) shows that the vertical (Q_b) and lateral loads (H) are applied to the rail segment by a wheelset. These forces were recorded by four loads cell, two to measure the vertical force (c_{b1}, c_{b2}), one for the

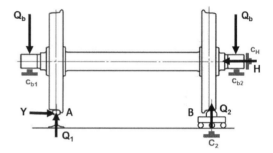

Figure 10. Technical scheme of the test bench.

vertical reaction (c_2) and another one for the lateral force (c_H). The rail vertical reaction (Q_1) was derived from the Q_2. To properly evaluate the rail reaction (Q_1 and Y) it was considered the standard inclination of the rail (1/20). In this conditions, the structural system is isostatic: at point A, where the wheel-rail contact point occurs, we could assume a theoretical revolute joint, whereas at point B a roller joint.

4.2 Experimental results

The results of laboratory tests are reported in Figure 11. The chart represents a generic load cycle. The highest vertical and lateral load applied by the test bench is about 45 kN. For each load curves, the signal is reported in function of time. Moreover, in this chart the rail foot strain and the value of constant K are shown together thanks to two different reference axes.

Analysing the results, the strain curve (ε_{tot}) has the same shape of the lateral load curve (H). This confirms the existence of an experimental constant (K), as done in the theoretical approach. Besides, when the lateral load (H) is applied, the vertical reaction (Q_2) decreases because of the isostatic equilibrium of the structure increases the rail reaction (Q_1), tilting the wheelset. Also, at the beginning of the load

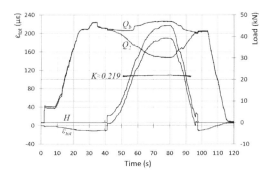

Figure 11.   Results of laboratory test.

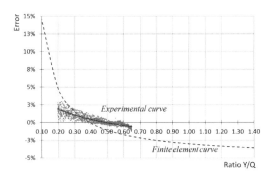

Figure 12.   Experimental and finite element error curve.

cycle, there is no lateral load (*H*) and the strain curve became negative (ε_{tot}): this is the effect of the action of the vertical load (Q_b) on the measure which represents a part of the experimental error, as seen in the finite element simulations (Fig. 7).

Considering the average value of constant *K* (0.219 kN/µε), which is quite the same value of simulations, the experimental and the finite element error curve has the same trend. Unfortunately, the technical specifications of the test bench have limited the *Y*/*Q* ratio to 0.65.

4.3   Extension of results

The structure of the test bench did not allow to perform experimental tests by moving the vertical and lateral load along the rail. Despite this limitation, in order to extend the results and determine the strain effects of contact forces on the entire length of the rail segment, the rail was shifted through the two fastenings (Fig. 13).

The term *d* is still the distance of the strain gauges, while the distance *m* is the traslation of the rail along the x-axis and the term *n* is the distance of the strain gauges from the application point of contact forces. Thanks to this procedure, the strains were measured in different sections of the rail,

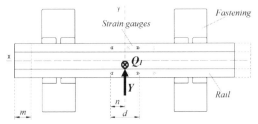

Figure 13.   Position of rail between the two fastenings.

Table 2.   Positions of strain gauges.

Rail position	Distance d mm	Distance m mm	Distance n = (d/2) − m mm	Constant K = Y/ε_{tot} kN/µε
P0	120	0	+ 60	0.219
P20	120	+ 20	+ 40	0.219
P80	120	+ 80	− 20	0.195
P140	120	+ 140	− 80	0.182

leaving the application point of vertical and lateral forces in the mid-section between the fastenings. The load cycle (Fig. 11) was repeated for different positions of the rail with the contact forces in the area between the V-shaped strain gauges (Table 2: position P0 and P20) and with the contact forces externally (Table 2: position P80 and P140).

The results of laboratory test (Figs. 14–16) confirm the previous outcomes: the same behaviour for the signal of the recorded strain (ε_{tot}) and the lateral load curve (*H*).

However, in this case the value of constant *K* is different for the forces out of the area between the V-shaped strain gauges (Table 2: position P80 and P140). This condition shows that it is not possible to obtain the magnitude of the lateral force (*Y*) using only one value of constant *K* when contact forces are applied out of the area between strain gauges.

In order to investigate this experience, additional finite element simulations, based on the previous experimental tests, were conducted. The contour plots of the rail foot surface (Figs. 17–20) show the extension and compression areas for the rail positions P80 and P140 under a vertical (*Q*) and a lateral (*Y*) load of 100 kN (*Y*/*Q* = 1.0). The extension and compression areas are calculated along the main axes of the two measuring grids of the four V-shaped strain gages (ξ and η).

When contact forces are completely external to the strain gauges' area (Figs. 19–20), the experimental results (Table 2) have shown that the combination of the measuring grids produces a different value of constant *K*. This effect occurs when the strain gauges are placed in compressed rail section

84

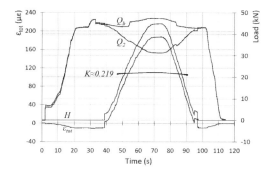

Figure 14. Results of laboratory test (Rail position P20).

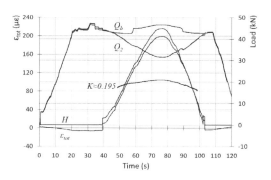

Figure 15. Results of laboratory test (Rail position P80).

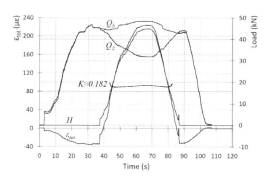

Figure 16. Results of laboratory test (Rail position P140).

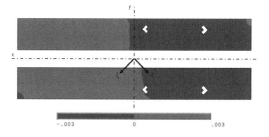

Figure 17. Contours plot of the rail foot surface along ξ axis (Rail position P80).

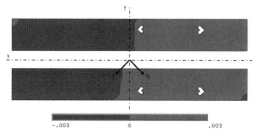

Figure 18. Contours plot of the rail foot surface along η axis (Rail position P80).

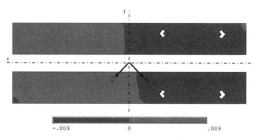

Figure 19. Contours plot of the rail foot surface along ξ axis (Rail position P140).

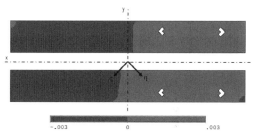

Figure 20. Contours plot of the rail foot surface along η axis (Rail position P140).

instead of stretched and vice versa: there is no longer the symmetrical arrangement of the measuring grids like for the rail positions P0 and P20. This situation occurs even if the forces are just outside the area of the strain gauges (Figs. 17–18) because of an uneven strains distribution.

This complementary finite element investigation confirms the experimental evidence: it is possible evaluate the lateral force (Y), starting from the measurement of the rail foot strains (ε_{tot}) and using the constant K, only when contact forces are applied in the area between the strain gauges.

5 CONCLUSION

In this paper, it was presented an experimental method to estimate the wheel-rail lateral contact force starting from the measure of strains on the rail foot surface. The experimental measurement system (SMCT) is composed by four V-shaped strain gauges, symmetrically arranged on the rail foot. A suitable combination of the recorded strains allows to reproduce the same continuous signal of the applied lateral force (Y), decoupling the effects of the vertical one (Q). The study, based on finite element simulations and laboratory tests, show how to find a constant ratio (K) between the applied lateral load (Y) and the recorded strains (ε_{tot}) on the rail foot surface. Thanks to this constant ratio and the low influence of the vertical force, it is possible to evaluate the wheel-rail lateral contact force with a percentage error below 4% for values of Y/Q greater than 0.3. Moreover, the experimental and numerical investigations have shown the feasibility of the measurement method only when wheel-rail contact forces are applied externally to the strain gauges' area.

In order to make operative a prototype of the measurement systems, further experimental investigations, both in laboratory and on a rail line, have been scheduled. The first aim of these analysis is to validate the experimental prototype of the SMCT measurement system applied on a longer rail. In conclusion, the research confirms that the rail can be used as a useful measurement system to ensure the safety of railways traffic and as a mean for the development of new track maintenance management.

REFERENCES

Accattatis, F.M.D. et al. 2014. Measurement of the vertical loads transferred to the rail. *Ingegneria Ferroviaria* 11: 1001–1041.

Ahlbeck, D.R. & Harrison, H.D. 1977. Technique for Measuring Wheel/Rail Forces with Trackside Instrumentation. *ASME Winter Annual Meeting, 27 November – 2 December 1977*. Atlanta, USA.

Ahlbeck, D.R. & Harrison, H.D. 1980. Techniques for Measurement of Wheel-Rail Forces. *The Shock and Vibration Digest* 12.

Ahlbeck, D.R. & Harrison, H.D. 1981. Development and evaluation of wayside wheel/rail load measurement techniques. *Proceedings of the International Conference on Wheel/Rail Load and Displacement Measurement Techniques, 19–20 January 1981*. Cambridge Mass., USA.

Bracciali, A. et al. 2001. Progetto e validazione di un sensore estensimetrico multifunzione per il binario ferroviario. *XXX Convegno AIAS, 12–15 September 2001*. Alghero, Italy.

Bracciali, A. & Folgarait, P. 2004. New Sensor for Lateral & Vertical Wheel-Rail Forces Measurements. *Railway Engineering Conference, 6–7 July 2004*. London, England.

Bruner, M. et al. 2015. Tecniche di misura sperimentali per la determinazione delle forze laterali di contatto ruota-rotaia. *IV Convegno Nazionale Sicurezza ed Esercizio Ferroviario: Soluzioni e Strategie per lo Sviluppo del Trasporto Ferroviario, 2 October 2015*. Roma, Italy.

Bruner, M. et al. 2016. The Rail Strain under Different Loads and Conditions as Source of Information for Operation—*Proceedings of the Third International Conference on Railway Technology: Research, Development and Maintenance, J. Pombo, Civil-Comp Press, Stirlingshire, 5–8 April 2016*. Cagliari, Italy.

Delprete, C. & Rosso, C. 2009. An easy instrument and a methodology for the monitoring and the diagnosis of a rail-*Mechanical System and Signal Processing* 23: 940–956.

Di Benedetto, L. et al. 2010. Diagnosi delle condizioni di marcia di un rotabile con un sistema di misura delle forze laterali e verticali al contatto ruota-rotaia. *XXXIX Convegno AIAS, 7–9 September 2010*. Maratea, Italy.

Malavasi, G. 2014. Contact Forces and Running Stability of Railway Vehicles. *International Journal of Railway Technology* 3(1): 121–132.

Milkovic, D. et al. 2013. Wayside system for wheel–rail contact forces measurements. *Measurement, Journal of the International Measurement Confederation* 46: 3308–3318.

Moreau, A. 1987. La verification del la securite contre le deraillement. *Revue Générale des Chemins de Fer* 4: 25–32.

Yifan, L. et al. 2011. Wheel-Rail Lateral Force Continuous Measurement Based on Rail Web Bending Moment Difference Method. *Applied Mechanics and Materials* 105–107: 755–759.

Transport Infrastructure and Systems – Dell'Acqua & Wegman (Eds)
© 2017 Taylor & Francis Group, London, ISBN 978-1-138-03009-1

Design and maintenance of high-speed rail tracks: A comparison between ballasted and ballast-less solutions based on life cycle cost analysis

M. Giunta & F.G. Praticò

University Mediterranea, Reggio Calabria, Italy

ABSTRACT: The increase of train speed and axle load in the European rail network is an essential goal to make the railway transport more and more competitive for passengers and freights. High speed trains call for a better structural and geometrical stability of the track. To this aim it is crucial to apply innovative track design and materials. In the last decades, various types of ballast-less track systems have been developed and put in service around the world. These systems seem to perform better than ballasted solutions especially when high-speed passenger trains share the track with freight trains. The main advantages of innovative slab systems are the following: low maintenance needs/costs, higher availability, increased service life (50–60 years), higher lateral stability, reduction of weight and height of the track, easier and more economic vegetation control. Weaknesses of slab tracks against ballasted tracks are as follows: higher construction cost, higher noise radiation. In the light of the above considerations, in the study presented in this paper a life cycle cost assessment of two competing track solutions (ballasted and ballast-less) has been carried out, considering short- and long-term perspectives. Cost analysis uses the present value of agency (construction, inspection, maintenance and renewal), environmental (CO_2 emission), and user (delays-related etc.) costs. The analysis of the trend of agency, user, and externality costs of the alternatives over the entire life cycle of the infrastructure allows recognizing the most sustainable option. Results show that solutions that are more affordable in the short term can yield maintenance and renewal processes which are unfavorable or less sustainable in the long term. Furthermore, in the long term, the difference between the (total) present values of the two solutions becomes too small to yield sound conclusions in favor of the ballast-less solution with respect to the ballasted one.

1 INTRODUCTION

1.1 Background

Track design and maintenance are nowadays affected by the increasing axle loads and train speed in railway lines. Higher loads and speed require greater structural and geometrical stability of the track. It appears crucial to withstand the stresses in the rails, fasteners, sleepers/slabs, ballast and subgrade due to: i) the static mass of the vehicles; ii) the dynamic actions, such as lateral centrifugal forces on curves, longitudinal acceleration and braking forces; iii) vertical inertial forces from the motion of the wheel-set and its suspension; iv) the vibrational forces induced from imperfections in the rail surface (corrugations, joints, welds, defects) and in the wheels (flats and shells); v) the dynamic response of the track components to the above actions (Tzanakakis, 2013).

Based on the abovementioned considerations, accuracy in design, application of enhanced maintenance concepts and new or improved construction methods are needed (Esveld, 1999; Esveld, 2001; Gautier, 2015).

Ballast-less tracks were developed to achieve the following objectives:

- Increase speed;
- Increase capacity;
- Reduce the number of track maintenance operations and thereby the costs for maintenance.

Many types of ballast-less systems have been developed in the last years: Shinkansen, Rheda, Sonneville-LVT, Züblin, Stedef and Infundo-Edilon, IPA.

In general, ballast-less track is a continuous slab of concrete in which the rails are usually supported directly on the upper surface (of the slab) by using resilient pads.

Ballast-less track exhibits the following benefits in comparison with the traditional ballasted track: low maintenance costs (approximately 20–30% less), higher availability, increased service life (50–60 years), higher lateral stability, reduction of weight and height of the track, and easier and cheaper vegetation control.

On the other hand, the main weaknesses of slab tracks are: higher construction cost, higher noise radiation due to the lack of noise absorption of the ballast bed (Bilow & Randich, 2000; Esveld, 2001; Darr & Fiebig, 2006; Lichtberger, 2005). The mitigation of noise and vibration may further increase the costs of the slab track construction (Di Mino, et al. 2009).

In the light of the above considerations, an analysis based on long time frames is needed. The Life Cycle Cost (LCC) Analysis is an appropriate tool to evaluate the suitability of different track solutions considering the level of traffic in the line, the maximum speed allowed and other boundary conditions.

1.2 *Objective of the work*

The objective of the study described in this paper is to set up a LCCA-based method to compare two competing rail track solutions (ballasted and ballast-less). Short-term and long-terms perspectives are considered. An LCCA-based model able to evaluate the total cost of competing track solutions is proposed and applied. By means of this model, traditional tracks (made up of ballast, sleepers, fastenings, baseplates, rail) and innovative tracks (pre-stressed concrete slabs, fastenings, pad and rails) are compared.

The present value of agency (construction, inspection, maintenance and renewal), environmental (CO_2 emission), and user (delays-related etc.) costs is considered.

2 METHODOLOGY

2.1 *Basic concepts of LCCA*

According to the ISO 15686-5, "Life Cycle Costing (LCC) is a valuable technique which is used for predicting and assessing the cost performance of constructed assets".

LCC includes construction, operation, maintenance, end of life, Environment (including energy and utilities) costs (ISO 15686-5).

The following definitions apply:

i. External Costs: costs associated with an asset which are not necessarily reflected in the transaction costs between provider and consumer (e.g. business staffing and productivity, user costs, etc). Collectively these elements are referred to as externalities.
ii. Environmental cost impacts: ...costs (or savings via rebates) to LCC depending on the effects on the environment. Examples could include cost premiums for the use of non-renewable resources or for green house gas emissions. Where these costs are external to the constructed asset they will form part of a WLC analysis.
iii. Whole Life Cost (WLC) elements. Typically the difference to LCC is that the elements of

WLC include a wider range of externalities or non construction costs, such as finance costs, business costs and income streams.

Best practice LCCA calls for including not only direct agency expenditures (for example, construction or maintenance activities) but also user costs. User costs are costs to the public resulting from work zone activities, including lost time and vehicle expenses. Another important class of costs to consider is the externality costs, namely the costs of the environmental impact. Usually, these costs are not considered in the LCC, because their estimate is very difficult. However it is very important to take into account also these intangible costs because they can significantly affect the total cost (Stalder, 2001; Zoeteman, 2001, Lee et al., 2008). The model proposed in this work, includes this class of costs.

LCCA Methodology include the following main steps:

- Establish alternative design strategies.
- Determine activity timing.
- Estimate agency costs.
- Estimate user costs.
- Estimate externality costs.
- Determine life-cycle cost.

The implementation of the LCCA has involved two main phases:

- The Life Cycle Inventory (LCI) refers to data collection. It is a portion of LCCA. It entails the detailed tracking of all the flows in and out of the system. Raw resources or materials, construction processes, energy, emissions to air (CO_2e), unit costs are included. It is crucial to take into account that if the scope is too wide then the LCI may become quite useless to help in decision-making. On the contrary, if the scope is too narrow then the results may be biased and the output partisan;
- The Life Cycle Cost Model (LCCM), which focuses on cost estimation and discounting, and allows determining the total cost pertaining to a given solution.

2.2 *The proposed LCCA model*

A model to evaluate the suitability in long-term perspective of different alternatives for high speed rail tracks has been set up (Praticò & Giunta 2016, Praticò & Giunta 2016a). Three main classes of cost have been associated to a given track solution:

– the agency costs (C_{ag}), which refer to the expenditures for construction, maintenance and rehabilitation. Consequently, these costs include the initial cost and the running costs. The latter is affected by the service life of the track structure and its components;

- the user costs (C_{us}), mainly related to the delays, in operational phase, caused by the maintenance (slowdowns) and/or rehabilitation (re-routing) works.
- the externality costs (C_{ext}) connected to the impacts on the environment (resources depletion, noise, air pollution, etc.) due to the construction/maintenance/renewal processes.

It follows:

$$C_{ag} = C_{cons} + C_{main} + C_{renw} \qquad (1)$$

where C_{cons} is the cost for track construction C_{main} the costs for maintenance and C_{renw} the costs of renewal.

Regarding the user costs, they can be calculated starting from the costs of delays for maintenance and renewals. According to Lovett et al., 2015, user costs can be calculated as:

$$
\begin{aligned}
C_{us} &= CD_{main} + CD_{renw} \\
&= ((CO + CF) \cdot NL + CW) \cdot TD_{main} \\
&\quad + ((CO + CF) \cdot NL + CW) \cdot TD_{renw}
\end{aligned} \qquad (2)
$$

where, CD_{main} and CD_{renw} are the costs (euro/train-hour) of delays caused by maintenance and renewal activities respectively, CO is the locomotive operating cost (euro/locomotive-hour), CF is the fuel cost (euro/locomotive-hour), NL is the average number of locomotives, CW is the crew cost (euro/train-hour), TD_{main} is the average length (hours) of delays for maintenance activity and TD_{renw} is the average length (hours) of delays for renewal activity.

The externality costs refer to environmental impacts produced during the construction and the life cycle of the track (CEX_{kj}). Each j-th impact produced by the k-th process (Q_{kj}) can be associated to a unit cost (UP_{kj}). Having in mind the symbols already defined, the externality costs can be calculated by means of this equation:

$$C_{ext} = \sum_{K}\sum_{j} CEX_{kj} = \sum_{k}\sum_{j} Q_{kj} \cdot UP_{kj} \qquad (3)$$

The quantification of these costs is a difficult task; in this work the quantity of CO_2 equivalent corresponding to the given process and material has been considered.

For a given mixture and amount of greenhouse gas, the carbon dioxide equivalent refers to the amount of CO_2 that would have the same Global Warming Potential (GWP) when measured over a specified timescale (generally, 100 years). Regarding the cost of a ton of CO_2e, it should be noted that it is extremely variable (Ian et al. 2009, Praticò et al, 2011) and this variability is a critical factor because it may imply different incidence of externality costs on total cost.

Consequently, in a long term perspective, the equilibrium between the sum of agency and user costs (tangible costs) and externality cost (intangible costs) is crucial. To this aim a balancing factor (v) has been defined as the minimum ratio of tangible to intangible costs (see equation 7).

In order to make all the discussed and defined costs comparable, they are discounted to a base year considering the interest (r) and inflation (i) rates. The present values (PV) for the quoted costs can be calculated by means of the equations (4) to (6):

$$
\begin{aligned}
PV_{ag} &= C_{cons} + \sum C_{main} \cdot \left(\frac{1+i}{1+r}\right)^{E_{main}} \\
&\quad + \sum C_{renw} \cdot \left(\frac{1+i}{1+r}\right)^{E_{renw}}
\end{aligned} \qquad (4)
$$

$$
\begin{aligned}
PV_{us} &= \sum CD_{main} \cdot \left(\frac{1+i}{1+r}\right)^{E_{main}} \\
&\quad + \sum CD_{renw} \cdot \left(\frac{1+i}{1+r}\right)^{E_{renw}}
\end{aligned} \qquad (5)
$$

$$PV_{ext} = CEX_0 + \sum_{k} CEX_k \cdot \left(\frac{1+i}{1+r}\right)^{E_k} \qquad (6)$$

where PV_{ag} is the present value agency costs, PV_{us} is the present value user costs, PV_{ext} is the present value externality costs, i is the inflation rate, r is the interest rate, CEX_0 and CEX_k refer to the externality costs at construction stage and at k-th maintenance/rehabilitation phase respectively, E_{main} is the expected life for maintenance activity, E_{renw} is the expected life for renewal, E_k is the expected life for the k-th activity of maintenance or renewal.

The total present value (TPV) of a given alternative is the sum of the present values of all costs related to them, as already defined. In order to take into account the issue related to the CO_2 cost fluctuation, which affects the externality costs, a balancing factor has been defined as follows:

$$v = \min_{j=1,2,..k} \frac{PV_{ag_j} + PV_{uc_j}}{PV_{ext_j}} \qquad (7)$$

where j is the j-th alternative and k is the number of the alternatives in comparison.

Having in mind this factor, the TPV for a given j-th alternative can be derived from the equation:

$$TPV_j = PV_{ag_j} + PV_{uc_j} + v \cdot PV_{extj} \qquad (8)$$

This allows determining the differential (gain) between two competing solutions:

$$G_{tot} = TPV_S - TPV_B \qquad (9)$$

where TPV_S and TPV_B are the discounted total cost pertaining respectively to ballasted and slab tracks.

In this way the gain produced by one alternative (i.e. Ballast) with respect another (i.e. Slab) can be evaluated in a short-term or/and long-term time frame.

3 APPLICATION AND RESULTS

3.1 Data acquisition

The proposed LCCA-based model was applied to two track systems: a traditional ballasted track and the Shinkansen type (Figure 1). The main characteristics of the track components and their average expected service life (Milford & Allwood, 2010) are reported in Table 1. A double track line with a length of 1000 m has been considered.

For ballasted track, the main components are: rails, sleepers, fastenings, ballast and sub-ballast.

The Shinkansen slab track consists of a sublayer stabilized using cement, cylindrical "stoppers" to prevent lateral and longitudinal movement,

reinforced pre-stressed concrete slabs measuring 4.93 m × 2.34 m × 0.19 m (4.95 m × 2.34 m × 0.16 m in tunnels) and asphalt cement mortar injected under and between the slabs. The slabs weigh approximately 5 tonnes (Esveld, 2001).

In order to estimate the costs, according to the formulas set up in the previous section, a preliminary inventory phase has been carried out. The data gathered to estimate each cost are shown in Table 2.

The sources of data have been Italian executed and ongoing projects and the scientific literature.

It should be noted that maintenance encompasses all those minor activities aiming at repairing (corrective maintenance) or preventing (preventive maintenance) rails and sleepers damage, including tamping, track stabilization, ballast injection.

Maintenance costs are affected by the traffic (typically Millions of Gross Tons, MGT) and speed: the higher the traffic the higher the maintenance costs. Furthermore, higher speeds correspond to higher maintenance costs (Baumgartner, 2001; Silavong et al., 2014; Thompson, 1986).

The maintenance costs in this work have been considered as annually running (Jimenez-Redondo

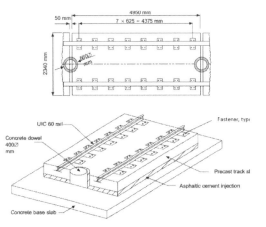

Figure 1. Slab track [*Source:* Esveld 2001].

Table 1. Track components and expected service life.

Component	Service life [Years]
Rail—*60 UNI*	28
Sleepers—*Pre-stressed mono-block*	40
Fastenings—*Elastic type Vossloh W14 AV*	40
Ballast—*Crushed stones, 500 mm. depth*	40
Subballast/Concrete road-bed—*Cement treated layer 200 mm depth*	40
Slab—*Pre-stressed concrete with cylindrical bollard*	60

Table 2. Life cycle inventory.

Costs	Inventory
C_{cons}	Components track
	Quantities
	Unit Prices
	Construction processes and costs
C_{main}	Traffic (Million of Goss Tons)
	Speed
	Service life of components
	Scheduling of maintenance
	Materials and components
	Unit price
	Maintenance processes and costs
C_{renw}	Traffic (Million of Goss Tons)
	Speed
	Service life of track
	Scheduling of renewal
	Components track
	Unit price
	Disposal process and costs
	Reconstruction processes and costs
C_{us}	Length of maintenance renewal activities
	Operation costs of train
	Train composition
	Fuel cost
	Crew cost
	Length of delays
C_{ext}	Materials
	Processes
	CO_2e emissions
	Cost of a ton of CO_2

et al., 2012) and have been calculated according to the following formula, which was set up and calibrated on the basis of data gathered from the Italian industry:

$$C_{main} = \left[2.2 \cdot \frac{(V-100)}{200} + 4 \right] \cdot GTK^{\left[-0.05 \cdot \frac{(V-100)}{200} + 0.63 \right]} \quad (10)$$

where GTK stands for Gross Tons × Kilometer, and V [Km/h] stands for speed.

Regarding the user costs, having in mind equation 2, the average delay for maintenance and renewal has been derived on the basis of operating costs (locomotive operating costs, crew cost for train, fuel/energy cost). Data were gathered from RFI (Italian Agency Owner of Railway network) and Trenitalia.

On the other side, the externality costs have been estimated considering the CO_2e emission associated to material production and processes (Milford & Allwood, 2010) and considering the unit price of the CO_2e.

3.2 Results of the application

The results of the study are reported in the following Figures 2–7.

Figures 2 and 3 illustrate the value the initial costs (construction), running costs (maintenance/renewal) user costs, and externality costs for the two track alternatives under comparison, and for a time frame of 0, 30, 60 and 120 years.

The comparison of the two figures highlights the higher construction and externality costs of the slab track at year 0 (construction stage) with respect to the ballast track. For the construction costs, the results comply with the literature

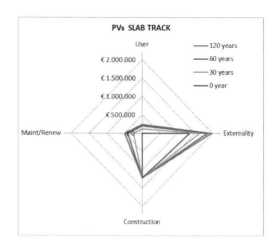

Figure 3.   Trends of PVs for slab track.

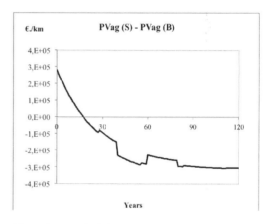

Figure 4.   Trend of the PVs of agency costs.

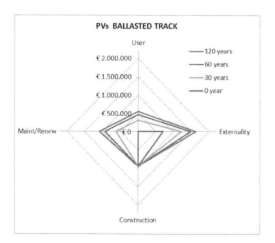

Figure 2.   Trends of PVs for ballasted track.

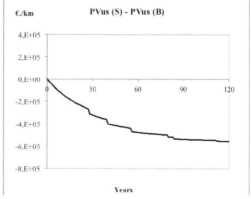

Figure 5.   Trend of differential of PVs user costs.

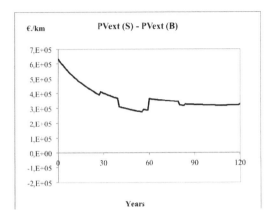

Figure 6.   Trend of the PVs of externality costs.

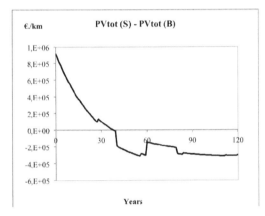

Figure 7.   Trend of differential of PVs total cost.

(Schilder & Diederich, 2007; Pichler & Fenske, 2013; Gautier, 2015). High externality costs are due to the great carbon footprint associated to the production of the cement.

Conversely, the slab track exhibits lower maintenance/renewal and user costs, due to the higher service life of the entire structure and of its components (i.e. concrete slab).

The trend of the user costs follows the one of running costs because the user costs, as already explained, depend on the delays due to the activities carried out on the track during the service life.

The following Figures 4–7 illustrate the trend over time (until 120 years) of the gain, evaluated for each considered cost.

The gain is calculated as the difference between the discounted costs (Present Value, PV) pertaining to the Slab track (S) and the one pertaining to the Ballasted track (B).

With this assumption, a positive value of the gain indicates that the ballasted track is preferable to the slab track and *vice versa*.

As for the gain of the agency costs (Figure 4), it can be observed that in the initial period the ballasted track alternative appears more convenient than the slab track.

However, the gain of this alternative decreases over time and reaches the value 0 after 16 years.

The discontinuities along the curve are due to the concentration of expenditures for the given solution (maintenance and/or renewal).

The gain of user costs (Figure 5) is negative during the reference period. This indicates the higher appropriateness of slab tracks with respect to ballasted tracks. This derives from the appreciable availability and service life of slab tracks.

Externality costs follow a reverse trend. In this case, the ballasted solution appears more "environmental friendly". Note that the embedded carbon of a slab track is 0.3–0.4 Kg CO_2/Kg. In contrast, for ballast and sleepers it is 0.005 Kg CO_2/Kg and 0.27–0.28 Kg CO_2/Kg, respectively (Milford, 2010, Praticò et al. 2013).

By referring to the total cost of the two solutions (Figure 7), the tendency of the gain may appear similar to the one observed for agency costs (Figure 4). Anyhow, some differences have to be considered.

In the short term the ballasted track is economically convenient, but this advantage becomes smaller and smaller over time.

After 36 years the break-even point is achieved. In the long term, due to the negative value of the gain, the slab solution emerges as convenient choice, even if the gain is somewhat small and constant.

A sensitivity analysis has been carried out in order to evaluate how the variability of the user and externality costs can affect the results. It has been found that the higher the user costs, the higher the difference of the present value (i.e. gain) between the ballasted and the slab track. User costs affect also the externality cost EX' due to the facts that different values of user costs result in different values of the parameter v (see eq. 7). By referring to the environmental costs, it is noted that thanks to equation 7, variation in CO_2e cost do not impact EX' value. For example if CO_2e cost is 11 €/ton it turns out EX' = 1.6 M€ (ballasted) or 1.9 M€ (slab). In contrast if Co2e cost is multiplied by ten it still results: EX' = 1.6 M€ (ballasted) or 1.9 M€ (slab).

4   CONCLUSIONS

In the present paper the comparison between ballasted and ballast-less track solutions has been carried out based on Life Cycle Cost Analysis.

A method to evaluate the costs of the two alternatives has been proposed and applied.

The method aims at offering to practitioners and researches a tool for evaluating the best strategy for the design and management of a railway track over time (maintenance and renewal). Long-term perspective, tangible and intangible costs are considered.

For each solution, technical features, traffic, speed of the railway line, costs of construction, maintenance and renewal, environmental impacts produced by the construction/maintenance/renewal activities, greenhouse emissions and related costs have been considered.

The trend of agency, user, and externality costs of the alternatives over the entire life cycle of the infrastructure allows recognizing the most sustainable solution from different standpoints.

Based on the data gathered, it can be reasonably observed what follows:

– solutions that are more affordable in the short time can yield maintenance and renewal processes which are unfavorable or less sustainable in the long term;
– the externality costs play an important role in the evaluation of the two alternatives;
– in the long term, despite the change of sign in total gains, the difference between the total present values appears quite negligible;
– the above observations imply that further analyses and studies are needed to yield rigorous conclusions in favor of the ballast-less solution with respect to the ballasted one or *vice versa*.

REFERENCES

Baumgartner, J.P., 2001. *Prices and costs in the railway sector*. École Polytechnique fédérale de Lausanne. Laboratoire d'Intermodalité de Transports et de Planification.
Bilow, D.N., Randich, P.E., 2000. *Slab track for the next 100 years*. Portland Cement Association. Skokie, IL.
Darr E. & Fiebig W. 2006. *Feste Fahrbahn: Konstrktion und Bauarten für Eisenbahn und Strassenbahn*, Second edition. Eurailpress. Germany.
Di Mino G., Giunta M. & Di Liberto M.C., 2009. Assessing the open trenches in screening railway groundborne vibrations by means of artificial neural network. *Int. Journ. Advances on Acoustics and Vibration*, Article ID 942787, 12 pages ISSN: 1687–6261.
Esveld C., Slab track: A Competitive Solution, 1999.
Esveld C., *Modern Railway Track*, Second Edition. Delft University of Technology 2001.
Gautier, P.E., 2015. Slab track: Review of existing systems and optimization potentials including very high speed. *Construction and Building Materials* 92: 9–15.
Ian, W.H., Parry, I.W. H & Timilsina, G.R., 2009. Pricing Externalities from Passenger Transportation in Mexico City, Policy Research Working Paper 5071, The World Bank, Development Research Group. Environment and Energy Team.
ISO 15686–5:2008 Buildings and constructed assets—Service-life planning—Part 5: Life-cycle costing.

Jimenez-Redondo, J., N. Bosso, N., Zeni, L., Minardo, A., Schubert, F., Heinicke, F., and Simroth, A., 2012. "Automated and cost effective maintenance for railway (acem-rail)". *Transport Research Arena* 48:1058–1067. doi: 10.1016/j.sbspro.2012.06.1082.
Lee, C.K., Lee, J.Y. & Kim, Y.K., 2008. Comparison of environmental loads with rail track systems using simplified life cycle assessment (LCA). *WIT Transactions on the Built Environment* 101:367–372.
Lichtberger B. *Track compendium*. First edition, Eurail Press. 2005.
Lovett, A.H., C.T. Dick, C.J. Ruppert, Jr. & C.P.L. Barkan, (2015). Cost and delay of railroad timber and concrete crosstie maintenance and replacement. *Transportation Research Record: Journal of the Transportation Research Board* Vol. 2476, pp. 37–44.
Milford R.L. Allwood, J.M, 2010. Assessing the CO_2 impact of current and future rail track in the UK. *Transportation Research Part D* 15: 61–72.
Pichler, D.; Fenske, J., 2013. Ballastless track systems experiences gained in Austria and Germany. *Proc. of AREMA Annual Conference*, Indiana Convention Center Indianapolis.
Praticò, F.G.; Vaiana, R.; Giunta, M.; Iuele, T.; Moro, A., 2013. Recycling PEMs back to TLPAs: Is that possible notwithstanding RAP variability? *Applied Mechanics and Materials* Vols. 253–255 (2013) pp 376–384.
Praticò, F.G.; Giunta, M., 2016. Assessing the sustainability of design and maintenance strategies for rail track by means life cycle cost analysis. In *Proceedings of COMPRAIL 2016 15th International Conference on Railway Engineering Design and Operation*, July 19–21 Madrid, Spain.
Praticò, F.G.; Giunta, M., 2016a. Issues and perspectives in railway management from a sustainability standpoint. In *Proceedings of International Conference on Transportation Infrastructure and Materials*, July 16–18, Xian, China.
Praticò F.G., Vaiana R., and Giunta M., Sustainable rehabilitation of porous European mixes, *ICSDC 2011: Integrating Sustainability Practices in the Construction Industry—Proceedings of the International Conference on Sustainable Design and Construction* 2011, pp. 535–541.
Schilder, R.; Diederich, D., 2007. Installation Quality of Slab Track—A Decisive Factor for Maintenance. *2007 RTR Special* 76–78.
Silavong, C.; Guiraud, L.; Brunel, J., 2014. Estimating the marginal cost of operation and maintenance for French railway network. *Proc. ITEA Conference–*Toulouse, France.
Stalder, O., 2001. The life cycle costs (LCC) of entire rail networks: An international comparison, Rail International, *International Railway Congress Association*, Vol. 32, No. 4, pp. 26–31, ISSN 0020-8442.
Thompson, L.S., 1986. *High-Speed Rail. Technology Review*, v. 89, pp: 32–43, 70.
Tzanakakis K. 2013. *The Railway Track and Its Long Term Behaviour. A Handbook for a Railway Track of High Quality*, Volume 2 of the series 2013 Springer Tracts on Transportation and Traffic pp 279–292.
Zoeteman, A., 2001. Life cycle cost analysis for managing rail infrastructure: Concept of a decision support system for railway design and maintenance *EJTIR*, 1, no. 4 pp. 391–413.

Transport Infrastructure and Systems – Dell'Acqua & Wegman (Eds)
© 2017 Taylor & Francis Group, London, ISBN 978-1-138-03009-1

Instability exposure and risk assessment of strategic road corridors in a geomorphologically complex territory

R. Pellicani, I. Argentiero & G. Spilotro
Department of European and Mediterranean Cultures, University of Basilicata, Matera, Italy

ABSTRACT: Risk and exposure of the road corridors of Matera Province (Basilicata Region, Southern Italy) to landslide phenomena was assessed. The provincial road network (1,320 km length) represents the main connection network among thirty-one urban centers due to the lack of an efficient integrated (road, railway and aerial) transportation system through the whole regional territory. The strategic importance of these roads consists in their uniqueness in connecting every urban center with the socio-economic surrounding context. The exposure was evaluated in terms of amount of traffic, as a function of population of each centers. The vulnerability was assessed in function of the presence of criticalities along roads. The exposure and vulnerability to landslides were combined in order to evaluate and map the risk. The classification of the road sections in terms of risk levels represent a support for decision making and allows to identify the priorities for designing appropriate landslide mitigation plans.

1   INTRODUCTION

Landslides affecting transportation corridors can cause direct and indirect consequences, respectively, in terms of traffic disruption and impact with vehicles.

In general, quantifying, in mathematical terms, the landslide risk can be very complicated, due to several aspects, related to the complexity in assessing the temporal probability of a specific landslide event with given intensity (hazard) and the probability of damaging a given element at risk, i.e. vulnerability (Glade 2003, Pellicani et al. 2014a). Assessing the landslide risk with regards to a mobile elements at risk can be even more difficult (Pellicani et al. 2016).

In the recent literature, quantitative risk assessment procedures, attempted to estimate the risk in terms of annual probability of direct impact, in terms of life loss of occupants of a vehicles, especially with regards to rockfall phenomena, have been developed (Corominas et al. 2005; Pantelidis 2011; Ferlisi et al. 2012; Budetta et al. 2015; Nicolet et al. 2016). Among the different typologies of landslide (Varnes 1984), rockfalls are generally characterized by small size, but by relatively high magnitude due to the high falling velocity of blocks and accordingly by greater impact and damaging potential on elements at risk, especially those mobile (vehicles). Other typologies of landslides, such as flows and slides, affecting road corridors are more widespread on the territory than rockfalls, as they affect hillslopes with small slope angle and characterized by different lithologies, unlike the rockfalls affect-ing mainly steep and rocky slopes. For these reasons, landslides have greater impact on the fixed element at risk (road and traffic). Nevertheless, the assessment of landslide risk along road corridors is poorly treated in literature.

In this paper, a procedure for assessing and mapping the landslide exposure and risk along the road corridors of Matera Province (Basilicata region, Southern Italy) is presented.

The road exposure was evaluated considering the amount of vehicular traffic on each road stretch. This estimation was carried out by using population data and by ranking in different orders the roads connecting several urban areas (nodes) according to the type of connection and number of linked nodes. Subsequently, the assessment of landslide risk was carried out by using a qualitative matrix approach. This procedure consists in overlaying the consequences and hazard maps and by combining in a matrix the relative classes. The consequences were derived by combining the vulnerability and exposure maps; while the hazard was evaluated in function of susceptibility and landslide intensity, depending on size and velocity of instability phenomena.

2   MATERA PROVINCIAL ROAD NETWORK

The road corridors of the Matera Province, in Basilicata region (Southern Italy), extend for 1,324 km, connecting 31 municipalities (Figure 1). This road network exerts an important role for the entire

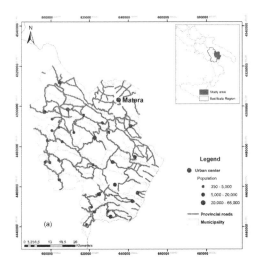

Figure 1. Study area: Matera provincial road network (Basilicata region, Southern Italy).

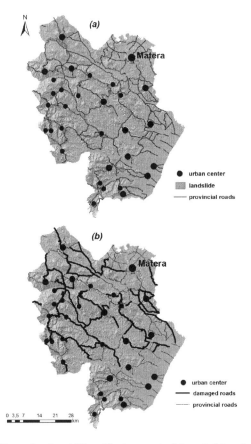

Figure 2. Landslides affecting provincial roads (a) and road segments subjected to landslide damages and reparation works.

transportation system of the provincial territory, of which represents the main connection network between among the urban centers, due to the lack of other type of effective infrastructures (railway, motorway, aerial, etc.). The strategic importance of these roads consists in their uniqueness in connecting every urban center with the socio-economic surrounding context. Consequently, without these roads, the urban centers would be isolated.

These road corridors and their relative vehicular traffic are continuously exposed to landslide processes (Figure 2a). In the last years, about the 44% (584 km) of the total length of roads was affected by damages and, subsequently, by reparation works (Figure 2b). These landslide phenomena are characterized both by high intensity and low frequency and by low intensity and high frequency. This last typology is particularly hazardous for the roads since it is widely distributed along the transportation network and its occurrence (depending by the return time) is connected to rainfall events.

3 METHODOLOGY: FROM EXPOSURE TO RISK

Landslide risk was defined by Varnes (1984) as *the expected number of lives lost, persons injured, damage to property, or disruption of economic activity due to a particular damaging natural phenomenon for a given area and period of time.* For a given category of elements at risk, the specific risk can be quantified as the product of hazard, i.e. the probability of occurrence of a specific hazard scenario with a given return period in a given area, vulnerability, intended as degree of loss to element at risk, and exposure, in terms of amount or economic value of elements at risk (Van Westen et al. 2006). In the following, the method for assessing landslide risk along the road network is explained, focusing on the procedure for exposure estimation.

3.1 *Road exposure estimation*

The exposure of the road corridors was evaluated in terms of amount of potential traffic. This choice was derived from the assumption that the economic value and the reconstruction costs of the entire road network are constant (exposure in monetary terms) and performing a probabilistic analysis of vehicles and passengers distribution along the road network is not possible, due to the lack of detailed input data in relation to the analysis scale.

A relatively simple model to calculate traffic volume on road segments (links) connecting several towns (nodes) is the Gravity model (McNally 2007, Hong & Jung 2016). It is based on Newton's Law and assumes that the trips produced at an origin and attracted to a destination are directly proportional to the total trip productions at the origin and the total attractions at the destination. In particular, the vehicular flow is assessed by means of a Four Step model, consisting in *trip generation* (depending on the activity system, represented by socioeconomic and demographic data), *trip distribution*, *mode choice* and *route choice* (depending on the transportation system, represented by road network typology and characteristics). The Gravity model application is not straight forward because depends on several factors complex to be determined, unless using survey data, such as total number of households and employees, trip purposes, traveler behavior, etc. Another factors is a calibrating term representing the reluctance or impedance of persons to make trips of various duration or distances.

For all these reasons it was choose to to prioritize and rank the links from busiest to least busy, without necessarily knowing the exact daily traffic volume. Therefore, the traffic along each road section, connecting two or more towns, was qualitatively assessed as a function of population of each of them. The road classification was carried out as following:

- first order road: direct link connecting only two towns;
- second order road: link connecting first order road segments;
- third order road: link connecting second order road segments.

The qualitative estimation of potential traffic volume on these three types of roads was carried out by calculating, for the 1st order roads, the arithmetic average of population of the interconnected urban centers, for the 2nd order roads, the average of traffic values of the connected 1st order roads, and, finally, for the 3rd order road, the average of traffic values of the connected 2nd order roads.

3.2 *Risk assessment along road network*

The spatial distribution of landslide risk along the road corridors of Matera Province was assessed and mapped using a qualitative matrix approach (Chowdhury and Flentjie 2003, AGS 2007), in which risk is obtained by combining in a two-dimensional table or matrix a set of hazard categories with a set of consequence categories (Figure 3). The first matrix regards the landslide hazard, i.e. the probability of a landslide of certain magnitude to occur, is a function of the return time, which depends on landslide typology and intensity. For a given type of landslide mechanism, the intensity changes according to the areal extension and velocity of landslide. Hazard matrix was obtained qualitatively through the following steps:

- Assessing the velocity range of landslide phenomena, by associating a velocity category to each type of landslide mechanism.
- Evaluating the areal extension of landslide phenomena and subdivision into four classes.

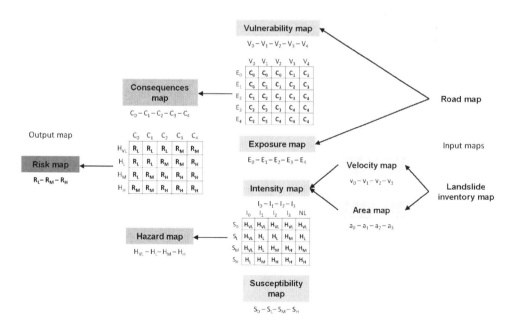

Figure 3.   Flow chart synthesizing the procedure for landslide risk assessment.

- Determining the landslide intensity by combining in a matrix the velocities and the areas with relative classes.
- Assessing the landslide susceptibility of the entire Matera provincial area through a polynomial heuristic-bivariate statitical model (Pellicani et al. 2014b). A set of thematic maps related to the predisposing factors was prepared and correlated with the landslide inventory map in order to obtain through a bivariate procedure (Van Westen 1993) the weights representing their importance on the instability process; finally, the factors were weighted and combined among them in a polynomial function.
- Combining the intensity classes with the susceptibility classes in a matrix and classifying these combinations in terms of five hazard classes: very low, low, medium, high. A column was added to the hazard matrix to consider the area not affected by existing landslides, in which future landslides could occur.

The landslide hazard map was obtained by overlying in GIS the susceptibility map and the landslide intensity map.

Consequences are generally defined as the outcome or potential outcome to an element at risk arising from the occurrence of a landslide of certain magnitude (Glade and Crozier 2005). Therefore, consequences are a function of the amount of elements at risk and the vulnerability of the affected elements.

The vulnerability of provincial road corridors crossing the Matera territory was assessed considering the previous criticalities and repair works carried out along road network. In particular, about the 44% (584 km) of roads was affected, in the last years, by instability and different types of works were carried out in order to repair it (Figure 2b). As the vulnerability is intended as the degree of damage to a given element at risk caused by an instability phenomenon, the road repair works, representative of the type and severity of damages, were ranked in five classes, to which a score ranging from 0 (no works) to 1 (severe damages and structural consolidation works) was assigned. As the road stretches affected by damages were repaired by performing different typology of work, the vulnerability values were derived by summing, for each road section, the several scores associated at the work categories and by normalizing the resulting values from 0 to 1. Five vulnerability classes were associated to the following score ranges: null (V_0) to zero value, low (V_1) to 0–0.25, medium (V_2) to 0.25–0.45, moderate (V_3) to 0.45–0.75 and high (V_4) to 0.75–1. At the same way, the exposure values, previously assessed, were normalized and subdivided into five classes: very low ($E_0 = 0$–0.1), low

($E_1 = 0.1$–0.25), medium ($E_2 = 0.25$–0.4), moderate ($E_3 = 0.4$–0.5) and high ($E_4 = 0.5$–1). The consequences map was obtained by overlying in GIS the vulnerability map and exposure map and by combining in a matrix the five vulnerability classes with the five exposure classes and ranking the combinations into five classes: insignificant, minor, medium, major and catastrophic. Finally, the risk matrix was produced by combining among them the hazard and consequence classes and by associating to each combination the following three risk classes: low, medium and high. The risk map was obtained by overlaying in GIS the hazard and consequence maps and associating the corresponding risk classes.

4 RESULTS AND DISCUSSION

The classification of provincial road segments was carried out considering the criteria defined in the paragraph 3.1 and some special conditions. In particular, border roads, which link directly towns of Basilicata with towns of Apulia (for examples Matera and Gravina di Puglia) or connect urban centers, belonging to Matera province, to major roads (state highway, for example SS106 *Ionica* and SS407 *Basentana*), were considered as first order roads. While road segments connecting towns to industrial areas, small villages and abandoned or relocated urban centers as damaged by landslides (for example, Craco and Alianello) or in presence of a better alternative roadway were classified as second order roads. Finally, road connecting secondary roads outside region or unpopulated sites were assumed of 3rd order. The resulting road classification is shown in Figure 4a. Out of the total length of road network, 1st order roads are about the 45% (600 km), 31% (407 km) and 24% (317 km) are, respectively, 2nd and 3rd order roads. By calculating for each road the amount of traffic, as indicated in the paragraph 3.1, the exposure values were achieved. These were normalized from 0 to 1 in order to obtain the exposure map (Figure 4b). Then the same were subdivided into five classes.

The exposure zonation has highlighted that the 15% of roads is characterized by moderate (E_3) amount of traffic, the 17% by both very low (E_0) and medium (E_2) exposure, the 20% by high exposure (E_4) and the 31% by low exposure (E_1).

The landslide risk analysis was carried out exclusively along the road corridors. For this reason, the spatial data contained in the input maps were made uniform, in order to obtain raster maps with the same spatial resolution, i.e. 20×20 m. The landslide inventory map was transformed into a raster map, classified in terms of landslide mechanism type, and then reclassified in terms of

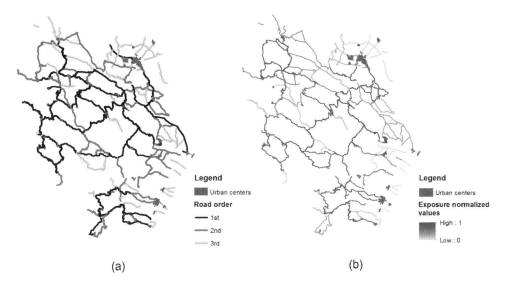

Figure 4. (a) Classification of road network into three orders and (b) exposure map.

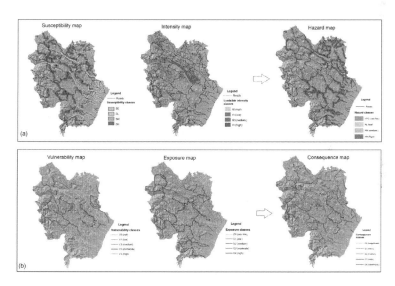

Figure 5. (a) Landslide hazard map obtained by overlying susceptibility and intensity maps; (b) Consequence map obtained by overlying vulnerability and exposure maps.

landslide velocity and area to obtain the intensity map. While, the vulnerability and exposure raster maps were obtained by rasterizing the road vectors and reclassified them, respectively, in terms of road repair work category and amount of traffic.

By overlying the landslide intensity map on the susceptibility map and by combining the corresponding classes into the hazard matrix, the hazard zonation was obtained (Figure 5a). Based on the hazard zonation, the 57% of the road corridors is affected by high hazard level, 3% is characterized by medium hazard and 40% by low or very low hazard.

The vulnerability zonation along road corridors has revealed that about the 56% of roads is free by damage, the 8% by low vulnerability, the 16% by both medium and moderate vulnerability and 4% by high vulnerability. By comparing the vulnerability and exposure maps (Figure 5b) it can be noted that road stretches affected by highest levels of vul-

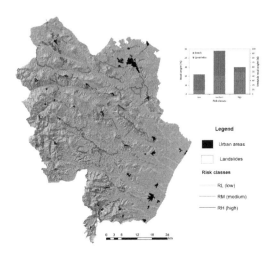

Figure 6. Landslide risk map of Matera provincial road network and graph showing the percentages of the entire road network and of the segments affected by landslides included in each risk class.

of risk corresponds to the following combinations: very low hazard with major and catastrophic consequences, low hazard with medium and high consequences, medium hazard with minor consequences and high hazard with insignificant and minor consequences. The graph in Figure 6 shows the percentages of the overall road network and of the stretches crossed by landslide bodies included in each of the three risk classes. The road sections subjected to low landslide risk are 22% (290 km) on the total length, while the remaining 48% (642 km) and 30% (392 km) of the examined road corridors is affected, respectively, by medium and high risk levels. The comparison between the risk map and the landslide inventory recognized along roads has also revealed that about the 10% (136 km) of roads are crossed by instability phenomena. In particular, on the total of instability phenomena, the 49.5% of landslides affected sections where the risk was evaluated high and 41% sections where the risk was assessed medium. The remaining 9.5% of landslides crossed road stretches where the risk was mapped as low.

nerability (i.e. V_4 and V_3) are conversely characterized by low amount of traffic. Indeed the 78% of roads free by damage (V_0) has moderate and high exposure levels. The distribution of consequence levels along roads (Figure 5b) was achieved by overlapping of vulnerability and exposure maps among them and from the combination of the corresponding classes into the consequence matrix. In particular, the assignment of consequence levels to each combination of vulnerability and exposure classes was carried out by assuming more prevalent the influence of exposure than vulnerability, since this last does not derive from a probabilistic analysis of potential damage degree but from the evaluation of distribution of past damages on the roads. Based on consequences matrix, the 29% of roads is affected by insignificant consequences, the 19% by minor, 36% by medium, 9% by major and 7% by catastrophic consequences.

Hazard and consequence maps were overlaid, the corresponding classes were combined in the risk matrix and the risk classes (low, medium, high) were associated to each combination, up to obtain the final landslide risk map along the 1,324 km of Matera provincial roads (Figure 6). Road sections affected by very low and low hazard and by insignificant and minor consequences, or by medium hazard and insignificant consequences or medium consequences and very low hazard are characterized by low risk. Conversely, medium and high hazard levels with major and catastrophic consequences, medium consequences with both medium and high hazard levels or catastrophic consequences with low hazard involve a high risk level. Finally, a medium level

5 CONCLUSIONS

A procedure for assessing and mapping the landslide exposure and risk along the Matera provincial road network (Basilicata region, Southern Italy) was explained. The main aim of this study is provide a landslide risk map as support to local governments for identifying the priorities for the design of appropriate mitigation plans. This study was carried out for the provincial roads as they represent the main network connecting the urban centers of the Matera Province and is affected for the 10% of its total length by landslide instability phenomena.

Road exposure to landslides was estimated in terms of amount of traffic, qualitatively obtained by ranking the road segments in three different order and by calculating, for 1st order roads (direct links among towns), the arithmetic average of population of interconnected urban centers and, for 2nd and 3rd order roads (connecting, respectively, 1st and 2nd order roads), the average of traffic amount on connected road stretches. The landslide risk was evaluated, in absence of detailed input data due to the scale of analysis, by a qualitative matrix-based approach. The proposed procedure has allowed to assess and map landslide risk, through the following three sequential steps: (i) producing landslide velocity and areal extension maps from landslide inventory map, according to the movement type, and combining them for obtaining the landslide intensity map; (ii) combining, pairwise, intensity and susceptibility maps and vulnerability and exposure maps for obtaining the hazard and consequences maps, respectively; (iii) overlying the hazard and consequences maps and combining the corresponding classes in a matrix in order to obtain the final risk map of the roads, subdivided in low, medium and high risk levels.

The comparison between the risk map and the distribution of landslides crossing the road stretches, has showed the majority of instability phenomena (about 91%) are located on road sections classified at medium and high risk. This is a significant result, as it shows that it is possible to use, for a regional scale landslide risk assessment, a simplified-qualitative procedure, based on a few data, easy to find and manage, and reliable in relation to the scale of analysis. Although the proposed methodology for the risk assessment did not consider the prediction of landslide run-out and temporal probability, the results obtained allow to rank the road stretches in terms of risk levels and, consequently, to identify the priorities for designing detailed field surveys and appropriate landslide risk mitigation plans.

ACKNOWLEDGMENTS

The Authors wish to point out that Dr. R. Pellicani has been responsible for the research methodological approach and Prof. G. Spilotro for research coordination.

REFERENCES

AGS 2007. Pratice note guidelines for landslide risk management 2007. Extraxt from *Australian Geomechanics Journal and News of the Australian Geomechanics Society* 42 (1): 63–114.

Budetta, P., De Luca, C., Nappi, M. 2015. Quantitative rockfall risk assessment for an important road by means of the rockfall risk management (RO.MA) method. *Bulletin of Engineering Geology and the Environment*: 1–21 DOI 10.1007/s10064-015-0798-6.

Chowdhury, R.N. & Flentje, P. 2003. Role of slope reliability analysis in landslide risk management. *Bulletin of Engineering Geology and Environment* 62: 41–46.

Corominas, J., Copons, R., Moya, J., Vilaplana, J.M., Altimir, J., Amigó, J. 2005. Quantitative assessment of the residual risk in a rockfall protected area. *Landslides*, 2: 343–357.

Ferlisi, S., Cascini, L., Corominas, J., Matano, F. 2012. Rockfall risk assessment to persons travelling in vehicles along a road: the case study of the Amalfi coastal road (southern Italy). *Natural Hazard*, 62: 691–721.

Glade, T. 2003. Vulnerability assessment in landslide risk analysis. *Die Erde* 134, 121–138.

Glade, T. & Crozier, M. 2005. The nature of landslide hazard impact. In: *Landslide hazard and risk*, edited by: Gladem T., Anderson M., and Crozierm M., Wiley, Chichester, 43–74.

Hong, I. & Jung W.-S. 2016. Application of the gravity model on the Korean urban bus network. *Physica A: Statistical Mechanics and its Applications* 462: 48–55.

McNally, M.G. 2007. The Four Step Model. Chapter 3 in Hensher and Button (eds). *Handbook of Transport Modeling*, Pergamon.

Nicolet, P., Jaboyedoff, M., Cloutier, C., Crosta, G.B., Lévy, S. 2016. Brief communication: on direct impact probability of landslides on vehicles. *Natural Hazards and Earth System Sciences* 16: 995–1004.

Pantelidis, L. 2011. A critical review of highway slope instability risk assessment systems. *Bulletin of Engineering Geology and the Environment* 70: 395–400.

Pellicani, R., Van Westen, C.J., Spilotro, G. 2014a. Assessing landslide exposure in areas with limited landslide information. *Landslides* 11 (3): 463–480. DOI 10.1007/s10346-013-0386-4

Pellicani, R., Frattini, P., Spilotro, G. 2014b. Landslide susceptibility assessment in Apulian Southern Apennine: heuristic vs. statistical methods. *Environmental Earth Sciences* 72 (4): 1097–1108. DOI: 10.1007/s12665-013-3026-3

Pellicani, R., Spilotro, G., Van Westen, C.J. 2016. Rockfall trajectory modeling combined with heuristic analysis for assessing the rockfall hazard along the Maratea SS18 coastal road (Basilicata, Southern Italy). *Landslides*: 1–19. DOI 10.1007/s10346-015-0665-3

Transport Infrastructure and Systems – Dell'Acqua & Wegman (Eds)
© 2017 Taylor & Francis Group, London, ISBN 978-1-138-03009-1

Safety in the III Valico tunnels

A. Focaracci
Foundation Fastigi President, CEO, Prometeoengineering.it srl, Rome, Italy

ABSTRACT: The Genoa-Milan High-Speed Railway Line (III Valico di Giovi) is characterized by the presence of several tunnel sections, including the double-barreled Valico tunnel with a length of 27 km (not including interconnections, making it one of the longest galleries in the national territory), and the Serravalle tunnel with a length of 7 km. In recent years the issue of tunnel safety has been the subject of specific laws, both at the national level, with Ministerial Decree 28.10.2005 on Safety of Railway Tunnels, and at the European level through Council Decision 2008/163/EC of 20 December 2007 on the technical specifications of interoperability relating to Safety of Railway Tunnels in the trans-European conventional rail system and to high-speed referred to as Technical Specifications for Interoperability or TSI. The III Valico di Giovi is part of the European Genoa-Rotterdam corridor and, therefore, falls within the before mentioned TSI introduced following the approval of the final design in 2005. The safety design, based on in-depth analyses carried out by means of fire and exodus models and probabilistic risk analysis, has led to the definition of the functional layout and performance specifications of safety systems, and the development of a final plan that is among the most advanced in Europe from an operational safety point of view.

1 INTRODUCTION

The Genoa-Milan High-Speed Railway Line ("III Valico di Giovi") is characterized by the presence of numerous tunnel sections, including the double-barreled Valico tunnel (not including interconnections, making it one of the longest galleries in the national territory), and the Serravalle tunnel with a length of 7 km. As part of the final design of the railway line, which was built in 2004–2005, the safety of the tunnels was addressed according to the relevant laws and RFI standards that were required at the time. In light of recent changes in the regulatory framework, which do not entirely overturn the security principles already used in the (III Valico) final design, but rather refine and integrate some of the concepts verifying the design choices through a security design methodology based on the risk analysis, it has been necessary to re-evaluate the design choices concerning safety of the III Valico galleries. Present document describes the safety measures (active and passive) adopted in the III Valico railway tunnels and the final project completion added in 2005 in order to adequalte the project to the new safety regulations for railway tunnels.

2 THE PROJECT

The III Valico Railway runs along a route of about 53 km characterized by the presence of several long tunnels, making the project particularly demanding. The type of the tunnels planned are consistent with the latest security standards, including the construction of two side-by-side, single-track tunnels with cross-connections, which enable each tunnel to be safe place from any events occurring in the other. The proposed route will start about 800 m before Bivio Fegino on the line originating from Genova Piazza Principe. The line passes under the Liguria Apennines with a tunnel, approximately 27 km in length, that opens up into the municipality of Arquata Scrivia where the railway line is expected to connect with the Libarna track, running along the Plain of Novi and, then, passing under the Serravalle Scrivia territory with a tunnel about 7 km in length. The route ends in Tortona where a grade-level connection with the Piacenza/Milano line is planned. The plan for the Valico Tunnel, which constitutes the most significant infrastructure work of the project, includes the construction of four windows, including two exploratory narrow tunnels, which were partially completed in 1996–98, for further development. Upon their completion these tunnels will constitute the Castagnola adit (Municipality of Fraconalto) and the Val Lemme adit (Municipality of Voltaggio). Following the general diagram of the route with the works planned for the new rail link and the associated plant equipment and a summary table of the salient features of individual tunnels are reported.

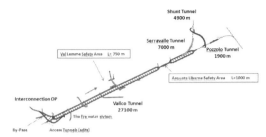

Figure 1.   Layout route.

3   SAFETY DESIGN

The method adopted for the safety design is the IRAM-RT methodology, already employed in the safety design of the recently opened to traffic tunnels on the Bologna-Florence High-Speed railway line. The safety design of a railway tunnel foresees the following operating steps:

– analysis of infrastructure vulnerability, starting from the acquisition of geometric, structural and equipment installation characteristics of the work, traffic and accident data;
– identification, structural and plant design of safety requirements that may prove to be necessary after the vulnerability analysis (Ministerial Decree of 10/28/05);
– risk analysis for the verification of safety targets achievements (Ministerial Decree 28/10/05);
– operating procedures and in particular the preparation of emergency management plans (Ministerial Decree 10/28/05).

4   UTILITIES SYSTEM EQUIPMENT

The key works for which the new regulations have required the design are:

– the construction of a safety area inside the Valico tunnel provided with suitable exit routes, smoke extraction and automatic shutdown system;
– the construction an open-air safety area between the Valico tunnel and the Serravalle tunnels equipped with an automatic shutdown system;
– the construction of new by-passes;
– the upgrading of the system services of the tunnel access;
– the construction of new ventilation shafts and adaptation of fire safety equipment.

4.1   Val Lemme safety area

In compliance with STI requirements, the construction of a safety area for passengers and freight trains is planned at a mid-point within the Valico gallery (about 27 km). This safety area will be accessible by an adit through which a conflagrant train can be driven and which will allow the controlled exodus of travelers and the intervention of rescue teams. The Val Lemme safety area consists of two evacuation tunnels, which extend 750 m from the axis of the adit, parallel to the tunnel axis, located 35 m between the even and odd track, respectively. The evacuation tunnels are accessible by the platform, through branches, placed at a center-to-center distance of 50 m and are connected by a walkway, placed over the two barrels, at the Val Lemme tunnel access connection. Access from outside the safety area takes place through the Val Lemme adit, 1592 m long; at the tunnel entrance and at the kp 0+700.00 two ventilation chambers are placed.

The tunnel safety area will be equipped with the following systems to efficiently and effectively contrast the tunnel emergencies:

– ventilation system/smoke control;
– fire water system;
– automatic extinguishing foam system;
– hazardous liquid collection system.

The ventilation system/smoke control is designed according to the engineering approach to fire safety with reference to the international standards as NFPA 92B and NFPA 130 and analysing similar systems as those designed for the Turin–Lyon railway.

Using a probabilistic approach to safety design, the flow rate of the ventilation system was assumed to range from 200 m^3/s to 400 m^3/s. The design of the system was carried out with reference to the maximum capacity and, considering the uncertainties related to possible dysfunctions, the complex geometry and the behavior of airflow circuits in the presence of hot smoke, a variability of about 100 m^3/s was estimated.

The system features a distributed fume extraction design whereby extraction is carried out by a set of extraction points located along the safety area, placed at the tunnel access and inside of six distributed bypasses. The scheme above described, allows optimizing the extraction of fumes in relation to the most likely locations where the fire may occur within the safety area and to geometric characteristics of the same area. Fumes, once drawn and channeled, are conveyed into a false ceiling inside the Val Lemme tunnel access to be expelled through the shaft foreseen in the project.

The ventilation control unit is located in a tunnel made ad-hoc before the shaft, designed to accommodate four two-stage axial fans able to extract up to 120 m^3/s each. The following figures show the ventilation control unit at Val Lemme, the location of the by-passes used for the fumes extraction, and an illustrative section showing the path fumes.

The bypass system connecting the barrels of the train tunnel with the safety area leading to the exodus

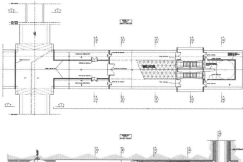

Figure 2. Layout to the ventilation control unit at Val Lemme.

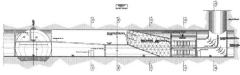

Figure 3. Layout odd platform for the fumes extraction.

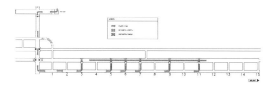

Figure 4. By-passes used for the fumes extraction.

shaft is equipped with pressurization system, which will create an excess of pressure in the safety area so as to prevent the entry of fumes from the compromised barrel. This is accomplished by means of a pair of fans (one backup) capable of preventing fumes from entering the safety area. Fresh air is drawn into the safety area from outside through a vent in the false ceiling along the adit, passing through a control unit located outside the tunnel access. In the case of malfunctioning fans, this control unit is also able to provide a minimum of excess pressure to the safety area. Moreover, the design of the external control unit includes a Saccardo ventilation installation, capable of pressurizing the entire tunnel access. In conclusion, the design includes an extraction system in the rescue vehicles parking area that picks up the fumes directly from vehicle exhausts.

The fire water system, in accordance with the provisions of the Ministerial Decree 28.10.2005 and reference TSIs, will be composed of two pressurization control units, a storage tank and a network of fire hydrants equidistant to 125 m. For fires extin-

guishing of flammable liquids and combustible fuels the design proposes fire protection by means of monitors to AFFF foam additives (Aqueous Film Forming Foam), cooling agent and the formation of a protective film on any liquid fuel (B class). The protection system provides a high foam flow of up to 3000 l/min directly to the fire location, inhibiting combustion on the surfaces and subsequently cooling them. The use of foam allows better coverage of the wet surfaces. In case of spillage and fire of hazardous liquids the AFFF additive will cause the rapid formation of an impermeable liquid film on the surface of the spilled liquid. The designed system, thanks to the high flows and possibility to concentrate their action at the fire location, allow for significant mitigation of the force of the fire, such higher as earlier the system is activated. The presence of an automatic extinguishing system at the Val Lemme area reduces the uncertainties of risk management, by permitting the reduction of the fire power to less than 100 MW, resulting in a significant improvement of tunnel system performance in terms of emergency management.

Along the entire length of the safety area there will be a collection system of potentially hazardous liquids. The spilled liquids and waters discharged from the automatic shutdown will be channeled into a tank located in the lower point of the safety area where the flammable liquids will be separated.

4.2 Arquata Libarna safety area

The definition of new safety standards has necessitated the creation of an external safety area, with a length of 1166 m, located in the vicinity of the Arquata Libarna PC. The above mentioned area is accessible by emergency vehicles through a special road. The safety area contains zones equipped with a Triage area, a technological building, a rescue helicopter pitch, and razed pathway for positioning the bimodal tack mechanism.

The area's external security systems are:

– fire water system;
– automatic foam extinguishing system;
– hazardous liquid collection system.

The fire water system, in accordance with the provisions of the Ministerial Decree 28.10.2005 and reference TSIs, has been designed based on what is already reported in the 2005 project considering the increase of pump capacity from 600 to 800 l/min. The automatic extinguishing system is the Monitors type similar to the one designed for the Val Lemme area but with monitors spacing equal to 50 m which can be active in groups of 3–6. The system facilitates the extinguishing operations and is able to handle high flow rates of foam. It is also effective against dangerous liquid fires with an AFFF foam additive able to extinguish class B fires. In addition,

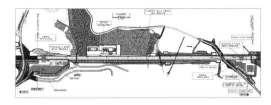

Figure 5. Layout of the Arquata Libarna outdoor safety area.

there will be a potentially hazardous liquid collection system. The Figure 5 illustrates the layout of the Arquata Libarna outdoor safety area.

4.3 Cross connections

The escape system of the main line consists of a series of by-pass link connections between the two single-track railway tunnels (odd and even) every 500 m approximately, in both the III Valico and Serravalle tunnels. By-passes are used for people escaping from one railway tunnel to a parallel railway tunnel; each bypass is compartmentalized into both galleries. The conceptual basis for the safety analysis is the consideration of safe place of the intact (unaffected) tunnel. The ventilation system foreseen (pressurization of the bypass links road) allow keeping the exodus ways free from the fumes produced in the compromised tunnel, with the following basic criteria:

– ensuring effective pressure in the link road with respect to the compromised tunnel both when the access doors (to the compromised tunnel and the unaffected tunnel) are open, or closed;
– guarantee, even in minimum load conditions, a suitable flow rate of air replacement to the considerable possible presence of persons inside the bypass;
– determine the air velocity in the exodus zones with values compatible with the emergency situation of the passengers, hit by substantial flow of air;
– reduced start-up times of the fans (less than 30 sec) in order to reach, in the shortest possible time (about 35 sec), the standard overpressure expected for the volumes involved. The system will still keep the by-passes free from any fumes present in the line gallery where the accident occurred.

The exodus system for the Interconnection lines consisted of two pedestrian by-pass crossing that are connected one to the other. Since at the starting point the crossings are rather long and of reduced section, a filter chamber beside the rail tunnel was created. This chamber is pressurized in a way similar to the "transition chamber" in the adit, with the same considerations in terms of safety analysis (intact/compromised tunnel) and the conditions assumed in calculations.

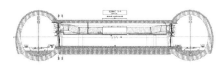

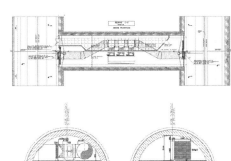

Figure 6. By-pass link connections.

4.4 Access tunnels (adits)

The exodus system with the adit allows for a widened area at the end of each side tunnel, which forms a space that is intended to allow for the reverse gear of rescue vehicles and to accommodate the beginning of the passenger flow from the tunnel towards the outside. Each of these areas, referred to as "transition chambers", is equipped with a series of doors (on the side railway tunnel and the on the side adit) and a ventilation system capable of keeping the same chamber in slight overpressure with respect to the tunnel. Furthermore, there is a second filter area, between the two rails tunnels (crossing the tracks) also equipped with a ventilation system, able to keep the same area in slight overpressure with respect to the tunnel. In case of tunnel fire, the ventilation system prevents fumes from entering the exodus adits, allowing the passengers to evacuate in safety. Finally, there is an extraction system designed for the adits in the emergency vehicle parking area, similar to that of the Val Lemme safety area, which directly picks up the fumes from the vehicle exhausts.

4.5 Ventilation shafts

In line with the ventilation strategies adopted for Italian railway tunnels and with the provisions of Annex II of the Ministerial Decree of 28/10/2005, the points of transition from a twin-tube tunnel with a single barrel tunnel will be designed with measures preventing the circulation of fumes from the compromised tube to the unaffected tube, by means of ventilations shafts. Intake grids positioned on top of the tunnel will suck fumes into a circular segment section plenum. As a results of the design specifications based on the analysis of train fire scenarios and risks (where a thermal power of 10 MW was used for potential passenger train fires and 50 MW for freight train fires), the adjustment intervention of the III Valico final design includes the addition of new ventilation shafts and the modification of the extraction flow rate for those already planned. The ventilation shafts have been designed based on the thermos-fluid dynamic simulations results to allow flow rates extraction of approximately 200 m³/s.

5 RISK ANALYSIS

The Valico tunnel system, considered together with the Compasso tunnel and the Voltri Interconnection as one single system, is greater than 9000 m in length. Therefore, in accordance with Annex III of the Ministerial Decree of 28/10/2005 the expanded risk analysis of the system was performed. The analyses conducted using the IRAM RT method have shown how the safety measures in the tunnel design allow for a level of risk which falls within the attention zone, primarily due to the high volume of freight trains predicted, as exemplified in the next figure.

The adoption of a training program aimed to limit the contemporaneity between freight trains and passenger trains would determine further risk reduction. In order to verify the functionality of the designed works in terms of both smoke management and exodus, numerous three-dimensional simulations for representative scenarios were carried out. The analysis made it possible to calibrate the statistical models adopted for the risk calculation, review the emergency management timelines, supporting the choices made with regard to safety systems such as power shutdown and ventilation. The execution of the simulations made it possible to reduce the uncertainties associated with the transportation system efficiency, in the occurrence of chaotic nature hazardous events such as fires. As an example, passenger exodus simulations and thermo-fluid dynamic simulations of smoke extraction carried out for the Val Lemme safety area are reported in the following sections.

5.1 Val Lemme safety areas—thermo fluid dynamic smoke extraction simulations

The analysis of accident scenarios, using the three-dimensional Fire Dynamics Simulator calculation code, was performed by means of the simulation of the smokes spread generated by a stationary train at the Val Lemme safety area. The simulations target is the functionality and performance of the smoke extraction system verification. The proposed smoke extraction system involves the construction of n. 6 extraction points to be assessed based on the maximum distance between the ventilation outlets. The identified scenarios are summarized below:

- passenger train with a potential thermal power of 20 MW and suction capacity of about 200 m³/s;
- freight train with a thermal power of 50 MW and a suction capacity of about 400 m³/s.

The results of the simulations, expressed in terms of the distribution of the fumes in the tunnel, profiles of temperature, concentration of carbon monoxide and visibility for different time intervals, showed that:

- in both the analysed scenarios the ventilation system ensures environmental conditions at the dock compatible with the timelines, taken as reference for the exodus of the passengers and the train crew;
- the optimum configuration for the extraction points has been detected in a mixed combination, which provides for a distribution of vents every 100 m with the three control units every 50 m. The vents gathered centrally, where it is easier to happen the train on fire, nevertheless ensure the protection of 450 meters of tunnel that includes an entire passenger train.

In the case of freight train, even if fire occurs at the train end, the simulations have shown that the vents at 100 m distance are sufficient to ensure that the ventilation system guarantees the protection of drivers. In the case of freight train fire and in the phases of Firefighter intervention, the opening of the extraction vent localized on the graft of the adit has been designed, in order to allow the extraction to the maximum system capacity. The simulations conducted show that the security plan as a whole, thanks to the presence of the combined distributed

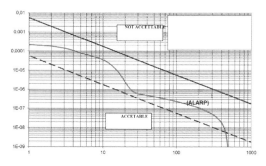

Figure 7.   Global FN curve.

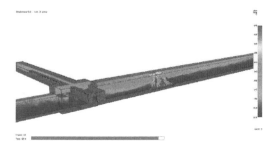

Figure 8. Temperature simulation.

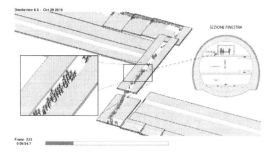

Figure 9. Exodus simulation.

extract ventilation system, guarantees a minimum level of security largely compatible with the current regulations at national and European level. The following figures summarize the three-dimensional analyses performed for the verification procedure of the tunnel access extraction system.

The simulation results have shown that the fume extraction zone is characterized by flows leading to high load losses in justification of the high fan performance necessary for the system operation phase.

5.2 Val Lemme safety areas—exodus simulations

The analyses of accident scenarios was performed using the three-dimensional Fire-Dynamics Simulator calculation code by the simulation of the smoke propagation generated by a passenger train stopped at the Val Lemme safety area, coupled with the simulation of the exodus process, conducted through the EVAC code.

The targets of the simulations consist of:

– simultaneous verification of the functionality and performance of the safety systems;
– verification of the escape times of the passengers in accident conditions;

The scenario analyses, with the primary purpose of verifying the emergency management, is characterized by a greater verisimilitude. The assumption adopted for the definition of reference fire, to check safety conditions of the exposed population,

was characterized by a generated maximum thermal power close to 20 MW with a gradual increase in a time equal to 10 min.

The simulations carried out on the exodus process of 500 persons aboard a passenger train stopped at the Val Lemme safety area, show that, the combined fire simulation with the exodus simulation localized at the dock, in presence of train accident, all the passengers leave the train and enter in the connecting branches with a total evacuation time of about 3 minutes.

5.3 Conclusions

The safety of travelers is a major issue, which in recent years has become increasingly important also because of a significant change in the regulatory framework (Ministerial Decree of 28/10/2005 and Technical Specifications for Interoperability).

Based on these recent changes, the union of the primary and most competent engineering companies in Italy has allowed the creation of a consortium of design, which by increasingly advanced and cutting-edge technologies, has contributed to the current project of the III Valico railway design. The approach adopted in the design led to the definition of a performance benchmark for facilities that allow for emergency management in the most likely scenarios supporting the risk analysis results provided by the Ministerial Decree of 28/10/2005.

The obtained results at the design level discussed in this article, backed up by detailed analyses conducted by means of fire and exodus models and the analysis of probabilistic risk, have led to the definition of the functional layout, and of the performance specifications of security installations, and to the development of a final project that is among the most advanced in Europe from the point of view of safety in the operation phase.

REFERENCES

Focaracci A., Lunardi. P. Silva C. (2001): "The Bologna to Florence Hight Speed Railways line: 92 km through the Appenines", Progress in Tunneling after 2000 – Bologna 2001.
Focaracci A., (2007): "Designing Safety"—Italian Risk Analysis Method—Le Strade n°4–2007.
Focaracci A., (2007): "Commission Tunnels Safety to route and railways"—Conference Safety in Tunnels—Genova, 27–28 March 2007.
Focaracci A., (2010): "Italian Risk Analysis method IRAM" 11th International Conference Underground Construction Prague 2010—Transport and City Tunnels. 837–845.
Focaracci A., (2010): "Safety in Tunnels: innovation and tradition" 11th International Conference Underground Construction Prague 2010—Transport and City Tunnels. 846–851.

AllBack2Pave: Towards a sustainable recycling of asphalt in wearing courses

D. Lo Presti & G. Airey
University of Nottingham, Nottingham, UK

M. Di Liberto, S. Noto & G. Di Mino
Universita' degli Studi di Palermo, Palermo, Italy

A. Blasl, G. Canon Falla & F. Wellner
Technische Universität Dresden, Dresden, Germany

ABSTRACT: Nowadays, asphalt plant technologies allow producing asphalt mixtures incorporating up to 100% reclaimed asphalt. Unfortunately policies are still behind technology and in order to suggest guidelines for a widespread use for surface courses, road managers feel the need of having a deeper understanding of optimised design strategies, information related to the handling in asphalt plants and on the performance of these mixes. This paper provides the summary of the main results and the details of the main idea behind "AllBack2Pave 2013–2015" a two-years, 500 K€ project funded by the CEDR Transnational Road Research project that evaluated the feasibility of going towards 100% recycling of asphalt pavements into surface courses. The project, coordinated by the Technische Universität Dresden in Germany, together with the University of Nottingham and University of Palermo, was structured so to provide an overview of the European panorama by involving 3 EU countries from the South (Italy), Centre (Germany) and North (UK). In order to facilitate the deployment of lean concepts and lean production practices, the investigation was implemented in close collaboration with the private sector, including asphalt mixing plants, chemical additives producers and waste material managers.

1 INTRODUCTION

Nowadays, the amount of recycling of Re-claimed Asphalt (RA) in new asphalt pavements has grown to the point that it is no longer a mere alternative when virgin materials are not easily available (i.e. The Netherlands), but a common practice in almost all of Europe. However, the dismantling and end of life strategies for these pavements in European countries are very divergent relating to the amount of RA recycled in new pavement layers, although, due to mainly policy's constrains, over the whole Europe the share of recycling of RA in new asphalt courses remains rather lower than it could be technically. In fact, some of the most recent European collaborative research projects, such Direct-MAT, 2011 and Re-Road, 2012, have shown that the durability of hot asphalt mixes with RA proved to be satisfactory, however the complete reuse of the reclaimed construction materials requires a precise assessment of the properties of the components.

On this wave, in 2012 the Technical Group Research of CEDR, the Conference of European Directors of Roads, initiated a programme in response to the common research needs of its National Road Administration members. This Transnational Research Programme called "Road construction in a post-fossil fuel Societyî had as overall aim to develop new concepts for road construction especially for pavements in the context of the finite nature of fossil fuels and other resources. Economic and environmental benefits are the driving forces behind research. In turns, the required solutions needed to be feasible, valid and cost-effective through increased recycling and consequent minimisation of the impact in the society.

"AllBack2Pave: towards 100% recycling of reclaimed asphalt in wearing courses" was one of the selected proposal. The project started in November 2013, lasted 2 years, and was led by the Technische Universität Dresden in Germany, together with the University of Nottingham in the UK and University of Palermo in Italy. The investigation was implemented in close collaboration with the private sector, including asphalt mixing plants and chemical additives producers and main idea behind it was evaluating the feasibility of going towards 100% recycling of asphalt pavements into surface courses by following the structure shown in Fig. 1 and aiming at the following objectives:

– establishing, through laboratory tests on binders and asphalt mixes, whether the use of high rates of RA is feasible in developing mixes with a high level of durability.
– developing the so-called "AllBack2Pave end-user manual" with best practices to produce cost-effective asphalt mixes with high RA content.
– assessing the sustainability performance of the technologies, at first through an evaluation of the design life of the mixtures and then with an holistic approach based on Life-Cycle based techniques such as Life-Cycle Assessment (LCA) and Life-Cycle Cost Analysis LCCA.

This paper illustrates the outcomes of the project and each of the following paragraph will summaries the main results of each of the following technical Work Packages:

• In Work Package 2 "Laboratory Design" different reclaimed asphalts and virgin materials were collected and characterized in Germany and Italy. Mix design of currently used asphalts for wearing courses in these countries were undertaken considering 0%, 30%, 60% and the closest feasible amount to 100% of RA.
• Work package 3 "Plant scale manufacturing and end-user manual" aimed at validating the laboratory design and led to recognize and solve issues related to the plant manufacturing.
• Work Package 4 "Performance prediction" aimed at assessing the durability of the high-content RA Warm asphalt mixes in terms of mechanical properties and evaluating design life of the mixes in typical traffic scenarios.
• Work Package 5 "Sustainability Assessment" involved carrying out a life cycle assessment, life cycle cost analysis in the context of a broader sustainability assessment of the investigated technologies. Additionally a state of the art review on sustainability rating systems was provided together with the development of a bespoken sustainability assessment methodology.

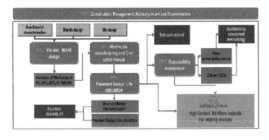

Figure 1.   AllBack2 Pave project structure.

2 LABORATORY-BASED DESIGN

Within this phase the partners aimed at designing two commonly used asphalt mixtures for surface courses, SMA and AC, incorporating the highest amount possible of RA (Fig. 2). These mixtures are characterized by a strong coarse aggregate skeleton that gives good resistance to permanent deformation, in case of the SMA obtained also thanks to a relatively high bitumen content (>6% and often a PmB), and providing enhanced resistance to fatigue. Furthermore, the advantages in terms of longer life and improved performance make these mixes economically viable, even if the SMA initial costs is around 20% higher than conventional asphalt concrete mixes. Therefore, being successful with maintaining these high standards, by replacing a high percentage of virgin materials with reclaimed asphalt, will then result in a more economic and potentially environment-friendly technology.

Within this laboratory exercise several key points were discovered in both case studies. Good results as well as discovering limitations have led the partners to highlight some key areas that deserves attention before undertaking the design of high-content reclaimed asphalts for wearing courses:

2.1 Sampling of RA

There is a common concern that the high quality requirements for asphalt wearing courses are not met if the major volumetric component of the mix comes from a recycled material. The background of this fear is the inherit heterogeneity of the RA, which properties depend in great extent on factors such as the technique used to reclaim the asphalt, the

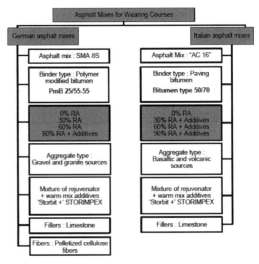

Figure 2.   SMA11S and AC16 case studies.

maintenance history of the road, the storage conditions, etc., which are not considered in standard mix design procedures. In fact, high quality aggregates with high resistance to wear/abrasion (polishing) are needed for wearing courses. Thus, properly milled (i.e. layer by layer) and stockpiled RA is a mandatory prerequisite in order to produce durable asphalt wearing courses with high content of RA. This is a major concern because in the majority of the cases the RA is produced and stockpiled by non-selective methods in which valuable high quality aggregates of wearing courses are milled/stockpiled together with lower quality materials of deeper layers. That means that the aggregates of the RA will automatically be downgraded losing economic value. One possible solution is to allow the use of high percentage of RA in wearing courses only if the RA material proceeds from the same location and layer where the new mix will be placed.

2.2 Characterization of the RA towards designing asphalt with the highest feasible content

The RA needs to be characterized before the actual mix design. This is because with ageing and oxidation some changes may occur in the mix: for the binder, this includes hardening and loss of ductility and for the aggregates the gradation may change due to degradation caused by traffic loads and environmental conditions. It is therefore important to characterize the RA with the same rigor used for the virgin aggregates except for the fact that family of materials is a complex agglomeration of stones that retain visco-elastic properties. Within the project the RAs were subjected to binder recovery procedures to obtain the so-called white RA. Then white RA was characterized according to the standard used for aggregates for road pavement.

Gradation: In particular, the gradation of the RA aggregates is one of the most important factors associated with the control of asphalt mixes with high RA content. At very high recycling rate, the gradation of the RA affects the overall performance of the final mix including stiffness, fatigue resistance, rutting and moisture damage. Aiming towards 100% recycling, the gradation of the selected RA must be as close as possible to the bands, but also the distribution of particle sizes in the aggregates of the RA must have just the right density so that the final mix will contain the optimum amount of asphalt binder and air voids.

In case the grain size distribution of the RA does not meet the control points of the target mix, as for the SMA11S in Fig. 3, it is necessary to add virgin aggregates in order to balance the grading. Using an iterative algorithm that decreases the amount of RA and increases the amount of virgin aggregates can do this. In our investigation, the results

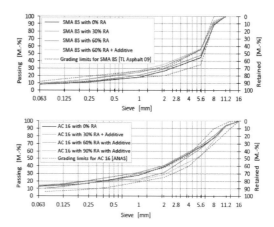

Figure 3. Grading of RA aggregates (white-curve) for the SMA and AC with limits respectively from TL09 and ANAS case study.

indicated that the maximum feasible percentage of RA was 70% for the SMA11S case, while for the AC16 case was over 90%.

Skid resistance: The aggregates ability to resist skidding is crucial when designing wearing course mixes. The Polished Stone Value (PSV) is a common evaluation method used for measuring skid resistance. Two skid tests were performed to determine the PSV of the RA aggregates by using the British pendulum. As example, in the SMA case study, the tests showed that the RA aggregates have an average PSV value of 51, which means that the polishing characteristics of the RA aggregates are good enough to be used for wearing courses of federal high volume roads in Germany (min. required PSV value of 51, acc. to TL Asphalt-StB 07).

2.3 Preliminary binder blend design

The binder selection for a specific mix design is usually done based on traffic and loading requirements expected in the wearing courses of the selected road. For instance, According to the German bitumen specification, for a federal high volume road with a surface layer made of a SMA material, it is required to use modified bitumen of the type 25/55–55. The contract specifications of this type of bitumen are presented in Figure 4.

Then, in order to obtain a high-content RA mixture comparable to a more standard mix, it is necessary that the binder blend of bitumen contained on the RA and the other additives/rejuvenators complies with the above-mentioned standard. In order to do that it is therefore necessary to perform an accurate binder blend design, which is specific to the selected RA, virgin bitumen and other components.

Attribute	Units	Specification	Requirements (PmB 25/55-55)
Pen @25°C	1/10mm	EN 1426	25 to 55
R&B Temp	°C	EN 1427	≥55
Tensile ductility	J/cm²	EN 13589	≥2@10°C
Flammable point	°C	EN ISO 2592	≥235
Fraass temp	°C	EN 12593	≤ -10

Figure 4. Contractual requirements for a PmB 25/55–55 (TL Bitumen-StB 07, 2007).

Within the project two very different RAs were selected: the RA used for the SMA in Germany was short-term aged (pen22) while the one used in Italy for the AC16 case study, was extremely aged (pen8). To determine suitable virgin bitumen, in both cases a comprehensive blend study was performed at the laboratory of UNOTT. This study allowed defining a procedure to assess whether the selected virgin bitumen and additives/rejuvenators would allow obtaining a final binder with the desired target properties even considering the variability of degree of blending (DoB) and binder content of the RA. The targeted properties were based on both conventional and rheological properties. A detailed explanation of this procedure has been published elsewhere (Lo Presti et al. 2016, Del Barco Carrion et al. 2015). As a result, for the SMA11S case study the properties of the virgin bitumen from the RA did not differ much from the properties of the target binder, therefore, it was decided to use as virgin binder a PmB 25/55–55 (i.e. the same type and grade as the target). Instead for the Italian case study, the RA was so much aged that each of the mixes (30%, 60% and 90%) needed the addition of a rejuvenator that was also supposed to act as warm-mix additive.

3 PLANT SCALE MANUFACTURING AND END-USER MANUAL

3.1 Selection of the asphalt mixing plant

Generally three types of batch plants are in use for which further modular machinery may allow higher recycling rates:

- 1) *The batch plant* (most widely used, charge addition—maximum RA content ~30%) where RA is added cold directly into the mixer and heated up indirectly through contact with superheated stone fractions into the mixer This processes may provoke an intense aging of the RA binder.
- 2) *The batch plant with RA heating drum* (maximum RA content of ~ 60% or counter-flow RA heating drum with maximum RA content of up to 100%): allows gentle heating of RA into the drum to avoid excessive binder aging.

- 3) *The batch plants with additional parallel drum* (maximum RA content up to 100% for counter-flow heating drum, ~60% for parallel RA heating drum): where RA and the virgin constituents are added in separate material flows to the mixer; this process allows the gentle heating of the RA. Then *continuous drum mixing plants* (maximum RA content ~25%) exist and this is where the RA is heated in the same drum mixer as the aggregates but this may provoke an intensive aging of the RA binder.

According to this classification, two plants were selected to produce the mixtures and with two different aims: 1) understanding production issue in a batch plant with a parallel drum capable of manufacturing 100% RA mixtures (SMA11S); 2) Evaluating strategies to maximize the amount of RA in a simpler batch plant with RA heating drum (AC16).

3.2 SMA and AC mixes manufacturing

SMA11S mixtures were produced at the mixing plant of Richard Schulz Tiefbau GmbH & Co that is located in Gilching, near Munich, Germany (Fig. 5). The mixing temperature were established at the design stage and were measured straight after the production from the silo (Table 1).

The AC16 mixtures were instead produced in Italy with a simpler batch plant with coaxial cylinder RA heating plant of Ferrara & Accardi near

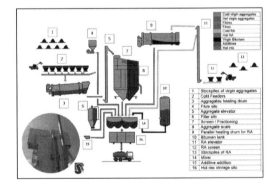

Figure 5. Scheme of the batch plant with a parallel drum used for the SAM11S case study.

Table 1. Mixing Temperatures in German Plant (SMA11S).

Mix	Mixing temperature [°C]
0% RA	170
30% RA	170
60& RA	170
90% RA + Add.	165

Catania (Figure 6). Due to the ageing of the RA, and also to try maximizing the RA amount, all the mixes were produced with the addition of the additive which was fed directly into the mixer. Despite several trials to try producing the 90% RA moisture, this plant allowed producing only the asphalt with up to 60% of RA. In fact, in order not to have an efficient filtering of harmful fumes, the maximum percentage of RA permitted in the production of high-content RA mixes is directly proportional to the amount of virgin aggregate content inside the drum. For this reason, the mixture with 60% of RA was produced by feeding the recycling ring of the drum with only 50% of the total weight of required RA, while the remaining cold part was directly sent to the mixer, once the virgin aggregate and hot RA were already inside. With regards to the mixture with 90% RA, because of the limits of the mixing plant system, most of the RA needed to be fed directly cold into the mixer which was not able to achieve a mixing temperature above 110°C. The too small amounts of virgin aggregate inside the drum did not allow a sufficient heat exchange to ensure the attainment of temperatures necessary for the blending; in addition, the shortage of virgin material within the drum did increase the temperature of the fumes, leading it abruptly to the threshold value permitted by the system. Thus all mixes were produced in plant except the one with 90% RA. The latter was then manufactured in the laboratory for the experimental survey. The mixing temperatures are reported in Table 2 provided for each of the blends produced.

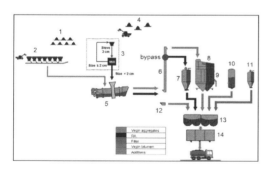

Figure 6.   Scheme of the batch plant with coaxial RA heating drum used for the AC16 case study.

Table 2.   Mixing Temperatures in the Italian Plant (AC16).

Mix	Mixing temperature [°C]
0% RA	160
30% RA + Add.	163
60& RA + Add.	165
90% RA + Add. (170 produced in lab)	–

3.3   End-user manual methodology

The scope is to define the guidelines of drafting of an End-Users Manual addressed to all stakeholders such as road agencies, production plants and construction enterprises. Two main technical goals can be summarised: 1) the consensus between mixtures produced in laboratory and in plant; 2) the achievement of the pavement performances targeted with the mix design.

Therefore the main concept behind the idea of the End-Users Manual lies on trying to maximize both the RA use within new road pavement without compromising the performance of the superstructure. Since the overall RA recycling process, from the dismantling of old pavements to the reuse within new pavements, is complex and several users have specific roles and objectives, it seemed most practical to use tailored check-lists for each user that summarise all actions to be implemented to achieve the purposes. The structure of the checklist provides for each operation, the Basic Rule, required in order to perform well each step of the process. The type of rule can be identified by a regulatory requirement (Regulation, R), a prescription contract (Specification, S), and a good practice (Best Practices, BP).

In order to evaluate the quality of the generic process path, a basic rule can be weighted depending on the category (Regulation, R; Specification, S; Best Practices, BP) to which it belongs. The weights may be set conventionally 3, 2, 1, by setting a descending hierarchy that goes from R to BP. To assess the reliability of the generic process, the REliability Measure parameter (REM) was defined as follows:

$$REM = \sum_{i=1}^{n} WV_{[BRi=YES]}$$

where:

- n is the total number of the basic rules of the given path;
- WV [BR$_i$ = yes] the value, varying from 1 to 3, of the ith basic rule put in place.

The reliability of the generic process is calculated by dividing REM to the maximum achievable REM, i.e. when the checklist is populated with all its positive answers. It then defines the coefficient of reliability "Reliability Ratio" (RR) varying from 0 to 1, as:

$$RR = \frac{\sum_{i=1}^{n} WV_{[BRi=YES]}}{\sum_{i=1}^{n} WV_{[AllBRi=YES]}}$$

where: WV [AllBR$_i$ = yes] is the maximum value of the REM.

Table 3. Check List applied to the Italian case (AC16).

Operation	Basic rule (Code)		Description	Done (Y/N)
Single Source Separate Stockpiles	Storage Area (4S-SA1)	BP	Minimum area should be not less than 1500 m². Area should be sloped (six degree is ideal)	Y
Single Source Separate Stockpiles	Storage Area (4S-SA2)	BP	Treatment of the surface area (no water, no clay)	Y
Single Source Separate Stockpiles	Storage Area (4S-SA3)	BP	Permeable to air roof with permeable to air membrane	N
Single Source Separate Stockpiles	Stockpiling (4S-S1)	BP	Stockpile must have conical shape with maximum height of 6 m	Y
Single Source Separate Stockpiles	Stockpiling (4S-S2)	R	Searching for deleterious materials (EN 12697-42) Sampling according to EN 932-1	N
Single Source Separate Stockpiles	Stockpiling (4S-S3)	R	Determining aggregate grading (EN-13043) Sampling according to EN 932-1	N
Single Source Separate Stockpiles	Stockpiling (4S-S4)	R	Determining binder content (EN 12697-1) Sampling according to EN 932-1	Y
Crushing & Fractioning	Preliminary Screening (CF-PS1)	BP	Preliminary screening of the finer particles by a suitable sieve (3/16 ASTM series or equivalent)	N
Crushing & Fractioning	Before & After analysis of RAP gradation CF-BAG1)	BP	Gradation control before & after the in ine crusher to determine the RAP aggregate size (within 1B&A analysis 2 samples for RAP source)	N
Crushing & Fractioning	Number of Screening (Sieves) Unit (CF-NSU1)	BP	The Plant must have 3 screening unit at least, typically 3/4, 3/8, 3/16 (ASTM series or equiv.)	N
Batch Plant	Drying RAP (BP-D2)	BP	The plant must have a rotary drum dryer with recycling ring (T = 110 ÷ 130°C)	Y
Warm Recycling	Mixing Time Control (WR-MTC1)	BP	Verifying the mixing time is within the range 25 ÷ 90 s	YES
Warm Recycling	Emission Control (WR-EC1)	R	Verifying the pollutant emission according to UE Directive 75/2010	YES

In order to evaluate the technical processes of both Italian plant and German one, the methodology described above, has been applied. According to the responses of the checklist, the "Reliability Ratio is equal to 0.52 and 0.85 respectively. Here the checklist for Italian plant is reported as example in Table 3

More details of the results of this section are reported elsewhere (Allback2Pave D3.1, 2016).

4 MECHANICAL CHARACTERISATION AND PREDICTED PERFORMANCE

On the basis of both the scientific literature and the availability of test devices in the laboratories involved, the mechanical properties of the eight produced wearing course asphalt mixes were investigated according to the following plan: (i) the stiffness modulus by both Four Point Bending Beam (4PBB) test and Indirect Tensile stiffness Test (ITT); (ii) the resistance to fatigue also by 4PBB test and ITT; (iii) the resistance to permanent deformation through Wheel Tracking Test (WTT) and Uniaxial Compression Test (UCT) and (iv) the resistance to moisture damage by means of Indirect Tensile strength Test.

In addition, the characterisation of the binders extracted from the final mixes has been carried out in order to understand whether performance-related tests on binders compare with mixture's results also to validating the laboratory mix design undertaken in WP2. This was done by assessing (i) rutting resistance by Multiple Stress Creep Recovery (MSCR); (ii) fatigue resistance by Time Sweep tests and analysis of (iii) the thermal cracking resistance plus the determination of critical temperatures. A part of the stiffness and fatigue related asphalt test results have been used as input data for design life calculations.

4.1 *Performance-related properties of binders*

Given the results of all the tests, different conclusions can be drawn: for each case study, with regards to the use of the additive and as a result of comparison with the design stage. In the SMA11S case, RA binder and virgin bitumen were known to have not very different properties. In this sense, results for high RA content binders, even without using rejuvenators, show that binder recovered from high-content RA mixes have similar properties to recovered virgin binder. Specifically, as the percentage of RA increases in the binders (30–60%), rutting resistance improved, while fatigue and low-temperature behaviour did not change in comparison to the scenario without RA. However, when the rejuvenator was added, rutting resistance slightly decreases compared to the virgin one, but fatigue and thermal cracking resistance improved, obtaining binders with good predicted-performance (Figs. 7, 8).

In the AC16 case, RA binder and virgin bitumen were known to be extremely different; therefore, some issues could be expected when high RA percentages were to be used. However, thanks to the use of rejuvenator, and in accordance to the design carried out in previous deliverables, rutting resistance and low-temperature behaviour improved with higher RA content. Regarding fatigue results, binders with "30% and 60% of RA + Additives" showed very similar behaviour to that of the "0% RA" content, and only in the case of "90% RA + Additive" fatigue life would be shorter, but still within acceptable limits. With these results we can confirm that the additive generally allows obtaining a good fatigue behaviour of the binders up to 60% RA. Some details are reported in Figure 7. The whole investigation is detailed elsewhere such the deliverable 4.1 of the project (Blasl et al. 2016, Allback2Pave D4.1, 2016).

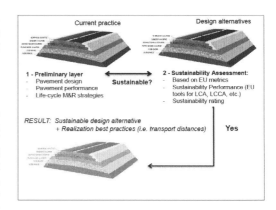

Figure 8. Suggestion for a performance-based CEDR sustainability assessment methodology.

4.2 *Performance-related properties of mixes*

- **Stiffness:** The results of the stiffness related indirect tensile tests and four point bending beam tests confirm a comparable dependence of the material behaviour of the amount of RA within the investigated eight asphalt mixes. Regarding the SMA mixes it can be concluded that, as expected, the stiffness increases with an increasing amount of reclaimed asphalt. However, the asphalt mixture with 60% RA + Additive shows a lower stiffness due to its higher binder content and maybe also due to the presence of additives. The fourth SMA mix has been designed as an experimental asphalt mixture, where originally the allowance of additives was not required by the results of the laboratory tests on recovered binders in line with the mix design procedure. For comparison, the AC mixes show almost no differences between the temperatures related stiffness. It seems that the additive has completely compensated the influence of the RA what confirm an optimal blend design procedure.
- **Rutting:** the results of the wheel tracking tests and the modified uniaxial compression tests do not lead to a consistent ranking of the asphalt mixtures. Regarding the SMA materials the plastic deformation results of the two different laboratory tests can be qualitatively compared, whereas the AC mixtures results show completely different rankings. Regarding the SMA 8 S mixes it can be concluded that the amount of reclaimed asphalt has no significant influence on the plastic deformation. The mixture with 60% RA and additives has a higher binder content in comparison to the other mixes, which leads to higher plastic strains. Therefore, a relation between additive and the development of higher plastic strains at the mixes with 60% RA cannot be concluded. In terms of the AC16 materials the accumulated plastic

Mixture	Design critical temperatures [°C]			Recovered binder critical temperatures [°C]			Relative error [%]		
	High	Int	Low	High	Int	Low	High	Int	Low
SMA 8S	80.1	22.5	-13.5	81.4	19	-16.8	1.6	-15.6	24.4
SMA 8S with 30% RA	79.5	19.8	-15.5	83.6	19.9	-16.6	5.1	0.4	7.1
SMA 8S with 60% RA	79.7	20.5	-15.0	85.6	21.8	-16.4	7.4	6.4	9.4
SMA 8 S with 60% RA + Additive	79.1	18.7	-15.5	76.5	17.1	-19.8	-3.3	-8.4	28.0

Figure 7. Performance related tests on recovered binder SMA11S case study: critical temperatures (top).

strains considerable decrease with an increasing amount of RA. The influence of temperature and stress level on the permanent deformation behaviour is significant. Therefore, the evaluation of the plastic deformations should take into account a wide range of the testing conditions. More details are reported in the deliverable

- *Fatigue:* The results of the fatigue related tests provide the conformation that the stiffer an asphalt material is the higher is the number of load cycles until failure. The outcomes of the fatigue related indirect tensile tests and four point bending beam tests partially show a different dependence of the material behaviour of the amount of RA within the investigated eight asphalt mixes. Clear conclusions and comparison of the results obtained by both test results very difficult to undertake.
- *Moisture damage:* Finally, with regards to the moisture damage resistance, all asphalt mixes show no substantial sensitivity to damage moisture.

Details will be reported elsewhere (Allback2-Pave D4.1, 2016).

5 SUSTAINABILITY ASSESSMENT

AllBack2Pave supports the philosophy that sustainability principles should be already considered at the design stage of a road pavement, as well as any other structures, so that a sustainable transport infrastructure could be defined as: *a construction that maximizes the recycling of waste/secondary materials within its structure and minimizes the environmental impacts, through the reduction of energy consumption,* natural *resources and associated emissions while meeting performance conditions, specification requirements and social needs throughout its entire life cycle.* With this concept as driver, within the WP5 led by the University of Nottingham, the consortium closely collaborated with industrial partners and sustainability professionals to collect data and sustainable practices for providing road authorities with tools to be used for further designs and to performing a broad sustainability assessment of the investigated technologies. This work package produced the following reports: D5.1: A state of the art review of existing sustainability assessment tools of the impact of road pavement infrastructures; D5.2: Evaluation of the environmental impact (through LCA technique) and economic impact (through LCCA techniques) of the defined technologies taking into account the European level of the project and adapted to real case studies; D5.3: Sustainability assessment of the AllBack2Pave technologies adapted to real case studies at European level, through a methodology

developed by this project and proposed in details for ease of use by CEDR members.

Each of these sections provided specific inputs such as an overview of sustainability assessment system, a screening on freely available tools to perform LCA and LCCA, practical example of environmental and economical impact assessment for road pavements and finally also a bespoke sustainability rating system for asphalt pavement which adapts to Europe the existing BE2ST tool developed by The University of Winsconsin and make uses of the ECORCE software developed from IFSTTAR (Dauvergne et al. 2014). A summary of the forecasted advised methodology for a sustainable decision related to new design, maintenance and rehabilitation of EU road pavements is indicated in the figure below and will be detailed elsewhere (Allback2pave D5.1, D5.2, D5.3, 2016).

6 CONCLUSIONS

AllBack2Pave is a 2 years project that involved a big efforts from three partners to provide CEDR with strategies, methodologies and tools to design, manufacture high-content RA asphalt mixes and assess the sustainability of their use within Europe. Here is a summary of the main findings:

RA is a family of materials, therefore a specific preliminary characterization of the RA is necessary. Furthermore, any mix design containing RA should first have a preliminary binder blend design (such as in WP2) that take into account concepts such as degree of blending of the RA, as well as allowing the use of additives. In WP3, the project provided also an End User manual to produce the investigated technologies. This provide best practices at the asphalt plants to maximize RA use within new road pavement without compromising its performance. Within WP4 plant-produced asphalt mixes have been investigated, as well as recovered binders, and indication on the validity of the undertaken mix design are given. At last, the WP5 provided CEDR an answer to the question: Is maximizing recycling of asphalt into pavement a sustainable practice? It was proved that maximizing reclaimed asphalt can be a sustainable option, however in some of the case studies other variables were more important than recycling. These were maximizing durability of the asphalt mixes and decreasing transport distances during construction and maintenance (100 Km is a good reference).

The authors are aware that most of the results not included here. This is because approval from CEDR is still pending. Results will be soon available in the project website and at the conference.

ACKNOWLEDGEMENTS

The research presented in this paper was carried out as part of the CEDR Transnational Road research Programme Call 2012. The funding for the research is provided by the national road administrations of Denmark, Finland, Germany, Ireland, Netherlands and Norway.

REFERENCES

Allback2Pave D2.1; D3.1; http://allback2pave.fehrl.org (accessed Sep 16). Report D4.x and 5.x will be available soon.

Blasl, A. Kraft, J. Lo Presti, D. Di Mino, G. Wellner, F. 2016. Performance of asphalt mixes with high recycling rates for wearing layers, EE congress 2016

Dauvergne, M., Jullien, A., Proust, C., Tamagny, P., Ventura, A., Coelho, C., et al. *ECORCE M User's Manual*. Nantes, FR: French Institute of Science and Technology for Transport Development and Networks (IFSTTAR, 2014).

Direct-MAT. 2011. Dismantling and recycling techniques for road materials. Available at: http://www.direct-mat.eu

Jiménez del Barco Carrión, A. Lo Presti, D. Airey, G., 2015 "Binder design of high RAP content hot and warm asphalt mixture wearing courses" - Road Materials and Pavement Design, Taylor&Francis,, Special Issue: EATA 2015

Lo Presti, D. Jiménez del Barco Carrión, A. Airey, G. Hajj, E. 2016. *Towards 100% recycling of reclaimed asphalt in road surface courses: binder design methodology and case studies*: Journal of Cleaner Production. 131, 43–51

Re-Road. 2012. End of life strategies of asphalt pavements. Available at: http://re-road.fehrl.org

RDO Asphalt 09. 2009. Richtlinien für die rechnerische Di

TL AG-StB 09. 2009. Technische Lieferbe-dingungen für Asphaltgranulat. Köln. FGSV Verlag GmbH.

TL Asphalt-StB 07. 2007 Technische Lieferbedingungen für Asphaltmischgut für den Bau von Verkehrsflächenbefestigungen. Köln. FGSV Verlag GmbH.

Transport Infrastructure and Systems – Dell'Acqua & Wegman (Eds)
© 2017 Taylor & Francis Group, London, ISBN 978-1-138-03009-1

Feasibility and preliminary design of a new railway line in the Dolomites area of Veneto Region

M. Pasetto, G. Giacomello, E. Pasquini & A. Baliello
Department of Civil, Environmental and Architectural Engineering, University of Padua, Padua, Italy

ABSTRACT: The current conformation of the railway line situated in the Dolomites area of Veneto Region does not represent an efficient network to fully connect the territory and serve the global mobility demand. In fact, the existing railway line lies along the Piave valley and connects cities on the Veneto plain with Belluno and the Cadore area, whereas public road transport links the train stations with the other towns in Belluno province. Given this background, a new railway line (from Calalzo to Cortina d'Ampezzo passing through Auronzo) could meet the province mobility needs, leading to a potential increase in traffic and thus improving economic and tourism development in the area. Indeed, historical as well as natural and tourist (ski) areas would benefit. Moreover, a new railway line is an attractive and valid alternative to the development of the existing road network, allowing a proper integration between rail and road transportation systems. In this sense, a preliminary design has shown noise levels at least halved with respect to a road context eliminating any level crossing, thus also proving to be a convincing solution under the environmental point of view. However, a more specific and detailed assessment of the environmental impact of the new railway line will be necessary taking into account that the region involved is one of the most scenic Dolomite areas (UNESCO Heritage Site), with complex morphology and geology. A 45 km long single-track railway layout has been designed hypothesizing two passenger trains per hour and assuming a minimum speed of 80 km/h, a minimum bend radius of 300 m, a maximum longitudinal slope of 20‰ and gauge of 1435 mm. A total cost of 500 million euros has been estimated for the realization of the new infrastructure.

1 INTRODUCTION

The territory of the province of Belluno has always been penalized by its distance from the industrialized Veneto plain because of the area's orographic and geomorphological characteristics. In particular, the dolomitic territory of Cadore and the Ampezzo Valley, in the central-northern part of the province, enjoys a limited north-south accessibility. In the other directions, it has a system of links and a viability typical of mountain areas (mainly via valleys).

The increase in traffic that has been registered all over Italy in recent years has also affected Cadore and the Ampezzo Valley, where there has been a rise in tourist numbers, an increase in the mobility of the residents for regular (home, work, school) and sporadic travel (free time) and the movement of goods. The road network carries steady flows of traffic along the valleys, which gradually decrease leaving the main valleys for the peripheral areas.

1.1 *The territorial setting: the province of Belluno, Cadore and the Ampezzo Valley*

The province of Belluno is in northern Italy, in the north of the Veneto Region, between the Trentino

Alto Adige and Friuli Venezia Giulia Regions and is a prevalently mountain territory. Although it is the largest province in Veneto (3678 km^2), it is the one with the fewest inhabitants (206,856) (ISTAT 2013). Compared with other not easily reached mountain areas, however, the area is fairly well populated and characterized by large municipalities uniting sparse hamlets.

The two most populated dolomitic areas of the province and the most important for tourism are Cadore and the Ampezzo Valley. The former is a well-defined morphological and historical unit corresponding to the upper basin of the river Piave upstream of the village of Longarone. The latter corresponds to the valley of the Boite torrent and historically has always been linked to Cadore.

The area of Cadore and the Ampezzo Valley lies in the Dolomites, characterized by two associated types of rocks: dolomitic and volcanic, which derive from totally different processes. The dolomitic rock is more resistant to atmospheric agents than the rocks of volcanic origin, which erode easily: the results are high dolomitic peaks overlooking deep valleys. Because of this morphology, in many parts of this area and along the valleys (especially that of the Boite) there are phenomena of instability of the mountainsides and surface subsidence. The distinctiveness of these mountain areas has led to their classification as "Sites of Community Importance" and "Special Protection Areas".

1.2 The economic-social context of the Dolomites area

The principal municipalities in the area are: Calalzo di Cadore (approx. 2,250 inhabitants, 43 km² in area), Domegge di Cadore (2,441 inhabitants, 50 km² in area), Lozzo di Cadore (1,383 inhabitants, approx. 30 km² in area), Auronzo di Cadore (approx. 3,350 inhabitants, 220 km² in area), San Vito di Cadore (1,857 inhabitants, approx. 61 km² in area) and Cortina d'Ampezzo (5,907 inhabitants, approx. 252 km² in area) (ISTAT 2013).

The main economic activities in the area are livestock rearing, forestry and wood processing, but the most important sources of wealth are the many companies that produce glasses and the tourist sector. Indeed, tourism (both summer and winter) is of enormous economic importance in these areas. According to the 2012 data (ISTAT 2013), the province of Belluno accounts for approximately 86% of tourist presences in the mountain zone of Veneto Region. In terms of tourist numbers, Auronzo di Cadore accounts for about 70% of the entire Cadore and about 7% of that of the province of Belluno. Cortina d'Ampezzo instead attracts around 25% of the annual tourist presences in the province of Belluno. Cadore alone accounts for 42.3% of the entire mountain tourist sector of the Veneto Region.

The large volume of tourists who visit the Dolomites travel in their own cars. It emerged from an ISTAT study that the majority of travel in the area is by private transport, with negative effects on the environment due to air and noise pollution and on traffic in general.

There is a growing need to upgrade the mobility of the territory, with a modern public transport system to complement and integrate use of the private car. However, this is a situation for which, at the moment, there is no easy solution in the short-medium term. This fact will probably be strongly dependent on a series of dynamics, already underway, but that will increasingly affect the situation in the territory of the province of Belluno. The demographic decline, ageing of the resident population, depopulation of the more marginal areas and concentration of the population in the most important centres, the rise in tourist flows and the increasingly difficult economic management of the public services.

In order to avoid the increase of these problems, the transport infrastructures must assume characteristics and dimensions that allow them to fully integrate into the territory with the possibility of offering places and landscapes from a different perspective, in which the natural and historical background that has led to the formation of this environment is clear.

1.3 The mountain railway in the context

Although major international railways lines exist in the north-east of Italy (like the Brennero line and the Pontebbana railway) that connect the plain areas and the Adriatic with the Alpine area and the regions of central Europe, the Dolomite areas of Cadore and the Ampezzo Valley are not directly crossed by these lines.

The province of Belluno is currently served by two lines that connect Belluno with Padova and Venice. The stretch of railway Belluno—Calalzo, built for military purposes in 1914, has the characteristics of a mountain line: single-track, not electrified, 25‰ maximum slope, bend radii of 200 m (Agostini 2014).

In the past (1921–1964) a railway line existed (known as the "Dolomites railway—*Ferrovia delle Dolomiti*"), which connected Calalzo di Cadore with Cortina d'Ampezzo and terminated at Dobbiaco. Nowadays, the stretch Dobbiaco—Cortina d'Ampezzo is a footpath, while the stretch Cortina—Calalzo has been almost entirely converted into a cycle path. The Dolomites railway involved a total rise of 810 m and was 65 km long, had a maximum slope of 35‰, a minimum bend radius of 60 m, a narrow gauge of 950 mm and electrification at 3000 V (Gaspari 2005).

In this context, the general rediscovery of rail transport in the last decades has also stimulated, in the municipalities of the Dolomites, the idea that the train could be the ideal solution to solve the problems of mobility and contemporarily enrich the tourist offering.

This rediscovery is supported by examples of transport systems integrated in the landscape and still functioning, such as: the Stubaital railway (1904) that unites Innsbruck with the district of the valleys of the Stubai Alps in Austria; the Bernina Express (1880) that covers the stretch between Tirano, Saint Moritz and Chur (Switzerland) in an itinerary of outstanding landscape with slopes of up to 70‰ (Wall 1987); the Trento–Malè–Marilleva railway (1909) in Trentino Alto Adige (Forni 1975, Serra 1989); the Zillertal railway (1902) that crosses the valley of the Ziller uniting Jenbach with Mayrhofen in the Austrian Tyrol; the Merano-Malles railway (1906) along the Val Venosta (Marseiler 2006), a railway closed through lack of use, then reopened and now growing strongly thanks to efficient management.

These examples demonstrate that a railway has the potential to provide an efficient and sustainable response to the mobility needs in the area of the present study.

2 FEASIBILITY STUDY OF THE PLANNED ROUTE

2.1 Project hypothesis

The hypothesized railway route runs for about 45 km in an area with complicated morphology that implies very high construction costs (for tunnels and bridges). Otherwise, costs may be compensated in a relatively short period of time if an increased tourist flow and the scenic beauty of the places traversed by the infrastructure are considered.

The tourism interest in this project is even clearer if a possible extension of the line to Misurina is considered, where the parking places for visitors to the Tre Cime di Lavaredo have become insufficient, or further north towards Dobbiaco, in Trentino Alto Adige.

It is hypothesized that the railway will be single-track, with trains travelling in opposite directions passing one another in the stations.

Given the orographic characteristics of the areas to be crossed, a railway line was designed, with numerous tunnels and bridges. It begins from the existing station at Calalzo di Cadore at an altitude of approximately 740 m. The route runs along the valley of the river Piave and then of the river Ansiei passing through Auronzo di Cadore. In the vicinity of the Somadida forest the line enters a tunnel, exiting at Chiapuzza (hamlet of San Vito di Cadore) and, traversing the valley of the Boite torrent, ends at Cortina d'Ampezzo (Fig. 1).

The fundamental elements that characterize a railway route are essentially the minimum radius of the bends and maximum slope of the gradients. These elements are set according to the maximum speed and characteristics of the trains that will use the line.

Were therefore defined: the stops and stations (Table 1), minimum bend radius (300 m) and maximum slope (not above 18‰).

Table 1. Stations and stops along the designed railway line.

Station name	Station or stop	Partial distance (m)
Calalzo di Cadore	station	0
Domegge di Cadore	stop	3,400
Lozzo di Cadore	station	3,000
Cima Gogna	stop	4,700
Auronzo Centro	station	4,500
Auronzo Impianti	stop	2,000
Somprade	stop	3,400
San Marco	station	6,000
Briglie-Somadida	stop	3,400
San Vito di Cadore	station	6,600
Acquabona	stop	4,600
Cortina d'Ampezzo	station	3,700

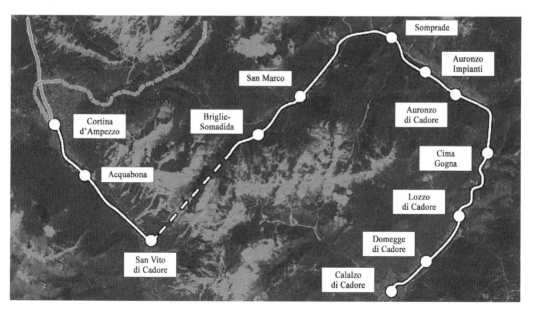

Figure 1. Designed railway layout: project (white), prolongation of the railway line towards Dobbiaco (grey) and prolongation of the railway line towards Misurina (dotted grey).

The maximum superelevation allowed on a bend is 160 mm for speeds less than 160 km/h. For bend radii, greater than those corresponding to the maximum superelevation, the non-compensated acceleration reduces in proportion to the superelevation.

The planned railway line, according to the schedule of the Italian Railway Network (*Rete Ferroviaria Italiana* - RFI), is of "type B" (secondary lines). The route has 43 straight stretches and 41 round bends linked by transition bends on entering and leaving.

The railway line leaves from the existing station at Calalzo with a single not electrified track. Its route is planned first along the right bank of the river Piave and then of the river Ansiei. The railway axis has been designed with the intention of maintaining an adequate distance between infrastructure and buildings and between infrastructure and the right bank of the Piave.

The station at Calalzo could also be used as terminus by the trains, because it already has at least three tracks to allow trains travelling in different directions to pass one another. From Calalzo, the line crosses the river Molinà over a bridge and continues in the open along the right bank of the Piave remaining at some distance from the hamlet of Vallesella (for which no stop is planned). The stop in the municipality of Domegge di Cadore is planned at some distance from its centre, so access will have to be arranged. Between the municipalities of Domegge and Lozzo di Cadore the railway line runs parallel to the Piave through a tunnel. The station in the municipality of Lozzo is planned slightly before the built-up area and, having two tracks, will allow the passing of trains travelling in opposite directions. Immediately after the station at Lozzo there are two critical points: the intersection with the state road "Alemagna" (which will be raised to allow the line to pass underneath and maintain the design slopes) and the Ruoiba landslide, on the right-hand side of the river Piave. In that point the valley becomes very narrow and enclosed, forming a "V" (the railway route will be shifted, immediately after the municipality of Lozzo, onto the left bank over a bridge).

The route will then return to the right bank just before the confluence of the river Ansiei. The route lies parallel to the valley of the river Ansiei along the left bank. The line runs to the west of Cima Gogna (hamlet of the municipality of Auronzo di Cadore), with a stop planned at the end of the hamlet (easily reached and close to the intersection with the road to Val Comelico).

After this the railway line, over a succession of stretches in the open and through tunnels, reaches the municipality of Auronzo di Cadore. Two stations are planned at Auronzo: one in the town centre ("*Auronzo Centro*") and one next to the ski lifts ("*Auronzo Impianti*"). The two stations are on the right shore of the Santa Caterina Lake and, in the project, are designed to be in tunnels: the "*Auronzo Impianti*" station is at an altitude of 850 m a.s.l., the same as that of the ski lifts.

From the station "*Auronzo Impianti*", the line passes over a bridge on the Val da Rin and another on the left bank of the river Ansiei and then arrives at the station of Somprade (Fig. 2). After this, the route passes over a small viaduct, an ample stretch in the open following the valley of the river Ansiei and a long tunnel, after which it arrives at the station of San Marco. The route then returns to the right bank of the river Ansiei over a bridge, entering the Somadida forest, where a station at Briglie—Somadida is planned. This station, conceived according to a model at low environmental impact, will serve tourists visiting the reserve situated in a splendid area of the Dolomites. The route then enters the longest tunnel (about 5 km) and exits in the valley of the Boite at Chiapuzza (hamlet of the municipality of San Vito di Cadore), where a station is planned. From here the line arrives at the station of Acquabona, in the lower part of the valley. The route then crosses the industrial zone of Zuel in a tunnel and, remaining on the left bank of the Boite torrent, terminates at the station of Cortina d'Ampezzo, situated next to the cross-country skiing trails in the south-east of the town.

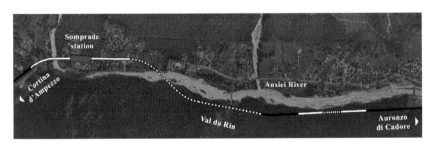

Figure 2.   Crossing of the Val da Rin and river Ansiei, arriving at the Somprade station (gallery in black and bridge in dotted white).

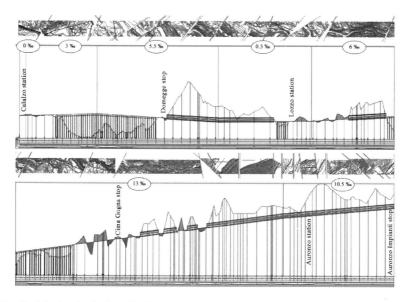

Figure 3. Detail of the longitudinal profile between Calalzo station and Auronzo Impianti stop.

Table 2. Steady gradients of designed railway line.

Steady gradients	Slope (‰)	Length (km)
1	0.0	0.1
2	3.0	1.7
3	5.5	2.9
4	0.5	2.2
5	6.0	2.0
6	13.0	6.2
7	10.5	2.6
8	18.0	14.2
9	0.9	5.0
10	12.0	1.5
11	18.0	4.1
12	12.0	2.9

From the altimetric point of view, the slopes assigned to the route are lower than the maximum prescribed value (18‰). Figure 3 shows a part of the longitudinal profile whereas Table 2 reports all the slopes utilized and the lengths of the gradients. More specifically, between Calalzo and Auronzo slopes are very gentle (the steepest is 13‰), while between Auronzo and Cortina the slope reaches 18‰ (in the tunnel between Somadida forest and Chiapuzza it remains below 1‰).

The construction of embankments, cuttings, bridges and tunnels is planned. The gradient assigned to the slopes depends on the angle of friction of the soils but is generally set at 2/3 on an embankment and 1/1 in a cutting.

To reduce the transversal dimensions of a section (in embankment or cutting), shallow slopes supporting the soil with scarp or counter scarp walls can be built (in masonry or simple or reinforced concrete). For the bridges, it can be planned that the crossing over a waterway, road or railway, will be at an angle of 90°, but the railway route often does not allow this so the bridge must be built diagonally. Bridges also exist on bends in the route. The choice of bridge structure to be built is directly linked to each individual situation. In the case of tunnels a profile limit is stipulated within which every vehicle including transported loads must remain.

2.2 The rolling stock and infrastructure

The rolling stock was chosen in order to meet service needs (times) and to define the design parameters of the infrastructure. The requisites identified are: travel times (maximum speed and high acceleration and braking performances), journey comfort (noise, vibrations, air-conditioning, passenger information and entertainment), accessibility (flush access, large doors for easy and rapid mounting and dismounting of the passengers, multifunctional spaces for the transport of bicycles or skis), environmental impact.

Among the many types of trains that meet the above requisites, the powered railway carriages produced by the Swiss Stadler Rail "GTW 2/6" and "GTW 4/12" were chosen. The first is a light articulated railway carriage useable for local transport, with two driving axles (located in the central section) on a total of six axles, and offers 111 seats. The second, with four driving axles (located in the central section) on a total of twelve axles, instead offers 245 seats. The GTW 4/12 trains are two GTW 2/6 permanently coupled via an intercommunicating passage accessible also to the passengers.

Once the type of train had been decided, it was possible to classify the railway line (the category determines the characteristics of the superstructure) on the basis of "maximum weight per axle" and "maximum load per metre allowed" (i.e. the ratio between the total weight of the loaded vehicle and its length including buffers). The planned railway line can be classified in category B2, with a maximum weight per axle of 18 tons and a weight per unit length of 6.4 tons per metre.

According to the RFI regulations (Mayer 2004), the line can also be classified depending on the intensity of traffic measured by the dummy load, expressed in gross tons hauled daily [gth/d]. RFI has divided the lines on the basis of the dummy load starting with group 1 (with a load above 102,000 gth/d) up to group 9 (with a load less than 1,000 gth/d). This line can be classified in group 9.

The characteristics of the infrastructure (Fig. 4) were hypothesized as being those of a secondary line according to RFI specifications (Mayer 2004): rails type UIC 60, with type R200 steel (type R260 on isolated bends with radius less than 600 m), gauge of 1,435 m, pre-compressed reinforced concrete sleepers (module 0.66 m, weight 2,500 g, length 2.3 m and trapezoidal section variable in height, with maximum dimensions at the heads and reduced in the centre), indirect joints between sleepers and rails, staggered joints between the rails, stone ballast (basalt, porphyry or granite rocks, with sharp edges and almost uniform size 25–30 mm in diameter, ballast depth 0.35 m), settlement plan (i.e. the support for the ballast that follows the planimetric trend of the route), useful life of the superstructure of 30 years.

The arrangement of the substructure for the construction of an embankment will consist of the prior removal of the vegetated layer of soil, to a depth of at least 20 cm. Analyses of the soils will be conducted to learn the carrying capacity and mechanical characteristics.

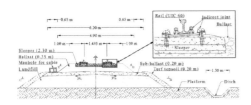

Figure 4.   Cross section of the designed line.

2.3   The timetable and transport service

The train timetable was drawn up on the basis of the localities to be served and the other data hypothesized for the project. The timetable defines the route, travel times, performances of the services (spatial accessibility, temporal accessibility and commercial speed), the attractiveness of the services, the capacity of the system to be competitive with other methods of transport, and the stops.

The presence of an optimal number of stops/stations:

– offers better cover of the territory with a greater number of departures and destinations;
– increases the attractiveness of rail transport also for travel from, to and between secondary destinations, guaranteeing higher profitability of the service (fuller trains), also in small places that lack attractors of "point-to-point" traffic;
– allows a larger swathe of the population to have access to the comfort of rail transport.

The presence of a railway stop signifies, especially in rural and mountain ambits, promotion of the territory, as it greatly modifies the temporal geography, i.e. the time necessary for journeys.

In this study, an estimate was made of the travel time necessary to reach the station at Cortina d'Ampezzo leaving from Calalzo di Cadore and also an estimate of the number of trains that can transit in an hour in both directions.

The inputs used in the calculation were: place of departure (Calalzo di Cadore), destination (Cortina d'Ampezzo) and vice versa, distance travelled between every stop/station, travel time of these distances, stopping times at the single stops/stations (4 minutes in the stations with more than one track, to allow trains coming from opposite directions to pass, and 2 minutes in the other stops), average speed.

The travel time between Calalzo di Cadore and Cortina d'Ampezzo results as being 62 minutes. Moreover, the time necessary for two trains to pass one another in opposite directions at the stations of Lozzo di Cadore and San Vito di Cadore is about 50 minutes whereas at Auronzo Centro and San Marco is about 35 minutes. Therefore, in 1 hour, two trains can transit in opposite directions, with a time separation of 28 minutes (Fig. 5).

It was hypothesized that the transport service could be divided on the basis of working and holiday periods: on weekdays, a service would connect all the stops (from Calalzo to Cortina and vice versa, focused on local people and tourists, with the above timetable), while at weekends and on holidays there would be a service to Venice (with a maximum of 3 trains per day that connect Cortina directly with Venice). This would create a tourist service for the passengers of cruise ships that arrive in Venice, offering the opportunity of a daytrip to Cadore and the Ampezzo Valley.

2.4 The estimate of the costs

The advantages of a railway line with respect to a road are the possibility of providing more transportation, the smaller surface area necessary (for the same amount of traffic), the greater comfort, less congestion of the infrastructure due to traffic and less influence of atmospheric agents.

Against these advantages it is necessary to take into account the economic and social magnitude of the investments and therefore conduct an evaluation of the choices with a careful analysis of the costs, impacts and benefits. The construction costs of a railway are always high, as are the fixed costs that must be sustained independently of the amount of traffic. It follows that a railway line can only be justified economically when traffic is sufficiently intense.

It was decided to insert in this feasibility study a comparison between the construction costs (Table 3), running costs and revenues that can be expected, also within the perspective of a wider territorial plan or growth of international and intermodal traffic.

Costs and revenues are difficult to determine in a uniform way as the former depend on the construction characteristics and the latter on the type of service. As the costs depend on the construction characteristics of the line, the localities to be served were decided first, then the number of tracks, type of line, type of rolling stock, type of station (simple stop or station with double the number of tracks for trains to pass one another).

The overall cost depends on the values of maximum slope, length of the route, number of structures, and in particular tunnels, that lead to the cost per kilometre. The running costs were calculated taking into account the capacity of the line (number of trains that can circulate in the unit of time and train speed).

The following construction cost was estimated: 9.5 million euro for each bridge (with a fixed cost equal to 160 million euro), 8.95 million euro for each tunnel, 0.1 million euro per km for train control system and other technologies, 4 million euro for each railway vehicle (there are a total of 5 trains), 1 million euro for each station/stop (there are 12 stations/stops).

Given its insertion in a delicate area, the total construction cost was recalculated utilizing a corrective factor of 1.4 (Law n. 443 of 2001) and the cost of the 12 stations (12 million euro), train control system (4.54 million euro) and trains (20 million euro) were added, thus obtaining a total construction cost of the entire infrastructure of 509.09 million euro.

The maintenance costs of the railway line were estimated utilizing a standard value of 7,000 euro/km that, for the total length of 46 km, is 322,000 euro/year. For maintenance of the trains a cost of 625,000 euro/train was hypothesized, which for 5 trains becomes a cost of 3.13 million euro per year.

To calculate the earnings a service was hypothesized, as indicated in the previous section, composed of: 3 trains from Venice to Cortina d'Ampezzo (and vice versa) on weekends and holidays, and 20 trains per day from Calalzo di Cadore to Cortina d'Ampezzo (and vice versa) on weekdays. Were also estimated: the average capacity (seats plus standing room) of a GTW 4/12 train equal to approxi-

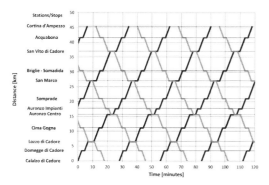

Figure 5. Timetable of designed railway line.

Table 3. Construction costs.

Railway line stretch	L*	E/T**	B***	Tu****	Total
	km	million euro			
Calalzo di Cadore – Auronzo di Cadore	17.7	10.5 €	49.9 €	78.5 €	138.9 €
Auronzo di Cadore – Briglie-Somadida	14.2	8.1 €	36.1 €	58.1 €	102.3 €
Briglie-Somadida – San Vito di Cadore	5.0	0.0 €	0.0 €	44.7 €	44.7 €
San Vito di Cadore – Cortina d'Ampezzo	8.5	6.3 €	11.8 €	33.5 €	51.6 €
Total	45.4	24.9 €	97.8 €	214.8 €	337.5 €

*Length, **Embankment/Cutting, ***Bridge, ****Tunnel.

mately 300 people, a one-way ticket from Venice to Cortina of about 25 euro per person and a one-way ticket from Calalzo to Cortina of about 5 euro. Differentiating the income by summer season, winter and during autumn/spring, an annual income was calculated for the passenger service of approximately 13 million euro. Subtracting annual cost for the maintenance of the line and trains from this, a value is obtained from which the number of years to amortize the initial construction costs of the line is derived, which in this case is about 50.

3 CONCLUSIONS

The railway connection between Calalzo di Cadore, Auronzo di Cadore and Cortina d'Ampezzo would respond to some needs, such as:

– improving connections between the municipalities of Cadore and the Ampezzo Valley;
– activating new tourist circuits to integrate with those already existing and that spread to nearby areas;
– creating an alternative to investments in the road sector in the Dolomites area, increasingly opposed by the local populations;
– encouraging a multimodal system of road/rail traffic for both people and goods.

This new stretch of railway would interest a pool of users corresponding to the resident population and a tourist sector that is busy in both summer and winter, which would constitute the basic prerequisite for the development of tourism and production in the areas involved. The railway would also encourage a reorganization of the existing network of infrastructures that would allow the improvement and enhancement of the historical centres, natural areas and ensure greater functionality of the new residential and productive areas.

The development of a railway line that connects Venice with Cadore and Cortina also aims to create a wider offering for the tourists who arrive in Venice, giving them the opportunity to make a daytrip to Cadore and the Ampezzo Valley.

It must anyway be taken into account that the zone involved in the study is one of the most striking areas of the Dolomites from the landscape-environmental point of view, characterized by a highly complex morphology and geology. The environmental impact caused by a railway line is certainly less than that of a new fast road because:

– a single-track railway line requires an overall width of around 4 metres with an impact comparable more to a cycle path than to a road;
– lower level of noise generated, less than that of a vehicle on a road and concentrated in the moments when a train passes 2 or 4 times per hour;
– no level crossings are planned as the railway line and the existing road never intersect.

Further studies are anyway required so that the railway can become an integral and sustainable part of the territorial context. Following the approach of other researchers (Keshkamat et al. 2009, De Luca et al. 2012), it will also be useful to conduct a study utilizing GIS and multi-criteria analysis.

The creation of the new railway line would be further stimulated if it was decided to build a railway connection between Cortina d'Ampezzo and Misurina or else to extend the planned railway line from Cortina d'Ampezzo to Dobbiaco (where the *Ferrovia delle Dolomiti* existed). In this way the number of passengers per kilometre could markedly increase, encouraging the development of new accommodation and sports facilities, with the creation of new jobs and an environment more conducive to demographic growth.

REFERENCES

Agostini, A. 2014. 1914-2014, *Belluno—Calalzo di Cadore: Una ferrovia nel cuore delle Dolomiti*. Belluno: Momenti AICS.
De Luca, M., Dell'Acqua, G. & Lamberti, R. 2012. High-speed rail track design using GIS and multi-criteria analysis. *Procedia—Social and Behavioral Sciences*. 54: 608–617.
Forni, M. 1975. Piccola storia della ferrovia Trento-Malè. *Ingegneria Ferroviaria* 30(11): 70–75.
Gaspari, E. 2005. La ferrovia delle Dolomiti: *Calalzo, Cortina d'Ampezzo, Dobbiaco, 1921–1964*. Bolzano: Athesia.
ISTAT. 2013. *Veneto*. Roma: ISTAT.
Keshkamat, S.S., Looijen, J.M. & Zuidgeest, M.H.P. 2009. The formulation and evaluation of transport route planning alternatives: a spatial decision support system for the Via Baltica project, Poland. *Journal of Transport Geography* 17:54–64.
Marseiler, S. 2006. *Cent'anni di ferrovia in val Venosta: 1906–2006*. Bolzano: Assessorato al Turismo ed alla Mobilità.
Mayer, L. 2004. *Impianti Ferroviari*. Roma: CIFI.
Serra, M. 1989. Il prolungamento della ferrovia elettrica Trento-Malè. *Ingegneria Ferroviaria* 44(1–2): 54–58.
Wall, H. 1987. *Bernina Express: in treno attraverso le Alpi*. Salò: ETR.

Transport Infrastructure and Systems – Dell'Acqua & Wegman (Eds)
© *2017 Taylor & Francis Group, London, ISBN 978-1-138-03009-1*

Investigation of the causes of runway excursions

N. Distefano & S. Leonardi
Department of Civil Engineering and Architecture, University of Catania, Catania, Italy

ABSTRACT: The risk of runway excursion dependent on multiple factors related to operating conditions. These include Runway Contamination, Adverse Weather Conditions, Mechanical Failure, Human Error.

A multivariate analysis of historical data on accidents on runways carried out in order to quantify the effect that various factors have upon runway excursions (landing and takeoff overruns and landing and takeoff veer-offs).

In this paper, an in-depth data study was conducted of all runway excursion accidents over the period spanning 2006–2015 in 8 geographical regions to investigate the causes of runway excursion accidents. All data in this study are from the Aviation Safety Database, published by Aviation Safety Network (ASN) by Flight Safety Foundation, and have been augmented by appropriate investigative reports when available.

The technique deployed in this work is the multiple logistic regression model, often used in recent literature and proved suitable to examine and quantify the effect of various factors on accidents risk. This technique revealed interesting relationships among the variables both for landing accidents and for take-off accidents. The goal of a logistic regression analysis is to predict correctly the outcome for individual cases using the parsimonious or least complex model.

The results of this paper show that the main cause of landing veer-off is the mechanical failure in most of the regions for any type of aircraft; the leading cause of landing overrun is human error in all regions especially for small aircrafts: the takeoff overrun mainly are caused by mechanical failure, especially for very large aircrafts. The weather conditions assume a dominant role, especially for small aircraft and for overrun accidents.

1   INTRODUCTION

A major initiative to improve safety is the increasing role of Safety Management Systems (SMS), intended as "an organized approach to managing safety, including the necessary organizational structures, accountabilities, policies, and procedures" (ICAO, 2012). ICAO has championed this initiative, which is now included in international standards for airline operations (ICAO, 2009). The basic structure involves four "pillars": identifying safety hazards; safety risk management through remedial actions to address safety risks; continuous monitoring and assessment of the safety level sought and achieved; and programs for continuous improvement in the overall level of safety.

Another major contributor to the improved safety record can be traced to the careful investigation of past accidents to determine what led to the accidents and what needs to be done to prevent such events from occurring again. This reactive approach to improving aviation safety has been enhanced by the thorough analysis of data from numerous accidents, which has aided in the identification of recurring patterns or risk factors that are not always apparent when individual accidents are investigated. More recently, proactive approaches to determining ways to improve safety have become increasingly popular. An example of such a proactive approach is the analysis of incident data to identify areas of increased risk that may lead to an accident (Oster et al. 2013)

A large number of air accidents occur during the take-off and landing phases. Most occur beyond the designated safety and protection areas, around the runway, when an aircraft overruns the runway-end during take-off or landing, or when it undershoots the runway, with regard to the threshold, during landing.

The most common accidents that occur during these flight phases are:

– Landing overrun (LDOR—LanDing OverRun).
– Landing veer-off (LDVO—LanDing Veer-Off).
– Landing   undershoot   (LDUS—LanDing UnderShoot).
– Take-off overrun (TOOR—Take-Off OverRun).
– Take-off veer-off (TOVO—Take-Off Veer-Off)
– Ground collision after take-off.

Runway excursions during take-off and landing continue to be the highest category of aircraft accidents and often exceed 25% of all annual commercial air transport accidents (IATA 2011).

Runway excursions can result in loss of life and/or injury to persons either on board the aircraft or on the ground. The effect of runway excursions can result in damage to aircraft, airfield or off-airfield installations including other aircraft, buildings or other items struck by the aircraft.

A runway excursion accident is defined as an accident where an aircraft on the runway surface departs the end of the runway or side of the runway surface during take-off or landing (IATA 2011).

It consists of two types of events:

– Veer Off: A runway excursion in which an aircraft departs the side of a runway
– Overrun: A runway excursion in which an aircraft departs the end of a runway

It excludes both accidents where the aircraft did not initially land on a runway surface, and take-off excursions that did not start on a runway (e.g., inadvertent take-offs from taxiways).

The current definitions of runway overrun do not specify any distance for arrivals and departures where an incident can not be considered to be an overrun or undershoot, because it is too far out from the runway. For example, the current definitions used by Eurocontrol are:

– Overrun on Take-Off: "A departing aircraft fails to become airborne or successfully reject the take-off before reaching the end of the runway".
– Overrun on Landing: "A landing aircraft is unable to stop before the end of the runway is reached".

The common taxonomy group of the ICAO commercial aviation safety team refers to: Runway excursion as veer off or overrun off the runway surface. This definition is only applicable during either the take-off or landing phase. The excursion may be intentional or unintentional (for example, deliberate veer off to avoid a collision, brought about by a runway incursion). This classification applies in all cases where the aircraft left the runway regardless of whether the excursion was the consequence of another event or not.

Arnaldo Valdés et al. (2011) developed probabilistic models of runway overruns and proposed risk models supported by historical data; they noted that runway overrun risk depends on multiple factors in relation to operating conditions including wind, runway surface conditions, landing or take-off distance required, presence of obstacles, runway distance available, existence and dimensions of runway safety areas.

Oster jr. et al. (2013) says there is still a role for careful accident investigation and there are still lessons to be learned from the few accidents that these carriers have. But with improvements in safety and major reductions in accidents, airline safety analysis will have to shift toward analysis of incident and operational data with the intent of identifying safety risks before accidents occur. There are two potential benefits from looking at these types of data. One is to address the question of why some sequences of events result in accidents while other sequences do not result in accidents. A better understanding of how potential accidents were avoided in some situations may lead to more such avoidances in the future. A second potential benefit is to identify trends in or the emergence of potentially hazardous sequences of events before they result in an accident. Here again, by identifying such trends, it may be possible to take corrective action before an accident occurs.

Distefano et al. (2014) proposes a risk assessment procedure in which the probability of each accident is proportional to the cumulative probability of the causes identified for the accident.

Čokorilo et al. (2014) has produced a model in order to estimate the frequency of aviation accidents in different environmental conditions and was based on the physical characteristics of the aircraft. The data were processed and then aggregated into groups, using cluster analysis based on an algorithm of partition binary 'Hard c means.'

Ayres jr. et al. (2012) shows that location models could be improved if greater attention was paid to causality but data difficulties exist, for example, on meteorological influences on overrun distance and that this is often mis-recorded in accident dockets so lateral deviations are more difficult to model; while in consequence modelling the variation in aircraft type, wingspan and speed ought to be included as well as pavement type variations that will affect deceleration.

Wagner et al. (2014) focuses his work on predicting if the excursion will generate fatalities: human errors are the strongest associated feature with fatal excursions, another feature strongly associated with fatal excursions is adverse weather conditions, fatal excursions occur more frequently on commercial flights than other categories of aircraft operation, and overruns are the most fatal category of runway excursions.

The risk of runway excursion dependent on multiple factors is related to operating conditions. The goal of this paper is to identify risk factors for type of accident of runway excursion in relation to certain elements such as the aircraft type and geographical region. The low accident rate of aviation means that no particular airport has sufficient accident occurrences in the recent past to support an accident frequency model with reasonable statistical confidence (Piers, 1994; Hale, 2001). Therefore, this study is based on a large database of relevant accident cases. The results of the current study can be used by a broad range of civil aviation organizations for risk assessment and cost-benefit studies of actions improvements.

2 RUNWAY EXCURSION DATA

The primary data source used in this work is a database created by Aviation Safety Network (ASN). The Aviation Safety Network is a private, independent initiative founded in 1996. On line since January 1996, the Aviation Safety Network covers accidents and safety issues with regards to airliners, military transport planes and corporate jets.

The ASN Safety Database contains detailed descriptions of some 15.800 incidents, hijackings and accidents to airliner, military transport category aircraft and corporate jet aircraft safety occurrences since 1921. Most of the information are from official sources (civil aviation authorities and safety boards), including aircraft production lists, ICAO ADREPs, and country's accident investigation boards.

For the purposes of this paper, it created a database containing solely runway excursion accidents, in a period between 2006 and 2015, for all categories of aircraft, and in all world regions.

Runway excursions are categorized into the following categories: Landing Overrun (LDOR), landing Veer Off (LDVO), Take-Off Overrun (TOOR), and Take-Off Veer Off (TOVO). The four runway excursion categories are mutually exclusive.

In this database for every event are recorded information about:

– date
– airport
– airport's country
– accident type
– phase of flight
– potential cause
– n° of fatalities
– n° of occupants
– aircraft type
– nature of flight
– aircraft damages and
– dynamic event.

Figure 1 shows the distribution of events contained in the database as a function of the type of accident.

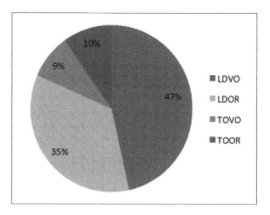

Figure 1. Distribution of events by type of accident.

4.1 Data preparation

In order to make statistically significant data to be analysed fields suitable to regroup together the elements of some records of the database have been created.

The following are the criteria used for the definition of new fields.

2.1.1 Geographical regions
The airport's countries where there was the runway excursion were grouped as recited in the IATA classification of the geographic regions (IATA 2011):

– AFI: Africa,
– ASPAC: Asia Pacific,
– CIS: Commonwealth of Independent States,
– EUR: Europe,
– LATAM: Latin America & the Caribbean,
– MENA: Middle East & North Africa,
– NAM: North American, and
– NASIA: North Asia.

Table 1 shows the number of events by type of accident in each geographic regions.

2.1.2 Risk factors
The following four classes of risk factors have been defined which includes all the causes of the events recorded in the database:

– Aircraft System Faults:
  – Engines
  – Brake (wheel brakes, spoilers or
  – reversers)
  – Hydraulic
  – Electric
  – Main gear
  – Tire
  – Other
– Human errors:
  – Incorrect Flight Planning

Table 1. Distribution of events by type of accident in geographic regions.

	LDVO	LDOR	TOVO	TOOR
AFI	18	17	3	6
ASPAC	43	21	5	2
CIS	8	5	3	6
EUR	25	16	6	3
LATAM	21	31	4	8
MENA	21	12	1	3
NAM	44	35	12	10
NASIA	2	0	0	0

Table 2. Distribution of events by type of accident and by risk factor.

	LDVO	LDOR	TOVO	TOOR
Aircraft System Faults	53	25	5	13
Human errors	49	42	12	8
Weather Conditions	38	15	6	4
Runway Conditions	12	11	1	5
Unknown	30	44	10	8

- Communication/Coordination
- Pilot error
- Visual Illusion
- Fatigue
- Excessive speed
- Loss of control
- Other
- Weather Conditions:
  - Low Visibility
  - Rain
  - Wind Shear
  - Tailwind
  - Crosswind
  - Ice
  - Low Ceiling
  - Strong Wind
  - Turbulence
  - Freezing Rain
  - Other
- Runway Conditions:
  - Wet
  - Contaminated—Standing water
  - Contaminated—Rubber
  - Contaminated—Oil
  - Contaminated—Ice
  - Contaminated—Slush
  - Contaminated—Snow
  - FOD
  - Wildlife Hazards
  - Down Slope

When the accident report does not mention the potential cause, this was referred to as unknown.

Table 2 shows the number of events by type of accident in relation to the class of risk factors.

2.1.3 Aircraft classes

The accidents were analysed according to the different classes of aircraft operations represented in: General Aviation (GA), Corporate Aircraft (CA), Commuter Aircraft (Com A) and Transport Aircraft (TA)—to address these different operational perspectives. The aircraft operation classes are defined as:

- General Aviation aircraft (GA): Typically these aircraft con have one (single engine) or two engines (twin engine). Their maximum gross weight is usually below 14.000 lb.
- Corporate Aircraft (CA): Typically these aircraft can have one or two turboprop driven or jet engines (sometimes three). Their maximum gross mass is up to 90.000 lb.
- Commuter Aircraft (Com A): Usually twin engine aircraft with a few exceptions such as the De Havilland DHC-/which has four engines. Their maximum gross mass is below 70.000 lb.
- Transport Aircraft (TA):
  - Short-Range (S-R): Their maximum gross mass is usually below 150.000 lb.
  - Medium-Range (M-R): These are transport aircraft employed to fly routes of less than 3.000 nm (typical). Their maximum gross mass is usually below 350.000 lb.
  - Long-Range (L-R): These are transport aircraft employed to fly routes of less than 3.000 nm (typical). Their maximum gross mass usually is above 350.000 lb.

Table 3 shows the number of events by type of accident in relation to the aircraft class.

2.1.4 Nature of flight

Nature of flight was reported in ASN database; the following are the items used for the analysis:

- Ambulance,
- Cargo,
- Domestic Non Scheduled Passenger,
- Domestic Scheduled Passenger,
- Executive,
- Ferry/positioning,
- International Scheduled Passenger,
- International Non Scheduled Passenger,
- Private,
- Other (Agricultural, Training, Test, Parachuting), and
- Unknown

Scheduled services are flights scheduled and performed for remuneration according to a published timetable, or so regular or frequent as to constitute a recognizably systematic series, which are open to direct booking by members of the public,

Table 3. Distribution of events by type of accident and aircraft class.

	LDVO	LDOR	TOVO	TOOR
GA	29	25	4	7
CA	39	31	14	14
Com A	48	33	6	10
S-R	45	34	8	3
M-R	14	11	2	1
L-R	7	3	0	3

Table 4. Distribution of events by type of accident and nature of flight.

	LDVO	LDOR	TOVO	TOOR
Ambulance	3	2	2	2
Cargo	27	15	4	5
Domestic Non Scheduled Passenger	15	5	2	0
Domestic Scheduled Passenger	72	48	10	14
Executive	11	14	3	4
Ferry/positioning	7	11	2	5
International Scheduled Passenger	19	13	1	1
Military	7	9	4	2
Private	11	11	2	2
Other	3	6	3	1
Unknown	7	3	1	2

A non-scheduled air service is a commercial air transport service performed as other than a scheduled air service. A charter flight is a non-scheduled operation using a chartered aircraft.

Table 4 shows the number of events by type of accident in relation to the nature of flight.

3 STATISTICAL APPROACH FOR EXCURSION DATA ANALYSIS

In many transportation applications discrete data are ordered, for example to analyse categorical frequency data (e.g., accident, serious incident, incident).

Although these data are discrete, application of the standard or nested multinomial discrete models does not account for the ordinal nature of the discrete data and thus the ordering information is lost. To address the problem of ordered discrete data, ordered probability models have been developed.

Ordered probability models (both probit and logit) are a form of discrete outcome models that relate dependent variables on an ordered discrete scale to a series of predictor variables. Ordered probability models are derived by defining a latent variable z as a basis for modelling ordinal ranking data. This unobserved variable can be specified as a linear function for each observation such that:

$$z = \beta X + \varepsilon \qquad (1)$$

where X is a vector of variables determining the discrete ordering for observation n, β is a vector of estimable parameters, and ε is a random disturbance. Using this equation, observed ordinal data, y, for each observation are defined as:

$y = 1$ if $z \leq \mu_0$

$y = 2$ if $\mu_0 < z \leq \mu_1$

$y = 3$ if $\mu_1 < z \leq \mu_2$

$y = \ldots$

$y = I$ if $z \geq \mu_{i-1}$,

where the μ are estimable parameters (referred to as thresholds) that define y, which corresponds to integer ordering, and I is the highest integer ordered response. Note that during estimation, non-numerical orderings such as never, sometimes, and frequently are converted to integers (for example, 1, 2, and 3) without loss of generality.

The μ are parameters that are estimated jointly with the model parameters (β). The estimation problem then becomes one of determining the probability of I specific ordered responses for each observation n. This determination is accomplished by making an assumption on the distribution of ε. If ε is assumed to be normally distributed across observations with mean = 0 and variance = 1, an ordered probit model results with ordered selection probabilities as follows:

$$P\ (y = i) = \Phi\ (\mu_i - \beta X) - \Phi\ (\mu_{i+1} - \beta X),$$

where μ_i and μ_{i+1} represent the upper and lower thresholds for response category i and $\Phi\ (\ldots)$ is the cumulative normal distribution.

Estimation is done using maximum likelihood methods. For this study, XLSTAT software was used to estimate separate ordered probit models in order to examine the risk factors of the runway excursions depending on the flight nature and in relation to different geographic regions of occurrence and the aircraft class involved. Runway excursions risk factors are assessed on a five-point scale.

When interpreting the results of the ordered probit models, a positive value for particular β implies that if the condition corresponding to that parameter is true (e.g., the runway excursions category is LDVO) the probability of the highest ordered discrete category (e.g., risk factor: aircraft system faults) will increase and the probability of the lowest category (e.g., risk factor: runway conditions) will decrease.

Assessing the impacts on intermediate categories is more difficult because these results are conditional based upon the threshold values (μ). As such, to examine the effects of a specific variable on one of the interior categories, marginal effects are computed as the change in the estimated probabilities when a specific indicator variable is changed from zero to one with all other variables held equal to their means. These marginal effects can be interpreted as the change in the probability of a particular preference category, $P(y = i)$, given a change in the respective variable.

Ordered probability models were developed in order to gain greater insight as to how differences in geographic regions, aircraft type, and other factors influenced the cause of the runway excursion.

4 RESULTS AND DISCUSSION

Table 5 presents the parameters of the ordered probit model that has been developed for the estimation of the probabilities of the causes that generate the runway excursions. Coefficient estimates are provided for this model, along with standard errors, as well as confidence intervals for each variable. The modelling results allowed to infer interesting considerations, both general and in relation to individual types of accidents.

The comments relating to each type of accidents will be performed with reference to a subdivision of accidents in relation to the different geographic regions of occurrence.

For each type of runway excursion in each geographic region has been identified the cause that has the highest probability of generating the event.

4.1 Generality

The weather conditions play an important role, although not predominant, for the overrun accidents (both in take-off and landing).

In NAM and EUR the probability that an overrun is due to bad weather conditions is about 16% (with 18% values for smaller aircraft); while the probability that a veer-off is due to unfavourable weather conditions is about 13% (with values 8% for larger aircraft).

In LATAM, AFI and ASPAC the probability that bad weather conditions cause all kinds of accidents is on average 17% for all types of aircraft. Only for a L-R aircraft such probability is reduced to 10% in LATAM and ASPAC.

Table 5. Ordered Probit Model (normalized coefficients).

Variable	Coeff.	Std. Error	95% Confidence interval
Event			
LDVO	0.000	0,000	
LDOR	−0.201	0,061	(−0.320, −0.081)
TOVO	−0.117	0,059	(−0.232, −0.002)
TOOR	−0.015	0,060	(−0.134, +0.103)
Airport's country			
NAM	0.000	0,000	
EUR	0.023	0,066	(−0.106, +0.151)
LATAM	−0.076	0,069	(−0.211, +0.058)
AFI	−0.190	0,067	(−0.321, −0.058)
ASPAC	−0.192	0,070	(−0.329, −0.055)
CIS	−0.116	0,062	(−0.237, +0.006)
MENA	−0.126	0,067	(−0.256, +0.005)
NASIA	−0.083	0,058	(−0.197, +0.030)
Aircraft class			
GA	0.000	0,000	
CA	−0.040	0,076	(−0.189, +0.109)
Com A	−0.026	0,082	(−0.187, +0.135)
S-R	0.028	0,084	(−0.137, +0.194)
M-R	0.034	0,071	(−0.105, +0.172)
L-R	0.115	0,067	(−0.016, +0.247)
Nature of flight			
Ambulance	0.000	0,000	
Cargo	−0.125	0,135	(−0.390, +0.139)
Domestic Non Scheduled Passenger	−0,101	0,098	(−0.294, +0.092)
Domestic Scheduled Passenger	−0.105	0,181	(−0.461, +0.250)
Executive	−0.118	0,110	(−0.333, +0.096)
Ferry/positioning	−0.011	0,102	(−0.211, +0.189)
International Scheduled Passenger	−0.064	0,120	(−0.299, +0.171)
Military	−0.102	0,101	(−0.300, +0.096)
Private	−0.043	0,103	(−0.244, +0.158)
Unknown	−0.117	0,083	(−0.281, +0.046)
Other	−0.010	0,085	(−0.177, +0.156)

In LATAM and AFI the probability that the various types of accidents are caused by unknown causes assumes a higher percentage than in countries of NAM and EUR, especially for small aviation. This consideration is more evident in the countries AFI even more than in those of LATAM.

In ASPAC the percentage of accidents involving large aircraft (M-R and L-R) is 5.6%. For large aircraft (L-R), the most likely cause is the mechanical failure for all types of accident; human error is manifested by a reduced rate compared to other types of aircraft.

In CIS occurred only 5.6% of total accidents; the predominant cause for all incident types and for all types of aircraft is human error (from 20 to 30%); the probability that accidents are caused by unknown factors is about 25%. The mechanical failure is relevant in the case of LDVO, while the weather conditions are the potential cause of 18% of accidents on average for all types of aircraft.

In NASIA there were only 2 LDVO accidents; for them the unknown cause is the most likely (43%).

4.2 LDVO (Landing Veer-Off)

In EUR and NAM, the main causes of this type of accident are those related to mechanical failure, regardless of the type of aircraft and the flight nature. For example, the LDVO for GA aircraft and "Private" flight nature, has a 42% probability of being caused by mechanical failure, 31% to have as a triggering event a human error, 13% to be caused by weather conditions and only 5% to be caused by the runway conditions.

In LATAM, AFI, ASPAC and MENA the leading cause of LDVO for small aircraft is human error, while for large aircraft in LATAM, MENA and ASPAC is the mechanical failure and in AFI is always human error.

4.3 LDOR (Landing Overrun)

In EUR and NAM, this accident has a higher probability of occurrence for causes related to human error (30%), regardless of the type of aircraft. It should however be pointed out that, as regards the nature of the flight, in the case of "Other" (Test, training, parachuting), the mechanical failure is the main cause of this accident type (over 30%).

In LATAM, human error is the main cause (around 27%) of LDOR for small aircraft and for aircraft of larger size (L-R) the main cause is mechanical failure (40%).

In AFI and ASPAC, the main cause of LDOR for small aircraft is human error (about 20%); the unknown causes, however, assume a high percentage (between 40 and 50%) for all types of aircraft.

In NEMA, the main cause of LDOR is human error (approximately 25%) for small aircraft, while the unknown causes contribute in a high percentage (between 30 and 40%) for all types of aircraft, with the exception of L-R mainly caused by mechanical failure (33%).

4.4 TOVO (Take-Off Veer-Off)

In EUR and NAM, the TOVO has an equal probability of occurring from causes related to human error and mechanical failure in the case of small aviation (GA and CA); for larger aircraft (com-A and S-R), the probability that the human error is the main cause grows slightly, while for large aircraft (M-R and L-R) this type of accident is infrequent.

In LATAM the main cause of TOVO is human error (around 27%), although the percentage of unknown causes is very high for all types of aircraft.

In AFI and ASPAC, accidents during take-off involve only small aviation (CA, Com A); for such aircraft, the most likely cause of TOVO is human error (20%) followed by the bad weather conditions (17%). Also in this case the unknown causes are in high proportion (45%).

4.5 TOVO (Take-Off Overrun)

In NAM and EUR accidents of this type are quite rare (9.9% in NAM and 6% in EUR); the mechanical failure is certainly the most likely cause, especially for very large aircraft.

Also in AFI and ASPAC the TOOR type accidents are not very frequent (13,63% in AFI and 12.4% in ASPAC) and involve only small aircraft (CA, Com A); for these aircraft the most likely cause of such accidents is human error (about 26%) followed by the bad weather conditions (18%). The probability of unknown causes is high in percentage (30%).

In CIS the incidents of this type are quite frequent (27,27%) and the likely cause is human error (25–30%) in the case of small and medium-sized aircraft. The large aircraft have not recorded any such incident.

In LATAM the frequency of these accidents is very low (2.8%) and the most likely cause is the mechanical failure (about 35%) for small aircrafts, and even 62% for aircraft type L-R. The probability that the TOOR is due to unknown causes is instead quite low.

5 CONCLUSION

Aviation safety analysis historically has emphasized accident data. For the most part, the aviation industry and government regulators have used data reactively to identify the causes of aircraft accidents and to take steps to prevent these types of accidents from recurring. But there are still lessons to be learned from the few accidents that these carriers have. With improvements in safety and major reductions in accidents, airline safety analysis will have to shift toward analysis of incident and operational data with the intent of identifying safety risks before accidents occur.

This paper shows how the same event may be caused by different risk factors in relation to different geographic regions or types of aircrafts.

This results represent a very proactive tool for future aviation safety analysis.

Future work includes efforts on improving predictive performance with statistic methods that address imbalanced data more effectively. Further, it will be considered the consequences of the events in terms of damage to the aircraft and/or persons.

REFERENCES

Arnaldo Valdés, R.M., Gómez Comendador, F., Mijares Gordún, L. & Sáez Nieto, F.J. 2001. The development of probabilistic models to estimate accident risk (due to runway overrun and landing undershoot) applicable to the design and construction of runway safety areas. Safety Science 49: 633–650.

Ayres, M. Jr., Shirazi, H., Carvalho, R., Hall, J., Speir, R. & Arambula, E. 2013. Modelling the location and consequences of aircraft accidents. Safety Science 51: 178–186.

Distefano, N. & Leonardi, S. 2014. Risk assessment procedure for civil airport. Journal International Journal for Traffic and Transport Engineering (IJTTE) 4 (1): 62–73.

Hale, A., 2001. Regulating airport safety: The case of Schiphol. Safety Science 37: 127–149.

IATA. 2011. Runway excursion analysis report 2004–2009 (RERR 2nd Edition—Issued 2011). Montréal, Québec.

International Civil Aviation Organization. 2009. Safety management manual (2nd ed.). Montreal, Quebec.

International Civil Aviation Organization, Co-operative Development of Operational Safety and Continuing Airworthiness Programme. (2012). Safety Management System (SMS)

Oster, C.V. Jr., Strong J.S. & Kurt Zorn C. 2013. Analyzing aviation safety: Problems, challenges, opportunities. Research in Transportation Economics 43: 148–164.

Piers, M., 1994. The development and application of a method for the assessment of third party risk due to aircraft accidents in the vicinity of airports. In: 19th International Council of Aeronautical Sciences Congress, Anaheim September ICAS proceedings. ICAS (19): 507–518.

Wagner, D.C.S. & Barkerb, K. 2014. Statistical methods for modeling the risk of runway excursions. Journal of Risk Research 7 (17): 885–901.

Aviation Safety Network (ASN Database), 2016. <http://aviationsafety.net/database/> (Giugno 2016).

Transport Infrastructure and Systems – Dell'Acqua & Wegman (Eds)
© 2017 Taylor & Francis Group, London, ISBN 978-1-138-03009-1

A review of sulfur extended asphalt modifiers: Feasibility and limitations

S.E. Zoorob
Energy and Building Research Centre, Kuwait Institute for Scientific Research, Kuwait

C. Sangiorgi & S. Eskandersefat
Department of Civil, Chemical, Environmental and Materials Engineering, University of Bologna, Bologna, Italy

ABSTRACT: Sulfur Extended Asphalt Modifier (SEAM) is currently being marketed as an additive for hot asphalt mixtures. Typically 40% by mass of the binder phase in the asphalt mix can be replaced by SEAM using conventional mix design and production techniques. A number of full scale trials using SEAM have been reported in the literature and, except for some minor concerns regarding rutting and moisture resistance, it appears that overall the SEAM modified mixes have the potential for improved mechanical performance compared to conventional asphalt. During the production stages of SEAM asphalt mixes in conventional hot mix plants, the sulfur component will exist in the liquid phase, which requires careful thermal management to control gaseous emissions. As a consequence, this paper includes a general review of exposure limits for sulfur dioxide and hydrogen sulfide emissions and their short and long term health effects on healthy and asthmatic individuals.

1 INTRODUCTION

In the early 1970s, due to concern over asphalt cement (bitumen) supply issues and an anticipated overabundance of elemental sulfur, several organizations in the US and Canada, began to evaluate the potential for sulfur to substitute for bitumen as binder extender, this became known as Sulfur Extended Asphalt (SEA). Indentation/hardness tests carried out on elemental sulfur-bitumen systems, showed that a large increase in resistance to indentation can be obtained in the range in which the continuous sulfur phase begins to form (i.e. at sulfur-bitumen ratios of 1.5–2.0) (Beaudoin & Sereda 1979).

In the late 1980s, a sharp rise in the price of sulfur in addition to the fact that the use of hot liquid sulfur during production generated a significant amount of fumes and odors brought its use in road paving to an end. However, the recent rises in bitumen prices coupled with the production of low sulfur fuels has once again made sulfur a marketable product in the asphalt industry.

To overcome the problems with hot liquid sulfur used in bituminous mixes, Sulfur Extended Asphalt Modifier (SEAM) in solid pellet form (patented process) was developed and marketed as an additive for hot mix asphalt under the brand name Shell Thiopave™ (Shell 2014, FHWA 2012, Tran et al. 2010).

The SEAM pellets are in a solid, non-sticky, non-melting form at ambient temperatures, which eases handling, storage and transportation. The pellets have melting point in the range of 93 to 104°C, which has been specifically designed to melt when in contact with a hot asphalt mix. The additive is intended to act both as a bitumen extender and as an asphalt mixture modifier. The manufacturers claim (backed up by a number of full scale investigations) that SEAM improves the performance of conventional bitumens by imparting toughness, strength and rut resistance. Furthermore, it is claimed that SEAM additives can provide cost savings when compared to other polymer modified asphalts (Rock binders Inc 2014).

2 SEAM IN ASPHALT PAVEMENTS

2.1 Composition of SEAM pellets, mix proportions and construction of asphalt layers

The key to the composition of SEAM (Thiopave™) is that liquid sulfur is plasticized by the addition of carbon black at a concentration of at least 0.25% (preferably 0.4 to 0.8%) and the plasticized sulfur is further treated with amyl acetate at a concentration of at least 0.08% (preferably between 0.2 and 0.4%) to produce an even more manageable plasticized sulfur additive. The carbon facilitates the plasticization reaction with the sulfur, additionally the carbon creates an ultraviolet light shield which helps to prevent ultraviolet degradation of the final bitumen plus aggregate product. On the other hand, the amyl acetate helps to reduce unwanted odors from the product and thereby improve its overall handling (Bailey et al. 2003).

SEAM may be used in asphalt mixes made with conventional bitumen in weight ratios from 20%/80% to 50%/50% SEAM/bitumen. SEAM melts readily into the bitumen and becomes a part of the binder in the asphalt mixture. Stiffness and Marshall stability are expected to increase with increased amounts of sulfur (Predoehl 1989).

Once a SEAM/bitumen ratio is selected, the equivalent volume replacement of binder (by% weight) can be determined. Since sulfur is about twice as dense as bitumen, a given weight of sulfur has therefore about half of the volume of the same weight of bitumen (Mahoney et al. 1982).

The sulfur pellets at ambient temperature are added to the pre-heated aggregate and bitumen during the asphalt mixing process rather than the sulfur being pre-blended with bitumen. For health and safety reasons, the manufactures stress that when incorporating SEAM, the asphalt mix discharge temperatures must be carefully controlled to never exceed the recommended range 135–146°C. At this temperature, the pellets melt quickly and the shear conditions in the mixer are high enough to disperse the sulfur into the asphalt mix in a very short time that is compatible with asphalt mix production.

Delayed delivery trucks arriving at the paving site with SEAM mix below a temperature of 115°C shall be rejected (Rock binders Inc. 2014). The manufacturers claim that SEAM may be used with any conventional bitumen grade, but not polymer-modified bitumen. Warm Mix Asphalts (WMA) that incorporate water foaming type additives in their compositions are also not recommended for Sulfur extended asphalt production (FHWA 2012).

2.2 Stiffness evolution of SEAM asphalts and general findings from full scale SEAM trials

In a recent investigation (Tran et al. 2010), 10 asphalt mix designs were laboratory manufactured (2 control and 8 Thiopave composites). Details of the mix compositions are listed in Table 1. Figure 1 shows the average and% increase in dynamic modulus (E*) for each of the 10 mixes tested at 10 Hz and at 21°C after 1 day and 14 days of curing.

When a PG 67–22 was used as the base binder in 4 Thiopave mixes, the average 1-day E* results of these Thiopave mixes were comparable or greater than the respective average 1 day E* values for the PG 67–22 control mix. In addition, except for the 30% Thiopave mix with 2% design air voids, the other Thiopave mixes yielded average 1-day

Table 1. Sample identification for data of Fig. 1.

Mix ID	Base Binder	% Thiopave	% Design Air Voids
Control-67	PG 67-22	0	4
Control-76	PG 76-22	0	4
Thio-67-30-2	PG 67-22	30	2
Thio-67-30-3.5	PG 67-22	30	3.5
Thio-67-40-2	PG 67-22	40	2
Thio-67-40-3.5	PG 67-22	40	3.5
Thio-58-30-2	PG 58-28	30	2
Thio-58-30-3.5	PG 58-28	30	3.5
Thio-58-40-2	PG 58-28	40	2
Thio-58-40-3.5	PG 58-28	40	3.5

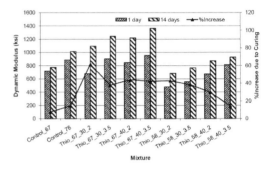

Figure 1. Average E* results at 10 Hz and 21°C for all mixtures after 1 and 14 days curing.

E* results comparable to those of the PG 76–22 control mix. The average 14-day E* results of the Thiopave mixes were comparable or greater than the respective average 14-day E* values for both the control mixes (Tran et al. 2010).

A more recent laboratory investigation has also shown that sulfur modified asphalt mixtures continue to increase in stiffness and reach a plateau after approximately 30 days. This increase in stiffness recorded at 30 days, amounted to about 50% of the initial measurements that were taken at 9 days following compaction (Cocurullo et al. 2012).

A large full scale investigation was conducted by the California DoT to determine whether the incorporation of sulfur with a soft grade of bitumen could change the temperature-viscosity relationship of the resulting binder, thereby making it more useful in cold and hot climates (Predoehl 1989). The two trial sections (a hot and a cold climate test sections) utilized the same sulfur/bitumen blends, i.e. 20% and 40% sulfur by weight of total binder, and each project used the same blending AR 2000 bitumen (i.e. RTFOT residue with viscosity of 2000 Poise at 60°C). Overall, based on the findings of this full scale study, SEAM binders were a viable alternative to conventional bitumens. Results of physical tests (Hveem system) on compacted briquettes and field cores revealed no significant differences in stability, cohesion and surface abrasion between SEAM and conventional asphalt mixes. The findings of this study raised one major concern, in that it appeared that SEAM blends with over 20% sulfur by weight (e.g. at a 40%/60% sulfur/ bitumen ratio) should not be utilized in overlays in colder climate areas due to an early thermal cracking potential. This was not an issue for the hotter climate test section (Predoehl 1989).

The US Federal Highway Administration also completed a field study to compare the performance of sulfur-extended asphalt pavements to conventional asphalt control pavements. A representative set of pavements from 18 States was chosen to provide a comprehensive evaluation. The sulfur/ bitumen ratios investigated were (10/90, 20/80, 25/75, 30/70, 35/65, 40/60) including surface & binder courses. The primary conclusion was that there was no difference in overall performance between the SEAM and control sections. A laboratory study complemented the field study, and cores were obtained from many of the pavements for testing. In general, the laboratory test results supported the results of the field study. Overall, sulfur did not increase or decrease most test properties, and often it had no effect on a given test property of a mixture. However, sulfur did decrease the resistance to moisture susceptibility in the laboratory (this is further discussed below). There were also minor trends indicating that with some mixtures, sulfur reduces the susceptibility to rutting. Additionally stress-controlled repeated load indirect tensile fatigue tests were carried out on cores obtained from the trial sections and when considering all projects there was an indication that sulfur decreased the fatigue life (Stuart 1990).

Nicholls (2012) summarised key findings from selected full scale trials covering the period 2004 to 2008 from a number of countries including: Canada, US, Saudi Arabia, Qatar (Al-Ansary 210) and China.

Cores taken from a wide range of sites have shown that the incorporation of sulfur pellets does not result in any impairment of the compaction, with the bulk density being marginally greater than the control in most cases. That marginal difference may be due to the higher density of the pellets than the bitumen. In general, Marshall stability of the sulfur-modified mixtures were similar to the control mixtures on day 1, however the stability values were observed to increase with time.

Al-Mehthel (2010) also presented excellent performance results from four full scale trials (using 30/70 sulfur/bitumen blends) laid in 2006 at various locations across Saudi Arabia that were monitored for signs of rutting and fatigue cracking for a period of up to four years. Nazarbeygi (2012) also

reported excellent pavement condition index (PI) and no wheel path depressions or cracking of a full scale trial monitored for 3 years made with a 35/65 (Sulfur/bitumen) hot mix asphalt.

2.3 Rutting susceptibility of SEAM mixes

In 2004, the City of Calgary undertook a series of full-scale demonstration projects aimed at evaluating the potential performance enhancement offered by several premium surfacing materials, including (Johnston et al. 2005):

1. Stone Matrix Asphalt (SMA), 12.5 mm nominal maximum size (NMS) using 6.0% Performance Grade (PG) 70-31 polymer modified binder (PMA).
2. 12.5 mm NMS SMA, modified with 3% manufactured shingle modifier (MSM), with PG 67-37 PMA, for a total binder content of 6.0%.
3. Superpave (12.5 mm NMS, Fine Graded) using the same PG binder and content as 1.
4. Superpave (12.5 mm NMS, Fine Graded) using the same percentage of PG binder and MSM modification as 2.
5. Superpave (12.5 mm NMS, Fine Graded) incorporating 3.9% 150/200 A bitumen and 2.6% SEAM.

An Asphalt Pavement Analyser (APA) (a modified version of a laboratory wheel tracker) was used to provide an assessment of the rutting potential of the various mixture types. Field cores were acquired from the various installations within several months of construction. Superpave gyratory compactor specimens were also fabricated from plant mixes produced for the projects. Figure 2 presents the average APA results, ranked based on laboratory fabricated samples. The average APA field core test results are shown for comparison. The results clearly show that with respect to rutting, the SEAM mixes were the worst performers. Although it is not entirely clear from this investigation whether the choice of bitumen grade for SEAM incorporation was optimal. Selecting a harder bitumen grade may have reduced rutting, but in turn this may have also narrowed down the

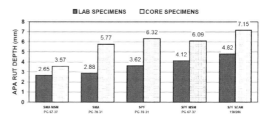

Figure 2. Comparison of lab prepared and core specimen asphalt pavement analyzer rut depths.

allowable temperature window for production and laying operations.

Other laboratory based rutting investigations showed much more promising results. For example, 45°C wheel tracking trials using Iranian 60/70 pen grade bitumen with sulfur replacement levels ranging from 25 to 45% showed progressively reduced wheel tracking depths with increased sulfur replacement levels (Johnston et al. 2005).

Nicholls (2012) also collated the results of laboratory conducted rutting tests of sample extracted from selected the full scale trials. Overall, the results were very positive, with rut depth values of sulfur-modified mixes being 20 to 55% less than the corresponding control mixes.

2.4 Moisture susceptibility of SEAM mixtures

In one large laboratory investigation carried out by the Washington State DoT, it was found that mixture moisture susceptibility (assessed using the Lottman moisture conditioning test) increases when increasing amounts of sulfur are added to the binder (i.e. increasing SEAM ratio). Samples were subsequently investigated using a SEM, and it became apparent that the sulfur crystals formed in the samples prior to moisture conditioning were often destroyed or broken during the conditioning process. Since strength (more specifically stiffness) seems to be imparted to the SEAM samples through establishment of a network of sulfur crystals in the voids and binder, it is likely that this strength will decrease when that network is damaged by the combination of moisture and freeze-thawing. The investigation found that, as expected, increasing amounts of added sulfur results in increased mixtures stiffness. At the same time, increasing amounts of added sulfur generally results in increased mixture stiffness loss following the moisture and freeze-thaw conditioning (Mahoney et al. 1982).

In a large scale US Federal Highway Administration investigation (Stuart 1990), the susceptibility to damage by moisture of SEAM pavements was evaluated on cores obtained from a number of trial pavements (representing 18 States). The testing procedure was in accordance with ASTM D 4867 (minor variation on the Lottman procedure) with resilient modulus and tensile strength values being determined before and following moisture conditioning. It was concluded that when considering all the projects, the effect of sulfur was to reduce both the Tensile Strength Ratios (TSR) and the resilient Modulus Ratios (MrR), but not the visual percent stripping. It was hypothesised that the lower ratios were related to a loss of cohesion rather than a loss of adhesion. Overall the results clearly indicated that the SEAM binders were weakened by the moisture conditioning processes (Stuart 1990).

In another recent investigation (Tran et al. 2010), moisture susceptibility testing was performed on six mix designs as shown in Figure 3 (details of the mix compositions were shown earlier in Table 1). Testing was conducted in accordance with ALDOT 361-88 and AASHTO T 283-07 after the specimens were allowed to cure for 14 days at room temperature.

For the ALDOT method, the conditioned specimens were vacuum saturated to the point at which 55% to 80% of the internal voids were filled with water.

The indirect tensile strength ratio at $25 \pm 0.5°C$ was calculated for each set by dividing the average tensile strength of the conditioned specimens by the average tensile strength of the unconditioned specimens (Tran et al. 2010).

For the AASHTO T 283 method, the conditioned specimens were vacuum saturated so that 70–80% of the internal voids were filled with water. These specimens were then wrapped in plastic and placed in a leak-proof plastic bag with 10 mL of water prior to being placed in the freezer at $-18 \pm 3°C$ for a minimum of 16 hours. After the freezing process, the conditioned samples were placed in a $60 \pm 1°C$ water bath for 24 ± 1 hours to thaw. All samples, conditioned and unconditioned, were brought to room temperature in a $25 \pm 0.5°C$ water bath to equilibrate the sample temperature for 2 hours just prior to testing. Calculation of the failure load, splitting tensile strength, and TSR value was conducted at $25 \pm 0.5°C$.

As shown in Figure 3, the Thiopave modified mixes had lower TSR values than the control mixtures tested by both the ALDOT and AASHTO methods. In this study, the TSR values for the Thiopave mixtures were found to be lower than the commonly accepted failure threshold of 0.8 (Tran et al. 2010).

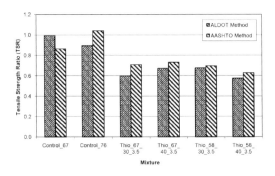

Figure 3. Summary of TSR results (after 14 days of curing).

In yet another more recent laboratory investigation (Cocurullo et al. 2012), asphalt mixtures composed of 40/60 pen grade bitumen modified with 30% SEAM by mass of total binder were slab compacted and cored. Indirect tensile stiffness modulus test results on dry and moisture conditioned (immersed in water at 85°C) sulfur modified specimens showed that after 9 days of water immersion approximately 50% of the initial dry modulus was retained. Interestingly, further tests revealed that following water immersion tests, long term dry recovery periods (minimum 2 weeks) caused the specimens to partially recover their stiffness values with time.

No definitive explanation has been put forward to justify the loss of mechanical property in moisture conditioned SEAM asphalt mixtures. The greatest impact of sulfur on bitumen properties and performance probably relates to air oxidation of the sulfide forms indigenous to crude oil to various sulfur oxide species. For example, conversion of aliphatic sulfides to the corresponding sulfoxide $(RS(O)R)$, occurs during oxidative aging tests for bitumen. The change in physical properties of sulfides with oxidation is dramatic. For example, methyl sulfide (CH_3SCH_3), is a light ($\rho = 0.846$ g/mL), water-insoluble, low boiling liquid (bp = 36°C); whereas methyl sulfoxide $(CH_3S(O)CH_3)$ is a heavy ($\rho = 1.48$), water miscible liquid boiling at 189°C. With regard to bitumens, the sulfoxide group exhibits a significant interaction with most aggregates which is fairly susceptible to water stripping, in contrast to the negligible adsorption of sulfides onto aggregate. Oxidation of sulfides to sulfoxides would thus be expected to increase asphalt hardness/viscosity, and to increase the concentration of amphoteric species. It would also potentially increase susceptibility to water stripping and strength of asphalt-aggregate adhesion (Green et al. 1993).

3 SEAM FROM A WIDER PERSPECTIVE

The sulfur industry is different from many other important modern mineral industries in that the disposal of excess supplies of sulfur is becoming a more important issue than that of how to maintain sustainable production. Worldwide recovered sulfur output is expected to increase significantly in the future. Sulfur surpluses were expected beginning in 2010 with acceleration thereafter as a result of increased production, especially from oil sands in Canada, natural gas in the Middle East, expanded oil and gas operations in Kazakhstan, and heavy-oil processors in Venezuela. Additional increases are expected to come from Russia's growth in sulfur recovery from natural gas and Asia's improved sulfur recovery at oil refineries.

Sulfur recovery from global petroleum refineries alone is predicted to reach 50 Mt/yr in 2025. Future gas production is also likely to come from deeper, hotter, and more sour deposits that would result in even more excess sulfur production. If non-traditional uses for elemental sulfur increase significantly, and alternative technologies for reinjection and long-term storage are not developed, the oversupply situation will result in tremendous stockpiles accumulating around the world.

The aforementioned discussion on the advantages and disadvantages of incorporating SEAM as a partial replacement of the bitumen fraction in asphalt mixtures is typical of the type of information readily available to road engineers and road material specifiers. In the opinion of the authors, this type of analysis is slightly restrictive in scope, in that when a new composite equals or even outperforms its rivals in terms of density, voids, stiffness, fatigue, or even price performance, there remains other equally important more holistic issues to consider, in particular when the new composite contains a non inert bulk waste material.

In the following sections, an attempt is made to expand the discussion by exploring the exposure to sulfur dioxide and hydrogen sulfide emissions and their short and long term health effects on healthy and asthmatic individuals.

3.1 Reaction kinetics of H_2S evolution

At temperatures above 130°C the sulfur ring molecules suffer partial decomposition and polymerize in to long diradical chains. When sulfur is blended with bitumen at high temperature these sulfur radicals react with some components of bitumen provoking two competitive reactions. These radicals may either extract hydrogen from hydrocarbon molecules with subsequent hydrogen sulfide formation by means of dehydrogenation reactions, or can be incorporated to bitumen structure as carbon-sulfur bonds by means of addition reactions. Which of these two reactions is predominant depends on temperature, sulfur content and heating rate. Generally, the dehydrogenation reactions predominate at temperatures above 150°C (Masegosa et al. 2012).

H_2S gas concentrations can gradually increase to high levels in the air voids of a loose paving mixture, during storage in silos and during truck delivery to the paving site. The "entrapped" gas is released when the air pockets in the mixture are opened up as the mixture is dumped from the delivery trucks or as the mixture is subjected to mechanical mixing. High paver screed temperatures can also cause unacceptable emissions (Saylak & Tex 1988).

The SEAM Materials Safety Data Sheet states that momentary levels of H_2S from 1 to 20 ppm have been recorded in the paver hopper as a new mix was being dumped, and at the paver screed level while it was being preheated. The same document also states that at processing temperatures below 150°C the sulfur-bitumen mix gives off very little H_2S, but that at temperatures above 150°C, the rate of H_2S evolution increases significantly and at very high temperatures emissions may not dissipate readily and are therefore considered hazardous (Rock binders Inc 2014).

In one recent investigation (Masegosa et al. 2012), blends of 40pen bitumen containing 5% and 10% sulfur by mass of bitumen were analyzed at elevated temperatures in order to determine the amount of volatile fractions evolved from the blends. The mass loss rates at 130°C and 140°C (composed of H_2S + evaporated S), averaged over the first 7 days of thermal exposure are shown in Table 2 (Masegosa et al. 2012). The results indicated that the percentage of sulfur evolved by the blends depended primarily on the storage temperature. No significant differences were found in the rate of volatiles evolved by the blends as a function of the blended sulfur content. It has been observed that the volatile compounds produced by the blends contain both sulfur gas as well as hydrogen sulfide. In this study, the authors concluded that it would be best to maintain the temperature as close as possible to 130°C during the stages of mixing, handling and storage of sulfur bitumen mixtures, in order to control emissions.

3.2 Toxicological profile of hydrogen sulfide

Hydrogen Sulfide (H_2S) is a colorless, flammable gas. H_2S is rapidly absorbed from the lungs following inhalation exposure. Absorption of the gas through the skin is minimal. Following absorption, H_2S is widely distributed around the body, primarily as undissociated H_2S or as HS—ions. H_2S binds reversibly to metalloenzymes, including those involved in aerobic cellular respiration such as cytochrome oxidase. H_2S is a metabolic poison, it blocks the electron transport chain in mitochondria, inhibiting cellular utilisation of oxygen (Costigan 2003). H_2S has a pungent odor reminiscent of rotten eggs, the rotten egg odor is recognizable up to 30 ppm. It has a sweet odor at 30 ppm to 100 ppm. The odor is detectable at

very low concentrations and the threshold for perception is between 0.02 to 0.13 ppm. Nonetheless, odor perception is unreliable as a warning of high exposures, olfactory fatigue may develop at concentrations of ≥ 100 ppm (Costigan 2003, Simonton 2007). During vigorous exercise, low level exposures (5 to 10 ppm) cause a shift to anaerobic respiration, leading to increased lactic acid formation and irritation to eyes, nose and throat, but evidence on eye irritation is inconsistent. The balance of evidence suggests that eye irritation is only likely to occur at high exposure concentrations, in region of hundreds of ppm (Costigan 2003, Simonton 2007).

In asthmatic individuals, levels as low as 2 ppm can cause bronchial constriction as well as spontaneous abortion. Between 50 to 200 ppm symptoms may include, severe respiratory tract irritation, eye irritation/acute conjunctivitis, shock, convulsions, coma and death in severe cases (Simonton 2007). Exposures for a few hours to concentrations of 500 ppm and above are likely to be fatal. Exposure for a few minutes to concentrations of 1000 ppm and above are likely to cause rapid unconsciousness and death (within minutes).

These findings may be of relevance to personnel who are continuously involved with the production and laying of SEAM mixtures for a number of years.

3.3 Other SEAM gaseous emissions to consider

Assuming a typical SEAM/Bitumen ratio of 40%/60%, with a sulfur density twice that of bitumen. For one tonne of asphalt mix manufactured at a presumed binder content of 6% (by mass of mix), the amount of sulfur encapsulated in the asphalt mix would be approximately 51 kg. In view of the substantial amounts of sulfur that can potentially be used, the authors decided to conduct a small laboratory trial to investigate emissions.

Several 10 kg samples (40%/60% SEAM/bitumen) asphalt mixes were manufactured. Different hot batches were kept in the loose state in metallic trays in laboratory ovens preset at various temperatures varying from 130°C to 210°C. The temperature of each mix and the gaseous emissions were continuously monitored for up to 6 hours. A Dräger X-am 5000 hand held gas detector with H_2S and SO_2 sensors was especially purchased for the trial. Although not an accurate or conclusive investigation by any means, nonetheless, the following trends were very prevalent:

- at 150°C and below, there was very little detectable H_2S emissions (all measured values were well below a concentration of 5 ppm). It was very clear that the H_2S suppressant was functioning

Table 2. Rate of mass loss through gaseous emissions of sulfur stored at high temperatures.

Sulfur in blend (%mass of bit.)	Storage at 130°C	Storage at 140°C
5%	0.52%/day	0.66%/day
10%	0.31%/day	0.60%/day

in line with the manufacturer's expectations. Above about 160°C, H_2S emissions can become a concern, and as expected, higher temperatures further intensify emissions.

• in the case of SO_2 emissions, things were quite different. At 150°C, concentrations were certainly higher than 10 ppm, with some readings even approaching 20 ppm.

The aforementioned results prompted the authors to investigate further the effects of SO_2 on human health and acceptable exposure limits. While health effects are documented at various concentrations by different researchers and organizations, a sampling of thresholds for health effects is provided here. Adverse effect on pulmonary function by an exposure at low concentrations (0.75 ppm) is reversible. At higher concentrations, airways can become severely obstructed and the effects can involve the lower airways. At 5 ppm, dryness of the nose and throat can be observed and resistance to bronchial airflow significantly increases. At 6–8 ppm tidal respiratory volume may noticeably decrease. At 10 ppm, sneezing, coughing, and wheezing may be observed, possibly accompanied by eye, nose and throat irritations. Nosebleeds may also be seen. At this level, asthmatics are likely to experience asthmatic paroxysm, lasting possibly for several days. At 20 ppm, bronchospasms tend to begin and eye irritation is very likely. At 50 ppm discomfort becomes extreme, but permanent injury is unlikely if exposure is less than 30 minutes duration. Above 50 ppm, reflex closure of the glottis can take place and last for a period of minutes. Exposure to sulfur dioxide at a concentration of 400 ppm will likely constitute an immediate danger to life (Miller 2004).

4 CONCLUSIONS

Sulfur Extended Asphalt Modifier (SEAM) in solid pellet form (primarily composed of sulfur that is plasticized by the addition of carbon black and amyl acetate) was marketed as an additive for hot mix asphalts. A typical SEAM/bitumen ratio recommended for most mixes is 40%/60% (sulfur/bitumen) by mass of total binder.

SEAM may be used with any conventional bitumen grade, but not polymer-modified bitumens and the use of free lime is not recommended. Conventional mix designs and production techniques are entirely suitable for SEAM mixes.

A substantial number of laboratory and full scale trials using SEAM have been reported in the literature, and except for some minor concerns regarding rutting and "more than minor" concerns with respect to moisture resistance, it appears that overall the SEAM modified mixes have the potential for improved mechanical performance (stability, stiffness, cohesion, surface abrasion) compared to asphalts manufactured with conventional bitumens.

When sufur and bitumen are heated and combined the sulfur can react chemically with the bitumen causing dehydrogenation which results in hydrogen sulphide gas being formed, causing also significant changes in the rheological properties of the bitumen.

Research on the reactions of elemental sulfur with various bitumens and residues has confirmed that H_2S evolution starts at temperatures as low as about 130–135°C. However, the activation energy and the reaction order are different for various types of bitumens. These findings may be of relevance to personnel who are continuously involved with the production and laying of SEAM mixtures for a number of years.

In the UK, for instance, control of inhalation exposure to H_2S in the workplace must comply with occupational exposure standards of 5 ppm (8 hour timed weighted average) and 10 ppm (short term exposure limit). With respect to SO_2 emissions, the US national ambient air quality standards for SO_2 recommends that repeated exposures to 5-minute peak SO_2 levels (≥ 0.60 ppm) could pose an immediate significant health risk for a substantial proportion of asthmatic individuals at elevated ventilation rates.

The aim of this paper was to highlight various examples of short and long term aspects that highway materials engineers must consider when assessing the viability of encapsulating waste materials, including SEAM or other sulfur based composites, in roads. We therefore all have a duty to pass on to our future generations road networks that also function as potential "linear quarries" and not as "linear waste dumps". The decision making process must weigh up all the short and long term factors on a case by case basis and should principally be based on sound engineering principles.

REFERENCES

Al-Ansary, M. 2010. Innovative solutions for sulphur in Qatar. *The Sulphur Institute's (TSI) Sulphur World Symposium*, Doha, Qatar.

Al-Mehthel, M., Al-Abdul, W., Al-Idi, S. & Baig M.G. 2010. Sulfur extended asphalt as a major outlet for sulfur that outperformed other asphalt mixes in the Gulf. *Sulphur Institute's (TSI) Sulphur World Symposium*, Doha, Qatar.

Bailey, W.R., Rock Binders Inc. & Shell Canada Energy. 2013. *International Patent Application no. WO 03/014231 A1, PCT/US02/25333*, Sulfur additives for paving binders and manufacturing methods.

Beaudoin, J.J. & Sereda, J. 1979. A two-continuous-phase sulfur-asphalt composite-development and characterization. *CAN. J. CIV. ENG.* Vol. 6, pp. 406–412.

Cocurullo, A., Grenfell, J., Yusoff, N.I. and Airey, G.D. 2012. Effect of moisture conditioning on fatigue properties of sulfur modified asphalt mixtures. *7th Rilem International Conference on Cracking in Pavements*, Vol. 2, Edited by Scarpas A. et. al., pp. 793–803.

Costigan, M.G. 2003. Hydrogen Sulfide: UK Occupational Exposure Limits. *Occup. Environ Med 2003*. 60: pp. 308–312.

FHWA-HIF-12–087 Tech Brief. 2012. An Alternative Asphalt Binder, Sulfur-Extended Asphalt (SEA), *US Dept. of Transportation, Federal Highway Administration*, Office of Pavement Technology.

Green, J., Yu, S., Pearson, C. & Reynolds, J. 1993. Analysis of Sulfur Compound Types in Asphalt. Strategic *Highway Research Program, SHRP-A-667*, Sept. 1993.

Johnston, A.G., Yeung, K. & Tannahill D. 2005. Use of Asphalt Pavement Analyzer Testing for Evaluating Premium Surfacing Asphalt Mixtures for Urban Roadways. *Annual Conference of the Transportation Association of Canada*, Calgary, Alberta, pp. 1–16.

Mahoney, J.P., Lary, J.A., Balgunaim, F. & Lee, T.C. 1982. Sulfur Extended Asphalt Laboratory Investigation, Mixture Characterization. *Washington State Department of Transportation*, Final Report WA-RD 53.2.

Masegosa, R.M., Canamero, P., Cabezudo, M.S., Vinas, T., Salom, C., Prolongo, M.G., Paez, M. Ayala. 2012. Thermal behaviour of bitumen modified by sulphur addition. *5th Eurasphalt Eurobitume Congress*, 13–15th June 2012, Istanbul.

Miller, V. 2004. Health Effects of Project SHAD Chemical Agent: Sulfur Dioxide, Prepared for the National Academies by The Center for Research Information, Inc., *Contract No. IOM-2794-04-001*, pp. 1–58.

Nazarbeygi, A.E. & Moeini, A.R. 2012. Sulfur Extended Asphalt Investigation, Laboratory and Field Trial. *5th Eurasphalt & Eurobitume Congress*, 13–15th June 2012, Istanbul.

Nicholls, J.C. 2012. Sulphur Extended Asphalt Modifier. *5th Eurasphalt & Eurobitume Congress*, 13–15th June 2012, Istanbul.

Predoehl, N.H. 1989. An evaluation of sulfur extended asphalt (SEA) pavements in cold and hot climates. Office of Transportation Laboratory, State of California Dept. of Transportation, *Final Report No. FHWA/CA/TL-89/01*.

Rock Binders Inc., Material safety data sheet *http//wrbailey.com/rockbinders/index.html*. Accessed May 2014.

Saylak, D. & Tex B. 1988. Sulfur-Coated Asphalt Pellets. *US Patent No. 4,769,288*, Sept. 6.

Shell Thiopave™. *www.shell.com/sulphur/thiopave*. Accessed May 2014.

Simonton, S., 2007. Human Health Effects from Exposure to Low-Level Concentrations of Hydrogen Sulfide. article published online in the Oct. 2007 issue of *Occupational Health & Safety*. http://ohsonline.com/home.aspx

Stuart, K.D. 1990. Performance evaluation of sulfur-extended asphalt pavements—laboratory evaluation. Office of Engineering and Highway Operations R & D, *US Dept. of Transportation, Federal Highway Administration*, Publication No. FHWA-RD-90–110.

Tran, N., Taylor, A., Timm, D., Robbins, M., Powell, B. & Dongre, R.. 2010. Evaluation of Mixture Performance and Structural Capacity of Pavements Using Shell Thiopave. *NCAT Report 10–05*.

Transport Infrastructure and Systems – Dell'Acqua & Wegman (Eds)
© *2017 Taylor & Francis Group, London, ISBN 978-1-138-03009-1*

In-plant production of warm recycled mixtures produced with SBS modified bitumen: A case study

A. Stimilli, F. Frigio & F. Canestrari
Università Politecnica delle Marche, Ancona (AN), Italy

S. Sciolette
Laboratorio Dottor Sciolette, Arriccia (RM), Italy

ABSTRACT: Asphalt mixtures produced at reduced temperatures through Warm Mix Additives (WMA) allow reductions of fuel consumption and harmful emissions, ensuring economic and environmental benefits. Considering that nowadays mixtures include more and more often reclaimed aggregates, the combination of recycling and WMA technologies represents a major challenge in road construction and needs further investigations for identifying drawbacks/advantages. Several concerns derive from regular in-plant productions since warm recycled mixtures have been mainly optimized through laboratory studies without evaluating possible issues of large-scale in-plant productions and lay-down processes. So far, few field constructions were realized, limited to small trial sections and without considering the use of modified bitumens. Given this background, the paper describes in-plant productions of warm recycled mixtures prepared with three chemical WMA additives for the construction of an extensive motorway segment. Dense-graded mixtures for binder and base courses as well as open-graded mixtures for wearing courses were produced. Three full-scale trial sections included warm mixtures whereas a further section, used as reference for comparison purposes, comprised analogous mixtures realized through hot recycling according to the current practice. The paper describes the construction steps and the controls carried out to verify technical standard requirements in terms of volumetric properties, compactability and Indirect Tensile Strength. The main objective was to attest the feasibility of large-scale productions to adequately reproduce the mix design previously implemented through laboratory studies when WMA technologies and recycling techniques are concurrently involved. Moreover, gas emissions monitoring at the asphalt plant during the real scale productions was conducted through Continuous Emissions Monitoring Systems in order to quantify the potential benefits in terms of pollution. Results demonstrate the suitability of WMA chemical additives to produce at low temperature mixtures with adequate performance, concurrently recording a significant reduction in pollutants without needing mix design modifications or implementation of expensive new technologies.

1 INTRODUCTION

1.1 *Warm mix asphalt*

Technologies to produce Warm Mix Asphalt (WMA) are gaining growing interest in the field of road construction due to the number of economic and environmental advantages deriving from the significant reduction in the production temperatures compared to traditional Hot Mix Asphalts (HMA) (D'Angelo et al. 2008, Frank & Prowell 2014). Nowadays, several WMA additives based on different operational mechanisms are available on the market (i.e. organic waxes, chemical additives, water-based products) (Rubio et al. 2012). Due to the recent implementation of such technologies, over the last years their effectiveness has been mainly investigated through laboratory studies

rather than field experimental projects (D'Angelo et al. 2008). The consequent lack of in situ experience determines low reliability on these production techniques and risk to compromise the complete exploitation of WMA technologies potential. This matter is even more accentuated by the more and more urgent need of combining warm technologies and recycling techniques, so dealing with warm mixtures incorporating Reclaimed Asphalt Pavement (RAP) material. Such a challenge introduces further variables that could determine significant variations on mixtures behavior (Dinis-Almeida & Lopes Afonso 2015, Dinis-Almeida et al. 2016, Lee et al. 2009). Additional concerns are related to the use of Polymer Modified Bitumens (PMB), whose use is currently increasing worldwide for the production of road mixtures. Polymers significantly improves bitumen properties, but usually require higher working temperatures to gain acceptable mixture workability. Therefore, the use of warm technologies could result not suitable when combined with PMBs. This aspect further complicates the overall analysis and WMA technology applicability. In this context, full-scale trial sections are essential to provide reliable performance data and appropriately assess warm mixtures suitability, concurrently optimizing possible concerns related to the operations of the asphalt plant during production and potential in-situ problems during lay-down.

1.2 Full-scale trial project

The research project presented in this paper, realized in April 2016, consists in the construction of four full-scale trial sections located along a stretch of an Italian motorway pavement (800 m along the motorway A1, segment Orte—Fiano Romano, sud direction, km 513+000 – km 513+800). The A1 motorway is a double carriageway high traffic volume road with three lanes per direction. The project involved the re-construction of the slow lane after milling of all existing bituminous courses. Three sections (each 200 m long) were reconstructed with bituminous pavement courses all realized with recycled warm mixtures, adopting in each case a different WMA additive. All the selected WMA products were chemical additives since, based on previous laboratory studies (Frigio & Canestrari 2016; Frigio et al. 2016), this kind of WMA technology was found the most promising in terms of performance when dealing with modified bitumens and RAP aggregates (for both open and dense graded mixtures). A further section was built with traditional HMAs and used as reference for comparison purposes. A Styrene-Butadiene-Styrene (SBS) polymer modified bitumen was used for preparing all produced mixtures. Trial sections were constructed along a continuous stretch of the motorway in order to have the same boundary conditions in terms of existing materials, courses thickness, road geometry (straight flat stretch), traffic volume and weather conditions.

1.3 Objectives

The main objective of the study was to evaluate the feasibility of WMA technologies for producing recycled bituminous mixtures prepared with a PMB and the related environmental consequences. Three different chemical WMA additives were selected in order to evaluate the effectiveness of different chemical compositions and operational mechanisms on mixture workability, volumetric properties and main mechanical performance.

Moreover, the study allowed the evaluation of possible issues related to the asphalt plant operation when producing at reduced temperatures in large scale amounts, as well as advantages and drawbacks during lay-down and compaction. Finally, thanks to the large quantity of material involved for the construction of the trial sections, the asphalt plant worked at reduced temperatures for long operational windows, so giving the opportunity to have appropriate durations for adequately monitoring pollutant emissions, key aspect to define the overall convenience of WMA technologies.

2 EXPERIMENTAL PROGRAM

2.1 Materials

In each trial section, the pilot project involved the construction of three bituminous courses prepared as follows:

- wearing course: open graded mixture (4 cm thick) with 15% by aggregate weight of selected RAP (fraction 8/16 mm, deriving only from old motorway porous courses), virgin basalt aggregate fractions combined with 70% cellulose-30% glass fibers (dosed at 0.3% by aggregates weight, added to prevent draindown issues) and 5.25% by aggregate weight of total bitumen content (virgin bitumen + reclaimed bitumen from RAP);
- binder course: dense graded mixture (10 cm thick) with 25% by aggregate weight of RAP (RAP 0/16 mm, unfractioned, deriving from old binder and base motorway courses), virgin limestone aggregate fractions and 4.8% by aggregate weight of total bitumen content;
- base course: dense graded mixture (15 cm thick) with 30% of RAP by aggregate weight (RAP 0/16 mm, unfractioned, deriving from old binder and base motorway courses) virgin limestone aggregate fractions and 4.5% by aggregate weight of total bitumen content.

For all courses, the virgin bitumen used was a SBS polymer modified bitumen (i.e. 3.8% of SBS by bitumen weight), the same typically employed for maintenance and construction activities in the Italian motorway network. Likewise, the aggregate grading curve (as well as the RAP percentage) was equal to the one usually adopted at the asphalt plant for each corresponding course in case of HMA productions. For determining the right amount of virgin bitumen content, RAP fractions were previously subjected to bitumen extraction. In each trial section it was used a different type of chemical WMA additive and, within the same section all bituminous courses were prepared with the same WMA product and at equal working temperatures (i.e. 130° and 120°C in case of warm mixtures, 170°C and 160°C in case of hot mixtures, for mixing and compaction, respectively). Each additive was dosed in accordance with the range prescribed by the producer, in quantity different for each course (Table 1) depending on the "working" bitumen content (virgin bitumen + reactivated RAP bitumen) in order to consider the different amount of RAP aggregates. This implies that the exact quantity of each WMA additive was determined taking into account a certain degree of RAP bitumen reactivation that, based on previous specific studies (Stimilli et al. 2015), can be assumed equal to 70% in case of HMAs. Due to the reduced working temperatures, this value was precautionary considered lower for WMA mixtures and equal to 60%.

WMA products were selected with different chemical composition and operational mechanisms. In particular, one trial section (section 1) was realized with an additive (coded as C1) mainly composed of ammine substances which act as surfactants and adhesion enhancers. It is a viscous liquid at 25°C with a density of about 1.0 g/cm³, a pour point of about −8°C and a flash point higher that 140°C.

In the second section, a chemical additive (coded as C2) containing alkylates and fatty acids, which act as viscous regulators, was added to asphalt mixtures. At ambient temperature, it is an amber-colored, inodorous liquid, insoluble in water. It is characterized by a density of 0.86–0.90 g/cm³ at 20°C, a flash point higher than 220°C and a boiling point between 300°C and 408°C.

The third section was realized with a chemical additive (coded as C3) that, analogously to C2 (even if characterized by a different chemical composition), acts as viscous regulator. C3 is liquid at ambient temperature with a density of about

1.00 g/cm³ at 20°C and a flash point higher than 200°C.

In each case, the additive was added to the virgin bitumen by means of a volumetric pump (equipped with a timer and an electronic control) connected with an external tank where the WMA additive is stored and that allowed to supply in continuous the bitumen with the additive according to the predetermined delivery capacity. Specialized technicians carried out the calibration of the pump few days before warm mixtures productions. The WMA agent injection happened right before adding each material component (i.e. virgin bitumen, virgin and RAP aggregates) to the mixing chamber. In this way, veery limited modifications to the asphalt plant were applied for producing at reduced temperatures without compromising the possibility to easily switch to standard HMA productions. Moreover, since WMA additives are characterized by a rapid decay curve, in order to fully exploit WMA additives potential, it is preferable to use the bitumen immediately after the addition of the chemical agent without insertion of rest times, as it happened in the present investigation. Apart the addition of WMA additives, the operational steps followed at the asphalt plant for producing warm mixtures were the same usually adopted for HMAs.

2.1 Working program

The construction activities lasted four days as following described. In the first working day, the original pavement was milled for a total depth of 34 cm. The milling activities, starting from section 1, were carried out in two subsequent steps in order to keep separate the RAP aggregate produced by milling the porous surface course from RAP aggregates of binder and base courses. Concurrently, the warm mixture with additive C1 for the base course (B_WC1) was prepared at the asphalt plant. After completion of the milling activities in the first section, the base mixture was laid-down and compacted. Right after, the base course of the second section (B_WC2), whose mixture was meanwhile produced at the asphalt plant (with the additive C2) after mixture B_WC1, was constructed.

The second working day started with the production and in situ application of the base course for section 3 (B_WC3) and the binder course for section 1 (BIN_WC1), both prepared at reduced temperatures. Afterwards, the asphalt plant was switched to produce the base course of section 4 (B_H prepared at standard temperatures). The latter was produced as last mixture of the day in order to optimize the working timetable and avoid long waiting periods needed to reach the desired mixing and compaction temperature when switching from

Table 1. Main mixture characteristics per each course.

Course	RAP [% by agg. weight]	Total bitumen [% by agg. weight]	WMA additive [% by virgin bitumen weight]
WEARING (OG: 4 cm)	15 (RAP 8–16 mm)	5.25	C1 => 0.42 C2 => 0.70 C3 => 0.45
BINDER (DG: 10 cm)	25 (RAP 0–16 mm)	4.80	C1 => 0.50 C2 => 0.80 C3 => 0.50
BASE (DG: 15 cm)	30 (RAP 0–16 mm)	4.50	C1 => 0.55 C2 => 0.90 C3 => 0.55

hot to warm productions. In fact, rising the asphalt plant operational temperature is much faster than the opposite. Moreover, in such way the warm production was continuous with a long duration that permitted the monitoring of pollutant emissions for a significant period. Concurrently, the asphalt plant had enough time to reach a full capacity at reduced temperature, so allowing the observation of potential critical issues in the mid-long term operation. The same plan was carried out during the third working day when mixtures for the binder course were completed in the following order: warm binder for section 2 (BIN_WC2), warm binder for section 3 (BIN_WC3) and hot binder for section 4 (BIN_H).

The last working day was dedicated to the production and subsequent construction of all wearing courses, starting from section 1 (OG_WC1) and finishing with section 4 (OG_H) in order to have a continuous warm production with all the advantages previously described.

2.2 Mixtures evaluations

After in-plant production, a certain amount of each asphalt mixture was collected and brought to the laboratory located right near the plant. By means of the Superpave Gyratory Compactor (SGC) set up with standard compaction parameters (vertical pressure 600 kPa; rotation speed: 30 revolutions/min; vertical mold inclination: 1.25°), this material was used to compact cylindrical specimens useful to check volumetric (i.e. air voids, Compaction Energy Index) and mechanical properties (i.e. Indirect Tensile Strength) of the corresponding mixture, in accordance with technical standard acceptance requirements. A minimum of four samples was prepared per each material (diameter = 100 mm, height = 63.5 mm). The number of gyrations was selected based on technical standards specifications for motorway pavements (i.e. 130 and 200 for OG and DG mixtures, respectively). It is worth noting that prior to compacting, each mixture was kept for one hour in a force-draft oven at a temperature equal to the in-situ compaction temperature (i.e. 120°C for WMA mixtures, 160°C for HMA mixtures) in order to properly simulate the aging effects underwent during lay-down and transportation. Concurrently, each mixture was subjected to solvent extraction in order to determine the actual bitumen content and the aggregate gradation. This evaluation was useful to verify weather in-plant produced warm mixtures complied with the original laboratory mix design or the reduced working temperatures caused problems at the asphalt plant in terms of material composition with consequences on volumetric and mechanical properties. Moreover, in situ visual inspections during and after lay-down

and compaction allowed the evaluation of in situ mixture workability to confirm what observed through the laboratory evaluation.

2.3 Emissions monitoring

Emissions of pollutants were monitored during in plant productions in order to evaluate the effective environmental benefits deriving from lower working temperatures. In compliance with current regulations, a specialized staff carried out the monitoring of gases and dusts emissions through direct in plant detections. An in-situ Continuous Emissions Monitoring System (CEMS) was adopted as a tool to monitor flue gases (Harrison & Kemmer 1996). The system consists of a sample probe, a filter, a sample line, a gas conditioning system, a calibration gas system, and a series of gas analyzers which reflect the parameters being monitored. Emissions were measured using a pollutant concentration monitor apparatus. A "hot dry" extractive method was adopted: the sample is carried along a sample line into a sample conditioning unit (without dilution) where it is dried and filtered to remove particulate matter and moisture (also the latter was measured through a specific control unit). Once conditioned, the sample enters a sampling manifold. Here, individual gas analyzers extract each investigated component and measure the corresponding concentration by means of various techniques (e.g. infrared and ultraviolet adsorption, chemiluminescence, fluorescence, chromatography and spectrography). Through a Data Acquisition and Handling System (DAHS), the signal output from each gas analyzer is recorded and automatically collected into an emissions database. One advantage of this method, compared to dilution extractive methods, is the ability to measure the percent amount of oxygen in the sample, which is often required in the regulatory calculations for emission corrections. Measurements were performed both during warm and hot productions to obtain a reliable comparison at equal boundary conditions (e.g. wind intensity and direction, air temperature, humidity, asphalt plant configuration) so avoiding the reference to historical data of generic HMA productions and effectively isolating the impact of working temperatures reduction. With the aim to obtain reliable data, the monitoring of emissions was extended for several hours, so guaranteeing that the asphalt plant was working at full capacity. Totally, four detections were realized, one per each type of WMA additive and one for the reference HMA mixture. In each case, the measurement point was the same and was chosen to detect the most significant emission rate (measurement point set in correspondence of the stack connected with the aggregate drying drum). Table 2 lists all fluid dynamic and chemical parameters monitored,

Table 2. Fluid dynamic and chemical parameters for monitoring pollutant emissions.

Parameters	Standard
Flow rate	ISO 16911:2013
Temperature	–
Humidity	–
Dusts	UNI EN 13284
Oxygen	Italian DM 25.8.2000
Carbon monoxide (CO)	Italian DM 25.8.2000
Nitrogen oxide (NO_x)	Italian DM 25.8.2000
Sulfur oxide (SO_x)	Italian DM 25.8.2000
Volatile Organic Compound (VOCs)	UNI EN 12619
Polycy. Aromatic Hydrocar. (PAHs)	Italian DM 25.8.2000

specifying the corresponding technical standard used to carry out the measurement.

3 RESULTS

3.1 Bitumen percentage and aggregate gradation

Table 3 summarizes the results for each mixture in terms of total bitumen content (Bt), expressed in percent by aggregate weight, and aggregate gradation, both determined after solvent extraction.

Regardless of the bituminous course considered, warm mixtures were characterized by a total bitumen content close to the design value. Also in terms of gradations, WMA mixtures were able to achieve the desired aggregate skeleton, without significant differences compared to the reference HMA and in total agreement with the grading envelope prescribed by the technical standards. The limited variability detected is mainly attributed to the production process and to the inherent heterogeneity provided by RAP aggregates. Such results suggest that in plant productions of WMA mixtures in large-scale amounts are feasible and reduced temperatures do not alter the material composition of the resulting mixture even without needing of significant modifications to the asphalt plant. Also, they validate the hypotheses assumed for calculating the total bitumen content in case of reduced temperatures (i.e. lower RAP bitumen reactivation). The similarity in terms of material composition among mixtures excludes the influence of further parameters except reduced temperatures on mixture behavior, so allowing a significant performance comparison.

3.2 Compactability and volumetric properties

One key aspect to judge the suitability of a warm mixture is surely its workability and the conse-quent volumetric properties. Although reduced production temperatures, the mixture must guarantee good compactability and an air void content within the limit range prescribed by the technical standards. In this sense, the Compaction Energy Index (CEI), calculated for each investigated mixture using SGC data recorded during laboratory specimens compaction, can provide an idea of the compaction efforts needed to achieve a certain target density (Mahmoud & Bahia 2004). For dense-graded mixtures, CEI was determined as area under the densification curve from the 8th gyration (considered to simulate the effort applied by the paver) to the 92% density; for open-graded mixtures, CEI was calculated as the area under the densification curve from the 8th gyrations until the final gyration. Concurrently, the air voids content (V_m) was determined for each specimen based on the maximum (EN 12697–05) and the bulk density of each material (EN 12697–06). Results are summarized in Figure 1 as average of four replicates.

In terms of compactability, mixtures prepared with WMA additives generally achieved good performance in line with the reference HMA mixture, so demonstrating that all WMA products were able to overcome the higher bitumen viscosity caused by reduced production temperatures.

In particular, the additive C2 appears the one able to guarantee the highest compactability aptitude, even higher than the reference material in the case of open graded mixtures. Such an outcome is consistent with the operational mechanism of this product. As viscous regulator, it should be able to modify bitumen viscosity, so allowing acceptable workability at reduced production temperatures.

The same principle should characterize the additive C3. However, mixtures prepared with the latter always achieved higher CEI values compared to the other materials, proving the important influence of the additive chemical structure.

The additive C1 exhibited very good performance in case of dense-graded mixtures, whereas it showed lower potential when added to porous asphalts. The different behavior dependent on the type of pavement course is reasonable considering that the product C1 mainly acts as surfactant: open-graded mixtures have a discontinuous aggregate skeleton with high air voids percentage. Therefore, the internal friction during lay-down and compaction is much higher and the final aggregate skeleton configuration is achieved earlier, so preventing the additive to express its entire potential in terms of workability. However, the contact between aggregate particles happens in few specific points. Considering that the product C1 is also composed by adhesion enhancers, it could result particularly efficient in terms of performance providing higher resistance

Table 3. Aggregate gradation and bitumen content.

Sieve [mm]	WEARING course						
	WC1	WC2	WC3	H	Design	Envelope	
20	100	100	100	100	100	100	100
14	87.6	89.8	88.1	89.0	90.3	85	94
10	46.9	48.4	50.3	52.0	45.8	38	53
6.3	24.3	26.2	25.0	25.2	24.9	22	32
2	15.2	16.4	15.8	16.1	16.0	14	22
0.5	9.8	10.7	10.1	10.7	9.9	9	15
0.25	7.7	9.2	8.8	8.2	7.7	7	12
0.063	5.8	6.1	7.2	6.8	4.6	4	8
Bt [% by agg. wt]	5.00	5.23	5.14	5.10	5.25	–	

Sieve [mm]	BINDER course						
	WC1	WC2	WC3	H	Design	Envelope	
20	95.8	97.1	94.9	95.4	91.2	85	98
14	82.7	84.0	78.4	80.9	78.5	70	87
10	66.0	62.2	64.2	61.6	61.4	58	78
6.3	49.3	47.3	48.9	47.4	48.2	46	66
2	27.8	29.9	28.5	29.8	28.9	25	38
0.5	12.0	12.2	13.8	15.4	14.1	11	21
0.25	8.2	8.5	9.3	10.3	9.1	7	17
0.063	6.4	6.0	6.2	5.8	4.7	4	8
Bt [% by agg. wt]	5.02	4.93	4.85	4.79	4.80	–	

Sieve [mm]	BASE course						
	WC1	WC2	WC3	H	Design	Envelope	
20	93.8	92.3	93.9	92.5	83	73	94
14	76.3	73.6	75.6	76.6	65.4	51	76
10	57.1	58.6	57.6	55.3	56.2	40	64
6.3	38.5	43.8	42.0	39.9	47.4	31	55
2	26.2	26.3	26.3	24.3	23.7	19	38
0.5	13.8	10.6	12.7	9.9	13.4	8	21
0.25	9.6	6.6	9.1	6.9	8.7	5	16
0.063	5.8	5.0	5.2	4.0	4.4	4	8
Bt [% by agg. wt]	4.47	4.68	4.45	4.15	4.5	–	

and balancing the lower workability effects. In fact, an improved adhesion in the very limited contact points is a key aspect for open-graded mixtures, especially when subjected to the action of water.

As far as the air voids content is concerned, it is possible to notice that all open-graded and binder mixtures were characterized by values within the limit range prescribed by the technical standards (i.e. open-graded mixture: $V_m \geq 14\%$ at 130 SGC gyrations; binder: $V_m \geq 2\%$ at 200 SGC gyrations). Moreover, generally the air voids trend was consistent with the CEI values trend: the higher the CEI, the higher the air voids.

On the contrary, for the base course, all WMA mixtures exhibited air voids contents lower than the limit prescribed by the standards (i.e. $V_m \geq 2\%$ at 200 SGC gyrations) differently to the HMA mixture that had an air voids content in line with what required. Considering that all base mixtures were characterized by similar aggregate gradations, such a result suggests that the WMA additive amount considered for designing the base course could be reduced. This overestimation can be related to a lower RAP bitumen reactivation than what theoretically hypothesized for the calculation (i.e. 60%). Probably, due to the higher percentage of RAP in the base course (i.e. 30%) compared to the other bituminous courses, the reclaimed bitumen release at reduced working temperatures is even lower than 60%. Since WMA additives were dosed based on the hypothetical effective

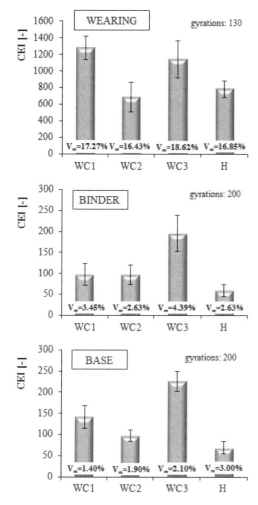

Figure 1. CEI and air voids.

needed for evaluating the acceptability of a mixture was based on the Indirect Tensile Strength (ITS) calculated as following specified:

$$ITS = \frac{2 \cdot P_{max}}{\pi \cdot t \cdot d} \qquad [GPa] \qquad (1)$$

where P_{max} = maximum load in kN; t = specimen height in mm; d = specimen diameter in mm.

Results are summarized in Figure 2 (as average of four replicates for all the investigated mixtures) along with error bars expressed in terms of standard deviation. All specimens were conditioned in air (dry conditioning) at the test temperature of 25°C for a minimum of four hours prior to testing. With the aim to verify the water susceptibility of the wearing course (crucial aspect for open-graded mixtures due to the high air voids content), the ITS test was performed on both dry (ITS_dry) and wet (ITS_wet) specimens. Wet specimens were conditioned in water at 40°C for a period of 72 hours prior to dry conditioning and testing (according to EN 12697-12 Method A). The ITS ratio (ITSR) between wet and dry specimens was calculated to quantify the effect of water in terms of performance (Fig. 2).

Regarding open-graded mixtures, all materials exhibited lower ITS values compared to the refer-

bitumen acting in the mixture (i.e. 60%), the additive effects in terms of workability were too high. Bearing in mind this aspect for future investigations, it is worth noting that the lower air voids of WMA mixtures could be partially attributed also to the slightly higher bitumen content detected with respect to the reference mixture (see Table 3).

In all cases, the in-situ visual inspection during lay-down and compaction confirmed the good results above described in terms of workability and volumetrics. No specific issues were detected by the operators during construction with respect to the reference HMA mixture.

3.3 Indirect tensile strength

In accordance with technical standards requirements, the preliminary performance analysis

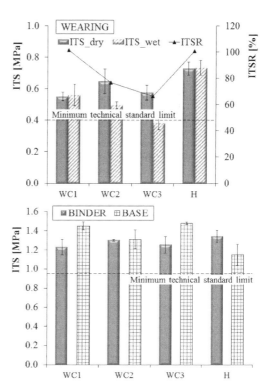

Figure 2. ITS results.

ence HMA, both in dry and wet conditions. However, data recorded in dry conditions were always higher than the limit prescribed by the technical standards regardless of the type of WMA additive considered. In wet conditions, additives C2 and C3 determined a significant loss in performance compared with the reference mixture as confirmed by the ITSR trend. On the contrary, the additive C1 was able to guarantee an optimum resistance also in wet conditions, demonstrating an ITSR ratio even higher than the reference material. Such consideration confirms what previously anticipated about the potential benefits deriving from the presence of adhesion enhancers within the chemical structure of the additive C1 that helps to improve the resistance developed at the few contact points between aggregate particles, so enhancing overall mixture performance especially against water action.

Regarding dense-graded mixtures (both binder and base course), no significant differences were detected with respect to the HMA material. WMA mixtures demonstrated ITS values equal or even higher than the reference ones, regardless of the additive type considered. This indicates that reduced production temperatures did not negatively affect mixture performance. It is worth noting that also for the base course, although the potential overestimation of WMA additive amounts previously discussed, WMA mixtures were characterized by ITS values generally higher than the reference mixture.

In conclusion, large-scale productions of warm mixtures are possible even at low production temperatures without need of modifying the production process, the asphalt plant assembly or the aggregate mix design. However, attention must be paid to the additive amount and the effective bitumen content, especially in presence of high percentage of RAP aggregates. Moreover, the selection of a WMA chemical additive with a chemical composition able to promote bitumen-aggregate adhesion appears fundamental in case of open-graded mixtures for developing good water resistance.

3.4 Emissions

As discussed in paragraph §2.4, concurrently to the assessment of mechanical/volumetric properties and the verification of the working process feasibility of the asphalt plant, the environmental impact of productions at low temperatures was also evaluated.

Table 4 summarizes the values obtained for those parameters considered more relevant in terms of environmental impact for both the reference HMA mixture and WMA mixtures (WMA results are expressed as the average of three independent detections). All data are quantified in terms of concentration of each substance identified within the investigated gas medium and are calculated in relation to the reference oxygen concentration established by the technical standards. Therefore, all measured values were corrected as follows:

$$E = \frac{21-O}{21-OM} \cdot EM \qquad (2)$$

where E = corrected concentration; EM = measured concentration; O = reference oxygen concentration; OM = measured oxygen concentration.

Table 4 also indicates the threshold acceptance values specified for each substance by the current regulations.

Although all emissions data were within the threshold limit regardless of the mixture production temperatures considered, it is worth noting that the average values measured for warm productions were generally lower than those recorded for HMA mixtures. In details, the following remarks can be drawn:

– a significant reduction (around 18%) in the nitrogen oxide (NO_x) concentration was detected in case of WMA productions (i.e. 52.4 mg/Nmc) with respect to HMA productions (i.e. 61.8 mg/Nmc). This result is direct consequence of the lower temperature measured at the fumes stack of the asphalt plant for WMA productions than for HMA mixtures (temperature reduction around 14 °C). Such a finding confirms that, as expected, a direct correspondance exists between NO_x concentration and plant operational temperatures.
– Also the sulfur oxide (SO_x) amount was lower for WMA mixtures than for the reference HMA. Although SO_x emissions depend on both the content of the volatile sulfur in the raw materials and the type of fuel used (equal for all mixtures in the present study), the temperature achieved during the combustion process is, also in this case, the main parameter to control SO_x emissions.

Table 4. Emissions data in terms of corrected concentration E.

Parameters	Standard limit [mg/Nmc]	Production temperatures	
		WMA	HMA
Dusts	20	11.5	11.8
CO		1081.0	1021.0
NO_x	500	52.4	61.8
SO_x	1700	113.2	129.0
VOC	–	8.4	10.4
IPA	0.1	< 0.067	< 0.067
Temperature at the stack [°C]		92.7	107.0

- As far as the VOC concentration is concerned, the same improving trend demonstrated by warm productions compared with hot productions in terms of NO_x and SO_x can be observed, corresponding to a COV concentration reduction around 23% (i.e. 8.43 mg/Nmc Vs 10.4 mg/Nmc for WMA and HMA mixtures, respectively).
- No significant changes were detected in CO and IPA concentrations between WMA and HMA productions.

The above-described data demonstrate that in-plant productions at reduced working temperatures can provide relevant benefits in terms of environmental impact, allowing a significant reduction in those substances considered more dangerous for air pollution. It is also worth noting that the monitoring was carried out only at the asphalt plant during production and did not involve lay-down and compaction phases, aspect that could further enlarge environmental benefits, with particular emphasis for the impact on operators' health.

4 CONCLUSIONS

The present paper describes a project concerning the in-plant production of warm recycled mixtures through WMA chemical additives. Volumetric and compactability properties as well as main mechanical performance parameters were assessed in accordance with the acceptance requirements prescribed by the technical standards. Laboratory results demonstrate the feasibility of large-scale productions when recycling techniques and WMA technologies are combined without any need of modifications to the asphalt plant or to the production process. No alterations in the mixture composition (i.e. aggregate gradation and bitumen content) were caused by lower production temperatures. Optimum performance were achieved by WMA mixtures, with minimum gap compared with the reference HMA both in terms of volumetrics (i.e. air voids), compactability (i.e. CEI) and Indirect Tensile Strength. Results also highlight the importance of the WMA additive chemical structure to balance the lower production temperatures. Particularly, for open-graded mixtures the use of an additive composed of surfactants and adhesion enhancers appears the most suitable solution to guarantee low water susceptibility. Attention must be paid to the additive amount considering the effective RAP bitumen reactivation degree. Specific study are recommended to investigate this aspect. Finally, the monitoring of pollutant emissions allowed an overall evaluation of WMA mixtures benefits. The findings in terms of pollutants concentration demonstrated that a tempera-

ture reduction up to 40 °C is able to successfully induce significant emissions reductions avoiding the expensive implementation of specific solutions (e.g. asphalt plant fairing) often required to obtain acceptable emissions.

In a context of more and more urgent environmental needs, the overall findings obtained through this experimental investigation represent an important reference to promote the use of WMA technologies for the production of new bituminous mixtures in combination with the use of reclaimed aggregates and modified bitumens. Future works will involve also the monitoring of emissions during lay-down and compaction to quantify the overall environmental impact of WMA mixtures. Moreover, the trial sections will be continuously monitored to check the performance evolution of each section and obtain a reliable evaluation of long-term performance.

REFERENCES

D'Angelo, J., Harm, E., Bartoszek, J., Baumgardner, G., Corrigan, M., Cowsert, J., Harman, T., Jamshidi, M., Jones, W., Newcomb, D., Prowell, B., Sines, R. & Yeaton, B. 2008. Warm-Mix Asphalt: European Practice. *Report FHWA*, USA, American Trade Initiatives.

Dinis-Almeida, M. & Lopes Afonso, M. 2015. Warm Mix Recycled Asphalt – a sustainable solution. *J Clean Prod* 107:310–316.

Dinis-Almeida, M., Castro-Gomes, J., Sangiorgi, C., Zoorob, S.E., Afonso, M.L. 2016. Performance of Warm Mix Recycled Asphalt containing up to 100% RAP. *Const Build Mater*, 112:1–6.

Frank, B. & Prowell, B. 2014. Method for calculating Warm Mix energy saving based on stack gas measurements. In *Proceedings of the ISAP 2014*, 1:49–58.

Frigio, F. & Canestrari F. 2016. Characterisation of warm recycled porous asphalt mixtures prepared with different WMA additives. *Eur J Envir and Civil Eng*, published online.

Frigio, F., Stimilli, A., Bocci, M. & Canestrari F. 2016. Adhesion properties of warm recycled mixtures produced with different WMA additives. In *Proceedings of the 4th CEW*, Delft (The Netherlands).

Harrison, R. & Kemmer, T. 1996. Preferred and alternative methods for estimating air emissions from hot-mix asphalt plants. In *Final Report of Emission Inventory Improvement Program* (EIIP), Vol. 2 Chapter 3.

Lee, S.J., Amirkhanian, S.N., Park, N.W., & Kim K.W. 2009. Characterization of warm mix asphalt binders containing artificially long-term aged binders. *Const Build Mater* 23(6): 2371–2379.

Mahmoud, A.F.F. & Bahia, H.U. 2004. Using the gyratory compactor to measure mechanical stability of asphalt mixtures. *Wisconsin Highway Research Program 05–02*.

Rubio, M.C., Martinez, G., Baena, L. & Moreno, F. 2012. Warm mix asphalt: an overview. *J Clean Prod*, 24:76–84.

Stimilli, A., Virgili, A. & Canestrari, F. 2015. New method to estimate the "re-activated" binder amount in recycled hot-mix asphalt. *RMPD* 16:442–459.

Transport Infrastructure and Systems – Dell'Acqua & Wegman (Eds)
© 2017 Taylor & Francis Group, London, ISBN 978-1-138-03009-1

Environmental and engineering performance assessment of biofilters and retention systems for pavement stormwater

M. Bassani, L. Tefa, E. Comino, M. Rosso & F. Giurca
Politecnico di Torino, Torino, Italy

A. Garcia Perez & R. Ricci
Biosearch Ambiente s.r.l., Torino, Italy

F. Bertola & F. Canonico
Buzzi Unicem s.p.a., Casale Monferrato (AL), Italy

ABSTRACT: This work is aimed at investigating the use of green infrastructure technologies for the infiltration of stormwater runoff from urban road pavements into soil, and as a means to support the evapotranspiration and harvesting processes. Porous concrete pavements and vegetated biofilters were considered to assess pollutant removal and hydraulic performances. The investigations were at bench-scale for single materials, and at column and box scale for composite structures. The pollutant reduction was determined on the basis of suspended solids, heavy metals and hydrocarbon concentration, with promising results that were found to be dependent on filter material characteristics, layer distribution, and permeability.

1 INTRODUCTION

Increasing urbanization worldwide is having a severe impact on the environment and quality of life. Available data confirm that 50% of the world's population already lives in cities, and forecasts indicate that this will rise to 67% by 2050 in developing countries, and 86% in developed ones. Cities will be home to about 6.3 billion inhabitants, while rural areas will not face this demographic growth (van Leeuwen, 2015).

The increase in urban land cover causes irreversible environmental changes, a major one being the loss in its ability to absorb rainwater. In urban areas, rain falls on impervious surfaces where it is rapidly directed into the drainage sewer system, often overloading it and sometimes flooding the roads. Furthermore, stormwater runoff carries waste material, bacteria, heavy metals, and other dangerous pollutants with the first flush responsible for the highest level of contaminants.

Ranging from 65% to 95% of the total surface of an urban area, the built cover predominates on pervious vegetated surfaces (Ferguson 2005). Road pavements, sidewalks, and parking lots occupy most of the urban land destined for industrial, commercial, and residential use. In intensely built-up areas, they cover more than 50% of all the land. As a result, road infrastructures are considered environmentally unfriendly, and there is universal agreement on the need to limit their impact especially in urban areas. Many commenters sustain that urban transportation infrastructures may help the achievement of a "Low Impact Development" (LID), by limiting as much as possible any damage to soil, vegetation, and aquatic systems (US EPA 2000, Prince George's County 1999). To alleviate their effects, existing and future urban road facilities should be more environmentally friendly and sustainable. One solution is to include new "Green Infrastructures" (GI) such as pervious pavements, planter boxes, bio-swales, and other solutions as part of an urban stormwater management system.

The development of GI in urban areas is strongly promoted by the European Commission (Dige 2011). Nevertheless, there is limited experience and a lack of knowledge of their potential benefits globally and, specifically, in Italy. To bridge this gap, a team of researchers from the Department of Environment, Land and Infrastructure Engineering of the Politecnico di Torino, Biosearch Ambiente s.r.l., and Buzzi Unicem s.p.a. investigated the use of alternative GI technologies. Specifically, the work aimed at the assessment of their capabilities to infiltrate, evapotranspirate, and/or harvest urban stormwater runoff, thus contributing to a reduction in the water discharged into the drainage system.

In the research program, two GI technologies, porous concrete pavements, and biofilter boxes to treat the water runoff pollutants from impervious pavements, were investigated under identical experimental conditions to derive ecological and hydraulic indicators useful to their design as urban drainage facilities. A comparison between the hydraulic performance and pollutant removal capacity of the two GI systems, can be of use to designers who wish integrate them into an urban stormwater management system.

1.1 *Pervious pavements*

The use of pervious pavements is well established in the LID approach (Dietz 2007). They include porous asphalt, pervious concrete, concrete blocks, plastic or concrete grid systems (Eisenberg et al. 2015). Their application in urban areas results in a reduction in stormwater volumes that would otherwise go directly into the storm-sewer system (US EPA 1999, Holman-Dodds et al. 2003). Aihablame et al. (2012) reported that pervious pavements can reduce the runoff by between 50% and 93% in comparison with traditional impervious pavements. An in situ investigation of permeable pavements by Abbott & Comino-Mateos (2003) demonstrated a reduction in peak flows and an extended duration of outflows. Collins et al. (2006) observed a reduction in the peak flow rate for pervious pavements compared to the peak runoff flows from usual asphalt surfaces: an 80% reduction for porous concrete, 72–84% for permeable interlocking concrete pavers, and 89% for concrete grid pavers.

Road pavements are also a source of pollutants, which are washed off by runoff and then contaminate water (Sansalone & Buchberger 1997, Drapper et al. 2000). However, numerous authors showed that porous pavements can mitigate this problem. Legret et al. (1996) compared the quality of runoff from porous pavements with that from conventional surface drainage systems, finding a reduction in pollution concentration in terms of suspended solids (64%), and heavy metals such as Pb (79%), Zn (72%), Cd (67%). Furthermore, Dierkes et al. (1999), Pagotto et al. (2000), Fach and Geiger (2005), Gilbert and Clausen (2006) calculated the capacity of different pervious pavements in lowering heavy metal concentration. These systems have also been recognized as being efficient in the retention of hydrocarbons (Pratt et al 1999). Coupe et al. (2003) tried to inoculate a permeable pavement surface with specifically hydrocarbon-degrading microorganisms, finding that the treatment did not improve the oil degradation capability in comparison with the naturally developed microbial communities. Newman et al. (2004) found that pervious pavements with

geotextiles are able to reduce the concentration of hydrocarbons in runoff.

Rahman et al. (2015) demonstrated that the use of construction and demolition waste (CDW) in combination with geotextiles increases pollutant removal. CDW was found to have geotechnical and hydraulic properties equivalent or superior to that of conventional granular materials from quarries.

The described advantages can be eroded by clogging phenomena that can occur over time, as well documented in Abbott & Comino-Mateos (2003), and Kayhanian et al. (2012). Thus, pervious pavements need regular maintenance (Scholz & Grabowiecki 2007).

1.2 *Biofilters*

Several studies have promoted the use of vegetation and soils for the interception and infiltration of stormwater runoff from road pavements, to improve water quality and hydrological response of a site. While it is clear that in urban areas and near rural roadways, stormwater treatments are useful to mitigate the effects of pollutants, there is a distinct lack of knowledge in the literature as regards the integration of full-scale stormwater GI technologies in the urban environment.

Some authors (Dietz, 2007; Davis et al., 2009) use the term "biofiltration" or "bioretention" to indicate soil and vegetation based systems used for stormwater pollution removal. The biofiltration process includes adsorption, plant uptake and removal of pollutants, plus biological transformation and degradation (Kadlec & Wallace, 2009). According to Zinger et al. (2013), the most dangerous pollutants in terms of ecological toxicity are Cd, Pb, Ni, Fe, Mg, Cu, Zn, Cr, suspended solids and hydrocarbons. They demonstrated that the level of concentration of pollutants in stormwater can vary greatly depending on the particular site and the time elapsed since the last rain event.

The removal efficiency depends on the characteristics of the materials used to filter stormwater (Lim et al. 2015). Compost had high removal efficiencies for heavy metals (>90%), while potting soil and commercial mix offered the best uptake for Zn and Cd (Reddy et al. 2015). However, no single filtering material was capable of removing all heavy metals (Li et al. 2016). Therefore, a combination of filtering materials should be investigated for the simultaneous removal of heavy metals. A contribution may also be provided by vegetation and the microbial community that can grow on geotextiles and/or inside the filtering materials (Mazer et al. 2001, Bayon et al. 2015). Most of the growth in biofilm responsible for degrading pollutants is concentrated in the geotextile layer. This is due to its retention capacity, which stabilises microorganisms and provides them with enough time to consume hydrocarbons.

Experiments on biofilters have been documented from the laboratory to the full scale. For example, Zinger et al. (2013) based their lab experiences on small columns, while Legret et al. (1996) and Kazemi et al. (2015) based their work on large boxes. A combination of biofilters with rainwater tanks was investigated by Demuzere et al. (2014), while Hurley & Forman (2011) contributed to the assessment of a detention basin.

2 MATERIAL AND METHODS

In this investigation, the pollutant retention capacity of several combinations of biofilters and pervious pavements was evaluated starting from the bench scale (small samples of single filtering material), to complete laboratory systems in which the materials were combined in different ways to form columns and boxes. At the same stage, hydraulic performances were evaluated to obtain fundamental parameters for design purposes. The investigation involved the following stages:

1. the development of a microbial biofilm on artificial supports;
2. the assessment of ordinary and recycled materials for the formation of pervious pavements and biofilters;
3. the formation of different system configurations;
4. their analysis in terms of hydraulic and pollutant removal properties.

According to Zinger et al. (2013), the laboratory chemical test methods on pollutant removal properties carried out on inflows and outflows included: (a) concentration of suspended solids (AP-AT CNR IRSA 2090 B Man 29: 2003), (b) concentration of heavy metals (UNI EN ISO 17294-2:2005), and (c) total hydrocarbons (AP-AT CNR IRSA 5160 B2 MAN 29: 2003).

2.1 Biofilm

A microbial biofilm inoculum was prepared to introduce active microorganisms into the two GI systems to favour organic pollutant abatement by biodegradation processes. A polyester needle-punched non-woven geotextile (400 g/m²) with no chemical binders, was used as support for biofilm development.

For biofilm development, the geotextile was introduced into a plastic container and irrigated continuously in a water recirculating system. Deionized water was added with 10 g/l of compost, incubated at room temperature for 5 days, and filtered through the same geotextile. Biofilm growth was evaluated through direct and microscopy observation.

After biofilm development, this geotextile was included in GI systems in some interfaces between layers. Its pollutant removal properties were evaluated in preliminary bench-scale tests, as described in Section 2.3.

2.2 Stormwater

Small volumes of input stormwater were collected from urban pavements in two steps, and then used to run tests at the bench and column scales. The differences in concentration of suspended solids, heavy metals (Cd, Cr, Fe, Mn, Ni, Pb, Cu, Zn) and hydrocarbons between the two stormwater samples are reported in Table 1 and Table 4.

A larger volume of stormwater was also synthetized in a tank, mixing tap water with a concentrated solution of heavy metals and diesel fuel. The blend of pollutants included was set to be quite different from the one collected from the pavement with more hydrocarbons and fewer heavy metals. The results for pollutant concentrations are listed in Tables 5, 6 and 7.

2.3 Preliminary pollutant removal assessment at the bench scale

A first materials assessment of the two GI technologies was carried out on the basis of the literature review, previous experience, and local availability.

The materials considered included: (a) natural sand, (b) organic soil, (c) compost, (d) expanded clay, (e) geotextile with biofilm, and (f) subgrade soil. All of them were submitted to a preliminary bench scale test to get information on their pollutant reduction capability (Figure 1).

The stormwater was poured into glass bowls containing the abovementioned materials and filtrated for 3 hours. Outflow filtered water was collected and tested for the presence of the same pollutants and pH.

As shown in Table 1, input stormwater had a high content of hydrocarbons and suspended solids. Conversely, metal content was very low. Filtration through a single media slightly altered the neutral pH of the outflow, and led to a significant reduction in hydrocarbon content.

The best performing substrate was the geotextile with biofilm, which decreased the initial amount of 2700 μg/l of hydrocarbons to non-detectable levels. The same level of reduction was recorded for sand, compost and subgrade soil, which showed an abatement percentage of 99%.

The inflow amount of 518 mg/l of suspended solids was adequately reduced by material samples: from the 99% abatement by subgrade soil to the

Table 1. pH and concentration of hydrocarbons, Suspended Solids (S.S.) and heavy metals in input stormwater and outflow water.

Pollutant [µg/l]	InSW	Sand	OS	C	EC	GTX	SS
pH	6.8	7.4	7.5	7.5	6.7	7.7	N/A
S.S. [mg/l]	518.0	34.0	73.0	128.0	145.0	37.0	3.0
Hydrocarbons	2700	33	110	27	930	5	29
Cadmium	<0.1	<0.1	<0.1	<0.1	<0.1	<0.1	<0.1
Chrome	<2.5	<2.5	<2.5	<2.5	<2.5	<2.5	<2.5
Iron	<25.0	<25.0	<25.0	<27.0	<25.0	<25.0	<25.0
Manganese	8.0	20.0	2.5	2.5	13.0	2.5	2.5
Nickel	1.2	1.2	1.0	2.3	5.8	1.0	1.0
Lead	<1.0	<1.0	<1.0	<1.0	<1.0	<1.0	<1.0
Copper	<10.0	<10.0	<10.0	<10.0	<10.0	<10.0	<10.0
Zinc	<25.0	<25.0	<25.0	<25.0	<25.0	<25.0	<25.0

Notes: InSW = input stormwater, OS = organic soil, C = compost, EC = expanded clay, GTX = geotextile, SS = subgrade soil.

Table 2. Density (γ) and coefficient of permeability (k) of materials.

Materials	γ [kg/m³]	k [m/s]
Subgrade soil	2135	8.28 · 10⁻⁹
Subbase material	2277	2.35 · 10⁻⁷
Gravel 5–15 mm	1449	9.07 · 10⁻⁵
Gravel 15–30 mm	1419	N/A
CDW aggregates	1184	N/A
Organic soil	594	1.00 · 10⁻⁴
Concrete mix 1 (4–16 mm)	1873	1.24 · 10⁻⁴
Concrete mix 2 (10–20 mm)	1919	1.24 · 10⁻⁴

Table 3. Total Layer Volume (LV), Water Storage Capacity (WSC) in the layers, Void content for single layers (V) and Average for the whole system (AV) in the five boxes.

Box of Figure 3	Layer	LV [m³]	WSC [m³]	V [%]	AV [%]
(b)	OS	0.106	0.029	27	33
	G5–15	0.053	0.019	36	
	G15–30	0.053	0.023	43	
(c)	OS	0.106	0.031	29	44
	CDW	0.106	0.062	58	
(d)	OS + CDW	0.106	0.025	24	29
	G15–30	0.106	0.036	34	
(e), (f)	CL	0.240	0.084	35	28
	Subbase	0.240	0.055	23	
	Subgrade	0.480	0.134	28	

Notes: OS = organic soil, CL = concrete layer, G = gravel.

72% by expanded clay. Metal content remained very low after filtration with no significant variations between inflow and outflow values.

Table 4. pH and concentration of pollutants [µg/l] in Input Stormwater (InSW), in outflow water from columns (a) and (e).

Pollutant	InSW	Vegetated biofilter (a)	Pervious pavement (e)
pH	7.4	6.9	7.5(*)
S.S. [mg/l]	1127.0	82.0	4.0
Hydrocarbons	12000	67	79
Cadmium	0.2	0.6	<0.1
Chrome	3.8	8.8	<2.5
Iron	1100.0	2200.0	<25.0
Manganese	210.0	505.0	295.0
Nickel	15.0	125.0	9.5
Lead	10.4	12.5	38.0
Copper	41.5	79.5	7.7
Zinc	130.0	N/A	N/A

Note: (*) in this test the concrete layer was not included in the column (e).

Table 5. pH and concentration of pollutants [µg/l] in Input Stormwater (InSW) and outflow from biofilter (a) and (b).

System	InSW	System (b)	
Layer	–	OS	G15–30
Depth [cm]	–	40	80
pH	6.4	6.8	6.8
S.S. [mg/l]	1.0	10.5	9.4
Hydrocarbons	117000	860	550
Cadmium	17.5	<0.1	<0.1
Chrome	7.0	<2.5	4.0
Iron	34.0	34.0	34.0
Manganese	107.0	33.0	12.0
Nickel	2.0	26.0	57.0
Lead	51.0	<1.0	<1.0
Copper	67.0	14.2	36.9
Zinc	272.0	52.0	54.0

Table 6. Concentration of pollutants [µg/l] in the Input Stormwater (InSW) and in outflow from biofilter system (c) and (d).

System	InSW	System (c)		System (d)	
Layer	–	OS	CDW	OS+CDW	G15–30
Depth [cm]	–	40	80	40	80
pH	6.4	6.8	7.2	7.1	7.0
S.S. [mg/l]	1.0	10.5	12.2	10.1	7.6
Hydro-carbons	117000	860	560	550	570
Cadmium	17.5	<0.1	0.1	0.1	0.1
Chrome	7.0	<2.5	54.0	6.0	6.0
Iron	34.0	34.0	32.0	45.0	32.0
Manganese	107.0	33.0	44.0	13.0	4.0
Nickel	2.0	26.0	23.0	11.0	30.0
Lead	51.0	<1.0	1.3	1.1	2.9
Copper	67.0	14.2	17.6	10.9	19.5
Zinc	272.0	52.0	31.0	49.0	73.0

Table 7. pH and concentration of pollutants [µg/l] in the Input Stormwater (InSW) in paved systems (e and f).

Pollutant	InSW	System (e)		System (f)	
Layers	–	Sub-base	Sub-grade	Sub-base	Sub-grade
Depth [cm]	–	40	90	40	90
pH	6.4	11.3	10.9	11.8	11.5
S.S. [mg/l]	1.0	14.8	43.3	16.3	14.0
Hydro-carbons	117000	10500	1500	740	670
Cadmium	17.5	0.1	0.2	<0.1	<0.1
Chrome	7.0	5.0	8.0	8.0	14.0
Iron	34.0	<25.0	192.0	<25.0	39.0
Manganese	107.0	3.0	17.0	3.0	4.0
Nickel	2.0	2.0	2.0	3.0	2.0
Lead	51.0	<1.0	5.8	<1.0	1.9
Copper	67.0	<10.0	<10.0	<10.0	10.0
Zinc	272.0	<25.0	39.0	<25.0	28.0

Figure 1. Bench scale tests on different media. On the left: grey coloured input stormwater and transparent filtered outcome water. On the right: water outflow samples.

2.4 Physical characterization of materials

In the first stage of the investigation, the organic soil, the subgrade soil, and the geotextile were considered suitable to be employed in the formation of the two GI technologies.

Additional granular materials were also included to build layers. Specifically, three natural unbound granular materials (a granular pavement subbase, and two gravels 5–15 and 15–30 mm), and a Construction and Demolition Waste (CDW) were employed. They are used in road pavements, embankments, and biofilters. CDW was submitted to a leaching test in accordance with Italian standards (UNI 10802) to assess the heavy metal content of the eluate. Results in µg/l are: Cd < 0.25, Cr = 12.5, Fe = 5.7, Mn < 0.14, Ni < 1.5, Pb < 4.2, Cu = 4.8, Zn = 4.5. All these values fall within Italian limits (D.Lgs 152/2006), so the CDW was considered suitable in the investigation.

Figure 2 exhibits the particle size distribution of the granular materials selected, with the exception of the organic soil. The subgrade soil had a well-graded particle distribution in a wide range of diameters (0–50 mm). The granular material for subbase layers presented a well-graded particle distribution in the 0–30 mm range, while the CDW particles and the two gravels (15–30 and 5–15) exhibit a discontinuous distribution in the same range, which is typical for materials selected through a sieve operation in a relatively small size interval. Their grain dimension indicates gravels which have high permeability. Figure 2 also shows the grain size distribution of the aggregates used in the two concrete layers described later.

The materials used to form layers in columns and boxes were subjected to density and permeability analysis. Loose materials were compacted into known volume moulds, weighed, and then tested in a permeameter under a constant hydraulic head.

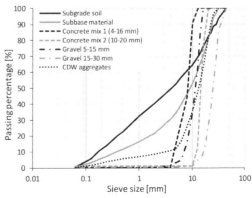

Figure 2. Particle size distribution of materials.

157

In the case of subgrade soil and subbase granular material, Table 2 reports the average dry density derived from the Proctor compaction method on samples prepared at the optimum moisture content and at this value ±2%.

Table 2 also contains the unit mass of compacted samples of the two gravels, the CDW and the organic soils. For the two concrete mixtures, the density was measure on compacted cylindrical samples.

The subgrade soil and the subbase material present the lowest coefficients of permeability, on account of their grain size distribution (Figure 2). Very high coefficients of permeability were observed for the other materials. In particular, the two concrete mixtures for porous pavement layers exhibited the same value of permeability coefficient ($1.24 \cdot 10^{-4}$ m/s), due to their high void content. The gravel with particles gradation 5–15 mm had a permeability of $9.07 \cdot 10^{-5}$ m/s. Due to its very low density and high porosity, the organic soil was the material with the highest coefficient of permeability.

2.5 Columns and boxes

Figure 3 reports the stratigraphy of cylindrical columns (20 cm in diameter) and boxes (rectangular 60×100 cm for pavements, and cylindrical 60 cm diameter for vegetated biofilters). Thicknesses of the layers were defined on the basis of literature review and current practices. Materials were laid to form compacted layers of uniform thickness. In particular, granular materials and concrete mixtures were manually compacted in columns with a Proctor hammer, while a falling weight acting on a rigid plate was used to compact layers in boxes.

Vegetated systems need to meet requirements to support the vegetation growth, although they were not planted during this first investigation. Pervious pavements require structural and functional characteristics in terms of quality of materials and installation. In the meantime, all they need to do is satisfy water infiltration and pollution reduction principles. When making a comparison, it should be noted that in pervious pavements the drainage layers are in the upper part of the system, whereas they are located in the lower part of vegetated biofilters.

The vegetated biofilter of Figure 3a was built in a column and is composed of a double layer of coarse (15–30 mm) and fine gravel (5–15 mm) divided by the geotextile previously treated with the microbial biofilm. The two granular layers contribute to drainage and water retention, and to promote physical processes for pollution removal. A 40 cm layer of organic soil was placed on top to create the substrate for vegetation rooting.

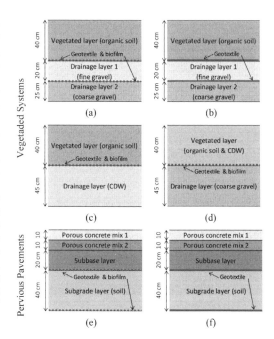

Figure 3. Layered systems used in columns and boxes: (a) vegetated column, (b) vegetated box with traditional gravel drainage layers and organic soil, (c) vegetated box with CDW drainage layer and organic soil, (d) vegetated box with traditional gravel drainage layer and a mixture of organic soil and CDW, (e) pervious pavement column and box with pre-treated geotextile with biofilm, (f) pervious pavement box with ordinary geotextile.

The same stratigraphy in terms of materials and layer thickness was adopted for the system placed in a box in Figure 3b, with the distinction that in this case the geotextile was not pre-treated with biofilm.

To evaluate the ecological and economic potential of vegetated systems with recycled materials, two additional systems containing CDW were set up. A first box (Figure 3c) was prepared with a bottom layer of 45 cm of CDW for water storage purposes, and a top layer of organic soil. A second box (Figure 3d) was prepared with a bottom layer of coarse gravel (15–30 mm) and a top layer made up of a mixture of two thirds organic soil and one third CDW.

Pervious pavement stratigraphy was defined according to ordinary solutions for low-volume roads and parking lots. Two Portland cements, CEM I 52.5 R for concrete mixture 1, and CEM II/A-LL 42.5R for concrete mixture 2, were used in a quantity of 350 kg/m³. To avoid the release of Cr VI, in the production of these two cements ferrous sulfate was added to the cement powder as a reducing agent.

Mixture 1 was made up of aggregates of sizes 4–16 mm, and with a Water/Cement (W/C) ratio equal to 0.31. To improve workability, a plasticizer was added, while the cohesion between particles was enhanced through the addition of a viscosity modifying admixture for self-compacting concrete. Aggregates of 10–20 mm were employed in mixture 2 with a W/C ratio equal to 0.38. Compressive strength on cubic samples at 28 days of curing was 7.1 MPa for mixture 1, and 8.4 MPa for mixture 2.

A column (Figure 3e) and two boxes (Figure 3e and 3f) were built with a subgrade of 40 cm, a sub-base layer of 20 cm, and two porous layers made up of concrete mix 1 and 2 with a combined thickness of 10+10 cm.

The pores of the upper (surface) layer were smaller to contrast the clogging produced by urban wastes (leaves and trash), making urban road cleaning operations easier. The lower concrete layer had a higher void content that resulted in a greater direct water storage capacity.

In systems (e) and (f) subbase and subgrade were separated by the geotextile to avoid layer contamination and, in the case of system (e) the geotextile was pre-treated to develop biofilm.

In order to prepare and clean cylinders and boxes, they were firstly saturated with tap water, allowed to dry for 3 days before being saturated with polluted stormwater for 3 hours. Outflow samples were collected in some interfaces between layers and at the bottom. The released water was stored in glass bottles and then analysed.

3 RESULTS AND DISCUSSION

3.1 Hydraulic assessment

The hydraulic capacity of boxes was evaluated in terms of Water Storage Volume (WSV), void content for each single layer (V), and weighted Average Void content (AV). Table 3 reports the values measured in the five boxes of Figure 3, also including the total Layer Volume (LV).

The vegetated boxes exhibited a higher water storage capacity with values ranging from 29 to 44% of voids. In particular, the vegetated biofilter (c) proved to be the best in water retention due to the considerably high void content (58%) of the CDW aggregate layer. The organic soil layers of vegetated boxes were able to store water from 25 to 31 dm³ in a layer with a total volume of 0.106 m³. Considering the two gravel layers of box (b), WSV reduces when particle size decreases.

The paved boxes showed a lower average void content equal to 28%, due to low WSC of subgrade and subbase layers. The pervious concrete layers functioned as a reservoir, collecting in total 35 dm³/m².

3.2 Pollutant removal assessment

3.2.1 Columns

Table 4 lists the concentration of hydrocarbons, suspended solids, and heavy metals in the input stormwater and in the outflow water of the two columns (a) and (e) in Figure 3.

These systems performed very well in hydrocarbon abatement, since they resulted in a reduction of 99% in the starting concentration of 12000 µg/l. In terms of suspended solids and heavy metals, the pervious pavement outperformed the vegetated one. The paved column (e) decreased the concentration of Cd, Cr, Fe, Ni and Cu, while water passing through the vegetated column (a) collected huge quantities of these metals from the filtering materials, since their concentrations resulted higher than those of the input stormwater.

3.2.2 Boxes

The concentration of pollutants in the vegetated box (b) in Figure 3, in comparison with the input stormwater is provided in Table 5. The hydrocarbons were strongly reduced (abatement of 99–100%), as well as heavy metals, with the exception of Fe that did not change its concentration, and Ni that seemed to be released from the system. Water passing through the filter obviously collected a huge amount of suspended solids, especially in the organic soil layer.

A comparison of the results for the two layers provides evidence that the gravel in the biofilter is able to reduce S.S., hydrocarbons and Mn. It does not significantly affect the pH and the concentrations of Fe, Cd, Cr, Pb and Zn, while it produces an increase in Ni and Cu.

The concentration of pollutants in outflow water from the biofilters (c) and (d) is shown in Table 6. Both performed well in hydrocarbon reduction (abatement of 99–100%), while the pH value did not change significantly remaining around the neutral value. Compared to system (a), the addition of a previously developed biofilm of bacteria did not have a predominant role in pollutant abatement.

As expected, in boxes (c) and (d) the level of suspended solids detected were higher in the outflow than in the inflow due to the presence of the organic soil layer. CDW materials released Cr in box (c) and Fe in box (d). Both systems, like that of Table 5, added Ni and Cu to the outflow. Other heavy metals experienced a significant reduction in their concentration levels.

When considering results at the same depth from the surface, the layer consisting of OS and CDW (system d) seems to perform better than the layer with OS only. This suggests that CDW are more performant when used in combination with organic soil.

Table 7 presents the results obtained from pervious pavements. At a depth of 40 cm, the box (e) had a low capability in hydrocarbon reduction (91%). But at 90 cm, the hydrocarbons in output water is reduced to 1500 μg/l, which corresponds to an abatement of around 99%. In the case of box (f), the reduction was greater than 99% at both depths. Water passing through the two systems collected suspended solids: their concentration rising (at a depth of 90 cm) from 1 mg/l to 43.3 mg/l and 14.0 mg/l in boxes (e) and (f) respectively.

As expected, water filtrated through concrete increased its pH from 6.4 to 11.3 and 11.8 in systems (e) and (f) respectively. However, the pH decreased when the water passed through the subgrade, reaching values equal to 10.9 and 11.5 respectively.

The levels of Cd, Mn, Pb, and Zn were greatly reduced by the two paved systems. Although the majority of heavy metals were removed, their increase in the subgrade layer revealed that this material released some of the pollutants into the outflow water.

4 CONCLUSIONS

Looking at the results, the following conclusions can be drawn:

1. pervious pavements are more effective in pollutant removal than vegetated biofilters made up of organic soils in combination with other materials;
2. this finding could be due to the differences in layer permeability, which determines the rate of stormwater filtration, with water moving fast in vegetated biofilters, and very slow in pavement systems;
3. this difference in filtration rate partially explains the differences in pollutant removal performance, since bacteria may have more time to treat the same stormwater when the water flows slowly;
4. furthermore, denser materials present a greater surface area for the biofilm development, the hydrocarbons and metal sorption (in which liquid substances are attracted by solid particles as a consequence of physical-chemical processes);
5. vegetated biofilters may benefit if an intermediate layer with low permeability coefficients is used at the bottom; this condition will certainly be investigated in the near future;
6. the pollutant reduction performance of CDW increases when it is combined with organic soil;
7. pollutant reduction is not affected by microbial film inoculated through the geotextile, while the geotextile by itself promotes the fixation of naturally developing microbial communities, thus confirming previous investigations in literature;
8. the differences observed in the pollutant removal capacity of systems formed in cylindrical columns and boxes require further investigation;
9. the outflow concentration values of hydrocarbons, suspended solids, and heavy metals for all the systems satisfy the acceptance limits of Italian regulations for wastewater.

ACKNOWLEDGEMENTS

The research performed by the Department of Environment, Land and Infrastructure Engineering of the Politecnico di Torino, the Biosearch Ambiente s.r.l., and the Buzzi Unicem s.p.a. was partly founded by Regione Piemonte in the frame of WIN_STREET Project ("Water IN STReet design with Environmental Engineering Technologies—for urbanized areas", code F.E.S.R. 2007/2013).

REFERENCES

Abbott, C.L. & Comino-Mateos, L. 2003. In-situ Hydraulic Performance of a Permeable Pavement Sustainable Urban Drainage System. *Water and Environment Journal*, 17(3), 187–190.

Ahiablame, L.M., Engel, B.A., & Chaubey, I. 2012. Effectiveness of low impact development practices: literature review and suggestions for future research. *Water, Air, & Soil Pollution*, 223(7), 4253–4273.

Bayon J.R, Jato-Espino D., Blanco-Fernandez E., Castro-Fresno D. 2015. Behaviour of geotextiles designed for pervious pavements as a support for biofilm development. *Geotextiles and Geomembranes*, 43, 139–147.

Collins, K.A., Hunt, W.F. & Hathaway, J.M. 2006. Evaluation of various types of permeable pavements with respect to water quality improvement and flood control. In *8th International Conference on Concrete Block Paving*. San Francisco, CA: American Society of Agricultural and Biological Engineers.

Coupe, S.J., Smith, H.G., Newman, A.P. & Puehmeier, T. 2003. Biodegradation and microbial diversity within permeable pavements. *European Journal of Protistology*, 39(4), 495–498.

Davis, A.P., Hunt, W.F., Traver, R.G., Clar, M., 2009. Bioretention technology: overview of current practice and future needs. *Journal of Environmental Engineering*, ASCE, 135, 109–117.

Demuzere M., Coutts A.M., Gohler M., Broadbent A.M., Wouters H., Lipzig van N.P.M., Gebert L., 2014. The implementation of biofiltration systems, rainwater tanks and urban irrigation in a single-layer urban canopy model. *Urban Climate*, 10, 148–170.

Dierkes, C., Holte, A. & Geiger, W.F. 1999. Heavy metal retention within a porous pavement structure. In *Proc.*

the Eighth International Conference on Urban Storm Drainage.

Dietz, M.E. 2007. Low impact development practices: A review of current research and recommendations for future directions. *Water, air, and soil pollution,* 186(1–4), 351–363.

Dige, G. 2011. *Green infrastructure and territorial cohesion. The concept of green infrastructure and its integration into policies using monitoring systems.* Technical Report 18. European Environment Agency, Copenhagen, Denmark.

Drapper, D., Tomlinson, R. & Williams, P. 2000. Pollutant concentrations in road runoff: Southeast Queensland case study. *Journal of Environmental Engineering,* ASCE, 126(4), 313–320.

Eisenberg, B., Collins Lindow, K. & Smith, D.R. 2015. *Permeable Pavements.* Reston VA: American Society of Civil Engineers.

Fach, S. & Geiger, W.F. 2005. Effective pollutant retention capacity of permeable pavements for infiltrated road runoffs determined by laboratory tests. *Water Science and Technology,* 51(2), 37–45.

Ferguson, B.K. 2005. *Porous pavements.* CRC Press. Gilbert, J.K. & Clausen, J.C. 2006. Stormwater runoff quality and quantity from asphalt, paver, and crushed stone driveways in Connecticut. *Water Research,* 40(4), 826–832.

Gilbert, J.K. & Clausen, J.C. 2006. Stormwater runoff quality and quantity from asphalt, paver, and crushed stone driveways in Connecticut. *Water Research,* 40(4), 826–832.

Holman-Dodds, J.K., Bradley, A.A. & Potter, K.W. 2003. Evaluation of hydrologic benefits of infiltration based urban storm water management. *JAWRA Journal of the American Water Resources Association,* 39(1), 205–215.

Hurley S.E. & Forman R.T.T. 2011. Stormwater ponds and biofilters for large urban sites: Modeled arrangements that achieve the phosphorus reduction target for Boston's Charles River, USA Ecological Engineering, 37, 850–863.

Kadlec & Wallace 2009. Treatment Wetlands, 2nd ed. CRC Press, Boca Raton, FL.

Kayhanian, M., Anderson, D., Harvey, J.T., Jones, D. & Muhunthan, B. 2012. Permeability measurement and scan imaging to assess clogging of pervious concrete pavements in parking lots. *Journal of Environmental Management,* 95(1), 114–123.

Kazemi F. & Kelly Hill K. 2015. Effect of permeable pavement basecourse aggregates on stormwater quality for irrigation reuse. *Ecological Engineering,* 77, 189–195.

Legret, M., Colandini, V. & Le Marc, C. 1996. Effects of a porous pavement with reservoir structure on the quality of runoff water and soil. *Science of the Total Environment,* 189, 335–340.

Li J., Jianga C., Leib T., Li Y., 2016. Experimental study and simulation of water quality purification of

urban surface runoff using non-vegetated bioswale. *Ecological Engineering,* 95, 706–713.

Lim W., Lim H.S., Hu J.Y., Ziegler A., Ong S.L. 2015. Comparison of filter media materials for heavy metal removal from urban stormwater runoff using biofiltration systems. *Journal of Environmental Management,* 147, 24–33.

Mazer G., Booth D., Ewing K., 2001. Limitations to vegetation establishment and growth in biofiltration swales. *Ecological Engineering,* 17, 429–443.

Newman, A.P., Puehmeier, T., Kwok, V., Lam, M., Coupe, S.J., Shuttleworth, A. & Pratt, C.J. 2004. Protecting groundwater with oil-retaining pervious pavements: historical perspectives, limitations and recent developments. *Quarterly Journal of Engineering Geology and Hydrogeology,* 37(4), 283–291.

Pagotto, C., Legret, M. & Le Cloirec, P. 2000. Comparison of the hydraulic behaviour and the quality of highway runoff water according to the type of pavement. *Water Research,* 34(18), 4446–4454.

Pratt, C.J., Newman, A.P. & Bond, P.C. 1999. Mineral oil bio-degradation within a permeable pavement: long term observations. *Water science and technology,* 39(2), 103–109.

Prince George's County 1999. Low-impact development design strategies: An integrated design approach. Prince George's County, MD Department of Environmental Resources.

Rahman Md. A, Imteaz M.A., Arulrajah A., Piratheepan J., Disfani M.M. 2015. Recycled construction and demolition materials in permeable pavements systems: geotechnical and hydraulic characteristics. *Journal of Cleaner Production,* 90, 183–194.

Reddy K.R., Xie T., Dastgheibi S. 2014. Removal of heavy metals from urban stormwater runoff using different filter materials. *Journal of Environmental Chemical Engineering,* 2, 282–292.

Sansalone, J.J. & Buchberger, S.G. 1997. Characterization of solid and metal element distributions in urban highway stormwater. *Water Science and Technology,* 36(8–9), 155–160.

Scholz, M. & Grabowiecki, P. 2007. Review of permeable pavement systems. *Building and Environment,* 42(11), 3830–3836.

US EPA (US Environmental Protection Agency) 1999. *Stormwater technology fact sheet. Porous pavement.* Washington, D.C: Office of Water, EPA 832-F-99-023.

US EPA (US Environmental Protection Agency) 2000. *Low impact development (LID), a literature review.* Washington, D.C: Office of Water, EPA-841-B-00-005.

van Leeuwen, K. 2015. Too little water in too many cities. *Integrated Environmental Assessment and Management,* 11(1), 171–173.

Zinger Y., Blecken G.-T., Fletcher T.D., Viklander M., Deletic A. 2013, Optimising nitrogen removal in existing stormwater biofilters: Benefits and tradeoffs of a retrofitted saturated zone. *Ecological Engineering,* 51, 75–82.

Transport Infrastructure and Systems – Dell'Acqua & Wegman (Eds)
© 2017 Taylor & Francis Group, London, ISBN 978-1-138-03009-1

Slope hazards and risk engineering in the Canadian railway network through the Cordillera

J.L. Rodriguez, R. Macciotta & M. Hendry
University of Alberta, Edmonton, Canada

T. Edwards & T. Evans
Canadian National Railway, Canada

ABSTRACT: The Canadian railway industry is a significant contributor to the Canadian economy, with safe and fluid operations being priorities for the industry. The presence of the Cordillera in Western Canada exposes railway corridors to numerous slope hazards with the potential to endanger railway personnel, operations, equipment and infrastructure. Railways in Canada have a long history of operation delays and infrastructure damage caused by slope instabilities. In this regard, slope hazards are a substantial aspect of the risk engineering approaches adopted by the railways crossing the Cordillera. This paper summarizes some engineering approaches applied to slope hazards along some sections in the Canadian Cordillera. This is illustrated with three landslide case studies with the potential to negatively affect railway operations. The risk engineering approaches adopted at each location are also presented, highlighting the importance of proactive investigation to increase the knowledge about the particular hazard with the aims of continuously enhancing the risk management strategies.

1 INTRODUCTION

Canadian railways have over 50,000 km of tracks connecting urban areas, production regions and ports across the country (Minister of Transport 2016), and are a significant contributor to the Canadian economy. Most of the operations consist of freight transportation by the two major railway companies in Canada: Canadian Pacific Railway (CP) and Canadian National Railway Company (CN).

The railway network in western Canada crosses the Cordillera (in British Columbia and western Alberta) and connects the cities of Vancouver and Prince Rupert with other provinces and markets in the US (Figure 1). The corridors through the cordillera are exposed to varied ground hazards, including slope instabilities, flooding, embankment failures, ground subsidence, frost heave, and snow avalanches. The most frequent ground hazards in these corridors are slope instabilities (rock falls, rock slides, soil slides, debris flows) (Macciotta et al. 2013). Examples of some of these are illustrated in Figure 2.

The large extent of these corridors and the environmental characteristics of the region (extreme weather events and complex geology) make ground

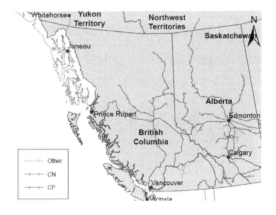

Figure 1. Canadian western railway network.

hazard control very challenging. In this context, adequate risk engineering strategies are necessary to maintain operations within tolerable risks. These strategies require a thorough understanding of the slope instability mechanisms in order to assess their likelihood and the potential impact to the railway infrastructure and operations.

The complexity of slope instability mechanisms is reflected on the uncertainty associated

a)

b)

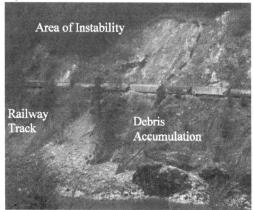

c)

Figure 2. Examples of slope instabilities along railway corridors through the Cordillera: a) soil landslides, b) rock slides, c) rock falls.

with assessing slope instability likelihood and consequences. Risk engineering measures under such uncertain conditions are best dealt with within a risk management framework. In this framework, slope instabilities can be identified, analyzed and evaluated, in order to take objective actions to control the hazards and minimize risk (Porter 2013), therefore minimizing traffic disruptions.

CN and CP have proactively been adopting risk engineering measures to control potential and existing slope instabilities in the Cordillera. These control measures include displacement monitoring and early warnings (slide fences), track deflection monitoring, slope stabilization, Protection measures (ditches, embankments, rock sheds, rock fall meshes), and operation strategies (i.e. patrols in front of trains, reduced speeds). Slope stabilization can involve water control (drainage and surface water management), earth movement, retaining structures, reinforcement (piles, anchors), or combinations of the above (USGS, 2000). Although these can prove effective, they are often cost-prohibited and technically challenging. In these cases, combinations of protection, warning and operation strategies provide feasible options.

This paper presents some examples of risk engineering applied to slope hazards along railway corridors crossing the Cordillera.

2 ROCK FALLS ALONG THE FRASER RIVER VALLEY AND ALONG CN'S SQUAMISH SUBDIVISION

Two corridors with long histories of rock fall occurrences are discussed. CN and CP main lines travel along one of the corridors, which runs parallel to the Fraser River valley between Vancouver and interior British Columbia. A particular section of this corridor is characterized by steep slope cuts to accommodate the railway alignment (Macciotta et al. 2013). The other corridor (CN's Squamish subdivision) runs along Anderson and Seton lakes, and links the city of Vancouver with the city of Lillooet, British Columbia. This corridor is characterized by steep slopes of sheared and weathered rock, where rock falls frequently originate (Figure 3a) (Macciotta et al. 2015b).

Rock falls are a frequent ground hazards along these corridors, where rock slides and debris flows are also present but less frequent. Proactive control strategies in place include: scaling loose rock blocks, slope reinforcement (rock anchors and shotcrete surfaces), rock fall catchment nets and drapery systems, and catchment ditches. Slide fence systems that detect material potentially blocking the tracks, and protection sheds have been particularly adopted along the section within the Fraser River corridor. Patrols in front of trains, and train speed constrains are characteristic of the section along the Anderson and Seton lakes. Examples of some

a)

b)

c)

Figure 3.   a) Steep rock slopes along mountainous railway corridors in western Canada, b) railway equipment
removing fallen debris, and c) patrol in front of train.

control approaches are illustrated in Figure 3, and
Figure 4.

These risk engineering approaches control the
slope hazards along these corridors. Furthermore,
CP and CN keep records of slope instabilities that
include the location, volume, date, and environmental conditions. These allow quantification of
the most hazardous sections of track, slope instability frequencies and their influence in the fluidity
of operations.

The extensive database on slope instabilities has
promoted research with the objective of enhancing the current risk engineering approaches
implemented by CP and CN. Macciotta et al.
(2015b) analyzed the trigger mechanisms of rock
falls along Anderson and Seton lakes to predict
periods of time when rock falls are more likely
to occur. These trigger mechanisms included
weather conditions, seismic events, and biological influences. The results from this study helped
propose a rock fall risk control chart based on
the freeze-thaw cycles and precipitation rates
along the study area (Figure 5). This chart is
meant to aid scheduling maintenance work under
nonhazardous periods, increase awareness on the
hazardous periods, and increase the fluidity of
operations.

a)

b)

c)

Figure 4.   Examples of rock fall hazard control: a) sheds
(for rock fall, debris flow and avalanche protection),
b) rock anchors, c) attenuator nets.

Weather-based criteria for periods of higher rock fall hazard - Squamish 124-157		3-day cumulative precipitation	
		≤ 2 mm	> 2 mm
a) Has there been a freeze-thaw cycle within 3 days?	No	Non-hazardous period	Hazardous period
	Yes	Hazardous period	Hazardous period
b) Within 2 first weeks of spring thaw?	No	Non-hazardous period	
	Yes	Hazardous period	

Notes:
90% of rock falls occur within hazardous periods
10% of rock falls occur within non-hazardous periods
An average of 50% of the time within a year is under a hazardous period warning

Figure 5. Weather base criteria for rock fall hazard along the Squamish Subdivision, (after Macciotta et al. 2015).

3 RIPLEY LANDSLIDE

The Ripley landslide was initially identified in 1951 (Hendry et al. 2015) and did not show significant signs of movement until 2005 when building a siding that required the construction of a retaining wall to increase the construction area.

The increase in landslide displacement rate was identified by shearing of two piezometers within the Ripley Landslide and the development of tension cracks and a back scarp in 2007. The landslide moves at a rate between 25 and 180 mm/year (Bunce & Chadwick 2012; Hendry et al. 2015). A risk assessment in terms of loss of life (train crew), environment, and operations concluded that the high costs associated with slope stabilization works were not reflected in significant reduction of the overall risk to operations (Bunce and Chadwick 2012). The risks associated with displacement of the landslide have been successfully managed through regular inspections and maintenance. The risk engineering approach adopted by CN and CP include displacement monitoring with a GPS system and continuous, in-place, down-hole, inclination measurements (Slope Inclinometers and recently Shape Accel Arrays) that provide early notice of excessive movement. Other approaches adopted at this site are groundwater monitoring, scheduled surveys using LiDAR imaging, regular track geometry surveys and visual inspections (Figure 6).

The Ripley Landslide has also served as laboratory for landslide research and assessment of new monitoring technologies. Technologies that have been deployed include an acoustic monitoring system, satellite radar technology, and a network of optical fiber to monitor deformations at the retaining wall (Hendry et al. 2015, Huntley et al. 2014). Ongoing research activities include investigating the relationship between slope displacements and groundwater pressures, and the response of the railway track to cumulative landslide displacements. Research on the landslide displacements

a)

b)

c)

Figure 6. Instrumentation at the Ripley Landslide: a) GPS station, b) reflector for satellite radar technology, c) location of piezometer, Shape Accel Array, reflector and data acquisition system.

and their effect on track geometry have allowed proposing an enhanced early warning system based on displacement rates, trends of deformation and track quality (Macciotta et al. 2015).

The correlations found in these studies provide the basis to assess other landslide movements in the area and optimize the risk engineering approaches adopted.

4 10-MILE SLIDE

The 10-Mile Slide is a landslide located in the province of British Columbia, near the town of Lillooet, North-East of Vancouver. The importance of the landslide is associated with the location of a

Figure 7. 10-mile Slide and location of the GPS monitoring system.

Figure 8. 10-mile Slide instrumentation and survey: a) GPS Unit, b) downhole slope displacement readings, c) geotechnical investigation and instrument installation, d) remote sensing (LiDAR imaging).

highway (Highway 99) and a railway line (now part of CN) within its boundaries (Figure 7). The landslide is approximately 200 m wide and 140 m high, with an approximate volume of 750 000 m³. The 10-mile Slide is sliding on a through-going shear surface at velocities up to 10 mm/day.

The landslide related risks to the railway have been managed through scheduled track geometry measurements, visual inspection of the track before each train passes (patrols), track deformation monitoring, and regular track maintenance. A retaining wall was installed immediately downslope from the railway tracks to prevent deformations caused by loosening of materials associated with the slope deformations and delay retrogression of the landslide. Construction of the retaining wall strengthened a cost-effective risk management strategy along the railway section.

Active research activities are also part of CN's monitoring efforts. Displacement monitoring of some of the piles that compose the retaining wall has allowed for the measurement of the response of the wall to retrogression of the landslide. Application of new technologies for landslide monitoring also include aerial LiDAR and photogrammetry, drone surveys, installation of a newly developed cost-efficient GPS system, and monthly ground-based LiDAR image acquisitions (Figure 8).

Currently, CN has decided to stabilize the section of the landslide to maintain the alignment of their track by supporting the slope with an array of deep piles.

5 CONCLUSION

This paper illustrates the adoption of different risk engineering approaches to slope instabilities along railway corridors in the Canadian Cordillera. These corridors are exposed to varied slope instabilities, including rock falls and soil landslides. The most suited engineering approach depends on the type of slope instability addressed, its volume, its particular mechanism and the resources available for risk control. Table 1 presents a summary of the approaches discussed.

The cases presented show the importance of a proactive approach towards slope management. Such proactive approach requires a thorough understanding of the slope instability mechanisms in order to assess their likelihood and the potential impact to the railway infrastructure and operations. Furthermore, promoting research activities has proven to enhance the risk engineering strategies through increased understanding of the slope hazards. In turn, this is reflected in increased fluidity of operations and resource optimization.

The results and risk management techniques presented in this paper are site specific and do not

Table 1. Risk Engineering Strategies applied in each case history.

Case	Risk Engineering Strategies
Rockfall along the Fraser River Valley	Includes: • Scaling loose rock blocks • Slope reinforcement (rock anchors and shotcrete surfaces) • Catchment nets and drapery systems • Catchment ditches • Slide fences • Protection shed • Records keeping
Rockfalls along CN's Squamish Subdivision	Includes: • Scaling loose rock blocks • Slope reinforcement (rock anchors and shotcrete surfaces) • Catchment nets and drapery systems • Catchment ditches • Patrol trucks • Speed constrains • Records keeping

Ripley Slide	Includes:
	• Down-hole surveys
	• Slope Inclinometers
	• Shape Accel Arrays
	• Groundwater monitoring
	• Ground Surveys
	  ○ GPS
	  ○ Satellite Radar
	  ○ Visual inspections
	  ○ LiDAR imaging
	• Acoustic monitoring system
	• Network of optical fiber to monitor

10-mile Slide	Includes:
	• Patrol trucks
	• Ground Surveys
	  ○ GPS
	  ○ Satellite Radar
	  ○ Visual inspections
	  ○ LiDAR imaging
	  ○ Photogrammetry
	  ○ Drone
	• Slope Inclinometers
	• Track maintenance

provide a guide that can be directly implemented on other slope instabilities. However, the case histories presented in the paper highlight the importance of implementation of adequate risk engineering, and the role of investigation and research to achieve safe and fluid operations.

ACKNOWLEDGMENTS

This research was made possible by the (Canadian) Railway Ground Hazard Research Program, which is funded by the Natural Sciences and Engineering Research Council of Canada (NSERC), Canadian Pacific Railway (CR), Canadian National Railway Company (CN); and the Canadian Rail Research Laboratory (CaRRL) (www.carrl.ca), which is funded by NSERC, CP, CN, the Association of American Railways—Transportation Technology Center Inc., Transport Canada, the National Research Council of Canada, and Alberta Innovates—Technology Futures.

REFERENCES

Cui, Y., Miller, D., Nixon, G., and Nelson, J., 2015. British Columbia digital geology. British Columbia Geological Survey, Open File 2015-2. Accessed August 25th, 2016. http://www.empr.gov.bc.ca/Mining/Geoscience/BedrockMapping/Pages/BCGeoMap.aspx

Hendry, M.T., Martin, C.D., Choi, E., Edwards, T. & Chadwick, I. 2013, "Safe Train Operations over a Moving Landslide.", 10th World Conference on Railway Research, November 25–28.

Hendry, M.T., Macciotta, R., Martin, C.D. & Reich, B. 2015, "Effect of Thompson River elevation on velocity and instability of Ripley Slide", Canadian Geotechnical Journal, vol. 52, no. 3, pp. 257–267.

Huntley, D., Bobrowsky, P., Zhang, Q., Sladen W., Bunce, C., Edwards, T, Hendry, M., Martin, D., and Choi, E. 2014. "Fiber optic strain monitoring and evaluation of a slow-moving landslide near Ashcroft, British Columbia, Canada". Proceedings of World Landslide Forum 3, 6 p. Beijing, China.

Macciotta, R., Cruden, D.M., Martin, C.D., Morgenstern, N.R., & Petrov, M. 2013. "Spatial and temporal aspects of slope hazards along a railroad corridor in the Canadian Cordillera". In P.M. Dight (Ed.) Slope Stability 2013: International Symposium on Slope Stability in Open Pit Mining and Civil Engineering, Brisbane, Australia. pp. 1171–1186.

Macciotta, R., Hendry, M., & Martin, C.D. 2015. "Developing an early warning system for a very slow landslide based on displacement monitoring". Natural Hazards, 1–21. http://doi.org/10.1007/s11069-015-2110-2.

Macciotta, R., Martin, C.D. & Cruden, D.M. 2015a, "Probabilistic estimation of rockfall height and kinetic energy based on a three-dimensional trajectory model and Monte Carlo simulation", Landslides, vol. 12, no. 4, pp. 757–772.

Macciotta, R., Martin, C.D., Edwards, T., Cruden, D.M. & Keegan, T. 2015b, "Quantifying weather conditions for rock fall hazard management", Georisk: Assessment and Management of Risk for Engineered Systems and Geohazards, vol. 9, no. 3, pp. 171–186.

Minister of Transport 2016, Transportation Canada 2011, Comprehensive Review, Transport Canada, Minister of Public Works and Government Services.

Porter M. & N. Morgenstern 2013, Landslide Risk Evaluation—Canadian Technical Guidelines and Best Practices related to Landslides: a national initiative for loss reduction, Geological Survey of Canada.

Spiker, E.C. & Gori, P.L. 2003, National Landslide Hazards Mitigation Strategy: A Framework for Loss Reduction, U.S. Geological Survey, USGS Publications Warehouse.

Computational model to conform aerodrome geometry to ICAO design standards

E.J. Silva & N.D.F. Gualda
Escola Politécnica, University of São Paulo, Brazil

ABSTRACT: The article presents a computational model to conform an aerodrome geometry to the ICAO Annex 14 standards. These standards impose geometric characteristics to runways, taxiways and aprons, as well as protection surfaces free of interferences according to the project aircraft. The model generates a KML file that can be read in a virtual globe such as Google Earth or in a GIS desktop as QGIS. It encompasses both CAD and GIS routines written in Python language. It allows automatic analysis of possible aerodrome geometry conflicts. Model functionality has been verified for both hypothetic and real world instances. The paper shows results of its application to the Viracopos Airport in Campinas, Brazil (SBKP). A DTM (Digital Terrain Model) of the SBKP site was achieved with subsequent generation of the runway and of the associated OFZ (Obstacle Free Zone) surfaces using SRTM (Shuttle Radar Topography Mission) data.

1 INTRODUCTION

ICAO Annex 14 to the Convention on International Civil Aviation (ICAO, 2004) sets forth mandatory design standards for international airports of 191 countries and is a main technical basis for domestic airports as well. Given a project aircraft, those standards require geometric characteristics for runways, taxiways and aprons, as well as protection surfaces free of interferences related to airport and surrounding objects. Airspace interference outside the aerodromes must be checked as buildings, antennas and other objects must not protrude a set of imaginary surfaces that start in the runway vicinity and slope upwards and outwards.

This article, developed as part of a doctoral re-search, recognizes the complexity involved in assessing the compliance of a given aerodrome design to applicable design standards, especially regarding its geometric aspects. After analyzing existing tools and the aerodrome geometric characteristics, a computational model was developed to analyze conformity of an aerodrome geometry to the ICAO Annex 14 standards (ICAO, 2004). The model incorporates and integrates CAD (Computer Aided Design) and GIS (Geographic Information System) routines. Besides the automatic generation of the related geometric entities, the proposed model performs geometric conflicts analysis, allowing the assessment of existing objects as to Annex 14 design standards. Results are shown for both real world and fictitious test instances.

1.1 *Paper structure*

Section 1 defines paper objectives and structure. In section 2, aerodrome components, their geometry and related design standards are discussed. Section 3 presents related existing software and their features. In section 4, the proposed model is described. Its application to a real world instance is presented on section 5. Section 6 presents the study conclusions.

2 AERODROME GEOMETRY

Whereas *airside* denotes the portion of an aerodrome intended for aircraft operations, *landside* stands for passengers and cargo facilities (e.g.: terminals and parking lots). Textbooks, such as Horonjeff, et al. (2010) and Kazda & Caves (2007), present a comprehensive approach to both the airside and landside design principles.

This research tackles only the airside, from the perspective of Annex 14 to the Convention on Inter-national Civil Aviation (ICAO, 2004). This document establishes Standards and Recommended Practices (SARPs) to be adopted by 191 signatory States in their international airports. While standards are mandatory, recommended practices are optional, but States may turn them obligatory, too. When standards cannot be met, an aeronautical study may be necessary to assure that the TLS (Target Level of Safety) is attained. Silva & Gualda (2013) show the application of design standards flexibilizations to São Paulo/Guarulhos

International Airport airside for A380–800 and B747–8 aircraft operation.

For domestic airports in Brazil, ANAC (National Civil Aviation Agency) adopts a design standards framework quite close to that of ICAO.

FAA—Federal Aviation Administration (FAA, 2014) in the USA, on the other hand, issues a design standards framework for domestic airports that substantially differs from the ICAO one.

As a result, the modelling proposed in this research fits most world international airports and domestic airports in countries where ICAO standards are adopted. For different standards, such as those imposed by FAA in the USA, model adaptations are required.

Aerodrome airside design encompasses both con-figuration analysis and sizing of physical components such as runways, taxiways and aprons, as well as their safety areas on the ground and in the air.

2.1 Airside components: geometry and standards

Several aerodrome geometry criteria are established according to ARC (Aerodrome Reference Code), a classification that accounts for aircraft dimensions and performance.

Besides physical components, such as runways and taxiways, the airside comprises protection areas on the ground and imaginary surfaces in the airspace.

2.1.1 Protection areas on the ground

Protection areas on the ground surround runways and taxiways.

Runway Strip is a rectangle that surrounds the runway with a width that depends on the ARC code number. It starts at a given distance before the threshold and extends after the runway end or the stopway end. Such given distances depend on the ARC code number and on whether it is an instrument runway or a non-instrument one.

The Runway Strip follows the runway centerline profile and constitutes the starting edge of some of the airspace protection surfaces. For this reason, a Runway Strip must be represented as a 3D polygon or surface rather than a simpler 2D geometry.

Runway End Safety Area (RESA) is an area on the ground longitudinally adjacent to a Runway Strip and has similar geometric characteristics to a Runway Strip.

Taxiway Strips exclude movable objects at given distances from the taxiway centerline. It is required for taxiways other than aircraft stand taxilanes. At curved sections, the path followed by aircraft critical points such as wingtips and stabilizers must be analyzed to assure proper clearances.

2.1.2 Protection areas in the air

In order to protect aircraft in the air, several imaginary surfaces apply, which exclude fixed objects. Additionally, during operations, temporary surfaces are created on the ground and in the air to exclude movable objects, such as other aircraft and service vehicles. Approach and take-off operations require their own protection surfaces.

Annex 14 (ICAO, 2004) obstacle limitation surfaces apply to aerodrome planning and design, whereas PANS-OPS (ICAO, 2006) concerns flight procedures design and obstacle assessment.

Figure 1 shows the basic geometry of obstacle limitation surfaces according to ICAO.

This set of surfaces control land use in the aerodrome neighborhood, especially the construction of buildings and other infrastructure. On precision approach runways, additional OFZ (Obstacle Free Zone) surfaces are required. Such surfaces are closer to the runways and represent significant criteria for positioning taxiways, bypasses, runway holding position marks and holding areas. The standards for aerodrome obstacle limitation surfaces depend on the ARC code number and on the approach procedures to be adopted – e.g. visual, non-precision, or precision one.

2.1.3 Airside geometry representation

The airside geometry representation should be as simple as possible, for model efficiency.

Runway and taxiway centerlines can be represented by lines or points with a set of x,y,z coordinates. Runways and taxiways, as well as other physical airside components, can be represented as polygons. This is also true for protection surfaces on the ground or in the air. At most, polygonised sur-faces are necessary to duly account for the complex intersections that occur between airspace protection surfaces. Fortunately, surfaces do not overlap, as the lower surface rules.

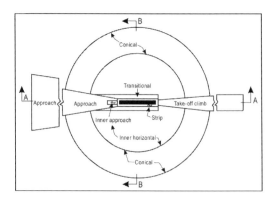

Figure 1. Geometry of obstacle limitation surfaces. Source: (ICAO, 2004, pp. 4–2).

In curves, wheels and other parts of an aircraft describe a mathematical complex pattern, even if the nose wheel follows an arc of a circle. This has implications for designing pavement fillets and for analyzing safety clearances during taxi, as well as on apron maneuvers. Such complex patterns may be well represented by a sequence of points, even though a larger number of points are necessary for higher precision.

As an analogy, an aerodrome can be represented by a non-continuous adjacent set S of surfaces $z_S = f_S(x,y)$. Different "z" surfaces can be represented by layers (e.g.: runways, protection areas on the ground, airspace protection areas).

Therefore, every layer of the aerodrome geometry can be represented by points, lines, polygons and polygonised surfaces, avoiding the use of solid representations.

2.1.4 Obstacle representation

When evaluating airspace infringement by a building or antenna, a single point representation seems reasonable. However, objects closer to runways will deserve a more complex procedure.

For movable objects, a movement simulation or an investigation of the critical position is necessary. TRB (2010) presents a case of airspace infringement, in which a crane operating over a 1,000 rail at Oakland Port became an obstacle in only some of its possible positions. Aircraft

are also movable objects of complex geometric representation.

3 AERODROME ANALYSIS AND DESIGN SOFTWARE

A web search has shown no open source software related to aerodrome design, but some related commercial software. Table 1 summarizes those types of software and their main features.

No mention was found that existing computational tools could be called by programming procedures, allowing the extension of their functions, multiple scenario evaluation or optimization.

4 MODEL DEVELOPMENT

This section presents key aspects of the model development.

4.1 Conceptual background

Geospatial and CAD (Computer Aided Design) concepts, of fundamental importance to the model, are briefly introduced in the following sub items.

4.1.1 Geospatial considerations

Computer handling of geospatial data can be performed through user implementations or with

Table 1. Aerodrome design and analysis software.

Feature	3DAAP	AeroTurn3D*	PathPlanner*	Obstacle Surface Planner	ArcGis for Aviation: Airports
License	Freeware, but requires Autodesk Map 3D	Proprietary	Proprietary	Proprietary	Proprietary. Requires ArcGis for Desktop
Aircraft docking analysis		✓	✓		
Jet Blast analysis		✓	✓		
Airspace protection surfaces infringement	✓			✓	✓
Fillets analysis and design		✓	✓		
Control Tower and runway line of sight analysis	✓				
ALP (Airport Layout Plan) support	✓				✓
Ground clearances and protection areas analysis		✓	✓		
Aerodrome optimization					
Source	(FAA & Planning Technology, Inc, 2005).	Transoft Solutions (2016)	Transoft Solutions (2016)	Simtra (2016)	ESRI (2016)

Source: Several sources mentioned on the table.
*: Succeeded by AviPLAN.

the aid of the so-called GIS software (Geographic Information System). For developers, GDAL and OGR free open source libraries deserve consideration, as they "are the backbone of many Desktop GIS software" (Leidig & Teeuw, 2015). For the interested reader, Chen et al. (2010) compares an extensive list of free open source GIS software.

Although civil engineering designs can often disregard the shape of the Earth, the greater the distance involved, greater the error introduced will be. An aerodrome extends over considerable distances, especially when airspace protection surfaces are considered. Moreover, terrain topography and obstacles are usually surveyed and stored over a geodesic coordinate system. In some cases, even the final designed geometry will be more conveniently represented on a geodesic system.

When geodesic data is transformed into a projection in the Cartesian coordinate system, two sources of distortions are expected: scale distortion and height distortion (Burkholder, 1993). Both distortions can be treated with proper adjusting factors or with a simplified approach, in which projection is properly selected and a maximum height imposed to keep errors within an acceptable range.

From the above considerations, the local topographic plan (ABNT, 1998), was considered as the projection system to be adopted. It is restricted to a squared area of 100 km X 100 km and can keep errors as small as 1:20,000. Adaptations were made to ensure complete numeric consistency on direct and inverse transformations using the matrix transformations described by Dal'Forno, et al. (2011) and Andrade (2003).

Another key component of this proposed modelling is a DTM (Digital Terrain Model) to bridge the gap between terrain topography data and automatic retrieval of height at a given x,y coordinate. For this, a TIN (Triangulated Irregular Network) was adopted. Delaunay triangulation is performed over terrain input data, considering only 2D (x,y) information. From this triangulation, a bilinear interpolation model is set, considering the elevation of each of the terrain input points. Li, et al. (2005) can be consulted for a deeper discussion on alternatives for DTMs.

4.1.2 *CAD considerations*

A CAD (Computer Aided Design) is a computational system for digital design and for draft generation in 2D and/or in 3D. A geometric design comprises a set of geometries and the underlying relationships. Consequently, a geometric design has at least two concerns, namely: i) geometry generation; and ii) model construction paradigm. These topics are here further analyzed:

i. Geometry generation: points, lines, polygons and mathematical functions are usually adopted as primitives for 2D geometry construction. By utilizing mathematical functions, fewer control points might be required, design traces can be smoothed and direct calculations of areas might be easier. For 3D geometries, the same aforementioned primitives can be used, in addition to cones, pyramids, spheres, particle systems and others. Choosing primitives directly influences the easiness and the flexibility to represent different shapes. As detailed by Hosaka (1992), once a geometry is generated, operations may be executed, such as translations, scaling, rotations, mirroring, bending, etc. In computational geometry, such operations are usually performed via matrix calculations.

ii. Model construction paradigm: Monedero (2000) contrasts variants programming, which is a static model, and the interactive one. In variants programming, the model is completely generated by an internal programming procedure, and in case altering part of it is necessary, an entire new model must be generated. This approach requires complete anticipation of the involved geometries and their relationships. On the other hand, an interactive model, sometimes called "parametric design" or "history-based design", allows the user to change the whole or parts of the model, employing an internal data structure, such as a graph tree that keeps track of the sequence used to build the model (Monedero, 2000). Additionally, a feature based approach (Bidarra & Bronsvoort, 2000) can be utilized in the model to represent entities based on their semantic structure (runways and their protection surfaces, for example). For further discussion on CAD construction paradigms refer to Shah (1998).

4.2 *Proposed model*

Figure 2 shows the proposed model basic structure.

Input data refer to ICAO design standards, critical aircraft (ARC), site available areas, site topographic data, and surrounding fixed and movable objects. With these data it is possible for the airport designer to propose an airside geometric configuration (number, position, length and orientation of runways, taxiways, airport fixed and movable objects). The model generates protection areas and surfaces and performs the analysis of conformity of the airside geometry and of existing objects to the Annex 14 standards.

Implementation considers that the model may be extended in the future, for simulation or optimization purposes, for example. Bearing this in

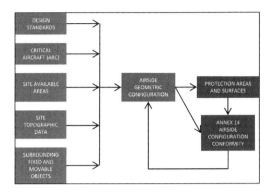

Figure 2. Proposed model basic structure.
Source: Prepared by the authors.

mind, the following criteria were set for choosing the most suitable programming language: availability of libraries to deal with geospatial and CAD manipulations; support to 2D and 3D visualization, as well as animations; possibility of developing an integrated GUI (Graphic User Interface); easiness of native data structures; existence of a wide collaborative community; commitment to openness; and independence from IDEs (command line execution).

The above criteria led the Python language to be selected. This is an interpreted, high-level, multi-paradigm, multi-platform programming language, promoted by the Python Software Foundation (PSF, 2016). Although computational performance is not deemed as a strong Python point, tools such as Cython allow speeding up execution, by means of a special compiling technique (Cython, 2016). Visualization can be generated by Python with the Matplotlib library and a GUI can be developed via wxPython toolkit, for example.

4.2.1 Geometry generation and model construction paradigm

Consideration of Python language indicated that points, lines and polygons would be the most suitable primitives, facilitating CAD operations, visualization with the Matplotlib library and exportation to KML (see item 4.2.2). Moreover, from the discussion in item 2.1, it can be assumed that geometries to be generated do not require complex representations, as design standards explicitly require the representation of given points and lines.

Note that using complex primitives can simplify and speed up geometry construction coding. Their pitfalls, however, lie on storage, manipulation, visualization, or other tasks. Nevertheless, if a complex

primitive can be transformed back into a simpler one, reliably and efficiently, this hybrid approach deserves consideration.

Regarding the model construction paradigm, the aerodrome geometry is strongly constrained by design standards so that different aerodrome alternatives differ only due to very objective and enumerable design criteria, such as orientation and number of runways, length of runways, runway/taxiway separations, etc. To facilitate revision of a given aerodrome geometry, it is convenient to represent the aerodrome components individually. Therefore, a feature-based approach (Bidarra & Bronsvoort, 2000) was utilized to represent these components.

For the above reasons, aerodrome geometry is generated by an internal programming procedure (static model), but its components are created based on a feature-based approach. The latter feature shall facilitate a future development of the model to account for simulation or optimization applications.

Obstacles are assessed against the infringement of a protection surface by a routine that: i) first compares both convex hulls; and ii) in case they overlay, height tests are performed. Protection surfaces are transformed into TINs (Triangulated Irregular Networks) and interpolated height is retrieved just as done with DTMs.

4.2.2 File formats

With the constant progress of GIS software and data acquisition technologies, it is interesting that designs can be translated into standard interoperable data formats. The comparison of different layers, such as DTMs, protection surfaces, noise contours, satellite and aerial imagery, soil occupation patterns, and roads, for example, might provide valuable insights and analysis power to decision makers. Airports are an example of infrastructure that keeps several relationships with the surrounding environment. It was found that the KML (Keyhole Markup Language) standard could be used to write files that can be easily read in GIS software such as the free open source QGIS and in a virtual globe such as Google Earth. For this task, Simplekml Python library, licensed by GNU standard, has been chosen. All other file exchange requirements have been kept as simple as supportable in CSV (Comma-Separated Values) format.

4.2.3 Pseudocode

A pseudocode of the proposed model is presented in Figure 3. Algorithm 1 generates the aerodrome geometric model, which is an input to Algorithm 2 that allows obstacle assessment.

Inputs:

Terrain_points ← topographic data

Limits ← site external boundaries and internal exclusion areas

Configuration ← defining parameters (# of runways, orientation,..)

Model:

DTM ← TIN(Terrain_points)

For each aerodrome component i:

   C_i ← *component_generation (DTM, Limits, Configuration)*

   P_i ←*protection areas_generation (C, DTM, Limits, Configuration)*

Outputs:

Kml_file ←model2kml (C,P)

Internal_file←(C, P, DTM, Limits)

End

Algorithm 2: obstacle assessment

Inputs:

P,C, DTM ← Algorithm 1

Objects ← surrounding objects

Fleet ← on site aircraft xy position and azimuth

Model:

For each aircraft i:

   Fleet_positioned ←positioning (Fleet_i,DTM)

For each object i in [Objects] and [Fleet_positioned]:

   For each surface j in P_j:

      If convex_hull (object_i) intersetcts convex_hull (Pj):

         If height (object_i) > height(P_j):

            Violations_{ij} ← True

Outputs:

Violations of object i over surface j

end

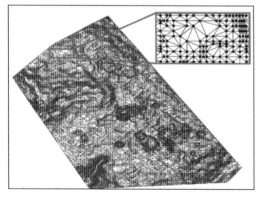

Figure 3.   SBKP aerodrome site TIN DTM. Graphic output generated by the Matplotlib Python library. Source: Prepared by the authors from USGS (2015) data.

5   MODEL APPLICATION

This section presents details of the model applications for both real world and fictitious instances.

5.1   Digital terrain modelling

As discussed in section 2.1, DTM (Digital Terrain Modelling) is essential for aerodrome geometry analysis and design. To assess the methodology described in item 4.1.1, SRTM (Shuttle Radar Topography Mission) elevation data were used to create a TIN model, for an arbitrary delimitation of the Airport of Viracopos (SBKP). SRTM data were obtained from USGS (2015) in GeoTIFF raster format (1 arc-second global, published on 2014). Data were pre-processed on QGIS software to generate a CSV input file with 23,893 height points, corresponding to 1-m contour lines. This DTM is shown in Figure 4, at WGS84 horizontal datum. Visualization was generated by the Matplotlib library.

5.2   Aerodrome geometry

Figure 5 depicts the Runway, the inner portion of the Runway Strip and the OFZs of the SBKP Airport. As SRTM data were used for DTM construction, differences to Google Earth data are apparent in the picture. It is also possible to visualize the irregularities of the Inner Transitional OFZ higher contour. As prescribed, 45 m height is attained at irregular orthogonal distances from the runway centerline, given the runway height profile. This visualization is possible by importing the generated KML file into Google Earth software.

5.3   Obstacles

Aircraft are movable objects situated inside the aerodrome and very close to protection surfaces.

Hence, after a geometric configuration is generated, aircraft must be also generated and positioned on taxiways/runways in the different situations they are allowed to operate. This is essential for obstacle assessment and for the subsequent conformity validation of a geometric configuration. Aircraft height is retrieved automatically from the DTM and cor-

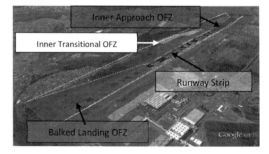

Figure 4.   SBKP aerodrome OFZ surfaces as viewed on Google Earth after importing a model generated KML file. Source: Prepared by the authors.

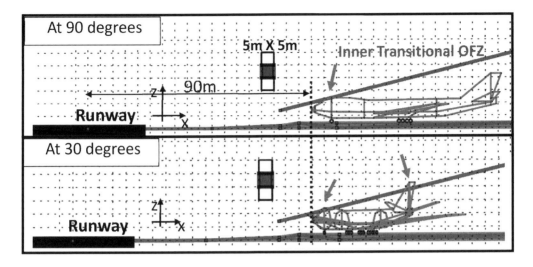

Figure 5. Obstacle assessment graphic output generated from the computational model for a fictitious instance. Graphic generated through Matplotlib Python library.
Source: Prepared by the authors.

responds to the maximum of two heights: DTM at nose wheel; and DTM at the main gear. Figure 6 depicts a fictitious test instance output automatically generated by the model. In this picture, the aircraft considered, a B747-8, waits to enter the Runway at 90 m from the Runway centerline. However, as the aircraft is positioned at different angles, wingtips and tail tips can protrude a protection surface. In Figure 6, while at 90° only aircraft nose is in the imminence of infringing the Inner Transitional OFZ, at 30° the vertical stabilizer infringes this OFZ surface. The model can perform obstacle assessment in a 3D environment and present a list of detected violations for the different combinations of objects and protection surfaces.

6 CONCLUSIONS

A clean-sheet computational model to conform an aerodrome to ICAO Annex 14 is presented, bridging a gap in the literature.

Modelling requirements of the aerodrome geometry were pinpointed. Simple geometric primitives were employed to represent aerodrome components: runways and taxiways, as well as on ground protection areas and airspace protection surfaces were represented by lines and polygons (2D and 3D).

Geospatial concerns were treated, as the distortions related to geographic projections, as well as the construction of a DTM (Digital Terrain Model), including the DTM height retrieval. A local topographic coordinate system and a TIN

paradigm for DTMs were also implemented and validated.

Along with GIS, CAD features regarding geometry generation and internal representation were considered. User interaction can be replaced by an internal programming procedure.

The model was implemented in Python language and applied to the case of Campinas International Airport in Brazil (SBKP), demonstrating its functionality. Graphic results were generated by the Matplotlib Python library. Additionally, model interoperability with the KML standard is provided, allowing results visualization on Google Earth.

Designers can benefit from the automation provided by this type of model, which favors analyses of design alternatives regarding their compliance with ICAO standards.

ACKNOWLEDGEMENTS

The authors acknowledge CAPES (Coordenação de Aperfeiçoamento de Pessoal de Nível Superior/Coordination for the Improvement of Higher Education Personnel) for a Doctoral Fellowship (Bolsa de Doutoramento), CNPq (Conselho Nacional de Desenvolvimento Científico e Tecnológico/ Brazilian National Council of Scientific and Technological Development) for a Research Productivity Fellowship (Process 307622/2015-0) and LPT/EPUSP (Laboratório de Planejamento e Operação de Transportes da Escola Politécnica da Universidade de São Paulo—Transportation Planning and Operation Laboratory of the School of

Engineering of the University of São Paulo) for technical support.

REFERENCES

ABNT, 1998. NBR 14166 - Rede de Referência Cadastral Municipal—Procedimento, Rio de Janeiro: ABNT.

Andrade, J. B. d., 2003. Fotogrametria. 2ed. Curitiba: SBEE.

Bidarra, R. & Bronsvoort, W., 2000. Semantic feature modelling. Computer-Aided Design, 32(3), pp. 201–225.

Burkholder, E. F., 1993. Design of a Local Coordinate System for Surveying, Engineering, and LIS/GIS. Journal of Surveying and Land Information Systems, 53(1), pp. 29–40.

Chen, D., Shams, S., Carmona-Moreno, C. & Leone, A., 2010. Assessment of open source GIS software for water resources management in developing countries. Journal of Hydro-environment Research, 4(3), pp. 253–264.

Cython, 2016. C-extensions for Python. [Online] Available at: http://cython.org. [Accessed 28 Jul 2016].

Dal'Forno, G. L., Aguirre, A. J., Hillebrand, F. L. & Gregório, F. d. V., 2011. Transformação de coordenadas geodésicas em coordenadas no plano topográfico local pelos métodos da norma NBR 14166/1998 e o de rotações / translações. A Mira: Agrimensura e Cartografia.

ESRI, 2016. ArcGIS for Desktop Extensions. ArcGIS for Aviation: Airports. [Online] Available at: http://www.esri.com/software/arcgis/extensions/aviation/airports/features. [Accessed 26 Jul 2016].

FAA, 2014. Federal Aviation Administration. AC 150/5300–13 A. Airport Design. Includes Change 1. Washington: U.S. Department of Transportation.

FAA, T. C. A. F. T. I. L. & Planning Technology, Inc, 2005. Three-Dimensional Airspace Analysis Programs. User Manual: Program Installation and Menu Commands: AutoCAD Map 3D 2005, s.l.: s.n.

Horonjeff, R., McKelvey, F., Sproule, W. & Young, S., 2010. Planning and Design of Airports. s.l.: McGraw-Hill Education, 5 Edition.

Hosaka, M., 1992. Modeling of curves and surfaces in CAD/CAM. s.l.:Springer-Verlag.

ICAO, 2004. International Civil Aviation Organization. Annex 14 to the Convention on International Civil Aviation. Volume I. Aerodrome Design and Operations.

ICAO, 2006. International Civil Aviation Organization. Procedures for Air Navigation Services: Aircraft Operations (PANS-OPS). Doc 8168.

Kazda, A. & Caves, R. E., 2007. Airport Design and Operation. Bingley: Emerald Group Publishing, 2nd Ed.

Leidig, M. & Teeuw, R., 2015. Free software: A review, in the context of disaster management. International Journal of Applied Earth Observation and Geoinformation, Volume 42, pp. 49–56.

Li, Z., Zhu, Q. & Gold, C., 2005. Digital terrain modeling. s.l.: CRC Press.

Monedero, J., 2000. Parametric design: a review and some experiences. Automation in Construction,9(4), pp. 369–377.

Planning Technology Inc., 2004. Three-Dimensional Airspace Analysis Programs (3DDAAP) process implementation: airport layout plan, navigational aid screening, web coordination, & photogrammetry standards. [Online] Available at: http://www.3daap.com/ [Accessed 18 Aug 2015].

PSF, 2016. Python Software Foundation. [Online] Available at: https://www.python.org/psf/. [Acessed 28 Jul 2016].

Shah, J. J., 1998. Designing with Parametric CAD: Classification and comparison of construction techniques. In: Geometric Modelling: theoretical and computational basis toward advanced CAD applications. Tokyo: Springer Science+ Business Media, LLC, pp. 53–68.

Silva, E. J. & Gualda, N. D. F., 2013. São Paulo/Guarulhos International Airport Airside Compatibility Analysis for A380–800 and B747–8 Aircraft Operation. Journal of the Brazilian Air Transportation Research Society, 9(1), pp. 9–22.

Simtra, 2016. A Transoft Solutions Company. Obstacle Surface Planner. [Online]. Available at: http://www.simtra.com [Accessed 26 Jul 2016].

Transoft Solutions, 2016. Transoft Solutions. [Online] Available at: http://www.simtra.com/. [Accessed 26 jul 2016].

TRB, 2010. Transportation Research Board. Airport Cooperative Research Program. Report 38. Understanding airspace, objects, and their effects on airports, Washington: s.n.

USGS, 2015. United States Geological Survey. Shuttle Radar Topography Mission (SRTM) 1 Arc-Second Global. [Online] Available at: https://lta.cr.usgs.gov/SRTM1 Arc [Accessed 29 Jul 2016].

Transport Infrastructure and Systems – Dell'Acqua & Wegman (Eds)
© 2017 Taylor & Francis Group, London, ISBN 978-1-138-03009-1

A speed model for curves of two-lane rural highways based on continuous speed data

Raul Almeida & Luís Vasconcelos
CITTA and Department of Civil Engineering, Polytechnic Institute of Viseu, Viseu, Portugal

Ana Bastos Silva
CITTA and Department of Civil Engineering, University of Coimbra, Coimbra, Portugal

ABSTRACT: The concept of design consistency refers to the conformance of road geometry with driver expectancy. An inconsistency in road design can be described as a geometric feature or a combination of features with unusual or extreme characteristics that may lead to unsafe driving behavior. A handful of methods and approaches have been trying to model design consistency. One of the most used criteria to evaluate design consistency is based on operating speed measurement. The present Portuguese design manual proposes a method to evaluate the geometric design consistency that is usually considered too simplistic, assuming that the speed on the curve is uniform and the values of the acceleration and deceleration are based on constant values. To investigate the limitations of this approach, field measurements were recorded using a vehicle equipped with a precision data logger that combines video footage with kinematic variables. The analysis of the resulting database has shown that some assumptions of traditional speed models do not hold and has motivated the development and calibration of a speed model on curves for two-lane rural highways. This model can differentiate the driving dynamics of a set of drivers on two-lane rural highways under a wide range of geometric conditions and can be integrated in a global methodology to assess design consistency on two-lane rural highways.

Keywords: Design consistency, Speed profile, Operating speed, Two-lane rural highways, Global positioning system

1 INTRODUCTION

The objective of transportation is usually described as the safe and efficient movement of goods and people. In the specific case of road transportation, this objective presupposes the existence of a road design that does not defraud the natural expectations of drivers. Thus, speed is normally used as one of the most relevant factors to assess the performance of a road infrastructure.

In recent decades, speed has been a major focus of interest for researchers in many fields, resulting in numerous operating speed models that can be applied to roads from different countries.

Design speed is set according to the desired functionality for a particular road. It establishes the starting point for the selection of most of the geometrical features of the road, which enables it to correspond to the expectations of the drivers in terms of travel time, economy and comfort. It should also be consistent with the expected speeds by drivers on the same road.

Operating speed is also a strong indicator of road performance, and is normally estimated by the 85th percentile of the distribution of observed speeds in real environment.

It is important to define road designs that do not require significant changes in the geometric characteristics of consecutive elements, and inherently abrupt changes to driver behavior. Therefore, the need to predict the behavior of drivers in these situations has resulted in various methodologies to evaluate the design consistency of two-lane rural highways.

2 LITERATURE REVIEW

The necessity to check the speed at individual elements of the road may, in many cases, prove to be overly restrictive and inadequate to ensure a comfortable and safe road design. There are many methods to ensure a good balance regarding contiguous elements of the road and design consistency for two-lane rural roads. Those methods can be categorized into four major groups: *a)* methods that use traffic behavior parameters (including operating speed profiles); *b)* methods that use indices related to road geometry; *c)* methods that use driver workload; *d)* methods that use checklists to fulfill design consistency criteria (Fitzpatrick et al. 2000a).

A consistent road design can be obtained when free-flow speed variations are made in a gradual way along the road. With this goal in mind, it is vital to have adequate models to estimate operating speed depending on the geometric characteristics of the road. With these models it is possible to define design consistency criteria with particular regard in establishing limits to speed differential between two consecutive geometrical elements, as recommended by the present Portuguese design manual.

The method used in the present Portuguese design manual to evaluate design consistency on two-lane rural highways consists first to determine operating speed in the various elements of the road and then calculate the operating speed profile. This method assumes that the acceleration and deceleration are constant, and that the operating speed is also constant along the curved sections of the road.

Between the various parameters that may be associated with operating speed, some are more significant than others, such as the maximum legal speed, number of lanes, road type, percentage of heavy vehicles, weather, visibility, period of the day, and factors related with driver behavior (Ye et al. 2001).

Although there is already a significant number of studies relating to operating speed, it continues to be justifiable the development of studies that allow to identify the parameters that affect operating speed, particularly in the development of continuous operating speed profiles on two-lane rural highways (Cafiso & Cerni 2012).

Typically, the design manuals of different countries establish the operating speed on road sections through a focused analysis on specific geometric elements of the road. In the United States, AASHTO (2011) proposes a number of recommendations regarding design speed and operating speed according to the type of road. This procedure is also found in the present Portuguese design manual regarding design speed and operating speed.

According to Lobo et al. (2013), the problem of speed modeling is an issue that concerns not only road infrastructure management entities but also researchers from academia, resulting in models developed for different geometric elements, type of vehicles and environmental conditions. These studies have led to various methodologies to evaluate design consistency on two-lane rural roads.

Road geometry has been identified as the most conditioning factor regarding operating speed, and can be characterized by various parameters such as the curve radius, the Degree of Curvature (DC), the Curvature Change Rate (CCR), and the deflection angle.

The model developed by Morrall & Talarico (1994) in their study to determine side friction on two-lane rural highways, consider the degree of curvature as the only feature that has a significant effect on operating speed.

Passetti & Fambro (1999) investigated the effect of the transition curves relative to operating speed on curves, considering the curvature $(1/R)$ as the only relevant feature to develop the their speed model.

The radius is considered by Misaghi & Hassan (2005) as the most statistically significant parameter for estimating the operating speed, according to the study conducted on several horizontal curves.

Other parameters, like road width have been studied by several authors using different approaches, as is the case of Lamm et al. (1988) that proposed several speed models for different categories of road width. Krammes et al. (1995) used the approaching speed on tangent in their speed model, seeking to translate the influence of the driver expectations through variables that characterize the upstream and downstream sections of the road, considered in the literature as being particularly related to the field of vision of drivers.

Operating speed, according to Gibreel et al. (1999), is the most widely used criterion to evaluate design consistency, and is usually defined as the speed corresponding to the 85th percentile of the speed distributions of a number of vehicles operating in free-flow conditions.

Several models have been developed to estimate operating speed on curves. However, the model format, the independent variables, and regression coefficients are substantially different from one model to another. This may prove that there is differences in driver behavior from one region to another, and highlights the lack of a model that is universally accepted by the scientific community.

The procedure to collect the data in most of this cases was performed using equipment that would necessarily have to be handled by an operator (e.g. LIDAR speed gun). These devices are likely to present several problems during field data gathering, such as human error and influence on the driver behavior. Another method is by the use of pavement sensors. Though, since it is a bulkier equipment, takes a longer time to install, and it remains visible to the driver, which normally tends to influence driver behavior.

Most of these methods allow the data to be collected at the medium point of the curve. Thus, the models that result from such methods assume that operating speed remains constant inside the curve, such as the model developed by Lamm et al. (1999). This model uses the CCR variable as the only explanatory variable for estimating the operating speed on curves, and assumes that the acceleration and deceleration occurs only at the tangent, and is equal to ±0.85 m/s^2. Similar values are considered by the present Portuguese design manual (±0.80 m/s^2).

Another model was developed by Ottesen & Krammes (2000), which is essentially based on the same assumptions that the model by Lamm et al. (1999), where the data is collected with speed guns in various spots on tangent and curves along two-lane rural roads.

The present study presents a method to collect continuous speed data based on Global Positioning System (GPS) that allows to collect a large amount of data without significant influence on the drivers, which facilitates the development of speed models.

3 OBJECTIVES

The objective of this study is to develop an operating speed model on curves that uses geometric characteristics of the road as explanatory variables. It can be used as a surrogate method to estimate the operating speed on curves to the present Portuguese design manual, and also be incorporated in a global methodology to evaluate design consistency.

The analysis is based on continuous speed data contrary to other models that use spot-speed data. This procedure allowed the development of the speed model and also the opportunity to analyze the speed variations from tangent to curved sections, and within the curved sections.

The research allowed to assess a new methodology to obtain the geometrical characteristics of the road based on the data collected and using specific software related to road design. Therefore, this paper also evaluates a new methodology for collecting and processing continuous speed data on two-lane rural highways.

A comparative analysis is also performed with the proposed model that is developed using continuous speed data, and other speed models that used different methodologies to collect speed data.

4 DATA COLLECTION METHODOLOGY

The analysis presented in this paper is not based on spot-speed data gathering using equipment placed in various locations of the road as observed in other studies, but based on continuous speed data collected over time and length. Consequently, the

model and the results will become more accurate in case of a similar sample of data.

Continuous speed profiles allow to study driver behavior namely the variations of operating speed throughout the various elements of the road, and the points where those variations occur. These phenomena could not be studied before the use of continuous speed data, mainly concerning speed variations within the curved sections.

According Pérez-Zuriaga et al. (2010) this technique has some limitations, particularly regarding the equipment used as it can influence driver behavior, though in a minimal way, since they are aware of being observed and tested.

However, this method has a great advantage because it uses a portable GPS device that collects a large amount of continuous data for further analysis and that allows to build continuous speed profiles for individual drivers.

4.1 Equipment and road characteristics

The equipment used for data gathering was a data logger from Race Technology Ltd (DL1 MK3 model). The data logger has a built in 6 g accelerometer with three axes and a 20 Hz GPS (20 recordings per second), enabling to reference the data for periods of time in relation to the position on the road segment.

This equipment has an accuracy of about three meters in terms of positioning and an accuracy in relation to speed of around 0.1 km/h. In addition to the GPS functionality, it is also possible to connect external sensors. This device was complemented with another device (VIDEO4 model) that allows video recording using a high definition camera placed inside the vehicle perfectly synchronized with the data logger.

The data was collected in two two-lane rural road segments in Portugal. The gathering and processing of data allowed to differentiate the driving dynamics of a set of drivers on two-lane rural highways.

The segments of road are characterized by a low traffic volume with an Annual Average Daily Traffic (AADT) of approximately 6500 vehicles. Since this study was focused on horizontal alignments, the sample data was limited to relatively flat stretches (grades less than $\pm4\%$). A total of 47 curves were analyzed for each direction of traffic. The radius of the circular curves varied between 40 m and 500 m.

The pavement was in good condition for all the extension of the road. The road segments presented a total length of about 13.5 km long with pavement shoulders throughout most of its extension. Road width had an average of 7 m. The data was collected between August and September during work days under dry weather. None of the road segments presented important intersections.

All the drivers had more than five years of driving experience and were allowed a testing period before each field session. All of the laps were recorded in real time with the support of the installed equipment and checked afterwards with the video footage to guarantee that the drivers were always in free-flow conditions.

4.2 Data analysis and modeling

The data obtained from the equipment allowed the modeling of the road segments with the accuracy needed to obtain the various geometric characteristics of the horizontal alignments. The coordinates taken from the data logger were exported with the use of the data review program provided with the equipment. Using the software AutoCAD Civil 3D it was possible to determine the beginning point and end point of each element of the road segments (tangents, curves, and transition curves). Consequently, all the geometric characteristics of each road segment were obtained (the radius was obtained from the trajectory but it can be considered to be very close to the real axis radius).

The software provided by the equipment offers an extensive analysis of the journeys made by the drivers on the road segments (Fig. 1), i.e. path view overlaid with satellite maps, speed and acceleration diagrams, synchronized video footage with the position on the road and the corresponding speed diagram, etc.

Since the equipment takes into account GPS, it is assumed that some positioning errors might occur, but since the recording are taken at intervals of 0.1 seconds those errors did not influence the modeling of the road segments.

With the road segments characteristics retrieved from the software analysis, there was the need to construct curvature diagrams of all the road seg-

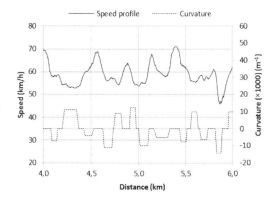

Figure 2. Example of a partial curvature diagram and speed profile of a single driver.

ments. These diagrams were combined with actual speed profiles of drivers to analyze speed variations and driver behavior on curves (Fig. 2).

5 OPERATING SPEED MODEL

Many of the methods used to evaluate design consistency on two-lane rural highways are based in the analysis of operating speed. The data gathering methodology used in this paper was chosen because it is the most representative in recent studies and one that allows to observe driver behavior concerning speed and geometric design of the road when compared to other methods of data collection.

This method is similar to that adopted by Pérez-Zuriaga et al. (2013) and other researchers that explored continuous speed data gathering, like the study of Cafiso & Cerni (2012), but with some differences, namely in obtaining and modeling the geometric characteristics of the road, and data reduction and analysis.

One of the main advantage of using this data gathering procedure was the possibility to collect a substantial amount of data which allowed the development of the operating speed model on curves. Although many observations were collected and validated under free-flow conditions, the sample is not representative of all the drivers that drive in a particular road under the same circumstances.

The operating speed model developed in this paper can only be used as a tool to evaluate design consistency when associated to operating speed profiles and consistency thresholds.

Taking into consideration all the parameters of the horizontal alignment on both segments of the road, the most important variables are the curve radius, curve length, and deflection angle.

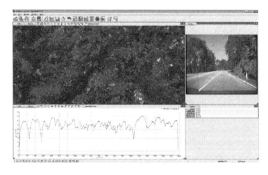

Figure 1. Display of the software analysis software provided with the data gathering equipment.

The curve radius varied from 40 m to 500 m, the curve length from 33 m to 232 m, and the deflection angle from 9.85° to 120.75°.

With the purpose to develop several regression models, the relations between these parameters were studied. The radius of the curve was the variable that revealed the highest correlation to operating speed, and the length of the curve revealed the least correlation. Therefore, the radius of the curve was used as the explanatory variable to develop different models.

To develop the operating speed model the minimum value of the observed speed data was used in each of the 47 curves for each driver. A total of 284 observations were obtained and validated.

The assumption that the operating speed on curve is constant was not observed. From the analysis of the continuous speed profiles, most of deceleration and acceleration length occur within the curve (Fig. 3). Only in very specific situations a constant speed on curves was observed, for instance, curves that are very close to each other and have a small tangent between them, or curves with similar radii. These situations do not allow the driver to increase or decrease the travel speed, taking the decision to maintain a constant speed between these elements (see rightmost curves in Fig. 3).

From a single regression analysis resulted the exponential model shown in Figure 4 with the radius as explanatory variable and $R^2 = 0.511$. It can be obtained by the following equation:

$$V_{85} = 81.29\, e^{(-3.86/\sqrt{R})} \tag{1}$$

where V_{85} = operating speed on curve (km/h); R = radius (m).

The model shows a high slope for small radii and decreases as the radius gets bigger. It indicates that

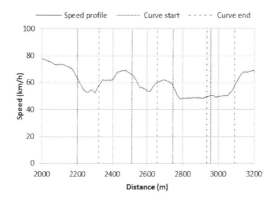

Figure 3. Example of a partial speed profile of a single driver.

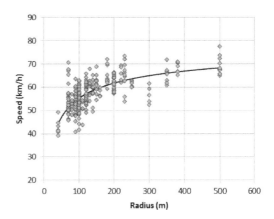

Figure 4. Proposed operating speed model on curves.

for curves with large radius this variable is not as significant in the choice of speed as for curves with small radius. This analysis is consistent with the model developed by Pérez-Zuriaga et al. (2010).

6 COMPARISON OF MODELS

The proposed model was compared with others in the literature. All models that were considered for the analysis were applied to curves with radii between 40 m and 500 m to comply with the proposed model in this study.

The model developed by Morrall & Talarico (1994) on two-lane rural highways in Canada was calibrated with observations from 9 curves with pavement shoulders. The curve radius varied between 290 m and 3490 m. The model presents the degree of curvature as the explanatory variable, and has an $R^2 = 0.631$. This model assumes that the speed on curve is constant, and it is represented by the following equation:

$$V_{85} = e^{(4.561 - 0.00586\,DC)} \tag{2}$$

where V_{85} = operating speed (km/h); DC = degree of curvature (°/100 m of arc length). The variable DC present in this model can be defined as one degree per 100 m of arc length (SI units), hence it is possible to relate this variable to the variable curve radius, i.e. $DC = 5729.578/R$, where R is the radius (m).

Another model was developed on two-lane rural roads in the United States by Passetti & Fambro (1999) with observations from 12 transition curves and 39 circular curves. The model uses the inverse of the radius as the explanatory variable, and has an $R^2 = 0.68$. This model also assumes that the speed on curve is constant, and it is represented by the following equation:

$$V_{85} = 103.9 - 3020.5 / R \qquad (3)$$

where V_{85} = operating speed (km/h); R = radius (m).

In another study conducted by Fitzpatrick et al. (2000b), the authors also developed a model to estimate the V_{85} for vehicles traveling on two-lane rural highways. The speed data were collected in 176 locations in the United States. The model uses the inverse of the radius as the explanatory variable. Two equations were developed for grades less than ±4%, but only the one that best fits in this analysis will be used in the comparison. Other grades were considered by these researchers to develop more equations. The model presents the inverse of the radius as the explanatory variable, and has an R^2 coefficient of 0.76. It can be represented by the following equation:

$$V_{85MC} = 104.82 - 3574.51 / R; \quad 0\% \le G \le 4\% \qquad (4)$$

where V_{85MC} = operating speed on curve midpoint (km/h); R = radius (m); G = grade (%). It was noted by the authors that operating speed on curves is very similar to the speed on tangents when the curve radius is equal or higher than 800 m. On curves with radii below 250 m operating speed decreases very rapidly.

Kanellaidis et al. (1990) developed a model in Greece that studied the relation between operating speed on curves and various geometric characteristics of the road. The data was collected in 58 locations, and the model can be represented by the following equation:

$$V_{85} = 129.88 - 623.1 / \sqrt{R} \qquad (5)$$

where V_{85} = operating speed (km/h); R = radius (m).

The inverse of the squared root is the independent variable in Equation 1, which is similar to the model developed by Kanellaidis et al. (1990). From other models shown on Table 1 that also use the radius of the curve as the explanatory variable, and performing a comparative analysis of those models with the proposed model, it can be concluded that this model tends to underestimate operating speed on curves (Fig. 5). That can be explained because the operating speed used to develop this model was always the minimum observed in each curve from the continuous speed data, when the other models used spot-speed collection methods, usually at the medium point inside the curve, which not always correspond to the geometrical point where the minimum speed occurs.

Table 1. Comparison of models.

Model	Equation
Kanellaidis et al. (1990)	$V_{85} = 129.88 - 623.1 / \sqrt{R}$
Morrall & Talarico (1994)	$V_{85} = e^{(4.561 - 0.00586DC)}$
Passetti & Fambro (1999)	$V_{85} = 103.9 - 3020.5 / R$
Fitzpatrick et al. (2000b) $0\% \le G \le 4\%$	$V_{85MC} = 104.82 - 3574.51 / R$
Proposed model	$V_{85} = 81.29 \, e^{(-3.86/\sqrt{R})}$

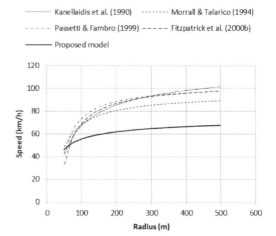

Figure 5. Comparison of operating speed models.

7 CONCLUSIONS

The necessity to have methods to evaluate design consistency holds to prevent unexpected changes to the geometrical characteristics of consecutive road elements and combinations of elements that do not comply with driver expectations. Some of the methods that exist today to estimate design consistency on two-lane rural highways are based on operating speed profiles that do not comply with actual driver behavior. Thus, it is necessary to have models that are able to predict operating speed concerning road characteristics and driver behavior.

The methodology adopted for data gathering allowed to model genuine operating speed profiles and thereby analyze in greater detail the behavior of drivers on several elements of the road. However, it was not conclusive if this type of equipment has less influence than other data gathering methods.

The speed models that result from this type of analysis will be more accurate that those that use data gathering on specific points of the road, such as the midpoint of the curve. Thus, this methodology allows to determine the beginning point and end point of both acceleration and deceleration phenomena, as well as the minimum speed observed for each driver inside the curve.

It was observed that the speed is not constant on the curve, as is assumed in various methods to evaluate design consistency on two-lane rural roads.

The use of the software AutoCAD Civil 3D for modeling the road segments, combined with the instrumented vehicle for field data gathering, resulted in an innovative methodology that allows in an expeditious way to obtain the geometric characteristics of the horizontal alignment, revealing a procedure that is essential to the development of operating speed models on two-lane rural roads.

REFERENCES

AASHTO. (2011). *A policy on geometric design of highways and streets*. 6th Edition. Washington, D.C., American Association of State Highway and Transportation Officials.

Cafiso, S., Cerni, G. (2012). *New approach to defining continuous speed profile models for two-lane rural roads*. Transportation Research Record - Journal of the Transportation Research Board, Vol. 2309. Washington, D.C., TRB—Transportation Research Board, pp. 157–167.

Fitzpatrick, K., Wooldridge, M.D., Tsimhoni, O., Collins, J.M., Green, P., Bauer, K.M., Parma, K.D., Koppa, R., Harwood, D.W., Anderson, I., Krammes, R.A., Poggioli, B. (2000a). *Alternative design consistency rating methods for two-lane rural highways*. Report No. FHWARD-99-172. Texas Transportation Institute, U.S. Department of Transportation, FHWA-Federal Highway Administration.

Fitzpatrick, K., Elefteriadou, L., Harwood, D.W., Collins, J.M., McFadden, J., Anderson, I.B., Krammes, R.A., Irizarry, N., Parma, K.D., Bauer, K.M., Passetti, K. (2000b). *Speed prediction for two-lane rural highways*. Report No. FHWA-RD-99-174. Texas Transportation Institute, U.S. Department of Transportation, FHWA-Federal Highway Administration.

Gibreel, G.M., Easa, S.M., Hassan, Y., El-Dimeery, I.A. (1999). *State of the art of highway geometric design consistency*. Journal of Transportation Engineering, Vol. 125(4). ASCE-American Society of Civil Engineers, pp. 305–313.

Kanellaidis, G., Golias, J., Efstathiadis, S. (1990). *Driver's speed behavior on rural road curves*. Traffic Engineering and Control, Vol. 31(7/8). London, England, Printerhall Limited, pp. 414–415.

Krammes, R.A., Brackett, R.Q., Shafer, M.A., Ottesen, J.L., Anderson, I.B., Fink, K.L., Collins, K.M.,

Pendleton, O.J., Messer, C.J. (1995). *Horizontal alignment design consistency for rural two-lane highways*. Report No. FHWA-RD-94-034. Texas Transportation Institute, U.S. Department of Transportation, FHWA—Federal Highway Administration.

Lamm, R., Choueiri, E.M., Hayward, J.C., Paluri, A. (1988). *Possible design procedure to promote design consistency in highway geometric design on two-lane rural roads*. Transportation Research Record, Vol. 1195. Washington, D.C., TRB—Transportation Research Board, pp. 111–122.

Lamm, R., Psarianos, B., Mailaender, T. (1999). *Highway design and traffic safety engineering handbook*. McGraw-Hill.

Lobo, A., Rodrigues, C., Couto, A. (2013). *Free-flow speed model based on Portuguese roadway design features for two-lane highways*. Transportation Research Record—Journal of the Transportation Research Board, Vol. 2348. Washington, D.C., TRB—Transportation Research Board, pp. 12–18.

Misaghi, P., Hassan, Y. (2005). *Modeling operating speed and speed differential on two-lane rural roads*. Journal of Transportation Engineering, Vol. 131(6). ASCE—American Society of Civil Engineers, pp. 408–418.

Morrall, J.F., Talarico, R.J. (1994). *Side friction demanded and margins of safety on horizontal curves*. Transportation Research Record, Vol. 1435. Washington, D.C., TRB—Transportation Research Board, pp. 145–152.

Passetti, K., Fambro, D. (1999). *Operating speeds on curves with and without spiral transitions*. Transportation Research Record—Journal of the Transportation Research Board, Vol. 1658. Washington, D.C., TRB - Transportation Research Board, pp. 9–16.

Pérez-Zuriaga, A.M., Camacho-Torregrosa, F.J., García-García, A.G. (2013). *Tangent-to-curve transition on two-lane rural roads based on continuous speed profiles*. Journal of Transportation Engineering, Vol. 139(11). ASCE—American Society of Civil Engineers, pp. 1048–1057.

Pérez-Zuriaga, A.M., García-García, A., Camacho-Torregrosa, F.J., D'Attoma, P. (2010). *Modeling operating speed and deceleration on two-lane rural roads with Global Positioning System data*. Transportation Research Record—Journal of the Transportation Research Board, Vol. 2171. Washington, D.C., TRB—Transportation Research Board, pp. 11–20.

Ottesen, J., Krammes, R. (2000). *Speed-profile model for a design-consistency evaluation procedure in the United States*. Transportation Research Record—Journal of the Transportation Research Board, Vol. 1701. Washington, D.C., TRB—Transportation Research Board, pp. 76–85.

Ye, Q., Tarko, A., Sinha, K. (2001). *Model of free-flow speed for Indiana arterial roads*. Transportation Research Record—Journal of the Transportation Research Board, Vol. 1776. Washington, D.C., TRB—Transportation Research Board, pp. 189–193.

Transport Infrastructure and Systems – Dell'Acqua & Wegman (Eds)
© 2017 Taylor & Francis Group, London, ISBN 978-1-138-03009-1

Rheological characterization of cold bituminous mastics produced with different mineral additions

C. Godenzoni, M. Bocci & A. Graziani
Dipartimento di Ingegneria Civile Edile e Architettura, Università Politecnica delle Marche, Ancona, Italy

ABSTRACT: Nowadays, cold bituminous mixtures produced with bitumen emulsion (CBEMs) represent one of the most attractive alternatives with respect to traditional hot mix asphalt. In particular, the use of CBEMs is increasing because they combine economical and environmental efficiency. CBEMs can be schematized as composite of coarse aggregate particles (virgin and/or reclaimed) covered and bound by a bituminous mastic (fresh bitumen and filler-sized particles), also incorporating part of the fine aggregate particles (sand). Therefore, the mechanical behavior of the whole CBEMs is strictly connected to the behavior of its associated Cold Bituminous mastic (CBm). The main objective of this study is to evaluate the effect of different mineral additions and volumetric concentrations on the LVE properties of CBms. Complex shear modulus (G^*) was measured on CBms prepared with Calcium Carbonate (CC) or cement (CEM) at 0.15 and 0.3 as mineral addition ratios. In addition, the influence of curing time (1 and 3 days) on LVE properties, was also investigated. The testing protocol included Brookfield viscosity measurements during mixing phase and Complex shear modulus testing, performed at temperatures ranging from 5°C to 60°C. The results showed that CBm composition strongly affects the viscosity and the mechanical response of CBms. In particular, at higher temperatures and higher concentration ratios, phase separation was observed. The linear viscoelastic behavior showed an evolutive trend at longer curing times, especially for CBms prepared with cement.

1 INTRODUCTION

The use of Cold Bitumen Emulsion Mixtures (CBEMs) for pavement construction, maintenance and rehabilitation constantly increased over the last 20 years [Liebenberg et al., 2004; Serfass et al., 2004; Santagata et al., 2009, Nassar et al., 2016a]. Economic and environmental benefits, with respect to both hot and warm mixtures, derive from the reduction of energy necessary for heating both bitumen and aggregates. Moreover, when emulsion-based materials are used for pavement recycling, consumption of virgin aggregate can be significantly reduced [Stroup-Gardiner, 2011]. The composition of CBEMs can be extremely variable and affecting their mechanical response. In particular, CBEM properties vary depending on the proportions in which binders, bitumen and cement, are mixed.

When comparing the composition and the behavior of CBEMs to that of conventional Hot-Mix Asphalt (HMA) mixtures, two aspects are of the outmost importance. First, water instead of heating is used in order to allow CBEMs laydown and compaction. Indeed, moisture content is the main factor controlling the rheological properties of CBEMs in the fresh state, in particular their ability to achieve suitable volumetric properties [Grilli et al., 2016]. Second, CBEMs can be defined as evolutive materials because their physical structure evolve over time. This process, known as curing, leads to the improvement of the mechanical properties (e.g. stiffness and strength) until reaching a log-term cured state [Jenkins and Moloto, 2008; Godenzoni et al., 2016a; Graziani et al., 2016].

Because of the simultaneous presence of both bituminous and hydraulic binders, the curing process of CBEMs is actually due to the interaction of different physical and chemical mechanisms [Brown and Needham, 2000]. Curing mechanisms related to the bituminous phase basically involve emulsion breaking and water expulsion.

Huang and Di Benedetto (2015) proposed a useful schematization of HMA as a multi-phase material:

– bituminous mastic, composed by bituminous binder and filler-sized particles that establishes the connection between bituminous binder and mixture;
– fine aggregate mixture or mortar, composed by bituminous mastic and fine aggregate particles, that fills voids between the coarsest aggregate particles;
– coarse aggregates skeleton.

The same representation can be adopted for CBEMs. In this case, cold bituminous mastic (CBm) is composed by bitumen emulsion (bitumen droplets dispersed in water) and filler-sized particles. Over time, water evaporates and emulsion breaking occurs; bitumen droplets coalesce forming a continuous thin film of bitumen that englobes filler-sized particles. In addition to filler-sized particles commonly used with HMA (e.g. recovered dust), other mineral additions such as cement or hydrate lime can be included in CBEMs [Nassar et al., 2016b].

CBEMs can also be analyzed at different time-scales because water content evolves over time as well as the physical state of the mixture, i.e. from the fresh state to the cured state.

This feature is noticeable at mixture level and at each smaller level of investigation (mortar or mastic-level).

The performance of CBEMs mainly depends on the rheological properties of its CBm. In this context, it is important to investigate the linear viscoelastic (LVE) properties in order to understand how the proportion of each constituent can affect the whole LVE behavior of the associated mixture. Different proportions, generating various microstructures, can produce a wide range of bituminous material behaviors [Elnasri et al. 2013].

In addition, considering LVE properties of CBm, mineral addition used as emulsion breaking regulator, material stiffener and filler, plays a fundamental role.

The main objective of this study is to evaluate the effect of different mineral additions and volumetric concentrations on the LVE properties of CBms. The complex shear modulus (G^*) was measured on CBms prepared with limestone filler, i.e. Calcium Carbonate (CC) or cement (CEM). In addition, the influence of curing time (1 and 3 days) was also investigated.

2 BACKGROUND ON RHEOLOGICAL PROPERTIES OF BITUMINOUS MATERIALS

The mechanical behavior of CBms can be analyzed using the same approach commonly adopted for bituminous binders.

Bituminous materials exhibit a thermo-viscoelastic behavior, i.e. dependent to applied temperature and loading frequency [Pellinen et al., 2007]. In this context, G^* is often used to identify their LVE response and it is defined as:

$$G^* = \frac{\tau^*(t)}{\gamma^*(t)} \tag{1}$$

with

$$\tau^*(t) = \tau_0 \cdot \left[\cos(\omega t) + i \cdot \sin(\omega t)\right] \tag{2}$$

$$\gamma^*(t) = \gamma_0 \cdot \left[\cos(\omega t - \delta) + i \cdot \sin(\omega t - \delta)\right] \tag{3}$$

where τ_0 and γ_0 are the amplitude of stress and strain respectively, ω is the angular frequency, t is the time and δ is the phase angle.

By substituting these expressions (periodic sinusoidal loads) in the previous Equation (1), the following form can be obtained:

$$G^*(i\omega) = \frac{\tau_0}{\gamma_0} e^{i\delta} = G_1 + iG_2 \tag{4}$$

As reported in Equation 4, G^* can be decomposed in two components

- $G_1 = G^* \cdot \cos(\delta)$, is the storage modulus, used as indicator of the elastic behavior of the material (elastic behavior);
- $G_2 = G^* \cdot \text{sen}(\delta)$, is the loss modulus, used as indicator of the dissipated energy when a load is applied to the material and, hence, represents the viscous component of the material behavior (viscous behavior).

The phase angle represents the relative amounts of recoverable and non-recoverable deformation, therefore, it concurs to identify the material behavior: when δ tends to 0° the bitumen behaves as an elastic material, whereas δ values close to 90° a viscous behavior is detected.

3 EXPERIMENTAL PROGRAM

3.1 Materials

A cationic overstabilized bitumen emulsion designated as C 60 B 10 (EN 13808) was adopted to prepare the cold bituminous mastics. This emulsion was selected because it is commonly used during cold recycling operations. Its breaking behavior is evaluated by measuring the mixing stability with cement (UNI EN 12848).

Bitumen emulsion properties were reported in Table 1.

Two mineral additions were selected:

- Portland limestone cement II A/LL, strength class 42.5 R (UNI EN 197-1);
- limestone filler, i.e. finely ground calcium carbonate ($CaCO_3$).

Portland cement and calcium carbonate were characterized determining the Rigden voids and the fineness (specific surface) using the Blaine method (Table 2).

Table 1. Characterization of mineral additions.

Characteristics of the bitumen emulsion

Property	Value
Water content (UNI EN 1428)	40%
pH value	3
Settling tendency @ 7 days	8%
Breaking value (UNI EN 13075-1)	180%
Mixing stability with cement (UNI EN 12848)	≤0.2 g
Application temperature	5–80°C

Table 2. Characterization of mineral additions.

Parameter	Cement	Calcium carbonate
Rigden voids [%] UNI EN 1097-4	40.18	31.14
Specific surface [m²/g] UNI EN 196-6	0.41	0.51

Table 3. Summary of CBms produced.

Mineral addition	Residual bitumen [g]	Concentration ratio	Curing time [day]	Sample code
None	100	0	1	BE_1d
			3	BE_3d
Cement		0.15	1	CEM0.15_1d
			3	CEM0.15_3d
		0.3	1	CEM0.3_1d
			3	CEM0.3_3d
Calcium carbonate		0.15	1	CC0.15_1d
			3	CC0.15_3d
		0.3	1	CC0.3_1d
			3	CC0.3_3d

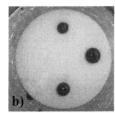

Figure 1. Mastic preparation: a) mixing by mechanical device; b) and c) pouring of liquid mastic in silicone molds (8 mm and 25 mm diameter).

Two different volumetric concertation ratios (volume of mineral addition over volume of bitumen) of mineral addition were considered for CBms preparation: 0.15 and 0.3. These values were calculated starting from the mass concentration ratio and considering the particles densities of 1.02, 3.15 and 2.69 g/cm³ for bitumen, cement and calcium carbonate, respectively. The adopted volumetric concentration ratios were intended to simulate the common composition of CBms, where mass ratio between bitumen and cement is usually around 1 [Godenzoni et al., 2016b].

CBms were coded with an alphanumeric code composed by letters referring to the types of mineral addition adopted (CEM or CC, for cement or calcium carbonate, respectively) and a number representing the volumetric concentration ratio (0.15 or 0.3).

For example, a CBm produced using cement at 0.15 of volumetric concentration ratio, was coded as CEM0.15. In addition, bitumen emulsion was also tested and coded as BE (without mineral additions).

3.2 Mastics production

A standardized procedure for the production of CBms is not available. In this experimental study 25 g of pre-wetting water was initially added to 100 g of calcium carbonate or cement and stirred by hand for one minute, in order to facilitate mixing and avert breaking of the emulsion (Table 3). Afterwards, bitumen emulsion was added and stirred by hand until obtaining a homogeneous compound without any lumps. Then a high-shear mixer was adopted for the second mixing phase

and the lower rotation speed rate equal to 489.3 rpm was selected in order to avoid the premature emulsion breaking (Fig. 1a).

CBms were mixed continuously for one hour at room temperature and sampled after 15, 45 and 60 minutes (end of mixing phase) of mixing in order to check the mastic homogeneity by measuring CBm viscosity and reject mastic when sedimentation phenomena occur [Santagata et al., 2016].

The homogeneity control was carried out by measuring material viscosity using a Brookfield rotational viscometer.

At the end of this phase, mastics were poured in silicone molds with different dimensions (8 and 25 mm) (Fig. 1b and c). Then, CBm specimens were cured for one and three days at 40°C, in a temperature controlled chamber.

In particular, the production of small size test specimens has avoided separation between the materials characterized by different densities, obtaining homogenous samples for rheological laboratory testing.

At the end, CBms codes were updated adding information about the curing time (1d and 3d, for one and three days of curing, respectively). For example, a CBm produced using cement at 0.15 of volumetric concentration ratio and tested after one day, was coded as CEM0.15–1d. The list of CBms produced is given in Table 3.

3.3 *Testing program and protocols*

3.3.1 *Brookfield rotational viscometer*

In order to quantitatively assess the occurrence of segregation phenomena, CBm samples were characterized in terms of dynamic viscosity (η). Measurements were performed by means of a Brookfield rotational viscometer operated at ambient temperature (about 25°C). About 10 ml of CBm were poured into the cylinder. The spindle (S21) was lowered into the samples and subsequently, the test was run at different shear rates γ (ranging from 50 up to 200 s^{-1}).

3.3.2 *Strain-sweep and frequency-sweep tests*

Rheological measurements (G^*) were performed using a stress/strain controlled Dynamic Shear Rheometer (DSR) device equipped with parallel plate-plate geometry. Strain-Sweep (SS) tests were preliminary performed with the aim to investigate the LVE limit and define a suitable range of strain level for both BE and CBms [Dondi et al., 2014; Frigio et al., 2016]. SS tests were performed at 0°C by using 8 mm parallel-plate-geometry and 2 mm gap, applying a constant frequency of 10 rad/s (1.59 Hz). A unique strain level of 0.01% was adopted as LVE limit for all materials, in order to simplify the testing procedure. This value was selected based on the LVE limit detected with CEM mastic, even if BE and CC mastic were characterized by higher LVE limits.

Frequency-Sweep (FS) test was conducted in a range of frequencies between 0.01 and 10 Hz, at the temperatures of 0, 10, 20, 30, 40, 50 and 60°C. The 8 mm plate-plate measuring system with 2 mm gap was adopted from 0 up to 40°C and 25 mm plate-plate measuring system with 1 mm gap was used from 30 to 60°C (Fig. 2).

4 RESULTS AND ANALYSIS

4.1 *Dynamic viscosity (η)*

Viscosity provides an evaluation of the flow characteristics of materials. In this study, measurements in terms of dynamic viscosity were carried out in order to verify the homogeneity of CBm samples

Figure 2. Dynamic shear rheometer with different measuring system: a) 8 mm and b) 25 mm parallel-plate-geometry.

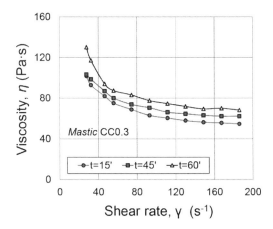

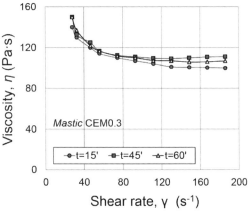

Figure 3. Dynamic viscosity measurements as a function of shear rates, at different sampling times: a) mastic CC0.3 and b) mastic CEM0.3.

and to evaluate segregation phenomena [Santagata et al., 2016; Wang et al., 2016].

In Fig. 3, an example of results obtained from CC0.30 and CEM0.30 mastics was reported.

As commonly observed for suspension, a typical shear thinning behavior was detected for both

CC0.3 and CEM0.3 mastics: viscosity decreases with the increasing in shear rate showing an asymptotic trend. As expected, viscosity values increase with mixing time for CC0.3 mastic; differently, a clear behavior over time cannot be distinguished for CEM mastic.

As it can be observed at higher shear rate (γ = 200 s^{-1}), viscosity reached a constant value of about 60 Pa·s and 100 Pa·s for CC0.30 and CEM0.30 mastics, respectively.

A similar behavior was also detected for CC0.15 and CEM0.15 mastics but viscosity measurements were characterized by lower values due to the lower volumetric concentration ratio of mineral addition.

According to previous results, it can be concluded that CBm samples can be considered homogeneously mixed, i.e. no segregation phenomena occur during the mixing phase and consequently, can be adopted for mechanical testing.

4.2 Complex shear modulus (G*)

In order to analyze G^*, graphs as *Black diagram*, where the norm of complex shear modulus ($|G^*|$) is reported as a function of phase angle (δ), were adopted.

The rheological parameters $|G^*|$ and δ of BE, CC0.15, CC0.30, CEM0.15 and CEM 0.30, cured 1 and 3 days are reported, in terms of Black diagram, in Fig. 4. Results represent the averaged values calculated on two replicates specimens.

Considering the CC mastic (Fig. 4a and Fig. 4c), it can be noted that at both curing times (i.e. 1 and 3 days) and volumetric concentration ratio (i.e. 0.15 and 0.30) did not affect the mechanical response of this material. In fact, the reported Black diagrams is practically superposed. At high temperatures, $|G^*|$ and δ were 85 Pa and 85°, respectively while at low temperatures, $|G^*|$ and δ were 131·10^6 Pa and 35°, respectively. In particular, results carried out on CC0.15 and CC0.30 were approximately the same of those obtained on BE.

The rheological behavior detected for CEM mastic was totally different: the influence of both concentration ratio (i.e. 0.15 and 0.3) and curing time (i.e. 1 and 3 days) was marked (Fig. 4b and Fig. 4d).

Considering CEM0.15-1d mastic: at high temperatures, $|G^*|$ and δ were 5000 Pa and 51°, respectively while at low temperatures, $|G^*|$ and δ were 93·10^6 Pa and 38°, respectively. Differently with CEM0.15-3d mastic, at high temperatures, the $|G^*|$ and δ were 64 Pa and 51°, respectively while at low

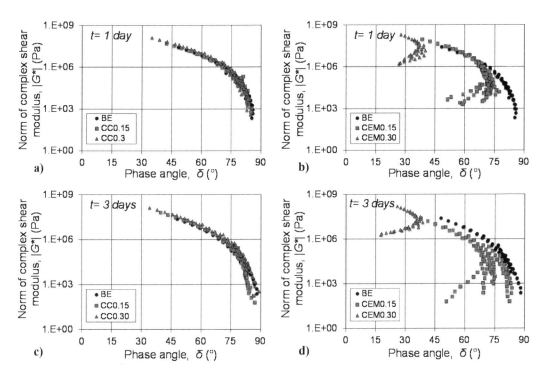

Figure 4. Black diagram, effect of concentration ratio: a) BE, CC0.15 and CC0.3 mastics at 1 day of curing; b) BE, CEM0.15 and CEM0.3 mastics at 1 day of curing; c) BE, CC0.15 and CC0.3 mastics at 3 days of curing; d) BE, CEM0.15 and CEM0.3 mastics at 3 days of curing.

temperatures, $|G^*|$ and δ were $21 \cdot 10^6$ Pa and $38°$, respectively.

Considering CEM0.30-1d mastic: at high temperatures, $|G^*|$ and δ are $1.9 \cdot 10^6$ Pa and $27.4°$, respectively while at low temperatures, $|G^*|$ and δ were $208 \cdot 10^6$ Pa and $27.7°$, respectively.

Differently with CEM0.30–3d mastic, at high temperatures, $|G^*|$ and δ were $2 \cdot 10^6$ Pa and $18°$, respectively while at low temperatures, $|G^*|$ and δ were $158 \cdot 10^6$ Pa and $26°$, respectively.

According to previous results, it is possible to assess that an increasing in concentration ratio led to an increase of $|G^*|$ and, consequently, a reduction in δ. Fig. 4b highlighted that the mechanical behavior of CEM mastic was strictly related to the amount of cement added in the material.

In fact, increasing cement content, stiffness properties move from those achieved with BE (viscoelastic fluid behavior) to those obtained with CBEM (viscoelastic solid behavior).

This observation is probably due to an increase in the development of cementitious products (at higher cement content and curing time) that stiffened the structure of CEM mastics.

Although, the effect of curing time on CEM0.15 mastic was evident, it was less marked with respect to that obtained analyzing CEM0.3.

As it can be noted from Fig. 4d, CEM0.15 mastic cured 3 days was not characterized by a unique trend in the Black diagram. This peculiarity can be related to phase separation phenomena that probably occur in the mastic, at high temperatures (from 40°C to 60°C).

In Fig. 5, the comparison between CC and CEM mastics at both volumetric concentration ratios (0.15 or 0.3), was reported considering one day of curing.

As it can be observed from Fig. 5a, at volumetric concentration ratio equal to 0.15, the rheological response at lower temperatures of both CC0.15 and CEM0.15 mastics were practically superposed. At temperature of 20°C, the Black diagram was characterized by two different trends for the considered CBms: G_0 values were almost the same for CC0.15 and CEM0.15 mastics but, the adoption of cement as mineral addition caused a strong reduction in δ when temperature increased (δ changes from 85° with CC0.15 to 50° with CEM0.15, at temperature of 60°C). This observation highlighted the cementitious microstructure of CEM0.15 mastic promoted a higher viscoelastic response with respect to that detected with CC0.15 mastic.

Moving towards higher concentration ratio (Fig. 5b), CC0.30 and CEM0.30 mastics exhibited a totally different mechanical behavior: at high temperatures, the addition of calcium carbonate promoted rheological properties more close to a viscoelstic liquid material (liquid-like behavior, $\delta \rightarrow 90°$).

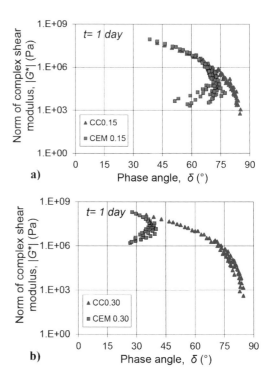

a)

b)

Figure 5. Black diagram, effect of type of mineral addition: a) CC0.15 and CEM0.15 at 1 day of curing; b) CC0.3 and CEM0.3 at 1 day of curing.

Whereas, the adoption of cement at high temperatures, promoted a strong reduction in phase angle, i.e. rheological proprieties more related to a viscoelastic solid material (solid-like behavior, $\delta \rightarrow 0°$).

5 CONCLUSIONS

The present research focuses on the rheological characterization of two cold bituminous mastics produced in laboratory with bitumen emulsion and two mineral additions: cement and calcium carbonate. The influence of both volumetric concentration ratio (0, 0.15 and 0.30) and curing time (1 and 3 days) on linear viscoelastic properties (i.e. complex shear modulus G^*) was evaluated. Considering the overall results, the following conclusions can be drawn:

– The adoption of Brookfield rotational viscometer for evaluating the homogeneity of CBm samples during production, was satisfactory.

– The influence of the selected curing times was negligible on CC mastics and slightly affected the LVE properties of CEM mastics at all volumetric concentration ratios. Shorter curing times (e.g. few hours) should be considered to better analyze the effects of curing time at mastic-scale analysis.

- The volumetric concentration ratio strongly affects the LVE properties of CEM mastics at both curing times. Increasing of cement volumetric concentration led to an increase of $|G^*|$ and a reduction in δ (viscoelastic behavior) probably related to the stiffening contribution of the developed cementitious structure.
- The effect of the type of mineral addition was comparable with that achieved increasing cement concentration ratio. Moving from calcium carbonate to cement as mineral addition, the LVE behavior evolved from that of liquid material (viscous behavior) to that of solid material (viscoelastic behavior).

REFERENCES

Brown, S., & Needham, D. (2000). A study of cement modified bitumen emulsion mixtures. Asphalt Paving Technology, 69, 92–121.

Dondi, G., Mazzotta, F., Sangiorgi, C., Pettinari, M., Simone, A., Vignali, V., & Tataranni, P. (2014). Influence of cement and limestone filler on the rheological properties of mastic in cold bituminous recycled mixtures. Sustainability, Eco-efficiency, and Conservation in Transportation Infrastructure Asset Management, 61.

Elnasri, M., Airey, G. & Thom, N. 2013. Experimental Investigation of Bitumen and Mastics under Shear Creep and Creep-Recovery Testing. The Airfield and Highway Pavement Conference. Los Angeles. California 9th-12th June.

Frigio, F., Ferrotti, G., & Cardone, F. (2016). Fatigue rheological characterization of polymer-modified bitumens and mastics. In 8th RILEM International Symposium on Testing and Characterization of Sustainable and Innovative Bituminous Materials (pp. 655–666). Springer Netherlands.

Godenzoni, C., Cardone, F., Graziani, A., & Bocci, M. (2016a). The effect of curing on the mechanical behavior of cement-bitumen treated materials. In 8th RILEM International Symposium on Testing and Characterization of Sustainable and Innovative Bituminous Materials (pp. 879–890). Springer Netherlands.

Godenzoni, C., Graziani, A., & Corinaldesi, V. (2016b). The Influence Mineral Additions on the Failure Properties of Bitumen Emulsion Mortars. In 8th RILEM International Conference on Mechanisms of Cracking and Debonding in Pavements (pp. 327–333). Springer Netherlands.

Graziani, A., Godenzoni, C., Cardone, F., & Bocci, M. (2016). Effect of curing on the physical and mechanical properties of cold-recycled bituminous mixtures. Materials & Design, 95, 358–369.

Grilli, A., Graziani, A., Bocci, E., & Bocci, M. Volumetric properties and influence of water content on the compactability of cold recycled mixtures. Materials and Structures, 1–14.

Huang, S. C., & Di Benedetto, H. (Eds.). (2015). Advances in Asphalt Materials: Road and Pavement Construction, Chapter 13 "Paving with asphalt emulsion". Woodhead Publishing.

Jenkins, K. J., & Moloto, P. K. (2008). Updating bituminous stabilized materials guidelines: mix design report. Phase II–Curing protocol: improvement. In Technical Memorandum Task 7.

Liebenberg, J. J. E., & Visser, A. T. (2004). Towards a mechanistic structural design procedure for emulsion-treated base lavers. Journal of the south african institution of civil engineering, 46(3), 2–9.

Nassar, A. I., Mohammed, M. K., Thom, N., & Parry, T. (2016b). Mechanical, durability and microstructure properties of Cold Asphalt Emulsion Mixtures with different types of filler. Construction and Building Materials, 114, 352–363.

Nassar, Ahmed I. and Thom, Nicholas and Parry, Tony (2016) Examining the effects of contributory factors on curing process of cold bitumen emulsion. In: Fourth International Chinese European Workshop on Functional Pavement Design, 29 June 6–1st July 2016, Delft, Netherlands. (In Press)

Pellinen, T., Zofka, A., Marasteanu, M., & Funk, N. (2007). Asphalt Mixture Stiffness Predictive Models (With Discussion). Journal of the Association of Asphalt Paving Technologists, 76.

Santagata, E., Baglieri, O., Tsantilis, L., & Chiappinelli, G. (2016). Storage stability of bituminous binders reinforced with nano-additives. In 8th RILEM International Symposium on Testing and Characterization of Sustainable and Innovative Bituminous Materials (pp. 75–87). Springer Netherlands.

Santagata, F. A., Bocci, M., Grilli, A., & Cardone, F. (2009). Rehabilitation of an Italian highway by cold in-place recycling techniques Vol. 2 (pp. 1113–1122). Proceedings of the 7th international RILEM symposium on advanced testing and characterization of bituminous materials, Rhodes, Greece.

Serfass, J. P., Poirier, J. E., Henrat, J. P., & Carbonneau, X. (2004). Influence of curing on cold mix mechanical performance. Materials and structures, 37(5), 365–368.

Stroup-Gardiner, M. (2011). Recycling and reclamation of Asphalt pavements using in-place methods (No. Project 20–05 (Topic 40–13)).

Wang, H., Yang, J., & Gong, M. (2016). Rheological Characterization of Asphalt Binders and Mixtures Modified with Carbon Nanotubes. In 8th RILEM International Symposium on Testing and Characterization of Sustainable and Innovative Bituminous Materials (pp. 141–150). Springer Netherlands.

Transport Infrastructure and Systems – Dell'Acqua & Wegman (Eds)
© 2017 Taylor & Francis Group, London, ISBN 978-1-138-03009-1

A comprehensive methodology for the analysis of highway sight distance

M. Castro & C. De Santos-Berbel
Department of Civil Engineering, Transport and Territory, Technical University of Madrid, Spain

L. Iglesias
Department of Geological and Mining Engineering, Technical University of Madrid, Spain

ABSTRACT: As one of the main elements of geometric design, sight distance must be considered carefully for the safe and efficient operation of highways. An application developed on Geographic Information Systems (GIS) was conceived for the three-dimensional estimation of sight distance on highways, as opposed to conventional two-dimensional techniques, which may underestimate or overestimate the actual visibility conditions. It is capable of computing the available sight distance of a highway section given the driver's eye height, the target height, the vehicle path and an elevation model. The outcome can be studied in detail with the aid of the tools and capabilities developed, including sight-distance graphs. The influence of the input features, such as the nature of the elevation model, its resolution and the spacing between path stations on the results accuracy was analyzed. The interpretation of results is also essential to explain sight distance deficiencies and provide insight into the effect of roadside elements on those results. In addition, the sight-distance graph permits the detection and characterization of sight-hidden dips, an undesirable shortcoming in the spatial alignment of highways. The versatility of GIS enables, moreover, an integrated research of highway safety. It allows the incorporation of diverse operational factors such as accident data, traffic volume, operating speed and design consistency to detect and diagnose potentially hazardous spots or, eventually, identify the factors involved in a particular accident. This paper describes the methodology utilized and reviews the main issues through case study examples.

1 INTRODUCTION

As one of the main elements of geometric design, sight distance must be considered carefully for the safe and efficient operation of highways. In response to this, highway geometric design standards in different countries set minimum sight distance threshold values (Ministerio de Fomento 2016, AASHTO 2011, FGSV 2012). In order to facilitate the geometric design of roads, some guidelines propose two-dimensional analytical procedures to estimate the available sight distance. Nevertheless, these procedures may not be practical since they consider separately horizontal and vertical alignment, which may lead to overestimate or underestimate the actual available sight distance (Ismail & Sayed 2007). It is more common instead, to develop algorithms based on line-of-sight loops in three dimensions (3-D). Such procedures retrieve the cross-sectional profile of the terrain below the line of sight between the observer and the target locations, detecting whether the vision is obstructed. Ismail and Sayed (2007) devised a precise algorithm to compute the available sight distance. Besides algorithms based on line-of-sight loops, procedures based on viewsheds were developed to study sight distance on highways (Castro et al. 2011, Jha et al. 2011).

Computer-aided applications for road design estimate and compare available sight distances to stopping sight distance and passing sight distance. They also include visualization tools that simulate the driver's perspective while travelling (Kühn et al. 2011, Castro 2012). Such visualization tools are utilized to supervise proper 3-D alignment coordination, although it requires this checking procedure is performed by experienced engineers (Larocca et al. 2011).

Methods based on line-of-sight loops enable the depiction of sight-distance graphs. These charts represent on the horizontal axis the stations where the driver is sequentially placed, and on the vertical axis the sight distance variables ahead each driver position (Kühn & Jha 2011, Castro et al. 2014). Besides the comparison available and required sight distances, such charts result advantageous to evaluate the 3-D alignment coordination (Roos & Zimmermann, 2004, Jha et al. 2011, Castro et al. 2015a). The German Road and Transportation Research Association provided a framework both for virtual perspective generation and sight-distance graphs on the design of rural highways (FGSV 2008). Campoy-Ungría (2015) proposed a procedure to estimate available sight distance on highways based on prismatic line-of-sight buffers launched directly on a high-density LiDAR cloud of points, not requiring any terrain surface.

A Geographic Information System (GIS) is useful to calculate sight distances because it gathers numerous advantages. Besides the 3-D treatment of the sight distance issue, it enables safety integrated analyses accounting for other factors such as geometrics, accident data, traffic volume, operating speed and design consistency at ease (Altamira et al. 2010; Castro & De Santos-Berbel 2015). Nowadays, affordable data sources are available to characterize the highway and the roadsides with higher accuracy, which GIS is capable to handle and exploit at ease. Khattak and Shamayleh (2005) assessed highway safety through GIS data visualization.

This paper reviews the GIS-based methodology developed to study sight distance on highways. The main issues are contemplated by means of case study examples. Following this introduction, the second section provides detailed description of the methodology. Particular attention is paid to inputs (driver's eye height, target height, vehicle path and elevation model) and output (sight distance graph). In the third part, case studies are presented. The first case illustrates the influence of roadside elements (vegetation) through the use of a Digital Terrain Model (DTM) and a Digital Surface Model (DSM). The second one describes how to detect and analyze shortcomings in the spatial alignment of highways through the use of sight-distance graph. The third one proposes a solution based on GIS tools (lines of sight) and multipatch datasets in order to represent overhanging roadside features (e.g. cantilever traffic signal) adequately. Finally, conclusions are presented.

2 METHODOLOGY

2.1 Sight distance algorithm

An application developed on ArcGIS was conceived for the 3-D estimation of available sight distance on highways and has already been validated by the authors (Castro et al. 2014). It is capable of computing the available sight distance of a highway section given the driver's eye height, the target height, the vehicle path and an elevation model.

The computational routine for available sight distance estimation launches a line-of-sight beam iteratively from every station on the vehicle path towards the stations ahead, determining whether each target is seen by the driver. Once the loop from a station has been completed, the virtual driver is moved forward to the next station, where an identical loop is launched. Available sight distance is defined as the distance, measured along the vehicle path, between the driver's position and the farthest target seen without interrupting the line of sight. According to this definition, the algorithm checks lines of sight from the driver position as shown in Figure 1. At station i, the available sight distance is determined by the station $i+2$ (line of sight in light grey), which is the furthest one seen before the first line of sight is blocked (station $i+3$, with line of sight in dark grey). Software stores the binary value of visibility of every line of sight.

Figure 2 shows the complete process to study sight distance on highways, including the input

Figure 1. Available sight distance estimation through lines of sight.

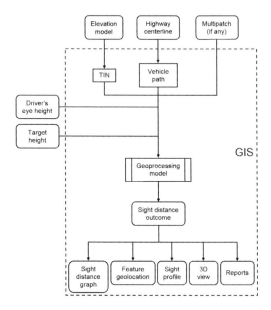

Figure 2. Flowchart of the sight distance study procedure.

and output data as well as the auxiliary tools developed.

2.2 Input data

Each particular highway requires two important datasets to study sight distance. Whereas a Digital Elevation Model (DEM) is necessary to recreate the highway environment, a file containing the points that define vehicle trajectory is needed.

With respect to DEMs, the two types mentioned in the previous section are available for sight distance modelling: DTM's and DSM's. Either of them is handled by the application as long as it is of the form of a Triangular Irregular Network (TIN). To choose one over the other is not a trivial matter owing to their influence in results. A DTM is a 3-D representation of the terrain surface which depicts exclusively the elevation of the bare ground. However, the reality contains many more elements influencing sight distance than the bare ground. Features such as vegetation, traffic signs, buildings and many other elements are not included in a DTM. These models comprise such additional roadside features, making available further information about features by the roadsides which could limit the available sight distance. However, where overhanging features are present by the roadsides, the intrinsic features of DSMs hamper sight distance analysis. These surfaces do not support two points on its surface having the same horizontal projection while their elevation values are different. This fact hinders a lifelike representation of overhanging features, which is particularly troublesome when they are partially located above the road, as occurs for tree crowns or cantilever signals.

Current techniques provide cost-effective high resolution DEMs. The remote sensing LiDAR (Light Detection and Ranging) devices emit a pulse beam. The pulse is received back if any surface is hit, which allows its geospatial location and its characterization (Topcon 2010). Usually, those data might have been collected by airborne surveying or terrestrial surveying. The latter ones are known as Mobile Mapping Systems (MMS). Whereas the airborne LiDAR is able to capture around one point per two m^2, the MMS is able of deliver more than 200 points per m^2 when closer to the sensor. However, the area covered from the airborne standpoint is much greater and the performance much higher, whilst the MMS are limited by elements that may create shadow areas beyond them. In both cases, raw data usually contain much noise. Items not forming part of the static landscape are captured, such as vehicles or even wildlife. Moreover, the abovementioned overhanging elements, i.e. aerial power lines, hamper the use of DSM. All those points are therefore entitled to be removed.

The shape of roadside elements is another fact to bear in mind. DSMs from airborne LiDAR can hardy model vertical roadside features, such as traffic signs or guard-rails, although they could represent considerably larger vertical devices (e.g. gantries) depending on the resolution. In general, roadside vertical equipment is better depicted by DMSs derived from terrestrial LiDAR.

Regarding the trajectory, points that compose driver's path could be obtained from several sources. If the highway geometrics are known, a theoretical vehicle path can be extracted. Otherwise, it might be deduced from inventories or precise-enough cartographic data. Moreover, the track of a GNSS receiver mounted on a car driving along the studied highway constitutes a reliable data source for this input when highway geometrics are unknown or not reliable. The application has tools that simplify its treatment. Points should have an attribute, namely station, which indicates their distance to the origin measured along this trajectory.

As generic inputs, driver's eye height and target height must be considered. Those values are usually taken from highway design standards (AASHTO 2011, Ministerio de Fomento 2016).

The accuracy and resolution of these models, along with the spacing between the path stations, come also into play. The effect of these factors has been studied by the authors (Castro et al. 2015b). This was tested by comparing the available sight distance results of elevation models of different resolutions and varying the spacing between stations on each model. The elevation model resolution ranged from 1 to 5 m, all built up of a squared mesh. The paths tested had stations no closer than 1 m and no further than 20 m. It was found that the DEM resolution has a larger effect on outcome than the spacing between stations.

To avoid the issues created by overhanging features in the DSM, the use of multipatch datasets can be contemplated. This supports the insertion of roadside elements such as cantilever signals, gantries or overpasses to achieve an adequate sight distance modeling.

2.3 Output

The results, stored by the application after the computational process, are plotted on the sight-distance graph. Also, software may retrieve the longitudinal profile between the observer and observed points at the request of the user. This feature permits the detection of the area that obstructs vision.

Due to the GIS geolocation capabilities, the outcome can be shown on the map. Mapped features facilitate the integrated analysis along with other factors. Furthermore, results may be exported in full detailed reports. The 3D visual inspection of the scene modelled is also possible in ArcSCENE to get a better understanding of modelling issues.

3 CASE STUDIES

For the purpose of illustrating the capabilities of the methodology described hereby, several case studies are presented. The first case illustrates the influence of roadside elements (vegetation) through the use of a DTM and a DSM. The second case describes how hidden dips are characterized throughout the use of sight-distance graph. The third one proposes a solution based on GIS tools (lines of sight) and multipatch datasets in order to calculate and analyze the influence of a cantilever traffic signal in the visibility of truck drivers. All cases correspond to two-lane rural highways located in the region of Madrid (Spain).

3.1 Roadside elements

Roadside elements such as vegetation, traffic signs or buildings may be taken into account in the sight distance studies when using a DSM instead of a DTM. Regarding the roadside vegetation, trees and plants may reduce available sight distances, especially when it comes to forests and densely wooded areas. A sub-section of highway M-611 was selected to illustrate this issue. The design speed is assumed to be 40 km/h. A horizontal curve of radius 23 m is flanked by respective spirals and tangents. Figure 3 shows the actual view of a vehicle approaching a right curve, where a densely wooded area is found close to the inner roadside. In this case, both an airborne DTM and an airborne DSM arranged in a 1-m square mesh were used. The vehicle path was derived from cartographic data and was discretized into stations spaced 5 m apart.

Figure 3. Real view of curve where sight distance is limited by vegetation.

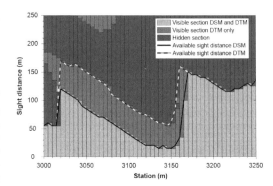

Figure 4. Sight distance graph comparing results using DTM and DSM.

Figure 4 shows the sight distance graph superposing the results of both a DTM and a DSM. When a DTM is used as input, the minimum available sight distance is 55 m (slashed line) whereas the corresponding to the DSM is 15 m (solid black line) around station 3150. The latter one is more in line with reality. This value would not comply with the stopping sight distance set at 50 m by the Spanish standard (Ministerio de Fomento 2016). Moreover, the available sight distance is reduced around 40 m all the way in front of the curve. This difference is highlighted in medium grey in Figure 4, which represents the sections that are seen when the input is the DTM whilst the study with DSM determined they cannot be seen. This example shows how significant is the influence of vegetation by the roadsides by means of the choice of DEM.

In addition, a study carried out by the authors found that sight distance results using airborne DTM, airborne DSM and MMS DSM were all statistically significantly different (Castro et al. 2016). The differences were particularly relevant in sub-sections where the available sight distance was shorter.

3.2 Hidden dips

Some combinations of horizontal and vertical alignments might produce shortcomings in the driver's perspective. Hidden dips are a common shortcoming on highways where the profile adjusts the terrain more strictly than the horizontal alignment. A hidden dip is produced where the driver is able to see two separate sections of the roadway while the stretch in between remains concealed. This typically occurs where a sag follows a crest curve on a rather straight horizontal alignment. It is essential to avoid potentially hazardous spots within the hidden section, such as intersections or unexpected changes in direction. Moreover, these alignments may mislead drivers at the beginning of a passing maneuver, since oncoming traffic remains unnoticed in the hidden section.

The sight-distance graphs generated by the application developed are suitable to detect and analyze hidden dips. Figure 5 shows a straight section of M-611 highway on a rolling profile where there is a hidden dip. The corresponding sight-distance graph is illustrated in Figure 6. Sections seen by the driver are colored in light grey, in medium grey stations not seen and in dark grey target stations in the dip where an object of 0.75 m height would not be seen. The reason for this value is explained later. The maximum dip depth is 3.9 m according to the longitudinal profile between the observer and the observed points retrieved by software.

The perspective shortcoming is first noticed at station 710. When the driver reaches station 725, the available sight distance is 380 m, a stretch of 240 m remains hidden, and a further segment is seen over again up to 905 m. The hidden dip ranges up to station 1060, totaling 350 m.

As this type of shortcoming may produce passing issues, the study of passing sight distance results interesting. In this section posted speed is 90 km/h. For this value, the current Spanish standard (Ministerio de Fomento 2016) demands a passing sight distance of 205 m along a section of 340 m or larger. However this threshold is exceeded along

Figure 5.   Real view of hidden dip on straight alignment.

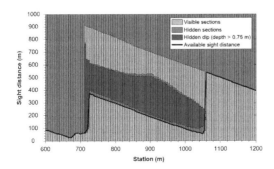

Figure 6.   Sight-distance graph of hidden dip.

165 m only. Thus passing should be prohibited all along the hidden dip range.

Furthermore, the German guidelines for the visualization of rural roads (FGSV 2008) describe the conditions under which a hidden dip is potentially hazardous for drivers, regardless of the design speed. Three conditions must be fulfilled simultaneously along a range of at least 60 m: The hidden sections must not spread out beyond 600 m from the driver, the hidden section has to cover more than 75 m and the depth of diving must exceed 0.75 m (hence the area in dark grey in Figure 6). These values are largely exceeded in the present case, reporting a measure of risk exposure during passing maneuvers. Hence this spot may be potentially hazardous.

Similarly, the Swiss standard (VSS 1991) determines the maximum distance to consider the reappearing section at 500 m for that speed. That occurs only from station 920, therefore the hazardous stretch would range 140 m.

3.3   Overhanging elements

In this case study, a cantilever traffic signal in a highway section was simulated. The aim is to analyze the effect of the location of this signal on drivers sight distance, avoiding the problems associated to the use of a DSM. As in previous cases, a DEM is needed. In this case an airborne DTM arranged in a 1-m square mesh was utilized. In addition, the cantilever traffic signal was modeled as a multipatch file taken from an open library (Trimble 2015). This multipatch was placed on 5 different locations (Table 1) along a section of highway M-104. The cantilever traffic signal location covers exactly the width of the traffic lane where signing applies.

The highway horizontal alignment is composed by a right horizontal curve, followed by a long tangent and a left curve. In the vertical alignment, there is a sag curve approximately on the middle of the tangent, between the two horizontal curves. The cantilever traffic signal locations 1 and 2 are supported by the roadside on different spots of the right curve. Location 3 is at the beginning of the sag curve and location 4 is at the lowest point of

Table 1.   Locations where cantilever signal was placed.

Location	Station (m)
1	5260
2	5305
3	5400
4	5450
5	5630

the sag curve. Location 5 is between the sag vertical curve and the left horizontal curve. Figure 7 shows a 3D view made using ArcSCENE, where the different locations of the cantilever traffic signal considered are depicted. The right and the left horizontal curves overlap approximately with two crest vertical curves.

The clearance height was 5.5 m, according to the current Spanish standard (Ministerio de Fomento, 2016) whereas the maximum height of the structure is 6.81 m. Moreover, such standard requires that the impact of gantries and cantilever signals on sight distance is checked.

In this study, geometric characteristics of a theoretical truck path were considered. Therefore the observer height was set at 2.5 m. According to the Spanish design standard (Ministerio de Fomento 2016), the path considered was that resulting of the parallel offset of 1.5 m from the highway centerline. Also, according to the same standard, target height was set at 0.5 m. Figure 8 shows the sight-distance graph when the calculation was launched without cantilever traffic signal. There is a first zone of shorter sight distance due to the first horizontal curve and the first crest vertical curve (minimum available sight distance of 115 m at station

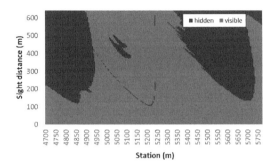

Figure 9. Sight-distance graph corresponding to cantilever signal at station 5260 (location 1).

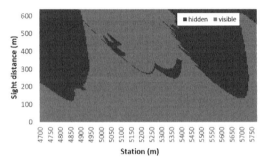

Figure 10. Sight-distance graph corresponding to cantilever signal at station 5450 (location 4).

4855). Then, the available sight distance increases until drivers approach the second horizontal curve and the second crest vertical curve (135 m of minimum sight distance at stations 5680–5715).

Figure 9 shows sight-distance graphs corresponding to location 1 of the cantilever traffic signal. In location 1, there are some lines of sight corresponding to driver location between stations 4940 and 5240 that intersect traffic signal post, but its effect is negligible. The cantilever signal itself has no effect on sight distance. Similarly, Figure 10 shows the sight-distance graph corresponding to location 4 of the cantilever traffic signal (near station 5450). Comparing Figures 9–10, it can be noticed that not only the cantilever traffic signal effect moves (due to the change of location) but also the non-visible area becomes larger. This latter effect is due to signal location at the lowest point of the sag curve. As a result, the cantilever signal intercepts much more lines of sight, producing a more relevant hidden area.

Figure 7. View in ArcSCENE of the cantilever traffic signal possible locations.

4 CONCLUSIONS

The proposed procedure demonstrates the potential and versatility of GIS in highway sight distance

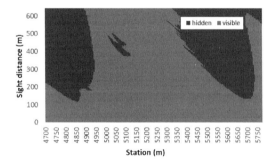

Figure 8. Sight-distance graph without traffic signal.

studies. The inputs needed for the study, namely the driver's eye height, the target height, the vehicle path and the elevation model were described. The importance of the resolution and nature of the DEM was particularly emphasized. To achieve precise results, it is desirable to use high resolution models (1 node per m^2). The outcome can be studied in detail with the aid of the tools and capabilities developed, including the sight-distance graph, line-of-sight profiles and mapped features. Sight-distance graphs permit a detailed analysis of roadside elements which limit sight distance.

Throughout three case studies on in-service highways, the strengths and capabilities of the methodology were proved. To take account of the effect of roadside elements or vegetation on sight distance, DSMs must be used instead of a DTM. In particular cases, the reduction of available sight distance is especially significant while considering these entities. The second case study showed how to detect and characterize sight-hidden dips. The parameters that may indicate the risk exposure of these sections, namely range, length of hidden section, dip depth minimum available sight distance and distance to reemerged stretch can be identified at ease with the tools provided.

A DSM leads to biased sight distance modelling where there are overhanging features because lines of sight are obstructed by the model surface even below the overhanging feature. The proposed GIS-based procedure may contemplate multipatch structures to model them, overcoming the difficulties inherent to the presence of overhanging elements. The third case study showed how to model properly a section with a cantilever traffic signal and its real impact on sight distance. The easiness to place multipatch objects is an additional advantage provided by this procedure. This simplifies the simulation of object location to evaluate its possible effects on sight distance. Also, due to the availability of multipatch datasets libraries, modelling effort is reduced.

Therefore the GIS based methodology presented is useful not only to study sight distance but also seek for potential safety issues though integrated analysis. Diverse operational factors such as accident data, traffic volume, operating speed and design consistency can be incorporated to locate and diagnose potentially hazardous spots or, eventually, to identify the factors involved in a particular accident.

ACKNOWLEDGEMENTS

The authors gratefully acknowledge the financial support of the Spanish Ministerio de Economía y Competitividad and European Regional Development Fund (FEDER). Research Project TRA2015-63579-R (MINECO/FEDER).

REFERENCES

Altamira, A.L., Marcet, J.E., Graffigna, A.B. & Gómez, A.M. 2010. Assessing available sight distance: an indirect tool to evaluate geometric design consistency. In *Proceedings of the 4th International Symposium on Highway Geometric Design*.

American Association of State Highway and Transportation Officials (AASHTO) 2011. *A Policy on Geometric Design of Highways and Streets*. Washington DC: AASHTO.

Campoy-Ungría, J.M. 2015. *Nueva metodología para la obtención de distancias de visibilidad disponibles en carreteras existentes basada en datos LiDAR terrestre*. Doctoral dissertation. Valencia: Universidad Politécnica de Valencia.

Castro, M., Iglesias, L., Sánchez, J.A. & Ambrosio, L. 2011. Sight distance analysis of highways using GIS tools. *Transportation Research Part C: Emerging Technologies*, 19(6): 997–1005.

Castro, M. 2012. Highway design software as support of a project based learning course. *Computer Applications in Engineering Education*, 20(3): 468–473.

Castro, M., Anta, J.A., Iglesias, L. & Sánchez, J.A. 2014. GIS-based system for sight distance analysis of highways, *Journal of Computing in Civil Engineering*, 28(3): 04014005.

Castro, M., De Blas, A., Rodriguez-Solano, R. & Sanchez, J.A. 2015a. Finding and characterizing hidden dips in roads. *Baltic Journal of Road and Bridge Engineering*, 10(4): 340–345.

Castro, M & De Santos-Berbel, C. 2015. Spatial analysis of geometric design consistency and road sight distance. *International Journal of Geographical Information Science*, 29(12): 2061–2074.

Castro, M., García-Espona, A. & Iglesias, L. 2015b. Terrain model resolution effect on sight distance on roads. *Periodica Polytechnica: Civil Engineering*, 59(2): 165–172.

Castro, M., Lopez-Cuervo, S., Paréns-González, M. & De Santos-Berbel, C. 2016. LIDAR-based roadway and roadside modelling for sight distance studies. *Survey Review*, 48(350): 309–315.

Forschungsgesellschaft für Straßen- und Verkehrswesen (FGSV) 2008. *Hinweise zur Visualisierung von Entwürfen für außerörtliche Straßen*. Bonn: FGSV Verlag.

Forschungsgesellschaft für Straßen- und Verkehrswesen (FGSV) 2012. *Richtlinien für die Anlage von Landstraßen*. Bonn: FGSV Verlag.

Ismail, K. & Sayed, T. 2007. New algorithm for calculating 3D available sight distance. Journal of Transportation Engineering, 133(10): 572–581.

Jha, M.K., Karri, G.A.K. & Kühn, W. 2011. New three-dimensional highway design methodology for sight distance measurement. *Transportation Research Record: Journal of the Transportation Research Board*, 2262: 74–82.

Khattak, A.J. & Shamayleh, H. 2005. Highway safety assessment through geographic information system-based data visualization. *Journal of Computing in Civil Engineering*, 19(4): 407–411.

Kühn, W., Volker, H. & Kubik, R. 2011. Workplace simulator for geometric design of rural roads. *Transportation Research Record: Journal of the Transportation Research Board*, 2241: 109–117.

Larocca, A.P., Da Cruz Figueira, A., Quintanilha, J.A. & Kabbach Jr, F.I. 2011. First steps towards the evaluation of the efficiency of three-dimensional visualization tools for detecting shortcomings in alignment's coordination. In *Proceedings of the 3rd International Conference on Road Safety and Simulation*.

Ministerio de Fomento 2016. *Norma 3.1-IC: Trazado*. Madrid: Ministerio de Fomento.

Roos, R. & Zimmermann, M. 2004. Quantitative methods for the evaluation of spatial alignment of roads, In *Proceedings of the 2nd International Conference of Società Italiana di Infrastrutture Viarie (SIIV)*.

Topcon 2010. IP-S2 specifications. Available from Internet: <http://www.topcon.co.jp/en/positioning/products/pdf/ip_s2.pdf>.

Trimble 2015. Sketchup PRO. Available from Internet: <http://buildings.trimble.com/products/sketchup-pro>.

Vereinigung Schweizerischer Strassenfachleute (VSS) 1991.*Linienführung; Optische Aforderungen (SN-640140)*. Zurich: VSS.

Transport Infrastructure and Systems – Dell'Acqua & Wegman (Eds)
© 2017 Taylor & Francis Group, London, ISBN 978-1-138-03009-1

Performance improvements of asphalt mixtures by dry addition of polymeric additives

M. Ranieri & C. Celauro
Department of Civil, Environmental, Aerospace, Materials Engineering, University of Palermo, Palermo, Italy

L. Venturini
Iterchimica s.r.l., Suisio, Bergamo, Italy

ABSTRACT: This paper shows the results of an experimental study concerning the development and optimization of asphalt mixtures for binder and base courses, improved with specifically engineered additives. The focus was on the mechanical improvements of the mixtures as achievable via dry modification with polymeric additives by making use of aggregate and bitumen of average quality, as locally available, in order to limit the consumption of virgin materials. The results allowed interesting conclusions to be drawn about the use of polymeric additives for these mixtures. In particular, the modified mixtures proved to have better performance in terms of both permanent deformation resistance and stiffness modulus. Moreover, it is possible to use lower binder contents, thus proving the economic feasibility of this modification, when considering the advantage of consuming mixtures with components of average quality for production of highly performing mixtures.

1 INTRODUCTION

In recent years growing traffic has demanded higher quality pavements. Binder and base courses must have structural characteristics such as not to allow degradations. A method that can improve the quality of the pavements, avoiding degradation, with benefits (and savings) regarding the need for maintenance, is the addition of polymers in the asphalt mixtures (Bense 1983, Serfass et al. 2000).

In addition awareness of environmental problems has led to research being focused on development of new technologies of additives that can improve the mechanical performances of asphalt mixtures. Among these specially engineered additives are interesting solutions, both to meet the related environmental issues, in Italy and in all other economically developed countries, like for example the productive reuse of secondary raw materials, and to solve specific technical requirements of pavements by improving the physical and mechanical properties of the mixtures.

The use of additives makes it possible to increase the performance of asphalt mixtures, by decreasing production and laying costs, and to reduce the environmental impact.

It also favors sustainable development because it makes it possible to reduce consumption of valuable natural resources (Venturini et al. 2016).

In this context plastics are the materials with the highest degree of diffusion in many different sectors. The polymer market shows a strong potential for development and diversification; the reasons of this success are clear: low weight, workability, versatility, hygiene, different selection options, recycling and recovery (Celauro et al. 2004).

There are two main processes for adding polymers to asphalt mixture: by modifying the bitumen (PMB—wet process); or by adding the polymers during the mixing phase (PMA—dry process) (Pettinari et al. 2014).

The wet process needs specific equipment (for mixing and to facilitate the reaction with bitumen at high temperatures), while this is not required for the dry process. Therefore, the dry process is easier to implement.

The aim of this study was optimization of mixtures that do not necessarily employ high-performance materials, making use of locally available stone aggregate and bitumen, also aiming to improve the traditional mixtures as made possible by dry addition of suitable polymers.

2 MATERIALS AND METHODS

The experimental plan was developed by using the following materials (aggregates, binders, additives, mixtures):

• aggregates (D_{max} = 16 mm);
• binder (neat bitumen 50/70);

- additive (plastomeric compound);
- mixtures:
  α. traditional mixture;
  β. mixture with plastic additive.

The mix design was achieved by carrying out:

1. Marshall test with 4 percentages of bitumen and CE = 75 blows per face (compaction effort), according to the EN 12697-34:2012 standard;
2. compactability test with a gyratory compactor (D = 150 mm), in accordance with the EN 12697-31:2007 standard.

Finally, tests on optimized mixtures were performed in order to assess the following properties:

○ rutting resistance, according to the EN 12697-22:2007 standard, method B;
○ complex stiffness modulus, according to the EN 12697-26:2012 standard, annex B;
○ fatigue resistance, according to the EN 12697-24:2012 standard, annex D.

2.1 The stone aggregates

For composition of mixtures it appears appropriate to refer to aggregates with a diameter between 0 and 16 mm, a suitable range for binder and base courses. The aggregates used in the mixtures were crushed limestone from a quarry whose composition and physical and mechanical properties are given respectively in Table 1 and Table 2.

The mixture target (a typical dense graded high modulus asphalt) was obtained for the composition of the fractions provided by implantation and by appropriate sub-fractions (see Fig. 1).

Table 1. Composition of the aggregates' available fractions.

Sieve mm	Passing % Fractions				
	a_2 (20/25)	a_3 (10/15)	a_4 (6/10)	a_5 (0/6)	Filler
32	100	100	100	100	100
24	100	100	100	100	100
20	94.73	100	100	100	100
12	18.33	99.96	100	100	100
8	0.77	85.67	99.91	99.70	100
4	0.51	38.3	72.62	97.78	100
2	0.49	15.66	41.54	70.60	100
0.4	0.44	6.62	16.84	26.09	99.31
0.18	0.41	5.28	11.21	17.19	93.78
0.075	0.33	4.01	6.42	10.57	74.25

Table 2. Physical and mechanical characteristics of the available aggregates.

Characteristics	Fractions				
	a_2	a_3	a_4	a_5	filler
Bulk specific weight, g/cm³ (EN 1097-7:2008)					2.85
Apparent specific weight, g/cm³ (EN 1097-6:2013)	2.82	2.83	2.84	2.85	
Los Angeles abrasion, % (EN 1097-2:2010)	22.10	20.19	20.64	20.12	
Sand equivalent, % (EN 933-8:1997)				91.38	90.41
Voids ratio (CNR 65:1978)	0.80	0.79	0.71	0.73	
Absorption coefficient (EN 1097-6:2013)	0.64	0.51			

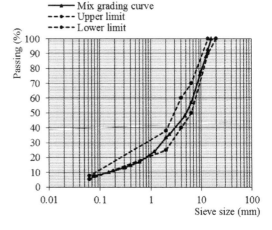

Figure 1. Mix grading curve.

The mixture of aggregates was then subjected to the pycnometer method for determination of the real specific weight and the value was 2.842 g/cm³, according to the EN 1097-3:1999 standard.

2.2 Bitumen

The bitumen used in this research was neat bitumen 50/70, more easily available in contexts such as in Italy; its characteristics are reported in Table 3.

2.3 Additive

The additive used was a particular polymeric compound of selected polymers, Low-Density

Table 3. Characteristics of the bitumen used in the mixtures studied.

Characteristic	Unit	Value	Standard
Specific weight at 25°C	g/cm³	1.033	EN 3838:2005
Penetration at 25°C	dmm	68	EN 1426:2015
Softening point Ring and Ball	°C	50.5	EN 1427:2015
Penetration Index		−0.21	EN 12591:2009
Temperature Fraas	°C	−12	EN 12593:2015
Ductility at 25°C	mm	>100	ASTM D113:2007
Viscosity at 60°C	Pa·s	255.5	EN 13302:2010
Viscosity at 100°C	Pa·s	3.917	EN 13302:2010
Viscosity at 135°C	Pa·s	0.435	EN 13302:2010
Viscosity at 150°C	Pa·s	0.222	EN 13302:2010
Mixing temperature ($\eta = 0.17$ Pa·s)	°C	155	EN 13302:2010
Compaction temperature ($\eta = 0.28$ Pa·s)	°C	145	EN 13302:2010
After RTFOT:			
Change in mass	%	−0.19	EN 12607-1:2015
Penetration at 25°C	%	44	EN 1426:2015
Softening point Ring and Ball	°C	64.5	EN 1427:2015
Viscosity at 60°C	Pa·s	668	EN 13302:2010

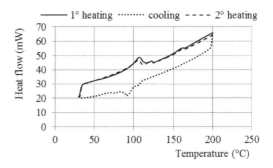

Figure 2. DSC test results on plastomeric compound.

Polyethylene (LDPE), Ethylene (EVA) as well as others with low molecular weight and average melting point, that is semi-soft and flexible granules. The plastomeric compound was not designed for modification of the bitumen, but rather to improve the mechanical performance and durability of the asphalt mixtures (Iterchimica S.r.l 2015). The physical properties are:

- aspect: granules;
- colour: shades of grey;
- dimensions: 2.00–4.00 mm;
- softening point: 160°C;
- melting point: 180°C;
- melt index: 1÷5;
- apparent density at 25°C: 0.40–0.60 g/cm³.
- specific weight: 0.934 g/cm³.

A DSC (Differential Scanning Calorimetry) test was carried out on this material in accordance with the ISO 11357-3:2011 standard, which makes it possible to characterize thermal behaviour because it provides temperature and enthalpy values corresponding to glass transition, melting point and crystallization through heating from −60°C to 180°C, cooling from 180°C to −60°C and again heating; the results are shown in Figure 2, where there are three peaks: the first is more or less at

106°C and it is usually for low-density polyethylene, the second one at 120°C is for the high-density polyethylene, while the last peak at 160°C is for polypropylene.

Since the area under the curve is proportional to the mass and the two areas at 120°C and 160°C are quite low, the quantity of polypropylene and High-Density Polyethylene (HDPE) is minimal.

The other peaks are not important because they refer to crystallization of polymers at low temperatures. This test confirms that the plastomeric compound is a low-density polyethylene with a small quantity of high-density polyethylene and polypropylene.

This additive has low affinity with the bitumen and therefore, for practical needs, it is much more advantageous to add additives in the asphalt mixture (known in the technical literature as DRY method): the additive is added to the hot aggregates, before mixing with the bitumen (Celauro et al. 2004). In view of this, the optimal process to make a mixture is the succession of the following components: aggregates, additives, bitumen and filler.

3 FORMULATION TESTS

3.1 Marshall Stability

The first step in the formulation of these mixtures was applying the Marshall Method only to the traditional mixture in order to assess the physical and mechanical characteristics typically considered in the Italian specifications.

Four percentages of bitumen were selected ($b'_1 = 4.8\%$, $b'_2 = 5.2\%$, $b'_3 = 5.5\%$, $b'_4 = 5.9\%$, by weight of the aggregates) and four specimens for each percentage were produced, for a valid repetition. The volumetric properties (v, air voids, and VFB, voids filled with bitumen) were determined according to the requirements of the EN 12697-8:2003 standard. The calculation of the maximum specific weight (γ_t) of the mixture was performed according to the "C" (mathematical) process specified by the EN 12697-5:2010 standard, while the calculation of the apparent specific weight (γ_{app}) was performed according to the EN 12697-6:2012 standard. The results of the Marshall test are reported in Table 4 and Figures 3–4, where S

Table 4. Marshall test results (2 × 75).

B %	v %	S KN	F mm	R KN/mm	VFB %	γ_{app} g/cm³
4.8	6.03	12.24	4.27	2.87	69.22	2.46
5.2	2.88	13.39	4.07	3.29	82.90	2.52
5.5	2.52	14.67	4.40	3.33	86.05	2.52
5.9	1.82	13.30	5.33	2.49	89.75	2.53

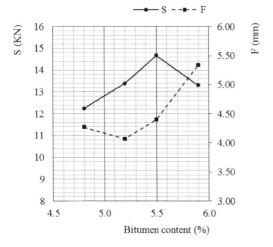

Figure 3. Marshall Stability and displacement, at different bitumen contents.

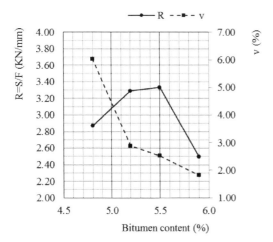

Figure 4. Marshall Ratio and Marshall voids, at different bitumen contents.

Table 5. Marshall test results in accordance with MIT specification.

Required results	Unit	Course Base	Course Binder
Marshall Stability, S	KN	8	10
Marshall Ratio	KN/mm	>2.5	3 ÷ 4.5
Residual voids	%	4 ÷ 7	4 ÷ 6

is the Marshall Stability, F is the Marshall Flow that is the corresponding displacement and R the Marshall Ratio S/F.

As can be seen from the graphs above, regarding the Italian specifications, such as the example in Table 5 (MIT 2001), the requirements are respected for bitumen contents between 4.8 and 5.1%.

3.2 Gyratory compactor

After some information obtained from the Marshall method the second step in the mix design of these mixtures was to consider some compactibility characteristics also considered by the Superpave method introduced in 1992 by the Strategic Highway Research Program SHRP (Cominsky et al. 1994).

The mixtures were subjected to gyratory compaction in such a way that a measure of the internal stability of the mixture during the compaction was obtained too, i.e. the gyratory shear ratio $\sigma = S/P$, where S is the shear stress, and P is the ram pressure. Acquisition of this feature was carried out on specimens compacted up to N = 200 rpm, in order to evaluate the behaviour of the mixtures in the different conditions of densification that affect it from the time of laying and throughout the design life years. In general, an increasing trend of the shear ratio σ in the initial stage of compaction (approximately, in the first 50 rpm) and stabilization even beyond the maximum value of N, together with fulfillment of the volumetric requirements (VMA, VFA), ensure a correct formulation and good stability during operation (Roberts et al. 1996).

This test gives a good idea of the job-site density values, according to course thickness. Conducted ahead of the other mechanical tests, this test is used to make a preliminary selection or screening of mixes, and for optimizing the asphalt mix composition.

The gyratory compactor test was initially carried out on two specimens with different bitumen contents (4.9, 5.1 and 5.4%) and three different plastomeric compound contents (P.0 = 0% of polymer, P.3 = 0.3% of polymer, P.6 = 0.6% of polymer, by weight of mineral aggregates). Later, the specimens with additives being subjected to bleeding due to the excessive binder content, it was also decided to carry out the test, for the percentage 0.3

of the plastomeric compound, on two specimens with two lower bitumen contents (4.3, 4.6). The densification curves recorded during the gyratory compaction made it possible to obtain parameters of the regression lines, K and C_1, that respectively define the workability and the self-densification of these mixtures (see Table 6).

Table 6. Values of workability and self-densification.

Mixture	b %	p %	% $G_{mm} = C_1 + k * \log (N)$		
			C_1	K	R^2
AC 4.3/P.3	4.3	0.3	0.8009	0.0757	0.9980
AC 4.6/P.3	4.6	0.3	0.8072	0.0739	0.9984
AC 4.9/P.0	4.9	0.0	0.7972	0.0753	0.9989
AC 4.9/P.3	4.9	0.3	0.8141	0.0788	0.9952
AC 4.9/P.6	4.9	0.6	0.8278	0.0748	0.9900
AC 5.1/P.0	5.1	0.0	0.8190	0.0710	0.9988
AC 5.1/P.3	5.1	0.3	0.8168	0.0786	0.9940
AC 5.1/P.6	5.1	0.6	0.8274	0.0755	0.9880
AC 5.4/P.0	5.4	0.0	0.8141	0.0788	0.9952
AC 5.4/P.3	5.4	0.3	0.8345	0.0741	0.9837
AC 5.4/P.6	5.4	0.6	0.8366	0.0742	0.9702

From the values reported in Table 6 it can be observed that for the same aggregate skeleton the workability does not depend on the bitumen or additive content. Instead, when the bitumen or additive content increases, the values of the initial densification C_1 and, consequently, the compactness at any number of revolutions also increases.

Shear ratio values, determined during the test as mentioned above and automatically recorded by the equipment, are reported in Figure 5.

The shear ratio lines in Figure 5 show that the specimens with a percentage of 0.3 of plastomeric compound and 4.3 or 4.6 of bitumen, and specimens without any percentage of plastomeric compound but with the percentage 4.9 or 5.1 of bitumen, mobilize shear ratio values that are maintained constant during the design life years. By contrast, a slight excess of bitumen and a slight excess of additive cause a fall of the shear ratio and therefore the content is not optimal.

Regarding the voids, Table 7 shows the average values of specimens prepared with the gyratory compactor that, considering for example the requirements on the air voids (Anas S.p.A 2009) at 10, 100 and 190 rpm, reported in Table 8, confirm the results seen in Figure 5.

Table 7. Average values of the air voids of the specimens at 10, 100 e 190 gyrations.

Mixture	b %	p %	No. of gyrations		
			10	100	190
AC 4.3/P.3	4.3	0.3	12.6	4.7	2.8
AC 4.6/P.3	4.6	0.3	12.1	4.4	2.6
AC 4.9/P.0	4.9	0.0	13.0	5.1	3.2
AC 4.9/P.3	4.9	0.3	11.1	2.7	1.0
AC 4.9/P.6	4.9	0.6	10.2	2.0	0.7
AC 5.1/P.0	5.1	0.0	11.2	3.8	2.0
AC 5.1/P.3	5.1	0.3	10.8	2.4	0.8
AC 5.1/P.6	5.1	0.6	10.2	1.9	0.6
AC 5.4/P.0	5.4	0.0	9.9	1.7	0.4
AC 5.4/P.3	5.4	0.3	9.6	1.4	0.3
AC 5.4/P.6	5.4	0.6	9.5	1.0	0.3

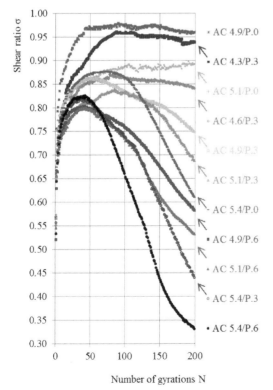

Figure 5. Shear ratio lines of the traditional mixture and mixture with plastomeric compound.

Table 8. Air void values with variation in the number of gyrations (Anas S.p.A requirements).

N° gyrations	% voids
10	11 ÷ 15
100	3 ÷ 6
190	≥2

4 TESTS ON OPTIMIZED MIXTURES

4.1 Permanent deformation resistance

The wheel tracking test was carried out considering two percentages of bitumen 50/70 as well as two percentages of plastomeric compound (i.e. P.0 the control mixture and P.3 the one with 0.3% of polymer), and making overall two slabs for each mixture, with dimensions $305 \times 305 \times 50$ mm and air voids content v = 4.5%. The average values are reported in Figure 6.

In Figure 6 it is possible to note that, for the same percentage of plastomeric compound, there is no substantial difference regarding rut depth; by contrast, at the same percentage of bitumen, the rut depth values of the mixture with the plastomeric compound are half of those of the mixture without the plastomeric compound.

Moreover, the introduction of the plastomeric compound leads to almost constant results with variation in the bitumen content in the mixture, for the contents studied, unlike what happens for traditional mixtures, which always have an increase in rut depth with an increase in the bitumen content. This leads to the conclusion that the introduction of the polymer, making permanent deformation resistance indifferent to variations in the bitumen content, is advantageous in the case of mixtures which require particularly high bitumen contents.

The additive used thus appears to be particularly advantageous regarding permanent deformation resistance, since it reduces rutting, increasing the cohesive ability of the bituminous mixture, as already demonstrated by many tests (Iterchimica S.r.l 2015). A confirmation can be obtained by calculating an additional parameter as the wheel-tracking slope in air (WTS$_{air}$), that is the average rate at which rut depth increases with the number of passages (see Table 9).

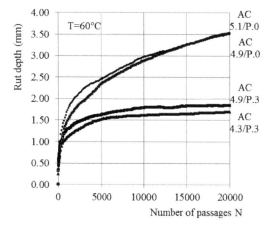

Figure 6.   Trend of rut depth in the wheel-tracking test.

Table 9.   Values of WTS$_{air}$ (Wheel-Tracking Slope in air) of the mixtures studied.

Mixture	WTS$_{air}$ mm/10^3 cycles
AC 4.9/P.0	0.10
AC 5.1/P.0	0.13
AC 4.3/P.3	0.01
AC 4.9/P.3	0.01

The fact is that this parameter increases when the percentage of bitumen increases, while it decreases when the percentage of additive increases.

4.2 Stiffness and fatigue cracking resistance

The loading configuration adopted in the fatigue tests (as well as in the complex modulus tests) was sinusoidal bending on prismatic specimens constrained at the two outer clamps and point loads at the two inner clamps (4 point bending beam test). The deformation was kept constant during the test (controlled strain).

The tests were conducted at 20°C and 10 Hz on beams with dimensions $400 \times 45 \times 50$ mm and obtained from a slab of dimensions $400 \times 305 \times 50$ mm. The deformation was 350 µε for short duration tests and 150 µε for long-term tests.

The fatigue criterion used was the classical one, referenced as N$_{f50}$. It corresponds to the number of cycles for which the modulus decreases to 50% of its initial value. The initial value was calculated at the 100th load cycle. The value of the strain amplitude leading to failure at one million cycles is hereafter called "$\varepsilon_{10}6$".

The complex modulus tests were carried out in the same way as the fatigue tests, with the only difference that the test did not finish when the beams broke or when the value was half its initial value (at 100th cycle), and was only carried out up to 150 cycles. The deformation was 25 µε and the frequencies were 1, 10, 30 and 1 Hz. The complex modulus tests and the fatigue tests were carried out on optimized mixtures, that is ones with a content of 4.9% of bitumen for the traditional mixture, 4.3% of bitumen and 0.3% of plastomeric compound.

The complex modulus test parameters are reported in Table 10 and show what it was reasonable to expect: the complex modulus values are highest at low temperatures and high frequencies, and are lowest at high temperatures and low frequencies; besides, the mixture with the plastomeric compound has higher values than the traditional mixture at the same temperature and frequency conditions.

Table 10. Values of complex modulus and phase angle of the mixtures studied.

Mixture	|E*|, Mpa			φ, Deg.		
	Frequency, Hz			Frequency, Hz		
	1	10	30	1	10	30
AC 4.9/P.0	3692	7282	8756	36	24	23
AC 4.3/P.3	4168	7528	9188	30	20	18

Table 11. Fatigue line parameters of the mixtures studied.

Mixture	A	B	R^2	ε_{10}^6
AC 4.9/P.0	6328	−0.262	0.9828	169.5
AC 4.3/P.3	4532	−0.240	0.9797	164.5

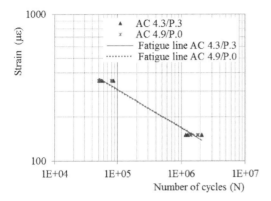

Figure 7. Fatigue lines.

Fatigue test results made it possible to obtain the regression lines (Wöhler curves) shown in Figure 7, i.e. Eq. (1).

$$\varepsilon = a*N^{-b} \tag{1}$$

For each fatigue lines the following parameters were calculated in order to make a judgment on performance:

• "a" is a constant and it depends on the physical and mechanical characteristics of the material, test temperature and frequency;
• slope of the fatigue lines (b);
• coefficient of determination (R^2);
• admissible strain level at $N = 10^6$ loading applications (ε_{10}^6) in order to characterize fatigue resistance.

Table 11 summarizes these parameters for the mixtures studied.

The fatigue lines have high values of the regression coefficients R^2 (see Table 11) and this means that the results are slightly dispersed and very reproducible. The fatigue lines being very close to each other, one can conclude that the mixtures studied offer comparable performances, in terms of fatigue resistance. Finally, the complex modulus was also evaluated by means of a triaxial cell. The loading configuration adopted for characterizing the stiffness of the bituminous mixture (dynamic modulus) was direct compression on cylindrical specimens. The tension was kept constant during the test (controlled stress), which was carried out according to the EN 12697-26 standard, annex D.

The tests were conducted at 10, 20, 30 and 40°C and six frequencies for temperature (20, 10, 5, 1, 0.5 and 0.1 Hz) on cylinders with dimensions 100 × 150 mm and obtained after compaction with a gyratory compactor. The stress levels applied were chosen in such a way that the strain response was kept within 50–150 με. The complex modulus tests were carried out on the optimized mixtures seen before, with two percentages of voids (2.5% and 5.5%).

The isotherms obtained from the triaxial cell were used for determination of the Master Curves. The values of the shift factor were calculated and optimized according to the formula of Arrhenius (Celauro et al 2010). In this way, by horizontal translation of the shift of isotherms relating to the test temperatures it was possible to construct the master curve at a reference temperature of 20°C for each bituminous mixture. Figures 8 and 9 show the master curves for the two mixtures with 2.5% and 5.5% voids.

From Figure 8 it is interesting to note that in the whole range of frequencies studied the mixture with the plastomeric compound provides higher dynamic modulus values than those of the traditional mixture.

Compared to the previous case, in Figure 9 it is interesting to note that at low frequencies the mixture with the plastomeric compound still provides higher dynamic modulus values than the traditional mixture, while at high frequencies it provides similar values.

For the mixtures studied it is observed that the dynamic modulus always increases with an increase in the percentage of the voids in the mixture (in the range of investigation).

However, while for the mixtures without additives this increase in the mechanical performance (manifested by an upward shift of the Master Curve of the dynamic modulus), is observable, both at low and high frequencies in the test (and hence both at high and low temperatures), in the case of the mixture with the plastomeric compound this increase is much more limited, the values of the dynamic modulus being very similar for the two different percentages of voids studied (the Master curve are close).

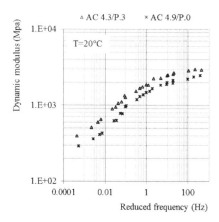

Figure 8. Master curves with v = 2.5%.

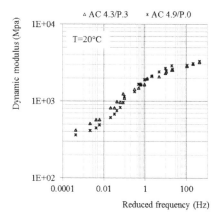

Figure 9. Master curves with v = 5.5%.

This leads to the consideration that the mechanical performances of the mixture are less dependent on the variation in air void content, with substantial insensitivity to this parameter.

5 CONCLUSIONS

Limited to the materials studied, the laboratory tests reported in can be considered adequate to indicate the possibility of using the plastomeric compound in order to improve mechanical performances of bituminous mixtures produced with component of average quality.

The optimized mixtures show good stability values and compaction and the positive influence that the plastomeric compound has on these mixtures regarding permanent deformation resistance (at the same bitumen content, the mixture with the plastomeric compound is more resistant than the traditional mixture, with rut depth values reduced to less than half).

Moreover, the presence of the plastomeric compound allows optimization of mixtures with the lowest binder contents and also makes it possible to obtain higher complex modulus values than traditional mixture without additive (35% more) and leads to the same performance regarding fatigue resistance. From triaxial cell tests it can be said that the dynamic modulus always increases with an increase in the percentage of voids in the mixture. However, while for the traditional mixture there is a slight increase in the mechanical performance, both at low and high frequencies of the test (and hence both at high and low temperatures), in the case of mixtures with the plastomeric compound, there is substantial insensitivity of this parameter characteristic to variation in the air void content.

A possible future development is a thorough statistical study considering several kinds of bitumen and sources of aggregate, as locally available in order to highlight the performance sensitivity to aggregate-bitumen combinations.

REFERENCES

Anas 2009. Capitolato Speciale d'Appalto per lavori stradali. Norme tecniche.

Bense, P. 1983. Enrobés armés par déchets de matières plastiques. Bull Liaison Lab Ponts Chauss, 99–106.

Celauro, B. et al. 2004. Effect of the construction process on the performance of asphalt paving mixes with Crumb Rubber from scrap tires. 3rd Eurasphalt & Eurobitume Congress, ISBN 908 0288 446, Vienna, 757–769.

Celauro, C. et al. 2010. Production of innovative, recycled and high-performance asphalt for road pavements. Resources, Conservation and Recycling, 54 (6): 337–347.

Cominsky, R.J. et al. 1994. The Superpave Mix Design Manual for New Construction and Overlays. Report SHRP-A-407, Strategic Highway Research Program-National Research Council-Washington, DC.

Iterchimica S.r.l 2015. SuperPlast. Conglomerati bituminosi ad elevate prestazioni. Libretto Tecnico. Iterchimica S.r.l, Bergamo.

MIT 2002. Studio a carattere pre-normativo delle Norme Tecniche di Tipo Prestazionale per Capitolati speciali d'Appalto. MIT Ministero delle Infrastrutture e dei Trasporti-Ispettorato per la Circolazione e la Sicurezza Stradale.

Pettinari, M. et al. 2014. The effect of Cryogenic Crumb Rubber in cold recycled mixes for road pavements. Construction and Building Materials 63: 249–256.

Roberts, F.L. et al. 1996. Hot Mix Asphalt Materials, Mixture Design, and Construction. NAPA Research and Education Foundation (2), Lanham-Maryland.

Serfass, J.P. et al. 2000. Enrobé bitumineux modifiés au polyethylene. Revue Génerale des Routes et Aérodromes 787: 47–57.

Venturini, L. et al. 2016. New technologies in road pavement design-Increase of the service life. Second Serbian Road Congress. June 9–10.

Transport Infrastructure and Systems – Dell'Acqua & Wegman (Eds)
© *2017 Taylor & Francis Group, London, ISBN 978-1-138-03009-1*

An application of the parallel gradation technique to railway ballast

M. Esposito, F.S. Capaldo, S. Andrisani & B. Festa
University of Naples "Federico II", Naples, Italy

ABSTRACT: Minimizing the decay phenomenon of ballast performances is essential to preserve functions and substructure over time.

The track design usually ignores the degradation of ballast and associated plastic deformations resulting from the passage of many wheels. This is due to a lack of understanding of failure mechanisms of this material and the absence of stress-strain constitutive models, which include plastic deformation and breakage of the particles under a large number of load cycles.

This limitation involves an oversimplified design of the ballast and technological deficiencies in the construction, which will impact on frequent and expensive maintenance. A good understanding of the ballast behavior and parameters that control its performance will help reduce ballast maintenance costs while at the same time preserving «effectiveness» and «efficiency».

In order to keep high standards related to ballast fulfillments, several railway standards select material based on physical and geometric requirements, but there is never a direct correlation with the deformation behavior of the track: using a stiffness parameter of subgrade would allow one to consider the track as a beam resting on an elastic foundation.

The aim of this study is to characterize the ballast by means of experimental tests, in order to assess the resilient modulus provided by dynamic triaxial tests, simulating the passage of the wheels.

According to UNI-EN13286–7 and ASTM-D2850–87 the specimen to be tested must have a diameter D greater than 5Dmax (maximum particle size of the material) and height H = 2D.

Since the ballast contains aggregates too large (D_{max} = 6–7 cm) to be tested in the common triaxial apparatus, this study applies the parallel gradation technique to railway ballast. The aim of this approach is to obtain a scaled down grading curve compatible with the available equipment, keeping shape of the particles, surface roughness and mineralogical nature.

1 THE BALLAST

1.1 Introduction

The railway track is placed on the ballast, which plays the role of elastic soil. Its functions are to:

– distribute vertical loads on the road body;
– ensure the track geometric design conditions (levels and alignments due to building and maintenance conditions);
– absorb stresses induced in the track by trains;
– confer elasticity to the track in order to allow the arming to react elastically to the stresses to which it is subjected;
– allow drainage of rainwater;
– absorb longitudinal stresses due to temperature changes;
– provide a filter between the track and environment against vibrational phenomena.

These properties can be affected by the presence of the operating loads that, due to the rolling process, determine the lifting of the sleeper that, beating the ballast during its descent, can generate dynamic forces stressing the ballast and causing scaling, size redistribution and rubbing; also, the lift of material from the foundation can occur in cases of a lack of substructure or in the presence of a poorly constructed substructure or even contamination of the crushed stone, generated by the loss of loads, vegetation residues or other atmospheric agents (Lichtberger, 2010).

Minimizing the decay phenomenon of ballast performances is essential to preserve its functions and the substructure over time.

1.2 Ballast fulfillments

There are no universal standards to this material's characteristics, such as size, shape, hardness, friction, texture, abrasion resistance and the particle size composition which can allows to obtain the optimum performance under all load types and with all types of subgrade; therefore, a large variety of materials are employed for the formation of the layer (i.e. basalt, limestone, granite, dolomite, gneiss and quartzite).

In order to ensure the functions which the ballast must carry out, and which have been defined in section 1.1, it must have characteristics such as size, shape, particle size, surface roughness, particle density, resistance, durability, hardness, resistance to attrition and water.

In Italy the activity of management and maintenance of the railway network, as well as the design, construction and start-up of the plants is provided by RFI (Rete Ferroviaria Italiana).

According to the Technical Standards provided by RFI, and in compliance with the UNI EN 13450, the stone used for a new railway ballast or for renewal and maintenance, must comply with special requirements. The UNI EN 13450 contains all the requirements that must be followed for the identification of the material; in Table 1, for instance, granulometric curve fulfillments are summarized.

In addition to the Italian requirements, different standards and technical specifications have been issued by different railway organizations around the world to meet their design requirements. In general, as already seen with Italian standards, the ballast has to be edgy, strong, hard and tough under the expected traffic and in difficult environmental conditions, and have a uniform particle size distribution.

For example, the Australian law is the AS 2758.7: "Aggregates and rock for engineering purposes. Railway ballast" (AS2758.7, 1996) (Table 2).

The AREMA (The American Railway Engineering and Maintenance-of-Way Association) is a North American rail industry group. Like RFI in Italy, it publishes technical specifications for the design, construction and maintenance of rail infrastructure, thus defining the requirements in the United States and Canada.

The material intended to be used as a railway ballast, is defined in terms of physical and geometrical characteristics, responding to grain size, elongation, specific weight, the friction resistance requirements, etc. The requirements which

Table 1. Granulometric curve UNI EN 13450.

Sieve n° mm	Passed percentage %
80.00	100
60.00	100
50.00	70–99
40.00	30–65
31.50	1–25
22.40	0–3

Table 2. Granulometric curve AS 2758.7.

Sieve n° mm	Passed percentage %
63.00	100
53.00	85–100
37.50	20–65
26.50	0–20
19.00	0–5
13.20	0–2
9.50	–
4.75	0–1
1.18	–
0.075	0–1

Table 3. Granulometric curve AREMA.

Sieve n° inch	Passed percentage %
2 1/2	100
2	90–100
1 1/2	60–90
1	10–35
3/4	0–10
3/8	0–3
No 4	0–0.5

the ballast must fulfill in terms of granulometric curve according to American standards (AREMA, 2010), are reported in Table 3 by way of example.

1.3 *Mechanical characterization of ballast*

As discussed above, various railway standards in the world (RFI, Australian and AREMA) select the ballast for railway use, based on physical tests to assess the correct form of aggregates, granulometric curve, abrasion resistance and climate changes, but the identified parameters never have direct correlation with the deformation behavior of the track and which would allow to schematize its behavior as a beam resting on elastic foundation, characterized by the parameter K. the triaxial test is probably the most appropriate test to assess the material's ability to withstand vertical forces.

The stress state of the ballast during the passage of the train can be, with a good approximation, well described by the compression phase in a triaxial test, limited to the area which is located below the sleepers, and by the traction phase with reference to the area between two successive sleepers (Kaya M., 2004), as shown in Figure 1.

The aim of the study is, therefore, to characterize the ballast by means of experimental tests, by referring to the Resilience Modulus M_R, which is

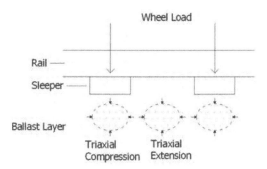

Figure 1. Stresses in ballast.

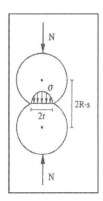

Figure 2. Contact stresses between two spheres.

used to represent the load subgrade bearing capacity when it is stressed by mobile loads, such as road or rail loads. M_R is defined as the ratio between the deviator stress q and the corresponding axial recovered deformation and is evaluated, for example, through a triaxial test.

1.4 Parallel gradation scaling technique

According to UNI-EN13286-7 and ASTM-D2850-87 the test specimen must have a diameter D greater than $5*d_{max}$ (maximum particle size of the material) and height H = 2D, in order to reduce the edge effect given by the plates during the test.

With reference to a railway ballast, the maximum size of the aggregates is less or more 6–7 cm, involving D ≥ 35 cm and H = 70 cm.

Since the ballast contains aggregates too large to be tested in the equipment that are located in the common test laboratories, the application of the parallel gradation scaling technique is proposed, in order to model this material. The objective of this technique is to retain the shape of the particles, the surface roughness and the mineralogical nature in order to obtain a reduction of the material that is characterized by a maximum dimension of the aggregates compatible with that of the available apparatus.

This technique was used for the first time by Lowe with reference to materials used for the dams and then recalled by other authors (M. Kaya, 2004) (Sevi, 2008) in following studies aimed at testing the ballast.

A model sample of perfect spheres (Figure 2), regardless of grain size, would closely duplicate the contact stresses and void ratio characteristics of a larger prototype gradation. Thus in models of coarse gradations, where the only difference between prototype and model sample is the difference in size of particles, the model sample should closely duplicate the behavior of the larger prototype.

The theoretical basis of the parallel gradation scaling modeling scheme is based on the Hertz formula for the maximum stress at the contact of two bodies.

The maximum contact stress, σ_{Nmax}, is located at r = 0, where the radius of the contact area is a. This calculation assumes the two objects are spherical and of the same radius, R. P is the compressive force acting on the particles and G is the shear modulus of the materials. This equation assumes that both materials exhibit the same elasticity and Poisson's ratio, ν (Hertz, 1956).

$$\begin{cases} \sigma_{N_{max}} = \dfrac{3P}{2\pi a^2} = \dfrac{4Ga}{(1-\upsilon)\pi R} \\ a = \left[\dfrac{3P(1-\upsilon^2)}{4E}\right]^{\frac{1}{3}} = \dfrac{1}{2}\left[\dfrac{3(1-\upsilon)PR}{G}\right]^{\frac{1}{3}} \end{cases} \quad (1)$$

This relationship for particles having a perfect geometric similarity shows that the values of contact stresses and strains are independent of particle size. Laboratory data of tests run on quartz, which is a highly elastic material, have found that the coefficient of friction is constant and independent of both contact area and normal load. Therefore, it appears that the deformational characteristics of elastic rock materials should not depend on the grain size of the material tested (Sevi, 2008).

However, other studies conducted by (Kaya et al. 2004; Kaya et al. 1997) on the same material, have surprisingly led to completely contrasting results than Sevi and shown in Figure 3:

Many predictive formulas have been developed to estimate the modulus of the material, taking into account the dependence on the stress state, but the hyperbolic model is the most widespread (Janbu, 1968):

$$E = Kp_r\left(\frac{\sigma_3}{p_r}\right)^n \quad (2)$$

where σ_3 = confining stress, p_r = reference pressure (atmospheric pressure, used to make conversion from one system of units to another), K and n are dimensionless numbers.

With reference to the Resilient modulus, hyperbolic model (Figure 4) gives the same result both the loading e unloading phases and becomes:

$$E_{ur} = K_{ur} p_a \left(\frac{\sigma_3}{p_a}\right)^n \qquad (3)$$

Two steps are involved in evaluating the modulus parameters K and n. The first is to determine the values of Ei for each test, and the second is to plot these values against σ_3 (on log-log scales) to determine the values of K and n. The value of K is equal to the value of E_i/p_a when $\sigma_3/p_a = 1$.

The value of n is the slope of the line on this plot, and may be determined graphically (Duncan *et al.* 1980).

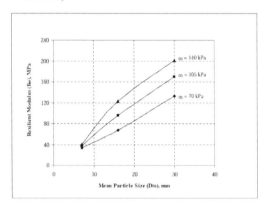

Figure 3. Resilient modulus versus mean particle size, varying confining pressure.

In this study two different scaled down granulometric curves have been studied and the hyperbolic model has been applied to both of them.

2 TRIAXIAL TEST

2.1 *Introduction*

The triaxial test is aimed to reproducing a given stress state on a sample and later, through the pore pressure measurement, even the evolution of the effective stresses up to breaking conditions.

Observing Figure 5, it is possible to get an idea of the trend of the total and residual vertical deformations deriving from a triaxial test at constant confining pressure and with the deviatoric stress that varies with a sinusoidal law as a function of time.

The total and residual deformation curves, which increase with the number of load applications, show parallel trends after a certain number of loading cycles; therefore, the resilient deformation first varies significantly with the number of cycles and then it assumes an almost constant value. M_R is then calculated as the ratio of q and the corresponding resilient deformation ε_{ra}, stabilized around a single value (Giannattasio *et al.* 1989).

2.2 *The equipment*

Triaxial tests were performed using the equipment INSTRON 8502, provided by the road laboratory LaStra of DICEA (Department of Civil, Architectural and Environmental Engineering; University of Naples Federico II) and showed in Figure 6.

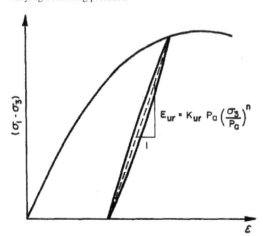

Figure 4. Loading and unloading modulus.

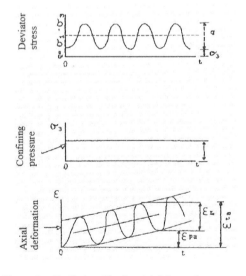

Figure 5. Vertical and horizontal deformations.

Figure 6.   INSTRON 8502 available at the LaStra.

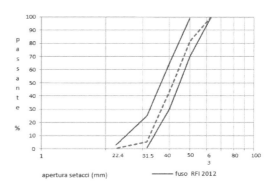

apertura setacci (mm)   ―――― fuso RFI 2012

Figure 7.   Adopted granulometric curve.

Table 4.   Converted sieves.

EN Sieve n° mm	Converted sieve mm	Passed percentage %
63.00	20.00	100
50.00	15.87	81.53
40.00	12.70	43.06
31.50	10.00	5.28
22.50	7.14	0.53
0.063	0.02	0.48

Table 5.   Granulometria 1.

Sieve n° mm	Passed perc. %	Retained percentage %
20.00	100.00	0.00
16.00	82.10	17.90
12.50	40.28	41.82
10.00	5.28	35.00
8.00	1.96	3.32
0.063	0.00	1.96

Table 6.   Granulometria 2.

Sieve n° mm	Passed perc. %	Retained percentage %
10.00	100.00	0.00
8.00	82.10	17.90
6.30	41.49	40.61
4.00	1.96	39.53
2.00	0.51	1.45
0.063	0.00	0.51

The press is equipped with a servo-hydraulic frame that allows the loading of samples for cyclic or static tests and connected to a suitably equipped computer with control software.

The diameter D of the sample is 10 cm and the height H is 20 cm.

2.3   The material

This study analyzes basaltic material taken from a quarry in Campania (Italy), selecting a granulometric curve compatible with RFI 2012 Specifications, as showed in Figure 7.

With reference to the INSTRON 8502, in a sample with D = 10 cm and H = 20 cm the maximum particle size d_{max} of the material to be tested is instantly assessable through the limit dictated by the specification UNI EN 13286-7: $d_{max} = D\backslash5 = 20$ mm.

The conversion coefficient c is, therefore, obtained through the ratio between the diameter of the maximum standard sieve $D_{setstand\ (max)}$ in the real granulometric curve and d_{max}:

$$c = D_{setstand\ (max)} \backslash d_{max} = 3.15.$$

The coefficient thus obtained allows the determination of the sieve sizes corresponding to the reduced grain size, obtained simply as the ratio between the size of the sieves of the real curve and c:

Finally, since the previous operation led to the determination of sieves that do not match to the real ones, then a linear interpolation must be carried out in order to derive the passing percentage trough a real sieve and determine the final reduced size curve, called "Granulometria 1".

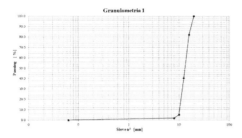

Figure 8. Granulometria 1.

Figure 9. Granulometria 2.

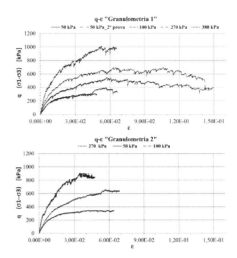

Figure 10. Static triaxial tests.

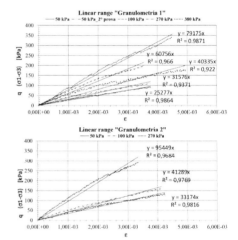

Figure 11. Elastic Modulus, versus confining pressure.

Reasoning similarly and setting the conversion coefficient c = 6.30 a second granulometric curve, called "Granulometria 2", has been obtained.

2.4 Static tests

In order to highlight the rheological model representative of the ballast behavior and to make a clear distinction between linearity and plasticity field, static triaxial tests were carried out.

Tests were conducted controlling the axial s and assuming a final axial strain of 20%, in order to leave unchanged mechanical characteristics and the resistance of the material; the corresponding advance speed of the test is 0.0002 cm/s.

Tests were performed with different confining pressure, both for Granulometria 1 and 2, but they were not always carried to term, because of systematic arrests of the apparatus. In any case, the post processing of the outputs showed that in almost all the tests the specimen reached the breakage, intended as the depletion of the resistance (Figure 10).

From the analysis of static triaxial tests it is possible to derive the modulus of elasticity of the ballast, obtained as the slope of the linear portion (Figure 11).

2.5 Cyclic tests

Cyclic tests were carried out controlling the load. Vertical stress was increased up to σ_3 in order to achieve the spherical state; then the deviator stress was varied cyclically, varying in a different range depending on whether the test was performed in the elastic or plastic range (defined by the output of static tests).

Since the train speed varies point by point, it is important to study the influence of the load on the frequency behavior of the ballast; hence, the tests were conducted with frequencies 1, 5, 10 and 20 Hz.

In this stage we were not interested in studying the fatigue behavior of the material, therefore the specimens were subjected to 500 load cycles, cycles from 1 to 5, from 149 to 155 and from 495 to 500 were acquired, with a resolution of 100 points per wave.

Tests in the elastic range were carried out on the same sample before and after tests in the plastic range, with the aim of verifying the influence of the stress state on the linear behavior.

An example of output of the dynamic tests is given in Figure 12.

M_R was evaluated on 151 load wave.

2.6 Results

Outputs were classified as a function of particle size and were divided in two classes, depending on whether or not they were the result of tests carried out in the linearity range.

By resorting to mathematical software Mathcad, the above-mentioned moduli were finally rep-

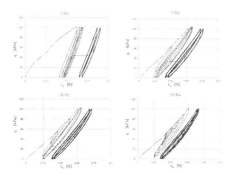

Figure 12. Cyclic test on "Granulometria 1", elastic range (q = 0–100 kPa), σ_3 = 8psi.

Table 7. K and n from linear regression.

Granulometric curve		K kPa	n	R^2
Gran. 1	Linear range	4.6×10^4	0.378	0.611
	Plastic range	9.2×10^4	0.327	0.792
Gran. 2	Linear range	1.4×10^4	0.648	0.960
	Plastic range	5.9×10^4	0.421	0.756

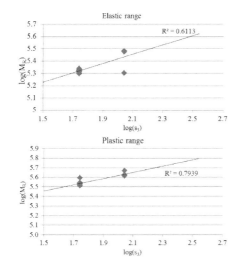

Figure 13. Linear regression (Granulometria 1).

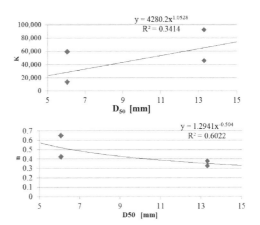

Figure 14. Trend of K and n.

Table 8a. M for Granulometria 1, Prova 2.

	A1	mean	st.dev.
		2,13E+05	9,45E+03
by Tests Analysis		μ−2σ	μ+2σ
f	Mr	1,94E+05	2,32E+05
Hz	kPa	test A1	
1	2,19E+05	ok	
5	2,18E+05	ok	
10	2,15E+05	ok	
20	1,99E+05	ok	

resented in a bi-logarithmic diagram (σ_3 on x-axis) reported in Figure 13, obtaining the intercept values of the regression line on the ordinate axis (K) and the corresponding slope (n).

Finally, the variability of K and n was correlated to the mean diameter of the aggregates of both granulometric curves.

Regressions (see Figure 14) show a growing trend of K function with increasing average diameter and a decreasing trend of the slope function.

3 CONCLUSIONS

This article proposed an approach to study the bearing capacity of the ballast.

In fact, the analysis of the literature shows that the ballast is not completely defined in terms of stiffness.

The Resilient modulus is the best parameter to schematize the bearing capacity of the ballast, because it is a dynamic parameter used to represent the load bearing capacity of the substrate when this is stressed by mobile loads (i.e. road or rail loads).

The above parameter was, therefore, studied by means of static and dynamic triaxial tests, varying

Table 8b. M for Granulometria 1, Prova 3.

A1		A2	mean	st.dev.
			2,08E+05	2,81E+03
by Tests Analysis			μ−2σ	μ+2σ
f	Mr	Mr	2,02E+05	2,13E+05
Hz	kPa	kPa	test A1	test A2
1	2,05E+05	2,10E+05	ok	ok
5	2,08E+05	2,10E+05	ok	ok
10	2,07E+05	2,12E+05	ok	ok
20	2,04E+05	2,07E+05	ok	ok

Table 9a. M for Granulometria 2, Prova 3.

A1		A2	mean	st.dev.
			1,82E+05	2,55E+03
by Tests Analysis			μ−2σ	μ+2σ
f	Mr	Mr	1,77E+05	1,87E+05
Hz	kPa	kPa	test A1	test A2
1	1,81E+05	1,85E+05	ok	ok
5	1,78E+05	1,84E+05	ok	ok
10	1,79E+05	1,83E+05	ok	ok
20	1,82E+05	1,83E+05	ok	ok

Table 9b. M for Granulometria 2, Prova 4.

A1		A2	mean	st.dev.
			1,95E+05	3,79E+03
by Tests Analysis			μ−2σ	μ+2σ
f	Mr	Mr	1,88E+05	2,03E+05
Hz	kPa	kPa	test A1	test A2
1	1,91E+05	1,98E+05	ok	ok
5	1,91E+05	1,99E+05	ok	ok
10	1,91E+05	1,98E+05	ok	ok
20	1,93E+05	2,00E+05	ok	ok

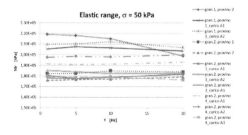

Figure 15. Mr versus f (Hz).

confining pressure, deviator stress and load frequency, for a scaled down granulometric curve, in order to adapt the test size to the available triaxial cell size.

Results seem to show that the modulus is independent from load frequency, as shown in Tables 8 (a and b) and 9 (a and b) and in Figure 15, with reference to the elastic range.

Furthermore, the regressions (elastic and plastic range, cfr Figure 13) allow to correlate, with good approximation, the resilient modulus to the stress state applied.

Additional tests should be conducted, introducing at least one other reduced grain size, in order to increase the sample size and improve the evaluations.

ACKNOWLEDGMENTS

This work would not have been possible without the dedication and constant support of the Las. Tra laboratory technicians Alfredo Caruso, Anna Formisano and Attilio Sannino, who are gratefully acknowledged.

REFERENCES

American Railway Engineering and Maintenance-of-Way Association (AREMA), 2010. *Part 2: Ballast*.
ASTM D 2850–87. (s.d.). *Standard Test Method for Unconsolidated-Undrained Triaxial Compression Test on Cohesive Soils*.
Australian Standard, 1996. AS2758.7: *Aggregate and rock for engineering purposes—Part 7: Railway Ballast*.
Duncan, J., Byrne, P., Wong, K., & Mabry, P. 1980. *Strenght, stress-strain and bulk modulus parameters for finite element analysis of stresses and movement in soil masses*. College of Engineering, Office of Research Services, University of California, Berkeley: Report N. UCB/GT/80–01.
Giannattasio, P., Caliendo, C., Esposito, L., Festa, B., & Pellecchia, W. 1989. *Portanza dei sottofondi*. Fondazione Politecnica per il mezzogiorno d'Italia; Napoli
Hertz, H. 1899. *The principles of mechanics presented in a new form*. New York: Dover Publications.
Janbu, N. 1968. *Soil compressibility as determined by oedometer and triaxial tests*. European Conference on Soil Mechanics and Foundation Engineering, (p. 19–25, Vol.I). Wiesbaden.
Kaya, M., Jernigan, R., Runesson, K., & Sture, S. 1997. *Reproducibility and conventional triaxial tests on ballast material*. USA: University of Colorado, Department of Civil Environmental and Architectural Engineering.
Kaya, M. 2004. *A study on the stress-strain behavior of railroad ballast materials by use of parrallel gradation technique*. Middle East Technical University.
Lichtberger, B. 2010. *Manuale del binario*. Hamburg: Eurail Press; ISBN: 978–3-7771–0408–9
RFI, 2011. *Technical Standards RFI DTCINC SP IFS 010 A*. Roma: Gruppo Ferrovie delle Stato S.p. A.
Sevi, A. 2008. *Physical modeling of railroad ballast using the parallel gradation scaling technique within the cyclical triaxial framework*. Missouri: Ph.D. thesis
UNI EN 13450, 2003. *Aggregati per massicciate ferroviarie*.
UNI EN 13286–7, 2006. *Miscele non legate e legate con leganti idraulici—Metodi di prova. Part 7: Prova triassiale ciclica per miscele non legate*.

Transport Infrastructure and Systems – Dell'Acqua & Wegman (Eds)
© *2017 Taylor & Francis Group, London, ISBN 978-1-138-03009-1*

Rolling resistance impact on a road pavement life cycle carbon footprint analysis

Laura Trupia, Tony Parry, Luis Neves & Davide Lo Presti
Faculty of Engineering, Nottingham Transportation Engineering Centre (NTEC),
University of Nottingham, UK

ABSTRACT: In terms of methodology, there is a continuing discussion related to the introduction of the impact of road pavement surface properties on rolling resistance in pavement Life Cycle Assessment (LCA).

The aim of this paper is to analyse if the current level of knowledge of this component is sufficient to be implemented in the pavement LCA framework. The study compares the CO_2 emissions, calculated with two different rolling resistance models in the literature and performs a sensitivity test on the pavement deterioration rate, for two UK case studies.

The rolling resistance models and the pavement deterioration rate significantly affect the LCA results.

The results show that the methods of modelling and the methodological assumptions need to be transparent in the analysis of the impact of the pavement surface properties on fuel consumption, in order to be interpreted by decision makers and implemented in the LCA framework.

Keywords: Rolling Resistance, Pavement surface properties, Life Cycle Assessment, Carbon Footprint

1 INTRODUCTION

The transport sector and particularly the road transport sector, is responsible for the emission of a large amount of Greenhouse Gases (GHG) contributing to global climate change. In 2009, transport accounted for around a quarter of UK GHG and road transport was the most significant source, accounting for 68% of total transport GHG emissions (UK Department of Energy & Climate Change 2015).

Life Cycle Assessment (LCA) is a standardized approach to quantify the direct and indirect environmental impacts associated with a product, process or service. In the road transport sector this approach is promising because it can estimate the impacts of infrastructure and operations on the environment and then assist highway authorities, companies and government institutions in evidence-based decision-making (Muench et al. 2014). Over the last years, the LCA approach has begun to permeate into the planning, construction, operation, and maintenance processes in highway asset management (Matute et al. 2014). However, the complete introduction of this approach in the asset management decision making process is not possible yet, due to an incomplete understanding and uncertainty regarding the impact of some relevant phases and components of a road pavement LCA. The use phase is one of the most relevant

and complex parts of the life cycle that requires investigation in different areas of scientific literature: rolling resistance, albedo, carbonation, lighting and leachate (Santero and Horvath 2009).

Rolling resistance is one of the forces resisting vehicle movement partly due to the energy loss associated with the Pavement-Vehicle Interaction (PVI), due to the physical interaction between pavement and tyre. Much of the rolling resistance can be tracked to tyre properties, but it is also affected by other parameters related to the characteristics of the pavement, such as the pavement surface properties. The impact of the pavement surface on rolling resistance has been an area of study for many years because of its effect on vehicle fuel consumption and emissions and the opportunity to reduce them with conventional maintenance strategies. The pavement surface properties that affect rolling resistance include roughness and macrotexture, usually represented by parameters International Roughness Index (IRI) and Mean Profile Depth (MPD) or Mean Texture Depth (MTD) (Sandberg et al. 2011)

Calculating the impact of pavement surface properties on the rolling resistance and then on vehicle fuel consumption is complex, although over the last years, some studies have been performed to estimate the emissions related to these components (Hammarström et al. 2012; Wang et al. 2014; Wang et al. 2012).

However, the different rolling resistance models and the incomplete knowledge related to the influence of specific variables and assumptions on the results, generates a high level of uncertainty in their interpretation.

In the UK, not only are there no significant studies involving national case studies on the impact of the rolling resistance on the LCA of a pavement, but there are is also a lack of general pavement deterioration models able to predict the change of unevenness and texture depth over time (deterioration rate of IRI and MPD). The change in these parameters may be different for each lane, since it depends on the traffic volume and type, on the surfacing type and on the regional climate. While, some empirical models to describe the deterioration rate of IRI and MPD have been developed (Lu et al. 2009; Tseng 2012), these models are calibrated for specific areas and maintenance treatments and are not applicable to each case study (in these models, the value of MPD tends to increase over time, which is not typical in the UK, where MPD may decrease over time).

This generates a further level of uncertainty concerning the introduction of PVI rolling resistance into the LCA approach. In order to introduce pavement LCA results into the decision making process of highway authorities, governmental institutions and companies, it is necessary that methods of modelling and the assumptions in LCA and carbon footprint studies are transparent and lead to consistent results.

This paper analyses the effect of the pavement surface properties on vehicle fuel consumption and the impact of some methodological assumptions related on the results. To do this, it compares the CO_2 emissions calculated with two different rolling resistance models and performs a sensitivity test on the pavement deterioration rate.

The main aim of this study is to evaluate if the existing approaches to incorporating the impact of PVI rolling resistance into the pavement LCA framework can be done with sufficient confidence for decision making in the UK.

In this paper, the authors will extend the analysis performed for a previous case study (Trupia et al. 2016 (in press)) to a second case study, with a higher volume of traffic, in order to consider the research questions: Are the rolling resistance models ready for implementation in pavement LCA? Can they be applied in the UK? How do pavement deterioration and the models used to describe it affect the results?

2 METHODOLOGY

Figure 1 shows the outline of the process used in this research. It involves the calculation of the

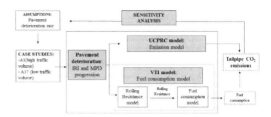

Figure 1.   Outline process.

Table 1.   Case study details.

Case study	A17	A1
AADT	15372	51502
Road type	A—road single carriageway	Motorway dual carriageway
Construction year	2009	2009
Analysis period	20	20
Surface materials	Hot Rolled Asphalt (HRA)	Thin surfacing

GHG emissions due to PVI rolling resistance, using two models from the literature and a sensitivity test performed on the pavement surface properties deterioration rates.

In this study, only the CO_2 emissions are considered for the GHG estimation, since this is the biggest component of the vehicle tailpipe CO_2e emissions (over 99.8%) (Wang et al. 2014).

The study is applied to two case studies, with different Annual Average Daily Traffic (AADT) and pavement design (see Table 1).

2.1 Calculation of the tailpipe CO_2 emissions with VTI and UCPRC models

The CO_2 emissions due to the effect of the pavement surface properties on vehicle fuel consumption for two different UK case studies were estimated using: the model developed at the University of California Pavement Research Center (UCPRC, Davis) (Wang et al. 2014) and the model developed by the Swedish National Road and Transport Research Institute (VTI), within the European Commission project Miriam (Models for rolling resistance In Road Infrastructure Asset Management systems) (Hammarström et al. 2012). Details related to the models are provided in the references, but the key elements are summarised here.

In the UCPRC model, the vehicle CO_2 emission factors are a continuous function of MPD and IRI, but the coefficients in the function are different for each combination of the categorical variables (pavement, road and road-access type, vehicle type). The CO_2 emissions for a specific vehicle type can be calculated directly, based on the analysed pavement segment's MPD and IRI values by using equation (1) and multiplying by the vehicle mileage travelled.

$$T_{CO_2} = a_1 \times MPD + a_2 \times IRI + Intercept \quad (1)$$

where T_{CO_2} is the tailpipe CO_2 emission factor, the terms a_1, a_2 and $Intercept$ are the coefficients derived from the linear regression, depending on surface type and access type, year and vehicle type, IRI is the road roughness (m/km) and MPD is the macrotexture (mm). The Intercept term identifies the total CO_2 emissions related to the total driving resistance, excluding the impact of the pavement condition, which is estimated from the other two terms. To develop this equation, the authors used two different software models: Highway Development and Management Model - version 4 (HDM-4) (PIARC 2002), an empirical - mechanistic model software tool to perform cost analysis for the maintenance and rehabilitation of roads, calibrated for US conditions, and MOVES (Motor Vehicle Emission Simulator), the US EPA highway vehicle emission model based on national data. HDM-4 was used to estimate the rolling resistance, while MOVES was used to model the vehicle emissions as a function of the rolling resistance

The VTI model includes a general rolling resistance model (equation (2)) to estimate the contribution of the rolling resistance to the total driving resistance and a fuel consumption model (equation (3)) to calculate the vehicle fuel consumption (Hammarström et al. 2012). Once the fuel consumption related to a specific type of vehicle was estimated using this model, it was converted to CO_2 emissions, assuming the conversion process proposed by International Carbon Bank & Exchange (ICBE) (2010).

$$F_r = m_1 \times g \times$$
$$(0.00912 + 0.0000210 \times IRI \times V + 0.00172 \times MPD)$$
$$(2)$$

$$F_{CS} = 0.286 \times$$
$$\left(\begin{pmatrix} 1.209 + 0.000481 \times IRI \times V + 0.394 \times MPD \\ + 0.000667 \times V^2 + 0.0000807 \times ADC \times V^2 \\ - 0.00611 \times RF + 0.000297 \times RF^2 \end{pmatrix}^{1.163} \right)$$
$$\times V^{0.056}$$
$$(3)$$

The rolling resistance model developed by VTI is mainly based on empirical data from coastdown measurements in Sweden; the fuel consumption model has been calibrated based on results

obtained from a software VETO, based on a theoretical model developed at VTI to calculate fuel consumption and exhaust emissions from traffic due to various characteristics of vehicles, roads and driving behavior (Hammarström and Karlsson 1987; Karlsson et al. 2012)

For both models, only the CO_2 emissions directly related to the pavement surface properties (IRI, MPD) are calculated; the other terms of the equations are considered equal to zero, since their estimation is not the aim of the study.

Applying the equations previously defined, it is possible to estimate the total CO_2 emissions related to the pavement condition in terms of IRI and MPD (see Figure 2), namely the total component (total area, representing the total CO_2 emissions related to the IRI and MPD). This total component can be considered as the sum of the basic component (dark grey area, representing the value of emissions if the IRI and MPD remain constant over time—no deterioration) and the deterioration component (light grey area, equal to the difference between the first two and representing the emissions due to the deterioration of the pavement properties during the study analysis period, in terms of IRI and MPD).

Pavement engineering studies tend to focus on the deterioration component because of the opportunity to reduce these PVI emissions, taking direct action on the road surface condition, through appropriate maintenance. Obtaining pavement condition improvements is in general more rapid and easy than other approaches to reduce rolling resistance emissions that involve technology improvements or traffic reduction.

In this paper, all the components were estimated, since they can provide a better understanding of the behaviour of the two rolling resistance models.

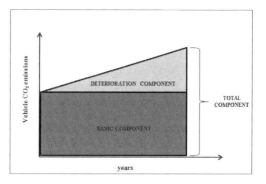

Figure 2. Total CO_2 emissions, divided into basic component (dark grey area) and deterioration component (light grey area) (from Trupia et al. 2016 (in press)).

Table 2. Pavement deterioration rate, in terms of IRI and MPD, during the analysis period.

	Scenario	MPD mm	IRI m/km
A17	Average deterioration	1.8–0.8	1.0–2.3
	Worst deterioration	1.5	1.0–5.0
	No deterioration	1.8	1.0
A1	Average deterioration	1.6–0.6	1.0–2.3
	Worst deterioration	1.3	1.0–5.0
	No deterioration	1.5	1.0

Table 3. CO_2 emissions due to pavement surface condition (average deterioration).

A17

	Emission of CO_2 (ton)		
Model	Basic	Deterioration	Total
VTI	10272	−634	9638
UCPRC	1170	217	1387

A1

	Emission of CO_2 (ton)		
Model	Basic	Deterioration	Total
VTI	109344	−4205	105139
UCPRC	18058	4586	22645

2.2 Sensitivity test on pavement deterioration rate

The use of these models, correlating pavement surface properties to vehicle fuel consumption and emissions, requires as an input parameter, the estimation of pavement condition deterioration rate with time (in terms of IRI and MPD).

As mentioned in the introduction, in the UK there are no models able to predict the deterioration rate of these parameters over the years. For this reason, the time progression of IRI and MPD on the assessed road segments over the analysis period (20 years) is generated according to literature data for specific maintenance strategies (Aavik et al. 2013; Jacobs 1982; Wang et al. 2014).

In order to take into account the uncertainty related to these parameters and the range of potential impact during the use phase, different scenarios of deterioration of IRI and MPD are considered for the two case studies (see Table 2) and compared in a sensitivity analysis.

The average deterioration values include an initial and final condition value and a linear change with time is assumed. This is also the case for the IRI values in the worst deterioration scenario. The MPD in the worst deterioration scenario and the MPD and IRI for the no deterioration scenario are held constant. Note that in the average deterioration scenario, the MPD falls with time from a high initial value; this is common in the UK were high MPD values are specified for new surfacing to assist in provision of high-speed wet skidding resistance.

3 RESULTS

Table 3 summaries the results obtained for the two case studies, using the average deterioration rate scenario (see Table 2). As explained above, the deterioration component is calculated as the difference between the total component and the basic component (represented by the No deterioration scenario).

For both case studies, the two models provide substantially different results; this difference amounts to one order of magnitude for all the components, with the exception of the deterioration component, where the two models provide a negative term.

The difference in results obtained for the basic component shows an interesting aspect related to the comparison of the two models; regardless of the deterioration in the IRI and MPD over the analysis period, the two models return considerably different results (10272 ton against 1170 ton in the A17 and 109344 ton against 18058 ton in the A1). This difference reflects the substantially different estimated total components. The different validation of the two models, together with the different approaches used, can be considered the main reason for this significant difference in the results: indeed, the models were calibrated for different countries with different input data, in terms of weather, vehicles, and roads.

Particularly significant are the results obtained for the deterioration component, which take into account the evolution of the pavement surface properties. The deterioration term for the VTI model is negative for both case studies and this means that overall the deterioration in pavement surface properties produces a reduction of the vehicle tailpipe CO_2 emissions over the years, rather than an increase as was expected. The negative term related to the deterioration component is due to the different impact given to the IRI and MPD terms by the two models (see Figure 3 and Figure 4).

The VTI model assigns to the MPD term a greater impact on the rolling resistance and on the emission estimate than for IRI (even at high speed, which increases the impact of the IRI); the opposite consideration is true for the UCPRC model, where the IRI term has a larger impact. The significance of this different behaviour becomes particularly relevant for pavement surfaces where the IRI tends to

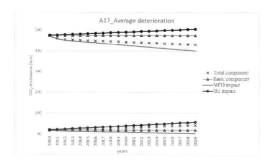

Figure 3. Impact of IRI and MPD in the VTI (above) and UCPRC (below) models for the A17 case study.

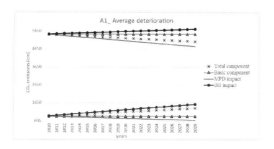

Figure 4. Impact of IRI and MPD in the VTI (above) and UCPRC (below) models for the A1 case study.

increase and the MPD tends to decrease, as in these case studies. The VTI model gives a negative value for the deterioration component, because the MPD term decreases faster than the IRI term increases. Therefore, the pavement surface properties deterioration and the models used to describe them have a significant impact on the emissions results. This consideration is confirmed by the sensitivity test performed on this input variable that shows how the IRI and MPD deterioration rate can change the rolling resistance results in a pavement LCA see (Table 4 and Table 5). The two tables provide the results of the sensitivity test obtained from the two models in terms of basic component, deterioration component and total component (see Figure 2), taking into account three different scenarios of deterioration in IRI and MPD during the analysis period, Average (A), Worst (W) and No deterioration (No) as defined in Table 2.

The results obtained performing the sensitivity test for the two case studies show:

– For both models, the range of potential impact due to the PVI is wide;
– The lowest emissions in the two models occur under different pavement deterioration rate scenarios (no deterioration in the UCPRC model and average deterioration in the VTI model). In the UCPRC model the deterioration compo-

Table 4. Sensitivity test results for the A17 case study.

A17

| | Emission of CO_2 (tonne) | | | | | |
| | VTI | | | UCPRC | | |
Scenario	B*	D**	T***	B	D	T
A	10272	−634	9638	1170	217	1387
W	10272	500	10772	1170	1134	2304
No	10272	0	10272	1170	0	1170
	% compared to the Average scenario					
A	100	100	100	100	100	100
W	100	−79	112	100	522	166
No	100	0	107	100	0	84

* Basic component
** Deterioration component
*** Total component

Table 5. Sensitivity test results for the A1 case study.

A1

| | Emission of CO_2 (tonne) | | | | | |
| | VTI | | | UCPRC | | |
Scenario	B	D	T	B	D	T
A	109344	−4205	105139	18058	4586	22645
W	109344	4716	114059	18058	19634	37693
No	109344	0	109344	18058	0	18058
	% compared to the Average scenario					
A	100	100	100	100	100	100
W	100	−112	108	100	428	166
No	100	0	104	100	0	80

nent increases over time, so the absence of deterioration minimizes the total emissions.
– In the VTI model, the deterioration component, under the average condition of pavement deterioration, tends to decrease, producing an overall reduction in the calculated emissions. This effect levels off under the "worst deterioration" pavement condition, when the IRI effect is larger than the MPD effect.
– This means that in both models, the CO_2 emissions are significantly higher in the case of the worst pavement deterioration scenario.
– The results shows the CO_2 emissions due to the pavement roughness are very sensitive to the pavement surface deterioration over time.

It is clear that the choice of model used to estimate the PVI rolling resistance CO_2 emissions and the surface deterioration model chosen both have a large influence on the results. As mentioned above, the differences in the models are due to the country where they were developed and hence validated and the modelling approach taken.

In the development of these models, different variables come into play affecting strongly the results obtained. In particular, some methodological choices (such as rolling resistance measurement methods, road surface measures, direct or indirect estimation of emissions) and site specific elements (weather, vehicle types and technology, type of roads, pavement design models and deterioration) can play a relevant role, producing a rolling resistance and fuel consumption model that is not suitable for every geographical location.

The two models assessed in this research are very different, both in terms of methodological approach and in terms of site specific elements. The UCPRC models was developed in California, by using two software tools (HDM-4 – calibrated for US conditions—and MOVES—a US EPA highway vehicle emission model based on national data) and implementing IRI and MPD data related to US roads.

The VTI rolling resistance model is based on empirical data from coastdown measurements in Sweden and the fuel consumption model was calibrated using calculated values from VETO, a theoretical model to estimate the fuel consumption and exhaust emissions of vehicles. Different climates, type of roads, pavement deterioration processes and models, traffic composition and vehicle technology characterize California and Sweden affecting the models developed and the results produced.

The two models consider the impact of the pavement surface properties, IRI and MPD, in different ways. In the UCPRC model the IRI has a larger impact on the rolling resistance than the MPD and the opposite is true for the VTI model. This difference is particularly significant in this case study, where the MPD falls over time, producing opposite results when the two models are used; in the UCPRC model, the deterioration component is positive, since the impact of the increase in IRI is larger than that due to the reduction in MPD; while for the VTI model the deterioration component is negative. Therefore, the pavement condition deterioration over time has a strong impact on the rolling resistance, significantly affecting the results. This is confirmed by the sensitivity test performed on the IRI and MPD deterioration rate that showed that the CO_2 emissions due to PVI rolling resistance are very sensitive to this factor.

4 CONCLUSIONS

The significantly different results obtained for both case studies show that rolling resistance models should be used carefully and the results interpreted with caution, especially considering the different weight given to the IRI and MPD terms by the two models.

The sensitivity test performed confirms that the results are sensitive to the pavement deterioration rate of IRI and MPD.

In the UK, currently there are no validated models able to predict the relationship between rolling resistance and pavement surface properties and the use of the models in the literature could lead to unreliable results.

In addition, there are no models to predict the deterioration of roughness and texture depth over time depending on maintenance treatments, traffic volume and type, surface properties and materials. This gap has to be filled before introducing the use phase in the pavement LCA framework.

As a result of this investigation, the authors believe that overall, further research is necessary to develop standardized procedures for PVI rolling resistance emission estimates, so to obtain comparable and reliable pavement LCA results beneficial in the decision making process. Furthermore, it is recommended to any national or regional road authority that wants to introduce the use phase into the pavement LCA framework, to at least carry out an exhaustive calibration and validation of the existing models presented in this work, if not developing their own models to accurately predict evolution of pavement surface properties of their road networks over time.

ACKNOWLEDGMENTS

This study was funded from an Engineering and Physical Sciences Research Council (EPSRC) Doctoral Training Grant (DTG) - Faculty of Engineering. The authors would like to thank staff at Lincolnshire County Council and Highways England for providing data for the case studies.

REFERENCES

Aavik A, Kaal T, Jentson M Use Of Pavement Surface Texture Characteristics Measurement Results In Estonia. In: XXVIII International Baltic Road Conference, Vilnius, Lithuania 26–28 August, 2013.

Hammarström U, Eriksson J, Karlsson R, Yahya M-R (2012) Rolling resistance model, fuel consumption model and the traffic energy saving potential from changed road surface conditions, VTI Rapport 748 A.

Hammarström U, Karlsson B (1987) VETO: Ett datorprogram för beräkning av transportkostnader som funktion av vägstandard, VTI meddelande, ISSN 0347–6049.

Jacobs F (1982) M40 High Wycombe By-pass: Results of a Bituminous Surface-texture Experiment. Transport and Road Research Laboratory, UK.

Karlsson R, Carlson A, Dolk E (2012) VTI - Energy Use Generated by Traffic and Pavement Maintenance: Decision Support for Optimization of Low Rolling Resistance Maintenance Treatments, MIRIAM SP3.

Lu Q, Kohler ER, Harvey JT, Ongel A (2009) Investigation of Noise and Durability Performance Trends for Asphaltic Pavement Surface Types: Three-Year Results, UCPRC-RR-2009–01.

Matute JM, Chester M, Eisenstein W, Pincetl S Life-Cycle Assessment for Transportation Decision Making. In: Transportation Research Board 93rd Annual Meeting, 2014. vol 14–1287.

Muench ST, Lin YY, Katara S, Armstrong A (2014) Roadprint: Practical Pavement Life Cycle Assessment (LCA) Using Generally Available Data. Paper presented at the International Symposium on Pavement LCA 2014, Davis, California, USA, October 14–16 2014.

PIARC (2002) Overview of HDM–4. The Highway Development and Management Series Collection.

Sandberg U, Bergiers A, Ejsmont JA, Goubert L, Karlsson R, Zöller M (2011) Road surface influence on tyre/road rolling resistance. Report MIRIAM SP1_04.

Santero NJ, Horvath A (2009) Global warming potential of pavements. Environmental Research Letters 4:034011.

Trupia L, Parry T, Neves L, Lo Presti D (2016 (in press)) Rolling Resistance Contribution To A Road Pavement Life Cycle Carbon Footprint Analysis The International Journal of Life Cycle Assessment doi:10.1007/s11367–016–1203–9.

Tseng E (2012) The construction of pavement performance models for the California Department of Transportation new pavement management system. Department of Civil and Environmental Engineering - University of California Davis.

UK Department of Energy & Climate Change (2015) 2014 UK Greenhouse Gas Emissions, Provisional Figures.

Wang T, Harvey J, Kendall A (2014) Reducing greenhouse gas emissions through strategic management of highway pavement roughness Environmental Research Letters 9:034007.

Wang T, Lee I-S, Harvey J, Kendall A, Lee EB, Kim C (2012) UCPRC Life Cycle Assessment Methodology and Initial Case Studies for Energy Consumption and GHG Emissions for Pavement Preservation Treatments with Different Rolling Resistance.

Transport Infrastructure and Systems – Dell'Acqua & Wegman (Eds)
© 2017 Taylor & Francis Group, London, ISBN 978-1-138-03009-1

Intermediate stage traffic technical solution of prince Branimir Street in Zagreb

M. Šimun
University of Applied Sciences Zagreb, Croatia

S. Mihalinac
ZG-projekt Ltd., Zagreb, Croatia

K. Rehlicki
Elipsa-S.Z. Ltd., Zagreb, Croatia

ABSTRACT: Traffic network of each city is determined by Urban Master Plan (UMP) which reflects in traffic corridors of specified width. The urban master plan is plan of the local level and in our case covers 220 km^2 for the city of Zagreb. The plan gives indicators for the construction, renovation and protection of space in the area of Plans scope and to ensure spatial plan preconditions for realization of traffic and infrastructure projects. On the area of a large city is a complex interdependence of traffic infrastructure. The quality of traffic infrastructure is manifested as accessibility and mobility of space and advancement of new traffic corridors as a basis for a sustainable development. In this paper, the traffic technical solution of prince Branimir Street in Zagreb, road section from Zavrtnica to Vjekoslav Heinzel Street, has been elaborated. Construction of this road section was initiated in 2012 through the I. phase of construction but only part of the width of the planned traffic corridor. As a result of that property-right issues were partially solved. A fully built, in a full corridor width, was planned for II. phase of construction. In the meantime the property-right issues were solved but still not in a complete width of the planned corridor. Because of that, it has been approached to preparation of technical solution which would keep the elements constructed in a I. phase, and built new ones in intermediate stage which would also be incorporated in the II. phase of construction. In intermediate stage, elements which improve traffic conditions, should be constructed, such as enhancement of traffic capacity and satisfying level of service. In the paper were analyzed all the components and traffic technical solutions of intermediate stage.

1 INTRODUCTION

1.1 *General*

Traffic network of the city of Zagreb is determined by Urban Master Plan (UMP)—Urban master plan of the city of Zagreb (2007) which reflects in traffic corridors of specified width for road categorized as urban highway, city avenue, the main street, city streets and corridors of public transportation areas. The plan gives indicators for the construction, renovation, modernize and upgrade existing streets, solve stationary traffic and improve bicycle traffic. Transport situation in the city of Zagreb is a direct consequence of Urban Master Plan (UMP) planned traffic corridors. Figures 1 and 2 showing Urban Master Plan (UMP) of the city of Zagreb—purpose of space and road network. Parts of the city in which they are built traffic corridors into full width, from the point of road capacity, function with satisfactory level of service. In other parts of the city is very low capacity of road network, reduced availability of the city center and there is a problem of structure and organization of traffic flows. Also, the development of transport network doesn't follow the development of the city. Planned traffic corridors in other parts of the city are characterized by no kind of construction (not started any construction in the planned corridor) or kind of contruction only in the part of the planned width or discontinuity of development corridors with the level of development of planned width or in full width of planned corridor.

The cause of this condition partly lies in the fact that it is an urban area where there are built residential business and commercial objects and any construction of traffic corridors is actually a "puncture" through the built area with unavoidable demolition of existing buildings. Prince Branimir

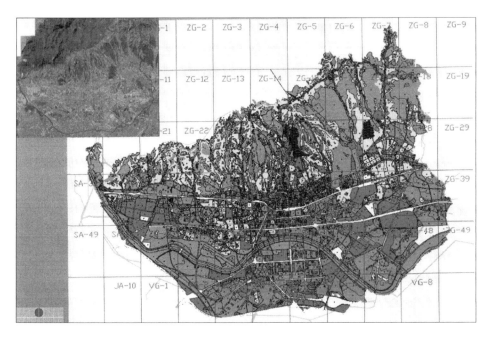

Figure 1.   Urban master plan of the city of Zagreb—purpose of space.

Figure 2.   Urban master plan of the city of Zagreb—traffic network.

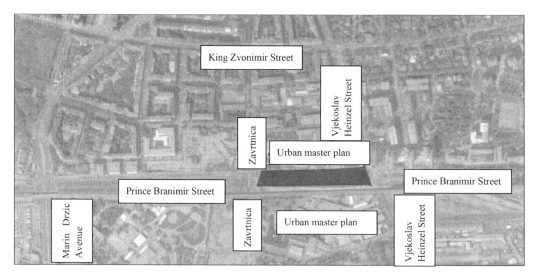

Figure 3. Grip of the project.

Street, as one of the main traffic corridors from the center to the eastern part of the city is an example of discontinuity corridor, where for many years the corridor was interrupted on the part of the Avenue Marin Drzic to the Vjekoslav Heinzel Street and all because of the residential business and commercial objects within the plan of the early traffic corridor.

Few years back on the part of Zavrtnica to the Vjekoslav Heinzel Street was built only one part of the planned width of the corridor as a kind of "junction" of the old and new Branimir Street. The construction of "junction" has been made the continuity of traffic corridor but due to insufficient road capacity as a result of partially development the junction is only a "bottleneck".

The quality of traffic infrastructure is manifested as accessibility and mobility of space and advancement of new traffic corridors as a basic for sustainable development—Book of regulations (NN 110/2001) & Book of regulations (NN 34/2005).

1.2 Position and role in city traffic network

Prince Branimir Street has been planned by the Urban Master Plan (UMP)—Urban master plan of the city of Zagreb (2007) to be a traffic corridor of main city street from main railway station to town district Sesvete and represents one of the main traffic corridors from the center to the east side of the town. Figure 2. showing prince Branimir Street in city road network. For years there were only two separated parts: part from main railway station to Marin Drzic Avenue and part from Vjekoslav Heinzel Street to Dubrava Street in city district Sesvete.

Between 2005 and 2008 it has been constructed a part from Marin Drzic Avenue to Zavrtnica in a section with dual two-lane carriageway each with two traffic lanes and pedestrian-bicycle way.

Continuity of traffic corridor couldn't be realized because there hasn't been built a part from Zavrtnica to Vjekoslav Heinzel Street in a length of 350 meters so that this part has been analysed in this paper.

In Urban Master Plan (UMP), in range of a grip, prince Branimir Street has been planned in a width of 50 meters. On the south, it borders on railway corridor and Urban Master Plan (UMP) Heinzel-Radnicka-railway line. Figure 3 showing grip of the project. On the north side corridor borders on Urban Master Plan (UMP) area bounded with streets Banjavciceva-Heinzel-prince Branimir-Zavrtnica.

2 DEVELOPMENT OF PHASES OF CONSTRUCTION

The building of the section initiated in 2012 through the I. phase of construction within partial width of the planned street corridor as a result of unsolved property-right issues—Decision 2006 (12). The building of entire width corridor was planned through the II. phase of construction. The I. phase of construction was finished in 2012. In years after achieving the I. phase of construction, additional property-right issues were solved. But since they were not solved entirely, it was not possible to build the planned II. phase of construction. That situation led to finding a new technical solution with the aim to improve the conditions of ever-increasing traffic: increasing capacity and achieving satisfying level of service. All components of road design build in the I. phase and planned in the II. phase of construction were analyzed and resulted in a traffic and road design of an intermediate stage of construction. All design components built in the I. phase of construction are preserved and upgraded with additional components, entirely compatible with the planned II. phase of construction—Decision 2006 (16).

3 TECHNICAL SOLUTION

Basic design of the I. phase of construction was single lane carriageway with no additional traffic lanes at Zavrtnica and Heinzelova intersections (Elipsa-S.Z. Ltd., 2012). Footpath in minimum width was designed only on south side of carriageway. At time when designed, it was not sufficient for existing traffic volume, but was only possible due to unsolved property-right issues. The I. phase resulted in connection of existing unlinked segments of prince Branimir Street but with congestions on Heinzelova Street intersection. Figure 4 showing technical solution situation of the I. phase of construction.

In the II. phase of construction was planned dual-two lane carriageway with additional traffic lanes for left and right turns at Zavrtnica and Heinzelova intersections—Elipsa-S.Z. Ltd. (2014). Footpath in minimum width was designed on south side of carriageway. On north side of carriageways footpath with cycle lanes was designed.

Intermediate stage is presented with combined carriageways from the I. and the II. phase of construction. Intermediate stage in fully use carriageway built in the I. phase of construction and build north side corridor carriageway as planned in the II. phase but with modification on east side necessary to form intersection. Main difference in design of intermediate stage and design of the II. phase of construction is on intersection with Heinzelova Street. Design of intermediate stage not allow direct left turn on intersection with Heinzelova Street. Left turn is achievable indirectly on intersection with Zavrtnica Street. Figure 5 showing technical solution situation of intermediate stage of construction.

The II. phase of construction planned to be activated as soon as property-right issues are solved. Figure 6 showing technical solution of planned the II. phase of construction.

Each component of road design built in the I. phase and planned in the II. phase of construction were analyzed and redesigned, resulting in new combined technical solution of intermediate stage—Rehlicki (2016). Cross-sections of a corridor defined through the I. phase, the II. phase and intermediate stage are shown in Figure 7.

3.1 Horizontal alignment

In the I. phase of construction was built one single lane carriageway, partially on south and north side of planned corridor. Horizontal alignment is defined with two curves 80 meters in radius connecting three straights. Design speed of 50 km/h is determined by curves radius.

In the II. phase of construction are planned two duo lane carriageways, one on south and one on north side of planned corridor. Horizontal alignment is defined with one axis. Design speed is 50 km/h and determined by traffic regulation.

Intermediate stage, according to drive direction, has two horizontal alignments. For drive direction west-east horizontal alignment completely match the I. phase of construction. For drive direction east-west horizontal alignment match north carriageway horizontal alignment of the II. phase of construction.

Traffic corridor is defined by an unique axis set in central median on which are based all the

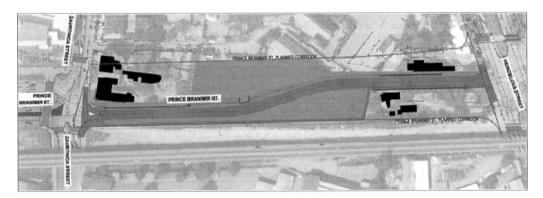

Figure 4. The I. phase of construction.

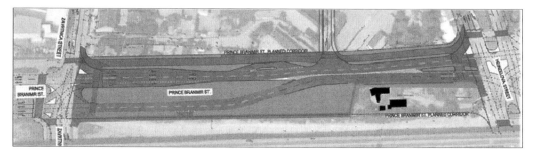

Figure 5.   Intermediate stage of construction.

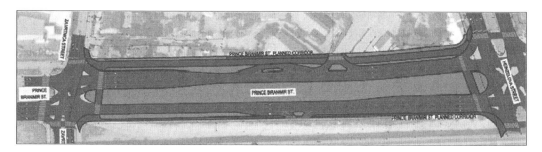

Figure 6.   Planned the II. phase of construction.

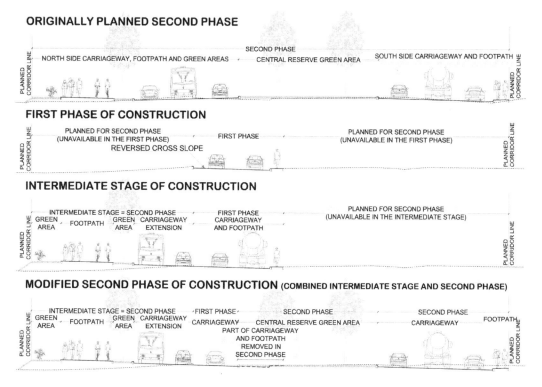

Figure 7.   Cross-sections of a corridor.

229

other ele-ments of cross-section across the width of the planned corridor—Korlaet (1995). Because of phase building in the I. phase and traffic technical solution, it was necessary to define separate axis that will best define the elements of the cross-section for the purpose of phase building. In a transport-functional sense, considering directions of the vehicle, traffic technical solution consists of two axis, different for every vehicle di-rection. Figure 9 showing defined axis.

3.2 Vertical alignment

Vertical alignments of the I. phase, the II. phase and intermediate stage are shown in Figure 8. Vertical alignment deviation in intermediate stage in regard to the II. phase of construction is result of keeping existing carriageway on north side of corridor, build in the I. phase of construction, and super elevation, changing grade from existing carriageway to carriageway planned through the II. phase of construction.

3.3 Pavement structure

The same pavement structure as defined in the I. phase is retained in the planned II. phase of construction. As shown in Figure 10 total thickness of existing pavement structure is 61 cm. Extension of existing carriageway while maintaining the same layer thickness, cross fall and subgrade fall requires, on extended side, pavement structure in total thickness of 51 cm. The frost-resistant pavement structure requires total thickness (depth) of 57 cm. Due to keeping existing pavement layer thickness and insuring frost-resistant pavement structure, it was necessary to reduce subbase layer thickness from 30 cm down to 26 cm and reduce subgrade fall from 4% down to 3%.

Layer thickness of pavement and subbase, based on layer thickness and subgrade fall, designed in the I. phase of construction are recalculated and redefined for intermediate stage of construction—Babic (1997). Layer thickness and subgrade fall of intermediate stage pavement structure is shown in Figure 10.

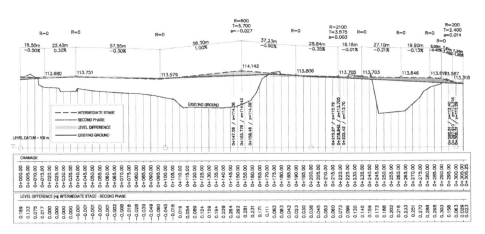

Figure 8. Vertical alignments.

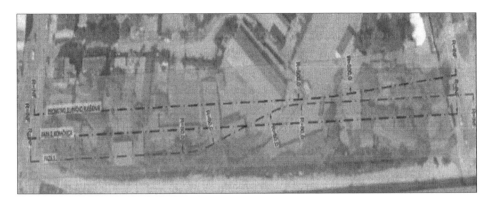

Figure 9. Defined axis.

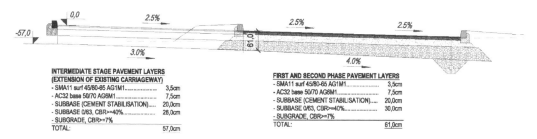

INTERMEDIATE STAGE PAVEMENT LAYERS
(EXTENSION OF EXISTING CARRIAGEWAY)
- SMA11 surf 45/80-65 AG1M1.................... 3,5cm
- AC32 base 50/70 AG6M1......................... 7,5cm
- SUBBASE (CEMENT STABILISATION)..... 20,0cm
- SUBBASE 0/63, CBR>=40%.................... 26,0cm
- SUBGRADE, CBR>=7%
TOTAL: 57,0cm

FIRST AND SECOND PHASE PAVEMENT LAYERS
- SMA11 surf 45/80-65 AG1M1.................... 3,5cm
- AC32 base 50/70 AG6M1......................... 7,5cm
- SUBBASE (CEMENT STABILISATION)..... 20,0cm
- SUBBASE 0/63, CBR>=40%.................... 30,0cm
- SUBGRADE, CBR>=7%
TOTAL: 61,0cm

Figure 10. Intermediate stage pavement structure.

Table 1. Comparative overview carriageway upgradeability through phases of construction.

	First phase	Intermediate stage	second phase
Carriageway			
built, planned [m²]	2304	4534	6820
to remove in second phase [m²]	1104	81	-
realized in year	2012	-	2016
cost in all* [million Kn**]	6.8	-	9.0

* Realized cost in all include: carriageway, footpath, traffic lights, traffic signs, road markings, sewer, lighting, telecommunication infrastructure, landscaping, urban equipment.
** 1 Kn = 0.133€.

3.4 Validity and sustainability of intermediate stage

Validity and sustainability of the intermediate stage technical solution is shown in Table 1. Almost 50% of carriageway built in the I. phase of construction should be removed to be upgraded to the II. phase technical solution. Unlike the I. phase of construction, less than 2% of carriageway built in intermediate stage should be removed to be able to upgrade to technical solution of the II. phase of construction.

4 CONCLUSION

Building of traffic corridors in a full width represent a complex project, which extends through long time period, require a coordinate work of large number of city and municipal services. It includes planning and harmonization of activities, solving property-right issues and insurance of financial resources, both in the design stage and building stage. A transportation system is characterized by incoherence of urban areas, low road capacity, reduced availability of town center, adverse traffic effect on environment and increased degree of motorization. Systematic construction of main traffic corridors should be a priority and the only measure of sustainability of existing traffic network. The level of construction should correspond to development stage of urban area. This is achievable only through multi phase of construction. The development of the corridor is achieving through construction phasing in a way that next phase is a suffix to a previous phase. It is unacceptable to tear down allready built part of infrastructure, except in case when there was a long period between two phases. Traffic system is inimitable so it is necessary to continuosly invest in it's development.

Today's low road capacity of traffic infrastructure and lowered availability of town center is caused not only by inappropriate building but also by absence of systematic traffic control. Only combination of measures based on urban planning can contribute to growth of motorization degree and increased road capacity.

As shown in this case, originally planned construction through two phases, due to the impossibility of solving property-right and legal issues in the planned time frame, led to new phase of construction—intermediate stage. Each phase corresponds to traffic needs but taking into account the spatial possibilities at its time of construction.

In May 2016 property-right issues were solved entirely. After four years in use, 48% of carriageway built in the I. phase was removed and in July 2016 was built full width corridor according to planned the II. phase of construction. So there was no need for application of intermediate stage. This paper was not intended to cover feasibility analysis of construction for each phase but to show that for each stage of development there is appropriate technical solution.

REFERENCES

Babic. B. 1997. *Design of pavement structures*, Zagreb, Croatian Society of Civil Engineers.

Book of regulations, NN 110/2001 www.nn.hr.

Book of regulations, NN 34/2005 www.nn.hr.

Decision 2006 (12) http://www1.zagreb.hr/slglasnik/.html#/home.

Decision 2006 (16) http://www1.zagreb.hr/slglasnik/index.html#/home.

Elipsa-S.Z: Ltd. 2012, *Prince Branimir Street project—the I. phase of construction.*

Elipsa-S.Z: Ltd. 2014, *Prince Branimir Street project—the II. phase of construction.*

Korlaet, Z. 1995. *An introduction to the design and constructions of roads*, Zagreb, Faculty of civil engineering University of Zagreb.

Rehlicki, K. 2016, *Technical solutions of prince Branimir Street in Zagreb*, Diploma thesis, University of Applied Sciences Zagreb.

Urban master plan (UMP) of the city of Zagreb, decision 2007 (16) http://www1.zagreb.hr/zagreb/slglasnik.nsf/7ffe63e8e69827b5c1257e1900276647/fc7bc5b1e3f36f97c12573a90032dc3d/$FILE/Odluka%20o%20dono%C5%A1enju%20Generalnoga%20urbanisti%C4%8Dkog%20plana%20grada%20Zagreba.pdf.

Transport Infrastructure and Systems – Dell'Acqua & Wegman (Eds)
© 2017 Taylor & Francis Group, London, ISBN 978-1-138-03009-1

Research and analysis of dolomite aggregate for resistance to fragmentation

L. Šneideraitienė & D. Žilionienė
Department of Roads, Vilnius Gediminas Technical University, Vilnius, Lithuania

ABSTRACT: Mechanical and physical properties of aggregates, used for producing asphalt mixture, influence the quality indices of pavement structure, i.e. functionality, reliability and durability. Investigation of various mechanical and physical properties of numerous aggregates by using test methods of EU standards Lithuania started in 2004. Requirements for resistance to fragmentation of aggregate, the technical specifications of mineral materials in Lithuania and in Latvia have been determined. The resistance to fragmentation of Lithuanian high quality dolomite aggregate meets Lithuanian and Latvian requirements according to *TRA ASPHALT 08 Technical Specifications for Road Asphalt Mixtures* and *Ceļu SPECIFICATIONS 2015* for course mineral materials, bituminous mixtures and surface treatments.

1   INTRODUCTION

According to Bulevičius *et al.* (2011) the principal cause of defects in road pavement is the impact of heavy vehicles which influence functionality of road pavement, driving comfort and traffic safety. In order to improve the properties of road pavement and the quality of traffic the scientists of Lithuania, Latvia and various countries carry out investigations of structural pavement layers, analyse the effect of their properties on pavement performance, mechanical and physical properties of asphalt mixtures used for structural pavement layers, etc. Aggregates (crushed stone, dolomite, gravel) differ in their size, shape and working conditions of the material that depend on the size and nature of loads, working temperature, environmental aggressiveness, etc. Mechanical and physical properties of aggregate, used for producing asphalt mixture, influence the quality indices of pavement structure, i.e. functionality, reliability and durability. Aggregate suitability to asphalt mixtures is dependent on its mechanical and physical properties. Aggregate properties determine the mode of production of asphalt mixture, the thickness of structural pavement layers and other peculiarities of road pavement structure. In each case, selection of aggregate shall be economically justified. The selected aggregate shall be cheap and easily available. When a stronger aggregate is used, the service life of road pavement structure is longer, the structure is more reliable, the individual structural pavement layers are thinner, the costs of aggregate and other materials are lower, etc. Often the aggregate is selected not because of its better properties but because it is cheaper.

Bulevičius *et al.* (2011) indicate that since 2004 Lithuania began to investigate mineral physical and mechanical indicators by EN standard test methods. Many tests of resistance to fragmentation done with numerous minerals from different manufacturers, e.g. *EN 1097-2:2010 Tests for Mechanical and Physical Properties of Aggregates—Part 2: Methods for the Determination of Resistance to Fragmentation* impact resistance (*SZ*) and the coefficient of Los Angeles (*LA*) indicator values provided. Regulations for resistance to fragmentation for Lithuanian rock determined too (Figure 1).

Rock processing takes place according to technological schemes for the manufacture of bulk products used in the production of other construction products (Fook *et al.* 1988; Labus *et al.* 2016).

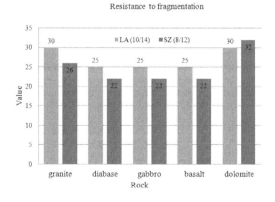

Figure 1.   Category value of rock resistance to fragmentation by TRA MIN 07 (Lithuania).

Dolomite is one of the most available sedimentary rocks in the territory of Lithuania and Latvia (Haritonovs *et al.* 2014, Skrinskas *et al.* 2010). Dolomite quarries in Lithuania contain hundreds million tons of this material. Either, high quality dolomite produced applying special extraction technology, and the mechanical properties of this material are similar to granite (Šernas *at al.* 2016).

Haritonovs *et al.* (2015) specify a reason to import dolomite aggregates from other countries because Latvian dolomite lacks the mechanical strength for motorways according to the *Latvian Road.* Latvian dolomite of its low quality, mainly *LA* value, cannot use for average and high volume roads (Haritonovs, Tihonovs 2014; Haritonovs *et al.* 2015). Therefore, mostly importable magmatic rocks (granite, diabase, gabbro, basalt) or dolomites are used and makes asphalt concrete expensive (Haritonovs *et al.* 2014).

According Šernas *et al.* (2016), it is a cost-effective to use local road construction materials (on that ground and dolomite) because the transportation increases price of ready to use materials about 30 percent and the mechanical and physical properties of the materials are practically comparable to these properties of importable mineral materials. Considering these facts and to avoid ruts on road pavements, there is very important question of using local high quality dolomite aggregates in producing asphalt mixes, which properties would be similar or better comparing to asphalt concrete mixes with granite aggregates. Šernas *et al.* (2016) used empirical and performance based tests for designed asphalt concrete mixtures with high quality dolomite aggregates. These tests showed that:

– asphalt concrete mix AC 11 VS with high quality dolomite aggregates could be used for the asphalt concrete pavement wearing course with ESALs ≤ 3.0 million during design period, and
– use of stone mastic asphalt mix SMA 11 S with high quality dolomite aggregates for asphalt concrete wearing course should be evaluated additionally, taking into account the change of mechanical and physical properties of these aggregates during road maintenance.

Respectively Vaitkus *et al.* (2015) state that dolomite aggregates commonly used in asphalt concrete pavement for the base course of surfacing and road foundation layers' mixtures.

This paper focus on experimental tests of Lithuanian high quality dolomite aggregates for resistance to fragmentation (*SZ* value) and Impact Test (*LA* value). Analysis of test results showed that dolomite meets Lithuanian *TRA ASPHALT 08 Technical Specifications for Road Asphalt Mixtures* and Latvian *Ceļu SPECIFICATIONS 2015* standards for coarse mineral materials, bituminous mixtures and surface treatments.

In addition, requirements for mineral materials indicated in *TRA MIN 07 Technical Specifications for Road Mineral Materials when resistance to fragmentation of aggregates is determined in accordance with LST EN 1097-2* presented in TRA ASPHALT 08 by asphalt quality and type.

2 TEST METHODS AND REQUIREMENTS

One of the most accurate tests to determine coarse aggregate strength is the test of resistance to fragmentation in accordance with *LST EN 1097-2:2010 Tests for Mechanical and Physical Properties of Aggregates – Part 2: Methods for the Determination of Resistance to Fragmentation.* The main method of this test lies in the determination of resistance to fragmentation by use in Latvia of the Los Angeles Test method. However, the alternative Impact Test method is the most frequently used in Lithuania. For this reason, the specimens tested by both methods to determine the strength of Lithuanian dolomite aggregates and to discover a relationship between these two tests. In consequence, *LA* and *SZ* values show the same property of tested material but the methods of tests are different.

In the Los Angeles test method of coarse aggregate, 10/14 mm size fraction (5000 ± 5) g specimen together with ten steel balls, each Ø 45–49 mm in diameter and weighing in total 4690–4860 g, rotated for 500 revolutions at a constant speed of 31–33 min[-1] in a closed drum. In the Impact Test times by a weight falling from 370 mm height. After the tests, the percentage mass loss of material, having passed the control sieve, is calculated.

Using the Los Angeles Test method, it is allowed by the standard to test also other aggregate fractions alternative to the standard 10/14 mm size, whereas, the Impact Test method allows to test. Therefore, when using the Los Angeles Test method a 10/14 mm size fraction of coarse aggregate selected.

The test results assessed according to their conformity to the categories. For example, LA_{20} when the mass loss is ≤ 20%, *SZ* shows the percentage mass loss of material having passed five test sieves, e.i. SZ_{18} when the mass loss is ≤ 18% (Table 1).

LA calculated by the formula:

$$LA = \frac{5000 - m}{50} \qquad (1)$$

where *m* – retained mass on a 1.6 mm sieve, g. *SZ* calculated as:

$$SZ = \left(\frac{M}{5}\right)\%, \qquad (2)$$

Table 1. Values and categories of Los Angeles coefficient (*LA*) and Impact Resistance (*SZ*).

Requirements	Los Angeles coefficient	Category *LA*	Value of resistance to fragmentation%	Category *SZ*
TRA MIN 07	≤ 20	LA_{20}	≤ 18	SZ_{18}
(Lithuania)	≤ 25	LA_{25}	≤ 22	SZ_{22}
	≤ 30	LA_{30}	≤ 26	SZ_{26}
Celu SPECIFICATIONS	≤ 20	LA_{20}	–	–
2015 (Latvia)	≤ 25	LA_{25}		
	≤ 30	LA_{30}		
	≤ 40	LA_{40}		

Note: According to *TRA MIN 07* requirements for surface dressing *SZ18 (LA20) and SZ22 (LA25), and* according to *Celu SPECIFICATIONS 2015 - LA20.*

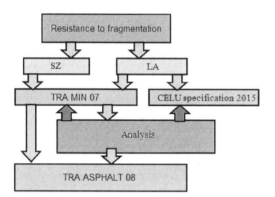

Figure 2. Experiment plan of dolomite aggregates for resistance to fragmentation.

where *M* – the sum of percentage mass passing five test sieves.

Table indicates Los Angeles coefficient (*LA*) and impact resistance (*SZ*) requirements of categories for course aggregates by *Celu SPECIFICA-TIONS 2015* and course aggregates for bituminous mixtures and surface treatments according to *TRA MIN 07* requirements.

3 EXPERIMENTAL RESEARCH

Figure 2 presents the experimental plan of the high quality dolomite aggregates of resistance to fragmentation by *SZ* and *LA* tests.

3.1 *Object and research methods*

High quality dolomite aggregates produced at AB "Dolomite" quarry "Petrašiūnai-2" in Lithuania under the special production technology. In 2014–2016 period, the experimental studies of resistance to fragmentation for this dolomite done in AB "Dolomite" laboratory.

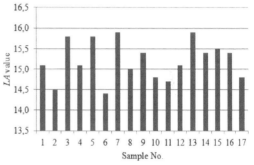

Figure 3. Results of Los Angeles coefficient *(LA)* of the dolomite aggregates.

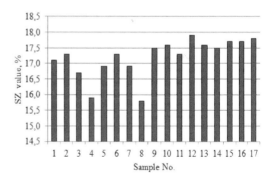

Figure 4. Results of resistance to fragmentation (*SZ*) of the dolomite break-stone.

Whatever test results of high quality dolomite obtained according to *EN 1097-2* after 17 tests by *LA* method and 17 tests by *SZ* method. Figure 3 and Figure 4 show results of these tests.

3.2 *Results*

The received values of the high quality dolomite aggregates for resistance to fragmentation according to *EN 1097-2:2010* by Los Angeles method showed that all 17 samples meet requirements of the Los Angeles coefficient category LA_{20} for the aggregates used in bituminous mixtures and for surface dressing according to *TRA MIN 07, TRA ASPHALT 08* and *Ceļu SPECIFICATIONS 2015*. Maximum value obtained of sample No. 7 and sample No. 13 is 15.9. The lowest value was 14.4 of the sample No. 6. The average value of this test was 15.2.

The received values of the high quality dolomite aggregates for resistance to fragmentation according to *EN 1097-2* by Impact Test method showed that all 17 samples meet requirements of the *SZ* value for the aggregates used in bituminous mixtures and for surface dressing according to *TRA MIN 07* and *TRA ASPHALT 08*. For SZ_{18} the maximum value was 17.9% (sample No. 12), and the lowest value was 15.8% (sample No. 8). Average value was 17.2%.

Positive results of the experiments shows necessity to continue a long-term research of the high quality dolomite aggregates.

4 CONCLUSIONS

Lithuanian high quality dolomite aggregates according to *EN 1097-2 Tests for Mechanical and Physical Properties of Aggregates—Part 2: Methods for the Determination of Resistance to Fragmentation by LA* method test results correspond the mineral materials, used for bituminous mixtures and surface dressings according to Lithuanian *TRA MIN 07 Technical Specifications for Road Mineral Materials, TRA ASPHALT 08 Technical Specifications for Road Asphalt Mixtures* and Latvian *Ceļu SPECIFICATIONS 2015* requirements of the Los Angeles coefficient category LA_{20}.

Lithuanian high quality dolomite aggregates by *SZ* method test results in the total value corresponds Lithuanian *TRA MIN 07 Technical Specifications for Road Mineral Materials* and *TRA ASPHALT 08 Technical Specifications for Road Asphalt Mixtures* requirements for mineral materials used in bituminous mixtures and surface dressings (impact resistance category SZ_{18}).

Lithuanian high quality dolomite aggregates meets the Lithuanian and Latvian requirements for bituminous mixtures and surface treatments according to Lithuanian *TRA ASPHALT 08 Technical Specifications for Road Asphalt Mixtures* and Latvian *Ceļu SPECIFICATIONS 2015* requirements.

Although the experimental results are positive and it is necessary to continue a long-term research and tests on the high quality dolomite aggregates what enable to use local high quality dolomite aggregates instead expensive importable granite aggregates in producing asphalt concrete mineral mixes.

REFERENCES

Bulevičius, M.; Petkevičius, K.; Žilionienė D.; Čirba, S. 2011. Testing of Mechanical-Physical Properties of Aggregates, Used for Producing Asphalt Mixtures, and Statistical Analysis of Test Results, *The Baltic Journal of Road and Bridge Engineering* 6(2): 115–123. http://dx.doi.org/10.3846/bjrbe.2011.16.

Fook, P.G.; Gourley, C.S.; Ohkere, C. 1988. Rock Weathering in Engineering Time, *The Quarterly Journal of Engineering Geology* 21: 33–57. http://dx.doi.org/10.1144/GSL.QJEG.1988.021.01.03.

Haritonovs, V.; Zaumanis, M.; Brencis, G.; Smirnovs, J. 2013. Performance of Asphalt Concrete with Dolomite Sand Waste and BOF Steel Slag Aggregate. *The Baltic Journal of Road and Bridge Engineering* 8(2): 91–97. http://dx.doi.org/10.3846/bjrbe.2013.12.

Haritonovs, V., Tihonovs, J., Smirnovs, J. 2015. High Modulus Asphalt Concrete with Dolomite Aggregates. In *IOP Conference Series: Materials Science and Engineering* 96(1): 1–10. IOP Publishing. http://dx.doi.org/10.1088/1757–899X/96/1/012084

Haritonovs, V., Tihonovs, J. 2014. Use of Unconventional Aggregates in Hot Mix Asphalt Concrete. *The Baltic Journal of Road and Bridge Engineering* 9(4): 276–282. http://dx.doi.org/10.3846/bjrbe.2014.34

Labus, M.; Such, P. 2016. Microstructural Characteristics of Wellbore Cement and Formation Rocks under Sequestration Conditions, *Journal of Petroleum Science and Engineering* 138: 77–87. http://dx.doi.org/10.1016/j.petrol.2015.12.010.

Skrinskas, S.; Gasiūnienė, V. E.; Laurinavičius, A.; Podagėlis, I. 2010. Lithuanian Mineral Resources, Their Reserves and Possibilities for Their Usage in Road Building, *The Baltic Journal of Road and Bridge Engineering* 5(4): 218–228. http://dx.doi.org/10.3846/bjrbe.2010.30.

Šernas, O.; Vorobjovas, V.; Šneideraitienė, L.; Vaitkus, A. 2016. Evaluation of Asphalt Mix with Dolomite Aggregates for Wearing Layer, *Transport Research Procedia* 14: 732–737. http://dx.doi.org/10.1007/s00603–011–0174.

Vaitkus, A.; Vorobjovas, V. 2015. Dolomito skaldos panaudojimas asfalto viršutinio sluoksnio mišiniams, *Lietuvos keliai* 1(34): 57–61. ISSN 1392–8678.

Transport Infrastructure and Systems – Dell'Acqua & Wegman (Eds)
© 2017 Taylor & Francis Group, London, ISBN 978-1-138-03009-1

Design vehicles and roundabout safety—review of Croatian design guidelines

Š. Bezina, I. Stančerić & S. Ahac
Faculty of Civil Engineering, University of Zagreb, Zagreb, Croatia

ABSTRACT: In this paper, review of Croatian national guidelines is given, with the emphasis on the geometric design and fastest path (speed profile) analysis. The review is presented through several theoretical examples of single-lane suburban roundabouts with various external radii, designed according to the recommendations of these guidelines. Design vehicles used in this analysis are truck with trailer, semi-trailer truck and tri-axle bus. These vehicles occupy different swept path widths when moving through the roundabout with the same external radii, which leads to differences in resulting entrance and exit widths and consequently the speed profiles on roundabouts. These differences, as well as the impact of analyzed design vehicles on the roundabout geometric design and speed profile, are also presented.

1 INTRODUCTION

Compared to the conventional intersections, roundabouts can provide numerous benefits, such as improved intersection safety and capacity, reduced maintenance costs and air pollution (Ahac et al. 2014; Hydén et al. 2000; Mandavilli et al. 2003). Their design process is highly iterative and usually consists of several steps:

– initial geometric design of elements,
– performance and safety checks (swept path analysis, fastest path analysis and sight distance tests), and
– final design of elements.

These steps are prescribed by national guidelines or standards. Each step is equally important; if any of them is ignored, previously mentioned benefits can be annulled.

In this paper, review of Croatian guidelines (HC 2014) for the design of roundabouts on national roads will be presented via 12 theoretical examples of single-lane suburban roundabouts with various external radii. These theoretical examples will be designed according to the analyzed Croatian guidelines (HC 2014), for following design vehicles: truck with trailer, semitrailer truck and tri-axle bus.

The emphasis of this review will be on the geometric design and fastest path analysis. Main objective of the study presented in this paper is to compare the Croatian guidelines (HC 2014) with the existing European (CROW 1998, FGSV 2006, RS 2011 & TSC 2011) and American (TRB 2010) guidelines, and to recommend possible improvements.

2 BACKGROUND

Main differences between roundabout design guidelines and standards are related to the design of roundabout elements (roadway edges, entrance and exit widths, islands, circular lane widths), fastest path analysis and sight distance tests (Ahac et al. 2014, 2016a,b; Bastos Silva & Seco 2005; Galleli et al. 2014; Montella et al. 2013; CROW 1998, FGSV 2006, RS 2011, SETRA 1998, TRB 2010, TSC 2011 & VSS 1999).

2.1 Geometric design elements

As a rule, the dimensions of the roundabout elements are determined on the basis of the design vehicle swept path width, with the addition of lateral protective widths. Design vehicles in European (FGSV 2001, RS 2011) and American guidelines (AASHTO 2011) are semi-trailer trucks, trucks with trailers and buses with the maximum allowable length. Those three types of design vehicles have different dimensions (length, wheel base) and they sweep different path widths when driving on the circular arc. Because of that, resulting dimensions of the roundabout elements differ, especially entrance and exit widths, circulatory road width, as well as entry and exit radii (Tables 1, 2).

2.2 Speed analysis

One of the most important roundabout design feature, which is affecting vehicle speed on roundabout, and consequently roundabout safety, is deflection around a central island (Bastos Silva & Seco 2005). The influence of deflection on

Table 1. Recommended lane widths (CROW 1998, FGSV 2006, TRB 2010, VSS 1999).

Country	Entrance m	Exit m	Circulatory roadway m
Germany	3.5–4.0	3.75–4.5	6.5–9.0
Switzerland	3.0–3.5	3.5–4.5	5.0–7.0
The Netherlands	3.5–4.0	4.0–4.5	5.0–6.0
USA	4.2–5.5	/	4.5–6.0

Table 2. Recommended entry and exit radii (CROW 1998, FGSV 2006, TRB 2010, VSS 1999).

Country	Entry radius m	Exit radius m
Germany	14–16	16–18
Switzerland	12	14
The Netherlands	8–12	12–15
USA	15–30	15–60

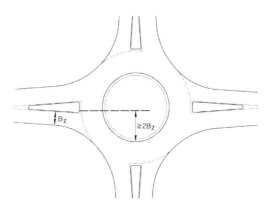

Figure 1. Deflection analysis according to (FGSV 2006).

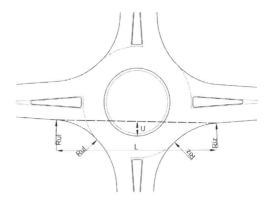

Figure 2. Geometry features used in determination of the fastest path radius according to (CROW 1998, TSC 2011).

roundabout performance can be evaluated by defining the radius (or radii) of the centerline of a vehicle traveling along the so-called fastest path through the roundabout and then calculating the vehicle speed (Galleli et al. 2014). This theoretical fastest path is representative of most used trajectory by drivers under free flow conditions when minimizing their driving discomfort (Bastos Silva & Seco 2005).

At first glance, this roundabout performance check seems straightforward: firstly, fastest path is defined, then the path radii is measured or calculated, and at the end, the vehicle speed is estimated based on speed-radius relationship. However, inconsistencies in design standards and practices concerning procedures for the definition of the deflection, construction of the fastest path, vehicle speed estimation and the speed profile requirements can be observed (Ahac, 2016a).

As mentioned above, speed analysis on roundabouts can be conducted using different approaches:

– by measuring the roundabout's geometry features and checking the achieved deflection (as described in German guidelines (FGSV 2006), and shown in Figure 1);
– by measuring the roundabout's geometry features and calculating the path radii and vehicle speed (as described in Dutch (CROW 1998) and Slovenian guidelines (TSC 2011), and shown in Figure 2);

– by constructing the fastest paths through the roundabout, measuring the path radii and then calculating the vehicle speed (as described in American (TRB 2010) and Serbian guidelines (RS 2011), and shown in Figure 3).

According to (CROW 1998, TSC 2011), the radius of the driving curve is calculated with the aid of the following equation:

$$R = \frac{(0.25 \cdot L)^2 + (0.50 \cdot (U+2))^2}{U+2} \quad (1)$$

where R [m] = radius of the driving curve; L [m] = distance between the tangent of entry radius and the tangent of the exit radius; and U [m] = distance between the edge of the central island and the right side of the lane of the access road, measured at the tangent point (Fig. 2). The design of the roundabout is correct if the radius of the driving path is between 22 and 23 m.

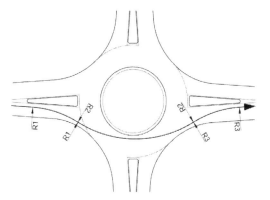

Figure 3. Fastest path radii according to (TRB 2010, RS 2011).

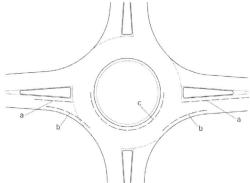

Figure 4. Fastest path lateral clearances.

The relation between the speed on the path curve and the radius of this curve is defined with the following equation:

$$V_i = 7.4 \cdot \sqrt{R_i} \qquad (2)$$

where V_i [km/h] = predicted design speed; and R_i [m] = radius of the driving curve. Acceptable vehicle speed through the roundabout should be in the range from 30 to 35 km/h (CROW 1998).

According to (TRB 2010, RS 2011), speed analysis procedure is composed from following steps: construction of the fastest path on the analyzed roundabout, measurements of path radii (R_i), and estimations of vehicle speed based on speed-radius relationship:

$$V_i = \sqrt{127 \cdot R_i \cdot (f \pm e)} \qquad (3)$$

where V_i [km/h] = predicted design speed; R_i [m] = radius of curve; f [–] = side friction factor; and e [–] = superelevation.

According to the analyzed documents, superelevation values are assumed to be +0.02 (TRB 2010) or +0.025 (RS 2011) for entry and exit curves and –0.02 (TRB 2010) or –0.025 (RS 2011) for curves around the central island.

According to (TRB 2010, RS 2011) fastest path should be drawn with the prescribed distances to the particular geometric feature: (a) painted edge line of the splitter island, (b) a right curb on the entrance and exit of the roundabout, and (c) a curb of the central island, as shown in Figure 4. American guidelines (TRB 2010) define these distances as (a) = 1.0 m, (b) = (c) = 1.5 m, while Serbian guidelines (RS 2011) define only (b) = 1.5 m and (c) = 2.0 m. According to (TRB 2010), acceptable entry vehicle speed is 40 km/h.

3 REVIEW OF CROATIAN GUIDELINES FOR ROUNDABOUT DESIGN

Latest Croatian guidelines for roundabout design (HC 2014) were published in June 2014, and their usage is mandatory for the intersections on the state roads. Geometric design of suburban roundabouts according to these guidelines is carried out in nine major steps:

- Selection of the external radius.
- Determination of the circulatory roadway width by using movement trajectory of two-axle design vehicle driving in a full circle.
- Determination of the central island truck apron width by using movement trajectory of heavy goods vehicle (truck with trailer or semi-trailer truck) driving in a full circle.
- Selection of the approach roadway lane width and splitter island form and length.
- Designing the outer roadway edge on entry and exit: selection of the entrance and exit width.
- Control of the entry angle and roadway widening severity.
- Swept path analysis on the roundabout for the design vehicle and for all movement directions.
- Determination of the fastest path and vehicle speed through the roundabout.
- Conducting the sight distance tests.

As in all abovementioned design standards, roundabout design process described in Croatian guidelines (HC 2014) is iterative: if the initial roundabout design does not meet the requirements for unobstructed design vehicle movement, travel speed and sight distance, it has to be re-defined. The design vehicle selection, swept path analysis and fastest path analysis given in these guidelines is briefly described below.

239

3.1 Design vehicles and swept path analysis

According to Croatian guidelines (HC 2014) round-about swept path analysis can be performed with two design vehicles: semitrailer truck (16.5 m long) and truck with trailer (18.75 m long). Dimensions of those vehicles are in compliance with (European Committee Directive).

Two-axle vehicles are also mentioned in the guidelines (HC 2014) because their swept paths are used for the determination of the circulatory roadway width. Namely, required widths of a circulatory roadway are specified for different wheel base and front overhang lengths of two-axle vehicles and external radii. Their choice should be harmonized with the needs of users at the proposed roundabout location and with the consent of the road administration.

Swept path analysis is conducted by drawing the design vehicle (body) movement trajectories in all possible directions on roundabout blueprint. It ensures that conditions for unobstructed vehicle movements on roundabout entry and exit are achieved. The minimum protective lateral width along the trajectories is 0.5 m (exceptionally 0.3 m) on all segments, except on the outside edge of the circulatory roadway where the minimum lateral width is 1.0 m.

3.2 Geometric design elements

According to the Croatian guidelines (HC 2014), on small and medium sized suburban round-abouts (with outer radius Rv from 11 to 25 meters) the splitter island is triangular (Fig. 5.). Recommended length of these islands (m) is between 15 to 50 meters.

Required speed profiles on roundabouts are achieved by proper design of the right curb on entrances and exits, i.e. the proper selection of entry radius (Rul), exit radius (Riz), and entrance width (e). Circulatory roadway width (u) is determined based on two-axle vehicle movement trajectory, while driving in a full circle. In order to determine the radius of this circle, roundabout's outer radius is reduced by the protective lateral width, which is 1.0 meter. Recommended and borderline dimensions of abovementioned design elements for small and medium sized suburban single-lane round-abouts are given in Table 3.

After the selection of basic design elements, entry angles (Φ) and roadway widening severity on entrance (S) are checked. Severity on entrance is the dimensionless parameter of roundabout design, which represents relationship between entrance width and lane width. The dimensions of both entry angle and severity on entrance, as well as entrance widths, depend on the design of the splitter island and applied entry and exit radii. Borderline dimensions for the entry angles are 0–77°, and recommended values are 20–40°, while both borderline and recommended values for the severity on entrance are between 0 and 2.9.

3.3 Speed analysis procedure

Speed analysis procedure described in Croatian guidelines (HC 2014) is based on the procedure described in Dutch (CROW 1998) and Slovenian (TSC 2011) guidelines. According to these guidelines, speed analysis on roundabout is conducted by measuring the roundabout's geometry features and calculating the path radii and vehicle speed.

However, there are some changes in Croatian guidelines (HC 2014) (Fig. 6).

– Geometry feature L that is used in calculation of the path radii (Eq. 1) and vehicle speed (Eq. 2) is measured via fastest path through roundabout—it is defined as the distance between the tangent of entry radius and the tangent of the exit radius on the fastest path.
– The vehicle path is determined by the clearance of 1.0 m from the roadway edges or the roundabout

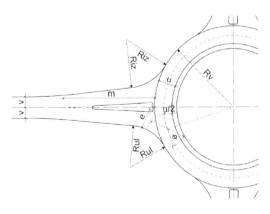

Figure 5. Design elements of suburban roundabouts.

Table 3. Recommended and borderline dimensions (HC 2014).

Element	Recommended m	Borderline m
Entrance width (e)	4.0–7.0	3.6–10.0
Lane width (v)	3.0–3.5	2.5–7.0
Entry radius (Rul)	8.0–20.0	6.0–25.0
Exit radius (Riz)	10.0–25.0	8.0–50.0
Circulatory roadway width (u)	4.5–6.0	4.0–9.0

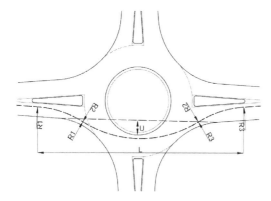

Figure 6. Fastest path construction elements (HC 2014).

design elements: (a) painted edge line of the splitter island, (b) a right curb on the entrance and exit of the roundabout, and (c) a curb of the central island.

- Acceptable vehicle speed through the roundabout should be in the range from 25 to 40 km/h.

The differences between Croatian guidelines (HC 2014) and aforementioned foreign standards and guidelines (CROW 1998, RS 2011, TRB 2010, TSC 2011) inspired the research described shown below. The aim of this research was to establish if and in what way these discrepancies influence the results of speed analyses.

4 STUDY

For the purpose of this study, theoretical examples of single-lane suburban roundabouts with various external radii were designed (Rv = 17.5, 20.0, 22.5, and 25.0 m). Initial design elements of these roundabouts were defined according to the analyzed Croatian guidelines (HC 2014).

Swept path analysis was performed for three design vehicles (truck with trailer 18.75 m long, semitrailer truck 16.5 m long, and tri-axle bus 15.0 m long). Because tri-axle buses play an important role in passenger transport on long-distance lines in Croatia, this type of vehicle was included in the research, even though Croatian guidelines (HC 2014) do not define buses as design vehicles.

For each design vehicle, four roundabouts with different external radii were designed. On those roundabouts fastest path analysis was performed, and vehicle speed was estimated according to Croatian (HC 2014), Dutch (CROW 1998) and American guidelines (TRB 2010), as described below.

4.1 Roundabout design

After the selection of the external radii (Rv), lane widths (v), and shape and length of the splitter islands (m), circulatory roadway widths (u) were defined as the sum of the lateral protective widths and width (sv + Δv), and rounded to 0.1 m (Table 4, Figs. 5, 7). Widths (sv + Δv) were determined on the basis of the two-axle vehicle movement trajectories, while driving in a full circle of radii Rv-1 (Table 4). All trajectories were drawn by software (Autodesk Vehicle Tracking 2015).

In this study, the two-axle vehicle used to determine (sv + Δv) widths was 12.0 m long bus from German guidelines (FGSV 2001), because dimensions of the two-axle vehicle needed for the construction of the movement trajectories are not specified in Croatian guidelines (HC 2014). Croatian guidelines (HC 2014) only give (sv + Δv) widths for a specific set of Rv-1 values (8.0, 9.0, 11.0, 12.5, 15.0, 18.0, and 21.5 m). Initial entrance widths (eul), exit widths (eiz) and outer edge radii (Rul and Riz) were also selected.

After an initial roundabout design was finished, swept path analyses were conducted for all design vehicles. The trajectories were drawn by a software

Table 4. Applied dimensions of roundabouts design elements.

Outer radius [m]	17.5	20.0	22.5	25.0
m [m]	30.0	30.0	30.0	30.0
v [m]	3.5	3.5	3.5	3.5
Rul [m]	13.0	13.0	13.0	13.0
Riz [m]	15.0	15.0	15.0	15.0
Rv-1 [m]	16.5	19.0	21.5	24.0
sv+Δv [m]	4.94	4.62	4.32	4.11
u [m]	6.5	6.2	5.9	5.7
u' [m]	1.0	1.0	1.0	1.0

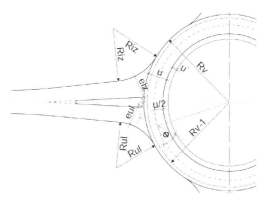

Figure 7. Suburban roundabout scheme.

(Autodesk Vehicle Tracking 2015) for all movement directions: straight, right, left, U-turn and circular. Based on these trajectories circular swept path widths (sv + Δv') were measured and following elements were defined: truck apron widths (u'), entrance widths (eul), exit widths (eiz), outer edge radii (Rul) and outer edge radii (Riz) (Tables 4–7).

Entry angles (Φ) and roadway widening severity on entrance (S) were also controlled. Obtained values of entry angles were within limit range, while severity on entrance values were within the recommended range (Tables 5–7).

Swept path analysis showed that selected three-axle bus occupies greatest turning width, compared to semi-trailer truck and truck with trailer, and that it requires the greatest entry and exit roundabout width. In addition, out of the 12 defined roundabouts some of them have the same elements dimensions (Tables 5–7).

- Roundabouts with the outer radius of 20.0 m dimensioned for the design vehicle truck with trailer (20TT) and semi-trailer truck (20ST).
- Roundabouts with the outer radius of 22.5 m dimensioned for the design vehicle truck with trailer (22.5TT) and semi-trailer truck (22.5ST).

Therefore, the vehicle speed estimation was conducted only on 10 examples of roundabouts.

4.2 Fastest path speed analysis

For each roundabout fastest path analysis and vehicle speed estimation were conducted in accordance

Table 5. Applied dimensions of roundabouts design elements for Truck with Trailer (TT).

Roundabout Id.	17.5TT	20.0TT	22.5TT	25.0TT
Outer radius [m]	17.5	20.0	22.5	25.0
sv+Δv' [m]	5.39	4.95	4.62	4.38
eul [m]	5.25	5.75	6.00	6.00
eiz [m]	5.75	6.25	6.50	6.50
Φ [°]	42.0	41.7	47.0	47.2
S [-]	0.44	0.40	0.44	0.56

Table 6. Applied dimensions of roundabouts design elements for Semi-trailer Truck (ST).

Roundabout Id.	17.5ST	20.0ST	22.5ST	25.0ST
Outer radius [m]	17.5	20.0	22.5	25.0
sv+Δv' [m]	5.72	5.13	4.84	4.56
eul [m]	5.25	5.75	6.00	6.00
eiz [m]	5.75	6.25	6.50	6.75
Φ [°]	42.0	41.7	47.0	47.2
S [-]	0.44	0.40	0.44	0.56

Table 7. Applied dimensions of roundabouts design elements for three-axle Bus (B).

Roundabout Id.	17.5B	20.0B	22.5B	25.0B
Outer radius [m]	17.5	20.0	22.5	25.0
sv+Δv' [m]	5.95	5.43	5.05	4.77
eul [m]	5.50	6.00	6.50	6.50
eiz [m]	6.00	6.25	6.50	7.25
Φ [°]	43.3	46.5	49.0	49.1
S [-]	0.33	0.33	0.35	0.38

Table 8. Speed analysis variant parameters.

Variant	Clearances m	Radius	Speed
CRO-1	a = b = c = 1.0	Eq. 1	Eq. 2
CRO-2	a = 1.0, b = c = 1.5	Eq. 1	Eq. 2
USA-1	a = b = c = 1.0	measured	Eq. 3
USA-2	a = 1.0, b = c = 1.5	measured	Eq. 3
NLD	–	Eq. 1	Eq. 2

with the Croatian (HC 2014), The Netherlands (CROW 1998) and American guidelines (TRB 2010). In these analyses, different fastest path clearances from the roadway edges were taken into consideration, to test the effect of their changes on the resulting speed (Table 8).

Following assumptions were also made.

- Superelevation values are +0.025 for entry and exit curves, and –0.025 for curves around the central island.
- Side friction factor varies with vehicle speed and is determined in accordance with (AASHTO 2011).

In order to investigate the effect of different approaches to the construction of the fastest paths and different equations for the vehicle speed calculation, speed analyses conducted according to (TRB 2010, HC 2014) included the definition of the fastest path radii on entrances (R1), around the central islands (R2), and on exits (R3).

5 DISCUSSION

The vehicle speed calculated in accordance with the Croatian guidelines (HC 2014) with clearances of 1.0 m from the central island and the roadway edges exceed the recommended value of 40 km/h. They are also higher than calculated speed on the fastest path with clearance of 1.5 m. The deviations are in range from 2 km/h to 4 km/h, and they

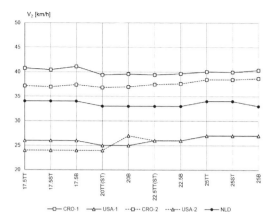

Figure 8. Vehicle speed around the central island.

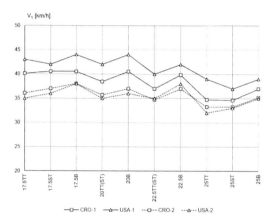

Figure 9. Entry vehicle speed.

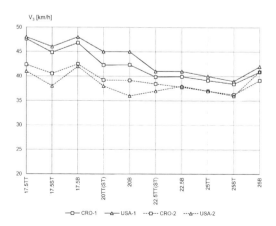

Figure 10. Exit vehicle speed.

decrease with an increase of an external radius of the roundabout.

The vehicle speed calculated in accordance with the Dutch guidelines (CROW 1998) is within the recommended limits (30 to 35 km/h), and is smaller than the speed obtained in accordance with the Croatian guidelines (HC 2014). This discrepancy was expected, due to differences in the definition of length L, and consequently the calculated radius of the driving curve. The vehicle speed around the central island (Fig. 8) calculated in accordance with the American guidelines (TRB 2010) is significantly smaller than the speed calculated in accordance with the Croatian guidelines (HC 2014). Average difference is 14 km/h for the 1.0 m path clearance, and 12 km/h for the 1.5 m path clearance.

Even though the entrance and exiting widths of roundabouts designed for three-axle buses are larger than those designed for semi-trailer truck and truck with trailer (0.25 and 0.50 m), speed deviation rate around the central island is negligible (1 km/h). These differences are larger for speed on entry and exit paths (Figs. 9, 10), where larger lane widths needed to accommodate three-axle buses led to the deviation of speed from 1 to 3 km/h.

6 CONCLUSIONS

Results from swept path analysis on theoretical examples of roundabouts showed that three-axle bus with the length of 15.0 meters needs largest dimension of the design elements. That is because this vehicle takes up a greater width when turning than the semi-trailer truck and truck with trailer. It is evident that the impact of these vehicles on the performance and safety of roundabouts must not be ignored, especially in tourist countries such as Croatia, where three-axle buses play an important role in the transport of passengers on intercity and international routes.

According to the results from fastest path analysis and vehicle speed estimations on the theoretical examples of roundabouts, it can be concluded that the vehicle path clearances have the greatest impact on vehicle speed. After that comes the impact of the design vehicle type, because they have an effect on the roundabout's entrance and exit width. It should be noted that 1.0 m clearance does not always ensure unhindered passage of a passenger car through the roundabout. Because of that, larger minimum clearances from the central island and right curb on the entrances and exits are recommended, for instance 1.5 meters. All this indicates to the need to revise the guidelines in terms of vehicle path clearances.

REFERENCES

AASHTO (American Association of State Highway Transport Officials). (2011). A Policy on Geometric Design of Highways and Streets. Washington DC.

Ahac, S., Džambas, T., Stančerić, I. & Dragčević, V. 2014. Performance checks as prerequisites for environmental benefits of roundabouts. In Stjepan Lakušić (ed.), *Road and Rail Infrastructure III, Proceedings of the Conference CETRA 2014, Split, 28–30 April 2014.* Zagreb: Department of Transportation, Faculty of Civil Engineering, University of Zagreb.

Ahac, S., Džambas, T. & Dragčević, V. 2016a. Review of fastest path procedures for single-lane roundabouts. In Stjepan Lakušić (ed.), *Road and Rail Infrastructure IV, Proceedings of the Conference CETRA 2016, Šibenik, 23–25 May 2016.* Zagreb: Department of Transportation, Faculty of Civil Engineering, University of Zagreb.

Ahac, S., Džambas, T. & Dragčević, V. 2016b. Sight distance evaluation on suburban single-lane roundabouts, *Građevinar* 68(1): 1–10.

Autodesk Vehicle Tracking. (Integrated swept path analysis software). (2015). Autodesk, San Rafael, CA.

Bastos Silva, A. & Seco, A.: Trajectory deflection influence on the performance of roundabouts, In European Transport Conference (ETC) Association for European Transport, Strasbourg, 2005.

CROW (Dutch Technology Platform for Transport, Infrastructure and Public Space). (1998). Eenheid in rotondes. CROW 126, Ede, Netherlands.

European Committee Directive. (2002). "Directive 2002/7/EC of the European Parliament and of the Council of 18 February 2002 amending Council Directive 96/53/EC laying down for certain road vehicles circulating within the community the maximum authorised weights in international traffic and the maximum authorised weights in international traffic." OJ L 67, Brussels, Belgium.

FGSV (Forschungsgesellschaft für Straßen und Verkehswesen). (2001). Bemessungsfahrzeuge und Schleppkurven zur Überprüfung der Befahrbarkeit von Verkehrsflächen, Forschungsgesellschaft für Strassen- und Verkehrswesen. Köln: FGSV.

FGSV (Forschungsgesellschaft für Straßen und Verkehswesen). (2006). Merkblatt fur die Anlage von Kreisverkehren. FGSV K 1000, Köln, Germany.

Gallelli, V., Vaiana, R. & Iuele, T.: 2014. Comparison between simulated and experimental crossing speed profiles on roundabout with different geometric features, *Procedia – Social and Behavioral Sciences* 11: 117–126.

HC (Smjernice za projektiranje kružnih raskrižja na državnim cestama). (2014). Hrvatske ceste d.o.o.

Hydén, C. & Várhelyi, A. 2000. The effects on safety, time consumption and environment of large scale use of roundabouts in an urban area: a case study, *Accident Analysis and Prevention* 32: 11–23.

Mandavilli, S., Russell, E.R. & Rys, M.J. 2003. Impact of Modern Roundabouts on Vehicular Emissions, In *Proceedings of the 2003 Mid-Continent Transportation Research Symposium, Ames, 21–22 August 2003.* Ames: Iowa State University.

Montella, A., Turner, S., Chiaradonna, S. & Aldridge, D. 2013. International overview of roundabout design practices and insights for improvement of the Italian standard, *Canadian Journal 0f Civil Engineering* 40: 1215–1226.

RS (Pravilnik o uslovima koje sa aspekta bezbednosti saobraćaja moraju da ispunjavaju putni objekti i drugi elementi javnog puta). (2011) Službeni glasnik RS, br. 50/2011

SETRA (Service d'Etudes sur les Transports, les Routes et leurs Aménagements). (1998). Aménagement des carrefours interurbains sur les routes principales, Carrefours plans—guide technique.

TRB (Transport research board). (2010). Roundabouts: An Informational Guide, 2nd Edition. Washington, D.C.

TSC (Krožna križišča, TSC 03.341: 2011). (2011). Ljubljana: Ministrstvo za infrastrukturo in prostor – Direkcija RS za ceste.

VSS (Vereinigung Schweizerischer Strassenfachleute). (1999). Schweizer Norm SN 640 263: Knoten mit Kreisverkehr. Zürich.

Transport Infrastructure and Systems – Dell'Acqua & Wegman (Eds)
© *2017 Taylor & Francis Group, London, ISBN 978-1-138-03009-1*

Laboratory investigation into the modification obtained of paving bitumen with different plastomeric polymers

E. Saroufim & C. Celauro
Department of Civil, Environmental, Aerospace, Materials Engineering, University of Palermo, Palermo, Italy

N. Khalil
Department of Civil Engineering, University of Balamand, Lebanon

ABSTRACT: The use of polymer/bitumen modified blends in road paving applications has been increasingly advancing from the technological point of view. Plastomeric polymers are known to be amongst the most widely used in bitumen modification. Therefore, this paper presents the influence and degree of modification of several plastomeric-modified bitumen in terms of rheological properties and morphology. Conventional tests such as penetration grade, Ring-and-Ball softening point and storage stability were performed. Rheological properties of the modified blends were characterized in terms of dynamic mechanical analysis via frequency sweep tests using the Dynamic Shear Rheometer (DSR). The morphology of the samples is investigated by means of fluorescent light optic microscopy. New processed and mixed polymers are introduced into the analysis. The results confirm that the fundamental properties and morphology of the modified bitumen are directly dependent on the type of polymer in use. Overall, the polymer-modified bitumen improves the original bitumen properties: remarkably stable blends were found with overall good performance especially for treated polymers. The microscope pictures provide useful information on the different types of interaction between polymers and the bitumen.

1 INTRODUCTION

Bitumen is known to be a viscoelastic material with properties suitable for pavement construction. But bitumen alone has poor mechanical performance, it is hard and brittle in cold weather and soft in hot environments (Blanco et al., 1996). With the increase in traffic load nowadays, pavement surfaces are highly stressed and affected by elevated traffic volume and temperature variations. Consequently the need to enhance pavement surfaces and minimize maintenance problems has allowed the field of bitumen/polymer modification to grow and become important in designing optimally performing pavements. Only small amounts of additives are sufficient to maintain bitumen's own advantage and improve the desired properties (Blanco et al., 1996). It has been recorded that pavements with polymer modification exhibit greater resistance to rutting and thermal cracking, and decreased fatigue damage, stripping and temperature susceptibility (Yildirim, 2007).

There is a wide variety of polymers that have been used as additives to bitumen for producing so-called PMBs (polymer modified bitumens). Normally, two main types of polymers are popular for bitumen modification: elastomers and plastomers.

The latter would be better classified as thermoplastic polymers, since the term "plastomer" simply refers to their processability, thus recalling some technological properties they offer.

As reported in the literature, elastomeric modification works by making the binder more elastic and reducing its stiffness (Robinson, 2005). Plastomeric modification of bitumen, in contrast, makes bitumen stiffer and reduces its temperature susceptibility, particularly at high in-service temperatures (McNally, 2011). This is important in Mediterranean regions where high temperatures affect long-term road conditions (Viola and Celauro, 2015). Thus, plastomers would be an appropriate modifier to sustain the needs for road construction in warm climates. They are also known to reduce the risk of rutting (Robinson, 2005, McNally, 2011), which is amongst the major distresses of flexible pavements and is particularly severe in hot regions.

Although significant research has been done in this area, plastomeric PMBs are still poorly understood scientifically, due to the complex nature of bitumen and its interaction with the polymer system (Brule, 1996). This paper highlights this category of modification.

Plastomers are known to offer several advantages beside some drawbacks in terms of PMB performance, depending on the polymer type taken into consideration (Zhu et al., 2014). For instance, polyethylene-based polymers, once added to bitumen, allow limited improvement in elasticity and potential storage stability problems, whereas Ethylene-Vinyl Acetate (EVA) in appropriate content gives relatively good storage stability to the final blends (Zhu et al., 2014). Consequently, the plastomeric modification process exposes challenges in this domain since pavement construction techniques require the binder to remain stable before and after application on the road (Navarro et al., 2004). Therefore, not only enhancement in rheological properties is required but also production of blends that will resist separation under high temperature storage, yet remain at low cost.

Modification of binders is a complex mechanism that requires simplification. Therefore, different types of polymers, including new compositions, were used in order to create a link between the rheological properties and stability of the PMBs, and their influence on morphology. In this context, conventional tests are carried out on different plastomeric polymer/bitumen blends. Apart from the typical empirical/technological tests on road bitumen, the rheological properties of the polymer/bitumen are investigated through dynamic mechanical analysis while the morphology is evaluated by assessing the state of dispersion of polymers into the bitumen matrix.

2 MATERIALS AND METHODS

2.1 Physical properties and classical tests

The physical properties of pure bitumen from vacuum distillation having a penetration grade 50/70 are shown in Table 1. Five different polymers were used for modification of one binder in order to highlight the effect of the polymer used on the final blend produced.

The polymers investigated in this research are Low-Density Polyethylene (LDPE), High Density Polyethylene (HDPE) and Ethyl-Vinyl-Acetate (EVA) copolymer supplied by Versalis. The latter contains 28% vinyl-acetate.

HDPE, EVA and LDPE were used as received. LDPE was processed (P-LDPE) in a Brabender mixer at 210°C for 30 min at a rotational speed of 64 rpm. Finally, LDPE and EVA were blended in the same mixer but in lower processing conditions, at 180°C for 10 min at a rotational speed of 64 rpm. The EVA content in the blend was 20% wt/wt.

The modified binders were prepared at 180°C using a Silverson High shear mixer. The polymer content for all the modified binders was kept constant and equal to 5% by weight as it has been seen in other studies to be an effective content for modification (Pérez-Lepe et al., 2003) and it is, therefore, one of the most common contents for practical applications. An additional mix was done with reduced polymer content (2.5%), only for HDPE, for comparison purposes regarding the final properties of the blend. Mixing was carried out for an extra 30 min after the entire polymer sample was fed, at a constant rate, into the heated bitumen to produce homogenous mixtures. The modified samples were then sealed in aluminum containers and stored for further testing.

Empirical tests such as penetration and softening point were run according to the following standards: EN 1426:2007 and EN 1427:2007, respectively. The results are shown in Table 2.

After mixing with the base bitumen, the modified binders underwent the storage stability test according to EN 13399:2004 Standard in order to monitor the phase separation between binder and polymers because it is considered a critical aspect of modification results.

Table 1. Conventional properties of the base bitumen.

Characteristics	Units of measurement	Values
Grade		50/70
Penetration	dmm	68
Softening point	°C	50.5
Fraas breaking point	°C	−12
Ductility	mm	over 100
Increase in softening point (after RTFOT)	°C	3.5
Increase in penetration (after RTFOT)	%	64.7

Table 2. Conventional tests on polymer modified binders.

	T	T
Requirements	Intermediate	High
Characteristics	Pen at 25°C	SP
Test method	EN 1426	EN 1427
Unit of measure	dmm	°C
PURE BITUMEN 50/70	68	50
LDPE	35.7	56
HDPE	25.7	58
LDPE/EVA	35	61
P-LDPE	31.7	57
EVA	40.1	61.1
2.5% HDPE	33	55.5

2.2 Dynamic Shear Rheometer

The viscoelastic properties of the polymer/bitumen blends are investigated by means of Dynamic Mechanical Analyses (DMA). Frequency sweep tests were carried out within the linear viscoelasticity range using a Dynamic Shear Rheometer (Physica MCR 1). Depending on the temperature range the tests were carried out with parallel plates of diameter 8 or 25 mm. The temperatures ranged from −10°C to 80°C. The rheological properties of the binders were measured in terms of their complex modulus G^* and phase angle δ in a controlled-strain loading mode.

2.3 Fluorescence microscopy

Fluorescent microscopy is the most valuable method for studying the phase morphology of polymer modified bitumen, as it allows direct observation of different microstructures of modified bitumen. This is because the polymer-rich domains appear yellow while the asphaltene-rich domains remain black (Lesueur, 2009). This allows qualitative evaluation of the homogeneity and the structure of the modified bitumen (Sengoz and Isikyakar, 2008a).

In this study, the preparation of samples implied pouring of the blends, which were thoroughly stirred to ensure homogeneity, into small elements subjected immediately to a cooling regime. Prior to fluorescent image analysis, the frozen samples were sharply cut and placed on a lamella to examine the fractured part. The images were taken with a magnification of 16:1 using a trinocular OPTIKA® microscope, N-400FL model.

3 RESULTS AND DISCUSSIONS

3.1 Conventional test results

It can be seen from Table 2 that the addition of the selected polymers to bitumen shows a decrease in the Penetration (Pen) and an increase in the Softening Point (SP) for all the modified blends. However, the modification does not significantly affect the softening point in relation to penetration: the increase in the softening point is relatively uniform for all the blends in the range of the 5–10% for all the polymers, while there is a considerably larger decrease in penetration values registered, especially for the HDPE at 5%. This roughly indicates that the bitumen in use becomes—in any case—harder and less susceptible to temperature variation.

3.2 Storage stability

Storage stability is one of the most critical aspects of asphalt modification. According to the EN 14023:2010 Standard, the specific storage stability requirement for PMBs is that the softening point difference between top and bottom ($\Delta T_{R\&B}$) is $\leq$ 5°C after 3 days of storage.

The stability results are shown in Table 3. The large increase in temperatures for LDPE and HDPE in both percentages indicates that the samples are unstable and macroscopic phase separation readily occurs during their storage at high temperatures, which leads to undesirable problems during transportation. Surprisingly, the processing of LDPE as well as the addition of EVA to the LDPE show significant stability. These results emphasize the possibility of transforming previously unstable polymer/bitumen blends into absolutely stable ones. Two methods are suggested in this paper, one through treatment of virgin polymer with other components and one through a mechanical process.

3.3 Stiffness master curve

The effect of polymer addition on the complex modulus loss and storage modulus of binders can

Table 3. Storage stability.

Requirements	Storage stability
Characteristics	$\Delta T_{R\&B}$
Method	EN 13399
Test	EN 1427
Unit of measurement	°C
LDPE	28
HDPE	20
LDPE/EVA	5
P-LDPE	0.4
EVA	1.75
2.5% HDPE	10

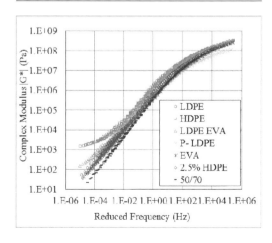

Figure 1. Complex modulus |G*| (Pa) master curves at T_{ref} = 30°C.

247

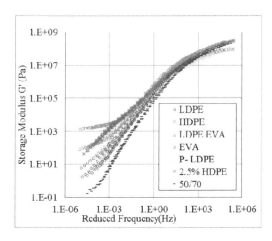

Figure 2. Storage modulus master curve G' (Pa), at T_{ref} = 30°C.

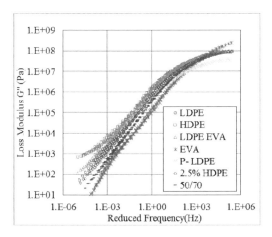

Figure 3. Loss modulus G" (Pa) at T_{ref} = 30°C.

be observed in Figures 1, 2 and 3 respectively. The reference temperature, T_{ref}, is set at 30°C for the application of the Time-Temperature Superposition Principle (TTSP). All the polymer/bitumen blends displayed higher complex modulus when compared to the pure binder. A clear increment of the results is evident in the low frequency (high temperature) zone. To a certain extent, this indicates an improvement in the mechanical behaviour of the base binder. On the other hand, an increase in |G*| in the low frequency range is very common in polymer modification. It can indicate that the modifier has made the base binder less susceptible to temperature (Asgharzadeh and Tabatabaee, 2013). The considerable difference in the degree of modification depends on the modifier type: this proves that the properties of the blend primarily depend on the rheological characteristics of the

polymer and its content. In the low frequency domain, which corresponds to high temperatures, a significant increase in stiffness for all the PMBs is spotted. This is clearly seen for 5% HDPE, with a value approximately equal to 1.5 kPa, which denotes solid-like behaviour and improvement of elasticity. Furthermore, the plateau region seen for this polymer is an indication of a predominant polymer network within the modified binder (Airey, 2002, Pérez-Lepe et al., 2003), while all the other PMBs, at very low frequencies, show similar asymptotic behaviour to each other.

Thus the high temperature stiffness is enhanced for all the PMBs, and particularly for 5% HDPE.

The presence of the plateau region at high temperatures (low frequencies) triggered the need to investigate the behaviour of HDPE in lower percentages for a better understanding of the material, and also to investigate whether this behaviour and the overall stability are replicated. At low frequencies, the 2.5% curve of HDPE has an asymptotic shape similar to that of the base binder but with higher stiffness values; while 5% HDPE tends to form a plateau. With lower polymer content, the behaviour of the modified bitumen remains close to that of the straight-run bitumen. This could be an additional indicator that the rheological response at 5% HDPE is strongly contributed by the polymer. Nevertheless, both blends are found to be unstable, as seen previously. Thus at a certain concentration of HDPE the blend receives its properties from the polymer itself.

At low temperatures (high frequencies), this is no longer the case, since all the mixtures exhibit similar trends, apart from those with P-LDPE and pure EVA which attain lower storage modulus values.

The results shown in Figures 2 and 3 for G' and G" comply with the complex modulus master curve. The addition of polymer ends up with an increase in the modulus values at high temperature, which is beneficial for limiting permanent deformation in road structures.

3.4 *Phase angle master curves*

Figure 4 represents the phase angle master curves for all the binders studied. A clear change in the shape of the phase angle curves is observed after polymer addition. This indicates an improvement in elastic behaviour after modification.

Once again, HDPE modification at 5% displays a small plateau region at intermediate frequencies followed by a reverse slope in the lower frequency range. This plateau means that the elastic and viscous complex modulus components vary in the same proportion so that the phase angle does not change (Da Silva et al., 2004). Such a plateau

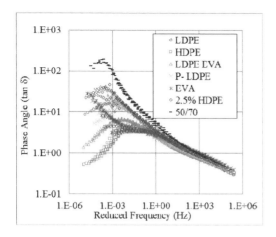

Figure 4. Phase angle (tan δ) master curves at $T_{ref} = 30°C$.

Table 4. Letter coding description according to EN13632.

Phase continuity	P: Continuous polymer phase
	B: Continuous bitumen phase
	X: Continuity of both (crosslinking)
Phase description	H: Homogeneous*
	I: Inhomogeneous
Size description	S: Small (< 10 μm)
	M: Medium (from 10 μm to 100 μm)
	L: Large (>100 μm)
Shape description	r: Round, cylindrical
	s: Elongated
	o: Other

* If no structure is visible but fine, homogeneous, slightly yellow particles are detected, the product should be called homogeneous.

region for the phase angle, together with a relatively high modulus (as much as possible insensitive to temperature variation) could be favourable to binder deformation resistance (Lu and Isacsson, 1997). In relation to this comportment, it is worth recalling that plateau rheological curves, in general, are a characteristic of the viscoelastic response of rubbers. Therefore, they imply a polymer-rich elastic network (Airey, 2003, Da Silva et al., 2004) emphasizing the polymer elastic properties which govern the overall behaviour of the blend.

3.5 Fluorescence microscopy

Bitumen is a complex material. Its components are typically simplified by arranging them into a colloidal structure (Lesueur, 2009). Addition of polymer makes it even more complex. Morphology facilitates the analysis of the internal structure, as it investigates the polymer phase dispersion in the bitumen matrix by means of a simple microscopic visualisation process.

In this study, the morphology of the produced PMBs is represented in Table 5 as a general characterization related to the nature of the continuous phase as well as a description of the phases and shapes in a letter coding format according to EN 13632: 2005 (Table 4).

It was not possible to take representative microphotographs of the blend with EVA, since the dispersion was visible only for couple of seconds before the frozen material quickly melted under the microscope lamp. This could be a result of the EVA concentration in the blend and its low melting temperature (around 40°C).

The morphologies of the PMBs vary in accordance with the differences in the characteristics of the polymers used (in terms of type, density, etc.). However, a desired morphology is illustrated in the form of a biphasic condition in which a polymer-rich phase and an asphaltene-rich phase coexist in a micro-scale metastable equilibrium stage (Polacco et al., 2006). From the different shapes depicted in Figures 5 (a) to (e) and listed in Table 5 it can be deduced that the morphology is correlated with the properties previously examined in this paper. An inception of a polymeric network can be roughly seen for P-LDPE, where a uniform dispersion of small irregularly shaped polymer particles is spotted with a smaller quantity of medium particles having non-uniform dispersion and a tendency to group. In any case, based on previous studies (Lu and Isacsson, 2000, Sengoz and Isikyakar, 2008b) polymeric phases are known to be strongly evident when at least 6% of polymer is used in the blend, i.e. at a slightly higher percentage than the one used in this study. The P-LDPE blend, which has acceptable mechanical performances and excellent storage stability, displays uniformly distributed small particles with some parts of polymers linked together in clear shapes. This morphology indicates a structured material, because of the interconnection in the dispersed phase. Thus, amongst the polymers studied in this research P-LDPE proves to be a promising modifier for this grade of bitumen. As a new approach in polymer modification, P-LDPE polymer gives the blend satisfactory mechanical performances and stresses the fact that available resources can be treated for enhancing the properties of the desired blends. The same applies to the LDPE/EVA stable blend, where elongated small particles are uniformly distributed in the bitumen matrix.

On the other hand, 5% HDPE, which showed an improvement in the stiffness modulus, has a similar

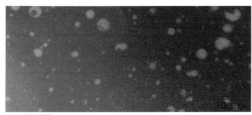

(a) LDPE

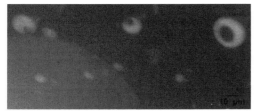

(b) 2.5% HDPE

(c) HDPE

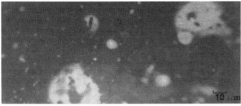

(d) P- LDPE

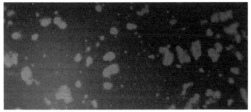

(e) LDPE/EVA

Figure 5. Microphotographs of the polymer/bitumen blends studied.

Table 5. Morphological characterization through letter abbreviation.

PMB	Phase continuity	Phase description	Size description	Shape description
LDPE	B	I	S	r/o
HDPE	B	I	M	r/o
P-LDPE	B	H/I	S/M	r/o
LDPE/EVA	B	H	S	s/r
EVA	–	–	–	–
2.5% HDPE	B	I	S	r/o

morphological phase separation, which is attributed to Brownian coalescence, followed by gravitational flocculation and later on leads to creaming (Habib et al., 2011). The difference here between HDPE and P-LDPE blends is in the degree of homogeneity of the distribution of these particles in the bitumen mass. Unlike P-LDPE, HDPE does not show any uniform distribution of smaller polymer particles. This indicates that the polymer is not well dispersed within the bitumen, implying that the blend, at high temperature ranges, primarily acquires polymer properties. The mechanism of the polymer-rich phase separation has also been proved to be due to creaming, preceded by coalescence, as seen in a previous study on HDPE (Pérez-Lepe et al., 2006). Finally, stabilization effects are seen in this research for the LDPE, either by mechanically process-ing the polymer or by adding another copolymer (EVA) known for its relatively good storage stabil-ity. In the same perspective, a similar approach is desirable when working with HDPE too.

As a result, successful modification should be reflected not only in the enhanced rheologi-cal properties and satisfactory storage stability at macroscopic level, but in the dispersed polymeric phase on the microscopic scale as well.

4 CONCLUSIONS

The addition of polymers to bitumen enhances the mechanical properties of the modified binder. Even though plastomers are known to be an effective and economical modifier for the bitumen, they reveal some major drawbacks, as mentioned in the intro-duction. The main disadvantages are typically seen in mechanical property deficiencies and also regard-ing the stability in storage and transportation of these PMBs. Hence overcoming their drawbacks is an advance in bitumen modification and a challenge for researchers in the field of paving materials.

In this paper, a variety of polymer/bitumen blends was studied, at typical polymer contents. The selected polymers were of the plastomeric type. The results in terms of storage modulus and

structure to that of the P-LDPE blend in terms of "large particle" morphology but with less intercon-nection between the bitumen and polymer. HDPE previously proved to be unstable. One of the main causes of this instability is the tendency of PMB to phase angle indicate an improvement in perform-ance for all PMBs. The highest enhancement of elastic behaviour is registered for the blend pro-duced with 5% HDPE. On the other hand, the same blend also showed the highest stiffness in penetration values, evident storage instability and large circular spheres with no homogenous distribution of smaller particles in micrographs. This indicates that the mechanism of separation of the polyethylene from the bitumen is preceded by the coalescence and flocculation phenomenon of the dispersed phase. On the opposite side, the more stable bitumen/polymers blends, represented by LDPE undergoing mechanical processing and LDPE mixed with EVA, have improved rheological properties. They show a homogenous dispersion of small particles in the bitumen. Thus micrographs, regardless of the type of polymer, can be a useful indicator of the rheo-stability of the blend.

Processing of LDPE and addition of EVA to LDPE showed an effective way of stabilizing the virgin polymer. Considering the good results obtained for the HDPE modifier, in terms of rheo-logical performances, further stabilization methods should be developed for the HDPE/bitumen blend in order to stabilize the blend and avoid filler type behaviour or, in other terms, the hardening (stiff-ening) effect of the polymer.

REFERENCES

Airey, G. D. 2002. Rheological evaluation of ethylene vinyl acetate polymer modified bitumens. Construc-tion and Building Materials, 16, 473–487.

Airey, G. D. 2003. Rheological properties of styrene butadiene styrene polymer modified road bitumens. Fuel, 82, 1709–1719.

Asgharzadeh, S. & Tabatabaee, N. 2013. Rheological master curves for modified asphalt binders. Scientia Iranica. Transaction A, Civil Engineering, 20, 1654.

Blanco, R., Rodríguez, R., García-Garduño, M. & Castaño, V. M. 1996. Rheological properties of styrene-butadiene copolymer–reinforced asphalt. Journal of Applied Polymer Science, 61, 1493–1501.

Brule, B. 1996. Polymer-modified asphalt cements used in the road construction industry: basic principles. Transportation Research Record: Journal of the Transportation Research Board, 48–53.

Da Silva, L. S., De Camargo Forte, M. M., De Alencas-tro Vignol, L. & Cardozo, N. S. M. 2004. Study of rheological properties of pure and polymer-modified Brazilian asphalt binders. Journal of materials science, 39, 539–546.

Habib, N. Z., Kamaruddin, I., Napiah, M. & Tan, I. M. 2011. Rheological properties of polyethylene and polypropylene modified bitumen. International Jour-nal Civil and Environmental Engineering, 3, 96–100.

Lesueur, D. 2009. The colloidal structure of bitumen: Consequences on the rheology and on the mecha-nisms of bitumen modification. Advances in colloid and interface science, 145, 42–82.

Lu, X. & Isacsson, U. 1997. Characterization of Styrene-Butadiene-Styrene Polymer Modified Bitumens-Comparison of Conventional Methods and Dynamic Chemical Analyses. Journal of testing and evaluation, 25, 383–390.

Lu, X. & Isacsson, U. 2000. Modification of road bitu-mens with thermoplastic polymers. Polymer Testing, 20, 77–86.

Mcnally, T. 2011. Polymer modified bitumen: properties and characterisation, Elsevier.

Navarro, F. J., Partal, P., Martínez-Boza, F. & Gallegos, C. 2004. Thermo-rheological behaviour and storage stability of ground tire rubber-modified bitumens. Fuel, 83, 2041–2049.

Pérez-Lepe, A., Martínez-Boza, F. J., Gallegos, C., González, O., Muñoz, M. E. & Santamaría, A. 2003. Influence of the processing conditions on the rheo-logical behaviour of polymer-modified bitumen. Fuel, 82, 1339–1348.

Pérez-Lepe, A., Martínez-Boza, F. J., Attané, P. & Gallegos, C. 2006. Destabilization mechanism of polyethylene-modified bitumen. Journal of Applied Polymer Science, 100, 260–267.

Polacco, G., Stastna, J., Biondi, D. & Zanzotto, L. 2006. Relation between polymer architecture and nonlinear viscoelastic behavior of modified asphalts. Current opinion in colloid & interface science, 11, 230–245.

Robinson, H. 2005. Polymers in asphalt, iSmithers Rapra Publishing.

Sengoz, B. & Isikyakar, G. 2008a. Analysis of styrene-butadiene-styrene polymer modified bitumen using fluorescent microscopy and conventional test methods. Journal of Hazardous Materials, 150, 424–432.

Sengoz, B. & Isikyakar, G. 2008b. Evaluation of the prop-erties and microstructure of SBS and EVA polymer modified bitumen. Construction and Building Mate-rials, 22, 1897–1905.

Viola, F. & Celauro, C. 2015. Effect of climate change on asphalt binder selection for road construction in Italy. Transportation Research Part D: Transport and Environment, 37, 40–47.

Yildirim, Y. 2007. Polymer modified asphalt binders. Construction and Building Materials, 21, 66–72.

Zhu, J., Birgisson, B. & Kringos, N. 2014. Polymer modi-fication of bitumen: Advances and challenges. Euro-pean Polymer Journal, 54, 18–38.

Transport Infrastructure and Systems – Dell'Acqua & Wegman (Eds)
© 2017 Taylor & Francis Group, London, ISBN 978-1-138-03009-1

Long-term monitoring of half warm mix recycled asphalt containing up to 100% RAP

José Manuel Lizárraga, Antonio Ramírez, Patricia Díaz & Miguel Martín
Sacyr Construcción, Madrid, Spain

Francisco Guisado
Repsol, Madrid, Spain

ABSTRACT: This paper presents the main results from a project-site investigation to compare the mechanical performance differences between half warm mix recycled asphalt (HWMRA) and a conventional mix (HMA) section in Spain. To this end, HWMRA was manufactured using total recycled contents up to 100% RAP at temperatures below 100ºC. This study was therefore performed in three phases. In the first phase, a preliminary laboratory testing analysis was conducted to ascertain the optimal mix design solution through the influence of different curing times in a forced-draft convection oven at 50°C. A second phase assessed the viability of manufacturing these mixes in a discontinuous asphalt mixing plant adapted with a secondary flow-parallel drying drum, while the final phase involved the construction of a road test section formulated with two emulsion contents. At the current time, little difference is seen regarding the performance of the HWMRA compared to the control HMA section.

1 INTRODUCTION

Environmental concerns about the preservation of the non-renewable resources and increasing damage to the environment from greenhouse gas (GHG) emissions have created increasing awareness and rising costs in raw materials (Tutu & Tuffour 2016). In this way, the reuse of recycled material is happening more than ever before in a sustainable technique practice, and cost-effective method able to generate significant economic savings in rehabilitation activities of up to 34% (Kandhal & Mallick 1997). These materials still retain high costs and good properties, which therefore should be reused in manufacturing and laying of new mixtures in road pavements (Botella et al. 2015) and also to favorably reduce environmental burdens by 23% (Chiu et al. 2008).

As regards cleaner production technologies, they can be classified according their manufacturing temperatures as Cold Mix Asphalt (CMA) spread usually at room temperatures, Half-Warm Mix Asphalt (HWMA) spread below 100°C, and Warm-Mix Asphalt (WMA) spread between 110 and 140°C. In this way, half-warm mixes in combination with reclaimed asphalt pavement (HWMRA) can be considered as an innovative concrete material produced at temperatures around 60–70ºC lower than those used in the production of conventional HMA and yet still have similar mechanical performance. The reduction in manufacturing and compaction temperatures entails significant advantages regarding environmental conservation and lowers energy consumption by 50%, thus saving significant costs compared to conventional HMA (Olard et al. 2008; Blankendaal et al. 2013). Moreover, the reduction of half-warm mix production temperatures implies significant emission reductions of 58% carbon dioxide and 99.9% sulfur dioxide (Rubío et al. 2013). Despite the environmental and money saving advantages that the use of these types of mixtures implies, there are some concerns and uncertainties about the mechanical behavior and life span of these mixes. For this reason, the reuse of high RAP content is still downgraded in other layers (base and binder courses) rather than replacing it within asphalt surface courses, thus laboratory testing analysis along with field investigations is required to demonstrate the performance and effectiveness of these mixes.

Recent studies have been focused on showing that reducing temperatures to half-warm production range and incorporating RAP should not compromise the performance of these asphalt mixtures (Dinis-Almeida et al. 2012; Swaroopa et al. 2015). In addition, most of the current asphalt mixing plants are not completely equipped neither for manufacturing nor to deal with total recycling production contents up to 100% RAP. In fact, the content of RAP to be added in new asphalt mixtures is still limited and downgraded to less than 20% RAP in asphalt batch plants (Zaumanis & Mallick 2014) because of the difficulties in mixing the RAP, together with the virgin aggregates. Nevertheless, there are some examples of reusing total contents up to 100% RAP in a modified asphalt mixing plant using an additional flow parallel dryer drum for RAP heating (García & Lucas, 2014). To validate this half-warm mix design procedure, a set of preliminary laboratory tests were performed to evaluate whether the use of high rates of RAP are feasible in developing recycled mixes with a high level of durability and performance (García et al. 2011). To this end, both formulations of HWMRA (2.5 and 3.0% by weight of the mix) were produced in the laboratory to accurately ascertain their volumetric and mechanical performance, and subsequently these asphalt mixes were reproduced in a modified discontinuous asphalt mixing plant capable of reaching up to 100% RAP production at temperatures below 100°C. A third phase involved the construction and deployment of HWMRA in a service road using two different emulsion contents (2.5 and 3%), respectively. These mixes were intended for use in binder courses for low traffic load categories. In short, performance is compared in the laboratory and after plant manufacture, laying and compaction in the field with the aim of increasing the confidence in using these cleaner production technologies.

2 HALF WARM MIX RECYCLED ASPHALT

2.1 Service road structure

The test section was built on a service road parallel to the A1 motorway, between km 203 and 204, located in the town of Lerma, province of Burgos in Spain. The service road is currently being subjected to T2 traffic loads with an average daily intensity of heavy vehicles equal to or greater than 200 and lower than 800. The service road consists of two lanes with one way traffic on a carriageway width of 10 m. The rehabilitation works consisted of milling a layer of 5 cm thickness, which was removed and replaced using half warm mix recycled asphalt for binder course with the following dimensions: 400 m long and a thickness of 5 cm over its entire carriageway width. The remaining 300 m were rebuilt with a conventional hot mix asphalt layer (AC16 D). In this way, the slow traffic lane was designed with 3% emulsion, whilst the fast traffic lane was formulated with 2.5% emulsion content. Finally, the asphalt concrete surface course (AC16 D) was paved in the total length of 700 m and with a thickness of 40 mm.

2.2 Material and methods

For the successful development of this research project, both HWMRA were characterized and

tested in the laboratory, specifically, two HWMRA with total recycled contents of up to 100% RAP and two emulsion binder contents (2.5% and 3.0%) and a conventional mixture with a binder 50/70 penetration grade. The characterization of the materials, as well as the mixture itself, was based on European Committee for Standardization (CEN) standards.

2.3 RAP characteristics

As was already previously mentioned in the 2.1 section, the RAP was obtained from damaged road pavement through a Wirtgen (W/1000) cold milling machine. The RAP was transported to the asphalt mixing plant to be classified into two fractions, that is, the fine fraction (0/5 mm) and coarse fraction (5/25 mm). This was done with the aim of ensuring gradation, optimal binder contents in the final mix design and also with the aim of avoiding excessive mix heterogeneity. For this purpose, both RAP fractions were homogenized, quartered, treated and characterized. This was done in terms of residual binder content per EN 12697–1, RAP binder recovery through the centrifuge extractor method per EN 12697–3, penetration test EN 1426, softening point per EN 1427, and gradation curves per EN 933–2, that is, black and white grading curves, where black curves are the RAP gradation before the extraction of the aged binder, while the white curves are the RAP gradation once the residual binder has been extracted. The average binder content value was 7.02% in the fine fraction and 3.66% in the coarse fraction by total weight of aggregates as collated in Table 1. Moreover, the recovered binder showed an average penetration value of 17 dmm and a softening point of 67.3°C. The physical binder characterization and particle grain size distribution of the RAP are collated in Tables 2–3.

The aggregate grading curves of both HWMRA fell within the limits stipulated for dense graded asphalt concrete mixtures (AC16 D) in accordance with the following proportion of RAP fractions: 40% fine fraction (0/5 mm) and 60% coarse fraction (5/25 mm), as illustrated in Figure 1.

Table 1. Determination of the binder content.

Fraction	Unit	0/5 mm	5/25 mm
binder/aggregate (b/a)	%	7.02 ± 0.1	3.66 ± 0.07

Table 2. Physical characterization of the binder.

Properties	Unit	Test method	Value
Penetration	0,1dmm	EN 1426	17
Softening point	°C	EN 1427	67.3

Table 3. Particle grain size distribution: white and black.

| Sieve size UNE (mm) | RAP | | | |
| | 0/5 mm | | 5/25 mm | |
	white	black	white	black
31.5	100	100	100	100
22.4	100	100	99.7	99.2
16	100	100	95.6	84.5
8	100	100	58.9	27.4
4	84.9	76.7	26.1	1.2
2	60.5	44.3	19.4	0.9
0.5	34.4	15.6	12.0	0.7
0.25	25.6	8.1	9.5	0.6
0.063	14.0	1.3	5.5	0.2

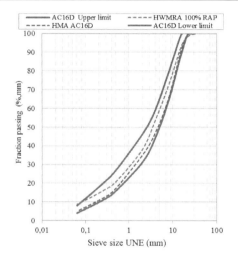

Figure 1. Aggregate grading curves of HWMRA and HMA.

Table 4. General technical specifications for cationic bitumen emulsion (CSS-1).

Property	Test method	Unit	Emulsion CSS-1
Viscosity at 25ºC	EN12846-1	S	23
pH	EN 12850	–	3.0
Water content	EN 1428	%	38.8
Sieving	EN 1429	%	0.01
Residual binder content	EN 1431	%	61.2
Penetration of residual binder at 25ºC	EN 1426	dmm	66

2.4 *Bituminous emulsion*

Both HWMRA have been manufactured with a cationic slow setting bituminous emulsion (CSS-1), which was selected as the asphalt binder to provide adequate workability during the compaction phase, coating with RAP aggregates and to guarantee a thick asphalt emulsion. In this regard, the bituminous emulsion was formulated with high residue content greater than 61% by weight of the emulsion. For emulsion production, a bitumen 50/70 dmm penetration grade was employed. The asphalt emulsion meets the current pavement recycling specifications according to EN 13808 standard. The characterization of the bitumen emulsion consisted of analyzing the viscosity, water content, sieving, residual binder content and penetration as collated in Table 4.

3 METHODOLOGY

This project investigation was divided into three main phases. The first phase consisted of a preliminary field investigation together with a laboratory testing program to accurately evaluate the optimal mix design solutions through a comprehensive accelerated ageing study (0, 24, 48 and 72 h) in a forced-draft convection oven at 50ºC. This study enables the reproduction of field pavement conditions one year after construction (Bowering 1970; Thenoux & Jamet 2002). The second phase assessed the feasibility of manufacturing these mixes in a modified discontinuous asphalt mixing plant equipped with a secondary flow parallel drum. Cores were extracted and tested after construction for determining density, Indirect Tensile Strength (ITS) at 15ºC, stiffness modulus (ITSM) at 20ºC and Indirect Tensile Strength Ratio (ITSR) through the water sensitivity test, in June 2012. The last phase involved the construction process (laying and compaction) of a test road section to validate both the short and long-term mechanical performance of these mixes under real conditions. Finally, a set of monitoring campaigns were also performed to control and assurance the quality of these mixes in the field. In April 2014, new pavement cores were taken and tested to verify the evolution of volumetric characteristics in terms of air void contents, resistance to Indirect Tensile Fatigue Test (ITFT) at 20ºC, stiffness modulus (ITSM), water sensitivity and Indirect Tensile Strength (ITS) two years after construction. Specific details are thoroughly shown in the following sections.

4 PRELIMINARY LABORATORY STUDY

4.1 *Mix design*

HWMRA samples were initially produced in the laboratory using RAP and bituminous emulsion, thus, the initial conditions began by heating the

RAP at temperatures between 90° and 100°C and bitumen emulsion at 50°C. The Superpave Gyratory Compactor (SGC) was chosen and employed for compaction assessment of the half warm mixes in accordance with EN 12697-31. The compaction of the samples was performed at a temperature of 70°C, thus the number of gyros to be applied on asphalt samples was adapted to obtain final air void contents between 2.5 and 4.0%, so it was necessary to apply, at least, 65 load cycles to obtain the required mix performance. The compaction began using the following conditions shown below:

i. Internal angle velocity: 0.82 °C
ii. Constant speed of rotation: 30 rpm
iii. Vertical contact pressure load: 600 kPa
iv. Number of load cycles: 65 gyrations
v. Mold diameter: 100 mm

The determination of the optimal mechanical performance of HWMRA was conducted according to the apparent density curves under saturated surface dry (ssd) conditions, as per EN 12697-6, maximum density per EN 12697-5, air void contents per EN 12697-8, water sensitivity test per EN 12697-12, and stiffness modulus per EN 12697-26, in comparison with different emulsion contents ranging from 2.0 to 3.5% (at 0.5% increments). The target properties of the mix design were set out in accordance with the resistance to water action, indirect tensile strength and air void contents. The retained strength ratio was fixed above 90%, air void contents should be between 2.5 and 4.0%, and Indirect Tensile Strength (ITS dry) at 15°C between 1.5 and 2.0 MPa according to the traffic load category (García & Lucas, 2014). In order to determine the retained strength ratio, asphalt samples were prepared using two-thirds (2/3) of the total compaction energy used for determining their mechanical and volumetric properties in the laboratory. The average retained strength values for all mixtures were higher than 90%, except for 0% emulsion content, as indicated in Table 5.

4.2 Influence of curing on HWMRA

After HWMRA had been characterized, a new set of laboratory specimens were prepared with the optimal emulsion content of 2.5%, to analyze the influence of curing as optimization of the mix design. In this way, HWMRA 2.5% were then assessed and tested after being subjected to an accelerated ageing process (0, 24, 48 and 72 h) through a forced-draft convection oven at a temperature of 50°C. This was done to evaluate both volumetric characteristics and mechanical behavior such as: (i) Indirect Tensile Strength (ITS) at 15°C and (ii) stiffness modulus at 20°C. Typically, these types of mixes manufactured at low temperatures and with bitumen emulsion tend to require

Table 5. Mechanical-volumetric properties of HWMRA.

| Properties | % Emulsion content | | | | |
	0%	2.0%	2.5%	3.0%	3.5%
Maximum density, (g/cm³)	2.481	2.428	2.407	2.389	2.377
Apparent density, ssd, (g/cm³)	2.268	2.327	2.328	2.338	2.339
Geometric density (g/cm³)	2.164	2.256	2.258	2.278	2.289
Air voids (%)	8.6	4.2	3.3	2.1	1.6
Modulus (MPa)	3754	3034	2891	2861	2364
ITSd, (MPa)	1.55	1.87	2.13	1.93	1.72
ITSw, (MPa)	1.08	1.82	2.08	1.89	1.69
ITSR (%)	69.5	97.3	97.6	98.1	98.3

a certain curing period to develop their ultimate mechanical properties such as: stiffness modulus and ITS (Godenzoni et al. 2016). For this reason, mechanical tests were carried out to ascertain whether it could have had a loss of water content by evaporation during the laboratory compaction phase; however, most of the water content was lost during the mixing process between 60 and 120 seconds. The development of ultimate mechanical performance of these mixes can be largely attributed either to emulsion breaking or moisture loss (Bocci et al. 2011; Serfass et al. 2004). In this way, a servo-pneumatic mechanical testing machine was used to determine both ITS and stiffness modulus. Therefore, ITS test consisted of applying a constant deformation rate of 50 ± 2 mm/min after a conditioning period of 2 hours, according to EN 12697-23. Thereafter, the laboratory specimens were compared along with specimens that still hadn't been subjected to an accelerated curing process, thus, the stiffness modulus and ITSdry values remained relatively constant within the first 24 hours. Nevertheless, ITS at 15°C showed a sharply increase of cohesiveness at the end of the curing process between 2.13 and 2.34 MPa, while the stiffness modulus at 20°C varied from 2891 to 3642 MPa, as collated in Table 6.

4.3 Wheel tracking test

To complete the preliminary laboratory study of the mechanical performance of HWMRA, the resistance to permanent deformation was performed using the wheel tracking test at temperatures of 50 and 60°C, respectively. This test was performed with duration of 10,000 load cycles, per procedure b in air, with a contact load pressure of 700 N, with a frequency of 26.5 ±1 load cycles/min, according to EN 12697-22.

Prior to the wheel tracking test, the prismatic specimens were manufactured with the following dimensions: 300 mm length · 300 mm width · 50 mm height, and thereafter compacted until reaching up to 98% of the benchmark density through the vibratory steel roller device in accordance with EN 12697-33.

The wheel tracking test was also conducted for HWMRA 2.5%, at 50°C, to evaluate the influence of the temperature on mix resistance to permanent deformation in asphalt binder courses, since these mixes are exposed to significant lower thermal gradients compared to surface courses. In this sense, the average deformation slope at 50°C, was 0.068 (mm/1000 load cycles) between 5000 and 10,000 load cycles, and at 60°C was 0.109 (mm/load cycles). Moreover, HWMRA 3.0% was also assessed and tested at 60°C, resulting in a slope deformation of 0.143 (mm/1000 load cycles) and with an average proportional rut depth of 5.3%. In relation to what has been mentioned above, HWMRA 2.5% meets the minimum strengths required for the potential T2 traffic loads. Table 7 shows the mean apparent density, deformation slope values and proportional rut depths.

5 ON-SITE ROAD TESTING

5.1 Reproducibility of the HWMRA in an modified asphalt plant

After the preliminary laboratory testing phase, it was decided to evaluate the feasibility of manufacturing these types of asphalt mixes in a modified discontinuous asphalt mixing plant used for conventional HMA production. However, it was necessary to make some adjustments to reach up to 100% RAP production at half warm temperatures. To achieve this, the batch plant was adapted with a secondary flow-parallel drying drum with a differentiated inlet ring for giving special treatment to each RAP fraction and for driving off the moisture content in the fine fraction. This batch plant features a delayed combustion chamber, wherein the RAP is heated through the hot gases before coming into contact the flame, thus, there is no direct contact either with the burner flame or at high temperatures zones. An additional hot gases recirculation circuit was installed for potential suction improvement, partial thermal energy recovery, and replacement of secondary fresh air for temperature control. The manufacturing proc-

Table 6. Curing of the HWMRA samples at 50°C.

Curing time	0 h	24 h	48 h	72 h
Apparent density (g/cm³)	2.313	2.308	2.300	2.315
Air voids, Vm, (%)	3.9	4.1	4.4	3.8
Modulus (MPa)	2891	2984	3077	3642
ITSdry (MPa)	2.13	2.11	2.17	2.34

Table 7. Wheel tracking test at 50°C and 60°C.

Testing	HWMRA 100% RAP		
	2.5% 50°C	2.5% 60°C	3.0% 60°C
Apparent density, ssd, (g/cm³)	2.302	2.328	2.330
Deformation at 5000 load/cycles, RDAIR, (mm)	0.52	2.11	2.48
Deformation at 10.000 load/cycles, RDAIR, (mm)	0.86	2.66	3.19
Wheel tracking slope, WTSAIR, (mm/10³ load cycles)	0.068	0.109	0.143
Proportional rut depth, PRDAIR, (%)	1.42	3.47	5.3

ess began by weighting the right proportion of RAP, heating it in the drum dryer at 100°C, and heating the bituminous emulsion at 50–60°C. All the components were then mixed together in the mixer, transported by the bucket elevators to the discharge hopper, and after which the resulting asphalt mix was unloaded to the delivery trucks at 100°C. Another crucial aspect to reach up to 100% half-warm mix production was the proper characterization of the RAP. To this end, the RAP was classified into two fractions, the fine (0/5 mm) and coarse fraction (5/25 mm). For this case study, the modified batch plant was fixed at a maximum theoretical output capacity of 120 t/h. Figure 2 shows a schematic overview of the pretreatment and processing of RAP, together with the installation of specific equipment. The facilities of the modified discontinuous asphalt plant are illustrated, as follows: 1. Grinding roller mill 2. Screening 3. Belt conveyor of the coarse fraction (5/25 mm) and 4. Fine fraction (0/5 mm) 5. Cold feeding bins 6. drum dryer 7. Vertical RAP elevator 8. Flow-parallel drum and 9. Vertical virgin aggregates elevator.

5.2 Spreading and compacting

For the final phase to validate this technology under real conditions, a pilot road test section was previously built near the asphalt mixing plant to define the optimal compaction procedure. The paving process of these mixes was conducted with the same machinery used for actual field pavement construction of conventional mixtures, namely, an asphalt paver, a pneumatic road roller and vibratory road roller compactor. Spreading and laying of the HWMRA mixes began using a Volvo (ABG 8820) asphalt paver in which the vibrating screed was fixed at a paving speed of 2.5 m/min and loading frequency of 1700 rpm to induce a high precompaction degree on the asphalt mixture.

A Dynapac (CC422) double-drum vibratory road roller with a maximum theoretical operating weight of 10.4 ton, together with a Dynapac (CP271) pneu-

Figure 2. Facilities of the discontinuous asphalt mixing plant using a secondary flow parallel drum (UM 260 t/h).

Figure 3. Monitoring process: a) manufacturing below 100°C and b) spreading at temperatures between 85 and 90°C.

Table 8. Volumetric and mechanical results of HWMRA.

Properties	HWMRA 100% RAP	
	2.5%	3.0%
Binder content (%s/m)	6.17	6.25
Density, ssd, (g/cm³)	2.364	2.365
Air voids, Vm, (%)	3.0	2.65
Modulus at 20°C (MPa)	5248	5057
ITS dry (MPa)	2.80	2.92
ITSR (%)	92.2	90.3

matic tyre road roller of 27 ton (mass with maximum ballast) were employed to achieve the benchmark density above 98% of the asphalt samples that have previously been prepared for determining their mechanical and volumetric properties in the laboratory. To this end, the compaction protocol adopted during the road construction consisted of applying at least four double passes with the vibratory roller compactor and four passes with the pneumatic road roller. In addition, from manufacturing until final paving were continually monitored through an infrared thermographic camera system (FLIR B360) to detect possible homogeneity defects generated during laying and compaction as illustrated in Figure 3.

6 RESULTS OF THE TEST ROAD SECTION

6.1 Asphalt samples

Both half-warm formulations (2.5 and 3.0%) were taken after in-plant manufacturing to evaluate the reproducibility of these mixes in a modified batch plant. Hence, more than six laboratory specimens were compacted using the gyratory compactor by applying the same standard conditions, which have previously been obtained in the preliminary mix design. This was performed for determining air void contents, stiffness modulus at 20°C, water sensitivity and ITS dry at 15°C. The average stiffness modulus values were above 5000 MPa, resulting in accordance with those obtained for the conventional mixtures. From the ITSR test, both HWMRA showed retained strengths above 90%, as collated in Table 8.

6.2 Cores performance

In June 2012, more than 32 pavement cores were extracted after each mix was compacted for determining thicknesses, density, air void contents, stiffness modulus at 20°C, ITS at 15°C. Moreover, in April 2014, almost two years after construction of the service road, and with mixtures being subjected to T2 traffic loads; more than twelve new pavement cores were extracted for each type of asphalt mixture to assess the long-term in-situ performance, specifically, the resistance to fatigue at 20°C. These scheduled coring campaigns were performed to both analyze the short-term performance and their evolution in terms of density, air voids, stiffness modulus, ITS, as collated in Table 9.

6.3 Particle size distribution

The control and quality assurance of these mixes was done in terms of grading, particle grain size distributions and binder contents in the final mix design. The final binder content for HWMRA (2.5%) was 6.17% and 6.25% (of the total weight of aggregates) for HWMRA 3%. The average binder content resulting from batches-in black didn't vary by a percentage greater than ± 0.3% by total weight of aggregates regarding the preliminary laboratory mix design. In this context, both half warm mixes fell within the limits stipulated for dense graded asphalt concrete mixtures (AC16 D), see in Figure 4.

Table 9. Performance of pavement cores extracted in 2012, and two years after construction in 2014.

| | HWMRA – 100% RAP | | | | |
| | Cores – 2012 | | Cores – 2014 | | HMA Cores |
Properties	2.5%	3.0%	2.5%	3.0%	2014
Height (mm)	53.6	55.2	54.2	57.0	50.9
Apparent density, (g/cm³)	2.309	2.250	2.277	2.307	2.315
Air voids, Vm, (%)	5.2	7.4	4.27	6.46	–
Modulus at 20°C (MPa)	4758	3431	3505	4990	5399
ITS in dry at 15°C (MPa)	2.64	2.29	2.54	2.43	2.85
ITSR (%)	90.2	94	79.2	88.4	94.5

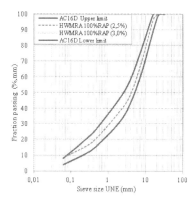

Figure 4. Aggregate grading curves of both HWMRA (2.5–3%).

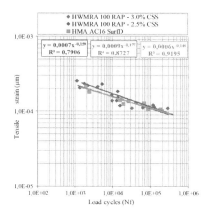

Figure 5. Fatigue laws of HWMRA 100% RAP and HMA.

6.4 Fatigue test

In April 2014, the resistance to fatigue cracking was carried out under controlled stress mode at 20ºC, using a harvesine load waveform with a loading

Table 10. Fatigue laws for HWMRA and HMA cores.

| | HWMRA 100% RAP | | HMA AC16D |
Regression constants	3.0%	2.5%	
a (μm/m)	0.0007	0.0009	0.0006
b (–)	0.159	0.177	0.149
R^2	0.7906	0.8727	0.9195
Modulus (MPa)	3284	3792	4287
ε_6 (μm)	63	69	75

frequency of 10 Hz and with a loading time of 0.1 seconds, according to EN 12697-24 (AENOR, 2003).

The failure criterion was defined using the classical approach as the number of cycles to 50%, (Nf), reduction in the initial stiffness modulus measured at 100th load cycle. The deformation slope and fatigue resistance of the HWMRA (3.0%) and HMA cores were practically identical, resulting in a very similar fatigue resistance behavior. For the stress levels tested, fatigue performance of HWMRA (2.5%) was slightly better than that of the control mixture. Nevertheless, the deformation slope of HWMRA (2.5%) was slightly steeper than that of the other two mixtures, which therefore indicates that for greater number of load cycles the fatigue performance could be somewhat worse than that of the other two mixtures, as depicted in Figure 5. The regression constants, stiffness modulus and microstrains (a, b, R^2, ε_6) are collated in Table 10.

7 CONCLUSIONS

This research project was focused on demonstrating the suitability of manufacturing half-warm mixes with total recycled contents of RAP at temperatures below 100ºC and using a cationic bituminous emulsion for use in asphalt binder courses. In this regard, this research project involved the mechanical characterization of the optimal mix design solution in the laboratory, in-plant manufacturing, laying, construction and monitoring of its performance over time. The optimal job mix formula was chosen and employed after an accelerated ageing process (0, 24, 48 and 72h) in a forced-draft convection oven. The main conclusions that can be drawn from this research project are the followings:

– Short and long term in-situ performance of the half-warm mixes with total recycled contents of RAP showed similar results to those obtained for conventional mixtures in terms of water sensitivity, stiffness modulus and indirect tensile strength.
– The aggregate grading curves of both HWMRA formulations fell within the limits stipulated for dense graded asphalt concrete mixtures (AC16 D) and the binder contents obtained in the field were quite similar than that of the job mix formula. Thus, these types of mixes were successfully

reproduced in a modified batch plant with a secondary flow-parallel drum.

- The average deformation slope (WTSair) for HWMRA 2.5% reached at a temperature of 50°C, was of 0.068 (mm/load cycles) and proportional rut depth (PRDair) of 1.42%, resulting in good results. However, the wheel tracking test conducted at 60°C can be considered hardly representative, for both base and binder courses since these mixtures are commonly exposed to lower heating temperatures compared to wearing courses.
- Cores extracted two years after construction, showed that both HWMRA (2.5% and 3%) had a similar mechanical performance to the conventional hot asphalt mixture, however, HMA cores showed higher stiffness modulus and ITSdry values than that of the other two half-warm mixes. Despite this, HWMRA (3.0%) cores displayed a significant increase of the stiffness modulus and ITS dry two years after construction of the service road, which therefore could be related to the evolution of their mechanical properties for these types of mixtures manufactured with emulsion.
- It appears the decrease of the stiffness modulus and ITS at 15°C was largely produced due to pavement cores extracted at center of the travel lane, so these cores showed lower density than those values obtained in 2012.

This project is a firm commitment for designing, manufacturing and applying new cleaner production technologies using total recycled contents of RAP at low temperatures. It is therefore concluded that these findings encourage continuing endeavors in developing these sustainable asphalt mixtures for road pavements. Current research is being carried out to further evaluate the long-term in-situ performance of this technology.

ACKNOWLEDGEMENTS

I would like to express my fully acknowledgement to European Union's Seventh Framework Programme for Research, Development and Demonstration under grant agreement number 607524, Marie Curie Initial Training Network actions (ITN) FP7-PEO-PLE-2013-ITN for its financial support. Moreover, we would also like to express our acknowledgement to Eduardo Ortega of Sacyr's R&D laboratory.

REFERENCES

AENOR. Asociación Española de Normalización y Certificación. Normas UNE EN 12697. Métodos de ensayo para mezclas bituminosas en caliente. Madrid.

Blankendaal T, Schuur P, & Voordijk H. (2013). Reducing the environmental impact of concrete and asphalt: a scenario approach. Journal of Cleaner Production, 66 (2014), 27–36. doi:10.1016/j.jclepro.2013.10.012

Bocci M, Grilli A, Cardone F, & Graziani A. (2011). A study on the mechanical behaviour of cement–bitumen treated materials. Construction and Building Materials, 25(2), 773–778. doi:10.1016/j.conbuildmat.2010.07.007

Botella R, Miró R, Díaz P, Ramírez A, Guisado F, & Moreno E. (2015). Sustainable Urban Surface Asphalt Layers. Springer, 11, 619–627. doi:10.1007/978–94–017–7342–3

Bowering RH. (1970). Properties and behavior of foamed bitumen mixtures for road buildings. Canberra, Australia. In Proceedings of the 5th Australian Road Research Board Conference (pp. 38–57).

Chiu CT, Hsu TH, & Yang WF. (2008). Life cycle assessment on using recycled materials for rehabilitating asphalt pavements. Resources, Conservation and Recycling 545–556. doi:10.1016/j.resconrec.2007.07.001

Dinis-Almeida M., Castro-Gomes J., & Antunes, M.D.L. (2012). Mix design considerations for warm mix recycled asphalt with bitumen emulsion. Construction and Building Materials, 28(1), 687–693. doi:10.1016/j.conbuildmat.2011.10.053

García JL, Guisado F., Paez A., & Moreno, E. (2011). Reciclado total de mezclas bituminosas a bajas temperaturas. Una propuesta para su diseño,caracterización y producción. ASEFMA, 1–12.

García JL, & Lucas, F. (2014). Mezclas templadas con reutilización del RAP con tasa alta y tasa total. Aplicación, Experiencias reales y resultados. Asfalto y Pavimentación, IV(14), 51–65.

Godenzoni C, Cardone F, Graziani A, & Bocci M. (2016). The Effect of Curing on the Mechanical Behavior of Cement-Bitumen Treated Materials. In 8th RILEM International Symposium on Testing and Characterization of Sustainable and Innovative Materials (pp. 879–890). doi:10.1007/978–94–017–7342–3

Kandhal PS, & Mallick RB. (1997). Pavement Recycling Guidelines for State and Local Governments Participant's Reference Book.

Olard F, Le Noan C, Beduneau E, & Romier A. (2008). Low energy asphalts for sustainable road construction. In Proceedings of the 4th Eurasphalt and Eurobitume congress (p. 12). Copenhaguen, Denmark: European Asphalt Pavement Association (EAPA).

Rubío MC, Moreno F, Martínez-Echevarría J, Martínez G, & Vázquez J.M. (2013). Comparative analysis of emissions from the manufacture and use of hot and half-warm mix asphalt. Journal of Cleaner Production, 41(2013), 1–6. doi:10.1016/j.jclepro.2012.09.036

Serfass J., Poirier J., Henrat J., & Carbonneau, X. (2004). Influence of curing on cold mix mechanical performance. Materials and Structures, 37 (April 2003), 365–368.

Swaroopa S, Sravani A, & Jain PK. (2015). Comparison of mechanistic characteristics of cold, mild warm and half warm mixes for bituminous road construction. Indian Journal of Engineering and Materials Sciences, 22, 85–92.

Thenoux G, & Jamet A. (2002). Foamed Asphalt Technology. Revista Ingeniería de Construcción, 17(2), 84–92.

Tutu KA, & Tuffour YA. (2016). Warm-Mix Asphalt and pavement sustainability: a review. Journal of Civil Engineering, 6, 84–93. doi:http://dx.doi.org/10.4236/ojce.2016.62008

Zaumanis M, & Mallick RB. (2014). Review of very high-content reclaimed asphalt use in plant-produced pavements: state of the art. International Journal of Pavement Engineering, 1–17. doi:10.1080/10298436.2014.893331

Transport Infrastructure and Systems – Dell'Acqua & Wegman (Eds)
© 2017 Taylor & Francis Group, London, ISBN 978-1-138-03009-1

Fiscal commitments to encourage PPP projects in transport infrastructure

N. Nose
World Bank, Washington, DC, USA

C. Queiroz
Claret Consulting LLC, Washington, DC, USA

M. Nose
International Monetary Fund, Washington, DC, USA

ABSTRACT: This paper aims at analyzing how different types of fiscal commitments to Public Private Partnerships (PPP) can be effectively utilized in the preparation and implementation of PPP projects in transport infrastructure. The instruments include risk mitigation instruments offered by international financial institutions and public financial support for infrastructure projects. In PPP projects in roads, it is important to identify risks and allocate responsibility for the identified risks between the public and private sectors. In particular, allocating revenue related risks is critical because it involves uncertainty for future demands. However, not all public sector agencies are capable to take those risks, especially in developing economies, due to their insufficient fiscal condition and limited track record of similar type of projects. In order to attract private sector investment to PPP road projects in those countries, the fiscal commitments discussed in the paper could be utilized to mitigate the risks.

1 INTRODUCTION

Managing risks is crucial for the preparation and implementation of PPP program in roads and other modes of transport. The main risks of PPP projects are those affecting gross revenue of the project, in addition to those leading to significant cost increases, such as changes in design during construction. Revenue related risks usually reflect uncertainty in both the predictability of future demand (e.g., traffic volumes) and the willingness of users to pay tariffs.

In order to manage risks associated with PPP programs, parties concerned should identify risks for each stage of the project and allocate responsibility for the identified risks between the public and private sector. The best approach is not to try to transfer all risks to the private sector, as this would result in less interest by the private sector or a much higher cost to the public sector. As a result, risk allocation is a very important component in the assessment of any PPP project. The general rule is that risks need to be allocated to the party that is best capable to manage them. This means that the government would need to take some risks that it can manage better or because the costs of the private sector assuming such risks would be too high.

However, not all public sector agencies are able to accept such risks. Especially in developing economies, the public partner may not be sufficiently creditworthy or does not have a proven track record in the eyes of private financiers to be able to attract private investments without external support. In those cases, several types of possible financial instruments can be used to facilitate the mobilization of private capital to finance transport PPP projects in a country where financing requirements substantially exceed budgetary or internal resources.

Several International Financial Institutions (IFIs) and other international or bilateral organizations can support those countries with financing instruments consisting of risk mitigation instruments and public sector financial support including, for example, the World Bank Group (International Development Association, International Bank for Reconstruction and Development, International Finance Corporation, Multilateral Investment Guarantee Agency) and the Asian Development Bank. Such organizations can provide resources in several areas, including private infrastructure financing products, such as viability gap funds, financial intermediary loans, equity, and partial risk guarantees (World Bank, 2014). Some of the IFIs can also finance Special Purpose

Vehicles (SPV) on a risk sharing basis. In the particular case of the World Bank Group, this can be done through its International Finance Corporation (Web-1, 2016). Another World Bank affiliate, the Multilateral Investment Guarantee Agency (Web-2, 2016) (Web-3, 2014) offers political risk insurance. Several types of possible financial support are discussed in the next paragraphs.

2 FINANCIAL INSTRUMENTS, RISK SHARING AND GUARANTEES

In the particular case of roads, a study (Queiroz and Kerali, 2010) suggests that toll road traffic forecasts are characterized by large errors and considerable optimism bias. As a result, financial specialists need to ensure that transaction structuring remains flexible and retains liquidity such that material departures from traffic expectations can be accommodated.

Risks should be identified for each stage of a project, and responsibility should be allocated for the identified risks.

As PPP are legally long-term contractual agreements, responsibilities should be clearly defined as they will determine the costs that the public and private partners will ultimately pay. For example construction risk is usually transferred to the private sector, which means that it will be responsible (and will not be able to claim additional compensation) for delays and cost-overruns in completing the works. The best approach is not to try to transfer all risks to the private sector, as this would result in less interest (or no interest, i.e., no bidders) by the private sector or a much higher cost to the public sector. As a result, risk allocation is a very important component in the assessment of any PPP project (Robert, 2009).

2.1 Risk matrix

A good practice in preparing risk matrices is to adopt the following structure for each potential risk of the project:

a. Description of the risk
b. Proposed allocation of the risk (usually two columns – 'Grantor' and Concessionaire'- and one of them gets checked for a particular risk)
c. Comments

The general rule is that risks need to be allocated to the party that is best capable to manage them. This means that the government would need to take some risks that it can manage better or because the costs of the private sector assuming such risks would be too high. The private sector will price the risk of the project based on how individual risks are

allocated, their likelihood of occurrence, and impact. If the private sector is transferred a risk that it cannot control (for example inflation being higher than forecast) it will either take a very conservative scenario (such as assuming a very high inflation rate) or simply not accept it (and therefore will not make any proposal, thus reducing competition). The risk allocation exercise requires a very good understanding of the market and project finance principles in order to allocate the risk in a way that balances the public and private sector concerns and interests (Queiroz and Lopez-Martinez, 2012).

The preparation of a risk matrix would help the government to decide which risk should be allocated to which party. The risk matrix should be prepared with a legal perspective in mind because it should provide the basis for drafting the PPP legal agreement/ concession agreement.

The risk allocation matrix should be updated and refined as project preparation evolves. It is usually prepared with the support of transaction experts and in consultation with potential bidders. Ultimately the risk allocation will determine if a PPP project is financeable (i.e., lenders will not finance it if they believe the risk allocation is not appropriate), so the public sector should remain flexible when designing such a matrix (Queiroz and Lopez-Martinez, 2012).

The Global Infrastructure Hub (GIH) provides good examples of risk matrices, showing the allocation of risks between the public and private sectors in typical PPP transactions, along with mitigating measures, within the transport, energy and water and sanitation sectors (GIH, 2016).

In road projects in developing economies, there may be a perception of a risky environment for private investment, and the use of risk mitigation instruments can help reduce the risk perception and facilitate private sector investment.

2.2 Risk mitigation

Several instruments can be used to facilitate the mobilization of private capital to finance transport infrastructure PPP projects in developing economies, where financing requirements substantially exceed budgetary or internal resources. Risk mitigation instruments are financial instruments that transfer certain defined risks from project financiers (lenders and equity investors) to creditworthy third parties (guarantors and insurers) that have a better capacity to accept such risks. These instruments are especially useful when the public partner is not sufficiently creditworthy or do not have a proven track record in the eyes of private financiers to be able to attract private investments without support. The advantages of such instruments are multifaceted (Matsukawa and Habeck, 2007):

a. The public sector is able to mobilize domestic and international private capital for infrastructure implementation, supplementing limited public resources.
b. Private sector lenders and investors will finance commercially viable projects when risk mitigation instruments cover those risks that they perceive as excessive or beyond their control.
c. Governments can share the risk of infrastructure development using limited fiscal resources more efficiently by attracting private investors rather than having to finance the projects themselves, assuming the entire development, construction, and operating risk.

Commonly used risk mitigation instruments include guarantees and insurance products. Guarantees typically refer to financial guarantees of debt that cover the timely payment of debt service. Procedures to call on these guarantees in the event of a debt service default are usually relatively straightforward. In contrast, insurance typically requires a specified period during which claims filed by the insured are to be evaluated, before payment is made by the insurer. Examples of risk mitigation instruments available that could apply to transport infrastructure PPP projects include (Matsukawa and Habeck, 2007):

a. Partial Risk Guarantees, currently known as project-based guarantees; and
b. Political Risk Insurance (PRI).

Project-based guarantees are designed to provide risk mitigation with respect to key risks that are essential for the viability of specific investment projects. Project-based guarantees generally cover government-related risks, which are risks within the control of the government and public entities. There are two main types of project-based guarantees: loan guarantees and payment guarantees. Loan guarantees cover defaults of debt service payments and could be granted for public sector or private sector projects. Loan guarantees could protect commercial lenders financing a private sector project from debt services defaults caused by government actions or inactions. This type of guarantee was previously known as a partial risk guarantee (PRG) (World Bank, 2016). They typically cover the full amount of debt. Payment is made only if the debt default is caused by risks specified under the guarantee. Such risks are political in nature and are defined on a case-by-case basis. Project-based guarantees are offered by multilateral development banks (World Bank, 2016) and some bilateral agencies. Figure 1 provides an illustration of how a project-based guarantee can apply to a highway concession contract. *Mutatis mutandis*, such

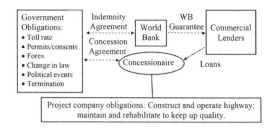

Figure 1. Structure of a highway concession contract and World Bank Project-based guarantee (Queiroz, 2006).

illustration also applies to other modes of transport infrastructure.

Political risk insurance typically requires a specified period during which claims filed by the insured are to be evaluated, before payment by the insurer is made. PRIs usually insure foreign direct investments against losses related to single or a combination of: Currency inconvertibility and transfer restrictions; Expropriation; War, civil disturbance, terrorism, and sabotage; Breach of contract; and Non-honoring of sovereign financial obligations (Web-5, 2014). Unlike project-based guarantees, PRIs do not require a counter-guarantee from a host government. Premium rates are decided on a per-project basis and vary by country, sector, transaction and the type of risk insured. PRIs are provided by private PRI providers, national export credit agencies (ECAs) and multilateral agencies such as MIGA. Figure 2 provides an illustration of how a PRI can apply to a highway concession contract; such illustration also applies to other transport infrastructure.

A recent case that utilized its guarantee instrument for a road project is MIGA's support to Vietnam's BT20 National Highway 20, providing non-honoring of sovereign financial obligations cover of $500 million to the project's lenders to cover a 15-year loan and associated costs to finance the road's construction works (Web-7, 2014). The project consists of the rehabilitation and upgrading of the National Highway 20 that extends from the Dong Nai to Lam Dong provinces of Vietnam, under a build-transfer contract between the BT20 Consortium and the Ministry of Transportation. BT 20, Cuu Long Joint Stock Company, is a consortium comprised of Dong Mekong Construction Production Trade Service Co., PetroVietnam Construction Joint Stock Company, Cuu Long Corporation for Investment, Development and Project Management of Infrastructure, and Building Materials Corporation No. 1 Co (Web-8, 2014). Goldman Sachs (loan arranger and lending syndicate member) has provided loans to BT 20 to support the construction financing of the phase

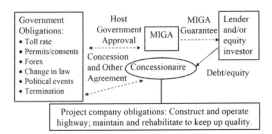

Figure 2. Structure of a highway concession contract and MIGA political risk insurance (Web-6, 2014).

Figure 3. Structure of MIGA political risk insurance to BT20 National Highway 20 Project.

1 of the BT20 National Highway 20 Project with the unconditional guarantee provided by Ministry of Finance. Figure 3 provides an illustration of MIGA's support to the BT20 National highway 20.

World Bank Group guarantee instruments (Web-9, 2016) have been relatively underused, partly due to policy constraints. Recent reforms, however, have made them more flexible and accessible helping the World Bank Group (IBRD/IDA, IFC and MIGA) to work together more effectively, tackling clients' needs and catalyzing private sector participation (World Bank, 2013). Accordingly, this should contribute to mobilizing support to potential PPP transport projects.

2.3 Public sector financial support

Road and other transport PPP projects in developing economies should seek to minimize the need for public financial support in order to maximize the benefits of a concession relative to its costs. However, public financial support may be appropriate if it helps ensure the mobilization of required amounts of private capital.

Overall, the type and level of government financial contribution to a PPP project should be limited to what is required to attract private financing and promote a successful project.

Following the recent global financial crisis, the higher cost and lower availability of loans and an increase in risk aversion, governments have had in several cases to resort to increased support to PPPs to enable them to go forward. This has taken the form of subsidies (or grants to the concessionaire) as well as increased risk-bearing.

A PPP subsidy is a direct government contribution or grant to pay for a portion of costs that is not repaid by the concessionaire. Governments can provide subsidies by making upfront cash contributions to pay for capital costs (i.e., construction subsidies). Alternatively, once a project has been built, governments can make regular payments to the private company based on the availability and quality of the service it is contracted to provide. A third option is for governments to pay a fee per user, such as the number of vehicles using a toll road (World Bank, 2012). The latter is known as shadow toll, a concept created for Design, Build, Finance and Operate—DBFO (Bain and Wilkins, 2003) roads in the United Kingdom, and has also been used in other countries such as Finland and Portugal (Web-10, 2014).

Subsidies to PPPs help make sure projects that will produce a net economic or social gain can be commercially financed. There are two broad reasons why an economically justified project may not be financially viable. First, infrastructure projects can create public benefits that are not reflected in the price consumers are willing to pay for the service, such as a toll road that creates third-party benefits by increasing mobility and lowering vehicle emissions. Second, user fees can be deliberately set at a low level to keep them at a socially acceptable level (Web-10, 2014).

When estimating the minimum required levels of subsidies and/or PPP guarantees that will make a PPP project attractive to private investors, countries should note that the provisions of subsidies and guarantees without proper assessment of the fiscal cost may lead to budget shortfalls in the future because of direct or contingent liabilities. In this regard, governments should establish a high quality Public Financial Management (PFM) system to select projects with high value-for-money and to ensure the long-term commitments will not overburden the public budget in future (Engel, Fischer, and Galetovic, 2014). Government officials in developing countries need a relatively simple, user friendly tool that assesses potential fiscal costs and risks in PPP projects quickly so several options can be tested in a short period at low cost. The financial models included in the Toolkit for PPP in Roads and Highways and/or PPP Fiscal Risk Assessment Model (P-FRAM) are useful tools that can be used for such purpose, with relatively small training required (Web-11, 2016; Web-13, 2016).

2.4 Risk sharing and forms of PPP

Different forms of transport infrastructure concession contracts, such as availability fee, shadow tolls, Build-Operate-Transfer (BOT), and Build-Own-Operate (BOO), provide increased risk transfers to the private sector. Under shadow tolls, BOT and BOO the demand risks are borne by the private partner, but under shadow tolls the concessionaire does not assume the risks associated with toll collection (e.g., sensitivity of demand to the level of toll rates).

The cost of public sector risk bearing is an important element to consider when requesting and evaluating PPP proposals. One of the key premises that should be considered using PPPs is the optimum allocation of project risks to the partner that is best able to manage them cost effectively. Consequently, to truly assess the impact of private sector involvement, governments need to adopt an approach to quantify the short-term impacts of the project on the public budget and the long-term potential cost of the risks the government chooses to retain (Aldrete et al, 2010).

Even countries more advanced in the implementation of PPP projects have resorted to financial instruments to attract private investors. This is the case, for example, of France and Spain that jointly launched the Perpignan—Figueras Rail Concession, which provides a link between French and Spanish rail systems, thus reducing travel times and transport bottleneck (Web-12, 2016). The project, which combines high-speed trains (travelling at up to 350 km) as well as freight convoys (moving at 120 km/h), received a state subsidy covering 57% of the construction cost, as well as bank guarantees (European Commission, 2004).

3 SUMMARY AND CONCLUSIONS

There are financial instruments that can help make transport infrastructure PPP projects more attractive to the private sector, potentially resulting in a more competitive selection of the private partner.

Even if infrastructure projects in developing economies are economically and socially justified, often they are not, *per se*, able to attract private investors. However, they may become feasible PPP projects if appropriate support is given to the project, particularly through financial instruments such as guarantees and subsidies. Potential PPP projects in developing economies may want to take advantage of one or more of the financial instruments discussed in the paper, such as capital subsidies or through proper involvement of multilateral partners.

Even more advanced economies, such as France and Spain, have granted subsidies to projects, turning them into successful PPP projects, such as the Perpignan-Figueras Rail Concession.

Developing countries tend to prefer PPPs to conventional public procurement in financing infrastructure to relieve their budget pressure. This highlights the importance of fiscal prudence in using PPPs for infrastructure investments. Countries assuming long-term liabilities (e.g., availability payments) and issuing PPP guarantees (e.g., project-based guarantees) should improve their PFM system to ensure the appropriate project appraisal and selection, for example by utilizing available user-friendly financial models.

REFERENCES

Aldrete R., Bujanda A. & Valdez-Ceniceros G.A. 2010. Valuing Public Sector Risk Exposure in Transportation Public-Private Partnerships, at: http://utcm.tamu.edu/publications/final_reports/Aldrete_08–41–01.pdf

Bain R. 2009. Error and Optimism Bias in Toll Road Traffic Forecasts. Transportation, 469–482.

Bain R. and Wilkins M. 2003. The Evolution of DBFO Payment Mechanisms: One More for the Road at: http://www.robbain.com/Evolution%20of%20DBFO.pdf

Engel, E., Fischer, R. & Galetovic, A. 2014. The Economics of Public-Private Partnerships: A Basic Guide, Cambridge University Press, NY. USA.

European Commission. 2004. European Commission Resource Book on PPP Case Studies Perpignan-Figueras Rail Concession, France & Spain, at: http://ec.europa.eu/regional_policy/sources/docgener/guides/pppresourcebook.pdf

Global Infrastructure Hub-GIH. 2016. Allocating Risks in Public-Private Partnership Contracts. Sydney, Australia. http://www.globalinfrastructurehub.org/content/uploads/2016/07/160610-GIHub-Allocating-Risks-in-PPP-Contracts-2016-Edition.pdf

Matsukawa T. & Habeck O. 2007. Review of Risk Mitigation Instruments for Infrastructure Financing and Recent Trends and Developments. Trends and Policy Options No. 4, Public-Private Infrastructure Advisory Facility (PPIAF) and the World Bank, at http://www.ppiaf.org/ppiaf/publications/Trends%20and%20Policy%20Options

Queiroz, C. 2006. The Potential of Private Financing to Enhance Road Infrastructure in the Baltic States. In 26th International Baltic Road Conference, Kuressaare (Saaremaa), Estonia, 28–30 August 2006. http://www.bjrbe.vgtu.lt/news/news002.php

Queiroz, C. & H. Kerali. 2010. A Review of Institutional Arrangements for Road Asset Management: Lessons for the Developing World. World Bank Transport Paper No. TP-32: at http://go.worldbank.org/6HDCYBMRT0

Queiroz, C. & A. Lopez-Martinez. 2012. Legal Frameworks for Successful Public Private Partnerships. in "The Routledge Companion to Public-Private Partnership", at: http://www.routledge.com/books/details/9780415781992/

World Bank. 2012. Best Practices in Public-Private Partnerships Financing in Latin America: the role of subsidy mechanisms, at: http://api.ning.com/files/JgnTHgt50ouqoEiQqvTYC-edJiYj9tuAn-qH*A6j39vd-fQ72wjVqJmhEM2-CjumJlv9ZKv9b7 s7rU0Obc2XfiYOsVHsnFZAA4/BestPracticesroleofsubsidiesmechanisms.pdf

World Bank. 2013. Enhancing the World Bank's Operational Policy Framework on Guarantees. http://consultations.worldbank.org/consultation/modernizing-operational-policy-guarantees-public-consultations

World Bank. 2014. World Bank Group Guarantee Products, at: http://treasury.worldbank.org/bdm/pdf/Brochures/WBG_Guarantees_Matrix.pdf

World Bank. 2016. World Bank Group Guarantee Products, at: https://ppp.worldbank.org/public-private-partnership/sites/ppp.worldbank.org/files/documents/wbg-guarantees_final_0.pdf

Web sites: Web-1: http://www1.ifc.org/, consulted September 2016.

Web-2: http://go.worldbank.org/3FAJ6 LS510, consulted September 2016.

Web-3: http://www.miga.org/whoweare/index.cfm?stid= 1792, consulted 5 August 2014.

Web-5: http://www.miga.org/investmentguarantees/index.cfm, consulted 10 August 2014.

Web-6: http://siteresources.worldbank.org/EXTGUAR-ANTE/Resources/structure.pdf, consulted 15 August 2014.

Web-7: http://www.miga.org/news/index.cfm?aid = 3660, consulted 15 August 2014.

Web-8: http://www.miga.org/documents/BT20_National_Highway_ESRS_April_24_2013.pdf, consulted 15 August 2014.

Web-9: http://web.worldbank.org/external/default/main?menuPK = 64143540&pagePK = 64143532&piPK = 6 4143559&theSitePK = 3985219, consulted 15 August 2016

Web-10: http://pppnetwork.ning.com/page/financing-public-private-partnerships-best-practices-in-latin-ame, consulted 15 March 2014

Web-11: http://www.ppiaf.org/sites/ppiaf.org/files/documents/toolkits/highwaytoolkit/index.html, consulted April 2016

Web-12 http://www.eiffage.com/en/le-groupe-eiffage/a65_motorway.html, consulted 2016

Web-13: http://www.imf.org/external/np/fad/publicinvestment/pdf/PFRAMmanual.pdf, consulted April 2016.

Transport Infrastructure and Systems – Dell'Acqua & Wegman (Eds)
© 2017 Taylor & Francis Group, London, ISBN 978-1-138-03009-1

Mix design and volumetric analysis of hot recycled bituminous mixtures using a bio-additive

E. Bocci
Università eCampus, Novedrate (CO), Italy

F. Cardone
Università Politecnica delle Marche, Ancona, Italy

A. Grilli
Università degli Studi di San Marino, San Marino, San Marino Republic

ABSTRACT: Nowadays, the use of Reclaimed Asphalt (RA) as a constituent material for hot Asphalt Concretes (AC), is gaining increasing interest because of the important technical and environmental advantages. In addition, the reuse of the old bitumen from the RA, allows the required amount of new bitumen to be reduced with clear economic returns. However, only a small percentage of RA can be recycled in new ACs (typically less than 30%), as the excess of oxidized aged bitumen may lead to a brittle behavior of the pavement layer. In order to avoid this issue, when high amount of RA are used, specific additives are recommended in order to restore the bitumen characteristics leading to a mixture with the expected mechanical properties. The present paper deals with the use of a bio-based Additive (A) for the production of AC for binder layer containing high quantities of RA. In particular, the study focuses on the mix design phase, highlighting the effects of RA and A on the volumetric properties of the mixtures. The experimental program included four ACs with no RA, 40% of RA, 40% of RA and additive A, 50% RA and additive A respectively. Results showed that the optimum bitumen content decreased when decreasing RA content or when adding A. Moreover, the use of A allowed obtaining the desired volumetric properties even with a significantly lower amount of virgin bitumen to be added to the mixture without penalizing its mechanical response.

1 INTRODUCTION

The ordinary road pavement maintenance procedure, consisting of milling of distressed asphalt layers before overlaying, determines the production of large amounts of Reclaimed Asphalt (RA) (Bocci et al. 2010). As this material includes precious components as bitumen and mineral aggregate, it should not be relegated to a waste product but should be advantageously exploited through hot and cold recycling techniques (Karlsson & Isacsson 2006, Al-Qadi et al. 2007). Particularly, hot recycling of RA allows important economic and environmental benefits to be achieved, through the reduction of aggregate and bitumen supplying and thus the preservation of natural resources (Grilli et al. 2015, Olard et al. 2008).

However, it is typically supposed worldwide that Asphalt Concretes (AC) with high RA content do not provide good performances because of the aged binder that RA contains. The ageing mechanism is complex and it is accompanied by negative effects as oxidation and chemical modifi-

cation (Petersen & Glaser 2011, Read & Whiteoak 2003). Consequently the binder becomes harder and more brittle (Stimilli et al. 2014), thus more prone to cracking, with significant disadvantages on the mix behavior (Karlsson & Isacsson 2006, Molenaar et al. 2010).

In addition, the maximum amount of RA to be reused in AC production depends not only on the ability to correct the physicochemical characteristics of the aged bitumen but also on the type of processing of the mix plant (Frigio et al. 2014). For these reasons, the amount of RA which is typically recycled in hot mixtures does not exceed 30%.

To enable an increase in RA content, the use of specific additives is strongly recommended. The specific additives are intended to mobilize and restore the mechanical properties of the aged bitumen, in order to achieve adequate blending with the fresh virgin binder and suitable workability and mechanical performance of the mix (Chen et al. 2007, Grilli et al. 2013, Shen et al. 2007, Zaumanis et al. 2014).

The present research project deals with the use of a specific bio-based Additive (A) to produce AC using a high amount of RA, without penalizing the mix performance and complying with the Italian specifications. In particular, the study focuses on the mix design phase, in order to highlight the effects of RA and A on the volumetric properties of the mixtures.

2 OBJECTIVE AND EXPERIMENTAL PROGRAM

This paper deals with hot recycling of RA using a bio-based Additive (A). This study provides the volumetric mix design of four ACs containing no RA, 40% RA, 40% RA plus A, 50% of RA plus A, respectively. The main objective was to determine the effect of RA content and A on the volumetric properties of the AC. In addition, the effectiveness of A to produce AC with high percentages of RA complying Italian specifications (Autostrade per l'Italia 2008, ANAS 2010) was evaluated.

To this aim, the experimental program included the mix design and testing of four mixtures including:

- no RA (00RA), as reference AC;
- 40% of RA (considering its aggregate weight) by granular mixture weight (40RA);
- 40% of RA and 6% of A by aged bitumen weight (40RA6A);
- 50% of RA and 6% of A by aged bitumen weight (50RA6A).

Table 1 summarizes the compositions of the mixtures.

Mixtures were compacted by means of a Gyratory Compactor (GC) according to EN 12697-31. The mix design procedure consisted in the volumetric analysis of the mixtures, carried out through the following steps:

- selection of appropriate constituent materials and aggregate gradation;
- production of four mixtures using different bitumen contents (4.3, 4.8, 5.3 and 5.8% by mix-

ture weight), according to the Italian technical specifications;
- compaction of 3 specimens for each mixture by means of a GC at 180 gyrations, in order to evaluate the overall compaction curve (EN 12697-10);
- calculation of the bitumen content (optimum bitumen content) that allows 4% of air voids content (V_m) to be obtained at 100 gyrations and checking of V_m at 10 and 180 gyrations to respect the Italian specification.

Finally, volumetric and mechanical characteristics were evaluated in detail on the design mixtures following the present protocol:

- production of the mixture using the calculated (optimum) bitumen content;
- compaction of 3 specimens for each mixture by means of a GC at 180 gyrations for the volumetric analysis and the study of the compaction curve (EN 12697-10);
- compaction of 3 specimens by means of a GC at 100 gyrations and determination of Indirect Tensile Strength ITS (EN 12697-23) and Indirect Tensile Coefficient ITC.

ITC is an empirical mechanical parameter which can be related to the stiffness properties of the mixture, and be calculated according to Eq. 1:

$$ITC = \frac{\pi \cdot D \cdot ITS}{2 \cdot \delta_{v,\max}} \tag{1}$$

where D is the specimen diameter and $\delta_{v,\max}$ is the vertical deformation corresponding to the load peak during ITS test.

3 MATERIALS

3.1 Aggregate

Virgin limestone aggregates, limestone filler and two fractions of RA were sampled in a selected mix plant and characterized in terms of gradation (Fig. 1). In particular, the RA after bitumen extraction (gradation of the aggregates in the RA) was sieved adopting the wet sieving method. The bitumen content in the RA, measured by mix weight, was 4.6% in the fine fraction and 3.2% in the coarse one. In order to optimize plant RA management it was used approximately 2/3 of fine RA and 1/3 of coarse RA for the mix design.

The bitumen recovered from RA according to the procedure defined by EN 12697-1 showed a penetration of 18 dmm (EN 1426) and a softening point of 75°C (EN 1427).

Table 1. AC composition.

Mixture	RA content [% by mix]	A content [% by aged bit.]
00RA	0	0
40RA	40	0
40RA6A	40	6
50RA6A	50	6

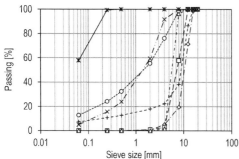

Figure 1. Grading of aggregates and RA (after bitumen extraction).

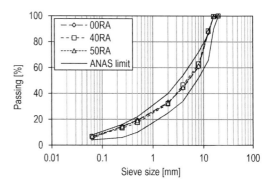

Figure 2. Mix gradations.

Table 2. Aggregate proportioning.

Fraction	% by weight		
	00RA	40RA	50RA
8/16 G_c 90–20	45	34	34
4/10 G_c 85–15	7	3	2
4/8 G_c 90–10	5	0	0
0/4 G_f 90	35	21	12
16 RA 0/12	0	13	16
12 RA 0/8	0	27	34
Filler	8	2	2

The different fractions of RA and virgin aggregate were proportioned in order to obtain a fixed gradation curve complying with the aggregate band for a binder course (ANAS 2010).

Table 2 shows the percentage of each fraction in the different ACs and Figure 2 plots the gradation curves.

3.2 Virgin bitumen

The virgin binder used in the present study was a 50/70 bitumen (EN 12591). The characteristics of the bitumen are shown in Table 3.

3.3 Bio-Additive (A)

The recycling Additive (A) used in the present study is a tall oil derived from the processing of pine wood in paper industry. It is a miscible, long-lasting and sustainable bio-based material able to mobilize oxidized RA binder and restore aged bitumen physical properties. The characteristics of A are shown in Table 4.

In order to identify the optimum dosage of A, the physical properties of Virgin Bitumen (VB),

Table 3. Virgin bitumen properties.

Property	Unit	Norm	Value
Penetration at 25°C	dmm	EN 1426	53
Softening point	°C	EN 1427	48
Loss of mass after RTFOT	%	EN 12607-1	0.08
Retained pen. after RTFOT	%	EN 1426	55
Increase of softening point	°C	EN 1427	16

Table 4. Bio-additive properties.

Property	Unit	Value
Flash point	°C	>295
Dynamic viscosity at 25°C	mPa·s	71
Kinematic viscosity at 0°C	mm²/s	355
Kinematic viscosity at 40°C	mm²/s	43
Kinematic viscosity at 100°C	mm²/s	9
Density	g/cm³	0.93

Table 5. Physical properties of the binders.

Property	VB	RB	RB5A
Penetration at 25°C [dmm]	53	18	32
Softening point [°C]	48	75	65
Dynamic viscosity at 100°C [Pa·s]	4.9	19.4	13.0
Dynamic viscosity at 135°C [Pa·s]	0.54	1.35	0.95
Dynamic viscosity at 160°C [Pa·s]	0.18	0.30	0.27

bitumen recovered from RA (RB) and bitumen recovered from RA plus 5% of A (RB5A) by weight have been determined (Table 5). Hence, on the basis on the results obtained with 5% of A and previous experiences using empirical tests, the optimum content of A was fixed as 6% by aged bitumen weight to achieve even better performance.

3.4 *Mixtures*

The mix production process simulated the operative procedure at the plant. In particular, for the 00RA mixtures, both virgin aggregate and bitumen were heated at 150°C. For the hot recycled ACs, 15% of cold RA was added to the mix, whereas the virgin aggregate and the remaining part of RA (25% for mixtures 40RA and 40RA6A, 35% for the mixtures 50RA6A) were heated at 170°C. The bio-additive A was directly sprayed on the RA particles (before the heating in the case of hot-added RA).

4 MIX DESIGN RESULTS

4.1 *Determination of the optimum bitumen content*

Figures 3–6 show the air voids content as a function of bitumen dosage in order to calculate the optimum bitumen content for the different mixtures. Bitumen introduced by RA was considered

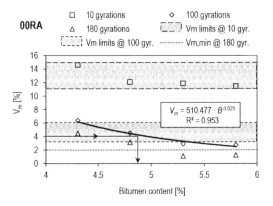

Figure 3. Air voids content vs bitumen content, mix 00RA.

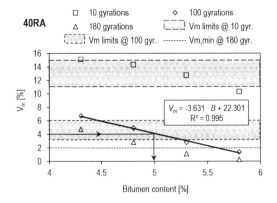

Figure 4. Air voids content vs bitumen content, mix 40RA.

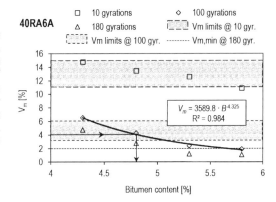

Figure 5. Air voids content vs bitumen content, mix 40RA6A.

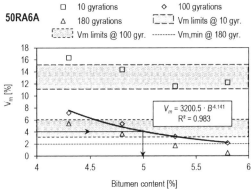

Figure 6. Air voids content vs bitumen content, mix 50RA6A.

entirely reactivated. Note that 40% and 50% RA carried 1.7% and 2.0% of aged bitumen in the mixture, respectively.

Establishing a target V_m of 4% at 100 gyrations, an optimum bitumen content equal to 4.8% by mix weight was identified for the mix 00RA (Fig. 3).

The mixture 40RA showed an optimum bitumen content equal to 5.0% by mix weight (Fig. 4), corresponding to 3.3% of virgin bitumen and 1.7% of aged bitumen from RA. This slight increase in total binder content with respect to the mix 00RA was probably due to the higher viscosity of the oxidized binder included in the RA.

For the mix 40RA6A (Fig. 5), the optimum bitumen content resulted equal to 4.8% by mix weight (3.1% of virgin bitumen), denoting that the bio-additive effectively mobilized the bitumen from the RA, improving the compactability of the mixture.

When increasing the RA content up to 50% (mix 50RA6A), the total bitumen content corresponding to 4% of V_m was found to be 5.0% (3.0% of virgin bitumen) by mix weight (Fig. 6). This indicates that the presence of more 10% of RA determined a decrease of the mix workability, related to the lower virgin bitumen/aged bitumen ratio.

As it can be verified from the Figures, the design bitumen content values also allowed the fulfilment of the volumetric limits provided by Italian Specifications (Autostrade per l'Italia 2008) at 10 gyrations (V_m between 11 and 15%) and 180 gyrations ($V_m \geq 2\%$).

4.2 Analysis of the volumetric properties

The compaction curves, which describe the reduction in V_m when increasing the number of gyrations N, were determined according to EN 12697-10 for all mixtures. In particular, the trend of these curves is linear in a semi-logarithmic plane:

$$V_m = V_{m,0} - m \cdot \log N \qquad (2)$$

where $V_{m,0}$ is the intercept of the y-axis and indicates the air voids at the beginning of the compac-

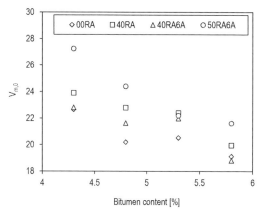

Figure 7. Coefficient $V_{m,0}$ vs bitumen content.

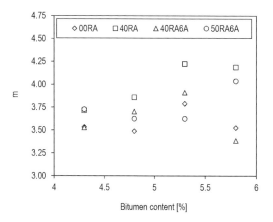

Figure 8. Coefficient m vs bitumen content.

Table 6. Volumetric properties of the design mixes.

Mixture	$V_{m,10}$ [%]	$V_{m,100}$ [%]	$V_{m,180}$ [%]	VMA_{100} [%]	VFB_{100} [%]
00RA	12.1	4.5	3.1	15.8	71.5
40RA	13.2	3.9	2.2	15.5	74.9
40RA6A	13.4	4.4	2.8	15.5	72.4
50RA6A	13.9	4.5	2.7	15.7	71.5

tion ($N = 0$), whereas m is the slope and represent the mix workability.

Figures 7 and 8 show the values of $V_{m,0}$ and m for the different ACs as a function of the bitumen content respectively. It can be observed that, for all the mixtures, the intercept $V_{m,0}$ tended to decrease when increasing bitumen content. This is due to the higher volume of binder that fills the voids between the aggregate particles of the loose AC.

Differently, the slope m showed an increasing trend with bitumen content up to a maximum value approximately in correspondence of $B = 5.3\%$. Over this value ($B = 5.8\%$), the coefficient m exhibited a slight decrease. This indicates that the compaction is promoted when increasing binder content, as it helps to lubricate the aggregate surface reducing the friction between the solid particles. However, this effect is not evident for high contents of binder, as the excess of bitumen probably hinders the efficient stress distribution through the aggregate skeleton.

5 CHARACTERIZATION OF THE DESIGN MIXTURES

The mixtures including the optimum bitumen content were produced in order to verify the volumetric and mechanical properties according to Italian Specifications.

5.1 Volumetric properties

Table 6 shows the V_m of the design mixtures at 10, 100 and 180 gyrations. Moreover, the Void in the Mineral Aggregate (VMA) and the Void filled with Bitumen (VFB) are presented.

It can be observed that, in general, the ACs showed comparable volumetric properties, despite the added virgin bitumen content was significantly different (4.8%, 3.3%, 3.1% and 3.0 by mix weight for the mixtures 00RA, 40RA, 40RA6A and 50RA6A, respectively). All the results fulfilled the requirements at 10 (V_m between 11 and 15%), 100 (V_m between 3 and 6%) and 180 (V_m higher than 2%) gyrations. In addition, the values of VMA and VFB proved to satisfy the requirements defined by SHRP (Cominsky 1994), i.e. VMA higher than 13% and VFB between 65 and 75%.

The average value of the coefficients m and $V_{m,0}$ of the design mixtures, determined through the linear interpolation of the compaction curves (in the semi-log plane), are presented in Figure 9.

From the graph it can be observed that the mixtures 40RA and 50RA6A showed higher m and $V_{m,0}$ values with respect to the mixtures 00RA and 40RA6A. It can be assumed that this result is related to the higher total bitumen content (5.0% for 40RA and 50RA6A vs 4.8% for 00RA and 40RA6A). Indeed, in the mix design it was observed that both m and $V_{m,0}$ increased when increasing bitumen content up to 5.3% (Figs. 8 and 9).

Comparing the results for the ACs with the same bitumen content (i.e. 00RA with 40RA6A and 40RA with 50RA6A), it can be noted that the presence of the bio-additive determined an increase in m denoting the capability of A to improve the mix workability.

5.2 *Strength properties*

The ITS and ITC of the designed mixtures, compared with the technical specification limits (ANAS 2010), are shown in Figure 10.

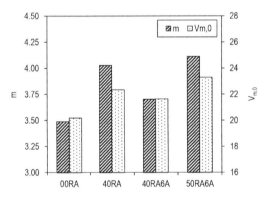

Figure 9.   Coefficients m and $V_{m,0}$ for the design mixes.

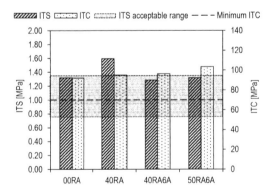

Figure 10.   ITS and ITC values for the design mixes.

The ITS of the mixtures 00RA, 40RA6A and 50RA6A were comparable and met the acceptance range defined by Italian Specifications. On the contrary the mixtures 40RA exceeded the maximum limit of ITS, denoting a not satisfying mechanical response. This indicates that, using the proper dosage of A (in this case 6% by aged bitumen weight), high percentages of RA, up to 50%, can be used attaining satisfying mechanical performance and comparable to those of the traditional AC (i.e. 00RA mixture). In addition, the 40RA and 50RA mixtures with A allow the highest advantages to be reached in terms of amount of recycled materials.

6   CONCLUSION

The present paper deals with the use of a specific bio-based Additive (A) to produce AC using a high amount of RA, without penalizing the mix performance and complying with the Italian specifications. In particular, the study provides the volumetric mix design of four ACs containing no RA, 40% RA, 40% RA plus A, 50% of RA plus A, respectively. The main objective was to determine the effect of RA content and A on the volumetric properties of the ACs and verify the design mixtures to comply with Italian specifications.

The analysis of the experimental results allowed the following conclusions to be drawn:

- an optimum total bitumen content of 4.8% by mix weight was identified for the mixes 00RA and 40RA6A, whereas 5.0% by mix weight was determined for the mixes 40RA and 50RA6A;
- as 40% and 50% RA carried 1.7% and 2.0% of aged bitumen in the mixture (which was considered as entirely reactivated), the virgin bitumen content resulted 3.3%, 3.1% and 3.0% respectively for the mixes 40RA, 40RA6A and 50RA6A;
- the slope m of the compaction curves showed a maximum value approximately in correspondence of $B = 5.3$%, indicating that the compaction is promoted when increasing binder content, as it helps to reduce the friction between the aggregate particles, but the excess of bitumen probably hinders the efficient stress distribution through the solid skeleton;
- the designed mixtures showed comparable volumetric properties and fulfilled the requirements by Italian Specifications and SHRP in terms of V_m, VMA and VFB;
- the slope m for the designed mixtures increased with bitumen content and in presence of A, denoting the capability of the bio-additive to improve the mix workability;

– using the proper dosage of A, good mechanical performances can be obtained even for high contents of RA (up to 50%).

Future investigations will regard the full scale validation of the hot recycled mixtures and the determination of the stiffness and fatigue properties.

REFERENCES

ANAS 2010. Capitolato speciale di appalto, Parte 2ª Norme tecniche per pavimentazioni stradali/autostradali.

Al-Qadi, I., Elseifi, M. & Carpenter, S.H. 2007. Reclaimed asphalt pavement—A literature review. Report No. FHWA-ICT-07-001, Illinois center of transportation.

Autostrade per l'Italia 2008. Capitolato speciale di appalto, parte seconda, opere civili.

Bocci, M., Canestrari, F., Grilli, A., Pasquini, E. & Lioi, D. 2010. Recycling techniques and environmental issues relating to the widening of a high traffic volume Italian motorway. International Journal of Pavement Research and Technology 3(4): 171–177.

Chen, J.S., Huang, C.C., Chu, P.Y. & Liu, K.Y. 2007. Engineering characterization of recycled asphalt concrete and aged bitumen mixed recycling agent. Journal of Materials Science 42: 9867–9876.

Cominsky, R.J. 1994. The Superpave Mix Design Manual for New Construction and Overlays. Strategic Highway Research Program, Washington DC, United States.

EN 12591. 2009. Specifications for paving grade bitumens. European Committee for Standardization, Brussels, Belgium.

EN 12697-10. 2007. Test methods for hot mix asphalt—Compactibility. European Committee for Standardization, Brussels, Belgium.

EN 12697-23. 2006. Test methods for hot mix asphalt—Determination of the indirect tensile strength of bituminous specimens. European Committee for Standardization, Brussels, Belgium.

EN 12697-31. 2007. Test methods for hot mix asphalt—Specimen preparation by gyratory compactor. European Committee for Standardization, Brussels, Belgium.

EN 1426. 2007. Determination of needle penetration. European Committee for Standardization, Brussels, Belgium.

EN 1427. 2007. Determination of softening point. European Committee for Standardization, Brussels, Belgium.

Frigio, F., Pasquini, E., Partl, M.N. & Canestrari, F. 2014. Use of reclaimed asphalt in porous asphalt mixtures: laboratory and field evaluations. Journal of Materials in Civil Engineering 27(7).

Grilli, A., Bocci, E. & Bocci, M. 2015. Hot Recycling of Reclaimed Asphalt using a Bio-based Additive. 8th International RILEM SIB Symposium, Ancona, Italy.

Grilli, A., Bocci, M., Cardone, F., Conti, C. & Giorgini, E. 2013. Laboratory and in-plant validation of hot mix recycling using a rejuvenator. International Journal of Pavement Research and Technology 6(4): 364–371.

Karlsson, R. & Isacsson, U. 2006. Material-related aspects of asphalt recycling: State-of-the-art. Journal of Materials in Civil Engineering 18(1): 81–92.

Molenaar, A., Hagos, E. & Van de Ven, M. 2010. Effects of Aging on the Mechanical Characteristics of Bituminous Binders in PAC. Journal of Materials in Civil Engineering 22(8): 779–787.

Olard, F., Noan, C., Bonneau, D., Dupriet, S. & Alvarez, C. 2008. Very high recycling rate (>50%) in hot mix and warm mix asphalts for sustainable road construction. Proceedings of the 4th Eurasphalt and Eurobitume Congress, Copenhagen, Denmark.

Petersen, J.C. & Glaser, R. 2011. Asphalt Oxidation Mechanisms and the Role of Oxidation Products on Age Hardening Revisited, Road Materials and Pavement Design 12(4): 795–819.

Read, J. & Whiteoak, D. 2003. The Shell Bitumen Handbook. Fifth edition. Shell UK Oil Products Limited.

Shen, J., Amirkhanian, S. & Miller, J.A. 2007. Effects of rejuvenating agents on superpave mixtures containing reclaimed asphalt pavement. Journal of Materials in Civil Engineering 19(5): 376–384.

Stimilli, A., Ferrotti, G., Conti, C., Tosi, G. & Canestrari, F. 2014. Chemical and rheological analysis of modified bitumens blended with "artificial reclaimed bitumen". Construction and Building Materials 63:1–10.

Zaumanis, M., Mallick, R.B., Poulikakos, L. & Frank, L. 2014. Influence of six rejuvenators on the performance properties of Reclaimed Asphalt Pavement (RAP) binder and 100% recycled asphalt mixtures. Construction and Building Materials 71: 538–550.

Transport Infrastructure and Systems – Dell'Acqua & Wegman (Eds)
© 2017 Taylor & Francis Group, London, ISBN 978-1-138-03009-1

A water curtain-based wayside protection system for tunnel fire safety

S. Terribile & G. Malavasi
Department of Civil Building and Environmental Engineering, University of Rome "La Sapienza", Rome, Italy

P. Firmi & A. Pranno
Rete Ferroviaria Italiana S.p.A., Technical Department, Infrastructure Standard, Rome, Italy

ABSTRACT: Tunnel fire safety plays a key role in the railway operations and many procedural and technological improvements have been realized in order to prevent and mitigate the fire risk. The fire prevention and protection procedures can be approached through two main different strategies: the deterministic/prescriptive based approach and the performance-based approach. In the study presented in this paper, the possibility of application of the water curtain-based wayside fire protection system, as one of the auxiliary systems for improving and upgrading the fire safety in the railway tunnels for the smoke compartmentation, has been studied and analysed. A finite element-based dynamic simulation model has been developed in order to simulate different fire scenarios and to assess the water curtain-based fire protection system performances in terms of efficiency against the toxic smokes and gases propagation into the railway tunnel. Some preliminary numerical results have been presented.

1 INTRODUCTION

In the context of the railway tunnel safety, derailment, collision and fire are considered critical scenarios (Kuesel et al. 1996). Indeed, in a closed environment (like a railway tunnel), the fire consequences can be more serious than ones that can happen in an open space: this is due to several factors such as the reached high temperature, the smoke and toxic gases production and propagation and the possible visibility lack (Malavasi & Rainoldi, 2013).

The fire prevention and protection procedures can be approached through two main different strategies: the deterministic/prescriptive based approach and the performance-based approach.

The prescriptive-based approach consists of normative and rules that impose to respect specific obligations to each kind of activities in order to be compliant with the fire safety constraints (minimum safety requirements).

The performance-based approach consists of the application of engineering principles, rules and expert judgments: in tunnel fire safety, they are based on the scientific assessment of the combustion phenomenon, the fire effects and the human behaviour, aiming at the human life, assets and environment protection, the fire risk and its effects quantitative assessment (Beard & Carvel, 2005). In the last years, many researches and studies were conducted on the fire dynamics and people behaviour evacuation, taking into account several

scientific techniques (e.g. engineering, chemistry, physics, computer science, etc.), social sciences (e.g. sociology, psychology, etc.), medical sciences (e.g. biology, physiology, etc.) and economical sciences (e.g. statistics, financial management, etc.) (D.M. 09/05/2009). The scientific feature of the performance-based approach is also related to the fact that several Computational Fluid-Dynamics (CFD) codes and software tools can be applied in order to simulate the fire development and the smoke and hot gases propagation.

In the last years, the legislative framework for the tunnel fire safety has been characterised by a continuous evolution. In the European context, the first legislation, recognised as reference guideline, was the UIC Codex 799-9 R "Safety in Railway Tunnels (SRT)" (UIC Fiche 799-9, 2002); then, the Technical Specification for Interoperability "Safety in Railway Tunnels" (TSI-SRT) was published in 2008 (first version) (TSI 07/03/2008) and in 2014 (second and current version) (TSI 12/12/2014): the TSI-SRT is the cross-cutting mandatory legislation because it defines minimum requirements for three strictly related railway subsystems (Infrastructure, Energy, Operations).

In the Trans-European railway Network (TEN) context, the orographic complexity of Italy is the main reason of a very high concentration of tunnels (Fig. 1). In Italy, nowadays, the Interministerial Decree 28/10/2005 "Sicurezza nelle Gallerie Ferroviarie" is identified as reference national normative: it defines a set of minimum requirements to

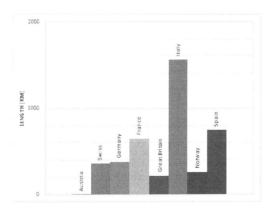

Figure 1. The tunnel distribution in the European railway network.

be respected through the adoption of prescriptive measures (only for railway tunnels with more than 1000 m of length); in certain conditions, it suggests a performance-based approach in order to identify the tunnel risk level (D.M. 28/10/2005).

2 WATER CURTAIN-BASED PROTECTION SYSTEMS

In the last applications, the water curtain-based protection systems have been used as one of the suitable solutions in order to mitigate the smoke and hot gases, released from a fire source in closed spaces. These systems have been recognized as a useful technique to mitigate major industrial hazards because it combines attractive features to different types of risks (such as pool fire, fuel release, etc.).

Nowadays, in the industrial context, three main typologies of water-based protection systems exist:

- Water spray (composed by one or more shaped sprays, placed side by side)
- Water curtain (composed by a continuous water wall)
- Water-vapor curtain (composed by one or more sprays, placed side by side, or by water vapor screen)

Generally, fixed installations allow rapid response to an emergency, especially when the water spray can be activated automatically by gas sensors; there are also clear advantages in achieving mitigation without bringing the lives of emergency response crews and site personnel in danger. The advantage of a mobile monitoring system is the flexibility of when, where and how to apply.

In case of accidental flammable gas releases, the spray curtain may be used as a direct-contact reactor exchanging heat, mass and quantity of motion with the dispersing gas cloud. Fluid curtains are also used to mitigate the consequences of flammable and toxic substances dispersion in the environment.

In order to reduce and/or eliminate the released gaseous and solid substances, the design criteria must pay attention to three main performance functionalities:

- Absorption (due to the chemical interactions among gas particles and liquid droplets)
- Dilution (due to the "air entrainment" effect, generated by the volumetric flow rate)
- Temperature reduction (due to energy exchange in terms of heat)

A water curtain spray directed vertically, horizontally or at different angles (e.g. 45°) through a smoke and hot gases can have a number of mitigation effects:

- Mechanical effects of acting as a barrier to the passage of smoke and gases;
- Mechanical effects of dispersion and dilution by air entrainment;
- Mechanical effects of imparting upward quantity of motion to smoke and gases;
- Thermal effects by cooling in case of smoke and hot or burning gases;
- Physic-chemical effects of absorption of smoke and gases (with or without chemical reaction)

From the operative point of view, the water curtain-based protection system functions are based on the fire source identification and isolation through a set of water-curtains, which compose a distributed system at fixed distance along the railway tunnel or in the cross-section (Fig. 2). As regards the railway context, a questionnaire-based survey has shown that in Europe the application of such systems is not

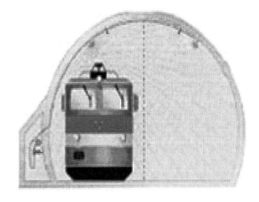

Figure 2. Example of a water curtain-based protection system installation inside a railway tunnel.

so diffused; only in Austria, the railway IM (OBB) has performed some studies for the possible installation of the water curtain-based systems along the passenger railway station platforms.

Up to now, in Italy, some applications of the water curtain-based protection systems have been realized into the underground railway passenger stations in order to create a physical water-based barrier between the fire source and the passenger evacuation path.

These surveys explains better that the application of this kind of systems along the railway tunnels is a possible innovative work in order to improve the fire safety levels (UpTun Project, 2008). Unfortunately, inside the railway tunnels, the application of the automatically activated protection systems is forbidden due to technical and procedural limitations (due to the presence of electric power supply systems), which have to be switched off during a fire emergency. For this reason, it is important to note that specific protection systems (activated by remote control centre) can be installed inside the railway tunnel in order to generate a physical barrier to the smoke and hot gases propagation and to facilitate the passenger evacuation.

3 METHODOLOGY

In order to study the fire characteristics, the smoke and hot gases development and propagation and the related water curtain-based protection system inside a closed environment (like a railway tunnel), the risk theory, the thermo-fluid-dynamics equations, the fire combustion concepts and water jet theory have been taken into account.

3.1 Equations of motion

The physical law of a thermo-fluid dynamic system can be represented by a set of six differential equation in six variables (also known as "Navier-Stokes" equations): the three velocity components (u, v, w) in x, y, z spatial directions, the temperature (T), the pressure (p) and the density (ρ). This set is very sensible to the spatial-temporal changes of the initial conditions in terms of solution convergence.

The Navier-Stokes equations consists of a time-dependent continuity equation for conservation of mass, three time-dependent conservation of quantity of motion equations and a time-dependent conservation of energy equation (Mc Grattan et al. 2015). In order to solve the Navier-Stokes equations, several mathematical techniques can be used: the main three are Large Eddy Simulation (LES), the Direct Navier-Stokes (DES) and the Raynolds Average Navier Stokes (RANS). A Large Eddy Simulation (LES) is a popular technique for simulating turbulent flows. A Direct Numerical Simulation (DNS) a simulation in computational fluid dynamics in which the Navier-Stokes equations are numerically solved without any turbulence model. This means that the whole range of spatial and temporal scales of the turbulence must be resolved. All the spatial scales of the turbulence must be resolved in the computational mesh, from the smallest dissipative scales (Kolmogorov scales), up to the integral scale L, associated with the motions containing most of the kinetic energy. RANS techniques are based on the theory that the turbulent fluid motion is described by averaged and time-varying motion; the parameters have to be averaged by time: this produces a reduction on the computational time because the average fluid motion scale are bigger than turbulent fluid motion ones. These techniques requires the application of other equations (k-ε model, k-ω model, etc.) in order to solve the problem.

3.2 Turbulence models

Turbulent flows of a single Newtonian fluid, even those of quite simple external geometry (such as a fully-developed pipe flow) are very complex and their solution at high Reynolds numbers requires the use of empirical models to represent the unsteady motions. It is self-evident that the addition of particles to such a flow will result in:

- Complex unsteady motions of the particles that may result in non-uniform spatial distribution of the particles and, perhaps, particle segregation. It can also result in particle agglomeration or in particle fission, especially if the particles are bubbles or droplets.
- Modifications of the turbulence itself caused by the presence and particles motions. The turbulence could be damped by the presence of particles, or it could be enhanced by the wakes and other flow disturbances that the motion of the particles may introduce.

The Kolmogorov scales are the smallest scales used in order to describe a turbulent flow. Basically, the Kolmogorov theory introduces the concept that the smallest scales have to be universal in order to describe the turbulence phenomena: in other words, the scales have to be similar for each turbulent flow and they are dependent to two parameters (ε is the mean rate of dissipation per unit mass of fluid in a time unit and v is the fluid kinematic viscosity). The three main Kolmogorov scales are "Length Scale", the "Time Scale" and "Velocity Scale".

3.3 Fire and smoke dynamics

The fire dynamics into a tunnel can be studied and analysed through the following main parameters:

- Heat Release Rate (HRR)
- Flame Length
- Natural Ventilation
- "Backlayering" Effect

During the fire spread, it produces energy (in terms of heat) that is time-dependent and it changes in function of the calorific power and the combustion speed.

The fire growth speed depends to the ignition process, the flames propagation and the combustion rate. Each object, subjected to the combustion process, is characterized by growth time and the combustion speed. During the initial phase, a fire can be modelled by the "*T-squared curve*" (Eq. 1):

$$Q = \alpha t^2 \qquad (1)$$

Where Q = heat release rate [kW]; α = fire growth coefficient [/]; and t = time [s].

Generally, fire can be classified into four main classes in function of the growth speed (slow, medium, fast and ultra-fast). In the fire safety engineering, the flame length is an important factor to be considered in order to better understand the fire diffusion inside the tunnel. The flame length is defined as the distance between the fire centre and the flame impact section (Braubaskas, 1980).

The natural ventilation is a very complex parameter to be studied in order to understand the fire behaviour. Inside railway tunnels, the natural ventilation is mainly longitudinal and it has influence on the oxygen flow with consequently fire alimentation and HRR increase.

During a tunnel fire, another important aspect to be considered is the "Backlayering" phenomenum. Even if the longitudinal ventilation has been generated, the flames and the combustion products can move towards opposite direction: this effect happens when the heat release rate and the air speed are equal to critical values.

4 COMPUTATION FLUID-DYNAMICS (CFD) SOFTWARE APPLICATION

Computational Fluid Dynamics (CFD) is fluid mechanics-related science that uses numerical algorithms and methods to solve and analyse the problems that involve fluid flows. Due to the high complexity of the physical and chemical phenomena that characterized the fluids flow (such as interactions liquids-gases, two-phases flow, turbulence, etc.), computers have to be used in order to perform the calculations required to simulate the interaction of liquids and gases with surfaces defined by boundary conditions. In all of these approaches, the same basic procedure has to be followed to design and develop the simulation model.

During the pre-processing phase, the standard procedure require the following steps:

- Definition of the problem geometry (physical regular or non-regular bounds);
- Division of the volume (occupied by the fluid) into discrete cells (the mesh): the mesh may be uniform or non-uniform;
- Definition of the physics (equations of motion, enthalpy, radiation, species conservation, etc.) and relative modelling;
- Identification of the boundary conditions: this involves specifying the fluid behaviour and properties at the boundaries of the problem.
- Definition of the initial conditions (only for transient problems).
- After that, the simulation has to be started and the equations are solved iteratively in a steady-state or transient way.
- Finally, a postprocessor has to be used for the analysis and visualization of the resulting solution.

The purpose of this research study was to investigate the possibility of installation of a water curtain-based protection system in railway tunnels and propose a new method to assess their performances in terms of efficiency (considered as radiation blockage, smoke propagation obstruction and visibility improvement) by using a CFD simulation model, developed with the free and open-source FDS and SMOKEVIEW software tools.

The FDS model has been applied for two main objectives:

- To compare the experimental data, collected during the tests, with the numerical results;
- To compare the following two main scenarios: the first based on the absence of the water curtains in order to study the fire growth and smoke and hot gases behaviour inside the railway tunnel; the second based on the activation of the water curtains during time (distributed in variable number along the tunnel length) in order to mitigate the effects of smoke and hot gases propagation and increase the passenger safety and evacuation facilitation levels and the rescue teams accessibility.

The main developments steps of the model, referring to commands of the FDS code, have been reported below:

- Geometry: "Obstruction"
- Computational Domain: "Mesh"
- Boundary Conditions: "Vent" (inlet and outlet)
- Fire: "HRR-time curve" (t-squared)
- Water curtain system: "Nozzle flow rate", "Droplet Initial Velocity", "Spray Pattern Shape", "Liquid Water Droplet"

The results of a case study have been reported in Section 5.

5 CASE STUDY

In the study presented in this paper, a finite element-based dynamic simulation model of a railway tunnel has been developed in order to assess the performance of the water curtain-based fire protection system for the mitigation of smoke and toxic gases propagation.

The simulation model of the railway tunnel has been defined in some different steps regarding the geometry, the fire source, the water curtain-based fire protection system and the boundary conditions inside the tunnel (Table 1). The simulated domain is a single-tube, double track railway tunnel: it includes the tunnel structural elements, the ballast, the passenger coach (as fire source) and the water-based protection system. This domain has specified overall dimensions equal to 200.0 m × 10.5 m × 7.85 m. The mesh of domain consist in 137632 cubic cells of side equal to 0.5 m.

As regards the mesh dimension, the average size of the discretization cell (computational grid) is related to the characteristic diameter of the fire: this parameter indicates the goodness of the resolution grid, according to the following expression (Mc Grattan et al. 2015):

$$D^* = (Q / c_\infty T_\infty \rho_\infty \sqrt{g})^{2/5} \qquad (2)$$

where D^* = characteristic fire diameter [m]; Q: total heat release rate [kW]; ρ_∞ = air density [kg/m^3]; c_∞ = air specific heat [kJ/kg * K]; T_∞ = air temperature [K].

As input, Q = 10000 kW, ρ_∞ = 1.205 kg/m^3, c_∞ = 1 kJ/kg * K, T_∞ = 293 K, $\sqrt{g}$ = 3.13 m/s^2. As output, the fire characteristic diameter (D*) has been calculated equal to 2.41 cm.

For the mesh dimension optimization, the results have been divided in three main sets:

- <u>Coarse</u>: suggested cell dimension (δx) is 60.22 cm and the total number of cells is 81000;
- <u>Moderate</u>: suggested cell dimension (δx) is 24.09 cm and the total number of cells is 1244160;
- <u>Fine</u>: suggested cell dimension (δx) is 15.06 cm and the total number of cells is 4665600;

At this first step of the research study, the cell dimension has been selected between the coarse and moderate classes (δx = 50 cm) in order to obtain the results compliant with the current available computational resources: this value has been set due to the fact that the railway tunnels are particular civil work in which the longitudinal dimension (x) is predominant to the others. The purpose is to study the fire and smoke behavior along the entire length of railway tunnel. A grid sensitivity analysis for the grid resolution verification will be performed. In the future simulations, a multi-mesh approach should be adopted in order to reduce the cell.

The protection system (located at 125 m from the fire source) has been composed by four single water shields: two of them displaced at the lateral sides (n°1 and n°2), one at the top side of the railway tunnel (n°3) and the last one (n°4) along the ballast in the middle of the railway trackside (Fig. 3).

The fire model has been represented by a single passenger coach with the flames, smoke and gases source located on the roof of the rolling stock. The fire source has been characterised by analytic function of Heat Release Rate (HRR) with a maximum value equal to 10 MW.

The simulations have been done for a time of about 26 minutes.

Table 1. Railway tunnel simulation model set-up.

Input parameters	Value u.o.m
Tunnel Length	200 m
Tunnel Width	10.5 m
Tunnel Height	7.25 m
Mesh Cell Size (x-direction)	0.5 m
Mesh Cell Size (y-direction)	0.5 m
Mesh Cell Size (z-direction)	0.5 m
Max Heat Release Rate (HRR)	10 MW
Fire Growth Coefficient (α)	0.01
Combustion Reaction	Polyurethane Foam
Water Shields Pressure	5 bar
Water Shields Flow Rate	383 l/min, 283 l/min
Liquid Particles Mean Diameter	1 mm
Longitudinal Wind Speed	0.5 m/s

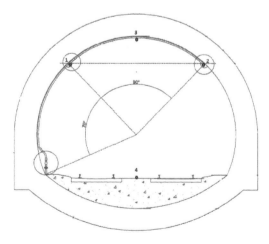

Figure 3. Water curtain-based protection system configuration: nozzles installation on the railway tunnel cross-section.

Four sets of monitoring devices (such as thermocouple, opacimeter, etc.) has been located at different distances from the water-curtain protection systems (space step equal to 10 m) at two different heights (+1.75 m and +7 m) in order to identify all the possible critical areas for the human life safety and sustainability and to monitor the parameters along the evacuation paths in terms of length and time. Moreover, two measurement devices has been located near the water curtain ("sensor x" at −10 m before and "sensor y" at +10 m beyond) in order to study the behaviour of smoke and hot gases with the presence of the protection system (Figs. 4–5).

5.1 *Performance key parameters*

The performance assessment of the innovative water curtain-based fire protection system has been done into the simulation model taking into account the following parameters:

- Temperature [°C]
- Visibility [m]
- Heat Flow [kW/m²]

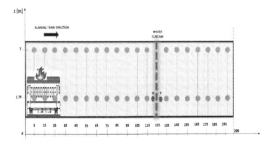

Figure 4. Measurement devices longitudinal configuration: sensors sets installation along the railway tunnel.

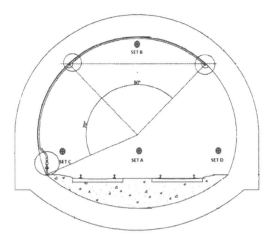

Figure 5. Measurement devices transversal configuration: sensors sets installation on the railway tunnel.

- Fractional Effective Dose—FED [/]
- Volume Fraction [mol/mol]:
  - Carbon Monoxide (CO)
  - Carbon Bioxide (CO₂)
  - Oxygen (O₂)
  - Soot

5.2 *Life safety and evacuation key parameters*

The assessment of the improvement provided by the innovative water curtain-based fire protection system in terms of life safety and evacuation has been done into the simulation model taking into account the "Human life safety critical conditions". The human life safety critical conditions are the limit conditions under which a person can be exposed in case of fire. The design verification consists to avoid the occurrence, through prevention and protection measures against fire, of conditions more severe than each of the limits (Fava & al. 2013).

5.3 *Preliminary simulation results*

The simulation model of the railway tunnel, developed under the input conditions described above, returns the following preliminary results. As regards the "Temperature" parameter (expressed in °C degrees), in the proximity of the water curtain protection system at 1.75 m and 7 m, it is possible to note a reduction due to the cooling effect of the water particles on the smoke and hot gases generated by the fire. Both at the railway tunnel ceiling (measurement devices set B) and on the lateral evacuation sidewalks (measurement devices sets C), the temperature decreases during time beyond the water curtain (Figs. 6–7): despite that for the railway tunnel fire safety there are no commonly recognised standard thresholds, the temperature and visibility values could be compared to fixed

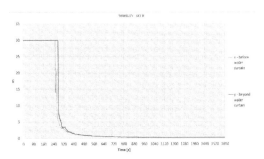

Figure 6. Comparison of temperature between the sensor x (−10 m before the water curtain) and sensor y (+10 m beyond the water curtain), located at the ceiling of the railway tunnel (height = 7 m—Set B).

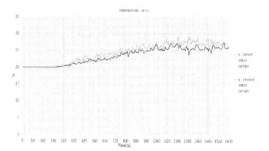

Figure 7. Comparison of temperature between the sensor x (−10 m before the water curtain) and sensor y (+10 m beyond the water curtain), located on the railway tunnel lateral sidewalk (height = 1.75 m—Set C).

Figure 8. Comparison of visibility between the sensor x (−10 m before the water curtain) and sensor y (+10 m beyond the water curtain), located at the ceiling of the railway tunnel (height = 7 m—Set B).

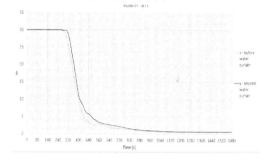

Figure 9. Comparison of visibility between the sensor x (−10 m before the water curtain) and sensor y (+10 m beyond the water curtain), located on the railway tunnel lateral sidewalk (height = 1.75 m—Set C).

thresholds as, for example, in the metro systems (ISO/TS 13571-2002).

As regards the "Visibility" parameter (expressed in m), in the proximity of the water curtain protection system at 1.75 m and 7 m, the first results have shown that the system doesn't produce effective improvements in terms of visibility conditions. It is important to note that the presence of the water particles provides a time delay effect on the degradation of visibility conditions, both at the railway tunnel ceiling (measurement devices set B) and on the lateral evacuation sidewalks (measurement devices sets C) (Figs. 8–9).

The numerical results have shown that the water curtains are capable to block the radiation generated by the fire and reduce the temperature.

The use of the Lagrangian approach in the FDS model can give the following advantages:

• The Lagrangian model is slightly faster to compute than the Eulerian one.
• The Lagrangian model enables to calculate the initial velocity of the droplets.
• The specification of the initial trajectories of the droplets is considerably simpler than with the Eulerian approach.
• With the Lagrangian approach, an accurate representation of the droplet size distribution can be modeled.

A limitation to be highlighted (in the modeling of the water droplets by using the Lagrangian approach) is that the FDS code enables the user to apply a two-phase model in which the smoke and hot gases are studied as a continuous pattern characterized by thermal energy, mass and velocity and, on the contrary, the discharging curtain not as a continuous water layer but a transient discrete pattern with air space inside the curtain; so, each water curtain is not capable to use for the blockage of smoke spreading, as the smoke particle would transmit through the transient air space. As regards

the chemical substances concentration, further simulations will be done in order to assess the efficiency of the system, but the expected results are not promising due to the same drag effects.

6 CONCLUSIONS AND FUTURE WORK

In the research study proposed in this paper, an innovative finite element-based dynamic simulation model has been developed in order to assess a water curtain-based fire protection system performances in terms of efficiency against the heat and combustion products propagation into the railway tunnel. This kind of systems have to be designed in order to perform some key functions:

• Dilution (due to the "air entrainment" effect, generated by the volumetric flow rate)
• Temperature reduction (due to energy exchange in terms of heat).

An interesting property is that the FDS simulation model can be applied in order to optimize the water curtain system configuration along the railway tunnel cross-section. The model can assess the system performances by varying the nozzles number per each water curtain (n°4, 6, 8), the 3D position of each single nozzle and the type/commercial model (in terms of orifice diameter). Some possible tests may be performed taking into account the variation of water flow [l/min] and pressure [bar].

As future works, a detailed research study will be done in order to assess the water curtain-based protection system performance in terms of combustion products and FED concentration in order to guarantee the life safe conditions for passenger evacuation. Moreover, the calibration of simulation model will be performed taking into account any collected experimental data. Then, a model sensitivity analysis will be done in relation to the assessment of the water curtain-based protection system performance characteristics, the comparison to life safe conditions thresholds (to be identified and quantitatively defined) and the effects of variation of the railway tunnel geometrical features. As last future development, a simple evacuation model will be integrated in the core one in order to study more in detail the smoke and hot gases propagation in relation to the people evacuation.

REFERENCES

Babrauskas, V. 1980. Flame Lengths under Ceiling, *Fire and Materials*, 4, 3,119–126.

Beard, A. & Carvel, R. 2005. *The Handbook of Tunnel Fire Safety*.

D.M. 09/05/2009. Direttive per l'attuazione dell'approccio ingegneristico alla sicurezza antincendio. *Decreto Ministeriale Italiano*.

D.M. 28/10/2005. Sicurezza nelle Gallerie Ferroviarie. *Decreto Ministeriale Italiano*.

Deliverable n°251. September 2008. Engineering Guidance for Water Based Fire Fighting Systems for the Protection of Tunnels and Sub Surface Facilities. *UPTUN Project*. WP 2 Fire development and mitigation measures.

Fava, P. & al. A.A. 2012–2013. Simulazione di un Incendio in Galleria tramite il Software FDS. *Tesi di Laurea Magistrale*. Università degli Studi di Padova.

ISO/TS 13571:2002. Life-Threatening Components of Fire-Guidelines for the Estimation of Time Available for Escape Using Fire Data. *ISO Guidelines*.

Kuesel, T.R. et al. 1996. Tunnel Engineering Handbook.

Malavasi, G. & Rainoldi, G. 2013. Il Rischio Incendio nelle Gallerie Ferroviarie. L'elaborazione di Piani di Emergenza Esterna. *IV Convegno Nazionale-Sicurezza ed Esercizio Ferroviario. Giugno 2015*. Roma, Italy.

McGrattan, K. et al. November 2015. Fire Dynamics Simulator Technical Reference Guide-Volume1: Mathematical Model. National Institute of Standards and Technologies (NIST) *Special Publication 1018-1-6th Edition*.

McGrattan, K. et al. November 2015. Fire Dynamics Simulator-User's Guide. National Institute of Standards and Technologies (NIST) *Special Publication 1019-6th Edition*.

TSI 07/03/2008. Technical Specification of Interoperability relating to Safety in Railway Tunnels in the trans-European Conventional and High-speed Rail System. *Technical Specification for Interoperability*.

TSI 12/12/2014. Technical Specification for Interoperability relating to "Safety in Railway Tunnels" of the Rail System of the European Union. *Technical Specification for Interoperability*;

UIC Fiche 799-9. September 2002. Safety in Railway Tunnels. *UIC Codex*.

Transport Infrastructure and Systems – Dell'Acqua & Wegman (Eds)

3D control of obstacles in airport location studies

D. Gavran, S. Fric, V. Ilić, F. Trpčevski & S. Vranjevac

Department for Roads, Railroads and Airports, Faculty of Civil Engineering, University of Belgrade, Serbia

ABSTRACT: While looking for a potential site of a new airport, meteorological and environmental analyses, as well as navigational analyses, are all of the ultimate importance. Potential airport location failing to pass any of these three checks is considered absolutely inappropriate for further airport development. Especially in mountainous regions, navigational analyses are concentrated on the control of obstacles which is synonym for the analyses of Obstacle Limitation Surfaces. Obstacle Limitation Surfaces are imaginary surfaces defining the volume of airspace in the vicinity of the airport that should ideally be kept free from obstacles so as to provide for safe aircraft operations either during an entirely visual approach or during the visual segment of an instrument approach. Also, establishment of such a protected volume of airspace prevents the uncontrolled growth of manmade obstacles in the vicinity of the airport. The paper presents contemporary 3D techniques of orienting the runway so as to minimize the extent of protrusions through the Obstacle Limitation Surfaces. These techniques are based on the triangulated 3D model of terrain surface and on the models of Obstacle Limitation Surfaces. The techniques in concern are demonstrated on an airport located in the mountainous Balkan region.

1 INTRODUCTION

Most of the airport projects are concentrated on expansions and reconstructions of the existing facilities. Even when working on airport master plans, these master planes are usually confined to existing airport locations. But, when working on a master plan of a new airport, the whole planning process starts with the, so called, location study (ICAO 1987, FAA 2007, Horonjeff et al. 2010). Practically, meteorological and environmental analyses, as well as navigational analyses, are carried out for all promising airport locations near the city in concern. Potential airport location failing to pass any of these three checks is considered absolutely inappropriate for further airport development. Especially in mountainous regions, navigational analyses are concentrated on the control of obstacles which is synonym for the analyses of Obstacle Limitation Surfaces. Obstacle Limitation Surfaces are imaginary surfaces defining the volume of airspace in the vicinity of the airport that should ideally be kept free from obstacles so as to provide for safe aircraft operations either during an entirely visual approach or during the visual segment of an instrument approach. Also, establishment of such a protected volume of airspace prevents the uncontrolled growth of manmade obstacles in the vicinity of the airport.

The paper presents contemporary 3D techniques of orienting the runway so as to minimize the extent of protrusions through the Obstacle Limitation Surfaces. The techniques in concern are based on the triangulated 3D model of terrain surface and on the models of Obstacle Limitation Surfaces. Officially, primary products of these analyses are The Aerodrome Obstruction Chart—Type A and The Aerodrome Obstruction Chart—Type B. In addition, the deployment of triangulated 3D surfaces introduces isopachyte plan, which further improves the control of obstacles.

These techniques are demonstrated on a Trebinje airport located in the mountainous Balkan region (Fig. 1). Trebinje is small city in southern Hercegovina with population of 25,000, the number expands to 70,000 in summer months. It is merely 25 km inland from the Croatian historical coastal

Figure 1. Future Trebinje international airport.

city of Dubrovnik. The vicinity of Dubrovnik creates an opportunity for the new Trebinje airport (Nikolić et al. 2009) to compete for passengers with the existing Dubrovnik airport.

Despite the fact that navigational, climatologic and environmental analyses together are crucial in the search for an optimal airport location, it immediately became apparent that the proper setting of the approach procedures in relation to the existing ground features would be decisive. The vicinity of existing airports in Mostar, Podgorica, Tivat and Dubrovnik, as well as the recently established borders between Croatia, Montenegro and Bosnia and Hercegovina, imposed further limitations.

2 MODELLING TERRAIN SURFACES

As a rule, planning or design process starts from the digital terrain model. The most widely adopted terrain model for civil engineering purposes is TIN—Triangulated Irregular Network model (Fig. 2). By definition, TIN model connects terrain points by using non-overlapping triangles tending to be as much equiangular in plan projection as possible (Green & Sibson 1978, Gavran 1996).

By simple editing (switching triangles' edges) it is possible to incorporate any kind of manmade or natural feature (ridge, escarpment, pavement edge) into the TIN, making the model identical to the natural surfaces. For rough examination of large areas of the terrain, the grid model could be quite appropriate (Fig. 3) (Gavran 1996, Petrie & Kennie 1987).

The generation of a grid model is much easier to program than that of a TIN model. But, for subsequent geometrical analyses, the TIN model is much simpler to work with. In fact, each triangular facet is a part of a simple plane (as the three triangle's vertices define the perfect plane), while the grid

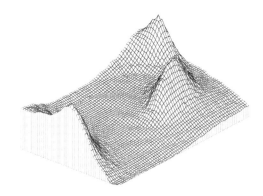

Figure 3. Grid terrain model.

cell is a part of a curved (twisted) surface. Since the cutting of the longitudinal profiles and cross sections, volume calculations and other geometrical analyses are much easier to execute (or to program) on simple triangular facets, even when the grid terrain model is at the disposal, at the start of the design process it is "exploded" into triangles. In fact, each "twisted" grid cell (defined with four points) could be easily exploded into two triangles (each one defined with three points).

3 MODELLING OBSTACLE LIMITATION SURFACES

Having the digital terrain model completed the control of obstacles moves to the modelling of obstacle limitation surfaces. The shapes and sizes of obstacle limitation surfaces are published by ICAO—International Civil Aviation Organization manuals (ICAO 2004). In general, these are imaginary surfaces constructed around a particular runway. These are approach and take-off surfaces (extending up to 15 km in front of each runway's threshold), the inner horizontal surface (circular surface with the radius of 4 km, 45 m above the lower threshold), the conical surface (climbing at the grade of 20% around the perimeter of an inner horizontal surface, and having the width of 2 km) and the transitional surface climbing from the runway strip up to the inner horizontal surface at the rate of 1:7 (7:1 in American format, or cca 14%). The entire set of surfaces is moved and rotated (together with the runway) in order to minimize terrain protrusions. Obstacle limitation surfaces are also checked against the natural (trees) and manmade (buildings, towers, power lines) features.

To proceed with the obstacle control, a triangulated model of obstacle limitation surfaces should be constructed (Fig. 4). In essence, standard surfaces (with the straight approaches) could be easily

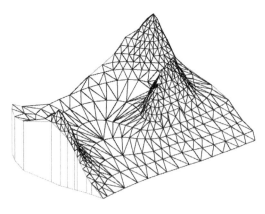

Figure 2. TIN terrain model.

Figure 4.   Obstacle limitation surfaces' model.

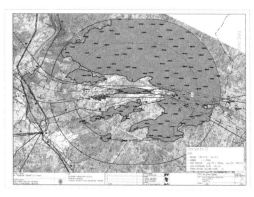

Figure 5.   Aerodrome obstruction chart—Type B.

modeled by using general purpose CAD systems, while curved approach or take-off paths could be modeled by using software solutions intended for road modeling. In fact, it is quite simple to program the construction of a triangulated surface following any reasonable curved centerline in plan or profile projection (Gavran 1996).

4   3D CONTROL OF OBSTACLES

3D control of obstacles is performed upon the triangulated terrain model, on one side, and the model of the obstacle limitation surfaces, on the other.

4.1   *Producing standard graphical documents*

The graphical documents representing the relation between the obstacles and the obstacle limitation surfaces are The Aerodrome Obstruction Chart—Type A and The Aerodrome Obstruction Chart—Type B. Aerodrome Obstruction Chart—Type B is more illustrative (Fig. 5) (ICAO 1983, 1989). This is the map representing obstacle limitation surfaces in plan projection, as well as all natural and man-made obstacles in the area. Apart from being a crucial element of the airport location study, the Type B map accompanies the flight crew on the route to a particular airport. The map informs the crew on the most prominent obstacles surrounding an airport. Based on these obstacles the crew decides upon the procedures (turns) to be performed in the case of the abandoned approach etc.

One of the most important features of The Aerodrome Obstruction Chart—Type B are the thick blue lines indicating terrain penetration through the obstacle limitation surfaces. While positioning the runway centerline, the model of these surfaces is moved and rotated along with the runway. For each promising position of the runway, the hidden line removal should be called in plan projection, thus approximately indicating areas where obstacle limitation surfaces sink beneath the terrain surface.

To sharply delineate the terrain penetration line, it is necessary to deploy specific (and rather sophisticated) tool. This is the tool dealing with the penetrating triangles: the terrain triangles and the triangles forming the model of the obstacle limitation surfaces.

In this particular case, our design team deployed the tool for decomposing penetrating triangles into the subtriangles that do not intersect any more, but touch each other along the lines of intersection. The software had been used for years while modeling intersecting cut and fill slopes (Fig. 6) (Gavran 1996, 2012).

The software is supposed to work on triangulated cut/fill slope models. After decomposing the fill slopes' triangles, the lower subtriangles (below the intersection lines) are to be removed, while modeling cut slopes, the upper triangles are the surplus triangles.

The algorithm that handles intersection of multiple triangles is a sophisticated one, because all of the newly created subtriangles deriving from one "explosion" (between two particular triangles) and touching each other perfectly, must be checked again for potential "explosions" with the rest of the starting triangles. To speed up the process, family relations are introduced between the triangles.

The pretriangles are the triangles belonging to the starting set of triangles, while the subtriangles created in the explosion of one particular triangle are brothers (or sisters). Besides the brothers and the sisters, each subtriangle has its mother and father: the triangle of origination and the triangle in relation to which the originating triangle was exploded. As the algorithm starts to dissipate the

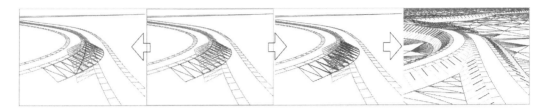

Figure 6.   Modelling intersecting cut and fill slopes.

triangles, the number of candidates for the "explosion" grows rapidly. Keeping track of family relations, unnecessary (impossible) "explosions" are skipped, making the software run faster. But, to cut the long story short, to delineate the intersection between the terrain triangles and the triangulated model of the obstacle limitation surfaces, only a small fraction of this algorithm should be deployed. Only the intersection lines between the pretriangles are to be generated and no pretriangles decomposition is needed (left process on Fig. 6).

The simplified algorithm of resolving triangles' explosion is given in Figure 7. General idea of how explosion goes is given in the upper left corner, while the main body of the program starts just below. The list of triangles to be exploded (*listri*) contains particular triangles, each one defined by its vertices (*t1,t2,t3*), name (*en*, which is important only within exact programming environment), color (*col*), layer (*lay*) and father's number within the list (*No*, pretriangles have no father). The length of the starting set of triangles is *len + 1* (as the list of the triangles is zero-based, the index of the first pretriangle is *0*, while the index of the last one is *len*). As the program keeps running, the list *listri* grows in size, while new triangles are added to the list.

Program takes triangles one by one and tries to explode them. Taken triangle, as candidate for explosion, is potential mother triangle and is drawn in black. At the start, variable *codx* takes "T" value. When (and if) the mother explodes, it changes to nil value.

The question is: in collision with which triangle the mother could explode? If the mother triangle is one of the pretriangles (if its index *no* is less than or equal *len*), it could be any of the pretriangles. But, if the candidate mother is one of the newly created triangles (resulting from previous explosions, *no > len*), then it could be exploded only when colliding with the pretriangles positioned in the list *listri* after her father. If the mother of the potential mother have not found a potential father before finding the actual one (father), neither the potential mother could collide with the triangle preceding the actual father—the potential mother (if resulting from the explosion and not being one

of the pretriangles) is only a fraction of its mother. Consequently, the counter for potential fathers (*no1*) starts from 0, if the candidate mother is one of the pretriangles, or from her father's index *No* increased by 1, if the mother results from a previous explosion. The program then cycles trough list *listri*, searching for a potential father. The fathers' cycle goes to the end of that portion of the list containing pretriangles only (*no1 < = len*), as pretriangles only could act as potential fathers. Taking newly created triangles (resulting from the explosions) for candidate fathers has no sense—a triangle not colliding with a particular pretriangle could not collide with its children either.

Unlike father, potential mother could be any of the pretriangles, or newly created triangles. Spatial triangle could collide with two or more triangles. In this case, particular triangle must be exploded in the relation to the first one (father). Its children (resulting from the explosion) will be touching the father's plane by their edges, but would still collide with some other triangles. Therefore, these new triangles must be given a chance to act as mothers and collide with the second (or even third) triangle. When a potential mother finds an appropriate father, it explodes, variable *codx* takes value nil and program takes the next triangle from list *listri* as a potential candidate mother.

Resolving a potential collision between a potential mother (black triangle) and a potential father (white one) starts in subroutine *incalc*. Actually, *incalc* subroutine determines in how many points father's edges penetrate mother's plane. Mother triangle resides in plane **P**, while father triangle resides in plane **G**. If these two planes are parallel, or coincide, there is no collision. Also, if some of the mother's edges reside within a father's plane, there is no collision either (the potential mother touches the potential father by its edge). In these cases, the program leaves subroutine *incalc*. Otherwise, the program proceeds with determining penetration points of father's edges through mother's plane. These points are stored in the list named *intpnt*. The rule is: no point could appear in the list twice. If one of the father's vertices resides exactly within a mother's plane, it is logical that this vertex would be added to the list *intpnt*

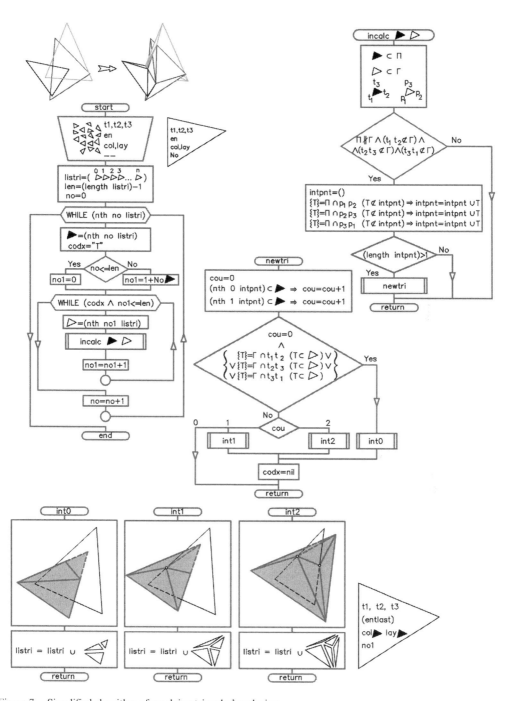

Figure 7. Simplified algorithm of resolving triangles' explosion.

twice—once for each of the two edges meeting at that vertex. But, this is prohibited and the maximum length of *intpnt* list is 2. If, after analyzing all of the three father's edges, the length of list *intpnt* stays at 1, it means that one of the father's vertices resides exactly within the mother's plane, while the remaining two vertices are located on the same side of the mother's plane. In that case, there is no collision. Also, there is no collision if the length of *intpnt* list is 0.

The collision between the potential mother and the potential father occurs (mothers' explosion takes place) only if the length of *intpnt* list equals 2. In that case, the program proceeds to subroutine *newtry* which executes the explosion. The subroutine first determines how many of the father's penetration points (points from *intpnt* list) reside within the mother triangle (the number of these points is *cou*). If no point from *intpnt* list resides within mother's triangle, but any of the mother's edges penetrates a potential father, then the collision exists and the explosion is executed by subroutine (*int0*). If no point from *intpnt* list resides within mother's triangle, but no mother's edge penetrates a potential father, then the collision does not exist and flow leaves subroutine *newtri*. If only one of the points from *intpnt* list resides within mother's triangle, the explosion is executed according to the scheme from *int1* subroutine. For two points residing within a mother's triangle, subroutine *int2* is executed. Each one of the three subroutines executing an explosion (*int0*, *int1* and *int2*) explodes the mother's triangle into the triangles that touch the father's triangle either by their edges, or by their vertices. Graphically, the mother entity (triangle) is erased, but it remains within *listri* list. If the mother's triangle belongs to the starting set of pretriangles, it could even serve as a potential father for some other triangles. Each newly created triangle (emanating from the explosion) is drawn by using mothers color and layer and is added to the end of the list *listri*. In fact, the entire group of newly created triangles (ranging in size from 3 to 5) is added to the end of *listri*. Within the list, newly created triangle is represented by its vertices, by its color, layer and father's index (*no1*).

The Aerodrome Obstruction Chart—Type A is a combination of plan and profile projection (Fig. 8). In the lower part of the document there is a relatively narrow plan depicting the approach surface, with all the obstacles marked with the symbols proposed by ICAO. The longitudinal profile resides in the upper part of the drawing. The profile spans the length of the approach (or take—off) path. All the obstacles marked in the plan projection are placed at their distinctive elevations in the profile. Runway and approach surface profiles are also superimposed.

The software for marking the obstacles and the correlation of obstacles in plan and profile projections is a rudimentary one (Gavran 1996, 2012). But the most interesting part of the obstacle profile is the terrain itself. The terrain profile is not a simple longitudinal profile cut along the extended runway centerline and following the approach path, nor a kind of combination of the profiles generated along the diverging edges of the approach path. The terrain profile is supposed to be a kind of a shadow profile (ICAO 1989). At each incremental step along the approach path, the maximum terrain elevation is taken from the terrain cross section, providing the cross section spans the exact width of the approach (or take–off) path at this particular location. The profile outlines the exact terrain shadow for the observer standing aside the approach path (Fig. 9).

To produce such a profile, we turned again to the existing software tool intended primarily for road design. This is the tool transferring 3D points labeled with station/elevation attributes into the longitudinal profile.

When setting the vertical alignment of the street, it is highly recommended to observe some specific points scattered in the vicinity of the centerline, such as entrances to nearby houses, shop windows etc. These points are marked with station / elevation attributes and transferred from 3D space into the longitudinal profile.

For the creation of a shadow terrain profile along the approach path, only a small automation is added to the existing tool. Points are now automatically attached to every vertex of each terrain triangle enclosed within the approach path and then labeled with station / offset (and elevation) assemblies in relation to the centerline of the approach path. The "cloud" of points generated in this manner is then transferred into the longitudinal profile generated

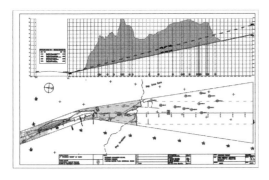

Figure 8.   Aerodrome obstruction chart—Type A.

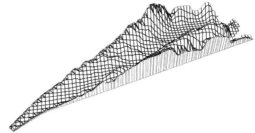

Figure 9.   Shadow terrain profile.

along the centerline of the approach path. When taken from the TIN model produced by exploding the grid model into triangles, this cloud nicely reflects the terrain morphology (Fig. 10). Finally, the outline of the shadow profile is redrawn manually, through the highest points within the profile.

4.2 *Producing non-standard graphical documents*

Besides the plan projection presented on The Aerodrome Obstruction Chart—Type B, some cross sections (perpendicular to the runway centerline) are always helpful. These cross sections usually contain terrain and the obstacle limitation surfaces. But, in the case of Trebinje airport we came to a conclusion that isopachytes' projection would give a much clearer picture than any set of cross sections. Till then, we had been using isopachytes only on resufac-ing and ground remodeling projects. Isopachytes are the contour lines delineating equal differences in elevation between the two triangulated surfaces (the proposed and the existing surface). At the location of each node (from both triangulated surfaces) the vertical difference between the two surfaces is determined and the new point, having the elevation equal to that difference, is set at this position. The TIN model generated from thus positioned points represents the thickness between the two surfaces. On grading projects, the model is negative in cut areas and positive in areas to be filled. Contours generated from such a TIN model are isopachytes (Fig. 11) (Gavran 2012).

For construction purposes, 1.00 m isopachytes are suitable for ground remodeling projects, while the interval of 1.00 cm is suitable for road resurfacing projects. On road resurfacing projects isopachytes may be used to represent the variable thickness of the leveling course (the course laid after the scraping of the existing pavement and beneath the newly applied wearing course).

In the case of Trebinje airport, apart from the terrain penetration line through the obstacle limitation surfaces, the idea was to somehow depict the sheer extent of this penetration. Therefore, the TIN model representing the "thickness" of the penetration was created and contours were generated from such a model. By definition, these contours were isopachytes. Bearing in mind the area to be covered, the scale of Type B map and the sole purpose of these isopachytes, the interval of 50 m was adopted (Fig. 12).

As the isopachytes resembled the general morphology of the terrain, the picture of the terrain penetration extent became quite clear and our design team was very satisfied with this graphical document. In essence, whenever one of the two surfaces to be compared is rather flat, then the general flow of the isopachytes resembles the contours of the opposing surface. In this particular case, obstacle limitation surfaces played the role of a "flat" surface, while isopachytes "imitated" contours of the terrain.

Unlike Trebinje airport project, in some cases we were not thoroughly satisfied with the isopachytes' application. When working on a dredging plan for

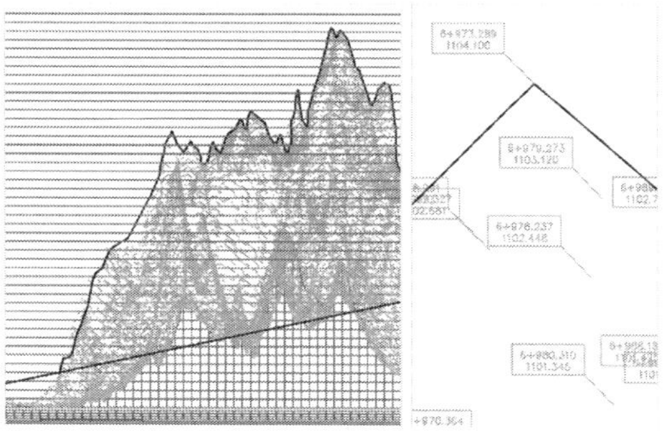

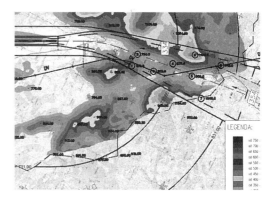

Figure 10. Transferring cloud of points into the profile.

Figure 12. Isopachytes' representing terrain protrusions.

Figure 11. Isopachytes' application on grading projects.

Kuwait harbor (Lukić et al. 2003), the isopachytes generated between the existing and the proposed bottom of the harbor were hard to follow even for the eye of a professional (Fig. 13). This happens whenever the vertical differences between the two surfaces are relatively small and when the surfaces frequently change sides in the vertical sense (between cut and fill).

Contours generated from the obstacle limitation surfaces are always welcome. In plan projection they give a general three-dimensional picture of the entire assembly of obstacle limitation surfaces. In municipality plans they impose vertical limits on the structures planned in the area surrounding the airport. With the triangulated 3D model of the obstacle lim-

itation surfaces completed, it is exceptionally easy to generate contours from such a model (Fig. 14).

5 CONCLUSION

The paper presents contemporary 3D techniques of obstacle control, while working on airport location studies. In general, 3D control of obstacles is based on comparison between the two triangulated models: the surface terrain model and the model of obstacle limitation surfaces.

Primary products of these 3D analyses are The Aerodrome Obstruction Chart—Type A and The Aerodrome Obstruction Chart—Type B, as required by ICAO. Generation of these graphical documents is supported by software tools primarily developed for terrain remodeling, grading and road design.

In addition, the deployment of triangulated 3D surfaces introduces new documents, such as isopachyte plan, which is not obligatory, but further refines the control of obstacles.

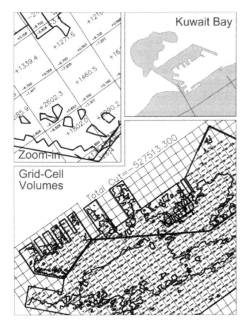

Figure 13. Kuwait harbor dredging plan.

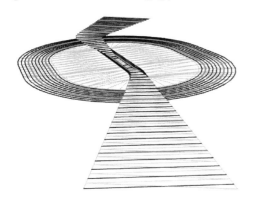

Figure 14. Generating contours from obstacle limitation surfaces.

REFERENCES

FAA, 2007. *Airport Master Plans*. Washington D.C.: U.S. Department of Transportation, Federal Aviation Administration, USA.

Gavran, D. 1996. *Razvoj metodologije i tehnoloških postupaka za prostorno projektovanje aerodroma* (in Serbian): doctoral thesis. Belgrade: Faculty of Civil Engineering University of Belgrade. 252 p.

Gavran, D. 2012. *GCM++ (Gavran - Civil Modeller)*. User Guide. Belgrade. Autodesk. 86 p.

Green, P.J. & Sibson, R. 1978. Computing Dirichlet Tessellations in the Plane. *The Computer Journal* 21(2).

Horonjeff, R., McKelvey, F.X., Sproule W.J., Young S.B. 2010. *Planning and Design of Airports. Fifth Edition.* New York: McGraw Hill.

ICAO, 1983. *Airport Services Manual—Part 6—Control of Obstacles*. Montreal: International Civil Aviation Organization.

ICAO, 1987. *Airport Planning Manual—Part 1*. Montreal: International Civil Aviation Organization.

ICAO, 1989. *Aeronautical Charts—Annex 4 to the Convention on International Civil Aviation*. Montreal: International Civil Aviation Organization.

ICAO, 2004. *Annex 14 to the Convention on International Civil Aviation*. Montreal: International Civil Aviation Organization.

Lukić, M. et al. 2003. *Kuwait Port Dredging Plan*. Belgrade/Kuwait: Al - Tawbad.

Nikolić, D. et al. 2009. *Ispitivanje lokacijskih uslova, potvrda lokacije i dopuna Master plana aerodroma Trebinje* (in Serbian). Belgrade: Faculty of Civil Engineering University in Belgrade and Public Enterprise "Airport Trebinje".

Petrie, G. & Kennie, J.M. 1987. Terrain Modelling in Surveying and Civil Engineering. *Computer Aided Design* 19(4).

Transport Infrastructure and Systems – Dell'Acqua & Wegman (Eds)
© 2017 Taylor & Francis Group, London, ISBN 978-1-138-03009-1

Environmental and economic assessment of pavement construction and management practices for enhancing pavement sustainability

J. Santos & V. Cerezo
IFSTTAR, AME-EASE, Route de Bouaye, Bouguenais, France

G. Flintsch
The Charles Via, Jr. Department of Civil and Environmental Engineering, Center for Sustainable Transportation Infrastructure, Virginia Tech Transportation Institute, Virginia Polytechnic Institute and State University, Virginia, USA

A. Ferreira
Road Pavements Laboratory, Research Center for Territory, Transports and Environment, Department of Civil Engineering, University of Coimbra, Coimbra, Portugal

ABSTRACT: Stakeholders in the pavement sector have been seeking new engineering solutions to move towards more sustainable pavement management practices. The general approaches for improving pavement sustainability include, among others, reducing virgin binder and virgin aggregate content in hot mix and warm mix asphalt, reducing energy consumed and emissions generated in the mixtures' production, applying in-place recycling techniques, and implementing preventive treatments. In this study, a comprehensive and integrated pavement Life Cycle Costing-Life Cycle Assessment (LCC-LCA) model was developed to investigate, from a full life cycle perspective, the extent to which several pavement engineering solutions (hot in-plant recycling mixtures, warm mix asphalt, cold central plant recycling and preventive treatments) are most efficient at improving the environmental and economic aspects of pavement infrastructure sustainability, when applied either separately or in combination, in the construction and management of a road pavement section located in Virginia, US. Furthermore, in order to determine the preference order of alternative scenarios, a multi-criteria decision analysis method was applied. The results showed that the implementation of a recycling-based maintenance and rehabilitation strategy where the asphalt mixtures are of type hot mix asphalt containing 30% RAP best suits the varied and conflicting interests of stakeholders. This outcome was found to be robust even when different design and performance mixtures and treatment type scenarios were considered.

1 INTRODUCTION

With the recent launch of the Build America Investment Initiative (White House 2014)—a US government-wide initiative designed to address the country's pressing infrastructure investment needs and promote economic growth—many Departments of Transportation (DOTs) are likely to renew their efforts both in the construction of new highway infrastructures and in the maintenance of those already built.

The activities central to the construction, operation and maintenance of highway infrastructures are notorious for the large quantities of natural materials and energy they consume, as well as for the considerable environmental impacts they generate. In addition, the environmental impacts of these activities, along with stringent environ-

mental regulations, have strengthened the commitment of DOTs to deliver infrastructures in a more environmentally friendly way, while also using funds in the most economically responsible way possible. These factors have motivated DOTs, and the pavement community in general, to investigate sustainable engineering solutions that reduce both environmental impacts and the costs of road pavement construction and maintenance. Some examples of solutions with the potential to improve pavement sustainability include (but are not limited to): (1) asphalt mixes requiring lower manufacturing temperatures, such as Warm Mix Asphalt (WMA), half-warm mix asphalt and cold asphalt mix technologies, (2) in-place pavement recycling, (3) pavement preservation strategies and preventive treatments, (4) long-lasting pavements, (5) Reclaimed Asphalt Pavement (RAP) materials,

(6) reclaimed asphalt shingles materials, (7) industrial wastes and byproducts, biobinders, etc.

Several studies have, to some extent, corroborated the environmental benefits with which the aforementioned solutions are a-priori associated. However, many of these studies have applied methodologies that disregard the environmental burdens of some processes and pavement life cycle phases. In addition, as the primary goal of transportation agencies is to provide maximum pavement performance within budgetary constraints, an environmentally advantageous solution might not be preferable to a technically equivalent solution if it is not economically competitive. Furthermore, there are still some questions about (1) the cost of such solutions throughout their life cycle, (2) which factors are the key drivers of their economic performance, and (3) which stakeholders benefit most from the application of those solutions.

Addressing this multifaceted issue and providing answers to the aforementioned questions requires multidimensional life-cycle modelling approaches, such as Life-Cycle Assessment (LCA) and Life Cycle Costing (LCC). These approaches enable long-term economic and environmental factors to be included in the decision-making process by providing a comprehensive and cumulative view of both the environmental and economic dimensions of a given technical solution. However, it is important to underline that life-cycle modelling approaches alone will not necessarily determine which solution is most suitable for a given purpose. Rather, the information that they provide should be used as one component of a more comprehensive decision making process, which, among other aspects, will allow the assessment of tradeoffs among the interests of the multiple stakeholders.

2 OBJECTIVES

The main objectives of this paper are as follows: (1) to investigate, from a life cycle perspective, the extent to which several pavement engineering solutions (i.e., hot in-plant recycling mixtures, Sasobit® WMA, cold central plant recycling [CCPR], and preventive treatments) best improve the environmental and economic dimensions of pavement infrastructure sustainability, when applied either separately or in combination, in the construction and management of a road pavement structure, (2) to raise awareness of the importance of considering materials and processes in extending the system boundaries of environmental and economic life cycle assessments, which may eventually reverse the sustainability of a solution and (3) provide designers, contractors, local and state agencies, and road users with an improved understanding

of how materials considerations, treatment typology, design, construction, and application timing can enhance pavement sustainability, and the related tradeoffs in these stakeholders' various requirements.

For this purpose, a comprehensive and integrated pavement LCC-LCA model, encompassing six pavement life cycle phases into the system boundaries, has been developed. They are as follows: (1) materials extraction and production, (2) construction and M&R, (3) transportation of materials, (4) work-zone traffic management, (5) usage, and (6) End-of-Life (EOL).

Finally, to account for the often conflicting interests of the multiple pavement management stakeholders, the pavement construction and maintenance scenarios considered in this paper were further analyzed using a Multi-Criteria Decision Making (MCDM) method.

3 METHODOLOGY

3.1 Principles of the integrated pavement life cycle costing and life cycle assessment model

The research work presented in this paper builds on the process-based LCA (P-LCA) and LCC models introduced by Santos et al. (2015a,c) and Santos et al. (2015b), respectively, to develop a comprehensive and integrated pavement LCC-LCA model. The proposed pavement LCC-LCA model follows a cradle-to-grave approach and relies on a hybrid inventory approach that allows the sub-models to connect with one another by data flows. Specifically, it connects the monetary flows associated with exchanges of the pavement life cycle system that are directly covered by the LCC model but for which specific process data are either completely or partially unavailable (e.g., interest on loan, taxes and insurance of construction equipment, etc.). In other cases happen that data are available, but their collection and subsequent analysis are highly demanding, either time- or resource-wise and were therefore disregarded in the previous P-LCA models (e.g., construction equipment manufacturing and maintenance, on- and off-road vehicles tires manufacturing, lubricant oil production, etc.) (Santos et al. 2015a,c). These data are combined with an input-output methodology for deriving the underpinning environmental burdens. Thus, by interactively integrating the strengths of Process-based Life Cycle Inventory (P-LCI) and Input-Out (I-O) LCI, the resources which are readily available can be used in a more efficient, consistent and rational way and with less effort, helping to reduce the "cutoff" errors and improving the consistency between the system boundaries of the pavement life cycle when analyzed concomitantly from the economic and environmental viewpoint. For this purpose, the pavement LCC-LCA model uses Carnegie Mellon University's Economic Input-Output Life Cycle Assessment tool (EIO-LCA) (CMUGDI 2010). This tool utilizes the Leontief's methodology to relate the inter-sector monetary transactions sectors in the US economy, compiled in a set of matrices by the Bureau of Economic Analysis (BEA) of the US Department of Commerce, with a set of environmental indicators (e.g. consumption of fossil energy, airborne emissions, etc.) per monetary output of each industry sector of the economy. The environmental burdens at sector level associated with a particular commodity under analysis is therefore calculated by multiplying its monetary value, previously adjusted to US dollars of the EIO-LCA model's year according to sector specific economic indices from the US Department of Labor, by the respective sectorial environmental multipliers obtained from the EIO-LCA model. The US 2002 EIO-LCA benchmark consumer price model for the US economy was preferred to the Producer Price Index model because the monetary values of the commodities whose environmental burdens the study aimed to quantify are better represented by retail price (e.g., construction equipment acquisition, tire acquisition, lubricating oil acquisition, etc.), which allows for further accounting of the environmental impacts associated with their distribution to wholesalers.

3.2 Scope of the study

3.2.1 Functional unit

The functional unit considered in this case study was defined as a 1 km one-way road pavement section of an Interstate highway in Virginia, with two lanes, each of which was 3.66 m wide. The Project Analysis Period (PAP) was 50 years (VDOT 2014), beginning in 2011 with the construction of the pavement structure. The annual average daily traffic for the first year was 20,000 vehicles, of which 25% were trucks (5% were single-unit trucks and 95% were combination trucks). The traffic growth rate was set equal to 3% per year.

The initial pavement structure was designed using the pavement structural design method AASHTO'93 (AASHTO 1993) for flexible pavements, as defined by the *Chapter V-Pavement Evaluation and Design* of the Virginia Department of Transportation's (VDOT's) Manual of Instructions for the Materials Division (VDOT 2014). The assumptions considered during the design process and the Hot Mix Asphalt (HMA) mixtures properties are presented in Santos et al. (2017). Based on those assumptions, a pavement structure was designed with a structural number of 7.38.

As far as the pavement maintenance and rehabilitation (M&R) is concerned, three main groups of alternative M&R strategies (scenarios) were considered to be available for application over the PAP of the pavement structure. The first two groups of alternative M&R strategies, hereafter named VDOT and Recycling-based VDOT strategy, respectively, were based on the M&R plan outlined by VDOT (VDOT 2014), in which functional and structural treatments and a major rehabilitation are applied in pre-established years (M&R activities ID 1, 2 and 3 in Table 2, respectively). Briefly, the functional and structural treatments consist of patching and asphalt concrete overlays with different areas of application and thickness, whereas the major rehabilitation comprises the milling and replacement of the existing bounded layers.

Table 1. Identification of the alternative M&R scenarios.

Type of scenario	Scenario ID	Scenario name
VDOT	1	HMA – 0% RAP
	2	HMA – 15% RAP
	3	HMA – 30% RAP
	4	Sasobit® WMA – 0% RAP
	5	Sasobit® WMA – 15% RAP
	6	Sasobit® WMA – 30% RAP
Recycling-based VDOT	7	HMA – 0% RAP
	8	HMA – 15% RAP
	9	HMA – 30% RAP
	10	Sasobit® WMA – 0% RAP
	11	Sasobit® WMA – 15% RAP
	12	Sasobit® WMA – 30% RAP
Preventive maintenance	13	Microsurfacing – 0% RAP
	14	THMACO – 0% RAP[a]

Table 2. M&R activities considered in each scenario, and respective application years.

M&R Scenario ID	M&R activity ID					
	1	2	3	4	5	6
1 to 6	12, 44	2	32	–	–	–
7 to 12	12, 44	22	–	32	–	–
13	9, 17, 25, 41, 49	–	32	–	7, 15, 23, 39, 47	–
14	10, 18, 27, 41 50	–	32	–	–	7, 16, 24, 39, 47

Furthermore, the two aforementioned M&R scenarios were considered to be different from each other in that only conventional asphalt materials and treatments were implemented in the first group, while in the second group the major rehabilitation (M&R activity ID 6 in Table 2) was carried out through the combination of the CCPR in-place recycling technique, and conventional asphalt layers. The recycling-based M&R activity was designed in such a way that it provided equivalent structural capacity to the non-recycling-based counterpart and took into account VDOT's surface layers requirements for layers placed over recycling-based layers (VDOT 2013). The third group of M&R strategies consisted of preventive maintenance treatments.

The VDOT and Recycling-based VDOT strategies were also further divided into HMA and Sasobit® WMA scenarios with three distinct RAP contents (0%, 15% and 30%). As for the preventive alternative maintenance strategies, two additional scenarios were considered depending on the type of preventive treatments adopted: microsurfacing and thin hot mix asphalt concrete overlay (THMACO) [M&R activities ID 4 and 5 in Table 2, respectively). A summary of all considered scenario names is provided in Table 1. For further details of the M&R activities and actions considered in the different scenarios the reader is referred to Santos et al. (2017). Table 2 shows the M&R activities considered in each scenario and the respective application years.

In order to determine pavement performance over time, VDOT's pavement Performance Prediction Models (PPPM) were used. VDOT developed a set of PPPM in units of Critical Condition Index (CCI) as a function of time and category of the last M&R activity applied. CCI is an aggregated indicator ranging from 0 (complete failure) to 100 (perfect pavement) that represents the worst of either load-related or non-load-related distresses. VDOT classifies M&R activities into five categories: (0) Do Nothing (DN), (1) Preventative Maintenance (PM), (2) Corrective Maintenance (CM), (3) Restorative Maintenance (RM), and (4) Reconstruction/Rehabilitation (RC). Using the base form corresponding to Equation (1), VDOT defines PPPM for the last three categories (Stantec Consulting Services and Lochner 2007). The coefficients of VDOT's load-related PPPM expressed through the Equation (1) for asphalt pavements of Interstate highways are presented in Table 3 (Stantec Consulting Services and Lochner 2007).

$$CCI(t) = CCI_0 - e^{a + b \times c^{\ln\left(\frac{1}{t}\right)}} \qquad (1)$$

where $CCI(t)$ is the CCI in year t since the last M&R activity (i.e., CM, RM or RC); CCI_0 is the

Table 3. Coefficients of VDOT's load-related PPPM expressed by the Equation (1) for asphalt pavements of interstate highways.

M&R activity category	CCI_0	a	b	c
CM	100	9.176	9.18	1.27295
RM	100	9.176	9.18	1.25062
RC	100	9.176	9.18	1.22777

critical condition index immediately after treatment; and a, b, and c are the load-related PPPM coefficients (Table 3).

VDOT did not develop individual PPPM for PM treatments. Thus, in this case study, the considered PM treatments (i.e., microsurfacing and THMACO) were respectively modelled as an 8-point and 15-point improvement in the CCI of a road segment that takes place whenever the CCI falls below the trigger value of 85 (Chowdhury 2011). Once the treatment is applied, it is assumed that the pavement deteriorates according to the PPPM of a CM, without reduction of the effective age. In the case of the application of CM, RM and RC treatments, the CCI is brought to the condition of a brand new pavement (CCI equal to 100) and the age is restored to 0 regardless of the CCI value prior to the M&R activity application.

To estimate the environmental impacts and costs incurred by road users during the pavement usage phase resulting from vehicles traveling over a rough pavement surface, a linear roughness prediction model, expressed in terms of International Roughness Index (IRI), was considered (Equation 2).

$$IRI(t) = IRI_0 + IRI_{grw} \times t \qquad (2)$$

where $IRI(t)$ is the IRI value (m/km) in year t, IRI_0 is the IRI immediately after the application of a given M&R activity and IRI_{grw}, which was set at 0.08 m/km (Bryce et al. 2014), is the IRI growth rate throughout time. It was assumed that the application of an M&R activity other than PM restored the IRI to the value of a brand new pavement (IRI equal to 0.87 m/km). The IRI reduction due to the application of a PM treatment was determined based on the expected treatment life and assumed that there was no change in the IRI_{grw} value after the PM application (the same assumption was also made in the case of the remaining M&R activities). By assuming treatment life cycle periods of three and five years (Chowdhury 2011), respectively, for microsurfacing and THMACO preventive treatments, the resulting reductions in the IRI value were found to be 0.24 and 0.40 m/km.

3.2.2 System boundaries, system processes, life cycle inventory data and main assumptions

The pavement LCC-LCA model includes six pavement life cycle phases but the environmental burdens associated with the EOL phase were disregarded based on the consideration of the "cut-off" allocation method. Also excluded from the system boundaries were the environmental burdens due to the transportation of the workforce to the workplace. Furthermore, only real monetary flows were accounted for in order to avoid double counting the environmental impacts (Swarr et al. 2011).

For the sake of brevity only the methodology used to model the materials production phase is presented in detail. The various models used for each component of the remaining pavement life cycle phases, as well as the main data required to run those models, can be found in Santos et al. (2017).

In this case study it was assumed that all asphalt mixes were produced through a natural gas-fired conventional drum-mix plant. In a conventional drum mix plant, to prevent additional aging of the RAP binder, RAP is not heated directly. Instead, the virgin aggregates are superheated before-hand so that when the RAP is introduced into the drum it is dried and heated by conduction. Such a superheated temperature is likely to cause additional energy consumption, which may eventually offset the RAP's associated economic and environmental benefits.

In order to assess these tradeoffs, along with the sensitivity of the air emissions due to the variations in composition and manufacturing temperature of the mixes and the moisture content of the raw materials, the heat energy required to produce the asphalt mixes was determined through an energy balance represented by Equation (3).

$$Q = \frac{\sum_{i=0}^{M} m_i \int_{T_{0i}}^{T_{fi}} C_i(T)dT + L_v \times \left(m_{wvf} - m_{wvo} \right)}{Heating\ Eff\ F} \quad (3)$$

where Q is the heat energy required to produce the asphalt mixture (J); m_i is the mass of material i (kg); M is the total number of materials, including water; T_{fi} is the final temperature of the material i (°C); T_{0i} is the initial temperature of the material i (°C); $c_i(T)$ is the specific heat capacity coefficient, as a function of temperature, of material i [J/(kg/°C)]; L_v is the latent heat required to evaporate water (2256 J/kg); m_{wvf} is the final mass of water vapor (kg); m_{wvo} is the initial mass of water vapor (kg); and $HeatingEffF$ is a factor that represents the casing losses.

To account for the fact that specific heat capacities of minerals and fluids increase substantially with temperature, the equations presented by Waples and Waples (2004a,b) were adopted, taking the temperature of 20°C as the reference temperature. The heating requirements for the aggregates applied in bound layers other than surface layers were modeled by considering the specific heat value of limestone (880 J/[kg/°C]). In the case of the surface layers, the value for quartzite (1013 J/[kg/°C]) and diabase (860 J/[kg/°C]) were taken to represent the aggregate used in the surface mixes and THMACO, respectively. With regard to binder and water, the third equation proposed by Gambill (1957) and the equation developed by Somerton (1992), both cited and displayed in Waples and Waples (2004b), were adopted, respectively. The initial moisture contents of fine and coarse aggregates were assumed to be 3% and 1%, whereas for RAP a value of 4% was assumed. As for the *HeatingEffF*, a value of 80% was adopted for the production of all mix types after calibrating the model with the data corresponding to the HMA production in the case study of Munster, Indiana, reported by West et al. (2014). The HMA mixing temperature was set at 160°C and the initial temperature of all raw materials other than bitumen was assumed to be equal to the ambient temperature of 15°C. Bitumen was assumed to be stored at 160°C in heated tanks located in the asphalt plant facility. The volume of natural gas required to heat the insulated storage tanks was calculated based on the total quantity of binder heated, the total time the bitumen spent in the tanks throughout the paving season and the heat capacity of the tanks. As for the WMA, whose mix design was considered the same as that of the homologous HMA, it was assumed that the addition of 1.5% of Sasobit® per mass of bitumen reduced the mixing temperature by 25°C in relation to the reference temperature of 160°C. This assumption was based on the range values of reduction of temperature of 20–30°C commonly referred to in the literature (D'Angelo et al. 2008). Moreover, it was also assumed that the RAP used in WMA could be blended with new asphalt binder at this lower temperature.

In order to determine the air emissions resulting from the mixing process of all mixes considered in this case study, a methodology was developed based on the Emission Factors (EFs) published by the AP-42 study of HMA plants (US EPA 2004) corresponding to a natural gas-fired filter-controlled drum-mix plan, and the thermal energy required to produce the asphalt mixes. Firstly, the average EFs referring to the production of an HMA with 0% RAP were taken as reference. Secondly, as CO_2 emissions primarily result from fuel combustion, the average emission of this Greenhouse

Gas (GHG) was combined with the fuel emission coefficient (53.1 kg/MMBtu) reported by US Energy Information Agency (EIA) to determine the quantity of natural gas whose combustion would release the same amount of CO_2 (US EIA 2013). Thirdly, for each mix, an EF multiplier was determined through the ratio between the thermal energy computed with Equation (3) and the thermal energy calculated according to the procedures previously described. Finally, GHGs and air pollutant EFs from the mix's production were derived by multiplying the EFs taken as reference by the EF multipliers. The values of the EF multipliers and the natural gas consumption requirements for producing all mixes considered in this case study can be found in Santos et al. (2017). Emissions and energy consumption due to the operation of the wheel loader at the asphalt plant facility were estimated based on the rate at which the wheel loader could move aggregates (Santos et al. 2015c) and the methodology adopted by the US EPA's NONROAD 2008 model (US EPA 2010).

In addition to the process-based components described throughout this section, the input-output life cycle inventory approach was adopted to estimate the environmental burdens associated with the manufacturing, repair, maintenance, interest on loan and insurance of the asphalt plant setup and auxiliary equipment. The environmental burdens were amortized by applying the portion of the asphalt plant's setup and auxiliary equipment depreciation that was allocated to the quantity of asphalt mixes consumed in a given construction activity and considering the average annual production of asphalt mixes. A similar approach was adopted in the construction, M&R, and transportation of materials phases for determining the environmental burdens associated with the construction equipment and hauling trucks, but taking as allocation factors the number of usage hours and hauling km traveled for a given construction activity.

As for the economic dimension, the proposed pavement LCC-LCA model accounts for the costs incurred by the highway agencies during the construction and M&R of the pavement, by the road users when facing a disruption of the normal traffic flow as a consequence of the constraints imposed by a work zone traffic management plan, and throughout the PAP as a consequence of the pavement's deterioration.

3.3 Life cycle impact assessment and life cycle costs computation

The US-based impact assessment methodology, the Tool for the Reduction and Assessment of Chemical and other environmental Impacts 2.0 – TRACI 2.0 (Bare et al. 2011) developed by the US EPA,

was adopted in this study to conduct the impact assessment step of the LCA. The TRACI impact categories used in the analysis include acidification air, eutrophication air, human health criteria pollutants, and photochemical smog formation. The time-adjusted characterization model for the Climate Change (CC) impact category that was proposed by Kendall (2012) was used, as opposed to the traditional time-steady International Panel on Climate Change model. Furthermore, three energy-based indicators were also included in the assessment: (1) primary energy obtained from fossil resources, (2) primary energy obtained from non-fossil resources and (3) feedstock energy. The feedstock energy was fully allocated to the virgin binder, with none attributed to RAP. This assumption aimed to avoid double counting, as it should have been accounted for in the previous pavement system.

In computing the life cycle costs, the concept of net present value was applied and a real discount rate of 2.3% was adopted (OMB 2013).

4 RESULTS AND DISCUSSION

4.1 Overall performance

In order to determine the preference order of alternative scenarios, the MCDM Technique for Order of Preference by Similarity to Ideal Solution (TOPSIS) method was applied. Three main criteria were considered: Highway Agency Costs (HAC), Road User Costs (RUC) and environmental impacts. The last criterion was further broken down into eight sub-criteria, each representing one environmental impact category. To inform decision-makers about the consequences of the weighting in the ranking of the alternative scenarios, a combinatorial weight assignment method was used for the main criteria, while the weights assigned to the environmental sub-criteria remained unchangeable and equal to those adopted by the US-based Building for Economic and Environmental Sustainability (BEES) software (Lippiatt 2007). Since the energy demand indicators considered in the proposed LCC-LCA model are not available in the BEES software, they were given a weight of five points each, which is the weight assigned to the Fossil Fuel Depletion impact category considered in BEES. All the weights assigned to the environmental sub-criteria were posteriorly rescaled, so that the sum of their values totaled 100 points. Thus, in the MCDM, the final weight of each environmental sub-criterion is the value resulting from multiplying the weight of the main environmental criterion by the weight determined, as explained above. The best scenario for all possible weighting combinations between the three main criteria is displayed in Figure 1 through

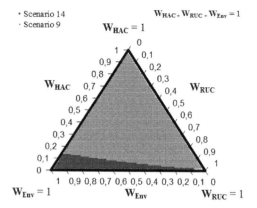

Figure 1. Best scenarios for all possible weighting combinations of the main criteria. (Note: W_{HAC} = weight assigned to highway agency costs; W_{RUC} = weight assigned to road user costs; W_{Env} = weight assigned to environmental impacts).

a triangular diagram. Each point in the triangle area corresponds to a specific weighting set and the relative weights always add up to a total weight of 1 (or 100%). From Figure 1, we can conclude that of the competing scenarios, scenarios 9 and 14 rank best. Of those, if the decision is exclusively based on either HAC or RUC, scenario 9 ranks best. Alternately, if the environmental performance is the only criterion taken into account, then scenario 14 outperforms the remaining scenarios.

In order to give the decision-maker a clearer picture of the drawbacks and advantages of each possible technological solution, expressed in terms of emissions, costs and consumption of natural resources, Table 4 presents for each alternative solution the values of the main criteria as well as the respective raking when they are considered to be equally important. For the sake of brevity, the CC impact category is taken as representative of the environmental criteria.

4.2 Sensitivity and scenario analysis

Due to their recent application, there is a lack of results obtained from comprehensive field studies about the long-term performance of road pavements that incorporate new pavement engineering solutions. Thus, it is pertinent to consider that new paving materials/solutions may not be as durable as conventional materials.

Having this aspect in mind, a sensitivity and scenario analysis were conducted to examine how variations across a set of parameters and assumptions affect the outcomes, and thereby, the relative merits of the alternatives being compared. For this study, the one-(factor)-at-a-time sensitivity analysis method was used (Pianosi et al. 2016). In this method, output variations are induced by varying one input factor at a time, while all others are held at their default values. Table 5 presents the triangular diagrams that display the best scenarios for

Table 4. Values of the main criteria for each M&R scenario, as well as the respective raking, when the main criteria are considered to be equally important.

M&R Scenario ID	CC (tonnes CO_2-eq)	HAC (K$)	RUC (K$)	Ranking
1	4 262	974	3 044	11
2	4 217	880	3 044	7
3	4 178	768	3 044	3
4	4 263	995	3 044	12
5	4 220	901	3 044	9
6	4 178	788	3 044	4
7	4 121	904	2 978	8
8	4 082	822	2 978	5
9	4 049	722	2 978	1
10	4 121	923	2 978	10
11	4 083	840	2 978	6
12	4 047	740	2 978	2
13	3 897	1 001	3 702	14
14	3 404	1 059	2 956	13

Table 5. Results of the sensitivity and scenario analysis expressed in terms of the best M&R scenarios for all possible weighting combinations of the main criteria.

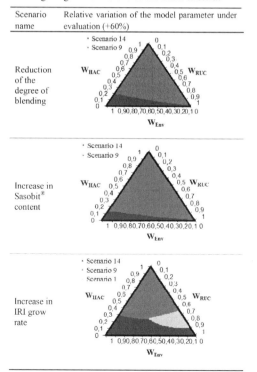

Scenario name	Relative variation of the model parameter under evaluation (+60%)
Reduction of the degree of blending	
Increase in Sasobit® content	
Increase in IRI grow rate	

all possible weighting combinations between the three main criteria when each model parameter is individually changed +60% with respect to the base scenario. Table 5 clearly shows that the position of scenarios 9 and 14 in the ranking of the best compromise solutions is robust. In general, only a slight change is seen when considering scenarios different from the baseline. The only exception to this outcome is observed when different IRI grow rates are considered. In these circumstances, for a given set of weights, scenario 1 was found to be the best compromise scenario. This happens when the weight values assigned to the RUC criterion are high while those assigned to the remaining criteria are low, and can be explained by the increase in the roughness-related user costs associated with higher IRI grow rates.

5 SUMMARY AND CONCLUSIONS

This paper presents a comprehensive and integrated pavement LCC-LCA model to investigate the potential environmental and economic benefits resulting from applying in-plant recycling mixtures, WMA, cold in-place recycling techniques and preventive treatments throughout the life cycle of a pavement structure.

For the conditions considered in this case study, the recycling-based VDOT M&R strategy, with asphalt mixtures of type HMA containing 30% RAP, has been shown to be more compatible with highway agency and road users' demands for affordable road maintenance and usage over its life cycle than the remaining technical solutions investigated.

Furthermore, a sensitivity and scenario analysis were undertaken to assess the robustness of the outcomes in response to variations in some of the most relevant input values. The analysis showed that variances to the key assumptions considered when assessing the life-cycle environmental and economic performances of multiple pavement construction and maintenance practices do not considerably alter the overall advantage of implementing a recycling-based VDOT M&R strategy, where the asphalt mixtures are of type HMA containing 30% RAP.

ACKNOWLEDGEMENTS

This work has been supported by the project EMSURE- Energy and Mobility for Sustainable Regions (CENTRO-07-0224-FEDER-002004) and by the Transportation Pooled Fund TPF-5(268) National Sustainable Pavement Consortium. João Santos wishes to thank the Portuguese Foundation of Science and Technology (FCT) for a personal research grant (SFRH/BD/79982/2011).

REFERENCES

American Association of State Highway and Transportation Officials (AASTO), 1993. Guide for design of pavement structures. 4th ed. Washington, DC: American Association of State Highway and Transportation Officials.

Bare, J., 2011. TRACI 2.0: The tool for the reduction and assessment of chemical and other environmental impacts 2.0. *Clean Technologies and Environmental Policy*, 13 (5): 687–696.

Bryce, J., Flintsch, G. & Hall, R., 2014. A multi criteria decision analysis technique for including environmental impacts in sustainable infrastructure management business practices. *Transportation Research Part D: Transport and Environment*, 32: 435–445.

Carnegie Mellon University Green Design Institute (CMUGDI), 2010. *Economic input-output life cycle assessment (EIO-LCA)*, U.S. 2002 Industry Benchmark model. Available from: http://www.eiolca.net

Chowdhury, T., 2011. *Supporting document for the development and enhancement of the pavement maintenance decision matrices used in the needs-based analysis*. Virginia Department of Transportation, Maintenance Division, Richmond, VA.

D'Angelo, J., Harm, E., Bartoszek, J., Baumgardner, G., Corrigan, M., Cowsert, J., et al., 2008. *Warm-mix asphalt: European practice*, Report No. FHWA-PL-08-007, Federal Highway Administration, U.S. Department of Transportation.

Gambill, W., 1957. You can predict heat capacities. *Chemical Engineering*, 64: 243–248.

Kendall, A., 2012. Time-adjusted global warming potentials for LCA and carbon footprints. *The International Journal of Life Cycle Assessment*, 17: 1042–1049.

Lippiat, B., 2007. *BEESRG 4.0: building for environmental and economic sustainability technical manual and user guide* (The National Institute of Standards and Technology Report No. 7423).

Office of Management and Budget, 2013. *Discount rates for cost-effectiveness, lease purchase, and related analyses*. Table of past years discount rates from Appendix C of OMB Circular No. A–94.

Pianosi, F., Beven, K., Freer, J., Hall, J., Rougier, J., Stephenson, D. & Wagener, T. 2016. Sensitivity analysis of environmental models: a systematic review with practical workflow, *Environmental Modelling & Software*, 79: 214–232.

Santos, J., Flintsch, G. & Ferreira, A. 2017. Environmental and economic assessment of pavement construction and management practices for enhancing pavement sustainability, *Resources, Conservation and Recycling*, 116:15–31.

Santos, J., Ferreira, A. & Flintsch, G., 2015a. A life cycle assessment model for pavement management: methodology and computational framework, *International Journal of Pavement Engineering*, 16(3): 268–286.

Santos, J., Bryce, J., Flintsch, G. & Ferreira, A., 2015b. A comprehensive life cycle costs analysis of in-place recycling and conventional pavement construction and maintenance practices. *International Journal of Pavement Engineering*, (available online). http://dx.doi.org/10.1080/10298436.2015.1122190

Santos, J., Bryce, J., Flintsch, G., Ferreira, A. & Diefenderfer, B., 2015c. A life cycle assessment of in-place recycling and conventional pavement construction and maintenance practices. *Structure and Infrastructure Engineering: Maintenance, Management, Life-Cycle Design and Performance*, 11(9): 119–1217.

Somerton, W., 1992. *Thermal properties and temperature-related behavior of rock/fluid systems.* Developments in Petroleum Science, 37, Amsterdam: Elsevier.

Stantec Consulting Services & Lochner, H., 2007. *Development of performance prediction models for Virginia department of transportation pavement management system.* Virginia Department of Transportation, Richmond, VA.

Swarr, T., Hunkeler, D., Klöpffer, W., Pesonen, H.-L., Ciroth, A., Brent, A. & Pagan, R., 2011. *Environmental life cycle costing: a code of practice.* Pensacola (FL): Society of Environmental Chemistry and Toxicology (SETAC).

United Sates Energy Information Administration (US EIA), 2013. *Carbon dioxide emissions coefficients.*

United States Environmental Protection Agency (US EPA), 2010. *Exhaust and crankcase emission factors for nonroad engine modelling - compression-ignition* (Report No. NR-009d).

United States Environmental Protection Agency (US EPA), 2004. *AP-42: Compilation of air pollutant emission factors* (Vol. 1: Stationary point and area sources, Chap. 11: Mineral products industry 11.1).

Virginia Department of Transportation (VDOT), 2014. *Manual of instructions for the materials division.* Virginia Department of Transportation Materials Division.

Virginia Department of Transportation (VDOT), 2013. *Project selection guidelines for cold pavement recycling.* Virginia Department of Transportation Materials Division.

Waples, D. & Waples, J., 2004a. A review and evaluation of specific heat capacities of rocks, minerals, and surface fluids, part 1: mineral and non-porous rocks. *Natural Resources Researches*, 13(2): 97–122.

Waples, D. & Waples, J., 2004b. A review and evaluation of specific heat capacities of rocks, minerals, and surface fluids, part 2: fluids and porous rocks. *Natural Resources Researches*, 13(2): 123–130.

West, R., Rodezno, C., Julian, G., Prowell, B., Frank, B., Osborn, L. & Kriech, T., 2014. *Field performance of warm mix asphalt technologies* (National Cooperative Highway Research Program Report No. 779). Transportation Research Board, Washington, D.C.

The White House, 2014. *FACT SHEET: Building a 21st century infrastructure: increasing public and private collaboration with the Build America investment initiative*, Office of the Press Secretary.

Transport Infrastructure and Systems – Dell'Acqua & Wegman (Eds)
© 2017 Taylor & Francis Group, London, ISBN 978-1-138-03009-1

Road route planning for transporting wind turbines in Europe

F. Autelitano, E. Garilli & F. Giuliani
Dipartimento di Ingegneria Civile, dell'Ambiente, del Territorio e Architettura—DICATeA, University of Parma, Parma, Italy

ABSTRACT: The upward trend in wind energy production had created a challenge for the roadway system. The road infrastructure had to accommodate a wide range of complex vehicle configurations for the transport of wind turbine components (nacelles, tower sections and rotor blades), which are often considered oversize/overweight loads by the transport authorities. The paper provided an overview of established difficulties encountered during the road transport of wind turbine components, examining how the growth in the size of them, had affected the transport industry and the route assessment process. The authors proposed an operational planning strategy, based on maximum swept path width, to facilitate the identification of optimal routes for rotor blades transport. The methodology, which does not substitute a rigid swept path analysis, could be used by the road managing authorities for checking planimetric restrictions and constraints on their road infrastructure and for a ready evaluation of possible abnormal road transport corridors.

1   INTRODUCTION

The use and development of renewable energy technologies plays an important role at European level to respond to climate change and move towards a low carbon future. Among renewable energy sources, wind power represents one of the most important in the world, with an extraordinary development in the last few years (EAWA, 2016; IEA-ETSP & IRENA, 2016; WWEA, 2016a).

Transport of wind turbine components (nacelle, tower sections and rotor blades) from the factory floor to the project site represents a critical part of the logistic and cost structure of a wind project, which requires specialized vehicles and suitable infrastructures (Hau, 2013). It involves handling sensitive and valuable components that can weigh several tonnes and be well over fifty meters in length (Cotrell et al., 2014). There are three potential stages of turbine component delivery: construction, operation and decommissioning. Initially, all turbine components are delivered to the site during the construction period for installation. Throughout the operational phase of the wind farm, it is possible that individual components may require replacement. The expected life of the wind farm is 20–25 years, after which the turbines will be decommissioned (Gash & Twele, 2012). Some efficient transportation solutions for delivering turbine components were improved. But, even when other modes of transport are employed (rail, water and air) for some or most of the journey, trucks are usually required for the last leg in order to carry turbine components in austere or landlocked territories.

Road transport of wind turbine components represents both an economic problem and a technical one (Cotrell et al., 2014; Neff & Bai, 2012). These components, which are considered abnormal loads (overweight/oversize) by transport authorities, do not fit easily with road infrastructure and conventional transport methods and vehicles. On the one hand, transport is an important capital cost component. Abnormal vehicles must obtain permits for every state through which they will travel and are frequently required to be accompanied by escort vehicles, with one escort in front and one behind. International literature has listed that transport cost is approximately the 6–7% of capital cost of the entire wind project (Cotrell et al., 2014; Ray, 2007; TPI Composite, 2003). On the other hand, the ever-increasing demand for energy and the growing popularity of green power has led designers and engineers to build larger turbine units, which provide better land utilization and maximization of electricity production (Abu-Rub et al., 2014). As turbines grow, it becomes more difficult to transport their components, which are already pushing the limits of what can be carried by trucks. In the European context these large and oddly shaped trucks have usually to pass through at least one major urban center and sometimes to travel through residential neighborhoods. Moreover, they often can not fit through small underpasses or around sharp corners and heavy shipments may not be able to pass safely over certain bridges and overpasses (IWEA, 2011; Shihundu & Morrall, 2011).

From that standpoint, the paper examined several established difficulties encountered during the road transport of wind turbine components, providing a critical analysis of the strategies and technologies used to overcome them. The authors wanted to propose theoretical and practical transport strategies, mainly based on swept path analysis, for a ready route assessment and the identification of abnormal road transport corridors already at planning level.

2 WIND POWER IN EUROPE

The wind power has raised worldwide dramatically more than other renewable energy resource, reaching a wind power generation capacity of more than 456 GW at the end of June 2016 (WWEA, 2016b). Europe was the undisputed global leader in wind energy technology until about the year 2010 and now represents about the one third of the global market with 150 GW of installed wind power capacity: 138 GW onshore and 12 GW offshore. Germany is the EU country with the largest installed capacity (47 GW), followed by Spain (23 GW), UK (14 GW), France (11 GW) and Italy (9 GW) (EAWA, 2016; Wind Europe, 2016; WWEA, 2016b). The European onshore wind market is still quite dynamic: some countries have a considerable market growth potential (Norway, Poland and Turkey) and others are still experiencing a modest growth (Germany, France and The Netherlands). But, there are some saturated markets (Denmark and UK), in which little or no new installed capacity will be added in the future (Henzelmann et al., 2016).

Regarding the markets and manufacturers, the FTI Consulting annual report (2016) on wind turbine original equipment manufacturers said that 2015 had been the first year in which a Chinese firm (Goldwind) took the top spot as the world's largest suppliers of turbine (Table 1).

These data were strongly influenced by the surging Chinese market that installed a record of 28.7 GW last year. Nearly all the capacity of Chinese suppliers was reserved for the home market, where foreign manufacturers accounted less than 5% of market share. The acquisition of French transport and electricity conglomerate Alstom, gave to General Electric (GE) a firm foothold in Europe. In the light of these considerations, the European wind service market is currently dominated by Vestas, GE, Siemens, Gamesa and Enercon, even though other suppliers, such as Nordex-Acciona (Germay-Spain) and Senvion (Germany) plays an active role. Some, like Vestas and Enercon, are wind pure-players, whereas others, such as GE and Siemens, count wind turbine as only part of their energy generation interest.

Table 1. World wind turbine market share in 2015.

Manufacturer	Nation	Market share (%)
Goldwind	China	12.5
Vestas	Denmark	11.8
General Electric	USA	9.5
Siemens	Germany	8.0
Gamesa	Spain	5.4
Enercon	Germany	5.0
United Power	China	4.9
Ming Yang	China	4.1
Envision	China	4.0
CSIC Haizhuang	China	3.4
Others	–	31.4

The functioning principle of wind turbine is to convert the kinetic energy contained in the wind into mechanical power which can be used for specific tasks or transformed into electricity via electromagnetic induction. The basic elements of the wind power system are the blades, the rotor hub, the nacelle and the tower (Fig. 1). Specifically, the blades capture and convert the wind's energy to rotational energy which is transferred by a rotor hub to the rotor shaft. The rotor shaft is also connected to the gearbox, which changes the low rotating speed from the blades to a high rotating speed for input to the generator. The rotor shaft, rotor brake, gearbox and generator components are housed within a nacelle. The rotor blades, rotor hub and nacelle are supported by and elevated on a tower. Most large wind turbines are up-wind horizontal-axis turbines with three blades (Gash & Twele, 2012; IEA-ETSP & IRENA, 2016).

For utility-scale (MW-sized) sources of wind energy, a large number of wind turbines are usually built close together to form a wind plant, also called wind farm. The onshore or land-based technology has evolved over the last 10 years to reduce the Levelized Cost of Energy (LCOE) (Abu-Rub et al., 2014; IEA-ETSAP & IRENA, 2016). Machines have become bigger with taller hub heights, larger rotor diameters and in some cases bigger generator, depending on the specific wind and site conditions. The growing trends of emerging turbine size and power from 1980 are shown in Figure 2 (EAWA, 2009; IEA-ETSAP & IRENA, 2016). Currently, manufacturers offer utility-scale turbines with rotor diameters ranging from 70 m to 140 m, generators from 1.7 MW to 5 MW and hub heights from 60 m to 150 m. In addition, some suppliers have started to offer 6–8 MW onshore platforms with a rotor diameter over 150 m. Table 2 provides an overview of the commercially available onshore solutions at 2016 offered by the most

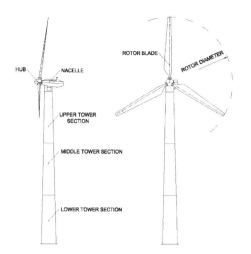

Figure 1. Basic components of an horizontal-axis turbine with three blades.

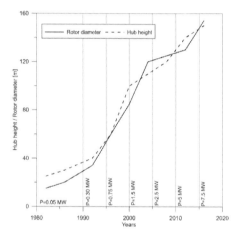

Figure 2. Growth of wind turbines.

important European manufacturers. Each turbine is characterized by an alphanumeric acronym: the letter is representative of the manufacturer, the number describes the rotor diameter in meters and the number in brackets gives the rated power in megawatts.

3 WIND TURBINE COMPONENTS AS ABORMAL LOADS

These days heavier nacelles, bigger tower section and longer blades, mean that wind turbine components are often considered abnormal or super loads by road transport authorities. The European Union had a clear legislation since 1996 (Directive 96/53), recently revised (Directive 2015/719), on allowed weights and dimensions (length, width, height and overhang) in road transport, even though there are big differences between the rules and procedures currently applied in the Member States. According to these standards, the abnormal loads are those that exceed maximal accepted weight (overweight) and/or dimensions (oversize). A load is considered overweight when it exceeds 40 t for articulated vehicles and 44 t for articulated vehicles loaded with 12.00 m containers. A load is defined oversize when the total length of the vehicle exceeds 12.00 m or 16.50 m in the case of an articulated vehicle (18.75 m for road train), or the overall width exceeds 2.55 m, or the overall height exceeds 4.00 m. Vehicle weight is a hard constrain, since excessive loads can damage pavements or exceed the bearing capacity of bridges and culverts. The dimensional constrain underlie different road transport problems. The principal factors limiting the permissible height are the clearances under any overhead bridges or power lines. Abnormally long and wide vehicles could have difficulty in

Table 2. Overview of the European commercially available onshore wind turbines at 2016.

VESTAS	GE	SIEMENS	GAMESA	ENERCON	NORDEX ACCIONA	SENVION
V90(1.8/2.0)	GE100(1.7)	SWT82(2.3)	G80(2.0)	E70(2.35)	AW70(1.5)	MM82(2.0)
V100(1.8/2.0)	GE103(1.7)	SWT93(2.3)	G87(2.0)	E82(2.35)	AW77(1.5)	MM92(2.0)
V110(2.0)	GE 82.5(1.85)	SWT101(2.3)	G90(2.0)	E92(2.35)	AW82(1.5)	MM100(2.0)
V90(3.0)	GE87(1.85)	SWT108(2.3)	G97(2.0)	E103(2.35)	N90(2.5)	M122(3.0–3.2–3.4)
V105(3.45)	GE116(2.0/2.3)	SWT120(2.3)	G114(2.0)	E101(3.05–3.5)	N100(2.5)	M114(3.2–3.4)
V112(3.45)	GE107(2.2/2.4)	SWT101(3.0–3.2–3.4)	G106(2.5)	E115(3.2)	N117(2.4)	M104(3.4)
V117(3.45)	GE88(2.5)	SWT108(3.0–3.2–3.4)	G114(2.5)	E126(4.2)	AW100(3.0)	M144 (3.4)
V126(3.45)	GE100(2.5–2.75)	SWT113(3.2)	G126(2.5)	E141(4.2)	AW116(3.0)	M126(6.2)
V136(3.45)	GE120(2.5–2.75)	SWT130(3.3)	G132(3.3)	E126(7.58)	AW125(3.0)	M152(6.2)
	GE103(2.75–2.85)	SWT154(6.0)	G128(5.0)		AW132(3.0)	
	GE103(3.2)	SWT154(7.0)	G132(5.0)		N100(3.3)	
	GE130(3.2)				N117(3.0)	
	GE137(3.4)				N131(3.0)	

travelling around the sharp curves, particularly in urban areas, in mountain passes, on highway ramps, over certain roads with short vertical curves and some bridges. Moreover, these loads can cause obstruction and danger to other road users.

The wind turbine components are usually transported separately as a single abnormal indivisible load with special vehicle and telescopic multi-axle flat-bed trailers and assembled on site. A single turbine can require up to 10 hauls, not counting the ones for the crane: 1 nacelle, 1 hub, 3 blades and 3 to 5 tower sections. The wind turbine nacelle represents an overweight load (Fig. 3). It can weigh between 70–120 t (3 MW) to 130–240 t (5 MW) depending on its drive train concept. Dimensions instead do not typically exceed standard construction trailer dimensions.

Single-piece towers are occasionally used, whereas in nearly all cases the steel or concrete towers are produced in 3 to 5 sections of 25–35 m (25–55 t). These sections are transported in a horizontal position by low-bed trailers (Fig. 4) or semi-trailer equipped with clamp lift adapters. The tower transport challenge is therefore twofold: each section can be considered as an overweight and overheight load. Tower sections vary in length

Figure 3. Transport of a nacelle.

and dimension depending on where they are in the tower. The towers have a conical shape, with a diameter that diminishes from the base up to the tower head, so that lower sections (base) are the shorter and heavier whereas higher sections (top) are taller and lighter. The most critical section with respect to its transport is therefore the bottom one. Bottom tower sections are generally limited to 4.3–4.6 m in diameter, but already in the case of a 100 m tower, the bottom section diameter is already over 5 m so that this segment can scarcely be transported by road in one piece. The only way out is manufacturing it in half-shells which must be welded together at the installation site (Gash & Twele, 2012; Hau, 2013).

The blade transport challenge is caused by the difficulty of transporting long blades around turns, through narrow passages, and beneath overhead obstructions. Some manufacturers have attempted to design turbine blades that can be broken into multiple pieces during transport and resemble on site. So far, these efforts have met with no success because segmented turbines are too fragile after reassembly and do not retain their structural integrity under strong winds (Hau, 2013). The rotor blades in the range of 35 and 40 m are mostly transported on edge in a pack of three in a specially designed long trailer (Fig. 5). Longer rotor blades are instead carried individually in vertical position (Fig. 6). When the maximum blade chord length exceeds 4 m the passage below most bridges (overhead height more of 5 m) is no longer possible. In this case, a special blade pitching mechanism is used to turn the blade into the horizontal position before passing below bridges.

Some extendible trailers are equipped with an extra moveable support beam on top of the load floor. This makes it possible to retract the vehicle during transport without moving the load, by temporarily creating a shorter wheelbase even the tightest bends can be negotiated. Recently, a new rotor blade adapter (hydraulic lifter) was designed

Figure 4. Transport of a tower segment.

Figure 5. Transport of a pack of three rotor blades.

Figure 6. Transport of a longer rotor blade.

Figure 7. Transport of a blade lifted with the blade adapter.

for lifting and tilting upwards the wind blade up to approximately 60° (Fig. 7). Thus, the position of the rotor blade can be adjusted to the road, guaranteeing a flexible transport.

4 ROUTE PLANNING AND SWEPT PATH ANALYSIS

The transport of wind turbine components provides a preliminary phase, defined route planning, which includes the research of fully possible delivery routes and the individuation of the most efficient path which best satisfies technical, financial and functional criteria (Bazaras et al., 2013; Neff & Bai, 2012). The route assessment process consists in finding, with GIS and on-site inspection, a route with adequate width and height clearances (especially in tunnel and bridge) and suitable turning radii (Petraška & Palšaitis, 2012; Ray, 2007). Once identified the most efficient solution, the recognition of critical points in the route is necessary. The dimensional requirements of abnormal vehicle/loads may require alterations to the existing road infrastructure (e.g. widening on corners

and barriers), accommodation of street furniture (e.g. lighting, traffic signals, power lines) and protection and upgrade of road related structures (e.g. bridges and culverts). Analysis of each constraint or group of constraints includes any concerns regarding vertical alignment and swept path analysis to assess horizontal alignment.

Vertical alignment issues are difficult to assess during a desk-based exercise. Theoretical limits on longitudinal gradients and cross falls are given in few turbine transport guidelines, but the limits vary significantly. For a 3MW wind turbine with a rotor diameter bigger than 90 m, Vestas and Nordex transport guidelines state that road longitudinal slopes and longitudinal radii (convex or concave) must be less than 6–8° and 200 m, respectively. Moreover, all roads must be clear of overhead obstructions to a minimum height of 5 m to allow the passage of high loads.

The horizontal alignment of the existing road layout is checked by undertaking vehicle swept path analysis, i.e. the evaluation of the envelope swept out by the sides of the vehicle body, or any other part of the structure of the vehicle (Pecchini & Giuliani, 2013).

These calculations generally require special software. Some authors (Flores et al., 2015; Godavarthy et al., 2016) have conducted field tests with a wind blade transporter vehicle and compared its swept path to that generated by the AutoTURN software, concluding that this software vehicle simulation can reasonably model the vehicle's turning characteristics. Figure 8 represents a swept path analysis of a blade transport (Nordex N117-2.4) elaborated with AutoTURN 10. This process that demonstrates if the vehicles can manoeuvre safely and efficiently within the site layout is very useful when it is contextualized in a specific area considering a preordained truck/trailer configuration.

For a ready planimetric verification at planning level should be more interesting to approach the problem from a theoretical point of view, considering the maximum vehicle swept path width, i.e. the amount of roadway width that a truck covers in negotiating a turn (AASHTO, 2011). Assuming a no skidding or tyre slippage and an infinitely small tyre contact, it may be demonstrated that the rear steering wheels of a vehicle follow circular paths concentric to those traced out by the front wheels while the steering angle is constant. Effect of driver characteristics are minimized by assuming that the speed of the vehicle for the turning radius is less than 15 km/h. The geometrical path of a vehicle with front and rear steering is showed in Figure 9. The swept path width $(R_o\text{-}R_i)$, is established by the outer trace of the tractor front overhang (R_o) and the path of the inside trailer rear wheel (R_i) (AASHTO, 2011). The R_o value is described by the relation:

$$R_0 =$$

$$\frac{1}{\sqrt{\left(\sqrt{\left(R_i + \frac{W_t}{2}\right)^2 + (WB_t)^2} + \frac{W_T}{2}\right)^2 + (WB_T + FO_T)^2}}$$

(1)

where WB_T and WB_t are the tractor and trailer wheelbases, W_T and W_t are the tractor and trailer widths and FO_T is the tractor front overhang.

Figure 9 shows that the most influencing parameter on the radius of the outer trace of the tractor

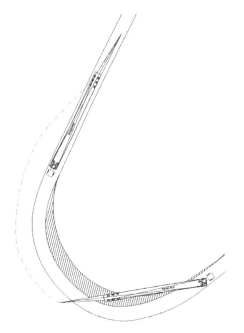

Figure 8. Example of a swept path analysis with Auto-TURN.

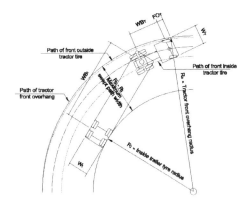

Figure 9. Turning characteristics of a typical tractor/trailer configuration truck.

front overhang is the trailer wheelbase, i.e. the distance from the king pin to the rearmost trailer axle group. For wind turbine components transport, this parameter depends on the longest elements, that are the rotor blades. Table 3 summarises the blade dimensions of some of the latest commercially avaliable onshore solutions of Vestas, which is the most important European wind turbine manufacturer.

Considering the maximum swept path width as principal parameter, a methodology and a corresponding operational planning tool could be developed to facilitate efficient preliminary route plans for wind turbine components. The Vestas turbines were chosen as reference. In parallel, different vehicle configurations were selected, considering the same tractor (Scania LA6 × 4HSZ) and several load floor extensions of a telescopic trailer (Nooteboom TELE-PX Super Wing Carrier) (Fig. 10). These configurations were chosen so that the trailer floor base length (L_t) almost coincided with the rotor blade length, as a common practice of transporting

According to equation 1, it was constructed a diagram (Fig. 11), which relates the inside trailer tyre radius, which can be considered in first approximation equal to the inside road curve radius, the maximum vehicle swept path width and the trailer length (blade length). Once fixed the inside road curve radius, the user can easily define the maximum swept path width for a predetermined blade length. Vestas transport guidelines gives a minimum inside turning radius ranging from 28 to 43 m for a blade length of 45 m and 55–60 m, respectively. The passage in a more binding path element with an assigned maximum swept path width can be satisfied through a reduction of the trailer length which requires a reduced inner radius. The extent of the vehicle

Table 3. Blade dimensions of onshore Vestas wind turbines at 2016.

Rated power		Blade dimensions	
		Length	Max. chord
kW	Wind turbine	m	m
1.8–2.0	V90	44.0	3.9
	V100	49.0	3.9
2.0	V110	54.0	3.9
3.0	V90	44.0	3.5
	V105	51.2	4.0
	V112	54.7	4.0
3.45	V117	57.2	4.0
	V126	61.7	4.0
	V136	66.7	4.1

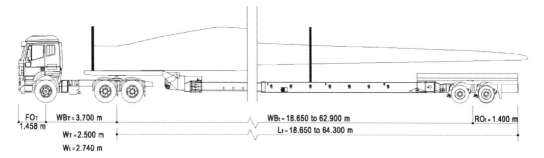

FO$_T$
1.458 m

WB$_T$ = 3.700 m

W$_T$ = 2.500 m

W$_t$ = 2.740 m

WB$_t$ = 18.650 to 62.900 m

L$_t$ = 18.650 to 64.300 m

RO$_t$ = 1.400 m

Figure 10. Tractor/trailer configurations for the rotor blade transport.

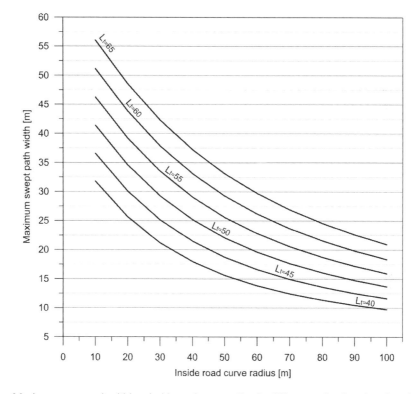

Figure 11. Maximum swept path width vs inside road curve radius for different trailer floor base lengths (L$_t$).

length reduction generates an equal value of the rear blade overhang increase. This overhang can be usefully compensated, where there are the conditions, by the lifting of the blade with the possible use of a rotor blade adapter (Fig. 7). By way of example, a truck/trailer transporting a rotor blade of 65 m and negotiating a turning maneuver in a local urban context (inside road curve radius of 50 m) can be considered. By plotting in the graph a vertical line from the x-coordinate value of 50 m, it intercepts the curve L$_t$ = 65 m (no blade rear overhang) identifying a maximum swept path width of about 33 m. At this point it is possible to easily evaluate a swept path width reduction assuming gradually greater rear blade overhangs, compatibly with the technical wind turbine transport guidelines. The user has to shift vertically downward considering the underlying curves, which represent shorter trailers or temporarily shortened configurations. Thus, the same blade should be transported, assuming for example a rear blade overhang equal to 20 m, by a trailer with a floor base length of 40 m which covers a maximum swept path width of 16 m.

5 CONCLUSIONS

The upward trend in wind energy production had created a challenge for the roadway system. The road infrastructure had to accommodate a wide range of complex vehicle configurations for the transport of wind turbine components (nacelles, tower sections and rotor blades), which are often considered oversize/overweight loads by the transport authorities. The paper provided an overview of nacelles, tower sections and rotor blades transport challenges, examining how the growth of wind energy in Europe, and especially the growth in the size of wind turbine components, had affected the transport industry and the route assessment process. The authors proposed an operational planning strategy, based on maximum swept path width, to facilitate the identification of optimal routes for rotor blades transport. The methodology, which does not substitute a rigid swept path analysis, could be used by the road managing authorities for checking planimetric restrictions and constraints on their road infrastructure and for a ready evaluation of possible abnormal road transport corridors. To effectively connect the European industrial centres, the identifications of these road corridors should provide an international cooperation and involvement, especially for strategic trans-European routes.

REFERENCES

AASHTO. 2011. *AASHTO Greenbook. A policy on geometric design of highways and streets*. American Association of State Highway and Transportation Officials, Washington, DC.

Abu-Rub, H., Malinowski, M. & Al-Haddad, K. 2014. *Power electronics for renewable energy systems, transportation, and industrial applications*. Chichester: John Wiley & Sons Ltd.

Bazaras, D., Batarliene, N., Palšaitis, R. & Petraška, A. 2013. Optimal road route selection criteria system for oversize goods transportation. *Baltic Journal of Road and Bridge Engineering 8*(1), 19–24.

Cotrell, J., Stehly, T., Johnson, J., Roberts, J.O., Parker, Z., Scott, G. & Heimiller, D. 2014. *Analysis of transportation and logistics challenges affecting the deployment of larger wind turbines: Summary of results*. Technical Report NREL/TP-5000-61063. National Renewable Energy Laboratory. Denver West Parkway Golden, CO.

EWEA. 2016. *Wind in power: 2015 European statistics*. European Wind Energy Association, Brussels.

EAWA. 2009. *Wind energy—the facts: a guide to the technology, economics and future of wind power*. London: Earthscan.

Flores, J., Chan, S., Homola, D. 2015. *A field test and computer simulation study on the wind blade*; Proceedings of the fifth International symposium on highway geometric design, Vancouver, 22–24 June 2015.

Gasch, R. & Twele, J. 2012. *Wind power plants: Fundamentals, design, construction and operation*. Berlin Heidelberg: Springer-Verlag.

Godavarthy, R.P., Russell, E. & Landman, D. 2016. Using vehicle simulations to understand strategies for accommodating oversize, overweight vehicles at roundabouts. *Transportation Research Part A 87*: 41–50.

Hau, E. 2013. *Wind turbines. Fundamentals, technologies, application, economics*. Third edition. Berlin Heidelberg: Springer-Verlag.

Henzelmann, T., Büchele, R., Hoff, P. & Wollgam, G. 2016. Onshore wind power. Playing the game by new rules in a mature market. *Think act. Beyond mainstream*, January, 1–19.

IEA-ETSAP & IRENA. 2016. *Wind Power. Technology Brief*. International Energy Agency, Paris and International Renewable Energy Agency, Abu Dhabi.

IWEA. 2011. *Transport of abnormal load to wind farm*. Irish Wind Energy Association, Naas.

Neff, M.R. & Bai, Y. 2012. *Developing a multi-modal freight movement plan for the sustainable growth of wind energy related industries*. Report 25-1121-0001-467. Mid-America Transportation Center. Lincoln, NE.

Nelson, V. 2014. *Wind energy. Renewable energy and the environment. Second edition*. Boca Raton, FL: CRC Press.

Pecchini, D. & Giuliani, F. 2013. Experimental test of an articulated lorry swept path. *Journal of Transportation Engineering 139*(12): 1174–1183.

Petraška, A & Palšaitis, R. 2012. Evaluation criteria and a route selection system for transporting oversize and heavyweight cargoes. *Transport 27*(3), 327–334.

Shihundu, D. & Morrall, J. 2011. *Oversize/overweight vehicles: An investigation into the safety and space requirements for alternative energy projects*. Proceedings of the 2011 Conference and Exhibition of the Transportation Association of Canada, Edmonton, 11–14 September 2011.

Ray, J.J. 2007. A web-based spatial decision support system optimizes routes for oversize/overweight vehicles in Delaware. *Decision Support Systems 43*(4), 1171–1185.

TPI Composites. 2003. *Cost study for large wind turbine blades: WindPACT blade system design studies*. Report SAND2003-1428. Sandia National Laboratories. Albuquerque, NM and Livermore, CA.

Wind Europe. 2016. *The European offshore wind industry. Key trends and statistics 1st half 2016*. Bruxelles.

WWEA. 2016a. *WWEA assessment report*. World Wind Energy Resource Assessment, Bonn.

WWEA. 2016b. *WWEA half-year report 2016*. World Wind Energy Resource Assessment, Bonn.

Transport Infrastructure and Systems – Dell'Acqua & Wegman (Eds)
© 2017 Taylor & Francis Group, London, ISBN 978-1-138-03009-1

The method of the friction diagram: New developments and possible applications

P. Colonna, N. Berloco, P. Intini & V. Ranieri
DICATECh—Department of Civil, Environmental, Building Engineering and Chemistry, Polytechnic University of Bari, Bari, Italy

ABSTRACT: This paper presents an updating and an application of a tool developed by the authors: the Friction Diagram Method. The model on which the Friction Diagram is based was explained in previous papers and here it will be briefly summarized. A sensitivity analysis has been performed in order to establish which input parameters mostly influence the percentage of Friction Capital that a vehicle is using traveling on a given road layout with different boundary conditions. Furthermore, some possible applications of the Friction Diagram to real cases will be shown. In particular, the examples carried out show how the Friction Diagram, together with the speed and visibility diagrams, could be considered an important tool for selecting the speed limits of a road, both in dry and in wet conditions. The implications of the method provide some guidance for both the design of new road infrastructure and the improvement of existing roads.

1 INTRODUCTION

Road friction is a crucial matter in road design, maintenance and operation since it is related to safety issues (Wallman & Astrom 2001, Li et al 2013). Indeed, if the vehicle requires more friction than the ability of the road to provide it, then a loss of friction between tire and road occurs. The phenomenon of road friction depends on several features belonging to different spheres: the road, the vehicle and the environment.

All these influences were considered in a new method developed by the authors: the Friction Diagram Method (FDM). This method is based on the estimate of the safety level of a road by calculating and comparing for each road section the friction force that the road can potentially provide to a driving wheel "FP" with the resultant force acting on it "FD". The quantity "FP" is a measure of the "Friction Capital" available to the vehicle, while the quantity "FD" represents the share of "Friction Capital" used by the vehicle. The model on which the Friction Diagram is based was explained in previous works (Colonna et al. 2014, 2016a, b, c) and it will be briefly summarized in next section.

In this paper, the possible applications of the Friction Diagram to real cases will be shown. Different vehicles are considered, with particular regard to the Design Critical Vehicles (DCVs), the vehicles presenting the worst performance for each combination of both the horizontal and vertical alignments. On the other hand, different environmental characteristics, speed motion conditions and road geometry features are considered as well. The impact of each variable on the output of the model, represented by the percentage of friction used (F_{USED} [%]) is shown through the use of a sensitivity-uncertainty matrix.

Furthermore, some considerations about the maximum safe speed with regard to the risk of skidding in each combination of road geometry and vehicles are given. Currently, the recognized method by the Italian Standard (D.M. 6792–2001) for controlling the speed of a road layout is the speed diagram. In the case of an existing road infrastructure, the speed diagram could not be verified if the horizontal alignment does not allow the imposed acceleration and deceleration distances (considering an acceleration rate: a = 0.8 m/s²) or if the speed of consecutive geometric elements exceeds the fixed thresholds. In these cases, the designer reduces the allowed speed limit in order to comply with the design speed range. Similar situations occur for the sight distance diagram. If it is not possible to change the road geometric layout or to eliminate the obstacles that limit the visibility, the designer reduces the speed limit in order to guarantee the stopping distances, or eventually the overtaking distances, with the new maximum allowed speed.

Therefore, the road speed limits do not consider the phenomenon of friction and, consequently, the numerous factors which can greatly influence its value according to the geometrical characteristics of the road, the type of the vehicle, its motion conditions and the boundary conditions.

It can occur that the speed limit (which is derived from the study of the visibility and the speed diagrams) definitely ensures the stopping distances and the correct deceleration and acceleration between the geometric elements of the road; but it could not guarantee the friction of all vehicles for all plano-altimetric combinations and for all possible environmental conditions.

The Friction Diagram Method provides a useful tool to prevent these dangerous situations, for both the design of new road sections and the improvement of safety of existing road infrastructures.

The contemporary and iterative use of the speed diagram, the sight distance diagram and friction diagram allows to obtain, for each section of a road, the speed which satisfies the three checks and ensures greater road safety, defined in this paper as the safe speed.

2 THE FRICTION DIAGRAM METHOD

2.1 Update of the 3rd criterion of Lamm

Among all the authors who studied the phenomenon of road friction, Lamm is one of the most important. The outcome of his studies is the Lamm's Third Criterion (Lamm et al. 1998, 1999, 2002) which determines the level of safety design of a curve road section (good, fair or poor) by comparing the "side friction assumed" (f_{RA}) with the "side friction demand" (f_{RD}).

The first coefficient "f_{RA}" is the maximum cross friction coefficient that the road can provide to the vehicle in the design conditions. The second coefficient "f_{RD}" represents the friction coefficient that a vehicle, considered as a material point, requires from the road when it navigates a curve at a constant speed equal to the operating speed "S_{85}".

Therefore, the Lamm's Third Criterion allows to measure the level of safety by considering the difference between f_{RD} and f_{RA}. These two terms abide by the following hypotheses:

- they are only function of speed, cross slope and radius of horizontal curvature;
- the motion condition is characterized by a speed constant, equal to the operating speed;
- the vehicle is modeled as a material point, without considering possible different geometric features which are able to influence the redistribution of the vehicle weight on the axles;
- the topographic factor "n" can represent by itself all the possible longitudinal actions.

The applicability of the framework given by Lamm about road friction was limited only to the horizontal curves. It does not provide any indication about other road geometric types and other important features:

- the rolling resistance of tires (see European Regulation No. 1222/2009);
- the influence of road geometry on the distribution of the vehicle weight on the road surface;
- the environmental and surface conditions;
- the forces acting on the vehicle along the road.

If the maximum friction force provided by the road to the vehicle (A_{LIM}) is greater or equal to the resultant (R) of the cross and longitudinal forces acting on the vehicle ($A_{LIM} \geq R$), then a vehicle can be considered as safe from the point of view of friction. Hence, the quantity A_{LIM} can be defined as the Friction Potential, the limit performance of the road in terms of friction.

Therefore, in order to correctly estimate the friction along a road layout, the comparison between the two above defined quantities in all the possible road geometric conditions can be made, once the parameters reported in Table 2 are known.

The Friction Diagram Method FDM (Colonna et al. 2016c) is a tool able to analyze the friction of a road layout considering at the same time physical, dynamics and environmental factors. It provides an evaluation of the skidding risk associated to a road layout by introducing a synthetic parameter, the Friction Used (F_{USED}). The Friction Used is plotted on the y-axis against the distances on the x-axis of a diagram. This diagram, the "Friction Diagram", allows to individuate high-risk road sections with respect to friction by a simple visual inspection.

The Friction Diagram can be plotted below the road elevation profile, by connecting the values of the F_{USED} obtained for each cross section.

Table 1. III Lamm's criterion.

Level of safety design	Recommended interval
Good	$f_{RA} - f_{RD} \geq 0.01$
Fair	$-0.04 \leq f_{RA} - f_{RD} < 0.01$
Poor	$f_{RA} - f_{RD} < -0.04$

Table 2. Factors influencing A_{LIM} and R.

Factors influencing A_{LIM}	Factors influencing R
Adherent weight	Road geometry
Friction coefficient	Vehicle speed
	Boundary conditions
	Motion conditions

The Friction Diagram Method FDM is based on three key concepts:

- Friction Potential "F_P";
- Friction Demand "F_D";
- Friction Used "F_{USED}".

2.2 The Friction Potential

The Friction Potential is the maximum friction force that a road surface can provide to the vehicle (Padmanaban & Pawar 2015). It varies with the adherent weight (Schaar 1993), the friction coefficient (Canale et al. 1998) and the road geometry. It is computed as follows:

$$|Fp| = P_a (g, s) \times f_a (c, s) \qquad [N] \qquad (1)$$

where:

- P_a (g, s) = adherent weight on the driving wheels, depending on the speed "s" and the road geometry "g";
- f_a (c, s) = friction coefficient in any direction (depending on the speed "s" and the surface conditions "c").

2.3 The Friction Demand

The Friction Demand is the friction force required by the vehicle to the road. It is equal to the resultant force of all the longitudinal and cross forces acting on the vehicle. It is computed as follows:

$$F_D = (L^2 + C^2)^{0.5} \qquad [N] \qquad (2)$$

where:

- "L", sum of all the longitudinal forces: rolling resistance, aerodynamic drag, slope resistance, curve resistance, longitudinal wind force;
- "C", sum of all the cross forces: centrifugal force, cross component of the vehicle weight, cross wind force.

2.4 The Friction Used

The parameter introduced, able to immediately define the skidding risk, is the Friction Used. It is defined as:

$$F_{USED} = F_D/F_P \times 100 \qquad [\%] \qquad (3)$$

If the Fused is less than 100% (if the Friction Demand is less than the Friction Potential), then the vehicle is in safety conditions. Therefore, safety conditions are defined by the distance of the F_{USED} from its limit value. In very safe situations, F_{USED} will be low; while in situations related to high skid-

ding risk, F_{USED} will be close to 100%. Moreover, if the vehicle stability is not guaranteed according to the FDM, then F_{USED} will be greater than its limit.

2.5 Methodological approach to implement the Friction Diagram Method

In order to compute the F_{USED} value in each cross section of a road layout and to plot the related Friction Diagram, the following steps have to be followed:

- analysis of road geometry and motion conditions;
- estimation of the Friction Potential;
- computation of the resultant of the longitudinal forces acting on the vehicle;
- computation of the resultant of the cross forces acting on the vehicle;
- computation of the resultant R of all the forces acting on the vehicle, equal to the Friction Demand, except for the sense of the vector;
- comparison between the Friction Demand and the Friction Potential, by computing the Friction Used.

The FDM has been implemented in a spreadsheet which allows the easy application of the method, once the "Road Layout Data" and the "Vehicle Data", reported in Table 3 are known.

In particular, the variables considered in the Friction Diagram Method for the calculation of the F_{USED} are the following: road friction coefficient (f_a), cross slope φ and longitudinal slope α_m, vehicle characteristics (wheelbase [m] (P_L) and front track [m] (P_C), height of the center of gravity [m] (h_G), longitudinal section [m²] (S_L) and cross section [m²] (S_C), longitudinal drag coefficient [–] ($C(N)_L$) and cross drag factor [–] ($C(N)_C$), mass (m)), resistance class of the tires, vehicle speed, acceleration, deceleration, wind speed and wind direction.

2.6 The design critical vehicle

The forces acting on the vehicle taken into account by the FDM are the following:

- weight force, depending on the vehicle mass;
- inertial actions, depending on the vehicle mass and the speed variation;

Table 3. Road layout data and vehicle data needed for the implementation of the method.

Road layout data	Vehicle data
Friction coefficient	Mass
Cross section	Height of center of gravity
Travel direction	Longitudinal and cross pitch
Design speed	Vehicle cross section
Allowed acceleration	Vehicle longitudinal section
Allowed deceleration	Drag coefficients

– centrifugal forces, depending on the vehicle mass, the vehicle speed, the horizontal and vertical radii of curvature;
– aerodynamic actions, depending on the wind speed and the longitudinal and cross sections of the vehicle.

All the actions above reported depend on vehicle characteristics (Jadav Chetan & Patel Priyal 2012, Genta 1983, Zagati et al. 1983). Therefore, for a given road layout, different vehicles correspond to different F_D and F_P values. Hence, since F_{USED} is the ratio between F_D and F_P, different vehicles are related to different F_{USED}, and to different Friction diagrams. The analysis of a road layout is not complete if only data belonging to a given type of vehicle are considered. In fact, in this case, the Friction Diagram is valid only for that type of vehicle. Preliminary analyses (Colonna et al. 2014; Colonna et al. 2016a) revealed that:

– it is not possible to individuate a unique type of vehicle associated to F_{USED} values always greater than the other vehicles;
– the difference between the F_{USED} values related to different type of vehicles is included between 0 [%] and 35 [%].

Therefore, the type of vehicle to be used for computing the F_{USED} have to be identified, in various boundary conditions, without underestimating the level of skidding risk. This means that the type of vehicle maximizing the F_{USED} value in different conditions has to be found.

For this aim, in the FDM, the "Design Critical Vehicle—DCV" is introduced: it is the vehicle to be considered for plotting the Friction Diagram in different road geometric conditions. Using the DCV allows road designers to control for skidding risk by using a unique Friction Diagram. Other studies by the authors revealed that (Colonna et al. 2016b, c):

– the vehicle associated to the maximum value of F_{USED} varies with the specific type of plano-altimetric road section;
– the value of F_{USED} varies with the type of traction (front or rear wheel drive) in uphill and downhill road sections;
– the motion condition (constant speed, acceleration, deceleration) largely influences F_{USED}.

Hence, a set of critical vehicles, rather than a unique vehicle, was considered. Each vehicle of the set is "critical" in a given combination of road geometry and motion conditions.

Twenty road geometric conditions were studied, obtained by combining horizontal alignment (4 conditions) and the vertical alignment (5 conditions).

Since cross slope is considered in the static equilibrium of a vehicle navigating a curve section, the spiral transition curve is divided into two sections: "ante level section" (a. l. s.) and "post level section" (p. l. s.). They represent the sections before and after the point where the cross slope of the external lane is equal to zero (level section).

Three motion conditions were considered: constant speed, acceleration and deceleration. In order to individuate the characteristics of the critical vehicle related to each of the 20 combinations of the geometric elements shown in Figure 1, the influence of the F_{USED} value on the 6 vehicle parameters shown in Table 4 and previously defined, were evaluated.

Among the 120 resultant combinations (20×6), 6 recurrent combinations representing theoretical critical vehicles were identified. Each theoretical critical vehicle was considered both having a front wheel drive and a rear wheel drive.

These critical vehicles should be considered merely theoretical, since each of them represents an unfavorable combination of parameters among all the possible combinations.

Further developments of the research led to select a set of real critical vehicles, in order to represent the real different types of vehicles commercially available. The mean characteristics of the chosen real critical vehicles are shown in Table 5. However, further research developments will eventually include other types of vehicle according to their performances with respect to the skidding risk.

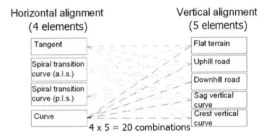

Figure 1.   Possible combinations of both the horizontal and vertical alignments.

Table 4.   Theoretical critical vehicle features.

Critical vehicle	$C(N)_L$ [–]	$C(N)_C$ [–]	h_G [m]	PL [m]	PC [m]	M [kg]
I	1.15	1.15	1.50	6	2.50	7,031
II	1.15	1.15	0.75	6	1.50	2,109
III	1.15	1.15	1.50	1.7	1.50	1,195
IV	1.15	1.15	0.75	1.7	1.50	598
V	1.15	1.15	0.75	6	2.50	3,515
VI	1.15	1.15	1.50	6	1.50	4,218

Table 5. Real critical vehicle features.

Critical* vehicle	$C(N)_L$ [–]	$C(N)_C$ [–]	h_G [m]	PL [m]	PC [m]	M [kg]
I	0.32	1	0.76	2.33	1.46	972
II	0.33	1	0.84	2.78	1.61	1,829
III	0.38	1	0.87	2.56	1.54	1,647
IV	0.60	1	1.65	6.30	2.23	15,750
V	0.31	1	1.28	3.52	1.76	3,432
VI	0.70	1	1.10	3.57	1.73	3,579

*I—City Car; II—Sport Car; III—Cross-country; IV—Bus; V—Van; VI—Light Truck.

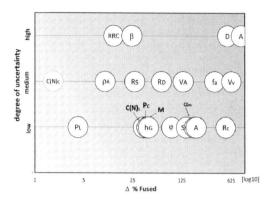

Figure 2. The sensitivity-uncertainty matrix.

The further applications shown in this study are based on the real critical vehicles. Therefore, for each cross section characterized by a given geometry, the maximum F_{USED} value among the six F_{USED} values computed for each real critical vehicle is chosen. The Friction Diagram is obtained by connecting the F_{USED} values obtained for each cross section.

3 SENSITIVITY ANALYSIS

3.1 Input parameters

A sensitivity analysis has been performed in order to establish which input parameters mostly influence F_{USED}. The several factors taken into account (Colonna 2016) belonging to different spheres of influence, are listed as follows.

– Vehicle features: Longitudinal drag coefficient $C(N)_L$ [–]; Cross drag coefficient $C(N)_C$ [–]; Wheelbase P_L [m]; Front track P_C [m]; Height of the center of gravity h_G [m]; Mass M [Kg]; Rolling resistance coefficient RRC [Kg/t];
– Road design features: Longitudinal slope $\mathbf{a}_m$ [rad]; Cross slope φ [rad]; Radius of the circular curve R_C [m]; Parameter of the spiral transition curve A [m]; Length of the horizontal curve s [m]; Radius of the convex vertical curve R_D [m]; Radius of the concave vertical curve R_S [m];
– Vehicle dynamics: Speed V_V [km/h]; Friction coefficient fa [–]; Acceleration rate A [m/s²]; Deceleration rate D [m/s²];
– Environmental conditions: Air speed V_A [km/h]; Air direction (with respect to the vehicle) $\mathbf{b}$ [°]; Air density $\mathbf{r}$ [Kg/m³].

3.2 The sensitivity-uncertainty matrix

For each analyzed parameter, the ranges of variability were identified, in consideration of the boundary conditions, the types of vehicles and the motion condition. In relation to the variability of each parameter and to the easy or difficult avail-

ability of data, for each parameter three degrees of uncertainty have been assigned qualitatively: low, medium and high.

Then, a sensitivity analysis was carried out considering:

– 21 parameters, characterizing each of them with three values (standard, minimum and maximum);
– 27 plano-altimetric combinations;
– 3 different motion conditions of the vehicle (constant speed, uniform acceleration or deceleration).

Therefore, it was possible to verify the variation of F_{USED} for each of the 5,103 cases analyzed, coming to the following matrix of uncertainty vs sensitivity.

The first five parameters that mostly influence the model, ranked by decreasing importance, are: acceleration, speed and deceleration of the vehicle, the horizontal curve radius and, finally, the friction coefficient.

The high degree of uncertainty related to acceleration and deceleration is caused by the value assumed by Italian standards, equal to ± 0.8 m/s². Realistically, these values can be appreciably greater. The medium degree of uncertainty related to speed is caused by the use of theoretical and not real or detected speeds. Similar considerations can be made for f_a since the state and age of the surface course and the tires are often not considered.

In order to get reliable results by applying the FDM, it is necessary to have as accurate as possible information about these parameters that are able to influence the model more than the others.

4 USING THE FRICTION DIAGRAM METHOD TO DEFINE THE SAFE SPEED

In case of new road infrastructures design, the friction diagram can be used as a "design tool", conditioning the horizontal and vertical layout of the road in order to always ensure the friction used percentage (F_{USED}) less than 100%, for each road section.

The friction diagram should be used in conjunction with the speed and the sight distance diagrams, being sure that, for each road section, a safe speed is assigned, within the design speed range, in order to verify the three checks, conducted in both directions of travel and for different boundary conditions.

In the case of safety improving of an existing road infrastructure, the friction diagram can be used as a diagnostic tool, useful to identify the road sections, characterized by a specific plano-altimetric combination, in which the friction used could exceed the friction potential. Also in this case, it should be used with the speed and the sight distance diagrams, for each travel direction and for different boundary conditions.

For improvement of existing roads, the safe speed, intended as the lowest speed value among the three speed values which satisfy respectively the visibility, the friction and the speed diagram, assumes an even more important value.

The two possible uses of the friction diagram, although apparently similar, are conceptually different from each other.

In the case of a new road, the designer can define the geometrical elements of the road in such a way that the three checks are satisfied. Depending on the function of the new infrastructure, on its interactions with the land use and the type of expected traffic, it will be possible to fix for each road section the minimum safe speed. The safe speed will influence the road geometry, as a result.

The higher is the strategic value of the infrastructure, the higher is the level of service to be guaranteed and, therefore, the higher is the value of the minimum safe speed to ensure in the different sections of the road.

In the case of safety improvement of existing road infrastructures, it is firstly necessary to distinguish three types of intervention that depend on the available budget, the function of the road under study and the boundary conditions:

– functional improvements on the geometry (high budget, strategic function of the road and not limiting boundary conditions);
– functional improvements that exclude geometric variations (medium budget, not important function of the road and limiting boundary conditions);
– minimal functional improvements (low budget, negligible function of the road, and limiting boundary conditions).

The diagnostic tools that it is possible to use simultaneously for the verification of an existing road infrastructure are the speed diagram and the sight distance diagram, both required by the existing Italian Standard.

In the first iterative loop, the existing road infrastructure is approximated to a type of road defined by the Italian Standard (even if the existing road was built certainly long before the entry into force of the Standard). Subsequently, the speed range of that type of road is identified and it enables the construction of the speed diagram (in both directions of travel). Once the speed diagram is known, it is possible to plot the visibility diagram (in both directions of travel).

In the road sections where the verification of one of the two diagrams is not satisfied, the two checks will have to be repeated, lowering progressively the speed in order to obtain a value able to satisfy both of them.

If this value is acceptable for the purpose of the level of service to ensure on the road infrastructure, for its strategic function and for the type of traffic, the recommended improvements will avoid variations on the road geometry and will be focused on other types of countermeasures. Otherwise, if at the same time, there is the availability of a high budget and there are no territorial limiting restrictions, functional improvements on the geometry will be implemented.

The above described procedure, however, does not eliminate the risk of skidding of a generic road vehicle because:

– the speed diagram does not take into account the vertical alignment of the road;
– the two tests do not take into account the different types of road vehicles;
– the two tests do not take into account the conditions of the road surface (new, dry, wet, slippery, smooth);
– the two tests do not take into account the different motion conditions of vehicles.

The friction diagram is therefore the tool that, contextually used with the speed and visibility diagrams, ensures higher road safety. The loop described above should be conducted simultaneously with the three procedures. For each road section, there will be a threshold value of the speed which satisfies the speed diagram, a threshold value of the speed which satisfies the visibility diagram and a threshold value of the speed which satisfies the friction diagram. The safe speed, for that section, will be the lowest speed among these three threshold values and it will be used for the subsequent plotting of the other two diagrams.

Once completed the diagnosis of the road, using the conditions previously explained, the improvement interventions can be better identified. If improvements on the road geometry are not feasible, it will be possible to implement actions to improve the visibility, i.e. with the possible elimination of obstacles in the vicinity of the road, and/or improve the friction, i.e. by remaking the surface course of the road pavement. Surely, even in the case of low budget available, it will be possible to operate on the speed limits, redesigning the road signs.

The new project of road signs should be based on the identification of the safe speed and joined by solutions that ensure compliance with the imposed speed limits (speed control) in order to guarantee the success of the countermeasure.

The example reported in this section refers to a rural road located in the District of Bari, in Italy (S.P. 239). Only a part of the road is shown for graphic reasons, selected in order to support the procedure previously described.

The road is not a strategic infrastructure in the national road network context and it is located in an environmentally protected area with a lot of ancient olive trees considered natural archaeological monuments. The section shown in the example has a constant longitudinal slope equal to 4.5%. In order to improve the road safety, interventions on the geometry are therefore excluded as well as the removal of olive trees for increasing the visibility.

Considering the cross section of the road, about eight meters wide, the road has been approximated to the road type—F2-defined by the Italian Standard, with a design speed ranging from 60 to 100 km/h.

Hence, considering a maximum design speed equal to 100 km/h, it is possible to note (Fig. 3) that none of the three checks is satisfied (for the entire length or for some parts of the section considered). Once the same checks were carried out with a maximum speed of 70 km/h, it is possible to note that the visibility and the speed checks are satisfied, while the friction check is not met in a small part of the road section and for only one direction of travel ("return" direction in the Figure 3).

Further calculations have been conducted iteratively in order to obtain the speed satisfying the friction check in every part of the road section. Its value is equal to 66.67 km/h. This speed is the safe speed for the road under study and it eliminates the skidding risk for any vehicle and for wet conditions.

In the re-design process of the road signs, the limit of 70 km/h must therefore be fixed on the entire road but not in approaching the curve for the return direction, where it will be equal to 60 km/h.

The study of the entire infrastructure leads to various safe speeds for each road section. The selection of speed limits will be carried out trying to uniform as much as possible these speed values and at the same time to ensure a higher speed in

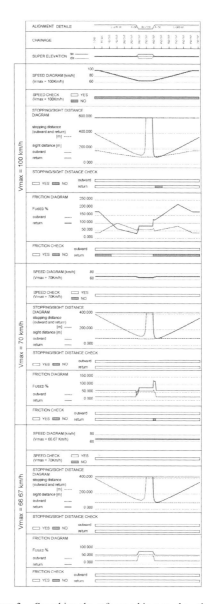

Figure 3. Searching the safe speed in a road section.

sections where it is possible to allow it. In the example carried out, the control of the new speed limits has been guaranteed with an automatic speed control system with the possibility to sanction users who do not comply with them.

5 CONCLUSIONS AND FURTHER RESEARCH

This paper presents an update of a tool developed by the authors: the Friction Diagram Method.

A sensitivity analysis has been carried out in order to identify the input parameters on which it is necessary to pay more attention for the proper implementation of the FDM.

At the same time, in the example shown before, the attention is focused on the road sections in which, while the speed and the visibility diagrams are verified, the risk of skidding is still present. The application of the method leads to identify the "safe speed", namely the speed able to verify all the three checks about speed, visibility and friction. Therefore, it provides a rational tool for improving existing road infrastructures and selecting safety countermeasures, including the re-design of the road signs and the speed limits, under dry and wet conditions. This can be useful since sometimes the speed limits are not selected in a rational way by the road Agencies or designers, setting speed limit values that could not eliminate the risk of skidding.

Regarding the critical vehicles considered in this paper, compared to six ideal vehicles considered in the previous studies, six real vehicles were identified. While the theoretical critical vehicles represent the maximum unfavorable conditions among all the possible combinations, the real ones represent the prevailing traffic composition on the road under study and, at the same time, represent the real different types of vehicles commercially available.

Future insights will deepen the analysis by dividing the critical vehicles into two categories: the light vehicles (city car, sport car, cross-country vehicles etc.) and the heavy vehicles (buses, vans, light trucks etc.). In this way, the safe speeds, and consequently the speed limits, could be differentiated for the two macro categories. With this purpose, the use of variable message signs would be appropriate, showing the speed limits in case of dry or wet conditions, for passenger cars and heavy vehicles.

Considering the current and future developments related to self-driving cars, the speed that averts the skidding risk could play an even more important role in the future. The technological equipment of future vehicles could autonomously:

- properly modulate the acceleration and deceleration between consecutive geometric elements of the road (with databases containing the design features of the traveled road);
- measure the sight distances (using a variety of techniques such as radar, lidar, GPS, odometry) and therefore modulate the speed to ensure the stopping distance;
- to measure in real time the friction coefficient and other environmental factors in order to obtain F_{USED} and therefore modulate the speed to avert the skidding risk.

In this case, it would be necessary to implement an algorithm able to apply the FDM on board and in real time. The proposed method should be diversified for roads where traditional and self-driving vehicles can circulate together and roads exclusively devoted to next-generation vehicles.

REFERENCES

Canale, S., Leonardi, S. & Nicosia, F. 1998. Nuovi criteri progettuali per una politica di sicurezza stradale, Atti del XXIII Convegno Nazionale Stradale dell'A.I.P.C.R., Verona, Italy.

Colonna, P., Berloco, N., Intini, P., Perruccio, A. & Ranieri, V. 2016a. Proposal of a new method for analyzing the road safety conditions related to friction. *International Journal of Systems Applications, Engineering & Development*, 10, 87–96. University Press, UK.

Colonna, P., Berloco, N., Intini, P., Perruccio, A. & Ranieri, V. 2016b. *Sicurezza stradale—un approccio scientifico a un problema tecnico e comportamentale*. WIP Edizioni, Bari, Italy.

Colonna, P., Berloco, N., Intini, P., Perruccio, A. & Ranieri, V. 2016c. Evaluating the skidding risk of a road layout for all types of vehicle. *Transportation Research Record: Journal of the Transportation Research Board*, 2591, 94–102.

Colonna, P., Perruccio, A. & Ranieri, V. 2014. Influence of vehicle characteristics on skidding risk, SCORE@POLIBA2014 Conference, Bari, Italy.

European Parliament Council. 2009. Regulation (EC) No. 1222 on the labelling of tyres with respect to fuel efficiency and other essential parameters.

Genta, G. 1983. *Meccanica dell'autoveicolo*. Levrotto & Bella.

Italian Ministry of Infrastructures and Transports. 2001. *Functional and geometric standards for roads construction*. D.M. n. 6792. Rome. Italy.

Jadav Chetan, S. & Patel Priyal, R. 2012. Parametric Analysis of Four Wheel Vehicle Using Adams/Car, *International Journal of Computational Engineering Research*.

Lamm, R., Psarianos, B., & Cafiso, S. 2002. Safety evaluation process for two-lane rural roads: A 10-year review. *Transportation Research Record: Journal of the Transportation Research Board*, 1796, 51–59.

Lamm, R., Psarianos, B., & Mailaender, T. 1999. *Highway Design and Traffic Safety Engineering Handbook*. McGraw-Hill.

Lamm, R., Psarianos, B., Choueiri, E. M. & Solilmezoglou G. 1998. A practical safety approach to highway geometric design International studies: Germany, Grece, Lebanon and USA. Transportation Research Circular. USA.

Li, Y., Liu, C., & Ding, L. 2013. Impact of pavement conditions on crash severity. *Accident Analysis & Prevention*, 59, 399–406.

Padmanaban, S. & Pawar, P. 2015. Estimation of Tire Friction Potential Characteristics by Slip Based On-Road Test Using WFT, SAE Technical Paper 2015-26-0225.

Schaar, A. 1993. Driving simulation with Innovative Tools, *Automobile Technical Journal* (ATZ), 95, 256–262.

Wallman, C.G. & Astrom, H. 2001. Friction Measurement Methods and the Correlation Between Road Friction and Traffic Safety, Swedish National Road and Transport Research Institute, VTI, Linkoping, Sweden.

Zagati, E., Zennaro, R. & Pasqualetto, P. 1983. *L'assetto dell'autoveicolo*. Ed. Levrotto & Bella.

Transport Infrastructure and Systems – Dell'Acqua & Wegman (Eds)
© *2017 Taylor & Francis Group, London, ISBN 978-1-138-03009-1*

The relationships between familiarity and road accidents: Some case studies

P. Intini, P. Colonna, N. Berloco & V. Ranieri
DICATECh—Department of Civil, Environmental, Building Engineering and Chemistry,
Technical University of Bari, Bari, Italy

E. Ryeng
Department of Civil and Transport Engineering, Norwegian University of Science and Technology,
Trondheim, Norway

ABSTRACT: Familiarity with the route can lead to distraction, inattention and more dangerous behaviors. Conversely, unfamiliar drivers could be unaware of possible dangers hidden in the road environment. In order to inquire in detail the relationships between accidents and familiarity, a database composed of 633 fatal and injury accidents (over the period: 2005–2014) related to 84 sections of two important two-way two-lane rural Norwegian highways (E6, E39) was investigated. Familiarity of drivers with the place of the accident was defined by considering a distance measure from the residence. Two sites characterized by high percentages of namely familiar and unfamiliar accidents (selected basing on distance of involved drivers from residence) were analyzed to a micro-scale level in order to find possible recurring patterns and related factors. Familiar drivers were found to be over-involved in hitting vehicles in rear-end accidents, while only some indications without clear patterns were found for the unfamiliar accidents.

1 INTRODUCTION

1.1 *The concept of familiarity with a road*

A very common situation for a driver is to be familiar with the road on which he/she is traveling. This happens because many travels from a given origin to a given destination are often repeated over time (during the same week, for the whole year and more). A typical example is the home to work travel, that is the commuting trip. It is reasonable to assume that, having to reach the same destination from the same origin, the path chosen is usually the same. In fact, the main aim of a driver is the maximization of the utility related to a travel (see e.g. Noland 2013). Once all possible routes by car from a given origin to a given destination have been explored, it is most likely that the driver chooses, from that moment on, the most useful solution for the same trip. Repeating the same travel on the same route several times makes the drivers likely familiar with that route.

Being familiar with a given route means, by definition of the adjective "familiar" (Oxford Univ. Press Dictionary 2016), having a good knowledge of its characteristics for long or close association. This implies that the driver has a good knowledge of the features of the roads (in general, including segments and intersections, rural and urban, with

different cross sections), which are part of the familiar route. The concept of route familiarity is a sort of "selective experience" independent from the driving experience in the strictest sense of the word (that is e.g. measurable in terms of years of driving license and annual mileage). The driving experience is related to the act of driving itself and so, it can influence the capability related to a given task (see e.g. Fuller 2005) to be accomplished by the driver. On the other hand, familiarity is related to the act of driving on a specific route and so, it can influence the behavior of the driver in a similar way like experience does. In fact, the expectations of the drivers, given other boundary conditions being constant, are most likely to meet reality because the driver already knows what to expect by the road and so he is "experienced" about it (but at the same time he can be more or less experienced with the act of driving itself).

Studies in literature have suggested that familiarity can influence driving behavior inducing inattention and changes in perception (Yanko & Spalek 2013; Martens & Fox 2007). This can happen because drivers focus on other thoughts different from driving, since the driving task could have become less demanding. However, parallel to the increased proneness to distraction, a familiar driver could be more skilled at more dangerous behaviors on the familiar route, like speeding and rule breaking (see e.g. Rosenbloom et al. 2007), especially the more aggressive drivers (Colonna et al. 2016).

On the other hand, it is clear that also unfamiliar drivers have some behavioral weak points. In fact, if the road system is not able to self explains (Theeuwes & Godthelp 1995), expectations of unfamiliar drivers could be different by the real road situation inducing errors or unsuitable maneuvers.

However, familiarity and unfamiliarity can be seen as two features of the same phenomenon if drivers' perception of risk is taken into consideration. Indeed, the subjective perception of risk can be able to influence driving tasks, but this can happen at two different levels: one internal to the driver and the other external to the driver (Colonna & Berloco 2011). In the internal sphere of risk (the driver's unconscious), the perceived risk is compared with a personal constant budget of safety that a driver is ready to take, and this difference (supposed to be constant as deduced from Wilde 1982) determines the behavior in the medium/long term period. Instead, in the external sphere of risk (the driver's consciousness, that is practically the interface driver-road-environment), the driver could have to suddenly face a real risk given by the road environment different from the perceived one: the more is the difference, the more demanding will be the task of avoiding the danger. All drivers (in this case both the familiar and the unfamiliar drivers) have a conscious and an unconscious sphere, and so they are exposed to both the internal and the external risks. However, for familiar drivers, it is possible that the acquired knowledge of the road environments can alter the equilibrium between the perceived risk and the target level of risk in the medium/long period, resulting in possible pitfalls. So, in this case, a higher danger could come from the internal sphere (unconscious). Instead, for the unfamiliar drivers, a higher hazard could come from the external sphere: they do not know the road and so, they could be completely unaware of possible sudden dangers hidden in the road.

1.2 Accidents and familiarity

Given that driving experience is normally considered a variable to be accounted while performing statistical analyses related to road accidents; one should expect that also drivers' familiarity with the road on which the accident occurred is taken into account. Actually, in some sources, the familiarity of drivers is contemplated as a factor taken in consideration (see e.g. Liu & Ye 2011; Baldock et al. 2005).

Establishing if a driver was familiar with the road on which the accident happened is a complex matter to address. It is even perplexed by the fact that often this information is required when some specific analysis is performed, since it is not normally included into accident databases as a variable. However, the inclusion of a "familiarity" variable related to a given road accident would be complex too. In fact, police officers should ask to drivers involved in the accident if they were familiar, barely familiar or not familiar with the place of the accident. The problem is that this is only a perception by the driver which is not an objective measure. Some drivers could state to feel confident with a given road only after some travels on it and vice versa. On the other hand, it should be considered that possible driving behavioral changes due to familiarity belong to the subjective sphere of the driver. Therefore, the more reliable relationship between accidents and familiarity should take into account the subjective feeling of being familiar with a given road on which the accident happened. Since this information cannot be found in databases, it could be achieved through surveys.

A survey asking about previous road accidents as a driver, over a short period of time (in order to acquire realistic information) could be widespread. Questions about the familiarity of drivers with the place of the accident (i.e. unfamiliar, barely familiar or familiar) should be included in this survey. Anyway, it is evident that also this strategy shows some weak points. First of all, a huge number of surveys should be widespread for this aim, considering the response rate and the detail of the information required (to be a driver, to have experienced some recent road accidents as a driver, to be able to report detailed characteristics about the accident). Furthermore, self-reported information could not correspond to reality in some cases.

Therefore, in order to find relationships between road accidents and drivers' familiarity, some indirect measures could be used. They have the advantage of being objective, but the disadvantage of being not easy to achieve or not being directly related to subjective familiarity. For example, one could measure drivers' familiarity with a road by using the frequency of traveling on it on a given time period (daily, weekly, monthly, yearly...). This is a common measure of familiarity (see e.g. Liu & Ye 2011; Brown et al. 2005) since a driver who travels daily or close to daily on a given road is most likely familiar with it. Anyway, again, this is not a measure that can be found in a common accident database.

Instead, an accident database could include information about the origin of the drivers (see e.g. Blatt & Furman 1998; Yannis et al. 2007). The zip code associated to the place of residence could be an indirect measure to be used for the aim of associating road accidents to familiarity. This measure is used in this study since it can be representative of the familiarity of drivers with the road. In fact, it is most likely that a driver whose residence is placed very far from the place of the accident, is not familiar with it. On the other hand, it is most likely the opposite concept: a driver who is traveling on a road close to the place of residence is most likely familiar with that road. This common sense rule could be violated by some exceptions: e.g. if the driver's permanent residence is different from the real residence placed very far from the permanent one or, if the driver is familiar with some roads very far from his residence for different reasons (e.g.: for traveling always to the same vacation place or for working reasons, as it happens to the professional drivers). However, since the aim is to find recurrent patterns at particular sites characterized by very high or very low accidents to familiar drivers and not the analysis of a single accident only, then errors due to the wrong classification of drivers based on the distance criterion could be tolerated, since the analysis considers several accidents and not single crashes.

1.3 Aim of the study

Hence, since relationships between road accidents and drivers' familiarity are not easy to describe, the aim of this study is to make a contribution in this sense. In particular, the main objective is the identification of recurrent accident patterns related both to unfamiliar and familiar drivers. This will be addressed by inquiring accidents at particular sites selected through the methods described in next section.

2 METHODS

2.1 Accident database

A part of an accident database provided by the Norwegian Public Road Agency (NPRA) was used. It regards 84 two-way two-lane rural road sections of two important Norwegian highways: the E6 (37 sections) and the E39 (47 sections). The Norwegian itinerary of the E6 connects the southern boundary with Sweden to Kirkenes in the North of Norway, for a length of 2628 km. Instead, the Norwegian itinerary of the E39 connects Trondheim to Kristiansand, for a length of 1140 km.

The database used for this study includes 10 years of accident data, from 2005 to 2014. In particular, the accident database is composed by 633 fatal and injury traffic accidents in which at least one vehicle was involved along the considered sections. It provides general information about accidents, including the exact localization of the accident along the road section, and specific information about the vehicles and the involved units (drivers and passengers).

2.2 Measures

As explained in the introductory section, the measure used to define familiarity of the drivers with the place where the accident occurred is their distance from the place of residence. In particular, this distance was computed by considering the road itinerary associated to the shortest travel time between the center of the city/town of residence and the exact place where the accident occurred.

Once the measure of distance from the residence has been defined, there is still the problem of defining a threshold distance after which a driver can be considered as unfamiliar with the road. Finding an exact distance boundary for the aim explained above is a not realistic approach. Instead, defining a short distance from residence below which the driver can be considered most likely familiar and, defining a very long distance above which the driver is supposed to be unfamiliar is a more realistic strategy. In this study, a distance of the place of the accident from 0 to 20 km from the place of residence was used to consider drivers as familiar with the road where the accident occurred. This distance was chosen according to a report in which average car commuting travels were computed for Norway: 15.8 km for car drivers and 21.7 km for car passengers (Hjorthol et al. 2014). A distance of the place of the accident greater than 200 km from the place of residence was used to consider drivers involved in the accidents as unfamiliar with the road where the accident occurred. This distance was set by weighting the definition of long trip (100 km) from the same Norwegian report used for the definition of commuting travels (Hjorthol et al. 2014) and the distance above which the plane is the preferred mean of transport (300 km according to Hjorthol et al. 2014; 400 km according to Thrane 2015). Based on these threshold distances two areas of familiarity can be defined, as summarized in Table 1.

2.3 Sites inquired in detail

For the aims of this study, some road sites have to be selected for a detailed analysis of the possible

Table 1. Ranges of distance used to define familiarity of drivers involved in the accidents.

Range	Distance* (km)		Drivers involved
	From	To	
1	0	20	Familiar
2	200	–	Unfamiliar

*Distance from the drivers' place of residence to the place where the accident occurred.

accident patterns for which drivers' familiarity can play a not negligible role. Therefore, sites in which large percentages of unfamiliar drivers were involved in accidents and sites in which large percentages of familiar drivers where involved in accidents were inquired. In order to individuate patterns, road sites where less than 10 accidents occurred were not considered. The definition of drivers' familiarity was based on the distance values shown in Table 1.

Road sites having experienced accidents in which at least one unfamiliar driver was involved for a percentage greater or equal than 70% of the total number of accidents were analyzed. They are here defined as the "unfamiliar" road sites. In the same way, road sites having experienced accidents in which all drivers were familiar for a percentage greater or equal than 70% of the total number of accidents were analyzed too. They are defined as the "familiar" road sites. In the latter procedure of sites' selection, the accidents in which at least one driver was defined as unfamiliar were considered, rather than requiring that all drivers were unfamiliar. This choice is based on the consideration that drivers distant more than 200 km from the residence can be considered as an "anomaly" in the traffic flow among the other drivers. Moreover, the percentage of 70% was chosen since it was the higher percentage allowing to individuate at least a road site in each of the two categories (unfamiliar and familiar road sites). The minimum of 10 accidents for each site was set in order to obtain meaningful percentages of familiar and unfamiliar accidents.

Characteristics of the two road sites selected by using these criteria are summarized in Table 2.

Moreover, both sites have posted speed limits included between 50 and 80 km/h. Road widths measured in correspondence of the accidents at the two sites according to the database, are included between 6 and 8.6 meters (maximum 8.3 meters for the site 2). The driveway density (number of driveways or minor intersections per kilometer) is 2.1 for the site 1 (corresponding to 5 total driveways) and 1.1 for the site 2 (corresponding to 6 total driveways). The site 1 is characterized by 8 curve sections along the segment (3.3 curves per km), while the site 2 is characterized by 24 curve sections (4.4 per km). Three out of the total curves belonging to the site 2 are near to 90-degree curves.

Table 2. Characteristics of road sites inquired.

Site	AADT (veh/day)	Road	Length (m)	Crashes	Vehicles
1 (Familiar)	6697	E39	2400	11	23
2 (Unfamiliar)	5165	E6	5500	10	19

Therefore, the sites seem comparable in terms of traffic, speed limits and widths. Instead, a noticeable difference in terms of driveway density and number of curves per km can be observed. Those differences were considered in the analysis of the results of the accident reconstructions.

3 RESULTS AND DISCUSSION

3.1 "Familiar" site

The "familiar" site inquired experienced eleven accidents in ten years, having involved 23 vehicles. Among these eleven accidents, eight of them were classified as "familiar". This means that, in more than 70% of those accidents (eight), all drivers being resident from 0 km to 20 km from the place of the accident, were involved. Among the 23 vehicles involved, missing zip codes associated to 6 vehicles were found. A summary of the accidents together with the indications about the vehicles and the drivers involved is given in Table 3.

Table 3. Details about the "familiar" site.

#	Type*	Vehicle/Drivers' familiarity**			
		A	B	C	D
1	Rear-End/ Angle	Car (ND)	Car (NC)		
2	Rear-End/ Angle	Car (ND)	Car (F)		
3	Rear-End/ Angle	Motorcycle (F)	Unknown (ND)		
4	Head-on	Car (ND)	Bicycle (F)		
5	Run-off	Car (F)			
6	Rear-End/ Angle	Car (F)	Car (F)	Car (F)	
7	Run-off	Truck (F)			
8	Rear-End/ Angle	Tractor (F)	Tractor (ND)	Car (NC)	
9	Rear-End/ Angle	Car (F)	Car (F)		
10	Rear-End/ Angle	Car (F)	Camper (ND)	Car (F)	Camper (F)
11	Run-off	Car (NC)			

*Rear-End and Angle accidents were grouped together since most of the crashes classified in the database as "while turning from the same direction" were described as rear-end crashes.
**Classification of familiarity: F = familiar (≤20 km), U = unfamiliar (≥200 km), NC = not classified (20–200 km), ND = missing data. The definition "camper" includes campers, vans, cars with trailer, light trucks.

320

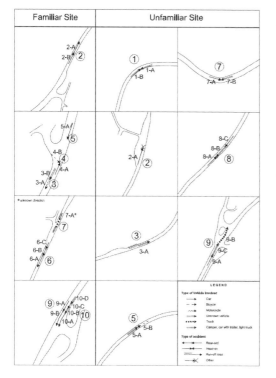

Familiar Site	Unfamiliar Site

Figure 1. Schemes of accidents occurred at the two road sites.

The accidents analyzed, for the aim of finding familiar accident patterns, were the number 2, 3, 4, 5, 6, 7, 9, 10 of the list shown in Table 3. Schemes representing collisions are reported in Figure 1.

The accident 2 happened on a straight road before a curve section in good weather, road and visibility conditions. The car A was rear-ended by the car B. The driver of the vehicle B was considered familiar while any measure was available for the vehicle A.

The accidents 3, 4 and 5 occurred in the same curve road section in correspondence of a driveway. The accident 3 happened in good weather, road and visibility conditions. The motorcycle A was rear-ended by the unknown vehicle B. The driver of the vehicle A was considered familiar while any information was available for the vehicle B. The accident 4 happened on wet road and in bad visibility conditions (but with presence of road lighting). The cyclist B, coming from the driveway, hit the car A traveling on the main road. The cyclist B was considered familiar, while any information was available for the vehicle A. The accident 5 occurred on wet road and in good visibility conditions. The car A, ran off the road when entering in the curve from the previous straight road section. The driver

of the vehicle A, considered familiar, fell asleep while driving as found out from the report.

The accidents 6 and 7 happened on a straight road section. The accident 6 occurred on wet road and in good visibility conditions. The cars A and B were stopped on the roadway, and the car B was rear-ended by the car C coming from behind. All the drivers were considered familiar. The accident 7 happened on icy road and in good visibility conditions. The driver of the truck A tried to overtake another vehicle but he lost control of its vehicle, and he hit the road barrier. The driver of the vehicle A was considered familiar.

The accidents 9 and 10 happened on a curve section (showing a very high radius of curvature) in correspondence of a driveway. The accident 9 occurred in good weather, road and visibility conditions. The car B is stopped in order to turn to the driveway, and it was rear-ended by the car A. Both drivers were considered as familiar. The accident 10 happened in good weather, road and visibility conditions. The vehicles A and B were stopped in order to turn to the driveway. The car C was successful in braking behind the vehicles A and B. Instead, the vehicle D coming from behind, failed in stopping and so, the preceding vehicles were rear-ended. All the drivers were familiar, except for the driver of the vehicle B, showing missing data.

The other three accidents (1, 8, 11) were two rear-end/angle crashes and a run-off road crash.

Some general considerations can be made on considering the results of the accident analyses. Most of the accidents were rear-end crashes (7 out of 11, and five of them were considered "familiar" accidents). The other recurrent type is the run-off road (3 out of 11, and two of them were "familiar"). Only one head-on crash was recorded (familiar).

At this road site, the percentage of missing zip codes (resulting in missing classification of the familiarity of drivers) is close to 25%, higher than the average (less than 15%) of all the sample of road sites inquired. This occurrence makes the interpretation of results more complex. Anyway, among the 8 familiar accidents analyzed, familiarity was not defined for only 4 drivers. Except for the accident 3, these drivers were rear-ended or otherwise hit. Therefore, even with missing data, it is possible to affirm that the driver who hit the cars in the rear-end accidents was considered familiar in 4 out of 5 of the familiar crashes (4 out of 7 total rear-end crashes). This is a recurring pattern at this site.

However, the argument of a recurring pattern in rear-end accidents related to familiar drivers have to be analyzed in more detail. In fact, the road site is characterized by a great number of driveways connecting nearby peripheral residential areas.

In some cases, they do not seem to be adequately signaled along the road or they are not provided by turning lanes to simplify the maneuver. It seems logical that, at this road site, a high share of rear-end accidents is present, since they can be related to the presence of driveways (and indeed, 5 rear-end crashes out of 7 were near a driveway). On the other hand, the presence of a nearby residential area can be likely related to a high share of familiar drivers in the traffic flow at this road site. This occurrence can be responsible for the over-representation of familiar drivers in the accidents happened there.

Nevertheless, in four out of five familiar rear-end accidents (among the seven total rear-end accidents), the crash started with a familiar driver hitting from behind other cars. Finding who were at fault was obviously not coherent with the aim of this paper and anyway, it is an information not derivable from the information present in the database. However, this information is useful since it can be likely related to drivers' speed. A rear-end crash is often caused by a car traveling at high speed that fails in braking promptly. The braking process can be perplexed by wet road or bad weather conditions (e.g. this was observed at this site too). Familiarity of drivers with the road site could have been responsible for higher speeds (as introduced in the first section), potentially related to the rear-end crashes. Moreover, as stated in the introduction, familiarity of drivers can be related to distraction (potentially linked to rear-end crashes too) and more dangerous behaviors. The accident 5 (familiar driver fell asleep) and the accident 7 (familiar driver who tried a likely dangerous overtaking maneuver on icy road) can be related to this tendency. However, given the small number of items, these should be considered only as indicators.

3.2 "Unfamiliar" site

The "unfamiliar" site inquired experienced ten accidents in ten years, having involved 19 vehicles. Among these ten accidents, seven of them were classified as "unfamiliar". This means that, in the 70% of those accidents, at least one driver being resident from more than 200 km from the place of the accident, was involved. Among the 19 vehicles involved, only one vehicle was not associated to a zip code. A summary of the accidents together with the indications about the vehicles and the drivers involved is given in Table 4.

The accidents analyzed, for the aim of individuating recurring unfamiliar accident patterns, were the number 1, 2, 3, 5, 7, 8, 9 of the list presented in Table 4. Schemes representing those collisions are reported in Figure 1.

Table 4. Details about the "unfamiliar" site.

#	Type*	Vehicle/Drivers' familiarity*			
		A	B	C	D
1	Head-on	Camper (U)	Camper (U)		
2	Other	Car (U)			
3	Run-off	Camper (U)			
4	Head-on	Motorcycle (NC)	Truck (NC)		
5	Head-on	Camper (ND)	Bicycle (U)		
6	Run-off	Car (NC)			
7	Head-on	Car (NC)	Camper (U)		
8	Head-on	Car (NC)	Camper (NC)	Car (U)	
9	Rear-End/ Angle	Car (NC)	Truck (NC)	Camper (U)	
10	Rear-End/ Angle	Car (NC)	Car (NC)		

* See footnotes to Table 3.

The accident 1 happened at a curve section on icy road, during rain and in good visibility conditions. The vehicle A invaded the opposite lane in curve resulting in a head-on crash with the vehicle B coming from the other direction. Both drivers were considered unfamiliar with the road.

The accident 2 occurred in a curve section in good weather and road conditions but without road lighting at night. The car A hit a moose crossing the roadway. The driver was classified as unfamiliar.

The accident 3 happened at a curve section in good weather, road and visibility conditions. The vehicle A ran off the road on the right and it tipped over. The driver was classified as unfamiliar.

The accident 5 happened at a curve section in good weather, road and visibility conditions. The bicycle B, invaded the opposite lane in curve resulting in a head-on crash with the vehicle A coming from the other direction. The cyclist was classified as unfamiliar, while no zip code was available for the other driver.

The accident 7 occurred at a curve section on icy road in good weather and visibility conditions. The car A lost control and invaded the opposite lane resulting in a head-on crash with the vehicle B coming from the other direction. The driver of the vehicle B was considered unfamiliar, while the driver of the vehicle A came from a distance between 20 and 200 km (neither familiar nor unfamiliar).

The accident 8 happened on a straight road section in good weather, road and visibility conditions. The car A invaded the opposite lane resulting in a head-on crash with the vehicle B coming from the other direction. Then, the car C coming from the same direction, was damaged by junks of vehicle B. The driver of the vehicle C was classified as unfamiliar, while the others were neither familiar nor unfamiliar.

The accident 9 occurred in a curve section (showing a high radius of curvature) in correspondence of a driveway, in good weather, slippery road and without road lighting. The truck B was moving on the left in order to turn to the driveway, while it was rear-ended by the vehicle C. Then, both the vehicles B and C were rear-ended by the vehicle A. The driver of the vehicle C was classified as unfamiliar, while the other drivers were neither familiar nor unfamiliar.

The other three accidents (4, 6, 10) were a head-on, a run-off and a rear-end/angle crash.

Some general considerations can be made on considering the results of the accident analyses. Most of the accidents were head-on crashes (5 out of 10, and in four of them there was at least one unfamiliar driver involved). Two run-off crashes (one of them with the presence of an unfamiliar driver) and two rear-end crashes (one of them with the presence of an unfamiliar driver) were recorded. Only one crash with an animal was found (unfamiliar).

It is important to note that even if they are considered as two divided categories, the head-on and the run-off crashes have some common traits to be evaluated for the aims of this study. In fact, the head-on accidents are often caused because a vehicle invades the opposite lane. The reasons for these invasions could be various, but except for the case of the driver who fall asleep or is distracted, they can be related to some maneuvering errors. The same errors can be related to the run-off road crashes together with the speed selection. Considering together the head-on and the run-off crashes, in five out of seven total accidents at least one unfamiliar driver was involved. In three out of five of those accidents, the unfamiliar driver invaded the opposite lane or ran off the road at curves. In only one of the inquired cases, the invasion of the opposite lane happened in a road section essentially straight (even if included between two curves). This case could be better related to distraction or fatigue, but in that case (accident 8) the driver was not classified as unfamiliar. Therefore, from this analysis, only some indications about unfamiliar drivers can be taken, without the identification of clear accident patterns.

Anyway, also in this case, it is important to analyze the context of the road site inquired. The road site is a segment of a rural rolling/mountainous road with no important city placed at a close distance. Given these conditions, it is expected that a significant part of the traffic flow can be composed of unfamiliar drivers. This can explain the over-representation of unfamiliar drivers in the studied accidents. Some of them were involved in the possible causation of accidents likely due to driving errors. This is coherent with the concepts stated in the introduction: if the road environment is particularly demanding, presenting sudden surprises, the unfamiliar drivers could be not able to correctly address the road demand, since their expectations do not meet reality. However, in this case, no recurrent patterns were found for the unfamiliar accidents.

4 CONCLUSIONS

A micro-analysis of the accidents occurred at specific sites selected for their noticeable high share of familiar and unfamiliar drivers (based on the distance between the accident place and the residence) was conducted.

In detail, two road sites were further analyzed: the site 1, showing a high concentration of crashes for familiar drivers and the site 2, showing a high concentration of crashes for unfamiliar drivers.

Some interesting results were highlighted through the reconstruction of accidents. In fact, some recurrent accident patterns were individuated, especially for site 1. There, most of accidents were rear-end crashes, likely due to the high density of driveways along the road segment. Familiar drivers were found to be in most cases the drivers hitting vehicles from behind, resulting in the rear-end crash. This is coherent with the possible tendency to speeding and distraction for familiar drivers. Instead, at the site 2, most of the accidents were head-on collisions, likely associated to the high demanding road geometry of the road segment (as well as run-off-road crashes). Unfamiliar drivers were found to be in some cases the drivers who lost control or who anyway invaded the opposite lane, resulting in crashes. This is coherent with the possible difference for them, between expectations and the surprising reality (Colonna & Berloco 2011). However, in this case, no clear patterns were individuated for the unfamiliar accidents.

Anyway, these results come from the analysis of a very limited number of case studies. Furthermore, these sites were selected by setting empirical threshold percentages of drivers' familiarity. Therefore, these interpretations should be mainly taken as a cause for further reflections about the topic, that should be analyzed more in depth. For example, at the road sites inquired, the most recurrent types of accident are multi-vehicle crashes. The influence of familiarity on the accident type, considering more in detail the interactions between familiar and unfamiliar drivers in multi-vehicle crashes should be analyzed by focusing on other specific sites. Moreover, it is evident that several other factors (related to road, driver, environment) are related to the accidents besides the familiarity of drivers. In this paper, they were considered only locally, as related to each accident. It is evident as well that, the highlighted accident patterns could be explained by the differences themselves in the road sites features, complementary to, or independent from drivers' familiarity. Therefore, a more comprehensive statistical analysis would be needed for further studies, in order to assess the importance of the familiarity as predictor variable of accident occurring and type, but considering other variables, such as for example drivers' age and gender, curvature rates, other geometric features, weather etc. (Russo et al. 2013).

However, some of the highlighted facts suggest that drivers' familiarity/unfamiliarity could be better considered for road design and safety-based maintenance. Demanding and surprising road features should be avoided on segments where unfamiliar drivers are likely to travel (i.e. touristic and rural mountainous highways). On the other hand, possible shift of familiar drivers to more risky behaviors could be taken into account at roads near residential areas (e.g. by modifying minor intersections into roundabouts for reducing speeds at intersections). However, further studies are needed to better address the possible practical applications.

ACKNOWLEDGMENTS

The authors acknowledge Damiano De Gennaro who, during his thesis, helped with the reconstruction of the accidents used throughout the paper.

REFERENCES

Baldock, M.R.J., Long, A.D., Lindsay, V.L.A. & McLean, J. 2005. *Rear end crashes*. Adelaide University. Australia.

Blatt, J. & Furman, S.M. 1998. Residence location of drivers involved in fatal crashes. *Accident Analysis & Prevention*, 30(6), 705–711.

Brown, J., Fitzharris, M., Baldock, M., Albanese, B., Meredith, L., Whyte, T. & Oomens, M. 2015. *Motorcycle In-depth Crash Study* (No. AP-R489-15). Austroads.

Colonna, P. & Berloco, N. 2011. External and internal risk of the user in road safety and the necessity for a control process. Xxiv PIARC World Congress. Mexico City.

Colonna, P., Intini, P., Berloco, N. & Ranieri, V. 2016. The influence of memory on driving behavior: how route familiarity is related to speed choice. *Safety science*, 82, 456–468.

Fuller, R. 2005. Towards a general theory of driver behaviour. *Accident Analysis & Prevention*, 37(3), 461–472.

Hjorthol, R., Engebretsen, Ø. & Uteng, T.P. 2014. *Den nasjonale reisevaneundersøkelsen 2013/14: nøkkelrapport*. Transportøkonomisk institutt. Norway.

Liu, C. & Ye, T.J. 2011. *Run-off-road crashes: an on-scene perspective* (No. HS-811 500). NHTSA's National Center for Statistics and Analysis. Washington, DC.

Martens, M.H. & Fox, M.R. 2007. Do familiarity and expectations change perception? Drivers' glances and response to changes. *Transportation Research Part F: Traffic Psychology and Behaviour*, 10(6), 476–492.

Noland, R.B. 2013. From theory to practice in road safety policy: Understanding risk versus mobility. *Research in transportation economics*, 43(1), 71–84.

Rosenbloom, T., Perlman, A. & Shahar, A. 2007. Women drivers' behavior in well-known versus less familiar locations. *Journal of safety research*, 38(3), 283–288.

Russo, F., Biancardo, S.A., Busiello, M., De Luca M. & Dell'Acqua, G. 2013. A statistical look at gender and age differences as related to the injury crash type on low-volume roads. *WIT Transactions on the Built Environment*, 134, 213–224.

Theeuwes, J. & Godthelp, H. 1995. Self-explaining roads. *Safety science*, 19(2), 217–225.

Thrane, C. 2015. Examining tourists' long-distance transportation mode choices using a Multinomial Logit regression model. *Tourism Management Perspectives*, 15, 115–121.

Wilde, G.J. 1982. The theory of risk homeostasis: implications for safety and health. *Risk analysis*, 2(4), 209–225.

Yanko, M.R. & Spalek, T.M. 2013. Route familiarity breeds inattention: A driving simulator study. *Accident Analysis & Prevention*, 57, 80–86.

Yannis, G., Golias, J. & Papadimitriou, E. 2007. Accident risk of foreign drivers in various road environments. *Journal of safety research*, 38(4), 471–480.

Transport Infrastructure and Systems – Dell'Acqua & Wegman (Eds)
© 2017 Taylor & Francis Group, London, ISBN 978-1-138-03009-1

Geotechnical asset management for Italian transport agencies: Implementation principles and concepts

P. Mazzanti
NHAZCA S.r.l. and Department of Earth Sciences, University of Rome "Sapienza", Rome, Italy

P.D. Thompson
Consultant, Bellevue, WA, USA

D.L. Beckstrand
Landslide Technology, Inc., Portland, OR, USA

D.A. Stanley
D.A. Stanley Consulting, Bellingham, WA, USA

ABSTRACT: Internationally, Transportation Asset Management (TAM) has been accepted practice for many years. National, regional or state statutes and policies encourage transportation agencies to adopt asset management principles, including performance- and risk-based management. Transport agencies also develop their own short- and long-term plans, objectives and goals to guide agency programs.

Italy recently reorganized the Ministry of Infrastructure and Transport (MIT) and developed ten key concepts to guide the Ministry. While all these concepts have application to TAM, several stand out, including utility, simplification, involvement, maintenance and safety. Development of TAM for transport systems can be a guiding force in improving development and management of transport systems.

Consideration of geotechnical assets is important in TAM development for Italy to meet national, regional and agency goals. Italy has a highly-developed multi-modal system of surface transportation and thousands of km of tunnels also due to the complex topography. The system ranges from roadways to airports to rail transport and ferry and ship transport infrastructure, all of which depend on support from geotechnical assets, whether embankments below pavement, earth retaining walls, rock slopes, etc. These type of assets are often overlooked, but the value of geotechnical assets is likely to be a substantial percentage of the total value of the system. The high geological and geomorphological activity of the Italian peninsula is a relevant factor of stress for the national transportation network and, especially, for geotechnical assets. Furthermore, many of these assets are in declining condition, because, like every transport agency in developed countries, there are not sufficient funds to do everything that is needed. Maintaining and improving the condition of geotechnical assets can be a key step for the achievement of the MIT goals.

Geotechnical Asset Management (GAM) can contribute to improvement to the condition, utility, safety and life cycle cost of a transportation network. GAM will assist transport agencies to understand the current and future condition of geotechnical assets, and through integrated life cycle cost and risk analysis, the alternative actions available to rehabilitate, preserve and eventually replace assets. Geotechnical asset management offers guidance in supporting decision-making to preserve and improve the transport system.

1 INTRODUCTION

1.1 *Role of agency goals and objectives and transportation asset management*

This paper explores potential processes for creating a Transportation Asset Management (TAM) program for the management of geotechnical assets associated with the Italian road system.

The purpose of a TAM program is: "to meet a required level of service, in the most cost-effective manner, through the management of assets for present and future customers." (Gordon, et al, 2011). TAM ties essential agency processes to a performance management system that includes target levels of service and performance metrics that when monitored over time, give a picture of how the agency is progressing toward its targets. TAM benefits include optimized costs for the system, improved communication within the agency and with outside stakeholders, reduced risk, and improved asset condition and system performance.

The Italian transport agency goals and objectives have already been created, as discussed below, and provide the basis for development of an asset management process. What remains is the development of a step-by-step roadmap to guide transport agencies through the process of self-assessment, setting agency targets and desired outcomes, incorporating TAM into transport system organizations, establishing performance standards, developing TAM Plans, developing TAM tools and processes (inventory procedures, monitoring and data collection, information systems, communicating with public and stakeholders, etc.) through established agency protocols.

Ideally, Geotechnical Asset Management (GAM) programs should develop alongside or trailing TAM programs. However, TAM systems may include all transportation assets and therefore may be much slower to move forward. But, even if the TAM system is not entirely developed, a GAM program can proceed apart from, but in alignment and compatible with the TAM system. This can be done by first making connections between GAM and the broader agency goals and objectives and then following a roadmap that mirrors the TAM roadmap. The steps in a TAM/GAM roadmap can vary considerably, but Figure 1 shows an example.

1.2 Why manage geotechnical assets?

Geotechnical assets have a rarely noticed importance in transport networks: all transport assets are supported by or protected by geotechnical assets, whether soil embankments below highway surfaces, rock slopes and rockfall mitigation devices above roadways and railroads, rock below bridge foundations or retaining walls that support roadways and tunnels, that could not otherwise be constructed.

The relevance of geotechnical assets is even greater in a geologically young country like Italy that is characterized by intense geological and geomorphological processes. Landslides and rockfalls are the most common events that interact with the transportation assets. However, erosion, floods, earthquakes and volcanic activities interact quite often with the transportation network.

Moreover, settlement and subsidence are frequent in low-lying areas, due to the compressibility of recently deposited soils. Hence, instability problems can frequently affect embankments, roads, runways/airstrips and tunnels. Due to the complex topography, the Italian transportation network includes thousands of kilometers of tunnels, a substantial percentage of all the tunnels in Europe.

However, despite the complexity and extent of geotechnical assets, they are often constructed with a "build and forget" mentality. But these assets must be maintained throughout their lifecycle, just as concrete or steel bridges and asphalt and concrete pavements must be maintained. GAM programs offer the ability and tools to make cost-effective decisions about geotechnical assets that contribute to the good performance, reduced risk and improved cost-effectiveness of the transport system as a whole.

1.3 What are geotechnical assets?

One of the first steps in management is to decide what geotechnical assets Italian transport agencies are responsible for and what can be managed. While taxonomic classifications have recently been developed (Anderson, 2016), geotechnical assets can generally be subdivided into tangible and intangible assets. Tangible assets are those that have been built to physically support the transportation system. Intangible assets include results of geological or geotechnical data gathering. Table 1 contains examples of both types. Poor management of either type results in decreased resilience, inefficient stewardship, and higher life cycle costs.

In the USA, only a few programs are operating or under development for geotechnical assets. Specific assets include: rock and soil slopes, retaining walls and sand and gravel sources/quarries. In addition, one GAM-based program focuses on managing geotechnical hazards, such as rockfall, landslides, debris flows, volcanic hazards, and floods. As stated above, Italy has an abundance of

Figure 1. Steps for Development of Geotechnical Asset Management for Italian Transport Agencies.

Table 1. Example geotechnical assets.

Physical assets	Data assets
Embankments	Geologic Maps
Soil and Rock Slopes	Geotechnical Reports
Tunnels	Road Closing Events/Costs
Retaining Walls	Boring Logs
Coastal Revetments	Performance monitoring data
Foundations	Laboratory Data
Geotechnical Instruments	Geotechnical Event Data
Rockfall Control Measures	Maintenance History
Geotechnical Instruments	As-built Plans
Material Sources	Photographs
Levees	Flood Records

each of these asset types and hazards. As part of a gap analysis, the agencies can determine which assets to manage.

1.4 High level transport agency goals and objectives

The Italian transport system is the responsibility of the Ministero delle Infrastrutture e dei Trasporti (Ministry of Transport and Infrastructure or MIT). The MIT recently reorganized and adopted a mission statement: "Italy does not need large or small projects, but useful ones." The term "Useful" applies to growth of the country, to modern and efficient logistics, for rapid mobility, and a safe, clean, convenient and widespread system of transport. Useful also relates to recovery and livability of urban centers and for the defense and maintenance of the country. (MIT website: http://www.mit.gov.it/ communicazione/news/nuovo-mit, accessed September 2, 2016) The MIT mission also includes modernizing the existing road and motorway network.

As part of their efforts to modernize the agency, the MIT has developed "Ten Key Concepts" to guide its operations and investments: Utility, Simplicity, Participation, Sustainability, Care, Integration, Safety, Iron, South, and Europe.

Since 1946, MIT has operated a government-owned company, Azienda Nazionale Autonoma delle Strade (the National Autonomous Roads Corporation or ANAS) whose mission is the "management of ordinary and extraordinary maintenance of streets and highways," including:

- progressive improvement of the network of roads,
- construction of new roads and highways,
- service of information to users,
- protection of the heritage of roads and highways,

- adoption of safety measures, and
- conduct of research and experimentation in the field of roads and traffic and implementation of the results.

(ANAS website http://www.stradeanas.it/, accessed September 2, 2016).

In addition, the MIT has engaged in strategic long-range planning with development of its July 2016 "Connect Italy—Strategies for transport infrastructure and logistics." Together the mission statement, Ten Key Concepts and the "Connect Italy" strategic planning effort provide ample high level guidance to create and execute a roadmap for a performance-based asset management system for geotechnical assets. The "Connect Italy" infrastructure strategy lays out the architecture of the transport system, but importantly defines the "objectives of transport policy in Italy." This critical section lists: sustainable and safe mobility, quality of life and competitiveness in urban and metropolitan areas, and support for industrial policies of the supply chain (MIT website: http://www.mit.gov.it/comunicazione/news/piste-ciclabili/connetterelitalia-pubblicata-sul-sito-la-nuova-strategia-del: accessed September 2, 2016). Specific strategic objectives outlined in the document include:

- Useful infrastructure, streamlined and shared;
- Modal and intermodal integration;
- Enhancement of existing infrastructure assets;
- Sustainable urban development.

MIT and ANAS work together to operate the Italian network of roads and highways within the framework of mission statements, adopted key concepts, and strategic planning objectives. Incorporation of transportation asset management principles into both agency's business and technical processes in accord with these objectives, concepts and strategies will help align the organizations, strengthen each agency and provide long term improvements to the road and motorway networks.

2 SELF-ASSESSMENT AND GAP ANALYSIS

The incremental process of advancement in asset management necessarily occurs in phases spread over many years. During that time, much can change in an agency's institutional and economic environment, in the needs of stakeholders, in the agency's delivery capability, and in technology.

A useful general approach to implement asset management is self-assessment. The American Association of State Highway and Transportation Officials has published an example (Cambridge et al, 2002) which addresses business processes, data, tools, documentation, and training in the following key areas:

- Policy guidance
- Planning and programming
- Program delivery
- Information and analysis

The strategic self-assessment is couched in very general terms in order to be applicable to all transportation agencies and all types of assets. Volume 2 of the AASHTO Asset Management Guide provides more detail (Gordon et al, 2011).

In the specific domain of geotechnical assets, it is useful to think about how information about these assets enters into each of the business processes addressed in the strategic assessment. For example:

- Policies and criteria to guide inventory and condition surveys, and the identification and prioritization of needs.
- Processes to ensure that geotechnical needs are considered in road and bridge projects, and in the maintenance budget.
- Reliable tracking of actual costs of preservation, reconstruction, risk mitigation, and incident recovery on geotechnical assets?
- Efficient and effective delivery of planned inspections, maintenance activities, and projects.
- Maintenance of a complete inventory with effective presentation tools, such as maps and charts.
- Capability to forecast future conditions and needs.
- Incorporation of geotechnical needs within agency tools for planning of pavement and bridge work.
- Consideration of geotechnical risk and long-term cost, including future maintenance cost, within the design of new facilities.

Clearly many of these concerns apply to all types of assets, so in many agencies it may be more cost-effective to consider all infrastructure assets in the same self-assessment. This would be a prelude to developing a coordinated multi-year program that can fill all the gaps.

3 GEOTECHNICAL ASSET MANAGEMENT PLAN

A Geotechnical Asset Management Plan (GAM Plan) is a written document, or a set of written documents and databases, which describes the processes and outputs of agency GAM activities. (Thompson 2016) The GAM Plan complements other agency planning documents such as strategic plans, service plans, and investment plans, but focuses on the preservation and performance of the agency's valuable infrastructure over a typical time frame of ten years. The GAM Plan serves several important functions:

- Documents the ways in which infrastructure performance contributes to agency success.
- Specifies an agreed means of measuring the contribution of asset performance toward agency success.
- Provides a commonly-understood set of criteria for evaluating and selecting investments.
- Establishes a basis for incentives for superior performance, internal and external to the agency.
- Engages stakeholders in a discussion of the tradeoff between funding and performance.
- Provides an objective method of allocating resources among preservation, risk mitigation, and service improvement goals.
- Specifies research requirements and analytical tools for relating current decisions to future outcomes.
- Provides the functional and quality assurance requirements for asset inventories and data collection processes.
- Provides the justification for capital and operating expenditures on preservation, maintenance, and risk mitigation.
- Documents the agency's commitment and accomplishments in the continuous improvement of its management processes.

All of the basic components of asset management and TAM Plans have been codified in various standards documents in recent years. In the United Kingdom, the authoritative source is Publicly Available Specification 55, volumes 1 and 2 (BSI 2008). In the United States, a basic framework is described in a financial management context in Government Accounting Standards Board Statement 34 (GASB 1999), and in a strategic planning context in Volume 1 of the AASHTO Guide for Asset Management (Cambridge et al 2002) as well as in ISO 55000. A more detailed adaptation of the same principles is New Zealand's International Infrastructure Management Manual (IIMM, NAMS 2006). AASHTO has built on this concept in great practical detail with the AASHTO Transportation Asset Management Guide, Volume 2: A Focus on Implementation (Gordon et al 2011).

A key aspect of successful asset management implementation, brought out in the IIMM and the AASHTO Guide, is the notion of continuous improvement. A variety of human and automated ingredients need to be improved in tandem. The amount of progress that can be made in asset management tools is limited by the human and organizational readiness to use the technology, and vice versa. In a more tangible sense, the technology to produce quality asset management information depends on management willingness to accept asset management information in decision-making (and to see the value and pay the cost of producing this information); and management acceptance, in turn, depends on the quality of information that can be produced. A small improvement in the decision making process must be matched by an incremental improvement in technology, which then spurs the next small improvement in decision making.

These same principles are widely used in the private sector, often taking the form of performance management frameworks such as the Balanced Scorecard and Six Sigma (Proctor et al 2010, Gordon et al 2011).

In 2015, the US Federal Highway Administration published a draft of its proposed TAM Plan requirements (FHWA 2015). The rule specifies that the plan shall cover at least a 10-year period, shall be made easily accessible to the public, and shall establish a set of investment strategies that improve or preserve condition and performance in support of national highway performance goals. The plan is to be tied in to existing programming processes and must contain the following minimum content:

1. TAM objectives, aligned with agency mission;
2. Performance measures and targets;
3. Summary of asset inventory and condition;
4. Performance gap identification;
5. Life cycle cost analysis;
6. Risk management analysis;
7. Financial plan;
8. Investment strategies.

The risk-based TAM Plan is required to identify the hazards affecting the movement of people and goods, assess the likelihood and consequences of adverse events, and evaluate and prioritize mitigation actions. The life cycle cost analysis is a quantitative network-level analysis that considers current and desired condition levels, asset deterioration, effects of adverse events, and treatment options over the whole life of assets.

The proposed US rules require that TAM Plans address only pavements and bridges, but encourage agencies to expand the scope to include any or all infrastructure assets within the right-of-way corridor. Alaska is the first state to develop a plan for geotechnical assets which includes all of the components required in the federal rules (Thompson 2016).

4 PERFORMANCE MEASURES AND TARGETS

Transportation Asset Management is a discipline that relies on performance measures that are tracked over time. Decisions regarding life cycle management, capital programming, and maintenance policy are evaluated according to predicted changes in measurable objectives. Research-based analysis methods are used to forecast future performance and to estimate the outcomes of agency actions.

Transportation agencies spend money constructing or acquiring geotechnical assets, and spend additional money over the lifespans of these assets to keep them functioning as intended. In general, the intended function is to maintain (or refrain from disrupting) desired levels of safety, mobility, environmental sustainability, and economic efficiency of transportation service. Through its maintenance forces and contractors the agency implements treatments that maintain or enhance the characteristics of its geotechnical assets to minimize the frequency of disruptions. These desirable characteristics make up a property called resilience. In general, resilience is defined as:

"… the capability of a system to maintain its functions and structure in the face of internal and external change and to degrade gracefully when it must." (Allenby & Fink 2005).

"Internal and external change" can be interpreted in the context of geotechnical assets as changes caused within the asset itself (i.e. normal deterioration) and change caused by external forces (natural extreme events). "Maintain its functions and structure" can be interpreted as the avoidance of transportation service disruptions. "Service disruptions," in turn, can be interpreted as unintended changes in the safety, mobility, sustainability, or economic performance of the roadway. Based on this reasoning, a slope may be considered to have high resilience (or low vulnerability) to the extent that it is sufficiently able to refrain from causing service disruptions due to normal deterioration or adverse events. Examples of measurable factors affecting resilience are:

Rock slopes:

- Ditch (or catchment) effectiveness: assesses how often falling rocks reach the roadway, combining the effects of all design, mitigation, and geometry concerns.
- Rockfall activity: assesses how active the slope is in producing falling rocks, combining the effects of all condition characteristics, geological character, climate, and hydrology.

Embankments and soil slopes:

- Roadway displacement or slide deposit: assesses the direct effect on the roadway surface of earth movement, combining the effects of all relevant condition characteristics and mitigation features.

- Length of affected roadway and roadway impedance: assesses the geometry of the site.
- Movement history: assesses the combined effect of geological character, climate, hydrology, and permafrost characteristics.

Retaining walls

- Vertical and horizontal wall alignment: assesses one aspect of condition (deformation), combining the effects of drainage, geometry, and foundation.
- Impacts to the roadway: assesses the effect of physical condition of the roadway as it relates to wall condition.
- Critical component health: assess all aspects of wall condition other than deformation, such as corrosion.

These characteristics are typically summarized into an index (often on a scale where 100 is best and 0 is worst) or a set of states (such as good-fair-poor) which can be recorded and tracked over time.

Acceptable asset performance is characterized by determining whether a given asset satisfies a set of level of service criteria, usually the same criteria used in separating Poor performance from Fair. These criteria may vary by class of road, with high-volume roads or essential lifeline routes subject to more rigorous criteria. Additional criteria may be used to identify opportunities for preventive maintenance or risk mitigation.

At the network level, performance is typically assessed in the form of average of asset performance index values, or as the percent of the inventory in Good or Poor condition. The agency can track changes in this performance, and will typically report changes publicly in order to establish mutual accountability for making progress toward shared goals.

Future performance goals are expressed as performance targets over a timeframe of 5–10 years. Aspirational targets can be established to indicate the desired direction of future performance trends. For example, many agencies have an aspirational safety objective of "zero fatalities." Aspirational LOS and performance measures have been considered for unstable slopes in Alaska (Stanley & Pierson, 2011).

However, there are more mature asset management processes with fiscally-constrained performance targets, developed from a research-based analysis of deterioration rates, intervention costs, action effectiveness, and the likelihood and consequences of service disruption, to indicate the expected performance level under a specific fiscal scenario. Such tools are common in the management of pavements and bridges, but are still in their infancy for geotechnical assets. Alaska's Department of Transportation and Public Facilities has developed an example set of analysis tools as Excel spreadsheets (Thompson 2016).

5 INVENTORY, ASSESSMENT, AND MONITORING

Transportation asset management relies on data and analysis to improve decision making. Thus the first step, and most costly investment, in GAM implementation is the creation of an accurate inventory of these assets, and establishment of a recurring process of field inspections to keep the inventory up-to-date and to monitor conditions over time.

5.1 Inventory and assessment approaches

Site-based geotechnical performance assessment focuses on a limited set of visual observations of the key physical properties that have most effect on deterioration rates or life cycle cost.

5.1.1 Focus on screening-level efforts

Often, agencies may not have a good grasp on the number, location, or type of geotechnical assets for which they are responsible. Inventories can start with plan sets and online street viewer systems carried out by entry-level geotechnical personnel. Follow-up, site specific condition assessments require the knowledge and judgement of more experienced geotechnical personnel. Condition assessments evaluate the asset according to a set of parameters where numerical scores are assigned to eight to twelve observable conditions then combined into a single metric enumerating its condition. Subsurface explorations should not be part of the effort. Field time for each condition assessment should take no longer than 20 to 30 minutes per asset. If longer times are required, such as for long linear assets such as levees, they can be broken up into smaller segments or evaluated to a more detailed reporting level.

5.1.2 Limit to observable criteria

There are many unknowns in the geotechnical field. At the screening level, the asset's condition is effectively communicated by its performance. For example, the performance of a rock slope is tied directly to its rockfall activity and the effectiveness of the roadside catchment ditch. Retaining wall performance is observable through its deformation, corrosion, and performance of the adjacent pavement. Initial observations of surface features of retaining walls is sufficient for screening level assessments. More in-depth investigation of buried reinforcement elements is possible through exhumation, use of retrievable reinforcing sample coupons, and remote electrochemical testing, but these methods are better suited for long-term monitoring activities for high-risk structures.

5.1.3 *Utilize existing systems and data*

Many systems have been developed to assess the condition of geotechnical assets. Systems such as the Rockfall Hazard Rating System (RHRS) (Pierson & VanVickle, 1993) have been implemented along select Italian motorways (Budetta, 2004; Budetta & Nappi, 2013). Other systems have been generated for the evaluation of landslide slopes and retaining walls by US transportation departments (Federal Highway Administration, states of Washington, Ohio, Oregon, and Alaska). These systems contain concepts utilized by the geotechnical community for decades and can be adapted for condition assessment and correlated to risk and mitigation costs while still containing concepts familiar to most geotechnical professionals (Beckstrand and Mines 2016).

5.1.4 *Data collection and storage*

Data collection and storage platforms have advanced in recent years. The advent of user-friendly mobile devices combined with GIS data collection applications, such as ESRI's *Collector* mobile application, has expanded opportunities for efficient data collection and storage. These systems facilitate field data collection and subsequent storage directly into an online GIS platform that integrates readily into planning activities and analysis.

5.1.5 *Simplify results*

Clear and concise communication of assessment results is critical. Results from existing systems, such as the RHRS, are clear to geotechnical personnel well versed in its use, but may be confusing to non-geotechnical personnel. GAM system output should be understandable to non-geotechnical personnel and use scores and descriptions that clearly communicate results. Descriptors such as 'Good', 'Fair', and 'Poor' describe the meaning of the scores as shown below. Scores that drop over time from 100 toward 0 through Good and Fair to Poor communicate deterioration.

5.2 *Monitoring*

Inventory and condition assessment provide a baseline for life cycle management. Uncertainty is a constant companion in geotechnical engineering. Each asset can be characterized by unpredicted events, and unforeseen findings. Risk management principles may be applied in GAM programs to reduce the effects of uncertainty. As noted above, observational methods (Terzaghi, 1937; Peck,

Table 2. Example Rock Slope Evaluation Approach.

Score Range	Descriptor	Rock Slope Condition
80 to 100	Good	Little to no rockfall activity with good ditch effectiveness
40 to 80	Fair	Moderate activity with moderate to limited ditch effectiveness
0 to 40	Poor	Frequent or constant activity with limited to no ditch effectiveness

1969) can be used to reduce the impacts of unpredicted events and to systematically update the assessment.

In the last few decades, a strong technological evolution has provided to the community several tools to improve observational abilities and perform effective geotechnical monitoring that can support GAM programs. Ranging from remote satellite systems to contact apparatus today it is possible to perform a multi-scale monitoring approach in space and time, thus supporting asset management and decision making.

Geotechnical monitoring tools can be classified in two main groups: contact systems and remote systems. Contact systems need the physical contact between the instruments and the ground while remote methods do not require such contact as they are mainly based on sensors receiving and, often, also emitting electromagnetic waves. A suitable combination of these tools may answer most of the needs of management of geotechnical assets.

Specifically, remote sensing systems like satellite InSAR, aerial LiDAR and Photogrammetry may provide considerable information at regional/national scale (thus controlling the entire transportation networks) while contact systems may allow for the continuous control of high-risk sites where an early detection of potential threats due to adverse events can reduce the risk.

In some cases, continuous monitoring can be considered also the most effective and ultimate solution for the management of high risk areas. Over the last years ANAS and RFI (Rete Ferroviaria Italiana) – the two main players of the Italian National Transportation networks that control more than 50,000 kms of roads—performed several successful applications that demonstrated the efficacy of the monitoring approach.

Hence, Italy could be a leading country in the development of a "Smart GAM", i.e. a management plan that makes extensive use of geotechnical monitoring tools.

6 COMMUNICATE PERFORMANCE

An important concept in all transportation asset management systems is the ability to conduct effective communication about the program internally within the agency, between related transport agencies (MIT, ANAS and railway agencies, e.g.), and with the public and stakeholders including government administration and legislative bodies. In Italy, there must be robust communication systems that will assist developers of geotechnical asset management programs in communicating about the process and to collect and disseminate data and information about geotechnical assets and their condition, performance, lifecycle costs, risk/safety issues, among others.

An Italian example of communication is the ANAS Road Management Tool, which provides surveillance through remote cameras, traffic management, data collection and real time event management capabilities through a central control room. Managers in the control room may dispatch resources as needed to roadway events.

Alaska DOT has developed an ArcGIS On-Line-based mapping product that shows a snapshot of the state's program for managing unstable slopes derived from databases of road network and geological data (Figure 2).

A more traditional TAM presentation provides the public and other stakeholders a quick view of status of high-level agency-wide performance attributes through the use of DOT website performance dashboards or performance report publications, such as attributes of congestion, safety, road surface condition, financial performance, management performance, customer service, and timely completion of projects.

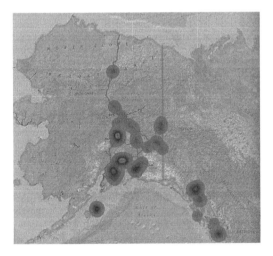

Figure 2. GIS heatmap for Alaska's rock slope performance.

7 CONCLUSION

As Italy moves forward to modernize and improve its transport network of roads, motorways and rails, it has an opportunity to follow the international pattern of implementing transportation asset management as a business process and as a set of tools to aid in management of geotechnical assets. Italy has substantial numbers of geotechnical assets at risk, such as rock and soil slopes and retaining walls throughout the country located in complex and active geological terrains as recently demonstrated by the August 24, 2016 earthquake in Central Italy that triggered more than 100 landslides intersecting with transportation roads (CERI website: http://www.ceri.uniroma1.it/index. php/2016/08/earthquake-central-apennines-italy/, accessed September 10, 2016).

Italy's current effort to effect change through reorganizing the Ministry of Infrastructure and Transport, utilizing "Ten Key Concepts" and the operations of the National Autonomous Roads Corporation, as well as comprehensive planning such as the "Connect Italy" all combine to provide the high-level goals and objectives necessary to formulate and implement an asset management plan for geotechnical assets. Conducting TAM and GAM programs to manage assets will further the agency goals and objectives.

Aging infrastructure elements such as unstable slopes and retaining walls are often overlooked until they fail, sometimes catastrophically, but they can be managed through TAM/GAM processes, starting with gap analyses to assess the weaknesses in the transport agency's management of geotechnical assets. Agencies can then proceed through inventory and assessment, beginning with screening level basic efforts utilizing simplified condition ratings.

Furthermore, the international leadership of Italy in the field of monitoring technologies and remote sensing can be a driving force for the development of a "Smart Geotechnical Asset Management" (SGAM) plan that includes the continuous collection and storage of monitoring data for the systematic update of the GAM in support of decision making. ANAS is one of the most advanced and innovative agencies at global level in the application of cutting edge monitoring and sensing technologies for the control of geotechnical assets. Technologies like satellite and terrestrial InSAR and LiDAR, for example, have been commonly used for the last several years for management plans of single assets under critical conditions (Mazzanti et al, 2015; Brunetti & Mazzanti, 2015). With this technology, transport agencies can then define standard procedures for the continuous geotechnical monitoring of high risk areas and assets.

As all of these pieces of the puzzle come together, transport agencies will develop the capability to provide analysis-based decision-making support to agency authority in planning and selecting projects and project elements best suited for improving performance and reducing life-cycle cost and risk for the transport network.

REFERENCES

Allenby, B. & Fink, J.. 2005. Toward inherently secure and resilient societies. *Science* 309:5737, pp. 1034–1036.

Anderson, S.A., Schaefer, V.R., & Nichols, S.C. 2016. Taxonomy for geotechnical assets, elements, and features. *Compendium of Papers*, TRB Annual Meeting, Washington, D.C., USA: Transportation Research Board.

Beckstrand, D. & Mines, A. 2016. Jump starting a geotechnical asset management program with existing data. Submitted for publication to *Transportation Research Record: Journal of the Transportation Research Board*, in preparation.

Brunetti A.& Mazzanti P., 2015. Monitoring an unstable road embankment for public safety purposes by terrestrial SAR interferometry. Proceedings of the 9th International Symposyum on Field Measurements in Geomechanics (Sydney, 9–11 September 2015) pp. 769–780.

BSI. 2008. *Asset Management Part 1: Specification for the optimized management of physical assets.* London: British Standards Institute, Publicly Available Specification 55–1 (PAS 55–1).

BSI. 2008. *Asset Management Part 2: Guidelines for the application of PAS 55-1.* London: British Standards Institute, Publicly Available Specification 55–2 (PAS 55–2).

Budetta, P. & Nappi, M. 2013 Comparison between qualitative rockfall risk rating systems for a road affected by high traffic intensity *Natural Hazards and Earth System Sciences* 13: 1643–1653.

Budetta, P. 2004 Assessment of rockfall risk along roads *Natural Hazards and Earth System Sciences* 4: 71–81.

Cambridge Systematics, Inc., Parsons Brinckerhoff Quade and Douglas Inc., Roy Jorgensen Associates Inc., & Paul D. Thompson. 2002. *Transportation Asset Management Guide,* Washington: American Association of State Highway and Transportation Officials.

FHWA. 2015. *Notice of Proposed Rule-Making on Transportation Asset Management Plans.* Federal Register 80:34, Page 9232 and 9236. Washington: US Federal Highway Administration.

GASB. 1999. *Basic Financial Statements — and Management's Discussion and Analysis — for State and Local Governments.* Government Accounting Standards Board Statement 34.

Gordon, M., Jason Smith, G., Thompson, P.D., Park, H., Harrison, F., & Elston, B. 2011. *AASHTO Transportation Asset Management Guide, Volume 2: A Focus on Implementation.* Washington: American Association of State Highway and Transportation Officials.

Mazzanti P., Bozzano F., Brunetti A., Esposito C., Martino S., Prestininzi A., Rocca A., Scarascia Mugnozza G., 2015. Terrestrial SAR Interferometry Monitoring of Natural Slopes and Man-Made Structures. G. Lollino et al. (eds.), Engineering Geology for Society and Territory, Volume 5, Springer International Publishing, Switzerland, 189–194.

NAMS Steering Group. 2006. *International Infrastructure Management Manual (IIMM).* Auckland: National Asset Management Steering Committee, New Zealand.

Peck, R.B. 1969. Advantages and limitations of the observational method in applied soil mechanics. Géotechnique, 19(2), 171–187.

Pierson, L & VanVickle, R. 1993. *Rockfall Hazard Rating System Participants Manual.* Washington: US Federal Highway Administration Report FHWA SA-93–057.

Proctor, G., Park, H., Varma, S., & Harrison, F. 2010. *Beyond the Short-Term: Transportation Asset Management for Long-Term Sustainability, Accountability, and Performance.* Washington: US Federal Highway Administration Report FHWA-IF-10–009.

Stanley, D.A. & Pierson, L.A. 2011. Geotechnical asset management performance measures for an unstable slope management plan. In *Proceedings: 62nd Highway Geology Symposium:* pp. 133–152. Lexington, KY, USA.

Terzaghi K., 1937. Settlement of structures in Europe and methods of observations. American Society of Civil Engineers. Proceedings, Vol. 63, pp. 1358–1374.

Thompson, P.D. 2016. *Geotechnical Asset Management Plan: Technical Report.* Juneau: Alaska Department of Transportation and Public Facilities.

Transport Infrastructure and Systems – Dell'Acqua & Wegman (Eds)
© 2017 Taylor & Francis Group, London, ISBN 978-1-138-03009-1

Evaluation of workability of warm mix asphalt through CDI parameter and air voids

D.M. Mocelin, L.A.T. Brito, M.G. Johnston, V.S. Alves, G.B. Colpo & J.A.P. Ceratti
Federal University of Rio Grande do Sul, Brazil

ABSTRACT: Warm Mix Asphalt (WMA) has grown in production over the past years due to its various advantages, such as reduced greenhouse gas emissions, lower energy consumption and reduced compaction temperature among others. In Brazil, it was not until few years ago that warm mixes set off. With typical temperature reductions of around 30°C below the Hot Mix Asphalt (HMA) production, the use of surfactant additives has proved effective also in enhancing bitumen coating and lubrication of the asphalt binder in the mixture—arguably, yielding and improvement in WMA workability at lower temperatures. This research evaluated the workability of warm mixes accounting the Construction Densification Index (CDI), obtained in the compaction curve of a Superpave Gyratory Compactor (SGC) and also by means of air voids control achieved in Marshall specimens. Torque measurements during the mixing process improved the discussion along with mineral particle distribution obtained from digital image processing. The experiment tested a standard dense HMA and equivalent WMA varying both mixes on their production and compaction temperatures at −15°C, −30°C and −45°C of the reference HMA (143°C), totalling seven analysed mixtures. The results showed no workability variation on the mixes production but an improvement on the compactability; the latter being highly sensitive to surfactant presence and to temperature decrease. In general, the mixtures at reduced temperatures only reach the proper compaction using WMA surfactant additive. The tests carried out allow the conclusions that the addition of WMA surfactant additive enables a decrease in the compaction temperature of about, at least, 30°C without significant loss in workability and improving particle homogeneity in the mixes.

1 INTRODUCTION

Environment awareness over the past decades motivated all engineering fields for more sustainable products and rational use of natural sources. Pavement engineering has followed such concern and the industry has made new products available to reduce impact in road maintenance. Amongst such developments, temperature reduction in asphalt mix production has gained space worldwide.

Several benefits can be associated to temperature reduction of Hot Mix Asphalt (HMA), such as the reduction in greenhouse gas emissions in-plant and during paving jobs, energy savings due to lower fuel consumption, healthier working environment.

In addition, benefits to asphalt mix performance due to reduced heat exposure during production may be considered. Standard temperatures for mixing HMA average 150°C; at this temperature, light volatile particles present in the asphalt binder can evaporate and oxidize the material, resulting in material's drop in performance.

Another immediate benefit of the so-called Warm Mix is the mix plant coverage; material can be supplied at longer distances; lower compaction temperature thresholds allow longer transports periods. Prowell et al. (2012) also highlight better compaction rates using WMA and the possibility of carrying out jobs in cold seasons or places of cold weather.

Different techniques can be used to allow temperature reduction in asphalt mixes; the majority uses some type of additive blended in binder. Surfactant additives proved to be effective and easy to use in warm mix production (Hurley & Prowell, 2006; Kvasnak, 2010; Bennert et al., 2010).

Surfactant additives have their origin from surface-active agents. These agents act in the interface between aggregates and the asphalt binder provoking better lubricity between both and, hence, enabling an easier binder coating over aggregates at lower than conventional temperatures.

Temperature of HMA preparation and compaction is a function of the binder viscosity. Because surfactant additives do not change significantly this property (Johnston *et al.*, 2015), temperature recommendations for WMA typically follows additive supplier, whose expertise recommends production and compaction temperature ranges.

This paper attempts to measure, yet in a qualitative fashion, how such additives improve asphalt mixes workability and what is the temperature reduction limit threshold that can be achieved with this WMA technique. A better understanding of the densification process makes a parallel to field compaction, while monitoring the torque during the mix process enables to determine the required effort in both HMA and WMA production.

Finally, an investigation of the homogeneity after compaction assesses the effect on temperature reduction of the produced mix.

2 MATERIALS

The study was carried out with asphalt mixtures with characteristics and materials commonly used in Brazilian highways; ergo, national standards were applied. Materials description are summarized below.

2.1 Aggregates

The aggregates used are of basaltic origin, from a quarry located in southern Brazil (Santo Antônio da Patrulha/RS). The particle size distribution used is Figure 1.

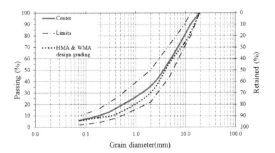

Figure 1. Particle size distribution.

Table 1. Results of binder characterization.

Properties	Brazilian Standard	Binder 50/70	Binder + Surfactant (WMA Additive)
Softening Point	46 min	49	48
Penetration	50–70	67	57
Specific Gravity	—	0.957	1.017
Brookfield Viscosity 135°C (cP)	274 min	408	480
Brookfield Viscosity 150°C (cP)	112 min	237	242
Brookfield Viscosity 177°C (cP)	57–285	90	91

2.2 Asphalt Binder

The asphalt binder used was a type 50/70 (pen gradation). Such binder is widely used in most of Brazilian highways. The benefit of using a standard asphalt binder—with no other modification—was also important for bias, allowing the effect of the surfactant additive to become evident.

Results in Table 1 summarize both used binders: a standard 50/70 used to produce the tested HMA and the same binder with surfactant additive for the tested WMA. The addition of surfactant did not change significantly the characteristics.

2.3 Surfactant additive

The surfactant used in this research is available in liquid form, allowing the additive to be mixed with the binder on terminals and, therefore, distributed using normal binder supplying process. Due to their improved convenience, provided no changes in the production line of asphalt mix is required, this type of additive are of growing use (NCHRP, 2011).

The surfactant additive was added to a ratio of 0.4% by weight of asphalt binder, as per supplier recommendation. Typically, such use allows a 30°C reduction in mix and compaction temperatures. The used surfactant was Ingevity Evotherm® M1.

2.4 Mixtures parameters

Two different mixes were designed using the Marshall method—HMA and WMA. Both used the same aggregate gradation and binder type, only varying incorporation of WMA surfactant additive.

The first step in the study was to determine design characteristics; to do it, HMA was mixed at 153°C and compacted at 143°C, while WMA used 123°C & 113°C, (−30°C to HMA), respectively, following supplier's recommendation.

The final characteristics for both mixes are shown in Table 2 and evidence no significant change in the parameters.

Important to notice that for WMA production, binder is pre-heated at the same temperature as HMA, regardless of the additive presence. The temperature reduction shall only be carried out for mixing and compaction.

Table 2. Final characteristics of the asphalt mixtures.

Properties	Standard DNIT – 031/2006	HMA no additive	WMA w/ surfactant additive
Binder Content (%)	—	5.7	5.7
Air Voids (%)	3–5	4.3	4.2
Voids w/asphalt (%)	72–82	76.9	78.2
Gmm (kN/m³)	—	25.06	25.04
Gmb (kN/m³)	—	23.98	24.04

3 METHODOLOGY

To investigate mixtures compactability, eight tests were performed with the superpave gyratory compactor (SGC): four tests using standard unmodified binder and the other four prepared with surfactant additive. Each group of tests varied mixing temperature only, at: 143°C, 128°C, 113°C and 100°C. This correlates to –0°C, –15°C, –30°C & –43°C of the standard HMA compaction temperature. Temperature reductions were always performed during mixing and compaction procedures.

Going below the maximum temperature reduction recommended by the additive supplier and extending the tests for the mixes with unmodified binder enable to compare the densification process via the Construction Densification Index (CDI) calculated during compaction with the SGC.

The mixtures were prepared using a laboratory mixer with torque measurement sensor. Torque was monitored on the blades moving mechanism throughout the mixing time. These measurements allowed the evaluation of the workability of all mixes during the production process, for the same temperature scheme.

CDI is defined as the area from the 8th gyration to 92% of Gmm in the densification curve (Figure 2). Theoretically, CDI represents the work applied by the roller to compact the mixture to the required density during construction. The number of eight gyrations is selected to simulate the effort applied by a typical paver during the process of laying down the mixture, while the 92% of Gmm is the density at the completion of construction and the pavement is open to traffic (Mahmoud and Bahia, 2004).

Because to construct CDI curves required high number of gyrations, the resulting samples have low air voids. Hence, controlling their final air voids is of little interest. Nonetheless, as the air voids represent one of the key parameters for a mix design, the same temperature scheme was also applied for a typical Marshall compaction, enabling a cross analysis between CDI from SGC and air voids from Marshall specimens, using standard compaction effort of 75 blows per specimen side.

Masad *et al.* (1999), Tashman *et al.* (2001) and Vasconcelos *et al.* (2005) used imaging techniques to investigate homogeneity in asphalt mixes. Aggregate scatter in both horizontal and vertical directions have direct effect in mechanical properties. Hence, homogenous distribution of materials during the compaction process is vital for an adequate behaviour of pavement layers.

In order to determine if the lower temperatures were affecting the homogeneity of the resulting mix, 112 mm high by 150 mm in diameter specimens, produced in the SGC to designed air voids were cut into 10 mm slices, producing from each side of the slice one image (Figure 3).

Images were then scanned and using FIGI/IMAGEJ software boundaries of aggregates were determined automatically by threshold limits on an 8-bit grey scale. The minimum size aggregate captured were those with and area of 22.15 mm², equivalent to material passing in a sieve #4 (4,75 mm).

Figure 4 illustrates, in simplicity, the sequence in the digital processing steps for particle detection. After aggregates were sized and orientation tagged, their occurrence was grouped into three concentric

Figure 3. Specimen slicing for image processing of particle sizing and orientation.

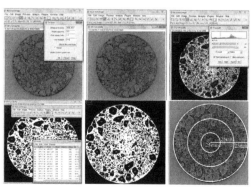

Figure 4. Image processing steps to obtain particle size and distribution (HMA in detail).

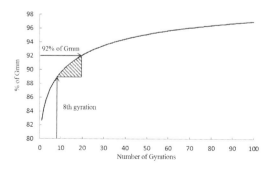

Figure 2. Hatched area corresponding to CDI determination.

radial zones of 25 mm enabling to verify materials homogeneity.

4 RESULTS

4.1 Construction Densification Index—CDI

The CDI is considered a volumetric parameter, related to compactability of asphalt mixtures. Higher CDI means a longer distance from the 8th gyration to 92% of Gmm in the densification curve, i.e., a difficulty to achieve a proper compaction (see Figure 2). Mixes with lower CDI values are deemed to have better constructability and, thus, of preference, while excessively low values of CDI could be an indication of a tender mixture and should be avoided (Mahmoud & Bahia, 2004).

Table 3 show the results of CDI obtained for all four tests carried out. The analysis of the results demonstrated that with greater temperatures, a reduction on CDI values suggests swifter compactability.

Despite the obvious expected results, for the viscosity-temperature characteristics of bitumen, it is evident that the addition of surfactant to the binder provoked an increase in compactability, here inferred from CDI reduction values at each tested temperature. Figure 5 presents the relation of CDI to temperature summarized.

The WMA mixtures at tested temperature of 113°C returned intermediate (27.9) CDI values to HMA mixtures at test temperatures of 143°C (25.4) and 128°C (28.8); this evidences clear improvement in the densification process with the incorporation of the additive. A variation of –30°C resulted in near CDI values of the standard mixing HMA temperature.

These values of CDI obtained are similar to low compared to values found in literature (Mahmoud &

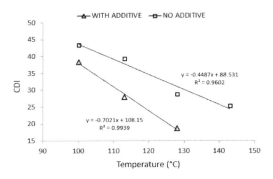

Figure 5. CDI to temperature variation.

Bahia, 2004; Nascimento, 2008; Soares, 2014), indicating good compactability of the mixes.

The CDI value is also dependent on granular matrix skeleton. As particle size distribution of both mixes used in this study have a high percentage of finer particles (55% smaller than 4.5 mm), this may have contributed to low CDI values.

Mahmoud & Bahia (2004) investigate a set of mixes with different materials and binder contents, finding a significant variation in densification behaviour of the mixtures tested. Results of that study confirms that CDI results are highly dependent on the mineral skeleton and the type of binder. Nonetheless, as all tests carried out in this study have the same particle size distribution and binder type, the comparative effect in unharmed.

4.2 Air Voids versus Compaction Temperature

Using standard Marshall compaction procedure to test resulting Air Voids (AV) at the same temperature test sets used for CDI tests, higher AV values at lower temperatures would be expected.

Mix design properties (Table 2) of both HMA and WMA shows a target value of 4.3% AV for HMA (compacted at 143°C) and 4.2% AV for WMA (compacted at 113°C). Brazilian standard DNIT 031/2006 for the asphalt wearing courses states desired air voids percentage between 3% to 5%, averaging a designed target of 4% AV. The experiment test results are recorded in Table 4. Figure 6 presents the results obtained for all of the analysed asphalt mixtures.

Sensibility of the mixtures to temperature and the effect of the additive are evident in the results. For the same temperatures, the mixtures with additive featured lower values of air voids than HMA. A higher level of compactability can be credited to an increase in lubrication provided by the surfactant.

The differences in the air voids were 1.5%, 1.6% and 1.1% for the temperatures of 100°C, 113°C

Table 3. CDI results obtained for the mixtures.

| Temperature (°C) | Construction Densification Index (CDI) | | | |
	No additive	Average	With additive	Average
100	43.40 *	43.40	38.41 *	38.41
113	51.72 26.95	39.34	24.22 31.61	27.92
128	28.95 28.66	28.81	15.68 21.69	18.69
143	25.40 *	25.40	– –	–

Average is a result of the two tested samples on each temperature. (*) means error has occurred during measurements and data was discarded. (–) no test at 143°C was carried out for WMA (w/additive).

Table 4. Air voids results obtained.

| Temperature (°C) | Air Voids (%) | | | |
	No additive	Average	With additive	Average
100	6.34	6.2	4.85	4.7
	6.31		4.65	
	6.07		4.71	
113	6.02	5.8	4.25	**4.2**
	5.58		4.19	
	5.92		4.27	
128	5.12	5.0	3.91	3.9
	5.08		3.98	
	4.87		3.82	
143	4.11	**4.3**	–	–
	4.45		–	
	4.32		–	

(–) no test at 143°C was carried out for WMA (w/ additive).

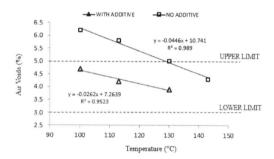

Figure 6. Air voids versus temperature results.

and 128°C, respectively. A steeper trend in AV increase with temperature variation—comparing angular coefficient of regressions (0.0262 for WMA & 0.0446 for HMA)—also corroborates to the CDI values, indicating a considerably smoother compactability.

For the HMA, a 15°C reduction shifted AV to outside the upper limit threshold. Conversely, for all three tested temperatures, WMA remained inside AV limits. Despite adequate AV values at 100°C for the WMA, caution is required to use such reduction in WMA production; CDI values in section 4.2 shows a closing pattern of WMA and HMA towards 100°C; below this point it is uncertain the surfactant additives improvement in compactability.

4.3 Torque measurements—mixtures workability

Measurement of workability in asphalt mixes is non-standardized; neither procedures nor standards for guiding the torque measurement method was found. Hence, preliminary tests using a torque-

monitored mixer were performed in order to establish certain guidance. Rotation speed of the blades were set to 20 rpm; the mixer's drum was set to rotate in reverse to the sense of the blades and the amount of material to be tested was limited to 20 kg with a mixing time of 500 s. The mixer drum relies on a controlled temperature environmental chamber, thereby allowing mixing temperature control.

Torque measurements for the asphalt mixes presented some level of variation throughout the tests. Such variations are possibly associated to particle size discontinuity, inherent to the material. Final torque values, for each mix to all same six temperature/mix combination afore mentioned, were averaged discarding initial ten seconds and when torque peaks exceeded mean value plus one standard deviation. This procedure avoided great variation observed when mixer blades met a coarse lump against the drum spiking torque results.

Figure 7 illustrates the torque readings of a mixture, depicting parameter variation and outlier values disregarded, as an example. The average torque results obtained for the mixes are presented in Figure 8.

Results show no significant change between the different temperatures and mixes with and without additive. Values ranged between a maximum of 22.5 N.m and a minimum of 20.5 N.m.

To test the torque apparatus responsiveness and also guide results assessment, two very homogene-

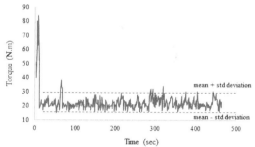

Figure 7. Results of torque measurements.

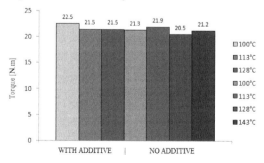

Figure 8. Torque measurement comparison for workability assessment.

ous and of lower kneading effort required—water and sand—were used to test the equipment using the same mixing parameters. Water, alone in the drum, resulted in a 1.67 N.m torque reading and sand, 9.78 N.m.

The tests revealed a good sensibility of the equipment and allowed a benchmark of the results, provided literature was not found about such measurements and, yet if found, could greatly vary according to equipment, amount of material, rotation speed, among others.

Little torque variation indicates that both HMA and WMA had similar workability under tested conditions. Torque for WMA at 113° (21.5 N.m) nears the torque required to mix HMA at 143°C (21.2 N.m); this could be expected as the surfactant additive ought to compensate the temperature loss and maintain workability of the mix. Notwithstanding, the little torque alteration within the HMA temperature variation suggests another secondary reasoning.

For all tests, the asphalt binder was heated up to 153°C; it was done so because for WMA production there is no alteration in temperature to the added binder. Because the mixing time is fairly short (500 s) it is arguable that the temperature decay from 153°C to one of the mixing temperatures (100°C to 143°C) – at which both environmental mixing chamber and added aggregates shall be—may happen at a steady low pace. The mixture workability can then be attributed to the asphalt binder at its added temperature, resulting in slight variation in the measured torque.

4.4 Homogeneity from imaging processing

The temperature reduction during mixing and compaction processes demonstrated not to have great impact in workability and granted WMA with increased compactability. With the purpose of analysing the effect of such variation in the specimens homogeneity, image processing of sliced samples enabled the evaluation.

This analysis was performed at standard temperatures: mixing at 153°C and compaction at 143°C for the HMA, while for WMA 123°C & 113°C respectively. After the image processing as described it section 3, results indicated a considerable more homogenous specimen for the warm mixes.

Figure 9 illustrates six slices of a HMA bottom half specimen, while Figure 10 illustrates an equivalent imagery for a WMA specimen. Figure 11 brings a comparison of the resulting frequency of aggregates greater than 4.75 mm in each of the three radial zones.

Results clearly demonstrates a more homogenous distribution of aggregates across radial section for the WMA. A concentration of coarser aggregates towards the edge of the specimen is

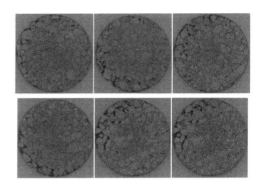

Figure 9. HMA Specimen 2 – Faces 7 to 12.

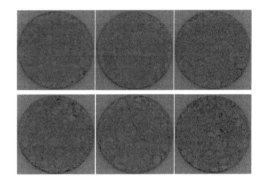

Figure 10. WMA Specimen 3 – Faces 7 to 12.

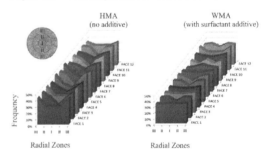

Figure 11. Image processing steps to obtain particle size and distribution in quadrants.

evidenced in the histogram and in the faces images in Figure 9.

5 CONCLUSIONS

This study aims to help understanding workability and compactability improvements of surfactant agents used for WMA. The later was observed with a better densification obtained via CDI measurements during compaction, whereas workability was found unchanged for all tested conditions.

Analysis of the SGC tests demonstrated that greater temperatures lead to a CDI reduction suggesting swifter compactability. Although obvious, as with rising temperatures asphalt binders have lower viscosity, the addition of surfactant to the binder led to an increase in compactability, inferred from CDI reduction values at each tested temperature.

WMA compacted at 113°C returned intermediate (27.9) CDI values compared to HMA at 143°C (25.4) and 128°C (28.8); this is a clear improvement in the densification process with the incorporation of the additive. A variation of −30°C resulted in near CDI values of the standard mixing HMA temperature.

Sensibility of the mixtures to temperature and the effect of the additive were evident in Marshall compaction air voids control. For the same temperatures, mixtures with surfactant (WMA) featured lower values of air voids than HMA, also indicating a higher compactability of the mix. This can be credited to an increase in lubrication provided by the additive.

Despite adequate air voids values were reached at 100°C for the WMA, caution is recommended to lower below this limit. CDI shows a matching pattern of WMA and HMA towards 100°C. Below this temperature air voids may be met, but densification starts to slow down what may result in unwanted collateral effects.

Regarding the workability tests from torque measurements, the little variation in results indicates that both HMA and WMA presented similar workability under tested conditions. Because the mixing time was short, it is arguable that the temperature binder decay from 153°C to one of the mixing temperatures (100°C to 143°C) may have happened at low pace. The mixture workability can then be attributed to the asphalt binder at its added temperature, resulting in slight variation in the measured torque.

Results in this study agrees with Prowell et al. (2012) findings, confirming that, despite the low temperatures, WMA more easily achieves the desired density due to an increase on lubrication provided by the surfactant. In addition, the digital image processing of specimen sliced into chops have clearly demonstrated a better homogeneity of particle distribution in WMA.

An additional conclusion is that CDI ought to be further investigated to act as reference to determine the amount of temperature reduction for WMA additives. The CDI value obtained in a standard HMA can be expected to be achieved at a lower temperature by means of using a warm mix additive.

ACKNOWLEDGEMENTS

The authors wish to express their gratitude for the products suppliers—Ingevity specialty chemical, Greca asfaltos & Triunfo Concepa—and all other laboratory partners for theirs technical contribution and financial support to the research group.

REFERENCES

Agência nacional do petróleo, gás natural e biocombustíveis. Cimentos Asfálticos de Petróleo. Resolução nº 19 de 11 de julho de 2005 da ANP – Regulamento Técnico ANP nº 19/2005.

bennert, T.; Reinke, G.; Mogawer, W.; Mooney, K., Assessment of Workability and compatibility of Warm-Mix Asphalt. Transportation Research Record: Journal of the Transportation Research Board, No 2180, Washington, DC., USA, pp. 36–47, 2010.

Departamento nacional de infraestrutura de transportes. DNIT 031: Pavimentos Flexíveis – Concreto Asfáltico – Especificação de Serviço. Rio de Janeiro, 2006.

Hurley, G. e Prowell B., Evaluation of EVOTHERM® for use in Warm Mix. National Center for Asphalt Technology (NCAT) - NCAT Report 06–02. Auburn, Alabama, USA, 2006.

Johnston, M. G.; Bock, A. L.; Brito, L. A. T.; Ceratti, J. A. P.; Ribeiro, R. Influência do período de condicionamento em estufa sobre os parâmetros de dosagem de misturas asfálticas mornas. XVIII Congresso Ibero Latinoamericano del Asfalto, Bariloche, Patagônia, Argentina, 2015.

Kvasnak, A. et al., Alabama Warm Mix Asphalt Field Study: Final Report. National Center for Asphalt Technology (NCAT) – NCAT Report 10-XX. Auburn University, Auburn, AL, USA, 2010.

Mahmoud, A. F. F.; Bahia, H. Using the gyratory compactor to measure mechanical stability of asphalt mixtures. Wisconsin Highway Research Program 0092–01–02, Madison, Wisconsin, EUA, 2004.

Masad, E.; Muhunthan, B.; Shashidhar, N.; Harman, T. (1999) Internal Structure Characterization of Asphalt Concrete Using Image Analysis. Journal of Computing in Civil Engineering, v. 13, Nº2, April.

Nascimento, L. A. H. (2008) Nova Abordagem da dosagem de Misturas Asfálticas Densas com Uso do Compactador Giratório e Foco na Deformação Permanente. Dissertação de Mestrado – COPPE/UFRJ, Rio de Janeiro.

Prowell, B. D.; Hurley, G. C.; Frank, B. Warm-mix asphalt: best practices. 3ª ed. Lanham: National Asphalt Pavement Association. Quality Improvement Series 125, 2012.

Soares, J. S. (2014) Investigação da Relação entre Parâmetros da Compactação Giratória e de Deformação Permanente em Misturas Asfálticas Densas. Dissertação de Mestrado – Programa de Pós-Graduação em Engenharia de Transportes, Universidade de São Paulo, São Paulo.

Tashman, L.; Masad, E.; Peterson, B.; Saleh, H. (2001) Internal Structure Analysis of Asphalt Mixes to Improve the Simulation of Superpave Gyratory to Field Conditions. Journal of the Association of Asphalt Paving Technologists, v. 70, p. 605–655.

Vasconcelos, K. L.; Evangelista JR., F.; Soares, J. B. (2005) Análise da Estrutura Interna de Misturas Asfáticas. In: XVII Congresso Brasileiro de Pesquisa e Ensino em Transportes, Recife, PE.

Transport Infrastructure and Systems – Dell'Acqua & Wegman (Eds)
© 2017 Taylor & Francis Group, London, ISBN 978-1-138-03009-1

A laser profilometer prototype for applications in road pavement management system

G. Cerni & A. Corradini

Università degli Studi di Perugia, Perugia, Italy

ABSTRACT: Correct planning of road maintenance should require intervention before road reliability conditions become unsustainable. To this aim, it is necessary to evaluate the rate of road degradation, which is strictly linked to road roughness, using high performance instruments like the laser profilometer. The International Roughness Index, obtained from profilometric survey, provides basic information regarding the general state of a road pavement, but does not take into account the characteristics of distresses. In this context, the University of Perugia (on behalf of the Province of Perugia) developed a new prototype of laser profilometer and an innovative post-processing software. This equipment enables global and local analysis. The global analysis provides the intervention priorities of a whole road infrastructure, while, with local analysis, the single distresses and the cross slopes of a limited road portion can be evaluated. Information about the type of distress are very useful in order to plan maintenance work.

Keywords: pavement, profilometer, maintenance

1 INTRODUCTION

Road maintenance is the complex of operations and activities aimed to conserve the functional and structural characteristics of pavements over the useful lifetime of road infrastructure (CNR 1988, Durango & Madanat 2002, Robinson et al. 1998). In particular, the purpose of road pavement maintenance is to guarantee correct function maintaining adequate safety levels while efficiently employing economic resources (Dekker et al. 1998, Canale et al. 1998).

Preventive diagnosis is guaranteed by the study and interpretation of deterioration curves, which need to be elaborated employing the principal indicators of pavement state, namely skid resistance, evenness, bearing capacity and noise levels. The monitoring over time of these parameters, combined with the use of appropriate threshold levels, permits the control of pavement damage. This approach allows the identification of the maintenance interventions at an early stage, so as to optimize costs and limit inconvenience to road users. The monitoring of the functional state of a road can be carried out using high performance devices, such as the laser profilometer.

A laser profilometer is a measuring instrument used to investigate road pavement surface, paying special attention to its longitudinal profile in terms of surface texture (Sayers & Karamihas 1998). As is well known, texture is defined by the irregularities on a pavement surface that deviate from an ideal, perfectly flat surface. Pavement texture has been categorized into three ranges based on the wavelength

(λ) of its components: microtexture ($\lambda < 0.5$ mm), macrotexture ($0.5 < \lambda < 50$ mm), and megatexture ($50 < \lambda < 500$ mm). Wavelengths longer than the upper limit of megatexture are defined as roughness (Henry 2000). Different sizes of texture will affect pavement surface characteristics in different ways (Sandburg & Ejsmont, 2002). Small texture affects friction, while large texture affects ride quality (Flintsch et al. 2002). Noise and rolling resistance are principally controlled by macrotexture and megatexture. In general, it can be stated that microtexture and macrotexture have good effects on ride quality, while greater texture are poorly accepted. Roe et al. report that increased macrotexture reduces total accidents, under both wet and dry conditions (Roe et al. 1998). Furthermore, this study shows that increased macrotexture reduces accidents at lower speeds than previously believed.

With the aim to convert a pavement evenness measurement into a single number, the highway industry introduced the International Roughness Index (IRI) as a quality indicator (Sayers et al.1986). Lin et al. (2003) stated that IRI may completely reflect pavement distress conditions. Thus, it is feasible to use IRI as a pavement performance index. Tighe et al. (2000) observed that IRI is significantly related to single-vehicle accident rate.

According to ASTM E 1926-08 (ASTM Standards 2000), IRI is obtained from measured longitudinal road profiles and is calculated using a quarter-car model (Fig. 1). A quarter-car system

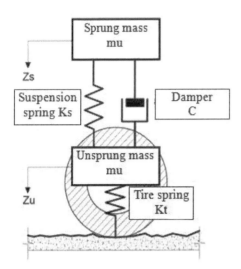

Figure 1. Quarter-car vehicle model.

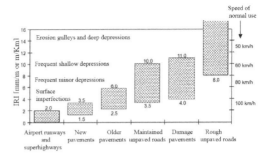

Figure 2. IRI ranges represented by different classes of road and velocity (replotted from Sayers & Karamihas 1998).

Figure 3. Profilometer prototype.

is composed of two solid masses m_s (sprung mass) and m_u (unsprung mass), which respectively represent ¼ of the body of the vehicle and one wheel of the vehicle (Jazar 2014). A suspension spring of stiffness k_s and a shock absorber with viscous damping coefficient c, connect the two masses, while the sprung mass is in contact with the real pavement surface by a tire spring of stiffness ks.

During the simulation, the quarter car system runs over the longitudinal profile at a constant speed (V) of 80 kilometres per hour (km/h). The passage over this profile induces dynamic excitation to the quarter car system, generating a certain level of movement in both the sprung and unsprung masses (Z_s e Z_u) in relation to surface irregularities. As a result, IRI value for a given section length (L) is computed according to Eqn 1.

$$IRI = \frac{1}{L} \cdot \int_{0}^{L/V} |\dot{Z}s - \dot{Z}u| \, dt \qquad (1)$$

The IRI is a dimensionless parameter (generally expressed as m/km or mm/m) and it is null for a perfectly smooth road surface. In addition, its typical range values change with pavement type and corresponding speeds (Fig. 2). Although this index provides basic information regarding the general state of a road pavement, it does not take into account the information related to the characteristics of the distresses that are associated with a specific level of unevenness. For this reason, two road segments, with different typologies of surface irregularities, could give comparable values of IRI.

2 OBJECTIVES

The state of degradation of a road surface represents a key parameter in the management pavement system. With the end to obtain data pertinent to the state of pavement surfaces, in order to determine objective criteria for the prioritization of maintenance works, the Provincia of Perugia decided to equip itself with a profilometer (Fig. 3).

As an alternative to the direct acquisition of the instrument, the Provincia chose to construct a prototype in collaboration with the Engineering Faculty, involving the Department of Civil and Environmental Engineering and the Department of Electronic and Computer Engineering. The principal motivation for this choice related to the possibility to take full advantage of the data recorded by the apparatus's sensors, while containing costs. To this end, particular attention was given to the creation of a post-elaboration software capable of providing, as well as IRI indexes, information relative to the typology, the gravity and the localization of irregularities. In addition, cross slope values can be obtained by elaborating data provides by the instrumentation.

Nowadays, the profilometer prototype is at an early stage of use and it is employed to monitor the major road networks within the Provincia of Perugia.

3 PROFILOMETER HARDWARE COMPONENTS

The profilometer was built using the following components (Fig. 4):

– a bar with a length equal to 2.20 m;
– 10 lasers installed on the bar every 20 cm;
– an inertial unit installed at the center of the bar;
– a control unit for analogue/digital conversion;
– a GPS device;
– an odometer;
– a computer inside the vehicle for data collecting.

Each laser measures the distance between the sensor and the pavement surface at rates up to 20 kHz, while the inertial unit, evaluating acceleration and angular velocities along three orthogonal axes, detects the vehicle movements and makes the profile independent from these contributes. The profilometer records the position using GPS technology, georeferencing both laser and inertial measurements. In particular, the Geographic Information System used is the World Geodetic System, established in 1984 (WGS84). In the last version of the prototype, an odometer was added with the aim to achieve greater precision and address, together with the inertial unit, the possible loss of data. The computer inside the vehicle records measurements in real time, which will then be transferred to a post processing software. Such technology provide both longitudinal profiles and cross slopes along the path of the vehicle.

Figure 4. Profilometer hardware equipment: bar (a); lasers installed on the bar (b); inertial unit (c); on-board computer (d).

4 PROFILOMETER POST-PROCESSING SOFTWARE ANALYSIS

The development of the post-processing software (Fig. 5) has represented the most difficult step of the project. It is used to analyse and elaborate the whole database in terms of heights (i.e. distances between road surface and sensors), accelerations, velocities and position.

4.1 *Software set up*

The first goal of this experimentation was to obtain longitudinal profiles (in terms of elevation) and cross slopes as close as possible to the real ones. In fact, during profilometric surveys laser measurements are generally affected by vehicle oscillations, but such contributions should not be taken into account in the final profile analysis. In this context, an already known profile, especially realized for this application, was studied in a preliminary investigation in order to correctly set up the instrumentation. In particular, a PVC profile was laid down over the pavement surface and speed bumps were positioned along the wheel paths, causing longitudinal and transversal oscillations to the vehicle. The intention was to implement an algorithm capable of providing, in these operative conditions, a profile as faithful as possible to the original one.

With the same aim, high-resolution measurements were carried out on eight road portions (with an overall length of 389 m) using a laser scanner. This data was than compared with those provided by the post-processing software of the profilometer. The comparison in terms of IRI showed an average error equal to 0.55 mm/m. The same evaluation with regard to cross slopes resulted in an absolute error of 0.85% and 0.65%, respectively before and after the introduction of the odometer in the prototype.

The introduction of the odometer not only improved the calculation of distances and velocities, but also enabled us to know the real

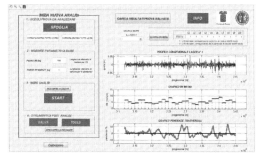

Figure 5. Main interface of the post-processing software.

path of the vehicle during GPS data loss. This was made possible by setting a procedure in which both velocities obtained from the odometer and angular velocities provided by the inertial unit, step by step were integrated. This technique allows the identification of the plan position at a time step j from the knowledge of the position at a time step j–1. In order to verify procedure reliability, simulated data losses with different durations were introduced into the GPS database (an example is shown in Figure 6a).

The integration procedure required the knowledge of the initial conditions, which consist of the coordinates for the starting point of the integration and the initial angle to the north point (initial direction).

Integrating step by step, the estimated path shown in Figure 6b can be obtained, but it is clearly affected by problems of drift. The problem of the drift is due to small errors that are added together during the integration phase, resulting in an error that is more consistent the longer the integration time. This problem is generally difficult to resolve because of its non-linear nature and although the use of more sophisticated instrumentation can reduced it, it cannot be eliminated.

However, a new post processing procedure was designed. This procedure takes into account the final conditions, both real and obtained by integration (in terms of positions and angles), and the total length covered, in a context of iterative solution. In fact, it was observed that the angular error, i.e. the angle resulting from the difference between the expected and the obtained by the integration, cannot be linearly distributed throughout the affected section because the coincidence between the "end integration point" and the "real end point" would not be obtained. Therefore, the strategy to proceed by attempts is to adjust two variables which are the length and the position of the stretch, within the affected section, where the angular error can

be distributed. The prototype program performs numerous attempts by shifting the error along the section subject to examination. All the resulting paths comply with the final direction. The criteria for choosing between all attempts consists in setting the minimum value of the distance between the "end integration point" and the "real end point". Figure 6c shows the result obtained applying the correction procedure to a period of simulation of no GPS signal equal to 96 seconds.

4.2 Location of critical point on the pavement surface

The decision to realize on our own the equipment enabled us to maximize all the information provided by the apparatus's sensors. In particular, through the data analysis carried out by the post-processing software, different types of pavement distresses could be highlighted and classified. Irregularities were divided into two main categories: structural distresses (deep) and surface distresses.

The structural distresses were associated with the depressions of the road surface caused by structural collapses. They could present significant extensions and induce longitudinal and transversal oscillations in the vehicle.

The surface distresses include localized singularities such as potholes, marked cracks and detachments of asphalt concrete, which caused sudden movements of the vehicle.

This division stems from the need to distinguish between the types of intervention required to restore the evenness of a damaged road. For example, if accentuated depressions are present, a simple road resurfacing to restore the evenness will not be sufficient, since the problem could mainly reside in the deeper layers of the pavement.

To this end, the post-processing software is based on two methods for the detection of road irregularities:

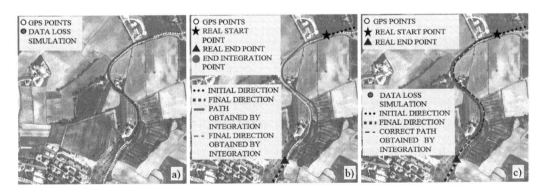

Figure 6.   Example of GPS data loss simulation (a); estimated path (b); final path corrected after iterative procedure (c).

1. SSD Method (Separation of Superficial Distress), to identify surface discontinuities.
2. SWVW Method (Shifting Windows of Variable Width) to identify wave irregularities.

For both the discontinuities, the software elaborates a summarizing index (structural and superficial) which provides additional information beyond those obtained by analyzing IRI values.

4.2.1 *The SSD method*

The method consists in a repeated application of a Separation of Superficial Distress routine to analyse and elaborate the peaks of the first derivative of the elevation profile. In synthesis the SSD routine calculates the first derivate of the profile, where the value is compared point by point with a predetermined threshold. In this way, the discontinuities, which present an accentuated slope in the profile, are identified.

Based on a prototype procedure implemented in the software, it is possible to separate the irregularities with reference to their length λ and depth.

With reference to the length, surface discontinuities are classified as followed (Table 1).

In addition, based on the maximum distress depth, three different classes can be established, i.e. low (from 10 to 20 mm), medium (from 20 to 30 mm) and high (greater than 30 mm).

At the end of the routine, by removing the superficial discontinuities and applying a joining polynomial, the SSD profile is obtained (Fig. 7). This produces a continuous profile without localized discontinuities, which may then be analysed with the second SWVW method as illustrated in the following paragraph.

4.2.2 *The SWVW method*

In the SWVW method, Shifting Windows of Variable Width are applied in steps, relative to the size of the window itself, moving along the length of the profile. In relation to their width, different wave irregularities were identified using 6 classes obtained by dividing the range of roughness in four portions and the range of megatexture into two portions. Each portion has a constant amplitude in the logarithmic scale. In this way, six bands are obtained as a function of wavelength λ, as reported in Table 2.

Fourier spectral analysis is carried out within each shifting window. The spectrum used is the Power Spectral Density (PSD) of slope and not that of elevation in order to obtain a more uniform spectral density graph so as to allow a more adequate comparison between the various wave irregularities (Sayers and Karamihas 1998). This graph is therefore the most indicated to assess which possible wave bands represent a more relevant part of the irregularities of the global profile. The area under the PSD of slope quantifies the irregularities of the longitudinal profile. Modifying the window width chosen to analyse the profile, it can be noted that, for the same irregularities, the spectrum change considerably and other bands are also influenced. It is therefore appropriate to use shifting windows, rather than fixed windows, which move along the scanned profile with a width slightly greater than the wavelength analysed. Essentially, the methods consist of 6 passages, one for each band analysed. The preliminary phase conducted through SSD method resulted to be necessary to ensure the elimination of noticeable irregularities, in relation to slope, which influence the spectrum in reference to all bands. To quantify each band irregularity, the areas under the PSD curves between fixed upper and lower band limits were calculated (Fig. 8). For each computed area, three different levels (low, medium and high) were defined in order to establish the quantity of irregularity within each band.

Table 1. Surface discontinuities.

Name	Wavelength [mm]
Small potholes	$50 < \lambda < 150$
Medium potholes	$150 < \lambda < 500$
Big potholes	$500 < \lambda < 1500$

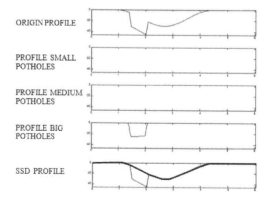

Figure 7. Profile obtained after SSD method application.

Table 2. Wave irregularities.

Band number	Band name	Wavelength [mm]
Band 1	Short microwave	$50 < \lambda < 158$
Band 2	Medium microwave	$158 < \lambda < 500$
Band 3	Large microwave	$500 < \lambda < 1581$
Band 4	Short macrowave	$1581 < \lambda < 5000$
Band 5	Medium macrowave	$5000 < \lambda < 15810$
Band 6	Large macrowave	$15810 < \lambda < 50000$

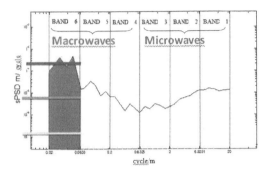

Figure 8. PSD of slope.

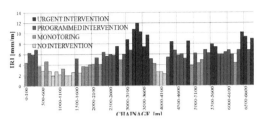

Figure 9. IRI index analysis.

Figure 10. Different road conditions in relation to damage.

The detailed information obtained from these two prototype methods supply plentiful data useful for local analysis. However, the quantity of information is so vast it is difficult to use to organize global maintenance. For this reason, the data is summarized into two main indexes. In particular, the SWVW procedure results provide the structural index as relevant to the macrowaves contribution, while the results obtained within the microwave ranges, together with those of the SSD method, determine the identification of the superficial index.

4.3 Software output

In the post processing software, the IRI values for hundred meter stretches are calculated in order to provide, in a simple and practical way, indications for maintenance interventions (Fig. 9). These interventions are defined based on IRI thresholds, which have been set with reference to existing state of a 95 km provincial roads sample. The study revealed the necessity to modify the IRI limits, with respect to those usually utilized for highways, according to the following categories:

– IRI <3.5 where the pavement is in satisfactory conditions (24% of the analysed roads)
– 3.5 < IRI < 5 where the pavement is slightly damaged therefore requiring frequent monitoring (35% of the analysed roads);
– 5 < IRI < 7 where the pavement is damaged requiring intervention (29% of the analysed roads);
– IRI >7 where the pavement is significantly damaged requiring urgent intervention (12% of the analysed roads).

In the actual version of the software, the elaborated results from the various sensors are synchronized with video footage. In the resulting video, the chainage is recorded in the lower left of the screen and the IRI value in the upper right (Fig. 10).

As well as the IRI index, the structural and the superficial indexes are represented in graphs (an example is shown in Fig. 11). These indicate the type of distresses present, globally quantified by the IRI index, and therefore provide detailed information in order to address the choice of intervention. When these index values are higher than prefixed thresholds, their labels are displayed in the video footage (Figure 10).

In addition, a graph of cross slope for every 10 meters is constructed to verify the presence of curves with unacceptable values (an example of counter slope is presented in Fig. 12).

The recorded data are georeferenced (using WGS84) on an orthophoto map to facilitate their geographical location (Figs. 13–15). In particular, the IRI index is recorded on the orthophoto map according to the new classification introduced.

With reference to distress localization, the validity of the orthophoto map identification was also verified by analyzing laser scanner data collected on a later investigation. In addition, after this second laser scanner survey, which was carried only in specific road stretches, it was decided to conduct site visits in the critical sections identified by the program. Both the site visits and the laser scanner survey confirmed the optimal performance of the post-processing software to identify actual distress (Cerni & Durantini 2013).

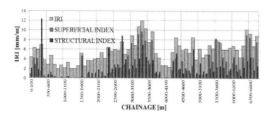

Figure 11. Structural and superficial index.

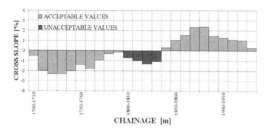

Figure 12. Cross slope analysis.

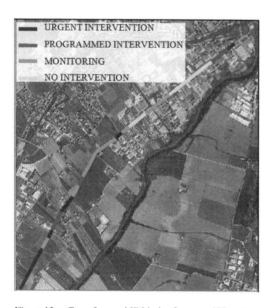

Figure 13. Georeferenced IRI index for every 100 meters.

4.4 Intervention priority

Distress analysis provides a structural and a super-ficial index for every 100 meters, which together with the IRI index permit the construction of an intervention priority program. From an opera-tional point of view, the identification of this intervention program is one of the most useful instruments present in the post-processing soft-ware in order to define correct road maintenance. Based on a decreasing classification in terms of

Figure 14. Georeferenced indication for superficial and/or structural distresses for every 20 meters.

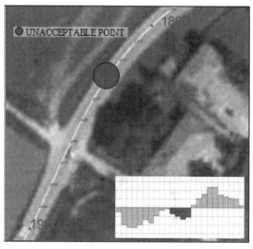

Figure 15. Georeferenced indication of critical cross slope for a curve.

index values, the necessary interventions can be objectively prioritized in table form.

5 CONCLUSIONS

Correct planning of maintenance activities is essential to preserve both the structural and func-tional characteristics of a road infrastructure, in consideration of the best use of available finan-cial resources. To this aim, an accurate analysis of the existing situation by investigating parameters such as evenness is necessary. The elaboration of the pavement profile can supply indications of not only the superficial conditions, but also reveal deeper structural problems.

Road profiles can be obtained by carrying out field investigations using high performance instruments, such as the laser profilometer. In this light, the Province of Perugia decided to adopt a profilometer prototype, which represents, together with the post-processing software, an important apparatus for the maintenance of its road network. In fact, the instrumentation provides high quality results, together with a high speed of data acquisition and an ease of use. Data reliability of both the longitudinal profile and the cross slope were verified through an initial trial set up survey, which certified the accuracy of the equipment settings.

The post-processing software enables both global and local analysis. The global analysis provides the intervention prioritization of a whole road infrastructure, while the local analysis allows a detailed evaluation of the single distresses and the cross slopes of a limited road stretch.

The post-processing software carries out road irregularities detection employing two different prototype methods: the SSD method classifies surface discontinuities, while the SWVW method identifies structural distress (deep). The surface distresses include localized singularities such as potholes and detachments of asphalt concrete, while structural distress are associated with road surface depressions caused by structural collapse.

The possibility of distinguishing different kinds of distress is very useful in order to identify the typology of intervention, avoiding a simple road resurfacing without evaluating the real cause of deterioration.

Furthermore, the post-processing software could be extended to take into account additional data available in the Provincia of Perugia database with reference to traffic, accident and friction as well as evenness, in order to more effectively prioritize interventions. The development of a pavement performance prediction model could also be a future objective.

ACKNOWLEDGEMENTS

The authors thank the Province of Perugia (Italy), and in particular the staff of the Traffic Management Department, for their support during the GPS field surveys. The help of Dr. Andrea Durantini in the software development has been greatly appreciated.

REFERENCES

ASTM Designation E 1926. 2000. Standard Practice for Computing International Roughness Index of Roads from Longitudinal Profile Measurements. *American Society of Testing and Materials*. 995–1012.

Canale, S. & Nicosia, F. & Leonardi, S. 1998. Programmazione degli interventi manutentivi in base ai limiti di budget. *Atti del XIII Convegno Nazionale Stradale dell'A.I.P.C.R.*, Verona.

Cerni, G. & Duranti, A. 2013. La realizzazione di un prototipo laser come strumento diagnostico per la manutenzione programmata delle pavimentazioni stradali. *Strade & Autostrade*, 6.

CNR n.125. 1988. Istruzioni sulla pianificazione della manutenzione stradale. *Consiglio Nazionale delle Ricerche*, Roma.

Dekker, R. & Plasmeijer R.P., & Swart, J.H., 1998. Evaluation of a new maintenance concept for the preservation of highways. *IMA Journal of Management Mathematics*, 9(2): 109–156.

Durango, P.L. & Madanat, S.M. 2002. Optimal maintenance and repair policies in infrastructure management under uncertain facility deterioration rates: an adaptive control approach. *Transportation Research Part A*, 36(9): 763–778.

Flintsch, G.W. & Al-Qadi, I.L. & Davis R. & McGhee K.K. 2002. Effect of HMA Properties on Pavement Surface Characteristics. *Proceedings of the Pavement Evaluation*, Conference, Roanoke, Virginia.

Henry, J.J., 2000. Evaluation of Pavement Friction Characteristics—A Synthesis of Highway Practice. *NCHRP Synthesis 291*, Washington, D.C.

Jazar R.N. 2014. *Vehicle Dynamics: Quarter Car Model*. 985–1026, New York: Springer.

Lin, J.D. & Yau, J.T. & Hsiao, L.H. 2003. Correlation Analysis Between International Roughness Index (IRI) and Pavement Distress by Neural Network, *Transportation Research Board,* Washington, D.C.

Robinson, R. & Danielson, U. & Snaith, M. 1998. *Road Maintenance Management-Concepts and Systems*. Basingstoke, UK: McMillan Press Limited.

Roe, P.G. & Parry, A.R & Viner, H.E. 1998. High and Low Speed Skidding Resistance: The Influence of Texture Depth, *TRL Report* 307, Crowthorne, U.K.

Sandburg, U. & Ejsmont J.A. 2002. *Tyre/Road Noise Reference Book*. Kisa, Sweden: Informex.

Sayers, M.W. & Gillespie, T.D. & Paterson, W.D.1986. Guidelines for the Conduct and Calibration of Road Roughness Measurements. *The World Bank*, technical paper n. 46, Washington, D.C.

Sayers, M.W. & Karamihas, S.M. 1998. *The little book of profiling*. The Regent of the University of Michigan. Courtesy of the University of Michigan Transportation Research Institute.

Tighe, S. & Li, N. & Falls, L. & C. & Haas R, 2000. Incorporating Road Safety into Pavement Management. *Transportation Research Record Journal of the Transportation Research Board*, 1699(1): 1–10.

Transport Infrastructure and Systems – Dell'Acqua & Wegman (Eds)
© 2017 Taylor & Francis Group, London, ISBN 978-1-138-03009-1

Speed management in single carriageway roads: Speed limit setting through expert-based system

N. Gregório, A. Bastos Silva & A. Seco
CITTA, Department of Civil Engineering, University of Coimbra, Coimbra, Portugal

ABSTRACT: A number of approaches has been developed and adopted to set speed limits in interurban roads. Among others, the Australian family of applications (*XLimits*), later also adopted in the USA, is a relevant example of this type of approach. This model enables the selection of adequate speed limits by taking into consideration a wide set of explanatory variables, aiming at describing the infrastructure, land use, local safety and operational characteristics. Hence, the major objective of this work is to provide a generalized decision-support methodology for speed limit setting in interurban single carriageway roads, crossing different types of road environments, considering a range of objective variables. This work thoroughly describes the methodological approach to be used in this phase, which involves the evaluation of experts and the collection of a set of variables related with the road environment. Some early results, regarding these evaluation, are presented and discussed.

1 INTRODUCTION

Speed is among the most important factors involved in road operation, influencing its users behaviour. Usually, the most noticeable effect of speed is its impact over road accidents, whether over occurrence risk or over severity, whose relationship has been demonstrated by several studies (Baruya 1998; Farmer et al. 1999; Kloeden et al. 2001; Taylor et al. 2002; Nilsson 2004; Aarts & van Schagen 2006; Elvik 2013). Speed also affects environmental conditions, such as emissions, fuel consumption, noise and the overall quality of life (ERSO 2006; Kockelman 2006; Austroads 2010).

Despite its widespread use and acceptance throughout the world, no consensus has so far been achieved among practitioners on the most reliable and effective method to set the speed limit in each road section. This is a major concern, since it leaves the technical community without definitive guidance in this field (Forbes et al. 2012). Furthermore, the ever growing urban developments around most roads has, in many countries, led to the existence of not only purely rural environments, but also disperse and non-consolidated built-up areas in their surroundings, with the boundaries between these zones very often difficult to identify. This has led to ever more complex road environments, where the traditional road design and management principles are no longer sufficient, with problems of coherence and homogeneity arising to speed limit setting strategies (Aarts & van Schagen 2006; Hauer 2009; Stuster et al. 1998).

Hence, the major objective of the current work is to provide an integrated decision-support methodology for speed limit setting in interurban single-carriageway roads, crossing different types of road environment, and taking into account a range of significant and objective variables.

A number of approaches have over the years been developed and adopted to set speed limits in interurban roads. Most of these methodologies usually give prevalence to geometric features of the road layout, both in their vertical and horizontal alignments, and especially in critical sections, such as curves, intersections or in stretches with higher slopes. Examples of this approach can be found in numerous studies and are usually the basis of official guidelines and statutory documents. However, given the road environment complexity, a wider set of factors related to the surrounding areas, safety, traffic and users need to be included. A robust methodology should consider the wider number of factors possible, weighting their significance in this process.

The current work aimed to deliver an accessible and easy to use methodology, based on data which can be collected remotely and is easily measurable. Therefore, a different approach is considered, which emphasizes factors related with the prevailing road environment, especially focusing on road integration into its surrounding areas. Previous work has already been done to develop a robust methodological approach, based on an expert-based system (Seco et al. 2008; Correia & Bastos Silva 2010; Correia & Bastos Silva 2011;

Bastos Silva et al. 2012; Bastos Silva et al. 2016; Gregório et al. 2016). However, additional development was still needed (I) to compose a new database, involving more itineraries crossing different zones, and thus broadening the model's scope, as well as a wider set of explanatory variables, representing more accurately the road environment; (II) to develop and to apply a systematic method of expert assessment in laboratory and data collection, enhancing its applicability and usability. This work intended to thoroughly describe the methodological approach that was composed to the overall process, and to present and discuss some early results of the experts' assessment to be included in the final model.

2 SPEED LIMIT SETTING METHODS

The scientific and technical community has been using several speed limit setting methods, which can be divided into four major types: engineering methods, economic optimization methods, harm minimization strategies and expert-based systems.

2.1 Engineering methods

The engineering methods usually set speed limits based on an analysis of traffic and the road environment on the section under study and its surrounding roads. This approach includes the Operating Speed Method, which has been widely used throughout the world, but especially in the United States, with examples in several states (such as Illinois, for instance) (Forbes et al. 2012). This method is based on the 85th percentile of speed (V85) and its usual procedure includes setting the speed limit in a value equal to or higher than V85, eventually adjusting it according to specific infrastructure and traffic conditions (Forbes et al. 2012). It takes into account that speed values equal to or near a standard deviation over the mean value (which is near that percentile) tend to correspond to a minimum accident risk for the driver (DfT 2013). It also takes into consideration that this kind of speed limit is in accordance to the perception of a vast majority of users about which is the adequate speed under specific traffic conditions, thus contributing to a more uniform speed regime (FHWA 2009; Forbes et al. 2012).

Nevertheless, this method presents several disadvantages: it unrealistically assumes that drivers select their travel speed taking an adequate and objective consideration of road safety issues; it is considered to be the cause of a gradual increase of the average operating speed; it produces an inadequate measure of speed consistency; it tends to be less effective the more residential

the surrounding environment is; and it cannot be considered as objectively rational, since it is based on an erroneous driver perception of speed impacts (Elvik 2010; Park & Saccomanno 2006; TRB 1998).

Another engineering method, the Road Risk Method, has been used in New Zealand and Canada. This method is similar to the Operating Speed Method, since both determine the recommended speed limit based on a previously defined value, which is then adjusted having into account several other factors. The difference between these two methods arises from the fact that, while in the former the base value corresponds to the 85th percentile of the free flow speed, this method uses the functional classification of the road instead (Forbes et al. 2012). Therefore, speed limit can be reconciled with the functional nature of the road, in a more adequate way to its prevalent use.

2.2 Harm minimization methods

The Harm Minimization strategies address the speed limit setting problem on a road safety perspective, considering that it is against ethics to allow situations in which accident occurrence is possible. Thus, they focus on the tolerance and integrity of the human body in accident situations to set the speed limit. It has been implemented in some countries, with Sweden (Vision Zero) and the Netherlands (Sustainable Safety) as the most representative cases.

Vision Zero takes into account three fundamental ethical imperatives: no individual can die or suffer chronic injuries as a consequence of accident; road safety cannot be considered as a mobility function, but rather as a road safety function; a monetary value or cost can never be attributed or associated to human life (Vaa 1999). Furthermore, this approach also usually forgives the driver in case of accident, since even the systematic non-observance of the established rules by the users supposes their inadequacy (Whitelegg & Haq 2006). However, the activity of particularly vulnerable users is not appropriately addressed (Vaa 1999), and its realism and rationality are also questioned (Elvik 1999; Rosencrantz et al. 2007).

The Sustainable Safety approach also considers that it is not ethically acceptable for a system to allow frequent accident occurrence with fatal or serious injuries (Vaa 1999), aiming to create a road system in which accident occurrence is strictly limited by an intrinsically safe road environment. Thus, speed limit setting must allow to influence both traffic homogeneity, and road layout and user behaviour predictability, and must be safe and credible (Wegman & Aarts 2006). Lack of credibility from the driver's perspective may lead them to question and ignore it, and, thus, need to be avoided (Goldenbeld & van Schagen 2007).

2.3 Economic optimization methods

The Economic Optimization methods intend to attribute a monetary value to all costs related to mobility, including those which are due to accident-caused damage. Among the various available methods, the Optimum Speed Limits approach has been the most disseminated. This methodology was initially proposed by Oppenlander (1962), and intends to regulate traffic speed from a general society's point of view, recognizing that individual users do not always select speed taking into account the risk imposed over the other road user individuals. This is due to the non-consideration by the traditional methods of the external costs associated with mobility—like those arising from fuel consumption, emissions, noise and accidents—, which produces a market imperfection (Elvik 2010). Therefore, curves of the cost function associated to each one of these factors must be developed to the various road sections, under different traffic conditions. The optimal speed value must be set as the minimum point of the total function, corresponding to the minimum transport cost from the society's point of view.

This method is not currently prescribed by any authority or agency, though a number of studies were carried out in countries like Sweden and Norway.

2.4 Expert systems

Expert-based systems are related with the engineering methods, as they intend to address their lack of consistency and uniformity, aiming at more realistic results (Austroads 2005). These systems are computational based programs which are used to solve complex problems recurring to decision algorithms and a database, allowing it to simulate the behaviour, reasoning, evaluation and decision process of experts (MnDoT 2012). Among all the components that compound such a system, the database—the knowledge base—is particularly relevant, since the system bases its decisions on it. This database includes information arising from the experts' knowledge and experience, structured in tasks to execute and decisions to take (TRB 1998).

The development of the model that establishes the relationship between the knowledge base and the factors shall be carried out by collecting information in a set of representative cases. The calibration of this model uses the speed limit values previously devolved by the experts, establishing the knowledge base. After the definition of the model function, the system is prepared to estimate the speed limit value, based on data related to each one of the considered factors.

In general, the advantages of using this type of approaches are evidenced by practice, which shows its comprehensiveness, consistence and reliability, as well as the fact of being easily reproducible in different contexts (Forbes et al. 2012).

This type of systems has been widely applied in several places in the last few decades, with Australia (XLIMITS family programs) and USA (USLIM-ITS) as the most representative cases. Although there is not any known case of its application in a European country so far, it is considered to be a relevant improvement of the engineering methods, due to its more realistic results, involving more easily available data and embracing a wider number of factors.

3 METHODOLOGICAL APPROACH

The model is an expert-based system, following the previously presented assumptions, as well as Seco et al. (2008), Correia & Bastos Silva (2010), Correia & Bastos Silva (2011), Bastos Silva et al. (2012), Bastos Silva et al. (2016) and Gregório et al. (2015). As it was previously stated, the fundamental component of such a system is the knowledge base, i.e. the database which incorporates the information provided by the experts' technical experience and knowledge, and whose content will allow the system to take decisions.

3.1 Data collection

A number of 4 experts are involved in this process. They were selected based on their experience and expertise in the field of traffic engineering, and above all in speed management. These experts intervene in two main phases: they were intended to analyse the final set of variables to include in the model; and they independently attributed a speed limit value to each analysed road section, based on their experience and expertise. Their evaluation must only be influenced by road functionalities and interaction with the surrounding environment, and the existing posted speed limits were disregarded.

Therefore, a number of road itineraries was selected, corresponding to road segments of inter-urban highways, in which data collection is carried out. These itineraries are representative of several characteristic situations of this type of roads: they cross several types of environment, namely urban areas, transition areas and rural environments (intra-section variability); some of them were selected in regions with dispersed urban occupancy (cases in most of northern and central

Portugal), and other ones in regions with a more concentrated and consolidated urban pattern, like in southern Portugal (inter-section variability); they do not present too irregular or heterogeneous layouts. Eventually, more than 130 km of roads will be included in the database.

The process of speed limit attribution and data collection is carried out in successive road sections of 200 meters, in which each itinerary is divided. This value is considered to be a sufficiently short length to guaranty a homogeneous level of both road physical features and surrounding environment characteristics. On the other hand, this distance is also considered to be long enough to enable experts to make stable assessments. The experts assisted, in laboratory, to video recorded in which one of itineraries, in both directions. This video was recorded in dry weather conditions.

Data related with each one of the variables included in this model is collected through the examination of satellite/aerial imagery and by using the videos recorded in the considered itineraries.

3.2 Model estimation

The model is estimated by using this data and the speed limit values proposed by each expert in each road section, which is the dependent variable. This model is a Multinomial Logit, presenting four alternatives, which are contemplated in the expert decision-making process. These alternatives are the following speed limit values:

- 30 km/h (18.6 mph), which has been associated with zones within urban areas with a strong residential, commercial or services function, with high volumes of pedestrians and cyclists (it even originated the concept of Zone 30, now widely used all over Europe);
- 50 km/h (31.1 mph), the statutory speed limit for urban roads in Portugal, i.e. roads crossing areas where there is a significant urban occupancy and presence of multiple functions, other than those related to motorized through traffic;
- 70 km/h (43.5 mph), which is considered as a representative value for a transition zone (suburban or peri-urban context, with a disperse urban occupancy);
- 90 km/h (55.9 mph), the statutory speed limit for interurban highways in Portugal, crossing rural areas where there is none or negligible urban activity.

The reference alternative must be 90 km/h, since it is the statutory speed limit for interurban highways in Portugal.

Taking into account that speed limit setting is essentially a discrete choice problem, the Multinomial Logit model was selected to estimate the

system's function, which establishes the relationship between the knowledge base and the included factors.

Discrete choice models are based on the theory of stochastic utility, in which the choice is carried out by the user aiming to maximize the utility function. The utility function is composed as a combination of known explanatory variables, the systematic part of the utility, and its random part, which is unknown (Ben-Akiva & Lerman 1985). Thus, this function has the following form:

$$U_{in} = V_{in} + \varepsilon_{in} \qquad (1)$$

where U_{in} represents the utility function, given by decision-maker n to alternative i, V_{in} its systematic part and ε_{in} the error between the systematic part and the true utility. The systematic part of the utility function given by user n to alternative i is, in its turn, represented by the following expression:

$$U_{in} = \beta_{0i} + \beta' X_{in} \qquad (2)$$

where β_{0i} is the specific constant of each alternative i, β' represents the vector of weights and X_{in} is a vector of attribute values for each alternative i, given by a decision-maker n.

The error between the systematic part of the utility and its true value can be regarded as the part of the utility which is unknown to the analyst, i.e. as its random part (this is the most widely accepted theory).

The hypotheses assumed about the statistical distribution of the error term of the utility function lead to the adoption of different types of discrete choice models. The Logit models are one of them. The Multinomial Logit model, which is part of this family, was developed as a generalization of a binary choice model in a context involving more than two alternatives.

This model is based on the assumption that the error terms of the alternatives are all independent and identically distributed (IID), according to a Gumbel distribution, also known as type I extreme value distribution (Ben-Akiva & Bierlaire 1999). This distribution directly implies that any difference between the error terms is logistically distributed. According to this hypothesis, the probability of choice of alternative i in this model is the following (Ben-Akiva & Lerman 1985):

$$P_n(i) = P(U_{in} \geq U_{jn}) = \frac{e^{V_{in}}}{\sum_{j \in C_n} e^{V_{jn}}} \qquad (3)$$

where C_n is the choice set.

Having into account that distributions are identically distributed, it is considered that the mean assumes the same value for all the errors. However, the mean is irrelevant, since only differences in the utility matter for changing the probability of choice. The best method to estimate this model is through maximum likelihood, using the following function:

$$L^* = \prod_{n=1}^{N} \prod_{j \in C_n} P_n(i)^{y_{in}} \qquad (4)$$

where y_{in} represents the binary variable which assumes value 1 if the decision-maker n chooses alternative I, and 0 otherwise. By using logarithms, the maximization of the likelihood function can be simplified in the following expression:

$$L = \sum_{n=1}^{N} \sum_{i \in C_n} y_{in} \log\left(P_n(i)_n\right) \qquad (5)$$

Finally, the obtained model must be validated, by carrying out other trials in one or more itineraries, which have not been involved in previous stages, and following a similar procedure.

3.3 Explanatory variables

The set of explanatory variables was selected according to previous work already developed and to some shortcomings previously identified. Examples of new variables now included are those related with lateral barriers, intersections, road lighting and parking. These variables were also selected having into account the experts suggestions, in a previous stage before their assessment.

Variables with a binary character represent the existence of the feature. Likewise, discrete variables represent the number of elements of that type which can be found along the section in analysis. Other variables, with a continuous character, represent a proportion of the road section where the feature exists.

On the other hand, edification density is expressed in number of edifications per 100 meters of road, while distance-related variables are measured in meters. To determine these variables, only edifications located within a range of less than 30 meters of the roadway are taken into account, according to Bastos Silva et al. (2012).

Every variable, except NCRO and ISLAND, are determined for both sides of the road–R for the nearside of the road; L for the offside.

Portugal is a country with right-hand traffic.

a. Intersections and motorized accesses:

– ROUND: Number of roundabouts along the section;

– INTCR: Number of intersections with 4 or more legs along the section;
– INTEN: Number of 3-leg intersections along the section;
– NATER: Number of lateral local accesses, both public and private;
– AGRO: Number of rural roads and paths, namely small local access roads without pavement.

Comparing with previous work, it was decided to create new variables regarding each type of intersection. A new variable was also created to small rural roads, given their specific location in rural zones.

b. Pedestrian accesses:

– NAPED: Number of public pedestrian paths along the section;
– NAPEH: Number of private pedestrian paths through buildings, such as doors and gates, which are directly connected to the road;
– NCRO: Number of formal pedestrian crossings along the section;
– SIDEW: Proportion of the section with pedestrian sidewalk.

A new configuration was given to variable SIDEW, which used to be a binary variable and now corresponds to a proportion of the section.

c. Parking:

– GARAG: Number of accesses to off-road parking, namely garages and other private parking situations;
– NAPAR: Number of on-road parking places along the section;
– IPAR: Existence of illegal parking along the section (binary variable);
– LOG: Number of accesses to open spaces separated from the road, both public and private, including off-road parking areas, petrol stations and private patios or terraces, among others.

A new variable was created to address the problem of illegal parking, which is widespread in various locations. Another variable was composed to include accesses to public and private open spaces near the road, and connected with it. As a result, previously considered variables, such as the number of petrol stations, were removed.

d. Infrastructure:

– ISLAND: Proportion of the section with the presence of a central island;
– RLIGH: Proportion of the section with the presence of road lighting;
– SIGN: Number of speed control traffic lights along the section;
– NBUS: Number of bus stops along the section.

A new variable was added to represent the existence of road lighting along the section, which was previously mentioned and identified as a relevant indicator of urban development. A new configuration were given to *ISLAND*, which now represents a proportion of the section, instead of the mere existence of this feature.

e. Density and visibility:

– *ED*: Edification density per 100 meters of road, within 30 meters off the carriageway;
– *MD*: Minimum distance between edifications and the carriageway, for edifications within 30 meters off the carriageway (in meters);
– *MED*: Median distance between edifications and the carriageway, for edifications within 30 meters off the carriageway (in meters);
– *OPAC*: Proportion of the section with the presence of a physical barrier on the roadside, including walls, fences, long barriers or moats.

The first three variables were previously defined as objective variables, to better represent the effect of buildings and other structures near the road over speed limit setting (Bastos Silva et al. 2012). These variables intended to substitute other subjective variables, and their adoption allowed to obtain more accurate results.

On the other hand, *OPAC* is a new variable added to represent the optical effect that physical barriers on the immediate roadside have over the driver's perception, namely regarding two main problems: these structures provide both a sense of opacity (which harms visibility) and of porosity (when there are access points) on the roadside. However, having into account the diversity of situations that may fit under this variable, a wide range of different cases were included, for the sake of simplicity and ease of use.

3.4 Expert assessment

A certain degree of disagreement is expected between the speed limit values proposed by each expert. Despite Multinomial Logit model being prepared to deal with these dissimilarities between expert evaluations, it tends to produce models representing average assessments, regardless of the fact that models would present probabilities for each alternative. Since in safety-related problems it is recommended to use conservative, rather than average, options when traffic operational conditions are to be selected, a *virtual conservative expert* is used, selecting for each road section the second most restrictive speed limit value from the choices made by each expert, as in Bastos Silva et al. (2012).

The trials are carried out in laboratory, in order to assure more homogeneous and replicable conditions, due to its more controlled environment and to the fact that each expert always analyse each road in the exactly same conditions of other experts (avoiding the influence of undesirable factors to this process, such as weather conditions). Laboratorial conditions also allow the expert to avoid the sense of continuity between two successive sections. The suitability of these methodology was already tested, showing a high degree of consistence (Gregório et al. 2015).

4 PRELIMINARY ANALYSIS OF RESULTS

An analysis with some early results obtained for the experts' evaluation was carried out, allowing to take some preliminary conclusions. This analysis included the assessment of two experts over 73 km of roads in both directions (a total of 146 km), corresponding to 730 road sections. These sections included urban, transition and rural zones, and were collected in different regions, as it was previously mentioned.

Although each expert was allowed to use each one of the four available alternatives, the 30 km/h alternative was not used in any section, which is mostly due to the characteristics of the road sections considered in this analysis. Thus, in case their assessments were not identical, experts could diverge by a value of 20 km/h (if their assessments were 50 and 70 km/h, or 70 and 90 km/h) or by 40 km/h (in case their assessments were 50 and 90 km/h).

To analyse the consistency of their assessments, a statistical test was performed. Since this analysis may be considered as a paired observations-like situation—two results with the same nature for each one of the cases, collected under the same conditions and assumptions, by analysing the same road sections –, a paired samples t-test, a parametric test, was performed. This test allows to compare two population means in case the corresponding two samples are correlated, and it assumes the following hypotheses (Washington et al. 2011):

$$H_0: \mu_d = 0 \text{ vs. } H_1: \mu_d \neq 0 \tag{6}$$

where μ_d is the difference between both sample means.

The obtained results are shown in Table 1. The SPSS software was used in this analysis.

A p-value of 0.030 was obtained in this test, which implies that the null hypothesis cannot be rejected for a significance level of 1%, but can already be rejected

Table 1. Results obtained for the paired samples t-test.

	N	df	t	p-value
Paired t-test	730	729	–2.171	0.030

for levels of 5% or over. This is a slightly inconclusive result, although it indicates that the null hypothesis can be rejected for some of the usual significance levels and, therefore, the difference in means between the two samples is not significantly closer to 0. This result implies that the two assessment sets are not significantly similar in statistical terms.

In order to better analyse these outcomes, those sections where the expert assessment diverged were identified. Thus, within the 730-cases set, divergence occurred in 192 sections, corresponding to 26.3% of the set. However, situations involving 50 and 90 km/h in the same section were not reported. Such a divergence would be excessive, and could indicate some inconsistency of the experts' assessment and of the overall methodology. Indeed, having all the divergent cases occurring in 50–70 and 70–90 km/h situations is an expected and reasonable outcome.

The obtained results are presented in Table 2.

All sections where divergent assessment occurred belong to transition zones. This was again an expected outcome, since in these zones characteristics of both urban and rural areas prevail very often in the same road section, under a peri-urban or suburban environment, sometimes coexistent with other non-residential or mixed uses. This is a particularly relevant issue for some variables, especially edification density, intersections and lateral accesses (all those variables usually express average values in these sections). Besides, in these sections usually problems related with speed limit inconsistencies are more recurrent and severe.

These results were expected and are entirely acceptable. In fact, similar experts' evaluations were reported in a wide number of sections (nearly 75%), in different road environments and regions, proving the adequacy of the method and of the conditions under which this process was carried out. On the other hand, experts diverged in their judgment in a number of cases that is big enough to justify the scope of this work, proving that the problem being addressed is relevant and widespread all over the road network.

The overall results proved that one of the experts (Expert 1) is slightly more conservative, which also justifies the use of the previously described *virtual conservative expert* method. In any case, it is expected that the assessment of the four experts, as a whole, will express the same proportion of divergence, since it is more likely to occur in these sections in transition zones, disregarding their propensity to a more or less conservative approach.

5 CONCLUSION

This work aimed to present an accessible and easy to use methodology to set speed limits in interurban single carriageway roads. Thus, a methodological approach involving an expert system was developed, including a knowledge base composition through expert assessment on a number of road sections, and data collection regarding a set of factors related to the road environment and its surrounding areas. The usual factors related with road geometry and layout were disregarded in this phase, given the short length (200 m) of the road sections. The experts' assessment was delivered in laboratory, by watching video recorded in each one of the analised roads. In this phase, only some results of this evaluation process were presented and discussed, allowing to conclude that the most problematic sections are those in transition zones. However, the overall results were consistent and showed the adequacy of the methodology. Further work shall now be carried out in order to develop the knowledge base and then to estimate the final model. Moreover, further work can still be developed about the expert assessment process, involving additional trials and, in a later stage, a thorough analysis in each one of the sections where divergent results were obtained, in order to identify the responsible factors.

Finally, in later phases, longer stretches of road must be considered, involving a number of adjacent sections, and particularly focusing on transitions between them, in terms of road consistency and layout homogeneity, and intending to correspond to drivers expectancy. Later on, this analysis will be performed on a route perspective, involving a longer road extension and focusing on issues related with safety, economic and environmental effects.

ACKNOWLEDGEMENTS

The authors would like to thank the Foundation of Science and Technology (Fundação para a Ciência e a Tecnologia—FCT), under the MIT-Portugal Program.

REFERENCES

Aarts, L., van Schagen, I. 2006. Driving Speed and the Risk of Road Crashes: A Review. *Accident Analysis and Prevention* 38(2): 215–224.

Table 2. Divergence cases in the experts' assessment, where each expert was more conservative.

	50, in 50–70 km/h		70, in 70–90 km/h	
	No.	%	No.	%
Expert 1	37	5.1%	74	10.1%
Expert 2	34	4.7%	47	6.4%

Austroads. 2005. *Balance between Harm Reduction and Mobility in Setting Speed Limits: A Feasibility Study.* Research Report AP-R272/05. Sydney: Austroads.

Austroads. 2010. *Impact of Lower Speed Limits for Road Safety on Network Operations.* Publication No. AP-T143/10. Sydney: Austroads.

Baruya, A. 1998. Speed-Accident Relationships on European Roads. MASTER Project. *9th International Conference Road Safety in Europe*; Bergisch Gladbach, 21–23 September, 1998.

Bastos Silva, A., Seco, A., Gregório, N. 2016. Setting Speed Limits in Interurban Single Carriageway Highways Using Experts Judgment. *Transport* 31(2): 282–294.

Bastos Silva, A., Seco, A., Santos, S. 2012. Setting Safe Speed Limits in Two-Lane Rural Highways Using Expert's Judgments. 91st Transportation Research Board Annual Meeting; Washington, D.C., 22–26 January, 2012.

Ben-Akiva, M., Bierlaire, M. 1999. Discrete Choice Methods and Their Applications to Short-term Travel Decisions. In *Handbook of Transportation Science, Vol. 23. International Series in Operations Research and Management Science.* Kluwer.

Ben-Akiva, M., Lerman, S. 1985. *Discrete Choice Analysis: Theory and Application to Travel Demand.* Cambridge: MIT Press.

Correia, G., Bastos Silva, A. 2010. Setting Speed Limits in Rural and Interurban Two-Lane Highways Using Expert Opinion Crossed with Measurable Road-Side Characteristics. *89th Transportation Research Board Annual Meeting; Washington, D.C., 10–14 January, 2010.*

Correia, G., Bastos Silva, A. 2011. Setting Speed Limits in Rural Two-Lane Highways by Modeling the Relationship between Expert Judgment and Measurable Roadside Characteristics. *Journal of Transportation Engineering* 137(3): 184–192.

DfT. 2013. *Setting Local Speed Limits.* DfT Circular 01/2013. London: Department for Transport.

Elvik, R. 1999. Can Injury Prevention Efforts Go Too Far? – Reflections on Some Possible Implications of Vision Zero for Road Accident Fatalities. *Accident Analysis and Prevention* 31(3): 265–286.

Elvik, R. 2010. A Restatement of the Case for Speed Limits. *Transport Policy* 17(3): 196–204.

Elvik, R. 2013. A Re-Parameterisation of the Power-Model of the Relationship between the Speed of Traffic and the Number of Accidents and Accident Victims. *Accident Analysis and Prevention* 50: 854–860.

ERSO. 2006. *Speeding.* European Road Safety Observatory. Retrieved on 31 August, 2016, from http://www.erso.eu.

Farmer, C., Retting, R., Lund, A. 1999. Changes in Motor Vehicle Occupant Fatalities after Repeal of the National Maximum Speed Limit. *Accident Analysis and Prevention* 31(5): 537–543.

FHWA. 2009. *Manual on Uniform Traffic Control Devices for Streets and Highways.* 2009 Edition. Washington, D.C.: Federal Highway Administration.

Forbes, G., Gardner, T., McGee, H., Srinivasan, R. 2012. *Methods and Practices for Setting Speed Limits: An Informational Report.* Report No. FHWA-SA-12-004. Washington, D.C.: Federal Highway Administration—Office of Safety.

Goldenbeld, C., van Schagen, I. 2007. The Credibility of Speed Limits on 80 km/h Rural Roads: The Effects of Road and Person (ality) Characteristics. *Accident Analysis and Prevention* 39(6): 1121–1130.

Gregório, N., Bastos Silva, A., Seco, A. 2016. Speed Management in Rural Two-Way Roads—Speed Limit Definition Through Expert-Based System. *European Transport Conference 2015; Frankfurt am Main, 28–30 September, 2015. Transportation Research Procedia* 13: 166–175.

Hauer, E. 2009. Speed and Safety. *Transportation Research Record* (2103): 10–17.

Kloeden, C.N., Ponte, G., McLean, A.J. 2001. *Travelling Speed and the Risk of Crash Involvement on Rural Roads.* Report No. CR 204. Canberra: Australian Transport Safety Bureau.

Kockelman, K. 2006. *Safety Impacts and Other Implications of Raised Speed Limits on High-Speed Roads.* Final Report. NCHRP Document 90 (Project 17–23). Washington, D.C.: Transportation Research Board.

MnDoT. 2012. *Methods for Setting Posted Speed Limits. Transportation Research Synthesis.* TRS 1204. Saint Paul: Minnesota Department of Transportation.

Nilsson, G. 2004. *Traffic Safety Dimension and the Power Model to Describe the Effect of Speed on Safety.* Doctoral Thesis. Lund: Lund Institute of Technology.

Oppenlander, J. 1962. A Theory on Vehicular Speed Regulation. 41st Annual Meeting of the Highway Research Board; Washington, D.C., 8–12 January, 1962. Highway Research Board Bulletin (341): 77–91.

Park, Y., Saccomanno, F. 2006. Evaluating Speed Consistency between Successive Elements of a Two-Lane Rural Highway. *Transportation Research Part A: Policy and Practice* 40(5): 375–385.

Rosencrantz, H., Edvardsson, K., Hansson, S.O. 2007. Vision Zero—Is It Irrational? *Transportation Research Part A: Policy and Practice* 41(6): 559–567.

Seco, A., Bastos Silva, A., Galvão, C. 2008. Speed Management Model Development Applicable to Regional and National Single Carriageway through Roads. *2nd European Road Transport Research Arena—TRA 2008; Ljubljana, 21–25 April, 2008.*

Stuster, J., Coffman, Z., Warren, D. 1998. *Synthesis of Safety Research Related to Speed and Speed Management.* Report No. FHWA-RD-98–154. Washington, D.C.: Federal Highway Administration.

Taylor, M., Baruya, A., Kennedy, J. 2002. *The Relationship between Speed and Accidents on Rural Single-Carriageway Roads.* Report TRL511. Crowthorne: Transport Research Laboratory.

TRB. 1998. *Managing Speed—Review of Current Practice for Setting and Enforcing Speed Limits.* TRB SR254. Washington, D.C.: Transportation Research Board.

Vaa, T. 1999. *Vision Zero and Sustainable Safety: A Comparative Discussion of Premises and Consequences.* Borlänge: Swedish National Road Administration.

Washington, S.P., Karlaftis, M.G., Mannering, F.L. 2011. Statistical and Econometric Methods for Transportation Data Analysis. Second Edition. Boca Raton: Chapman & Hall/CRC.

Wegman, F., Aarts, L. 2006. *Advancing Sustainable Safety: National Road Safety Outlook for 2005–2020.* Leidschendam: SWOV—Institute for Road Safety Research.

Whitelegg, J., Haq, G. 2006. *Vision Zero: Adopting a Target of Zero for Road Traffic Fatalities and Serious Injuries,* for the UK Department for Transport. Stockholm Environment Institute. Norwich: HMSO.

Transport Infrastructure and Systems – Dell'Acqua & Wegman (Eds)
© 2017 Taylor & Francis Group, London, ISBN 978-1-138-03009-1

Effects of mineral fillers on bitumen mastic chemistry and rheology

R.M. Alfaqawi, G. Airey, D. Lo Presti & J. Grenfell
Nottingham Transportation Engineering Centre, University of Nottingham, Nottingham, UK

ABSTRACT: Age hardening of bitumen is one of the key factors determining the lifetime of an asphalt pavement. This study attempts to investigate the effect of different fillers including hydrated lime on mastic ageing. The Thin Film Oven Test (TFOT) was used for short-term ageing, and the Pressure Ageing Vessel (PAV) was used to simulate long-term ageing. The changes due to the ageing are measured by the changes in complex modulus |G*| of the mastics. Chemical changes in the binder were evaluated by the means of Fourier Transform Infrared Spectroscopy (FTIR). Results indicated that different filler mineralogy and properties can significantly affect the rate of oxidative ageing, and this depends mainly on the filler type and concentration. It also showed that hydrated lime reduces the hardening rate more than limestone and granite filler and this correlated with the chemical changes measured by FTIR in terms of rate of change of the carbonyl oxidative products.

1 INTRODUCTION

Age hardening of bitumen has long been considered one of the main factors that can significantly affect the durability of pavements. When bitumen is age hardened, this process is usually accompanied by stiffening and embrittlement of the binder, thus asphalt pavements become brittle and their ability to support traffic-induced stresses and strains may significantly reduce. In addition, excessive hardening can also weaken the adhesion between the bitumen and aggregate, which contributes to a reduction of the durability of pavements and eventually increases the maintenance cost (Wu, 2009). Even though it is only one component of asphalt mixtures, the overall performance of the asphalt pavement is largely influenced by the viscoelastic properties of the bitumen. Generally, bitumen ageing takes place in two stages; short-term ageing at high temperatures during asphalt mixing and paving and long-term ageing at ambient temperatures during in-service. In bitumen ageing, two mechanisms can be involved; the main one is irreversible, which is characterised by chemical changes of the binder where two processes take place, oxidation and loss of volatile components. The other mechanism is a reversible process called physical hardening that can be attributed to molecular restructuring.

The oxidation of the hydrocarbon components in bitumen is largely accepted as the major factor of the hardening and stiffening of asphalt pavements Petersen et al., (1974a). According to Moraes & Bahia (2015), even though binder ageing in pavement always occurs when the binder is in contact with aggregates and mineral filler, in most ageing studies, asphalt binders are individually aged without accounting for aggregate-induced interactions.

This is also the same in the current specifications. It is important to recognise that the most intimate contact between bitumen and aggregate in a paving mixture involves the fine aggregate and filler. Anderson et al. (1992) stated that the properties of the fine fraction dominate in terms of physico-chemical interactions between the bitumen and the mineral surface, as the fines are embedded in the bitumen and, thus, the majority of the surface area is generated by the fine minerals. Therefore, the mechanisms of asphalt mastic ageing are influenced by the characteristics of the asphalt binder and the mineral filler properties, in addition to the molecular interaction between them. The chemical and rheological changes associated with ageing are well understood for neat bitumen as the result of much research on this subject (Petersen 2009, Read & Whiteoak 2003). However, for the ageing of asphalt mastics, many topics in this research area still need to be investigated Moraes & Bahia (2015).

Little and Petersen, (2005) investigated the unique effect of hydrated lime on the performance-related properties of asphalt pavements. They compared the effect of hydrated lime to a similar size filler comprised of limestone. They performed extensive laboratory testing, on mastics and mixtures. The results confirmed that when HL is added to bitumen, its effect will range between that equal to other mineral fillers to a considerably greater effect, depending on the physicochemical interaction developed between the bitumen and the HL. This interaction has proven to be different when different binders are used.

In order to develop a better understanding of the way by which the mineral fillers affect bitumen ageing, an ageing study on the bitumen-filler mastics including DSR and FTIR testing after different ageing stages was carried out. Mastic ageing has its advantages compared to asphalt ageing. As mentioned above, the surface area of fillers is much larger than the coarse aggregate, which will make the adsorbing and catalysing phenomenon more obvious. In addition, as the fillers in mastics can be viewed as being embedded in the binder, the effects of 'void content' and 'binder film thickness', which were concerns during the mixture ageing studies, could be avoided in the mastic ageing.

2 MATERIALS AND EXPERIMENTS

2.1 Materials

Three types of fillers were used in this study, Granite filler (G), Limestone filler (LS), and hydrated lime (HL). All fillers were substantially finer than 63μm in particle size.

A 40/60 Pen bitumen with a penetration of 45 dmm and a softening point of 50°C was used to prepare mastics.

Granite or limestone fillers were mixed with the 40/60 bitumen to produce mastics with 50% filler by mass. Then the filler (Granite or Limestone) was replaced by hydrated lime at percentages of 10% and 20% by mass.

2.2 Filler tests

2.2.1 Geometrical properties

The morphology and geometrical properties (size and shape) of the selected fillers for this study were assessed by a grading test and a scanning electron microscopy imaging technique. A Beckman Particle Size Analyser (LS200) was used for measuring the particle size distribution of fillers. In this analyser, a flux curve is generated which is a composite of the flux curves for all the particles that pass through a laser beam during a sample run. The LS 200 system converts the composite flux curve into a particle size distribution. This device uses water as a liquid dispersant, so it was not suitable for testing the Hydrated lime fillers as there is a reaction between water and the hydrated lime-Ca(OH)$_2$. So, the particle size distribution for hydrated lime was given by the supplier as it was tested using a Malvern Mastersizer.

2.2.2 Particle density

The particle density (SG) was measured by the pycnometer method in accordance to BS EN 1097-7:2008. This test is required for the Rigden voids calculation.

2.2.3 Specific surface area

The Specific Surface Area (SSA) was determined to evaluate the stiffening power relation with the fineness of the filler. It is well known that there is an increase of the SSA with the increase of the fine particle content. Also for larger values of SSA more bitumen is expected to be adsorbed by the filler. Consequently, the filler stiffening effect should increase with SSA. The BET method is used for the calculation of SSA.

2.2.4 Fractional voids

The voids in dry compacted filler or Rigden Voids (RV) were determined with the Rigden equipment in accordance with BS EN 1097-4:2008. A filler sample is first compacted in a mould with the defined compaction system, and then the void content is calculated considering the compacted volume and the filler particle density. Several studies have concluded this test value to be closely related to the filler stiffening power (Moraes & Bahia 2015, Bahia et al 2010).

2.3 Mastic tests

2.3.1 Mastic ageing procedures

The Thin Film Oven (TFOT) Test is a standard binder test used for measuring the combined effects of heat and air on a film of bituminous binder. The purpose of the test is to simulate the effect of ageing on binders due to short-term ageing during production in an asphalt mixing plant. The test was carried out according to BS-EN 12607-2:2000. In this test, mastic samples with the same volumes are placed in a TFOT pan to form a thin layer of the same thickness, which are then held in the TFOT oven at 163°C for 5 hours.

The Pressure Ageing Vessel (PAV) ageing procedure is used to simulate long-term field oxidative ageing of asphalt binders in accordance with BS-EN 14769:2012. In this procedure, samples are aged in a vessel pressurised to 2.1 MPa and heated to 90°C in order to simulate in-service ageing over an estimated 7 to 10 year period (Bahia 2014). The mastics in this study are aged for 20 h at 90°C in the PAV after the short-term ageing in TFOT oven.

2.3.2 Dynamic shear rheometer

The rheological properties of the mastics were characterised in this study with a Bohlin Gemini model DSR. A frequency sweep is performed with a 0.3% strain amplitude over a range of 0.1 to 10 Hz and temperatures from 0 to 50°C. The resulting complex shear modulus |G*| data is used to produce master curves using the time-temperature superposition principle, with shift factors following the Williams-Landel-Ferry (WLF) equation. The selected reference temperature for the master curves is 30°C.

2.3.3 Ageing index

In order to show the effects of ageing on the mechanical and rheological properties of mastics, an ageing index was calculated after each stage of ageing: short-term and long-term ageing. The Ageing index was calculated using |G*| at 20°C and 0.4 Hz according to Equation 1.

$$Aging\ Index = \frac{G^*_{aged\ mastic}}{G^*_{un\ aged\ mastic}} \qquad (1)$$

2.3.4 Chemical analysis of bitumen and mastics

Fourier transform infrared FTIR spectroscopy is the most powerful means of detecting and identifying chemical bonds (functional groups) in either organic or inorganic materials. The asphalt binder's main functional groups can be quantified by using the FTIR technique. Therefore FTIR is widely used to characterise the oxygen-containing functionalities found in bitumen, particularly those produced during oxidation Wu, (2009). Typical wave numbers for bitumen ageing products are sulfoxides S = O at 1030 cm^{-1} and C = O in carbonyls at 1700 cm^{-1}. The FTIR test was performed on the recovered bitumen from mastics before and after ageing. Carbonyl C = O ageing index was used to evaluate the effect of ageing on different mastics and it was calculated as follows: Carbonyl index (C = O): $A_{1700}/\Sigma A$.

The sum of the area ΣA represents: $A_{1700} + A_{1600} + A_{1460} + A_{1376} + A_{1030} + A_{864} + A_{814} + A_{743} + A_{724} + A_{(2953, 2923, 2862)}$.

3 RESULTS AND DISCUSSION

3.1 Filler

shows the gradation curves of the used fillers; it can be seen that hydrated lime and limestone filler almost have the same gradation, but granite filler is significantly coarser. However more than 95% of the filler particles are smaller than 63 μm. The properties of the fillers such as the Specific Gravity, Rigden Voids SSA are shown in Table 1.

These results are expected to show a strong correlation with the filler particle shape and morphology. The results show that, although the hydrated lime and limestone have almost the same grading and particle size, HL has significantly higher Rigden Voids and specific surface area compared to both the limestone and granite fill-

ers. In Figure 2 it can be seen from the microscopic morphology of the different filler particles by SEM imaging that, hydrated lime has mostly granular type particles with very rough texture. However Granite and Limestone fillers have mostly angular type particles with smooth to slightly rough texture. The SEM imaging results can be correlated with the other filler tests. As the larger value of the specific surface area SSA and RV of hydrated lime are consistent with the granular shape of particles. The higher SSA of hydrated lime could be related to the internal pores within the hydrated lime particles and its rough surface compared to limesone and granite fillers.

3.2 Mastics

3.2.1 Dynamic Shear Rheometry (DSR)

The master curves of the complex modulus G* of bitumen and mastics are shown in Figure 3 a to c, it can be noticed that the oxidative ageing distorted the shape of the original 40/60 bitumen master curve by decreasing the slope, and thus decreasing the time-dependency of the material response in this frequency range. This effect for bitumen is in line with previous research by Moraes & Bahia (2015).

On the other hand, the addition of mineral filler to bitumen shifted the curve toward higher frequencies without changing the slope, and thus a more elastic response, which is expected to be due to the addition of elastic filler particles.

An important observation is that ageing of the mastics, regardless of filler mineralogy, also resulted in a rotation of the master curve in the reverse direction of that caused by ageing of the binder alone. The result of which is aged mastic

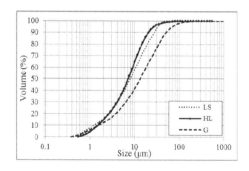

Figure 1. Gradation curves of fillers.

Table 1. Filler properties.

Filler type	Specific gravity	Rigden voids (%)	Surface area (m²/g)
Granite (G)	2.66	47.15	1.26
Limestone (LS)	2.65	39.82	1.58
Hydrated Lime (HL)	2.22	61.62	2.24

Figure 2. SEM imaging of fillers a) Granite, b) Limestone, c) Hydrated lime.

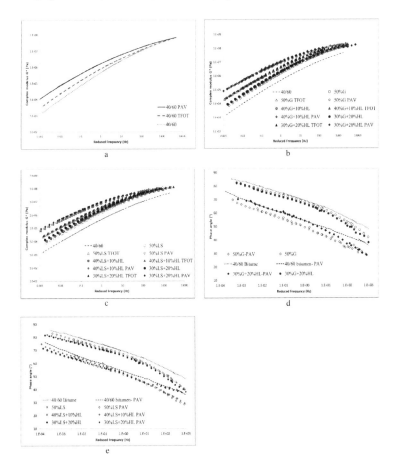

Figure 3. Oxidative ageing effect on master curves of bitumen and mastics: a) complex modulus of 40/60 bitumen, b) complex modulus of granite mastics c) complex modulus of Limestone mastics, d) phase angle of granite mastics, e) phase angle of limestone mastics.

master curves that are only shifted in comparison to the neat, without the distortion and rotation observed for ageing of binder alone.

The master curves of the phase angle of original and PAV aged bitumen and mastics are shown in Figure 3 (d and e), it is clear that the addition of the mineral fillers to bitumen decreases the phase angle in all cases. However after PAV ageing the effect of hydrated lime replacement is clear. It

reduces the change in phase angle or increases the phase angle in the range of frequencies tested. This is a relative measure of the ability of mastics with hydrated lime to dissipate shear stress. The increase of phase angle by hydrated lime should be beneficial to reduce cracks in aged pavements.

The effect of hydrated lime on the phase angle was more significant with the granite mastics than the mastics with limestone.

362

3.2.2 Ageing Index calculation

As can be seen from Figure 3, the comparison between the filler effect on stiffening and ageing of mastics is not clear from the master curves. In order to make the comparison more quantitatively, the complex modulus (G*) values at 20°C, 0.4 Hz for the 8 mm testing geometry were selected for the ageing index calculations It can be seen in

Figure 4(a) that all mastics complex modulus values increase with ageing.

In all cases after ageing, mastics containing granite and hydrated lime fillers are stiffer than those containing limestone and hydrated lime filler with the same percentages. It could also be noticed, that by replacing with 20% of hydrated lime does not make significant additional effect compared to the 10% of hydrated lime with granite fillers.

(a)

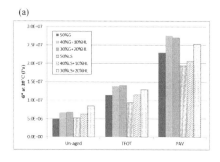

(b)

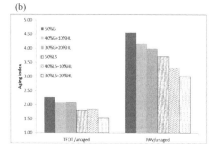

Figure 4. Complex modulus (G*) ageing index of filler mastics. (a) Mastics stiffness (G*), (b) Ageing index.

It also can be seen that the combination of 30%LS+20%HL is a stiffer mastic before ageing, however after short-term and long-term ageing, mastics with LS+HL are softer than mastics with G+HL fillers, thus meaning this combination reduces the rate of ageing and oxidation within the mastics. The initial stiffening effect of HL could be related to its higher RV and the larger surface areas of hydrated lime and limestone filler compared to the granite filler.

3.2.3 FTIR results

The infrared absorption spectrum between 500 cm^{-1} and 4000 cm^{-1} has been recorded for bitumen re-covered from granite mastics, and the absorption bands belonging to asphalt functional groups that are affected by ageing, such as carboxylic acids (C = O) were used to evaluate the increase in oxidative ageing products, which can be determined through FTIR, utilising a very strong band around 1700 cm^{-1}.

Figure 5 shows the FTIR spectrum for the recovered binder from mastics before and after different ageing stages. It can be noticed that the carbonyl area increases with the ageing of the mastics. In addition, according to this scan result, the carbonyl index was calculated and presented in Figure 6. It can be seen that the binder recovered from the mastic of hydrated lime with granite contains less oxygenated products after both short—and long-term ageing. These results can explain the lower ageing index (G* ageing index) for mastics including hydrated lime which means a slower rate of increase of the heavier oxygenated products in mastics containing hydrated lime with ageing.

In order to show the effect of hydrated lime on the rate of stiffening of the mastics after ageing, the mastics containing 50%G, 40%G+10HL, and 30%G+20HL were aged in the TFOT oven for different ageing times up to 20 hours, and the complex moduli of the aged mastics were measured at 20°C and 0.4 Hz and presented in Figure 7(a). From these results the ageing index was calculated as mentioned before and the results are presented in Figure 7(b). It can be seen that, mastics containing hydrated lime are initially stiffer than the granite only mastic.

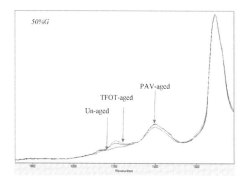

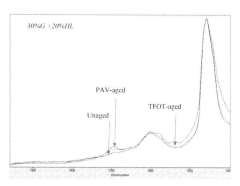

Figure 5. FTIR spectrum for recovered bitumen of granite mastics.

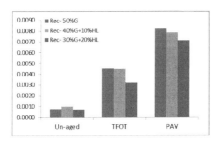

Figure 6. Change of Carbonyl of recovered binder from mastics with ageing.

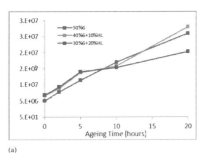

(a)

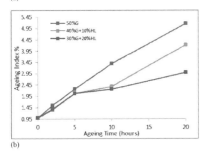

(b)

Figure 7. (a) Change of G* at 20°C with ageing time, (b) G* ageing index.

However with ageing the rate of increase in stiffness of the granite only mastics is higher than the mastics with hydrated lime. And it is clear that replacing 10% or 20% of Granite filler by hydrated lime reduces the ageing index by almost 50% compared to the mastic without hydrated lime.

These results prove that hydrated lime is a more effective filler than limestone and granite in terms of reducing the rate of ageing and the production of oxidative ageing products such as carbonyls. Thus, this proves that there is a physicochemical interaction between hydrated lime and the bitumen which is not present with limestone or granite fillers.

4 CONCLUSIONS

In this study the effect of different mineral fillers of the ageing of bitumen mastics were investigated. The bitumen mastics were aged for short—and long-term ageing periods in the lab and the change in rheological and chemical properties of the mastics and the recovered binders of the mastics were evaluated by means of the DSR and FTIR respectively. The following conclusions could be drawn:

– The ageing of bitumen mastics is different from the ageing of pure bitumen as can be seen from the complex modulus master curves.
– The mineralogy and the properties of the fillers can significantly affect the stiffening ratio of the bitumen mastics and that depends mainly on the filler type and concentration in the asphalt mastic.
– The effect of the hydrated lime replacement on the phase angle change after PAV ageing was measured. It was clear that hydrated lime increase the phase angle after the PAV ageing. This effect could be beneficial to reduce the cracks of lime treated pavement after ageing.
– Replacing 10% or 20% of granite or limestone fillers in the bitumen mastics by hydrated lime filler is more effective in terms of reducing the rate of ageing on the mastics. It can reduce ageing by up to 50% after long-term ageing and this could be used to produce more durable asphalt pavements.

REFERENCES

Anderson, D.A., Bahia, H.U. and Dongre, R. (1992) Rheological Properties of Mineral Filler-asphalt Mastics and Their Relationship to Pave-ment Performance, American Society for Testing and Materials, Philadelphia.
Bahia HU, Faheem A, Hintz C, Al-Qadi I, Reinke G, Dukatz E. Test methods and specification cri-teria for mineral filler used in HMA. In: Program NCHR, editor. NCHRP Project 9–45. USA: Trans-portation Research Board; 2010.
BS EN 1097-4:2008, Tests for mechanical and physical properties of aggregates. Determination of the voids of dry compacted filler
BS EN 1097-7:2008, Tests for mechanical and physical properties of aggregates. Determination of the particle density of filler. Pyknometer method
Moraes, Raquel, and Hussein Bahia. "Effect of Mineral Fillers on the Oxidative Aging of Asphalt Binders: Laboratory Study with Mastics." Trans-portation Research Record: Journal of the Transportation Research Board 2506 (2015): 19–31.
Petersen, J.C., Barbour, F.A. and Dorrence, S.M.(1974a) Catalysis of Asphalt Oxidation by Mineral Aggregate Surface and Asphalt Compo-nents, Proc. Association of Asphalt Paving Tech-nologists, Vol. 43, pp 162–177.
Petersen, J. C. A Review of the Fundamentals of Asphalt Oxidation - Chemical, 9 Physicochemi-cal, Physical Property, and Durability Relation-ships. Transportation 10 Research Circular E-C140, Transportation Research Board, 2009.
Read, J., and D. Whiteoak. The Shell Bitumen Handbook, Fifth Edition, 2003.
Wu, Jiantao. "The influence of mineral aggregates and binder volumetrics on bitumen ageing." Nottingham: University of Nottingham (2009).

Transport Infrastructure and Systems – Dell'Acqua & Wegman (Eds)
© 2017 Taylor & Francis Group, London, ISBN 978-1-138-03009-1

Functional and durability properties evaluation of open graded asphalt mixes

V.M. Kolodziej, G. Trichês, J.S. Ledezma, G.C. Carlesso, L.M. Jardín & R.M. Knabben
Universidade Federal de Santa Catarina, Santa Catarina, Florianópolis, Brasil

ABSTRACT: Open graded asphalt mixes are one solution to the current demand for roads that offer quality, safety and comfort to users. However, these mixtures have problems of durability and mechanical strength. In this context, the aim of this research is to obtain a mixture that presents a balance between the permeability, acoustic and durability properties. This study evaluated mixtures with two different aggregate gradations with a maximum aggregate size of 9.5 mm and a polymer modified binder, through a series of laboratory testing. One of the two aggregate gradations was studied in Brazil (G1), and another gradation was based on a work done in the United States (G2). The research concluded that G2 mixture shows satisfactory performance related to acoustic, permeability and durability properties.

1 INTRODUCTION

From a roadway users perspective, the more interested functions of a pavement are more related to its surface performance, including smoothness, safety, low noise, and good visibility of markings at night and during raining.

Traffic noise pollution has become a growing problem because of the continuous increase of traffic volume. Traffic noise is a primary source of noise pollution that affects human health and life quality of bordering residents (WHO, 2011). Sources of noise produced by vehicle traffic are divided into motor noise, aerodynamic noise and tire/pavement noise (Rasmussen et al, 2007). According to Bernhard and Wayson (2004), in light vehicle traffic situations and speeds above 50 km/h, the amount of tire/pavement noise exceeds motor and aerodynamic noise.

Sound absorption of a road surface is considered one of the parameters that influence the road noise levels, which are generated by the contact tire/pavement (Li et al., 2015). Existing studies show that the sound absorption is effective to reduce tire/pavement noise in the frequency range between 800 and 1000 Hz (Sandberg & Ejsmont, 2002).

Sound absorption coefficient is the ratio of the incident acoustic energy and acoustic energy absorbed by the surface. This value is always positive and can vary between 0 and 1, where 1 represents a purely absorbing material (Gerges & Arenas, 2010). Factors such as air-void content, texture patterns, layer thickness and the structure of voids influence the sound absorption capability of surfaces (Liu et al., 2016).

Presence of water on road surfaces involves safety problems for users by the loss of tire/pavement adhesion, the risk of aquaplaning and the decreased visibility, which increases the number of traffic accidents in rainy days (Meurer Filho, 2001). A surface with high air-voids contents has the ability to eliminate the water depth of pavement surfaces, which is formed during a rain. According to Oliveira (2003), pores of mixtures can be classified as effective pores (which contribute to passage and to water storage), semi-effective pores (that not contribute to water passing, but allow water storage) and ineffective pores (that not contribute to storage or passage of water).

Both the noise caused by tire/pavement contact as the presence of water can be reduced or minimized by optimizing characteristics of pavement surface. Open graded asphalt mixtures have the capacity to absorb the tire/pavement noise and allow a passage of water through its internal structure (PIARC, 2013).

2 EXPERIMENTAL DESIGN

2.1 Materials and mix designs

This experiment includes two aggregates gradations and one SBS polymer modified asphalt binder type 60/85. One granitic nature aggregate was used for all mixtures. The chosen aggregate gradation G1 was previously studied by Guimaraes (2012) in Brazil, and this mixture showed good resistance to abrasion loss and permeability. Aggregate gradation G2 was previously studied by Lu et al. (2009) in the United States. In this study, however, there was a modification of the gradation G2, inserting a larger percentage of coarse aggregate (material retained on 4.75 mm sieve), in order to improve its permeability. The aggregate gradations G1 and G2 used in this study are showed in Table 1.

In order to obtain the optimum binder content of the mixtures were evaluated four binder contents (4.5%, 5.0%, 5.5% and 6.0%), being shaped three specimens in Superpave Gyratory Compactor for each binder content. For each specimen was determined the air-void content, porosity and abrasion loss. The optimum binder content was the one corresponding to air-void content higher than 18%; porosity higher than 10% and abrasion loss less than 20%.

2.2 Test methods

The mix properties that are critical to pavement surface performance were evaluated in this study including air-void content, porosity, resistance to raveling, moisture sensitivity, permeability and acoustic absorption.

The Superpave Gyratory Compactor was used to fabricate specimens of 100 mm diameter for the Cantabro test, moisture susceptibility and acoustic absorption tests. The specimens were fabricated using 50 gyrations of the SGC, applied pressure of 600 kPa, and rotation angle of 1.25°. The LCPC (*Laboratoire Central des Ponts et Chaussées*) compactor was selected to compact slabs specimens of dimensions 50 cm length, 18 cm width and 5 cm thick for permeability test.

2.2.1 Air-void content

The air void content of specimens was calculated from the theoretical maximum specific gravity (G_{mm}) measured in accordance with ASTM D2041 and the bulk specific gravity (G_{mb}) measured using the DNER-ME 117/94 (Equation 1).

$$V_v = \left(\frac{G_{mm} - G_{mb}}{G_{mm}} \right) \times 100 \qquad (1)$$

Table 1. Aggregate gradations used in the study.

Sieve size mm	Percentage Passing (%)	
	G1	G2
9.5	100	100
6.3	43.07	–
4.75	12.51	55
2.36	12.51	14
1.18	12.51	12
0.6	12.51	10
0.3	12.51	7
0.15	7.14	6
0.075	4.08	5

2.2.2 Porosity

In order to obtain the porosity of mixtures was measuring the amount of water that penetrates inside the specimen for the upper face, when the side and bottom faces are waterproofed by adhesive tape and paraffin remaining on its upper face a constant film of water for ten minutes. This procedure was performed as recommended by AFNOR-NF-P-98-254-2, 1993.

2.2.3 Resistance to raveling

Resistance to raveling was evaluated using the Cantabro test on specimens, following the procedure described in DNER-ME 383/99. Specifically, compacted specimens were put inside a Los Angeles Abrasion machine drum without steel balls, and the drum was turned for 300 revolutions in 10 minutes. The percentage of mass loss during this process was calculated by Equation 2.

$$D = \left(\frac{P_i - P_f}{P_i} \right) \times 100 \qquad (2)$$

where P_i = initial sample weight of the sample; P_f = final weight of the sample; and D = Cantabro abrasion loss.

2.2.4 Moisture Susceptibility

Moisture susceptibility of the mixtures was determined using the AASHTO T 283 test method with some modifications as specified in ASTM D 7064. For moisture conditioned specimens, they were first saturated at a vacuum of 87.8 kPa for 10 minutes, and then submerged in water during the 16 hrs freeze cycle. After, in the 24 hrs thaw cycle, specimens were submerged in the 60°C water bath. Instead of five freeze/thaw cycles as specified in ASTM D 7064, only one cycle was applied.

2.2.5 Permeability

The permeability of compacted slabs was determined with LCS permeameter in accordance with Spanish standard NLT-327/00. In this test is measured the time required for the percolation of a column of water in the asphalt mixture layer. The mixture permeability characteristics were determined as a function of flow time in accordance to Equation 3.

$$\ln K = 7.624 - 1.348 \ln T \qquad (3)$$

where K = permeability coefficient; and T = water flow time in LCS permeameter.

2.2.6 Acoustic Absorption

The acoustic absorption coefficient of a material is the ratio of acoustic energy not reflected by the surface material for a normal incidence plane wave. The sound absorption coefficient as a function of frequency was measured in accordance with ISO 10534-2:1998, using an impedance tube of 100 mm diameter.

3 RESULTS AND DISCUSSION

3.1 Air-void content

Results of determination of air-void content for the three specimens for each asphalt binder content and each aggregate gradations G1 and G2 are represented in Figure 1 and Figure 2, respectively. It is noted that air-void content in the mixtures decrease as it is increased the binder content. All samples had air-void contents greater than 18%, which is the minimum required for open graded friction course mixtures.

3.2 Porosity

In regard to test results determination of porosity, all specimens had higher than 10% (minimum required to open mixtures), and their values decrease as it is increased the binder content. In general, higher porosity is related to higher permeability of mixture. In Figure 3 and Figure 4 are represented the results of porosity of G1 and G2, respectively.

3.3 Resistance to raveling

One of the main failure modes of an open mixture is the raveling, which is caused by vehicle traffic and high temperatures. This failure mode was evaluated for each specimen by the Cantabro test. The results are shown in Figure 5 and Figure 6 for G1 and G2 mixtures, respectively. Analyzing this results it can be seen that the resistance to raveling increases as the binder content is increased. The mixture G1 has lower resistance to raveling than the mixture G2, and the reason is explained by the higher air-void contents and porosity identified in the first one that decreases its resistance. It can also be seen in Figure 5, only for 6.0% binder content the mix G1 meets the specification of less than 20% Cantabro abrasion loss. In the case of mixture G2, the specification of less than 20% abrasion loss holds for 5.5% and 6.0% binder contents.

3.4 Optimum binder content

Based on the results obtained for the air-void contents, porosity and abrasion loss for G1 and

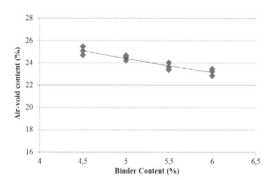

Figure 1. Air-void content of G1.

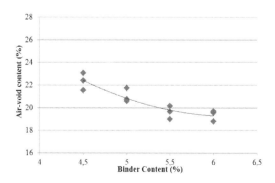

Figure 2. Air-void content of G2.

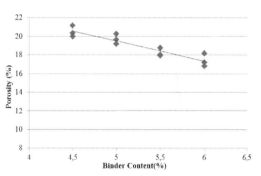

Figure 3. Porosity of G1.

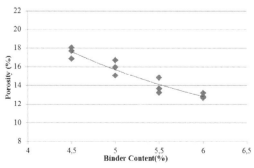

Figura 4. Porosity of G2.

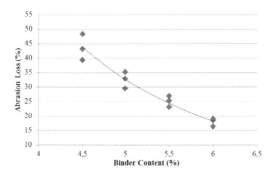

Figure 5.   Cantabro abrasion loss of mixture G1.

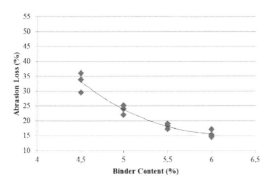

Figure 6.   Cantabro abrasion loss of mixture G2.

G2 mixtures, were selected the optimum binder contents that met both the following conditions: (1) air-void contents higher than 18%; (2) porosity greater than 10%; and (3) abrasion loss less than 20%. As can be seen in Figures 1, 2, 3 and 4; all binder contents met the air-void contents and porosity specifications. However the results of the abrasion loss of the mixture G1 (Fig. 5) show that only for a binder content of 6.0% was achieved a abrasion loss lower than 20%. The mixture G2 met the specification with 5.5% and 6.0% percentages of binder. The optimum binder content selected for G1 was 6.0% and for G2 was 5.5% based on the criteria of abrasion loss.

3.5   Moisture susceptibility

Figure 7 shows the indirect tensile strength for each mix (G1 and G2) in both dry and wet (moisture-conditioned) conditions. The minimum value for the indirect tensile strength is 0.55 MPa in accordance with DNER-ES 386/99 specification to open mixtures in Brazil. Only G2 mixture reaches values above 0.55 MPa for both wet and dry conditions, while G1 mixture showed values below the minimum specified for the two conditions (wet and dry).

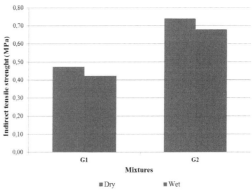

Figure 7.   Indirect tensile strength test results.

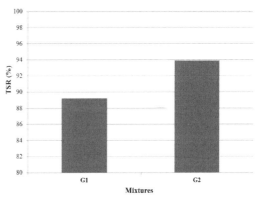

Figure 8.   TSR test results.

Figure 8 shows the results of tensile strength retained (TSR) for two mixtures (G1 and G2). It is conclude that all mixtures showed values above 80% that is the minimum required by ASTM D 7064. The G2 mixture showed better performance with TSR value of 93.1% and the mixture G1 showed greater damage caused by moisture with a value of 89.2%.

3.6   Permeability

The permeability measurements were made at three points of each slab (right edge, left edge and center). The test results of permeability using the LCS permeameter are shown in Figure 9. From the analysis of results it was found that the mixture G1 has a permeability coefficient greater than G2. The presence of a higher air-void contents and porosity in the mixture G1 allows passage of a greater amount of water through interior mixture. The K values for two mixtures exceeded the minimum value of 0.12 cm/s suggested by NCAT (National

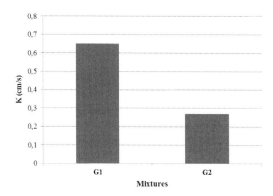

Figure 9. Permeability of mixes measured by LCS permeameter.

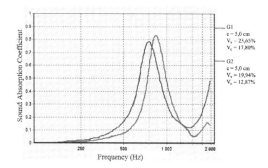

Figure 10. Spectra of sound absorption coefficients of mixes G1 and G2.

Center of Asphalt Technology) for the new generations of open mixtures (Alvarez et al. 2006).

3.7 Acoustic absorption

Figure 10 shows the spectrum of sound absorption coefficient versus frequency for each specimen compacted in SGC with 5 cm height and 10 cm diameter for each mixture (G1 and G2). It was observed that the maximum absorption coefficients are similar for the two mixtures, but the peak sound absorption of the mixture G1 (0.84) is higher than that of the G2 mixture (0.78). This can be explained by the greater air-void contents and porosity in the mixture G1, which allow a greater passage of sound wave into the same. The frequencies where the peaks have been reported (resonance frequencies) were in the range between 750 Hz and 1000 Hz. The results showed that the greater air-voids contents and porosity is related with greater peak sound absorption coefficients.

The mixture G1, which contains approximately 23% of air-void and 18% of porosity, is the best sound attenuation material because showed better

absorption performance in the frequency range of 800 to 1000 Hz, where the noise tire/road is more intense.

4 CONCLUSIONS

This paper presents a laboratory study about the performance (permeability, noise reduction and durability) of two small aggregate size open-graded asphalt mixes (G1 and G2). From the results of a series of tests, the following conclusions were obtained:

- The two mixes G1 and G2 with optimum binder percentages of 6.0% and 5.5%, respectively, met the specifications of an air-void content higher than 18%, porosity greater than 10% and abrasion loss less than 20%.
- In relation to results of indirect tensile strength, G1 didn't reach the minimum resistance value required of 0.55 MPa, while G2 exhibited satisfactory indirect tensile strength values. The results of TSR of the two mixes were higher than 80%, meeting specification.
- The permeability of the mixture G1 was higher than the permeability of the mixture G2, due to the presence of a greater percentage of air-voids and porosity. Both mixes showed a satisfactory performance for the permeability test, presenting results greater than 0.12 cm/s.
- G1 and G2 mixes showed high values of sound absorption coefficient within the range of frequencies 800 and 1000 Hz, where the intensity of noise generated by the tire/pavement contact is critical.

The mixture G1 showed better performance than the mixture G2 from the point of view of functional properties (permeability and sound absorption), but didn't reach the minimum value of indirect tensile strength and had high abrasion loss.

Finally, the mixture G2 (American experience) met all limits and specifications for all tests showing a satisfactory performance in relation to acoustic, permeability and durability properties.

REFERENCES

Alvarez, A.E.; Martin, A.E.; Estakhri, C; Button, J.; Glover, C.J.; Jung, S.H. 2006. Synthesis Of Current Practice On The Design, Construction, And Maintenance Of Porous Friction Courses. FHWA/TX-06/0–5262–1. Texas Transportation Institute. Austin, Texas.

ASTM D2041. 2011. Standart Test Method for Theoretical Maximum Specific Gravity and Density of Bituminous Paving Mixtures.

ASTM D7064/D7064M – 08. 2008. Standard Practice for Open-Graded Friction Course (OGFC) Mix Design.

Bernhard, R.J. & Wayson R.L. 2004. An introduction to tire/pavement noise of asphalt pavement. Asphalt Pavement Alliance. USA. 27 p.

DNER-ES 386/99: Pavimentação—Pré-misturado a quente com asfalto polímero—camadaporosa de atrito. 1999. Departamento Nacional de Estradas de Rodagem. Rio de Janeiro, RJ.

Gerges, S.N.Y. & Arenas, J.P. 2010. Fundamentos y Control del Ruído y Vibraciones.2. ed. Florianopolis: Nr Editora, 787 p.

Guimaraes, J.M.F. 2012. Concreto Asfáltico Drenante em Asfaltos modificados por polímero SBS e borracha moída de pneus. Dissertação (Mestrado) – Programa de Pós-graduação PPGEC-UFSC, Universidade Federal de Santa Catarina. Florianópolis.

ISO 10534–2 E: 1998 Acoustics—Determination of sound absorption coefficient and impedance tubes—Part 2: Transfer-function method.

Li M.; Van Keulen, W.; Tijs, E.; Van De Ven, M. 2015. Molenaar, A. Sound absorption measurement of road surface with in situ technology. *Applied Acoustics* 88: 12–21.

Liu, M.; Huang, X.; Xue, G. 2016. Effects of double layer porous asphalt pavement of urban streets on noise reduction. *International Journal of Sustainable Built Environment* 5: 183–196.

Lu, Q.; Fu, P.C.; Harvey, J.T. 2009. Laboratory Evaluation of the Noise and Durability Properties of Asphalt Surface Mixes. Research Report: UCPRC-RR-2009–07 University of California Pavement Research Center UC Davis, UC Berkeley. EEUU.

Meurer Filho, E. 2001. Estudo de Granulometria Para Concretos Asfálticos Drenantes. Dissertação (Mestrado) – Programa de Pós-graduação PPGEC-UFSC, Universidade Federal de Santa Catarina. Florianópolis.

NLT-327/00. 2000. Permeabilidad in situ de pavimentos drenantes con el permeámetro LCS.

Oliveira, C.G.M. 2003. Estudo de propriedades mecânicas e hidráulicas do concreto asfáltico drenante. Dissertação de mestrado em Geotecnia. Universidade de Brasília.

PIARC (World Road Association). 2013. Quiet Pavement Technologies. Technical Committee D.2 Road Pavements.

Rasmussen, R.O.; Bernhard, R.J.; Sandberg, U.; MUN, E.P. 2007. The Little Book of Quieter Pavements. FHWA, USA. FHWA—IF-08–004. 37 p.

Sandberg, U.; Ejsmont J. 2002. Tyre/road Noise Reference Book. Informex, Kisa, Sweden.

WHO. 2011. Burden of disease from environmental noise—Quantification of healthy life years lost in Europe. World Health Organization. – Regional office for Europe and EC-JRC.

Transport Infrastructure and Systems – Dell'Acqua & Wegman (Eds)
© 2017 Taylor & Francis Group, London, ISBN 978-1-138-03009-1

Re-use in asphalt pavements of fillers from natural stone sawmilling sludge

E. Santagata, O. Baglieri, L. Tsantilis, G. Chiappinelli & P.P. Riviera
Politecnico di Torino, Turin, Italy

ABSTRACT: The experimental investigation reported in the paper focused on the use in asphalt pavements of fillers originated from natural Stone Sawmilling Sludge (SSS). Performance characteristics of bituminous mastics and mixtures prepared with two types of SSS fillers, differing in mineralogical origin of stones and in stone cutting techniques, were compared with those of reference mastics and mixtures containing standard mineral filler. In the case of mastics, the testing program included the evaluation of their rheological properties at different temperatures and loading conditions. In the case of corresponding mixtures, laboratory characterization was based on the evaluation of their workability, volumetric characteristics, stiffness, and rutting potential. Experimental results indicated that SSS fillers can be conveniently used in asphalt pavements yielding satisfactory levels of performance.

1 INTRODUCTION

The construction of flexible pavements requires large volumes of aggregates to be employed in bitumen-bound layers (Zoorob & Suparma 2000), leading to a massive exploitation of natural resources and to significant environmental concerns (Drew et al. 2002). In such a context, re-use of alternative materials such as recycled aggregates allows a serious waste management problem to be solved and simultaneously contributes to reduce the depletion to which available natural materials are exposed.

The manufacturing process of natural ornamental stones (including marble and granite) generates huge quantities of sawmill sludge (Akbulut & Gurer 2007, Ribeiro et al. 2008, Yilmaz et al. 2010). This by-product derives essentially from the cutting phase to which stone is subjected and, once dried, it is considerably fine (Sarkar et al. 2006).

Due to its variable chemical composition and particle size, Stone Sawmilling Sludge (SSS) can be used as aggregate replacement in several construction materials, such as ceramic and clay composites (Acchar et al. 2006, Saboya et al. 2007) and cement concrete (Hwang et al. 2008, Pereira et al. 2008, Topcu et al. 2009, Yilmaz et al. 2010, Corinaldesi et al. 2010, Aliabdo et al. 2014). Furthermore, SSS has also been considered as a possible stabilizer of plastic soil since its use leads to a reduction of plasticity index and optimal moisture content, with a corresponding increase of unit weight and CBR index (Okagbue & Onyeobi 1999, Agrawal & Gupta 2011, Sivrikaya et al. 2014).

The study presented in this paper focused on the performance evaluation of two types of SSS, supplied from two Italian quarries and derived from the cutting of natural stones, added as a substitution of the filler fraction in asphalt concrete mixtures. Such an application has already been documented in literature, although limited information was gathered with respect to potential field performance (Chandra et al. 2002, Akbulut & Gurer 2007, Karasahin & Terzi 2007, Choudary & Chandra 2008). Thus, this specific aspect was addressed in the investigation described in the following, which included rheological tests carried out on mastics in several aging conditions and performance-related tests performed on bituminous mixtures.

2 MATERIALS AND METHODS

2.1 *Bituminous mastics*

Bituminous mastics considered in the experimental investigation were prepared by using two SSS fillers that were supplied from two different Italian quarries, and a standard calcareous filler (FR), used for comparative purposes. The two SSS fillers differed not only in terms of stone mineralogy, but also with respect to employed stone cutting techniques. In particular, one was obtained from diamond sawing (FD), while the other one derived from bucksaw blade operations (FT). Particle size distribution and bulk density (ρ_f) of the three fillers are presented in Tables 1 and 2, respectively.

A single 70/100 penetration grade Bitumen (BB) was used for the preparation of mastics and

Table 1. Particle size distribution of fillers.

Sieve size [mm]	% Passing		
	FT	FD	FR
1.000	100	100	100
0.500	100	96.0	100
0.250	100	90.0	99.0
0.125	97.0	85.0	97.0
0.063	80.0	80.2	89.8

Table 2. Bulk density of fillers.

	FT	FD	FR
ρ_f [g/cm^3]	2.72	2.88	2.72

Table 3. Penetration and softening point of bitumen.

pen$_{25}$ [dmm]	$T_{R\&B}$ [°C]
83	45.6

Table 4. Calculated filler-to-bitumen ratios.

Mastic code	Filler code	$(P_{0.063})_F$ (%)	$(P_{0.063})_M$ (%)	B_M (%)	$R_{F/B}$
MT	FT	80.0	8.0	5.6	1.14
MD	FD	80.2	8.0	5.6	1.14
MR	FR	89.8	8.0	5.6	1.28

mixtures. It was subjected to basic characterization tests for the determination of penetration at 25°C (pen$_{25}$) and softening point temperature ($T_{R\&B}$, with the ring and ball apparatus) according to standards UNI EN 1426 (2007) and UNI EN 1427 (2007), respectively. Results are reported in Table 3.

Bituminous mastics were prepared in the laboratory by adopting a predefined filler-to-bitumen ratio (by weight) ($R_{F/B}$). Target values of such a parameter were calculated by means of Equation 1:

$$R_{F/B} = \frac{(P_{0.063})_M \cdot (P_{0.063})_F}{B_M \cdot 100} \tag{1}$$

where $(P_{0.063})_M$ is the filler fraction dosage adopted in the job mix formula of the bituminous mixtures (set at 8%), $(P_{0.063})_F$ is the percentage of filler fraction passing through 0.063 mm sieve, and B_M is the design bitumen content of the bituminous mixtures (by weight of mineral aggregates).

Input data and filler-to-bitumen ratios obtained from Equation 1 are listed in Table 4, where codes used to identify mastics are also indicated. On account of the negligible differences existing between the calculated $R_{F/B}$ values of the three materials, in the experimental investigation a single value was adopted for all mastics, fixed at 1.2.

Mastics were considered in their original state and after Rolling Thin Film Oven (RTFO) and Pressure Ageing Vessel (PAV) treatments, carried out to simulate short-term and long-term aging conditions, respectively (AASHTO T 240-2009, AASHTO R 28-2009).

Mastics were subjected to the following tests:

• softening point tests (UNI EN 1427–2007), with the ring and ball apparatus, for the determination of filler "stiffening power";

• viscosity tests (UNI EN 13302-2010), with the Brookfield viscometer at temperatures comprised between 125 and 190°C, in order to gather information on filler-bitumen affinity;

• multiple stress creep recovery tests (AASHTO TP 70-2010), with the Dynamic Shear Rheometer (DSR) on short-term aged mastics at high in-service temperatures, for the evaluation of rutting resistance;

• flexural creep tests (AASHTO T313-2010), with the Bending Beam Rheometer (BBR) on long-term aged mastics at low temperatures, for the assessment of thermal cracking resistance.

2.2 Bituminous mixtures

Bituminous mixtures were obtained by combining four mineral aggregate fractions with the same fillers (FT, FD, FR) and Bitumen (BB) employed for the preparation of mastics. Their job mix formula, which was the same for all filler types, is shown in Table 5. Particle size distribution curves of the four aggregate fractions and of the target mixture are displayed in Figure 1.

Preparation of the bituminous mixtures was performed at 165°C by means of a BBMAX80 mixer (certified standard NF P 98-250-1).

The investigation performed on the bituminous mixtures focused on the following performance-related characteristics:

• workability and volumetric parameters determined from the analysis of gyratory compactor densification curves and resulting specimens (UNI EN 12697-31-2004UNI EN 12697-6-2008-Proc. B; UNI EN 12697-5-2010; UNI EN 12697-8-2003);

• master curves of the complex modulus determined from the results of triaxial tests performed

Table 5. Job mix formula of bituminous mixtures.

Aggregate size fraction	Mass percentage
Coarse gravel	10.0
Medium gravel	29.0
Fine gravel	39.0
Sand	14.0
Filler	8.0
Bitumen	5.6

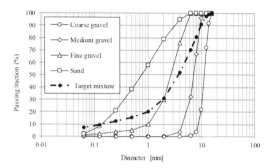

Figure 1. Particle size distribution curves of mineral aggregate fractions and target mixture.

with the asphalt mixture performance tester on cores obtained from gyratory compactor specimens (UNI EN 12697-31-2004; AASHTO TP 79-2010; AASHTO PP61-2010);
- semi-circular bending tests performed on portions of gyratory compactor specimens at an intermediate in-service temperature (20°C) for the evaluation of the potential for crack propagation (UNI EN 12697-31-2004; UNI EN 12697-44-2010);
- wheel tracking tests carried out on roller compactor slabs at a high in-service temperature (60°C) for the assessment of anti-rutting potential (UNI EN 12697-33-2007; UNI EN 12697-22-2004).

3 ANALYSIS AND RESULTS

3.1 Bituminous mastics

3.1.1 Softening point tests
Stiffening power represents an important characteristic that can be taken into account in order to gain an insight into the role played by filler in the mastic phase.

As shown in Table 6, this parameter was evaluated by comparing the softening points of mastics with that of base bitumen.

It can be observed that for all mastics relative increments of the softening point were greater than

5%, thus complying with the minimum requirement that is indicated in Italian specifications in the case of filler-to-bitumen ratios of 1.5 (ANAS S.p.A, 2010). Moreover, it can be noticed that the two SSS fillers exhibited a stiffening power which was significantly higher than that recorded for the reference calcareous filler, with the greatest stiffening potential associated to the use of the filler obtained from bucksaw blade operations (mastic MT).

3.1.2 Viscosity tests
Viscosity tests were performed at several temperatures and shear rates in order to yield information on the affinity between filler and bitumen.

Results are synthetically presented in Figure 2, which shows viscosity values obtained for the base bitumen and the three mastics at all temperatures adopted for testing (125, 135, 150, 165, 175, 190°C) at a single reference value of shear rate, equal to $6.8s^{-1}$.

As expected, mastics showed viscosities which were higher than those recorded for neat bitumen in the entire range of investigated temperatures. Coherently with the outputs of softening point tests, mastics containing the two SSS fillers yielded viscosity values which were significantly higher than those obtained for the mastic prepared with the reference calcareous filler. However, in contrast with the outcomes of softening point tests, the SSS filler which affected binder viscosity to the greatest extent was that obtained from diamond sawing

Table 6. Softening points of mastics ($T_{R\&B}$) and their increments with respect to bitumen ($\Delta T_{R\&B}$).

Mastic code	$T_{R\&B}$ [°C]	$\Delta T_{R\&B}$ [°C]
MT	63.4	17.8
MD	57.9	12.3
MR	54.0	8.4

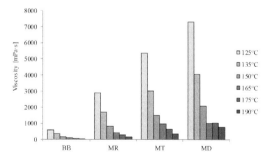

Figure 2. Viscosity values of Base Bitumen (BB) and mastics (MR, MT, MD).

(mastic MD), thus suggesting its greater affinity with bitumen. In any case, it should be emphasized that such a difference is not surprising since the two test procedures differ significantly in terms of strain conditions imposed to the test specimens and explore different temperature ranges.

3.1.3 *Multiple stress creep recovery tests*

Multiple stress creep recovery tests were carried out on RTFO-aged bitumen and mastics in order to obtain information on various aspects of their response under repeated loading, including non-recoverable creep compliance (parameter J_{nr}) and elasticity (R).

Values of non-recoverable creep compliance are reported in Figures 3 and 4, where results obtained at two different levels of stress applied in the creep phase, 0.1 kPa and 3.2 kPa, are presented at the four investigated temperatures (58, 64, 70, 76°C).

The stiffening effect of filler is clearly highlighted by the substantial difference existing between the results obtained for neat bitumen and for the three mastics. Moreover, mastics containing SSS filler (MT and MD) showed lower values of non-recoverable creep compliance with respect to reference mastic MR, regardless of temperature and stress level. This outcome seems to confirm the superior anti-rutting performance of mastics

prepared with SSS fillers, suggested by softening point tests. However, no significant differences were found when comparing the two mastics containing different types of SSS filler.

The same observations can be made when considering elasticity values obtained at 0.1 and 3.2 kPa. As shown in Figures 5 and 6, from the analysis of both $R_{0.1}$ and $R_{3.2}$ parameters, it can be inferred that when compared to the reference mastic, mastics MT and MD exhibited a higher capability to recover damage at all investigated temperatures.

Outcomes of the above described MSCR tests are in line with the results of rheological tests reported by Choudhary and Chadra (2008) on mastics prepared with granite dust.

3.1.4 *Flexural creep tests*

The rheological response of mastics at low temperatures was assessed by carrying out flexural creep tests on PAV-aged materials at -12 and -18°C. In particular, flexural creep stiffness (S) and creep rate (m) were evaluated at a standard loading time of 60 seconds, thus obtaining parameters S_{60} and m_{60} (Figures 7 and 8).

As expected, the presence of filler led to an increase of S_{60} and a decrease of m_{60}, thus indicating a higher flexural stiffness and a lower relaxation capability of mastics with respect to

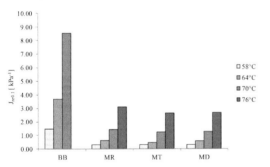

Figure 3. Non-recoverable creep compliance at 0.1 kPa ($J_{nr0.1}$).

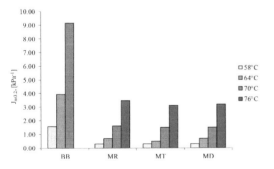

Figure 4. Non-recoverable creep compliance at 3.2 kPa (Jnr3.2).

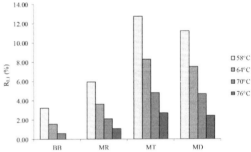

Figure 5. Elasticity values at 0.1 kPa (R0.1).

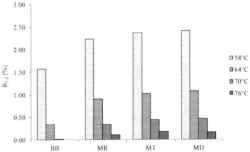

Figure 6. Elasticity values at 3.2 kPa ($R_{3.2}$).

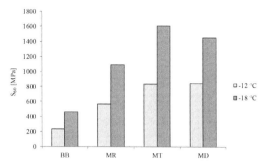

Figure 7. S_{60} values at –12 and –18°C.

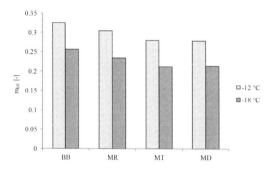

Figure 8. m_{60} values at –12 and –18°C.

bitumen. These effects were slightly more pronounced in the case of SSS fillers, thus suggesting that their use can lead to a reduction of thermal cracking resistance.

3.2 Bituminous mixtures

3.2.1 Workability and volumetric characteristics
In order to gather information on the workability of bituminous mixtures, cylindrical specimens with a diameter of 150 mm were prepared by means of a gyratory compactor. A summary of the main compaction parameters is presented in Table 7.

The effect of filler type on the compaction properties of the mixtures was evaluated by referring to densification curves that were modelled according to the following Equation (2):

$$C = C_1 + k \cdot \log\left(N_x\right) \qquad (2)$$

where C is the compaction level reached after N_x gyrations, C_1 is the initial value of compaction (self-compaction) and k is the slope of the curve, that is related to the workability of the mixture.

Results collected for the three mixtures prepared with fillers FT, FD and FR, marked in the study as CT, CD and CR, respectively, are synthesized in Table 8, where each value is the average of three runs.

Table 7. Compaction parameters.

Compaction parameter	Reference value
Mold diameter	150 mm
Angle of inclination	1.25°
Speed of rotation	30 rpm
Vertical pressure	600 kPa
Mass of mixture into mold	4440.8 g
Number of gyrations	100
Temperature	155°C

Table 8. Workability parameters.

Parameter	CT	CD	CR
C_1 (%)	76.4	76.6	76.4
K	9.2	9.5	9.0

Obtained results clearly indicate that the three materials exhibit a quite similar behavior in terms of both self-compaction and workability. With respect to workability, only a slight increase of k can be observed for mixtures CT and CD, in contrast with results obtained on mastics in terms of stiffening power and viscosity.

Volumetric tests were performed to determine maximum density ρ_{mc} on loose materials, bulk density ρ (saturated surface dry) of compacted specimens, and void characteristics at 100 and 10 gyrations, V_{100} and V_{10}, respectively. Average results of three determinations are listed in Table 9.

Data listed in Table 9 indicate that the influence of filler type on the final volumetric properties of mixtures is quite limited. However, it can be noticed that, in agreement with the information on workability, the mixture containing the reference filler exhibited the highest V_{100} value and the lowest V_{10} value.

3.2.2 Complex modulus master curves
Mechanical characteristics of bituminous mixtures were assessed via dynamic tests performed in the triaxial configuration by means of the Asphalt Mixture Performance Tester (AMPT). For the determination of the complex modulus (E*), tests were carried out by applying a sinusoidal load at frequencies comprised between 0.01 and 25 Hz at three test temperatures (4, 20 and 40°C).

Test specimens were prepared by means of the gyratory compactor (diameter 150 mm, height 170 mm) and then cored and cut by sawing to obtain diameters and heights of 100 and 150 mm, respectively. The mass of the mixture used for compaction was specifically selected to obtain geometrical percentages of voids equal to those of the specimens compacted at 100 gyrations (Table 9).

Master curves of the complex modulus (E*) versus reduced frequency (f_r) were built at the reference temperature of 20°C by modelling raw data according to the Hirsch model (Christensen et al., 2003). Results obtained for the three mixtures are presented in Figure 9.

From the analysis of master curves obtained for mixtures CT, CD and CR it can be observed that in the high frequency (i.e. low temperature) range the three curves follow the same trend, thus revealing a negligible influence of filler type. However, in the low frequency (i.e. high temperature) range the three mixtures exhibited a different behavior under repeated loading. In particular, mixture CD exhibited the highest complex modulus values, followed by CT and CR. This finding, that can be related to the volumetric peculiarities of each mixture and to the intrinsic characteristics of their mastics, is in good agreement with the outcomes of multiple stress creep recovery tests. Hence, obtained master curves suggest that the two bituminous mixtures containing SSS fillers can provide better anti-rutting performance with respect to the reference material.

3.2.3 Semi-circular bending tests

Semi-Circular Bending (SCB) tests were carried out at 20°C to assess the potential for crack propagation of bituminous mixtures.

A half cylinder test piece (diameter 150 mm, height 50 mm) with a center notch of 10 mm was loaded in the three-point bending configuration by imposing a constant deformation rate of 5 mm/min until failure. Average values (from six replicates) of strain at maximum force (ε_{max}) and

fracture toughness (K_{IC}) determined from experimental data are summarized in Table 10.

Fracture toughness values indicate that the mixture containing the calcareous reference filler exhibited the highest resistance to crack propagation. By contrast, values of the strain at maximum force seem to indicate that materials CT and CD showed a more ductile behavior than reference mixture CR. This last result is consistent with those of fatigue tests performed by Chandra et al. (2002) on bituminous mixtures containing marble dust.

3.2.4 Wheel tracking tests

The resistance of bituminous mixtures to permanent deformation under loading was evaluated by means of wheel tracking tests performed at 60°C.

Rectangular slabs were produced with a roller compactor by imposing a target geometric density equal to the geometric density obtained at 100 gyrations on cylindrical specimens.

Results of wheel tracking tests are graphically presented in Figure 10, where rut depth (ε_r), expressed in percentage, is plotted versus the number of wheel passes (N). Table 11 contains the average values (from 2 repetitions) of the maximum permanent deformation (ε_{pmax}) and of the slope of rutting curves ($\Delta P_r / \Delta \log N$).

Slope values of rutting curves shown in Table 11 suggest that mixtures containing the reference calcareous filler (CR) and the filler derived from bucksaw blade operations (CT) exhibit a better anti-rutting behavior. Moreover, it can be observed that the two mixtures containing SSS filler lead to the highest values of maximum permanent deformation. These findings are in partial

Table 9. Volumetric properties of mixtures.

	CT	CD	CR
ρ_{mc} [kg/m³]	2550	2538	2577
ρ [kg/m³]	2414	2425	2527
V_{100} (%)	5.3	4.5	5.8
V_{10} (%)	14.4	14.0	13.5

Table 10. Results of SCB tests.

	CT	CD	CR
ε_{max} (%)	2.10	1.93	1.83
K_{IC} [N/mm³/²]	10.3	10.2	13.1

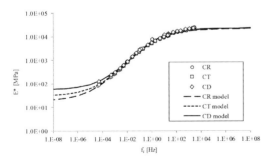

Figure 9. Master curves of the complex modulus E*.

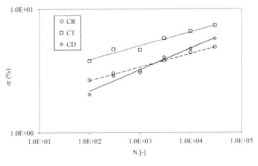

Figure 10. Results of wheel tracking tests.

Table 11. Permanent deformation parameters derived from wheel tracking tests.

	CT	CD	CR
$\Delta P_i/\Delta logN$ [-]	0.11	0.18	0.11
ε_{pmax} (%)	7.34	5.76	4.93

disagreement with the outcomes of multiple stress creep recovery tests performed on mastics and of dynamic tests carried out in the triaxial configuration on mixtures. However, it should be considered that notwithstanding their simulative character, wheel tracking tests are typically associated to a greater dispersion of results. These may be significantly affected, especially in terms of maximum deformation, by local biasing effects related to the initial contact between test slab and loading wheel.

4 CONCLUSIONS

Based on the experimental results presented in this paper, it can be concluded that fillers retrieved from Stone Sawmilling Sludge (SSS), obtained from both diamond and bucksaw blade operations, can be used for the preparation of bituminous mixtures for road paving applications.

With respect to their anti-rutting potential, it was found that the use of SSS fillers can provide improvements in the performance properties of mixtures at high temperatures, as indicated by several performance-related parameters analyzed at both the mastic and mixture scale. On the other hand, the use of these alternative fillers may not provide beneficial effects with respect to the resistance to crack-related distresses. Thus, use of SSS fillers should be carefully considered depending upon environmental and traffic scenarios.

Further research should be performed in order to validate the conclusions of this preliminary investigation by taking into account a wider array of base materials.

ACKNOWLEDGEMENTS

The study reported in this paper was funded by the CCIAA (Chamber of Commerce, Industry, Crafts and Agriculture) of Verbano Cusio Ossola.

REFERENCES

AASHTO PP 61–2010. Standard practice for developing dynamic modulus master curves for Hot Mix Asphalt (HMA) using the Asphalt Mixture Performance Tester (AMPT).

AASHTO R 28-2009. Accelerated aging of asphalt binder using a pressurized aging vessel.

AASHTO T 240-2009. Effect of heat and air on a moving film of asphalt binder (Rolling Thin Film Oven Test).

AASHTO T 313-2010. Determining the flexural creep stiffness of asphalt binder using the Bending Beam Rheometer (BBR).

AASHTO TP 70–2010. Multiple Stress Creep Recovery (MSCR) test of asphalt binder using a dynamic shear rheometer.

AASHTO TP 79–2010. Standard method of test for determining the dynamic modulus and flow number for asphalt mixtures using the Asphalt Mixture Performance Tester (AMPT).

Acchar W., Vieira F.A., Hotza D. 2006. Effect of marble and granite sludge in clay materials. Materials Science and Engineering A 419: 306–309.

Agrawal V. & Gupta M. 2011. Expansive soils stabilization using marble dust. International Journal of Earth Sciences and Engineering 4: 59–66.

Akbulut H. & Gurer C. 2007. Use of aggregates produced from marble quarry waste in asphalt pavements. Building and Environment 42: 1921–1930.

Aliabdo A.A. Elmoaty. A.E.M.A., Auda E.M. 2014. Re-use of waste marble dust in the production of cement and concrete. Construction and Building Material 50: 28–41.

ANAS S.p.A. 2010. Capitolato special d'appalto—Norme tecniche.

Chandra S., Kumar P., Feyissa B.A. 2002. Use of marble dust in road construction. Road Materials and Pavement Design 3(3): 317–330.

Choudhary R. & Chadra S. 2008. Granite and marble dusts as filler in asphalt concrete. Proceedings of the First International Conference on Transport Infrastructures, Beijing (China), 24–26 April 2008.

Christensen, D.W., Pellinen T.K. & Bonaquist R.F. 2003. Hirsch model for estimating the modulus of asphalt concrete. Journal of the Association of Asphalt Paving Technologists, vol 72, pp 97–121.

Corinaldesi V., Moriconi G., Naik T.R. 2010. Characterization of marble powder for its use in mortar and concrete. Construction and Building Material 24: 113–117.

Drew L.J., Langer W.H., Sach J.S. 2002. Environmentalism and natural aggregate mining. Natural Resources Research 11(1): 19–29.

Hwang E.H., Ko Y.S., Jeon J.K. 2008. Effect of polymer cement modifiers on mechanical and physical properties of polymer-modified mortar using recycled artificial marble waste fine aggregate. Journal of Industrial and Engineering Chemistry 14: 265–271.

Karasahin M. & Terzi M. 2007. Evaluation of marble waste dust in the mixture of asphaltic concrete. Construction and Building Material 21: 616–620.

Okagbue C.O. & Onyeobi T.U.S. 1999. Potential of marble dust to stabilise red tropical soils for road construction. Engineering Geology 53: 371–380.

Pereira F.R., Ball R.J., Rocha J., Labrincha J.A., Allen G.C. 2008. New waste based clinkers: belite and lime formulations. Cement and Concrete Research 38: 511–521.

Ribeiro R.C.C., Correia J.C.G., Seidl P.R. 2008. Use of aggregates produced from marble and granite quarry waste in asphalt pavements: a form of clean technology. Proceedings of the 2008 global symposium on recycling, waste treatment and clean technology, Cancun (Mexico), 12–15 October 2008.

Saboya F., Xavier G.C., Alexandre J. 2007. The use of the powder marble by-product to enhance the properties of brick ceramic. Construction and Building Material 21: 1950–1960.

Sarkar R., Das S.K., Mandal P.K., Maiti H.S. 2006. Phase and microstructure evolution during hydrothermal solidification of clay-quartz mixture with marble dust source of reactive lime. Journal of the European Ceramic Society 26: 297–304.

Sernas O., Strumskys M., Skrodenis D. 2014. Possibility of use of granite fines in asphalt pavements. Proceedings of the 9th International Conference "Environmental Engineering", Vilnius (Lithuania), 22–23 May 2014.

Sivrikaya O., Kiyildi K.R., Karaca Z. 2014. Recycling waste from natural stone processing plants to stabiliseclayey soil. Environmental Earth Sciences 71: 4397–4407.

Topcu I.B., Bilir T., Uygunoglu T. 2009. Effect of waste marble dust content as filler on properties of self-compacting concrete. Construction and Building Material 23: 1947–1953.

UNI EN 12697-22-2004. Bituminous mixtures—Test methods for hot mix asphalt—Part 22: Wheel tracking.

UNI EN 12697-31-2004. Bituminous mixtures—Test methods for hot mix asphalt—Part 31: Specimen preparation by gyratory compactor.

UNI EN 12697-33-2007. Bituminous mixtures—Test methods for hot mix asphalt—Part 33: Specimen prepared by roller compactor.

UNI EN 12697-44-2010. Bituminous mixtures—Test methods for hot mix asphalt—Part 44: Crack propagation by semi-circular bending test.

UNI EN 12697-5-2010. Bituminous mixtures—Test methods for hot mix asphalt—Part 5: Determination of the maximum density.

UNI EN 12697–6-2008 Bituminous mixtures—Test methods for hot mix asphalt—Part 6: Determination of bulk density of bituminous specimens.

UNI EN 12697-8-2003. Bituminous mixtures—Test methods for hot mix asphalt—Part 8: Determination of void characteristics of bituminous specimens.

UNI EN 13302-2010. Bitumen and bituminous binders—Determination of dynamic viscosity of bituminous binder using a rotating spindle apparatus. Yilmaz H., Guru M., Dayi M., Tekin I. 2010. Utilization of waste marble dust as an additive in cement production. Material and Design 31: 4039–4042.

UNI EN 1426-2007. Bitumen and bituminous binders—Determination of needle penetration.

UNI EN 1427-2007. Bitumen and bituminous binders—Determination of the softening point—Ring and Ball method.

Zoorob S.E. & Suparma L.B. 2000. Laboratory design and investigation of the properties of continuously graded asphaltic concrete containing recycled plastics aggregate replacement (Plastiphalt). Cement and Concrete Composites 22(4): 233–242.Clayey Soil", Environmental Earth Sciences 71, 4397–4407, 2014.

Transport Infrastructure and Systems – Dell'Acqua & Wegman (Eds)
© 2017 Taylor & Francis Group, London, ISBN 978-1-138-03009-1

Risk compensation in a changing road environment

A. Oikonomou, P. Tafidis, P. Kyriakidis, S. Basbas & I. Politis
Aristotle University of Thessaloniki, Thessaloniki, Greece

ABSTRACT: The objective of this paper is to identify risk compensation effect i.e. the tendency of drivers to adjust their behavior in changing conditions of the road environment. For this reason a questionnaire survey was conducted in order to find out whether the effect takes place as a result of changes in the road safety levels and to address any possible connection between aggressive driving and driver behavioral adaptation. The survey was conducted after the beginning of road works in Egnatias' Street in Thessaloniki. The analysis revealed that males comprise the majority of drivers not to adapt their driving behavior while being the more aggressive at the same time.

1 INTRODUCTION

According to World Health Organization (2009) road accidents is one of the biggest problems globally and the main reason for the death of children aged between 10 and 19 (WHO 2008). The urge to deal with the problem in a more sufficient manner has led to the study of drivers' psychological aspects, that is, research and interpretation of their driving behavior. Risk compensation, the tendency of drivers to adjust their driving behavior when they perceive changes in the road environment, the vehicle or themselves, is among the most popular explanations about drivers' behavior.

The notion of risk compensation was introduced in early 70's and was highlighted as one of the most prominent parameters to affect driver behavior (Summala 1996). Lave and Weber (1970) and Peltzman (1975) proposed risk compensation as an answer to the technological approach in road safety research. Peltzman (1975) introduced the term as the drivers' tendency to take higher risks whenever safety conditions improve, the result being having more accidents despite the improvements in road safety. Naatanen and Summala (1974) pointed out that when it comes to the driver safety it is not skill but the driver's preferred driving pace that matters. Risk compensation as a term is getting more common to indicate changes in driver's behavior towards road safety improvements (Levym & Miller 2000).

A confirmation of risk compensation theory comes from a study from McCarthy (1986), where it is demonstrated that the drivers' decision to use a safety belt resulted in them undertaking higher risks while driving. Singh and Thayer (1992) found out that the risk compensation effect is more likely to occur to drivers that are not against getting exposed to higher levels of risk. Traynor (1993) concluded that drivers react to changes in safety levels by a variable which measures their aggressiveness in driving and another one dependent on dangerous driving situations.

Aggressive driving behavior is a key factor in reducing road safety. Aggressive driving is usually characterized by the influence of disorders and leads the driver to impose its own desired level of risk on others. The three main categories of emotional disorders which affect driving behavior according to James and Nahl (2000) are impatience and inattentiveness, power struggle, recklessness and road rage.

Typical traffic violations by aggressive drivers are: red and yellow light running, ignoring stop signs, excessive lane changing, speeding, tailgating, blocking intersections, driving while drunk and bad behavior.

According to Vaa (2007) humans like every other living organism display the tendency to adapt to the constant environmental changes. Road environment is a special case of an environment that demands the driver's constant adaptation so accidents are avoided. Risk compensation effect also represents a special case of driver's behavioral adaptation and concerns involuntary decisions made by the drivers driving an ABS equipped car in situations where the distances between vehicles are getting shorter (Sagberg et. al 1997). This process is risk compensation because despite the introduction of a system that reduces road risk, drivers compensate by choosing to drive faster or closer to the car ahead.

Studying the effectiveness of the enforcement in the use of seat belts in New Jersey, Asch et al. (1991) noticed that while road fatalities were reduced by 18.9%, the overall number of road accidents increased by 16.5%. Assum et al. (199), in their study about risk compensation in regard with road lighting, found out that when road lighting is sufficient drivers tend to drive faster, sometimes even faster than during daylight, while when lighting is insufficient or non-existent they tend to driver more carefully.

The direct observation and accurate estimate of risk compensation is difficult due to the limited capacity to instantly monitor drivers' actions when they perceive changes on road safety levels. In this study risk compensation is detected indirectly, through the actions drivers stated they perform in a changing road environment. The approach is to compare drivers' stated behavior before and after the changes and to examine possible factors affecting the results. Finally, it is important to highlight that risk is not the only parameter affecting driving behavior, although its better observation and understanding could lead to a more efficient implementation of road safety measures.

2 METHODOLOGY

2.1 Questionnaire survey

A questionnaire survey was conducted in order to study whether risk compensation effect manifests as a result of changes in road safety levels and in order to investigate the relation between aggressive driving and drivers' behavioral adaptation. The area of the study was Egnatias Street in Thessaloniki, Greece, where the subway is being constructed. The street is one of the major streets of the city's road network and an area of mixed land uses like residence, commerce, city services, urban green etc. The reason of this choice was the observed worsening road safety levels that increase accident risk as a side effect of the ongoing works. Drivers were interviewed in two open-air parking sites nearby (Ancient Agora Square, Fillipou Street), securing that the drivers' majority were frequent users of Egnatias Street.

The questionnaire comprises of three sections. In the first section there are general question items regarding the driver profile and trip habits. These are gender, age group, overall driving years, driving days per week, driving hours per day and most usual trip purpose. The second section concerns the adaptation of drivers' behavior. The questionnaire items are about how much affected were the drivers by the special traffic conditions the ongoing works brought about. In particular lane curves, limited visibility at intersections, locally narrowed street lanes, increased incidents of unruly pedestrian crossing behavior, fragmented bus lanes and sporadic traffic pole roadways separation. The

third section concerns the detection of drivers who adopt an aggressive stance when they deal with traffic conditions that occur as a result of the ongoing works. In particular whether they are speeding at the affected area, yellow light running at intersections with limited visibility, their stance towards unruly pedestrian behavior, maneuvering in narrowing road sections, entering congested intersections while vehicles ahead are already trapped and their stance towards buses that are getting to move again after a stop. The questionnaire survey consisted of one hundred (100) interviews. The interviewees were drivers that had just parked their cars or were about to leave or waiting inside their cars at the parking sites. The questionnaire items of the survey were coded into specific variables (Table 1 and 2).

2.2 Descriptive and inferential statistics

Variables of the survey were processed through IBM SPSS 21. Table 3 and 4 presents descriptive statistics for the variables along with normality tests for skewness and kurtosis. In a reliability test for internal consistency the group of variables concerning drivers' behavioral adaptation has a Cronbach's Alpha of .825, while the group of variables concerning aggressive driving .633. In table 2 it is shown that the sample consists mostly by males. Most prevalent age group is 25–54. On average they drive 5 to 6 days per week, 2.5 hours a day with cars of 1468 cc. Most usual trip purpose is commuting. We can notice the variable AggrDr1 for speeding through the works-affected area has a

Table 1. Variables.

Variables	Description
Gender	Interviewee gender
Age Group	Interviewee age
DrivYrs	Interviewee driving years
DaysWeek	Days per week driving
HrsDay	Hours per day driving
carCC	Engine displacement in cc3
TripPurpose	Interviewee trip purpose
Adapt1	Adaptation due to lane curves
Adapt2	Adaptation due to limited visibility
Adapt3	Adaptation due to narrower traffic lanes
Adapt4	Adaptation due to pedestrian crossings
Adapt5	Adaptation due to fragmented bus lanes
Adapt6	Adaptation due to lane poles
AggrDr1	Speeding through the affected area
AggrDr2	Behavior at intersections
AggrDr3	Behavior towards pedestrian crossings
AggrDr4	Maneuvering in narrowing traffic lanes
AggrDr5	Behavior at congested intersections
AggrDr6	Behavior towards buses

Table 2. Scale of measure.

Variables	Scale of measure
Gender	0: Male 1: Female
Age Group	0: 18–24 1: 25–54, 2: 55–64 3: 65+
DrivYrs	0: 0–5 1: 6+
DaysWeek	Open type
HrsDay	Open type
carCC	Open type
TripPurpose	0 = Commuting 1 = Professional 2 = Education, 3 = Personal, 4 = Recreational 5 = Shopping 6 = Other
Adapt1	0: Permanent adaptation 1: Non-permanent adaptation
Adapt2	0: Permanent adaptation 1: Non-permanent adaptation
Adapt3	0: Permanent adaptation 1: Non-permanent adaptation
Adapt4	0: Permanent adaptation 1: Non-permanent adaptation
Adapt5	0: Permanent adaptation 1: Non-permanent adaptation
Adapt6	0: Permanent adaptation 1: Non-permanent adaptation
AggrDr1	0: Non-aggressive driver 1: Aggressive driver
AggrDr2	0: Non-aggressive driver 1: Aggressive driver
AggrDr3	0: Non-aggressive driver 1: Aggressive driver
AggrDr4	0: Non-aggressive driver 1: Aggressive driver
AggrDr5	0: Non-aggressive driver 1: Aggressive driver
AggrDr6	0: Non-aggressive driver 1: Aggressive driver

Table 3. Descriptive statistics.

Variables	Σ	min	max	Mean	StdDev
Gender	100	0	1	0,30	0,461
Age Group	100	0	3	0,90	0,628
DrivYrs	100	0	1	0,63	0,485
DaysWeek	100	1	7	5,69	1,710
HrsDay	100	0,5	8	2,30	1,573
CarCC	100	1000	2500	1468	306,456
TripPurpose	100	0	6	1,33	1,589
Adapt1	100	0	1	0,50	0,503
Adapt2	100	0	1	0,27	0,446
Adapt3	100	0	1	0,52	0,502
Adapt4	100	0	1	0,29	0,456
Adapt5	100	0	1	0,44	0,499
Adapt6	100	0	1	0,54	0,501
AggrDr1	100	0	1	0,04	0,199
AggrDr2	100	0	1	0,20	0,401
AggrDr3	100	0	1	0,70	0,261
AggrDr4	100	0	1	0,27	0,446
AggrDr5	100	0	1	0,26	0,440
AggrDr6	100	0	1	0,39	0,491

Table 4. Normality tests.

Variables	Skewness		Kurtosis	
	Stat.	Std. Error	Stat.	Std. Error
Gender	0,886	0,241	−1,240	0,478
Age Group	0,825	0,241	2,548	0,478
DrivYrs	−0,547	0,241	−1,736	0,478
DaysWeek	−1,223	0,241	0,534	0,478
HrsDay	1996	0,241	4,285	0,478
carCC	0,634	0,241	0,053	0,478
TripPurpose	0,718	0,241	−1,070	0,478
Adapt1	0,000	0,241	−2,041	0,478
Adapt2	1,052	0,241	−0,912	0,478
Adapt3	−0,081	0,241	−2,034	0,478
Adapt4	0,940	0,241	−1,140	0,478
Adapt5	0,245	0,241	−1,980	0,478
Adapt6	−0,163	0,241	−2,014	0,478
AggrDr1	4,714	0,244	20,641	0,483
AggrDr2	1,541	0,246	0,381	0,488
AggrDr3	3,338	0,246	9,334	0,488
AggrDr4	1,060	0,250	−0,896	0,495
AggrDr5	1,125	0,245	−0,750	0,485
AggrDr6	0,451	0,245	−1,835	0,485

big kurtosis with a value of 20.641. Also AggrDr3 for aggressive driving towards unruly pedestrian crossings has a kurtosis value of 9.334.

2.3 Analysis

Figure 1 presents the relation between gender and driver behavioral adaptation for the six individual changes in the road environment that were included in the study. As it is seen there is a statistically significant difference between males and females, with males appearing not to adjust their driving behavior as a result of the changes in the road environment.

Another parameter that seems to have statistically significant relation with the preservation of driving behavior is the driver experience. Figure 2 shows the differentiation in driver behavioral adaptation (cases 1, 5 and 6) depending on the driver inexperience (up to 5 years of driving) or experience (more than 6 years of driving). Engine displacement seems to have a significant impact in driver behavioral adaptation. In Figure 3 we notice that driver behavioral adaptation due to fragmented bus lanes tends to remain unchanged the more the engine displacement increases. The same

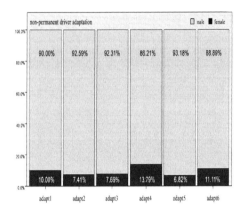

Figure 1.   Driver behavioral adaptation by gender.

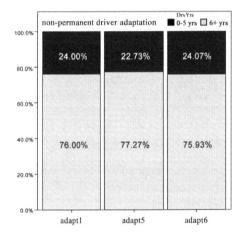

Figure 2.   Driver behavioral adaptation by driving years.

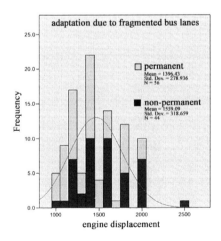

Figure 3.   Driver behavioral adaptation due to fragmented bus lanes.

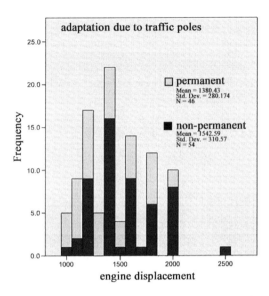

Figure 4.   Driver behavioral adaptation due to traffic
poles.

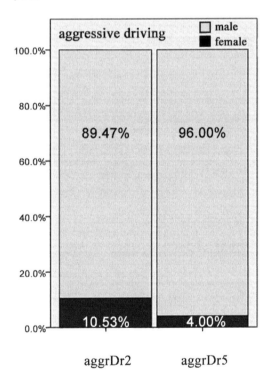

Figure 5.   Aggressive driving by gender.

tendency is present in the case of traffic poles roadways separation (Figure 4).

As it was expected males appear to drive more
aggressively than females. Figure 5 shows the two

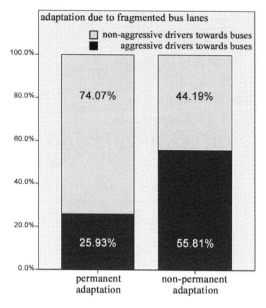

Figure 6. Adaptation and aggressive driving due to bus lanes.

out of the overall six instances in the survey with a differentiation by gender that is statistically significant. These are yellow light running at intersections of limited visibility (AggrDr2) and entering congested intersections while vehicles ahead are jammed (AggrDr5).

Figure 6 shows that the majority of those who adapt their driving behavior when they deal with fragmented bus lanes drive more cautiously towards a bus that is about to rejoin traffic after a stop. Similarly, the majority of drivers who do not adjust their driving when they deal with fragmented bus lanes act more aggressively towards buses that rejoin traffic after a stop.

3 CONCLUSIONS

Risk compensation effect was studied extensively in the past mostly when safety technologies, that were new at the time, were introduced. Subsequently, risk compensation does not get much attention despite being present in drivers either when new technology innovations are introduced or like in this study the road environment changes. Analysis in the present study reveals that risk compensation effect is active

for the drivers in Thessaloniki that need to drive through Egnatias street where the ongoing subway works alter the road conditions frequently. From the analysis it is derived that the effect is weaker in males and in those who drive more aggressively in general. However it is necessary to emphasize that to study risk compensation effectively would require observing the actual driver behavior over a long time, a thing not possible in the present study where the effect was identified

REFERENCES

Asch, P., Levym, D.T., Shea, D. & Bodenhorn, H. 1991. Risk compensation and the effectiveness of safety belt use laws: A case study of New Jersey. *Policy Sciences* 24(2): 181–197.

Assum, T., Bjernskau, T., Fosser, S. & Sagberg, F. Risk compensation - The case of road lighting. *Accident Analysis and Prevention* 31(5): 545–533.

James, L. & Nahl, D. 2000. Aggressive driving is emotionally impaired driving. University of Hawaii.

Lave, L. B. & Weber, W. E. 1970. A Benefit-Cost Analysis of Auto Safety Features. *Applied Economics* 2(4): 265–275.

Levym, D.T. & Miller, T. 2000. Risk Compensation Literature - The Theory and Evidence. *Journal of Crash Prevention and Injury Control* 6(2): 82–89.

McCarthy, P. 1986. Seat belt usage rates: A test of Peltzman's hypothesis. *Accident Analysis and Prevention* 18(5): 425–438.

Naatanen, R. & Summala, H. 1974. A model for the role of motivational factors in drivers' decision-making. *Accident Analysis and Prevention* 6(3–4): 243–261.

Peltzman, S. 1975. The effects of automobile safety regulation. *Journal of Political Economy* 83: 677–725.

Sagberg, F., Fosser, S., Sætermo, I.V. 1997. An investigation of behavioural adaptation to airbags and antilock brakes among taxi drivers. *Accident Analysis and Prevention* 29(3): 293–302.

Singh, H. & Thayer, M. 1992. Impact of seat belt use on driving behavior. *Economic Inquiry* 30(4): 649–658.

Summala, H. 1996. Accident risk and driver behavior. *Safety Science* 22(1–3): 103–117.

Traynor, T. 1993. The Peltzman hypothesis revisited: An isolated evaluation of offsetting driver behavior. *Journal of Risk and Uncertainty* 7(2): 237–247.

Vaa, T. Modelling Driver Behavior on Basis of Emotions and Feelings: Intelligent Transport Systems and Behavioural Adaptations. In C. Cacciabue (ed), *Modelling Driver Behaviour in Automotive Enviroments*: 208–232. London: Springer.

World Health Organization 2009. Global Status Report on Road Safety: Time for Action.

World Health Organization 2008. The Global Burden of Disease: 2004 Update. Geneva.

Transport Infrastructure and Systems – Dell'Acqua & Wegman (Eds)
© *2017 Taylor & Francis Group, London, ISBN 978-1-138-03009-1*

A PSO algorithm for designing 3D highway alignments adopting polynomial solutions

G. Bosurgi, O. Pellegrino & G. Sollazzo
Department of Engineering, University of Messina, Messina, Italy

ABSTRACT: Intelligent optimization algorithms for highway alignments have produced good results so far. However, considering the numerous constraints and factors directly implied in the infrastructure design, the researchers' efforts usually focus only on simplifying the alignment choice, supporting engineers in the design phase. Implementing strategic considerations regarding comfort and safety would be also very important. In this paper, the authors propose a method for designing improved 3D highway alignments using a specific optimization algorithm, based on a Swarm Intelligence technique, adopting an innovative polynomial transition curve as the unique horizontal curvature element, called PPC (Polynomial Parametric Curve). This geometric solution assures higher levels of comforts for users than the traditional ones (clothoid—circular curve—clothoid), because the PPC shows more gradual trends of the main dynamic variables involved while driving, and defines each whole curve through a unique element, simplifying the design procedure. The authors provide technical and operational details for improving a Swarm optimization model through the adoption of the PPC and prove the efficacy of the proposed procedure through a specific significant example.

1 INTRODUCTION

In recent years, various researchers have continually developed intelligent techniques for automatically defining road alignments and optimizing their geometric design. Generally, the researchers' aim is to help designers in the choice of the optimal alignment in a specific study area and in compliance with some of the several constraints. Each specific design requires a very deep examination of the area characteristics and of the related constraints. Then, many innovative tools and algorithms have been produced during the last years, based on various and effective optimization techniques. However, there have been no attempts to define and propose innovative optimization algorithms producing advanced alignments, which can improve comfort levels for drivers. All the existing optimization methods consider traditional solutions in the design of the highway alignments.

Traditionally, the clothoid is assumed as the geometric element for the transition curves. Considering the increasing levels of traffic and vehicle speed, recent studies based on the evaluation of numerical tests and practical experiments have proposed the adoption of different curves, assuring a smoother gradation of the steering manoeuvre in accordance with actual drivers' behaviour. A proper composition of the different geometric elements forming roads, considering transition curves, eliminates abrupt variations of driving behaviour. This also guarantees to handle in a proper way the dynamic variables that influence safety and comfort. For this reason, different kinds of polynomial curves and spirals have been studied and the related results seem very significant.

Therefore, the authors propose a different and improved optimization algorithm for defining 3D road alignments. This procedure not only takes into consideration several kinds of constraints (economical, geometric, environmental, etc.), but draws a road with higher levels of users' comfort, adopting the PPC (Polynomial Parametric Curve) as the unique curve element. The specific mathematical formulation of this curve guarantees a smoother and more gradual variation of the main variables generally considered for checking comfort along the roads (Bosurgi and D'Andrea, 2012; Bosurgi et al., 2015). This paper suggests technical and operational solutions for introducing the PPC in an existing Artificial Intelligence (AI) algorithm, based on the Particle Swarm Optimization (PSO) method, for optimizing highway 3D alignments and minimizing total costs. Then, the algorithm designs alignments made up of a composition of tangent sections and PPC on the horizontal plane, while a succession of tangent sections and parabolic curves describes the vertical profile of the road.

2 ALIGNMENT OPTIMIZATION METHODS

The idea of optimizing the design of road alignments showed a slow but constant development after 1970. The most famous approaches exploit modern AI techniques and, in particular, the Genetic Algorithms (GA), derived from Darwin's theory on genetic evolution (Jong and Schonfeld, 1999; 2003; Jong et al., 2000; Jha and Schonfeld, 2000; 2004; Kang et al., 2012; Kim et al., 2007; Maji and Jha, 2009; Samanta and Jha, 2012). Their outcomes have been very encouraging and represented great advances in this topic. For instance, some papers introduced considerations concerning road interconnections, bridges, or sight distances into the analysis, producing effective algorithms. The Swarm Intelligence (SI) and, in particular, the PSO method resulted as the best alternative to the GA, especially in terms of execution time; various papers have presented the excellent results of the related algorithms (Angulo et al., 2012; Shafahi and Bagherian, 2013; Bosurgi et al., 2013).

Both GA and PSO start from some random initial solutions and, after each cycle, improve and correct them in order to minimize costs (defined through a specific fitness or cost function), and observe the several constraints. The main differences concern the analytical methods adopted for modifying and optimizing the various solutions. Since GAs consider an evolutionary theory in which only the best ones survive and reproduce after each cycle, the solutions of each cycle are rearrangements and combinations of the previous ones. The PSO, on the contrary, moves in the search space the different solutions, as it derives from the observation of flocks of birds moving in the space and searching for food. In detail, the SI is an AI technique introduced by Kennedy and Eberhart (1995), based on the social interaction typical of particular simple groups of animals, that, without a leader and through simple and coordinated behaviours, can perform very complex global tasks. Even if the individuals do not have a real collective awareness, the group is able to achieve exceptional goals.

The hypothesis of the PSO is that each bird (particle) does not know the real food position, but only its distance from it; then, it modifies its speed, step by step, and thus its position, in order to minimize this distance. Similarly, each particle symbolizes a possible solution, moving in the space of the solutions, searching for global optimum. Each particle changes its position according to its own experience and by imitating other particle movements and experience. The position is evaluated through a specific fitness (or cost) function to be maximized (or minimized). The equations to calculate the position x and the speed v after each iteration are very simple:

$$x_{ij}(t+1) = x_{ij}(t) + v_{ij}(t+1) \tag{1}$$

$$v_{ij}(t+1) = w_{ij}v_{ij}(t) + c_1 r_1 \left[g_{b,j}(t) - x_{ij}(t) \right] + c_2 r_2 \left[p_{b,ij}(t) - x_{ij}(t) \right] \tag{2}$$

$$v_{ij}(t) = \begin{cases} v_{ij}(t) & if \ v_{ij}(t) \leq \pm V_{\max j} \\ \pm V_{\max j} & if \ v_{ij}(t) > \pm V_{\max j} \end{cases} \tag{3}$$

where i is i-th particle (variable), j the j-th dimension, g_b and p_b are the best positions achieved by the swarm and by the i-th particle respectively, w the inertial coefficient (used to consider the influence of previous speed), c_1 and c_2 are acceleration constants, both r_1 and r_2 are random numbers in [0,1] range, t is the time, and $V_{\max}$ represents the limit value of the speed, for better performance.

Finally, since GA and PSO show different advantages, various papers proposed hybrid methods for obtaining better results and overcoming some operational problems (Eberharth and Shi, 1998; Thangaraj et al., 2001; Juang, 2004; Shafahi and Bagherian 2013). This combination may assure a more extensive exploration of the searching space, typical of the GAs, and, due to the PSO, guarantees a higher calculation speed and improves the model convergence.

3 BRIEF NOTES ABOUT PPCS

The transition curve is an essential element for causing a gradual influence of the radial acceleration in infrastructure alignments, guaranteeing a better visual perception of the curve, and assuring a more realistic gradation of the steering speed according to the driver's behaviour. Traditionally, for improving the comfort for users while driving along curves, the clothoid represents the most used transition curve in the design of road alignments. However, different authors examined the effects of more powerful geometric solutions that, according to their results, may improve the levels of comfort for users along curves (Changping, 2007; Can and Kuscu, 2008; Easa and Mehmood, 2008). Among the various solutions analysed, the application of splines, polynomial curves, and multispirals provided the best outcomes (Sànchez-Reyes and Chacòn, 2003; Habib and Sakai, 2009; Jha et. al, 2010; Kobryń, 2011; Ziatdinova et al., 2012). Fifth degree polynomial curves showed high geometric and computational versatility (Baykal et al., 1997; Tari, 2004; Baykal and Tari, 2005), also in resolving some design cases with very complex geometry. Bosurgi and D'Andrea (2012)

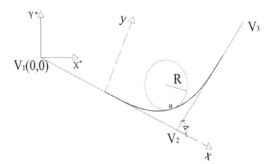

Figure 1. PPC for transition between tangents.

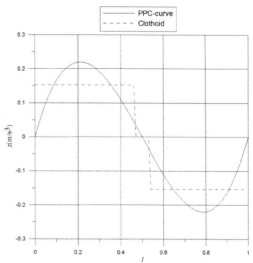

Figure 2. Rate of change of radial acceleration (Bosurgi and D'Andrea, 2012).

introduced a parametric version of a fifth degree polynomial curve and named it "Polynomial Parametric Curve" (Fig. 1). It represents one of the most promising solutions for improving comfort along roads (Bosurgi et al, 2014, 2015).

The expression of the curvature κ of the PPC, represented by a fifth-degree polynomial as evidenced in equation (4), may properly adapt to every different geometric situation by changing the specific boundary conditions (for the curvature itself and for its derivative) that modify the curve characteristics.

$$\kappa(l) = \left[\frac{1}{R}(al^5 + bl^4 + cl^3 + dl^2 + el + f) \right]^{\beta} \; [1/m] \; (4)$$

where $l = s/L$ is the normalized abscissa (L = total length of the curve and s = curvilinear abscissa), R represents the minimum value of the radius of the osculating circle [m], the coefficients a, b, c, d, e, f can be obtained by imposing the boundary conditions for the curvature and its derivative, and β is a parameter useful for modifying the shape and the length of the curve.

Practically, the PPC is defined using two different parameters, called α and β, that influence its geometric shape and, thus, the trends of the dynamic variables involved while driving. The first parameter, α, represents the normalized abscissa of the minimum curvature point and influences the symmetry of the PPC. The other one, β, is a scale parameter: it amplifies the length of the curve and, for a fixed value of the minimum radius, it provides an exponential geometric flexibility of the PPC in the space.

The PPC assures better performances especially in terms of users' comfort and seems to be more appropriate than the clothoid in the design of roads. In order to prove this assessment, some considerations regarding the dynamic aspects of the curve are very explanatory. Regarding the checks on the motion and on the effects over the

driver, numerous numerical examples confirmed the advantages of the PPC, assuming as references the main dynamic variables (rate of change of radial acceleration, steering speed, and roll speed). The particular geometric characteristics of the PPC guarantee the continuity of these variables involved while driving that directly influence the driver's behaviour and the comfort. In general, a more gradual variation of the curvature produces smoother trends for the dynamic variables, and this helps users to drive better and in a more comfortable way. Fig. 2 provides a comparison of the trend of the rate of change of radial acceleration between the PPC and the traditional solution (clothoid-circular curve-clothoid). Although in this particular example the PPC shows a higher peak due to the specific adopted parameters—however lower than the physiological limit -, the related smooth curve provides higher level of comforts to the users rather than the traditional solution. Moreover, the PPC does not imply discontinuities, producing very significant increases in the driving comfort. The other dynamic variables present similar smooth trend, confirming the high quality of this geometric solution.

4 OPTIMIZATION PROBLEM MODELLING

The optimization problem concerns the choice of an optimum backtracking 3D highway alignment linking two known points in the 3D space in compliance with the various considered constraints.

In this context, the optimum solution is the one minimizing the total costs, defined through a proper cost function. In this specific case, the horizontal alignment is defined using the PPC. The position of the vertices defining the alignment may represent the independent variable, according to the research goals. Since three coordinates describe this position in a 3D space, each solution is represented by 3 n_p independent variables to optimize (Eq. 5), where n_p is the number of the internal vertices of the alignment preliminarily fixed.

$$s_i = \left[x_{i1}, y_{i1}, z_{i1}, x_{ik}, y_{ik}, z_{ik}, x_{in_p}, y_{in_p}, z_{in_p} \right] \qquad (5)$$

A Digital Terrain Model (DTM) reproduces the region topography with a very high detail level (Fig. 3). In the study region, it is possible to evidence some areas characterized by various constraints (economic, historical, environmental, etc.). This characterization is handled through two specific rectangular grids easily adaptable to all conditions. The first grid ("Gridval") provides the unit cost for each square meter of road built in a selected cell, due to the economic (and natural or historical) value of the area and to the soil conditions. The second one ("Gridrisk"), that may have different sizes and characteristics, relates to the environmental hazard of the study area (as well as seismic, hydraulic, and geomorphological hazard).

The optimization algorithm can be divided into various steps: 1) data input and numerical check; 2) initial representation of the region DTM and of the study area; 3) solution initialization; 4) geometric analysis of the straight segments; 5) PPC and parabolic curves fitting; 6) solution evaluation and evolution using the swarm core; 7) operator application; 8) stopping criteria checking and final representation.

The initial solutions are very essential to solve the problem properly and, in general, there are various alternatives for choosing them. The easier method is a random definition of the several initial solutions for favouring a whole exploration of the study area; then, a particular algorithm randomly selects the x and y coordinates of the initial alignment vertices. However, this could provide improper solutions with critical loops and intersections between following straights. For overcoming this limitation, other algorithms can be adopted for performing this task, such as specific methods for selecting non back-tracking alignments (Jong, 1998; Bosurgi et al., 2013).

Regarding the vertical alignment, the algorithm chooses the z coordinate of each vertex. Since an appropriate query of the DTM provides the ground elevation for each vertex, the algorithm slightly modifies these values, in order not only to maintain as close as possible the road and the ground elevations, but also to define a smooth and correct vertical profile, in compliance with the adopted road standards. For simplifying the analysis, the vertices of the horizontal path are vertices of the vertical profile also. Concerning the radius design, the authors adopted particular methods, determining random radius values within a given range for both alignments. The aim of assuring the geometric continuity of the alignment significantly conditions the choice of the radii. In detail, in order not to induce intersections between the curves, maintaining also if needed a certain minimum straight segment, the algorithm must check the radii of following curves. Special penalty functions will charge additional costs to improper solutions, in which the difference between following radii is too high or the radius values exceed the Road Standard boundaries.

In general, the cost function to be minimized is the key element of the optimization procedure (Eq. 6). In this case, the total cost of each solution, C_T, is the sum of the real cost, C, and various penalties, P, useful to discard incorrect alignments by increasing the related cost in proportion to the related violation.

$$C_T = C_{len} + C_{loc} + C_E + P_{R\min} + P_R + P_{ret} \\ + P_G + P_{Rv} + P_{dz} + P_E \qquad (6)$$

where C_{len} is the length-dependent costs, C_{loc} the location-dependent costs, C_E the earthwork costs, $P_{R\min}$ the minimum radius of circular curves penalty, P_R the ratio of following circular curves radii penalty, P_{ret} the minimum and maximum tangent segment length penalty, P_G the maximum grade penalty, P_{Rv} the minimum radius of parabolic curves penalty, P_{dz} the maximum height (depth) of fill (cut) sections penalty, and P_E the environmental penalty.

To guarantee a rapid convergence and a larger exploration of the search space, some specific operators, derived from the GA theory, have been introduced in the model. There are two different kinds of operators: stochastic and corrective

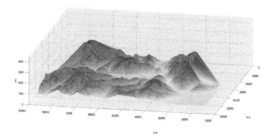

Figure 3. DTM representation of the region.

ones. The stochastic operators facilitate the moving from local optima and the solution optimization, through a more exhaustive exploration of the search space. The corrective operators, on the contrary, modify specific infeasible solutions, characterized by inappropriate values of deviation angles or intersections between tangents.

5 PPC IMPLEMENTATION

Adopting the PPC in the horizontal alignment of the road, each curve is described by only one element and, as a result, design, calculation, and representation are simpler than adopting the traditional composition. Even if there is not an explicit expression for defining both the clothoid and the PPC, considering the latter, the total number of geometric elements decreases to the minimum and, thus, from a computational point of view, the advantages are very clear and important. In the following, some operational and technical solutions to actually adopt the PPC in the PSO algorithm are provided.

5.1 *PPC fitting and design*

This phase involves the choice of the minimum radius of the osculating circle for each PPC, in compliance with comfort and geometric constraints, avoiding intersections between following curve paths and maintaining, if needed, a minimum residual straight. A specific procedure assigns to each radius an initial value, chosen for favouring high service levels for users as the one corresponding to the maximum transversal grade and the maximum design speed, in compliance with the Italian Road Standards (2001).

After the radius values are assigned, it is possible to process the PPCs (considering typical values of the parameters - $\beta = 1.0$ $\alpha = 0.5$ - for simplifying the analysis and reducing the computational time). Each curve is numerically calculated and defined, according to the number of reference points properly chosen. Obviously, a reduction in the number of points describing the PPC decreases the quality of the representation, but increases the performance of the numerical evaluation (Fig. 4).

Then, it is time to check the geometric consistency of the alignment, looking for critical point of intersections between following curves that could happen if the tangents of the curves are too long compared to the available distance between the vertices. If all the residual straights are sufficient, the horizontal alignment is complete. If some straights are shorter than the minimum value, the critical segments are further processed and corrected, by reducing in sequence all the critical radius values.

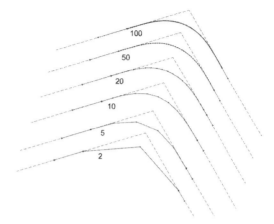

Figure 4. Same PPC drawn for different number of reference points.

Whether the procedure reduces some radii beyond the limit value, the related improper solutions are later penalized, using an appropriate penalty function. In general, for all the penalty functions related to the curves, the authors have assumed the same expression adopted for the previous models with circular curves, since it is easy to assess a certain similarity and correlation between their radius in a traditional approach and the minimum radius of the osculating circle in the PPC.

5.2 *Location costs and environmental penalties*

Evaluating location costs and environmental penalties represents a critical task for the algorithm, because of the specific analytical formulation of the PPC. As said, since there is not an explicit expression describing the relations between the x and y coordinates of the curve points on the horizontal plane, only a point-by-point numerical calculation permits to evaluate and represent the PPC. Each cell of the reference grids has a specific unit value for both economic and hazard classifications. The objective of this sub-step is to estimate the length of road passing through each cell.

Although for straights this could be easily evaluated through a linear system of equations, the PPC formulation does not permit to follow this approach. Then, the authors proposed a different but equally efficient procedure. Since the grid characteristics and the x and y coordinates of the PPC points are known, it is simple to locate the indices of the cell in which each point is. The algorithm counts how many points are located in each cell and the segments that are one less. Moreover, as the curve is practically defined point by point, the distance between two following points along the PPC is constant (Eq. 7). For more clarity,

Table 1. Points and segments for cells according to Fig. 5.

Cell	Points	Segments
i, j – 1	3	2
i, j	3	2
i, j + 1	2	1
i + 1, j + 1	1	0
i+1, j+2	2	1

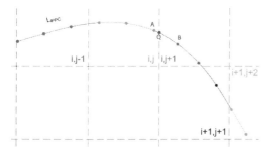

Figure 5. Location costs using PPC.

Table 1 provides the number of points and segment for each cell, according to Fig. 5.

$$L_{sPPC} = \frac{L_{PPC}}{n_{PPC}} \tag{7}$$

where L_{sPPC} is the length of each segment, L_{PPC} is the total length of the PPC and n_{PPC} is the number of the segments.

Regarding the boundary elements (i.e. segments between points located in different cells), the authors have assumed to consider the curve segment as linear. This approximation produces a very little error that may become irrelevant (10^{-3}) considering the whole length of the PPC, especially if the number of segments used for defining it is large enough (>100). In detail, if Q is in the intersection point with the grid, A is the point in cell {i – 1, j}, and B is the point in cell {i, j}, the length of the two segments (AQ for cell {i – 1, j} and QB for cell {i, j}, Fig. 5) can be evaluated through expressions (8) and (9).

$$l_{i,j} = \overline{AQ} = \sqrt{\left(x_Q - x_A\right)^2 + \left(y_Q - y_A\right)^2} \tag{8}$$

$$l_{i-1,j} = \overline{QB} = \sqrt{\left(x_B - x_Q\right)^2 + \left(y_B - y_Q\right)^2} \tag{9}$$

Then, in this way it is possible to estimate how long are the segments passing through each cell and, for a generic cell i, j, the related length of the PPC can be calculated using equation (10).

$$[L_{PPC}]_{i,j} = rl_p + t \cdot L_{sPPC} + rl_f \tag{10}$$

where rl_p is the residual length in the i-th j-th cell of the previous boundary segment, t is the number of segments in the i-th j-th cell (equal to the number of points in the cell minus one), L_{sPPC} is the length of each PPC segment, and rl_f is the residual length in the i-th j-th cell of the following boundary segment.

5.3 Earth-costs

Regarding the earth-costs evaluation, the main complications derive again from the specific analytical formulation of the PPC. In this context, the earthwork volumes are evaluated using the "average-end-area" method. In order to properly evaluate these volumes, it is important to define a method to locate the reference station useful for reproducing the ground profile. Although both the tangent points of the PPCs always constitute station points, there are two possible approaches:

– all the PPC points are assumed as station points;
– if the station points are equally located along the whole alignment, only few selected points of the PPCs are considered as station points.

In the first case, it is easy to evaluate the ground elevation for each station point (Fig. 6a), but the computational costs in the calculation procedure may increase widely. Considering the second scenario, the number of points is limited, but it is necessary to define a specific rule for selecting the points along the PPC. In this study, the authors have applied the second approach. In detail, since after the stations are equally located along the alignment (St*) the related progressive distances are known as well as those of the tangent points, the algorithm can easily select the closest PPC points (St) to the station target point (Fig. 6b). Then, these become the selected station points and the DTM model of the

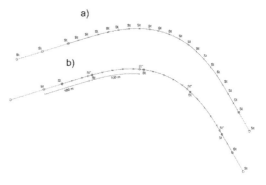

Figure 6. PPC stations for vertical profile calculations.

study area provides the related ground elevations, for calculating the earthwork volumes and costs.

6 NUMERICAL EXAMPLE AND DISCUSSION

For evaluating and practically using the proposed method, the authors coded an original script. The results of a specific example are useful for showing the effectiveness of the entire procedure. The reference maps considered in this example have rectangular shape and measure 4000 m on X side and 3000 m on Y direction. The study area topography is reproduced through a particular DTM. In order to get a proper representation of the region characteristics, the authors defined the DTM using more than 15.000 points with various elevation values (Fig. 3). The reference grids for economic and environmental hazard constraints have the following resolution: the cell size is 175 m in direction X and 120 m in direction Y.

The authors assigned four different unit costs (useful to estimate location-dependent costs, from C1 low to C4 very high) to each cell of the Gridval matrix, in order to define an ideal representation of the study area. Four different hazard levels (from H1 low to H4 very high) describe the environmental hazard map. They are representative of seismic, geomorphologic and flood hazard levels for each cell, and relate to four different unit penalties. The greater the hazard level, the higher the probability that a critical event happens (or its magnitude is greater), the greater the unit cost related to this hazard level must be. Since these examples represent only a test of the algorithm efficient, for characterizing the test area, the authors have defined ideal distribution for location-dependent unit costs and hazard levels.

The elaboration started after all calculation variants (population size: 20, max cycles number: 500, $V_{max} = 500$ m, $V_{maxz} = 10$ m, w = 0.5, $c_1 = 1.5$, $c_2 = 2.5$), unit costs (pavement, maintenance, environmental and earthwork costs), geometric and environmental unit penalties had been specifically defined through appropriate input boxes. The final solution is made up of the coordinates of 4 internal vertices ($n_p = 4$). Moreover, the algorithm analysed 50 sections for describing the vertical profile of the road. The maximum number of cycles is 500, and all the PPCs count 50 building points. In order to underline the model efficiency, the best alignments found by the algorithm at cycles 5, 25, and 500 are presented below: Fig. 7 provides an evolution of the horizontal alignment, while Fig. 8 confronts the vertical profiles for the different cycles.

Although the new algorithm includes the PPC and the related modifications, an observation and

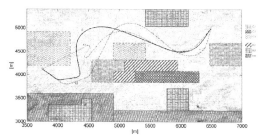

Figure 7. Alignment evolution.

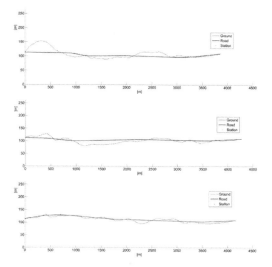

Figure 8. Vertical profile evolution.

comparison of the previous figures proves the correct and effective evolution of the road in the study area. In the first phase of the analysis, the solution is widely incoherent and improper from a technical point of view: Fig. 7a represents an alignment with abrupt and not correct changes of curvature passing also through some of the high cost areas. Furthermore, the selected road does not fit the territorial characteristics, as confirmed by the vertical alignment provided in Fig. 8a, in which the ground and the road profiles are clearly inconsistent, with very high height differences.

Fig. 7b, 7c, 8b and 8c permit to follow the efficacious optimization of the alignment. Step by step, the road is effectively improved and the curve characteristics are more homogeneous. Furthermore, the critical areas are avoided everywhere, and the road elevation is properly defined, in compliance with terrain height and features. The final solution is acceptable and shows an effective optimization of the alignment. In particular, both the horizontal path and the earthworks are widely optimized. The algorithm drew a correct

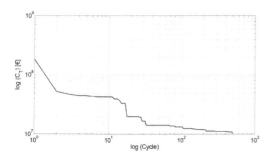

Figure 9. Cost evolution of the best solution on advancing cycles.

alignment, reducing both geometric and economical penalties and reducing the peaks difference between the ground and the road profiles.

Moreover, plotting the total cost evolution of the best solution as function of cycles on a logarithmic scale (Fig. 9) evidences the solution improvement and optimization after each single computational cycle. Cycle after cycle, the algorithm actually modifies the different solutions reducing the cost and improving the quality of the alignment. It should be noticed that, probably, more iterations should have produced a more accurate solution with excellent geometric features. However, since the aim of the paper is to evidence the introduction of the PPC in this kind of optimization method and other proofs of the effectiveness of the PSO approach are provided in literature, the authors stopped only after 500 cycle the optimization procedure. This number of cycles is enough to assure that the technical solutions adopted for handling the PPC in this PSO optimization method are efficient and the alignment optimization is properly performed. The natural geometric characteristics of the PPC assure better performance in terms of comfort for users, assuring a smoother variation of the dynamic variables involved while driving.

7 CONCLUSIONS

In this paper, the authors have proposed an innovative optimization algorithm, based on the PSO method, for designing highway alignments considering the PPC as unique curvature element. The adoption of the PPC guarantees higher levels of comfort for users than the traditional transition curves. In particular, the paper presented various operational solutions for introducing the PPC in the optimization method. In general, the optimization algorithm leads to define an efficient procedure useful for preliminary locating the most appropriate path for a road in a specific study

area. The resulting algorithm proposed in this paper guarantees to not only minimize the related costs and design an alignment in compliance with the main constraints, but also improve the users' comfort along the road, as a direct consequence of the PPC dynamic advantages. Moreover, the PPC permits to solve the curve and define the transition between two straights using only a single element, producing benefits also in terms of design, evaluation, and drawing of the curve.

This procedure, as also proved by the numerical example provided by the authors, is very efficient and can help engineers in the planning phase. The cost function may be further modified and improved, in order to keep into considerations other important factors conditioning the alignment design. Future researches may introduce comfort considerations in the cost function, in order to try to optimize the road characteristics considering also the peak values of the rate of change of radial acceleration. The modification of the vertical curves might be equally interesting, defining 3D alignments increasing the comfort of user both on the horizontal and vertical planes.

REFERENCES

Angulo, E., Castillo, E., Garcia-Ròdenas, R. and Sànchez-Vizcaino, J. (2012). Determining Highway Corridors. Journal of Transportation Engineering, 138, 5, 557–570.

Baykal, O. and Tari, E. (2005). A New Transition Curve with Enhanced Properties. Canadian Journal of Civil Engineering, 32(5), 913–23.

Baykal, O., Tari, E., Coskun, Z. and Sahin, M. (1997). New Transition Curve Joining Two Straight Lines. Journal of Transportation Engineering, 123(5), 337–45.

Bosurgi, G. and D'Andrea, A. (2012). A Polynomial Parametric Curve (PPC-Curve) for the Design of Horizontal Geometry of Highways. Computer-Aided Civil and Infrastructure Engineering, 27, 303–312.

Bosurgi, G., Pellegrino O. and Sollazzo G. (2013). A PSO Highway Alignment Optimization Algorithm Considering Also Environmental Constraints. Advances in Transportation Studies an International Journal, Section B, 31, 63–80.

Bosurgi, G., Pellegrino, O. and Sollazzo G. (2014). "Using Genetic Algorithms for Optimizing the PPC in the Highway Horizontal Alignment Design." Journal of Computing in Civil Engineering, 10.1061/(ASCE) CP.1943–5487.0000452, 04014114. 30 (1), Jan. 2016.

Bosurgi, G., Pellegrino, O. and Sollazzo G. (2015). An Algorithm Based on the PPC (Polynomial Parametric Curve) for Designing Horizontal Highway Alignments. Advances in Transportation Studies an International Journal, Section A, 36, 5–20.

Can, E. and Kuscu, S. (2008). Investigate of Transition Curves with Lateral Change of Acceleration for Highways Horizontal Geometry. Integrating Generations

FIG Working Week 2008, Stockholm, Sweden 14–19 June 2008.

Changping, W. (2007). Polynomial Curve Fitting Method for the Design of Highway Horizontal Alignment. CNKI Journal, 03, 30–2.

Easa, S.M. and Mehmood, A. (2008), Optimizing Design of Highway Horizontal Alignments: New Substantive Safety Approach. Computer-Aided Civil and Infrastructure Engineering, 23(7), 560–73.

Eberharth, R.C. and Shi, Y. (1998). Comparison between Genetic Algorithms and Particle Swarm Optimization. Lecture Notes in Computer Science Volume 1477, 611–616.

Habib, Z. and Sakai, M. (2009). G2 Cubic Transition between Two Circles with Shape Control. Journal of Computational and Applied Mathematics, 223, 133–144.

Italian Road Standards (2001). "Ministerial Decree n. 6792 (11/05/2001) of the Ministry of Infrastructures and Transports. Functional and geometric standards for building roads" Rome (in Italian).

Jha, M.K. and Schonfeld, P. (2000). Integrating Genetic Algorithms and GIS to Optimize Highway Alignments. Transportation Research Record, 1719: 233–240.

Jha, M.K. and Schonfeld, P. (2004). A highway alignment optimization model using geographic information systems. Transportation Research, Part A, 38(6): 455–481.

Jha, M.K., Kumar Karri, G.A. and Kuhn, W. (2010). Selection of 3 D Elements for Different Speeds in the 3D Modelling of Highways. 4th International Symposium on Highway Geometric Design, June 2–5, Valencia, Spain.

Jong, J.-C. (1998). Optimizing highway alignments with genetic algorithms. Ph.D. Dissertation, University of Mariland, College Park.

Jong, J.-C. and Schonfeld, P. (1999). Cost Functions for Optimizing Highway Alignments. Transportation Research Record, 1659: 58–67.

Jong, J.-C. and Schonfeld, P. (2003). An evolutionary model for simultaneously optimizing 3-dimensional highway alignments. Transportation Research, Part B: Methodol., 372: 107–128.

Jong, J.-C., Jha, M.K. and Schonfeld, P. (2000). Preliminary Highway Design with Genetic Algorithms and Geographic Information Systems. Computer-Aided Civil and Infrastructure Engineering, 15(4): 261–271.

Juang, C.F. (2004). A hybrid of Genetic Algoritghm and Particle Swarm Optimization for Recurrent Network Design. IEEE Transactions on systems, man and cybernetics—Part B: Cybernetics, Vol. 34, No. 2, 997–1006.

Kang, M.-W., Jha, M.K. and Schonfeld, P. (2012). Applicability of highway alignment optimization models. Transportation Research, Part C, 21: 257–286.

Kennedy, J. and Eberhart, R.C. (1995). Particle Swarm Optimization. Proceeding of IEEE International Conference on Neural Networks, 1942–1948.

Kim, E., Jha, M.K., Schonfeld, P., and Kim, H.S. (2007). Highway Alignment Optimization Incorporating Bridges and Tunnels. Journal of Transportation Engineering, 133, 2, 71–81.

Kobryń, A. (2011). Polynomial Solutions of Transition Curves. Journal of Surveying Engineering, 137(3), 71–80.

Maji, A. and Jha, M.K. (2009). Multi-objective Highway Alignment Optimization using a Genetic Algorithm. Journal of Advanced Transportation, 43(4), 481–504.

Samanta, S. and Jha, M.K. (2012). Applicability of Genetic and Ant Algorithms in Highway Alignment and Rail Transit Station Location Optimization, International Journal of Operations Research and Information Systems, 3(1), 13–36.

Sànchez-Reyes, J. and Chacòn, J.M. (2003). Polynomial Approximation to Clothoids via S-Power Series. Computation Aided Design, 35, 1305–1313.

Shafahi, Y. and Bagherian, M. (2013). A customized Particle Swarm Method to solve highway alignment optimization problem. Computer-Aided Civil and Infrastructure Engineering, 28: 52–67.

Tari, E. (2004). The New Generation Transition Curves. ARI Bulletin of the Istanbul Technical University, 54(1), 34–41.

Thangaraj, R., Pant, M., Abraham, A. and Bouvry, P. (2001). Particle swarm optimization: Hybridization pespectives and experimental illustrations. Applied Mathematics and computations. Vol. 217, 12, 5208–5226.

Ziatdinova, R., Norimasa, Y. and Kima, T.-W. (2012). Fitting G2 Multispiral Transition Curve Joining Two Straight Lines. Computer-Aided Design, 44, 591–596.

Transport Infrastructure and Systems – Dell'Acqua & Wegman (Eds)
© 2017 Taylor & Francis Group, London, ISBN 978-1-138-03009-1

QEMS. ANAS quality evaluation management system

Eleonora Cesolini & Stefano Oddone
ANAS S.p.A., Rome, Italy

Sonia Gregori
Anas International Enterprise S.p.A., Rome, Italy

ABSTRACT: The ANAS Quality Evaluation Management System (QEMS) is an innovative tool that enables the collection, the analysis and the evaluation of all the information related to road pavement for new construction projects and for maintenance of existing roads. The system provides on-time quality reports which include information about the quality of the activities involved in road construction and maintenance, and data about the performances of the Key Suppliers (KS). Based on this, with the QEMS you can check the quality performance of each activity performed by KS. In addition, the QEMS may be used by each KS, who is allowed accessing only to its own results. This interactive tool is proposed as a support to enhance the quality of road construction by providing a synthetic and on-time analysis of performances, all the results are showed with different colors on geographical maps in order to facilitate the user to read them.

In the QEMS, each project is divided into more audit areas each characterized by a unique pavement design. Then, in each Audit Area, the activities of each KS (e.g., laying, compaction), are audited acquiring selected information (laboratory test data, on site audit data, management quality audit data), defined as parameters; each parameter is evaluated according to defined evaluation criteria based on weights, optimal values and acceptable values. Within the same audit area all the activities are organized into a hierarchical structure, called Activity Tree.

The QEMS is completely configurable and you can simulate more scenarios with various weight factors in order to get reports with different evaluation criteria.

1 INTRODUCTION

The Quality Evaluation Management System (QEMS) is a software tool wholly conceived, designed and developed by ANAS. The tool analyze and evaluate the pavement from the designing phase to the construction and maintenance phase.

The software uses a Geographical interface to show the results of analysis; it intends to be used by all stakeholders involved in pavement design, construction and maintenance (commissioner, auditors, general contractors, supervision consultants, material suppliers and production facilities,…).

The final aim of this interactive tool is to enhance the quality of road construction by:

- providing synthetic and on-time analysis of performances, highlighting any problem and allowing putting in place fast and optimized solutions;
- supporting activities of performance evaluation, allowing each involved KS, to access and to monitor his own quality performances;
- allowing timely put in place of corrective actions, helping in the improvement of the Quality Levels (knowledge allows improvements);
- providing tools for communication between different parties (Public Works Authority, General Contractors, Supervision Consultants, etc.).

 In summary, the main characteristics of QEMS are:

- flexibility: it allows to easily define and modify, over the time, the performance evaluation model (weighting factors, parameters, activities, layers, mixtures, materials, user profiles, etc.), making possible to easily adapt QEMS to new requirements or new technologies;
- accessibility by all the stakeholders involved in the road pavement construction projects: each user accesses only his data;
- friendly interface, with an Intuitive Graphical User Interface, including reports displaying results on geographical maps.

2 KEY CONCEPTS & DEFINITIONS

The analyses of performances are based on the QEMS configuration and on all information acquired during audit activities on pavement construction projects (the results of the audits).

To illustrate the QEMS, it is necessary to introduce some preliminary concepts and definitions, described in the following sections.

2.1 Project

Set of the overall information related to a specific construction or maintenance process of one or more roads. The definition of a Project includes:

• The geographical location of the Project,
• List of the KSs involved in the delivery of the final road.

An example of a geographical location of a Project is displayed in Fig. 1.

2.2 Audit Area

In QEMS Audit Area is a part of the Project containing road sections with the same pavement design. A project may have one or more Audit Areas.

Each Audit Area is composed by one or more Road Sections, with a start and end chainage. Each road section is represented on the map with a polyline (Fig. 2).

2.3 Key Supply Chain Partners (KSs)

Key Supply Chain Partners (KSs) are the subjects responsible for the successfully delivery of the Projects. The Key Supply Chain Partners are classified in categories according their role (Pavement Design Consultant, Contractor, Supervision Consultant, Materials Supplier and Production Facilities, Laboratories, etc.).

2.4 Sources and plants catalogue

Production affects significantly the quality and the performances of all types of mixtures. The engineering properties and behaviors of materials used in the construction of pavement layers influence the overall performance of the pavement structure over its design life. In QEMS also plants have to be defined and they are audited and evaluated to ensure that plants are capable of producing materials according to the physical and engineering properties as specified in the required standards.

2.5 Activities

Activities are the operations necessary to accomplish the successfully deliver of the road construction projects whose phases and processes are carefully audited.

Road pavement constructions are complex sets of multiple processes (or Activities) interconnected referred to as "Supply Chain". The pavement construction (complex multiple processes), is so divided in several single activities that are audited.

The Main Activity (which includes all the activities involved in the project of road Construction) is named: "Overall Road Construction". This Activity includes sub-Activities (carried out by different KSs):

Figure 1. Example of a Project.

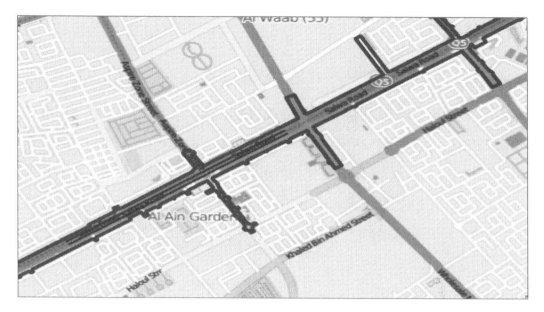

Figure 2. Example of Audit Area.

- Pavement Design
- Quality Management System
- Pavement Construction

These sub-Activities articulate into further sub-sub-Activities, because they are performed by different KSs or at different times.

For example:

- Pavement Design includes sub-activities: Geotechnical Investigations, Subgrade Layer Design, Traffic Design Estimation, etc.;
- Pavement Construction includes a set of sub-activities for each pavement layer: Construction of Surface Layer, Construction of Base Layer, etc.;

For this, the Activities have been structured in a hierarchical tree structure.

2.6 Activity tree template

In QEMS the complex relations between activities are organized into a hierarchical structure defined as Activity Tree.

Development of the three template stop when an activity can be measured through parameters. The evaluation of Parameters could be through laboratory tests or specific audits.

The evaluation of Activities at higher level occurs through the evaluation of activities at lower level.

For instance, the performance evaluation of a HMA Layer depends on:

- Construction: Laying, Compaction, etc.;

- Production: Mixing Temperature, Compaction Temperature, etc.;
- Materials and Mixture Properties.

Therefore, the evaluation of a single layer occurs auditing a set of sub-activities.

Weighting factors are assigned to each activity and each parameter to establish the performance evaluation criteria.

3 CONFIGURATION DATA

3.1 Materials and Pavement Layers

In order to define the pavement structure of each Audit Area, in the QEMS it is necessary to configure: Raw Materials, Material Types, Mixtures and Pavement Layers. For each pavement layer must be specified the materials and mixture used. Therefore, the evaluation of pavements quality needs to go through the quality assessment of the used materials. Follows an example list of layers with their mixtures and materials.

3.2 Parameters

Parameters are the basic items directly audited and used as evaluation criteria to assess the quality of the Activities.

For each Leaf Activity of the Activity Tree Template it is necessary to define the Parameters to evaluate it.

Parameters are for example, the temperature of the asphalt, the thickness of a layers, etc.

397

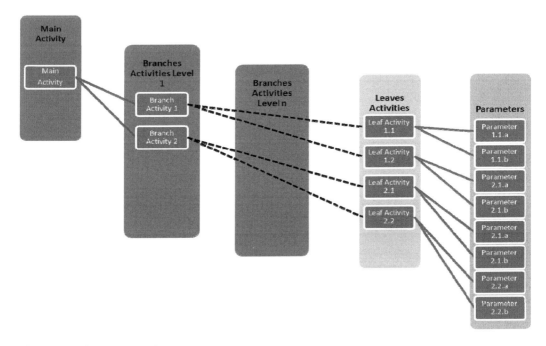

Figure 3. Activity Tree Template.

Table 1. Example of Pavement Layers, Mixtures and Materials.

Pavement Layers	Mixtures	Materials
Asphalt Surface Course (SC)	Hot Mix Asphalt	Neat Asphalt Binder; PMB, Coarse Aggregates, Crushed Aggregates, Fine Aggregates, Filler
Asphalt Base Course (BC)	Hot Mix Asphalt	Neat Asphalt Binder; PMB, Coarse Aggregates, Crushed Aggregates, Fine Aggregates, Filler
Sub base / road base (SB)	Sub base Mixture	Crushed Aggregates, Fine Aggregates, Cement, Filler

In the QEMS a Parameter can be single (eg. the temperature of the HMA during compaction activities) or composed by a group of Sub-Parameters (eg. the Parameter is the grading curve and the sub-Parameters are the percentages of passing for the different diameters).

Some Parameters are related to specific Pavement Layers, so QEMS enables the definition of condition of existence of each Parameter depending on the characteristics of the layer (i.e. Layer, Mixture, Raw Material, and Material Type). These conditions must be compatible with those set for the Leaf Activity containing the Parameter. The Parameters so configured, added only once in the Activity Tree Template, will be automatically available on the specific Activity Tree of each Audit Area, depending on the pavement structure of the Audit Area.

4 AUDITS

In each Project, Activities are audited to assess their quality upon specified criteria, therefore, in the QEMS the audit results coincides with Parameters results. Different types of Parameters require different types of audit with different type of results. In the QEMS, three types of Audit are possible:

- Laboratory Testing Results Audit—Laboratory test, to assess materials properties;
- On Site Audit—Audits performed by auditors on construction as well as production sites, to assess Construction Practice and Production Facilities.
- Documental Audit—Audits of documents, manuals and procedures adopted by the KSs.

5 EVALUATION

5.1 *Normalized Scores and Quality Indicators*

In QEMS the evaluation process begins with the evaluation of Parameters, all the values of them are converted in Normalized Scores. Normalized Scores are real numbers from 0 to 100 indicating the quality level of the result of the Parameter (100 = max quality).

Combining the results obtained for a Parameter in a defined time interval (i.e. combining more Normalized Scores using specific algorithms), we obtain a result that expresses the quality of the Parameter in the defined time interval; such result is named: Parameter Quality Indicator (PQI).

Combining the PQIs of the Parameters of a Leaf Activity, we obtain a result that expresses the quality of the Leaf Activity in the defined time interval; such result is named: Activity Quality Indicator (AQI).

Both PQIs and AQIs (as well as the Normalized Scores) are expressed with real numbers from 0 to 100 indicating the quality level (100 = max quality); this criterion provides values combinable and comparable one each other.

5.2 Evaluation of the Normalized Scores

The criterion for the evaluation of the Normalized Scores from Parameter Results depends on the typology of Parameter input data that may be:

- Selection from a data list (e.g. on-site and documental audits),
- Input of a Numerical value (e.g. lab test results).

If the Parameter input data is based on the selection from a data list, the criterion is based on the direct assignment of a Normalized Scores to each item of the data list, an example of such criterion is shown in Table 2. Example of criterion to assign Normalized Scores to Data List type Parameters.

If the Parameter input data is a numerical value, the criterion for the evaluation of the Normalized Scores is based on the definition for each Parameter (using the QEMS configuration functions), of threshold limits. In this case, two ranges of Normalized Scores are considered:

- Optimal Range—Parameter within standard threshold limits (Score 100);
- Acceptable Range—Parameter within tolerance limits; in this case, the tolerance limits depend

Table 2. Example of criterion to assign Normalized Scores to Data List type Parameters.

Parameter's Value	Normalized Score
Yes	100
Observation—Acceptable	75
Observation—Partial Satisfactory	50
Observation—Poor	25
No	0

on testing procedures and accuracy of the test method; where tolerance was not clearly specified, a Coefficient of 5% was generally used to set tolerance limits.

Within the Optimal Range, the Normalized Score is 100 while within the Acceptability Range the Normalized Score varies linearly from 100 to 0. Outside of the Optimal and the Acceptability Range, the Normalized Score is zero.

5.3 Evaluation of Activities in the Activity Tree template

The main activity "Overall Road Construction" is organized in Sub Activities providing braches Activities at different intermediate levels and Leaf Activities at the lowest level of the hierarchical structure. Beside Activities layout, in the computational model, the definition of Activities requires also to define 2 characteristics:

- Weighting Factor—For each Activity, a weight factor has to be defined depending on its own importance; it is used to evaluate the AQI of the Activity at a higher level.
- Percentage of Worst Results—for each Activity has to be defined the percentage of worst results to be considered in the computation; such value is used to calculate the AQIs of the Activity using a subset of QIs values (the worst) of his Sub-Activities or Parameters.

The quality of the Leaves Activities is measured upon Parameters (evaluation criteria). In light of this, once all the Activities were configured, then in each Leaf Activity the Parameters selected as its own evaluation criteria were defined.

The evaluation of the AQI of the *Leaf Activities* depends on:

- the values of the PQI of the Parameters of the Leaf Activity,
- the Weighting Factors of the Parameters of the Leaf Activity,
- the defined Percentage of Worst Results defined for the Leaf Activity.

The evaluation of the AQI of the *Activities* with child activities depends on:

- the values of the AQI of its child Activities,
- the *Weighting Factors* of its child Activities,
- the defined Percentage of Worst Results.

The configuration of the *Activity Tree Template* requires the definition, for each Activity, of his repeatability depending on the pavement structure (eg. the layer construction activity has to be repeated for each layer). The QEMS system automatically generates a specific Activity Tree for each

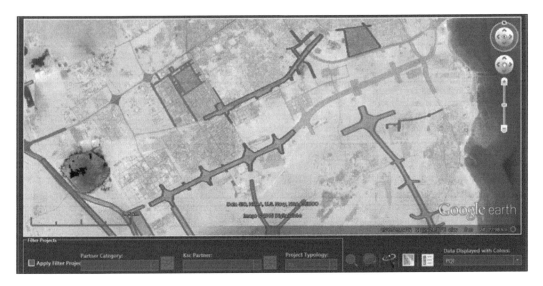

Figure 4. Example of Report.

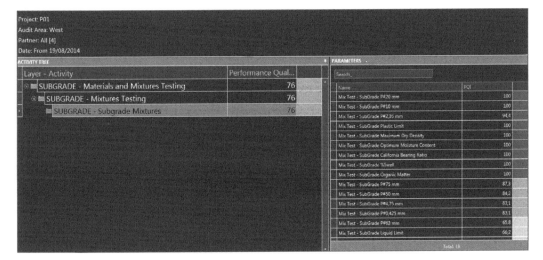

Figure 5. Example of Activity Quality Indicator.

Audit Area depending on the defined pavement structure.

For this reason, the audit results have to be collected in each audit area; despite this, the validity of such results may be extended to the entire **Project** or to multiple **Projects**. For this reason was introduced in QEMS the concept of "scope" to be defined for each Parameter. This aspect permits to introduce as the geographical validity of the audit results in the computation of the Quality Indicators.

In synthesis:

• for each Activity with Parameters (Leaf Activities), the Activity Quality Indicator Measured (AQI_M) is evaluated on the base of the Parameter QIs; this

indicator is said "Measured" because his value derives from the results obtained directly from the measurements of the Parameters of the Activity;

• for each Activity with child Activities, is calculated the Activity Quality Indicator Derived (AQI_D) on the base of the Activity QIs of his child Activities; this indicator is said "Derived" because his value derives indirectly from the results of other Activities.

6 REPORTS

QEMS provides some powerful reporting functionalities, including the following:

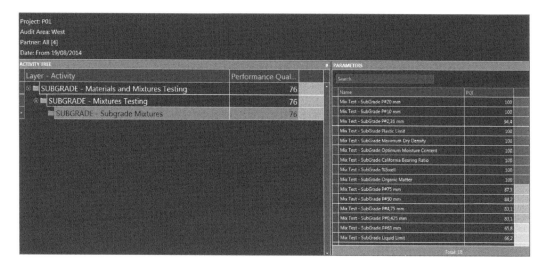

Figure 6. Example of Parameter Quality Indicator.

- Performance Reports of Activities
- Performance Reports of KSC Partners

The results of the evaluations are displayed on geographical maps: The polygons of a Project or the polylines of an Audit Area are shown with different colors depending on the Performances of the Activities (AQI).

Within the map it is possible to navigate obtaining reports; for example the Partner Quality Indicator Report or the Activity Quality Indicator Report.

It is possible to browse the reports obtaining detailed information on QIs and audit results.

REFERENCES

Cesolini, Oddone & Cuciniello (2015) 3rd MESAT2015 Middle East Society of Asphalt Technologists—The ANAS Quality Evaluation Management System (QEMS) Applied to Road Pavement Construction.

Cesolini, Oddone & Gregori (2016) Transport Research Arena TRA2016—ANAS Quality Evaluation Management System.

Transport Infrastructure and Systems – Dell'Acqua & Wegman (Eds)
© 2017 Taylor & Francis Group, London, ISBN 978-1-138-03009-1

Innovative approaches to implement road infrastructure concession through Public-Private Partnership (PPP) initiatives: A case study

M. Arata & M. Petrangeli
Anas International Enterprise S.p.A., Rome, Italy

Francesco Longo
ANAS S.p.A., Rome, Italy

ABSTRACT: During recent decades, major public road authorities showed an increasing trend in seeking for alternative ways to establish and implement road infrastructures concession systems, with the main aim to make it doable in a reliable, socio-economically sustainable and cost effective manner.

This phenomenon is mainly due to increasingly higher risks related to the establishment of road concession system and risk related to the introduction of road toll payment requirements.

During the last three years, ANAS International Enterprise S.p.A., a subsidiary company of the Italian Highway Agency ANAS S.p.A., has been appointed for assessing the feasibility and comprehensively structuring road concessions, through the implementation of a Public-Private Partnership (PPP) initiative, for a large part of road network of the Republic of Colombia.

The main activities consisted in (i) the development of a financial model, consistent with the rules and economic parameters related to the Colombian market; (ii) the assessment of the economic and financial feasibility; (iii) carrying out studies and simulations for several concessions scenarios implementation, including analysis of risks and technical, financial and legal assessment; (iv) preparation of tender documentation and delivery of technical support to the National Infrastructure Agency during the public-private partnership awarding process.

Along with these activities, the implementation of an innovative road concession system allowed the transfer and application of a number of best practices in various sectors. Indeed, improvement on the following aspects have been achieved:
• *Innovations in construction*, including alternative solutions for road embankment slope stabilization; barriers against rock falls; asphalt concrete mixtures modified with polymers;
• *Technological innovations*, including intelligent transport systems, telecommunications and information infrastructure, subsystem tolls, transport operating centers, subsystem video surveillance cameras hotspots, emergency call subsystem, weighing subsystem, lighting systems in urban areas and tolls;
• *Innovations in financial structuring*, including tailor-made regulating financial planning and monitoring tools, introduction of innovative solutions for the use of appropriate discount rates aimed to ensure investor profitability, assessment tool to manage public -and private-sides risks, assessing tools to ensure value for money, implementation of alternative solutions for public participation in Project Financing initiatives, development and implementation of alternative solutions for private funding in the financing process (including structuring infrastructure bonds), adoption of PPP structure type contracts provisions;

In this paper, the Authors describe in detail the concession implementation process, by particularly focusing on criteria adopted in selecting the contract type and related aspects, and highlighting main criticalities identified throughout the concession setting up process.

Aiming to give food for thought for further development of road infrastructure concession practice implementation, this paper aims to provide the Reader with an analysis of an actual successful case study, according to the trust demonstrated by the financial closing approved by the banks, and accurate and well-balanced comments on advantages, disadvantages and criticalities faced.

1 ALTERNATIVES FOR ROAD INFRASTRUCTURE CONCESSIONS

In the PPP acronym the most important P is the third one: Partnership. It is the meaning we want to give to this Partnership, and the equilibrium of the sharing of the burdens and responsibilities between the Public and the Private sectors, the key of success of such initiatives.

This has also been the lesson learned in Colombia, where ANAS has been awarded two contracts for the comprehensively structuring of around 3.000 km of roads, divided in 9 lots, of the fourth generation of concessions, the so called 4G highway program, after the first, second and third generations failed precisely for the lack of balance. In fact in the first generation all risks were borne by the public, ending to be much more costly for the government.

In the second generation the situation has been inverted, with the result that the private partner have either failed or fled the contracts. In the third one a better equilibrium has been found, but still lack of clear rules ended up creating delays, low levels of services and complaints.

Therefore it is important that the public administration proposing the concession properly plan the project and arrange a fair contract, financially attractive for the private sector, but clear and detailed in defining not only its obligations, but the level of service required (in the referred projects ANAS has expanded from 12 to 22 the service level return index to be satisfied by the concessionaires during their operation and maintenance period).

We must start from the analysis of the economic viability of the project. To that end we must have an accurate analysis of traffic flows with a careful projection, integrating future development of the territory crossed. Being this difficult and uncertain, given the long duration of road concessions, it is advisable not to oversize the infrastructure at the beginning (e.g. designing it for the first 10-year period), that would result in high CAPEX and related interest costs, but to foresee, both in the design and in the contract, an optional second phase of works for road widening.

But if cost optimization concern mainly the engineering of the construction, in terms of the proceeds must be financially considered that a road is not only a road. A road is a service corridor where run all networks, from communication and power lines to oil and gas, enabling their maintenance with easy access. A road is an avenue where users are clients with needs to be satisfied (petrol, food stations, shops and garages) and beneficiaries of advertisings. A road opens new commercial benefits to developing areas.

Therefore, another aspect that is worth considering are this "secondary incomes" that might result even in being self-sustaining of the project feasibility without the need for tolling, for example advertising, service areas, etc. A useful reflection where there are social problems of tolling acceptance. Also in this case some measures can be taken as widening the expropriation width to accommodate other services (paying use), arranging service areas with suitable structures and advertisement

spaces (to rent) and supplying the legal assessment for arrangements that capture part of the value added from the infrastructure to the productive activities of the area.

Once finalized the feasibility study, the Administration can decide the concession's duration and whether keeping it fixed or variable, according to the necessity of government grants (in the referred project some corridors needed from 30% up to 70% of public funding), researching and selecting the best international practices regarding the financing of infrastructure.

To this end ANAS had interviews with the management of banks and other lenders and analyzed major bond issues in the field of infrastructure, in order to understand the current potentiality of project financing and related conditions so as the current situation of the capital market in Colombia.

2 RISK ANALYSIS

Risk analysis is a fundamental step in the study of a PPP project, which allows to evaluate the feasibility of the project at 360 degrees. Within the foreseeable risk, the estimate is to assess the probability of occurrence of the risks that have been exemplified and, considering their level of impact, to evaluate their effects. In the specific Colombian experience, ANAS has developed the analysis in four steps:

- Risk identification: at this stage the possible causes and effects of risk factors were identified and allocated to the party (public and/or private) with more capacity to manage it;
- Risk evaluation: at this stage the analysis seeks to assess the materiality of the risks. The process includes an evaluation of the probability of occurrence and the related impact on the costs. The result from these first two phases is the risk matrix, where any risk is identified, described, assigned and evaluated qualitatively with respect to probability and impact;
- Risk valuation: at this stage, taking into account the qualitative assessment of the previous stage, it developed a risk map where, for the risks mapping in the Significant and Serious area, it is performed a quantitative evaluation, incorporating the criteria assessed in the qualitative stage a numerical approximation;
- Risk mitigation: solutions and mitigation tools have been proposed to optimize the management of risks.

The result applying this methodology, led to the definition of a risk financial plan with annual provisions that the public entity will use both to hedge against project's risks and to get a better idea of the real cost of the project and its feasibility. The methods used for the quantitative evaluation are the main methods of statistics, such as Chi Square and Bootstrap. In some specific cases, when historical information about the event under consideration was not available, experts meetings on the specific issue were organized, analyzing with statistical methods their answers.

In order to obtain a correct balance of risks between the public and private entities, it was considered a form of shared risks that was beneficial to the two parties involved. For example for the risks associated with extra cost arising from land acquisition or extra costs for environmental offsets: the private entity will assume the extra cost of up to 120% of the initial value; the extra cost between 120% and 200% of the initial value will be shared between the two parties, to the extent of 70% for the public entity and 30% for the private sector; the extra cost over 200% will be entirely supported by the public entity.

The correct balance of risks is crucial to the success of a PPP operation and for this analysis have taken into account the point of views of all involved parties and evaluated the market acceptance of the distribution designed. The discussion and pre-analysis phase therefore is essential.

It is important to point out that one of the success factors of the risk analysis is the level of advancement of the detailed design. The more you deepen the technical studies, the more it reduces the uncertainty about the estimates. In Colombia, the development requested of the technical studies for the fourth generation of highway concessions is very high, which allowed a more accurate risk as well as projects assessment, in general. With this fourth generation of PPP projects has been possible to improve the distribution of the risks of the concessions, solving the critical points of previous generations who had a distribution of risks which in some cases proved to be critical. In the 4G concessions in Colombia, for example, the risk for lower toll revenues is now borne by the public body. However the traffic study has been developed to an advanced level, thus reducing the uncertainty of the risk. The same consideration can be done on all risks for which the analysis carried out has decreased the uncertainty.

3 INNOVATIONS IN CONSTRUCTION

Main constructive innovations concerned road embankment slope stabilization and asphalt mixtures.

3.1 Slope stabilization

The traditional works type (reinforced concrete retaining walls on direct foundation or on piles,

with or without ground anchors) are characterized by the use of materials as concrete and steel. A plus point of these solutions is the fact that they can solve situations of high criticality, while among the weaknesses there are:

- Need of careful selection of materials and high quality controls;
- High costs and complex organizations, especially during construction;
- Some types of intervention require specialized technologies and skilled labor;
- The works are poorly adapted to future additions, for example in the case of road widening, and eventual demolition is expensive, as it is the landfill disposal of materials
- Works require maintenance (typically drainage interventions), but material's degradation is inevitable.

The proposed innovative solutions include applications of "bioengineering" and they can be alternative or integrative to traditional solutions. Among the plus points are noted:

- The required materials are available locally. Other materials are metal nets, cables, rods and screws, geomats, geogrid and geosynthetics. Such materials are very low volume. This reduces transportation costs;
- Environmental integration is optimal;
- The execution of the works is simple, small in size, with less impact on traffic deviations;
- Specialized supervision is required, but unskilled labors are sufficient for their erection;
- Interventions are flexible and easy to extend or integrated in the future;
- With proper maintenance durability is significant.

Material used in bioengineering reduces the costs of works: based on our experience, savings over traditional interventions range between 25% to 30%.

Other bioengineering solutions used on the Colombian projects have been rock slope protection,

Figure 1.   Example of actual concrete retaining wall.

Figure 2. Example of (a) Rock fall nets (b) Rock slope protection.

ground slope reinforcing (soil nailing) and Rock fall nets.

3.2 *Asphalt mixtures*

It is believed that one of the oldest methods of paving was the so called "Roman road". Created to facilitate communications within the empire, this road was developed in various stages and some of its stretches are still in good condition. Nowadays asphalt mixtures and concrete are the most usual pavement construction materials because they provide a good support and allow passage of vehicles without major damages.

An asphalt mix is usually a combination of asphalt and aggregate minerals. The relative proportions of these minerals determine the physical and performance properties of the mixture. Conventional asphalt mixtures have properties that meet the normal requirements of pavements; however, when severe climate conditions and high traffic volumes occur, it is necessary to seek appropriate and economically viable technical solutions that enable the concessionaire to meet service levels required, both at the beginning and at the end of operating time, making it vital for the maintenance cost of the road.

Since a large percentage of car accidents that occur every day are related to the deterioration of the roads, it is very important to prolong the life of pavements. This is achieved by studying potential changes in the designs. The benefits of these developments would impact both road safety and project economy.

Polymer modified asphalts are conventional mixtures which are added with a polymer that enhances the properties, producing:

- Increased adhesion properties even in presence of high humidity;
- Increased resistance to deformation and fatigue;
- Reduced susceptibility to thermal variations.

According to the polymer applied and the amount thereof, can be obtained increases in pavement service life up to 100%, against an increase cost of the mixture up to 25%.

Analyzing graphs, it is observed that for a polymer modified asphalt, the life cycle is much greater than that of conventional binders. Defined the Load Equivalence (LE) as the structural performance of an asphalt layer compared to another, for modified asphalt mixtures with Reactive Elastomeric Terpolymer (RET), this is about 1.5 compared to a conventional asphalt. Economic evaluations that have been conducted show that the use of polymers with a LE greater than 1.2, represents a saving on the final cost of the project.

Consequently:

- Polymers are the economical choice when traffic volumes are high or climatic conditions are tough and the project needs to include the maintenance to allow the return of capital invested;
- Requested service levels of the road can be preserved for longer, requiring less maintenance interventions more spaced in time, which means better service index and a greater availability of the road by users;
- The polymer modified asphalt mixtures also allow us to optimize the use of natural resources. In our case, we reduced the environmental impact reducing the thickness of the asphalt, using quality stone materials. In the first corridors delivered, the use of the Dense Hot Mix modified polymers, represented a 6% savings in the construction cost. Ate

4 TECHNOLOGICAL INNOVATIONS

Traffic, especially in metropolitan areas, has become an everyday problem difficult to solve. An issue that causes undesirable effects as increased travel time or failure times in public transport, intolerable air pollution and noise levels that seriously affect health and has its counterpart in significant economic losses. A problem accentuated in recent decades by the increasing use of the automobile, a diffuse decentralization process and an increasing number of trips and amplitude thereof.

As regards the environment, the situation is also worrying. Transportation is, in this area, responsible for over 60% of carbon monoxide emissions, 50% of nitrogen oxides and 33% of hydrocarbons. Moreover, the consequences of pollution on the health of European citizens represent a cost of 0.5% of EU GDP.

Colombia is no stranger to this problem, due also to the effects of population growth with economic, environmental and quality of life consequences.

The purpose of implementing an Intelligent Transportation System (ITS) is to allow the collection, storage, processing, analysis and distribution of information related to road operation, in order to improve traffic management and achieve a more efficient and safe operation of the infrastructure, generating more information to users, reducing travel time, reducing the cost of fuel consumption and environmental impact, among others.

Any technological platform achieves its true operating profit, and therefore the return on investment, when it come across both the technical components (equipment) as the operation and maintenance processes, needed for an operating strategy and continuous improvement. Technology alone does not achieve the goals without processes and people involved.

The major subsystems required in the design of road concessions are:

• Telecommunications infrastructure: consisting of hybrid solutions with fiber optic as backbone, different wireless systems and mobile technology solutions, to connect all elements along the corridor;
• Electronic Toll Collection System (ETC);
• Operational Control Center (OCC): room where centralized management of all subsystems is made;
• Variable (or Dynamic) Messaging Systems VMS (DMS), integrated with information screens to provide useful information for travelers;
• Video surveillance cameras (high-speed) for vehicle identification, counting and speed control systems;
• Weighing system;
• Intelligent lighting systems in critical areas.

Perceived benefits in projects integrating ITS are: better information in real time, faster response to emergencies, greater fluency in circulation, increased road safety, automatic vehicle and freight location, environment amelioration, and, at least but not at last, development of related industries.

Studies indicate a clear contribution of ITS to develop a more sustainable mobility in the sense of increasing services without increasing impacts. It has been found that ITS not only contribute to increased productivity of transport systems, with the positive impact this has on the economy, but that is in itself a market with high economic potential and interesting future projection. The trend is to make the ITS design by subsystems, allowing proper integration of the different elements. An important role to meet the functional requirements is to establish a governance model and organization with appropriate policies.

Due to the diversity of manufacturers and systems today existing on the market, both for the collection and the transport of data, is always advisable to make preliminary pilots projects, led by a severe test plan, where it is verified that the product complies with all features offered. The resulting information can be used to explore other business alternatives.

The concession contract imposes the contractor to use the mentioned technologies.

5 INNOVATION IN FINANCIAL STRUCTURING

According to ANAS experience, it is proposed, as a new alternative, to use the Italian practice in the financial model, called Financial Regulatory Plan (FRP), where toll and additional revenues remunerate only admitted project's costs: return on capital invested, depreciation costs, operation and maintenance costs. In this way the FRP seeks to determine if the project's costs are balanced, at the end of the concession, by the income received from tolls and eventual government grants, discounted at WACC (Weighted Average Cost of Capital).

The pursuit of this balance allows to review what was executed annually and to determine pricing adjustment policies. In addition, FRP allows to calculate the value that the contracting entity should pay to the operator in case of early termination of the concession's contract in a given year. This FRP is an official document that is part of the concession contract and every five years it is revised to allow small adjustments ensuring regulatory balance, without waiting to have unexpected concerns at the expiration of the concession.

To calculate the financial balance should be used a discount rate in accordance with market conditions and projects with similar levels of risk and technical features. This rate corresponds to the WACC, with which the projected cash flows are updated to calculate the IRR (Internal Rate of Return) of the project. Thereby it is considered that companies are financed by two mechanisms: through financial resources, which are denominated debt or through direct investment, denominated equity.

In infrastructure projects worldwide equity ratio to investment value ranges from 20% – 30% while the rest is financed by third parties. The financial leverage can be of 3 types, multilateral agencies, international banks and local banks.

Multilateral agencies promote infrastructure development in different countries and make long term development loans, compared to that of commercial banks, that can have the same duration of the concession, which makes it much more profitable for the concessionaire. But this type of loan has a much longer and difficult process to be obtained than a commercial bank credit.

International banks have a capacity of debt much higher than local banks. Moreover, because its desire for risk diversification, a project like this is framed within their target investment. Unfortunately these banks do not usually pay in local currency, and given the length of payback, create a currency exchange risk very high to the project.

Local banks are a source of leverage in local currency, avoiding exchange risk. This source has lower time limits than multilateral banks and generally do not gather the entire loan individually.

There is the possibility of creating a structure of debt leverage through a hybrid of the above, making an international bank to grant a loan in foreign currency at a good rate at a local bank, and this use these currencies for their international operations and fit a loan in local currency to the concessionaire with the guarantees on the project.

In the following are described various financial instruments used by private subjects to finance infrastructure projects.

In large PPP projects, especially in the UK, it is not unusual that the Administration actively contribute to finding the best financing conditions thereof, requiring the opening of proceedings for the election of debt providers, known as "debt funding competition". The competition may not be adequate under conditions of limited financial liquidity.

The Project bonds are specific bonds issues related to a project, with return of capital depending on the cash flows generated by the project itself. In order to make the Project Bond an instrument with market appeal, encouraging the financing of public infrastructure by the private sector, have to be introduced amendments to legislation, as:

- Companies incorporated in order to build and manage an infrastructure project can issue bonds, with the approval of the local supervisory bodies, including during the construction phase of infrastructure. These bonds are nominative and cannot be transferred to people who are not qualified as institutional investors;
- Until the beginning of the operation of the infrastructure by the concessionaire, obligations can be guaranteed by the financial system;
- To introduce a tax incentive to the signing of Project Bonds, in Italy a tax system like the taxation of government bonds is foreseen;
- Project bond issue can also be used to allow refinancing credit operations or senior debt acquired previously.

Another possible tool to limit the risks associated with PPPs could be a public guarantee fund in order to provide part of the bank debt of the Licensor.

6 TENDERING PROCESS

ANAS, classified among the leading Italian contracting authority, has experienced since many years the concessions structure using the project financing tool for large infrastructure projects. It has designed and optimized tender procedures aimed at involvement of private partners, in compliance with the Italian legislation on public works and developed with reference to the international scenario in the field of PPP contracts.

This type of procedure has fostered not only the financial support of private sector in conjunction with the public one for the execution of large infrastructure projects, but also allowed the insertion of project proposals made by private subjects within the National Development Plan framework. In this way the role of "Private Promoter" has been fully integrated in the governmental planning process, contributing concretely to its definition and to the implementation of its action plans.

ANAS experience in the field of PPP has been successful also at international level. It is the case of Colombia, where ANAS has supported the local contracting authority to the development of a PPP concession system of new generation, which the government is now using for the implementation of its important infrastructure investment plan.

Referring in particular to the characteristics of the contract experienced in Colombia, it should be stressed that ANAS, on the basis of the Italian model, has contributed to the adoption of a methodology which provides the use of a single model of Cotract in all concessions tendering, consisting of a general part, such as applicable to each procedure, and a special one provided with specific attachments for each project.

The procedure involves a first phase of pre-qualification, targeted to the selection of the potential concessionaires that satisfy the technical and financial requirements provided in the tender instructions. Such kind of restricted procedure, well consolidated in Italy, guarantees a qualified participation for the benefit of the contracting authority and the subsequent production process.

To note also the great importance, in this context, of an extensive promotional plan, aimed to attract the world of investors both in business and banking. In the Colombian case, the success of such promotional initiatives, implemented through specific "road show", has contributed significantly to accelerate the procurement and awarding process of the first concessions of "fourth generation".

7 CONCLUSIONS

In the PPP acronym, the most important P is the third one: Partnership. It is important that the public administration proposing the concession properly plan the project and arrange a fair contract, financially attractive for the private sector. Also the risk analysis is a fundamental step in the study of a PPP project. The definition of a Risk Financial Plan with annual provisions that the public entity will use to hedge against project's risks is suitable.

To analyze the economic viability of the project, we must have an accurate analysis of traffic flows with a careful projection, integrating future development of the territory crossed. However, it is advisable not to oversize the infrastructure at the beginning, but to foresee, both in the design and contract, an optional second phase of works.

Another aspect that is worth considering when structuring a PPP initiative are the "secondary incomes" that might result even self-sustaining of the project feasibility without the need for tolling. Some examples of secondary incomes are incomes by rents, by advertising, by sub concession fees, etc.

Cost optimization concern mainly the engineering of the construction. Main innovations concerned slope stabilization and asphalt mixtures. For the first aspect, applications of "bioengineering" can be alternative or integrative to traditional solutions. Material used in bioengineering reduces the costs of works, saving between 25% to 30% over traditional interventions. Concerning the asphalt mixtures, polymer modified asphalts increase pavement service life up to 100%, against an increase cost of the mixture up to 25%. Polymers are the economical choice when traffic volumes are high or climatic conditions are tough, requiring less maintenance interventions.

Regarding the ITS technologies applied to road network, studies indicate a clear contribution of these in developing a more sustainable mobility. It has been found that ITS not only contribute to increased productivity of transport systems, with positive impact on the economy, but that is in itself a market with high economic potential and interesting future projection for the development of related industries.

Finally it is proposed at the contracting authority, but it isn't compulsory, to use the Financial Regulatory Plan (FRP), where toll and additional revenues remunerate only the return on capital invested costs, depreciation costs, operation and maintenance costs. FRP allows to calculate the value of the concession's contract in a given year, ensuring regulatory balance. In infrastructure projects worldwide equity ratio to investment value ranges from 20% to 30%. The financial leverage can be of 3 types, multilateral agencies, international banks and local banks. There is the possibility of creating a structure of debt leverage through a hybrid of the above. Project Bond can also be a sound instrument with market appeal, but amendments to legislation might be needed.

It is important, for the success of a PPP tendering procedure, to foresee a prequalification phase and an adequate promotional plan to attract investors.

REFERENCES

2020 European Fund for Energy, Climate Change and Infrastructure (Marguerite).

Ahmed, Priscilla, Project Finance in developing countries: IFCb lessons of experience/International Finance Corporation, Washington, D.C., 1999;

Balzer, L.A. (2001). "Investment risk: A unified approach to upside and downside returns".

Beato, p. (2007). Road Concessions: Read Lessons from the Experience of Four Countries rned. Substainable Development Department uses.

Bent Flyvbjerg, Mette Skamris Hola and Søren Buhl (2002) "Underestimating costs in public works projects: Error or lie?".

CAF: "public infrastructure and private equity—concepts and experiences and in"America and Spain", 2010;

CEPREDENAC—UNDP: the local risk management. Notions and clarifications on the concept and practice. 101 pg. 2003 (digital Version).

Chu, j. (1999). "The BOOT approach to energy infrastructure management: a means to optimize the return from facilities". Facilities. Vol. 17, no. 12–13.

Dams (2000) World Commission "dams and development: A new framework for decision-making".

Dragutin Nenezić, Branko Radulović Analysis of Finance Options and Models, and Financial Support Measures for Public-Private Partnerships in Serbia;

Essinger, J. & Rosen, J. (1991). Using technology for risk management. London: Woodhead—Faulkner.

Fitch Ratings, "Report update issuance of straight bonds—concessionaire of the West", July 2012;

Flanagan, R., & Norman, G. (1993). Risk management and construction.

Grimsey, D., & Lewis, M. (2002). "Evaluating the risks of public-private parnertships for infrastructure projects". International Journal of Project Management. Vol. 20.

Guida al PPP—Manuale di buone prassi, European PPP Expertise Centre and technical Unità di Finanza di Progetto, 2011;

http://ec.europa.eu;
http://financialtribune.com;
http://www.aipcr.it;
http://www.ani.gov.co;
http://www.ambteheran.esteri.it;
http://www.caf.com;
http://www.eib.org;
http://www.fitchratings.com;

http://www.hm-treasury.gov.uk;
http://www.ice.gov.it;
http://www.imf.org;
http://www.infomercatiesteri.it;
http://www.PIARC.org;
http://www.sace.it;
http://www.simest.it;
http://www.smbcgroup.com;
http://www.stradeanas.it;
http://www.utfp.it;
http://www.worldbank.org;
Infrastructure, Project Finance: "Financing of infrastructure and project bonds", 2010;
Jefferies, M. (2006). "Critical success factors of public-private sector partnerships". Engineering, Construction and Architectural Management. Vol.13, N ° 5.
Juan Benavides, reforms to attract private investment in road infrastructure;
Kahneman & Tversky, a. (1979). "Prospect theory: An analysis of decisions under risk", econometric. Vol. 47.
Lanfranco Senn, Il finanziamento delle Infrastrutture di Trasporto;

Lara, (February 2003). Risk Management in Toll Road Concessions. Massachusetts Institute of Technology.
Leonardo Freitas, Project Finance and Public-Private Partnerships: A Legal and Economics View from Latin America Experience, Business Law International, September, 2010;
Mills, A. (2001). "A systematic approach to risk management for construction". Structural Suvey. Vol. 19, N ° 5.
Private toll roads: lessons from Latin America, Samuel Carpenter and Raphael Barcham, Polytechnic University of Madrid (Spain), and Norbridge Inc., USA;
Thomson Reuters Eikon (various analysis);
Tool Roads Latin America Special Report, FitchRatings 2009;
Toolkit for Public-Private Partnerships in Roads and Highways, PPIAF—World Bank (2009).
UNDRO: Natural Disasters and Vulnerability Analysis in Report of Expert Group Meeting (9–12 July, 1979). Geneva.

Transport Infrastructure and Systems – Dell'Acqua & Wegman (Eds)
© 2017 Taylor & Francis Group, London, ISBN 978-1-138-03009-1

Highway functional classification in CIS countries

V.V. Silyanov
State Technical University—MADI, Russia

J.I. Sodikov
Tashkent Automobile Road Institute, Uzbekistan

ABSTRACT: Road infrastructure are not only costly to build but also expensive to maintain in order to adequately meet public expectation. Managing costly asset requires systematic approach which assures adequate decision in each step of project life-cycle namely planning, designing, building and managing. Presently, highway functional classification is widely used in developed countries however in CIS (Commonwealth of Independent Countries) countries planning, designing and managing of highways merely based on technical (or administrative) classification. Authors are suggesting highway functional classification which takes into account not only technical (or administrative) classification but also functional classification. Recommended functional classification requires justification in parameters such as speed, safety, riding quality and environment. The parameters were estimated by utilizing Analytical Hierarchy Process. Optimum spending was defined as a function of road network size and importance weights of highway functional classification.

1 INTRODUCTION

1.1 *Brief statistics*

There are 9 member states in CIS which are Armenia, Azerbaijan, Belarus, Kazakhstan, Kyrgyzstan, Moldova, Russia, Tajikistan, and Uzbekistan. The extent of public road network is about 1 232 850 km, land area 22 045,1 thousand km sq, number of bridges and overpasses is 95 593. Total number of vehicle fleet is 359 720,4 thousand units. There are about 1 million people work in road sector.

1.2 *Perception functional classification*

The review of literature revealed that developed countries implemented functional highway classification for planning, designing, and managing whereas CIS (Commonwealth of Independent Countries) countries utilize based on technical classification. Functional classification serves for which purpose the roadway is served such as for mobility or accessibility. Technical classification serves for technical definition of roadway such as geometric parameters. There have been a number of attempts to implement functional classification in CIS countries. It was recommended to review existing Russian classification of roads, especially in urban areas by taking into account land use, traffic accidents spots, vehicles and pedestrians flows, traffic safety issues, including traffic speed

regulation in order to develop functional classification. In a proposed new classification one of the important principles is to provide road use properties, which would be defined by the rate of speed provision, comfort levels and traffic safety by road classes, Devyatoe & Vilkovay (2006). In the Republic of Belarus road classification has functional assignments. There are four classes, namely: highways, high-speed roads, ordinary roads and low class roads. Functional assignments are for highways—has purpose to serve long distance commute without serving near by territories, for high speed roads—has purpose to serve local commutes with high speed, ordinary roads—has purpose to serve for general purposes, and finally low class roads serve for small town and villages, Design norms (2016). In Uzbekistan road classification has road type, road function, road class (technical category), and administrative class. Road network consist of following types: highways, city access highways, main, local and internal roads, SHNK 2.05.02–07 «Automobile Roads» (2008). According to intergovernmental standard of CIS countries, road technical classification according to road use properties and access function is divided into the following classes: highways, high speed roads, partial speed roads, and ordinary roads, GOST 33382–2015 (2015). One of the fist works on road classification in USSR was carried out by Babkov F. He suggested that there were need in national and functional classification of highways.

However, he proposed to utilize national classification, but he mentioned that highway functional classification should be based on the traffic flow, which may not necessarily be coincident with the national one, Babkov & Zamakhayev (1967). Silyanov (2004) reviewed classification of Russian highways and there are two separate standards for rural (SNIP 2.05.02–85) and urban roads (SNiP II-60–75). Both standards don't have highway functional classification.

Paraphantakul (2014) analyzed worldwide road classification and found that number of road classes vary from 4 classes in Greece to 12 classes in South Africa. Author combined the relationship structure and the classification categories from various literature and produced eight general road classification themes: access control, road surface, usage, administration, link role, place status, transport mode, and function. In European, North American and Australian classification themes have similarities such as function, access and link role are the most recognized themes. In the study, there is overlapping between function, access and link role themes, which may lead to a biased conclusion.

2 PRESENT ROAD CLASSIFICATION

Intergovernmental standard GOST 33382–2015 on road classification among CIS countries divide highways into the following classes: highways, high speed roads, partial high speed roads, and ordinary roads. Highways are dedicated to safe and uninterrupted traffic flow with high speed and volume, which serve for long distance commute. Access is not allowed to a certain types of transport means, pedestrians and bicyclists. Measures are taken to avoid wild and domestic animals on the roads. There are requirements to which highways must satisfy:

– Division opposite direction by installing median strip or separate design of carriageway for each direction;
– At least two traffic lanes for each direction;
– Avoid intersection at grade with roads, bicycle and pedestrian lanes, and also wild animal migration routes, domestic animal crossings;
– Access provision by grade separation;
– Prohibit bicycles, motorcycles, tractors, and also other types of transportation which may disrupt traffic flow, except machinery for road repair or emergency situations;
– Prohibit access heavy weight, dangerous and oversized vehicles;
– Limit access points in grade separation.

High speed roads are allowed to design at grade level but under condition that it would not cross direct traffic flow and satisfy other requirements of highways. Partially high speed roads have multilane without median strip; access allowed at grade and grade separation. Ordinary roads serve for low volume traffic therefore they have one or two traffic lanes; access allowed at grade and grade separation (see Table 1).

As a matter of fact, CIS countries inherit the same history therefore standards and norms are based on research during that period with some modifications to fit todays realities such as increase of axle load and change in vehicle composition. Table 2 shows road network length by road category. Majority of roads are IV and V category which make up about 63% and 74% for Uzbekistan and Russia respectively.

The problem with existing road classification is that it doesn't fully satisfy to manage adequately

Table 1. Highway Classification according to GOST 33382–2015.

Class	Category	Design traffic volume*	Number of traffic lanes
Highways	IA	$\dfrac{>7}{>14}$	More than 4
Speed roads	IБ		
Partial speed roads	IB		3 or 4
	IIA		3
	IIБ	$\dfrac{3-7}{6-14}$	
	III	$\dfrac{1-3}{2-6}$	
Ordinary roads	IV	$\dfrac{0.1-1}{0.2-2}$	2
	V	$\dfrac{<0.1}{<0.2}$	1

*– Thousand vehicle per day (in numerator—actual number of vehicles, in denominator—passenger car equivalent).

Table 2. Road network length comparison by road category between Uzbekistan and Russia.

Road category	Uzbekistan		Russia	
	Length, km	Percentage	Length, km	Percentage
I	1246	3%	4800	1%
II	2363	6%	27800	5%
III	9800	23%	109800	20%
IV	21333	50%	320200	59%
V	7788	18%	81600	15%
Total	42530	100%	544200	100%

road assets. Road classification basically served for planning and designing purposes. While road maintenance takes into account road category for budget allocation and repair. But key question is whether the road section corresponds to the category which was assigned during design or not, what kind of purpose does certain road section serve for? In order to properly answer these questions, the authors are proposing functional highway classification.

3 PROPOSED HIGHWAY FUNCTIONAL CLASSIFICATION

3.1 Highway functional classification

Two primary transportation functions of roadways, namely mobility and access, and describes where different categories of roadways fall within a continuum of mobility-access. In addition to mobility and access, other factors that can help determine the proper category to which a particular roadway belongs—such as trip length, speed limit, volume, and vehicle mix. The concept of functional classification defines the role that a particular roadway segment plays in serving this flow of traffic through the network. Roadways are assigned to one of several possible functional classifications within a hierarchy according to the character of travel service each roadway provides. Planners and engineers use this hierarchy of roadways to properly channel transportation movements through a highway network efficiently and cost effectively, Highway Functional Classification Concepts (2013). In Serbia, functional classification was proposed as primary and secondary functions due to the fact, that unique basic function of a certain rural road section is virtually impossible to achieve and, in any case, would be spatially, economically and environmentally unacceptable. Therefore, certain mixing of functional tasks is necessary and requires the definition of primary and secondary function of a certain road section, Maletin & Tubić (2015). In other words, it's logical that certain road section may serve different purposes such in Serbian case, local road (direct access road) has primary function as direct access to property, and secondary function as collection/distribution of traffic flow. To avoid misunderstanding in practical usage and to simplify data management the authors came up with simplified highway functional classification. Based on literature review, the relationship between number of classification and level of country development the following highway functional classification was proposed for CIS countries:

- Arterial:
  - Major Arterial;
  - Minor Arterial;

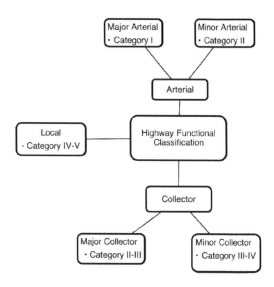

Figure 1. Highway Functional Classification taking into account road category.

- Collector:
  - Major Collector;
  - Minor Collector;
- Local

Figure 1 shows highway functional classification in relationship with road category. As a road category is primary principle for designing and maintaining roads it would be appropriate to introduce functional classification in line with it.

Highway functional classification serve not only for identifying what function is playing certain road section but also importance of that section. As one of the performance indicators AADT (Annual Average Daily Traffic) or VKMT (Vehicle Kilometers Traveled) plays significant role in budget spendings. In order to assign importance factors to each functional classification analytical hierarchy process was utilized.

3.2 Analytic Hierarchy Process

The Analytic Hierarchy Process (AHP) proposed by Saaty (1980) is a very popular approach to multi-criteria decision-making that involves qualitative data. It has been applied during the last several decades in various decision-making problems. The method uses a reciprocal decision matrix obtained by pairwise comparisons so that the information is given in a linguistic form. The pairwise comparison method was introduced in 1860, Fechner (1965). Based on pairwise comparison, Saaty proposes the analytic hierarchy process as a method for multi-criteria decision-making. It provides a way of breaking down the general method into a hierarchy of sub-problems, which are easier to evaluate. It utilizes multiple criteria, and provides a simple process for weighting portions of the hierarchy that is difficult to enumerate directly. This method describes a general multi-factor decision problem in way that decision tree in each of the hierarchy levels include some types of criteria. The idea behind making decisions is to compare the relative importance of each criterion in adequate manner so that it can fit into general concepts. They are some of the concerns regarding application of AHP in making pair-wise comparisons between alternatives, the more alternatives the more problems occur regarding the consistency of the comparisons, and the assignments of the scores. An expert makes decision regarding the relative importance of each criterion and then specifies a preference, which is rated on a scale from 1 to 9, for each decision alternative. If there are n alternatives, then n*(n–1)/2 pair-wise comparisons are needed. Obviously, for practical application of AHP, the number of alternatives must be reasonable limited. The result of AHP is a prioritized ranking that indicates the overall preference of each alternative. The expert responsible for adequateness of assigned that they reflect the importance of the issues. The consistency of the judgments of the expert can be measured with a Consistency Ratio (CR). The CR is calculated as follows:

$$CI = \frac{(\lambda_{max} - n)}{n - 1} \quad (1)$$

$$CR = \frac{CI}{RI} \quad (2)$$

where, CI – consistency index, λ max = the eigenvalue corresponding to the principal eigenvector n = the number of alternatives or criteria being compared RI = the random index, a dimensionless value that is a step function of n.

CR of 0.1 or less is considered acceptable. If a decision-maker's responses fail the consistency test, then the analyst must repeat iteration until consistent responses are obtained. Applications of AHP highlighted in various transportation projects evaluation such as Kang & Lee (2006), Azis (1990), Hagquist (1994), Kim & Bernardin (2002), Masami (1995), Tabucanon & Lee (1995), Kengpol (2002). A popularity of AHP method in solving multicriteria problems in infrastructure projects, clearly related to easy to use and the way it prioritize alternatives. The logic behind prioritization is based on expert's judgment. Besides in decision-making the method can use quantitative and qualitative data. Despite of advantages there are some disadvantages such as scaling alternatives from 1 to 9, weakness of priorities estimation

methods, large number of comparisons when many alternatives exists, Ramanathan (2001), Taslicali & Ercan (2006).

4 RESULTS AND DISCUSSIONS

One of the key purposes of introduction highway functional classification is to achieve optimal road asset management. There are two main factors which affect road asset management. One is financial constraint, the other one is time. To manage efficiently road assets at network level importance weights were assigned to each highway functional classification by analytical hierarchy process. First, importance was assigned for each functional classification such as arterial, collector, and local. Intensity of importance varies from 1 to 9; 1 – equal importance, 3 – somewhat more important, 5 – much more important, 7 – very much more important, 9 – absolutely more important, 2,4,6,8 – intermediate values used when compromise is needed. Tables from 4 to 6 describe alternatives and parameters values and pairwise values.

In appendix 1 is given detailed description of pairwise matrix and R code for calculating analytical hierarchy process for highway functional classification. The result is shown in Table 3.

Table 3. Assigning importance weights.

	Weight	Arterial	Collector	Local	CR
Weight	1.00	0.57	0.27	0.15	0.09
Safety	0.47	0.27	0.13	0.07	0.00
Speed	0.33	0.22	0.08	0.03	0.02
Riding Quality	0.14	0.08	0.04	0.02	0.01
Environment	0.06	0.01	0.02	0.04	0.00

Table 4. Alternatives and parameters values.

	Speed (60–120)	Safety (0–10)	Environment (0–10)	Riding Quality (0–100)
Arterial	120	10	6	90
Collector	80	6	8	70
Local	60	4	10	50

Table 5. Parameters pairwise values.

	Speed	Safety	Environment	Riding Quality
Speed	1	1/2	5	4
Safety	2	1	5	4
Environment	1/5	2	1	1/4
Riding Quality	1/4	1/4	4	1

Table 6. Alternatives pairwise values.

	Speed		
	Arterial	Collector	Local
Arterial	1	3	6
Collector	1/3	1	3
Local	1/6	1/3	1
	Safety		
	Arterial	Collector	Local
Arterial	1	2	4
Collector	1/2	1	2
Local	1/4	1/2	1
	Environment		
	Arterial	Collector	Local
Arterial	1	1/3	1/6
Collector	3	1	1/2
Local	6	2	1
	Riding Quality		
	Arterial	Collector	Local
Arterial	1	2	5
Collector	1/2	1	2
Local	1/5	1/2	1

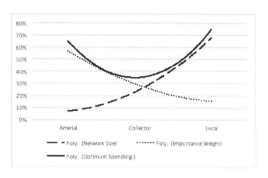

Figure 2. Optimum spending depending on functional classification and network size.

Based on Table 2 and Table 3 the relationship between importance weight and network size were found. It's obvious that network size increases from arterial to local. Whereas importance factor decreases from arterial to local. At the network level and multiyear programming, it can be assumed that optimal spending under limited budget should be primarily focused on major and minor arterial, major collectors and secondarily on minor collector and local roads (see Fig. 2).

Highway functional classification plays indispensable role in not only in planning, designing, and maintenance but also in budget allocating. In above-mentioned research works shows that there is overlapping in road function between arterial and collector roads, and between collector and local roads. Along with road inventory and condition data collection there is a need in functional classification survey.

The present classification in CIS countries, mainly based on technical category, which doesn't satisfy up-to-date requirements of road asset management. The research proposes that functional classification in line with technical classification would efficiently manage road assets. Proposed highway functional classification takes into account two functions such as mobility and access. The five type of functional classification, namely major and minor arterial, major and minor collector, and local roads were recommended. In order to assign

importance weight, analytical hierarchy process was utilized. Based on importance weights and network size, optimum spending on road maintenance was recommended. Previously, the authors proposed simplified road asset management system for developing countries in which one of key factors was traffic, Sodikov & Silyanov (2015). Traffic volume changes over time and doesn't precisely reflect the importance of road section. On the other hand, highway functional classification has direct relationship with traffic flow, besides it has functional purpose with importance weight. There are number of recommendations proposed:

- Highway functional classification not only divide roads into classes like in technical category division, but also each class has functional purpose. Based on which road agencies would be able to adequately and efficiently manage their assets.
- Continuously surveying road network for identification and homogeneous sectioning under highway functional classification, would lead to more appropriate decision-making.
- Based on optimal spending curve one can build decision tree model to predict prioritization road maintenance program at network level.

Further research will focus on defining detailed requirements for each functional classification, actual division road network according the new classification and development GIS map.

REFERENCES

Azis, I.J., 1990, "Analytic Hierarchy Process in benefit-cost framework: A post-evaluation of the Trans-Sumatra highway project", *European Journal of Operational Research. Volume 48, Issue 1.*

Ali Kamil T. & Sami E., 2006. The analytic hierarchy & the analytic network processes in multicriteria decision making: a comparative study. *Journal of aeronautics and space technologies,* Volume 2, Number 4.

Babkov V & Zamakhayev M.1967, Highway Engineering, Mir Publishers, Moscow.

Chutipong P. 2014, Review of Worldwide Road Classification Systems, *Conference: National Transportation Conference*, At Bangkok, Volume: 9. https://www.researchgate.net/publication/284371945_Review_of_Worldwide_Road_Classification_Systems. Last accessed: 1/11/17

Devyatoe M.M. & Vilkovay I.M 2006, Methodology of functional classification of roads for the purpose of their modernization. *Proceedings of roads and bridges*, Issue 16/2, Moscow 2006. http://www.complexdoc.ru/ntdpdf/541958/dorogi_i_mosty_sbornik_vypusk_162.pdf. Last accessed:1/11/17

Design norms 2016, TKP 45–3.03–19–2006 (02250), Ministry of architecture and Building of the Republic of Belarus. http://www.npmod.ru/3/Perechen_1/tkp30319.doc. Last accessed: 1/11/17

Fechner G.T., 1965, Elements of Psychophysics. Volume 1, (Holt, Rinehart & Winston, New York.Informational-analytical publication, 2012. *Status quo and development of roads in CIS countries*. http://www.e-cis.info/foto/pages/19824.doc. Last accessed: 1/11/17

GOST 33382–2015, Public domain automobile roads. Technical Classification.

Hagquist R.F., 1994, "High-precision Prioritization Using Analytic Hierarchy Process Determining State HPMS Component Weighting Factors", *Transportation Research Record* 1429, pp 7–14

Highway Functional Classification Concepts, 2013. Criteria, and Procedures. Federal Highway Administration.

Kengpol, A., 2002, "The Design of a Support System (DSS) to Evaluation the Investment in New Distribution Centre Using The Analytic Hierarchy Process (AHP)", *Capital Investment Model and Transportation Model.*

Kim, K. & Bernardin, V., 2002, "Application of an Analytical Hierarchy Process at the Indiana Department of Transportation for Prioritizing Major Highway Capital Investments", *Journal of the Transportation Research Board* 1816, pp 266–278

Masami S., 1995, "Analytic Hierarchy Process (AHP)-Based Multi-Attribute Benefit Structure Analysis of Road Network Systems in Mountainous Rural Areas of Japan", *International Journal of Forest Engineering* Vol. 7, No 1.

Mihailo M. & Vladan T., & Marijo V., 2015. Functional Classification Of Rural Roads In Serbia, *International Journal for Traffic and Transport Engineering, 5(2): 184–196*

Ramanathan R. 2001, A note on the use of the analytic hierarchy process for environmental impact assessment. *Journal of Environmental Management*, 63, 27–35

R source code for assigning importance weights, 2016, at github.com https://github.com/JamshidSod/HFC. Last accessed: 1/11/17

Saaty T.L. 1980. The Analytic Hierarchy Process, McGraw Hill International.

Seunglim K., Seongkwan M.L., 2006. AHP-based decession-making process for construction of public transportation city model: case study of Jeju, Korea. *Joint International Conference on Computing and Decision Making in Civil and Building Engineering.*

SHNK 2.05.02–07 «Automobile Roads», 2008. Goskomarhitektstroy the Republic of Uzbekistan, Tashkent.

Silyanov V.V., Review of the Roads Appraisal Process in Russia, technical paper, 2004. http://www.transport-links.org/transport_links/filearea/documentstore/331_Valentin%20Silyanov%20Paper%20-%202.12.04.doc. Last accessed: 1/11/17

Sodikov J.I. & Silyanov V.V. 2015, Road Asset Management Systems In Developing Countries: Case Study Uzbekistan, *Science Journal of transportation*, Moscow-China-Vietnam, 2015, Special Issue 6.

Tabucanon, M.T. & Lee, H., 1995, "Multiple Criteria Evaluation of Transportation System improvement projects: the case of Korea", *Journal of Advanced Transportation*, 29(1), pp 127–134.

APPENDIX 1

Analytical Hierarchy Process in R

Four parameters such as Speed, Safety, Environment and Riding Quality were used to assign importance weight to each highway functional classification namely, arterial, collector and local.

```
## First it's needed to load libraries:
library(ahp)
library(data.tree)
## Then, loading input file hfc.ahp and look to the file:
ahpFile <- system.file("extdata", "hfc.ahp", package = "ahp")
cat(readChar(ahpFile, file.info(ahpFile)$size))
## Version: 2.0
## Alternatives: &alternatives
## # Here, speed is assigned (max-120 kmph, in-60 kmph), safety(max - 10, min-0),
## # Environment(max-10, min-0), Riding Quality (max-100, min-0).
## Arterial:
## Speed: 120
## Safety: 10
## Environment: 6
## Riding Quality: 90
## Collector:
## Speed: 80
## Safety: 6
## Environment: 8
## Riding Quality: 70
## Local:
## Speed: 60
## Safety: 4
## Environment: 10
## Riding Quality: 50
## #
## # End of Alternatives Section
```

```
## ##############################
##########
## # Goal Section
## #
## Goal:
## # A Goal HAS preferences (within-level
comparison) and HAS Children (items in level)
## name: Assigning Importance Weight
## preferences:
## pairwise:
## # preferences are defined pairwise
## # 1 means: A is equal to B
## # 9 means: A is highly preferrable to B
## # 1/9 means: B is highly preferrable to A
## - [Speed, Safety, 1/2]
## - [Speed, Environment, 5]
## - [Speed, Riding Quality, 4]
## - [Safety, Environment, 5]
## - [Safety, Riding Quality, 4]
## - [Environment, Riding Quality, 1/4]
## children:
## Speed:
## preferences:
## pairwise:
## - [Arterial, Collector, 3]
## - [Arterial, Local, 6]
## - [Collector, Local, 3]
## children: *alternatives
## Safety:
## preferences:
## pairwise:
## - [Arterial, Collector, 2]
## - [Arterial, Local, 4]
## - [Collector, Local, 2]
## children: *alternatives
## Environment:
## preferences:
## pairwise:
## - [Arterial, Collector, 1/3]
## - [Arterial, Local, 1/6]
## - [Collector, Local, 1/2]
## children: *alternatives
## Riding Quality:
## preferences:
## pairwise:
## - [Arterial, Collector, 2]
## - [Arterial, Local, 5]
## - [Collector, Local, 2]
## children: *alternatives
## #
## # End of Goal Section
## ##############################
#########
## Next is to calculate priorities (eigenvalues)
and visualize our data
hfcAhp <- Load(ahpFile)
Calculate(hfcAhp)
Visualize(hfcAhp)
## And finally is to present resuls in table
AnalyzeTable(hfcAhp)
```

Transport Infrastructure and Systems – Dell'Acqua & Wegman (Eds)
© *2017 Taylor & Francis Group, London, ISBN 978-1-138-03009-1*

Efficient practices in railway ballast maintenance and quality assessment using GPR

L. Bianchini Ciampoli, A. Calvi & A. Benedetto
Roma Tre University, Rome, Italy

F. Tosti & A.M. Alani
University of West London, London, UK

ABSTRACT: The need for effective and efficient railway maintenance is always more demanded all over the world as the main consequence of aging and degradation of infrastructures. Primarily, the filling of air voids within a railway ballast track-bed by fine-grained materials, coming up from the subballast layers by vibrations and capillarity effects, can heavily affect both the bearing and the draining capacity of the infrastructure with major impacts on safety. This occurrence is typically referred to as "fouling". When ballast is fouled, especially by clay, its internal friction angle is undermined, with serious lowering of the strength properties and increase of deformation rates of the whole rail track-bed. Thereby, a detailed and up-to-date knowledge of the quality of the railway substructure is mandatory for scheduling proper maintenance, with the final goal of optimizing the productivity while keeping the safety at the highest standard. This paper aims at reviewing a set of maintenance methodologies, spanning from the traditional and most employed ones, up to the most innovative approaches available in the market, with a special focus on the Ground Penetrating Radar (GPR) Non-Destructive Testing (NDT) technique. The breakthrough brought by the application of new processing approaches is also analyzed and a methodological framework is given on some of the most recent and effective maintenance practices.

1 STATEMENT OF THE PROBLEM

Freights, bulk goods and commuters travel every day between cities, terminals and production poles upon ballasted rail tracks, all around the world. As a result of society development and technological enhancement, the speed of trains is increasing, as well as the requested performance of the track-bed. On the other hand, to ensure safety of the transport and highest standards of productivity, the railway network needs to be effectively and timely maintained, due to expected increased deformation rates by the passing loads.

A railway track-bed can be broadly divided into superstructure and substructure (Figure 1). The former includes steel rails, fastening system and sleepers, whilst the latter is composed of granular

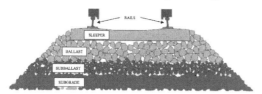

Figure 1. Typical railway track-bed cross-section.

layers laying upon the subgrade, namely, ballast and subballast. In more detail, ballast is a homogeneously hard-rock-derived graded material. Its main functions are to resist to the stresses imposed by passing loads and transmit the vertical forces, properly attenuated, to the subgrade. Thereby, ballast holds the primary roles of retaining the track in its correct position, absorbing the acoustic waves within the air-filled voids, and ensuring proper drainage of meteoric water (Benedetto et al., 2016a).

Differential track settlements are prevalently related to the substructure, whose mechanical response, is the major concern of designers and maintenance managers (Indraratna, 2016). Indeed, whilst the superstructure components react to forces elastically with negligible deformations, the stress cycles imposed by the passing loads induce relevant deformation into the granular layers composing the substructure (Chrismer, 1985). This cyclic loading determines a breakage of the sharp corner of ballast grains and a fragmentation of the weaker particles. Thereby, the grading of the substructure changes (Selig & Waters, 1994; Ebrahimi et al., 2012) with implications on the mechanical behaviour and drainage capacity of the material.

Furthermore, the air-filled voids can get polluted by the deposit of concrete and steel dust produced at the ballast-sleeper and wheel-rail contact (Indraratna et al., 2014), respectively, by the pouring of coal dust from the passing freights (Huang et al., 2010; Tennakoon et al., 2015), and by the upward migration of fine clay particles from the subgrade along with capillary water (Al-Qadi et al., 2010). Mostly, regardless of the nature of the process, this filling of ballast voids by fine particles is referred to as 'fouling'. When the ballast is fouled, especially by clay, its internal friction angle is undermined, with serious lowering of the strength properties and increase of deformation rates of the whole rail track-bed.

In view of the above, a detailed and up-to-date knowledge of the quality of the railway substructure is mandatory for scheduling proper maintenance, with the final goal of optimizing the productivity while keeping the safety at the highest standard.

This study evaluates a set of maintenance methodologies, spanning from traditional and most employed ones, up to the most innovative approaches available in the market, with a special mention to the Ground Penetrating Radar (GPR) Non-Destructive Testing (NDT) technique, whose application in railways is increasingly attracting the attention of researchers, companies and administration. In Chapter 2, an overview of the up-to-date methodologies for maintaining railway ballast is given; hence, the GPR method, pros and cons are discussed in Section 3; then, in Section 4, a scheme of efficient maintenance management is proposed; conclusions and research open issues are finally presented in Section 5.

2 RAILWAY MAINTENANCE

With regards to the aforementioned railway substructure, ballast plays a crucial role in maintaining serviceability of the track and ensuring its structural capacity (Al-Qadi et al., 2008; Al-Qadi et al., 2010). The fouling process causes degradation of the original substructure, and can compromise the performance of the railway track system. At worst, such an occurrence can increase the risk of derailment (De Chiara et al., 2014).

Thereby, monitoring and maintenance stages of track-beds are relevant to ensure high performance of the railway system over its service time, and granting the highest safety standards. At the same time, they are challenging tasks to be carried out due to the significant longitudinal extent of these infrastructures.

Traditional methods for the monitoring and assessment of ballast degradation are visual investigations, punctual drillings and diggings, which are carried out all along the track at discrete intervals. Nevertheless, these methods have several disadvantages. Indeed, visual inspections do not allow for subsurface investigation and understanding of the deep causes of the damage. On the other hand, drillings and diggings do not provide continuous data on the health conditions of the track, although they can return highly reliable information (Al-Qadi et al., 2008; Al-Qadi et al., 2010). Furthermore, these are labour intensive and time consuming (Shao et al., 2010).

Thereby, the recent focus is to move towards the use of time- and cost-effective technologies, capable of performing rapid and non-destructive inspection of the track-bed.

Amongst these technologies, two-dimensional (2D) (Cho et al., 2006) and three-dimensional (3D) (Sun et al., 2014) laser scanners, based on an optical method of investigation, are currently used for railway inspections. In more detail, 2D and 3D imaging methods have been used for assessing critical features such as size and shape of ballast particles.

In the last decades, GPR has attracted great interest in the assessment of railway ballast conditions. To that effect, there is a plenty of literature studies proving the capabilities of such a non-destructive tool to overcome the limits of the traditional investigation methodologies (Hyslip et al., 2003; Shao et al., 2010).

3 GROUND PENETRATING RADAR

3.1 Working principles

GPR is a geophysical inspection technique based on the transmission of Electromagnetic (EM) waves towards a medium, e.g., soil, and the reception of the transmitted or reflected signal. The behaviour of the traveling wave depends on the dielectric properties of the medium itself, according to the EM field theory. In more detail, the wave velocity is influenced by the dielectric permittivity ε, whereas signal loss and attenuation are mainly affected by the electric conductivity σ. The magnetic permeability μ is equal to the free space magnetic permeability μ_0 for all the non-magnetic materials, and does not affect the propagation of the EM waves.

In practical terms, when the EM impulse emitted by a source encounters a dielectric contrast in the medium, it generates a partial reflection and transmission of the signal. By collecting such reflections through a receiving antenna, it is possible to image the subsurface features, in both two or three dimensions.

Nowadays, several different GPR systems are available in the market. Main differences are related to the type of the antenna and centre frequency of investigation as well as to the working system. A comprehensive presentation of the theoretical and practical features of GPR is given by Daniels (2004) and Jol (2009).

3.2 *Railway applications*

The use of GPR for the investigation of railway substructures has mostly developed over the last 20 years. Originally, GPR surveys were carried out using ground-coupled antennas, with centre frequencies usually below 500 MHz. Although these antennas do not provide high resolution, they ensure deep signal penetration. In the last decade, low-frequency antennas have been replaced by higher frequency (1 GHz÷2 GHz) air-coupled systems, bringing many advantages in terms of ballast degradation assessment (Roberts et al., 2006). Indeed, air-coupled antennas are installed on track inspection vehicles, capable to monitor the track conditions continuously and without any interference with rail traffic (Al-Qadi et al., 2008; Fontul et al., 2014). On the other hand, higher frequencies provide greater resolutions and they allow detecting changes in voids volume of ballast, which are likely linked to fouling phenomena (Roberts et al., 2006).

Different features of the GPR signals collected in railway investigations have been taken into account by experts and related to several substructure quality parameters. In Hugenschmidt (2000), a ground-coupled 900 MHz system mounted onto a cart and towed by a locomotive was employed for surveying a railway section. A comprehensive data post-processing allowed the author assessing the ballast thickness and identifying several zones of clay migration upwards. Olhoeft & Selig (2002) used an instrumented vehicle equipped with several air-launched antennas with different centre frequencies to perform a visual multi-frequency analysis of GPR signals collected on both the centreline and sides of the track. The authors argued that the highest performances were reached with the 1 GHz antenna. They also showed how different results coming from tests performed at the sides of the track may indicate stability issues.

A comparison between data from 1 GHz and 2 GHz air-coupled antennas was performed by Roberts et al. (2006). It was noticed how the higher frequency was capable to detect differences in the scattering of EM waves, due to the pollution of air voids within the ballast.

More recently, frequency-based studies on railway fouling detection can be found in the literature. In Al-Qadi et al. (2008), the authors presented a multi-method approach to assess the ballast condition, based on direct analysis, scattering amplitude envelope, and Short Time Fourier Transform (STFT). A 2 GHz centre frequency antenna was used for the surveys.

By means of complex experimental and theoretical efforts, Shao et al. (2010) proposed an automatic algorithm for the classification of ballast conditions, based on the application of Support Vector Machine (SVM) onto processed GPR frequency spectra. The method relies on the assumption that the frequency spectrum is affected by the fouling conditions, in a predictable way, which involves the possibility to recognize the ballast quality by analyzing the relevant frequency spectrum.

Fewer laboratory activities can be found in the literature, although they provide essential information to calibrate railway-based models through a proper characterization of the surveyed materials (Shangguan et al., 2012; De Chiara et al., 2014; Fontul et al., 2014; Benedetto et al., 2016a; Tosti et al., 2016).

Also in view of the above studies, a relatively new railway-related research branch began to capture scientists' interest over the last years (Benedetto et al., 2016b), namely, the numerical simulation of both physical and dielectric properties of ballasted track-beds. Indeed, the simulation of the ballasted structure in virtual environment leads to undeniable advantages. One of the most outstanding is the possibility to replicate and extend the experimental conditions, in terms of both frequencies of investigation and track-bed characterization, by properly setting the numerical parameters, which is a massive save of time and economical resources.

3.3 *Advantages and limitations*

Although GPR and NDTs in general often allow to overcome the limits of traditional approaches, a deep knowledge about their limitations is required. This is a simple rule to avoid overestimation of the methods and misinterpretation of the results, which may lead to detrimental effects. Undeniably, the employment of air-coupled system mounted onto rail vehicles grants a traffic-speed continuous data collection, which involves no disturbance to the railway traffic and a comprehensive monitoring of the track. Furthermore, a regularly scheduled GPR survey is definitely much more cost-effective, if compared to traditional tests. It also holds very few concerns about safety issues, which cannot be said for traditional track-bed drilling and trenching.

In turn, GPR is not a direct method, as it measures the potential effects of fouling on dielectric properties of the whole track-bed. Since such effects can be produced by multi-sources, specialist expertise in the interpretation of the data is mandatory, even in the case of automatic methods. Furthermore, as the EM behaviour of the ballast is highly influenced by size distribution and mineralogy of the aggregates, a proper characterization is always required for calibration purposes. In such a framework, numerical simulation can represent a useful mean of integration for GPR.

In general terms, GPR appears as a powerful tool for reducing costs and simultaneously increasing the effectiveness of railway ballast monitoring. Nevertheless, due to its limitations it cannot be employed as a stand-alone technology at the current state of the art, and it grants the highest performance when integrated with other methods.

4 SCHEME OF EFFICIENT MAINTENANCE

In this Section, a procedural GPR-based scheme for performing efficient railway maintenance is proposed.

Once taken into account the complex environment of railways and the relevant features of GPR, a prompt for an early-stage and non-destructive fouling detection can be listed hereafter (Figure 2).

1. *Physical characterization of ballast.* The first step necessarily consists in a comprehensive knowledge of the physical and dielectric properties of the rock-derived aggregates. Such information can be derived from "ad hoc" laboratory tests performed on materials excavated in situ, or it can be provided directly by the railway network administration. In addition, if the mineralogical nature of the ballast is known, shape and size of the aggregates can be assessed by 2D or 3D laser scan tests.

2. *Numerical simulation.* Once the aggregates and the subgrade are characterized, it is possible to reproduce the survey environment within a virtual scenario, numerically simulated. Thereby, synthetic GPR signals related to different fouling conditions can be generated.

3. *GPR survey.* Hence, GPR data can be collected along the railway section or network. In view of the above, most effective radar configurations are equipped with air-coupled antennas, with centre frequencies ranging between 1 GHz and 2 GHz.

4. *Signal processing.* The collected signal requires now a proper processing in order to remove noise and clutter and, more in general, to reduce the ambiguities and uncertainties. The processing technique must be chosen with reference to the specific model that is planned to be used.

5. *Model calibration.* The parameters used in the selected model need to be calibrated with regard to the physical properties of the tested material.

6. *Model application and fouling detection.* The model can now be applied to the database collected along the track. The expected output is an overall assessment of the track-bed, and the identification of the most fouled areas.

7. *Crosscheck.* The unambiguousness of the results can be significantly raised through a comparison between the data collected on-site and the simulated ones.

8. *Definition of the maintenance priorities.* If the results are confirmed by simulation, a map of ballast deterioration throughout the whole track is now available, and therefore it is possible to prioritize maintenance actions.

5 CONCLUSIONS

This paper aims at reviewing existing practices for the maintenance of railway assets and it introduces a new methodology scheme for the application of effective GPR-based maintenance actions, based on the most recent and up-to-date processing approaches developed in this area of expertise. The first Section introduces the problem, which is tackled more thoroughly in the second Section. Insights on the potential of GPR are given in Section 3, along with a short review of the most acknowledged scientific evidences by literature. In view of the above, several assumptions have been made. Firstly, the most efficient and effective GPR configuration for railway applications is composed of air-coupled pulsed systems, mounted onto rail vehicles and equipped with a high frequency antenna. Secondly, accuracy and reliability of GPR can reach proper standards when integrated

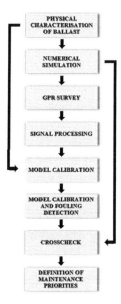

Figure 2. Scheme of effective maintenance using GPR.

with complementary methods. Accordingly, in the maintenance scheme proposed in Section 4, it is worthwhile noting that the application of a GPR model requires proper calibration by a physical characterization of the materials (e.g., performed by laser scanner), as well as a crosscheck by numerical simulation, to validate the results. To that effect, an integration of different methodologies would seem the best solution.

ACKNOWLEDGEMENTS

This work has benefited from the network activities carried out within the EU funded COST Action TU1208 "Civil Engineering Applications of Ground Penetrating Radar."

REFERENCES

Al-Qadi, I.L., Xie, W. & Roberts, R. 2008. Time-frequency approach for ground penetrating radar data analysis to assess railroad ballast condition. In Research in Nondestructive Evaluation, 19, pp. 219–237.

Al-Qadi, I.L., Xie, W., Roberts, R. & Leng Z. 2010. Data analysis techniques for GPR used for assessing railroad ballast in high radio-frequency environment. In Journal of Transportation Engineering, 136(4), pp. 96–105.

Benedetto, A., Tosti F., Bianchini Ciampoli L., Calvi A., Brancadoro M.G. & Alani M.A. 2016a. Electromagnetic characterization of clean and fouled railway ballast through gpr signal processing and numerical simulation. In Construction and Building Materials, In press.

Benedetto, A., Tosti, F., Bianchini Ciampoli, L., Pajewski, L., Pirrone, D., Umiliaco, A. & Brancadoro, M.G. 2016b. A simulation based approach for railway applications using GPR. In proc. of 16th International Conference on Ground Penetrating Radar (GPR 2016), Hong Kong, June 13–16.

Cho, G.C., Dodds, J. & Santamaria, J.C. 2006. Particle shape effectson packing density, stiffness and strength of natural and crushed sands. In Journal of Geotechnical and Geoenvironmental Engineering, 132(5), pp. 591–602

Chrismer, S. M. 1985. Considerations of factors affecting ballast performance. In American Railway Engineering Association (AREA), Bulletin 704, AAR Research and Test Department Report No. WP-110, pp. 118–150.

Daniels, D.J. 2004. Ground Penetrating Radar, The Institution of Electrical Engineers, London.

De Chiara, F., Fontul, S. & Fortunato E. 2014. GPR laboratory tests for railways materials dielectric properties assessment, In Remote Sensing, 6(10), pp. 9712–9728.

Ebrahimi, A., Tinjum, J.M. & Edil T.B. 2012. protocol for testing fouled railway ballast in large-scale cyclic triaxial equipment, In Geotechnical Testing Journal, 35 (5), pp. 1–9.

Fontul, S., Fortunato, E. & De Chiara F. 2014. evaluation of ballast fouling using GPR, In proc. of 15th International Conference on Ground Penetrating Radar (GPR 2014), Brussels, June 30 – July 4.

Huang, H., Tutumluer, E., Hashash, Y., & Ghaboussi, J. 2010. Laboratory validation of coal dust fouled ballast discrete element model, In Paving Materials and Pavement Analysis, pp. 305–313.

Hugenschmidth, J. 2000. Railway track inspection using GPR, In Journal of Applied Geophysics, 43, pp. 147–155

Hyslip, J.P., Smith, S.S., Olhoeft, G.R. & Selig E.T. 2003. Assessment of railway track substructure condition using ground penetrating radar, in Proc. Annu. Conf. AREMA, Chicago, NY.

Indraratna, B. 2016). 1st Ralph Proctor lecture of ISSMGE. Railroad performance with special reference to ballast and substructure characteristics, In Transportation Geotechnics, 7, pp. 74–114.

Indraratna, B., Nimbalkar, S., Coop, M. & Sloan S.W. 2014. A Constitutive model for coal-fouled ballast capturing the effects of particle degradation, In Computer and Geotechnics, 61, pp. 96–107.

Jol, H. 2009. Ground penetrating radar: theory and applications, In Elsevier ed., Amsterdam, The Netherlands.

Olhoeft, G.R. & Selig, E.T. 2002. Ground penetrating radar evaluation of railway track substructure conditions, In 9th International Conference on Ground Penetrating Radar (GPR 2002), Santa Barbara, California, April 29–2 May.

Roberts R., Rudy J., Al-Qadi I.L., Tutumluer E. & Boyle J. 2006 Railroad ballast fouling detection using ground penetrating radar – a new approach based on scattering from voids, In 9th European Conference on NDT.

Selig E. T. & Waters J. M. 1994. Track geotechnology and substructure management. In Thomas Telford ed., London.

Shangguan, P., Al-Qadi, I.L. & Leng Z. 2012. Groundpenetrating radar data to develop wavelet technique for quantifying railroad ballast-fouling conditions, In Transportation Research Record, 2289, pp. 95–102.6

Shao, W., Bouzerdoum, A., Phung, S.L., Su, L., Indratatna, B. Rujikiatkamjorn C. 2010. Automatic classification of ground-penetrating-radar signals for railway-ballast assessment, In IEEE Transactions on Geoscience and Remote Sensing Vol. 49(10), pp. 3961–3972.

Sun, Y., Indraratna, B. & Nimbalkar S. 2014. Three-dimensional characterisation of particle size and shape for ballast, In Géotechnique Letters, 4 (3), pp. 197–202.

Tennakoon, N., Indraratna, B., Nimbalkar, S. & Sloan S.W. 2015. Application of bounding surface plasticity concept for clay-fouled ballast under drained loading, In Computer and Geotechnics, 70, pp. 96–105.

Tosti, F., Benedetto, A., Calvi, A., & Bianchini Ciampoli, L. 2016. Laboratory investigations for the electromagnetic characterization of railway ballast through GPR. In Proc. 16th International Conference of Ground Penetrating Radar (GPR 2016), Hong Kong, June 13–16, 2016.

Transport Infrastructure and Systems – Dell'Acqua & Wegman (Eds)
© 2017 Taylor & Francis Group, London, ISBN 978-1-138-03009-1

BIM experience in infrastructural large projects: Doha Metro— Al Jadeda Station al Matar B case study

A. Ingletti, M. Scala & L. Chiacchiari
3TI PROGETTI ITALIA SpA, Rome, Italy

ABSTRACT: Civil engineers and architects typically outline projects and construction data using technical drawings. At the beginning of the XXI Century, the development of new modeling technologies has led to introduction of a new intelligent approach to 3D design: the Building Information Modeling (BIM). BIM enables to improve the planning, design, construction and management of infrastructure or buildings for the entire life cycle. 3TI is involved in Doha Metro Red Line South Underground main stations project, as consultant in architectural finishes and MEP works of Al Jadeda Station al Matar B. BIM process has been efficiently applied. This paper wants to underline that all project phases have been studied using this new technology, from the analysis of the two-dimensional drawing, through the construction of a 3D model, achieving an high detailed level for every single designed part. All specialists involved in the project have the knowledge of the process and have to manage their updating in order to reach a global understanding. Design coordination is the responsibility of all parties (meant as specialists of different sectors) as an ongoing process until the completion: the outputs are not simple representations but a set of information. Definitely, this paper discusses the importance of BIM technology in Civil Engineering and Architectural sectors, as an important mean for the time optimization compared to the traditional approach and the reduction of costs, thanks to the prevision and the resolution of the issues, during the planning through the interface checking. The analysis of the case of study highlights as the coherence of BIM procedure application enables to adopt a maintenance strategy leading to a longer life-cycle of Al Jadeda Station.

Keywords: BIM process, interface checking, maintenance strategy

1 INTRODUCTION

1.1 *BIM System*

The Building Information Modeling technology represents a powerful process of development and use of a computer software model able to integrate data from several users in one Building Information Model. This "Model Based" methodology, as called in AGC (2006), encloses a three-dimensional, geometric, object-oriented representation of the project and has data attributes making the model "intelligent".

Collaboration between professionals and all the figures involved in the project is ensured, thanking to the flexible environment of work.

Indeed, design and project data are well-structured by the use of several software for analysis (such as Risa 3D, RAM, STAAD, ETABS), design (Revit and Architectural Desktop, Dynamo Studio, Bentley's MicroStation…), coordination, detailing and rendering software (Navisworks, Tekla Structure, and Graphisoft).

The design and construction practices can be digitally managed by BIM tools, thus the whole life-cycle of the project can be monitored.

The composition of all the items and data of a project represented by the System Implementation phase gives an integrated and structured model, threedimensional, smart and ultra-detailed, able to cover several topics:

– Evaluation
– Collaboration
– Implementation strategy
– Cost and Time
– Quality and Safety
– Deliverables
– Design and Documentation
– Contract administration
– Commissioning.

As highlighted by Succar (2008), BIM process assures interconnections within different fields: procedures in BIM, players and deliverables are interlocked, so that guidelines and best practices, professionals' activities and technologies (for software, networks and communication) do not represent self-contained compartments, but mutually enrich each other.

The above mentioned mechanism of "knowledge transferring" within sectors leads to relevant technical, operational and business advantages, thanks to a greater control on the projects components. Particularly, a correct and clear implementation of all the information ensures a successful development of the project and a reduction of modification and reissues, which results in efficiency maximization and time saving. In this way BIM can be defined a *model checker*, since the first phases of project implementation up to the operational phase, and more, as for example manufactured product information, can be uploaded into the BIM database too, and this allow even to budget and schedule maintenance.

Essentially the BIM is not simple software but a complex management system in planning steps of a building or infrastructure. The 3D modelling, in fact, includes dimensional, quality, timing and coordination data, and along with outputs provided, such as work order flows, costs, accounts, and financial elements, it supports the decision-making processes. Therefore, the key business drivers can now rely on Just-In-Time (JIT) delivery of materials and equipment, which is an outcome realizable by complete coordination prior to the start of construction, very early during planning stage. In addition, the possibility to simulate maintenance procedures allows to asset a thought plan or verify the impact of retrofits or maintenance works in advance, besides aligning the decisions, with the goal of wanted high maintainability-levels. Lifecycle benefits regarding maintenance are guaranteed, and a robust management strategy can be then construct for the adoption of BIM.

1.2 *BIM experience in large projects*

3TI boasts numerous valuable experiences in the preparation of projects of major structures and infrastructures in BIM system.

A large project carried out with BIM technologies is the new 520-bed hospital in La Spezia. This fan-shaped building opens up to the sea, in order to not interfere with the panoramic view. Public spaces have been designed, along with a building of radiation therapy and about 750 parking spaces carefully hidden by a series of terraces.

Furthermore, 3TI has developed five stations within the development plan of the metro lines at Doha (Qatar), as detailed in the next paragraph.

On these and other projects, all the complexity of BIM technology has emerged, together with its great functionality.

Particularly, the consolidated experience of 3TI enables to claim that BIM technology is actually a pervasive focus on information, not limited to a representation. Thus, as per the Company's understanding, in order to optimize the power of BIM, it is important that model, workflow, family and design settings are set up at the beginning, in order to converge toward the desired central point of clear management of the complexity of the project.

All design phases have to be thought in BIM philosophy and of course, all specialists involved in the project have to know and manage BIM technology in order to achieve the best benefits.

Rightly related to the latter point, of advantages can be listed a few:

- designing only once;
- simultaneous updating of plant, sections and elevations;
- functional capability on improving cost analysis;
- smaller project team necessity;
- advanced understanding of design constructability;
- interference checking, with finding "mistakes" before they are made;
- revisions in much less time.

In this sense, the integration within parties is important and 3TI successfully created a collaborative BIM environment. This approach has been adopted within both the Professionals hierarchy and technical staff, enhancing also the staff-tasks in using the BIM solution day by day. The design models have been processed through an iterative activity of model-based design-coordination-training, to insure the constructability with greater focus on clashes.

2 DOHA METRO PROJECT

2.1 *Overview on the project*

Doha Metro is a singular and prestigious project of the new passenger rapid transit system in Qatar's capital city, which is actually under construction and in the near future will establish easier and more convenient connections within the city's central areas as well as the outskirt areas of Doha (Figure 1). The aim of the project is to promote public transportation as a valid alternative to private means of transport.

The planned construction stage is divided into two phases: the first will see the construction of three of the four lines (Red, Gold, and Green line) and 37 stations. These lines are expected to be open to the public in late 2019. The second phase will be completed by 2026, and will involve the expansion

Figure 1. A panoramic view of Doha (Jon Bowles photo-gallery).

of the Phase 1 lines, and the construction of an additional one, the Blue Line. Others 72 stations will also be built.

Within the Qatar Integrated Railway Project (QIRP) the appointed Architect UNStudio has developed the new Doha Metro Network design for the 35 stations in package Phase 1, and around 60 stations in package Phase 2.

Doha metro project requires a careful planning to respect the strict timing and to minimize the environmental impact. Thus, in order to guarantee flexibility at the project implementation, UNStudio has drafted an Architectural Branding Manual, which includes a set of design guidelines, architectural details and material outlines. The Architect has produced this manual also for assuring the spatial quality and clarity of the network. All the appointed Design&Built contractors have to refer to it, and respect the architectural of the stations, which reflects the heritage of the country in shape and colours.

Likewise, the manual establishes the phased project timetable, so that the entire Project life cycle is divided into four stages. Specifically every phase of the project design is named:

- M1 – Concept Phase
- M2 – Preliminary Design
- M3 – Detailed Design
- M4 – Construction Design.

This staging is reflected on the items of which the each station is composed.

2.2 Branding Manual

As basis for the design phases UnStudio has provided a detailed manual, which includes guidelines and practical specifications to be followed and respected for the project.

The Manual represents the guidance on the production of various elements which compose the station units, up to the more elementary detail.

Every single element, is included in the book described by means of schedules composed by designs and detailed descriptions, this finalized to realize an efficient construction site, with a high level of quality control. Modular forms are described, along with all the elements' features, such as shelters, ceilings, walls, floors, and the layout configurations, equipment, wayfinding and furniture.

2.3 The 3TI BIM Design Stations

3TI has been involved in the project of metro Doha for the design of many stations of red line and green line. The red line, called "Cost Line", runs from the north to the south for 40 km, has 17 stations and connects the city center to the Hamad International Airport. The green line, called "Education Line" runs from the east to the west, has 8 stations and passes through the Education City.

The Company has developed five stations of the Doha metro project, four of the red line (Al Doha Al Jadeda station—AMB, Oqba Ibn Nafie station—AME, Al Matar station—AMD, Umm Ghuwailina—AMC) and one of the green one (Al Mansoura station—Al Khubaib, Figures 2–3). On the basis of stated criteria and defined elements' characteristics by means of the Branding Manual, 3TI Progetti has developed phases M2, M3, M4 for each station, falling within its assignment (Tables 1–2).

Al Doha Al Jadeda station is the *pilot station* and any other stations have been developed on the basis of this major project, by reaching in the end the same level of detail and definition of the M4 phase.

Figure 2. Doha Metro Project—Red Line.

Figure 3. Doha Metro Project—Green Line.

427

Table 1. Red line stations.

Station name	Al Doha Al Jadeda (AMB)	
	Item	Phase
	Landscaping	M3–M4
	Architectural detailed design and finishes services	M3–M4
	Entrances and Pop ups	M2–M3–M4
	Signage	M3–M4
	BOQ	M3–M4
	Video animation	

Station name	Oqba Ibn Nafie station (AME)	
	Item	Phase
	Landscaping	M2
	Architectural detailed design and finishes services	M4
	Entrances and Pop ups	M2–M3–M4
	Signage	M3–M4
	BOQ	M3–M4

Station name	Al Matar (AMD)	
	Item	Phase
	Landscaping	M2
	Architectural detailed design and finishes services	M4
	Entrances and Pop ups	M2–M3–M4
	Signage	M3–M4
	BOQ	M3–M4

Station name	Umm Ghuwailina (AMC)	
	Item	Phase
	Signage	M3–M4
	BOQ	M3–M4
	Video animation	

Table 2. Green line stations.

Station name	Al Mansoura (Al Khubaib)	
	Item	Phase
	Landscaping	M2
	Architectural detailed design and finishes services	M4
	Entrances and Pop ups	M2–M3–M4
	Signage	M3–M4
	BOQ	M3–M4

For better and more effective management of projects and design files, each station has been splitted into 3 items: Item A—Station Box, Item B - Entrances and Pop ups and Item C—Landscape, as can be seen in Figure 4.

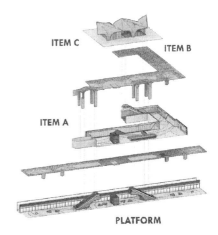

Figure 4. The general simplified Station Components (Branding Manual) - Landscape around.

2.4 *Al Jadeda Station Al Matar B case study*

Al Doha Al Jadeda station is placed on the junction Al Doha Al Jadeda Street—Bring Road. The station serves a mixed-use area, that includes a Civil Defence building, a main police station, hotels, retail and residential buildings.

3TI has been awarded for the design finishes, structures and Mechanical, Electrical and Plumbing (MEP) Engineering services for items B—Entrances and Pop ups; the same services have been carried out for Item A - Station Box, except structures. In both cases have been developed signage and bill of quantity (BOQ).

Al Doha Al Jadeda station is on four levels and has four public entrance pods with vertical circulation, single lift, stairs and two escalators; these elements connect the station box to the concourse level, which includes staff and public facilities as prayer rooms, toilets, etc. and opens vertical circulation by stairs and escalators to the platform level. The station box results in a platform level (island type) with single tracks and three sections of subways.

There are four main entrances. Entrance N°1 is adjacent to the station concourse and it is connected to Entrance N°2 trough a subway under Al Matar Street to North East, and Entrance N°2, in turn, is connected by a subway to North West to Entrance N°3. The Entrance N°2 has a short length and will be connected with Entrance N°4 after a subway extension. This last entrance will be constructed in a different time, but this solution will not have negative effects on the pedestrian area or affect emergency escape conditions.

Materials are fundamental components of architectural fit-out work materials specified within the Architectural Branding Manual. Modularization and prefabrication is the base of each single element defined in the Manual. This strategy aims to facilitate construction operations, developing dimensionally and functionally flexible products. Prefabrication means to realize functional parts of a work in advance and outside construction site, giving the opportunity to reduce costs or complexity.

This approach is applied to different stages of the Al Jadeda station project and the BIM technology represented the best way to design and pre-assembly integrated system units. BIM permits to create parametric elements to repeat endlessly: the structural pieces are controlled by the parametric structure within properly called *families*, allowing their forms to be automatically generated and later fabricated (Figure 5).

In Al Jadeda Station, petal panels fill all areas around the ceilings' vaults, which are the main characteristic elements of the local architecture sinking its roots in the traditional heritage. They are modelled for example by modular elements in the shape of petals (Figure 5–6).

The architectural design is not the only technical sector which has been developed by means of BIM applications. MEP systems in fact, have been designed in BIM: ducts for HVAC systems, piping runs for water and gas supply and disposal; routing trays and control boxes for electrical and communication systems and all the related activities have been carry out by BIM technology.

Figure 5. Shape and families characterization (3TI Progetti).

Figure 6. Doha vault-game (UNStudio website).

In addition to technical drawings, many rendering and videos, to easier and clearer explain the developing of the station project to the Client, have been recorded. These videos were also used to publicize the project to a non-technical audience.

2.5 BIM Environment

Designing in "BIM oriented" way means communicate, through the exchange of a unique model, with colleagues and partners, without loss of quality. BIM allows indeed collaboration between designers and also interoperability between software tools.

In the presented Al Jadeda Station case study several software for analysis, design, coordination, detailing and rendering (mostly Revit, Naviswork and Dynamo) have been used. By means of these tools, it is possible to build a unique big model, by linking many objects and sources together, also attaching attributes and merging different file types. Follows an in-depth examination about software mentioned before.

The entire project has been developed in Revit. This station is a very complex architectural and engineering object, and designing it using traditional software would have been certainly much more complicated in planning, time consuming control, reports production and technical management.

Revit is a software to design with parametric modelling that, unlike Autocad, works not with simple lines, but with well-defined objects of complex shapes. It works in three dimensions and allows multiple users to work simultaneously on the same document through an internal system called worksets. This system consists on a server-based method of work-sharing which allows team members to work on a local file (authorized based on its role). Every update of secondary file is synchronized on the main file. This process allows a parallel constant updating. The whole concept of BIM includes Revit Architecture for planning and architectural design, Revit Structure, for structural management of the project and Revit MEP for the plants, all the three active simultaneously on the same project and, therefore, on the same file. Once an amendment is made on the 3D model, the software modifies directly and simultaneously plans, sections, elevations and details' sheets. The design simplification and flexibility is facilitated by using templates and libraries that are the basis for definition of an object model and its assembly. "Families" are the parametric components that make up the graphics system for designing complex sets (furniture and interior design) and simple ones (walls and pillars). Even the printing process of documents is more simple and rapid (automatic scales both on the basis of line weights both according to the scale of views).

The BIM new approach reduces planning management time during the whole design process. In this way, areas and volumes, energy analysis, lighting and automatic material take-offs doable can clearly be defined at the earliest stages of the project. The software is intended for all experts involved in architectural and engineering planning and technical designers of construction companies, because its main objective is to control the project evolution from concept to detailed design, keeping control on time and costs.

Naviswork is then a software able to integrate the design itself with the as-built world. It enables to have control of the project from a global point of view: the holistic view of the project undoubtedly allows managing models, identifying and clashing the models for interference, and virtually constructing a building using a construction timeline. Within the many potentialities and resources of the tool, it can be mentioned that it offers the possibility of live interactive walkthroughs, to be used as demonstration for clients or contractors.

Al Jadeda Station project has been coordinated also by means of the use of this software, particularly for the revision, simulation and analysis of the modelling. In the studied case Naviswork has been of great importance in detection, identification and management of interferences in a more effective way. The main file contains external links of all different technical files (drawing files, MEP files, structural files, etc.), on which the experts constantly and simultaneously work. In this way Navis "annotates" every updating of the files and allows locating and predicting problems before building (imperfection detected between a structural component and airducts can be seen in Figure 7).

When it identifies interference, under a key-parameter finding, a circle in red color is showed (Figure 8) around the clashing area (also comments and annotations can be added, which will be visible in every file of the projects too). In the Al Jadeda Station, for example, Naviswork has helped to identify interference between the ceiling and an utility duct: during the modelling phase, MEP engineers had updated the size of duct in a larger one, and it was interfering with ceiling at the moment of analysis; consequently the height of ceiling have been lowered to solve the defect.

BIM tools have been useful also for determining the right positions of catwalk accesses for inspection: maintenance of plants is guaranteed by means of this 90 cm high technical floor, and the panels of finishes are interrupted at wanted intervals which respect the fusion of functionality of Architecture-Structure-MEP.

Figure 7. Structure-Duct interference detection (3TI Progetti).

Figure 8. Plumbing fixtures (floor drain)-Structural plinth floor clash detection (3TI Progetti).

Dynamo is finally a software able to develop complex parametric geometries, and to make the using of all kind of data-processing more effective and accurate than before. It provides faster solution of problems regarding geometry and models, and is able to transform simple drawings into interoperable workflows for analysis, documents and simulation. This software defines various text-based programming languages, creates relationships and sequences of actions that make custom algorithms.

As regards the case study, 3TI has used Dynamo software to construct the basic geometry of the modular element "vault", both in the platform and in subway level, taking advantage of the software's ability in processing parametric elements that can repeat endlessly. As known, the delivery of the final documents of a project involves the preparation of a huge number of documentations and drawings. The use of this software has enabled 3TI to significantly reduce time in the preparation of title blocks of the drawings, as once defined the scheme, the patterns, matrices or images positions, then automatically and rapidly, drawings' layouts in all deliverable are completed.

2.6 BIM—More than 3D

One of the most important innovations offered by BIM in the planning process is the introduction of the "fourth dimension" (4D—the time): with specific tools, in fact, BIM will track all temporal stages of the project from design to construction, allowing the interaction with the actual "step".

Once concluded the phase M4 of 3D modelling, up to the level of detail of shop drawings, it is possible to define times and ways for site and construction management, by adding to each element, already modelled, many detailed parameters, agreed with the construction company. In this way, quantity of all elements in the construction site and costs can be controlled. BIM is able to indicate all information recorded by users and more: ranging from number of walls to number of furniture, even neon maintenance or bulbs life before next replacement.

As to each elements of the project it is associated a precise value, the insertion or deletion of any object from the model involves a change in the overall construction cost, which can be controlled minute by minute in the design phase. Therefore, BIM includes also a fifth phase: 5D—cost control, which can be managed before the construction site setting-up.

As regard to Al Jadeda Station project, these stages have not started yet.

2.7 Benefits and disadvantages

As regards the different disciplines, a list of benefits can be analysed, noticing that it is reflected in the outcomes of each BIM-structured-model, the high contribution in work effectiveness.

In the architectural field, BIM users make the difference, as the greater efficiency in graphical interface and communication within informatics tools and areas of work are the winning components.

In order to guarantee that the available functions used are fully and correctly tamed, staff has to be qualified and able to quickly transfer the model, between various tools, without loss of quality.

For those who still works with "no BIM" computer packages, separated files (Figure 9) or in bulk documentations have to be controlled, each single element designed: for instance ones useful for the definition of bill of quantities, ones related to the development of photorealistic images, once for land register interaction, once to carry out the energetic studies and produce certification activities etc.

Modeling the Al Jadeda Station directly in 3D has allowed to have the full awareness of the evolution of each element of the project and to identify any design flaws, which have been immediately resolved. This has been possible thanks to BIM technology, as described at paragraph 2.5, and also their easier dialogue mode in the integration within all the parent tools. In particular, in those models that involve significant calculations, structural analysis, dynamic assessment, the computational

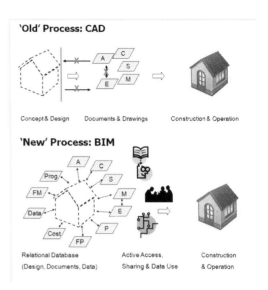

Figure 9. Conventional CAD vs BIM Approach, Azhar et al. (2008).

burden can be reduced due to less linking time between modelling packages and simulating calculation programs. The transferring phase of sections' characteristics, used materials' features or manufacturing components units is cancelled, by eliminating any potential human error, as the user always working on updated model.

Another important benefit using BIM is reduction of processing time of bill of quantity, as indicated in the BIM Handbook (2008). In case of study from the model tables containing any data of project items like space areas, number of different element and their technical characteristic, etc... have been taken out automatically. These sheets have hugely reduced time of calculation as well as human errors. Furthermore making use of parametric elements directly linked with their main outputs in automatic tables, it was possible to quickly suppose the costs of the other station, in which the same objects with same technical characteristic were used.

Extremely significant after the BIM birth, has been the reduction in time spent for the identification of complications between networks and plants for services (such as passage of electric cables, pipes and ventilation, gas or smoke ducts interfering with structural elements).

It is clear that designing in BIM means to be expert to control all the details, communicate with colleagues and partners (which potentially are using other software), sharing all data and project information. For this reason the Company needs more qualified resources with deep knowledge of this technologies, the ones who can be called *bimmers*. This latter point conduct to the arising of the production team costs at first step.

The direction that is taking the actual designing mode evolution is represented by the adoption of techniques ever more modern. The evidence of the quality BIM-based design, and the recognised benefits will lead to the standardization of the processes and technologies. In the meanwhile in technical specifications is noticeable that innovative solutions will experience a smooth transition to BIM, as per Art. 23 of the Legislative Decree 18 April 2016, n. 50 "New Code of the public contracts".

2.8 *Economic advantages*

The Designer Company 3TI during the design of Al Jadeda Station (and other stations) have experienced and appreciated that BIM gives a significant reduction in designing times and running calculations. An example is identified in the possibility that BIM tools allow to have the bill of quantity as an output of the software: it is precise and correct (provided that all the elements inserted in the model are right and not doubled), not a business of BOQ Expert anymore (even though the Expert of course controls the output of the software). Furthermore the Design Engineer Company may rely on the experience gained over years of working in the BIM Environment particularly for the reduced time of implementation of drawings on only one model platform instead of on a set of different software's decks. Less time is translated into a better undertaking and greater productivity from the workers, which lead in minor commitment of human resources.

Moreover the Designer can guarantee to the Construction Company anytime the awareness of the dimension of the project. Every step of planning process can be monitored and controlled and the company can have information at all moments on technical specifications, sizes and characteristic of each element, cost's estimate, time schedule, location of materials' storage, etc. thanks to the use of BIM by the Designer. Obviously cost benefits for construction company comes after phase M4, from phase 4D—site management, as better anticipated in previous paragraphs. Finally, the most important thing is that the global vision of the project guarantees the higher standard of security: in an interactive way, banned areas are 'marked out' by sound emitted from transceivers when a danger is signalled.

Stanford University Center for Integrated Facilities Engineering in the studies listed in the annual technical report CIFE (2008) based on 32 major projects using BIM, indicates benefits such as studied by Azhar et al. (2008):

- up to 40% elimination of unbudgeted change;
- cost estimation accuracy within 3%;
- up to 80% reduction in time taken to generate a cost estimate;
- a savings of up to 10% of the contract value through clash detections;
- up to 7% reduction in project time.

2.9 *Conclusions*

Having gambled in Building Information Modeling (BIM) as a design tool and management of the entire life of an infrastructure, 3TI has been found many advantages in Italy as well as abroad.

"The Italian engineering companies, in fact, in terms of quality and technical skills (of which it is definitely now part of the BIM), can boast strong competitiveness abroad, and assuming that critical mass is sufficiently high for addressing major projects, they are able to excel in all areas" (interview to the 3TI Chairman).

This paper aims to emphasize how great and powerful has been found the use of BIM system in modelling and design projects, as per 3TI experience, and how it has been needful for the design of case study.

Control and validation of models is possible in every step of the project: indeed, in the BIM workflow verification of data does exist consistency, whether they come from the same company as well as by external professionals working on/with the same model. The interoperability is surely the attribute that allows the proper communication among all workers involved, avoiding economic waste.

Certainly it is essential a control highlighting beforehand the design and planning inconsistencies, as well as real interference between the construction components, as has been experienced in the petal vaults' ceiling of the Al Jadeda Station design. The model checking allows through customized rules (internally agreed by users), to optimize the process and make it easily communicable to stakeholders.

Although the advantages offered by a technology such as BIM are obvious, and appear increasingly clear to those who use it, the difficulties to its spread are many. For example, the effort of design firms or companies for transition from traditional CAD to BIM is considerable (both in terms of software implementation and in staff's recruitment and training).

It is also necessary to say that if it is true that BIM represents an excellent solution for quality guarantee, it is also true that the *bimmers* have to produce a clear and "clean" model. So, while a well-structured BIM model, created from the beginning following quality standards, would be certainly manageable in any aspect and without any significant difficulties, a model poorly controlled will lead to complications: the re-editing of a model constructed for example with overlapped lines or components' layers not right placed, would be time-consuming and costly.

Progressive use of specific methods and electronic tools, such as electronic modelling and information for the construction industries, infrastructure companies and engineering firms, is moving the Italian Legislator to insert notes in the Standards accordingly.

ACKNOWLEDGEMENTS

The authors are thankful to Arch. Franca Francescucci for her diverse support and valuable advises, excellent cooperation in the content of the paper and suggestions for improvements, and to Eng. Luca Mezzadri for his technical point of view and images.

REFERENCES

Associated General Contractors of America (AGC) 2006. *AGC Contractors' Guide to BIM*. Available on line at Whole Building Design Guide website: https://www.wbdg.org/bim/bim_libraries.php?l = a.

Azhar, S.; Nadeem, A.; Mok, J.Y.N.; and Leung, B.H.Y. 2008. Building Information Modeling (BIM): A New Paradigm for Visual Interactive Modeling and Simulation for Construction Projects. Proceedings of the First International Conference on Construction in Developing Countries (ICCIDC-I), August 4–5, Karachi, Pakistan.

Eastman, C., Teicholz, P., Sacks, R. & Liston, K. 2008. BIM Handbook: A Guide to Building Information Modeling. Canada. John Wiley & Sons

Gao J. and Fischer M. 2008. Framework & Case Studies Comparing Implementations & Impacts of 3D/4D Modeling Across Projects - Technical Report #TR172 March 2008. Center for Integrated Facility Engineering (CIFE), Stanford University. Available online at http://cife.stanford.edu/sites/default/files/TR172.pdf.

Legislative Decree 50/2016. Regulation of public procurement regarding works, services and supplies implementing Directives 2014/23/EC, 2014/24/EC and 2014/25/EC

Paloscia, L., Interview to 3TI Chairman Ingletti Alfredo, available on line at the Strade e Autostrade on line website: http://online.stradeeautostrade.it/notizie/2016-05-25/fatturato-in-crescita-per-3ti-progetti-53179/

Succar, B., Building information modelling framework: A research and delivery foundation for industry stakeholders. *Automation in Construction,* 2009, 18, 357–375.

UNStudio website: http://www.unstudio.com/projects/qatar-integrated-railway-project-qirp

Transport Infrastructure and Systems – Dell'Acqua & Wegman (Eds)
© 2017 Taylor & Francis Group, London, ISBN 978-1-138-03009-1

The pavement management system for improving airports and roads infrastructures

G. Battiato & L. Cosimi
RODECO Group, Voghera, Italy

C. Stratakos
NAMA LAB S.A., Athens, Greece

ABSTRACT: The PMS (Pavement Management System) provides a set of objectives and well-organized procedures for determining areas having high maintenance priorities and work plans, assigning resources and budgeting for pavement M&R (Maintenance and Rehabilitation). It also can be used to quantify information and furnish detailed recommendations for works required to maintain a set of pavements at a satisfactory level of service and safety while minimizing the cost of M&R. By the PMS it's possible not only to evaluate the actual condition of a pavement network, but also to predict their future condition through the use of pavement condition parameters, such as residual life and overlay, IRI and PCI ect. Since the late '80 s RODECO Group has developed RO.MA® (Road Management) PMS & Asset Management software, in order to increase the quality of the Roads and Airports Infrastructures in the short, medium and long term, for improving pavement durability and safety.

1 INTRODUCTION

The purpose of a PMS is to improve the quality of airport infrastructures and plan maintenance and rehabilitation measures in the short, medium and long term, maximizing available resources that are not always adequate to the customer needs.

RO.MA.® PMS satisfies these requirements as asset management tool, easy to use and can be customized following recommendations of the Road and Airport Agency.

The survey of functional and structural characteristics of the pavements, using high-performance systems, is the basis of PMS: the most important parameters, measured in the infrastructure and used by RO.MA.® PMS software, are the bearing capacity (measured with HWD) for ACN-PCN evaluation, roughness and planimetric profiles, surface distresses and PCI (Pavement Condition Index), skid resistance.

RODECO Group has recently developed and introduced new technologies for automatic measurement of the PCI, IRI and distress; in particular the ADE System (Automated Distress Evaluation) which allow more extensive and widespread use of high-performance systems for surveying and quality control of road and airport infrastructures.

The RO.MA.® PMS has been continuously updated relating to the evolution of high-performance systems, the new needs expressed

by users and, in particular, the international experience gained by RODECO in the field of Pavement Evaluation and Management System.

Particular attention was engaged in software development, simplification and flexibility of data entry procedures, in order to reduce the amount of information necessary to obtain the results of the PMS.

The main requirements of the software are the following:

- provide a methodology to assess the pavement condition (surface and structural parameters), using the Pavement Quality Index (PQI);
- provide a device to define an optimum plan of scheduled maintenance and rehabilitation of airport and road pavements, with a priority list of interventions;
- estimate the optimal time to apply maintenance measures, using appropriate decay models to predict pavement condition in the future;
- propose planned maintenance activities optimizing the cost/benefit ratio, with some budget constraints introduced by the user.

Some recent RO.MA.® applications (as Larnaka and Sarajevo Airports) have specifically highlighted as an efficient PMS system can optimize the maintenance on a multiannual basis, substantially reducing the overall cost of management of airport infrastructure.

2 PAVEMENT EVALUATION

RO.MA.® PMS software was designed to process data coming from the analysis of the results of the Pavement Evaluation phases (field tests). For the evaluation of structural and surface characteristics of an airport pavement network, there are different types of high-performance systems, which allow to quickly record all the parameters required for a proper evaluation of the PMS.

According to ICAO Annex 14 and ENAC "Italian Civil Aviation Authority", the typical pavement surveys, that it's possible to conduct, are the following:

* definition of the pavement bearing capacity with HWD (Heavy Falling Weight Deflectometer) and calculation of ACN and PCN parameters;
* survey of longitudinal and transverse profiles using Laser Profilometer;
* measurement of pavement surface condition and evaluation of PCI (Pavement Condition Index);
* investigations with GPR (Ground Penetrating Radar) and coring to define the pavement stratigraphy;
* survey of the skid resistance.

2.1 HWD (Heavy Falling Weight Deflectometer)

The structural characteristic of an airport pavement network are analyzed using the Heavy Falling Weight deflectometer (HWD), which can adequately simulate the load condition of an aircraft.

For each HWD measuring point, through the software RO.ME. (Road Moduli Evaluation, developed by RODECO), are estimated:

* the values of the E1, E2, E3 moduli (asphalt layers, subbase and subgrade) under test condition;
* the value of the E1 modulus (asphalt layer), referred to 20°C;
* residual fatigue life of the pavement in years;
* the critical layer;
* the calculated theoretical reinforcement necessary to support the project traffic, in mm;
* calculation of ACN/PCN in accordance with ICAO.

2.2 Laser profilometer

The Laser Profilometer is used for the survey of the pavement roughness, longitudinal and transverse profiles and the definition of the index IRI (International Roughness Index), measured in mm/m.

The IRI is a standardized index that contains the information required to establish the regularity of a pavement surface, as defined by the World Bank Technical Paper No. 45.

Figure 1. RODECO Laser Profilometer during Malta International Airport surveys.

The irregularities of a pavement are the result of an infinite number of wavelengths, componing the longitudinal profile of a pavement.

The profilometer can detect the actual profile of the pavement in the XY coordinates (relative), where X is the distance measured by the odometer and Y represents the elevation profile of the pavements.

Generally, the system is capable of storing the actual average profile every 100/200 mm of each section.

The knowledge of the amplitude to the shortwaves (1–3.3 m), medium (3.3 m - 13 m) and long (13 m - 60 m) is very important to identify the cause of the irregularities.

Where the irregularity is related to the short waves, the problem can be found in the surface layers of the pavement (surface distresses), while the irregularities related to the long and medium waves may be due to problems of subsidence in the bottom layers.

The software allows, through simulation, to analyze the filtered results obtaining the values of irregularities at the wavelengths desired.

Starting from the real profile, the following parameters are estimated for sections of 25 m:

* IRI (International Roughness Index) averaged in mm/m on 25 m sections;
* irregularities filtered to short waves from 1 to 3.3 m, in mm/m, on sections of 25 m;
* irregularities filtered to medium waves from 3.3 to 13 m, in mm/m, on sections of 25 m;
* irregularities filtered to long waves from 13 to 60 m, in mm/m, on sections of 25 m;
* simulation of a 3 m straightedge for calculating the maximum deflection, as required by ICAO standards;
* the maximum deflection on sections of 45 m and the number of irregularities that in this

section exceed 20–30 mm as required by ICAO standards;

- cross slope (%).

The equipment of RODECO Laser Equipment includes:

- various lasers, for a width of about 3 meters, with a sampling frequency of 5 mm;
- n. 3 accelerometers and n. 2 high-precision gyroscopes;
- software for raw data processing, to calculate cross-slope, rutting, longitudinal profiles (IRI) and irregularities for short, medium and long waves.

2.3 *Distress Survey: PCI (Pavement Condition Index) analysis*

The survey of the surface distress for rigid and flexible pavements is necessary to:

- control the surface degradations and their evolution over time;
- identify degraded areas to plan emergency actions;
- provide preventive maintenance and rehabilitation to slow or halt the degradation process and prolong the service life of the pavements.

RODECO Group has developed advanced technologies for the automatic detection of surface distress as the ADE (Automated Distress Evaluation).

Digital images, using Videocar system, are acquired each 4 meters, on all airport pavements (runways, taxiways, aprons), and simultaneously different types of surface distress (classification using 3 level of severity) are detected each 25 m section.

The main surface distresses that ADE system can process are the following:

- longitudinal cracks;
- transverse cracks;
- alligator cracking;

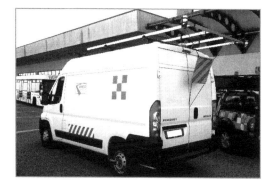

Figure 2.   RODECO Videocar-ADE System.

- raveling;
- depressions.

Each type of surface distress should be evaluated by using 2 indices:

- quantity index: % of the area affected by the distress;
- quality index: L (low) = low severity—M (medium) = Moderate severity—H (high) = high severity.

The images of the pavement, as detected by the ADE system, are post-processed with a specific software called "Automated Distress Evaluation", through it's possible to recognize and codify, on a database, the different types of distress, in a completely automatic way.

From ADE surveys, PCI (Pavement Condition Index) is calculated for each sample unit, in accordance with the ASTM D5340-12 (airport pavements) and ASTM D6433 (road pavements).

Starting from Videocar survey, global quantity of possible preventive maintenance/rehabilitation is defined, together with needs and priorities of maintenance, like crack sealing or other type of preventive treatments.

The PCI is a numerical index, ranging from 0 for a failed pavement to 100 for a pavement in perfect condition. PCI is divided into three classes:

- $70 \leq PCI < 100$ good
- $55 < PCI < 70$ fair
- $PCI \leq 55$ poor

Using this classification, a detailed map of all airport or road pavements is realized, that permits to identify promptly the critical sample unit (red color in the map).

2.4 *Ground Penetrating Radar (GPR) Survey*

By GPR technology, it is possible to continuously detect the pavement stratigraphy and estimate the thickness of the different layers. The output of this survey consists of numerical tables and/or graphics that provide the thickness of the different layers of the existing airport pavements. The data obtained from the GPR surveys are used to process the HWD data, to estimate the elastic moduli of different layers, and for the PMS drawing.

RODECO Group has a radar system, that may mount up to 3 antennas operating between 200 MHz and 2 GHz, and that can investigate the pavement airport infrastructures up to 3 m of depth.

The antennas are installed on a vehicle properly equipped with the control unit, a computer process, a high-resolution color video and an encoder.

The collected data can be displayed on screen in real time and then processed to produce tables and graphs.

2.5 Skid Resistance Survey

The skid data collection has a very important rule to assess the adequacy of air traffic safety level.

The GripTester, approved by ICAO, is a very simple device, consisting of a trailer towed to a vehicle; the trailer has two side wheels and a central wheel, braked during movement, used to measure the friction coefficient.

The braked wheel is constantly sprinkled with water by a distributor during the tests; water comes from a tank installed on the driving vehicle.

The water flow is regulated with an electronic pump to ensure the desired thickness (eg. 1 mm of water) between the tire and pavement; this water thickness don't depend of the speed measurement.

During the trailer movement, two longitudinal strain gauges measure the strain "Fo" (which is opposed to the travel) and the vertical load "Fv", given by the weight of the trailer; "Fv/Fo" ratio is the Grip Number (GN).

The strain gauges are connected to an electronic system that records data value of the friction coefficient measured every 10 meters on an onboard computer. The tests are performed on airport pavements according to the requirements of the legislation (ICAO Annex 14—Aerodromes Volume I—Aerodrome Design) at a speed of 65 and 95 km/h.

Measurements are performed along 4 lines, at ±3 and ±6 meters from the runway centerline.

Processed data can be provided in charts and graphics for every test point values, and the relative average values every third of the runway length, as required by ICAO standards, are reported.

The Skid Resistance of road pavements is also measured by SCRIM equipment.

3 DEFINITION AND EVALUATION OF HOMOGENEOUS SECTIONS

One of the most important step in the interpretation and processing data is related to choice airport or road sections with similar characteristics; an homogeneous section is characterized by having virtually constant parameters throughout its considered length. The procedure to calculate the homogeneous section permits to define the pavement section having similar features: once identified, they cannot be further divided into subsections having significantly different behavior. The subdivision of an airport or road network element into homogeneous sections may be based on various parameters, whose considered are the following:

- IRI;
- PCI;
- deflection basis parameters (D0, SCI "Surface Curvature Index" and BCI "Base Curvature Index");

- traffic;
- number of layers used for backanalysis.

There are various statistical techniques available for sharing data in a series of homogeneous sections. One technique is the cumulative sum method, introduced in the '50 s for the industrial quality control, and resumed in the "AASTHO Guide for Design of Pavement Structures guides".

This method is the best among existing in the identification of homogeneous sections, but at the same time presents a difficult to translate in an algorithm. It was therefore used another method, inside the RO.MA.® PMS software, based on dichotomization of the measures.

This method considers a probability of change of an r—parameter at a specific significance level; r—parameter is the ratio between the quadratic distance between two successive data and variance of data.

The method uses the Carré Moyen test of consecutive differences for the study of a value's randomness and Gaussian characteristics in a section. In addition, it uses a dichotomist technique to perform the segmentation. Through this, the method verifies whether data behaves under a normal probabilistic distribution before performing the segmentation.

The segmentation method searches for the point where the decision function g(i) is maximized, as shown in the following equation:

$$g(i) = \frac{n}{(n-i)i} \cdot \left[\sum_{j=1}^{i} (x_j - \bar{x}) \right]^2 \qquad (1)$$

$$g_k = \max_{1 \le j \le k} S(i) \qquad (2)$$

where n is the total number of data contained in the analyzed section, x_j is the j-value of the measured point and $\bar{x}$ is the mean of the data in the analyzed section. The point where g(i) is maximum within the section, represents the point where the mean changes. It is also the first element of the second segment of the same section. Later, the procedure is repeated in both segments, obtaining consecutive maximum points of g(i). The segmentation process ends when the g(i) function contains no more maximum points within either analyzed segment, or when the homogeneous section length is less than a length value fixed by the user.

4 DESCRIPTION OF RO.MA.® PMS SOFTWARE

The RO.MA.® PMS software, developed for road networks, and specialized for airports, is a tool that can provide the optimal maintenance and rehabilitation strategies, identified through clear and simple procedures. The software, starting from a database relating to the pavement condition and its further division into homogeneous sections, evaluates alternative strategies in a reference period of 10 years, considering the one with the highest effectiveness/cost ratio, for each homogeneous section.

The software flow chart starts from the survey data (residual life, IRI and PCI), then proceeds to the definition of homogeneous sections, with the method previous described, for all considered parameters.

As the database was created, and enriched over the time by surveys, the software works following four main steps, which will subsequently be detailed separately:

• database of pavement parameters referred to the current condition of the pavement, definition of the homogenous sections and priority levels set for the network elements.
• prediction models of the IRI, PCI, and RL (Residual Life) over the time and parameter range for specific maintenance treatments;
• maintenance measures and their characteristics;
• calculation of the optimal strategy for each homogeneous section, considering budget constraints or free budget.

4.1 Definition of database and priority levels

The first step is to create a database, which shall contain all the features of the surveyed pavements. First, it's necessary to identify, inside the entire airport or road network, the branches which will be assigned a priority for maintenance management; this priority decreases with the importance of the branch itself. It will be necessary to report the following data for each branch:

• ID code;
• traffic flow: traffic can be divided into three categories (high, medium, low), where the lower priority level is associated with high traffic level, and so on;
• number of layers of the backcalculated model, type and thickness, and average moduli E1, E2, E3 (MPa), respectively, of the asphalt concrete layer at 20° C, subbase and subgrade;
• average IRI, PCI and RL values, calculated for the homogeneous sections;
• geometrical characteristic, such as width, length and area of the homogeneous sections, and parameters measured by the laser profilometer, such as cross slope, and straight-edge;
• average value of the PQI parameter (Pavement Quality Index), which expresses the overall pavement condition. This variable is calculated as the sum of the various analyzed parameters (IRI, PCI, and RL); each of them has an associated multiplier w_i such that:

$$\sum w_i = 1 \qquad (3)$$

The PQI formula, instead, is the following:

$$
PQI = w_1 \cdot \frac{100}{IRI_{max} - IRI_{min}} \cdot (IRI_{max} - IRI)
$$
$$
+ w_2 \cdot \frac{100}{PCI_{max} - PCI_{min}} \cdot (PCI_{max} - PCI)
$$
$$
+ w_3 \cdot \frac{100}{RL_{max} - RL_{min}} \cdot (RL_{max} - RL) \qquad (4)
$$

Once the database is created, the priorities for maintenance will be calculated for each section, needed to define the list of maintenance strategies during the PMS time period.

4.2 Prediction models of pavement parameters (IRI, PCI and RL) and selection of maintenance level

Prediction models for IRI, PCI and RL parameters have been provided inside the software as a function of the possible traffic class (of the airport or road network). For each parameter, basing on technical experience, a decay curve was calculated for each traffic class. These models depends on time, so the value of a particular parameter can be calculated at any time during the service life of the pavement.

For each parameter it is necessary to identify different ranges, when a type of maintenance treatment can be used, according to the three maintenance classes:

A. Preventive maintenance: required treatments that restore mainly the PCI parameters;
B. Partial reconstruction maintenance: required measures that restore PCI, IRI and RL parameters not in a complete way;
C. Rehabilitation maintenance: required treatments that restore PCI, IRI and RL parameters to the project condition.

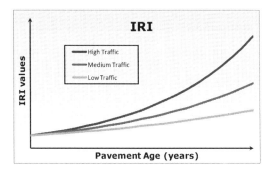

Figure 3. Prediction model for IRI parameter, depending on the traffic level.

4.3 Types of maintenance and rehabilitation treatments and their characteristics

Within the RO.MA.® PMS software, it was decided to use a set of 9 maintenance treatments, divided into the three maintenance classes (see classification above-mentioned). For each treatment, it was necessary to define:

- the maintenance class (preventive, partial reconstruction or rehabilitation);
- what parameters are repaired and how it restores (the recovery of each parameter is expressed as a percentage of the existing value);
- the unit cost (depending of the type of treatment);
- the expected life for each maintenance and rehabilitation treatment (the required time of pavement to decay in the condition before applied maintenance).

4.4 Definition of the optimum strategy for each homogeneous section considering budget constraints or free budget

Once defined maintenance treatment and their application range, the last step of the software is to identify the best possible strategy for each homogeneous section during the PMS time.

Strategy refers to a plan of action that includes one or more specific maintenance operations designed to restore and improve the performance characteristics of the pavements.

Starting from the current value of PQI, the software calculates the best possible maintenance strategy, combining the various types of treatments. This calculation is performed considering the possibility that the strategies will be not applied to the first year of analysis, but they are deferred over the time, allowing the decay of all parameters, following their prediction model.

This approach is used because some interventions have a useful life lower than the period of analysis on which the strategy is evaluated, then calculating strategies that involve the application of multiple treatments it's possible to cover the entire period of analysis.

The strategies are established from the decision tree, starting from the set of maintenance treatments listed in the previous chapter: a strategy will therefore consist of one or more branches of the tree, depending on the duration of the proposed maintenance period. At the end of the analysis, the results, for each strategy, will be a vector having the following features:

- ID code, related to the homogeneous section;
- maintenance treatments types and year of application.

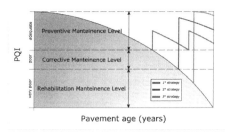

Figure 4. Definition of different strategies for an homogeneous section.

Each section will have different maintenance strategies, because managing authority could decide to take action on the section at different time, as shown in the Fig. 4: this is linked closely to the range values of individual parameters representing the pavement condition.

The next step is to define the effectiveness of each strategy, its cost and the effectiveness/cost ratio, in order to choose which is the optimal strategy for each section. In this case it's necessary to consider two possible cases: the case where there is a free budget, and if there are economic constraints.

4.4.1 Free budget Analysis
For each strategy it's necessary to identify:

- his effectiveness, calculated on the PQI parameter;
- total costs of maintenance treatments considered.

About the calculation of the effectiveness of single-strategy (E), it will be identified graphically by the included area between the curve that represents the strategy after his application and the PQI curve before. About the overall costs of each strategy, they will be referred to the initial year of the analysis period (CCA: actualized construction costs), depending on the discount rate r, that expresses the amount of interest paid/earned as a percentage of the balance at the end of the annual period.

The ratio between the effectiveness of single strategy and total cost (E/CCA) will be calculated for all possible strategies of each homogeneous section; the strategy, having the higher value of the previous ratio, will be chosen. At the end of this process, which will be repeated for each homogeneous section, it will be possible to estimate the total budget to be invested for the implementation of the best identified strategies.

4.4.2 Restricted budget analysis
This method, used to identify the best strategies when a preset budget is fixed in the time of analysis (10 years), consists of the following steps:

a. definition of the budget to invest;
b. definition of maintenance strategies during the period of analysis, for all homogeneous sections;
c. calculation of effectiveness "E" and total cost "CCA" of each maintenance strategy;
d. calculation of the ratio effectiveness/cost of each strategy (E/CCA);
e. selection of the strategy (referred to a k homogeneous section) having the higher value of E/CCA: it will be used as reference for the beginning of the iterative calculation;
f. calculation, for the k section, the marginal cost-effectiveness ratio MCE of all other expected strategies on the same k section. It is expressed in terms of effectiveness and cost of each i-th strategy identified for the section k, as compared with the selected strategy in step 5;
g. if MCE is negative or $E_i < E_k$, the *i-th* strategy is eliminated and will not be considered in further steps, otherwise the combination replaces the selected strategy in step 5, and the available budget is updated;
h. the process is repeated until no other strategy can be selected for each section in each year of analysis, and when the allocated budget is fully committed.

This method allows, in addition to the definition of the best strategies to be applied to all homogeneous sections, to perform an analysis at different budget levels: inserting a different value for the estimated budget, strategies choice will change, in accordance with the most appropriate capital to invest.

5 IMPLEMENTATION OF AIRPORT PAVEMENT MANAGEMENT SYSTEM: THE EXAMPLES OF LARNAKA AND SARAJEVO

5.1 Larnaka International Airport

The software, for the Larnaka Airport, was applied at the project level: the customer has identified different areas where HWD test has been conducted, in order to investigate the pavements having different distresses on the surface.

Tests have been performed with a step of 20 m, collecting data useful for the realization of a maintenance project such as to meet the institution's needs and ensure the necessary characteristics of bearing for the next 15 years.

The various areas have been divided into homogeneous sections, based on the following parameters:

- D0, which is the measured deflection at the center of the loading plate;
- SCI (Surface Curvature Index), which characterizes the surface layers of the pavement;

- BCI (Base Curvature Index), which characterizes the base layer.

For each section, the following parameters have been calculated, starting from the backanalysis process:

- Modulus of elasticity of the bituminous layer at 20°C (E1), and its thickness;
- Modulus of elasticity of the base layer (E2), and its thickness;
- Modulus of elasticity of the subgrade layer (E3);
- Modulus E0: surface modulus calculated under the load plate;
- Residual fatigue life of the most critical layer;
- Theoretical overlay required to reinforce the weak pavement;
- ACN/PCN coefficients.

The output of the software was the following:

- Identification of the homogeneous sections and application of the maintenance measures;
- Year of the rehabilitation for the weakest homogeneous sections;
- Type of maintenance;
- Total surface area of the rehabilitation;
- Unit cost (€/m²);
- Total Cost of the maintenance for each homogeneous section;
- Total cost for all the PMS period (there were no budget constraints) and divided by year.

The adopted solutions, added to preventive and curative maintenance, consisting in the application of a special fiberglass grid. This type of maintenance has been optimized by the software, capable of detecting the thickness of the pavement to mill, strictly related to the depth in which to insert the grid; the aim was to reduce the time and costs of the intervention, obtaining the same final result of a new resurfacing.

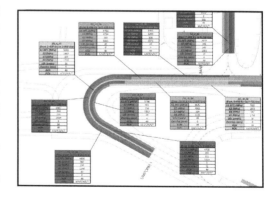

Figure 5. Definition of the homogeneous section for the Larnaka Aiport.

5.2 Sarajevo International Airport

RODECO Group performed a series of high performance surveys on the Sarajevo Airport pavements, in order to plan a pavement management system for the next 10 years. Using all the technologies described before, the actual pavement condition has been detected.

RO.MA.® PMS software was used to define the strategy of preventive maintenance and rehabilitation, once made the surveys using high-performance systems (IRI, PCI, HWD etc.). The software output was a list of maintenance measures per year and for each homogeneous section located within the airport network. In the chart obtained by the software the following parameters were shown:

- year of implementation of maintenance measures belonging to the chosen strategies for each homogeneous section;
- type of maintenance measures to apply, their dimension and location;
- unit cost ($€/m^2$) and total cost of maintenance per year ($€$).

The primary objective of the PMS for this airport was to plan and especially "optimize" the pavement maintenance and rehabilitation strategy, i.e. to maintain the highest level of functional and structural characteristics of the pavement with minimal cost, assuring a significant savings for the managing authority.

The main results of the PMS application on Sarajevo Airport was to demonstrate that, even if the very critical condition of the runway pavement, it was feasible to project a total rehabilitation by a reconstruction limited to the surface layers, applying special materials and technologies, with a huge saving of economic resources and time.

6 CONCLUSIONS

Some practical examples of the use of the RO.MA.® PMS were presented with a brief description of the pavement evaluation and PMS phases of the RO.MA. method. The pertinence of the proposed method has been discussed in terms of technical and economic aspects. The implementation of a database on the user's PC containing PE and PMS data allows the user to prepare more realistic annual maintenance budgets based on a scientific approach to maintenance rehabilitation problems. Moreover, the use of new software for graphic representations of the PMS results increases the information available for the agency to manage its network in terms of efficiency and economy, and the graphic presentation of the data allows also the user to easily understand the results of the RO.MA. analysis.

The use of the system in all reported cases in this paper has allowed the optimization of the maintenance budgets for all agencies. This was accomplished through the use of RO.MA.® PMS techniques to identify specific areas of the airport network that are in need of maintenance; at the same time, RO.MA.® PMS software has identified areas of the airport that did not need maintenance. The users have benefitted from this analysis by knowing exactly where implementing maintenance techniques, saving time and money.

REFERENCES

AASHTO Guide for design of pavement structures, 1993, Appendix J: J1–J5.
ASTM D5340-12, Standard Test Method for Airport Pavement Condition Index Surveys, 2012.
ASTM D6433-11, Standard Practice for Roads and Parking Lots Pavement Condition Index Surveys, 2011, 1–2.
Battiato, G., 1990. Evaluation of PCN (Pavement Classification Number) of airfields by means of the F.W.D. In *Proceedings for Third International Conference on Bearing Capacity of roads and airfields*, Trondheim, Norway, 3–6 July 1990.
Battiato, G., 1998. Prove e controlli non tradizionali sulle pavimentazioni bituminose. La valutazione della qualità delle pavimentazioni stradali mediante l'impiego dei sistemi ad alto rendimento. In *Proceedings for SITEB 7° corso tecnico di base su leganti e conglomerati bituminosi*, S. Donato Milanese, Italy, 25–27 November 1998.
Battiato, G., 2000. Valutazioni e controlli di qualità delle Pavimentazioni Aeroportuali. In *Proceedings for Le pavimentazioni aeroportuali*: 90–108, Rome, Italy, 30 Novembre 2000.
Battiato, G.; Al Emadi K.M.I.; Liberati P., 2003. Description et application du systeme de gestion routier (PMS) developpe pour l'Etat du Qatar. In *Proceedings for XXII Congres Mondial Routier—C6 Comite sur la gestion routiere*, Durban, South Africa, 19–25 October 2003.
Battiato, G.; Larsen B.K., 1990. Description and Application of RO.MA.® (Road Evaluation and Pavement Management System). In *Proceedings for III International Conference on Bearing Capacity of roads and airfield*, Trondheim, Norway, 3–6 July 1990.
Battiato, G.; Amé E.; Wagner T., 1994. Description and Implementation of RO.MA.® for Urban Road and Highway Network Maintenance. In *Proceedings I for Third International Conference on Managing Pavements*, 43–50, S.Antonio, Texas, USA, May 22–26 1994.
Battiato, G.; Larsen B.K.; Ullidtz P.,1987. Verification of the analytical – empirical method of Pavement Evaluation based on FWD testing. In *Proceedings for 6th International Conference on Structural Design of Asphalt Pavements*, Ann Arbor, Michigan, USA, 1987.
ENAC Numero: 3/2015-APT, Ed. n. 1 del 1 Ottobre 2015, Airport Pavement Management System - Linee guida sulla implementazione del sistema di gestione della manutenzione delle pavimentazioni.
ICAO Annex 14 to the Convention of International Civil Aviation – Aerodromes – Volume I – Aerodrome Design and Operations, Sixth Edition, July 2013: ATT A4-A10, A28–A29.

Transport Infrastructure and Systems – Dell'Acqua & Wegman (Eds)
© 2017 Taylor & Francis Group, London, ISBN 978-1-138-03009-1

The use of porous asphalt for the improvement of the grading plan geometry and drainage of pavement surfaces on urban roads

V. Ilić, M. Orešković & D. Gavran
Department for Roads, Railroads and Airports, Faculty of Civil Engineering, University in Belgrade, Serbia

I. Pančić
Transportation Institute CIP, Belgrade, Serbia

ABSTRACT: Porous asphalts are applied all over the world to minimize the effect of noise caused by everyday traffic. Beside the effect of noise reduction, thanks to their open texture and improved drainage characteristics, porous asphalts decrease the impact of spraying water behind a moving vehicle. Since drainage capability of porous asphalt as a surface permeable layer is much higher compared to a conventional dense-graded asphalt surface course, the values of cross grades of pavement wearing course on the road sections with porous asphalts should be additionally considered. Hence, the primary topic of this paper is how the application of porous asphalt could affect the geometric design of grading plans of pavement surface on urban roads, especially in intersection zones, where different longitudinal and cross grades of intersecting directions have to be mutually aligned. Due to a higher drainage capacity of porous asphalt layers, layout plan of storm water inlets can be changed if pavement cross grades are reduced.

1 INTRODUCTION

Correct design and construction of drainage system of urban roads is one of the most responsible tasks of civil engineers engaged in the design of transport infrastructure in cities. Design of drainage elements of the largest number of urban road sections is based on the city sewage systems, which, in addition to their drainage function, play an important role in protection against pollutions. Due to intensive urban development and construction of buildings right next to R.O.W. (right of way) boundaries, drainage systems of urban roads have to be dimensioned with certain capacity reserves in order to be able to accept rainwater run-on from neighboring building roofs and facades (Maletin 2009).

Key elements of the whole pavement drainage system are storm water inlets which accept rainwater from gutters and divert it to the storm drain pipes. Also, they represent a membrane which retains larger debris and waste materials washed from the pavement surface (Maletin 2009). Efficient drainage concept implies that storm water inlets have to be placed in the direction of major flows of rainwater collected along the road curb (Despotović 2009). Except directly next to the curb, storm water inlets should be placed at the lowest elevations of pavement surface so that their full capacity could be reached.

Particularly sensitive areas in terms of stormwater drainage of urban roads are intersections, where pavement surfaces of different intersecting roads have to be accurately graded and fitted into each other in geometrical sense. Grading plan of pavement surface is used by hydrotechnical professionals as an initial background drawing for determining the exact locations of storm water inlets.

In this paper, the recommendations for proper setting of storm water inlets in intersection zones, as well as the analysis of the impact of porous asphalt application as a wearing course of pavement structure, are given. In the case study of real four-leg intersection located in Belgrade it was shown how the reduction of pavement cross grades affects the change of grading plan geometry of pavement surface. Also, through the analysis of simplified cross sections of urban road, the impact of modified cross grades on elevations heights of outside pavement edges and curb elements was discussed in detail. In additional case it was shown how the grading plan of pavement surface on parking lot is changed due to the application of porous instead of standard dense-graded asphalt pavement.

2 GENERAL PRINCIPLES OF GEOMETRIC DESIGN OF GRADING PLAN

In order to design and plot a grading plan of pavement surface at intersection the basic design elements such as longitudinal grades, vertical curves, grades of all cross section elements of intersecting roads, as well as the methods of attaining superelevation (superelevation development length, superelevation rate and axis of pavement rotation), have to be defined. In general, longitudinal grades of road depend on the rank of a road according to functional classification and the objective location requirements. All of the above mentioned elements must be mutually harmonized in intersection zones, in order to get merged and uniquely designed pavement surfaces in 3D.

Outside pavement edges, which are usually followed by the gutters that collect rainwater flowing along the streets, must be designed with required longitudinal grades which guarantee adequate conditions for pavement drainage. For standard conditions and normal roughness of pavement whose flatness lies within the construction tolerance the minimun value of longitudinal grade is $i_N = 0.5\%$. The adoption and implementation of longitudinal grades below this limit deteriorate drainage conditions, and this requires a detailed analysis of runoff parameters and, in certain circumstances, the specific allocation of storm water inlets along urban road section, or even the application of non-standard drainage elements. Maximum longitudinal grade does not depend on drainage requirements, because its upper limit is conditioned by the functional rank of road section within the urban road network and the general infrastructure requirements defined in urban development plans (Maletin et al. 2010). Also, the maximum values of longitudinal grades can be used only on the sections between adjacent intersections, while in the intersection zone a certain reduction of longitudinal grades is needed.

Based on a shape and pattern of contour lines drawn in grading plans road designers can perceive uniformity of grading changes and notice possible deficiencies of suggested design solutions. This particularly applies to the horizontal distances between contour lines generated next to the pavement edges, so for example, variation of distances between contours lines indicates some irregularities, and if contour interval changes inconsistently the pavement surface is improperly modeled in 3D. Those areas on pavement surface, where contours do not look like smooth lines, require further corrections. Some elevation corrections of grading plan are also necessary in cases when outside pavement edges followed by the gutters and curbs do not have longitudinal grades required for efficient drainage, or their longitudinal grades have excessive values due to high superelevation rates of pavement.

Bearing in mind a simple fact that water always runs perpendicular to the direction of contour lines, in order to make a storm water inlet layout plan, it is necessary to determine the specific drain area per one storm water inlet on the basis of previously collected hydrotechnical parameters such as designed storm event, rainwater runoff conditions and maximum flow. For the exact definition of these parameters a constant monitoring of precipitations and research in the field are essential.

It is very important to avoid the positioning of storm water inlets at pedestrian and / or bicycle crossings. In addition, storm water inlets should be positioned directly before or after pedestrian crossing in order to collect as much water as possible, and prevent rainwater runoff across the pedestrian crossings. In real urban road network, however, is not so rare that storm water inlets are placed wrongly, and because of that the required amount of water can not be cached, or the rainwater that pours down the pavement surface just misses incorrectly installed storm water inlets.

3 BASIC CHARACTERISTICS OF POROUS ASPHALT

Porous asphalts are bituminous mixtures characterized by high percentage of air voids. Pavements with porous asphalt mixtures must have adequate bearing capacity and satisfactory drainage of storm water. Infiltrated water further flows to the layer which contains reservoir for collected rainwater, or runs through all layers of pavement structure to the poorly compacted sub-grade. Percentage of air voids, even up to 25%, is achieved due to an open gradation curve. During the design process of porous asphalts, it is not enough to take into account only the characteristics of the material, traffic load, etc., but also many other parameters that can contribute to the reduction of overall noise levels and improve their drainage capabilities such as the size of aggregate grains, discontinuity of mixture and porosity. The size of aggregate grains in the mixture affects the permeability, durability and stability under the traffic load. By reducing the grain size, noise caused by the interaction of tire with the road surface will be reduced. If the porosity is increased, the absorption of noise and the drainage of water through porous layers are also increased.

Due to the high content of voids in the mixture, the use of porous asphalt significantly reduces the occurrence of "aquaplaning" effect, if these voids are not clogged (Bendtsen 2002). There seems to be a tendency that porous pavements on highways and rural roads keep the porous structures open during their lifetime. This is explained by the existence of "self-cleaning mechanism", where the vehicle tires press water down in the voids of the porous pavement under high pressure during the periods of rain. On the other hand, on low speed urban roads water pressure, due to slower rotation of vehicle's wheels, is not high enough to ensure a continuous "self-cleaning" effect of the pavement, causing the clogging of voids in the pavement on such roads. This was revealed first by the engineers from Copenhagen in Denmark, after their experimental research was conducted on real test sections (Bendtsen 2005, Bendtsen et al. 2005).

Friction resistance of porous asphalt is an important characteristic, as well as drainage. At lower speeds, the friction resistance of porous asphalt is smaller in comparison with standard asphalt mixtures because the exposed surface of aggregate at the tire pavement contact is smaller. Consequently, the microtexture of a surface porous layer and the roughness of exposed aggregate particles become main features. Lack of "friction" at lower speeds is not considered the major disadvantage, since low friction values are usually dangerous at high speeds that are common on motorways. Therefore, the porous asphalts are suitable for urban roads paving where traffic speed is slower.

Uncontrolled traffic growth has a negative impact on the environment. Except noise and harmful exhaust gases, a significant source of pollution is a large amount of particles generated by friction and tire wear. Porous asphalts have a positive impact on the improvement of the quality of water that flows on the road surface and infiltrates through pavement layers. A large amount of dangerous particles is retained in the upper layers. Quality control of water that runs through the whole pavement structure was done in several countries around Europe and similar results were obtained (Gopalakrishnan et al. 2014). The water that is collected at the bottom of the reservoir is diverted via a perforated tube out and analyzed. It has been noticed that the concentration of the adsorbed metals zinc, lead, and others is only 20% to 30% relative to their concentration in surface layers. In order to prevent that dangerous particles penetrate into natural soil and pollute groundwater, a regular vacuuming and maintenance of porous pavements are recommended.

Porous asphalts are prone to clogging if they are not properly maintained. Proper maintenance of porous asphalt extends their lifetime and retains their functional characteristics. Porous pavements are cleaned by using water jets under high pressure, after which a vacuum is needed to remove liquid and retained solid pieces. It is recommended that first cleaning should be performed three months after construction, and then after every six months. Recent experiences in the field have shown that the water pressure around 20 MPa gives satisfactory results (Fletcher & Theron 2011).

The biggest drawback of porous pavements is their high cost and limited durability compared to standard dense-graded asphalt mixtures. Due to the relatively short exploitation period (maximum 8 to 10 years) (Ferguson 2005) and high investment costs, porous asphalts are rarely applied in poor countries. Also, a major disadvantage of these pavements is high maintenance costs, especially during winter.

4 APPLICATION OF POROUS ASPHALT FROM THE ASPECT OF PAVEMENT DRAINAGE

One of the most important properties of porous pavements is their ability of pavement draining, i.e. reduction of surface runoff from pavement surfaces. In order to better understand the hydraulic regime, and a new hydrologic balance that is established by the construction of porous pavement, all components, inflow (surface run-on, precipitation), as well as runoff of water in the system have to be analyzed (Fig. 1). The main inflow (filling) sources of system are primarily atmospheric precipitation (rainwater) and surface run-on from the adjacent pavement and other surfaces. One part of the inflow immediately infiltrates through wearing course of pavement, and the other part evaporates into the atmosphere or further flows along certain hydraulic gradient according to the grading characteristics of a surface course. A certain portion of the infiltrated water inflow in the pavement can be collected by drainage pipes laid on the subgrade surface (formation level), and all the rest of the

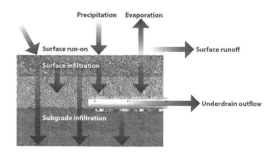

Figure 1. Water balance variables for permeable pavements (source: Eisenberg et al. 2015).

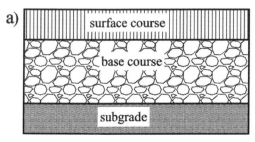

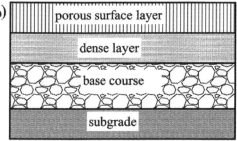

Figure 2. a) Porous asphalt paved over water permeable base course with open-graded base reservoir; b) Porous asphalt paved over the existing pavement- Overlay (source: Ferguson 2005).

infiltrated inflow is further drained through the subgrade material into the subsoil.

Because of its open texture, porous pavement reduces the effects of spraying and splashing water, and this phenomenon was reduced by more than 95% after construction a new porous asphalt layer (Rungruangvirojn & Kanitpong 2009). This feature changes over time due to clogging of the pores.

Overall, in terms of drainage of porous pavements, two types of pavement structures with wearing porous asphalt layers can be distinguished:

1. Porous asphalt layers with infiltration base (reservoir). With this type of pavement structure, porous asphalt layers are constructed over the permeable base layer, through which the infiltrated water inflow is drained to the so-called "reservoir" from unbound aggregates, and further to the uncompacted subgrade of pavement structure all the way down to the natural ground (subsoil) (Fig. 2a). A reservoir layer should consist of pure, uniformly granulated aggregates with 30%–40% of voids (Eisenberg et al. 2015). The thickness of the layer containing reservoir is determined on the basis of the required capacity for storage of infiltrated water, structural requirements, seasonal changes of high groundwater levels in the soil, and the frost depth. This type of porous pavement is mostly used for parking areas in residential or commercial complexes, then for the construction of pedestrian and bicycle paths, driveways, and for other

asphalt pavements which are not subjected to high traffic loads.

2. Porous asphalt layers paved as wearing (top) course over the previously constructed, waterproof flexible or rigid pavement. This type of porous asphalts is often called "Overlay" in the literature, while in the USA "Permeable Friction Course—PFC" is a widely used term (Fig. 2b). Unlike porous asphalts with the reservoir in the base course, which allows water to infiltrate into the subgrade, this type of porous asphalt has entirely different drainage mechanism. All infiltrated water in "Overlay" layer is transversally diverted through voids to specially designed gutters or hollow curbs laid along pavement edges. Most problems with the drainage of "Overlay" layer are caused by the water that remains trapped at the contact surface between the new layer of porous asphalt and the old waterproof pavement course. Trapped water can cause serious problems during the winter at low temperatures because there is not enough space to expand due to the snow and ice in the cavities of the porous surface layer.

5 MODIFICATION OF THE GRADING PLAN GEOMETRY AFTER THE APPLIACTION OF POROUS ASPHALT—CASE STUDY

In order to clearly understand the impact of porous asphalt application on the changing of pavement cross grades, and consequently on the modification of grading plan geometry, two practical cases from site are shown. The first case refers to the standard four-leg intersection (crossroad), and the second to the application of porous asphalt as a surface layer of the porous pavement structure at parking lots.

In Figure 3a the grading plan of the intersection of Severna tangenta and Zage Malivuk streets is presented. The plotted drawing of grading plan represents the existing situation where the wearing top course of pavement was made from standard dense-graded hot mix asphalt. Severna Tangenta Street was considered as a major road (arterial) and Zage Malivuk Street as a minor road (cross street). All sections where changes of pavement cross grade occur, as well as all vertical points of intersection of longitudinal alignments, for both major and minor road directions are precisely labeled. Figure 3a shows that in this case cross grades of the major road are equal to 2.50% and have crown shape along the entire intersection zone, and that adjacent longitudinal grades of the cross street are also 2,50% in order to fit into the cross grades of the major road. At the same time, the longitudinal grade of the major road of 1.00%

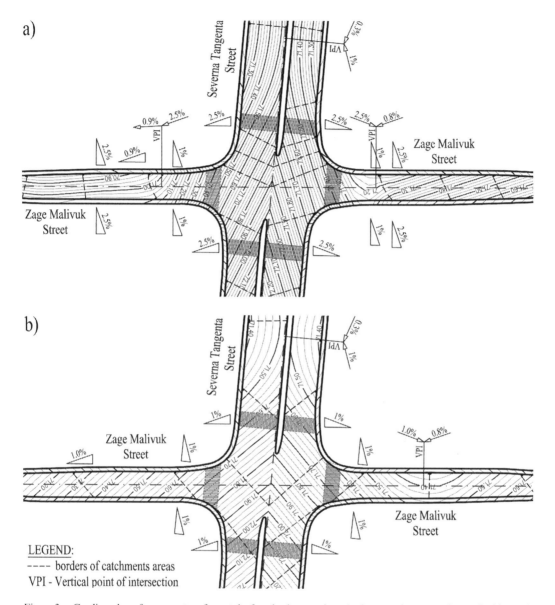

Figure 3.   Grading plan of pavement surface at the four-leg intersection: a) when wearing course is paved with standard dense-graded asphalt mix; 4b) when wearing course is paved with porous asphalt.

represents the pavement cross grade of the minor road in the intersection zone. The shape of contour lines on the grading plan clearly indicates that vertical alignment of minor road accurately follows cross grades of pavement of the major road. Equidistance or contour interval is 2.50 cm. On the grading plan of pavement surface every fourth contour line is labeled in ascending order, and the positions of pedestrian crossings and storm water inlets are displayed too. Taking into consideration the designed storm event (designed rainfall) and

other hydrological parameters of the site in Belgrade where the intersection is located, the specific drain area for one storm water inlet is estimated as 250.00 m^2.

The grading plan of the same intersection, with changed values of pavement cross grades due to laying of porous asphalt as a wearing course of pavement structure, is plotted again in Figure 3b. Cross grades of the pavement on the major and minor road are reduced from 2.50% to 1.00%. The longitudinal grade of the major road remained unchanged, since its correction is very rarely, almost never, done. However, due to the reduction of pavement cross grades of the major road the longitudinal grades of the minor road had to be changed radically. As a consequence of reduced longitudinal grades of the minor road in the intersection zone, the positions of vertical points of intersection (VPI) significantly shifted in the vertical alignment of the minor road compared to the previous case. Although the grading plan in Figure 3b was plotted using the same equidistance as in the grading plan in Figure 3a, due to lower values of pavement cross grades, horizontal distances among contour lines are wider. It is evident that contour lines in the grading plan in Figure 3b seem more harmonious and are smoother compared to the contour lines in the grading plan plotted in Figure 3a.

After the construction of porous asphalt run-off from the pavement surface reduce substantially due to the infiltration of one part of surface run-on and precipitation through porous layers. Therefore, the specific drain area, which was used for determining the number of storm water inlets, has to be estimated again. If it is assumed that the runoff coefficient of porous wearing course is 0.50 (for dense-graded asphalt mixtures its value is usually between 0.90 and 0.95), it can be easily concluded that the specific drain area in comparison with the one which was used for the positioning of storm water inlets in Figure 3a, should be increased almost two times, or to the value of approximately 450.00 m². For this reason, the layout plan of storm water inlets in Figure 3b is different from the one which is displayed in Figure 3a. Hence, for the drainage of the identical pavement surface a smaller number of storm water inlets is needed.

The reduction of pavement cross grades of primary urban roads and city arterials with separate carriageways can be performed in two ways, as shown in simplified drawings of cross sections in Figures 4a and 4b. In densely populated urban areas where buildings are placed right next to the sidewalk edges, changing of the pavement cross grades can extensively impact on other elements of public infrastructure and facilities. Two specific cases from road design practice are discussed:

1. Cross grade decreases by revolving a traveled way about the outside lateral edges of the traveled way, and the axis of rotation coincides with the contact edge between the surface wearing course of pavement and adjacent side curbs. From the cross section drawing in Figure 4a it can be seen that the elevation levels and vertical position of outside curb elements next to the ends of the traveled way stayed the same, while the elevation heights of the curbs along median

edges decreased. The advantage of this method of traveled way superelevation is that elevation heights along the outside pavement edges remain the same, and the elevations of existing storm water inlets do not change either. In addition, this method of attaining superelevation is more suitable for urban roads that pass through city centers (downtowns) among densely built structures, because the elevations of entrance steps in the surrounding buildings do not have to be changed. Major disadvantage of this superelevation method is that most of the layers of the existing pavement structure, including a wearing course, have to be destroyed and removed later in order to construct a stable and flat base course for the new layers of porous asphalt. The entire median has to be lowered down vertically which may cause additional problems if some underground utility lines have been already installed below the median surface. These complex construction works are inevitably followed by enormous investment costs.

2. In the second case pavement cross grade is reduced by revolving a traveled way about median edges. Using this method of attaining supereleva-tion all structures and facilities which were built next to the pavement edges are functionally disrupted due to the rising of the elevations of side curbs along the outside edges of traveled way (Fig. 4b). This further leads to the uplifting of sidewalks and elevations heights of entrance steps of neighboring buildings, which can be very inconvenient for all residents of these buildings. Not only functionality of buildings, but also their aesthetics and architecture suffer because of the entrance steps uplifting. Also, vertical positions of storm water inlets and other elements of drainage system that are located right next to the side curbs along the outside pavement edges have to be corrected. For primarily ranked urban roads whose edges are not encroached with houses and buildings, this superelevation method for the decreasing of pavement cross grades represents better engineering solution compared to the superelevation method illustrated in Figure 4a. Existing layers of pavement structure together with the wearing course of dense-graded asphalt mix can be kept, but in that case, due to reduced cross grades, a new wearing course of porous asphalt must be paved in a layer of variable thickness.

Changing of the pavement cross grade at parking lots whose pavement structure was founded on permeable subgrade or embankment is much easier to implement, as shown in the example in Figure 5. Since all surface run-on is infiltrated through porous layers to the subgrade surface and

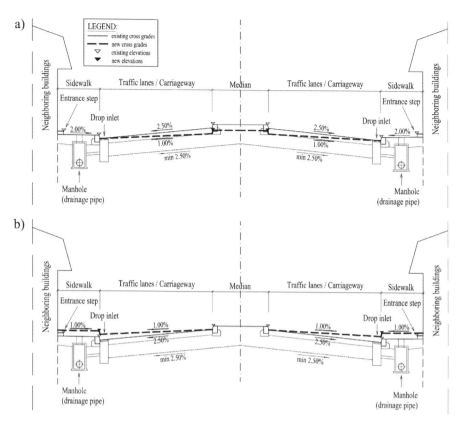

Figure 4. Simplified cross sections of urban road witch illustrate the effects of cross grade reduction by revolving a traveled way: a) about the outside lateral edges of pavement surface; b) about median edges.

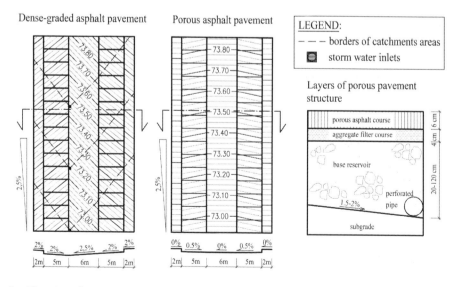

Figure 5. Changing of pavement cross grades at parking lot after the construction of wearing course of porous asphalt over water permeable base with reservoir.

further into natural soil, with this type of parking lots there is no need for setting storm water inlets and the construction of the whole system for pavement drainage, as it is necessary to do for parking areas with standard dense-graded asphalt pavement. The biggest drawbacks for a wider application of this type of pavement structure for the construction of parking lots are high cost of used materials, very rigorous quality requirements during construction and relatively expensive maintenance treatments.

6 CONCLUSION

The key reason for writing this paper was to investigate how the construction of new porous asphalt surface layers can affect the existing geometry of grading plans of urban roads. In order to clearly understand the impact of porous asphalt application on the changing of pavement cross grades, and consequently on the modification of grading plan geometry, two practical cases from site are shown: four-leg intersection and parking lot.

The reduction of pavement cross grades of primary urban roads and city arterials with separate carriageways can be done in two ways: by revolving a traveled way about the outside lateral edges of pavement surface or about median edges. In the first case elevation levels of the outside curb elements next to the ends of the traveled way stayed the same, while the elevation heights of the curbs along the median edges decreased. The advantage of this method of the traveled way superelevation is that elevation heights along outside edges of pavement remain the same, as well as elevations of existing storm water inlets. However, in order to construct a stable and flat base course for the new layers of porous asphalt most of the layers of existing pavement structure, including a wearing course, have to be removed. In the second case, by revolving a traveled way about median edges, all structures which were built next to the pavement edges are functionally disrupted due to the rising of elevations of side curbs along the outside edges of traveled way. For primarily ranked urban roads whose edges are not encroached with houses and buildings, this superelevation method for decreasing pavement cross grades is a better solution compared to the superelevation method applied in the first case.

Unlike complex modifications of grading plans of intersections, changing of cross grades of pavement in parking lots, due to construction of new permeable porous layers, does not represent a hard task for road designers. If new porous pavement structure is founded on a permeable subgrade there is no need for setting storm water inlets and the construction of the whole system for pavement drainage.

Is it rational to change pavement cross grades due to the implementation of a new porous asphalt layer as a wearing course on urban roads remains the basic dilemma. If we only take into account increased investment costs then this proves an unreasonable decision, but after comparing the shapes of contour lines of grading plans plotted for pavements with and without porous asphalts, the issue of ride comfort, besides noise and pollution reductions, as well as the elimination of "splash and spray" effect, are gaining in importance. Also, it should not be forgotten that application of porous asphalts, especially on urban roads with higher operating speeds, contributes to traffic safety.

REFERENCES

Bendtsen, H., Larsen, L.E. & Greibe, P. 2002. *Development of noise reducing pavements for urban roads*, Report No. 4. Copenhagen: Danish Transport Research Institute.

Bendtsen, H. 2005. *The DRI-DWW Noise Abatement Program - Project description*. Issue No. 24. Copenhagen: Danish Road Institute & Danish Road Directorate.

Bendtsen, H. et al. 2005. *Two-layer porous asphalt for urban roads*, Report No. 144. Copenhagen: Danish Road Institute & Danish Road Directorate, p. 5–16.

Despotović, J. 2009. Kanalisanje kišnih voda (In Serbian). Belgrade: Faculty of Civil Engineering University of Belgrade.

Eisenberg, B. et al. 2015. *Permeable Pavements*. Reston: American Society of Civil Engineers-ASCE, Virginia, USA.

Ferguson, B.K. 2005. *Porous Pavements*. Boca Raton: CRC Press, Taylor & Francis Group, Florida, USA.

Fletcher, E. & Theron, A.J. 2011. *Performance of open graded porous asphalt in New Zeeland*. NZ Transport Agency research report 455. Wellington: NZ Transport Agency.

Gopalakrishnan, K., Steyn, W.J. & Harvey, J. 2014. *Climate Change, Energy, Sustainability and Pavements*. Berlin: Springer-Verlag, Germany.

Maletin, M. 2009. *Gradske saobraćajnice* (In Serbian). Belgrade: Orion Art.

Maletin, M., Anđus, V. & Katanić, J. 2010. *Technical guidelines for intersection design (PGS-PR / 07)*. Belgrade: Faculty of Civil Engineering University of Belgrade.

Rungruangvirojn, P. & Kanitpong, K. 2009. Measurement of Visibility Loss due to Splash and Spray: the Comparison between Porous Asphalt, SMA, and Conventional Asphalt Pavements. *Transportation Research Board, 88th Annual Meeting. 11-15 January 2009*. Washington D.C.

Transport Infrastructure and Systems – Dell'Acqua & Wegman (Eds)
© 2017 Taylor & Francis Group, London, ISBN 978-1-138-03009-1

Urban tracks for Light Rail Transit (LRT). Dublin experience

M. Corsi
Track Design Manager, Transport Infrastructure Ireland, Republic of Ireland

ABSTRACT: Luas Cross City (LCC) is Dublin new 5.9 km light rail extension (Luas is the commercial name for Dublin LRT), being built as part of the overall development, since 2001, of a new, modern and efficient LRT network for the Irish Capital.

One of the major challenges for what is essentially a city central, partly on-street LRT, has been the design and construction of the embedded track system. This design was influenced by the experiences of previous Dublin LRT projects and by the outcomes of benchmark analyses with other European networks and trial track panels.

TII (Transport Infrastructure Ireland), is delivering LCC through a D & B contract, based on very strict Employer Technical Requirements. The resulting innovative track design is described in detail in this paper.

1 LRT EMBEDDED TRACK SYSTEMS

Urban light rail track is referred to as "embedded track" because the rails (generally grooved type) are embedded into the road surfaces.

Embedded track international standards do not exist, and each Country (and even every city in the same Country) applies different requirements, tolerances, design principles and construction philosophies.

Among the various National standards or guidelines, the French CERTU, and German VDV would be the most commonly referred to, with the English ORR and Irish GDRIRS being more functional requirements.

Together with city specific systems, mostly based on long lasting experience (Zurich, Freiburg for example), there are off-the-shelf or turn-key proprietary systems by the largest track manufacturers available.

Among all of them, the main differentiator parameters are the form of gauge control (with or without sleepers, direct fixation, no fixation at all), concrete levels and reinforcement (low, middle or high concrete, and traditional reinforcement, no reinforcement, fibre-reinforcement), the type of construction (top-down or bottom-up) rail insulation (rubber profiles, rubber filler blocks, chambers, number and materials of the components, etc.), casting system (precast or in-situ), rail-road joint type, finishes (exposed concrete, asphalt, modular elements, grass).

Within this frame, the following sections describe the track system specifically developed and adopted for Dublin Light Rail system.

2 THE LUAS (DUBLIN LRT) NETWORK EMBEDDED TRACK TYPES

2.1 Overall Luas network

The overall Luas network is formed by two lines, Green and Red and it was built over a decade starting in 2001, with the northern section of the vertical line (Green Line) currently under construction, the "Luas Cross City" (LCC).

2.2 Initial lines A-B-C

Luas Green and Red Lines (Lines A-C and Line B) were opened to passenger traffic in 2004. The two lines were not connected in the City Centre and measured a total length of approx. 25 km.

Few sections of these lines were on-street running, with embedded track built with grooved rail embedded in up to three stages concrete structural slab.

The original track system had the following features:

- RI59 N rails installed with no sleepers (*EN 14811 Railway applications—Track—Special purpose rail—Grooved and associated construction*)
- ALH type factory-made rail encapsulation.
- Concrete shoulders 150 mm wide on each side of the rails for asphalt or granite setts finish.
- 150 mm subbase (3% cement bound) +300/350 mm reinforced concrete structural slab +100 mm reinforced infill slab (or 100 mm asphalt or granite sett). Total depth 550/600 mm from bottom of subbase.

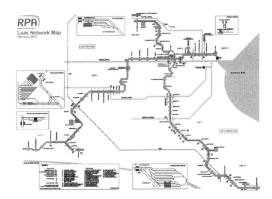

Figure 1. Schematic of the Luas (Dublin LRT) network at completion of Luas Cross City.

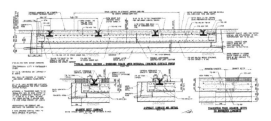

Figure 2. Typical cross sections of Embedded Track for lines A-B-C.

- Top down construction with temporary supports.
- Stray current active protection (ALH encapsulation) and passive protection (current collection cage around each rail). (*EN 50122: Railway applications—Electric safety, earthing and the return current*)

2.3 Luas extensions A1-B1-C1

The two lines were subsequently extended between 2007 and 2011, with the construction of three new sections, A1-B1-C1 for an additional total length of approx. 13.5 km.

Each of those extensions was delivered with a different D&B or Employer's design contract, with the result of three new interoperable but conceptually different embedded track systems, with different structural design, sleepers, concrete shoulders and rail encapsulation systems.

2.4 Special type A1

For Line A1 a special type of embedded track was tested over some of the busiest road junctions, to experiment the durability of embedded track with no shoulders. The selected solution was to limit the

Figure 3. Embedded Track for line A1 under construction.

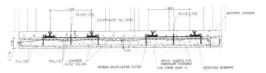

Figure 4. Typical cross sections of Embedded Track for line B1.

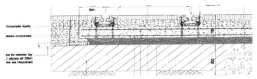

Figure 5. Typical cross sections of Embedded Track for line C1.

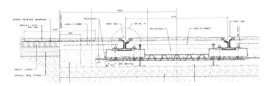

Figure 6. Typical cross sections of Special Embedded Track for line A1 at road crossings—no concrete shoulders.

Figure 7. Special Embedded Track for line A1 at road crossings—no concrete shoulders.

concrete height to 40 mm below the rail head, and use polymer modified asphalt, of the type TSCS (Thin Surface Course System), (*NRA National Road Authority Specification for Road Works. Series 900 Road Pavements—bituminous bound materials*).

Also, to seal the joint with the rail encapsulation at the surface, a hot bituminous filler was poured on both rail sides (40 mm wide on running side to cover for the wheel tread width in the worst scenario).

This track has been in operation for approximately five years and no issues have manifested so far.

3 ORIGINAL EMBEDDED TRACK ISSUES

3.1 General

In the course of the last 12 year's operation, the various track types aged differently. In general there were no major issues, to the satisfaction of the Infrastructure Manager and the Maintainer.

Despite a general positive outcome, some recurring defects were noted, as listed below.

3.2 Early Back Flange Contact (BFC)

Early Back Flange Contact (BFC) appeared on some of the tighter curves where no gauge control bar or sleeper was provided (Lines A-B-C). The absence of sleepers and active gauge control may have contributed to gauge spread under lateral load, exacerbated by the high lateral flexibility of rail encapsulation (factory applied rail coating), also due to the single encapsulation casting process (same mechanical characteristics in both vertical and lateral planes). Early side wear of the keeper, friction noise, and potential for derailment risk were the consequences.

3.3 Rail replacement

Guiding rails on tighter curves (R < 50 m) started reaching side wear maintenance limit in 2013.

Figure 8. Shoulder cut-out with holes for the additional fastenings. New rail yet to be inserted and grouted.

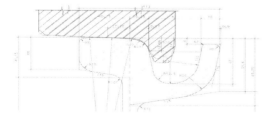

Figure 9. Wheel-Rail interface on Dublin Luas.

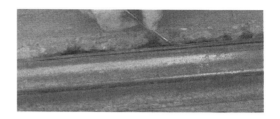

Figure 10. Concrete shoulders "V" shape initial crack.

Figure 11. A and B. Examples of concrete shoulders structural collapse and failed repair works.

It was soon evident that the existing track systems presented two shortcomings in relation to side wear.

The first was the adoption of R260 rail steel hardness for the grooved rail on all alignments (*EN 13674 Railway applications—Track—Rail standards*) presenting high carbon medium hardness steel, with medium side wear, and low chances of successful gauge corner repair welding at low temperature.

The second was the challenge associated with rail replacement in a full rail concrete embedment scenario. The shoulders were initially designed to provide additional rail lateral stability in the lack of sleepers, a clear demarcation line between road and track (maintenance responsibility between LRT operator and City Council), and a protection/support for rail encapsulation and road surface

In terms of rail replacement, all the "shoulder-based" designs indicated a proposed "cut line", next to each side of the rail to free it up prior to grouting-in the new encapsulated rail.

As it emerged at first trial, the remaining part of the shoulder would not be structurally sound and therefore the new inserted rail would have to be chemically anchored in the main slab via an alternative fastening prior to re-grouting the missing part of the shoulder.

This made the rail replacement very complicated and expensive. Also, the vertical cold joint within the shoulder made it fragile and prone to failure, particularly under road traffic in shared running sections.

Dimension, structural soundness and level of reinforcement of the shoulders were highlighted as critical factors.

Figures 12. A. Absence of rail-road joint and ingress of water under the rail foot—imprinted concrete finish with stains from water pumping. – B. Damages at the exposed encapsulation joint, with repair works (cut-out slot, filled-in with polymer based pourable joint sealant).

3.4 Concrete shoulders deterioration

Concrete shoulders deteriorated as a consequence of several different factors.

External edge of wheel-tread runs over the brittle concrete surface, damaging the joint. Luas wheels are 110 mm wide to negotiate S&C with deep flange-way, and this makes the tread up to 40 mm wider than the rail head.

These cracks developed into deeper failures for water ingress, rail pumping and winter freezing, exacerbated by road traffic at shared sections.

In 2013, 60% of the embedded track at road crossings showed structural deterioration with 18% requiring shoulder replacement (once or even twice).

3.5 Non-sealed rail-shoulder joint

With the exception of Luas A1 road crossings, all other solutions had the rail encapsulation exposed to the surface with no sealant.

That arrangement contributed to the development of several issues for the ingress of water, debris, dirt often deteriorating into cracked shoulder and damage to the encapsulation or to its joints (collars).

4 NEW DEVELOPMENTS

Following the analysis of previous installations and building on the lessons learned from the maintenance outcomes, TII undertook a comprehensive benchmark analysis of existing European systems (France, Germany, Italy, UK and Switzerland) to establish a new embedded track standard for the future lines in Dublin.

As part of this thorough analysis, a 5 m long single track trial test panel was also built to test constructability, resistance and durability of four different track arrangements and finishes. Following the outcomes of that real-scale test, a new idea of embedded track was developed and used for the construction of the *Rosie Hackett Bridge* track in Dublin city Centre.

4.1 Track trial panel

The track trial panel was delivered in 2013, with the ongoing condition monitoring regime still in place.

It was divided in two parts, one finished with asphalt, and one with modular elements (granite setts). The asphalt section was divided in two (one per rail), with one full depth asphalt as required by the City Council (100 mm in total, 60 binder and 40 wearing), and one thin wearing course with 40 mm PMSMA (Polymer Modified Stone Mastic Asphalt).

The setts section was also divided in two, with and without rail concrete shoulders.

Two different pourable polymer-based joint sealants were also tested, one on each rail.

While the evaluation of the long term durability is ongoing, the following initial outcomes were noted:

– Polymer based joints are working well and holding on to rail and concrete in the "with shoulder configuration".
– 40 mm thin layer asphalt is also working well with no joint sealant (asphalt laid against the rail) with no sign of deterioration along the rail or de-bonding from concrete base.
– granite setts in the without-shoulder configuration are getting dislodged for the poor vertical stability under traffic loads, due to the ineffective bedding mortar (40 mm thickness) possibly due to the poor bedding preparation and water content control. This stresses the importance of site quality control in setts laying.
– Setts along the rail with no shoulder do not present instability issues, but the wider bedding problem jeopardized the full panel.

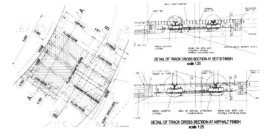

Figure 13. Test panel plan and cross sections.

Figure 14.   Test panel completed.

Figure 15.   Rosie Hackett Bridge—Embedded Track with granite setts and derailment containment channel.

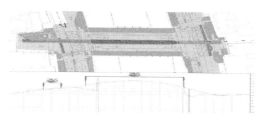

Figure 16.   Rosie Hackett Bridge—Plan profile of the track alignment.

– 100 mm hot rolled asphalt (composed by 60–40 mm courses) is performing well with the exception of the joint with the polymer sealant showing signs of deterioration along the joint itself. These have developed initially in the first 12 months and seem to have stabilized since.

4.2   *Rosie Hackett Bridge track*

The first section of Luas Cross City track (100 m) was installed as part of the construction of a new public transport dedicated bridge over the River Liffey, the Rosie Hackett Bridge, in 2014.

The bridge was designed and commissioned by the City Council, while TII had the responsibility for the design, technical specifications and site supervision for track and LRT systems installation.

River navigability and vertical track alignment led to a very slim deck with extreme span to depth ratio, imposing significant challenges to the structural design of the bridge and the track itself.

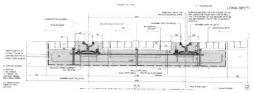

Figure 17.   Rosie Hackett Bridge—Embedded Track Cross Section over the bridge.

Figure 18.   Rosie Hackett Bridge—Embedded Track with asphalt finish across the Quays.

Figure 19.   Rosie Hackett Bridge—Embedded Track with asphalt finish across the Quays.

The architectural requirement (granite setts finish) further called for very shallow track slab, with the adoption of a proprietary slim gauge control system, in the form of flat steel sleepers at 1.5 m interaxis and structural concrete shoulders for additional rail lateral support. Additionally, for the first time in Dublin, a "low profile" rubber encapsulation and pourable polymeric sealant was used on a real project on both rail edges.

As part of the bridge works, track was also built across both Quays for approximately 25 m on each side. In those sections, no shallow track requirement and heavily trafficked road junctions called for a different track system with no concrete shoulders, heavier bi-block sleepers, thicker track slab and a polymer based joint sealant.

In this case, while the initial TII recommendation was for 40 mm Thin Wearing Course (TWC) system, similar to Line A1 road crossings and the most successful finish of the track trial panel, a strong opposition of a key stakeholder led to a design change towards the use of the standard 60+40 mm binder-wearing Hot Rolled Asphalt (HRA).

Soon after opening the road to traffic, some sections developed minor rail-to-asphalt joint failures, particularly at bus turning movements.

A mix of causes appears to have contributed: type of asphalt and large aggregate, lack of proper compaction along the rail, asphalt milling prior to sealant laying (dislodging larger aggregate), low bitumen quantity, and its evaporation for the use of a flame torch to dry up the joint prior to sealant pouring.

5 NEW EMBEDDED TRACK FOR LUAS CROSS CITY (LCC)

5.1 Background

Following agreements with the various stakeholders, all the lessons learned from the previous installations were considered for the new track design.

Essentially, the new track had to fulfill all the railway requirements (*GDRIRS Guidelines by CRR —Irish Commission for Railway Regulation, BoStrab SpR, 2004, EN 50122: Railway applications— Electric safety, earthing and the return current*), including rail stability, ease of maintenance, joint stability, stray current among others, but also the road requirements, as the shared track was part of the urban road network (*NRA Specification for Road Works. Series 700-800-900*).

That led to the track design being structurally based on DMRB (*Irish Design Manual for Road and Bridges by National Road Authority HD24-25-26/10*) in what referred to the pavements and structures (capping, foundation and structural slab).

A typical example is the integration of the concrete shoulders into what is essentially a semi-rigid road pavement design, with the requirements in terms of skid resistance to be achieved with exposed aggregate treatment of the shoulder's surface.

This design was carried out in a joint effort between the awarded D & B Joint Venture (JV) for LCC and TII, within the strict rules dictated by a D & B contract. Based on these, the Contractor has full design responsibility within stringent Employer Requirements (ER) produced by the tendering Authority (TII in this case) and the Authority carries out compliance checks to verify the adherence to the ER.

5.2 Design objectives for LCC embedded track

Most of the design objectives have already been described in the course of this paper.

The fundamental requirement was the capacity to withstand heavy road traffic for a design life of 60 years in all its components, with the exception of the rails in curves with radius less than 50 m.

The track had to be highly standardized, light and easy to install, easy and cost effective from maintenance viewpoint (like for example allowing easy rail replacement and gauge corner welding

repair for tight curves), and provide an overall low Life Cycle Cost.

The proposed embedded track had to facilitate its installation within constrained central areas and stringent traffic management conditions, while ensuring track laying geometry and tolerances (*RPA (Railway Procurement Agency) 2007. Track alignment—Design Handbook—absolute H&V tolerance 5 mm, relative 3 mm/3 m, gauge +–2 mm, twist at 3 m 1:250*).

5.3 LCC embedded track design

– RI59 N rails with higher steel grade. (*EN 13674 Railway applications—Track—Rail standards*)
– 18 m long rubber encapsulation profiles (to minimize joints), low profile, with optimized joints to sleeper fasteners interface.
– Double side rail-to-road polymer based pourable joint filler, (40–25 mm wide).
– Light gauge bar (flat steel sleepers) at 3 m centers to facilitate construction and maintenance (rail replacement as described below).
– Variable height foundation (depending on capping CBR values, min 200 mm) in C8/C10 pourable lean concrete mix + 300 mm reinforced (twin layer plus top mesh for asphalt finish only) concrete structural slab + 100 mm infill asphalt finish. (*EN14227-1 Hydraulically bound mixtures. Part 1. Cement bound granular mixtures/EN 206-1:2002/A2:2005: Concrete—Part 1: Specification, Performance, Production and Conformity*)
– For granite setts finish, (setts are 120 mm deep, 100 mm square, over 40 mm bedding), the required insert pocket between shoulders is 160 mm deep instead of 100 mm, making the main slab only 240 mm thick. This is to keep foundation levels constant irrespective of the track finishes.
– Total single track structure width 2200 mm for a standard gauge of 1435 mm, with chamfered slab edges for improved road pavement interface.
– Track depth 600 mm from bottom of foundation.
– Top down construction with lateral bracing as temporary supports.
– Stray current active protection (rail encapsulation) and passive protection (reinforcement). In excess of 10 Ohm-km measured on site pre and post pour testing.

5.4 LCC embedded track innovations

– Higher rail steel grade for curves below 50 m. The 290GHT (heat treated) steel grade offers higher wear resistance, longer side wear life, less corrugation (longer grinding intervals), vibrations and ground-borne noise. Reduced carbon content from HB260, facilitates weld-repair at low temperature. Rail life extended by approx. 7 to 8 years.

- Improved rail rubber encapsulation profile with longitudinal joints staggered from the sleeper fasteners position (to avoid clamping damages), low profiles (25 mm below top of rail and keeper) and optimized top edge overlap with the pourable joint.

- Light gauge bar provided at 3 m centers for construction and future rail replacement purposes only. Self-regulating spindles to support track during construction and facilitate the lateral adjustments through push-pull bars anchored to

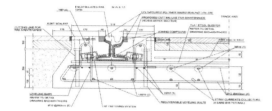

Figure 20. Luas Cross City Embedded Track with asphalt.

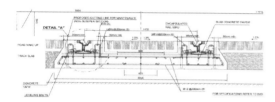

Figure 21. Luas Cross City Embedded Track with granite setts.

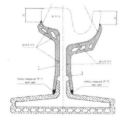

Figure 22. Rail encapsulation made of three profiles and with low type side profiles with 45 degree upper edge.

Figure 23. Flat steel sleepers.

the rail heads. Sleeper's fasteners are protected with a plastic cap to facilitate future rail replacement, and their position is indicated by a notch in the rail keeper.

- Standardized cross section for different finishes with standard reinforcement cages and shoulder stirrups to speed up construction process in narrow sites and between operational traffic lanes.

- Rail concrete shoulders to avoid rail-road joint issues was not a new feature in Dublin. But to prevent shoulder structural issues, two measures were adopted: more structurally sound design, with a wider footprint (up to 210 mm wide) and higher reinforcement (with hook type stirrups every 200 mm and longitudinal 16 mm bar), and the sealing joint at the running edge of the rail to prevent wheel contact and seal rail jackets from water ingress.

- Larger shoulder with the reinforcement set back 40 mm from the rail foot to facilitate rail replacement by saw-cutting at a constant rail offset, yet leaving its structural vital part intact. Sleeper's fasteners re-use for rail replacement.

- High density pourable polymer-based sealing joint on both rail sides for the reasons mentioned above.

- Track slab edges chamfered arrangement to mitigate the risk of adjacent flexible road surface reflective cracking and to reduce transversal footprint of the track while avoiding overlapped stress-relieving membrane otherwise required by the DMRB. Track slab width only 2,200 mm, with significant savings and improved buildability.

Figure 24. Construction detail of the shoulder reinforcement.

Figure 25. Construction detail of the rail arrangement with concrete shoulders and joint filler on both sides. To note the precision in relative levels of the three elements and the pigmented and exposed aggregate texture of the shoulders.

5.5 Construction sequence and photos

Figure 26. C8/C10 lean concrete foundation.

Figure 27. Main slab reinforcement cage and shoulder stirrups.

Figure 28. Shoulder stirrups details. Their precise levelling will be completed after the rail-sleeper panel is in place (sitting on sleeper spindles) and at final levels.

Figure 29. Sleepers and rails are laid, assembled over the reinforcement cage, welded, encapsulation profiles are joined over the welds and final track alignment adjustments are made. Formworks are assembled with chamfered sides.

Figure 30. Main slab concrete pour.

Figure 31. Shoulders are poured between shutters installed over the slab pour. Timber profiles hanging from the rail head to form the joint recesses. Retardant is sprayed over the shoulder surface to expose aggregate.

Figure 32. Joint sealant application. The joint surfaced are dried and prepared for better adhesion of the flowable polymer based material.

Figure 33. Tack coat and base-wearing courses applied and rolled for asphalt finish sections.

Figure 34. A-B. Setts laying.

Figure 35. Track laying in College Green (Trinity College in the background).

Figure 36. Double track laying in St. Stephen's Green North.

458

6 CONCLUSIONS

Following 15 years of track construction experience in Dublin and benchmark analyses with the major European networks, with the additional information gathered from real scale testing, significant innovations were introduced in the new embedded track design for Luas Cross City, as described in the paper. Monitoring regime is already in place to evaluate the benefits of those innovations in terms of rail-road joint, but tram operation and full monitoring will not start until late 2017. It is expected that the new track will deliver higher performances in terms of noise and vibration, corrugation development, rail and road surfaces durability, stray current resistance, and overall low maintenance cost.

REFERENCES

CRR (Irish Commission for Railway Regulation) — GDRIRS (Guidelines for the Design of Railway Infrastructure and Rolling Stock) Section 7 "Tramways" (doc. number RSC-G-008-B)

ORR (UK Office of Rail Regulation) "Design Requirements for Street Track". Tramway Technical Guidance Note 1.

CERTU LROP Tome 1 and 2 "Plate-formes de tramway". AITF.

RPA (Railway Procurement Agency) 2007. Track alignment—Design Handbook.

RPA (Railway Procurement Agency) 2001. Track alignment—Tramway Clearances.

UIC 719. Earthworks and Trackbed construction for railway lines.

UIC 15273. Railway applications. Gauges.

NRA (Irish National Road Authority) DMRB (Design Manual for Road and Bridges). HD25–26/10: Pavement and Foundation Design.

NRA DMRB. HD24/06: Pavement Design and Maintenance: Traffic Assessment.

NRA Specification for Road Works. Series 700 Road Pavements—General.

NRA Specification for Road Works. Series 800 Road Pavements—Unbound and Cement Bound Mixtures.

NRA Specification for Road Works. Series 900 Road Pavements—bituminous bound materials.

BoStrab SpR, 2004. German Federal Regulations on the Construction and Operation of Light Rail Transit Systems, Guidance Regulations (SpR).

IS EN14227-1 Hydraulically bound mixtures. Part 1. Cement bound granular mixtures.

IS EN 206-1:2002/A2:2005: Concrete—Part 1: Specification, Performance, Production and Conformity.

IS EN 13674 Railway applications—Track—Rail standards.

IS EN 14811 Railway applications—Track—Special purpose rail—Grooved and associated construction.

EN 50122: Railway applications—Electric safety, earthing and the return current—Fixed installations. Part 1: 2011 and Part 2: 2010.

Transport Infrastructure and Systems – Dell'Acqua & Wegman (Eds)
© 2017 Taylor & Francis Group, London, ISBN 978-1-138-03009-1

Study on paving factors of stone path in campus based on characteristics of pedestrian walking

Yang Yanqun, Huang Qinwei, Xu Meiling & Yu Sheng
College of Civil Engineering, Fuzhou University, Fuzhou, China

ABSTRACT: Stone path is usually seen in places such as campus, park, scenic spots, etc.; And this study is about stone path with gradient smaller than 1%. The pavement of stone path is closely related to the characteristics of pedestrian walking. Therefore, the author collects 306 valid samples by studying the stone paths and the characteristics of the students in a university, and analyze their influencing factors including the gender, height, single/multi-walker, weight-bearing status, personality analysis, etc., obtaining the reasonable pedestrian stride and its main factors affecting it by variance analysis. According to the reasonable pedestrian stride, the author also designs the pavement parameters such as the slate width, spacing of slates, and the center-to-center spacing of two adjacent slates, etc. At last, the author verifies the conclusion of the paper by Electroencephalograph (hereinafter EEG) experiment, proving that these pavement parameters are worthy of reference.

1 INTRODUCTION

In university campus, the open spaces between two adjacent buildings are usually filled with plants for the greening of environment, and lawn is the most common one. To prevent the pedestrians between the two adjacent buildings from trampling on the lawn, the campus management will usually pave a path on the lawn. Among various pavement materials, stone path itself is of ornamental functions and can get people through without affecting the landscape (Han, 2014). Therefore, stone path is widely used in campus; it can be seen in places like teaching building, library, experimental building and small park, etc.

In the college, as a supplement to cement concrete road and sidewalk paving road, etc., stone path is basically used by students. Compared to the users of the cement concrete road and sidewalk paving road, stone path users are often in a leisurely walking state. Most stone paths in campus are paved on flat lawn (gradient <1%), and only seldom of them are paved on area with steep slope, thus this study is about stone path with gradient smaller than 1%.

Due to different causes, the pavement of stone path is of comparatively high casualty. In the current code, Except the code in *Code for the Design of Urban Green Space (GB50420-2007)* that the width of green lane should less than 0.8 m, other related codes including *Code for the Design of Urban Road Engineering (CJJ37-2012)* and *Code for the Design of Park (CJJ 48-92)* have not stipu-

Figure 1. Stone path.

lated any pavement elements such as slate width and the spacing of two slates, etc. The design of stone path does not take into account the walking characteristics of pedestrians because of the casualty of the stone path construction, which causes discomfort to the pedestrians walking on it. First of all, pedestrians will easily miss their steps and step into the gap between the two slates, causing accidental injury and damages to the shoes; secondly, during peak hour of pedestrian traffic flow, the discomfort leads to the bad traffic capacity of stone path, increasing the congestion degree rather than easing the traffic pressure; and finally, due to poor comfort of walking caused by improper pavement, some pedestrians may give up walking on it and choose to walk on the roadside lawn, which greatly goes against its original intention. Therefore, it's necessary to take walking characteristics of pedestrians into the consideration of the pavement elements of stone path in university campus.

Compared with the research in motor vehicles, the research in walking characteristics of pedestrians is later and more complex. One main cause is the diverse forms of walking, which brings about no fixed pattern in walking. The earliest relevant research was conducted by Henderson L F in Department of Mechanical Engineering in Sydney University. After investigating the walking speed of 693 university students, 628 pedestrians in Sydney and some children aging from 6 to 8, he put forward that the walking speed of one single pedestrian is presented as Gauss distribution, the average speed of pedestrians is 1.34 m/s with 0.26 m/s standard deviation (Henderson, 1971, 1972, 1974). In 1975, the research of Palmer A et al. stated clearly that gender is an influencing factor of walking characteristics. Variance analysis demonstrated that gender and age are very important independent factors.

Apart from the study of basic section, in recent years many scholars from home and abroad have studied the walking characteristics in different road sections including transportation hubs, pedestrian crosswalk, mountain ladder section, etc. (Ma et al, 2009, Alhajyaseen et al, 2011, Virkler et al, 1994, Sisiopiku et al, 2003, Laplante & Kaeser, 2004, Zhang et al, 2012, Zhang, 2014, Zhu & Lin, 2015, Xu, 2013, Yang, 2009, Zhang, 2009, Zhao et al, 2006). Meanwhile, they have also studied the walking characteristics in different areas (Tanaboriboon et al, 1986, Chen & Dong, 2005, Lam et al, 2002, Rastogi et al, 2013, Laxman et al, 2015).

To probe into the relations between pedestrian's walking characteristics and the pavement elements of stone path, the author has done researches in the walking characteristics of pedestrians in university campus in their normal walking conditions, and analyzed the factors which influence the pedestrian stride, stride frequency, and the walking speed.

2 MATERIAL AND METHODS

There are many factors influencing pedestrian's stride, stride frequency and walking speed, such as the height, gender, age, etc (Ma et al,2009). According to the make-up of the pedestrian traffic flow in college, the author analyzes pedestrians' walking characteristics based on their gender, height, single/multi-walker, weight-bearing state, characteristics analysis, etc., obtaining the walking characteristics of the pedestrians. Then, integrated with the feet size of the students, the author gives suggestions on the pavement elements of stone path in campus.

2.1 Design and procedure

1. Personality Test
The personality of the participant is defined by filling out the psychological assessment questionnaire before collecting the walking characteristics. According to the result, the participants will be divided into two group: A and B.

People in group A are more confident, aggressive, easy to be nervous, and have a sense of achievement.

People in Group B are comparatively at ease and calm, and remain unruffled to anything.

2. Parameters Measurement of Pedestrians' Walking Characteristics
Test personnel laid down the tape on the cement concrete pavement, and invited the participants to walk on the road paralleled to the tape (walking length is about 20 m). There were 6 video cassette recorders (VCR) next to the tape to make sure they can record the foot path and scale of the tape. The test personnel timed the walking of participants by stopwatch; and after the test, they analyzed these videos (Fig. 2).

The stride frequency is the walking steps of the subject divided by walking time. The walking speed is the total walking length divided by walking time. The stride is obtained by intercepting the stride of each walking and getting the mean of it. The subjects' height and feet size were registered through inquiry before the experiment.

2.2 Participants

To ensure the reliability of the experimental result, the experiment invited 310 college students to attend the experiment at random. After collecting data and eliminating invalid samples, the paper carried out statistical analysis of the 306 valid samples. The distribution of the 306 valid samples is shown in Table 1.

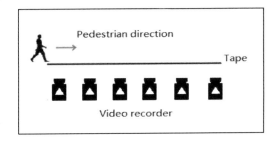

Figure 2. Date collection of pedestrian walking.

Table 1. The Distribution of 306 samples.

Classification		The number of people	Percentage (%)
Gender	Male	171	55.88
	Female	135	44.12
Height (cm)	155–165	116	37.91
	165–175	141	46.08
	175–185	49	16.01
Personality	A	134	43.79
test result	B	172	56.21
Single walking		60	19.61
Multi walking		246	80.39
Weight-bearing		105	34.31

3 RESULTS

3.1 Preliminary analysis

Table 2 lists the statistics of the stride frequency, walking speed and stride of the 306 valid samples.

Seeing from the statistical analysis, the stride frequency, walking speed, and stride of university students are presented in Gaussian distribution.

The main construction elements of stone path are slate width b, center-to-center spacing of two adjacent slates c, and the edge spacing of two adjacent slates d. In addition, b+d = c (Fig. 6). The main influencing factor of b is the feet size of students. According to the investigation data, the mean of feet size of male subjects is 25.95 cm, while female 23.48 cm. Besides, stride fluency has a great impact on d, while stride frequency and walking speed barely have relations with b and c. Consequently, this paper put emphasis on the analysis of the stride.

Table 2. The statistics of stride frequency, speed, and stride.

	Stride frequency	Walking speed (m/s)	Stride (m)
Mean	1.76	1.22	0.68
Standard deviation	0.21	0.22	0.12

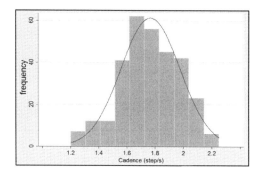

Figure 3. Statistical analysis of stride frequency.

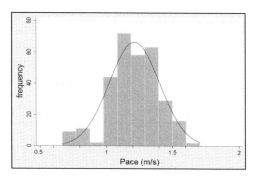

Figure 4. Statistical analysis of walking speed.

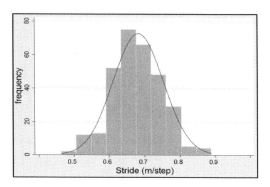

Figure 5. Statistical analysis of stride.

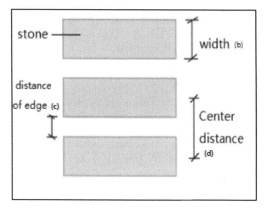

Figure 6. Construction elements of stone path.

3.2 Influnce of gender

The distribution figure of male and female stride is drawn by software Stata. We can see from the Fig.7 that male and female stride are basically presented in Gaussian distribution.

To explore the influence of gender on pedestrians' stride and do significance test on the mean of male and female stride, we build original hypothesis as follow:

H0: $\mu a = \mu b$

where a = male; b = female.

The variance analysis of the samples was done by using software Stata; significant level $\alpha = 0.05$.

On the basis of the table, we can see that in the null hypothesis H0, while $p < 0.01$, it can be considered that the mean of the two groups are of significant difference. The result of Bonferroni's statistical inspection suggests that the male stride is observably higher than female stride ($p < 0.005$).

3.3 Influence of height

Difference heights may also cause the variation on walking characteristics. There is a difference

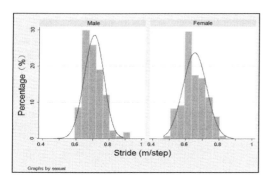

Figure 7. The distribution of male and female stride.

Table 3. Variance analysis.

Source	SS	df	MS	F	Prob > F
Between groups	0.217	1	0.217	51.99	0.000
Within groups	1.271	304	0.004		
Total	1.488	305	0.004		

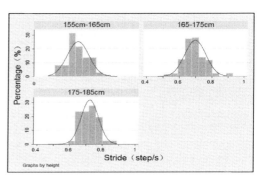

Figure 8. Stride distribution in different heights.

Table 4. Bonferroni statistical test table.

Row mean-Col mean	160	170
170	0.48684	
	0.000	
180	0.75230	0.25646
	0.000	0.042

of stride between a tall and big stature pedestrian and a short and small stature person. It can be seen from the statistical data of Chen et al (Chen & Dong, 2005). that foreign pedestrians have a faster walking speed than Chinese pedestrians due to height, etc. Integrated with the height of the students in campus, this paper divides the height into three groups as 155–165 cm, 165–176 cm, and 175–185 cm.

The distribution of stride in different heights was drawn by using software Stata. We can see from the Fig. 8 that the stride in different heights is presented as Gaussian distribution.

To look into the influence of gender on pedestrian's stride, we had done the significance test on the mean variation of stride in different heights, and analyzed the variance of these samples by Stata. The result illustrates that the stride of 155–165 cm is remarkably lower than that of 165–176 cm and 176–185 cm (p < 0.005), while there is no significant difference between the mean of stride of 165–175 cm and that of 175–185 cm. Due to space limitations, we only list the Bonferroni statistical test table.

3.4 Influence of single/ multi-walker

Besides the personal and environmental factors, the walking speed is also influenced by whether the pedestrian is walking alone or not. As shown in Fig. 9.

The significance test and variance analysis were done on the difference in stride mean of single/multi walkers. Result indicates that while the

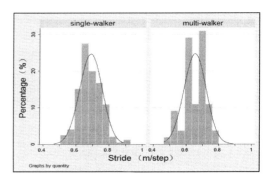

Figure 9. The stride of single-walker and multi-walker.

corresponding p tested in the null hypothesis H0 is less than 0.01, we can regard that the two group of means are of great variation. In line with the result of Bonferroni statistical test, the stride of single-walker is obviously bigger than that of multi-walkers (p < 0.005).

3.5 Influence of weight-bearing and A&B types personalities

The result of Bonferroni statistical test shows that there is no significant difference between A-type personality pedestrians and B-type personality pedestrians (p < 0.005), nor is there any difference whether the pedestrian bears weight or not (p < 0.005). Hence the paper doesn't develop the exposition and argumentation about it.

Table 5. Different classes of KPI in pavement. Unit: centimeter.

Class 1 KPI	Class 2 KPI
Slate width (b)	Feet size
Edge spacing of two slates	(c) c = d-b
Center-to-center spacing of	height
two slates (d)	gender
	single/multi—walker

3.6 The pavement elements analysis of stone path based on KPI

KPI (Key Performance Indicators) is a kind of targeted quantitative management indicators which measures the process performance through setting values in, sampling, calculating, and analyzing the key parameters of the input and output end of one process in an organization (Fu & Xu, 2001). It is a tool which can change the strategic target into an operational vision. Associated with the above experimental analysis, the author gives the different classes of KPI in stone path pavement.

Feet size affects the slate width, thus the width of peddles should at least holds the length of a foot (275 mm). Besides, it should reserve a 10 cm space (Wu, 2002). Meanwhile, considering the cost, the length of the slate should not be too long. After analyzing the samples, the author builds the following formula between slate width b and gender:

$$b = 0.03x + 0.32$$

where x is the proportion of male in total number; Unit: meter.

The indicators that influence the center-to-center spacing of two slates are height, gender, and single/multi-walker. Since the height and single/multi-walker are hard to estimate in different colleges and using phases, the KPI of them is set as 0. According to the stride data, the author builds the following formula between center-to-center spacing of two slates d and gender:

$$d = 0.05x + 0.65$$

Due to c = d-b, thus the relation of the edge spacing of two slates c is as follow:

$$c = 0.02x + 0.33$$

4 VERIFICATION AND DISCUSSION

According to the analysis of experimental data, we can receive the following conclusions:

1. The main factors affecting pedestrian's stride in campus are gender, height and single/multi-walker;
2. Pedestrian's stride is generally distributed in 0.6–0.8 m/step, thus irrespective of artistic, etc., the center-to-center spacing of two slates should conform to pedestrians' stride range;
3. Among all the students, the mean of male feet size is 25.95 cm, while that of female is 23.48 cm, the male average stride is 0.70 m/step, while female is 0–65 m/step.

On the basis of the above conclusions, this paper puts forward the suggested value of stone path pavement elements.

We simulated a temporary stone path in campus according to the suggested value in Table 6. In order to explore whether this simulated one is more comfortable than others, we conducted the verification experiment by using the EEG equipment.

The verification includes experiments on 15 groups of male and 15 groups of female. Each participant conducts 3 sets of comparative experiments with head-mounted wireless EEG device (E2015U908-14). Participants in the 1st set walk on stone path paved according to the parameters in Table 6; participants in the 2nd set walk on the real stone path in campus (slate width b = 28 cm, Edge spacing of two slates c = 27 cm, center-to-center spacing of two slates d = 55 cm, all of them are smaller than the suggested value in Table 6);

Table 6. The suggested value of stone path pavement elements.

Pavement site	b	c	d
The institute with more male students (such as civil engineering institute and mechanics institute)	35	34	69
The institute with more female students (such as school of arts and humanities and school of foreign languages)	33	32	65
Institutes of equal proportion of male and female, the library, etc.	34	33	67

Figure 10. The simulated stone path (left).

Figure 11. The real stone path in campus (right).

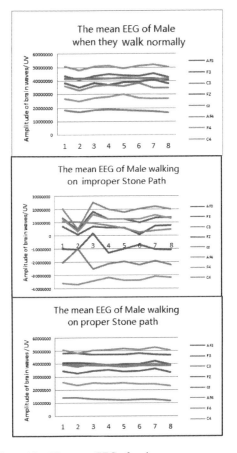

Figure 12. The mean EEG of male.

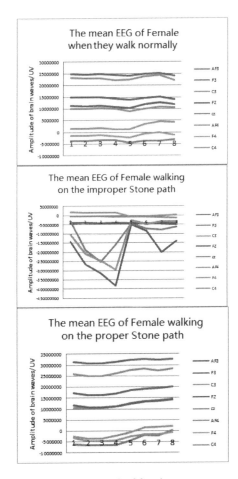

Figure 13. The mean EEG of female.

participants in the 3rd set walk normally on the ordinary road. The walking distance is 8 m, and participants have 2 minutes to rest before they taking each experiment, ensuring there is no interference between experiments. After the test, we analyzed the changes in EEG of these participants.

The means of male and female walking on the two different stone paths and walking normally are listed hereinafter:

The result indicates that both male and female will have an increase in EEG when they walk on stone path. But when they walk on a stone path paved as suggested values and a regular road, the EEG remains a comparatively stable state. This explains that there is scarcely any difference between walking on a stone path built according to the suggested values in Table 6 and walking normally, but participants might feel uncomfortable when they walk on the improper stone path. Comparing the EEG changes of male to that of female, we can see that although both of them show big wave in EEG when walking in the real stone path, female's EEG fluctuation is obviously bigger than male's. The probable reason of this is that the stride of female is smaller than male's, and the feet size of female is also shorter than that of male. Therefore, the stone path paved in line with the parameters in table 6 is more suitable for pedestrians to walk as compared with other stone paths. These parameters are worth referring to.

5 EXPECTATION

The research results of the paper can provide reference to the short stone path in campus. In consideration of the widely usage of stone path and fully taking its advantages, we need do further studies in the following aspects:

1. The selected road section in this experiment is 20 m long, which can meet the need of lawn width linking the two adjacent buildings in the college. However, if the stone path is taken as a long recreational trail, it needs to be further explored and studied integrated with pedestrian's physiological and psychological characters (including EEG, Electrocardiogram, Skin Conductance, etc.).
2. In the analysis, the difference in personalities has relatively small impact on the stride, but it has great influence on the stride frequency and walking speed. In the research of the pavement elements of long stone path, further study of stride frequency and walking speed is needed.
3. In the field research of sites like park, etc., we find out that environmental factors such as longitudinal slope, pedestrian volume, weather, etc. will also affect the pavement elements of stone path. Thus we need to consider the impact of environment as well in the further study of pavement elements of stone path.
4. The collected samples of this experiment are only for college students. In the future it can be considered to extend to faculties, other people in parks and communities, etc., so as to provide reference for the pavement elements of stone path.

REFERENCES

Alhajyaseen, W.K.M. et al. 2011. Effects of Bi-directional Pedestrian Flow Characteristics upon the Capacity of Signalized Crosswalks [J]. *Procedia - Social and Behavioral Sciences* 16(1):526–535.

Chen R. & Dong L. 2005. Observations and Preliminary Analysis of Characteristics of Pedestrian Traffic in Chinese Metropolis (in Chinese) [J]. *Journal of Shanghai University: Natural Science* 11(1):93–97.

Fu Y. & Xu Y. 2001. *Performance Management* (in Chinese) [M]. Shanghai: Fudan University Press.

Han, G. 2014. On the Construction Techniques of Garden Road Pavement (in Chinese) [J]. *Architectural Engineering Technology and Design* (14).

Henderson, L.F. 1971. The statistics of crowd fluids. [J]. *Nature* 229(5284):381–3.

Henderson, L.F. & Lyons, D.J. 1972. Sexual differences in human crowd motion. [J]. *Nature* 240(5380):353–355.

Henderson, L.F. 1974. On the fluid mechanics of human crowd motion[J]. *Transportation Research* 8(6):509–515.

Lam, W.H.K. et al. 2002. A study of the bi-directional pedestrian flow characteristics at Hong Kong signalized crosswalk facilities [J]. *Transportation* 29(2):169–192.

Laplante, J.N. & Kaeser, T.P. 2004. The continuing evolution of pedestrian walking speed assumptions [J]. *Ite Journal* 74(9).

Laxman, K.K. et al. 2015. Pedestrian Flow Characteristics in Mixed Traffic Conditions [J]. *Journal of Urban Planning & Development* 136(1):23–33.

Ma Y. et al. 2009. Impact Analysis of Pedestrian's Characteristics on Walking Behavior [J]. *Traffic & Transportation: Academic Edition* (1).

Mavros, P. et al. 2016. Geo-EEG: Towards the Use of EEG in the Study of Urban Behaviour[J]. *Applied Spatial Analysis & Policy* 9(2):1–22.

Virkler, M.R. & Elayadath S. 1994. Pedestrian speed-flow-density relationships[J]. *In Transportation Research Record: Journal of the Transportation Research Board* 1438 (1):51.

Palmer A. 1975. Sex differences and the statistics of crowd fluids[J]. *Behavioral Science* 20(4):223–227.

Rastogi R. et al. 2013. Pedestrian flow characteristics for different pedestrian facilities and situations[J]. *European Transporttrasporti Europei* (53):5.

Sisiopiku V.P. & Akin D. 2003. Pedestrian behaviors at and perceptions towards various pedestrian facilities: an examination based on observation and survey data [J]. *Transportation Research Part F Traffic Psychology & Behaviour* 6(4):249–274.

Tanaboriboon Y. et al. 1986. Pedestrian Characteristics Study in Singapore[J]. *Journal of Transportation Engineering Asce* 112(3):229–235.

Wu T. 2002. Urban Detail: Analysis on Design principles and Methods (in Chinese) [D]. Chongqing University.

Xu H. 2013. Research on Pedestrian's Traffic Characteristics in the Ladder section of Mountain City (in Chinese) [D]. Chongqing Jiaotong University, 2013.

Yang L. 2009. Research on the Traffic Characteristics of Pedestrian Comprehensive Passenger Transport Terminal (in Chinese) [D]. Jilin University.

Zhang G. 2009. Research on the Pedestrian Behavior and Microscopic Simulation on the Square of Passenger Transport Terminal (in Chinese) [D]. Beijing Jiaotong University.

Zhang J. 2014. Analysis on Traffic Characteristics of Pedestrian at Signalized Intersection (in Chinese) [D]. Chang'an University.

Zhang Y. et al. 2012. Analysis of Walking Road System in University Campus from the Perspective of Environmental Behavior (in Chinese) [J]. *Journal of Xi'an Aviation College* (5):60–63.

Zhao G. et al. 2006. Study on Pedestrian Flow Traffic Characters of Special Games (in Chinese) [J]. *Road Traffic & Safety* (2): 19–21.

Zhu D. & Lin Y. 2015. Characteristics, Models and Methods– A Preliminary Exploration of the Microscopic Simulation of Pedestrian Connection Behavior in Influenced Urban Realm around Rail Transit Stations (in Chinese) [J]. *Architectural Journal* (3):24–29.

Transport Infrastructure and Systems – Dell'Acqua & Wegman (Eds)
© *2017 Taylor & Francis Group, London, ISBN 978-1-138-03009-1*

Performance evaluation of regenerated HMA with very high RAP content

N.A. Alvarado Patiño & J. Martinez
Laboratoire de Génie Civil et Génie Mécanique, INSA de Rennes, France

E. Lopez
Fenixfalt SAS, Buros, France

ABSTRACT: This research presents the thermomechanical performances of Hot Mix Asphalt (HMA) containing conventional and regenerated Reclaimed Asphalt Pavement (RAP). Different mixtures containing 40 to 70% RAP are evaluated and compared to a control mix (without RAP). The material properties are determined through an extensive experimental program including workability, rutting resistance, susceptibility to water, stiffness, fatigue resistance and performances at low temperature. The results show that, compared to the control mixture, the use of RAP significantly increases the stiffness of the mixes without a significant loss of fatigue resistance; however, RAP does reduce the performance of HMA at low temperatures. The use of an additive allows the mixtures to partially recover properties such as workability, fatigue resistance and ductility at low temperatures.

1 INTRODUCTION

1.1 *Objectives and context*

The objective of this research work is to increase the rate of Reclaimed Asphalt Pavement (RAP) in Hot Mix Asphalt (HMA) without important loss of the thermomechanical performances by using an additive.

An experimental laboratory program is developed involving two main steps: (i) composition and production of different mixtures with variable RAP content with and without the additive; (ii) testing of the general and thermo-mechanical properties of the mixes. The material considered is a typical class-4 dense French asphalt known as grave-bitume (GB4). For all the mixtures, the grading curve of the aggregates and the final binder content are kept constant and are in accordance with French standards.

1.2 *Previous results*

Using high rates of RAP in HMA gives rise to different limitations. First, the optimization of the manufacturing process for mixing RAP and virgin components is necessary in order to guarantee correct homogeneity between virgin and aged binders (Olard et al. 2008, Navaro 2011, Valdés et al. 2011; Vislavičius & Sivilevičius 2013). Moreover, differences between laboratory and plant production must be controlled (Reyhanegh & Daniel 2016).

Secondly, a sufficient level of mechanical performances like cracking at low temperature and fatigue resistance must be guaranteed. Current observations show that when using mixtures with high percentages of RAP, a reduction in workability, an increase in rutting resistance and in elastic stiffness occur. This is due to the ageing of the RAP binder and the consequent hardening of the final binder (Shu et al. 2008, Olard et al. 2009, Hajj et al. 2009, Silva et al. 2012, McDaniel et al. 2012, Mangiafico et al. 2015).

The effect of RAP content on the fatigue resistance of the mixtures is observed differently, depending on the authors and on the conditions. In some cases, up to a certain content of RAP, the fatigue resistance of the material is not reduced and is even sometimes improved (Huang et al. 2005, Tabakovic et al. 2010, Mangiafico et al. 2015, Perez-Martinez et al. 2016). When the RAP content is increased, the fatigue resistance may even deteriorate compared to the control mixture without RAP (Shu et al. 2008, Hajj et al. 2009).

For virgin HMA, the performances at low temperatures are mainly controlled by the type of binder (Arand 1990, Pucci 2001, Maia et al. 1997, Soenen & Vanelstraete 2003, Olard 2004). Information on the cracking resistance of RAP at low temperatures is relatively recent, and supposes a loss in performance (Huang et al. 2004, Hajj et al. 2009).

In order to compensate for performance loss in RAP mixtures, special binders and additives are introduced allowing an increase in RAP content.

As a consequence, an increase in workability has been obtained (Silva et al. 2012, Lopez et al. 2015). Moreover, the additives have enhanced the cracking resistance of RAP (Lopez et al. 2011, Largeaud et al. 2015, Porot et al. 2016).

2 MATERIALS AND TEST METHODS

2.1 Characterization of the material components

The components are characterized exhaustively to reduce the heterogeneity of the final mixtures (Vislavičius & Sivilevičius 2013). The stocks of RAP and virgin aggregates are mechanically homogenized in piles. A second laboratory homogenization is required for each sample (one control test per tonne). The gradation results are shown in Figure 1.

All virgin aggregates used in this work come from massive rocks (sandstone quartzite), while the filler is a limestone. The average density of the aggregates is 2630 kg/m³.

The main properties of RAP are summarized in Table 1. Properties can be considered as homogeneous according to the RAP classification. Penetration and softening temperatures are representative of an aged binder from surface layers.

Different grades of the virgin binder (20/30, 35/50, 50/70 and 70/100) are used to ensure the same class (20/30) of the final binder for all the mixtures.

The additive employed for this research (REGE-FALT®) has the function of reducing the oxidized hydrocarbon components included in the RAP binder and to facilitate the interpenetration of the particles of old and new binders, which leads to the integration of the components of the new bitumen and to an increase in the performances of the mixtures (Lopez et al. 2011, Lopez et al. 2015).

2.2 Tests performed for characterizing the mixes

The manufacturing conditions of the mixtures are in agreement with standard specifications (NF EN 12697-35+A1 2007) and constant for all RAP

Table 2. Tests performed for the formulation of the mixes.

Measured properties	Test	N_R**
Workability	Gyratory shear press (GSP)	3
Sensitivity to water	Duriez	5×2
Rutting resistance	Wheel tracking test	2
Complex modulus	2PB-TR	4
Fatigue resistance	2PB-TR	6×3
Low temperature resistance*	UTST	3×4
	TSRST	3

*According to (NF EN 12697-46 2013).
**Number of replicates.

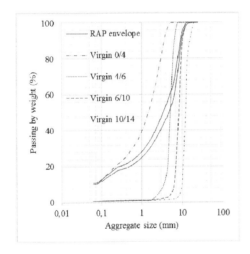

Figure 1. Grading curves of the different aggregates.

Table 1. Main properties of RAP.

Properties	Test results
Type*	12.5 RA 0/10
Petrography	Quartzite-eruptive
Density of particles (coated)	2472 kg/m³
Soluble binder content	5.11%
Penetration at 25°C	16.4 (0.1 mm)
Softening point (Ring and Ball)	65.5°C
Category*	$TL_1 - B_1 - G_1 - R_1$

*According to (NF EN 13108-8 2006).

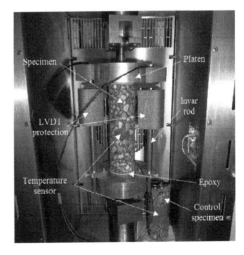

Figure 2. Equipment for the low temperature tests in climate chamber.

contents: pre-heating of virgin aggregates and RAP at 190°C and 180°C respectively. In order to guarantee correct homogeneity between components, a mixing duration of 5 mn is adopted. The preheating temperature of the virgin binder depends on its penetration grade.

Standard tests performed on the mixes agree with French specifications (Table 2). Three different equipment are used for the compaction of the samples. The GSP specimens have a diameter of 160 mm and are tested at a temperature of 180°C. The Duriez specimens, of 120 mm diameter, are compacted statically with application of a vertical force of 180 kN during 5 mn. The specimens for the remaining tests are compacted with a plate compactor in rectangular molds of different sizes (length and width in mm): 500*180 for the rutting tests and 600*400 for the others. In the rutting tests all the mixtures are compacted with the same energy, as for the following tests, the compaction energy is adapted to the required voids content.

Two types of low temperature tests are performed using the equipment shown on Figure 2: the Uniaxial Tension Stress Test (UTST) at four different temperatures (+20°C, +5°C, −10°C and −25°C) and the Thermal Stress Restrained Specimen Test (TSRST) with an initial temperature of +10°C and a temperature gradient of −10°C/h.

The temperature is measured using thermocouples fixed to the surface of the specimen and both on the surface and inside a control sample for temperature correction (Figure 2).

3 MATERIALS COMPOSITION

Three proportions of RAP content are considered (40%, 55% and 70%, with and without additive) and a control mixture containing only virgin aggregates and virgin binder. A constant grading curve for all the above mixtures is obtained by first mixing the 70% RAP formula with different proportions of the elementary fractions of virgin aggregates to obtain a conventional grading curve of GB4 material. For all the mixtures, the RAP grading curve is considered in total without separation into different fractions; thus the constant grading of the different mixes is obtained by adjusting the fractions of virgin aggregates (Table 3).

Constant total binder content is fixed for all the mixtures by modifying the amount of the virgin binder content (Table 3). Similarly, the final binder penetration (class 20/30) is obtained by adjustment of the virgin binder grade using the usual mixing law (NF EN 13108-8 2006):

$$a \lg pen_1 + b \lg pen_2 = (a+b) \lg pen_{mix} \qquad (1)$$

where pen_{mix} = calculated penetration of the binder in the mixture with RAP; pen_1 = penetration of the RAP binder; pen_2 = penetration of virgin binder; a and b = portions by mass of the RAP binder (a) and of the virgin binder (b). ($a + b = 1$).

The additive content Q (‰) is determined from the proportion of RAP (R) in the mixture and the RAP binder content (X) by the following empirical relation: $Q = 36\ R\ X$.

4 GENERAL PROPERTIES

All mixtures are manufactured according to the requirements of French standards for the GB4 (NF EN 13108-1 2007). The grading curve and the binder content are verified systematically for each test.

4.1 Workability (GSP test)

The workability tests are performed with the Gyratory Shear Press (GSP type 2). Figure 3 shows the results are shown in Figure 3 for the different mixtures in terms of voids content versus the number of gyrations. All mixtures are in agreement with the French specifications (voids content at 100 gyrations $V_{100} < 9\%$). Compared to the control

Table 3. Main properties of the mixes.

Id	RAP (%)	Grading mixes (passing by weight)						Additive (‰)	Virgin binder		Total binder (%)	Penetration at 25 °C (1/10 mm)	Softening point (°C)	Density (kg/m³)
		0.063 mm (%)	2 mm (%)	6.3 mm (%)	10 mm (%)	14 mm (%)			Grade	(%)				
0R	0	8.4	30.9	59.6	81.3	97.0	–		20/30	4.90	4.90	29	61.6	2449
40R	40	8.4	31.0	59.2	81.4	97.1	–		35/50	2.85		30	58.0	2461
40R′	40	8.4	31.0	59.2	81.4	97.1	0.7		35/50	2.85		30	58.0	2461
55R	55	8.4	31.0	59.8	81.9	97.1	–		50/70	2.09		29	58.6	2465
55R′	55	8.4	31.0	59.8	81.9	97.1	1.0		50/70	2.09		29	58.6	2465
70R	70	8.4	31.0	59.3	81.8	97.1	–		70/100	1.32		24	60.9	2470
70R′	70	8.4	31.0	59.3	81.8	97.1	1.3		70/100	1.32		24	60.9	2470

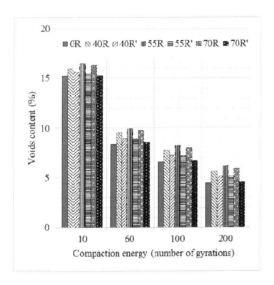

Figure 3. Workability of the mixes (GSP test).

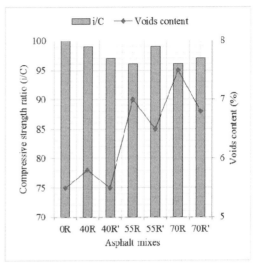

Figure 4. Compressive strength ratio (i/C) and voids content (Duriez test).

mix (0R), the mixtures containing RAP become less workable. The additive, however, increases the workability, especially for the mix with the highest rate of RAP (70R'), producing a voids content equal to that of the control mix.

4.2 *Sensitivity to water (Duriez test)*

Susceptibility to water is determined using the Duriez test which measures the ratio i/C of the compressive strength in dry and saturated conditions at 18°C. A slight reduction in the i/C ratio is observed for the mixtures containing RAP (Figure 4). Nevertheless, the ratio remains above 95%, which is far higher than the specifications for this type of material (i/C > 70%). Moreover, the voids content is observed to increase with the ratio of RAP, so too does the favorable effect of the additive: as in the GSP test.

4.3 *Rutting resistance*

The rutting resistance is measured by the mlpc® wheel tracking device at a temperature of 60°C. The voids content V of the compacted plates follow the same variations displayed in the results above, showing the favorable effect of the additive (Figure 5). The values are in agreement with French specifications (8% < V < 5%). The rutting percentages at 30,000 cycles ($P_{30,000}$) decrease with an increase in RAP content, and are slightly higher when using the additive, but the variations remain low (about 1%). For all the mixes the rutting values stay small compared to the allowable values (<10%).

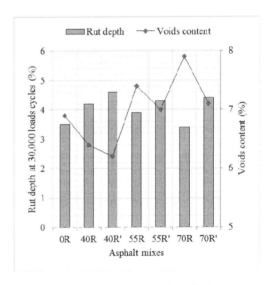

Figure 5. Initial voids content and rut depth measured at 30,000 cycles.

5 STIFFNESS AND FATIGUE RESISTANCE

The stiffness and the fatigue resistance are measured on standard trapezoidal specimens at 4.5% voids content, using strain-controlled two-point bending tests (2PB-TR). Stiffness tests are performed on four samples at a temperature of 15°C, a frequency of 10 Hz and a strain level of about 39 μm/m. Fatigue tests are carried out according to the standards (10°C; 25 Hz) with three levels of deformation and six samples per level.

472

5.1 Complex modulus and phase angle

Due to the hardness of the aged binder, the mixes exhibit a complex modulus that increases with the RAP content and is much higher (about 50%) than that of the control mix (Figure 6). For the same reason, the phase angle decreases with the RAP content (from 13 to 7 degrees). Similar results have been obtained by Mangiafico et al. (2015) and by Perez-Martinez et al. (2016) signifying that RAP increases the elastic component of the stiffness and reduces the viscous component. On the other hand, the additive very slightly reduces the complex modulus without affecting the phase angle.

5.2 Fatigue resistance

The fatigue results (Figure 7 and Table 4) show that for all the mixtures, the fatigue resistance is quite high ($\varepsilon_6 > 147$ μm/m) and that, compared to the control mix, the addition of RAP leads to a small degradation of ε_6 (less than 8 μm/m).

Moreover, an increase of ε_6 is observed with the additive, especially for the mix 70R', which is above the control mix 0R. On the other hand, the scatter $\Delta\varepsilon_6$ is slightly higher when using additive but it remains of the same order as that of the control material. Another effect of the additive which was observed is the decrease in the slope b of the Wöhler curve compared to the mixes without additive. This leads the life duration of the pavement to be less susceptible to the variations of the strain level, when using the additive.

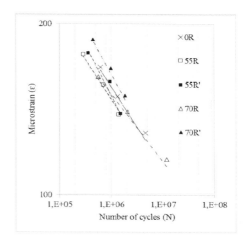

Figure 7. Fatigue resistance of the mixes (10°C; 25 Hz).

Table 4. Synthesis of stiffness and fatigue results.

Mixture	E* (MPa)	φ (°)	ε_6 (μm/m)	$\Delta\varepsilon_6$ (μm/m)	b
0R	12,285	13.1	154.8	5.6	−0.14
55R	17,751	8.9	147.2	3.9	−0.14
55R'	17,105	8.9	149.5	6.0	−0.18
70R	19,093	7.3	150.4	3.6	−0.12
70R'	18,844	7.3	163.0	5.6	−0.17

6 LOW TEMPERATURE PERFORMANCES

The specimens are cored from plates compacted in the laboratory and have a cylindrical shape: 160 mm height by 50 mm diameter (Figure 2), with a voids content of $5.5 \pm 0.5\%$.

6.1 UTST results

The strain rate of the test ($4{,}2 \pm 0.8 \ 10^{-6} \ s^{-1}$) is set at a lower value to the one recommended by the standards, but this parameter has been shown to have no effect on the results at low temperatures (Olard, 2004, Steiner et al. 2016). Typical stress-strain curves show two types of fracture (Figure 8): fragile fracture for low temperatures (−25°C; −10°C) and ductile for higher ones (+5°C; +20°C). In the case of ductile fracture, the tensile strength β_t and the failure strain $\varepsilon_{failure}$ are taken at maximum stress according to the standards.

For all the mixtures the tensile strength is maximum for intermediate temperatures (Figure 9 and Table 5) corresponding to the fragile-ductile transition temperature T_{fde} (Olard 2004). Compared to the control mix, this temperature is increased by the presence of RAP: from −15°C for 0% RAP up to 0°C for 70% RAP. The effect of the binder

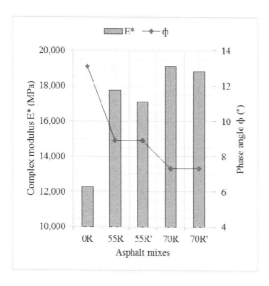

Figure 6. Complex modulus and phase angle of the mixes (15 °C; 10 Hz).

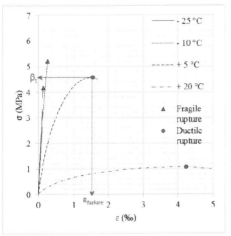

Figure 8. Typical stress-strain response from UTST tests 40R'.

hardness on the tensile strength at low temperatures has been shown by Olard (2004). For the mixtures containing 70% RAP, the additive has a favorable effect as it allows the enhancement of the transition temperature (from 0°C to −3°C). In the ductile domain the tensile strength increases significantly with RAP content. This effect is due to a higher mobilization of the aged binder through its elastic stiffness (Olard 2004, Arand 1990).

The failure strain (Figure 10 and Table 5) decrea-ses with RAP content due to the ageing of the RAP binder, and the differences increase with the temperature. On the other hand, the additive increases the failure strain, allowing the recovery of some of the ductility.

6.2 TSRST results and strength reserve

The TSRST test simulates the thermal stress σ_{cry} produced in a longitudinally restrained pavement caused by an environmental temperature drop.

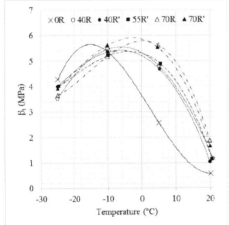

Figure 9. Tensile strength versus temperature.

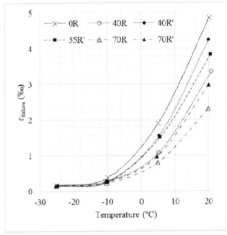

Figure 10. Tensile failure strain versus temperature.

Table 5. Synthesis of UTST/TSRST results.

	UTST									TSRST			Reserve	
	βt (at T °C)				$\varepsilon_{failure}$ (at T °C)								$\Delta\beta t$ (at T °C)	
Mix	−25 (MPa)	−10 (MPa)	+5 (MPa)	+20 (MPa)	−25 (‰)	−10 (‰)	+5 (‰)	+20 (‰)	T_{fde} (°C)	$\sigma_{cry,failure}$ (MPa)	$T_{failure}$ (°C)		−20 (MPa)	+10 (MPa)
0R	4.28	5.32	2.57	0.57	0.166	0.382	1.899	4.854	−15.3	3.74	−28.2		2.83	1.60
40R	3.52	5.46	4.79	1.14	0.134	0.272	1.068	3.343	−6.1	3.52	−24.0		1.34	3.95
40R'	4.00	5.34	4.66	1.03	0.143	0.265	1.549	4.236	−6.3	3.67	−26.5		1.86	3.86
55R'	4.01	5.21	4.87	1.14	0.155	0.261	1.503	3.821	−3.9	3.82	−24.0		1.15	4.16
70R	3.66	5.18	5.60	1.87	0.148	0.231	0.803	2.299	+0.3	3.66	−21.8		0.57	4.94
70R'	3.92	5.61	5.52	1.65	0.161	0.263	0.963	2.971	−2.9	3.95	−23.9		1.02	4.68

The test results are characterized by the maximum stress reached $\sigma_{cry,failure}$ (cryogenic failure stress) and by the corresponding temperature $T_{failure}$. For all the tests, the initial temperature ($T_0 = +10°C$) is chosen such that it has no effect on the final result (Tušard et al. 2014).

The results (Figure 11 and Table 5) show that the failure stress does not depend on the RAP content, but that the failure temperature increases significantly, which leads to a reduction in the cracking resistance of RAP mixtures at low temperatures. Similarly to the UTST results, the additive allows the decrease of the failure temperature.

The strength reserve $\Delta\beta_t(T)$ at a given temperature T is defined as the difference between the tensile strength $\beta_t(T)$ and the cryogenic stress $\sigma_{cry}(T)$ derived respectively from the UTST and the TSRST tests (NF EN12697–46 2013). It corresponds to the allowable stress afforded to traffic when the pavement is submitted simultaneously to cryogenic stress.

Figure 12 shows that both the maximum strength reserve and the corresponding temperature increase with RAP content.

Moreover, the reserve curves allow us to differentiate two domains depending on temperature (Table 5).

For temperatures higher than −5°C the RAP mixtures exhibit significantly higher strength reserves than the control mixture (up to +110%).

Inversely, for temperatures lower than −5°C the strength reserve of the control mix is higher than that of the RAP mixtures. In this domain, the additive increases the strength reserve up to 0.6 MPa.

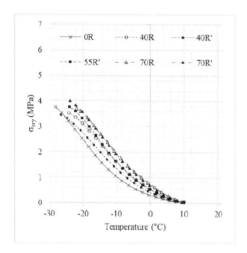

Figure 11. Cryogenic stress from TSRST test.

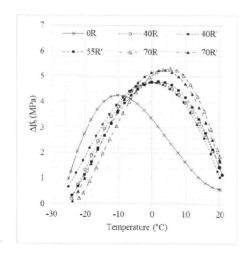

Figure 12. Tensile strength reserve of the mixes.

7 CONCLUSIONS

The extensive set of tests performed on a dense HMA with different RAP contents and the presence or absence of an additive lead to several conclusions.

The increase in RAP content in the mixes shows a reduction in the workability and a small effect on the sensitivity to water and on the rutting resistance. The elastic stiffness is strongly increased and the fatigue resistance is slightly reduced. In terms of behavior at low temperatures, the performances are significantly reduced.

The use of the additive allows the partial or total recovery of the performances deteriorated by the RAP; such as workability and fatigue resistance. Furthermore it has only a small effect on the elastic stiffness. At low temperature, an increase of the thermomechanical properties is observed when using the additive.

ACKNOWLEDGEMENTS

This research project is part of the Ph.D. thesis prepared by the first author under a contract CIFRE 2013/0334 between the French ANRT (Agence Nationale de la Recherche et de la Technologie) and the Fenixfalt company. The authors gratefully acknowledge the following companies for their participation in the project: Le Foll TP, Roger Martin and Braja-Vésigné. The authors are also grateful to C. Garand from the LGCGM laboratory of INSA—Rennes, for his help in the layout and in the organization of the low temperature tests.

REFERENCES

Arand, W. 1990. Behaviour of asphalt aggregate mixes at low temperatures. *Mechanical Tests for Bituminous*

Mixes, *Proceedings of the 4th RILEM Symposium*: 67–81. Budapest.

Hajj, E.Y., Sebaaly, P.E., & Shrestha, R. 2009. Laboratory Evaluation of Mixes Containing Recycled Asphalt Pavement (RAP). *Road Materials and Pavements Design* 10(3): 495–517.

Huang, B., Kingery, W. & Zhang, A. 2004. Laboratory Study of Fatigue Characteristics of HMA Mixtures Containing RAP. *International Symposium on Design and Construction of Long Lasting Asphalt Pavements*: 501–522. Auburn.

Huang, B., Li, G., Vukosavljevic, D., Shu, X. & Egan, B.K. 2005. Laboratory Investigation of Mixing Hot-Mix Asphalt with Reclaimed Asphalt Pavement. *Transportation Research Record* (1929): 37–45.

Largeaud, S., Faucon-Dumont, S., Eckmann, B., Hung, Y., Lapalu, L. & Gauthier, G. 2015. Caractérisation du comportement à basse température des liants bitumineux. *Revue générale des routes et de l'aménagement (RGRA)* (928): 70–77.

Lopez, E., Gasca-Allue, C., Chifflet, P. & García-Serrada, C. 2011. La nouvelle vie du bitume. Une innovation récente de régénération des bitumes vieillis des enrobés. *Congrès mondial de la Route (AIPCR)*. Mexico.

Lopez, E., Willem, M. & Mabille, C. 2015. Bétons bitumineux recyclés et régénérés à fort taux d'AE. *Revue générale des routes et de l'aménagement (RGRA)* (928): 22–27.

Maia, A.F., Marciano, Y., Archimastos, L., Corté, J.-F., Gourdon, J.-L., Piau, J.-M. & Faure, M. 1997. Low-temperature behaviour of hard bitumens: experiments and modeling. *Mechanical Test for Bituminous Mixes (MTBM), proceeding of the 5th. International RILEM Symposium*. Lyon.

Mangiafico, S., Sauzéat, C., Di Benedetto, H., Olard, F., Pouget, S, Dupriet, S. & Van Rooijen, R. 2015. Effet de l'ajout d'agrégat d'enrobé sur le module complexe et la résistance en fatigue des enrobés bitumineux. *Revue générale des routes et de l'aménagement (RGRA)* (928): 35–40.

McDaniel, R., Shah, A., Huber, G.A. & Copeland, A. 2012. Effects of reclaimed asphalt pavement content and virgin binder grade on properties of plant produced mixtures. *Road Materials and Pavements Design* 10(3): 495–517.

Navaro, J. 2011. Cinétique de mélange des enrobés recyclés et influence sur les performances mécaniques. *Thèse de doctorat: École Nationale Supérieure d'Arts et Métiers ParisTech*.

NF EN 12697-35+A1 2007. *French standard. Bituminous mixtures. Test methods for hot mix asphalt*. Part 35: Laboratory mixing. AFNOR.

NF EN 12697-46 2013. *French standard. Bituminous mixtures. Test methods for hot mix asphalt. Part 46: Low temperature cracking and properties by uniaxial tension test*. AFNOR.

NF EN 13108-1 2007. French standard. Bituminous mixtures. Material specifications. Part 1: Asphalt Concrete. AFNOR.

NF EN 13108-8 2006. French standard. Bituminous mixtures. Material specifications. Part 8: Reclaimed Asphalt. AFNOR.

Olard, F. 2004. Comportement thermomécanique des enrobes bitumineux à basses températures. Relations entre les propriétés du liant et de l'enrobé. *Thèse de doctorat: Institut National des Sciences Appliquées de Lyon*.

Olard, F., Bonneau, D., Dupriet, S., Beduneau, E. & Seignez, N. 2009. Formulation spécifique en laboratoire des enrobés à très fort taux de recyclés (jusqu'à 70%), à chaud ou en technique E.B.T.®. *Revue générale des routes (RGRA)* (872): 62–66.

Olard, F., Le Noan, C, Bonneau, D., Dupriet, S. & Monnier, Y. 2008. Le recyclage à très fort taux (>50%) des enrobés chauds et des enrobés tièdes pour une construction routière durable. *Revue générale des routes (RGRA)* (866): 56–62.

Perez-Martinez, M., Marsac, P., Gabet, T., Hammoum, F., de Mesquita Lopes, M. & Pouget, S. 2016. Effects of Ageing on Warm Mix Asphalts with High Rates of Reclaimed Asphalt Pavement. In Chabot et al. (eds), *Proc. 8th RILEM International Conference on Mechanisms of Cracking and Debonding in Pavements, Nantes, 7–9 June 2016*: 113–118. Springer.

Porot, L. & Vuillier, B. 2016. SYLVAROAD™ RP1000 Additive, un régénérant BIO pour AE à forts taux. *Revue générale des routes et de l'aménagement (RGRA)* (933): 54–57.

Pucci, T. 2001. Approche prévisionnelle de la fissuration par sollicitation thermique des revêtements bitumineux. *Thèse de doctorat: École Polytechnique Fédérale de Lausanne*.

Reyhanegh, R.-R. & Daniel, J.S. 2016. Mixture and Production Parameters Affecting Cracking Performance of Mixtures with RAP and RAS. In Chabot et al. (eds), *Proc. 8th RILEM International Conference on Mechanisms of Cracking and Debonding in Pavements, Nantes, 7–9 June 2016*: 307–312. Springer.

Shu, X., Huang, B. & Vukosavljevic, D. 2008. Laboratory evaluation of fatigue characteristics of recycled asphalt mixtures. *Construction and Building Materials* 22(7): 1323–1330.

Silva, H., Oliveira, J.R. & Jesus, C.M. 2012. Are totally recycled hot mix asphalts a sustainable alternative for road paving?. *Resources, Conservation and Recycling* 60: 38–48.

Soenen, H. & Vanelstraete, A. 2003. Performance indicators for low temperature cracking. *6th International RILEM Symposium on Performance Testing and Evaluation of Bituminous Materials*: 458–464. Zurich.

Steiner, D., Hofko, B., Dimitrov, M. & Blad, R. 2016. Impact of Loading Rate and Temperature on Tensile Strength of Aspahalt Mixtures at Low Temperatures. In Chabot et al. (eds), *Proc. 8th RILEM International Conference on Mechanisms of Cracking and Debonding in Pavements, Nantes, 7–9 June 2016*: 69–74. Springer.

Tabakovic, A., Gibney, A., McNally, C. & Gilchrist, M.D. 2010. Influence of recycled asphalt pavement on fatigue performance of asphalt concrete base courses. *Journal of Materials and Civil Engineering* 22(6): 643–650.

Tušar, M., Hribar, D. & Hofko, B. 2014. Impact of characteristics of asphalt concrete wearing courses on crack resistance at low temperatures. *Transport Research Arena*. Paris.

Valdés, G., Pérez-Jiménez, F., Miró, R., Martínez, A. & Botella, R. 2011. Experimental study of recycled asphalt mixtures with high percentages of reclaimed asphalt pavement (RAP). *Construction and Building Materials* 25: 1289–1297.

Vislavičius, K. & Sivilevičius, H. 2013. Effect of reclaimed asphalt pavement gradation variation on the homogeneity of recycled hot-mix asphalt. *Archives of Civil and Mechanical Engineering* 13(3): 345–353.

Transport Infrastructure and Systems – Dell'Acqua & Wegman (Eds)
© 2017 Taylor & Francis Group, London, ISBN 978-1-138-03009-1

Finite element modelling of flexible pavement reinforced with geogrid

G. Leonardi & R. Palamara
"Mediterranea" University of Reggio Calabria, Reggio Calabria, Italy

ABSTRACT: The always increasing patterns of vehicle traffic on roads, in all weather, sets that the maintenance and rehabilitation of infrastructures among the engineering key tasks for transport infrastructures. Furthermore, the need to guarantee high performance wilts increasing pavement life, has led research and industry to focus a greater attention on the use of pavement reinforcements. In recent years, numerous kinds of geosynthetic grids have been introduced on the market, which can be used for pavement reinforcement, extending extend pavement life, guaranteeing high performance and reducing costs of service and maintenance. The use of geosynthetics can produce several benefits, such as drainage, reinforcement, filter, separation and proof. In this paper, the effectiveness of geogrids as reinforcement was investigated. The study proposes a numerical investigation using a three-dimensional Finite Element Method (FEM) to analyze the importance, in terms of rutting at the top of pavement system, of geogrids in the behavior under wheel traffic loads on pavement. In the model, a multi-layer pavement structure was considered with a geogrid reinforcement and the model dimensions, element types and meshing strategies are taken by successive attempts to obtain desired accuracy and convergence of the study. FEM results show that a geogrid reinforcement can provide lateral confinement at the bottom of the base layer by improving interface shear resistance and reduce rutting at the top of pavement system.

1 INTRODUCTION

In recent years, the need to increase pavement service life and to guarantee high performance has turned a greater attention on the use of pavement reinforcements. During the life period of pavement structure, it is vulnerable to different kinds of distresses. In fact, under the traffic load the asphalt mixtures are submitted to elastic, plastic, visco-elastic and visco-plastic deformations. Permanent deformation (rutting) is one of the serious distresses in which pavement structure may be involved.

Rutting can be defined as the permanent deformation of pavement along the wheel path caused by load traffic. To estimate permanent rut depth of the asphalt layer, in general nonlinear visco-plastic Finite Element Analysis (FEA) is commonly performed using different programs like ABAQUS, ANSYS or PLAXIS. A lot of research has been conducted to reduce the pavements rutting phenomenon. One of the recent methods is related to reinforcing pavement structures by means of the use of geogrids and in general by the use of different geosynthetic types.

The use of geogrid reinforcement in the construction of road pavement started in the 1970s. Then, the technique of geogrid reinforcement has been increasingly used and many experimental and analytical studies have been performed to assess geogrid behavior in the flexible pavement [Howard, 2009].

Herbst et al. [10] illustrated an interesting set of data from an experimental site in Austria, where the comparative benefits of geogrids and geotextiles could be directly assessed. Penman & Hook [14] described how glass-fiber based geogrids had successfully used as interlayers to extend the design life of asphalt pavements on airport runways, taxiways and aprons.

The results obtained by Chehab et al. [15] revealed a partial improvement in reflective cracking resistance due to the incorporation of fiber-reinforced interlayer. It is clear from these studies that many practitioners see significant benefit in using asphalt geogrid reinforcement.

Designing a flexible pavement reinforced with glass fiber grid and evaluating the effectiveness of reinforced pavement performance is a complex problem requiring considerable research and study [Buonsanti et al, 2012; Leonardi, 2015]

In this study, a three-dimensional finite element model, using ABAQUS software, was developed to investigate the importance of geogrids reinforcement, in terms of rutting. Furthermore, the influence of the position of geogrid reinforcement has been investigated considering three different positions within the pavement. The geogrid reinforcement layer was first placed at the HMA—base interface, then it was placed in the middle of the base course and finally at 2/3 depth of the pavement surface.

2 ASPHALT REINFORCEMENT AND PERMANENT DEFORMATION

2.1 *Flexible pavement improvement by geosynthetics: functions and mechanisms*

The geosynthetics used in flexible pavement are essentially fiber glass grids and geogrids. Geogrids have traditionally been used in three different pavement applications: (a) mechanical sub-grade stabilization, (b) aggregate base reinforcement, and (c) asphalt concrete (AC) overlay reinforcement. Geogrids used like reinforcement in transport application perform two primary functions: separation and reinforcement, in which the geogrid improves the mechanically property of the pavement system [Cardile et al 2014, Calvarano et al. 2016]. Three fundamental reinforcement mechanisms have been identified involving the use of geogrids to reinforce pavement materials: lateral restraint (Fig. 1), improved bearing capacity (Fig. 2), and tensioned membrane effect (Fig. 3).

Lateral restraint refers to the confinement of the aggregate material during loading. The geogrid minimizes lateral movement of aggregate particles and increases the modulus of the base course, which leads to a wider vertical stress distribution and consequently to a reduction of vertical deformations [Calvarano et al. 2016]. The second mechanism, improved bearing

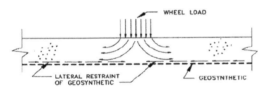

Figure 1. Lateral restraint.

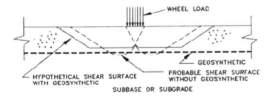

Figure 2. Bearing capacity increase restraint.

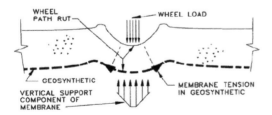

Figure 3. Membrane tension support.

capacity, is achieved by shifting the failure surface of the pavement system from the relatively weak subgrade to the relatively strong base course material.

As the of permanent deformations increase the tension membrane mechanism develops. If the geosynthetic has a sufficiently high tensile modulus, tensile stresses will be mobilized in the reinforcement, and a vertical component of this tensile membrane resistance will help to support the applied wheel loads.

2.2 *Permanent deformation (Rutting)*

Permanent deformation in the form of rutting is one of the most important failure mechanisms in asphalt pavements.

Rutting is a serious issue for road users for two reasons: driving in the rut is difficult and the water collecting in these depressions cannot drain freely off the pavement surface and may cause aquaplaning, making rutting a potential safety hazard. Rutting in flexible pavement develops gradually with increasing numbers of load applications, usually appearing as longitudinal depressions in the wheel paths accompanied by small upheavals to the sides. In addition, as a consequence of the increased tire pressure and axle load, the surfacing asphalt layer is subjected to increased stresses, which result in permanent (irrecoverable) deformations. The permanent deformation accumulates with increasing number of load applications. The permanent deformation in the surface layer, thus, accounts for a major portion of rutting on flexible pavements subjected to heavy axial loads and high tire pressures.

The main cause of rut initiation is shear strains in asphalt. Generally, there are three causes of rutting in asphalt pavements: accumulation of permanent deformation in the asphalt surfacing layer, permanent deformation of subgrade, and wear of pavements caused by studded tires. These three causes of rutting can act in combination, i.e., the rutting could be the sum of permanent deformation in all layers and wear from studded tires. In the past, subgrade deformation was considered to be the primary cause of rutting and many pavement design methods applied a limiting criterion on vertical strain at the subgrade level. However, recent researches indicate that most of the rutting occurs in the upper part of the asphalt surface layer.

3 METHODOLOGY

3.1 *Finite Element Model*

The present work aim to model reinforced and unreinforced flexible pavement using 3-D analysis with ABAQUS software, by modelling its dimensions, layers' characteristics and different meshing. The objective is to compare an unreinforced flexible pavement section with a reinforced section in terms of surface rutting, considering for the last case different positions in the system of layers.

Despite the fact that it requires considerably more computational time and computer memory, the 3D analysis is still considered superior to the 2D. The necessity for adopting the 3D analysis arises from the following advantages:

1. it better reflects the complex behavior of the composite pavement system materials under different configurations traffic loads,
2. it is preferred when verifying numerical model results with laboratory or field test results,
3. it allows the simulation of the loaded wheel rectangular footprint [Bassam Saad, 2005].

3.2 Model and materials

The flexible pavement system (Fig. 4) used in 3-D ABAQUS software consisted of asphalt concrete (HMA) surface layer, base layer, sub-base layer and sandy subgrade layer, subjected to traffic load with 0.01 second loading period.

To model the flexible pavement 5 parts were instanced, simulating 5 m wide and 5 m long layers with variable thickness, in particular: HMA surface course thickness is 0.076 m, base course thickness is 0.216 m, sub-base course thickness is 0.254 m and sandy sub-grade layer thickness is 1.454 m. The geogrid has been modelled as 5 m wide layer with a thickness of 0.003 m and with open meshes (Fig. 5).

Model dimensions have been selected to reduce any edge effect error, keeping the elements' sizes within acceptable limits.

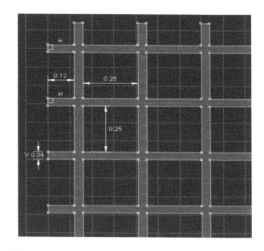

Figure 5.   Open meshes dimension.

All pavement layers were modelled by using 3D deformable solid homogeneous elements except for the geogrid, modelled like a membrane.

The materials data required for the simulations have been assumed to represent realistic material properties published in previous studies.

An elastoplastic material model for the asphalt concrete layer is necessary to allow this layer to deform permanently. Permanent deformations in layers with only elastic properties are zero when the load is removed. Based on previous analyses, creep, Drucker Prager, and modified Drucker Prager material models have been used for HMA, base, and subgrade layers, respectively.

For example, the model used in the study conducted by Zaghloul and White (1993) employed 3D dynamic finite element was a visco-elastic model for the asphalt concrete, an extended Drucker-Prager model for granular base course and Cam Clay model for the clay subgrade soils.

Zaman et al. (2003) successfully developed a viscoelastoplastic creep model representing the time-dependency of asphalt mixtures to evaluate their rutting potential and to identify factors having a significant effect.

Therefore, as regards the HMA layer a mechanical model of Creep type has been chosen, for the geogrid and the base layer a linear elastic behavior, for the remaining layers a mechanical elastic-elastic behavior, according to the Drucker-Prager model of hyperbolic [Leonardi, 2015]. Table 1 shows the mechanical characteristics of the various layers.

After defining the geometry and the mechanical properties of the various parts the features interface assignment and the mesh creation have been done. The interface mechanical characteristics are very important because they significantly affect

Figure 4.   Flexible pavement system.

Table 1. Mechanical characteristics.

Landslide susceptibility	HMA	Base	Subbase	Subgrade	Geogrid
Modulus of elasticity [Pa]	1 e⁹	316 e⁶	41 e⁶	10 e⁶	28 e⁹
Poisson's ratio	0.35	0.35	0.3	0.3	0.3
Friction angle [°]			15	10	
Initial tension [kPa]			10000	10000	
A	9 e⁹				
m	0.67				
n	−0.5				

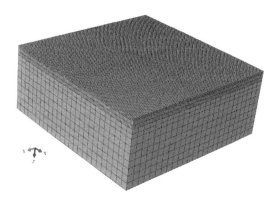

Figure 6. Mesh model.

the response of the model depending on the type of contact that is considered.

As regards the behavior at the interface between the geogrid and the layer of HMA, the interaction is constituted by the tangential and normal component. For the normal component, the hard contact type behavior has been considered while in the tangential direction a penalty type behavior with a coefficient of friction equal to 1.5 has been chosen. The same features have been used for the geogrid and the base layer interface while, to simplify the model, the other contact surfaces has been defined as tie constraint, able to limit both the displacements and the relative rotations.

The mesh model dimensions have been selected so to reduce any edge effect errors. The generated mesh has been designed to give an optimal accuracy. To improve the convergence speed, linear brick elements 8 nodes reduced integration (C3D8R) have been used for all layers (Fig. 6), whereas for the geogrid the mesh membrane type M3D4 (4-node quadrilateral membrane) has been used (Fig. 7).

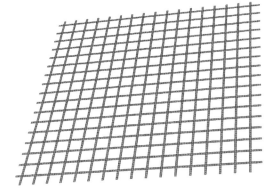

Figure 7. Geogrid mesh model.

3.3 Load and boundary conditions

The traffic load is considered as the main factor in designing the pavement system. This includes axial loads, configuration of axles, tire contact areas, number of load repetitions, vehicle speed. Heavy vehicles, such as trucks, significantly influence pavement distresses and failure. For loading area, wheel load can be applied on the wheel area, on the entire wheel path, or on an equivalent circular/semi-circular area. In this study the traffic load has been applied in correspondence of the central part of the model, simulating the passage of a twin wheel of a heavy vehicle having an axle of 80 kN. The load has been simulated as a pressure of 550 kPa acting on a 17 × 19 cm rectangular area (Fig. 8).

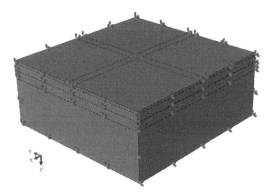

Figure 8. Load area and boundary conditions.

For the loading duration equivalent, pulse, and moving loading are the most common methods.

In this analysis, the application of the load impulse time, modelled as a truck speed of about 50 km/h, has been considered and calculated using the formula proposed by Brown (1973):

$$\log t = 0.5 \cdot h - 0.2 - 0.94 \cdot \log V$$

where t is the time in seconds of application, h is the thickness of the surface layer in meters, V is the speed in km/h.

Since the boundary conditions have a significant influence in predicting the response of the model and given the size adopted, the model has been constrained at the bottom (U1 = U2 = U3 = UR1 = UR 2 = UR3 = 0); X-Symm (U1 = UR2 = UR3 = 0) on the sides parallel to y-axis; and Y-Symm (U2 = UR1 = UR3 = 0) on the sides parallel to x-axis.

4 ANALYSIS AND RESULTS

4.1 Geogrid position

The present study has been undertaken to investigate the optimum position of the geogrid in a flexible pavement system. Therefore, a series of finite element simulations have been carried out to evaluate the benefits of the different geogrid positions. Three different configurations have been studied, namely the HMA-base layer interface (Fig. 9), in the middle of base course (Fig. 10) and inside the HMA layer at a height of 1/3 of its thickness from the bottom (Fig. 11).

Figure 9.   First configuration: geogrid at the HMA-base layer interface.

Figure 10.   Second configuration: geogrid inside the base layer.

Figure 11.   Third configuration: geogrid inside the HMA layer.

The pavement behavior has been simulated and the displacements have been considered as a response of traffic load applied. The final magnitude of the displacements, U, evaluated at the load center at the end of loading, resulted to be affected by the position of the geogrid in flexible pavements. The final magnitude of the surface displacements U was 1.76 mm (Fig. 12) for the first configuration, 1.75 mm (Fig. 13) for the second configuration and 1.4 mm (Fig. 14) for the third configuration, while for the unreinforced test section the surface displacement was 2.01 mm (Fig. 15).

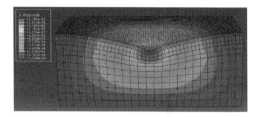

Figure 12.   Displacements for the first configuration (Fig. 10).

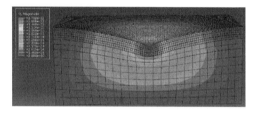

Figure 13.   Displacements for the second configuration (Fig. 11).

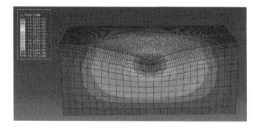

Figure 14.   Displacements for the third configuration (Fig. 12).

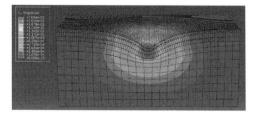

Figure 15.   Displacements for the unreinforced test section.

481

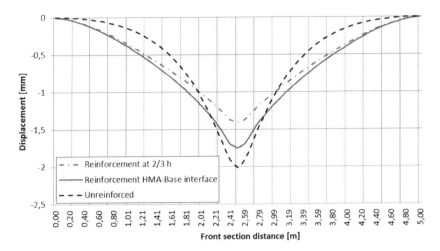

Figure 16. Rutting comparison.

Table 2. Displacements and improvement.

Section	Displacements [mm]	Improvement [%]
reinforcement inside the HMA layer at a 1/3	1.40	30.00
reinforcement in HMA-base layer interface	1.76	12.40
reinforcement in the middle of base course	1.75	12.90
Unreinforced test section	2.01	

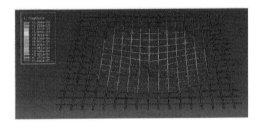

Figure 17. Displacement in correspondence of the geogrid.

The examination of the permanent surface deformation after the traffic load showed that the results for the geogrid-reinforced test sections are lower than the results obtained for unreinforced test section (Fig. 16).

The results show the improvement offered by the reinforcement in flexible pavement at the different geogrid positions. The presence of the geogrid implies an improvement in term of surface displacement reduction, corresponding to the 12.4%, 12.9% and 30% for test section with the geogrid placed at HMA-base layer interface, in the middle of base course and inside the HMA layer at a distance of 2/3 from the bottom respectively (Table 2). Results confirm that there is higher efficiency when the geogrid is placed inside the HMA layer at a distance of 2/3 from the bottom. In this test session, the displacement of the reinforcement geogrid following the application of the traffic load was equal to 1.3 mm (Fig. 17).

5 CONCLUSION

Several researches have been conducted to reduce the rutting phenomenon in flexible pavements. One of the prevention methods is the geosynthetic reinforcement. The analysis conducted in this paper with regards to the flexible pavement reinforcement, showed that the geogrid reinforcement positively influences the permanent deformation of asphaltic surface; in fact, it may be observed a significant decrease in the vertical displacement obtained for the reinforced pavement section compared to the unreinforced pavement section.

Based on the 3-D ABAQUS software outputs applied on pavement structure to evaluate the benefits of reinforcing pavement with geogrid at three positions, the following conclusions can be drawn:

1. A significant improvement in the pavement behavior is obtained by placing the geogrid layer inside the HMA layer at a distance of 2/3 of thickness. Vertical displacement and effective stress responses are significantly lower for

reinforced pavement system in comparison with unreinforced pavement.

2. Moderate improvement in pavement system behavior is gained by adding geogrid at the top of sub-base layer.

3. The best location of adding geogrid results to be in the vicinity of the point of tire pressure application within the asphalt concrete layers.

4. The use of geogrid significantly enhances the asphalt concrete resistance to the deformation and to the rutting failure development.

REFERENCES

Buonsanti, M., Leonardi, G., and Scopelliti, F. (2012), Theoretical and computational analysis of airport flexible pavements reinforced with geogrids. RILEM Bookseries, Volume 4, 2012, Pages 1219–1227.

Calvarano, L.S., Gioffrè, D., Cardile, G., Moraci, N., (2014), A stress transfer model to predict the pullout resistance of extruded geogrids embedded in compacted granular soils, in proceeding of 10th International Conference on Geosynthetics ICG 2014, Berlin.

Calvarano L. S., Palamara R., Leonardi G., Moraci N. (2016). Unpaved road reinforced with geosynthetics. Procedia Engineering, 158 (2016) pp 296–301, ISSN: 1877-7058- DOI:10.1016/j.proeng.2016.08.445, Ed. Elsevier.

Chehab, G., Chaignon, F., Thompson, M., and Palacios, C. (2008), Evaluation of fiber reinforced bituminous interlayers for pavement preservation. In: Pavement Cracking, CRC Press.

Herbst, G., Kirchknopf, H., and Litzka, J. (1993), Asphalt overlay on crack-sealed concrete pavements using stress distributing media. In: 2nd Int. RILEM Conf. Reflective Cracking in Pavements, Liege, CHAPMAN & HALL.

Hook, K. and Penman, J. (2008), The use of geogrids to retard reflective cracking on airport runways, taxiways and aprons. In: Pavement Cracking, CRC Press.

Howard, I. L. and Warren, K. A. (2009). Finite-element modelling of instrumented flexible pavements under stationary transient loading. J. Transportation Eng. ASCE, 135 (2): 53–61.

Leonardi, G., (2015), Finite element analysis for airfield asphalt pavements rutting prediction. Bulletin of the Polish academy of Sciences: Technical Sciences, Volume 63, Issue 2, 1 June 2015, Pages 397–403.

Saad, B., Mitri, H., and Poorooshasb, H., (2005), Three-Dimensional Dynamic Analysis of Flexible Conventional Pavement Foundation. In Journal of Transportation Engineering.

Zagh, S., and White, T., (1993) Use of a Three-Dimensional, dynamic finite element program for analysis of flexible pavement, Transportation Research Record.

Zaman, M., Pirabaroban S., and Tarefder, R., (2003) Evaluation of rutting potential in asphalt mixes using finite element modelling, in Annual Conference of the Transportation Association of Canada.

Transport Infrastructure and Systems – Dell'Acqua & Wegman (Eds)
© *2017 Taylor & Francis Group, London, ISBN 978-1-138-03009-1*

Critical success factors for PPP infrastructure projects in India

R.P. Pradhan, A. Verma, S. Dash, R.P. Maradana & K. Gaurav
Indian Institute of Technology Kharagpur, West Bengal, India

ABSTRACT: Public Private Partnership (PPP) mode is a strategic step to boost infrastructure development in an economy. However, the achievement of this strategy depends upon the stress of its Critical Success Factors (CSFs). This research investigates the CSFs for PPP infrastructure projects in India. Using Principal Component Analysis (PCA), the study finds the factors- lack of community participation, lack of political will and absence of good governance- that delay the projects and make them financially unviable.

1 INTRODUCTION

Infrastructure is a key to development. However, the association between these two varies from place to place, time to time, and sector to sector. India is no exception to this issue. Infrastructure problems in India are very acute and range from poor condition of roads to shortage of energy. The airports as well as seaports of India need to be upgraded to meet international standards. With rapid urbanization for its ever-growing population, there is an additional demand for all kinds of infrastructure, such as electricity, drinking water, transportation and other basic amenities. The growth of these infrastructure requires enormous investment in the Indian economy. However, due to lack of budgetary finance, government encourages the Public Private Partnership (PPP) mode of infrastructure development. Still after more than two decades of globalization and privatization, the PPP mode of infrastructure development is not so common in India. Consequently, the objective of this paper is to examine the critical success factors for PPP infrastructure projects in the Indian economy.

The remainder of this paper is organized as follows. Section 2 describes methodology. Section 3 describes the empirical results. Section 4 discusses the results. The final section concludes with policy implications.

2 RESEARCH METHODOLOGY

The investigation of this problem is a two-step process. First, the identification of Critical Success Factors (CSFs) for PPP infrastructure projects; and second, pointing out the most suitable CSFs for successful implementation of PPP infrastructure projects. In the first phase, on the basis of existing literature, we identify twenty six critical success factors for PPP infrastructure projects (Akintoye et al., 2003; Birnie,

1999; Boynton and Zmud, 1984; Brodie, 1995; Chan et al., 2010; Doloi, 2009; Grant, 1996; Li et al., 2005; Mohr and Spekman, 1994; Ng et al., 212; Qiao et al., 2001; Sanvido et al., 1992; Scandizzo, 2010; Tiong, 1996; Verma et al., 2016; Verma, 2014; Yeo, 1991; Zhang, 2005). All these factors are categorically placed in four heads, namely technical, financial & economic, political & legal, social & risk-oriented factors. We develop a questionnaire to know the importance of these factors in each category. About four hundred respondents were contacted to gather this information. The respondents are mostly executives of PPP infrastructure projects in India.

The study deploys Principal Component Analysis (PCA) to identify the most critical success factors. PCA is a data driven technique and is mostly used to reduce the original factors into a relatively lower number which, can be used to represent relationships among sets of interrelated variables. It can be done by employing any one of the following rotations: orthogonal rotation (*e.g. varimax*), or oblique rotation (*e.g. promax*).

3 EMPIRICAL RESULTS

Before proceeding with PCA results, the appropriateness of this analysis was evaluated by employing sampling adequacy tests, Kaiser-Meyer-Olkin (KMO) test, Bartlett's test, and correlation matrix of responses. We do not report these results due to space restriction. We report the PCA results below in the form of component extraction and component loadings.

Four components are extracted in this analysis and they explain 51% of total variance in the original dataset. This is further verified by analyzing component correlation matrix. Correlations among all the four components are found to be low (i.e., below 0.3) and hence, this justifies the selection of four components (see Table 1).

In response to above identified factors, we use scree plot (see Fig. 1) to authenticate these results.

Table 2 provides the loadings of all four factors, which have been obtained on the basis of PCA.

Table 1. Component correlation matrix.

Component	C1	C2	C3	C4
C1	1.000			
C2	0.051	1.000		
C3	0.049	0.070	1.000	
C4	−0.027	0.145	0.039	1.000

Note: C1-C4 are components of CSFs.

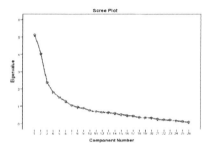

Figure 1. Scree Plot of Eigen Values over Components.

Table 2. Pattern matrix with component loadings.

Component	F1	F2	F3	F4
P3	0.826			
F5	−0.763			
S5	−0.738	−0.301		
R5	0.659			
S4	0.857			
S1	0.635			
P5	0.597		0.398	
R6	−0.546			0.451
F4		0.712		
T2	0.396	0.690		
T3		0.648		
P4		0.632		
R1		0.542		0.340
T1	0.318	0.539	0.311	
F3	0.456			−0.415
F6	0.358	0.539		
S3	−0.320		0.613	
P1			0.610	
R4		−0.405	0.578	
F1	0.313		0.561	
P2			0.474	−0.438
F2			0.448	
R2				0.702
R3	−0.508			0.616
T4				0.549
S2				−0.395

Note: F1-F4 are factors of component matrix; and P3-S2 are corresponding components for factor loadings.

To avoid repetition in loadings, a choice was made between items of highly similar content on the basis of loading size.

From PCA results, the 26 CSFs are finally reduced to 22 and these are categorically clustered into 4 major factors. These include factor 1- stable project environment-political and social, factor 2- shared role of public and private parties in project execution, factor 3- favorable environment for service delivery by private consortium, and factor 4- government control over risk allocation. The sub-success factors under each of these four factors (Factors 1–4) are illustrated in Table 3.

Table 3. Final component loadings.

Component	Factors and description			
	F1	F2	F3	F4
P3	0.826			
R5	0.659		Stable	
			project	
S4	0.857		Environ-	
			mental-	
S1	0.635		Political	
			and	
P5	0.597		Social	
F4		0.712	Shared	
			role of	
			public	
			and	
T2		0.690	private	
			parties	
			in project	
T3		0.648	execution	
P4		0.632		
R1		0.542		
T1		0.539		
F6		0.539		
S3	Favorable for		0.613	
	Environment			
P1	Service		0.610	
	Delivery			
	by Private			
R4	Consortium		0.578	
F1			0.561	
P2			0.474	
F2			0.448	
R2	Government			0.702
	Control			
	over			
R3	Risk			0.616
	Allocation			
T4				0.549
R6				0.451

Note: F1-F4 are factors of component matrix; and P3-S2 are corresponding components for factor loadings.

4 DISCUSSION

We intend to highlight the key critical success factors that can enhance the performance of these four attributes. We deploy a few additional tools, such as rank agreement factor, maximum rank agreement factor, percentage disagreement and percentage agreement factors, to examine the relative importance of these critical success factors and attributes. This study brings out the relative importance index and significance index to address these CSFs in the Indian economy.

4.1 Relative importance index

We have used the Relative Importance Index (RII) to get the ranks of these four attributes in terms of their criticality as perceived by respondents. This is based on the findings of Iyer and Jha (2005) and Verma et al. (2016). The study declares five most important factors for successful implementation of PPP infrastructure projects in India. These include *need for community participation, undue political interruption during project implementation, experience of government in packaging PPP projects, financial viability of project*, and *flexibility of contract for change in output specification and renegotiations* (see Table 4).

4.2 Significance index and agreement analysis

A significance index is deployed to analyze the CSFs' ranking across various attributes. Responses from different groups, viz, government sector, private player, consultants are separately indexed and their results are compared to each other to identify CSFs from each respondent group. Agreement analysis is applied to check the rank agreement among various respondents. The results of this analysis are presented in Tables 5–6. A strong Percentage of Agreement (PA) has been observed between government and private players over CSFs. The analysis identified five key success factors amongst a total of 26 CSFs. The basic characters of a PPP infrastructure project, such as large sunk investment, high sensitivity to demand variations, great exposure to financial markets and vulnerability to political instability are to be kept in mind before analyzing the key success factors as well as proposing relevant mitigation measures for it. Indian PPP scenario needs to be redefined with a more mature and strong institutional environment with a willingness and commitment to implement the partnership. Each CSF is further analyzed and recommendations are made to formulate an institutional environment which shall allow for better implementation of PPP infrastructure projects in the Indian economy.

Table 4. Relative importance index.

Component	Success attributes	RII	Ranking
S2	Need for community participation	0.863	1
P2	Political interruption & pressure	0.833	2
T1	Government sector experience	0.757	3
F6	Financial viability	0.749	4
T3	Contract flexibility	0.742	5
S4	Acceptable toll	0.737	6
P1	Stability in political system	0.709	7
P4	Favorable legal framework	0.678	8
F2	Adequate local financial market	0.676	9
F3	Promising economy	0.653	10
F4	Availability of long-term cash flow	0.643	11
S1	Community response during implementation	0.643	11
R5	O & M agreement risk allocation	0.643	11
S5	Job opportunities for local community	0.641	12
P3	Environment sensitivity of project	0.618	13
T2	Private consortium capability	0.613	14
P5	Statutory and institutional arrangements	0.605	15
R4	Supply agreement	0.597	16
R1	Concession agreement risk allocation	0.582	17
F1	Long-term demand	0.575	18
T4	clarity in service quality delivery	0.562	19
F5	Competition from other PPP projects	0.542	20
R2	Shareholder agreement risk allocation	0.542	20
R3	Loan agreement/insurance agreement	0.494	21
R6	Guarantees/comfort letters/support	0.484	22
S3	Reliable service delivery	0.473	23

Note: S2-S3 are components of critical success factors; and RII is relative importance index.

Table 5. Ranking and agreement between government and private sector.

Critical success factor	Government index	Sector rank	Private index	Sector rank	Agreement analysis
T3	88.18	1	88.57	1	RAF = 06
S2	84.55	2	79.05	3	RAF Max = 3.4
F6	83.64	3	76.19	4	PA = 82.3%
T1	80.91	5	73.33	6	
P2	80.00	6	73.33	6	

Note: F1-F4 are factors of component matrix; and P3-S2 are corresponding components for factor loadings.

Table 6. CSF rankings and agreement between consultants to government and private sector.

Critical success factor	Government index	Sector rank	Private index	Sector rank	Agreement analysis
S2	88.18	1	86.67	1	RAF = 2
P2	87.27	2	86.67	1	RAF Max = 3
T1	72.73	5	74.29	2	PA = 33.33%
F6	69.09	7	70.48	3	

Note: F1-F4 are factors of component matrix; and P3-S2 are corresponding components for factor loadings.

4.3 *Important factors for successful PPP projects implementation*

In this section, we highlight the importance of the previously mentioned five attributes in order to justify their implementation in PPP infrastructure projects.

4.3.1 *Flexibility in contracts*

Flexibility in contractual arrangement calls for trade-off between capturing efficiency gains and contractual rigidity. It is generally infeasible to prepare a contractual document between parties to the PPP transaction that could comprehensively mitigate all project risks—present and future—and thereby be used to govern a PPP project until completion. It is rather suggested that PPP contracts attempt to be more flexible and dynamic as opposed to being comprehensive and static (Orr, 2005). A contract can go under renegotiations in two ways, viz., discretionary and contractual. The former one is followed in Indian industry where there is no prior established rule or provision for negotiation in the contract and both the parties try to establish a common understanding on new terms. This method is vulnerable to transparency issues like lobbying.

To tackle the drawbacks, the consortium facilitates availability of objectives, restrictions and outcomes of the renegotiation process, to the public. It can also conduct public hearings before new terms are signed. Second method of renegotiations,

which is followed in the Latin American countries under Economic & Financial Re-equilibrium model, is that a set of priority rules are established in the contract, which provide a guideline for the negotiation process. Renegotiations, an inherent feature of PPP contracts, can be facilitated as per the changing project environment. Instead of preparing a rigid guideline for project delivery, a set of strong conditions for monitoring of project should be fixed and initial design should be allowed to adapt as it fits the actual scenario. There should be ample space for modification in scope, investment plan, price or condition of service provided to the concessionaire by government and in price of services charged by concessionaire. Furthermore, agreements can have a clause to modify financing structure and specific conditions for refinancing and/or conversion of debt into equity. Flexibility in technology selection to private party without compromising on quality of service delivery should also be given. Lastly, the consortium should note that provision of flexibility should be at all levels of project implementation, i.e., strategic, tactical and operational.

4.3.2 *Need for community participation*

For any successful PPP project, it is recommended that the public authority shall consult with all key stakeholders at the initial stages of project preparation and project structuring. These stakeholders include public or project users, community groups and associated Non-Governmental

Organization (NGOs), private operators, financial institutions, political representatives, and other government organizations (ICRA, 2011).

Absence of a buy-in from the people at large has led to significant hurdles at various stages of a PPP project in recent years. For instance, disturbance caused by displaced people during land acquisition; or resistance to collection of revenues in the form of toll, charges or tariffs during operational stage. So, public support is critical to any project. This helps in sustaining PPPs through innovative organizational arrangements and incentives. The local community participation can be encouraged by involving users directly or indirectly into legal and organizational arrangements for supervising PPP infrastructure projects. During the initial stage, proper communication of project benefits to various stakeholders helps in mobilizing public support. Structured ICE (information, communication, and education) and public consultation activities also help to bring out the concerns, apprehensions and acceptance of various stakeholders on the project. The public or private party can involve local NGOs for this purpose. The NGOs can use existing contacts with the municipality and other influential bodies to ensure maximum support of the people for unhindered execution of project (ICRA, 2011; Verma et al., 2016).

4.3.3 *Financial viability*
Infrastructure projects are usually financed by a combination of various forms, like private equity, debt, governmental funds and user charges. In Indian market, there are usually no user charges and the services are subsidized for political reasons. Hence, the PPP gets more exposed to the volatility of capital market. The financial meltdown and scarcity of credit availability may impose significant financial risk. To overcome it, innovative financing schemes can be applied which will provide support for project development and structuring arrangements that stand the test of time. This shall render the project less dependent on volatile markets by cutting government financial support in the form of Viability Gap Fund (VGF) or Project Development Fund (PDF). Such innovations in financing can be done by employing cost effective technical measures to generate funds and effective resource utilization to avoid time and cost over-run.

Concessionaire can utilize the annuity payments received during operations stage from public authority by securitizing it and thus generating more funds for future. Some projects like the Vadodara-Halol toll road have used deep discount bonds, take-out financing, cumulative convertible preference shares and long term loans to avert financial risks. In order to convince Indian users to pay for the services, an efficient marketing and

structuring of project needs to be done. Additionally, optimal tariffs and concession periods can be designed to replace high tolls over a long period by ensuring better monitoring projects and holding contractors accountable. Cost recovery mechanism should also be strengthened instead of subsidizing the services for political benefits. Strong institutions can be set-up which focus on lending to different phases of a project based on the risk profile to help projects continually restructure themselves such that the risk of their debt matches with the risk profile of the project at any stage (Mahalingam and Kapur, 2009).

4.3.4 *Government experience in packaging PPP projects*
Good governance is the key requirement to provide transparency, fair treatment and open competition. Absence of governance or weak authority makes the potential investors and lenders apprehensive about the increasing cost of money. Thus, many good players are lost, which reduces competitive pressure on bidders, resulting in increased costs and reduced quality. Cost and time over-runs are also caused due to negative attitude of rent seeking and other forms of corruption (Delmon, 2011). Good governance requires putting into place the enabling institutions, procedures and processes surrounding PPPs in order to fully benefit from the arrangement. Authorities should take initiatives for better structuring of PPP projects. Technical and regulatory capacity should be added to public institutions and officials, to manage the PPP process so as to maximize returns for all shareholders. There is an absence of institutionalized mechanism to do traffic volume forecasting in India, not only for 25 years, but also for 5–10 years. To meet this, there can be a framework within government body to carry out forecast studies, so that cost-recovering risks can be mitigated at pre-bid stage. This will attract parties to submit higher bids for feasible projects. Dedicated and cross-sectoral professional units need to be installed to support project implementation (Nataraj, 2007). Streamlining of approvals and clearances is the most critical requirement to expedite project implementation. Reforms in the functioning of land, labor, and financial markets, as well as the removal of restrictive regulation are required which will enable private sector to become a capacity builder (Verma, 2013).

4.3.5 *Political interruption and pressure*
The primary objective of PPP can only be reached when government agencies are a willing partner with active participation to make the project viable for private players. The main reasons of private sector failure to deliver are lack of law and order support from administration side leading to delay

and unwanted hurdles in project execution. There is a lot of political syndicate pressure to give contracts to influential elements. Moreover, the prices are invariably based on political pressures or considerations. Absence of credible government commitment and presence of bureaucratic red tape makes capital more expensive resulting in higher tariffs. To address issues of transparency and corruption, the government should instill reforms to structure and support PPP arrangements by publishing mandatory disclosures and fair practices to be followed by all projects. The possibility of setting up a web based PPP market place can be explored to add transparency. Dedicated grievance redressal cell/dispute resolution mechanism should be set-up. A single window clearance for high priority projects should be done to safeguard private players from red-tape delays. New market based products can be developed such as independent pre-bid rating, to assist investors in identifying well-structured PPP projects. Most importantly, management information systems should be employed to continuously monitor the performance of the PPP projects over the project life cycle. MIS can help in evaluation of PPP projects by tabulating and summarizing various stage experiences as a database for future (see, for more details, Verma et al., 2016).

5 CONCLUSION

This paper provides an overview of critical success factors of PPP infrastructure projects in India. On the basis of principal component analysis, we recognize four factors, namely stable project environment-political and social, shared role of public and private party in project execution, favorable environment for service delivery by private consortium, and government control over risk allocation. The success of these factors are further predicted by some sub-factors. These include community participation, undue political interruption during project implementation, experience of government in packaging PPP projects, financial viability of the project, and flexibility of contract for change in output specification and renegotiations. It can be noted that this study has been principally focused towards overall PPP infrastructure scenario in the Indian economy. There can be further research to examine this issue sector-wise, such as housing, transportation, water supply and policy recommendations for each infrastructure sector can be made. This study does not provide any meaningful outcome in terms of understanding the clustering effects of the similar attributes and their predictive capacity. This can be possible by using some advanced statistical methods, which is beyond the scope of this research.

ACKNOWLEDGEMENTS

We thank the RCG School of Infrastructure Design and Management and Vinod Gupta School of Management at Indian Institute of Technology Kharagpur (India) to carry out this research. We also grateful to the anonymous referees and conference convener for their constructive comments and suggestions.

REFERENCES

Akintoye, A.; Hardcastle, C.; Beck, M.; Chinyio, E.; Asenova, D. 2003. Achieving Best Value in Private Finance Initiative Project Procurement, *Construction Management and Economics*, 21(5): 461–470.

Birnie, J. 1999. Private Finance Initiative (PFI)—UK Construction Industry Response, *Journal of Construction Procurement*, 5 (1): 5–14.

Boynton, A.C.; Zmud, R.W. 1984. An Assessment of Critical Success Factors, *Sloan Management Review*, 25 (4): 17–27.

Brodie, M.J. 1995. Public/Private Joint Ventures: The Government as Partner—Bane or Benefit? *Real Estate Issues*, 20 (2): 33–39.

Chan, A.; Lam, P.; Chan, D.; Cheung, E.; Ke, Y. 2010. Critical Success Factors for PPPs in Infrastructure Developments: Chinese Perspective, *Journal of Construction, Engineering & Management*, 136 (5): 484–494.

Delmon, J. 2011. Public-Private Partnership Projects in Infrastructure, Cambridge University Press, Cambridge.

Doloi, H. 2009. Analysis of Pre-qualification Criteria in Contractor Selection and their Impacts on Project Success, *Construction Management and Economics*, 27 (12): 1245–1263.

European Investment Bank. 2000. The European Investment Bank and Public Private Partnerships. The Newsletter of the International Project Finance Association, 1: 3–4.

Grant, T. 1996. Keys to Successful Public-Private Partnerships, *Canadian Business Review*, 23(3): 27–28.

Hambros, S.G. 199). Public-Private Partnerships for Highways: Experience, Structure, Financing, Applicability and Comparative Assessment, Canada.

ICRA 2011. Toolkit for Public Private Partnership Frameworks in Municipal Solid Waste Management Volume I –Overview and Process. Public Private Partnerships Knowledge Series, Government of India, New Delhi.

Iyer, K.C.; Jha K.N. 2005. Factors Affecting Cost Performance: Evidence from Indian Construction Projects, *International Journal of Project Management*, 23 (4): 283–295.

Li, B.; Akintoye, A.; Edwards, P.J.; Hardcastle, C. 2005. Critical Success Factors for PPP/PFI Projects in the UK Construction Industry, *Construction Management and Economics*, 23 (5): 459–471.

Mahalingam, A.; Kapur, V. 2009. Institutional Capacity and Governance for PPP projects in India. Lead 2009 Conference, CA, USA.

Mohr, J.; Spekman, R. 1994. Characteristics of Partnership Success: Partnership Attributes, Communication Behaviour, and Conflict Resolution Techniques, *Strategic Management Journal*, 15 (2): 135–152.

Nataraj, G. 2007. Infrastructure Challenges in South Asia: The Role of Public-Private Partnerships. Discussion Paper, No. 80. Asian Development Bank Institute, Tokyo.

Ng, S.T.; Wong, Y.M.W.; Wong, J.M.W. 2012. Factors Influencing the Success of PPP at Feasibility Stage- A Tripartite Comparison Study in Hong Kong, *Habitat International*, 36 (4): 423–432.

Orr, R. 2005. Proceedings of the First General Counsels Roundtable, Collaboratory for Research on Global Projects, Working Paper Series, Stanford, CA, USA.

Qiao, L.; Wang, S.Q.; Tiong, R.L.K.; Chan, T.S. 2001. Framework for Critical Success Factors of BOT Projects in China, *Journal of Project Finance*, 7 (1): 53–61.

Sanvido, V.; Grobler, F.; Parfitt, K.; Guveris, M.; Coyle, M. 1992. Critical Success Factors for Construction Projects, *Journal of Construction, Engineering and Management*, 118 (1): 94–111.

Scandizzo, P.L.; Ventura, M. 2010. Sharing Risk through Concession Contracts, *European Journal of Operational Research*, 207 (1): 363–370.

Tiong, R.L.K. 1996. CSFs in Competitive Tendering and Negotiation Model for BOT Projects, *Journal of Construction, Engineering and Management*, 122 (3): 205–211.

Verma, A. 2013. Critical Success Factors in Indian Scenario for PPP Projects in Infrastructure Development, M. Tech. Thesis, Indian Institute of Technology Kharagpur, India.

Verma, A.; Pradhan, R.P.; Bele, S.K.; Gaurav, K. 2016. Critical Success Factors for PPP Projects in Infrastructure Development: The Indian Scenario, *Indian Journal of Regional Science*, (forthcoming).

Yeo, K.T. 1991. Forging New Project Value Chains- a Paradigm Shift, *Journal of Management in Engineering*, 7 (2): 203–211.

Zhang, X. 2005. Critical Success Factors for Public-Private Partnerships in Infrastructure Development, *Journal of Construction, Engineering and Management*, 131 (1): 3–14.

Transport Infrastructure and Systems – Dell'Acqua & Wegman (Eds)
© 2017 Taylor & Francis Group, London, ISBN 978-1-138-03009-1

An on-ramp metering and speed control application on O1 freeway in Istanbul

H.G. Demir & Y.K. Demir
Faculty of Engineering, University of Niğde, Niğde, Turkey

ABSTRACT: In this study, to make improvements in freeway traffic flow, speed limits and ramp metering have been applied on on-ramps of the European side of 2.5 km Istanbul O1 Freeway during peak hours. Speed limits from 10 to 70 km per hour in multiplies of 10 km per hour and local ramp metering control ALINEA were applied on on-ramps of the study area. Simulation software CORSIM was used to test control results. The comparison of the results with the current condition shows that 20 km/h constant speed limit that was applied to on ramps of Beşiktaş reduce the average travel time per vehicle and the effect of shock waves on the study area. It shows as an alternative for local ramp-metering. The total travel time of vehicles on mainstream results %5 decrease.

1 INTRODUCTION

Traffic management measures can increase the efficiency of road transport infrastructure. These applications are made to regulate traffic and improve traffic flow. Freeway management application are also made for these reason. With the development of traffic technology and control theory freeway management are also developed alongside. Ramp metering and variable speed limits are proven technique to improve traffic flow on freeways. Ramp metering is a equipment which use a traffic signal placed at the end of a ramp to allows traffic to enter the freeway under control. Variable Speed Limit Signs (VSL) systems consist of Variable Message Signs (VMS) equipped along a freeway to display the current speed limits advisory or regulatory under prevailing traffic condition, weather conditions, construction or maintenance activities and other factors. In this study, speed limits and ramp metering were applied on on-ramps. The aim of this application was to enter the vehicles from on ramp to main road at a lower rate. In this way, it is expected to reduce congestion at the entrance on the mainstream. With the reduction of congestion, it is expected to prevent the formation of shock waves and queues. With the application of ramp metering, the effects of speed limit application were compared. Controls were applied to the on ramps of 2.5 km segment of O1 Freeway, which is located on the European side of Istanbul, near the Bosphorus Bridge (Figure 1). The study segment has two on ramps Beşiktaş and Balmumcu. Congestion occurs during evening rush hours in direction from Europe to Anatolian side. Geometrical irregularities are available such as a horizontal curve and a bottleneck.

Figure 1. Working Zone of O-1 freeway (Demir, 2012).

2 CORSIM

For testing the estimated control effects on the zone, CORSIM (McTrans, 2012) microscopic simulation software was used. CORSIM RTE has the capability to intervene in the system with the VSL control device. As part of the way for the creation of CORSIM simulation model, the road was coded into CORSIM network format by aid of the maps (Google, 2007). Traffic volume data are obtained by video recording for calibration of CORSIM model. The record was taken from 16:23 to 17:27 for evening peak period in 16 October 2007. Then for examining the records, observation stations (OS1 and OS2 in Fig. 2) were set using the maps and on site measurement by GPS devices. Speed and volume data were measured in one-minute interval. Total 65 minutes of surveys were used for

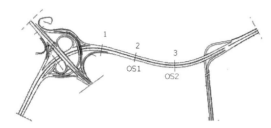

Figure 2. Observation stations (OS1 and OS2) in the working zone.

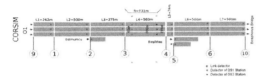

Figure 3. Schematic Figure of of the study site including two on on-ramps and detectors (Demir, 2012).

calibration (Demir, 2012). Calibration was confirmed by Theil's test comparing observed speed and flow with CORSIM outputs at two observation stations (Demir, 2012).

3 APPLICATION OF SPEED LIMITS ON ON-RAMPS

Because of disturbing effects of the on ramps on main flow, variable speed limits were applied from 10 km to 70 km per hour in multiplies of 10 km per hour and in matrix form on amp of Beşiktaş and Balmumcu (Figure 2). The study site of freeway segment was divided into six sections (Figure 3) so that the effects on the on-ramp traffic could be analyzed more efficiently. After simulation of control application it could be seen that for 20 km per hour speed limit on Beşiktaş on ramp minimum queue was built (R02 represents 20 km per hour speed limit applied on Beşiktaş on ramp). For the control application control tools in CORSIM software was accessed with the intervention of RTE that was written with Visual C.

4 RAMP METERING CONTROL APPLICATION FOR ON RAMPS OF BESIKTAS USING ALINEA ALGORITHM

ALINEA (Asservissement Lineaire d'entree Autoroutiere) was the first local ramp metering control strategy to be based on straightforward application of classical feedback control theory (Papageorgiou

et al. 1991). ALINEA has had multiple successful field applications (Paris, Amsterdam, Glasgow, and Munich) (Bogenberger and May 1999). ALINEA algorithm use the system output as input for the next iteration. Ramp flow is regulated to keep the downstream measured occupancy (O_{out}) below critical occupancy (O_{cr}). Figure 4 illustrates the sketch of an on ramp. Occupancy "O" is the portion of time that a vehicle is detected at a certain location. It has a unique value for traffic condition, as seen in Figure 5. That has an advantages for the ALINEA metering algorithm. Papageorgiou (Papageorgiou et al. 1991) calculated ramp metering rate, r(k) (veh/h) in discrete time interval k = 1, 2, ... as shown Equation 1 below:

$$r(k) = r(k-1) + KR [O - Oout (k-1)] \qquad (1)$$

The metering rate r(k–1) at previous time interval k–1, $K_R > 0$ is a regulator parameter. Acceptable result could be found if this value was selected 70 veh/hour. This value is not sensitive (Papageorgiou et al. 1991). O is desired occupancy which is selected slightly lower than the critical occupancy O_{cr} (occupancy corresponding to the capacity flow). O_{out}: measured occupancy measured downstream on the on ramp during time interval k, r(k) metering rate is updated between 40 s and 5 min (Chu and Yang 2003). Metering rate is calculated

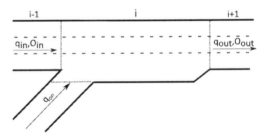

Figure 4. Sketch of an on-ramp.

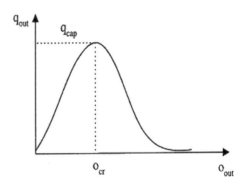

Figure 5. Fundamental diagram (May, 1990).

by measuring the difference between the measured occupancy, and the desired occupancy.

The metering rate r(k) is between the range [r_{min}, r_{max}]. Where r_{max} is the flow capacity of on ramps (for single lane uncontrolled on ramp equal between 1800 and 2200 veh/hour). $r_{min}>0$ is minimum admissible ramp flow (200 400 veh/hour) (Papageorgiou et al. 2007). ALINEA control was applied on Beşiktaş on ramp (A01 represent ALINEA control application). K_R value was selected 70 veh/hour. Desired occupancy value was selected %20. Measured occupancy value was measured on the loop detector by CORSIM which was placed 40 meter far.

5 THE EFFECTS OF CONTROL APPLICATIONS ON WORKING ZONE

For determine the effects of control application on freeway segment, graph and charts have been generated. To evaluate of performance criteria, mean speed, mean travel time, queue and supply of working area had been considered. The necessary data for the generation of chart and graphs were taken from CORSIM. With using of a function inside the CORSIM RTE interface the detector and link data for the study site were recorded in a database every ten seconds. At the end of simulation which was 120 min data was converted into chart and graphs. The graphs and charts which were used described below:- Speed Graphs: This graphs shows the speed variation on the links according to time. Shock waves can be easy observed. Speed is represented in gray color on the graph. The dark color defined low speed whereas the light color defined high speed level.- Density Graphs: This graphs shows the variation of density according to time. The density value are derived from link values. On this graphs lighter colors represent lower density whereas darker color defined higher density value. Demand-Supply Graphs: On the horizontal axis, the time is shown, and on the vertical axis, the vehicles (included on ramps) demand on working zone against vehicles outflow in link 6 are given.

- Average Travel Time Table: This represent the average travel time per vehicle before and after shock wave occurred. The average travel time is the sum of the vehicles travel time exit on each link per 10 sec interval divided by the number of total vehicles per 10 sec interval exit on each link. The last columns on table represent variation of the average travel time of the vehicles according to current situation (%). Queue Graphs: This graph shows the vehicles waiting on the virtual link that could not entry to the working zone.

R00 symbolizes the current state of the study zone. When current situation is examined in

Figure 6, it can be seen that an upstream prorogating shockwave starting on Link-6 at 35 minutes. Table 1 summarize speed variation before and after shock wave on links of study site. Average speed and standard deviation of speed are the two parameters pointed in Table 1. After the occupation of shockwave, standard deviations of speed on Link-1, Link-2 and Link-3 increased. Contrastively then, standard deviations of speed on Link-5 and Link-6 decreased. But in total, standard deviation of speed increased in the study zone.

Table 2 lists average travelling time for each links before and after shockwave occurred. Total average travelling time for study zone increased from 102.41 seconds to 365.55 seconds after formation of shock-wave. The highest increments in travelling time is observed on Link-3 and Link4. This increments are result of a bottleneck effect. Because of high demand of vehicles joined from Beşiktaş ramp, density increased in Link-5.

The shockwave formation can be seen in Figure 7. The Figure points out that densities fluctuations along simulation on Link-4 and Link-5. Especially after 50th minutes density spreads upstream. Figure 8 represents vehicles discharge against the total demand. After 30th minutes, the discharging curve diverges from demand curve, although discharge flow is 7560, demand is 8362 at 120th minutes of simulation. End of the simulation, study zone could discharge 14,076 vehicles

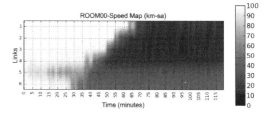

Figure 6.   R00 speed graph.

Table 1.   R00 mean speed before and after the shock wave.

S.T.	(km/h)	L1	L2	L3	L4	L5	L6	A.O.	V(%)
0-35	m.	101.51	100.36	98.52	91.03	74.71	96.23	93.73	-0.37
	s.d.	2.22	2.21	3.24	8.94	8.69	4.09	10.77	-6.51
35-120	m.	53.39	42.42	35.11	26.03	37.62	41.22	39.30	6.94
	s.d.	39.65	36.75	31.05	17.60	9.23	8.18	28.16	4.03
0-120	m.	67.46	59.37	53.65	45.04	48.47	57.31	55.22	3.18
	s.d.	39.91	40.64	38.95	33.42	19.16	26.05	34.76	-1.19

S.T.:Simulation Time m.:mean s.d.:standart deviation

Table 2.   R00 mean travel time before and after the shock wave.

S.T.	(km/h)	L1	L2	L3	L4	L5	L6	A.O.	V(%)
0-35	m.	9.78	19.24	10.49	25.70	24.27	12.92	102.41	0.00
	s.d.	1.95	6.17	2.63	9.83	5.08	3.80	12.47	0.00
35-120	m.	36.11	84.92	49.03	115.43	52.62	27.44	365.55	0.00
	s.d.	22.11	43.09	20.95	34.15	7.26	2.80	108.21	0.00
0-120	m.	28.41	65.71	37.76	89.18	44.33	23.19	288.59	0.00
	s.d.	22.14	47.09	24.90	50.20	14.53	7.30	150.53	0.00

S.T.:Simulation Time m.:mean s.d.:standart deviation

totally. The inadequate discharging causes queuing on entering sections as seen on Figure 9. End of the simulation, number of vehicles waiting for entering working zone, at main road (O1), Balmumcu on ramp and Beşiktaş on ramp were 400, 500 and 150 vehicles respectively. The inadequate discharging causes queuing on entering sections as seen on Figure 9. End of the simulation, number of vehicles waiting for entering working zone, at main road (O1). Balmumcu on ramp and Beşiktaş on ramp were 400, 500 and 150 vehicles respectively.

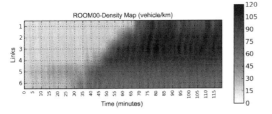

Figure 7.   R00 density graph.

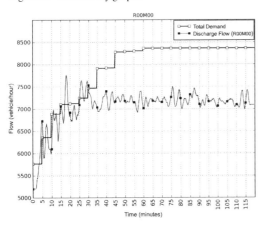

Figure 8.   R00 demand-supply graph.

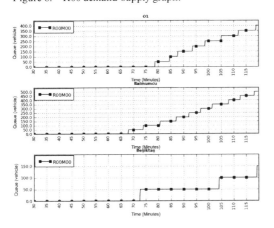

Figure 9.   R00 queue graph.

5.1   *Application 1 (R02)*

With R02 control strategy 20 km per hour speed limit was applied on Beşiktaş ramp. R02 raised average speed of whole simulation up to 3.18% relative to uncontrolled state (Table 3). Also standard deviation of speed and average travelling time (Table 4) decreased 1.19% and 5.06% respectively relative to uncontrolled state. This control created a speed pattern as shown Figure 10. It explains that, shockwave propagation could not be prevented but postponed to 40th minutes of the simulation. Figure 11 shows that the control succeeded reducing and postponing queuing at O1 main road and Balmumcu on ramp. At the end of the simulation, 1000 vehicles were observed at on ramps and O1 entrance. The number of vehicles discharged from working zone along whole simulation is 14,070 vehicles.

Table 3.   R02 mean speed before and after the shock wave.

S.T.	(km/h)	L1	L2	L3	L4	L5	L6	A.O.	V(%)
0-35	m.	101.51	100.36	98.52	91.03	74.71	96.23	93.73	-0.37
	s.d.	2.22	2.21	3.24	8.94	8.69	4.09	10.77	-6.51
35-120	m.	53.39	42.42	35.11	26.03	37.62	41.22	39.30	6.94
	s.d.	39.65	36.75	31.05	17.60	9.23	8.18	28.16	4.03
0-120	m.	67.46	59.37	53.65	45.04	48.47	57.31	55.22	3.18
	s.d.	39.91	40.64	38.95	33.42	19.16	26.05	34.76	-1.19

S.T.:Simulation Time m.:mean s.d.:standart deviation

Table 4.   R02 mean travel time before and after the shock wave.

S.T.	(km/h)	L1	L2	L3	L4	L5	L6	A.O.	V(%)
0-35	m.	9.75	19.31	10.50	27.19	24.86	11.48	103.10	0.67
	s.d.	1.87	6.53	2.70	10.27	4.65	1.82	11.85	-4.97
35-120	m.	33.96	79.13	45.42	109.13	50.04	26.96	344.63	-5.72
	s.d.	22.10	43.61	20.43	32.89	8.00	3.65	111.05	2.62
0-120	m.	26.88	61.64	35.20	85.16	42.67	22.43	273.99	-5.06
	s.d.	21.63	45.81	23.45	46.75	13.52	7.75	144.36	-4.10

S.T.:Simulation Time m.:mean s.d.:standart deviation

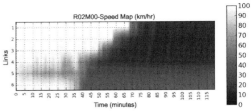

Figure 10.   R02 speed graph.

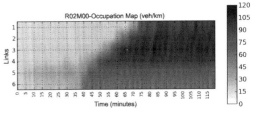

Figure 11.   R02 density graph.

5.2 Application 2 (A01)

Figure 11 and Figure 13 clearly show that ALINEA control strategy prevents congestion on main road which free flow condition was prevailed. Hence average speed increased while its standard deviation decreased over whole simulation time. With improvement of speed, average travelling time drops from 288 second to 96 seconds (Table 6). The adverse effect of this control was long queues that was 3950 vehicle observed on Beşiktaş on-ramp. From the result of AO1 application it can be said that Beşiktaş on ramp cause traffic congestion on main road. Ramp metering control on AO1 did not cause effective solution for traffic jam. The number of vehicles discharged from working zone along whole simulation is 11671 vehicles which is the lowest value of the application.

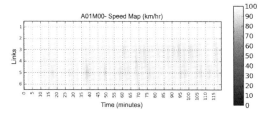

Figure 12. A01 speed-time graph.

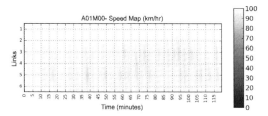

Figure 13. A01 density-time graph.

Table 5. A01 mean travel time before and after the shock wave.

S.T.	(km/h)	L1	L2	L3	L4	L5	L6	A.O.	V(%)
0-35	m.	101.66	100.57	98.48	98.29	97.16	98.17	99.06	5.29
	s.d.	2.02	2.21	3.15	2.23	3.60	2.55	3.10	-73.09
35-120	m.	101.21	99.85	95.75	96.37	95.14	96.94	97.54	165.41
	s.d.	2.20	2.37	4.48	2.90	4.29	2.57	3.95	-85.41
0-120	m.	101.34	100.06	96.55	96.93	95.73	97.30	97.99	83.09
	s.d.	2.16	2.35	4.32	2.86	4.20	2.62	3.78	-89.26

S.T.:Simulation Time m.:mean s.d.:standart deviation

Table 6. A01 mean travel time before and after the shock wave.

S.T.	(km/h)	L1	L2	L3	L4	L5	L6	A.O.	V(%)
0-35	m.	9.73	19.12	10.50	25.46	20.51	12.31	97.63	-4.67
	s.d.	1.83	6.03	2.63	9.91	8.56	6.82	12.89	3.37
35-120	m.	9.71	18.74	10.61	24.97	19.98	11.69	95.69	-73.82
	s.d.	1.58	4.18	2.05	6.69	5.56	3.41	8.54	-92.11
0-120	m.	9.71	18.85	10.58	25.11	20.14	11.87	96.26	-66.64
	s.d.	1.66	4.80	2.24	7.78	6.58	4.68	10.04	-93.33

S.T.:Simulation Time m.:mean s.d.:standart deviation

Change on standard deviation of the mean speed during total simulation time. According to current condition R02 application decreased the travel time, shifted shock wave, decreased queues on O1 main road and Balmumcu on ramp, but increased queue on Beşiktaş on ramp. A01 ALINEA application on Beşiktaş on ramp increased the mean speed on main road during simulation time and increased the travel time after shock waves. It has been seen at the 50th min. queued up 1500 vehicles, at the end of simulation 3950 vehicles on Beşiktaş on ramp. R02 application was effective in reducing the travel time and shifting the shock wave and could be considered as an alternative.

6 EVALUATION AND RESULTS

In this study 2.5 km of O1 Freeway, disturbing effects of the ramps on main flow speed limits and ALINEA algorithm were applied at on ramp to improve freeway traffic flow at peak hour. For testing the estimated effects on the zone, COR-SIM microscopic simulation software was used. The looking at the current state of the traffic flow within the working zone, at the 35th minutes of the simulation time, shock wave was emerged propagating from link 5 to upstream. After the formation of the shock wave, the average speed reduce from 94 km/h to 37 km/h, mean travel time per vehicle increase from 102 seconds to 365 seconds. From the results of simulation, it can be seen that the traffic problem arise from the bottleneck due to high demand of Beşiktaş on ramp. The VSL impacts were criticized under different indicators such as travel times per vehicle, shock wave effects, mean travel speed, standard deviation of mean speed. R02 has the same properties with the current applications before shockwave occupation. After formation of the shock wave it can be seen that the mean speed increase according to current condition. There was no with 20 km per hour. It should be emphasized that vehicle speed up on ramps as joining on the mainroad. However, this type of control application needs testing before. Travelling speed with 20 km per hour is thought to be possible with intelligent vehicle and transportation technology. The provision of speed is the subject of a different investigation. However, it is thought that the spread of electric vehicles in the future, more precise speed limit could be implemented. The primary results show that Application 1 is capable of reducing the total travel time per vehicle as %5 percent against increasing queue length on Beşiktaş. But total queue is 50 vehicles less than uncontrolled case. In addition it shifted the beginning of the shock wave without making changes on average speed. It can be seen that there is a trade-off between main-stream and ramps queues. For this reason the control problem should be optimized. Dynamic speed limits application on ramps can be suggested as further work because of changing condition of the freeway traffic flow.

ACKNOWLEDGMENT

This study was generated from the thesis of Hatice Göçmen Demir which completed on the structure of thesis by Istanbul Technical University. Authors wish to express a sincere thanks to all person and companies for their contributions.

REFERENCES

Bogenberger, K. & A. May (1999). Advanced coordinated traffic responsive ramp metering strategies. Working Paper.

Chu, L. & X. Yang (2003). Optimization of the alinea ramp-metering control using genetic algorithm with micro- simulation. Transportation Research Board 82nd Annual Meeting, Washington, DC.

Demir, H.G. (2012). Flow Control in Urban Freeways by Applying Variable Speed Limits in A Model-predictive Control Framework. Ph. D. thesis, Istanbul Technical University, Istanbul, Turkey.

Google (2007).Satellite data. Web Page. http://www. googleearth.com.

Papageorgiou, M., H. Hadj-Salem, & J. Blosseville (1991). Alinea: A local feedback conrol law for on-ramp metering. In Freeway operations, highway capacity, and traffic flow, pp. 58–64. National Research Council, Washington, D.C.

Papageorgiou, M., H. Hadj-Salem, & F. Middelham (1991). Alinea local ramp metering- summary of field results. In Freeway operations, highway capacity, and traffic flow, pp. 58–64. National Research Council, Washington, D.C.: TRB.

Papageorgiou, M., E. Kosmatopoulos, L. Papamichail, & W. Yibing (2007). Alinea maximises motorway throughput:an answer to flawed criticism. Traffic Engineering and Control 48, 271–276.

Transport Infrastructure and Systems – Dell'Acqua & Wegman (Eds)
© *2017 Taylor & Francis Group, London, ISBN 978-1-138-03009-1*

Transport infrastructure capacity calculations using green travelling planner

I. Celiński, M. Staniek & G. Sierpiński
Faculty of Transport, Silesian University of Technology, Katowice, Poland

ABSTRACT: The article provides a discussion on the chosen approaches to the calculation of the traffic capacity of road network cross-sections and movements at intersections in line with the idea of sustainable development of transport. The methods of calculating traffic capacity used until now have emphasised exclusively the physical aspect of the respective measure. The said measure is related to the number of vehicles covering a specific road cross-section over a unit of time. Under such an approach, the characteristics being significant for individual means of transport are mainly their physical properties, rather than their potential in terms of allowing transport to develop in a sustainable manner. The authors argue that a certain specific parity should be taken into account in calculations of road network cross-section capacity, ensuring that such cross-sections are used efficiently from the point of view of transport sustainability. The said parity regulates the manner in which individual means of transport access the network. The article presents a methodology for calculation of the parity based on the features of a tool known as Green Travelling Planner (trip planner). The discussion is supplemented with findings of original research into multimodal cross-sections in the Upper Silesian urban area in Poland. There is a perfect correspondence between the methodology discussed in this article and the extensive framework of actions undertaken by the EU with regard to transport (European Commission. 2011, Janic M. 2016).

1 INTRODUCTION

Within the road network, the theoretical traffic capacity of a road cross-section or movement at an intersection depends on a large number of variable parameters. The characteristics taken into account in the evaluation of the traffic carrying capacity of infrastructure elements are the following: average vehicle speed, level of service, safe gap between vehicles, other distances between vehicles, directional distribution of traffic, share of heavy goods vehicles, type of intersection, geometrical parameters of the road, and many others. Essentially, the capacity calculation procedures are related to a single, global measure expressed by the number of vehicles travelling through a specific road cross-section over a unit of time. Traffic capacity is measured in vehicles per time unit. The measurement is usually carried out at intervals of 15, 20 or 30 minutes, or of one hour. If traffic is highly heterogeneous, the measurement interval is reduced to an order of tens of seconds. In this approach, the capacity calculation procedures are of a technical nature, taking into account exclusively the physical aspect of the traffic processes taking place within the transport network. Above all, however, they fail to take into account sustainable development, which is a critical aspect when it comes to the correct shaping of

road (transport) networks. It needs to be pointed out that the capacity of road network cross-sections and of movements at intersections is a certain time- and space-limited resource. As in the case of any resource that is limited, certain conditions need to be determined for its utilisation. Road traffic is a self-organising process, so the related resource is utilised, as it were, naturally on a random basis. This randomness can be assumed to represent an unfavourable phenomenon. Traffic management can determine the conditions of utilising capacity only to a limited extent. Combining traffic organisation and control in an accurate manner may help regulate to the full extent the way in which the resource is utilised.

The authors argue that a specified parity should be taken into account in the calculations of capacity of transport network cross-sections. This parity should be understood as a reasonable division of the capacity resource into appropriate parts, sectioned off and allocated on an exclusive basis to different means of transport, in such a way as to assure their efficient utilisation in terms of sustainable development of transport. The process of sectioning off and allocating the resource may take place in the same way as in ICT networks.

In this article, the said parity will be described as a rationally determined proportion of the amount of traffic in private transport to the amount of traffic in public transport (PrT: PuT, Private Transport to Public Transport). The parity components are referred to respectively as PPrT and PPuT. The parity component related to public transport vehicles indicates, in the given network cross-section, how important the relevant means of transport is from the perspective of transport sustainability (PPuT). At the same time, it indicates the value of the parity component for private transport. This approach has already been implemented to a certain extent in transport networks, for instance, by the provision of bus lanes for public transport vehicles. The problem in this respect consists in the determination of reasonable and clear criteria for the division of traffic carrying capacity as a highly limited resource. This would make it possible to set forth precise conditions for the provision of bus lanes and other similar solutions. Furthermore, one needs to bear in mind that capacity is a highly heterogeneous resource across the entire transport network. The drivers of change in traffic stream distribution include mainly suburbanisation processes (Aguiléra, A. et al. 2009, Handy, S. et al. 2005, Khattak, A. J. & Rodriguez, D. 2005, Limanond T. et al. 2011, Lu, X. and Pas, E. I. 1999, Millward, H. & Spinney, J. 2011, Sáez A. E. & Baygents J. C. 2014). Consequently, the criteria for the division of capacity must be universal for the entire area it concerns, and apply 24 hours a day.

Following the ICT network analogy, a reasonable capacity division might lead to a greater diversity of rates charged for access to the road network, also for different public transport operators. What this means is that the quality of transfer within the transport network for various means of transport is proportional to the rights to use it which one has acquired. If this approach were adopted, private users would be able to acquire rights to use bus lanes at prices making it possible to bring the idea of transport sustainability into life.

The fundamental issue addressed in this article is related to the method of calculating the parity components (PPuT, PPrT = 1-PPuT) for any road network cross-section or movement at an intersection. The capacity measure should take into account the value of the parity for the given means of transport in the given road network cross-section, if calculation of the components becomes technically feasible. Consequently, the parity and capacity thus calculated should impose the methods of traffic management and control within the specific road network cross-section. Any restrictions imposing certain forms and methods of traffic management and control resulting from the parity adopted can be bypassed only if operators of public transport services are unable (for various reasons) to take advantage of the parity (component, share in

traffic) determined for them. In that case, the unused resource can be resold to private vehicle or heavy goods vehicle users in quantities that do not put transport sustainability at risk.

The methodology of calculating capacity in road network cross-sections accounting for the parity discussed in this article is relevant mainly for transport networks characterised by strong diversity of means of transport and roads, and ones in which roads have minimum two lanes in each direction within the network area. Moreover, it is valid for networks with a large number of cross-sections, where more than one means of transport operates. Taking the foregoing into account, the article defines the multimodal cross-section and describes research into road assets of this kind, performed in the Upper Silesian conurbation in Poland.

As described in this article, the parity component for Public Transport (PPuT) was calculated using a specialised travel planner known as Green Travelling Planner (GT Planner) (Esztergár-Kiss D. & Csiszár Cs. 2015, Sierpiński G. et al. 2014, Sierpiński G. et al. 2016, Sierpiński G. 2017). The PuT value can be determined by way of basic calculations using data acquired from GT Planner. The data stored in GT Planner describe actual and planned transport behaviours of the inhabitants of the area covered by GT Planner's operation. GT Planner is usually built for the entire transport network area of a single urban agglomeration.

The approach suggested in the article also requires a modification of the measures used in the calculations of capacity, from object-related ones (vehicles) to subject-related ones (number of travelling persons).

2 MULTIMODAL CROSS-SECTION

The share of vehicles of a specific type in a traffic stream determines the traffic structure, which in turn influences capacity (measured as before, in vehicles). In classical methods of capacity calculation, this influence results, however, only from the physical properties of infrastructure elements and vehicles (lane width, vehicle dimension, vehicle speed etc.). This approach to capacity calculation usually fails to take into account the differences in vehicle occupancy rates within the specific cross-section. A bus carrying 50 passengers on average is the equivalent of only two passenger cars, more or less, from the point of view of classical capacity measures. In reality, from the perspective of both transport sustainability and the size of the physical road transport processes, the minimum difference in share for this resource corresponds to 8:50, while the maximum one to 2:50 (for different passenger vehicle occupancy rates). Therefore, if the share in the resource in the road network cross-section is calculated in travels, each two private vehicles registered in the road network cross-section cause a theoretical loss of at least 42 travels. Estimating the loss of cross-section capacity, it is assumed that 2 passenger cars occupy approximately the road surface, which can by use by one bus. It was also assumed equal 50 passengers in the bus at rush hour. Number of the passengers in each of the two cars was set as 4, although based on extensive research authors in the Silesian conurbation in 2009 it was 1.4. In this manner specified minimum loss of capacity resulting from the existing (observed) modal split. Estimatrion was realized for rush hours due to the fact that during this period traffic volume is close to capacity. In realistic situation, the variance filling buses in rush hour on different lines modifies these losses (from 42 trips), but according to the authors mostly up (>>42 trips). In cross-sections where the average daily traffic is around 50,000 vehicles per 24 hours, this means a loss of around 1,050,000 travels per day, per road network cross-section. These figures show the essence of the problem related to the sustainable development of transport.

Traffic structure is usually studied within a precisely determined road cross-section making it possible to measure the value effectively. The percentage share of individual means of transport in the network cross-section is referred to as modal split. This modal split can be expressed with the following equation:

$$MS_{cs} = STM_1 : STM_2 :,...,: STM_n \qquad (1)$$

where MS_{cs} = modal split in the cross-section (relation); STM_i = share of the ith means of transport; n – number of means of transport;

$$\sum_{i=1}^{n} STM_i = 100\% \qquad (2)$$

The value expressed in the form of equation (1) can be defined very accurately only for individual traffic streams (RMS $\cong$ 0). Except in the case of individual traffic streams, the modal split structure changes to a greater or smaller extent as a result of changes in the traffic flow management plan, different movements of vehicles within the road network, and local interaction between vehicles.

Figure 1 illustrates the dynamics of the changes of modal split in one of the measurement cross-sections in the Upper Silesian conurbation (Poland). The figure shows that the process is highly heterogeneous. Nevertheless, even approximate determination of modal split with regard to its distribution in time and space of the road network

makes it possible to determine the parity components for public transport. Moreover, knowledge of the modal split makes it possible to build better traffic models, streamline transport task allocation, plan lines and perform routing for means of public transport, set up PuT stops in appropriate places, parameterise the prioritisation process in road traffic control systems, and sometimes also make rates for transport services uniform.

Figure 1 expresses the idea of sustainable development of transport in the form of a parity point (PPuT). In this case, it consists in the balancing of the share of private vehicles and other vehicles in the traffic structure. If the traffic carrying capacity measures are expressed in travels rather than in vehicles, the largest capacity reserves will be available with regard to the number of public transport vehicles that can be allowed to enter the road network. Currently, this reserve is hidden in the number of private vehicles within the transport network used by travelling persons capable of changing their transport-related behaviour patterns in a conscious manner. As it has already been mentioned, the new procedures suggested in the article for the calculation of capacity in the road network are reasonable in cross-sections in which more than one means of transport has been identified. These are referred to as multimodal cross-sections characteristic from the point of view of the entire road network (Celiński, I. et al. 2015). Due to the temporal and spatial irregularities in road traffic distribution, the authors of the article suggest that such cross-sections should be determined deliberately in the space of the road network analysed (Figure 2 shows a heterogeneous spatial distribution of those cross-sections). The criterion used to determine whether a cross-section is multimodal, which the authors adopted arbitrarily, is that it must be possible to register at least four different types of means of transport in that cross-section. This criterion means that cars, buses, heavy goods vehicles and trams need to be registered in the modal split. The research conducted by the authors in the road network of the

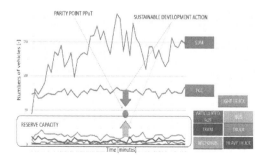

Figure 1. Modal split dynamics and parity point value.

Upper Silesian urban area shows that the number of points meeting the above criterion is not very high. For instance, in the main city of the urban area (Katowice), the number of such cross-sections ranges at around 20 measurement points.

Figure 2 shows the spatial distribution of 100 points where multimodal cross-sections are located, studied by the authors in the area of the Upper Silesian conurbation in Poland.

The pilot study has made it possible to understand how important for road network organisation it is to take into account various types of modal split in capacity calculations: observed modal split (presented in the pie charts in Figure 3) and hidden modal split, discussed further on in the article.

Figure 3 shows selected results of the measurement discussed in this article in multimodal cross-sections. A significant heterogeneity of the modal split can be observed in the transport network space. The points where multimodal cross-sections have been determined should account for cases of cross-sections of roads with different numbers of lanes. If trams are excluded from the criterion, the number of multimodal cross-sections increases.

Knowledge of the traffic structure in multimodal cross-sections allows one to determine the actual breakdown of capacity (conventional parity, observed parity, modal split) for a specific infrastructure ele-

ment by the individual means of transport (2). This is analogous to the division of bandwidth in ICT networks with unrestricted access for all users with equal (random) access rights. The modal split shown in Figure 3 is a certain random set of values imposed by the network structure and its organisation, and consequently expressed with demand channelled through actual use. In the authors' opinion, a different modal split value (obligatory parity) can be determined in transport network cross-sections. Obligatory parity, as opposed to conventional parity, pursues goals related to sustainable development of transport. Obligatory parity is a set of values for individual means of transport taking into account suppressed demand in the transport network (Szarata, A., 2010a, 2010b, 2012, 2013). Consequently, parity components can be determined for the individual means of transport. A parity determined in this way will make it possible to manage and control road networks in a spirit of transport sustainability (Beckerman, W. 1994., Ionescu G. 2016., Kanister, D. 2008, Monzon-de-Caceres A. & Di Ciommo F. 2016, Okraszewska, R. 2008, Stanley, J. & Lucas, K. 2014).

3 CROSS-SECTION PARITY

One method of determining the parity component for public transport and its other types in a road network cross-section or intersection movement involves using a travel planner, i.e. GT Planner (Fig. 4). This planner makes it possible to reproduce the transport behaviour patterns of people living in a specific area.

The application shown in Figure 4 allows users to plan their travels. The starting point and the destination need to be indicated on the map, and the means of transport should be chosen together with several other significant travel parameters. Consequently, a transfer route is mapped out in the transport network covered by GT Planner operation. The route can be parameterised in the form of data describing transport-related behaviour:

Figure 2. Multimodal cross-section in the Upper Silesian urban area.

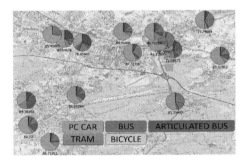

Figure 3. Modal split in cross-sections in the Upper Silesian urban area (POLAND).

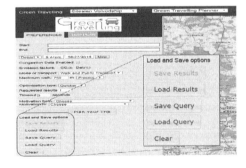

Figure 4. GT Planner—demand parametrisation option.

$$TB_i = (GP, AP, T, TM, TP, \Delta t, R, MOT) \qquad (3)$$

where TB_i = transport behaviour of the ith user; GP = travel generation points set; AP = travel absorption point set; T—start time; TM—means of transport; TP—indirect point set, dt—start time shift window; R—space resistance; MOT—travel motivation.

The set of transport behaviours determined in accordance with equation (3) should constitute a certain sample of behaviours observed in the actual road network. This sample, after the data have been processed, can provide a good estimate of the values of modal split distribution in actual road network cross-sections. In GT Planner, however, it is possible to store all route planning activities, also those which will not be converted into actual travels. For instance, a GT Planner user may plan a travel which is impossible at the given time given the set of data included in the GTFS (no bus or train is about to depart at the moment, or the waiting time to change to another means of transport is too long; no public transport line is available between the starting point and the destination, or the combination of multimodal connections does not fit within the travel time determined earlier; one cannot leave one's urban bike at the point of destination, etc.) Therefore, in each transport network cross-section, a value can be defined determining the number of planned transport behaviours that were impossible for various reasons and consequently were not converted into actual behaviours. The said value describes what is referred to as transport-related preferences, which to a certain extent constitute what is referred to as suppressed demand in the transport network. Knowledge of the suppressed demand characteristic and of actual transport behaviours makes it possible to determine the parity based on the demand characteristic in the specific transport network cross-section instead of the observed parity (actual modal split):

$$| MS_{cs} - DMS_{cs} | \geq 0 \qquad (4)$$

where DMS_{cs} = desired modal split;

The difference in value described by equation (4) indicates how different traffic in the specific transport network cross-section (specified parity) is from the idea of transport sustainability (obligatory parity). The said value describes the relationship between satisfied and unsatisfied demand. Balancing demand at that point also consists in an attempt to channel suppressed demand. At the same time, when switching to capacity measures expressed in travels rather than vehicles, applying the obligatory parity makes it possible to indicate by how much the capacity value can increase in the specific road network cross-section and with regard to specific means of transport.

Figure 5 illustrates the idea of how to proceed when determining obligatory parity in road network cross-sections based on modal split calculated for known characteristics of observed demand (referring to demand observed in network cross-sections in the form of moving vehicles or GT Planner data) and suppressed demand (travels recorded only in GT Planner). The figure shows a multimodal cross-section in the transport network of the city of Katowice (Poland), in the direct vicinity of a large transport hub (with buses, trams and trains being the available modes).

It is clear that three out of the six travels planned between the same point of start and destination (neighbouring cities of Mikołów and Katowice) are not present in the relevant measurement cross-section (marked with the horizontal red line). There may be various reasons for this particular distribution of travel in the network's space. Travels that should appear in a single measurement cross-section spread out over a much wider area of the transport network for instance as a result of the failure to match the public transport offering to suppressed (hidden) demand, or of poor access to the network. Demand for a specific transport network cross-section is therefore composed of actual and suppressed transport behaviours:

$$D_{cs} = \sum_{i=1}^{m} TB_i + \sum_{j=1}^{l} STB_j \qquad (5)$$

where D_{cs} = overall demand in the cross-section; TB_i = transport behaviour of the ith user actually displayed (observed) in the cross-section; STB_i = suppressed transport behaviour of the jth user (not observable in any real cross-section of the road network—they can be observed in GT Planner); m—number of actual transport behaviours, l—number of suppressed travel behaviours.

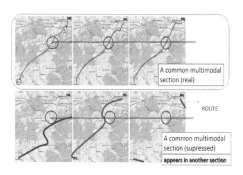

Figure 5. Comparison between real modal split and suppressed modal split (data from GT Planner).

The gap in Figure 5 left by the three travels which did not take place using means of public transport is probably filled immediately by private means of transport.

4 PARITY-BASED CALCULATION

In line with the idea of transport sustainability and with the methodology suggested by the authors, capacity in a road network cross-section can be expressed/calculated taking the obligatory parity for access to that cross-section into account. The relevant parity can be calculated using a tool such as GT Planner. Therefore, the general capacity of a road network cross-section can be expressed in the form of a sum:

$$C'_{cs} = PPuT * C_{CS} + (1 - PPuT) * C_{CS} \tag{6}$$

where C'_{cs} = sustainable cross-section capacity; PPuT—cross-section public transport parity component; C_{cs}—simple capacity calculation; (1-PPuT) = modal split for other vehicles, parity component.

The expressions written as equation (6) make it possible to divide capacity in the given cross-section of the road network into the part intended for private vehicles and freight-carrying vehicles, and the part intended for public transport vehicles (the sum can be composed of parts related to all means of transport). Moreover, expression (6) can be calculated, as far as the measures are concerned, on the basis of the number of passengers (or into travels), rather than of the number of vehicles. In this case, capacity for the given road cross-section is expressed with the number of travels broken down into two kinds of means of transport. If there is more than one lane in the given cross-section of the road, traffic can be managed reasonably in the lanes. If the number of travels taking place in the given road cross-section using public transport is close to or larger than 50%, it needs to be assumed that a single lane should be sectioned off exclusively for the purposes of means of public transport. The key aspect consists in answering the question about the minimum percentage of these travels that should lead to a bus lane for PuT being sectioned off. The obligatory parity is calculated taking into account the suppression of some travels and acknowledging the fact that they need to be shifted to other transport network cross-sections. The capacity of a road with two lanes can be calculated for instance using the HCM equation:

$$C_{CS} = C_0 * (v/c)_i * f_d * f_w * f_{HW} \tag{7}$$

where C_{cs} = overall classical capacity in the cross-section; C_0 = constant input capacity; $(v/c)_i$ = volume-to-capacity ratio for level of service i; f_d = adjustment factor for directional distribution of traffic; fw = adjustment factor for narrow lanes and restricted lateral clearance; f_{HW} = adjustment factor for the presence of heavy vehicles in the traffic stream.

Using the above equation, one can determine separately (in a breakdown) the capacity for private vehicles and public transport vehicles:

$$C_{cs}^{PPuT} = C_0 * (v/c)_i * f_d * f_w * f_{HW} * (PPuT/100) \tag{8}$$

where C^{PPuT} = public transport share of classical HCM capacity in the cross-section based on the obligatory parity component for PuT;

$$C_{cs}^O = C_0 * (v/c)_i * f_d * f_w * f_{HW} * (1 - PPuT/100) \tag{9}$$

where C^O = non-public transport share of classical capacity in cross-section based on obligatory parity component for other vehicles;

Capacity expressed in the form of equations (8) and (9) can be written as a unit convenient from the point of view of transport sustainability analysis, in the following form:

$$C_{cs}^{SD} = C^{PPuT}{}_{cs} * \bar{n} + C^O{}_{cs} * \bar{p} \tag{10}$$

where C^{SD}—sustainable development capacity in the cross-section; n—average occupancy of means of public transport; p—average occupancy of other vehicles. If the expressions describing the components of the sum total of equation (10) are defined as capacity for travels made by means of public transport $C^{PPuT(trip)}$ and capacity for travels using other means of transport, $C^{O(trip)}$ is a comparison of these values makes it possible to make reasonable changes to the traffic distribution by lanes in the given road network cross-section. The relation, written as the following inequality:

$$\frac{C^{PPuT(trip)}}{C^{O(trip)} \cdot \lambda} \geq 1 \tag{11}$$

where λ = sustainable development capacity scale indicator, means that a reasonable criterion is obtained for sectioning off one or more lanes exclusively for the purposes of public transport (e.g. bus lanes).

The scale indicator λ should be defined as 2 (physical proportion between private vehicles and public transport).

In the case of capacity calculations for movements at point elements of infrastructure, the methodology followed will be similar. The HCM equation for the determination of capacity in a minor stream is as follows:

$$C_r = V \frac{e^{v*t_c/3600}}{1-e^{v*t_f/3600}} \qquad (12)$$

where V = overriding (major) stream volume; t_c—critical gap in the major stream [sec], t_f—follow up time in the minor stream [sec].

According to the transformations applied in equations (7) and (8), the sustainable capacity of a stream at a road intersection will be written in the following manner:

$$C_r^{SD} = V \frac{e^{v*t_c/3600}}{1-e^{v*t_f/3600}} * (PPuT/100) \qquad (13)$$

where: C^{SD} —sustainable stream capacity for PuT transport; V^r = overriding stream volume;

5 RELIABILITY OF THE PARITY VALUE

The value of the obligatory parity component for means of public transport calculated in the manner described above is obtained on the basis of data stored in the GT Planner archive (concerning actual and suppressed transport behaviours). The reliability of these data can be estimated on the basis of measurement data obtained from recorders (visual, inductive and others) installed in the transport network cross-sections. The proportion expressed as equation (13) determines, therefore, the level of accuracy of mapping demand to the transport network in different groups of means of transport.

$$CR_{PPuT} = \frac{t_r}{t_p} \qquad (14)$$

where CR_{PPuT} = reliability of the parity component value; t_r—number of actual travels in the observed cross-section; t_p—number of planned/requested travels in the cross-section;.

The relationship expressed in the form of equation (13) is not linear across the whole range of the variable on the x-axis, therefore the behaviour of the relationship needs further studies. This, however, can only happen after GT Planner has been openned for public use (at present, it is at the stage of functional tests). As for the calibration of the number of planned travels with those observed in

the transport network in the given cross-section, the authors expect high consistency of the results. In the case of suppressed travels, the reliability of the parity component may be much lower. This results from the fact that travels that were routed (i.e. attempted) using the planner account for an undetermined part of the general number of suppressed travels. The assumed high consistency between the actual transport behaviours in the transport network and those planned results from the fact that according to the authors' data, the group that uses planners is representative of the society at large. For the purposes of studying transport behaviour patterns, this number exceeds by 500% the required representative sample sizes in large agglomeration areas. The inhabitants like to plan their travels, just like 42% of the general public (Gretzel, U. et al. 2007).

6 HIDDEN RESERVE OF CAPACITY AND PARITY TRADE

The fact that a conventional parity is observed in road network cross-sections resulting from the self-regulation of road traffic streams leads to hidden capacity reserves appearing in each cross-section, expressed with the number of travels that can be translated into means of public transport. These hidden reserves are visible only when capacity is converted from units expressed in the number of vehicles into the number of travels using the parity component for means of public transport. Each replacement of a private vehicle with a public one leads to an increase in capacity of the relevant transport network cross-section, expressed with the number of travels. This change, however, should not exceed the number of travels by public transport declared by the previous users. A similar aspect concerns switching between means of public transport operating in the specific network. In this case, the prevalence of one operator may disrupt the sustainable development of transport.

In the authors' opinion, there may be cases in which selected transport network cross-sections, with a determined obligatory parity for the individual types of means of transport, might remain unused. If the PPuT parity component is unused by public transport operators, a part of it can be sold to other transport network users. It must be assumed that in transport networks, including dense ones, every part of the resource will be utilised. It is highly likely that particularly affluent users of private means of transport and fleet operators will be interested in resources from the transport network being resold so that they can use them for their own vehicles. Nevertheless, these resources should be guaranteed to public transport operators. The total obligatory parity in a road network cross-section can be written as the following equation:

$$TPTN = (PPuT) + PPrT + PFHV + ... \qquad (15)$$

where: $TPNT$ = total parity in the transport network; $PPrT$ = parity component for individual users in the transport network; $PFHV$ = parity component for freight veh. in the transport network.

The parity component for public transport is composed of the following elements:

$$PPuT = \sum PPuT_A^{op1} + \sum PPuT_{AP}^{op2} + \sum PPuT_T^{op3} .... \qquad (16)$$

where: $\sum PPuT_A^{op1}$ = parity component for op1 number operator in buses in the transport network; $\sum PPuT_{AP}^{op1}$ = parity component for op2 number operator in articulated buses.

7 CONCLUSIONS

Using a travel planner with the features GT Planner offers makes it possible to determine in a reasonable manner the various components of the obligatory parity in transport network cross-sections. This comment also concerns the constituents of that parity with reference to various types of means of transport and to the operator.

The problem addressed in the article is simple only in relation to the idea of founding a mandatory modal split in selected sections of the transport network and the determination on the basis of their capacity. In fact, calculations of the value of the parity problem is very complex. Evidence of this even the fact that there are currently no tools to implement the proposed concept. According to the authors, such as GT Planner tool enables closer parity optimal value from the point of view of sustainable transport development. This value should also take into account the conditions of transport infrastructure and its organisation in the various sections of the transport network. These issues are described in various Authors publications.

Another issue for discussion is the question of whether the current method for capacity calculating are adequate for appearing in the road network, significant percentages of public transport. The authors, in the Silesian conurbation identified sections where this share is approaching to 25%. Assuming further steps towards the sustainable development of the transport this share should continue to grow.

The actions addressed here with regard to the determination of the components of the obligatory parity can be supplemented with education regard to the culture of mobility (Okraszewska R. et al. 2014).

REFERENCES

Aguiléra, A., Wenglenski, S., Proulhac, L. 2009. Employment suburbanisation, reverse commuting and travel behaviour by residents of the central city in the Paris metropolitan area, *Transportation Research Part A* 43: 685–691.

Beckerman, W. 1994. *Sustainable Development: Is it a Useful Concept?* Environ. Values.

Celiński, I. Sierpiński, G., Staniek, M. 2015. *Przekroje multimodalne w sieciach transportowych.* Logistyka 4: 2714–2723.

Esztergár-Kiss D., Csiszár Cs. 2015. Evaluation of multimodal journey planners and definition of service levels. *International Journal of Intelligent Transportation Systems Research* 13: 154–165.

European Commission. 2011. *White Paper: Roadmap to a Single European Transport Area — Towards a competitive and resource efficient transport system.* Com(2011) 144.

Gretzel, U., Yoo, K.H., Purifoy, M. 2007. Online travel Reviever Study, *Laboratory for Intelligent Systems in Tourism*: 4–7.

Handy, S., Cao, X., Mokhtarian, P. 2005. Correlation or causality between the built environment and travel behavior? Evidence from Northern California, *Transportation Research Part D* 10: 427–444.

Ionescu G. 2016. *Transportation and the Environment: Assessments and Sustainability.* CRC Press Taylor & Francis Group.

Janic M. 2016. *Transport Systems: Modelling, Planning, and Evaluation.* CRC Press Taylor & Francis Group.

Kanister, D. 2008. The sustainable mobility paradigm. *Transport Policy* 15: 73–80.

Khattak, A.J., Rodriguez, D. 2005. Travel behavior in neo-traditional neighborhood developments: A case study in USA, *Transportation Research Part A* 39: 481–500.

Limanond T., Butsingkorn T., Chermkhunthod Ch. 2011. Travel behavior of university students who live on campus: A case study of a rural university in Asia, *Transport Policy* 18: 163–171.

Lu, X., Pas, E.I. 1999. Socio-demographics, activity participation and travel behavior, *Transportation Research Part A* 33: 1–18.

Millward, H., Spinney, J. 2011. Time use, travel behavior, and the rural–urban continuum: Results from the Halifax STAR project, *Journal of Transport Geography* 19: 51–58.

Monzon-de-Caceres A., Di Ciommo F. 2016. *CITY-HUBs: Sustainable and Efficient Urban Transport Interchanges.* CRC Press Taylor & Francis Group.

Okraszewska R. 2008. *Przestrzenne sytuacje konfliktowe wywołane rozwojem systemu transportowego w warunkach równoważenia rozwoju*, Gdańsk University of Technology.

Okraszewska R., Nosal K., Sierpiński G. 2014. The Role of the Polish Universities in Shaping A New Mobility Culture — Assumptions, Conditions, Experience. Case Study of Gdansk University of Technology, Cracow University of Technology And Silesian University of Technology. *Proceedings of ICERI2014 Conference*: 2971–2979.

Sáez A.E., Baygents J.C. 2014. *Environmental Transport Phenomena*. CRC Press Taylor & Francis Group.

Sierpiński G., Staniek M., Celiński I. 2014. Research And Shaping Transport Systems With Multimodal Travels — Methodological Remarks Under The Green Travelling Project. *Proceedings of ICERI2014 Conf.*: 3101–3107.

Sierpiński G., Staniek M., Celiński I. 2016. Travel behavior profiling using a trip planner. *Transportatrion Research Procedia* 14C: 1743–1752.

Sierpiński G. 2013. Revision of the MoCHCH05dal Split of Traffic Model. Activities of Transport. *Communications in Computer and Information Science* 395: 338–345.

Sierpiński G. 2017. Technologically advanced and responsible travel planning assisted by GT Planner. *Lecture Notes in Network and Systems* 2: 65–77.

Silva Cruz, I., Katz-Gerro, T. 2015. Urban public transport companies and strategies to promote sustainable consumption practices, *Journal of Cleaner Production,* http://dx.doi.org/10.1016/j.jclepro.2015.12.007

Stanley, J., Lucas, K. 2014. Workshop 6 Report: Delivering sustainable public transport, *Research in Transportation Economics* 48: 315–322.

Szarata A. 2010a. Modelowanie symulacyjne ruchu wzbudzonego i tłumionego, *Transport Miejski i Regionalny* 3:14–17.

Szarata A. 2010b. Modelowanie ruchu tłumionego w ujęciu symulacyjnym, *Modelowanie podróży i prognozowanie ruchu* 94: 169–282.

Szarata A. 2012. *Badania ankietowe dotyczące zjawiska ruchu wzbudzonego w podróżach transportem zbiorowym*, 9th Scientific-Technical Conference Logitrans, Szczyrk.

Szarata, A. 2013. The simulation analysis of suppressed traffic. *Advances in Transportation Studies*: 29: 35–44.

Transportation Research Board. 2010. *Highway Capacity Manual.*

Transport Infrastructure and Systems – Dell'Acqua & Wegman (Eds)
© 2017 Taylor & Francis Group, London, ISBN 978-1-138-03009-1

The influence of road marking, shape of central island, and truck apron on total and truck accidents at roundabouts

Jwan Kamla, Tony Parry & Andrew Dawson
Faculty of Engineering, Nottingham Transportation Engineering Centre (NTEC), University of Nottingham, UK

ABSTRACT: For in-service roundabouts where traffic volume has changed since design, road marking is an important factor affecting the safety and capacity of roundabouts. The principal aim of this study is to investigate the influence of road markings, truck apron, and shape of the central island on total and truck accidents. The results indicate that, the highest rate of total and truck accidents for the selected roundabouts were recorded for concentric-spiral marking, followed by spiral, partial- concentric, concentric, and roundabouts with no markings. The majority of the selected roundabouts have oval shape and they have higher rate of total accidents and truck accidents, than circular shape roundabouts. In addition, only three of the selected roundabouts have truck aprons and the rate of truck accidents are high in these locations.

1 INTRODUCTION

Road transport authorities are responsible for promoting the safety of road networks, and their aim is to reduce fatalities and injuries arising from accidents on road networks. The main cause of many accidents is driver error, and trucks in particular are a type of vehicle whose effect on the safety of the road network should be taken into account, as they cause many fatalities and serious injuries, because of their size, the freight they carry, and the different and difficult maneuvers that they require compared to cars and other types of vehicles (Carstensen et al. 2001). Roundabouts have become popular in developed countries; in the United Kingdom (UK) roundabouts are widely used instead of other junction types. Roundabouts are considered safer than other intersection types because; the number of conflict points decreases, they lead drivers to reduce their speeds, they regulates turning movement of other vehicles, and they are considered to give better operational performance (Kennedy 2007; Design Manual of Roads and Bridges (DMRB) TD 16/07 2007).

Geometric layout, operational analysis, and safety evaluation are significant requirements for the roundabout design process. Small modifications in geometry can lead to considerable changes in the safety and/or operational performance of roundabouts. Any sudden change in geometric design leads the roundabouts to be less safe (Kennedy 2007). Truck rollover accidents are common at roundabouts (Kemp et al. 1987). Weber et al. (2009) stated that issues with trucks at roundabouts mainly include accommodating trucks within the available geometry.

For a well-designed roundabout with balanced traffic movement and efficient operations, no additional road markings at approaches and within the circulatory carriageway are required. For in-service roundabouts where traffic volume has changed since design, road marking is considered to be an important factor affecting the safety and capacity of roundabouts. When the circulatory carriageway is wide, this may confuse drivers if there are no markings within the circulatory lanes. Weber et al. (2009) indicated that bigger roundabouts are better for trucks and other large vehicles. They stated that the use of road markings within the circulatory makes the roundabout safer for truck drivers, as they can stay in their own lane. Using road markings will reduce three types of accidents: "side to side collisions on the circulatory lanes, drivers being forced on to the central island, and collisions between entering and circulating vehicles" (DMRB TA 78/97 1997, p. 2/1). Using road markings within the circulatory lanes at grade-separated roundabouts increases the efficient use of the circulatory lanes, as drivers can choose the right path (DMRB TD 16/07 2007). According to DMRB TA 78/97 (1997) there are four types of markings: concentric, partial-concentric (for wide circulatory width), concentric-spiral, and spiral (more suitable for large roundabouts).

Regarding the shape of the central island, the majority of the roundabout design guidelines advise the use of circular roundabouts; the majority of other types of intersections are converted to non-circular roundabouts and these intersections can have poor accident records (Kennedy 2007). Alphand et al. (1991b) illustrated that oval roundabouts have higher accident rates than circular ones, however Rodegerdts et al. (2010) reached the conclusion that the latter are safer because they encourage a constant speed with the circulatory lanes, while oval roundabouts increase the speed in the straight line then induce speed decrease when the vehicle approaches the arc, which precipitates loss of control accidents within circulatory lanes.

Another geometric parameter that can be used within the small roundabout is a truck apron, which is "an over run area (a raised low profile area around the central island)" which can be important for trucks using small roundabouts (DMRB TD 16/07 2007, p.7/4). Gingrich and Waddell (2008) in a study of a roundabout during morning and evening peak, recorded that 624 trucks travelled within the roundabout, 77% of which did not use the truck apron, and of those that did, 67% did so to prevent other cars in the adjacent lane from travelling beside them.

The principal aim of this paper is to investigate the influence of road markings, truck apron, and shape of the central island on total and truck accidents.

2 DATA SOURCES AND DESCRIPTION

2.1 Description of the selected locations

For this study, 70 roundabouts in the UK including 284 approaches were selected. The selected roundabouts comprise nine roundabouts on the motorway M1, ten roundabouts on the M6, six roundabouts on the M5 and nine roundabouts on the M4, with the others located on different motorways and A-class roads. They include signalized, un-signalized, or partially signalized roundabouts. They are either two lane, or three lane, and the majority of roundabouts are grade-separated. Note that in this study, a roundabout is considered partially signalised when one or more of the approaches and circulatory lanes are signalised, but not all.

The characteristics of the whole 70 roundabouts include:

– Legs
  • 3 legs = 12,
  • 4 legs = 39,
  • 5 legs = 12, and
  • 6 legs = 7,
– Traffic signals
  • Yes = 20,
  • No = 28, and
  • Partial = 22,

– Lanes
  • 2 lanes = 39 and
  • 3 lanes = 31,
– Grade type
  • At grade = 19 and
  • Grade separated = 51.

Despite these differences, the selected roundabouts are similar with respect to design standards (all are on modern motorways) and traffic levels. Individual lane widths, traffic signage and maintenance regime will all be similar and other features such as sight lines will meet the minimum required standards.

2.2 Accident data

For the selected locations STATS19 data (UK accident statistics) were acquired within a 350 m radius from the center of the roundabouts, for 11 years (2002–2012). The accident data from STATS19 includes information about collision circumstances (accident reference code, year, accident severity, time and date, weekday, weather and lighting condition, road type and road surface condition, vehicle and casualty numbers, easting/northing coordinates, etc.), vehicle details (accident reference, type of vehicle, etc.) and reported contributory factors.

2.3 Traffic data

Average Annual Daily Traffic (AADT) was acquired from the Department of Transport, Traffic Counts (DFT 2016).

2.4 Geometric information

Road marking, shape of central island, and truck apron information, was acquired from aerial photographs of an on-line mapping site.

3 METHOD

In order to identify the locations that triggered accidents, the Earth Point program, Excel to KML, was used, which displays Excel files on Google Earth and can be found at (Earth Point *2016*).

Firstly, the accident position coordinates were converted from grid easting/northing to latitude/longitude using Grid InQuest Version 6.6.0 available as free download from Ordnance Survey Ireland: "The Grid InQuest software provides a means for transforming coordinates between The European Terrestrial Reference System 1989 (ETRS89) and the National coordinate systems of Great Britain, Northern Ireland, and the Republic of Ireland" (Quest Geo Solutions Ltd 2004). Then the converted points were uploaded to Google Earth and checked to see whether they were located in the selected study roundabouts. Once they had all been checked, then the number of accidents were counted manually.

In order to separate truck accidents from total accidents, so as to upload them separately to Google Earth, the STATS19 Excel sheet for vehicle details was used, one of the columns of which indicates the types of vehicles. Each truck accident has a reference code, which was used in casualty details to find and highlight truck accidents. With truck accidents highlighted, they were then filtered from other casualty details, then easting and northing, by the same process, was converted to latitude and longitude and uploaded to Google Earth. After uploading them, the numbers of truck accidents for each circulatory and approach of the roundabouts were counted manually. Note that the definition of truck accidents in this study is any accidents involving a truck.

Note that DMRB TD 16/07 (2007) uses 100 m from the entry line as a measurement guide for the design of roundabouts including speed limit within 100 m on approach, for maximum flare length, and for maximum exit kerb radius. Therefore, the ruler from Google Earth was used to compute the percentage of total and truck accidents within 100 m distance, within the circulatory lanes, and beyond 100 m distance, so as to understand how far the accidents occurred away from a given approach's entry line.

Road markings, shape of central island, and truck aprons for the selected locations were investigated using Google Earth. Then for each type of marking, and for each shape of roundabout central islands, the number of accidents was recorded and the rate of accidents per roundabout was identified.

4 RESULTS

4.1 Overview

In the UK, roundabouts are commonly used as a high traffic volume junction. This has led to the construction of large roundabouts with high Inscribed Circle Diameter (ICD) and results in a high circulating speed. Two points are important and should be considered by design organizations during roundabout rehabilitation and safety improvement (DMRB TD 16/07 2007):

- the need to consider the geometry of each part, and
- the need to review the existing roundabout marking

Therefore, for the selected roundabouts accident rate based on different geometric characteristics were identified and the results are given in the following sections.

4.2 General total and truck accident trends

According to STATS19 data, 5,520 casualties in all categories of accidents around the selected locations (entry, exit, and circulatory lanes) were recorded during the 11 years from 2002 to 2012. Of these, 26.6% of accidents include trucks. It was found that more fatalities occur in truck accidents (2.10%) than in accidents involving only other types of vehicle (1.7%). This shows that truck accidents are more dangerous because of their size, weight, and manoeuvrability as identified by previous studies (DFT 2014; Trucks V 2013; US Department of Transportation 2014; Carstensen et al. 2001; Grygier et al. 2007; Kennedy 2007).

It was found that 60% and 57% of total and of truck accidents occurred within 100 m of the entry line, 32% and 36% within the circulatory lanes, and 7% of total and truck accidents happened on approaches at a distance greater than 100 m from the entry line. The locations that recorded higher number of accidents beyond 100 m distance are very busy and big roundabouts. In Italy, Montella (2007), in a study for 15 urban roundabouts (55 approaches) of three- and four-leg types, found that 65% of accidents occurred at roundabout approaches, with 15% and 20% in the circulatory and exit lanes, respectively.

4.3 Road marking

Accident rates were identified for each type of marking and the results are reported in Table 1.

It was found that the highest rates of total and truck accidents were recorded in roundabouts having concentric-spiral markings; followed by concentric, partial-concentric, spiral, and roundabouts with no markings.

Five of the roundabouts with concentric markings are five and six-arms, with accident rates of 78, and the other eleven roundabouts with concentric marking are three and four-arm roundabouts with accident rates of 56. And because concentric-spiral and spiral marking is more

Table 1. Road marking type and the rate of total accident, and truck accidents for the selected locations.

Marking type	Roundabout no.	Total accident		Truck accident	
		No.	Rate	No.	Rate
Concentric	16	1006	63	218	14
Partial-concentric	15	916	61	188	13
Concentric- spiral	16	1420	89	403	25
Spiral	11	676	61	150	14
None	12	216	18	28	2

suitable for big roundabouts (DMRB TA 78/97 1997), it is necessary to re-assess the big roundabouts that have concentric markings and consider changing these markings in order to make the path within the roundabouts more efficient for the users, and thereby reduce accidents that might occur because of insufficient marking within the roundabouts.

Roundabouts with concentric spiral markings are associated with a higher rate of total and truck accidents. The rates of total and truck accidents for concentric and spiral type marking are similar, followed by partial concentric marking. Note that five of the roundabouts that have no markings within the circulatory lanes are grade-separated and the rate of total and truck accidents are 25 and 3.4, respectively; while in the other seven at-grade locations with no marking the rate of total and truck accidents is lower (13 and 1.57, respectively). This indicates that these grade-separated roundabouts may benefit from marking within the circulatory lanes.

Regarding the roundabouts that have spiral markings, four of them are small at-grade roundabouts. Based on DMRB TA 78/97 (1997) spiral markings are suitable for big roundabouts. Comparing these three at-grade locations to the other eight roundabouts with spiral marking, a similar rate of total accidents (59 relative to 62) was recorded; and higher truck accident rate (16 relative to 13) was recorded. These three at-grade roundabouts probably require re-assessment.

4.4 Shape of central island

In this study, there are 43 oval-shape central island roundabouts, and 27 circular shape roundabouts. Table 2 reports the rates of total accidents, truck accidents per roundabout, and AADT with respect to oval shape and circular shape roundabouts. It is clear that the rates of total and truck accidents are higher in oval shaped rather than circular shaped roundabouts. The probable reason for having this result is 42 out of 43 of the oval shape roundabouts are grade-separated. The rate of total and truck accidents at grade-separated roundabouts is higher relative to these rates at at-grade roundabouts (a rate of 71.9 relative to 29.9 for total accidents per roundabouts, and a rate of 17.3 relative to 5.4 for

Table 2. Central is land shape type.

Shape type	Roundabout no.	Total accident No.	Total accident Rate	Truck Accident No.	Truck Accident Rate	AADT per roundabouts
Oval	43	3315	77	811	19	60543
Circular	27	919	34	176	7	35388

truck accidents per roundabouts). In addition, they have higher AADT as reported in Table 2; oval shape roundabouts have higher AADT by 41% relative to circular shape roundabouts. This result is in line with Alphand et al. (1991b) although their result was for total accident rates.

4.5 Truck apron

The set of roundabouts considered in this study included 19 at-grade roundabouts and they were considered small compared to grade-separated roundabouts. For the 19 selected locations, only three roundabouts have truck aprons. In these locations, the rates of total and truck accidents are 90, and 20, respectively; while for the other 16 locations without truck apron the rate is 16, and 2.0, respectively. This indicates that availability of truck aprons in the three small roundabouts has not reduced accidents to the general level and could even be somewhat causative of the much higher accident rates observed; given that both truck and all-vehicle accident rates are increased relative to other roundabouts. Note that the percentage of trucks in these locations is high (9–10%) and, probably because these locations have enough space (circulatory width of >10 m) for trucks to negotiate the roundabouts, they may not use the truck aprons. Gingrich and Waddell (2008) have found that during morning and evening peak periods 77% of trucks did not use truck aprons.

5 CONCLUSIONS

This paper reported the total and truck accident numbers within 70 roundabouts in the UK, and considered them in light of their road markings.

Based on DMRB RD 16/07 (2007) big roundabouts require spiral and spiral-concentric markings, while small roundabouts require concentric markings. In this study regarding road marking, some big roundabouts have simple concentric marking and recorded high total and truck accident rates. In addition, spiral markings are recommended for big roundabouts but some of the small roundabouts have spiral markings and recorded high rates of total and truck accidents and these locations may require a re-assessment regarding marking in order to make the movement path easier for the roundabout users and provides safety at these roundabouts.

Oval shaped roundabouts have a higher rate of total and truck accidents relative to those of circular shape but all the oval shapes are big roundabouts with higher AADT.

For the selected locations, only three of the roundabouts have truck aprons and the rate of truck accidents are high in these locations.

Addition of truck aprons in these roundabouts has not reduced accident rates to the general level.

REFERENCES

Alphand F, Noelle U and Guichet B, 1991b. Roundabouts and road safety; state of the art in France. Intersections without traffic signals II. *Proceedings of an International Workshop* 18–19 July 1991 in Bochum, Germany, pp126–140.

Carstensen A Gitte, Hansen W, Hollnagel V, Højgaard H, Jensen I, Kines P, Klit L, Kofoed P, Mikkelsen J, and Petersen O, 2001. Analysegruppen for Vejtrafikuheld (*AVU*), *Lastbiluheld- en dybdeanalyse af 21 uheld.*

Department For Transport (DFT) 2014. *Reported road casualties in Great Britain*, Annual Report, Statistical Release.

Design Manual for Roads And Bridges, TA 78/97, 1997. *Design of Road Markings at Roundabouts*, London, UK.

Design Manual for Roads And Bridges, TD 16/07, 2007. *Geometric Design of Roundabouts*, London, UK.

DFT 2016. *Department for Transport, Traffic Counts.* Available at: http://www.DfT.gov.uk/traffic-counts/area.php/ [Accessed 10 Aug. 2016].

Earth Point 2016. *Excel To KML - Display Excel files on Google Earth.* Available at: https://www.earthpoint.us/ExcelToKml.aspx/ [Accessed 10 Aug. 2016].

Gingrich, M. and Waddell, E. 2008. Accommodating trucks in single and multilane roundabouts. *In: Transportation Research Board, National Roundabout Conference*, Kansas City, 12–18.

Grygier, P.A., Garrott, W.R., Salaani, M.K., Heydinger, G.J., Schwarz, C., Brown, T. and Reyes, M. 2007. Study of heavy truck air disc brake effectiveness on the national advanced driving simulator. *In: The 20th ESV Conference Proceedings.*

Kemp, R., Chinn, B. and Brock, G. 1978. Articulated vehicle roll stability: methods of assessment and effects of vehicle characteristics. *In Transport and Road Research Laboratory: TRRL Laboratory Report 788.*

Kennedy, J. 2007. *International comparison of roundabout design guidelines*, TRL.

Montella, A. 2007. Roundabout in-service safety reviews: safety assessment procedure. *Transportation Research Record: Journal of the Transportation Research Board,* 2019, 40–50.

Quest Geo Solutions Ltd. 2004. Grid InQuest DLL User Manual (Version 6).

Rodegerdts, L., Bansen, J., Tiesler, C., Knudsen, J., Myers, E., Johnson, J., Moule, M., Persaud, B., Lyon, C., Hallmark, S., Isebrands, H., Crown, R., Guichet, B., and O'Brien, A., 2010. NCHRP Report 672: Roundabouts: An informational guide, Transportation Research Board, Washington, DC.

Trucks, V. 2013. *European Accident Research And Safety Report.* Available at: http://www.volvotrucks.com/SiteCollectionDoments/VTC/Corporate/Values/ART%20Report%202013.pdf/[Accessed 29 Sep. 2016].

US Department Of Transportation 2014."*Traffic Safety Facts", National Highway Traffic Safety Administration*, DOT HS 812 101, Washington, DC 20590.

Weber, M. Button, N. 2009. Accommodating small and large users at roundabouts. *In: Annual Conference of the Transportation Association of Canada: Sustainability in Development and Geometric Design for Roadways*, Vancouver, British Columbia, Canada.

Transport Infrastructure and Systems – Dell'Acqua & Wegman (Eds)
© *2017 Taylor & Francis Group, London, ISBN 978-1-138-03009-1*

A runway veer-off risk assessment based on frequency model: Part I. probability analysis

L. Moretti, G. Cantisani, P. Di Mascio & S. Nichele
Dipartimento di Ingegneria Civile, Edile e Ambientale. Sapienza, University of Rome, Rome, Italy

S. Caro
Departamento de Ingenieria Civil y Ambiental, Universidad de Los Andes, Bogota, Colombia

ABSTRACT: This two-part paper presents a comprehensive airport risk assessment methodology for veer-off accidents of aircraft runways. Veer-off is a lateral runway excursion, and it occurs when an aircraft leaves the runway during a movement. The most frequent causes for this type of accidents are inappropriate pilot performance and aircraft condition; while other contribution factors include transversal wind, contamination of the runway, and poor visibility. The consequences of these events involve both damage to the aircrafts and potential health effects on passengers and crew members. Statistical data collected throughout the world by the authors were used as a reference to conduct frequency and risk assessments of veer-offs. The objective of the first part of the paper is to assess the probability of a veer-off accident. To accomplish this objective, a cumulative probability distribution was used to represent the phenomenon. It was found that the exponential curve described by the Poisson distribution could be properly used to describe this type of accident. The results show that the average frequency of a veer-off accident is 1.44 in ten million movements for commercial flights over 30 Mg (ton) and that veer-offs are more frequent to occur during landing than during take-off procedures. The proposed analysis permits to compute the probability of veer-offs at any airport, after considering its specific conditions (e.g. number and type of plane, kind of movement, bearing capacity of the subgrade, etc.).

1 INTRODUCTION

Risk management represents a fundamental aspect of Air Traffic Management (Eurocontrol 2001). Several studies reported in the literature have identified some of the main factors involved in airport accidents: weather; crew technique decision or performance; and airport system factors (e.g. malfunctioning of assistance instruments) (Flight Safety Foundation 2000, Flight Safety Foundation 2009).

The concept of safety requires a detailed analysis accounting for different perspectives of the subjects involved. According to Kirkland et al. (2004), there are several objective or subjective types of safety levels of a system: the real level of safety, the target level of safety and the achieved or the deemed level of service. A good-quality safety management design for airport infrastructures needs an accurate methodology to evaluate the risk and prevent accidents (Attaccalite et al. 2012, Di Mascio et al. 2012, Canale et al. 2005, Wong et al. 2009a, b, Kirkland et al. 2004, Di Mascio & Loprencipe in press). Therefore, to reduce risk, service providers should carry out a risk assessment (Spriggs 2002), which includes to calculate both the probability of occurrence of certain hazard, as well as the magnitude of the consequences of such an occurrence (ISO 1995, International Civil Aviation Organization 2010).

A runway veer-off is an operational risk present in airports, defined as the excursion in which an aircraft leaves the side of the runway. ICAO Annex 14 provides recommendations for acceptable consequences in case that an aircraft runs off the side of the runway. Specifically, the Aerodrome Design Manual Part 1. 5.3.22 specifies the following regarding the function of runway strips "...., *it should be graded in such a manner as to prevent the collapse of the nose landing gear of the aircraft. The surface should be prepared in such a manner as to provide drag to an aircraft and below the surface, it should have sufficient bearing strength to avoid damage to the aircraft.....*" (International Civil Aviation Organization 2013a).

In many countries, national specifications require special risk evaluations (Moretti et al. in press). In Italy, for example, the licensing regulations of aerodromes, issued by the Italian Civil Aviation Authority, ENAC, have transposed the

ICAO standard and, in case of insufficient bearing strength of the strip, ENAC requires a runway veer-off risk assessment (Ente Nazionale per l'Aviazione Civile 2014).

It should be noticed that lateral offset from the travel lanes represents a critical operational aspect not only in airport fields. Downgrade roadways, for example, are required to have adjacent truck escape ramps. These ramps act as emergency areas and provide a location for out-of-control vehicles (National Cooperative Highway Research Program 1992, American Association of State Highway and Transportation Officials 2001). As for the road context, a systematic methodology has determined the need for truck escape ramps and their locations (Abdelwahab et al. 1996): likewise, airport design should reduce the risk of veer-off events to its least. For this reason, the materials and geometrical configuration used in the lateral airport runway safety areas should guarantee technical, environmental and economic sustainability, in order to balance the stakeholders' competing interests (Di Mascio et al. 2012, Di Mascio et al. 2014, Moretti 2013, Moretti et al. 2014, Transport Canada Civil Aviation 2011). In the case of a veer-off event, the materials used for the runway strip should contribute to reduce the risk of damage to an aircraft, the health impact on the occupants of the airplane and the air traffic interruption (Transport Canada Civil Aviation 2011).

This research assesses the risks of veer-offs by evaluating their likelihood and severity. The analysis takes into account existing methodologies to assess risk of plane crashes (Wong et al. a, b, Kirkland et al. 2004), and it considers that the level of risk has to be As Low As Reasonably Practicable (ALARP) (Airport Cooperative Research Program 2011, Airport Cooperative Research Program 2014, UK Civil Aviation Authority 2014). This analysis considers available statistical data and it proposes a new reference system to describe this type of accident.

2 METHODOLOGY

The present research relies on statistical data obtained from several international sources (Agenzia Nazionale per la Sicurezza del Volo 2016, Airdisaster.com 2016, Airsafe 2013, Australian Transport Safety Bureau 2012, Aviation safety network 2005, Boeing 2000, 2002, 2003, 2006, 2010, 2012, 2013, 2014, 2015, 2016, General Aviation Manufacturers Association 2005, Flight Safety Foundation 2014, International Civil Aviation 2013b, National Transportation Safety Board 2014). Based on this information, 3500 veer-off records that occurred between 1953 and 2015 and

that involved several types of flights (passenger aviation, cargo, training, military and agricultural flights) were analysed.

At last, this study used a reduced number of records to get a more homogeneous database. This is due to the fact that accident reports of general aviation are not always complete and clearly recorded in the literature, so the overall database would not have been consistent. Thus, the analysis was conducted on 301 events (accidents or serious incidents), related to passenger/cargo flights within the period 1953–2015 (Fig. 1). The number of events reported until the 1970s was relatively low, most likely due to underreporting.

The date of the accidents allowed calculating the average interval between two consecutive events. Considering all accidents involving passenger and cargo flights from 1953 to 2015, the number of days between two successive events is 78. However, since data corresponding to accidents that occurred many decades ago are scarcely available, as observed in Figure 1, if we consider only the events since the 1980s, the frequency of the recorded veer-offs levelled off at approximately 8.6 events per year, resulting in an average number of days between two veeroff events of 42. Figure 2 represents the frequency of the observed times between two consecutive events.

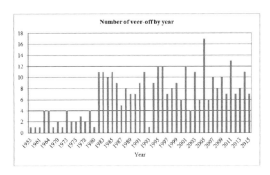

Figure 1. Number of veer-offs by year.

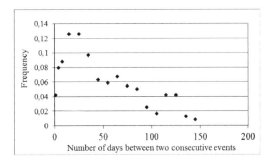

Figure 2. Frequency of observed time between two consecutive events.

Taking into account data collected over the last years, the authors estimated the frequency of these accidents against the total number of movements over the same period of time.

The analysis considered the frequency of veer-offs that occurred to commercial flights over 30 Mg (ton). The results from this analysis show that the average veer-off probability is equal to $1.44 \cdot 10^{-7}$, whereas the average frequency of an aircraft accident is $30 \cdot 10^{-7}$. The order of magnitude of the veer-off rate computed taking into account the evaluated databases coincides to the one reported in the "Final Report on the Risk Analysis in Support of Aerodrome Design Rules", prepared by the Norwegian Civil Aviation Authority (AACN), which reports a frequency of occurrence of a veer-off of $2.2 \cdot 10^{-7}$ (Norwegian Civil Aviation Authority 2001).

A runway veer-off may occur during four different phases of a flight: 1) landing, 2) take-off, 3) taxiing, or 4) during other operational condition. Ac-cording to the collected data, 98% of veer-offs occurred during landing and take-off, and veer-offs during landings were found to be approximately 3 times more likely to happen than during take-offs (223 veer-offs during landings, 78 during take-offs), as confirmed by AACN (2001). The ratio of incidence between the two types of accidents is 2.8 (Fig. 3).

Table 1 summarizes ten common causes of veer-off events with their associated frequencies for the same set of data. As shown, the total of all percentages exceeds 100 percent because of the multiple contribution factors associated with the same veer-off event.

The causes for veer-offs were categorized into five general classes: 1) aircraft conditions, 2) pilot performance, 3) weather conditions, 4) conditions of the runway surface, and 5) airport management.

A transfer-in procedure within a fault tree analysis (FTA) was applied to calculate the likelihood of occurrence of each factor, as listed in Table 2 (Ale 2002).

Table 1. Common causes of veer-offs.

Causes	Frequency %
Pilot performance	63
Wind characteristics	25
Aircraft landing gear system	14
Aircraft maintenance	7
Aerodrome management	6
Weather conditions	6
Runway surface (e.g. presence of contaminants)	6
Engine failure	6
Collision with obstacle(s)	3
Other	9

Table 2. Causing factors classified by category.

Factors	Frequency %
Aircraft	46
Pilot performance	39
Weather conditions	31
Runway surface characteristics	17
Airport management	11

The Ishikawa diagram in Figure 4 categorizes all potential causes that were found as a result of this analysis.

Consequences of a veer-off include damage to the aircrafts and potential health impact on occupants. Regarding mechanical damages, Table 3 summarizes eight types of possible consequences, with their associated frequencies, which were obtained from the analysed data.

Statistics on health effects show that there were 383 fatalities that occurred in 32 veer-offs. The average number of fatalities per veer-off with deaths was 11.90; whereas the average number of deaths per veer-off was 1.27 (calculated considering 301 veer-offs collected by the authors). The analysis of the available data also shows that the accidents recorded in the case of passengers and cargo flights were more serious but less frequent than in other types of flights.

Three categories of accidents were considered in this study:

• accidents that resulted in deaths,
• accidents that caused injuries but did not involve deaths,
• accidents that did not cause injuries or deaths.

Figure 5 shows the distribution of health effects caused by veer-offs.

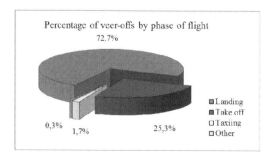

Figure 3. Percentage of veer-offs by phase of flight.

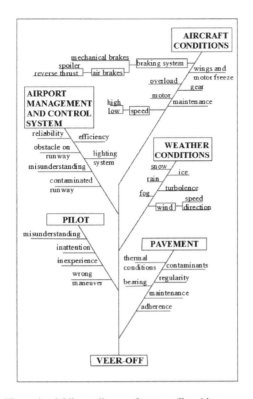

Figure 4.   Ishikawa diagram for veer-off accident.

Table 3.   Veer-off consequences.

Consequences	Frequency %
Mechanical damage	50
Collision with obstacle(s)	32
Damage to landing gear system	32
Crash into an embankment or in a drainage channel	13
Fire	11
RWY, taxiway (TWY) or apron crossings	5
Crossing of airfield perimeter fence	4
Other	3

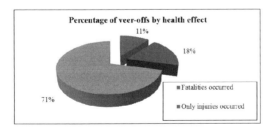

Figure 5.   Percentage of veer-offs by health effect on aircraft occupants.

3   RESULTS

The collected records highlight that the severity of a veer-off accident largely depends on the type of flight. As a matter of fact, injuries suffered by occupants due to veer-offs are more likely to occur in commercial aviation (passenger and freight flights) than in other types of flights. Indeed, for veer-offs occurring in these flights, the ratio between fatalities and occupants on board is equal to 36%, which proves the severity of these events.

The values in Tables 4 to 6 correspond to the statistical parameters pi, calculated for each j category of the total accidents, according to the following equations:

$$p_{1,j} = \frac{n_{events,j}}{N_{events,TOTAL}} \tag{1}$$

$$p_{2,j} = \frac{n_{pax,j}}{N_{pax,TOTAL}} \tag{2}$$

$$p_3 = \frac{n_{deaths}}{N_{pax,TOTAL}} \tag{3}$$

$$p_4 = \frac{n_{deaths}}{N_{pax,j}} \text{ with } j = 1 \tag{4}$$

$$p_{5,j} = \frac{n_{injuries,j}}{N_{pax,TOTAL}} \text{ with } j = 1,2 \tag{5}$$

$$p_{6,j} = \frac{n_{injuries,j}}{N_{pax,j}} \text{ with } j = 1,2 \tag{6}$$

Table 4.   Accidents with deaths.

Indicator	Amount %
p_1	11
p_2	25.5
p_3	9.2
p_4	36
p_5	0.1
p_6	1.3

Table 5.   Accidents with injuries (without deaths).

Indicator	Amount %
p_1	18
p_2	25.5
p_5	2.6
p_6	10.5

Table 6. Accidents without injuries or deaths.		
	Amount
Indicator	%
p_1	71
p_2	49

where $n_{events, j}$ is the number of observed accidents with j category, j is a code to representing the type of accident (with fatalities $j = 1$, injuries $j = 2$ or without them $j = 3$), $N_{events, TOTAL}$ is the total number of observed accidents, $n_{pax, j}$ is the number of passengers involved in accidents with j category, $n_{pax, TOTAL}$ is the total number of passengers involved in veer-offs, and n_{death} is the total number of fatalities involved in veer-offs.

As observed, fatal accidents account for approximately 11% of the total amount of accidents, but involve 25.5% of the total people on board of the flights evaluated. Also, the overall percentage of deaths of occupants during veer-offs was found to be 36%. Considering all the passengers involved in all the examined veer-offs, the percentage of deaths was 9.2% (Table 4).

The cases of accidents with injuries, but without deaths, account for approximately 18% of the total, and involve 25.5% of people on board (Table 5).

Finally, the results show that there were no health effects in 71% of the events, which involve 49% of the total amount of people on board of the affected planes (Table 6).

As mentioned previously, veer-offs may be due to two main reasons: a misalignment at the beginning of a take-off or landing procedure, or a loss of the directional control by the pilot during the landing, after touchdown, or during the take-off roll.

As reported in the literature, accident rates also depend on the type of movement, as demonstrated in Figure 3. Indeed, the accident rate of veer-off depends strongly on the instrumentation available during the movement: on non-instrument runways or with non-precision instruments, the frequency of occurrence is higher than the one recorded on runway with precision instruments. With reference to landing veer-offs, the analysis conducted by AACN found that 59% of the events occurred after a precision approach, while the remaining 41% occurred as a result of a non-instrument or non-precision instrument approach (Boeing 2015). With respect to take-off veer-offs, the ratio between flights with precision instruments and flights with no-instruments or with non-precision instruments is on average 10.2. This value allowed calculating the frequency of occurrence of a veer-offs as a

function of the type of movement. According to the collected data, the frequency of occurrence as a function of the instrumentation of the runway and the type of movement was calculated, obtaining the following results:

- instrument landing: $1.37 \cdot 10^{-7}$,
- non-instrument landing: $9.72 \cdot 10^{-7}$,
- instrument take-off: $6.96 \cdot 10^{-8}$,
- non-instrument take-off: $6.82 \cdot 10^{-9}$.

The statistical analysis of the veer-off events show that these accidents could be properly described using a Poisson distribution because of the following characteristics of the phenomenon (Scozzafava, 1995):

- the event is something that can be counted in whole numbers,
- occurrences are independent, so that one occurrence neither diminishes nor increases the chance of another,
- the events occur randomly in a given interval of time (or space),
- the probability of occurrence of an event in a small time interval Δt is proportional to Δt and it can be estimated as proportional to the product $\lambda \Delta t$, where λ is the mean number of events per interval T_a (i.e., the average interval between two consecutive events), as expressed by Equation 7, and

$$\lambda = \frac{1}{T_a} \qquad (7)$$

- it is possible to count how many events have occurred, but it is meaningless to ask how many of such events have not occurred yet.

Equation 8 gives the probability of observing x events within a time interval t:

$$p_x = P(X_\lambda = x) = \frac{(\lambda t)^x}{x!} e^{-\lambda t} \qquad (8)$$

where λ is the mean, variance and prevision values of distribution.

Veer-off accidents have been modelled also by evaluating the path related to the event. Regarding the spatial occurrence of the veer-offs, the initial configuration representing the departure of the air-craft in the runway affects the outcome and the final stopping location of the aircraft. This research evaluated the veer-off path by defining three geometrical characteristics: 1) the local drift angle, 2) the average drift angle, and 3) the drift radius. The initial two cases refer to a polar coordinate system with a pole located at the runway threshold, whereas the former is determined by a pole at the initial point of the veer-off path, as shown in Figure 6.

The data herein evaluated show that in more than 66% of veer-offs the average drift angle was up to 5 degrees, as shown in Figure 7.

The obtained average drift angles are represented in Figure 8.

The final point of the drift path leads to determine the lateral probability distribution, as depicted in Figure 6: it describes the probability that at the end of a veer-off the aircraft travels beyond a certain distance from the centreline of the runway. The curve trends of the final positions resulted in a negative exponential distribution. They were compared with the curves obtained by the Norwegian Civil Aviation Authority in a veer-off event risk analysis (Norwegian Civil Aviation Authority 2001) and it is possible to conclude that the results are comparable.

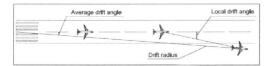

Figure 6.   Location of accidents system.

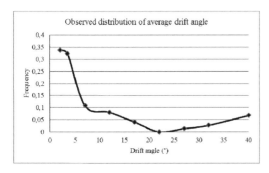

Figure 7.   Observed distribution of average drift angles.

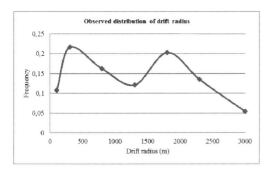

Figure 8.   Observed distribution of drift radius.

At the end of a veer-off accident, there are three main scenarios depending on the final position of the aircraft:

- veer-off over the runway,
- veer-off over the Cleared and Graded Area (CGA), with the external gear over the CGA, but the wing tip inside the strip,
- veer-off over the CGA, with the wing tip over the strip.

The final position of the aircraft allows calculating the transversal and longitudinal distribution probability curves related to the centreline of the runway. The position of the aircraft after the veer-off from the longitudinal and transverse edges of the runway could be described by a distribution function of the cumulative probability. Considering the final position and the size of the runway, the possibility of describing the phenomenon using an exponential distribution function was examined. The probability distribution depends on a 2D geometrical configuration and, in order to simplify the model, the problem was divided in two 1D scenarios mutually perpendicular. The overall curve of probability consists of four separate branches. Two branches have the abscissa axis orthogonal to the longitudinal axis of the runway and two branches have the abscissa axis coincident with the longitudinal axis of the runway. The two transversal branches of the cumulative probability distribution are not necessarily symmetrical in relation to the geotechnical and anemometry local conditions. Likewise, the two longitudinal branches with the origin in the midpoint of the runway are not necessarily symmetrical in relation to the runway length required by traffic.

Figure 9 represents the average cumulative probability distributions related to veer-off events during both take-off and landing procedures. As observed, the data does present an exponential trend.

Taking into account the movement type, the lateral deviation was found to be larger during take-offs than during landing procedures considering equal levels of veer-off probabilities (Fig. 10).

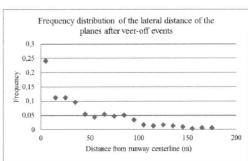

Figure 9.   Frequency distribution of the lateral distance of the planes after veer-off events.

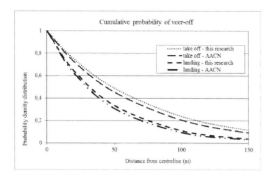

Figure 10. Cumulative probability of veer-off.

Results of this research are close to the ones obtained by AACN.

4 CONCLUSIONS

The first part of this paper presents the methodological advancements achieved in the development of veer-off accident frequency models.

The authors worked using a single comprehensive database of relevant accidents to quantify the probability of veer-off events during take-off and landing, both instrument and non-instrument or with non-precision instrument operations. Veer-offs were clustered according to their frequencies, causes and con-sequences. Inappropriate pilot performance, aircraft damage, meteorological conditions and runway contamination were determined to be the main causes of the accidents that involved areas lateral to the run-way and, rarely, also areas across the boundaries of the airports. Statistics of accidents and operations show that since 1990 the frequency of a veer-off is 1.44 in ten million movements, whereas the average frequency of an aircraft accident is 30 in ten million movements.

According to the available data, it was concluded that 97% of the total veer-offs occurred during landing and take-off procedures, and that veer-offs during landing are approximately 2.8 times more likely to occur than during a take-off. Concerning the consequences of a veer-off, it was determined that the mechanical damage of the aircraft is the most likely consequence (50%). On the contrary, the least likely consequence is the crossing of the airfield perimeter fence (5%). Regarding the health effects of these accidents on occupants, statistics show that at least one fatality occurred in 11% of the cases, whereas injuries were reported in 18% of the total veer-offs. Besides, there were not health effects in 71% of the events.

With respect to the movement, this type of accident is more frequent during landing than during take-off procedures, as mentioned previously, but the average final position of the aircraft during take-off is more distant from the runway centreline than during landing.

The probability distributions of the final transversal location of the aircrafts after an incident were obtained from the analysis of the veer-off reports. The proposed probabilistic model for the frequency of the accidents uses a multivariate logistic regression according to the Poisson distribution. This probabilistic density distribution was observed to properly describe the phenomenon, since it has an exponential qualitative trend and because 90% of the veer-off events are included within the strip width.

Also, the analysis show that the final spatial location of an aircraft involved during veer-off may be represented through a polar coordinate system. The drift angle and the drift radius allow identifying the risk areas in and around airport runways. The findings obtained in the study demonstrate the importance of veer-offs and the need for mitigating these accidents.

REFERENCES

Abdelwahab, W.M. & Morral, J.F. 1996. Determining need for and location of truck escape ramps. *Journal of Transportation Engineering*, Vol. 123, No. 5, pp. 350–356. doi: http://dx.doi.org/10.1061/(ASCE)0733-947X(1997)123:5(350).

Attaccalite, L., Di Mascio, P., Loprencipe, G. & Pandolfi, C. 2012. Risk Assessment Around Airport. *Procedia: Social & Behavioral Sciences*, pp. 852–861, 53. doi:10.1016/j.sbspro.2012.09.934.

Agenzia Nazionale per la Sicurezza del Volo 2016. Available from www.ansv.it.

Airdisaster.com 2016. Available from www.airdisaster. com.

Airport Cooperative Research Program (ACRP) 2011. ACRP Report 50: Improved models for risk assessment of runway safety areas. Transport Research Board.

Airport Cooperative Research Program 2014. ACRP Report 107: Development of a runway veer-off location distribution risk assessment model and reporting template. Transport Research Board.

Airsafe 2013. Available from www.airsafe.com.

Ale, B.J.M. 2002. Risk assessment practices in The Netherlands. *Safety Science*, vol. 40, pp. 120–126.

American Association of State Highway and Transportation Officials 2001. A policy on geometric design of highways and streets. Green book, pp. 259–269.

Australian Transport Safety Bureau 2012. Available from www.atsb.gov.au.

Aviation safety network 2005. Airliner accident statistics 2004. Statistical summary of fatal multi-engine airliner accidents in 2004.

Boeing 2000. Statistical summary of commercial jet airplane accidents - worldwide operations 1959–1999.

Boeing 2002. Statistical summary of commercial jet airplane accidents - worldwide operations -2001.

Boeing 2003. Statistical summary of commercial jet airplane accidents - worldwide operations 1959–2002.

Boeing 2006. Statistical summary of commercial jet airplane accidents - worldwide operations -2005.

Boeing 2010. Statistical summary of commercial jet airplane accidents - worldwide operations 1959–2009.

Boeing 2012. Statistical summary of commercial jet airplane accidents - worldwide operations 1959–2011.

Boeing 2013. Statistical summary of commercial jet airplane accidents - worldwide operations 1959–2012.

Boeing 2014. Statistical summary of commercial jet airplane accidents - worldwide operations 1959–2013.

Boeing 2015. Statistical summary of commercial jet airplane accidents - worldwide operations 1959–2014.

Boeing 2016. Statistical summary of commercial jet airplane accidents - worldwide operations 1959–2015.

Canale, S., Distefano, N. & Leonardi, S. 2005. A risk assessment procedure for the safety management of airport infrastructures. Proceedings at III International Congress of Società Italiana Infrastrutture Viarie, SIIV.

Di Mascio, P., Cardi, A., Di Vito, M. & Pandolfi, C. 2012. Distribution of Air Accidents Around Runways. *Social and Behavioral Sciences*, vol. 53, pp. 862–871, Proceedings, V International SIIV Congress Sustainability of Road Infrastructures, Rome, Italy, 29th-31st October 2012. doi: 10.1016/j.sbspro.2012.09.935.

Di Mascio, P., Moretti, L. & Panunzi, F. 2012. Economic Sustainability of Concrete Pavements. *Procedia - Social and Behavioral Sciences*, vol. 53, pp. 125–133, Proceedings, V International SIIV Congress Sustainability of Road Infrastructures, Rome, Italy, 29th-31st October 2012. doi: 10.1016/j.sbspro.2012.09.866.

Di Mascio, P., Loprencipe, G. & Moretti, L. 2014. Competition in rail transport: methodology to evaluate economic impact of new trains on track. ICTI2014 – *Sustainability, Eco-efficiency and Conservation in Transportation Infrastructure Asset Management*, 669–675, Losa & Papagiannakis (Eds)-Taylor & Francis Group.

Di Mascio, P. & Loprencipe, G. in press. Risk analysis in the surrounding areas of one-runway airports: a methodology to preliminary calculus of PSZs dimensions. *ARPN Journal of Engineering and Applied Sciences* (ISSN 1819-6608).

Ente Nazionale per l'Aviazione Civile (ENAC) 2014. Regolamento per la costruzione ed esercizio degli aeroporti.

Eurocontrol 2001. ESARR 4 – Risk assessment and mitigation in ATM.

European Aviation Safety Agency 2014. Annual Safety review 2013.

Flight Safety Foundation 2000. FSF ALAR Briefing note 8.1 - Runway excursions and runway overruns. Available from www.aviation-safety.net.

Flight Safety Foundation 2009. Reducing the risk of runway excursions. Available from www.aviation-safety.net.

Flight Safety Foundation 2014. Available from www.aviation-safety.net.

General Aviation Manufacturers Association 2005. General aviation statistical databook 2004.

International Civil Aviation Organization 2010. Annex 13: Incident reporting, data systems and information exchange.

International Civil Aviation Organization 2013a. Annex 14: Aerodrome Design and Operations.

International Civil Aviation Organization 2013b. Safety Management Manual, SMM. ICAO Doc 9859.

International Air Transport Association 2015. Safety report.

ISO 1995. ISO 8402:1995/BS 4778. Quality management and quality assurance. Vocabulary.

Kirkland, I.D.L., Caves, R.E. & Humphreys, I.M. 2004. An improved methodology for assessing risk in aircraft operations at airports, applied to runway overruns. *Safety Science* 42, pp. 891–905.

Moretti, L., Di Mascio, P. & D'Andrea, A. 2013. Environmental Impact Assessment of Road Asphalt Pavements. *Modern Applied Science*, 7(11), 1–11. http://dx.doi.org/10.5539/mas.v7n11p1.

Moretti, L. 2014. Technical and economic sustainability of concrete pavements. *Modern Applied Science*, 8(3), 1–9. DOI: 10.5539/mas.v8n3p1.

Moretti, L., Cantisani, G. & Caro, S. (in press). Airport veer-off risk assessment: an Italian case study. *ARPN Journal of Engineering and Applied Sciences*. (1819–6608).

National Cooperative Highway Research Program 1992. Truck escape ramps. NCHRP, n. 178.

National Transportation Safety Board 2014. Available from www.ntsb.gov.

Norwegian Civil Aviation Authority 2001. Final report on the risk analysis in support of aerodrome design rules. Rules, AEAT/RAIR/RD02325/R/002 Issue 1, 2001

Scozzafava, R. 1995. Primi passi in probabilità e statistica. Zanichelli Editore.

Spriggs, J. 2002. Airport risk assessment: examples, models and mitigations. 10th safety-critical systems Symphosium, Southampton.

Transport Canada Civil Aviation 2011. Airport pavement bearing strength reporting. Advisory Circular AC302-011.

UK Civil Aviation Authority, UK CAA 2014. Licensing of aerodromes.

Wong, D.K.Y., Pitfield, D.E., Caves, R.E. & Appleyard, A.J. (2009a). The development of a more risk-sensitive and flexible airport safety area strategy: Part I. The development of an improved accident frequency model. *Safety Science* 47, pp. 903–912. doi:10.1016/j.ssci.2008.09.010.

Wong, D.K.Y., Pitfield, D.E., Caves, R.E. & Appleyard, A.J. (2009b). The development of a more risk-sensitive and flexible airport safety area strategy: Part II. Accident location analysis and airport risk assessment case studies. *Safety Science* 47, pp. 913–924. doi:10.1016/j.ssci.2008.09.011.

Transport Infrastructure and Systems – Dell'Acqua & Wegman (Eds)
© 2017 Taylor & Francis Group, London, ISBN 978-1-138-03009-1

A runway veer-off risk assessment based on frequency model: Part II. risk analysis

L. Moretti, G. Cantisani, P. Di Mascio & S. Nichele
Dipartimento di Ingegneria Civile, Edile e Ambientale. Sapienza, University of Rome, Rome, Italy

S. Caro
Departamento de Ingenieria Civil y Ambiental, Universidad de Los Andes, Bogota, Colombia

ABSTRACT: This two-part paper presents a comprehensive airport risk assessment methodology for aircraft runway veer-off accidents around the world. Veer-off is a lateral runway excursion, and it occurs when an aircraft gets out the runway during a movement. The most frequent causes for this type of accidents are inappropriate pilot performance and aircraft condition; while influential factors include transversal wind, contamination of the runway, and poor visibility. The consequences of these events involve both damage to the aircrafts and potential health effects on passengers and crew members. Statistical data collected by the authors were used as reference for conducting frequency and risk assessments of veer-offs. A successful probabilistic model of veer-off accidents was developed in the first part of the paper. This second part assesses the level of safety guaranteed by the Cleared and Graded Area (CGA), which is a part of the runway strip that enhances the deceleration of an aircraft in case of an excursion. This area has to be dimensioned according to the requirements provided by the International Civil Aviation Organization Standards (ICAO) and the regulations issued by the Italian Civil Aviation Authority. Therefore, a model of veer-off accidents was developed to quantify the risk of an event. The results from the risk analysis showed that there are some cases for which the Real Level of Safety (RLS) was higher than the proposed Target Level of Safety (TLS), meaning that the level of damage expected is higher than the proposed one. An application example of the model complying with the requirements of the ICAO Regulation for non-instrument Code A runways is also presented. Overall, the results obtained from this study show the proposed model contributes to the implementation of Safety Management Systems at airports, as stipulated by the ICAO.

1 INTRODUCTION

A high-quality safety management design of airport infrastructures needs an accurate methodology to evaluate the risk and prevent accidents (Attaccalite et al. 2012, Wong et al. 2009a,b, Kirkland et al. 2004, Di Mascio & Loprencipe in press).

The initial portion of this paper focused on the development of a statistical model to assess the frequency of aircraft veer-offs during landings and take-offs, including both instrument and non-instrument or non-precision movements. Data available since 1990 for commercial flights over 30 Mg (ton) showed that the average frequency of a veer-off accident was 1.44 in ten million movements.

Also, data collected by the authors suggested that veer-offs are more frequent to occur during a landing (probability d_l equal to 73%) than during a take-off (probability d_{to} equal to 27%), as confirmed by the Norwegian Civil Aviation Authority (2001).

Currently, the reference code of runway plays a decisive role in the definition of the extension of lateral runway surfaces (International Civil Aviation Organization, 2013). Available specifications vary according to the different designations of runway (i.e. instrument or non-instrument runways).

The methodology proposed in this study is a useful tool to evaluate the veer-off risk of a specific air-port (Moretti et al. in press) and, as such, it offers the airport management body the possibility of implementing appropriate measurements to achieve specific safety requirements.

2 METHODOLOGY

The probability that an aircraft travels beyond certain distance from the runway centreline was defined by combining the probability of veer-off events and the probability of the lateral location distribution after the accident. Statistics show that veer-offs are more frequent during non-instrument runway operations, although these movements are less than 10% of all operations. Indeed, the probability of a veer-off in a non-instrument operation, p_{non-I}, is 0.09, which results in a probability of precision operation, p_p, of 0.91.

In order to represent more effectively the actual frequencies of veer-off events, the authors divided the veer-off probability distribution into four different situations, taking into account landings and take-offs, both on instrument or non-instrument runways. Using the statistical analysis of the data presented in the first part of this article, the following distributions were defined (Equation 1, 2, 3 and 4):

$$f = 1.37 \cdot 10^{-7} \cdot e^{-0.0219x} \qquad (1)$$

$$f = 9.72 \cdot 10^{-7} \cdot e^{-0.0219x} \qquad (2)$$

$$f = 6.96 \cdot 10^{-8} \cdot e^{-0.0143x} \qquad (3)$$

$$f = 6.82 \cdot 10^{-9} \cdot e^{-0.0143x} \qquad (4)$$

where Eq. 1 is valid for landings on instrument runways, Eq. 2 is valid for landings on non-instrument runways, Eq. 3 is valid for take-offs from instrument runways, Eq. 4 is valid for take-offs from non-instrument runways, and f is the frequency of an aircraft running beyond a certain distance x measured from the runway (RWY) centreline.

Six cumulative frequencies were defined by calculating the expected average interval between two subsequent events, and by associating the consequences of an event with its frequency. Highly frequent events are associated with mechanical damages of the aircrafts, whereas unlikely events cause the aircraft to crash through the airfield perimeter fence, as observed in Table 1.

On the other hand, the consequence of a veer-off event with reference to the potential health effects on people is defined through the coefficient D. This quantity corresponds to the ratio between the number of fatalities and injuries and the total number of occupants involved in the veered off aircraft (Equation 5):

$$D = \frac{\sum_{event,j}(N_{fatalities} + N_{injuries})}{\sum_{event,j} N_{occupants}} \cdot 100 \qquad (5)$$

For each mechanical or physical available consequence, a statistical analysis was performed using the veer-off data presented in the first part of the paper. Table 2 shows the average frequencies for damage D obtained.

Substantial health damages, on the other hand, are related to the ramming through the airfield perimeter fence ($D = 63.1\%$), whereas minor damages were observed to occur when the landing gear system crashes or when the airplane crashes into an embankment aside the runway ($D = 5.5\%$); the average value of D is equal to 18%.

Then, according to the levels of D, five categories of severity were defined (Table 3).

Table 2. Health effects, D, for each consequence.

Consequence	Damage $D\%$
Crash against an embankment or against a drainage channel	5.5
Landing gear damage	5.6
TWY, RWY, apron crossings	18.8
Mechanical damages	20.7
Collision with obstacle(s)	22.7
Fire	36.6
Crossing of airfield perimeter fence	63.1

Table 1. Consequences based on the likelihood of occurrence of veer-off events.

	Probability	Observations	Consequences
A	Highly frequent	< 3 months	Mechanical damages
B	Frequent	> 3 months	Landing gear damage Collision with obstacle (s)
C	Occasional	> 2 years	Fire Crash against an embankment or against a drainage channel
D	Remote	> 4 years	RWY, taxiway (TWY) crossing
E	Improbable	> 10 years	Apron crossings
F	Extremely improbable	> 20 years	Crossing of airfield perimeter fence

Table 3. Categorization of severity.

Severity	Health effects on occupants		$D\%$	Consequence
	Injury	Fatality		
1 Negligible	No	No	0	No damage
2 Minor	Yes	No	$0 < D$ and $D < 15$	Landing gear damage; Crash against an obstacle
3 Major	Yes	No	$15 < D$ and $D < 25$	TWY, RWY, Apron crossings; Collision with obstacle (s)
4 Hazardous	Yes	Yes	$25 < D$ and $D < 50$	Fire
5 Catastrophic	Yes	Yes	$50 < D$ and $D < 100$	Crossing of the airfield perimeter fence; hull loss

3 RESULTS

Using the definition of the frequency interval and the categories of damage, it is possible to compute the level of risk as "the combination of the harm probability category with the consequence of a specific hazard being realized" (Airport Cooperative Research Program, 2014), which is described by the elements of the risk assessment matrix (Equation 6):

$$R_{ji} = p_j \cdot D_i \tag{6}$$

Thus, for each kind of potential consequence, the cumulative frequency of a particular veer-off is the product of the probability of veer-off events and the j damage of those veer-off events, as expressed by Equation 7:

$$p_j = p_{veer-off} \cdot p_{event,j} = 1.44 \cdot 10^{-7} \cdot p_{event,j} \tag{7}$$

where $p_{event,j}$ derives from the data presented in the first part of this paper.

Figure 1 shows the computation of the current risk levels of aircraft damage with respect to each result and category. Each curve is a series of points of equal risk and it represents, for each damage category, its level of risk.

For each consequence and corresponding damage level listed in Table 2, the authors calculated the actual level of risk. The results are presented in Table 4.

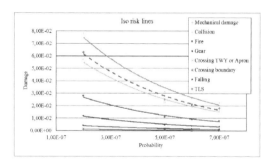

Figure 1. Iso-risk lines for each consequence.

Table 4. Level of risk for each consequence.

Consequence	Actual level of risk
Crossing of airfield perimeter fence	$0.273 \cdot 10^{-9}$
Crash against an embankment or against a drainage channel	$0.840 \cdot 10^{-9}$
TWY, RWY, apron crossings	$0.948 \cdot 10^{-9}$
Landing gear damage	$2.42 \cdot 10^{-9}$
Fire	$5.59 \cdot 10^{-9}$
Collision with obstacle(s)	$11.5 \cdot 10^{-9}$
Mechanical damages	$15.6 \cdot 10^{-9}$

The iso-risk lines for each consequence, listed in Table 4, are represented in Figure 1. These results highlight the need for actions that could help reducing the risk in certain conditions.

Combining the values in Table 4, it is possible to calculate the Real Level of Safety (RLS). This quantity is obtained by combining the mean current veer-off probability ($1.44 \cdot 10^{-7}$) and the weighted sum of recorded damages for each consequence, according to the following expression (Equation 8):

$$RLS = p_{veer-off} \cdot D_{average} = 1.44 \cdot 10^{-7} \cdot \frac{\sum D_j \cdot p_{event,j}}{\sum p_{event,j}} \tag{8}$$

For the purposes of this research, a Target Level of Safety (TLS) equal to $1.26 \cdot 10^{-8}$ was assumed. This value implies to have a veer-off with the mean current veer-off probability and the damage D equal to half of the actual level of 9%. This value of damage is reasonably achievable with the application of requirements for safety areas that are both width and the bearing capacity of the soil located in the CGA. These requirements guarantee a stopping performance of the CGA, as they have been tested and validated for various weather conditions (Australian Government Civil Aviation Safety Authority, 2003).

According to the statistical data collected by the authors, it was found that the RLS is higher than TLS, as showed in Figure 2.

According to Figure 2, some safety actions should be taken in airports in order to comply with the mini-mum desired TSL. As for some of the consequences (involving collision with obstacle(s) and mechanical damages), the level of risk exceeds the imposed threshold. In these cases, a proper re-design of the runway strip is necessary to close gaps and to accomplish the imposed requirements. Available actions involve, among others, dimensional and geotechnical features on CGA and on the strip.

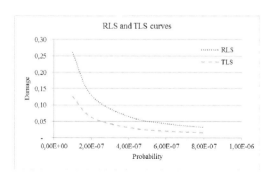

Figure 2. Comparison between RLS and TLS curves.

RLS and TLS should be compared not only considering the consequences of veer-offs, but also assessing these values for each runway code. Reported results have shown that the problem of veer-offs prevention has geometrical implications to be considered. For each type of runway code (RW_{code}) defined by the ICAO Annex 14 (International Civil Aviation Organization 2013), the risk of the main landing gear departing beyond the CGA (Ente Nazionale per l'Aviazione Civile 2014) (R_{CGA_GEAR}) was calculated using equations (1), (2), (3) and (4). The obtained results are listed in Table 5.

The results show that the RLSCGA_GEAR level of safety, which considers the geometrical configuration of the CGA, is typically lower than the pro-posed TLS, although for more than 25% of the runway runway codes the RLS is higher than TLS.

The ICAO regulation states that the CGA should be "(...) *constructed as to minimize hazards arising from differences in load bearing capacity to aeroplanes which the runway is intended to serve in the event of an aeroplane running off the runway*" (International Civil Aviation Organization 2013, 2006). Many national aeronautical authorities have interpreted this statement by defining a minimum value of a bearing capacity index of the soil composing the CGA, and a maximum value of the subsidence of the soil. For example, a minimum CBR (Californian Bearing Ratio) of 15% and a subsidence less than 15 cm are currently required in Italy (Ente Nazionale per l'Aviazione Civile 2008). If these values are not achieved, the Italian Authority states that a specific veer-off risk assessment is necessary (Ente Nazionale per l'Aviazione Civile 2008). However, no indication about the allowable risk level is provided.

Table 5. Veer off risk of the main landing gear departing beyond the CGA with D = 18%.

| RW_{code} | R_{CGA_GEAR} | | | |
	Landings on precision instrument runways	Landings on non-instrument runways	Take-offs from instrument runways	Take-offs from non-instrument runways
1 A	1.09E-08	9.58E-08	7.35E-09	8.31E-10
2 A	1.09E-08	7.70E-08	7.35E-09	7.20E-10
3 A	2.62E-09	3.58E-08	2.90E-09	4.37E-10
1B	1.10E-08	9.74E-08	7.42E-09	8.40E-10
2B	1.10E-08	7.83E-08	7.42E-09	7.28E-10
3B	2.66E-09	3.64E-08	2.93E-09	4.41E-10
1C	1.14E-08	1.01E-07	7.59E-09	8.58E-10
2C	1.14E-08	8.09E-08	7.59E-09	7.44E-10
3C	2.75E-09	3.76E-08	2.99E-09	4.51E-10
4C	2.75E-09	3.76E-08	2.99E-09	4.51E-10
3D	2.90E-09	3.97E-08	3.10E-09	4.67E-10
4D	2.90E-09	3.97E-08	3.10E-09	4.67E-10
4E	2.90E-09	3.97E-08	3.10E-09	4.67E-10

Table 6. Veer off risk of the main landing gear departing beyond the CGA with D = 9%.

| RW_{code} | R_{CGA_GEAR} | | | |
	Landings on precision instrument runways	Landings on non-instrument runways	Take-offs from instrument runways	Take-offs from non-instrument runways
1 A	9,00E-09	7,94E-08	6,08E-09	6,88E-10
2 A	9,00E-09	6,38E-08	6,08E-09	5,96E-10
3 A	2,17E-09	2,96E-08	2,40E-09	3,62E-10
1B	9,14E-09	8,07E-08	6,15E-09	6,96E-10
2B	9,14E-09	6,48E-08	6,15E-09	6,03E-10
3B	2,20E-09	3,01E-08	2,43E-09	3,65E-10
1C	9,45E-09	8,34E-08	6,28E-09	7,11E-10
2C	9,45E-09	6,70E-08	6,28E-09	6,16E-10
3C	2,28E-09	3,11E-08	2,48E-09	3,73E-10
4C	2,28E-09	3,11E-08	2,48E-09	3,73E-10
3D	2,40E-09	3,29E-08	2,57E-09	3,87E-10
4D	2,40E-09	3,29E-08	2,57E-09	3,87E-10
4E	2,40E-09	3,29E-08	2,57E-09	3,87E-10

Therefore, this study could be used to provide reference regarding allowable values of risk. For ex-ample, the current average risk level might be considered the maximum allowable risk, so that the values listed in Table 5 could be used as references.

The authors also calculated the level of risk R_{CGA_GEAR} with a damage D equal to half of the current damage ($D = 9\%$). Table 6 shows the results.

In this case, for some cases of RW_{code} the results of R_{CGA_GEAR} are higher than the selected TLS value. are higher than the selected TLS value. Also, as observed, the most critical results were obtained for landings on non-instrument runways. These results may depend on the geometrical characteristics of the CGA, mostly for 1 and 2 non-instrument runway categories, whereas the width of the CGA of instrument runways can guarantee the required TLS. The enlargement of the CGA should not be overlooked as a protection strategy.

4 CONCLUSIONS

This study has focused on the development of a risk-based methodology to mitigate the severity of runway veer-offs. It involved the analysis of selected events that occurred between 1953 and 2015 throughout the world. Collected data included frequency of the events and causes and consequences of these accidents. The proposed risk analysis considered commercial flights over 30 Mg (ton), and it demonstrated that the veer-off accident requires a rigorous and thorough evaluation of the current operative conditions to mitigate these type of accidents.

The statistical analysis of the available data showed that the current veer-off probability is equal to $1.44 \cdot 10^{-7}$ and that the damage (defined as the ratio between the number of fatalities and injuries that occur to the occupants of the veering off aircraft with respect to the total number of occupants) is equal to 18%. Consequently, the Real Level of Safety (RLS) was calculated to be equal to $2.6 \cdot 10^{-8}$.

Using the statistical analysis of the data presented in the first part of this article, the frequency curves of an aircraft running beyond a certain distance x measured from the runway centreline were computed, and the level of risk at different distance from the runway was determined. For each runway reference code, the level of risk was computed considering both the RLS and the Target Level of Safety, TLS. In the analysis, the value of TLS was assumed to be equal to half of the current damage (i.e. for the damage $D = 9\%$, the TLS was equal to $1.26 \cdot 10^{-8}$). The most critical results from these results were found for landings on non-instrumented runways, where the risk of the main landing gear departing beyond the CGA presented the higher values (R_{CGA_GEAR} equal to $1.01 \cdot 10^{-7}$ during landings on non-instrument 1C runways).

Many national aeronautical authorities demand a risk assessment when the minimum value of the bearing capacity of the soil composing the CGA is too low, but generally there is not any indication regarding allowable levels of risk. Within this context, this study might be used to provide a reference for allowable risk values. For example, current or target average risk values might be considered as the maximum allowable risk levels.

REFERENCES

Airport Cooperative Research Program (ACRP) 2011. ACRP Report 51: Risk Assessment Method to Support Modification of Airfield Separation Standard. Transport Research Board.

Attaccalite, L., Di Mascio, P., Loprencipe, G. & Pandolfi, C. 2012. Risk Assessment Around Airport. *Procedia: Social & Behavioral Sciences*, pp. 852–861, 53. doi:10.1016/j.sbspro.2012.09.934.

Australian Government Civil Aviation Safety Authority 2003. CASR 139 Manual of Standards.

Di Mascio, P. & Loprencipe, G. in press. Risk analysis in the surrounding areas of one-runway airports: a methodology to preliminary calculus of PSZs dimensions. *ARPN Journal of Engineering and Applied Sciences* (ISSN 1819-6608).

Ente Nazionale per l'Aviazione Civile (ENAC) 2008. Circolare 07/02/2008 Linee guida per l'adeguamento delle strip aeroportuali.

Ente Nazionale per l'Aviazione Civile (ENAC) 2014. Regolamento per la costruzione ed esercizio degli aeroporti.

International Civil Aviation Organization 2006. Aerodrome Design Manual – part I.

International Civil Aviation Organization 2013. Annex 14 – Volume I Aerodrome Design and Construction.

Kirkland, I.D.L., Caves, R.E. & Humphreys, I.M. 2004. An improved methodology for assessing risk in aircraft operations at airports, applied to runway overruns. *Safety Science* 42, pp. 891–905.

Moretti, L., Cantisani, G. & Caro, S. in press. Airport veer-off risk assessment: an Italian case study. *ARPN Journal of Engineering and Applied Sciences*. (1819–6608).

Norwegian Civil Aviation Authority 2001. Final report on the risk analysis in support of aerodrome design rules. Rules, AEAT/RAIR/RD02325/R/002 Issue 1, 2001.

Wong, D.K.Y., Pitfield, D.E., Caves, R.E. & Appleyard, A. J. 2009a). The development of a more risk-sensitive and flexible airport safety area strategy: Part I. The development of an improved accident frequency model. *Safety Science* 47, pp. 903–912. doi:10.1016/j.ssci.2008.09.010.

Wong, D.K.Y., Pitfield, D.E., Caves, R.E. & Appleyard, A. J. 2009b). The development of a more risk-sensitive and flexible airport safety area strategy: Part II. Accident location analysis and airport risk assessment case studies. *Safety Science* 47, pp. 913–924. doi:10.1016/j.ssci.2008.09.011.

Transport Infrastructure and Systems – Dell'Acqua & Wegman (Eds)
© 2017 Taylor & Francis Group, London, ISBN 978-1-138-03009-1

Risk factors that increased accident severity at US railroad crossings from 2005 to 2015

E. Dabbour

Abu Dhabi University, Abu Dhabi, United Arab Emirates

ABSTRACT: This study identifies and quantifies the effects of different risk factors that increase accident severity at railroad-highway grade crossings. The research is based on utilizing binary logit to analyze all vehicle-train accidents that occurred at railroad-highway grade crossings in the United States from 2005 to 2015. The study investigates the temporal stability of the identified risk factors throughout the analysis period to identify the most significant risk factors that are temporally stable. Age and gender of the driver were found to be the most-significant temporally-stable risk factors identified. It was also found that darkness and adverse weather conditions may reduce accident severity, but they were not temporally stable throughout the analysis period. Accidents related to young drivers were found to be less severe. The findings of this research have the potential to help decision makers develop policies and countermeasures that reduce the severity of injuries at railroad-highway grade crossings by focusing on risk factors that consistently exhibit significant effects on the severity of accidents.

1 INTRODUCTION

Accidents at railroad-highway grade crossings (or simply railroad crossings) are usually among the most severe accidents due to the large ratio between the weight of a typical train and that of a typical motor vehicle. That ratio is estimated to be approximately 4,000 to 1 (Hao & Daniel 2016; Yan et al. 2010). According to data obtained from the Federal Railway Administration (FRA 2016), there was a total of 25,556 reported accidents at railroad crossings in the United States during the period from 2005 to 20015. Those accidents included 9,727 serious accidents that resulted in 3,204 fatalities and 10,685 injuries.

Several research studies attempted to identify risk factors that increase the frequency of accidents at railroad crossings so that countermeasures may be implemented to reduce the number of those accidents (Saccomanno et al. 2007; Oh et al. 2006; Saccomanno et al. 2004; Austin and Carson 2002). Other studies focused on analyzing risk factors that increase accident severity at railroad crossings, given that an accident has already occurred. Hao et al. (2016) used ordered probit modeling to analyze all accidents that occurred at railroad crossing in USA during the period from 2002 to 2011 where they found that injury severity increases during AM and PM peak hours as well as during PM off-peak period. Hao and Daniel (2013) used ordered probit modeling to determine the significant factors influencing the severity

of drivers' injuries at railroad crossings in USA where they found that female and older drivers were more likely to be involved in severe accidents as compared to male and younger drivers, respectively. Eluru et al. (2012) developed a latent class model to identify risk factors that may increase the severity of driver injuries at highway–railway crossings by using accident data from 14,532 crossings in USA for the period from 1997 to 2006. The factors that were found to significantly influence driver's injury severity include driver's age, time of the accident, and weather conditions. Hu et al. (2010) used logit modeling to identify risk factors that increase the severity of accidents at railroad crossings using data from Taiwan for the period from 1995 to 1997. Variables such as daily number of trains, daily number of trucks, and the use of obstacle-detection devices were found to influence the severity of accidents. McCollister and Pflaum (2007) used binary logit to develop models that predict the probability of accidents, injuries, and fatalities at railroad crossings. The risk factors that were found to increase injury severity include train speed, number of trains, traffic volume, and accident history. The proportion of heavy vehicles in the traffic volume and the proximity of traffic control devices were also found to impact the severity of accidents.

It must be noted that most of the previous studies that investigated the factors influencing the severity of drivers' injuries at railroad crossings were based on compiling data from different years into one dataset that was used for the analysis. This approach disregards the fact that the factors identified in those studies might be temporally unstable so that they might have been significant during the period being analyzed and then their significance might change. The change in the significance over time might be due to the ongoing changes in vehicle and train technologies, drivers' attitudes, traffic volumes, road conditions, conditions of railroad crossings, and policy implementation. Understanding the temporal trends of the factors affecting injury severity would help researchers assessing the effectiveness of implementing different safety treatments so that they could identify whether any safety improvements are attributed to the specific treatment or simply attributed to the temporal instability of the factors being addressed. Therefore, the purpose of this research study is to investigate temporal stability of the factors affecting the severity of drivers' injuries resulting from accidents at railroad crossings. The study is based on using binary logit modeling to analyze all accidents that occurred at railroad crossings in USA for the period from 2005 to 2015. A separate model is developed for each analysis year, and the models are compared together to determine whether the effects of the identified factors are temporally stable throughout the analysis period that extends for 11 years from 2005 to 2015. The findings of this analysis have the potential to assist decision makers identifying the most significant risk factors that are temporally stable so that different resources may be allocated to reduce the impacts of those risk factors. This research study also provides more-recent quantitative measures that may be used to determine the feasibility of implementing different countermeasures in reducing the severity of accidents at railroad crossings.

2 DATA COLLECTION

The dataset used for the research is obtained from the Federal Railway Administration (FRA 2016) and it covers all accidents that occurred at railroad crossings in the United States for eleven years starting from 2005 until 2015. Incidents related to suicide, or attempted suicide, were excluded from the analysis since they are actually deliberate actions and not accidents. Using accident records that extend for eleven years provides the benefits of capturing year-to-year temporal changes over that period so that countermeasures may be developed to reduce the impacts of risk factors that are found to be temporally stable. During the period from 2005 to 2015, there was a total of 25,556 reported accidents at railroad crossings, from which 23,767 accidents were between trains and motor vehicles. The remaining 1,789 accidents were between trains and either pedestrians or non-motorized vehicles.

3 METHODOLOGY

The most common modeling methods used in investigating the factors influencing injury severity in traffic accidents include:

a. Binary logit modeling (e.g. Wang et al. 2016; Ye et al. 2016; Wong and Chung 2008; Chang and Yeh 2006);
b. Ordered logit, or probit, modeling (e.g. Hao et al. 2016; Hao and Daniel 2014; Haleem and Abdel-Aty 2010; Gray et al. 2008); and
c. Mixed logit modeling (e.g. Naik et al. 2016; Wu et al. 2016; Behnood & Mannering 2015; Kim et al. 2013).

In binary logit modeling, the response variable must be dichotomous (e.g., injury accidents vs. no-injury accidents). Ordered logit, and probit, modeling allows for ordinal response variables (e.g., different severity levels of accidents), but it is based on the assumption that the impacts of contributing factors on injury severities are consistent, or proportional, across all observation in all the ordered classes. This assumption may not reflect the heterogeneity nature found in the data being analyzed. Mixed logit models have the flexibility to account for individual heterogeneity of the impacts of different factors on injury severities by allowing parameters to follow a distribution and to randomly vary across observations. However, they may disregard the ordinal nature of the outcome variable.

Based on the above brief discussion, binary logit modeling was selected as the tool utilized in this research study to analyze accident data. Binary logit is a generalized linear model that predicts the probability of occurrence of an event by fitting data to a logit function in the form of (McCullagh and Nelder, 1989):

$$f(z) = e^z/(1 + e^z) \tag{1}$$

where z is the logit function, which is a measure of the total contribution of all the explanatory variables (risk factors) used in the model; and $f(z)$ is a dichotomous variable that is assumed to follow Bernoulli distribution. It represents the probability of an accident to be serious (i.e. resulted in either a fatality or an injury), given that the accident has already occurred. Based on that, the variable $f(z)$ takes the value of "1" if the accident is serious, or "0" if the accident is minor (i.e. resulted in property-damage only with no injuries). Selecting this binary structure for the outcome variable (i.e. modeling the accident as either severe or minor) has the potential to overcome the inconsistency in reporting accident severity. For example, an injury accident in the United States is defined as an accidents that resulted in an injury for a person (e.g. vehicle driver) that required medical treatment (FRA 2016). In Australia, according to the Australian Transport Safety Bureau (ATSB 2012), an injury accident is defined as an accident that resulted in an injury that required hospitalization for less than 30 days. Using binary logit modeling will overcome the inconsistency in the thresholds used to define injury accidents by simply classifying an accident to either severe (that resulted in death or any injury) or minor (that did not result in any injury).

The logit function has the following form:

$$z = \beta_0 + \beta_1 x_1 + \beta_2 x_2 + \beta_3 x_3 + \cdots + \beta_k x_k \quad (2)$$

where β_0 is the intercept and $(\beta_1, \beta_2, \beta_3, ..., \beta_k)$ are the regression coefficients of the explanatory variables $(x_1, x_2, x_3, ..., x_k)$, respectively. The explanatory variables (risk factors) that were investigated include:

d. Heavy vehicle (*HV*): takes a value of "1" if the vehicle that collided with the train was a heavy vehicle (i.e. a truck or a bus), and a value of "0" otherwise;
e. Freezing air temperature (*FR*): takes a value of "1" if the air temperature at the time of the accident was below the freezing point (0°C or 32°F), and a value of "0" otherwise;
f. Darkness (*DA*): takes a value of "1" if the crossing was not lighted at the time of the accident (by either daylight or illumination), and a value of "0" otherwise;
g. Adverse weather (*AW*): takes a value of "1" if the weather at the time of the accident was adverse (i.e. rain, snow, sleet, or fog), and a value of "0" otherwise;
h. Obstructed view (*OB*): takes a value of "1" if the view of the vehicle driver was obstructed (by a natural object, a built object, or another train on an adjacent track), and a value of "0" otherwise;
i. Female driver (*FE*): takes a value of "1" if the vehicle driver was female, and a value of "0" otherwise;
j. Younger driver (*YO*): takes a value of "1" if the age of the vehicle driver was below the 21 years, and a value of "0" otherwise; and
k. Older driver (*OL*): takes a value of "1" if the age of the vehicle driver was above 65 years, and a value of "0" otherwise.

The speed of the train (for accidents when a train hits a vehicle) and the speed of the vehicle (for accidents when a vehicle hits a train) were not investigated since they are obvious risk factors that have been investigated in an enormous number of previous research studies (e.g. Hao et al. 2016; Haleem and Gan 2015; Liu et al. 2015; Hao & Daniel 2014; Fan & Haile 2014; Russo & Savolainen 2013; Eluru et al. 2012; Miranda-Moreno et al. 2009).

A Wald test is used to test the statistical significance of each of the estimated coefficients ($\hat{\beta}_1$, $\hat{\beta}_2$, $\hat{\beta}_3$, ..., $\hat{\beta}_k$). The Wald statistic of asymptotic chi-square distribution is the squared value of the Z statistic and is computed as:

$$Wald = [\hat{\beta}_i / SE(\hat{\beta}_i)]^4 \quad (3)$$

where $\hat{\beta}_i$ is the ith estimated coefficient and SE ($\hat{\beta}_i$) is the standard error of that coefficient. In this study, the Odds Ratio (*OR*) is used to interpret the significance of different explanatory variables (risk factors) where an estimate of the odds ratio of a certain explanatory variable is exp ($\hat{\beta}_i$) while holding the other explanatory variables unchanged. Finally, the 95% confidence interval (*CI*) is also used in this study to describe the upper and lower limits of the odds ratio with 95% confidence level and is given by $|\hat{\beta}_i \pm Z_{0.95} \cdot SE(\hat{\beta}_i)|$.

4 RESULTS AND DISCUSSION

4.1 *Descriptive statistics*

The distribution of all accidents between trains and motor vehicles at railroad crossings, for the period from 2005 to 2015, is shown in Table 1. For the purpose of this study, accidents are classified as either minor (no injury) accidents or serious (injury or fatal) accidents. As shown in the table, there was a total of 23,767 accidents between trains and motorized vehicles during the analysis period. They include 15,462 minor accidents, 6,394 injury accidents, and 1,911 fatal accidents. The mean and standard deviation values of all categories are also shown in the table. It is noted that the percentage of serious accidents during that period, as compared to all accidents, is 34.94% with an average mortality rate of 0.1348 deaths per accident. When comparing this mortality rate with the overall mortality rates for all vehicular accidents at highway intersections in USA, as obtained from the National Highway Traffic Safety Administration (2016), it was found that the mortality rate for vehicle-train accidents is significantly higher than that for vehicle-vehicle accidents. This is due to the significantly large weight of a typical train as compared to the weight of a typical motor vehicle (Hao and Daniel 2016; Yan et al. 2010).

Table 1 also shows a significant decrease in the number of total accidents between the period from 2005 to 2008 and the period from 2009 onward. By applying statistical significance two-tail t-test for the difference between the two means, at 5% significance level, the *t*-stat is found to be 6.83 [*df* = 9 and *p* < 0.001]. The decrease in the number of serious accidents was also highly significant with

Table 1. Number (and percentages) of accidents at railroad crossings in USA from 2005 to 2015[a].

| Year | No Injuries | Serious Accidents | | | Total Accidents | Total Injuries[b] |
		Injury	Fatal	Total		
2005	1,966 (67.51%)	710 (24.39%)	236 (8.10%)	946 (32.49%)	2,912 (100%)	1,293
2006	1,892 (67.43%)	666 (23.73%)	248 (8.84%)	914 (32.57%)	2,806 (100%)	1,343
2007	1,755 (66.73%)	658 (25.02%)	217 (8.25%)	875 (33.27%)	2,630 (100%)	1,292
2008	1,483 (65.53%)	608 (26.87%)	172 (7.60%)	780 (34.47%)	2,263 (100%)	1,148
2009	1,184 (66.37%)	454 (25.45%)	146 (8.18%)	600 (33.63%)	1,784 (100%)	880
2010	1,164 (62.18%)	559 (29.86%)	149 (7.96%)	708 (37.82%)	1,872 (100%)	991
2011	1,181 (62.79%)	549 (29.18%)	151 (8.03%)	700 (37.21%)	1,881 (100%)	1,166
2012	1,119 (61.99%)	534 (29.59%)	152 (8.42%)	686 (38.01%)	1,805 (100%)	1,095
2013	1,172 (61.91%)	573 (30.27%)	148 (7.82%)	721 (38.09%)	1,893 (100%)	1,075
2014	1,362 (65.89%)	556 (26.90%)	149 (7.21%)	705 (34.11%)	2,067 (100%)	975
2015	1,184 (63.86%)	527 (28.43%)	143 (7.71%)	670 (36.14%)	1,854 (100%)	1,106
Total[c]	15,462	6,394	1,911	8,305	23,767	12,364
Mean	1,406	581	174	755	2,161	1,124
SD[d]	305.16	70.05	37.97	105.18	406.32	138.20

[a] For accidents between trains and motor vehicles after excluding suicide/attempted suicide.
[b] Including fatalities.
[c] For 11 years (2005 to 2015).
[d] Standard deviation for the population.

t-stat 5.84 [df = 9 and $p < 0.001$]. Furthermore, the decrease in the total number of injuries and fatalities was significant with t-stat value of 3.91 [df = 9 and $p = 0.004$].

4.2 Binary logit models

The binary logit coefficients (and standard errors) of the risk factors that were investigated are shown in Table 2. The corresponding odds ratios, along with their 95% confidence intervals, are shown in Table 3. In both tables, risk factors that are statistically significant (with $p < 0.05$) are shown in bold. As shown in the two tables, the most significant risk factor that is temporally stable throughout the entire analysis period is when the driver is over the age of 65 years (OL). This finding may be explained by the possible deterioration in health conditions of older drivers, which make them more susceptible to injuries. Furthermore, older driver may also have slower reaction times and therefore may be slower than other drivers in reacting to emergency situations by taking the necessary actions needed to reduce the severity of an accident, such as steering, accelerating, or braking. This finding is consistent with the findings of other research studies that found that older drivers are more likely to be inured in accidents at railroad crossings (Zhao & Khattak 2015; Haleem and Gan 2015; Hao and Daniel 2013; Eluru et al. 2012). This finding is also consistent with other research studies where older drivers were found to be more likely to be injured in vehicle-vehicle

highway accidents (Teftt, 2008; Williams and Shabanova, 2003).

Another significant risk factor is associated with female drivers (FE). This risk factor is significant in 9 years out of the 11-year analysis period. More important, it has consistently been significant since 2011 onward. This finding is also consistent with other previous research studies (Haleem and Gan 2015; Zhao & Khattak 2015; Hao and Daniel 2013; Eluru et al. 2012). This finding may be attributed to one of the following two explanations:

a. The differences between male and female drivers in terms of body size and structure; or
b. Possible cognitive differences between male and female drivers in taking last-second corrective measures to reduce the severity of an accident (given that the accident is unavoidable).

To test the two hypotheses shown above, a binary logit analysis was conducted on a sample data of all drivers involved in single-vehicle collisions in North Carolina between 2007 and 2013, where the outcome was whether the collision led to rollover (Dabbour 2012). Based on that analysis, it was found that driver gender was not a significant factor with p-value found to be 0.127. Since avoiding rollover in a single-vehicle collision usually requires last-second corrective measures, it can be concluded that female drivers do not lack those last-second corrective measures; and therefore, the increased probability for female drivers to be injured in an accident at railroad crossing may

Table 2. Binary logit coefficients (and standard errors) [a].

Year	HV	FR	DA	AW	OB	FE	YO	OL
2005	0.078	**−0.281**	**−0.275**	**−0.539**	−0.002	0.180	**−0.412**	**0.429**
	(0.079)	**(0.121)**	**(0.097)**	**(0.141)**	(0.161)	(0.094)	**(0.093)**	**(0.137)**
2006	−0.027	−0.112	**−0.394**	−0.253	−0.125	**0.333**	**−0.512**	**0.777**
	(0.081)	(0.151)	**(0.099)**	(0.137)	(0.196)	**(0.096)**	**(0.098)**	**(0.135)**
2007	0.007	−0.201	**−0.279**	−0.265	−0.040	**0.386**	**−0.619**	**0.670**
	(0.083)	(0.124)	**(0.103)**	(0.140)	(0.181)	**(0.095)**	**(0.104)**	**(0.140)**
2008	−0.082	−0.249	**−0.226**	−0.083	−0.342	**0.305**	**−0.628**	**0.537**
	(0.089)	(0.129)	**(0.106)**	(0.138)	(0.193)	**(0.103)**	**(0.112)**	**(0.148)**
2009	0.167	−0.048	**−0.393**	−0.131	−0.408	**0.268**	**−0.891**	**0.416**
	(0.100)	(0.138)	**(0.126)**	(0.159)	(0.222)	**(0.115)**	**(0.133)**	**(0.161)**
2010	**0.201**	−0.191	**−0.632**	**−0.567**	0.097	0.175	**−0.711**	**0.371**
	(0.097)	(0.133)	**(0.120)**	**(0.158)**	(0.209)	(0.111)	**(0.127)**	**(0.147)**
2011	−0.034	−0.180	**−0.387**	**−0.677**	−0.039	**0.345**	**−0.970**	**0.892**
	(0.095)	(0.137)	**(0.114)**	**(0.166)**	(0.204)	**(0.111)**	**(0.129)**	**(0.154)**
2012	0.037	−0.154	**−0.369**	**−0.440**	**−0.772**	**0.471**	**−0.704**	**0.444**
	(0.097)	(0.182)	**(0.112)**	**(0.179)**	**(0.266)**	**(0.111)**	**(0.131)**	**(0.154)**
2013	−0.091	−0.132	−0.173	−0.100	−0.067	**0.317**	**−1.011**	**0.410**
	(0.095)	(0.133)	(0.104)	(0.149)	(0.202)	**(0.107)**	**(0.138)**	**(0.147)**
2014	−0.072	0.011	**−0.231**	**−0.284**	−0.114	**0.412**	**−0.929**	**0.489**
	(0.093)	(0.121)	**(0.102)**	**(0.142)**	(0.197)	**(0.104)**	**(0.133)**	**(0.142)**
2015	−0.102	−0.243	−0.018	**−0.452**	**−0.462**	**0.285**	**−0.857**	**0.562**
	(0.097)	(0.148)	(0.106)	**(0.157)**	**(0.230)**	**(0.111)**	**(0.130)**	**(0.147)**

[a] Numbers in bold indicate significant risk factors.

Table 3. Odds ratios (with 95% lower and upper Confidence Intervals) [a].

Year	HV	FR	DA	AW	OB	FE	YO	OL
2005	1.08	**0.76**	**0.76**	**0.58**	1.00	1.20	**0.66**	**1.54**
	(0.93–1.26)	**(0.60–0.96)**	**(0.63–0.92)**	**(0.44–0.77)**	(0.73–1.37)	(0.99–1.44)	**(0.55–0.79)**	**(1.17–2.01)**
2006	0.97	0.89	**0.67**	0.78	0.88	**1.39**	**0.60**	**2.18**
	(0.83–1.14)	(0.67–1.20)	**(0.56–0.82)**	(0.59–1.02)	(0.60–1.29)	**(1.16–1.68)**	**(0.49–0.73)**	**(1.67–2.83)**
2007	1.01	0.82	**0.76**	0.77	0.96	**1.47**	**0.54**	**1.95**
	(0.85–1.19)	(0.64–1.04)	**(0.62–0.93)**	(0.58–1.01)	(0.67–1.37)	**(1.22–1.77)**	**(0.44–0.66)**	**(1.48–2.57)**
2008	0.92	0.78	**0.80**	0.92	0.71	**1.36**	**0.53**	**1.71**
	(0.77–1.10)	(0.61–1.01)	**(0.65–0.98)**	(0.70–1.21)	(0.49–1.04)	**(1.11–1.66)**	**(0.43–0.66)**	**(1.28–2.29)**
2009	1.18	0.95	**0.68**	0.88	0.67	**1.31**	**0.41**	**1.52**
	(0.97–1.44)	(0.73–1.25)	**(0.53–0.86)**	(0.64–1.20)	(0.43–1.03)	**(1.04–1.64)**	**(0.32–0.53)**	**(1.11–2.08)**
2010	**1.22**	0.83	**0.53**	**0.57**	1.10	1.19	**0.49**	**1.45**
	(1.01–1.48)	(0.64–1.07)	**(0.42–0.67)**	**(0.42–0.77)**	(0.73–1.66)	(0.96–1.48)	**(0.38–0.63)**	**(1.09–1.93)**
2011	0.97	0.84	**0.68**	**0.51**	0.96	**1.41**	**0.38**	**2.44**
	(0.80–1.17)	(0.64–1.09)	**(0.54–0.85)**	**(0.37–0.70)**	(0.64–1.44)	**(1.14–1.75)**	**(0.29–0.49)**	**(1.80–3.30)**
2012	1.04	0.86	**0.69**	**0.64**	**0.46**	**1.60**	**0.49**	**1.56**
	(0.86–1.26)	(0.60–1.23)	**(0.56–0.86)**	**(0.45–0.91)**	**(0.27–0.78)**	**(1.29–1.99)**	**(0.38–0.64)**	**(1.15–2.11)**
2013	0.91	0.88	0.84	0.91	0.94	**1.37**	**0.36**	**1.51**
	(0.76–1.10)	(0.68–1.14)	(0.69–1.03)	(0.68–1.21)	(0.63–1.39)	**1.11–1.69**	**(0.28–0.48)**	**(1.13–2.01)**
2014	0.93	1.01	**0.79**	**0.75**	0.89	**1.51**	**0.39**	**1.63**
	(0.78–1.12)	(0.80–1.28)	**(0.65–0.97)**	**(0.57–0.99)**	(0.61–1.31)	**(1.23–1.85)**	**(0.30–0.51)**	**(1.24–2.15)**
2015	0.90	0.78	0.98	**0.64**	**0.63**	**1.33**	**0.42**	**1.75**
	(0.75–1.09)	(0.59–1.05)	(0.80–1.21)	**(0.47–0.87)**	**(0.40–0.99)**	**(1.07–1.65)**	**(0.33–0.55)**	**(1.31-2.34)**

[a] Numbers in bold indicate significant risk factors.

be mainly attributed to the differences between male and female drivers in terms of body size and structure. This explanation is consistent with the explanation provided in another research study (Eluru et al. 2012) where it was concluded that the higher physiological strength of a male driver (as compared to a female driver) might result in a less severe injury for male drivers.

Darkness conditions at railroad crossings (*DA*) were found to reduce the severity of collisions at railroad crossings in 9 years out of the 11-year analysis period. However, the effect was not temporally stable in the most recent years since it was not significant in years 2013 and 2015. A possible explanation for this finding is that drivers are usually more alert in darkness conditions and therefore they may take immediate evasive measures to reduce the severity of an accident with a train given that such an accident was unavoidable. This finding is consistent with the findings of other research studies (Yau 2004; Krull et al. 2000) where it was found that darkness conditions may reduce the severity of traffic accidents at highway intersections. However, previous research studies found that darkness conditions may contribute to increased severity of accidents at railroad crossings (Hao et al. 2016; Zhao & Khattak 2015; Hao & Daniel 2014). Other research studies found that darkness conditions may also increase the severity of vehicle-vehicle accidents at highway intersections (Kim et al. 2013; Zhang et al. 2011; Abdel-Aty 2003). In a research study by Plainis and Murray (2002) it was suggested that the reaction time for drivers generally increases when driving at low luminance levels; and therefore those drivers would have less time to take any evasive measures to reduce the severity of accidents. Further research may be needed to clarify those conflicting findings regarding the effect of darkness conditions on the severity of accidents.

Adverse weather conditions (*AW*) were found to be a significant factor in reducing the severity of collisions at railroad crossings in only 6 years out of the 11-year analysis period; and therefore this factor may be considered as temporally unstable. However, several research studies found that adverse weather conditions may reduce the severity of accidents, either at railroad crossings or at highway intersections (Behnood and Mannering, 2015; Haleem and Gan 2015, Fan & Haile 2014; Behnood et al., 2014; Krull et al., 2000). A possible explanation is that drivers are usually more alert during adverse weather conditions and therefore they would travel at lower speeds and would also be ready to take immediate evasive measures to reduce the severity of an accident with a train given that such an accident was unavoidable. However, few other studies suggested that adverse weather conditions may actually increase the severity of accidents at railroad crossings (Hao et al. 2016; Hao and Daniel 2014). Further research may be needed to clarify those conflicting findings.

An interesting finding is that young drivers (*YO*) were found to be consistently less likely to be injured in vehicle-train accidents at railroad crossings. This factor is temporally stable throughout the analysis period. This finding is consistent with the findings of previous research studies that analyzed accident severity at either railroad crossings or highway intersections (Haleem and Gan 2015; Eluru et al. 2012; Wong & Chung 2008). No temporally-stable significance was found for the effects of freezing temperature (*FE*), heavy vehicles (*HV*), and obstructed view of the driver (*OB*). Therefore, it may be concluded that those factors have no temporally-stable correlation with the severity of accidents at railroad crossings.

5 CONCLUSIONS

In this paper, binary logit modeling was used to investigate the temporal stability of the factors that increase accident severity at railroad-highway grade crossings (or simply railroad crossings). This objective was achieved by analyzing all accidents that occurred in USA at railroad crossings during the period from 2005 to 2015. It was found that the most temporally-stable factor was the old age of vehicle driver (if the age of the driver is above 65 years). This may be explained by the possible deterioration in health conditions, which make older drivers more susceptible to injuries. Furthermore, older drivers may also have slower reaction times and therefore may be slower than other drivers in reacting to emergency situations by taking the necessary actions needed to reduce the severity of an accident, such as steering, accelerating, or braking.

Female drivers were also found to be more likely to be injured in accidents at railroad crossings, which is mainly attributed to their smaller body size and more fragile body structure.

Darkness and adverse weather conditions were both found to reduce the probability of being injured in an accident at a railroad crossing, which may be explained by the extra attention paid by drivers under those conditions so that drivers may take the immediate actions needed to reduce the severity of an accident at a railroad crossing. However, both factors were not temporally stable since darkness was significant in only 9 years out of the 11-year study period, and adverse weather was significant in only 6 years out of the 11-year study period. Furthermore, the findings of previous research studies regarding these conditions are mixed since some research studies confirm the findings provided in this research study while other previous research studies suggest that darkness and adverse weather conditions actually increase accident severity at railroad crossings. Further research may be needed to investigate the effects of these two factors on accident severity at railroad crossings.

It was also found that drivers below the age of 21 years are less likely to be injured in accidents at railroad crossings. Finally, it was found that freezing temperature, heavy vehicles, and obstructed view of the driver are not significant factors in increasing accident severity at railroad crossings.

It must be noted that train and vehicle speeds were not investigated since these two risk factors have been previously investigated in an enormous number of research studies and they were found to be obvious risk factors that significantly increase accident severity at railroad crossings.

Decision makers may use the findings of this research study to identify the more temporally-stable significant risk factors that increase the severity of accidents at railroad crossings so that they can allocate the resources needed to reduce the effects of those risk factors. Furthermore, the binary logit models developed in this paper provide more-recent quantitative measures that may be used to determine the feasibility of implementing different countermeasures in reducing the severity of accidents at railroad crossings.

REFERENCES

Abdel-Aty, M. (2003). Analysis of driver injury severity levels at multiple locations using ordered probit models. *Journal of Safety Research* 34: 597–603.

Austin, R.D., and J.L. Carson (2002). An Alternative Accident Prediction Model for Highway–Rail Interfaces. *Accident Analysis and Prevention* 34(1): 31–42.

Australia Transport Safety Bureau (2012). Australian Rail Safety Occurrence Data: 1 July 2002 to 30 June 2012. *Final Report RR-2012-010*. Australia Transport Safety Bureau, Canberra, Australia.

Behnood, A. & Manne ring, F. (2015). The temporal stability of factors affecting driver-injury severities in single-vehicle crashes: Some empirical evidence. *Analytic Methods in Accident Research* 8: 7–32.

Behnood, A.; Roshandeh, A.; and Mannering, F. (2014). Latent class analysis of the effects of age, gender, and alcohol consumption on driver-injury severities. *Analytic Methods in Accident Research* 3–4: 56–91.

Chang, H. & Yeh, T. (2006). Risk Factors to Driver Fatalities in Single-Vehicle Crashes: Comparisons between Non-Motorcycle Drivers and Motorcyclists. *Journal of Transportation Engineering* 13(3): 227–236.

Dabbour, E. (2012). Using Logistic Regression to Identify Risk Factors Causing Rollover Collisions. *International Journal for Traffic and Transport Engineering* 2(4): 372–379.

Eluru, N., M. Bagheri, L.F. Miranda-Moreno, and L. Fu (2012). A Latent Class Modeling Approach for Identifying Vehicle Driver Injury Severity Factors at Highway–Railway Crossings. *Accident Analysis and Prevention* 47: 119–127.

Fan, W. & Haile, E. (2014). Analysis of severity of vehicle crashes at highway-rail grade crossings: multinomial logit modeling. *Proceedings of 93rd Annual Meeting, Transportation Research Board*, Washington, D.C.

Federal Railway Administration (2016). Highway-rail accidents statistical data. Federal Railway Administration, Office of Safety Analysis. Available online from https://safetydata.fra.dot.gov/OfficeofSafety/publicsite/query/query.aspx (date accessed September 23, 2016).

Gray, R.; Quddus, M.A.; and Evans, A. (2008). Injury severity analysis of accidents involving young male drivers in Great Britain. *Journal of Safety Research* 39: 483–495.

Haleem, K. & Abdel-Aty, M. (2010). Examining traffic crash injury severity at unsignalized intersections. *Journal of Safety Research* 41: 347–357.

Haleem, K. & Gan, A. (2015). Contributing factors of crash injury severity at public highway-railroad grade crossings in the US. *Journal of Safety Research* 53: 23–29.

Hao, W. & Daniel, J. (2016). Driver injury severity related to inclement weather at highway–rail grade crossings in the United States. *Traffic Injury Prevention* 17(1): 31–38.

Hao, W. & Daniel, J. (2014). Motor vehicle driver injury severity study under various traffic control at highway-rail grade crossings in the United States. *Journal of Safety Research* 51: 41–48.

Hao, W., & Daniel, J. (2013). Severity of injuries to motor vehicle drivers at highway-rail grade crossings in the United States. *Transportation Research Record: Journal of the Transportation Research Board* 2384: 102–108.

Hao, W., Kamga, C., and Wan, D. (2016). The effect of time of day on driver's injury severity at highway-rail grade crossings in the United States. *Journal of Traffic and Transportation Engineering* 3(1): 37–50.

Hu, S.R., C.S. Li, and C.K. Lee (2010). Investigation of Key Factors for Accident Severity at Railroad Grade Crossings by Using a Logit Model. *Safety Science* 48(2): 186–194.

Kim, J., Ulfarsson, G., Kim, S., Shankar, V. (2013). Driver-injury severity in single-vehicle crashes in California: A mixed logit analysis of heterogeneity due to age and gender. *Accident Analysis and Prevention* 50: 1073–1081.

Krull, K., Khattak, A., Council, F. (2000). Injury Effects of Rollovers and Events Sequence in Single-Vehicle Crashes. *Proceedings of the 80th Annual Meeting, Transportation Research Board*, Washington, D.C.

Liu, J., Khattak, A., Richards, S., and Nambisan, S. (2015). What are the differences in driver injury outcomes at highway-rail grade crossings? Untangling the role of pre-crash behaviors. *Accident Analysis and Prevention* 85: 157–169.

Miranda-Moreno, L.F., L. Fu, S.V. Ukkusuri, and D. Lord (2009). How to Incorporate Accident Severity and Vehicle Occupancy into the Hot Spot Identification Process? *Transportation Research Record: Journal of the Transportation Research Board* 2102: 53–60.

McCollister, G.M., and C.C. Pflaum (2007). A Model to Predict the Probability of Highway Rail Crossing Accidents. *Journal of Rail and Rapid Transit* 221: 321–329.

McCullagh, P. & Nelder, J. A. (1989). *Generalized linear models, 2nd Edition*. Chapman and Hall, New York, NY.

Naik, B.; Tung, L.; Zhao, S.; and Khattak, A. (2016). Weather impacts on single-vehicle truck crash injury severity. *Journal of Safety Research* 58: 57–65.

National Highway Traffic Safety Administration (2016). Fatality Analysis Reporting System (FARS). Available online at http://www-fars.nhtsa.dot.gov (date accessed: September 15, 2016).

Oh, J.; Washington, S.; and Nam, D. (2006). Accident Prediction Model for Railway–Highway Interfaces. *Accident Analysis and Prevention* 38(2): 346–356.

Plainis, S. & Murray, I. (2002). Reaction times as an index of visual conspicuity when driving at night. *Ophthalmic and Physiological Optics* 22(5): 409–415.

Russo, B. & Savolainen, P. (2013). An examination of factors affecting the frequency and severity of crashes at rail-grade crossings. *Proceedings of 92nd Annual Meeting, Transportation Research Board*, Washington, D.C.

Saccomanno, F., Fu, L., and Miranda-Moreno, L. (2004). Risk-Based Model for Identifying Highway–Rail Grade Crossing Blackspots. *Transportation Research Record: Journal of the Transportation Research Board* 1862: 127–135.

Saccomanno, F., Park, P., and Fu, L. (2007). Estimating Countermeasure Effects for Reducing Collisions at Highway–Railway Grade Crossings. *Accident Analysis and Prevention* 39(2): 406–416.

Tefft, B (2008). Risks older drivers pose to themselves and to other road users. *Journal of Safety Research* 39(6): 577–582.

Wang, X., Liu, J., Khattak, A., Clarke, D. (2016). Non-crossing rail-trespassing crashes in the past decade: A spatial approach to analyzing injury severity. *Safety Science* 82: 44–55.

Williams, A. & Shabanova, V. (2003). Responsibility of drivers, by age and gender, for motor-vehicle crash deaths. *Journal of Safety Research* 34(5): 527–531.

Wong, J. & Chung, Y. (2008). Comparison of methodology approach to identify causal factors of accident severity. *Transportation research Record: Journal of the Transportation Research Board* 2083: 190–198.

Wu, Q.; Zhang, G.; Zhu, X.; Liu, X.; and Tarefder, R. (2016). Analysis of driver injury severity in single-vehicle crashes on rural and urban roadways. *Accident Analysis and Prevention* 94: 35–45.

Yan, X., Han, L., Richards, S., and Millegan, H. (2010). Train–vehicle crash risk comparison between before and after stop signs installed at highway–rail grade crossings. *Traffic Injury Prevention* 11(5): 535–542.

Yau, K. (2004). Risk factors affecting the severity of single vehicle traffic accidents in Hong Kong. *Accident Analysis and Prevention* 36: 333–340.

Ye, X., Poplin, G., Bose, D. Forbes, A. Hurwitz, S., Shaw, G., Crandall, J. (2016). Analysis of crash parameters and driver characteristics associated with lower limb injury. *Accident Analysis and Prevention* 83: 37–46.

Zhang, Y., Li, Z. B., Liu, P., & Zha, L. (2011). Exploring contributing factors to crash injury severity at freeway diverge areas using ordered probit model. *Procedia Engineering*, 21: 178–185.

Zhao, S. & Khattak, A. (2015). Motor vehicle drivers' injuries in train–motor vehicle crashes. *Accident Analysis and Prevention*, 74: 162–168.

Transport Infrastructure and Systems – Dell'Acqua & Wegman (Eds)
© 2017 Taylor & Francis Group, London, ISBN 978-1-138-03009-1

Development of a road asset management system in Kazakhstan

G. Bonin & S. Polizzotti
SPT Srl, Rome, Italy

G. Loprencipe
Sapienza Università di Roma, Rome, Italy

N. Folino & C. Oliviero Rossi
Università della Calabria, Rende (CS), Italy

B.B. Teltayev
KazDorNII (Kazakhstan Highway Research Institute), Almaty, Kazakhstan

ABSTRACT: The paper describes the development of a custom Road Asset Management System (RAMS), that will take care of about 23'500 km of roads that make the Kazakhstan main network of roads (Republican roads), that also includes 6 international corridors. The system has two main roles, (1) organizing the asset information (road cadaster or "road passport" according to the Kazakh standard) in a modern digital database and (2) managing the maintenance of the network, optimizing through economic analysis the budget allocation for maintenance works. The system development takes also care of organizing the data collection procedures for both roles, that will be done using automated devices installed on mobile laboratories. The system will also use data from other sources, such as the growing Intelligent Transport System (ITS) equipment (mainly weather stations, cameras, Weigh In Motion (WIM) devices and traffic counters for the purposes of this system). The system is organized as a web based service and it is accessible through any internet connected device, offering the operators the possibility to browse the database or update it in any place with an internet connection available. One of the key element of the system is its ability to make analysis and forecasts: the system is developed to measure periodically condition data across all the network, to have a clear understanding and control on the status of the roads. This module uses Highway Development and Management Model (HDM-4) to make pavement maintenance analysis and optimization of resources. The system will start its operation with the first complete data collection, that will be calibrated over the years by the repetition of condition analysis, allowing to improve reliability and quality of analysis forecasts. The system will also serve for other analysis, such as the control of Asset Value, analysis on the effect of new road projects over the network.

1 INTRODUCTION

Road networks may be still considered, as a whole, one of the more expensive assets for each Country to develop and maintain; nevertheless, there is usually not enough attention to the management and optimization of the expenditures related to it.

The reason for this comes from multiple factors, including the higher attention to the development of the road networks (opposed to rational management and maintenance) and to the overall complexity of the technical and economic evaluation.

In the past 40 years, several attempts to rationalize the management of such complex systems (asset management) has been started, but only in the last 20 years, with the widespread availability of cheap and powerful computing tools allowed to efficiently manage the problem.

2 ASSET MANAGEMENT SYSTEMS

Roads assets include a wide range of items, whose most valuable is usually the road pavement, followed by the artificial structures and tunnels (depending on terrain characteristics): these assets are managed through dedicated systems named, according to the core item Pavement Management Systems (PMS), Bridge Management Systems (BMS) and Tunnel Management Systems (TMS).

Modern road network systems are getting more and more complex, in particular higher categories highways that nowadays include many Intelligent Transport System (ITS) systems, as well as road furniture (e.g. roadside safety devices and vertical road signs) that require dedicated and continuous monitoring and maintenance. The management of such complex system require a new holistic approach, that should take into account all the parts of the road: these new systems go under the name of Global Management Systems (GMS) (Loprencipe et al., 2012, Bonin et al., 2007).

3 INFORMATION ON KAZAKHSTAN

Kazakhstan became an independent republic after the fall of Soviet Union, in 1991; over the last 25 years the Country progressed steadily reaching the upper-middle income status.

Kazakhstan has a wide territory (2'724'900 sq. km) and a very small density (6.5 people/sq.km with a population of less than 18 million people); this is mostly due to its location and climatic conditions.

Kazakhstan is located in the middle of Central Asia, just South of Siberia; the climate is continental but, due to its size, it may be further divided into 4 climatic zones. In most of its Central-North and North-Eastern regions it suffers from very cold winters, where the soil may be continuously frozen for more than 4 months.

3.1 Road network

Kazakhstan, due to its position in the Central Asia and to its mostly flat land, it is a natural location for international routes, such as the corridors that connects South (Kyrgyzstan, Uzbekistan and others through these) and North (Russian Federation) and mostly East (China) and West (towards Europe through Russian Federation).

For this reason, the World Bank funded a large project on the "Western Europe-Western China international transit corridor (CAREC[1] 1B and 6B)", worth more than US$ 2 Billion (in the references). This large project started in 2009 and mainly targets the construction or improvement of these international corridors, but also includes a specific fund for the improvement of Public Administration and Management procedures, including the development of a Road Asset Management System for the Republican Roads of Kazakhstan.

---

[1]CAREC: Central Asia Regional Economic Program (http://www.carecprogram.org/), a partnership of central Asian Countries and multilateral institutions (funds and investment banks) to promote development through cooperation.

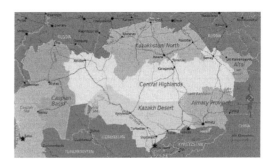

Figure 1. Kazakhstan main road network and neighbour Countries.

Table 1. Kazakhstan Republican Road network pavement type.

Pavement type	km	Network share, %
Asphalt concrete Warm Mixed Asphalt (WMA) or Hot Mixed Asphalt (HMA)	10'001	42.6%
Cold Mixed Asphalt (CMA)	11'491	48.9%
Cement concrete	97	0.4%
Gravel and cold asphalt (transition)	1'797	7.6%
Unpaved	107	0.5%
Total	23'493	100%

The Republican road network consists of about 23'500 km of roads and is the main network of roads in the Country; it includes (i) international corridors (ii) main national road network (iii) the toll road network.

Most (91.5%) of the Republican roads have an Asphalt concrete pavement, although about 8.1% are gravel or unpaved roads (see Table 1).

3.2 Review of the current situation—road organizations

Kazakhstan is still moving in its transition from the old centrally planned economy to the market economy: while many transformations have already been made, these are not yet fully consolidated and many changes in the road organizations, management procedures and standards are still in progress.

The Road Asset Management System is a modern tool that may serve to support the transition, not necessarily toward the market economy, but certainly to a more modern and efficient management.

The road sector is now controlled by a governmental authority, the Committee of Roads (CoR) that is under the Ministry of Investment and

Development (MID); the CoR is surrounded by some other publicly owned road organizations:

- Road manager
  National Company (NC) KazAutoZhol is a Joint Stock Company (JSC) that has been established recently (resolution 79/2013/CoR and Government Resolution of December 26, 2013 № 1409).
  The CoR realizes some of its main objectives through KazAutoZhol, such as (i) realization of a state policy on development of a network of highways and (ii) maintenance of roads and their structure for ensuring uninterrupted and safe traffic.
- Maintenance operator
  Republican State Enterprise (RSE) KazakhAvtoDor is the road organization that, through its local regional branches, conducts routine repair, maintenance and landscaping of the republican road network. The Committee for Roads acts as a client for such works through KazAutoZhol.
  Wider scope repairs are usually performed through bidding processes opened to private contractors, although (Oblast) branches of KazakhAvtoDor may participate, as a rule, for republican roads sectors that belong to their region.
- Research, technical and scientific advisory
  JSC KazDorNII is the Kazakh road research institute, whose stocks are owned by MID. The most general direction of the institute's scientific activity is enhancement of roads operation reliability, such as (i) investigating road safety conditions in order to improve normative base of roads designing and reconstruction and (ii) investigating opportunities to use traditional materials as well as local and new materials, including composite and industrial waste materials for construction of roads.
  KazDorNII has also a full set of testing equipment for road materials as well as special testing vehicles (Mobile Laboratories, ML) for the data collection of roads.

3.3 Review of the current situation—road standards and practice

The current Kazakh standards are developed on the basis of the Soviet Union standards, continuously updated and coordinated with the other Countries of the Commonwealth of Independent States (CSI).

Regarding the management of road assets, 3 standards are mostly relevant to the process:

- PR RK 218–28–03 (currently updating) on Certification of roads (Road Passport), that includes technical details for the definition of the "rad

passport". The contents of this "road passport" is completely defined within the standard and represents the official inventory of road assets, which goes under the name of "road cadastre" in other Countries.
- PR RK 218–27–14 (currently updating) on Diagnostics of road assets, that includes the standards and procedures for the testing of roads assets, mainly road pavement and road structures (bridges and culverts).
- PR RK 218–19–01 on Spring and Autumns survey, that describes the procedures for the semi-annual road condition assessment and the procedures for the calculation of the synthetic Global Road Quality Score.

Other standards define specific tasks and topics, such as traffic measurements, material specifications and testing, quality control.

3.4 Current maintenance and repairs estimates

It is important to notice how the terminology slightly differs in Kazakhstan from the international common practice: this may lead to inconsistencies and misunderstandings in the analysis and reports when international consultants operate in the Country, however the terminology in use is rather clear and consistent if it is well exposed and clarified.

Under the term "maintenance", in the Kazakh practice, are included: winter maintenance, landscaping (mowing, trimming of vegetation), drainage cleaning and the current maintenance (other small maintenances excluding road repair).

The repairs on road pavement or structures are divided into:

- "current repairs", with smaller scope, including patching potholes, crack sealing, repair of edge breaks, etc. (performed by public maintenance company KazakhAvtoDor);
- "mid-term repairs", with intermediate scope, including any pavement overlay—that is not common practice in Kazakhstan—or recycle/reconstruction of top layer of pavements, under 50 mm (performed by public or private contractor);
- "major repairs" or "reconstruction", with the wider scope: the definition differs mainly on being the "reconstruction" a set of road pavement works that involve an upgrade of road category.

The current practice for road maintenance is driven by the "spring and autumn survey" standard, that is a road maintenance assessment that is performed twice a year, using the autumn survey results as a basis for the budget estimate for the maintenance and repairs.

There are, although, several shortcomings in the current practice:

1. Subjective assessment: the autumn and spring surveys on the road conditions are carried on by a commission that includes inspectors from road manager organization, road works/repair organization and Police; the reports are based on visual assessment and expert's judgement of the inspectors rather than quantitative, objective, measurements.
2. Cost estimate without prediction of pavement deterioration: he current method, regarding the road pavement maintenance, is essentially a cost estimate of the repair works, based on the previously detailed visual assessment results. Such cost estimate, (a) being the assessment made during the autumn and (b) without the use of a road pavement deterioration model, always results in out of date cost estimates, when they are applied: the repair works, after the discussion and approval process inside the Authority, usually start more than 18 months after the assessment.
3. Optimization (i): reports are further used to calculate, by a set of formulas that "weighs" the asset conditions to calculate a global quality index, used to assign a "score" to each road; such score is used to prioritize the budget and to represent the status to the Authorities (ministries) involved in the budget allocation for the next years. The synthetic single score including all the maintenance needs may result in unclear identification of the issues.
4. Optimization (ii): the current cost estimate process does not use any optimization strategy and this is made worse by the fact that the budget allocation are usually lower than the necessary need; the latter, that is likely in any Country, without a budget reallocation strategy through optimization, causes maintenance inefficiencies and may carry also to an overall lower care of the cost estimate process (there is a general lack of confidence/interest into making it accurate, since no one will take advantage of it).
5. New technologies: this lacking process of assessment/analysis/optimization/control makes it very difficult to take advantage of new technologies, either on the materials, machinery, management and execution of works.

4 DEVELOPMENT OF THE KAZAKH ROAD ASSET MANAGEMENT SYSTEM (RAMS)

Some earlier projects identified the need to support reorganization of the road sector in Kazakhstan with the development of a modern Road Asset Management System (RAMS).

The main goal of the project described in this paper is to address the shortcomings identified in the current process for road maintenance management in Kazakhstan.

This is largely supported by international organizations and funding agency, because it has been identified as a key element to provide an assessment on the real conditions of roads, to evaluate the best strategies to solve the issues and to optimize the available budget allocated for the purpose of improving road transport.

The RAMS main achievements will be:

- The creation of a digital inventory, through a structured database, of data regarding the Republican road network, including asset data, conditional data (mainly regarding road pavement), information regarding road maintenance expenditures and other data.
- The establishment of improved methodologies for the collection of data (road asset inventory and road pavement condition), through the use of advanced testing vehicles (mobile laboratories), whose availability in the Country, already sufficient to start the activities, will be further extended through a parallel project.
- The establishment of a new budget estimate and optimization procedure, through the use of HDM-4 analysis. This will also allow the possibility to run iterative budget analysis, to expand the sensibility on the optimal budget allocation by considering different scenarios.
- The new HDM-4 based analysis allows also to integrate the maintenance analysis with other projects, such as new roads or reconstruction: specific longer term analysis will allow to tune the budget allocated for maintenance with future new expansion of the network.

Following this, it is to remark that a complete Road Asset Management System is not only the software and equipment used to collect, store, analyse and deliver information and results, but it is, indeed, extended to the whole process of managing the Road Asset, including, to a higher level, the road sector organization and, to a more practical level, all the rules (laws, technical standards, operating methodologies) and the experience of all the people involved.

The completion of such complex system will take some years and will need adjustments and adaptation, as well as continuous support from the government, the stakeholders and the operators.

The transition towards the completion of the system and the reform of the supporting standards and maintenance approaches will be done in 3 stages:

1. First stage: foundation/development
   includes the development of the core system (database, software and Information

Technology (IT) architecture), the arrangement of equipment for data collection and definition of methodologies and operating procedures. During this phase it should be also detailed the strategic plan for the further implementation of the system, that should also include the support to Performance Based Maintenance Contracts.

2. The second stage: initialization

It is foreseen an initial minimum 3/5-year time frame for this, that will focus on completing the reference data collection, the adjustment of standards, methodologies and procedures for the system operations. This will be done through the feedback from the actual use of the system. During this stage the system management will be maintained within the Ministry, with the direct involvement of the Road Manager, to allow a more comfortable transition to the new procedures.

3. The third stage: system complete (fully operational)

This starts when (i) the system is ready enough to complete every year the cycle of data collection-analysis-budget forecast for all the Republican road network predicting reliably the road maintenance expenditures and (ii) the new performance standards and all the related procedures are set properly and all the stakeholders are ready to take full advantage of the new procedures.

The readiness in stage 3 represents the maturity, intended as the capacity to provide reliable forecasts, as well as allowing further improvements through research.

The improvements include the whole process, starting from the general rules and standards, the correct setup of the road sector management process, the data collection and analysis procedures, and also to take full advantage of any new technology.

4.1 *RAMS database*

The RAMS database is the core of the system, as it stores all the data in a structured digital database. This, itself, represents a great step forward for any administration that is still based on traditional practices, often relying on paper documents and reports.

This situation is also common, yet, in many Countries. The typical situation is the "incomplete transition" to modern, fully digital systems: although data are exchanged daily through the offices using email or by exchanging data files, the data are usually not structured or stored, in the final revision, in some easy to reuse and reanalyse format, such as any database format.

Too often the final elaboration of the data are stored in a PDF format or as a text documents

(such as Microsoft Word), making it very unpractical for any reuse.

Kazakhstan has a dedicated organization and some State protocols to enhance the use of digital technologies that goes under the name "Digital Kazakhstan 2020"; part of this program, that is now in an advanced status, is the transfer of all the Governmental Servers and Systems to a modern Cloud based infrastructure, that will assure the best level of security for data, maintenance standards for hardware and availability (in terms of fast communication channels and redundancy/failure tolerance).

The Kazakhstan RAMS, that is a natural candidate to became a Governmental system, also uses this cloud based architecture, with a dedicated MS SQL server for the DBMS (Data Base Management System). This assures the availability of data (accessible through dedicated Virtual Private Network (VPN) channels or through Internet, depending on the level of access) and a reliable and powerful platform for the analysis needs.

The database is structured according to the data needed, that is divided into 3 main categories:

1. Asset data (road passport data in the local standard).
2. Road condition data (mainly through the mobile testing vehicles)
3. Other data, including those transferred from other systems (such as traffic data, weather condition data, etc.) and data input by road organizations (operations data—road open/closed/roadworks, emergency data—road accidents/severe weather conditions, management data—e.g. actual expenditure data for roadworks and maintenance).

Data needed for the yearly analysis through HDM-4 (that comes from all the 3 categories) is a subset of all the available data fields (in particular regarding 1) asset data and 2) road condition data): the local technical standards foresees a wider group of data, that is also included in the RAMS database and may be useful for additional detailed analysis.

4.2 *RAMS workflow*

The general workflow of an Asset Management system, shown in Figure 2, has been used as a reference for the development of Kazakhstan RAMS; it shows clearly that there are several cycles to control the implementation and the efficiency of the system, that shall be carefully enforced and managed.

In addition to the items in the diagram, it is important to pursue the continuous improvement of the system, that includes:

Figure 2. General workflow of an Asset Management system (FHWA Asset Management Primer, 1999).

- calibration of the analysis tool (HDM-4), through the availability of feedback data from the implementation of the maintenance works and comparison with the previous predictions;
- refinement in data collection procedures, that may be enhanced from availability of new equipment and technologies, but it is mainly due to the increased experience of the operators, the data quality assurance procedures and the better understanding of the road network;
- critical analysis of the whole process performance is necessary, to identify shortcomings and inefficiencies; this may be done through an internal evaluation of the system performance, with the improvement of regulatory documents and standards and through an expert's external supervision and dedicated research projects.

4.3 Data collection

Although the system is flexible to store data for different needs, it is important to clarify that the main scope of this Project, the development of the Kazakh RAMS, is to assure the possibility to store information and execute analysis for the strategic budget allocation.

For this purpose, the Ministry of Investment and Development (MID) of the Republic of Kazakhstan (RoK) invested in the purchase of highly specialized mobile testing vehicles for its National Research Institute for Roads (KazDorNII) and for the OblZhollaboratory road organization.

KazDorNII has 2 mobile testing devices for this purpose:

- A fully equipped Multi Function Vehicle (MFV®) with Falling Weight Deflectometer (FWD), Ground Penetrating Radar (GPR),

IRI and rutting measurement systems, cameras, Global Positioning System (GPS) and fully automated Laser Cracking Measurement System (LCMS®), manufactured from Dynatest®.
- A vehicle with cameras and GPS, with specific software dedicated to asset recognition and positioning (Road Passport data collection), manufactured from NPO Region®.

KazDorNII usually serves as the reference scientific institute for the RoK, including the preparation of technical standard and advisory for technical and quality control issues. Their vehicles are currently used regularly for the quality control of the newly built roads. KazDorNII will likely be directly involved in the data collection of RAMS and as technical and scientific advisor for the RAMS management.

The OblZhollaboratories are the regional local units directly controlled by the MID, consisting in 14 Oblast (regional) units, plus 2 newly established for the main cities of Kazakhstan and their metropolitan region, Astana and Almaty.

OblZhollaboratories will be the main source of pavement condition data, due to the large availability of equipment and to its dislocation in the Country: the MID is completing the fleet of 16 mobile testing vehicles (one per each branch), including the refurbishment of some previously owned vehicles.

These vehicles are fully equipped with Falling Weight Deflectometer (FWD), Ground Penetrating Radar (GPR), International Roughness Index (IRI) and rutting measurement systems, cameras, GPS and a semi-automated pavement defect measurement system.

The data will be collected during the end of summer or in autumn, before the winter snow; it is foreseen that, given the availability of vehicles, the surveys will be repeated every year during the initial phase of RAMS operation. This will assure a repeated feedback from the roads and the improvement of the analysis prediction accuracy. When the system will be mature enough, in terms of road condition prediction accuracy, it will be possible to reduce the frequency of surveys.

Survey data, after initial processing with on board equipment and software, will be transferred (stored) to the RAMS database for further processing.

The data processing is performed by trained operators and, currently, is greatly affected by the recognition of asset data and from the analysis of the pavement defects, that are not fully automated. Additional statistic strategies (such as those described in Krawczyk B., Szydło A, 2013) are adopted to reduce the amount of surveyed area (subsample) to be processed; this approach is also confirmed by other pavement distress methods (such as the Pavement Condition Index (PCI) survey methodology described in the ASTM-D6433). This approach will be further improved through specific research projects, when data will be available, and improved based on the availability of new software to assist the image recognition process.

Before the data will be fully published in the RAMS, therefore made available to the users and for the analysis, it will be performed a quality assurance procedure.

Since the main data available through the use of the mobile testing vehicles are those regarding the pavement condition and considering that the input data for the analysis tool (HDM-4) also includes other data, it is necessary also to provide them.

HDM-4 does not allow a detailed analysis of assets like artificial structures (bridges, culverts, retaining walls, tunnels etc.), drainage and technological systems (such as Intelligent Transportation System (ITS) devices), so the conditional data regarding these asset categories is collected using other methods, mainly through specific surveys and through the analysis of the reports form the road manager (KazAvtoZhol) and the road maintenance road organization (KazakhAvtoDor). It is possible that the RAMS will be further expanded with additional modules dedicated to these asset categories (Bridge and Systems in particular).

The drainage conditions, from the engineering point of view, are particularly important in some areas, because the flat terrain and the thawing of winter snow and ice causes big problems, with persistent floodings and soil softening issues.

The other main categories that have to be taken into account are:

- climate data: this has to be assessed and monitored periodically, as it is not necessary to track the continuous temperature changes for the analysis, but the Country has severe climate conditions, with hot summers and extreme winters, resulting in high annual excursion (may result in much more than 100°C if measured on the road surface, close to 90°C as standard air temperature, (see KazHydroMet.kz) and daily excursions (typical around 20°C as air temperature).
- Traffic data and vehicle fleet composition: this is one key parameter for any pavement analysis; this is currently measured using traditional methods, but there is an increasing number of ITS devices that will allow continuous monitoring of traffic and axle loading data (trough Weigh In Motion (WIM) ITS devices): data from the continuous monitoring will allow better understanding of the specific loading conditions, that shall be performed as a separate, analysis before the input in HDM-4.

4.4 Analysis

The Performance modelling and Alternative evaluation shown in Figure 2 are performed in the Kazakh RAMS through HDM-4.

HDM-4 (Highway Development and Management Tools collectively referred to as HDM-4), is the last revision of the Highway Design and Maintenance Standards Model (HDM), developed by the World Bank's Transportation Department for the analysis, appraisal and optimization/prioritization of road management and investment alternatives. HDM-4 is currently distributed and updated by HDMGlobal, an international consortium of academic and consultancy.

The software includes a Pavement Management System (PMS) based on a Pavement deterioration model (engineering module) and an Economic model, which are used to perform an integrated analysis.

The pavement performance models used in HDM-4 are based on a deterministic approach (se details in Watanatada et al., 1987; ISOHDM, 1995a,b).

The input parameters include, traffic, climate conditions, vehicle fleet consistency, road geometry data, safety information (accident rates) and details on the road pavement, that is the main engineering item. The pavement is detailed through its structural and construction characteristics (layer's thicknesses), conditional data (roughness (in terms of IRI), cracks, potholes, edge breaks, etc.) and surface characteristics (friction and texture depth).

The analysis consists in a two integrated sub-analysis:

- an evaluation of the pavement performance, predicting its deterioration over time, with or without the application of maintenance strategies
- an economic analysis, performed through the evaluation of the Vehicle Operating Costs (VOC).

With this approach, it is possible to consider the effects on the users, rather than the sole agency cost and asset value.

The agency costs, essentially represented in the following Figure 3 as Road Work Costs (RWC) may be added to the User's Cost (UC, or VOC) to calculate the Total Road Transport Cost and select the best option.

This tool will greatly enhance the analysis capabilities: in the current maintenance evaluation process, the budget allocation decision is based on the Cost Estimates calculated using the data collected using by mean of the visual assessment method described before; the current process, also, is unable to predict the pavement condition over time, therefore the budget allocations are decided on the basis of conditional data that are usually more than

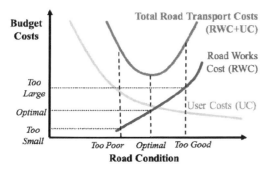

Figure 3. Total cost, Road User's Cost and Road Works Cost.

Table 2. Strategic, network, and project-level decisions (Zimmerman et al., 2011 and FHWA, 1999).

Decision level	Decision maker	Types of decisions/ Activities
Strategic	Legislator Commissioner Chief Engineer Council Member	Performance targets Funding allocations Pavement preservation strategy
Network (Tactical)	Asset Manager Pavement Management Engineer District Engineer	Project and treatment recommendations for a multi-year plan Funding needed to achieve performance targets Consequences of different investment strategies
Project (Operational)	Design Engineer Construction Engineer Materials Engineer Operations Engineer	Maintenance activities for current funding year Pavement rehabilitation thickness design Material type selection Life cycle costing

1 year old (including 2 winter seasons) when the maintenance and repair works actually start.

The HDM-4 allows the execution of project level analysis, road work programming under constrained budgets, and for strategic planning of long term network performance and expenditure needs. It is designed to be used as a decision support tool within a road management system.

This allows to serve Strategic and Network (tactical) decision levels, according to the definition described in the previous Table 2. This will allow a direct involvement in the analysis of the high level decision makers (usually Ministries) and mid-high level decision makers (typically the road management agency, KazAvtoZhol in RoK).

The analysis, after its completion, validation and acceptance, it is loaded into the RAMS database and made available to the authorized users.

4.5 *Reporting and information to stakeholders*

A very important function of the RAMS is the possibility to provide information to the stakeholders and to the road users, with specific tools and contents.

The system, through its data base core, may directly provide data, through reports (predefined or custom made), queries and through a GIS interface.

The data are provided to each user category on the basis of the user's needs and according to the RAMS manager authorization.

The higher level of authorization is granted to the Authorities and to the system administrators.

Reports are usually targeted to specific needs, including periodic reports for official purposes, evaluation of performance, verification.

Queries may be used to perform specific data analysis, including specific asset verification; the large amount of structured data will allow research projects, that are encouraged to further improve the system.

Direct browsing of data may be performed on the raw data (tables, data fields) or through the GIS mapping tool. This will also allow further development of the system, by allowing the professional road operators to browse the system on the road through mobile devices.

The GIS mapping tool is also the main medium of data publishing to the road users, the population. The data will be prepared through the main GIS tool within the RAMS, then the data will be published on an additional dedicated service (web server) on the basis of public mapping services (such as Google maps or Yandex maps). This will allow the integration of proprietary data (such as a synthetic 3 level pavement condition index or the information on the road service—road open/closed/with reduced lanes) with publicly available data through an interface familiar to users.

5 CONCLUSIONS

The development of a new Road Asset Management System is a complex task, that involves the knowledge of the state of the art for the equipment and procedures, as well as the understanding of the stakeholders' needs.

It is very important to understand the current procedure in use and to be able to make the system "useful" to all the involved players, to make its use really fruitful.

It is, indeed, really important to communicate the system capabilities and needs, to involve all the stakeholders to use it for their everyday duties and, therefore, continuously improving it.

The biggest advantages of the use of the RAMS in the RoK will be the availability of a structured, fast, always available source of data for the road assets and the possibility to produce more reliable, detailed and useful analysis, including the optimization of budget allocations, that are nowadays not possible.

REFERENCES

ASTM D 6433–11. 2011. Standard Practice for Roads and Parking Lots Pavement Condition Index Surveys. West Conshohocken, PA, USA.

Bonin, G., Cantisani, G., Loprencipe, G., Ranzo, A. (2007). Dynamic effects in concrete airport pavement joints [Effetti dinamici nei giunti delle pavimentazioni aeroportuali in calcestruzzo]. *Industria Italiana del Cemento*, 77 (834), pp. 590–607.

FHWA Asset Management Primer, 1999.

ISOHDM. 1995a. *International Study of Highway Development and Management Tools Modelling Road Deterioration and Maintenance Effects in HDM-4*. Asian Development Bank, Manila, Philippines.

ISOHDM. 1995b. *International Study of Highway Development and Management Tools Modelling Road Users Effects in HDM-4* (Asian Development Bank, Manila, Philippines).

Krawczyk, B., & Szydło, A. (2013). Identification of homogeneous pavement sections. *Roads and Bridges-Drogi i Mosty*, 12(3), 269–281.

ISO 690.

KazHydroMet.kz, historical records of temperatures in the main cities of Kazakhstan see http://www.kazhydromet.kz/en/inforeg/

Loprencipe, G., G. Cantisani & P. Di Mascio 2014. Global assessment method of road distresses, *Fourth International Symposium on Life-Cycle Civil Engineering IALCCE* 2014.

Pavement Design, Construction and Management: a digital Handbook, 2015. AASHTO. Washington, DC.

South West Roads: Western Europe-Western China international transit corridor (CAREC 1B and 6B) Project. World Bank loan no. 7681-KZ, retrieved on http://www.worldbank.org/projects/P099270/south-west-roads-western-europe-western-china-international-transit-corridor-carec-1b-6b?lang = en&tab = overview

Watanatada, T., Harral, C., Paterson, W., Dhareshwar, A., Bhandari, A. & Tsunokawa, K. 1987. The Highway Design and Maintenance Standards Model—Description of the HDM-III Model. The World Bank, Washington, DC. Vol. 1.

Zimmerman, K.A., Wolters A.S., K. Schattler, A. Rietgraf. 2011. Implementing pavement management systems for local agencies.

Transport Infrastructure and Systems – Dell'Acqua & Wegman (Eds)
© 2017 Taylor & Francis Group, London, ISBN 978-1-138-03009-1

Building Information Modeling (BIM): Prospects for the development of railway infrastructure industry

M. Leone, A. D'Andrea, G. Loprencipe & G. Malavasi
Dipartimento di Ingegneria Civile, Edile e Ambientale. Sapienza, University of Rome, Rome, Italy

L. Bernardini
Italferr S.p.A., Gruppo Ferrovie dello Stato Italiane (Italian State Railways Group), Rome, Italy

ABSTRACT: The last decade has seen a profound change of information representation tools with which the Infrastructure Manager must necessarily interact with the other parties involved, to explore all the essential decision-making processes for railway development and operation. In all phases of infrastructure life (project, construction, operation, adjustment, etc.), the digital representation can greatly facilitate the organizational model that always puts the information content of all railway components (geometry, size, location, materials, design and proper maintenance, stress state, etc.) at the center of any decision-making process. On account of its strong multidisciplinary nature, a railway infrastructure causes numerous interactions that need to be coordinated properly to meet the performance levels now required in any transportation system. In this sense, the development of BIM processes for punctual building has now reached a level of maturity able to satisfy also the recent guidelines of the European Standards (2014/24 / EU) adopted to promote, first of all, the transparency of public work contracts and design contests. However, as regards linear construction, the full application of these BIM processes with an adequate level of maturity still remains rather complex considering that the current processes representation in use has achieved an adequate level of development. In this case study, applications have been developed in BIM mode with the aim of highlighting the main potential of this innovative process including some improvements in order to promote the full implementation of the BIM technology for rail infrastructure.

1 INTRODUCTION

Currently railways, especially High-Speed trains, are considered one of the best nationwide means of transport to solve the problem of mobility.

However, building, maintaining and operating HSR lines is expensive; it involves a significant amount of sunk costs and may compromise the transport policy of a country.

The full lifecycle development of a railway infrastructure includes several separate main stages, each supervised by the technical expert in charge.

A railway construction project is made up of the following parts:

- Concept;
- Design/Planning;
- Construction;
- Maintenance;
- Adjustment/Implementation;
- Disposal (demolition or other use).

The planning and construction of railway lines involve a number of operating branches. For example: civil engineering, electrification, signaling,

telecommunication, etc. In order to avoid discordant dynamics and a contrasting technical project development it is fundamental to bring in a technical coordination of the various stages belonging to railway construction projects (Project Manager as regards project planning/design).

Furthermore, decisions as to how and to what end construction stages are handled, may have to be revised constantly. This may even concern already well-established information.

As a matter of fact, the transition from the construction stage to that of maintenance often turns out to be conflicting or unrelated. This is due to the use of two different responsible control bodies as well as two software computer programs for data processing with a consequential loss and discordance of information.

In addition, software undergoes a continuous and unforeseeable upgrading. This makes it almost impossible to maintain its application the same throughout the lifecycle of a railway network.

Nowadays the means of controlling the entire process is offered by BIM. The common definition of BIM is: "*the use of shared digital representation of a built object (including buildings, bridges, roads, process plants, etc.) to facilitate design, construction and operation processes and to form a reliable basis for decisions*" (ISO 29481-1:2016).

On these grounds the EU has introduced the norm 2014/24/EU art. 22 c.4 for the use of BIM when constructing public buildings and transport systems. For public work contracts and design contests, Member States may require the use of specific electronic tools, such as of building information and electronic modeling tools or similar.

So far the BIM has often been applied for built objects like buildings and bridges. Instead projects of infrastructural networks have found very little BIM application both in literature and construction.

Despite this, there are some case studies using the BIM mode for a railway line section with the aim of highlighting the main potential of this innovative process.

This was made possible by the cooperation between Italferr S.p.a. (Italian State Railways Group) and University of Rome with the developing of some master degree theses.

2 REVIEW

The intended function of Digital and Information technology, when employed by BIM, is to integrate all activities concerning projects and constructions of public works.

On account of this, scientific literature (Nurain, 2013) evidences three main categories:

- theoretical perspectives;
- integration;
- implementation.

Theoretical perspectives explore studies that have offered key theoretical perspectives on digital technology use.

What is more, researchers have explored the application of digital technology in the Architecture, Engineering and Construction (AEC) disciplines. Researchers described their behavior as regards their development and use (Taylor & Bernstein, 2009). This takes into account the four main elements of BIM:

- visualization;
- coordination;
- analysis;
- integration.

Instead, visualization, integration and automation are the three stages belonging to virtual planning/design and development/construction.

Finally, when proceeding with BIM, priority is given to automation, information and transformation.

The concept of *integration* reveals having two different aspects: the first one regards people and processes and the second one concerns the integration of interoperability systems.

The integration of some stages and disciplines is extremely useful from an early stage of project. This allows the virtual construction of the work in its definitive configuration and the possibility to see the integration of more disciplines.

Still the BIM process seems to be too complicated for many and initially adopted in a limited manner (Howard & Björk, 2008).

Nevertheless, literature is full of examples dealing with stage integration and the coexistence of various disciplines (Bonin et al., 2009).

The evolution from design 2D to 3D has improved the structure work and the conflict management of structural aspects and Mechanical, Electrical and Plumbing (MEP). Moreover, the use of 4D has brought together the modeling of the stages concerning the construction site like planning and scheduling, data analytics and management (time related information) and those regarding safety. Lifecycle integration represents a further important element dealt with by the BIM. This allows its modeling from the very beginning of the project. The emerging advantages can be summarized in the following three points (Aranda-Mena et al., 2009):

- the possibility of exchanging models with the technical project advisors (technical ability);
- the ability to project in 3D during the entire design process (operative ability);
- the ability to complete important projects of big dimensions more and more efficiently (business ability).

A further outstanding aspect of integration concerns the information exchange among the various technicians belonging to the planning and constructing process, thus forming real centralized 3D data bases. In spite of the many proposals presenting standard models and data service, interoperability (integration of systems) is still a problem of the BIM building process (Grilo & Jardim-Goncalves, 2010). There are exchange systems like the Industry Foundation Classes (IFC), the most advanced, that allow the exchange of information and models. There are other exchanges that adopt interoperability only in some sectors. A great effort has to be made to find a single standard exchange suitable for all operators and digital platforms. However, there are different types of *implementation*:

- referring to BIM technology with procedures already carried out;
- including the main advantages and problems dealt with;

- with the users' perceptions and expectations;
- with the suggested technical standards for implementations;
- with activated ecological engineering.

The approach (Arayici et al., 2011) of bottom-up implementation results being more efficient for the BIM implementation when linked with a pull-technology strategy. The latter reduces the resistance to change and the risk of substituting the already existing well performing processes. Positive aspects can include:

- the opportunity to produce design and documentation directly from the BIM model (technical result);
- 3D design during the entire process (operative result);
- minor risk of wrong interaction between structural aspects and MEP (business results).

Other studies show how the BIM process can upgrade all application disciplines and that users recognize improvement in all construction indicators (Singh et al., 2011). Expectations, instead, concentrate on the development of norms, contents, software, formation and certification as well as the use of the BIM in ecological projects (Young et al., 2008, Cantisani et al., 2011, Moretti et al., 2016). Major problems concern integrated digital and technological implementations, costs, more time and effort for 3D models, resources and proper coordination. The guidelines for implementation focus on the following themes:

- project making;
- development and respect for consolidated practices;
- BIM monitoring and support supply for users.

Other recommended rules address the matter of technical evolution, procedure and organization. The BIM can play an important role in eco-sustainable projects (Zhu, 2010) by identifying its best option beforehand. However, some positive examples aside, the ECO-BIM is still underestimated.

3 DESCRIPTION OF CASE STUDY

The project refers to a quadruple-track railway belonging to the Milano Rogoredo—Pavia network. Its construction consists in two 26 km new rail tracks alongside the two already existing rails tracks from Pieve Emanuele to Pavia.

In this case the BIM process has been applied for a 6 km long section starting from km 21+646.880 to km 27+241.060. By doing so, it has been possible to verify the full potential of BIM application development. This regarded the pros and cons of layout interference with other built objects. The studied track presents 24 interferences, (18 box culverts, two road underpasses, one electrical substation, one bridge crossings a channel, two box structures for crossings on foot and by bike). In fact, this section represents a railway network project for densely populated urban centers with a lot of interference. This project has used the support of the integration software owned by the Bentley family.

MICROSTATION: It is a 3D modeler used by most software of this platform. It allows modeling any kind of frame in a parametric mode providing the user with immediate programmable report, for example the estimate. In addition, this program allows assigning the kind of material and other specific properties (service, mechanical capacity, general hyperlink, etc.). It also allows the publication of the I-model to be used during the whole project by all final users and types of computer even without the Bentley software.

POWERCIVIL: It is modeler that makes planning linear infrastructure easier for civil linear project. It can be used to elaborate a Digital Terrain Model (DTM), horizontal and vertical road alignment and their coordination. It is also able to model box structures, pipes, retaining walls, etc. Furthermore, it can create libraries for the parametric and programmable cross sections to solve interference with DTM and other punctual built object.

POWER RAIL TRACK: It is a specific railway project modeler that allows to insert cants, transition curves and rail switches and to develop railway station layouts.

PROJECTWISE: It is the managing software. It coordinates all software and users of the project. It saves all project information and authorizes modification and visualization by technicians. Modifications are communicated to other users by alert notice. In addition, this software memorizes all project changes and the users who made them. Basically, it is the managing platform for user interoperability in design stage and it guarantees operation simultaneity.

This railway line project has been developed and elaborated in the following three steps:

1. The available cartographic data have been used for a DTM to insert the above-given railway section. Before starting, a preliminary operation of data reprocessing and cleaning is always needed to obtain a good definition of the terrain model (Fig. 1).
2. The realization of the railway solid in 3D. This point has three steps.
   a. The horizontal alignment design (Fig. 2). This was done by following in parallel the two already existing railways, imposing a distance

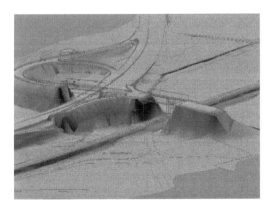

Figure 1.   Digital Terrain Model (DTM).

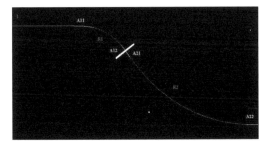

Figure 2.   Plane view of the horizontal alignment.

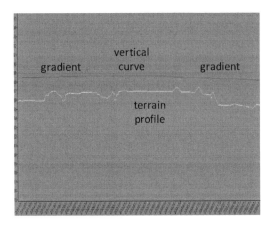

Figure 3.   Vertical view of the elevation profile.

the tracks, requiring it to maintain a constant vertical distance between the old and the new railway. The design of the vertical curves was based on the project of railway speed.

c. The setting-out of the cross section in case of either embankment or trenches and its closure rules with the terrain, as well as the extrusion of this section on the alignments. (Fig. 4).

The railway 3D solid (Fig. 5) needs minor and major built objects to solve interference with the terrain. The railroad equipment (ballast, rails, and sleepers) and other civil and technologic railway installations complete the infrastructure (electrical, traction, signaling, cable duct, etc.).

This part of the case study describes the building of two objects used to avoid interference with the territory. The first one is a box culvert structure to restore the natural water flow when interrupted by the embankment construction. The second one is a rail crossing solved by a new road underpass.

The box culvert structure design needed a preliminary horizontal placement (Fig. 6) and the choice of the reference height and the longitudinal slope (Fig. 7).

In this case, the extrusion of the section along the design axis considers the subtraction of the

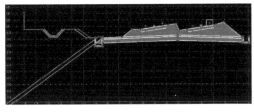

Figure 4.   Cross section of the project: definition of all rules.

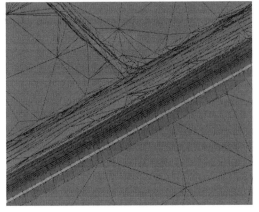

Figure 5.   The railway 3D solid.

between track centers of 7.60 m. The layout was then continued by using the new coordinates and referring only to the odd track (Italian practice). Its geometry consisted of a straight (L = 4500 m) and two opposite horizontal curves (R1 = 904 m; R2 = 1500 m).

b. The determination of the elevation profile and the insertion of the gradients and the vertical curves (Fig. 3) considered the quadrupling of

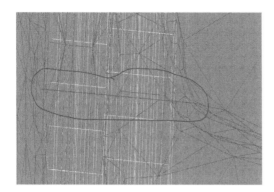

Figure 6.   Box culvert structure horizontal placement.

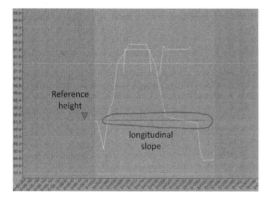

Figure 7.   Box culvert structure vertical placement.

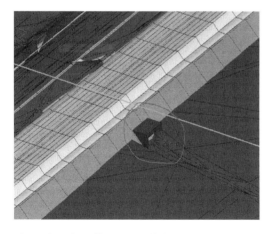

Figure 8.   The railway 3D solid in correspondence of the box culvert structure.

embankment volume to avoid having a volume redoubling (Fig. 8).

This process makes it possible to assign the characteristic properties to each material in order to draw up an estimate subsequently. This method enables the complete control of costs and timing during design phase as well as the workflow regarding the construction stages.

The modeling of the road underpass follows the same operating procedure used for other built objects. This means definition of road horizontal layout first and vertical profile after, followed by cross section design with which the solid of the built object has been obtained. Also in this case, the subtraction of the embankment volume concerning the overlapping part of the underpass is indispensable to avoid a wrong estimate and visualization problems.

The underpass design requires the definition of various cross sections that vary throughout the corridor (Fig. 9).

Fig. 10 shows some of the defined cross sections used for the underpass modeling.

The underpass has been completed with road pavings and road furniture (parapets, barriers and so on.) and draining pipes for rain water discharge (Fig. 11).

The hydraulic system has been placed in the fitting spaces of the cross section. Additionally, this 3D multi-disciplinary project works together with

Figure 9.   The 3D model of the underpass.

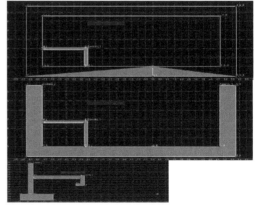

Figure 10.   The defined cross sections used for the underpass modeling.

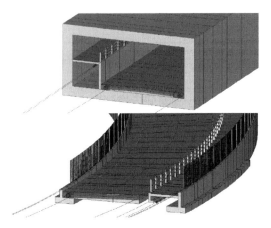

Figure 11. The 3D model of the underpass completed with road pavings and street furniture.

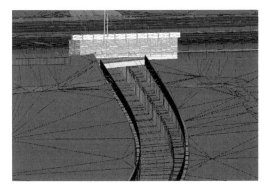

Figure 12. The railway 3D solid in correspondence of the underpass and the noise barrier.

a unified and integrated view to limit design errors. In case of road underpasses, design errors may give rise to high-risk situations in the event of adverse climatic conditions (flooding).

Fig. 12 represents the complete underpass model placed in the railway project.

Besides, in correspondence to the underpass (Fig. 12) there is a noise barrier model. On account of its 3D modeling and BIM process It has been possible to place it correctly in the most suitable foundation system for the given substrate (foundation piles for the embankment, lag bolts in correspondence of built objects) this choice of design is directly defined in the model in terms of materials, technical characteristic, geometry, etc.

4 CONCLUSIONS

All BIM mode applications developed in this case study have been the result of intensive research in order to find the most suitable solution among those offered by computer technology.

The above-cited applications met the expected and acknowledged design requirements of the engineering company in charge as well as the technical specification required by the customer. Furthermore, the models developed for this case study can be reused with minor modifications for other projects benefitting from one of the major advantages of the BIM process. Finally, the BIM process has the potential to overcome procedural delays that have slowed down the construction industry development in many countries especially in Italy, due to shortage of the design. In conclusion, the implementation of the BIM process in railway infrastructures needs steady improvement. This regards library development and available models for all users so as to encourage the development of this methodology and therewith its usability of information throughout the lifecycle of an infrastructural work.

ACKNOWLEDGEMENTS

We thank our colleagues from Italferr SpA (Italian State Railways Group), Eng Tascione P. and from Bentley SpA Eng. Cristallini C. and Magalotti E. who provided insight and expertise that greatly assisted the research.

REFERENCES

Aranda-Mena G., Crawford J., Chevez A. and Fröese T. (2009). "Building information modelling demystified: Does it make business sense to adopt BIM?" *International Journal of Managing Projects in Business*, 2(3), 419–434.

Arayici Y., Coates P., Koskela L., Kagioglou M., Usher C. and O'Reilly K. (2011). "Technology adoption in the BIM implementation for lean architectural practice". *Automation in Construction*, 20(2), 189–195.

Bonin, G., Cantisani, G., Ranzo, A., Loprencipe, G., & Atahan, A.O. (2009). Retrofit of an existing Italian bridge rail for H4a containment level using simulation. *International Journal of Heavy Vehicle Systems*, 16(1–2), 258–270.

Cantisani, G., Loprencipe, G., and Primieri, F. (2011). The integrated design of urban road intersections: A case study. In *The International Conference on Sustainable Design and Construction* (pp. 722–728) ICSDC 2011: Integrating Sustainability Practices in the Construction Industry, Kansas City, Missouri.

Directive 2014/24/EU of the European Parliament and of the Council of 26 February 2014 on public procurement and repealing Directive 2004/18/EC.

Grilo A. and Jardim-Goncalves R. (2010). "Value proposition on interoperability of BIM and collaborative working environments". *Automation in Construction*, 19(5), 522–530.

Howard R. and Björk B.-C. (2008). "Building information modelling - experts' views on standardisation and industry deployment". *Advanced Engineering Informatics*, 22(2), 271–280.

ISO 29481-1:2016. Building information models—Information delivery manual—Part 1: Methodology and format.

Moretti, L., Cantisani, G., & Di Mascio, P. (2016). Management of road tunnels: Construction, maintenance and lighting costs. Tunnelling and Underground Space Technology, 51, 84–89.

Nurain Hassan Ibrahim (2013) "Reviewing the evidence: use of digital collaboration technologies in major building and infrastructure projects" Journal *of Information Technology in Construction*.

Singh V., Gu N. and Wang X.Y. (2011). "A theoretical framework of a BIM-based multi-disciplinary collaboration platform". *Automation in Construction*, 20(2), 134–144.

Taylor J.E. and Bernstein P.G. (2009). "Paradigm trajectories of building information modeling practice in project networks". *Journal of Management in Engineering*, 25(2), 69–76.

Young Jr. N.W., Jones S.A. and Bernstein H.M. (2008). "Building information modelling (BIM): Transforming design and construction to achieve greater industry productivity." McGraw Hill Construction.

Zhu S.T. (2010). "Sustainable building design based on BIM". Paper presented at the *EBM 2010: International Conference on Engineering and Business Management*, Vol 1–8.

Transport Infrastructure and Systems – Dell'Acqua & Wegman (Eds)
© 2017 Taylor & Francis Group, London, ISBN 978-1-138-03009-1

Dissipated energy approach for the fatigue evaluation of RAP asphalt mixtures

M. Pasetto
University of Padua, Padua, Italy

N. Baldo
University of Udine, Udine, Italy

ABSTRACT: The results are discussed of a laboratory investigation on the fatigue resistance of Base Courses Asphalt (BCA) mixes, for road pavements, evaluated by means of the four-point bending test. The experimental analysis was performed on bituminous mixtures with Reclaimed Asphalt Pavement (RAP), up to 40% by weight of the aggregate. Both conventional and polymer modified bitumen was used in the investigation. The mix design was based on volumetric (gyratory) and indirect tensile tests, on both wet and dry samples. The fatigue behaviour was evaluated at 20°C and 10 Hz, under strain control mode, by means of the conventional approach, based on the reduction in the initial stiffness modulus, as well as using a dissipated energy method focused on the Plateau Value concept. Between the different approaches used for the fatigue data analysis, it has been verified a qualitative consistency in the mixtures ranking, with a higher fatigue resistance presented by the polymer modified mixtures produced with RAP aggregates.

1 INTRODUCTION

In the last decades one of the most relevant research themes within the road scientific community has been the reuse of discarded materials in the pavement structure, in order to improve the sustainability of the road constructions (Kavussi & Qazizadeh 2014, Oluwasola et al. 2015, Pasetto & Baldo 2012, 2013a, 2014a, Xue et al. 2009, Wu et al. 2007).

Among the several type of wastes considered, the Reclaimed Asphalt Pavement (RAP) aggregate is one of the most investigated, due to its strategic importance in the road construction industry (Celauro et al. 2009, Oliveira et al. 2013, Pereira et al. 2004, Silva et al. 2012, Stimilli et al. 2016). Nevertheless, the influence of the RAP aggregate on the fatigue life of the asphalt concretes is still not fully clarified.

Some studies reports a lower fatigue resistance (Colbert & You 2012, Miró et al. 2011, Pereira et al. 2004, Silva et al. 2012, Stimilli et al. 2016), primarily due to a brittle response of the mix, as a consequence of the stiff aged binder of the RAP particles (Colbert & You 2012). In order to minimize such problem, a soft bitumen may be used as new binder to be added to the RAP mixtures (Celauro et al. 2009, Oliveira et al. 2013, Pereira et al. 2004, Silva et al. 2012).

Rejuvenator agents represent also an effective option useful to improve the fatigue response of mixtures containing RAP materials (Celauro et al. 2009, Huang et al. 2005). However, an increment of the fatigue resistance of RAP mixtures was documented by few researchers who used a low penetration grade bitumen, by means of a proper mix design method (Miró et al. 2011, Valdés et al. 2011). Eventually, a comparable fatigue resistance between RAP hot mix asphalts and high modulus bituminous mixtures has been recently observed.

Therefore, the fatigue behaviour of RAP mixes represents a topic which has to be furthered studied, in order to improve the knowledge of the mechanical response of such materials.

In this experimental study, the fatigue behaviour of polymer modified asphalt concretes, for base courses of road pavements, prepared with Reclaimed Asphalt Pavement (RAP), was investigated using the four-point bending test (4PBT), under strain control mode.

The conventional stiffness reduction method and a dissipated energy approach were used in order to analyse and elaborate the fatigue data.

The main goal of the study was to design high performance recycled bituminous mixtures, combining RAP aggregates and polymer modified bitumen, in order to improve the fatigue behaviour of conventional polymer modified mixes.

2 MATERIALS

2.1 *Aggregates and bitumen*

The RAP considered in the investigation derives from the demolition of highway asphalt pavements, characterized by a similar composition and located in the Northern Italy. Conventional crushed limestone, sand and filler were also utilized in order to produce the bituminous mixtures studied in the experimental trials. Two different binders have been used: a bitumen modified with SBS (styrene–butadiene–styrene) polymers and a conventional one. All the materials have been supplied by private companies located in the North-eastern Italian area. The conventional aggregates (LS—limestone) were made available in four different fractions: 0/4, 4/8, 8/12 and 12/20 mm. The grading curves of the natural aggregates (EN 933–1), along with that of RAP are illustrated in Figure 1. The RAP aggregate has been tested for the grading curve determination, after the extraction of the bitumen (white curve, centrifugation method; EN 12697–1).

A bitumen content of 5.3% by weight of the aggregate was determined, by means of the cold extraction test performed on the RAP. The bitumen of RAP was recovered through the ABSON method (EN 12697–3); the extracted binder has been subsequently tested following the conventional test methods: penetration at 25 °C (EN 1426), softening point (Ring & Ball Method; EN 1427), Fraass breaking point (EN 12593) and viscosity at 160°C (EN 13702–2). The results are reported in Table 1; the experimental data determined for the conventional bitumen and the polymer modified binder are also presented. On the basis of the experimental data shown in Table 1, the RAP binder results really aged and hard. A lower penetration, as well as higher values for the softening point and the viscosity, have been observed for the polymer modified binder with respect to the conventional bitumen.

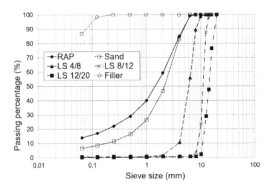

Figure 1. Grading curves of the aggregates.

Table 1. Bitumen characterization.

Properties	RAP binder	Conventional bitumen	Polymer modified
Penetration (dmm), 100 g, 5 s at 25°C	18	52	43
Softening point (°C), R&B method	64	48	98
Fraass breaking point (°C)	-	−18	−18
Viscosity at 160°C	0.256	0.13	1.33

2.2 *Hot mix asphalts*

Six asphalt concretes have been analysed and optimized in the experimental trials, with regards to the acceptance requisites of bituminous mixtures for road base courses.

The conventional bitumen and the modified binder were utilized for the preparation of two different types of asphalt concretes, associated to the code AC (Asphalt Concrete) and PMA (Polymer Modified Asphalt), respectively.

For each one of the two types of hot mix asphalts, three different bituminous mixtures, characterized by an increasing RAP content and therefore by a different lithic matrix, were considered. The mixtures AC/R0, AC/R2, AC/R4 have been prepared with standard bitumen and a RAP content of 0% (R0), 20% (R2) and 40% (R4), respectively. Similarly, the mixtures PMA/R0, PMA/R2 and PMA/R4, have been produced with a RAP quantity of 0%, 20% and 40%, but using the modified binder.

In other laboratory researches it has been verified the possibility to optimize asphalt concretes with higher RAP contents, up to 50% (Celauro et al. 2009, Colbert & You 2012, Miró et al. 2011, Pereira et al. 2004) 60% (Valdés et al. 2011) and 100% (Oliveira et al. 2013, Silva et al. 2012); however, rejuvenator agents were used in order to improve the mechanical response of such mixtures.

In the present experimental study, such agents were not considered, in order to satisfy a direct demand of the local road agencies (North-eastern area of Italy); hence, the maximum quantity of RAP was assumed equal to the 40% by weight of the aggregate.

However, in this research the attention was primarily focused on the improvement of the fatigue resistance of asphalt concretes produced with significant quantities of RAP, rather than on pushing the RAP dosage up to very high percentages.

3 METHODS

3.1 *Mix design phase*

A trial and error method (Pasetto & Baldo 2014a) has been adopted to design the grading curves of the asphalt concretes. The different particle sizes of the aggregates have been combined assuming as a reference grading envelope the one utilized in the North-eastern Italian area, for road base courses (Veneto Strade 2012). Figure 2 illustrates the design grading curves of the asphalt concretes and the reference envelope adopted. The control asphalt concretes (AC/R0 and PMA/R0) were made only with limestone aggregates, to properly compare the mixtures prepared with RAP material.

Cylindrical specimens (150 mm diameter and 60 mm height), have been produced using a gyratory compactor, in order to support the mix design method. The fundamental gyratory parameters, that are pressure, speed and angle of rotation, were assumed equal to 600 kPa, 30 revs/minute and 1.25°, respectively. The ITS tests have been conducted at 25°C, following the principal specifications of EN 12697–23 Standard. In order to evaluate the moisture resistance of the mixes, both dry and wet specimens have been tested. The wet specimens have been submerged for 15 days in water, by means of a thermostatic bath. The ratio between the ITS of wet (ITS_{wet}) and dry (ITS_{dry}) samples expresses the Tensile Strength Ratio (TSR); such parameter was determined for each asphalt concrete considered.

On the basis of the CIRS design approach (CIRS 2001), the volumetric and mechanical acceptance thresholds presented in Table 2 have to be accomplished to design the optimum bitumen percentage. The volumetric characteristics, expressed in terms

of the residual air voids content were checked at three different compaction conditions (10, 100 and 180 revs); instead, the mechanical strength has been evaluated at the design number of revolutions (100 revs).

The assumption of total blending between the binder of RAP and the virgin bitumen was taken in consideration in the mix design method (Al-Qadi 2007). The OBC corresponds to the virgin binder identified in the asphalt concrete optimization; hence, for the mixtures produced with RAP materials, the total binder content has to be computed as the sum of the OBC and the binder of RAP.

The RAP aggregates were preheated for 2 hours at 90°C before the mixing phase. On the basis of the CIRS specifications (CIRS 2001), the mixing temperatures have been fixed at 150°C and 170°C for the conventional binder and the hard modified one, respectively.

3.2 Fatigue tests

The four-point bending apparatus has been used for the fatigue tests, following the most relevant specifications reported in Annex D of the European EN 12697–24 Standard; hence, a continuous sinusoidal waveform, with a frequency of 10 Hz, was used for all the trials. The bending tests were carried out under constant strain conditions, adopting three different strain values, within the range 200–600 μm/m, to adequately capture the fatigue response of each mix. The fatigue study has been performed at 20°C, that is a significant temperature for the fatigue evaluation of hot mix asphalts in the north-eastern part of Italy. The fatigue data, namely stress and strain, the phase angle and dissipated energy, were obtained by the apparatus for each loading repetition. For each hot mix asphalt, slabs (300 × 400 × 50 mm) were compacted by means of a laboratory compacting roller, according to the main specifications of the EN 12697–33 Standard. The beam specimens (400 × 50 × 60 mm), to be used in the fatigue trials, have been obtained from the cut of the slabs.

Most of the times, the fatigue resistance of asphalt concretes is evaluated by means of the classical stiffness reduction method (Celauro et al. 2009, Miró et al. 2011, Oliveira et al. 2013, Pereira et al. 2004, Silva et al. 2012, Valdés et al. 2011). Such method is based on the empirical assumption that at a 50% reduction of the initial stiffness modulus, the beam specimen achieves the structural failure. In the current investigation the fatigue resistance was evaluated using such empirical method, but also applying the dissipated energy method originally introduced by Carpenter & Shen (2005, 2006), which has been developed by means of more rational hypotheses.

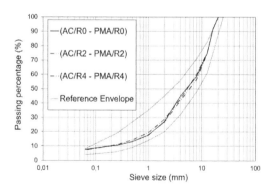

Figure 2. Design grading curves of the mixtures.

Table 2. Mix design acceptance requisites.

Parameter	Threshold
V_a at 10 revs	10–14%
V_a at 100 revs	3–5%
V_a at 180 revs	>2%
ITS_{dry}	>0.6 MPa
TSR	>75%

3.3 Stiffness reduction approach

For a fatigue test carried out under the strain control mode, the strain is maintained at a constant value; as a consequence, the stress decreases until the failure of the specimen, that is conventionally defined by a 50% reduction of the initial stiffness modulus. Such criterion has been assumed also in relevant experimental investigation reported in the literature (Ghuzlan & Carpenter 2003, Khalid 2000).

In the present laboratory study, the conventional fatigue curves have been developed combining the initial value of strain ε_0 and the number of loading cycles N_f, at which a 50% reduction of the initial stiffness is achieved. According to previous studies (Artamendi & Khalid 2005), the strain value observed at the 100th cycle was used as initial strain (EN 12697–24, Annex D).

For the interpolation of the fatigue data, a power law model has been used:

$$\varepsilon_0 = aN_f^b \qquad (1)$$

where a and b represent the interpolation coefficients; their values depend on the type of asphalt concretes.

3.4 Dissipated energy approach

The fatigue response were also analysed with the energy method of Carpenter & Shen (2005, 2006), recently also utilized in other investigations (Pasetto & Baldo 2013b, 2014b, Yoo & Al-Qadi 2010). In such method, the Plateau Value (PV) of the Ratio of Dissipated Energy Change (RDEC) represents the most important parameter for a rational evaluation of the fatigue behaviour of asphalt concretes. The RDEC is computed using the formula:

$$RDEC = \frac{DE_{n+1} - DE_n}{DE_n} \qquad (2)$$

where DE_n and DE_{n+1} represent the dissipated energy produced in load cycle n and n+1, respectively.

It was argued that the RDEC is not influenced from other dissipated energy types, related to mechanical work or heat development (Carpenter & Shen 2005, 2006). Hence, the RDEC can be considered a rational parameter, useful to characterize the internal damage developed within the asphalt concrete, during the fatigue trial.

The trend of RDEC in a fatigue test, can be subdivided in three different phases. During the first phase there is a sharp reduction of RDEC, whereas a plateau zone, characterized by a constant RDEC value, is typical of the second phase;

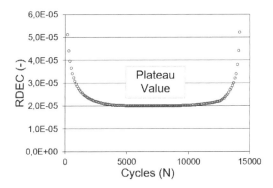

Figure 3. Determination of PV for mix AC/R2 at 400 μm/m.

in such zone the quantity of input energy which undergoes a transformation into internal damage remains constant. In the third and last phase, the RDEC increases fastly and eventually the specimen achieves the failure condition. The RDEC value that remains constant during the second phase, has been named Plateau Value (PV) of RDEC (Carpenter & Shen 2005, 2006). Such value results really useful for a comparison of the fatigue resistance of different hot mix asphalts, from an energy based point of view; basically, the lower the PV, the greater the fatigue resistance of the mixture will be (Carpenter & Shen 2005, 2006, Pasetto & Baldo 2013b, 2014b, Shu et al. 2008, Yoo & Al-Qadi 2010). Figure 3 illustrates an example of the determination of PV for the mix AC/R2 at 400 μm/m.

In this research, different strain levels have been used in order to obtain the damage curves (PV curves—PVC), expressed as a function of N_f (loading cycles involved for a 50% stiffness reduction) and PV.

The interpolation of the PV—N_f data pairs, has been performed using a power law model:

$$PV = aN_f^b \qquad (3)$$

where a and b are regression coefficients; their values depend on the type of hot mix asphalt.

4 RESULTS AND DISCUSSION

4.1 Mix design results

For both the binder utilized in the trials, three different aggregate structures were optimized. Each asphalt concrete was designed with an equal quantity of the coarser limestone material, that is the fraction 12/20 mm. Conversely, for the other two limestone fractions (4/8 and 12/20 mm), the quantities were different, in relation to the RAP content.

The aggregate structure of the control asphalt concretes (AC/R0, PMA/R0), was characterized by the 47% of sand, 7% of the limestone fraction 4/8, as well as the 7% of the fraction 8/12, plus 5% of filler.

The asphalt concretes containing 20% RAP (AC/R2, PMA/R2) were made with the 25% of sand, 9% of both the limestone fractions 4/8 and 8/12. The filler was fixed at 3%.

The introduction of 40% RAP in the aggregate composition (mixes AC/R4 and PMA/R4), allowed to reduce the filler and the sand contents to 1% and 8% respectively; instead, for the limestone, the fraction 4/8 was utilized at 12% and the 8/12 at 5%.

As it was expected, the greater the RAP content, the lower the filler amount required to design the mixtures.

Tables 3–4 report the results of the mix design phase. For each mix the most important design properties have been reported: OBC (by weight of the aggregate); Air Voids (Va) at 10, 100 and 180 revs; bulk density at 100 revs; Indirect Tensile Strength (ITS) for dry and wet conditions, at 100 revs.

The bitumen content reported in Table 3 is the virgin binder that resulted to be necessary for the optimization of the asphalt concretes.

For all the hot mix asphalts, the air voids thresholds fixed in the CIRS mix design method (CIRS 2001), were completely accomplished, depending on the particular OBC. Each of the asphalt concrete studied has presented strength properties, namely dry ITS and TSR, much higher than the

Table 3. OBC and volumetric properties.

Asphalt concretes	OBC %	Va, 10 revs %	Va, 100 revs %	Va, 180 revs %
AC/R0	5.00	13.9	4.9	3.4
AC/R2	3.94	13.6	4.5	3.3
AC/R4	2.88	11.2	3.5	2.3
PMA/R0	5.00	14.0	5.0	3.8
PMA/R2	3.94	13.8	4.8	3.7
PMA/R4	2.88	11.3	3.8	2.6

Table 4. Physical and mechanical properties.

Asphalt concretes	Bulk density Mg/m³	ITS dry MPa	ITS wet MPa	TSR %
AC/R0	2.457	1.04	0.89	86
AC/R2	2.473	1.54	1.36	88
AC/R4	2.484	1.95	1.78	91
PMA/R0	2.451	1.62	1.46	90
PMA/R2	2.465	2.08	1.92	92
PMA/R4	2.477	2.32	2.16	93

CIRS thresholds. Furthermore, the wet ITS values resulted also higher than the dry ITS threshold for all the mixes; hence, it has been verified the excellent moisture resistance of the asphalt concretes investigated.

The total binder percentages identified by the mix design method were within the range typically adopted for the base courses, that is 4.5–5.5% by weight of the aggregate (CIRS 2001), for all the mixes. According to the results outlined in previous investigations, reported in the literature (Al-Qadi et al. 2007, Ghuzlan & Carpenter 2003), the new bitumen content necessary to satisfy the acceptance requisites resulted a decreasing function of the RAP quantity; the greater the RAP content, the lower the virgin binder to be added to the mixes. Such experimental evidence is congruent with the assumption of a total blending between RAP binder and new bitumen.

No workability issues were observed during the mixing of the components, namely bitumen and aggregates, neither for the compaction of the mixes produced with RAP, for both the types of bitumen.

4.2 Fatigue analysis based on the stiffness reduction approach

Figure 4 shows the conventional fatigue curves, obtained by means of the stiffness reduction method, for the asphalts concretes considered. Table 5 reports the interpolation coefficients along with the coefficient of determination R^2. The tensile strain ε (10^6) was determined following the specifications reported in the Standard EN 12697–24, Annex D, for a fatigue life of 1,000,000 loading repetitions.

Accordingly to the results presented in Table 5, the polymer modified mixes were characterized by a greater fatigue resistance, compared to the asphalt concretes produced with the conventional bitumen, in terms of tensile strain ε (10^6). This was

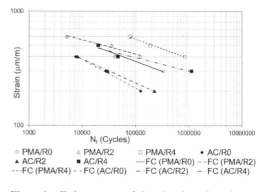

Figure 4. Fatigue curves of the mixes investigated.

Table 5. Regression coefficients of the fatigue curves and ε (10^6) values.

Mixes	a µm/m	b –	ε (10^6) µm/m	R^2 –
AC/R0	3687.1	−0.248	120	0.9972
AC/R2	2606.3	−0.209	145	0.9991
AC/R4	1501.8	−0.117	298	0.9378
PMA/R0	2655.9	−0.173	243	0.9562
PMA/R2	1779.3	−0.125	316	0.9665
PMA/R4	4009.9	−0.170	383	0.9911

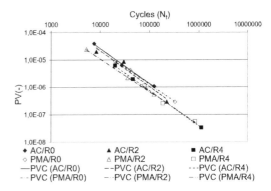

Figure 5. Damage curves of the mixes investigated.

Table 6. Regression coefficients of the damage curves and PV_6 values.

Mixes	a µm/m	b –	PV_6	R^2 –
AC/R0	3.3201	−1.278	7,13101E-08	0.9976
AC/R2	3.5992	−1.306	5,25056E-08	0.9587
AC/R4	2.0827	−1.289	3,84261E-08	0.9999
PMA/R0	1.9145	−1.234	7,55189E-08	0.9984
PMA/R2	0.6936	−1.203	4,19864E-08	0.9999
PMA/R4	1.0138	−1.231	4,16824E-08	0.9793

verified also for the mixtures with RAP, especially for the greater RAP content (40%). Indeed, the fatigue curves of the polymer modified mixtures are shifted towards higher loading cycles values; moreover, they present a lower slope (Table 5), with respect to the control mixes. The improvement of the fatigue performance due to the polymer modified bitumen was clearly expected, whereas the effect of the RAP on the fatigue life represents a new contribution on the characterization of the RAP mixes. Moreover, according to the results of Table 5, it has been verified that the combined use of a polymer modified bitumen along with RAP materials allows to achieve the highest fatigue life, namely 383 µm/m for HM/R4, with an increment of 58% with respect to the control mix prepared with limestone aggregates (HM/R0). It is worth of mentioning that the increase in the fatigue life, due to the use of RAP materials, resulted nonlinear with the RAP quantity; indeed, the increment was higher with the 40% of RAP rather than the 20%.

4.3 Fatigue analysis based on the dissipated energy approach

Figures 5 shows the damage curves, whereas Table 6 reports the coefficients of interpolation and determination (R^2); the fitting of the experimental data resulted very good for all the mixes studied. In similarity to the conventional approach, the PV for a 1,000,000 loading cycles (PV^6) has been determined, in order to synthetically quantify and compare the fatigue performance of the mixes.

With respect to the control asphalt concretes (AC/R0 and PMA/R0), the mixes made with RAP aggregates were characterized by lower PV_6 values, namely, by a damage reduction. Depending on the RAP content, the PV_6 reductions achieved the 42% for the mixes prepared with conventional bitumen and the 36% in case of PMB.

It has to be observed that the higher the RAP content, the better the fatigue resistance, for both the binders. Such results could be justified by the chemical affinity between the bitumen of the RAP

and the virgin binders, which probably allows the development of a tough adhesion between the thin film of the virgin binders and the RAP grains. From another point of view, the thin film of aged binder that covers the RAP particles could be considered as an intermediate stiff layer between the stiffer mineral grains of the RAP and the softer virgin bitumen, so admitting a layered configuration within the RAP asphalt concretes. On the basis of such structural configuration (Huang et al. 2005; Oliveira et al. 2013), the binder of the RAP contributes to reduce the stress concentration within the hot mix asphalt, with a consequent beneficial effect on the fatigue behaviour.

However, the lowest PV_6 values were obtained for PMA/R4 and PMA/R2, namely the mixes produced with both RAP and polymer modified bitumen. The improved fatigue response shown by the polymer modified asphalt concretes with RAP, could be due to the flexibility associated to the polymers of the modified binder.

Currently, for road base asphalt concretes, the Italian technical specifications (CIRS 2001) prescribe a maximum RAP amount equal to 30%; furthermore, for such mixes, the use of polymer modified binders is not allowed. Hence, the feasibility to improve the fatigue performance of polymer modified asphalts characterized by a conventional aggregate structure, by means of the integration of RAP, up to 40%, allows to meet the demand for sustainability and high performance of the road constructions.

Focusing the attention on the relative ranking of the mixes (Table 6), the analysis performed with the dissipated energy approach, confirms the trend already outlined with the conventional method. However, the dissipated energy approach relies on rational principles and not on an empirical failure assumption as the conventional method.

5 CONCLUSIONS

The experimental study described and discussed in this paper concerns the fatigue performance evaluation of hot mix asphalts for road base courses, produced with RAP, up to 40% by the weight of the aggregate, and polymer modified bitumen.

The fatigue data obtained by four point bending fatigue tests, carried out at 10 Hz and 20°C, have been analysed using the conventional stiffness reduction approach and the dissipated energy method, originally proposed by Carpenter.

With respect to the reference mixture, produced using only natural aggregates, all the hot mix asphalts with RAP materials have shown a better fatigue behaviour; the improvements were even greater for the mixes prepared with polymer modified bitumen, for both the fatigue data analysis method adopted.

Based on the results obtained with this experimental study, a positive effect of the RAP materials on the fatigue resistance of the hot mix asphalts considered has been outlined.

In qualitative terms, the comparative ranking of the mixes identified by the empirical approach and that established by the dissipated energy method was totally analogous.

The results obtained in the present paper represent a further validation of the dissipated energy method based on the Plateau Value concept and contribute to clarify the importance of a rational approach for the fatigue evaluation of road bituminous mixtures.

REFERENCES

Al-Qadi I.L., Elseifi M. & Carpenter S.H. 2007. Reclaimed Asphalt Pavement—A Literature Review. Technical Report FHWA-ICT-07-001, Federal Highway Administration (FHWA), Washington, DC, USA.

Artamendi I. & Khalid H. 2005. Characterization of fatigue damage for paving asphaltic materials. *Fatigue & Fracture of Engineering Materials & Structure* 28: 1113–1118.

Carpenter S.H., Shen S. 2006. Dissipated energy approach to study hot-mix asphalt healing in fatigue. *Transportation Research Record* 1970: 178–185.

Celauro C., Celauro B. & Boscaino G. 2009. Production of innovative, recycled and high-performance asphalt for road pavements. *Resources, Conservation and Recycling* 54(6): 337–347.

CIRS – Ministero delle Infrastrutture e dei Trasporti. 2001. Capitolato speciale d'appalto tipo per lavori stradali. Italy (in Italian).

Colbert B. & You Z. 2012. The determination of mechanical performance of laboratory produced hot mix asphalt mixtures using controlled RAP and virgin aggregate size fractions. *Construction and Building Materials* 26(1): 655–662.

Ghuzlan K.A. & Carpenter S.H. 2003. Traditional fatigue analysis of asphalt concrete mixtures. *Proc. Transportation Research Board 2003 Annual Meeting, Washington DC, 12–16 January 2003*.

Huang B., Li G., Vukosavljevic D., Shu X. & Egan B.K. 2005. Laboratory investigation of mixing hot-mix asphalt with reclaimed asphalt pavement. *Transportation Research Record* 1929: 37–45.

Kavussi A. & Qazizadeh M.J. 2014. Fatigue characterization of asphalt mixes containing electric arc furnace (EAF) steel slag subjected to long term aging. *Construction and Building Materials* 72: 158–156.

Khalid H. 2000. A comparison between bending and diametral fatigue tests for bituminous materials. *Materials and Structures*, 33: 457–465.

Miró R., Valdés G., Martínez A., Segurac P. & Rodríguez C. 2011. Evaluation of high modulus mixture behaviour with high reclaimed asphalt pavement (RAP) percentages for sustainable road construction. *Construction and Building Materials* 25: 3854–3862.

Oliveira J.R.M., Silva H.M.R.D., Jesus C.M.G. & Abreu L.P.F. 2013. Pushing the Asphalt Recycling Technology to the Limit. *International Journal of Pavement Research Technology* 6(2): 109–116.

Oluwasola E.A., Hainin M.R. & Aziz M.M.A. 2015. Evaluation of asphalt mixtures incorporating electric arc furnace steel slag and copper mine tailings for road construction. *Transportation Geotechnics* 2: 47–55.

Pasetto M. & Baldo N. 2012. Fatigue Performance of Asphalt Concretes with RAP Aggregates and Steel Slags. In: Scarpas A. et al. (ed.), *Proc. 7th RILEM intern. Conference on Cracking in Pavements, RILEM Bookseries, 4: 719–727, Delft, 20–22 June 2012.* Dordrecht: Springer Netherlands.

Pasetto M. & Baldo N. 2013a. Resistance to permanent deformation of road and airport high performance asphalt concrete base courses. *Advanced Materials Research* 723: 494–502.

Pasetto M. & Baldo N. 2013b. Fatigue Performance of Asphalt Concretes made with Steel Slags and Modified Bituminous Binders. *International Journal of Pavement Research and Technology* 6(4): 294–303.

Pasetto M. & Baldo N. 2014a. Influence of the aggregate skeleton design method on the permanent deformation resistance of stone mastic asphalt. *Materials Research Innovations* 18(3): S96-S101.

Pasetto M. & Baldo N. 2014b. Fatigue performance and stiffness properties of Stone Mastic Asphalts with steel slag and coal ash. In: Kim Y.R. (ed.), *Asphalt*

Pavements—Proc. of the intern. Conference on Asphalt Pavements, ISAP 2014, 1(82): 881–889, Raleigh, 1–5 June 2014. Boca Raton: CRC Press, Taylor & Francis Group.

Pereira P.A.A., Oliveira J.R.M. & Picado-Santos L.G. 2004. Mechanical characterisation of hot mix recycled materials. *International Journal of Pavement Engineering* 5(4): 211–220.

Shen S., Carpenter S.H. 2005. Application of the dissipated Energy concept in fatigue endurance limit testing. *Transportation Research Record* 1929: 165–173.

Shu X., Huang B. & Vukosavljevic D. 2008. Laboratory evaluation of fatigue characteristics of recycled asphalt mixture. *Construction and Building Materials* 22: 1323–1330.

Silva H.M.R.D., Oliveira J.R.M. & Jesus C.M.G. 2012. Are totally recycled hot mix asphalts a sustainable alternative for road paving?. *Resources, Conservation and Recycling* 60: 38–48.

Stimilli A., Virgili A., Giuliani F. & Canestrari F. 2016. In plant production of hot recycled mixtures with high reclaimed asphalt pavement content: A performance evaluation. In Francesco Canestrari & Manfred N. Partl (ed.): *Proc. 8th RILEM intern. Symp. on Testing and Characterization of Sustainable and Innovative Bituminous Materials, RILEM Bookseries, 11: 927–939, Ancona, 7–9 October 2015*. Dordrecht: Springer Netherlands.

Valdés G., Pérez-Jiménez F., Miró R., Martínez A. & Botella R. 2011. Experimental study of recycled asphalt mixtures with high percentages of reclaimed asphalt pavement (RAP). *Construction and Building Materials* 25: 1289–1297.

Veneto Strade. 2012. Capitolato speciale d'appalto tipo per la manutenzione e la costruzione delle infrastrutture stradali. Italy (in Italian).

Wu S., Xue Y., Ye Q. & Chen Y. 2007. Utilization of steel slag as an aggregate for stone mastic asphalt (SMA) mixtures. *Building and Environment* 42: 2580–2585.

Xue Y., Wu S., Hou H., Zhu S. & Zha J. 2009. Utilization of municipal solid waste incineration ash in stone mastic asphalt mixture: pavement performance and environmental impact. *Construction and Building Materials* 23: 989–996.

Yoo P.J. & Al-Qadi I.L. 2010. A strain-controlled hot-mix asphalt fatigue model considering low and high cycles. *International Journal of Pavement Engineering* 6: 565–574.

Transport Infrastructure and Systems – Dell'Acqua & Wegman (Eds)
© *2017 Taylor & Francis Group, London, ISBN 978-1-138-03009-1*

Correlation between risk level and operating speed of road section

D. Bellini & M.C. Iaconis
Regione Toscana—Progettazione e Realizzazione Viabilità Regionale Pisa-Siena-Pistoia, Italy

M. Rossi
Dipartimento di Ingegneria Civile e Industriale, Università di Pisa, Pisa, Italy

ABSTRACT: The use of accident data has always been a challenge for the management of road networks in Italy. On one hand, there were many attempts to find theoretical models to interpret the data of accidents, while on the other arose the difficulty of finding quality data for targeted studies, especially for the rural road networks. In recent years however, the possibility of finding accident data was used to develop a useful procedure at a planning level to identify the road sections on which further investigation should be carried on, in order to improve road safety conditions. The results obtained with the application of this procedure to some roads under the jurisdiction of the Tuscany Region Road Authority made possible to build some bubble diagrams, which allow an immediate and complete view of all homogeneous sections of the road network. For those sections for which was obtained a high risk level, deeper investigations were performed by measuring the operating speed, in order to evaluate a relationship between those detected discrepancies between the design conditions and the real operating condition of those homogeneous sections (correlated with the perception of the infrastructure by drivers) and the risk of these sections themselves, allowing the planning of those actions necessary in order to increase road and traffic safety level.

1 INTRODUCTION

The use of accident data has always been a difficult challenge for the Road Authorities in Italy. On the one hand many attempts to find theoretical models to interpret the data of accidents were carried on, on the other it came up the difficulty of finding quality data for targeted studies and, especially for the secondary road networks, the difficulty to obtain updated data on traffic flows.

In recent years however raised up the possibility to obtain a set of data that, combined, permitted to define a useful procedure in the planning level in order to identify those road sections on which further investigation must be carried on, in order to improve road safety conditions, as also stated into the Italian Road Safety Guidelines (D.M. n. 182/2012), from which the entire study moved on.

This was possible thanks to SIRSS project: it consists in the normalization of information about the location of the accidents in accordance with a road graph, in addition to the georeferencing of each accident.

The analysis here performed is the starting point in the developing of a new procedure: data along 14 homogeneous section on regional road S.R.68 "della Valdicecina" within the Province of Pisa were available. Georeferencing of accidents, as well

as the construction of a road graph for homogeneous road sections, made it possible to determine the average annual rate of accidents for each homogeneous section, on the basis of some indicators, as described in the Final Report "Criteria for the classification of the network of existing roads", drawn up and approved by the CNR with DP CNR n. 13465 of 11/09/1995 and published on 13/03/1998, as described in the following paragraphs.

With all these data available, a correlation between risk level and speed data of homogeneous section was found.

2 ROAD SAFETY CONDITION ANALYSIS AT A NETWORK LEVEL

2.1 *C.N.R. procedure to evaluate accident rate and to classify each homogeneous sections within a road*

The accident rate analysis is part of the characterization of each road within a network in terms of road safety, which enables the identification, in relation to a statistically significant comparison threshold, the accident rate involving the entire road.

The analysis was here performed for 14 homogeneous section on regional road S.R.68 "della Valdicecina" within the Province of Pisa. In addition to calculating the accident rate following the CNR report quoted above, the analysis was deepened with the calculation of the density of accidents.

In order to classify the entire road on the base of the accident rate, the procedure indicated in the CNR report requires the following information:

– georeferenced accidents data, including the date, number and type of vehicles involved; only accidents with people injured were considere. All these data were provided by SIRSS in a specific database;
– division into homogeneous section of the entire road;
– TGM of each hgomogeneous section, referred to each specific year.

The accident rate of each homogeneous section is evaluated as the number of accident in relation to one million vehicles per kilometer, and is given by:

$$T_i = \frac{10^6 \cdot N_i}{365 \cdot l_i \cdot \sum_t TGM_{i,t}} \quad (1)$$

with:

N_i number of accidents with injuries occurred in the tth year within the analysis period considered on the ith homogeneous section;

L_i length (km) of the ith homogeneous section;

$TGM_{i,t}$ Average daily traffic on the ith homogeneous section in the tth year within the analysis period.

The CNR report provides also instructions about the identification of a statistically significant reference threshold in order to classify homogeneous section as, low, medium and high accident rate, based on the results obtained by the accident rate evaluation as stated above:

$$T_{inf}^{\bullet} = T_m - K \cdot \sqrt{\frac{T_m}{M_i} - \frac{1}{2 \cdot M_i}} \quad (2)$$

$$T_{inf}^{\bullet} = T_m + K \cdot \sqrt{\frac{T_m}{M_i} + \frac{1}{2 \cdot M_i}} \quad (3)$$

with:

K Poisson's probability distribution constant, assumed with the value 1.645 in order to have an error probability less the 10%;

M_i traffic momentum of the ith homogeneous section within the entire analysis period, evaluated as:

$$M_i = 365 \cdot l_i \cdot \sum_t TGM_{i,t} \quad (4)$$

Each homogeneous section is classified as with low, medium or high accident rate on the bases of the T_i value evaluated with the equation (1) compared with the thresholds evaluated with equations (2) and (3).

The obtained results made it possible to build some bubble diagrams, which allow an immediate overview on each homogeneous section within a road or an entire network, highlighting those with higher risk level. These diagrams have, for each homogeneous sections, on the abscissa the values of the accident density D_i of each homogeneous section, measured in number of accidents per km per year, and in ordinate the V_i values of the speed (operating speed of project speed) of each homogeneous section, while the area of the bubbles is the accident rate T_i of the homogeneous section.

The choice to build this type of representation raised by the necessity to evaluate the risk level of each infrastructure which the homogeneous section belongs apart from the TGM, that represents the road users risk level component. In particular, with this type of representation it was tried to summarize the information regarding the contribution of inhomogeneity of radius of curvature within paths with many bends.

On each diagram it is possible to identify 4 different areas, on the base of the speed of 70 km/h (speed over which the italian standard D.M. 2367/2004 confirms the mandatory installation of crash barriers, identifying this speed value as a discriminating value above which it can be considered that the damage resulting from an impact is more than proportional to the speed itself) and on the base of the average accident density D_m evaluated on a reference base (national, regional, network, single road):

– Low risk level area: $V_i \leq 70$ km/h, $D_i \leq D_m$
– Medium risk level area: $V_i > 70$ km/h, $D_i \leq D_m$
– High risk level area: $V_i \leq 70$ km/h, $D_i > D_m$
– Very high risk level area: $V_i > 70$ km/h, $D_i > D_m$

The bubbles area is proportional to the accident rate of the road section: bigger is a bubble, higher is the accident rate of the section to which bubble refers.

It was also necessary to identify three different ranges for the bubbles area, to make it easier to read the graph by assigning to each of them a different color.

In the following Figure 1 an example of the bubbles graph is reported.

This division is based on the evaluation of the average accident rate T_m on the same basis used for the evaluation of D_m:

– Low accident rate: GREEN color, $T_i \leq 2 \cdot T_m$
– Medium accident rate: YELLOW color, $2 \cdot T_m < T_i \leq 3 \cdot T_m$
– High accident rate: RED color, $T_i > 3 \cdot T_m$

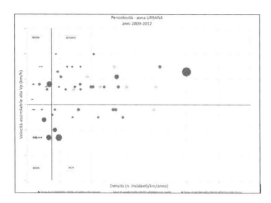

Figure 1. Example of a bubbles graph.

The analysis of the bubbles diagrams make it possible to identify those situations marked by higher anomaly than the average of the entire network (or single road) analyzed.

In most case, the anomaly higher than the average of the entire network is caused by the wrong perception of the path by the road users, with the results that it is hard for them to drive at a "safe" speed along the entire road or on its some specific section.

For these situations it is necessary to deeper the investigation in order to identify those factors, such as the geometry of the homogeneous section, the composition of traffic flows, environmental conditions prevailing, etc. responsible of the anomaly.

The benefit is twofold: not only to go to primarily analyze situations with higher absolute numbers regarding the amount of people involved but, among these, the ones that have the greatest chances to be improved because suffering from abnormalities of some kind.

In particular, it seems useful to consider primarily the situations relating to red bubbles in the very high risk level area of the diagram as representative of road sections characterized by high number density of accidents, high accident rates and high speed.

In general, for the yellow and red bubbles some actions must be planned in order to improve the road safety conditions, like reinforcement of road signs, control activities by traffic police organs, awareness campaigns, new design of the road, etc.

2.2 Correlation between accident rate and geometrical characteristics of the road path

Once the accident rate analysis is completed and bubbles graphs are built, it is possible to identify those sections for which some action should be planned.

This research focused to find a possible correlation between the accident rate and some characteristics (geometrical, mostly) of the road path: in this way, it could be possible to plan some design changes in the road path and to evaluate how much the safety condition will improve.

The research starts from the assumption that the right perception of the path (and then the speed maintained by users along the road) is recognized as one of the main reason at the base of high accident rates.

It was assumed that each user drive along a bend at a constant speed, and he accelerates and decelerates along the following straight and before the following bend as he recognizes the layout of the road path.

Then, a correlation between the accident rate and the difference between the speed theoretically attainable at the end of a straight and the design speed of the following bend, under the hypothesis of linear deceleration assumed equal to *0.8 m/s²* as stated in the Italian Standard D.M. 6792/2001 ("Norme funzionali e geometriche per la costruzione delle strade").

The first step was to find a model to evaluate the design speed of each bend. This is a theoretical speed and it can be evaluated by taking in account what is reported in the Italian Standard D.M. 6792/2001 ("Norme funzionali e geometriche per la costruzione delle strade). It can be evaluated for each bend on the road path from:

$$\frac{V_p^2}{R \cdot 127} = q + f_T \qquad (5)$$

with:

q   transversal slope
f_T maximum transversal grip coefficient, inversely proportional to the design speed
R   radius of curvature
V_p design speed.

Knowing the radius of each bend and the length of each straight along a road path, both evaluable from technical maps C.T.R. or using special vehicles, the theoretical speed V_{end} at the end of each straight was evaluated using the equations of uniformly accelerated motion and under the following hypothesis:

– if the stop distance L_{STOP} (evaluated with the procedure reported in "Il tracciato stradale: dinamiche di percorrenza e visibilità", F.S.Capaldo, 2005) is lower than the length L_{ST} of the straight, it was assumed that the driver will start to brake when he will be at a distance of L_{STOP} from the following bend;
– if the stop distance L_{STOP} is higher than the length L_{ST} of the straight, it was assumed that

the driver will start to brake when he will be at a distance of L_{ST} from the following bend.

Once the speed V_{end} at the end of each straight is evaluated, it can be compared with the speed of the following bend:

$$\Delta V = V_a - V_{end} \qquad (6)$$

with:

V_a design speed of the bend following the straight

V_{end} speed evaluated at the end of the straight.

From the comparison, three different cases are possible:

1. The theoretical speed at the end of the straight is lower than the design speed of the following bend: in this case, the user has the time to drive along the straight for a certain time before starting to decelerate, in order to enter the following bend at the right speed;
2. The theoretical speed at the end of the straight is equal to the design speed of the following bend was looked: it is a limit condition, where the user must start to decelerate immediately after he enters on straight, otherwise he will enter the following bend at a speed higher than the design speed;
3. The theoretical speed at the end of the straight is higher than the design speed of the following bend: it is the worst condition, because even if the driver start to decelerate just after he enters on the straight, he will not have the possibility to enter the following bend at a safe speed.

Of course, the case n. 3 is the worst and the most risky, and it is the one that should be analyzed and correlated with accident rate.

The research of the correlation started from the analysis of available data. Actually, accident rate in the period 2010–2013 data and speed data on S.R.68 "della Valdicecina" (within the Province of Pisa) are available.

Previously from analysis were excluded those homogeneous section within urban areas: here, there are too many variables to be considered to evaluate a risk level, as pedestrian crossings, private access on the road, public lighting, etc.

At the end, data on 24 homogeneous section were available: excluding the urban section, the analysis was performed on 14 sections, where the accident rate was evaluated.

Then, according to what stated above, all the positive speed differences (previous point n. 3) were summarized along a homogeneous section and normalized it on standard length of 1,000 m.

In the first step, it was investigated the correlation considering all the 14 sections: evaluating the Pear-

son correlation coefficient r (where the X series was the ΔV, and Y series the accident rates), it was found a value of 0.317. The critical value under two variables and n-2 degrees of freedom, at a significance level of 95%, is 0.532: this means that it seems no correlation is existing between ΔV and accident rate.

Then a second step in the analysis was performed. Since 6 homogeneous sections were newly designed and built in the period 2004–2009, a new analysis was performed considering two different groups: one with those sections newly designed, and one with all the other sections (where only ordinary maintenance was performed).

An F-test was run considering the accident rate of the two different groups, to investigate if the two groups must be treated as different independent groups of if they can be treated as a single group.

The result obtained was $F_{eval} = 101.382$.

Entering the F_{crit} critical value table for two data series, each one with n-1 degrees of freedom and with a significance level of 95%, it was found the value of 4.88. Since $F_{eval} > F_{crit}$, it can be affirmed that the two groups are totally independent (at least from the point of view of accident rate), and they must be treated and analyzed independently.

Then, the Pearson correlation coefficient r was evaluated again separately for both groups, even considering the ΔV as the X series and the accident rate as the Y series. The results obtained are reported here below:

– r_{ND}: 0.81
– r_{old}: 0.29

where r_{ND} is the coefficient evaluated for the group constituted by the newly designed sections and r_{old} the group of all the other sections.

The results show that there is a really good correlation between ΔV and the accident rate for those newly designed sections: within this group, at lower ΔV values correspond lower accident rate values.

Some comments on this results are reported in chapter 3.

3 CONCLUSIONS AND FURTHER DEVELOPMENT

In this study authors tried to provide a procedure to manage accidents data available (and georeferenced) in order to make it a useful instrument to plan actions along specific sections of a road (or a network), in order to consistently improve the road safety condition; from another point of view, such a procedure could make it also available to forecast the entity of the road safety condition improvement.

The authors investigate the correlation between geometrical characteristics of a homogeneous section along a road path (here taken in account by considering a normalized speed difference ΔV between the speed theoretically attainable at the end of a straight and the design speed of the following bend) and the accident rate evaluated on the same section, in order to develop a possible procedure to help road authorities to focus and to plan on actions on some homogeneous sections of their network. The correlation, was investigated by analyzing data available for the 14 homogeneous sections of S.R.68 "della Valdicecina".

The result from an F-test, run onto accident rate data, shows and confirm that those sections newly designed and built (according the Italian Standard) must be treated as independent group.

The Pearson correlation coefficient values, evaluated for both groups of sections (newly designed and old), are here reported:

– r_{ND}: 0.81
– r_{old}: 0.29

where r_{ND} is the coefficient evaluated for the group constituted by the newly designed sections and r_{old} the group of all the other sections.

These results shows that many parameters affect the accident rate within a certain road section: indeed, the low r_{old} value means that there is no correlation between ΔV and accident rate within old sections of S.R.68, or from another point of view, the accident rate must be correlated *also* to some other characteristics (e.g. visibility distance, road path altimetry, the presence of side entrances on the road, the presence of dangerous intersections, etc.). Instead, the high r_{ND} value means that there is a good correlation between ΔV and accident rate within newly designed sections of S.R.68, because of all other "dangerous" characteristics are no more present along the road path (the aim of a new design of one or more road section should be to remove as much as possible all the dangerous characteristics listed above): according to the results obtained, on newly designed road section only geometrical data are correlated to accident rate data.

It is also relevant to note that the accident rates used within this research are data that already take account of the traffic volume, major factor in accidents government. The main factor contributing to accident rates, the traffic volumes, is therefore already considered within the parameter chosen for analysis, for which the highest correlation in the accident rate is certainly with the volume of traffic:

traffic volume is, indeed, a key factor contributing to accident rates according to the statistical CNR model.

However, to find a mathematical relationship able to forecast the lowering of accident rate as a consequence of a new design, on the basis of the correlation found here, it is necessary to deep the research by developing a new step: indeed, it is necessary to extend the analysis on more samples, and—more important—also to develop a model able to consider all those parameters listed above.

Once such a model will be available, it will be a powerful help for public administration to understand how much specific actions could lower the accident rate on road sections, or how much specific action could raise the safety condition.

REFERENCES

Bellini D. et al. (2016), *Migliorare la sicurezza delle reti extraurbane secondarie,* Strade & Autostrade n. 116, Gennaio/Febbraio 2016.

Bordin M. & M. Stefanutti (1997), *Analisi dinamica delle norme tecniche CNR sulle curve stradali,* Strategie e strumenti dell'ingegneria delle infrastrutture viarie—Convegno SIIV, Pisa.

Busi R. & L. Zavanella (2002), *La classificazione funzionale delle strade,* EGAF, Forlì.

CNR (1998), Criteri per la classificazione della rete delle strade esistenti ai sensi dell'art. 13, comma 4 e 5 del nuovo codice della strada, Roma.

Figueroa Medina A. (2004), *Reconciling Speed Limits with Design Speeds,* JTRP Technical Reports, INDOT Division of Research, West Lafayette, Indiana.

Francesco Saverio Capaldo (2005), Il tracciato stradale: dinamiche di percorrenza e visibilità.

Giuffrè O. et al. (1997), *Contributo alla definizione dei criteri di classificazione della viabilità esistente,* Strategie e strumenti dell'ingegneria delle infrastrutture viarie—Convegno SIIV, Pisa.

Marchionna A. et al. (2011), *Application of design consistency evaluation tools for two-lane rural roads: a case study from Italy,* Transportation Research Board 90th Annual Meeting, Compendium of Papers No. 11-0689, Washington.

Ministero delle Infrastrutture e dei Trasporti (2012), *Monitoraggio del PNSS: Linee Guida per la valutazione dei risultati degli interventi di Sicurezza Stradale,* a cura di L. Persia, D. Usami, Direzione Generale per la Sicurezza Stradale, Roma.

Rossi M. (2014), Key elements to draw up a traffic plan for rural roads, PhD thesis, Pisa.

Transport systems

Transport Infrastructure and Systems – Dell'Acqua & Wegman (Eds)
© 2017 Taylor & Francis Group, London, ISBN 978-1-138-03009-1

Implementing innovative traffic simulation models with aerial traffic survey

A. Marella, A. Bonfanti & G. Bortolaso
Trafficlab (Progectolab Group), Alba, Italy

D. Herman
RCE systems s.r.o., Brno, Czech Republic

ABSTRACT: Traffic data and human driver behaviours simulation are two of the most important parameters for the proper implementation of a traffic simulation. In our case study, we implemented an innovative method to obtain both a complex set of data of OD matrix and detailed human driver behaviours data in order to set a specific scenario simulation. Firstly, we recorded an aerial video and, subsequently, we conducted an advanced traffic analysis of aerial video data using DataFromSky service. The result is a complex complete set of traffic parameters. Data collection is based not only on classic OD matrix (volume data, vehicles classification, hourly rates, etc.), but also on innovative dynamic vehicle database parameters (speeds, lateral and tangential accelerations, travel times and distances, trajectories, etc.). All of these parameters are supplied to each vehicle detected. For this case study, our team provided as well a set of specific simulation parameters, such as gap time and follow-up time, for each entry/exit analysed. It is clear that most of this data were used to the calibration and improvement of our simulation network. In particular, we inserted specific values for each vehicle class (dimensional and dynamic values) and a detailed calibration of our behavioural models in Krauss settings. The resulting traffic simulation scenario shows the highest correlation value between real and simulated driver behaviours, probably, never obtained before.

1 INTRODUCTION

The dataset used for this case study come from a survey that took place in Sheffield (UK), on the A6178, called Sheffield Road, that connects the city with Rotherham (Figure 1). The road within the investigated section is one lane per direction,

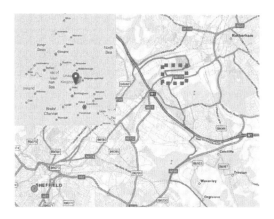

Figure 1. Study area localization.

separated by a painted traffic divider between them, slightly curved in the middle of the two straight segments. The survey area was recorded by means of a UAV by Vertex Access, for an interval time of nearly 10 minutes, on a segment of almost 250 meters placed between Temple Road and Grange Lane enter onto the main road. In fact, the actual distance that was considered for the data extraction is less than that, since it depends on the exact place where the entry and exit gates were positioned. In particular, the examined length was about 229 meters.

Once the video has been recorded from above the study area, DataFromSky team allows to collect a large amount of data for each of the detected vehicle passed within it: this tool allows not only to know exactly how many vehicles entered and exited the area, but also at which time and how many seconds it needs to cover the distance, and so the travel time. The dataset doesn't restrict to that, providing also the speeds and accelerations measured for each timestamp (from second up to milliseconds) between an entry and an exit gate.

The entire set of parameters become useful in order to calibrate and implement the behavioural car-following model, that defines the speed of a vehicle in relation to the vehicle ahead: a modification of the Krauss model is used as default by the simulation software SUMO (Open Source Licensed). The idea on which the models work, both the original and the modified one, is letting vehicles drive as fast as possible while maintaining perfect safety, being always able to avoid a collision if the leader starts braking.

Through the software SUMO the network were loaded from OpenStreetMap, and four detectors were placed on lanes in the same position as the ones used with DataFromSky, so that the compared trajectory, the real and the simulated ones, have the same length. The comparison among different scenarios was implemented by the analysis of the travel times, that provide an overall measure of vehicles, alone or in a group, driving into the network. The term of comparison (scenario no.0) was represented by the real investigation of the travel time, provided by the automatic calculation of DataFromSky.

Finally, two different simulations were developed with SUMO—*Simulation of Urban Mobility*: the scenario no.1, with the default values assigned by the simulator to the 'car' vehicle (only this type of vehicle has been considered in this case study), and the scenario no.2, where the aggregated values of speed and acceleration got from the DataFromSky analysis was used to better define the vehicle driving behaviour, showing improvement in the output simulation.

2 WHAT IS AND WHAT DOES DATAFROMSKY

Every year, traffic jams and accidents inflict huge costs worldwide. The most effective strategies for dealing with these notorious problems include improvements in traffic management, signalizing systems and the quality of road network. However, to design and implement the changes successfully, a large amount of data is needed. Conventional approaches to data collection offer simple measurements of multiple quantities, often disjointed. DataFromSky offers an innovative approach to the traditional problems—the use of aerial video which brings numerous advantages.

DataFromSky (Apeltauer *et al.*, 2015) is a service based on automatic extraction of time-space trajectories of vehicles from aerial video. A wide range of traffic parameters, such as speed, accelerations and gate counting, are derived through an advanced computer vision analysis of the image. There is no need for expensive equipment, sophisticated sensor and time-consuming installation. The video can be collected using any small aerial platform, be it an Unmanned Aerial Vehicle (UAV) or an aerostatic balloon, or simply shot from high-rise.

Thanks to long-term development of specialized algorithms for detection and tracking, DataFromSky is able to provide comprehensive information about the behaviour of road users merely from aerial video data taken by an ordinary camera. Indeed, unlike in many commonly used systems for traffic monitoring, the only requirements for detailed analysis of road user behaviour by DataFromSky tool is an aerial video and a description of the recorded scene.

Our detection algorithms are capable of recognizing vehicles as small as 16×16 pixels to allow for coverage of a large area by one device only (more than 30,000 m²). As high-quality video stabilization algorithms are an integral part of DataFromSky tool, the sensing platform is not required to be 100% stable during the recording.

The overall system can be divided into three main parts:

– Pre-processing
– Vehicle detection
– Tracking

In the pre-processing step, the acquired image gets undistorted and geo-registered against a user-selected reference frame. The efficiency and robustness of vehicle detection methods in aerial images have been addressed several times in the past. DataFromSky developers designed a highly optimized multi-layer detector and trained it on a hand annotated training dataset with 80,000 positive and 80,000 negative samples. The detector is capable of learning from its errors and improving over time. The tracking part is an extremely challenging task due to the presence of noise, occlusion, dynamic and cluttered backgrounds, and changes in appearance of the target. A subtype of sequential importance sampling filter has been employed to deal with these challenges. Due to the outstanding performance of detection and tracking algorithms, DataFromSky team can provide high-quality data and guarantee a hit rate of over 96%.

Each vehicle is labelled by a unique ID and all its positions, speeds, and accelerations are recorded during its passage through the monitored section. Thanks to extraction of spatiotemporal information about each vehicle in the scene, DataFromSky offers unlimited possibilities in data interpretation and inspection.

The service offered by DFS is thus articulated into the following steps:

– The data collection is performed by external partners operating drones/UAVs commercially in the target areas.
– The recorded videos are automatically analysed by the Extractor tool developed by

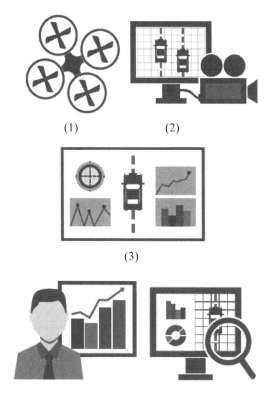

(1)   (2)

(3)

Figure 2.   DataFromSky analysis steps.

Figure 3.   DataFromSky output video data example.

Figure 4.   DataFromSky analysis data example.

DataFromSky. It is used to extract trajectories and classifications.
- The output is delivered to the client together with software for further analysis and data interpretation—DataFromSky Viewer.
- DataFromSky developers can modify the software according to any needs and requirements.

DataFromSky also includes a tool for advanced data analysis. A virtual counting gate may be defined anywhere in the input video to establish the number of vehicles in this place. Further, analysists can also obtain detailed information about traffic lane utilization, sectional measurements, average speeds, observed anomalies, etc. (Figures 3 and 4). The tool also provides advanced functions for traffic engineers such as *Tg* (*time gap*) and *Tf* (*time to follow*) estimations, and computation of vehicle lateral acceleration.

Thanks to the advanced technology of DataFromSky, application and research fields are very wide.

3   WHAT IS AND WHAT DOES SUMO

"Simulation of Urban Mobility" (SUMO) (Jakob *et.al.* 2012) is an open source traffic simulation package including the simulation application itself as well as supporting tools, mainly for network import and demand modelling. Through a three steps elaboration process, the micro simulator reproduces and then forecasts how a given traffic demand distributes and behaves within a road net, on the basis of the geometric characteristics and intersection regulation of the roads, considering the interaction among different types of vehicle.

An example of the procedural phases needed to perform a traffic simulation is shown in the following graph (Figure 5): the road network can be either generated using *netconvert* or *netgenerate*, whereas od2trips, *duarouter*, *dfrouter*, *jtrrouter* are the applications that allow to describesontoment of vehicles, each of which is defined at least by a unique identifier, the departure time and the vehicle's route through the network.

Additional variables may add further devices and components to the simulation, or more accurately define specific part in order to refine and improve the quality of the simulation.

By means of these inputs, the application sumo/sumo-gui performs a time-discrete simulation, whose model is space-continuous and where each

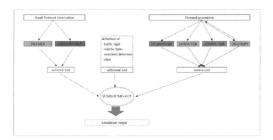

Figure 5. Common Network and Demand generation procedures.

vehicle's movement through the network is computed using a car-following model.

Car-following models usually compute a vehicle's speed by looking at the maximum allowable speed, its distance to the leading vehicle and the leader's speed. In particular, SUMO uses an extension of the stochastic car-following model developed by Krauss (1998) as default, because of its high execution speed and simplicity.

Indeed, speed, acceleration and deceleration of every single vehicle can't be the ones desires, as if each of them was alone and free to choose its own free flow travel speed. On the contrary, the behaviour of each driver is influenced by the other's driving, even more as the flow increases, and most of all by the leader's.

Assuming two cars following each other on the same road, the acceleration profile of the following car is just a function of the relative speeds between the two cars, the follower (f) and the leader (l):

$$\frac{\partial v_f(t-\tau)}{\partial t} = \lambda \cdot [(v_f(t) - v_f(t)]$$

v_f = follower's speed
v_l = leader's speed
λ = sensibility
τ = reaction time
t = time of analysis

In the modified Krauss model working on SUMO the correlation among vehicles is defined by considering for the following vehicle the maximum allowable speed, always ensuring it to avoid collision with the vehicle ahead, even though in case of hard braking. The desired speed describes the will of a vehicle to reach it, while still taking into account the safety limits.

$$v_{safe} = v_l(t) + [g_n(t) - v_l(t)\tau] / \left[\frac{v_f(t) + v_l(t)}{2b} + \tau \right];$$

where:

$$v_{safe} = v_l(t) + [g_n(t) - v_l(t)\tau] / \left[\frac{v_f(t) + v_l(t)}{2b} + \tau \right];$$
$$g_n(t) = x_l(t) - x_f(t) - \min Gap$$

v_{safe} = safety speed
g_n = distance between follower and leader
b = deceleration
x_l = leader position
x_f = follower position

The main input parameter for the Krauss model's specification in SUMO, associate to every defined type of vehicle, are:

- *Acceleration* [m/s²]: the acceleration ability of vehicles of this type that a driver chooses;
- *Deceleration* [m/s²]: the deceleration ability of vehicles of this type that a driver chooses;
- σ: the driver imperfection (between 0 and 1); for values above 0, drivers with the default car-following model will drive slower than would be safe by a random amount;
- τ [s]: the driver's reaction time; drivers attempt to maintain a minimum time gap of τ between the rear bumper of their leader and their own front-bumper + minGap;
- *minGap* [m]: empty space after leader;
- *maxSpeed* [m/s]: the vehicle's maximum speed;
- *SpeedFactor* and *SpeedDev*: the first defines the vehicles expected multiplicator for lane speed limits, whereas the second represent its relative standard deviation, so the ratio of the standard deviation to the mean; they are used to sample a vehicle specific chosen speed from a normal distribution, with mean the one calculated with the SpeedFactor and a deviation speedDev. A vehicle keeps from the distribution its chosen SpeedFactor for the whole simulation and multiplying it with edge speeds can compute the actual speed for driving on each edge. Thus vehicles can exceed edge speeds, although vehicle speeds are still capped at the vehicle type's maxSpeed. Using speed distributions is highly advisable to achieve realistic car following behaviour: if all vehicles have the same maximum speed on any given road, they will not be able to catch up with their lead vehicle causing unrealistic large headways.

4 ANALYSIS

One of the main problem in comparing simulated to real data is the collection of the data themselves. As a matter of fact, it's generally not easy to be able, in terms of cost, devices and concrete capability, to extract some sort of data from the real study environment.

The service proposed by DataFromSky can overcome lots of limitation in the road traffic field, but not this only, providing sufficient amount of data to enable vehicle location and movement monitoring. From here we started to develop our data: having a video from above the study area (watch it at *https://vimeo.com/117601820*), we automatically analysed the traffic activity within an entry and an exit gate we set as shown in Fig. 6. Moreover, tables Table 1, Table 2 and Table 3 exhibit an example of the main part of the whole extracted dataset needed for the following processing steps.

For each of the two directions, data coming from table Table 1 have been reported into a graph, with entry time [s] on the x-axis, and actual travel time on the y-axis, calculated as the difference between the exit and entry time. Although not many factors can influence the short chosen road section (only 229 m), since the covered distance is almost straight and there are no turning vehicles and no intersection, except for two accesses that were not used within the recorded time interval, the curve is not flat at all. This means, as we expected, that even if the network is very simple, maybe overly, drivers have different behaviour: the curves have mean travel time equal to 16,7 s and 17,1 s, in the East-to-West and West-to-East direction respectively, but the curves trend vary between 13 s and 21 s in the first case, 11 s and 21 s in the second.

On the other hand, both Table 1 and Table 2 were useful for investigate and collect the aggre-

gated value needed to represent the vehicle characteristic within the SUMO network, shown in Fig. 7. In particular, two types of vehicle have been defined, one per direction, and the following Krauss parameter of Table 4 have been adopted instead of the default ones, reported in Table 5.

In particular, Speed Factor represent the ratio between the actual speed on the road, taken the 85th percentile of the mean speeds of all vehicles as reference, and its speed limit, whereas the Speed Deviation comes from the calculated standard deviation over the mean speeds on the study segment.

The maximum acceleration and deceleration taken for the simulation correspond to the 95th percentile of the actual maximum values (positive and negative ones) for each vehicle, as shown in Fig. 8 (a) and (b).

On the contrary, the following default values are commonly used in SUMO

The simulations with the two different definitions of vehicle, the default one for scenario no.1 and the one coming from DFS for the scenario no.2, were run. In both case, vehicle were introduced at the beginning of the road at the time corresponding to time recorded data from DataFromSky, so that at the entrance were almost equal the distances, in time and space, between following vehicles.

In order to evaluate the quality of the simulation done with SUMO software and, in particular, to do a benchmarking of the analysis performed with default and real parameters, in comparison with the observed behaviour, data sets have been shown on the graphs below, and the mean values reported on Table 6.

It can clearly be seen that the results obtained from the scenario no.1 can't describe with sufficient degree of accuracy the reality of the phenomenon, since they do not provide a variance in the simulation. The curve appears rather flat, which means that the travel time doesn't significantly change and every vehicle driving on the network spend almost the same time between the entry and the exit of the lane. Besides, the mean value around which the curve is built differs from the mean of

Figure 6. Position of the entry and exit gates with DFS on the examined lanes.

Table 1. Example of DataFromSky output data: id, type, gate, time, distance, speed.

Vehicle id	Vehicle type	Entry gate	Entry time (ms)	Exit gate	Exit time (ms)	Travelled dist. (m)	Avg. speed (m/s)
24	Car	14	8320	20	26440	228,6	12,61
28	Car	14	40120	20	57360	228,2	13,24
29	Car	14	54640	20	69040	228,5	15,87
...	...						

Table 2. Example of DataFromSky output data: tangential acceleration for every second.

	Vehicle id			
Travel time (s)	24	28	29	...
1	0.10	−0.50	0.97	...
2	0.66	0.34	0.22	...
3	0.77	0.38	−0.26	
4	0.69	0.43	0.01	
5	0.27	−0.02	−0.38	
6	0.19	0.29	−0.87	
7	0.19	0.38	0.07	
8	0.24	−0.17	−0.23	
9	−0.01	−0.16	−0.50	
10	−0.03	−0.22	−0.58	
11	0.11	0.32	−0.32	
12	−0.10	0.06	0.19	
13	−0.13	−0.03	−0.32	
14	−0.30	−0.16	−0.31	
15	−0.21	−0.19	−0.31	
16	−0.19	−0.21	0.00	
17	−0.31	−0.21	0.00	
18	−0.28	−0.44	0.00	
19	−0.25	0.00	0.00	
20	0.00	0.00	0.00	
a max [m/s^2]	0.77	0.43	0.97	
a min [m/s^2]	−0.31	−0.50	−0.87	

Table 4. Krauss Model real car parameters (scenario2).

	n cars	Speed factor	Speed dev.	a max (m/s^2)	d max (m/s^2)
East to West	30	1.13	0.11	1.85	−0.90
West to East	48	1.11	0.13	1.12	−3.27

Table 5. Krauss Model default car parameters (scenario1).

	n cars	speed factor	speed dev	a max (m/s^2)	d max (m/s^2)
East to West	30	1.0	0.0	2.9	−7.5
West to East	48	1.0	0.0	2.9	−7.5

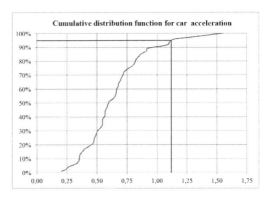

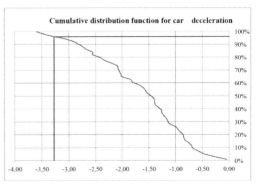

Figure 8. Example cumulative distribution function of acceleration (a) and deceleration (b) for West to East direction.

Table 3. Main aggregated output values from DataFromSky.

	n cars	Mean distance (m)	Mean speed (m/s)	St. dev. (σ)	85ile mean speed (m/s)	Mean travel time (s)
East to West	30	228.41	13.86	1.46	15.69	16.65
West to East	48	235.97	13.99	1.76	15.41	17.10

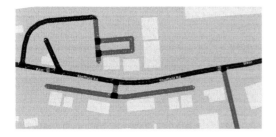

Figure 7. Sumo network representation with Instant Induction Loops.

the scenario no.0: the first is 1,4 s higher in the East-to-West direction, and 1 s higher in the West-to-East direction (Figures 9 and 10).

On the contrary, for both considered directions, it is clearly visible the greatest similarity of the scenario no.2 to the observed data set's curve. The divergence between their relative mean is limited

Table 6. Comparison between travel time on different scenarios.

	Mean travel time (s) Scenario no.0 (with DataFromSky survey)	Mean travel time (s) Scenario no.1 (with SUMO default values)	Mean travel time (s) Scenario no.2 (with SUMO survey values)
East to West	16.65	18.06	16.52
West to East	17.10	18.09	16.82

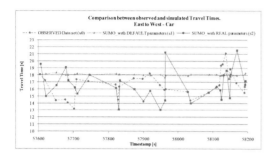

Figure 9. Comparison between observed and simulated Travel Times, direction East to West.

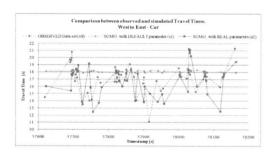

Figure. 10. Comparison between observed and simulated Travel Times, direction West to East.

to 0,13 and 0,28 seconds, and the trend fits very well with the one of the scenario no.0. There is still lack of precision in the position of the maximum and minimum peaks of the curve, but this is due to the fact that we implemented an overall type of vehicle, got by aggregate parameter consideration rather than the single vehicle dynamic characteristics, just like commonly happens when compute a traffic simulation. By means of that, vehicles were entered the network at the same time they were observed to do it, but the simulation keep randomness in choosing its dynamic parameter among the distribution we set. That is the reason why in this type of analysis it is not necessary to have the complete overlaying of the curves, since that's not what we expected from the simulation.

5 CONCLUSIONS

The value of the current work should be identified not only in the appreciable precision degree achieved through the processing, but primarily in the methodological approach proposed, that starting from the existing models has turned to a formulation that fits the particular dataset considered, with an elaboration and a contextualization of the parameters.

It represents only a preliminary attempt to implement a commonly used simulation tool, with the aid of a full and not easily obtainable data set, in order to better reproduce what is the common approach of vehicle in that type of environment, with that type of drivers.

In future work all the aspect that have been considered should be increased, from the network to number and types of sampling vehicles over which extract the information. That upper analysis should even more highlighted the differences between the default set of vehicles parameters and the one coming from a real analysis of the drivers' attitude in a specific context.

REFERENCES

Apeltauer, J. & Babinec, A., & Herman, D. & Apeltauer, T. 2015. Automatic vehicle trajectory extraction for traffic analysis from aerial video data. *The International Archives of the Photogrammetry, Remote Sensing and Spatial Information Sciences*, Volume XL-3/ W2, 2015 PIA15+HRIGI15 – Joint ISPRS conference 2015, 25–27 March 2015, Munich, Germany.

Jakob, E. & Behrisch, M. & Bieker, L. 2012. Recent Development and Applications of SUMO - Simulation of Urban MObility; Daniel Krajzewicz. *International Journal On Advances in Systems and Measurements*, 5 (3–4):128–138.

Krauß, S. 1998. Hauptabteilung Mobilität und Systemtechnik des DLR Köln 1998. Microscopic Modeling of Traffic Flow: Investigation of Collision Free Vehicle Dynamics ISSN 1434–8454.

Vertex Access, 296 Sheffield Road, Rotherham, S60 1DX Phone: +44 (0) 1709 379 453 Web: www.vertexaccess.com

Transport Infrastructure and Systems – Dell'Acqua & Wegman (Eds)
© 2017 Taylor & Francis Group, London, ISBN 978-1-138-03009-1

Social exclusion and high-speed rail: Some evidence from three European countries

F. Pagliara, L. Biggiero & F. Menicocci
Department of Civil, Architectural and Environmental Engineering—University of Naples Federico II, Naples, Italy

ABSTRACT: Very few contributions are present in the literature dealing with the issue of social exclusion related to High-Speed Rail (HSR) services. The objective of this study is to compare the current situation of non-HSR users in Italy, Spain and England and the factors preventing them from choosing this service. Three countries with very different HSR networks have been chosen in order to get evidence of the different behaviour. For this purpose, three surveys have been delivered to Italian, English and Spanish users of the transport systems for long distance journeys. Data about their socioeconomic characteristics and their perception of social exclusion have been collected based on seven principles. The main result of these surveys has been that a relationship between social exclusion and HSR is evident, especially in terms of economic and geographical exclusion. Moreover, these factors are perceived differently based on the different "shape" of the HSR networks and the service provided.

1 INTRODUCTION

According to Levitas *et al.* (2007) social exclusion is *"The lack or denial of resources, rights, goods and services, and the inability to participate in the normal relationships and activities, available to the majority of people in a society, whether in economic, social, cultural or political arenas. It affects both the quality of life of individuals and the equity and cohesion of society as a whole."*

In the literature different approaches have been proposed to solve the problem of social exclusion related to transport systems. Indeed, transport systems planning should be integrated with the urban and social policies. One first step towards the reduction of social exclusion might be that of promoting activities to increase accessibility.

The objective of this paper is to analyse whether High Speed Rail (HSR) systems can increase social exclusion for long-distance trips, taking into account that other transport alternatives are available to users. Specifically, considering that the trend in future transportation systems investments is represented by these services, authors would like to analyse this phenomenon and its impacts on social exclusion. For this purpose, three case studies have been analysed, i.e. Spain, UK and Italy.

This paper is organised as follows. Section 2 presents a literature review on the link between HSR and social exclusion. Section 3 introduces the Spanish, English and Italian HSR systems. The methodology is proposed in section 4. Conclusions and further perspectives are presented in section 5.

2 SOCIAL EXCLUSION AND HSR SYSTEMS

Among the very few studies present in the literature on the link between HSR systems and their impacts on social exclusion, the statistical analysis of surveys carried out by Cass *et al.* (2005) reports interesting results. It indicates that HSR has both positive and negative social impacts. The positive social impact is represented by the increased accessibility and activities for commuting HSR users. The concept of accessibility represents the relationship between the system of activities in a given territory and the transport system serving it. The study carried out in Spain by Monzón *et al.* (2010) shows the role played by the selection of the commercial speed. Indeed, an increase from 220 km/h to 300 km/h in a given corridor results in significant negative impacts on spatial equity between locations with and without a HSR service.

The same authors propose an assessment methodology for HSR projects following a twofold approach, i.e. addressing issues of both efficiency and equity. The procedure uses spatial impact analysis techniques and is based on the computation of accessibility indicators. Efficiency impacts are evaluated in terms of increased accessibility resulting from the HSR project, with a focus on major urban areas. Likewise, spatial equity implications are derived from changes in the distribution of accessibility values among these urban agglomerations (Monzón *et al.* 2013).

The paper by Chen & Wei (2013) reports the case study of Hangzhou East Rail station in China. This area is undergoing a rapid industrialization and thus workers' incomes are increasing significantly. However, HSR is still not affordable for the majority of the population. In the contribution by Shi & Zhou (2013), the aim is that of analysing transportation equity change in terms of accessibility change experienced by cities served by the HSR line in China. The main research findings, from the equity assessment, reveal that investments in HSR systems do not have a strong impact in fostering social exclusion in terms of being excluded from the use of the new high speed infrastructure.

From these contributions it is clear that elements of social exclusion from HSR are represented by the accessibility and the income. Therefore high income users can afford HSR but also a high accessibility to HSR station can contribute to increase its choice.

3 HIGH SPEED RAIL SYSTEMS IN SPAIN, UK AND ITALY

Since January 2016, Spain has the world's second longest high-speed network, after China, and the longest in Europe (Pagliara *et al.* 2016a), with around 3,100 km of HS lines in operation. The service of HSR in Spain—known as AVE, Alta Velocidad Española—is operated by RENFE Operadora, the Spanish national railway company. Since 2005, AVE trains run on a HSR network owned and managed by ADIF, the public company in charge of the management of most of the Spanish railway infrastructure. Although RENFE Operadora is the only company operating the high-speed trains nowadays, private companies may be allowed to operate trains in the future, in accordance with the EU legislation. It is envisaged that the Madrid-Valencia corridor will be the first case to introduce competition in the HSR services in the country.

The first HS line was opened in 1992, connecting the cities of Madrid, Cordoba and Seville. It was designed according to the technical standards of the French high-speed TGV. In the following years, the network was extended towards the northern part of the country, with the aim to create a connection to France and thus to the European HSR network (see Figure 1).

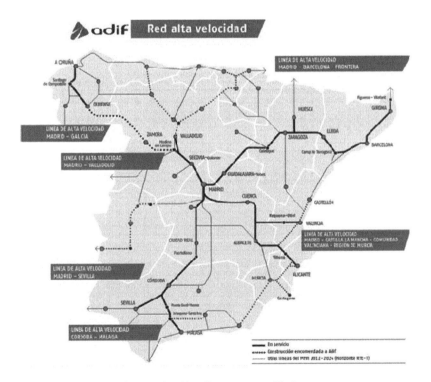

Figure 1. Spain's HSR network as of November 2015 (Source: www.adif.es).

Great Britain has been one of the last nations to introduce HSR systems. High Speed 1 was inaugurated in 2007 and it connects London with the Channel Tunnel. The travel time between London and Paris is only two hours and a half.

At its opening the line was positively welcomed by travelers and this was one of the first reasons for the design of another line, the High Speed 2, whose works should start in 2017. It will link London Euston to Birmingham (2026), and then, following a 'Y' shape, it will connect Manchester (2027) on the west and Leeds (2033) on the east. The line will be 540 km long, with speeds of almost 400 km/h (see Figure 2).

In Italy, the first HSR line was inaugurated in 1992 between Florence and Rome with the so called "Direttissima", which allowed trains to run at 230 km/h covering the 254 km between Rome and Florence in about two hours. However, this project dated back to 1970.

The new generation of HSR (i.e. with trains running at 300 km/h) started in December 2005 between Rome and Naples and Milan and Bologna. Later, in December 2009, the project was extended with the Milan-Turin and the Bologna-Florence lines,

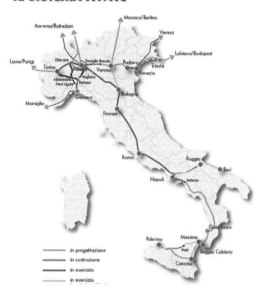

IL SISTEMA AV/AC

Figure 3. Italy's HSR network as of 2014 (Source: Delaplace *et al.* 2014).

as well as with the urban penetration into the cities of Rome and Naples. In 2015 the Italian HSR network was operational for more than 1400 km (see Figure 3).

He national Italian network and operations are all owned by FS (State Railway) Holdings, a fully government owned company. It has three key operating subsidiaries: Trenitalia operates all freight and passenger trains, including the high-speed trains, RFI (Rete Ferroviaria Italiana) manages the infrastructure, and TAV (Treno Alta Velocità SpA) is responsible for the planning and construction of the new HS infrastructure.

The introduction of the new private operator Nuovo Trasporto Viaggiatori (NTV) in April 2012, competing with the Trenitalia on the national HSR network, represents the first case of competing HSRs operating on the same line (i.e. multiple operators on a single infrastructure). NTV represents the first private society to benefit of the European liberalisation of the High-speed train networks.

4 THE METHODOLOGY

This contribution is based on the framework of factors that may limit the mobility of socially excluded people, proposed by Church *et al.* (2000). In their paper, the categories of exclusion connected to transport, applied for urban trips, are the following ones:

Figure 2. UK's HSR network as of June 2016.

1. Physical exclusion: physical barriers, i.e. lack of disabled facilities or timetable information, limiting accessibility to transport services.
2. Geographical exclusion prevents people from accessing transport services, especially those living in rural or peripheral urban areas.
3. Exclusion from facilities, concerning the low accessibility connected with facilities, like shops, schools, health care or leisure services.
4. Economic exclusion represents the high monetary costs of travel preventing or inhibiting access to facilities or employment and thus having an impact on incomes.
5. Time-based exclusion refers to other demands on time, like combined work, household and child-care duties, reducing the time available for travel.
6. Fear-based exclusion concerns to the fears for personal safety precluding the use of public spaces and/or transport services.
7. Space exclusion is the security or space management preventing given groups having access to public spaces, like first class waiting rooms at stations.

In this paper, these categories have been adapted to medium-long distance trips.

Starting from this premise three Revealed Preference (RP) surveys were carried out. The questionnaires were created on web platforms and 414 useful ones were collected in Spain, 359 in the United Kingdom, 968 in Italy. Due to the survey method used, based on the web platform, the sample needed to be weighted (Pagliara *et al.* 2016a, b, Pagliara & Biggiero 2016). The percentages of gender and age classes, based on the Spanish census (INE 2015), the UK census (ONS 2014) and the Italian census (ISTAT 2015) have been considered to adjust the sample. Then, those observations with a trip length lower than 60 km have been removed from the sample since they typically correspond to regional trips, not operated by HSR services. In this case authors tried to avoid any bias present in the data set used to make inferences. In Table 1 the socioeconomic and trip characteristics of the non-HSR users for the three cases are reported.

It is interesting to notice some differences and similarities among the case studies. Concerning the age, there are not significant differences, just for the case of Italy it is possible to see almost 80% of non HSR-users are less than 55 years old. This result can be justified considering economic factors or work or study engagements. Spain and Italy have a similar trend concerning the non-HSR users nationality (almost 100% of the sample), while in UK non-HSR users are a bit more than 50% and this can be supported by the multiethnic image of the country. Concerning the occupation, non-HSR users have a

Table 1. Socioeconomic and trip characteristics for the three countries.

Characteristics	Levels	Spain non-HSR users [%]	UK non-HSR users [%]	Italy non-HSR users [%]
Age	18–23	13.23	13.41	10.40
	24–34	12.88	16.61	22.22
	35–55	32.91	34.06	44.47
	>55	40.98	35.92	22.21
Gender	M	48.02	50.02	58.11
	F	51.98	49.98	41.89
Nationality	National	97.21	56.61	99.90
	Other	2.79	43.39	0.10
Education	Degree or more	73.48	38.11	46.11
	Other	26.52	61.89	53.89
Occupation	Full time/ part time worker	56.71	33.64	62.59
	Student	12.50	20.12	15.10
	Unemployed	2.00	0.84	3.70
	Freelance	1.97	10.11	6.87
	Retired	26.83	27.64	11.45
	Other	–	7.65	0.29
Monthly income	<€1500	24.90	12.42	48.20
	€1500–€3000	53.16	13.06	28.62
	>€3000	21.94	74.52	23.17
Trip purpose	Work	4.81	14.01	30.39
	Study	0.94	9.66	13.23
	Holiday	33.00	18.38	42.25
	Personal activities	61.24	57.94	14.13
Travel type	Alone	31.67	43.29	39.25
	Partner	39.58	18.21	13.03
	Colleagues	–	8.39	14.68
	Friends	15.07	8.30	16.15
	Relatives	13.68	21.83	16.89

partial or full time job, even if in UK there are a lot of students and retired. In Italy and in Spain almost 80% of non-HSR users have a household monthly income less than 3,000€ while the opposite can be found in UK (this result can be justified considering that non-HSR users represent the majority of the sample (see Figure 4). Almost 60% of non-HSR users in Spain and in UK have travelled for personal purposes, while in Italy only 14% have travelled for the same purposes. On the contrary in this country non-HSR users have travelled for holidays and work purposes. Almost 40% of non-HSR Spanish users have travelled with the partner, while the same percentage in UK and in Italy have travelled alone. The mode choice is represented in Fugure 4.

It is evident in Italy and in Spain that HSR is the most chosen transport alternative for long-distance trips.

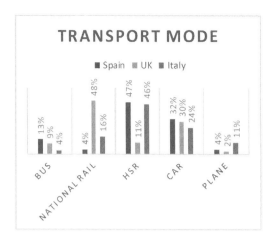

Figure 4.   Transport mode choice (all samples).

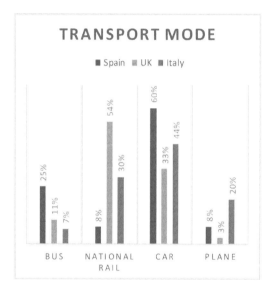

Figure 5.   Transport mode choice for non-HSR users.

Car has been chosen by 32%, 30% and 24% in Spain, UK and Italy respectively. In UK, as stated before, HSR has been chosen by the 11% of the respondents, the same for the bus, while 48% of the respondents have chosen train and 30% car.

Non-HSR users are represented in Figure 5.

In Table 2, the choice among the seven Church *et al.*'s categories of social exclusion has been analysed and summarized in Figure 6.

It follows that in Italy, geographical exclusion is the most perceived, followed by the economic and time-based. This is justified by the fact that the HSR system connects only big cities and therefore it is not considered as a real alternative transport mode.

Table 2.   Church *et al.*'s categories of social exclusion (non-HSR users)—Comparison.

Categories of exclusion	Case studies		
	Spain [%]	UK [%]	Italy [%]
Economic	39.85	48.73	27.20
Time-based	10.59	11.07	6.30
Spatial	6.27	1.38	–
Fear-based	1.23	–	0.90
Geographical	23.87	24.04	58.70
Physical	9.75	9.90	5.60
Facilities	8.44	4.88	0.30
Total	100.00	100.00	100.00

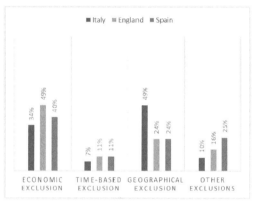

Figure 6.   Church *et al.*'s categories of social exclusion (non-HSR users)—Comparison.

Even if the presence of two companies competing on the same HSR line has reduced travel costs, there are still a lot of users who cannot afford it. In UK, on the other hand, the first two factors of exclusion are inverted. Geographical exclusion is less felt with respect to the economic one. The limited extension of the HSR network in UK could lead to think that geographical exclusion could have been highly felt. However this has not been the result, considering that a limited network with low accessibility to the stations does not make HSR a real alternative with respect to the conventional rail system, which is capillary and efficient. For Spain, it is possible to note that the result is very similar to that obtained in UK, even for different reasons. Indeed, the Spanish network, as reported before, is very extended but the economic factor is still an element of exclusion.

It is interesting to highlight that in all the three cases time-based exclusion is the third factor. The other elements of exclusion are less perceived by non-HSR users.

In the following Tables, from 3 to 8, the factors of exclusion are reported on the basis of the trip purpose and the hosuehold monthly income. In Tables 3 and 4 the results for Spain are reported.

It results that Spanish non-HSR users perceiving the factors of exclusion are those who have travelled for personal purposes and have a low-medium income. Specifically, those travelling for personal purposes are free to plan the trip, taking

Table 3. Church *et al.*'s categories of social exclusion vs trip purpose (non-HSR users)—Spain.

| Categories of exclusion | Trip purpose | | | | |
	Work [%]	Study [%]	Holiday [%]	Personal activities [%]	Total [%]
Economic	4.47	35.23	1.00	59.29	100.00
Time-based	–	11.69	2.85	85.47	100.00
Spatial	–	24.07	3.01	72.92	100.00
Fear-based	–	32.41	–	67.59	100.00
Geographical	5.49	44.51	–	50.00	100.00
Physical	–	23.50	2.65	73.85	100.00
Facilities	–	22.84	3.06	74.10	100.00

Table 4. Church *et al.*'s categories of social exclusion vs monthly income (non-HSR users)—Spain.

| Categories of exclusion | Monthly income | | |
	Low-medium [%]	High [%]	Total [%]
Economic	77.96	22.04	100.00
Time-based	93.01	6.99	100.00
Spatial	67.40	32.60	100.00
Fear-based	57.23	42.77	100.00
Geographical	73.85	26.15	100.00
Physical	69.37	30.63	100.00
Facilities	55.52	44.48	100.00

Table 5. Church *et al.*'s categories of social exclusion vs trip purpose (non-HSR users)—UK.

| Categories of exclusion | Trip purpose | | | | |
	Work [%]	Study [%]	Holiday [%]	Personal activities [%]	Total [%]
Economic	11.25	11.09	18.45	59.21	100.00
Time-based	31.25	1.45	10.74	56.57	100.00
Spatial	13.55	11.69	24.04	50.73	100.00
Fear-based	–	–	–	–	–
Geographical	9.43	17.04	11.06	62.47	100.00
Physical	9.64	7.45	41.40	41.52	100.00
Facilities	36.40	–	–	63.60	100.00

Table 6. Church *et al.*'s categories of social exclusion vs monthly income (non-HSR users)—UK.

| Categories of exclusion | Monthly income | | |
	Low-medium [%]	High [%]	Total [%]
Economic	53.32	46.68	100.00
Time-based	21.39	78.61	100.00
Spatial	47.35	52.65	100.00
Fear-based	–	–	–
Geographical	53.88	46.12	100.00
Physical	52.53	47.47	100.00
Facilities	46.31	53.69	100.00

Table 7. Church *et al.*'s categories of social exclusion vs trip purpose (non-HSR users)—Italy.

| Categories of exclusion | Trip purpose | | | | |
	Work [%]	Study [%]	Holiday [%]	Personal activities [%]	Total [%]
Economic	34.16	10.90	36.05	18.90	100.00
Time-based	–	4.05	91.22	4.73	100.00
Spatial	–	–	–	–	–
Fear-based	88.89	–	11.11	–	100.00
Geographical	29.72	10.34	40.06	19.88	100.00
Physical	9.21	15.79	55.26	19.74	100.00
Facilities	–	–	100.00	–	100.00

Table 8. Church *et al.*'s categories of social exclusion vs monthly income (non-HSR users)—Italy.

| Categories of exclusion | Monthly income | | |
	Low-medium [%]	High [%]	Total [%]
Economic	70.06	29.94	100.00
Time-based	97.30	2.70	100.00
Spatial	–	–	–
Fear-based	88.89	11.11	100.00
Geographical	80.12	19.88	100.00
Physical	38.16	61.84	100.00
Facilities	100.00	–	100.00

into account the economic and time availability, but also the trip origin and the destination.

In Tables 5 and 6 the results for UK are reported.

The same considerations made for Spain can be adapted to the case study of UK. Indeed, also in this case, those suffering of social exclusion re the ones travelling for personal purposes.

Concerning the income, those suffering of geographical and economic exclusion are the ones of low-medium income. On the other hand high income English non-HSR users highly perceive the time-based exclusion.

In Tables 7 and 8 the results for Italy are reported.

The latter are very similar to the case study of Spain with the difference that the only category of exclusion is represented by those who have travelled for holidays. Also in this case, this category was the most represented in the sample, confirming that trips fo toursim purpose, but also for personal purposes, are represented by a highly elasticity of choice (e.g. destination, time, budget etc.). Therefore the same comments can be applied to this case study.

5 CONCLUSIONS AND FURTHER PERSPECTIVES

In this paper the relationship between HSR and social exclusion has been analysed. Following the framework proposed by Church *et al.* (2000). The results of three Revealed Preference surveys have shown that only some criteria are perceived by the users and among them only some of them are the most relevant. For those who have not chosen HSR, the main reason for that, is the economic exclusion, i.e. the cost of the HSR ticket. It follows the geographic exclusion, i.e. the low accessibility to the departure/arrival station. The fact that both criteria are greatly perceived by low income classes can be interpreted by the location of residences of these classes of travellers. For the higher cost connected with the use of the residences, it is clear that those having higher incomes live in city centres, which, in general, are served by a good public transport and by taxi as well. Indeed a good public transport system can allow an easy access to the departure/arrival station.

Further perspectives will consider the collection of a larger data set which can support these findings and the specification and calibration of some mode choice models.

Indeed, a quantitative approach should be taken into account with the aim of better evaluating the perception of social exclusion related to mode choice.

REFERENECES

Cass, N., Shove, E., Urry, J. 2005, Social exclusion, mobility and access. *The social review* 3: 539–555.

Chen, C-L., Wei, B. 2013, High-Speed Rail and urban transformation in China: the case of Hangzhou east rail station. *Built Environment* 39: 385–398.

Church, A., Frost, M., Sullivan, K. 2000, Transport and social exclusion in London. *Transport Policy* 7: 195–205.

Delaplace, M., Pagliara, F., Perrin, J. e Mermet, S. 2014, Can High Speed Rail foster the choice of destination for tourism purpose? *Procedia Social and Behavioral Sciences* 111: 166–175.

Levitas, R., Pantazis, C., Fahmy, E., Gordon, D., Lloyd, E., Patsios, D. 2007. *The Multi-dimensional Analysis of Social Exclusion*. London: Department for Communities and Local Government (DCLG).

Monzón, A., Ortega, E., López, E. 2010. Social impacts of high speed rail projects: addressing spatial equity effects. *Proceedings of the 12th WCTR, July 11–15, Lisbon, Portugal*.

Monzón, A, Ortega, E., López, E. 2013. Efficiency and spatial equity impacts of high-speed rail extensions in urban areas. *Cities* 30: 18–30.

Pagliara, F., Menicocci, F., Gomez, J., Vassallo, J.M. 2016a, Economic, geographical and time-based exclusion as main factors inhibiting Spanish users from choosing High Speed Rail. *Paper submitted to Transport Policy*.

Pagliara, F., De Pompeis, V., Preston, J. 2016b, Travel cost: not always the most important element of social exclusion. *Paper submitted to Transport Policy*.

Pagliara, F., Biggiero, L., 2016, Social exclusion from High Speed Rail systems: An exploratory study. Paper accepted for publication by Ingegneria Ferroviaria.

Shi, J., Zhou, N. 2013. How Cities Influenced by High Speed Rail Development: A Case Study in China, *Journal of Transportation Technologies* 3: 7–16.

http://www.adif.es
http://www.ine.es
http://www.istat.it
https://www.ons.gov.uk

Transport Infrastructure and Systems – Dell'Acqua & Wegman (Eds)
© 2017 Taylor & Francis Group, London, ISBN 978-1-138-03009-1

Modeling airport noise using artificial neural networks and non-linear analysis

M. De Luca
University of Naples "Federico II", Naples, Italy

D.C. Festa & G. Guido
University of Calabria, Rende, Italy

ABSTRACT: The noise generated by air traffic is an issue debated very much in the scientific community. It is a long history of consciousness, because it is not an easy topic to understand. From 1966 in Europe, this matter was recognized as a source/cause of pollution, whose effects are dangerous to human health (particularly about hearing and extra-hearing damage). In this work two models are built to estimate the level of aircraft noise (*Lva*) around populated areas. In particular the study evaluates the effects of this phenomenon on people and buildings. Data of the level of noise pollution (Lva), were detected on the airport of *Lamezia Terme* (IATA: SUF, ICAO: LICA), during the period 2006–2008, through 7 receptors located close to the airport. To the air traffic, reference was made to the data provided by the "post-holder" office at the airport of *Lamezia Terme*. The data were processed with the Multivariate Analysis and Artificial Neural Network technique. Two models were obtained: Model 1 (Model MVA) and Model 2 (Model ANN). Both models showed good predictive ability in terms of Lva. In particular, model 2 (ANN model) was better than model 1 (MVA Model) in terms of residual. In addition, to test the simulative capabilities of the two models (MVA model and ANN model), an experiment is conducted on a sample data not used for the construction of the two models.The comparison showed that the ANN model is the most reliable because it has the lowest residual.

1 INTRODUCTION

The noise generated by air traffic is an issue debated very much in the scientific community. It's a long history of consciousness, because it is not an easy topic to understand. Indeed, since 1966 in Europe, this matter was recognized as a source/cause of noise/pollution, whose effects are dangerous to human health (particularly about hearing and extra-hearing damage). In the last years, many researchers studied this issue, providing support tools for the management and control of this phenomenon, especially in places close to populated areas, to control the effects on people and on the buildings (buildings, artifacts, etc.).

Ozkurt et al. (2015), studied the results of noise in Izmir Adnan Menderes Airport (Turkey). The results showed that about 2% of the resident population was exposed to noise levels of 55 dB(A) or higher during day-time in İzmir.

Hua-Kun Yan, et al. (2013), in the Dalian International Airport compared the aircraft noise pollution and the cost-risk effects. The findings showed that the aircraft noise pollution of the offshore airport was lesser than that of the expanded inland port was lesser than that of the expanded inland airport; the land-use cost, noise reduction charges and other risks of the offshore airport were also lesser; the creation of the offshore airport may be more favorable to the city's development.

Sadr et al. (2014) studied land use planning around airports, by employing Remote Sensing (RS) and Geographic Information Systems (GIS), in conjunction with an optimization algorithm using an Integrated Noise Model (INM) software, to establish the potential effects of aircraft noise at Imam Khomeini International Airport (IKIA) in Tehran (Iran). The results indicated that developing IKIA together with the residential development will increase airport noise.

Maldonado et al. (2013), studied airport noise exposure around Viracopos International Airport quantifying the proportion of highly annoyed people in surrounding zones using simulations, integrated noise models and geographic information systems.

Licitra et al. (2014), simulated the present and future scenarios and they evaluated the health endpoints produced by Galileo Galilei Airport of Pisa (Italy).

Sari et al. (2014), developed an aircraft noise model for the İstanbul Atatürk Airport (Turkey).

It was found that 1.2% of the land area of İstanbul City exceeded the threshold of 55 dB(A) during daytime.

Heleno et al. (2014), made an analysis of airport noise based on LAeqD and LAeqN methodology. The goal of the study was to make an analysis of airport noise through the LAeq metric and to propose new alternatives based on Day Equivalent Sound Level (LAeqD) and Night Equivalent Sound Level (LAeqN) noise metrics.

Black et al. (2007) made a study in residential neighborhoods near Sydney Airport with high exposure to aircraft noise. Noise aircraft measurements were analyzed. The study showed that those who have been chronically exposed to aircraft noise were more likely to report stress and hypertension.

2 TECHNIQUES USED IN DATA ANALYSIS

Two different types of techniques are used for the analysis in this study: MultiVariate Analysis (MVA) and Artificial Neural Network (ANN). For the first the description is omitted because it is present for many years in the technical literature. For the second, much more recent, the following basic principles are described (*Čokorilo et al, 2014*).

2.1 *The Artificial Neural Network multilayer approach*

Inspiration for the structure of the ANN is taken from the structure and operating principles of the human brain. It is made up of neurons linked by connections that represent the connections between the biological synoptic neurons. The function of a biological neuron is to add its input and produce an output. This output is transmitted to subsequent neurons, through the synoptic joints, only if the transmitted signal is high (i.e., greater than a predetermined value), otherwise, the signal is not transmitted to the next neuron. In the network, therefore, a neuron calculates the weighted sum, using (Eq. 1), (considering the input xi and weights wi) and compares it with a threshold value; if the sum is greater than the "threshold" value, the neuron "lights up" and the signal is transmitted. Otherwise, the neuron does not turn on and the flow stops.

$$I = \sum_{i=1}^{n} w_i \cdot x_i \qquad (1)$$

The activation value "ui" rather than "uj", connected to weight Wij, is a function of the weighted sum of the input. This function may take various forms. In this study, a function of type (Eq. 2) was used.

$$u_j = \frac{1}{1 + e^{-\left(\sum_{(i)} w_{ij} \cdot u_i + \theta_j\right)}} \qquad (2)$$

where θj is the bias unit uj (i.e. the degree of sensitivity of uj when it receives an input signal from ui).

2.2 *Multi Layer Perceptron (MLP) and the Back Propagation (BP) algorithm*

In this study, a neural network with MLP architecture was used. Training was carried out using the Back Propagation (BP) algorithm. The neurons (or units) that comprise this type of network are organized into layers: an input layer, an output and a number of intermediate layers between input and output referred to as hidden, defined by the user. Initially the weights are assigned random values normalized in the range [0,1] or [−0.5, +0.5] and initially there is a pattern "p" input to the network: Xp = (X0, X1, X2, ..., Xn − 1) with X0 = 1 and a vector consisting of the output values Tp = (T0, T1, T2, ..., Tm − 1). In this way, the network will consist of (n − 1) input neurons and (m − 1) output neurons. The "weighted sum of the inputs" for each layer is calculated using Eq. 1 and its value of activation, i.e. output, using Eq. 2. Then, the weights must be changed so that the output of the network (i.e. the output of the last layer of neurons) increasingly approximates the target set by the user.

It was defined a function error (Eq. 3) proportional to the square of the difference between the output and target for all output neurons:

$$E_p = \frac{1}{2} \sum_j \left(T_{pj} - O_{pj}\right)^2 \qquad (3)$$

Subsequently, Back Propagation is applied i.e. the weights are varied so that error Ep tends towards zero (starting from the last layer to the first). We define, for the current pattern p, a variation Δwij of weight wij between the neuron i and j that given by (Eq. 4).

$$\Delta_{p \cdot w_{ij}} = -\alpha \frac{\partial E_p}{\partial w_{ijp}} + \beta \left(\Delta_{p-1}\right) w_{ij} \qquad (4)$$

where α is the learning coefficient (learning rate), β is momentum, and Δp − 1wij is the variation of the same weight calculated according to the previous model. The new weights are given by Eq. 5.

$$w_{ij}^{new} = w_{ij}^{old} + \Delta_p \cdot w_{ij} \qquad (5)$$

The variation of the weights is calculated starting from the layer of output neurons and backward toward the first hidden layer. The derivatives can be calculated using Eq. 6.

$$\Delta_p \cdot w = \alpha \cdot A_i \cdot \delta_j + \beta \cdot \Delta_{p-1} \cdot w_{ij} \qquad (6)$$

where A_i is the value of the i-th neuron of the layer being considered; δ_j is given by Eq. 7 if we are considering the output layer.

$$\delta_j = (T_j - O_j) \cdot O_j \cdot (1 - O_j) \qquad (7)$$

It is given by Eq. 8 for all other intermediate layers.

$$\delta_j = I_j (1 - I_j) \sum_k w_{jk} \cdot \delta_k \qquad (8)$$

To train a network, this process must be run many times (at least 1,000) with different patterns, each of which features a different weight. This process is performed until the error is less than a predetermined value (the value is set by the user). When the process converges, the network is ready to classify a new input with an unknown target. The parameters α and β are chosen by the user with values between 0 and 1; in the present study α was assumed equal to 0.5 and β equal to 0.4. In particular, α is linked to the convergence of the network *(De Luca et al., 2016)*.

3 DATA COLLECTION AND INSTRUMENTATION USED IN MONITORING

The investigation interests the runway of the international airport of Lamezia Terme (IATA: SUF, ICAO: LICA) whose layout is shown in Fig. 1. In the last few years the airport has highly incremented passengers traffic and handling activities *(De Luca & Dell'Acqua, 2013)*. Airports belonging to a local air transportation system where competition is strong exploit their inputs lesser than airports with local monopoly power. In 2011, more than 2 million of passengers were recorded at the airport of Lamezia Terme.

The sound level measurements are performed following the European Directive 2002/30/CE, and they are measured with the instrument shown in Figure 1 from October 1 2006 to October 2, 2008 h 24.

Figure 1. Overview of Lamezia Terme airport.

There are 7 survey stations, whose location and georeferencing are reported in Table 1 and Figure 2.

Specifically, in phonometric reference (Fig. 3) we use the Lva parameter, whose expression is shown below (see equation 9), as recommended by European Directive 2002/30/CE on noise monitoring resulting from air traffic).

$$L_{va} = 10 \log \left[\frac{1}{N} \sum_{j=1}^{N} 10^{L_{vaj}/10} \right] \qquad (9)$$

Table 1. Location of "Sound level meter".

Label "Sound level meter" station	Latitude	Longitude	Distance runway center [m]
REC-1	16°15,940'00"	38°54,161'00"	1687
REC-2	16°14,915'00"	38°54,168'00"	382
REC-3	16°13,531'00"	38°54,450'00"	1634
REC-4	16°14,133'00"	38°54,317'00"	707
REC-5	16°14,724'00"	38°54,571'00"	470
REC-6	16°15,239'00"	39°54,426'00"	735
REC-7	16°13,917'00"	38°54,282'00"	238

Figure 2. Location of "Sound level meters".

Figure 3. Instrumentation.

589

Table 2 shows data collection; in particular for each "sound level meter" (indicated by acronym SL) the daily maximum values of noise levels detected are illustrated.

Table 2. Lva detected by Sound level meters.

Date	SL1 Lva [db]	SL2 Lva [db]	SL3 Lva [db]	SL4 Lva [db]	SL5 Lva [db]	SL6 Lva [db]	SL7 Lva [db]
01/01/2006	98	90	67	65	80	89	131
02/01/2006	99	90	67	65	80	89	131
03/01/2006	100	90	67	65	81	89	131
04/01/2006	100	91	67	65	81	89	131
05/01/2006	101	91	68	65	81	89	132
06/01/2006	102	92	68	65	81	90	132
07/01/2006	102	92	68	66	81	90	132
08/01/2006	103	92	68	66	81	90	132
09/01/2006	103	93	68	66	82	90	132
....	....	....	....	....	....	....	....

Table 3. Data collection.

Date	Company	Type Aircraft	T, Thrust [N]	V_{max} [km/h]	$V_{takeoff}$ [km/h]	P, Power [Kw]
1/1/2006	Helvetic	F 100	67	845	241	16159
1/1/2006	Austrian	CRJ	78	860	245	19062
1/1/2006	A.Berlin	F 100	67	845	241	16159
1/1/2006	Airone	B 737	64	946	270	17280
1/1/2006	Alitalia	M 82	186	811	231	42938
1/1/2006	Airone	B 737	64	946	270	17280
1/1/2006	Airone	B 737	64	946	270	17280
1/1/2006	Alitalia	A 319	115	850	243	27899
......	....	.....	......	....	....	......

Table 4. Aggregated/Organized data in *Lva* classes.

Lva classes [dB]	D, Distance [m]	P, Power [kW]
20–30	1122	619001
30–40	965	574497
40–50	866	640848
50–60	934	566705
60–70	864	503595
70–80	950	564299
80–90	828	503451
90–100	884	547246
100–110	549	568255
110–120	607	489310
120–130	439	486237
130–140	358	460735
140–150	377	553864
150–160	389	620240
160–170	470	652417
170–180	470	698177
180–190	470	739234
190–200	470	704856

In addition (for each movement, from October 1 2006 to October 1 2008) in Table 3 other information is reported (*De Luca et al, 2016*).

In particular, the Vmax (maximum speed) and the Thrust T are defined by the aircraft manufacturers. The takeoff speed ($V_{takeoff}$), i.e. the speed with maximum energy production of noise on the airport structure, is assumed equal to 0.285*Vmax. The Power P is derived using the following equation (10).

$$P = T \times V_{takeoff} \qquad (10)$$

Subsequently data contained in Tables 2 and 3 are aggregated into classes of *Lva* (class amplitude equal to 10 dB). For other variables, see Table 4, the average value in each class is considered.

4 MVA APPLICATION (MODEL 1)

Model 1 (Eq. 11) was obtained using MVA (shown in Chapter 2) on data set contained in Table 4; the structure of the model and the variables used are the following:

$$L_{va} = a_1 \cdot P^{1/2} + a_2 \cdot d^2 \qquad (11)$$

Lva, (dependent variable);
D, Distance (Predictor);
P, Power (Predictor).

Model 1 (Eq. 11) is characterized by a coefficient of determination $\rho^2 = 0.88$ and a significance greater than 95% (see Table 5).

It is immediate to observe that:

- a_1 presents positive value therefore if the Power, P, increases Sound Level (Lva) increases;
- a_2 presents negative value therefore if the distance, D, increases the Sound Level (Lva) decreases.

5 ANN APPLICATION (MODEL 2)

Model 2 is obtained using the ANN technique shown in Chapter 3 using the same variables in the previous Chapter (MVA application). In particular,

Table 5. Model parameters.

Par.	Est.	Std. Error	Interval Lower Bound	Interval Upper Bound	Sign.
a_1	0.232	0.010	0.210	0.254	>95%
a_2	−1.31E-4	1.28E-5	−1.58E-4	−1.04E-4	>95%

Model 2 is obtained using 70% of dataset for training and 30% for testing. Fig. 4 shows the ANN architecture while the parameters of the ANN estimates are shown in Table 6.

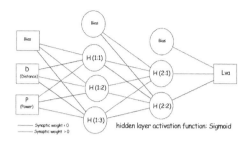

Figure 4.   ANN Architecture.

Table 6.   Parameters of the ANN estimates.

		Predicted				
		Hidden Layer 1			Hidden Layer 2		Output Layer
Predictor		H (1:1)	H (1:2)	H (1:3)	H (2:1)	H (2:2)	Lva
Input Layer	(Bias)	−0.200	−0.287	−0.287			
	D	−0.352	−0.713	−0.713			
	P	−0.100	0.718	0.718			
Hidden Layer 1	(Bias)				0.724	−0.249	
	H (1:1)				0.227	−0.837	
	H (1:2)				0.293	2.244	
	H(1:3)				0.316	−1.044	
Hidden Layer 2	(Bias)						0.738
	H (2:1)						2.495
	H (2:2)						−5.480

Table 7.   Comparison between ANN and MVA and Residual.

Lva Observed	Lva MVA	Lva ANN	D Distance	P Power	Res. 1	Res. 2
25	17	36	1122	619001	8	11
35	54	50	965	574497	19	15
45	87	65	866	640848	42	20
55	60	55	934	566705	5	0
65	67	78	864	503595	2	13
75	56	53	950	564299	19	22
85	75	83	828	503451	10	2
95	69	65	884	547246	26	30
105	135	116	549	568255	30	11
115	114	114	607	489310	1	1
125	137	125	439	486237	12	0
135	141	129	358.6	460735	6	6
145	154	132	377	553864	9	13
155	163	155	389.4	620240	8	0
165	158	164	470.3	652417	7	1
175	165	181	470.3	698177	10	6
185	170	185	470.3	739234	15	0
195	166	182	470.3	704856	29	13
			Total Residual		**257**	**165**

6   ANN VERSUS THE MVA

Table 7 and Figure 5 show the comparison between the two models and they denote that Model 2 is better than Model 1 because the residual has a lower total sum (see last two columns of Table 7).

To test the simulative capabilities of the two models (MVA model and ANN model), an experiment is conducted on a sample data not used for the construction of the two models. Table 8 shows the results obtained.

Even in this case, the ANN model is more reliable than the MVA model. Table 9 and Figure 6 show the results of the comparison (*Dell'Acqua, 2016*).

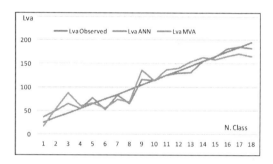

Figure 5.   Graphical Comparison: ANN Vs MVA.

Table 8.   Data used to validation models.

P Power	D Distance	Lva Observed
476200	734	80
395404	470	100
402424	237	120
521579	470	140
672155	470	160
888617	470	180
952087	470	200

Table 9.   Comparison between ANN and MVA (validation phase).

P Power	D Distance	Lva Observed	Lva MVA	Lva ANN
476200	734	80	89	83
395404	470	100	117	116
402424	237	120	140	125
521579	470	140	139	132
672155	470	160	161	164
888617	470	180	190	181
952087	470	200	197	185

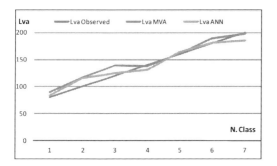

Figure 6. Graphic Comparison between ANN and MVA (Validation phase).

7 CONCLUSIONS

In this work, two models are built to estimate the level of aircraft noise (*Lva*) around populated areas.

Data of the level of noise pollution (Lva) are detected on the airport of Lamezia Terme (IATA: SUF, ICAO: LICA), during the period 2006–2008, through 7 receptors (SL) situated close to the airport. The traffic data, for the same period, are provided by the "post-holder" office of the airport of *Lamezia Terme*.

The data organized in Table 4 are processed through the MVA technique and ANN technique (see Chapter 2) and two models have been obtained: Model 1 (MVA model) and Model 2 (ANN model). Both models show good predictive ability in terms of Lva. In particular, model 2 (ANN model) is better than model 1 (MVA Model), in terms of residual (see Table 8).

For a first procedure of validation, the models are tested on a sample of data that have not been used to construct the two models. The sample comes from the same airport (*Lamezia Terme*) and the same period (2006–2008). Even in this case, both models show a good predictive capacity and the comparison between them denotes that Model 2 is better than Model 1.

REFERENCES

Ozkurt, N., Hamamci, S., F. & Sari, D. 2015. Estimation of airport noise impacts on public health. A case study of İzmir Adnan Menderes Airport. *Transportation Research Part D: Transport and Environment,* 36: 152–159; doi: 10.1016/j.trd.2015.02.002.

Hua-Kun Yan, Nuo Wang, Liao Wei & Qiang Fu, 2013. Comparing aircraft noise pollution and cost-risk effects of inland and offshore airports: The case of Dalian International Airport, Dalian, China. *Transportation Research Part D: Transport and Environment*, 24: 37–43; doi: 10.1016/j.trd.2013.05.005.

Sadr, M. K., Nassiri, P., Hosseini, M. & Monavari, M, 2014. Assessment of land use compatibility and noise pollution at Imam Khomeini International Airpor., *Journal of Air Transport Management*, 34: 49–56; doi:10.1016/j.jairtraman.2013.07.009.

Maldonado F., Heleno, T. & Slama, J. 2013. Noise mitigation action plan of Pisa civil and military airport and its effects on people exposure. *Journal of Air Transport Management*, 31: 15–1; doi: 10.1016/j.jairtraman.2012.11.001.

Licitra, G., Gagliardi, P., Fredianelli, L. & Simonetti, D. 2014. *Applied Acoustics*, 84: 25–36; doi:10.1016/j.apacoust.2014.02.020.

Sari, D., Ozkurt, N., Akdag, A., Kutukoglu, M. & Gurarslan, A. 2014. Modeling of noise pollution and estimated human exposure around İstanbul Atatürk Airport in Turke., *Science of The Total Environment*, 482–483: 486–4921; doi:10.1016/j.scitotenv.2013.08.017.

Tarcilene A., Slama J. & Maldonado, F. 2014. Analysis of airport noise through LAeq noise metrics. *Journal of Air Transport Management*, 37, Pages 5; doi:10.1016/j.jairtraman.2014.01.004.

Black, D. A., Black, J. A., Issarayangyun T. & Samuels S.E.2007. Aircraft Noise Exposure and Resident's Stress and Hypertension: A public Health Perspective for Airport Environmental Management. *Journal of Air Transport Management*, 13: 264–276. http://dx.doi.org/10.1016/j.jairtraman.2007.04.003.

Čokorilo, O., De Luca, M. & Dell'Acqua, G. 2014. Aircraft safety analysis using clustering algorithms. *Journal of Risk Research*, 17(10). DOI: 10.1080/13669877.2013.879493.

De Luca, M. 2015. A Comparison Between Prediction Power of Artificial Neural Networks and Multivariate Analysis in Road Safety Management, *Transport*, 1–7, ISSN: 1648–4142, DOI: http://dx.doi.org/10.3846/16484142.2014.995702.

De Luca, M.; Abbondati, F.; Pirozzi, M.; Zilioniene, D. 2016. Preliminary study on runway pavement friction decay using data mining. *Transportation Research Procedia*, 14: 3751–3670. DOI: 10.1016/j.trpro.2016.05.460.

De Luca, M. & Dell'Acqua, G. 2013. Runway surface friction characteristics assessment for Lamezia Terme airfield pavement management system. *Journal of Air Transport Management*, 34: 1–5, ISSN: 0969–6997, doi: 10.1016/j.jairtraman.2013.06.015.

De Luca, M., Abbondati, F., Yager, T.J. & Dell'Acqua, G. 2016. Field measurements on runway friction decay related to rubber deposits, *Transport*, 31(2): 177–182. DOI: 10.3846/16484142.2016.1192062.

Dell'Acqua, G., De Luca, M., Prato, C.G., Prentkovskis, O. & Junevičius, R. 2016. The impact of vehicle movement on exploitation parameters of roads and runways: a short review of the special issue. *Transport* 31(2): 127–132, DOI: 10.3846/16484142.2016.1201912.

Transport Infrastructure and Systems – Dell'Acqua & Wegman (Eds)
© 2017 Taylor & Francis Group, London, ISBN 978-1-138-03009-1

A new safety performance index for speed-related crashes

G. Guido, V. Astarita, A. Vitale & V. Gallelli
University of Calabria, Rende, Italy

F.F. Saccomanno
University of Waterloo, Waterloo, Canada

ABSTRACT: The use of GPS-equipped smartphones by drivers can facilitate the tracking and monitoring of vehicle operations in real-time. This kind of solution would be implemented to obtain estimates of individual vehicle operating speeds from the traffic stream. However, before GPS data can be used in safety performance analysis, tracking errors caused by en-route satellite signal disruptions need to be taken into account to accurately reflect real-world traffic conditions. The main aim of this paper is to demonstrate how, when adjusted for errors, GPS-based estimates of instantaneous speeds can be used to highlight locations where safety is compromised due to poor road geometry. A vehicle-specific Track Safety Performance Index (TSPI) is developed to measure the difference between individual vehicle operating speed and the design speed for the location. The results from two case studies support the use of error-adjusted GPS probe data for identifying sites where safety at a given location is compromised by driving too fast for the underlying geometric restrictions. Locations with higher TSPI values were found to correspond closely to sites with a higher number of speed-related crashes as reported over a period of five years. The results from the two case studies were found to be consistent in linking GPS probe estimates of TSPI to locations of higher crash risk.

1 INTRODUCTION

One of the most common methodologies to estimate safety makes use of inferential statistics applied to crashes databases therefore being considered a reactive approach to the problem. Although this method seems to intuitively link the causes to effects, a good knowledge of the dynamics of the events preceding the crash may provide a more useful support to the implementation of appropriate countermeasures.

Safety performance indicators represent a useful tool for evaluating road safety conditions on the basis of objective parameters deducible from the vehicle kinematics. Most of these indicators provide a causal or mechanistic basis for explaining interactions between different pairs of vehicles belonging to the traffic stream that can compromise safety (Hayward 1971, Minderhoud & Bovy 2001, Huguenin et al. 2005, Saccomanno et al. 2008, Guido et al. 2011, Astarita et al. 2011).

However, some accidents, as well as some potential conflicts, are caused by the aggressive behavior of users who adopt speeds unsuitable for the road geometry causing isolated vehicle crashes. Indeed, safety is influenced by a number of traffic and geometric factors, such as driver features and conditions (experience, stress, tiredness, etc.), road characteristics (type of road, road surface, geometric features, etc.), traffic conditions (volume, speed, density, etc.), vehicle attributes (maneuverability, braking capability, stability, etc.), and environment (weather conditions, light conditions, etc.). Among these factors speed is universally recognized as the main factor for the risk of accidents. The World Health Organization states that "speed has been identified as a key risk factor in road traffic injuries, influencing both the risk of a road crash as well as the severity of the injuries that result from crashes".

According to the AASHTO Green Book (AASHTO 2001), one of the most important factors in roadway design is the design speed. When roads are designed encouraging drivers to travel at higher desirable speeds several safety problems may occur if the differences between vehicle design speeds, operating speeds and speed limits are considerable at a given location (Stamatiadis et al. 2009).

A number of empirical studies have explored the relationship between vehicle operating speeds and crash occurrence (Aarts & van Schagen 2006, Taylor et al. 2000, Hassan & Abdel-Aty 2011, Nilsson 2004). The American Association of State Highway and Transportation Officials Green Book (AASHTO 2011) recommends different "safe design speeds" or standards for different types of roads and traffic conditions, taking into account features such as, super elevation, radii of curvature, sight distances, lengths of crests and sags, etc. Higher design speeds reflect safer road geometries, such as, sweeping curves, steeper banking, longer sight distances, and more gentle hills and valleys. These permit drivers to increase their operating speeds without incurring an increased crash risk. To fully appreciate the speed-safety relationship, we need to understand how individual drivers at a given location select their operating speeds based on what they perceive to be safe (Walton & Bathurst 1998, Haglund & Åberg 2000, Sümer et al. 2006, Wallén Warner & Åberg, 2008, Mannering 2009). Hence, setting operating speeds below safe design thresholds ensures a degree of safety for individual vehicles in the traffic stream along a given route.

One of the challenges in this analysis is to obtain accurate empirical estimates of individual vehicle operating speeds over time. GPS smartphones offer a potentially low-cost means to acquire time-dependent estimates of instantaneous speeds in a nonintrusive reliable manner, without the need for stringent setup controls for videotaping and image detection (Guido et al. 2012, Guido et al. 2013, Guido et al. 2014, Bierlaire et al. 2010, Herrera et al. 2010). The focus of this research is on the use GPS probe vehicle estimates of instantaneous speeds for input into a Track Safety Performance Index (TSPI) for a given location.

Two case study applications are introduced to demonstrate how uncertainty in GPS-based speed estimates can affect the resultant measure of safety performance. When estimates of speed in TSPI are properly adjusted for GPS errors, the case studies indicate that they can provide a reliable method for highlighting potential safety problems along the route. Logically we would expect that if crashes occur consistently at a given site that their location along the route would correspond to sites with high TSPI measures. Factors affecting the accuracy of the GPS-based TSPI are explored empirically for the case study highway segments.

The paper is organized according to the following structure. Section 2 provides the formulation of the safety performance index introduced in this work. Section 3 describes the model adopted for the safe design speed used as benchmark in safety evaluation. In section 4 a brief description of a benchmarking exercise is presented that is addressed to estimate the accuracy of instantaneous operating speeds obtained from GPS probes. Section 5 introduces a case study application. Section 6 provides the results for a second case study application highlighting the transferability of the methodology. Section 7 summarizes the major conclusions of the study.

2 SAFETY PERFORMANCE INDEX FORMULATION

Safety performance in this paper for a given vehicle is expressed as the difference between instantaneous vehicle operating speed and recommended safe design speed for the location. For uncongested conditions instantaneous vehicle operating speeds are assumed to reflect drivers "desired" speed at a given location subject to prevailing weather conditions (in this paper assumed ideal). Equation 1 provides a mathematical formulation of a Track Safety Performance Index (TSPI) for an individual vehicle at a specific point in its trajectory (or distance increment i), such that:

$$\text{TSPI} = \alpha \cdot \left(\frac{v_i^{op} - v_i^t}{v_i^t} \right) \quad (1)$$

where:
  v_i^{op} = mean operating speed of sample of vehicles at location i.
  v_i^t = safe design speed at location i.
  α = scaling factor.

v_i^t in Equation 1 represents the safe design standard for the geometric conditions at location i, as provided in the AASHTO Green Book. v_i^{op} represents the mean instantaneous operating speed of a sample of GPS vehicle probes traversing location i, such that for a sample of m_0 probes:

$$v^{op}_i = \sum_{m=1}^{m_0} v_{i,m} / m_0 \quad (2)$$

In Equation 2, $v_{i,m}$ is the instantaneous speed of the mth vehicle traversing location i, as extracted from GPS output.

The underlying premise in the study is that each driver sets his/her operating speed based on a perception of what is safe for the location. An overestimation of "what is perceived to be safe" will inevitably contribute to increased crash risk, especially as it relates to single vehicle speed-related crashes. Safe design speed is essentially established from geometric restrictions on mobility at a given location. For this paper, we have assumed that all drivers in the probe sample have the same instantaneous operating speed (sample mean) and that they face the same safe design speed for the location v_i^t.

TSPI from Equation 1 is location specific, but it can also be expressed for the entire segment by weighting each distance increment (or location) i by the mean speed of vehicles in the sample at that location, such that:

$$TSPI = \frac{\displaystyle\sum_{i=1}^{n}\left[TSPI_i \cdot \frac{d_i}{v_i}\right]}{\displaystyle\sum_{i=1}^{n}\frac{d_i}{v_i}} \qquad (3)$$

Equation 3 reflects the safety performance index of the entire route segment based on differences between mean vehicle operating speeds and safe design standards. Two inputs in Equations 1 and 3 need elaboration: 1) procedure for establishing the safe design speed v_i^t along the route, and 2) procedure for treating uncertainty in the GPS speed estimates at location i for vehicle m ($v_{i,m}$).

3 SAFE DESIGN SPEED MODEL

The AASHTO Highway Safety Manual protocols define safe design speeds to consider "… a number of important factors, such as, underlying topography, anticipated operating speed of traffic stream over the longer tract of road, adjacent land use and development density, and the functional classification of the highway" (AASHTO 2010).

In this paper, the design speed is calculated based on the V85 model introduced by Lamm (2002).

For a two-lane highway and uncongested traffic conditions, the safe design speed was established by Lamm by first estimating the V85 speed along the route. V85 reflects the operating speed that is exceeded by 15% of vehicles in the traffic stream, thus providing a measure of the upper range of "risk behavior" in the driver sample. For two-lane undivided highways, Lamm suggested a V85 safe design speed of 105.3 Km/h for tangents under ideal uncongested conditions.

The tangent design speed is reduced accordingly to account for restrictions to mobility caused by sharper horizontal curvature. For such curves, Lamm suggested a safe design speed expression of the form:

$$v^t(km/h) = 105.31 + 2*10^{-5}*CCR_s^2 - 0.071*CCR_s \quad (4)$$

Where CCRs reflects the curvature change rate and is used to tailor the safe design speed for tangents to account for differences in horizontal alignment. For circular and transitional curves, CCRS is expressed as a function of curve length and radius, such that:

$$CCR_s = \frac{\left(\dfrac{L_{cl1}}{2R} + \dfrac{L_{cr}}{R} + \dfrac{L_{cl2}}{2R}\right)*63.7}{L} \quad (5)$$

where:
 L_{cr} is the length of circular curve (m),
 L_{cl1} and L_{cl2} are the lengths of clothoids preceding and succeeding the circular curve (m),
 R is the radius of circular curve (m),
 L = L_{cr}+L_{cl1}+L_{cl2} (km).

4 UNCERTAINTY IN GPS PROBE SPEEDS

The major problem with estimating instantaneous operating speeds from GPS probes is their lack of precision as affected by real-world road and traffic conditions commonly encountered along the route (Chalko 2007). This concern was addressed experimentally by Guido et al. (2014), with respect to a GPS benchmarking traffic study. This study yielded GPS confidence intervals for instantaneous vehicle operating speeds, subject to varying satellite signal disruptions en-route. A brief description of this benchmarking exercise has been presented in this section of the paper.

The level of precision associated with GPS output is related to the number of satellites in view and their relative position. This is normally reported by the GPS receiver in terms of its Circular Error Probable (CEP) range. A device calibration exercise is carried out to establish CEP in terms of different radii centered on a known fixed ground control point. Since this is a result of a "fixed" ground control test, CEP may not be able to account for the myriad of factors affecting error under dynamic road and traffic conditions.

Guido et al. (2014) obtained a sample of smartphone probe vehicle operating speeds for varying satellite signal reception and compared these speeds to benchmark values obtained using a high frequency video V-Box (Race Logic ®) mounted on the same vehicle. Lack of precision in the speed estimates was expressed as the signal Root Mean Square Percentage Error (RMSPE).

$$RMSPE = \sqrt{\frac{\displaystyle\sum_{i=1}^{N}\left[\frac{S_i - \hat{S}_i}{\hat{S}_i}\right]^2}{N}} \qquad (6)$$

where
 S_i = speed of probe,
 $\hat{S}_i$ = V-Box benchmark speed,
 N = number of probes.

In order to obtain estimates of RMSPE, a relationship was established between the reported signal CEP range and its corresponding RMSPE for a sample of probe vehicle passes. Instantaneous speeds obtained from four GPS-enabled smartphones were compared to benchmark values

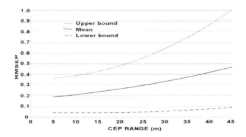

Figure 1. RMSPE mean and 95th percentile confidence intervals for different CEP values.

obtained from the high frequency V-Box receiver. Confidence intervals for probe operating speeds in the sample were subsequently obtained experimentally for varying road and traffic conditions.

Figure 1 illustrates the relationship between mean RMSPE and CEP, along with corresponding RMSPE confidence intervals (95%).

From Figure 1, we note that the width of the 95% confidence interval becomes decidedly more pronounced with increasing CEP values, suggesting increased uncertainty in the GPS speed estimate with increased loss of satellite signal strength.

A series of CEP-related speed adjustments for mean and 95% confidence intervals ("lower bound" and "upper bound") were assessed based on this analysis and then applied to the case study exercise.

5 ANALYSING SAFETY PERFORMANCE FOR CASE STUDY HIGHWAY SEGMENT

A case study based on GPS-based instantaneous speeds is used to yield the TSPI for a two-lane rural highway segment. This case study seeks to provide insights into several practical safety questions: Which sites along the highway segment are more prone to crashes? How well do these sites correspond to locations of reported crashes (total and single vehicle speed-related)? How is the assessment of safety along the highway affected by uncertainty in GPS instantaneous speed estimates? Do GPS-based instantaneous speeds provide a reliable metric for assessing road safety performance?

The SS106 is a two lane undivided highway in southern Italy serving about 5,000 to 20,000 vehicles per day (De Luca 2015). Historically the number of crashes along this highway have been high and it's considered to be one of the most dangerous two-lane roads in the country. The crash mortality rate for this highway is also high, averaging 8.5 fatalities per 100 crashes, as compared to a national average of 4.9 fatalities per 100 crashes. The major cause of crashes along this highway is attributed vehicles travelling too fast for the conditions (geometric).

As illustrated in Figure 2, the segment of the SS106 used for this case study consists of a 2.4 Km tangent, bounded by two circular curves of radius 660 m (for Curve 1) and 325 m (for Curve 2), respectively.

Figure 3 illustrates the mean operating speeds along the SS106 case study segment as obtained from GPS output, with corresponding 95% confidence intervals reflecting uncertainty in the GPS signal from Guido et al. (2014). These profiles are juxtaposed on the safe design speed of the segment from Lamm's formulation for tangents and curves. The mean instantaneous operating speeds were estimated for 32 GPS vehicle passes along the segment in increments of 100 m. About 3,500 of instantaneous speed data were acquired by the probe vehicles (GPS) along the highway stretch. Mean operating speed of this sample of vehicles was then calculated for each 100 meters segment as in Equation 2. Sites where operating speeds exceed safe design thresholds can be viewed as posing potential safety problems.

For this case study, the mean operating speeds were found to be close to the safe design thresholds for curve 1 and tangent portions, but exceeding these thresholds in the sharper curve 2 portion of the segment. The upper bound speed profiles (dotted) suggests that there is a 2.5% chance that operating speeds will exceed safe standards as recommended by AASHTO based on the GPS sample. This suggest that although "on average" operating speeds can be viewed as being safe, these is a chance that these results may be biases due to errors in the GPS receiver. This gives a kind of statistical "safety margin" that we may wish to consider as a basis for possible intervention along the case study highway segment.

Figure 2. Case study highway segment.

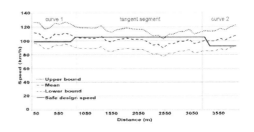

Figure 3. 95% confidence interval for the v^{op} speed profile and safe design speed threshold.

596

Figure 4 illustrates the upper bound profile for TSPI for the SS106 case study, based on higher risk driver behavior. Let's consider two safety intervention thresholds (A and B). Threshold A suggests intervention for elements where TSPI ≥ 2 and B (less tolerant of risk) where TSPI is greater than zero. Threshold A would suggest intervening in the sharper Curve 2 areas; while Threshold B recommends intervention along the entire case study segment (curves and tangent). This latter threshold would obviously come at a higher intervention cost. Presumably, threshold B would be adopted for a less restrictive safety budget and a reduced tolerance to crash risk. A similar comparison can be carried out considering other driver risk sensitivities, as would be reflected in a different TSPI measure.

The safety index TSPI can be used directly to guide intervention strategies along the case study segment. For this application TSPI has been aggregated for the tangent and curve elements, and the results are summarized in Table 1 confirming the previous results that the major safety concern along the case study segment is in Curve 2. For this portion of the case study, TSPI has been found to be significantly higher than either Curve 1 or the tangent. Based on these results, we could recommend Curve 2 be given priority in safety intervention.

TSPI has been normalized using a feature scaling of the form:

$$t = \frac{(x - x_{min})}{(x_{max} - x_{min})} \quad (7)$$

where t is the normalized TSPI value, x is the observed TSPI value, and x_{min} and x_{max} are the minimum and maximum TSPI values, respectively.

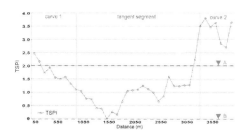

Figure 4. Track safety performance indices that exceed thresholds for intervention.

Table 1. TSPI for curve 1, tangent section and curve 2.

ID	Section	Length (m)	Radius (m)	TSPI (U.B.)	t
1	Curve 1	850	660	1.714	0.450
2	Tangent	2,400	–	0.973	0.256
3	Curve 2	600	325	3.384	0.890

Conceptually crashes can be viewed from two perspectives: a) crashes that result from safety problems at a given location (consistent), or b) random "acts of God" or crashes that result from difficult-to-predict events. It is reasonable to assume that if crashes occur with consistency, their location would correspond to sites with higher values of TSPI or risk. As defined in this paper, TSPI focuses on deviations between vehicle operating speeds and safe design thresholds. This measure of TSPI reflects risk of crashes related to excessive speed, normally involving a single vehicle. Presumably other types of crashes would be picked up by different forms of TSPI. Hence, for the index introduced in this paper, our intent is to compare the location of single vehicle speed-related crashes to TSPI as presented in Equation 1 (i.e. difference between instantaneous operating speed and safe threshold). If crashes are consistent, we would expect the location of single vehicle speed-related crashes to correspond to sites of higher TSPIs.

The Centro di Monitoraggio Sicurezza Stradale in the Province of Crotone reported a total of 40 crashes for the SS106 case study segment over the period 2007–2011. These resulted in 4 deaths and 111 injuries. Of the 40 crashes, 19 (47.5%) took place on sites with high TSPI values (> 2), and 21 (52.5%) on sites with moderate TSPI (between 1–2). As illustrated in Figure 5, twenty-eight (28) of these crashes (70%) were single vehicle and attributed by the police to excessive speed, 57.1% of these occurring on sites with TSPI > 2. This suggests that for the first SS106 case study segment, speed-related crashes took place at locations with higher TSPIs, where drivers presumably overestimated safe design speeds for the underlying road geometry.

The kinds of intervention that are likely to be most effective in reducing crashes are improvements in highway geometry along the curves, or an improvements in driver speed advisory and enforcement such that drivers are more aware of geometric/traffic mobility restrictions along the segment. The nature of the relationship between different treatments and changes in TSPI and the number of likely crashes is not considered to be within the scope of this paper.

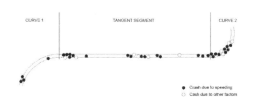

Figure 5. Locations of crashes due to speeding on the SS106 segment.

6 TRANSFERABILITY OF RESULTS

One of the major issues in using mobile GPS probe data for input into safety performance analysis is the transferability of the results. To address this issue a second 10 Km stretch of two-lane highway (SS107) was selected for analysis (Fig. 6).

For this application 85th percentile of TSPI profiles were obtained based 45 mobile GPS probes. As illustrated in Figure 6, the SS107 case study

Figure 6. Locations of speed-related and totally crashes for 10 Km SS107 case study route.

Table 2. TSPI and crash rate for the SS107 segment (2009–2013).

ID	Curve or Tangent	Length (m)	Radius (m)	TSPI	t	Crashes[1]	Rate[2]
1	C1	48	220	−0.114	–	0	0
2	T1	86	0	−0.122	–	0	0
3	C2	58	186	−0.126	–	0	0
4	T2	94	–	−0.149	–	0	0
5	C3	54	141	−0.174	–	0	0
6	T3	74	–	−0.168	–	0	0
7	C4	320	117	−0.120	–	0	0
8	T4	299	–	−0.052	–	0	0
9	C5	355	181	−0.030	–	0	0
10	T5	194	–	−0.090	–	0	0
11	C6	236	211	−0.060	–	0	0
12	T6	195	–	−0.023	–	0	0
13	C7	73	171	−0.031	–	0	0
14	T7	116	–	−0.029	–	0	0
15	C8	403	408	−0.033	–	0	0
16	T8	156	–	−0.052	–	0	0
17	C9	578	618	0.033	0.246	1	0.173
18	T9	337	–	0.122	0.911	3	0.889
19	C10	220	920	0.086	0.638	2	0.908
20	T10	205	–	0.089	0.665	0	0
21	C11	170	658	0.134	1.000	2	0.076
22	T11	393	–	0.115	0.856	3	0.764
23	C12	358	287	0.084	0.628	2	0.559
24	T12	128	–	0.055	0.408	0	0
25	C13	73	428	0.008	0.063	0	0
26	T13	316	–	0.029	0.213	1	0.316
27	C14	372	169	−0.133	–	0	0
28	T14	195	–	−0.041	–	0	0
29	C15	366	184	−0.015	–	1	0.273
30	T15	138	–	−0.058	–	0	0
31	C16	246	243	−0.038	–	1	0.407
32	T16	604	–	−0.092	–	0	0
33	C17	286	176	−0.269	–	0	0
34	T17	376	0	−0.308	–	0	0
35	C18	122	529	−0.101	–	0	0
36	T18	453	0	0.077	0.574	2	0.442
37	C19	122	224	0.084	0.626	0	0
38	T19	189	0	0.078	0.579	0	0
39	C20	100	172	0.051	0.377	0	0
40	T20	132	0	0.067	0.496	0	0
41	C21	434	187	0.017	0.126	1	0.230
42	T21	274	0	0.008	0.056	0	0

[1]Number of speed related crashes; [2]Crash rate per 100 m.

route consists of 42 distinctive geometric sections (21 curves and 21 tangents).

GPS probe operating speeds were adjusted for satellite signal reception errors as discussed above from the previous benchmark exercise. Upper bound speed profiles were compared to the design thresholds for the route to produce 42 mean section-specific TSPI values. The design speeds for each section were obtained based on Lamm (2002), as discussed previously.

A total of 64 crashes were reported by police for these road sections over the period 2009–2013, of which 30% (13) were single vehicle crashes attributed by the police to excessive speed. The location of both speed-related and total crashes are shown in Figure 6. For each of the 42 geometric elements, estimates were obtained for TSPI, number of speed-related crashes and crash rate and, these have been summarized in Table 2. The crash rate is expressed in terms of the number of crashes per 100 m over the 2009–2013 period.

The purpose of this case study is to test the transferability of the previous SS106 results, i.e. that for the SS107 higher TSPIs correspond to sites with a higher number of reported speed-related crashes. A close match would attest to the consistency of the findings between the SS106 and the SS107, and provide stronger evidence supporting the use of mobile GPS probes for targeting potential speed-related safety problems along a highway.

Given that TSPI and crash rates may not be normally distributed, a nonparametric Spearman rank correlation was obtained to test similarities between TSPI values and reported speed-related crash rates for the 42 geometric elements along the SS107. A Spearman ranking index is of the form:

$$\rho_s = 1 - \frac{6 \cdot \sum\limits_{i=1}^{n} d_i^2}{n \cdot (n-1)} \qquad (8)$$

where:
d_i = rank difference between two samples.
n = sample size (no. of sections).

Higher values of ρ_s reflect reduced differences between TSPI and speed-related crash rates.

As summarized in Table 3, the correlation parameter (ρ_s) in the Spearman test was found to be significant at the 5% level. Hence it can be concluded that TSPI provides a good metric for the location of speed-related crashes along the SS107 case study route. This supports the findings that were obtained from the first case study the SS106.

The extraction of GPS mobile probe data from the SS106 and SS107 has demonstrated a significant degree of consistency for estimating instantaneous operating speeds and TSPI values. Both case studies provide a convincing link between GPS-based TSPI values and speed-related crashes, i.e. the higher the TSPI, the higher the number of speed-related crashes.

7 CONCLUSIONS

This paper has presented a road safety performance study in which safety is analyzed comparing average operating speed profiles to a recommended safe speed thresholds along a given road segment. Operating speeds have been obtained from a smartphone probe vehicles sample, while safe speed thresholds are based on v85 Lamm model mainly accounting for geometric conditions. The assumption is made that each driver sets his individual operating speed based on his perception of the safe design speed. The safe design speed can be overestimated by drivers, and this poses potential crash risks at specific road locations. Hence, safety performance is expressed for a given road segment as the degree to which operating speeds at a given point in time exceed safe design standards.

A safety performance index (track safety performance index or TSPI) was formulated to assess lack of safety over the case study segment, accounting for the difference between the probes speed profiles and selected safe design thresholds. The TSPI metric introduced in this paper only reflects speed-related crashes caused by restrictive road geometry. Other safety concerns could be considered using different expressions for TSPI from that introduced in this paper.

Two case studies were analyzed for two-lane highways in which instantaneous operating vehicle speeds were obtained from a GPS equipped smartphones. Operating speed profiles from the probes sample were then adjusted to account for uncertainty in the speed estimates under dynamic road and traffic conditions. The results of this analysis show that TSPI is affected by the uncertainty in speed estimation, as well as by the thresholds used to guide intervention.

The case study applications have confirmed that GPS mobile probe measures of TSPI closely match sites of single vehicle speed-related crashes for several years of reported data. The two case studies yielded consistent results concerning the relationship between TSPI and speed-related crashes, i.e. as the TSPI at a given location is increased the

Table 3. Spearman test for the SS107 sections.

N	ρ_s	d.o.f.	t	p-value
42	0.60	40	4.694	<0.001

observed number of speed related crashes at this location is also increased.

Notwithstanding the preliminary limitations of the smartphone sample used in this study, the proposed methodology has yielded some promising results. The approach has been shown to yield reliable traffic (instantaneous speed) attributes for individual vehicles in the traffic stream. These can be used as inputs into safety assessment and in developing cost-effective safety treatments. Of course, uncertainty in safety performance indicator needs to be taken into account to ensure that decisions concerning safety performance from smartphone probe estimates are not based on incorrect unreliable measures of speed and other traffic attributes.

A comparison among TSPI measures along the segment closely reflects historical crashes sites along the case study segment. This provides an interesting dimension to the analysis. If historical crashes occur where risks have been identified by TSPI as being high, then we would expect these crashes to exhibit a high degree of consistency and hence reflective of potential safety problems. On the other hand, if historical crash sites don't match high TSPI sites, then crashes along the segment can be viewed as being essentially random events and difficult to predict and resolve.

REFERENCES

Aarts, L., & van Schagen, I. 2006. Driving speed and the risk of road crashes: A review. *Accident Analysis and Prevention*, 38: 215–224.

AASHTO 2011. *A Policy on Geometric Design of Highways and Streets*. 6th Edition.

AASHTO, 2001. *A Policy on Geometric Design of Highways and Streets*. 4th Edition.

AASHTO, 2010. *Highway Safety Manual*. First Edition.

Amundsen, F.H. & Hyden, C. 1977. *Proceeding of First Workshop on Traffic Conflicts. Institute of Transport Economics*, Oslo/Lund Institute of Technology, Oslo, Norway.

Astarita V., Giofré V., Guido G., & Vitale A. 2011. Investigating road safety issues through a microsimulation model. *Procedia Social and Behavioral Sciences*, 20: 226–235.

Bierlaire, M., Chen, J., & Newman, J. 2010. *Modeling Route Choice Behavior From Smartphone GPS data*. Report TRANSP-OR 101016, Transport and Mobility Laboratory, Ecole Polytechnique Fédérale de Lausanne.

Chalko, T.J. 2007. *High accuracy speed measurement using GPS (Global Positioning System)*. Melbourne, Australia.

De Luca, M. 2015. A Comparison Between Prediction Power of Artificial Neural Networks and Multivariate Analysis in Road Safety Management. *Transport*, 1–7, DOI: http://dx.doi.org/10.3846/16484142.2014.995702

Guido G., Astarita V., Giofré V., & Vitale A. 2011. Safety performance measures: a comparison between microsimulation and observational data. *Procedia Social and Behavioral Sciences*, 20: 217–225.

Guido G., Vitale A., Saccomanno F.F., Festa D. C., Astarita V., Rogano D., & Gallelli V. 2013. Using smartphones as a tool to capture road traffic attributes. *Applied Mechanics and Materials*, 432: 513–519.

Guido, G., Gallelli, V., Saccomanno, F.F., Vitale, A., Rogano, D., & Festa, D. 2014. Treating uncertainty in the estimation of speed from smartphone traffic probes. *Transportation Research Part C*, 47: 100–112.

Guido, G., Vitale, A., Astarita, V., Saccomanno, F.F., Giofré, V.P., & Gallelli, V. 2012. Estimation of safety performance measures from smartphone sensors. *Procedia Social and Behavioral Sciences*, 54: 1095–1103.

Haglund, M. & Åberg, L. 2000. Speed choice in relation to speed limit and influences from other drivers. *Transportation Research Part F*, 3: 39–51.

Hassan, H. & Abdel-Aty, M. 2011. Exploring visibility-related crashes on freeways based on real-time traffic flow data. *Proceedings of the 90th Annual Meeting of the Transportation Research Board*, Washington, D.C.

Hayward, J. 1971. *Near misses as a measure of safety at urban intersections*. PhD Thesis, Department of Civil Engineering, The Pennsylvania State University, Pennsylvania.

Herrera, J.C., Work, D.B., Herring, R., Ban, X., Jacobson, Q., & Bayen, A. 2010. Evaluation of traffic data obtained via GPS-enabled mobile phones: the Mobile Century field experiment. *Transportation Research Part C*, 18: 568–583.

Huguenin, F., Torday, A., & Dumont, A. 2005. Evaluation of traffic safety using microsimulation. *Proceedings of the 5th Swiss Transport Research Conference – STRC*, Ascona, Swiss.

Lamm, R., Psarianos, B., & Cafiso, S. 2002. Safety Evaluation Process for two-lane rural roads: a 10 year review. *Transportation Research Record*, 1796: 51–59.

Mannering, F. 2009. An empirical analysis of driver perceptions of the relationship between speed limits and safety. *Transportation Research Part F*, 12: 99–106.

Minderhoud, M. & Bovy, P. 2001. Extended time to collision measures for road traffic safety assessment. *Accident Analysis and Prevention*, 33: 89–97.

Nilsson, G. 2004. *Traffic safety dimensions and the power model to describe the effect of speed on safety*. Bulletin 221, Lund Institute of Technology, Lund.

Saccomanno, FF., Cunto, F., Guido, G., & Vitale, A. 2008. Comparing safety at signalized intersections and roundabouts using simulated traffic conflicts. *Transportation Research Record*, 2078: 90–95.

Stamatiadis, N., Grossardt, T., & Bailey, K. 2009. How driver risk perception affects operating speeds. *Advances in Transportation Studies an International Journal*, 17(A): 17–28.

Sümer, N., Özkan, T., & Lajunen, T. 2006. Asymmetric relationship between driving and safety skills. *Accident Analysis and Prevention*, 38: 703–711.

Taylor, M.C., Lynam, D.A., & Baruya, A. 2000. *The effects of drivers' speed on the frequency of road crashes*. TRL Report, No. 421, Transport Research Laboratory TRL.

Wallén Warner, H., Haglund, M., & Åberg, L. 2008. Drivers' beliefs about exceeding the speed limits. *Transportation Research Part F*, 11: 376–389.

Walton, D. & Bathurst, J. 1998. An exploration of the perceptions of the average driver's speed compared to perceived driver safety and driving skill. *Accident Analysis and Prevention*, 30(6): 821–830.

Transport Infrastructure and Systems – Dell'Acqua & Wegman (Eds)
© 2017 Taylor & Francis Group, London, ISBN 978-1-138-03009-1

Monitoring the transport sector within the sustainable energy action plan: First results and lessons learnt

I. Delponte
DICCA, Department of Civil, Chemical and Environmental Engineering, University of Genoa, Genoa, Italy

ABSTRACT: The Sustainable Energy Action Plan (SEAP), promoted by the Covenant of Mayor (CoM), is a key tool for policies aimed at reducing fossil fuel consumption and GHG emissions, in accordance with the Kyoto protocol and its updates. It is implemented at the municipal level and is constituted by a Baseline Emission Inventory (BEI) and a set of measures to be implemented (Action Plan, AP). In particular, the paper will investigate the "weight" of the transport sector in terms of emissions within the BEI and the relevance and the variety of its actions inserted in the AP.

To achieve an actual implementation of the SEAP and to obtain its expected targets, monitoring is a crucial component. SEAP monitoring has to look at both the progress of each single action and its global environmental effect, which requires more than one level of evaluation and control. Economical crisis and conjunctural effects mainly influenced the transport Co2 decrease, making AP less important, or quite indifferent in terms of reducing. Is the plan still effective? Or a change of approach can be provided?

According to these statements, the author proposes an analysis of the case of Genoa SEAP (2010) and its monitoring process, recently concluded (2015), particularly focusing on transport sector and its goals. The activity implemented in the city of Genoa is representative of that challenge and is a helpful test case for other cities addressing the same issues.

1 INTRODUCTION

1.1 General background

Climate change concern has brought several different approaches to manage and reduce greenhouse gas emissions connected with energy generation and consumption, at both global and local scales (Wilbanks & Kates 1999, GHG Protocol 2015). In this trend, a leading role has been certainly played by the European Union, which from the first years of this century has been implementing environmental policies to face climate change scenarios and favorite low emission actions (Mertens 2011, EU Climate, 2015).

Summing up briefly the steps of the engagement process by EU in the energy sector, a particularly meaningful moment was when, in 2005, was explicitly expressed the need of a shared policy at the UE level around these topics. The first result was the publication, in 2006, of the Green Paper Energy "A European Strategy for Sustainable, Competitive and Secure Energy", anticipating the exigent of a common planning on energy efficiency and RES exploitation.

In 2007 the Action Plan for Energy Efficiency for the 5-years period 2007–2012 was drawn up, containing the targets of 20% of reducing and the definition of the fields of intervention for achieving the score of reducing energy demand. In the same year, the so-called SET plan (Strategic Energy Technology plan) was promoted, a strategy dealing with the energy field new technologies to be implemented. With the 2008, the engagement of EU achieved meaningful pillars by means of fundamental instruments, such as the first was the Climate Action, which promotes strategies for all the involved stakeholders at long and short terms. Afterwards we have the 2nd Strategic Energy Review, which introduces the well-known "20–20–20" strategy.

More recently, the European Commission presented the "Roadmap for moving to a low-carbon economy in 2050" (2015). This Roadmap aims at a reduction of GHG emissions in the EU 27 by at least 80% in 2050 vis-a-vis emissions in 1990.

1.2 The Sustainable Energy Action Plan (SEAP)

After the European Directives, the Member States adopted the targets, drawing up national Action Plans for the emissions reducing, since the first Two-Thousands. But in consequence of the adoption of the Renewable Energy and Climate Change Package in 2008, the European Commission reckoned to launch, at local entities scale, the initiative of the Covenant of Mayors

(CoM), with the aim of sparking and support the efforts by Municipal Administrations, as the basic unit of the public administration and citizens, in the process of actualization of energy and climate change policies (Alberti & Marzluff 2004, Derissen et al. 2011). In this way, the decisive role of municipalities in the mitigation of the main causes (and consequent effects) of climate change was acknowledged, above all taking into account that the 80% of the energy consumptions and production of CO_2 is associated to urban activities.

The CoM initiative, launched on 29th January 2008 by the European Commission, and the planning tool it promotes, the Sustainable Energy Action Plan (SEAP), is located within this framework and aimed to promote the implementation of EU commitments on the Kyoto Protocol, by means of an unilateral and voluntary participation of European cities.

The SEAP is based on the results of the "Baseline Emission Inventory" (BEI), which quantifies the energy consumption of the territory for the adopted reference year, and identifies several long (LT) and short term (ST) actions in different priority areas, in order to get the expected GHG reduction. Drawing up, implementation and monitoring are the three integrated phases by which the goals of the SEAP can be achieved, through a coordinate initiative at Municipal level involving public institutions, private stakeholders and simple citizens. In this way, the SEAP turns out to be a key document for the authorities that defines the municipal energy policies and, at the same time, a dynamic tool to be upgraded and optimized on the obtained results and the benefit/cost ratio of every action in view of the compliance of the EU objectives.

In order to quantify the emission reduction targets and the achievements of the SEAP actions within 2020, an overall view as a reference situation in terms of final energy flows and CO_2 emissions inside the Municipality territory is needed. In this way, the Baseline Emission Inventory tool has the function to quantify the amount of CO_2 emitted due to energy consumption in the territory of the Covenant signatory in a given baseline year. Thus, the Baseline Emission Inventory is considered a strategic instrument because it provides the quantification and the sources areas of main CO_2 emissions to set the priorities of the future reduction interventions and to assess the impact and the progress of the SEAP actions. The final energy consumptions are divided in two sectors: "Buildings, equipment/facilities and industry" and "Transport". The baseline data collection is to be implemented in order to be representative of the considered local situation; the BEI must be accurate, complete and, most of all, available in the future to be compared in the following years during the monitoring activities through 2020. In order to quantify the CO_2 emissions, EC in the SEAP guidelines gives two methodologies. Each local authority can select between the "Standard" emission factors method, in line with the IPCC (Intergovernmental Panel on Climate Change) principles which consider the CO_2 production due only to energy consumption within the territory of the Municipality, or the LCA (Life Cycle Assessment) approach in order to assess, on the other side, the impact of the whole cycle process of energy production.

After the data collection and the compilation of the BEI the analysis of the results leads to the identification of the main criticalities and the main areas by which improvements are to be attained.

In response to those, the Action Plan (AP) is the operative tool by which getting the definition of sustainable development strategies, regulations, and actions, in accordance with the environmental policy adopted by the Municipality.

Hence, the aim of the Action Plan is to set out actions and to define priorities in respect to the different fields for the reduction of CO_2 emissions, both in public and in private sector. Actions are outlined with different time-horizon: short-term, for intermediate targets within at most 4/5 years from the commencement of the SEAP, and long-term, referring to the strategy towards 2020.

The priority setting process to rank the foreseen measures in order to arrange their relevance has to be defined by means of different criteria, involving the stakeholders so as to sort the most effective actions. In this document every single action is to be precisely described and, moreover, all its details must be clearly defined (such as objectives, expected saving and CO_2 emission reduction as well as the responsible in charge for the action, timetables and deadlines, involved players and stakeholders, budget and risk analysis) in a form: this precise work reveals to be fundamental in the following implementation of the SEAP, since the forms become a key reference document during all the subsequent phases of the CoM.

It is a shared idea that only through accurate monitoring activities and tracking progress can real SEAP implementation can be achieved. In 2014, the Covenant of Mayors Office (CoMO), in collaboration with the Joint Research Centre (JRC) of the European Commission, released the "Reporting Guidelines on Sustainable Energy Action Plan and Monitoring" document a tool aimed to control and check the progress of SEAPs. For the implementation of a Sustainable Energy Action Plan in fact, after planning, has to take into account the changing and updating needs, the knowledge scenario and the related administration initiatives; simultaneously, the territory feedback and the economic and regulatory framework also need to be considered. In this sense, monitoring activities are supposed to be the way to control processes and to recalibrate objectives and instruments of implemented measures. The assessment phase deriving from the monitoring should be able to refine the approach in light of the needs and difficulties. Thus, through a multi-stage strategy, we are able to develop virtuous tools for the implementation of actions.

The SEAP, thus, is a concrete working tool to be implemented and, meanwhile, strictly monitored, through which CoM's Signatories are committed to approach and to achieve the final objectives established in the strategic vision. In this sense, a forward-looking vision of the local administrations is a key-point to a successful approach to the SEAP, like to the CoM from which it follows.

From the operative standpoint, as stated in SEAP Guidelines, CoM signatories are committed to producing two documents after the SEAP submission. The first one, to be submitted every two years, is an implementation report containing qualitative and quantitative information on interventions to evaluate, monitor and verify the status of the Action Plan (SEAP Implementation Status) and its effect; the second one is an update of the CO_2 emission inventory, named the Monitoring Emissions Inventory (MEI), to be compared with the previous BEI for monitoring the progress in terms of emission reductions every four years. CoM provides a monitoring template for the SEAP Implementation Status, in which every measure presents new fields to be filled in such as staff capacity allocation, overall budget spent so far and, where possible, main barriers encountered during SEAP implementation.

From the operative viewpoint, MEI appears to be an updated version of the BEI referring to a specific monitoring year in order to easily compare the two documents and to track the progress of SEAP implementation. As the Baseline Emissions Inventory, the MEI calculates the current amount of final energy consumption and the associated CO_2 emissions in terms of energy carrier and sector such as building equipment facilities, industries, transport, agriculture, forestry and fisheries. Indicating the fuel emission factors makes it then possible to automatically evaluate the associated CO_2 emissions. Furthermore, information on energy supply (municipal green energy purchases, local/distributed electricity and heat/cold production generated from renewable energy sources and CHP plants) must be included in the related parts of the MEI template.

2 THE CASE OF GENOA

2.1 The SEAP of 2010

The Municipality of Genoa joined the initiative on February 2009 and its Sustainable Energy of Action Plan (SEAP), with the aim of a 23,7% CO2 reduction by 2020, has been the first to be officially published by the EU Commission.

For the Municipality of Genoa, the reference baseline year was set in 2005 (first year with the whole data available) and the CO_2 inventory was based on the overall consumptions of energy, except those coming from the industrial sector. The "Standard" method was used to evaluate GHG emissions, by following IPCC guidelines and particularly the "bottom-up" approach, based on the fuel end-use in the several fields. The results of the data analysis for year 2005 of the energy flows in Genoa are classified as required in the JRC template not including energy consumption data of industries and of long-distance transports (railway, highways, sea and air transport). The total CO_2 emissions have been calculated by means of the National/European Emission Factors as the total of the contributions from each energy source. In compliance with SEAP guidelines, the presented data do not reflect the whole energy flow of the city of Genoa but only the emissions generated by the sectors addressed by the SEAP, namely the civil sector and local transports. Energy consumption scenario in Genoa shows that the civil sector represents the by-far prevailing use of energy (77%) in respect to the transport sector. Particularly, natural gas is revealed as the main fuel source being used in Genoa city and electricity consumption results significant, both of them primarily used in tertiary and residential sectors.

Within the two sectors the SEAP focuses on - civil field and local transport-, CO_2 emissions from electricity consumption (35%), from natural gas consumption (35%) and from fuel (gasoline + Diesel) consumption for local transports (27%) result to be very similar. From the BEI, the overall CO_2 emissions in the baseline year 2005 resulted of 2.142.484 tons: this value has been considered as the 100% reference target to be compared in order to evaluate and quantify the effectiveness of the mitigation SEAP actions (Schenone et al, 2015).

2.2 The monitoring of 2015

Due to the delay in publishing the monitoring template by CoM, original deadline for submitting the Monitoring Template will be set in 2015 (instead of 2014), in order to become familiar with the reporting framework.

Considering that Genoa was the first city to attain the SEAP approval, it was among the firsts as well (with Glasgow and Goteborg for instance, to cite a few with similar characteristics) to submit the report, so it is quite impossible to make in-depth comparisons with other results coming from the monitoring phase. A wide literature on the field is supposed to be ready and consistent in a couple of years.

By the way, as a frontrunner in the drafting of the Monitoring Emission Inventory, Genoa Municipality used data collection and processing tools according to a logic of governance and coordination with local organizations and individuals to seize opportunities resulting from initiatives of local, regional and international level.

Just as with the BEI, the MEI compilation is also quite complex because of the difficulty in collecting consistent and coherent data. In many cases, in fact, the availability of complete data sources with the same level of granularity (or aggregation) is almost impossible, thus making statistical processing or adoptions of other indicators necessary.

Based on these considerations and on the governance process, it was decided to report the MEI to 2011, the year for which the Regional Environmental Information System of the Region of Liguria (where Genoa is located) is able to produce complete energy balances at the regional, provincial and municipal levels, and from which it is possible drawing information on final energy consumption of Genoa territory.

The sector with the highest energy consumption and from where the most of the GHG emissions derives is the civil one (public administration, commercial, residential and public lighting), representing the 77% of the total, against the 23% incidence of the transport sector.

In the civil sector the leading cause of fuel consumption and emissions is due to the residential area affecting the total with 42% in the case of emissions; then the tertiary, with a weight of 31%, the public administration (5%), and public lighting (4%).

Comparing with the BEI, for the consumption of 2011 it was observed a reducing effect of –9,5% thanks to the civil sector and –2,5% to transport, which produced an average decrease of CO_2 emissions of the –9,4%.

After calculating the MEI, the SEAP actions monitoring has been conducted with an approach designed to define both the progress of single actions and its environmental monitoring. In particular, the action progress analysis consisted in for the improvement verification updated to 2015 of the SEAP interventions, both qualitatively through the definition of eight classes (not started yet, in definition phase, started, in progress, advanced, completed, postponed, canceled) and quantitatively through progress percentages.

Table 1. BEI 2005 and MEI 2011 GHG emission in Genoa Municipality territory (tCO_2).

SECTOR	BEI 2005	MEI 2011
Municipality	114.801	100.889
Tertiary	606.851	609.117
Residential	934.340	831.458
Public Lighting	18.257	17.293
Transport	468.235	422.226
TOTAL	2.142.484	1.980.982

On the other side, the environmental monitoring concerned the estimated energy savings and the associated reduction of CO_2 to 2015 due to each action, as required by the JRC Monitoring Guidelines. In addition, where possible, values of energy production from renewable sources have been shown.

It should be underlined that the percentage allocation of CO_2 savings was conferred by assessments differing from case to case, in order to preserve the uniqueness of cases and the specificity conducted estimates. This allowed highlighting the presence of critical issues affecting interventions performance and, by means of the comparison between the progress and the environmental monitoring, to better understand the nature of the criticalities and then identify possible corrections or incentives.

The variety of monitored actions can be attributed in general to some typical situations. In some cases, actions progress can appear at an advanced stage, but the CO_2 reduction imputation is still equal to 0, because of the expected benefits will be activated with the commissioning. This is typical of infrastructural interventions, which also require the convergence of other actions to work at full performance. Another case is when CO_2 allocation is corresponding to the progress percentage, i.e., with regard to plants intervention actions, where the implementation immediately leads to the savings achievement. Moreover, in the case of actions to be redefined, points not equal to 0 have been assigned in the progress status where some alternative hypotheses have already emerged in the internal comparison to the administration. This has been done in the belief that preserving the sustainability objective of the plan, it is possible and adequate, in some cases, proceed with a relocation of the initiatives on the occasion of funding or new perspectives not yet present in the early stages of the SEAP preparation.

It is noted that the allocation of the CO_2 saving, in many cases, is the result of deterministic algorithms (even if calculated from estimates and approximations), in others a mixture of deterministic methods, also through tools simulation, and in others on the basis of examples from the literature or derived from observation of urban governance dynamics.

According to the innovative nature of the SEAP tool, the carried on monitoring activities stands out for its experimental nature, requiring a continuous monitoring in terms of scientific content and methods proposed.

Regarding the progress of Genoa's SEAP actions, almost all of the interventions have been completed, started or, at least, defined. During the monitoring activities critical issues emerged, related to interventions no longer valid due to updates of administrative policies or extreme difficulty in the implementation, making necessary for some actions to be deleted. Plus, four actions related to local electricity production at the end of 2015 still not started mainly due to technological hurdles and realization extreme difficulties (Delponte et al. 2013).

3 FOCUS ON TRANSPORT

The SEAP comprises numerous municipal competencies to attain a meaningful reduction of CO_2 emissions and an energy policy of strong impact directly addressed to citizens and urban transport being one of these. To reduce urban GHG emissions from transport, it is important to examine all parameters contributing to emissions, which are usually very strictly related to city shape and settlement location (Kennedy et al. 2009). There are many actions that could be copied from one city to another to optimize urban movements. However, the majority of these actions must be tailored to the geomorphology and transportation attitude of citizens (Van de Coeveringa and Schwanenb 2006, Rickwood et al. 2008).

The complex nature of connections between land-use transformation, climate change, and energy requires new investigations from global to local levels.

This paper mainly focuses on the third sector concerning municipal transport measures, although all of the above-mentioned aspects are closely interrelated and involve other urban planning issues, such as density, the presence of an appropriate mix of activities (reducing local mobility), and the role of green areas (mitigating summer heat) (Steemers 2003, Wilson 2006, Banister and Anable 2009, Wilson and Piper 2010, Romerolankao 2012, Pasimeni 2014).

Geographic peculiarities, combined with the role of Genoa as a port and seafaring city, has influenced both its urban development and infrastructure (which is typically linear along the coast). Thanks to a simulation using the software tool VISUM (for preliminary studies) and MTCP (for plan scenarios) and calculation support from Transport Energy Environment (TEE, a GIS-based decision support system software), the plan included several actions addressed to sustainability, strictly verified for their estimation.

Regarding private and commercial vehicular circulation, two types of estimation were executed, one related to the urban fleet and the other to traveled paths. In the first case, compared with the percentage distribution by national estimate, we found that the fleet was not far from the readings, although some categories were more precise because of the obvious standardization of national conditions. Some algorithms were run to configure actual metropolitan fluxes (Delponte et al. 2013). After extracting the Genoa fleet and fluxes, to every vehicular typology was assigned features regarding fuels of the circulating fleet (parameters of consumption owing to Euro 0,1 and other categories). Using not average but actual travel patterns (and their lengths) avoids an estimation error about 25% of emissions caused by approximation, owing to a downscaling effect that overestimates current metropolitan conditions relative the rest of the region.

The simulation also determines foreseen actions concerning the public sector, in which it is convenient to invest because of great difficulties in using private automobiles (lack of parking and space devoted to infrastructure). In the development of the tool, it was planned to implement a system of urban mobility to enable easier access to and movement around the city, thanks to alternative means of transport, surface and underground local public transport, cycle paths, pedestrian isles, intermodal use of public elevators, funiculars, and introduction of increased water-based transport (Navebus, as in Table 2). The program regarding rationalization of the municipal fleet and greater use of a car-sharing system for internal needs were also sorted for attaining the residual percentage of reduction. This reduction is not significant from a numerical standpoint but is important to the leading role of public administration as an example for citizen behavior.

Actions for private and vehicular circulation were simulated together and then assessed for the pursuit of the general target. Establishing a unique simulation, the foreseen percentages were recalculated with consideration of the multiplier effect of contemporaneous realization of the interventions. The concurrent actions were allowed in such a way to maximize repercussions on the integrated transport system.

From the quantitative point of view, the more effective actions related to transport emissions reducing are the new parking system provided (high tariffs for the city-centre and an annual fee for residents in the boroughs) and promotion of cycling and pedestrian modes (environmental islands).

Table 2. List of local transport reduction expectations on the total BEI (in percentage, ST short term, LT long term).

Actions	Reducing
Municipal Fleet	
Rationalization of use of the municipal fleet_ST	0,036%
Renewal of the municipal fleet_LT	0,010%
Local public transport	
Eco-friendly fleet transition plan_ST	0,20%
Eco-friendly fleet transition plan_LT	0,18%
Strengthening of the local railway system_LT	0,50%
Elevators and funiculars_ST	0,20%
Elevators and funiculars_LT	0,40%
Navebus_LT	0,03%
Protected axes_ST	0,60%
Protected axes_LT	0,80%
Extension of the Subway line_ST	0,30%
Extension of the Subway line_LT	0,30%
Private and commercial transport	
Resident parking policy: Blue Areas_ST	4,20%
Resident parking policy: Blue Areas_LT	4,20%
Environmental islands_ST	1,40%
Environmental islands_LT	1,40%
Interchanging hubs_ST	0,30%
Interchanging hubs_LT	0,30%
Infrastructures_ST	3,00%
Large-scale infrastructures_LT	1,00%
City Logistics_LT	0,40%
Other	
Car sharing_LT	0,40%
Soft mobility – Cycling facilities_ST	0,10%
Soft mobility – Cycling facilities_LT	1,50%

Table 3. BEI 2005 and MEI 2011 GHG emission due to transport in Genoa Municipality territory (tCo2).

	BEI 2005	MEI 2011
Municipality	9.838	8.836
Public transport	32.765	43.595
Private transport	425.631	369.795
TOTAL	468.235	422.226

4 LESSONS LEARNT FROM MONITORING

As showed in Table 3, transportation weighted totally 468.235 tCo2 in the 2005, with a certain decrease in 2011. Going deeply, in the transport sector the percentage of consumption related to private transport (89%) is higher with respect to public transport (9%) and municipal utilities (2%); this percentage results being almost unchanged also within CO_2 emissions.

For analyzing the consumptions' trends related to metropolitan travels, the differences (2005–2011) about the performances of the private fleet were at first considered. The second step was taking into the account the divergence on traffic fluxes. The traffic congestion monitored along the incurred years showed a reduction of −5,6% on the urban area, while the renewal of the vehicular fleet is not so far from the National statistics (Euro 0 and Euro 1, most pollutant, have almost disappeared).

Matching the two previous aspects and considering also that the network was not strongly changed, the results showed 369.795 tCO_2 for the private transport sector, namely a −13% in respect of the BEI.

As far as the public side was concerned, what already said about privates fits as well, regarding general criteria. Nevertheless, in those six years, simultaneously there was an introduction of new more performing public means as well as an increase of energy consumptions. But this result was due only to the more frequent use of the air conditioning on board and the relevance of less pollutant vehicles has already starting to be observed since 2013.

Looking at the monitoring phase, after 5 years, Transport Actions (TRA) generally present a discrete progress (Figure 1): most of the actions are in fact started, ongoing, advanced and completed (above all the short-term ones –ST- which conclusion was expected for 2015). Three actions have been postponed and two deleted, because of the extreme difficulties in the realization of the interventions. It should however be specified that the total mobility system emitted, in recent years, smaller quantities of carbon dioxide in respect of foreseen trends and also less than implied on the basis of actions contents implementation.

As shown by Genoa's SEAP general progress in 2015, in terms of both action progress in percentage and CO_2 emissions reduction, very few sectors correspond to the expected values.

In absolute terms, TRA and EDI (Buildings) are running well, but considering that they are the decisive ones, the most efforts must be devoted to them. This acknowledgement should determine a consequent governance strategy for Genoa in implementing "core" actions. Is this already the case? Are they easy to implement, and what is the level of feasibility of the interventions they

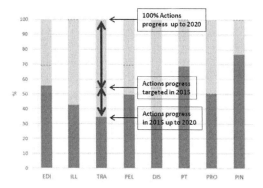

Figure 1. Genoa's SEAP actions progress in 2015 (in percentage) compared to the expected reduction by 2020 for each sector (focus on TRA action progress).

promote? Are there any "enhancement factors" that can be crucial for their realization (which implies a large part of the overall efficiency)?

Hence, the main outcomes here shortly represented underline a certain decrease in transportation consumptions. They were also compared with not local trends: differences and analogies were verified in order to demonstrate the features of the Genoa case. This could significantly due to:

– The bigger performances of the fleet represents a generalized environmental advantage, which reflects upon consumptions and emissions, likely the international trends;
– As come up from the Italian economical framework of those years, after decreasing the citizens' buying power, the mobility demand and the resources engaged on transport sector diminish (fuel litres, travels, fluxes...);
– Economical crisis (which decreasing trends are getting more and more visible in the years after 2011) influenced a lot the population life-style and, specifically in transportation, was decisive for the ownership of the second-cars (fleet is currently decreasing only for cars, not for other types of vehicles) and regarding the use, less intensive, of the existent vehicles (fluxes in decrease, too). In many cases, a mentality change clearly shifts to more sustainable habits (for instance the doubling of the car-sharing system members, even if also that has been influenced by a shortage of travels);
– About local context, huge efficacy has been registered for the new parking system, because the introduction of the tariffs in the city-centre and the annual fee foreseen for residents disincentivized the use of individual car for daily and short travels; (and, so, reducing kilometres and fluxes)

– Moreover, in Genoa, an increasing percentage of people uses bike for daily commuting; the modal shift is partly due to the municipal initiatives, but above all for a higher sensibilization towards sustainable alternatives in favour of less pollutant and cheaper means of transport, as an indirect effect of the crisis (and of the increasing cost the the fossil fuels);
– At the end, the remote accessibility to public services can be considered as a life-style change in the urban areas: where ICT's offer prevents from physical moving, it has a notable influence on mobility phenomena.

Some other conclusions, with a general methodological significance, can be drawn up as well.

– By the comparison between BEI and MEI, it comes out that the reducing, due the field of transport, registered in 6 years is about −2,5% of CO_2 tons on the total of BEI, but is quite −10% taking into account only the comparison BEI-MEI in the sector. It means that the decrease of 2005–2011 emissions is more relevant in the case of transportation than for the other sectors of SEAP.
– Regarding the Action Plan, the percentage of initiatives' realization is about 35%; the result is not so satisfactory, considering that the monitoring was made in the middle of the validity of the SEAP (the expected outcomes for 2015 were about 55% of realization, see Fig. 1).
– But, at the same time, the emissions of the MEI show that the sector is reducing its potential pollution of almost 10%, that corresponds to the half of the predicted reduction for all the actions' realization (and the trend for the upcoming years, so far, is not foreseen moving away).

This is due, as told before, to conjunctural aspects that are not related to the soundness of municipal initiatives, but, largely, to vehicles performances and reduction of mobility demand.

This is good for the air we inhale, because a lesser pollutant transport system is good for everybody, but a concluding remark can be pinpointed about the incisiveness of municipal planning actions in the urban transportation field; their importance in the agenda can be absolutely underestimated in the future, if results even better can be achieved by no-local effects. The Genoa case shows that a certain efficacy in carrying out governance actions and achieving environmental goals has been registered for parking tariffs actions (which weakens more and more the mobility demand; others are more difficult to be realized, from the spatial planning point of view, but at the same time, they are less incisive for the partial contribution they bring.

Moreover, many recent reports say that the transport sector is one of the most difficult to be treated in order to reach a substantial "energy transition", because of the strong dependence on fossil fuels, which in other sectors is decisively reducing. So, transport planning has to regain its legitimacy whereas it does not seem to be an evident urban priority.

Urban areas are worldwide responsible of almost 80% of the total GHG emissions but, at the same time, municipalities can be considered as favourite areas that play a crucial role in the safeguard of the environment and in the setting out of more resilient initiatives. So, the Covenant of Mayor represents a relevant part of this strategy: through the related commitments and the SEAP, cities are called to work hard for facing the challenge of GHG reduction. At this point, future research should investigate the role of the transport sector in the SEAP tool, namely how to deal with it, respecting the identity and the general aim of the plan and considering the increasing municipal role. Maybe that lack of solutions by transport planners observed in many cases can underline, by the contrary, the opportunity to involve not only environmental and technological experts in solving problems, but also sociological ones, taking more into account urban livability's complexity. A more comprehensive approach could give nourishment to the current impasse of ideas.

REFERENCES

Alberti M.; Marzluff J., 2004. Ecological Resilience in urban ecosystems: Linking urban patterns to human and ecological functions. *Urban Ecosystems*, 7, 241–65.

Banister, D., & Anable, J., 2009. Transport Policies and Climate Change. In S. Davoudi, J. Crawford, & A. Menhood (Eds.), *Planning for Climate Change. Strategies for Mitigation and Adaptation for Spatial Planners*, 55–69. London: Earthscan.

Delponte, I., Pittaluga, I., and Schenone, 2013. C., Develop, implementation and monitoring of the Sustainable Energy Action Plan of the Municipality of Genoa, *Proceedings of 31st UIT Heat Transfer Conference*, 599–608.

Derissen S., Quaas M.F., Baumgärtner S., 2011. The relationship between resilience and sustainable development of ecological-economic systems. *Ecological Economics 2011*, 70, 1121–8.

EU Climate. Available online: http://ec.europa.eu/dgs/clima/acquis/index_en.htm.

Genoa Sustainable Energy Action Plan Summary Available online: http://www.urbancenter.comune.genova.it/sites/default/files/archivio/allegati/SEAP%20summary_0.pdf.

International Local Government Greenhouse Gas Protocol. Available online: http://www.ghgprotocol.org/city-accounting.

Kennedy, C., Steinberger, J., Gasson, B., Hansen, Y., Hillman, T., Havranek, M., Pataki, D., Phdungsilp, A., Ramaswami, A., and Mendez, G.V, 2009. Greenhouse gas emissions from global cities, *Environmental Science & Technology* 43, 7297–302.

Mertens K., 2011. Recent Developments of EU Environmental Policy and Law J. *Eur. Env. Plan. Law*, 8, 293–298.

Pasimeni, M.R., Petrosillo, I., Aretano, R., Semeraro, T., De Marco, A., Zaccarelli, N., Zurlini, G., 2014. Scales, strategies and actions for effective energy planning: A review. *Energy Policy* 65, 165–174.

Reporting Guidelines on SEAP and Monitoring Available online: http://www.covenantofmayors.eu/Library,84.html.

Rickwood, P., Glazebrook, G., and Searle, G., 2008. Urban Structure and Energy. A Review, *Urban Policy and Research* 26 (1), 57–81.

Roadmap for moving to a low-carbon economy in 2050. Available online: http://ec.europa.eu/clima/policies/roadmap/index_en.htm.

Romero-lankao, P., 2012. Governing Carbon and Climate in the Cities: An Overview of Policy and Planning Challenges and Options. *European Planning Studies*, 20(1), 7–26.

Schenone, C., Delponte, I., Pittaluga, I., 2015. The preparation of the Sustainable Energy Action Plan as a city-level tool for sustainability: The case of Genoa. *Journal of Renewable and Sustainable Energy*, 7 (3).

Steemers, K., 2003. Energy and the city: density, buildings and transport. *Energy and buildings*, 35(1), 3–14.

Van de Coevering, P., and Schwanenb, T., 2006. Re-evaluating the impact of urban form on travel patterns in Europe and North-America, *Transport Policy* 13 (3), 229–239.

Wilbanks J., Kates R.W., 1999. Global changes in local places: How scale matters. *Climate Change*, 43, 601–28.

Wilson, E., 2006. Adapting to Climate Change at the Local Level: The Spatial Planning Response. *Local Environment*, 11(6), 609–625.

Wilson, E., & Piper, J., 2010. *Spatial Planning and Climate Change*. Routledge, Ed., Oxon.

Transport Infrastructure and Systems – Dell'Acqua & Wegman (Eds)
© 2017 Taylor & Francis Group, London, ISBN 978-1-138-03009-1

A model to evaluate heavy vehicle effect in motorway traffic flow

R. Mauro, S. Cattani & M. Guerrieri
DICAM, University of Trento, Trento, Italy

ABSTRACT: Highway operating conditions are also determined by interactions between passenger cars and heavy vehicles in the flow. Therefore, a suitable model which describes and predicts the heavy vehicle effect can be useful to achieve more realistic criteria to evaluate the Level of Service (LOS) for infrastructures in the uninterrupted flow. After reviewing briefly the models available in literature to take account of the effects on operating conditions due to heavy vehicles, the paper examines some traffic data analysis of the Italian A22 motorway and later presents a closed-form traffic model for heavy vehicles via the queuing theory. Some applications of this model are displayed for operational purposes and a critical evaluation of the results is also made in comparison with some experimental evidence.

1 INTRODUCTION

The plano-altimetric geometry, the vehicular flows and the heterogeneity in the vehicular traffic mix composing the traffic streams are the most important factors affecting the Level of Service (LOS) of motorways and highways.

Despite the proportion of heavy vehicles percentage (trucks, buses, recreational vehicles, etc.) is generally smaller with respect to the total flow, its impact on highway performances is remarkable.

The effect of heavy vehicles on the MOE (measures of effectiveness) is mainly ascribed to two factors: their larger dimensions and the lower performance (acceleration, maximum speeds, etc.) compared to passenger cars (Sun, et al., 2008).

In addition, the performances of heavy vehicles may vary under different traffic and geometric conditions, and are related to the following factors:

- traffic regime: unsaturated or saturated conditions;
- traffic flow rate for unsaturated conditions;
- terrain type: level, rolling and mountainous.

Since the first versions the Highway Capacity Manual (HCM) has provided the methodology to evaluate the Passenger Car Equivalents (PCEs) for the capacity and Level of Service (LOS) analysis.

PCEs were first formally introduced by the HCM (1965) but the algorithms used for their evaluation have been modified in the following versions of the manual.

The HCM 2000 defines PCE as "*the number of passenger cars displaced by a single heavy vehicle of a particular type under specified roadway, traffic and control conditions*". By means of PCEs, traffic flows can be expressed in terms of Passenger Car

Units (PCUs). Therefore, PCEs have an important role in highway design and functional analysis (Webster & Elefteriadou 1999; Elefteriadou 2014).

Nowadays many different methods are available to estimate the PCEs as function of one or more traffic flow parameters (Shalin & Kumar 2014):

- PCEs based on Flow Rates and Density;
- PCEs based on Headways;
- PCEs based on Speed;
- PCEs based on Queue Discharge Flow;
- PCEs based on Delays;
- PCEs based on volume-to-capacity ratio;
- PCEs based on Vehicle-Hours;
- PCEs based on Travel Time;
- PCEs based on HCM method.

Although the evaluation of LOS is generally carried out by means of HCM 2010 methodology, many other alternative methodologies can be used for this purpose (Wu, 2016).

Furthermore, the HCM 2010 doesn't allow evaluating the LOS in the case of particular types of traffic control strategies, implemented to increase safety and/or capacity, and cannot also be used for highway performance analysis in *real-time*.

In the light of previous critical issues, this research introduces a novel closed-form traffic model, based on the P-K theory (Pollaczek 1930; Kleinlock 1975), for the evaluation of MOE in *real-time*.

The proposed model does not need to estimate the PCEs.

First applications were done to evaluate the effect of trucks' Overtaking Prohibition (OP) in terms of LOS.

The OP is a very common strategy utilized to increase the passing lane capacity and motorway safety, although it can generate the formation of platoons of heavy vehicles. In addition, increases of rear-end collisions may occur in the right lane (Mauro & Guerrieri 2016).

The case study of the Italian motorway A22 (Autostrada del Brennero) was examined.

The sampling of traffic flows, platoons, speeds was performed in four observation sections. The proposed traffic model has been used to evaluate delays, queues and level of service in real-time.

2 NOVEL CLOSED-FORM TRAFFIC MODEL FOR EVALUATING HIGHWAY MOE

The queuing theory can be used as the basis for considering the influence of heavy vehicles on the level of service of highway segments.

The novel closed-form traffic model presented in this section is based on the following general assumptions under which the model works:

- heavy vehicles travel only on the right lane, with speed v_{vl} (as in the case of trucks' overtaking prohibition);
- the passenger cars travel with a speed v, except for the case in which they must queue behind heavy vehicles before starting overtaking maneuvers (while in waiting for overtaking maneuvers the speeds of light and heavy vehicles are the same $v = v_{vl}$);
- in the right lane the passenger cars, before reaching heavy vehicles, decelerate with a constant value of deceleration d. Instead, overtaking maneuvers take place with a constant acceleration a;
- denoting with γ the percentage of heavy vehicles $\gamma = Q_{vl}/Q_{tot}$ (Q_{vl} = traffic flow of heavy vehicles and Q_{tot} = total flow) and with β the percentage of passenger cars travelling along the right lane ($Q_{mn} = \beta(Q_{tot}-Q_{vl})$). The passenger cars travelling on the overtaking lane are: $Q = Q_{tot}-Q_{vl}- Q_{mn}$;
- for the passenger cars that have to make the overtaking manouvres, the right lane is regarded as a hypothetical acceleration lane in which the service time w is equal to the interval time spent in queuing before the overtaking manoeuvres;
- the proposed model defines LOS for motorway and highway as a function of the average speed.

In the light of the above assumptions the average speed can be estimated with the following steps:

1. estimation of γ, β, Q_{tot}, Q_{mn}, Q_{vl} (by means of sampling traffic flows);
2. calculation of critical gaps by means of the following relationships:

$$T = \frac{v - v_{vl}}{2a} + 2\delta \tag{1}$$

In which δ is the *perception-reaction time* (δ = 1 s)

3. in the event of Erlang headway distribution (with a parameter K, Mauro & Branco 2012) for vehicles travelling on the overtaking lane, any service time s and vehicle headways, according to P-K relationships, the mean b of the service time s, b = E[s], and the variance $\sigma^2(s)$ can be calculated as follows:

$$b = E[s] = T + \frac{e^{kQT} - \sum_{i=0}^{k} \frac{(kQT)^i}{i!}}{Q \sum_{i=0}^{k-1} \frac{(kQT)^i}{i!}} \tag{2}$$

$$\sigma^2(s) = \frac{(k+1)\left[e^{kQT} - \sum_{i=0}^{k+1} \frac{(kQT)^i}{i!}\right]}{kQ^2 \sum_{i=0}^{k-1} \frac{(kQT)^i}{i!}} + (E[s] - T)^2 \tag{3}$$

4. the time spent in queuing E[w]:

$$E[w] = b + \frac{Q_{vl}(b^2 + \sigma^2(s))}{2(1 - Q_{vl} \cdot b)} \tag{4}$$

5. the average speed of passenger cars (V_{pc}) is

$$V_{pc} = \left(\beta \cdot E[w] \cdot v_{vl} + \beta \cdot t_a \cdot \left(\frac{v + v_{vl}}{2}\right) + \right.$$
$$\left. (t_r - \beta \cdot t_a - \beta t_d) \cdot v + \beta \cdot t_d \cdot \left(\frac{v + v_{vl}}{2}\right)\right) / (\beta \cdot E[w] + t) \tag{5}$$

where:

$$t_r = \frac{v_{vl}}{(v - v_{vl}) \cdot Q_{vl}} \tag{6}$$

is the time taken by a passenger car with a speed v to travel in the space comprised between two successive heavy vehicles, whose speed is v_{vl};

$$t_a = \frac{v - v_{vl}}{a} \tag{7}$$

is the time taken by a passenger car to accelerate from v_{vl} to v (equal to speed of passenger cars in the overtaking lane)

$$t_d = \frac{v - v_{vl}}{d} \tag{8}$$

is the time required by a passenger car to decelerate from v to v_{vl}.

Figures 1÷3 summarize the correlation between average speed V_{pc} and total flow Q ($V_{pc} = V_{pc}$ (Q)) for different traffic conditions and for $\gamma = 10\%$ (see Fig. 1), $\gamma = 20\%$ (see Fig. 2), $\gamma = 30\%$ (see Fig. 3).

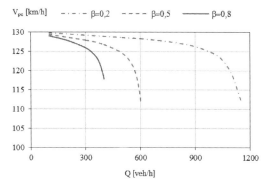

Figure 1. $V_{pc} = V_{pc}$ (Q) relationships for different β (a = 0.6 m/s², d = 2 m/s², v = 130 km/h, δ = 1 s, v_{vl} = 90 km/h, γ = 10%).

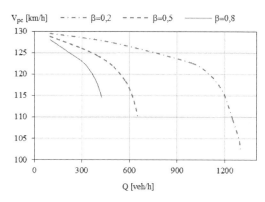

Figure 2. $V_{pc} = V_{pc}$ (Q) relationships for different β (a = 0.6 m/s², d = 2 m/s², v = 130 km/h, δ = 1 s, v_{vl} = 90 km/h, γ = 20%).

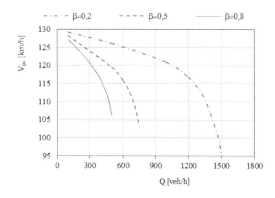

Figure 3. $V_{pc} = V_{pc}$ (Q) relationships for different β. (a = 0.6 m/s², d = 2 m/s², v = 130 km/h, δ = 1 s, v_{vl} = 90, γ = 30%).

3 THE A22 MOTORWAY CASE STUDY

The A22 motorway, "Autostrada del Brennero", links the Po Valley and the A1 Freeway with Austria and Germany. The A22 is one of the principal axes of the Italian highway network and belongs to the TEN-T Network (Trans-European Road Network, corridor Helsinki—La Valletta).

The A22 overall length is about 313 km.

Fig. 4 shows the layout and overtaking prohibitions along the A22. Instead, Fig. 5 shows the typical semi-cross section.

The overtaking prohibition for heavy vehicles is managed as follows:

- vehicles with weight > 7.5 t, from hour 0:00 to 24:00, from 0 to 85 km;

Figure 4. A22 motorway layout and overtaking prohibitions.

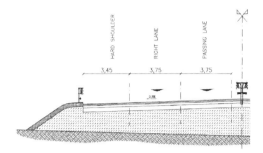

Figure 5.   A22 Typical semi—cross Section.

- vehicles with weight > 12 t, from hour 6:00 to 22:00, from 85 to 313 km.

The Annual Average Daily Traffic (AADT) is between 41,907 vehicles per day (section "Brennero—Vipiteno") and 62,646 vehicles per day (section "Ala Avio—Affi").

3.1 Estimation of capacity and distributions of heavy vehicles between lanes

The capacity analysis was done for two observation periods: 5÷11 May 2014 and 8÷14 December 2014. The observation sections are given in Table 1.

The macroscopic flow parameters (i.e. flow "q", speed "v" and density "k") have been calculated for intervals of 5 minutes and 15 minutes.

Traffic flows were estimated in terms of passenger car unit (PCU), by means of the following PCE relationship:

$$\text{PCE} = \frac{(1-\gamma)\cdot(\overline{d}_{AL}+\overline{d}_{LA}-\overline{d}_{AA})+\gamma\cdot\overline{d}_{LL}}{\overline{d}_{AA}} \quad (9)$$

In which:

- γ: percentage of heavy vehicles
- d_{AA} : average headways of couples of passenger cars;
- d_{AL} : average headways of couples of vehicles in which the leader vehicle is a passenger car and the follower vehicle is a heavy vehicle;
- d_{LL} : average headways of couples of heavy vehicles;
- d_{LA} : average headways of couples of vehicles in which the leader vehicle is a heavy vehicle and the follower vehicle is a passenger car.

PCEs relationships have been tabulated as function of the slope p (see Table 1) in Table 2.

In all, N5' = 24,192 couples (v; k), (q; k), (v; q). have been carried out.

For each section the flow diagrams are obtained by means of the Drake model (May 1990; Wang et al. 2009) in which the relationship between speed (v) and density (k) is given by the following equation:

Table 1.   Observation sections.

Observation section	Location	Horizontal alignment	Vertical alignment; slope (%)
Kofler	063+500	Tangent	Curve: R = 10,000 m; slope = 0.41%
S. Michele	123+960	Tangent	Tangent, slope = 0.03%
Portale Affi	205+500	Curve R = 1000 m	Curve: R = 10,000 m; slope = 0.23%
Mantova Sud	271+900	Tangent	Curve: R = 15,000 m; slope = 0.00%

Table 2.   PCEs relationships.

p	Right lane	Passing lane	Carriageway
> 0	PCE = $1.166\cdot(\gamma\%)^{-0.126}$	2	$1.126\cdot(\gamma\%)^{-0.205}$
> 0	PCE = 1.35	2	$1.220\cdot(\gamma\%)^{-0.132}$
≈ 0	PCE = $1.186\cdot(\gamma\%)^{-0.112}$	2	$1.161\cdot(\gamma\%)^{-0.172}$

$$v = v_f \cdot e^{-\frac{1}{2}(\frac{k}{k_{jam}})^2} \quad (10)$$

where v_f is the free-flow speed and k_{jam} the jam density. The typical fundamental diagrams (with LOS limits, see HCM 2010 values) for the right lane, passing lane and carriageway are depicted in Figure 6 (Guerrieri & Mauro 2016). Instead, Table 3 illustrates the flow relationship parameters for the years 2003 (Mauro et al. 2013) and 2014.

The Level of service has been calculated for all the sections (considering an hourly volume VHP = 0.1·AADT, see Table 5). Figures 7–8 show the frequency distribution histograms of the percentage of heavy vehicles in the traffic stream on the right lane and on the overpassing lane at weekdays (section San Michele). It is immediately verifiable that overtaking prohibitions are generally respected by heavy vehicles over 7.5 t, though this leads to the formation of many platoons.

Figure 9 shows the flow distribution between lanes in the northbound roadway for the section San Michele.

The analysis of the speed and flow processes shows that on the A22 motorway the observed average speed of passenger cars (passing lane) is close to 130 km/h (see Fig. 10), instead the observed average speed of heavy vehicles (v_{vl}) is close to 80 km/h (right lane).

In addition, sampling of platoons has been carried out at observation sections shown in Table 2 on 7th May 2014 (Wednesday). The flow parameters of platoons, with 2÷20 heavy vehicles for each platoon, have been investigated by collecting data

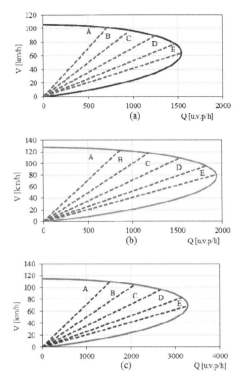

Figure 6. Speed—Flow diagrams for the right lane (red), passing lane (b) and carriageway (c), A22, northbound roadway.

Table 3. Traffic flow parameters (A22).

Lane/ carriageway	v_f [km/h]	k_{jam} [pcu/ lane/km]	C [pcu/h]	v_{jam} [km/h]
Year 2014				
Right lane	106	24	1552	65
Passing lane	128	25	1916	77
Carriageway	115	47	3254	70
Year 2003				
Right lane	107	24	1534	65
Passing lane	130	25	1983	79
Carriageway	118	49	3490	71

regarding vehicle speeds, headways and mass. The following parameters have been analyzed:

- frequencies of the number of platoon sizes of "i" heavy vehicles (i = 1,2,...20);
- minimum headway (τ_{min}), mean headway (τ_{mean}), and maximum headway (τ_{max}) between vehicles of each platoon;

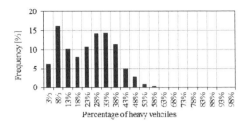

Figure 7. Frequency distribution of the heavy vehicle percentage. Section San Michele, northbound roadway, right lane, weekdays.

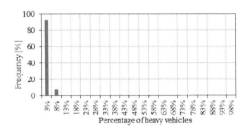

Figure 8. Frequency distribution of the heavy vehicle percentage. Section San Michele, northbound roadway, passing lane, weekdays.

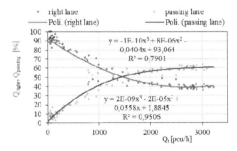

Figure 9. Relationship between lane flow rate and total flow rate Q_t (San Michele, northbound roadway).

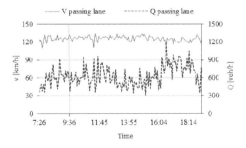

Figure 10. Observed average speed of passenger cars and flow processes. Section S. Michele, northbound roadway (weekday).

- minimum speed (V_{min}), mean speed (V_{mean}), and maximum speed (V_{max}) of vehicles of each platoon;

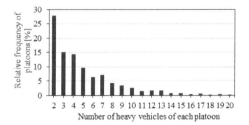

Figure 11. Frequency of platoons as function of the number of vehicles for each platoon. Section San Michele, northbound roadway.

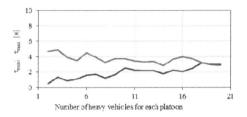

Figure 12. Relationship headway—platoon size (red line: τ_{min}; blue line: τ_{mean}). Section San Michele, northbound roadway.

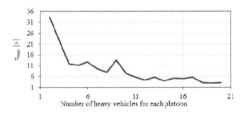

Figure 13. Relationship maximum headway–platoon size. Section San Michele, northbound roadway.

- the number of platoons whose vehicles travel with headways comprised within predetermined values (from $0 < \tau \leq 1$ seconds, to $1 < \tau \leq 20$ seconds).

Fig. 11 shows the relative frequencies of the number of platoons with "i". In all the sections of the A22 the size of platoons rarely exceeds ten vehicles ($i > 10$).

Fig. 12 and Fig. 13 show respectively the trends of minimum headway (τ_{min}), mean headway (τ_{mean}) and maximum headway (τ_{max}).

The mean headway is characterized by low fluctuations; instead, the minimum headway increases monotonously as function of the platoon size.

The number of platoons whose vehicles travel with headways comprised within predetermined classes ($\tau_I \div \tau_j$) are given in Table 4.

Through the traffic model explained in Section 2 the mean of passenger cars speed has been estimated, in real-time, using the traffic sampling of 7 May 2014 (Wednesday). In order to implement the algorithm, the following input data were used: a = 0.6 m/

Table 4. Number of platoons whose vehicles travel with headways comprised within classes of headway. Section San Michele, northbound roadway.

τ_i [s]	τ_j [s]	$\tau^* = (\tau_i + \tau_j)/2$ [s]	N	V_m [km/h]
0	1	0.5	15	87.28
1	2	1.5	90	85.07
2	3	2.5	225	84.6
3	4	3.5	179	85.51
4	5	4.5	109	86.25
5	6	5.5	71	88.06
6	7	6.5	33	90.45
7	8	7.5	27	92.45
8	9	8.5	18	89.86
9	10	9.5	13	94.04
10	11	10.5	6	90.31
11	12	11.5	6	87.36
12	13	12.5	4	94.25
13	14	13.5	5	89.47
14	15	14.5	1	81.5
15	16	15.5	3	88.22
16	17	16.5	2	90.92
17	18	17.5	0	–
18	19	18.5	0	–
19	20	19.5	0	–

s^2; d = 2 m/s^2; δ = 1 s, v = 130 km/h, v_{vl} = 90 km/h. The mean value of the coefficients β is 0.58, instead γ = 11%–18% depending on the section.

The first results of the traffic model are given in Fig. 14; further phases of the research require a more accurate calibration of the model in order to reduce the deviations between observed and estimated values (Fig. 15).

However, in the hypothesis of uninterrupted flows, the proposed model allows the real-time estimation of the motorway density (k) and Level of Service (LOS), as shown in Fig. 16 for the case of the right lane. Finally, the speed levels carried out by means of the model presented above can be used as input data for the calculation of the reliability ϕ. As well known, the flow instability is a random event whose occurrence depends on the speed level process. The probability that no instability will occur in the interval time T defines the reliability ϕ of the vehicle flow (for the period T). The ϕ value for a lane traffic flow can be estimated with the expression (Ferrari 1988; Mauro et al. 2013):

$$\phi = 1 - 19,80 \cdot \left(\frac{Q}{10,000}\right)^{8,82} \cdot T^{1,983} \cdot M^2 \quad (11)$$

where Q = flow rate constrained along the off-side lane (veh/h); T = period of time (minutes); M = absolute value of "b", angular coefficient of relationship $\sigma_v^2 = a + b \cdot \ln(k)$ (m$^2 \cdot$km$\cdot$s^2). The coefficient σ_v^2 is the standard deviation of speed level.

Table 5. Level of service estimation for each segment of A22.

Initial	Final	Section Name	AADT (Carriageway) [veh/day]	VHP [veh/h]	LOS
0 + 000	15 + 870	Brennero—Vipiteno	23,049	2305	C
15 + 870	38 + 030	Vipiteno—Bressanone	24,287	2429	C
38 + 030	47 + 600	Bressanone—Bress.ne Z.I.	26,613	2661	D
47 + 600	53 + 070	Bressanone Z.I.—Chiusa	25,793	2579	C
53 + 070	77 + 470	Chiusa—Bolzano nord	25,181	2518	C
77 + 470	85 + 330	Bolzano nord—Bolzano sud	26,828	2683	D
85 + 330	101 + 800	Bolzano sud—Egna Ora	31,764	3176	E
101 + 800	121 + 450	Egna Ora—S. Michele	32,346	3235	E
121 + 450	131 + 440	S. Michele—Trento nord	31,831	3183	E
131 + 440	136 + 460	Trento nord—Trento centro	28,855	2886	D
136 + 460	142 + 000	Trento centro—Trento sud	29,081	2908	D
142 + 000	157 + 850	Trento sud—Rovereto nord	33,586	3359	F
157 + 850	166 + 740	Rovereto nord—Rovereto	33,201	3320	F
166 + 740	179 + 125	Rovereto sud—Ala Avio	34,141	3414	F
179 + 125	206 + 670	Ala Avio—Affi	31,323	3132	E
206 + 670	225 + 370	Affi—Verona nord	26,519	2652	C
225 + 370	228 + 000	Verona nord—int. aut. A4	33,237	3324	F
228 + 000	243 + 670	int. aut. A4—Nogar. Rocca	34,471	3447	F
243 + 670	256 + 180	Nogar. Rocca—Mantova nord	33,580	3358	E
256 + 180	265 + 000	Mantova nord—Mantova sud	33,658	3366	F
265 + 000	276 + 710	Mantova sud—Pegognaga	33,918	3392	F
276 + 710	285 + 630	Pegognaga—Reggiolo Rolo	30,050	3005	D
285 + 630	302 + 175	Reggiolo Rolo—Carpi	30,492	3049	D
302 + 175	312 + 150	Carpi—Campogalliano	32,817	3282	F
312 + 150	313 + 085	Campogalliano—Autosole	33,388	3339	F

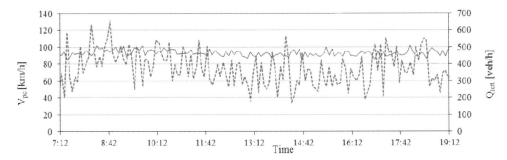

Figure 14. Average speeds (V_{pc}) of the passenger cars (blue line) in the right lane as function of the flow rates (Q) on the overpassing lane (red line), calculated with the traffic model explained in the section 2. Section San Michele northbound roadway (7/5/2014).

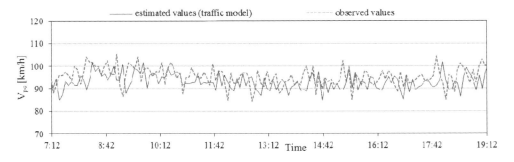

Figure 15. Estimated and observed values of the average speeds of passenger cars on the right lane.

615

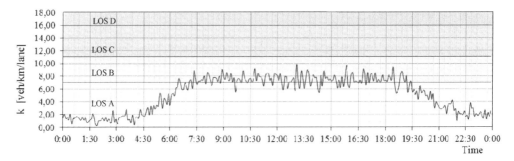

Figure 16. Density (k) and LOS estimated with the proposed traffic model, Section San Michele northbound roadway (7/5/2014) (right lane).

4 CONCLUSIONS

The presence of heavy vehicles is particularly interesting in traffic study, and more generally in Highway Engineering, with reference to many technical applications, especially to topics like capacity, level of service, road safety and air pollution emissions.

The research proposes a closed-form traffic model based on the P-K queuing theory that allows to estimate in real-time (under predetermined assumptions) many traffic flow parameters, like the mean speed of the passenger cars travelling in the right lane (V_{pc}), the lane density (k) and the Level of Service (LOS). The novel traffic model does not require the use of Passenger Car Equivalents (PCEs) with regard to LOS analysis.

The model has been employed to investigate the functionality of the Italian motorway A22 (Autostrada del Brennero). Along this motorway, heavy vehicle overtaking prohibition is in force. The first part of the research was devoted to the calculation of the macroscopic flow parameters. The traffic survey has shown that overtaking prohibitions are generally respected by heavy vehicles; this implies that many platoons travel on the motorway right lanes.

In order to estimate the average passenger cars, this research has employed the following values of acceleration (a), deceleration (d), perception-reaction time, (δ), speeds (v and v_{vl}): a = 0.6 m/s^2, d = 2 m/s^2, δ = 1 s, v = 130 km/h, v_{vl} = 90 km/h.

In the hypothesis of uninterrupted flows, the proposed traffic model allows to estimate the motorway density and level of service in real time (the right lane was examined). In addition, the results can be used as input data for the calculation of the reliability of the vehicle flow ϕ, in real-time.

Although the new traffic model requires a more accurate calibration, the first results show that the procedure is characterised by a reliable and rapid analytical approach in passenger cars speed and in LOS evaluation.

REFERENCES

Elefteriadou, L. 2014. An Introduction to Traffic Flow. Springer.

Ferrari, P. 1988. *The reliability of motorway transport system*. Transp. Res. Part B, 22(4), 291–310.

Guerrieri, M. & Mauro, R. 2016. *Capacity and safety analysis of Hard-Shoulder Running (HSR). A motorway case study*. Transportation Research Part A: Policy and Practice, Vol. 92, pp. 162–183.

HCM (1965). TRB, National Research Council, Washington, DC.

HCM (2000). TRB, National Research Council, Washington, DC.

HCM (2010). TRB, National Research Council, Washington, DC.

Kleinlock, L. 1975. *Queuing Systems, I*. New York: John Wiley and Sons.

Mauro, R. & Guerrieri, M. 2016. *Safety and capacity analysis of* A22 *motorway (in Italian)*. Internal Report, DICAM, Unitn.

Mauro, R., Branco, F. 2012. *Two vehicular headways time dichotomic models*. Modern Applied Science, 6 (12), pp. 1–12.

Mauro, R., Giuffrè, O., Granà, A. 2013. *Speed stochastic processes and freeway reliability estimation: Evidence from the A22 freeway, Italy*. Journal of Transportation Engineering, 139 (12), pp. 1244–1256.

May, A.D. 1990. *Traffic Flow Fundamentals*, Prentice-Hall.

Pollaczek, F. 1930. *Über eine Aufgabe der Wahrscheinlichkeitstheorie*. Mathematische Zeitschrift.

Shalini, K., Kumar, B. 2014. *Estimation of the Passenger Car Equivalent: A Review*. International Journal of Emerging Technology and Advanced Engineering Volume 4, Issue 6, pp. 97–102

Sun, D., Lv, J., Paul, L. Calibrating Passenger Car Equivalent (PCE) for Highway Work Zones using Speed and Percentage of Trucks. 2008. TRB 2008 Annual Meeting CD-ROM.

Wang, H., Li, J., Chen, Q. Y,. Ni, D. 2009. *Speed-Density Relationship: From Deterministic to Stochastic*. TRB 88th Annual Meeting at Washington D. C.

Webster, N; Elefteriadou, L.. 1999. *A Simulation Study of Truck Passenger Car Equivalents (PCE) on Basic Freeway Sections. Transportation Research. Part B: Methodological* 33(5), 323–336.

Wu, N. 2016. *New Features in the 2015 German Highway Capacity Manual*. Transportation Research Procedia 00 (2017) 000–000.

Transport Infrastructure and Systems – Dell'Acqua & Wegman (Eds)
© 2017 Taylor & Francis Group, London, ISBN 978-1-138-03009-1

Quality and energy evaluation of rail-road terminals by microsimulation

A. Carboni & F. Deflorio
Department of DIATI, Politecnico di Torino, Torino, Italy

ABSTRACT: The aim of this paper is to present a method to evaluate the quality and energy performance of inland freight terminals, using a quantitative approach based on traffic microsimulation models. The transport of goods should meet requirements of environmental sustainability and combined transport can be an eco-friendly option for medium/long distance connections, in which the railway mode can provide its efficiency if properly accessible. The analysis of the terminal aims at quantifying Key Performance Indicators (KPIs) by traffic microsimulation. The relevant features of the typical phases of the internal process are represented and the traffic flow data of arrivals are disaggregated by specific service needs. The model is used to compare the chosen indicators in different scenarios, varying the arrival rates for trucks and reducing the duration of check-out and cranes operations. Results of quality and energy are also reported disaggregated for the different types of service lines.

1 INTRODUCTION

The intermodal transport is defined as a transhipment of goods through the same transport unit using two or more transport means and without the manipulation of the freight themselves (UNI/CE). To be sustainable from the environmental and economic points of view the intermodal transport must be efficient for all the actors involved in the process. The crucial point is to take advantage of the operational benefits—as cost, capacity and flexibility—of transport modes, then merging them into a single transport chain increasing the attractiveness of rail freight transport.

The Directive 92/106/CEE defines the "combined transport" as "the transport of goods between Member States where the lorry, trailer, semi-trailer, with or without tractor unit, swap body or container of 20 feet or more uses the road on the initial or final leg of the journey and, on the other leg, rail or inland waterway or maritime services where this section exceeds 100 km as the crow flies and make the initial or final road transport leg of the journey".

The UE, in the White Paper, has reiterated the need of reduce drastically the greenhouse gas emissions worldwide with the goal of maintain the global warming under 2°C. In total, by 2050, Europe must reduce emissions by 80–95% compared to 1990 levels (European Commission 2011). The White Paper further states that freight transport over short and medium distances (roughly below 300 km) will continue to be carried, in large measure, with trucks. In longer distances, options for road decarbonisation are more limited, and freight multimodality has to become economically attractive for shippers. Thirty per cent of road freight over 300 km should shift to other modes such as rail or waterborne transport by 2030, and more than 50% by 2050, facilitated by efficient and green freight corridors.

Figure 1 shows that only in Estonia, Latvia and Switzerland the road transport mode has a contained share in refer to total percentage of freight transport, whereas in the other countries it still largely, probably due to different geographical contexts or policies adopted.

The transport by railway, sea or inland waterways is preferred on long distances for economic scale reasons to reduce the impact of road transport, whereas the initial and final haulage are

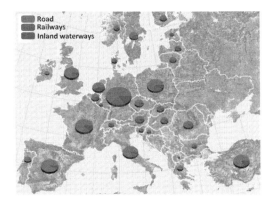

Figure 1. Inland freight transport modal share (from EU transport in figures 2014, elaboration by Castaldo (2014)).

managed in road transport mode, because it provides more flexibility and accessibility (Carreira et al. 2012).

The transfer between the two modes is operated in intermodal terminals that deal both of the transfer operations and the network change. More specifically, the International Union of Railways defines the rail-road terminals as interchange hubs between rail and road traffic. They are fitted with all the equipment required to handle and tranship loading units from one transport mode to the next in a rapid and efficient manner: gantries and mobile cranes, modern computer systems integrating tracks, storage areas, transshipment areas and connections to roads and motorways.

The intermodal transport chain must guarantee a good level of speed, efficiency and safe transhipment of ITUs (Intermodal Transport Units, such as containers, semitrailers and swap bodies) between the transport modes, also to consider the specific features of the territory. In fact, in UE the concentration of intermodal terminals is not uniform, with a relevant number of infrastructures in the north regions and around mountain (Alps, Pyrenees).

2 STATE OF THE ART

2.1 Principal equipment and processes in the high capacity terminals

In Figure 2 there is a synthetic scheme of a typical rail-road combined transport terminal with the main processes, in detail:

- *Check-in operations* for the trucks incoming at the terminal, including the inspection procedures and the documents management for goods and drivers. These two operations can also be performed in two distinct phases and places.
- *Loading or unloading operations* under cranes, from truck to railway wagon or vice versa, or

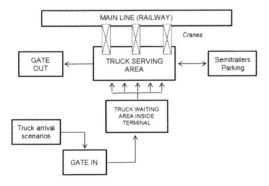

Figure 2. Key elements in a typical inland terminal.

even in special areas in case of technical stops or in parking lots for semitrailers.
- *Check-out operations* for the trucks leaving the terminal.

Although there are other procedures inside the terminal, these specifically involve operations from the railway side and therefore not directly involved in this study. The main elements included in rail-road terminal are: rail siding for wagon both to transhipment that manoeuvres and other operations, buffer lanes for ITUs, loading and driving lanes for the trucks, gates and internal road network (Ballis & Golias 2004). In general, the terminal operations in real-life environment are very complex and their detailed description is out of the scope of this paper.

The focus of this research is on the road part of the terminal and its functional interactions with rail, crane sub-systems and gate operations. The traffic microsimulation model later described in the paper is then built to represent these features of the terminal.

2.2 Modal transfer energy

The energy analysis in intermodal transport for freight regards principally the following aspects (Zumerchik et al. 2011):

- Line Haul Energy is the fuel/energy needed to transport goods from origin to destination through different modes.
- Modal Transfer Energy is the fuel/energy used in the terminal for modal transfer by cranes, drayage trucks, yard tractors, service vehicles, as well as energy use for switching.
- Storage Energy refers to deposits and storing.

The interest of this paper is on the second point, therefore the energy spent to transhipment operations inside an inland terminal. In particular, the focus is on the energy consumption by trucks within the terminal on the road side, which is in charge of the various transport operators using the terminal. This estimation has been performed with a traffic microsimulation approach, where each vehicle can be monitored along the simulation period with a sub-second step time resolution. The other modal transfer energy components are not considered for the following reasons. The energy use due to the crane operations for the ITUs handling depends on the type of crane, ITU weight and dimension, and these data are not available, although they can be easily measured by the terminal operators. For containers operations, an average value of 4.4 kWh/TEU is considered in IFEU et al. (2014). The energy profile of the internal road tractors cannot be independent by the operations scheduled in the terminal. Therefore many vehicle random interactions can be detected, during the terminal operations and at the queues in relevant phases of the process. Also, this contribution can be measured by the terminal operators, since usually they manage the fleets.

2.3 *KPIs*

Ballis & Golias (2004) underline that the importance of the quality of service offered by the combined transport terminals is recognized as an important factor associated to the attractiveness of the combined transport.

Different actors are involved in the terminal process and this creates several points of view to evaluate the efficiency of terminal services and its performance. In this study two main stakeholders are taken into account: users who want to ship goods and terminal operators who have to manage the process. The first one can be considered those customers who use the terminal, then in practice represented by truckers. Their main benefits are the energy saving inside the terminal and total time spent between the check-in and check-out operations (turnaround time). Both of these aspects can be considered part of the quality assurance of the terminal, which is in charge of the second actor involved.

The Key Performance Indicators must be clear, coherent, compatible, controllable, complete, pertinent and feasible. The intermodal terminals have a fundamental role, therefore an impact, on entire logistic chain of combined transport both on economic, quality and efficiency point of view. Currently the European intermodal terminals are not subject to standard quality evaluations or some Level of Service (LOS) as in other transport sectors.

At global level the performance measure of railroad terminal essentially takes place by means of the turnaround time, which is the full time between the entrance of a truck in the terminal for the delivery or pickup of ITU and its exit from the same, but also the degree of utilization and productivity. The last one is function of the average time of ITU permanence in intermodal terminal and it has obviously relevant implications on transport efficiency: the lower the residence time, the greater productivity (Zumerchik et al. 2011).

Ballis (2003) has correlated standard values of selected quantifiable indicators with six possible Level of Service, letter A represent the excellence, the ideal situation, whereas letter F is for an unacceptable state. The set of selected indicators are the following:

- Waiting time for users, it is the turnaround time included the initial queue.

- Reliability, measure the congruency of effective transport service and the scheduled time.
- Flexibility, i.e. as the system is able to respond to contingencies without changing the level of service.
- Qualification, the terminal ability of responding to more complex logistical requirements.
- Safety and Security, related to the loss or damage ITUs and their documents.
- Accessibility, definable through the actual working hours of the terminal.

3 METHODOLOGY

A typical terminal for the transhipment of ITUs between railway and road mode is simulated taking into account the specific road infrastructure and vehicles, as well as their interaction at relevant points of the internal process. Also, the energy consumption estimation for vehicles using the terminal is considered. All of these requirements can be modelled by simulation. Although the relevant events of the whole process could be represented by a discrete-event simulation model, such as in Ricci et al. (2016), if the focus is on the traffic interactions along the connecting roads, vehicle queues and their energy consumptions, microsimulation tools can provide a more effective modelling. They are based on a time-sliced approach and widely used in traffic engineering studies. The terminal layout was then modelled with *Aimsun®* and applied to a large-sized Italian terminal (*Hupac Busto Arsizio*) on the basis of the Open Street Map information.

3.1 *Model and hypotheses*

The typical activities in a rail-road combined transport terminal, by a truck driver point of view are the following:

- physical check-in (code, damages, labels, irregularity)
- documents exchange and verification
- loading and unloading operation with cranes or semi-trailers
- check-out.

The first two sub-processes may take place simultaneously or in two different areas, manually or automatically, for example with particular Optical Character Recognition Cameras, or Barcodes, RFID or other security technologies. In the scenario these operations are performed in different places manually.

The trucks arrival can be divided into three categories, based on their mission: delivery, collection or both.

The terminal layout is assumed with check-in and check-out connected to the same external road infrastructure. Internal roads are composed by section, nodes and roundabouts with specific, lanes, traffic rules. In Figure 3 the basic terminal layout is shown with the truck activities locations used for the simulation; in particular the screenshot represent the path of line 1 (described in more detail below) in which the codes indicate:

- Ck_in_phy1 = physical check-in
- Ck_doc5 = documents exchange and verification
- Barrier = automatic barrier for the entrance to the terminal
- Crane 1 = loading and unloading operation
- Ck_out = check-out.

The method proposed in this paper consists in the traffic simulation of the trucks flow as public transport lines with specific stops along the route inside the terminal. Any stop model one of the activities listed above, characterized by different times due to the usually expected operations that can take place.

In particular, eight service lines were chosen to create possible situations based on the characteristics of the mission:

- LINE 1–2, the vehicles should carry out the *delivery* operation. They would have to do the physical check-in at the entrance and then will be served by a crane;
- LINE 3–4, these vehicles should do only the *collection*, this means that they would not have to the physical check-in, so can go directly to the documents point. One crane tranships the unit and then the activity of check-out would require more time in order to verify the load;
- LINE 5–8, dedicated *both for delivery and collection* which are supposed at two different cranes. The duration for documents exchange and veri-

fication doubles to take into account the two activities requiring separate procedures.

The reference scenario (S0) is set assuming the assignment of the 8 lines to the 6 cranes as reported in Table 1 and an average service time duration for any crane equal to 3 minutes, so the unitary service rate μ is equal to 20 vehicles per hour. To consider the random phenomena linked to this operation, due to the position, the type of ITU or weather conditions. A normal distribution is assumed with 60 s for the standard deviation. The total average service rate is then 120 vehicles per hour, if we consider all the six cranes operating.

To consider the system in under-saturation conditions, for this reference scenario, the arrival rate has been set subsequently. The frequency of the 8 lines is then calculated assuming a number of "equivalent lines" equal to 12, because 4 lines are for a single service (delivery or collection) and 4 lines with both of them. The single arrival rate λ was considered roughly equivalent to 10 vehicles per hour and consequently the mean time interval between departures was set to 6 minutes.

Each line has its timetable, such as reported for the line 8 in Figure 4, and the frequency has been assumed identical for all the lines. To consider random phenomena, also the headway has been modeled with a normal distribution with 30 seconds of standard deviation. To avoid the concentration of vehicles in short time intervals the lines departures are scheduled every 45 seconds, by dividing equally the mean headway among the 8 lines.

The stops are fundamental elements for the building of the model in which the vehicles perform some typical activities in an inland terminal. As aforementioned, each line includes specific stops during the route that may also vary their duration depending on the character-

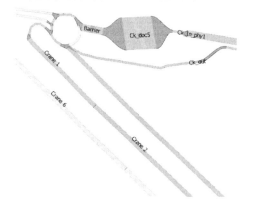

Figure 3. Positions of truck activities and service line 1.

Table 1. Operation order* for cranes and relation with lines.

Line	Crane					
	1	2	3	4	5	6
1			I			
2						I
3		I				
4					I	
5	II		I			
6		I			II	
7	II			I		
8				I		II

* I = first; II = second.

620

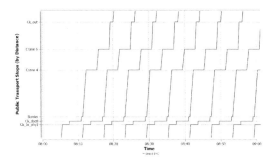

Figure 4.   Example of timetable for the line 8.

istics of the service line itself. The time required for the documents exchanging and verification is set at 100 s (standard deviation = 30 s) for each operation, whereas 200 s is for lines with double service (delivery and collection). The six cranes were represented as fixed stops in specific zones of the road sections. This is an assumption which relates to the way the cranes are used for operations and it would need to be confirmed case by case. In fact, even with electric gantry cranes on fixed tracks, they can cover larger areas, maybe overrunning the action space of the nearby crane, if required.

The vehicles type that use the lines previously described are trucks. In particular their characteristics are defined with some dimension variability of to distinguish the possible different truck types (e.g. trailer or articulated trucks) and the related space used on the road.

3.2   Emission and energy consumption model

The emission model used and also implemented in Aimsun® was propose by Panis et al. (2006). It is chosen in this study for its rich data set used during the calibration process. The traffic emissions are estimated on an instantaneous emission model integrated in a microscopic traffic simulation model. These vehicle emissions are calculated in relation to the type, the instantaneous speed and acceleration. Panis et al. (2006) have focused on some pollutants, for their potential health impacts and external costs: Nitrogen Oxides (NO_x), Volatile Organic Compounds (VOC), and Particulate Matter (PM). Carbon dioxide (CO_2) is also estimated for its effect on global climate change and can be useful if converted in fuel: 1 kg of CO_2 corresponds to 0.4 l of fuel, with negligible differences between diesel and petrol.

The emission functions for each vehicle are derived using non-linear multiple regression techniques (Eq. 1).

$$E_n(t) = \max[E_0, f_1 + f_2 v_n(t) + f_3 v_n(t)^2 + f_4 a_n(t) + f_5 a_n(t)^2 + f_6 v_n(t) a_n(t)] \quad (1)$$

where $v_n(t)$ and $a_n(t)$ are the instantaneous speed and acceleration of vehicle n at time t. E_0 is a lower limit of emission (g/s) specified for each vehicle and pollutant type, and f_1 to f_6 are emission constants specific for each vehicle and pollutant type determined by the regression analysis. To derive the emission functions, in Panis et al. model, for heavy vehicle the measurements were carried out from two type of vehicles (Iveco Eurocargo and Volvo FH12–420) respectively in number of 1638 and 4514.

3.3   Performance indicators

The KPIs chosen for this paper can be classified in two following groups: quality and energy.

As for the quality, the delay and queue length are useful from the terminal management point of view in order to identify possible bottlenecks of the system. Furthermore, the turnaround time is important for several actors. The truck driver and the client are interested because is the total time spent for terminal operations. The terminal operator can use this value to be more competitive. The policy maker can chose this to compare similar terminals for quality evaluation. Finally, less time is required in inland terminals, means less time in the logistic chain, increasing the competitiveness of the rail-road combined transport.

As for the energy estimation, the fuel consumption refers to the road vehicles travelling inside the terminal from the first road section before the check-in point and the final section after the check-out point, during the simulation period. The fuel consumption, in our case is estimated from the emission data, provided by the Panis et al. model, as explained in §3.2.

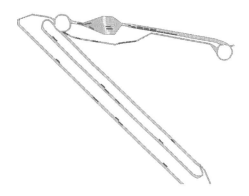

Figure 5.   Screenshot of vehicle simulation along roads inside the intermodal terminal.

4 APPLICATION AND RESULTS

4.1 Baseline scenarios and experimental conditions

The scenario S0 was used as baseline to build the model with characteristics described in the section 3.1. In this scenario an equilibrium condition has been set for services and arrivals.

This baseline scenario has been tested with two complementary scenarios (S1 and S2), simulating respectively the decrease and increase of the arrivals at terminal. For sake of simplicity a uniform variation of the headways for arrivals has been applied for all the lines, assuming a value of 5 minutes for scenario S1 and 7 minutes for scenario 2.

Any microsimulation experiment is composed by ten replications for a 1 hour simulation period and the final indicators are evaluated on the basis of the average values. A warm up period equal to 30 minute before the simulation period has been also used to obtain significant statistical indicators, avoiding the empty network state at the initial time.

4.2 Results for quality indicators in baseline scenarios

A global view of the terminal performance can be observed in Figure 6 by the queue indicator at the various sections during the whole simulation period, including the six cranes where the value is approx. 1 and the check in/out points with a high value at the exit.

The travel time for the 8 service lines provides a disaggregate estimation of the turnaround time, highlighting the different quality levels experienced in the terminal, which is in a range between approximately 12 and 32 minutes (Table 2). As

expected, the higher values are for the double service trucks. Also, a different value is observed for collection and delivery services, since the first one requires less time at the physical check-in point.

The light increasing trend over time for all the values shows that the system is just over the equilibrium. This is also confirmed by the values in Table 3.

Indeed, observing the values related to the scenario 1, obtained increasing the time intervals for arrivals from 6 to 7 minutes, they are quite constant, as should be in stationary conditions.

On the other hand, when the arrivals increase, the quality performance of the terminal, operated in simulation with the same service level, is affected more dramatically and the model estimates, in the worst case, a turnaround time of 45 minutes (Table 4).

Table 2. Turnaround time for service line in S0.

Turnaround Time [min]	6.15	6.30	6.45	7.00
Line 1 Delivery 1	17,0	19,7	23,0	25,2
Line 2 Delivery 6	17,4	18,7	21,7	24,4
Line 3 Collection 2	12,1	13,5	14,4	17,3
Line 4 Collection 5	12,3	14,0	14,2	16,1
Line 5 D+C	20,5	23,7	26,6	29,2
Line 6 D+C	24,8	27,3	30,0	33,0
Line 7 D+C	23,2	26,8	29,2	31,6
Line 8 D+C	23,2	26,8	28,7	32,4

Table 3. Turnaround time for each service line in S1.

Turnaround Time [min]	6:15	6:30	6:45	7:00
Line 1 Delivery 1	14.4	15.6	15.6	16.6
Line 2 Delivery 6	15.4	16.7	16.3	16.7
Line 3 Collection 2	11.8	13.0	13.3	13.3
Line 4 Collection 5	11.7	12.2	13.2	13.4
Line 5 D+C	20.1	21.0	21.0	22.3
Line 6 D+C	22.3	22.6	23.6	24.3
Line 7 D+C	22.2	23.4	25.9	25.0
Line 8 D+C	23.5	23.9	23.9	23.5

Table 4. Turnaround time for each service line in scenario S2.

Turnaround Time [min]	6.15	6.30	6.45	7.00
Line 1 Delivery 1	20,6	25,6	30,8	33,2
Line 2 Delivery 6	21,2	26,3	32,1	35,6
Line 3 Collection 2	13,3	16,0	18,9	23,0
Line 4 Collection 5	13,0	15,2	18,4	21,3
Line 5 D+C	23,3	29,4	33,6	38,5
Line 6 D+C	27,8	34,9	39,8	45,3
Line 7 D+C	25,6	30,7	35,2	39,9
Line 8 D+C	26,0	32,0	37,6	41,6

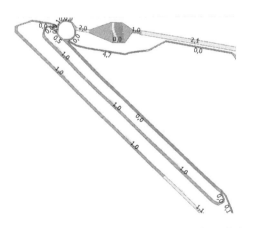

Figure 6. Maximum queue at various section of the terminal.

4.3 Results for quality indicators in improved scenarios

To test the model in assessing possible improvements of the terminal operations, two further scenarios have been simulated (S3 and S4).

In S3 a reduction of the service time in the last phase of the process is generated, whereas the arrivals are as in S0. This can be obtained by modifying the check-out operations for the loaded vehicles with a collected ITU, with the support of security technologies. In Table 5 the higher variations are observed for collection vehicles and at the end of the period, when the interactions between vehicles are more relevant. However, also for vehicles not involved in the improvement at check-out point, there is an observed reduction of the turnaround time, since there is a unique exit gate and interaction phenomena occur.

In S4 the simulation is focused to assess a possible improvement of the quality level of the S2, by assuming a better positioning system for the crane areas, which may generate a reduction of the service time. In particular, this value was set up to 2 minutes, in place of 3 minutes, to balance the increased time interval of arrivals.

The technical solution tested in S4 is not sufficient to reduce the turnaround time for each line. Therefore, a further scenario (S5) is simulated to combine the decrease of the cranes service time and the introduction of security technologies in the check-out area (as S3). Table 7 shows that the

Table 5. Turnaround time variation [%] for service line in S3.

Turnaround Time	6:15	6:30	6:45	7:00
Line 1 Delivery 1	−5%	−9%	−15%	−18%
Line 2 Delivery 6	−5%	−2%	−11%	−14%
Line 3 Collection 2	−10%	−8%	−18%	−25%
Line 4 Collection 5	−15%	−15%	−16%	−19%
Line 5 D+C	−3%	−8%	−7%	−11%
Line 6 D+C	−5%	−8%	−6%	−10%
Line 7 D+C	−6%	−7%	−7%	−6%
Line 8 D+C	−5%	−10%	−8%	−12%

Table 6. Turnaround time for each service line in scenario S4.

Turnaround Time [min]	6:15	6:30	6:45	7:00
Line 1 Delivery 1	20.4	25.3	31.0	34.8
Line 2 Delivery 6	20.3	27.2	31.2	36.2
Line 3 Collection 2	12.2	15.7	19.2	23.4
Line 4 Collection 5	12.7	15.6	18.0	23.3
Line 5 D+C	22.4	26.7	33.1	37.4
Line 6 D+C	24.3	30.1	34.8	39.7
Line 7 D+C	24.6	29.9	34.8	39.6
Line 8 D+C	24.9	29.8	34.4	39.2

Table 7. Turnaround time for each service line in scenario S5.

Turnaround Time [min]	6:15	6:30	6:45	7:00
Line 1 Delivery 1	17.7	21.2	24.5	27.4
Line 2 Delivery 6	18.0	22.2	25.9	30.1
Line 3 Collection 2	8.8	9.0	10.3	12.8
Line 4 Collection 5	9.6	9.9	10.3	12.2
Line 5 D+C	19.6	23.5	26.2	29.7
Line 6 D+C	22.7	26.4	30.4	35.3
Line 7 D+C	21.7	26.7	29.1	33.2
Line 8 D+C	20.6	26.3	28.9	33.7

Table 8. Summary of the scenarios explored.

Scenario	Rate of Arrivals	Service rate for cranes	Service rate for check-out	Average turnaround time [min]
S0	BASE	BASE	BASE	22.4
S1	−	BASE	BASE	18.7
S2	+	BASE	BASE	28.3
S3	BASE	BASE	+	20.3
S4	+	+	BASE	27.3
S5	+	+	+	22.0

combination of these two improvements better responds to the increase in demand.

Before reporting the energy results, a summary of the explored scenarios and their average turnaround time is reported in Table 8 to compare the assumptions and the actions simulated.

The global results on turnaround time indicate that the increase of demand simulated in S2 and the consequent decrease of the quality level can be managed improving cranes and check out operations. However, to maintain the quality level as observed in S0, both of the improvements should be applied as in S5, since the only improvement of the crane performance is not enough (S4).

4.4 Energy results

A disaggregate estimation of the energy used in the terminal by road vehicles has been collected during simulation for each line (Table 9). As expected, their variation is in line with the level of congestion within the terminal. However the variation is not linearly dependent with the traffic: at the same headway variation for arrival (S1 and S2) a different variation of fuel consumption is estimated, respectively −16% and +19%. In S3 the improvement of the quality has a related reduction in energy, although its value is not relevant (−4%). The S4 and S5 scenarios show the expected reduction in total fuel used respect their reference scenario (S2).

Table 9. Total fuel [l] for the different investigated scenarios.

	S0	S1	S2	S3	S4	S5
Line 1	24.0	20.1	29.7	22.8	28.8	27.2
Line 2	25.1	20.6	30.4	23.5	29.9	27.3
Line 3	21.2	17.7	24.5	20.2	23.6	20.5
Line 4	22.1	18.3	25.5	20.4	24.6	22.7
Line 5	26.3	21.3	31.7	25.3	30.5	28.7
Line 6	34.5	28.9	40.3	33.7	38.3	36.0
Line 7	33.0	29.1	39.3	31.9	37.3	35.7
Line 8	28.8	25.1	35.1	27.8	33.4	30.3
Total	215.1	181.0	256.5	205.6	246.3	228.5
		−16%	19%	−4%	15%	6%

Table 10. Time and fuel estimation in idle phases for S2.

S2	Travel Time [s]	Stop Time [s]	% Time	Total Fuel [l]	Fuel Idle [l]	% Fuel
Line 1	932	648	70%	29,73	7,10	24%
Line 2	978	657	67%	30,37	6,90	23%
Line 3	772	489	63%	24,52	4,41	18%
Line 4	757	436	58%	25,45	4,20	17%
Line 5	1268	985	78%	31,74	8,36	26%
Line 6	1392	911	65%	40,29	7,26	18%
Line 7	1445	964	67%	39,32	7,27	18%
Line 8	1421	1097	77%	35,10	8,37	24%
			68%			21%

These values can be converted in energy per vehicle and are in a range between 2 to 6 litres.

This estimation being related to the global process in the terminal, include also the fuel consumption during the stop phases of the vehicles, assuming the engine is on. An approximate estimation of the fuel consumed during these phases can be performed calculating the running time of the vehicle travel time. Table 10 shows the comparison between the variation in terms of time and fuel for the most congested scenario (S2). The average contribution of idle phases to the global time spent inside the terminal is 68% in terms of time, which is higher than the 21% of total fuel.

5 CONCLUSIONS

The model is built to simulate the terminal with a traffic microscopic approach and has provided an estimation of the required indicators for its quality, expressed as the turnaround time, and energy, expressed as the total fuel consumed by the road vehicles. Also the flexibility of the simulated terminal can be easily assessed, since the demand variation for arrivals is modifiable in the model.

The model can then be used as a decision support tool for terminal operators to provide a view on the current state of the system and explore possible improvements aimed at improve its efficiency, for example by removing bottlenecks. All data used, including the terminal layout, although describe realistic scenarios, are assumed mainly to test the ability of this method to reproduce the required process operations and not to provide an appraisal of the terminal performance. In this first phase, the main focus was then to evaluate the effectiveness of the simulation model, by monitoring the variation of the chosen indicators and their consistency with expectation for the established scenarios.

Further research can be carried out to apply the model with real data provided by a terminal operator and better tune the parameters used during simulations. Also, pollutant emissions inside the terminal can be estimated with the simulation tool to investigate on possible actions aimed at improving the air quality at specific critical areas. A future research can be devoted to study other alternative simulators to compare their ability in reproducing the required performance for terminals.

REFERENCES

Ballis, A., 2004. Introducing Level of Service Standards for Intermodal Freight Terminals. In *TRB*. Washington: Transportation Research Record: Journal of the Transportation Research Board, pp. 79–88.

Ballis, A. & Golias, J., 2004. Towards the improvement of a combined transport chain performance. *European Journal of Operational Research*, 152(2), pp.420–436. Available at: http://www.sciencedirect.com/science/article/pii/S0377221703000341 [Accessed February 8, 2016].

Carreira, J.S., Santos, B.F. & Limbourg, S., 2012. Inland intermodal freight transport modelling. In *European Transport Conference 2012*. Glasgow: ETC Proceedings.

Castaldo, M., 2014. È sufficiente ridurre i costi di produzione per rilanciare il trasporto ferroviario? In *Mercintreno 2014*. Roma: Trenitalia.

European Commission, 2011. *WHITE PAPER*, Bruxelles.

IFEU et al., 2014. EcoTransIT World (Ecological Transport Information Tool for Worldwide Transports). Methodology and Data Update., (December 2014), p.106. Available at: http://www.ecotransit.org/download/ecotransit_background_report.pdf.

Panis, I.L., Broekx, S. & Liu, R., 2006. Modelling instantaneous traffic emission and the influence of traffic speed limits. *Science of the Total Environment*, 371(1–3), pp.270–285.

Ricci, S. et al., 2016. Assessment methods for innovative operational measures and technologies for intermodal freight terminals. *Transportation Research Procedia*, 14, pp.2840–2849. Available at: http://dx.doi.org/10.1016/j.trpro.2016.05.351.

Zumerchik, J., Sr, J.L. & Rodrigue, J., 2011. Incorporating Energy-Based Metrics in the Analysis of Intermodal Transport Systems in North America. *Journal of Transportation Research Forum*, pp.97–112.

Transport Infrastructure and Systems – Dell'Acqua & Wegman (Eds)
© 2017 Taylor & Francis Group, London, ISBN 978-1-138-03009-1

New features of Tritone for the evaluation of traffic safety performances

V. Astarita, V.P. Giofrè, G. Guido, A. Vitale, D.C. Festa, R. Vaiana, T. Iuele, D. Mongelli, D. Rogano & V. Gallelli
Dipartimento di Ingegneria Civile, University of Calabria, Rende (CS), Italy

ABSTRACT: Recent research papers have confirmed that traffic simulation can identify near crashes events and establish a good base for the estimation of real crashes risk. The objective of this paper is to present the new features of a microsimulation model originally developed to estimate road safety performance.

The presented microsimulator has many new features that can be useful to engineers and researchers such as:

– Dynamic calculation of traffic and road safety indicators.
– Simulation of satellite location data obtained by GPS and smartphones.
– Simulation of adaptive traffic lights activated by FCD data.
– 20 different acoustic emission models.
– Possibility of taking into account "instrumented" vehicles to assess new Intelligent Transportation System performances.

The paper describes the above listed new features of TRITONE that combined with the calculation of road safety indicators evaluations allow planners to benefit from the availability of a useful and innovative tool for traffic simulation.

1 ORIGINS OF TRITONE TRAFFIC SIMULATOR

Currently there are several commercially microsimulation packages for road traffic, which provide detailed information on its likely evolution. However, only recently some simulators have evolved to reproduce road safety performance indices (Guido et al., 2011a, Astarita et al, 2011). For this reason, in 2009 the University of Calabria developed TRITONE, a traffic simulation package (Astarita et al., 2012) able to provide, not only the evolution of road traffic and all usual performance indices but also road safety performance indices. In detail, TRITONE is able to represent accurately the possible areas of a road network that have a higher risk of accidents. TRITONE can, in fact, reproduce dynamic traffic evolution evaluating risky situations by taking into account the geometrical aspects of road infrastructure and the actual behavior of drivers considering in detail all the characteristics of vehicles and drivers. Road safety parameters indicators produced by TRITONE allow the prediction of dangerous situations and plan preventive solutions already in the design phase.

TRITONE was conceived with the aim to become a leading tool in the field of road safety.

This is because in addition to being functionally efficient and user-friendly and allowing the user to evaluate directly and dynamically many road safety parameters such as Deceleration Rate to Avoid the Crash (DRAC) and Time to Collision (TTC) indicators (Guido et al., 2011b), TRITONE has been developed with new features that are described in the following parts of this paper.

The equations describing these two indicators are determined by the kinematic interactions between Following Vehicle (FV) and Lead Vehicle (LV) and are the following:

$$DRAC_{FV}(t+\Delta t) = \frac{\left(V_{FV}(t) - V_{LV}(t)\right)^2}{2*\left[\left(X_{LV}(t) - X_{FV}(t)\right) - L_{LV}(t)\right]}$$

$$TTC_{FV}(t+\Delta t) = \frac{\left(X_{LV}(t) - X_{FV}(t)\right) - L_{LV}(t)}{V_{FV}(t) - V_{LV}(t)}$$

where,

t = time interval (s)
X = position of the vehicles (m)
L = vehicle length (m)
V = speed (m/s)

Moreover, TRITONE was conceived as a university research tool, and is therefore widely used in this area since it is available to researchers and scholars in the field. It allows researchers to validate their mathematical simulation models through appropriate standardized libraries or direct implementations.

2 NEW FEATURES OF TRITONE

In recent years, many new features have been added to this traffic simulation package. In most cases new algorithms have been added that reflect current development trends of road traffic processes.

The new features that this paper intends to present are:

- Tools for the dynamic calculation of road safety indicators.
- Simulation of GPS sensors (smartphones or other equipment) in vehicles
- Simulation of adaptive traffic lights triggered by mobile devices.
- Noise pollution rating.
- Auto-generation of road networks from GIS maps

3 NEW FEATURES IN THE DYNAMIC CALCULATION OF TRAFFIC AND ROAD SAFETY INDICATORS

TRITONE can be used to find solutions to a large number of issues, TRITONE can, in fact, carry out the traditional dynamic analysis on the evolution of traffic circulation allowing users to compare their various design scenarios (which may include normal intersections, roundabout intersections, signalized intersections, junctions split-levels, etc.). Moreover, it can be used to evaluate and optimize road networks in cases of exceptional events (emergencies, construction sites, accidents, evacuation plans, etc.). In all cases, it is possible to use a network of roads corresponding as much as possible to reality (lane width, slopes, narrow passages, etc.) allowing users. that accept and use this tool. to obtain reliable results.

The original purpose of TRITONE was to evaluate in advance the risk of road accidents occurrence, this task can be performed now with new multiple representation tools offered by new versions of TRITONE. New tools allow the user to locate easily and quickly the critical safety issues that arise on the network. Figure 1 shows an example of road sections where it is possible to pinpoint possibly unsafe situations with a study of road safety indicators, such as DRAC and TTC.

DRAC and TTC indicators are evaluated directly in TRITONE allowing the user to identify

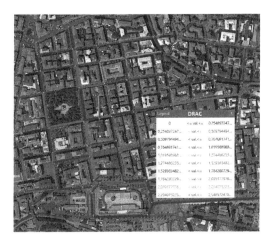

Figure 1. Road sections with a high probability of accidents.

on which link of the urban road network the probability of accident occurrence is higher.

TRITONE, unlike the Surrogate Safety Analysis Model project (SSAM) from the Federal Highway Administration (FHWA, 2008), calculates and evaluates the safety indicators (DRAC and TTC), dynamically during the simulation without post-processing.

TRITONE identifies the risk areas of the road network by imposing thresholds to DRAC (< 3.40/3.35 m/s^2) (AASHTO, 2004, Archer, 2005) and TTC values (> 1.5 s) (Van der Horst, 1991) measured on link.

In each simulation time step, TRITONE dynamically calculates the DRAC and TTC values for each vehicle. These are then averaged over time and network links. In a specific interval, each link has an average value of the DRAC and TTC that is dynamically compared with a given thresholds. In this way TRITONE has the possibility of pinpointing critical areas or links during the simulation.

With a few steps, it is possible to model the road network, adding the traffic demand and after calculation of the circulating traffic identify the network links that present critical values.

4 SIMULATION OF SATELLITE LOCATION DATA OBTAINED BY GPS AND SMARTPHONES

GPS technologies are now available for most of the vehicles on our roads. These sensors are often supplied by the insurance companies to verify the good behavior of the drivers, by security companies to prevent theft or simply through smartphones or any other mobile communications device in the vehicle. Satellite location technology has allowed the Intelligent Transportation System (ITS) industry to provide a myriad of new systems that are based on smartphones or other instrumentation in the vehicle.

Many papers have been recently presented on smartphone applications as a solution to many traffic problems (Astarita et al., 2014, Festa et al., 2013), moreover, there have been many researches on obtaining traffic data from instrumented vehicles as traffic probes (Floating Car Data), among them Astarita and Florian, (2001), Rose (2006), Astarita et al., (2006), Sohn and Hwang (2008), Dailey and Cathey (2002), Axer and Friedrich (2014). Some researchers have been studying in particular traffic safety issues: Biral et al. (2005), Vaiana et al. (2014), Guido et al. (2014), Kerner et al. (2015).

The use of smartphones has been studied to verify the penetration rate required to obtain positive outcome in traffic information systems: Ygnace et al., (2000) and Ferman et al., (2003). Some researchers have been exploring directly on the field such as in the Mobile Century Project directed by Alex Bayen (Herrera et al.,2010) and in Guido et al. (2012, 2013).

There are now many systems that exploit these devices, but they have a big drawback, as they can only really be useful (and tested) after a substantial percentage of users adopts them. In fact, the costs for testing some of these systems in the design phase on a whole road network has a very high cost. For this reason in TRITONE a new module has been developed that can reproduce the presence of GPS sensors or other Global Navigation Satellite System (GNSS) on simulated vehicles.

TRITON is able to simulate Floating Car Data (FCD). TRITONE can reproduce the presence of a GPS signal in a vehicle, whether it originates from the insurance black box or from a smartphone. This makes it possible to study in detail the evolution and develop algorithms for complex ITS systems on entire urban networks. TRITONE performs this task with an algorithm that can simulate both the presence of the devices and the position error in GNSS (Fig. 2). Moreover, a map-matching algorithm is added to obtain data that are similar to those that would be obtained in the real implementation of such systems. With this methodology (Lou et al., 2009) it is possible to assess the performance of ITS systems that are based on Floating Car Data (FCD).

At the end or during a simulation it is possible to know the location track of vehicles carrying a GPS (GNSS) sensor, this allows the behavior of users in an ITS system to be predicted and/or the performance of the entire simulated road network under a certain ITS control strategy to be evaluated.

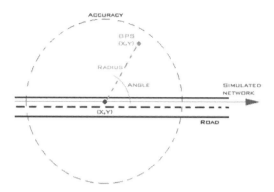

Figure 2. Simulation of GNSS error.

5 SIMULATION OF ADAPTIVE TRAFFIC LIGHTS ACTIVATED BY FCD DATA

The article "A Cooperative Intelligent Transportation System for Traffic Light Regulation Based on Mobile Devices as Floating Car Data (FCD)" (Astarita et al., 2016) for the first time introduces the possibility of regulating traffic light systems through the use of FCD coming from common smartphones that are on board vehicles circulating on the road network. A simple explanation of how it works is contained in Gordon (2016).

Previous academic works have been dedicated only to priority control systems of traffic lights for emergency vehicles or public transport (Ekeila and other, 2009), dedicated radio links or dedicated localization systems and/or extraction of information from FCD. No other academic seminal paper on using a smartphone application data to regulate traffic lights is known to the authors.

TRITONE, having the ability cited above to simulate FCD has allowed the development and testing of algorithms useful to obtain a better regulation of adaptive traffic lights. In fact, with TRITONE it has been possible to evaluate algorithms with different values of traffic demand and different percentages of vehicles equipped with GNNS. Figures 3–4 show how the arrival of an instrumented vehicle, with a GPS sensor, near a traffic light, changes the intersection rules. The adaptive traffic light gives priority to this vehicle (always ending the current phase with a smooth transition of green, yellow and then red). The system estimates the time required for this vehicle to perform the maneuver and adjusts the green phase in a congruent manner.

Subsequently, at the end of the maneuver of this vehicle the traffic light cycle returns to standard values always respecting the phase rotation.

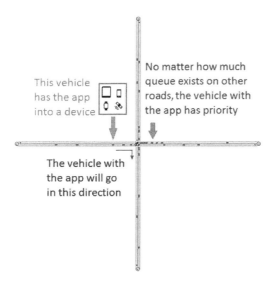

Figure 3. Example of precedence of the green traffic light.

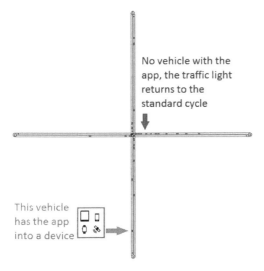

Figure 4. Example of precedence of the green traffic light.

6 EVALUATION OF NOISE POLLUTION

To cope with increasing noise pollution many nations like the United States (FHWA), the EU (2002/49/EC), Russia (ODM 218. 013–2011), Japan (ASJ RTN-Model 2008) and China (HJ2 4–2009) have started to impose clear limits for the protection of human health. Over the years, studies have been devoted to the possibility of coupling noise emissions models for road traffic with the current micro-simulation techniques. The development of noise models began in the 50 s, taking into account mainly average values of traffic flows, both light and heavy vehicles.

Many models have been developed for traffic noise prediction (see Barone et al. 2011, Crocco et al. 2011 and Mongelli 2013 for more details). There are now more complex solutions that assess amplification and attenuation of sound also taking into consideration additional features such as the type of road surface, the topography of the terrain, the elements present on the roadside such as anti-noise walls or weather conditions, etc.

TRITONE also has been expanded to take into account the calculation of noise emissions in a more detailed way. The introduction of a noise emission module in TRITONE has provided multiple formulations to be coupled with the microsimulation tools, currently in TRITON there are included 20 different noise simulation models, based on the following formulations:

- CSTB in non-built-up areas (NMPB-Routes-96, 1995)
- CSTB in built-up areas (NMPB-Routes-96, 1995)
- Josse (Quartieri et al., 2009)
- Alexandre (Quartieri et al., 2009)
- Burgess (Burgess, 1977)
- Griffiths and Langdon (Griffiths & Langdon, 1968, Bertoni et al., 1988)
- Garcia and Bernal (Grippaldi et al., 199)
- Johnson and Saunders (Johnson & Saunders, 1968)
- OMTC (Quartieri et al., 2009)
- Lamure (Lamure, 1965)
- Corriere and Lo Bosco (Corriere & Lo Bosco, 1991)
- CoRTN (Anon, 1975, HMSO, 1988)
- EMPA (Quartieri et al., 2009)
- C.N.R. (Canelli et al., 1983, Cocchi et al., 1991)
- RLS 90 (RLS, 1990, RLS, 1981)
- C.E.E. (Quartieri et al., 2009)
- CETUR (CETUR, 1980)
- Brambilla
- NMPB (NMPB-Routes-96, 1995, NMPB-Routes-2008, 2009, Directive 49/EC, 2002, SETRA, 2009)

The implementation of most current models of noise emission calculation, allows users to have a wide choice for various traffic simulation scenarios. TRITONE allows users also to perform easily a comparison between different sound emission models. The module structure of the TRITONE simulator is described in Figure 5. It is necessary first to build the network with geometric and other properties of the roads and the definition of the O/D matrices, a choice then can be made on the behavioral models for drivers (car following, lane changing, etc.), it is then possible also to choose the sound emission model that has to be used. The simulator will simulate the movement of vehicles on the net and show, in addition to the traditional output graphs common of all micro-simulators also those of sound levels in thematic maps.

For each instant of simulation, TRITONE provides the calculation of traffic noise levels as a function of the noise emission model chosen and the traffic flow present in each moment on each arc of the road network.

The simulation then allows the variation of the noise level with time to be obtained, for example assessing both what happens in the hours in which there is intense traffic and in those in which there is a scarce presence of vehicles. Also it allows the making of design decisions based on the actual behavior of the vehicles with respect to changes in the type of infrastructure.

To assess the quality of the TRITON simulator, some tests were carried out in different parts of the urban area of the city of Cosenza. In the real site, for comparison with the simulated values, traffic noise over time also was detected by a sound level meter. An example of the results obtained are shown in Figure 6.

These experimental results show that it is possible to extend the functionality of existing road traffic microsimulators by adding more and more features that might be required by international standards. By coupling results of researches on

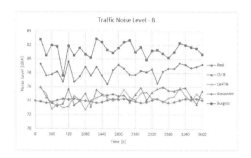

Figure 6.   Result of sound micro-simulation for experimental site.

noise emission models with microsimulation it seems possible to analyze driving behavior, traffic flows and their effects on the road adjacent areas.

7   AUTO-GENERATION OF TRANSPORTATION NETWORKS FROM GIS

To obtain a good quality in the results from traffic simulations a high quality of input parameters is necessary. For this reason, to avoid mistakes and to increase the quality of input information in TRITONE a module has been developed that can automatically extract simulation parameters from GIS systems. This module of TRITONE allows the reconstruction of the transportation network of an entire city automatically, taking into account also rail lines. The input data can be taken from one of the most common GIS systems on the Internet (OpenStreetMap). Data are then filtered, according to the user's needs, and used by TRITONE to reconstruct the roadway or railway network.

The roads are drawn precisely, taking into account details of curved sections. The module can assign to every road automatically its geometric (number of lanes, width, etc) and functional characteristics (speed limit, etc.).

One example of a full network reconstructed with this module is the network of Sioux Falls (South Dakota—USA), shown in Figure 7.

As can easily be seen in the picture the access ramps are modeled precisely allowing to have reliable results on the movements of vehicles and their position in space (feature that can be useful in ITS modelling).

This new feature results in increased accuracy of the coded road information, but can also still result in significant savings of human resources associated with network coding tasks. For example, a person needs around 35 hours to model the 4105 road sec-

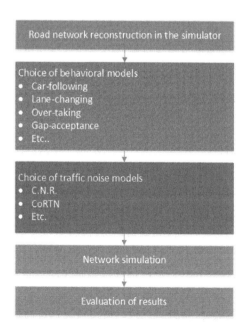

Figure 5.   TRITONE structure.

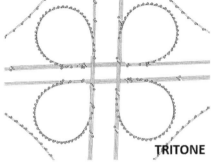

Figure 7.   Example of GIS reconstruction.

tions of the Sioux Falls network (about 116 km2), while with this feature needs only 5.20 minutes.

8   CONCLUSION

In conclusion, the idea of pairing the common standardized simulation techniques to new concepts such as Safety Performance evaluations or the simulation of ITS systems based on satellite localization has led to the creation of a useful and innovative tool for traffic simulation.

New future development of TRITONE, to keep pace with future technologies in the transportation field, include the real time use of TRITONE for the real time implementation of traffic control strategies.

REFERENCES

American Association of State Highway and Transportation Officials (AASHTO), 2004. *A Policy on Geometric Design of Highways and Streets*, AASHTO, Washington, D.C.

Anon 1975. *Calculation of Road Traffic Noise*, London, United Kingdom Department of Environment and welsh Office Joint Publication, HMSO.

Archer, J., 2005. *Methods for the assessment and prediction of traffic safety at urban intersection and their application in microsimulation modeling*. PhD Thesis, Department of Infrastructure, Royal Institute of Technology, Sweden.

Astarita, V. and Florian, M. 2001. *The use of mobile phones in traffic management and control*. In Intelligent Transportation Systems Proceedings IEEE, 2001, pp. 10–15.

Astarita, V., Bertini, R., D'Elia, S. and Guido G. 2006. *Motorway traffic parameter estimation from mobile phone counts*. European Journal of Operational Research, Vol. 175 (3), pp. 1435–1446.

Astarita, V., Festa, D.C., Mongelli, D. W. E., Tassitani, A. 2014. *New methodology for the identification of road surface anomalies*, Service Operations and Logistics, and Informatics (SOLI), 2014 IEEE International Conference.

Astarita, V., Giofrè, V. P., Guido, G., Vitale, A. 2011. *Investigating road safety issues through a microsimulation model*. Procedia-Social and Behavioral Journal, Vol. 20, pp. 226–235.

Astarita, V., Giofrè, V. P., Vitale, A. 2016. *A Cooperative Intelligent Transportation System for Traffic Light Regulation Based on Mobile Devices as Floating Car Data (FCD)*, 2016, American Scientific Research Journal for Engineering, Technology, and Sciences (ASRJETS), Vol. 19–1, pp. 166–177.

Astarita, V., Guido, G., Vitale, A., Giofrè, V. P. 2012. *A new microsimulation model for the evaluation of traffic safety performances*, European Transport, 2012, pp 1–16, Paper N° 1, ISSN 1825–3997.

Axer, S. and Friedrich, B. 2014. *Level of service estimation based on low-frequency floating car data*. Transportation Research Procedia, Vol 3, pp 1051–1058.

Barone, V., Crocco, F., Mongelli, D.W.E. 2011. *A mathematical model for traffic noise prediction in an urban area* in Proceedings of the 4th WSEAS International Conference on Urban Planning Transportation (UPT '11), pp. 405–410, Corfu Island, Greece.

Bertoni, D., Franchini, A., Magnoni, M. 1988. *Il rumore urbano e l'organizzazione del territorio*, Pitagora Editrice, Bologna.

Biral, F., Lio, M. D., Bertolazzi, E. 2005 *Combining safety margins and user preferences into a driving criterion for optimal control-based computation of reference maneuvers for an ADAS of the next generation*. In Proceedings of Intelligent Vehicles Symposium, IEEE,, pp. 36–41.

Burgess, M.A. 1977. *Noise prediction for Urban Traffic Conditions, Related to Measurement in Sydney Metropolitan Area*, Applied Acoustics, vol. 10, pp 1–7.

Canelli, G. B., Gluck, K., Santoboni, S. A. 1983. *A mathematical model for evaluation and prediction of mean energy level of traffic noise in Italian towns*, Acustica, 53, 31.

CETUR, 1980. *Guide du bruit des transports terrestres*, fascicule prévision des niveaux sonores.

Cocchi, A., Farina, A., Lopes, G. 1991. *Modelli matematici per la previsione del rumore stradale: verifica ed affinazione del modello CNR in base a rilievi sperimentali nella città di Bologna*, Acta di 19° Convegno Nazionale AIA, Naples 10–12 April, 1991.

Corriere, F., Lo Bosco, D. 1991. *Valutazione previsionale dell'inquinamento acustico nella variabilità urbana*, Autostrade, n. 1.

Crocco, F., D'Elia, S., Mongelli, D. 2011. *An integrated prediction model to define the level of noise in urban areas*, in Proceedings of the 4th WSEAS International Conference on Urban Planning Transportation (UPT '11), pp. 399–404, Corfu Island, Greece.

Dailey, D. J and Cathey, F. W. 2002 *"Virtual speed sensors using transit vehicles as traffic probes.* In Proceedings of The IEEE 5th International Conference on Intelligent Transportation Systems, pp. 560–565.

Directive 2002/49/EC of the European parliament and of council of June 25 2002 relating to the assessment and management of environmental noise, official journal of the European communities, 1189/12–25, 18.7.2002.

Ekeila, W., Sayed, T, Esawey, M., 2009. *Development and comparison of dynamic transit signal priority strategies,* Transportation Research Record Journal of the Transportation Research Board 2111(2111):1–9, December.

Ferman, M. A., Blumenfeld D. E. and Dai X., 2003. *A simple analytical model of a probe-based traffic information system.* In IEEE Transactions on Intelligent Transportation Systems,Vol. 1, pp. 263–268.

Festa, D.C., Mongelli, D.W.E., Astarita, V., Giorgi, P. 2013. *First results of a new methodology for the identification of road surface anomalies,* Service Operations and Logistics, and Informatics (SOLI), 2013 IEEE International Conference.

FHWA, 2008. *Surrogate Safety Assessment Model (SSAM),* HWA-HRT-08-049.

Gordon J. 2016. *Talking to traffic signals.* Traffic technology international, October/November 2016, pp52–55.

Griffiths, I.D., Langdon, F.J. 1968. Subjective *Response to road traffic noise,* Journal of Sound and Vibration 8, 16–32.

Grippaldi, V., Barbaro, S., Cosa, M. 1996. *Modelli di previsione per il rumore da traffico in ambiente urbano: un quadro di confronto.*

Guido, G., Astarita, V., Giofrè, V. P., Vitale, A., 2011a. *Safety performance measures: a comparison between microsimulation and observational data.* Procedia-Social and Behavioral Journal, Vol. 20, pp. 217–225.

Guido, G., Saccomanno, F. F., Vitale, A., Astarita, V., Giofrè, V. P., 2011b. *Un algoritmo per l'acquisizione delle traiettorie veicolari da immagini video.* Strade & Autostrade, 2011, Vol. 1, pp. 132–135.

Guido, G., Vitale, A., Astarita, V., Saccomanno, F, Giofrè, V. P. and Gallelli, V. 2012. *Estimation of safety performance measures from smartphone sensors.* Procedia-Social and Behavioral Sciences, Vol. 54, pp. 1095–1103.

Guido, G., Vitale, A., Saccomanno, F, Festa, D.C., Astarita, Rogano D. and Gallelli, V. 2013 *Using smartphones as a tool to capture road traffic attributes".* In Applied Mechanics and Materials, Vol. 432, pp. 513–519.

Guido, G., Gallelli, V., Saccomanno, F., Vitale, A., Rogano, D. and Festa, D. C. 2014. *Treating uncertainty in the estimation of speed from smartphone traffic probes.* Transportation Research Part C: Emerging Technologies, Vol. 47, pp. 100–112.

Herrera, J. C., Work, D. B., Herring, R., Ban, X. J., Jacobson Q. and Bayen, A. M. 2010. *Evaluation of traffic data obtained via GPS-enabled mobile phones: The Mobile Century field experiment.* Transportation Research Part C: Emerging Technologies, 18(4), 568–583.

HMSO Department of Transport, 1988. *Calculation of Road Traffic Noise,* United Kingdom.

Johnson D.R. & Saunders E.G., 1968. *The evaluation of noise from freely flowing road traffic,* J. Sound. Vib., 7(2):287–309.

Kerner, B.S., Demir, C., Herrtwich, R. G., Klenov, S. L., Rehborn, H., Aleksić, A. and Haug A. 2015 *Traffic state detection with floating car data in road networks.* In Intelligent Transportation Systems, 2005. Proceedings. pp. 44–49.

Lamure, C. 1965. *Niveaux de bruit au voisinage des autoroutes,* Proc. Fifth International Congress on Acoustics.

Lou, Y., Zhang, C., Zheng, Y., Xie, X., Wang, W., Huang, Y. 2009. *Map-matching for low-sampling-rate GPS trajectories,* Proceedings of the 17th ACM SIGSPATIAL International Conference on Advances in Geographic Information Systems. ACM.

Ministère de l'Ecologie, de l'Energie du Développement durable et de l'Aménagement du terroire, 2009. *Road noise prediction, 2 -Noise propagation computation method including meteorological effects,* Methodologic guide (NMPB2008). Service d'études sur les transports, les routes et leurs aménagements Éditions Sétra.

Mongelli, D. W. E. 2013. *An integrated prediction model for traffic noise in an urban area,* Service Operations and Logistics, and Informatics (SOLI), 2013 IEEE International Conference.

NMPB-Routes-96 (SETRA-CERTU-LCPC, CSTB), the French national computation method, referred to in 'Arrêté du 5 mai 1995 relatif au bruit des infrastructures routières, Journal Officiel du 10 mai 1995, Article 6' and in the French standard 'XPS 31–133'

Quartieri, J., Mastorakis, N. E., Iannone, G., Guarnaccia, C., D'Ambrosio, S., Troisi A. and Lenza, T.L.L. 2009. *A Review of Traffic Noise Predictive Models,* Recent advances in applied and theoretical mechanics, ISBN: 978–960–474–140–3, pp. 72–80.

Rose, G. 2006. *Mobile phones as traffic probes: practices, prospects and issues.* Transport Reviews, Vol. 26(3), pp. 275–291.

RLS, 1981. *Richtlinien für den Lärmschutz an Strassen.* BM für Verkehr, Bonn.

RLS, 1990. *Richtlinien für den Lärmschutz an Strassen.* BM für Verkehr, Bonn.

SETRA, 2009. *Prévision du bruit routier, 1 - Calcul des émissions sonores dues au trafic routier,* juin 2009, 124p, 978-2-11-095825-9.

Sohn, K and Hwang K. 2008. *Space-based passing time estimation on a freeway using cell phones as traffic probes.* Intelligent Transportation Systems, IEEE Transactions on, Vol. 9(3), pp. 559–568.

Ygnace, J. L., Drane, C., Yim Y. B. and De Lacvivier R. 2000. *Travel time estimation on the san francisco bay area network using cellular phones as probes.* California Partners for Advanced Transit and Highways (PATH).

Vaiana, R., Iuele, T., Astarita, V., Caruso, M. V., Tassitani, A., Zaffino C. and Giofrè, V. P. 2014 *Driving behavior and traffic safety: an acceleration-based safety evaluation procedure for smartphones.* Modern Applied Science, Vol. 8(1), 88.

Van der Horst, A. R. A., 1991. *Time-to-collision as a Cue for Decision-making in Braking.* In: Brown, I. D., Haselgrave, C.M., Moorhead, I. and Taylor, S. (eds), Vision in vehicles 3: 19–26.

Transport Infrastructure and Systems – Dell'Acqua & Wegman (Eds)
© 2017 Taylor & Francis Group, London, ISBN 978-1-138-03009-1

Genetic algorithm-based calibration of microscopic traffic simulation model for single-lane roundabouts

O. Giuffrè, A. Granà & M.L. Tumminello
DICAM, University of Palermo, Palermo, Italy

A. Sferlazza
DEIM, University of Palermo, Palermo, Italy

ABSTRACT: A calibration procedure for microscopic simulation models based on a genetic algorithm is proposed. Focus is made on single-lane roundabouts for which many random factors such as gap-acceptance affect operations. A comparison is performed between the capacity functions based on a meta-analytic estimation of critical and follow up headways and simulation outputs of a roundabout built in Aimsun microscopic simulator. Aimsun parameters were optimized using the genetic algorithm tool in MATLAB® which automatically interacted with Aimsun through a Python interface. Results showed that applying the genetic algorithm in the calibration process of the microscopic simulation model, a good match to the capacity functions was reached with the optimization parameters set. By this way, automation of the calibration process results effective for analysts which use traffic microsimulation for real world case studies in the professional sphere.

1 INTRODUCTION

1.1 *Background*

Microscopic traffic simulation models have become increasingly useful tools for the advanced analysis of transport systems and have proven to be an active field of research in computer science and transportation engineering (Barceló 2010). Advances in research and current application to road and highway planning and design over the last few years have outlined their great potential to assess operational performances and safety effects, since they can support the evaluation of road policy and infrastructure changes before implementing them in the real world (see e.g. Papathanasopoulou et al. 2016, Giuffrè et al. 2016a, Vasconcelos et al. 2013). Differently from analytical approaches, microscopic traffic simulation models capture road traffic interactions through a combination of complex algorithms which take into consideration car following, lane changing and gap acceptance describing real-world driving behavior. Thus, microsimulation enables the analyst to develop increasingly higher levels of complexity and uncertainty in operations of road networks and single installations (Ayyub 2011). However, concerns are often expressed by practioners about the possible misuse of traffic microsimulation. In simulation studies, indeed, model calibration is a very crucial task, since reliable results must be obtained from the analysis that we made.

Results of several applications, as the technical literature in the road engineering sector refers, show that simulation-based optimization methods and, among these, the genetic algorithms can be usefully applied in the calibration process of microscopic traffic simulation models (see e.g. Chiappone et al. 2016, Essa & Sayed 2016, Rubio-Martín et al. 2015, Nayeem et al. 2014, Menneni et al. 2008). Incorporating the optimization problem within the model calibration, the iterative process of manually adjusting the model parameters, that users and practioners are required to perform, is now automatized. However, in order to automate the iterative process of manually adjusting the model parameters, microscopic traffic simulation models have to be still enhanced with custom models and/or have be equipped with various APIs (when available) to remotely control the simulation. Now for practical implications, the question is how to provide integrated software solutions, user-friendly for practioners and transportation engineers which use microsimulation for real world case studies in the professional sphere.

Despite several encouraging results already obtained from studies on the specific topic, it is noteworthy that commercial traffic simulation software, specially formulated on the operating principles of genetic algorithms are not yet available. Thus engineers often need an analytical aid to develop their own codes and adapt the objective function to the specific requirements of the problems that are often encountered in professional practice. However, it is now possible to use software packages with specific optimization tools as for example MathWorks's® proposes. On this regard, Chiappone et al. (2016) applied the genetic algorithm tool of MATLAB® to reach convergence of the outputs from Aimsun microscopic simulator to the empirical data (that is to minimize the differences between the field measurements observed in the speed-density diagram and the simulator's outputs obtained for a selected freeway segment); the automatic interaction with Aimsun software was achieved through an external Python script, so that the data transfer between the two programs could automatically happen.

The theory and background of the genetic algorithms are not presented in this paper since they are not the objective of this work; instead, we are interested on the applications of genetic algorithms to the calibration of traffic simulations with Aimsun. However, the reader is referred to the large number of online manuals for further knowledge of genetic algorithms as for example the online MathWorks's® website proposes.

1.2 *The objectives of the paper*

The paper proposes a calibration procedure for microscopic simulation models based on a genetic algorithm. Focus is made on single-lane roundabouts for which many random factors such as vehicles arriving and gap acceptance can affect operations. Capacity functions were calculated by specifying the well-known Hagring model for a single-lane roundabout (Hagring 1998). Based on the results obtained in a previous work by Giuffrè et al. (2016b), a meta-analytic estimation of the critical and the follow up headways was used to calculate the capacity functions; they represented the empirical target values of entry capacity to which we compared the simulated output. For pursuing the above stated objective, the calibration parameters—that cause the model to best reproduce empirical capacities—were preliminarily identified using sensitivity analysis for a single-lane roundabout built in Aimsun. A comparison was performed between the empirical capacity functions and the simulation outputs obtained with the default values for the selected parameters of Aimsun. The genetic algorithm tool in MATLAB® was then applied in order to minimize the differences between the two data set and the automatic interaction with Aimsun was implemented through an external Python script that we wrote.

2 PRELIMINARY ANALYSIS

2.1 *The roundabout geometric configuration*

Based on current design applications, a scheme of single-lane roundabout was selected for this study.

Single-lane roundabouts are designed for low-speed operations and represent one of the safest treatments for at-grade intersections (see Rodegerdts et al. 2010). Figure 1 exhibits the sketch of the at-grade single-roundabout, features of which comply with the instructions by the Italian standards on geometric design of compact roundabouts (2006).

The geometric design of the roundabout included: a) the outer diameter (that is the basic parameter used to define the size of a roundabout and is measured between the outer edges of the circulatory roadway) of 39,0 m; b) the circulatory roadway width of 7,0 m; c) the entry width of 3,75 m; d) the exit width of 4,50 m; e) the length of legs reaching the roundabout of 35 m without parking possibilities for vehicles from 20 m up to the approach zones. By this way, based on the selected roadway widths, the resulting speeds resulted moderate enough to accommodate mixed traffic flows of passenger cars and heavy vehicles in whatever context, also in the built environment.

2.2 *Behavioural parameters and capacity model*

Traffic operations at roundabouts are typically ruled by the gap acceptance process, which specifically for single-lane sites is also facilitated by speeds moderated by the particular geometric design. In gap acceptance process the critical headway and the follow-up headway are two key factors in determining entry capacity which, in turn, depends on circulating flow under a specified arrival headway distribution (see Mauro 2010). Many methods have been developed to estimate these parameters from gap data; as a matter of fact, technical literature proposes several empirical research studies focusing on estimation of critical and the follow-up headways at roundabouts. However, the decision whether a hypothesis is valid or a statistical method is appropriate for the estimation of the population parameters, or in general for the study to be performed, cannot be based on the

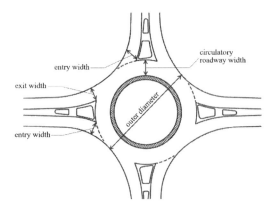

Figure 1.   The sketch of the single-lane roundabout.

results of a single empirical research, since results can vary from one research to the next. Rather, analysts need a mechanism to synthesize data across studies (see Borenstein et al. 2009). In this view—finding that the application of quantitative methods to summarize the results of several empirical studies is now widespread in several research fields (see e.g. Elvik 2011)—in a previous work, we performed a meta-analysis, that is the analysis where each (primary) study yields an estimate of statistical mean values of the critical headway and the follow-up headway (i.e. the effect sizes), we assessed the dispersion in these effects and then we computed a summary effect for each parameter. The reader is referred to Giuffrè et al. (2016b) for a more effective presentation of the results of the meta-analysis of effect sizes, performed as part of the literature review through the random-effects model.

Table 1 shows the summary effects (i.e. the random quantitative meta-analytic estimate) for each parameter, the 95% lower and upper limits for each summary effect, the results of the Cochran's Q test (or a measure of heterogeneity that is the sum of the squared deviation of each effect size from the mean, weighted by the inverse-variance for each study) and the Higgin's index I^2 (the ratio of true heterogeneity to total observed variation). According to Borestein et al. (2009) the p-value close to zero and I^2 less than 25% for both headways, confirmed the absence of heterogeneity in the data set examined for the single-lane roundabout. At last, the meta-analytic estimates for the two headways were found nearly consistent across all studies and, compared to the single studies, provided a more reliable result for the parameters of interest.

Further activities involved the construction of empirical functions for entry capacity; for this purpose, the general Hagring's model for multi-lane intersections was used (Hagring 1998). The Hagring's model was specified for the single-lane roundabout where a Cowan's M3 headway distribution (Cowan 1975) was assumed for the circulating stream and the meta-analytical values estimated for the critical and follow up headways were used or:

$$C_e = Q_c \left(1 - \frac{\Delta\, Q_c}{3600}\right) \frac{\exp\left[-\dfrac{Q_c}{3600}(T - \Delta)\right]}{1 - \exp\left(-\dfrac{Q_c}{3600} T_f\right)} \quad (1)$$

where C_e = entry-lane capacity [pcu/h]; Q_c = circulating traffic flow [pcu/h]; T = critical headway [s]; T_f = follow–up headway [s]; Δ = minimum headway of major flow [s] that, according to literature, was assumed equal to 2.10 s.

Thus the entry capacities derived from the Hagring model for each entry of the single-lane roundabout, were used in the following as reference for calibration purposes.

2.3 Aimsun modelling

The microscopic simulation package AIMSUN (version 8.1) was used for microscopic modelling of the single-lane roundabout; no priority was created for the legs approaching the roundabout, but priority to vehicles moving anticlockwise on the ring was established. Detectors were located so that they could replicate the possible location of field detectors, that is at each entry/exit and upstream/downstream of each entry. The model building produced a single-lane roundabout scheme that was reproducible and verifiable; at the same time, this scheme reproduced as accurately as possible the geometric intersection pattern of a possible real world counterpart (Dowling et al. 2004). Before applying the GA-based procedure that automatically runs to test the model validity, traffic conditions on the roundabout were reproduced in Aimsun with default settings; indeed, we considered the default parameters of Aimsun version 8.1. As far as the traffic demand, we used O-D matrices with due consideration to the direction of turn. Passenger cars were only considered at this stage; the attributes of passenger cars were the default values given by Aimsun. In order to reproduce the traffic demand and represent realistic traffic conditions on the roundabout, we assigned O/D matrices such as reproducing the circulating flow Q_c, from 0 veh/h to 1400 veh/h with a step of about 200 veh/h; these matrices guaranteed saturation conditions at entries, so that the number of vehicles entering the roundabout was the capacity for the specific entry each time considered. It should be noted that at the time when the analysis presented in this research was carried out, we found that no single-lane roundabout having a modern design was installed in our City. However, different existing unsignalized and

Table 1. The meta-analytic estimates for critical and follow-up headways at single-lane roundabouts (Giuffrè et al. 2016a).

Statistics	Headway [s]	
	Critical	Follow-Up
Random estimate	4.27	3.10
Standard error	0.11	0.07
95% lower limit	4.05	2.96
95% upper limit	4.49	3.25
Z-value	37.46	41.82
p-value	0.00	0.00
Q	23.28	33.90
I^2	1.19	20.37

signalized intersections, having size comparable to the roundabout that we built, had been identified in the urban road network and some of them were likely to be converted into the single-lane roundabout as that chosen for calibration purposes (see e.g Giuffrè et al. 2011; Giuffrè et al. 2014).

Calibration parameters were preliminarily identified by using a sensitivity analysis; it required that some model parameters were changed and adjusted in an iterative way until model outputs were close to empirical data, based on a predefined level of agreement between the two data sets. The comparison between the set of model parameters manually derived for calibration purposes and the best combination of the values for the GA-optimized parameters is reported in sections 3 and 4. Thus in the context of microsimulation based modelling, the capacity functions derived from Equation 1 were compared to the simulation outputs obtained using Aimsun.

In order to automate the iterative process of manually adjusting the model parameters, the calibration of the microscopic traffic simulation model was then formulated as an optimization problem which searches for an optimum set of model parameters through an efficient search method as will be explained in the following section 4.

3 STRUCTURE OF THE GA-BASED METHOD

Now we propose the formulation and the solution of the calibration problem: the formal interpretation of the problem is given and, subsequently, the solution by applying genetic algorithms is described.

Let $\{u_k\}_{k=1...N}$ and $\{y_k\}_{k=1...N}$ be two input-output sequences of observed data acquired during suitable traffic measurements; they are the "experimental surveys". We want to reproduce the same output sequence corresponding to the same input sequence by simulation. In order to obtain the simulated output, denoted with $\{\hat{y}_k\}_{k=1...N}$, the model has to be calibrated; this means to find values for the model parameters such that the simulated output $\{\hat{y}_k\}_{k=1...N}$ is as close as possible the observed output $\{y_k\}_{k=1...N}$ given the same input $\{u_k\}_{k=1...N}$.

The cost function is defined as:

$$j(\beta)=\frac{1}{N}\Sigma_{i=1}^{q} w_i\left[\Sigma_{k=1}^{N} g\left(y_{i,k}-\hat{y}_{i,k}(u_k,\beta)\right)\right]$$

(2)

where k = the discrete time instant; N = the number of measures (each one at each time instant); q = the number of outputs considered for the identification procedure; w_i = the weight associated with the error on the i-th variable (the generic i-th variable

will be specified for the problem under study in the next section); $g(\cdot)$ = either the square or the absolute-value function; $y_{i,k}$ = the experimental value of the i-th variable at the instant k; $\hat{y}_{i,k}(u_k,\beta)$ = the corresponding simulated value that is a function of the input u and the parameter vector β.

The solution of the calibrating problem will be the parameter vector β^* that minimize the objective function (see Equation 2), or:

$$\beta^* = \arg\min_{\beta} j(\beta)$$

(3)

Problem expressed in Equation 3 can be solved iteratively; however, two problems have to be solved. The first problem involves the initial condition to be chosen, whereas the second problem is represented by the stopping criteria. According to Davis (1991), the problem of the initial condition should not be undervalued. It should be noted that most algorithms only search for local minima; when we face multiple minima (non-convex problem), the algorithm usually converges only if the initial guess is already somewhat close to the final solution. However, this problem is avoided if genetic algorithms are used, since they are evolutionary optimization algorithms robust with respect to the initial condition (Davis 1991).

The second problem, instead, can be easily solved by selecting a maximum number of iterations or, the algorithm can be stopped when:

$$\left|\frac{j(\beta)_k - j(\beta)_{k-1}}{j(\beta)_{k-1}}\right| < \varepsilon$$

(4)

where ε = the error stop quantity; $j(\beta)_k$ and $j(\beta)_{k-1}$ = the values of $j(\beta)$ computed at the iterations k and $k-1$, respectively. This stopping criteria means that the algorithm will be stopped when the objective function variation, between two consecutive instants, is less than a quantity ε freely chosen. The reader is referred to the online MathWorks's® website, or to the large number of online manuals, for further details on genetic algorithms and how to perform genetic algorithms optimization.

In our application the "experimental surveys" consist of entry capacity values calculated by using the Hagring model as a function of the meta-analytic estimation of the critical and the follow up headways (see previous subsection 2.2). The estimated output was generated by means of Aimsun software which ran with a fixed model corresponding to the model under study, and tuned with a suitable set of parameters. It is obvious that if the selected parameters are incorrect, then the estimated capacity values do not coincide with the experimental survey. For this reason, let us select the objective function in Equation 2 as follows:

$$J(\beta) = \frac{1}{N} \Sigma_{k=1}^{N} \left[\left(C_k - \hat{C}_k(\beta) \right)^2 \right] \qquad (5)$$

where N = 32, since we have considered observations (see subsection 2.3) distributed over eight hours, in each of them we ran four simulations (one every fifteen minutes).

In the case of Aimsun traffic simulation model, the model behaviour is depending on a wide variety of model parameters. In general, if we consider the model to be composed of entities, (i.e. vehicles, sections, intersections, and so on), each of them described by a set of attributes, (i.e. parameters of the car following, the lane change, gap acceptance, speed limits and speed acceptance, and so on), the model behaviour is determined by the numerical values of these parameters.

As part of the calibration process we evaluated beforehand the calibration parameters by using a sensitivity analysis, or recalculating the simulation outputs under different alternative assumptions to determine the impact of the explored variables, and then we were able to establish for the particular model the influence of some important parameters on the capacity of the entries. In this way, we pursued the objective of finding the values of the parameters which were able to produce a valid model. Thus, we choose as parameters for the optimization the following:

$$\beta = \left[R_T, H_{min}, S_{acc} \right] \qquad (6)$$

where R_T = the reaction time; H_{min} = the minimum headway; S_{acc} = the speed acceptance.

More specifically the driver's reaction time or more easily the reaction time, as used in the car-following model of Aimsun, is the time in seconds that it takes a driver to react to speed changes in the preceding vehicle. The reaction time assigned to a vehicle, moreover, is a global parameter of Aimsun, that means during each trip is constant. In each simulation run, we set the parameter as fixed and equal to simulation step, that is the same value for all vehicles. Moreover, the reaction time may influence the computing performance and some simulation outputs, such as the section capacities: in general, the lower the reaction time is, the higher capacity values can be obtained. The reason for this is that the drivers are more skilful, as they have shorter reaction times; they can drive closer to the preceding vehicles, they can find gaps more easily, they have more opportunities to enter the network, etc. The minimum headway is primarily a lane-changing model parameter; setting this parameter ensures the minimum headway (minimum time in seconds) between the leader and the follower. Aimsun traffic simulation model includes the minimum headway between leader and follower

as a restriction of the deceleration component in the car following model and applies this constraint before updating the position and the speed of the vehicle (i.e. leader) respect to its follower.

At last the speed acceptance ($S_{acc} \geq 0$) represents the level of goodness of the drivers or the degree of acceptance of speed limits: $S_{acc} \geq 1$ means that the vehicle will take, as maximum speed for a section, a value greater than the speed limit, whereas $S_{acc} \leq 1$ means that the vehicle will use a lower speed limit. In the Aimsun car-following model the speed acceptance, together with other parameters such as the target speed and the section speed limit, helps to define the desired speed for each vehicle on each section. The speed acceptance may, moreover, influence the behaviour of the gap-acceptance model; several vehicle parameters (i.e. speed acceptance, turning speed, desired speed and so on), influencing all vehicles of a particular type when driving anywhere in the network, also have an influence on the output of the model. Now using Equations 5 and 6, the problem in Equation 3 can be solved implementing the genetic algorithm tool in MATLAB®. This tool has been applied in order to minimize the differences between the two set of empirical and simulation data, and the automatic interaction with Aimsun was implemented through an external Python script.

Starting from a generic initial condition, the genetic algorithm generates a set of parameters β, and then the Aimsun software is running with the parameters β. Aimsun is attached to MATLAB® via a python subroutine that allows the data transfer between the two programs. Thus Aimsun provides a set of estimated outputs (one for each β) and the algorithm computes the objective functions 5 associated with each β. At last, the algorithm selects the best parameter β and generates a new set of parameters β that is the new generation. The cycle goes on and on until the specified stopping criterion occurs. We chose the stopping criterion with a specified, fixed, maximum number of iterations (50 generations). In our case, moreover, the initial population used to seed the genetic algorithm is given by 20 individuals, by using the default setting of Aimsun as first individual. Based on these choices, the computational time is about 4 hours by using an Intel(R) Core (TM) 2 Quad CPU Q9300 2.50GHz and 8Gb of RAM. We stopped the algorithm after 50 iterations so that the value of the cost function in Equation 5 reached a steady-state and the algorithm could be stopped. An upper bound β'' and a lower bound β' for β were introduced to restrict the search domain and the condition $\beta' \leq \beta \leq \beta''$ was then established to avoid that parameters without a physical meaning were generated (i.e. negative reaction time, negative distance among vehicles, etc...). The best β^* obtained by solving the optimization problem can only represent the best value of reaction time, the

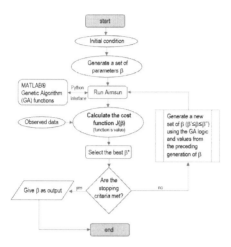

Figure 2. Block diagram representation of the GA.

minimum headway and the speed acceptance such that the simulated capacity values track as well as possible the empirical ones. This gives an efficient automated calibration procedure for simulation with Aimsun.

Figure 2 shows the outline of GA which summarizes how the genetic algorithm works and executes the processing steps, whereas the results are presented in the next Section.

4 SIMULATION RESULTS

The optimization problem was solved by applying the algorithm illustrated before. According to the convergence condition, after 50 generations (approximately 4 h of computing time), the algorithm converged on the optimal solution. The best combination of the values for the simulation parameters included:

- the reaction time of 0.86 s instead of the Aimsun default value of 0.80 s;
- the minimum headway of 1.58 s instead of the Aimsun default value of 0 s;
- the speed acceptance of 1, instead of the Aimsun default value of 1.1.

These values were different from values derived from manual calibration or: the reaction time of 0.85 s; the minimum headway of 1.6 s, the speed acceptance of 1.

The resulting parameter set returned by the genetic algorithm with the least error was chosen as the best parameter set (or the best combination of the model parameters), and then used for predicting the simulated capacity.

Figure 3 shows the Aimsun simulation outputs which were gained by using both the best output parameters and the parameters derived

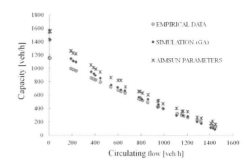

Figure 3. Entry capacity: simulation vs Hagring data.

from manual calibration ("Aimsun parameters"), and the empirical data which were based on the Hagring model where the meta-analytical estimation of the behavioural parameters was introduced for capacity calculation. We can see in Figure 3 that the simulation which used the best parameter set, as returned by GA application, generated a better match to Hagring data.

In addition to the better match to the empirical data obtained for the simulation with the optimization parameters as shown before, we also analysed the positive effects of the proposed optimization procedure by calculating the cost function in both cases, i.e. for simulation with the default and optimization parameters. Figure 4 shows the values of the fitness function—that is the cost function $J(\beta)$ in Equation 5 - during the optimization period, and the corresponding value of the same function computed with the default parameters. The cost function resulted equal to $J(\beta) = 126.8$ for optimized parameters and equal to $J(\beta) = 135$ for parameters derived from manual calibration. We want to highlight that the benefit in tuning the model parameters would have been greater if we had used Aimsun default parameters as the initial condition ($J(\beta) = 159.04$ for Aimsun default parameters).

This figure shows the positive effect derived from the application of the proposed genetic algorithm-based calibration, since a better matching to the field data is clear in comparison with the simulation that used the default parameters. The graphs in Figure 4 also plots the mean score of the population at every generation: the mean best fit is the mean value of the cost functions calculated for all individuals of the same generation, whereas the best fit is the cost function of the best individual within the generation. Because the genetic algorithm function minimizes the fitness function, the best fitness value for a population is the lowest fitness value for any individual in the population. To further test the model validity despite it would have been accepted, we used the GEH index as criterion for acceptance, or otherwise rejection, of the model. This index is a global indicator widely used in practice for validating traffic microsimulation models, especially when only aggregated

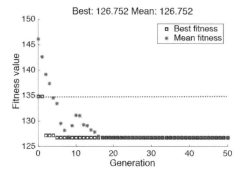

Figure 4. Values of the cost function $J(\beta)$ during the optimization period.

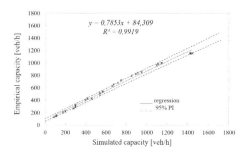

Figure 5. Simulated and empirical relationship with 95% confidence limits.

values are available as flow counts at detection stations aggregated to the hour and entry capacity values (see Barceló 2010, pp. 46–47). Since the deviation of each simulated value with respect to the measurement for each entry was lower than 5 in 88% of the cases (however, lower than 8 in the remaining cases), the model could be considered "calibrated" in terms of its ability to reproduce the empirical entry capacities at the examined single-lane roundabout and has been accepted as significantly able to reproduce local conditions and traffic behaviour in a satisfactory way. Note that for the examined roundabout the root mean squared normalized error—which provides information on the magnitude of the errors relative to the average measurement—resulted less than 0.10, whereas the mean absolute percent error—also calculated as supplemental parameter to measure the size of the error in percentage terms—resulted less than 5%. Figure 5 depicts an example of scattergram analysis developed to compare empirical versus simulated capacities at entries of the examined single-lane roundabout. The regression line of empirical versus simulated capacity was plotted along with the 95% Prediction Interval (95% PI). Based on the R^2 value of 0.9919 and the fact that most of points were within the confidence band of the regression lines, we reached the conclusion which the model could be accepted as significantly close to the reality.

5 CONCLUSIONS

This paper proposed a procedure based on a genetic algorithm to calibrate a microscopic traffic simulation model. Microscopic traffic simulation models are now widely used for real-world case studies in transportation engineering, since they are faster and safer than actual on-field implementations. However, without a proper calibration the results of the analysis could be misleading and unreliable.

We were interested on the applications of the genetic algorithm-based calibration to traffic simulations with Aimsun. Focus was made on roundabouts for which many random factors such as gap-acceptance can affect operations. In order to automate the iterative process of manually adjusting the model parameters, the calibration of the microscopic traffic simulation model was formulated as an optimization problem which searches for an optimum set of model parameters through an efficient search method. Calibration parameters were preliminarily identified using sensitivity analysis, whereas the optimum values for these parameters were obtained by minimizing the error between the simulated and empirical capacity by using a genetic algorithm.

The proposed calibration procedure was, indeed, applied to the Aimsun software by using a genetic algorithm to systematically modify the model parameters and then to fit the capacity values obtained from simulations to the capacity functions based on a meta-analytic estimation of the critical and the follow up headways. The genetic algorithm tool in MATLAB® was applied in order to minimize the differences between the two data set and the automatic interaction with Aimsun was implemented through an external Python script that we wrote so that the data transfer between the two programs could automatically happen.

The genetic algorithm-based approach appeared to be effective in the calibration of Aimsun; indeed, the results showed a better match to the empirical data than simple manual calibration. Despite this research is a first exploratory analysis for roundabouts, it should be noted that the comparison between the empirical data and the simulation results obtained with the default and optimized parameters, gave insights into the performance of the calibration procedure that we proposed. The sensitivity analysis performed to identify the calibration parameters with greater effect on output, indeed, confirmed to us that no additional benefits to the calibration process would be given by a higher number of calibration parameters. Thus, given the number of model parameters here used, we are confident that the procedure can be applied to other case studies. Based on the results of this research—summarized by the values of the cost function $J(\beta)$ during the optimization period in Figure 4—it is evident that benefits resulting from the use of a genetic algorithm can compensate the computational efforts deriving from the application of an optimization technique which automates the iterative process of manually adjusting the simulation parameters. Anyhow, if we had selected more parameters (a situation which can occur when more complex problems have to be managed), probably we would have had to deal with other issues such the nature of the obtained optimum (i.e. if it is a local minimum or the absolute minimum) and/or how well the local minimum approximates the absolute one, and so on.

At last, it should be said that the importance and the advantages of this procedure are also to improve the efficiency of the model to predict different scenarios (that may occur in reality) by varying different initial conditions. Moreover, the application of this procedure to more complex scheme of multilane roundabouts should be designed to address further problems that practioners and transportation engineers, which use microsimulation for real world case studies in the professional sphere, are often called to solve. For these future developments, more consideration is needed in order to specify the calibration procedure of microscopic simulation models through automated processes which incorporate the latest optimization techniques.

REFERENCES

Ayyub, B.M. 2011. *Vulnerability, Uncertainty, and Risk: Analysis, Modeling, and Management*. Reston, Virginia: American Society of Civil Engineers (ASCE).

Barceló, J. 2010. Fundamentals of Traffic Simulation. Londono: Springer, 442 p.

Borenstein, M., Hedges L.V., Higgins, J.P.T. & Rothstein, H.R. 2009. *Introduction to Meta-analysis*. Chichester, UK: John Wiley & Sons, Ltd.

Chiappone, S., Giuffrè, O., Granà, A., Mauro, R. & Sferlazza, A. 2016. Traffic simulation models calibration using speed-density relationship: An automated procedure based on genetic algorithm. *Expert Systems with Applications* 44: 147–155.

Cowan, R.J. 1975. Useful Headway Models. *Transportation Research* 9(6): 331–375.

Davis, L.D. 1991. *Handbook of genetic algorithms, 1st edition*. New York: Van Nostrand Reinhold.

Dowling, R., Skabardonis, A. & Vassili, A. 2004. *Traffic Analysis Toolbox Volume III: Guidelines for Applying Traffic Microsimulation Software. Report No. FHWAHRT-04-040*. Washington DC: Federal Highway Administration-United States Department of Transportation.

Elvik, R. 2011. Publication bias and time-trend bias in meta-analysis of bicycle helmet efficacy: a re-analysis of Attewell, Glase and McFadden, 2001. *Accident Analysis and Prevention* 43(3): 1245–1251.

Essa, M. & Sayed T. 2016. A comparison between PARAMICS and VISSIM in estimating automated field-measured traffic conflicts at signalized intersections. *Journal of advanced transportation* 50: 897–917.

Ghods, A.H & Saccomanno, E.F. 2010. Comparison of carfollowing models for safety performance analysis using vehicle trajectory data. *Annual Conference of the Canadian Society for Civil Engineering 2010 (CSCE 2010), Winnipeg, MB, Canada, June 9–12, 2010*. Volume 2: 13391348.

Giuffrè, O., Granà, A., Roberta, M. & Corriere, F. 2011. Handling underdispersion in calibrating safety performance function at urban, four-leg, signalized intersections. *Journal of Transportation Safety and Security* 3(3): 174–188.

Giuffrè, O., Granà, A., Giuffrè, T., Marino, R. & Marino, S. 2014. Estimating the safety performance function for urban unsignalized four-legged one-way intersections in Palermo, Italy. *Archives of Civil Engineering* 60 (1): 41–54.

Giuffrè, O., Granà, A., Marino, S. & Galatioto, F. 2016a. Microsimulation-based passenger car equivalents for heavy vehicles driving turbo-roundabouts. *Transport* 31 (2): 295–303.

Giuffrè, O., Granà, A. & Tumminello, M.L. 2016b. Gap-acceptance parameters for roundabouts: a systematic review. *European Transport Research Review* March 2016, 8:2.

Hagring, O. 1998. A further generalization of Tanner's formula. *Transportation Research Part B: Methodological* 32(6): 423–429.

Mathew, T.V. & Radhakrishnan, P. 2010. Calibration of microsimulation models for nonlane-based heterogeneous traffic at signalized intersections. *Journal of Urban Planning and Development* 136 (1): 59–66.

MathWorks[R], 2016. Online documentation on GA, URL: https://www.mathworks.com/help/gads/geneticalgorithm.html

Mauro, R. 2010. *Calculation of Roundabouts*. Berlin Heidelberg: Springer.

Menneni, S., Sun, C., & Vortisch, P. 2008. Microsimulation calibration using speed-flow relationships. *Transportation Research Record* 2088(1): 1–9.

Nayeem, M.A., Rahman, Md.K. & Rahman M.S. 2014. Transit network design by genetic algorithm with elitism. *Transportation Research Part C: Emerging Technologies* 46: 30–45.

Norme funzionali e geometriche per la costruzione delle intersezioni stradali [*Functional and geometric standards for the construction of road intersections*]. 2006. Ministero Infrastrutture e Trasporti [Minister of Infrastructure and Transport], in Italian.

Papathanasopoulou, V., Markou, I. & Antoniou C. 2016. Online calibration for microscopic traffic simulation and dynamic multi-step prediction of traffic speed. *Transportation Research Part C: Emerging Technologies* 68: 144–159.

Rodegerdts, L. et al. 2010. Roundabouts: An Informational Guide, 2nd edition. NCHRP REPORT 672. Washington DC: Federal Highway Administration-United States Department of Transportation.

Rubio-Martín, J.L., Jurado-Piña, R., Pardillo-Mayora, J.M. 2015. Heuristic procedure for the optimization of speed consistency in the geometric design of single-lane roundabouts. *Canadian Journal of Civil Engineering* 42(1): 13–21.

Vasconcelos, L., Seco, Bastos Silva, A. 2013. Safety analysis of turbo-roundabouts using the SSAM technique. *CITTA 6th Annual Conference on Planning Research*: 1–15.

Transport Infrastructure and Systems – Dell'Acqua & Wegman (Eds)
© 2017 Taylor & Francis Group, London, ISBN 978-1-138-03009-1

More accessible bus stops: Results from the 3iBS research project

M.V. Corazza & A. Musso
Sapienza University of Rome, Rome, Italy

M.A. Karlsson
Chalmers University of Technology, Gothenburg, Sweden

ABSTRACT: Although often perceived as poorly attractive and unreliable, buses are central in the development of sustainable mobility options in urban areas. Therefore, 3iBS - *the Intelligent, Innovative, Integrated Bus Systems* (a research project funded by the European Commission) promoted the research on a new generation of vehicles and facilities. Specific emphasis was placed on the design of bus stops as crucial elements to improve the quality and accessibility for all of bus services. The paper describes the conflicts to solve and stresses the need of univocal directions to design inclusive bus facilities; specific recommendations for implementing innovations in this field are eventually provided.

1 INTRODUCTION

Buses are central in the everyday life of many citizens. Nevertheless they are often perceived as unappealing (Stradling et al 2007, Beirao & Sarsfield Cabral 2007, NRTBI 2009, Dobbie et al. 2010). Therefore, 3iBS - *the Intelligent, Innovative, Integrated Bus Systems*, a research project funded by the European Commission (EC), moved from this assumption to propose solutions to revamp this mode of transport across Europe. More specifically, the project aimed to promote the research on bus systems with a special focus on a number of key topics and, among these, the improvement of accessibility for all. The results evidenced how accessibility has become a mainstream issue on the agendas of transit stakeholders. Therefore, this paper describes the main outcomes in this field, with a special focus on problems related to bus infrastructures, resulting from a specific activity dedicated to issue guidelines for accessibility and safety concepts in the future design of bus systems.

2 DEFINING ACCESSIBILITY AND DISABILITY

Accessibility can be broadly defined as the ability to travel between different activities (Vuchic 1999), it can be linked to convenience, as the ability to reach goods, services, destinations (Littman 2014), or the ease to do it (Sinha & Labi 2007). Engwicht defined accessibility as "the ease with which exchange opportunities can be accessed" (1999:167). Lynch stressed the importance of ease through the concept of immediacy, since accessibility is defined as "the general proximity in terms of time of all points ⋯ to a given kind of activity or facility" (1995:49); he also identified three sub-dimensions of accessibility: diversity (of things to be accessed), equity (of access for the different social groups) and control (over the access systems) (1981). For Grava (2003), accessibility is also a measure of the quality and operational effectiveness of a community. Therefore, the acknowledgment that transit has to serve all leads to the concept of fully accessible transit (Vuchic 2005). Hence, it is not uncommon to assess the accessibility of a given transit facility according to the level of ease by which one can reach it, typically by location-based criteria, or more generally short distance. For example, Banister observes that transport's "primary aim is to maintain a high level of accessibility with trip lengths being as short as possible" (2005:10). This shifts the focus on how to measure accessibility in general (Handy & Niemeier 1997, Geurs & van Eck 2001, Church & Marston 2003). Accessibility can be also associated with terms such as "within walking distance" or "walkable". Linking transit accessibility with walking can be, on the one hand, a way to emphasize the (supposed) modest physical efforts required to reach a given destination, and on the other a way of underestimating the efforts of those who are not able to autonomously walk. This may also imply that spatial accessibility is equal to temporal accessibility, but efforts spent to reach a given destination may differ according to the travellers' physical and perceptive conditions. For examples, the "walking distance" to a bus stop can be less challenging for an elderly person than the time spent standing in line, waiting for boarding, with no possibility to rest.

Accessibility can be also linked to perception. Hansen (1959) pioneered this concept by defining accessibility as the "intensity of possibility of interaction". Ropoport observed that: "people react to environments in term of the meanings the environments have for them" (1990:13). A bus stop and its surroundings can become the weakest link in the journey chain if they provide users with negative meanings (too distant, uncomfortable, unsafe, etc.).

The approach within the 3iBS project was based on the notion that "a Public Transport (PT) system is accessible to people who are able to use it". The emphasis is placed on the ability of a given PT system to provide appropriate meanings or, in other words, equal conditions of exchange by meeting the requirements of all, rather than on the specific abilities of the users. If a transit system is not, or poorly, accessible this means that a number of factors prevent passengers from using it, not the passengers themselves. Preventing factors may result from inappropriate design criteria for vehicles, infrastructures and communications. They occur because they fail to meet users' physical and/or cognitive requirements. However, users' needs are many and even indistinct, and so also the definition of disability. Actually, the line between a disabled individual and a non-disabled one is rather thin, since a person performing everyday activities (walking with prams, suitcases, shopping trolleys, etc.) requires the same space as does a wheelchair user. At the same time, other sensory or cognitive impairments do not entail such kind of spatial requisites. One might argue that the so-called "able-bodied" person does not even exist.

Therefore, physical, sensory, cognitive or developmental impairments give rise to a vast range of requirements, in some cases contrasting, and solutions to meet them all can only result in a kind of "relative optimum" for a majority of passengers. Ramps are a typical example of such a contrast: designed for wheelchair users, they can be challenging for blind users, who cannot perceive the length, unless trained or helped, and therefore prefer grades, easily detectable by walking sticks (Lauria 1994).

The design of inclusive solutions for transport infrastructure is based on the knowledge of users' needs, but given the variety of disabilities, some of them are still neglected. An example is dyslexia: although 8% of the world population is affected by this impairment, dyslexia is far from being considered in the provision of information in public areas. Difficulties in processing numbers, reading, and/or spelling, make the comprehension of basic travel information (i.e timetables, variable messages, announcements, etc.) a demanding task for dyslexic travellers who develop creative solutions such as the use of satellite navigation systems (Lamont 2010).

Disability is therefore difficult to define from a design perspective. However, Carmona et al. observed that there are two main models of disability: the "medical model" and the "social model". The former defines disability in terms of medical conditions, and the impairment factors are placed solely on the individual; the latter concerns the barriers "imposed by a disabling society/environment unable to make adjustments: this rather than personal impairment, is the disability factor. In this model, people have impairments, environments are disabling" (2003:127). Therefore, shifting the focus from single individuals with disabilities to places and the need to shape them according to everybody's requirements, represents, in terms of design, a more productive basis to start from.

3 THE REGULATORY FRAMEWORK TO PROMOTE ACCESSIBILITY FOR ALL

According to all of the above, accessibility and disability seem to be concepts under continuous development, and so are the legal and regulatory frameworks behind and the related financial supports.

3.1 The European strategy and regulations

Since the 1980s concepts such as accessibility and safety have been progressively included in many regulations in this field, both at supranational and national levels, although with different approaches. The European institutions have continuosly developed strategies to increase inclusiveness and equity according to general principles; currently, the most important course of actions are the Disability Action Plan 2006–2015 (CoE 2006) and European Disability Strategy 2010–2020 (European Commission 2010), the EU Disability Action Plan 2003–2010 (CSES 2009), all in line with the 2006 United Nations Convention on the Rights of Persons with Disabilities. Strategic sectors contemplated in these documents such as medical care, education, employment, and social safeguard are considered key areas to promote participation and equality of rights for people with disabilities. Accessibility is, thus, meant to ensure full fruition of goods, services and assistive devices, facilities, and communication technologies, in the same way as for "able-bodied". The importance of having full access to the built environment by applying Universal Design criteria is also stressed, as well as the need to provide fully accessible transport services. These comprehensive directions are primarily meant to steer national regulations towards a common vision and therefore can be considered expressions of general principles in safeguarding people with disabilities.

More specific design criteria can be found at national levels; as already stated, from the 1980s' onwards, European Countries have progressively enforced more and more specific design criteria to create barrier-free environments, with a special focus on dwellings. Initially, design criteria concerned mainly dimensions and layouts to accommodate physically challenged users, and especially those using wheelchairs. The philosophy behind such an approach was strongly affected by the General Product Safety Directive n. 92/59/EEC, according to which products available for the market should be safe, so as to prevent risks for the end-users: but even though this principle is imperative, its misinterpretation can lead to design concepts merely based on passive safety rather than active involvement of people with disabilities. The repetition of this approach, i.e. conceiving barrier-free environments mostly for wheelchair users resulted in a number of problems.

First, *it cast a shadow on the requirements of the remaining part of the population with disabilities*, which still affects the way in which many transport facilities are designed or upgraded. It is undeniable that especially for what concerns cognitive disabilities, accessibility is far from being fully addressed. As a result, although ramps and lifts abound nowadays, provisions for visually or hearing-impaired still do not, and even less do the aids for "less considered" disabilities.

Secondly, designing barrier-free environments mostly for wheelchair users, being mainly focused on barriers which can be overcome by support devices or equipment (typically, walking aids), *affected the standardization programs towards the developments of specifications of single components rather than to develop common process (Criteria and procedures) to apply univocally and systematically across Europe*. Currently, standardization covers a wide range of Assistive Technology (AT) products; for instance, the European Committee for Standardization (CEN) issued specifications on: technical aids for disabled persons (CEN 2007); Man-Machine Interface for users with special needs (CEN 2012); and medical informatics (CEN 2010), but those which can evolve into a standardization of inclusive solutions for the built environment are still in the making (as CEN/CLC/JWG5 and 6 respectively on Design for All, and Accessibility in the Built Environment).

Eventually, this approach foster again and again *the misconception that inclusive design is an* *"adaptation to special needs"* in the attempt to increase the passive safety of users with disabilities, whereas it is a concept to shape the built environment according to the needs of all and with the aim to improve their active and autonomous participation in social life. Adaptive design is, therefore, justifiable and/or admissible in historic or natural environments only (cultural heritage sites, green areas), where physical obstacles are irremovable, thus restricting the access options.

Needless to say, to overcome these problems, the criteria of Design for All and AT technologies must be applied throughout the standardisation process and mandatorily included in the design of new facilities and in the rehabilitation and retrofitting of those already operational, according to the lesson learnt from the Americans with Disabilities Act (ADA). ADA enforces a comprehensive regulatory system, complemented by a vast range of technical specifications and standardized solutions including detailed norms for transportation facilities, as it considers standards as a mean to overcome discrimination and exclusion.

3.2 *Political support*

The regulatory framework described earlier is affected by the political support. Unlike the US, in many European Countries minorities started rather recently to lobby against exclusions, mostly thanks to the commitment of associations such as the European Disability Forum, the European Blind Union, or the European Union of the Deaf, just to mention a few of the most influential. Liaising with the European Union institutions is one of the tasks of these bodies, in order to keep the awareness high at political level.

Along with such poltical commitment, at the beginning of the 2000s, the European istitutions promoted a number of studies in this field. More specifically, the Conférence Européenne des Ministres des Transports (CEMT) and the EU developed several comparative analyses: on differences in regulations enforced (Législation pour améliorer les moyens d'accès. rapport CEMT/CS/TPH(2000)7/REV1), on the best practice (Améliorer l'accessibilité des transports—Rapport CEMT/CM(2004)27), and on the the quality of transit services in 49 cities (Citizens Network Benchmarking Initiative—2004), so to have a sound knowledge basis and raise the awareness on the problems faced by people with disabilities (Thomas et al. 2007). It is rather difficult to assess to what degree the needs and expectations of the people with disabilities have been met over the years, but since the proposal for a European Accessibility Act (EAA) is still to be published by the European Commission, more can be achieved in this field. The lesson learnt from ADA on standardization as a mean to improve social inclusion clearly shows the importance of enforcing a tool as the EAA and the need for a paradigm shift: it is necessary not only to have common mandatory inclusive design standards but also to enforce compulsory accessibility master plans (as for land use) to have a common quality of barrier-free environments across Europe.

However, at a European level, an indirect, but strong, political support is represented by the many research projects that the European Commission have funded over the years to improve the quality of life of people with disabilities, in line with the CEMT approach. The outcomes of these projects constitute a comprehensive set of guidelines, recommendations and best practice examples in the fields of safety and accessibility, but are far from being applied as consolidated practice across Europe, which again leads to contemplate the need for a tool as EAA and its implementation Europewide.

3.3 *Funding to improve accessibility*

Aside from those granted by the European Union, specific funds are usually linked to retrofit programs to refurbish major facilities, typically railways stations, and to comply with what the national regulations mandate. A good example is the "Access for All" station improvement scheme, launched recently in the UK. According to the national Railways Act 2005, it allowed more than 80 facilities to be rehabilitated and an additional fund was announced to have more stations transformed by 2019. It is also estimated that for every 1 GBP spent in this rehabilitation program 2.92 GBP is actually gained from additional revenues coming from the enlarged group of passengers (DfT 2012).

Other major railways operators in Europe have adopted similar initiatives, often issuing their own guidelines (for example RFI 2012) to renovate buildings and rolling stock, and solve typical problems, such as the access to platforms or the refurbishment of rest and waiting areas. Nevertheless, while this might result in improved accessibility of a majority of larger urban stations, mainly for physically-challenged and visually/hearing impaired passengers, the access continuity with the surrounding environment is usually not ensured, or of poor quality.

In the case of railways, one operator in charge of managing one national network is a happenstance that facilitates this kind of rehabilitation process. But different, local, smaller operators who manage bus fleets and facilities in just one area are more likely to adopt their own solutions for layouts, materials, textures, colours, etc, according to their own funding availability. Again, common universal design standard criteria and accessibility master plans, when mandatory across Europe, would facilitate rehabilitation processes compelling local operators to adopt common solutions. Examples of common funding encompassing more local areas are few but an interesting one is represented by the German Lower Saxony administration that, in 2013, launched a 400,000 Euro program to renovate bus stops in different cities, planning to upgrade 50 stops yearly (Vogt 2013).

Different players and areas also make difficult to prioritize plan interventions and above all raise appropriate funds. Usually, funding opportunities for upgrading transit facilities come from public, local (municipal) investment programs, seldom including some earmarked funds for improving accessibility and even more seldom from private financing. One exception is the German *Programme 233 and 234—Barrierearme Stadt* which enables banks to fund investments for barrier-free infrastructure, and especially transit facilities. This is a good example of the successful involvement of private bodies in a public accessibility improvement processes, thus supplementing public funding with private financing.

4 THE BUS STOP: A MULTI-REQUIREMENT ENVIRONMENT

"Bus stops are often dreary because they are set down independently, with very little thought given to the experience of waiting there, to the relationships between the bus stop and its surroundings" (Alexander 1977:452). Not much has changed since this remark. Therefore, the emphasis placed within 3iBS on bus stops, as elements to redesign to meet the requirements of all is two-pronged. On the one hand, according to the definitions of accessibility and disability reported earlier, it stresses the crucial role played by the environment when addressing problems related to access and impairment. On the other hand, it highlights the need for constant, extra funding and strong political support to increase momentum on these issues. The former calls for a common vision shared by planners, designers, operators and the largest part as possible of passengers; the latter requires a common regulatory framework to develop univocal design criteria and standards across Europe.

This is not an easy task, since a bus stop environment concentrates multiple functions, requirements, performance levels and planning and design criteria. In fact, the problem is multifaceted. Since a majority of the different phases of a journey may take place at the bus stop

(Figure 1), the design of this facility must provide different solutions to meet a number of common requirements and functions (walkability; comfort while waiting or boarding/alighting; ability to autonomously perform other travel functions as purchasing tickets, getting information, resting; readability and comprehension of signs and directions; feeling of inclusiveness, etc.); the travel experience may occur under different circumstances (good vs adverse weather, daytime vs nighttime, peak vs off-peak times, frequent vs novice travellers, solo travelling vs accompanied, beginning vs end-leg of a journey, etc.), and take place in different environments (outdoor vs indoor, secluded vs frequented, friendly vs hostile—perceived, etc.).

Therefore, it is relevant to conceive the bus stop as a system in which the main activity, i.e. boarding/alighting, is strictly interrelated with additional or associated activities occurring in the surroundings; each activity has its own meaning for the users, which makes them perceive the bus stop as a specific environment and react differently to it.

5 FROM CONCEPTS TO PROBLEMS

Once users' needs are analysed according to good practice and the vast scientific literature, and bus stops are designed consequently, the requirements for accessibility and appropriate meaning should be met. But, even when urban transit operators issue their own inclusive design guidelines for the network facilities, full accessibility across the whole transit network is a goal far from being accomplished. Unlike the railways sector, planning accessibility for bus systems results in different outcomes due to a series of factors, which can be synthesized into: *too many managing actors, too many facilities* and *too many end-users in different environments*.

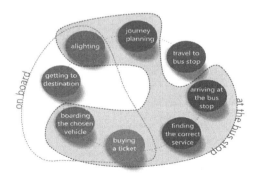

Figure 1.   Main activities of a bus journey.

5.1   *Multiplicity of actors*

In terms of management of bus stops, a number of different situations may occur: a) one operator managing the whole transit network and facilities; b) several operators, each managing a part of the local transit network; c) several operators, each managing either the transit network or the ground facilities.

For cases b) and c), the earlier mentioned lack of compulsory guidelines result in a lack of common vision which becomes a hindrance in the development of common design criteria for fully accessible stops for the whole urban area. Moreover, when bus stops are managed by a third party such as advertising companies, matching commercial functions with the needs of all may become difficult. For instance, excessive advertisement on the shelter walls may result in reduced space for information for visually-impaired users (who need areas to accommodate large prints/tactile information), and for the hearing-impaired ones (who need full visibility of the approaching buses). Poor availability of transparent surfaces may induce a perception of reduced security since the principle of "see and being seen" is strongly diminished, especially when inside the shelter.

The problem is not just a matter of conflicts between accessibility and commercial purposes. Other factors may contribute to inappropriateness. For instance, O'Neill & O'Mahony (2005) reported how, according to the Disabled Persons Transport Advisory Committee in the United Kingdom, disabled people were more likely to rely on taxi services. However, in the case study analyzed, 42% of the people with disabilities were happy with buses. Although a useful travel option, taxis may not be the best way towards inclusiveness, as it: a) misses the opportunity to make conventional transit and its facilities more accessible and b) increases somehow the dependency on private travels. As a result, if it is assumed at political level that disabled people are more likely to be satisfied with taxis, no attention will be paid to make bus stops more accessible for all.

An additional factor is the dominating attention paid to products rather than processes. Specific products for specific disabilities improve constantly, but Carmona et al. stress that: "Inclusive design is about attitudes and processes as much as about products" (2003:127). It is indisputable, then, that even the most advanced products (as commercial shelters), if not included in a process based on the integration of the needs of all, do neither "produce" inclusiveness, nor contribute to create less disabling environments. This may affect decisions concerning other elements of the travel chain or the decision process. For example, the conversion of obsolete bus fleets into more environmentally-friendly ones, with low-floor vehicles and on-board equipment to accommodate the largest part of passengers with special needs may be thwarted if the environment surrounding the bus stop is not accessible, not to mention that the related investment for such eco-conversion can, then, appear as an effort for inadequate return. Therefore system approaches might not, *per se*, result in integrated solutions.

5.2 *Limits in the feasibility*

Among the different management situations, in the case of situation a), i.e. one operator for the whole network, the task of a total rehabilitation might be too demanding. Single operators face the problem to *upgrade too many facilities*, i.e. a huge number of stops with different functions and size: from simple bus markers to shelters, hubs, and terminals. The approach in this case is rather common, since the upgrading process is linked to capacity: priority is given to larger facilities (hubs, terminals) for which a wide range of solutions is deployed, to remove what are considered the most common barriers (grades and gaps); then a minor provision of equipment is assigned to shelters (usually to improve ease of use such as armrests at seats, real time information displays and occasionally tactile aids), whereas stops with simple bus markers are left not upgraded. However, basic adaptations of bus markers are not difficult (Lauria 1994; Rickert 2007). Along with different ways of management and vast amounts of facilities to deal with, a third element of conflict is the *different end-users* that each stop facility attracts and *the difference in the surrounding environments* where it is located. Guidelines for bus stops design are available in grey and scientific literature (some of which are referenced in this paper) all providing well-known criteria for locating (usually commercial) shelters and their basic equipment in indistinct urban environments.

This approach may apply to the construction of a new infrastructure of urban relevance, such as a light railway, with dedicated platforms, which entails a total rehabilitation of the streetscape, including additional areas to accommodate pedestrians. Not the same can be obtained when locating a basic bus stop (typically a shelter) on the sidewalk, where passenger flows accessing/egressing the waiting area cannot be separated by the main pedestrian flows on the sidewalk and the space to accommodate passengers with disabilities is scarce (as in Figure 2, where a layout designed for all is described, including typical pedestrian movements around a bus stop, according to different Level of Services—LOS, and some basic requirements

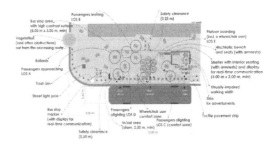

Figure 2.   A bus stop environment.

to accommodate a wheelchair user and a visually impaired one).

Still this approach becomes more difficult to adopt in historic centers, where appropriate location can be demanding, due to narrow sidewalks. As a result layouts such that of Figure 2 are unfeasible. In this case the built environment becomes disabling and the conflict has virtually no resolution: adaptive planning and design remain the only resources.

Likewise, allocating the proper space requirement for each user may be challenging and generally regulations for street furniture and for the use of sidewalks (typically Highway Codes) increase the amount of restrictions. Specifications for street furniture usually associate size of objects/obstructions (benches, bollards, vendors, etc.) to a given amount of pavement areas (GMUAM 2001, Winnipeg Transit 2006), but seldom include LOS and areas to accommodate users with special needs. The approach followed by Highway Codes and regulation on street functions is quite the reverse: LOS and service areas are provided, but for a very restricted group of street furniture. Usually, none includes systematic specifications for the areas required to accommodate users with special needs, or recommends functional layouts or adaptations criteria to the type/quality of the environment. The lesson learnt is that there is no general "recipe" to design a bus stop or to rehabilitate this kind of facility when already existing in a sensitive context (historic or consolidated areas) to improve accessibility; planning and design must equally consider requirements of all and constraints from the environment, in the awareness that only that kind of "relative optimum" can be obtained.

6   RECOMMENDATIONS FOR MORE ACCESSIBLE BUS STOPS

Conflicts and limits reported can be of no easy resolution but certainly can goad on to progress towards more inclusive adaptation of the built environment. Therefore, within 3iBS, Key Directions (KD) for the implementation of innovations regarding accessibility for future bus systems have been provided (Karlsson 2014). These KDs are:

a. *Process approach for implementing innovations;*
b. *Initial assessment of the local level of accessibility;*
c. *Long-term vision;*
d. *Strong leadership;*
e. *Organizational culture;*
f. *Political and public support;*
g. *Consideration of regulatory aspects;*
h. *Possibilities of innovation procurement;*
i. *Secure and long-term financing;*
j. *Sufficient and relevant resources;*
k. *Full analysis of existing solutions;*
l. *Stakeholder analysis;*
m. *Cooperation and collaboration;*
n. *User studies;*
o. *Evaluation;*
p. *Market of the innovation*

The translation of such recommendations into more accessible future bus stops involves a new vision based on the in-depth comprehension of what makes an environment "disabling", and where innovation of system and products are aimed at increasing inclusion, according to three main lines of action: 1) *Assessment of resources to increase accessibility*; 2) *Inclusion of less-studied disabilities* and 3) *Integrated proactive design for all.*

6.1 *Assessment of resources to increase accessibility*

The previously reported analyses of current political support and availability of funding evidenced the inadequacy of resources to progress towards more inclusive transit environments. Required resources must be regulatory and economic. For the former, specific universal design criteria for transit facilities must be standardized at EU-level and enforced at national level. Consideration of innovative legal and regulatory aspects to support the standardization process and its implementation (KD *g)* and especially, innovative, specific technical solutions for the design of new bus stops will help transit companies to initially assess the status of the facilities (KD *b*) and solutions (KD *k*), and plan the upgrading process, according to a process approach (KD *a*). The analysis of the existing solutions is of the utmost importance, as it indicates the compatibility between the status quo and the enforcing standards, and therefore the entity of the upgrading process. This entails a systematic enforcement of the previously mentioned standards across the whole bus stop network, starting from the very basic facilities.

Such a network-scale plan calls for a thorough assessment of the available resources (KD *j*) to refurbish facilities according to the enforcing standards, along with the application of a specific business plan targeting changes in perception of the service by the enlarged community of users (thus including those previously excluded). All of the above is possible only if secure and long-term financing is available (KD *i*). Funding is crucial: due to the multiplicity of managing actors and involved third parties, contributions from all of them are required and fundraising may involve non-conventional bodies such as sponsors, donors, and communities, etc. The possibility to use innovation procurement (KD *h*) to support the introduction of innovative devices (e.g. AT products for advanced communication systems at bus stops) must be contemplated. Needless to say, plans and funding rely on long-term visions (KD *c*) and strong and durable political support (KD *f*), at supranational and national levels for what concerns the issue of standardized solutions and innovations, and at local levels to launch such plans.

6.2 *Inclusion of less-studied disabilities*

All of the above relies on an in-depth knowledge of the requirements of the community to serve.

Traditionally, studies on disabilities have focused on walking, hearing and visually-impaired users respectively and as a result, transit facilities solutions have been usually (re)designed to meet the requirements of each category, separately. However, transit is used by passengers with all kinds of health conditions and personal travel experiences and often their impairments do not relate to physical barriers as grades or gaps. Meeting requirements associated with mental or intellectual impairments, specific conditions due to age or personal perception helps reduce vulnerability, disadvantage and exclusion. More specifically, for the design of the bus stops this calls for continuous analyses of the demand to serve (KD *n*), both ex ante and ex post (KD *o*), including demands of users with "conventional" as well as "less known" impairments. Prior to the upgrading process, studies of the community should be aimed at assessing how the requirements are met and the compatibility of the proposed solutions among the different categories of passengers. Conflicts may arise and the more the requirements to meet, the more likely this is going to happen. Moreover, ex ante assessments can highlight specific incompatibilities (if any) between the enforcing standards and specific disabilities to deal with (footnote: standards are based on the uniformity of technical criteria and the possibility of missing some specific requirement must be thus contemplated). Nevertheless, ex ante analyses of users' needs should result in "snapshots" of the bus stop requirements, to steer design of facilities towards a performance specifically targeted to increase accessibility. Dealing with enlarged communities of passengers (including those affected by less "common" impairments) includes not only the usual customer care process. Long-term observation studies on the fruition of the upgraded facilities by the different groups of users are also required to assess whether the whole built environment has gradually become less "disabling". Again, funding and political commitment are essential to ensure appropriate methodologies for the analyses of the users, and especially for those with less well-known disabilities. It is also essential that the panel of stakeholders involved in the analyses and in the assessment of resources includes representatives of all categories of disabled users and, further, that cooperation is ensured (KD *m*). In this way the risk of conflicts may be mitigated and the interests of all equally considered (KD *l*). A strong supporting organizational culture is obviously required (KD *e*).

6.3 *Integrated proactive design for all*

To improve bus stop accessibility, practice has so far relied on the provision either of *ex novo* facilities, with the limits earlier mentioned (inclusion of requirements from "conventional" disabilities only, capacity as a priority criterion, conflict of interests unsolved, etc.) or of existing facilities, with "tacked-on" adjustments to make them more barrier-free, more or less with the same limits and the additional restriction that in some case the built environment (for example historic centers) could thwart such rehabilitation attempts due to lack of proper space.

These situations are far from acceptable as, in both cases, the design approach is reactive: commercial components (shelters, seats, displays, etc.), designed for "able-bodied" users, are adjusted to meet the requirements of the considered majority of impaired users. Subsequent additions of ramps and tactile tiles represent "adjustment for all" *vs* "design for all". To have the latter winning the action is two-pronged. On the one hand, the design approach must become proactive, thus it must: i) include requirements from less well-known disabilities to develop more accurate design criteria, and ii) rely on standardized universal design criteria for transit facilities, resulting in the production of innovative systems and components enforced by sound and durable regulatory supports and by the compulsory adoption of accessibility plans for transit. On the other hand, once again the political support must become unrestricted and compelling. Transit companies must be required to

compulsory provide accessible infrastructures, vehicles, and communication systems to all passengers and control possible conflicts. This requires a strong commitment from the transit companies (KD *d*) to comply. Users must become aware that accessibility is a right and informed of the availability of most suitable travel options. Marketing of the innovative solutions (KD *p*) should be considered to disseminate the information to the different groups of passengers, according to different communication tools. Marketing can help increase awareness and lessons learned from environmental care can be applied to accessibility: *mutatis mutandis*, transit components, when compliant with standards and operating within accessibility plans, may be, in the near future, labelled or "tagged" as "accessibility-mindful", thus increasing the attractiveness of public transport.

7 CONCLUDING REMARKS

Alexander described the epitome of bus stops as: "easy to recognize, and pleasant, with enough activity around them to make people comfortable" (1977:453). To achieve this, there are still many avenues to explore, and accessibility is the main one, especially in political, regulatory and operational terms. Issues such the EAA, the inclusion of less well known disabilities, the compulsory adoption of common Universal Design criteria when planning and designing transit facilities, the mandate to enforce accessibility plans for transit companies, and the possibility to award fully-compliant products and components as "accessibility-mindful" are therefore necessary steps to make urban travels really for all.

REFERENCES

Alexander, C. 1977. *A pattern language*. Oxford University Press, New York.
Banister, D. 2005. *Unsustainable Transport. City transport in the new century*. Routledge, London.
Beirao, G. & Sarsfield Cabral, J.A. 2007. Understanding attitudes towards public transport and private car: A qualitative study. *Transport Policy*, Vol. 14, Issue 4, July, pp. 478–489.
Carmona, M., et al. 2003. *Public places Urban Spaces*. Architectural Press, Oxford.
CEN 2007. *Business Plan—CEN/TC 293. Assistive products for persons with disability*. http://standards.cen. eu/BP/6274.pdf
CEN 2010. *Business Plan—CEN/TC 251. Health informatics*. http://standards.cen.eu/BP/6232.pdf
CEN 2012. *Business Plan—CEN/TC 224. Personal identification, electronic signature and cards and their related systems and operations*. http://standards.cen. eu/BP/6205.pdf

Church R. & Marston, J. 2003. Measuring Accessibility for People with a Disability. *Geographical Analysis*, Vol. 35, n. 1, pp. 83–96.

CoE—Council of Europe (2006). *Disability Action Plan 2006 2015*. www.preventionweb.net/english/professional/policies/v.php?id = 35144

CSES 2009. Mid-term Evaluation of the European Action Plan 2003–2010 on Equal Opportunities for People with Disabilities. CSES, Sevenoaks.

DfT—Department for Transport 2012. *High level output specification 2012 – Railway Act 2005 Statement*. www.gov.uk/government/publications/high-level-output-specification-2012

Dobbie F. et al. 2010. *Understanding Why Some People Do Not Use Buses*. www.scotland.gov.uk/Resource/Doc/310263/0097941.pdf

Engwicht, D. 1992. *Reclaiming our cities and town*, New Society Publishers, Philadelphia.

European Commission 2010. *European Disability Strategy 2010–2020: A Renewed Commitment to a Barrier-Free Europe*. COM/2010/0636 final. http://eur-lex.europa.eu/legal-content/en/NOT/?uri=CELEX:52010DC0636

Geurs, K.T. & van Eck, J.R., 2001. *Accessibility measures: review and applications*. RIVM report 408505 006, National Institute of Public Health and the Environment, Bilthoven.

GMUAM—Gerencia Municipal de Urbanismo, Ayuntamiento de Madrid 2001. *Instruccion de Via Publica*. Nilo, Madrid.

Grava, S. 2003. *Urban Transportation Systems*. Choices for Communities. McGraw-Hill, New York.

Handy, S.L. & Niemeier, D.A., 1997. Measuring accessibility: an exploration of issues and alternatives. *Environment and Planning A*, Vol. 29, n. 7, pp. 1175–1194.

Hanson, W. 1959. How accessibility shapes land use. *Journal of the American Institute of Planners*, n. 15, pp. 73–76.

Lamont, D. 2010. The Effects of Dyslexia upon Personal Travel: Empirical Results of a Qualitative Study. *Proceedings of the 12th WCTR*, Lisbon.

Lauria, A. 1994. *Pedonalità urbana*. Maggioli, Rimini.

Littman, T. 2014. *Accessibility*, in: TDM Encyclopedia. www.vtpi.org/tdm/tdm84.htm

Karlsson, M.A., (ed.). 2014. *3iBS—Del. 1.1.2. Guidelines for the introduction of) accessibility and safety concepts in future design of bus systems* (restricted document).

Lynch, K. 1995. *City Sense and City Design:* MIT Press, Cambridge.

Lynch, K. 1981 *A theory of good city form*. MIT Press, Cambridge.

NRBTI 2009. *Quantifying the Importance of Image and Perception to Bus Rapid Transit*. www.nbrti.org

O'Neill, Y. & O'Mahony, M. 2005. Travel Behavior and Transportation Needs of People with Disabilities. *Transportation Research Record*, No. 1924, pp. 1–8.

Rapoport, A. 1982. *The meaning of the built environment*. The University of Arizona Press, Tucson.

Sinha, K.C. & Labi, S. 2007. *Transportation Decision Making*. Wiley and Sons, Hoboken, NJ.

Stradling, S. et al. 2007. Passenger perceptions and the ideal urban bus journey experience. *Transport Policy*, Vol. 14, n. 4, pp. 283–292.

Thomas, R., et al. 2007. *L'accessibilite des reseaux de transport en commun en Europe*. ÉNS d'Architecture, Grenoble.

RFI 2012. *Linee guida per la progettazione—Accessibilità nelle stazioni*. http://www.webstrade.it/pescara/peba/RFI_2011_LG_accessibilita-stazioni.pdf

Rickert, T. 2007. *Bus rapid transit accessibility guidelines*, World Bank, Washington, D.C.

Vogt, S. 2013. Region baut Haltestellen behindertengerecht um. *Hannoversche Allgemeine*, 14.2.2013.

Vuchic, V.R. 1999. *Transportation for livable cities*. CUPR, New Brunswick, NJ.

Vuchic, V.R. 2005. *Urban transit. Operations, planning and economics*. Wiley and Sons, Hoboken, NJ.

Winnipeg Transit 2006. *Designing for Sustainable Transportation and Transit in Winnipeg*. www. winnipegtransit.com/inside-transit/transitreports

Transport Infrastructure and Systems – Dell'Acqua & Wegman (Eds)
© *2017 Taylor & Francis Group, London, ISBN 978-1-138-03009-1*

Shaping the future of buses in Europe: Outcomes and directions from the EBSF and EBSF_2 projects

M.V. Corazza & A. Musso
Sapienza University of Rome, Rome, Italy

U. Guida & M. Tozzi
UITP, Brussels, Belgium

ABSTRACT: EBSF—*European Bus System of the Future* and its follow-up EBSF_2 are two research projects funded by the European Union and led by UITP with the aim of developing a new generation of buses across Europe. Within EBSF the accomplished goals were to increase the bus attractiveness by testing innovative vehicles. Coherently, EBSF_2 is now committed to raise the image of the bus through solutions for increased efficiency of the system. The environmental concern and the need to save energy are also behind the majority of both projects' innovations. These are tested through demonstrators, i.e. on vehicles operating in several European cities, and performance are independently evaluated. The paper describes the methodology and the test process adopted in both projects, and reports the results so far achieved and the new challenges ahead, with the research objective to provide advanced knowledge for further applications beyond the European projects field.

1 THE CONTEMPORARY SITUATION

Passenger cars are still the dominant mode in the everyday mobility of many European urban areas, even though the recent economic developments have started to orient mobility demand towards transit, and more specifically toward increasing bus ridership. Car-based lifestyles are still fostered by many behavioral patterns and recurring practice, among these: the assumed personal convenience of this mode, continuously growing urbanization and increased tolerance to distance and congestion (the average delay in minutes for one-hour journey driven in peak periods is 29 min, as monitored by TomTom 2013). Air and noise pollution, space consumption, climate change and environmental concerns are typically associated to the phenomena above described, and integrated strategies to fight congestion in urban areas have become more and more mainstream issues on local, national and supranational political agendas in Europe. Among the many proposed countermeasures, the effectiveness of (even cleaner) bus systems has been long acknowledged worldwide.

This is not surprising since buses remain the most widespread public transport mode with around half of all public transport passengers (30 billion per year) in the EU, reaching 100% in smaller towns and medium-sized cities (UITP 2011). Moreover, within transit, the most recent generations of buses are a very appropriate mode to meet sustainability constraints in terms of energy efficiency, emissions and space occupancy, as well as operational effectiveness, since they can be more easily adapted to different requirements of passengers and do not require heavy (re)design of infrastructures. Nevertheless, the modal switch from passenger cars to buses can be challenging in urban areas where inadequate performance such as poor regularity, punctuality, comfort, design and obsolete technologies contribute to the general modest attractiveness of this mode, if compared to other transit options, especially in Central, Eastern and Southern Europe as stated by the European Commission (2013), and generally reported in scientific literature (Stradling et al 2007, Beirao & Sarsfield Cabral 2007, NRTBI 2009, Dobbie et al. 2010).

The European Union has a major role in advocating for a momentous shift in transport planning and operating, especially by promoting more sustainable urban mobility policies and the several research projects it funded since the 1990s represent continuous opportunities to test, in real urban scenarios, innovative approaches. Great emphasis is placed on promoting studies and implementation of solutions to meet more sustainable requirements in local transport systems, and among the research projects in this field, a series of them is specifically dedicated to buses, such as EBSF—

European Bus System of the Future (2008–2013), *3iBS—the Intelligent, Innovative Integrated Bus System* (2012–2015), *ZeEUS—Zero Emission bUs System* (2013–2017), along with the more recent EBSF_2 (2015–2018) and *ELIPTIC— ElectrIfication of Public Transport In Cities* (2015–2018). More specifically, EBSF and EBSF_2 have a common, two-pronged task: developing innovative solutions to operate more environmentally-friendly and efficient vehicles and, at the same time, increasing the attractiveness of this mode of transport.

The paper first describes the measures tested in EBSF and those to be tested within EBSF_2, then it reports the results achieved in the former and outlines those expected to accomplish in the latter and, eventually, critically analyze the relevance of both in the field of the current practice for energy management across Europe.

2 ACHIEVED RESULTS AND THE CHALLENGES AHEAD FOR THE BUS OF THE FUTURE

2.1 *Innovation and demonstrations*

Both EBSF and EBSF_2 rely on the synergy between research and demonstration activities under real operational conditions. Research involves six main key working areas, also called Priority Topics (PTs) within EBSF_2; these are: i) energy strategy and auxiliaries, ii) green driver assistance systems, iii) IT standards introduction in existing fleet, iv) vehicle design (in terms of capacity, accessibility, and modularity), v) intelligent garage and predictive maintenance, vi) interface between the bus and urban infrastructure. The theoretical vision behind the PTs is that innovation for buses is a multi-comprehensive concept, synergically involving the study of requirements and performance of vehicles, infrastructures, operations and the relevant actors (i.e., passengers, operators, drivers, staff, manufacturers, suppliers, decision makers, etc., all equally considered).

Both projects reflect this comprehensiveness as they have been conceived as an open platform for dialogue on bus system efficiency that includes key representatives of all stakeholders involved in urban bus services. As already for EBSF, within EBSF_2, a multiplicity of partners, in this case over 40 bodies of international relevance, are involved in the projects (Figure 1), covering all areas of expertise (from manufacturing, to operations and research) and geographical regions in Europe. Their mutual collaboration, not limited by commercial relations, enables the development of solutions that represent the best technological answer to real operational requirements. To be noted that the projects

Figure 1.   EBSF_2 Consortium partners.

consortia involve four leader European bus manufacturers, along with many others as associated members. Together they represent more than the 70% of the European bus market and are therefore able to boost the introduction of key innovations on a large scale in Europe and beyond. In terms of Public Transport Authorities and Operators, this corresponds to operations by a fleet of more than 45,000 vehicles and more than 6 billion passengers per year. They are complemented by suppliers and research organizations as well as national and international associations that will ensure the highest dissemination and networking of the project results.

Central in both projects are the demonstration activities. The innovative measures tested in such real demonstrators are selected according to their technological maturity, in order to ensure a short step for commercialization after the end of the project. The use of simulators and prototypes is conceived as a preliminary step for the validation of the innovations in real operational scenarios, or as a necessary task to prove the potential of more futuristic solutions currently implemented at early stage of development (e.g. adaptation of bus segments to actual passenger demand through coupling systems).

Several urban areas across Europe provide test fields for the demonstrators (within EBSF: Bremerhaven, Brunoy, Budapest, Rome, Rouen, Gothenburg, Paris city and Madrid; the latter three also involved in EBSF_2, along with London, Paris area, Barcelona, San Sebastian, Dresden, Ravenna, Lyons, Helsinki and Stuttgart).

Even though the measures are very different, environmental concern and energy management are behind the majority of them. This also explains, for some PTs, the major emphasis placed on the vehicles' more challenging components, when the goal is to achieve a more balanced energy management: auxiliaries, with a special attention to Heating,

Ventilation and Air Conditioning (HVAC) and driver assistance systems. Currently auxiliaries account for 15% to 25% of the total energy budget of an Internal Combustion Engine (ICE) bus, but they can rise up to 50% in the case of electric propulsion with no excess heat energy to exploit. Thus, requirements for energy management for different propulsion technologies can vary, and so do the optimal solutions. A combination of measures can increase the efficiency of air conditioning units as well as reduce heat loss via the interior and exterior design of buses. Effective circulation of air could also help maintain a comfortable temperature inside the bus with less energy input. Fresh air is required to control moisture/condensation on window surfaces which is imperative especially during snowy conditions. A large number of driver assistance systems, including eco-driving systems, are already available for passenger cars and trucks, but not for public transport operations where the driver needs to keep to a set schedule and start/stop frequently. "After-market" eco-driving systems designed generically for vehicle fleets provide limited access to data other than vehicle accelerations and braking. Thus, they neither take into account several factors which influence energy consumption, nor differences between the driver's need for feedback in e.g. diesel compared to electric buses. Moreover, in eco-driving, like in other driver assistance systems, the efficiency of the feedback relies on the information transmitted to the driver but also on the design of the Human Machine Interface (HMI). Present interfaces typically include displays showing the driving performance in terms of basic feedback which may suffice for the experienced driver but not for the less experienced ones. At the same time excessive information will be difficult to perceive and may even distract the driver (Tozzi et al. 2016).

Innovations which seem to be specifically designed to improve operations or attractiveness have also positive consequences in terms of energy savings. The design of bus doors may serve as case in point: within EBSF, prototypes with larger door blades or with a fifth door were respectively tested in Gothenburg and Budapest with the primary aim to speed up boarding and alighting operations. But the possibility of reducing overall fuel consumption due to the consequent shortened dwell time at stops may be not negligible. In light of the EBSF results (synthesized in Musso & Corazza 2015a), energy efficiency proved to be a major issue and therefore became a specific area of investigation within EBSF_2 as well, where shortened dwell time can be associated to controlled door openings for less heat exchange or to a totally indoors bus stop, to provide a controlled, single thermal environment.

2.2 Assessing the innovation: the methodological approach

According to the dedicated evaluation methodology adopted in both projects, the assessment of each measure is a classical "before-vs-during-the-implementation" comparison of results, with Key Performance Indicators (KPIs) measuring the performance variations at each single case study and cross-case. However, to have a more accurate assessment within EBSF_2, quantitative Performance Targets have been introduced to assess (in percentage), at the end of the project, whether the achieved results met the expectations. Demonstrators can select one or more Performance Targets, each of them includes one or more KPIs. Among the 65 Performance Targets thus far selected, those specifically targeted to assess a more balanced energy management are: i) Improving the overall energy efficiency of fleets; ii) Reducing the consumption of conventional fuels or electric energy; iii) Increasing the uptake of fully electric and hybrid options, iv) Improvement of auxiliaries management; v) Decrease harsh decelerations/accelerations; vi) Reduction in HVAC power requirement; vii) Reducing noise and air emissions; viii) Reduce the vehicle's energy consumption; ix) Expected increase in retrofitting programs, x) Energy savings in auxiliary systems. Expectations are high for all of them, with improvement targets varying from 5 up to 30%, according to the different demonstrators (Musso & Corazza 2015b).

The performance comparative analyses are also fostered by additional activities, such as modelling and surveys. Among these, a study on the theoretical possibility to transfer the good results achieved elsewhere in Europe (the so-called Transferability Exercise) and a survey on the energy management of the bus fleet (the 2013 EBSF "Survey on Energy Efficiency", specifically aimed at identifying critical issues in public transport companies concerning energy efficiency, submitted to 34 European companies, all accounting for a total annual production of 2.3 billion vkm) stressed also how the perception of innovation may be affected by economic concerns, as reported in the next section.

2.3 Lessons learnt from EBSF

Although EBSF_2 is in progress and tests for the majority of the measures just started, the EBSF ones evidenced some interesting results, namely:

- Ecodriving which led to a reduction of fuel consumption in the two UCs where it was tested (respectively by –17% and –8%);
- New internal layouts, which sped up boarding and alighting operations, thus reducing dwell times by –13% and –6% in the two test sites where

this performance was surveyed, increased capacity (respectively +3% and +8%), and improved punctuality up to 52% in one out of the three test sites where this parameter was monitored, whereas in the other two +3% and +6% were recorded, respectively;

- Predictive, remote maintenance systems, which strongly reduced the amount of days in the workshop of the tested vehicles (respectively –60% at one test site and –75% at another), and contributed to decrease the ratio of non-working vehicles by 25% at one test site.

In other test cases, some parameters for instance like fuel consumption, passenger demand, operating costs, commercial speed and service efficiency provided negligible variations (±1% if compared to the baseline performance prior to EBSF) or did not vary at all. Such results, shortly defined as of "no change" were considered positive, as well, given the short testing periods, the limited amount of testing vehicles or extent of operations.

An example of the quality of the "no-change" situation is provided by the test scenario for a predictive, remote maintenance system. In this test site (a small size city), the "no-change" situation had to be ascribed to the relatively short duration of the "during" phase and, partly, to the gist of remote maintenance system itself. As a matter of fact, the primary goal of this test was to develop a reliable system to detect failures and prevent breakdowns whose efficacy in terms of saving resources (both fuel and money) or improving performance should require implementation periods far beyond the 8-month duration of the test. Moreover, as the demonstrator was applied on a small scale (just 10 vehicles on a single line, out of the local fleet of 152 vehicles, operating 22 lines), it was not surprising that it resulted uninfluential, especially for what concerns the operating costs on their whole. But, at the same time, the "no-change" situation proved that this remote maintenance system, even on such a short-range basis, worked properly: had it not worked so and problems strictly linked to the installation and fine-tuning of the system had arisen in a reiterated way or/and with a high level of severity, the test operations would have been inevitably affected and extra resources to manage the service required, since the not-working vehicles involved in the test should have been replaced, anyhow, with consequent, unavoidable negative changes, especially in the service performance levels.

Finally, passengers' appreciation of the innovations, according to the surveys run during the test activities, was high especially for that which concerned improved quality of service (up to 13%), on-board comfort (up to 25%) and image of the bus (up to 16%) according to Musso & Corazza (2015a).

At the same time, the lesson from the Transferability Exercise (in which a total of 35 European stakeholders took part) was clear: the EBSF measures, although successful and innovative, might be transferred if rewarded by reduced fuel consumption or improved service efficiency, but in any case they do not have to increase operating costs. Energy consumption or improved driving styles affecting fuel consumption are therefore considered just for their saving potentials, and not in terms of the prospective environmental benefits. But this result can be also interpreted in light of the lower percentage usually represented by fuel costs in the overall expenditure for operations: being this a minor item, it is important that remains unaffected by the introduction of innovations.

The Transferability Exercise highlighted one more issue: demonstrators more appreciated by the respondents, thus more likely to be theoretically transferable are those where a mix of measures were contemporarily tested (for instance, cases studies including new HMI driving devices along with energy management and ITS-based support systems), rather than a very specifically-focused trial of a single measure.

The EBSF "Survey on Energy Efficiency" provided, in turn, a snapshot of companies willing to switch to more environmentally friendly propulsion systems, but still anchored to conventionally-fuelled fleets, mostly focused on reducing costs, and moderately interested in the use of less polluting fuels, also because the majority of the surveyed fleets were old and doomed to obsolescence within a very short term (Carbone & Proia 2013).

2.4 The contribution of 3iBS

These results can be analyzed in synergy with those from a complementary, but independent, survey carried out within 3iBS (the 2013 "European Bus Systems in Europe: current fleets and future trends" survey), involving a panel of more than 70 stakeholders, from 63 cities of different size, corresponding to a total fleet of around 70,000 vehicles (buses and trolleybuses) serving a population of over 100,000,000 inhabitants in 24 European countries. Some core questions addressed the contemporary and future situations of propulsion systems. As in the previous EBSF survey, diesel-fuelled was the most common option (79%), followed by biodiesel (9.9%), CNG (7%), and electricity (1.2%); other fuels (hydrogen, bioethanol, liquefied petroleum gas, etc.) accounted only for 3%. Looking at the future plans of the stakeholders considering different propulsion systems, the majority of respondents are willing to change the current situation towards less "traditional" propulsion systems.

In fact, electric traction seems to become the real option of the future as more than 40% of the respondents want to move towards more electric vehicles. More specifically, respondents, who could select more options, seemed to favor fully electric buses with batteries (45.5%) or hybrids (69.7%). Among the "no-electric" options, the increased preference for CNG (28.3%), biogas (13.2%) and biodiesel (18.9%) was one more proof of the willingness to change, but about one respondent out of three still relied on diesel (as it totaled up to 34%) as reported by Tozzi, Guida & Knote (2014). This can be interpreted as the respondents' awareness that diesel-based propulsion cannot be totally avoided and is coherent with the EBSF survey result of the 70% of companies stating their aim to purchase (or having already purchased) Euro VI vehicles.

2.5 More goals ahead, the EBSF_2 tasks

The EBSF accomplishments opened the way for further developments within EBSF_2.

More specifically, the EBSF_2 vision and expected outcomes can be considered able to provide a paradigm shift if compared to the more conservative positions of the Transferability Exercise and the two surveys respondents. At the same time, the trend to change towards more eco-friendly propulsion systems observed in the two surveys cannot be ignored if the goal is to develop advanced energy management systems. This is the reason why greatest importance is attached to the role played by eco-driving, saving strategies and the enhancement of Zero Emission Vehicle (ZEV) operations, which became central in a series of case studies. Therefore it is expected that:

- The effective strategies for energy and thermal management of buses, in particular auxiliaries may contribute significantly to an overall reduction in energy consumption for both ICE and battery-powered buses. The proposed demonstrators on HVAC will implement advanced technologies for the various components and will combine them individually. Improved HVAC systems that can use about 20% less energy are tested in different climate conditions.
- Specific driver assistance technologies, encouraging an energy saving driving style, may help drivers contribute substantially to the overall goals for energy efficiency. The EBSF_2 solutions aim to reduce the fuel/energy consumption between 5 and 8% (depending on the demonstrator activities). These technologies not only aim to save fuel/energy, but also to increase passenger comfort and hence the attractiveness of the bus system. The assisted driving will also help the ICE-powered buses in complying with the real driving emissions limits set by the Euro VI emission regulations, because of smoother driving and less transient duty-cycle of the vehicle, and resulting into a more efficient way to manage the crucial interoperability of the engine and the emission control system.

- The test of driver assistance system and HMI technologies compliance with IT standards will contribute to move from vertical solutions to fully interoperable ones, both at on-board and back-office level, for an easier integration of such systems in the existing fleets.
- Electric buses (thanks to clean, high tech appearance, low noise, zero emissions) can provide a challenging and appealing opportunity to design concepts and new functions for bus stops as well as address interactions between passengers, vehicles and urban infrastructures. The highest innovation potential identified in EBSF_2 is the indoor stop for electric buses implemented in Gothenburg which will be analysed in terms of users' perception and interplay between vehicle and infrastructure, such as the passageway between outdoors and indoors; this will be also an opportunity to measure how to ensure an 'indoor' climate and atmosphere, and study accessibility issues.

2.6 Selected case studies within EBSF_2

The activities of the demonstrators to pursue all of the above are complex, but a few examples are useful to describe the EBSF_2 testing scenarios.

2.6.1 Barcelona, boosting the efficiency of full-electric buses

The Barcelona demonstrator is one of the bus networks selected in EBSF_2 to test and validate technological solutions to boost the efficiency of full electric buses. Two 12 meters electric buses are already operating on routes crossing the city centre as part of the ZeEUS Project (Figure 2). The results already achieved within ZeEUS paved the way for more efficient operations currently about to be tested in EBSF_2, based on extended drive ranges and reduced overall energy consumption thanks to a multitasking demonstrator which includes two technological innovations: an energy predictive system and an eco-driving system.

The former is a system able to achieve up to 20% energy-savings for the auxiliaries, by implementing a "self-learning" approach (i.e., by learning and predicting energy demands in order to optimize the management of energy flows between the auxiliary systems and the energy storage unit).

Figure 2. The electric bus under test in Barcelona.

Figure 3. The hybrid bus under test in Lyons.

The consideration behind the test is that driving cycles of urban buses on specific routes are usually similar, therefore they might successfully put into operation an intelligent and adaptive system to optimize usual energy requirements. This is supported by the awareness that frequent loading and discharge of energy to and from the storage system increases the consumption and the energy demand for cooling system. Moreover, auxiliaries on full electric buses are powered and run independently from the driving styles.

Thanks to the predictive system, several online vehicle-parameters (including the position and the driving styles) will be collected, sent to the self-learning algorithm which in turn will provide revised parameters to steer the vehicles towards more efficient functioning of, among the others, the steering pump and the air compressor. This will also provide the opportunity to improve the efficiency of more components (sensors, compressor, touch panels, etc.). During the test, a driver assistance system will assess in real time the driving style so as to have the driver always informed about his/her driving efficiency. This means that quantitative data will be automatically collected by a control unit and transferred to the back-office. A small group of drivers will be selected and efficiency improvements will be measured for each of them, due to the high potential impact of driver behavior on energy consumption (Musso & Corazza 2015b).

2.6.2 Lyons, exploiting ZEV-mode options

The demonstrator in Lyons includes two main test fields still based on the binomial "energy management and ecodriving": the former relies on an energy management system on a hybrid bus (Figure 3) modified to operate a 2 km ZEV-mode extension on the route, and to mitigate the consequent detriment in the batteries lifetime.

The latter relies on a new approach to ecodriving, based on the use of a mathematical model which calculates in real-time a dynamic "fuel economy performance" index, taking into account the vehicle operating conditions (Blain 2015).

Figure 4. The testing environment in Lyons.

The test to exploit a 2 km ZEV mode (corresponding to the bus ride across the Croix Rousse tunnel opened in 2013 for buses, pedestrians and cyclists, as in Figure 4), manages the batteries State of Charge (SOC) efficiently and assess the impact on the durability of the batteries started in the second quarter of 2016.

More specifically, the test relies on a hybrid vehicle (Euro 6 Urbanway Hybrid) currently operating in the so-called "Arrive & Go" mode, which enables operations for 50 meter in ZEV mode. Modifications on the SOC management are required to force batteries charging operations ahead of the ZEV strip and perform the no emission ride accordingly. Bus stops along the route are simulated. The testing activities also require the preliminary training of the drivers and at least 5 running tests to have an average performance. Data are collected daily, in real time, to assess the most relevant parameters (for example, distance, speed, acceleration, amps, SOC and voltage of batteries, ICE parameters, vehicle weight, profile, position, weather conditions). One might argue that a ZEV 2 km strip would appear a rather modest performance if compared, for instance, to the virtually trebled ZEV mileage currently tested in Cagliari by a trolleybus line, within ZeEUS (Gasparini 2015). As a matter of fact, the focus of the test is not on the distance to cover but on the possibility to investigate how extended ZEV mode operations on hybrid vehicles can make them competitive with the full electric option, or eligible as a reliable alternative for operations in restricted areas of the city. One more field of investigation will include the assessment of the environmental potential: fuel consumption and air and noise emissions of the tested vehicle will be compared with those recorded by a regular hybrid bus.

Unlike Barcelona where the demonstrator is tested on the same vehicles and routes, or the local ZEV mode trial which involve just one bus, for the ecodriving system in Lyons a much larger fleet is deployed, due to the specific test features of this technological innovation. This ecodriving system is based on a real time algorithm which assesses fuel consumptions, and identifies driving styles which may cause excessive consumption. This requires three different test categories: a trial in a controlled environment (in which performance from a hybrid bus "as-it-is" will be compared to those coming from a test vehicle equipped with the ecodriving system); a test in real operational conditions (where the performance of a fleet of 40 diesel buses equipped with this technological innovation will be compared with those of a fleet of 40 vehicles operating without such equipment); and lastly, the validation of this technological concept, by comparing performance with and without the design of the selected IT architecture compliant with the EN13149 standard and EBSF project results which are now integrated into the IT×PT (Information Technology for Public Transport) platform (Tozzi, Guida & Knote 2014, Blain 2015).

2.6.3 *Gothenburg, improving energy management*

For the Gothenburg case, the newly-opened line 55 will be the arena to test performance due to three technological innovations, more specifically: a new air-to-liquid heat pump and integrated air conditioner on the roof, driven by electricity and bio-fuels that will improve the efficiency of heating by 30%; new buses (fully electric) with internal layout features such as large double doors, low-floor, and flexible/folding seats to increase accessibility, as well as new and light interior design and on-board WiFi; and an advanced design concept for bus stops including traditional outdoor facilities and an indoor one, previously mentioned.

Regarding energy strategies and auxiliaries, the demonstration will include simulation, laboratory tests, and a field trial. The simulations and laboratory tests will deal with investigations of the impact of vehicle designs on energy use by taking into consideration the number and type of windows, insulation, heated surfaces rather than heated air in compartment, as well as the effects of the number of passengers on-board. Comparisons

will be made between different solutions to assess the possibilities for storing and distributing heat, for combining heat and cold storage as well as testing solutions to achieve a distribution of temperature which is perceived as comfortable by passengers and drivers. A new heating system driven by electricity and biofuels (instead of diesel) will be tested (on one of the buses) in the field trial. It will consist of this innovative heat pump and integrated air conditioner on the roof of the vehicle. In addition the coolant liquid is to be heated with a heater combining 16 kW heating capacity from bio-fuel with 7 kW electrical heating while driving and 9.2 kW electrical heating in depot. A key to the improved performance is that the bus is pre-heated while standing in depot, either by hot coolant water or by electric power. Such multifold test plan will certainly enable a comprehensive measurement of the energy use for auxiliaries.

But in the case of Gothenburg, innovative findings are also expected to come from the research questions on how design of vehicles and facilities might affect the energy management. As said (and also already proved within EBSF) larger bus doors contribute to reduce dwell times, but in this case the interplay between bus design and bus stop is also expected to provide more accurate suggestions on how to speed up boarding and alighting operations (thus reducing idle times and preventing unnecessary periods with opened doors which affect the overall thermal comfort levels on-board, both causing excessive energy consumption) according to the direct observation of the passengers' behaviors. These issues will be assessed both by collecting data based on a number of participants boarding and alighting at bus stops and on-board, and by questionnaires (or short structured interviews) to drivers. Along with that, different bus stops with new layouts will be tested at different locations, including indoor areas (Figure 5).

The tests for these facilities will rely on real trials and field studies. Quantitative data (for instance the number of passengers accessing/

Figure 5.   The indoor bus stop in Gothenburg.

egressing each stop; dwell time calculations and average speed) will be collected through technical counting systems. Observations of travelers' behavior at bus stops (activities, positions, clustering, etc) will complete the collection of data and information. The questionnaires will also include queries aimed at assessing the passengers' appreciation due to improved on-board travel comfort, appearance, and increased quality of service and satisfaction (Tozzi et al. 2016, Musso & Corazza 2015b).

3 LEARNING FROM THE RESULTS

If results from the surveys and the Transferability Exercise are compared, the emphasis placed by the respondents on the economic side of operations seems to be a stronger and stronger criterion for the renewal of the fleets in Europe, and an observed poor attention to environmental issues in the EBSF Transferability Exercise may corroborate this.

But the reiterated interest within EBSF_2 for measures such ecodriving or to manage energy relies on the need to address the economic issue (i.e. reduce the operating costs) not by simply saving, but by innovating. Innovation, therefore, may be a driver affecting the overall cost structures from many sides. The EBSF_2 measures within the Energy Management and Driver Assistance PTs are directly targeted to reduce the fuel consumption and control the energy balance by acting on auxiliaries or operating ZEV modes. But improving the quality of maintenance operations, for example by making parking operations speedier at depots, or progressing with the IT back-office and on-board interoperability strongly contribute to improve the overall efficiency of the system. Last but not least, improved ergonomics and design of facilities and vehicles may speed up boarding and alighting operations, and more in general travel times with not negligible effects on the energy saving potentials.

That these innovations may equally address both the economic and the energetic issues is also evidenced by the progress in the assessment process. The evaluation process is on-going and a set of comprehensive KPIs is available to detect the performance variations. Evaluation categories mostly are focused on operations and labour, costs, energy efficiency, passengers' and drivers' perception. Demonstrators are currently selecting the most suitable ones, and it is not accidentally that the trend seems to assess the majority of the measures under the energetic and the economic points of view. A selection of some recurring KPIs, thus far selected, is reported in Table 1.

Table 1. A selection of Performance Targets and KPIs.

Performance Target	KPI definition	Unit
Improve energy efficiency of fleets	Average vehicle energy consumption per km	kWh/km
	Energy efficiency (energy used per vehicle)	MJ/vkm
Reduce energy costs per vehicle	Average energy costs per vehicle per 10,000 km	€/(vehicle* 10,000 km)
Improve auxiliaries management	Total average energy demand by auxiliaries per km	kWh/km
Mitigate air emissions	CO_2 emissions	g/vkm
Assess passengers' perception of on-board noise	Noise perception by passengers	Scale 1–10 (survey)
Decrease vehicle dwell time	Average dwell time / total travel time during the test	%

4 CONCLUSIONS

EBSF first and EBSF_2 now are continuously building a comprehensive knowledge of the drivers and barriers to increase the attractiveness and the efficiency of bus systems, by completing the development, test and validation of a set of promising technological solutions in real operational scenarios, and evaluating to which extent such innovations can contribute to improve cost and energy savings. Both also contribute to the progress in the development of some key solutions that would have a great impact on bus systems, but require more steps before wide acceptance on the market (one example for all is modularity). The success of EBSF is expected to be replicated by EBSF_2. To conclude, a set of Guidelines and Tools for the introduction of innovations will be derived from the experience of the demonstrators and it will pave the way to the introduction of the EBSF_2 technologies and concepts beyond the test fields.

ACKNOWLEDGEMENT

The authors wish to thank all the participants in the EBSF and EBSF_2 projects who actively contributed to the results above described.

REFERENCES

Beirao, G. & Sarsfield Cabral, J.A. 2007. Understanding attitudes towards public transport and private car:

A qualitative study. *Transport Policy*, Vol. 14, Issue 4, July, pp. 478–489.

Blain, L. 2015. *EBSF_2 Deliverable 5.1. Demo description and implementation plan*. Restricted document.

Carbone, D. & Proia, E. (ed). 2013. *EBSF Deliverable—Report on the Energy Efficiency of Bus System*. Restricted document.

Dobbie F. et al. 2010. *Understanding Why Some People Do Not Use Buses*. www.scotland.gov.uk/Resource/Doc/310263/0097941.pdf

European Commission, 2013. *Quality of life of cities*. Publications Office of the European Union, Luxembourg. http://ec.europa.eu/regional_policy/sources/docgener/studies/pdf/urban/survey2013_en.pdf.

Gasparini, P. 2015. *ZeEUS Deliverable 27.1—Cagliari: Demo Description and Implementation Plan*. Restricted document.

Musso, A. & Corazza, M.V., 2015a. Visioning the bus system of the future: a stakeholders' perspective, *Transportation Research Record: Journal of the Transportation Research Board*, No. 2533, pp. 109–117.

Musso, A. & Corazza, M.V., (ed.), 2015b. *EBSF_2 Deliverable 2.2. Report on test scenarios and validation objectives*. Restricted document.

NRBTI 2009. *Quantifying the Importance of Image and Perception to Bus Rapid Transit*. www.nbrti.org

Stradling, S. et al. 2007. Passenger perceptions and the ideal urban bus journey experience. *Transport Policy*, Vol. 14, n. 4, pp. 283–292.

TomTom International 2013. *TomTom Congestion Index 2013*. http://www.tomtom.com/lib/doc/congestionindex/2013–0322-TomTom-CongestionIndex-2012-Annual-EUR-mi.pdf

Tozzi, M., et al. 2016. A European initiative for more efficient and attractive bus systems: the EBSF_2 Project, *Transportation Research Procedia*, Vol. 14, pp. 2640–2648.

Tozzi, M., Guida, U. & Knote, T., 2014. *3iBS—the Intelligent, Innovative Integrated Bus Systems*. Proceedings of the Transport Research Arena 2014, Paris.

UITP—Union International des Transports Public 2011. *Position Paper – A comprehensive approach for bus systems and CO2 emission reduction*. http://www.uitp.org/sites/default/files/cck-focus-papers-files/20111122_UITP%20EU%20position%20paper_comprehensive%20approach%20for%20bus%20systems%20and%20CO2%20emission%20reduction.pdf

Transport Infrastructure and Systems – Dell'Acqua & Wegman (Eds)
© 2017 Taylor & Francis Group, London, ISBN 978-1-138-03009-1

Waterfront and sustainable mobility. The case study of Genoa

P. Ugolini, F. Pirlone, I. Spadaro & S. Candia
DICCA, University of Genoa, Genoa, Italy

ABSTRACT: The paper analyses European port cities. Port cities are cities which grow up in close connection with their ports. Over the years, ports influenced cities development becoming the main driver for urban sprawl. Nowadays many port areas are no more exploited for port's trade. Several cities used these spaces designing modern waterfronts for leisure, culture and tourism activities. Waterfront revitalisation is also fundamental for urban mobility. Port cities can use these areas to develop new transport infrastructure promoting sustainable mobility. In a densely built up area it is increasingly difficult to find space for bicycle lanes or pedestrian zones, waterfront revitalization projects can be the perfect occasion to solve this problem. This paper analyses the case of one important port city in the Mediterranean: Genoa. This analysis is necessary to define new forms of mobility and transport promoting sustainability inside port cities' centres taking often advance of port's abandoned areas.

1 PORT-CITIES

Port cities are fascinating. They are a combination of many different cultures, architectural styles, and people. Port and city interact in many ways and one is essential for the other. The intrinsic complexity of the port city has drawn attention from a vast number of scholars belonging to a variety of scientific fields (Ducruet C., 2011).

Port cities are cities which grow up in close connection with their ports. These two entities have been developed together over the years. The presence of a natural harbor or bay attracts boats and sailormen, and merchandises and good attracts merchants: in this way the first urban settlement is established around the main market.

Different from port cities, it is the case of a city with a port established a long time after the city foundation. This port is the result of the necessity of a city to trade and it is normally settled far from the old town centre. On the contrary port cities are all built around a natural harbor.

The Mediterranean Basin has been the cradle of world civilization since the first settlements in Jericho (Israel West Bank) and port cities were fundamental for Europe development. In the Meddle age port cities were centers of primary importance for trade of merchandises and knowledge. Some of the most flourishing and prosperous towns of the XIII century were port cities, such as Venice, Istanbul, Pisa and Genoa. All around the main harbor were established new small ports strictly connected to the first trade center. Also the hinterlands, in the proximity of a big port city, reported a fast rate of development. The majority of activities carried out in these villages were related to port trades. The opening of the Mediterranean trading area coincides with the first crusade which revived trade with the Near East interrupting the Muslim predominance on the eastern Mediterranean. Crossing the Persian Gulf and the Red Sea, Arab caravans brought luxury goods to Egypt and Israel and from the port of Alexandria, Acre and Joppa sailors from Venice, Genoa, and Pisa transported the goods to Italy on their way to the markets of Europe. The easiest way to reach the north from the Mediterranean was by Marseilles, getting back on the top of the Rhone valley. Thank to this affluence of goods the *Italian Repubbliche Marinare—* Maritime Republics, established their own "State" in the Mediterranean shores. In each port city the merchants organized themselves into guilds, becoming richer and more influential in social and political every day life. The extent of trade and urban life were really connected. At the same with the ending of Viking and Barbarian attacks a north trading route was opened, from the British Isles to the Baltic Sea. The center of this northern maritime traffic was the county of Flanders.

Medieval port cities were really compacts: the center with the main market was always established immediately behind the principal harbor and the built-up area was surrounded by high walls to protect its inhabitants against enemy raid. There were three characteristics that marked ports: in the first place, ports had harbors that were the center of the movement of people and products; secondly, the urban morphology of ports always had particular buildings or spaces that dominated the city, such as dockyards, warehouses, customs houses, open markets, inns and pubs; finally, ports could also be identified by the particular socio-economic groups that they sheltered such as merchants, bankers, bookkeepers, shopkeepers, shipbuilders and foreigners (Antunes C., 2010).

The Mediterranean supremacy started to crack with the discovery of America. The main trade moved on the Atlantic cost facilitating the establishment of new port cities in Portugal and in Great Britain. The Mediterranean Sea became a peripheral basin of commerce, because the main maritime routes where no more between Europe and the close Middle-East, but between Spain, Portugal, France and England and the new colonies settled in America, Africa and Asia. European colonial powers would be the first to establish a true global maritime trade network from the 16th century: most of the maritime shipping activity focused around the Mediterranean, the northern Indian Ocean, Pacific Asia and the North Atlantic, including the Caribbean (J.P. Rodrigue, 2013). With the opening of the Suez Canal the second half of the 19th century saw an intensification of maritime trade to and across the Pacific and inside the Mediterranean Sea but here boats normally were directed to England crossing the basin.

Port cities have played a pivotal role in each Industrial Revolutions; ports have thrived because all Industrial Revolutions up until now relied on the trade of raw materials: coal and iron ore since the 18th century, whereas oil and gas were added to the commodity mix since from the late 19th century (Jansen, M. 2016). After the second industrial revolution, many port cities renovated their structure to host new factories handling coal, ore, oil, oil products, and chemicals. As for the Middle Age, ports were the driver for urban development. Many people moved to port cities from the countryside to work in the factories. Port cities significantly incremented their population and it was necessary to establish new settlements for workers and their families outside the city wall. Since the 80 s of the XXI century the spread of container transport obliged port cities to build new and modern docks for container ships and cargo containers. These port infrastructures require much more space than the previous docks to storage both loaded and empty containers. Normally loaded containers are stored for short periods then they are sent to northern Europe by train, trucks and barges, on the contrary empty containers can remain on the docks for longer periods awaiting the next use. The container terminals were often set up far away from the original/historic port with good rail or highway connection. Because of containers trade many old ports were abandoned or underexploited and their infrastructures (merchandise stores, cranes, quays…) began to degrade. But the

city and the port were no more a single entity but two different realities developing one close to the other with few points of contact. Many barriers—fences, rails, walls, port police, and customs service … -were placed between the port and the city dividing even physically port cities. The port and the city were facing away from each other developing their infrastructure separately. Huge highway and rail crossroads were also built to get easier the container loading and unloading and to facilitate the access to port for wagons and trucks. All these elements contributed to the widespread of the idea that ports are modern walled city that make the access to the sea impossible for port cities' citizens. This physical separation compounded the negative effects of isolation and of insecurity also linked to the development of criminality. -the common belief considers ports as danger zone were criminality is not sufficiently controlled. This felling of insecurity is another element which contributed to port and city split.

The situation has completely changed at the end of the 20th century: different old ports were renovated and revitalized and many efforts have been made to re-establish the relationship between the port and the city.

In the wide and complex panorama of urban transformation, waterfront revitalization is one of the most interesting phenomena of urban renewal of the last decades, bringing 'cities on water' around the world to a new leadership (Giovannazzi O., Moretti M., 2010). Many big cities worked towards water to offer a better quality of life regenerating people's space and respecting the environment. Waterfront and riverfront after New York and San Francisco's experiences are now of central importance for urban planning in many international cities such as Buenos Aires, Shanghai, London, Barcelona and Porto. These port cities are good examples, best practices of urban renewal increasing the quality of life starting from degraded port areas. This was possible paying particular attention to the design of new public spaces, transports, residential area respecting the environment. The cruise industry had more benefits than other economic sectors, from port regeneration offering new activities for tourists near the sea. The food chain (local products and tradition) also had its place in the renewal process. Many historical markets closed to old ports were renovated thanks to waterfront revitalization projects and now are points of interest for tourists and citizens.

As maritime trade flows within and between ports change, the recognition of new values of the waterfront, including through tourism, have spurred projects to attract attention, investment, and people: Opportunities to improve quality of life and environment and enhance economic vitality have led to the revitalization of degraded and underutilized port regions, often through strategies that incorporate their rich historical and cultural heritage to create unique new public spaces (Ivinis C., 2013).

Waterfront revitalization is also fundamental for urban mobility. Port cities can use old ports' areas to develop new transport infrastructures promoting sustainable mobility. In a densely built up area it is increasingly difficult for urban planner to find space for bicycle lanes or pedestrian zones. At many locations, it might appear complicated to provide for on-road cycling in a way that satisfies the needs of bicycle riders and other road users and waterfront revitalization projects can be the perfect occasion to solve this problem. The European White paper of transport—Roadmap to a single European transport area, towards a competitive and resource-efficient transport system (2011) – claimed the necessity of a competitive and sustainable transport system across EU Countries reaching the 60% emission reduction target. Waterfront renewal can contribute to reach this goal. One of the first port cities which decided to renovate its port in favor of sustainable transport was Portland in Oregon. In 1974 Portland tore down a portion of the Route 99 W freeway to reclaim the waterfront for a pedestrian/cycling seawall and park. Recently the city of Boston even defined the "South Boston Waterfront Sustainable Transportation Plan" (2015). The Sustainable Transportation Plan lays out a blueprint for transportation and public realm improvements for the South Boston Waterfront over the next two decades (Boston Redevelopment Authority, 2015). The main goals of the plan are: Improve Access and Mobility for All (improve multimodal access to/from/through and mobility within the South Boston Waterfront for residents, workers, maritime-related commerce, and visitors); Support Economic Growth and Vitality (deliver the transportation infrastructure needed to support a world-class economy); Reinforce Sustainable Policies and Programs (align programs and policies to support more sustainable transportation choices and demand management to and within the South Boston Waterfront); Enhance the Public Realm (contribute to enhancing the attractiveness and quality of the urban character through ongoing transportation investment); Contribute Environmental and Health Benefits (Realize the positive environmental effects and health benefits that result from a more sustainable transportation plan. Invest Smartly for the Future); Advance strategic investment to ensure the long-term financial and operational sustainability. The plan for Boston waterfront is good example, a best practice, which could be transferred to other port cities in Europe. This paper analyses the port city of Genoa proposing new solutions for sustainable mobility inside the waterfront area and the old town centre.

2 THE PORT CITY OF GENOA

Genoa is one of the most famous examples of port-city. The port of Genoa is beautifully located between the Ligurian Sea and the Apennine Mountains. It is the major Italian seaport with a trade volume of 51.6 million tons of cargo.

Genoa is one of the main port cities of the Mediterranean; in truth, the northern most port of the Mediterranean Sea. This position can provide remarkable advantages for the city, which can aspire to become a juncture between Europe, Africa and Asia Minor (Gabrielli B., 2010).

Genoa has strongly been connected with its port since the beginning of its history. For years the city and the port lived together in a state of interdependence and the local community based their identity on the relationship with the sea and sailing, in a prosperous union.

The extraordinary shape of seaside cities—that we can better appreciate from the sea—is the result of the synthesis between urban culture and maritime culture; this harmonious union of maritime and urban cultures offers a different point of view that is full of charm and semantic values (Clemente M. 2013).

Genoa is a typically port city. The first settlement, dated 6th and 5th centuries BC, was established on a hill (called Castello) controlling a natural harbor. The port was created according to the nature of the site and the city developed around the bay. Thanks to the prosperous commerce with the Middle East, in the XIst century Genoa emerged as an independent city-state. In the Middle Age, Genoa held one of the largest and most powerful navies in the Mediterranean financed with trade, shipbuilding, and banking. The commercial routes along the Mediterranean coast enriched enormously the town which became one of the most influent port cities in the Mediterranean with Venice, Pisa, and Amalfi, the so-called Maritime Republics. In the XIIsd century was built the first port according to an established project innovating the pre-existing settlement. Legal and administrative means guaranteed the maintenance and the development of port's infrastructures. The old town centre was characterized by the presence of copious markets and there were also entire streets or arcades (for example the present Sottoripa—ex Ripa Maris) consacrated to the trading of goods. For this reason, Genoa was called both a port and a market city. The main markets of S. Giorgio, Banchi and Soziglia were placed on the principal streets which connected the city with the port. The markets were the real point of contact between the city and the port, the physical place where merchandises, people, ideas and culture melted together. S. Giorgio and Banchi are located perpendicularly to Sottoripa on the same axis; Soziglia is rearward in comparison to the first two in the middle of an artisanal district.

Every day were sold textile products coming from the north of Europe, spices from the Middle East and the refined silk coming from china. During the Middle Age Genoa could count on its colonies settled in Africa (Tabarca), in Crimea (Caffa, Cherson, Cembalo…), in the Caucasus region (Copa, Mapa, Casto…) and in Anatolia (Amasra, Carpi…) to trade metals, stones and other precious goods. The urban fabric reflected port configuration, indeed the roads were built in correspondence to the main quay to get easy the loading and unloading of merchandises.

Since the 1970s the predominance of the iron and steel industry and of the containers traffic separated the port of Genoa from the city. Huge highway crossroads and fences were built separating port activities from Genoa daily life. In particular, it was impossible for Genoa citizens to have any visual contact with water. Maritime traffic affects only container terminals set up in the west part of the city far from the old port that was progressively abandoned. In 1985 the remaining activities finally moved and the Municipality of Genoa formed a Commission, in cooperation with the Region Liguria and the Port Authority, to study all the possible solutions to connect again the old port with the old town centre. In 1992 Genoa has a unique opportunity to renovate its waterfront thanks to an International Exhibit hosted inside the old port as part of the Columbus Celebration (for the 500 years since the discovery of America). State investments helped the city to afford the restoration of the majority of the old port area. Waterfront revitalization included the recovery of almost all the historic docks, the rearrange of the open space in a big square and promenade along the sea and the realization of a big aquarium, a panoramic lift and an ice rink. The old port was permanently converted to urban uses, in particular for leisure, culture and tourism activities. The waterfront regeneration processes have also concerned the relationship between the city and the port.

The restoration of the waterfront and the maintenance operations in the historic centre have led residents to a change in their perception of their own city; the re-ownership of the waterfront and the notable quantity of historical places recovered, but also the reconversion of large disused industrial sites through the introduction of new uses, have been translated in terms of growing sense of belonging and have reinforced elements of identity

through the rediscovered use of collective and representative functions at a metropolitan level. In addition, the overall image of the city is seen from the outside not only as more linked to the harbor and the industry (often seen in a static or decadent way), but also as a new element linked to art culture, architectural beauty, and new form of acceptance (in a dynamic and more propulsive vision) (Gastaldi F., 2014).

Other international events (the G8 meeting in 2001 and the European City of Culture in 2004) provided resources for regenerating other portions of the old port and of the old town centre. Nowadays the working port of Genoa is almost all set in the districts of Sampierdarena and Voltri, far away from the city centre. It covers an area of about 700 hectares of land and 500 hectares on water with 30 km of operative quays. It is incredible wide covering around 22 kilometers of coastline. The cruise terminal instead is located immediately westward of the Old Port. This is a privileged position because cruise tourism has a direct access to waterfront leisure activities. Further interventions restored other port abandoned areas and buildings: on the site of the medieval Arsenals were built the Sea Museum and the new Faculty of Economics. Actually the urban waterfront covers an area of 117.000 sq m—of which 86.000 sq m are open spaces—and it is made up of: 23% Culture, permanent exhibition and museums; 20% Congress Centre; 17% Leisure, education; 15% Facilities and parking lots; 12% Offices; 7% Restaurants and bars; 6% Retail.

In the last months, the Municipality of Genoa has decided to open an international ideas competition to renovate the last portion (Figure 1) of the port inside the city center starting from a project donated to Genoa by the Renzo Piano Building workshop.

The overall project—called "Bluprint for Genoa"—outlines the profile and the future functions of the part of the city located near the east

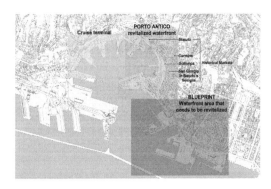

Figure 1. The waterfront of Genoa. The port, the city and the historical markets.

port entrance: the challenge is to combine tourism with the industrial activities situated in the neighborhoods, where companies of excellence work on ship repairing industry and refitting and where are foreseen other important projects.

The part of the waterfront designed by the Blueprint shows some strong features: a central position; the presence of two docks with more than 400 moorings; a big pavilion for events designed by Jean Nouvel; the historical "Palasport"—a structure with a great potential—and a close promenade, "Corso Italia", for leisure activities and bathing. The requalification of areas of the former Trade Fair gives to Genoa the possibility of realizing an essential tile to complete the waterfront restyling started in 1992 by Renzo Piano with the renovation of Porto Antico, today one of the main touristic attraction and livable places in Genoa. The guiding principle of the project is defined by Piano as a "mending" of the urban pattern, the win back of the sea by the city and the reopening to the water of part of land reclaimed from the sea (the canals).

3 SUSTAINABLE MOBILITY FOR THE WATERFRONT OF GENOA

In this paragraph are reported different researches on the waterfront of Genoa. In particular, innovative solutions are here analyzed for a sustainable mobility inside the old port area. A better urban mobility Plan is of paramount importance to overtake the remaining barriers between the port and the city. Port cities can use these areas to develop new transport infrastructure promoting sustainable mobility. It is increasingly difficult to plan bicycle lanes, new bus lines or pedestrian zones in a densely built up area. For this reason, waterfronts are an extraordinary opportunity for urban planners to have new areas to dedicate to sustainable mobility.

3.1 *Genoa as port-market city*

Genoa is not only a port city but can be considered as a port-market city. This is a result of the fact that the urban pattern is marked by the presence of many markets and a central port. The authors have developed a detailed study on market cities within the European project "Marakanda Historical markets" (2012–2015) funded with the European Neighbourhood and Partnership Instrument—ENPI CBCMED (scientific responsible: P. Ugolini and F. Pirlone). Marakanda brought together 10 direct partners from six different Mediterranean Countries: Municipality of Florence (Italy, Toscana); Municipality of Genoa (Italy, Liguria); Local Authorities Union of Xanthi District (Greece,

Anatoliki Makedonia—Thraki); Municipality of Limassol (Cyprus); University of Genoa—Research Centre in Town Planning and Ecological Engineering (Italy, Liguria); Municipal Institute of Markets of Barcelona (Spain, Cataluña); PLURAL—European Study Centre (Italy, Liguria); Municipality of Favara (Italy, Sicilia); National Research Centre (Egypt); Souk El Tayeb (Lebanon).

The project collected and exchanged best practices to enhance competencies and capacities of small entrepreneurs operating in city markets. The partnership promoted the integration of productive chains strategies at Mediterranean level with focus on high quality agro-food and handicraft products. Local authorities and private stakeholders were both involved in the revitalization process to guarantee a sustainable management of historic city markets. The participative approach proposed by the project firstly involved the local dimension in order to ensure the effective participation and territorial relevance of the initiatives carried out. Marakanda involved essentially local communities by working with them on common aspects of the day to day life and ordinary running of the economy and markets.

Marakanda established also common quality standards for the efficient management of historic city markets. Particular attention was given to the connection between markets, ports and old town centers. Sustainable mobility was preferred and the project supported pedestrian walkways and cycle routes. A better connection was of fundamental importance to revitalize city markets and to reestablish the historical relationship between the city with its markets and port. These elements lived together in a state of interdependence and markets renewal was necessary to restore a prosperous union. A good connection is also important for tourism. Markets can be a veritable tourist attraction promoting local foods and traditions. Pedestrian walkways which connect old town centres, waterfront and markets are for this reason of paramount interest.

In figure 2 is reported a study which analyses every possible sustainable connection between the historic city markets, the waterfront and the old town centre of Genoa.

New pedestrian walkways are identified to get easy for citizens and tourists to move around the city. Cruise tourism could also benefit from these interventions: from cruise terminal tourist can follow a safe and pleasant itinerary to discover Genoa historical buildings.

3.2 *The old town centre and the port of Genoa: identification of new itineraries*

The aim of this research is to enhance the relationship between the port and the city of Genoa

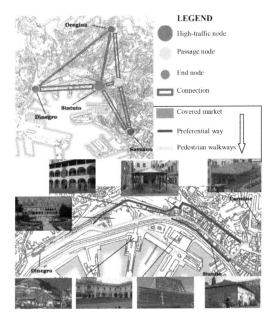

LEGEND

- High-traffic node
- Passage node
- End node
- Connection
- Covered market
- Preferential way
- Pedestrian walkways

Figure 2. Identification of the main roads and ways between the historical markets of Genoa: Dinegro, Statuto, Carmine and Sarzano.

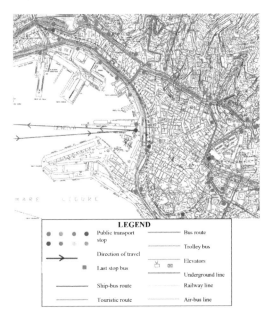

LEGEND

Public transport stop	Bus route
Direction of travel	Trolley bus
Last stop bus	Elevators
Ship-bus route	Underground line
Touristic route	Railway line
	Air-bus line

Figure 3. Analysis on public transport inside the waterfront area.

supporting sustainable mobility. Waterfront revitalization is analyzed to a larger scale including the old town centre and the historic markets inside a more complete reclamation project. This is because throughout history, port, old town centre and city markets were a part of a single productive system. Now these elements could interact together again preferring the realization of public spaces for leisure, culture and tourism activities and directional centers.

The authors have done a detailed study on urban mobility inside the new waterfront and the old town centre of Genoa (from the cruise terminal—Piazza Fanti d'Italia—to Mura della Marina) in favour of sustainable mobility. In particular, it has been analysed the public transport network. In this area there are 3 bus lines, a trolleybus, different public elevators, one underground line and the main railway station (Figure 3).

On the whole the public transport system is sufficient and it covers the entire area. Different problems are caused instead by car traffic. In this case the introduction of interchange station parking could push drivers to use public transportation. There are different users that walk inside the study area, mainly students, tourists and workers. There are many pedestrian walkways both inside the old town centre and the waterfront. These paths are generally safe inside the waterfront, but the same

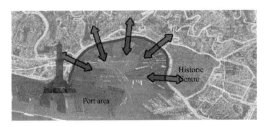

Figure 4. The outline of the possible connections between the port and the city.

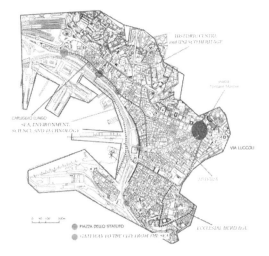

Figure 5. Touristic routes (pedestrian and cycling areas) inside Genoa waterfront and old town centre.

cannot be said for many pedestrian streets inside the old town centre that, especially at night, can be dangerous.

For this reason of primary importance is to create new safe vertical connections between the waterfront area and the historical centre (Figure 4) to get stronger the relationship port—city.

The most critical area is the old district of Prè. Here the waterfront is completely separated from the city by a physical barrier, the "*sopraelevata*" an elevated road build in the '60 s. This isolation condemned the main road of the district—via Prè—to become a degraded and unsafe place. This road, if better connected to the rest of the old town and the port, can become a market street to promote local products with a touristic interest. In Figure 5 are reported different sustainable ways that could connect the study area preferring pedestrian walkways and cycle tracks.

All the old town centre's districts can attract different users according to their vocations: night life, culture, wine and food shops and markets, ... all moving in a sustainable way.

3.3 *Conclusions*

In conclusion, waterfront revitalisation is of paramount importance to relocate in the international context port cities and to improve the urban quality of life. It is preferable a larger-scale reasoning which includes inside the project area appropriate connections between the port, the city and its historical markets in favour of sustainable mobility.

The concept of sustainability has been applied to port services and infrastructures revitalizing historic port areas. Several port cities such as Barcelona, Genoa and Marseille are reclaiming, or reclaimed, their waterfront from freeways and industry. These cities are using ports' abandoned areas designing modern waterfronts for leisure, culture and tourism activities because they have realized that public access to water—for the simple pleasures of walking, cycling, beaches, boating, etc. –is a necessary asset for any city that calls itself livable.

Urban planners and policymakers, pursuing waterfront projects, conceive innovating approaches for interlinking socio-economic and environmental dimensions. Public and private engagement is a precondition for project success, so it is preferred a joint cooperation since the planning phase. Waterfront renewal has in many cases a catalytic effect becoming the starting point for the revitalization of the entire city. Ports are a perfect place for urban experiments because they are located in strategic areas, close to city centers and transports routes. Sometimes this restoration process transforms an unused port in one of the most dynamic parts of the city, a driver for new interventions and for future restoration projects. This is because many town planners consider old ports as a driver for economic development capable of attracting the interest of international investors. Particular attention is given to public spaces and sustainability. The aim is to crate open and permeable places, lived in and felt to be a part of the city. New waterfronts are more enjoyable, fostering relations, displays, development and enhancement in the fields of leisure, sport, mobility and culture. Port cities look to the future transforming post-industrial areas and buildings into new production areas.

Genoa is an evident example of this revitalization process. Its renovated waterfront—the old port—interacts with the city and it is a magnet for resources and fluxes capable of exploiting opportunities to generate new economies. But there is still a lot of work to do: this paper suggests new itineraries and mobility solutions to carry on the waterfront revitalization enhancing the current relationship between the city, the port and its historical markets and supporting sustainable development.

REFERENCES

Artuso, N. & Salvetti, L. 2013. Verso il completamento del waterfront di Genova, dagli anni '90 ai programmi futuri, *PORTUS: the online magazine of RETE*, n. 26, November 2013, Year XIII, Venice.

Boston Redevelopment Authority 2015, *South Boston Waterfront Sustainable Transportation Plan*, VHB, USA: Boston.

Busi, R. & Pezzagno, M. 2006. *Mobilità dolce e turismo sostenibile: un approccio interdisciplinare, Gangemi Editore*, Rome.

Cátia Early, A. 2010, *Modern Ports: 1500-1750*, Institute of European History (IEG), ISSN 2192-7405, Germany: Mainz.

Carnevali, G. & Delbene, G. & Patteeuw, V. 2003. *Developing and rebooting a waterfront city*, NAi Publishers, Nederland: Rotterdam.

Clemente, M. 2013. Sea and the city: maritime identity for urban sustainable regeneration, *TRIA International Journal on city planning culture11* (2/2013) 19–34/print ISSN 1974-6849, e-ISSN 2281–4574.

European Commission 2011. *The European White paper of transport—Roadmap to a single European transport area, towards a competitive and resource-efficient transport system*. Commission services, Belgium: Bruxels.

European Commission 1992. *Green Paper: Towards a new culture for urban mobility*, Commission services, Belgium: Bruxels.

Fera S., Minella M. 1999. *Genova porto e città*, Sagep, Genoa.

Gabrielli B. 2010. The Renaissance of Cities: the Case of Genoa, *PORTUS: the online magazine of RETE*, n.16, June 2010, Year XIII, Venice.

Gabrielli B., Ghiara H. 2013., Genova, dal waterfront alla città portuale del XXI secolo, *PORTUS: the* online *magazine of RETE*, n. 25, June 2013, Year XIII, Venice.

Gastaldi F. 2014. Waterfront redevelopment and gentrification in the inner city of Genoa, *PORTUS: the online magazine of RETE*, n. 30 Year XIV, Venice.

Genoa Port Authority 1999. *Piano, porto, città. L'esperienza di Genova.* Skyra Editore, Milan.

Genoa Urban Lab 2008. *Quaderno 1*, ECIG Edizioni Culturali Internazionali, Genoa.

Genoa Urban Lab 2011. *Quaderno 2*, ECIG Edizioni Culturali Internazionali, Genoa.

Ivins C. 2013. *Waterfront Revitalization in Port Cities: Risks and Opportunities for Emerging Economies*, Brics Policy Centre, Policy brief publication, Brazil: Rio de Janeiro.

Jansen M. 2016. The Fourth Industrial Revolution and the future of ports, World Economic Forum, Switzerland: Geneva.

Municipality of Genoa 1995. *White book on Genoa mobility*, Mobility and Transport department, Genoa.

Municipality of Genoa 2012. *Urban mobility plan for the city of Genoa*, Mobility and Transport department, January 2010 updated with C.C. n. 28/2012, Genoa.

Ribalaygua B., Latorre E.M. 2014. Port and City. An approach to the study of new urban port role from its perception as node and place, *PORTUS: the online magazine of RETE*, n. 27, May 2014, Year XIV, Venice.

Rodrigue J.P. 2013. *The Geography of transport system*, ISBN 978-0-415-82254-1, Routledge Edition, USA: New York.

Schubert D. 2013. L'ultima frontiera del waterfront rigenerazione urbana: Nord Europa, *PORTUS the online magazine of RETE*, n. 25, giugno 2013, Anno XIII, Venice.

Troin J.F. 1997. *Le metropoli del Mediterraneo. Città di frontiera, città cerniera*, Jaca Book, Milan.

Notes

Selena Candia is responsible for the researches on waterfronts and on cycle mobility as preferred sustainable mean of transport inside the port city of Genoa.

Francesca Pirlone is responsible for port cities general description and she carried out different researches on sustainable mobility for the city of Genoa.

Ilenia Spadaro is responsible for a detailed study on the historical evolution of port of Genoa.

Pietro Ugolini is responsible for the researches on Mediterranean port cities and on Mediterranean historical markets.

Transport Infrastructure and Systems – Dell'Acqua & Wegman (Eds)
© 2017 Taylor & Francis Group, London, ISBN 978-1-138-03009-1

New forms of mobility for an alternative territorial fruition: The rediscovery of tourist footpaths

R. Papa & R.A. La Rocca
Department of Civil, Architectural and Environmental Engineering (DICEA), University Federico II of Naples, Naples, Italy

ABSTRACT: The search for modalities of movement inspired by the sort of "soft mobility" that can be an alternative to the use of a car represents an opportunity to promote sustainability in territorial use and its fruition. In this sense, the definition of methods and tools aimed at aiding decision-makers in identifying policies and technical interventions for both the promotion of tourism and the safeguarding of the existing system of resources is one of the main targets of town planning. The enhancement of existing cultural paths, as well as the realization of new foot routes, constitutes an opportunity for tourism development and territorial sustainability. In the context of these premises, this study represents a proposal to characterize a network of "cultural tourist paths," also considering the application of GIS technologies in supporting both the planning and use of such tourist routes.

1 CULTURAL TOURIST PATHS AS "COMMON HERITAGE"

The use of soft mobility is currently one of the primary agents of territorial fruition, especially for those who put particular emphasis on environmental issues and sustainability. The focus on a developmental model of conscious tourism that favors a "slow" model has concentrated attention on the current national and European policies which aim for the revaluation of cultural paths, as well as a renewed interest in historical, environmental, cultural and territorial identity.

In Italy, the recent directive from the Ministry of Heritage, Culture and Tourism (2016) in establishing the *year of the routes*, underlines the role of historical-cultural paths both for the enhancement of territorial resources (even economic ones) and for the promotion of sustainable tourist typologies.

The Directive, in particular, mainly emphasizes two aspects. On the one hand, it points out the need for in-depth examination of the state of the existing routes capable of being adapted for the soft mobility manner of moving, in order to create an Atlas of Routes. On the other hand, it highlights the exigency of the different bodies (Region, Universities, Research Institutes, Associations, etc.) engaged in territorial development operating in a collaborative vision in order to promote "slow tourism" as a user typology compatible with the valorization of historical and cultural identities of the territories through which the paths cross.

The Directive, then, also wants to be a tool for supporting and fostering the less well-known regions by encouraging tourist typologies different from the mass tourism model. For these reasons, "cultural tourism" (WTO 1985; ICOMOS 2002, Richards 2003) has been identified as the form of tourism most adequate for the use and promotion of this kind of territorial resource.

Even though there is not yet a uniform definition of cultural tourism, it is considered to be the form of tourism that favors territorial cultural aspects, in its rediscovery of the primitive educational function of the journey, intended as an experience, as it was for the nineteenth-century model of the *Grand Tour*.

If the identification of cultural tourism as a sustainable typology of tourism is a recent acquisition, the European project to activate a cooperative network by the revaluation of cultural routes dates back to the Sixties, when the European Council certificated the first two historical routes: the Way of Saint James and the Rural Habitat.

The first example is perhaps the best known among European routes; it has been the object of literary and cinematic interest, as well as historical, cultural and religious. Set up in the ninth century, as a route for pilgrims to reach the grave of Saint James, at present, the Way of Saint James represents an icon for cultural tourism. Data on fluxes in tourism, in fact, show how cultural motivation has increased in recent years (Table 1), despite religious motivation remaining the main impetus for

Table 1. Types of motivation for traveling the Saint James Way 2012–2015.

Motivations	2012	2013	2014	2015
Cultural and religious	101.171	117.715	120.412	141.969
Only religious	79.490	86.291	101.013	99.680
Only cultural	11.827	11.804	16.461	20.809
Total	192.488	215.810	237.886	262.458

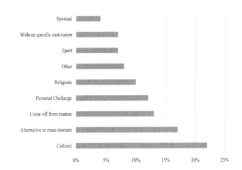

Figure 2. Percentage of types of motivation for walking the Via Francigena in Italy (elaboration from TCI 2015).

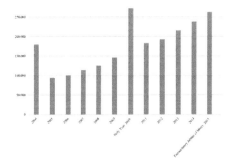

Figure 1. Number of pilgrims in the period 2004–2015 (elaboration from www.prgrinossantiago.es consulted on 08/08/2016).

the flow of tourists, with numbers growing significantly on occasions of holy events (Fig. 1).

A recent survey carried out by the Italian Touring Club in 2015, using a sample of 400 tourists, gives the main reasons that users indicate for having decided to retrace the medieval Via Francigena (Fig. 2).

In particular, it can be noted that religious motivation inspired only 10% of the sample, who were mainly driven by a cultural reason (22%). On the other side, the Habitat Rural paths refer to a system of routes characterized by the presence of particular, predominantly rural ecosystems identifiable on the basis of specific local features (special craft production and/or gastronomic interest, specificity of the natural landscape, etc.).

The Program of the Cultural Routes of the Council of Europe is based on the idea that the valorization of leisure through culture can activate cooperation among different countries. The wandering journey, then, can represent the means to rediscover heterogeneous heritages still interconnected within common historical and cultural roots.

In this sense, the path can be assumed to be a shared "common cultural heritage," able to link different places through the physical and immaterial resources of the lands across which it runs.

Since 1997, the European Institute of Cultural Routes (EICR) has been the technical body of the Council of Europe. The Institute evaluates already

existing cultural routes and supports new projects for enhancing territorial resources. At present, there are thirty-two certified routes representing different, interesting geographical regions and referring to different themes inspired by European history and culture.

In particular, the classified routes can be clustered into nine macro-categories:

- pilgrimages (the Way of Saint James since 1987; the Via Francigena, the Saint Olav);
- European celebrity (European Mozart Ways; the Schickhardt Route; the Saint Martin of Tours Route; the Huguenot and Waldensian trail; the Via Habsburg; in the Footsteps of Robert Louis Stevenson; Destination Napoleon; European Routes of Emperor Charles V);
- monasteries (European Route of Cistercian abbeys; the Cluniac sites in Europe);
- historical and architectural heritage (Routes of El legado andalusí, Transromanica; European Cemeteries Route; ATRIUM; Réseau Art Nouveau Network; the European Route of Megalithic Culture);
- rural landscapes (Routes of the Olive Tree; Iter Vitis)
- trade lines (Phoenicians' Route; the Viking Routes; La Hanse);
- industrial and craft lines (The Pyrenean Iron Route; The European Route of Ceramic);
- prehistoric art (Prehistoric Rock Art);
- specialized towns (European Route of Historic Thermal Towns).

The Cultural Routes of the Council of Europe express the will of a shared and cooperative project involving all political and administrative levels (from local to national to transnational), as well as the local population, in development of the program. They also represent a project of cooperation between cultural, educational, and tourism interests aimed at promoting European values.

Among the European countries, Italy is one of the most engaged in the Program of routes (Fig. 3). In fact, the relevance of the historical routes for tourism in Italy has also been recognized at the institutional level, particularly in the current government's program focused on the enhancement of the Via Francigena and the Route of the Phoenicians.

It should be noted that the Program of Cultural Routes, rather than focusing on economic objectives, aims to promote sustainable and non-invasive forms of tourism by proposing alternative modes of visitation and transportation. The promotion of sustainable tourism, in fact, is one of the criteria necessary to obtain European certification. The need to promote culture in order to propose different models for utilizing territorial resources has also been adopted in the cultural itineraries defined by the Italian Ministry that focus on "soft mobility" as a way of travelling along the routes with opportunity for territorial knowledge.

This framework, on one side, highlights the awareness (at the national and pan-European levels) that cultural routes represent an important means for promoting lesser-known destinations and contribute to the sustainability of tourism development, while on the other side, it underlines the need to make these resources available through appropriate actions of territorial planning and design. Such actions should provide for urban and territorial policies targeted to the coordination of the differ-ent aspects involved. In particular, as regards the public component such actions should include:

- the coordination between the target of the development plan and those of the town planning;
- the dimensioning of the infrastructure (physical and technological) in order to support the tourism fruition of the territorial resources;
- the implementation of facilitations (tax incentives and so on) in order to improve the involvement of the private sector and to increase the supply services for the better use of the routes (accommodation, catering, management, etc.).

As regards the private component, actions should include:

- the collaboration among the different interests involved in the development project;
- the promotion of entrepreneurial initiatives aimed at enhancing the local identities and territorial resources;
- the compliance with minimum standards of quality that have to be defined and monitored;
- the sharing of information to avoid redundancy;
- the coordination among actions of promotions in order to ensure the brand awareness.

The synergy between these two components (public and private) strongly involved is a prerequisite for the realization of a shared, balanced and sustainable project of territorial development. The support for the realization of this project can came also from the correct use of technology (Papa 2013).

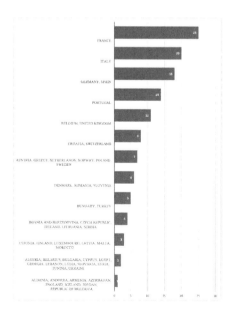

Figure 3. Country involved in the European Cultural Route Program (elaboration from www.camminideuropa.eu, consulted on 08/08/2016).

2 CULTURAL PATHS AND TOURIST DEMAND

Over the past few years, there has been a radical change in tourist behavior and habits: personal use of tablets and smartphones has, in fact, exceeded the purposes exclusively connected to business needs, with that use spreading to recreation and entertainment, too. In this change, tourism can have an active role in promoting sustainable life style in this contributing to the opening of new perspectives of the recent concept of urban smartness (Papa & Fistola 2016).

In truth, our entire interpersonal system of communication has changed quite rapidly. Referring to tourism, in particular, the whole information supply system is modifying itself and a multiple "virtual dimension" (Fistola & La Rocca, 2001) is being established beside the traditional supply of tourists.

Some scholars (Orlando 2013; Garau 2014) identify at least three types of use for new technologies in the fruition of cultural heritage as applied to tourism.

The first is a *communicative* type in which the use of technologies refers to the diffusion of information to a wide audience (audio-guides, QR codes, smart plaques, etc.) using narration.

The second is an *educational* type in which the use of technologies refers to the transmission of complex information to a specialized audience (students, scholars, researchers, teachers, etc.) using augmented reality, visual ambience, interactive reconstructions of virtual environments, immersive virtual environments, etc.

The third type can be defined as *customized*, as it refers to the use of mobile devices (smartphones, tablets, cameras, etc.) capable of combining geographical data with cultural information to be used on site (via an app, for example).

The development of these technologies is widely used in the cultural tourism sector and is increasingly linked to the composition of the supply-side, oriented to this type of tourist demand.

The most frequent applications include integration with geolocation tools present in the device; these provide, for example, for proposing a territorial area as a special route, characterized by a strong central theme (food, history, legend, wine, etc.) and allows the users to personalize the path. These applications, generally speaking, merge the use of GPS (Global Position System) and Web GIS (Geographic Information System): in other words, through the GPS, users can visualize their localization on a geo-referenced map and choose their own path according their own exigencies (disposition of time, difficulty, interests, curiosity, etc.).

The segmentation of the information is particularly significant in allowing users to customize their path, as the afore-mentioned exigencies can vary deeply by gender, age, composition, cultural level, etc. In this sense, it is important that information is organized in order to satisfy all the different typologies of user. Nevertheless, it is also necessary to introduce information regarding the basic service supply system (accommodation, food services, leisure activities, etc.) that might be present along the path. In fact, the density, variety, quality and organization of this supply system can deeply influence choices. Furthermore, this information allows the users to know the path and, thus, to plan their travel according to their needs (time available, difficulty of the path, personal objectives and motivations, etc.).

For these reasons, the planning of cultural tourist routes has to be based on the presence of an efficient supply system able to support the choices of the users; the supply system has to be oriented along at least three orders of objective, in particular:

- to indicate safety standards for the fruition of the paths, both those existing and those to be planned;

- to promote the territorial potentialities according to sustainable modalities;
- to enhance tourist typologies that generate low environmental impacts (eco-tourism).

The realization of these objectives assumes the definition of tools and rules mainly aimed at:

- verifying the feasibility of the valorization project or program;
- defining criteria for the award of the paths at local, regional and national levels;
- outlining the use and walkability of the paths;
- delineating the rules for the management, maintenance and usability of the paths;
- supervising the relationship between the private and public sectors.

The last point, in particular, underlines the need for cooperation among all the subjects that are interested in the development of the project of valorization. Moreover, cooperation is an essential condition if we consider that the paths run across different counties, beyond regional and administrative borders. On the other side, tourism development involves different sectors and different stakeholders, and partnership among these subjects can foster economic development through tourism (Briedenhann & Wickens 2004).

In this sense, the planning of paths can be assumed to be an occasion to define networks (physical and immaterial) among all the different subjects and levels involved (Meyer 2004; Mitchell & Hall 2005). The importance of the involvement of all the levels (administrative, economic sector, local population) and the positive effect derived from cooperation can be indicated in five main points:

- the establishment of a network among poles that might not hold the same attraction individually;
- the enhancement of lesser-known destinations;
- the increasing of employment and the training of workers in new professional skills;
- the diffusion of such new tourist typologies as slow tourism;
- the decreasing of negative impacts on natural environments.

These conditions should improve the distribution of benefits derived from the development of tourism, including for those work categories not directly engaged in the tourism sector.

3 PRINCIPLES FOR THE PLANNING OF CULTURAL TOURIST PATHS

The planning of a cultural tourist path is primarily a matter of landscape and territory, a project that has to consider both the physical and immaterial components of a territorial heritage (Berti 2013).

The complexity of this heritage needs a holistic and systemic approach in order to achieve the valorization of its potentialities and, at the same time, to assure the safeguarding of its local identities. Such an approach, furthermore, allows us to consider not only the physical links among the different poles that a path can connect inside a region, but also the immaterial relations that the paths can establish among their components.

According to this systemic vision, attention has to focus on the relations generated between the different poles that the path connects and the territories it runs across. In this vision, three main dimensions can characterize the project of the path:

• immaterial: the path can create an immaterial link among attractions located in different regions, connected by a unique theme;
• material: the path creates physical links connecting attractions along defined directions;
• uniqueness: the path itself represents an attraction on the basis of the elements that make it unique.

In the first case, attractions are not physically connected to each other; this allows users to create their own route.

In the second case, the path constitutes a physical link between attractions and has a strong relation to the territory.

In the third case, attractions are part of the path, and relations are generated inside the elements that constitute it.

With reference to these relations, we can propose a first classification of paths as:

• linear paths;
• paths "in stages";
• "closed track" paths;
• multipolar network paths.

The first type draws a unique and recognizable path through the territory, generally retracing ancient historical ways, proposing a main theme (natural, historical, religious, etc.). In this case, the relationship between the path and the territory is the result of temporal phases that have characterized the territorial history. Normally, the path leads towards a node-pole that physically represents the final destination of the route, and from the point of view of perception, the main goal for users walking the path. In the most common cases, the path connects two different poles of attraction (a source pole and a destination pole).

The Way of Saint James is perhaps the best known path in the world, and its particularity stems from the fact that it connects different origin-poles that all lead towards the same final destination point, located in the city of Santiago de Compostela.

The Via Francigena represents another famous case of a linear path that connects England (Canterbury) to Italy (Rome). The Via Francigena, in particular, is an example of the effects that a path can trigger on territorial resources that would otherwise have not been considered.

The "in stages" type of path is a subtype of the first one, in the sense that it is composed of a network of leg-poles that defines its geometry. These poles can be either attractions or facilities strategically located along the path (Figs. 4–5) and they must respect specific requirements, such as:

• locations at a fixed distance from one another (no more than 20 km);
• the presence of auxiliary services and facilities along the path;
• the availability of these services all year long;
• the provision of a network of low-cost or free accommodations (hostels, privates rooms, monasteries, etc.);
• the outlining of procedures aimed at labelling the accommodation services;
• the definition of agreements to identify users of the path and collect data on their characteristics and preferences.

In the case of a "closed track" path, the origin-pole and the destination-pole coincide and the path itself can constitute an attraction for its structure alone. With this type of path, even more so than with others, it is important to install an efficient system of signage in order to guarantee safety and beneficial conditions of fruition for users. The question of signage, in fact, represents a difficulty in the planning of paths, especially in regard to the necessity of defining standards for homologating information (Polci 2009; Zabbini 2012).

The "multipolar network" path can be considered to be a combination of the previous types and is probably the most suitable model for the study of existing routes and the design of new routes. In this type of path, the poles are connected to each

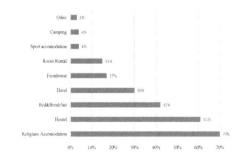

Figure 4. Percentage of preference for the accommodation structures along the Via Francigena. (Elaboration from TCI 2015).

Figure 5. Tourist Representation of a path "in stages" in Molise where some leg-poles coincide with attractions. (Source: Regione Molise Progetto Leader 2007–2013).

other according to complementary relations, even among poles that are hierarchically different.

In other words, all the elements composing the network are indispensable and complete each other. The hierarchy among poles depends on the role that each pole plays within the network, rather than on their physical dimensions. It depends on the ability that each pole has in triggering cultural, environmental and economic exchanges inside the network.

Within the network, we can individualize four main levels among the different poles:

- the first level corresponds to the origin pole (base pole);
- the second level corresponds to the set of attractive intermediate-poles;
- the third level corresponds to scattered-poles;
- the fourth level corresponds to the destination pole (magnet-pole).

The origin pole is the physical place where the path begins and it plays a strategic role, especially for all the activities regarding the knowledge and planning of the path. This pole has to be characterized by good accessibility and it has to be specialized in order to provide dedicated services and facilities, for instance: meeting point; presentation and advertisement of the path; accommodation services; information services; user monitoring facilities; enrollment of users; specialized equipment, etc.

The "intermediate-poles" identify the physical nodes that are equipped to support the use of the path. These may coincide with smaller urban centers, whose attractiveness stems from their very well-equipped supply system of facilities and adjunctive services or in a particular monument localized within the node.

The "scattered poles" indicate the network of possible stages in order to split the length and difficulties of the path. These nodes will be properly marked and equipped so as to ensure their function within the network.

The "magnet pole" corresponds to the end of the path (destination) and it plays a special and symbolic role within the network, as it represents the purpose of the journey.

It will need to be equipped with adequate accommodation services in order to host and manage the incoming fluxes of users.

4 BIG DATA, GIS AND TOURIST PATHS

In recent years, new technologies for collecting, sharing and managing Big Data represent a powerful backing for the identification and the design, in real time, of routes aimed at different uses and not least at the use related to leisure activities.

In order to manage and to analyze these data GIS technology (Geographical information System) permit to geolocalize the data.

In order to be able to manage and to analyze these databases, there is the need to use the GIS (Geographical Information System technology, that compared to the traditional software of alphanumeric data management, offers the possibility of reading the alphanumeric data and to localize them in the space (geolocalize), through the assignment of spatial coordinates. Being able to analyze the data and to localize it position, offers the benefit of improving the knowledge of the phenomenon both in quality and in speediness (Maguire et al. 2005).

The GIS approach and the growing availability of sophisticated software allows us to develop a Decision Support System able to identify and to select the most suitable and realizable solution to satisfy the users and to preserve the original environmental and cultural characteristics.

The GIS application to the fruition of routes allows the users to choose the path that will be the more adapt to theirs needs and aspirations.

This can be achieved by integrating the information developed in GIS with the "Mobile App" that can be easily downloaded on the personal mobile devices as additional multimedia data.

Data and information that these applications offer refer to leisure as well as to the augmented reality and so on. They are largely used for tourism as they allows users to integrate information and to localize the attractions within a territorial area.

Data and information, that these applications offer, refer to leisure as well as to the augmented reality. They are largely used for tourism as they allow users to integrate information and localize the attractions within a territory (Angelaccio et al.).

These technologies have radically changed the current way of communicating and living the tourist experience; they certainly integrate cultural information but especially they play a strategic role in the improvement of safety and reliability of the paths.

In the context of these considerations, the planning of paths can be developed by the adoption of a GIS methodology based on different geometries.

On one side, the previous typologies of poles can be structured according to a punctual geometry linked to descriptive information about amenities and facilities located at each pole. On the other side, a linear geometry can be generated by connecting the data about the characteristics of the track (slope, track fund, travel time, etc.) to each arc of the path.

In a successive phase, the GIS can be implemented through additional information that can be both graphical and alphanumeric, made available on an appropriate web site. This allows users to select information through their smartphone or tablet according their own needs.

As we have already stated, routes represent complex and dynamic systems generating relations among and through different territorial elements and levels. Even though they cannot be considered as exhaustive needing further in-depth research development, for the planning of routes compatibles both with territorial characteristics (physicals as well as socio-economical) and the exigencies of users some phases can be indicated:

- to determine the typology of the route in order to understand the potential and critical issues related to its geometry;
- to define the extend of the route;
- to delimit the intervention considering all the elements (physical, social, cultural, economic, etc.) engaged in the planning of the routes;
- to consider the possibility of involvement of local populations and stakeholders;
- to individualize actors and applicable forms of funding (cooperation, project financing, etc.).

5 POTENTIAL DEVELOPMENTS AND RESEARCH APPLICATIONS

5.1 A technological platform as tool for shared solutions

Designing tourist routes aimed at promoting the territorial knowledge through sustainable forms of mobility necessarily involves different aspects that are inherently included in the relationship between tourism and land use.

Tourism, in fact, is characterized by being a cross activity which involves multiple sectors (public and private). As such, tourism is an ideal field for testing models of territorial development based on the integration between the different interests involved.

By the use of innovative techniques and tools it is possible to envisage solutions able to drive tourism and territorial development towards some shared conditions and to reduce environmental, social and economic impacts. In this sense, the construction of a technology platform can represent an operational tool to deal with the "dimension" of a network or an integrated cooperative system.

In particular, the platform can overcome the lack of coordination and cooperation between the

different countries that the paths run across, generating new balance and positive effects.

The technology platform, then, could support the physical design of the tourist footpaths, serving the choices of different tourism typologies, on one hand, and the implementation of the supply of goods and services connected to the fruition of the paths, on the other hand.

In a starting phase, the technological platform could allow to both the public actors and private investors the optimal management of the footpaths. At the same time, it could allow the final users to appreciate the whole system of the present resources (territorial, social, cultural, historical).

Technologically, the e-services can be located on a specialized Web Portal aimed at integrating different elements of the present supply system of tourist services. The architecture of the technological platform can be articulated into four macro-areas (Fig. 6) that, also by using specialized apps, can integrate information for the fruition of the footpaths network. In particular, the macro-areas refer to:

- territorial promotion,
- territorial safety,
- institutional cooperation,
- local participation.

The macro-area of the promotion guarantees interactivity between the supply and the demand components by using functions aimed at optimizing the territorial fruition. In the framework of this context this macro-area should contain telematics functions able to ensure interactivity between the supply operators and the different typologies of tourist user (family, senior, adventure, young, etc.). In particular, in this macro-area, services should be oriented at supporting the users' choices with additional information that can be displayed directly on the users' mobile devices.

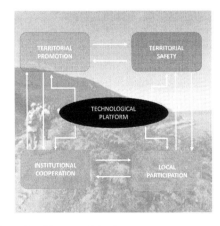

Figure 6. Diagram of the four macro-areas composing the platform.

The safety macro-area should include a monitoring system for the tracking of users along the path. In this way, users can be assisted in case of need and can be supported in the selection of the difficulties that characterize the paths.

The institutional cooperation macro-area is targeted to the coordination among public and private actors involved into the tourist development project.

The area of local participation should be targeted to provide information and services to improve the involvement of the resident and of all the social levels concerned.

The architecture of the platform would require further details that could constitute possible research developments. In this context, it seemed appropriate to underline that the platform could represent a technical tool to define a "dialog space" where cooperation between different interests can be realized also to improve local identities.

6 CONCLUSION

The demands of tourism are evolving along with the rise in demand for more conscious forms of use of territorial and environmental resources through the modification of behaviors and lifestyles.

This change is also linked to the social structure composing the present demand within tourism that prefers the cultural aspect of the journey. On the basis of this change, the rediscovery of cultural paths can be an occasion to promote alternative forms of fruition of territorial resources through the use of soft mobility (La Rocca 2010).

This paper has tried to underline the strategic role that Italy and, in particular, the regions of southern Italy can play, considering both the variety and significance of its heritage and the recent interest of the political and administrative sectors in promoting cultural routes. Nevertheless, cooperation between the private and public sectors has been individualized as a structural condition and a factor of success in the planning of territorial path networks.

With reference to a holistic approach, the planning of a network of cultural paths has to consider the role that they can play in the integration of different territories. In this sense, paths are both an immaterial tool of sharing projects and a physical tool of territorial development according to sustainability and resilience models.

REFERENCES

Akerkar, R. 2012. Big Data & Tourism. In TMRF Report edited by Technomathematics Research Foundation.
Angelaccio, M., Buttarazzi, B. & Onorati, M. 2014. Modelli di Turismo e tecnologie Smart Cities per la valorizzazione di borghi, giardini, itinerari e territori "intorno" alle città intelligenti. *Smart City and Digital Tourism*. Roma: Universitalia.
Berti, E. 2013. Itinerari culturali del consiglio d'Europa: nuovi paradigmi per il progetto territoriale e per il paesaggio. *Alma Tourism*, 2013 (7): 1–12, www.alma-tourism.unibo.it.
Briedenhann, J. & Wickens, E. 2004. Tourism Routes as a Tool for the Economic Development of Rural Areas – Vibrant Hope or Impossible Dream? *Tourism Management*, I, 25: 71–79.
Fistola R. & La Rocca R.A. 2001. The virtualization of urban functions. In H. Bakis & W Huh (eds), *Geocyberspace: Building territories on the geographical space of the 21th century*, NETCOM, 15 (1–2): 39–48, Montpellier, France.
Garau C. 2014. From Territory to Smartphone: Smart Fruition of Cultural Heritage for Dynamic Tourism Development. *Planning Practice & Research*, 29 (3): 238–255. DOI: 10.1080/02697459.2014.929837.
International Council on Monument and Sites 1999. *International Cultural Tourism Charter Managing Tourism at Places of Heritage Significance*. Adopted by ICOMOS at the 12th General Assembly in Mexico, October 1999. www.icomos.org.
La Rocca, R.A. 2010. Soft Mobility and Urban Transformation. *TeMA. Journal of Land Use, Mobility and Environment* 3 (SP): 85–90. Naples: University of Naples Federico II. DOI:10.6092/1970–9870/125.
Maguire, D.J., Batty, M. & Goodchild, MF. 2005. *GIS, spatial analysis, and modeling*. Esri Press.
Meyer, D. 2004 *Routes and Gateways: Key issues for the development of tourism routes and gateways and their potential for Pro-Poor Tourism*, background paper, output of the ODI project: Pro-poor Tourism Pilots in Southern Africa, http://www.odi.org.uk.
Mitchell, M. & Hall, D. 2005. Rural tourism as sustainable business: key themes and issues. In D. Hall, I. Kirkpatrick & M. Mitchell, *Rural Tourism and Sustainable Business*, Clevedon: Channel View.
Orlando, M. 2013. Didattica e turismo 2.0—Nuove tecnologie per la divulgazione del patrimonio culturale. *Storia e Futuro* 32 June 2013 http://storiaefuturo.eu
Papa R. & Fistola R. 2016 (eds). *Smart Energy in the Smart City. Urban Planning for a Sustainable future*. Switzerland: Springer International Publishing.
Papa R. 2013. Editorial Preface Smart Cities: Researches, Projects and Good Practices for the City. TeMA Journal of Land Use, Mobility and Environment 1 (2013): 3–4 DOI 10.6092/1970–9870/1544
Polci S. 2010. *La valorizzazione della Via Francigena. I percorsi, l'accoglienza, l'offerta culturale*. CIVITA, Roma.
Richards, G. 2003. What is Cultural Tourism? In A. van Maaren, (ed.), *Erfgoed voor Toerisme*. Nationaal Contact Monumenten.
Touring Club Italiano 2015. Il turismo sulla Via Francigena. Touring Club Italiano, Roma. "Survey sul turismo lungo la via Francigena" Roma: Centro Studi TCI, ottobre 2015.
World Tourism Organisation 1985. The State's Role in Protecting and Promoting Culture as a Factor of Tourism Development. Madrid: WTO.
Zabbini, E. 2012. Itinerari culturali e patrimonio intangibile. In *Proceeding of the 32nd National Conference of The Italian Association of Regional Studies (AISRE)*, 1–12, www.aisre.it.

Transport Infrastructure and Systems – Dell'Acqua & Wegman (Eds)
© 2017 Taylor & Francis Group, London, ISBN 978-1-138-03009-1

Terminal operator liability

Francesca d'Orsi
Studio Legale d'Orsi—Neri, Roma, Italy

ABSTRACT: The figure of the terminal operator, which can best be defined as the main player in the complex of port operations within the port area, is still one of the most controversial figures in the context of maritime law, and more specifically as regards the juridical qualification to be given to the contracts stipulated with them for the execution of the aforementioned operations and the consequent regime of liability in terms of the fulfilment of their obligations. Given that the law does not give any definition of a terminal operator, it does not discipline the contractual circumstances in which the juridical relations between them and the user who is benefiting from their activities should be included either.

1 INTRODUCTION

The figure of the terminal operator, which can best be defined as the main player in the complex of port operations within the port area, is still one of the most controversial figures in the context of maritime law, and more specifically as regards the juridical qualification to be given to the contracts stipulated with them for the execution of the aforementioned operations and the consequent regime of liability in terms of the fulfilment of their obligations.

2 DEFINITIONS

Neither the Navigation Code nor Law no. 84 of 28 January 1994, reforming the port system under our legal system, give a definition, even brief, of the terminal operator.

A contribution in this sense could be that of the 1991 UNCITRAL Convention (the United Nations Convention on the liability of Operators of Transport Terminals in International Trade) which, however, did not enter into force for lack of the necessary ratifications for that purpose, and it is hardly likely to enter into force in the near future. The definition given by the UNCITRAL under Article 1 "Operator of a transport terminal" means a person who, in the course of his business, undertakes to take in charge goods involved in international carriage in order to perform or to procure the performance of transport-related services with respect to the goods in an area under his control or in respect of which he has a right of access or use".

As mentioned, the Navigation Code does not even mention this figure, whereas Law no. 84 of

28 January 1994 has introduced it into our legal system, at least at the level of statutory regulations, clearly distinguishing between, on one side, the legal system of port operations (Article 16), including all operations regarding goods carried out in a port, such as the loading, unloading, tranship-ment, deposit and handling of goods in general, which can be performed by specialised companies if authorised by the Port Authority or, if there is no Port Authority, by the Maritime Authority; and on the other by the successive Article 18, which is a public law ruling governing the case of the concession, always granted by the Port Authority or, otherwise, by the Maritime Authority, of areas and wharves (terminals) for the execution of port operations on the concessionaire's own behalf or for third parties, in which case the concession itself refers precisely to the specific figure of the so-called "terminal operator".

The holding of an authorisation for the performance of port operations can, in some cases, be accompanied by the right to dispose exclusively of port spaces and areas. And in practice, the holding of a concession for the use of state land in a port means a port terminal. The company which holds both the authorisation and a concession is thus defined as a terminal company or operator, and is, in practice, the subject on which all the other possible authorised companies hinge, in a system which can be defined as reticular.

Therefore, a qualifying element of terminal operators, under Law no. 84 of 28 January 1994, is the fact that they perform port operations within state-owned areas and at state-owned wharves the use of which is granted to them exclusively. Normally, the exclusive use of the areas is connected with the need to use fixed wharf equipment, the utilisation of which requires the exclusive availability of port areas, but with increasing frequency port terminals are assigned to companies whose activity does not involve the use of this type of equipment, but which nevertheless requires the exclusive use of areas for storage or for the parking of vehicles.

Precisely because of this exclusivity, Article 16, paragraph 4, letter a), in fact requires the authority appointed for the purpose to assess every authorisation application, first and foremost on the basis of the requisites of a personal and technical-organisational nature, and the financial capacity and professional standing of the applicant operators and companies, which must be adequate for the activity to be performed, and they must also present an operational programme and the determination of a staff of workers directly employed by the same. Which, in other words, implies an assessment aimed at ensuring that only the companies with greater capacities are authorised to use the port spaces, also taking into account the fact that the port is, in any case, a limited area in which only a limited number of companies can reasonably operate. Consequently, it becomes essential for the port authority or the maritime authority to adopt selective criteria to assess all the possible applicant companies, and to guarantee access to the services market and to allow for the execution of port operations only on the part of those with the best credentials, i.e. those with special characteristics and entrepreneurial capacities, as well as a development and investment programme suitable for guaranteeing the return of economic benefits for the port and for the ancillary businesses that depend on the port for their survival.

Therefore, in view of the above, the law attributes to the port or maritime authority also the power to establish a maximum number of companies which can be authorised to operate at a port (Article 16, paragraph 7) as well as the duration of the relative authorisations, which will depend on the operational programme proposed by the company, always without prejudice to the port or maritime authority's power to periodically check on respect for the programmes on the part of the authorised companies, naturally as well as the power to revoke the authorisation also in the case of failure to respect the programme (Articles 16, paragraph 4, letter b) and 6).

The terminal operators are, in fact, characterised by the fact that the subjects which operate in the port are simultaneously concessionaires of state-owned port spaces and companies authorised, pursuant to Article 16, paragraph 3, of Law no. 84/94, to carry out port operations.

Over time, this has led to some perplexity also regarding the possibility of the creation of monopolies, and in fact, before this situation was remedied by the law (Law no. 186/200), the subject of the monopoly of the technical-nautical services was a hot topic as regards the letter of the law and case law, based mainly on the distinction between port operations and technical-nautical services.

The law, intervening on the problem of the execution of technical-nautical services, has probably brought this knotty question to an end, finding a definitive solution with the introduction of Article 1 of Law no. 186 of 30.06.2000, which has substantially amended Article 14, 1 bis, of Law no. 84/1994. In fact, the law now specifies that: the technical-nautical services of pilotage, towage, docking and inshore services are services of general interest for the purpose of guaranteeing the safety of navigation and berthing in the ports where they have been instituted. With the introduction of this provision into the legal fabric of Law no. 84/1994, therefore, also the legislator has pointed out, once and for all, bringing to an end the knotty question of the distinction between port operations contemplated by Law no. 84/1994 and the technical-nautical services (the so-called navigation ancillary services), that port operations are in fact different from the technical-nautical services (pilotage, inshore services, towage and docking), emphasising that only the execution of these latter, because of the close connection between their performance and the fulfilment of specific public service obligations, serves to guarantee the safety of the ports and the prevention of damaging events in the ports, thus justifying, albeit implicitly, a control of the prices of the same. Furthermore, in order to determine the binding criteria for regulating and disciplining port services (not to be confused by the afore-mentioned technical-nautical services) by the port and maritime authorities, pursuant to the amended Article 16 of Law no. 84/1994, Ministerial Decree no. 132/2001 has been issued (see Council of State opinion of 29 January 2001).

Having fully considered questions of a general nature, this paper will continue with a discussion of the terminal operator, as a legal person, and its liability.

3 LIABILITIES AND RESPONSABILITIES

First and foremost, the terminal operator cannot undoubtedly be considered either as merely a dispatcher or simply a mere depository for the goods delivered, but it should rather be considered as the subject around which the main activities within the port are carried out, the real protagonist of the port operations (Carbone & Funari 1994; Carbone 1995; Midoro 1995).

Since the law does not give any definition of a terminal operator, it does not discipline the contractual circumstances in which the juridical relations between the same and the user who is benefiting from its activities should be included either.

Doctrine has highlighted the mixed nature of contracts between the owner of the goods and the terminal operator, which includes within its scope both the circumstance of depositing goods (deposit in favour of third parties) and the contracting of services, consisting of the performance of port operations on behalf of third parties by the terminal operator (Vincenzini 1989).

It has also been maintained that the terminal operator could be included within the sphere of multi-mode transport, in which the obliged party supplies not only the transport by various means, but also the activities ancillary to the same and strictly inherent to the main performance, namely that of the transport (Riccomagno 1998).

Independently of the exact juridical qualification to be given to contracts stipulated by the terminal operator, as regards the relations between the latter and the third party recipients of the goods, the use of so-called Himalaya Clause in the contracts in use in international maritime traffic implies that the specific regulations on transport contracts are applicable, with specific regard to exoneration from liability and compensation limits, to the terminal operator which has supplied its services on behalf of the dispatcher, almost as if it was acting as its auxiliary/delegate (Piras 2008, Potenza 2002).

In fact, the relationship between the terminal operator/terminal company and the maritime shipping company should be examined in depth and therefore it should be verified whether it is an agent of the maritime shipping company or a contractor independent of the maritime shipping company; this distinction, far from being a question of mere nomenclature, is important on the legal level, regarding the applicability of the waivers and of the limitations of liability, contemplated in favour of international maritime shipping companies, of certain provisions of international law (the Hague Visby Rule on bills of lading), even if, in the case of the above-mentioned Himalaya Clause, this problem is practically annulled by the application of the clause (Dani, 1986).

Coming to the terminal operator's liability, an important contribution has been given by the Uncitral Convention, whereby the legal discipline of the terminal operator's liability is strictly connected to the definition of the terminal operator/company given by the Convention.

With regard to the liability for loss or damages, liability begins when the goods are taken in charge, therefore it is strictly connected to the effective availability, and therefore involving full control, of the goods on the part of the operator; it is a presumed liability similar to that of the carrier (road or maritime), except for the exonerating circum- stances for the terminal operator when it is demonstrated that the subject, its employees and/or representatives or other subjects of which it takes avail in the execution of its obligations, have taken all measures deemed reasonably necessary in order to avoid the event or its damaging consequences (art 5. Uncitral).

With regard to delays, the terminal operator/ company bears liability if it does not make the goods available to the assignee within the term expressly indicated in the contract, or within a reasonable time in respect of the assignee's legitimate request for the redelivery.

Substantially, the convention adopts the standard principles of international law; this tendency can also be found in the drafting of the terminal contracts, and in fact an examination of the contracts most commonly used in this specific sector allows for listing the main obligations of a terminal operator/company, which are represented by activities that are typical of various contracts: deposit contracts; mandate contracts; services and transport contracts, as well as a series of additional activities linked to the main activities, such as the supply of cranes and structures, and the rental of spaces (Carbone 1995).

Thus, there are several performances to which a specific discipline is applied, also as regards liability. However, the purpose of the terminal contract is certainly unique since, in view of the fact that it is extremely common, it can be defined as a socially typical contract, in consideration of the fact that there is a certain uniformity of content between the various terminal operators/companies and that it is based on a (typical) services contract, the discipline of which, however, derives from the application of the law to the single performances, but within the framework of a complex works contract (Antonini 2008, Riguzzi 2006, Lefebvre et al. 2008 Gaggia A., 2008).

One doubt of a practical nature is that which regards the identification of the moment at which the prejudice is caused in order to render applicable the discipline of one or another type of contract of which the actual contract is composed, also from the viewpoint of the evidence. According to reason, adhering to the general principle of the so-called "proximity of the evidence", it must be maintained that the subject bearing the burden of proof is the terminal operator/company, also because this latter is certainly the party most interested in proving, for example, the applicability of a specific discipline rather than others (e.g. the law on transportation, for the compensation limits and the term of lapse); however, if it is not actually possible to identify the precise moment in time when the prejudice occurred, the general discipline of Article 1218 of the Civil Code must be applied.

Always from analysing the general nature of the contracts, one can maintain that it is possible to establish compensation limits contractually, valid within the limits contemplated by Article 1229 of the Civil Code, but only in the case of slight negligence. In the same way, the exoneration for so-called indirect damages, such as loss of profit, is established.

4 CONCLUSION

In conclusion, therefore, from the above analysis, one can maintain that the terminal operator's liability is subjective for presumed negligence, disciplined by Articles 1218 and 1693 of the Civil Code. Consequently the exonerating circumstances follow the same discipline, without prejudice to the fact that the terminal operator bears the burden of proving the specific moment in time at which the damaging fact occurred. Similarly, compensation limits and terms of lapse, if contemplated, are those laid down by law for the single cases.

With regard to the position of the third party which receives the goods, however, a party which is extraneous to the contract with the maritime shipping company, in view of the Himalaya Clause this party benefits from the typical system of the maritime contract, with the consequent applicability of the relative discipline, whether it bases its action on contractual or non-contractual liability, qualifying the terminal contract as a contract in favour of the third party.

From the insurance viewpoint, considering the very large variety of obligations imposed on terminal operators/companies, the insurance market has not drawn up standard policies; however, PORT OPERATORS' AND TERMINAL OPERATORS' LIABILITY POLICIES can be found on the market, specifically designed to cover the civil liability of the insured party and holding the same harmless for what it may be held to pay on such grounds, in relation to and for damages occurring within the sphere of its professional activity, for:

material loss or damage caused to the goods while they are under the care, custody and responsibility or within the sphere of control of the insured party;

direct damage to containers or other equipment for transport, ships or other means of transport owned by third parties, caused by the insured party while the goods are in practice under its custody and responsibility;

indirect damages deriving from or due to material loss or damage caused to the goods or by the equipment used to transport the goods; without prejudice to the activities and/or risks excluded by agreement.

The guarantees provide can be extended to errors and omissions, to the activity of the custodian and/or the warehouse, to damage to the insured party's own equipment, and third party civil liability for damages to people and property.

REFERENCES

Antonini A., Benelli G. Carretta M., Gaggia A., Ingratoci C., Lobietti C., Morandi F., Pellegrino F., Pruneddu G., Rizzo M.P., Sia A.L., Tassinari G. 2008. Trattato breve di diritto marittimo, Milano.

Carbone S.M. & Funari F., 1994. Gli effetti del dirtitto comunitaio sulla riforma portuale, in Dir. Mar.

Carbone S.M., 1995. Contratto di trasporto marittimo di cose, Editore Giuffrè Editore.

Dani A. 1986. Persone che possono beneficiare del limite di debito, Diritto Marittimo 1986.

Gaggia A., 2008 L'operatore terminalista, in Trattato breve di diritto marittimo (a cura di Alfonso Antonini) Trattato breve di diritto marittimo, Milano.

Lefebvre D'Ovidio A. – Pescatore G. Tullio L. 2008. Manuale di diritto della Navigazione, giuffrè Milano).

Midoro R. 1995. Riflessioni sull'impresa terminalistica portuale, in Economia e diritto del terziario.

Piras M., 2008 La clausola Himalaya, in Trattato breve di diritto marittimo, II, Milano, 281 ss.

Potenza M., 2002. La clausola Himalaya al vaglio della giurisprudenza italiana: verso l'estensione a favore del terminal operator del regime di polizza di carico? Diritto dei Trasporti 2002, 245 Trib. Genova 20 novembre 2000.

Riccomagno M. 1998 La risoluzione dlle controversie delle controversie nel Trasporto multimodale, Trasporti.

Riguzzi M. 2006. Il contratto di Trasporto Giappichelli Torino.

Vincenzini E. 1989. Ina realtà portualee una figura giuridica nascente: il terminal operator in Dir. Trasp. 1989 II.

Transport Infrastructure and Systems – Dell'Acqua & Wegman (Eds)
© *2017 Taylor & Francis Group, London, ISBN 978-1-138-03009-1*

Information provision in public transport: Indicators and benchmarking across Europe

M. Pirra & M. Diana
DIATI, Corso Duca degli Abruzzi, Politecnico di Torino, Torino, Italy

A. Castro
Epidemiology, Biostatistics and Prevention Institute (EBPI), University of Zurich, Zurich, Switzerland (affiliation during METPEX project: ZHAW—Zurich University of Applied Sciences—Institute of Sustainable Development, Winterthur, Switzerland)

ABSTRACT: The measurement of the quality of transport services has been the object of an intensive research activity in the last decades. The EU project METPEX aims at advancing the state of the art in this crucial research area through a targeted survey proposed in eight European cities. The dataset is then analyzed to identify latent constructs that can measure the quality of the traveler experience. Several different indicators are proposed to represent detailed aspects related to the perceived quality of transport services. In this paper, we concentrate on those indicators capturing the quality of information provision services in different travel stages: before starting the journey, when passing through stations or stops and while travelling. The results show the added value of separately considering the quality of information provision at different stages of the journey experience, for which different entities could be responsible (e.g. infrastructure managers versus service operators).

1 INTRODUCTION AND LITERATURE REVIEW

Previous scientific work has attempted to measure the quality of transport services. The EU project METPEX (A Measurement Tool to determine the quality of the Passenger Experience) sets the goal of increasing our knowledge on this topic by focusing on current research gaps and less explored issues. Namely, METPEX studies how quality is perceived for the whole traveler experience, not just focusing on the journey made with a specific service but taking into account all phases, from the pre-trip information acquisition process to the final leg to get to destination. Within such framework, the ambition is to assess how the quality of individual trip legs is combined to form an overall quality assessment of multimodal journeys. Rather than an industrial and managerial viewpoint, which is usually focusing on single services (or on services provided by a given operator) and is adopting methodologies adapted from marketing research, a transport policy and planning perspective is considered, where a more holistic view of the system has to be taken. Beyond the much investigated quality issues in public transport, the project considers active transport means (walk, bike) and the specific viewpoints of special user groups, including commuters, women and physically challenged individuals.

The present work reports some preliminary analyses done to build and test quality indicators according to the above research perspective (Diana et al. 2016), especially focusing on those that measure the quality of information provision processes in public transport systems (METPEX 2015). These analyses have been carried out during the first phase of the research, where the METPEX team tried to deepen the understanding of the different facets that concur in forming the quality perception of a transport service without, at this step, taking into account the travelers' specific profiles and features. Therefore, the following analyses are targeted at defining some latent variables related to the perceived quality of information provision, rather than considering any objectively measurable quantity.

The definition of both the measurement tool and of the subsequent synthetic indicators, including those presented in this paper, was also based on past studies in this field. Many articles are found in literature which analyze several different factors influencing perceived service quality, as also information provision. In the following, we shortly review those more relevant according to this and the above introduced METPEX perspective. For example, the relation between the overall satisfaction for a travel trip and the satisfaction for specific aspects composing the journey itself has been analyzed by de Oña et al. (2013), while their relative importance has been explored by Stradling et al. (2007).

In many works the perceived quality of a wide range of components is analyzed through different methods and techniques (de Oña et al. 2014, Eboli & Mazzulla 2011, Tyrinopoulos & Aifadopoulou 2008). In these works, the contribution of both subjective and objective aspects is considered. Among the different aspects that users have to evaluate, information provision is certainly an aspect affecting transit service quality. In fact, "passengers need to know how to use transit service, where the access is located, where to get off in the proximity of their destination, whether any transfers are required, and when transit services are scheduled to depart and arrive" (Cirillo et al. 2011). The importance of the information provision is found in the interesting qualitative study proposed by Beirão & Sarsfield Cabral (2007) to understand the travelers' attitudes towards transport. In fact, several respondents think that the bus system, in this case that of the metropolitan area of Porto, is difficult to use and information is difficult to obtain. In specific, bus users notice that problems occur when the bus company changes timetables or routes and no information is provided to travelers. In recent years, many transit agencies have realized that that service information is important for travelers and can be effectively used to increase ridership potentially attracting new riders to the transit system (Eboli & Mazzulla 2012).

In an ideal perspective, the user should be able to access information at every stage of its journey. This is fundamental in the case of multimodal travel and it is a key point affecting customers' modal choice (Grotenhuis et al. 2007). Through pre-trip information the user could plan routes and connections, knowing the location of the nearest bus stop, fares, routes to the desired destination, possible transfer locations, time of departure and approximate journey duration. Schmitt et al. (2015) observe that this aspect has a strong influence on the customer' mode choice, mainly during unfamiliar transit travels.

However, the need of an efficient information provision is also relevant for habitual trips. During the displacement, the traveler should be able to know the correct vehicle to board and where to transfer and proper information should be given while onboard in different ways (i.e., by audio announcements, by monitors, by personnel, on mobile devices...). These good practices are important for passengers since they "decrease uncertainty and develop psychological strategies to reduce the adverse effects of uncertainty and to raise feelings of security" (Cheng & Chen 2015). Moreover, return trip information should be available at the destination (i.e., departure times and changes in route numbers). The availability of a wide range of mobile apps equipped with real time transit information is triggering many new challenging issues, concerning for example the influence they could have on travelers' behavior (Brakewood et al. 2014, 2015). On the whole, the information required by passengers should be provided during the trip, at stops, and beforehand to plan the journey properly (pre-trip information) (Eboli & Mazzulla 2012). Consistently with such framework, the indicators defined in the following sections will be based on all these aspects too.

Some examples of quantitative indicators that are based on the evaluation of the information provision can be found in literature. In Nathanail (2008), an indicator of the passenger information during the trip is defined. It is graded by a trained checker and it is based on the type and quality of information provided at five, arbitrarily selected, stations along the train itinerary. Then, the itinerary is evaluated according to the average grade of all five stations. In Eboli & Mazzulla (2011), a more objective indicator of the attribute "availability of schedule/maps on bus, and announcements" is proposed. It is defined as the ratio of the number of vehicles with functioning information device on board to the total number of vehicles sampled in a certain time period. The functioning of these information devices on different days is monitored by a trained checker. The attribute "availability of schedule/maps at bus stops" is evaluated too. A score in a scale [0–10] is assigned to each stop of a line, where 0 goes to a stop without any kind of information device on the place and 10 to one with schedule and maps. The indicator is then the average value of the scores assigned to all the line stops.

However, beyond such examples, most of works on literature concentrates on the subjective evaluation given by travellers on some specific quality issues, which is also a specific focus of METPEX. In de Oña et al. (2014) a three step approach, which aims at monitoring the transit service quality of the metropolitan area of Granada over the years, is developed. Passengers are asked to evaluate 14 rather general service features, such as information, frequency and punctuality, cleanliness of the vehicle, or speed of the trip and a Service Quality Index is calculated from the passengers' perceptions.

A further key point in understanding the users' perception according to the METPEX perspective is the consideration of the whole journey experience, from the origin to the destination of the displacement. This means taking into account elements that could provide more satisfaction to the passenger, such as a good provision of information before the departure, beyond the information both during the trip and at public transport stations or stops, the latter being called "land-side" to contrast it with the former one. Many researchers have addressed these aspects (Tyrinopoulos & Antoniou 2014, Carreira et al. 2013, 2014, Lu et al. 2009) and related findings were useful to define the structure and the contents of the survey that is presented in the following section. Section 3 will then describe the analyses that lead us to the definition of an initial set of indicators capturing the underlying dimensions of the quality experience, and will particularly focus on three indicators related to information provision in different phases of the journey. Section 4 will present some benchmarking exercises based on such indicators, while the concluding section summarizes the results and offers some perspectives on the next steps of this work in progress.

2 EXPERIMENTAL FRAMEWORK

Different group of travelers have different needs and viewpoints, thus different variables describing service aspects that are relevant for different type of users and travel modes have been formulated. This involved the definition of more than 500 satisfaction questions for the measurement tool related to a wide range of aspects of the journey experience. Only a selection of these questions was presented to each user, namely around 60 of them, according to both the characteristics of the traveler and of the kind of journey under investigation. Unlike most of previously reviewed studies, the measurement tool is in fact referred to a specific journey rather than to a more or less generic opinion about a service.

Specifically, a multi-level questionnaire was set up, where the evaluation and the satisfaction with different aspects is checked. The structure of the questionnaire can be found in Diana et al. (2016, p. 1166). Firstly, some baseline characteristics of the user are analyzed, as socio-demographic ones, together with some attributes related to the last journey done (origin, destination, travel time, travel means etc.) that is the focus of the remainder questionnaire. Then, the attention goes to the overall journey itself and to the satisfaction with 21 more general quality components. The corresponding 21 questions are named "Tier 1" (T1_x in the following, where x ranges between 1 to 21) and take the form of a satisfaction rating exercise using a 5 points ordinal scale, ranging from 1, not satisfied, to 5, really satisfied. Such T1 questions are presented in the first two columns of Table 1. Some of these indicators are usually found in the literature, as highlighted in the previous section. However, some innovative aspects are presented and investigated. For example, the evaluation of intermodality topics, that is the use of different forms of transport during same journey, is introduced thanks to the definition of questions such as T1_5 and T1_15. This is a cross-cutting concept in contemporary transport research and policy, since it is related to a variety of strategic issues, ranging from the minimization of transport-related impacts through the use of an optimal "modal mix", till the possibility of successfully launching new transport services that are enabled by new technologies whenever potential customers do not necessarily stick to more traditional means (Diana 2010). Also the "design" concept, related to the physical infrastructures encountered by the

Table 1. Tier-1 categories description.

Question ID	T1 category name
T1_1	Design of stations was adequate for my needs
T1_2	Design of transport interchanges (main terminals) was efficient
T1_3	Design of transport stops was adequate for my needs
T1_4	The city supported my mobility needs
T1_5	The different modes of transport I used worked well together
T1_6	My passenger rights (e.g. able to access all transport services) were respected
T1_7	The overall accessibility of my journey was adequate for travellers with additional needs
T1_8	Provision of information on arrivals and departures was adequate for my needs
T1_9	Public Transport Staff were receptive to my needs
T1_10	The quality of travel information available during journey was good
T1_11	The quality of pre-trip information before I started my journey was good
T1_12	The quality of transport infrastructure (e.g. whole transport service) during my journey was good
T1_13	The quality of my ride was good
T1_14	My safety and security while travelling was good
T1_15	Support for intermodal (e.g. different forms of transport during same journey) travel was provided
T1_16	Recognition of the needs of motorised vehicle users
T1_17	Ticket purchasing process was easy to follow
T1_18	Time the journey took was as promised
T1_19	Transport availability was adequate for my needs
T1_20	Vehicle design was suitable for my needs
T1_21	Value for money of services was good

travelers in their journeys, has not been analyzed so much until now. Among T1 questions, instead, it is considered through T1_2 but also in T1_1, in T1_3, in T1_20 and, in a certain way, in T1_12 (Table 1). The passenger experience, center of the METPEX project, comes out mainly when it is asked to evaluate the meeting of mobility needs (T1_4) and the respect of travelers' rights (T1_6). Niche users opinions are investigated too, mainly thanks to T1_7 and T1_16 which focus on travelers with additional needs and motorized vehicles users.

Finally, and most importantly for the purpose of the present paper, information provision aspects are considered in three distinct questions, namely T1_8, T1_10 and T1_11, respectively related to land-side, on-trip and pre-trip phases. Information to travelers is in fact supplied in different phases of the journey experience, and it is important to distinguish perceived quality issues at different stages, for which different entities could be responsible. For example, due to the vertical separation between infrastructure managers and public transport operators, information systems at stops and on-board the vehicles might be designed and provided by different entities.

In the following sections of the questionnaire, it is asked to the user to concentrate on the longest trip leg among those composing the journey under investigation. In this part, the interviewee must evaluate some aspects of his/her displacement according to the same 5 points scale to have uniformity with the previous section. The set of asked questions was personalized according to both the mode used in this leg and on the kind of user, on the basis of the answers given in the baseline part. Overall, 305 questions were thus organized in 10 different mode-specific sets and 11 different user group-specific sets, each set comprising around 15 questions.

The last part of the questionnaire is named "Tier 2" (T2 in the following) and its questions are organized in 21 sets, each set representing a deeper assessment of one of the above introduced "T1" quality aspects. The interviewee had to answer to only one set, being selected depending on the previous answer to a control question. This structure is intended to enable the analyst to acquire both general information about the journey, as well as very detailed information about specific aspects whilst maintaining a practical limit on the number of questions a participant is asked, thereby responding to the concern about developing an overlong survey tool. The number of T2 questions asked for each T1 components is shown in Table 2 (second column).

The surveying period began on September 15th, 2014 and ran until October 29th, 2014, in eight different European cities: Bucharest (Romania), Coventry (UK), Dublin (Ireland), Grevena (Greece), Rome (Italy), Valencia (Spain), Vilnius (Lithuania) and Stockholm (Sweden). Additionally, the questionnaire was distributed by FIA (Fédération Internationale de l'Automobile) through their national networks of motorists to reach a wider target of both transit and private means users. Overall, 6,360 responses have been collected through a variety of surveying protocols. After the data had been cleaned, the total number of valid observations was 6,276: 1,472 responses from paper-and-pencil method, 3,805 responses from on-line web survey method, 289 responses from the navigator app, 420 responses from game app and 290 responses from focus group method. More information concerning the data acquisition can be found in Susilo et al. (2016), while the questionnaire contents and the implementation of the survey are presented in METPEX (2013) and in Diana et al. (2016, 2017).

For the sake of briefness, we do not present here descriptive statistics on the sample; those are available in METPEX (2014). In the following, we concentrate on the data gathered through the

Table 2. Cronbach's Alpha of the Tier-2 questions within the 21 Tier-1 categories.

Question ID	N of Items (T2 questions)	Sample size	Cronbach's Alpha
T1_1	12	67	0.950
T1_2	13	67	0.949
T1_3	13	67	0.958
T1_4	14	139	0.965
T1_5	14	96	0.951
T1_6	14	73	0.967
T1_7	14	45	0.968
T1_8	10	64	0.963
T1_9	11	48	0.946
T1_10	17	110	0.980
T1_11	14	119	0.965
T1_12	13	107	0.957
T1_13	7	192	0.918
T1_14	12	94	0.933
T1_15	13	100	0.968
T1_16	12	109	0.940
T1_17	14	45	0.965
T1_18	12	96	0.965
T1_19	13	126	0.961
T1_20	13	45	0.913
T1_21	13	71	0.947

T1 and T2 questions and we analyze the relationships and the latent factors that lie underneath the answers given by the users. This is a preliminary work to the definition of proper indicators that could help in understanding the real feeling of users about the whole journey experience.

3 ASSESSMENT OF THE INDICATORS THROUGH CRONBACH ALPHA AND PRINCIPAL COMPONENT ANALYSIS

In a first explorative test, the consistency of the hypothetical summated rating scales defined by jointly considering all variables within each of the 21 T2 groups has been tested by using the Cronbach's Alpha analysis (Cronbach 1951). This analysis helps to check if a certain group of variables can be considered as a unique indicator of an underlying unidimensional construct, or if it is representative of more than one latent variable. In order to have a sufficient number of observations, the whole dataset comprising data collected in all demonstration sites was jointly considered for this analysis. Data groups with Cronbach's Alpha values above 0.7 are considered acceptably consistent (Cureton & D'Agostino 1983). As the last column of Table 2 shows, all T2 groups are reliable and the variables are consistent within the groups since the Cronbach's Alpha value is around 0.9 in all groups. Therefore, according to this test, no item can be removed in order to increase the consistency of the data.

As further step, in order to reduce the number of T2 questions representing the T1 categories, a

Principal Component Analysis (PCA) is performed. The PCA is a variable reduction technique that aims to minimize the number of representative variables (components) that show the highest possible variance of the observations (Bishop 2006). An Explorative Factor Analysis (EFA), an alternative variable reduction technique, might also provide interesting information since it identifies "underlying constructs that cannot be measured directly" (Suhr 2003). However, EFA has been rejected for this reduction of variables because it requires additional statistical assumptions concerning the correlation of factors and error terms that might not be fulfilled in the METPEX database, due to the small number of observations that are available for some of the variables.

As a side note, the METPEX data of T1 and T2 questions are ordinal, since they are expressed through rating scales. Although PCA is originally designed for interval data, this variable reduction technique can also be applied to ordinal data (Kolenikov & Angeles 2004). Furthermore, since no multicollinearity was found among the T2 questions, a varimax rotation was selected for the PCA, in order to ease the interpretation of the factors.

When performing the PCA, the first coefficient checked is the Kaiser-Meyer-Olkin measure, which varies from 0 to 1. Values close to 0 indicate that the sample is inappropriate for the factor analysis due to diffusion in correlation patterns, values around 0.5 can be considered acceptable sample conditions and values close to 1 indicate suitability (Cureton & D'Agostino 1983). Table 3 summarizes the results for the Kaiser-Meyer-Olkin measure in the 21 T2 groups. Thanks to the fact that values goes from 0.848 to 0.946, the sample is said to be suitable for factor analysis.

Furthermore, eigenvalues of an item analyzed in a factor analysis represent the variance associated to this item, while communalities indicate the common variance. The so-called Kaiser's criterion states that the number of items that actually represent the whole group after a reduction corresponds to the number of factors with eigenvalues above 1 if three conditions are met: 1) there are less than 30 variables, 2) average communality is higher than 0.7 and 3) the sample size is higher than 250 (Cureton & D'Agostino 1983). According to this rule, as the last column of Table 3 shows, six groups could be reduced to 2 items while the other fifteen could be reduced to just 1 item. However, only four groups report average communality higher than 0.7, ranging in general from 0.562 to 0.778. Results of eigenvalues in groups with average communality between 0.6 and 0.7 could be considered if sample size is higher than 300 (Cureton & D'Agostino 1983), which is the case in only one group (T1_4). Thus, sample size is relatively small in many T1 categories to give a clear-cutting answer, while requirements of Kaiser-Olkin measures and average communality are met partially or totally

(Table 2). By looking at the overall values in Table 3, we can therefore conclude that PCA can provide valuable insights if applied to our dataset.

From this preliminary analysis we can conclude that there could possibly be more than one underlying dimensions for six out of the above 21 groups, namely T1_5, T1_6, T1_7, T1_12, T1_17 and T1_20. On the other hand, each of the three sets of questions pertaining to the quality of information provision can be represented by a unique latent variable. We present those three sets in Tables 4–6. It can be noticed that indicator T1_10 is more heterogeneous than the others, since Table 5 contains items pertaining to both on-board the vehicle and at stations and stops. Yet a unique component can represent all questions also in this case, according to both Cronbach alpha and Principal Component Analysis.

4 CORRELATION AMONG INFORMATION QUALITY INDICATORS AND BENCHMARK ACROSS DIFFERENT CITIES

In T1 categories whose corresponding T2 representative questions are represented by only one factor according to Table 3, summated rating scales are created by summing up the satisfaction scores of the T2 questions pertaining to the same T1 category (Spector 2006). This process can therefore be applied also to the three indicators reported in

Table 3. Main parameters of the factor analysis within 21 groups of Tier-2 questions.

T1 Category	Sample size of T2 questions	Kaiser-Meyer-Olkin measure	Average communality	Components with eigenvalue above 1
T1_1	89–112	0.931	0.610	1
T1_2	88–110	0.925	0.658	1
T1_3	84–103	0.920	0.639	1
T1_4	240–311	0.946	0.660	1
T1_5	150–180	0.924	0.695	2
T1_6	89–142	0.866	0.696	2
T1_7	59–81	0.848	0.694	2
T1_8	83–103	0.910	0.699	1
T1_9	60–95	0.932	0.665	1
T1_10	159–201	0.937	0.677	1
T1_11	146–231	0.935	0.650	1
T1_12	166–196	0.934	0.711	2
T1_13	209–242	0.911	0.684	1
T1_14	149–281	0.930	0.605	1
T1_15	125–149	0.940	0.707	1
T1_16	152–191	0.909	0.562	1
T1_17	72–105	0.906	0.670	2
T1_18	124–200	0.936	0.713	1
T1_19	172–218	0.938	0.646	1
T1_20	89–113	0.871	0.591	2
T1_21	110–180	0.933	0.778	1

Table 4. List of rating items included in indicator T1_8 (land-side information provision).

T1_8: Provision of information on arrivals and departures was adequate for my needs

Design of timetable information
Knowledge of staff about your transport needs
Lighting of written information
Public address announcements for arrivals
Public address announcements for departures
Quality of information on mobile devices
Reliability of real-time information on arrivals at transport stops
Reliability of real-time information on departures at transport stops
Reliability of real-time information on station arrivals
Reliability of real-time information on station departures

Table 5. List of rating items included in indicator T1_10 (on-trip information provision).

T1_10: The quality of travel information available during journey was good

Adequacy of information for onward journey planning
Clarity and ease of use of audio information
Clarity and ease of use of information displays
Clarity and ease of use of map information
Clarity of directional information
Clarity of warnings and hazards during journey
Clearness of traffic information announcements in stations/transport stops
En route information on mobile devices
Information updates which give sufficient time to change journey (e.g move to new platform)
Reliability of information provided by transport staff
The accuracy of fare information at stations
The accuracy of fare information at transport stops
The accuracy of timetable information at stations
The accuracy of timetable information at transport stops
Clarity of travel information
Provision of information in different languages
Provision of service information in multiple formats

Table 6. List of rating items included in indicator T1_11 (pre-trip information provision).

T1_11: The quality of pre-trip information before I started my journey was good

Accessibility of pre-trip information
Accuracy of fare information on the web/apps
Accuracy of route information on the web/apps
Accuracy of timetable information on the web/apps
Comprehensiveness of information provided
Ease with which seat reservations can be made
Ease with which you can speak to transport staff in advance of your journey
Level of information enabling you to get to the station or transport stop
Level of information enabling you to make connections
Understandability of information provided
Provision of information in different languages
Provision of service information in multiple formats
Quality of service provided by call centre staff
Advanced Information on planned service disruption

Tables 3–5 to come up with three different scales representing the perceived quality of information before starting the trip, when arriving at stations or public transport stops and while travelling. We are aware of the potential shortcomings of such data treatment, due to the fact that ordinal rather than metric variables are considered here; however we notice that scales are commonly developed through such process in social sciences (Spector 2006). Additionally, the process is robust when using scales with at least five points (Bollen & Barb 1981). More complex aggregation methods, tailored on ordinal data, have also been proposed and already applied in the transport sector (Diana et al. 2009).

A correlation analysis is performed between these summated scales and the values of the corresponding T1 question. Since the variables are ordinal, Spearman's rho is calculated instead of Pearson's rho. Spearman correlation rather than Pearson correlation is selected for the calculations because although factor scores are metric variables, T1 questions are ordinal. As explained previously, T2 questions have been designed to deepen the knowledge of the corresponding T1 quality component, therefore we would expect high correlation values. Regarding the three indicators of interest here, correlations are higher for T1_8 (0.638) than for T1_10 and T1_11 (0.387 in both cases) and they are all statistically significant at the 0.01 level. Concerning T1_10, lower correlations could be an indication of the higher heterogeneity of questions composing the scale that was already previously noted. The finding related to T1_11, instead, points at the inherent complexity of aspects pertaining to pre-trip information, that can be provided through a variety of channels and for different purposes, ranging from fares to schedules, from directions to reach service points to the management of unexpected events.

It is finally interesting to conduct a benchmarking exercise to check the mean values taken by these three indicators in some of the aforementioned cities. We preliminarily notice that on one hand the three indicators have a different number of items, on the other some ratings are missing due to non-response. In order to make the values of different indicators comparable and to consider also observations where only part of the responses are available, the values of the scales has been normalized on the 0–1 range, where 0 indicates that the respondent has given the minimum score to all questions s/he answered and 1 the maximum. Additionally, we do not consider observations where less than three items from those composing an indicator have been answered.

It is also important to stress that, due to the previously introduced survey design, no respondent

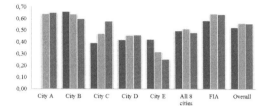

■ T1_8 (land-side info quality) ■ T1_10 (on-board info quality) ■ T1_11 (pre-trip info quality)

Figure 1. Normalised mean scores for three information quality indicators in different cities.

answered to more than one T2 questions set, therefore ratings for the three different indicators are coming from different individuals. Additionally, given the high number of T2 questions sets, sample sizes in each one are relatively small, as reported in the second column of Table 3. Therefore, in the following we are presenting mean values and standard deviations only for those five cities, among the above listed eight, where at least 10 observations for each indicator are available. With such threshold, only two out of three indicators could be computed for one of those five cities.

The results are showed in Figure 1. The first five groups of bars report the standardized mean values of the three indicators for such five cities, whereas the sixth group reports the mean values over all eight cities where the survey was deployed. The last two groups of columns report the mean scores for FIA motorists and the overall values, averaged both over the eight cities and the FIA observations. It can be noticed that FIA motorists were relatively more satisfied than those that answered in cities, so that the sixth group of bars would probably represent a correct term of reference to evaluate the performances in individual cities.

It is insightful to compare indicator mean values pertaining to the same city, but also to make comparisons among cities concerning the different satisfaction patterns. In particular, cities "B" and "E" score relatively higher concerning the quality of land side information and lower concerning pre-trip information quality, while city "C" is showing the opposite trend. Too detailed comments on such aspects would infringe the anonymity of this benchmarking exercise, however we can comment that this is clearly reflecting the different organization of public transport in such cities, particularly concerning the amount of staff that is present on-board the vehicles and the efforts put in setting up a modern information provision system through the web. In summary, by using such indicators, policy makers can therefore realize how they should priorities their efforts in order to improve customer satisfaction concerning information provision.

5 CONCLUSIONS

The main target of the European project METPEX is the study on how quality is perceived for the whole traveler experience. The main goal is to build proper indicators that could provide useful information on users' perception and that could help policy makers in providing inclusive, passenger-oriented integrated transport systems that are accessible by all citizens. More than 6300 interviews have been collected in late 2014 in eight European cities to gain a rich dataset that could help in creating and understanding the latent factors underlying the quality components.

In this paper, our attention was on the 21 quality issues asked in the survey (Tier-1 part) and on the questions representing a focus on a particular sub-set of travel satisfaction aspects (Tier-2 part). We searched for latent factors hidden in them that could more thoroughly represent the corresponding T1 categories. Cronbach alpha followed by Principal Component Analysis allowed us to find the most important components, which are two for 6 cases over the 21 quality issues presented in T1. Among those components, we focus then on three indicators that can measure the perceived satisfaction of information provision services in different phases of the journey experience, namely before starting the journey, when going through public transport stations or stops and while travelling.

The correlation between the three indicators and the corresponding synthetic "T1" questions is high but far from perfect: this shows us that such indicators are giving us richer information that cannot be gathered with a simple question, even if some of the underlying rating items are very detailed and maybe overlapping. Finally, a benchmarking exercise has shown the advantages of considering three separate rather than a unique indicator. The provision of public transport services in most countries is by now the outcome of a complex process where several different entities interact; similarly, informing both prospective and actual passengers might be a task falling under the responsibility of several different actors in the system. An analysis of the patterns of the values of the three indicators across some of the eight METPEX cities has shown the added value for policy makers of the proposed sets of indicators. Some cities perform relatively better in the provision of information before starting the journey, whereas others receive higher satisfaction ratings in the travelling phase.

Future research work will more systematically explore the complete set of indicators stemming from the Tier-2 questions, beyond information provision issues, and will study how such issues are viewed by individuals with different socio-demographic background. Transport quality profiles are also being developed to more effectively communicate the results of the analysis to stakeholders (Diana et al. 2017). A measurement model that is linking each indicator with the overall satisfaction rating for the whole journey experience is also foreseen through the use of Structural Equation Models (SEM). This would allow understanding the relative importance of different aspects studied during the project to shape the overall satisfaction for a journey experience.

ACKNOWLEDGMENTS

This study is part of the METPEX (a MEasurement Tool to determine the quality of the Passenger EXperience) project (www.metpex.eu), which has received funding from the European Union's Seventh Framework Programme for research, technological development and demonstration under grant agreement no 314354.

REFERENCES

Beirão, G. & Sarsfield Cabral, J.A. 2007. Understanding attitudes towards public transport and private car: A qualitative study. *Transport Policy* 14(6): 478–489.

Bishop, C.M. 2006. *Pattern Recognition and Machine Learning, Information Science and Statistics*. New York: Springer-Verlag.

Bollen, K.A. & Barb, K.H. 1981. Pearson's R and coarsely categorized measures. *American Sociological Review* 46(2): 232–239.

Brakewood, C., Barbeau, S. & Watkins, K. 2014. An experiment evaluating the impacts of real-time transit information on bus riders in Tampa, Florida. *Transportation Research Part A* 69: 409–422.

Brakewood, C., Macfarlane, G.S. & Watkins, K. 2015. The impact of real-time information on bus ridership in New York City. *Transportation Research Part C* 53: 59–75.

Carreira, R., Patrício, L., Jorge, R.N. & Magee, C. 2014. Understanding the travel experience and its impact on attitudes, emotions and loyalty towards the transportation provider - A quantitative study with mid-distance bus trips. *Transport Policy* 31: 35–46.

Carreira, R., Patrício, L., Jorge, R.N., Magee, C. & Eikema Hommes, Q.V. 2013. Towards a holistic approach to the travel experience: A qualitative study of bus transportation. *Transport Policy* 25: 233–243.

Cheng, Y. H. & Chen, S.Y. 2015 Perceived accessibility, mobility, and connectivity of public transportation systems. *Transportation Research Part A* 77: 386–403.

Cirillo C., Eboli, L. & Mazzulla, G. 2011. On the Asymmetric User Perception of Transit Service Quality. *International Journal of Sustainable Transportation* 5(4): 216–232.

Cronbach, L. 1951. Coefficient alpha and the internal structure of tests. *Psychometrika* 16: 297–334.

Cureton, E.E. & D'Agostino, R.B. 1983. *Factor Analysis: An Applied Approach*. Lawrence Erlbaum Associates.

de Oña, J., de Oña, R., Eboli, L. & Mazzulla, G. 2013. Perceived service quality in bus transit service: A structural equation approach. *Transport Policy* 29: 219–226.

de Oña, R., Eboli, L. & Mazzulla, G. 2014. Monitoring Changes in Transit Service Quality over Time. *Procedia - Social and Behavioral Sciences* 111: 974–983.

Diana, M. 2010. From mode choice to modal diversion: a new behavioural paradigm and an application to the study of the demand for innovative transport services. *Technological Forecasting & Social Change* 77(3): 429–441.

Diana, M., Duarte, A. & Pirra, M. 2017. Transport quality profiles of European cities based on a multidimensional set of satisfaction ratings indicators. In *Transportation Research Board 96th Annual Meeting, Washington DC, USA, January 2017.*

Diana, M., Pirra, M., Castro, A., Duarte, A., Brangeon, V., Di Majo, C., Herrero, D., Hrin. G.R. & Woodcock, A. 2016. Development of an integrated set of indicators to measure the quality of the whole traveller experience. *Transportation Research Procedia* 14: 1164–1173.

Diana, M., Song, T. & Wittkowski, K.M. 2009. Studying travel-related individual assessments and desires by combining hierarchically structured ordinal variables. *Transportation* 36(2): 187–206.

Eboli, L. & Mazzulla, G. 2011. A methodology for evaluating transit service quality based on subjective and objective measures from the passenger's point of view. *Transport Policy* 18(1): 172–181.

Eboli, L. & Mazzulla, G. 2012 Performance indicators for an objective measure of public transport service quality. *European Transport* 51: 1–21.

Grotenhuis, J.-W., Wiegmans, B.W. & Rietveld, P. 2007. The desired quality of integrated multimodal travel information in public transport: Customer needs for time and effort savings. *Transport Policy* 14: 27–38.

Kolenikov, S. & Angeles, G. 2004. The Use of Discrete Data in PCA: Theory, Simulations, and Applications to Socioeconomic Indices. In *Proceedings of 2004 Joint Statistical Meeting, Toronto, Canada, August 2004.*

Lu, A., Aievoli, S., Ackroyd, J., Carlin, C. & Reddy, A. 2009. Passenger environment survey: representing the customer perspective in quality control. *Transportation Research Record* 2112: 93–103.

METPEX. 2013. Development of standard format for the measurement instruments. Project Deliverable D3.1.

METPEX. 2014. Report on survey results and behavioural analyses from each location. Project Deliverable D4.2.

METPEX. 2015. A comprehensive set of quality and accessibility indicators for transport services. Project Deliverable D5.2.

Nathanail, E. 2008. Measuring the quality of service for passengers on the Hellenic railways. *Transportation Research Part A* 42: 48–66.

Schmitt, L., Currie, G. & Delbosc, A. 2015. Lost in transit? Unfamiliar public transport travel explored using a journey planner web survey. *Transportation* 42:101–122.

Spector, P.E. 2006. Summated rating scale. In Jupp, V. (ed.), *The SAGE Dictionary of Social Research Methods*: 295–297. London: SAGE Publications.

Stradling, S., Anable, J. & Carreno, M. 2007. Performance, importance and user disgruntlement: a six-step method for measuring satisfaction with travel modes. *Transportation Research Part A* 41(1): 98–106.

Suhr, D.D. 2003. Principal Component Analysis vs. Exploratory Factor Analysis. *SUGI30, Statistic and Data Analysis*: 203–230.

Susilo, Y.O., Woodcock, A., Liotopoulos, F., Duarte, A., Osmond, J., Abenoza, R., Anghel, L.E., Herrero, D., Fornari, F., Tolio, V., O'Connell, E., Markucevičiūtė, I., Krtharioti, C. & Pirra, M. 2016. Deploying traditional and smartphone app survey methods in measuring door-to-door travel satisfaction in eight European cities. *WCTR 2016. Shanghai, China, July 2016.*

Tyrinopoulos, Y. & Aifadopoulou G. 2008. A complete methodology for the quality control of passenger services in the public transport business. *European Transport* 38: 1–16.

Tyrinopoulos, Y. & Antoniou, C. 2014. Public transit user satisfaction: Variability and policy implications. *Transport Policy* 15: 260–272.

Transport Infrastructure and Systems – Dell'Acqua & Wegman (Eds)
© 2017 Taylor & Francis Group, London, ISBN 978-1-138-03009-1

Public transport resilience during emergency: A simulated case in Torino

F. Deflorio, H.D. Gonzalez Zapata & M. Diana
Department of DIATI, Politecnico di Torino, Torino, Italy

ABSTRACT: Nowadays detailed Public Transport (PT) data are available thanks to open standards such as the General Transit Feed Specification (GTFS), introduced to enable information services usually provided via web. In this paper this dataset has been used, after some conversion and filtering operations, for accessibility analysis, based on some indicators to measure how the zones of a city are well connected, in terms of capacity, number of transit options and their efficiency.

The study focuses mainly on the assessment of the consequences of transport resources disruption, considering a large flooding event as risk scenario, which could generate also the closure of bridges and then limit the PT coverage. The selected case study is in Torino (Italy), to measure the impact on the various zones of the city, in terms of variation in their transit service accessibility. Relevant capacity reductions (more than 90%) are observed at the north-east of the city.

1 INTRODUCTION

In recent years a number of analyses focused on understanding the behavior of a transport network either public or private, under extraordinary working conditions. The goal is generally to measure the degree of reactivity and the strategic role of certain connections within the network, thus strengthening them in order to minimize service disruptions.

The type of failures that may occur over a network due to external agents can be quite varied, ranging from natural disasters such as floods, forest fires and landslides, to causes of anthropic origin, such as transit incidents or terrorist attacks within the system. This work will focus on the total or partial disruption of public transport lines, without considering the possibility of still operating them through detours.

The main objective is to understand the performance and behavior of the transit transport network under normal and anomalous conditions. According this aim, a base network model without considering disruptions on the system is built and analyzed, in order to create a general view of the transport infrastructure. In the second phase, service disruptions are introduced and some performance indexes are proposed and monitored. The following section will lay out the theoretical foundations to define such indexes, based on accessibility and resilience concepts.

2 MEASURING NETWORK ACCESSIBILITY AND RESILIENCE

Accessibility is a concept difficult to define, even if it is widely used in different contexts. One of the most common definitions in the literature was proposed by Hansen (1959), according to whom accessibility is "the potential of opportunities of interactions", where the concept is associated with the spatial distribution of activities, considering the preference of people or companies to overcome spatial separation. In that way, the accessibility is not a completely objective parameter, since it depends on the ability and desires of the users to cover longer or shorter distances to access to certain service.

An accessibility measure may be analyzed from different perspectives depending on the application. Different categorizations can be found in literature, which provide an overview of components of accessibility. For example, Geurs & van Wee, 2004 distinguish four basic perspectives on accessibility:

1. Infrastructure-based measures, which analyze the performance of transport infrastructure.
2. Location-based measures, based on indicators related to the spatial distribution of activities.
3. Person-based measures at the individual level, considering the individual requirements and limitations.
4. A utility-based measure, which considers the benefits that people derive from level of access to the spatially distributed activities.

Geurs & van Wee, 2004 also define four basic and mutually interacting accessibility components:

1. The land-use component, given by the spatial distribution of opportunities.
2. The transportation component, considering the supply and demand of transport.
3. The temporal component, reflecting the availability of opportunities.
4. The individual component, reflecting the needs, abilities and opportunities of individuals.

In general those components interact between them, but the emphasis on some of them depends on the study purpose. For example transport studies concentrate on the transport component of accessibility, using transport models based on factors, such as travel time, reliability, comfort, but ignoring the land-use and individual components of accessibility. On the contrary, urban planners typically focus on the land use component and less on the transport component and differences between population segments (Geurs & Östh 2016).

Beyond accessibility issues, studies related to infrastructure and network resilience have been popularized in recent years, in part due to the increase of natural disasters either in the number and intensity, as well as terrorist attacks in transport networks (Mcmanus et al. 2007; Henry & Ramirez-Marquez 2012; Barker et al. 2013). The concept of resilience is related with the capacity of a system to recover after an external disruption; this capacity determines the time to return at the initial state, a more resilient system will have a shorter recovery time compared to one less resilient (Henry & Ramirez-Marquez 2012).

Barker et al., 2013 illustrates the system behavior in function of time, where the states through which the entity passes before returning to the initial or desired state are the following:

– Reliability, governing the system in the absence of disruptions; it contemplates the availability of service (Zhang et al. 2011).
– Vulnerability, representing the adverse effect on the network performance caused by any interference (Nagurney & Qiang 2007).
– Survivability, i.e. a mitigation approach to the vulnerability pursuing a minimization of the original impact (Westmark 2004).
– Recoverability, considering the speed at which the system returns to the original or desired state.

In terms of transport, network resilience requires robustness to minimize the impact due to disruptions as well as a rapid recovery back to normal operations and performance (Barker et al. 2013). The impact depends on safety margins established during the service planning and on system operation

parameters, such as network connectivity, reserve capacity of not affected links or residual capacity of the affected arcs, influence the network reactivity and survivability. The system recovery speed depends on the spread/absorption capacity of the network and the complementary management tools available for the system.

3 APPLIED METHODOLOGY

The analysis of public transit performance is mainly based on the methodology proposed by Mamun et al. 2013, including some modifications in the definition of the indexes that will be later discussed.

3.1 Trip alternatives definitions

The selection of the alternative transit paths for any Origin Destination pair is made for a specific time of the day, considering only the lines which present trips during the analysis period. In this case, both direct connections and the alternatives which include one line transfer are included. These alternatives are classified as follows:

– Direct connections, which include the trips made using only one transit line from the origin to the destination.
– Direct Transfers, related to the trips made with one transfer; in this case the end stop of the first track represents the initial stop of the second.
– Indirect Transfers, which refer to the trips with one line transfer, but in this case, the walking distance between the transfer stops is also considered; the final stop of the first track and the initial stop in the second line are different.

3.2 Transit accessibility analysis

Accessibility is first considered to evaluate the existing transit service, and then its variation is assessed to analyze the impact generated by disruptions on the public transport network. Traditionally, accessibility was calculated as the combination of spatial coverage (related to the proximity of people to stop) and the service characteristics (frequency and capacity); however, the transit connectivity (trip coverage) is included in order to take account of the journey comfort (Mamun et al. 2013). Therefore, transit accessibility (A_{ijl}) is measured as the combination of the spatial coverage score at the origin zone and the total hourly service capacity (temporal coverage) between the origin and destination.

3.2.1 Spatial coverage
Spatial coverage (R_{il}) is the proportion of the area in the origin i served by the transit line l. It is

computed using the ratio between the spatial coverage area of a transit line ($B_{i,l,buffer}$) and the total area ($B_{i,total}$) (Mamun et al. 2013).

$$R_{ij} = \frac{B_{i,l,buffer}}{Bi_{i,total}} \qquad (1)$$

The spatial coverage area is here considered as the area covered by a particular route within a walking distance of 300 meters from the stops. This calculation differs from the original methodology, where the buffer is calculated around the line and not around the stop. The area covered by each stop is calculated in a first step, then, the coverage area of the line in the origin is determined as the sum of the areas covered by each stop belonging to the line found in the analyzed zone.

$$B_{i,l,buffer} = \sum_k B^k_{i,l} \qquad (2)$$

A limitation of this approach may be the overestimation of the spatial coverage of those lines where stops are very close between them.

3.2.2 Temporal coverage

The temporal score refers to the available seats per hour from the origin i to the destination j through line l; this parameter is calculated as the product of the hourly public transport runs of l from i to j (service frequency) and the capacity of the vehicles on the line.

$$S_{ijl} = V_{ijl}U \qquad (3)$$

The capacity of the line is a function of the total number of seats by vehicle (U) and the frequency of service (V_{ijl}); this information is generally provided by the transportation agencies.

3.2.3 Transport accessibility

The spatial and temporal coverage are combined to obtain the transport accessibility (A_{ijl}) for each origin/destination pair by transit line l connected to the origin:

$$A_{ijl} = R_{il}S_{ijl} \qquad (4)$$

Then the total accessibility for each i-j pair is calculated by considering the aggregation of the possible trip alternatives between the zones. It is important to note that in this case the selected alternatives are those resulting after the grouping done in function of the line connected to the origin:

$$A_{ij} = \sum_l A_{ijl} \qquad (5)$$

In real situations, the level of service is conditioned by the network congestion. Therefore, it might be useful to consider a demand parameter linked to the use of the transport service. Since this information is not available in all cases, the original methodology proposed by Mamun et al. 2013 uses the population of zones. However, this requires the hypothesis that the population is uniformly distributed and that all zones are populated, presenting problems in areas with a large number of transient population (industrial zones, business districts or even parks). Since the focus is on the loss of travel opportunities in emergency events, it was considered more appropriate not to use any parameter related to service demand, in order to avoid distortion of results.

3.3 Travel time estimation

The total door-to-door travel time between the origin i and the destination j is estimated as the aggregation of different components, where components change depending on the alternative analyzed. In all cases the access and egress time are considered and taken equal to 5 min each one, while in-vehicle time is calculated through the service data available.

Additionally, it is included a waiting time at the stops (entry and transfer stops) assumed as one-half of the scheduled headway. If the scheduled headway of an entry stop is greater than 10 min the waiting time is assumed as 10 min. This default value is taken from the Mamun et al. 2013 methodology and it represents the average waiting time from the National Household Travel Survey (NHTS) data in the United States.

In that way, the travel time (T_{ijl}) calculated for a direct alternative considers: access time, waiting time, in-vehicle time and egress time:

$$T_{ijl} = T_{access} + T_{wait} + T_{in-vehicle} + T_{egress} \qquad (6)$$

The travel time for transferred trips includes an additional in-vehicle time for the second line and a waiting time at the transfer stop equals to one-half of the schedule headway of the second route, even if the waiting time turns out to be more than 10 minutes.

An additional walking time from the ending stop on the first track to the initial stop on the second one, is calculated considering the air distance between the involved stops and a walking speed of 0,7 m/s.

Finally, the concept of connectivity decay factor (f_{ijl}) is introduced in order to take into account the perception of people when selecting the route to travel from one place to another (Eq. (7)).

$$f_{ijl} = \frac{L}{1 + \alpha e^{-\beta l}} \qquad (7)$$

where α and β are coefficients available in Mamun et al. 2013, estimated using information related with the transport demand, and L is equals to 1.

3.4 *Definition of performance indicators*

The public transport network behavior is assessed under three different components: capacity, network flexibility and connection efficiency. All of them related with the zones accessibility and the performance of network under unusual conditions. Each parameter has an associated indicator: TOI, CONX, EFF respectively.

This study does not consider the origin/destination demand matrix. Therefore, in order to evaluate the performance of the network, each zone is considered both as an origin and as a destination Thus, a square matrix with the results of each index is generated, whose dimension correspond to the number of zones studied. Additionally, the analysis is made hourly, thereby measuring the variation of parameters during the day.

3.4.1 *Transit Opportunity Index (TOI)*

The Transit Opportunity index measures the possibility that people have to move from one zone to another in terms of the total seats per hour available at a specific time of day (service capacity), modified by the perception of journey alternatives in function of the travel time. According to the definition of the decay function, trips with a travel time grater that 60 minutes contribute very little in the index result (Eq. (8)).

$$TOI_{ij} = \sum_l A_{ijl} f_{ijl} \qquad (8)$$

The TOI index is calculated by the combination of spatial and temporal factors, summarized as the multiplication of the Transit access and Transit connectivity. First, the index for each trip alternative is independently computed, then, the total TOI for each Origin-Destination pair is calculated by adding all single alternatives between the zones. The Transit Opportunity Index is also useful to measure the level of opportunity of any origin or destination into the city; this value can be interpreted as the possibility of entering or leaving a specific zone in a specific moment of the day (e.g., hospitals, schools, universities, railways stations, etc.).

3.4.2 *Connectivity Index (CONX)*

The connectivity index measures the flexibility level of the public transport network in terms of the possible trip combinations between each Origin-Destination pair. The index represents the total number of trip options considering both the possible combinations of routes (direct or indirect changes)

as well as the transfer alternatives between each pair of lines. In that way, each direct alternative represents just one connection between the zones, while in the case of options which includes a line change, either directly or indirectly, each possible combination of routes is taken as a link alternative, and in the cases of two partially overlapping routes, which share more than one transfer stop, each one of them would represent an additional alternative linkage. According to this, it is assumed as more flexible a transfer option which allows diverse change alternatives between the routes associated with a specific trip choice.

3.4.3 *Efficiency Estimation (EFF)*

The transit efficiency is function of the total travel time between two zones and the distance between their centroids. It is used to measure how convenient or expensive is a journey in terms of time, since travel time is perhaps one of the most influential parameter in the selection of routes within an urban context, even more important than the distance between places (Balijepalli & Oppong 2014). In the case of public transport, the time employed is not directly proportional to the physic distance between the places as in the case of private transport (considering equal speed), where in general the shortest or the fastest path is selected.

Public transport lines are generally designed to pass through the areas of greatest interest for the population, but in certain cases, some regions which are rather close among them require a relatively higher travel time. Therefore, the proposed index considers a relative coefficient, which is function of the minimum physical distance between the interest points (centroid zones), determined using the Dijkstra's method on the road network graph, and the minimum travel time between them measured on the public transport network.

4 SIMULATION OF SERVICE DISRUPTIONS

The main objective of the transport analysis presented in this work is to understand the performance and behavior of any public transport network under special work conditions. First, the simulation in normal conditions without considering disruptions on the system is performed to create a general view of the transport infrastructure. Then, the calculation process considering the disruptions into the network provides further results for comparison. This procedure has been implemented in Python, using QGIS to manage and represent geographical information.

All the transit data used to perform the analysis are generally available through the General Transit Feed Specification (GTFS), an open standard widely adopted to publish PT data on internet. This information includes the stops and routes, frequency of service, vehicle classification, etc. (More information can be found on: https://developers. google. com/transit/gtfs/)

4.1 Modeling of network failure

The model considers the network as a graph, described by a sequence of nodes (stops) and arcs, linked between them through different data structures, which connect the nodes with transport routes, and these in turn with the origin and destination zones. This organization, where there is not a unique data structure which includes the graph information, makes difficult the removal of an element within the data set, therefore, an "ignore it" approach is chosen during the calculation, without removing it from the graph.

In this order of ideas, the introduction of system disruptions to the model is through as an established format shown in Table 1.

– The field "ID" just represents an identification of the disruption (it is not employed in the model)
– The "Type" column specifies the disruption classification, which indicates how to consider the line. There are two interruption types: 0 when the line is completely disabled; 1 if the line has a partial failure (in this case the two segments of lines are assumed as independent lines not connected between them).
– In the "Shape" field the line codes are introduced; one route is composed at least by two shapes, one for each line direction. In some cases, there are more than two shapes, some of them available only in specific hours.
– In the column "From" the last enabled stop before the onset of the failure must be indicated. Depending on the route direction analyzed, each direction will have a different "beginning" stop. This field is required only if the disruption type is 1.
– Finally, the column "To" represents the first stop enabled after the disruption. In a similar way, it depends on the route direction and is only required when disruption type is 1.

In the network simulation two different types of failure analysis are contemplated (namely, specific or general analysis), depending on the disruption extension or the study aims. In both cases, the above introduced classification of failure types and the standard structure presented in Table 1 are considered.

Table 1. Established format for disruption data.

ID	Type	Shape	From	To
1, 2, ..., n	0/1	Shape code	Stop code	Stop code

The specific approach to simulate disruptions can be feasible if only a network failure point is analyzed, or if the study focuses on few lines of the system, because in this case the input file is manually set, then, any input data verification can be managed by the analyst.

On the other hand, the general analysis considers a wider disruption area represented by a shapefile. In this case, the algorithm considers all the public transport stops and lines within the affected area, filters them and generates the input disruption file automatically following the format in Table 1. Unlike the specific analysis, the classification of failure type (0/1) is made following some parameters, in order to generate a more realistic analysis:

– If the affected percentage of the line is greater than 80%, the line is assumed as completely disrupted (Type 0).
– If the line is not completely disrupted, it is initially assumed as partially damaged and the residual segments are verified. If the residual segments have less than 4 stops (3 arcs), the segment is not considered useful (very short) and it is removed; after that, the percentage of damage is verified again according the previous condition.

The model makes three validations before considering a line as potentially available after the disruption; the sequence is shown in Figure 1.

4.2 Measurement of disruption effects

The effects on the public transport network are estimated through the relative variation in the indices defined (TOI, CONX, EFF and EFFAV, the latter to be later introduced). These differences allow visualizing the effects for the various zones, in terms of loss of capacity, variation on travel time and reduction of network flexibility. The variation is calculated as a percentage respect to the base scenario (Eq. (9)).

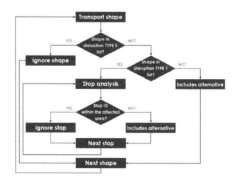

Figure 1. Decision making structure for the alternatives chosen (disruption analysis).

$$index_{variation}[\%] = \frac{index_{base} - index_{disrupted}}{index_{base}} * 100 \qquad (9)$$

4.2.1 Transit Opportunity Index Variation

The Transit Opportunity Index (TOI) variations are related to the transport capacity modification and the changes on the total travel time. This variation can take both positive and negative values due to the way in which its calculation is structured.

The negative variations in the TOI are associated for example, with the loss of a transit line with a high hourly capacity. This negative variation helps to determine the most critical links in the network. By contrast, the positive changes are more related with trip time variations. Indeed, if the sample connection is selected based on the hourly capacity, there is the possibility that the alternative chosen is not the best, considering the travel time, and the result for TOI in the base analysis is penalized by the time decay function. For that reason, if this connection is removed, it may penalize the accessibility from the capacity point of view, but improve it in terms of travel time. If the benefit due to the travel time improvement is greater than the variation due to the loss of capacity, the TOI is larger compared to the base scenario result.

4.2.2 Connectivity index variation

This index is directly related to the number of travel alternatives and it is mainly sensitive to those including any line change. The importance of a public transport line for a specific city zone is measured by the number of line connections lost when the transit route is disturbed. In general, higher hourly capacity lines (e.g. subway lines or trams) have a larger impact compared to low capacity lines, since they have a large number of line changes along its route.

4.2.3 Efficiency indices variation

The variation of the efficiency indices (both the efficiency of the fastest route and the average efficiency) is directly related to any change in travel times due to the elimination of one or more connections between the studied areas, and it can be associated with the loss of a less efficient or a more efficient route.

A first index (EFF) estimates the efficiency of the fastest route; a variation in this parameter means that the affected public transport line is part of the shortest path (in terms of time) between the two analyzed zones. On the other hand, the index of average efficiency (EFFAV) measures the global variations in the travel time between zones. This coefficient might take either positive or negative percentage values, because it is based on an average travel time, calculated considering all possible alternatives between a specific origin-destination pair.

5 APPLICATION TO THE TORINO CASE STUDY

As a case study, the city of Torino was chosen, due to its morphological features that were considered relevant for the proposed analysis, such as the confluence of four rivers within the urban area of the city, which interact at different levels with the public transport infrastructure, representing a potential risk factor of the system.

Considering the position of these rivers, the most interesting to analyze is perhaps the Dora river, which crosses the city near to the center dividing it in two parts. The river is mostly channelized along its route in the city and it is crossed by a large number of bridges, many of them built centuries ago, with relatively low clearances, therefore, more susceptible if floods occur. This river is also connected to the Mont Cenis dam, for which inundation maps related to its failure are available (Fig. 2).

5.1 Simulation data

For this analysis, the inundation area shown in Figure 2 is considered as the affected zone, according to the emergency plan available. The simulation was performed in normal conditions and in the emergency scenario after the dam failure. In both cases, the same city zoning employed by urban demographic studies was used, thus considering 94 zones.

The needed information concerning the Turin public transport system is provided on the web by the *"Gruppo Torinese Trasporti (GTT)"* and *"5T S.r.l"* transportation agencies, through the above introduced GTFS standard format. The complete transport network is composed by 1264 shapes (grouped in 212 transit lines) and 6965 stops.

5.2 Base scenario results

The model uses the travel times included within the GTFS database which is not considering real time traffic dynamics, but it rather takes into account the experienced vehicle travel times used to set the timetable. The indices are calculated on an hourly

Figure 2. Inundation map in Turin for the scenario of Mont Cenis dam failure.

basis for a whole week, in order to observe both their hourly and daily variations. However, Tuesday values on the morning peak hour (7:00 am) are only presented here for briefness.

The algorithm calculates the four indices for each origin-destination pair and, to visualize them on the map, it aggregates the results for each zone as origin or destination. In the case of the Transit Opportunity Index (TOI), the results is measured in terms of transit capacity and it is shown in terms of relative values (Fig. 3), calculated respect to the maximum absolute value of TOI reached in the week, equals to 59517seats per hour per zone. In Figure 3, it is apparent that the transport supply is concentrated around the city center.

According to the map shown in Figure 4, the travel efficiency is much higher on the peripheries of the city, having an opposite behavior pattern respect to that shown by the transit opportunity analysis.

This greater efficiency in the outskirts of the city is associated with a decrease of travel time per kilometer, usually due to a longer distance between stops as well as more direct routes and possibly to both less congestion and to the presence of road infrastructures with higher free flow speed, all elements enabling higher travel speeds.

5.3 Variations due to PT disruption

The disruption analysis tries to estimate the effects on the public transport network due to a failure in the Mont Cenis dam, using as input data the inundation area shown in Figure 2. The analysis is made in general mode (see section 4.1) and the disruption lines are selected automatically. During the simulation, 335 affected shapes were found (174 partially affected and 161 completely damaged).

In Figure 5 the variation of TOI is presented after the disruption: there is an evident general impact in the city, with a greater impact in the direct affected zone, but also in the norther part of the city, where the variation rises more than 90%, whereas in the south, the reduction is little (less than 20%).

In the case of efficiency the trend is the same, but the difference between the north and the south of the city is much more visible.

According to the information presented on the maps, the impact on the public transport network generates a complete disconnection of the zones at the north of the city, affecting both production and residential zones (Figs. 6–7).

A particular situation is presented in the zones located on the hills (on the right side of the maps). Those zones are not directly affected by the inundation, but the estimated level of impact suggests that the related public transport lines pass through the north of the city to reach the city center. This can be clearly confirmed by analyzing the connection loss shown in Figure 7, which reflects a high vulnerability of those zones.

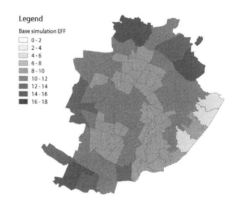

Figure 4.   EFF by origin—base scenario analysis (Km/h).

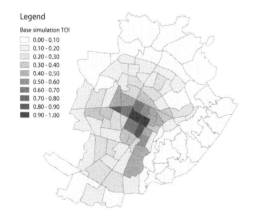

Figure 3.   TOI by origin—base scenario analysis (%).

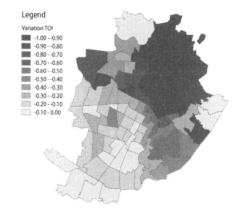

Figure 5.   Variation of TOI by origin (%).

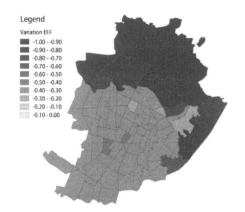

Legend

Variation EFF

- ■ -1.00 - -0.90
- ■ -0.90 - -0.80
- ■ -0.80 - -0.70
- ■ -0.70 - -0.60
- ■ -0.60 - -0.50
- ■ -0.50 - -0.40
- ■ -0.40 - -0.30
- ■ -0.30 - -0.20
- ■ -0.20 - -0.10
- ☐ -0.10 - 0.00

Figure 6.   Variation of EFF by origin (%).

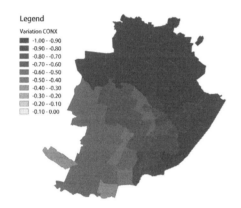

Legend

Variation CONX

- ■ -1.00 - -0.90
- ■ -0.90 - -0.80
- ■ -0.80 - -0.70
- ■ -0.70 - -0.60
- ■ -0.60 - -0.50
- ■ -0.50 - -0.40
- ■ -0.40 - -0.30
- ■ -0.30 - -0.20
- ■ -0.20 - -0.10
- ☐ -0.10 - 0.00

Figure 7.   Variation of CONX by origin (%).

6   CONCLUSIONS

The methodology proposed and applied in this work offers an alternative to traditional computing accessibility models for transportation systems based on both supply and demand data. Considering that not always detailed and reliable demand data are available, this model only uses information related to public transport service characteristics, as well as a city zoning to perform the analysis. The study does not consider the analysis of the passengers' distribution within the system, but it determines the weight of the alternatives in terms of their travel time, which penalizes options that have higher values. This can be directly associated to the users' perception, since passengers tend to prefer faster alternatives. The model quantifies three main parameters: the access to public transport and its capacity (TOI), the degree of network connectivity (CONX) and the efficiency of its connections (EFF).

Concerning the scenario with public transport network interruptions, as expected, the obtained variations of the three indices show that the impact is mainly concentrated in the direct influence area

of the affected lines. More remote affected areas are nevertheless identified, mainly in the eastern part of the city. This means that among all travel alternatives available, at least one of them crosses the disrupted area. The most relevant values were found in the north-eastern quadrant, where the reduction is of more than 90% in terms of transport capacity (TOI). The other indices indicate a total disruption in the north of the city and in some eastern zones, representing, in some cases, a risk of total disconnection, even if the zone is not included into the inundation area.

Finally, the model was built based on the public transport system of Torino, but the way in which the different algorithm's modules were built makes the model easily transferable to other cities, provided that the input data are available in the GTFS format.

REFERENCES

Balijepalli, C. & Oppong, O., 2014. Measuring vulnerability of road network considering the extent of serviceability of critical road links in urban areas. *Journal of Transport Geography*, 39, pp.145–155. Available at: http://dx.doi.org/10.1016/j.jtrangeo. 2014.06.025.

Barker, K., Ramirez-Marquez, J.E. & Rocco, C.M., 2013. Resilience-based network component importance measures. *Reliability Engineering and System Safety*.

Geurs, K.T. & Östh, J., 2016. Advances in the Measurement of Transport Impedance in Accessibility Modelling. *EJTIR Issue*, 16(2), pp.294–299.

Geurs, K.T. & van Wee, B., 2004. Accessibility evaluation of land-use and transport strategies: Review and research directions. *Journal of Transport Geography*.

Hansen, W.G., 1959. How Accessibility Shapes Land Use. *Journal of the American Institute of Planners*, 25(2), pp.73–76. Available at: http://www.tandfonline.com/action/journalInformation? journalCode = rjpa19.

Henry, D. & Ramirez-Marquez, J.E., 2012. Generic metrics and quantitative approaches for system resilience as a function of time. *Reliability Engineering and System Safety*, 99, pp.114–122. Available at: http://dx.doi.org/10.1016/j.ress.2011.09.002.

Mamun, S.A. et al., 2013. A method to define public transit opportunity space. *Journal of Transport Geography*, 28, pp.144–154. Available at: http://dx.doi.org/10.1016/j.jtrangeo. 2012.12.007.

Mcmanus, S. et al., 2007. Resilience Management A Framework for Assessing and Improving the Resilience of Organisations Executive Summary, New Zealand.

Nagurney, A. & Qiang, Q., 2007. A Transportation Network Efficiency Measure that Captures Flows, Behavior, and Costs with Applications to Network Component Importance Identification and Vulnerability. In *POMS 18th Annual Conference*. Dallas, Texas, U.S.A., pp. 1–22.

Westmark, V.R., 2004. A Definition for Information System Survivability. In *Proceedings of the 37th Hawaii International Conference on System Sciences*. Honolulu, Hawaii.

Zhang, C., Ramirez-marquez, J.E. & Rocco Sanseverino, C.M., 2011. A holistic method for reliability performance assessment and critical components detection in complex networks. *IIE Transactions*, 43, pp. 661–675.

Transport Infrastructure and Systems – Dell'Acqua & Wegman (Eds)
© *2017 Taylor & Francis Group, London, ISBN 978-1-138-03009-1*

The "form" of railways: Between construction and regeneration of contemporary landscape

F. Viola
DICEA, Università degli Studi di Napoli "Federico II", Napoli, Italy

C. Barbieri
DiARC, Università degli Studi di Napoli "Federico II", Napoli, Italy

ABSTRACT: Transport infrastructures—in particular railways—have a "special beauty", punctual and systemic at the same time. On the one hand, infrastructural architecture can be exemplary works of technical merit which express adherence to practical functionality and the wise use of materials according to their nature and costs. On the other, each infrastructural work has an "added value": it belongs to a complex landscape system of greater scale through which it affects the visual, physical and cultural quality of the inhabited landscape. The recent closure of many obsolete railway lines makes it necessary to redefine the landscape character using a new design strategy more attentive to the preservation of identity characters, to avoiding the homologation of the contemporary landscape. This kind of design strategy proposes the enhancement of railway architectural characters, thus avoiding a nostalgic reconstruction of their original form. The goal is to implement a new design paradigm which works with surviving elements and signs, produced by time, and looking for to ensure successful integration between old and new. When the engineering works are abandoned they gain a unique character—becoming aesthetically interesting. Residual elements are also used in many fields of Contemporary Art to give them new life and new social functions through post-production processes and formal contaminations. These artistic techniques can be applied to reuse abandoned railways too: hybridization, repetition and serial fittings can generate surprising and expressive new projects.

1 INTRODUCTION

Over the past decades the railways network went through a huge phase of restructuring in all industrialized countries due to the changing mobility needs of goods and people, to the reallocation of productive activities in the territory—such as raw materials extraction or shipbuilding industry—to which the railroad was bound from birth.

The disposal of a large part of the railways affected equally both the major metropolitan areas and the peripheral areas where there is a strong competition with the road transport system—cheaper and widely disseminated—forcing the closure of many kilometers of "deadwood", lines with a limited number of users and unsustainable management costs. A precious legacy of areas and architecture made of tracks of rails, depots, workshops for the repair of trains, a loto of small stations and surveillance tolls, has suddenly been abandoned and, in many cases, is still waiting to know their fate. It is a set of artifacts with an extraordinary economic, urban and cultural value, not only for the great opportunities that the conversion could offer to the redevelopment of urban and suburban areas—originally marginal compared with regional dynamics, but now in a strategic position -, but also for the intrinsic qualities of infrastructural places characterized by an unexpected and amazing mix of archaeological finds and wild nature (VIOLA 2004). Industrial buildings, domestic and rural artifacts, slender pylons, hardy steel portals suspended on the tracks and massive brick sheds, are one beside the other in these places with no apparent relationship. The connections between the repertoire of architectures and tracks, on the one hand, and the set of formal, perceptive and symbolic relations, on the other, are the leitmotif of our study and the qualifying aspect according to which we intend to suggest some inputs for the project of transformation of these places.

On the issue of brownfield infrastructure generally two different positions have been taken. On the one hand there were those who only seize the opportunity of building on abandoned areas. On the other, there were those who, by agreeing to compete without bias with pre-existing conditions, tried to redeem the qualities of the places through the redevelopment

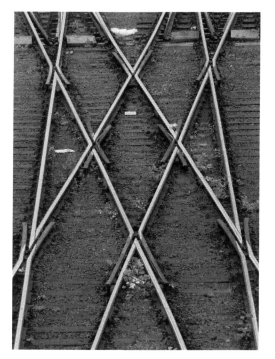

Figure 1. Intersection of tacks.

project. But along with an extraordinary heritage of architecture that have marked the history of modern transportation, the disposal of the railways questioned also the ways of crossing the space and the perception of the landscape from these places, developed over time. A deep transformation of the areas and buildings often followed the closure of railways: within a few years these areas and buildings have been homologated to many anonymous other that make up the contemporary landscape.

All the traces of a previous infrastructural function have been deleted and where previously ran the trains were opened new roads, built other houses, shopping centers, industrial warehouses.

In Italy the railway network restructuring process started later than in other European nations—even for the decreased availability of resources—and the conversion of railway areas can still be seized as an opportunity to put in place interventions more careful to identity of places and useful to trigger virtuous recovery actions of the many abandoned sites around the sediments of tracks. The recovery of abandoned railroads can be an opportunity to introduce a new architectural and environmental quality in those marginal places among the city, the infrastructure and the countryside, where signs of degradation are more evident. Just this filed, so far without significant experiments in planning, is the topic of this paper.

Figure 2. Elements and legacy of railways.

The hypothesis from which we started is that the reuse of abandoned railways is an important opportunity to implement regeneration processes of degraded or marginal areas compared to regional dynamics, provided that they respect their own specific identity, an identity made of recurrences and singularity.

2 BETWEEN ENGENEERING AND ARCHITECTURE

The railways have traditionally been a specific competence of the engineers and the only occasions when architects had a role in the infrastructural design process were those in which it was necessary to give a new form to values expressed by the infrastructure themselves: i.e. the emblematic case of stations, that was a favored theme in the last century because of the experimentation of new typologies related to the experience of the trip. It was established the belief that the construction of the railways has been exclusively determined by technical and economic reasons, indifferent to the framework conditions in order to make the transport as fast and economical as possible (MUMFORD 1961).

After years of unconditional faith in the domain of reason over reality, there is today more awareness that vagueness and contradictions are intimately part of the architect's work, creating opportunities for his progress. In fact, the condition of permanent instability is a urge to explore new paths, to experience more and more advanced expressive approaches, up to reach those extreme limit points beyond which the initial borrowings principles are contradicted and replaced by new ones. Compared to many studies of the recent years about railways, in this paper we will focus especially on architectural built form that links the railway to the territory and on the role that it had—and may still have—in landscape construction. From our point of view, the attention to the relationship that architecture establishes with the reality of places is important when "reality" is not interpreted as a condition to simply join or undergo, but as something to positively transform, giving the project an active role and not simply a decorative one.

On the contrary, many recent architectural and engineering studies about transport infrastructures seem to have overlooked the complex constructive and type-morphological relationships which traditionally have bind the artifacts to places, focusing on increasingly evanescent and impalpable aspects. In Architecture, these studies transcend the purpose of projects' realization to make them only producer of non-figurative quality; in Engineering, they split the skills in increasingly specialized fields

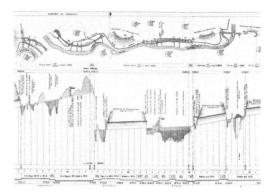

Figure 3.   Railway Parma-La Spezia, 1879–1894.

and give priority to the theoretical and virtual simulations instead of field testing (BRANZI 2006).

In recent years, among architects and urban planners it has spread the belief that the city and the territory of the future will be characterized by increasingly weak urbanization systems and intangible and enzymatic architectures, without defined formal features and material sediment. Furthermore, engineers who deal with infrastructures are increasingly attracted by resources and traffic flows management rather than—as is their tradition-, the creation of artefacts able to leave tangible traces in the territory.

3 THE RE-USE OF RAILWAYS: TECHNICAL FORMS AND LANDSCAPE

The topic of the relationship between railway infrastructure and territory, in its most recent meaning of the re-use and environmental restoration, is today in a complex scenario, in which the architecture skills appear wider and less defined than in the past: a dilation of the field of work involving both the issues addressed and the methods and tools used in the project. Overcoming the traditional opposition between the consolidated city and suburbs, between the built and natural space—that have long dominated the theory and practice in the last century -, the opportunities of transformation have become much more frequent and important than the new constructions for architects. The contemporary territory is no longer considered as a neutral space, a void to fill with new objects, but it is conceived as a complex place in which traces of human presence, heterogeneous in consistence and identity, are stratified layer upon layer. And in this emphatic fragmentation scenario—in which the individuality appears prevalent than the overall design -, the railways are the only elements that characterize the sites, according to the clear identity of their architectures, the autonomy of the tracks compared to the surroundings, and the perceptual opportunities of the crossed space.

The strategy of disused railways recovery, wishing to protect the quality of places and infrastructural architectures, is part of the vast opinion movement that opposed, in the recent years, to environmental desertification policies according to which humanity seems to have no more place to live. The need for protection of places identity has become increasingly strong since the crisis of Reason, in its various forms ranging from *ecological thinking* - that in architecture fueled researches for environmental sustainability—to the *regionalist claims* in the political field, up to the formulation of a *Geophilosophy* in contemporary thought.

The *Geophilosophical* reflection, as written by Luisa Bonesio, in front of «a landscape ravaged by a careless use of technology, where identities are deleted from the effects of a senseless logic of production and consumption», interrogates the territory in view of «a new responsibility and a full awareness of thinking, of acting and living» (BONESIO 1996). The *Geophilosophy* is also a Physiognomy because it recognizes in every place a particular aspect, an individual expression of its unique character. It is an appreciation of the character inherently spiritual and symbolic of the inhabited space that has many points of contact with the «soul-searching of places» elaborated by Norberg-Schulz in *Genius Loci*.

When faced with the trend currently prevailing in making indistinct places, the reuse of abandoned railways is an opportunity to assert a new and more advanced idea of environmental restoration, able to align the transformation of the territory with the enhancement of its identity.

On the contrary, the prevailing attitude over landscape, especially in our country, systematically rejects any suggestion of change, demanding to match the reality of places with an ideal and abstract image of them.

On the other hand, it's necessary to define precisely what constitutes the aesthetic identity of a place to go beyond generic statements of principle, and it's necessary to identify the shared characters, though not exactly "objectives", in order to base on them the hypothesis of transformation.

Especially for railways, the landscape's identity is made up of recurrences and singularity. Apparently similar tracks and architectures, with similar typological, formal and constructive features in different and distant places, but all the more different from each other the stronger the bond that unites them to the context and to local memory.

The intertwining of repertoires and individuality, on the one hand, and formal, functional and

Figure 4.   Güterbahnhof Pankow, Berlin.

symbolic relations between railways and territories, on the other, may therefore be the main thread of the same strategies of transformation.

The issue of reuse of abandoned railways is often addressed in Industrial Archeology, a recently born discipline and full of inner contradictory aspects because of the short temporal distance between the artifacts of the first industrial revolution and our time, unlike other protected historical archeologies. Railways, built mainly during the last century, have not yet acquired that status because of they belong to an age culturally too close to ours to let us make critic judgments.

The term "Archeology" is undoubtedly effective to describe the nature of railways' finds: as many old artifacts, they are objects bearing clear traces of physical consumption due to the use, to the time and to the weather, that are very important features especially in a design perspective.

As opposed to what happens for other kinds of architecture that draw from Beauty the reasons for their survival over time, the abandoned railway buildings gain this privilege thanks to a renewed functional vitality.

4 WORKING WITH THE RESTS

Forms determined by the technical do not have a secondary role in transport architectures. Generally Engineering is not pleased with its own beauty and it never worried about having to persuade the public with appearance. However, even engineers know how to grasp the difference between a beautiful bridge and one less successful. The attribution of a different value to build works is not as important for Engineering as a discipline in itself, as it is for our culture and for how we assign a different value to the artifacts that surround us, distinguishing between objects that characterize the environment of our lives (MANTEROLA 2011).

The principle that railways architectures and manufacts must be incorruptible, unalterable, emblematically reflected in oversizing the constructive elements, with generous use of solid materials—such as steel, stone and brick -, often ostentatious to enhance the solid character of the buildings in solids details of the stations' furnitures and railway coaches. Some kinds of architectures have been invented specifically for railways, such as tunnels, stations, locomotives remittances, coal and water tanks, workshops for maintenance of rolling stock, cabins for traffic surveillance suspended on the rails.

These are the elements that have contributed the most to build the image of a unified system both for their strong formal characterization both fot the ability to give life to a complex network able to communicate—through the serial repetition of elements along the lines—with landscape in its broadest extension.

The serial logic used to build railways architectures often prevents us from recognizing their aesthetic value, traditionally attributed to individual pieces rather than the products appearing as the result of cultural environments or extensive technical expertise. In industrial production, objects can be replicated in thousands of specimens without losing the original quality of the prototype, while for men it is not possible to reproduce the same objects equal to each other: human life is made of experiences and the repetition is, in a sense, the end of the experience. The artisan object shows the secret of the trade and its tradition through its forms: it is the emblem of the experience gained over a period by human existence. For this we are spontaneously attracted by craft objects with a sentimental attitude and it is difficult to love in the same way industrial products because they are the effect of a mechanical action. The abandonment and the degradation by time, however, often transform the serial elements present along the tracks (machines, electrical equipment, sleepers and rails, metal trellises, lights, etc.), assigning them a new individual character and making them less perfect, unique and aesthetically interesting.

Considered in longer timescale, the current phase of railways infrastructures' disposal can be seen as the latest expression of the troubled relationship that architecture has established with the industry over the past two centuries (JÜNGER 1991). What we are experiencing in recent years is the phase in which the abandoned industrial artifact is judged regardless of the use for which it was designed, the choice of materials, its solidity and resistance. The industrial product has turned into an object to contemplate or to rethink in its appearance through post-production and assembly operations—as in the works of Kurt Schwitters. After him, many others—like Fernandez, Arman, Rosenthal, Barry, Chamberlain—used the scraps, giving them a new life and a social function through art. The new assembly techniques used, from "serial assemblies" to "accumulations", offer an extensive repertoire of compositive solutions that can be models for the design of railways recovery.

Beyond the proposed objectives and the personal artistic career, the use of anonymous industrial materials by modern architects has always been functional to create unique and original works, in a time when architectural production—from computerized procedures used in the design to industrialized building components—seems to refer to a mechanical assembly operation.

Working "in" and "with" the remains, however, become in recent years an ordinary condition in the architect's work and, perhaps, the only possible strategy to offer an adequate response the complexity of the contemporary. Many argue that our age is not destined, as in the past, to produce contents but post-productions, collecting and regenerating all that is already available. This is particularly

701

true for the transformation of abandoned areas in which the role of the objects is prevalent than the faded figures of the background.

The contemporaneity space appears structurally transient, a construction site animated by continuous processing activities in which the forms of architecture lose stability.

As written by Ernst Jünger, in the present landscape nothing is made to last forever and with that character of permanence that we appreciate in ancient buildings, nor in the spirit in which art tries to establish a formal language: everything, however, is provisional and made for short-term use.

According to this situation, our territory looks like a landscape in transition: in it there is no stability of shapes, continuously created by a dynamic restlessness.

The unpredictability of the initial conditions characterizing the nature of abandoned territories is a key factor in the development of the project, introducing the opportunities of subversion of traditional approaches as surrealist artists did: looking for an art that could overcome the logic and shared conventions, they artfully created unforeseen situations to go beyond reality immediately perceived. As in the contemporary projects of transformation, the surrealists tried to create an artistic space in continuous change, including simultaneously the past, the present and the future, contradictory conditions and conflicts.

But designing with wastes requires a complex approach to reality. On the one hand, it is necessary to look at the pre-existing conditions with no preconceived ideas that may lead to exclude what instead could be useful to the project. Look without preconceptions—also paying attention to seemingly worthless materials—means being able to look tendentiously at reality, being interested in the transformation of the existing rather than in its preservation, with a "creative eye" and not with a "contemplative" one. In this sense, the repertoire of references for the reuse project of railways is very broad and it also includes the trivial forms with no author, the engineering works and even the objects of Nature without any aesthetic feature, like those "objects with poetic reaction" that Le Corbusier loved to collect and draw in his notebooks to find their "secret form". These are objects that can also be found "by chance" in the territory, confused among many other worthless materials, abandoned wrecks and figures worn by time. In the railway context, the designer defines some criteria of selection of the objects left along the tracks: formal affinities, analogies mnemonic, emotional suggestions, etc. These criteria make the design process less accidental, as opposed to what happens in "automatic procedures" that, by relying on mere chance, submit the materials to various treatments, removing the artist's control of the creative process.

The use of fragments as materials for the reuse project has, in some ways, a role similar to the one of quotations used by the architects in their works. In both cases, it is to affirm the idea of architectural continuity over time against the belief that history is made of fractures and that the past cannot be in any way related to the contemporary.

The relationships between objects, thier sequences and rhythms, are more valuable than the objects themselves and them spaces they define. In these conditions, an effective design strategy can only be implemented through a set of discontinuous elements, creating a system made up of geometric references between new and existing space, of triangulation based on benchmarks and emerging discrete points in the territory rather than relying on the continuity of the parties.

It comes to design interventions of inter-scalar transformation that do not develop, as normally happens, according to a concatenation of choices—from the general to the particular-, but according to systems of elements, even partial and discontinuous, able to create meaningful places through perceivable syntactic orders.

In this case, the additive process—which consists in arranging different parts in a horizontal sequence—seems the most suitable because the individual elements already have their own configuration and their order is already dictated by the linear track of rails.

5 GENERALIZATION OF THE RESEARCH

A condition to safeguard in the re-use project of abandoned railways is the public use of the tracks. On the contrary, often it happened that, once closed the lines, the property was split and sold to settle debts of the railway companies.

In Italy, in 1991, it was created a company called Metropolis with the aim to "enhance" the abandoned manufacts of 'Ferrovie dello Stato' by selling the buildings and the relative areas. In other European countries, more careful to the preservation of historical heritage, such as France, Spain and Belgium, the divestment was followed by the establishment of companies with public capital that reused the closed lines for tourism and cultural purposes. The case of the Spanish Vias Verdes—Fundación de los Ferrocarriles Españoles (FFE), is exemplary from this point of view: in Spain, today there are more than 1,700 kilometers of rail infrastructure in disuse that have been converted into cycle routes and green public walks.

Equally important it is to be able to preserve the public use of the path so that the knowledge experience of the territory by train can survive, although through other forms of transport (SCHWARZER 2004). It would be desirable to use new locomotion

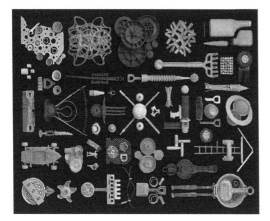

Figure 5. Barry Rosenthal, Grid, 2014.

Figure 6. High Line, New York, 2009.

systems with low environmental impact on the old tracks, as well as in many European countries, to replace the old technically and economically not sustainable train.

The reuse project has to face also with unscheduled and spontaneous aspects, using them as a new resource: the natural vegetation wedged between the rails, creating long green lawns and large patches of trees and wild shrubs, graffiti and inscriptions that covered the walls, gaps and breaks created from the ruins of buildings.

The green is a constant presence in infrastructure places, closely linked to the real identity of the railway. The relationship is obvious in the case of tracks that develop within the natural territory, less obvious in urban lines. Although initially the lines were outside the cities, the constructions have gradually surrounded the tracks because of the suburbs growth, changing the character of places: inside and above the city, in the interstices left free from buildings, deep green corridors raised along the train paths right into the city. These are elements of continuity with the countryside that introduced forms of biodiversity in the city (RITTER 1994). Once the railways were closed, this unforeseen nature grew on the edge of the tracks is one of the most significant legacy to be valued in the reuse project, as happened emblematically in the 90's transformation of the Görlitzer line in Berlin or in the recent recovery project for the High Line in New York.

Berlin and New York cases teach that the reuse of abandoned railways requires a flexible strategy of interventions, able to adapt to the large-scale, with typical and repeatable solutions, but also able to identify specific solutions for each place.

Once lost the original unity of the infrastructure, the project has to ensure a permanent condition of transformation, favoring the spontaneous evolution of the parts and the physiognomic change over time according to their use. As well as in archaeological sites, where the task of excavations is not to bring to light separate pieces of a buried reality but—through a careful reconstruction of the various levels—to restore the meaning of layered sediments, in the same way the primary goal of the reuse project for abandoned railways should be to integrate the complexity and contradictions of current condition by giving the lines a new life (VIOLA 2013).

To define the physical dimensions of the interventions, their articulation, their hierarchies, their components and the different times of their mutations, is the primary task of the recovery project of tracks and railway architectures, linking the past and a possible future of a particular territory. Therefore, a project designed for recovering a linear complex system and its different elements, "rebuilding" the contemporary landscape through an intervention that is a new "work of art" itself.

REFERENCES

Bonesio, L. 1996. Appartenenza e località. L'uomo e il territorio. *Atti degli incontri di geofilosofia*, Milano: SEB.
Branzi, A. 2006. Modernità debole e diffusa: il mondo del progetto all'inizio del 21° secolo. Milano: Skira.
Budoni, A. 2014. Catturare il valore del suolo per sviluppare reti di trasporto locale su ferro. *Ingegneria Ferroviaria* (n.5): p. 431.
Jünger, E. 1991. L'operaio. Dominio e forma, Parma: Guanda.
Manterola, J. 2011. Ingegneria come opera d'Arte. Milano: Jaca Book.
Maternini, G., Riccardi, S., Cadei, M. 2014. Trasformazione a tramvia di un sistema ferroviario. Il caso studio tram-treno nell'area metropolitana di Brescia. *Ingegneria Ferroviaria* (n.3): p. 225.
Mumford, L. 1961. The city in history: Its origins, its transformations, and its prospects. New York: Brace & World.
Ritter, J. 1994. Paesaggio. Uomo e natura nell'età moderna. Milano: Angelo Guerini e Associati.
Schwarzer, M. 2004. Zoomscape. Architecture in Motion and Media. New York: Princepton Architectural Press.
Viola, F. 2004. Ferrovie in città. Luoghi e architetture nel progetto urbano. Roma: Officina.
Viola, F. 2013. Pietra su Pietra. La storia come materiale di progetto. Salerno: Cues

Transport Infrastructure and Systems – Dell'Acqua & Wegman (Eds)
© 2017 Taylor & Francis Group, London, ISBN 978-1-138-03009-1

Modelling ambulance and traffic behaviour using microsimulation: The LIFE project application

F. Galatioto, V. Parisi, E. McCormick & C. Goves
Transport Systems Catapult, Milton Keynes, UK

ABSTRACT: This paper presents the preliminary results of the application of a microsimulation model (VISSIM, PTV Group) to replicate the behaviour of ambulances in urban area and how different reactions of general traffic can impact on the travel time of an ambulance. The work is part of an Innovate UK collaborative funded project, namely LIfe First Emergency Traffic Control (LIFE) with the aim to develop an innovative application for an intelligent transport system that operates in real-time to enable ambulances to reach life threatening emergency cases quicker by integrating ambulance route finder applications with traffic management systems.

The microsimulation model setup within the project has been developed to understand and evaluate the impacts and the best scenarios to improve ambulance response time and gains in cost-saving, whilst on balance mitigating adverse impacts such as residual congestion.

1   INTRODUCTION

1.1   *Context*

Given the aging demographics and rapid urbanisation, cities need to be equipped to respond to emergency (eg. 999 calls) more quickly.

By 2050, over 25% of the UK's population will be over 65. This has implications on the overall health services as well as the NHS Trust to cope with anticipated rise in ambulance call outs amidst worsening urban congestion. Presently, the Government has set out targets for ambulance services to reach 75% of emergency calls within 8 minutes. For this reason, there is a growing need to develop new and innovative applications for an even more intelligent use of the existing transport system that will support in real-time emergency vehicles to reach life threatening emergency cases quicker.

Hence, this paper will discuss the methodology and the preliminary results of the modelling implementation of a "LIfe First Emergency Traffic Control" or "LIFE" system seeking to identify the best solution to reduce the time to respond to emergency calls, whilst operating a resilient service with a cost and fuel efficient fleet.

1.2   *Background*

A recurrent problem that affect emergency vehicles is reaching their destination on time. This is caused by increasing traffic and congestion in modern cities. Especially in urban areas, the mix of road users and the presence of traffic lights, makes driving at a higher speed without causing any harm to anybody even more difficult for the emergency vehicle drivers itself.

Currently, the issues of ambulance services to meet the Government's target of responding to 75% of life threatening calls within 8 minutes (Guardian, 2015), is becoming more complex year on year. Although critical emergency calls are time-sensitive (i.e. heart attacks, strokes etc.), the BBC in 2015 reported a 116% rise in ambulance delays over the past year alone, proving the situation is becoming increasingly critical.

An international best practice review suggested that the response time should be approx. 5 min. to increase the survival rate by 12% (NHS, 2014).

Aside from operational issues, ambulance services currently face (e.g. 10% of 999 calls are genuinely life-threatening conditions, while categorising 40% of calls, resulting in dispatched vehicles before having determined the exact nature of the problem), LIFE will approach the poor ambulance response performance from a transport and city level. This is further exacerbated that ambulances are not legally allowed to travel 10 mph above the speed limit, as this could potentially cause more accidents. On average, four ambulances are involved in crashes each day resulting in a total cost for compensation and repair bills of £300 K every month (EveningStandard, 2008).

In Plano, Texas, the Emergency Vehicle Preemption (EVP) system has dramatically reduced the number of emergency vehicle crashes from an average of 2.3 intersection crashes per year to less than one intersection crash every five years (US DOT, 2006), whilst in Fairfax, Virginia, the EVP system has been shown to save anywhere from a few seconds to a few minutes. Ambulance drivers cited savings of 30–45 seconds at a single intersection (US DOT, 2006).

England has 10 Ambulance Service NHS Trusts all of which have similar needs and same pressure to meet the Government's 75% target (NHS, 2015). Of 1.1 m emergency calls last year, approx. 68% require emergency transport (NWAS, 2015).

Across the UK, the Diagnostic and Ambulance Services Market (D&AS) has an annual growth of 2.7% (2010–2015) with a revenue of £6bn, driven by UK's ageing population. Total demand of the £800 m a year UK ambulance service industry is expected to grow significantly in the next 10 years demonstrated by the double digit growth in spending on private ambulances by the NHS over the past two years (Plimsoll, 2013).

Crashes are a significant problem for ambulances and all emergency vehicles, in the US it was found that 25% of all accidents involving emergency vehicles occurred at signalised intersections especially where line of sight was blocked by buildings or vegetation (Viriyasitavat & Tonguz, 2012). In the UK cost for compensation and repair bills of ambulance crashes totals £300 K every month (EveningStandard, 2008).

Also the bus priority applications that can be seen in many cities worldwide use pre-emption systems to save time especially when they are beyond schedule. (CORDIS, 2016)

An application that is a precursor of the LiFE concept is the EViEWS system, deployed in Texas since 2014. Through GPS signals transmissions and a wireless communication, speed and position data of emergency vehicles are sent to traffic lights, that are then turned green to provide prioritisation. (Lloyd, 2014)

The latter however is not the first emergency vehicles pre-emption technology used so far. Global Traffic Technologies (GTT) introduced the Opticom system is 1968, which again makes use of GPS signals and wireless communication, in combination with the siren activation, to trigger the green phase according to the ETA to the signal (ITS Int, 2015).

The pre-emption technique has proved through the different on site applications to actually reduce journey times, however also traffic modelling studies demonstrate that it easies the movement of emergency vehicles in urban areas. (Sharma, et al., 2013) Algorithm based approaches have been carried out to determine whether enabling priority at intersections would actually reduce travel times for emergency vehicles and it has been found out that since the green wave allows the emergency vehicles to go faster, they do not experience slowdowns at the intersection and additionally crashes are prevented (Viriyasitavat & Tonguz, 2012). Not many are the experiments carried out so far using traffic microsimulation software. An example though is represented by (Wang, et al., 2013) where a Vissim model has been used to validate a mathematical model that estimates travel times for vehicles under pre-emption control. The approach used here sees the deployment of an emergency lane that is activated at random times, from where vehicles move to the outside lane to let the EV through. Other studies focused on the emergency vehicles routing decisions, where assignment models that take into account the demand and travel costs variability on the network have been used to develop a travel time function able to forecast emergency vehicles routing. (Musolino, et al., 2013).

1.3 Research gaps

There is a clear need for the traffic systems of the city to support faster response times through prioritisation. In the US Emergency Vehicle Pre-Emption (EVP) at signalled crossings has been taken up widely across the country with improved ambulance journey times.

This is not the first time a concept like this takes place: other systems, such as the European Compass 4D, looked at prioritising vehicles at traffic lights through V2I/I2V communication, as well as warning drivers of possible road hazards (V2V communication). However, the system that has seen only a trial version of it, is not deployed yet and even though it has been proved to be efficient in terms of reducing emissions at intersections and journey times. For these reasons a test with emergency vehicles should be run and is the focus of this paper.

Moreover, current and traditional traffic microsimulation modelling tools, have been used using default functions which are definitely not designed to model an ambulance and the behaviour of traffic when an ambulance is going through it.

Unlike EVP, the LIFE project idea goes beyond traffic lights responding to an ambulance's approach but integrates route finder applications with traffic management systems. The proposed application will be developed so it can be integrated with existing systems in ambulance vehicles and will therefore compliment current solutions. Also, it is going to be designed using ethnographic research into ambulance driver behaviour to ensure LIFE is fit-for-purpose for the end-user with feedback loop mechanisms throughout the project.

As the project has begun 10 months ago, this paper will focus mainly on the initial results of the implementation of the microsimulation modelling methodology and in the steps to model the improvements of the response time of ambulance services through a predefined corridor.

1.4 *Main aims*

In order to overcome the above identified issues, the LiFE project aims at reducing the travel time of emergency vehicles in a safely manner, namely prioritising them at traffic lights or rerouting them where the combination of traffic and road conditions would allow a shorter drive.

Key and unique aspect of the work undertaken in this paper is the process that has allowed the microsimulation tool to be adjusted in order to enable an existing microsimulation model to replicate traffic conditions when an ambulance on blue light (highest emergency code) is going throughout a representative urban corridor.

1.5 *Objectives*

The real-world datasets used to calibrate and validate the microsimulation tool have been made available by the North West Ambulance Service (NWAS) based in Liverpool, UK.

Within this paper several objectives will be achieved:

– For the first time a traffic microsimulation model will be used to reproduce realistic traffic behaviour in presence of ambulance;
– the modelling framework and API module settings used within the microsimulation tool will be presented;
– a real study area will be reproduced using real-world ambulance data, and
– preliminary results of selected traffic management scenarios will be presented alongside the benefit of implementing the new modelling software.

The reduction of the frequency and severity of collisions with ambulances significantly decreases the cost to the fleet as well as public liability associated with fatalities, injury and damage to property. 3) Managing vehicle intelligently could have huge environmental impacts for cities with benefits such as reduction of CO_2 emissions, improved air quality as well as cost savings resulting from less fuel consumption and wear-and-tear on emergency vehicles.

2 THE PROPOSED MODEL

2.1 *The microsimulation model*

The traffic microsimulation software used to analyse the LiFE system is the German Vissim 8.12 commercially available through PTV Group.

The reason why it has been chosen among the other commercial software available is the more realistic behaviour of the vehicles, with the possibility for them to overtake both in the same lane and in the opposite lane. Vehicles that have enough lateral space (defined by the user) and a desired speed higher than those in front, may overtake in the same lane, as well as in the opposite lane if not upcoming traffic is detected.

Since the main purpose of the application of a microsimulation traffic models to analyse trips and travel time of emergency vehicles (EVs) was also to reproduce and monitor the reactions of standard drivers in urban areas that traditionally in presence of ambulance react in different ways depending on the road layout and speed, therefore the feature required was from one hand the ability for the EV to overtake standard traffic, but also for the standard traffic to implement manoeuvres that are not by default implemented in a microsimulation software.

In fact, by default the overtaking feature in Vissim allows all the classes of vehicles defined by the user to be overtaken by vehicles with higher desired speed, but vehicles of same defined classes may overtake as well depending on the speed they desire to travel at, as mentioned earlier. In other words, it was not possible for EVs only to overtake all the other vehicles on the road network by default.

Hence, in order to also achieve a realistic behaviour of the traffic in presence of an EV, *an external driver model* able to control vehicles' behaviour on the Vissim network has been identified as the only solution to model realistically ambulances and traffic affected by EVs.

2.2 *The external driver model*

By designing and implementing an external driver model, the intention is to override most of the vehicles' behaviour. In this way, Vissim vehicles do not follow anymore the behaviour embedded in the software, but instead the instructions the external model send to them in certain circumstances (approaching traffic lights, EVs behind or in front, etc.).

Emergency services are mostly operating in critical conditions; thus an efficient journey planning is vital in improving time-response to a distress call. Modelling and analysing travel time in different traffic conditions and road types could prove very helpful in decision making and better allocation of existing resources.

Vissim is an established software package in microscopic traffic simulation, capable of providing a realistic model of road users while allowing the flexibility to inject external models for vehicle behaviour. VISSIM has an External Driver Model for as DLL interface which provides the option to replace the default driving model with a user defined one as long as the choice is activated for that particular vehicle type.

The possibility of incorporating the external driver model into the software has made it feasible to reproduce realistic conditions in the traffic simulator, where standard vehicles are made aware of EVs approaching and react accordingly.

2.3 General Implementation Guidelines

The DLL is implemented in C++ and has to follow VISSIM structural guidelines.

The DriverModel.h is a pre-defined header file containing all the definitions necessary during the simulation, the DriverModel.cpp represents the main source file which must implement three functions: DriverModelSetValue, DriverModelGetValue and DriverModelExecuteCommand.

During the simulation run, VISSIM calls the DLL every time step for each vehicle type actively controlled by the external model in this order: SET, GET and EXECUTE COMMAND.

The *Set* method passes the current vehicle values referring to time, vehicle id, lane and link information, velocity, acceleration, width, height, maximum acceleration, turning indicator, x and y coordinates, information regarding lane change and lane target, road information, coupled with data about adjacent vehicles. At any time, the known neighbouring vehicles in the current link are two ahead, two behind, two left and two right. If the signal DRIVER_DATA_VEH_CURRENT_LINK is active, the model also gets information about adjacent links.

The *ExecuteCommand* method handles the data received about a particular vehicle in Set and makes the necessary adjustments to change the default behaviour. If no correction is needed, the data is passed back to VISSIM without alterations.

In the *Get* method, VISSIM retrieves the value associated with the vehicle id specified in Set and applies the new behaviour. The currently available commands are driver init, create driver, move driver and kill driver.

A minority of parameters are updated per vehicle type or once per DLL initialization and are not called every tick.

2.4 Specifications

In order to implement the behaviour of ambulances and the affected vehicles, a certain degree of traffic disturbance was implemented.

Every vehicle has a different degree of ambulance awareness distance which is a Gaussian distribution of 70 meters mean and 20 metre

dispersion. In congested areas, the awareness falls to 10 meters.

When an ambulance is approaching, every vehicle checks if the distance between the two is within the awareness threshold, if it is not, the DLL collects the values suggested by VISSIM and passes them back, without any change. If the distance is in the awareness range and the vehicle has a velocity greater than 5 m/s, it changes lanes in order to allow the emergency vehicle to pass. If the velocity is small, the vehicle is caught in congestion and the ambulance is behind, the vehicle is moving outside the road while decelerating with an exponential function of lane lateral position, allowing the ambulance to overtake.

Without coordinated commands, vehicles might decide to take inconsistent decisions such as, some move left while some move to the opposite direction, thus creating a situation where unrealistic scenarios could arise. In order to avoid this, the vehicle which leads the decision making is the one directly ahead of the ambulance. Any other driver affected by the ambulance firstly checks the specified car action and then applies the same behaviour if the conditions are favourable.

The first limitation experienced was the fact that at any time a vehicle could interrogate the ID of the neighbours but not their types, thus a car would be unaware of who is the ambulance in the proximity.

The problem was solved by defining a 'Vehicle' structure holding all the necessary parameters updated during 'Set' calls and a matrix with the neighbours' unique identification numbers as seen in the Figure 1.

Every Set, ExecuteCommand and Get calls are dealing with data embedded within the current vehicle object.

In parallel, there is a hash map which holds an array of 'Vehicle' objects and uses as a key the vehicle unique identification number. The map contains all the vehicles active in the simulation.

If a vehicle knows its neighbour's id it can check on the map the type of that particular id alongside other vehicle specific parameters. Furthermore, in order to improve efficiency there could be two structures, one holding regular vehicles and one holding the emergency vehicles.

		2 lanes left	1 lane left	current lane	1 lane right	2 lanes right
		0	1	2	3	4
2 ahead	0	id (0,0)	id (0,1)	id (0,2)	id (0,3)	id (0,4)
1 ahead	1	id (1,0)	id (1,1)	id (1,2)	id (1,3)	id (1,4)
				Current Vehicle		
1 behind	2	id (2,0)	id (2,1)	id (2,2)	id (2,3)	id (2,4)
2 behind	3	id (3,0)	id (3,1)	id (3,2)	id (3,3)	id (3,4)

Figure 1.   Matrix for the neighbours vehicle' ids.

The latter category contains less elements and a query takes considerably less resources.

The ambulance has the ability to identify gaps in the upcoming traffic, to evaluate if the vehicle ahead is offset lane by checking the lane lateral position and accelerate past and to ignore the red traffic warnings.

The scenario when the non-emergency behaviour vehicle has no ambulance behind in the current lane is described in the Figure 3.

The scenario when the ambulance is two cars behind is described in Figure 4.

The scenario when the ambulance is more than two cars behind in the current lane is described in the Figure 5.

The scenario when the ambulance is one car behind is described in Figure 6.

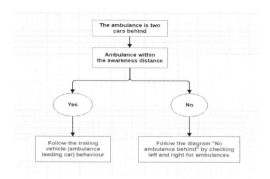

Figure 4.   Ambulance two cars behind.

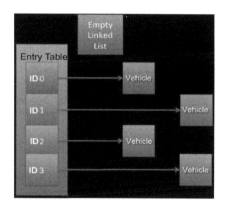

Figure 2.   Vehicle Hash map.

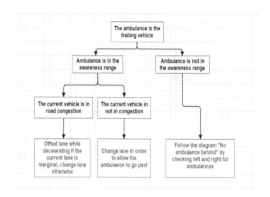

Figure 5.   The ambulance is more than two cars behind in the current lane.

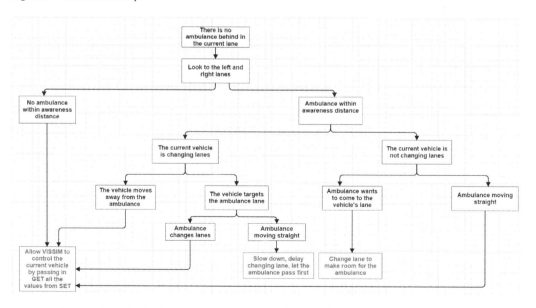

Figure 3.   No ambulance behind in the non-emergency's vehicle current lane.

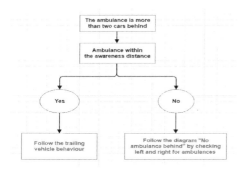

Figure 6. The ambulance is one car behind in the current lane.

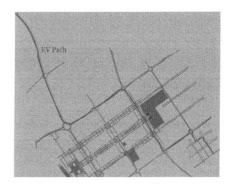

Figure 7. Traffic network layout and ambulance trip (red).

3 RESULTS

3.1 The study area

For the application and calibration of the newly developed external driver model an urban network extended on an area of 3.5 km^2 in the vicinity of the Transport Systems Catapult premises have been used. The network (Figure 7) was previously calibrated and validated for a different project, hence the testing could rely on an existing proven tool.

Since red calls should arrive at destination within 8 minutes, the ambulance journey (in red) was chosen of a length of 5 km which in normal traffic or overnight takes up to 6 minutes, but during peak hours can easily take over 10 minutes.

The time modelled for this exercise was between 7:30 am and 10:00 am, with 30 minutes warm up period. This is because the network is particularly congested at that time. However, to stress test the external model, higher level of flows were also considered.

3.2 The modelling results

As a result of the novel external driver model developed, several scenarios of ambulance going through traffic both on links and junctions have been replicated and realistic behaviour have been possible to replicate in the microsimulation model (Figure 8).

For each scenario result, 10 different unique simulations (using different seeds) were carried out.

In Figure 9 are reported the results of running the simulation with 3 to 5 ambulances inserted in the network every 10 minutes, this was done to limit multiple ambulances to proceed on the same corridor at the same time, even though that is a real possibility.

It can be observed that the external model produces a real benefit of around 4 minutes for traffic level which can be classified as busy, while for free flow, as expected, the time saving is minimal (mainly associated with the ability of EVs to go

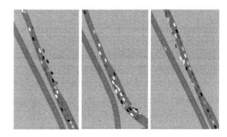

Figure 8. Screenshots of three different modelling scenarios of ambulance (long yellow vehicle), using the novel external driver model.

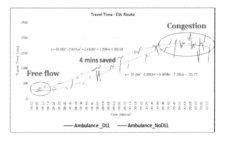

Figure 9. Travel time of Ambulances (EVs) through the simulation period with and without implementation of the external model (DLL).

through red light at signalised junctions). Similarly, for highly congested period EVs cannot take advantage of either the external model, since surrounding vehicles are not able to move easily, and the ability to go through red light.

This, as highlighted in Figure 10, which refers to results from the same network, but implementing reduced flow levels, provides a useful perspective on when both temporally and geographically a LIFE system application will make a real difference in enabling the ambulance to reduce the travel time to meet the Travel Time (TT) target.

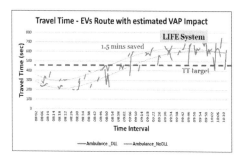

Figure 10. Traffic network layout and ambulance trip (red).

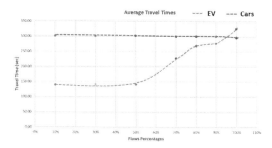

Figure 11. Preliminary results of Emergency Vehicle speed/flow curves (orange) compared with standard traffic (red).

4 DISCUSSION AND CONCLUSIONS

This paper has presented the methodology to develop an external driver model to enable a commercial microsimulation model to accurately model ambulance (or generically Emergency Vehicle) trips across a transport network and the preliminary results of the application of proposed driver model.

The external driver model has been designed and developed to work independently from any microsimulation network configuration and geometry, this is because it overrides the vehicle behaviour according to the dynamics on the surrounding environment (eg. speed limits, speed of the surrounding vehicles, position of the EV in respect to each individual vehicle).

The results obtained show that the application of the new driver model developed provides the expected benefit that in real-world condition an ambulance going through an urban area and for different traffic levels is expected to achieve, however the model results also highlight that with congested traffic levels, ambulance travel time cannot be currently reduced beyond a certain limit, which unfortunately is not enough to achieve the set target of 8 minutes. This reinforce the need to implement a new system which is currently under development as part of the LIFE project.

A simple link based pre-emption strategy has been tested using the new external driver model and benefit up to a certain congestion level have been observed, however a more structured and coordinated strategy seems to be required and this will be part of the future steps of the project.

5 FUTURE WORK

Since this paper presents the preliminary results of the application of an external driver model to enable a commercial microsimulation model to replicate ambulance trip accurately and also to model the behaviour of the traffic when an emergency vehicle is going through it, the next step the LIFE modelling group is currently developing is the extension of the modelling functionalities to model new scenarios where intelligent priority scenarios can be designed and tested on any traffic network.

Future steps are, a) to develop a traffic control strategy that by detecting the ambulance and its route in advance will automatically develop a new traffic light planning, dependent on traffic condition and congestion on the network; b) an algorithm to calculate in real-time the ambulance TT. For the latter, specific speed flow curves for ambulance will be derived using the new modelling tool developed and presented in this paper. This is currently under development and full results may be presented at the conference.

In anticipation, Figure 11 presents the result of an initial test performed using the external driver model algorithm in a non-congested urban transport network. The different points have been obtained by changing the percentage of total flow (standard vehicles) hence, the travel time of the ambulances injected in the network at 10 minutes intervals have been averaged and reported in the graph.

It is still an early stage result and needs calibration and tuning, as well as a validation with the AVL data available, however the preliminary results show encouraging pattern, which is definitely different from the general traffic as expected.

This will help in developing the pre-emption strategies as well as to accurately predict the Ambulance travel time in the future.

REFERENCES

BBC, 2015. http://www.bbc.co.uk/news/health-30742817
CORDIS, 2016. http://cordis.europa.eu/result/rcn/4102 9_en.html.

EveningStandard, 2008. http://www.standard.co.uk/news/ambu-lances-in-crashes-four-times-every-day–6653861.html.

Guardian, 2015 https://www.theguardian.com/society/2015/jan/16/englands-ambulance-services-trial-new-system-cut-response-times.

ITS Int., 2015. Priority management saves time, money and lives. ITS International. September October 2015. http://www.gtt.com/1521–2/.

Lloyd, R., 2014. Next-Generation Preemption System Clears the Way for Harris County, Texas, First Responders http://www.fireapparatusmagazine.com/articles/print/volume-19/issue-3/features/next-generation-preemption-system-clears-the-way-for-harris-county-texas-first-responders.html.

Musolino, G., Polimeni, A., Rindone, C., Vitetta, A. 2012. Travel time forecasting and dynamic routes design for emergency vehicles. Procedia—Social and Behavioural Sciences 87 (2013) 193–202. SIDT Scientific Seminar 2012.

NHS, 2014. Ambulance Services, England 2013–14. Workforce and Facilities Team, Health and Social Care Information Centre. Version 1.0. 31 July 2014

NWAS, 2015. North West Ambulance Service NHS Trust Annual Report 2014/2015.

Plimsoll, 2013, Ambulance Services Market Report

US DOT, 2006. Traffic Signal Pre-emption for Emergency Vehicles. A CROSS-CUTTING STUDY.U.S. Department of Transportation. Federal Highway Administration. January 2006.

Viriyasitavat, W., Tonguz, O.K., 2012. Priority Management of Emergency Vehicles at Intersections Using Self-organized Traffic Control. Carnegie Mellon University. 9–2012.

Wang, J., L. Yu, and F. Qiao. 2013. Micro Traffic Simulation Approach to the Evaluation of Vehicle Emissions on One-way vs. Two-way Streets: A Case Study in Houston, Downtown. Proc., 92nd Annual Meeting of the Transportation Research Board, Washington, DC.

Transport Infrastructure and Systems – Dell'Acqua & Wegman (Eds)
© 2017 Taylor & Francis Group, London, ISBN 978-1-138-03009-1

Development and application of national transport model of Croatia

Gregor Pretnar
PNZ Svetovanje Projektiranje d.o.o., Ljubljana, Slovenia

Uwe Reiter
PTV Transport Consult GmbH, Karlsruhe, Germany

Igor Majstorović
Gradevinski Fakultet Sveučilišta u Zagrebu, Zagreb, Croatia

Ana Olmeda Clemares
Ineco, Madrid, Spain

ABSTRACT: The Croatian Ministry of Maritime Affairs, Transport and Infrastructure commissioned the development of the National Transport Model, including collection of all available data, carrying out necessary surveys, developing networks models and demand models for freight and passenger demand for the base year, calibrating and validating the models, and developing forecast models. Model has 985 internal and 267 external transport zones and over 360.000 km of modelled links. The purpose is to identify shortcomings, bottlenecks and issues in the current and the planned future transport systems. The model is used to identify specific measures and projects for the different transport modes and their integration supporting the selected strategies. The model produces quantitative results allowing to determine impacts of the strategy alternatives and of the measures on traffic conditions, on social and environmental impacts. The availability of a National Model guarantees that similar approaches are used at the regional level, improving the general transport planning approaches all over the country.

1 ABOUT THE PROJECT

The National traffic model for the Republic of Croatia was co-financed by the EU from the European Regional Development Fund under Transport Operational Programme 2007–2013 within the project "Support for the preparation of the Republic of Croatia's Transport Development Strategy and designing of the national Traffic Model for the Republic of Croatia—National Traffic Model for the Republic of Croatia". Beneficiary of the project and contact information:

Ministry of Maritime Affairs, Transport and Infrastructure; Directorate of EU Funds
Krležin Gvozd 1a, 10000 Zagreb, Croatia www.mppi.hr; www.promet-eufondovi.hr

The contents of this publication are the sole responsibility of Consortium led by PTV Transport Consult GmbH and do not necessarily reflect the opinion of the European Union.

2 INTRODUCTION

The intention of the Croatian Ministry of Maritime Affairs, Transport and Infrastructure in commissioning the model development was to obtain a quantitative tool that could support the development of the National Transport Strategy, help to analyse current conditions and forecast future conditions, provide the basis to identify necessary strategies and measures and finally being able to calculate the impacts of strategies and measures on the future transport system and the influencing processes, like social, economic and environmental processes. A National Transport Model is the necessary tool to plan the sustainable development of the transport system.

3 MODEL DEVELOPMENT

The model was developed using the software suite **PTV VISION**. It follows the classical 4-step approach. National Model of Croatia is a synthetic model using network data, socio-economic data and behavioural data as its foundation. Only a synthetic model, calibrated and validated to actual empirical data, is capable of scientifically and correctly forecasting future developments and of calculating impacts of changes in influencing conditions (exogenous factors like economic and social development) and of changes within the transport system itself, e.g. implementation of strategies and measures (endogenous).

For passenger transport, a trip generation model was developed for the resident population and for the tourists, information of the actual destinations of trips of residents and visitors was used for trip distribution. Similarly, for freight transport, data on import/export, production, processing and consumption of numerous commodity types was used to develop the freight generation and distribution. For both models, mode choice and assignment were based on costs including actual travel times.

The socio-economic data was collected from different sources at the level of model zones mainly from National Statistics. The basis for the behavioural data was a household survey with more than 3,000 interviews and a survey of freight operators, both carried out within the project. Empirical data was complemented by traffic counts and public transport passenger counts. The empirical data from external sources and from the own surveys was used to calibrate and to validate the model. To represent differences between the summer season with high numbers of tourists visiting the country and the rest of the year, two different models were produced, an off-season model and a seasonal model.

Model development consisted in the development of a base year model and of forecast models for the horizon years 2020, 2030 and 2040. Forecast was based on the available and accepted official data of future socio-economic development of Croatia and the surrounding countries. For all forecast horizon years, a so-called do-minimum scenario was developed that is used as a reference scenario, including only those projects and measures that are already under development or that are planned and financed. The do-something or strategy scenarios include additional projects and measures being part of the national transport strategy.

3.1 Network and zoning

The network system was composed of road and rail systems, airports, maritime and inland ports. The national and international connections were mainly established by primary road network (motorways, trunk roads and state roads), railway network and ports. In addition, roads in major urban areas have relevant functions for the national network and a significant impact on connectivity and accessibility. These network elements also serve as public transport routes and as alternative routes. Therefore, urban main roads were also included in road network.

As result the following network elements were included in the model:

– Road network with all relevant levels (motorways; state, county, local and urban roads) with design standard and condition characteristics;
– Railway lines of importance to international, regional and local transport with design standard, traction, gauge, interoperability, conditions—restrictions and reliability characteristics, strategic national and international connections, capacity utilization, rolling stock, operators and safety;
– Public transport routes for rail, boat (ferry), tram, city and intercity bus lines with timetables;
– Seaports and inland ports with their characteristics, purpose and capacities;
– Airports for which the land-side traffic was modelled (by car, public transport and freight traffic);
– Intermodal facilities for passenger and freight transport.

Republic Croatia was divided into 985 traffic zones. Basic level for zoning were cities and municipalities. Traffic zones with population greater than 8.000 persons were broken down to lower level of territorial unit, i.e. city and municipality to settlement, settlement to statistical circle.

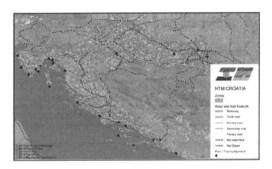

Figure 1.   Road and rail network.

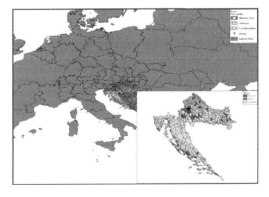

Figure 2.   Zoning.

3.2 Passenger demand model

Passenger demand model consists of first three steps in traditional four-step model:

- generation (production and attraction),
- distribution and
- modal split.

Production and attraction primarily depend on the gross domestic product, motorization rate and spatial socio-economic structure as well as of behaviour patterns, whilst attraction also partly of the traffic supply. Distribution and mode choice are calculated simultaneously and consider both attraction (e.g. number of shopping area) and transport supply (accessibility). Distribution approximately equally depends on the spatial socio-economic structure and behaviour patterns on one hand and on the traffic supply on the other. Modal split particularly depends on the traffic supply and to a considerable extent also on the land-use.

Last step in the development of passenger demand model is assignment of demand on the multimodal network. Static stochastic learning procedure (Lohse) was used. Public transport was assigned by the intermodal method based on timetables. Method takes into account journey, transfer and waiting times, based on actual time-tables. Both, internal and external transport, were assigned simultaneously.

This approach enables that the forecast calculation considers changes in spatial structure, gross domestic product, motorization rate, residents, jobs, etc., as well as transport supply while calibrated parameters of the model remain unchanged.

3.2.1 Trip generation

Production and attraction were calculated by the method of origin-destination groups (13 origin-destination groups were considered for passenger transport). These 13 origin-destination groups actually represent 5 trip purposes (work, school, shopping, leisure and vacation, other) in conjunction with the home-bound trips and combinations between them. The trip purpose of business was specifically modelled as a separate demand model.

From the household survey (performed within the project itself) it was found that different parts of Croatia have different travel behaviour patterns. Therefore, the model of passenger transport production and attraction on the average workday was specially developed for Continental and Adriatic regions of the country.

For each of the thirteen origin-destination groups a mobility rate (number of trips per day per person concerned) was determined. In most of these groups, the person concerned is a resident. However, in groups home-work, work-home, work-other and other-work, the concerned person is an employee, or in school trips, the person concerned is a pupil, and for secondary school or university a student. Based on surveys across households in the Republic of Croatia certain mobility rates were determined for all origin-destination groups.

Development of the seasonal traffic was based on the same basis as the average workday model. Thirteen origin-destination groups were taken into account to calculate the generation of passenger traffic, but the destination groups home-school and school-home were replaced by the destination groups home-vacation and vacation-home. Modified destination groups include trips from home to holiday and private facilities intended for vacation use (home-vacation) and trips in the opposite direction (vacation-home); there are no school trips during the holidays. Therefore, the purpose of (daily) leisure was extended for a special holiday subcategory of leisure (vacation).

3.2.2 Trip distribution and mode choice

Distribution and modal split were calculated simultaneously. That is to say, at the same time the destination and the transport mode by which the trip is done were chosen. The calculation was carried out on the basis of the EVA probability function (developed by prof. Lohse from TU Dresden and included in PTV Vision software), for the average workday traffic and for the traffic during the tourist season.

Input data for the distribution and modal split sub-model were:

- productions and attractions,
- generalized prices or generalized times for the road motorized and public transport,
- EVA model parameters.

Basic parameters of the model were set based on the stated researches, recommendations of the software manufacturer and previous experiences in modelling of the national and regional models. Based on these data and the EVA functions, within the multiple iterations, the trip matrices for passenger car transport, public transport, cycling and walking were calculated.

3.2.3 Assignment

Road assignment was carried out in several iterations. Based on the information obtained in the previous iteration, users find a new optimal route in the next iteration. Therefore, in the iterative process search for optimal route runs until the network equilibrium and the appropriate impedance matrix convergence were reached.

Network assignment was based on the function of generalized price or generalized time. Toll was incorporated through the function of generalized time, where the monetary values were converted into the equivalent of time.

In seeking the optimal routes, also the effects of traffic congestions and jams were taken into account, i.e., the effects of driving speed reductions. BPR function was used for this, the most common volume-delay function, which reflects the travel time, depending on the volume and road capacity. It was useful both for modelling of non-urban and urban roads.

Intermodal method for public transport assignment allowed the entire public transport network to operate as a unified system that includes rail, bus and maritime lines of various levels. The method, based on timetables, requires precise arrival and departure times of vehicles or trains at stations and stops to be set for all public transport lines. The network was therefore modelled in that way.

For each origin-destination pair of zones a favourable connection was found or calculated. It was assumed that passengers are aware of the time-table, and would take the first available line of public transport offering a favourable route. Among various combinations of routes, more favourable routes were chosen. The most favourable routes were determined on the basis of the whole chain of route segments, including ticket price, which was included in the function of generalized time.

3.2.4 Input data

The main input data of the passenger model is listed below:

- Transport network data for all modes (road, public transport),
- Socio-economic data, i.e. population and employment data, disaggregated to traffic zone level,
- Behavioural data from household survey (n = 3.000),
- Transport cost parameters (distance related, time related, toll costs) per transport mode.

3.3 Freight demand model

Freight transport as a whole is a very complex and heterogeneous process. The multi-modal freight model follows a highly disaggregated approach to calculate the freight volumes based on origins and destinations of homogenous commodity types. This includes both, domestic freight flows and external freight flows (import/export/transit). The freight model considers all transport modes relevant from a national perspective, i.e. road (HGV and LGV), rail and vessel. The same network as for the passenger model is used, enhanced with additional mode-specific freight parameters and

transhipment infrastructure for intermodal handling of goods.

The modal freight trip matrices were calculated with a commodity based multi-modal model using an enhanced 4-step approach. As a big advantage of this synthetic multi-modal approach, the proposed methodology guarantees:

- the adequate consideration of commodity-specific affinities regarding different transport modes,
- the ability to reflect multi-modal transport and inter-modal transport chains,
- a realistic calculation of future freight demand based on socio-economic changes and/or network modifications (e.g. new links or transhipment hubs),
- the consideration of all possible transportation modes for route choice.

The different commodity types, which are considered, range from agricultural goods (e.g. cereals, fruits, vegetables), raw materials (e.g. coal, raw wood, ores), oil products, industrial products (e.g. steel and metal products, chemical products) to construction materials and consumer goods.

3.3.1 Freight generation

In practice, the reasons for the transport of goods are the different locations of production and consumption of a certain good and the resulting need of exchange.

Hence, as a first step of the demand calculation, the generated volumes per traffic zone are determined for each commodity. This is done both for the production side (also referred to as origin side) and the consumption side (also referred to as destination side).

In general, the determination of origin and destination vectors is conducted in 2 steps:

1. Determine production and consumption volumes on national level for Croatia and the countries of the region
2. Break down of these national volumes to traffic zone level

On national level, there is the constraint that the total generated volume of all origin zones must equal the total generated volume of all destination zones. While the origin volumes consist of local production volumes and import volumes, destination volumes are the sum of local consumption and export.

The production and consumption volumes on national level are broken down to traffic zone level by the distribution of the decisive land uses for each commodity. Depending on the type of commodity, decisive land uses can be for example population, employees by economic sector as well output/production capacities of production facilities.

Thus, calculating the local production and consumption per zone and adding the import/export volumes, for each commodity two vectors are generated. One includes the generated origin volumes per zone, the other vector the attracted destination volumes.

3.3.2 Freight distribution

Like traffic generation, distribution calculation is applied successively and separately for each commodity. Using a gravity model, the generated origin and destination volumes per zone are distributed, resulting in yearly ton flows between the traffic zones. The trip distribution calculation is carried out in 2 steps:

1. Calculation of evaluation matrix based on a skim matrix including the impedances between traffic zones
2. Calculation of trip matrix (yearly ton flows) based on the evaluation matrix and the origin and destination volumes.

A monetary impedance matrix calculated from the VISUM network model is used as skim matrix. The matrix values in [€] are calculated with the following impedance function.

$$w_{ij} = \sum C_{handling} + \sum (Length \cdot c_{km}) + \sum (Time \cdot c_h) \qquad (1)$$

where w_{ij} = impedance between traffic zone i and traffic zone j [€]; $C_{handling}$ = handling costs [€]; c_{km} = distance cost rate [€/km]; and c_h = time cost rate [€/h].

For all commodities, different impedances (costs) result from different cost rates for required logistic system (e.g. bulk, container) and different requirements for transport speeds due to varying urgencies (loss of value).

3.3.3 Mode choice

The route and mode choice were conducted simultaneously in an interim assignment step for the ton flows. The ton flow matrices of each commodity were assigned to the multi-modal network with an equilibrium assignment procedure.

The mode choice decision for a certain commodity and origin-destination relation was based on the transport costs. It is always the most cost efficient route and transport mode that is chosen. This can be either a direct transport with one transport system or a multimodal chain with a combination of transport modes and transhipments in between.

The total transport costs, which are the determining factors for the route and mode choice during the assignment, consist of:

– Time costs (mode specific time costs + commodity specific loss in value),

– Distance costs (mode specific distance related transport costs),
– Handling costs (costs for loading/unloading and transhipment).

The modal transport costs were allocated to all network links. Different sets of transport costs consider:

– The link attributes (permitted transport mode; transfer link) and
– The logistic system, the commodity is allocated to.

As noted before, the ton flow matrices of each commodity were assigned to the multi-modal network with an iterative equilibrium assigning procedure, resulting in the most cost-efficient route and mode choice for each OD relation. From the multimodal ton flow assignment, ton flow matrices by mode can be derived.

The modal ton flow matrices were converted to vehicle trip matrices by average loading factors (depending on distance classes) and a factor for the unloaded drives. Finally, the generated HGV and LGV trip matrices were assigned to the network together with the trip matrices from the passenger demand model.

3.3.4 Input data

The main input data of the freight demand model are listed below:

– Socio-economic data, i.e. population and employment data, disaggregated to traffic zone level,
– Data on national production (production volumes, location of major production facilities) for each commodity,
– Import/export data from trade statistics, aggregated to about 50 freight model commodities,
– Transit data from UN COMTRADE statistics in, aggregated to the freight model commodities,
– Transport cost parameters (distance related; time related; transhipment costs) per transport mode for the route and mode choice calculation,

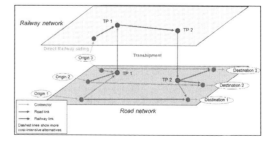

Figure 3. Mode and route choice for freight transport.

– Operational parameters (e.g. average loading factors and share of empty trips by commodity and vehicle type),
– Growth rates and macro-economic parameters are required for forecasting internal and external freight flows.

4 MODEL CALIBRATION AND VALIDATION

In modern society the transport model represents one of the key bases for decision-making on transport and spatial policy, on the investments in the infrastructure demanding time and funds, on the form and dimensions of roads and railways, their impacts, etc. It is therefore important that the model results are reliable. Reliability and credibility are the key characteristics of good and useful transport models. The necessary precision of the model is achieved by the calibration, whilst the reliability and credibility required are proved by the validation.

The growing role of the transport policy has initiated that transport models are becoming increasingly complex. Validation became an obligatory part of a model and the only way to justify its quality. Increasing complexity of models also led to greater complexity of the validation procedures.

Validation procedure for the verification of the adequacy of the National traffic model for the Republic of Croatia (NTMC) was based on following documents:

– JASPERS Appraisal Guidance (Transport): The Use of Transport Models in Transport Planning and Project Appraisal, August 2014
– Design Manual for Roads and Bridges, Volume 12, 1997.
– Variable Demand Modelling—Convergence Realism and Sensitivity, TAG Unit 3.10.4, 2010.
– Critical opinions on these documents and examples of good practice have also been taken into account.

For the freight model, the calibration and validation process comprises three levels:

– total national freight volumes by mode [tonnes],
– port and border crossing volumes [tonnes],
– freight vehicle volumes at 480 count locations.

4.1 Passenger demand validation

Validation criteria for demand is less standardized than for validation of the assignment. Mostly due to lack of independent statistical data (as it is case with traffic counts and assignment). Nevertheless, few quantitative and qualitative tests were done to prove the model.

First step was checking the input data for demand that are socioeconomic and behavioural data. Socioeconomic data (number of population, workplaces, school places…) were taken from the official databases where they have already been submitted to various checks and should be reliable. There were some issues that required additional analyses (e.g. number of workplaces assigned to company's seat, no data available on shopping areas…).

Results of household survey were compared with existing data from previous studies and practice. It was established that most important indicators lie within expected benchmark values (e.g. between 2.5–3 daily trips per person, cca. 40% of work and school trips, characteristic morning and afternoon peak hours, average trip length and duration…).

First validation was done for trip duration distribution. Although such distribution is also input data, it serves also for checking results of the model. Modelled distribution is not only direct result from the survey, but also considers impedance (travel time, length…) between all pairs of zones.

4.2 Transport flow validation

Most traditional type of validation is validation of transport flows.

Considering previously mentioned guidelines and recommendations we suggested following criteria to be accepted by the client for the National Transport model for the Republic of Croatia:

– $R2 > 0.9$
– 65% of GEH <5
– 85% of GEH <10
– difference in transport work < 3%.

In the Table 1 goodness of fit is presented.

Only two indicators do not match expected criteria, but this is for season traffic, where input data were less reliable (no household survey, data about overnight stays, count data). With the use of matrix estimation for internal (Croatian) traffic numerical results would be much better and would fit all criteria. We do not encourage it, as this would decrease quality of forecasting models.

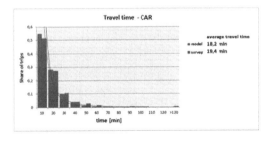

Figure 4. Validation of travel times distributions.

Table 1. Goodness of fit.

	Correlation	GEH < 5	GEH < 10	Difference [veh*km]
Offseason weekday	0.94	67%	88%	< 1%
Offseason peak hour	0.91	65%	91%	< 1%
Season weekend	0.96	70%	82%	< 3%
Season peak hour	0.92	61%	84%	< 1%

Table 2. Recommended elasticities of demand to a change in supply.

Mode	Travel time change	Recommended elasticity	Result [change in car]	Result [change in public transport]
car	± 20%	<−2.0	−0.29 to −0.38	0.58–0.66
public transport	± 20%		0.28–0.35	−1.42 to −1.66

In freight demand model for all validation levels, a satisfying model quality was achieved. The correlation factor R2 is 0.92 and 95 percent of the count locations have a GEH < 5.

4.3 Realism test

If a model adequately reproduces the existing situation, it does not yet mean that it is appropriate for traffic forecasts. The demand model should also behave realistically. A change in the traffic supply should lead to a realistic change in the demand. A change in the demand should be consistent with general experiences.

A change in travel time or trip price (cost in one word) particularly affects the mode choice and trip distribution. This impact must be in realistic limits.

Acceptability of the model response is determined by the elasticity of demand. The equation of arc elasticity is as follows:

$$e = (\log (T1) - \log (To))/(\log(C1) - \log (Cc \quad (2)$$

where e = elasticity, T1 = demand after change, To = demand before change, C1 = changed travel cost and Co = original travel cost

Recommended elasticities for 20% changes in costs are presented in Table 2.

According to the WebTAG recommendations, the elasticity on travel time must be significantly greater than on monetary costs, but not greater than −2.0. This means that a change in travel time has a significantly greater impact on the demand than a change in monetary cost. However, also the later influences to some extent. But with the increasing value of time, travel time gains importance, whilst the impact of direct monetary costs decreases.

Sensitivity of the model must be within the recommended limits. For passenger cars the sensitivity is tested to a 20-percent change in travel time. Sensitivity for cars is between 0.29 and 0.38. This is consistent with the WebTAG recommendations.

For public transport the sensitivity is also tested to a change in travel time. Sensitivity to the change in travel time is greater than for passenger cars and is 1.42 to 1.66, but still well within benchmark.

Based on the sensitivity analysis we concluded that the model responds realistically on systemic and developmental changes and is therefore suitable to offer realistic traffic predictions. This test demonstrates an ability for a real change in modal split.

5 APPLICATION AND USE OF THE NATIONAL TRANSPORT MODEL

After development, calibration and validation of the transport model, it can be used to assess current and future conditions, to identify potential strategies and measures to improve conditions and to assess the impacts of these strategies and measures on the transport system itself and on influenced processes, like environmental impacts, accessibility and social inclusion, social impacts and impacts on economic development.

5.1 Analysis of current conditions

The model was used to analyse current conditions on the Croatian road network and on the National public transport system. Examples are described below.

Of course the highest traffic flows can be observed on the motorway and trunk road networks from Zagreb to all parts of the country and around the larger metropolitan areas in Croatia, mainly Zagreb, Zadar, Rijeka, Split, Varaždin, Osijek and Dubrovnik. The traffic flow analysis is the basis for identifying the major OD relations in the country and to determine external impacts like environmental impacts emission of pollutants and noise.

For understanding the internal impacts and to identify bottlenecks in the network and potential shortcomings, it is necessary to relate the actual flows to the provided capacities. The Figure 5 shows the volume/capacity ratio, again for the off-season conditions and the conditions within the summer season.

Differences are clear; while in the off-season situation, high volume/capacity ratios can be observed mainly around the major cities, and here more in the continental cities of Croatia, in the summer season high volume/capacity conditions occur also on the motorway network towards the Adriatic coast and in and around the Adriatic cities. Levels above 75% volume/capacity ratio are critical, potentially bearing the risk of congestion and traffic breakdowns at peak hours.

Detailed analyses can be carried out for public transport. To display supply/service quality, the Figure 6 to the right shows the accessibility of the city centres of the major cities from the whole territory of the country in 30 minute steps. This could also be compared to similar accessibility plots for car traffic.

The accessibility analysis shows that there are service gaps between the cities, mainly between Zagreb and Zadar and Zagreb and Osijek, with very long access times or even no public transport at all. However, these are remote areas with low population densities.

More graphical and quantitative analyses are possible and have been carried out for the current conditions represented in the base year model.

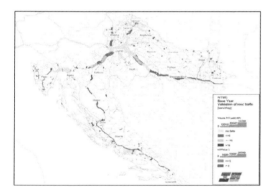

Figure 5.   Validation of link volumes.

Figure 6.   Road traffic—volume/capacity ratio for annual average (ADT) and for seasonal traffic (ASDT).

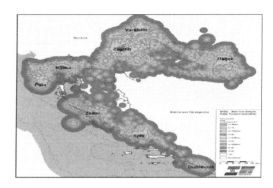

Figure 7.   Accessibility of city centres by public transport.

Model results can also be used for a more complex analysis, for example for defining so-called "Functional Regions". Functional regions are areas with a high frequency of internal regional interaction. The concept of functional regions is used worldwide to understand and define functionally connected areas that need to manage the transport system across administrative borders. The most commonly approach to define functional regions is using data on population commuting to work and school because the pattern of daily commuting rule is a good approximation for staging other types of interaction.

The National Transport Model has been used to determine commuter trips into and out of the major cities as a basis to define functional regions for Croatia (see Figure 7).

5.2   Analysis of future conditions

Since the National Transport Model of Croatia not only represents current conditions but was designed also for the forecast horizon years 2020, 2030 and 2040, the analyses can be carried out also for the future transport conditions. To identify future bottlenecks and shortcomings, a so-called do-minimum scenario is used. In the do-minimum scenario, future demand is calculated based in forecasts of socio-economic and behavioural development, whereas the transport networks represent current conditions only enhanced with those measures and projects that are already under construction now or are planned and fully financed.

5.3   Identification of strategies and measures

Analyses of current and forecasted future traffic conditions, accessibility and transport impacts are an important input to develop strategies and measures to alleviate the identified shortcomings and to

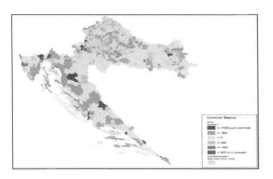

Figure 8. Display of commuter flows for the definition of functional regions.

transform the transport system in accordance with more general transport visions and objectives.

The development of strategies and measures is based on the overall Vision and the defined objectives of the National Transport System. These objectives include:

– Ensure economic, social and environmental sustainability,
– Provide accessibility and social inclusion,
– Increase traffic safety,
– Increase transport efficiency and quality of services,
– Reduce negative impacts on the natural, manmade and social environment.

6 DEVELOPMENT OF REGIONAL AND URBAN TRANSPORT MODELS

Numerous regional and urban transport and mobility master plans are now under development throughout Croatia. All regions and metropolitan areas are carrying out these master plans. One important component of these regional and urban master plans is the quantitative analysis of current and forecasted conditions and the impact assessment of proposed measures and changes. Regional and urban transports model are needed for these

exercises. The National Transport Model forms the basis for the development of these models representing smaller areas.

Technically, the National Model can be used to extract so-called sub-network models, then forming the basis for the implementation of further details, like more detailed zoning structure in the respective area of interest, more detailed allocation of socio-economic data based on the refined transport zones, adding details to the road network and adding local public transport lines and services.

7 SUMMARY

The Ministry of Maritime Affairs, Transport and Infrastructure has commissioned the development of the National Transport Model in preparation of the development of the National Strategy. The consultant team have finalised the work on developing this powerful tool in form of a synthetic 4-step model, by developing the base year model with network model, passenger and freight demand models, have calibrated this model with survey data and have validated it against empirical data. The result is a robust model capable of forecasting future transport conditions and calculating impacts of exogenous changes (like political and socio-economic conditions) and endogenous changes in form of transport strategies and measures.

REFERENCES

Dieter Lohse: Ermittlung von Verkehrsströmen mit n-linearen Gleichungssystemen unter Beachtung von Nebenbedingungen einschließlich Parameterschätzung (Verkehrsnachfragemodellierung: Erzeugung, Verteilung, Aufteilung), Dresden, 1997.
European Commission—Directorates General for Energy, for Climate Action and for Mobility and Transport: EU Energy, Transport and GHG Emissions—Trends to 2050, ISBN 978–92–79–33728–4, 2013.
NTM Croatia Consortium: NTM Croatia Model Development Report, 2016.

Transport Infrastructure and Systems – Dell'Acqua & Wegman (Eds)
© 2017 Taylor & Francis Group, London, ISBN 978-1-138-03009-1

The effects of the different uses of mobile phones on drivers: Experimental tests on an extra-urban road

S. Zampino

AIIT Puglia e Basilicata, Italy

ABSTRACT: It is well known that the alteration of the cognitive load due to the use of mobile phones influences the driving behavior.

The development of smartphones, which allow mobile phone calls, driving navigation, text messaging, etc., introduces more variables in terms of effects on the driver behavior and, consequently, on road safety: the misuse of these phone devices is one of the most serious causes of accidents.

In order to develop an easy method to evaluate the relationships between cause and effect, a direct experimentation on the road was performed, using a GPS navigation system enabled to monitor the driving behavior on a four lanes rural road.

Three groups of drivers were selected by age: their driving performances were evaluated while making a phone call, writing a text message, setting a navigation route.

During the driving tests, different parameters were analyzed (i.e. acceleration/deceleration, lateral position of the vehicle, use of the braking system).

The results of the driving tests highlighted dangerous behaviors during the most complex tasks, with particular reference to the lane-keeping.

1 INTRODUCTION

1.1 Driving behavior and mobile phones

In normal conditions, driver behavior is not easily predictable and when motorists are engaged in multiple tasks, brain concentration can be dangerously diverted from the main driving tasks. As a consequence, the driving parameters (i.e. time of reaction, speed, acceleration, lateral position) are significantly altered.

These effects are properly explained by the theory of the "single channel bottleneck" in human information processing, with specific reference to high speed tasks (Craik 1948, Welford 1967).

From this perspective, the impairment caused by the use of mobile phones while driving, has to be considered a serious health problem. In many countries, the legislation makes it illegal the use of handheld cell phones while driving, although the legislation does not always impose specific sanctions related to the increasing complexity and to the multiple possible uses of the most modern phone devices.

Although people overestimate their ability to multitasking, the high level of impairment caused by modern smartphones is perceived by drivers (Wogalter & Mayhorn 2005) and road agencies (Fig 1), but to which extent the use of smartphones is dangerous and which kind of effects the misuse of the "smart" devices can have on the vehicle speed and trajectories is not yet well known.

1.2 The multi-tasking interference

It is widely acknowledged that the effects on driving due to the interference of multiple tasks, affect the brain performance, as confirmed by functional Magnetic Resonance Imaging (fMRI) experiments conducted by the Carnegie Mellon University (Pittsburgh, Pennsylvania). The experiments show that the effects of cell phone speaking reduce the activity of the parietal lobe related with driving (Just et al. 2008).

Figure 1. Example of ITS message on a National Italian road.

In order to comprehend the mental processes of the human brain, many studies about the effects of the use of mobiles have been conducted with specific reference to phone calls.

These studies prove that the consequences due to the use of mobile phones are more complex than it was initially supposed. The driving behavior has been deeply investigated, in terms of sensitive cognition and cognitive workload, analyzing the different resources and the way in which the hierarchical control, which characterizes the human mind works.

In more detail, the effects of the mental workload on visual search and decision making, studied by visual detection and discrimination tests (performed while driving) cause different levels of impairment to spatial gaze concentration, visual detection and discrimination (Recarte & Nunes 2003).

The four dimensions multiple resources model (Wickens 2002) states that there are four dimensions, which have influence on time-sharing performances. According to this theory, two tasks demanding the same performance (i.e. visual tasks) interfere with each other more than tasks demanding separate levels of perceptual modalities (i.e. visual and auditory tasks).

More recently, the hierarchical model proposed by Logan and Crump (2009) states that there are two separate control loops: the outer loop and the inner one. The inner loop is automated and operates without knowing what the outer loop is doing.

According to this theory, the ordinary driving tasks, with practice, need a progressively reduced level of control: with extensive practice, the driving tasks are gradually encapsulated in automatic operations (Medeiros-Ward et. al. 2014).

When performing different tasks, such as driving and speaking, the human brain switches between primary (or automated) tasks and secondary tasks: consequently, the automated operations are partially impaired. While the effects on the "drained attention" are not perceived during ordinary driving tasks, the switching process reduces the time of reaction whenever the driver has to deal with external not ordinary events or more demanding extra tasks. Other unperceived effects are related to the visual information, which can be altered, affecting the so-called "intentional blindness".

The extreme unpredictability of the driver behavior and the different impacts related to the different kinds of simultaneous extra-tasks are demonstrated by experiments, which prove that the increase of external workload due to the most ordinary use of handheld phones does not necessarily affect vehicles lateral position. The reason of such a behavior can be differently explained, on the basis of both the theories reported: low-impact auditory tasks have no influence on visual tasks,

or, in a different perspective, the effects of an ordinary phone call are limited to the external loop of the human brain by reducing the effects of the eye movements (Cooper et al. 2013).

In other words, the reasons why the lane-keeping performance appears to be unaffected during ordinary phone calls can be due to the way in which the hierarchical control operates: driving, while making or receiving a low-concentration phone call, leaves less residual attention to other tasks directly related to eye movements which affect the lateral positon, but does not impair the driving automatic operations.

When the auditory tasks became more complex or are combined with more complex visual tasks, their effects on the automated operations become progressively more significant.

The different ways in which the auditory tasks affect the driving behavior are demonstrated by further experiments about the difference between the use of handheld and hands free cell phones, which, contrarily to common expectations, considerably influence the not automated driving operations, in case of prolonged use (National Safety Council 2012).

1.3 *The advent of the smartphones technology*

If the effects of the use of mobile phones while driving have been deeply investigated with regard to phone conversations, the most recent widespread use of smartphones has introduced a new perspective.

As a matter of fact, modern smartphones have progressively changed the way of using the traditional phone devices and the diffusion of the social networks combined with the introduction of innovative applications (i.e. *Facebook, Instagram, Whattsapp, Google maps*) has progressively stimulated a very different approach to the phone use.

The use of the most modern phone devices is no more simply based on auditory tasks, but it implies more demanding brain tasks and the touch-screen technology involves the direct observation of the device screen not only for text messaging (Reed & Robbins 2008)

The effects of such a different use are much more dangerous than the traditional use of mobiles for the following reasons:

– tasks which imply visual activities or text messaging are much more complex than simply speaking, and can affect the information processing stages, more than simple auditory tasks;
– performances are less predictable and looking at the phone screen inevitably affects visual channels;
– focal vision is diverted from the road and from the car instruments, while unnatural eye

movements, or the abnormal use of the peripheral vision alter the ordinary space perception (Maples et al. 2008, Robbins & Jenkins 2015).

The problem is emphasized by the constant increase of the misuse of the "smart" devices.

According to a recently published AT&T research (http://www.digitaltrends.com/mobile, May 2015), 70% of drivers declare to use their smartphones while operating vehicles. While texting and sending e-mails are the most common activities, an unexpected 27% of the interviewed drivers declared of checking Facebook, 14% of Tweeting, 14% admitted of using Instagram, 11% of using Snapchat, not to mention the use of smartphones to take "selfies" while driving.

2 GENERAL METHOD

2.1 Objective of the study

The aim of this paper is to give a contribution to the comprehension of the actual effects of the increasingly widespread and diversified use of mobile smartphones while driving, through a direct investigation, under normal but strictly controlled and pre-defined driving conditions.

In particular, the goal of the study was accomplished in a series of direct driving experiments in which a group drivers, selected by age and experience, performed different tasks involving different uses of their smartphones while driving one a pre-defined road section.

2.2 Description

In order to collect the data, rather uniform driving conditions were pre-defined: the tests were conducted on the same road section, driving the same car in the same hours of the day, under similar weather conditions (Temperature 27–32°C, Wind Speed 2–5 m/s) and under uniform conditions of light traffic.

The different tasks to perform during the tests were standardized, while the drivers had to use their personal devices, supposing a higher level of familiarization and a better knowledge of the Operating System.

Participants were selected on condition of declaring to be regular users of smartphone apps, to avoid dangerous consequences during the visual tasks.

Before starting each experiment, participants were carefully instructed about the tasks and the security warnings.

Their standard behavior (without using any device) was tested on the road and recorded, to allow a preliminary evaluation and to highlight

the peculiarities of the general driving attitudes of each motorist (i.e. standard lateral deviation, use of the braking system).

In order to monitor the driver behavior during the driving tests, it was made use of three devices: a *WI-FI Action-camera* (VTIN—mod EyPro VOD001B), a standard HD camcorder and a common phone device (Apple I-Phone 5 s) abled to record the driver behavior with reference to speed, acceleration, lateral position (detected with the *GPS* app—*Sygic Navigation system—All regions 2016*) (Fig. 2).

The choice of using ordinary devices and a common application to record the data was made in the prospect of reproducing the tests in other contexts.

2.3 Overall design

The tests were performed over the course of short repetitive drives.

The car used for the tests was a common model of Peugeot (mod. 206-1.1, 60 HP), equipped with the system of cameras and *GPS* previously described (Fig. 3).

In particular, the phone *Dash Camera* was positioned in correspondence to the central axis of the vehicle, while the *Action-camera* was positioned on the left of the driver, in proximity to the windshield.

The driving experiments were conducted on a dual carriageway (4 lanes) rural road (Fig. 4), on a pre-selected straight road section (S.P. n. 364 – Lecce-S. Cataldo, Eastbound – km 1,200–km 2,640–length 1440 m–lane width 3.50 m–GPS coordinates of the initial point 40°22'0.48"N – 18°13'10.56"E).

Three groups of participants (16 males, 5 females, 21 drivers in total), were selected according to the following age and Driving Experience (*DE*) criteria:

Figure 2. Devices and SW used during the tests.

Figure 3. Cameras and GPS devices positioned inside the car.

Figure 4. The straight section of the road where the driving tests were performed.

Group 1 – young drivers aged between 18 and 27 years (YD—*DE* < 10 years – 7 drivers);
Group 2 – expert drivers aged between 28 and 54 years (MD—11 < *DE* < 25 years - 7 drivers);
Group 3 – elder drivers aged between 55 and 65 years (ED—*DE* > 26 years – 7 drivers).

Participants, who previously reported having normal or corrected-to-normal vision, were asked to interact with their phone, performing three separate tasks, according to the following standardized sequence:

A. Firstly, the drivers were asked to make a hand-held phone call (with a pre-defined content);
B. Secondly, the selected drivers were asked to write and send a pre-defined text message with the *Whattsapp* application;
C. Finally, each driver was asked to select a route to a predetermined destination, using the *Google Maps* application.

After the first drive without performing any extra task (Full control), the pre-defined "smartphone tasks" were performed in three different and consecutive drives along the selected road section, each task starting after returning at the same point (km 1,400) of the road section and never exceeding the final point (km 2,400) of the same road section. A digital chronometer was used to record the duration of every single task.

The driving tests were conducted at a reference speed between 55–75 km/h, according to the trial procedure reported below.

10' *Welcome and general instructions*
5' *Familiarization with the car*
6' First Drive (baseline drive—full control) *return to the starting point*
6' Second Drive (Phone call) *return to the starting point*
6' Third Drive (Text messaging) *return to the starting point*
6' Fourth drive (Google maps route) *end of the test*

For each participant the instructions to interact with the smartphones were repeated at the same point of the route by an instructor, according to a general standardized procedure

The continuous presence of the instructor during the tests was necessary in order to prevent the risk of accidents.

No overtaking was allowed during the tests.

During the experiments, the basic data and the driving parameters were simultaneously detected and saved in the phone device (enabled to measure the speed and acceleration parameters) in a *cloud* directory and in the memory card of the action camera (abled to measure the lateral displacements with the use of optical *viewfinder* positioned upon the car windshield). The standard HD camcorder was used to record the general data of each test, inside the car.

The devices and the apps used to collect the data were activated by the instructor seated on the front passenger seat of the car.

With regard to the SW app used to record the test data, the *Sygic* application is not only a GPS navigation software, but it also allows memorizing a digital movie of each route by a *Dash camera SW* connecting the digital camera of the electronic device and recording the main kinematic data of each route.

2.4 Definition of the tasks

The tasks assigned to each motorist were pre-defined, in terms of sequence and complexity, in order to analyze and compare different behaviors under different circumstances.

The smartphone exercises that participants were required to perform while driving are showed in the following Table 1.

The phone call message was defined in order to evaluate the effects of a brief phone call, although the visual task of observing the speedometer to check the speed added a slightly more complex activity contemporary to the auditory task.

The following (visual) task was much more complex and demanding than the previous one and was defined in relation to the length of the road section (1000 m).

Table 1. Pre-defined messages/address for each task.

Task 1	Phone call	Ciao, come stai? Io mi trovo sulla strada Lecce San Cataldo, e sto andando a Sancataldo. (After looking at the speedometer) Attualmente viaggio alla velocità di ... chilometri all'ora.
Task 2	Wapp Message	Mi trovo a Lecce, arriverò tra 10 minuti.
Task 3	Gmaps Route	Piazza G. Mazzini, 10, Lecce.

Figure 5. Frames of the videos captured with the *Dash Camera*.

The final task was different from the second, shorter and less demanding in terms of brain effort, but more demanding in terms of visual application.

2.5 *Data acquisition and processing*

The movies recorded with the *Dash camera* application during the diving tests, contain the main data of the route and, precisely, the *GPS* coordinates and the instantaneous speed. Contemporary, the SW application recorded a set of kinematic parameters resulting in a series of diagrams (*speed vs distance, speed vs time, acceleration vs distance, acceleration vs time, etc.*) collected in a "digital travel book" instantaneously saved in a specific external *cloud* directory.

The videos of each test, simultaneously recorded with the three cameras, allowed a double check of the vehicle trajectories and were used to schedule each drive.

While, according to similar experiences, the speed data collected by the phone *GPS* antenna and by the device internal accelerometer can be assumed to be precise with a margin of error of almost 1% (Fazeen et al. 2012), the *GPS app* does not allow precise lateral position measurements. For this reason, a simple method to calculate the vehicle lateral position relative to the axis of the road was implemented.

In particular, two optical viewfinders were positioned upon the car windshield (exactly in front of phone camera and in front of the *Action-camera*).

The videos collected during the driving tests with the phone *Dash camera SW* and with the *Action-camera* were post-processed with *DVDVideosoft Free Studio* software (Fig. 5).

Each frame extracted from the videos (with an average rate of 0.5 fps) was analyzed with *C Thing Meazure* software, which (assuming as a reference the horizontal road markings and the marks on the viewfinder) allowed the calculation of the transverse displacements relative to the central lane axis.

In order to analyze and compare the driving performan-ces, the main data of each drive were extrapolated and reported in a digital worksheet.

3 RESULTS AND DISCUSSION

3.1 *The analysis of the driving performances*

The analysis of each driver behavior was processed through the extraction of the driving data collected from the GPS app and from the video camcorders.

In particular, for each participant and for each test, the following key measures were collected:

- Mean/maximum speed
- Speed variation
- Maximum deceleration/acceleration
- Left side displacements
- Right-side displacements
- Maximum left side displacements
- Maximum right-side displacements

The mean speed was evaluated with reference to the four drives, while the speed deviation was evaluated with reference to the second, third and fourth drive.

Likewise, the standard lane-keeping behavior of the motorists was evaluated during the first drive. The mean and maximum lateral displacements recorded during the following drives were compared with the data of the first drive. The driving parameters were analyzed for each group of drivers with particular reference to speed and lateral position.

Further data were evaluated with regard both to the longitudinal trajectories of the drivers and to the relationship between speed and lateral position during the tasks performed while using the smartphones.

3.2 *Speed*

The first aspect to be evaluated was the speed variation during the phone tests.

Assuming as a reference the baseline drive, all the drivers decreased their speed while performing the following three driving tests.

With regard to the mean speed, the data show that the speed reduction was related to the complexity of the task.

In particular, while the frequency of speed reduction was not significant during the second drive, 53% of the motorists reduced their speed by 10 km/h during the third and fourth drive, when the tasks regarded the visual use of the smartphones.

The effects of the visual tasks were more significant if related to the mean speed of each motorist during the tests (Fig. 6).

In particular, during the 3rd and 4th drive, more than 90% of drivers reduced their speed by more than 10% than the average speed.

The average speed reduction during the second drive was 12.70%, but it increased to 15.30% and to 17.26% during the third and fourth drive.

3.3 Lateral position

The most significant results of the tests regarded the lane-keeping performances.

For each driver, the lateral displacements recorded during the driving runs were compared to the standard behavior recorded during the baseline drive.

In particular, assuming as a reference the central axis of the lane, the mean fluctuation range between the left and the right vehicle positions, measured during the baseline drive, was 0.28 m (mean displacement to the left 0.17 m, mean displacement to the right 0.11 m).

During the following drives, the mean fluctuation range between the left and the right positions increased to higher values (second drive: 0.34 m, third drive: 0.44 m, fourth drive: 0.48 m).

The most significant result was that while the mean range increased during the second drive (phone call) by 0.06 m, it increased during the third and fourth drive (*Whattsapp* and *Google maps* exercises) by 0.16 and 0.20 m, respectively.

In more detail, 60% of participants increased their mean displacement to the left by more than

0.10 m during third drive, 64% of drivers increased their mean displacement to the left by the same measure, while during the second drive (phone call) the same increase of the mean fluctuation was limited to 18% of participants (Fig. 7).

The mean of the maximum driver displacements to the left were much higher during the third and fourth drives, as shown in Figure 8.

During the third and fourth drives the mean of the maximun driver displacements to the left increased to more than 0.6 m.

While, during the second drive, for 52% of drivers the maximum displacement to the left was more than 0.30 m, during the following drives, the percentage of drivers whose maximun displacment to the left was more than 0.30 m grew up to 82% (more than 0.50 m, for 44% of drivers).

The displacments to the right during the second, third and fourth drives were markedly lower (Fig. 9).

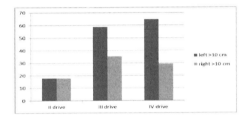

Figure 7. Mean displacement to the left and to the right, during the different tasks (driver behavior).

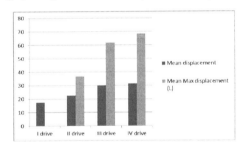

Figure 8. Comparison of the mean displacements to the left.

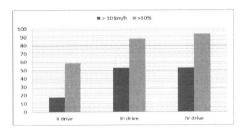

Figure 6. Mean speed: comparison with the baseline drive.

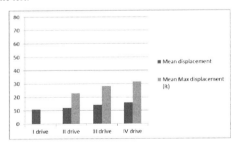

Figure 9. Comparison of the mean displacments to the right.

In particular, for 52% of drivers the maximum deviation to the right was more than 0.30 m during the third and fourth drives (for 33%, more than 0.50 m).

No significant difference was observed between female and male drivers, except during the phone call exercise (second drive): in that case, the deviation to the left was slightly lower for female drivers.

The more significant differences were detected between the first group of drivers (Young Drivers) and the third group of drivers (Elder Drivers).

The mean of the maximum displacements to the left was sharply higher during third and fourth drives for elder drivers (0.45 – m—YD; 0.91 m—ED).

In comparison, the difference between the mean of the maximum displacements to the right was more reduced (0.15 m—YD; 0.32 m—ED—IV drive).

The mean of the maximum left-right fluctuation recorded during the second, third and fourth drives show how the effects of the use of the smartdevices during the three pre-defined tasks were lower for young drivers than for elder drivers (Fig. 10).

All the data collected about the lane-keeping performances, showed oscillating trajectories with a prevailing trend to occupy the left side of the lane.

This trend was not exclusively related to the difficulty of the task (writing 3rd test more complicated the 4th exercise), but it was more probably influenced by the negative effect of the diversion of the focal vision from the road.

The interruption to focal information and the consequent inadequacy of the sole peripheral vision to define the correct spatial perception and the vehicle trajectory (Land & Horwood, 1995) induced the motorists to move away from the right edge of the carriageway (perceived as more dangerous). For elder drivers, whose visual field is much more reduced than young drivers, the tendency to occupy the adjacent lane was more significant.

3.4 Longitudinal trajectories and kinematic data

The tests conducted were used to study the longitudinal trajectories of the vehicles while driving and simultaneously performing smartphones extra tasks.

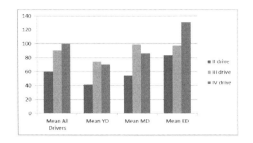

Figure 10. Mean of the maximum fluctuations for category.

The following diagrams (Figs. 11, 12. 13) show the maximum lateral displacements recorded for each motorist.

In more detail, while the maximum transverse fluctuations to the left and to the right were mostly comparable during the second drive (phone call), the maximum fluctuation range increased during the the text messaging test and during the Google maps exercise, more significantly (Figs. 12, 13)

Table 2 reports the data of the statistical (normal) distribution of the maximum lateral displacements, assuming as a reference the left edge of the lane (0.00 m left edge, 1.75 m central axis, 3.50 m right edge).

The prevailing tendency to occupy the left side of the lane during the third and fourth drives is confirmed by the percentile curves (Fig. 14).

As reported above, during the third and fourth drives a more noteworthy speed reduction was registerd.

Figure 11. Maximum displacements during the second drive.

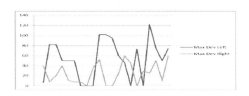

Figure 12. Maximum displacements during the third drive.

Figure 13. Maximum displacements during the fourth drive.

Table 2. Statistical distribution of the maximum deviation.

	Mean	St dev.	N. Dist. 1.75
2nd drive	168,11	36,45	0,57
3rd drive	158,23	51,59	0,63
4th drive	157,95	61,78	0,61

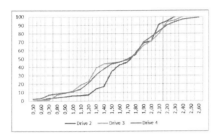

Figure 14. Maximum displacements percentile curves (axis 1.75 m).

Figure 15. Example of diagram of the trajectories (driver n. 16). (2nd M 0.02, SD 0.04; 3rd M 0.07, SD 0.14; 4th M 0.03, SD 0.21).

However, the maximum deceleration (1.3–1.8m/s^2) demonstrates a limited use of the braking system and for each test it was observed in correpondence to the most significant lateral displacments.

For each drive, the longitudinal trajectories of the drivers were examined by diagrams deducted by the video processing (Figure 15).

For the vast majority of drivers, the analysis of the trajectories showed a recurrent first displacement to the left followed by sudden trajectory corrections to the centerline and to the right edge of the lane. Single fluctuations of variable entity, around the central axis of the lane, followed (displacements to the left of more than 1.5–2 m were observed).

The trajectory Standard Deviation (SD) of each driver increased during the 3rd an 4th drives (Mean SD 2nd drive, 0.07 m, Mean SD 4th drive 0.27 m—reference central axis) and was higher for elder drivers.

4 CONCLUSIONS

The results of the tests confirm the very dangerous effect of the visual use of smartphones while driving.

In particular, the data collected show how the impairment due to the use of modern phone devices is not exclusively evaluable in terms of mental workload but it is strongly influenced by the effects of the visual tasks. The concentration of the focal vision on the smartphone, diverts the look from the road and produces the consequent negative effect of an incorrect reliance on the peripheral vision, with the further drawback of the loss of the correct space perception.

Drivers who rely on the sole guidance of the peripheral vision, perceiving of not checking exactly the space in front of the car, assume fluctuating trajectories. They mostly tend towards the left side of the lane (unconsciously perceived as safer than the right side), increasing the risk of accident in case of multiple vehicular interactions (i.e. frontal impacts in case of single carriageway roads).

To prevent such a risky behavior, a suggestion for smartphone manufacturers could consist in implementing a GPS based application to inhibit the activation of the most visual demanding smartphone apps, whenever the detected speed and acceleration exceed a default maximum value.

REFERENCES

Cooper J.M., Medeiros Ward N., Strayer D.L. 2013. The impact of eye movements and cognitive workload on lateral position variability in driving. In Human Factors and Ergonomic Society (ed). *Human factors Vol. 55 No. 5*: 1001–1014.

Craik K.J.W. 1948. Theory of the human operator in control systems II. Man as an element in a control system, *British Journal of Psychology, 38*, 142–148.

Fazeen M., Gozick B., Dantu R., Bhukhiya M., Gonzalez M.C., 2012. Safe Driving Using Mobile Phones. In IEEE Transactions on Intelligent Transport Systems.

Just M.A., Keller T.A., Cynkar J.A., 2008 - A decrease in brain activation associated with driving when listening to someone speak. In CMU Research Showcase. *Brain Research:* 70–80.

Land M., & Horwood, J., 1995. Which parts of the road guide steering? *Nature, 377* (6547), 339–340.

Logan G.D., & Crump, M.J.C., 2009. The left hand doesn't know what the right hand is doing. *Psychological Science, 20*, 1296–1300.

Maples W. C, De Rosier W., Hoenes R. Moore S., 2008. The effects of cell phone use on peripheral vision. In Optometry Journal of the American Optometric Association, 79.

Medeiros-Ward N., Cooper J.M., Strayer D.L 2014. Hierarchical Control and Driving. In American Psychological association (eds). *Journal of experimental psychology: General: Vol 143 No 3*: 953–958.

National Safety Council, 2012. Understanding the distract brain—Why driving while using hands-free cell phone is risky behavior–*White paper.*

Recarte M.A., Nunes L.M. 2003. Mental Workload while driving: effects on visual search discrimination and decision making. In American Psychological association (eds). *Journal of experimental psychology applied, Vol 9 No 2*: 119–137.

Reed N., Robbins R., 2008. The Effect of Text Messaging On Driver behavior—A simulator Study. In TRL – *Published Project Report—PPR 367.*

Robbins R. & Jenkins D., 2015. Eyes on the road, a review of literature and an in-car study of driving whilst navigating–RAC Foundation—London.

Welford A.T. 1967. Single channel operation in the brain, *Acta Psychologica, 27*, 5–21.

Wickens C.D. 2002. Multiple resources and performance prediction. In Taylor and Francis Ltd (ed). *Theoretical issues in Ergonomic Science, Vol 3, No 2*: 159–177.

Wogalter M.S., Mayhorn C.B. 2005. Perceptions of driver distraction by cellular phone users and nonusers. In Human Factors and Ergonomic Society (ed). *Human Factors Vol.47 No. 2*: 455–467.

Transport Infrastructure and Systems – Dell'Acqua & Wegman (Eds)
© 2017 Taylor & Francis Group, London, ISBN 978-1-138-03009-1

Back-propagation neural networks and generalized linear mixed models to investigate vehicular flow and weather data relationships with crash severity in urban road segments

L. Mussone
Politecnico di Milano, Milano, Italy

M. Bassani & P. Masci
Politecnico di Torino, Torino, Italy

ABSTRACT: The paper deals with the identification of variables and models that can explain why a certain Severity Level (SL) may be expected in the event of a certain type of crash at a specific point of an urban road network. Two official crash records, a weather database, a traffic data source, and information on the characteristics of the investigated urban road segments of Turin (Italy) for the seven years from 2006 to 2012 were used. Examination of the full database of 47,592 crash events, including property damage only crashes, reveals 9,785 injury crashes occurring along road segments only. Of these, 1,621 were found to be associated with a dataset of traffic flows aggregated in 5 minutes for the 35 minutes across each crash event, and to weather data recorded by the official weather station of Turin. Two different approaches, a back-propagation neural network model and a generalized linear mixed model were used. Results show the impact of flow and other variables on the SL that may characterize a crash; differences in the significant variables and performance of the two modelling approaches are also commented on in the manuscript.

1 INTRODUCTION

Road crashes are events that depend on a variety of factors characterising human behaviour, weather, road pavement, vehicle stability and performance. Crash events show different magnitudes when evaluated with respect to the effects on road users (crash severity), and the knowledge of the contributing factors that affect the severity should be used to improve road safety through the action of transport policy makers, designers and road agencies.

Traffic volume, weather conditions, and road characteristics affect crash severity in a multifaceted way (Wang et al. 2013). Specifically, Theofilatos & Yannis (2014) pointed out that the few papers available mainly deal with roads operating under uninterrupted flow conditions, and recur mainly to logit modelling (Al-Ghamdi 2002, Golob et al. 2008, Christoforou et al. 2010, Jung et al. 2010, Xu et al. 2013, Yu & Abdel-Aty 2013). Earlier, Shankar et al. (1996) stated that crash severity investigations had been historically limited to the localization of fatalities, even though the estimation of the other severity levels (i.e., property damage only—PDO –, possible injury, non-incapacitating injury) could help in understanding the benefits of safety-improvement projects. Seventeen years later, Xu et al. (2013) again underlined that most

of the research studies have been focused on the likelihood of a crash without considering the crash outcome severity.

One of the main obstacles to investigations of this type is the limited availability of comprehensive crash databases, and associated robust weather and traffic databases. Nowadays, however, a continuous flow of environmental and traffic data is collected by local road agencies with sensors of increasing quality and performance (Chong & Kumar 2003, Nekovee 2005). Contrary to what happened in the past, data are now frequently collected in intervals of shorter duration. Hence, available databases can be associated and merged with others containing data collected over several years of observations, thus supporting new robust inferences (El Faouzi et al. 2011).

To obtain reliable models and convincing results, the availability of high quality data representing the characteristics of drivers, together with traffic, weather, and pavement conditions is fundamental. Unfortunately, data describing every significant factor affecting crashes needs work to associate the available contrasting information that could be long, arduous, and sometimes unproductive.

The paper aims to bridge these gaps by providing knowledge on factors contributing to crash severity in an urban road network, considering only those influencing crashes along road segments. Data on crashes, traffic and the weather database of Turin's road network (Italy) were collected and used to calibrate and validate predictive models of crash severity. The Back Propagation Neural Networks (BPNN), a robust tool used to investigate complex phenomena without assuming any preliminary hypotheses on the model, was used. But BPNN cannot give an analytical formulation of the mathematical functions linking the variables that significantly affect a certain phenomenon, thus only a sensitivity analysis of the model can be performed. A Generalized Linear Mixed Model (GLMM) was also used (with its analytical formulation) to compare and assess results with those obtained by BPNN.

2 DATABASE FORMATION

2.1 Crash classification

Crash data were provided by the *Istituto Nazionale di Statistica* (National Statistics Institute, ISTAT). The ISTAT database contains details on crash dynamics and location, on the vehicles, and on gender and age of people involved, in accordance with current Italian legislation, specifically articles number 582, 583 and 590 of the Italian Penal Code 2015 (Repubblica Italiana 2015). The Italian law considers a road accident to be a crash when it results in at least one injury, and crash consequences are classified into the following five Severity Levels (SL):

- Very Slight Injuries (VSI), when the most seriously injured person has a prognosis of less than 20 days;
- Slight Injuries (SLI), when the prognosis is between 21 and 40 days;
- Severe Injuries (SEI), if the event causes an illness that endangers the life of the injured party, and/or the event results in permanent damage to the brain or any body organ;
- Guarded Prognosis (GPR), if the doctor cannot determine the disability, and issues a report of "guarded prognosis" (pending resolution of prognosis, the road crash must be considered and treated as a determining factor); and
- fatalities (FAT), including injured persons who die within 30 days of the crash.

The dearth of information in the ISTAT database was overcome by including crash data collected by Turin's Municipal Police (TMP). All the records of the ISTAT database were matched up to the TMP database and implemented with the following information: (a) historical data (time, nearest minute, day, month and year of the crash event); (b) locality data (street name, house number); and (c) generic information concerning crash SL.

Table 1 shows the number of crashes per year and SL, and evidences the decrease in all the SL classes between 2006 and 2012. Assuming the year 2006 as a reference point, the following years witnessed a decrease in crash occurrence across all severity classes.

2.2 Traffic data

Traffic data were provided by the 5T Company (Telematics, Technologies for Traffic and Transport in Turin), which monitors and controls over 300 urban traffic lights in Turin, and collects traffic data. 5T uses induction-loop traffic sensors located along the exiting lanes of the monitored intersections to collect vehicle flow data at 5 minute intervals. It is worth noting that from 2006 to 2012, the number of traffic sensors available varied from 662 to 1051 due to the installation of new ones and the elimination of some of the damaged ones. Figure 1a shows the portion of the road network monitored by 5T in 2006.

2.3 Weather data

The Environmental Protection Agency of the Piedmont Region (ARPA Piedmont) provided data on weather conditions. The Turin weather station considered in the paper is located in the city centre (238 m a.s.l., 1.5 m off the ground, latitude 45°.066667, longitude 7°.683333), and collects temperature, atmospheric pressure, wind speed and direction, solar radiation, and rainfall intensity data on an hourly basis. The maximum distance between the weather station and the farthest crash location included in the database was 9.6 km. Each crash record was associated with the weather data recorded at the time of the crash.

2.4 Database formation

Only crashes that occurred along segments provided with valid and reliable traffic data were extracted from the main database and used. The database adopted for modelling is a subset (and

Table 1. Injury crash records along segments of the urban road network of Turin (Italy, 2006–2012).

Year	VSI	SLI	SEI	GPR	FAT	Total
2006	1317	173	36	33	22	1581
2007	1303	205	69	42	29	1648
2008	1128	152	41	33	12	1366
2009	1061	160	34	27	21	1303
2010	1198	169	46	31	19	1463
2011	1037	171	51	18	18	1295
2012	917	141	44	13	14	1129
2006-'12	7961	1171	321	197	135	9785

(a)

(b)

(c)

Figure 1. (a) Turin's traffic monitoring network operated by 5T in 2006 (highlighted in black); (b) spatial distribution of road crashes that occurred in 2006; and (c) crashes which were associable to 5 min traffic flow (487 in total).

a random sample) of the total number of crashes that occurred and were recorded in the official database. Figure 1b shows all the crash records in 2006, while Figure 1c shows only crashes associated with traffic flow data.

3 DATA ANALYSIS AND TREATMENT

3.1 Variables

Table 2 lists the independent variables, their numbering and labels, the type of variable, the unit of measurement, and the range. The variables referring to the road are:

– road type (C1), which indicates the organization of the carriageways and the directions served (0 = unknown; 1 = one carriageway, one way; 2 = one carriageway, two ways; 3 = two carriageways, two ways; 4 = more than two carriageways, two ways);

– pavement conditions (C2), which have been distinguished with a numerical variable indicating the presence of water, snow or ice (0 = unknown, 1 = dry, 2 = wet, 3 = slippery, 4 = icy/frozen, 5 = snowy); and

– the road signage (C3), which indicates if it was absent (0), if it was composed of vertical signs only (1), horizontal markings only (2), if both were present (3), or if a temporary construction signage was present (4).

The variables that reflect the characteristics of vehicles A and B involved in the crash are:

– vehicle type (C4 and C6), ranging from 0 (passenger cars) to 20 (quad), also including the case of vehicles that fled the crash scene (19); and

– vehicle category (C5 and C7), from 0 to 8, in which 1 represents cars, 2 buses, 3 trams, 4 heavy vehicles, 5 industrial vehicles, 6 bikes, 7 motorcycles, 8 vehicles that fled the crash scene and 0 unclassified vehicles.

In modelling, variables describing roads and vehicles were assumed as categorical. The variables describing drivers involved in the crashes were assumed as numerical. They are:

– age (C8 and C11), which ranges from a minimum of 10 (driver B) to a maximum of 89 (driver A); this variable also assumed the null value in cases of unknown/unrecorded age;

– age class (C9 and C12), which groups the ages into 6 intervals ranging from 0 to 5: 0 in the case of unknown/unrecorded data, 1 for very young drivers (15–19 years old), 2 for young drivers (20–24 years old), 3 for adults (25–64 years old), 4 for elderly drivers (from 65 to 79), and finally 5 for very old drivers (over 80); and

– sex of drivers (C10 and C13), which assumes the value 0 in cases of unknown/unrecorded data, 1 for males, and 2 for females.

In Table 2, the lowest values for 'age of driver A' refer to scooter drivers, while those for driver B refer to pedestrians or cyclists. Air temperature (C14), wind speed (C15), solar radiation (C16), and rainfall precipitation (C18) were assumed as numerical with values that correspond to the measured values. The lighting condition (C17) was assumed as a Boolean variable (0 = dark, 1 = light). The Traffic Flow (TF) variables (C20 ÷ C25) are numerical and represent the volume of vehicles per hour (veh/h) measured every 5 min across the crash event, according to the time scale reported in Figure 2. Finally, the standard deviation (C26) for the seven flow values was added to the list to take into account flow fluctuations for the 35 min period before and after the crash. Finally, the output variable indicating the SL (C27) was assumed numerical and ranging from 2 (VSI) to 6 (FAT).

Table 2. Number and labels of variables.

#	Code	Description	Type	u.m.	Range min	max
1	C1	Road type	C	_	0	4
2	C2	PC	C	_	0	5
3	C3	Road signage	C	_	0	4
4	C4	Veh. A type	C	_	0	20
5	C5	Veh. A cat.	C	_	0	8
6	C6	Veh. B type	C	_	0	20
7	C7	Veh. B cat.	C	_	0	8
8	C8	Dr. veh. A age	N	_	16	89
9	C9	Dr. veh. A cl. age	N	_	0	5
10	C10	Dr. veh. A sex	N	_	0	2
11	C11	Dr. veh. B age	N	_	10	86
12	C12	Dr. veh. B cl. age	N	_	0	5
13	C13	Dr. veh. B sex	N	_	0	2
14	C14	Air temp.	N	°C	−7.5	+35.3
15	C15	Wind speed	N	m/s	0	9.95
16	C16	Light radiation	N	W/m²	0	996
17	C17	Light/dark	B	_	0	1
18	C18	Rainfall	N	mm/h	0	12.8
19	C19	TF1 (*)	N	veh/h	0	730
20	C20	TF2 (*)	N	veh/h	0	750
21	C21	TF3 (*)	N	veh/h	0	765
22	C22	TF4 (*)	N	veh/h	0	775
23	C23	TF5 (*)	N	veh/h	0	570
24	C24	TF6 (*)	N	veh/h	0	565
25	C25	TF7 (*)	N	veh/h	0	494
26	C26	Flow st. dev.	N	veh/h	0	193
27	C27	SL	N	–	2	6

Notes: PC = pavement conditions, TF = traffic flow, Dr. = driver, veh. = vehicle, cl. = class, N = numerical, B = Boolean, C = categorical.

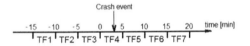

Figure 2. Time scale used to aggregate Traffic Flows (TF) across the crash event.

A criticism may be made of the use of all seven flow values (TF1-TF7). In fact, some of these values (in particular TF5-TF7) refer to intervals after the crash and therefore were caused by the crash itself. Nevertheless, the reason for using them is that they belong to the same time series and the interrupted nature of flow in urban roads makes each interval (though not as long as 5 minutes) a story apart, even when there is no crash. In addition, results will show that they play a different role in the models.

3.2 Database information content

The Principal Component Analysis (PCA) (Lebart et al. 1977) was used to investigate the information content of the database. Table 3 reports the variance explained by the first eight components. They account for about 81% of the total variance for both databases, while the first two components account for about 61%. The variables most linked to the first component are road type, road signage, and age of driver A, whereas those linked to the second component are light/dark, light radiation, and air temperature. Finally, traffic flows (TF1-TF7) are mainly linked to the third component. This means that the set of variables relating to road and driver can explain about 45% of the variance; those related to meteorological conditions about 16%, and those related to flow about 8%.

3.3 Data treatment

According to Table 1, the five SLs contained in the database were not equally represented and the dataset resulted imbalanced. This is not a problem for modelling approaches such as logistic regression, but it is for machine learning tools, and especially Artificial Neural Networks (ANN). With imbalanced datasets, ANN could not find the correct relationships between input and output for all categories present in the dataset. Over-sampling (with data duplication) or under-sampling (with data cancellation) techniques present advantages and disadvantages: under-sampling can remove important data and over-sampling can lead to over-fitting problems. Studies on imbalanced datasets have shown over-sampling to be more advantageous and useful than under-sampling (Chawla 2010).

The "focused re-sampling" method proposed by Japkowicz (2000), which consisted of an oversampling of those examples that occurred in the minority classes (specifically FAT, GPI, and SEI), was used. This approach implies duplication of the entire subset of data until their count is of the same magnitude as the most populated class. This approach avoids other possible biases in re-sampling data.

Another task performed was data normalization. Feature scaling, also called unity based normalization, was used for its simplicity. Let X_{min} and X_{max} be the two extreme values (minimum and maximum) of a variable X, the normalized variable X' (according to feature scaling) is:

Table 3. Percentage of variance explained by the first eight components in PCA.

Component	Simple value	Cumulative value
1	44.72	44.72
2	16.34	61.06
3	8.32	69.38
4	6.72	76.10
5	4.91	81.01
6	4.38	85.38
7	3.49	88.88
8	2.28	91.16

$$X' = \frac{X - X_{min}}{X_{max} - X_{min}} \qquad (1)$$

All the input variables were then normalized according to eq. 1, hence with values falling within the range [0,1]. The output variable (the output) is numerical, ranging from 2 to 6.

4 DATA MODELLING AND RESULTS

4.1 Back-Propagation Neural Networks (BPNN)

The BPNN used in this work is an example of an Artificial Neural Network (ANN) model, ANN models have a classical multilayer topology with feed-forward connections. Cybenko (1989), and Hornik (1991), described the capability of ANN in approximating any function belonging to the Lebesgue two space (L^2 space) with minimum error. Applications regarding transport, planning, control fields, and crash analysis are numerous starting from the 90's (Dougherty 1995, Mussone 1999, Mussone et al. 1999). Other contributions have faced the problem of crash prediction or severity (Abdelwahab & Abdel-Aty 2001, Chong et. al. 2004, Delen et al. 2006, Baluni & Raiwani 2014).

The downside in using the BPNN approach is that the relationships between variables are in a black box (the hidden layer of Figure 3), and no analytical formulation between input and output can be directly obtained. The effects of independent (input) variables can be interpreted only through a sensitivity analysis of the model.

The BPNN models were calibrated and validated with the Levenberg-Marquardt training algorithm. Performances were evaluated according to Mean Squared Errors (MSE) through the three phases of train, test, and validation. The model was constructed with an input layer including the 26 independent variables listed in Table 2, the hidden layer, and the output layer corresponding to the SL, tested with one neuron. All categorical variables are coded in binary format to reduce connection between their values. Finally, the best model was found to be made up of 25 neurons in the hidden layer for the model. It has a MSE lower than 0.08, which means that there are only 8 errors to each 100 classifications.

4.2 Generalized Linear Mixed Model (GLMM)

For the analysis of multilevel data, random clusters and/or subject effects should be included in the regression model to account for the correlation of data. The resulting model is a mixed model including fixed and random effects. Mixed models for continuous normal outcomes have been proposed for non-normal data and are generically classified as Generalized Linear Mixed Models (GLMMs). The extension of the methods from dichotomous responses to ordinal response data was actively pursued in the reviews of Agresti & Natarajan (2001).

The GLMM model is a regression model of a response variable that contains both fixed and random effects and comprises data, a model description, fitted coefficients, co-variance parameters, design matrices, residuals, residual plots, and other diagnostic information. Fixed-effects terms usually refer to the conventional linear regression part of the model. Random effects terms are associated with individual experimental units taken at random from a population, and account for variations between groups that might affect the response. The random effects have prior distributions, whereas the fixed effects do not.

The GLMM model structure is:

$$y_i|b \approx \mathrm{Distr}\left(\mu_i, \frac{\sigma^2}{w_i}\right) \qquad (2)$$

$$g(\mu) = \beta X + bZ + \delta \qquad (3)$$

where, y_i = the i-th element (dependent variable) of the y response vector, b = the random-effects vector (complement to the fixed β), Distr = a specified conditional distribution of y given b, μ = the conditional mean of y given b, and μ_i is its i-th element, σ^2 = the variance or dispersion parameter, w = the effective observation weight vector (w_i is the weight for observation i), $g(\mu)$ = link function that defines the relationship between the mean response μ and the linear combination of the predictors, X = fixed-effects design matrix (of independent variables), β = fixed-effects vector, Z = random-effects design matrix (of independent variables), and δ = model offset vector (residuals). The model for the mean response μ is:

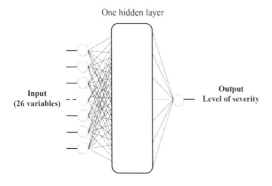

Figure 3. Back-propagation Neural Network structure for SL modelling adopted in this investigation.

$$\mu = g^{-1}\left(\hat{\eta}\right) \qquad (4)$$

where g^{-1} = inverse of the link function $g(\mu)$, and $\hat{\eta}$ = linear predictor of the fixed and random effects of the generalized linear mixed-effects model:

$$\eta = \beta X + bZ + \delta \qquad (5)$$

According to the Wilkinson notation, the GLMM model has the following structure:

$$y \sim \text{fixed} + (\text{random1}|\text{group1}) + ... + (\text{random N}|\text{group N}) \qquad (6)$$

where, the terms "fixed" and "random" are associated with independent variables and contain fixed and random effects, N = number of grouping variables in the model. Grouping variables are utility variables used to group, or categorize, observations, and are useful for summarizing or visualizing data by group.

For the SL output, a log link function and the Probability Mass Function (PMF) for the Poisson distribution was used. The fit method was the 'Laplace' one. Finally, the best performance was calculated through the minimization of the log-likelihood index; other indexes (i.e., Akaike's information criterion—AIC, Bayesian information criterions—BIC, and the Deviance parameter) were also estimated to control the minimization process.

According to eq. 6 notation, the GLMM model that gives the best performance was:

$$C27 \sim 1 + C8 + C9 + C11 + C14 + C16 + C18 + C21 + C24 + C25 + (1 | C1) + (1 | C2) + (1 | C3) + (1 | C4) \qquad (7)$$

where Cx identifies the x-th variable (Table 2). In Table 4 the fixed effect coefficients are drawn with a 95% confidence interval. The p values are lower than 0.001, with the exception of C9 (driver vehicle A age) and C18 (rainfall intensity) which are lower than 0.05. The Standard Error of estimates (SE) is generally much lower than the estimates, and lower and upper bonds of CI never include zero. Table 4 also reports the estimates for random parameters.

The effect of flows (C21, C24, C25) on SL has a different sign, positive for C21 (TF4), which anticipates the crash event, and for C24 (TF6), and negative for C25 (TF7). When flow after the crash (C25) increases, it is more likely that the SL decreases. The grouping variables are road type (C1), pavement condition (C2), road signage condition (C3) and vehicle A type (C4). The driver vehicle A age (C9) as well as the class age of vehicle B driver (C11) are negatively related to the SL. Furthermore, also light radiation (C16) and rainfall intensity (C18) are inversely related to the SL.

5 DISCUSSION

5.1 Sensitivity analysis of BPNN

In the case of the BPNN model, a sensitivity analysis was carried out to assess how output changes by varying input normalized variable values in the range [0,1] one by one. With this aim in mind, a first set of scenarios, referring to a particular set of input variables, was prepared. In addition to basic scenarios where variables are all zero or 1, another six scenarios were considered to study particular combinations of variable values (in Figure 4, from 4a to 4f). These scenarios aim to consider some possible and typical crash situations involving male or female drivers, during daytime or at nighttime, with rainy weather or dry road surface, or with elder drivers.

The effect of flow varies a lot for scenario and flow itself. When TF4 is high (Figure 4a), the SL is generally high for most of the scenarios, except for elderly drivers. On the other hand, the effect of TF7 (Figure 4b) depends very much on the particular scenario, though, generally, a higher SL is related to a higher flow. Low light radiation (nighttime to dawn) has a strong effect on young male drivers with dry pavement, while middle radiation has a strong effect on young female drivers in rainy conditions. Generally, a low radiation is more related to high SL than a high radiation.

5.2 Model output comparison

According to Powers (2011), the two model outputs were evaluated by confusion matrixes representing, for each output, the number of predicted cases (a_{ij}) on the reference databases. In this case, the output coincides to the SL, and a_{ij} are calculated on the resampled databases obtained according to what reported in Section 3.3.

There is s also interest in the measurement of its precision (the percentage of correct data predicted in respect of the total predicted) and its recall (the percentage of corrected data predicted in respect of the total to be predicted) capability. The main goal of learning is to improve the recall measurement without hurting the precision one. Tables 5 and 6 include the percentage of the predicted crashes to the total predicted for each SL, the "a priori" rate (PR), which expresses the complement of the recall rate, and the "a Posteriori" rate (PO) which is the complement of the precision rate, according to the following equations (n is the matrix dimension):

$$PR_i = 1 - a_{ii}/(a_{i1} + ... + a_{in}) \qquad (8)$$

$$PO_i = 1 - a_{ii}/(a_{1i} + ... + a_{ni}) \qquad (9)$$

Furthermore, comments on results are supported by the estimation of their accuracy (A):

Table 4. Fixed effects coefficients estimates and Random effects covariance parameters at 95% CIs) for the GLMM segment model.

Variable	Estimate	SE	p-value	Lower	Upper
Intercept	1.24210	0.2019	$<10^{-3}$	0.84629	1.63790
C8	0.00594	$<10^{-3}$	$<10^{-3}$	0.00428	0.00759
C9	−0.06953	0.018	$<10^{-3}$	−0.10540	−0.03367
C11	−0.00620	$<10^{-3}$	$<10^{-3}$	−0.00707	−0.00534
C14	0.00497	$<10^{-3}$	$<10^{-3}$	0.00318	0.00675
C16	−0.00016	$<10^{-4}$	$<10^{-3}$	−0.00023	−0.00010
C18	−0.07152	0.0203	$<10^{-3}$	−0.11142	−0.03163
C21	0.00056	$<2·10^{-4}$	0.002	0.00020	0.00092
C24	0.00111	$<3·10^{-4}$	$<10^{-3}$	0.00062	0.00161
C25	−0.00242	$<3·10^{-4}$	$<10^{-3}$	−0.00284	−0.00199

Group variable	Estimate
C1 (Intercept)	0.13646
C2 (Intercept)	0.38224
C3 (Intercept)	0.19893
C4 (Intercept)	0.11375

Indexes	
LogLikelihood	−11465
AIC	22959
BIC	23053
Deviance	22931
R^2 adjusted	0.3107

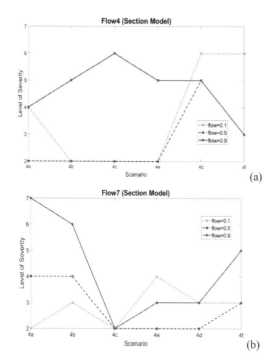

Figure 4. Effect of TF4 (a) and TF7 (b) (variables 22 and 26) on SL (BPNN model) for different flow values (0.1, 0.5, 0.9).

$$A = (a_{11} + a_{22} + \ldots + a_{nn})/\Sigma a_{ij} \qquad (10)$$

Table 5 reports the confusion matrices for the model calibrated through the BPNN. SL values lower than 2 (corresponding to the PDO crash type) and greater than 6 (which are unrealistic values) have also been included in the tables considering that the model output can fall outside the range of numerical values associated with each SL. The accuracy of 90% is certainly very high for BPNN model. PR and PO rates are low with the exception of SL 2 and 3. SL 2 is the more difficult to predict while SL 3 has the largest number of wrong cases assigned to it.

Table 6 contains the confusion matrices for the GLMM model. In this case, the capacity of SL prediction is significantly lower than the one for BPNN as indicated by the accuracy of 33%. GLMM has a superior capacity to provide results within the SL limits of 2 and 6, as confirmed by the absence of values that fall outside of the two limits. PR and PO rates are lower than the corresponding values for BPNN, showing a greater difficulty in predicting SL than the neural network modelling approach.

Comparisons with GLMM show a marked superiority of BPNN modelling as regards performance measured through confusion matrices. GLMMs, on the other hand, clearly show what variables are significant and their effect (sign and value of coefficients) though this is limited to the

Table 5. BPNN model confusion matrix for segments (row percentage values in brackets), and "a Priori" (PR) and "a Posteriori" (PO) rates.

Real SL	Predicted SL							CD	PR (%)
	<2	2	3	4	5	6	>6		
2	21	859	257	71	39	60	7	1314	35%
	2%	65%	20%	5%	3%	5%	0%		
3	18	150	1056	6	0	0	0	1230	14%
	1%	12%	86%	1%					
4	0	0	0	1250	0	0	0	1250	0%
				94%					
5	0	0	0	0	1287	0	0	1287	0%
					100%				
6	0	0	0	0	0	1311	0	1311	0%
						100%			
PO	–	15%	20%	6%	3%	4%	–	–	–

Notes: CD = crash data in the resampled database.

Table 6. GLMM model confusion matrix (row percentage values in brackets), and "a Priori" (PR) and "a Posteriori" (PO) rates.

Real RSL	Predicted PSL							CD	PR (%)
	<2	2	3	4	5	6	>6		
2	0	79	534	593	100	8	0	1314	94%
		52%	41%	45%	8%	1%			
3	0	72	444	588	102	24	0	1230	64%
		48%	36%	48%	8%	2%			
4	0	0	300	850	100	0	0	1250	32%
			24%	68%	8%				
5	0	0	393%	702	507	39	0	1287	61%
				55%	39%	3%			
6	0	0	0	483	621	207	0	1311	84%
				37%	47%	16%			
PO	–	48%	66%	74%	65%	26%	–	–	–

linear effect of variables without considering their possible reciprocal interaction.

6 CONCLUSIONS

The paper aims to achieve two goals: the evaluation of the crash Severity Level (SL) on urban road segments using environmental variables (some of which, like short-term flow, are innovative for this type of research), and the comparison of two different techniques for calculating SL, the Back-Propagation Neural Network model (BPNN) and the Generalized Linear Mixed Model (GLMM).

The results presented here provide new insights into urban roads and fill a gap in the knowledge acquired from the number of studies on rural freeways and expressways reported in literature.

From the use of the confusion matrixes technique, BPNN models evidenced their superiority in the prediction of the SL when compared to the GLMMs. This is attributable to their greater capability of accurately approximating any continuous and non-linear function. On the other hand, GLMMs (like any analytical model) allow a readier interpretation of model results. Other pros and cons in their use derive from the intrinsic characteristics of statistical and neural network methods, as clearly underlined by Karlaftis and Vlahogianni (2011). The authors suspect that the most significant limit of GLMMs for these applications is related to the constrained linearity of their functions. In addition, missing data may have contributed to the fact that the BPNNs, which are known to be capable of overcoming this problem, achieved better results.

However, both approaches (BPNN and GLMM), though with significant differences, indicate that flows have a relevant role in predicting severity: this role is not limited to the flow when the crash occurred (TF4), but also involves other flow data recorded before (TF1-TF3) and after (TF5-TF7) the crash. GLMM model shows the relevance of TF3, TF6, and TF7 only, but the BPNN model evinces more complex relationships for all seven variables. Weather variables also (i.e., rainy condition and light radiation) show a strong relation in some scenarios.

In future research, generalized non-linear models will be used to consider the higher order effects and the interaction between variables. Moreover, a mixed approach using both short-term flow and AADT values will be investigated to derive models over a mid-to-long term period and investigate the relationship between them.

ACKNOWLEDGEMENTS

The authors thank *Polizia Municipale di Torino*, and the *Città Metropolitana di Torino* for having provided crash data. Thanks are also due to *Consorzio 5T s.r.l.* for providing short-term flow data, and to the Environmental Protection Agency of the Regione Piemonte (*ARPA Piemonte*) for providing weather data.

REFERENCES

Abdelwahab, H.T. & Abdel-Aty, M.A. 2001. Development of artificial neural network models to predict driver injury severity in traffic accidents at signalized intersections. *Transp. Res. Rec.* 1746, 6–13.

Agresti, A. & Natarajan, R. 2001. Modelling clustered ordered categorical data: a survey. *Int. Stat. Rev.* 69, 345–371.

Al-Ghamdi, A.S. 2002. Using logistic regression to estimate the influence of accident factors on accident severity. *Accid. Anal. Prev.* 34, 729–741.

Baluni, P. & Raiwani, Y.P. 2014. Vehicular accident analysis using neural networks. *Int. J. of Emerg. Tech. and Adv. Engin.* 4 (9), 161–164.

Chawla, N.V. 2010. Data mining for imbalanced datasets: an overview. *Data Mining and Knowledge Discovery Handbook*, Springer US, 875–886.

Chong C-Y. & Kumar, S.P. 2003. Sensor networks: evolution, opportunities, and challenges. *Proceedings of the IEEE* 91(8), 1247–1256.

Chong, M.M., Abraham, A. & Paprzycki, M. 2004. *Traffic Accident analysis using decision trees and neural network*. arXiv preprint cs/0405050.

Christoforou, Z., Cohen, S. & Karlaftis, M. 2010. Vehicle occupant injury severity on highways: an empirical investigation. *Accid. Anal. Prev.* 42, 1606–1620.

Cybenko, G. 1989. Approximation by superpositions of sigmoidal functions. *Math. Control Signals Syst.* 2(4), 303–314.

Delen, D., Sharda, R. & Bessonov, M. 2006. Identifying significant predictors of injury severity in traffic accidents using a series of artificial neural networks. *Accid. Anal. Prev.* 38, 434–444.

Dougherty, M. 1995. A review of neural networks applied to transport. *Transp. Res. Part C: Emerg. Tech.* 3(4), 247–260.

El Faouzi, N-E., Leung, H. & Kurian, A. 2011. Data fusion in intelligent transportation systems: progress and challenges—a survey. *Inform. Fus.* 12(1), 4–10.

Golob, T.F., Recker, W.W. & Pavlis, Y. 2008. Probabilistic models of freeway safety performance using traffic flow data as predictors. *Saf. Sci.* 46, 1306–1333.

Japkowicz, N. 2000. The class imbalance problem: significance and strategies. *Proc. of the 2000 Intern. Conf. on Art. Intel.* (IC-AI'2000), Las Vegas, Nevada.

Jung, S., Qin, X. & Noyce, D.A. 2010. Rainfall effect on single-vehicle crash severities using polychotomous response models. *Accid. Anal. Prev.* 42, 213–224.

Karlaftis, M.G. & Vlahogianni, E.I. 2011. Statistical methods versus neural networks in transportation research: differences, similarities and some insights. *Transp. Res. Part C: Em. Tech.*, 19(3), 387–399.

Lebart, L., Morineau, A. & Tabard, N. 1977. *Techniques de la description statistique: méthodes et logiciels pour l'analyse des grands tableaux*, Dunod, Paris.

Mussone, L. 1999. A review of feedforward neural networks in transportation research, *e&i Elektrotechnik und Informationstechnik* 116(6), 360–365.

Mussone, L., Ferrari, A. & Oneta, M. 1999. An analysis of urban collisions using an artificial intelligence model. *Accid. Anal. Prev.* 31, 705–718.

Nekovee, M. 2005. Sensor networks on the road: the promises and challenges of vehicular ad hoc networks and vehicular grids. *Proc. Work. on Ubiq. Comp. e-Res.*, Edinburgh, Scotland, UK.

Powers, D.M. 2011. Evaluation: from precision, recall and F-measure to ROC, informedness, markedness and correlation. *J. Mach. Learn. Technol.* 2, 37–63.

Repubblica Italiana, 2015. Codice Penale (in italian). Testo coordinato del Regio Decreto 19 ottobre 1930, n. 1398, aggiornato con le modifiche apportate dalla L. 28 aprile 2015, n. 58, dalla L. 22 maggio 2015, n. 68 e dalla L. 27 maggio 2015, n. 69.

Shankar, V., Mannering, F. & Barfield, W. 1996. Statistical analysis of accident severity on rural freeways. *Accid. Anal. Prev.* 28 (3), 391–401.

Theofilatos, A. & Yannis, G. 2014. A review of the effect of traffic and weather characteristics on road safety. *Accid. Anal. Prev.* 72, 244–256.

Wang, C., Quddus, M.A. & Ison, S.G. 2013. The effect of traffic and road characteristics on road safety: a review and future research direction. *Saf. Sci.* 57, 264–275.

Xu, C., Tarko, A.P., Wang, W. & Liu, P. 2013. Predicting crash likelihood and severity on freeways with real-time loop detector data. *Accid. Anal. Prev.* 57, 30–39.

Yu, R. & Abdel-Aty, M. 2013. Using hierarchical Bayesian binary probit models to analyze crash injury severity on high speed facilities with real-time traffic data. *Accid. Anal. Prev.* 62, 161–167.

Transport Infrastructure and Systems – Dell'Acqua & Wegman (Eds)
© 2017 Taylor & Francis Group, London, ISBN 978-1-138-03009-1

The influence of pedestrian crossings features on driving behavior and road safety

A. Bichicchi, F. Mazzotta, C. Lantieri, V. Vignali, A. Simone & G. Dondi
Department of Civil, Chemical, Environmental and Material Engineering, University of Bologna, Italy

M. Costa
Department of Psychology, University of Bologna, Italy

ABSTRACT: Traffic safety depends upon the integrated and complex relationship between various components: driver psychology, traffic, vehicles and road infrastructures. According to statistics, the aspect that seems to be the most important, as it is responsible for most accidents, is the behavioral component and therefore the psychology of the driver of the vehicle. When considering the interaction between drivers and road infrastructure, pedestrian crossings play a significant role because theirs configuration, in terms of traffic signs and road signage, implicates different behavior responses. For this reason, it is important to identify which situations would be safer for vulnerable users. In the present study, a sample of 24 drivers performed a trial route of 59.4 km to investigate the totality of non-signalized pedestrian crossings located along the roadways. GPS kinematic parameters and mobile eye tracking data were registered to evaluate their approaching behavior. Pedestrian crossings were classified according to the configuration of road elements present in the each intersection. For each single crossing were evaluated the approach speed, the distance of pedestrian crossings perception and the fixation time to specific elements. The frame-by-frame video analysis shows that at least one element is visualized by 84% of drivers and the presence of certain elements and their configuration determine significant differences in sight distances and speed adaptation. Overall, driver behavior is significantly influenced by crossings complexity, determining remarkable effects on the incidence of fatal pedestrian accidents.

1 INTRODUCTION

1.1 Vulnerable road users

The pedestrian is one of the most vulnerable user involved in road accidents. In Italy, statistics confirm that the number of deaths among pedestrians is showing an overall decrease but it is likely that this is only due to the overall decline of fatal motor vehicle crash (–9.8%). Most of the crashes involving pedestrians occur on rural roads (4.63 deaths per 100 accidents), with a reduction on motorways and urban roads (3.46 and 1.04, respectively). As regards the roadway carriage, the highest mortality rate is recorded in single carriageway two-lane roads (5.03 deaths per 100 accidents) (ISTAT 2014, 2015). The most interesting data is, however, that about ¼ of fatal pedestrian accidents occurs at intersections with crosswalks.

Research on this topic is therefore more than actual. Recent interest on vulnerable users' protection is also encouraged by the European Commission that proposes to maintain the target of halving the overall number of road deaths in the EU decade 2011–2020.

Even premised that the risk of injury caused by running over is especially high for the elderly population, accident causes are multiple, ranging from infrastructures critical issues to improper driver. More specifically, at crosswalks pedestrians believe to be safe and they reduce their level of attention, compromising the overall safety of the intersection. Besides, it is clear that safety at the intersections depends on many factors and one most important is represented by the behavior related to drivers.

1.2 Drivers' behavior

While approaching pedestrian crossings, drivers visually incorporate multiple information that may affect the perception of vulnerable users present within the area of operation. It is hence important identifying how boundary conditions, in terms of road signs, road markings, geometrical elements of the road, can capture the drivers' attention and influence their driving behavior (Mazzotta et al. 2014). A recent naturalistic driving study by Habibovic focused on pedestrian accidents captured by onboard cameras and found that, at intersections, drivers failed to recognize the presence of the pedestrian conflict due to visual obstructions and/or because their attention was allocated towards something other than the pedestrian (Habibovic et al. 2013). For this reason, road design must take into account not only traffic factors, but also the gaze behavior of road drivers. From one side, the environment at the intersection should be designed avoiding any source of potential distraction; on the other hand, road elements and their configuration have the important role to alert drivers about the presence of the pedestrian crossing, raising their level of attention. The mechanism in which the user recognize and decide heavily affects the risk of accident and, for this reason, influencing the perception of the road users can induce precise behaviors and make driving role safer (Bucchi et al. 2012). When action needs to be taken, the driver seems to degrade the allocation of visual attention when it is overly general, but optimizes it when it is specific to a manoeuvre (Eyraud et al. 2014).

The recent interest of road engineers concerns the study of which elements of the road environment actively attract the attention of users.

The users' attention during the driving role focuses mainly on some specific areas of the intersection, ignoring other elements. In particular, attention is increased in more attention to more complex intersections, compared to those with less elements (Werneke & Vollrath 2012). More specifically, horizontal markings are more looked at than the vertical signals (Costa et al. 2014): the latter, in fact, have an important role, because of its advantage to be placed at a certain height but force drivers to turn their gaze at the right or left side with exception of central overhead signs. Horizontal markings are thus more visible because located in the direction of the user gaze, allowing drivers to keep the gaze straight ahead. Nevertheless, literature lacks of tangible evidence of horizontal markings effectiveness: a research by Herms that regarded pedestrian accidents at 400 intersection, demonstrates that approximately twice as many pedestrian accidents occur in marked crosswalks as in unmarked crosswalks (Herms 1972).

A similar research assessed the design features that affected signalized pedestrians crossings. The developed methodology suggested design modifications and operational features to improve both the safety level and the level of service of the intersection (Steinman & Hines 2004).

Concerning the travel speed in approach to intersections, other researchers found that users usually approach intersections with excess of speeding and an aggressive driving style (Papaiannou 2007). Anderson also showed that small changes in the travelling speeds of vehicles could have a remarkable effect on the incidence of fatal pedestrian accidents and, in addition, a key role in those accidents that could be completely avoided (Anderson et al. 1997).

For these reasons, roadway capability to attract or distract drivers may be thus considered as the direct cause of speed variations, a sort of an action-reaction process that should be deeply investigated to find measures for improve the safety level of intersections.

This study aimed to quantify visual (with eye-tracking system), and speed behavior of drivers as a function of constituent elements of pedestrian crossing with different design.

2 METHODOLOGY

Field tests have been carried out on C-type roads according to the Italian Road Code (1992), which are constituted by a two-lane carriageway with a variable section width. Routes were chosen according on their high rate of car-pedestrian collisions. The first was on a segment of the roadway SP610, known as the "Selice-Montanara", which connects Emilia and Toscana regions in the North of Italy. Similarly, the second segment was located on roadway SP26, known as the "Valle del Lavino", which develops from Bologna city area to Monte San Pietro. Overall, the experimental route was 59.4 km long, and intersected 15 small urban centers, in hilly terrain, with speed limit of 50 km/h. Because of the conspicuous presence of urban areas, a high number of pedestrian crossings was encountered (89) along the roadway.

The sample consists of twenty-four drivers, 15 men and 9 women (mean age: 30 ± 10 years) with normal vision, and without glasses or contact lenses that interfered with the eye-movements recording system. Their mean driving experience was 12 ± 10 years and the mean kilometers driven per year were 9858 ± 10212. None of the drivers had previous experience of the route and they were not informed about the study true aims. Participants were Psychology and Engineering undergraduate students and nobody of them was paid.

2.1 Devices

Participants drove a BMW 1 Series car equipped with a Video VBOX PRO data logger. Two cameras and a GPS antenna, positioned on the top of the car, and connected to the Video VBOX system, recorded the scene seen by the driver and their video outputs were synchronized with data about acceleration, speed and GPS coordinates.

Visual fixations were assessed using a mobile eye tracker, a device that has been extensively used in road safety research (Land & Tatler 2009; Underwood 1998; Van Gompel et al. 2007). The system used in this study was an ASL Mobile Eye-XG equipment constituted by two digital high-resolution cameras attached to lightweight eyeglasses. One camera recorded the scene image and the other the participant's right eye, creating a video for each participant, where a red cross showed the eye-fixations. This cursor, superimposed to the video of the scene, allowed researchers to detect what specific point of the scene the participant was looking at. The ASL Mobile Eye-XG equipment has a declared precision of 0.5°–1°, and a recording sample rate of 30 Hz. Its reliability in targeting looking behavior has been recently tested at a distance up to 150 meters from target point (Costa et al. 2014), by correlating the theoretical angle and the Interfix degree parameter obtained from the ASL Results software (r = 0.993 p <0.001).

The eye tracking equipment along with a computer and the VBOX PRO equipment were kept on the back seat and were monitored by one of the experimenters, who was instructed not to interact to the driver except for instructions or request for assistance.

Before the test, a calibration procedure was carried out for each participant. This procedure took place in a parking lot with the car being stationary and involved participants to look at least 15 visual points spread across the whole scene. Each driver was given a two kilometers trial run to get used to the car before starting along the experimental route.

2.2 Pedestrian crossings

As shown in Table 1, pedestrian crossings have been divided into 4 classes, depending on the configuration of four different elements: "zebra" marking, vertical signs, overhead gateway and central island (Figures 1a, 1b, 1c). For the reliability of this study, only pedestrian crossings within urban areas, and visible at more than 100 m were considered.

Considering the whole sample, 742 unobstructed intersections (cases with the presence of pedestrians were excluded) were considered. According to the above-mentioned classification, their distribution was: 29% of cases occurred on C1 configuration (214 on the total), 27% on C2 (203), 22% on C3 (163), and 22% on C4 (162).

Table 1. Typological configurations of the pedestrian crossings considered in the study.

	N° of crossings	Zebra markings	Vertical sign	Gateway	Central Island
C1	24	X	X		
C2	24	X		X	
C3	19	X	X	X	
C4	22	X		X	X

Figures 1a, 1b, 1c. Example of elements: gateway and zebra markings (a), vertical sign and zebra markings (b), central island and zebra markings (c).

3 RESULTS

A frame-by-frame analysis allowed noticing the visual fixations to the above-mentioned elements, which were considered "looked at" when the red cursor was superimposed on the specific element for at least one frame (33.33 ms). In addition, in each case the approaching speed and the distance at which the elements were looked at were considered and entered in the analyses. A pedestrian crossing was considered "perceived" when at least one of the mentioned elements was seen.

Based on this criteria, in 114 cases (15% of the total investigated) drivers did not looked at any element and did not consider the presence of the pedestrian crossing. The correlation between visual behavior and drivers characteristics denoted that pedestrian crossing perception was influenced by age (Pearson correlation r = 0.70) but not by gender (r = 0.021, p = 0,572).

3.1 Visual behaviour

A pedestrian crossing can be reasonably considered "well designed" when an approaching driver detect it sufficiently in advance, in order to assess the presence of pedestrians and eventually decrease speed in safe conditions. It is also assumed that drivers "safely" perceive a pedestrian crossing when the gaze at the first element occurs at an equal or greater distance than 50 m. This distance is required to stop a vehicle travelling at an average speed of 50 km/h with an average longitudinal slope of 4% (Ministerial Decree n. 6792, 2001).

A detailed data analysis permitted to determine the percentage of perceptions by each configuration and, in particular, if it occurs more of less than 50 m from the pedestrian crossing (Table 2). Thus, the percentage of not perceived configurations (none element gazed) was determined.

Central islands and overhead signs were the most looked elements. C4 was the pedestrian crossing configuration more looked at (in more than 90% of crossings) whereas the C1 configuration, composed by zebra markings and vertical signs only, was the less perceived (almost one crossing on four was ignored).

Regarding the property of pedestrian crossings to be perceived at a greater distance, it resulted that the C2 and C3 configurations had the highest percentages, due to the presence of the overhead sign. These configurations were respectively perceived more than 50 m in advance in 22% and 27% of the totality of cases. It is also possible to suppose that the central island, even if ensures the crossing an high probability to be perceived, in more than the 80% of cases is seen very close to the pedestrian crossing (<50 m).

At the same time, considerations emerge from the investigation of first element seen for each configuration. From Figure 2 it is evident that zebra

markings were the first element that was looked at in C1, C2 and C4. When present, in fact, horizontal signs are seen before other elements (vertical sign and gateway); when central island was present, zebra markings decrease their possibility to be seen first.

For each configuration it was computed the visualization of specific elements and then it has been related to the approaching distances. Gaze duration was computed multiplying the number of frames for the fixations to specific elements to the duration of each frame (33.33 ms). C4 resulted the configuration whose elements were gazed for longer time (M = 724 ms), followed by C1 and C3 (M = 296.48 ms and M = 296.54 ms, respectively). C2 was the configuration less looked at (M = 281.91 ms). Average of total fixation times to the four pedestrian crossing configurations are shown in Figure 3.

The time spent gazing elements increased considerably when the driver was closer to the intersection, with the exception of C3 configuration where the peak of time average was reached between 100 and 50 m in advance (see Table 3). Is therefore possible to suppose that the simultaneous presence of gateway and vertical sign anticipates the caption of the attention level. Similarly, considering each specific element, it resulted that zebra marking was the element most looked at (Figure 4) in every configuration, followed by the central island when present.

Considering the relationship between specific elements and sight distance:

- the quantity of frames focusing on zebra markings, which occur in all configurations, significantly increased as the driver approached the intersection, reaching a peak in the last 50 m before the pedestrian crossing (4.96 frames corresponding to 165.31 ms);
- the other elements (vertical sign, overhead sign and central island) got the attention of drivers before (from 150 m to 50 m before the pedestrian crossing) and, consequently, the number of frames slightly decrease in the last 50 meters;

Table 2. Perception of the 4 pedestrian crossing configurations as a function of distance.

Pedestrian crossing configuration	>50 m	<50 m	Not perceived
C1	17.29%	58.88%	23.83%
C2	21.67%	62.07%	16.26%
C3	26.99%	63.19%	9.82%
C4	11.11%	80.25%	8.64%

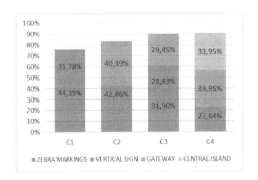

Figure 2. First element that was looked at in the different pedestrian crossing configurations.

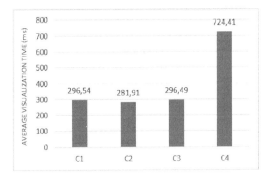

Figure 3. Mean total fixation duration for the four pedestrian crossing configurations.

Table 3. Average number of fixation frames for the different element of the four pedestrian crossing configurations.

	Zebra Markings			Gateway			Vertical Sign			Central Island		
	150–100	100–50	50–0	150–100	100–50	50–0	150–100	100–50	50–0	150–100	100–50	50–0
C1	4.8	67.9	165.2				17.4	27.5	13.5			
C2	8.5	68.7	142.5	23.1	27.5	11.3						
C3	11.4	77.5	92.8	18.6	14.7	7.3	17.5	39.0	17.3			
C4	5.5	114.8	242.7	38.0	38.8	9.4				17.4	134.9	122.4

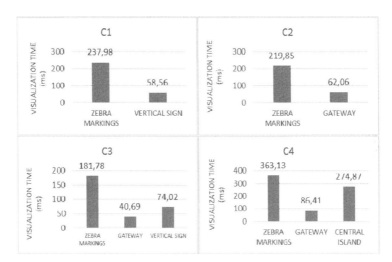

Figure 4. Average total fixation durations to the different elements of the four pedestrian crossing configurations.

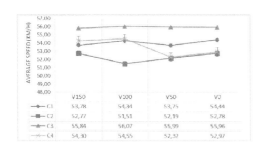

Figure 5. Average speed at 150 m, 100 m, and 50 m before the pedestrian crossing.

– overhead signs are visualized before the other elements, confirming their key-role in the improvement of pedestrian crossings in terms of safety (Ariën et al. 2013; Lantieri et al. 2014);
– central islands, were mainly fixated in the last 100 m.

3.2 Driving behaviour

Previous studies in literature suggest to analyze the drivers' speed approaching to the intersection, starting from 150 m before the zebra markings (Bella & Silvestri 2015). Speed data were recorded with the Video V-Box equipment. In particular, were considered the speeds detected in correspondence of 150, 100 and 50 m before the pedestrian crossing. Mean speeds are shown in Figure 5.

The overall peak of average speed occurs in the C3 configuration 50 m before the pedestrian crossing (56 km/h) while at a distance of 50 m the average speed is always higher than 50 km/h. Instead, V150 and V0 (average speed at 150 m and in correspondence of the crossing) are very similar for every configuration; interesting data are the significant speed decreasing in C2 and C4 (−1.26 km/h and −2.23 km/h).

4 DISCUSSION AND CONCLUSIONS

The described experimental field permitted to quantify how the configurations of road furniture influence drivers' behaviour. Horizontal, vertical signs and geometric elements as central islands influence, indeed, the pedestrian crossing sight distance, the observation time, and speed modulation. In particular, assuming that pedestrian safety and, more in general road safety, are assured when a crossing is perceived largely in advance, it has been demonstrated that the presence of gateways is a safety indicator, independently of boundary conditions. Similarly, overhead signs resulted to be the element visualized first with respect to other elements.

In addition, the central island resulted the element that induces the highest speed reduction, probably a cause of the reduction in lane width. The increasing observation time resulted slightly related to speed reductions. For this reason, configurations that are more complex do not distract drivers but induce them to increase their attention and to be more careful (decrease of average speed) confirming previous studies (Werneke & Vollrath 2012). On the contrary, almost 25% of the totally ignored pedestrian crossings, had a simple configuration (zebra markings and vertical sign only).

The aim of this study have been suitably reached, in particular:

1. Have been provided useful insights for a better comprehension of the elements attracting drivers' attention while approaching the crossings;
2. Regarding drivers' speed behaviour approaching crossings, a comparative evaluation of configurations' effectiveness have been performed.

The findings of this research offer interesting application for the design and adaptation of pedestrian crossings, in order to increase the level of safety offered to users. High-rate accidents roadways, once upgraded and adjusted following those criteria, should be studied in order to realize adequate before-after analyses.

Continuing this research, the sample for experimental trials could be incremented and clustered into different users' categories; moreover, the evaluation of further configurations e.g. provided with signaling lamps and back-colored markings could be considered for future trials. Future research could also focus on the pedestrian crossings accident analysis, in order to search for a direct correlation between elements or configurations and pedestrian injuries.

REFERENCES

ACI-ISTAT, 2015. Rapporto ACI—ISTAT sugli incidenti stradali, anno 2014 (in Italian).

Anderson, R.W.G., McLean, A.J., Farmer, M.J., Lee, B.H. & Brooks, C.G., (1997) Vehicle travel speeds and the incidence of fatal pedestrian crashes, Accident Analysis and Prevention, 29, 5, 667–674.

Ariën, C., Jongen, E.M.M., Brijs, K., Brijs, T., Daniels, S., & Wets, G. (2013). A simulator study on the impact of traffic calming measures in urban areas on driving behavior and workload. *Accident Analysis and Prevention*, 61, 43–53.

Bella, F., & Silvestri, M. (2015). Effects of safety measures on driver's speed behavior at pedestrian crossings. *Accident Analysis and Prevention*, 83, 111–124.

Bucchi, A., Sangiorgi, C., & Vignali, V. (2012). Traffic psychology and driver behaviour, in *Procedia—Social and Behavioral Sciences*, 53, 973–980.

Costa, M., Simone, A., Vignali, V., Lantieri, C., Bucchi, A. & Dondi, G. (2014). Looking behavior for vertical road signs. *Transportation Research Part F*, 23, 147–155.

Decreto Legislativo 30 aprile 1992, n. 285 *Nuovo codice della strada* (in Italian).

Decreto Ministeriale n. 6792 del 5 novembre 2001: *Norme funzionali e geometriche per la costruzione delle strade. G.U. n. 3, 04/01/2002* (in Italian).

EU Commission (2010). Road Safety Programme *2011–2020*.

Eyraud, R., Zibetti E. & Baccino, T. (2014). Allocation of visual attention while driving with simulated. *Transportation Research Part F*, 32, 46–55.

Habibovic A., Tivesten E., Nobuyuki U., Bärgman J. & Aust M.L. (2013). Driver behavior in car-to-pedestrian incidents: An application of the Driving Reliability and Error Analysis Method (DREAM). *Accident Analysis and Prevention*, 50, 554–565.

Herms, B.F. (1972). Pedestrian crosswalk study: accidents in painted and unpainted crosswalks. *Highway Research Record*, 406, 1–13.

Land, M.F. & Tatler, B.W. (2009). *Looking and acting: Vision and eye movements in natural behaviour.* Oxford; Oxford University Press.

Lantieri, C., Lamperti, R., Simone, A., Costa, M., Vignali, V., Sangiorgi, C. & Dondi, G. (2015). Gateway design assessment in the transition from high to low speed areas. *Transportation Research Part F: Traffic Psychology and Behaviour*, 34, 41–53.

Mazzotta, F., Irali, F. & Simone, A. (2014). La valutazione della sicurezza delle utenze deboli negli attraversamenti pedonali non semaforizzati tramite Mobile Eye Detector. In: *L'utente Debole Nelle Intersezioni Stradali, Parma, 27–28 Marzo 2014*, 181–190, EGAF EDIZIONI.

Mazzotta, F., Vignali, V. & Irali, F. (2014) Evaluation of the readability of road signs and roadside elements using Mobile Eye tracking device. *InBo*, 7, 253–260.

Papaioannou, P. (2007). Driver behaviour, dilemma zone and safety effects at urban signalised intersections in Greece. *Accident Analysis and Prevention*, 39, 147–158.

Steinman, N. & Hines, D.K. (2004). Methodology to assess design features for pedestrian and bicyclist crossings at signalized intersections. *Transportation Research Record*, 1878, 42–50.

Underwood, G. (Ed.) (1998). *Eye guidance in reading and scene perception.* Amsterdam: Elsevier.

Van Gompel, R.P.G., Fischer, M.H., Murray, W.S., & Hill, R.L. (Eds.) (2007). *Eye movements: A window on mind and brain.* Amsterdam: Elsevier.

Werneke, J. & Vollrath, M. (2012). What does the driver look at? The influence of intersection characteristics on attention allocation and driving behavior. *Accident Analysis and Prevention*, 45, 610–619.

Transport Infrastructure and Systems – Dell'Acqua & Wegman (Eds)
© *2017 Taylor & Francis Group, London, ISBN 978-1-138-03009-1*

The relationship between driving risk levels and drivers' personality traits, physical and emotional conditions

L. Eboli, G. Mazzulla & G. Pungillo
University of Calabria, Rende, Italy

ABSTRACT: This paper presents a study aimed to investigate the relationship between driver personality traits and the level of driving risk taken by driver during a trip, and also the relationship between driving risk levels and driver's physical and emotional conditions while driving. To this end, a survey was conducted to gain an objective measure of the risk level, which gives an indication of the driving conditions (safe or unsafe), and subjective judgements of the drivers about their personal characteristics. Objective measures were obtained from kinematic parameters (instantaneous speed, longitudinal and lateral accelerations) recorded along each trajectory followed by the driver. For this aim, smartphones equipped with GPS and accelerometer were used. On the other hand, drivers were asked to complete a questionnaire about their personality traits, and physical and emotional conditions while driving, which can influence their driving style. Each driver covers the same path several times in different days, in order to capture different physical and emotional conditions and driving styles. The sample of drivers belongs to a population of car drivers between 25 and 50 years old who every day drive by their private car for reaching their destination of work or study. Overall, drivers made about 170 tests covering about 1,100 km. We proposed a correlation analysis between the drivers' characteristics and the risk level obtained by using acceleration and speed data. As an example, we found that characteristics such as patience and meticulousness positively influence the level of safety.

1 INTRODUCTION

The phenomenon concerning road accidents is ever more relevant; every year about five millions of people in the world dies a violent death and a quarter of these are traffic accident victims. Driver's behavior is the primary cause of traffic accidents. In fact, road safety problems are more frequently related to certain drivers' attitudes and their physical and emotional conditions when driving; for this reason, it is necessary to study carefully the human factor in road system. The aim of this paper is to investigate the relationship between the level of driving risk taken by driver during a trip and his/her personality traits. We also analyze the relationship between driving risk levels and driver's physical and emotional conditions while driving. In order to collect this kind of information, a survey was conducted to gain an objective measure of the driving style, and drivers' opinions about their personal characteristics. The objective measures are represented by the kinematic parameters (instantaneous speed, longitudinal and lateral accelerations) recorded along the trajectories followed by the drivers through smartphones equipped with GPS and accelerometer. The acquisition of this type of data is increasingly spreading thanks to the

precision that we can obtain at low cost. In fact, this kind of measures were used also to study the bus comfort on board (Eboli et al., 2017, *in press*) and to study the design consistency of a road segments by drawing the speed profiles (Eboli et al., 2015). On the other hand, we conducted a survey for collecting the judgments of drivers about their personal characteristics. Each participant involved in the survey was asked to complete, for each trajectory, a questionnaire including some questions according to which he/she has to express a rating about some personality traits, and physical and emotional conditions while driving.

The database is interesting because all the drivers involved in the survey registered the same path run in different days, and complete the questionnaire for each path. This permits to capture the different conditions of the drivers and observe the possible changes of their driving style as a function of their physical and emotional states.

In this paper we propose a correlation analysis between the risk level calculated based on the kinematic parameters registered through smartphones, and the judgements expressed by the users through the questionnaire, about some personality traits, physical and emotional conditions which can influence the driving style.

In the following we report a brief literature review of the studies dealing with this kind of issues. Then, we propose the methodology, by describing the sample, the objective measure of the risk level, and the collection of the data concerning drivers' characteristics. After, we propose the results of the correlation analysis and the main conclusions about the work.

2 LITERATURE REVIEW

The major part of the literature studies dealing with the relationship between psychophysical factors and driving style comes from psychologists, pedagogues, and so on. They have generally analysed personal traits, driving behaviour, and habit.

Most studies on the causes of road crashes focus on the relationship between specific traits and abilities of drivers, and crash risk. Petridou and Moustaki (2000) proposed a review where they tried to delineate behavioural factors that collectively represent the principal cause of three out of five RTCs (Road Traffic Crashes) and contribute to the causation of most of the remaining. The personality factors most often selected for that purpose include the Big Five personality dimensions: openness, conscientiousness, extraversion, agreeableness, neuroticism. The studies proposed by White and Dahlen (2001), Dahlen et al. (2005), and Dahlen and White (2006) investigated the utility of combining the Big Five personality factors, which have been studied independently, in the prediction of self-reported driving anger expression and the frequency of aggressive and risky driving behaviours.

Much attention has been paid to older drivers in research on driving behaviour and safety. The study by Adrian et al. (2011), in fact, investigated how executive functions and personality traits are related with driving performance among older drivers. Significant correlations were found between poor driving performances and low scores on tests assessing shifting and updating functions. In addition, extraversion had a negative relation with driving performance and made the only contribution, among the psychological factors, to the prediction of driving performance.

Arnau-Sabatés et al. (2012) proposed a study to analyze the relationship between emotional abilities, such as adaptability, stress management or affectivity, and the influence of this relationship on self reported drivers' risky attitudes. The risky driving attitudes and emotional abilities of driving instructors were measured. The results demonstrate that risky attitudes correlate negatively with emotional abilities.

Driving anxiety that has developed following crashes has been studied relatively frequently, but anxiety per se and its effects on driving has not as yet garnered much attention in the literature. The study proposed by Dula et al. (2010) found that higher levels of general anxiety were related to a wide variety of dangerous driving behaviours.

The research of Lu et al. (2013) explores how and why anger and fear influence driving risk perception.

There are also studies where kinematic parameters were registered in addition to the information collected through the questionnaire addressed to the drivers. As an example, Harder et al. (2008) conducted a study to better understand the psychological and roadway correlates of aggressive driving. Survey data was used to investigate the relationship between personality, emotional, and behavioural variables and self-reported driving behaviour. Participants were asked to give information about their personality characteristics including hostility, anger expression, physical and verbal aggressiveness, competitiveness, and empathy; their emotional state including anger-provoking experiences, anger states, and negative thought patterns; and their behavioural tendencies with regard to driving. In addition, a part of the participants drove on a section of a simulated four-lane freeway.

In a simulator study proposed by Roidl et al. (2014), some participants took part in certain traffic situations which each elicited a different emotion. Each situation had critical elements (e.g. slow car, obstacle on the street) based on combinations of the appraisal factors. Driving parameters such as velocity, acceleration, and speeding, together with the experienced emotions, were recorded. Results indicate that anger leads to stronger acceleration and higher speeds. Anxiety and contempt yielded similar but weaker effects.

From this brief literature review it emerges that the major part of the studies analysed personality traits that do not vary over the time. On the contrary, our analysis focuses on personal conditions changing from day to day, that is the physical and emotional conditions of the drivers at the time of the driving.

3 METHODOLOGY

3.1 *Sample*

The drivers supporting this research belongs to a population of car drivers who every day drive by their private car for reaching their destination of work or study, by running always the same path. This definition of the population is due to our desire to investigate on the influence of psychophysical factors on driving style; so, it is important to collect data concerning the same path run in different days in order to capture different conditions of the drivers, and observe the possible changes of their driving style as a function of their psychophysical state.

The drivers involved in the survey are five licensed drivers who are between 25 and 50 years old. The drivers can be grouped into two categories. The first group is composed of three experts in the sector of transportation engineering, two females and one male, who know very well the objectives of the research and the proposed methodology. However, they become simple drivers who travel with their own cars from home to their place of work. The choice to be components of the sample arises from the desire to test the survey and personally experience the planned work. Two experts in driving compose the second group of drivers; they are two police officers working for a long time.

Each driver registered through a smartphone the information of thirty paths approximatively. Participants drove under good weather daytime conditions with light traffic. The differences in driver age, gender and driving experience were treated as random effects, and were not considered in this research.

The survey interested some parts of the Italian National road n.107 (S.S. 107) which is a rural two-lane road 138 km long connecting the Tyrrhenian with the Ionian Sea coast of the Calabria region, in the Southern Italy. The high flow, combined with heavy rains and frequent snowfalls, has gradually increased the danger of the road, especially in some tortuous and difficult parts. In fact, the road was in the list of the 10 most dangerous roads of Italy in 2007, occupying the 7th place (ACI, 2007). Some measurements were also carried out on one part of the Italian National road n.18 (S.S.18) that is an important arterial road connecting Campania with Calabria on the Tyrrhenian Sea coast.

The analysed road segments have a length of about 6 km. Each segment has a roadway with two lanes, one for each direction, having a width between 3.25 meters and 3.5 meters.

Overall, drivers involved in the survey make about 170 tests covering about 1,100 km.

With the aim of record the kinematic parameters along each trajectory followed by the driver, drivers ran repeatedly the analysed road segments. Smartphones equipped with GPS and accelerometer were used; by means of a free app, the kinematic values were recorded with a frequency of 1 hertz, together with the instantaneous vehicle position (latitude and longitude). Today, these devices are widely used because they allow the investigation on kinematic parameters with a low cost and at good levels of reliability. As reported in Eboli et al. (2016), it is proved that probe estimates obtained from smartphones provide accurate real-time traffic data for safety analysis.

3.2 Objective measure of driving risk level

The level of risk is identified on the basis of the percentage of external points at borderline of a safety domain. As a safety domain was chosen the one identified by Eboli et al. (2016) by means of a methodology that describes the relationship between lateral, longitudinal accelerations and speeds, and represents a tool for classifying driving behaviour of the car drivers. Each point, in the reported diagram, represents the kinematic conditions of the vehicle in a certain instant, and it is representative of a safe or an unsafe driving behaviour. Specifically, if the point is inside the safety domain the driver is in safety conditions, else the driving behaviour is unsafe.

As an example, Figure 1 shows a survey carried out by a driver having an unsafe behaviour. It shows that there is a high percentage of external points at borderline of the safety domain; the percentage is 9.97.

3.3 Collection of data concerning drivers' personal and psychophysical characteristics

Each participant involved in the survey was asked to complete a questionnaire aiming to collect judgments concerning personal traits and psychophysical conditions that can influence driving style and risk level (Eboli et al., 2016, in press). The final version of the questionnaire has been obtained after a deep and wide research on character aspects that could influence people driving style. We designed and realized a preliminary survey with the aim to investigate on various aspects and to verify the suitability for our objectives (Salandria, 2013). The analysis of the results of the preliminary survey helped us in the selection of the definitive list of aspects for the final survey, and in the observation of different categories of aspects. In fact, by observing some results of the preliminary survey we succeeded in identifying three different categories: personality traits, physical conditions, and emotional conditions.

Specifically, drivers made a self-evaluation by expressing a level according to a numerical rating scale about five pairs of adjectives representing general peculiarities of their character. We adopted a scale on five levels, from −2 to +2. The "−2" and "+2" levels are labelled "extremely", while "−1" and "+1" are labelled "quite". After a deep research and analysis of the attitudes, behaviours, characters, emotions of people, we selected five pairs of personality traits that we retain possible determinants of the driving style of car drivers.

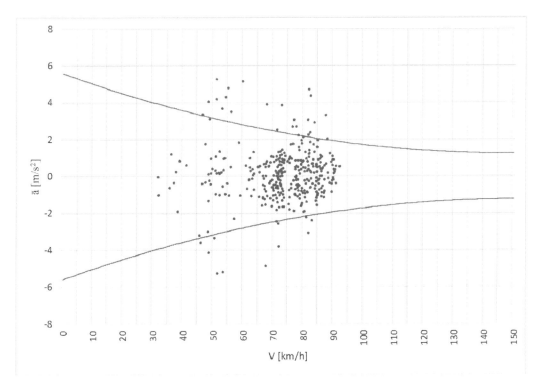

Figure 1. Graphical representation in the (V, ā) plane of the points inside and outside the safety domain, referred to a generic path.

The five pairs of adjectives according to the layout reported in the questionnaire are the following:

– lazy/dynamic
– irresolute/resolute
– impatient/patient
– impulsive/reflective
– superficial/meticulous.

The core of the questionnaire is the part aiming to collect information about the psychophysical conditions of the driver before starting the analysed path.

Specifically, the conditions were divided in two categories: physical and emotional. Also in this case drivers express a level from –2 to +2 about a series of adjectives representing physical and emotional conditions.

This part differs from the previous one because in this case drivers provide for information referred to their state in a specific moment, while in the first part they express opinions about general personality traits that however we retain as affectingtheir driving style. We selected three physical conditions that we consider possible determinants of drivers' driving style:

– tired/fresh
– sleepy/vigilant
– sick/healthy

We decided to analyse also five emotional conditions by considering the following five pairs of adjectives:

– gloomy/happy
– worried/carefree
– nervous/calm
– bored/interested
– angry/serene.

Also in this case the emotional conditions refer to the time before starting to drive the car.

4 CORRELATION ANALYSIS BETWEEN OBJECTIVE RISK LEVEL AND DRIVERS' CHARACTERISTICS

We propose a correlation analysis between the personal and psychophysical characteristics of the drivers and the risk level, in terms of safe or unsafe condition, obtained by using acceleration and speed data.

The aim of this kind of analysis is to understand which are the personal characteristics of the drivers mostly affecting drivers' risk level. In other words, we want to determine the personality traits, and the psychophysical conditions that influence the driving style, or the safe or unsafe drivers' behavior.

We have to make some preliminary remarks before starting the comments on the correlation analysis. The major part of the correlation coefficients show values lower than 0.3, indicating a low correlation. Only for two coefficients, belonging to the group of the personality traits, we found a medium correlation, being the value of the coefficient around 0.5. However, we retain that the obtained results can be considered for making a preliminary analysis, with the intent to expand the survey and improve the work.

Table 1 shows the correlation of the risk level (objective measure) with the drivers' personality traits. We can observe that the pair of adjectives mostly correlated with the objective measure of the risk level is "impatient/patient", while the personality trait having the lowest influence on the risk level concerns the laziness/dynamicity of the person. There are some coefficients showing a negative signs, and others with a positive sign of correlation. The positive sign indicates that when the value of the subjective judgement goes to the positive judgement (and therefore towards the adjective on the right), the risk level increases, or the behavior becomes more unsafe; *vice versa*, the negative sign means that if the value of the subjective judgement goes to the positive judgement driver behavior is safer. More specifically, if the driver is patient, the risk level decreases, as well as if the driver is dynamic and also meticulous. On the contrary, the risk level of people who are more resolute and more reflective increases.

These results can be interpreted by saying that laziness, superficiality, and impatience are personality traits that negatively influence the risk level and the safeness of the driving. In other words, an impatient person has a more unsafe driving, as well as a superficial person. We registered the highest values of the correlation coefficients just for these two personal characteristics. We can say also that a lazy person has an unsafe behavior, although the correlation coefficient is lower than the others.

On the other hand, two personality traits have a very different influence on driving risk level, that is the resoluteness and the reflectiveness; the correlation analysis shows that an irresolute person has a safer driving, as well as an impulsive person. It seems that to be a resolute and reflective person does not help in having a safe behavior, or in decreasing the driving risk level.

Table 2 shows the correlation coefficients of the risk level with the drivers' physical conditions. We observe lower values than the personality traits, due to the large variability of these conditions as regards the personality traits which are fixed over the time; however, we can define the physical conditions that mostly affect driving risk level and behavior. More specifically, we can see that there are two coefficients with a negative value. We can state that if the driver is fresh, the risk level decreases, as well as if the driver is vigilant. On the contrary, if the driver is healthy the risk level increases.

In Table 3 the correlation coefficients of the risk level with the drivers' emotional conditions are reported. As for the physical conditions, because of the variability of the data, also in this case the coefficients are quite low. The conditions mostly affecting driving risk level are being "gloomy-happy" and "nervous-calm". More specifically, if driver is gloomy or nervous, he has a safer driving, as well as he is bored. It seems that being gloomy or nervous is not a problem during the driving, but it can be an element that induces the driver to maintain a safer behavior, maybe because the driver is conscious of his condition and for this reason, he tries to drive with more cautiousness.

The value of the coefficients with a negative sign are very low, so we cannot provide a very reliable interpretation of the results; however, we can say that it seems that being carefree and serene helps to drive with a lower risk level.

Table 2. Correlation of the risk level with the drivers' physical conditions.

Physical condition	Correlation coefficient
tired/fresh	−0.101
sleepy/vigilant	−0.085
sick/healthy	0.061

Table 1. Correlation of the risk level with the drivers' personality traits.

Personality trait	Correlation coefficient
lazy/dynamic	−0.095
irresolute/resolute	0.273
impatient/patient	−0.477
impulsive/reflective	0.112
superficial/meticulous	−0.445

Table 3. Correlation of the risk level with the drivers' emotional conditions.

Emotional condition	Correlation coefficient
gloomy/happy	0.140
worried/carefree	−0.060
nervous/calm	0.108
bored/interested	0.046
angry/serene	−0.019

5 CONCLUSIONS

The aim of this paper was to analyze the relationship between drivers' personality traits, physical and emotional conditions and the risk level to be involved in a traffic accident calculated based on the kinematic parameters registered through smartphones during the driving. Specifically, we proposed a correlation analysis of the drivers' judgements about their personal characteristics (collected through a questionnaire) and an objective measure of the risk level obtained from kinematic parameters. This kind of analysis allows to determine the drivers' personal characteristics which more affect driving risk level.

The main and more interesting results suggest that characteristics such as patience and meticulousness positively influence the level of safety, because the level of risk decreases with the increase of these two human virtues. A finding apparently unexpected is the influence of the conditions "gloomy-happy" and "nervous-calm". Specifically, we found that being gloomy or nervous has a positive effect on the level of risk. A reasonable interpretation could be that gloom and nervousness induces the driver to maintain a safer behavior just because the driver is conscious of his condition and he tries to drive with more cautiousness in order to avoid to occur in a traffic accident.

As above specified, a future development of the work is surely the increase of the sample dimension, in order to obtain more solid and reliable results from the analysis of the data. We retain that the results reached in this paper are however interesting given that the main aim of the work was to test a new methodology and survey to combine subjective and objective data concerning driving style.

REFERENCES

ACI (2007). Incidenti stradali in Italia, 2007. Italian Bureau of Statistics.

Adrian, J., Postal, V., Moessingerc, M., Rascleb, N., & Charlesb, A. (2011). Personality traits and executive functions related to on-road driving performance among older drivers. *Accident Analysis and Prevention* 43: 1652–1659.

Arnau-Sabatés, L., Sala-Rocab, J., & Jariot-Garcia, M. (2012). Emotional abilities as predictors of risky driving behavior among a cohort of middle aged drivers. *Accident Analysis and Prevention* 45: 818–825.

Dahlen, E.R., &White, R.P. (2006). The Big Five factors, sensation seeking, and driving anger in the prediction of unsafe driving. *Personality and Individual Differences* 41: 903–915.

Dahlen, E.R., Martin, R.C., Ragan, K., & Kuhlman, M. (2005). Driving anger, sensation seeking, impulsiveness, and boredom proneness in the prediction of unsafe driving. *Accident Analysis and Prevention* 37: 341–348.

Dula, C.S., Adams, C.L., Miesner, M.T., & Leonard, R.L. (2010). Examining relationships between anxiety and dangerous driving. *Accident Analysis and Prevention* 42: 2050–2056.

Eboli, L., Guido, G., Mazzulla, G., & Pungillo, G. (2015, published on line). Experimental relationship between operating speeds of successive speeds of successive design elements in two-lane rural highways. *Transport.* (http://dx.doi.org/10.3846/16484142.2015.1110831)

Eboli, L., Mazzulla, G., & Pungillo, G. (2016). Combining speed and acceleration to define car users' safe or unsafe driving behaviour. *Transportation Research Part C* 68: 113–125.

Eboli, L., Mazzulla, G., & Pungillo, G. (2017, in press). Measuring Bus Comfort Levels by using Acceleration Instantaneous Values. *Transportation Research Procedia.*

Eboli, L., Mazzulla, G., & Pungillo, G. (2016, in press). The influence of physical and emotional factors on driving style of car drivers: a survey design. *Journal of Traffic and Transportation Engineering.*

Harder, K.A., Kinney, T.A., & Bloomfield, J.R. (2008). Psychological and Roadway Correlates of Aggressive Driving. Report CTS 08–30. Center for Human Factors Systems Research and Design, University of Minnesota.

Lu, J., Xie, X., & Zhang, R. (2013). Focusing on appraisals: How and why anger and fear influence driving risk perception. *Journal of Safety Research* 45: 65–73.

Petridou, E., & Moustaki, M. (2000). Human factors in the causation of road traffic crashes. European Journal of Epidemiology, 16, 819–826.

Roidl, E., Frehse, B., & Höger, R. (2014). Emotional states of drivers and the impact on speed, acceleration and traffic violations-A simulator study. *Accident Analysis and Prevention* 70: 282–292.

Salandria, A. (2013). L'influenza dei fattori psico-fisici sui comportamenti di guida dei conducenti delle autovetture (Thesis). University of Calabria, Italy.

White, R. P., & Dahlen, E. R. (2001). The role of personality and emotional factors in the prediction of crash-related conditions, aggressive driving, and risky driving behavior. Paper presented at the Mississippi Psychological Association Convention, Gulfport, MS.

Transport Infrastructure and Systems – Dell'Acqua & Wegman (Eds)
© 2017 Taylor & Francis Group, London, ISBN 978-1-138-03009-1

The impact of a bus rapid transit line on spatial accessibility and transport equity: The case of Catania

N. Giuffrida, G. Inturri, S. Caprì, S. Spica & M. Ignaccolo
Department of Civil Engineering and Architecture, University of Catania, Catania, Italy

ABSTRACT: Accessibility is a key issue to address spatial equity when planning for sustainable mobility. Accessibility indicators can be used to measure the performance of public transport as basic strategy to cause modal shift from private transport and reduce car dependence and urban sprawl. The purpose of this paper is to verify if the realization of a set of Bus Rapid Transit (BRT) lines with high level of service can provide an equitable access of residents to workplaces, when compared with a light improvement of the commercial speed of conventional bus lines with low level of service but high spatial coverage. To this aim we use a relative accessibility measure between private and public transport, weighted by socio-economic data of population. A high spatial resolution spatial analysis is used to capture the relevance of different stop density for walking access impedance, through a GIS transport modeling software. The methodology is tested for the city of Catania (Italy).

1 INTRODUCTION

In the last several years, in the field of planning and management of transport systems, a process that can be described as the transition from a mobility oriented to an accessibility oriented planning is taking place. While planning for mobility uses performance indicators mainly based on travel time and level of congestion, planning for accessibility takes into account a broader set of elements, in order to integrate the transport strategies with urban and territorial ones. In other words, there has been a gradual shift towards theories, approaches and support instruments to an integrated planning of land use and transport system that gives increasing importance to accessibility.

This work has multiple aims: on one hand it defines a new method for modeling public transport based on a high resolution spatial analysis with the use of GIS; on the other hand, it identifies a suitable accessibility index that allows a comparison with private transport and that includes social issues, in order to support an equitable transit planning. These goals are pursued by modeling the impact of two different transport scenarios on urban accessibility and testing the model in the multimodal transport network of Catania (Italy).

The paper provides at first a literature review regarding accessibility and its relation with social inclusion; the successive section shows the methodology by focusing on the network model, the GIS approach and accessibility measures used. Afterwards the case study is explored considering its context, the application of the methodology and the results obtained. Finally, conclusions are presented.

1.1 *Accessibility definitions and measures*

The concept of accessibility, especially in transport planning, plays an increasingly important role in the process of policy decisions and it is a useful tool to analyze the degree of access to an area above the surrounding territory. Accessibility measures may be valid social indicators, as they show the level of difficulty with which various categories of individuals can reach the economic opportunities or social interaction throughout the area.

However, drawing up a strict and unambiguous definition of accessibility is a complex task. One of the first scholars which considered its importance in the context of spatial planning was Hansen, who defined accessibility as "the potential of interaction opportunities" (Hansen, 1959).

A recent definition that highlights the mutual interaction between land use and transport systems has been provided by Geurs and van Wee (2004). According to the authors, the accessibility can be considered as the measure with respect to which the use of the territory and of transport systems allow groups of individuals to reach activities or locations by a combination of modes of transport.

From these and other definitions in the literature, four major accessibility components can be identified: land use, the transport system, the time factor and the individual dimension (Geurs & Van Wee, 2004).

A classification of accessibility measures depending on land use can be done considering the place in question as the origin or destination of the travel. We can therefore distinguish the origin accessibility or active accessibility, and the destination accessibility or passive accessibility (Cascetta, 2009):

– Active accessibility refers to the need to carry out the activities located throughout the area by a user that is in a particular place and it measures the ease with which the user can reach various destinations from an origin. It is useful in locating settlement decisions.
– Passive accessibility refers to the need for the various opportunities that are located in a certain area of the territory, to be achieved by the various users scattered throughout the study area. In other words, it measures the ease with which individuals, business and the services of a target area of the displacements can be reached by the users concerned. It is useful in the location decisions of public services and economic activities.

Most of the formulations in the literature refer to an urban active accessibility, whose indicator, in analytical terms, is generally a function of the number of spatial opportunities and of the generalized transport cost to reach them. In particular, the accessibility indices based on gravitational models provide a measure of the continuous type which weighs the value of the opportunities with respect to a spatial impedance function. The impedance function reflects the effect of decreasing accessibility due to the increase of distance, travel time, or in general of the generalized cost of shipping.

The first application of the gravity model to accessibility measures is attributed to Hansen (1959), which suggested that accessibility across regions was directly proportional to the attractiveness factors (jobs, shops, sports centers, etc.) and inversely proportional to the travel time between the zones, which represents the cost of moving. The Hansen's index has the following form:

$$A_i = \sum_j^n O_j \cdot f(C_{ij}) \tag{1}$$

where O_j is the number of opportunities in the zone j and $f(C_{ij})$ the impedance function among zones i and j. A negative exponential impedance is often used, such as:

$$f(C_{ij}) = e^{-\beta C_{ij}} \tag{2}$$

With C_{ij} generalized cost of travel among i and j zone and β is a parameter related to the cost, estimated by choosing a destination model.

The generic measure of cumulative opportunities can be considered a special case where $f(C_{ij})$ is equal to 1 if C_{ij} is less than the predetermined threshold; it is equal to 0 otherwise.

This type of indicators offers the advantage of requiring a relatively small amount of data (ease of processing and calculation), allowing to differentiate the areas of study and to derive the accessibility indices for each of them. They are particularly useful for assessing the potential of suburban residential areas in allowing access to activities such as shops, schools, workplaces, health care and other services.

1.2 Accessibility and social inclusion

Social exclusion is a condition of poverty combined with social marginalization, which causes the separation of certain groups or individuals from institutions and community, preventing full participation in joint activities of the society in which they live.

With regard to the transport field, social exclusion concept is closely linked to the accessibility to activities, goods and services. In this case, the transport system plays an important role, since limited accessibility to transportation, by public or private transport, prevents a person to reach job places, health centers and entertainment venues.

Public transport plays a pivotal role in worsening or alleviating social exclusion of vulnerable groups, since it affects their access to basic services as well as employment positions and social relations. It's also likely that the negative impact of the transport system on the environment, safety and public health would fall disproportionately on the most disadvantaged groups.

To develop an inclusive transport system is necessary for accessibility, safety and comfort of modes of transport to become the first priorities of transport policy. According to Litman (2009) this is strictly related to the improvement of:

– all phases of the path, including the environment related to the pedestrian access, in a way that also disabled people could reach and use public transport services;
– planning of transport infrastructure, considering the specific needs of vulnerable groups;
– safety of public transport, a key issue that affects mainly women and elderly people;
– frequency and punctuality of services;
– dissemination of maps and information guides.

Public transport may be able to reduce this mobility gap, by the promotion of social inclusion. In fact, when it's not accessible by the weakest population groups and it's unable to break down the barriers that do not allow the participation to social activities, public transport fails its primary goal: to give access to employment or educational opportunities, medical care services and entertainment venues. In summary, public transport should offer everyone the ability to move and therefore it's a determining factor for social inclusion policies.

2 METHODOLOGY

2.1 TransCAD model

In the present study it used a software for the management and analysis of transport data that is called TransCAD, produced by Caliper Corporation (USA). TransCAD combines a GIS (Geographic Information System) with analytical models and transport planning in one integrated environment.

2.2 Hansen's index computation

In order to evaluate Hansen Index an impedance matrix among zones has been calculated.

The impedance matrix tells us how difficult is to travel from an origin to a destination and considers the generalized cost of transport. It also translates into monetary cost factors not monetized ones, as the travel time (that the user tends to minimize), physical effort, comfort, safety and user stress.

The matrix has been realized considering impedance as the generalized cost of transport and taking into account parameters such as travel time, the cost of travel time, the number of transfers and travel fare.

Hansen function used by TransCAD to perform the calculations is:

$$A_i = \frac{\sum_{i=0}^{n} O_j \cdot e^{-\lambda C_{ij}}}{\sum_{i=0}^{n} O_j} \quad (3)$$

where A_i is accessibility of zone i, O_j are the opportunities in the zone j, C_{ij} is transportation cost between zones i and j and λ is the deterrence parameter. λ value should be calibrated fin each context according to the situation, but in this case it is used the value provided by default for each transport mode and trip purpose.

Results will be then normalized with the normalizing function:

$$A_{Ni} = \frac{A_i - A_{min}}{A_{max} - A_{min}} \quad (4)$$

where A_{Ni} is the normalized value of accessibility for zone i, A_i is the generic value of accessibility for zone i, and A_{min} and A_{max} are respectively the minimum and the maximum value of accessibility among all the case study's zones.

2.3 Relative accessibility measure

We define the Relative Accessibility Index as the relationship between the accessibility guaranteed by public transport and accessibility guaranteed by private transport. This concept can be expressed by the equation (1):

$$A_R = \frac{A_i^{TPb}}{A_i^{TPr}} \quad (5)$$

where A_R is the relative accessibility, A_i^{TPb} is the accessibility provided by public transport and A_i^{TPr} is the accessibility by private transport.

According to the aim of this work and in order to include social issues, we define the Relative Weighted Accessibility (RWA) index as follows:

$$RWA = \frac{A_i^{TPb(t)}}{A_i^{TPr(t)}} \cdot \left(1 - \frac{Pop_i^{Car}}{Pop_i^{Tot}}\right) \quad (6)$$

where Pop_i^{Car} is the population that has access to private transport (specifically people over 18 years old and under 70 years old) and Pop_i^{Tot} is the total population at the origin. People that do not have access to private transport are assumed to be highly correlated with the number of disadvantaged people, (such as children and elderly are) i.e. people who have lower car availability and therefore will benefit more largely of higher transit accessibility.

In this way, the RWA index is decreased for those zones where a high ratio of disadvantaged people is present.

3 CASE STUDY

3.1 Territorial framework and transport system

Catania is a city of about 300.000 inhabitants and it is located in the eastern part of Sicily; it has an area of about 183 km2 and a population density of 1.754,54 inhabitants / km2 (Istat, 2015b). It's part of a greater Metropolitan Area (750.000 inhabitants), which includes the main municipality and 26 surrounding urban centers, some of which constitute a whole urban fabric with Catania. The main city contains most of the working activities, mixed with residential areas. With reference to the urban area, the transport service is provided by 51 bus lines, a Shuttle line (ALIBUS) connecting the city center with the airport and a second fast bus (called BRT1) connecting the parking Due Obelischi with Stesicoro Square. BRT1 is the first of three lines provided by the City of Catania with equipped lanes protected by curbs on the majority of their path and was promoted commercially as Bus Rapid Transit. In Catania it is also present an urban subway line that currently connects the station "Porto" with the station "Borgo" from which continues as a surface long-distance line.

By 2016 it is expected the undergrounding of the line until the station "Nesima" and it's also planned the opening of a branch linking the station "Galatea" to Piazza Stesicoro.

3.2 *Transport model and scenarios*

Transport model has been implemented in Tran-CAD software, a software which combines Geographic Information System and analytical models and transport planning in one integrated environment. The zonation used for the city is the one given by ISTAT, which divides the study area in 2480 CENSUS sections (Fig. 1).

Methodology has been applied with reference to the trips made by the citizens of Catania for work purpose.

Three different scenarios have been considered:

– Base scenario: 51 urban public transport bus lines; it is assumed for these bus a commercial speed of 15 km/h.
– Project scenario 1: insertion of three rapid transit lines (L-EX, BRT1 e BRT2), with a commercial speed of 20 km/h.
– Project scenario 2: increase of commercial speed of all bus lines to 18 km/h.

3.2.1 *Base scenario*

Hansen accessibility has been evaluated for the first scenario, through the use of TransCAD; results are showed in the thematic map in Figure 2. The map reveals how the high-resolution zoning can represent a significant difference in accessibility index even for adjacent cells, which would have been 'flattened' by a lower resolution approach.

Relative Weighted Accessibility has been then evaluated for the base scenario by taking into account Equation 6 and results have been mapped (Fig. 3).

Weighting the accessibility index as indicated before generally produces lower values for zones with a low number of resident with access to private transport. So it can be noticed that the weighted index seems more adequate to support transit planner choices when the spatial distribution of disadvantaged social groups has to be taken into account properly.

Through the normalized accessibility values it is possible to derive a histogram showing the frequency distribution and the cumulative curve for traffic areas defined.

On the x-axis accessibility classes are shown, while on the y-axis we can see the frequencies. The cumulative establishes the percentage of areas with a certain accessibility value. The histogram can be seen in Figure 4.

3.2.2 *Project scenario 1*

In this case we tried to understand how accessibility to workplaces varies when three lines of rapid buses, L-EX, BRT1 and BRT2 are introduced into the network. It is assumed for these buses a commercial speed of 20 Km / h, superior to that of all other buses (15 Km / h).

As you can see in Figure 5 the three buses cover the whole territory of Catania and in particular: the BRT1 moves from north to south; the BRT2 from west to east and L-EX follows a route in the southeast.

The calculation of **RWA** for the Project Scenario 1 shows little improvements of accessibility for all the zones in the study area; a thematic map of the improvements can be seen in Figure 6.

Figure 1.   Transport model.

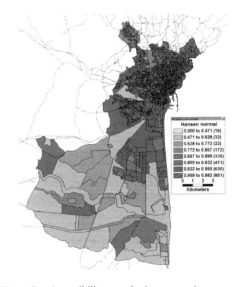

Figure 2.   Accessibility map for base scenario.

756

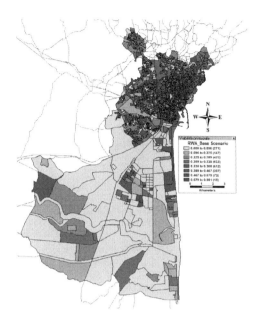

Figure 3. RWA map for base scenario.

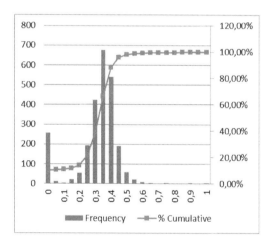

Figure 4. Histogram of base scenario's RWA.

Once again we present the results in a histogram (Figure 7) that shows the frequency distribution and the cumulative curve for the zones defined.

3.2.3 Project scenario 2

In this second scenario we tried to understand how accessibility to workplaces varies if we raise the commercial speed of all city buses already presented in the scenario base. This speed was increased from 15 km/h to 18 Km/h.

The calculation of RWA for the Project Scenario 2 shows big improvements of accessibility for all the zones in the study area; a thematic map of the improvements can be seen in Figure 8.

Figure 5. BRT lines.

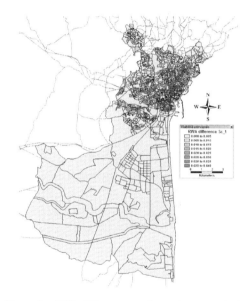

Figure 6. RWA difference map for project scenario 1.

Once again we present the results in a histogram (Figure 9) that shows the frequency distribution and the cumulative curve for the zones defined.

3.2.4 Results

Comparing the three scenarios we see how in all the cases we have a low accessibility in the industrial area (located in the south) and this is due to the geographical spread of the central areas served by buses and the industrial zone, and the low capillarity of the public transport network in that area. It can be noticed, however, an increment in accessibility as we get closer to the center or we move toward the northeast, because they are areas served by public transport lines. In any case

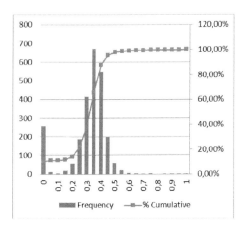

Figure 7. Histogram of RWA for scenario 1.

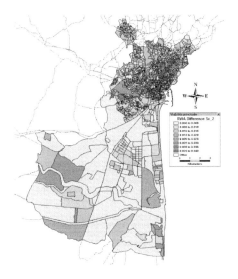

Figure 8. RWA difference map for project scenario 2.

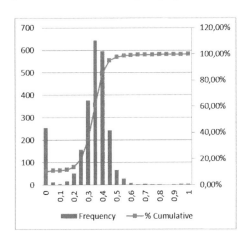

Figure 9. Histogram of RWA for scenario 2.

it is clear that between the baseline scenario and project scenarios we have a total increase of accessibility to workplaces.

4 CONCLUSION

In this paper new transit accessibility measure and indexes has been presented and their use has been experimented to evaluate the impact of introducing BRT lines in the transit network of the city of Catania. They have proved a good sensitivity to: the walking trip segment along the pedestrian network to access the transit system, the land use opportunities at destinations and the social groups using the transit system.

The relative weighted accessibility measure can be considered a useful tool to address a new approach in transport planning, when the main goal is to provide people with a high potential to reach the goods and services necessary for daily life, regardless their social status and car availability. A high transit accessibility compared to car accessibility is a precondition for economic development as it enables and empowers the opportunities for people interaction (workers and consumers), a performance indicator of the availability of less energy-intensive and lower impact modes of transport, a measure of equity in terms of distribution of transport options among citizens with different social and economic power.

In addition, the relative weighted accessibility can be a useful transport planning indicator to address changes in the transit network design and operation, in order to avoid the risk of uncritical increase in transit accessibility of zones where people already have a high car ownership and land use accessibility, thus increasing the social exclusion of people who have less access to private car and transit services.

REFERENCES

Cascetta E. (2009), *Transportation System Analysis Models and Applications*. Springer.
Geurs K. T. and Van Wee B., 2004. Accessibility evaluation of land-use and transport strategies: review and research directions. *Journal of Transport Geography*, Delft.
Hansen W. G., 1959. How Accessibility Shapes Land Use, *Journal of the American Institute of Planners*. Vol. 25, 73–76, USA.
Litman, 2009. Evaluating Transportation Land-use Impacts, *Victoria Transport Policy Institute*, Canada.

Transport Infrastructure and Systems – Dell'Acqua & Wegman (Eds)
© 2017 Taylor & Francis Group, London, ISBN 978-1-138-03009-1

Shared infrastructures: Technique and method for an inclusive social valuation

Saverio Miccoli
Department of Civil, Building and Environmental Engineering, La Sapienza University of Rome, Rome, Italy

Fabrizio Finucci
Department of Architecture, Roma Tre University, Rome, Italy

Rocco Murro
Department of Civil, Building and Environmental Engineering, La Sapienza University of Rome, Rome, Italy

ABSTRACT: In Italy, the implementation of many infrastructures is stuck because of ongoing conflicts between political and administrative decision-makers, operators and users. Aware of the current situation, a recent Legislative Decree introduced the obligation of the Public Debate between the stakeholders in order to reduce the likelihood of infrastructure projects not being implemented due to the lack of necessary social consensus. Consensus-building activities require the use of valuation approaches that actively involve the community. To this end, the paper proposes an inclusive valuation technique aimed at the direct and conscious deliberation of individuals involved in an infrastructure project combining deliberative methods, based on community debates, with a social multi-dimensional evaluation.

1 INTRODUCTION

In the Italian regulatory framework, the two main investment-planning tools are a transportation plan and a programme for strategically important infrastructure projects. In 1984 with Law n. 245, the General Transport Plan (GTP) of Italy was introduced with the aim of ensuring the existence of a single national transport policy. The Strategic Infrastructures Programme (SIP) was introduced in Italy's national legislation more recently, in compliance with Law 443/2001 (the so-called Objective Law). This measure provided for the identification of the public infrastructures and production plants of particular national interest, in full respect for the regions' autonomy, through a programme established by the Ministry of Transport and Infrastructure (MTI), in agreement with the competent ministers, after consultation with the Italian Inter-ministerial Committee for Economic Programming (ICEP). Furthermore, it was decided to include this programme in the Economic and Financial Planning Document (EFPD) specifying the relative funding.

The need to coordinate the infrastructural interventions linked to national requirements with the main transport policies at Community level also implied the thorough revision of Italy's legislation regarding both project planning and prioritis-

ing. To ensure funds were used efficiently, in the Community Programming 2014–2020, Member States were asked to guarantee a series of ex ante conditionalities of a regulatory, administrative or organizational nature. Within the framework of Community programming, the concept of conditionality are prerequisites Member States have to fulfil in advance to guarantee they have the regulatory and organisational resources necessary to ensure full adhesion to EU strategies and the effective feasibility of the investments supported with allocated funds.

The updating of national regulations and the adjustments introduced to fulfil the prerequisites of the *ex ante* conditionalities, resulted in the Infrastructure Appendix (IA) to the Economic and Financial Document (EFD), the single document that provides the framework for the planning and investments of national interest in the transport sector. The IA of 2015 was the first to recognise the infrastructure planning reform. The first part of the document provides a general overview of transport infrastructure planning based on an analysis of transportation demand (goods and passengers) and a SWOT analysis of the entire national infrastructure sector. In the second part of 2015 IA, strategies to address the critical issues identified by extending and improving the entire infrastructural network (railways, roads, ports

759

and airports) are proposed. To make this objective operational a number of implementing tools are used: a) the Strategic Infrastructures Programme (SIP); b) the ANAS Programme contract related to the restructuring of the motorway sector; c) the RFI Programme contract to extend and improve the national railway network; d) the National Plan for Ports and Logistics; e) The Airport Plan.

The implementing tools are linked to specific financial tools: a) the Connecting Europe Facility (CEF) and the European Fund for Strategic Investments (EFSI); b) the European Structural and Investment Funds; c) the Development and Cohesion Fund 2014–2020, for the achievement of strategic objectives related to national thematic areas. The document also specifies 25 top priority projects to include in the SIP; the selection is based on their integration with European networks, the progress of ongoing works, and the possibility of their being funded mainly with private capital. As regards the other projects included in the SIP (called Other Interventions), the document suggests proceeding following a preliminary consultation with the Regions. The other new developments envisaged in the IA to the EFD 2015 include the launching of a new IT system to monitor the progress of ongoing works (called Opencantieri) by the MTI.

The IA to the EFD 2016 is defined by the MTI itself as a new Planning process for public works centred on the planning of interventions that are useful from an intermodal and sustainable viewpoint and that enhance of existing infrastructures. It introduces two new tools: the first is the General Plan for Transportation and Logistics (GPTL) that will contain strategic guidelines for policies regarding the mobility of people and goods, and for the infrastructural development of the nation. The second is the Multiannual Planning Document (MPD) that will contain the interventions related to the transport and logistics sector, the feasibility studies of which must have received a positive *ex ante* evaluation to show the projects deserve funding. Furthermore, the scheme envisages the possibility for the MPD to be integrated with proposals from the Managing Bodies, the Regions and the Local Authorities involved. As provided for in Legislative Decree n. 228/2011, the MTI publishes a detailed annual report annexed to the EFD on the state of implementation of the MPD.

After representing the framework of reference and context of the EFD 2015 and summarising the ongoing activities, the IA to the DEF 2016 establishes a roadmap for the drawing up of the first MPD. In this process, the public works are divided into three main groups: a) the 25 top priority strategic works (Annexed to EFD 2015); b) the works included in the SIP divided into two typologies—

those with Binding Legal Obligations (BLO) and those without; c) the new projects proposed by local authorities or institutions. The 25 strategic projects along with those included in the SIP without BLO will be included in the Project Review after which they will be inserted in the first MPD 2017–2019. The SIP projects without BLO and any new projects proposed by local authorities or institutions will be subject to an evaluation related to infrastructural needs. This evaluation, of a quantitative nature, will consist in identifying the shortage of elements in the infrastructure system necessary to satisfy the demand. Once identified, feasibility projects will be carried out to find the most suitable solutions to remedy the shortage. Subsequently, the feasibility projects will be subjected to an *ex ante* evaluation, and only after they have undergone this procedure can they be included in the MPD.

The infrastructure planning and programming process is innovative also thanks to several procedural changes introduced with Legislative Decree 50/2016, or the New Procurement Code (NPC). The most important innovations include: a) the use of the Project Review; b) the feasibility projects; c) Public Debate in the decision-making process for the creation of infrastructures.

The Project Review is a tool provided for under Art. 202 of the NPC to verify if a project selected with previous planning processes has meantime become oversized due to variations in the general state of affairs or new market conditions. Depending on the outcome of the verification process, performed mainly with evaluation tools and procedures, the functional characteristics of the project in question can be redefined or funding may be revoked. The purpose of the feasibility project introduced and regulated in Arts. 23 and 202 of the NPC is to improve the quality of infrastructural planning, programming and design. The objective of design at this level is to verify the technical, economic, environmental and territorial conditions for the creation of infrastructures by selecting the solution with the best cost-benefit ratio for the Community. Lastly, the forms of participation are recognized in national regulations with the introduction of a structured process that involves communities in the decision making regarding infrastructure projects, i.e. Public Debate pursuant to Art. 22 of the NPC. Although timing, modalities and techniques have still to be established through future implementing decrees, the NPC clearly states that local administrations have to make public the feasibility projects regarding infrastructural and architectural interventions of social importance with a significant impact on the environment, the city and the territory. The outcomes of the public consultations including the minutes of meetings held and debates with the stakeholders also have to be published. Currently, open conferences with the participation of administrations, stakeholders and citizen committees are foreseen. This process will be mandatory for large infrastructure projects and an ad hoc MTI decree will establish size thresholds and what kind of project will undergo Public Debate. The debate will focus on the feasibility project to give the proposing institutions the possibility to listen and receive information and suggestions on every aspect of the project, and to add new ideas and further clarifications if necessary.

After outlining the regulatory framework, the paper: a) deals with the topic of inclusion in infrastructure planning; b) searches for shared decisions by proposing a deliberative multidimensional evaluation procedure; c) ends with a synthesis of the advantages of the proposed technique and highlights the main obstacles to its full implementation.

2 EVALUATION AND INCLUSION IN THE PROCESSES FOR THE CREATION OF INFRASTRUCTURES

Over the last two decades, the topic of conflict and consensus when it comes to creating large infrastructures, and public choice in general, has influenced Italy's political agenda. Albeit later compared to other countries, Italy has also put in place inclusive procedures to cope with the main decisional stumbling blocks linked to public works. Initially, a few regional laws were introduced, e.g. those of the Regions of Tuscany and Emilia Romagna, and local council regulations such as those introduced in Rome and Milan that regulate participatory practises. The adjustment process for the adoption of these practises at national level was finally regulated in the recent NPC with the introduction of Public Debate. The IA to the EFD 2016 also states that, "often, the projects handed down from on high, which do not take into account the micro-restrictions of the territory in question, give rise to opposition while work is in progress and this in turn gives rise to uncertain and approximate implementation costs and deadlines". In literature, this kind of opposition is called the NIMBY (Not In My Back Yard) syndrome. As the local community does not identify itself in the choices of the decision makers and believes the consequent transformation will prove detrimental to local interests, it begins to protest against the implementation of the project in question (Miccoli et al. 2015a). Currently in Italy, according to the Nimby Forum, there are 355 ongoing projects affected by the protests of local residents. By subdividing these conflicts according to the category of project in question,

it is possible to see that approximately one in every ten is due to an infrastructural intervention.

There are ongoing conflicts linked to 5 of the 25 strategic projects: the Turin-Lyons high-speed train link, the Valico dei Giovi railway, the Pede-montana Lombarda motorway, the Tangenziale Est in Milan a the Experimental Electromechanical Module MOSE. The lack of social consensus is affecting all these public works in terms of delays and increased costs. In addition, the progress of work on the 25 strategic projects does not mean that further protests will not arise. The overall situation regarding the progress of work on the 25 strategic projects is the following: a) under 15% of the work needed to complete seven works (28%) has been carried out (four have still to be started); b) under 50% of the work needed to complete almost half (12 projects) has been carried out; c) under 2/3 of the work needed to complete sixteen projects has been carried out. Italy's infrastructural modernisation process is also linked to many public works provided for in the SIP: a fragmented series of interventions of different sizes regarding ports, airports, interchange junctions, and stretches of railways and road.

In Italy's new infrastructural programme, the evaluation and inclusion of citizens in the decision-making process is considered crucial to the success of the interventions. Evaluation plays a pivotal role in various phases: the drawing up of feasibility studies for new projects proposed by local authorities or institutions and in the updating of the GPTL and MPD. During the implementation of the first MPD, evaluation is the main activity as far as decision making is concerned and includes selecting interventions, identifying priorities and Project Review. Each phase has to be supported by evaluation procedures that are adequate for the problems peculiar to the infrastructure project in question. These evaluation procedures must be capable of dealing with the innovations related to scientific approaches and to the regulations that from a structural point of view guide and reflect technological evolution. From an evaluation viewpoint, the MTI is preparing standardised guidelines for the evaluation of investments in public works with the aim of ensuring that all works are evaluated homogeneously and selected with transparency. As regards social inclusion, the Ministry itself is working on the choice of modalities for Public Debate as provided for in the NPC.

As far as this paper is concerned, these developments are irrelevant as our intention is to propose guidelines for an operational tool and the method to implement it capable of providing effective solutions in situations where the objective is not to help administrators-politicians who are the legitimate representatives of citizens to find out about their opinions through Public Debate, but to allow the citizens themselves to make informed, debated and shared choices on the infrastructural problems in question.

Therefore, preference regarding the two afore-mentioned decision-making pathways is not called into question; the same goes for their possible complementarity or coexistence, and the possibility to merge them into a single procedure.

In our opinion, in situations where it is necessary to know what a community thinks about alternative infrastructural solutions, the formalisation of an assessment pathway carried out directly by the members of the community with techniques and methods suited to dealing with the complex factors at stake, can result in an efficient choice because it is characterised by responsibility, transparency, rationality, and full awareness of the possible distributive effects. Thereby the valuation approach makes it possible to abandon the context of opinions elicited during a public debate and embrace that of certainties produced by scientific approaches. Substantially, we are speaking about a realistic choice that does not arise from theoretical or reductionist scenarios, but from situations associated with action and the effective adaptation of the members of a community. We firmly believe that the use of an evaluation technique of a dialogic-deliberative nature, combined with multi-criteria approaches, makes it possible to devise a more evolved and binding form of Public Debate, a tool otherwise destined to be used as a mere consultation mechanism with no totally inclusive decision-making value. Taking note of the opinions stated during citizens' assemblies undoubtedly provides important information for public decision-making, but it is a far cry from knowing the measure of opinions regarding choices based on analyses and on qualitative-quantitative data stated by those who for various reasons are interested in an infrastructure project. In the evaluation technique described below, the community itself, with different modalities and roles, fixes criteria, establishes priorities, and states evaluations with the aim of reaching a shared choice.

3 AN INCLUSIVE SOCIAL MULTIDIMENSIONAL EVALUATION

With the aim of providing an evaluation model that not only takes into account the complexity, risk and uncertainty attached to infrastructure projects, but also supports public choice and actively involves the community in order to reach a shared choice, we propose a procedure based on an inclusive social multidimensional evaluation. The purpose of a multi-criteria based multidimensional evaluation is

to take into consideration a number of alternative projects, various criteria and their order of priority, and the multiple impacts of an environmental, social and economic nature that the infrastructure in question would have on the territory. With a view to foreseeing problems linked to distributive justice, the evaluation approach is also social, capable of taking into account the viewpoints of all the stakeholders—be they single, collective, private or public—who are directly or indirectly affected by a complex project. As the procedure is based on dialogue and negotiation, as well as mediation between diverse interests and values, the evaluation model envisages the systematic participation of the community in question. It goes beyond the more traditional social consultation models and tries to directly involve the community through techniques based on information, communication, opinions and debate between stakeholders.

In recent decades in many countries, the involvement of citizens in decision-making processes has been consolidated through a variety of methods and tools that are indispensable for making public choice successful. The inclusive approach, intended as a tool to manage and bring together all the interests at stake in a given territory, combines the development and safeguard of the local community with a view to making sustainable decisions (Miccoli et al. 2015b). To this end, the action taken is based on transparent information, consultation, dialogue and negotiation, in a bid to ensure the community's participation in a decision-making process the aim of which is to make a shared choice based on consensus and full legitimacy. Participatory processes, therefore, imply a democratic value as they ensure that all the stakeholders even when their interests differ have the right to express their opinion and the possibility to influence and determine a choice. Another aspect of the utmost importance is a strategic value because public choice, which is founded on the democratic, transparent and rational principles of inclusion, is shared consensual and less conflicting, and has more chance of being successful.

An inclusive process can be developed through a number of different techniques, which can be classified according to the share of decision-making power that is transferred from the institutional decision maker to the participants. Some techniques simply disseminate information among the public, some involve the community in the consultations, some seek the cooperation of all the stakeholders to define possible scenarios and identify solutions, while others leave the decision-making entirely to the citizens.

In the most updated processes, the directly deliberative procedures based on assemblies, groups or citizen juries are becoming more common.

Although born in the '70 s, it was only in the '90 s that these procedures began to be widely used first in Germany and then in England, the USA and Australia. The various deliberative techniques developed over the years include the Plannungszelle (Dienel, 2002), Ned Crosby's Citizen Jury (Smith & Wales, 2000), the Opinion Deliberative Polls (Fishkin, 1991) and the Consensus Conference (Andersen & Jaeger 1999). The outcome of all these techniques is a preference based on aware and shared informed choices, obtained after all the stakeholders have stated and discussed their points of view, and possibly changed their opinions after discussing them (Miccoli et al. 2014b). Citizens are no longer the receiving end of the decision-making process but are full protagonists in the underlying debate (Miccoli et al. 2014a).

3.1 Technique and method for inclusive social evaluation

The evaluation technique we propose is divided into the following phases (Fig. 1):

– Defining the key elements on which the evaluation is to be based;
– The first participatory evaluation step with the community;
– Constructing the input matrix according to the criteria and sub-criteria identified through the community's evaluation;

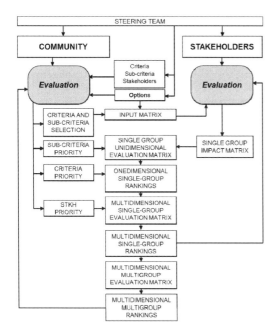

Figure 1.   Phases of the evaluation procedure.

– The second participatory evaluation step with the stakeholders and the construction of multiple impact matrixes;
– Defining partial rankings of the single stakeholder categories—according to the impact matrixes (opinions stated by the stakeholders) and of the priority vectors (opinions stated by the community)—and an overall multi-stakeholder ranking.

In the evaluation procedure, the stakeholders do not represent the entire community. The stakeholders are subjects, at times of a collective nature, (current and future, real and potential) who are directly or indirectly interested in an infrastructure project; they are asked to evaluate the alternatives according to their personal interest in it. On the other hand, the community is represented by all the subjects that share a territory, but unlike the stakeholders may have no interest whatsoever in the project, although they have the right to provide guidance to safeguard common interests. After consulting the community by means of a random representative sample, the scenario of the evaluation is obtained and it is the synthesis of the possible points of view present in a given territory.

During the initial phase of the procedure we propose, the course of actions and the elements to prepare the ground for the actual evaluation are established through a) identifying all the alternatives to be evaluated; b) suggesting the set of criteria and sub-criteria for the evaluation of the alternatives; c) specifying the categories of stakeholders interested in the infrastructure project, defined according to the characteristics of the alternatives to be evaluated and whether they impact at local, urban or regional level.

In the first step of the participatory evaluation, the community states its point of view through a deliberative approach by selecting among the criteria and sub-criteria submitted and adding others (where necessary), specifying their priorities and establishing the importance to attach to each stakeholder category. Once the criteria and sub-criteria to be used for the evaluation have been defined, it is possible to move on to the construction of the input matrixes by establishing for each sub-criterion the relative indicators (qualitative and/or quantitative) and to measure them.

The initial phase of the second step of the participatory evaluation consists in identifying the stakeholder categories present on the territory the members of which are called on to evaluate the matrix inputs by stating whether, in their opinion and according to their particular interests, the alternative would be capable of satisfying the sub-criterion/criterion taken into consideration. Once this phase has been completed, it is possible to define multiple impact matrixes for each stakeholder category according to the points of view stated. Lastly, by taking into account the outcomes of the second step and the evaluation of the community (priority vectors for the criteria, sub-criteria and stakeholder categories) different evaluation matrixes for each stakeholder category are formed.

By applying a two-tier multi-criteria based evaluation method (Miccoli et al. 2013), it is possible to presume multiple partial priority rankings of the alternatives related to single stakeholder categories. From a purely operational standpoint, the evaluation procedure is subdivided into two phases:

– the first consists of a series of unidimensional evaluations performed according to the sub-criteria, in accordance with the general criteria (based on the impact matrixes and priority vectors among the sub-criteria), which makes it possible to obtain a mono-dimensional ranking of the alternatives for each stakeholder category;
– the second, provides the multidimensional evaluation of the general criteria for each stakeholder category, by taking into account the results obtained from the unidimensional evaluations and from the priority vectors among the general criteria (indicated by the community), as well as defining the multidimensional rankings of the alternatives for each stakeholder.

The rankings obtained in this way make it possible to clarify the extent of conflict between the various stakeholder categories and can be a starting point for further debate and possibly rethinking the alternatives under examination with the aim of minimising conflict on one hand and harmonising the various interests at stake in the infrastructure project on the other. At a later stage, it is possible to define a multi-stakeholder evaluation matrix, based on the preferences stated by the single categories (indicated by the partial rankings), and appropriately weighed by the community (priority vector of the stakeholder categories). By applying once again the two-tier evaluation method, an overall multi-stakeholder ranking of the alternatives is obtained. To simplify the community's evaluation, ordinal scales for the measurement of priorities have to be used. Therefore, during the final evaluation, techniques that make it possible to employ the weights of the criteria expressed in the form of order of priority will be preferred.

3.2 The community's evaluation

The first participatory step is an evaluation based on the approaches of Deliberative Democracy. The underlying principle is that of presenting a problem to a group of citizens who, after listening to information provided by experts and the often-contrasting viewpoints of the various actors involved, suggest a common shared position. One of the many techniques developed over the years is Deliberative Opinion Polls, an idea pioneered by James Fishkin. In this technique, the community's point of view regarding a particular problem is stated only following a debate between a group of citizens chosen by lot and a group of experts on the issue under examination. These new forms of surveys differ greatly from the traditional ones as the latter do not gather opinions but something else, because people often have no clear idea and respond without being informed. On the other hand, the result of a Deliberative Opinion Poll is an aware and informed choice obtained "through debate, stating opinions and the possibility to change them after discussing and comparing them with those of others" (Fishkin et al. 2014).

In the evaluation procedure we propose, the community is asked to select and list the various factors by a) prioritising the general criteria proposed and, if necessary, single out new criteria; b) prioritising the sub-criteria proposed divided by the general criteria that can also be integrated; c) prioritising the categories of stakeholders involved in the evaluation process. The outcomes will be used to evaluate the impact matrixes defined during the stakeholders' evaluation. The preparatory phase of this participatory step consists in deciding on a work schedule for the duration of the procedure and forming a Steering Team whose task is to supervise the single discussion groups, ensure that everybody states his or her opinion, and provide the groups with unbiased information. In this phase, the individuals who interact with the community (experts and stakeholders) are identified. The experts are people capable of clearing up any doubts the participants may have regarding the criteria and sub-criteria under discussion; the selection of the stakeholders depends on their presence and activity on the territory, on the characteristics of the alternatives to evaluate and at which level they affect the territory. Once the preparatory phase has been completed, the participants who will form the sample are selected by means of the most suitable random sampling techniques. The techniques chosen must take into account the size and socio-demographic characteristics of the target population, to ensure the sample is fully representative of the communities involved in the project.

Schedules for two kinds of meetings are drawn up—group sessions during which the participants compare their ideas, and plenary sessions when the participants address the experts and stakeholders. At the opening of the first session, the experts explain the purpose of the evaluation, the main regulations for each daily session and the project to be evaluated. At the end of the introduction, the participants are asked to fill in a questionnaire; their answers will be based on their knowledge of the project. The participants are also asked to fill in a questionnaire at the end of the evaluation process after which the outcomes are compared and the participants' preference shifts at the end of the deliberative process—i.e. after exchanging information and opinions with each other—are analysed.

The community's evaluation is carried out in a number of sessions, one for each priority that has to be defined. Each single session is based on a model session and comprises two group meetings and two plenary meetings. A coordinator is chosen for each group with the task of guiding and moderating the debate to ensure participants total freedom of expression and opinion, to encourage everyone to state their opinion and to prevent the more active participants from monopolizing the debate. During the first sessions, the participants have to agree on and draw up a list of questions to put to the experts and the issues to discuss with them; during the subsequent plenary session the experts and stakeholders will clear up any doubts and answer the questions that arose from the group meetings. A new group meeting during which the answers provided are discussed, and a conclusive plenary session will bring the procedure to an end. At the end of each Model session, the participants fill in a questionnaire that enables them to formulate new criteria and sub-criteria and to prioritise them. At the end of this first participatory evaluation step, the opinions of the people called upon to represent the community are no longer based on superficial notions but on aware informed knowledge stemming from the debates, exchange of information and a deliberative pathway.

3.3 The stakeholders' evaluation

Before proceeding to the second participatory evaluation step, the Steering Team identifies the stakeholders. It is crucial for the territory and the project to be attentively examined before identifying the participants, and to do everything possible to ensure that everybody involved has the opportunity to state their opinion and that nobody is excluded a priori (Abelson et al. 2003). The preparation of informative material for the stakeholders involved is also indispensable. A wide range of existing techniques for public consultation and participation can be used to carry out the evaluation (questionnaires, interviews, market research, mobile territorial observatories etc.). For our case, we propose the use of three different methods a) Individual Evaluation; b) Compared Evaluation; c) Co-valuation. These methods produce impact matrixes directly assessed by the stakeholders;

by combining these matrixes with the priorities expressed during the community's evaluation the final step of the proposed procedure can take place. The choice of the method depends on the type of project, the specific characteristics of the territory, to what extent the participation of the possible stakeholders is expected to be dynamic, and on the relationship between the various individuals.

Individual Evaluation: With this method, the stakeholders do not interact during the evaluation process. They are contacted directly and provided with informative material, which they have to complete and hand in with their evaluation. This procedure requires a well-organized information and communication campaign on the territory for the entire duration of the evaluation process. The coordinators of this procedure set up an information point where the stakeholders can pick up the material, ask for information and clear up any doubts, hand in the material once filled in, and receive assistance regarding the evaluation itself. The duration of the evaluation depends on whether the coordinators of the procedure or the Public Administration involved have spread correct information to all the stakeholders, and on the reaction and response of the stakeholders themselves. The outcome is an evaluation on the part of the stakeholders who can state their opinion without having to discuss their point of view or compare it with those of others.

Compared Evaluation: Initially, this method foresees an individual evaluation as described above, followed by a meeting-assembly to which all the stakeholders who have performed the evaluation are invited. During the first phase of this public event, the outcomes of the first evaluation and the tendencies of each stakeholder category are stated. Subsequently, everybody in turn has the possibility to explain his/her intentions and the reasons behind his/her preferences. At the end of the assembly (or assemblies depending on the number of people involved), the stakeholders can rethink or modify their original evaluation—stated in the first phase—and another session can be called if deemed necessary. This method gives the stakeholders the possibility to compare, and possibly modify their original evaluation, after listening to the evaluation of others. This type of procedure can be repeated until the most widely shared position possible is achieved.

Co-valuation: Lastly, this method makes it possible to obtain a negotiated weighted evaluation from the dialogic interaction between the various stakeholders. One, or if necessary a number of meetings based on the fundamental steps of participatory techniques are organized. During these meetings, the participants are asked to evaluate the matter under examination and to discuss their evaluation with the other stakeholders. Once all the possible stakeholders have been identified, they are asked to participate in a workshop. In this phase, they acquire specific knowledge of the matter under examination, and they state their interest and compare it with and relate it to that of the other participants. During the workshop, alternating sessions with the participation of homogeneous stakeholder groups are held, as well as plenary sessions during which the evaluations of each group are stated, debated and discussed.

The work pattern is similar to that used for EASWs (European Awareness Scenario Workshops) (Andersen & Jaeger 1999) but in the procedure we propose the stakeholders work in homogenous (single category) and not heterogeneous groups; in this way the evaluations and opinions are always representative of a specific category and its interests. After a brief introduction and illustration of the criteria and sub-criteria, the group sessions during which the stakeholders formulate their evaluation are held. Once this phase is completed, two plenary sessions are held during which each group briefly illustrates their evaluation and the positions stated by the various categories are discussed. At the end of the debate, the participants are once again divided into groups so they can re-formulate their evaluation based on the positions stated by the other stakeholder categories. During the conclusive plenary session, the participants decide if the outcomes are exhaustive evaluations shared by every single participant, or if a new debate and a new evaluation are called for. This iterative process makes it possible to obtain evaluations that are the result of negotiated and shared interests.

4 CONCLUSIONS

The relatively recent updating of regulations regarding the infrastructure sector and contracts for public works prompted by the EU provides the opportunity to introduce cutting-edge techniques in the field of evaluation for public choice. These techniques are capable of indirectly generating innovation both in institutional administrative management and the enhancement of people's citizenship. The underlying consideration that supports the methodology we propose in this paper is oriented in this direction. We are aware that today the infrastructure sector is deemed to be closely linked to the dynamics of national economy. This belief is substantiated firstly by the fact that infrastructural projects considered strategic for the development and modernisation of the nation have been included in Italy's national economic programming, and secondly by the fact that along with the

EFD, the IA is published annually. This demonstrates the decisive role played by infrastructures in driving Italy's economic growth, but also the urgent need to realign the country's infrastructural conditions with the more advanced European standards. In fact, over the last few decades, the infrastructure gap between Italy and the rest of Europe has gradually widened with considerable repercussions on the overall performance of the country's system. Bridging this gap by programmatically focusing on priorities in infrastructure investments implies a pressing need to correlate their creation with the principles of sustainable development considering the dimensions and effects of works of this kind. Above all, however, citizens' interpretation of the impact these projects have on the environment, the economy and on the quality of social life needs to be taken into consideration.

In order to ascertain if conditions of this kind are included in the implementation plan, a preliminary evaluation of the feasibility of these projects with regard to established prerequisites of sustainability must be carried out. This is necessary because of both the multifaceted, heterogeneous and often conflicting nature of the interventions, and the expectations of those participating in the decision-making who for different reasons are interested in their implementation and who would not hesitate to protest and state their opposition if they were not consulted. It was partly due to this last consideration that a decision was taken to include the practice of Public Debate in the NPC. Therefore, the evaluation proposed in this paper is complex as it is necessarily multidimensional, social and inclusive given the problems associated to the sphere in which it is used.

Within the framework of public evaluation, those of a social and inclusive nature are radically different from ordinary institutional evaluations because of the perspective in which they are conducted. They are likely to become an important element of democratic life over the next few years as they could become the basis to forge collective choices and mould new, different, non-standardized scenarios of public interest. The in-depth, multifaceted nature of this kind of evaluation, and the freedom with which it is conducted, definable as civic evaluation, extols all forms of pluralistic democracy as it enables the members of a community to give voice to their fears, needs, and desires that otherwise would have remained unknown. It enables them to compare differing and dissonant convictions, opinions that were latent or potential until revealed. It makes it possible to achieve adhesion and consensus, and lastly, to reach a shared view of alternative solutions to collective problems. What becomes important from a political and cultural viewpoint is that a civic evaluation helps to bridge the increasingly wider and more frequent gap between the preferences stated by citizens and the decisions taken by their institutional representatives. It becomes the tool to implement the social, cultural and economic changes that are taking place, because if the self-determination of citizens is disregarded, the very foundations of liberty and democracy risk being distorted.

Many factors in decision-making methods may hamper the general use of inclusive evaluation techniques. They include the pride and arrogance of certain politicians and bureaucrats who are inclined to disparage the popular will expressed by majority votes; the fact that more time and money is required for deliberative evaluations; the reluctance to launch an experimentation program of these techniques, a precondition for their efficient and habitual application. However, most of these obstacles could be removed simply by increasing social culture on the most advanced forms of democracy.

REFERENCES

Abelson, J. et al., 2003. Deliberations about deliberative methods: issues in the design and evaluation of public participation processes. *Social Science & Medicine*, 57(2), pp. 239–251.

Andersen, I.-E. & Jaeger, B., 1999. Scenario workshops and consensus conferences: towards more democratic decision-making. *Science and Public Policy*, 26(5), pp. 331–340.

Dienel, P.C., 2002. *Die Planungszelle: der Bürger als Chance*, Wiesbaden: Westdeutscher Verlag.

Fishkin, J.S. et al., 2014. Deliberating across Deep Divides. *Political Studies*, 62(1), pp. 116–135.

Fishkin, J.S., 1991. Democracy and Deliberation: New Directions for Democratic Reform. *Canadian Journal of Political Science*, 26(3), pp. 596–597.

Miccoli, S., Finucci, F. & Murro, R., 2015a. A direct deliberative evaluation method to choose a project for Via Giulia, Rome. *Pollack Periodica*, 10(1), pp. 143–153.

Miccoli, S., Finucci, F. & Murro, R., 2014a. A Monetary Measure of Inclusive Goods: The Concept of Deliberative Appraisal in the Context of Urban Agriculture. *Sustainability*, 6(12), pp. 9007–9026.

Miccoli, S., Finucci, F. & Murro, R., 2013. La valutazione partecipata per la valorizzazione del paesaggio. Evaluation-Sharing in Landscape Enhancement. In M. Crescimanno, L. Casini, & A. Galati, eds. *Evoluzione dei valori fondiari e politiche agricole*. Bologna: MEDIMOND S R L, pp. 149–157.

Miccoli, S., Finucci, F. & Murro, R., 2015b. Measuring Shared Social Appreciation of Community Goods: An Experiment for the East Elevated Expressway of Rome. *Sustainability*, 7(12), pp. 15194–15218.

Miccoli, S., Finucci, F. & Murro, R., 2014b. Social Evaluation Approaches in Landscape Projects. *Sustainability*, 6(11), pp. 7906–7920.

Smith, G. & Wales, C., 2000. Citizens' Juries and Deliberative Democracy. *Political Studies*, 48, pp. 51–65.

Transport Infrastructure and Systems – Dell'Acqua & Wegman (Eds)
© 2017 Taylor & Francis Group, London, ISBN 978-1-138-03009-1

Automatic calibration of microscopic simulation models for the analysis of urban intersections

Sara Oliveira & Luís Vasconcelos
CITTA and Department of Civil Engineering, Polytechnic Institute of Viseu, Campus de Repeses, Viseu, Portugal

Ana Bastos Silva
CITTA and Department of Civil Engineering, University of Coimbra, Polo II, Coimbra, Portugal

ABSTRACT: Microscopic simulation models are being increasingly used in traffic engineering applications, but various issues concerning the extent to which its outputs reproduce field data still need to be addressed. In this perspective a proper calibration of the model parameters has to be performed so as to obtain a close match between the simulated and the actual traffic measurements. This paper aims to highlight the importance of calibration process, as the adjustment stage of the microsimulation models' parameters, applied to the analysis of urban, at-level, intersections. After the selection of the case studies (one roundabout and one intersection with traffic signal control), field observations were made, allowing the creation of a database that supported both the models' development and calibration. The next phase focused on the application of an Aimsun microscopic simulation model to the selected intersections. This involved the development of an optimization based calibration methodology, coupled with a sensitivity analysis. The optimization framework was implemented in Matlab using the pre-defined genetic algorithm in the optimization extension. With the models properly calibrated and validated, the performance indicators were obtained and conclusions about their approximation to reality were drawn. The recommended calibration methodology easily allows the replication of the observed conditions, revealing however poor adaptation to other intersections and a general lack of representativeness. The results obtained by the genetic algorithm are very sensitive to small changes in the initial set of parameters. The calibration methodology allows the replication of the observed conditions, revealing however poor adaptation to other intersections and a general lack of representativeness.

Keywords: Microsimulation; Roundabout; Signalized Intersections; Performance indicators; Capacity; Aimsun; Sensitivity analysis; Genetic algorithm

1 INTRODUCTION

Microscopic simulation models emulate realistically the traffic flow on a road network, which makes them appropriate tools for evaluating complex traffic facilities and control strategies. Simulation models include components related to the infrastructure, the demand, and associated behavioural models, which have complex data requirements and numerous model parameters. Calibration, as the adjustment of model parameters such that the model's output closely represents field conditions, is vital for appropriate decision making process.

A particular problem is the accurate modelling and calibration of unsignalized intersections, including roundabouts or traffic light controlled intersections. This is not a trivial task since several behavioural microscopic models are involved, such as car-following, gap-acceptance and lane-changing. In the literature several studies can be found addressing this topic but most of them are focused in uninterrupted traffic, namely in freeways (e.g. Punzo e Tripodi, 2007; Rakha et al. 2007; Vasconcelos et al. 2014b).

1.1 *Objectives*

The main objective of this research is to develop a sensitivity analysis to identify the role of several parameters in the simulation of urban intersections, and then develop and apply an automated calibration procedure to estimate optimal values for the most relevant parameters. The Aimsun microsimulation package (Casas et al. 2010) was used in this application, however the methodology is transversal to similar simulation models.

1.2 *Procedure*

This research followed the following analysis procedure:

1. Selection of the urban at-level intersections and data collection based on field observations;
2. Selection of the most relevant parameters through a previous selection, followed by a sensitivity analysis, involving a comprehension of the Aimsun sub-models, such as car-following, gap-acceptance and lane-changing;
3. Calibration of the selected intersections models using an optimization process by applying a Genetic Algorithm (GA), selection of performance indicators and analysis of the results.

2 DATA COLLECTION

Two intersections were studied: a roundabout, located in Viseu, and a signalized intersection, located in Coimbra, both in Portugal. The two-lane west entry of the roundabout, "Nelas Roundabout" (Fig. 1) was selected for the study due to the long queues that can be observed during the peak periods. In the off-peak, the traffic is fluid with very few stops. This roundabout has an inscribed circle diameter of 57 m and two circulatory lanes.

The traffic light approach selected for this study is the east entry to the "Casa do Sal" junction, a roundabout that is now operating with traffic lights (Fig. 2). The approach has three lanes and operates under heavy demand levels during most part of the morning and evening.

The data collection was based on video recordings, and the data treatment was supported by the use of LUT|VP3 software (Vasconcelos et al. 2013). The application log contains a list of records, each corresponding to a vehicle, with the corresponding time stamp and a code that indicates the respective lane. In the roundabouts the vehicles were observed during 121 minutes in three sections: approaching

Figure 1. Nelas roundabout (and the selected west entry).

Figure 2. Casa do Sal, signalized roundabout (east entry).

Table 1. Summary of the vehicle counts at the roundabout (121 minutes).

Entry lane		Circulatory lane	
Left	Right	Inner	Outer
683	1290	1274	534

Table 2. Summary of the vehicle counts at the signalized intersection (120 minutes, 28 cycles, green time: 33 s, yellow: 3 s).

Approach lane				
Period	Left	Middle	Right	Total
Green	404	400	388	1192
Yellow/Red	50	42	47	139
Total	454	442	435	1331

vehicles (at approximately 450 m before the entry), at the entry (yield line) and in the circulatory ring, in front of the entry. In the signalized intersection, vehicles were recorded, cycle by cycle, also during 120 minutes, as they passed the stop line during the green, yellow and red periods. A summary of vehicle counts is presented in the tables.

3 SENSIVITY ANALYSIS

A sensitivity analysis was performed with three main objectives: a) to provide a better insight about the influence of the different parameters in the simulation results; b) to facilitate the selection of parameters to calibrate; c) to reduce the range of values to use in the optimization procedure.

In a first approach, and based on previous studies (Vasconcelos et al. 2014a; Vasconcelos et al. 2014b), eight parameters were selected. Five of these parameters are defined at the vehicle type level assuming a normal truncated distribution of values. This way, the user must provide, for each parameter, the mean, the standard variation, the minimum and maximum values. These parameters are (Aimsun, 2012):

- Maximum acceleration, a (m/s^2) – is the maximum acceleration a vehicle can achieve under any circumstances. This parameter is required by the Gipps car-following model but is also used by the gap-acceptance and lane-change models. A high value means that a vehicle can achieve the desired speed faster and is able to change lanes and cross traffic streams using short intervals, thus resulting in a higher throughput;
- Normal deceleration, b (m/s^2) – is the maximum deceleration that a vehicle can use under normal driving conditions;
- Minimum spacing between vehicles, s (m) – is the distance that a vehicle keeps between itself and the preceding vehicle when stopped. A vehicle with a short spacing can seize short intervals and contributes to reduce the queue lengths;
- Give way time, gwt (s) – this parameter is used when a vehicle is in a give-way situation (e.g.: lane changing or yield/stop sign in a junction). When a vehicle has been at a standstill for more than gwt (in seconds), it will reduce the acceptance margins;
- Speed acceptance, SA – the speed acceptance parameter can be interpreted as the acceptance level of maximum legal speed section (v_{max}), i.e. the desired speed the driver will be $v_{max} \cdot SA$, which is achieved if the geometric and operating conditions permit.

The remaining three parameters are usually defined at the experiment level (same value for all vehicles) and reflect the reaction time of drivers under three different situations:

- Reaction time, rt (s) – is the time it takes a driver to react to speed changes in the preceding vehicle;
- Reaction time at stop, rts (s) – is the time it takes for a stopped vehicle to react to the acceleration of the vehicle in front;
- Reaction time at traffic light, $rttl$ (s) – is the time it takes for the first vehicle stopped after a traffic light to react to the traffic light changing to green.

The sensitivity analysis is based on two simple models (Figs. 3–4) that reflect the base conditions of the case studies: a roundabout and a traffic light junction. To get a better understanding of the way each parameter influences the simulation results, the driver-vehicle population was assumed uniform, only with cars. In both cases a very high demand value at the approach lane was set (5000 veh./h) to force a permanent queue at the entry.

Figure 3.   Roundabout simplified model.

Figure 4.   Traffic light approach simplified model.

The main characteristics of each model are, concerning the roundabout: single lane approach to a roundabout with only a circulatory lane with priority flow in the circulatory lane: 500 veh./h. For the traffic light approach the characteristics are: one single lane with a cycle of 90 s, 30 s of green and 3 s of yellow.

The next step consisted of evaluating the sensitivity of all parameters for the two models. The intersection capacities were measured for different values of each parameter, maintaining the default value on the remaining ones. The results are presented as the ratio between the current capacity and the reference capacity (obtained when all parameters take the default value) – Q/Q_{ref}.

3.1   *Roundabout*

With all the default values, the roundabout approach has a capacity of 948 veh./h. The ratio between the actual and the reference capacity for different parameter values is shown in Figure 5. The graphs also indicate the default value for the cars (circle marker) and truck (box marker). The following conclusions can be drawn:

- The maximum acceleration, the minimum spacing and the reaction time have a considerable effect on the capacity in the whole range of tested values;
- The effect of the normal deceleration and giveway time is very low when its value is near the Aimsun defaults (cars and truck) but is

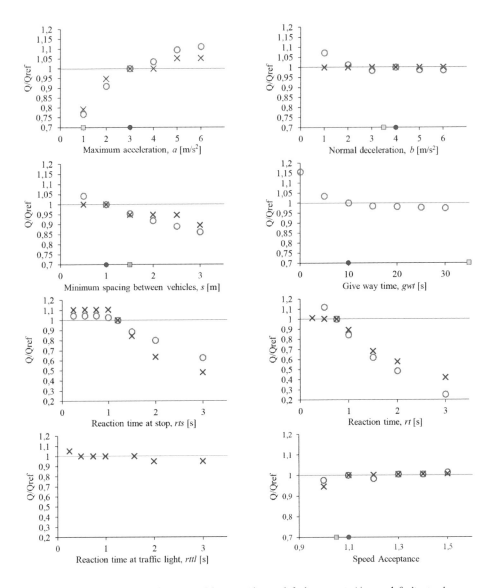

× Traffic light approach   ○ Roundabout   ● Aimsun default – cars   ▪ Aimsun default – trucks

Figure 5.   Sensitivity analysis for the traffic light approach and roundabout.

significant for values bellow 2 m/s² (deceleration) and 10 s (give-way time);
– The reaction time at stop only has a significant effect when its value exceeds the reaction time;
– For the simulated conflicting flow (500 veh./h), it was not possible to clarify the influence of parameter speed acceptance due to the existence of continuous queue at the entrance that prevents drivers from reaching its desired speed, and there is therefore one significant variability in the graph corresponds.

3.2 *Traffic light approach*

With all the default values, the traffic light approach has a capacity of 751 veh./h. The ratio between the actual and the reference capacity for different parameter values is presented in Figure 5, from where the following conclusions can be drawn:

– The maximum acceleration has a very significant effect on the capacity, particularly for values bellow 4 m/s²;

- Both the normal deceleration and the give-way time do not have any effect on the capacity, whatever the value tested;
- Increasing the minimum distance between vehicles tends to decrease the capacity;
- It was not possible to classify the influence of parameter speed acceptance factor due to the existence of continuous queue at the entrance that prevents drivers from reaching its desired speed;
- The capacity decreases for higher values of the reaction time and reaction time at stop;
- The reaction time at traffic light has a very subtle effect on the capacity.

Considering the modest role of the normal deceleration and the reaction time at traffic light in the model, it was decided to exclude these parameters from the automated calibration process.

Likewise, the give way time parameter was used only in the roundabout model.

4 CALIBRATION PROCEDURE

The automated calibration procedure takes the form of an optimization problem based on a Genetic Algorithm (GA). This is a widely used technique that employs ideas from natural evolution to find good solutions for combinational parametric optimization problems.

Building on the results of the sensitivity analysis, the parameter calibration problem was formulated in the following optimization framework:

$$
\min_{a,s,GWT,RT,RTSF,SA} f\left(M_{obs}, M_{sim}\right)
$$
$$
\begin{aligned}
\text{s.a.} \quad & 1 < a < 5\,m/s^2, \\
& 0,5 < s < 5m, \\
& 0 < GWT < 30s, \\
& 0,5 < RT < 1,5s, \\
& 1 < RTSF < 3, \\
& 0,9 < SA < 1,7
\end{aligned}
\tag{1}
$$

where $rtsf$ is the ratio between the reaction time at stop and the reaction time ($rtsf = rts/rt$) and f is the goodness of fit function that measures the distance between the observed and simulated measurements, M_{obs} and M_{sim}.

This function varies according to the intersection: in the roundabout, we were particularly interested in understanding how Aimsun was able to predict the variation of the density (and indirectly the queue lengths) during the simulation period and so the vector M was defined as the vehicle density between the upstream and downstream sections at the approach (veh./km), at each one minute observation—simulation period i. At the signalized intersection, and given that the permanent queues were

observed during the whole data collection period, it simply indicates the difference between the observed and simulated throughput (or entry capacity).

To measure the goodness of fit at the roundabout, as well as the signalized approach, the RMSE statistic was used. Its definition is:

$$
RMSE = \sqrt{\frac{\sum_{t=1}^{n}\left(M_{obs,t} - M_{sim,t}\right)^2}{n}}
\tag{2}
$$

The optimization framework was implemented in Matlab using the built-in genetic algorithm tool. The algorithm starts by generating an initial population (20 individuals, each of which corresponds to a set of 6 parameters in the roundabout and 4 parameters in the signalized junction).

For each individual, a Python script modifies the corresponding parameters in Aimsun, simulates the model in console mode and compares the observation and simulation outputs to compute the fitness value. The default Aimsun parameters that were subject to optimization (for the roundabout entry and traffic light approach) are presented on the following table.

When all individuals are evaluated, the GA generates a new population: besides elite children, who correspond to the individuals in the current generation with the best fitness values, the algorithm creates crossover children by selecting vector entries from a pair of individuals, and mutation children, by applying random changes to a single individual (MathWorks 2005).

After 73 generations and approximately 2 hours of computing time, the algorithm reached a convergence condition and returned the optimal solution for the roundabout analysis (see Fig. 6). The optimal parameters obtained are presented in the following table.

The effectiveness of the optimization method to fit the simulated time-series of densities to the observation is presented on Figure 7, which highlights the proximity between the density obtained in the field observations with the density simulated with the Aimsun default parameters and with the optimal parameters obtained.

It is noticeable that the difference between the default and optimal parameters has some discrepancy, revealing that the optimization process does not always returns values that can be considered as realistic; however the solution of all the optimal

Table 3. Default Aimsun parameters.

a (m/s^2)	s (m)	gwt (s)	rt (s)	rts	SA
3	1	10	0,75	1,20	1,1

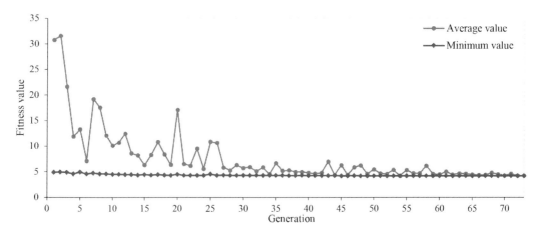

Figure 6. Convergence of the algorithm (roundabout entry analysis).

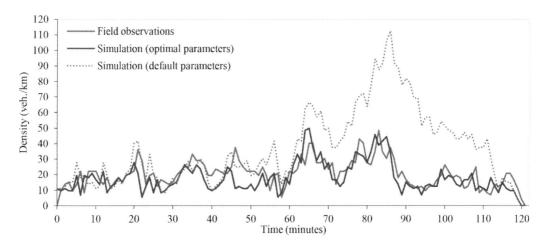

Figure 7. Time-series of densities (observations, simulation with default and optimal parameters) for the roundabout entry.

Table 4. Optimal solution for the roundabout.

a (m/s²)	s (m)	gwt (s)	rt (s)	$rtsf$	SA	RMSE
2,433	0,937	2,041	0,581	1,584	0,977	4,172

Table 5. Optimal solution for the signalized approach.

a (m/s²)	s (m)	rt (s)	$rtsf$	RMSE
2,340	0,940	0,994	1,669	0,008

parameters obtained combined leads to a fitness value that approximates in a very reasonable form the indicators in analysis (in the case of the roundabout: density).

In the signalized intersection, the objective was to understand if Aimsun was able to predict the intersection capacity.

Using the same optimization procedure, with the following indicated optimization framework, the parameters were obtained (Table 5).

$$\min_{a,s,RT,RTSF} f\left(\mathbf{M}_{obs}, \mathbf{M}_{sim}\right)$$
$$s.a. \quad 1 < a < 5 \; m/s^2,$$
$$0,5 < s < 5 \; m,$$
$$0,5 < RT < 1,5 \; s,$$
$$1 < RTSF < 3$$

(3)

The optimization resulted in a RMSE of 0,008, indicating that those parameters reproduced accurately the entry capacity. The optimal solution obtained is presented in the following table.

The value adjustment indicator shows a great fit between the field observation data values and those obtained after calibration. This is an analysis "simpler" than the one performed for the roundabout entrance involving a smaller number of parameters, where the approximation of the approach capacity, as the number of vehicles that entry the intersection per cycle can be considered excellent, again demonstrating the importance of the optimization process applied and validating its results. The value of the capacity obtained by optimization for the approach was 47,54 veh./cycle, coinciding with the resulting value of the field observations.

5 CONCLUSIONS

For the two cases studied, roundabout entry and traffic light approach, the calibration procedure based on optimization is validated and adequate to study specific urban intersections. The proximity of the calibrated simulated models to the observed data is significantly better than those without any parameter calibration, demonstrating the importance of the sensitivity analysis and posterior parameter calibration. However, the optimization may lead to unphysical parameter values and therefore it is recommended to define acceptable variation ranges for each parameter.

Simulation models must be carefully applied to the analysis of intersections, since its outputs are strongly related with the quality of inputs (e.g. traffic demand, the geometric design coding) and have high sensitivity to the values of the behavioural parameters adopted.

This way, future works should focus on the application of automated calibration procedures that can be applied to large networks and on the identification of simple methodologies that allow to quantify some of the most important parameters (eventually using instrumented vehicles) and to generalize the results to a population of drivers.

It would also be relevant to apply the optimization procedure at intersections with different layouts, confirming the applicability of the parameters for a particular type of intersection to other intersections of the same type.

REFERENCES

Aimsun, 2012. Aimsun Version 7 User's manual, TSS - Transport Simulation Systems, Barcelona, Spain.

Casas, J., Ferrer, J., Garcia, D., Perarnau, J., Torday, A., 2010. Traffic Simulation with Aimsun, in: Barceló, J. (Ed.), *Fundamentals of Traffic Simulation*, 145. Springer New York, pp. 173–232.

MathWorks, I., 2005. *Genetic Algorithm and Direct Search Toolbox for Use with MATLAB®: User's Guide*. MathWorks.

Punzo, V., Tripodi, A., 2007. Steady-State Solutions and Multiclass Calibration of Gipps Microscopic Traffic Flow Model. *Transportation Research Record: Journal of the Transportation Research Board*, 1999, pp. 104–114, doi:10.3141/1999-12.

Rakha, H., Pecker, C., Cybis, H., 2007. Calibration Procedure for Gipps Car-Following Model. *Transportation Research Record: Journal of the Transportation Research Board*, 1999, pp. 115–127, doi:10.3141/1999-13.

Vasconcelos, A.L.P., Seco, Á.J.M., Bastos Silva, A., 2013. A comparison of procedures to estimate critical headways at roundabouts. *PROMET—Traffic & Transportation*, 25(1), pp. 43–53.

Vasconcelos, L., Neto, L., Santos, S., Bastos Silva, A., Seco, Á., 2014a. Calibration of the Gipps car-following model using trajectory data. *Transportation Research Procedia*, 3, pp. 952–961, doi:10.1016/j.trpro.2014.10.075.

Vasconcelos, L., Seco, Á., Silva, A.B., 2014b. Hybrid Calibration of Microscopic Simulation Models, in: Sousa, J.F., Rossi, R. (Eds.), *Computer-based Modelling and Optimization in Transportation*, 262. Springer International Publishing, pp. 307–320.

Transport Infrastructure and Systems – Dell'Acqua & Wegman (Eds)
© *2017 Taylor & Francis Group, London, ISBN 978-1-138-03009-1*

On-street parking management: Bridging the gap between theory and practice

C. Piccioni, M. Valtorta & A. Musso
DICEA—Department of Civil, Building and Environmental Engineering, "Sapienza" University of Rome, Rome, Italy

ABSTRACT: More and more cities in Europe are adopting parking schemes targeted at improving urban accessibility, without increasing car dependency and the related negative externalities. However, possible links between such parking systems and contexts within they work are generally not investigated. This paper summarizes the findings of a study aimed at exploring, from a quantitative viewpoint, the gap between theory and practice also showing that, at practical level too, there is not yet a common understanding in selecting specific parking measures. Starting from an in-depth analysis of parking policies implemented by a sample of European cities, a set of indicators, specially defined for this analysis, were calculated. The statistical variability of the related independent variables was also examined, in order to find probable cause-effect links between local context features and selected measures as well as to provide a first methodological approach to explore the effectiveness of the applied parking strategies.

1   INTRODUCTION

Parking in urban environments is increasingly becoming a crucial issue. Cities have to cope with high rates of parking demand but the on-street parking provision appears very often inadequate. Such a supply likely results as a component of built environments designed many years ago, at a time when ownership and use of cars had not yet reached present values and parking was not considered a land-consuming activity.

Awareness that land is a precious and scarce resource, especially in high-density environments, has gradually grown over time; thus the main static and dynamics activities featuring mobility patterns, and their impacts on city users, are being revised in light of this concept (Musso & Piccioni 2012). On-street parking really provides direct accessibility to services/activities but shortage of accessible car spaces or, even more, their inefficient use (i.e. low turnover) can negatively affect local business also decreasing the livability of urban setting. On-street parking affects street capacity and road users' safety; maneuvering vehicles as well as illegally parked cars can create disruptions to traffic circulation, by increasing pollutant emissions. On-street parking also reduce public transport supply and performances; it actually can occupy portions of carriageway that otherwise would be used by specific traffic components (e.g. reserved bus lanes, cycling or walking paths).

More and more cities in Europe are adopting parking schemes targeted at improving accessibility to urban spaces, while limiting car dependency and the related negative externalities. Nevertheless, the maturity level of what experienced is quite uneven and measures are sometime applied at very local level. Besides, possible links between parking schemes and contexts within they work are generally not recognized at all. Finally, there is no objective evidence of which measures are preferred with respect to other ones.

To the Authors' knowledge there are no applications explicitly interpreting, from a quantitative viewpoint, parking management strategies applied at European level. To this end, such a paper aims at contributing to the current debate, by providing a methodological approach through which investigate and critically discuss whether and to what extent the strategies put in place differ from the theoretical framework supporting parking management policies.

2   THEORY BEHIND PARKING MANAGEMENT: ACCESSIBILITY AND ROLE OF INDICATORS

2.1   *Framing parking management evolution*

Many studies and researches have developed different parking management approaches, over the last decades, in order to meet the changing dynamics affecting urban mobility and sustainability (Shoup 2005; Litman 2008, 2013; Bates & Leibling 2012, Guo et al. 2012). Such analyses allowing marking how the paradigm shift is progressing, by passing from a "predict and provide" traditional approach (Owens 1995; Vigar 2001) up to a more contemporary and comprehensive strategy focused on a set of management levers (Litman 2013, 2016).

The former approach can be thought as borrowed by the transport planning theories explaining the concept of induced traffic demand (Næss et al., 2014). By transferring such a concept to parking (i.e. induced parking demand), the predict and provide approach sets wider minimum requirements aimed at providing as much parking lots as possible for free, or with the lowest price, in order to offer parking options easily accessible at every destinations. According to the main reference literature (Acutt & Dodgson 1997, Shoup & Manville 2005, Chatman 2008) the availability of parking at destinations seems to be a key issue influencing the users' choice by increasing car usage frequency and mileage. A generous parking provision also affects land use planning in a long-term perspective by increasing risk to use valuable land improperly while encouraging urban sprawl (Litman 2008). The latter focuses on the improved efficiency of parking facilities, by applying a sort of "predict and prevent" approach (Owens 1995) without increasing minimum requirements. This implies to rethink planning of parking supply, with a special attention to the existing one, by adopting pricing and/or regulations, creating a user's information system as well as by improving overall walkable conditions in the parking lot catchment areas (Litman 2016b). Some evidences stress how synergy resulting by the application of a set of parking measures allows to achieve increasing benefits, because total impacts are greater than the sum of their single ones (Litman 2016a). Such a paradigm shift represents thus a sort of watershed: it changes the way to conceive the "parking problem" and, so, the way to search for more proper parking solutions.

2.2 Evaluating accessibility for parking systems

Several definitions can be traced to the accessibility concept. Among them, that proposed over time by Dalvi (1976) as "the ease with which any land-use activity can be reached from a location using a particular transport system", by Burns (1979) as "the freedom of individual to decide whether to participate or not in different activities" as well as by the U.S. Department of Environment (1996) as "the ease and convenience of access to spatially distributed opportunities with a choice of travel", all well suited to the purpose of the present study. Nevertheless, the quantitative assessment of such "ease and convenience" is quite complex as it is a function

of varying types of trips and activities and most likely varies across people according to their tastes and preferences (Dong et al. 2006). By focusing on private individual transport is almost immediate understanding the influence of parking on accessibility in urban areas. The ease with which a driver is able to reach a particular activity affects not only the actual travel time but also the time required to find an available car lot as well as the monetary cost associated with its use (also including the payment ways). Accordingly, the accessibility concept can be declined both in physical and economic terms.

Physical accessibility, as practical availability of parking spaces, directly affects probability of finding an available car spot as well as time spent for parking. In congested areas such time is likely related to the "cruising for parking" (Shoup, 1997) phenomenon, affecting those drivers who, once realized that parking is at capacity in the destination zone, continue to drive at low speed all the time required to locate an available spot. In the most critical cases, drivers leave the area without being able to park their cars and then, unable to carry out the activities for which they were moved.

Economic accessibility is directly linked to the parking fee and, therefore, to the monetary transport cost. Over the years, an increasing number of Municipalities is implementing parking policies, by using just price as a management lever for achieving a better equilibrium between parking supply and demand. A key target is to improve the supply efficiency, by increasing the related turnover index (i.e. the number of cars served by a single spot during the reference period, generally calculated on a daily basis), limiting at the same time the related negative externalities.

2.3 Building accessibility indicators

According to the purposes of this study, the indicator is intended as a simplified measure, usually expressed in quantitative form, composed by one or more variables, able to describe in a plausible and reproducible way the trend of a phenomenon or changes affecting the system to which this refers (Musso & Piccioni 2010). Scientific literature suggests several indicators performing accessibility measures, most of which can be traced to 2 macroclasses such as: (1) indicators based on the utility and (2) indicators based on opportunities. The former type of accessibility indicators, generally pertaining the gravity and random utility models. The latter type is used in isochronic measures, representing the cumulative count of opportunities reachable within a given travel time or distance (Cervero 2005) as well as in the opportunities perception measures (Cascetta et al. 2013). Such measures, however, generally return aggregated values not consistent with the purpose this research. Thus, specific *ex novo* indicators suitable for the study have been defined, in order to recognize which elements featuring the physical and/or economic accessibility can likely be related to parking management practices.

3 METHODOLOGY

3.1 *Methodological approach*

This research was structured according to a main pillar, consisting of a comparative analysis involving a cities sample, aimed at characterizing accessibility from a quantitative viewpoint. More properly, the methodological approach was developed by performing a set of subsequent macro-activities such as: (1) cities selection and data collection; (2) accessibility indicators design; and (3) statistical analysis of indicators.

Such steps paved the way for the further analysis of correlation between indicators thus allowing applying the Principal Components Analysis (PCA), through which to explore the gap between the theoretical approach and the practical one in implementing parking management strategies. Finally, a clustering process was performed in order to recognize cities having an attitude towards one or more specific parking strategies as well as atypical contexts.

3.2 *Data collection*

The study focused on 24 cities belonging to 13 European countries (Fig. 1). It is worth mentioning that the heterogeneity of the selected urban environments—in terms of spatial extent, urban form,

Figure 1. Framework of sample cities.

socio-economic and transport aspects—increases the descriptiveness of this analysis, which thus does not exhaust its usefulness to the only characterization of large cities.

An extensive technical review of databases and statistical information (Urban Audit by Eurostat, Knoema Dataset, Official websites of parking operators and Municipalities) were carried out, in order to reach relevant data on cities sample according to the physical (e.g. size and spatial distribution of on-street, off-street and P&R) and economic (e.g. parking rates, pricing approaches) parking supply dimensions. A preliminary outlook of collected data shows how on-street paid parking supply is extremely variable. Moreover, on-street supply is always higher by at least an order of magnitude compared to the off-street parking.

In central urban areas, parking price is presumably higher than that applied in peripheral zones, so special emphasis was given to the maximum hourly rate of on-street parking and its related indicators. The minimum on-street parking fees is between € 0.24/h (Middlesbrough) and 2€/h (Vienna), the maximum hourly rate ranges from 1.20€/h (Rome) to 5€/h (Amsterdam). Except for Middlesbrough covering the last place in the ranking, Rome applies the lower rate of the whole cities sample. Naples is the most expensive, the applied parking rate follows an incremental logic based both on zoning and occupancy time: 1 hour costs 2€ but from the second onwards, the unit price is increased by 25%. In France, e.g. in the central area of Lyon and Nantes, by increasing the occupancy time by 50%, the parking fee rises by almost 150%. The rationale of such pricing policy, widespread in many European cities, is aimed at increasing the turnover rate, thus improving the efficiency of the overall parking system. The economic dimension of the P&R supply generally refers to the cost incurred by regular users of Public Transport (PT). 61% of the sample cites encourage P&R use, offering the service for free; the maximum cost is recorded in Barcelona (daily price is about 10€/8 hours).

3.3 *Accessibility indicators*

22 ad hoc indicators were defined and classified according to dimension and typology of parking supply (on-street, off-street, P&R). Concerning the physical aspects, a set of descriptive indicators was designed in order to investigate overall supply in relation to the main features of the selected cities (Table 1).

With reference to the economic accessibility, the selected indicators allow featuring the parking offer according to the disposable income (Gross Domestic Products) and two cost items relating to the use of private car and PT (Table 2).

Table 1.	Physical parking supply indicators.		
Code	Indicator	Unit	Descrip-tiveness
PION* 1	Pay parking lots/1000 inhabitants	Lots/1000 inhabitants	Potential Parking demand
PION 2	Pay parking lots/100 registered cars	Lots/100 cars	Parking demand
PION 3	Pay parking area/Urban area	%	Spatial features
PIOF **1	Parking lots/1000 inhabitants	Lots/1000 inhabitants	Parking demand
PIOF 2	Parking lots/100 registered cars	Lots/100 cars	Parking demand
PIOF 3	Parking lots/Pay off-street parking lots	Lots/lots	On-street parking
PIPR***1	P&R lots/1000 inhabitants	Lots/1000 inhabitants	Parking demand
PIPR 2	P&R lots/100 registered cars	Lots/100 cars	Parking demand
PIPR 3	P&R lots/Urban rail transit	Lots/km	Urban rail system
PIPR 4	P&R lots/Pay on-street parking lots	Lots/lots	On-street parking

*PION Physical Indicator On-street,
**PIOF Physical Indicator Off-street,
*** PIPR Physical Indicator Park & Ride.

Existence of different parking fees in the same city has been also considered by calculating respectively EIOF4, EIPR4 as well as EION4. Such an indicator is also useful to detect the propensity of Municipalities to manage parking demand by using a sort of zoning approach consistent with the attractive potential of the different city areas. Summarizing, a high number of indicators was intentionally chosen to do not miss interesting information on the phenomenon for the purposes of the next PCA.

3.4 Statistical analysis of indicators

As far as physical accessibility indicators is concerned, PION2 returns an average value of 9 lots/100 cars, which equals 11 cars/lot. The maximum value (28 lots/100 cars) is recorded by Copenhagen followed by Bologna (18 lots/100 cars) and Pisa (12 lots/car). The 2 Italian cities have higher PION1: Bologna reports the value of 96 lots/1000 inhabitants followed by Pisa (75 lots/1000 inhabitants). PIOF3 highlights the attitude of Municipalities for the on-street parking type, thus shrinking costs

Table 2.	Economic parking supply indicators.	
Code	Indicator*	Descriptiveness
EION** 1	On-street max. hourly parking fee/GDP per capita	City disposal income
EION2	On-street max. hourly parking fee/PT ticket	PT cost
EION3	On-street max. hourly parking fee/Fuel price	Car use cost
EION4	On-street max. hourly parking fee/On-street min. hourly parking fee	Pricing zoning
EIOF***1	Off-street average hourly parking fee/GDP per capita	Parking demand
EIOF2	Off-street average hourly parking fee/PT ticket	PT cost
EIOF3	Off-street average hourly parking fee/Fuel price	Car use cost
EIOF4	On-street max. hourly parking fee/Off-street average hourly parking fee	On-street parking
EIPR****1	P&R average daily parking fee/GDP per capita	Context prosperity
EIPR2	P&R average daily parking fee/PT ticket	PT cost
EIPR3	P&R average daily parking fee/Fuel price	Car use cost
EIPR4	P&R average daily fee/On-street max. hourly parking fee	On-street parking

*Such indicators are dimensionless as computed by the ratio of two items having the same measurement unit (€),
**EION Economic Indicator On-street,
***EIOF Economic Indicator Off-street,
**** EIPR Economic Indicator Park & Ride.

for infrastructure, service and maintenance. Barcelona, Brescia, Lisbon and Zurich recorded a ratio higher than 1, showing a tendency to limit land consumption caused by parking.

In terms of economic accessibility, Naples, among all, is the city applying the most expensive fee compared to its disposable income. Rome and Zurich, respectively, in the last and second to last position, offer a very affordable rate compared to their real purchasing power. On the sample basis, the average parking hourly cost exceeds 1.6 times cost of a PT ticket. Birmingham, Leeds and Antwerp confirm to use the economic leverage to restrict car access to the city center, e.g. in Leeds the hourly rate is 4.6 times higher than 1 bus ticket cost. In Italy, Naples and Brescia offer the highest hourly rates, about 2 times the ticket cost. A total of 19 cities record a value higher than 1 (i.e. 1 hour of parking costs more than 1 bus/metro ticket); Vienna offers both services at the same cost; as opposed in Monaco (0.95), Rome (0.80) and Zurich (0.77) 1 hour is cheaper than a PT ticket. EION4 shows certain homogeneity between pricing policies implemented in British cities where there is a high rates difference (average value is close to 5). Besides, Copenhagen (9.7), Antwerp (7) and Amsterdam (5.5) show high rate variability. The off-street average rate records a cost equal to the on-street maximum rate; Zurich, Barcelona, Munich and Lyon report a value lower than 1. In cities where such data is available, off-street parking is competitive when compared to on-street parking. With regard to P&R rates, two types of cities were identified, depending on which they offer free or pay parking. 7 cities belong to the second group, among them Barcelona offers the most expensive fare (daily cost is about 5 times the PT ticket), even if a monthly subscription is available.

3.5 Data scattering and heterogeneity of parking strategies

In order to increase the descriptiveness of the accessibility indicators, the main statistical variables were calculated. As known, mean describes the sample through a single value, which is much more representative as the sample has low data scattering. In the study context, a low data dispersion indicates a good homogeneity level of parking management approaches among different cities. In contrast, high dispersion ratios show lack of common practices or shared approaches; in such a case, the sample mean does not lend itself to represent the sample. Thus, data scattering was estimated by the Coefficient of Variation (CV) it allowing comparing elements having different measurement units. According to the above accessibility indicators, the main statistical variables are summarized in Table 3 and Table 4.

The economic indicators (CV = 0.42) showed a data dispersion minor than that calculated for the physical ones (CV = 0.82).

Such result shows how there was a greater consistency in implementing similar pricing policies (pricing lever is a recurrent element) if compared to approach managing physical supply. The higher heterogeneity of physical/spatial-related variables values can be likely traced to:

– a greater ability in managing price-related variables. Pricing plans can be changed in short term;

Table 3. Physical accessibility: statistical variables.

Code	Sample	Mean	Std dev.	CV	Min	Max	Max/ Min
PION1	18	38.6	21.3	0.6	11.9	96.2	8.1
PION2	18	9.0	6.1	0.7	3.5	27.9	7.9
PION3	18	1.8×10^{-5}	73.2×10^{-5}	0.7	33.1×10^{-5}	29.9×10^{-5}	9.0
PIOF1	15	27.9	27.5	0.98	2.2	87.5	40.4
PIOF2	15	6.4	6.6	1.0	0.4	23.8	58.6
PIOF3	14	0.8	0.8	0.97	0.04	2.6	64.5
PIPR1	17	9.6	8.1	0.8	1.5	28.3	18.4
PIPR2	17	2.0	1.6	0.8	0.4	5.4	12.5
PIPR3	14	173.9	211.8	1.2	10.9	771.95	70.8
PIPR4	13	0.3	0.2	0.7	0.1	0.83	7.9

Table 4. Economic accessibility: statistical variables.

Code	Sample	Mean	Std dev.	CV	Min	Max	Max/ Min
EION1	24	0.00	0.00	0.38	0.00	0.00	4.45
EION2	23	1.58	0.57	0.36	0.77	2.73	3.56
EION3	24	1.59	0.53	0.33	0.67	2.75	4.11
EION4	24	3.39	2.02	0.60	1.00	9.73	9.73
EIOF1	15	0.00	0.00	0.41	0.00	0.00	3.97
EIOF2	15	1.22	0.60	0.50	0.53	3.14	5.97
EIOF3	15	1.20	0.50	0.42	0.56	2.12	3.79
EIOF4	15	1.48	0.61	0.41	0.68	2.72	4.00
EIPR1	18	0.55	1.02	1.87	0.00	3.98	*
EIPR2	18	0.72	1.33	1.84	0.00	4.98	*
EIPR3	18	0.95	1.90	1.99	0.00	7.03	*
EIPR4	18	0.55	1.02	1.87	0.00	3.98	*

*Denominator = 0.

vice versa a physical supply expansion, where allowed, requires a long-term perspective;
– reliability and accuracy of available data. Data on physical parking supply are not always collected and/or updated by single Municipalities;
– different approaches. In some cities, parking management is performed by ignoring possible changes occurred in the functional mix/attractive potential of certain zones. In doing so, parking supply remains static without looking for a balance between on-street and off-street proximity supply.

Among the economic indicators, those related to P&R recorded the highest data scattering; such inhomogeneity can be likely attributed to:

– diversity of the proposed parking rates (e.g. daily, monthly, zooning-related subscriptions, etc.);
– different function assigned to P&R (data includes both on/off-street lots or only P&R supply for road-rail interchange nodes);

– potential transport side effect. Really, P&R allows a reduction in the private cars use for trips to the central areas; however, recent empirical studies (Clayton et al. 2015, Noel 1988) stress how a wide and free P&R supply incentives car use for trips between the origin and the interchange node, by shifting congestion from the city center to peripheral areas. A perception of economic effects by Municipalities also plays an important role, because they may fear losing revenues from parking located in central areas through introduction of P&R (Dijk & Montalvo 2011).

It is appropriate to note that for the other economic indicators descriptive of on-street component, data are concentrated in relative small-scale ranges, showing a certain spread of recurring practices.

4 CONTEXTUALIZING PARKING MANAGEMENT PRACTICES: THE PROPOSED APPROACH

4.1 Principal Component Analysis

The proposed accessibility indicators provides a good characterization of parking management strategies, according to the features of the single cities; however, just as they are do not allow a clear and concise picture of the main European trends and attitudes. For this reason, also because of the correlation between some indicators, the Principal Components Analysis (PCA) was applied. This technique enables the explanation of new latent factors as a linear combination of the original variables, limiting loss of information and maximizing the variability explained through a limited number of components.

The application of the PCA has requested 2 preparatory steps. The former involved the missing data management, so for cities reporting only one missing data by parking type, an unconditional mean imputation was adopted; cities with two or more missing data were excluded from the sample for that specific parking category. The latter step concerned the standardization of collected data, needed for managing variables having different unit of measurements; to this end the following equation was used: $Z = (X - \mu)/\sigma$, where Z = standardized variable; X = original variable; μ = sample mean, σ = standard deviation.

PCA was applied to each of the 6 groups of indicators (PION, PIOF, PIPR, EION, EIOF, EIPR) and the components selection was made by using the "Kaiser criterion", then choosing only the components with eigenvalues greater than 1. This implies to consider only a Principal Component

(PC) in case this is capable to explain a greater proportion of the total variance of that explained by a single variable.

Accordingly, 6 aggregate indicators were identified, one for each group of indicators:

– on-street (CPON), off-street (CPOF) and P&R (CPPR) physical components;
– on-street (CEON), off-street (CEOF) and P&R (CEPR) economic components.

Table 5 summarizes for each component, its eigenvalue, the accounted variance percentage and the equation expressing the PC based on the original variables belonging to the respective group.

CPON, CPOF, CPPR express the attitude (i.e. supply-oriented approach) of Municipalities in facilitating physical accessibility to urban areas by private car. Vice versa CEON, CEOF, CEPR quantify their propensity (demand-oriented approach) in managing the potential demand by using pricing, thus reducing economic accessibility. The accounted variance for all components exceeds 60% of the total variance, the minimum value refers to the P&R physical component (61%) and the maximum one (99%) to the P&R economic component. So, almost the totality of the data variance is explained by a single component and the loading (equation coefficients) recorded the same value (0.50) for each of the four (secondary) indicators, thus contributing with the same weight to the component structure. Such a result underline that CEPR does not depend on the links between parking measures and the context variables, but likely on the political choice to provide P&R car spaces for free.

Table 5. PCA relevant outcomes.

Code	Eigen-value	Accounted variance	Equation
CPON	1.92	63%	CPON = 0,67PION1 + 0,52PION2 + 0,45PION3
CPOF	2.8	93%	CPOF = 0,59PIOF1 + 0,58PIOF2 + 0,56PIOF3
CPPR	2.37	61%	CPPR = 0,59PIPR1 + 0,59PIPR2 + 0,31 PIPR3 + 0,44 PIPR4
CEON	2.57	64%	CEON = 0,55EION1 + 0,42EION2 + 0,53EION3 + 0,43EION4
CEOF	2.5	62%	CEOF = 0,48EIOF1 + 0,42EIOF2 + 0,56EIOF3 − 0,52EIOF4
CEPR	3.97	99%	CEPR = 0,50RIPR1 + 0,50EIP2 + 0,50EIPR3 + 0,50EIPR4

The estimated values for the 6 main components show the propensity of Bologna, Barcelona and Brescia—recording the higher values for CPON, CPOF, CPPR—to ensure high physical accessibility to urban areas by private car through provision of on-street (Bologna), off-street (Barcelona) and P&R (Brescia) car spaces. On the contrary, Birmingham, by recording the highest value for CEON, denotes a high propensity to shrink on-street parking demand. CEOF and CEPR refer the higher values to Barcelona, where low economic accessibility (high fee) may reflect need to cover high investment/operating costs.

In order to identify possible trends at European scale, 5 levels (very low, low, medium, high and very high) related to Municipalities' propensity in implementing parking policies were defined (Fig. 2).

78% of cities indicate a low or very low propensity to improve on-street physical parking supply. This figure manifests a clear policy to reduce cars dependency by constraining physical parking accessibility to private vehicles. The on-street economic component showed a low (34%) and medium (29%) propensity to manage supply by using pricing for controlling demand; besides a fairly high percentage (21%) amounted to very high values. 66% of

the sample shows a low or very low propensity in supporting off-street physical supply, although compared to on-street type increases (+12%) cities showing a medium or high propensity. Pricing policies supporting off-street parking are quite heterogeneous: 47% of cities prefer to keep low economic accessibility; other 40% provides service at a relatively low price. With reference to the P&R physical supply, 47% of cities show a very low propensity, thus preferring to restrain availability of parking spaces; vice versa 35% shows an attitude to ensure a high level of physical accessibility. The majority of the sample (61%) also provides service for free by highlighting a policy encouraging use of P&R system that is a very low propensity to disregard P&R as part of the total urban parking supply.

4.2 Cluster Analysis

A Cluster Analysis (CA) was carried out, aimed at identifying groups of cities applying similar parking management practices and possible factors affecting the decision-making choice on parking measures. A hierarchical clustering aggregative methodology was applied, by measuring clusters similarity through the "complete-link proximity" function: $D(X;Y) = max_{x \in X, y \in Y} d(x, y)$, where $D(X;Y)$ = distance between cluster X and Y, $d(x, y)$ = distance between point x belonging to the X set and point y belonging to the Y set. So, this function calculates the Euclidean distance between two specific clusters as the maximum distance between two elements belonging to them.

Dendogram of Figure 3 shows outcomes of CA; by choosing a similarity threshold value of 66.7, suitable for achieving a representative and significant amount of homogeneous groups, 6 clusters were identified.

First cluster includes Naples, Antwerp and Edinburgh that, although different in demographic, economic and transport terms, are all

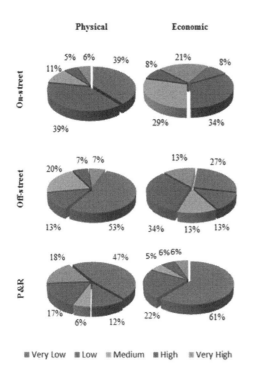

Figure 2. Parking management attitude of the sample cities.

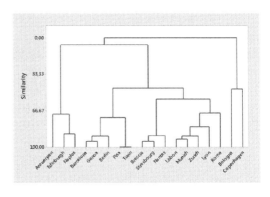

Figure 3. City clustering.

characterized by a low propensity to meet demand (low physical accessibility) and a high propensity to manage it by pricing. In line with theoretical guidelines, such a practice aims at reducing car use in urban areas through control measures both supply and demand-oriented, that is by increasing direct monetary costs for road users. Regardless of the level of physical accessibility, Northern European cities (Antwerp, Edinburgh, Amsterdam, Copenhagen) and even more the British ones (Leeds, Birmingham, Middlesbrough) show high propensity in using pricing policies. The cost of parking is not very convenient both when compared to the degree of richness of the context whether it towards the cost of PT. This suggests that attitude to control demand by pricing tool does not depend on the demographic/economic features, but mainly by cultural and social factors.

Second cluster includes Rome, Lyon, Munich, Zurich and Lisbon, all performing low physical accessibility and high economic accessibility that is few on-street parking lots offered at low price. Such cities are all highly urbanized in central areas, with consequent high functional mix also attracting commuters and visitors from their surrounding metropolitan areas. The clear unfeasibility to increase physical accessibility by building new on-street parking has been addressed in the single realities through different practices. Lisbon and Zurich,[1] e.g. compensate lack of on-street car spaces by offering extensive off-street supply at a relatively low price (respectively a high and average value for CPOF was recorded). The resulting increase in public land availability has been used to encourage use of more sustainable transport modes.[2] Munich, given the high extension of its urban rail network (174 km, it is the 3rd highest sample value), increased physical P&R supply (very high value for CPPR) also offering affordable service to PT users (low value for CEPR). Rome and Lyon do not show particular strategies to compensate their low physical accessibility. Rome is characterized by a low propensity to control on-street parking demand (very low value for CEON) and the related high economic parking accessibility (low price and a single fee rate in all zone), favors the motorized mobility, thus contributing to congestion.[3]

Third cluster, composed by Barcelona, Berlin, Pisa and Turin, record values close to the sample mean both for physical and economic component. This indicates a willingness to find a better balance between supply and demand by using pricing leverage.[4] Fourth cluster, including medium-size cities (Brescia, Strasbourg and Nantes) shows a high propensity for managing and limiting on-street parking demand (very low value for CPON and medium for CEON). Brescia integrates on-street parking with a wide off-street and P&R supply, both characterized by a high affordability (low values for CEOF and CEPR).[5] The French cities address parking demand mainly towards P&R facilities, by offering free parking. Finally, Copenhagen and Bologna belong to the last two clusters, both presenting a significant physical accessibility (respectively high and very high values for CPON). Both cities still turn away from the conventional paradigm by acting on price to manage on-street parking (CEON is medium for Bologna, very high for Copenhagen).

5 CONCLUSIONS

The study provides an analysis of contemporary approaches to parking management implemented at European level, by interpreting them according to some main parameters (evident and latent) affecting performances of the urban parking system. Findings confirm a decline of the pure conventional paradigm albeit some cities have not completely abandoned the typical "predict and provide" approach.

Pay parking is a common practice with varying prices depending on demo/socio-economic aspects and on the vision pursued by single Municipalities. Especially the Northern European cities show a clear willingness to implement a "car limiting" policy, by emphasizing use of parking pricing, also *ad hoc* for specific valuable areas. There is also an increasing propensity in encouraging use of off-street parking. In cities dealing with a conventional approach, the off-street typology is used to expand parking supply in built environments, where land is a scarcer resource. As opposed, "greener" cities aim at redeveloping part of the urban space by replacing on street parking lots with off-street ones, keeping constant the urban parking capacity.

---

[1] In Zurich this results from the "Historic parking compromise", implemented since 1996 aimed at shifting car parking from road to parking facilities.
[2] Zurich records highest cycle paths density (1.8 km/km²).
[3] Rome, with the highest motorization rate (734 cars/1000 people), is the 2nd congested city: on average a trip takes 38% of more time when compared with a free-flow condition.

---

[4] Barcelona integrates the on-street parking supply with a generous off-street provision (the highest value in the sample) to a higher but still competitive price if compared to on-street.
[5] Brescia is the first, among the Italian cities, for density of cycle paths (1.3 km/km²) and the fourth of sample as a whole.

Besides, there is a common understanding in ensuring high affordability for P&R supply (less common in cities covered by wide urban rail network) although this may reduce PT's ridership in peripheral city areas.

A lesson learned from this study suggests need to identify common guidelines, also proposing "light" measures, as follows:

- to adjust parking fees also according to PT service coverage and its real performances, mainly in car-oriented cities;
- to improve efficiency of urban parking systems by encouraging on-street parking turnover, also defining upper standards limits in central areas;
- to increase P&R functionality, since it is designed to improve urban intermodality without replacing road–based PT.

REFERENCES

Acutt, M. & Dodgson, J. 1997. Controlling the Environmental Impacts of Transport: Matching Instruments to Objectives. Transportation Research Part D: Transport and Environment 2(1): 17–33.

Barter, P. 2014. On-street parking Management, Sustainable Urban Transport Technical Document. Federal Ministry for Economic Cooperation and Development.

Burns, L. D. 1979. Transportation, Temporal and Spatial components of accessibility. Lexington Book, Toronto.

Cascetta, E., Carteni, A., Montanino M. 2013. A New Measure of Accessibility based on Perceived Opportunities. Procedia—Social and Behavioral Sciences, 87: 117–132.

Cervero, R., 2005. Accessible Cities and Regions: A Framework for Sustainable Transport and Urbanism in the 21st Century, Working Paper UCB-ITS-VWP-2005-3, UC Berkeley Center for Future Urban Transport.

Clayton, W., Parkhurst, G., Ben-Elia, E. & Ricci, M. 2015. Where to park? A behavioral comparison of bus Park & Ride and city center car park usage in Bath, UK. in 17th EPA Congress, Berlin, Germany.

Chatman, D. 2008. Deconstructing Development Density: Quality, Quantity and Price Effects on Household Non-work Travel, Transportation Research Part A 42.

Dalvi, M. & Martin, K. 1976. The measurement of accessibility: some preliminary results. Transportation, 5:17–42.

Dijk, M., & Montalvo, C. 2011. Policy frames of Park-and-Ride in Europe. Journal of Transport Geography, 19:1106–1119.

Dong, X., Ben-Akiva, M. E., Bowman J. L. and Walker, J. L. 2006. Moving from trip-based to activity-based measures of accessibility Transportation Research Part A: Policy and Practice, 40 (2):163–180.

Eurostat. Cities (Urban Audit). http://ec.europa.eu/eurostat/ web/cities. 25 February 2016.

Knoema. World Data Atlas. https://knoema.com/atlas. 18 February 2016.

Litman T. 2008. Recommendations for Improving LEED Transportation and Parking Credits, Victoria Transport Policy Institute.

Litman, T. 2013. Parking Management: Strategies, Evaluation and Planning. Victoria Transport Policy Institute.

Litman, T. 2016a. Parking Management—Comprehensive Implementation Guide. Victoria Transport Policy Institute.

Litman, T. 2016b. Land Use Impacts on Transportation, How Land Use Factors Affect Travel Behavior. Victoria Transport Policy Institute.

Lois, D. & Lopes-Saez, M. 2009. The Relationship Between Instrumental, Symbolic and Affective Factors as Predictors of Car Use: A Structural Equation Modelling Approach, Transportation Research Part A 43.

Musso, A. & Piccioni, C. 2010. Lezioni di Teoria dei Sistemi di Trasporto. Edizioni Ingegneria 2000, Roma.

Musso, A. & Piccioni, C. 2012. Tourist Coach Mobility Plans: current practices and operational criteria, in Cappelli, A. Libardo, A & Nocera S. (edited by) Environment, land use and transportation systems, Franco Angeli: pp.145–160.

Næss P., Andersen J, Nicolaisen M. S., Strand A. 2014. Transport modelling in the context of the 'predict and provide' paradigm. European Journal of Transport and Infrastructure Research, Issue 14(2): pp. 102–121.

Noel, E. 1988. Park and Ride: Alive, Well, and Expanding in the United States. Journal of Urban Planning and Development, 114:1(2), 2–13.

Owens, S. 1995. From 'predict and provide' to 'predict and prevent'?: pricing and planning in transport policy. Transport Policy, 2 (1): 43–49.

Shoup, D. 2005. The high cost of free parking. APA Planners Press.

Shoup, D. & Manville, M. 2005. Parking, People and Cities. Journal of Urban Planning and Development.

U.S. Department of Environment. 1996. Policy and Procedure Guidelines, Planning policy guidance: Town centres and retail developments, Revised PPG6.

Vigar, G., 2001. Reappraising UK transport policy 1950–1999: the myth of 'monomodality' and the nature of 'paradigm shifts'. Planning Perspectives 16, 269–291.

Transport Infrastructure and Systems – Dell'Acqua & Wegman (Eds)
© 2017 Taylor & Francis Group, London, ISBN 978-1-138-03009-1

Personality and driver behaviour questionnaire: Correlational exploratory study

J.F. Dourado
CITTA, Faculty of Science and Technologies of the University of Coimbra, Coimbra, Portugal

A.T. Pereira & V. Nogueira
Department of Psychological Medicine, Faculty of Medicine of the University of Coimbra, Coimbra, Portugal

A.M.C. Bastos Silva & A.J.M. Seco
CITTA, Faculty of Science and Technologies of the University of Coimbra, Coimbra, Portugal

ABSTRACT: The aim of our research is to explore the association between personality traits and dimensions of driver behaviour in a vast sample of the Portuguese population.

A community sample composed of 747 participants [417 (55.8%) women; mean age = 42.13 ± 12.349 years; mean; mean driving license years = 21.30 ± 11.338; mean years of regular driving = 20.33 ± 11.328] participated in an online survey. To evaluate personality traits the NEO-Five Factor Inventory-20 and the Impulsive Sensation Seeking scale were used. The Driver Behaviour Questionnaire composed of 24 items, assessing Infractions and Aggressive Driving (IAD), Non-Intentional Errors (NIE) and Lapses, was used.

IAD significantly and moderately correlated with age, gender (male), years of driving licence, Impulsivity and Sensation seeking; NIE and Lapses with Impulsivity. Other personality traits and socio-demographic variables presented significant but lower correlations with driving behaviour dimensions.

These results reinforce that personality traits should be accounted when studying driver behaviour for road safety assessment.

1 INTRODUCTION

Road trauma can be seen as one of the most significant diseases of the industrialized societies and it is an increasing public health and economic issue in developing countries. Organisation for Economic Co-operation and Development reports that every year approximately 1.24 million people die on the roads. In Portugal, the National Authority for Road Safety reported 30.604 road traffic accidents with injuries in 2014, with a total of 482 deaths, a number that may increase in a near future.

Driving can be defined as a "complex process involving individual factors, expressed within a social exchange among drivers, passengers, and pedestrians, which is ultimately impacted by contextual and environmental stimuli found inside and outside the vehicle" (Hennessy, 2011).

Oppenheim and Shina (2011) describe the driver behaviour as what the drivers do given their limitations, constraints, needs, motivation, level of alertness and personality. They also emphasize the importance of considering the context in which drivers drive to analyse the type of actions they take every moment. Additionally, Hennessy (2011) argues that any discussion of factors that impact driving outcomes should include personal factors.

Following this line of thought, traffic psychology researchers have been studding personal traits that are believed to have positive or negative correlation with aggressive and/or risky driver behaviours. They include: sensation seeking, having a positive correlation with risky behaviours (Yang et al., 2013); anger with positive correlation to violations (Lajunen et al., 1998), to loss of concentration (Dahlen et al., 2005) and to poor vehicle control (Deffenbacher et al., 2001); altruism with negative correlation to aggressive driving (Dahlen et al., 2006; Yang et al., 2013); and less conscientiousness that can be related to normlessness behaviour, which has been proven to be related to violations and risky behaviour (Iversen and Rundm, 2002; Oltedal and Rundmo, 2006; Yang et al., 2013).

Based on his taxonomy of human error Reason (1990) developed an Accident Causation Model,

787

further developed by (Verschuur and Hurts, 2007), as well as a Driver Behaviour Questionnaire (Reason et al., 1990) based on self-report questions to measure behaviours, the Manchester Driver Behaviour Questionnaire, to serve as tool for driver behaviour analysis. Previous studies have demonstrated the validity of the DBQ (Reason et al., 1990; Parker et al., 2000; Lajunen et al., 2004), and it has been amply used by researchers, with Wahlberg et al. (2011) mentioning fifty-four studies that have included at least part of the questionnaire is their studies. Yet, the relationship between the results of the self-report scales and objective driving data of the behaviours in question (e.g. speeding, close following, abrupt lane-changing, etc.), need more validation (Helmand and Reed, 2014), as for the relationship between the scales and crashes (Wahlberg et al., 2011).

The aim of this study is to explore the association between personality traits and dimensions of driver behaviour in a vast sample of the Portuguese population. This is part of a bigger research project which aims to analyse these relationships with real driving behaviours.

2 METHODS

2.1 Procedure

A community sample was recruited and invited to participate in an online survey on the relationship between personality and driver behaviour. Inclusion criteria were: driving license and regular driving for at least three years and age lower than 75 years old.

All the participants answered self-reported questionnaires, containing socio-demographic questions, personal accident history, and validated instruments to evaluate personality traits and driver behaviour.

2.2 Instruments

To evaluate the personality traits the researchers used the Big Five model (McCrae & John, 1992), in order to conciliate a global overview of the personality and the ability to differentiate specific traits. To assess the Big Five personality traits (Neuroticism, Extraversion, Openness, Agreeableness and Conscientiousness), the Portuguese version of the NEO-Five Factor Inventory-20-item (Bertoquini & Ribeiro, 2006) was used. This scale divides the items in five dimensions, four items each, on a 5-level Likert scale. It is a short version of the original NEO-PI-R (Costa & McCrae, 1992), validated for the Portuguese population by Bertoquini & Pais Ribeiro (2006). The NEO-FFI-20 has shown a clear factorial structure and an acceptable internal consistency (alphas ≥ 0.7) for the five dimensions

and an excellent adjustment by the confirmatory factorial analysis.

To assess Impulsivity and Sensation seeking, which have long been associated with a tendency to engage in behaviours without consideration of potential negative consequences (Caspi et al., 1997), the principal researcher in collaboration with a psychiatrist and a psychologist (both experienced in personality research and in the translation, adaptation and validation of psychological assessment instruments) translated the 19-item Impulsive Sensation Seeking Scale (Zuckerman, 1994). The ImpSS is part of the larger ZKPQ (Zuckerman et al., 1993), showing an acceptable internal consistency (alphas > 0.70) in heterogeneous samples (Mc Daniel & Mahan III, 2008; De Leo et al., 2010). This scale is composed of two dimensions, Impulsivity (eight items) and Sensation seeking (11 items), in binary item' score "true" = 1 or "false" = 0.

The driver behaviour evaluation was made using the Portuguese translation of the Driver Behaviour Questionnaire (Reason et al., 1990; Reimer et al., 2005; Correia, 2014) which consists of 24 items describing a variety of errors and violations during driving. The questionnaire divides the items in three dimensions—errors, lapses and violations—eight items each, on a 6-point scale, ranging from 0 to 5, where higher scores indicate more frequency in engaging risky driving behaviours.

2.3 Sample characterization

The total sample comprised 747 participants [417 (55.8%) women; mean age = 42.13 ± 12.349 years; mean driving license years = 21.30 ± 11.338; mean years of regular driving = 20.33 ± 11.328].

2.4 Statistical analysis

To verify if these self-reported questionnaires were valid and reliable tools to evaluate the dimensions in which the researchers were interested in, the analysis started with the study of the psychometric properties of the three instruments.

The total sample was randomly divided in two sub-samples, being sample A composed of 373 and sample B of 374 participants, who did not significantly differ in relation to socio-demographic and accident history variables (Qui-square tests and t-Student tests, with p-values > 0.05). Sample A was used to Exploratory Factor Analysis and sample B was used to Confirmatory Factor Analysis (Figure 1).

SPSS version 23.0, StatsToDo and AMOS software were used to Exploratory Factor Analysis, Parallel Analysis and Confirmatory Factor Analysis respectively.

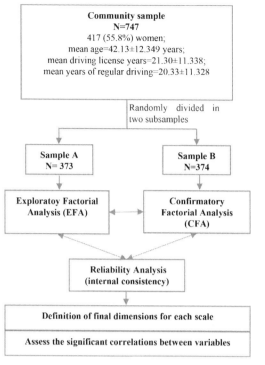

Figure 1. Statistical analysis methodology.

The Exploratory Factor Analysis was performed using the principal components method, with varimax rotation (Pereira et al., 2013). The suitability of data for factor analysis was assessed with the Kaiser-Meyer-Oklin (KMO) and the Barlett's Tests of Sphericity (Kline, 1994). The data was considered suitable for factorial analysis when KMO > 0.50 (Sharma, 1996) and the null hypothesis (H_0) was rejected (p-value ≤ 0.05). To help establishing the correct number of factors to extract from the factorial analysis, the criteria used were: i) the Kaiser criteria, to retain factors with eigenvalue >1 (Kaiser, 1958); ii) Cattel Scree Plot criteria, which implies the retention of all components in the sharp descent part of the plot before the eigenvalues start to level off, where line changes slope (Cattel, 1966; Kline, 2000). The selection of the items for each factor consisted of retaining items that showed strong factor loadings. As other authors have done (Elal et al., 2000), items with factor loadings >0.30 were chosen.

Using the Parallel Analysis, with the Monte Carlo Simulation, the number of significant factors for each scale was confirmed (Ledesma & Valero-Mora, 2007). Even though, different possible dimensional structures were obtained for each scale.

Following the exploratory analysis, the Confirmatory Factorial Analysis (CFA) was performed (Kline, 2010). The fitness of the previously obtained factorial dimensions from each scale to the data from the sample B was assessed. A series of indexes were calculated for the different possible dimensional structures for each scale, including the original factorial structures adopted by the authors of the scales. From the analysis and comparison of the results of the CFA, the most fitted models were chosen and then their internal consistency was assessed using the total sample (n = 747) and finally the structures for each scale were established. Details from this analysis will not be presented in this article, only the final factorial dimensions. These factorial dimensions were considered as variables in the present study.

The dimensions' internal consistency was measured by Cronbach's coefficient (α). High values of α indicate consistent and reliable measures (0.65 ≤ α <; 0.70, acceptable; 0.70 ≤ α < 0.80, good; 0.80 ≤ α < 0.90, very good; DeVellis, 1991). To analyse the ability of a single item to measure the attribute supposed to be assessed by the scale/factor dimension, Pearson correlations coefficients between each item and the total score (excluding the item) were examined. In other words, this coefficient allows to interpret the magnitude of the correlations between the item and its dimension and/or total scale. The Values of Pearson's Correlation Coefficient r ≥ 0.20 were considered to be acceptable, r ≥ 0.30 to be good and r ≥ 0.5 to be high (Cohen, 1992).

Also, the Cronbach's alphas excluding each item were computed and compared with the coefficient α of the total score. When this individual coefficient α was lower than the coefficient α of the total score, it meant that the item was contributing to the internal consistency of the scale, as its absence would decrease the total reliability of the scale.

Finally, the Pearson correlation coefficients between personality traits, the dimensions of the driver behaviour, and socio-demographic and personal accident history variables were analysed.

3 RESULTS

3.1 NEO-FFI-20

After performing the statistical analysis described in the chapter 2.4 and analysing in a qualitative way all the items, it was concluded that the best factorial structure for the analysed sample was the original structure, excluding two items (14 and 16).

The factors were:

Factor 1 - *Neuroticism*—tendency to experience such feelings as anxiety, anger, envy, guilt, and depressed mood.

Factor 2 - *Extraversion*—to enjoy human interactions and to be enthusiastic, talkative, assertive, and gregarious.

Factor 3 - *Openness to experience*—reflects the degree of intellectual curiosity, creativity and a preference for novelty and variety.

Factor 4 - *Agreeableness*—tendency to be compassionate and cooperative rather than suspicious and antagonistic towards others.

Factor 5 - *Conscientiousness*—a tendency to be organized and trustworthy, show self-discipline, act dutifully, aim for achievement, and prefer planned rather than spontaneous behaviour.

Each factor revealed to have an acceptable and good internal consistency ($0.60 \leq \alpha < 0.80$) (Table 1). Pearson correlation coefficients between each item and the dimensional score of its respective dimension (excluding the item) ranged from 0.359 (item 6—*feeling helpless many times and wishing someone to solve his/her problems*) to 0.619 (item 8—*feeling great emotions when reading a poem and observing a work of art*), with nine items having very good correlations and eleven good correlations. Cronbach's alphas excluding the item, except for item 6, were all equal or smaller than the alpha from the respective dimension. The increase of the alpha when eliminating the item 6, from the dimension Neuroticism, was very slight. Additionally, eliminating this item would make the dimension too small, and consequently inducing the decrease of its internal consistency. So one decided to maintain this item.

3.2 *Impulsive Sensation Seeking (ImpSS)*

Following the same methodological approach, the authors concluded that the best factorial structure for the analysed sample was the two factors' structure, resulting from the exploratory factor analysis, excluding items 4, 6, 7 and 10.

Table 1. Dimensions of NEO-FFI-20 and its internal consistency (Cronbach's α).

	Dimensions	Number of items	Internal Consistency (α)
NEO-FFI-20	Neuroticism	3 (excluded item 16)	0.68
	Extraversion	4	0.62
	Openness	4	0.74
	Agreeableness	3 (excluded item 14)	0.70
	Conscientiousness	4	0.74

The factors were:

Factor 1 - *Sensation Seeking*—defined by the search for experiences and feelings that are novel and intense, and by the readiness to take physical, social, legal, and financial risks for the sake of such experiences.

Factor 2 - *Impulsivity*—a tendency to act on a whim, displaying behaviour characterized by little or no forethought, reflection or consideration of the consequences.

Both factors revealed to have good internal consistency ($0.70 \leq \alpha < 0.80$) and the total score presented very good internal consistency with a coefficient $\alpha = 0.82$ (Table 2). The correlation strength between each item and the total score of its respective dimension (excluding the item) showed to be good and very good, being the minimum the item 15, $r = 0.329$. Concerning correlation coefficients between each item and the total score (excluding the item), they ranged from 0.305 (item 2 – *usually tending to think before taking an action*) to $r = 0.539$ (item 14 – *sometimes doing crazy things just for fun*), having 13 items good correlations and two items very good correlations. When analysing the total scale, Cronbach's alphas excluding the item were all smaller than the alpha of the total score (0.82).

3.3 *Driver Behaviour Questionnaire (DBQ)*

The best factorial structure for the DBQ was obtained as a result of the exploratory factor analysis, being composed of three factors, excluding items 1 and 24.

The factors were:

Factor 1 - *Infractions and aggressive driving*—deliberate deviations from practices believed necessary to maintain safe driving.

Factor 2 - *Non intentional errors*—to fail planned actions, which may result in dangerous outcomes.

Factor - *Lapses*—failures related to distractibility which can cause embarrassment but unlikely to have an impact on driving safely.

All factors revealed to have good internal consistency ($0.70 < \alpha \leq 0.80$) and the total scale a very

Table 2. Dimensions of ImpSS and its internal consistency (Cronbach's α).

	Dimensions	Number of items	Internal Consistency (α)	
ImpSS	Sensation seeking	10	0.79	0.82
	Impulsivity	5	0.76	

Table 3. Dimensions of DBQ and its internal consistency (Cronbach's α).

	Dimensions	Number of items	Internal Cons. (α)	
DBQ	Impulsive and aggressive driving	7	0.77	0.84
	Non intentional errors	9	0.73	
	Lapses	6	0.71	

good internal consistency (α > 0.80) (Table 3). Pearson correlation coefficients between each item and the dimensional score of its respective dimension (excluding the item) ranged from 0.315 (item 2) to 0.639 (item 12). When analysing correlation coefficients between each item and the total score (excluding the item) they were all higher than r = 0.311 (item 7—*forgetting where the car is parked*), The Cronbach's alphas excluding the item were all lower than the alpha of the total scale (0.84).

3.4 Correlations

Pearson correlation coefficients were estimated between socio-demographic variables, personality traits and the DBQ dimensions—Infractions and aggressive driving, Non intentional errors and Lapses (Table 4).

The strongest significant correlations (r > 0.2 and p-value < 0.05) with the dimension Infractions and Aggressive Driving (IAD) were: age (r = −0.239), gender (r = 0.254, male gender with higher scores on IAD), Impulsivity (r = 0.259), Sensation seeking (r = 0.301) and the ImpSS total scale (r = 0.337). Years of driving license, Extraversion and Agreeableness showed also significant but lower correlations with this driving behaviour dimension.

As for the dimension Non Intentional Errors (NIE), the variable with the strongest correlation was Impulsivity (r = 0.212). Conscientiousness, Neuroticism and the ImpSS total scale correlated with lower strength, but also significantly.

The Lapses variable presented stronger correlation with Impulsivity (r = 0.227, p-value < 0.001). This dimension also presented correlations with Neuroticism, Conscientiousness and the total ImpSS scale, but with lower strength.

The total DBQ score correlated significantly to Impulsivity (r = 0.299), Sensation Seeking (r = 0.246) and ImpSS (r = 0.312).

The variable accidents per driving license years (in which the respondent had been identified as responsible for the accident) correlated significantly with IAD (r = 0.214), NIE (r = 0.221), DQB (r = 0.260), and age (r = −0.211).

Table 4. Pearson Correlation Coefficient (r) between social demographic, personality traits, driver behaviour dimensions and accidents per year of driving licence variable.

		IAD	NIE	Lapses	DBQ	Accid/ driving lic year
IAD	r					**0.214**
	sig.					**<0.001**
	N					747
NIE	r	0.431				**0.221**
	sig.	<0.001				**<0.001**
	N	747				747
Lapses	r	0.326	0.540			0.176
	sig.	<0.001	<0.001			<0.001
	N	747	747			747
DBQ	r	0.792	0.802	0.764		**0.260**
	sig.	<0.001	<0.001	<0.001		**<0.001**
	N	747	747	747		747
Age	r	**−0.239**	−0.050	−0.092	−0.176	**−0.211**
	sig.	**<0.001**	0.173	0.012	<0.001	**<0.001**
	N	747	747	747	747	747
Gender	r	**0.254**	0.014	−0.133	0.079	0.046
	sig.	**<0.001**	0.697	<0.001	0.030	0.743
	N	747	747	747	747	747
Education	r	0.16	0.110	0.168	0.188	0.012
	sig.	<0.001	0.003	<0.001	<0.001	0.743
	N	747	747	747	747	747
Years of driving license	r	−0.174	−0.036	−0.072	−0.130	
	sig.	<0.001	0.323	0.048	<0.001	
	N	747	747	747	747	
Years of frequent driving	r	−0.166	−0.044	−0.080	−0.132	
	sig.	<0.001	0.225	0.029	<0.001	
	N	747	747	747	747	
Neuroticism	r	0.092	0.188	0.166	0.182	0.058
	sig.	0.012	<0.001	<0.001	<0.001	0.112
	N	747	747	747	747	747
Extraversion	r	0.138	−0.038	−0.071	0.027	−0.001
	sig.	<0.001	0.301	0.052	0.462	0.984
	N	747	747	747	747	747
Openness to experience	r	−0.017	−0.015	0.113	0.031	−0.013
	sig.	0.636	0.686	.002	0.400	0.731
	N	747	747	747	747	747
Agreeableness	r	−0.142	−0.038	0.010	−0.082	−0.017
	sig.	<0.001	0.296	0.779	0.026	0.640
	N	747	747	747	747	747
Conscientiousness	r	−0.026	−0.188	−0.157	−0.146	−0.071
	sig.	0.474	<0.001	<0.001	<0.001	0.053
	N	747	747	747	747	747
Impulsivity	r	**0.259**	**0.212**	**0.227**	**0.299**	0.064
	sig.	**<0.001**	**<0.001**	**<0.001**	**<0.001**	0.081
	N	747	747	747	747	747
Sensation seeking	r	**0.301**	0.115	0.131	**0.246**	0.047
	sig.	**<0.001**	0.002	<0.001	**<0.001**	0.202
	N	747	747	747	747	747
ImpSS	r	**0.337**	0.177	0.195	**0.312**	0.062
	sig.	**<0.001**	<0.001	**<0.001**	**<0.001**	0.089
	N	747	747	747	747	747

4 DISCUSSION AND CONCLUSION

Personality traits are of paramount importance for traffic psychology, since they are believed to have positive or negative correlation with aggressive and/or risky driver behaviours. It is also of great relevance to understand which factors and how they influence the driver behaviour in order to be taken into account when analysing the traffic safety conditions of a given infrastructure.

With this research the authors aimed to explore the association between personality traits and dimensions of driver behaviour in a vast sample of the Portuguese population (N = 747), contributing for a clearer understanding of the role of personality in this specific context.

Exploratory and confirmatory factorial analysis were performed to verify which factorial structure best suited the sample. These psychometric analyses represent a strength of the study. For the NEO-FFI-20 and ImpSS scales the dimensional structures obtained were similar to the original ones (Bertoquini & Ribeiro, 2002; Zuckerman, 1994), with some adjustments, to better fit the scales to the sample of this research. The excluded items were: 14 and 16 in the NEO-FFI-20 scale; 4, 6, 7 and 10 in the ImpSS scale; and items 1 and 24 in the DBQ. The number of factors obtained in the DBQ was the same as in the original versions (Reason et al., 1990; Reimer et al., 2005; Correia, 2014), yet the distribution of items in the factorial dimensions was slightly different in relation to the dimensions Non intentional errors and Lapses. In each scale the internal consistency results were acceptable, good and in one case very good.

Regarding Infractions and aggressive driving, younger and male individuals were the ones with higher scores in this behaviour.

Our correlational analysis emphasizes the relevance of the Impulsivity trait, which significantly correlated with all the driving behaviour dimensions. These results make sense, since people who are more impulsive, do not anticipate the consequences of their behaviours. Additionally, people who like to search for new experiences and intense feelings, and at the same time are less compassionate and cooperative, may have more tendency to commit infractions and to have an aggressive behaviour.

As for the Non intentional errors dimension, higher tendency to be impulsive and to be less self-disciplined and less organized is associated with the tendency to commit errors during the driving task, without intention, like fail to check the rear-view mirror before pulling out and changing lanes, for example. Furthermore, one can speculate that people who score high in the Neuroticism trait, may be more prone to Non intentional errors, due to their tendency to worry and ruminate, which may consume their attention, particularly in stressful situations.

The results also showed that people with lower scores in the Conscientiousness trait, that tend to be less cautious and focused, scored higher in NIE and Lapses.

For the behaviour Lapses younger people and women got the higher scores. People with higher scores in Neuroticism, Openness to experience and Impulsivity and with lower scores in Conscientiousness had a significant relation to failures related to distractibility, but that are unlikely to have an impact on driving safety.

It was also interesting to found a gender pattern in the correlations. Being female correlated with Lapses and male with IAD. In future studies it will be useful to analyse if personality traits still explain the DBQ levels after controlling for important demographic variables, such as age, gender and years of driving license.

Analysing the correlations related to accidents per driving license year, higher number of accidents/year have significant and moderate correlation with higher scores in IAD and NIE dimensions and with lower ages.

The obtained results are in line with studies in traffic psychology (Correia, 2014; Yang et al., 2013; Dahlen et al., 2006) and seam to reinforce the idea that the personality traits should not be ignored when studying the behaviour of drivers. Thus, other personal variables like socio-demographic variables as age, gender and driving licence years, which also showed to be significant, should be considered when analysing the driver behaviour.

The dimension of the sample (N = 747) is a good strength of this work. On the other hand, having used the NEO-FFI-20 instead of the more complete and detailed NEO-PI-R (240 items) or the NEO-FFI (60 items) may be a limitation of this work. However, the use of shorter versions is increasingly recommended in observational survey research, as it decreases the probability of response biases and increases the response rates.

Analysing the results described in this article it is possible to establish the hypothesis that people can be clustered by specific profiles that reflect their personality traits and correspond to different behaviour patterns on the road, and consequently to distinct risk situations. It is our intention to test this hypothesis in the near future. By using different personality and driving profiles in an experimental design, we hope to increase the knowledge of human behaviour on the road to better calibrate and validate tools (e.g. microsimulation models) for road safety assessment.

ACKNOWLEDGMENTS

This work was done with the support of the MIT-Portugal Program and the Foundation for Science and Technology. The authors are deeply grateful to all the participants in this study and to all the professionals involved.

REFERENCES

Bertoquini, V., & Pais-Ribeiro, J.L. 2006. Estudo de Formas Reduzidas do NEO-PI-R. *Psicologia: Teoria, Investigação e Prática* 11(1): 85–101.

Caspi, A., Begg, D., Dickson, N., Harrington, H.L., Langley, J., Moffitt, T.E., & Silva, P.A. 1997. Personality differences predict health-risk behaviors in young adulthood: Evidence from a longitudinal study. *Journal of Personality and Social Psychology* 73: 1052–1063.

Cattell R.B. 1966. The scree test for the number of factors. Multivariate Behavioral Research 1(2): 245–76.

Cohen J. 1992. A power primer. *Psycholl Bulletin* 112: 155–159.

Correia, J.P. 2014. Traços de personalidade, estados emocionais e condução: um estudo comparativo entre condutores de ambos os sexos. Doctoral Thesis, Faculty of Medicine, University of Lisbon, Portugal.

Costa, P.T., & McCrae, R.R. 1992. Revised NEO Personality Inventory (NEO-PI-R) and NEO Five-Factor Inventory (NEO-FFI) professional manual. Odessa, FL: Psychological Assessment Resources.

Dahlen, E.R., & White, R.P. 2006. The Big Five factors, sensation seeking, and driving anger in the prediction of unsafe driving. *Personality and Individual Differences* 41: 903–915.

Dahlen, E.R., Martin, R.C., Ragan, K. & Kuhlman, M.M. 2005. Driving anger, sensation seeking, impulsiveness, and boredom proneness in the prediction of unsafe driving. *Accident Analysis and Prevention* 37: 341–348.

De Leo, J.A., Van Dam, N.T., Hobkirk, A.L. & Earleywine, m. 2010. Examining bias in the impulsive sensation seeking (ImpSS) Scale using Differential Item Functioning (DIF) – An item response analysis. *Personality and Individual Differences* 50.

Deffenbacher, J.L., Lynch, R.S., Oetting, E.R. & Yingling, D.A. 2001. Driving anger: Correlates and a test of state-trait theory. *Personality and Individual Differences* 31: 1321–1331.

DeVellis FR 1991. Scale development - Theory and applications. London: SAGE Publications.

Elal, G., Altug, A., Slade, P., & Teckcan, A. 2000. Factorstructure of the eating Attitudes Test (EAT) in a Turkish university sample. *Eating Weight Disorders* 5: 46–60.

Hennessy, D. 2011. Chapter 12, Social, Personality and Affective Constructs in Driving. Handbook of Traffic Psychology. London: B. Porter Ed, Academic Press in an imprint of Elsevier.

Iversen, H., & Rundmo, T. 2002. Personality, risky driving and accident involvement among Norwegian drivers. *Personality and Individual Differences* 33: 1251–1263.

Kaiser, H.F. 1958. The varimax criterion for analytic rotation in factor analysis. *Psychometrika* 23:187–200.

Kline, P. 1994. An easy guide to factor analysis. London: Routledge.

Kline, P. 2000. The handbook of psychological testing (2nd ed.). London and New York: Routledge.

Kline, R.B. 2010. (3rd ed) Principles and practice of structural equation modeling. New York: Guilford Press.

Lajunen, T., Corry, A., Summala, H., & Hartley, L. 1998. Cross-cultural differences in drivers' self-assessments of their perceptual-motor and safety skills: Australians and Finns. *Personality and Individual Differences* 24-539-550.

Lajunen, T., Parker, D., Summala, H., 2004. The Manchester driver behaviour questionnaire: a cross-cultural study. *Accidents Analysis and Prevention* 36 (2): 231–238.

Ledesma, R.D. & Valero-Mora, P. 2007. Determining the Number of Factors to Retain in EFA: an easy-to-use computer program for carrying out Parallel Analysis. Practical Assessment, *Research & Evaluation* 12 (2).

McCrae, R.R. & John, O.P. 1992. An introduction to the five-factor model and its applications. *Journal of Personality* 60(2): 175–215.

McDaniel, S.R. & Mahan III, J.E. 2008. An examination of the ImpSS scale as a valid and reliable alternative to the SSS-V in optimum stimulation level research. Personality and Individual Differences 44: 1528–1538.

Oltedal, S. & Rundmo, T. 2006. The effects of personality and gender on risky driving behaviour and accident involvement. *Safety Science* 44: 621–628.

Oppenheim, I. & Shina, D. Chapter 15, Human Factors and Ergonomics. Handbook of Traffic Psychology. London: B. Porter Ed, Academic Press in an imprint of Elsevier.

Parker, D., McDonald, L., Rabbitt, P., Sutcliffe, P. 2000. Elderly drivers and their accidents: the aging driver questionnaire. Accidents Analysis and Prevention 32(6): 751–759.

Pereira, A.T., Bos, S., Maia, B., Soares, M.J., Valente, J., Nogueira, V., Pinto de Azevedo, M.H. & Macedo, A. 2013. Short forms of the Post-partum Depression Screening Scale: as accurate as the original form. *Archives of Women's Health* 16 (1): 67–77.

Reason, J., Manstead, A., Stradling, S., Baxter, J., & Campbell, K. 1990. Errors and violations on the roads: A real distinction? *Ergonomics* 33: 1315–1332.

Reimer, B., D'Ambrosio, L.A., Gilbert, J., Coughlin, J.F., Biederman, J., Surman, C., Fried, R. & Aleardi, M. 2005. *Accident Analysis and Preevention* 37: 996–1004.

Sharma, S. 1996. Applied Multivariate Techniques. New York: John Wiley & Sons.

Verschuur, W.L.G. & Hurts, K. 2007. Modeling safe and unsafe driving behaviour. *Accident Analysis and Prevention* 40: 644–656.

Wahlberg, A.E., Dorn, L. & Kline, T. 2011. The Manchester Driver Behaviour Questionnaire as a Predictor of Road Traffic Accidents. Theoretical Issues, *Ergonomics Science* 12(1): 66–86.

Yang, J., Du, F., Qu, W., Gong, Z. & Sun, X. 2013. Effects of Personality on Risky Driving Behavior and Accident Involvement for Chinese Drivers. *Traffic Injury Prevention* 14(6): 565–571.

Zuckerman, M., Kuhlman, D.M., Joireman, J., Teta, P., & Kraft, M. 1993. A comparison of three structural models for personality: The big three, the big five, and the alternative five. *Journal of Personality and Social Psychology* 65: 757–768.

Zuckerman. 1994. Behavioral expressions and biosocial bases of sensation seeking. New York: Cambridge University Press.

Transport Infrastructure and Systems – Dell'Acqua & Wegman (Eds)
© 2017 Taylor & Francis Group, London, ISBN 978-1-138-03009-1

Motorways of the sea: An outlook of technological, operational and economic tools

R.G. Di Meglio & G. Mainardi
Livorno Port Authority, Livorno, Italy

S. Ferrini
Scuola Superiore Sant'Anna and Consorzio Nazionale Interuniversitario Per Le Telecomunicazioni, Pisa, Italy

I. Toni
Polo Sistemi Logistici—Università di Pisa, Livorno, Italy

ABSTRACT: The paper addresses recent trends in the Motorways of the Sea (MoS) through an analysis of actions available to support and improve intermodal transport of people and goods among European and Mediterranean ports. It highlights public policy tools to back and finance MoS services in the European and Italian legal framework. It assesses how new technologies can be successfully deployed within the MoS related logistic chains, analyzing RFID, e-seals, port monitoring platforms, in the light of market needs for ITS. It discusses the design of port facilities in relation to MoS traffic. The paper deepens operational issues both on port side as well as on port inland relations, in economical and in technical terms. It deals with the Port of Livorno case, giving the chance to consider, in the real life of the leading Italian Ro-Ro port, the implementation of technological, operational and economic tools for MoS.

1 MOS IN THE POLICY AND FINANCING FRAMEWORK

1.1 Relevance of MOS for EU transport policy framework

EU policies in the field of MoS have targeted infrastructural, environmental and financial issues. Bearing in mind that the EU coastline is some 70,000 km long and that waterborne transport accounts for about 40% of intra EU trade (Suarez-Aleman, 2015). MoS linkages have been set to play a role both in terms of enhanced connectivity of transport network, as well as of improved sustainability of shipments and economic viability of services.

The European Commission White Paper on Transport (COM 2001) acknowledged that European funds, notably grants, were needed to ensure start up and make these services commercially attractive, as it was stated that "these lines will not develop spontaneously" (Commission, 2001).

In this respect, EU policy was intended to tackle the growth of road transport, which, despite all efforts made in the recent decades, is on the rise (Juan, 2016).

In fact, road transport proves to be still competitive, if we look at costs as well as at flexibility, while investments in ports do not necessarily turn to incentive modal shift, as technical barriers between maritime and inland transport (like incompatibility for loading units, missing last mile links etc) still exist (Baindur, 2011). Promoting MoS linkages, without solving inland connections, can thus result in greater road congestion in the port hinterland and in the vicinity of port areas, and ultimately lead to minor modal shift. Moreover, as far as MoS concept is concerned, there is lack of clarity on what is MoS and is therefore fundable with European schemes and what, on the contrary, is pure (Short Sea Shipping) SSS service. In this respect, Regulation EU 1315/2013 laying down the Guidelines of Trans-European Network of Transport (TEN-T), stated that MoS represent the maritime dimension of TEN-T, and thus recognized them as TEN-T corridor (Parliament, 2013). They must link two ports located in two different EU countries, either core or comprehensive network ports, and possibly connect European network with that of Third countries.

If we look at figures on EU funding leverage, the picture is quite positive, as 450 million EU grants for 45 MOS projects in the period 2007–2013 yielded 2 billion euros investments. Leverage factor is then over 4, below the new target of 15–20 set for

CEF funding but high when it comes to the relatively low funding rates, which hinder a real leverage effect (EU Parliament, 2016) (Expert Group 5, 2010). Presently, CEF Funding grants 30% of total eligible costs for works and 50% for studies, percentages that are considered low to exert the potential of EU funding. As a consequence, only projects that would have carried out anyway can find attractive the application for European funding, whereas riskier projects, or projects with a bigger need for initial ramp-up financial aid, are more likely to be disregarded by EU financing schemes.

European funding has been increasingly earmarked to support sustainability of maritime linkages, and the Connecting Europe Facility programme (CEF) now in force has steadily focused on Liquefied Natural Gas (LNG) as more enviromental-friendly fuel for transport. If we look at 2015 selected CEF proposals for the MoS priority, four out of twelve concerned LNG promotion in shipping and installations in ports, other three were linked to sustainability while only two were more a "traditional" MoS project, financing a new maritime linkage and the related works along with IT application. The remaining three proposals addressed navigation safety issues and data exchange among IT platforms. In 2014 MoS selected proposals were 27, of which 10 supported the upgrade of Maritime services or initiatives in ports or in the hinterland instrumental to carry-on the proposed MoS, while five dealt with LNG distribution in ports and six targeted goals of sustainability through enhanced MoS chains. This picture highlights how important are the environmental concerns in connection with MoS linkages and how the focus shifted from the economic and financial viability of project towards sustainability of MoS: within the TEN-T programme framework, a total of 11 projects in the field of LNG have been funded out of 45, reaching 103 million out of 369 million euros.

As a consequence, the environmental side of European support to MoS is gaining room, while more commercially focused actions or ICT projects somewhat lag behind. This trend can be partly due to the new goals set forth by the 2011 White Paper, in particular those goals referred to the use of more environmental-friendly fuels, which foresee horizontal priorities of EU transport that can be found not only for MoS, but obviously for all transport modes.

1.2 *Challenges and setbacks of financing programmes*

The impact of previous relevant programmes on modal shift, what has been stated as major target for MoS, seems to be rather scarce or even negligible in comparison with the massive growth of road transport in the meantime. European ports. The reason behind these setbacks is rooted in the instable shipping industry, in the challenging integration of international supply chains and the competitiveness of road transport, as European freight is carried mostly on roads even if a maritime solution exists. Another issue to be considered is the unbalanced trade flows for MOS which, in contrast with other branches of the shipping industry, have a greater need for balanced demand of transport for starting the connection (Baindur, 2011). The present 16 MoS operating services (Juan et al. 2015) face therefore relevant constraints as for financial and economic issues, even after operations have started. A public support scheme also available is State Aid. In the lack of EU funding to back a MoS project, European Commission allows Member States to support the project for a maximum duration of five years and up to 35% of total operational costs (European Commission, 2008).

National financing schemes can be also envisaged. In this respect, the Italian Marebonus deserves mention not only as example of national financing scheme for backing modal shift through the implementation of new MoS services, but also in the light of a global transport policy laid down by the Government (Ministero Infrastrutture e Trasporti, 2015). In fact, the 2015 unveiled National Strategic Plan for Logistics and Ports sets forth the improvement of accessibility as well as maritime and inland connections. The financing scheme, totaling 138 million euros between 2016–2018, is structured as incentive for the shipping company operating a MoS service, which has contributed to shift freight from road to sea transport; this incentive is then reversed to the haulier for each shipped unit, which has been diverted from all road transport to MoS service. The Marebonus incentive is the follower of the Ecobonus, established with Law 265/2002, which managed to divert 5% of total road traffic, not only by financially support MoS services, but also by granting aid to measures in the field of ICT and to bringing together small and medium haulers (Rete Autostrade del Mare, 2014). It has been acknowledged that intermodal solutions through national MoS services have somewhat contributed to sustaining the coastal navigation demand, spreading the offer of transport solutions along the Italian peninsula. Marebonus is still waiting for EU Commission's approval.

Despite the attained levels of modal shift, financing schemes such as Marebonus and Ecobonus clash with traditional setbacks of MoS, which have been partly explained above. First of all, although relevant players of the transport chains, hauliers account for only 5% of global costs, while ports operations costs and dues play by far a bigger role (Isfort, 2014); furthermore inland transport costs are much more likely to raise as consequence of poor coordination among players. Different business models and unequal distribution of costs are to be blamed for bias in loading and unloading of freight time, delays and inefficient planning of capacity (Van De Horst, 2015), resulting in higher costs for the whole MoS chain. Only truly pan-European schemes would therefore reduce congestion in Core European regions and foster intra-European trade, for the benefit of more peripheral regions too. As a consequence, to be more effective, MOS supporting policies should look towards better inter-firm cooperation, global planning of capacities and investments in inland terminals and eventually fine-tuning of operations of players through closer integration or agreements. In the following section, we explore how ICT tools can effectively contribute to the viability of MoS service since the ramp-up phase, which has proven to be the most critical.

2 HOW THE TECHNOLOGIES CAN BE SUCCESSFULLY DEPLOYED WITHIN THE MOS LOGISTICS CHAIN

2.1 ICT in MoS, introduction and regulations

Seaports are considered as intermodal points in complex logistics chains, enabling to connect different nodes of transport (i.e. other seaports, inland nodes) in terms of both physical flows (i.e. freight, cargo and passenger handling) and information flows. In this context, the Motorways of the Sea are not only a matter of maritime transport between two ports, but are rather the establishment of efficient and complete logistics chains among origin and destination points, through upgraded transport nodes of a fully integrated network. The vision of a fully integrated network combines: the removal of existing barriers between nodes, the overcoming of obstacles for fast exchange of information (for every transport node, independently of the concerned country), and the possibility to track, trace, monitor and control cargo, passengers and events. To do that, the upgrading of Information Communication Technologies (ICT) tools is as important as the upgrading of handling equipment and physical facilities.

Regulation framework (at European and national level) greatly emphasizes keywords as *Intelligent Transport Systems* (ITS) and *Digitalization*. European Directive 2010/40 defines ITS as *advanced applications which without embodying intelligence as such aim to provide innovative services relating to different modes of transport and traffic management and enable various users to be better informed and make safer, more coordinated and 'smarter' use of transport networks* (European Commission, 2010). ITS are part of ICT and they integrate telecommunications, electronics and information technologies with transport engineering, in order to plan, design, operate, maintain and manage transport systems.

In particular, the White Paper 2011 looks towards a competitive and resource efficient transport system, breaking down barriers to the completion of the internal market of transport, by developing the concept of *e-freight*. E-freight consists in the *wide deployment of information technology tools, to simplify administrative procedures, provide for cargo tracking and tracing, and optimize schedules and traffic flows* (European Commission, 2011). In addition to this, the Maritime Transport Strategy 2018 highlights the importance to deploy *e-maritime* services at both European and Global level. The *e-maritime* concept aims at promoting the competitiveness of the European maritime transport sector and a more efficient use of resources through the widespread use of ICT tools (European Commission, 2009).

The *e-freight* concept, the *e-maritime* services and the digitalization of all information related to cargo and passengers pave the way for the creation of digitalized environments as *Digital Single Market* (European Commission, 2015) and *Digital Administrations* (MIT Transport, 2015).

Already in 2010, European Commission provided the basis for the creation of digitalized environments, within the concept of *Maritime Single Window* (MSW). Indeed, Directive 2010/65 aims to establish MSW in the EU Member States, allowing the development of a platform to exchange data between EU Member States. MSW contributes to the development of a European information structure. Possibly, this would also provide the European maritime transport industry "a place to stand on". Directive 2010/65 supports the ONCE paradigm, based on data single electronic transmission and a single control declaration formality for ship departing/arriving from/in member states' ports (European Commission, 2010).

2.2 ICT in MoS, current practices

To track, trace, monitor and control cargo, passengers and environments, and to facilitate the communication and the exchange of information among two (or more) nodes of transport involved in MoS, each of these nodes have to be upgraded and innovated. The upgrading process consists in equipping each node of harmonized info-structure, fostering the use of standards (in terms of data exchange and communication). On the other hand, the innovation process consists in testing and setting-up advanced solutions, in order to both facilitate and accelerate the real-time data collection, communication and dissemination of information.

The first step towards upgraded seaports is the design of a coherent ICT infrastructure, in order to create more interconnected Seaports within the connection of the node both in itself (i.e. setting up a LAN of the communities that are based on the Port landside) and with other transport nodes along the MoS (i.e. setting up Internet wide-corridors of the member states communities) (Pagano et al, 2016). The ICT infrastructure enables to gather data from humans and machines (i.e. sensors and vehicles), in order to create more interconnected and smart environments and to facilitate the communication inside and between seaports placed in different geographical areas, making way to new paradigms as *Internet of Things* and *Big Data*.

The use of cutting-edge technologies in the transport field is the first step towards the Innovation process. Indeed, after creating an ICT infrastructure ashore, seaports should consider which are data to be gathered, and why and how to do it they can do it; the more seaports can collect and manage data, the more MoS increase their competitiveness.

Data generated in the transport and maritime field don't come exclusively from the documents accompanying cargo and passengers, but also from transporting and handling activities and from the port and logistics environment. In fact, users are interested in tracking, tracing, monitoring, and controlling goods, people and events, knowing their position in real-time, for different purposes (i.e. administration bodies are interested in detecting hazardous events in port areas and speeding-up control activities, while importers are interested in knowing the position and the status of their cargo during the shipment). In so doing, seaports should employ smart devices and technologies used to enable customized ICT and ITS applications, in order to both monitor port areas and gather information related to efficiency level of MoS (in terms of safety, security, speed, costs, times, etc.). Key enabling ICT technologies includes vehicle detection technologies (i.e. radar, laser, video image processing, infrared.), vehicle monitoring and tracking technologies (i.e. GPS, Radio frequency identification technologies—RFID), distributed sensors network, video cameras, connected vehicle technologies (V2X technologies) and so on. The proper integration among these technologies enables to obtain heterogeneous and real time information. Information, if rightly processed, allows identifying value-added services, increasing the competitiveness of seaports as gateways of MoS and, broadly speaking, of logistics chains.

Two of the most interesting and widespread wireless technologies are RFID and Wireless Sensors Network (WSN). RFID is the main low cost automatic identification technology used for tracking and tracing cargo and vehicles, using simple tags attached or embedded into objects that are to be identified or tracked (i.e. containers, trailers, pallets, track and so on). The information related to the objects are stored into the tags and they are read by readers placed in strategic points (i.e. gates, berths.). Sensors Network is composed of large number of sensor nodes that can be deployed on the ground, in vehicles, inside buildings, on the berths, on the quays, on the gates, on port equipment (i.e. cranes) (Jain et al, 2012). Each of the distributed sensor node has the capability to collect data and route them to a specific gateway, in order to facilitate the data processing and the communication tasks with application layer. The WSN are usually used to detect and monitor the environment. In MoS, integrating RFID with WSN can provide both identity and location of an object (thanks to RFID technology) and information regarding the condition of the object carrying the sensors (i.e. the status of the cargo should be detected as quickly as possible, and alarms should be triggered when temperature gradients cross a threshold). Sensors can "give expression" to containers, vehicles, freight, cranes and other objects, promoting the concept of *Internet of Things* or (as usually referred to in the field of transport) Internet of Goods.

Sensors, RFID, other smart devices, legacy systems enable the digitalization of a huge amount of data, introducing the *Big Data* paradigm. Gartner defines *Big Data* as *high-volume, high-velocity and/ or high-variety information assets that demand cost-effective, innovative forms of information processing that enable enhanced insight, decision making, and process automation* (Gartner, 2016). Data could be investigated using semi-automatic and automatic methodologies (business intelligence technologies), as OLAP, Data Mining (and its related techniques), for gathering, extracting, transforming, loading and reporting data from different data sources aforementioned (Bhavin et al, 2004). The design of distributed intelligent tools for MoS management could be used for gaining competitive advantage through real-time collaboration with trading partners, and offer a new way to rapidly plan, organize, manage, measure and deliver new services.

2.3 Case study: ICT in the port of livorno

The Port of Livorno is one of the largest multipurpose Port in the Mediterranean Sea, and it is one of the major Italian Port gateway belonging to core network of TEN-T (Scandinavian-Mediterranean Corridor).

In order to speed-up the information flows inside the Port Areas and among the other transport nodes, since 2000 Livorno Port is equipped with an optical fibre network. Thanks to this broadband backbone in terms of connectivity, Livorno Port is pervasively monitored and centrally controlled. The good connectivity inside the Port has established the basis to create the smart environment, by making Livorno Port as interesting inter-modal point of interest, where to develop and validate Intelligent Transport Systems and innovative ICT technologies. The Port is working to set-up a broadband backbone in the surrounding areas (with particular attention to the link with Guasticce Freight Village), to create controlled, monitored and fast transfers.

To support the design and implementation of the ICT Technical Agenda, in 2015 Livorno Port Authority has created the "Joint Laboratory of Advanced Sensing Networks & Communication in Sea Ports", that provides a continuous and effective presence of CNIT (National Inter-University Consortium for Telecommunications) researchers at the Port.

The objective of Livorno Port and the mission of the Joint Lab is to create smart and interoperable Port environments, within which people, vehicles, objects, equipment, and cranes can communicate with both each other and a central monitoring and control system. For this purpose, Livorno Port Authority has developed the monitoring and control system of the Port of Livorno, *MONI.C.A.* (Monitoring and control application). *MONIC.A.* collects and integrates all data coming from distributed sensors network, information systems (i.e Port Community System, Harbor's systems, Operators' systems, dangerous goods system, etc.), tracking and tracing devices (i.e. RFID) and other smart and intelligent devices. In addition to wireless sensors network, Livorno Port is testing the vehicular communication, to gather information coming from vehicles passing through the Port (i.e. destination, speed, hazardous events, etc.), becoming, by so doing, the new test-bed for European Telecommunications Standards Institute Plugtests 2016. To identify smart solutions for MoS, Livorno Port has participated in different European Commission and National Initiatives as *Business to Motorways of the Sea* (B2MoS 2012-EU-21020-S), *Monitoring and Operation Services for Motorways of the Sea* (MoS4MoS 2010-EU-21102-S), *Mediterranean Information Traffic Application* (MED.I.T.A. Med Programme 2007–2013), towards the creation of intelligent smart MoS and smart logistics chains. RFID technology and the digitalization of transport documents have had central role in these initiatives.

In particular, RFID technology has been used: i) to improve customs procedures and port security in terms of cost and time, speeding-up cargo controls (active RFID), and ii) to track and trace trailer and their cargo along both the MoS and the logistics chain (passive RFID). The creation of digital and smart ports is the real challenge to increase the competitiveness level of the MoS, considering that nowadays, users are interested in obtaining real time information about their cargo and the events occurred during the path.

3 INFRASTRUCTURE PRIORITY FOR EFFICIENT MOS IN SEAPORTS

3.1 Introduction

Infrastructure development of the country, both on the maritime and earth side, by the Maastricht Treaty onwards must comply with decisions taken collectively and move in synergy with the European mark. Thus, the decisions taken by the EU in 2010 with the Regulation no. 913/2010 are influencing the development of the national port system-but until 2020 since that regulation has established the criteria and rules for the construction of a European logistics network whose international access points will be the main port systems continental. (Commission, 2010).

The way to relaunch SSS is to build real sea motorways within the framework of the master plan for the trans-European networks. This requires better connections (accessibility) between ports and rail together with improvements in the quality of port services, port facilities, port community and port monitoring systems, shipping services (frequency, regularity, safety, security). (Directorate-General for Internal Policies, 2014).

Port infrastructures include not only facilities related to the delimitation of areas, but also port gates, road and rail connections to the hinterland and dedicated facilities.

The intermodal transport quality is a critical factor in the organization of the intermodal chain based on maritime transport. In general, studies on the intermodal quality, and in particular, those on maritime chains, are the pillars towards the optimal planning of intermodal transport and the removal of bottlenecks. (Isfort, 2015).

3.2 Terminal layout

Motorways of the Sea area is one of the MoSt complex to organize, implement and manage; must face the growing demand for mobility of passengers and cargo, matching the needs of businesses, ship owners, tourists and operators, aiming at optimize the connections of the port with users and with the surrounding area. In this scenario, relevant issues are related to port security, customs management of goods and passengers, infomobility, the management of operational yard, the use of technologies and of advanced sensors.

The main idea for the organization of area MoS is based on traffic flow management and the balance of incoming and outgoing flows of vehicles, making the distinction between heavy and light ones, already from the port access area. Even travelers without car or truck should benefit from a specialized connection through a mechanized pedestrian path which allow them a direct link to the boarding dock. MoS area also needs all the auxiliary services for workers, passengers and users such as: customs offices, control areas, local duty free, landing and boarding hall, shopping center, bar. It is also possible, to make modern, environmentally friendly, sustainable and technologically advanced terminal, to equip it with photovoltaic panels. (Mare, 2014).

MoS must have a modern platform of infrastructure and specific requirements: it must have an adequate surface for the movement of goods, including docks and maneuvering areas for loading and unloading, customs stations, the long stock areas for trucks and other means, buffer area for motor vehicle accumulation, yards accumulation of heavy vehicles; the viability input and output must be designed and managed not only for the standard conditions of operation in the winter months, when the passenger flow is considerably lower, but especially consider summer influx of conditions in which there is combination of heavy and light vehicles. The MoS platform need for a commercial area for passengers and port operators, easily accessible from both categories of users. This area can be connected directly with the operational areas for the best usability of services for heavy vehicles drivers. Buffer areas must be flexible and specialized by category of traffic to facilitate the movement of vehicles and customs operations. The same concept should be applied to ship buffer area. The viability for light and heavy vehicles must be provided not only both for different types of means but also for different streams, load/unload. It is possible to provide the realization of pre-boarding area where manage customs operations, with dynamic and flexible accumulation lanes that lead directly to the appropriated dock. The unloading flow is disposed in advance with loading one without overlapping. The management of the flow and of the relative accumulation and outflow are the main aspects for a MoS terminal, because these aspects directly related to passengers and port operators transit time. Another important feature of a functional MoS terminal is the creation of a single access gate. The gate must be equipped with technological components and sensors of the latest generation for monitoring, data warehousing and terminal security. The core facilities should be located centrally within the MoS terminal. The core of a MoS terminal can be identified in the boarding terminal, which should be conceived as a dynamic element. Inside are placement offices, bar, shop, information desk, companies, customs, duty-free, waiting rooms, toilet facilities. It must in fact be easily reached both by external users and passengers.

3.3 MoS and railway

Since MoS areas are created to move cargos from the road in order to limit the negative impacts generated by the intensive use of road transport in terms of pollution, congestion and safety, terminals need to develop a functional railway system to handle the amount of traffic coming in and out. Catching flows to be transferred from road to sea are those with a distance greater than 500 km. Such movements represent a marginal share of heavy road transport. But the environmental, social and economic impacts of this traffic part are much more significant than those generated by the short-range transport. The movement of goods by road is mostly concentrated over distances below 100 km, however, from the point of view of the negative and energy consume impacts, one ton of goods carried over 600 km generates negative impacts and consumption energy six times higher than those required to transport the same quantity of goods to a trip of only 100 km.

Intermodal transport is hindered by the situation of the national railway network which cannot guarantee the flow of any type of vehicle or goods across the country. At European level, the Italian railway network, from the infrastructural point of view, is the most backward: ports of central/southern Italy are not able to receive cargos (trailers, trucks, swap bodies) because of rail loading gauge. (Directorate-General for Internal Policies, 2006).

A MoS railway terminal provides specific infrastructure and equipment as a freight railway station, railway yard for handling wagons/goods, lifting means and cranes.

3.4 Extra Schengen area

The Extra-Schengen area is an area in which they are handled cargo and passengers, coming from countries outside the Schengen Agreement.

Within this area there must be space for the movement of vehicles and passengers, spaces for checks by police and agents of Customs, a viability provides the unloading/loading without intersection of the flows.

The design of a functional Extra-Schengen area provides the division of the different flow: entry and exit. There is the need to maintain separate the viability not only as it regards the flow of vehicles from internal countries of Schengen area than those with external origin, but also of commercial vehicles by private ones, or users that are to undergo inspections by customs. Separating vehicle flows is essential for a complete functionality of the area, allowing safety and reliability controls. An improper localization of different functions would imply traffic jams and queues at terminal gates.

In addition to roads and yards, for the monitoring of vehicles and passengers areas shall be equipped for additional services to users and operators.

3.5 Case study: Livorno port

The current MoS area in the port of Livorno is located in the north of the city, within Porto Nuovo zone. Analyzing the area from the point of view of infrastructural layout, it has good connections to the road network but the railway sector still has expansion capabilities.

The road network does not differentiate between incoming and outgoing flows, creating promiscuity between commercial and private vehicles.

The rail network presents railways tracks out of service that should be reactivated, even in relation to recent demands of logistics operators, interested in having dock directly linked to the railway.

The access gate of the vehicles is the Varco Galvani, which has two in/out lanes per way, differentiated for trucks and cars. The main problem is the placement of the gate, that does not have a buffer area sufficient for the vehicles exiting from the terminal.

Presently, all rail traffic in/out the port of Livorno goes through the Livorno Calambrone freight station, which lies in the Genova-Roma railway line and where all shunting operations are performed. The port railway network suffers from poor connectivity, which currently hinders the rail operations as a result of slow shunting operations and low quality accessibility to the shore.

As a consequence, Livorno Calambrone represents a bottleneck of the Core rail network, as time-consuming and expensive shunting operations are required to forward freight trains from the port to the hinterland. Works are already ongoing to directly link the largest container terminal of the port, 'Darsena Toscana' to the rail network.

It expects to achieve some interventions to develop the rail network so as to make more effective links between the port and the industrial hinterland, reducing the problems created by the Calambrone station. An important intervention, that represents the first step towards the enhancement of better rail connectivity of the port to the hinterland, is the realization of the rail overpass on the Genova-Roma railway line to avoid the Calambrone Station and put the train on the network. This infrastructure will be a source of development not only in a geographical perspective, but also from the point of view of fully integrated logistics systems. In fact, the rail overpass is needed to ensure fast and efficient transfer of goods between the port and the freight village of Guasticce. In this sense, the expected benefits from the realization of the rail overpass can be split into local and network ones, since the realization of the overpass will enhance significant improvements both for the logistics within the Livorno Logistic Node and the rail transport along the corridor as well.

The current layout of roads and infrastructure generates the need to make a short-term intervention to increase the functionality of the entire area. The implementation of buffer areas for commercial vehicles would avoid traffic jams and the flows in/out would not suffer slowdowns.

Within MoS terminal, temporary storage area (TC) of goods are essential for current trades and the lack of these zones creates considerable problems for operators. In case of short-term changes of the terminal, the creation of TC area is essential.

During 2015 it was approved the Port Masterplan for the Port of Livorno, which includes large infrastructure investment for the construction of an expansion to the sea insisting on the embankment in part already realized and partly newly built.

One of the main issues for the design of the *Darsena Europa* was the need to offer not only docks, quays and yards but also an intermodal equipment and efficient logistics tools that will lay the foundations to achieve the realization of the whole port-logistics platform as a single entity (Livorno, 2015).

In this sense it was necessary to characterize the port areas with specific logistics functions: rationalize and make efficient the different activities.

The project involves the port expansion in the sea with the construction of new yards and new docks. The functional areas are divided into *Molo Nord, Molo Sud, Darsena Petroli* and *Darsena Fluviale*. These will be equipped with all the necessary infrastructure for connecting the piers with the hinterland, areas for handling and movement of goods and direct connections to the internal markets (Livorno, 2015).

Within this allocation, it also planned to build a new area set to the Motorways of the Sea, in the *Molo Nord*. This area has been designed according to the expected traffic volumes and into consideration that the Port of Livorno is the first Italian port for RO/RO traffic, as shown by the statistics for the year 2015 (Autorità portuale di Livorno, 2015).

The plan intends to develop the road link (with motorway features) and the main railway lines, such as the Tyrrhenian axis and connecting via Pisa-Florence to the Adriatic coast. Concerning the organization of the new terminals of the port of Livorno, it was identified a solution for railway infrastructure considering the new configuration of the areas, *Molo Sud* and *Molo Nord*. The electrified railway track, 750 m length, in accordance with the European interoperability standards, will split with two new connections to two operative beams of different docks, one for the new container terminal (*Molo Sud*) and the other for the new MoS terminal in RO/RO traffic service (the ferroutage terminal).

Besides railway tracks, the new terminal will set up by operational yard for the handling of goods and vehicles, lifting storage area, storage technical warehouses and depots for terminal functions.

4 CONCLUSIONS

MoS often conjure up trucks loading and unloading from ferries within terminals, featured as slightly more than parking slots. Actually MoS are a rather complex world where onboard ICT tools communicate with infrastructures and smart devices placed in ports and within other logistic facilities. Such a framework is then subject to the challenges of the shipping industry, that is volatile freight rates, the availability of cargo and the seasonality of traffic. Moreover, MoS face the competition of all road cheap services, what is on the contrary hardly ever the case for deep sea connections. Viable MoS have to lay on financing schemes that have been often provided through public funding, especially when it comes to the testing and implementation of ICT and the launch of new maritime linkages. The paper shows how important is to ensure the sustainability and affordability of infrastructural and infostructural solutions, through a greater involvement of private operators and better integration of all these hardware and software components. As a consequence, MoS can have an impact on port communities and transport operators, broadening the logistic offer of ports and port related logistic systems. MoS services call for global solutions, involving shipping industry's stakeholders, public administrations such as Port Authorities and inland transport operators, fine-tuning private investments and public policies. In this respect, we pointed out how much important is the proper planning of port areas by competent public authorities for smooth and efficient working of MoS and MoS transport related services. Administrations and public research bodies are also essential in the start-up phase of new technological tools, as the industry may not be yet capable of identifying which IT solutions are most suitable for its needs. Public support, via European programmes, may still be needed in the future too, especially in the light of developing innovative tools and testing systems to trace, monitor and control MoS related cargo. In so doing MoS chains and ports can better compete with traditional all road transport through enhanced sustainability (both economical as well as environmental) and security of goods. Ports dedicated infrastructure can therefore become more efficient and better exploited, achieving greater return on investment and thus attracting the needed private funding.

REFERENCES

Autorità Portuale di Livorno 2015. Relazione generale—Piano Regolatore del Porto di Livorno.
Baindur, D. & Viegas, J. 2011. Challenges to implementing motorways of the sea concept-lessons from the past *Maritime policy and Management*. Vol. 1. - p. 673–690.
Bhavin, B. & Vipul, J. 2004. Data Mining: Knowledge Discovery in Databases *International Journal of Research in Computer Science and Management*. Vol. 1(2): p. 74–76.

Commission European 2001. Transport White Paper.
Commission European 2008. Communication from the Commission providing guidance on State aid complementary to Community funding for the launching of the motorways of the sea.
Commission European 2009. The European Union's Maritime Transport Policy for 2018: Brussels.
Commission European 2010. Directive 2010/40/EU.
Commission European 2010. Regulation (EU) No 913/2010 of the European Parliament and of the Council of 22 September 2010 concerning a European rail network for competitive freight.
Commission European 2011. Transport White Paper, COM(2011) 144 final.
Commission European 2015. Digital Single Market Priority.
Commission European Directive 2010. 2010/65/EU.
Directorate-General for Internal Policies European Parliament 2014. Improving the concept of "Motorways of the Sea".
Directorate-General for Internal Policies European Parliament 2016. Motorways of the Sea—Modernising European short sea shipping links.
European Parliament-Directorate General for Internal Polcies 2016. Assessment of Europe Connecting Facility-An in-depth analysis.
Expert Group 5-TEN-T Policy Review 2010. Funding Strategy and Financing Perspective.
Gartner IT Glossary 2016. http://www.gartner.com/it-glossary.
Giurini, A., La Tegola, O. & Miranda, L. 2012. La sicurezza sul lavoro nei porti *I Working Papers di Olympus.—09*.
Isfort Confindustria Italia 2014. *Autostrade del Mare 2.0, Risultati, Criticità, Proposte per il rilancio*.
Isfort Confindustria Italia 2015. *Sviluppo dell'intermodalità—Autostrade del Mare 2.0 e combinato marittimo*—Rapporto finale.
Jain, P.C. & Vijaygopalan, K.P. 2012. RFID and Wireless Sensor Networks.
Juan C., Olomos, F. & Pérez E. 2016. Decision support system to design feasible high-frequency Motorways of the Sea: a new perspective for public committment. *The Engineering Economics* Vol. 61 (3): p. 163–189.
Ministero delle Infrastrutture e dei Trasporti 2015. Piano Nazionale della Portualità e della Logistica.
MIT Transport Italian Ministry of Infrastructure and Transport 2015. Ministerial Decree n. 287.
Pagano, P., Falcitelli, M., Ferrini, S., Papucci, P., De Bari, F. & Querci, A. 2016. Complex Infrastructures: the benefits of ITS services in seaports in Pagano P. (eds), *Intelligent Transport Systems: from good practices to standards*. CRC Press.
Parliament European 2013. Regulation of the Council and the European Guidelines of the TEN-T.
Rete Autostrade del Mare, Sviluppo Italia 2014. Autostrade del Mare: Il Master Plan nazionale-Infrastrutture.
Suarez-Aleman, A. 2015. Short Sea Shipping in today's Europe. A criticial review of maritime policy *Maritime Economics and Logistics*. p. 1–21.
Van De Horst, M. & De Langen, P. 2015. Coordination in hinterland transport chains: a major challenge for seaports communities in Haralambides H. (eds), *Port Management* Palgrave Mac Millian.

Transport Infrastructure and Systems – Dell'Acqua & Wegman (Eds)
© 2017 Taylor & Francis Group, London, ISBN 978-1-138-03009-1

Assessing sustainability of road tolling technologies

D. Glavić, M. Milenković, A. Trpković & M. Vidas
The Faculty of Transport and Traffic Engineering, University of Belgrade, Belgrade, Serbia

M.N. Mladenović
Aalto University, Espoo, Finland

ABSTRACT: Facing a choice between a range of road tolling technologies, decision-making has to account for the effects that specific road tolling technology causes. For example, road tolling technology has direct impact on increase of vehicle operating costs, increase of greenhouse gas emissions, increase in travel time costs, as well as decrease of road users' level of service and safety. Thus, selecting the road tolling technology becomes a question of managing the technology's effects at an optimum level. To this end, this paper evaluates both wide-spread as well as new emerging road tolling technologies, including RFID, GNSS/CN, DSRC, smartphone, infrared, ANPR, smartcard, vignette, and ACM. The methodology for analyzing these road tolling technologies focuses on technical, financial, efficiency, environmental, and social aspects. As a result of this analysis, different technologies are comparatively ranked through a multi-criteria analysis.

1 INTRODUCTION

An important element in the delicate task of road pricing is the selection of tolling technology. This task is very complex and depends on many factors. The conventional road tolling technology involves toll plazas with manual toll collection in the form of cash, tokens, or other physically transferable material. This system of toll collection is inherently inefficient for many reasons. Consequently, transportation planners and road managers face the challenge of selecting Toll Collection Technology (TCT) for upgrading the existing system or when building new motorways.

A series of EU directives is trying to put some order in the area of road tolling technology, both in terms of technology and price systems as well as on other outstanding issues. The main objective of the EU is to achieve interoperability by policy called one market, one contract, one OBU (On Board Unit). Other objectives relate to the unification of prices and vehicle categories (Glavić 2016). EU based its policy on toll in following two documents:

– Directive 2004/52/EC. This Directive prescribes the conditions necessary to ensure interoperability of road tolls in the EU. This applies to the electronic collection of all types of road fees, on the entire road network, urban and intercity, highways, high-speed roads, and various infrastructure and facilities, such as tunnels, bridges and even ferries.

– Commission Decision 2009/750 / EC. This decision provides a description of the technical systems and interfaces necessary for the EETS (European Electronic Toll Service).

In accordance with Directive 2004/52/EC and the Decision 2009/75/EC European Union creates the toll policy in Europe according to the principle of "one OBU one Contract all EU".

Besides the many and diverse interests at stake, there is a multitude of TCT currently in use. Nowadays, the TCT used worldwide include vignettes, Dedicated Short Range Communication (DSRC) with barriers, DSRC Multi-Lane Free Flow (MLFF), barcode TCT, Radio-Frequency Identification (RFID) TCT, infrared TCT, smart card TCT, Automated Number Plate Recognition system (ANPR) TCT, Global Navigation Satellite System/ Cellular Network (GNSS/CN) TCT, tachograph-tolling, Automated Coin Machine (ACM) and smartphones tolling (Glavić & Milenković 2016).

Taking into account the diversity and conflict of interests as well as a multitude of TCT, the decision-making is further complicated having in mind that each TCT has certain advantages and drawbacks from the standpoint of road managers and users. For example, different TCT respond differently to the need to maximize the toll revenue, which is often an important criterion for road managers. On the contrary, different TCT cater differently to the user needs for reducing delays and costs, or improving safety. Moreover, TCT needs to achieve important sustainability goals from environmental, social, and economic aspect.

Similar decision-making issues have been observed before in selecting transport technology, e.g., traffic signal controllers (Abbas et al. 2013, Mladenovic et al. 2013, Mladenovic & Abbas, 2013). In order to deal with a range of decision factors and a multitude of TCT, this paper uses Multi-Criteria Analysis (MCA). In order to cater for all the dimensions of sustainability, the decision criteria include Level Of Service (LOS), environmental, social and economic impact of TCT. The aim is to identify a group of TCT that satisfy sustainability criteria, that have satisfying LOS for road users, and can fulfill tolling manager's objectives. The focus of the analysis is on the groups of TCT instead of individual TCT, aiming to test the decision-support framework and normalize the impact that different groups of TCT might have on the users, environment, and society. The grouping of TCT is additionally reasonable taking into account that there are more than 13 TCT worldwide, constraining the collection of necessary input data.

2 TOLL COLLECTION TECHNOLOGY OVERVIEW

This section provides an overview of all the TCT present worldwide at the moment of evaluation. Furthermore, this section will focus in particular on the dimensions of environmental sustainability, system architecture, system efficiency, and user friendliness.

2.1 Description of TCT

Vignette is a sticker which, once bought, pays the toll for a specific time period. The vignette must be applied to the car windshield (Glavić, 2013).

DSRC with barriers is a type of non-contact TCT where a vehicle does not have to stop at the toll plaza. The vehicle only needs to reduce its speed in order to establish the contact and recognition through the On-Board Unit (OBU), in order to receive the permission to pass through the gate (Glavić, 2013).

DSRC Multi-Lane Free Flow (MLFF) system, have antennas above particular locations that detect traffic flow and record the use on the OBU. The technology used in MLFF system is designed so that the vehicles can maintain their speed and change lanes while passing below the toll collection portal (Glavić 2013).

ANPR TCT system uses a stationary camera for recording and identifying the registration plate numbers of vehicles passing through the toll plaza. The identified registration plates are paired in the database and a particular amount of money is deducted from the user's account. If the recorded plate number is not read out properly or not found in the records, an enforcement violation alarm is generated to alert the authorities.

Barcode TCT is a subcategory of ETC. Within this system, the barcode is a sticker applied to the windshield of the vehicle which is read by a laser scanner while the vehicle passes through the toll plaza. It is the simplest as well as the oldest technology (Sharma 2014).

RFID TCT system contains the OBU installed on the front windshield of the vehicle. At toll plazas, this system is read by a RFID frequency reader or by an antenna. It can be either prepaid or postpaid, with or without a gate (Blythe 1999, Sorensen & Taylor 2005).

GNSS/CN technology includes a global navigation satellite system incorporated with the communication mechanism. It functions using a global positioning sytem unit (GPS/Gallileo/Glonas) mounted on the OBU which stores the coordinates of the vehicle and sends the transaction information to toll collection authorities via GSM/3G/4G (global mobile communication system) (Blythe 1999, Catling 2000, Charpentier & Fremont 2003).

Infrared TCT is similar to RFID and ETC DSRC systems; the only difference is that it has an active infrared unit installed in the vehicle which contains all the information. (Shieh et al. 2005, Staudinger & Mulka 2004, Tropartz et al. 1999).

Tachograph TCT records the mileage driven by the user through an OBU connected electronically to the vehicle´s odometer (the instrument that measures the mileage). A tachograph system which is in place in New Zealand requires manual, rather than electronic, data collection (Technology options for the European Electronic Toll Service 2014).

ACM is a machine with a slot for inserting coins and paper money. Automated coin machine sometimes has a basket in which drivers insert coins in order to pay toll. This machine based on the dimensions as well as weight of the coin recognizes the value which should be paid.

Smart card TCT represents a memory card in which the details of a particular person and certain amount of money are stored. The smartcard based toll-gate automated system functions on the basis of contact communication between the smart card and the reader. Whenever a smartcard is inserted into the smartcard reader, the reader will read the data stored in the card (Sridhar & Nagendra 2012).

Smartphones TCT is still in its early stages of development. An example of mobile and Smartphone ETC integration is the m-Toll project. The m-Toll project relies on the use of Smartphone WiFi connection to authenticate, validate and charge road users without the need for any dedicated hardware for the end user. Servers installed at the toll plaza can detect smartphones from a distance of 600 meters and deduct a particular amount of money from the connected account through NFC (Technology options for the European Electronic Toll Service 2014).

2.2 TCT aspects of environmental sustainability

One of the aspects by which the toll technologies are different is the environmental aspect. Some technologies, which require a speed change and stop the vehicle, require more fuel consumption and therefore a higher emissions of pollutants. For example, Coelho et al. (2005) quantified emission impacts of toll facilities on urban corridors in Lisbon, Portugal, and found that 61–80% CO_2 reduction could be achieved by entirely switching MTC to ETC. Also, according to the data collected between 1999 and 2005 on the New Jersey Turnpike in New Jersey, USA, Bartin et al. (2007) conducted a before-and-after study to examine the benefits of ETC on reducing air pollution. Results indicated that the overall emissions reduction of Carbon monoxide (CO), Hydrocarbon (HC), Nitrogen Oxide (NOx) and particulate matter (PM10) was 10.1%, 11.2%, 3.4% and 13.0%, respectively, after the deployment of ETC. Perez-Martinez et al. (2011) evaluated the energy consumption and the associated CO_2 emissions at three toll systems in Spain and found that the type of toll collection is a major factor affecting energy efficiency.

2.3 TCT aspects of SMART system architecture

It is to be noted that sometimes apart from the main objective the same technology may also serve in several other manners. Utilization of ETC systems, which represent smart technology, for navigation, theft prevention of automobiles and traffic surveillance are well examples of such secondary objectives.

2.4 TCT aspects of system efficiency

Different technologies have different toll collection efficiency parameters. The efficiency of a particular toll can be measured by the various parameters, by the capacity of a toll lane, travel speed, number of stops, the time spent waiting and others.

ETC lanes improve the speed and efficiency of traffic flow and save drivers time. Manual toll collection lanes handle only about 350 Vehicles Per Hour (vph), and automated coin lanes handle about 500 vph. An ETC lane can process 1200 vph, with ORT lanes allowing up to 1800 vehicles per hour (Tri-State Transportation Campaign 2004).

Since ETC allows toll transactions to be completed while vehicle travels at a higher speed, they may significantly reduce travel delay. Time saving is apparent from the difference between times for passing an MTC lane and an ETC lane (Tseng et al. 2014). According to the results of The Institute of Transportation in Taiwan (2001), the average toll transaction time for the ETC lane is 2.06 and for the MTC lanes it is 7.72 s.

2.5 TCT aspects of user friendliness and mobility

Users usually prefer technologies which are user friendly, that having high level of service and comfort during driving, which do not require stops or speed reduction due to enforcement or fee reasons, and for which travel delays do not exist.

The technologies which require minimum stopping or do not require stopping at all and where the capacity of the toll lane is higher provide greater mobility of the users especially in peak periods where queuing can be common on toll booths.

3 METHODOLOGY

Considering the semi-structured nature of the TCT selection problem, there is a need for a decision-support framework. MCA refers to the process of making decisions between a number of alternatives, by defining the criteria and their weights. The application of MCA thus results in the ranking of alternatives, from the most to the least favorable, thus allowing comparison of alternatives.

3.1 PROMETHEE decision-support framework

The MCA method selected for this research is Preference Ranking Organization Method for Enrichment Evaluations (PROMETHEE) was used. The PROMETHEE method is one of the most recent MCDA methods developed by Brans (1982) and expanded by Brans et al. (1986). PROMETHEE is primarily useful as the outranking method contrast with other multi-criteria analysis methods (Barns et al. 1986, Behzadian et al. 2010). One of the reasons for using PROMETHEE beside mathematical model is the fact that this method has excellent visualization of MCA process and results.

3.2 Definition of alternatives

All toll technologies can be grouped into three groups (three alternatives), depending on whether they require stopping, speed reduction or free flow driving of vehicles without any changes in vehicle speed. Alternative for MCA are:

Alternative 1: Manual tolling, Automated coin machine toll, vignette stickers, smart card tolling, bar code tolling and tachograph tolling.

Alternative 2: DSRC with barriers, INFRARED with barriers, RFID with barriers, ANPR with barriers

Alternative 3: DSRC MLFF, GNSS/CN, RFID ORT, SMARTPHONES, ANPR.

The following figures describe speed time dependences of each of 3 alternatives. The first group, Alternative 1, require speed reduction from about 120 km/h to 0 km/h and increase to around 120 km/h. This case is described by speed-time curve (Figure 2). The second group, Alternative 2, require speed reduction from about 120 km/h to around 30–50 km/h and increase to around 120 km/h. In this case exists delays only because of speed changes (Figure 3). In the case of Alternative 3 there was no delays because this technology do not require speed changes (Figure 4).

The first group, Alternative 1, includes technologies which require stopping with possible waiting

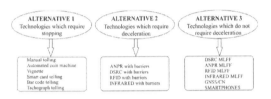

Figure 1.   Group of the alternatives.

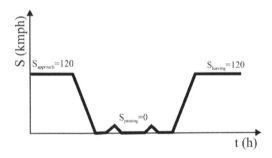

Figure 2.   Speed-time profile for the alternative 1.

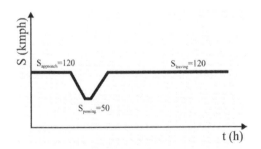

Figure 3.   Speed-time profile for the alternative 2.

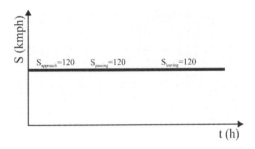

Figure 4.   Speed-time profile for the alternative 3.

in queue and stop and go conditions while queuing. This technology does not require enforcement for controlling toll users. Speed decrease to zero causes significant increase of VOC, TTC, GHG emission, lower LOS and decreased traffic safety.

The second group, Alternative 2, consist of the technologies which require deceleration to lower speed. This technology also do not require enforcement for controlling toll users. The decrease of speed to 30–50 km/h causes an increase of VOC, TTC, GHG emission, lower LOS and traffic safety compared to free flow speed.

The third group includes the technologies which do not require changes in the vehicles speed. This technology does require possible stopping due to enforcement for controlling toll users. Speed decrease can only be in the case of enforcement for controlling of toll users. This can increase VOC, TTC, GHG emission and lower LOS for users that are controlled.

3.3 Selection of criteria

As Table 1 below shows, the list of criteria consists of the following criteria: interoperability, efficiency, Enforcement, level of service, traffic safety, number of stops, an increase in travel time, visual impact of defacing the ambient environment, air pollution, noise, fair pricing for all users, the user friendliness and aesthetic effect to the vehicle.

3.4 Criteria weights

Determination of the weights is an important step in most MCA methods. Modified Digital Logic (MDL) method of subjective evaluation of criteria weights was used to determine the criteria weights. Each expert individually evaluated weights of the criteria and assigned points to the alternatives in a chart. Then, mean values of the obtained individual weights were used. For applications in which the number of design criteria is fairly large, assigning the importance weights among multiple criteria simultaneously may be very difficult for the decision maker. The Modified Digital Logic (MDL) method is used to address this issue

Table 1. The list and description of criteria.

Criteria	Description
Interoperability	Interoperability with other toll collection systems (with other countries).
Efficiency	This criterion includes the capacity of a lane, data rate and system complexity.
Enforcement	Control of the vehicles.
Level of service	The impact of the system on the level of service of traffic flow.
Number of stops	The number of stops during the tolling.
Traffic safety	The impact of the system on the traffic safety.
Delays	The increase in travel time that each of the TCT causes.
Visual impact ambient	The effect of TCT on visually defacing the environmental aesthetics.
Air pollution	To what extent the systems contributes to the pollution of the environment.
Noise	To what extent some systems influence the increase of noise.
Fair pricing	Whether the system is fair to all users by ensuring payment for what is used.
User friendliness	The easiness of using the system.
Aesthetic effect to the vehicle	The effect on the aesthetics of the vehicle.

by suggesting pair-wise comparisons of criteria (Dehghan-Menshadi et al. 2007). The decision makers use digital scoring scheme of {1, 2 and 3} to represent the less (1), equal (2), or more important (3) criteria. After all pair-wise comparisons are made, the MDL weights can be calculated as:

$$w_j = \frac{\sum_{k=1}^{n} C_{jk}}{\sum_{j=1}^{n} \sum_{k=1}^{n} C_{jk}}, \quad j \text{ and } k = \{1,\dots,n\} \text{ and } j \neq k$$

If two criteria j and k are equally important, then $C_{jk} = C_{kj} = 2$, otherwise $C_{jk} = 3$ and $C_{kj} = 1$ if the criteria k is more important than the criteria j. If the criteria k is less important than the criteria j, then $C_{jk} = 1$ and $C_{kj} = 3$.

The final values of the weights were obtained by determining the mean value of individual MDL weights for all 10 experts (Table 2).

3.5 Preference functions

Preference function are selected by analysing each criterion with indifference and preference thresh-

olds. Type of function and indifference and preference thresholds are presented in Table 2.

3.6 Scoring of alternatives

In the PROMETHEE the qualitative scale is defined by

- a number of ordered levels (from worst to best),
- numerical values associated to these levels,
- whether the numerical values should be minimized or maximized (scale orientation).

For purpose of this paper criteria are determined by min/max orientation, mostly scale 1–5 or 1–9 are used to evaluate alternatives by each criterion. For each criteria expert from that field conducted scoring (Table 2).

Table 2. PROMETHEE base matrix.

Criterion	Unit	Min/Max	Weight	Preference Fn.	Thresholds	Q: Indifference	P: Preference	S: Gaussian	Minimum	Maximum	Average	Standard Dev.	Toll with stopping	Toll with speed d...	Toll with free flow
Interoperability	9-point	max	8,00	Level	absolute	0,03	0,25	n/a	1,00	2,00	1,67	0,47	VB-8	VB-8	VB
Efficiency	5-point	max	8,00	Usual	absolute	n/a	n/a	n/a	2,00	3,00	2,67	0,47	bad	average	average
Enforcement	impact	min	8,00	Level	absolute	0,03	0,25	n/a	3,00	4,00	3,33	0,47	moderate	moderate	high
Level of serv...	6-point	max	8,00	V-shape	absolute	n/a	3,07	n/a	2,00	4,00	3,33	0,94	2,00	4,00	4,00
Number of st...	5-point	min	8,00	V-shape	absolute	n/a	2,00	n/a	0,00	2,00	0,67	0,94	2,00	0,00	0,00
Traffic safety	5-point	max	8,00	Level	absolute	1,00	2,00	n/a	2,00	3,00	2,67	0,47	bad	average	average
Delays	impact	min	8,00	Level	absolute	1,14	2,67	n/a	2,00	4,00	3,00	0,82	high	moderate	low
Visual impact...	impact	min	8,00	Usual	absolute	n/a	n/a	n/a	2,00	4,00	3,00	0,82	high	moderate	low
Air pollution	impact	min	8,00	Usual	absolute	n/a	n/a	n/a	2,00	4,00	3,00	0,82	high	moderate	low
Noise	impact	min	7,00	Usual	absolute	n/a	n/a	n/a	2,00	4,00	3,00	0,82	high	moderate	low
Fair pricing	5-point	max	7,00	Usual	absolute	n/a	n/a	n/a	5,00	5,00	5,00	0,00	very good	very good	very good
User friendli...	5-point	max	7,00	Usual	absolute	n/a	n/a	n/a	2,00	4,00	3,00	0,82	bad	average	good
aesthetic effe...	5-point	max	7,00	Level	absolute	1,00	2,00	n/a	2,00	3,00	2,67	0,47	bad	average	bad

807

3.7 PROMETHEE base matrix

All necessary input data regarding alternatives, criteria's, preference functions are shown in Table 2.

4 RESULTS

4.1 PROMETHEE I & II ranking

4.1.1 PROMETHEE I Partial Ranking

The PROMETHEE I Partial Ranking is based on the comparison of the leaving flow (Phi+) and the entering flow (Phi-). The PROMETHEE II Complete Ranking is based on the net flow Phi. Figure 5 shows PROMETHEE I Partial Ranking.

The left column corresponds to the Phi+ scores and the right column to the Phi- scores. They are oriented such that the best scores are upwards. The alternatives are ranked:

- according to the Phi+ (A3, A2, A1);
- according to the Phi- (A2, A3, A1).

Interesting is to notice that according to Ph-PROMETHEE I Partial Ranking (first place is A2) and Ph- PROMETHEE I Partial Ranking (first place is A3) we have different ranking.

4.1.2 PROMETHEE II Complete Ranking

According to the PROMETHEE II Complete Ranking i.e. Ph net flows given in Table 3 it can be concluded that PROMETHEE ranking is A3 (Phi = 0.226) followed by A2 (Phi = 0.186) and A1.

Table 3. Ranking of the toll alternatives using PROMETHEE model.

Rank	Action	Phi
1	A3-Toll with free flow	0.266
2	A2-Toll with speed decrease	0.186
3	A1-Toll with stopping	−0.452

4.1.3 GAIA visual analysis

The objective of GAIA is to describe the major features of the decision problems graphically:

- How much are actions different or similar to each other?
- Which criteria are conflicting with each other? Are there strong conflicts to solve? Are there groups of criteria expressing similar preferences?
- What is the impact of the weighing of the criteria on the PROMETHEE rankings?

The GAIA plane is a descriptive complement to the PROMETHEE rankings. GAIA starts from a multidimensional representation of the decision problem with as many dimensions as the number of criteria. A statistical method called the Principal Components Analysis is used to reduce the number of dimensions while minimizing the loss of information.

In Visual PROMETHEE three dimensions are computed:

- U is the first principal component, it contains the maximum possible quantity of information,
- V is the second principal component, providing the maximum additional information orthogonal to U,
- W is the third principal component, providing the maximum additional information orthogonal to both U and V.

Analyzing the GAIA plane given in Figure 6 one can conclude that tolling technologies with free flow are closest to decision stick and in same direction, closely followed by A2, while A1 is farthest and in the opposite direction from the decision stick.

4.2 Sensitivity analysis

Table 4 shows the final results of sensitivity analysis through possible range of criteria weights needed to keep the complete ranking unchanged and in order to retain the first place on the list.

From the above sensitivity analysis, the main conclusions are:

- Regarding sensitivity of the 1st place one can conclude that 1st place is stable according to most of criteria

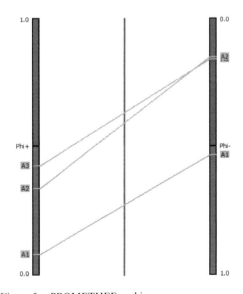

Figure 5. PROMETHEE ranking.

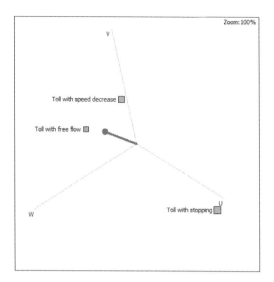

Zoom: 100%

Figure 6. Ranking of the toll alternatives using GAIA visual analysis.

Table 4. Sensitivity analysis for weight of criteria regarding the complete ranking and the first place.

Criteria	First place remains unchanged	Complete ranking remains unchanged
Interoperability	11.5%–100%	39.6%–100%
Efficiency	0%–100%	0%–43.2%
Enforcement	0%–100%	0%–100%
Level of service	0%–100%	0%–53.7%
Number of stops	0%–100%	0%–43.2%
Traffic safety	0%–100%	0%–100%
Delays	0%–83.3%	0%–69.2%
Visual impact ambient	0%–100%	0%–100%
Air pollution	2.1%–100%	2.1%–100%
Noise	1.1%–100%	1.1%–100%
Fair pricing	0%–100%	0%–100%
User friendliness	1.1%–100%	1.1%–100%
Aesthetic effect to the vehicle	0%–100%	0%–100%

– Regarding sensitivity of the complete ranking one can conclude that complete ranking is having lower stability then 1st place ranking stability but again according to most of criteria complete ranking is stable.

5 CONCLUSION

The A2 is having advantage over A3 according to PROMETHEE I ranking with negative flow Ph-. According to PROMETHEE I ranking with positive flow Ph+ A3 is having advantage over A2.

Analyzing the results according to PROMETHEE II complete ranking, one can see that the A3 (TCT which enable free flow of vehicles) are considered as the optimal solution, especially having advantage in ecological, social, level of service and traffic safety criterion. The results also indicate that the A3 is superior to A1 group of TCT.

The lack of analytically determined data in the scoring procedure for each of the alternatives per criterion is a limitation of this paper.

However, it is extremely difficult to analytically determine all values of the alternatives per criteria. In the situations where real data did not exist or was not available, the problem was overcome by including the experts from various fields who have the knowledge and experience and are therefore competent to subjectively evaluate the alternatives per each of the criteria. Thus, the MCA objectivity was realized through the expert evaluation using the scales 1–5 or 1–9 (Table 2).

The selection of the optimal toll collection system represents an important decision considering the fact that an inappropriate decision related to the selection of the optimal toll collection system can lead to economic, environmental and social problems both for the present and future generations.

The selection of the optimal toll collection system is significant for the road managers as well as for the users and society as a whole.

In addition, this paper provides a procedure for the proper selection one of the toll collection technology from an optimal alternative. The overall conclusion is that the Smartphone TCT, GSNN/CN TCT, DSRC MLFF, with RFID ORT are greenest and most efficient TCT.

5.1 Future research

There are several potential areas for further research. First, considering the fact that the smartphone technology is not yet a mature technology and it is in its development stage, there is a need for further testing and data collection on using smartphone-based TCT.

Second, relation between self-driving vehicles and TCT need to be investigated, as emerging technology. Currently there is no available data about this issue.

Third, when selecting the optimal toll collection technology, it is also extremely significant to consider the attitudes of the highway and toll users. In order to achieve this, a survey should be conducted to determine the attitudes of toll users.

Surveying toll users would provide the answers about the demands and expectations of the toll users. In other words, a series of opinions would be obtained which would significantly contribute to the ranking of toll collection systems from the users' point of view. The authors of this paper intend to make this the next step in their research on the topic.

REFERENCES

Abbas, M., Mladenovic, M., Ganta, S., Kasaraneni, Y., & McGhee, C. 2013. Development and Use of Critical Functional Requirements for Controller Upgrade Decisions. *Transportation Research Record: Journal of the Transportation Research Board*, (2355), 83–92.

Bartin, B., Mudigonda, S. & Ozbay, K. 2007. Impact of electronic toll collection on air pollution levels-estimation using microscopic simulation model of large-scale transportation network. *Transp. Res. Rec.* 2011: 68–77.

Behzadian, M., Kazemzadeh, R.B. Albadvi, A. & Aghdasi, M. 2010. PROMETHEE: A comprehensive literature review on methodologies and applications. *European Journal of Operational Research* 200: 198–215.

Blythe, P. 1999. RFID for road tolling, road-use pricing and vehicle access control. *Proceeding of the IEE Colloquium on RFID Technology*, London, UK, October 1999: 8/1–8/16.

Brans, J.P. 1982. Lingenierie de la decision. Elaboration dinstruments daide a la decision. Methode PROMETHEE. In: Nadeau, R., Landry, M. (Eds.), *Laide a la Decision: Nature, Instrument s et Perspectives Davenir*. Presses de Universite Laval, Quebec, Canada: 183–214.

Brans, J.P., Vincke, Ph. & Mareschal, B. 1986. How to select and how to rank projects: The PROMETHEE method. *European Journal of Operational Research* 24 (2): 228–238.

Catling, I. 2000. Road user charging using vehicle positioning systems. *Proceedings of the 10th International Conference on Road Transport Information and Control*, London, UK, 4–6 april 2000: 126–130.

Charpentier, G. & Fremont, G. 2003. The ETC system for HGV on motorways in Germany: first lessons after system opening. *Proceedings of the European Transport Conference (ETC)*, Strasbourg, France, 8–10 october 2003: 8.

Coelho, M.C., Farias T.L., & Rouphail, N.M. 2005. Measuring and modeling emission effects for toll facilities. Transp. Res. Rec. 1941: 134–144.

Dehghan-Manshadia, D., Mahmudib, H., Abediana, A., Mahmudic, R. 2007. A novel method for materials selection in mechanical design: Combination of non-linear normalization and a modified digital logic method. *Materials & Design* 28 (1): 8–15.

Glavić, D. & Milenković, M. 2016. Comparative analysis of road tolling technologies. *Second Serbian Road Congress*, Belgrade, Serbia, 9–10. June: 562–568.

Glavić, D. 2013. SWOT analysis of toll systems in Europe. *Put i saobraćaj* 59 (4): 21–30.

Glavić, D. 2015. Analiza sistema naplate putarine u Srbiji sa predlogom mera unapređenja. *Put i saobraćaj* 61(4): 56–62.

Glavić, D. 2016. Analiza naplate putarine u Srbiji i uporedna analiza sa novim trendovima u EU i svetu. *Second Serbian Road Congress, Belgrade*, Serbia, 9–10. June: 554–561.

Institute of Transportation. 2001. *Highway Capacity Manual in Taiwan*, Institute of Transportation: Taiwan.

M. Mladenovic, M. Abbas, S. Gopinath, & G. Faddoul. 2013. Decision-Support System for Assessment of Alternative Signal Controllers using Expert Knowledge Acquisition, *92nd Annual Meeting of Transportation Research Board*, Washington, D.C.

M. Mladenovic, M., Abbas. 2013. Improved Decision-Support Making for Selecting Future Traffic Signal Controllers using Expert-Knowledge Acquisition. *16th International IEEE Annual Conference on Intelligent Transportation Systems*, Delft, Netherlands.

Perez-Martinez, P.J., Ming, D., Dell'Asin, G. & Monzon, A. 2011. Evaluation of the influence of toll systems on energy consumption and CO2 emissions: a case study of a Spanish highway. *J. King Saud Univ. Sci. 23* (3): 301–310.

Sharma, P. & Sharma, V. 2014. Electronic toll collection technologies: A state of art review. *International Journal of Advanced Research in Computer Science and Software Engineering* 4 (7): 621–625.

Shieh, W.Y., Lee, W.H., Tung, S.L. & Ho, C.D. 2005. A novel architecture for multilane-free-flow electronic-toll-collection systems in the millimeter-wave range. *IEEE Transactions on Intelligent Transportation Systems* 6 (3): 294–301.

Sorensen, P.A. & Taylor, B.D. 2005. Review and synthesis of road-use metering and charging systems. Report Commissioned by the Committee for the Study of the Long-Term Viability of Fuel Taxes for Transportation Finance, *UCLA Institute of Transportation Studies.*

Sridhar, V. & Nagendra, M. 2012. Smart card based toll gate automated system. *International Journal of Advanced Research in Computer Engineering & Technology* 1 (5): 203–212.

Staudinger, M. & Mulka, E. 2004. Electronic vehicle identification using active infrared light transmission. *Proceedings of the ITS America 14th Annual Meeting and Exposition*, San Antonio, Texas, USA, 26–28 April: 357–376.

Tri-State Transportation Campaign. 2004. *The Open Road. The Region's Coming Toll Collection Revolution.*

Troupartz, S., Horber, E. & Gruner, K. 1999. Experiences and results from vehicle classification using infrared overhead laser sensors at toll plazas in New York City. *Proceedings of the IEEE/IEEJ/JSAI International Conference on Intelligent Transportation Systems*, Tokyo, Japan, 5–8 October 1999: 686–691.

Tseng, P.H., Lin, D.Y. & Chien, S. 2014. Investigating the impact of highway electronic toll collection to the external cost: A case study in Taiwan. *Technological Forecasting & Social Change* 86: 265–272.

Transport Infrastructure and Systems – Dell'Acqua & Wegman (Eds)
© *2017 Taylor & Francis Group, London, ISBN 978-1-138-03009-1*

Traffic impact of automated truck platoons in highway exits

R. Mesa-Arango & A. Fabregas
Florida Institute of Technology, Melbourne FL, USA

ABSTRACT: This paper assesses the impacts on travel time and travel time reliability associated with the operation of Autonomous Truck Platoons (ATPs) at freeway diverge areas. ATP technologies are mature, and commercial deployments are expected in the next few years (earlier than autonomous passenger cars). However, there is insufficient information about the impacts that ATPs will generate on the traffic stream, especially around exit lanes. This paper proposes an interdisciplinary framework to integrate ATPs into a microscopic traffic simulator. A combinatorial experiment is performed to test the impact of four experimental variables on travel time and reliability, i.e., (i) traffic volume projections, (ii) ATP penetration rates, (iii) ATP sizes, and (iv) ATP gaps. Two performance metrics are employed and statistically tested to quantify the impact of these variables on (a) though- and (b) divergent-traffic. Numerical results demonstrate the significance and impact of experimental variables on travel time and reliability.

1 INTRODUCTION

This research investigates the impact of Automated Truck Platoons (ATPs) on the traffic stream in the vicinity of a freeway diverge area. The impacts on average travel time and travel time reliability are considered for freeway users that drive through and turn right onto a right-side off-ramp. Truck platoons are groups of multiple trucks traveling close behind each other to reduce the power required to overcome aerodynamic resistance, which improves gas consumption and, hence, reduces operational costs, and undesired emissions. Although closer inter-truck gaps represent higher savings, small gaps are unsafe and inconvenient for human drivers. However, new automation technologies facilitate the implementation of ATPs with shorter gaps and higher financial benefits for the trucking companies. Simulations and testbeds report the significant energy savings that can be achieved with ATP systems, i.e., between 14% and 25% (Tsugawa et al. 2011, Vegendla et al. 2015, Bevly et al. 2015, Nowakowski et al. 2015).

Within the last three decades, several ATP-related projects have been developed around the world, i.e., in Europe, Japan, and the United States (U.S.), to test the feasibility of this technology and measure its benefits (Tsugawa et al. 2011, Nowakowski et al. 2015, Bishop et al. 2015, Shladover 2012, Kunze et al. 2009, Kunze et al. 2011, Janssen et al. 2015, Shladover 2010, Nowakowski et al. 2015). Furthermore, in the last two years several truck manufacturers have demonstrated their capabilities to develop and operate ATPs in real-world settings (Markoff 2016, Abt 2016, Alkim et al.

2016). ATP demonstrations, like the European Truck Platooning Challenge (ETPC) (Alkim et al. 2016), have been motivated by public agencies to understand the interdisciplinary challenges associated with this new technology. Given the imminent deployment of ATPs for commercial purposes, public agencies in the U.S. are energetically conducting projects to demonstrate their capabilities in connected freight corridors (Transportation Pooled Fund (TPF) Program 2016), and improve their understanding on the system (Bevly et al. 2015, National Cooperative Highway Research Program (NCHRP) 2015).

Many researchers have developed studies to anticipate the impact of ATPs from different perspectives, i.e., operations research (Saeednia & Menendez 2016, Liang et al. 2016, Van De Hoef et al. 2015, Larsson et al. 2015, Larson et al. 2015), game theory (Farokhi & Johansson 2015), human factors (Zheng et al. 2015, Skottke et al. 2014), and business/industry (Bevly et al. 2015, Shladover et al. 2015). However, few is known about the traffic impacts associated with the insertion of ATPs into the traffic stream. Although initial works suggest that increased ATP penetration rates will improve traffic flow stability, highway capacity, and critical/jam density (Bevly et al. 2015, Arem et al. 2006, Schakel et al. 2010, Arnaout & Bowling 2014), there is uncertainty about these results because a clear microscopic traffic simulation framework to model the interaction between traffic and ATPs under different conditions, e.g., ATP size, gap, penetration rate, among others, has not been formally presented. Developing a framework that facilitates the design of scenarios for different ATP conditions is a fundamental requirement to confirm the traffic benefits claimed in previous studies and test scenarios related to the new challenges that ATPs will impose on decision makers, i.e., traffic operation managers, and transportation planners. One of this scenarios correspond to the conditions that traffic users will face at freeway diverge areas. This is important because when large number of ATPs with multiple coupled trucks and short inter-truck gap drive close to off-ramp exits, the other users might experience reduced gaps that will generate dangerous conditions or delays in order to take the right exit at the right time. On the other hand, large penetration rates might reduce the number of uncoupled trucks on the road and hence, balance the negative vision stated above. Given the complexity of this interaction, simulation experiments are required to assess the impacts of ATPs on travel time and travel time reliability for multiple users under these conditions.

This paper is one of the few addressing the traffic impacts of ATPs and contributes to related literature by (1) proposing a novel interdisciplinary framework to integrate the complex operation of ATPs into a microscopic traffic simulator and conduct an efficient combinatorial experiment to assess the validity of experimental variables, and (2) providing meaningful insights about the operational aspects of ATPs in the vicinity of freeway diverge areas.

The paper is organized as follow. The introductory section illustrates and motivates the current work. Then, a comprehensive literature review on the state of research and practice related to ATPs is presented. Afterwards, the methodological framework to incorporate ATPs into a microscopic traffic simulation is proposed. Next, a combinatorial experiment is performed to assess the impact of experimental variables on travel time and reliability. Finally, conclusions and future research directions are provided.

2 LITERATURE REVIEW

This section presents a comprehensive literature review on ATPs, which includes previous and current implementations and projects, benefits and challenges of ATPs, and interdisciplinary research initiatives. Subsequently, the few works studying the traffic impacts of ATPs are summarized, which shows the gap on appropriate simulation frameworks that integrate ATPs and regular traffic, and indicates the lack of research to quantify the traffic impacts of ATPs on freeway diverge areas.

The ATP concept has evolved from preliminary tests—about two decades ago—to real world implementations in the last few months. The works

by Bishop et al. (2015), Nowakowski et al. (2015), and Shladover (2012), present a detailed review of previous ATP projects around the world, which are summarized next. ATP research efforts in Europe were initiated and promoted by a number of ATP projects, e.g., CHAUFFEUR, SARTRE, KONVOI (Kunze et al. 2009 and 2011), and COMPANION. Likewise, the Netherlands Organization for Applied Scientific Research (TNO) (Janssen et al. 2015) and the Eindhoven University of Technology (TU/e) have consistently promoted research on Cooperative Adaptive Cruise Control (CACC) for trucking operations in the Netherlands. In Japan, the Energy ITS project (Tsugawa et al. 2011) was developed to demonstrate the benefits of ATPs—among other Intelligent Transportation Systems (ITS)—for energy savings and global warming prevention. Moreover, intensive research on ATPs has been conducted by the California Partners for Advanced Transportation Technology (PATH) in the U.S. for about 2 decades (Shladover 2010), Nowakowski et al. 2015). These global efforts have demonstrated the benefits, feasibility, and technical requirements of ATPs. Likewise, models and real-world demonstrations have motivated recent implementations of the system.

Truck manufacturers and governments are engaged in the implementation of autonomous trucks and ATPs. In 2015, Daimler—one of the larger truck manufacturers in the world (Mercedes-Benz, Freightliner Trucks, Western Star Trucks, among others brands)—introduced the first licensed autonomous commercial truck to operate in the U.S. (Nevada), i.e., the Freightliner Inspiration (Markoff 2016). Later on, In March 2016, Daimler demonstrated its ATP capabilities by running a platoon of three heavy-duty Mercedes-Benz Actros trucks on a public stretch of the Autobahn highway system in Germany (Abt 2016). In the next month (April 2016), Rijkswaterstaat (traffic agency in the Netherlands), organized the ETPC to assess the current operations and requirements related to ATPs. As part of the challenge, several truck manufacturers (i.e., DAF Trucks, Daimler, Iveco, MAN Truck & Bus, Scania, and Volvo) drove their own ATPs from Sweden, Germany and Belgium to Rotterdam (Netherlands). The experiment collected and analyzed relevant information about ATPs in the real world (Alkim et al. 2016). Additionally, companies in the U.S. are actively developing business based on ATP concepts. For example, Otto is a start-up company led by former Google employees and engineers that develops technologies to provide autonomous upgrades to regular trucks. Likewise, Peloton offers ATP services and technologies to the trucking industry. The U.S. government is also very interested in the future of ATPs. Five jurisdictions allow to test and operate ATPs (i.e., California, Nevada, Michigan, Florida, and the District of Columbia), and there are several projects studying their opportunities. For example, the California, Arizona, New Mexico, and Texas DOTs are planning to demonstrate ATP capabilities for the I-10 Western Connected Freight Corridor Project in the next few years (TPF (Transportation Pooled Fund (TPF) Program 2016)). Similarly, the National Cooperative Highway Research Program (NCHRP) is conducting a project to understand the challenges of connected/autonomous vehicles in truck freight operations (NCHRP 2015). Additionally, Bevly et al. (2015) recently released the first phase of a comprehensive study on the commercial feasibility of driver assistive truck platooning, part of the Federal Highway Administration (FHWA)'s Exploratory Advanced Research Program. The substantial interest of researchers, governments, and companies on ATP systems is explained by the prominent benefits that this technology offers to business and society, which are summarized next.

When trucks drive close next to each other, i.e., with small inter-truck gaps, the power required to overcome the aerodynamic resistance (drag) generated by the air flow around the vehicles is less than the power needed by trucks that are driven in isolation, which generates fuel savings and emission reductions that are more critical at higher speeds. Based on the results from the Energy ITS project in Japan, Tsugawa et al. (2011) indicate that a platoon of three trucks driving at 80 km/h with a 10 m. gap has a 14% average fuel consumption improvement. Likewise, an ATP penetration rate of 40% along an expressway is related to a 2.1% reduction in CO_2 emissions for 10 m. gaps, and 4.8% reduction for 4 m. gaps. Similarly, Vegendla et al. (2015) study the aerodynamic influence of five different platooning configurations and show with simulations that trucks obtain the highest team fuel savings (up to 23%) when they travel one behind another (savings are reduced as they drive farther). Moreover, Bevly et al. (2015) demonstrate that the closer the platoon distance, the better the fuel savings. According to Nowakowski et al. (2015), shorter following gaps at highway speeds can generate energy savings as high as 20% to 25% for large trucks. Other ATP benefits include traffic safety improvements (Kunze et al. 2011), traffic flow benefits (Bevly et al. 2015), and congestion mitigation (Farokhi & Johansson 2015).

Furthermore, the interdisciplinary nature of ATP systems have captured the attention of researchers from multiple perspectives. For example, interesting problems related to the operation and optimization of ATPs have been approached by several authors. Saeednia & Menendez (2016) study acceleration/deceleration strategies for platoon formation and find that a hybrid approach is the best strategy. Liang et al. (2016) propose an iterative pairwise coordination method that sequentially joins the closest trucks and platoons to improve fuel efficiencies on the fly. Van De Hoef et al. (2015) propose optimization methods for platoon coordination based on pairwise clustering. Larson et al. (2015) propose a distributed network of controllers to enhance platooning and maximize fuel savings. Larson et al. (2015) combine platooning and vehicle routing to minimize the total fuel consumption of trucks dispatched between different origins and destinations in a network. Similarly, Farokhi & Johansson (2015) use game theoretical approaches to study the strategic interaction of passenger cars and trucks under truck platooning incentives to improve congestion. Most of these works focus on the optimization of individual firms and do not account for social benefits/impacts. Although Farokhi & Johansson (2015) approach this perspective, their games are broad and do not account for specific traffic interactions. Other considerations include driver discomfort and unexpected behavior related to small gaps (Zheng et al. 2015, Skottke et al. 2014), adoption willingness by trucking and logistics firms (Shladover et al. 2015, Bevly et al. 2015), and human factors important for successful implementations (Nowakowski et al. 2015). However, scarce attention have been placed to the impact of ATPs on the surrounding traffic.

Public agencies need to anticipate the traffic impacts associated with ATPs. Janssen et al. (2015) suggest that appropriate conditions need to be provided for the interaction between ATPs and other drives. Nowakowski et al (2015) indicate that several variables of the ATP system will have impacts on traffic, e.g., number of trucks allowed in an ATP, determination of exclusive ATP lanes, ATP coupling/decoupling procedures, among others. Likewise, the ETPC demonstrates the requirements to enhance the ATP-traffic interaction (Alkim et al. 2016). Few works have studied the impact of ATPs, and related technologies, in the traffic stream. Arem et al. (2006) use microscopic simulations of mixed traffic (passenger cars and trucks) with CACC. Although they conclude that CACC improves traffic efficiency, the authors acknowledge that extensive research is needed to anticipate traffic impacts. Schakel et al (2010) highlight the benefits of CACC for traffic flow stability but warn that fast shockwaves can be problematic for human drivers. Arnaut & Bowling (2014) develop a traffic simulation to assess improvements related to mixed-traffic using CACC at different penetration rates, and conclude that CACC improves traffic at penetration rates above 40%. Most of the optimization methods proposed in previous literature do not consider the interaction between traffic and ATPs e.g., Saeednia & Menendez (2016), Larsson et al. (2015). Only two studies investigate the impacts of ATPs on the traffic volume. Bevly et al. (2015) use CORSIM to show that two-truck platoons improve traffic at penetration rates above 60%. Secondly, Deng & Burghout (2016) compare the impacts of short gaps on the overall traffic flow performance. However, the specific characteristics of the ATPs considered in their derivations are unclear.

In order to develop accurate models and experiments to determine traffic impacts, it is very important to understand the specific features that differentiate ATPs from the rest of the traffic. For example, the expected platoon size (number of trucks clustered in an ATP) is uncertain. Typical ATP sizes include two to three trucks (Tsugawa et al. 2011, Bevly et al. 2015, Janssen et al. 2015, Shladover 2010, Zheng et al. 2014) but are envisioned to be as large as five to ten trucks (Abt 2016, TPF Program 2016).

Similarly, the magnitude of the gap between consecutive trucks is still unclear. Implementations and experiments have considered inter-truck gaps that range from 3 m. 15 m. (Tsugawa et al. 2011, Shladover 2010, Abt 2016, C Nowakowski et al. 2015, Zheng et al. 2014). Nowakowski et al. (2015) indicate that drivers feel uncomfortable 15 m. gaps and unsafe at less than 7 m. gaps. However, optimal fuel savings are achieved around 6 to 8 m.

The next section describes the methodology followed to study the traffic impact of ATPs on travel time and travel time for drivers in the vicinity of highway exits based on the lack of previous studies approaching this problem.

3 METHODOLOGY

The analysis framework used traffic microsimulation with automated scenario generation to model and analyze the segment near a freeway exit. This section provides an overview of the technical aspects involved in ATP simulation. The microsimulation tool chosen in the study is VISSIM, which gives the possibility to control individual vehicles via Dynamic Link Library (DLL).

3.1 *ATP control*

The overview of the control process is described in Figure 1. The process starts with an arrival generation (flow 1 in Figure 1) or a time step event from VISSIM. Connected vehicle applications use the dedicated short range (DSRC) frequency with a communication range about 300 ft. It was assumed that if the trucks get separated beyond the DSCR range connection is lost and control is returned

to the Heavy-Duty Truck (HDT) driver and the HDT will behave as regular simulated vehicle. In the algorithm, this will yield vehicle control back to VISSIM. ATP control was performed using a proportional controller taking into consideration acceleration, speed and gap between ATPs in the platoon. These data are passed from VISSIM to the DLL at the selected time step which was chosen as 1/20 seconds (flow 3, Figure 1). The ATP control action runs the gap control algorithm for execution in VISSIM (flow 2, Figure 1) When a vehicle reaches the end of the simulated segment is removed automatically by VISSIM.

3.2 *Traffic pattern generation*

Traffic pattern for ATPs was created using collected traffic counts from the analysis segment. The traffic pattern generation process is depicted in Figure 2-a. The process starts with an ATP generation pattern base on inter-platoon times and platoon size on an exclusive modified lane. The arrival creation lane allowed the ATPs to quickly form a platoon and move onto the traffic stream using a parallel merge area, i.e., platoon formation stage in Figure 2-a. Such initial process guarantees that the platoon will form and stay formed as part of the initial traffic pattern. The input traffic pattern is then formed and given a road segment to achieve stability. Once

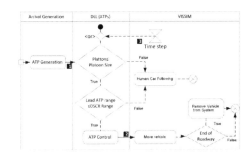

Figure 1.   Control framework for ATPs.

(a) Traffic pattern generation process

(b) Simulated ATP near a freeway exit

Figure 2.   ATP traffic pattern generation process and analysis segment.

the traffic contains the ATPs and it is in steady conditions, it enters the analysis segment. The analysis segment was chosen as two miles upstream of the exit ramp. An illustration of a truck platoon near the exit ramp is presented in Figure 2-b.

3.3 Computational experiment

A microscopic simulation experiment is proposed based on several scenarios generated from a baseline calibrated with real-time traffic data.

The main variables used to test the impact of ATPs on travel time and travel time reliability are (i) traffic volume projections (Passenger Cars (PCs) and HDTs), (ii) ATP penetration rates, (iii) ATP distance gaps, and (iv) ATP sizes. Furthermore, the following pseudocode describes the execution of the simulation-based numerical experiment.

Step	Description
0-01	Define analysis segment;
0-02	Compile traffic data from upstream/downstream sensors;
0-03	Calibrate baseline traffic simulation;
0-04	Develop and integrate ATP behavior in traffic simulator;
0-05	For all traffic flow projections
0-06	For all ATP penetration rates
0-07	For all ATP distance gaps
0-08	For all ATP sizes
0-09	Run traffic simulation;
0-10	Compile travel times and delays;
0-11	End;
0-12	End;
0-13	End;
0-14	End;
0-15	Return travel time & delays.

Intuitively, a representative highway section with and off-ramp access is selected on the study region. Then traffic counts are compiled from downstream and upstream sensors. Such data is used to calibrate the baseline in the traffic simulator with only trucks and passenger cars. Furthermore, the ATP driving behavior needs to be developed and implemented in the traffic simulator. Subsequently, a number of scenarios are tested under multiple combinations of the experimental variables. Finally, outputs associated with travel time and delays are collected for subsequent analyzes. Delays are used to quantify reliability, i.e., the higher the delays, the most unreliable the operation.

As recognized by Bevly et al. (2015), developing the ATP driving behavior and implementing it into the traffic simulator is a major challenge for ATP research. However, such challenge is properly addressed in this paper and meaningful insight are obtained from the numerical experiment described in the next section.

4 RESULTS AND DISCUSSION

This section describes the data inputs required to implement the traffic simulation experiment, provides a case study to test its implementation, and discusses meaningful insights obtained from the computations.

A 2.5-mile 2-lane segment of State Road 528 (Martin Andersen Beachline Expressway) approaching Exit 20 is selected for the analysis. Data from traffic counts reported by the Florida Department of Transportation (FDOT 2015) are collected to develop the Directional Design Hourly Volume ($DDHV$) for the base scenario, where $DDHV = AADT \times K \times D$, $AADT$ is the average annual daily traffic, K is the design hour factor, and D is the directional design hourly factor. The corresponding truck $DDHV$ ($DDHV_T$) is computed as $DDHV_T = DDHV \times T$, where T is the percentage of trucks. FDOT (2015) indicates that for this intersection $AADT = 20,000$ vehicles per day, $K = 0.094$, $D = 0.632$, and $T = 0.08$. Therefore, $DDHV = 118$ vehicles per hour, and $DDHV_T = 95$ trucks per hour. Additionally, FDOT (2015) reports that 34.01% of the traffic takes the Exit 20 off-ramp.

The previous inputs are used to generate a base scenario calibrated in VISSIM, where traffic variables are fine-tuned to replicate realistic traffic behavior. Likewise, it is assumed that the ATPs only drive through, i.e., do not diverge on the off-ramp.

Four sets of experimental variables are selected to analyze the changes in travel time and travel time reliability (delays) as a function of ATP deployments, i.e., (i) traffic volume projections $X_i = \{100, 150, 200, 250\}(\%)$, (ii) ATP market penetration rates $X_{ii} = \{20, 60, 100\}(\%)$, (iii) ATP sizes $X_{iii} = \{2, 3, 5, 10\}(trucks)$, and (iv) ATP gaps $X_{iv} = \{5, 10, 15\}(m)$. Thus, a total of $M = 144$ scenarios are considered in the experiment. The computational experiment is designed in Visual Basic, and it takes about 12 hours to run all the cases. A total of 137,367 observations are collected. Each observation describes a simulated vehicle trip in a specific simulated scenario with the corresponding travel time and delay. Three vehicle types are considered, i,e, PCs, HVTs (not in ATP formation), and HDTs operating in ATPs.

The dataset is used to estimate four Ordinary Least Squares (OLS) models that test the impact of experimental variables on travel time and reliability (described by delays) for off-ramp users (Table 1) and through traffic (Table 2).

The numerical results in Table 1 show how different ATP attributes impact travel time and delays for multiple off-ramp users. The average travel time for off-ramp PCs in the experiment is 87.22 second, and in average divergent HDTs spend 8.14 more seconds in the system, which is associated with the operational characteristics of each user type, i.e., HDTs

Table 1. OLS model for off-ramp travel time and delays.

Type	Variable	Travel Time Coeff.	Travel Time t Stat	Delay Coeff.	Delay t Stat
	Constant	87.22	516.46	5.07	21.83
PC	ATP Size	0.04	4.05	0.04	2.92
PC	ATP Gap	0.01	1.86	0.01	1.35
PC	Traffic Intensity	2.83	45.32	2.91	33.98
PC	ATP penetration	−2.06	−20.50	−2.23	−16.12
HDT	Constant	8.14	10.03	−6.34	−5.68
HDT	ATP Size	0.03	0.56	0.03	0.39
HDT	ATP Gap	0.01	0.34	0.01	0.24
HDT	Traffic Intensity	1.53	5.13	2.23	5.44
HDT	ATP penetration	−4.06	−4.90	−1.17	−1.03
	Observations	44487		44487	
	Multiple R	0.277		0.228	
	Adjusted R Square	0.076		0.052	

Table 2. OLS model for through-traffic travel time and delays.

Type	Variable	Travel Time Coeff.	Travel Time t Stat	Delay Coeff.	Delay t Stat
	Constant	81.43	632.80	−1.07	−8.60
PC	ATP Size	0.02	2.77	0.02	2.85
PC	ATP Gap	0.02	2.58	0.02	2.61
PC	Traffic Intensity	4.01	84.74	3.62	79.06
PC	ATP penetration	−1.40	−18.18	−1.43	−19.13
HDT	Constant	13.27	21.34	−0.33	−0.55
HDT	ATP Size	0.02	0.64	0.02	0.66
HDT	ATP Gap	0.01	0.20	0.01	0.21
HDT	Traffic Intensity	1.17	5.22	1.53	7.01
HDT	ATP penetration	−1.71	−2.67	1.48	2.38
ATP	Constant	5.04	10.66	4.64	10.14
ATP	ATP Size	0.16	6.10	0.07	2.55
ATP	ATP Gap	0.01	0.73	0.01	0.74
ATP	Traffic Intensity	2.48	16.14	2.34	15.68
ATP	ATP penetration	−0.63	−2.09	−1.07	−3.68
	Observations	92879		92879	
	Multiple R	0.352		0.291	
	Adjusted R Square	0.124		0.084	

are overall slower than PCs. Furthermore, the average delay is 5.07 for PCs and 5.07–1.17 = 3.90 seconds for trucks, which can be attributed to the lower desired speeds associated with HDTs.

Traffic growth has differential impacts on travel time for each user type, i.e., in average 2.83 seconds per 1% traffic increment for PCs and 1.53 seconds for HDTs, which is expected from classic traffic theory. By looking at delays, a 1% increment in traffic volume in average represents 2.91 seconds of delay for PCs and 2.23 seconds for HDTs.

Interestingly, the most critical ATP attribute impacting travel time is its market penetration. In average a 1% increment in ATP market penetration reduces off-ramp-related travel time by 2.06 seconds for PCs, and 4.06 seconds for HDTs. Such results highlight the potential benefits that will be perceived when platooned HDTs free additional highway space that can be efficiently used by other vehicle classes. The benefit might be larger for single HDTs (not in ATP formation) because their speeds will be similar to the speeds of the ATPs and time losses associated with maneuvers within ATPs are smaller. In general they might just follow leading ATPs until reaching their exits. On the other hand, the higher speed of PCs will generate more circumstances in which they will have to decelerate to change lanes, overcome ATPs, and break through ATPs to reach highway exits on time. Still the benefits obtained from the additional space gained after re-organizing HDTs into ATPs are significant. Furthermore, a 1% increment in market penetration in average reduces delays for PCs by 2.23 seconds and trucks by 1.17 seconds, and aligns with the insights presented for travel time.

The previous claims are also supported by the results associated with ATP size and ATP gap. In average each additional truck in ATP formation increases PC travel time and delays by 0.04 seconds. Likewise, each meter of inter-truck gap distance increases PC travel time and delays by 0.01 second. As ATPs cluster more trucks, the interaction between the platoons and divergent PCs will generate more delays because it will be harder to break through the ATP as more HDTs will have to adjust their speeds. Larger gaps provide more opportunities for PCs to change lanes through ATPs, which might promote risky behaviors, and reduces overall travel times for PCs and even ATPs. On the other hand, very small gaps (unacceptable for lane changing) might demotivate PCs to pass through the ATP and encourage drivers to anticipate their exit maneuvers taking advantage of the additional longitudinal space generated by the ATPs. These results motivate further investigations to assess specific instances where the PC-ATP interaction will generate delays, inefficiencies, and unsafe conditions. Although current results suggest similar conclusions for trucks, the low significance of ATP size and gap on travel time and delays (assessed by their low positive t-stats) indicate that these variable might have a negligible effect on travel times and delays for regular HDTs, which is expected as they are expected to follow ATPs rather than passing through them to take highway exits.

Furthermore, Table 2 shows how different ATP attributes impact travel time and delays for the through traffic. The average travel time for PCs in the experimental environment is 81.43 seconds. In general HDTs and ATPs are slower than PCs. While HDTs that drive through in average spend 13.27 seconds more than PCs, ATPs in average spend just 5.04 more seconds. This results highlight the overall time-savings achieved by HDTs engaged in ATPs. Reduced travel times for HDTs are amplified because each vehicle needs to account for specific behavioral characteristics of the human driver. On the other hand, only one driver will lead an ATP, and, therefore, the travel time for the followers will be enhanced. However, from the classic traffic-flow-theory perspective, ATPs will face higher delays because each follower will not have the possibility to improve their current speed as they are forced to follow the leader. In the experiment, each vehicle in an ATP experiences an average delay of $4.66–1.07 = 3.59$ seconds. However, in average PCs reduce delays by 1.07 seconds and HDTs obtain an additional 0.33-second reduction because they can move freely, are more flexible, and have better opportunities to reduce delays based on their independent decisions. Furthermore, traffic volume growth increments travel times for all user types, i.e., a 1% growth in overall traffic increases travel time and delays by 4.01 and 3.62 seconds for PCs, 1.17 and 1.53 seconds for HDTs, and 2.48 and 2.34 seconds for ATPs. The impact is more dramatic for PCs because they tend to drive faster, so increased traffic intensity will generate more congestion with higher travel time and delays. On the other, travel times and delays will be higher for ATPs than for HDTs because they are forced to travel as a group and have low flexibility to avoid undesirable circumstances generated by increased traffic. From the market penetration perspective, the overall through-traffic travel time will be reduced when larger fractions of HDTs operate as ATPs. The average travel time reductions per 1% increment in ATP market penetrations are 1.40, 1.71, and 0.63 seconds for PCs, HDTs, and ATPs respectively. In average, delays are mainly reduced for PCs and ATPs by 1.43 and 1.07 seconds per 1% increment in ATP market penetration. PCs will have more longitudinal space to maneuver and flexibility to reduce their delays and vehicles in ATPs will benefits by driving closer without human driver inefficiencies. Although HDTs will perceive lower travel times, in average their delays will be increased by 1.48 seconds per 1% increment in ATP market penetration. This might happen because as PCs gain more mobility, they might slightly affect HDTs. In general, additional entities in an ATP and larger inter-truck gaps will generate low increments on the travel for all users.

These numerical experiment provides interesting insight on future issues and opportunities related to the operation of ATP in freeway diverge areas. A summary of the work and conclusions are provided next.

5 CONCLUSION

This paper studies the impact of ATPs on travel time and reliability (delays) on the vicinity of a freeway diverge area.

An interdisciplinary method is proposed to incorporate ATPs into a microscopic traffic simulator and a combinatorial experiment assesses the impacts of (i) traffic volume projections, (ii) ATP market penetration rates, (iii) ATP sizes, and (iv) ATP gaps, with four OLS models for travel time and reliability (delays).

Traffic volume growth deteriorate travel time and delays for all user/movement types. Increased usage of ATPs improves travel time for all user/movement types. As the number of vehicles in the ATPs grows, travel times and delays are slightly deteriorated due to impedance on lateral movements and a resulting loss on automated efficiencies of the ATP. Although inter-truck ATP gap has the lowest impact on travel time and delays, it is especially important for PCs in highway exits. Larger gaps will represent a dilemma for users that might riskily accept them for lane changing. On the other hand, smaller gaps might force traffic to plan ahead and anticipate maneuvers taking advantage of the additional longitudinal space freed by ATPs.

The paper is one of the first efforts to study the impact of ATPs on freeway traffic. The main contributions of this research include: (1) providing a novel interdisciplinary framework to integrate the complex operation of ATPs into a microscopic traffic simulator and conduct an efficient combinatorial experiment to assess the validity of experimental variables, and (2) providing meaningful insights about the operational aspects of ATPs in the vicinity of freeway exits.

Future research directions include running multiple simulations associated with each scenario of the combinatorial experiment, which is achievable with multiple random seeds for each instance of the experimental variables; generalizing the results with additional experiments over different geometries, traffic behaviors, and traffic volumes; enhance ATP behavior to allow lane changing; incorporating Vehicle-to-Vehicle (V2V) and Vehicle-to-Infrastructure (V2I) communication and coordination for PCs, HDTs, and ATPs; and developing a systems-of-systems framework to understand the interaction between technical, behavioral, institutional, and physical components encompassing the ATPs.

REFERENCES

Abt, N., 2016. Daimler's Autonomous Platoon Debuts on Germany's Autobahn. *Transport Topics*. Available at: http://www.ttnews.com/articles/basetemplate.aspx?storyid = 41377.

Alkim, T. et al., 2016. *European Truck Platooning Challenge 2016. Creating next generation mobility. Lessons Learnt*, Rijkswaterstaat, RDW, and the Ministry of Infrastructure and the Environment. Netherlands. Available at: https://www.eutruckplatooning.com/PageByID.aspx?sectionID = 131542&contentPageID = 529927.

Arem, B. van, Driel, C.J.G. van & Visser, R., 2006. The Impact of Cooperative Adaptive Cruise Control on Traffic-Flow Characteristics. *IEEE Transactions on Intelligent Transportation Systems*, 7(4), pp.429–436.

Arnaout, G.M. & Bowling, S., 2014. A Progressive Deployment Strategy for Cooperative Adaptive Cruise Control to Improve Traffic Dynamics. *International Journal of Automation and Computing*, 11(1), pp.10–18. Available at: http://dx.doi.org/10.1007/s11633-014-0760-2.

Bevly, D. et al., 2015. Heavy Truck Cooperative Adaptive Cruise Control: Evaluation, Testing, and Stakeholder Engagement for Near Term Deployment. *Technical Report - Federal Highway Administration*, pp.1–135.

Bishop, R. et al., 2015. White Paper: Automated Driving & Platooning - Issues and Opportunities. *ATA Technology & Maintenance Council, Future Truck Program, Automated Driving and Platooning Task Force*, (2015), pp.1–48. Available at: http://orfe.princeton.edu/~alaink/SmartDrivingCars/ITFVHA15/ITFVHA15_USA_FutureTruck_ADP_TF_WhitePaper_Draft_Final_TF_Approved_Sept_2015.pdf [Accessed July 29, 2016].

Deng, Q. & Boughout, W., 2016. The impacts of heavy-duty vehicle platoon spacing policy on traffic flow. In *Transportation Research Board 95th Annual Meeting*. pp. 16–0403.

Farokhi, F. & Johansson, K.H., 2015. A Study of Truck Platooning Incentives Using a Congestion Game. *IEEE Transactions on Intelligent Transportation Systems*, 16(2), pp. 581–595.

Florida Department of Transportation (FDOT), 2015. Florida Traffic Online. *Traffic Data Shapefiles - Truck Traffic Volume*, (2015). Available at: http://www2.dot.state.fl.us/floridatrafficonline/viewer.html [Accessed July 29, 2016].

Janssen, R. et al., 2015. Future of Transportation Truck Platooning, (2015), pp.1–36. Available at: https://www.tno.nl/en/about-tno/news/2015/3/truck-platooning-driving-the-future-of-transportation-tno-whitepaper/ [Accessed July 7, 2016].

Kunze, R. et al., 2009. Organization and operation of electronically coupled truck platoons on German Motorways. *Second International Conference on Intelligent Robotics and Applications*, 5928, pp.135–146.

Kunze, R. et al., 2011. Automated Truck Platoons on Motorways—A Contribution to the Safety on Roads. In S. Jeschke, I. Isenhardt, & K. Henning, eds. *Automation, Communication and Cybernetics in Science and Engineering 2009/2010*. Berlin, Heidelberg: Springer Berlin Heidelberg, pp. 415–426. Available at: http://dx.doi.org/10.1007/978-3-642-16208-4_38.

Larson, J., Liang, K.Y. & Johansson, K.H., 2015. A distributed framework for coordinated heavy-duty vehicle platooning. *IEEE Transactions on Intelligent Transportation Systems*, 16(1), pp.419–429.

Larsson, E., Sennton, G. & Larson, J., 2015. The vehicle platooning problem: Computational complexity and heuristics. *Transportation Research Part C: Emerging Technologies*, 60, pp.258–277.

Liang, K.-Y., Mårtensson, J. & Johansson, K.H., 2016. Heavy-duty vehicle platoon formation for fuel efficiency. *IEEE Transactions on Intelligent Transportation Systems*, 17(4), pp.1051–1061.

Markoff, J., 2016. Want to Buy a Self-Driving Car? Big-Rig Trucks May Come First. *The New York Times*, p.B1.

National Cooperative Highway Research Program NCHRP, 2015. Challenges to CV and AV Application in Truck Freight Operations. *Transportation Research Board*, (2015). Available at: http://apps.trb.org/cmsfeed/TRBNetProjectDisplay.asp?ProjectID = 3936 [Accessed July 29, 2016].

Nowakowski, C. et al., 2015. Cooperative Adaptive Cruise Control (CACC) for Truck Platooning: Operational Concept Alternatives. *eScholarship, University of California*, pp.1–37. Available at: http://www.escholarship.org/uc/item/7jf9n5wm.

Nowakowski, C., Shladover, S.E. & Tan, H.-S., 2015. Heavy Vehicle Automation: Human Factors Lessons Learned. *Procedia Manufacturing*, 3, pp. 2945–2952.

Saeednia, M. & Menendez, M., 2016. Analysis of Strategies for Truck Platooning: A Hybrid Strategy. In *Transportation Research Board 95th Annual Meeting*. pp. 16–5380.

Schakel, W.J., Arem, B. van & Netten, B.D., 2010. Effects of Cooperative Adaptive Cruise Control on traffic flow stability. *Intelligent Transportation Systems (ITSC), 2010 13th International IEEE Conference on*, pp.759–764.

Shladover, S., 2010. Three-Truck Automated Platoon Testing. *Intellimotion*, 16(1), pp.7–9.

Shladover, S.E. et al., 2015. Industry Needs and Opportunities for Truck Platooning., (2015), pp.1–85. Available at: http://www.escholarship.org/uc/item/6723k932 [Accessed July 27, 2016].

Shladover, S.E., 2012. Recent International Activity in Cooperative Vehicle--Highway Automation Systems. *Federal Highway Administration*, pp.1–81.

Skottke, E.-M. et al., 2014. Carryover Effects of Highly Automated Convoy Driving on Subsequent Manual Driving Performance. *Human Factors: The Journal of the Human Factors and Ergonomics Society*, 56(7), pp.1272–1283. Available at: http://hfs.sagepub.com/content/56/7/1272.abstract.

Transportation Pooled Fund (TPF) Program, 2016. I-10 Western Connected Freight Corridor Project. *National Cooperative Highway Research Program (NCHRP)*, (2016). Available at: http://www.pooledfund.org/Details/Study/599 [Accessed July 13, 2016].

Tsugawa, S., Kato, S. & Aoki, K., 2011. An automated truck platoon for energy saving. *2011 IEEE/RSJ International Conference on Intelligent Robots and Systems*, pp.4109–4114.

Van De Hoef, S., Johansson, K.H. & Dimarogonas, D. V, 2015. Coordinating truck platooning by clustering pairwise fuel-optimal plans. In *2015 IEEE 18th International Conference on Intelligent Transportation Systems*. pp. 408–415.

Vegendla, P. et al., 2015. Investigation of Aerodynamic Influence on Truck Platooning. *SAE Technical Paper*.

Zheng, R. et al., 2014. Study on Emergency-Avoidance Braking for the Automatic Platooning of Trucks. *IEEE Transactions on Intelligent Transportation Systems*, 15(4), pp.1748–1757.

Zheng, R. et al., 2015. Biosignal analysis to assess mental stress in automatic driving of trucks: palmar perspiration and masseter electromyography. *Sensors (Basel, Switzerland)*, 15(3), pp.5136–50. Available at: http://www.mdpi.com/1424-8220/15/3/5136/htm [Accessed July 12, 2016].

Transport Infrastructure and Systems – Dell'Acqua & Wegman (Eds)
© 2017 Taylor & Francis Group, London, ISBN 978-1-138-03009-1

The potential of energy-efficient driving profiles on railway consumption: A parametric approach

Mariano Gallo, Fulvio Simonelli & Giuseppina De Luca
Dipartimento di Ingegneria, Università del Sannio, Benevento, Italy

ABSTRACT: This paper proposes a parametric approach to evaluate the effects of optimising driving profiles on railway consumption. To this aim, a kinematic optimisation model is formulated and solved for different case studies, also allowing for the energy recovered in the braking phase. The numerical results were used to build tables and abacuses to provide a preliminary evaluation of effectiveness of energy-saving strategies for different sections and train characteristics. The approach was tested on regional rail services in the Italian region of Campania, showing that energy-efficient driving strategies could produce significant reductions in energy consumption.

1 INTRODUCTION

Managing rail systems from different perspectives is an important part of transportation planning and operation. As regards the real-time operation of rail systems, some recent issues based on simulation of rail convoys have included: managing rail disruptions (D'Acierno et al., 2016) and eco-driving (Gallo et al., 2015). Further, in recent decades attention to energy efficiency has increased in all sectors both for economic and environmental reasons. The transport sector consumes about 30% of energy in Italy (ENEA, 2015) and 27% worldwide (IEA, 2014). Although energy consumption in the transport sector is due mainly to road transportation, energy efficiency in public transport and a policy for increasing public transportation use represent crucial objectives from the sustainability angle. In this context, the challenge to reduce rail energy consumption involves several aspects. Such aspects are both strictly technological (designing low-consumption electrical engines, designing energy-efficient braking recovery systems, etc.) and organisational (optimising rail service schedules and driving styles so as to minimise consumption on a railway line). The importance of energy-efficient driving strategies for reducing consumption was highlighted by Strobel et al. (1974) and relevant theoretical references for identifying optimality conditions are provided in Howlett (1988). More recently, Liu & Golovitcher (2003) proposed a solution algorithm to optimise trajectories using optimal control theory, Ke & Chen (2005) analysed optimal driving models under fixed block and mobile block conditions, while energy-saving methods for freight trains were studied by Lukaszewicz (2004) and Ke & Chen (2005).

Other papers (De Martinis & Gallo, 2013; Dominguez et al., 2012) dealt with the utilisation of energy recovery systems, while Gallo et al. (2015) and Zhao et al. (2015) proposed a general optimisation framework based on simulation approaches.

In the following, we consider a suburban railway line partitioned into sections. Each section links two stations and is protected by a signalling system; the departure of a train from a station is allowed only if the section is entirely free. The scheduled timetable is known and indicates the departure time of the trains from the stations. Scheduled timetables are designed so that the starting time from a station is equal to the starting time from the previous station plus the minimum running time, T_{mr}, between stations, plus the dwell time (necessary for the boarding and alighting of passengers), T_d, plus the reserve time, T_r. The reserve time is provided in order to recover an eventual delay of a train so that the starting time from the next station is respected (if the delay is lower than the provided reserve time).

If a train starts late, the driving style has to ensure the minimum running time, i.e. maximum acceleration, maximum cruising speed and maximum deceleration (compatible with comfort standards, speed limits and available power). This driving style is also known as *time-optimal* (or *all-out driving*) and is the driving style with the maximum energy consumption. If the train is on time, instead, the reserve time can be used for optimising the driving style in order to minimise energy consumption. Indeed, reduction of the cruising speed and/or the introduction of a coasting phase (the propulsion system is turned off and the train runs using kinetic energy, reducing its speed) make it possible to reduce energy consumption.

In this paper we use a model for optimising train speed profiles, with a view to minimising energy consumption, based on kinematic variables proposed by Simonelli et al. (2015). Our aim was to ascertain the effectiveness of energy-efficient driving and scheduling for different railway sections and train characteristics.

The paper is organised as follows: section 2 describes the optimisation model; section 3 introduces the parametric approach; section 4 focuses on the case study; finally, section 5 concludes the paper.

2 OPTIMISATION MODEL

The overall train trajectory along a section encompasses four motion phases: acceleration, cruising, coasting and braking. Under the assumption that acceleration and deceleration are at the maximum (comfort) feasible rate (a and a'), consistent with the optimality condition found in Howlett (1988), the problem of minimising mechanical energy can be set as an optimisation problem in three variables: cruising speed, v_{cr}, cruising duration, t_{cr}, and initial braking speed, v_b. Two constraints of the problem are: (a) the constraint of compatibility with the train schedule, expressed by imposing that the total space covered during the given travel time T equals the distance L between the two stations; (b) the inequality constraints related to the section speed limit and minimum initial braking speed. By introducing these constraints the model can be formulated as follows (Simonelli et al., 2015):

$$
\begin{aligned}
&[\hat{v}_{cr}, \hat{v}_b] \\
&= Arg_{v_{cr}, v_b} \min\{E^{acc}(v_{cr}) + E^{cr}(v_{cr}, t_{cr}) + E^b(v_b)\}
\end{aligned}
$$

s.t.

$$
s^{acc}(v_{cr}) + s^{cr}[v_{cr}, t^{cr}] + s^{cs}(v_{cr}, v_b) + s_b(v_b) = L
$$

$$
0 \le v_{cr} \le v_{max}
$$

$$
v_{b,min} \le v_b \le v_{cr}
$$

$$
t_{cr} \ge 0
$$

with:

$$
\begin{aligned}
E^{acc} = &M\frac{1}{2}v_{cr}^2 + \frac{Mg}{1000}\left[\frac{1}{2}r_0\frac{v_{cr}^2}{a} + \frac{1}{3}r_1\frac{v_{cr}^3}{a} + \frac{1}{4}r_2\frac{v_{cr}^4}{a}\right] \\
&+ Mgi\frac{v_{cr}^2}{2a}
\end{aligned}
$$

$$
E^{cr} = \frac{Mg}{1000}[r_0 v_{cr} + r_1 v_{cr}^2 + r_2 v_{cr}^3]t^{cr} + Mgiv_{cr}t_{cr}
$$

$$
\begin{aligned}
E^b = &-M\frac{1}{2}v_b^2 + Mgi\frac{v_b^2}{2|a'|} \\
&+ \frac{Mg}{1000}\left[\frac{1}{2}r_0\frac{v_b^2}{|a'|} + \frac{1}{3}r_1\frac{v_b^3}{|a'|} + \frac{1}{4}r_2\frac{v_b^4}{|a'|}\right]
\end{aligned}
$$

$$
s_{cr} = v_{cr}\, t_{cr} \qquad s_{acc} = v_{cr}^2/2a \qquad s_b = v_b^2/2|a'|
$$

$$
\begin{aligned}
s_{cs} = &\frac{D}{Z}\ln\frac{\left|1 + \left(\dfrac{v_{cr}+k}{D}\right)^2\right|}{\left|1 + \left(\dfrac{v_b+k}{D}\right)^2\right|} + \\
&- \frac{K}{Z}\left[\arctan\frac{v_{cr}+k}{D} - \arctan\frac{v_b+k}{D}\right]
\end{aligned}
$$

where the space and mechanical energy consumed in acceleration (E^{acc}, S^{acc}), in cruising (E^{cr}, S^{cr}), in coasting (E^{cs}, S^{cs}) and in the braking phase (E^b, S^b) are calculated by assuming the resistances to the motion, expressed in N over kN of weight, as a polynomial function of train speed through parameters r_0, r_1 and r_2. Obviously, in the absence of braking energy recovery systems the term E^b will be equal to zero.

In the expression of S^{cs}, parameters K, Z and D are related to the resistance parameters and to the slope of section i. In particular:

$$
K = \frac{r_1}{2r_2} \qquad D = \sqrt{\frac{r_0 + 1000i}{r_2} - K^2}
$$

$$
Z = \frac{r_2 D g}{1000}
$$

Importantly, the previous expression of distance in the coasting phase cannot be used for downhill slopes where:

$$
i \ge -\frac{4r_0 r_2 - r_1^2}{4000 r_2}
$$

When slope i does not respect this inequality, the motion equation during the coasting phase changes. Nevertheless, the case of steeply descending slopes will not be considered in the following analysis.

The optimisation model defined above, referring to homogeneous sections, can be extended to multiple homogeneous subsections by considering the same kinematic variables for each subsection and by introducing further variables, time and speed related to the crossing points between subsections.

With regard to the parameters of resistance to motion, although they should be evaluated experimentally for each train and truck, several formulations have been proposed in the literature and a sensitive analysis of mechanical equations to track and train parameters is provided in Boschetti & Mariscotti (2012). Resistance to motion will be calculated below through the Strahl formula for traction units:

$$R(v) = \frac{g}{1000}(3.3 * M_t + 0.03v^2)$$

where $R(v)$ is expressed in kN, v is the train speed (km/h) and M_t is the mass of traction unit in tons.

For passenger wagons, the Sauthoff formula will be used:

$$R(v) = \frac{gM_w}{1000}\left(1.9 + 0.0025v + 0.0048F_e\frac{n_w + 2.7}{M_w}v^2\right)$$

where $R(v)$ is expressed in kN, v is the train speed (km/h), M_w is the train mass in tons, F_e is a coefficient related to the train front area characteristic, usually taking the value 1.45, and n_w is the number of wagons.

3 PARAMETRIC APPROACH

By applying the model described above, the sensitivity analysis of the effectiveness of timetable scheduling and trajectory optimisation is carried out. In particular, our aim was to identify the kinds of services suitable for the implementation of an optimised travel time and subsequent optimised trajectory in order to obtain significant energy savings. Indeed, long distances or very short suburban services do not appear so sensitive to the optimisation of travel time and trajectory, due to the preponderance of cruising or acceleration/deceleration phase respectively. Below we therefore focus on regional rail services with section lengths between two successive stations from 1 to 55 km and train mass from 45 to 600 tons. Train departure/arrival times can be scheduled as a function of the minimum travel time, usually defined through a trajectory—known also as all-out in the literature—encompassing an acceleration phase at the maximum feasible acceleration rate, a_{max}, a cruising phase at the maximum allowed speed, v_{max}, and a braking phase at the maximum feasible deceleration rate, $|a_{min}|$. Nevertheless, scheduled timetables are usually designed by adding a reserve time to minimum travel time, fundamental for service regularity. Thus the analysis will be carried out for different reserve times, Tr, expressed as percentages of the minimum travel time.

In Figure 1 the energy saving percentage is plotted against train mass and section length varying in the range specified above with reference to a reserve time of 5% of the minimum travel time. The

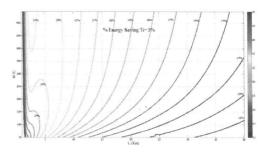

Figure 1. Analysis of energy saving (% w.r.t. total energy) with respect to train mass and section length with a reserve time equal to 5% of minimum travel time.

figure plots the isolines of the percentage of energy saving as a function of train mass and section length for a fixed percentage of reserve time. To build the abacus plotted in Figure 1, steps of 5 tons were considered for train mass and 200 metres for section length. Hence the number of combinations tested was 27,270.

The optimisation model for each combination was solved by a MatLab optimisation tool.

It is worth noting that the energy saving is always significant, varying from a minimum of 10% for long sections and low train mass to a maximum of 28% for very short sections. Apart from where sections are shorter than 5 km, the trend is quite regular and indicates that the percentage of energy saving increases with train mass and decreases with section length. Moreover, the effect of mass tends to decrease for increasing mass and decreasing length (the slope of the isolines grows). Similarly, the distance between the isolines increases with section length, showing that the maximum variability occurs for section lengths below 15 km, while for longer sections the sensitivity of the energy-saving percentage appears less significant.

In Figure 2 the same analysis was carried out considering a braking energy recovery system with an efficiency of 60%. Obviously, in the latter case energy saving is greater on all sections. However, it is worth noting that the differences between the optimisation of trajectories with and without braking energy recovery systems are more significant in the case of shorter sections: when the length is less than 10 km the percentage of energy saved is almost always more than 30% and it tends to become irrespective of train mass. This aspect is due to the relative importance of the kinetic energy collected during the acceleration phase which, for short sections, represents a great percentage of total energy spent and can be partially recovered through the braking energy recovery system.

In the previous figures the slope of the sections was assumed equal to zero and speed limit equal

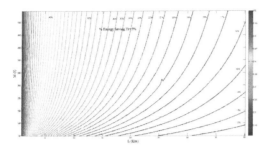

Figure 2. Analysis of energy saving (% w.r.t. total energy) with respect to train mass and section length with a reserve time amounting to 5% of minimum travel time and a braking energy recovery system.

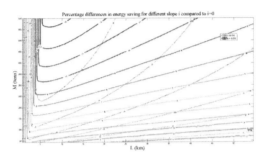

Figure 3. Percentage differences in energy saving for ascending and descending slopes of sections compared to the case of slope i = 0 with a reserve time equal to 5% of minimum travel time.

to 100 km/h, while the train mass was assumed the same for a traction unit and passenger wagons (50% each of total mass), and the number of wagons was assumed equal to 5. To assess the sensitivity of energy saving to the section slope and speed limit, the same calculations were carried out for several values of these parameters. Generally, uphill slopes involve a reduction in energy saved while downhill ones involve an increase consistent with the reduction in duration and effectiveness of the coasting phase.

The results are reported in Figure 3, where the percentage differences in energy saving are represented, through a contour plot, as a function of train mass and section length, for a descending slope (continuous isolines) and ascending slope (dashed isolines) of 5‰. Importantly, energy saving is more sensitive to descending than ascending slopes. Indeed, the percentage increase in energy saving always outweighs the decrease due to the same, in absolute value, ascending slope. This allows preliminary analysis of potential energy saving by using parametric evaluation without a slope. Indeed, in real networks the timetables usually

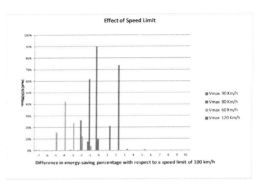

Figure 4. Effect of speed limit.

present services organised through round-trip runs, such that not considering the effect of the slope can lead, in the worst case, to underestimating the potential energy saving. Moreover, the effect of the slope is more sensitive to the train mass and section length for short sections (less than 5 km) and high train masses.

In Figure 4 the differences in energy saving are reported for several speed limits (V_{max}) compared to the base with a speed limit of 100 km/h. Generally, energy saving increases as the speed limit increases, although the effect of the speed limit is not so significant. For differences in speed limits of 10%, energy saving is nearly equal to the base case in 90% of the combinations of mass and length analysed.

4 CASE STUDY

The results obtained with the parametric analysis can be useful to provide a preliminary evaluation of the effectiveness of energy-saving strategies on real networks.

In Figure 5 the regional railway network of Campania and the distribution of section lengths are plotted. Looking at the railway network topology, independently of the actual services, over 97% of the station-station distances are less than 10 km. Given the results reported in Figures 1 and 2, in Campania most of the regional station-station sections are suitable for energy savings between 20% and 25% through optimisation of timetables and trajectories. These percentages rise to over 30% in the case of braking recovery systems. Starting from this preliminary consideration related to infrastructural and topological characteristics, our analysis was enhanced by considering the actual services defined from timetables and the actual train composition for each line. To this aim 3,175 station-station runs were extracted from timetables and for each service the actual train

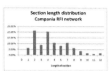

Figure 5.   Regional rail network in Campania.

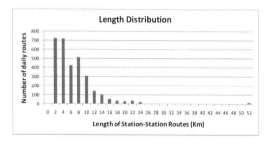

Figure 6.   Distribution of the lengths from stop to stop of actual rail regional services in Campania.

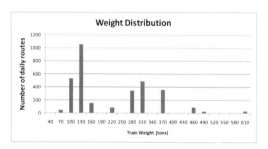

Figure 7.   Mass distribution of rolling stock used in regional services in Campania.

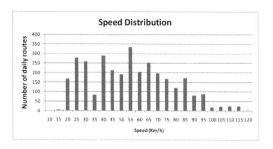

Figure 8.   Cruising speed distribution of regional services in Campania.

Table 1.   Rolling stock in Campania.

Train	Number of traction units	Total weight of traction units (tons)	Number of trailers	Total weight of trailers (tons)
Ale 126(1R+2M)AL	2	48	1	31
Ale 126(2M)AL	2	96	0	0
Ale 126(1M+1R)AL	1	48	1	31
Ale 724(2M+2R)	2	110	2	60
Ale 125(1M+1R)AL	1	48	1	31
Ale 125(2M+1R)AL	2	96	1	31
Ale 501/502(2M+1R)	2	92	1	31
Ale 506/426(2M+2R)	2	125	2	88
Ale 663(1101-1204)(1M)	1	40	0	0
Ale 663(1101-1204)(3M)	3	120	0	0
Ale 663(1101-1204)(2M)	2	80	0	0
Aln 668(2M)AL	2	74	0	0
E464	1	72	6–15	200–410

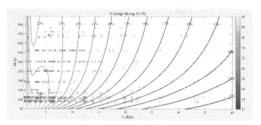

Figure 9.   Energy saving (% w.r.t. total energy) with respect to train mass and section length with a reserve time of 5% of minimum travel time and scatter plot of train mass and stop-stop length of services in Campania.

composition was considered. The distribution of section length, cruising speed and train weight are reported in Figures 6, 7 and 8 while Table 1 reports the trains with their mass characteristics.

The distribution of rail service characteristics show that 85% of daily routes cover less than 10 kms, while in terms of train mass two main categories may be considered: medium (nearly 130 tons) and high weight (and 330 tons). The cruising speed shows greater dispersion and the average is 60 km/h, although, as described above, this parameter does not significantly affect the effectiveness of trajectory optimisation. In Figure 9 the scatter plot of train mass and length of services from stop to stop of the regional rail services in Campania is positioned inside the contour plot of energy saving. The figure shows that,

given the length and mass characteristics for each route, most rail services may suitably achieve over 20% energy saving. In detail, the evaluation of potential energy saving based on the rail services database and parametric analysis, neglecting the slope, with the cruising speeds, provides an estimate of the total saving of 18.5% with respect to a reserve time of 5%. The same analysis can be carried out more accurately by optimising all trajectories, taking into account the characteristics of rail services such as train mass, section length, slope and cruising speed determined from schedules. The total energy saving arising from this further evaluation amounts to 20.5%, very close to the estimate provided by the simplified parametric analysis.

5 CONCLUSIONS

In this paper the potential of railway energy saving strategies was studied, highlighting the ranges of train masses and railway section lengths that are more promising in terms of consumption reduction.

Using a parametric approach applied to the rail network in the region of Campania (Italy) it was shown that up to 20% of total energy could be saved with proper optimisation of speed profiles, and up to 35% by combining speed optimisation with a braking energy recovery system. In general, the speed limit slightly affects the optimisation process, and the abacus in Figure 1, where the slope is also neglected, can be used for preliminary evaluation of potential energy saving for round trip rail services, leading at worst to an underestimation of energy saving.

ACKNOWLEDGEMENTS

Partially supported under research project PON—SFERE grant no. PON01_00595.

REFERENCES

Boschetti, G. & Mariscotti, A. 2012. The parameters of motion mechanical equations as a source of uncertainty for traction systems simulation. *XX IMEKO World Congress.*

D'Acierno, L., Placido, A., Botte, M. & Montella, B. 2016. A methodological approach for managing rail disruptions with different perspectives. *International Journal of Mathematical Models and Methods in Applied Sciences* 10: 80–86.

De Martinis, V. & Gallo, M. 2013. Models and methods to optimise train speed profiles with and without energy recovery systems: a suburban test case. *Procedia - Social and Behavioural Science* 87: 222–233.

Dominguez, M., Fernandez-Cardador, A., Cucala, A.P. & Pecharroman, R.R. 2012. Energy savings in metropolitan railway substations through regenerative energy recovery and optimal design of ATO speed profiles. *IEEE Transactions on Automation Science and Engineering* 9: 496–504.

ENEA 2015. Rapporto Annuale Efficienza Energetica RAEE 2015. ENEA Agenzia nazionale per l'efficienza energetica.

Gallo, M., Simonelli, F., De Luca, G. & De Martinis, V. 2015. Estimating the effects of energy-efficient driving profiles on railway consumption. *Proceedings of IEEE EEEIC 2015–15th International Conference on Environment and Electrical Engineering*: 813–818.

Howlett, P. 1988. *Existence of an optimal strategy for the control of a train.* School of Mathematics Report 3, University of South Australia.

IEA 2014. Key World Energy Statistics. IEA.

Ke, B.R. & Chen, N. 2005. Signalling block layout and strategy of train operation for saving energy in mass rapid transit systems. *IEEE Proceedings: Electric Power Applications* 152: 129–140.

Ke, B.R. & Chen, N. 2005. Signalling block layout and strategy of train operation for saving energy in mass rapid transit systems. *IEEE Proceedings: Electric Power Applications* 152: 129–140.

Liu, R. & Golovitcher, I.M. 2003. Energy-efficient operation of rail vehicles. *Transportation Research Part A* 37: 917–932.

Lukaszewicz, P. 2004. Energy saving driving methods for freight trains. *Advances in Transport* 15: 885–894.

Placido, A. & D'Acierno L. 2015. A methodology for assessing the feasibility of fleet compositions with dynamic demand, *Transportation Research Procedia* 10: 595–604.

Simonelli, F., Gallo, M. & Marzano, V. 2015. Kinematic formulation of energy-efficient train speed profiles. *Proceedings of AEIT International Annual Conference 2015*, Naples, Italy.

Strobel, H., Horn, P. & Kosemund, M. 1974. Contribution to optimum computer-aided control of train operation. 2nd IFAC/IFIP/IFORS Symposium on traffic control and transportation systems: 377–387.

Zhao, N., Roberts, C., Hillmansen, S. & Nicholson, G. 2015. A Multiple Train Trajectory Optimization to Minimize Energy Consumption and Delay. *IEEE Transactions on Intelligent Transportation Systems* 99: 1–10.

Transport Infrastructure and Systems – Dell'Acqua & Wegman (Eds)
© 2017 Taylor & Francis Group, London, ISBN 978-1-138-03009-1

The importance of railways for developing countries. The feasibility study for the rehabilitation of the Mchinji to Nkaya rail line in Malawi

Antonio Placido
D'Appolonia S.p.A., Transport Engineering—Mobility and Logistics, Napoli, Italy

Marco Marchesini
D'Appolonia S.p.A., Technical Engineering Development IT—Railway Systems, Roma, Italy

Vincenzo Cerreta
D'Appolonia S.p.A., Transport Engineering—Mobility and Logistics, Napoli, Italy

ABSTRACT: Malawi is a landlocked country in South-Central Africa which ranks among the world's least developed countries. The economy is predominantly based on agriculture and depends on substantial inflows of economic assistance from the International Monetary Fund (IMF), the World Bank, and individual donor nations. Because of poor reliability of rail infrastructure, import and export goods are mainly moved by road from/to neighbouring countries, with South Africa being the main partner. In particular, more than 95% of current freight is transported by road resulting in high transport costs as well as high negative impacts, such as pollution and car accidents. However, the Malawi Railway Network has the potential to be a vital asset for the entire southern region of Africa, as it is part of an international corridor connecting the Indian Ocean through the port of Nacala in Mozambique to the inner regions of the Continent. Indeed, besides Malawi, other landlocked countries like Zambia and Democratic Republic of Congo (DRC) are extremely interested in finding a competitive outlet to lower the cost of import and export trades. Currently most exports from these countries are being transported by road over a distance of more than 3,400 km to Durban port in South Africa, instead of less than 2,000 km by rail to Nacala port passing through Malawi. This paper provides a brief description of the traffic forecast, the infrastructure design, the operation and maintenance action plans, the environmental and social impact assessment, the multicriteria analysis as well as the cost benefit analysis culminating in a feasibility study report which identifies an optimal investment solution for the rehabilitation of the Mchinji to Nkaya rail section. The study also highlights the strategic importance of the railway to support the fragile economy of Malawi and that of the neighbouring countries.

1 INTRODUCTION

This paper summarizes the main activities carried out by D'Appolonia SpA in joint venture with the company TEAM Engineering SpA, as part of the feasibility study for the rehabilitation of the railway infrastructure from Mchinji in the Centre to Nkaya in the South of Malawi (Fig. 1). The Feasibility Study is expected to inform the Malawi government of the best possible investment solution to meet the current and future traffic demands on the line as well as a balanced maintenance plan, on the basis of the declared national and specific objectives. In doing so the railway line will play a vital role in national development of Malawi in particular by reducing the time it takes to move goods and people between locations in Malawi as well as from outside the country. Furthermore, it

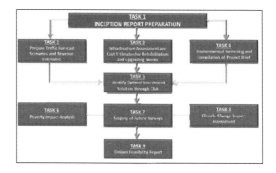

Figure 1. Proposed methodology for the Feasibility Study.

provides evidence of the importance of transport infrastructure for developing countries and gives some useful indications about rail service costs (e.g. unitary cost of service, cost of rolling stock, cost of signalling system) in that region of Africa, which can be used as references for other academic and/or professional studies.

2 BACKGROUND INFORMATION

The Malawi Railway Network is single track, with narrow (1,067 mm – "Cape") gauge. The railway was constructed between 1908 and 1980 and extends for approximately 797 km, running from Mchinji at the Zambian border in the west, via Lilongwe, to Blantyre and Makhanga in the South, up to the border with Mozambique (Fig. 2).

At Nkaya Junction, it links with the "Nacala Corridor" line, running eastwards via Nayuchi to Mozambique's deep water port at Nacala on the Indian Ocean. The link south from Makhanga to Mozambique's Beira corridor has been closed since the Mozambique Civil War, with plans for reconstruction not yet implemented. There is no direct link with neighbouring Tanzania, as there is a difference of gauge (1,067 mm/1,000 mm). An extension from Mchinji to Chipata in Zambia opened in 2010, and there is a proposal to eventually link up from there with the TAZARA railway at Mpika or Serenje (i.e. to the heart of the continent). Direct linkage is available with Mozambique, however, which has the same gauge track.

Between 2000 and 2012, the Malawi Railways network was concessioned to Central and East African Railways (CEAR), a private company, which made losses during every year of the concesssion. The very poor traffic performance and the recurrence of derailments/accidents are further evidenced in several technical/programmatic documents issued by various Authorities (SADC, 2012) according to which it is stated that CEAR's derailment rate results as being "one of the highest worldwide".

3 TRAFFIC FORECAST ANALYSIS

The aim of the Traffic Forecast Analysis is to assess the future traffic levels which will presumably affect the Mchinji-Nkaya railway line in 15 years' time and, consequently, estimating the revenues associated with the forecasted traffic scenarios.

As regards the first task, in order to simulate the future consistent traffic which will likely influence the circulation of goods within the entire Southern African region, an appropriate transportation model is used to assess the performance of transportation infrastructures (both road and rail) within Malawi taking into account its relations with neighbouring countries and ports (Fig. 3). In particular, the freight demand is estimated by means of a gravitational model (Cascetta, 2009) with mass variables corresponding to the employed people of each origin and destination.

After calibrating and validating the models on present traffic conditions (i.e. 2016), an in-depth analysis of the future economic growth of Malawi and Zambia is carried out so as to estimate the future freight demand. Indeed, both these landlocked countries are extremely interested in the rehabilitation of the Mchinji-Nkaya line, which would represent a cheaper and faster corridor to the Indian Ocean than those currently used. At the moment, most of the products to Europe, the United States and China are exported via Durban port in South Africa.

The economic projections are thus based on estimated GDP trends[1] in future years and four possible scenarios are considered:

Figure 3. Transportation model adopted for the simulation of the rail and road freight traffic.

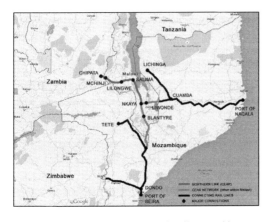

Figure 2. CEAR Railway Network railway and its connections with Mozambican Corridors.

[1]GDP trends for both Malawi and Zambia have been estimated according to data provided by The World Bank (www.theworldbank.com).

Scenario 1: in this scenario, both Malawi and Zambia are not expected to experience a massive growth, since it is assumed that trading levels will increase following a linear trend. This has to be considered as the most pessimistic scenario;

Scenario 2: Malawian economy is considered to grow slowly, while Zambia is expected to increase its trade exponentially;

Scenario 3: in this case, it is hypothesised that the Malawian GDP and import/export trade will develop considerably, in contrast to the Zambian economy, which is not expected to grow rapidly;

Scenario 4: in this scenario, it is assumed that the economy of both countries will run at full speed, growing according to an exponential trend. This has to be considered as the most optimistic scenario.

Based on these assumptions, the adoption of macroscopic simulations provide important indications about the amount of goods affecting the rail section between Mchinji and Nkaya in 15 years' time (i.e. time horizon mentioned in the ToR), as well as the resulting financial revenue. These values are required for the revenue estimation which depends on the cost that customers have to pay for moving goods. To this purpose, after carrying out a specific Sensitivity Analysis on fare structure, the rail transport cost considered is 0.089 USD/ton*km.

Among the four different scenarios, the one which foresees an exponential increase of the Zambia trading and, at the same time, a more linear growth of the Malawian economy (i.e. Scenario 2) is identified, according to the actual situation of both countries, as being the most likely and, therefore, is analysed in detail.

In particular, results highlight that the amount of tons travelling per year on the Mchinji-Nkaya section is 2,706,290, which corresponds to 3,484,329 ton*km per day. This value should guarantee a total annual revenue of 86,841,949 USD to the rail operator. The majority of this traffic is for transit flows and concerns Zambia trades through the Nacala corridor. However, the simulations demonstrate that the rehabilitation of the rail line is attractive even for Malawi. In fact, the main commercial hubs within the country (i.e. Nkaya and Kanengo station near Lilongwe, the capital of Malawi), consistent freight flows are attracted from road transport, resulting in lower negative externalities. A comparison between the current

Table 1. Results of the macroscopic simulations at 2031.

	Tons/year	Tons*km/day	$/year
Scenario 1	1,143,635	1,584,398	$ 39,483,198
Scenario 2	2,706,290	3,484,829	$ 86,841,939
Scenario 3	1,984,374	2,575,270	$ 74,135,739
Scenario 4	3,168,679	4,075,692	$ 101,566,245

Table 2. Results of the macroscopic simulations.

	Modal split at 2015 tons	Modal split at 2031 tons
Total OD freight	3,021,198	15,111,572
Total rail traffic	271,298	2,824,732
% Rail	8.98%	18.69%

situation at 2015 and the most likely traffic forecast at 2031 gives an idea of the benefits provided by the rail line to all the Sub-Saharan area (Table 2).

As can be seen, almost 3 mtpa will travel on rail, corresponding to 18.69% of the total OD freight matrix. This is even more significant if it is considered that railways in Malawi are not densely distributed, and some areas (e.g. the Northern region and the Southern region) can be reached only by road vehicles. Therefore, the traffic captured by rail could be probably even higher if further investments in rail infrastructure on other corridors are made (e.g. Sena corridor—from Blantyre to Beira Port).

In addition, it is necessary to consider that, besides freight transport, railways can be extremely strategic for passengers. The Mchinji-Nkaya rail section has proved to increase the accessibility of the Southern and the Central regions, within a country where transport mobility at the moment is very difficult. The rail line is, therefore, a valid alternative to bus transport, enabling a faster and a more comfortable connection between Blantyre and Lilongwe, as well as between all the other important cities on the corridor. Therefore, even though the main profits for the rail concessionaire are generated by freight traffic, the social benefits provided by the introduction of a passenger rail service are noteworthy.

4 INFRASTRUCTURE ASSESSMENT

In order to correctly define the type and level of interventions necessary for the rehabilitation and upgrading of the Mchinji-Nkaya line section, a site inspection of the line enabled a detailed assessment of the current conditions of the line, which resulted to be extremely bad.

According to this analysis and to the technical characteristics of the existing railway system, four different options for the rehabilitation and upgrading of the railway infrastructure have been considered:

Option 1: Rehabilitation of infrastructure aimed to return the railway to its original capacity. Under this option the line would have an estimated 15 years life for the railway trackage, but would require an ordinary programmed maintenance. The forecasted capacity is eight (8) trains per day with 1,836,800 tons/year. The estimated cost of this option is about 146,000,000M$ and the estimated duration of the infrastructure rehabilitation is 18 months based on the assumption of four construction sites, with an extension of approximately 100 km each;

Option 2: Rehabilitation and upgrading of the infrastructure to increase the capacity of the line by extending crossing loops at the main stations and constructing new triangles at Mchinji and Nkaya stations. The estimated cost of this option is about 164,000,000 M$ and the line would have an estimated 15 years life for the railway trackage, with an ordinary programmed maintenance. The forecasted capacity is eight (8) trains per day amounting to 2,663,360 tons/year. The estimated duration of the infrastructure rehabilitation is 20 months based on the assumption of four construction sites;

Option 3: Rehabilitation and upgrading of the infrastructure as for Option 2, plus increase of the axle load up to 18t. The estimated cost of this option is about 261,000,000 M$. Under this option the line would have an estimated 20 years life for the railway trackage, but would require an ordinary programmed maintenance. The forecasted capacity is eight (8) trains per day with 3,442,880 tons/year. The estimated duration of the infrastructure rehabilitation is 24 months, based on the assumption of four construction sites of approximately 100 km extension.

Option 4: Rehabilitation and upgrading of the infrastructure with a complete renewal of the permanent way and increase of the axle load up to 21 tons. The estimated cost of this option is about 507,000,000 M$. Under this option the line would have an estimated 30 years life for the railway trackage, but would require an ordinary programmed maintenance. The forecasted capacity is ten (10) trains per day with 5,278,000 tons/year. The estimated duration of the infrastructure rehabilitation is 36 months based on the assumption of four construction sites.

5 IDENTIFICATION OF OPTIMAL INVESTMENT SOLUTION THROUGH MULTI-CRITERIA ANALYSIS (MCA)

The elaboration of the MCA has been based on the best recognised international guidelines (Dogdson et al. 2009). The different MCA criteria adopted can be summarized as follows:

A. Transport criterion: "how will the different options respond to the transport demand, improving the transport capacity, reliability and safety in comparison with the present transport infrastructure?" This criterion has been given a 30% weight.
B. Financial criterion: "which is the most efficient way to rehabilitate the railway, with the minimum investment cost, sustainable recurrent costs and adequate revenues?" This criterion has been given a 50% weight.
C. Environmental criterion: "which is the rehabilitation option that is less impacting on the surrounding environment, in terms of air quality, noise, soil, water and biodiversity?" This criterion has been given a 10% weight.
D. Social criterion: "which is the most profitable option from the point of view of the local communities?" This criterion has been given a 10% weight.

For each of the above mentioned criteria, specific sub-criteria have been used in the MCA. More in detail:

A. Technical (transport) criterion: This technical criterion has been represented by the following sub-criteria:
  A.1. Railway Freight Capacity between Mchinji and Nkaya, calculated as the maximum annual ton*km that can be moved for each upgrading option, obtained by an optimal operational service. The measurement of freight capacity is given in number of tons per year.
  A.2. Transport Level of Service, expressed in terms of ratio between the expected traffic flow (tons/year) and the capacity offered by each option.
  A.3. Railway Passenger Services, expressed in terms of average speed (km/h) along the line.
  A.4. Improved Road Operations, expressed in terms of volume of freight traffic (tons) deviated from the road.
B. Financial criterion:
  B.1. Investment cost, Total financial investment cost, represented in million USD
  B.2. Operational costs, expressed in USD/ton*km.
  B.3. Traffic Revenues, expressed in total yearly traffic revenues in USD.
C. Environmental criterion:
  C.1. Synthetic parameter, ranking each option impact on air quality, noise & vibrations, soil & landscape, waste, water resources and biodiversity.
D. Social criterion: The social criterion will be represented by the following sub-criterion:
  D.1. Synthetic parameter, ranking each option impact on employment, economic development, poverty reduction, community health & safety and land acquisition.

The results of the MCA base case indicate that the preferred option is Option 3. In particular, the Transport criteria would favour Options 4, 3 & 2, which ensure higher values of railway capacity and level of service. Only Option 4, with higher

Table 3. Values of each sub-criterion for the different upgrading options.

	Opt. 1	Opt. 2	Opt. 3	Opt. 4
Cr. A1 [M tons]	1.5	2.7	3.4	**5.3**
Cr. A2 [f/Cap]	1.84	1.02″	0.79	**0.52**
Cr. A3 [km/h]	33.48	33.48	33.48	**44.66**
Cr. A4 [%]	12.15	18.40	**18.69**	**18.69**
Cr. B1 [M $]	**146**	164	261	507
Cr. B2 [$/ton*km]	0.054	0.054	**0.047**	**0.047**
Cr. B3 [M $]	47.2	85.5	**86.9**	**86.9**
Cr. C1	**30**	29	27	24
Cr. D1	42	44	**45**	44

Table 4. Distribution of weights for the different criteria.

	Trans.	Finan.	Envir.	Social
Weights case 1	30%	50%	10%	10%
Weights case 2	35%	45%	10%	10%
Weights case 3	40%	40%	10%	10%
Weights case 4	25%	45%	15%	15%
Weights case 5	30%	40%	15%	15%

Table 5. Results of the MCA for each distribution of weights adopted.

	Opt. 1	Opt. 2	Opt. 3	Opt. 4
Case 1	40.0	62.4	65.4	37.8
Case 2	37.0	60.2	62.9	37.1
Case 3	34.0	57.9	60.6	37.5
Case 4	42.0	64.6	67.1	37.3
Case 5	39.0	62.3	64.7	37.1

speed would guarantee a substantial attraction of passengers from Lilongwe and Blantyre. Options 2, 3 and 4 would all attract freight traffic from the parallel road, improving the road safety conditions. The Financial criteria, considering the construction cost (investment), operating cost and revenues, gives advantages to Options 3 & 2. Option 1 is the most advantageous for investment costs, while Options 3 & 4 for operation costs and revenues. The Environmental criterion (impact on air, noise, soil, water and biodiversity) presents the most advantageous synthetic parameter for Option 1, followed by Option 2. The Social criteria (impact on jobs, health, poverty and land acquisition) shows the advantage of Option 3, with a difference with the other options.

In order to analyse the stability of the results of the MCA base case, a Sensitivity Analysis has been conducted for a different distribution of weights (Table 4). The results of the Sensitivity Analysis in comparison with the base case (i.e. case 1) are collected in Table 5. The conclusion of the MCA

is that Option 3 (restructuring and upgrading the railway infrastructure, increasing the capacity of the line and stations, increasing the axle load to 18t and installing welded rails) can be considered as the best rehabilitation option according to the selected criteria.

6 ENVIRONMENTAL SCREENING AND POTENTIAL IMPACTS

The Environmental Brief has been developed for Project Option 3 and provides a description of the Project and of the selected Project option focussing on the analysis of alternatives, national policy and legal framework, environmental and social setting of the area and, finally, on the identification of the main environmental and social impacts (positive and negative) associated with the rehabilitation of the railway line, both during the construction and operation phases.

The main components likely to be affected are:

1. Natural environment: this includes air quality, noise & vibrations, soil, subsoil, land use and landscape, waste, water resources, biodiversity and sensitive habitats;
2. Socio-economic environment: this includes labour, local and national economy, poverty reduction, community health and safety, land access, work related safety, transport.

The Environmental Project Brief concludes that the Project has very positive impacts. In particular, it has the potential to contribute to the economy of the country through improved movement of exports/import goods, job creation, business promotion, enhancement of Community Health and Safety, and support to the agriculture industry, among others. From an environmental point of view, the Project will be responsible for the improvement of air quality as a consequence of diverting passenger and freight transport from road to rail, as well as reducing soil erosion. Limited negative impacts are expected to arise from the Project and are mostly limited to the construction phase (namely the interventions for the railway rehabilitation), thus of a temporary nature. They mainly refer to limited impacts on air quality due to particulate emissions from construction engines, the increase of the noise level, extraction and use of earth materials, harmless solid waste generation, potential work related accidents, limited loss of land and properties, risks of vandalism. It is important to consider that most of the railway sections cross uninhabited areas, thus limiting the impacts on human receptors. Moreover, through the implementation of appropriate mitigation measures, negative impacts will be minimised with no unacceptable residual impacts.

7 OPERATION MODEL AND DIMENSIONING OF TRAIN FLEET

On the basis of the results of the macroscopic simulations concerning the selected scenario (i.e. Scenario 2), it has been possible to design plausible Operation Plans to obtain indications about the requirements in terms of infrastructure equipment, the signalling system and rolling stock for the rehabilitation of the Mchinji-Nkaya rail line.

Basically, at the moment, the rail track would support a load of at most 15 tons per axle, and the actual status of the infrastructure enables a limit of about 0.25 mtpa. As far as the signalling system is concerned, train operations are currently managed by means of a telephonic paper order system. This method can be used safely with low traffic levels. For denser timetables (as expected), systems that provide train movement detection, as well as failsafe interlocking, should be considered. Furthermore, single-track line capacity is limited by the need for trains to decelerate, stop, and accelerate out of sidings to allow other trains to use the intermediate single-track sections. As a consequence, the need of frequent crossings of trains running in opposite directions consumes a lot of capacity and can become a critical factor. In addition, meeting at these sidings is the largest cause of train-interference delays on single tracks. Therefore, in order to best increase the capacity of the Mchinji-Nkaya line, for the Operation Plan no crossings between trains travelling in opposite directions are considered. As a consequence, it is hypothesised that on odd days trains run from East to West (e.g. from Mozambique to Zambia) and from West to East on even days. This solution has the great advantage of reducing travel times, avoiding inefficient stops at stations for crossings and, above all, minimising safety risks. Obviously, due to the alternate circulation, since trains travelling in one direction are allowed every two days, the number of runs per day has to be doubled.

Regarding the signalling system, Manual Block Signalling (MBS) or Automatic Block Signalling (ABS) could be adopted. The cost for signalling system has been estimated as 46,304,560 for the whole line (i.e. almost 8,900 USD/km).

By means of simple calculations, it has been possible to assess that, the upgrading Option 3 would meet the expected freight demand (i.e. Scenario 2) considering 6 freight runs from Nkaya to Mchinji on odd days and 6 runs travelling in the opposite direction on even days (Fig. 4). In addition to the freight service, a daily train connecting Lilongwe and Blantyre is included so as to provide also a passenger service.

Regarding the dimensioning of the fleet, the actual rolling stock available is out of date

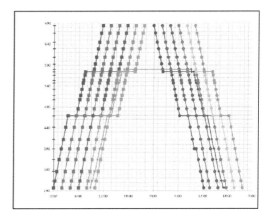

Figure 4. Hypothesized timetable of the Mchinji-Nkaya rail line.

Table 6. Average unitary cost for purchasing and revamping the rolling stock fleet.

Cost of new locomotive	$ 3,500,000
Cost of new freight wagons	$ 57,500
Cost of new passenger wagons	$ 700,000
Cost of revamping locomotives	$ 1,750,000
Cost of revamping freight wagons	$ 14,375
Cost of revamping pax wagons	$ 280,000
Cost of new shunting locomotives	$ 1,000,000
Cost of revamping shunting locomotive	$ 500,000

(average fleet age is 49 years) and has to be refurbished and or increased to meet the expected demand. In order to estimate the investment cost, the following assumptions have been taken into account:

- for freight service, double-headed trains coupled "in multiple" are considered, which means that at least two locomotives are necessary to haul each train;
- Besides the total locomotive fleet required to perform the service, an additional 30% has to be dedicated as a reserve for ordinary and extraordinary maintenance programmes or in case of disturbances to the ordinary conditions (e.g. spare locomotives);
- For the same reasons, a surplus of 40% of the total freight wagon fleet might also be sufficient to balance the empty wagons which have to be moved each day;
- The supplementary rolling stock (both locomotives and wagons) could also be used to meet possible peak demand periods (i.e. more than 6 runs per day needed);
- for passenger service, each train is composed of one locomotive and 8 passenger wagons for a total capacity of 600 passengers;

- Additional rolling stock for maintenance and reserve purposes is expected to be 30% of the actual fleet.
- The fleet belonging to other rail operators, especially CDN from Mozambique will contribute to moving the expected freight, thus reducing the resources needed.

According to these considerations, after a brief analysis of the average unitary costs (Table 6), the estimated investment costs for rolling stock amount to 52.09 M USD.

8 COST BENEFIT ANALYSIS

The CBA methodology is based on the EU Guide to Cost-Benefit Analysis of Investment Projects, Economic Appraisal Tool for Cohesion Policy 2014–2020 (European Commission, 2014), which combines financial, economic and risk analyses.

The Financial Analysis takes into consideration Investment costs and Operation and Maintenance (O&M) costs. Investment costs have been evaluated through Bills of Quantities and Cost Estimates regarding Option 3 and resulting from the infrastructure assessment (14-1397-H2, 2016a). The financial discount rate adopted for this study was 7.5% and the time extension of the cash flow has been taken equal to 30 years after the finalisation of the upgrade, i.e. until year 2050 included. The traffic flows that were considered are those estimated for the Scenario 2.

O&M costs have been estimated in the "with" and "without project" scenarios, grouped into the following categories:

- infrastructure operations, e.g. repairs, current maintenance, materials, energy, train operations, etc;
- services operations, e.g. staff cost, traffic management expenses, energy consumption, materials, consumables, rolling stock maintenance, insurance, etc.;
- services management, e.g. services management itself, fare/tolls collection, company overheads, buildings, administration, etc.

The unitary freight train operation cost was estimated as 0.0534 USD per ton × km.

Maintenance costs by contrast, should cover (and are usually distinguished into):

- routine maintenance: yearly work required to keep the infrastructure technically safe and ready for day to day operation as well as to prevent deterioration of the infrastructure assets;
- periodic maintenance: all activities intended to restore the original condition of the infrastructure.

No significant differences in terms of ordinary maintenance costs were considered between the hypotheses "with" and "without project". The assumption is that there will be a rough compensation between the differences that will occur among the two situations: a) increase of maintenance costs due to increased volumes of traffic for the situation "with project", and b) increased costs due to the deteriorated existing conditions in the "without project" case.

A periodic maintenance intervention is supposed for both scenarios every ten years. The cost of such maintenance was assumed by experience to be 10% of required cost for the upgrade of the line, i.e. 23.84 M USD. Differences between the two scenarios are only due the different phasing, with the first extra maintenance starting at year 2017 and 2026 for the "without project" and at five years later for the "with project" case.

Financial inflows are represented by the proceeds from the charges applied to users for the sale of transport services. The estimation of revenues has been based on the traffic volume forecasts, taking the charge system (i.e. 0.089 USD/ton*km) as invariant.

The financial CBA, as expected for this kind of investment, does not reach profitability, but the relevant indicators are still positive:

- the IRR is 6.45%, below the discount rate;
- the NPV is –31.61 M USD;
- the B/C ratio is 0.94, just below 1.

In the Economic Analysis, the main direct benefits related to changes of the following measurable have to be considered:

- The consumer surplus, defined as the excess of users' willingness to pay over the prevailing generalised cost of transport for a specific trip. The generalised cost of transport expresses the overall inconvenience to the user, computed as the sum of monetary costs borne (railway tariff) plus the value of the travel time, calculated in equivalent monetary units. Any reduction of the generalised cost of transport for the movement of goods and people determines an increase in the consumer surplus. The main items to be considered for the estimation of the consumer surplus in the railway project are: reduced railway travel time and reduced road users Vehicle Operating Costs (VOC) & accident costs;
- The producer surplus, defined as the revenues accrued to the transport operators and owners (railway and road) minus the costs borne. The change in the producer surplus is calculated as the difference between the change in the producer revenue (e.g. rail ticket income increase) and the change in the producer costs (e.g. train

operating & maintenance costs, road maintenance costs). The main items to be considered for the estimation of the producer surplus are the variation of the railway producer operating & maintenance costs and the parallel road maintenance costs.

The analysis is therefore carried out under the point of view of the general society, and this is the reason why this approach is frequently referred also as "socio-economic" approach. Excluding transfers of money between the parts involved in the analyses means to take into account also taxes, excise duties, levies that are collected by the State and by local administrations and used in order to deliver public services and, at the same time, means to exclude from the expenditures those implicit benefits that are linked to the new job creation. These are the notions that are at the base of the so-called "shadow prices" that are used inside economic CBAs and that from the practical point of view bring to the reduction of some costs considered by the financial approach with the application of financial/economic conversion factors. In our case a unique financial/economic conversion factor of 0.7 was adopted. A discount rate of 8% was also considered, as indicated in a previous pre-feasibility study (Gibbs & Parsons Brinckerhoff, 2014) on the same rail line. As a consequence, for the economic analysis, the different cost items are based on the same values adopted for the financial analysis, but they were reduced by transforming these costs into economic values. In addition, the "with project" scenario is bound to produce a modal shift from road to rail for both freight and passenger traffic. Such modal shift will produce a reduction in terms of operating costs, due to the better efficiency of the rail mode as a whole (and especially in terms of reduced energy consumptions). The calculation of the overall benefits was carried out by considering the tons × km and the passengers × km saved by moving traffic from the road mode to the rail mode and by comparing the relevant reduction of costs.

The following unitary economic operating costs were adopted in order to estimate such aspect:

- Truck operating cost: 0.0690 USD per ton × km;
- Bus operating cost: 0.0260 USD per pass × km;
- Freight train operating cost: 0.0374 USD per ton × km;
- Passenger train operating cost: 0.0395 USD per pass × km.

Values for trucks and buses have been assumed on the basis of a comparison with similar studies carried out by the Consultant, while values for rail are the result coming from the economic/financial conversion of the operating costs, as considered for the financial CBA.

The modal shift from the road to the rail mode produces some additional socio-economic benefits that have a significant impact on the final result.

These benefits are mainly represented by:

- The environmental impact of different transport modes;
- The benefits due to the reduction of road accidents;
- The reduction of road maintenance costs passing from the road mode to the rail mode (especially as regards the freight component);

The environmental impact of the different transport modes was estimated on the basis of international standards. This because the cost of pollution cannot be differentiated on the basis of the characteristics of local economies, giving that pollution has a global impact and cannot be estimated at local level. The values adopted in this study are therefore:

- Truck environmental cost: 0.0335 USD per ton × km;
- Bus environmental: 0.0215 USD per pass × km;
- Freight train environmental cost: 0.0122 USD per ton × km;
- Passenger train environmental cost: 0.0335 USD per pass × km.

The above values are the result of a Consultant's adaptation of data from other relevant European studies. Environmental values for rail services are relatively high because they are referred to trains hauled by diesel locomotives only. On the other side, values for road transport can be considered overestimated and therefore quite prudential for an African scenario, given that the environmental quality of the local vehicle fleet still does not match that one of the European vehicle fleet at year 2011.

The basic cost for road accidents was estimated at 0.01 USD per ton × km (Gibbs & Parsons Brinckerhoff, 2014) and a comparable value was adopted for passenger road traffic (0.01 USD per pass per km). Finally, following again the indications of Gibbs & Parsons Brinckerhoff (2014), the cost for road maintenance was also estimated at 0.01 USD per ton × km. The results of the economic analysis are:

- the IRR is 13.63%, that is over the discount rate (8%);
- the NPV is 170.77 M USD, therefore positive;
- the B/C ratio is 2.09, therefore much higher than 1.

Due to the difficulty in forecasting values in the future, besides CBA, it is good practice to assess the project risks by carrying out a sensitivity analysis and a risk analysis. In the Risk Assessment, risks are analysed through the probability distributions indicating the likelihood of the actual value of a variable falling within stated limits, while uncertainties have been analysed through defined variations of the main variables of the project (sensitivity analysis). The Sensitivity analysis allows the determination of the 'critical' variables or parameters of the model. Such variables are those whose variations, positive or negative, have the greatest impact on a project's financial and/or economic performance. The analysis has been carried out by varying the investment cost or the Benefits/Revenues values, one element at a time and also as a combined scenario, determining the effect of that change on EIRR/FIRR and on NPV. The results of both the above-mentioned analyses support the convenience of the investment. In fact, there is a 95% probability that the upgrading Option 3 economic IRR will be greater than 11.35% and the minimum value (11.90%) is reached if the investment cost is 20% higher and simultaneously the benefits are 20% lower than estimated in the CBA.

9 CONCLUSION

In this paper, the main activities carried out as part of the feasibility study for the rehabilitation of the Mchinki-Nkaya rail line are briefly described. Results have demonstrated that the project would help the Malawian economy to grow up supporting the GDP and reducing costs of import/export trades. In addition, the rail line would provide several social benefits which make the investment worthwhile especially from the economic point of view.

REFERENCES

Cascetta, E. 2009. Transportation systems analysis: Models and applications. Springer: New York (NY), USA.

Dodgson, J.S., Spackman, M., Pearman, A. & Phillips, L.D. 2009. Multi-criteria analysis: a manual. Department for Communities and Local Government: London.

European Commission 2014. EU Guide to Cost-Benefit Analysis of Investment Projects, Economic Appraisal Tool for Cohesion Policy 2014–2020.

GIBB, & Parsons Brinckerhoff 2014. Scoping study for the railway line in Malawi from Mchinji to Nkaya—including the section between Mchinji and Chipata in Zambia". Final Report.

SADC 2012. Southern African Development Community. Regional Infrastructure Development Master Plan.

Transport Infrastructure and Systems – Dell'Acqua & Wegman (Eds)
© *2017 Taylor & Francis Group, London, ISBN 978-1-138-03009-1*

Identifying driving behaviour profiles by using multiple correspondence analysis and cluster analysis

D.S. Usami, L. Persia, M. Picardi & M.R. Saporito
Sapienza University, Rome, Italy

I. Corazziari
ISTAT, Rome, Italy

ABSTRACT: Many studies investigated the relationship between accident risk factors related to driving behaviour (e.g. distraction, aggressive driving) and the involvement in a road accident. However less attention has been given to aspects related to driver insecurity and the effects of the overall driving patterns. The main target of this research were twofold: 1) The identification of driving behaviour profiles taking into account also insecurity. 2) The analysis of the association between the identified profiles and their accident involvement. A survey was undertaken among a sample of Italian drivers to assess driving distraction, aggressiveness, indiscipline and insecurity. The items of the used questionnaire were mostly derived from the literature (e.g. Driver Behaviour Questionnaire). Behavioural tendencies of drivers to distraction, aggressiveness, indiscipline and insecurity were studied through Multiple Correspondence Analysis. By using Cluster Analysis seven groups of drivers with similar behaviours were identified. A significant association between the seven groups and road accident involvement was found. This statistical approach allows the identification of driving behaviour profiles that could be used for driving training purposes. The answers to the questionnaire can for instance highlight an aggressive and/or insecure driving behaviour thus tailoring the theoretical and practical driving exercises to the specific driver needs, especially for novice drivers.

1   INTRODUCTION

1.1   *Background*

The prevalence of human factors in the causation of road accidents has been demonstrated by several studies (e.g. Usami et al. 2015). This led many researchers to focus on human related predictors of road accident involvement.

A review of the literature indicates that various socio-demographic factors like gender and age have been found to be related to road accident involvement (Abdel-Aty et al., 1998). Also information processing (e.g. selective attention), cognitive ability and personality traits (e.g. sensation seeking, anger, respect of authority, locus of control) have been found to be related to involvement in road accidents (e.g. Arthur et al., 1991).

It is also recognize that different personality traits lead to different driving styles. Many studies examined the association of a potential predictor alone. Dahlen et al. (2005) investigated the combined effect of driving anger, sensation seeking, impulsiveness, and boredom proneness on driving behaviour; also highlighting the importance of the use of multiple predictors in understanding unsafe driving behaviour.

Extensive research investigated the relationship between personality traits and driving behaviour, referring to the ways drivers choose to drive or habitually drive (e.g. speeding, driving aggressively, complying to norms, etc.).

Methods used within these studies to measure driving behaviour are mostly based on questionnaires, data from simulators or number of traffic offences.

Several questionnaires have been proposed in the last decades to measure driver behaviour. Some of them, like the Driver Behavior Questionnaire (DBQ, Reason et al., 1990) consider more aspects of driving behaviour. For example DBQ examines errors made while driving, deliberate violations of normal safe driving practice, and harmless mistakes that result from inattention. Other questionnaires are more specific examining for instance aggressive driving (Deffenbacher et al., 1994).

Based on a review of several existing questionnaires, Taubman-Ben-Ari et al. (2003) initially proposed four driving styles: (a) reckless and careless driving style, (b) anxious driving style, (c) angry and hostile driving style, (d) patient and careful driving style. After a factor analysis they found eight main factors, each one representing a specific driving style: dissociative, anxious, risky, angry, high velocity, distress reduction, patient and careful driving style. They found that these factors significantly predict self-reports of accident involvement and traffic offenses.

However, a driver hardly would fall on a single driving style category, his driving profile would probably fit more factors. This information is important especially from a driving school perspective. Knowing in advance the profile of a driver would help a driving instructor in defining a customized course, aimed at addressing the issues of a person way of driving.

1.2 Objective

Based on the existing knowledge, the present study combines four predictors of driving behaviour with two specific objectives: 1) The identification of driving behaviour profiles. 2) The analysis of the association between the identified profiles and their accident involvement.

The basic assumption is that a not balanced and patient driving behaviour is positively related to accident involvements, and also to a lacking knowledge of traffic rules and laws.

The driving behaviour is a complex and multidimensional construct and it has been measured considering four main dimensions: reckless, aggressive, careless and anxious driving behaviour. Each of the four dimension is complex and multidimensional in turn, so it is described by a set of indicators, measured by the corresponding questions (categorical variables) in the developed questionnaire.

To summarize and measure the four behavioural dimensions, a Multiple Correspondence Analysis (MCA) has been performed to each of the corresponding set of categorical variables.

The drivers' scores on the derived factors (principal components) have then been analysed by a Cluster Analysis (CA) to identify profiles of driving behaviors more related to accidents occurrences.

2 MATERIAL AND METHODS

2.1 Survey questionnaire

By applying to the field of driver education the European principles of Lifelong Learning, a questionnaire was developed able to identify different types of driving behaviour, so that we can better address initial and post-license training.

The aim is to get targeted training to obtaining a driving profile that would be safe, responsible, beyond a preparation aimed at passing an examination to obtain a driver's license.

The questionnaire is divided into 5 sections: driving experience, reckless (indiscipline), aggression, careless (distraction) and anxiety (insecurity).

The "driving experience" section is characterized by a number of questions related to driving frequency, type, mode of travel etc. Questions were derived from technical meetings held in collaboration with the EFA (European Federation of driving schools and driving instructors) and in EU studies (e.g. NovEV Project of CIECA, the International Commission of driving examiners).

The "reckless" section has been drawn from the Driver Behaviour Questionnaire (DBQ, Reason et al., 1990) by selecting questions on deliberate choices of the driver. Estimating the consequences of deliberate manoeuvres is difficult for a common driver and has different perceived consequences. For instance, parking a vehicle in second line is perceived differently from running a red traffic light.

The section of "aggressive behaviour" has been addressed both by means of the Driver Anger Scale (Deffenbacher et al., 1994), and the Driver Behavior Questionnaire (DBQ, Reason et al., 1990). Among the standards of human sciences used by the American Association of Psychology it includes the Driver Anger Scale, containing different traffic situations attributable to the behaviour of others (disrespectful), presence of other types of users (co-existence among different) or external situations (speed radar) and the consequences of these on your driving attitude.

"Distraction" section is once again mostly derived from the DBQ. It addresses the difference between distraction and inattention, where the first may be attributable to the conduct of a series of activities in parallel at driving (using the navigator or mobile phone) and the second to a mental state of disconnection (i.e. absorbed in thoughts) from the main task.

The "insecurity" section makes use of two studies (Ma et al., 2010; Ulleberg, 2001) in addition to the support received from the research of technical workshops in collaboration with the experience of the training carried out within the team EFA. The risk perception, the assessment of the situation, the sharing of the traffic environment, the decision making and past experiences are the factors that characterize this part of the questionnaire.

2.2 Sample description

Several drivers, clients of Italian driving schools from all the country, were contacted by e-mail and invited to take part to the survey.

More than 200 drivers initially agreed to answer to the questionnaire. However, only a part of them were correctly completed and were selected for the analysis.

The final sample was 104 participants (41% women) from all over Italy (mean age 40 years).

The demographic characteristics and the percentage of involvement in a road accident in the last two years are presented in Table 1.

2.3 Statistical methods

2.4 Multiple Correspondence Analysis and cluster analysis

Multiple Correspondence Analysis (MCA) is a factorial analysis method allowing summarizing a set of categorical variables by one or few quantitative and not correlated (orthogonal) ones. The main aim of MCA is to describe the association among the original variables, and to measure the underlying latent factor not directly observable, represented by the set of observable variables.

Separate MCA were performed for the four set of variables relating to the corresponding four predictors of driving behaviour.

To perform MCA, every item of each categorical variable has been coded in a binary form, 1 indicating the presence of the item, 0 otherwise, only one item presence allowed for each categorical variable. Performing an MCA on the resulting matrix (a complete disjunctive one) is equivalent to a principal component analysis on the same matrix (Jobson JD 1992).

The method provides a set of coordinate values which allow to analyse and possibly to graphically display on a Cartesian plane, the association among the original variables (among their categories). MCA also provide coordinate values (scores on the principal components) for individuals (drivers) allowing to evaluate drivers' profiles in term of their similarity referring to the considered predictor (their proximity can be graphically displayed on the same Cartesian plane where variables are displayed).

The drivers' scores on the derived factors (no more than two principal components for each of the four predictors of behaviour) can be analysed by other multivariate quantitative methods. In the present study an ascending hierarchical classification analysis was performed (Everitt BS 1993), to determine the most plausible number of drivers groups, homogeneous in term of their four predictor profiles. The variable created by the clustering process indicating which cluster every individual belongs to, represents an indirect measure of the driving behaviour, to be related to accident occurrences. The cluster method was the Ward's minimum variance method applied to the whole set of principal components obtained in the four MCA analysis. The Ward's method is normally used to aggregate cases when euclidean distances make sense, as it is the case with quantitative variables, and so in the case of the MCA scores. With this method, the sum squared distances of points from their cluster centroids are minimized. The description of clusters, that means the definition of driver profiles, is made through the analysis of the proportions of units with a specific category in the cluster, compared with the corresponding proportion of units reporting the same category in all the sample.

After having defined the clusters, new units can be classified ex post in one of the groups, by a discriminant analysis (Anderson TV 1984). The first step is to calculate the new units scores, projecting as supplementary units the new ones on the factorial planes obtained in the MCA analysis. Then a nonparametric discriminant analysis can be performed on these scores, to classify units in the given clusters, minimising distances between the new units and the clusters barycentres.

3 RESULTS

3.1 Factors contributing to driving reckless, aggressive, anxious and careless behaviour

The first MCA performed on the reckless indicator variables is summarized in Figure 1, displaying the first two principal components: behaviours respecting the rules are identified by negative values on the first (horizontal) principal components axis, with a tolerant attitude towards the other drivers. Negative values also on the second axis show less respect when the car ahead drives too slow, so that the risk to overtake other vehicles, even on the right when driving on the motorway, is not implausible. Such profile is associated with drivers aged more than 55. Positive values on the second axis identify drivers that do not always respect the public buses right of way. It is associated with younger female drivers.

Positive values on the first axis represent more reckless behaviours. When associated to negative values on the second axis, more extreme behaviours are described in term of risks: driving after alcohol

Table 1. Characteristics of the sample.

	Male	Female
Age		
<25	9,3%	11,5%
25–34	25,6%	11,5%
35–44	48,8%	45,9%
45–54	9,3%	19,7%
55–64	4,7%	6,6%
>65	2,3%	4,9%
Accident involvement in the last 2 years		
0 crashes	88,4%	85,2%
1 crash	9,3%	11,5%
2+ crashes	2,3%	3,3%

consumption, no speed limit respect, risky over-taking, no tolerance towards the other drivers, car races. Such reckless profile is associated with males.

To summarize the reckless predictor, only the first principal component is taken for the following CA.

The second MCA performed on the aggressive indicators provide two principal components axis, identifying four profiles of drivers (Figure 2). Negative values on the first axis describe the less aggressive profile, yet willing to show who is right (with negative values on both the axis), mainly resident in the North-east of Italy, with a primary school educational level, and (with positive values on the second axis) the no aggressive profile except in case of being forced to reduce its own driving speed due to ahead bikers or ahead camions limiting the view (drivers aged 25–34). With positive values on the first axis the more aggressive profiles: from drivers who feel nervous if forced to reduce their speed due to speed controls, traffic, bikers, camion limiting the views or loosing sends

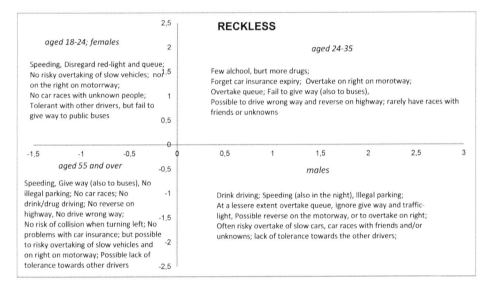

Figure 1.   Map of reckless behaviour associations (first and second principal components).

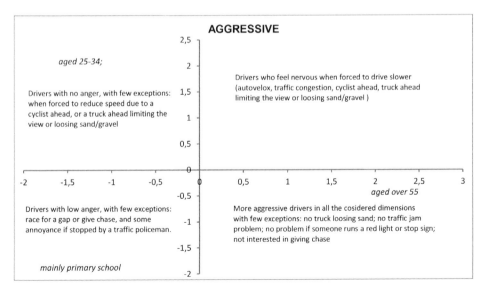

Figure 2.   Map of aggressive behaviour associations (first and second principal components).

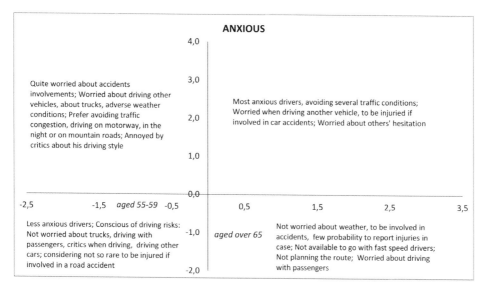

Figure 3. Map of anxious behaviour associations (first and second principal components).

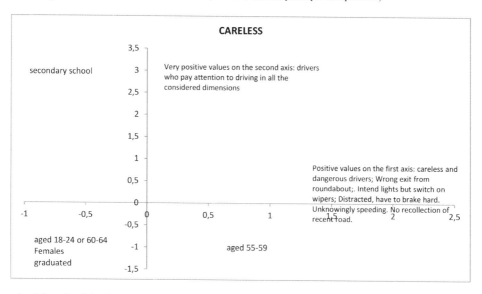

Figure 4. Map of reckless behaviour associations (first and second principal components).

(positive values on the second axis), to more aggressive drivers feeling nervous in all the considered dimensions, but not interested in showing their own reasons (aged more than 55, negative values on the second axis).

The third MCA refers to the anxious style of driving. Again two axes have been considered, the second one more clearly identifying anxious drivers (with positive values) with avoiding behaviours due to anxiety, against less anxious drivers with negative values on the same axis. Drivers showing higher consciousness of actual risks have negative values on the

first axis, while with positive values on the first one, drivers are characterized by less worrying about accident involvements, injuries and weather conditions.

The fourth and last MCA has been applied to the careless indicators variables. The two orthogonal axis are also conceptually orthogonal, as the first axis identified with positive values less care drivers, in every all the considered dimensions, while the second axis' positive values identifies very care drivers.

Negative values on both the dimensions identify more care drivers' profiles.

3.2 Driving behaviour profiles

To identify the overall driving behaviour in term of balance and patience, the first reckless axis discriminating between more respectful drivers and less ones, and the first two axis for the other three predictors, has been analysed by a cluster analysis. An ascending hierarchical classification has been adopted, using the Ward's algorithm. Starting from each driver being a single cluster, the method aggregates more and more couples of units, or a cluster with a unit, or two clusters, according to the minimization of the variance of the new cluster formed. The process stops when some criterion are met, considering the gain in minimizing the within clusters variance, and maximizing the between clusters one (Everitt BS 1993).

Seven clusters were identified with this method. The study of the characteristics of the drivers belonging to each cluster, allows identifying seven driving profiles in terms of balanced and patience behaviours as described by the four predictors, and other demographic characteristics.

In each driving profile does not exist a pure driving style. Driving profiles differ in the prevalence of a driving style (reckless, aggressive, careless and anxious driving) on another. A driver could act for instance aggressively (honks easily, discusses with other road users), disregarding traffic rules but confident in what he is doing (not being afraid of the complexity of the driving task).

The cluster medians and the median for each factor are reported in Table 2.

Drivers in Cluster 1 appear to be very aggressive but observing traffic rules.

In Cluster 2 there are drivers who show scores close to the median on most variables. However, they seem to be anxious (e.g. worried about driving other vehicles, in bad weather conditions, etc).

Cluster 3 includes drivers with moderate aggressiveness and recklessness.

The individuals in Cluster 4 are characterised by very high anxiety. They also reported high values in aggressive driving. They respect the traffic rules and pay attention while driving.

Reckless but mild behaviour mostly characterizes Cluster 5.

Drivers in Cluster 6 are anxious even if they appear to pay a lot of attention when driving. All the other variables have scores close to the median.

Drivers in Cluster 7 appear to be less anxious and conscious of driving risks. They are also aggressive and careless drivers.

Drivers in Clusters 2 seem to be those closer to a balanced average driving profile.

3.3 Driving behaviour profiles and road accident involvement

Finally, an attempt was made to assess the association between the seven driving profiles and self-declared accident involvement.

A significant association (95%) between the seven groups and road accident involvement was found (see Table 3).

Figure 5 shows the percentage of drivers reporting to be involved in a crash in the last two years by cluster. Drivers belonging to cluster 4 (aggressive and anxious, but observing rules), 6 (careful but anxious) and 7 (aggressive and careless) show the highest crash involvement. While drivers of clusters 1 (aggressive but observing rules), 2 (anxious but mild behaviour) and 5 (reckless but mild behaviour) seem to be the safest.

Even if the sample is limited and general conclusions cannot be stated, it seems that the combination of different risky driving styles leads to a higher accident involvement. The exception is Cluster 6 for which the level of anxiety seems to lead these drivers to be more aware avoiding distractions. However even the higher attention appears to not compensating the increased risk due to anxiety.

Table 2. Cluster medians.

	Reckless1	Aggress.1	Aggress.2	Careless1	Careless2	Anxious1	Anxious2
Cl. 1	−0,53	1,12	−0,37	−0,31	−0,22	0,05	−0,19
Cl. 2	−0,19	−0,45	−0,06	−0,44	−0,29	0,19	−0,01
Cl. 3	0,08	−0,09	0,85	−0,43	−0,24	−0,48	−0,04
Cl. 4	−0,73	1,47	−0,52	−0,43	1,01	−0,15	1,56
Cl. 5	1,13	−0,58	−0,61	−0,01	−0,23	0,15	−0,26
Cl. 6	−0,11	−0,44	−0,02	−0,19	0,78	0,36	−0,23
Cl. 7	0,11	−0,08	0,61	1,08	−0,05	−0,57	−0,21
Median	−0,11	−0,18	−0,06	−0,29	−0,14	0,14	−0,15

Table 3. Margin settings for A4 size paper and letter size paper.

Statistic	DF	Value	Prob
Chi-square	12	218.124	0.0397
Chi-square MH	1	77.900	0.0053
Phi coefficient		0.4580	
Contingency Coefficient		0.4164	
Cramer's V		0.3238	

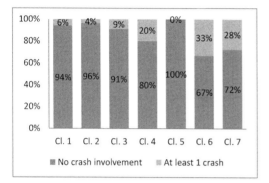

Figure 5. Percentage of drivers involved in a crash by cluster.

4 CONCLUSIONS

The paper identified seven driving profiles by clustering drivers according to four different driving styles: reckless, aggressive, anxious and careless driving behaviours. Three profiles were mostly characterised by a single (risky) driving behaviour (e.g. reckless, aggressive and anxious). While the other four profiles were mostly a combination of two risky driving styles. Careless driving was usually found in combination with other behaviours.

The analysis of the identified profiles with accident involvement shows a significant association (i.e. some clusters show a significantly higher level of accident involvement than others).

This statistical approach allows the identification of driving behaviour profiles that could be used for driving training purposes. The answers to the questionnaire can for instance highlight an aggressive and/or insecure driving behaviour thus tailoring the theoretical and practical driving exercises to the specific driver needs, especially for novice drivers. Answers to the questionnaire can be used to classify new drivers in one of the above clusters. A non-parametric discriminant analysis can be the statistical tool to attribute drivers to one of the seven individuated profiles, requiring different safe driving trainings.

The analysis has shown that among the traffic situations that can create insecurity fall, for example, driving in congested traffic conditions or on roads with many trucks.

In these situations, the theoretical lessons should provide information about the related traffic rules, road signs, the safety distance, but also aspects aimed to provide greater driving confidence such as understanding of perception/reaction time at different speeds, the importance of visual search, the importance of the rear mirrors etc. The practical exercises should then allow the driver to experience the concepts learned on the road. The driver will be able to tackle those situations that caused anxiety and insecurity with a better understanding and a higher confidence.

REFERENCES

Abdel-Aty M.A, Chen C.L, Schott J.R. 1998. An assessment of the effect of driver age on traffic accident involvement using log-linear models. Accident Analysis & Prevention, 30, Issue 6, November 1998, Pages 851–861.

Anderson, T.W. (1984). *An Introduction to Multivariate Statistical Analysis.* 2nd ed. New York: John Wiley & Sons.

Arthur Jr., W., Barrett, G.V., Alexander, R.A., 1991. Prediction of vehicular accident involvement: A meta-analysis. Human Perform. 4, 89–105.

Dahlen, E.R., & White, R.P. 2006. The Big Five factors, sensation seeking, and driving anger in the prediction of unsafe driving. Personality and Individual Differences, 41, 903–915.

Deffenbacher, J.L., Oetting, E.R., Lynch, R.S. 1994. Development of a driving anger scale, Psychological Reports, 74: 83–91.

Everitt BS (eds), Cluster Analysis. London: Arnold, 1993.

Jobson JD. Principal components, factors and correspondence analysis: In: Fienberg S, Olkin I (eds), Applied Multivariate Data Analysis. Volume II: Categorical and Multivariate Methods, Paris: Dunod, 1992: 343–482.

Ma M., X. Yan, H. Huang, M. Abdel-Aty. 2010. Occupational driver safety of public transportation: Risk perception, attitudes, and driving behavior. Proceedings of the Transportation Research Board 89th Annual Meeting.

Reason, J.T., Manstead, A., Stradling, S., Baxter, J.S., Campbell, K., 1990. Errors and violations on the roads: a real distinction? Ergonomics 33, 1315–1332.

Sanders, N. & Keskinen, E. (eds.) (2004). EU NovEV project; Evaluation of post-licence training schemes for novice drivers. Final Report. International Commission of Driver Testing Authorities CIECA, Rijswijk.

Taubman-Ben-Ari O. Mikulincer M, Gillath O. 2003. The multidimensional driving style inventory—scale construct and validation. Accident Analysis and Prevention 952: 1–10.

Ulleberg, P., 2001. Personality subtypes of young drivers. Relationship to risk-taking preferences, accident involvement, and response to a traffic safety campaign. Transport. Res. Part F: Traffic Psychol. Behav. 4, 279–297.

Usami, D.S., Giustiniani, G., Persia, L., Gigli, R. 2015. Aggregated analysis of in-depth accident causation data. International Journal of Injury Control and Safety Promotion. Article in Press.

Transport Infrastructure and Systems – Dell'Acqua & Wegman (Eds)
© 2017 Taylor & Francis Group, London, ISBN 978-1-138-03009-1

Solution to a capacity problem using non-conventional intersections

M. Vsetecka, M. Novak & T. Apeltauer
Brno University of Technology, Czech Republic

ABSTRACT: When designing the road network for the urban tender for new southern part of Brno city center, the authors of this paper came across a problem of heavily loaded intersection, which due to capacity reasons could not be designed as one-level—due to architectural issues and because of a nearby small river it was also not possible to use a multi-level design. Focus was therefore put on designing a non-conventional intersection, which diverts left turns to a neighboring intersection, not to nearby streets. The nearby streets can therefore be designed as purely service roads. In contrast to a one-way design or a design with only right turns permitted, this does not lead to lengthening the routes and is also important for the environment. The traffic at the traffic light-controlled complex of intersections has been tested using a microscopic traffic simulation, which proved the higher capacity of this solution. Sufficient capacity margin is important not just for the vehicles, but also pedestrians, because otherwise there would be a pressure to reduce the number of crossings. Several parameters of the results of the model were compared with conventional one-level and multi-level intersections.

1 INTRODUCTION

Many intersections, which could be labeled as non-conventional, have been described (Esawey 2012). Some moves, especially left turns, are dealt with using various non-standard ways in order to increase capacity, by separating the moves into more intersections.

A good example where such solution has been used is the intersection of the streets Aachenre Straße and Innere Kanalstraße in Köln/Rhein, Germany (50°56'11.9"N 6°55'28.7"E). This

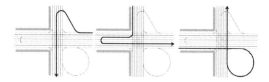

Figure 1. Non-conventional left turns at the intersection in Köln. From the left: semi-direct ramp, turns across separating lane and reverse ramp.

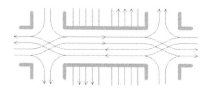

Figure 2. Diverging diamond interchange near Seclinu—left-side traffic on the bridge.

intersection includes a reverse semi-direct ramp and turns across the middle separating lane.

Another example is the multi-level interchange A1 and D549 near Seclin, France (50°32'41''N 3°3'21''E), a type of so-called diverging diamond interchange (Maji 2013, Hu 2014, Leong 2015) – left side traffic is used at a short stretch, which creates free left turns and thus increases the capacity and reduces delay time.

Nevertheless, use of non-conventional intersections is rather rare and so the authors welcomed the opportunity to co-operate on an urban study, which required use of such non-conventional intersection. The goal was to design a non-conventional intersection, including traffic control and then compare this layout with conventional intersection types.

2 SOURCE OF THE PROBLEM

The city of Brno has in long term been expanding quite unevenly—the southern part of the city center has not been very interesting for investors for several decades because of the unresolved issue of the main railway junction placement and the risk of floods from the nearby Svratka River. The area therefore currently consists of mostly brownfields. One challenge of building a new residence area here is designing an intersection that would connect the main city ring with the major road to Bratislava, which in 2014 had a traffic load of 73 thousand vehicles/day. Based on traffic model that takes into account the new housing development, the traffic load will increase to 89 thousand vehicles/day and after the planned roads on the city outskirts are finished, it should drop to 65 thousand vehicles/day. During the worst period (assuming the standard ratio of 8% between peak and daily traffic in Brno) the traffic load of the four-way intersection will therefore be over 7 thousand vehicles/hour, which is close to the upper limit of the capacity of a traffic-light controlled four-way intersection, especially if the intersection is also supposed to include pedestrian crossings. In addition, the urban design assumes a denser housing development compared to the traffic model.

In general, the following solutions are possible:

- Multi-level interchange, which however is not suitable due to architectural and urbanistic reasons in an above-ground layout and very complicated to build in underground layout due to the nearby river and high levels of groundwater.
- Distributing the traffic load to nearby roads, for example by separating the traffic into one-way system in parallel streets, which instead of one heavily loaded two-way intersection, would result in four less loaded intersections and non-problematic left turns. This solution was not chosen because it did not satisfy the requirement of concentrating the traffic corridors in one place, so that the negative effects of the heavy traffic impact as small area as possible (see Figure 4).
- No left turns, which would be replaced by traffic in minor streets (see Figure 4). This solution is not suitable because it would place a load on the nearby minor streets.
- Non-conventional intersection, which would thanks to its specific layout, allow handling such heavy traffic load required.

3 NON-CONVENTIONAL INTERSECTION

The requirement was to concentrate all traffic to just one corridor and also not increasing the driving distance. This limited the possible types of non-conventional intersections to those, which partly use (in continental Europe quite unusual) left-side traffic that allows left turns without having to give right of way to oncoming traffic.

Cross section ahead of the intersection is from the perspective of a driver entering the intersection: On the left a lane for left turns, in the middle driving lanes for oncoming traffic, on the right lanes for driving straight and turning right. Vehicles turning left therefore do not have to give right of way to the oncoming vehicles and thus do not reduce the intersection capacity.

Crossover of the position of the left-turning vehicles, i.e. the point where they cross the oncoming traffic, is located ahead of the intersection. Instead

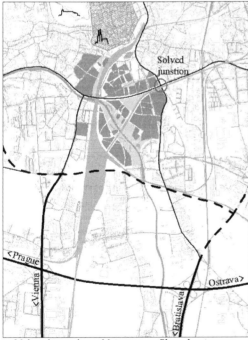

Figure 3. Map of the location of interest in the context of the city traffic system.
Map layer—urban design by architect Jana Kaštánková.

Figure 4. Parallel road for opposite-direction traffic (left), no left turns, which are replaced by minor roads (in the middle with just right turn, but with a diversion affecting four blocks, i.e. approximately 700 m, on the right with left turns at a different intersection and with no diversion).

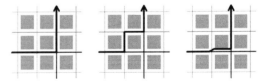

Figure 5. Possible solutions to left turns: 1. for capacity reasons, left turns are not possible at the transverse intersection; 2. left turns are performed at a previous intersection using minor roads; 3. left turns at a previous intersection is maintained, but realized using the main road and thus creating a non-conventional intersection.

of one collision area in the intersection, there are therefore additional, partial collision areas outside the actual intersection. In dense housing development with lots of intersections, it is possible to use nearby intersections for the crossover.

Non-conventional intersection therefore consists of several intersections—one, which will be called the central, and others, which will be called satellite. Central intersection is the actual intersection of the two major roads. Satellite intersections are used for the crossover of vehicles to the opposite side and also for connecting minor roads with the major one, although some turns can be limited, and also for pedestrian and bicycle crossings. The number of satellite intersections corresponds to the number of left turns that have to be dealt with non-conventionally to satisfy the capacity based on traffic load.

3.1 Traffic control

In order to maintain smooth traffic it is necessary to guarantee coordination of traffic control of the whole complex of non-conventional intersection (green waves), in other words mutual coordinated control of the central and satellite intersections. If this was not the case, apart from increased delay time (and the associated increased ecological footprint resulting from additional braking and

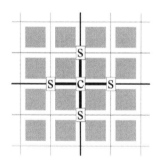

Figure 6. Non-conventional intersection complex: central (c) and satellite (s) intersections, stretches with partial left-side traffic highlighted in bold.

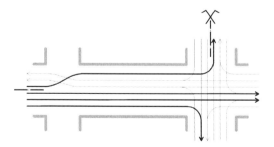

Figure 7. Driving scheme at one part of the intersection.

accelerating), there would be a risk of complete blockage of the non-conventional intersection if the traffic jams ahead of the individual traffic lights extended to the previous intersection. Another risk associated with non-functional coordination is the so-called stop-and-go effect, which arises as a result of gradual stopping and accelerating of vehicles, during which congestion arises, which moves in the opposite direction of traffic and reduces the traffic capacity of the road.

One of the factors affecting the design of coordinated traffic control is the distance between the individual intersections. The size of the housing blocks (i.e. the distance between the intersections) in the area of concern is 170 to 180 m. The almost perfectly symmetrical structure makes the traffic control coordination significantly easier and allows achieving much better results.

For maximal use of the capacity it is desirable that at the central intersections, the opposite direction traffic drives at the same time, and in contrast, at the satellite intersections, in order to allow the crossover of direction, that the opposite direction traffic does not drive at the same time. One shortcoming of this solution is that transverse movements are not possible, which is not ideal especially for pedestrians.

Assuming the standard speed of 50 km/h and the above mentioned distance of approximately 175 m, the cycle time necessary for coordination is 50 s. This is unusually short for such heavily loaded intersection. It is obvious that doubling the cycle time could be achieved by placing the crossover point one block further away (assuming same block size), which would create one additional atypical intersection between the central and satellite one, which would allow transverse movements.

Taking into account the surrounding traffic in the city of Brno, where the cycle time is unified to 100 s, the solution with 50 s cycle time was abandoned. The alternative with placing the crossover point to the intersection further away and creating one additional intersection was also abandoned, because it is desirable that the stretch where the vehicles are driving on the left side of the road is as short as possible. It was decided that an interval of 100 s will be used and the crossover performed

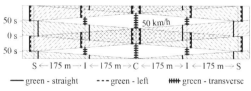

Figure 8. Coordination (green wave) of Control plans of central (C), additional intermediate (I) and satellite (S) intersections for the variant with 100 s cycle time and reduced capacity of the crossover.

at the nearest intersection, which however meant a reduction in capacity of the crossover.

3.2 *Pedestrians*

Due to urbanistic and architectural reasons, it is required that the pedestrians cross the intersection on the same level, even though designing the pedestrian crossings using different levels would increase the capacity and also since the pedestrians would not have to wait for green light, reduce the delay time.

One-level pedestrian crossing offers in general two possibilities. One possibility is concurrent (partially collisional) crossing and vehicle turning, where the drivers of turning vehicles must make sure they do not threaten the crossing

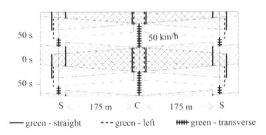

Figure 9. Coordination (green wave) of Control plans of central (C) and satellite (S) intersections for the variant with 100 s cycle time and reduced capacity of the crossover.

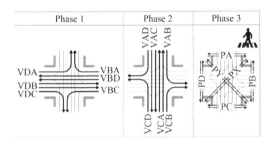

Figure 10. Phases of the control plan of the central intersection for the variant with non-collisional pedestrian crossing.

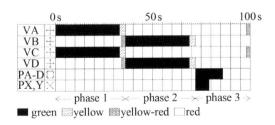

Figure 11. Control plan of the central intersection for the variant with non-collisional pedestrian crossing.

pedestrians (see Figure 12 and 13). Second possibility is non-collisional—the green phase for the crossing pedestrians and turning vehicles is not concurrent (see Figure 10 and 11). If the

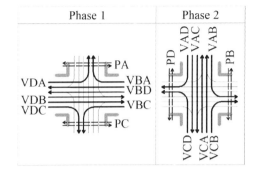

Figure 12. Phases of the control plan of the central intersection for the variant with partially collision pedestrian crossing.

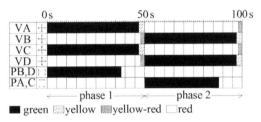

Figure 13. Control plan of the central intersection for the variant with partially collisional pedestrian crossing.

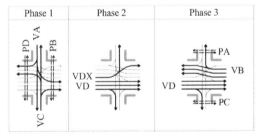

Figure 14. Phases of the control plan of the satellite intersection for the variant with transverse movement possible (see also Figure 9).

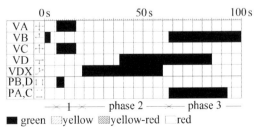

Figure 15. Control of the satellite intersection for the variant with transverse movement possible.

ratio of turning vehicles is high, there is no point in shortening the green phase for turning and leaving green phase for straight direction, especially if there are several driving lanes available.

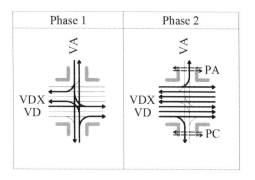

Figure 16. Phases of the control plan of the intermediate intersection (see also Figure 8).

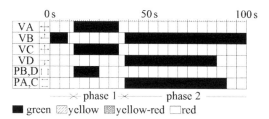

Figure 17. Control plane of the intermediate intersection.

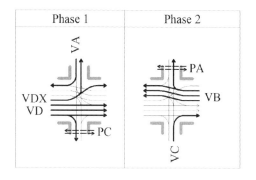

Figure 18. Phases of the control plan of the satellite intersection in variant that the intermediate intersection is used (see also Figure 8).

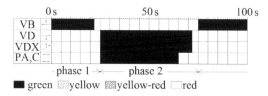

Figure 19. Control plan of the satellite intersection in variant that the intermediate intersection is used.

A possibility is then to create a green phase for all pedestrian crossings at the same time and thus also allow diagonal crossings.

3.3 Safety and orientation of drivers and pedestrians

Vehicles driving in the opposite direction obviously raise the question of safety and orientation.

Ahead of the satellite intersection will be signs that will clearly navigate the drivers to the correct driving lane. Using horizontal and vertical traffic signs and road markings, the driver will be guided to perform the crossover.

Traffic on the left side of the road will be physically separated—depending on the final architectural solution of the street either by verge posts or dividing belt.

Turns at the central intersection create a problem that stems from the (Czech) traffic regulations. Drivers turning right are not obliged to drive in the lane most on the right and instead can choose a lane. Drivers on the left therefore cannot smoothly turn to the left lane and instead have to give way to the oncoming right-turning vehicles. Solution to this problem is road markings, which would converge the right-turning vehicles. Physical separation is not an option because it would block the vehicles driving straight.

Pedestrians would cross the road with partial left-side traffic only at traffic-light controlled crossings, without collisions with vehicles coming from the unusual direction. In addition, a situation where vehicles drive in the opposite direction also arises in one-way streets or intersections with traffic islands and the authors did not encounter any problems with respect to the behavior of pedestrians in such situations.

4 ASSESSED VARIANTS OF THE INTERSECTION

Several variants of intersection configuration were selected for multi-criterial assessment:

- Non-conventional, described in previous section
- Conventional four-way intersection with left turns possible
- Conventional four-way intersection with no left turns possible
- Multi-level intersection

These variants are then further subdivided depending on the number of driving lanes (and thus the required width of the road) based on the possible crossings. One-level crossing is required in the given location, but for general comparison also variants with no one-level crossings were created.

All variants use traffic light control, which is designed uniformly for all variants based on the Czech technical standards TP 81 and TP 235. The cycle time used was 100 s (standard cycle time of the nearby road network) and for non-conventional intersection variant also 50 s (see section 3.1).

Multi-level intersection considered is that of the so-called "diamond" type, i.e. with one bridge/tunnel object for one of the pair of two-way traffic roadways. Taking into account the expected width of the road, only one driving lane for each direction is directed via underpass/overpass, which is fully sufficient for the desired capacity, but could cause problems for intersections nearby. For 50% green

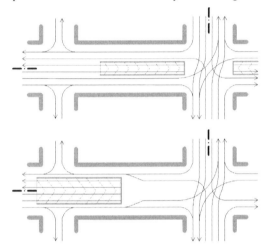

Figure 20. Configuration of multi-level intersection with an underpass (alternatively overpass) for one of the pairs of two-way traffic—at the top 4 lanes in underpass, at the bottom 2 lanes in underpass.

Table 1. Assessed intersection variants.

Var.	Type	Lanes*	Left turn	Ped**	$\frac{C}{s}$
A1	non-conventional	1,2,1	yes	no	50
A2	non-conventional	1,2,1	yes	yes	50
B1	non-conventional	1,1,1	yes	no	50
B2	non-conventional	1,1,1	yes	yes	50
C1	non-conventional	1,2,1	yes	no	100
C2	non-conventional	1,2,1	yes	yes	100
C3	non-conventional	1,2,1	yes	yes***	100
D1	conventional	1,2,1	yes	no	100
D2	conventional	1,2,1	yes	yes	100
E1	conventional	1,1,1	yes	no	100
E2	conventional	1,1,1	yes	yes	100
F1	no left turn	0,2,1	no	no	100
F2	no left turn	0,2,1	no	yes	100
G1	multi-level	1,-,1	yes	no	100
G2	multi-level	1,-,1	yes	yes	100

* Number of lanes for driving left, straight and right, **Pedestrians, ***All-green phase for pedestrians.

time and two getting lanes at the neighboring intersection, the calculated capacity is the same, however given the short distance between the intersections, there would be significant disruption of smooth traffic and also reduced capacity due to the zip and stop-and-go effect. An alternative would be to move the ramps for the underpass/overpass further away from the intersection, in other words extending the bridge or tunnel and using four-lane configuration (see Figure 20). The length of the ramp, assuming 8% slope, 1000 m radius of concave gradient deflection, 700 m radius of convex gradient deflection and 6 m vertical alignment difference, is 143 m.

5 ASSESSMENT CRITERIA

Traffic-engineering, architectural-urbanistic and economic criteria were selected for the assessment of the created variants. Architectural-urbanistic criteria cannot be objectively expressed and cannot be transformed directly into costs. Traffic-engineering criteria can usually be objectively described, but directly expressing them as costs is very complicated and questionable due to estimating the costs of human resources. For these reasons, objective multi-criterial assessment based on one parameter (costs) was not used and instead an approach was used where the individual criteria are assessed proportionally and anyone can interpret them based on their own personal preference.

5.1 Traffic engineering: Capacity

Capacity calculation was done using the standard procedure and equation 1:

Table 2. Capacity of the intersection variants.

Var.	left veh/h	direct veh/h	right veh/h	sum veh/h
A1	780	1760	780	13300
A2	640	1760	640	12160
B1	780	880	780	9780
B2	640	880	640	8640
C1	840	1880	840	14200
C2	680	1880	680	13000
C3	590	1320	590	9970
D1	390	880	840	8430
D2	320	880	680	7540
E1	390	440	390	6670
E2	320	440	680	5780
F1	0*	1880	840	10860
F2	0*	1880	680	10260
G1	390	880**	840	11560
G2	320	880**	680	10500

Values rounded to the nearest ten, *Left turns realized using minor roads, **Multi-level traffic has capacity equal to the saturated flow, i.e. 2000 veh/h.

$$K = \frac{G}{C}SL \qquad (1)$$

where K = capacity, G = green time, C = cycle time, S = saturated flow and L = number of getting lanes. The saturated flow is assumed to be 2000 vehicles/ hour for straight direction, for turns, depending on the radius (8 to 12 m) it is reduced to 1690 or 1780 vehicles/hour respectively. If there is a collision of turning vehicles and pedestrians, the saturated flow is reduced in accordance with the Czech technical standards to 1460 vehicles/hour—in reality of course, this value can be significantly different depending on the number of pedestrians and the ratio between the green time for pedestrians and green time for turning vehicles. The value used is based on the Czech standards and corresponds to 300 pedestrians per hour for one pedestrian crossing. The actual pedestrian load is very hard to predict—in this particular case however it is not so important, because all the variants assume the same number of pedestrians or the same coefficient is used so the results are comparable.

To make the capacity calculation a bit more clear, Table 3 also includes a more detailed calculation for one of the variants (A1).

5.2 Traffic engineering: Delay time of pedestrians

Calculating delay time for transverse movement (delay at one crossing) was done using Equation 2, which can be derived from weighted average delay (0 s for green light, half of the red time for red light):

$$D = \frac{G*0 + R*\dfrac{R}{2}}{C} = \frac{R^2}{2C} \qquad (2)$$

where D = pedestrian delay time, G = green time, R = red time and C = cycle time. Calculation of

Table 3. Capacity of the individual movements for the selected intersection (variant A1).

Turn*	L	$\dfrac{G}{s}$	$\dfrac{R}{m}$	$\dfrac{S}{veh/h}$	$\dfrac{K}{veh/h}$
VAB	1	22	12	1778	782
VAC	2	22	0	2000	1760
VAD	1	22	12	1778	782
VBA	1	22	12	1778	782
VBC	1	22	12	1778	782
VBD	2	22	0	2000	1760
VCA	2	22	0	2000	1760
VCB	1	22	12	1778	782
VCD	1	22	12	1778	782
VDA	1	22	12	1778	782
VDB	2	22	0	2000	1760
VDC	1	22	12	1778	782
total					13298

*Labeling—see Figure X.

the delay time for diagonal movement (delay at two crossings) was performed using Equation 3, which is derived from the assumption that if the imaginary green time on first crossing is 1 s (i.e. the last second), then the delay time corresponds to the sum of delay time based on Equation 2 and (fixed) green time between the end of green time at the first crossing and beginning of green time at the second crossing.

$$D = \frac{R_1^2}{2C} + I \qquad (3)$$

where W = pedestrian delay time, R_1 = imaginary red time at first crossing (in reality the cycle time minus one second) and I = interval between end of green time at first crossing and beginning of green time at second crossing.

5.3 Architectural-urbanistic: Street area

Based on discussions with architects and urbanists, the following is considered to be undesirable:

Table 4. Pedestrian delay time.

Var.	C	G	R	D1	D2
			s		
A2	50	12	38	14	37
B2	50	12	38	14	37
C2	100	37	63	20	62
C3	100	14	86	37	43*
D2	100	12	88	39	87
E2	100	12	88	39	87
F2	100	37	63	20	62
G2	100	12	88	39	87

C = cycle time, G = green time for pedestrians, R = red time for pedestrians, $D1$ = delay time for transverse crossing (1 cross-road), $D2$ = delay time for diagonal crossing (2 crossroads). * Crossing via 1 diagonal crossing.

Figure 21. Current pedestrian overpass closed to solved intersection. With this experience is clear, why overpass is unacceptable.

849

Table 5. Summary of intersection assessment.

Var.	Capacity	Delay	Urbanism	Costs
A2	88%	100%	100%	100%
B2	40%	100%	100%	100%
C2	100%	59%	100%	100%
C3	58%	61%	100%	100%
D2	24%	0%	100%	100%
E2	0%	0%	100%	100%
F2	62%	59%	0%	100%
G2	65%	0%	0%	0%

*Number of getting lanes for driving left, straight and right.

- Distributing (partly) traffic load to minor streets, which are only to be used by residents
- Multi-level intersection as a foreign element, which does not fit into the inner city area
- Multi-level pedestrian crossings—overpasses and underpasses. Pedestrians should be able to cross the road on just one level.

Based on the above it is quite obvious, which variants can be considered as suitable and which ones not, from the architectural-urbanistic perspective.

5.4 Economic: Costs

Cost estimates depend on the extent of construction considered. Non-conventional intersection does not include just the central, but also the satellite intersections, which would however be built either way in one form or the other, because the rebuilding of the traffic skeleton in the location in question is not limited to just this problematic intersection.

The variants assuming multi-level intersections are quite different in terms of their costs. Similar object was already built in Brno previously for approximately 100 mil. CZK (approximately 3.7 mil. EUR).

5.5 Assessment summary

Table 5 shows a comparison of the four assessed criteria; all expressed using the same unit (%) – 100% being the best, 0% being the worst possible result.

6 MICROSCOPIC SIMULATION

The purposed configuration of the traffic signals, including traffic control and calculation of non-conventional intersection capacity, was realized using microscopic simulation and the Aimsun 8.1.3 software. The simulation confirmed that intersection traffic control has a major impact on the intersection capacity. Very close attention must be paid to this in order to prevent stop-and-go effect, which would mean the theoretical capacity estimated based on green time would not be fully used because the entrance to the intersection is not fully saturated as a result of this stop-and-go effect.

7 CONCLUSION

Non-conventional intersection which uses direction crossover of left-turning vehicles at a neighboring intersection is a suitable design for the location of interest from traffic-engineering, architectural- urbanistic and economic perspective. The capacity of the crossroad would be 50 to 70% higher compared to a conventional four-way intersection because crossing over left-turning vehicles at a neighboring intersection allows better and more efficient design of traffic control. The average delay time for pedestrians would also shorten. In the meantime, the architectural-urbanistic requirement for one-level intersection and concentrating the traffic to main roads (not affecting the nearby minor roads), would be satisfied. In addition, the construction costs will be lower compared to the variant with multi-level intersection.

Non-conventional intersection can be a suitable solution for other locations as well, if they satisfy the following criteria:

- Heavy traffic load (8000 to 12000 vehicles/h/intersection)
- Partially distributing the traffic also to nearby minor roads is not possible or not desirable
- Multi-level intersection is not suitable because of the urbanistic situation or because the costs of building one are too high.

ACKNOWLEDGEMENT

This paper has been worked out under the project No. LO0108 "AdMaS UP—Advanced Materials, Structures and Technologies", supported by Ministry of Education, Youth and Sports under the "National Sustainability Programme I" and under the project FAST-J-14-2360 supported by Brno University of Technology.

REFERENCES

Esawey, M. E., and Sayed, T. 2012. Analysis of unconventional arterial intersection designs (UAIDs): State-of-the-art methodologies and future research directions. *Transportmetrica*

Hu, P., Tian, Z., Wu, X. & Xu, H. 2014. A Proposed Signal Operation and the Effect of ITS Cycle Length on Diverging Diamond Interchanges *Institute of Transportation Engineers. ITE Journal.*

Leong, L. V., Mahdi M. B. & Chin K. K.. 2015. Microscopic Simulation on the Design and Operational Performance of Diverging Diamond Interchange. *Transportation Research Procedia.*

Maji, A., Mishra, S., & Jha, M. 2013. Diverging Diamond Interchange Analysis: Planning Tool. *J. Transp. Eng.*

Transport Infrastructure and Systems – Dell'Acqua & Wegman (Eds)
© 2017 Taylor & Francis Group, London, ISBN 978-1-138-03009-1

A cycling mobility study case: The European cycling challenge 2015 in Naples

C. Aveta
Reconstruction Special Office, L'Aquila, Italy

M. Moraca
Freelance Engineer, Naples, Italy

ABSTRACT: In May 2015, the City of Naples and forty others European cities have joined the fourth edition of the European Cycling Challenge. This event has been created to promote the bicycle as a sustainable transport mean in urban areas, and lasted for all the month of May.

The event organizer, SRM Reti e Mobilità, provided an app where all citizens/cyclists could enroll and track their cycle journeys; the app was tracking, with a time interval of five seconds, the cyclist position (using GPS) and some journey details (journey name, length, speed, besides day and schedule).

These data were registered by the app and saved in a database. Several months after the event, each database has been sent to the participating cities.

Naples' database has been utilized to evaluate, for the first time, the cycling mobility in the city. Indeed, the City of Naples developed a cycle network longer than twenty kilometers in the recent years. Therefore, this evaluation aimed to understand how, when and where the cyclists have been using these paths.

The evaluation required the development of a methodological framework to analyze the database, with operations in Gis, Excel and Access environment.

The final product has been disaggregated in two categories, the territorial evaluation and the temporal evaluation. The territorial evaluation contains O-D matrices, an analysis involving Naples' districts, and a flow count analysis for road. The temporal evaluation includes an analysis for each day of the month, for each day of the week (Monday, Tuesday, and so on), and for two time slots (7,30–9,30 A.M. and 4,30–6,30 P.M.).

The overall results registered over 7961 kilometers covered and 1308 registered trips, with an average journey length of 6,07 kilometers and a massive use of the Waterfront cycle path (Via Francesco Caracciolo and Via Partenope).

1 EXECUTIVE SUMMARY

In the last ten years Italian cities extended considerably their cycle route networks. Indeed, the "Urban Mobility Report" (ISTAT, 2013) reveals that the length of the urban cycle routes is increased by 40%, from a value of 13,5 km per every 100 km^2 of urban surface in 2008 to a value of 18,9 km per every 100 km^2 of urban surface in 2013.

Then, the analysis of the cycle route use is important for cities, because of the need of quantifying the transiting flows and getting information useful for future planning: having useful tools is necessary.

The aim of this paper is to analyze the results extracted from the 2015 European Cycling Challenge (ECC) edition in Naples—utilized as a case study—defining a methodology for the data processing and showing the obtained results.

The methodology foresees progressive steps: in this way, the repeatability to all future surveys based on a similar georeferenced database

was guaranteed. This analysis represents the first cycling mobility survey carried out in the City of Naples.

1.1 *The European Cycling Challenge—event description*

The ECC (www.cyclingchallenge.eu)—whose first edition was held in 2012—is a challenge among urban cyclists which starts every May 1st and ends up in May 31th. The event foresees a virtuous competition among the citizens of involved cities based on the total mileage that the active cyclists/citizens of each city will have traveled in that month. To participate, is necessary to register all journeys by bicycle through the free App "Cycling365", or input own journeys manually via the event website.

All journeys are permitted and included in the database except journeys with an average speed greater than 30 km/h, a maximum speed higher than 40 km/h, or a length bigger than 30 km.

At ECC can participate all citizens using the bike within the boundaries of the involved city, including also residents out of the city who move themselves into the city by bicycle. For the 2015 Edition, 99 citizens of Naples (and from the neighboring urban centers) have registered their journeys in the month of May.

2 THE CYCLE ROUTE NETWORK IN NAPLES

The City of Naples has an area of 117,27 km², a road network of approximately 1200 km long, and it is divided in 10 Districts (Figure 1). Naples owns a population of 974.074 inhabitants—reference to 31/12/2015—and every District has a population grouped in a range from 80.000 to 140.000 inhabitants.

According to the data provided by the "2011 Italian Census" (ISTAT, 2011), the City of Naples is interested—on a typical working day—by 574.916 journeys; approximately the 59% (342.109) of journeys are within the city, while the remaining 41% are classified internal-external or external-internal journeys. Concerning this 41%, 193.928 journeys are towards Naples, while 38.880 depart from Naples; these numbers confirm the importance of the city of Naples as the principal

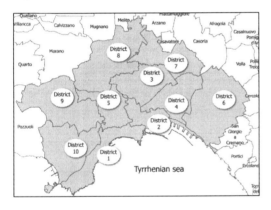

Figure 1. The City of Naples, its Districts and the neighboring cities.

Table 1. Working daily journeys in Naples (for transport modes).

	Internal	To Naples	From Naples	Total
Railway/ Underground	38.019	59.914	2.793	244.176
Bus	60.135	29.482	4.585	100.726
Car	115.081	98.660	30.435	135.813
Pedestrian/ Bicycle	128.874	5.872	1.067	574.916

regional aggregator. These data have been disaggregated for transport modes in Table 1.

In 2011, the City Administration started the construction of the first cycle route—from Bagnoli to Castel dell'Ovo (Via Nazario Sauro), with a length of 7,7 km—ended up in 2012. The City Administration was well aware of the benefits of the cycling in an urban contest highly congested by traffic like Naples (Van Hout, 2008) and the construction of a new cycle route wanted to increase the modal split between private vehicles and the bike in favour of the second.

The cycle route network in Naples was finally established with the Major order n.1233 of the 9th November 2012. The cycle route network goes through 4 Districts, and has a length of about 19 km.

The existing cycle routes in the city of Naples can be classified in 3 categories:

- Type 1: separated cycle routes. The routes are separated from the road, having a width of 1,25 m for lane (total width 2,50 m) and defined by delineator strips of width 12 cm. Plus, they have a separator curb from the road—in concrete—with a height of 20 cm and a width of 50 cm, and are separated from the footpath by the verge (variable width between 30 and 50 cm);
- Type 2: routes on footpaths. They are indicated through delineator strips having a width of 12 cm, the orange coloring the intermediate zone and an overall width of 2 m (the width of the single cycle lane is 1 m). Pedestrians have priority over cyclists;
- Type 3: routes in pedestrian areas or Restricted Traffic Zones ("RTZs"). The cyclists follow the rules of the other road users, and have indications through markings drawn on the road/pedestrian area (the bicycle symbol).

2.1 The cycle route network in District 10

The cycle route falling into the District 10 is unique and has a length of about 3,6 km. The cycle route

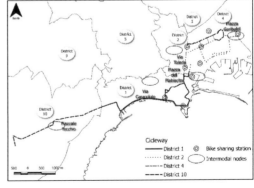

Figure 2. The cycle route network in Naples.

starts from Via Nuova Agnano and ends up in Via Caio Duilio. However, the path is not interrupted but continues towards Piazza Sannazzaro—through Galleria delle Quattro Giornate—in District 1.

The path falling in District 10 has Type 1 stretches (Viale Augusto and Via Caio Duilio) for a length of over 1,2 km, Type 2 stretches (Via Nuova Agnano and Viale J.F. Kennedy) for almost 1,8 km, and Type 3 stretches (Piazzale Tecchio and Largo Lala, pedestrian areas). There are several intermodal points that intercept the cycle route: Underground Line 2, Underground Line 6 (under construction), the railway line called "Cumana" and the bus station located in Piazzale Tecchio.

2.2 The cycle route network in District 1

The overall length of the cycle routes falling into District 1 is about 7 km. In District 1 there is a main cycle route and two secondary routes that are linked to the main. The main route carries on from Galleria delle Quattro Giornate and ends up in Via Toledo, entering into Historical Centre. The secondary paths start from Piazza Vittoria, crossing the historic area of Chiaia neighborhood (Piazza dei Martiri, Via Morelli, Via Chiaia) and Via Toledo, through Via Santa Brigida and Via Paolo Emilio Imbriani, ending up in Piazza Matteotti.

In this District there are Type 1 stretches (Galleria delle Quattro Giornate and the "Waterfront" from Via Francesco Caracciolo intersection Via Sannazzaro to Via Nazario Sauro intersection Via Raffaele De Cesare), for a total length of 3,5 km, while the rest is Type 3 in pedestrian areas (Via Toledo, Via Chiaia, Piazza Plebiscito) or in promiscuous with vehicles (the "30 km Zones" in Via Santa Lucia and Via Cesario Console). Again the intermodal points are numerous along the cycle network: Underground Line 1, Underground Line 2, Underground Line 6, the Cumana railway, and the Central and Chiaia funiculars.

From 2015, bike-sharing stations (currently suspended service, but in reactivation) are present near Castel dell'Ovo, Piazza Vittoria and Via Partenope.

2.3 The cycle route network in District 2

The cycle routes crossing the District 2 are connected to the Historic Centre of Naples (UNESCO World Heritage), have a total length of 5,2 km and the characteristic of being only in a promiscuous way with vehicles and pedestrians (cycle routes Type 3). The main cycle routes are two: the first, defined as "Decumano del Mare", originates from Piazza Garibaldi, and ends up in Via Roma after crossing Via Benedetto Croce. This cycle route is developed both in pedestrian areas and on footpaths, 30 km Zones and the RTZ so-called "Mezzocannone".

The second cycle route starts from Via Toledo, continuing the cycle route coming from District 1,

and ends up in Piazza Dante, foreseeing ramifications in Via dei Tribunali and Via Benedetto Croce, and has been developed in pedestrian area (Via Toledo) and on footpaths (except for the section near Piazza Dante, RTZ). The intermodal points located in the District are several, but the principal is Piazza Garibaldi—the main intermodal hub of the city—where converge: Underground Line 1, Underground Line 2, the railway called "Circumvesuviana" (which connects all the cities around Vesuvius, starting from Naples) and the bus hub, starting point of the main city bus lines. In the area it is also possible the connection with the ferry area starting and arriving in the Port of Naples (Calata Porta di Massa). From 2015, there are stations of bike-sharing located in Piazza Bovio, Via Brin, Via Toledo corner Via Diaz and Piazza Dante.

2.4 The cycle route network in District 4

The cycle routes located in District 4 transit both into the Historical Center and ends up in Piazza Garibaldi, following two parallel routes. Indeed, the two cycle routes (exclusively Type 3) move through the pedestrian area of Via dei Tribunali and Via San Biagio dei Librai. The cycle route network in District 4 has a total length of 3 km, and it has been developed in pedestrian areas and into RTZ of the Historical Center.

The first cycle route, called "Decumano Maggiore cycle route" with the historical reference to the road on which the route stands, starts from Via Benedetto Croce, and ends up in Piazza Garibaldi. The second cycle route, named "Decumano Superiore cycle route" (similarly to the previous), starts from Piazza Dante, and ends up in Piazza Garibaldi too. Unlike other Districts, intermodal connections are scarce because of the size of the streets and high population density: indeed, the only exception is the Duomo station of Underground Line 1, next to activation. However, in this District were located since 2015, the bike-sharing stations of Largo Donnaregina, Via Benedetto Brin and Piazza Garibaldi (side District 4).

3 THE METHODOLOGICAL FRAMEWORK

The App "Cycling 365" saved, for each journey registered and with a frequency of 5 seconds, the cyclist position (using GPS) and some journey details (journey name, length, besides day and schedule) in a database from the start to the end of the journey. Several months after the event, each database was sent to the participating cities.

The extraction of results from the European Cycling Challenge database has required the development of a methodology to define all the stages of the elaboration process, from the survey objectives to the specific type of investigation to be carried out, in order to ensure the replicability and comparison of results, obtaining understandable results by an unskilled audience. In detail, interested stakeholders have been divided in two main categories:

– Municipal technical offices, having the task of assessing the current status of the situation, according to the objectives identified, and investing resources in the development of the urban cycle network;
– Decision makers and carriers of public interests: subjects that, for institutional reasons or role, should guide the choices of the public administrators in the sector.

The methodological framework foresees the following steps, and is based on the typical scheme of transport surveys (Richardson & Ampt & Meyburg, 1995):

– Definition of the objectives;
– Definition of the key factors;
– Phenomenon analysis according to the identified key factors;
– Data processing;
– Evaluation of the results.

The survey objective is to provide an assessment of the satisfaction degree and the use of existing cycle routes in the City of Naples, providing information on the city areas for developing future cycle routes. These outputs require the definition of a set of "key factors", which allow the analysis to achieve the proposed objectives. The database provided from the European Cycling Challenge has a structure in which each row show information about the registered journey (journey's identity code, single point identification code belonging to the specific journey), time information (day of detection, hour, minute and second of detection) and spatial information (latitude and longitude detected by GPS). For this reason, the key factors identified are 2: "Space" and "Time". The key factors have been disaggregated in different "layers", to evaluate different aspects of the phenomenon at different levels:

– Time key factor:
– 0 24 time slot layer;
– 7,30–9,30 time slot layer ("Morning" time slot);
– 16,30–18,30 time slot layer ("Afternoon" time slot);
– Space key factor:
– City layer;
– District layer;
– Matrix Origin/Destination (O/D), layer obtained using the start and end points of the registered journeys;
– Street layer.

Each Time layer is referred to the entire month: for example, the 0–24 time slot layer is

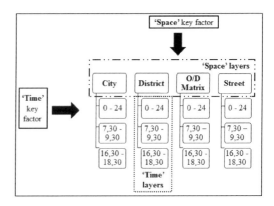

Figure 3. Methodology framework.

the overall analysis of the month of May, from 1st to 31th.

Classifications introduced allow to obtain a better detail level for the analysis. Joining one single layer belonging to each key factor allows to get a 3×4 matrix, in which each layer of the Space key factor is analyzed according to 3 different layers of the Time key factor (Figure 3).

The join permits to make a progressive detailed analysis and assesses whether there are particular phenomena, such as home-work journeys, home-school journeys, and how the movements are distributed over a time period and on the territory.

Data processing operations have been developed with the aid of different software tools (GIS, spreadsheet, SQL database). Indeed, the cycling flow analysis in a GIS can provide information to evaluate where cyclists choose to move and to predict where they could move in the future (Aultman-Hall & Hall & Baetz, 1997).

Operations have been performed starting from a first cleaning job of the database in GIS environment to eliminate some redundant information in the survey, such as the elimination of registered journeys outside Naples, or registered journeys on railway lines or motorways. The database thus obtained has been divided in accordance with predicted slots of the Time key factor, representing the mobility demand. The three sub-databases have been combined with the GIS containing information about the City of Naples (which represents the supply model of this procedure), using as joining factor the geographical coordinates of individual points (the database rows) presents in the sub-databases. The join of three sub-databases with the supply model, declined in the various detail scales of Space key factor, has allowed to obtain the 3×4 matrix displayed in Figure 3.

The evaluation of the results foresees the analysis of the information obtained during the data processing in order to provide to all stakeholders the information necessary for its own needs.

4 SURVEY RESULTS

The ECC database registered more than half a million points (equal to rows), which represent the "photographies" of all journeys made in May, with a time interval of 5 seconds.

The operations performed on the database, necessary to extract the data, have been predominantly of "spatial join" type, and various query types using the SQL environment that is present in the GIS software. Extracted data have been finally exported to a calculation sheet to create the tables available in the successive paragraphs.

The accuracy of the GPS data varies in some cases in a 5–10 meters range from the street. This is a limit also evidenced in literature (Lindsey & Hankey & Wang & Chen, 2013) because of it is impossible to understand if cyclists used cycle route or went in promiscuous with vehicles, nevertheless it allows to understand in which streets they passed.

The results have analyzed the most significant data for the survey: the registered journeys and the covered mileage.

4.1 Overall analysis

The total number of journeys registered in the month of May 2015—during the ECC—is 1.293, while the total mileage was 7.960 km, for a 6,16 km average mileage per journey.

The day with the largest number of registered journeys was on May 11th, with 84 journeys and a total mileage of 426,39 km (5,08 km per journey). The day with the most mileage covered was on May 19th with 448,37 km, 70 journeys and of 6,41 km average mileage per journey.

The analysis of the database by time slots shows that the recorded journeys are distributed during the day, without recording special spikes referable to home-work journey or home-school journey. Indeed, only 16% of the total journeys were recorded in the "Morning" time slot, while 19,4% of those registered in the "Afternoon" time slot. Furthermore, the morning journeys are longer than the afternoon journeys (7,59 km per journey in the morning time slot, compared to the 5,59 km on average registered in the afternoon time slot).

4.2 District analysis

The second level of analysis provides a count of the registered journeys inside the 10 Districts.

In Table 3 is shown that in the District 1 were recorded 713 journeys, while in the District 10 were recorded 542 journeys. Surprisingly, also the District 5 registers a high number of registered journeys, considering is a hilly terrain. The same trend was true also for the other time slots: however, in the District 5 journeys in the afternoon time slot are almost the same than in District 1, and higher than those recorded in the District 10. Plus, the sum

of registered journeys of Districts 3, 6, 7, 8, and 9 is less than 10% of the registered overall journeys.

The journeys shown in Table 3 are higher than those counted in Table 2: this is because a single journey can go through in more than a single District.

The District with the highest mileage registered is the District 10 with about 2.080 km, while the District 3 has registered the lowest mileage with almost 64 km. The sum of the registered mileage in District 1 and in District 10 is equal to the 50% of the overall mileage registered in the ECC, symptom that a separated cycle route is more attractive for cyclists. Results are similar for the other time slots, with prevalence of the District 10 in the morning time slot and District 1 in the afternoon time slot.

Table 2. Overall journeys and mileage about the ECC 2015.

Time slot	Journeys	Mileage (km)	Km per journey
0–24	1.293	7.960,94	6,16
Morning	213	1.616,78	7,59
Afternoon	251	1.409,01	5,59

Table 3. Registered journeys inside Districts.

District	Journeys per time slot		
	0–24	Morning	Afternoon
1	713	104	110
2	257	41	35
3	28	8	1
4	293	36	36
5	422	80	87
6	118	27	33
7	27	0	5
8	47	3	5
9	38	8	2
10	542	110	95
Total	2485	417	409

Table 4. Mileage per registered journeys.

District	Mileage per time slot (km)		
	0–24	Morning	Afternoon
1	1.765,50	192,04	327,36
2	680,55	101,47	164,19
3	63,76	31,36	–
4	778,83	25,84	159,79
5	1.414,50	441,20	189,51
6	478,91	64,90	195,76
7	78,09	–	6,47
8	149,04	9,67	26,38
9	108,28	34,61	–
10	2.080,01	667,94	267,25
Total	7.597,46	1.569,04	1.336,72

4.3 O/D Matrix analysis

The third level of detail has analyzed the origin and the end of journeys among the Districts (considered as internal centroids) and between the Naples Districts and the neighboring cities (considered as external centroids). Over 90% of the total journeys registered have been done within the city of Naples (with over 7200 km counted), with a small fraction of journeys to and from Naples (internal-external journeys and external-internal journeys).

The journeys registered in the morning and afternoon time slot followed this trend (Table 6 and Table 7).

The preferential directions for external-internal journeys related to the 0–24 time slot (Table 8 and Table 9) are the Pozzuoli-Naples directrix (15 journeys and 113 km covered) and the Volla-Naples

directrix (17 journeys and 93 km covered). Journeys from neighboring cities to Naples (47) have generated about 355 km (Table 9).

Journeys starting from Naples to neighboring cities have generated about 343 km (Table 10). Also in this case the Naples-Pozzuoli and the Napoli-Volla directrixes are the most traveled directions, the first with 143 km and the second with 80 km.

Table 5. Overall O/D Matrix, 0–24 time slot.

	Internal Destination	External Destination
Internal Origin	1193	51
External Origin	48	1

Table 6. Overall O/D Matrix, Morning time slot.

	Internal Destination	External Destination
Internal Origin	204	5
External Origin	4	0

Table 7. Overall O/D Matrix, Afternoon time slot.

	Internal Destination	External Destination
Internal Origin	197	5
External Origin	10	1

Table 8. Journeys starting from neighboring cities towards Naples.

Origin City	Journeys per time slot		
	0–24	Morning	Afternoon
Arzano	2	1	–
Casoria	2	–	–
Cercola	2	–	–
Marano	1	–	1
Melito	1	–	–
Mugnano	6	–	–
Pozzuoli	15	3	1
Quarto	1	–	–
Volla	17	–	8
Total	47	4	10

Table 9. Mileage per journeys starting from neighboring cities towards Naples.

Starting City	Mileage per time slot (km)		
	0–24	Morning	Afternoon
Arzano	25,93	21,85	–
Casoria	27,68	–	–
Cercola	11,81	–	–
Marano	5,05	–	5,05
Melito	23,20	–	–
Mugnano	47,06	–	–
Pozzuoli	113,68	25,89	9,65
Quarto	7,15	–	–
Volla	93,71	–	50,97
Total	355,27	47,75	65,66

Table 10. Journeys starting from Naples towards neighboring cities.

Destination City	Journeys per time slot		
	0–24	Morning	Afternoon
Arzano	2	1	0
Casoria	2	–	–
Cercola	2	–	–
Marano	1	–	1
Melito	1	–	–
Mugnano	6	–	–
Pozzuoli	15	3	1
Volla	1	–	–
Total	17	–	8

Table 11. Mileage per journeys starting from Naples towards neighboring cities.

Destination City	Mileage per time slot (km)		
	0–24	Morning	Afternoon
Arzano	20,58	2,83	2,54
Casoria	4,75	–	–
Cercola	22,40	–	–
Marano	9,90	–	9,90
Melito	21,25	–	–
Mugnano	40,16	7,34	11,58
Pozzuoli	143,73	32,18	14,10
Volla	80,92	–	–
Total	343,70	42,35	38,12

From the internal-internal component of the matrix O/D, journeys with origin and destination falling inside the same District have been extrapolated. The District where have been registered the largest number of internal journeys is the District 1, followed by District 10 and District 5. The journeys shown in Table 12 cover the 25% of the total mileage.

In Table 13 are presented data on total mileage divided by District and by time slots. The District 10 and District 5 registered almost the same mileage.

4.3.1 District 1 and District 10 detail

From the internal-internal component of the O/D matrix the journeys starting from District 1 and District 10 have been extrapolated, because of the data registered the highest number of journeys in that zones. For District 1 were recorded 167 journeys per 1033,73 km.

The mileage counted for the District 1–District 4 connection was 250 km, 245 km was the mileage for the District 1–District 10 connection and 181 km for the District 1–District 2 connection.

Table 12. Internal journeys per District.

District	Journeys per time slot		
	0–24	Morning	Afternoon
1	167	23	26
2	87	9	21
3	9	4	0
4	98	0	22
5	121	45	10
6	30	4	14
7	10	0	0
8	8	1	1
9	20	8	0
10	145	46	21
Total	695	140	115

Table 13. Mileage per internal District journeys.

District	Mileage per time slot (km)		
	0–24	Morning	Afternoon
1	1.033,73	155,03	139,67
2	542,70	78,07	162,54
3	55,76	31,36	–
4	605,82	–	129,88
5	841,90	289,69	79,51
6	344,42	37,92	188,53
7	69,96	–	–
8	85,49	7,28	16,48
9	108,28	34,61	–
10	1.437,33	568,87	157,66
Total	5.125,38	1.202,83	874,28

Table 16 and Table 17 show the results about journeys starting from District 10, with 144 journeys and 1433 km. The connection between District 1 and District 10 is the most evident (449 km total), then the connection with District 2 (378 km), and District 5 (264 km).

4.4 Street analysis

The ECC database permitted an evaluation of the cycling flows transiting in the streets of the City of

Table 14. Journeys starting from District 1 towards the other Districts.

Arrival District	Journeys per time slot		
	0–24	Morning	Afternoon
10	53	11	5
2	35	5	7
3	6	–	1
4	38	1	3
5	20	4	9
6	5	2	–
7	1	–	–
9	9	–	1
Total	167	23	26

Table 15. Mileage per journeys starting from District 1 towards the other Districts.

Arrival District	Mileage per time slot (km)		
	h24	7.30–9.30	16.30–18.30
4	312,89	81,06	23,29
10	181,69	27,29	33,99
2	63,53	–	14,87
5	257,04	3,94	15,51
3	121,90	20,18	46,70
6	50,98	22,56	–
9	4,07	–	–
7	41,63	–	5,30
Total	1.033,73	155,03	139,67

Table 16. Journeys starting from District 10 towards the other Districts.

Arrival District	Journeys per time slot		
	0–24	Morning	Afternoon
1	57	18	7
2	35	13	3
5	26	–	9
6	15	14	–
8	4	–	–
4	3	1	1
9	3	–	1
3	1	–	–
Total	144	46	21

Table 17. Mileage per journeys starting from District 10 towards the other Districts.

Arrival District	Mileage per time slot (km)		
	h24	7.30–9.30	16.30–18.30
1	449,52	164,95	62,63
2	378,98	171,76	17,61
5	264,86	–	64,47
6	238,90	221,93	–
8	47,15	–	–
4	28,86	10,23	9,70
9	13,14	–	3,26
3	12,04	–	–
Total	1.433,45	568,87	157,66

Table 18. The 15 busiest streets in Naples per time slots.

Street	Journeys per time slot		
	0–24	Morn.	After.
Via F. Caracciolo (1)	324	54	65
Via Nazario Sauro (1)	235	40	46
Via Caio Duilio (10)	185	44	26
Via Partenope (1)	185	30	38
Galleria Laziale (10–1)	182	41	22
Via Nuova Agnano (10)	171	47	27
Via Toledo (1–2)	159	41	25
Corso Umberto I (2)	156	22	23
Viale J. F. Kennedy (10)	141	37	22
Corso V. Emanuele (1)	137	56	27
Via Bagnoli (10)	133	45	19
Via Salvator Rosa (2)	120	36	15
Viale Augusto (10)	91	22	18
Via Torquato Tasso (1)	68	31	7
Via Domenico Fontana (5)	63	11	18

Naples utilizing the single journey characteristics (with geographic coordinates) and the map (in GIS environment) of the City of Naples, to get flows as the sum of journeys falling in a single street. Table 18 highlights the top 15 busiest streets, with in parentheses the belonging District.

Seven streets on the list are Type 1 cycle route, and fall in District 1 and District 10. Except Via Toledo—pedestrian area—the others are all main traffic streets.

5 CONCLUSIONS

The results showed that the cyclists preferred Districts where there are separated cycle routes (District 1 and District 10), or in close proximity. Flows drop progressively with the increasing distance from these areas, a sign that is necessary to increase the trust of citizens in the outlying areas, possibly creating new protected routes.

The results prompted the City Administration at approving the project of a cycle route in Corso Umberto, one of the busiest streets used by cyclists according to the data available. In addition, in 2013 the City Administration had already approved the construction of new cycle routes in own way in the stretch adjacent to the Port (Via Marina, Via Amerigo Vespucci and Via Alessandro Volta, District 2), and in the Industrial Area (Via Emanuele Gianturco, District 4), projects currently under construction, for further another 5 km of cycle routes in the east area.

The success of the 2015 edition of the ECC was confirmed by the results of the 2016 edition, in which there were recorded 14.701 kilometers (nearly double of the 2015 edition), and where from a first analysis of flows (available on the event website), emerges that the favorite routes are the same as the 2015 competition. In the future, travel demand models could be developed using information taken by this kind of survey.

REFERENCES

Aultman-Hall, L., Hall, F. & Baetz, B. 1997. Analysis of Bicycle Commuter Routes Using Geographic Information Systems. Implications for Bicycle Planning. *Transportation Research Record 1578.*
Istituto nazionale di statistica (ISTAT), 2011. 15° Censimento della popolazione e delle abitazioni 2011.
Istituto nazionale di statistica (ISTAT), 2013. Urban Mobility Report.
Lindsey, G., Hankey, S., Wang, X. & Chen, J. 2013. Feasibility of Using GPS to Track Bicycle Lane Positioning. *University of Minnesota, Center for Transportation Studies.*
Richardson, A., Ampt, E. & Meyburg, A. 1995. Survey Methods for Transport Planning. *Eucalyptus Press.*
Van Hout, K. 2008. Literature search bicycle use and influencing factors in Europe. *Instituut voor Mobiliteit (IMOB).*

Transport Infrastructure and Systems – Dell'Acqua & Wegman (Eds)
© 2017 Taylor & Francis Group, London, ISBN 978-1-138-03009-1

Secondary or no longer used rail infrastructures: A new tool for the regional railway and land planning

A. Cappelli
Department of Architecture and Arts, IUAV University of Venice, Venice, Italy

G. Malavasi & S. Sperati
Department of Civil, Building and Environmental Engineering, Sapienza University of Rome, Rome, Italy

ABSTRACT: To reach a sustainable development of the territory, it's implied that the choice of transportation methods must be related to the city form: compact or sprawl.

On opening or upgrading railway infrastructures or stations, have increased the real value (m^2) of both existing and new buildings located next to the station of the railway lines.

This research aims to demonstrate how the process of locationing and the development of the rail transportation supply could re-organize the growth of the demand in exploited areas by providing an innovative contribution on transportation planning using the Smart Regional Railway System.

Analyses were performed on two railway lines: Civitavecchia-Orte and Civitavecchia-Roma.

For the research purposes, it has been created an index, called "Cross Section Spatial Index". The innovation of this index is the way real estate value was considered, namely depending on the distance from an efficient transportation system.

1 RESEARCH OVERVIEW

The research aims to provide an innovative contribution to the regional transport planning, through a modeling approach defined as "Smart Regional Railway" [SRR] calibrated on two case studies: the lines Civitavecchia-Orte and Civitavecchia-Roma.

The effectiveness of the existing regional transport is part of a sustainable vision of collective transport (Ricci S. 2013), where the environmental consciousness and the need to modify the territory have consolidated the importance of a connection between land use and transportation planning (Maternini G. 2014).

The focus on the study of the existing rail network rather than on expanding or building new infrastructure stems from the interpretation of the current policies that are driving their intervention on modernizing and upgrading the existing (Ministry of Economy and Finance, 2016).

The scientific literature confirms the link between land use and transport planning, moreover it underline show correct urban and transportation planning decisions can affect the method of transport, the modal choice and the resulting quality of life of the citizen.

The urban planning decisions made in these years have defined the territorial form of today, while future urban form is overwhelmed by the specialization of the transport system, still strongly tied to the car and, consequently, to the private sphere (Brovarone E. 2010).

The reviewed studies have generally said and confirmed that:

- Intermodal transport is the essential approach to transport-related issues, as it allows to contain the external costs of transport and make it more efficient (Cappelli A. 2006);
- The individual transport may be discouraged, in favor of public transport, through an innovative territorial planning mindful ofcompact spatial distribution of the assets (Cappelli A. 2009);
- The geographical dispersion does not create competitiveness and convenience to the public transport operator, as it produces movements over long distances through private means of transport (Cappelli A. et al 2013);
- A higher density (Fig. 1) creates more profitability for public transport as it collects more movements and increases the transportation demand (Laconte P. 1996);
- A reduction in the use of private cars can only be sustained by a range of efficient and competitive public transport (Cascetta E. et al 2013);
- The sustainability of the transport system descend from a balance between the economic, environmental and social component (Cappelli A. et al 2015).

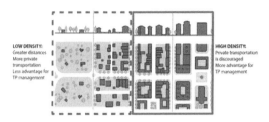

Figure 1.   Difference between Low-High density.

The analyzed literature confirms, therefore, as the sitting choices of residences and businesses can affect the development and success of public transport and vice versa.

A design aimed at the sustainable development of public transport, with particular reference to rail, must be based not only on the performance of the transportation system and on operating costs; even of different factors, such as: population size, territorial density and the sustainable and orderly development of the territory.

In summary, it can be said that the choice of mode of transport is related to the form of the city, meant as an urban or suburban area, diffuse or compact. In compact areas, the organization and management of public transport are more profitable. Therefore they represent a valid alternative to the private car (Abbadessa C. 1996).

Each area has its own characteristics, opportunities and problems, which affect the user's preferences: for example, in the case of a low density area, the choice generally falls on the private car, probably in view of a lack of effective alternatives.

Studies on new or upgraded high-speed lines have shown that, on opening or upgrading a rail infrastructure, or a station, the property value of new or existing buildings located near stations or railway lines (Martinez F. et al 2013) is increased (m²).

The bibliography (Leonetti M. 2015) deals with the reuse of brown field or abandoned rail infrastructure from different points of view: in the various examples, the plans are used for different purposes from the reactivation of the line (tourist paths, cycle paths and greenways).

The guideline of the European Union is clear (European Commission, 2011): the need to reduce transport externalities, the land consumption and emissions into the atmosphere by 2050 is a priority.

In 1996, at the European Conference Habitat II, the debate was on demographic urban development. The discussion underlined how metropolitan areas grew both in size both of inhabitants, while rural areas were increasingly subject of abandonment.

The growth of urban areas had been so fast that the services (including transport) couldn't follow the urban development. The areas with the greatest growth were the border areas, the ones between the old town and the countryside.

These areas had the feature required for that era: low density and greater comfort. In these areas, created quickly and supported by urban planning strategies, appropriate means of support were not always present, for instance an adequate public infrastructure network.

Therefore this implied that most of the movements were with private vehicles (Laconte P. 1996).

Even nowadays, the regional public transport presents different complications depending on the territory: for example, in "third world countries", the system is threatened by lack of funds, while in developed countries is declining because of the predominant use of private cars.

2   EVALUATION INDEX

2.1   Methodology

"The soil is an asset of common interest for the Community, although it is mainly privately owned, and which, if not protected, will undermine the sustainability and long-term competitiveness of Europe."(European directive, 2014).

The soil, as a resource, produces public benefits (transport, environmental and social terms).

Nowadays, however, the accessibility to land use information is not entirely guaranteed: the regions have taken the lead in preparing databases on soil uses, but only few of them are now able to provide accessible data in a simple and free way.

This research deals with the study of the relationship between transport planning and land use planning, placing a constraint on a better use of the existing rail infrastructure at a regional level (Italy) and defining economically sustainable development plans.

To understand the structure of the territory in ex-ante and ex-post stage, indicators represent an ideal key-role tool to read urban phenomena.

The indicators, since aim different objectives, are fundamental. For this reason, it's important to choose them carefully to prevent them from being ineffective and harmful in decision-making (Meadows D. 1998) or analysis.

In this context, the choice of the indicators was important (Pileri P. 2011): simplicity of application, statistical significance and consolidated application were used.

The interpretation is another delicate and crucial phase of the research: the indicator use is not automatically true. In fact although it reduces uncertainty, the indicator could be effected from the context in which it applies (Greeuw S. et al 2000).

To analyze the changes of land use compared to rail infrastructure and to conduct comparative analysis, it's important to rely on indicators (or indices) which are normalized on reference areas (eg. m^2) or on demographic variables (eg. The number of inhabitants). This will make possible to properly compare different territorial units.

2.2 Indicator choice and measures

In order to perform the analysis of the territory affected by railway lines Civitavecchia-Orte and Civitavecchia-Roma, paying particular attention to the relationship between spatial density (residential, production, etc.) and regional railway system performance (expected or possible), the following indicators were measured (OSDDT 2014).

2.3 Territorial indicators

a. "Territorial population density" and "intensity of land use" indicators [ab./ha or km^2]:
   - Territorial population density: (permanent residents)/(total land area).
   - Intensity of land use or territorial net population density: (permanent residents)/(urban land area).
b. Composition use index/land cover or coverage ratio composition (CLi):
   Relationship between the surface of a certain class/land use and the entire area considered as the study-unit (eg. Common). Composition indicators are also used to measure the heterogeneity or, on the contrary, homogenization of the landscape.
   $CLi = (LC_i / LC_{tot}) * 100$ where: LC = land cover/surface LC_{tot} = total surface study-area and i = type of land cover.
c. Used land per capita indicator:
   Relationship between the used land surface and the resident population in the area, expressed in m^2/hab—ha/hab.
   Cab = CS/hab where: Cab: used land per inhabitant; CS: used land [m^2; has]; hab: number of inhabitants [num].
d. Building number and built surface average:
   These indicators are useful to understand the quality and the main features of the built area.

2.4 Transport indicators

a. Coefficients and equipment index of the structures/human infrastructure:
   Infrastructures equipment or infrastructure density of the transport network (rail and road) for the reference surface [m/km^2].

2.5 Real estate value indicators

An additional indicator is the variation of the real estate value compared to the efficiency of the railway infrastructure.

The research, carried out on British study (Martinez et al 2013) cases which concern the trend of the real estate market at the time of the construction and opening of a new AV line (using various parameters such as the proximity to the station and the hedonic pricing), shows that the investments in transport infrastructures have often played a central role in the renovation of urban fabric. Two types of effects can be observed: one concerning the territory, mainly linked to the localization of residences and economic activities; the other one regarding economy and the price variation of properties in proximity to the station due to the increase in accessibility.

The study shows that the opening of a new railway infrastructure contributes to the development of the area and that people chose residences in the closest proximity to it even if they have to pay more.

For example, in the case of the Stratford station opened in 2010, during the first years of construction (2004) there was a +8,1% rise in the local real estate value, which then slightly decreased between 2005 and 2008, when the build process started; it reached a negative peak of –13,5% with the advent of the recession, but then grew to +4,56% in 2010 when the station was first opened.

Given the proper fiscal measures, the price variation resulting from the improvement of the transport services performance could support collective transport by reaching the value (Budoni A. 2014) (once common procedures are defined) which however would not cover the investment entirely and would still need public financing.

The efficacy of indicators originates in its availability, but in the case of studies on real estate value it's rather complex to detect its variations over time.

Because of that, an indicator named "Cross Section Spatial Index [CSSI]" was created for research purposes: this indicator is based on measuring a real estate value in relation to its distance from an efficient transport system offer.

However, even if the variation of real estate value overtime is unknowable, we can still measure it through space and location; namely, we can observe how it changes in different territorial environments when it's close or far from the local transport systems. This way it is possible to evaluate a real estate value in relation to the distance from an efficient transport offer. The indicator was applied with a 1–2–3–5 km buffer from the station involved.

The Spatial Cross Section substitutes for Time Cross Section and allows us to create an economic evaluation model and improve our knowledge in a field that is currently lacking contribution.

2.6 Calibration and application of indicators in study cases

Once picked the most suitable indicators, territorial and infrastructural analysis have been carried out and will be carried out on three regional railway lines: the unproductive line of Civitavecchia-Orte (Fig. 2), the exploited line of Civitavecchia-Rome (Fig. 3), focusing the efforts at locating the critical points linked to territorial density and to the correlation with the railway transport system.

These analysis were carried out using desktop open source QGIS software with which each railway line was studied, transposing every building (regular or not) of the relevant district into planimetry and tracing the existing road and railway infrastructures. Furthermore, the green areas and the most frequently reached destinations were depicted as well.

The area of study was divided into 1 km quadrants of territory crossed by the railway comparing that to a pedestrian section which could be easily traversed (Winkler B. 1990).

The choice of the routes is not random: the Civitavecchia-Orte line is a dismissed regional railway which shows how the presence of a railway infrastructure can transform and guide territory development; Civitavecchia-Rome is redeveloped line, from an organizational point of view.

The purpose of the research program is to fully optimize the existing regional railway transport system by renovating and enhancing it (namely increasing both the quality and the quantity of the offer), in areas where nowadays the demand would not seem to be enough to justify such a work, but which could be driving forces for the local development in the near future.

Fundamental elements for the project are the improvement of accessibility and the creation of an

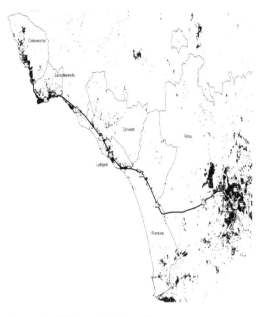

Figure 3. Civitavecchia-Roma line.

innovative and inter-modal transport system, which through a precise regional planning could allow to obtain an improvement in the transport quality, a change in the modality selection and finally an enhancement of life quality for the citizens.

The research also aims to show the possibility of the opposite process as well; namely that the improvement and enhancement of the public urban railway transport system could bring about the increase of demand in areas which are not been taken advantage of for whatever reason (comfort, accessibility, travel duration and frequency). Public transport has to adapt to the environment in order to increase its market share over the private means of movement.

These studies will result in the creation of a simulation model which could depict the situation and simulate its changing, thus providing support to decisions concerning territory planning with the purpose of improving collective public transport.

The originality of this model consists in the intent to increase the collective transport offer as a contribute to a territory development that could justify and represent a potential demand for mobility.

3 RESULTS AND FUTURE PROCESSING

To sum up, from the attached tables (Tables 1–2) we can conclude how the application of the above-mentioned indicators allows to define the main features of the two examined railway lines.

Figure 2. Civitavecchia-Orte line.

Table 1. Civitavecchia-Orte indicators.

Municipality	Quadrants N	Population	Territorial Density (inhabitant/Km²)	Average CLI (%)	Average Cab (m⁻¹)	Railway equipment Tot (m)	Road equipment Tot (m)	Number of buildings (n)	Built surface average (m²)
Civitavecchia	11	53069	719	9,6	119	9948	79528	925	1063
Tarquinia	5	16516	59	0,1	44	6214	8285	31	129
Allumiere	11	4059	44	0,2	67	8966	30689	72	313
Tolfa	6	5227	31	0,03	11	6009	11783	21	92
Blera	20	3385	36	0,6	171	15628	66803	464	163
Barbarano Romano	10	1091	29	1,1	369	9960	29911	287	266
Vejano	4	2282	51	0,14	26	3040	10589	30	220
Capranica	12	6554	160	2	103	11978	45171	569	301
Ronciglione	9	8741	167	2	139	7348	24896	468	269
Caprarola	5	5480	96	0,5	50	4179	13490	110	202
Carbognano	3	2021	116	0,3	24	2775	6828	36	219
Fabrica di Roma	6	8440	243	2	82	5466	22660	473	247
Corchiano	5	3907	119	0,4	37	4142	12433	97	233
Gallese	6	2934	79	0,5	68	4989	14380	64	456
Orte	13	8982	129	0,7	51	12813	30404	190	278

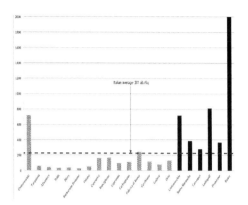

Figure 4. Comparing the average population density (Italy, Civitavecchia-Orte line and Civitavecchia-Roma line).

Table 2. Civitavecchia-Roma indicators.

Municipality	Quadrants N	Population Tot	Territorial density (inhabitants per km²)	Average CLI (%)	Average Cab (m⁻¹)	Railway equipment Tot (m)	Road equipment Tot (m)	Number of buildings (n)	Built surface average (m²)
Civitavecchia	7	53069	719	13,4	136	9482	84281	1653	474
Santa Marinella	30	18769	384	5,4	112	54352	235182	3796	750
Cerveteri	7	37441	279	4,8	172	9749	44527	727	448
Ladispoli	12	21078	812	10,4	129	14933	110000	2106	545
Fiumicino	22	78304	366	2,1	75	27140	78261	724	614
Roma	1	2864731	2225	0,5	2	3014	5126	24	312

Figure 5. Hotspot.

The information refer to 2015 ISTAT sources and allows to draw an up-to-date outline of the situation.

Regarding the single track line of Civitavecchia-Orte, how shown in the table (Table 1), the average residential density in the involved districts is 139 ab/km², compared to the national average of 201 ab/km². (Fig. 4)

Peculiar situations are represented by the Civitavecchia district, in which the average residential density is 719 ab/km², and by Fabrica di Roma district, in which it is 243 ab/km²; being the only two districts above the national average.

The territory coverage index (CLI) shows a really low percentage for the Civitavecchia-Orte line with values which go from 0% (regarding the buildings far from the station) to 30% (buildings near the station): these values show how the framework of buildings is distributed in hotspots, meaning that tiny built up areas are spread throughout the territory.

To support the information brought about by the first indicator, the CAB indicator shows an limited built up areas spread on a wide area.

Lastly, the dominant infrastructural outfit is road network.

The double track line of Civitavecchia-Roma (Table 2) has a population density of 719 ab/km², far above the national average.

The CLI indicator spreads out evenly on the territory, as we can deduce by observing the map (Fig. 3), in which the distribution of the buildings along railway tracks is depicted.

In this line the special shape is dictated by the linear distribution of the buildings alongside the existing railway line, differently from the hotspot pattern of the previous line (Fig. 5).

Leaving out the case of Rome, the CAB varies between 175 m² and in 75 m² in all the districts, which displays similar land use per inhabitant.

Lastly, like in the previous case, the infrastructures shape stresses a predominance of roads network.

Concerning the new CSSI indicator, the results shown in tables 3 and 4 confirm how the station acts as an attracting agent in the territorial context.

For the Civitavecchia-Roma line, it was more problematic to acknowledge the same role to the station due to land features that made it ambiguous case: within the four analyzed buffers, the value of real estate sales for an 85 m² apartment are mostly similar in some cases a further distance from the station resulted in an increase of value, due to the proximity to the sea.

In this particular context the territorial structure has two main attracting agents, the station and the shoreline represents the two major poles of interest.

The Civitavecchia-Orte line instead, is the bet: given the hotspot distribution of the line, some of the stops display very limited built up areas nearby; furthermore, the wider the distance from the station the lower the real estate value will be.

Table 3. Real estate sales Civitavecchia-Orte line.

Stop	Buffer 1000 m.	Buffer 2000 m.	Buffer 3000 m.	Buffer 5000 m.
Civitavecchia	279.000	175.000	108.000	90.000
Civitavecchia Porto Tarquinia	139.000	187.000	108.000	90.000
Aurelia	165.000	160.000	.	.
Mole del Mignone	.	.	.	.
Allumiere	.	.	.	.
Monteromano	.	.	.	.
Le Pozze	.	.	.	.
Civitella Cesi	.	.	.	.
Blera	128.000	93.000	.	.
Barulau di Barbarano	120.000	90.000	.	.
Barbarano Romano - Velano	.	.	.	145.000
Madonna del Piano	119.000	90.000	.	114.000
Ronciglione	130.000	110.000	.	.
Caprarola	.	109.000	90.000	80.000
Fabrica di Roma	75.000	.	.	42.000
Corchiano	.	.	55.000	.
Gallese	.	.	.	62.000
Castel Bagnolo di Orte	.	.	.	.
Orte	88.000	52.000	89.000	90.000

Table 4. Real estate sales Civitavecchia-Roma line.

Stop	Buffer 1000 m.	Buffer 2000 m.	Buffer 3000 m.	Buffer 5000 m.
Civitavecchia	279.000	175.000	108.000	90.000
Santa Marinella	155.000	179.000	240.000	150.000
Santa Severa	.	.	235.000	390.000
Marina di Cerveteri	139.000	195.000	.	129.000
Ladispoli-Cerveteri	189.000	195.000	.	190.000
Torre in Pietra-Palidoro	180.000	154.000	.	168.000
Maccarese-Fregene	170.000	189.000	.	.

The presented results are to be considered partial since new studies are being carried on in order to reach further developments and more in-depth analysis is with which to build the model and apply it in the final phase.

The research turned out to be very difficult, both for the collecting of the information to be analyzed and for the related numerical and cartographic processing.

Therefore more in depth studies and generalizations are being carried out. However, the already accomplished results allow us to confirm the estimation method and to make some first interesting considerations about the potential development of railway transport systems in not densely populated areas.

4 FINAL REMARKS

With the methodology of the study defined and on the basis of the elaboration of data and analyses undertaken, it is possible to express considerations which will allow for the continuation of the research.

In the first instance, in the opinion of the authors, the use of the parameter "Cross Section Spatial" will allow for the evaluation of the effects of the aperture or closure of one railway line on the value of real estate and the consequent effects on territorial development.

At present it is possible to affirm that the relative effects are "strong", in particular in the medium/low earnings areas where the accessibility to transport is strongly required and the cost of private transport impacts significantly on the families' budget.

The situation differs in the urban areas among the medium/high earnings bracket where accessibility to transport is far better and the presence of the railway line (barrier effect and acoustic impact) is more significantly felt.

The research, therefore, will carry out an in-depth study of these differences highlighting that the principal objectives of the study are the regional systems and railway lines in disuse or in various states of abandon and not densely populated, urban areas. It is in these areas that the problem appears more relevant and in need of further study inasmuch as in the dense urban areas the railway services, even if inadequately served, maintain their role as a valid public transport.

During the research it became apparent that the information pertinent to real estate values was lacking with regards to specific territorial settlements and above all historic trends.

This lack of information does not allow for a historic data bank that can establish a direct link between the evolution of the transport system and the presence or absence of railway connections.

For the reasons mentioned above the "Cross Section Spatial" parameter was suggested as a means to investigate the property market trend in relation to the distance of efficient transport (in particular railway in areas of medium/low earnings).

Obviously, were historic data banks available over a sufficiently extended period of time it would have proved simpler to study the various correlations of the role of railway transport. It is the authors' belief that such information be collected and stored and that the proposed indicators may consent to a sufficient in-depth study as to warrant an evaluation of a strategic nature.

REFERENCES

Abbadessa C. September 1996. *L'integrazione dei sistemi di trasporto nell'area metropolitana diffusa veneta.* Railway Engineering.

Brovarone, E. March 2010. *Pianificazione urbana e comportamenti di viaggio.* TeMa.

Budoni A. May 2014. *Catturare il valore dl suolo per sviluppare reti di trasporto locale su ferro.* Railway Engineering.

Cappelli A., Libardo A., Nocera S., Sardena A. & Antognoli M. 2015. *Accessibilità e qualità dei nodi di interscambio per lo sviluppo dei servizi di trasporto ferroviario regionale.* Proceedings of the IV conference Safety and Exercise Rail: solutions for the Rail Transport Development (SEF).

Cappelli, A. & Libardo A. & Nocera S. November 2013. *Teorie, strategie ed azioni per uno sviluppo efficiente del trasporto regionale.* Railway Engineering.

Cappelli, A. 2006. *Il nuovo corridoio ferroviario Torino-Lione.* Railway Engineering.

Cappelli, A. September 2009. *Il costo sociale del trasporto e della logistica in Italia.* Il Mulino.

Cascetta E., Carteni A. & Carbone A. March 2013. *La progettazione quality-based nel trasporto pubblico locale.*

Il sistema di metropolitana regionale della Campania. Railway Engineering.

Coppola E., February 2012. *Densificazione vs dispersione urbana.* TeMa.

European Commission, 2011. White Paper on transportation. Roadmap to a Single European Transport Area-for a competitive transport system.

European directive n° 2004/35/CE.

Greeuw, S., M. van Asselt, J. Grosskurth, C. Storms, N. Rijkens-Klomp, D. Rothman and J. Rotmans. 2000. *Cloudy crystal balls. An assessment of recent European and global scenario studies and models.* Environmental issues series No17. European Environmental Agency, Copenhagen.

Laconte, P. November 1996. *Regional Rail in Low-Density Areas.* Japan Railway & Transport Review.

Leonetti M. July 2015. *Reti ferroviarie secondarie: analisi e prospettive per uno sviluppo sostenibile dei territori marginali.* Thesis, La Sapienza.

Martinez F., Pagliara F., Tramontano A., 2013. *Valore edonico dell'accessibilità relativo agli immobili ad uso residenziale: processo di offerta casuale ed applicazione ad un nuovo collegamento ferroviario.* Railway Engineering.

Maternini G. March 2014. *Trasporti e Città: mobilità e pianificazione urbana,* EGAF Edizioni, Transportation Engineering Editorial series.

Meadows, D. September 1998. *Indicators and information systems for sustainable development.* The Sustainability Institute.

Ministry of Economy and Finance, Economic and Financial Document 2016, attachment: *Strategie per le infrastrutture di trasporto e logistica.*

OSDDT. 2014. *Come calcolare gli indicatori.*

Pileri P. 2011. *Misurare il cambiamento. Dalla percezione alla misura delle variazioni d'uso del suolo.* Land use in Lombardia region in the last 50 years. Lombardia region.

Ricci S. December 2013. *Ingegneria dei Sistemi Ferroviari: tecnologie, metodi ed applicazioni.* EGAF Edizioni, Transportation Engineering Editorial series.

Winkler B. Marzo-Aprile 1990. *Piano della mobilità per la città di Bologna.* Parameter.

Transport Infrastructure and Systems – Dell'Acqua & Wegman (Eds)
© *2017 Taylor & Francis Group, London, ISBN 978-1-138-03009-1*

Delay estimation on a railway-line with smart use of micro-simulation

F. Cerreto, S. Harrod & O.A. Nielsen
Department of Management Engineering, Technical University of Denmark, Kongens Lyngby, Denmark

ABSTRACT: This paper formulates a delay propagation model that estimates total railway line delay as a polynomial function of a single primary delay. The estimate is derived from a finite series of delays over a horizon that spans two dimensions: the length of the railway line and the number of trains in the service plan. The paper shows that the total delay estimate is a cubic relation for small primary delays. A probabilistic approach is presented to combine the total delay functions of primary delays given to different trains. The final estimate is the total delay on railway lines, after a random incident has occurred. The model can be integrated in railway timetable analysis to reduce the number of necessary simulations, and can be used when the computation speed is an issue, such as on-line rescheduling algorithms. The model is demonstrated with an analysis of a Danish suburban railway.

1 INTRODUCTION

Operational stability and robustness are crucial for railway transport. Not only are the passengers or users of the service sensitive to these measures of quality, but railways are usually integrated systems or networks, and failures at one location of the system affect other locations and services, sometimes quite catastrophically. Railway network planners are faced with many decisions about what quality of service to provide and what resources to allocate to deliver this service. Much of the literature demonstrates that there are frequently multiple feasible alternatives to allocate resources, and each alternative has a unique performance profile with characteristic statistics, especially with regards to punctuality and robustness. The analysis of these alternatives frequently requires laborious and inconclusive modeling with simulation software.

This paper contributes to the literature with a closed form function estimate of aggregate railway line delay propagation in response to a primary delay. Many railway and transit services are of the form of a single terminating railway line, and this function may supplement or replace the application of simulation for exploration of alternatives. On many railway lines passenger traffic is distributed over the line destinations, and aggregate delay is an appropriate measure of system performance and customer service.

This formulation is closed form under a set of assumptions that is later shown to be robust to variance. The formulation is derived from a finite series of deviations from the service plan (secondary delays) caused by a singular initial disruption (primary delay). The total delay generated by

disruptions on a railway line depends on the interactions between the trains, and a different total delay function is derived for each scheduled train. The probabilistic approach presented in this paper allows to estimate the contribution of the individual trains to the general function of the total delay on a selected railway line.

Using microsimulation, the model can be shown to be robust to deviations in assumptions, and the results may be used to establish bounds of expected performance of simulation models, and thus reduce the use of simulation models in preliminary, exploratory studies. Railway microsimulation is known for its heavy computational requirement, and the models proposed in this paper introduce new estimation of the total delay on railway lines with a very limited used of microsimulation, restricted to the initial calibration phase.

1.1 *Literature survey*

Prior literature on operational stability and delay propagation may be classified as proposing parametric methods, providing analytical methods, or demonstrating applications of simulation.

Parametric measures are functional relations fitted to empirical or experimental data. In these measures, the cause and effect relation may not be clearly understood, or it may be strongly limited to selected environments. Krueger (1999) presents many capacity estimation functions proposed for and validated on North American long distance railways. International Union of Railways (2004) defines procedures using timetable compression to estimate the capacity of European high density railway lines. Gorman (2009) fits linear multiple regression functions to large data sets of train operations on a North American railway to estimate delay as a function of train planning decisions.

Analytical methods derive system performance measures from known or presumed cause and effect relations in the railway service plan. Among these, Hasegawa et al. (1981) applies a hydrodynamic analogy to model railway traffic. The study models the delay propagation as a shock wave in a compressible fluid and finds the total delay as a cubic function of the primary delays by means of propagative velocity. Harker and Hong (1990) estimate the delay on a mixed single and double track railway where the train path is not defined in advance and is subject to a stochastic dispatching decision en route. Higgins et al. (1995) formulate decision rules for operation on a single track railway, and then calculate in closed form the expectation of system delay given a traffic pattern and defined probabilities of delay for trains, track segments, and terminals.

Railway delay models often lead to innovations in mathematics, such as Meester and Muns (2007) application of phase-type distributions. Meester and Muns derive the net delay distribution on connected railway network segments given the distribution of primary delays on each segment. The derivation asserts that recursive calculation of the solution may be attained with just three operations: sum, nonnegative excess beyond a bound, and maximum. The paper states that a phase-type distribution, a distribution of the absorption time of a continuous time Markov chain, can be contained in the three operations in closed form. However, the method depends on the assertion of independence of the primary delays. The method is demonstrated for a sample network of 24 directional line segments with seven transfer points.

Goverde (2010) presents an efficient delay propagation algorithm where timetables are modeled as timed event graphs (using max-plus algebra) and initial delays are known. The algorithm is very fast and in a few seconds can calculate the delay propagation over a large network consisting of many interdependent services, such as the Dutch national railway timetable. However, the model offers no functional relationship, and results must be calculated for each scenario separately. Kroon et al. (2007) proposes a stochastic linear program for the optimal allocation of supplement time along the route of a train path and finds that in a variety of realistic scenarios the supplement time should not be allocated uniformly along the train path. Finally, and most closely related to this paper, Landex (2008) proposes a delay propagation model computing the transfer of delay between trains through the scheduled buffer times. This model is used to study the relation between capacity consumption and the development of the disruptions, but does not take into account the recovery of train delays according to the timetable allowance. Cerreto (2016) extends Landex's delay propagation model to include the timetable allowance. The total delay on a railway line is described as a composite polynomial function of the primary delay generated at a station, which is cubic for small primary delays. The model allows to calculate the total delay with a limited use of microsimulation, but it returns a different total delay function depending on the first train delayed.

Simulation is widely used, experimentally and in practice. Relevant publications include Lindfeldt (2015), which extensively applies RailSys commercial railway simulation software to a variety of capacity and delay propagation topics. In particular, Lindfeldt simulates 336 timetable scenarios and then applies linear regression to determine the significance of many common heterogeneity measures in predicting aggregate secondary delay. Lindfeldt finds that the *mean pass coefficient*, a measure of the frequency of meets and overtakes, is the most significant indicator. Mattson (2007) uses microsimulation to study the interferences between trains under different capacity utilization values: Mattson finds this to be the most precise way to analyze secondary delays, but it is also demanding for very detailed input and the process is very time-consuming. Lastly, Cerreto (2015) applies Open-Track commercial railway simulation software to the analysis of a 21 km. line (nine stations) in the Netherlands with four configurations ranging from double track to quadruple track. Cerreto investigates methods to reduce the computation necessary for simulation based analyses and limits the number of simulation runs required with a heuristic process called the *skimming method*. Instead of simulating all combinations of trains and delays, a composite profile of train delay is estimated from an initial simulation analysis, and this composite delay function is used to calculate the aggregate system delay. The results demonstrate that capacity utilization is not strongly correlated with aggregate secondary delay, which contradicts the findings of some other literature.

2 INCIDENT, PRIMARY DELAY PROBABILITY AND TOTAL DELAY

Delays are positive deviations between the realized times and scheduled times of activities. In the literature, different classifications of delays are available. Most of the classifications distinguish between delays that are due directly to the variability of process times and delays that are originated by the subsequent conflicts in the actual operation (Goverde &

Hansen, 2013). The *primary delays* are unexpected extensions of the planned times of the individual processes scheduled. For instance, equipment failures and large passenger flows generate primary delays. The *secondary delays*, on the other hand, are delays generated by operation conflicts, which are due to primary delays themselves. When a train is delayed, it needs to use infrastructure elements at different times than planned. A conflict arises when two or more trains request to use the same element at the same time: they will be queued by dispatching decisions, since only one train at time can use one element or track section. The delay that generates from the queuing is called secondary delay.

The *cumulative delay*, or *total delay*, on a railway line is the sum of all the total positive deviations registered for all the trains at all the time measurement points.

The delay generation process begins with a disruption or incident. A primary delay generates when the failure intersects a scheduled event in the timetable, and secondary delays evolve from the interaction between different scheduled events in the timetable.

The model presented in this paper translates the probability density distributions of incidents on a railway line into the probability densities of primary delays and of secondary and cumulative delays.

The section below describes the probability of generation of primary delays to a selected train, given the characteristics of the incident and the timetable.

2.1 *Probability of primary delay to one train*

We consider those incidents that prevent trains from moving. Such events can be, e.g. failures at signal boxes, extended boarding times at stations, failures at other ground or onboard systems.

In several cases, it is possible to describe an incident by the distributions of its starting time and duration. The distribution of the starting time could be assumed uniform in preliminary studies, when detail information is not available. The distribution of the incident duration is still subject of studies in the railway field. Meng and Zhou (2011) propose the Normal distribution to model the disruption duration on single track lines, while the Exponential distributions is used by Schranil & Weidmann

Figure 1. Delay generation process: Primary delays happen when incidents cross scheduled events. Secondary delays generate from delayed scheduled event crossing other events in the timetable.

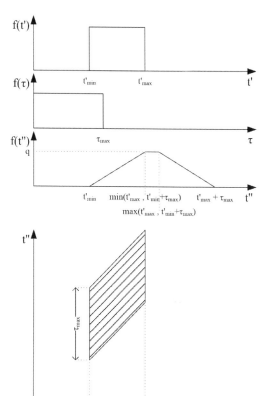

Figure 2. Probability density functions of the starting time, the duration, and the ending time of an incident. The last graph shows the joint probability density domain of t' and t''.

(2013) in Switzerland; finally Zilko et al (2016) propose an on-line model to predict the duration of a failure, based on the available knowledge at the beginning of the failure. The model uses the Copula Bayesian Networks to estimate the contribution of given influencing factors, based on historical data. In early studies the information available on the incidents may be insufficient to estimate these distributions, so we take the relaxed assumption of uniform distribution also for the incident. The formulation is, thus, simplified, but could be integrated with specific distributions fitted to the given incidents. Our model translates the probability of incidents into the probability of primary delays by integration of the probability densities. The structure of the model would not be affected choosing different distributions of the incident durations.

We define t' the starting time of an incident, t'' its ending time, and τ its duration, so that $t'' = t' + \tau$. Both t' and τ are assumed uniformly distributed, on independent ranges:

$$t' \in \mathcal{U}\left(t'_{min}, t'_{max}\right)$$

$$\tau \in \mathcal{U}\left(0, \tau_{max}\right)$$

Consequently, t'' follows a trapezoidal distribution in $[t'_{min}, t'_{max} + \tau_{max}]$ (Figure 2).

The central segment of the distribution spans from $\min\{t'_{max}, \tau_{max}\}$ to $\max\{t'_{max}, \tau_{max}\}$, and its constant value is $q = \dfrac{2}{t'_{max} - t'_{min} + \tau_{max} + |\tau_{max} - t'_{max}|}$.

We define L_i the event "*Train i experiences a primary delay*". The departure time of train i from the considered station is named θ_i, and the time separation between the train $i-1$ and the train i is the headway $h_i = \theta_i - \theta_{i-1}$. The incident generates a primary delay to the train i if it starts between the departures of trains $i-1$ and i, and it ends after the scheduled departure of the latter, θ_i.

$$L_i = \left(t' \in \left(\theta_{i-1}, \theta_i\right)\right) \cap \left(t'' > \theta_i\right)$$

The intersection probability is expressed by means of the conditional probability:

$$P\left(L_i\right) = P\left(t'' > \theta_i \,|\, t' \in \left(\theta_{i-1}, \theta_i\right]\right) \cdot P\left(t' \in \left(\theta_{i-1}, \theta_i\right]\right) \quad (1)$$

t'' depends on t'' through τ, and the conditional probability is derived hereunder.

The conditional probability density of $(t'' | t' \in (\theta_{i-1}, \theta_i])$ has a trapezoidal shape in the range $[\theta_{i-1}, \theta_i + \tau_{max}]$, with a central constant segment in the range $[\min\{\theta_i, \tau_{max} + \theta_{i-1}\}, \max\{\theta_i, \tau_{max} + \theta_{i-1}\}]$, and height $q = \dfrac{2}{h_i + \tau_{max} + |\tau_{max} - h_i|}$. The joint conditional probability corresponds to the striped area in Figure 4.

In the following formulation, equation (1) is split into two factors for a simpler explanation. We name the conditional delay probability $P(E1_i) = P(t'' > \theta_i \,|\, t' \in (\theta_{i-1}, \theta_i])$ and the event probability $P(E2_i) = P(t' \in (\theta_{i-1}, \theta_i])$ that corresponds to the start of the incident between trains $i-1$ and i. The probability of $E1_i$ depends on the relation between τ_{max} and h_i and is described by the following:

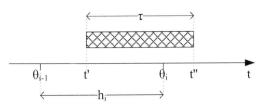

Figure 3. A train receives a primary delay if the incident begins (t') in the previous headway (h_i) and it ends (t'') after the scheduled departure time (θ_i).

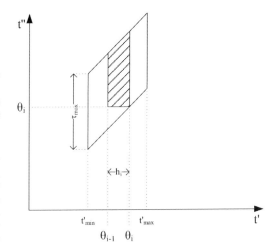

Figure 4. Joint conditional probability $P(t'' > \theta_i \,|\, t' \in (\theta_{i-1}, \theta_i])$.

$$P\left(E1_i\right) = \begin{cases} 1 - \dfrac{h_i}{2\tau_{max}} & \text{for } h_i < \tau_{max} \\[2mm] \dfrac{\tau_{max}}{2h_i} & \text{for } h_i > \tau_{max} \\[2mm] \dfrac{1}{2} & \text{for } h_i = \tau_{max} \end{cases}$$

$P(E2_i)$ is proportional to the headway of the train in the timetable cycle:

$$P\left(E2_i\right) = \frac{h_i}{\sum_i h_i} = \frac{h_i}{c}$$

where c is the timetable cycle, and is given by the sum of all the headways. The probability of every train in the cyclic timetable to experience a primary delay is given by the following:

$$P\left(L_i\right) = \begin{cases} \dfrac{h_i\left(2\tau_{max} - h_i\right)}{2c \cdot \tau_{max}} & \text{for } h_i < \tau_{max} \\[3mm] \dfrac{\tau_{max}}{2c} & \text{for } h_i > \tau_{max} \\[3mm] \dfrac{h_i}{2c} & \text{for } h_i = \tau_{max} \end{cases} \quad (2)$$

Note that the probabilities of the individual trains to receive primary delays do not sum up to 1. We denote $P(0)$ the probability that no train is delayed, that is

$$P\left(0\right) = 1 - \sum_i P\left(L_i\right). \quad (3)$$

2.2 Combined total delay functions

A total delay function describes the relation between primary delays given to a train and their cumulative effect on the railway line. Different train paths in a timetable are characterized by different stopping patterns, running time supplements and headway buffer times towards the following trains. Therefore, at every train path scheduled corresponds a characteristic total delay function of the primary delay.

We combine the characteristic total delay functions of different trains in a general total delay function that represents the effect of a primary delay to any of the trains in the timetable. The general function is a weighted average of the individual curves, where the weights are proportional to the individual probabilities of the trains to receive a primary delay.

The general total delay function is expressed by

$$d = \sum_i w_i \cdot d(i) \tag{4}$$

with weights

$$w_i = \frac{P(L_i)}{\sum_j P(L_j)} \tag{5}$$

Equation (4) allows the estimation of the general total delay given by an aleatory incident through the combination of total delay functions generated by selected trains. The estimation of individual total delay functions is relatively simple. In the following section we describe a delay propagation model to calculate the total delay $d(i)$ as a cubic function of the primary delay given to train i.

We reduce considerably the simulations necessary to estimate the general total delay combining the model described below and the probabilistic approach.

2.3 A finite series model of the total delay as a function of the primary delay

Previous literature demonstrates that the total delay on a railway line can be described as a cubic function of the primary delays given to a train. Cerreto (2016) models the total delay from the service timetable at all measurement points, as a function of timetable supplement, timetable buffer, and a single initial delay to one train. The model is summarized in this section.

The total delay model has a two dimensional analysis domain, namely the length of line and the number of trains included in the cumulative delay statistic. Trains on a single line with a single direction of movement are considered, which

is a common operating plan in Europe and urban North America. The time horizon of the model then begins with the departure of the first train at the beginning of the line and ends with the arrival of the last train at the end of the line.

The total delay d represents the unweighted utility loss experienced by the railway service due to a disruption. It is the sum of all individual delays at measurement points in the timetable over the analysis horizon, and is presented in (6).

$$d = \sum_{j,s}(d_{j,s} \mid d_{j,s} \geq 0) \tag{6}$$

with $d_{j,s}$ being the delay of train j registered at station or timing point s (difference between real and scheduled time).

The individual train delay $d_{j,s}$ is a combination of the hindrance from previous trains and the residual delay from the previous station. The delay is transferred to following trains due to a lack of buffer time, while a train keeps a residual delay from the previous station due to a lack of running time supplement. Equation (7) expresses the delay propagation on the two dimensions of the model, under the relaxed assumption of equal running time supplement a for all the trains between any pair of stations and equal buffer time b between any pair of trains.

$$d_{j,s} = p - (s-1)a - (j-1)b \tag{7}$$

Subject to non-negativity constraint: $d_{j,s} \geq 0 \ \forall \ j,s$.

p is the primary delay, which corresponds to the first train's delay at the first station $d_{1,1} = p$.

The total delay is derived summing up the individual train delays at all the stations. It results in (8).

$$d = \sum_{j,s \mid d_{j,s} > 0} d_{j,s} = \sum_{s=1}^{\frac{p}{a}} \sum_{j=1}^{\frac{p-(s-1)a}{b}} p - (s-1)a - (j-1)b$$
$$= \frac{(a^2 + 3ab)}{12ab}p + \frac{a+b}{ab}p^2 + \frac{1}{6ab}p^3 \tag{8}$$

The equation is valid for small values of primary delay that expire before the last train and before the last station.

Cerreto validates the model using microsimulation on a Danish suburban railway line with heterogeneous timetable. The model is robust and holds valid when the assumptions of equal running time supplement and buffer times are removed. The total delay on the line can be regressed to a cubic polynomial function. The application to a heterogeneous timetable, though, returns a different cumulative delay function for each train that receives a primary delay.

We introduce the index i to identify the total delay function $d(i)$ resulting from a primary delay given to train i.

The general total delay function is derived in section 2.2 combining the individual functions through the probability of each train to receive a primary delay.

3 CASE STUDY: THE NORDBANE IN COPENHAGEN

We simulated the operation of a suburban railway line in Denmark to validate the combination of different polynomial functions to describe the total delay against the primary delay. The suburban railway network in Copenhagen is a very densely occupied network with 2 minutes headway in the busiest section. Six different lines operate on the network, five running on the same central section. The suburban line is operated by uniform rolling stock in cyclic timetable. The selected section of the suburban network is the line from Hellerup to Hillerød. Overtakes in this section are prevented. Though it is theoretically possible at selected stations, it hardly occurs in real operation, due to the very high frequency of the train service.

The micro-simulation software OpenTrack by OpenTrack Railway Technology Ltd. and the Swiss Federal Institute of Technology (ETH Zurich) was used for the simulation. This micro-simulation uses continuous computation of train motion equations and simulates the interaction between trains through discrete processing of signal box states (Nash & Huerlimann, 2004). Given user defined infrastructure, rolling stock, and timetable databases, it is possible to calibrate the train paths defining the running time supplements; moreover, different driving behaviors can be modelled for on time trains and delayed ones. The strength of the micro-simulation models is the higher accuracy than the analytical models, and their flexibility to represent different contexts. Changes in the infrastructures and operating rules can easily be implemented and tested. The accuracy comes, though, at the cost of much longer computation time, as well as set-up time. Other micro-simulation software is available on the market, like RailSys by Rail Management Consultants GmbH (RMCon). Despite some differences in the approach, both the mentioned software suffer from long time needed to compute such detailed models (Landex, 2008).

Two different train paths run every ten minutes on the line between Hellerup and Hillerød with two different stopping patterns:

- Line A: runs throughout the entire line, skipping 5 stops in the first stretch
- Line E: only runs the first stretch, stopping at all the stations.

The line stationing and the schedules are summarized in Table 1.

The defined set of {1,...,10} minutes of primary delay was assigned separately to each train departing from Hellerup. The individual total delay functions were regressed from the corresponding total delay measured in the simulation, independently for line A and line E. The general total delay function of the line is calculated by the weighted average of the individual total delay functions of the trains.

For the model validation, we Monte Carlo sampled $n = 200$ failures at the departure signal from Hellerup, starting at a random time independent of the timetable. We identified the first impacted train and measured the primary delay generated by the failures. We measured the related total delay developed on the line and regressed it to individual cubic functions for lines A and E. The starting time of the disruption was extracted from a uniform distribution between 0 and 80 minutes, spanning over 8 consecutive timetable cycles. The duration of the failure was extracted from a uniform distribution between 0 and 10 minutes.

Table 2 compares the cases of primary delay experienced by each train line and the calculated

Table 1. Line stationing and scheduled.

Station		Stationing	Schedule*	
Name	Code	km	A	E
Hellerup	Hl	7,8	05	07
Bernstorffsvej	Btf	9,3			09
Gentofte	Gj	10,9			11
Jægersborg	Jæt	12,6			14
Lyngby	Ly	13,9	11	16
Sorgenfri	Stf	15,9			19
Virum	VG	17,7			21
Holte	Hot	19,0	16	23
Birkerød	BG	23,8	21	
Allerød	LG	29,3	26	
Hillerød	HG	36,5	32	

*Departure minutes of the hour reported. Each train path repeats every 10 minutes. | = pass-through.

Table 2. Cases of primary delay registered in the simulation and probabilities modeled.

Course	Cases of recorded primary delay		Model probability of primary delay	Weight
(i)	#	%	$P(L_i)$	w_i
0	76	38.0%	34.0%	
A	91	45.5%	48.0%	0.73
E	33	16.5%	18.0%	0.27

872

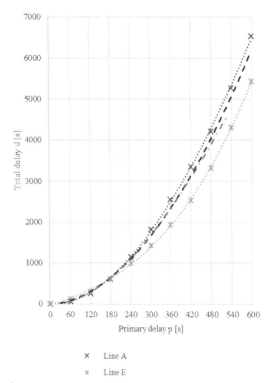

Figure 5. Total delay on the line as a function of the primary delay given to Line A (dotted dark gray line) and Line E (dotted light grey line). Modelled (dashed black line) and measured (dot-dashed line) general total delay.

probability. The weights for the general total delay function are calculated from the modelled probabilities.

The total delay general function of the railway line was regressed from the whole set of simulations. and compared to the combination of the individual delay functions.

Figure 5 compares the modelled general total delay on the line and the measured general total delay from the simulation.

4 RESULTS AND DISCUSSION

The total delay on a railway line can be regressed to a cubic function of the primary delay. Every train that receives the primary delay returns a different function, due to different interaction with the following trains, i.e. different buffer time.

The weighted average total delay function reflects the total delay function given by the joint simulation of failures independent of the timetable.

In this case study, a series of 200 microsimulation of a random failure at a signal box was well approximated by a reduced series of 20 microsimulations of primary delays to individual trains.

The modeled total delay function and the measured total delay hold tight up to 500 s of primary delay. This is due to a higher number of trains from line E that received smaller primary delay. As opposite, trains from line A tended to be the first delayed trains with higher values of primary delay. For this reason, the joint regressed total delay function is closer to line E for small primary delays and closer to line A at higher values of primary delay. Instead, the modelled general total delay function keeps the distance ratio between the two individual total delay functions throughout the entire primary delay range. A solution to this issue could be to cluster the distribution of the failure duration. In this way, different probabilities to be delayed can be calculated for different ranges of primary delays. The averaging weights would be calculated for individual clusters and the general total delay function would adapt to the probability to be delayed of individual trains.

5 CONCLUSIONS

This paper derives the total delay on a railway line as a closed function of the primary delay, under the assumption of equal buffer time between trains and equal running time supplement over the line.

We turn the estimation of the total delay given by an aleatory incident into the combination of the total delay functions of different trains. We determine each function's contribution to the general total delay with a probabilistic approach. The individual total delay function of each train is regressed from microsimulation. The result is a combined total delay function that does not depend on what train receives the primary delay. It is now possible to estimate the consequences of given incident, simulating independent primary delays on the individual trains instead. This allows broader timetable analyses without increasing the number of simulations needed.

The model allows to calculate the total delay on a railway line with high accuracy from the microsimulation, reducing the amount of simulation runs needed. Using this model, we needed only one tenth of the microsimulations used to estimate the total delay from the incident distributions. The number of microsimulations needed for the analysis may be further reduced, taking advantage of the good regressions of the individual total delay functions.

This is relevant for railway planners, because it allows timetable accurate analyses with a limited computational power, or on extended railway networks. At the same time, the accuracy of the model, together with the reduced computation needs, allows new applications in real-time rescheduling models, based on the total delay estimation.

The model accuracy could be further improved in the future clustering the distribution of the incident duration and introducing more complex distributions of the incident duration and starting time.

REFERENCES

Cerreto, F. (2015). Micro-simulation based analysis of railway lines robustness. In *6th International Conference on Railway Operations Modelling and Analysis* (pp. 164-1–164-13). Tokyo: International Association of Railway Operations Research. Retrieved from http://orbit.dtu.dk/fedora/objects/orbit:140777/datastreams/file_112408001/content

Cerreto, F. (2016). A Cubic Function Model for Railway Line Delay. In E. Spessa, M. Biscotto, & G. Smyrnakis (Eds.), *TRAVISIONS 2016 Contest book* (pp. 68–69). Warsaw. Retrieved from http://orbit.dtu.dk/en/activities/travisions-competition-2016(4d4a77a0-0541-4fb2-8548-e2ad7bd8269b).html

Gorman, M. F. (2009). Statistical estimation of railroad congestion delay. *Transportation Research Part E: Logistics and Transportation Review*, 45(3), 446–456. http://doi.org/10.1016/j.tre.2008.08.004

Goverde, R. M. P. (2010). A delay propagation algorithm for large-scale railway traffic networks. *Transportation Research Part C: Emerging Technologies*, 18(3), 269–287. http://doi.org/10.1016/j.trc.2010.01.002

Harker, P. T., & Hong, S. (1990). Two Moments Estimation of the Delay on a Partially Double-Track Rail Line with Scheduled Traffic. *Journal of the Transportation Research Forum*, 31(1).

Hasegawa, Y., Konya, H., & Shinohara, S. (1981). Macro-Model on Propagation-Disappearance Process of Train Delays. *Railway Technical Research Institute, Quarterly Reports*, 22(2), 78–82. Retrieved from http://trid.trb.org/view.aspx?id = 180725

Higgins, A., Kozan, E., & Ferreira, L. (1995). Modelling delay risks associated with train schedules. *Transportation Planning and Technology*, 19(2), 89–108. http://doi.org/10.1080/03081069508717561

Kroon, L. G., Dekker, R., & Vromans, M. J. C. M. (2007). Cyclic railway timetabling: A stochastic optimization approach. In F. Geraets, L. Kroon, A. Schoebel, D. Wagner, & C. D. Zaroliagis (Eds.), *Lecture Notes in Computer Science* (Vol. 4359 LNCS, pp. 41–68). Springer. http://doi.org/10.1007/978-3-540-74247-0_2

Krueger, H. (1999). Parametric modeling in rail capacity planning. In *WSC'99. 1999 Winter Simulation Conference Proceedings. "Simulation - A Bridge to the Future" (Cat. No.99CH37038)* (Vol. 2, pp. 1194–1200). IEEE. http://doi.org/10.1109/WSC.1999.816840

Landex, A. (2008). *Methods to estimate railway capacity and passenger delays*. Technical University of Denmark (DTU). Retrieved from http://findit.dtu.dk/en/catalog/2185768953

Lindfeldt, A. (2015). *Railway capacity analysis - Methods for simulation and evaluation of timetables, delays and infrastructure*. KTH Royal Institute of Technology. Retrieved from https://www.kth.se/polopoly_fs/1.613049!/15_002PHD_report.pdf

Mattsson, L.G. (2007). Railway Capacity and Train Delay Relationships. *Critical Infrastructure: Advances in Spaital Science*. Berlin, Heidelberg: Springer Berlin Heidelberg. http://doi.org/10.1007/978-3-540-68056-7_7

Meester, L. E., & Muns, S. (2007). Stochastic delay propagation in railway networks and phase-type distributions. *Transportation Research Part B: Methodological*, 41(2), 218–230. http://doi.org/10.1016/j.trb.2006.02.007

Meng, L., & Zhou, X. (2011). Robust single-track train dispatching model under a dynamic and stochastic environment: A scenario-based rolling horizon solution approach. *Transportation Research Part B: Methodological*, 45(7), 1080–1102. http://doi.org/10.1016/j.trb.2011.05.001

Nash, A., & Huerlimann, D. (2004). Railroad simulation using OpenTrack. *Computers in Railways IX*, 45–54. article.

Pellegrini, P., Marlière, G., & Rodriguez, J. (2014). Optimal train routing and scheduling for managing traffic perturbations in complex junctions. *Transportation Research Part B: Methodological*, 59, 58–80. http://doi.org/10.1016/j.trb.2013.10.013

Schranil, S., & Weidmann, U. (2013). Forecasting the Duration of Rail Operation Disturbances. In *TRB 92nd Annual Meeting Compendium of Papers* (pp. 1–20). Washington, D.C.: Transportation Research Board of the National Academies. Retrieved from https://trid.trb.org/view.aspx?id = 1241000

UIC. (2004). Leaflet 406 - Capacity, (June).

Zilko, A. A., Kurowicka, D., & Goverde, R. M. P. (2016). Modeling railway disruption lengths with Copula Bayesian Networks. *Transportation Research Part C: Emerging Technologies*, 68, 350–368. http://doi.org/10.1016/j.trc.2016.04.018

Transport Infrastructure and Systems – Dell'Acqua & Wegman (Eds)
© 2017 Taylor & Francis Group, London, ISBN 978-1-138-03009-1

Smart cyber systems incorporating human-in-the-loop towards ergonomic and sustainable transport systems

F. Galatioto
Transport Systems Catapult, Milton Keynes, UK

C. Salvadori
New Generation Sensors, Pisa, Italy

M. Petracca
Scuola Superiore Sant'Anna, Pisa, Italy

J. Garcia Herrero
University Carlos III de Madrid, Madrid, Spain

M.J. Santofimia
University Castilla—La Mancia, Ciudad Real, Spain

A. Pollini
BSD, Milano, Italy

ABSTRACT: This paper aims to introduce and explain a new framework that using standardised engineering principles and methods for Human-in-the-Loop (HiL) when human interactions and feedback are considered in smart Cyber-Physical Systems (CPS). This comes from an industrial need to better understand and predict human behaviour when interacting with CPS, which so far has been mainly addressed for and from ad-hoc solutions. The proposed framework objectives (fully described in the paper) will be achieved integrating in an adaptive Internet of Things system existing and new low-cost technologies, both on street and in-vehicle, with personal experience data coming from psychological, behavioural and physiological processing. This new approach is expected to enormously enhance the effectiveness of the proposed platform enabling a new focus on emotional states (anxiety, stress) and collective/shared feelings (eg. panic) that go substantially beyond the state of the art in urban management delivering new capabilities to the ITS sector.

1 INTRODUCTION

1.1 *Context*

The 2025 European City transport scenario is expected to be heavily affected by huge flows of pervasive information generated by people and vehicles, at all scales from local to global. The multitude and heterogeneity of participants, each with their own cultural models, needs and expectations must be considered as a key component in the EU City transport innovation process to address the longstanding problem of poor air quality, congestion in urban areas, and achieve greenhouse gas emission target reductions. Deployment of smart-city sustainable systems aspire to provide the right information to optimise "behaviours" in real towns. Most of the effort so far has been devoted to technological solutions, with less focus on their use by people and communities. In particular, the issues and linkages between traffic system status perception and decision making processes, at individual up to collective level, has not been properly investigated within the Intelligent Transport Systems (ITS) sector.

1.2 *Background*

The so-called "Internet of Things" has been defined as the network of physical objects or devices, independently of their nature (vehicle, entire buildings or other object), embedded with electronics, software, sensors and network connectivity that enables these objects to collect and exchange data on the web.

The main idea behind the IoT concept is to have worldwide interconnected objects (sensors), each one individually discovered and addressed as a resource in the network.

As sensors spread in the environment, and the capabilities of the so-called wireless sensor networks (the considered building blocks of Cyber-Physical Systems, a.k.a. CPS) are enhanced, several communication protocols must be considered. The first standard protocol, specifying both the Physical (PHY) and Medium Access Control (MAC) sub-layers of the ISO/OSI communication model, is the IEEE 802.15.4 (IEEE, 2003). The standard has been released in its first version in 2003 with the aim of enabling energy-efficient communications in Low-Rate Wireless Personal Area Networks (LR-WPANs). Kushalnagar et al. (2007) proposed the adaptation of the IPv6 over Low-Power Wireless Personal Area Networks (6LoWPANs), thus specifying a Network (NET) layer for Internet-like communication in IEEE 802.15.4-based networks. The 6LoWPAN concept comes from the idea that "the Internet Protocol could and should be applied even to the smallest devices" (Mulligan, 2007), and that low-power devices with limited processing capabilities should be able to participate in the envisioned IoT.

The Routing Protocol for Low-power and Lossy networks (RPL) (Winter, 2012) is the state-of-the-art routing algorithm developed by the networking community to enable routing capabilities in 6LoWPAN networks.

However, RPL is a "best-effort" routing protocol and does not satisfy the real-time requirements of several distributed applications for networks of smart cameras (e.g., tracking and action recognition).

Recent routing protocols permit (He et al. 2003) and (Bocchino, 2011) are characterised by highly efficient and scalable solutions able to tackle the needs and requirements of next generation sensor networks in which cameras are going to be part of the IoT scenario. Other protocol solutions such as the Constrained Application Protocol (CoAP) (Shelby, 2013), are advanced standards working at the Application (APP) layer to guarantee a seamless integration of IoT smart objects in the Internet world. In fact, CoAP is designed to enable the RESTful architecture in the constrained environment, thus providing an easy stateless mapping with HTTP in order to reach a full Machine-2-Machine (M2M) interaction.

This paper presents an ambitious concept which is the development of a new framework to embed more generic and standardised engineering principles and methods for Human-in-the-Loop (HiL) when human interactions and feedback are considered in smart CPS, generating what has been defined as Human-in-the-Loop CPS (HiLCPS). This comes from an industrial need to better understand and predict human behaviour when interacting with CPS, which so far has been mainly addressed for and from ad-hoc solutions, and the need for more robust and transferable approaches.

In particular, the human today, through the use of smart and mobile technologies can be considered as another sensor/actuator which brings new opportunities, as a result of the new capability to gather real-time context and human behaviour information. Human beings have therefore become an inherent element of the CPS, in what has been referred to as HiLCPS.

Future HiLCPS not only consider humans as operators affecting the loop control, but more importantly, as the source of the observation-analysis-adaption loop (Nunes, 2015).

1.3 Research gaps

Transport innovation cannot happen without an extensive sharing of information among all involved stakeholders. Ensuring interoperable transport data and services, where possible based on existing standards and technology, is one of the main components of the proposed conceptual platform called SMART H-LOOP. Currently, multi-modal information services across Europe lack interoperability and are fragmented in terms of what they offer including modal and geographical coverage, real-time information and quality levels. The project will investigate how to reduce the costs and exploit the benefits of sharing information in a context where different players do not always have the same goals.

Thus, SMART H-LOOP is expected to contribute towards EU-wide continuity and harmonised delivery of multimodal travel information services. This in turn is expected to encourage a positive modal shift to sustainable modes of transport and therefore improve the efficiency of Europe's transport network management.

SMART H-LOOP will specifically aim to go beyond current "informational" MultiModal Travel Information and Planning services (MMTIPs) which allow travellers to plan their journey, comparing different travel options and combining different variations of transport modes. Even the most modern MMTIPs includes a combination of several transport modes only (e.g. air, rail, waterborne, coach, public transport, demand responsive transport, walking and cycling), while SMART H-LOOP will integrate also structural user profile data with information coming from their cultural models, emotional states, risk and social cohesion perception. A possible solution to achieve that will be through a ubiquitous and adaptive CPS based on integrated ecosystem of innovative technologies (IoT, Common Sense Reasoning and Data Fusion, Big Data analysis) capable of solving different problems in different environments where the human-machine interaction is essential. This will be used as sources of information for real-time transport decision-making, even in contingency plans, as well as for medium to long-term trip planning.

In this scenario, the SMART H-LOOP sustainable city-user MMTIPs will allow the traveller to receive personalised routing results according not only to their specific travel preferences or needs, including the fastest or cheapest route, the fewest number of connections, the most environmentally friendly or the most accessible option for persons with reduced mobility, but also their decision will be supported taking into account emotional states (e.g. anxiety and stress) or feelings (vulnerability or exposure to risk).

1.4 Vision

The vision for the proposed SMART H-LOOP platform is disruptive in the Internet paradigm. In fact, it aims to completely distribute the intelligence at every level of the CPS chain, expanding the concept of smart everything and centred on the human factor. In fact, the idea is to propose a system capable of understanding and reacting to different type of events. Thus, the approach consists of three different points of view:

1. The "*dT point of view*", capable of understanding isolated events occurred at a certain time instant and the consequent immediate and contingent reaction. Time is considered as a discrete variable.
2. The "*ΔT point of view*", capable of understanding both the cause and effects of events and their impact in undergoing situations. Situations are considered to take place in a time interval. This perspective deals with managing situations in an accurate and specialised manner.
3. The "*T domain*", in which time is considered in a broader way, capable of understanding trends, attitudes, and behaviours. This perspective aims at maintaining the system in a continuous upgrade, capable of automatically reconfigure its models to the contingent reality and environments.

The above reported points of view can be mapped in a three-layer based system. In the lowest layer, the *dT point of view*, decisions are taken in a short time by using very few and local data. Systems in this layer have a local visibility of a particular event. The layer above performs a *ΔT point of view*, and takes decisions considering multiple events in a wide, but still local, area. In the last layer, the *T domain*, a global view of the system is maintained. For instance, by considering the ITS use case in which vehicles are connected to the infrastructure, the *dT point of view* can be realized by smart sensors installed inside the car taking local decisions, e.g., a pedestrian is crossing the road outside the crosswalk, then the car stop and such event is sent towards an ITS base station covering a specific area. The same base station,

by receiving events from other cars, e.g., warning messages or simple position notifications, can decide to notify other cars to slow down or change direction. The messages collected from multiple base stations can be used in a urban mobility control centre, performing the *T domain*, for statistic related to safety and traffic flow, or traffic management strategies in particular areas of the city. In this three layer-based architecture, a fast and local reaction to events can be performed, e.g., inside the car, in a road covered by an ITS base station, thus distributing the logic of the system from the root, the urban mobility control centre, to the leaves, smart sensors deployed in roads and vehicles.

1.5 Main aims

Through the implementation of the SMART H-LOOP platform the global aims are to provide a holistic framework for Human-in-the-Loop Cyber-Physical System (HiLCPS) that addresses the systematization of the design, implementation, and quality assurance processes as well as the formalization of model and programming interfaces.

This framework is aimed to be devised and tested through the implementation of use cases where smart dynamic reconfiguration of services and service composition are essential capabilities. For example, extreme event management in smart cities and safety in smart transportations may be two use cases that could be develop and tested in the future. The idea behind these scenarios is to prompt the system to deal with unexpected situations and involve human beings on the basis of the system's available means (services and devices) and according to a systemic view on real world including the community, the groups and the personal experience.

The proposed approach will offer the possibility to establish an engineering process for the construction of instructed knowledge-based system to be demonstrated in hot areas such as smart cities and smart transportation, where the interactions between human beings and machines are very strong.

1.6 Scientific and technological objectives

The following five scientific and technological objectives have been identified:

1. To *investigate and define* a methodology to implement a systematic design, implementation, and quality assurance of HiLCPS. This methodology will be enhanced with a toolchain, pattern designs, and reusable human behavioural models to ensure a cost-effective and delimited-time-to-market construction process. The methodology will be developed according to an innovative and foundational interdisciplinary approach trying to achieve an individual

877

psychological comfort optimisation balanced with network optimisation, following a cognitive and emotional ergonomics conceptual framework integrated with methods from operations management and modelling.

2. To contribute to *standardisation* efforts on HiLCPS designs and protocols (networks and Human Machine Interfaces).

3. To *develop an advanced high level understanding platform*, here referred to as Distributed Reasoning Platform (DRP). This platform enhances with intelligence the server and gateway layer, by combining data fusion and common-sense reasoning techniques to model and semantically enrich IoT devices, their sensing values, and their programming interfaces. This is the core of the "ergonomic reaction", since it has the knowledge of both the real world event and human models.

4. To *define* a modular self-adaptive, self-reconfigurable architecture compliant with already established cooperation (e.g., *FI-WARE*). This architecture, guided by the DRP, will articulate the process how humans devise clever uses of things they have at hand to overcome the shortage of a certain item (e.g., switch your cell phone during a blackout to illuminate). This architecture along with the reasoning framework and the big data processing platform comprise the core of the cognitive framework enabling a smart context. Quantitative aspects such as easiness to integrate new devices in the system and align with other sources will be evaluated.

5. To *build an open-source* integrated set of individual, social and collective *human behaviour models* in the framework of the socio-technical systems. When studying complex real-life phenomena, they need to be represented as networks of activity systems. Human actions will be described following embodied cognition theory (Wilson, 2014) (Bagnara, 2015). The achievement of this objective will be evaluated over the generated models, their accuracy, completeness, consistency, clarity, and adaptability.

This platform will play a double role. Firstly, it will accept general requests and feedback from humans in an almost-natural language, such as "*illuminate a certain physical space*", which will be translated into specific services activation. Secondly, context events and sensed magnitudes will be modelled and combined with domain specific and common-sense knowledge in a fully distributed architecture. This combination, along with big data analysis (spatio-temporal pattern analysis and activity recognition), will support the understanding of the ongoing situation(s), the elaboration of behavioural models, data visualisation, etc. The target will be jointly considered with O2, since

the adaptation capability of the system is going to depend on its understanding of what it is happening. Also, quality metrics will be used to validate the output of the reasoning framework in terms of correct alignment and interpretation of diverse sensor inputs and context events, such as detection of humans, traffic elements, track continuity, etc.

2 THE PROPOSED PLATFORM

2.1 *The three-layered architecture*

The proposed **SMART H-LOOP** is a three-layered architecture where key elements are individually addressed, as depicted in Figure 1.

At the top level, the sensing and actuators layer is characterized by its heterogeneity and dynamism. Sensor and actuator devices need to be seamlessly integrated as knowledge producer or consumer. SMART H-LOOP will allow bringing reliable, effective, and personalised information to the end users through the implementation of web services that can be easily accessed from personal mobile devices as well as embedded actuators in the real context of use. Particularly, an extensive search will be carried out to look for: (i) the most appropriate sensors to represent as real as possible the traffic environment, and (ii) the most effective actuators for showing a warming signal, from mobile phones, to street panels, following the cognitive ergonomics guidelines. At the intermediate layer, based on an *intelligent gateway* (re-named as "*ser-way*" for its improved processing and understanding capabilities), the Distributed Reasoning Platform or DRP is capable of understanding the evolution of monitored events and immediately reacting to them (see section 2.3). The DRP is intended to identify, model, and collect the domain specific knowledge involved in the considered context, including that related to human behaviour, and combine it in real-time with

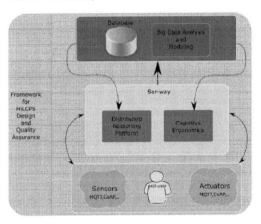

Figure 1. The proposed SMART H-LOOP framework layers architecture.

context events. Additionally, transversal knowledge, also known as common-sense knowledge, will be provided to support the understanding of context events. This knowledge is mainly devoted to describing and analysing the understanding tasks required as a primary need for both the decision making process and the direct ergonomic reaction exploiting the CPS actuators.

The upper layer, Big Data Analysis and Modelling layer, (see section 2.4) will enable the analysis and modelling of mental states or propositional attitudes of the system (beliefs, intentions, desires, hopes, etc.) exploiting the big data analysis on the data gathered from the sensors. Eventually, the system behaviour should be driven by these mental states since the functional requirements of HiLCPS cannot extensively captured, in advanced, as it happens with traditional software systems. Thus, Big Data Analysis and Modelling layer is thought to deeply understand behaviours, trends and attitudes typical of a contingent scenario and to feedback and adapt the models residents in the ser-way.

An example of how these layers will jointly work will be the following. Given traffic light sensors and induction loops, both geo-positioned sensors, it could be possible to detect cars jumping a traffic light (at the sensing and actuator layer) by combining the knowledge of the traffic light in red and the induction loop detecting a car passing by. However, it might be the case in which one car has jumped all traffic lights, in a row, of a certain avenue. To get to that conclusion and to achieve to that understanding, the intermediate layer (exploiting the DRP installed inside the ser-way) will model the jumping-light in a spatio-temporal manner. A causal explanation for those events can therefore be a car breaking the speed limit (distance-time between two traffic lights) and breaking traffic rules (hypothesis). From the big data analysis, these events match the behavioural pattern of a reckless high-speed driver. This hypothesis can be used to update the ser-way resident models improving the HiLCPS toward the nearby pedestrian accident prevention by such dangerous behaviours.

2.2 Smart sensors and actuators layer

SMART H-LOOP will adopt and extend the Internet of Things paradigm, by enabling the complete seamless interoperability among different networks (e.g., IEEE802.15.4/g, LoRa, etc.). Interoperability support is provided through the realisation of gateways (re-named as "ser-way" for its improved processing and understanding capabilities) and sensors nodes that embed one or more communication interfaces. Additionally, the exploitation of the RESTful paradigm will enable:

1. Sharing a common semantic model, not only among devices at this layer but also with upper layers. This supports the distributed processing and reaction implementation, thus holding the "dT point of view".
2. The dynamic re-configurability and re-programmability of sensors and actuators, thus exporting parameters, constants, and processing code as CoAP resources (Alessandrelli, 2013).

In this scenario, the development and deployment of smart and connected actuators is necessary to support the Situation Awareness (SA) (USDoT, 2003) and the cognitive and emotional ergonomics guideline described in the following subsection.

To better support SA applications, new class of embedded visual sensors (e.g. smart cameras), can be used for advanced applications. Smart cameras organized as IoT smart objects can be used to execute advanced low-complexity computer vision algorithms able to extract mobility related data (Salvadori 2012, 2014), to send images or video stream as a function of a detected event (Salvadori, 2013; Maggiani, 2015), and more in general to extract advanced information from the real world (Figure 2).

2.3 The "Ser-Way" a Distributed Reasoning Platform (DRP) layer

This is the core of the proposed architecture and what mainly distinguishes SMART H-LOOP from the other state-of-the-art proposals. In IoT systems, gateways usually perform a basic packet forwarding operation toward both, the nodes and/or the servers.

This layer is mainly intended to support the modelling, recognition, and understanding of ongoing activities from in the interpretation of sensor information and then, dictate based on the inference of that understanding with the SA and cognitive ergonomics guidelines. In this sense, two dimensions can be identified for the DRP: the understanding and acting. The *understanding* process is conceived here as a seek for *causal explanation*

Figure 2. Image processing of a camera tracking pedestrians.

to sensed information. According to Woodaward (2003) a causal explanation is *"any explanation that proceeds by showing how an outcome depends (where the dependence in question is not logical or conceptual) on other variables or factors counts as causal"*. For example, an explanation for a big crash noise and a sudden stop of the traffic flow can be the occurrence of a car accident. Explanations should be considered as preliminary hypothesis since there might be many different explanations catering for the same sensorial information, and it cannot be obviated that events are neither always clear nor precise due to sensor malfunctioning or low precision.

For that reason, a first stage will be intended to associate sensor measures (as captured effects of events) to *actions* that might have caused them.

2.4 Big data server layer

It is an important research activity to provide a high level understanding of both, the event causes, evolution and impact (e.g., degree of risk of the different sections of the road) and the human behavioural patterns that characterize the response of the citizens in such situations. Here, we hypothesize the role of citizen sensing in extracting relevant event and situational information, therefore proposing innovative algorithms based on the semantically enriched data gathered by DRP, historical registries, and user information (e.g., crowdsourced, smartphone, social networks).

SMART H-LOOP will integrate, in an advanced computing environment, different massively parallel-processing technologies which can acquire, store and elaborate a relevant and heterogeneous quantity of mobility, events, behavioural and contextual data. These data will draw from asynchronous user-generated information (e.g. mobility patterns), synchronous user-generated data (e.g. position in real time, crowdsourced information such incidents or traffic status), and historic databases (e.g. weather reports and trends).

2.5 Innovation components

We consider that the innovation at the three levels identified resides in the following components.

Perception of important elements is fundamental for forming a correct picture of the situation. Understanding is beyond simple perception; this level promotes an accurate current understanding of the environment through the integration of multiple pieces of information and a determination of their relevance to the person's goals. Finally, projection refers to the ability to foresee accurately future situation events and their dynamics.

SMART H-LOOP will implement a semantic composition approach based on the deep analysis capabilities, held at this level, about device, service, and system versatility. This analytical capability will be driven by a planning scheme that given a desired result (illuminate a room) it is going to look for a reconfiguration or composition setup through the whole space of available resources. To enable such a setup, a common semantic model should be shared among the different layers of the architecture.

We plan to use Scone[1] to model and hold explicit and implicit knowledge. Scone is also particularly well suited to manage possible worlds, through its implementation of the multiple-contexts mechanism (Santofimia, 2011).

On the functionalities themselves, the "brain" of the server, that is, the different algorithms, data analytics methods and techniques to be implemented in order to process the huge expected amount of data collected and infer valuable information regarding the identification and estimation of the evolution and impact of event and non-recurring incidents, accident prediction models and the relations between the citizen behaviour and all the possible influencing factors.

Specifically, long term data analysis will allow planners and managers to locate or reconfigure sensors and/or actuators especially for accident-prone "hotspots" or risk level assignation of a certain section of the road. On the other hand, our solution goes beyond current approaches, providing methods as predictive analytics aimed at discovering behavioural patterns and correlations and aggregation techniques, useful to evaluate the goodness of the DRP configuration and quantitatively guiding its parametrization.

3 SPECIFICATIONS

3.1 The FI-WARE compliance

FI-WARE constitutes an investment to promote the innovation of ICT sector thanks to European technologies; the architecture is comprised of a set of software pieces, all of them open source, known as Enablers, tackling different aspects of ICT technologies such as big data, IoT, security, cloud computing, etc. FI-WARE proposes a set of open standards defined through APIs which offer different technological capacities for developers. From a practical point of view, it is possible to define it as a set of open source software tools based on existing solutions and standards which have been designed to facilitate the creation of innovative ICT tools and services by developers.

Particularly, the functionalities defined in SMART H-LOOP have the ambition to be raised with the aim of creating a new model of smart city.

---

[1]http://www.cs.cmu.edu/sef/scone/

Figure 3. Transport System Catapult Virtual Reality Environment.

3.2 *The Virtual Reality environment*

SMART H-LOOP will offer a unique opportunity to study user reaction, interaction, and behaviour. It is intention of the authors to develop further the concept using and testing scenarios initially by implementation in a Virtual Reality (VR) experimental environment. Specifically, the use of VR will be across different use cases. However, before the implementation in a VR environment it is expected that an extensive use of simulations through which dynamic and iterative shifts from stepwise real-world implementations and VR experiments will be implemented. Such methodology will allow SMART H-LOOP to achieve an individual psychological comfort optimisation balanced with HiLCPS optimisation following the cognitive and emotional ergonomics conceptual framework integrated with methods from crisis management and smart transport operations management and modelling.

Moreover, the VR suite owned by the Transport Systems Catapult (UK) has been recently enhanced by integrating the existing Omnideck platform with a driving and a bike simulator. This will provide, for the first time, the opportunity to test simultaneous and synchronised simulation for the different VR suite users that will be able to interact to each other like in the real-world environment.

VR users can be equipped with smart sensors, including eye tracking, body sensors, etc., reproducing the different scenarios identified as the use cases for SMART H-LOOP.

4 DISCUSSION

The proposed platform originates from the needs of industrial stakeholders with common problems in relation to the lack of methods and tools supporting the process of building *HiLCPS*, highly distributed and connected digital technologies. This is because those companies seek common solutions through the implementation of predictive engineering methods and tools for reducing costs and time to market.

Additionally, the key market application proposed by SMART H-LOOP consists in a comprehensive framework for smart environments, provided as a FI-WARE enabler, in which services and devices can be automatically reconfigured and combined, if required, under special circumstances such as extreme event management, or device unavailability. The project will integrate a full CPS system from the sensor and actuators layer up to the server level in to demonstrate the feasibility of the approach and validate the methodologies and tools developed during the project. One ambition is to experiment the designed architecture at Technology Readiness Level (TRL) 5 through collaborative opportunities and funding streams such as H2020.

SMART H-LOOP plans to advance state-of-the-art software engineering by providing principles, methods, and tools that support the software lifecycle for systems such as HiLCPS that are mainly characterized by working under open world assumptions. This implies that all the interactions cannot be predefined nor prescribed, but the system is thought to be adaptive to different cultures and environments, exploiting the knowledge generated from Big Data analysis. Thus, the proposed methodology will aim at supporting the complex SMART H-LOOP socio-technical systems integration and adaption in a seamless way. The modelling and design process of such systems is co-evolutionary since hardware design (e.g., modularity, scalability), software design (e.g., programmability, real-time computing) and Interaction design (e.g., products and services) are carried out in parallel (divergence phase) and then integrated (convergence phase) in form of outputs.

By developing a newly emerging approach of embodied cognition to HiLCPS, SMART H-LOOP recognises that there is no clear-cut separation between body and cognition; and between body in action into the environment, and between the human being as a whole and the technologies she is going to adopt to transform its cognition and action. Such an action-perception loop resembles the humans-CPS circularity expanding further the notion of integrated and complex socio-technical systems as a whole.

SMART H-LOOP approach to ergonomics deals with this level of complexity, as one of its hallmarks is focusing on interactions among components, and on avoiding considering elements in isolation. Such analytic and design approach is systemic by nature; its focus being the study of systems (as opposed to individual elements), context, and complex interactions (in order not to isolate elements), holism, and emergence (in order to capture the various levels of explanation of one phenomena) (Wilson, 2014; Bagnara, 2015).

From the point of view of the physical environment, the proposed methodology plan to support the topological infrastructure modelling, deployment, and test. The challenge here is address the process of quality assurance, when interactions cannot be predefined nor prescribed. Approaches like Test-Driven Development (Beck, 2002) are based on writing small pieces of code specifically design to pass a test. Tests are intended to cover all system's functionalities. In this sense, this methodology that has gain credit over the last years, would fail to test the system reaction to unforeseen situations. Preparing a test for it means that the situation has already been foreseen. The novelty here consists in considering a mixed real-world and Virtual Reality (VR) experimental methodology that will consider the real SMART H-LOOP use context and the virtual world as a continuum.

Finally, the ambition behind the proposed SMART H-LOOP platform is to develop a methodology and tools enabling the design and the optimization of CPS in their integrity reinforcing Europe's know-how on a diversity of connected subject. The CPS in smart cities is envisaged as one of the case studies. The whole architecture will be based on IoT for the sensors and actuators which lead to the use of distributed intelligence and finally to analytics (BIG DATA) to cope with the data to be analysed. This data will be elaborated throughout the different layers of the CPS architecture. The aim is to be able to understand past events, monitor the present situation and predict future behaviours and ways of improving user lives. In order to involve individuals with these CPS, the SMART H-LOOP platform, through the use cognitive ergonomics will put the user at the centre of the CPS.

REFERENCES

Alessandrelli, D., Petracca, M. Pagano, P. 2013. 'T-RES: Enabling Reconfigurable in-network Processing in IoTbased WSNs', Distributed Computing Sensor Systems (DCOSS), 2013 IEE International Conference.

Bagnara, S, Pozzi, S. 2015. Embodied Cognition and Ergonomics. Journal of Ergonomics.

Beck. 2002. Test Driven Development: By Example. Addison-Wesley Longman Publishing Co., Inc., Boston, MA, USA.

Bocchino, S. et al. 2011. "SPEED routing protocol in 6 LoWPAN networks". 16th IEEE Conference on Emerging Technologies & Factory Automation (ETFA), pp.1–9, Sept. 2011.

He, H. et al. 2003. "SPEED: A Stateless Protocol for Real-Time Communicationin Sensor Networks". Proceedings of the International Conference on Distributed Computing Systems, pp. 56–55, May 2003.

IEEE, 2003. Computer Society, Wireless Medium Access Control (MAC) and Physical Layer (PHY) specifications for Low-Rate Wireless Personal Area Networks (LR-WPAN). The Institute of Electrical and Electronics Engineers.

Kushalnagar N., Montenegro G., & Schumacher C., 2007. IPv6 over Low-Power Wireless Personal Area Networks (6LoWPANs): Overview, Assumptions, Problem Statement, and Goals, RFC 4919.

LoRa. https://www.lora-alliance.org/What-Is-LoRa/Technology

Maggiani, C., Salvadori, C., Petracca, M., Madeo, S., Bocchino, S., Pagano, 2013 "Video Streaming Applications in Wireless Camera Networks: a change detection based approach targeted to 6LoWPAN" in Journal of Systems Architecture, vol. 59, no. 10, pp. 859–869, May 2013.

Mulligan, G. 2007. "The 6lowpan architecture". Proc. 4th IEEE Workshop on Embedded Networked Sensors (EmNets), pp. 78–82, June.

Nunes, D.S., Zhang, P., & Sá Silva, J. 2015. "A Survey on Human-in-the-Loop Applications Towards an Internet of All," in IEEE Communications Surveys & Tutorials, vol. 17, no. 2, pp. 944–965.

Salvadori, C, Petracca, M., Ghibaudi, M., Pagano, P. 2012 "On-board Image Processing in Wireless Multimedia Sensor Networks: a Parking Space Monitoring Solution for Intelligent Transportation Systems" in book Intelligent Sensor Networks: Across Sensing, Signal Processing, and Machine Learning", CRC Press, December 2012, pp. 245–266.

Salvadori, C., Petracca, M., Bocchino, S., Pelliccia, R., Pagano, P. 2014 "A Low-cost Vehicle Counter for Next Generation ITS", in Journal of Real-Time Image Processing, pp. 1–17, March 2014.

Salvadori, C., Petracca, M., Madeo, S., Bocchino, S., Pagano, P. 2013 "Video Streaming Applications in Wireless Camera Networks: a change detection based approach targeted to 6LoWPAN", in Journal of Systems Architecture, vol. 59, no. 10, pp. 859–869, May 2013.

Santofimia, M.J., Fahlman, S.E., Moya, F., & Lopez, J.C. 2011. Possible-world and multiplecontext semantics for common-sense action planning. In Ambient Intelligence and Smart Environments Ebook, Volume 14.

Shelby, Z., Hartke, K., Bormann, C., & Frank, B. 2013. "Constrained Application Protocol (CoAP)," IETF Draft Version 18.

USDoT, United States Department of Transportation. 2003. Bicycle and pedestrian detection, Minnesota Department of Transportation-Office of Traffic Engineering/ITS Section, SRF Consulting Group, Inc. (2003).

Wilson, J.R. 2014. Applied Ergonomics, Volume 45, Issue 1, January 2014, Pages 5–13.

Winter, T., Thubert, P., Brandt, A., Hui, J., Kelsey, R., Levis, P., Pister, K., Struik, R., Vasseur, J.P. & Alexander, R. 2012. RPL: IPv6 Routing Protocol for Low-Power and Lossy Networks. RFC 6550.

Woodward, J. 2003. "Making Things Happen: A Theory of Causal Explanation". Oxford University Press (2003).

Transport Infrastructure and Systems – Dell'Acqua & Wegman (Eds)
© *2017 Taylor & Francis Group, London, ISBN 978-1-138-03009-1*

Development, analysis and prediction of efficiency of bulk carriers in the world market

S. Galić & T. Stanivuk
Faculty of Maritime Studies, Split, Croatia

A. Marušić
Herzegovina University, Mostar, Bosna and Herzegovina

ABSTRACT: Today, a significant part of the world trade is performed by sea, of which a large share refers to bulk cargo. Bulk cargo shipping industry makes an essential part of the international shipping with ocean-going vessels that are the most effective and sometimes the only way to transport large quantities of bulk commodities. Transportation of bulk cargo by sea can be defined as transport of homogeneous cargo with vessels following irregular schedules of sailing. This paper provides the statistical analysis of bulk carrier vessels and further forecast of supply and demand of bulk cargo by type. The analysis also includes the key factors that encourage or discourage the growth of bulk cargo industry. One of them is the fulfilment of the new prescribed rigorous regulations on reduced discharge of greenhouse gases from these types of vessels, because the shipping industry is currently the only industrial sector which is already covered by a legally binding global agreement on the reduction of CO_2 emissions through technical and operational measures adopted by the International Maritime Organization (IMO).

1 THE FLEET OF THE SHIPS WORLDWIDE AND THE SEABORNE TRADE

At present, the global number of merchant ships exceeds the figure of 80,000. Research results indicate that in the period of 12 months the world's merchant fleet increased by 3.5% until 1 January 2015. That is the lowest annual growth rate in more than a decade (UNCTAD, 2015). At the beginning of 2015, the world's commercial fleet consisted of 89,464 vessels with a total tonnage of 1.75 billion DWT. The largest share in the global fleet is dry bulk that reached a share of 43.5% of the total seaborne shipping capacity at the beginning of 2015. The result is an increase of 4.4% between 2014 and 2015 and even greater expansion from 2010 to 2013. Bulk carriers are making one third of the world fleet, and are therefore the most common type of ships (Table 1). According to research results (Equasis, 2014) the total percentage of vessels by type and size are:

- Small and medium sized ships: tugs (19.6%), general cargo ships (19.1%), oil and chemical tankers (14.5%) and bulk carriers (12.9%) are the most common ship types by number, making about two thirds of the global fleet.
- Large and very large ships: bulk carriers (30.7%), oil and chemical tankers (29.5%) and container ships (22.9%) amount to about 85% of the fleet.

In the category of large and very large ships, bulk carriers (30.7%), oil and chemical tankers (29.5%) and container ships (22.9%) make about 85% of the fleet.

In terms of Gross Tonnage (GT), the large and very large size ships represent 80% of the fleet (Table 2), with oil and chemical tankers, bulk carriers and container ships dominating both categories at 86.1% (large) and 83.1% (very large).

The average age of the world fleet increased slightly according to data from 2014. Taking into consideration the delivery of several new buildings combined with reduced scrapping activity, higher tonnage cannot compensate for the natural cycle of aging fleets.

The world economy embarked on a slow-moving recovery led by uneven growth in developed economies and a slowdown in developing countries and economies in transition. In 2014, the world's Gross Domestic Product (GDP) increased by 2.5% compared to 2.4% in 2013. Meanwhile, the world trade increased by 2.3%, which was below the pre-crisis levels. According to the analyses and statistics, the global maritime shipments increased by 3.4% in 2014, which is the same growth rate as in 2013. This situation may be the result of several factors:

- a slowdown in development of major emerging economies;
- the price level of oil and the development of new refinery capacity; and
- uneven recovery in developed economies.

Table 1. Total number of ships in the world by size and type.

Ship type	Small (GT < 500)		Medium (500GT < 25000)		Large (25000GT < 60000)		Very Large (GT ≥ 60000)		Total	
General Cargo Ships	4356	13.90%	11650	30.90%	212	1.90%			16,218	19.10%
Specialized Cargo Ships	8	0.00%	201	0.50%	56	0.50%	2	0.00%	267	0.30%
Container Ships	17	0.10%	2255	6.00%	1619	14.80%	1193	22.90%	5084	6.00%
Ro-Ro Cargo Ships	30	0.10%	653	1.70%	619	5.70%	180	3.50%	1482	1.70%
Bulk Carriers	320	1.00%	3700	9.80%	5374	49.20%	1602	30.70%	10,996	12.90%
Oil and Chemical Tankers	1815	5.80%	6597	17.50%	2414	22.10%	1537	29.50%	12,363	14.50%
Gas Tankers	39	0.10%	1070	2.80%	216	2.00%	378	7.30%	1703	2.00%
Other Tankers	315	1.00%	531	1.40%	5	0.00%			851	1.00%
Passenger Ships	3657	11.70%	2528	6.70%	271	2.50%	156	3.00%	6612	7.80%
Offshore vessels	2531	8.10%	5227	13.90%	115	1.10%	157	3.00%	8030	9.40%
Service Ships	2405	7.70%	2361	6.30%	23	0.20%	6	0.10%	4795	5.60%
Tugs	15,747	50.40%	946	2.50%					16,693	19.60%
TOTAL	31,240	100%	37,719	100%	10,924	100%	5211	100%	85,094	100%

*Source: Equasis statistics—world merchant fleet in 2014.

Table 2. Total number of ships in the world according to gross tonnage.

Ship type	Small (GT < 500)		Medium (500GT < 25000)		Large (25000GT < 60000)		Very Large (GT ≥ 60000)		Total	
General Cargo Ships	1455	17.60%	49,812	22.80%	6942	1.70%			58,209	5.00%
Specialized Cargo Ships	2	0.00%	1596	0.70%	2107	0.50%	153	0.00%	3858	0.30%
Container Ships	7	0.10%	26,425	12.10%	62,925	15.20%	116,771	22.20%	206,128	17.70%
Ro-Ro Cargo Ships	11	0.10%	6306	2.90%	29320	7.10%	11696	2.20%	47,333	4.10%
Bulk Carriers	125	1.50%	54,518	25.00%	19,8021	47.90%	157,251	29.90%	409,915	35.10%
Oil and Chemical Tankers	586	7.10%	39,595	18.10%	88,677	21.50%	168,038	31.90%	296,896	25.50%
Gas Tankers	15	0.20%	6349	2.90%	9477	2.30%	40,813	7.80%	56,654	4.90%
Other Tankers	94	1.10%	1350	0.60%	162	0.00%			1606	0.10%
Passenger Ships	922	11.10%	10,579	4.80%	9649	2.30%	15,397	2.90%	3,6547	3.10%
Offshore vessels	719	8.70%	13,334	6.10%	5258	1.30%	15,374	2.90%	34,685	3.00%
Service Ships	595	7.20%	7510	3.40%	850	0.20%	992	0.20%	9947	0.90%
Tugs	3750	45.30%	931	0.40%					4681	0.40%
TOTAL	8281	100%	218,305	100%	413,388	100%	526,485	100%	1,166,459	100%

* Source: Equasis statistics—world merchant fleet in 2014.

Table 3 Ships world fleet (%)—according to the size of the ships and gross tonnage.

Total number of ship's world fleet by number

Very large	6%
Large	13%
Medium	44%
Small	37%

Total number of ship's world fleet by gross tonnage

Very large	45%
Large	35%
Medium	19%
Small	1%

* Source: Equasis statistics—world merchant fleet in 2014.

Table 4. Seaborne trade worldwide in 2014.

Crude oil	17%
Container cargo	15%
Minor bulks	15%
Iron ore	13%
Coal	12%
Other dry cargo	9%
Petroleum products	9%
Gas and Chemicals	6%
Grain	4%

* Source: United Nations Conference on Trade and Development—UNCTAD: Review of maritime transport 2015.

It is expected that the growth of world GDP and seaborne trade will continue to grow at a moderate pace over the next few years. However, odds remain uncertain and are subject to many risks, including the continuation of moderate growth in global demand and trade, the fragile recovery in Europe, geopolitical tensions and potential faster slowdown of the developing economies.

China achieved tremendous growth in the world seaborne trade in 2009. The challenge for shipping is to ensure that the dynamics of trade continues to be generated through the expansion of China and to replicate elsewhere. Therefore, the greater integration of China is the key driver in global production.

2 DRY CARGO TRADE

The demand for imports in developing countries, particularly China and India, remains the key driver of bulk carrier cargo growth in 2016. During 2014, the increase in world seaborne dry bulk shipments was estimated at 5%, and that was lower than the previous four years (Dry Bulk Trade Outlook, 2015). The growth was driven by strong expansion in iron ore trade (12.4%), which accounted for about 30% of the entire bulk cargo and reached 1.34 billion tons. In contrast, the transport of coal was estimated to be increased by 2.8%, which was significantly slower than in 2012 (UNCTAD, 2015).

Exports of dry bulk cargo such as bauxite, nickel ore, iron ore and coal were constrained by, among other factors, restrictions on exports, weather patterns, bans on mining activities, regulatory measures and policies seeking to promote national producers and industries.

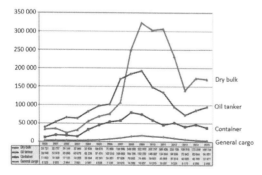

Figure 1. World tonnage order for propelled seagoing merchant vessels of 100 GT and above (from 2000 to 2015).
* Source: United Nations Conference on Trade and Development UNCTAD based on data supplied by Clarckson Research.

According to the statistics from 2014, the world merchant fleet grew up to 42 million GT. This was the result of the new construction of 64 million GT and 22 million GT ships that went for the scrap.

However more than 91% of GT delivered in 2014 was built in just three countries: China (35.9%), South Korea (34.4%) and Japan (21.0%), where China delivered dry bulk carriers followed by container ships and tankers, while Japan delivered mostly bulk carriers (UNCTAD, 2015).

3 BULK CARRIER FREIGHT RATES AND MARITIME TRANSPORT

The Capesize market has a cyclical nature. Such market circulates from depression into expansion, depending on a number of economic and political variables. This kind of market is very complex and reacts easily to various variables. It is not easy to predict a shipping cycle due to irregularities. Even the slightest fluctuation in freight rates can have a major implication on profits. The ability to accurately predict the Capesize market has potential advantage of allowing ship owners to better plan how to maximize profits or minimize losses or risks incurred on the shipping market. It is not possible to know how many years the market will be in expansion or recession. The cycles can last for several years. However, every cycle obviously depends on the law of supply and demand. If demand is strong and supply is low, then the market will be in expansion, but if the situation is reversed, then the market will be in depression.

According to the data of Baltic and International Maritime Council—BIMCO from 2016, on 10 February, the Baltic Dry Index (BDI) reached 290. At that moment, one bulk carrier, regardless of its size, age and fuel-efficient qualities earned a time charter average between 2417 to 2776 USD per day.

After that period, in the next few months the earnings for Capesize vessels were very small, but in August and September they achieved greater profits. Despite the fact that the earnings of Capesize vessels doubled in the next two months, these gains remained below the OPEX level for the largest part of the fleet.

During 2015, China exported 112 million tons of crude steel, which led to a decrease in the price of scrap steel. In January and February 2016, the export of steel from China decreased by 1.6% (17.85 million tons). In March 2016 exports amounted to 10 million tons (BIMCO, 2016).

According to data of Worldsteel Association, global crude steel production for January and February 2016 was 5.6% lower than in the same period of 2015, but at that time the production of

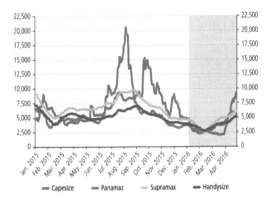

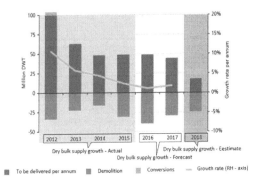

Figure 2. Baltic Exchange time charter averages USD/day (from 2015 to 2016).
*Source: BIMCO, Baltic Exchange, Clarksons Research.

Figure 3. The supply growth for 2016–2018 contains the existing orders only and is estimated under the assumptions that the scheduled deliveries would falls short by 10% due to various reasons and that 50% of the remaining ordered vessels are delayed/postponed.
* Source: BIMCO.

crude steel in China decreased by 6.5%. There are three key main elements to watch out for in 2016, including the amount of the dry bulk tonnage to be demolished.

Over the first two months in 2016, imports of iron ore to China increased by 6.4% to 155.8 million tons compared to the same period in 2015. On the other hand, imports of coal in China decreased by 10% to 28.8 million tons compared to the same period in 2015.

In March 2016, imports of Chinese iron ore amounted to 85.8 million tons, while imports of coal increased to 19.7 million tonnes. Strong imports of coal in March equalized the decline of coal in January and February.

In the first quarter of 2016, import of coal was smaller by 1.2% on an annual basis (BIMCO 2016).

3.1 Efficiency and opportunities for bulk carrier fleet on maritime market

Over the first three months of 2016, the number of demolished dry bulk shipping capacity reached a very high number, but regardless of that, the overall total dry bulk fleet continued to increase. Specifically the 16.7 million DWT of new capacity entered the fleet, and 14 million DWT went to scrap.

However, not all dry bulk sub-segments grew evenly. For example, the Capesize fleet, which has doubled in the last 6.5 years, was reduced to 7 vessels as well as the capacity which decreased by 0.2% in the first quarter of 2016.

According to research results (BIMCO 2016), a growth of the fleet by 1.1% or 10 million DWT is expected in 2016. Shareholders and investors work hard to postpone the delivery of new ships

in the freight market. By mid-March, there were only four new contracts signed, three in Japan and one in China. During March and April, the long-awaited orders for 30 VLOCs with a capacity of 400,000 DWT were confirmed. In March 2016, China COSCO Shipping Corporation and Vale signed a 27-year-long agreement that will allow the Chinese corporation to carry 16 million tonnes of iron ore per year for the Brazilian multinational corporation (BIMCO, 2016).

It is estimated that each VLOC can handle 1.6 million tons of iron ore from Tubarão (Brazil) to Baoshan (China) per year. This new series of VLOC carriers will open the market of 48 million tons of iron ore, and that is rather bad news for the international owners and operators. The delivery of 30 new VLOC ships has been ordered for the period 2018–2019 (BIMCO, 2016). In 2015, Brazil exported 191.6 million tons of iron ore to China; the existing and the new valemax ships (VLOC) will be able to carry more than half of Brazil's current annual exports of iron ore.

The export of grains and soybeans volume from Argentina and Brazil is expected to increase in 2016. According to research results (BIMCO 2016), there will be no massive increase in freight for handysize, supramax and panamax ships, because these ships are already waiting around the main loading areas. Another positive indicator for China is the rise of domestic steel prices from mid-February 2016. In mid-April 2016, iron ore rose from 41 USD per MT to 56 USD per MT on the international market, which is also positive indicator. However, the concern for the sustainability of freight rates in the years to come remains, because there is a fear that the activity of destroying ships could slow down if the BDI starts to grow.

If ship owners significantly slow down the destruction of vessels, the fleet of ships will inevitably increase and this will lead to a fundamental imbalance, because the market analyses predicted that the demand would grow slowly in the coming years. In order to change the trends in fleet enlargement, a trend of negative growth in the fleet on multi-year basis should be developed. It is estimated that the current utilization rate of dry bulk fleet approximates the state as it existed at the end of the 1970s (BIMCO, 2016).

What can be changed regarding the import/export are the coal imports into India, because India is planning to increase domestic production of coal and to completely stop importing coal in the next two to three years. India imported 176 million tons of thermal coal according to data from 2014. In 2015, this country imported 171 million tons of thermal coal. According to the analyses (Simpson Spence Young's Dry Cargo brokers—SSY), it is expected that India will import about 170 million thermal coal in 2016. It can also be expected that the domestic production of coal in India will continue to be high (Drewry Maritime Research 2016).

It seems that the ship owners will seek to recover their costs by reducing the proportion of the fleet so the market may return to a profitable situation in 2017 (Drewry Dry Bulk Forecaster research). However, the recovery of market to such an extent, where the ship owners will start earning profits, shall remain elusive for another year.

Key drivers for the shipping fleet will remain the steel industry and imports of iron ore and steam coal to China. In addition to iron ore, import of steam coal to China will ensure future employment of the dry bulk tonnage.

Increase in trade and contracting of supply will support the recovery of charter rates on major dry bulk shipping routes. There is a possibility that China will import more coal and iron ore (Drewry Dry Bulk Forecaster research). Macroeconomic policy decisions of the Chinese authorities had an impact on the overall dry bulk market increasing the demand for ships on major routes, which resulted in the recovery of charter rates. China's effort to revive its economy has helped to revive the trade of iron ore, so the further increase in ton/mile demand could be expected where most of the cargo is coming from Brazil.

Stricter conditions of financing options and restraint in new orders have helped the dry bulk market to keep supply under control so far. A high number of slippages and cancellations will impede any substantial growth in deliveries over the next few years. Demolition of old vessels is expected to result in a slowdown in growth and the fleet size in 2016.

Moreover, a similar rate of demolition and delivery in the next year will further help to deflate the oversupply.

The growth of demand, which did not seem to be likely to occur during the last year and early part of 2016, has helped the market dry bulk cargo that is slowly recovering.

It is expected that in 2016, the domestic production of coal in India will continue to be high.

If La Nina forms it would bring above average winter spring rainfall over central and Western Australia which may affect the transport of cargo and disrupt rail traffic and port activities. Such an event could result in increased slowdown at ports in the country, which will be reflected in the available supply of coal from other sources.

Freight rates for iron ore routes to Brazil-China and Australia-China for Capesize vessels will strengthen in the next two quarters, but the return of laid-up ships to transport cargo could disrupt the expected progress according to analyses (Drewry Dry Bulk Forecaster research) for August 2016.

Demand for coal-carrying vessels will increase in the coming quarters because Japan has decided to increase its coal-fired power generation, while China plans to cut domestic coal production which will increase import demand. Also, China plans to reduce domestic production of coal, which will increase the demand. It is expected that the grain trade will remain strong as recent heavy rains have enhanced the prospects for grains and Europe, CIS and North America, especially for maize and barley crops.

4 ENVIRONMENTAL CHALLENGES FOR BULK CARRIERS AND OTHER SHIPS

4.1 Implementation of the existing technologies

According to the 3rd International Maritime Organization—IMO greenhouse gas GHG study, Maritime transport emits about 1,000 million tons of CO_2 per year, and is responsible for about 2.5% of global greenhouse gas emissions. The latter is expected to reach around 5% by 2050, and this is not in accordance with the internationally agreed objective of maintaining global temperature rise below 2°C.

The international agreements require that global emissions of gases into the atmosphere should be at least halved by 2050. Although seaborne shipping, in most cases, is more fuel-efficient than other types of transports, greenhouse gas emissions from ships grow more every day and, unless adequate measures are taken, greenhouse gas emissions from ships will increase more than double by 2050.

This will happen because of the expected growth of the world economy and the associated demand for maritime transport.

Nowadays in the seaborne shipping there is a growing pressure in reducing greenhouse gas emissions and the industry asks for bulk carrier manufacturers to respond. With this in mind, statistics from 2013 predict that the new building bulk carriers will reduce the average carbon dioxide emissions by 40% by 2040 (Fig. 4).

According to 2nd IMO GHG Study, ship energy consumption and CO_2 emissions could be reduced up to 75% by applying operational measures and implementing the existing technologies (Fig. 5). There is a great potential in reducing emissions from ships through fuel-saving techniques provided by modern technologies (Table 5). This could be achieved with relatively little cost. The implementation of such techniques and measures would also significantly reduce ship running costs. However, not all of these solutions can be applied for every type of the vessel. Studies show that there is potential for reducing fuel consumption in the shipping sector up to 55%. This result is excellent compared to other forms of transport and economic sectors. Further reduction of emissions from ships can be achieved by introducing new innovative technologies.

Eco-ships have been considered as a solution to the dry bulk market for some time. However, as the research results show, not all stakeholders in the maritime shipping industry share the view that the eco-ships are worth paying attention to, and that they may affect the general situation. The benefits accruing from eco-efficient vessels on freight rates seem to be marginal with 43% of the respondents saying these ships have no significant impact at the moment. This can partly be attributed to the current low bunker prices that limit the fuel savings that modern tonnage gives (Platts, 2015).

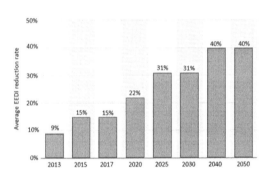

Figure 4. Statistic of estimated CO_2 emission reduction rate for new builds of bulk carriers between 2013 and 2050.
*Source: http://www.statista.com/statistics/216058/world wide-bulk-carrier-co2-emission-reduction-rate/.

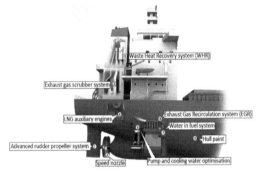

Figure 5. Existing technologies that can be implemented in the low emission concept—study for the 8,500 TEU container carrier and the 35,000 DWT bulk carrier.
*Source: http://www.greenship.org/lowemissionconcept study/lowemissionbulkcarrierstudy/technologiesused/.

Table 5. The most effective existing technical and operational measures to reduce CO_2 emissions from ships.

SOLUTION	RELATIVE CO_2 SAVINGS	SAVINGS/COSTS PER TONNE CO_2	TAKE-UP 2007–2011	
Speed reduction	17–34%	−280€/t	0%	50%
Propeller and rudder upgrade	3–4%	−150€/t	0%	0%
Hull coating	2–5%	−280€/t	0%	50%
Waste heat recovery	2–6%	+60€/t	0%	0%
Optimisation of trim and ballast	1–3%	−200€/t	0%	50%
Propeller polishing	1–3%	−280€/t	75%	75%
Hull cleaning	1–5%	−200€/t	75%	75%
Main engine tuning	1–3%	−250€/t	75%	75%
Autopilot upgrade	1–1.5%	−280€/t	75%	75%
Weather routing	1–4%	−280€/t	75%	75%

*Source: http://ec.europa.eu/clima/policies/transport/shipping/index_en.htm.

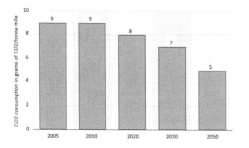

Figure 6. The statistic of current and estimated future transport efficiency improvement associated with bulk carriers between 2005 and 2050, by level of CO_2 used per tonne/mile.
*Source: http://www.statista.com/statistics/216033/tran sport-efficiency-improvement-associated-with-bulk-carriers/.

4.2 Current situation

Significant efforts to reduce emissions from ships have appeared through the IMO and the United Nations Framework Convention on Climate Change (UNFCCC) over the recent years. After years of international discussions, the IMO's Marine Environment Protection Committee set some important milestones including the adoption of the Energy Efficiency Design Index (EEDI) and the Ship Energy Efficiency Management Plan (SEEMP). Unfortunately, these measures alone cannot lead to an absolute reduction of harmful emissions from the ships. However, international discussions have yet to bring agreement on global market-based measures that would cut emissions from the sector as a whole, including the existing ships.

Table 6. SWOT analysis of supply and demand for bulk carriers and environmental challenges in bulk cargo shipping industry.

STRENGTH	WEAKNESSES	POSSIBILITIES	THREATS
The ability to accurately predict the market has potential advantage of allowing ship owners to better plan how to maximize profits or minimize losses or risks incurred in the shipping market.	The Capesize market has a cyclical nature. Such market circulates from depression into expansion, depending on a number of economic and political variables.	The ability to accurately predict the Capesize market has potential advantage of allowing ship owners to better plan how to maximize profits or minimize losses or risks incurred on the shipping market.	It is not easy to predict a shipping cycle because it is irregular. Even the slightest fluctuation in freight rates can have a major implication on profits.
It is expected that the world GDP and seaborne trade will continue to grow at a moderate pace over the next few years.	The recovery of the market to the point that ship owners will start earning profits will remain elusive for some time.	Increase in trade and supply will support the recovery of charter rates on major dry bulk shipping routes.	If ship owners significantly slow down the destruction of ships, the fleet will increase and this may lead to a fundamental imbalance.
Today, seaborne shipping is, in most cases, more fuel-efficient than other types of transport.	According to the 3rd IMO GHG study, Maritime transport emits about 1,000 million tons of CO_2 per year, and is responsible for about 2.5% of global greenhouse gas emissions.	According to 2nd IMO GHG Study, ship energy consumption and CO_2 emissions could be reduced by up to 75% through the application of operational measures and implementation of the existing technologies.	Greenhouse gas emissions from ships are getting larger and, unless appropriate measures are taken, greenhouse gas emissions from ships may increase more than double by 2050.
There is a great potential in reducing emissions from ships through fuel-saving techniques ensured by modern technologies.	Not all of fuel-saving techniques can be applied for every type of the vessel.	Studies show that there is potential for reducing fuel consumption in the seaborne shipping sector up to 55%.	The benefits accruing from eco-efficient vessels on freight rates seem to be marginal with 43% of respondents who say that these ships have no significant impact at the moment. This can partly be attributed to the current low bunker prices that limit the fuel savings that modern tonnage gives.
New ideas on technical and operational measures that are based on a gradual approach to monitoring, reporting and verification of emissions from ships.	According to the 3rd IMO GHG study, maritime transport emits about 1,000 million tons of CO_2 per year, and is responsible for about 2.5% of global greenhouse gas emissions.	Reliable information on the effectiveness of technologies to improve energy efficiency reduces the financial risk to investors.	Requirements imply that global greenhouse gas emissions should be at least halved by 2050.

* Source: author.

New ideas on technical and operational measures have been proposed by the European Commission and the IMO recently. The measures are based on a gradual approach to Monitoring, Reporting and Verification of Emissions (MRV) as the first step of further measures of efficiency for the existing ships.

The MRV regulation was adopted in 2015. It creates a EU-wide legal framework for monitoring, reporting and verification of CO_2 emissions from maritime transport. The regulation requires all ships over 5000GT, calling at EU ports from 1 January 2018, to collect and eventually publish verified annual data on CO_2 emissions and other relevant information. Statistics show that the transport efficiency measures will reduce emissions by seven grams of CO_2 per tonne/mile in 2030 (Figure 6).

5 CONCLUSIONS

Every year the number of merchant ships in the world increases, and the largest share of the global fleet belongs to dry bulk tonnage. Taking into consideration the delivery of several new buildings combined with reduced scrapping activity, higher tonnage cannot compensate for the natural cycle of aging fleets. It is expected that the growth of world GDP and seaborne trade will continue to grow at a moderate pace over the next few years. However, odds remain uncertain and are subject to many risks, including the continuation of moderate growth of global demand and merchandise trade, the fragile recovery in Europe, geopolitical tensions and the potential deceleration of economies in developing countries.

Rebalancing of the Chinese economy can significantly reshape maritime transport, changing the overall image of shipping and maritime trade. Import of steam coal to China will be the future challenge for dry bulk transport. The reduction of coal import to India can also affect the dry bulk transport, because India is planning to increase domestic production of coal and entirely cease importing coal in the next two to three years. The key drivers for the bulk fleet will remain the steel industry and imports of iron ore and steam coal to China.

These days there is a growing pressure to reduce greenhouse gas emissions from ships and the seaborne shipping industry is asking manufacturers of bulk carriers to respond to the situation. With that in mind, eco-ships have been considered as a solution to the dry bulk market for quite some time. Analyses predict that future new buildings will be able to reduce carbon dioxide emissions by 40% by 2040. Today, there is a great potential in reducing carbon dioxide emissions from ships by

applying the saving fuel techniques through the existing technologies. However, such solutions cannot be applied to all types of vessels, but studies show that there is potential for reducing fuel consumption in the maritime shipping sector up to 55%, while further reduction of emissions from ships can be achieved through implementation of new innovative technologies.

The stricter environmental regulations as well as low freight and charter rates should encourage further demolition of older ships. Also, these economic and regulatory initiatives would contribute to environmental protection and help to reduce overcapacity of the maritime shipping market.

Reliable information on the effectiveness of technologies to improve energy efficiency reduces the financial risk to investors. However, shipping companies are likely to take measures to improve the energy efficiency of their vessels if they are based on accurate and reliable information, as such information on the effectiveness of technologies to improve energy efficiency reduces the financial risk for investors.

REFERENCES

Ančić, I. 2015—*Influence of the required EEDI reduction factor on the CO2 emission from bulk carriers*, Energy Policy, article.

Baltic and International Maritime Council—BIMCO *https://www.bimco.org/*

Clarkson Research https://sin.clarksons.net/

Drewry Dry Bulk Forecaster research 2016 *http://www.drewry.co.uk/publications/view_publication.php?id=314*

Equasis and European Maritime Safety Agency EMSA Statistics. 2014—*The world merchant fleet in 2014*, study.

European Commission http://ec.europa.eu/clima/policies/transport/shipping/documentation_en.htm

Green Ship of the Future http://www.greenship.org

International Maritime Organization IMO. 2009 — *Second IMO GHG Study Executive Summary and Final Report*, London.

International Maritime Organization IMO. 2014 — *Third IMO GHG Study Executive Summary and Final Report*, London.

International Maritime Organization IMO *http://www.imo.org/en/MediaCentre/PressBriefings/Pages/01-2016-MTCC-.aspx*

Kyong, H. 2013 — *Forecasting the capesize freight market*, World Maritime University, Sweden, dissertation.

Platts-McGraw Hill financial. 2015: *Special Report Dry Bulk Market Survey 2015*, study.

Simpson Spence Young's Dry Cargo brokers — SSY *https://www.ssyonline.com/services/dry-cargo.*

Statista *http://www.statista.com*

United Nations Conference on Trade and Development UNCTAD: *Review of maritime transport 2015*, United Nations publication, ISSN 0566-7682.

Worldsteel Association *https://www.worldsteel.org/*

Transport Infrastructure and Systems – Dell'Acqua & Wegman (Eds)
© *2017 Taylor & Francis Group, London, ISBN 978-1-138-03009-1*

Enhanced financial analysis to evaluate mass transit proposals in terms of contribution on resilience increasing of urban systems

A. Spinosa
Cityrailways Engineering, Rome, Italy

ABSTRACT: The prevailing model of development of urban areas is increasingly oriented to a territorial isotropy, notwithstanding the evidences of a considerable literature. This is due to the absence of mathematical models able to translate this awareness into appropriate financial models. Nowadays a city no longer exists as mere physical built place, but as a dematerialized "cloud" of flows: people, goods, information. The vicious circle is all here: the car has allowed the urban sprawl; mass transit does not allow the widespread use of a city. This is the time of urban sustainability. If each urban element (home-work-facilities) is framed dynamically within the daily circadian rhythm that connects them, it will be possible to develop a sustainable solution that is not partialized on a specific sector. A balanced city produces social wealth: that turns into a financial one because it can be measured. The paper presents a methodology that combines the most modern epidemiological knowledge and the energy aspects of mobility. The Enhanced Financial Analysis (EFA) approach considers the Public authority as investor and manager of a planned infrastructure. Revenues come from fare and savings on current health expenditure. The model explores the morbidity of some peculiar diseases of which there is an identified cause-and-effect pollution (direct and indirect) and air quality. The avoided health expenditure is measured by health services that will not be paid out in the medium and long term compared with the do-nothing scenario. EFA allows introjecting some aspects that usually are relegated to the economic analysis, but are part of the current spending of the funder. This realizes a comprehensive financial analysis of the possible alternative technologies for a mass transit corridor, correcting the bias that may arise on the electrical transport—even on the rail transport—compared with apparently less expensive choices.

1 INTRODUCTION

The objective of social and environmental sustainability requires to look at the costs of the processes that animate the city every day from a broader perspective. If until now we have focused on the sustainability of housing on one side and on the reduction of transport emissions on the other, it is time to approach the theme of mobility from a joint perspective.

It is the time of urban sustainability. A city is a set of buildings but also a complex network of relationships in which homes, offices and transport are only the most visible part: 60% of CO_2 emissions is not in the car that produces it, but it is dematerialized in the choices that are upstream of the decision to take that car.

If there is a territorial sustainability there is also an effective indicator of risk in case of unsustainability? The answer is in the affirmative. Territorial sustainability is not only a virtuous exercise but an investment that a territory does in resources (human and environmental), productivity (how these resources become or generate usable wells) and stability (preserving these resources and their productivity over time). Summarizing in a word, it means increasing the resilience of a territory.

Although this is the information age, it does not mean that the displacements are decreasing. The cities are being renovated but the per capita movement of people and goods will increase by 30% at 2050 (The World Bank). This means that the mass transport will be essential for the sustainability of cities.

2 MOTIVE OF THE PAPER

This is the time of urban sustainability. If each urban element (home-work-facilities) is framed dynamically within the daily circadian rhythm that connects them, it will be possible to develop a sustainable solution that is not partialized on a specific sector. A balanced city produces social wealth: that turns into a financial one because it can be measured. The paper presents a methodology that combines the most modern epidemiological knowledge and the energy aspects of mobility to support administrators and designers in the choice of the most effective technology for a mass transport project.

3 METHODS

For each technology, the operation and maintenance costs are derived from the figures reported by 122 European operators for the period 2004–2014.

Unit prices are derived from the actual selling of orders as of the value communicated by the contracting authorities.

The figures related to energy costs are homogenized assuming as reference the Italian value (=100).

Similarly, the figures related to labor and maintenance costs are homogenized taking as reference the Italian value (=100). The reference indices for each country are provided by Eurostat.

The health care cost is assessed in terms of health services provided (i.e. avoided) on the useful life of the project.

All evaluations refer to the maximum daily demand i.e. the average weekday, identified as the day in which an urban transit system set the maximum trip demand. Weekdays equivalent per year (Istat) are derived from the ratio between the measured average displacements:

The daily average weekday is obtained by the annual demand dividing by 290 and vice versa.

To estimate the load in a particular day (e.g. summer, holiday or weekend) is necessary to apply the reduction factors shown in the second column of Table 1 (Istat*).

4 SHORT-TERM EXPOSURE TO AMBIENT FINE PARTICULATE MATTER

A large number of epidemiological studies have shown that short-term exposure to ambient Particulate Matter (PM) in outdoor air is associated with total and cause-specific mortality. This led to outdoor air pollution standards for PM10 in the European Union for daily and yearly averaging times. PM10 was seen as the most health relevant PM size fraction and is therefore measured on a regular basis in specified locations in all EU countries. Evidence is increasing, mainly based on studies from the US, that the smaller PM2.5 fraction is more consistently associated with health outcomes, and that this fraction is more specific for anthropogenic combustion-related emissions. EU registration therefore decided to put, following regulation in the US, a PM2.5 standard value in force starting 2015 with an anticipated limit value of 25 µg/m³ (yearly average). Contrary to the large amount of US studies on health effects of PM2.5, European studies on PM2.5 are sparse and studies in the literature on mortality effects of PM2.5 could only be identified for 10 European cities (Anderson et al., 2001, Atkinson et al., 2010, Branis et al., 2010, Garrett and Casimiro, 2011, Halonen et al., 2009, Mate et al., 2010, Mallone et al., 2011, Meister et al., 2012, Ostro et al., 2011 and Stolzel et al., 2007).

In addition to effects of PM2.5, there is increasing evidence that coarse particles (PM2.5–10) may play a role in generating adverse health effects (Brunekreef and Forsberg, 2005 and Sandstrom et al., 2005). In a national, multi-city time-series study of the acute effects of PM2.5 and PM2.5–10 in the US, Zanobetti and Schwartz (2009) found that both PM2.5 and PM2.5–10 were significantly associated with increased mortality for all and specific causes. Less evidence of effects of PM2.5 and PM2.5–10 is available from European studies. Several European studies failed to demonstrate significant effects of PM2.5 on all cause or cause specific mortality (Branis et al., 2010, Garrett and Casimiro, 2011, Halonen et al., 2009, Mallone et al., 2011 and Stolzel et al., 2007). However, these studies generally comprised relatively small populations (≤1 million) (Garrett and Casimiro, 2011, Halonen et al., 2009 and Stolzel et al., 2007) or short study periods (1 year) (Branis et al., 2010). In a recent study that included 6 years of data on particulate matter air pollution and mortality in London, UK, PM2.5 was not statistically significantly associated with all cause or cardiovascular mortality, and the effect on respiratory mortality was only statistically significant for one of the seven lags studied (Atkinson et al., 2010). Even less information is available for PM2.5–10, especially for North-West Europe.

PM10 and PM2.5 levels were statistically significantly (p < 0.05) associated with all cause and cause-specific deaths. For example, a 10 µg/m³ increase in previous day PM was associated with 0.8% (95% CI 0.3–1.2) excess risk in all-cause mortality for PM2.5 and a 0.6% (CI 0.2–1.0) excess risk for PM10. No appreciable associations were observed

Table 1. Number of equivalent weekdays (day in which an urban transit system set the maximum trip demand) per year.

Type of day	Value	Number	Weekday equivalent
Weekday	100%	203	203
Saturday	62%	48	30
Sunday	48%	48	23
Holyday	32%	7	2
July weekday	68%	22	15
July w-e	33%	8	3
August weekday	56%	22	12
August w-e	27%	7	2
		365	290

---

* Italian National Statistics Center.

for PM2.5–10. Effects of PM10, and PM2.5 were insensitive to adjustment for PM2.5–10, and vice-versa. PM10 and PM2.5 were too highly correlated to disentangle their independent effects.

Li MH et al. (2016) indicated that short-term exposure to a 10 μg/m³ increment of ambient PM2.5 is associated with increased COPD hospitalizations and mortality (N.A.H. Janssena et al., 2016).

In this analysis, health expenditure is assessed in terms of morbidity and health care costs by an increase of the PM10 concentration (in μg/m³, yearly average). Vice versa a reduction of PM10 concentration measures the avoided health care expenditure. The health services are calculated on the national tariff of health services of the Italian Ministry of Health (SSN, 2016).

The assessment of a rapid transit line health effects must be related to local conditions of the project corridor. The proposed methodology requires the following characteristic data:

- Average daily PM10 concentration (γ in μg/m³)
- Traffic average PM10 emission rate (ε_R)
- Daily road trip (R_T)
- Daily road mileage (R_M)
- Public transport rate
- Larger urban area resident population
- Population density (inh. /km²)

The Rapid Transit (RT) project corridor has a length (km) and a weekday average demand i.e. a weekday average mileage (trip km).

The new corridor subtracting a share of traffic by the road:

ΔR^T = Road subtracted trip = RT demand · (1-Public transport rate)

On safe side:

ΔR^M = Road subtracted mileage (trip km) = Road subtracted trip · corridor length

Finally:

ΔPM10 = Average daily PM10 concentration red.

$$= \varepsilon_R \cdot \frac{R_M}{\Delta R_M} \cdot \gamma$$

Table 2. Morbidity and health care costs by an increase of the PM10 concentration (in μg/m³, yearly average).

Pathology	Morbidity each 1.0 μg/m³	Annual cost of assistance (EUR)
Adenocarcinoma	6.4E-04	69,400
Lynphoma	2.7E-05	54,800
Asthma	9.2E-03	810
Chronic obstructive pulmonary Disease (COPD)	8.8E-03	4,440

5 SHORT-TERM EXPOSURE TO AMBIENT NOISE

Noise is pervasive in everyday life and can cause both auditory and non-auditory health effects. Noise-induced hearing loss remains highly prevalent in occupational settings, and is increasingly caused by social noise exposure (e.g. through personal music players). Evidence of the non-auditory effects of environmental noise exposure on public health is growing. Observational and experimental studies have shown that noise exposure leads to annoyance, disturbs sleep and causes daytime sleepiness, affects patient outcomes and staff performance in hospitals, increases the occurrence of hypertension and cardiovascular disease, and impairs cognitive performance in schoolchildren.

In principle, the noise/stress hypothesis is well understood: Noise activates the pituitaryadrenal-cortical axis and the sympathetic-adrenal-medullary axis. Changes in stress hormones including epinephrine, norepinephrine and cortisol are frequently found in acute and chronic noise experiments. The catecholamines and steroid hormones affect the organism's metabolism. Cardiovascular disorders are especially in focus for epidemiological studies on adverse noise effects. However, not all biologically notifiable effects are of clinical relevance. The relative importance and significance of health outcomes to be assessed in epidemiological noise studies follow a hierarchical order, i.e. changes in physiological stress indicators, increase in biological risk factors, increase of the prevalence or incidence of diseases, premature death (Basner et al., 2014).

Common noise related problems are interference with communication and sleep disturbance (Griefahn et al., 2000). Decreased quality of sleep is considered to be a major health outcome of environmental noise (Berglund and Lindvall, 1995). Noise exposure can also cause other nonauditory effects such as annoyance, changes of behavior and deterioration in performance. Long-term effects of road traffic noise on psychosocial health and wellbeing are also described (Ohrstrom et al., 1998). Studies have shown that people living near streets with busy traffic or airports close their windows, spend less time in their gardens, and have less visitors than people living in more quiet areas (Griefahn, 2000).

Primary physiological effects of noise exposure are vegetative reactions such as increases in blood pressure, heart rate and finger pulse amplitude, cardiac arrhythmia and changes in respiration and body movements (Berglund and Lindvall, 1995). Secondary effects are reduction in perceived sleep quality, increased fatigue, decreased mood or wellbeing and deterioration in performance. Cardiovascular effects of traffic noise such as hypertension and ischemic heart disease have also been described but epidemiological evidence is still limited in this respect (Babisch, 2000). In a recent study a relationship between aircraft noise exposure and hypertension has been reported (Rosenlund et al., 2001).

Noise policy largely depends on considerations about cost-effectiveness, which may vary between populations. Kim et al., (2012) has shown that the effect of traffic can cause annoyance in 11% of the population living within 500 meters from a major road; sleep disorders in 2% of the residents. For Bluhm G., Nordling E., Berglind N. (2004), frequent annoyance was reported by 13% of subjects exposed to Leq H24 > 50 dB(A) compared to 2% among those exposed to as lower of 50 dB(A), resulting in a difference of 11% (95% Confidence Interval 7%, 15%); sometimes or frequently occurring sleep disturbance was reported by 23% at Leq H24 > 50 dB(A) and by 13% at levels <50 dB(A), a difference of 11% (95% CI 4%, 18%).

In this analysis, health expenditure is assessed in terms of morbidity and health care costs by an increase of 1 dB (A) of the average noise level Leq H24 (or Lden, day-evening-night). Vice versa a reduction Sound Noise Level measures the avoided health care expenditure. The health services are calculated on the national tariff of health services of the Italian Ministry of Health (SSN, 2016).

The proposed methodology requires the Lden average value along the RT project corridor: health effects are calculated only on the resident population within a distance of 250 meters from each side from the corridor.

The reduction is proportional to the transit reduction of private vehicles and buses. Specific noise emissivity is given by the following values (current state of technology, UITP and TRCP**):

6 POLLUTION AND NOISE HEALTH EFFETCS CUMULATIVE EVALUATION

Measuring the energy demand of a RT corridor means verify its operation efficiency also in terms of induced morbidity. As said, this analysis counts only health spending that in the current state of the literature can be estimated with a robust confidence level.

Table 3. Morbidity and health care costs by an increase of 1 dB(A) of Lden average value.

Pathology	Morbidity each 1 dB(A)	Annual cost of assistance (EUR)
Annoyance	1.1E-01	533
Hypertension	3.3E-01	1,905

With regard to electric traction technologies are considered the emissions on the source (the power plant). In Italy, the national energy mix (GSE, 2014–2015) is 42.3% renewable source; 19.3% carbon; 28.9% natural gas; 4.9% nuclear; 4.6%, petroleum products and other sources. Power plants are located outside urban centers; for this it is assumed a reference density for the health impact of 100 inh. / km^2.

Health impact is assessed by including both the fine particulate that nitrogen oxides. Considering the specific effects on health (Medparticles, 2001–2011), the relation is:

$$PM10_{EQ} = 25 \cdot NOX + PM10 \text{ [mg/kWh]}$$

Imagining a project RT corridor of 10 km in an urban area of 100,000 inhabitants (density of at least 1000 inh. /km^2) is:

In the same example, it is possible to calculate the overall cost per energy unit. Purchase cost are the current ones at 2016 in Italy.

Table 4. Noise emission [in dB(A)] per mode of transport at normal travel conditions.

Technology	Specific emission per kW	Average speed (km/h)	Average standard emission per vehicle
Car	1.7E+00	30	83
Bus	5.8E-01	24	87
Trolleybus	4.2E-01	24	63
Tram (streetcar)	2.6E-01	24	66
Light Rail Transit	2.4E-01	27	72
Metro	1.9E-01	35	77
Commuter railway	2.0E-01	55	79

Table 5. Specific emissions per technology in mg per offered seat km.

Technology	NOX Specific emission per seat km	PM10 Specific emission per seat km	Average speed (km/h)
Car Euro V—diesel	4.50E-02	1.25E-03	30
Car Euro VI—diesel	2.00E-02	1.25E-03	30
Bus Euro V—diesel	1.36E-01	1.36E-03	24
Bus Euro VI—diesel	2.72E-02	6.80E-04	24
Trolleybus	1.80E-03	2.20E-03	24
Tram (streetcar)	1.50E-03	1.70E-03	24
Light Rail Transit	1.40E-03	1.53E-03	27
Metro	1.10E-03	1.22E-03	35
Commuter railway	1.30E-03	1.35E-03	55

** International Organization for Public Transport Authorities and Operators (UITP); Transit Cooperative Research Program (TCRP), USA.

Table 6. Specific emissions per technology in mg of PM10$_{EQ}$ per offered seat km.

Technology	PM10$_{EQ}$ specific emission per seat km
Car Euro V -diesel	1.13E+00
Car Euro VI -diesel	5.01E-01
Bus Euro V - diesel	3.40E+00
Bus Euro VI - diesel	6.81E-01
Trolleybus	4.72E-02
Tram (streetcar)	3.92E-02
Light Rail Transit	3.65E-02
Metro	2.87E-02
Commuter railway	3.39E-02

Table 8. Purchasing versus healthcare cost in urban area, Table 1–2.

Technology	Energy cost Wh per seat km	Healthcare cost		
		cent € per kWh	€ per liter of fuel	cent € per kWh from grid
Car (gasoline)	204	583	70.5	–
Car (diesel)	240	486	49.6	–
Bus (diesel)	68	79	8.1	–
Trolleybus	23	190	–	190
LRT	19	67	–	67

Table 7. Example of a PM10 related health impact calculation of a rapid transit line of 10 km, in an urban area of 100,000 inhabitants with a density of 1000 inh. /km².

NO-BUILT SITUATION

Figure	Value
Average daily PM10 concentration (µg/m³)	40.0
Traffic emission rate	35.0%
Public transport rate	9.0%
Larger urban area resident population	100,000
Population density (inh. /km²)	1,000
Corridor length (km)	10
Weekday average trip	139,400
Weekday average mileage (trip km)	1,672,800

RAPID TRANSIT (RT) PROJECT

Figure	Value
RT project demand	10,000
Road subtracted trip	9,100
Weekday subtracted mileage (trip km)	91,000

MEDIUM TERM PROJECT EFFECT

Figure	Value
Average daily concentration (µg/m³)	39.2
Expected reduction (µg/m³)	−0.8
Expected noise reduction in dB(A)	−1.0

The result is (all values are in euro):

SAVED HEALTHCARE SPENDING PER YEAR

Figure	Value
Adenocarcinoma	−3,382,711
Lymphoma	−112,686
Asthma	−567,542
Chronic obstructive pulmonary disease	−2,975,713
Annoyance	−117,260
Hypertension	−1,257,300
	−8,413,212

Table 9. Purchasing versus healthcare cost in urban area, Table 2–2.

Technology	Purchase cost cent € per seat km	Healthcare cost cent € per seat km	Purchase / healthcare cost ratio
Car (gasoline)	2.53	141.01	55.8
Car (diesel)	3.32	118.75	35.8
Bus (diesel)	0.91	5.41	6.0
Trolleybus	0.27	4.27	15.8
LRT	0.23	1.27	5.6

Table 10. Purchase versus healthcare cost, national average.

Technology	Purchase cost cent € per seat km	Healthcare cost cent € per seat km	Purchase / healthcare cost ratio
Car (gasoline)	1.50	8.39	5.6
Car (diesel)	1.41	5.90	4.2
Bus (diesel)	1.36	0.96	0.7
Trolleybus	0.27	0.51	1.9
LRT	0.23	0.15	0.7

The relationship that estimates the progression of health spending is:

$$C_H = k \cdot AVT \frac{1}{10} \ln(\delta \cdot L)$$

k is a monetary constant that, assuming PM10 background level equal to 20 µg/m³ is equal to EUR 3.85; AVT is the average daily traffic on the corridor; δ is the density along the corridor; L is the length of the corridor.

The characteristic values calculated above by mode of transport refer to an area with spatial density of not less than 1,000 km². It is possible to extend the calculation to the entire national

Table 11. Case of study: main figures.

RT CORRIDOR PROJECT EXAMPLE

line length	10.0	km
stops	20	
weekday demand	10,000	passenger
peak-hour demand	1,275	passenger
max load equivalent day per year	290	equivalent days
standard ticket (one ride)	1.50	EUR per trip
net fare	0.45	EUR per trip
fuel cost (diesel)	1.275	EUR per lt

Table 12. Saved healthcare spending (in euro cents).

Technology	Saved healthcare spending (15 years) pax.km	Saved healthcare spending per year per pax. km
Bus	−7.1914	−0.4794
Articulated bus	−8.3187	−0.5546
Bus RT	−7.3127	−0.4875
Articulated-Bus RT	−8.0521	−0.5368
Urban Tbus	−8.8628	−0.5909
Articulated Tbus	−9.8109	−0.6541
Trolleybus RT	−8.5775	−0.5718
Articulated-Tbus RT	−9.7789	−0.6519
Urban bty-Tbus	−8.8628	−0.5909
Articulated bty-Tbus	−9.8109	−0.6541
Battery-Tbus RT	−8.5775	−0.5718
Articulated bty-Tbus RT	−9.7789	−0.6519
Urban ebus	−8.8628	−0.5909
Articulated urban ebus	−9.8109	−0.6541
eBus Rapid Transit	−8.5775	−0.5718
Articulated-eBus RT	−9.7789	−0.6519
Compact tram	−9.5873	−0.6392
Standard urban tram	−10.4447	−0.6963
Compact vehicle-tramway	−9.4823	−0.6322
Standard vehicle-tramway	−10.1584	−0.6772
XL vehicle-tramway	−11.0191	−0.7346
LRT	−10.8945	−0.7263
XL vehicle-LRT	−11.4511	−0.7634
Light train metro	−13.0596	−0.8706
Heavy train metro	−13.0596	−0.8706
Commuter	−11.4737	−0.7649
XL train-commuter	−12.1008	−0.8067

territory. The urban space (Istat, 2015) sees 32.4 million people concentrated in 34,511 km² according to an average density similar to the reference one assumed in the previous calculation. On the other hand, the rural space (28.3 million) covers 266,829 km² with an average population density of 106 inh. / km².

Table 13. Total earnings per 1,000 passengers per useful life year. A compact tram is 24 m long; a standard, 30 m; a XL-tram, 40 m. A standard LRT is 40 m long; a XL-LRT, 55 m. A light-train metro is 60 meters long; a heavy-metro, 110. A commuter train is 100 m long; a XL-commuter, 200 m.

Bus	-1,974
Articulated bus	-1,456
Bus RT	685
Articulated-Bus RT	847
Urban Tbus	-1,417
Articulated Tbus	-920
Trolleybus RT	696
Articulated-Tbus RT	1,448
Urban bty-Tbus	-2,182
Articulated bty-Tbus	-1,654
Battery-Tbus RT	424
Articulated bty-Tbus RT	1,222
Urban ebus	-2,109
Articulated urban ebus	-1,452
eBus Rapid Transit	471
Articulated-eBus RT	1,282
Compact tram	1,189
Standard urban tram	1,798
Compact vehicle-tramway	1,616
Standard vehicle-tramway	1,964
XL vehicle-tramway	2,380
LRT	2,207
XL vehicle-LRT	2,456
Light train metro	-7,277
Heavy train metro	-1,816
Commuter	-862
XL train-commuter	-599

The reduction factor is a function of the relationship between density and the average length of the displacements:

Urban: 939 inh. / km²; 13 km
Rural: 106 inh. / km²; 38 km

The ratio rural / urban is

$$\frac{\ln(\delta \cdot L) \, rural}{\ln(\delta \cdot L) \, urban} = \frac{8312}{9447} = 11.9\%$$

Extending the calculation at a national level is:

7 RESULTS

The analysis is applied to an example case described in the following table:

The geographical area is Italy; all costs are expressed in euro (EUR).

The energy cost is a function of annual consumption (rate of the GSE, the Manager of national Electrical Services).

In the analysis are considered a total of 5 main alternatives (bus, trolleybus, ebus, tram, LRT, metro and commuter railway) and several variations (roadway, equipment, etc.) for a total of 27 technological alternatives.

The calculation model—organized in a detailed spreadsheet—is available at the following link: https://doc.co/yo3 L2T

The following tables exposes the outputs resulting from the application of the model to the various technologies.

The model allows to introject part of externalities costs in the financial analysis. In this way it is possible to absorb some of the thermal engine—bias in the evaluation of alternative technologies for a public transport corridor.

REFERENCES

Bibliography on operating costs and maintenance of urban transport

Barrero R., Van Mierlo J. and Tackoen X., 2008, Energy savings in Public Transport, IEEE.

Bruun, E., 2005, BRT versus LRT operating costs: comparisons using a parametric cost model. Transportation Research Record No. 1927. Transportation Research Board of the National Academies, Washington, DC, pp 10–20.

Grava, S., 2003, Urban transportation systems: choices for communities, McGraw Hill, New York.

Hess, D., Taylor, B., Yoh, A., 2005, Light rail lite or cost-effective improvements to bus service? Evaluating costs of implementing bus rapid transit. In: Transportation Research Record: Journal of the Transportation Research Board, No. 1927. Transportation Research Board of the National Academies, Washington, DC, pp 22–30.

Kittelson & Associates, 2007, Bus rapid transit practitioner's guide. TCRP Report 118. TRB, National Research Council, Washington, DC.

Levinson, H., Zimmerman, S., Clinger, J., Gast, J., Rutherford, S., Bruun, E., 2003, Bus rapid transit, volume 2: implementation guidelines. TCRP Report 90. TRB, National Research Council, Washington, DC.

Levinson, H., Zimmerman, S., Clinger, J., Rutherford, S., Smith, R., Cracknell, J., Soberman, R., 2003, Bus rapid transit, volume 1: case studies in bus rapid transit. TCRP Report 90. TRB, National Research Council, Washington, DC.

Moccia, L., Laporte, G., 2016, Improved models for technology choice in a transit corridor with fixed demand, Transportation Research Part B: Methodological Volume 83, January 2016, Pages 245–270.

Mwambeleko J.J., Kulworawanichpong T., Greyson G.A., 2015, Tram and Trolleybus Net Traction Energy Consumption Comparison, IEEE.

Punpaisan S. and Kulworawanichpong T., 2014, Dynamic Simulation of Electric Bus Vehicle, Standard International Journals (SIJ), vol. II, pp. 99–104, May 2014.

TCRP (Transit Cooperative Research Program), 2005, Innovations in bus, rail and specialized transit. Research Results Digest 70. Transit Cooperative Research Program, National Academy Press, Washington, DC.

Vuchic, R.V., Stanger, M.R., Bruun, E., 2013, Bus Rapid Versus Light Rail Transit: Service Quality, Economic, Environmental and Planning Aspects in M. Ehsani et al. (eds.), Transportation Technologies for Sustainability, DOI 10.1007/978-1-4614-5844-9, Springer Science+Business Media.

Vuchic, R.V., Urban Transit Systems and Technology, John Wiley & Sons.

Vuchic, V., 2005, Planning and selection of medium and high performance transit modes. In: Urban transit: operations, planning and economics. Wiley, Hoboken (Chap. 12).

Bibliography on ambient noise health effects

Basner, M., Babisch, W., Davis, A., Brink, M., Clark, C., Janssen, S., Stansfeld, S., 2014, Auditory and non-auditory effects of noise on health, The Lancet, Volume 383, Issue 9925, 12–18 April 2014, Pages 1325–1332.

Bluhm, G., Nordling, E., Berglind, N., 2004, Road traffic noise and annoyance-an increasing environmental health problem, Road traffic noise and annoyance-an increasing environmental health problem. Noise Health 2004; 6:43–9.

De Paiva Vianna, K.M., Alves Cardoso, M.R., Calejo Rodrigues, R.M., 2015, Noise pollution and annoyance: An urban soundscapes study, Noise pollution and annoyance: An urban soundscapes study. Noise Health [serial online] 2015; 17:125–33.

Ising, H., Kruppa, B., 2004, Health effects caused by noise: Evidence in the literature from the past 25 years, Noise Health (serial online), 6:5–13.

Kim, M., Chang, S.I., Seong, J.C., Holt, J.B., Park, T.H., Ko, J.H., Croft, J.B., 2012, Road traffic noise: annoyance, sleep disturbance, and public health implications, American Journal of Preventive Medicine 2012 Oct; 43(4):353–60.

Lercher, P., Evans, G.V., Meis, M., Kofler, W.W., 2002, Ambient neighbourhood noise and children's mental health, Occup Environ Med 2002;59:380–386 doi:10.1136/oem.59.6.380.

Stanfeld, S.A., et al., 2005, Aircraft and road traffic noise and children's cognition and health: a cross-national study, The Lancet, Volume 365, Issue 9475, 4–10 June 2005, Pages 1942–1949.

Bibliography on urban pollution health effects

Adar SD, et al., 2014, Ambient coarse particulate matter and human health: a systematic review and meta-analysis. Curr Environ Health Rep 1(3):258–274 (2014); doi: 10.1007/s40572-014-0022-z

Anderson JO, et al., 2012, Clearing the air: a review of the effects of particulate matter air pollution on human health. J Med Toxicol 8(2):166–175 (2012); doi: 10.1007/s13181-011-0203-1

Anderson, H.R., R.W. Atkinson, J.L. Peacock, L. Marsten, K. Konstantinou, 2004, Meta-analysis of time-series studies and panel studies of particulate matter (PM) and ozone (O3), WHO regional Office for Europe, Copenhagen, Denmark (2004) [report EUR/04/5042699 E82792 in *http://www.euro.who.int/__data/assets/pdf_file/0004/74731/e82792.pdf*]

Atkinson, R.W., G.W. Fuller, H.R. Anderson, R.M. Harrison, B. Armstrong, 2010, Urban ambient particle metrics and health. A time series analysis, Epidemiology, 21 (2010), pp. 501–511.

Baccini, M., Grisotto, L., Catelan, D., Consonni, C., Bertazzi, P.A., Biggeri, A., 2015, Commuting-Adjusted Short-Term Health Impact Assessment of Airborne Fine Particles with Uncertainty Quantification via Monte Carlo Simulation, Environ Health Perspect; DOI:10.1289/ehp.1408218

Chen, Y.-C., Weng, Y.-H., Chiu, Y.-W., Yang, C.-Y., 2015, Short-Term Effects of Coarse Particulate Matter on Hospital Admissions for Cardiovascular Diseases: A Case-Crossover Study in a Tropical City, Journal of Toxicology and Environmental Health - Part A: Current Issues.

Cheng, M.-H., Chiu, H.-F., Yang, C.-Y., 2015, Coarse particulate air pollution associated with increased risk of hospital admissions for respiratory diseases in a Tropical city, Kaohsiung, Taiwan, International Journal of Environmental Research and Public Health.

Correction Basagaña, X., Jacquemin, B., Karanasiou, A., Pascal, L., Pascal, M., 2015, Short-term effects of particulate matter constituents on daily hospitalizations and mortality in five South-European cities: Results from the MED-PARTICLES project, Environment International.

Garrett, P., E. Casimiro, 2011, Short-term effects of fine particulate matter (PM2.5) and ozone on daily mortality in Lisbon, Portugal, Environ Sci Pollut, 18 (2011), pp. 1585–1592.

Halonen, J.I., T. Lanki, T.Y. Tuomi, P. Tiittanen, M. Kulmala, J. Pekkanen, 2009, Particulate air pollution and acute cardiorespiratory hospital admissions and mortality among the elderly, Epidemiology, 20 (2009), pp. 143–153.

Heo, J., Schauer, J.J., Yi, O., Kim, H., Yi, S.-M., 2014, Fine particle air pollution and mortality: Importance of specific sources and chemical species, Epidemiology.

Janke, K., 2014, Air pollution, avoidance behavior and children's respiratory health: Evidence from England, Journal of Health Economics.

Janssen, N.A.H., G. Hoek, M. Simic-Lawson, P. Fischer, L. van Bree, H. ten Brink, et al., 2011, Black carbon as an additional indicator of the adverse health effects of airborne particles compared to PM10 and PM2.5, Environ Health Perspect, 119 (2011), pp. 1691–1699.

Khan, S.A., 2015, Atmospheric concentration of fine particulate matter and its impact on daily hospital admissions for respiratory disease in the prairies of Canada, International Journal of Environmental Studies.

Li MH, Fan LC, Mao B, Yang JW, Choi AM, Cao WJ, Xu JF., 2016, Short-term Exposure to Ambient Fine

Particulate Matter Increases Hospitalizations and Mortality in COPD: A Systematic Review and Meta-analysis, Chest. 2016 Feb;149(2):447–58. doi: 10.1378/chest.15–0513. Epub 2016 Jan 12. Review.

Linares, C., Carmona, R., Tobías, A., Mirón, I.J., Díaz, J., Influence of advections of particulate matter from biomass combustion on specific-cause mortality in Madrid in the period 2004–2009, 2015, Environmental Science and Pollution Research.

Mallone, S., M. Stafoggia, A. Faustini, G.P. Gobi, A. Marconi, F. Forastiere, 2011, Saharan dust and association between particulate matter and daily mortality in Rome, Italy, Environ Health Perspect, 119 (2011), pp. 1409–1414.

Meister, K., C. Johansson, B. Forsberg, 2012, Estimated short-term effects of coarse particles on daily mortality in Stockholm, Sweden, Environ Health Perspect, 120, pp. 431–436.

Mesa-Frias M, et al., 2013, Uncertainty in environmental health impact assessment: quantitative methods and perspectives. Int J Environ Health Res 23(1):16–30 (2013); doi: 10.1080/09603123.2012.678002

N.A.H. Janssena, P. Fischera, M. Marraa, C. Amelinga, F.R. Casseea, 2016, Short-term effects of PM2.5, PM10 and PM2.5–10 on daily mortality in the Netherlands, Science of The Total Environment, Volumes 463–464, 1 October 2013, Pages 20–26.

Pascal, M., Falq, G., Wagner, V., Pascal, L., Larrieu, S., 2014, Short-term impacts of particulate matter (PM10, PM10–2.5, PM2.5) on mortality in nine French cities, Atmospheric Environment.

Perez, L., M. Medina-Ramon, N. Kunzli, A. Alastuey, J. Pey, N. Perez, et al., 2009, Size fractionate particulate matter, vehicle traffic, and case-specific daily mortality in Barcelona, Spain, Environ Sci Technol, 43 (2009), pp. 4707–4714.

Shah, A.S.V., Lee, K.K., McAllister, D.A., Newby, D.E., Mills, N.L., 2015, Short term exposure to air pollution and stroke: Systematic review and meta-analysis, BMJ (Online).

Talbott, E.O., Rager, J.R., Benson, S., Bilonick, R.A., Wu, C., 2014, A case-crossover analysis of the impact of PM2.5 on cardiovascular disease hospitalizations for selected CDC tracking states, Environmental Research.

Tobias, A., L. Perez, J. Diaz, C. Linares, J. Pey, A. Alastruey, et al., 2011, Short-term effects of particulate matter on total mortality during Saharan dust outbreaks: a case-crossover analysis in Madrid (Spain), Sci Total Environ, 412–413, pp. 386–389.

Viana, M., Pey, J., Querol, X., De Leeuw, F., Lükewille, A., 2014, Natural sources of atmospheric aerosols influencing air quality across Europe, Science of the Total Environment.

Transport Infrastructure and Systems – Dell'Acqua & Wegman (Eds)
© *2017 Taylor & Francis Group, London, ISBN 978-1-138-03009-1*

A reading key of motorisation trend in Italy

E. Garilli, F. Autelitano & F. Giuliani
Dipartimento di Ingegneria Civile, dell'Ambiente, del Territorio ed Architettura—DICATEA, University of Parma, Parma, Italy

A. Guga & G. Maternini
Dipartimento di Ingegneria Civile, Architettura, Territorio, Ambiente e di Matematica—DICATAM, University of Brescia, Brescia, Italy

ABSTRACT: Since the beginning of motorisation, the rate between passenger cars to 1000 inhabitants (motorisation rate) has represented a significant socio-economic and productive development-index of a country. In the 70 s it was shown that the trend of the motorisation rate could be well represented by sigmoid functions. This study aims to present an analysis of the analytical expressions of the motorisation rate trend in Italy. In the current continental framework, Italy is one of the country with the greatest motorisation rate but also a country for which the interpretation of this simple statistical parameter necessarily requires a deepening. The motorisation rate is still well fitted by sigmoid function but with a gap between Northern and Southern Italy due to an infrastructure delay, highly noticeable in Southern Italy, but especially to a different management policy of the public transport and the related infrastructure investments.

1 INTRODUCTION

Since the beginning of motorisation, especially with the reorganization of infrastructure network within Europe after the Second World War, the rate between passenger cars to inhabitants has represented a significant socio-economic and productive development-index of a country. For that reason this rate has often been related through mathematical relationships to income per capita.

The analysis conducted in Italy by highway engineers exploited the growth in car ownership as a forecasting tool in order to plan investments and to provide an indispensable tool in the design of roads and intersections (Maternini & Caracoglia, 1966; Tocchetti, 1965). The analysis of traffic increase in the short- and medium-term, mainly associated with the dimensions of the various elements of the road, were based on linear functions. More complex functions, typically the sigmoid ones, are instead used for long-range (more than twenty years) forecasting. The sigmoid functions, widely used in the study of population growth, are characterised by an upper and a lower asymptote and by an inflection point. The use of such functions in prediction of car ownership trend is justified by its initial slow-growing, due to the limited availability of both passenger cars and road infrastructures for their circulation, followed by an

high-growing, caused by an increase in per capita income following the economic boom, and an ending with a slow-growing, sign of market and/or utility saturation. Car ownership in Italy increased from 0.0245 passenger cars per inhabitant (pc/inh) in 1955 to 0.116 pc/inh in 1965, during the economic boom, when the per capita income grew more than double in the same given time period. Yesterday's researchers predicted a car ownership saturation value in Italy between 0.30 and 0.35 pc/inh (Tocchetti, 1965). This projection framework has been redefined in the second half of the 70 s when, although significant macroeconomic factors mainly related to the oil market were intervening, the logistic curve was sill the reference, with estimated asymptotic values close to 0.32 (Maternini & Caracoglia, 1966). Even today, the approach of the logistic curve or of other curves falling in sigmoidal models can be a convenient and easily communicable reading key for transport and socio-economic investigations. But, the easy availability of more limited geographical areas disaggregated data allows to combine a long period analysis with an even easier reading of the short term trends, characterized by cause-effect processes of few years. The authors propose below a progressively scaled reading from the national statistical data to the regional and urban-metropolitan ones, recognizing time after time the most immediately significant aspects and prefiguring some development scenarios of territorial mobility needs.

2 EUROPEAN FRAMEWORK

The current motorisation rate in Europe (ACEA, 2016; EUROSTAT, 2016), defined by the rate between passenger cars to 1000 inhabitants (pc/1000 inh), is about 500 pc/1000 inh with an increase of 7% over the last 10 years. The European motorisation rate is represented in Figure 1: Italy has a fourth position with a value of 616 pc/1000 inh, that is greater than European average. Only Luxemburg, Liechtenstein and Iceland show higher vale; however the data of Luxemburg and Liechtenstein may be influenced by cross-border workers using company cars registered in the countries; whereas with regard to Iceland the data are strongly influenced by the low population.

In the past, a great value of the motorisation rate was an indication of economic well-being; today this is no longer the case in fact, the goal of the European politicians, as well as the Italian ones, is to reduce the motorisation rate, balancing the demand for increasing personal mobility and economic growth, with the need to respect the environment and provide an acceptable quality of life for all citizens (Black et al., 2016; European Commission, 2011).

3 MOTORISATION RATE TREND IN ITALY

The motorisation rate in Italy since 1952 is shown in Figure 2. Population and passenger cars data has been extrapolated from the national statistical offices (ACI, 2016; ISTAT, 2016).

The motorisation rate regression model is well described by sigmoid function as the Logistic function and the Gompertz function whose general expressions are given by Equation 1 and Equation 2 respectively:

$$y = \frac{a}{1 + b \cdot e^{-c \cdot x}} \tag{1}$$

$$y = a \cdot e^{-e^{b - c \cdot x}} \tag{2}$$

In this case a represents the motorisation rate that corresponds to the saturation limit; b and c parameters depend on the orographic-, geo-economic- and demographic-characteristics and x is the progressive analysis period expressed in years. Both functions presents a high correlation coefficient as can be seen from Table 1. Current national data (2015) are close to the asymptotic value a that reaches 630 pc/1000 inh in the logistic function and 676 pc/1000 inh in the Gompertz one. There is now ample evidence that the forecasts of the 70 s were disregarded: this is not due to the growth model used but to the steady state value. The curve shape and the experimental data indicate that has been achieved a glut of car ownership as a consequence of infrastructure and income per capita settlement.

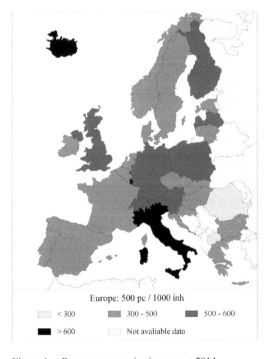

Figure 1.   European motorisation rate at 2014.

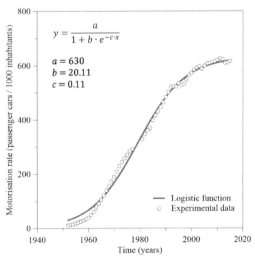

Figure 2.   Italian motorisation rate trend (1952/2015).

Table 1. Parameters of Logistic and Gompertz functions and percentage change of population (ΔP%).

Geographical area	ΔP * [%]	Logistic function					Gompertz function				
		a	b	c	r	r^2	a	b	c	r	r^2
Italy	6.66	630	20.11	0.11	0.997	0.994	676	1.48	0.07	0.999	0.997
Northern Italy	11.04	624	25.06	0.13	0.996	0.992	646	1.75	0.08	0.994	0.988
Piemonte	4.09	654	17.78	0.12	0.995	0.990	679	1.53	0.08	0.994	0.988
Lombardia	12.62	612	26.83	0.13	0.994	0.989	631	1.79	0.09	0.991	0.983
Veneto	11.76	617	31.60	0.13	0.998	0.996	643	1.90	0.08	0.997	0.994
Emilia Romagna	14.20	634	27.03	0.14	0.996	0.992	651	1.83	0.08	0.994	0.989
Southern Italy	0.59	641	36.24	0.10	0.998	0.996	740	1.78	0.05	0.996	0.993
Campania	2.77	637	45.46	0.10	0.995	0.991	723	1.91	0.06	0.992	0.984
Puglia	0.82	615	27.43	0.09	0.998	0.995	707	1.64	0.05	0.997	0.995
Basilicata	−5.30	702	31.25	0.09	0.997	0.994	843	1.69	0.05	0.999	0.998
Calabria	−4.14	706	35.97	0.09	0.998	0.996	861	1.74	0.05	0.998	0.997

* referred to 2004–2015.

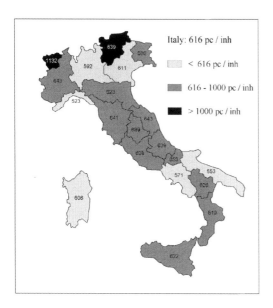

Figure 3. Regional Italian motorisation rate in 2015.

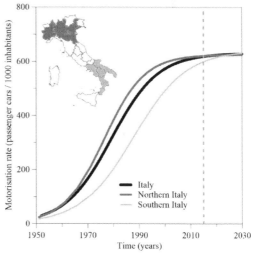

Figure 4. Italian motorisation rate trend (1952–2030).

Figure 3 show the regional Italian motorisation rate at 2015. It was decided to model the motorisation rate regression considering the Northern and Southern Italy data separately. Figure 4 and Figure 5 emphasizes how the national trend is strongly influenced by the motorisation rate value of Northern Italy (four sample regions were considered in this study: Piemonte, Lombardia, Veneto and Emilia Romagna) where the population and the consistency of passenger car fleet are prevalent (Fig. 4 and Table 1). In this analysis some regions,

like Valle d'Aosta and Trentino Alto Adige, were excluded, because they do not necessarily reflect the actual number of passenger cars per inhabitant due to their specific tax arrangement. In the same way, for the identification of the Southern Italy data (four sample regions: Calabria, Campagna, Puglia e Basilicata), Sicilia and Sardegna were excluded because they present transport peculiarity linked to their island status. The data reveal that the motorisation rate trend and values are greatly dissimilar from those of Northern Italy. These values suggests a continuing strong growth of motorisation rate which is still far from the asymptotic value of forecasting model (Fig. 5). This situation

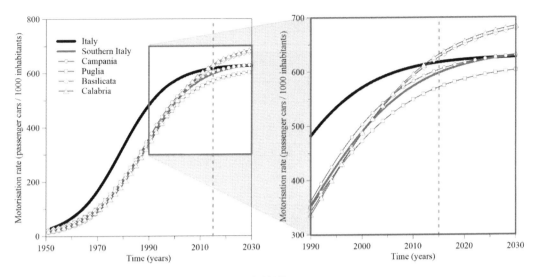

Figure 5.   Southern Italy motorisation rate trend (1952–2030).

is common to both the global and the regional scale with a decrease in resident population and a passenger car fleet far from those of Northern Italy (Table 1): this is mainly due to failure in completing primary road network as well as planned change of both existing rail network and high-speed railway line. This infrastructure delay, highly noticeable in Southern Italy, places the motorisation rate on upward part of the function, with substantial growth forecast suggestive of an unavoidable use of private motor cars transport in the absence of alternative regional and inter-regional public transport provision.

4   LOCAL-SCALE OF MOTORISATION RATE TREND OF TEN-YEAR TIME SERIES

The goodness of fit of these prediction curves is however undermined by the technological evolution of the automotive industry, and from the no longer consolidated direct relationship between the number of vehicles and the country's wealth or the income per capita. Existing evidences point out that economic and environmental benefits arise from sustainable mobility implementation. Several European programs (EVIDENCE, CIVITAS etc.) set out to provide a substantial change towards sustainable urban transport system (clean vehicles, urban freight, demand and mobility management strategies and collective passenger transport) (Black et al., 2016; Buehler & Pucher, 2011). Numerous European areas of advanced socioeconomic development, like Freiburg and Nurem-

berg (Germany), Ghent (Belgium), Bristol (UK), Strasbourg (France), etc. (Black et al., 2016; Buehler & Pucher, 2011; Buehler et al., 2016; European Commission, 2004) quantify the quality of life and the sustainability of the infrastructure system with the reduction of the motorisation rate, with the loosening of relationship ties between a passenger car and the demand for mobility.

The progression of motorisation rate of the last decade in Italian cities, distinguished by size and geographical location, should be read also according to this perspective (Busi & Maternini, 1995). Some Italian cities with more than 150,000 inhabitants were chosen by way of example (Figs. 6 and 7 and Table 1). Although all the data highlighted a reduction of the motorisation rate due to the global economic crisis (2011) and the consequent drastic reduction in the continental vehicles sales, in the northern Italy city it is possible to identify a systematic rate reduction which began before the crisis. Several and contributing factors are influencing this systematic reduction: targeted regional programs to promote both the replacement of old and highly polluting vehicles (Euro 0, Euro 1) and the reduction of the private car dominance, access and parking restriction to the private car in the city centre and improvements in public transport services (Black et al., 2016; European Commission, 2011). About thirty cities in Italy have developed the Sustainable Urban Mobility Plan (SUMP). This is the case for example of Milan and Turin, highly attentive to the development of local public transport with major investments, but also of Brescia, which has seen a drastic reduction in the motorisation rate from extremely high values (630

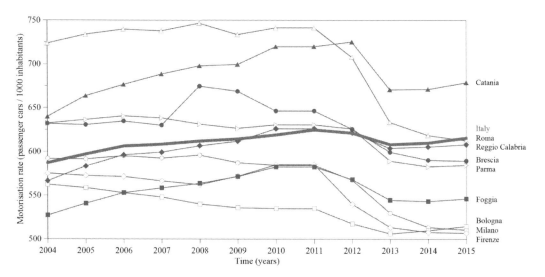

Figure 6.   Motorisation rate progression of some Italian cities (2004–2015).

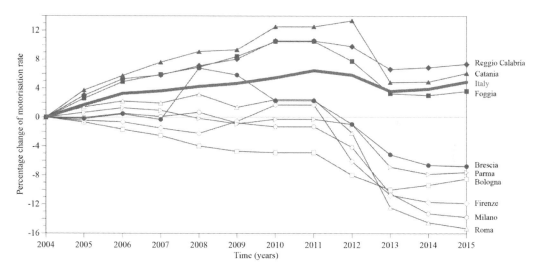

Figure 7.   Percentage change of motorisation progression of some Italian cities (2004–2015).

pc/1000 inh) to about 590 pc/1000 inh also with the introduction of the metro line and major sustainable mobility strategies (pedestrianisation, public transport and cycle lanes, car- and bike-sharing, park and ride, organised collective walking etc.). On the contrary, in the southern Italy none of this is noticeable in view of the marginal disincentive policies, which are unworkable for the lack of a real alternative to public transport and the almost total absence of efficient services and incisive policy for urban mobility.

5   CONCLUSIONS

In the current continental framework, Italy, with reference to the number of inhabitants and the territorial extensions, is one of the country with the greatest motorisation rate but also a country for which the interpretation of this simple statistical parameter necessarily requires a deepening. The national trend curves constructed on historical data are significant to permanently locate the asymptotic rate of growth which is no longer expected to vary appreciably in the future. The volatility of the relationship between motorisation rate and economic framework but, above all, between transport sustainability and economy, typical aspect of the last ten years, highlighted the gap between the Northern and Southern Italy trends, not only for the number of passenger cars but also for the slope of the growth curves. The focused trend analysis on region and on the most significant cities, showed clear differences between the relative growth curves. The car-use reduction for personal trips is strongly influenced by those policies which promote alternatives such as a comprehensive public transport system, especially in the urban areas and for the recurring trips between home and work. It is quite obvious that the management policy of the public transport and the related infrastructure investments implemented in the Northern Italy have led to a reversal trend, which indicates a close relationship between the motorisation rate and a possible and alternative mobility. The Sothern Italy data, in the simplicity with which they arose, can only be an incentive to reflect on the definition of a new and adequate reassessment of the effectiveness of infrastructural facilities and urban transport system services, apparently not credible as alternative to the car.

REFERENCES

ACEA. 2016. The automobile industry pocket guide. Bruxelles.

ACI. 2016. *Annuario statistico 2016*. Roma.

Black, C., Parkhurst, G. & Shergold, I. 2016. The EVIDENCE Project: Origins, Review Findings and Prospects for Enhanced Urban Transport Appraisal and Evaluation in the Future. *World Transport Policy and Practice 22*(1/2).

Buehler, R. & Pucher, J. 2011. Sustainable transport in freiburg: Lessons from germany's environmental capital. *International Journal of Sustainable Transportation 5*(1): 43–70.

Buehler, R., Pucher, J., Gerike, R., & Götschi, T. 2016. Reducing car dependence in the heart of Europe: lessons from Germany, Austria, and Switzerland. *Transport review:* 1–25.

Busi, R. & Maternini, G. 1995. Some considerations comparing metropolitan city motorization trends and public transport systems. In *Energy, environment and technological innovation. III International Congress—Vol. 1* (pp. 15–20). Caracas.

European Commission. 2004. Reclaiming city streets for people—Chaos or quality of life? Luxembourg.

European Commission. 2011. *WHITE PAPER—Roadmap to a Single European Transport Area—Towards a competitive and resource efficient transport system.* Bruxelles.

EUROSTAT. 2016. Passenger cars in the EU—statistics explained. Retrieved from http://ec.europa.eu/eurostat

ISTAT. 2016. Popolazione residente. Retrieved from http://dati.istat.it/

Maternini, M. & Caracoglia, S. 1966. Considerazioni sul futuro sviluppo del traffico di autovetture. *Le Strade* 6: 239–252.

Tocchetti, L. 1965. *Lezioni di costruzioni di strade, ferrovie ed aeroporti—Volume 2*. Napoli: Pellerano & Del Gaudio.

Transport Infrastructure and Systems – Dell'Acqua & Wegman (Eds)
© 2017 Taylor & Francis Group, London, ISBN 978-1-138-03009-1

Competition between European Mediterranean containers transhipments ports

Elen Twrdy & Milan Batista

Faculty of Maritime Studies and Transport, University of Ljubljana, Portorož, Slovenia

ABSTRACT: The phenomenon of increasing size of container ships calling at European ports is explained by economies of scale. Ports, as one of the most important links in the logistics chain, are forced to follow this fast progress of ship size and to adjust all technology and process to the new environment. Not all ports are able to do this because they do not have enough container throughput and in this condition the transhipment ports become important players.

This study present models to evaluate container transhipment ports in Mediterranean Sea (Med). Models are prepared based on available containers transhipment data and they describe the dynamic of container throughput in selected Med transhipment ports. We present a detailed analysis on dynamics of containers by using portfolio analysis, market share analysis and shift-share analysis. Then we used a simple Markov chain method to predict behaviour of this ports with respect to the identified trends in containers traffic and Lotka-Volterra dynamical model to identify possible competition/cooperation relationship between ports.

1 INTRODUCTION

According to UNCTAD the global seaborne shipments have increased by 3.4 per cent in 2014, and reached the total of 9.84 billion tons. In the same year containerized trade increased by 5.6 per cent and reached 171 million TEUs.

Trends in maritime container transport give preference to big container ships due to positive effects of economy of scale. Increase of container traffic and increase of ships size have resulted in decisions of shipping companies to directed as much as possible on a limited number of ports of call. That is clearly showed in growth of container throughput in European ports, where ports with the developed transhipment services recorded the highest containers volumes than other ports in Europe. The typical example are ports in Mediterranean (Med), where we can find a port with big transhipment incidence.

Container throughput in Med was significant in last 20 years and even during the global financial and economic crisis of 2008, some ports (Valencia and Malta) didn't had decline in overall container throughput or it was really small and insignificant. In the last period trade between Europe and the Far East has increased and become even more important than trade between Europe and the US.

For this reason, also the importance of Med ports has increased.

According to Notteboom (Notteboom, 2010, Notteboom et al., 2014) we can divide containers ports to the gateway ports, transhipment ports and mixed ports. Roughly speaking mixed ports has transhipment incidence, that is percent of share of transhipment in total throughput, over 50% and transhipment ports over 75%. There are some ports with the share of 95% of their total throughput.

A transhipment port must have a good geographical position, maritime condition and equipment for accepting big container ships, called mother ships. Containers are unloaded from the mother ship to a transhipment terminal and in a few days (as soon as possible) loaded to a smaller feeder ship that transports them to the final port. After unloading the containers, feeder ship is loaded with new containers that need to be transported to the transhipment terminal, and from terminal loaded on a mother ship.

Nowadays there are five major transhipment ports (Table 1, Figure 1) in Med: Algeciras, Valencia (Spain), Gioia Tauro (Italy), Marsaxlokk (Malta) and Piraeus (Greece). Historically Ports of Algeciras, Marsaxlokk and Piraeus are in first generation of transhipment ports started before 1990, in second generation, started in the mid/late

Table 1. Characteristic of main European Mediterranean transhipment ports.

Port	Transhipment incidence (%) in 2008	2012	Market served	Main player
Algeciras	95,2	91.1	Relay and interlining	Maersk, Hanjin
Valencia	43,9	51.0	West Mediterranean	MSC
Marsax-lokk	93	95.5	West Mediterranean, interlining	CMA-CGM
Gioia Tauro	92,9	93.6	Central and East Mediterranean	Maersk
Piraeus	8,2	80.0	East Mediterranean	Cosco Group

(Notteboom et al., 2014)

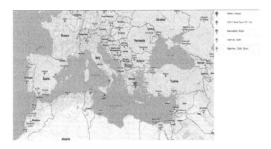

Figure 1. Location of transhipment ports in Mediterranean.

1990s, there is Italian port of Gioia Tauro, while in third generation, started in early/late 2000s, is port of Valencia. In our study we didn't include ports on African coast even though there are some important ports such as Port Said, which was in the first generation of transhipment ports.

2 THE DATA

In this study we use publically available containers total throughput data throughput for each analysed port for period 1995–2015. However as was noted by (Notteboom et al., 2014) the transhipment share in total throughput is hard to find. Some estimations are given in Table 1.

The total containers throughput data we use in this study were weighted by average percentages, except for port of Valencia, where we use natural growth function to model transition from low transhipment share of 18% to current 50% transhipment share in total containers throughput.

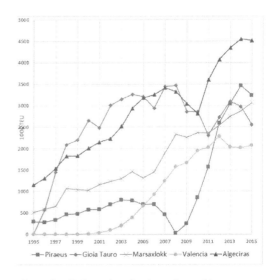

Figure 2. Estimated evaluation of transhipment containers throughput (in TEU).

3 ANALYSE OF THE DATA

In this section we will use a common market metrics to analyse time evaluation of transhipment containers traffic (Farris, 2009). There are three characteristic of a market of interest: its growth, its concentration and its instability.

3.1 Market share

Market share $S_{i,k}$ for port i in time k is defined as ratio between its actual port throughput $x_{i,k}$ at time k and total throughput x_k of port system at time k

$$s_{i,k} \equiv \frac{x_{i,k}}{x_k} \quad (i = 1,...,n; k = 0,1,2...) \qquad (1)$$

Market share is thus characteristics of port within system. From the definition follows that

$$\sum_{i=1}^{n} s_{i,k} = 1 \qquad (2)$$

The time evaluation of market share of discussed ports is shown on Figure 3. On this figure we can see that from the 1995 Algeciras lost a big part of market but is still a dominant player, while Pireaus made a big step from the bottom in 2008 and in 2015 obtain a 22 per cent of market.

3.2 Market concentration

Common measures for market concentration within a business system are the Herfindahl-Hirschman

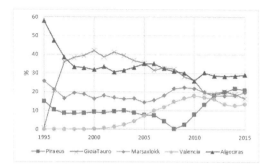

Figure 3. Evaluation of market shares.

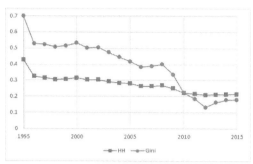

Figure 4. Evaluation of HH index and Gini index.

Index (HHI) and Gini index. HHI index is calculated using following formula

$$H_t \equiv \sum_{i=1}^{n} s_{i,t}^2 \qquad (3)$$

It can be shown that $1/n \leq H_t \leq 1$ where $H_t = 1/n$ if all ports has same market share and $H_t = 1$ if one port took all the market. The Gini index as is defined as "relative mean difference" of market shares (Santos and Guerrero, 2010)

$$G_t \equiv \frac{\sum_{i=1}^{n}\sum_{j=1}^{n}\left|s_{i,t}-s_{j,t}\right|}{2(n-1)} \qquad (4)$$

We normalize the index by $n-1$ rather than by n, and we have $0 \leq G_t \leq 1$. When all ports have equal market share then $G_t = 0$ and if one ports take all throughput then $G_t = 1$.

Time evaluation of HHI and Gini index is shown on Figure 4. We can observe that during last 20 years Gini index falls from about 0.7 to about 0.18. Also HHI falls from about 0.42 to about 0.2. This indicate tendency of containers transshipment deconcentration in Med. Roughly speaking HHI says that at present time all ports within the business system are equitable i.e. have shame market share. This is however not quite true (see Figure 3).

3.3 Market instability

A measure for instability is instability index (Hymer and Pashigian, 1962, Mazzucato, 1998)

$$I_t \equiv \tfrac{1}{2}\sum_{i=1}^{n}\left|s_{i,t}-s_{i,t-1}\right| \qquad (5)$$

From the Figure 5 we can see that containers transhipment was relatively instable business at the end of last millennium since about 20% of containers throughput was shifted between ports. Now the market is relatively stable since only 2 do 4% of containers are shifted between ports.

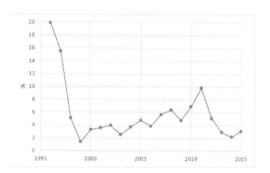

Figure 5. Evaluation of instability index (relative containers shift).

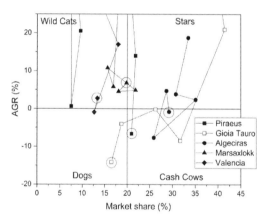

Figure 6. BCG matrix for 1998 to 2015 (in cicle). AGR is annual growth rate.

3.4 Growth-share matrix

On Figure 6 we present the time evaluation in Growth-share matrix (BGC—matrix) of Med ports. Using this matrix, we can describe dynamics of particular port as follows: only the port of Gioia Tauro is in a square 'dogs' but in 1995 it was in a square 'stars'; ports of Piraeus and Algeciras have a position of 'cash cow', while ports of Valencia and Malta are 'wild cats'.

In the language of BCG matrix no future large investments should be spend into ports of Piraeus and Algeciras because of their low growth, while port of Gioia Tauro should be sold off. But such recommendations should be taken with care.

4 FORECASTING WITH MARKOV MODEL

In this section we will use a simple Markov chain model (Twrdy and Batista, 2016) which could help us to answer a simple question about future evaluation of containers traffic in Med such as: will throughput grow, will AGR grow or drop etc.

Consider a sequence $y_0, y_1, ..., y_N$ where y is any time series data. To this sequence we assign a binary sequence $b_1, b_2, ..., b_N$ in the following way

$$b_i = \begin{cases} 0 & y_{i-1} \ge y_i \\ 1 & y_{i-1} < y_i \end{cases} \quad (i = 1, ..., N) \quad (6)$$

Now assume that this sequence is realization of Markov process (Kemeny and Snell, 1976) with the transition probability matrix P with two possible states:

$$\begin{array}{c} \text{next state} \\ \begin{array}{cc} 0 & 1 \end{array} \end{array} \quad (7)$$

$$\text{current state} \begin{array}{c} 0 \\ 1 \end{array} \begin{bmatrix} p_{00} & p_{01} \\ p_{10} & p_{11} \end{bmatrix}$$

The Markov process is described by

$$\pi_n = \pi_{n-1} P \quad (8)$$

where $\pi_n = \begin{bmatrix} p_0^{(n)} & p_1^{(n)} \end{bmatrix}$ is state vector and $p_i^{(n)}$ is probability that the process will after n steps be in state i. The steady state vector is given by

$$\lim_{n \to \infty} \pi_n = \begin{bmatrix} \dfrac{p_{10}}{p_{01} + p_{10}} & \dfrac{p_{01}}{p_{01} + p_{10}} \end{bmatrix} \quad (9)$$

Transition probabilities p_{ij}, $(i, j = 0, 1)$ are estimated from the realized bit-sequence using maximum likehood estimates which leads to formula (Guttrop, 1995)

$$\hat{p}_{ij} = \dfrac{n_{ij}}{\sum_k n_{ik}} \quad (i, j = 0, 1) \quad (10)$$

where n_{ij} is number of transitions from state i to state j in given sequence.

Before proceed with examples we note that Markov process is assumed to be random. Therefore, we must establish if patterns of bits in the bit-sequence. This is done by Run test (NIST, 2013) where the null-hypothesis is that the sequence is random. Practically the assumption require that the business conditions does not change over the years.

With described model we analyze Market Share data (MS) and Annual Growth Rate data (AGR). Results of calculations are given in Tables 2 and 3.

Table 2. Markov model for market shares.

	Row	Piraeus	GioiaTauro	Marsaxlokk	Valencia	Algeciras	Total
	p00	0.667	0.727	0.500	0.750	0.692	0.667
	p01	0.333	0.273	0.500	0.250	0.308	0.333
	p10	0.300	0.500	0.444	0.067	0.500	0.429
	p11	0.700	0.500	0.556	0.933	0.500	0.571
	ps00	0.474	0.647	0.471	0.211	0.619	0.563
	ps01	0.526	0.353	0.529	0.789	0.381	0.438
Run Test	Reject H0	1	0	0	1	0	0

Table 3. Markov model for AGR.

	Piraeus	GioiaTauro	Marsaxlokk	Valencia	Algeciras	Total
p00	0.615	0.643	0.300	0.583	0.545	0.455
p01	0.385	0.357	0.700	0.417	0.455	0.545
p10	0.833	1.000	0.667	0.571	0.625	0.750
p11	0.167	0.000	0.333	0.429	0.375	0.250
ps00	0.684	0.737	0.488	0.578	0.579	0.579
ps01	0.316	0.263	0.512	0.422	0.421	0.421
Reject H0	0	0	0	0	0	0

From Table 2 we can see that on a long run port of GioiaTauro and port of Algeciras will (about 60% of the time) lose their market shares, among other three ports. Port of Valencia will (about 80% of time) increase their market share.

Prediction about AGR is different. Except of port of Marsaxlokk, all other ports will on a long run note increase their AGR. These observations are graphically shown on graph in Figure 7.

According to all presented methods and models we prepare (Figure 7) a long run probability values for growth of market share and annual growth rate AGR. We can see that port of Marsaxlokk is the only one with the AGR and MS higher than 0,5.

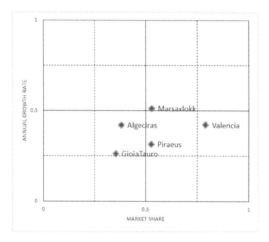

Figure 7. Long run probability values for growth of market share and annual growth rate.

5 COMPETITION-COOPERATION RELATIONSHIPS BETWEEN MED PORTS

In this section we use Lotka-Volterra predator-prey dynamical model to identify possible competition/cooperation relationships between Med tranship-ment ports (Twrdy and Batista, 2016). The system has the following form

$$\frac{dx_i}{dt} = a_i x_i + \sum_{j=1}^{n} b_{ij} x_i x_j \quad (i = 1, ..., n) \tag{11}$$

where x_i is total throughput of port i, t is time and n is the number of ports in the system. In our case $n = 5$ the coefficients a_i as natural decay/growth rates; off-diagonal coefficients b_{ij} $(i \neq j)$ are called interacting coefficients while diagonal terms b_{ij} are self-interacting coefficients.

For estimating of the coefficients of the system (5) from empirical data, we minimize the sum of squares of residuals:

$$\sum_{i=1}^{n} \sum_{t=1}^{N} \left(x_{i,t} - \hat{x}_{i,t} \right)^2 = \min, \tag{12}$$

where $x_{i,t}$ are throughputs at time t and $\hat{x}_{i,t}$ are solutions of (11) at time t. For more details see (Twrdy and Batista, 2016). The results of calculation are present in Tables 4, 5 and on Figures 8 and 9.

We have identified the following relationship evolve in last 20 years:

- we have a win-win situation between ports Marsaxlokk and Valencia, Marsaxlokk and

Table 4. Calculated interacting coefficients and growth coefficient for Lotka-Volterra model.

	Piraeus	GioiaTauro	Marsaxlokk	Valencia	Algeciras	Growth	R^2
Piraeus	−5.88E-05	−1.37E-05	8.19E-04	−3.89E-05	−5.45E-04	0.444	0.966
GioiaTauro	−1.01E-05	−8.39E-04	6.93E-05	−1.03E-04	6.05E-05	2.233	0.849
Marsaxlokk	−2.21E-05	5.84E-07	−1.78E-04	7.94E-05	7.08E-05	0.125	0.945
Valencia	−1.28E-05	−2.20E-04	1.45E-04	−4.52E-04	−2.88E-05	1.308	0.994
Algeciras	2.74E-05	1.34E-05	1.98E-04	−1.91E-04	−4.40E-05	−0.014	0.947

Table 5. Competition/cooperation matrix. Relations are read in rows. For example, Piraeus is predator for Marsaxlokk.

	Piraeus	GioiaTauro	Marsaxlokk	Valencia	Algeciras
Piraeus		lose	predator	lose	prey
GioiaTauro	lose		win	lose	win
Marsaxlokk	prey	win		win	win
Valencia	lose	lose	win		lose
Algeciras	predator	win	win	lose	

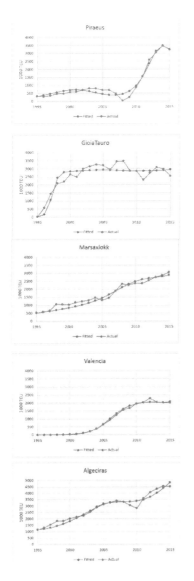

Figure 8.   Fitted containers throughput data with LVM.

Algeciras, Marsaxlokk and GioiaTauro and GioiaTauro and Algeciras.
- We have no win situation between GioiaTauro and Valencia, GioiaTauro and Piraeus, Valencia and Piraeus and Valencia and Algeciras
- We have competition between Marsaxlokk and Piraeus where Marsaxlokk is prey, and between Algeciras and Piraeus where Piraeus is pray.
- These last competition relationships are most likely spurious since Algeciras, Malta and Piraeus covers different hinterlands (see Table 1 and Figure 1).

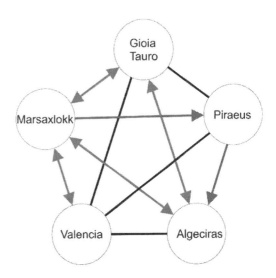

Figure 9.   LVM competition relationships between ports. <-> win-win, _--_ no-win, --> prey-predator.

6   CONCLUSIONS

Through Mediterranean Sea passes the main maritime road between the Far East and Europe and in the last few years' container traffic on this route has increased. Because of economy of scale new demands on market give preference to big ships. Those ships are limited with the number of ports that can accept them. In the Med there are few transhipment ports which have a good geographic location and a good feeder connection with smaller ports. We have analysed five ports (Algeciras, Gioia Tauro, Marsaxlokk, Piraeus and Valencia) and tried to find the relations between them. We found out that to four of those ports the container throughput has increased in the last year, while in the port of Gioia Tauro has decreased for 16 per cent. There aren't big differences in market share between those ports. We calculated the HH index and saw that the concentration is low (around 0,22) which means that all those ports are important players. BCG matrix shows us that no future large investments should be spend into ports of Piraeus and Algeciras because of their low growth while port of Gioia Tauro is in a very bad position with a low growth and a low market share. In the end we used Lotka-Volterra predator-prey dynamical model to identify possible competition/cooperation relationships between ports. In the long run we calculated that port of Marsaxlokk in Malta have a best possibility to become a major transhipment port in Med.

REFERENCES

BAIRD, A.J. 2007. The development of global container transhipment terminals. *In:* WANG, J. (ed.) *Ports, cities, and global supply chains.* Aldershot: Ashgate.

FARRIS, P. 2009. *Key marketing metrics: the 50+ metrics every manager needs to know,* Harlow, England; New York, Financial Times Prentice Hall.

GUTTROP, P. 1995. *Stochastic Modeling of Scientific Data,* Dordrecht, Springer Science+Business Media.

HYMER, S. & PASHIGIAN, P. 1962. Turnover of Firms as a Measure of Market Behavior. *Review of Economics and Statistics,* 44, 82–87.

KEMENY, J.G. & SNELL, J.L. 1976. *Finite Markov Chains,* New York, Springer-Verlag.

MAZZUCATO, M. 1998. A computational model of economies of scale and market share instability equal. Structural Change and Economic Dynamics, 9, 55–83.

NIST 2013. Engineering statistics handbook. Gaithersburg, Md.: National Institute of Standards and Technology (U.S.), International SEMATECH.

NOTTEBOOM, T. 2010. Concentration and the formation of multi-port gateway regions in the European container port system: an update. *Journal of Transport Geography,* 18, 567–583.

NOTTEBOOM, T., PAROLA, F. & SATTA, G. 2014. Partim transhipment volumes. PORTOPIA Consortium.

SANTOS, J.B. & GUERRERO, J.J.B. 2010. Gini's concentration ratio (1908–1914). *StatistiqueElectronic Journ@l for History of Probability and Statistics* [Online], 8.

TWRDY, E. & BATISTA, M. 2016. Modelling Of Container Throughput in Northern Adriatic Ports over the Period 1990–2013. *Journal of Transport Geography,* Article in Press.

Transport Infrastructure and Systems – Dell'Acqua & Wegman (Eds)
© *2017 Taylor & Francis Group, London, ISBN 978-1-138-03009-1*

Using the DataFromSky system to monitor emissions from traffic

V. Adamec & B. Schullerova
Institute of Forensic Engineering, Brno University of Technology, Brno, Czech Republic

A. Babinec & D. Herman
RCE Systems, Ltd., Brno, Czech Republic

J. Pospisil
Faculty of Mechanical Engineering, Brno University of Technology, Brno, Czech Republic

ABSTRACT: The issue of transport emissions and their impact on human health is becoming increasingly topical. One of the possible approaches to solving this issue is using data obtained from the DataFromSky software for modelling of intensities and emission from traffic. This is the software tool that uses real traffic information about the monitored traffic area, such as the type of the passing vehicles, their speed and acceleration profile. These data are obtained on the basis of monitoring the traffic area with the application of the drone technology or camera system. The software is then able to analyse the monitored area and the traffic intensity. Besides, the data obtained may then be used for modelling the emission with a suitable software tool. The paper introduces one of the possible approaches in the assessment of emissions from transport and subsequent application of these results to the transport management and planning, not only in urban conurbations, fully consistent with the concept of the Smart Cities.

1   INTRODUCTION

Transport is an indispensable part of everyday life of the society. However, it also has a negative impact such as air pollution and environmental pollution. This may cause damage to human health from serious illnesses to untimely death. The pollutants also affect vegetation and may cause reduction in agricultural production. They even cause damage to materials and buildings of historical significance.

The Sperling and Gordon study (2008) assumes that by the year 2030, there will be about 2 trillion cars driven in the world. Transport brings risks to human health and the environment in the form of air pollution from exhaust fumes in the form of CO, NOx and volatile organic substances. What also increases the emission load in cities significantly is congestion. 70–80% of the European population currently live in cities, where they also meet and move around. (Nieuwenhuijsen and Khreis, 2016). Air pollution, temperature increase, noise, accidents, and the immobility of the inhabitants are connected with increased untimely death rate and serious illnesses, especially in cities (Bhalla et al., 2014, WHO, 2015, Forouzanfar et al., 2015, Mueller et al. 2016). According to Zhao and Zhu (2012), about 30% of traffic congestion is caused when drivers are trying to find a free parking space. For example, the results of a study in Barcelona show that every day around a million vehicles spend twenty minutes searching for a parking space, which produces 2,400 tons of CO_2. The INFRAS study (2000) says that traffic congestion, pollution and road accidents in EU countries cost 502 trillion euros a year. Especially the economically quickly developing countries are trying to develop strategies for reducing air pollution not only in cities. Nieuwenhuijsen and Khreis (2016) give an example, which monitors current strategies for introducing cities without cars and promotes more pedestrian and bicycle zones for a healthy city life. It compares these strategies with, for example, the cities of Hamburg, Madrid or Oslo, which have been implementing their plans in the public transport and personal transport is already partly limited here. It also mentions cities like Brussels, Copenhagen, Dublin, Paris or Bogota, where various measures have been taken to limit vehicular traffic by, for example, supporting cycling, offering benefits when using public transport or reducing the number of parking spaces in the city centres (Cathkart-Keays, 2015). Other strategic measures include Carsharing, which, thanks to connecting with applications in smart mobile devices, allows an increase in efficiency and use of this service (CarSharing, 2016). The same applies to using the so called Mobypark applications, which allow sharing information about free parking spaces in cities, near public buildings, hospitals, in public parking houses or hotels (Mobypark, 2016).

The EU Commission is preparing a strategy for clean transport, which should become effective after 2020 and it is in compliance with the strategy proposal for low-emission transport. One of the significant impulses is the ever increasing air pollution in cities and a high share (up to one third) of road transport in the creation of greenhouse gases (EurActiv, 2016). The total share of transport in the creation of greenhouse gases is around 23%, as Figure 1 shows based on the EUROSTAT data (2016).

The Eurostat (2015) statistical data show that there is a slight decrease in greenhouse gas emissions, specifically CO_2, which is shown in Figure 2.

Decreasing the emission load in cities is one of the main goals of the Smart Cities concept (Pradeep, 2015, Adamec et al., 2016), which looks for tools which would not only help decrease the emissions and prevent them but also allow precise monitoring with fast and precise evaluation for specific sections and areas. Transport and infrastructure

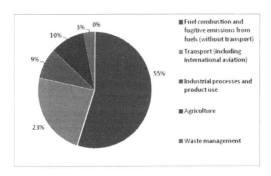

Figure 1. Greenhouse gas emissions, analysis by source sector, EU-28 in 2014 (Eurostat, 2016).

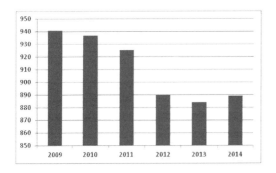

Figure 2. Greenhouse gas emission from transport [mil. tkm] (Eurostat, 2015).

are, therefore, a promising field, in which the latest technologies are used aiming to ensure traffic flow, its reliability and the decrease in the emission load already mentioned.

Monitoring emissions is one of the necessary parts of the measures whose goal is to increase the quality of the lives of inhabitants not only in cities. In the Czech Republic, transport emissions are monitored either based on real measurements, calculation or modelling. The following text briefly introduces these approaches.

2 MONITORING THE EMISSION LOAD FROM TRAFFIC IN CITIES

Stationary and mobile sources emitting pollutants into the air are monitored in the whole of the Czech Republic within the Register of Emissions and Air Pollution Sources (MZP, 2015). The balance of mobile air pollution sources (so called REZZO 4) includes emissions from road, railway, air, and water transport as well as emissions from non-road sources (agricultural, forest and building machinery, army vehicles etc.). As part of the air pollution monitoring, there are specialized stations marked as traffic "hot spots" focusing specifically on information about the air quality in areas which are significantly burdened by traffic. These places meet the criteria for locating the sampling equipment focusing on traffic according to the government regulation no. 597/2006 Coll. on monitoring and assessing the air quality. In pre-determined time intervals, various substances of organic and inorganic nature, especially those with a negative impact on human health, are measured and analysed here. The calculation criteria applied to the air quality assessment are based on Appendix i of the 2008/50/EC Directive and Appendix IV of the 2004/107/EC Directive (CHMI, 2014).

From the point of view of modelling, various software tools supporting the Intelligence Transport Systems (ITS), which allow creating models of line sources, are currently used to monitor the emission load from transport in cities. These software tools have an additional ability to calculate emissions based on the composition of the vehicle fleet (the share of vehicles from each emission group based on their frequency in real traffic) and the properties of the monitored area (corridor). Some of these tools are able to calculate the emission load for each type of vehicle drive. In the Czech Republic, software tools like MEFA, which contains emission factors including a great number of modelled pollutants and an extensive vehicle fleet, are used. It allows a fully automatic emission calculation for any number of line sources (ATEM, 2013). Another example is the ATEM air pollution model, which is a comprehensive tool for air quality assessment on a regional scale. It allows calculation of concentration, when there is a significant terrain obstacle between the pollution source and the reference point, which partly eliminates the influence of the source and allows calculation of pollution from line sources elevated above the terrain at the mouth of the tunnel portal (ATEM, 2015). In 2008, A Methodical Instruction to Reduce Dust Emissions from Traffic (Adamec et al., 2008) was published, which focuses on solid Particles Emissions (PM). This methodology uses the TIMIS software tool, which contains a vast database of concentration maps and it is commonly available as a Microsoft Excel module. The emission factors used in this programme are based on studies carried out between 2004 and 2008. However, other modelling software tools are also used in the Czech Republic, which are commonly applied all over the world such as EMME 4.2 (INRO, 2016), Pramics (Quadstone Paramics, 2016), EnViVer, which is based on VERSIT+ (PTV Group, 2016). The advantage of these software tools is that they can be used not only for calculating the emission load in cities. However, most of them are simulation tools or tools modelling situations based on given data. They do not allow transferring a real situation with the analysis and assessment of this data and displaying not only in the map data. On the contrary, they are able to make calculations and models in regular time intervals or evaluate the long-term situation of the emission load statistically.

In accordance with the Smart City concept, strategies which require demanding monitoring of the air quality are introduced in cities. The Smart Air Quality System is one example which is used in India. It evaluates data from the Closed-Circuit Television (CCTV) such as the total number of vehicles and the number based on the type of vehicle occurring in a specific area and considering the temperature, speed or density is very important (Mehta et al., 2016). Short-term monitoring (in the range of hours) is possible thanks to software tools available in the form of a web application which shows the development of the emission situation graphically and numerically Mehta et al., 2016, Narashid and Mohd, 2010).

It is the possibility of short-term or continuous monitoring of the emission load from traffic in cities which allows more efficient traffic planning and control and thus dealing with undesirable situations in which traffic congestion or other collisions occur and the concentration of vehicles in the particular section increases. That is why it is beneficial to apply this measure already in the traffic planning stage, if possible before the renovation or construction of new roads, intersections, buildings etc. A complex problem to deal with is the

historic city centres and old built-up areas which were originally designed for completely different traffic capacities which are no longer sufficient. That is why measures such as diversion routes, entry of vehicles only with permission, reinforcement of the public transport system etc. are taken. However, these measures are not sufficient in all cities. Therefore, there is also a possibility of monitoring the traffic online with immediate evaluation and if there is a connection with the traffic centre, traffic control is also possible, for example by changing the time intervals at traffic lights intersections, traffic control by the police, changing the driving direction in some streets in critical daytime periods or banning the entry of vehicles in these time periods etc. One possible way of implementing a software tool in the Smart Cities concept is using online monitoring of the traffic and its analysis in selected sections which are, for example, often affected by high levels of traffic and traffic congestion. That is why the following text introduces the DataFromSky software tool (DataFromSky, 2016), which was developed in the Czech Republic.

3 THE DATAFROMSKY SW

This software uses data obtained from aerial photographs and videos which are taken with cameras installed on high-storey buildings, on drones or other devices with the view of the selected area (Fig. 1). The system analyses and processes the data in two stages:

1. Georegistration: Establishing correspondence and mapping between video sequence frames and real world coordinate system.
2. Detection, localisation and tracking of objects of interest (vehicles) in the georegistered video sequence.

The georegistration stage applies distortion correction to every video sequence frame. Subsequently, it uses video stabilisation techniques of image feature matching and perspective transformation estimation, coupled together with known correspondence of landmarks in video sequence, to derive projection of video frames into real world coordinate system. For such georegistered video sequence, it is possible to calculate real world position of any given point on ground surface from its image in the video.

In the second stage, the vehicles are detected and localised in video sequence by application of coupled weak and strong vehicle classifiers, across the whole area of the video frames to produce sets of detections. To improve robustness and performance, the detection candidates are prefiltered by expected area of road surface and results of moving object detection. The detections are then matched and tracked through the video sequence to form vehicle trajectories in the inspected area. To reduce the effect of localisation noise, the trajectories are filtered to form smoothed curves based on vehicle dynamics models.

Based on this data, it is possible to monitor traffic density as well as collision situations or the time for which a vehicle remains still or how long it takes it to drive through an intersection. The collected data then can be used to directly derive trivial traffic flow properties, such as origin-destination matrices, dynamics of vehicles and counts. Additionally, the results can be applied to improve or inspect accuracy of traffic behaviour models by deriving their parameters or confronting the results of simulations, as have been done by CITILABS in their study of Dynasim model accuracy (DataFromSky Study, 2016).

The advantage of the software tool is using real data which does not need to be modelled using other software tools. The data obtained may be evaluated not only at intersections but also for specific road sections. Here, it is possible to monitor vehicles changing lanes, the time of a vehicle driving through intersections at a set speed or the time for which a vehicle remains in the monitored section in traffic congestion (Fig. 4, 5). With regard to the form of the videos obtained, which is from an aerial view, faces and vehicle licence plates are not visible so there is no possible security violation. If an infrared camera is used, records can even be made in poor lighting conditions such as at non-lit intersections, at dusk or at night. What may become a disadvantage is the distinguishing ability of the software tool, when it distinguishes the type of the vehicle (a passenger car/a lorry) but it cannot determine whether the vehicle uses petrol, diesel fuel or, for example, the CNG, LPG fuel. Using the drone technologies also has some limitation, which is the flight time, when videos can be made. That is why other options of camera

Figure 3. Identifying the speed of vehicles in a selected section.

Figure 4. An example of an analysis of vehicles driving through an intersection and pedestrian movement.

Figure 5. An analysis of vehicle speed and the intensity of traffic flows.

placement for monitoring roads and intersections are currently being explored, for example on high-storey buildings. However, these may not be present at each section to be monitored. That is why the possibility of using high-lift equipment is also being considered.

However, in or near cities, congestion occurs especially in specific daytime intervals. Therefore, we can presume that this system may also be used during these time intervals to deal with problematic situations in the section monitored.

4 CONCLUSIONS

The software currently allows an analysis of real data which can be used for traffic control and planning not only in cities but for any road section, intersection, parking place etc. Its application may be useful in monitoring, analysing and evaluating the traffic, when information on vehicle counts, flow gap time, follow time and Origin Destination matrix is offered. It is also able to measure macroscopic traffic flow characteristic at any point or region by analysing all vehicle trajectories in the place, speed and acceleration records for all vehicles tracked in the video. It is able to classify every tracked vehicle according to its visual appearance. Using the data mentioned above, application of the DataFromSky software to measuring the emission load in cities is currently being dealt with. The model area is the city of Brno. The research focuses on a possible connection of the software tool with the TIMIS software which is currently used (Adamec et al., 2016). Another possibility is extending the Data-FromSky SW functions by modelling the emission load during the analysis of the real data obtained. Ensuring the compatibility with another software tool or carrying out direct calculation modelling increases the efficiency and the possibility of evaluating the immediate situation in real time in the section monitored. The aim is to create a supporting tool for traffic control which is in accordance with the Smart Cities concept and leads to reducing the emission load from traffic. This software is currently used not only in the Czech Republic but also at universities in Palermo or Genova.

REFERENCES

Adamec, V. et al. 2016. Issues of Hazardous Materials Transport and Possibilities of Safety Measures in the Concept of Smart Cities. In *EAI Endorsed Transactions on Smart Cities*. Italy: EAI, ICST. org. Extended version.

Adamec et al. 2016. Metodický pokyn ke snižování prašnosti z dopravy. Ministry of Transport of Czech Republic. pp. 19 (In Czech).

Apeltauer, J. et al. 2015. Automatic vehicle trajectory extraction for tradic analysis from video data. Int. Arch. Photogramm. Remote Sens. Spatial Inf Sci,. XL-3/W2, pp. 9–15.

ATEM. 2013. MEFA. *Studio of Ecological Models* [online]. [2016-06-12]. Available on the Internet: http://www.atem.cz/mefa.php

ATEM. 2015. ATEM. *Studio of Ecological Models* [online]. [2016-06-12]. Available on the Internet: http://www.atem.cz/atem.php

BEIS. 2016. Department for Business Innovation & Skills. *GOV.* [online]. [2016-08-08]. Available on the Internet: https://www.gov.uk/government/organisations/department-for-business-innovation-skills

Bhalla et al., 2014 Transport for Health: The Global Burden of Disease From Motorized Road Transport. *World Bank Group*: Washington, DC. (2014). http://documents.worldbank.org/curated/en/2014/01/19308007/transport-health-global-burden-disease-motorized-road-transport

CarSharing. 2016. *The Carsharing Association* [online]. [2016-07-08]. Available on the Internet: http://carsharing.org/

CHMI, 2014. Graphic Yearbook 2014. *Czech Hydrometeorogical Institute* [online]. [2016-06-08]. Available on the Internet: http://portal.chmi.cz/files/portal/docs/uoco/isko/grafroc/14groc/gr14cz/I_uvod_CZ.html

DataFromSky. 2016. Applications. RCE Systems [online]. [2016–08–08]. Available on the Internet: http://datafromsky.com/

DataFromSky Study. 2016. Dynasim model improvement using our data. RCE Systems [online]. [2016-08-08]. Available on the Internet?: http://datafromsky.com/news/dynasim-model-improvement-using-data/

EurActiv 2016. *Businessinfo* [online]. [2016-07-07]. Available from Internet: http://www.businessinfo.cz/cs/clanky/uspornejsi-auta-a-dodavky-ale-i-nakladaky-komise-chysta-strategii-pro-cistou-dopravu–80815.html. (in Czech)

European Comission, 2015. Smart Cities. *Europa.* [2016-07-08]. Available on the Interner: https://ec.europa.eu/digital-agenda/en/smart-cities

Eurostat, 2016a. Transport statistics at regional level. Eurostat Statistic Explained [online]. [2016-08-05]. Available on the Internet: http://ec.europa.eu/eurostat/statistics-explained/index.php/Transport_statistics_at_regional_level

Eurostat, 2016b. Greenhouse gas Emission statistics. Eurostat Statistics Explained [online]. [2016-09-04]. Available on the Internet: http://ec.europa.eu/eurostat/statistics-explained/index.php/Greenhouse_gas_emission_statistics

Eurostat 2015. Greenhouse gas emissions from transport. European Environment Agency [online]. [2016-09-09]. Available on the Internet: http://ec.europa.eu/eurostat/tgm/table.do?tab=table&init=1&language=en&pcode=tsdtr410&plugin=1

IEEE Smart Cities. 2016. About Smart City. *IEE Smart Cities* [online]. [2016-07-08]. Available on the Internet: http://smartcities.ieee.org/about.html

INRO, 2016. EMME 4.2. *INRO* [online]. [2016-06-08]. Available on the Internet: https://www.inrosoftware.com/en/products/emme/

Metha, Y. et al. 2016. Cloud enable Air Quality Detection, Analysis and Prediction—A Smart City Application for Smart Health. In 3rd MEC International Conference on Big Data and Smart City. pp. 272–278. DOI 10.1109/ICBDSC.20167460380.

Mobypark. 2016. Le parking des voyaguers. *Mobypark* [online]. [2016-07-08]. Available on the Internet: https://www.mobypark.com/fr. (in French).

Mueller et al., 2016. Urban and transport planning related exposures and mortality: a health impact assessment for cities. *Environ. Health Perspect.* (in print)

MZP. 2015. Znečištění ovzduší z dopravy. *Ministry of Environment, Czech Republic* [online]. [2016-07-07]. Available on the Internet: http://www.mzp.cz/cz/znecisteni_ovzdusi_dopravy. (in Czech).

Narashid, R.H. and Mohd, W.M.N.W. 2010. Air quality monitoring using remote sensing and gis technologies. In *Proceedings of IEEE International Conference on Science and Social Research*, December, pp. 1186–1191.

Nieuwenhuijsen, M.J., Khreis, H. 2016. Car free cities: Pathway to healthy urban living. *Environment International*, 94, pp. 251–262.

Olofsson, Z., et al. 2016. Development of a tool to assess urban transport sustainability: The case of Swedish cities *International Journal of Sustainable Transportation*, 10(7), pp. 645–656.

Pillai, D., et al. 2016. Tracking city CO2 emissions from space using a high-resolution inverse modelling approach: A case study for Berlin, *Germany Atmospheric Chemistry and Physics*, 16 (15), pp. 9591–9610.

Pradeep, K. What's the really mean of Smart City?. *Smart City Projects* [online]. [2016-06-08]. Available on the Internet: http://www.smartcitiesprojects.com/whats-the-real-mean-of-smart-city/

PTV Group. 2016. *PTV Vision* [online]. [2016-05-08] http://vision-traffic.ptvgroup.com/en-us/products/ptv-visum/

Smart Cities Council. 2016. Information Center. *Smart Cities Council* [online]. [2016-07-08]. Availlable on the Internet: http://smartcitiescouncil.com/smart-cities-information-center/definitions-and-overviews

Taefi, T.T., et al. 2016. Supporting the adoption of electric vehicles in urban road freight transport—A multi-criteria analysis of policy measures in Germany. *Transportation Research Part A: Policy and Practice*, 91, pp. 61–79.

WHO, 2015. *Global health Observatory* [online]. [2016-09-02]. Available on the Internet: http://www.who.int/gho/road_safety/mortality/en/ (2015) accessed 26/5/2016

Zhao, H. and Zhu, J. 2012. Efficient Data Dissemination in Urban VANETs: Parked Vehicles Are Natural Infrastructures. *International Journal of Distributed Sensor Networks*. 8(12).

Transport Infrastructure and Systems – Dell'Acqua & Wegman (Eds)
© 2017 Taylor & Francis Group, London, ISBN 978-1-138-03009-1

Structuring transport decision-making problems through stakeholder engagement: The case of Catania metro accessibility

M. Ignaccolo, G. Inturri, N. Giuffrida, M. Le Pira & V. Torrisi
Department of Civil Engineering and Architecture, University of Catania, Catania, Italy

ABSTRACT: This paper presents a procedure for the structuring of a problem hierarchy by involving key stakeholders, rep-resenting a first step towards a participatory decision-making process. The case study regards the building of a new metro station in Catania (Italy), which will be the closest station to a high-demand district where healthcare and university services and a park-and-ride facility are located. Due to the distance and the high slope between the station and the district, a dedicated transit system linking the two nodes is under study, and four different alternatives have been proposed. Key stakeholders have been identified and involved via in-depth interviews. A questionnaire, a GIS map and a SWOT-like graph have been used to present them the problem and capture their preferences and opinions. From the results of the interviews, a first hierarchy of the problem has been built, that can be used for stakeholder-driven multicriteria analysis.

1 INTRODUCTION

1.1 *Stakeholder engagement in transport decisions*

Transport planning and decision-making is not a simple task, since decisions affect multiple actors with conflicting interests, i.e. the users of the transport systems, citizens, transport operators and, in general, all the stakeholders. Today is widely recognized the importance of engaging stakeholders and citizens in the decision-making process, to improve the quality and equity of the decisions made and to limit protests afterwards. A participatory approach becomes fundamental to find an alternative that should be the best trade-off between the "most shared" solution and the "optimal" one. With regards to transport planning, the EU strongly encourages the Member States to adopt innovative plans such as Sustainable Urban Mobility Plans (SUMPs), where participation is considered as a key issue of success for the decision-making process and for the implementation of the plan itself (Wefering et al., 2014).

Stakeholders involved in transport decisions are de-fined as "people and organizations who hold a stake in a particular issue, even though they have no for-mal role in the decision-making process" (Cascetta et al., 2015).

There are different levels of growing involvement, as represented by Arnstein (1969) in the so called "lad-der of citizen participation", from the lowest step of "Nonparticipation" to the highest one of "Citizen Power". The five Public Engagement levels proposed

by Kelly et al. (2004) ("Stakeholders identification", "Listening", "Information giving", "Consulta-tion", "Participation") have been integrated into the framework of transport planning by Cascetta et al. (2015). Le Pira et al. (2015a) propose a simple scheme to summarize and link the transport plan-ning process with monitoring and participation. The pro-posed decision-making process identifies three main actors and their related roles: planners and experts in charge of analyzing and modelling the transport system by defining the plan structure for the final technical evaluations; stakeholders and citizens that are involved in all the planning phases for the definition of objectives, evaluations criteria and alternatives; decision-makers in charge of the final decision supported by a performance-based ranking and a consensus-based ranking of plan alternatives.

In order to implement an effective participatory approach, it is necessary to understand what kind of tools and methods can help to design and speed up the process of taking a public decision, starting from the first essential phases of stakeholder iden-tification and analysis.

1.2 *Methods and tools for stakeholder engagement*

In general, participation processes require time and money and they are often regarded as compul-sory and quite formal steps of the decision-making process. A modelling approach can be used to sup-port the planning and designing of participation processes aimed at consensus building. In fact, knowing in advance the possible results of different scenarios of interaction among stakeholders can be helpful to plan effective participation processes. In this respect, Agent-Based Modelling (ABM) is suitable to repro-duce participation processes involving stakeholders linked in social networks, understanding the role of interaction in finding a shared decision, and investigating some important parameters such as stakeholder influence, degree of connection, level of communication for the success of the interaction process (Le Pira et al., 2015b, 2016).

There are several techniques that can be used and guidelines for stakeholder involvement in transport decisions (Kelly et al., 2004; Quick and Zhao, 2011; Wefering et al., 2014). Multi-criteria decision-making/aiding (MCDM/A) methods, typically used in traditional planning with a single decision-maker, can be extended for group decisions, proving their usefulness in structuring the problem to include dif-ferent criteria of judgments and points of views and dealing with the complexity of decisions regarding transport planning (Piantanakulchai and Saengkhao, 2003; De Luca, 2014, Le Pira et al., 2015a). MCDM/A in transport can largely benefit from the support of Geographic Information System (GIS), due to the intrinsic spatial nature of transport systems and the capability of GIS maps to easily visualize the impacts of transport choices on land use, environment and communities. Public Participation GIS (PPGIS) or Participatory GIS has been developed as powerful tools for supporting non-experts' involvement in transport decision-making process because of the power of visualization, which increases the awareness about the decision to be made (Sarjakoski, 1998; Tang and Waters, 2005; Zhong et al., 2008).

Therefore, a transport plan should be built with the help of quantitative methods to make a transparent, participatory decision-making process. These methods must include the stakeholders' perspectives and judgements in all the phases of the planning process. Besides, it is necessary to integrate different tools, i.e. (i) MCDM/A methods, (ii) engineering and simulation models, (iii) participatory GIS.

Based on this premise, this paper presents a procedure for the structuring of a problem hierarchy by involving key stakeholders, representing a first step towards a participatory decision-making process. The case study regards the decision about a new transit system in Catania (Italy), which should connect a new metro station with a high-demand district. Four different alternatives have been proposed and the key stakeholders have been identified and involved via in-depth interviews. A questionnaire, a GIS map and a SWOT-like graph have been used to present them the decision problem and capture their preferences and opinions. From the results of the interviews, a first hierarchy of the problem has been built, to be (eventually) used for analysis via appropriate MCDM/A methods, representing a first step of a stakeholder-driven decision-making process.

The remainder of the paper is organized as follows. Section 2 presents a short introduction to the expert-based approach used by the authors. Section 3 investigates the case study and illustrates the structure of the questionnaire, showing the application of the in depth interview method. Section 4 comments the survey's results and shows the procedure used to connect stakeholders' opinion to the construction of a decision problem hierarchy.

2 METHODOLOGY

A hierarchy is a stratified system of ranking and organizing people, things, ideas, etc., where each element of the system, except for the top one, is subordinated to one or more other elements. Though the concept of hierarchy is easily grasped intuitively, it can also be described mathematically. Diagrams of hierarchies are often shaped roughly like pyramids, but other than having a single element at the top, there is nothing necessarily pyramid-shaped about a hierarchy. At each step, the focus is on understanding a single component of the whole, temporarily disregarding the other components at this and all other levels. Through this process, the global understanding of the complex decision problem increases. By using the hierarchy, it is possible to integrate large amounts of information into the understanding of the situation, and with this information structure, to form a better and better picture of the problem as a whole.

The structure of the hierarchy consists of an overall goal, a group of options or alternatives for reaching the goal, and a group of factors or criteria that relate the alternatives to the goal. The criteria are further broken down into sub-criteria, sub-subcriteria, and so on, in as many levels as the problem requires.

Through an expert-based approach, involving via in-depth interviews the key stakeholders, it is possible to present the decision problem and to capture stakeholders' preferences and opinions, in order to define the hierarchy model of the problem. In doing this, stakeholders explore the aspects of the problem at levels from general to detailed and then they express it in the multileveled hierarchy. As they work to build the hierarchy, they increase their understanding of the problem, of its context, and of each other's thoughts and feelings about both.

The use of a Web GIS map presenting some of the main spatial characteristic of the project solutions gives a further contribution: it provides to the stakeholder a first quantitative analysis of the main impacts of the different options. This can be considered as a first step towards a tied approach (such as a WebGIS based MCDA), that can facilitate stakeholders in the understanding of a technical evaluation of the project characteristics.

The task of representing the evaluation problem as a network of interdependent elements distributed can be decomposed into the following steps. The first step concerned the evaluation of the criteria. Secondly, the associated sub-criteria have been identified for each criterion. They are chosen depending on the characteristics of the alternatives. Once the evaluation criteria and sub-criteria have been distinct, they have been to organised in a hierarchy.

The use of this kind of method in the solution of MCDA problems would overcome the unsuitability of traditional methods: in this way in fact the final decision taken by the stakeholder will be transparent and participated but it will also ensure a high technical level of the project solution chosen.

Figure 1. Territorial framework.

3 CASE STUDY

3.1 Territorial framework and problem presentation

Catania is a city of about 300.000 inhabitants and it is located in the eastern part of Sicily; it has an area of about 183 km^2 and a population density of 1.754,54 inhabitants/km^2 (Istat, 2015b).

It's part of a greater Metropolitan Area (750.000 inhabitants), which includes the main municipality and 26 surrounding urban centers, some of which constitute a whole urban fabric with Catania.

The main city (Fig. 1) contains most of the working activities, mixed with residential areas. With reference to the urban area, the transport service is provided by 51 bus lines, a Shuttle line (ALIBUS) connecting the city center with the airport and a second fast bus (called BRT1) connecting the parking Due Obelischi with Stesicoro Square. BRT1 is the first of three lines provided by the City of Catania with equipped lanes protected by curbs on the majority of their path and was promoted commercially as Bus Rapid Transit. In Catania it is also present an urban subway line that currently connects the station "Porto" with the station "Borgo" from which continues as a surface long-distance line.

By the end of 2016 it is expected the extension of the line until the station "Nesima" and it's also planned the opening of a branch linking the station "Galatea" to Piazza Stesicoro. In a few months a new subway station, named *Milo*, will be inaugurated. Its position will make it the closest station to an area where some education and health services and a park—and—ride facility (*S. Sofia* parking) are located (Fig. 2). Due to slope and distance between Milo station and S. Sofia parking, the implementation of a transit service linking the two transport nodes has been proposed.

3.2 Transit alternatives

3.2.1 Minimetro

People mover are modern automatic transport systems on rails, operation is automatic and vehicles are usually equipped with rubber tires circulating on metallic or concrete guides. Main characteristic of this kind of system are a segregated right of way, integral automation, a capacity that can go from very low values up to 3000–4.000 pass/h, a frequency of less than 1 minute and the possibility to overcome slopes of 15%. Between the various typologies of people mover, we are taking into account a particular one, called Minimetro, constructed by the Leitner Ropeways S.p.A.

3.2.2 Bus

Urban buses are the most popular means of collective public transport in our cities and their technology is now firmly established.

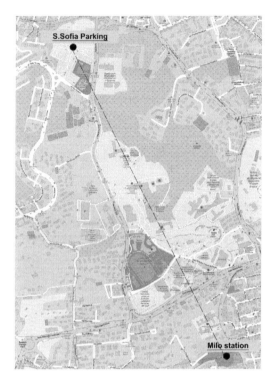

Figure 2. Area of intervention.

Taking into account existing arteries and various boundary conditions along the route Milo—Città Universitaria—S. Sofia, the bus line has been designed: a "tangential" line, which does not enter the heart of the university campus, but rather touches it and has its stops close to the main entrances. Given the fairly linear path (with forward and return lines side by side) and the large spacing between the stops, the system gets close in characteristics to a modern Bus Rapid Transit.

3.2.3 Monorail

Etna Rail is the name of a project of monorail designed for the metropolitan area of Catania, including 3 different lines, with the green one tangential to our study area. It is a modular system, designed for the prefabrication of all major components; it can overcome a maximum slope of 12%; the guide can be manual or with a fully automatic system "driverless" from a remote control room; the speed can reach up to 160 km/h.

3.2.4 Ropeway

Ropeway, which falls into the category of cable transport, is the most common technology used for public transport services on lines with very high slope (from 10% up). It is a shuttle which

consists of a pair of cars (or possibly groups of cars joined together), an ascending and a descending unit, permanently attached to the two ends of the same cable (a steel cable). Vehicles are built according to the transportation needs and characteristics; for the transport of people highly variable capacity cars are employed (from a few people to over one hundred), on the base of suitability to the service (citizens' movements, summer or winter tourism, etc.).

3.3 Stakeholders involved

Five stakeholders, representing the main interest groups affected by the intervention, have been involved:

1. University of Catania, in the person of Rector's delegate for mobility management;
2. Urban bus transit company (Azienda Metropolitana Trasporti—AMT Catania), represented by one of its transport planners;
3. University Students' Council, in the person of a representative student of the students' council;
4. Municipality of Catania, represented by an administration consultant;
5. Metro rail company (FCE—Ferrovia Circumetnea), in the person of its general manager.

3.4 Stakeholder survey

Before the beginning of the questionnaire, a Web GIS map of the main impacts of the four different alternatives has been shown to the stakeholders.

The map has been composed with the aid of the Open Source software QGIS (2016) by using the free plugin qgis2web. The base map references the live tiled map service from the OpenStreetMap (OSM) project. It's made available under the Open Data Commons Open Database License (ODbL). The base map already shows the future subway line track with a dotted line.

Four different layers have been added to the map, showing for each alternative:

– The design track;
– Value of hypothetical capacity (pass/h);
– A noise map
– The location of stops and stations.

Moreover, two more layers show the main interest point of the zone and the Traffic Analysis Zones from Urban Traffic Plan of Catania (PGTU, 2013).

Each stakeholder has been asked with some general information about his company/body: his company position; company's objective and specific abilities; company's main interests in relation to the project; collaboration with other bodies in relation to the project. The stakeholder has then been asked if he would consider the intervention necessary and the reasons of his opinion. Always in relation to the project and to the company's objectives, he has been asked to assign an importance level to four different impacts:

– Transportation impact, related to the ability to meet the demand, considering characteristics related to system performance, such as frequency, speed, etc.;
– Economic impact, related to construction, operation and maintenance costs;
– Social impact, relating to system security, ease of access, acceptability;
– Environmental impact, in terms of air pollution, noise, visual intrusion, etc.

The importance level has been evaluated by the stakeholder with a numeric scale going from 1 (*not important*) to 5 (*very important*).

The WebGIS map with an example of the survey proposed to the stakeholders can be found at the following website:

http://transportmaps.altervista.org/LinkMiloS-Sofia/index.html.

In the final part of the interview the stakeholder has been asked to conduct a SWOT analysis for each alternative, in order to consider for all the transit option:

– its strengths, i.e. the characteristics of the project providing an advantage over others; those should be internal features of the project;
– its weaknesses, i.e. the inherent characteristics that place the project at a disadvantage relative to others;
– its opportunities, meaning the external issues that the project could benefit of;
– its threats, the external elements in the environment that could cause failures of the project.

4 SURVEY'S RESULTS AND HIERARCHY CONSTRUCTION

In this section there are described survey's results and the hierarchy model construction associated to the case study, to be (eventually) used for analysis via appropriate multicriteria decision-making methods.

From the surveys, the fundamental aspects taking into consideration to build the hierarchy have emerged. For the first interviewed stakeholder (University of Catania), the transportation impact results the most uppermost one in terms of accessibility. Furthermore, the environmental and economic impacts were considered important, respectively in terms of pollution and visual intrusion, economic risks and costs distinguishing

in implementation and operation costs for each system.

The second stakeholder (AMT Catania), similarly to the first one, considered the transportation impact one of the most important aspect which have to be taken in consideration, especially in terms of capacity and frequency related to each transport system alternative. The stakeholder also considered the social impact associated to the realization of this transit service as the possibility of an urban redevelopment of the metropolitan area. The third involved stakeholder, University Students' Council, has given more attention to transportation impacts, such concern two main aspects of travel time reliability and comfort of the system. As regards economic impacts, the stakeholder referred to general costs. The fourth stakeholder, (Municipality of Catania) assigned the highest level of importance to the social impact in terms of acceptability and perceived security. The last stakeholder, Metro rail company (FCE), focused his attention on transportation and environmental impacts, like the other stakeholders, also taking into account the operational aspects of the system.

Stakeholders' answers about intervention importance and impacts comparison are summarized in Tables 1 and 2.

The following SWOTs (Tables 3, 4, 5, 6) show the main strengths, weakness, opportunities and threats associated to each transport alternatives.

As regards the first alternative, the economic aspect has an important role. Infact, bus has low

Table 1. Impacts level assigned by the stakeholders.

Do you think the intervention is necessary? Why?	
University of Catania	The intervention is necessary not only because it would be a benefit for University but it would fully explain the realization of metro station, solving the "last mile problem".
AMT	The intervention is necessary because there's a high demand level and a poor service supply.
Students' council	The intervention is necessary because students have high difficulties to find a parking lot in University campus and they need a fast transit service to get to the campus.
Municipality of Catania	The intervention is necessary because it's quite impossible to go through the path by walking or bike.
FCE	The intervention is necessary because of company's objective and also to achieve the final goal of traffic reduction and making the community more livable.

Table 2. Impacts level assigned by the stakeholders.

How do you rate those effects? (1 not important, 5 very important)

	Trasportation	Economic	Social	Environmental
University of Catania	5	4	3	4
AMT	4	2	3	2
Students' council	5	5	4	4
Municipality of Catania	4	3	5	3
FCE	5	3	3	4

Table 3. SWOT analysis for the alternative Bus.

ALTERNATIVE 1: BUS

Strengths	-Low costs of implementation -Immediate implementation -Flexibility
Weakness	-Pollution -High operation and management costs -High costs -Travel time unreliability
Opportunities	-Memorandum of understanding among University, AMT and Municipality of Catania -Possibility of higher level of transit service's performance (increased supply)
Threats	-Financial critically of urban bus transit company -Service punctuality because of road traffic

Table 4. SWOT analysis for the alternative Minimetro.

ALTERNATIVE 2: MINIMETRO

Strengths	-Capillarity -Good interchange with metro -Frequency
Weakness	-High operation costs -Visual impact -Rigidity of the system -Environmental and landscape impacts
Opportunities	-Element of attraction: innovation and modernity -Urban redevelopment of the area -Greater number of enrollees -Low operation costs
Threats	-Non-consolidated technologies -Uncertainty of ministerial authorizations for safety issues -Fixed supply -Dealing with a different transit company

Table 5. SWOT analysis for the alternative Monorail.

ALTERNATIVE 3: MONORAIL

Strengths	-Higher capacity
Weakness	-High implementation and operation costs -Visual intrusion -Rigidity of the system -Environmental and landscape impacts
Opportunities	-Potential to be part of a wider track -Urban redevelopment of a wide area of the metropolitan region -Good interchange with metro
Threats	-Financial uncertainty on realization -Dependence on foreign technologies -Political and economic interests -Higher number of stakeholder involved to reach the consensus

Table 6. SWOT analysis for the alternative Ropeway.

ALTERNATIVE 4: ROPEWAY

Strengths	-Moderate costs of implementation
Weakness	-Poor capillarity -Rigidity of the system -Environmental and landscape impacts -Sensitivity to atmospheric conditions -Uncertainty on use by users -Difficulty of increasing extension and capacity
Opportunities	-Exploitation for landscape tourism purposes -Innovation and modernity image -Architectural interest
Threats	-Fixed supply -Access and egress rigidity

cost of implementation, but at the same time high operation and management costs. Furthermore, the travel time unreliability due to mixed right of way must be taken into account.

The second alternative, represented by the minimetro is considered an element of attraction in terms of innovation and modernity. Conversely to the bus, it is a rigid system and if the urban situation changes, the system cannot be adapted.

It is a well-share opinion for the stakeholders, that the third alternative, the monorail represents a good opportunity for an urban redevelopment of a wide area of the metropolitan region, because it could be part of a wider track. It has also a good interchange with metro. Both minimetro and monorail have the economic problem of high implementation and operation costs, and the environmental problem in terms of visual intrusion and landscape impacts. The same considerations are also valid for the last alternative, with the only difference that the ropeway has moderate costs of implementation.

The hierarchy model for this case study is pyramid-shaped. The goal to be reached is represented by the choice of the best transport system to connect Milo metro station with S. Sofia parking. Four alternative ways of reaching the goal, and four evaluation criteria with three or four evaluation sub-criteria for each criteria were incorporated in the hierarchy. Table 7 shows a scheme of the four macro criteria with the associated sub-criteria, associated to a code in order to facilitate the graphic representation.

Fig. 3 schematically illustrates the developed hierarchy model for the choice of the best transport system, with the goal at the top, the four alternatives at the bottom, and the four criteria with their sub-critera in the middle. The one-way arrows indicate the influence between each element of the hierarchy.

Table 7. Criteria and sub-criteria definition for the hierarchy model.

Criteria	Code name	Sub-criteria	Code name
Transport	C_t	Accessibility	S_{t1}
		Travel time	S_{t2}
		Frequency	S_{t3}
		Comfort	S_{t4}
Economic	C_{ec}	Implementation cost	S_{ec1}
		Economic Risk	S_{ec2}
		Management cost	S_{ec3}
Environmental	C_{en}	Air pollution	S_{en1}
		Noise pollution	S_{en2}
		Visual intrusion	S_{en3}
Social	C_s	Acceptability	S_{s1}
		Urban requalification	S_{s2}
		Perceived security	S_{s3}

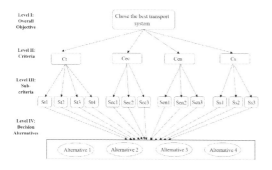

Figure 3. Tree illustrating the hierarchy of criteria and sub-criteria identified in the process.

Once the hierarchy has been constructed, it would be possible to use this hierarchy for analysis via appropriate multicriteria decision-making methods and to get to a final decision based on the results of this process.

5 CONCLUSION

Today is widely recognized the importance of engaging stakeholders and citizens in the decision-making process, to improve the quality and equity of the decisions made and to limit protests afterwards.

This paper presents a procedure for the structuring of a problem hierarchy by involving key stakeholders, representing a first step towards a participatory decision-making process.

The method presented in this paper have been applied to the case study of Catania metro accessibility, specifically regarding the realization of a dedicated transit service linking the two nodes Milo metro station and S. Sofia parking. Four alternatives have been proposed: bus, minimetro, monorail and ropeway.

The decision is based on different criteria, including non-monetary ones, and it should be able both to solve the mobility needs of the district and to improve the quality of life of the whole city.

The method consists on a creation of a stratified system organized in hierarchy in order to focus on understanding a single component of the whole, temporarily disregarding the other components at this and all other levels.

In this view, the key stakeholders belonging to the metro company, the public transport company, the University and the municipality of Catania were identified and involved via in-depth interviews. A questionnaire, a GIS map and a SWOT-like graph have been used to present them the decision problem and capture their preferences and opinions.

The hierarchy model for this case study is pyramid-shaped. The goal to be reached is represented by the choice of the best transport system to connect Milo metro station with S. Sofia parking. Four alternative ways of reaching the goal, and four evaluation criteria with three or four evaluation sub-criteria for each criteria were incorporated in the hierarchy, against which the alternatives need to be measured.

In conclusion, this methodology constitutes a very promising future research line in the field of transport planning and decision-making. From the results of the interviews, Hierarchy of the problem has been built, to be (eventually) used for analysis via appropriate multicriteria decision-making methods, representing a first step of a stakeholder-driven decision-making process.

Next studies could be addressed to the application of an AHP analysis by using the build hierarchy, in order to evaluate alternative solutions and to help decision makers find the alternative that best suits their goal. Moreover, a complete online platform of the survey could be implemented in order to address to the public, get a wider sample of answers.

REFERENCES

Arnstein, S.R., 1969. A ladder of Citizen Participation. *Journal of the American Planning Association*, 35, 216–224.

Cascetta, E., Carteni, A., Pagliara, F. & Montanino, M., 2015. A new look at planning and designing transportation systems: A decision-making model based on cognitive rationality, stakeholder engagement and quantitative methods. *Transport Policy* 38, 27–39.

De Luca, S., 2014. Public engagement in strategic transportation planning: An analytic hierarchy process based approach. *Transport Policy* 33, 110–124.

Kelly, J., Jones, P., Barta, F., Hossinger, R., Witte, A. & Christian, A., 2004. Successful transport decision-making—A project management and stakeholder engagement handbook, Guidemaps consortium.

Le Pira, M., Inturri, G., Ignaccolo, M. & Pluchino, A., 2015a. Analysis of AHP methods and the Pairwise Majority Rule (PMR) for collective preference rankings of sustainable mobility solutions. *Transportation Research Procedia*, 10, 777–787.

Le Pira, M., Inturri, G., Ignaccolo, M., Pluchino, A. & Rapisarda, A., 2015b. Simulating opinion dynamics on stakeholders' networks through agent-based modeling for collective transport decisions. *Procedia Computer Science* 52, 884–889.

Le Pira, M., Ignaccolo, Inturri, G., M., Pluchino, A., Rapisarda, A., 2016. Modelling stakeholder participation in transport planning. Case Studies on *Transport Policy* 4, 230–238.

Piantanakulchai, M. & Saengkhao, N., 2003. Evaluation of alternatives in transportation planning using multi-stakeholders multi-objectives AHP modelling. *Proceedings of the Eastern Asia Society for Transportation Studies*, Vol. 4, October, 2003.

QGIS Development Team, 2016. QGIS Geographic Information System. Open Source Geospatial Foundation Project. <http://www.qgis.org/>

Quick, K. & Zhao, Z.J., 2011. Suggested Design and Management Techniques for Enhancing Public Engagement in Transportation Policymaking. Report No. CTS 11–24, Center for Transportation Studies, University of Minnesota, Minneapolis, Minnesota.

Sarjakoski, T., 1998. Networked GIS for public participation—emphasis on utilizing image data. *Comput., Environ. and Urban Systems* 22 (4), 381–392.

Tang, K.X. & Waters, N.M., 2005. The internet, GIS and public participation in transportation planning. *Progress in Planning* 64, 7–62.

Wefering, F., Rupprecht, S., Bührmann, S. & Böhler-Baedeker, S., 2014. Guidelines. Developing and Implementing a Sustainable Urban Mobility Plan. Rupprecht Consult—Forschung und Beratung GmbH.

Zhong, T., Young, R.K., Lowry, M. & Rutherford, G.S., 2008. A model for public involvement in transportation improvement programming using participatory Geographic Information Systems. *Computers, Environment and Urban System* 32 (2008) 123–133.

Transport Infrastructure and Systems – Dell'Acqua & Wegman (Eds)
© 2017 Taylor & Francis Group, London, ISBN 978-1-138-03009-1

A demand based route generation model for transit network design

N. Oliker & S. Bekhor
Technion—Israel Institute of Technology, Haifa, Israel

ABSTRACT: This paper develops a route generation model for the transit network design problem. The generation of a large candidate route set is a preliminary step in the optimization of network design, which has a significant influence on its quality. Existing methods for route generation are widely based on variations of shortest path algorithms, generating fast routes between high demand centers. The idea of the proposed model is to form routes on corridor links, with high overall demand, in order to strengthen the service attractiveness. The suggested model form two types of routes: the commonly used shortest path and path that pass through high demand corridors. The model is compared to a shortest path method, and show a higher coverage of the network, producing a wider variety of routes, including links that combine multiple demands.

1 INTRODUCTION

The Transit Network Design Problem (TRNDP) has been widely addressed in the literature. The problem is to find the best transit network for a given infrastructure network topology and fixed public transport demand represented by an Origin-Destination (O-D) matrix.

In order to solve the problem, there is a need to generate and evaluate candidate transit routes. As the TRNDP is highly complex and comprise enormous solution space, the route generation is commonly performed as a preliminary step to a meta-heuristic solution method; a large set of candidate route is generated, followed by a meta-heuristic optimization that selects routes and assign frequencies or a schedule.

Existing methods for route generation are widely based on variations of shortest path algorithms. Such routes may accommodate fast and efficient lines between high demand O-D pairs. However, these methods may yield routes that miss high demand corridors.

The idea of the proposed model is to form routes on corridor links, with high overall demand, in order to strengthen the service attractiveness. These routes may accommodate trunk lines with high frequency that serves a large volume of passengers.

2 LITERATURE REVIEW

Route set generation is a procedure of layout transit routes in the network. Since the TRNDP is complex, most methods perform the route set generation as a stand-alone procedure, preliminary to the optimization. The approaches to route generation are heuristic, typically based on shortest path algorithms (Kepaptsoglou and Karlaftis, 2009). As travel time estimation is the widely accepted criteria for paths evaluation, the shortest path is the one with the minimal travel time.

Cipriani et al. (2012) found shortest paths between transit centers and terminals selected by the designer. Nayeem et al. (2014) simply found the shortest path between different pairs of nodes with high entering/leaving demand.

Pattnaik et al. (1998) found the shortest path between terminals chosen by the designer, and successively disable the use of one link of the shortest path to find alternative routes, under constrains of maximal length and overlapping ratio with shortest path.

Baaj and Mahmassani (1991) found k'th shortest paths connecting pairs of high demand or arbitrarily selected terminals. This process was repeated for different terminals.

Fan and Machemehl (2006) found all feasible routes that meet minimal and maximal length constrains. Their method produces a variety of routes but limited in the size of the solved network.

Bagloee and Ceder (2011) allocated stops and terminals by clustering demand centers. They lay down routes in shortest paths considering closeness to transit centers by modifying links' travel time. Their approach achieved convincing network coverage for large-scale networks.

3 METHODOLOGY

The proposed model includes a procedure for the generation of a candidate route set for transit network design. The model form routes by two methods: (a) shortest paths by travel time and (b) paths that pass through high demand corridors. Terminals are randomly selected among nodes with high entering/leaving demand.

The route set generation procedure is presented in Figure 1 and described as follows:

The inputs of the model are the network topology and fixed demand for each O-D pair.

Three preliminary steps are performed prior to the route set generation process. First, the transit demand is assigned over the network by a stochastic network loading. Dial's algorithm (Dial 1971) is a well-known method, which does not require path enumeration. This procedure produces the estimated desired distribution of passengers in the road network, without considering any transit service. The loading on each link represent the 'link weight', which is used to calculate the likelihood that a link will be included in a route.

Second, node weights are defined according to the demand entering and leaving the node. Two different weights are assigned to each node: an outgoing weight, equals to the total demand leaving the node, and an ingoing weight, equals to the total demand entering the node. These weights are later on used for the selection of terminals.

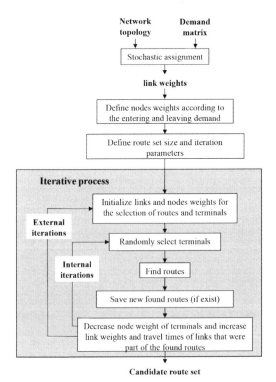

Figure 1. Route set generation scheme.

Third, the desired number of the routes in the set and the iteration parameters are selected. The parameters include: (1) the number of internal iterations; (2) the weight change in each iteration, specified in standard deviation units; and (3) the maximal number of iterations with no new route generation.

The iterative process includes external and internal iterations. In each external iteration the link and node weights are initialized to their original values found in the preliminary steps.

Each internal iteration includes a selection of terminals, by a random roulette wheel selection. The probability of each node to be selected as the origin terminal correspond to its outgoing weight divided by the total outgoing weights in the network. Identically, the probability of a node to be selected as the destination terminal corresponds to its proportional ingoing weight.

After terminals are selected two paths are found: (1) the shortest travel time path between the terminals and (2) the shortest path where the arc values correspond to the complementary link weights. The complementary link weights are the inversed values of the weights found by the stochastic loading. As paths are found in a shortest path method, and the path is aimed to pass through the maximal link weight, the values are inversed (i.e., the lowest arc value corresponds to the link with the maximal loading).

The two generated routes are then added to the route set. If these routes already exist in the list, their repetition is counted.

At the end of each internal iteration, the weights and travel times of the links that were part of the additional routes are increased (regardless if they are new to the route set). The weights of the nodes that were selected as terminals are decreased. In this way, the probability of the terminals and links to be included in routes in the next iteration is reduced. The size of change conducted to the weights in each iteration is pre-determined (in the third preliminary step) and measured in standard deviation separately for each scale of values (i.e., ingoing node weight, outgoing node weight, link travel time and link weight). After a pre-determined number of internal iterations are performed, an external iteration initializes the link weight, travel time and node weight to their original values.

The iterative process terminates when the desired number of routes is reached or alternatively when the maximal number of repetitions with no new route generation is exceeded.

The route generation methods are exemplified in Figure 2. The example network consists of 4 nodes, 3 links with given travel times and 2 O-D demands. The stochastic assignment produces the

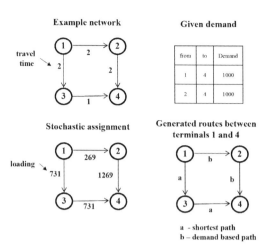

Figure 2. Route generation example.

Figure 3. Winnipeg road network (INRO, 1999).

loading presented in Figure 2. According to the suggested model two types of routes are generated, shortest travel time path and demand based path, pass through high demand corridors. For the terminals 1 and 4 the shortest path is 1–3–4. However, the demand based path is 1–2–4. The last produces a path that has longer travel time, but pass through links with higher overall demand, according to the stochastic loading. In this small example, the generation of routes by the shortest path method would avoid link 2–4, that centralize the highest demand in the network.

The methodology is aimed to produce both fast routes between demand centers and routes that pass through high demand corridors.

4 APPLICATION

The developed methodology was tested on the well-known Winnipeg network provided in the EMME/2 software (INRO, 1999). The network, presented in Figure 3, comprises 154 zones, 2975 links, 903 nodes and 5394 transit O-D pairs, with 18210 total passenger demand.

Runs were conducted with the following parameters: route set size was set between 100 and 10,000. The internal iteration number was set to 10, the weights change was set between 0.5 to 2.5 standard deviation and the maximal number of iterations with no new found route is 100.

The model was implemented in Matlab software, and the running time for a set of 10,000 routes was approximately 4 min. As the generation of a candidate route set is a one-time procedure in the transit network design process, we found the running time satisfactory.

5 RESULTS

The suggested model generate routes by two methods: the common shortest path and demand based path. In order to evaluate the suggested model, it was compared to a shortest path model. The compared model partially applied the suggested model, generates routes on the shortest path between different terminals. The procedure of terminals selection was identical to the suggested model and thereby the comparison enabled to evaluate the isolated impact of the demand based route generation procedure.

The evaluation was conducted by two measures, link coverage and demand coverage, given by:

$$\text{Link coverage} = \frac{\text{Nunber of links included in routes}}{\text{Nunber of network links}}$$

$$(1)$$

$$\begin{aligned}&\text{Demand coverage}\\&= \frac{\text{Total loading in links included in routes}}{\text{Total loading in network links}}\end{aligned} \quad (2)$$

The link coverage, expressed in (1) is the ratio of links included in the route set, out of all network links. The demand coverage, given in (2) express the ratio of loading found by the stochastic assignment, included in the routes. The first measure is the simple topological coverage of the network, and the second express the desired links coverage.

The results of the suggested model and the shortest path model are presented in Figures 4 and 5, showing the performance of the models for the generation of different sized route set. Each model was run 5 times for each size of route set, sum to a total of 25 runs for each model. The average values in each category are presented in the figures.

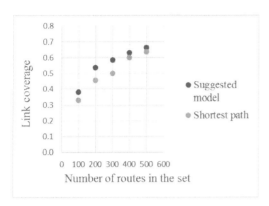

Figure 4. Link coverage results of the suggested and shortest path models.

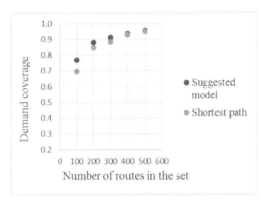

Figure 5. Demand coverage results of the suggested and shortest path models.

The comparison is presented for route set size of no more than 500 routes, as for larger route set the coverage values were very high for both models and the comparison was not significant.

The suggested model showed higher coverage of the network, both in the link coverage (Figure 4) and in the demand coverage (Figure 5) measures. These results can be explained by the two methods generate routes in the suggested model compared to the single method in the shortest path model. The two methods create higher variety of routes, produces by two different decision rules.

6 CONCLUSION

This paper presents a developed methodology for transit route generation as a preliminary step to a transit network design procedure. The model generates routes iteratively by two methods: shortest travel time path and path that pass through high demand corridors. The terminals of the routes are randomly selected among high demand centers.

The method is compared to the application of a shortest path method, and show a higher coverage of the network. The application of the demand based route generation produces a wider variety of routes and form routes that includes links that are desired by multiple demands.

Future work aims to incorporate the suggested route generation model into the transit network design model.

REFERENCES

Baaj, M.H. and Mahmassani, H.S., 1991. An AI-based approach for transit route system planning and design. Journal of advanced transportation, 25(2), pp. 187–209.

Bagloee, S.A. and Ceder, A.A., 2011. Transit-network design methodology for actual-size road networks. Transportation Research Part B: Methodological, 45(10), pp. 1787–1804.

Dial, R.B., 1971. A probabilistic multipath traffic assignment model which obviates path enumeration. Transportation research, 5(2), pp. 83–111.

Fan, W. and Machemehl, R.B., 2006. Optimal transit route network design problem with variable transit demand: genetic algorithm approach. Journal of transportation engineering, 132(1), pp. 40–51.

Inro Consultants, Inc. 2000. EMME/2 User's Manual, Release 9.2, 2000.

Kepaptsoglou, K. and Karlaftis, M., 2009. Transit route network design problem: review. Journal of transportation engineering, 135(8), pp. 491–505.

Nayeem, M.A., Rahman, M.K. and Rahman, M.S., 2014. Transit network design by genetic algorithm with elitism. Transportation Research Part C: Emerging Technologies, 46, pp. 30–45.

Pattnaik, S.B., Mohan, S. and Tom, V.M., 1998. Urban bus transit route network design using genetic algorithm. Journal of transportation engineering, 124(4), pp. 368–375.

Transport Infrastructure and Systems – Dell'Acqua & Wegman (Eds)
© 2017 Taylor & Francis Group, London, ISBN 978-1-138-03009-1

Multi-objective network design problem considering system time minimization and road safety maximization

I. Haas & S. Bekhor
Technion—Israel Institute of Technology, Haifa, Israel

ABSTRACT: This paper discusses the Network Design Problem (NDP) solved for travel time minimization and road safety maximization. Similarly to other NDP models, the model is formulated as a bi-level multi-objective optimization problem. The estimation of the system time benefit and safety benefit is performed using previously developed methods, and the optimal solutions are found using the multi-objective genetic algorithm. The proposed method is demonstrated on a real-size network, using a large set of candidate projects.

1 INTRODUCTION

The increasing demand for transport constitutes a huge challenge for transportation agencies around the world. Scholars and practitioners alike are constantly developing innovative concepts and initiatives for providing a tolerable service-level to all road users. In spite of great progress in this regard, and many different new promising research directions, it seems that traditional methods involving infrastructural investments are not likely to be abandoned in the upcoming future. This is largely due to the substantial impact such investments have on road network performance.

The large impact infrastructure projects have on road network performance on the one hand, and the great expenditures they incur on the other are only part of the reasons that make the selection of the optimal projects an intriguing problem among researchers. This problem, which has been known as the Network Design Problem (NDP), has been drawing much attention in recent decades.

The need to solve NDPs for large instances, i.e. for large networks and a large set of candidate projects, is yet another reason that keeps the NDP relevant even today. The combination of a large network with a large candidate project set raises the need to develop new solution methods for the problem. This is because a large set of candidate projects increases substantially the size of the solution space of the problem, and an increase of the network size, affects the computational effort required for the estimation of the objective sought. As a result, traditional methods involve great computational complexity, which can be reduced only by developing alternative approaches.

In addition, in contrast to the early NDP models that concentrated on travel time minimization only, today the tendency is to try and integrate several different objectives into NDP models. This adds another dimension of complexity to the problem, especially when the models are applied on large networks.

The focus of this paper is on the development of a multi-objective NDP model. The considered objectives in this model are travel time minimization and road-safety maximization. The model is applied on a real-size network with a large set of candidate projects. Although the general problem is not new in the literature, relatively few studies solved the problem using a real-size network for these two criteria with a large set of candidate projects.

This paper is organized as follows. First, a review of the NDP literature is presented including common solution methodologies. Next, the problem is formulated and the solution approach is described. Then, a case study applying the proposed model is presented together with the solutions obtained by implementing the proposed solution approach. The main findings are then discussed, and eventually conclusions are drawn and possible further research directions are outlined.

2 LITERATURE REVIEW

2.1 *The Network Design Problem*

The Network Design Problem (NDP) discusses the optimal selection of projects aiming to improve a given network, under a given cost. Due to its general definition the problem is relevant to various disciplines including energy systems, telecommunication networks, distributed computer networks and transportation networks (Minoux 1989).

The first transportation-related models of the NDP were developed during the 70's (LeBlanc 1975; Abdulaal & LeBlanc 1979). However, the common bi-level formulation of the problem was introduced only later (LeBlanc & Boyce 1986; Yang & Bell 1998). In this bi-level model, the upper level of the problem represents the perspective of the decision makers, striving to optimize a certain goal (minimize travel time or environmental impact, maximize road safety, etc.) and the lower level represents the users perspective, wishing to minimize their travel time, given a network configuration.

The NDP is commonly classified based on the type of the decision variables used in the problem. This type is associated with the sort of projects considered. When discrete decision variables are used, the problem is referred to as the Discrete Network Design Problem (DNDP) and the considered projects are addition of sections or lanes to the network, changing the allocation of lanes per direction or any other project that can be represented using discrete variables (Farvaresh & Sepehri 2013; Fontaine & Minner 2014). The Continuous Network Design Problem (CNDP) on the other hand, focuses on projects which can be represented using continuous variables as changing signal settings, determining toll charges, or adding capacity to existing lanes (Cantarella et al. 2012). The Mixed Network Design Problems (MNDP) involves both types of projects, and has also been applied in several studies (Cantarella & Vitetta 2006).

The early models of the NDP focused on the minimization of travel time only (Abdulaal & LeBlanc 1979). However, in recent years, additional objectives are considered with respect to the NDP. In many cases these have to do with environmental issues, and strive to minimize the total emission (Sharma & Mathew 2011; Miandoabchi et al. 2015). In less frequent times road safety considerations are also integrated into the problem as in Jiang and Szeto (2015).

2.2 Road safety estimation

When the NDP focuses on travel time minimization, one of several common approaches is used to estimate the travel time benefit; early studies concentrated on the estimation of the system optimum (Dantzig et al. 1979), later on the focus was shifted towards finding the user equilibrium (Farvaresh & Sepehri, 2013). This approach is still very common toady alongside other methods which estimate the travel time benefit as the consumer surplus (Jiang and Szeto, 2015). In spite of their differences, one principal unites all mentioned methods. In all of them the total travel time benefit (also referred to as system time benefit) is calculated based on the total travel time—the travel time of all the vehicles in all sections.

However, when road safety is considered, no consensus exists with respect to the estimation method that should be used. In cases were safety considerations were nonetheless integrated, this was performed using various methods. In some cases the total accident cost or number is estimated based on the vehicle miles traveled (Xu et al. 2013). In other cases the number of crashes is estimated based on the change of speed on the links as a result of implementing a certain combination of projects (Jiang and Szeto, 2015).

The lack of uniformity with respect to the estimation of road safety is not unique to NDP models but is also evident when examining road safety literature in general. Several models take the number of crashes, or more precisely, its decrease, as a proxy to the safety level of a network, while others put a greater emphasis on the reduction of the number of fatalities or injuries (Elvik et al., 2009).

One of the models that have gained popularity in recent years, as means of evaluating the number of road crashes in a network is the Crash Prediction Model (CPM). This model is generally location-specific, and associates different characteristics of the examined area with the occurrence of crashes. Those characteristics generally include traffic flows, supplemented by additional measures such as: density or volume/capacity ratio, speed, speed variation along the road section, and geometric characteristics such as shoulder width or the number of lanes and the length of the sections (Saccomanno et al., 2001; Lord et al., 2005).

The most suitable functional form for representing the occurrence of crashes is the negative binomial one. That is because crashes are discrete, non-negative, rare events, and therefore they cannot be adequately captured using linear models. Poisson models, which can represent discrete events, are also less suitable here, since crashes are over-dispersed, i.e. their variance is greater than their mean (El-Basyouny and Sayed, 2006). Using CPM, one can model crashes in sections (Saccomanno et al., 2001; Lord and Persaud, 2004), in intersections (Oh et al., 2004) or interchanges (Lord and Bonneson, 2005).

In spite of the extensive research concerning the development of CPM models, not many applications exist which apply these models for evaluating road safety at the macro-level, and they are also rarely integrated into NDP models. However, such a use seems to be self-evident, since when integrated with traffic assignment model, CPM can assist in determining the safety effect caused by changes in the network. In one such study, which concentrated on minimizing crashes in intersections, the optimal locations for implementing safety projects were chosen based on the use of a CPM model (Mishra, 2013). In this case, an assumption was made that all projects are mutually exclusive. Thus, allowing for the summation of their safety benefits, in order to maximize the total safety benefit obtained.

Another NDP-related study considered both road safety and the change of delay in sections and intersections as a result of implementing safety and operational improvements (Banihashemi, 2007). For this purpose, a CPM was used to evaluate the safety level, and the delay of each section and intersection was evaluated independently, based on traffic simulation and empirical models.

The current study makes use of a CPM calibrated for the Israeli network (Lotan et al., 2014). This CPM is a negative binomial model, which focuses on the estimation of crashes on the sections, and examines each travel direction separately. This model is based on the following explaining variables: area location, section length, traffic flow, standard deviation of the speed in the section, speed limit and number of lanes in the section. The model takes a negative binomial form as follows:

$$sec\_length - 0.282 \cdot flow + 0.045 \cdot std\_speed \quad (1)$$

where TCR = the total number of crashes in a section for a given direction in the rush-hour; is_ north = indicates whether the section is in northern Israel; is_multilane = states whether the section is of multilane type; sp_limit = the allowed speed limit of the section (in kph); sec_length = the length of the section (in km); flow = the traffic flow in the section (in vec./hours). and std_speed = the standard deviation of the speed in the section. Since the volume taken here is the rush-hour volume, the model estimates the number of crashes in the rush-hour accordingly. When this model is applied for every direction of every section in the network, a total estimation of the number of crashes in all network sections can take place.

Note that for a complete estimation of the safety level of the network, the expected number of crashes in intersections and interchanges should also be considered. This is especially important in light of the fact that these road elements account for the majority of crashes. Once a CPM is developed for the Israeli network that considers intersections and interchanges, the presented model can be extended to accommodate also these additional crashes and present a more accurate picture of the safety level of the network.

2.3 Solution space aspects

When discussing the NDP, one of the main challenges associated with this problem has to with its large solution space. When N candidate projects are considered, the total number of project combinations is 2^N. Therefore, depending on the size of the candidate project set, the solution space can grow very rapidly.

To overcome the challenge posed by the large solution space, many choose to solve the NDP using meta-heuristics. To this end, many different meta-heuristic methods have been used including among others the Genetic Algorithm (GA) (Cantarella & Vitetta 2006) and the ant colony (Poorzahedy & Rouhani 2007). Others developed methods for finding the exact solution of the problem (Fontaine & Minner 2014). However, these are usually more suitable for relatively small instances of the problem.

When concentrating on the system time for example, the difficulty to use a large set of candidate projects stems from the need to estimate the system time for many different project combinations. This estimation requires solving the user equilibrium by performing traffic assignment. Traffic assignment itself, can be a relatively time-consuming process, depending on the size of the network used.

In order to facilitate the travel time estimation of different project combinations, a heuristic was developed to approximate the system time of a given project combination, based on the calculation of a limited number of traffic assignments (Haas & Bekhor 2016). This heuristic is also used in this paper, and therefore for clarity purposes it is briefly described using a simple example:

Given a road network with 7 candidate projects, each of them is intended to improve the system time when compared to the base state of the network, where no projects are implemented, we are interested in estimating the system time of a combination of 3 projects (1, 4 and 5). Suppose that each project decreases the system time of the network, when compared with the base case, the maximal system time when implementing all 3 projects is at most the system time obtained using the project that delivers the lowest system time of all 3 projects (project 4 in our example). The minimal system time, on the other hand, is obtained when all candidate projects are implemented in the network. Figure 1 presents the lower and upper bounds of the system time for the described example.

Then, in order to determine the system time of each combination, we assume that each additional project in addition to the one that determined the upper bound, further improves the system time. Using interpolation, we can then estimate the overall system time of each combination.

In order to further increase the accuracy of the presented heuristic, by performing additional traffic assignments, we calculate the system time of each possible pair of candidate projects. Then

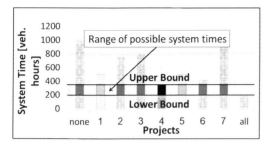

Figure 1. The upper and lower bounds of the system time based on single projects.

we set the bounds as described above, just that this time instead of using single projects to set the upper bound, we use pair of projects. Then, using interpolation we find the system time of a project combination, as follows:

$$T_c = \min_{i \in c} T_i - \frac{y-2}{N-2} \cdot (\min_{i \in c} T_i - T_N) \qquad (2)$$

where, T_c = the system time when combination C is implemented; y = the number of projects in the examined combination; T_i = the system time when implementing the most promising pair of projects in the combination C out of the projects considered; and T_N = the system time when all N projects are implemented. Eq. (2) defines the system time of a certain project combination, where this value is bounded by two values: the system time of the combination which includes all the projects; and the system time of the pair of projects delivering the lowest system time with respect to other projects in the combination.

A review of the literature reveals that in spite of their unprecedented importance for transport systems, road safety and travel time considerations are rarely combined in the NDP. Models that do combine both these considerations usually do not make use of CPM, but use more general modeling tools. In the next section we present a NDP model that combines both these considerations based on an existing CPM and the system time heuristic presented above.

3 METHODOLOGY

3.1 Model formulation

Similarly to the general bi-level formulation of the NDP (Yang & Bell 1998), the formulation of our model also follows a bi-level structure:

Upper level:

$$\text{Maximize } Z_1(x_i, f) \qquad (3)$$
$$\text{Maximize } Z_2(x_i, f) \qquad (4)$$

s.t.

$$\sum_{i=1}^{I} c_i \cdot x_i \leq B \qquad (5)$$

$$x_i \in \{0,1\} \qquad \forall i \in I \qquad (6)$$

Lower level:

$$\text{Minimize } H = \sum_{a \in A} \int_0^{f_a} t_a(\theta) d\theta \qquad (7)$$

Subject to

$$\sum_{k \in K} h_k^{rs} = q_{rs} \qquad \forall r, s \in D \qquad (8)$$

$$h_k^{rs} \geq 0 \; \forall r, s \in D, \forall k \in K \qquad (9)$$

$$x_i \leq \delta_{aki}^{rs} \; \forall a \in A; \forall r, s \in D; \forall k \in K; \forall i \in I \qquad (10)$$

$$f_a = \sum_{r \in D} \sum_{s \in D} \sum_{k \in K} \sum_{i \in I} h_k^{rs} \cdot \delta_{aki}^{rs} + \sum_{r \in D} \sum_{s \in D} \sum_{k \in K} h_k^{rs} \cdot \delta_{ak}^{rs} \; \forall a \in A \qquad (11)$$

$$\delta_{ak}^{rs} + \delta_{aki}^{rs} \leq 1 \; \forall a \in A; \forall r, s \in D; \forall k \in K; \forall i \in I \qquad (12)$$

where, $G(N, A)$ = transportation network, with N and A being the sets of nodes and links, respectively; Z_1, Z_2 = estimation of the upper level objective functions of a certain network configuration; c_i = the cost of project i; B = the given budget; i = candidate projects (i=1,..., I); $x_i = 1$ if project i was selected for implementation and 0 otherwise; f_a = the flow on link a; f = network flow; t_a = the travel time on link a; h_k^{rs} = the flow on path k connecting origin node r and destination node s; δ_{aki}^{rs} = a binary variable indicating whether path k includes link a, which is part of project i, connecting origin node r and destination node s; δ_{ak}^{rs} = a binary variable indicating whether path k includes link a, connecting origin node r and destination node s; D = the set of origin and destination pairs; r, s = origin and destination nodes ($r, s \in$D); and q_{rs} = total demand of origin-destination pair rs ($r, s \in$D).

The upper level, representing the decision makers' perspective striving to maximize two objectives in this case: maximizing the system time benefit, and the road safety benefit obtained by selecting a certain project combination (3,4). The only constraint of the upper problem is the budget constraint (5). The lower level represents the solution of the deterministic user equilibrium. The objective function value H is given in (7). Constraints (8) and (9) ensure flow conservation and positive flows on each path k. Constraint (10) determines the group of feasible links, for links which are included in the candidate project set. Constraint (11) defines the flow on each link to be the sum

934

of flows on the relevant link in each path, both for links that are part of the base network, and for links that are part of the completed project i. Constraint (12) ensures that a link can be considered at most once, either in its original state, when no project is implemented on the link, or as part of a project.

The objectives of the upper level represent the difference between two cases: the current case, where a set of projects is implemented in the network and the base case, where no projects are implemented in the network. Both these objectives can be formally stated as follows:

$$Z_1 = TT_{base} - TT_{current}$$ (13)

$$Z_2 = CN_{base} - CN_{current}$$ (14)

where, Z_1 is the travel time benefit of implementing a certain combination, which is defined as the difference between the travel time in the base case, TT_{base}, and the travel time in the current case, $TT_{current}$, where a certain project combination is used. In the same manner, Z_2, the safety benefit is defined as the difference between the number of crashes in the base case, CN_{base}, and the number of crashes in the current case, $CN_{current}$.

3.2 *Algorithm development*

In order to estimate the objective functions (12,13), we used the system time heuristic developed by Haas & Bekhor (2016), and for the crashes number we used the CPM developed by Lotan et al. (2014), estimating the crashes on road sections. The CPM was developed for the Israeli network, and that is why this was also the network used in the current study. Both these functions were integrated in the upper level of the NDP. In order to solve the multi-objective optimization model formulated we used a variation of the Niched-Pareto Genetic Algorithm II (NSGA-II) (Deb et al. 2002), implemented in MATLAB.

Using NSGA-II, we concentrated on optimizing both objectives simultaneously. Using the heuristic proposed by Haas and Bekhor (2016) and the CPM developed by Lotan et al. (2014), we could calculate the system time benefit and the safety benefit of every project combination. Later, using this data we found the Pareto optimal solutions.

4 CASE STUDY

4.1 *Network and project data*

The road network used for the case study is the interurban road network of Israel, which was used in previous studies (e.g. Bekhor et al. 2013). This

Table 1. Summary of the different project types.

Project type	1. New road section	2. Additional lanes	3. Interchange
No. of projects	12 (No. 1–12)	39 (No. 13–51)	24 (No. 52–75)
Additional capacity [veh./hr]	220,000	1,328,000	–
Free flow travel time improvement [min]	–	–	39
Total length [km]	110	467	90
Total cost [million is]	6,380	12,347	4,963

network includes 8334 links and 680 centroids. For the purpose of this study, the rush-hour demand with 614,565 private-vehicle travels is used (based on the year 2013). 75 projects were selected from the transportation master plan of the year 2040 (Israel Ministry of Transport and Road Safety, 2013) as the initial set of candidate projects, which affect 479 links. The cost of each project is fixed and known in advance and their total cost is 23,690 million Israeli Shekels (IS) (approx. 5,922. million US dollars). The features of the projects are presented in Table 1, and their distribution in the network is presented in Figure 2.

4.2 *Preliminary analysis of system time and safety*

The system time and the total number of accidents in the network were calculated based on the results of a set of traffic assignments, which were performed using the candidate project set. We first analyze the results for the case where a single project is added to the network. The values of the system time and the total number of accidents are shown in Figure 3 and Figure 4.

Figure 3 reveals that the majority of the projects contribute only slightly to the decrease of the system time (less than 100 vehicle hours for the whole network). However, there are several projects that cause more significant changes (up to 800 vehicle hours). Most of these projects involve the construction of new road sections (projects 1–12), which are likely to deliver higher benefits with respect to the system time. As can be deduced from the figure, other types of projects usually deliver lower benefit with respect to the system time. In addition, the system time when all the projects are implemented was also examined and it equals 194,925 vehicle hours, 5,362 hours less than the system time in the base case.

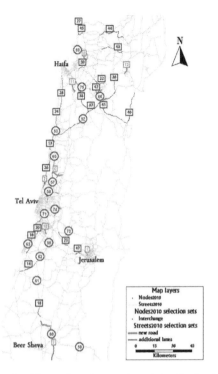

Figure 2. The used road network and the distribution of the projects.

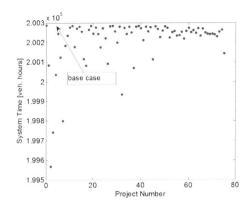

Figure 3. The system time for the base case and for every single project in the network.

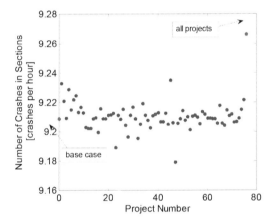

Figure 4. The total number of expected crashes for the base case and for every single project in the network.

In contrast to Figure 3, which shows that every project increases the travel time benefit, Figure 4 shows that there are projects that may increase the expected number of crashes, and therefore decrease the safety benefit. Note however that the number of crashes on the sections does not change dramatically. This holds true also when comparing the number of crashes in the base case with the case where all the projects are implemented. In addition, the implementation of all projects increases the overall expected number of crashes in the network. This finding is in accordance with the fact that the selected projects do not focus on safety maximization, but were proposed based on their expected benefit with respect to congestion relief. According to the model applied to estimate the expected number of crashes (1), the decrease in travel time (and consequently increase in travel speeds) causes an increase in the expected number of crashes.

Next, an analysis of the interdependencies between the different projects was performed with concern to both the system time benefit and the road safety benefit. This analysis approved that interdependency between projects exists to some extent with respect to the system time benefit, while for the road safety benefit no similar interdependency could be confirmed (Haas, 2016).

4.3 Solution of the optimization model

Next, we focus on solving the multi-objective problem. Our goal is to find the set of optimal projects that maximizes simultaneously the system time benefit and the safety benefit in the network. This set of projects should be included in the annual transportation investment plan. For this purpose, we set the budget constraint to 5,000 million IS (approx. 1,250 million US Dollars), which complies with the annual budget allocated for infrastructure investments in Israel (Israel ministry of Finance 2014).

According to the preliminary analysis conducted, the projects were found mutual exclusive with respect to the total safety benefit they deliver. Therefore, the total safety benefit was calculated by summing the individual benefits gained by each of the implemented projects in a combination. Note

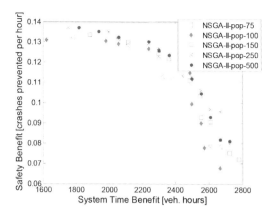

Figure 5. The Pareto front based on NSGA-II.

Table 2. Run times based on NSGA-II.

The used method	Run time [hr]	Number of solutions found
NSGA-II (population size of 75)	0.60	9
NSGA-II (population size of 100)	0.70	10
NSGA-II (population size of 150)	1.29	11
NSGA-II (population size of 250)	2.64	12
NSGA-II (population size of 500)	4.34	11

that as mentioned earlier, not all the safety benefits are positive, and several projects may deliver negative benefits as well.

The same approach could not be applied to the time benefit, as the projects may be interdependent with respect to the time benefit. Therefore, for the estimation of the system time benefit of a given project combination, we used the heuristic suggested by Haas and Bekhor (2016), which takes interdependencies into consideration. To this end, using traffic assignment, we calculated the system time of every possible pair of candidate projects and for the case where all projects are implemented. Then using these results we were able to estimate the system time of every possible combination in the network.

Next, we solved the problem using NSGA-II, with several different populations sizes (75, 100, 150, 250 and 500), in order to examine the sensitivity of the solution to different population sizes. In addition, the problem was solved 5 times for each population size, in order to strive and improve the quality of solution, due to the stochastic character of the GA. The Pareto front using NSGA-II

is presented in Figure 5. The run times together with the number of solutions found are presented in Table 2.

Figure 5 shows that no clear improvement is noticeable with the increase in population size. However, it can still be argued that the solutions based on the population size of 500 are of good quality when compared with smaller population sizes. Note also that as expected, larger population size take longer time to run, as Table 2 shows. The different magnitude of the values obtained in Figure 5 when compared with the values presented in Figure 3 and Figure 4 has to do with the definition of the benefit as presented in eq. (12–13). The results presented in Figure 5 show the obtained benefit, when compared with the base case, where no projects are implemented. Therefore, the values are expected to be smaller when compared with the "absolute" benefit obtained from an implementation of a single project, as presented in Figure 3 and Figure 4.

5 CONCLUSIONS AND FURTHER RESEARCH

This paper presented a solution for the multi-objective NDP. The objective criteria were minimizing the system time and maximizing the safety in the network. In the presented case, benefit interdependency was found with respect to the system time, but not with respect to the road safety. Therefore, when analyzing the safety in the network, the projects were considered mutually exclusive.

In order to find the Pareto front, the multi-objective optimization NSGA-II was used. No major differences were found using different population sizes.

An integration of road safety issued using a CPM is usually not performed when solving the NDP, and therefore the model presented here represents a novelty in this regard. In addition, the presented model was applied for a large set of candidate projects on a real-size network, which demonstrates its ability to provide solutions for realistic problems. It should be noted however that since CPM is location-specific, implementation of the model on different networks will necessitate the use of alternative CPM.

Several further research directions can be identified with respect to this work. For example, extending the used CPM to accommodate also intersections and interchanges, in order to increase the accuracy of the safety evaluation performed. Additional research directions include adding stochastic elements to the problem as the costs or the benefits of the projects. Another option is to add objectives to the problem as environmental considerations or to extend it to several time periods. Thus, consider scheduling issues as well, in addition to the selection of the projects.

REFERENCES

Abdulaal, M. & LeBlanc, L.J. 1979. Continuous equilibrium network design models. *Transportation Research Part B:* 13: 19–32.

Banihashemi, M. 2007. Optimization of highway safety and operation by using crash prediction models with accident modification factors. *Transportation research record* 2019:108–118.

Bekhor S., Cohen Y. & Solomon C. 2013. Evaluating Long-Distance Travel Patterns in Israel by Tracking Cellular Phone Positions. *Journal of Advanced Transportation* 47:435–446.

Cantarella, G.E. & Vitetta, A. 2006. The multi-criteria road network design problem in an urban area. *Transportation* 33:567–588.

Cantarella, B.E., Velonà, P. & Vitetta, A. 2012. Signal setting with demand assignment: global optimization with day-to-day dynamic stability constraints. *Journal of Advanced Transportation* 46: 254–268.

Dantzig, G.B., Harvey, R.P., Landsowne, Z.F., Robinson, D.W. & Maier. S.F. (1979). Formulating and Solving the Network Design Problem by Decomposition. *Transportation Research Part B* 13: 5–17.

Deb, K. Pratap, A., Agarwal, S. & Meyarivan, T. 2002. A fast and elitist multiobjective genetic algorithm: NSGA–II. *IEEE Transactions on Evolutionary Computation* 6: 182–197.

El-Basyouny, K. & Sayed, T. 2006. Comparison of two negative binomial regression techniques in developing accident prediction models. *Transportation Research Record* 1950: 9–16.

Elvik, R., Hoye, A., Vaa, T. & Sorensen, M. 2009. The Handbook of Road Safety Measures. Emerald Group Publishing, Bingley.

Farvaresh, H. & Sepehri, M.M. 2013. A branch and bound algorithm for bi-level discrete network design problem. *Network and Spatial Economics* 13: 67–106.

Fontaine, P. & Minner S., 2014. Benders Decomposition for discrete-continuous linear bilevel problems with application to traffic network design. *Transportation Research Part B* 70:163–172.

Goldberg, D.E., Deb, K. & Clark, J.H. 1991. Genetic algorithms, noise and the sizing of populations. *Complex Systems* 6: 333–362.

Haas, I. 2016. Developing models for the optimal selection of transportation projects (PhD thesis). Technion—Institute of Technology, Haifa, Israel.

Haas, I. & Bekhor, S. 2016. A parsimonious heuristic for the discrete network design problem. *Transportmetrica A: Transport Research* 12: 43–64.

Holland, G.H. 1975. Adaptation in natural and artificial systems, MIT press, Cambridge, MA.

Israel Ministry of Finance, 2014. State budget—proposal for the fiscal year 2015. (In Hebrew)

Israel Ministry of Transport and Road Safety, 2013. Israel Transport Master Plan for Land Air and Sea. (In Hebrew)

Jiang, Y. & Szeto W.Y. 2015. Time-dependent transportation network design that considers health cost. *Transportmetrica A: Transport Science* 11: 74–101.

LeBlanc, L.J. 1975. An algorithm for the discrete network design problem. *Transportation Science* 9: 183–199.

LeBlanc, L.J. & Boyce, D.E. 1986. A Bilevel Programming Algorithm for Exact Solution of the Network Design Problem with User-optimal Flows. *Transportation Research Part B* 20: 259–265.

Lord, D. & Persaud, B.N. 2004. Estimating the safety performance of urban road transportation networks. *Accident Analysis and Prevention* 36: 609–620.

Lord, D. & Bonneson, A. 2005. Calibration of predictive models for estimating safety of ramp design configurations. *Transportation Research Record* 1908: 88–95.

Lord, D., Manar, A. & Vizioli, A. 2005. Modeling crash-flow-density and crash-flow-V/C ratio relationships for rural and urban freeway segments. *Accident Analysis and Prevention* 37: 185–199.

Lotan, T., Bekhor, S., Toledo, T., Gitelman, V., Doveh, E., Morik, S. & Zatmeh S. 2014. A spatial analysis of travel speeds in Israel and their relation to road accidents. The Ran Naor Center for Road Safety Research and the Transportation Research Institute, Technion. (in Hebrew)

Miandoabchi, E., Daneshzand, F., Farahani, R. Z., & Szeto, W. Y. 2015. Time-dependent discrete road network design with both tactical and strategic decisions. *Journal of the Operational Research Society* 66: 894–913.

Minoux, M. 1989. Networks synthesis and optimum network design problems: Models, solution methods and applications. *Networks* 19: 313–360.

Mishra, S. 2013. A synchronized model for crash prediction and resource allocation to prioritize highway safety improvement projects. *Procedia-Social and Behavioral Sciences* 104: 992–1001.

Oh, J., Washington, S. & Choi, K. 2004. Development of accident prediction models for rural highway intersections. *Transportation Research Record* 1897: 18–27.

Poorzahedy, H. & Rouhani, O.M. 2007. Hybrid metaheuristic algorithms for solving network design problem. *European Journal of Operational Research* 182: 578–596.

Saccomanno, F.F, Fu, L. & Roy, R.K. 2001. Geographic information system–based integrated model for analysis and prediction of road crashes. *Transportation Research Record* 1768: 193–202.

Sharma, S. & Mathew, T.V. 2011. Multiobjective network design for emission and travel-time trade-off for a sustainable large urban transportation network. *Environmental and Planning B* 38: 520–538.

Xu, X., Chen, A. & Cheng, L. 2013. Stochastic network design problem with fuzzy goals. *Transportation Research Record* 2399: 23–33.

Yang, H. & Bell, M.G.H. 1998. Models and algorithms for road network design: a review and some new developments. *Transport Reviews* 18: 257–278.

Transport Infrastructure and Systems – Dell'Acqua & Wegman (Eds)
© 2017 Taylor & Francis Group, London, ISBN 978-1-138-03009-1

The impact of road networks in urban distribution spatial

L.M. Jardín, C. Loch & V.M. Kolodziej
The Federal University of Santa Catarina, Florianópolis, Santa Catarina, Brasil

ABSTRACT: The proposal of this article is to make an análisis of incidence of the road network in urban growth processes, using as parameter one of the most important indicators of urban dynamics, the Basic Service Infrastructure Network. Evolution and growth trends around the roads were evaluated based on the Multipurpose Cadastre and integration of dynamic models based on Multi-criteria and Cellular Automata evaluation, to generate Spatio-Temporal simulations and representations of present/future urban transformations near main communication roads.

1 INTRODUCTION

Road networks are vital in communication between cities, also are the main inductor of urban growth and territory structuring, originating transformations in the spatial organization of cities, delineating the urban morphology.

These routes and main avenues, have a close relationship with urban growth, because goods, services and main activities of a city happen in it or pass through. It doesn't mean this is a negative factor, but must be a balance between the ability to densify the areas near, the purpose and tracks use, with the legal instruments, urban and territorial planning that must exist in that territory.

In that way, the changes and transformations that occur in these areas near the tracks, present a greater intensity than the rest of the territory, so to analyze the territorial dynamics from the perspective of these areas's basic service infrastructure network, measuring them at the level of the cadastral parcel, gives us an opportunity to monitor urban growth, predicting possible densification scenarios around the road network and a more frequently reading in urban areas, which supports the territorial management of the institutions involved.

The infrastructure network is basically configured from the public space to the particular property, which demand for their services. In this sense, understanding the public and private space help to trace the premises of the spatial distribution, being Cadastre an essential tool of the space survey, in charge of base mapping, and appointed the official and systematic public record of the territory, based on the survey of the boundaries of each plot.

The study of cadastral plot linked to road and infrastructure networks brings us to the of the multipurpose cadastre application field, a needed concept to understand the links between the responsible institutions of providing, maintaining and creating the infrastructure networks.

The implementation of applied geotechnologies in multipurpose cadastre and the incidence of road network, provides the possibility, together with the dynamic models, to create future scenarios that achieve the ability to prevent some actions of the human need's logic.

2 LITERATURE REVIEW

The International Federation of Surveyors—FIG in 1998 defines Cadastre as a public inventory, methodically ordered, of data concerning all legal territorial objects from certain country or district, based on the measurement of its limits. Such legal territorial objects are systematically identified by means of some distinctive designation. This is defined by law, corresponding to the public or private law. The delimitation of the property, the identifier code, together with descriptive information, show for each separate land object the nature, size, value, and legal rights or restrictions associated with the land object.

For Blachut (1974), Multipurpose Cadastre simply means that the cadastral information, and in particular the cadastral maps, should have the quality and format that cadastral applications require, to be used by various agencies or user entities. The most important requirements for the existence of multipurpose, numeric data and maps must be based on the control system to ensure uniformity and accuracy, all tied to a coordinate system. Finally, the map, as basic component of the system should contain the main natural and artificial features of the land, beyond the limits of the property.

Polidori and Krafta (2005) point out that the city is in constantly change and must be analyzed considering these transformation processes over a given time interval. These changes together alter the city and the countryside, as the process of space-production consumes resources, produces new places and generates externalities, causing a change in the shape of the city, taking some features due to the growth process.

Scarassatti (2007) states that the main road system functions are to provide access, and ensure the free movement of its members. Also, adds that existence of a close link between the major urban projects, land use patterns and the transport system and transport, since the mobility assurance goals and accessibility to these same projects are fully linked to a need of conduct a capacity of prediction of road infrastructure.

For Portugal and Goldner (2003), the process of urbanization occurred over time and the consequent consolidation of activities, land use and the use of vehicular mobility medium, becomes more palpable the constraints and shortage of space in cities, which potentiates, consequently, possible larger installation impacts of infrastructure, particularly the road and transport.

3 APPLICATION MODEL

3.1 Study area

The study area is the Vila Nova neighborhood, located in the city of Joinville, State of Santa Catarina, Brazil (Figure 1). Joinville was founded in 1851, with the arrival of the first germans, swiss and Norwegians inmigrats. This city has a population in the 2010 census of 515.288 hab, and urbanization rate of 96.6%.

From its begginings the occupation of the territory occurred in dispersed character, and along paths that were based on the initial core, towards the route of the current main roads that reach periferics neighborhoods.

The growth of the urban fabric was basically supported by a large shaft, streets that still form the axis of the main urban roads. The configuration of this structure, it can be said that was due to the fact that the entire road link between Curitiba and Florianópolis, passed through the city center to the construction of the BR-101 in the early 1970s.

The first attempt to think more neatly the road structure and its systems were through the Urban Basic Plan (UBP) of 1965. The road guidelines were consolidated later in 1972. Trying to implant a mesh designed reticulated with main and secondary roads in counterpoint to the fabric of the "fish spine" deployed throughout the history of Joinville (Figure 2).

The identification of the pattern of organization of the road network by the PEU and the

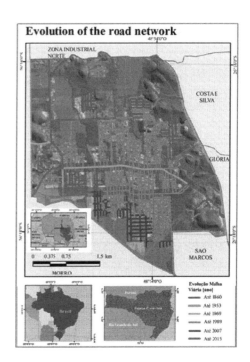

Figure 1. Evolucion de la road network in the neighborhood Vila Nova.

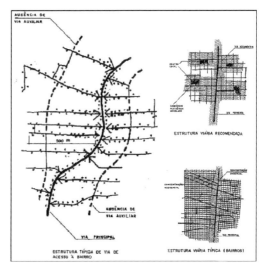

Figure 2. Models for occupation of the neighborhoods of the "Plan of Urban Structure" 1987. Font: Adaptation Joinville (1987).

suggestion for land, proposed in Vila Nova neighborhood, a model with main routes of access to the neighborhood, but also with auxiliary parallel pathways (Figure 2) to respond the demands that come from the perpendicular routes (secondary) many of them with no other connection to the main street (loop road type fish spine). Allowing connection or closure of roads with two or more alterative access to lots. This makes it possible to improve the mobility and displacement of people through the particular transportation as public.

3.2 Method

The study focuses on the evolution of the territory by means of sectoral records influenced by urban growth and road networks (Figure 3). The initial proposal was born to observe the federal road network, state and municipal presenting inducing characteristics of the development of the area they serve, generating a series of changes in the landscape configuration nearby, powered by urban growth, this will produce doubtless changes in network basic infrastructure. What will be the main element of study.

For the application of the methodological model (Figure 4) in the study area, firstly present an alphanumeric-cartographic database processed in GIS, after the multi-criteria assessment of factors and constraints, as well as maps with discrete intervals of Uses of Basic Infrastructure Network,

in this research was used the potable Water consumption 2005–2015, belonging to each neighboorhood's catastral parcel and by the stochastic projection module of Markov's chain, the transition matrixes were determined. Then, proceed to the application of the Automata Cellular function, the validation of participant parameters and preliminary results, and finally execute the simulation of the dynamic model in different time periods.

Neighboorhood's cadastral parcel base of the district in the year 2015 are analisadoos are conformed by 7821 parcels in which found still large portions with rural origin, showing remarkable depopulated spaces that translate in a neighborhood that is still in a rural to urban transformation, as also is the occupation consolidation in the urban voids with access to infrastructure networks.

The consumption information used is the "Media Annual Monthly Consumption" of Potable Water, worked for a single given annual consumption of the parcel which was soon separated into classes of intervals (Table 1).

The factors used were 15, related to basic infraestrurura networks, roads and demographics. By the proximity factors of urban road system and transport (Figure 5) are of great importance in promoting urban development, accessibility and flow to different parts of the neighborhood. The greater importance of a way, due to volume of traffic vehicles or passenger carrying, greater potential urban placement on them. In this group the way was separate by accessibility hierarchy. Also added the "collective transport line" that are within the neighborhood, being more attractive to the concentration of jobs, trades, services and residential areas benefited from this service.

Once obtained the maps with the public areas of influence (Fuzzy module), the weights of each

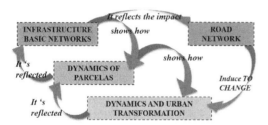

Figure 3. Deductive reasoning of the method adopted for research.

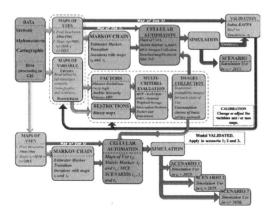

Figure 4. Synthetic method flowchart.

Table 1. Class range used for the maps of "Water consumption" (Consumption Categories).

Class	Mensal annual average consumption m³
1	0,00 m³
2	0,001 m³ a 5 m³
3	5,001 m³ a 10 m³
4	10,001 m³ a 15 m³
5	15,001 m³ a 20 m³
6	20,001 m³ a 25 m³
7	25,001 m³ a 50 m³
8	50,001 m³ a 100 m³
9	100,001 m³ a 500 m³
10	> 500 m³
11	Not apt

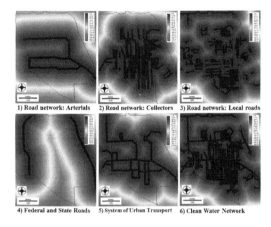

1) Road network: Arterials 2) Road network: Collectors 3) Road network: Local roads

4) Federal and State Roads 5) System of Urban Transport 6) Clean Water Network

Figure 5. Some of the factors used in the application of dynamic model.

Figure 6. Map Raster Aptitude potential of consumption resulting from the evaluation by Multiple Criteria (MCE) with OWA method in Bairro Vila Nova.

factor (Weight-AHP module), and restrictive variables maps (Boolean map), they are added to the multicriteria evaluation analysis by MCE routine (Multi-Criteria evaluation) with OWA (Ordered Weighted Average), which allows you to control the overall level of compensation between the factors and the level of risk in the analysis.

From this process, the suitability map or transition probability of consumption classes (Figure 6) is generated. This map expresses the potential evolution of fitness for all categories without any discrimination.

The simulation of scenarios of water consumption in 2015 is achieved by applying the Automata Cellular module (CA_MARKOV) in Idrisi software, where the combination of independent dynamic variable, the more probabilistic projection of the dependent variable allow emulate future scenarios. The 2015 projection was chosen because has given the real consumption of this year (Figure 7), and will serve to subsequently perform comparison and validation of the simulation model.

An important stage of dynamic modeling consist in comprobation and validation of Automata Cellular Markov's model simulation. To determine their quality and power of agreement ensures predictions of future scenarios with greater certainty of approaching reality.

The validation of the model is based on the comparison of Projected Uses Map in 2015 and the Real Uses Map in 2015 (reference map), using as a comparative criteria the Kappa Index, which gives a percentage of concordance between the maps (Table 2). Agreement that is based on the percentage of pixels correctly classified and also to evaluate the quality of the pixels with respect to the location of the changes.

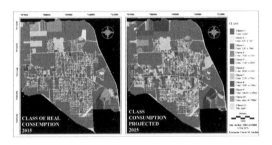

Figure 7. Comparison of the maps of 2015.

Table 2. Kappa index.

Kappa Index—Projected 2015 vs Reality 2015

Coefficients	Valuation	Qualitative interpretation*
%Correct	93,43%	Excellent
Kstandard	69,63%	Very good
Kno	74,42%	Very good
Kloc	76,11%	Very good
KlocationStrata	76,11%	Very good

*Quality Rango associated with Kappa value of Landis & Koch (1977).

Overall, the test had high rates of agreement between the 2015 year Consumer Categories of real and projected maps, supported in consumption data for 2005 and 2010 years, added to the explanatory dynamic array of independent variables presents the settings and adopted at different stages for this model.

After validating the model, it was made the simulation of categories maps of Water Consumption for the years 2020, 2025 and 2030. The simulations take into account the same independent variables (15) and treated them with Multi-criteria Evaluation.

4 ANALYSIS OF RESULTS

4.1 Demand infrastructure basic network

In general, considering that the Vila Nova neighborhood it is an urban neighborhood of residential character, made in his highways in mixed areas of trade, services, public facilities and other, changes in demand for basic infrastructure network can be considered mainly for residential purposes.

In this regard, Figure 8 shows the evolution of the average consumption of water service throughout the neighborhood, adding the projection of elaborate consumption for the years 2020, 2025 and 2030. Furthermore, it exposes the population evolution in the neighborhood, values they are also increasing in time.

It means that consumption displayed in 2005 (64428 m^3) in relation to the year 2015 (121147 m^3), shows an increase in annual average consumption of 88%. Likewise, we compare the district's population in 2005 (18587hab.) In relation to the year 2015 (24127hab.Taxa of Cresc. IBGE Media in Joinville) we see an increase of 29.8% of the population. This means that the increase in consumption was higher compared with the population over the same period.

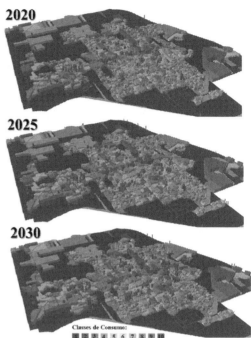

Class: 1 (Cons. 0 m^3); 2 (Cons. 0,01-5m^3); 3 (Cons. 5,01-10m^3); 4 (Cons. 10,01-15m^3); 5 (Cons. 15,01-20m^3); 6 (Cons. 20,01-25m^3); 7 (Cons. 25,01-50m^3); 8 (Cons. 50,01-100m^3) ; 9 (Cons. 100,01-500m^3); 10 (> a 500m^3).

Figure 9. Map projection of the categories of the annual average of monthly consumption of drinking water.

The simulation showed excellent results in the area of urban area where lots with reasonable proportion with respect to the occupation (size of lots not exceeding 2000 m^2). Finally, the stochastic prediction based on Markov chain, the artificial intelligence algorithms of cellular automata and factors developed by analyzes multiple criteria, which shaped the dynamic model employee, be concluded that the tool used to predict the quantity and space demands that infrastructure networks suffer over time is adequate (Figure 9).

4.2 Urban road system

The most emblematic example of public power intervention in the expansion of the road network observed in the review period was the conformation of the Binary system, a corridor for vehicular traffic, which improves mobility in the neighborhood, structuring the territory and changes inducer in corto plazo (Figure 10).

The binary structure allowed the neighborhood with two east-west highways, altering the typical structure of access to the neighborhoods of roads in the form of "fish spine" that lacks auxiliary or parallel tracks. Inducing, thus the changes in the

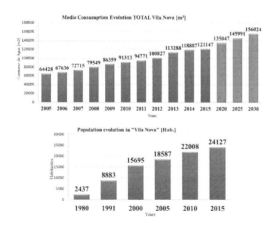

Figure 8. Evolution and projection of water consumption and the Population Evolution of the Vila Nova neighborhood.

Figure 10. Road System, Binary deployment in Vila Nova neighborhood.

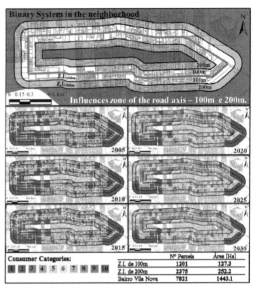

Figure 11. Evolution of "Water consumption" around the system torque in Vila Nova, Euclidean distance to 100m and 200m.

landscape, first on the tracks in the main streets, has an impact on the movement of trade and influenced the real estate appreciation of the area's lots esto was reflected in the water consumption in lots.

4.3 Impact area of binary

Evaluate the relationship between the demand of water service and the existing distance of lots to the road axis. This information is contextualized in the period 2005–2015 with the actual data of average consumption, and from 2015 to 2030 information from projections prepared by dynamic modeling.

The objective of this analysis in particular, is to show the features of consumer classes and spatial link. Important to clarify that the binary system, as previously mentioned, begins to function fully in 2014, so from that date to realize its impact on the environment.

The analysis considered tracks with I-clidianas distances to 100 m Binary System and 200 m, conforming areas around the streets where they assess the demands of the service infrastructure (Figure 11).

With regard to the areas that are within the "zone of influence" of the torque, first, that equidista 100 m, has a 127.3 Ha surface with 1201 lots involved, but the second, far shaft 200 m, has a 252.2 Ha area of 2375 lots, which shows an area-ratio lots 1061 m² /batch for the two cases, means that the concentric rings in 100 m and 200 m road axis have the same ratio of batches, allowing better tune analysis of the consumption.

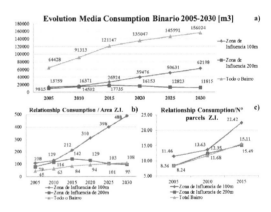

Figure 12. Development of consumption and consumption on the Binary.

Complementing maps the evolution of consumption 2005–2030 in torque area, the graphs annexed consumption growth in the same time interval, with the particularity that relate to the total consumption of the entire neighborhood, allowing visualizaar trends consumption (Figure 12).

In the period 2015 notes that 100 m influence area has an annual average of monthly consumption of 22.2% of the total respondent in the neighborhood, however consumption in the 200 m track consumption was only 14.6% of the total consumed in Vila new. That is, almost the same amount of area in each ring (Z.I.100 m = 127,3 Ha;

944

Z.I.200 m = 252,2 Ha-127,3 Ha), but the concentration of the consumption of lots abutting the torque are increased. The same analysis for the period 2010 showed consumption in the 100 m ring of 17.9%, and 200 m ring of 15.9% with respect to the total consumption of the neighborhood, with very close results. What proves that the period 2010–2015 increased consumption in areas abutting the binary system in the range of 100 m distance, more than the 200 m.

The results observed in Figure 12b shows the relationship of consumption (m³) and the interference area pudendal considerarse consumption density parameter. So, it is evident further growth of the 100 m range, indicating an upward trend in future projections. The reference to the ratio of consumption to the lots involved in each ring (Figure 12c) if you can see the 200 m ring has a similar behavior to the middle of the neighborhood, that is, consumption per lot area formed between the Euclidean distance 100 m and 200 m is the average (m³/lot) of what happens in the neighborhood. However this relationship (m³/lot) in the 100 m influence zone is clearly higher than in the 200 m, and in the 2015 period more than 48% compared to the consumption of a lot in 200 m ring. That is, the index of the year 2015 is 22,42 m³ / lot in 100 m ring, but will be 15,11 m³/lot in 200 m ring and 15.49 m³/lot the entire neighborhood.

In Figure 13, by extrusion of the consumption of lots of 2005 and 2015, one can see the increase in or the water service demands and put the densification of urban occupation. It is also observed as were used in 2005 lots were empty (no consumption).

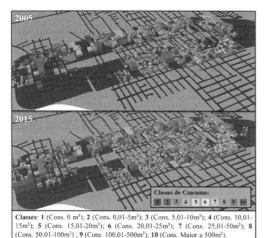

Classes: 1 (Cons. 0 m³); 2 (Cons. 0,01-5m³); 3 (Cons. 5,01-10m³); 4 (Cons. 10,01-15m³); 5 (Cons. 15,01-20m³); 6 (Cons. 20,01-25m³); 7 (Cons. 25,01-50m³); 8 (Cons. 50,01-100m³) ; 9 (Cons. 100,01-500m³); 10 (Cons. Maior a 500m³).

Figure 13. Consumption of lots represented by 3D extrusion in the influences zone binary.

In summary, the impact of the road system has on the lot is decisive, it generates changes in land value, use or purpose motivates the densification in areas close to the main axes, etc., as well, hence the dynamics of these impacts produze one growth in demand of basic infrastructure networks. In other words, the improvement and the interconnection of the road network of the neighborhood lead to a phenomenon of urban growth and the demands of infrastructure.

5 FINAL CONSIDERATIONS

The application of the model that converges in the simulation of scenarios involving several steps, such as software tools and databases used, where most of the results are previous to the dynamic model, they allow to show the trend lines of some of the activities that drive growth and create the features of the urban morphology of the neighborhood.

The results of all processes and applications made the crossing of information, maps and criteria, allow demonstrate the direct relationship of urban growth with the demands of urban infrastructure networks, as are the basic service and the road network. In addition to the ability to obtain predictions of future demands that will experience the urban population allowing contribute to the strategic planning of the analyzed areas and resorting to support the territorial management of the institutions involved in infrastructure networks.

Finally, the main contributions of this research was the linking of information from different institutions account a municipality for its functionality through the technical language register, managing demonstrate multipurpose data that produze each institution in their everyday territorial management. Information that institutions linked to infrastructure networks generate with greater temporary frequency (monthly) and focused on the batch, this allows reading and remains confident of the changes and transformations that take place in the first batch and mark a trend to urban growth. Being the great virtue of the research, the use of these readings sectoral entries for modeling the spatial dynamics, identifying key drivers of urban growth and transformation of the territory.

REFERENCES

Blachut, T. J. Cadastre carious functions characteristics techniques and the planning of land record system. 1974. Ottawa: Ca: National Research Council.

Joinville. 1987. Joinville Plano de Estruturação Urbana 1987: Analises e Recomendações. Joinville: Department of Planning and Coordination, City of Joinville.

Kaufmann, J. & Steudler, D. 1998. Cadastre 2014: A Vision for a Future Cadastral System. Switzerland: FIG - Working Group 1 Commission 7.

Polidori, M. C. & Krafta, R. 2005. Simulando crescimento urbano com integração de fatores naturais, urbanos e institucionais. Geofocus, v. 5: p.156–179.

Portugal, L. & Goldner, L.G. 2003. Estudo de Polos Geradores de Trafego: e de seus impactos nos sistemas viários e de transporte. São Paulo: Edgard Blücher Ltda.

Scarassatti, D. F. 2007. Modelagem Dinâmica na Projeção de Uso do Solo em função da Rede Viária de Transportes. Campinas/SP: Unicamp.

Transport Infrastructure and Systems – Dell'Acqua & Wegman (Eds)
© 2017 Taylor & Francis Group, London, ISBN 978-1-138-03009-1

Validation of pedestrian behaviour scale in Belgrade

B. Antić & D. Pešić
The Faculty of Transport and Traffic Engineering, University of Belgrade, Belgrade, Serbia

N. Milutinović & M. Maslać
The Higher Education Technical School of Professional Studies Kragujevac, Kragujevac, Serbia

ABSTRACT: Pedestrians, as the vulnerable road users, belong to the group of the most vulnerable categories in traffic. Considering the fact that the level of traffic safety mostly depends on the behaviour of all road users, it is very important to carry out the research on pedestrians behaviour, in order to recognize the differences between genders, age groups and other characteristics that can influence their behaviour. The current version of Pedestrian Behaviour Scale (PBS) was conducted among the participants in Belgrade, the capital city of Serbia, on a sample of 403 participants. Factor analysis showed that the data best fit into the five-factor solution (violations, errors, and lapses, aggressive and positive behaviours). The results showed that the greatest number of errors is made by the persons who forced walking, and most number of violations and lapses are made by the youngest and by the oldest age group of the pedestrians.

1 INTRODUCTION

Measurements of risky behaviour of traffic participants are conducted using tools that are based on their self-reported behaviour. The first such tool (Driver Behaviour Questionnaire—DBQ) was developed by Reason et al. (1990), and it was related to the self-reported behaviour of the driver. This was the beginning of the research and conceptual framework of a large number of studies on the behaviour of traffic participants.

In contrast to the large number of studies that are related to the behaviour of the driver (according to De Winter and Dodo (2010), 174 such studies have been conducted), a much smaller number of studies have examined the behaviour of pedestrians. Different versions of the self-reported behaviour were carried out in: Chile (Moyano Diaz, 1997), the UK (Elliott and Baughan, 2004), Turkey (Yildirim, 2007), France (Granié, 2008; Granié et al., 2013), New Zealand (Sullman and Mann, 2009), Brazil (Torquato and Bianchi, 2010), Spain (Sullman et al., 2011), and Serbia (Antić et al., 2016).

Due to the applied PBS version and social and cultural differences between the stated countries, the authors obtained diverse results. However, the most significant conclusion was the fact that the factor structure of the questionnaire best fitted into the three-factor and four-factor solutions. In his research, Moyano Diaz (1997) found the difference between violations, errors and lapses. After that, Yildirim (2007) developed a version of this

tool, differentiating between violations, aggressive behaviours and errors. In both cases, the results showed that more violations were made by men. Furthermore, more errors were noticed among young pedestrians (17–25 years) than among older pedestrians (25–49 years). In contrast to them, Granié et al. (2013) developed a Pedestrian Behaviour Scale PBS in France. This study validated the PBS for all ages, made a difference between violations, errors and lapses, and also provided an understanding of aggressive and positive behaviours by pedestrians toward other road users.

Personal traits and characteristics are related to the differences in behaviour of pedestrians crossing the street. Personal characteristics should be considered to be an important factor which influences pedestrians' attitudes and decisions in risky situations while crossing the street (Zhou et al., 2009). Several studies, conducted worldwide, have surveyed gender and age differences in pedestrians' behaviour. Men tend to make more traffic transgressions and cross the streets in risky situations (Rosenbloom et al., 2004; Díaz, 2002). Also, the young and adolescents make transgressions more frequently than older pedestrians (Díaz, 2002), while older traffic participants show more positive behaviours than younger pedestrians (Bernhoft et Carstensen, 2008).

In 2015, 37 pedestrians were killed in traffic accidents in Belgrade, which represents 36.6% of all killed people in that year in traffic accidents. In the same year, the total number of injured pedestrians was 927 (Road Traffic Safety Agency, 2016).

With this in mind, the main goal of this research is to obtain knowledge about the behaviour of pedestrians in Belgrade, and to attain other objectives related to this:

– to determine specific risky behaviours of pedestrians by applying the PBS,
– to determine the connection between risky behaviour with gender, age, reasons for walking, daily distances walked and the possession of driving license.
– to validate the PBS in Belgrade.

2 MATERIAL AND METHODS

The PBS conducted in Belgrade was based on the previously confirmed versions of the questionnaire on the behaviour of pedestrians (Moyano Diaz, 1997; Yildirim, 2007; Torquato and Bianchi, 2010; Granié, 2013) conducted in Chile, Turkey, Brazil and France. The PBS contained items which had the highest factor loadings in the previous research. In addition to these items, it also included the items relating the use of mobile phones when crossing the street, which were first used in the Serbian version of PBS (Antić et al., 2016).

In Serbia, on the basis of the Law on the Road Traffic Safety, Article 96 (paragraph 2) was defined and reads as follows: "When crossing the street pedestrians must not use a mobile phone or use the headset on their ears". Bearing this in mind, the items related to mobile phone use while pedestrians are crossing the street are grouped with violations. The items refer to talking on mobile phones, reading the contents (text messages, the Internet) and listening to music.

The method of collecting the data was a questionnaire filled in by 403 participants. The questionnaires were distributed to random pedestrians at several locations in Belgrade. Items were presented in random order to avoid bias in participants' answers. The questionnaire had two parts. The first part contained the questions about the social and demographic characteristics (gender and age), and the questions about the most frequent reasons for walking, the daily distance that a pedestrian walked and the possession of driving licence. All the questions were closed-type and participants circled one of the given answers.

The second part of the questionnaire was about the behaviour of pedestrians in traffic, and it contained twenty five items, divided into five groups (violations, errors, lapses, aggressive behaviour and positive behaviour). The first group contained the questions about violations in traffic. A violation was defined as an intentional deviation from the legal rules guiding pedestrian behaviour (Moyano

Diaz 1997, e. g. "I cross the street even though the pedestrian light is red") and the use of mobile phones while crossing the street (e.g. "I read the contents (text messages, the Internet)". Errors were defined as making decisions that put the pedestrian in danger, without disobeying the legal rules (Granié 2013, e.g. "I start street crossing at the marked pedestrian crossing but I finish outside of it"). Lapses were defined as ill-suited behaviours related to the lack of concentration on the task (Moyano Diaz 1997, e.g. "Before crossing the street I did not look left and right because I was in a hurry"). Aggressive behaviours were defined as conflicting behaviours with other road users (Yildirim 2007, e.g. "When I get angry at the driver who did not give me priority at the marked pedestrian crossing, I insult him/he"). Positive behaviours were defined as behaviours that appease social interactions (Granié 2013, e.g. "While crossing the street I give priority to vehicles even though I am at the marked pedestrian crossing"). In this part we used the Likert scale with answers ranging from 1 to 6, 1 being "Never" and 6 "Very often".

The data were analysed in the statistical software package IBM SPSS v. 22. The normality of distribution was tested using the Kolmogorov—Smirnov test. The internal consistency of the questionnaire was assessed using Cronbach's alpha statistic. A Principal Component Analysis (PCA) using the Kaiser's criterion for factor extraction and the orthogonal Varimax rotation method was performed to investigate the underlying structure of the questionnaire and to obtain dimensional aggregated measures of the behaviours of interest. The threshold of the statistical significance has been set to the conventional level of $p \leq 0.05$.

3 RESULTS

The research included 403 participants, out of which 218 were men (54%) and 185 were women (46%). The participants were between 15 and 71 years of age (M = 41.11, SD = 10.15). The reason why participants in Belgrade walked was predominately by necessity (58.3%).

The distances pedestrians would walk during the day were usually in the range of 300–800 meters (42%) or 800–1300 meters (32.0%). When it comes to the possession of a driving licence, the highest number of respondents possessed a driving licence (72.7%). The paper examined the association of gender, age, reasons for walking, daily distances walked, possession of a driving licence and the behaviour of pedestrians. The description of the sample is presented in Table 1.

This paper examined internal consistency and principal component structure of the questionnaire.

Table 1. Description of the sample.

		N	%
Gender	Men	218	54.0
	Women	185	46.0
Age	15–25	148	36.7
	26–35	81	20.1
	36–45	76	18.9
	46–55	65	16.1
	>56	33	8.2
Reasons for walking	Necessity	235	58.3
	Pleasure	186	41.7
Daily distances walked	300 m	55	13.6
	<300–800 m	170	42.2
	800–1300 m	129	32.0
	>1300 m	49	12.2
Possession of driving licence	Yes	293	72.7
	No	110	27.3

The questionnaire had an exceptional internal consistency (Cronbach's alpha 0.72). The internal consistency (Cronbach's alpha) was calculated for ordinary violations (.70), errors (.66), lapses (.72), aggressive behaviour (.71) and positive behaviour (.76). Cronbach's alpha test showed an acceptable internal consistency for all groups of items, except for errors. Principal Component Analysis (PCA) with orthogonal Varimax rotation was carried out on all 25 items of the scale. The scree plot indicated that the data best fitted a five-factor solution, which accounted for 60.6% of the total variance. The Kaiser–Meyer–Olkin measure of sampling adequacy was satisfactory (0.76) and Bartlett's test of sphericity was significant (0.0001). The factor loadings responded well for twenty items, while five items had factor loadings <.33, but they were expelled from further analysis.

Factor 1 "lapse" explained 14.8% of the variance. The lapse was defined by 3 items. In the classification by Moyano Diaz (1997) lapses were defined as ill—suited behaviours related to the lack of concentration on the task (e.g. "Before crossing the street I did not look left and right because I was in a hurry", factor loading: .872).

Factor 2 "positive behaviours" explained 13.2% of the variance. Positive behaviours were defined by 4 items. Positive behaviours were defined as behaviours that appease social interactions, (e.g. "While crossing the street I give priority to vehicles even though I am at the marked pedestrian crossing", factor loading: .751).

Factor 3 "violation", explained 12.6% of the variance. The violation was defined by 4 items, three of which related to the way of crossing the street (e.g. "I cross the street even though the pedestrian light is red", factor loading: .859), and three to the

use of mobile phones while crossing the street (e.g. "I read the contents (text messages, the Internet)", factor loading: .821). The items loading on this axis all had in common the intentional nature of the dangerous behavior or a deliberate offence contrary to the legal rules.

Factor 4 "error" explained 10.3% of the variance. The error was defined by 4 items. An error is a consequence of wrong pedestrians' decisions when crossing the street (e.g. "When I want to overtake a slow-moving person and I do not have enough space to do it on the sidewalk, I go onto the road", factor loading: .710).

Factor 5 "aggressive behaviours" explained 9.7% of the variance. Aggressive behaviours were defined by 3 items. Aggressive behaviours were defined as conflicting behaviours with other road users, in this case towards the driver (e.g. "When I get angry at the driver who did not give me priority at the marked pedestrian crossing, I wave my hand", factor loading: .548).

To examine the correlation between all the variables used in the study, a multiple correlation of coefficients is presented in the paper (see Table 2). Using Pearson's correlation, Table 2 shows the results obtained in the correlations between gender, age, reasons for walking, daily distances walked, possession of a driving licence and PBS scores.

Males walked significantly more out of necessity and crossed shorter distances during the day compared to females. The results also showed that males made significantly more violations and expressed more aggressive behaviours compared to females. Females showed significantly more positive behaviours, while in terms of making lapses and errors there were no significant differences between genders.

Age was positively associated with lapses and positive behaviour, while negatively associated with violations, errors and aggressive behaviour. When it comes to reasons for walking, it was found that persons who walked out of necessity made significantly more errors and expressed significantly more aggressive behaviours compared to those who walked for pleasure.

When it comes to making lapses, violations and expressing positive behaviours, there was no significant difference between persons who walked out of different reasons. Daily distances walked were positively associated with all types of behaviour, except with aggressive behaviour. Persons who possessed a driving licence showed significantly more positive behaviours, while persons who did not have a driving licence showed significantly more violations, errors and aggressive behaviours.

When it comes to the association between different types of behaviour obtained on the scale of behaviour, there are both positive and negative

Table 2. Correlation coefficients between gender, age, reasons for walking, daily distances walked, possession of a driving licence and PBS scores.

		1	2	3	4	5	6	7	8	9	10
1	Gender	–									
2	Age	.003	–								
3	Reasons for walking	.239**	−.106	–							
4	Daily distances walked	−.131*	−.011	−.136	–						
5	Possession of dri. lic.	.050	−.125*	.025	−.058	–					
6	Lapses	−.090	.135**	.003	.109*	.058	–				
7	Positive behavior	.133**	.239*	.048	.173*	−.154*	.038	–			
8	Violation	−.174*	−.494*	−.067	.100*	.222*	.230*	−.391*	–		
9	Error	−.031	−.333*	−.222*	.159*	.214*	.159*	−.351*	.536*	–	
10	Aggressive	−.253*	−.133*	−.143*	.015	.321*	2.82*	−.352*	.328*	.265*	–

*p < 0.05; **p < 0.01.

correlations. Lapses were positively associated with positive behaviour, violations and errors. Positive behaviours were negatively associated with violations, errors and aggressive behaviour. Violations were positively associated with aggressive behaviour and errors, while errors were positively associated with aggressive behaviour.

4 DISCUSSION

Pedestrian Behaviour Scales (PBS) were validated for the pedestrians in Belgrade. The scree plot indicated that the data best fit a five-factor solution, which accounted for 60.6% of the total variance. The factor analysis identified five axes: violations, errors, lapses, aggressive and positive behaviours.

The results of some previous studies, conducted in the UK and New Zealand best fit the three factor solution (Elliott and Baughan, 2004; Sullman and Mann, 2009) with somewhat lower percentages of explained variance (43.8% and 32.1%). However, the research conducted in France (Granié et al., 2013) showed that the results best fitted into the four factor solution (38.2% variance explanation).

The paper examined the association of gender, age, reasons for walking, daily distances walked, possession of a driving licence and the behaviour of pedestrians. Males reported significantly more violations and significantly more aggressive behaviour in comparison with females. These results confirm the results obtained in other studies (Moyano Diaz, 2002; Granié et al., 2013) that men were more prone to violations and expressed more aggressive behaviour than women who were distinguished by their positive behaviour. These results were expected having in mind gender differences in the society. These data tell us that women are more careful, they respect regulations, they care about other persons, while men are prone to competing

and proving themselves, which is reflected in their behaviour (Antić et al., 2016).

The association of age with the behaviour shows that as age increases, pedestrian lapses and tendencies to display positive behaviour increase. Older pedestrians show greater responsibility and more patience than younger pedestrians (Torquato and Bianchi, 2010), which is directly associated with a greater degree of expressing positive behaviour towards other road users. Older pedestrians also have reduced psychophysical abilities, which is directly reflected in the increased number of lapses. In contrast, the increase of age reduces the number of violations, errors and aggressive behaviour. Almost the same results have been obtained in other studies (Moyano Diaz, 1997; Yildirim, 2007; Torquato and Bianchi, 2010). The comparison of the pedestrian behaviour in relation to the reason for walking was also conducted in this study. Pedestrians who were forced to walk, i.e. did not have the ability to use other means of transport to meet their needs, made more errors and expressed more aggressive behaviours than people who walked for pleasure.

The results of studies conducted worldwide are similar, almost identical to our results. Such results are explained by the fact that pedestrians who are forced to walk make more errors, because they have more trepidation in urban environment, avoid interaction with other road users, focus on the goal of their journey and choose the shortest way (between parked or stopped vehicles). On the other hand, pedestrians who enjoy walking feel more comfortable in urban environment, devote time for walking and accept the interaction with other road users more easily.

Daily distances walked were positively associated with all types of behaviours except with aggressive behaviour. This result was expected given that several studies (Torquato and Bianchi, 2010; Antić et al., 2016) had confirmed that the increase in the distance travelled was directly related to the increase in deviant behaviour of pedestrians.

The possession of a driving licence, regardless of the category, showed important differences in behaviour between the examined groups. Namely, the respondents who had a driving licence made fewer violations and errors than those who did not possess a driving licence. Persons who possess a driving license are obviously aware of their vulnerability when they participate in traffic as pedestrians, and are therefore less exposed to dangerous situations.

The significant difference was observed in the behaviour towards other road users. The respondents who did not possess a driving licence were more prone to aggressive behaviours, and the respondents who had a driving licence were prone to positive behaviours. This result was expected, given the fact that persons who possess a driving licence are often found in traffic as drivers, and therefore have more frequent interaction with other road users.

5 CONCLUSION

Measurement of behaviour using a questionnaire is recognized as a valid measurement in social sciences (Corbett, 2001). This is particularly the case when studying risky behaviour, in this case of pedestrians, and when it is necessary to study the psychological factors that can explain the behaviour of pedestrians. This study is based on self-reported behaviours of pedestrians, and differentiates between several groups of behaviour: violations, errors, lapses, aggressive and positive behaviour. This capability of measuring positive and aggressive behaviour, along with the deviant behaviour of pedestrians, provides more detailed understanding of the behaviour of pedestrians as well as connecting psychological factors and factors mobility (Granie et al., 2013).

The results obtained in Belgrade are very similar to those obtained in Chile (Moyano Diaz, 1997), Turkey (Yildirim, 2007), Brazil (Torquato and Bianchi, 2010) and France (Granie et al., 2013). The similarities are related to almost all comparisons made between groups (gender, age, reasons for walking, daily distances walked and possession of a driving licence).

One of the problems in relation to pedestrian traffic safety is the use of mobile phones while crossing the street. The results of the studies in the field have shown that pedestrians make much more unsafe street crossings when talking over the mobile phone than when they are unobstructed (Hatfield & Murphy, 2007; Nasar et al., 2008). The results

of the studies conducted in Belgrade have shown that the use of mobile phones by pedestrians when crossing the street at the intersection amounts to between 11.5% and 13.4% (Pešić et. al., 2015). This was the main reason why the PBS conducted in the city of Belgrade included the items relating the use of mobile phones while crossing the street. The results showed that pedestrians reported violations which were related to the use of mobile phones while crossing the street. This confirms the assumption of a large number of researchers that pedestrians are increasingly using mobile phone in traffic every day.

Future research aimed at the safety of pedestrians in traffic should go further in this direction; the questionnaires should be improved and the questions should be clear and concrete. This version of the PBS may help the researchers of pedestrian risky behaviours in terms of the basis for the future versions of PBS. It is necessary to form shorter versions of PBS which would contain items related to the use of mobile phones while crossing the street.

REFERENCES

Antić, B., Pešić, D., Milutinović, N. & Maslać, M. 2016. Pedestrian behaviours: Validation of the Serbian version of the pedestrian behaviour scale. *Transportation Research Part F: Traffic Psychology and Behaviour* 41(2016): 170–178.

Bernhoft, I. M. & Carstensen, G. 2008. Preferences and behaviour of pedestrians and cyclists by age and gender. *Transportation Research Part F: Traffic Psychology and Behaviour* 11(2): 83–95.

Corbett, C. 2001. Explanations for "understating" in self-reported speeding behavior. *Transportation Research Part F: traffic Psychology and Behaviour* 4: 133–150.

De Winter, J.C.F. & Dodou, D. 2010. The Driver Behavior Questionnaire as a predictor of accidents: a meta-analysis. *Journal of Safety Research* 41: 463–470.

Elliott, M.A. & Baughan, C.J. 2004. Developing a self-report method for investigating adolescent road user behavior. *Transportation Research Part F: Traffic Psychology and Behaviour* 7(6): 373–393.

Granié, M.A. 2008. Influence de l'adhésion aux stéréotypes de sexe sur la perception des comportements piétons chez l'adulte. *Recherche Transports Sécurité* 101: 253–264.

Granié, M. A., Pannetier, M. & Guého, L. 2013. Developing a self-reporting method to measure pedestrian behaviors at all ages. *Accident Analysis and Prevention* 50: 830–839.

Hatfield, J. & Murphy, S. 2007. The effects of mobile phone use on pedestrian crossing behaviour at signalized and unsignalized intersections. *Accident Analysis and Prevention* 39(1): 197–205.

Moyano Díaz, E. 1997. Teoría del comportamiento planificado e intención de infringir normas de transito en peatones. *Estudos de Psicologia (Natal)* 2(2): 335–348.

Moyano Díaz, E. 2002. Theory of planned behavior and pedestrians' intentions to violate traffic regulations. *Transportation Research Part F: Traffic Psychology and Behaviour* 5(3): 169–175.

Nasar, J., Hecht, P. & Wener, R. 2008. Mobile telephones, distracted attention, and pedestrian safety. *Accident Analysis and Prevention* 40(1): 69–75.

Pešić, D., Antić, B., Glavić, D. & Milenković, M. 2015. The effects of mobile phone use on pedestrian crossing behaviour at unsignalized intersections – Models for predicting unsafe pedestrians behavior. *Safety Science* 82(2016): 1–8.

Reason, J.T., Manstead, A.S.R., Stradling, S., Baxter, J.S. & Campbell, K. 1990. Errors and violations on the roads: a real distinction? *Ergonomics* 33(10/11): 1315–1332.

Road Traffic Safety Agency 2016. Statistical report on the state of traffic safety in the Republic of Serbia for the year 2015.

Rosenbloom, T., Nemrodov, D. & Barkan, H. 2004. For heaven's sake follow the rules: Pedestrians' behavior in an ultra-orthodox and a non-orthodox city.

Transportation Research Part F: Traffic Psychology and Behaviour 7(6): 395–404.

Sullman, M.J.M., Gras, M.E., Font-Mayolas, S., Masferrer, L., Cunill, M. & Planes, M. 2011. The pedestrian behaviour of Spanish adolescents. *Journal of Adolescence* 34(3): 531–539.

Sullman, M.J.M. & Mann, H.N. 2009. The road user behaviour of New Zealand adolescents. *Transportation Research Part F: Traffic Psychology and Behaviour* 12: 494–502.

Torquato, R.J. & Bianchi, A.S.A. 2010. Comportamento de Risco do Pedestre ao Atraversaar a Rua: Um Estudo com Universitarios. *Transporte: Teoria e Aplicacao* 2(1): 19–41.

Yildirim, Z. 2007. Religiousness, Conservatism and their Relationship with Traffic Behaviours. Middle East Technical University.

Zhou, R., Horrey, W. J. & Yu, R. 2009. The effect of conformity tendency on pedestrians' road-crossing intentions in China: An application of the theory of planned behavior. *Accident Analysis and Prevention* 41(3): 491–497.

Transport Infrastructure and Systems – Dell'Acqua & Wegman (Eds)
© 2017 Taylor & Francis Group, London, ISBN 978-1-138-03009-1

The architecture of an information system for the power management system on ship

M. Krčum & A. Gudelj
Faculty of Maritime Studies, University of Split, Split, Croatia

ABSTRACT: The object of this paper is to present the design of an information system for the management of power system on a ship using software agents. In shipboard power system a large number of electric components are tightly coupled in a small space and when a fault happens in one part of the system may affect other parts of the shipboard power system.

The Power Management System is a critical part of the control equipment in the ship. It is usually distributed on various control stations that can operate together and share information between each other or independently in case of special emergency situations in which ship have to operate. The system becomes more complex by applying renewable energy system due to special rules implemented by International Maritime Organization (IMO).

Safe, secure and efficient shipping on clean ocean, suggested by IMO, require the development of appropriate design, operational knowledge and assessment tools for energy efficient design and operation of ships.

Based on an analysis of the information flow of the data processing for making appropriate decisions, a control architecture for power distribution, which has to be hierarchical, is proposed. This control architecture is implemented as multi-agent system. It is modelled by Colored Petri Net.

1 INTRODUCTION

Marine power system represents a more complex management challenge. Compared to onshore power system, intercom system has a wider range of frequencies and cable lengths are much shorter, which contributes to reduced electricity losses and significantly lower voltage drops. Such a system has some specificity, the design of management solutions are not confined to download ready-made solutions from the mainland. The system of Power Management System (PMS) aims to optimally manage all energy resources on board as well as consumption of electricity and other energy. The aim of the system is also reducing operating costs, which is achieved by minimizing failures (Häkkinen, 2003.). Production management system and distribution of electricity is a critical part of the managerial level on board. Usually the distribution is done at various sampling stations which can work together by sharing information or independently in the event of emergencies in which the ship must carry out their functions. The system becomes more complex use of renewable energy sources, because of special rules adopted by the International Maritime Organization (IMO).

The aim of this paper is to present the design guidelines and requirements which are subject to the design of the system in general. The specific aim is to introduce a new management system using Coloured Petri nets (CPT). In addition to the technical requirements, recently strengthened and other requirements that affect the design of PMS. On one side are the economic requirements for increasing efficiency set by shipping companies and ship owners. On the other hand daily strengthen regulatory pressures aimed at reduced environmental pollution and limit global warming. The regulatory influences come primarily through changes to the IMO MARPOL Convention development or her with Annex VI. Last changes relating to ships that were built after 1 July 2015 and do not have a hybrid or diesel-electric drive including a new chapter four. This chapter introduces guidelines for improving the energy efficiency of design and operational measures. These new requirements affect the need to introduce new management solutions as well as new sources of energy to the ship.

Due to the complexity of management problems that arise by increasing the share of renewable energy sources, which relate primarily to the stability of the system, this paper proposes a model of PMS for managing the electric power system with low share of renewable energy sources. Renewable part of the system consists of solar panels and batteries.

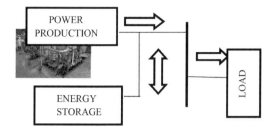

Figure 1. Hybrid power system.

2 MULTI-AGENT BASED SYSTEM FOR ENERGY MANAGEMENT

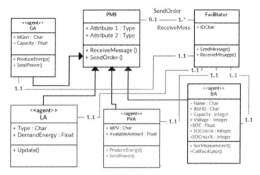

Figure 2. System infrastructure.

2.1 Multi-Agent System

A Multi-Agent System (MAS) is a distributed system consisting of multiple software agents, which work together to achieve a global goal. [1]. MAS has a great potential for modeling of autonomous decision making entities, which can be used to model and operate a power plant system. Agents have their own control over their behavior and internal states in any possible environment. They exhibit the following characteristics [2]:

– The agent must represent a physical entity so as to control its interactions with the rest of the environment.
– The agent must sense changes in the environment and take action accordingly.
– The agent must communicate with other agents in the power system via some kind of agent communication language with minimal data exchange and computational demands.
– The agent must exhibit a certain level of autonomy over the actions that it takes.
– The agent has minimally partial representation of the environment.

2.2 Agent model of the proposed hybrid energy system

Following the power plant modelling discussion found in Section 1, the platform consists of physical agents that are assigned to each type of physical entity found in a ship power plant. The essential data associated to agents are collected, organized, and stored in a database. Figure 2 shows a Unified Modelling Language (UML) class diagram of following types of agents: Photovoltaic array Agent (PVA), Battery Agent (BA), AC Generator Agent (GA) and Load Agent (LA).

Agent LA: The load agent LA is responsible for monitoring the evolution of the load. It manages these load to make it a controllable energy resource.

Agent PVA: This agent is dedicated to control the energy generated by the solar source. The agent PVA contains a DC/DC converter with MPPT which enables the PV sources to work at the maximum power point in a highly fluctuated environment. Indeed, the PVA agent normally uses a Maximum Power Point Tracking (MPPT) technique to continuously deliver the highest power through the converter to the load when there are variations in irradiation and temperature (Raju et al. 2015).

Agent BA: The agent BA, installed at each battery unit, can monitor the State of Charge (SOC) of the battery and manage the charge and discharge of the batteries. Obviously, the battery bank has two statuses (charging and discharging corresponding to the renewable energy source and the load, respectively). In fact, when the power, sent from the PV sources is insufficient to supply the load, the battery bank is discharged to meet the load demand as an energy supplier. In the opposite case, when the supply from the solar sources exceeds the load demand, the battery bank is charged and viewed as the load. BA has attributes such as State of Charge (SOC), SOCmax, SOCmin the capacity and the status (charge/discharge).

AC *Generator Agent (GA)*: This agent, installed at each diesel generator, has attributes such as the generator minimum/maximum output value.

The information stored in the database includes:

– available renewable energy generator power production capacity
– hourly solar insolation
– hourly demand load profile of the ship
– diesel generator capacity
– battery bank capacity
– hourly scheduling of different renewable energy generators
– unit cost of generation of different energy sources
– minimum and maximum SOC of battery bank.

The main objective is to fulfill the load at a particular time by minimizing the use of diesel generators. To accomplish this, the agents are required to cooperate and coordinate so that they make efficient use of the power supplied by other sources at the time of power shedding.

The Power Management System (PMS) controller is informed about the demand and producer availability in one negotiation cycle. At the timestamp of power shedding, if the electrical power from renewable resources is insufficient, demand is fulfilled by the diesel generators. PMS sends a message of announcement of tender to all agents in the ship's grid.

All agents reply a message of purchase of electrical power. The message includes the amount of electrical power i.e. load demand. All power producer agents in the grid reply a message of electrical power. The message includes the generator available power and states of charges for the battery storage.

In the next step, controller makes a decision on operation of different producers based on generated load demand. The demand is checked with the power available with the producer. If the demand is not fulfilled by first PV resources, PMS sends operation command to next producers (batteries and AC generators) until the demand is fulfilled. Finally, generation load pattern is displayed for a given time period and next negotiation cycle begins.

2.3 Concept of operations for modeled scenario

The concept of operations for modeled scenario uses four kinds of agents: Battery Agents, AC Agents and PV Agents. The most important work is done by the PMS controller. It uses the control algorithm, called the dispatch strategy for a hybrid energy system, for calculating the energy flows from the various sources like diesel generator and different types of renewable generators, towards the loads, including the charging and discharging of the battery system, on a time scale of minutes to hours, in such a way as to optimize system performance in terms of operating cost.

The solar power, load, state of charge (SOC) of the batteries and loads are monitored hourly. Based on these data, the agent takes best possible actions for energy management of the hybrid energy system on the ship. Considering all the possible options available for the solar resources, a flow chart is drawn as shown on Fig. 3.

By observing hourly operation of proposed hybrid energy system (see Fig. 1), there are four possible dispatch strategies to meet the load (Gupta et al. 2011).

1. *Battery charging strategy*: The use of only the battery to absorb the surplus power. The absorption of energy continues until:

    – Maximum battery SOC is reached.
    – The renewable power is not sufficient to meet the load.

2. *Battery discharging strategy*: Battery energy may be used to meet the load in a timestamp. The battery discharging continues until:

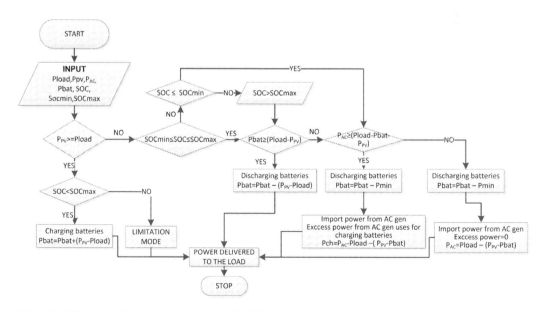

Figure 3. Flow chart of power management on the ship.

- Minimum battery SOC for discharge is reached.
- The renewable power is sufficient to meet the load.
- The renewable power is sufficient to meet the load as well as continue to charge the batteries.

3. *Load following strategy:* The diesel generator is running to follow the load, no charge to the battery and no discharge from the battery.
4. *Cycle charging strategy*: The diesel generator is running to cover the load demand and charge the battery. The diesel continues running for its prescribed minimum run time; after that, the diesel continues running until one of the conditions is met:

- The prescribed SOC set point has been met, or
- The renewable power is sufficient to meet the load.
- The renewable power is sufficient to meet the load as well as continue to charge the batteries.

Considered system has three operation modes to supply the load, which are given below:

Mode 1 (Stand Alone Mode): In this mode there is no sufficient solar power needed by the load and the state of charge (SOC) of the battery is also very low. All renewable sources are disconnected from the grid and the whole load will be supplied by the AC generator. In this mode the charging strategy is generally used in combination with the load following strategy. The PV power is not useful for the grid and is stored in battery storage units.

Mode 2 (normal mode): The total electrical power from renewable energy source is less than the power needed by the load, the energy deficit is covered by the battery source and the PMS controller puts the battery in the discharge condition (discharging strategy). If the storage cannot supply the whole, the rest will be supplied by the AC generator. If the generate power greater than the cycle charging strategy can be selected, else the load following strategy can be selected.

Mode 3: For this mode the photovoltaic panels produce the electric power more or equal than power references of the grid and the batteries are available. All sources are connected to the grid and the inverter delivers the electric power to meet grid power references. If the total power generated by the renewable energy is greater than the power needed by the load, the energy surplus is stored in the batteries (charging strategy). If the produced electric power from PV panels is less than power references of the grid, the batteries can be used to compensate this difference (discharging strategy).

In the case of lack of energy during 1 hour the energy is obtained either from the batteries or from the AC generator.

3 THE USE CPN MODEL OF THE INFORMATION SYSTEM

3.1 *Colored Petri Net*

Colored Petri Net (CPN) is a language for the modelling and validation of systems in which concurrency, communication, and synchronization play a major role. CPN is a discrete-event modelling language combining Petri nets with the functional programming language Standard ML. A CPN model of a system is an executable model representing the states of the system and the events (transitions) that can cause the system to change state. The CPN language makes it possible to organize a model as a set of modules, and it includes a time concept for representing the time taken to execute events in the modelled system.

Since PN simulates all of the system states and all transition judgments by token passing in a quite straightforward manner, the graphical representation for a moderate system shows very complex configuration. In CPN a place node owns several colors to represent different states and base on the colors the judgment functions in a transition node checks the states of the incoming place nodes. These characteristics dramatically simplify the graphical representation of the traditional PN and improve the execution efficiency too.

3.2 *Agent modeling with CPN approach*

By combining the MAS framework, algorithms and rules in CPNs, the CPNs model for the switching process of PMS operating mode is depicted in Fig. 4. The model is designed as the CPN tools (Version 4.0.1). In contrast to low-level Petri nets (such as Place/Transition Nets), each of these tokens carries a data value, which belongs to a given type. As an example, place BA has one token in the initial state. The token value belongs to the type *SOC* and represents state of charge for batteries. The detail declaration of all colors and variables is shown on the Fig. 5. The descriptions of all the places are illustrated in Table 1. Hourly intervals are considered for the design control strategy, where all the involved variables are assumed to be constant throughout these intervals. Accordingly, the place Next_hour represents daily hour and it has one token. Initially this value is 0, and it is updated for one hour.

The place BA is used to model the battery agent which controls states of the batteries. Three states for the batteries are considered and they

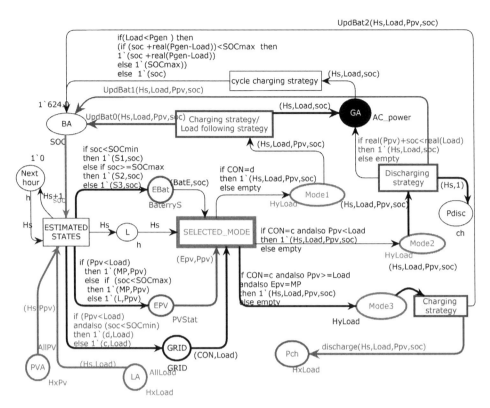

UpdBat2(Hs,Load,Ppv,soc)

if(Load<Pgen) then
(if (soc +real(Pgen-Load))<SOCmax then
1`(soc +real(Pgen-Load))
else 1`(SOCmax))
else 1`(soc)

cycle charging strategy   (Hs,Load,soc)

UpdBat1(Hs,Load,Ppv,soc)

1`624.0

UpdBat0(Hs,Load,Ppv,soc)   Charging strategy/
Load following strategy

BA

(Hs,Load,soc)

GA   AC_power

SOC

if soc<SOCmin
then 1`(S1,soc)
else if soc>=SOCmax
then 1`(S2,soc)
else 1`(S3,soc)

(Hs,Load,Ppv,soc)

if real(Ppv)+soc<real(Load)
then 1`(Hs,Load,soc)
else empty

1`0

Next
hour

h

Hs+1
soc

EBat

(BatE,soc)

if CON=d
then 1`(Hs,Load,Ppv,soc)
else empty

Mode1

Discharging
strategy

(Hs,1)

BaterryS

HyLoad

(Hs,Load,Ppv,soc)

Pdisc

Hs

ESTIMATED
STATES

Hs

L

Hs

SELECTED_MODE

if CON=c andalso Ppv<Load
then 1`(Hs,Load,Ppv,soc)
else empty

Mode2

ch

h

(Epv,Ppv)

HyLoad

(Hs,Load,Ppv,soc)

if (Ppv<Load)
then 1`(MP,Ppv)
else if (soc<SOCmax)
then 1`(MP,Ppv)
else 1`(L,Ppv)

if CON=c andalso Ppv>=Load
andalso Epv=MP
then 1`(Hs,Load,Ppv,soc)
else empty

(Hs,Ppv)

EPV

Mode3

Charging
strategy

PVA

PVStat

HyLoad

if (Ppv<Load)
andalso (soc<SOCmin)
then 1`(d,Load)
else 1`(c,Load)

GRID

(CON,Load)

AllPV

(Hs,Load)

GRID

Pch

discharge(Hs,Load,Ppv,soc)

HxPv

LA

AllLoad

HxLoad

HxLoad

Figure 4.   The CPN model for the cooperation process of the MAS.

Figure 5.   The color settings and variable declarations.

are represented on the CPN model by three token colors (P1, P2, P3) of colors set *batType* with following declaration:

colst batType = with P1 | P2 | P3.   (1)

It means that places having this colors set will have the value P1 or P2 or P3 as their token color. For the first color (P1), the battery is empty and this state is reached when its SOC (State of Charge) becomes equal or inferior to a minimum value (SOCmin). This condition is expressed as:

Table 1.   The meaning of places in CPN.

Places	Description
BA	Battery Agent
PVA	PV Agent
LA	Load Agent
GA	AC Agent
GRID	Grid connection
Ebat	Estimated states for batteries
EPV	This place represent in which mode PV is working
Mode1	Selecting mode of the power station
Mode2	
Mode3	
Pch, Pdis	Values of the surplus power (Pch), or the value of the load that has not been met (Pdis)
NextHour	This place is daily hours

$SOC <= SOCmin,$ (2)

where SOC is the estimated value of the state of charge.

For the second color (P2), the battery is fully charged and this state is reached when its SOC becomes equal or higher to a maximum value:

957

$$SOC >= SOCmax. \qquad (3)$$

For the third color (P3), the battery is in an intermediate state if remaining conditions are satisfied:

$$SOCmax. < SOC < SOCmax. \qquad (4)$$

The place PVA on Fig. 4 is used to model PV agents. Photovoltaic panels can work in the well-known MPPT mode (token color ML in color se PVstates) or in a power limitation mode (color L) when more power are available than required by the loads (Lu et al. 2010). In CPN model it is defined as color set PVstates.

To describe the connection with the ship's grid the place GRID is modelled. Two states have been defined. The first state (color d on Fig. 4 and 5) corresponds to the disconnection of the renewable sources from the grid. When the PV production is smaller than the required grid power and the storage units are fully discharged, the priority is given to charge the storage units in order to make available as soon as possible the power station in a safety operation. The second state (color c) corresponds to the grid connection.

The transition SELECTED MODE is the switcher and its aim is to switch from one mode to another according to the climate condition, the state of charge of the battery and the load. When this layer receives the information from the application layer in places Ebat, EPV and GRID, this layer selects the next node as aforementioned mode. Each mode is represented by a single place: Mode1, Mode2 and Mode3.

The transition SELECTED_MODE is enabled for all modes, but it can only occur for one mode at a. This situation is called a conflict (because the binding elements are individually enabled, but not concurrently enabled). This transition is in conflict with itself. The switching between the modes is determined by evaluating the corresponding arc expression given according to the rules of dispatching strategies described in Section 2. In presented CPN these rules are defined as arc functions to places Mode1, Mode2, Mode3.

4 CONCLUSION

The defining property of a shipboard power system is a large number of electric components that are tightly coupled in a small space and when a fault happens in one part of the system, it may affect other parts of the system. Safe, secure and efficient shipping on clean oceans, suggested by IMO requires the development of appropriate designs, operational knowledge and assessment tools for energy efficient design and operation of ships. The design of a future ship will require the development of new and increasingly sophisticated methods for modelling and simulation of complex systems that must be integrated in order to produce the total energy consumption of a ship. Control architecture for power distribution systems has to be hierarchical, distributed and easy to adapt. A complete logistic chain of this control architecture will be modelled by Colored Petri Net (CPN), which connects effective agents for autonomous control of complex distributed systems with agents for the control of power management systems.

REFERENCES

Colson, C.M. & Nehrir M.H. 2013. Comprehensive Real-Time Microgrid Power Management and Control with Distributed Agents-*IEEE Transactions on Smart Grid*, 4(1): 617–627.

CPNTools, CPN Group, Department of Computer Science, University of Aarhus, Denmark. http://www.daimi.au.dk/CPnets/

Gudelj, A & Krčum, M. 2013. Simulation and Optimization of Independent Renewable Energy Hybrid System. Transactions on Maritime Science 2(1): 28–35.

Gupta, A. Saini, R.P. & Sharma, M.P. 2011. Modelling of hybrid energy system-Part II: Combined dispatch strategiesand solution algorithm. Renewable Energy, 36: 466–473.

Häkkinen, P., Reliability of Machinery Plants and Damage Chains, World Maritime Technology Conference San Francisco USA, 17–20 Lokakuuta, US, 2003.

IMO, www.imo.org

Lu, D., Fakham, H., Zhou, T. & François, B. 2010. Application of Petri nets for the energy management of a photovoltaic based power station including storage units. *Renewable Energy*, 35: 1117–1124.

Raju, L., Shankar, S. & Milton, R.S. 2015. Distributed Optimization of Solar Micro-grid Using Multi Agent Reinforcement Learning. *Procedia of Computer Science. Science Direct, Elsevier*, 46(1): 231–239.

Transport Infrastructure and Systems – Dell'Acqua & Wegman (Eds)
© 2017 Taylor & Francis Group, London, ISBN 978-1-138-03009-1

Multimodal accessibility and the interest towards inter-urban carsharing services: A behavioural approach

S. de Luca, R. Di Pace, L. Elia & F. Martire
Department of Civil Engineering, University of Salerno, Fisciano, Italy

ABSTRACT: The underlying idea of the paper is that a generic traveller is first interested in carsharing, then he/she subsequently decides to join the service and to use it. Moreover, unlike urban carsharing services, inter-urban carsharing can be more significantly affected by the transport modes available and by the level of service supplied. In this conceptual context, the present paper investigates the most effective attributes able to explain and measure the interest towards an inter-urban carsharing service. In particular, the paper aims to demonstrate that users' interest significantly depends on the multimodal accessibility currently supplied and on the users' mode choice behaviour. To this aim, the interest in carsharing was investigated through an "ad-hoc" stated preferences survey and modelled through the calibration of the random utility theory. Different accessibility indicators were specified and tested, from simple ones (distance, travel time, generalised cost) to more complex and behaviourally coherent ones (Expected Maximum Perceived Utility).

1 INTRODUCTION

The choice to opt for a carsharing service may be interpreted as a choice process in which the users, before choosing carsharing as an alternative mode of transport, (i) develop an interest in the carsharing solution (as it is), (ii) acquire information on the carsharing service, (iii) make their choice based on the service's characteristics.

Existing scientific literature on carsharing focuses, in particular, on operational issues (the reader may refer to recent contributions by Correia and Antunes; 2012; El Fassi et al., 2013; Krumke et al., 2013), or on the analysis and/or modelling of users' behaviour.

In particular, studies on users' behaviour on carsharing are mainly concentrated in North America, and are focused primarily on the feasibility of carsharing programs and on the impact of carsharing on car ownership and vehicle usage. Interesting overviews may be found in Litman (2000), Haefeli et al. (2006), Shaheen et al. (2006), Barth et al. (2006), Shaheen and Cohen (2007), Shaheen et al. (2009), Shaheen (2013), Kent and Dowling (2013), de Luca and Di Pace (2014), Cartenì et al. (2016).

As regards modelling approaches, the representation of membership behaviour, frequency of usage and other choice dimensions have been dealt with by several authors. In particular, the use of carsharing has been modelled through Logistic regressions (Shaheen, 1999; Nobis, 2006), by binary Logit models (Cervero, 2003), by switching

models (de Luca and Di Pace, 2015), Multinomial Logit models (Fukuda et al., 2005; Cervero et al., 2007; Zhou et al., 2008, Zheng et al. 2009) or by hierarchical means-end approach (Shaefers, 2013).

As regards the probability of being an active member, the frequency of usage and/or membership duration have been investigated through regressive or econometric static models (Habib et al., 2012; Morency et al., 2012; Stillwater et al., 2009; Shaefers, 2013), through dynamic econometric models (Morency et al., 2009) and through activity-based microsimulation methods (Ciari, 2010).

In this context, no contributions have been founded on the analysis of the determinants that drive the users' interest to join/use the service.[1]

The idea is that a generic traveller is first interested in carsharing (as a potential solution), then he/she subsequently decides (choose) to join the service and to use it. The interest is independent of the main characteristics of the service itself (e.g. station, costs, type of vehicles, etc.) and can be interpreted as a measure of the service/alternative perception.

Modelling the interest may permit behavioural interpretation of the phenomenon, making it possible to forecast the potential demand of users willing to join/use the service and, finally, may allow for a more reliable estimate of the market

---

[1]A pilot study, with some first insights, was presented by the same authors in 2014 at the Euro Working group on Transportation.

potential if combined with a model able to estimate the choice of the service.

The present paper investigates the most effective attributes able to explain and measure the interest in an inter-urban carsharing system.

In particular, unlike urban carsharing services, inter-urban carsharing can be more significantly affected by the transport modes available and by the level of service supplied. Indeed, the paper aims to demonstrate that users' interest significantly depends on the multimodal accessibility currently supplied and on the users' mode choice behaviour.

To this aim, different accessibility indicators were specified and tested, from simple ones (distance, travel time, generalised cost) to more complex and behaviourally coherent ones (Expected Maximum Perceived Utility—EMPU).

The interest in carsharing was investigated through an "ad-hoc" stated preferences survey and modelled through the specification and the calibration of discrete choice models founded on the behavioural paradigm of the random utility theory. Different EMPU indicators of various complexities were investigated by specifying and calibrating specific mode choice models on users' actual mode choice preferences.

The paper is organised as follows: the methodological framework is proposed in section 2; the case study, the survey and some descriptive results are proposed in section 3; the estimation results and the sensitivity analysis are reported in sections 4 and 5; conclusions are drawn in section 6.

2 METHODOLOGY

The interest in carsharing was modelled through a binomial Logit model, where the alternatives were: "I would be interested" or "I would not be interested".

As previously mentioned, the main focus was on the investigation and estimation of the most effective set of attributes able to explain and measure the interest in a carsharing service. Together with socio-economic and activity related attributes, it has been assumed that the interest to join a carsharing service may be related to the current (active) accessibility provided by the available transportation system.

As is widely acknowledged, a carsharing service does not substitute existing modes, it is a viable solution if a transit system exists and it is complementary to existing transport modes. In a multimodal context, the higher the current multimodal accessibility from an origin to a destination is, the smaller the interest toward carsharing will be.

If transit services are reliable and competitive, the interest decreases (carsharing does not substitute transit services); if several alternatives exist, the interest decreases; if the travel costs decrease,

Table 1. Investigated accessibility measures.

Travel-related	Mode choice behaviour-related
Travel distance	Expected Maximum Perceived
Monetary cost	utility (simplex)—
Travel time by bus	EMPUsimplex
Travel time by car	
Generalised cost by car	Expected Maximum Perceived
Number of available	utility (complex)—
transport modes to	EMPUcomplex
reach the destination	
Access time to bus stop	

the interest is expected to decrease. Similarly, if users travel between origin-destination pairs with different accessibility, they may be characterised by different interest towards carsharing.

To this aim, different accessibility measures, of different complexities, were hypothesised and investigated (see Table 1).

– *Travel-related accessibility measures.*
  They are an aggregate measure of the spatial impendence between trip origin-destination pairs.
– *Mode choice Behaviour-related accessibility measures.*
  Such an approach makes it possible to take into account the factors that currently affect the transport mode choice process: the existing transport modes, the corresponding level of service supplied and the users' socio-economic and travel characteristics. Indeed, two different origin-destination pairs may be connected by the same number of transport modes, but the level of service supplied may be quite different. In other words, the level of service supplied could be perceived differently as the users' characteristics (travel or socio-economic) change.

Behaviour-related accessibility measures have been derived from the behavioural paradigm of the Random Utility Theory (RUT).

The RUT makes it possible to estimate the Expected Maximum Perceived Utility (EMPU) of the actual choice set. The EMPU attribute derives from the assumption that the generic decision-maker chooses from the available choice set, the alternative j with maximum perceived utility U_j, where the perceived utilities are modelled as random variables.

As the generic user associates a utility to each alternative, in the same manner he/she associates a utility to the given choice context, I, that he/she faces. Such a utility, usually called Expected Maximum Perceived Utility (EMPU) or satisfaction, is defined as the expected value of the maximum U_j over the alternatives available in the choice set, vector **U**.

The EMPU is a function of the expected values of the utilities of all the alternatives, E[**U**], and it depends on the joint probability density function of **U**, as well as on the composition of the choice set *I*. It can be formally expressed as:

$$s = s \, (E[\mathbf{U}]) = E[\max(\mathbf{U})] = \int ... \int ... \int \max(\mathbf{U}) \, f(\mathbf{U}) \, d\mathbf{U}$$

Depending on the assumptions on f(**U**), the EMPU variable can be expressed in a closed-form or can only be estimated through simulation techniques.

In our case study, the EMPU was estimated with regard to the choice set which users habitually face when travelling between the considered origin-destination pairs. To this aim, a specific set of random utility models was calibrated and, in particular, closed-form random utility choice models were preferred in order to obtain a closed-form expression for the expected maximum perceived utility.

In the case of the Multinomial Logit model (MNL), perceived utilities (random errors) are assumed independently and distributed as Gumbel variables (with the same scale parameter θ). Moreover, the maximum of a set of such random variables is also distributed as a Gumbel variable with parameter θ and the mean may be expressed in the following closed-form:

$$EMPU_{MNL} = \theta \ln \sum_{j \in I} \exp(E[U_j]/\theta) = \theta Y_I$$

where Y_I is the corresponding, well known, logsum variable. Evidently, the scale parameter cannot be separately estimated (due to the identification problem), but its estimation is embedded in the calibration procedure of the Binomial Logit which is calibrated to model the interest in the carsharing service.

In the case of the Hierarchical Logit model (HL), a similar formulation can be easily obtained. In the HL it is assumed that some of perceived utilities are correlated with each other and structured in groups (*G*). As it is possible to assign a systematic utility to each alternative belonging to each group, it is possible to assign a systematic utility to each group (*g*). The group-specific systematic utility (Y_g) can be derived from the systematic utilities of the alternatives belonging to each group and has the same formulation of the logsum variable introduced for the MNL model. Finally, the EMPU attribute can be expressed as a combination of the logsum variables related to the groups of alternatives that have been identified.

$$EMPU_{HL} = \theta_0 \ln \sum_{g \in G} \exp\left(\frac{\theta Y_g}{\theta_0} \right) \in$$

θ_0 and θ are two scale parameters, θ/θ_0 is usually estimated in the calibration stage of the HL model,

θ_0 cannot be separately estimated (due to the identification problem) and its estimation is embedded in the calibration procedure of the Binomial Logit which is calibrated to model the interest in the carsharing service.

Within the proposed methodological framework, two EMPU variables of different complexity were estimated:

– $EMPU_{simplex}$,
– $EMPU_{complex}$.

$EMPU_{simplex}$ was estimated from a binomial Logit choice model with a simple systematic utility function specification.

$EMPU_{complex}$ was estimated from a choice model calibrated on a much more realistic choice set and from a survey which was specifically carried out on the same case study.

The aim was threefold: (i) understanding if different types of EMPU variables may be able to explain the interest in the carsharing service; (ii) understanding if and how a more realistic (from a behavioural viewpoint) EMPU variable, derived from a more detailed choice model, may significantly affect the modelling of said interest; (iii) understanding the trade-off between a simple formulation (and economic survey), compared with a more realistic formulation that could be obtained through more expensive surveys.

Finally, it should be pointed out that the proposed considerations and the obtained results may be extended to multimodal contexts in which a transit system exists.

3 CASE STUDY, SURVEY AND DESCRIPTIVE RESULTS

The case study consisted of 500 interviews of randomly selected residents from the municipality of Salerno (Campania Region, Southern Italy) and from the four main municipalities belonging to the metropolitan area of Salerno.

Salerno is the capital city of the Salerno Province, it is located 55 km from Naples, it has about 140,000 residents, and it is characterised by 10,000 daily commuters. The four considered municipalities belonging to the metropolitan area of Salerno are: (1) Pontecagnano (25,000 inhabitants and 15 km from Salerno), (2) Baronissi (20,000 inhabitants and 10 km from Salerno), (3) Cava dè Tirreni (53,000 inhabitants and 12 km from Salerno), (4) Nocera Inferiore (46,000 inhabitants and 20 km from Salerno).

Only residents travelling for work between Salerno and the four municipalities (and vice versa) were considered, and respondents were sampled from each municipality with regards to the number of residents and coherently with census data ratios. Each respondent was presented with the same questionnaire, and he/she was asked to describe his/her usual travel habits (transport, travel cost, travel time, trip frequency, etc.); he/she was then introduced to the service and to the main qualitative characteristics (one way, fee typologies, types of car, dedicated parking slots, etc.). Users' socio-economic characteristics were then collected, as well as users' trip characteristics (mode, activity duration, trip frequency, etc.), their potential interest in opting for the service (as it is, without knowing the fees, the type of car or the parking location) and their main motivations.

Descriptive results indicated 75% of the intercepted users would be interested in the proposed carsharing service.

Interested users were mainly influenced by the inefficiencies of the public transport system (41%) and by the non-availability of the car transport mode (30%). It is interesting to note the obtainable financial gain by opting for the carsharing service is not the main determinant in the decision process (21%), whereas the access time to the bus stop is the main determinant for 7% of the users. Non-interested users are satisfied with the usual transport mode (30%) or they do not want to use an "unknown" car. Around 18% do not want to book in advance, 8% are not confident in the carsharing service, while 14% travel using carpooling with other users, thus, they already participate in carsharing to some extent.

With regard to the current mode chosen to travel, among the interested users, 73% make use of public transport, 17% travel by car and only 9% take part in carpooling. Non-interested users are mainly carpooling users (15%) and car users (54%).

In analysing the effect of the number of cars per household, it is interesting to note that respondents with a ratio less than 0.75 show a percentage of interest greater than 70%; for ratios between 0.75 and 1 the percentage decreases to 58%; for ratios greater than 1, the percentage is less than 50% (48%).

Finally, it is interesting to note that the interest in the carsharing service significantly decreases as the weekly trip frequency increases. As a matter of fact, a trip frequency which is less than three trips per week shows a percentage of interest equal to around 60%. This percentage decreases by 7% for four trips per week, and by more than 15% for weekly trips which are smaller than 5. This result confirms that a carsharing program can also be a viable solution for systematic users, however, the interest in the service depends on the number of systematic trips that each user has to do.

4 EXPECTED MAXIMUM PERCEIVED UTILITY (EMPU) ATTRIBUTE ESTIMATION

As previously introduced in section 2, two EMPU variables were investigated: $EMPU_{simplex}$, $EMPU_{complex}$. Each variable was estimated through two different mode choice models based on two revealed preferences (RP) surveys carried out in the same geographical area but in different periods.

The $EMPU_{simplex}$ was estimated from a simple mode choice model calibrated on a survey carried out on a regional level (de Luca and Cartenì, 2013).

The $EMPU_{complex}$ was estimated from a choice model specified and calibrated on a more detailed survey which made it possible to adopt more detailed systematic utility functions.

4.1 The $EMPU_{simplex}$ model

The $EMPU_{simplex}$ was estimated from a model which was specified and calibrated from a revealed preferences survey designed and conducted to that end in the winter of 2013.

The main aim of the survey was to understand some features of the systematic travel behaviour occurring within the province of Salerno. In particular, the trip purpose taken into account was home-to-work; the choice context consisted in two alternatives: car and public transportation (bus and/or rail).

The survey was by phone, it involved a sample of 1400 residents in the province of Salerno (which includes the case study) randomly selected according to stratified sampling based on gender, professional status and age.

The binomial-Logit mathematical formulation was adopted and only level of service, simple socio-economic and geographical attributes were introduced. The specification of the proposed modelling solution is shown below (see Table 2).

The value of time equal to 4 Euros can be observed and is consistent with estimated values in similar works by de Luca and Papola (2001), Cantarella and de Luca (2005). The male gender influences the utility of the car mode positively.

Age has a positive impact on the utility of the car; in particular, as the age attribute is a binary attribute segmenting users into classes of users who are older than 40 or less than 40, it appears that younger users associate greater utility to the car alternative.

The geographical attributes used are binary and assume a unitary value if the destination or the origin of the trip belongs to a specific geographical area. Although different types of aggregations of geographical origins and/or of possible destinations

Table 2. Mode choice model for EMPU$_{simplex}$ estimation.

Number of interviews: 470
Number of attributes : 7
pseudo-ρ^2: 0.74
pseudo-ρ^2 correct: 0.72

car   transit

Systematic utilities

$V_{car} = \beta_1 \cdot C + \beta_2 \cdot T_{car} + \beta_3 \cdot ASA + \beta_4 \cdot orig\_SA + 3 + \beta_5 \cdot dest\_SA + \beta_6 \cdot Gen + \beta_7 \cdot age_{<=40}$

$V_{transit} = \beta_1 \cdot C + \beta_2 \cdot T_{transit}$

Calibration results		
C	-1.06	monetary cost (€)
$T_{car} / T_{transit}$	-4.76	total travel time (h)
$orig\_SA$	1.31	binary attribute equal to 1 if the travel origin is the city of Salerno
$dest\_SA$	-0.73	binary attribute equal to 1 if the travel destination is the city of Salerno
Gen	1.20	binary attribute equal to 1 if the user is male gender, 0 if female gender
$age_{<=40}$	0.24	binary attribute equal to 1 if the user is less than 40 years old
ASA	0.24	alternative specific attribute

(* All coefficient estimates were significantly different to zero with P>95%

have been experienced, two of these were the most significant aggregations: *dest_SA*, if the destination is the city of Salerno, *orig_SA*, if the origin is the city of Salerno. The estimated coefficients showed that extra utility exists for all users moving from the city of Salerno associated with the *car* mode.

4.2 The EMPU$_{complex}$ model

The EMPU$_{complex}$ was estimated from a model which was specified and calibrated from a revealed preferences survey designed and conducted in the winter of 2014.

The main aim of the survey was to understand and effectively model the mode choice in a small part of the province of Salerno (the same case study of section 2). In particular, the trip purpose taken into account was home-to-work; the choice context consisted in four transport modes: car as driver, car as passenger, car-pool and bus.

In this case, the most effective modelling formulation was the hierarchical Logit model.

Calibrated parameters (see Table 3) show signs which are consistent with expectations.

The mutual ratios between the coefficients of representative attributes of travel time and the monetary cost attribute coefficient (willingness to pay in €/hour—value of time, V.O.T.) are equal to about 4 Euros for the time on board and 8 Euros the access/egress time to the bus stop.

Moreover, the incidence of attributes such as the activity duration at the destination and the frequency of weekly trips can be noted. As it was logical

Table 3. Estimation results for EMPU$_{complex}$ model.

Number of interviews: 200
Number of attributes: 12
pseudo-ρ^2: 0.35
pseudo-ρ^2 correct: 0.33

car   bus   car pool   pax

Systematic utilities

$V_C = \beta_1 T_{CAR} + \beta_2 C + \beta_6 Tstop_{(1,4)} + \beta_7 Tstop_{(>4)}$

$V_P = \beta_1 T_{CAR} + \beta_3 PAX$

$V_{CP} = \beta_1 T_{CAR} + \beta_2 C + \beta_9 Freq. + \beta_{10} CARAV + \beta_8 Tstop_{(<=1)} + \beta_4 CAR POOL$

$V_B = \beta_1 T_{BUS} + \beta_{12} T_{ACCESS} + \beta_2 C + \beta_9 Freq + \beta_{11} Gen + \beta_8 Tstop_{(<=1)} + \beta_5 BUS$

Calibration results		
T_{CAR}	-1.22	travel time by car (h
T_{BUS}	-1.22	travel time by Bus (h)
C	-0.23	monetary cost (€)
T_{ACCESS}	-1.83	access time to bus stop (h)
Gen	0.97	binary attribute equal to 1 if the user is of female gender
$Freq$	0.33	binary attribute equal to 1 if the trip frequency is smaller than 3
$CARAV$	1.66	car mode availability (number of cars / number of family components)
$Tstop_{(1,3)}$	-0.46	binary attribute equal to 1 if the stop time is less or equal to 1 hour
$Tstop_{(3,5)}$	0.25	binary attribute equal to 1 if the stop time is between 1 and 4 hours
$Tstop_{(>5)}$	0.32	binary attribute equal to 1 if the stop time is greater than 4 hours
ASA-Passenger	-2.05	alternative specific attributes
ASA-Bus	-1.97	
ASA-Car-pool	-2.62	
$\theta / \theta_0 = \delta$	0.73	model parameter, index of the correlation level between the alternatives

(*) All coefficient estimates are significantly different to zero with P> 95%

to expect, durations under 60 minutes penalise the *bus* and *car-pool* modes; very long layovers (several hours) increase the utility of the *car* mode. In particular, the attributes $T_{stop(1,3)}$, $T_{stop(3,5)}$ and $T_{stop(>5)}$ define temporal thresholds and their effect is not linear. As a matter of fact, longer stays increase the utility of the *car* mode, and the maximum utility of using the *car* is for time intervals ranging between 3 and 5 hours. The *frequency* makes it possible to understand how a high weekly trip frequency tends to encourage trips in organised forms (*car-pool*) or rather the use of public transport.

Finally, also socio-economic attributes were statistically significant such as the availability of cars and the gender.

5 MODELLING THE INTEREST: ESTIMATION RESULTS AND SENSITIVITY ANALYSIS

Starting from the methodological framework introduced in the previous section, homoscedastic (BNL) and heteroscedastic binomial choice (BMNL) models were specified and calibrated.

The obtained estimation results could be interpreted in terms of the attributes that proved to be statistically significant, but also as regards those attributes that did not show any significance.

5.1 Not significant attributes

First of all, it should be noted that all the socio-economic attributes—such as age, gender and income—did not prove to be statistically significant. This result is, however, interesting since most of the existing analyses on carsharing behaviour have showed that male, young and high income users are the demand segments who are more likely to opt for/use the service (Cervero et al., 2003 and 2007; Firnkorn et Müller, 2011; Lane, 2005; Shaheen et al., 2007; de Luca and Di Pace, 2015; Carteni et al., 2016). It can be concluded that users' socio-economic profile plays a role when the users have to pay for the service, but does not affect the interest.

Moreover, all the travel-related accessibility measures founded on simple attributes (time, distance, generalised cost) did not prove to be statistically significant. Therefore, the sole use of spatial impedance measures did not lead to any interpretation and to any significant modelling solution.

5.2 Significant attributes

Among the statistically significant attributes, the EMPU attribute proved to be statistically significant in both of the proposed formulations (Table 4).

EMPUs are the most statistically significant attributes and show, as expected, a negative effect on being interested. In fact, as EMPU increases, users perceive a smaller gain in a carsharing service. This result is interesting, since it allows us to understand that the perception of carsharing (and the consequent potential demand) is affected by the actual mode choice behaviour.

Though both EMPU attributes were statistically significant, the EMPU$_{complex}$ allowed for a better goodness-of-fit (BNL[c]) and made it possible to specify a more detailed systematic utility function.

Moreover, the relative weight of the alternative specific constant decreased, highlighting that the EMPU$_{complex}$, allowed for a more robust and reliable modelling (and interpretation) of the choice phenomenon, which reduced the need for the alternative specific constant.

Table 4. Modelling the interest: estimation results.

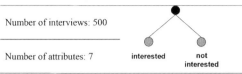

| Number of interviews: 500 | | | |

| Number of attributes: 7 | interested | not interested |

Systematic utilities

$V_{interest} = \beta_1 \cdot EMPU_x + \beta_2 \cdot KNOW + \beta_3 \cdot BUS + \beta_4 \cdot DIST + \beta_5 \cdot Freq$

$V_{not\text{-}interested} = \beta_{asc} \cdot ASC$

Calibration results

		BNL[s]	BNL[c]	BMNL[c]
pseudo-ρ^2		0.613	0.732	0.758
pseudo-ρ^2correct		0.584	0.694	0.716
$EMPU_{simplex}$		-0.56***	-	-
$EMPU_{complex}$		-	-0.86***	-1.18**
KNOW	equal to 1 if user is aware of what a carsharing service is and how it works	+0.37***	+0.41**	+0.83***
BUS	equal to 1 if user usually travels by Bus	+0.42***	+0.73***	+0.65***
DIST	travelled distance (km)	-	-0.23***	-0.18***
DIST (s.d.)		-	-	0.08***
Freq	number of weekly trips towards the declared destination	-0.30**	-0.62**	-0.71**
ASC	it is the alternative specific constant	+5.84***	+2.57***	+2.23***

*** results statistically significant with 95% confidence ** with 90%

In conclusion, the accessibility supplied by the multimodal transportation system plays a significant role even if estimated through simpler (or more aggregate) behavioural choice models. Obviously, as the behavioural accessibility measure is more realistic, the model effectiveness increases. This result indicates that inter-urban carsharing implementation should be accompanied by preliminary behavioural analysis on the current mode choice behaviour.

In analysing the other statistically significant attributes, it is noteworthy that travelling by bus greatly affects the interest in a carsharing service. In fact, the bus alternative offers less comfort than all the other existing alternatives. Moreover, this transit system is less reliable, often crowded and characterised by low frequencies. By contrast, carsharing offers the same flexibility and reliability as the car transport mode.

The weekly travel frequency increases the probability of not being interested. The interpretation is threefold: (i) travel costs become incomparable with bus and car transport modes; (ii) the need for booking several days in advance represents a mental obstacle; (iii) users that travel systematically are less interested in an alternative transport mode solution.

The distance represents a sort of spatial impedance and allows for the segmentation of the users involved. In fact, as distance increases, the probability of being interested decreases, independently from the level of service supplied by the available transport modes. Users that are farther from the final destination are less interested in a carsharing service, meaning that the perception of the advantages of this travel alternative decreases with the distance.

The interpretations might be manifold: for longer distances, users prefer to always use the same transport mode; for car users who travel for longer distances, users prefer to drive the same car. Finally, bus users travelling for longer distances prefer to travel with the same line, the same bus, the same driver and the same timetable.

However, it should be said that the distance may seem similar to the EMPU attribute, but it should be remembered that the EMPU attribute does not include only level of service attributes, but also socio-economic and activity-related attributes that play a significant role in determining the perceived utilities. As a matter of fact, when the $EMPU_{simplex}$ was tested, DIST did not prove to be statistically significant; whereas the use of the $EMPU_{complex}$ showed DIST to be significant. Moreover, correlation analysis between the $EMPU_{complex}$ and DIST showed a value smaller than one (0.43). Therefore, it can be concluded that the spatial distance allows for a geographical segmentation which is particularly useful for the interpretation and simulation of the catchment area of an inter-urban carsharing service.

The dummy attribute equal to 1 if user is aware of what a carsharing service is and how it works (*KNOWN*) showed a positive value highlighting that interest, and, thus, the carsharing demand is affected by the level of awareness of the service itself and could be greatly increased through specific marketing policies.

Together with the homoscedastic formulations, heteroscedastic solutions were also investigated. Both random coefficients and error components were estimated, but only random coefficients were statistically significant. In particular, only the travelled distance attribute (*DIST*) proved to be normally distributed. This result is logical since users belong to different municipalities and the kilometres travelled are different, thus, the perception of travelling distance may significantly vary among users. However, the goodness-of-fit slightly

Table 5. Sensitivity analysis.

Attribute	Variation of each attribute			
	+10%	+20%	+30%	+40%
EMPU	74.6	73.9	72.9	70.6
Travel distance	74.7	73.7	72.3	70.8
Trip frequency	74.1	71.7	67.9	64.5

improved, whereas all the previously discussed attributes continued to be significant and to play a similar role.

The $BMNL_{[c]}$ were applied to the calibration sample and a sensitivity analysis was carried out with respect to the EMPU, distance and trip-frequency attributes.

In the following table (see Table 5) the probability of being interested is shown with reference to the variation of each attribute. In particular, 10%, 20%, 30% and 40% variations were simulated.

First of all, it should be remembered that more than the 75% of the intercepted users stated that they were interested in the service. Thus, starting from this value, it can be noted that as EMPUs increased, the probability of being interested decreased by about 5%. A similar result was obtained by reducing the spatial distance. Both results highlight that the service's catchment area may change by about 10% for short distances and when there is good accessibility between the involved trip origin and destination.

A much greater decrease may be observed by varying the weekly trip frequency. Indeed, if trip frequency increases by 30% then the interest decreases of about 12% can be observed. In particular, the more the trips become systematic, the more the carsharing service decreases in its appeal.

6 CONCLUSIONS

Although carsharing has become a consolidated transport alternative, carsharing behaviour has been mainly investigated through ex-post analysis and in urban contexts. In this paper the interest in joining a carsharing service was investigated with regard to an inter-urban context and through an ex-ante approach based on a stated preferences survey.

The obtained results make it possible to draw several conclusions.

First of all, a statistically significant interpretation of the phenomenon and a mathematical formulation were obtained. The goodness-of-fit was satisfactory and the attributes composing the systematic utilities were coherent with the expectations.

Secondly, maximum utility attributes (EMPU) were statistically significant. This result is noteworthy since it allows for a behavioural interpretation of the phenomenon and indicates that the interest in carsharing does not only rely on the level of service of the available transport modes, but it also depends on the overall perceived utility of the available transport modes.

Thirdly, other significant attributes were: the weekly trip frequency and the travel distance. Both reduce the interest in carsharing and, therefore, systematic users who travel for longer distances are a challenging segment to attract. Moreover, the level of interest increases for those users who are already familiar with the service and for those who usually travel by bus.

Unlike existing insights in literature on carsharing choice behaviour, the interest in the service is not directly affected by age, gender or income attributes. Such attributes most probably play a role only when the user has to decide whether to opt for the service. However, on the other hand, it should be noted that such attributes concur within the EMPU variable.

Finally, the sensitivity analysis showed a greater sensitivity to the trip frequency variation and the EMPU attribute. As the EMPU attribute and the trip frequency change, the level of interest may increase/decrease by 20%.

In conclusion, some research perspectives seem worthy of interest: simultaneously and explicitly modelling the interest and the decision to opt for the carsharing service; investigating the role of accessibility on an urban scale.

REFERENCES

Barth, M., Shaheen, S.A., Fukuda, T., Fukuda, A. 2006. Carsharing and Station Cars in Asia. Journal of the Transportation Research Records 1986: 106–115.

Cantarella, G.E., de Luca S. 2005. Multilayer feedforward networks for transportation mode choice analysis: an analysis and a comparison with random utility models, Transportation Research part C, 13(2), 121–155.

Cartenì, A., Cascetta, E., & de Luca, S. 2016. A random utility model for park & carsharing services and the pure preference for electric vehicles. Transport Policy, 48, 49–59.

Cascetta, E. 2009. Transportation Systems Analysis: Models and Applications. Springer Verlag, US.

Cervero, R & Tsai. Y. 2003. City CarShare in San Francisco: Second-Year Travel Demand and Car Ownership Impacts. In Transportation Research Record: Journal of the Transportation Research Board, No. 1887, 117–127.

Cervero, R., Golob, A. & Nee, B. 2007. City CarShare: Longer Term Travel Demand and Car Ownership impact. In Transportation Research Record: Journal of the Transportation Research Board, No. 1992, 70–80.

de Almeida Correia, G., Antunes, A.P. 2012. Optimization approach to depot location and trip selection in one-way carsharing systems, Transportation Research Part E: Logistics and Transportation Review, 48(1), 233–247.

de Luca, S., Cartenì, A. 2013. 'A multi-scale modelling architecture for estimating of transport mode choice induced by a new railway connection: The Salerno-University of Salerno-Mercato San Severino Route, Ingegneria Ferroviaria, 68(5), 447–473.

de Luca, S., Di Pace, R. 2015. Modelling approaches for simulating the impact of carsharing on mode choice behaviour, Transportation Research A 71, 59–76.

de Luca, S., Papola, A. 2001. 'Evaluation of travel demand management policies in the urban area of Naples', Advances in Transport, 8, 185–194.

El Fassi, A., Awasthi, A., Viviani, M. 2012. Evaluation of carsharing network's growth strategies through discrete event simulation, Expert Systems with Applications, 39(8), 6692–6705.

Firnkorn, J. and Müller, M. 2011. What will be the environmental effects of new free-floating car-sharing systems? The case of car2go in Ulm. Ecological Economics 70, 1519–1528.

Fukuda, T., Fukuda, A. & Todd, M. 2005. Identifying potential market of carsharing users through modal choice behaviour and socio-economic perspective: a case study on Bangkok Metropolis. Presented at the 84th Annual Meeting of Transportation Research Board, Washington DC.

Habib, N. K., Morency, C., Tazul, I.M. & Grasset, V. 2012. Modelling users' behaviour of a carsharing program: Application of a joint hazard and zero inflated dynamic ordered probability model. Transportation Research Part A: Policy and Practice, Vol 46(2), 241–254.

Haefeli, U., Matti, D., Schreyer, C., Maibach, M. Evaluation 2006. Car-Sharing. Federal Department of the Environment, Transport, Energy and Communications, Bern.

Kent, J.L. 2013. Dowling, R. Puncturing automobility? Carsharing practices Original Research Article Journal of Transport Geography, 32, 86–92.

Krumke, S.O., Quilliot, A., Wagler, A.K., Wegener, J-T. 2013. Models and Algorithms for Carsharing Systems and Related Problems, Electronic Notes in Discrete Mathematics, 44(5), 201–206.

Lane, C. 2005. PhillyCarShare: First-year social and mobility impacts of carsharing in Philadelphia, Pennsylvania. In Transportation Research Record: Journal of the Transportation Research Board, No. 1927, 158–166.

Lane, C. 2005. PhillyCarShare: First-year social and mobility impacts of carsharing in Philadelphia., Pennsylvania. In Transportation Research Record: Journal of the Transportation Research Board, No. 1927, 158–166.

Litman, T. 2000. Evaluating carsharing benefits. Journal of the Transportation Research Board 1702, 31–35.

Morency, C., Habib, N. K., Grasset, V. & Tazul, I. M. 2012. Understanding members' carsharing (activity) persistency by using econometric model. Journal of Advanced Transportation, vol 46(1), 26–38.

Morency, C., Habib, N.,K., Grasset, V. & Tazul, I. M. 2009. Application of a dynamic ordered probit model for predicting the activity persistency of carsharing member. Presented at the 88th Annual Meeting of Transportation Research Board, Washington, DC.

Morency, C., Trepanier, M. & Basile, M. 2008. Object-Oriented Analysis of Carsharing System. In Transportation Research Record: Journal of the Transportation Research Board, No. 2063, 105–112.

Nobis, C. 2006. Car Sharing as a Key Contribution to Multimodal and Sustainable Mobility Behaviour—The Situation of Carsharing in Germany. In Transportation Research Record: Journal of the Transportation Research Board, No. 1986, 89–97.

Shaefers, T. 2013. Exploring carsharing usage motives: A hierarchical means-end chain analysis. Transportation Research Part A 47:69–77.

Shaheen, S. & Cohen, A. 2007. Growth in Worldwide Carsharing: An International Comparison. In Transportation Research Record: Journal of the Transportation Research Board, No. 1992, 81–89.

Shaheen, S. & Cohen, A. 2013. Carsharing and Personal Vehicle Services: Worldwide Market Developments and Emerging Trends. International Journal of Sustainable Transportation 7(1), 5–34.

Shaheen, S. 1999. Dynamics in Behavioral Adaptation to a Transportation Innovation: A Case Study of CarLink-A Smart Carsharing System. Institute of Transportation Studies UCD-ITS-RR-99-16.

Shaheen, S., Cohen, A., Chung, M. 2009. North American Carsharing: 10-Year Retrospective. In Transportation Research Record: Journal of the Transportation Research Board, No. 2110, 35–44.

Stillwater, T., Mokhtarian, P.,L. & Shaheen, S. 2008. Carsharing and built environment: a GIS-based study of one US operator. Presented at the 87th Annual Meeting of Transportation Research Board, Washington, DC.

Zheng, J., Scott, M., Rodriguez, M., Sierzchula, W., Platz, D., Guo, J. & Adams, T. 2009. Carsharing in a university community: Assessing Potential Demand and Distinct Market Characteristics. In Transportation Research Record: Journal of the Transportation Research Board, No. 2110, 18–26.

Zhou, B., Kockelman, K. M. & Gao, R. 2011. Opportunities for and Impacts of Carsharing: A Survey of the Austin, Texas Market. International Journal of Sustainable Transportation 5 (3), 135–152.

Transport Infrastructure and Systems – Dell'Acqua & Wegman (Eds)
© *2017 Taylor & Francis Group, London, ISBN 978-1-138-03009-1*

Sustainable transport—parking policy in travel demand management

D. Brčić, M. Šoštarić & K. Vidović
Faculty of Transport and Traffic Science, Zagreb, Croatia

ABSTRACT: The global urbanisation process considers mobility as a high priority in today's urban agglomerations, such that the mobility is the basic condition of the sustainable urban development. The living trends in urban agglomerations and the high motorisation rate increase, together with considering space rationality, demand a necessity for transport and other experts to manage entire travel demand, so the modal shift of mobility becomes an imperative. The parking policy and parking demand management is an unavoidable tool which is used to make changes in current modal split due to the excessive private car usage in the urban environment. The goal of this paper is to show the fundamental reasons for using parking policy and parking demand management in the City of Zagreb, and to show an analysis of the current state of policy with the basic advantages and disadvantages. The analysis is used to make conclusions on defining future approach to parking demand management policy in the City of Zagreb as a part of the comprehensive travel demand management approach.

Keywords: travel demand management, parking policy, City of Zagreb

1 INTRODUCTION

Urban areas in global, and especially in European environment, face a number of challenges today: the economic crisis, climate change, dependence of transport system on fossil fuels, as well as health risks caused, directly or indirectly, by the transport system. Transport demand in urban areas is growing, due to a series of economic and social factors, with number of motorized trips growth by cars in particular.

The excessive use of cars, as a product of economic prosperity has created a global problem in commuting urban spaces: insufficient capacity of transport infrastructure. Therefore, transport and other experts have to answer the question how to satisfy the need for mobility with space, energy, environmental and economic rationality, which is requirement on sustainable urban transport system.

There is an urgent need to implement the modal distribution, with a new political vision of redefining the transport strategy of urban space.

The transport demand management is a key strategy for cities trying to solve the modal distribution in order to reduce the excessive use of cars, and accordingly encourage non-motorized modes of travel and public mass transportation, as a function of the energy, environmental, ecological and economic rationality that is creating the preconditions for human-oriented city tailored for citizen's needs ("liveable city"), (Penelosa, 2005), (Bohler-Beadeker et al. 2014).

Entire transport policy should support economic prosperity of the urban environment. Therefore, the transport policy should include a range of strategies and measures in order to complementary achieve the set of objectives; increased mobility and availability, energy efficiency, GHG emissions reduction, noise reduction, the number of traffic accidents reduction and reduced external costs of transport system (Lopez-Ruiz, et al. 2013).

Parking demand management strategy is proven and reliable tool for reduction of transport demand generated by usage of cars (Lopez-Ruiz, et al. 2013), (Brčić 1999). The parking management strategy uses a variety of measures such as the supply of on street and off street parking areas, along with the administrative arrangements of availability of parking, as well as and economic measures trough charging of parking (Brčić 1999), (Litman 2011), (Slavulj 2013).

This paper will present the main objectives for the implementation of parking management strategies in the City of Zagreb, which is based on a Transport study of the City of Zagreb (MVA Consultancy 1999). Based on the analysis of the implemented measures and the state of parking management strategies in the period from 2000th to 2015th, together with advantages and disadvantages, comments and conclusions will focus on future guidelines and recommendations for redefining the future goals of transport policy of the City of Zagreb.

2   CITY OF ZAGREB TRANSPORT POLICY

Zagreb is the economic, cultural, historical, administrative, and transportation centre of the Croatia. It is also the largest city in Croatia with a population of 790,017 (2011), which is located on an area of 641 km² with a population density of 1232.48 inhabitants/km² (Statistical Yearbook—City of Zagreb, 2012). Transport policy of the City of Zagreb is based on the Transport study that was prepared and published in late 1999 by the consulting firm MVA from Great Britain (MVA Consultancy 1999). The temporal extent of the study is 2000–2020, with a target time horizon for 2005, 2010 and 2020.

The objectives of the General Transport Plan were:

1. Improvement of economic efficiency of the transport system
2. Protection of the environment from the adverse effects of traffic
3. Increase of the safety of passengers
4. Increase of the availability of transport facilities

These main goals were further complemented by the City of Zagreb with additional requirements— guidelines on the development of the transport system:

5. Improvement of the capacity of public transport and service levels
6. Better access to transport networks and transport facilities
7. Reduction of the use of cars in the city centre,
8. Increase of the number of parking areas
9. Improved conditions for non-motorized transport
10. Increase of traffic safety
11. Noise and air pollution reduction

Spatial coverage included the administrative borders of the City of Zagreb with 872,400 inhabitants that lived in 306,300 households. The degree of motorization was 213 vehicles/1000 inhabitants. The modal distribution of motorized modes of travel accounted for 49.6% on travel by car in the city of Zagreb (MVA Consultancy 1999).

In order to reduce the use of cars in the central part of the city and for the specific purpose of travel, as well as achieving other objectives in the study, the proposed measures for transport management among others, was the parking supply management.

Research has discovered a deficit of about 7000 street parking spaces (with 3587 existing) in the central part of the city. It should be noted that the system of on street and off street parking spaces in the City of Zagreb in 1999 was very modest with a modest effect on the overall transport policy. For example, offer of public parking-garage spaces was modest, amounting to only 678 parking-garage spaces with a very minor impact on the overall transport policy.

Already mentioned above Transport study also planned and the locations for Park & Ride system, close to the terminals and public transport stops.

Intention of such concept of parking policy and management of parking space in supply was change in the future modal distribution of travel, discouraging users from using cars to for trips to work, especially in the central part of the town, and to encourage the use of public transport (MVA Consultancy 1999). This would lead to reduction of the pressures of arrival by car in city canter. Administrative limit of parking time was introduced in order to reduce the volume of trips with the purpose of migrating to work, which dominates by the share of peak loads (morning and afternoon peak load), and in long-term reduce the offer of street parking places.

3   ANALYSIS OF PARKING POLICY IN THE CITY OF ZAGREB

Zagreb has registered a strong increase in transport demands for the period from 2001 to 2015 which is reflected by increasing level of motorization. Growth trend in transport demand is shown through a comparative analysis of the degree of motorization in observed period (Statistical Yearbook—City of Zagreb, 2001–2015), (Statistical Yearbook Republic of Croatia, Croatia, 2001–2015) as shown on Figure 1.

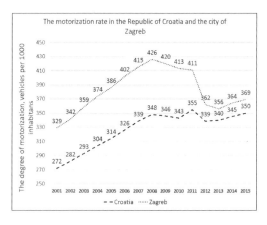

Figure 1.   The motorization rate in the Republic of Croatia and the City of Zagreb.
Source: Author in prep.

Analysis indicates that strong increase of degree of motorization has been noticed at the national level and in the City starting from 2001 until the beginning of global economic crisis in 2008, and in particular it should be noted that it has almost 30% higher. Registered decrease of the degree of motorization in the years 2008–2015, both in the country and in the city of Zagreb is largely the product of the economic crisis than the systematic implementation of an integrated transport policy at the national level. This downward trend in the degree of motorization travel in Zagreb, although unfortunately extensive research after traffic studies has not been performed.

In accordance with the objectives of Zagreb Transport study, strategic and management measures for parking supply are set to achieve the reduction of travel to work by car in the city centre, which were gradually imposed during the period 2000–2015.

The analysis shows that on street parking system has grown from 3587 street parking spaces available in 1999 to 10,000 available parking spaces in 2006, followed by the introduction of concept of administrative restrictions of parking and parking charges. The initial goals set by Transport study in segment of parking policy were met.

According to the proposal of the Study (MVA Consultancy 1999), the central part of the city was divided into three zones in which the time-limited on street parking is introduced with following restrictions: 1 hour (red zone), 2 hours of parking (yellow zone) and 3 hours of parking (green zone).

Demand management, charging and enforcement is responsibility of company Zagrebparking—Zagreb Holding, which is 100% owned by the City of Zagreb. According to the solution from the Transport study, the price per unit of time (1 hour) is created on the principle of non-linear relations 3–2–1, following the presumption that in the first zone (red) price of parking hour is three times higher than in the green zone, and in the second zone (yellow) is twice as high as in the green zone.

For tenants who have a residential address in the zone, and the businesses that have a business address in the parking zone, a special ticket (preferential ticket) is provided, which exempts from the time limit. The price of such special ticket is several times lower than the regular price of a unit of parking. The measure is favouring residential parking and businesses operating in the zone, with the intention to preserve the normal function of a mixed-use area in the central part of the city.

Although the Transport study (MVA Consultancy 1999) planned in the central part of the city introduction of around 10,000 on street parking spaces. The analysis for the paper has concluded

that after 2006 the significant increase in the number of on street parking spaces has occurred, and that they are spread in accordance with the spread of charging zones further from the central part of the city Figure 2. Analysis is presented in Figure 3, where timeline of the increase and expansion of on-street parking zone and the wider area of the City is visible.

In 2006 the planned number on street parking spaces was reached (11,613), and it was doubled by 2008 (approximately 20,200 spaces). The offer in the first zone has not increased (limited available space), but yellow and green zones has spread together with a special zone of off street parking spaces that where not introduced in a time limited parking.

The very next year (2009) on-street parking supply is significantly changed, and has almost tripled in first (red zone) due to the expansion of the zone (yellow zone was turned to red). That was

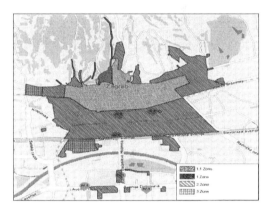

Figure 2.   Parking zones in City of Zagreb.
Source: Zagrebački holding.

Figure 3.   Number on street parking spaces per zone.
Source: Authors in prep.

the consequence of the increase of the degree of motorization (double) compared to 1998 and pressure of cars in the zone of the city centre.

Although as the result of the economic crisis, the level of motorization was in significant decline until 2013 (Figure 1), city authorities increase the capacity to 25,600 (2010), and finally to 29,500 places (2015). The first (red) zone is not increased, but supply of on street places in yellow zone has grown significantly and has approximately doubled compared to 2009 (by expanding the zone area), when the change of relationship between the first (red) and the other (yellow) zone has occurred.

For insight regarding the systematic implementation of management strategies of on street parking supply in the city, the analysis of pricing policies was made, together with its fluctuations during the observation period (Slavulj 2014).

Traffic study of the City of Zagreb, apart from administrative restrictions supply on street parking spaces, conceptually designed and parking pricing policy, which is conceived as the economic pressures on discouraging the parking in city canter.

Nonlinear relationship of parking prices in the area, as already stated, was 4 kn (red), 2 kn (yellow) and 1 kn (green) in 2000. Figure 3 presents the analysis of price of street parking.

The analysis revealed that the measures of economic deterrence of parking were extremely restrictive ending with 2008, and after that the trend was reduction of price for parking. This approach is not in line with the increase in supply of street parking spaces, and it is not in a function of deterring the arrival in the central part of the city by car and parking. Analysis of ticket sales for street parking for the period from 2013 to 2015 (SMS, parking machines, kiosks). Figure 5 shows that the number of tickets sold has followed consistent growth trend: 2013/2014 is 14.3%, and 2013/2015 to 19%, which indicates an increase in demand for parking in the zones, and increase in transportation demand of cars in the street parking charging zones.

For the purpose of consistent analysis, it is necessary to note that Traffic study planned exclusion of residents and businesses in the area of the time limit on-street parking, with the aim of social and economic prosperity of the immediate urban area.

Figure 4 shows the number and ratio of preferential parking cards in relation to the offer of on-street parking spaces.

Presented figure, according to the analysis, indicates that the number of preferential cards issued is almost equal to the total number of offered street parking spaces. In particular, the disproportion is particular noticeable in the number of preferential parking spaces and offer in places in the first zone (red) zone, which exceed the offer of street parking spaces for about 50%. This fact is particularly

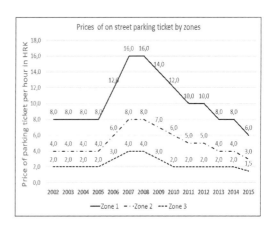

Figure 4. Prices of on street parking ticket by zones. Source: Authors in prep.

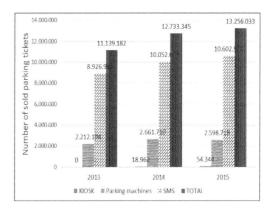

Figure 5. Analysis of ticket sales for the period from 2013 to 2015. Source: Author in prep.

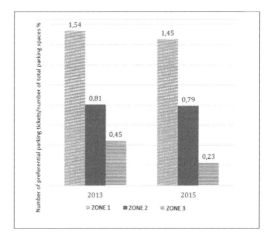

Figure 6. Number of preferential parking tickets issued with the number of parking spaces. Source: Autor in prep.

important for regulations on preferential tickets which allows ticket from the first (red) zone to be applied in other areas.

Offer of parking-garage spaces in the central part of the City of Zagreb has changed from a modest offer of 678 places in 1999, to 6356 places in 2015, which represents a tenfold increase. Prices per hour of parking in the garages under the administration of the company Zagrebparking is in general 4 kn/hour (can vary from 2 to 5 kn/hour), and linearly depends on the number of hours of use, while in private garages is dependent on the garage management, and its evaluation of supply and demand, and therefore is generally much higher.

It should be noted that the total supply of parking-garage spaces is open to public, but only 38% is owned and managed by the company Zagrebparking—Zagreb Holding Company, while the rest of the offer of 62% is owned by private investors. This ratio of ownership is a barrier for the city administration for implementing consistent pricing policy, given that fact that it has an impact on only 1/3 of the offer.

4 CONCLUSION

Following the analysis of situation in the transport demand and the extent and effects of demand management for street and off street parking in the city of Zagreb, for the period 2000 to 2015, which are based on a Traffic study, following conclusion can be made:

1. The motorization rate in the City of Zagreb in the past 15 years has more than doubled.
2. The supply of street parking spaces and the very concept of parking management in a traffic study is well defined, with a function of deterring the use of cars in the central part of the city, especially for trips with the purpose of going to work.
3. The offer of street parking spaces is conceptually divided into three zones, with the administrative limit on the length of parking for 2 or 3 hours depending on the zone.
4. The offer of street parking spaces increased from the planned 10,000 to 29,500 spaces in 2015 (increase of 3 times)
5. Analysis of the tariffs in certain zones shows that by 2008 it had significant economic function of deterring the use of cars. After 2008, the policy of tariffs of parking in the zones continuously decreases, and this is not a function of deterring the use of cars.
6. Number of preferential tickets used by residents and commercial entities in the area is almost equal to the supply of on street parking spaces in total in all zones. The first (red) zone has a

particularly large number of preferential tickets in excess of about 50% of offer of street parking spaces in the zone, which is nonsense.
7. In fact, preferential tickets in the total supply of on street parking spaces voids administrative limit parking on 2 or 3 hours of parking, and accordingly the economic pressure to discourage long-term parking, due to the fact that preferential tickets have preferential price. The Figure 6 shows that the number of preferential tickets exceeds overall supply of parking in the first zone, that is nonsenses.
8. Number of parking garage spaces in the central part of the city increased almost tenfold. Demand management of public parking garage spaces is mainly (62%) owned by private investors, while a smaller portion (38%) is under management of Zagrebparking. This fact is reflected in the different approach in tariff policy.

Taking into account all mentioned facts, and although there was no comprehensive research after a traffic study on modal distribution of travel it can be concluded that management strategies of on street parking spaces supply does not significantly affect deterrence of the use of cars in the central part of the city, especially regarding travels for the purpose of going to work (Engels, et al. 2012).

Therefore, it is necessary to state that the target set by Transport study for discouraging excessive use of personal cars using supply control strategies of on street parking spaces did not give results for several key reasons (MVA Consultancy 1999):

Unplanned expansion of zones and volume of on street parking spaces,

1. Inconsistent approach in issuing preferential tickets for charging zones
2. Construction of garages which are mostly owned by private investors.
3. Non-selective tariff policy in zones and garages.

We believe that it is necessary to redefine the strategy of street parking management using following measures:

1. Review of zones and volume of on street parking spaces with the aim of defining a number of smaller zones.
2. Urgent revision of tariff policy.
3. Revision of preferential tickets and limitation of the validity of preferential cards to micro zones (local neighbourhood).
4. Encourage creation of garage parking spaces in order to create conditions for reducing of supply of street parking.

It can be concluded that it is possible to set out the expert reasoned arguments and steps to achieve the goal of a significant change in modal distribution of trips, clearly with the application of a series of complementary strategies (public transport improvements, expansion of pedestrian zones, the implementation of infrastructure for the use of bicycle traffic and similar.)

REFERENCES

Bohler-Beadeker, S., Kost, K. & Merforth, M. 2014. Urban Mobility Plans, National Approaches and Local Practice, Bonn, GIZ, Germany.

Brčić, D. 1999. Contribution to the study of the impact of parking policy on the model of managing demand for transport in cities, PhD dissertation. Faculty of Transport and Traffic Sciences, University of Zagreb, Zagreb, Croatia.

Brčić, D., Šoštarić, M. & Pilko, H., 2013. Parking Policy Management in the City of Zagreb, IV International Conference "Towards a Humane City", 13–14 October 2013. Novi Sad, Republic of Serbia, pp. 103–111.

Bulletin Ministry of Internal Affairs 2015. Zagreb, Croatia.

Data Zagreb Holding Company 2015. Zagrebparking, Zagreb, Croatia.

Engels, D., Kontić, D., Matulin, M., Mrvelj, Š., Van Cauwenberge, B., Valkova, J., Vilarinho, C., Pedro Tavares, J. & Van Aken, E. 2012. CIVITAS ELAN Final Evaluation Report, ELAN Delivarable No. 10.11. European Commission, Brussels.

Lopez-Ruiz, H., Christidis, P., Demirel, H. & Kompil, M. 2013. Quantifying the Effects of Sustainable Urban Mobility Plans, EC Joint Research Centre, EU.

Litman, T.A. 2011. Parking Management, www.vtpi.org.

MVA Consultancy 1999. Transport study City of Zagreb, Croatia, Zagreb.

Penalosa, E. 2005. The Role of Transport in Urban Development Policy, Eschbom: GTZ GmbH, Germany.

Slavulj, M. 2013. Urban mobility planning with transport demand management measures, PhD dissertation. Faculty of Transport and Traffic Sciences, University of Zagreb, Zagreb, Croatia.

Statistical Yearbook—City of Zagreb, 2001–2015, Croatia, Zagreb.

Statistical Yearbook Republic of Croatia, Croatia, 2001–2015.

Transport Infrastructure and Systems – Dell'Acqua & Wegman (Eds)
© 2017 Taylor & Francis Group, London, ISBN 978-1-138-03009-1

Traffic flows merging cooperation analysis

D. Nemchinov
CNS Corporation, Moscow, Russia

A. Mikhailov
Irkutsk University, Irkutsk, Russia

P. Pospelov, A. Kostsov & D. Matiyahin
Moscow Auto and Road University, Moscow, Russia

ABSTRACT: The object of our current research is the traffic flows interaction in the merging zones of on-ramps. First of all we took into consideration that the traffic density increase and change of the vehicles dynamic greatly influence on the driver's behavior. The second is that on-ramp merging zone is a source of increase of traffic flow local density causing reduced speeds, capacity drop and congestion. The field data collection had been collected on the multi-lane highways (2 to 5 lanes in each direction) in 2015. The results of data processing are presented in this article.

1 INTRODUCTION

The interaction of merging traffic flows in conjunction areas of interchange on-ramps with a major highway is a subject of this study.

The main purpose of this work is data collection and studying of related task of traffic flow distribution on multilane highways with the number of lanes from 2 to 5 in each direction. Mathematical methods describing the process of traffic flows merging are developed on the grounds of results received from traffic flows cooperation analysis, and by which typical designs for merging areas and borders of their application depending on traffic volume and density of the major highway and the connecting interchange ramp, speed-change lane length related to traffic flows density and spacing distribution between them, will be developed. Further, this method will become a part of *code of practice for highway designs.*

Respectively the main target is to develop calculation technique of merging zones capacity estimation, which should be included into the new Highway Capacity Manual. There were many reasons to start this research. The capacity estimation techniques which are currently in use had been developed in the period of 1960–1980 (Drew, 1967, Roess, 1993) and had been included into Highway Capacity Manual 1982. Since HCM 1982 was published a lot of traffic flows characteristics have changed (Elefteriadou, 1995, Sang Gu Kim, 2003, Laval, 2006, Treiber, 2006, Ning, 2011). The development of calculation technique of merging zones capacity estimation was stopped in Russian Federation 30 years ago.

First of all we had taken into consideration that the traffic density increase and change of the vehicles dynamic greatly influence on the driver's behavior. The second is that on-ramp merging zone is source of increase of traffic flow local density causing reduced speeds, capacity drop and congestion.

Based on analysis of normative and methodological documentation and studying of traffic flows merging areas in locations of interchange ramps conjunction, which have been held in different countries, tasks of experimental part of analysis have been defined—the analysis of spacing between vehicles in traffic flow and their effect on vehicles cooperation, driver's choice of gap for entering into a through traffic. Method for carrying out the analysis also has been developed, adequate equipment has been defined and also assessment for minimum number of measurements has been exercised.

On the assumption of aims and work composition while making measurements the following data should be considered:

- Traffic volume and density for each lane of multilane highways in joining areas of interchange on-ramps;
- Traffic volume and density on interchange ramps;

- General through and merging traffic volumes (density) and their ratio when the vehicle behavior starts to change considerably (deceleration or movements to the left lane);
- Gaps (headways) distribution statistics for main roadway with different traffic volumes (densities);
- Main through and merging traffic volume (density) and their ratio when congestions appear in right and adjacent lanes of the major roadway and in the speed-change lane;
- Through traffic speed in the merging and adjacent areas;

While making measurements only considerable traffic factors such as ramp geometry and spacing between successive ramps should be considered.

As a result, eventually, information of spacing distribution in traffic flows for each lane on multilane roads will be derived. Analysis of traffic flows interaction in merging area will be a base to develop typical designs for merging areas and define borders for their application depending on traffic volume and density of the main highway and the connecting interchange ramp, speed-change lane length related to traffic flows density and spacing distribution between them.

The field data collection has been conducted on the multi-lane highways (2 to 5 lanes in each direction) in 2015.

2 PROCESSING OF TEST DATA

Based on test data, the following traffic flow characteristics were defined:

- Interval between vehicles of the secondary flow on the ramp t_f (Figure 1);

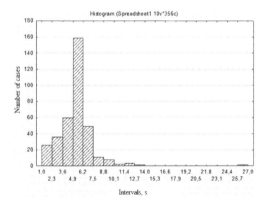

Figure 1. The example of gained intervals distribution on the ramp (Junction from IKEA, Himki).

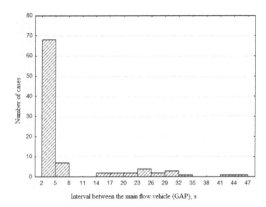

Figure 2. The example of gained intervals distribution between the secondary flow vehicle and the main flow vehicle (Junction from IKEA, Himki).

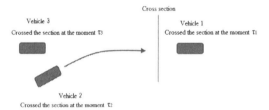

Figure 3. Estimation of the critical interval and its components.

- Interval between vehicles of the secondary flow following the main flow vehicles while merging h (Figure 2);

The value of interval between vehicles of the secondary flow (i.e. minimal intervals) is used while calculating the capacity of the ramp.

The critical interval t_f is the most significant design characteristic while determining the capacity of the merging area. The critical interval values were defined by the data processing gained from special video in which survey areas were in video camera lens, located in 4–5 m in excess along its entire length.

In determining critical intervals, those main flow intervals were selected, which were used by the only one vehicle from the secondary flow. The scheme of determining the critical intervals is shown in Figure 3.

The evaluation of the traffic flow characteristics was made by the following procedure:

1. Choose intervals in main flow which were used by the only vehicle $t = \tau_3 - \tau_1$
2. Draw the histogram of the used intervals t
3. Determine the statistics of distribution for the intervals used. The resulting median value—is the critical interval tc.

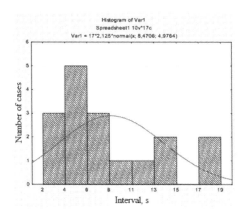

Figure 4. Critical intervals histogram. Average value – 8,5 s.

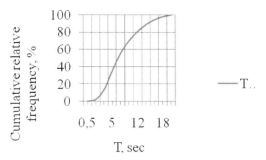

Figure 5. Cumulative relative frequency curve for the critical intervals. The 50th percentile value – 6,5 sec.

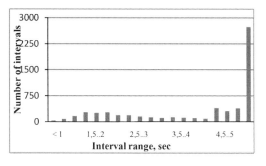

Figure 6. Intervals distribution with free traffic flow in the right lane of 3-lane carriage way.

The evaluation results are shown in Figures 4 and 5.

The evaluation of main traffic flow characteristics was carried out using ranges of intervals between the vehicles in the traffic flow presented below. As a result of work performed, data of intervals distribution in traffic flow for 4–8 lane roads in each direction was received.

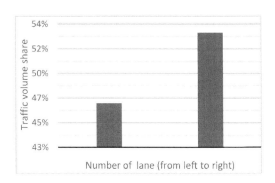

Figure 7. Traffic volume distribution on 2-lane carriage way.

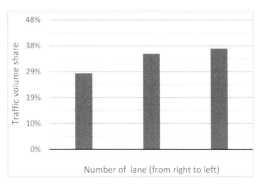

Figure 8. Traffic volume distribution on 3-lane carriage way.

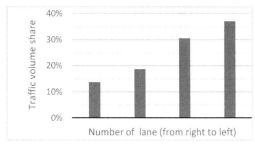

Figure 9. Traffic volume distribution on 4-lane carriage way.

The example of received data for one of examined cases is shown in Figure 6.

The received data also allowed defining the distribution of traffic volume on the lanes. The received data is presented in Figures 7–9.

Traffic volume calculations for traffic flows merging areas are carried out in accordance with methods and conditions set forth by Drew (Drew, 1967).

The calculation results are shown in Figure 10.

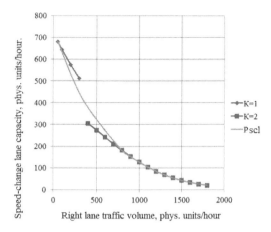

Figure 10. Speed-change lane capacity depending on the right lane traffic volume.

3 CONCLUSIONS

Conducted measurements have allowed revealing regularities of intervals distribution in the traffic flow composition for automobile roads with different number of lanes and roads with medians for dividing directions of travel. On the basis of the reveled regularities, traffic intervals were defined, taken be drivers on the main travelled way for making the departure maneuver from the interchange ramp.

Conducted on the basis of experimental data analysis of traffic volume distribution (flow density) for multilane roads has revealed that with a number of lanes more than 4 (more than 2 in each direction), the right lane loading is significantly less than it is in other lanes. This loading disproportion greatly exceeds similar disproportion revealed in foreign surveys and is likely concerned with the wide-spread application in domestic practice interchange junction ramps with short speed-change lanes or without them, with nonoptimal junction angle not providing drivers on the ramp visibility of the main travelled way vehicles in the rearview mirror. (If the ramp joins at an angle significantly less than 90 degrees, than the needed angle of head rotation exceeds physical abilities of human organism).

In the designing traffic interchanges for creating a traffic capacity reserve which will be required in the period of the development of urban area, it seems appropriate for the total coerced traffic volume in the right lane and on the ramp of the interchange not to exceed 1600 vehicles per hour.

An auxiliary lane in case of the number of lanes 2 or more for each direction should be added after interchange when the coerced traffic volume

per hour of one lane exceeds the value shown in nomogram (Figure 10). In case of exceeding the mentioned value the lane should be added after interchange instead of speed-change lane.

With 2 lanes on the ramp, as a rule, minimum one lane should be added after interchange or, if necessary, two. Having justification on the basis of the capacity calculation for the lack of necessity for adding auxiliary lane, the number of lanes on the combined ramp should be decreased before the speed-change lane beginning of junction on such combined ramp with the main travelled way, or the lanes of speed-change lane should be successively tapered.

Application of the developed proposals will let provide the increase in efficiency of traffic interchanges functioning. The performed work corresponds the scientific and technical level of similar works, being performed in other countries, and herewith let consider the particularity of traffic flows in the roads of Russian Federation.

REFERENCES

Drew, D.R. & LaMotte. L.R. &Wattleworth, J.A. & Buhr, J.H. 1967. Gap Acceptance in the Freeway Merging Process. *Highway Research Record* 208: 1–36.

Drew, D. R. & Buhr, J. H. & Whitson, R.H. 1968. Determination of Merging Capacity and Its Applications to Freeway Design and Control. *Highway Research Record* 244: 47–68.

Elefteriadou, L. & Roess, R.P. & McShane, W.R. 1995. Probabilistic Nature of Breakdown at Freeway Merge Junctions, *Transportation Research Record* 1484: 80–89.

Highway Capacity Manual 2010. Transportation Research Board, Washington, D.C., 2010.

Laval, J.A. & Daganzo, C.F. 2006. Lane-changing in traffic streams, *Transportation Research Part B* 40(3): 251–264.

Ning, W. & Lemke, K. 2011. A new Model for Level of Service of Freeway Merge, Diverge, and Weaving Segments. 6th International Symposium on Highway Capacity and Quality of Service Stockholm, Sweden, June 28–July 1, 2011.

ODM 218.2.020-2012. Branch road methodical document. Guidelines on the assessment of road capacity, Federal Road Agency, Russia, 2012.

Roess, R. & J. Ulerio 1993. Capacity of Ramp-Freeway Junctions, Final Report. NCHRP Project 3-37. Polytechnic University, Brooklyn, N.Y., Nov.

Sang, G.K. & Young, T.S. 2003. Development of a New Merge Capacity Model and the Effects of Ramp Flow on the Merge Capacity. TRB 2003 Annual Meeting 1–19.

Treiber, M. & Kesting, A. & Helbing, D. 2006. Understanding widely scattered traffic flows, the capacity drop, and platoons as effects of variance driven time gaps, *Physical Review E—Statistical, Nonlinear, and Soft Matter Physics* 74(2): 1–10.

Transport Infrastructure and Systems – Dell'Acqua & Wegman (Eds)
© 2017 Taylor & Francis Group, London, ISBN 978-1-138-03009-1

The relaunch of branch lines as a territory project for the inner areas

F. Alberti

Dipartimento di Architettura, Università degli studi, Florence, Italy

ABSTRACT: In the Italian National Strategy for Inner Areas, included in the Partnership Agreement 2014–2020, improving accessibility is one of the priority axes of intervention, either as a lever of territorial cohesion, or a driver of economic development. Therefore, the presence of a railway line, even if underused or abandoned, may represent for these areas a strategic asset to activate revitalization processes, through the integration of regional and mobility policies. Good practices demonstrate how innovative management models in rail service, together with an approach to regional planning oriented to public transport, may have significant multiplier effects on the demand side. On this basis, the paper presents two project researches for inner areas in Tuscany, both characterized by the presence of an old railroad now in decline, whose "smart" reuse is the key element for a "territory project" aimed at the enhancement of the local conspicuous territorial capital.

1 INNER AREAS IN ITALY: FROM WEAKNESS TO OPPORTUNITY

The Partnership Agreement for Italy 2014–2020 adopted by the EC has highlighted the importance for the country's development of "inner areas", which include a large number of minor centres, settled in different historical phases and in various regions. Therefore, it relates to a very wide range of territorial areas, which since the 50s of last century have been cut out of the process of Italy's industrialization and economic growth.

We know that this process followed very different trajectories between Centre-North regions on one side, and Southern Italy and the islands on the other. But even in regions where it was quicker and more intensive (such as in Emilia Romagna, Piedmont and Lombardy), it generated significant differences between the "central" areas, corresponding to cities and urban systems of any size directly affected by the economic development, and the "peripheral" ones, mainly, but not exclusively, located in high hilly and mountain areas, marked by more or less evident signs of economic de-growth, population decline and an increase of aging population.

In addition to being located at a significant geographical distance from supply centres of essential services—such as, in particular, education, health-care and mobility—inner areas, as described by the dedicated National Strategy set out in the Partnership Agreement, are characterized by a largely underused "territorial capital", concerning environment, cultural heritage, local knowledges, etc.: a waste of resources that results in high social costs, for example in terms of hydrogeological instability, due to the lack of slope maintenance, or deterioration of the historic heritage and landscape. But, at the same time, it is also a measure of the "economic development potential" of these areas, that cover about 60% of the total national territory, where about a quarter of the Italian population still lives split into over 4,000 municipalities. Therefore, local development policies—refers the Strategy—have to be, "first and foremost, policies for activating latent local capital".

If reduced accessibility to basic services is for locals a strong limit to their citizenship rights, the objective difficulty of moving from a place to another, due to the geomorphological features of the territory, the distance from major infrastructure networks, the poor conditions of local roads, and the inadequacy of public transport, represents the main obstacle to the economic development of inner areas and may be a cause of their "desertification"—a situation that we find not only in the regions of South Italy, but in the whole southern Europe: Portugal, Spain, France, Greece (Camagni, 2011).

The ultimate objective of the National Strategy for the inner areas, which sums up all other objectives, is to reverse the demographic trend, both in numbers and in terms of generational change (Figure 1); then, beyond the advantages that may be obtained through the development of immaterial infrastructures and the remote access to a growing range of services provided by ICT's, it is easy to understand that the improvement of material links, through the enhancement of transport infrastructure and services, is the *sine qua non* for achieving the result.

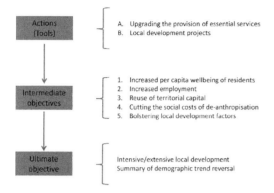

| Actions (Tools) | A. Upgrading the provision of essential services |
| | B. Local development projects |

Intermediate objectives	1. Increased per capita wellbeing of residents
	2. Increased employment
	3. Reuse of territorial capital
	4. Cutting the social costs of de-anthropisation
	5. Bolstering local development factors

| Ultimate objective | Intensive/extensive local development |
| | Summary of demographic trend reversal |

Figure 1. Objectives of the Italian Strategy for inner areas.

2 BRENCH RAILWAYS AS THE "INNER AREAS" OF TRANSPORT INFRASTRUCTURE

Although not explicitly mentioned in the National Strategy, among the underused resources of the territorial capital that are spread all over the country, a large number of branch railways must be included. The range of the lines is varied, with regard either to operating and maintenance conditions (totally abandoned lines, closed but still functional lines, closed railroads occasionally used as a tourist attraction, operating lines with reduced numbers of train rides and stops, etc.) or to the technical characteristics of the infrastructure (track gauge, power, presence of viaducts and tunnels, etc.).

The stories of these railways largely reflect the transformation process that in the second half of the twentieth century led to the concentration of activities and population in some parts of the country and to the formation of inner areas. Despite the disastrous damages occurred to the national rail network during the Second World War—at the end of the conflict, in 1945, 7,000 km of track, equal to 30% of the total track line length, had been destroyed, as well as about the same rate of railway bridges (Maggi 2012)—by 1955 the restoration works are completed for approximately 22,000 km of track, with a loss of 1,000 km compared with the pre-war network.

Between 1955 and 1972, going hand in hand with the explosion of individual motorisation and the development of the motorway network, other 2,100 km of lines (including 1,500 km of lines under concession) became unproductive and closed. The new geography of development, drawn by the routes of private mobility, that receive most of the public investment (in the 60's road transport already absorbs 80% of the total budget allocated

to infrastructure in Italy), has, as a side effect, the concentration of 95% of passenger traffic and freight on just a half—11,000 km—of the total operating tracks, threatening the survival of train services on the other half.

In 1985, a decree attached to the Budget Law, signed by the former Minister of Transportation Claudio Signorile, contains a list of 57 lines, some of which still used daily by a not insignificant number of commuters, that should be immediately closed as "not included in the railway network of general interest". The measure, presented as an act of rationalization, actually reveals the indifference of the State to seeking a way, through the reorganization of services, to take as much advantage as possible from existing rail infrastructure as an alternative to the spread of car mobility.

After the protests that followed the decree, only 6 lines were actually closed at the time. Nevertheless, it has become the first step towards the closure in the following years of an increasing number of secondary railways, as an inevitable outcome of the vicious circle between poor services and demand reduction: a situation that the transfer of power on local railways from the State to the regional authorities (2001) has been able to correct only in few cases.

In Apulia and in the autonomous Provinces of Trento and Bolzano, a different approach to extraurban mobility in areas with a high tourist vocation and an efficient use of European funds have, in fact, led to the successful reactivation of railway lines that had been abandoned in previous decades, with new railroad equipment, rolling stock and mode of operation.

On the contrary, other regional authorities have continued to channel all available resources provided by the State, more and more reduced because of public spending cuts, only to the enhancement or even maintenance of the stronger lines, playing at local level the same policy of depletion and closure of secondary railways started in the 80's by the central Government. Significant, in this sense, the case of the Piedmont Region, which in 2012 decreed the closure of twelve railways, used every day by about 6,000 people: in the lack of demand-side policies in favour of sustainable modes of transportation on the commuter routes, the operating cost per passenger of the lines had gradually grown becoming unjustified in relation to the service standards.

From the post-war period to the present day the number of disused railway branches across the country amounts to 162 tracks, for a total length of 5,800 km (www. ferrovieabbandonate. it). More difficult is quantifying the lines with extremely low traffic density, that over the years have seen a constant reduction of train rides and stops: a prelude to possible further closures in the coming years.

2.1 Secondary railroads as a resource for the accessibility of inner areas

The recognition of the inner areas as a "national issue" of strategic importance leads naturally to identify the improvement of regional accessibility and mobility as one of the priority axes of intervention, either as a lever of territorial cohesion (in order to support the maintenance of residential functions, land conservation and the protection of cultural heritage), or a driver of economic development (especially with regard to tourism, agriculture, local manufactures, etc.).

This gives a new perspective also on secondary railways, which can be taken into account on the basis of different criteria than those which have led to the gradual decimation of lines and services.

The methodology itself used for the definition of inner areas identifies in the presence of a "silver class" train station (that is, according to the classification adopted for the Italian railway network, a small-medium station with a daily flow of about 2,500 people, including passengers, employees and users of complementary services) one of the criteria characterizing a municipality or a group of neighbouring municipalities as a "service provision centre"; the other criteria are the presence of a hospital with emergency room, diagnostic services and short stay ward (DEA level 1), as well as of the whole range of secondary schools. Inner areas are therefore defined by the time required for reaching in the quickest way the nearest service provision centre, distinguishing the so-called "intermediate" areas (located at a distance of 20 to 40 minutes), from the "peripheral" areas (40 to 75 min.) and the "ultra-peripheral" ones (over 75 min.).

In coherence with this approach, especially with regard to those territories with a valley in the middle, where a road and a railway linked to a service hub run parallel to each other, we can look at the latter, even if considered inefficient according to the usual operating standards, as a strategic resource of the territorial capital, an asset to be valued in the logic of integration of regional and mobility policies, actions for social inclusion and actions for growth.

As far as transport efficiency is concerned, the above mentioned regional good practices highlight how innovative management and operating models can have a significant multiplier effect on the demand even in places with low-density settlement, which is especially true if new services are conceived as part of a multi-modal transport offer, capable of meeting the inhabitants' mobility needs, as well as of enforcing the supply of tourism services.

The experience of the Province of Bolzano, where a single public transport company supplies integrated services by road, rail and cable car, is from this point of view very significant. Among the railway lines, Merano-Malles (the Venosta Railway Line) and Soprabolzano-Collalbo (Ritten Railway), both inaugurated at the beginning of last century, are very interesting examples, not only at national level, of historical infrastructure, that have been renovated in recent times in order to accommodate ordinary train services.

After being inserted in 1985 in Minister Signorile's list of "dead branch lines", the Venosta Railway Line was left to languish by the national operator (RFI) until it closed in 1991. In 1999 the abandoned railway was acquired by the Province of Bolzano. Six years later, it will reopen with its 60 km of track line and nineteen stations, after an investment of 130 million Euros, used for a complete renovation of its equipment, advanced traffic management and control systems, the restoration of bridges and other structures, the purchase of trains with high comfort and performance and the building of road underpasses, architectural station canopies and areas for multi-modal interchange from train to bus or private car. Along with these interventions new services have been targeted at new kinds of users, like the transport of bike by train and bike rental at the stations. Thus transformed, from a "dead branch" Merano-Malles has established itself in a short time as the flagship line of South Tyrol, with over 18,000 trains operating during the year and 3 million passengers (two-thirds are locals, one-third tourists), becoming the model for the adaptation of railways in the other valleys.

In the municipality of Ritten, lying on a plateau of 110 sq. km, 1,000 m height above the city of Bolzano, a 4.5 km narrow-gauge and single track railway has come back to life, with a train frequency of 30 minutes, after the opening in 2009 of one the most modern cable-cars in Europe, connecting the terminal station of the railway at Soprabolzano to the city centre of the Province capital with a frequency of 4 minutes and a capacity of 550 people per hour. Such an intermodal combination of sustainable transport vehicles has made of this territory, where live 7,600 people, a unique case of an inner area that has suddenly ceased to be marginal as a consequence of a single act of modernization.

In addition to improving passenger services on the main railway lines, it is important to reflect on how to make easy, cheap and safe moving from the innermost areas towards the access points to the railway. On the other hand, regional planning can play a decisive role in promoting public transport accessibility to inner areas, making of the existing tracks the backbone of the functional reorganization of settlements. The principle of "transit-oriented developments" (Calthorpe, 1993), postulated with reference to the model of the compact city, translates in this case into the idea that rail stations, located within very small centres or even in open country, should anyway play a role of "functional milestones" of the territorial system, accommodating in their immediate surrounding (with a preference, wherever possible, for the re-use of railway buildings and areas that are no longer necessary for transport) basic public facilities, services for tourists, areas for multi-modal interchange, etc., as well as the access points to the network of nature trails.

3 RETHINKING THE ROLE OF BRENCH RAILWAY LINES IN TUSCANY. TWO RESEARCH PROJECTS

Moving from these assumptions, the next part of the paper presents two research projects recently undertaken by the Research Unit SUP&R (Sustainable Urban Projects & Researches) at the Department of Architecture of Florence, focused on the "smart" reuse of old railroads now in decline—by means of innovative services on the line and towards the stations—as the key element for "territory projects" aimed at the enhancement of local territorial capital (heritage, landscape, agriculture, etc.) in the inner areas. The field of both studies is Tuscany, a Region that before the unification of the Kingdom of Italy, in the Grand Ducal period, and even more so in the decades immediately following, occupied an important position in the development of rail transport in the peninsula, so that about 3/4 of 1,500 km operating tracks in the region date back to the nineteenth century. Today, the situation in the sector of rail transport has peaks and troughs. On the one hand, it must be recognized the commitment of the Region to maintain or increase train service levels on the busiest commuter lines, despite the reduction in state funding: an effort rewarded by a progressive increase of the total number of passengers in recent years (234,000 passengers in 2015, +11,4% in comparison to 2008). On the other, the present situation also shows the stagnation or depletion of weaker lines, reflected in the cut of train rides and stops and in the increase of bus services overlapping to trains or competing with them.

The studies relate to this kind of situation, explored in two very different local contexts, that is the Cecina Valley, in the hinterland of the Tyrrhenian coast, and Garfagnana, in the Appennine Mountains. Both of them reached a first level of maturity as thesis works, followed by Prof. Francesco Alberti and carried out respectively by Fabrizio Baroncini at the master degree course on Urban and Regional Planning and Design (2015) and by Elisabetta Mennucci at the 2nd level master "Designing the smart city" (2016) of the University of Florence.

3.1 *The rail link from Cecina to Volterra*

The territory object of study, straddling the Provinces of Livorno and Pisa, lies along the river Cecina, between the homonym town (28,000 inhabitants) and Volterra (10,5000 ab.): that is, on one side, the main urban hub of the Etruscan Coast (one of the most beautiful Tuscan stretches of coastline), served by the Tyrrhenian railway line Rome-Pisa-Genoa, and, on the other side, one of the most outstanding examples of fortified city in Tuscany, rich in remains ranging from the Etruscan period to the Medicis, located about 40 km from the sea, on a hill overlooking the Cecina Valley. In the middle, lies an area of great interest for its geology and landscape, dotted with ancient villages located on the top of the hills as well as small mining centres on the plains; the last ones include in particular Saline di Volterra, that during last century asserted itself as an important location of salt industry.

The two main centes were directly connected from 1912 to 1958 by one rail line, with four intermediate stops placed either at the beginning of the roads leading to the villages on the hills, or at the edge of the lowland centres. The last 4 km-stretch of the line was a rack railway with a gradient of 100‰, in order to climb the slope between the valley and the foot of the walls of Volterra (Figure 2), outside of which, not far from one of the gates of the city, the terminal station was built.

In 1958 the rack section closed, outclassed by the speed of road vehicles, with the consequence

Figure 2. Volterra. The old rack train at the foot of the city walls (Photo Archive Albertini).

that the line was reduced to the track Cecina Saline di Volterra. Although in 1985 this one was in turn in the list of the "lines not included in the railway network of general interest", it has survived until this day, with a traffic volume that has been gradually reduced to the current 4 pairs of trains running only on school day. The agony of the railroad is reflected either in the loss of population in the villages or in the loss of jobs at the salt-plant of Saline di Volterra, whilst at the coast, where seaside tourism is flourishing and full accessibility by road and train is provided, Cecina has in the meantime constantly grown and developed.

In 2014, following the approval of the Regional Landscape Plan of Tuscany (Piano Paesaggistico Regionale della Toscana, PPRT), the Cecina-Saline di Volterra becomes topical once again. The Plan includes in fact a "Project for the slow enjoyment of Regional landscape", aimed at improving slow accessibility to some regional places with special landscape qualities, by using the tracks of totally abandoned or underused railways: in the first case, the track structure should be converted to a nature trial; in the second one (that includes the Cecina-Saline di Volterra), new trains should be put in operation on the line to provide a tourist rail service. In addition to that, a concept plan attached to the Project shows some complementary interventions to be implemented along the Cecina Valley in favor of "slow mobility", such as hiking paths, riverside cycleways, mountain-bike and horse trails.

In respect to the Regional proposal, the objective of the research project carried out at the Department of Architecture of Florence is to define a strategic scenario for the Cecina Valley, focused not only on the relaunch of the currently running line between Cecina and Saline, but also on the reactivation, with modern technology, of a rail service in the section Saline-Volterra. A direct, uninterrupted rail link between Cecina and Volterra is seen in fact as an invaluable option to get the most out of the synergy between two strengths of the Region—such as Volterra and the Etruscan coast—and to reconnect Volterra to the Tyrrhenian corridor, not only to support tourism but all economic activities as well, with the ultimate aim to reverse the demographic trend of the innermost areas.

The study revolves around three key issues, two of them conditioning the feasibility itself of the proposal, which were therefore dealt with at first:

– The availability on the rolling stock market of a train with adequate comfort and performance, able to run the whole track, including the slope from Saline to Volterra, on the same path of the old rack railway; through the analysis of case studies, possible solutions have been identified in different railcars by Stadler, running in combined adhesion and rack railways in Switzerland (Brig-Zermatt, Zermatt-Gornegratt and the Monte Generoso Line), as well as on the line between the seaside and the city centre at Catanzaro, in the Region Calabria;
– How to solve, by means of urban design, the problem of a new terminal station at Volterra (exact location of the stop and adjacent interchange area, connections with the old city centre, etc.), since the re-use of the old station is no longer practicable, due to recent urban transformation.

The third theme is defining a framework of possible interventions, in addition to those set out by the PPRT, aimed at strengthening the role of the railway line as the guide-element of the functional reorganization of the territory. According to a first draft of the strategic scenario, interventions should apply the following principles:

– Concentrate in the areas adjacent to the railway stations the building capacity still resulting from the local planning instruments of all concerned municipalities, using abandoned freight yards for new developments; all new industrial sheds should however be concentrated in few locations (such as the existing industrial zones of Cecina and Saline laying along the rails);
– Locate additional service activities—like small camper areas, sports yards, public facilities, etc. – in the same areas close to the train stops: they will be easy to reach by train either from the coast or the hinterland;
– Add (few) new stops along the line, where necessary for a better service to the territory and making railway more competitive to cars (for example, to serve the populous neighborhood of San Pietro in Palazzi at Cecina; or at the exit Cecina Nord of the motorway Aurelia, where a park-and-ride could be realized to stop cars at the edge of the town; or west of Saline, where the medieval castle of Montegemoli and many holyday farms are located within 2 km from the line).

3.2 The reorganization of mobility in Garfagnana

Garfagnana is an inner area in the Apennines, in the north of the Province of Lucca, on its border with Liguria and Emilia, characterized by widespread settlement of ancient small centres (sometimes just groupings of buildings), inside a natural environment and landscape of highest quality. The total population is about 28,400, spread over 16 municipalities. The main centre is Castelnuovo di Garfagnana (pop. ca. 6000), at the confluence of the River Serchio and its tributary Turrite Secca, built around a remarkable medieval fortress (the Fortezza Ariostesca, after Ludovico Ariosto, author of *Orlando Furioso*, who stayed there as a governor on behalf of the Duchy of Este).

Transport infrastructure of Garfagnana consists of a corridor in the valley of Serchio, covered by the Lucca-Aulla railway (90 km) as well as the State roadway n. 445, and a secondary pattern of mountain roads engraving the territory. The construction of the railway, begun in 1884, had a troubled history and was completed only 75 years later (1959) with the inauguration of the gallery Lupacino (7.5 km), which connects Garfagnana with Lunigiana, that is the territory of the high valley of the River Magra, in the Province of Massa Carrara. The line is connected at one end (Aulla) with the railway La Spezia-Parma, and at the other end (Lucca) to the regional lines towards Florence, Pisa and Viareggio.

Although railways actually link Garfagnana to many regional and trans-regional destinations, rail transport has lost attractiveness hand-in-hand with the increase of traffic flows on the road network. Public transport itself is today mainly provided by bus services, including bus lines partially overlapping rail service, that are both inefficient and barely used, and targeted services such as school bus.

Given the need of re-modulating public transport supply both in terms of efficiency and sustainability, we must be aware that orography and spread settlement, which makes Garfagnana representative of many other inner areas in Italy and abroad, make it anyway impractical to apply here the typical "urban" approach to sustainable mobility, based on the integration between soft mobility and conventional public transport.

The objective of the research project is therefore to set out a different model of mobility based on the optimization of the existing railway, that actually runs on the route which collect all traffic flows affecting the area. This shall be accompanied by a complete reorganization of services by road as feeder services to the main route (Figure 3), resultant of a process of public participation. A very interesting precedent, in this sense, is the participation process carried out in 2015 by the municipality of San Casciano Val di Pesa, in the Chianti area, with funds from the Region Tuscany: a wide consultation that has involved local associations, voluntary groups, transport operators and private citizens, which has finally brought in summer 2016 to test a new shuttle service, having rather flexible schedules and routes, in order to connect the small hamlets scattered in the territory to the main centre.

The key elements of the project, that got the endorsement of the Union of the Municipalities of Garfagnana, are therefore:

– the study of more affordable operating modes for the railway line, to ensure on one hand an adequate service for locals, and support on

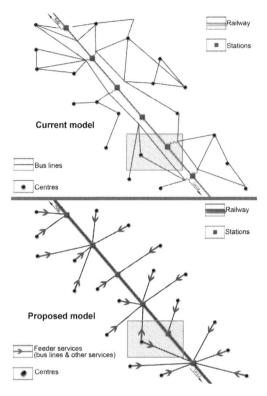

Figure 3. Current and proposed models for public mobility in Garfagnana.

the other the development of quality tourism, potentially attracted by the rich cultural heritage and environmental resources of the area together with its closeness to places of interest like Florence, Lucca, Cinque Terre;
– the optimization of feeder services by road to the railway stations, that should combine more rational and flexible public transport services (including as a possibility share taxis, on-call services, school bus used for additional services, etc.) with innovative forms of collective or shared transport at low-cost, that see the active involvement of the community, according to the principles of pooling economy;
– the development of a smart ICT platform for moving in Garfagnana, to provide citizens with an interactive information tool aimed at facilitating the integrated use of fixed and flexible services, so that to meet the mobility needs of each user.

4 CONCLUSIONS

Distant, by definition, from the urban hubs providing the main services to citizens and businesses, but "with resources that are missing in the central areas, with demographic problems but with high potential appeal" (Barca, 2012), the inner areas are a major challenge for the sustainable development of Italy and many other European countries.

Assuming that rural and environmental resources, cultural heritage, local traditions, etc., are the strengths that can be used to leverage opportunities for future development in the inner areas, difficult accessibility is at one time the most obvious weakness and the main threat to their possible economic revival. On the other hand, it is precisely due to this limitation and consequent lack of competitiveness with other regional areas, that they have been until now marginalized from the economic life and development of the nation.

Although isolation produces similar effects in very different contexts (such as unemployment, migration, aging, decay of buildings and the environment, etc.), it's evident that there is no ready-made solution to overcome this condition. As far as accessibility and transport are concerned, any action carried out with a sectoral approach, in areas that from the start are very weak in terms of demand, is condemned to failure. On the contrary, any sectoral contribution is essential in the formulation of an overall strategic scenario using a place-based approach.

In hilly or mountain territories, with a valley acting as a transport infrastructure corridor, the presence of a railroad, especially if connected directly to an urban hub, can represent a key input in the building up of this strategy, even if it is abandoned or underused. The railway, in addition to being an environmentally friendly mode of transport, can be improved in performance (in terms of capacity, transport safety, etc.) simply working on the service to be supplied, with a minimum of new works and little impact: two aspects that are particularly relevant in areas with high environmental quality.

The placement of the railway line in the valley, that is on the main axis of territorial distribution, makes it easily accessible from the outside, encouraging the development of sustainable tourism; moreover, it allows to intercept all commutes from inside outwards (and vice versa) and most internal trips.

The case studies of Cecina Valley and Garfagnana highlight two important aspects of the enhancement of secondary railways aimed at improving the accessibility of inner areas, which, depending on the circumstances, may have different weight in setting the strategy:

– on one side, the quality of the service (in terms of frequency, regularity, comfort, speed, ease of use) performed along the main axis, which should be the fixed component of local transport; to this aspect are also related all interventions on and along the line, aimed at optimizing the interactions between railway and territory, with particular attention to the role of the stations, seen as functional milestones of the settlement system;

– on the other side, the way to connect the innermost centres to the railway stations. With dispersed settlements and a weak demand, the solution to the problem can only be found with a totally new approach to mobility, that exceeds the distinction between public/private or collective/individual transport.

In this way, under the signs of pooling economy and smart city, the issues of social and technology innovation become fully part of the territory project for giving shape to the flexible component of local transport, complementary to the rail service. Such an integrated offer, in addition to encouraging a reduction of car dependency in the inner areas, responding to a universally valid objective of sustainability, can be an important enabling factor and a social inclusion driver as well, by ensuring access to mobility to non-motorized citizens. Furthermore, it can have positive effects as a catalyst of micro-entrepreneurial activities related to the provision of transport services, supplementary or alternative to the traditional ones, according to the idea of a community that finds within itself the answers to its own needs.

REFERENCES

Barca F., 2012. *Metodi ed obiettivi per un uso efficace dei Fondi Comunitari* 2014–2020. Document of the Minister of Regional Cohesion, in agreement with the Ministers of Labour, Social Policies and Food Agriculture and Forestry Policies.

Calthorpe P., 1993. *The Next American Metropolis. Ecology, Community and the American Dream*, New York: Princeton Architectural Press.

Camagni R., 2011. Coesione territoriale: quale futuro per le politiche territoriali europee? In L. Resmini & A. Torre (eds), *Competitività territoriale: determinanti e politiche*: 33–52. Milano: FrancoAngeli.

Maggi S., 2003, *Le ferrovie*, Bologna: Il Mulino.

Maggi S. & Giovani A., 2005. *Muoversi in Toscana. Ferrovie e trasporti dal Granducato alla Regione*. Bologna: Il Mulino.

Pucci P., 2008. Infrastrutture come progetti di territorio: con quali progetti e conquali strumenti. In A. Belli et al. (eds) *Territori regionali e infrastrutture. La possibile alleanza*: 226–276. Milano: FrancoAngeli.

Transport Infrastructure and Systems – Dell'Acqua & Wegman (Eds)
© 2017 Taylor & Francis Group, London, ISBN 978-1-138-03009-1

Total management tool oriented to carbon footprint reduction in terminals of containers

F.E. Santarremigia, G.D. Molero & M.D. Esclapez
Department de Proyectos de Investigación, AITEC, Parque Tecnológico, Paterna, Valencia, Spain

S. Awad-Núñez
Transportation Department, Universidad Politécnica de Madrid, Madrid, Spain
Escuela Técnica Superior de Ingeniería de Caminos, Canales y Puertos, Ciudad Universitaria, Madrid, Spain

ABSTRACT: Port terminals, together with inland terminals, must assume as their own the strategic need for environmental sustainability of the facilities, without losing sight that sustainable development should be based on other two fundamental foundations: safety and cost efficiency. There is wide research on calculating the carbon footprint and methodologies aimed at their reduction, but global tools that propose solutions and operational changes, taking into account process efficiency and safety are lacking. This paper shows the development and implementation of a methodology addressed to managers of terminals of containers and port authorities that allows the selection of Container Handling Equipment (CHE) that best fits the characteristics and needs of each terminal, removing the not viable alternatives according to its specific boundary conditions. The methodology is based on the fact that the kind of CHE used in the terminal defines the design of the layout, with a direct incidence on the environmental impact, the safety and the cost efficiency associated to the facility. This methodology evaluates by means of Analytic Hierarchy Process (AHP) the layouts linked to straddle carriers, forklifts, reach stackers, platforms and gantry cranes. To make easy the implementation of the methodology, a Graphical User Interface (GUI) has been developed considering four different scenarios: (i) port terminals, (ii) port terminals storing dangerous goods, (iii) inland terminals and (iv) inland terminals storing dangerous goods. This GUI may be applied to both newly constructed facilities and for the management of change of operative ones. Furthermore, it allows the extension of the expert board required, constituting a constantly updated database.

1 INTRODUCTION

The European Union has called for the need to drastically reduce world Greenhouse Gas (GHG) emissions, with the goal of limiting climate change below two Celsius degrees (EC, 2011a). Commission analysis (EC, 2011b) shows that by 2030, the goal for transport will be to reduce GHG emissions to around 20% below their 2008 level. Transport has become cleaner, but increased volumes mean it remains a major source of noise and local air pollution. Current challenges facing the logistics sector leads to the need to develop new methodologies for sustainable management of infrastructures, taking into account social, economic and environmental factors.

Like other sectors, initially container terminals focused their efforts on improving the operational performance to reduce the operating costs per container and to meet the service level requirements imposed by their direct customers. However, current interpretation of the customer concept includes a set of stakeholders such as carriers, shipping companies, Administration but also society in general. Therefore, managers of terminals of containers have reformulated their strategy and value proposition so that accommodate the new required, incorporating strategic objectives of safety, security and environmental sustainability. Port terminals, together with inland terminals, must assume as their own the strategic need for environmental sustainability of the facilities, without losing sight that sustainable development should be based on other two fundamental foundations: safety and cost efficiency.

The improvement of the design and the management of intermodal infrastructures, both seaport and inland terminals, will smooth the pressure on coastal roads and congestion of ports, reducing emissions that affect people health and ecosystems, by diminishing the excessive share of goods transported by road. The transport trough the terminal, port operations and industrial activities at terminals are contributors to GHG emissions too.

To reduce the amount of pollutants emitted, ports are beginning to retrofit CHE with emissions control systems, replace older equipment with newer cleaner equipment (WPCI, 2016). According to Geerlings & van Duin (2011), the most effective measure for CO_2 reduction in container terminals is undoubtedly the adaptation of the terminal layout. Those authors remarked too that it is not possible to make an optimal decision able to satisfy all the objectives. Otherwise, the problems may be structured in a hierarchical structure.

The main purpose of this paper is the development of a decision making tool addressed to managers and designers of container terminals, integrating practices reducing carbon footprint without affecting the operational and economic efficiency of the facilities. The methodology is devoted to bridge the gap between academic research and the end user needs by means of a Graphical User Interface (GUI) that allows the selection of criteria that best fit the characteristics and needs of each terminal. The layout design was studied by the prioritisation of CHE alternatives used in the terminal yard, provided the univocal relationship between CHE and the layout design (Monfort et al, 2011). The mathematical model of multicriteria decision theory of Analytic Hierarchy Process (AHP) was applied to prioritize the alternatives defined based on operational and cost efficiency, safety, security and environmental criteria that will influence the selection process.

2 METHODOLOGY

The Analytical Hierarchy Process (AHP) used in this paper, introduced by Saaty (1980; 2016), is a useful procedure to be applied when there are multiple criteria to take into account and it is necessary to prioritise them in a decision making process. One of the main advantages of the methodology is that it allows the comparison between measurable criteria with other intangible ones. In AHP methodology, after considering the opinions of an expert board, the importance of several criteria involved in a decision making process are established. Those criteria, considered as standards on which a judgment or decision may be based, are weighted by the expertise panel using a methodology of comparison of pairs of criteria. Then, the possible alternatives for the goal achievement are evaluated considering all the weighted criteria in order to obtain a prioritisation of the alternatives. The process followed in this paper started with an analysis stage, including the problem formulation and the establishment of criteria involved in the decision-making process (Molero, 2016). To prioritise the criteria involved in the design of terminals of containers, an expertise board was configured,

constituted by professional experts with wide technical experience and recognized prestige in their areas.

The kickoff of the methodology was carried out through an expertise panel constituted by expert technicians of Port Institute of Studies and Cooperation (FEPORTS) together with experts in transport and logistics of dangerous goods (AITEC). The novelty of the methodology proposed is that among the creation of a GUI, further opinions of experts will be incorporated periodically to the process after a carefully study of the profile of the candidates to join the expert panel. Within the starting panel of experts, the model development and the identification of criteria were carried out in a synthesis stage. The definition of the hierarchical model and the prioritisation of the criteria were supported by *Super Decisions* software. The board of experts completed surveys of criteria comparisons that resulted in comparison matrices that considered four different scenarios: (i) port terminal without dangerous goods and dangerous goods area, (ii) port terminal with dangerous goods or dangerous goods area, (iii) inland terminal without dangerous goods and dangerous goods area, (iv) inland terminal with dangerous goods and dangerous goods area. The distinction between those different cases is based in the fact that the need and requirements for port areas and inland terminals may be different, provided the high environmental, safety and social impact related to the sensitive location of the seaports. Among the differences between inland terminals and ports regarding the requirements of the yard CHE is that in maritime environments machines must be resistant to corrosion in a higher extent. Furthermore, the requirements of the law involving hazardous goods storage are very restrictive in port terminals. Some of the drawbacks related to the storage of dangerous goods in ports may overcome by means of inland terminals as strategic facilities of support to port terminals.

2.1 *Criteria definition*

We established the criteria implied on the design of terminals of containers driven by a thorough study of the state of the art, the existent legal rules and the good practices in the environmental, safety and security field. Three main areas of first level criteria were identified by means of this research: Equipment, Safety and Security and Environmental Care. Within these areas (or first level criteria), 21 criteria second level and 50 criteria of third level were identified.

2.1.1 *Criteria for Equipment area*
Five criteria of second level and 20 criteria of third level were identified according to literature (Lee &

Kim 2012; Monfort et al. 2011; Golbabaie et al. 2012; Guo and Huang 2012; Gobierno de España, 2013; Junliang et al. 2015; Yang et al. 2014; Kaysi & Nehme 2015):

Second level criterion B1 (Economic) includes several criteria of third level:

- C11 Automation cost of CHE.
- C12 Ground cost.
- C13 Personnel cost for the operation of the CHE.
- C14 Maintenance cost of CHE to ensure green, safe, secure and resilient facilities.
- C15 Expansion of the terminal related costs.

Second level criterion B2 (Equipment Performance), including:

- C21 Amount of Twenty-foot Equivalent Unit (TEU) containers moved in an hour when operating at full capacity.
- C22 Average time between a truck (or train) arrival and departure to the yard terminal.
- C23 Percentage of time a door is serving container traffic.
- C24 Equipment inactivity rate.
- C25 Average time the containers remain without manipulation.

B3 criterion (Capacity of the terminal) includes:

- C31 Number of TEU that can be stored.
- C32 Number of possible simultaneous inputs or outputs through the door.
- C33 Number of cranes.
- C34 Number of containers moved per hour by the machine.

B4 criterion (Expansion of the terminal):

- C41 Expansion possibility of the CHE.
- C42 Expansion complexity.
- C43 Expansion time.

Criterion B5 (Functionality of the CHE) involves:

- C51 Automation level, estimated by the number of operators required to work properly.
- C52 Level of simplicity of the automation.
- C53 Adaptability of the machine to work under different scenarios.

2.1.2 Criteria for Safety and Security area

In this case, a second level criterion was identified. It was, B6 (called Safety and Security too) that was divided in seven criteria of third level according to relevant research in the issue (FHWA 2012; Peilin et al. 2012; Bernechea & Arnaldos-Viger 2013; Li et al. 2014; Argenti et al. 2015; Assadipour et al. 2015; Zhang et al. 2015; Antao et al., 2016; Axelsen et al. 2016; Nogal et al. 2016; Ocalir-Akunal 2016):

- C61 Danger level of the dangerous goods.
- C62 Dangerous goods amount.
- C63 Distance to the urban.
- C64 Equipment reliability.
- C65 Evacuation time during an emergency.
- C66 Density of surrounding population.
- C67 Climatic conditions.

2.1.3 Criteria for Environment Care area

This area involves 5 criteria of second level and 23 criteria of third level (Filbrandt 2008; Geerlings & van Duin 2011; Awad-Núñez et al. 2015; He et al. 2015 Canbulat et al. 2015; Yang et al. 2013):

B7, related to the location of the facility includes the following criteria of third level:

- C71 Industrial ground availability.
- C72 Flood risk.
- C73 Available water resources.
- C74 Acoustic impact.
- C75 Landscape impact.

B8, regarding the environmental concern directly linked to the design of the terminal includes:

- C81 Energy efficiency (consumption) involving day lighting and thermal insulation to fight climate change through environmental excellence practices.
- C82 Waste system management avoiding the hazardous consequences of accidental fluid leaks.
- C83 Protection of groundwater and surface water plan.
- C84 Conditioning of the hazardous materials storage area.
- C85 Containers per waste fraction.

B9 is related to the environmental management of the facility, including:

- C91 Energy efficiency (emissions), favouring the use of equipment with lower GHG emissions.
- C92 Waste minimisation policies.
- C93 Scheduling of product transportation and the loading and unloading of materials in the yard.
- C94 Preventive measures against noise pollution.
- C95 Sewerage network maintenance.

B10 is related to the construction process of the terminal, including:

- C101 Management of construction and demolition waste.
- C102 Minimization of water consumption during the construction of the terminal.
- C103 Environmental management of the equipment, vehicles and facilities during construction.
- C104 Recovery of topsoil layer plan.

Finally, criterion B11 considers emergency situations by:

- C111 Means to address spills in storage in case of managing malfunctions in the terminal and disposal of preventive equipment.
- C112 Procedures to be applied in case of risk, including handling procedures and management of abnormal situations and emergencies.
- C113 Staff training to avoid accidents and manage emergency situations.
- C114 Natural events influence.

2.2 Criteria prioritisation

The expert panel considered that the three criteria of first level, or general areas, had the same importance in the design of the layout for the scenario considered, remaining all three with a local normalized weight of 0.33 (w_{ci}), but this value can be modified by the users in the GUI developed according to their specific needs. To prioritise the second and third level criteria, in importance terms, expertise panel compared them by pairs on their clusters or areas. The expert board filled the questionnaires following Delphi methodology (Linstone and Turoff 1975). The data collected constituted matrices of comparison that allowed the calculation of the global normalized weights for each criteria of third level (W_{CG}) as described by Saaty (1980, 2016) elsewhere. The consistency rates allowed for the comparison matrices were lower than 0.1. The global normalized weights for each criteria of third level (W_{CG}) were calculated as follows (1):

$$W_{CG} = w_{ci} \cdot w_{cj} \cdot w_{ck} \tag{1}$$

Where w_{ci} is the local normalized weight for a criterion of first level or main area, w_{cj} is the local normalized weight for a criterion of the second level and w_{ck} is the local normalized weight for a criterion of third level. The collection of W_{CG} data for all the criteria of third level allowed a proper comparison of those criteria even coming from different areas.

2.3 Alternatives of layout definition and prioritization

For the design of the layout we considered five different alternatives according to the kind of CHE to be used in yards of facilities of this nature, as far as CHE define the configuration of a port container terminal (Monfort et al, 2011). We considered one kind of CHE operating in the terminal yard and five possible alternatives of design of the terminal were consequently established (Table 1). Those alternatives, together with the three levels of criteria and the main goal constitute the hierarchical model to apply AHP.

Table 1. Layout alternatives considered according to the kind of yard equipment used for TEU containers.

Alternative	TEUs per row/ column	Length between batteries m	Transversal corridor length m
A1 Straddle Carrier	1/3	1.2–2.0	≥10
A2 Forklift	≤2/5–7	≥10	≥10
A3 Reach stacker	3/5–7	≥10	≥10
A4 Platform*	1/1	≥10	≥10
A5 Gantry crane	~8/~5	≥15	≥5

* Platform requires interconnexion equipment for TEUs loading and unloading tasks.

The methodology allows obtaining values of normalized local weights of the alternatives when compared with criteria of third level (wak). Moreover, it is especially relevant to obtain a value of the global normalized weight for each alternative versus the main goal that allows addressing the issue in a holistic way (WAG).

W_{AG} was calculated as follows (2) for each criteria of third level:

$$W_{AG} = \sum_{i=1}^{3} \left[\sum_{i=1}^{n_i} \left(\sum_{k=1}^{m_{ij}} \left(w_{ck} \cdot w_{ak} \right) \cdot w_{cj} \right) \cdot w_{ci} \right] \tag{2}$$

Weights of the alternatives in relation with criteria of first level (w_{ai}) and second level (w_{aj}) were calculated as shown in (3) and (4),

$$w_{ai} = \sum_{j=1}^{n_i} \left(\sum_{k=1}^{m_{ij}} \left(w_{ck} \cdot w_{ak} \right) \cdot w_{cj} \right) \tag{3}$$

$$w_{aj} = \sum_{k=1}^{m_{ij}} \left(w_{ck} \cdot w_{ak} \right) \tag{4}$$

$n_i = [5,7,5]$; $m_{1j} = [5,5,4,3,3]$;

$m_{2j} = [1,1,1,1,1,1,1]$; $m_{3j} = [5,5,5,4,4]$

2.4 Evaluation of the economic viability of the resulting alternatives

The preferred design of layout obtained through this methodology should be also economically viable to be considered useful. That is why we defined the following boundary conditions to determine the economic viability of a layout alternative. The total incomes of the terminal (I_T, in €/year) were considered as the sum of the incomes from TEUs storage (I_s, in €/year) and incomes from movements of TEUs (I_m, in €/year) (5). Movement incomes (I_m) may be calculated as (6), considering m_y as the average moved TEUs per year in the facility and M_p as the average price per movement of TEU in €.

Movement incomes (I_a) were calculated as (7), were S_p is the storage average price per day and TEU (€/year·TEU) and d is the average number of storage days per TEU (days/year). Considering the average data obtained of prices consulted to CHE manufacturers for the five alternatives, the GUI we created is able to calculate the number of years to amortize the cost of the chosen CHE, n_i, (8). In this equation P_i is the price of the equipment in €. Therefore, we considered that the alternatives of design are viable if them only if the cost of the related equipment can be amortized in a period equal or less than 15 years.

$$I_T = I_s + I_m \tag{5}$$

$$I_m = m_y \cdot M_p \tag{6}$$

$$I_S = \frac{m_y}{2} \cdot S_p \cdot d \tag{7}$$

$$n_i = \frac{P_i}{I_T \cdot 0.3} \tag{8}$$

Therefore, in the GUI developed, users have the chance of introducing the values of M_p, S_p, d, m_y, and the area (A) of the specific terminal to be studied. Within these data, the program will provide not only the alternatives of layout ranked against the main goal (or the distinct defined criteria), but also can filter only the economically viable alternatives. For instance, a gantry crane layout may be the preferred alternative for a terminal of containers, but it would not be reasonable to use this kind of equipment in a terminal with a low flow of container movements. The program is able to identify inconsistent data entered by the user too. This is due to the fact that m_y and d are related by the area of the yard, that should be introduced by the user.

3 RESULTS AND DISCUSSION

3.1 Criteria prioritisation

The W_{CG} values achieved by the AHP methodology for the criteria of third level were ranked in importance order, according to the opinions of the expertise board. It is important to have in mind that for the situation exposed in this paper we have considered that the three criteria of first level have the same importance, it means, the same value of w_{ci}. Otherwise, the manager or designer of a concrete terminal may consider that, for a specific situation of interest, those values of w_{ci} may be different. In the GUI developed, the user can vary these values according to the requirements, and observe how the preferred layout alternative may change too.

Maintaining the values of 0.33 for w_{ci} for the three criteria of first order, we obtained that the 20 more relevant criteria suppose more than the 80% of the sum of all the W_{CG} values achieved for the four scenarios considered. The results achieved for the distinct scenarios were consequently different too, but some common trends can be extracted from the results. Therefore, when analysing the 10 most influential criteria for all the scenarios involved (Tables 2–5) we found that criteria from Safety and Security rea are very relevant for all scenarios. For all the situations there are 4 safety criteria among the 10 most relevant criteria. The only Safety and Security criterion that was not considered in this higher part of the rank was climatic conditions (C67). In fact, in all the situations studied, the most relevant criterion is related to this area, being the distance to urban core the most relevant criteria for scenarios regarding

Table 2. Ranking of the main criteria of third level according to its global importance in the decision making process, where W_{CG} means the global weight of criterion versus the goal in scenario (i), port terminals without dangerous goods.

Position	Criterion	W_{CG}
1	C65	0.1160
2	C101	0.1099
3	C13	0.1060
4	C64	0.0862
5	C31	0.0474
6	C91	0.0463
7	C62	0.0441
8	C63	0.0415
9	C103	0.0357
10	C51	0.0314

Table 3. Ranking of the main criteria of third level according to its global importance in the decision making process, where W_{CG} means the global weight of criterion versus the goal in scenario (ii), ports terminals with dangerous goods.

Position	Criterion	W_{CG}
1	C64	0.1263
2	C13	0.1054
3	C63	0.0845
4	C71	0.0764
5	C31	0.0566
6	C61	0.0539
7	C81	0.0392
8	C74	0.0375
9	C65	0.0336
10	C14	0.0331

Table 4. Ranking of the main criteria of third level according to its global importance in the decision making process, where W_{CG} means the global weight of criterion versus the goal in scenario (iii), inland terminals without dangerous goods.

Position	Criterion	W_{CG}
1	C63	0.1282
2	C13	0.0957
3	C64	0.0897
4	C71	0.0877
5	C101	0.0548
6	C61	0.0538
7	C31	0.0493
8	C74	0.0447
9	C14	0.0389
10	C65	0.0295

Table 5. Ranking of the main criteria of third level according to its global importance in the decision making process, where W_{CG} means the global weight of criterion versus the goal in scenario (iv), inland terminals with dangerous goods.

Position	Criterion	W_{CG}
1	C63	0.1293
2	C12	0.0937
3	C71	0.0862
4	C61	0.0756
5	C65	0.0435
6	C62	0.0418
7	C24	0.0382
8	C102	0.0382
9	C11	0.0364
10	C72	0.0312

inland terminals. For scenario (i) the most important criterion was evacuation time (C65), this criterion is highly relevant for the other scenarios, but for scenario (ii) equipment reliability (C64) was considered even more important. The other six positions for the ten most relevant criteria area almost equally divided between Equipment performance and Environmental Care criteria for all the situations.

For the Environmental Care area, the availability of industrial floor (C71) is the most relevant criterion for scenario (ii), (iii) and (iv). While considering port terminals without dangerous goods, experts consider that the most relevant Environmental care criterion is that related to the management of the constructions and the demolition waste, using an economical and environmental viable plan. The criteria directly related to GHGs, C91 appears among the 20 most relevant criteria in all the scenarios, but also criteria C81 and C103

remains in important positions (linked to this kind of emissions too). Regarding the Equipment performance area, those criteria related to direct costs seems to be the most relevant for the experts for choosing the CHE, specially C11 (automation cost of CHE), C12 (ground cost), C13 (personnel cost) and C14 (maintenance cost), together with C31 related to the storage capacity of the facility.

3.2 Alternatives prioritisation

While studying the results it was pointed out that platform is perceived as the preferred reliable CHE alternative for all the situations studied, according its W_{AG} values (Figs. 1–2). This is to be expected due to the fact that platform as storage and disposal system of the terminal ensures the no accumulation of high risks in small areas of surface, being the most valuable alternative for the Safety and Security area in all cases too. Moreover, even having in mind the higher surface demanding characteristic of platforms, it is considered the most inexpensive device to be used in terms of automation cost (C11), personnel cost (C13), technical maintenance cost (C14) and expansion related cost (C15).

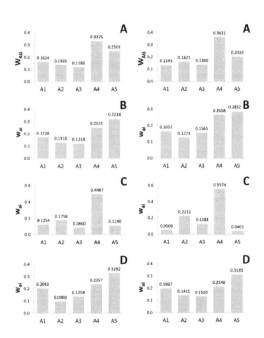

Figure 1. Prioritisation of layout design in terms of W_{AG} (A) or w_{ai} for Equipment area (B), Safety area (C) and Environment Care area (D) for scenario (i) port terminals without dangerous goods in left side and (ii) port terminals with dangerous goods in right side, considering $w_{ci} = 0.33$ in all cases.

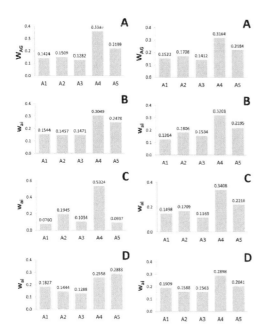

Figure 2. Prioritisation of layout design in terms of W_{AG} (A) or w_{ai} for Equipment area (B), Safety area (C) and Environment Care area (D) for scenario (iii) inland terminals without dangerous goods in left side and (iv) inland terminals with dangerous goods in right side, considering $w_{ci} = 0.33$ in all cases, considering $w_{ci} = 0.33$ in all cases.

If we centre the study in the case of port terminals we can observe that gantry crane is the preferred machine to be used, in terms of Equipment performance and Environment Care. Gantry cranes do not require any other auxiliary equipment for loading or unloading containers tasks. On the other hand, gantry crane option is opposite to platforms in terms of the surface of floor needed for the containers disposal. The preference of gantry cranes for those two areas in ports is mainly due to the high degree of influence of the criterion cost of the floor (C12) and availability of industrial floor (C71). In fact, the high value of the criteria C71 makes gantry crane option to be considered the preferred one in the Environmental care area in most of the scenarios except in scenario (iv). This is due to the fact that in inland terminal it is easier to find industrial available floor, with lower associated environmental risks. Furthermore, the presence of dangerous goods makes that stacking the containers implies risks. Those risks may be avoided using cheap platforms without stacking capacity. The option of gantry cranes is nowadays used terminals which have a high capacity

of investment, generally public terminals as ports areas, when the Equipment performance area of criteria is considered as the most relevant. In our case, when considering Safety and Security and Environment care with the same importance that Equipment performance, this situation is compensated with the characteristics of platforms.

4 CONCLUSION

The kind of CHE used in terminals of containers has a direct impact on its carbon footprint. This paper shows that the application of a decision making methodology for choosing the most suitable kind of CHE has a direct impact on its carbon footprint and determine the design of the layout of the terminal. For this study, not only environmental concerns have been considered, but also equipment performance, safety and security issues were studied in a holistic approach by means of AHP methodology. A hierarchic structure of criteria was constituted by 3 criteria of first level, 11 criteria of second level and 50 criteria of third level. Four different scenarios were considered; (i) port terminals, (ii) port terminals storing dangerous goods, (iv) inland terminals and (iv) inland terminals storing dangerous goods, given their different inherent characteristics. Those differences were manifested in the different order in the hierarchy of the criteria considered for the different situations. However, despite the differences, criteria related to safety and security were the most relevant in all scenarios. When the three criteria of first level had the same local normalized weight, platform was perceived as the most suitable yard CHE, mainly due to safety and security concerns.

The main novelty of the methodology is that we have developed a friendly user GUI with the purpose of bridging the gap between academic research and market application. This tool for low carbon emission management system includes safety and cost efficiency parameters in a holistic way. This application allows:

- Finding the most suitable layout to be applied to a terminal of containers in any of the scenarios considered, for the global objective or even for all the individual criteria considered.
- Comparing between the different deign alternatives.
- Customize the tool according to the specific needs of the facility, changing the value of the local normalized weight of (w_{ci}) for the three criteria of first level.
- Discarding the design alternatives that are not viable in terms of costs, according to the specific characteristics of the terminal.

- The tool allows the users to fill up their own questionnaires, acting as experts. Those opinions are studied and included, or not, in the whole program according to the curriculum vitae of the candidate to become an expert. Therefore, the tool is periodically updated and checked.

The developed GUI that is addressed to (i) engineers, designers and researchers in the field of container terminals; (ii) managers of existing terminals that want to validate the current layout design, or manage changes CHE and layout design in an appropriate way; (iii) managers of new container terminals.

Further developments of the research implies including the interactions between the yard equipment and the different options of equipment for loading and unloading operations

ACKNOWLEDGEMENT

Authors thank the Valencian Institute of Business Competitiveness (IVACE) for its financial support under research project with reference IFIDTA/2015/8.

REFERENCES

Antao P., Calderón M., Puig M., michail A., Wooldridge C.; Darbra R.M., 2016. Identification of Occupational Health, Safety, Security (OHSS). *Safety Science*, 85, 266–275.

Argenti, F., Landucci, G., Spadoni, G., Cozzani, V., 2015. The assessment of the attractiveness of process facilities to terrorist attacks. *Safety Science*, 77, 169–181.

Assadipour G., Ke, G.Y., Verma, M., 2015. Planning and managing intermodal transportation of hazardous materials with capacity selection and congestion. *Transportation Research Part E* 76, 45–57.

Awad-Núñez S., González-Cancelas N., Soler-Flores F., Camarero-Orive A., 2015. How should the sustainability of the location of dry ports be measured? A proposed methodology using Bayesian networks and multi-criteria decision analysis. *Transport* 30 (3), 312–319.

Axelsen C., Grauert M., Liljegren E., Bowe M., Sladek B., 2016. Implementing climate change adaptation for European road administrations. *Proceedings of 6th Transport Research Arena, April 18–21, 2016.* Warsaw, Poland.

Bernechea E.J. & Arnaldos-Viger J., 2013. Design optimization of hazardous substance storage facilities to minimize project risk. *Safety Science* 51 (1), 49–62.

Canbulat, O., Aymelek, M., Kurt, I., Koldemir, B., Turan, O., 2015. The green sustainable performance comparison of the three biggest container terminals in Turkey. In *Proceedings of International Association of Maritime Economists (IAME) 2015 Conference.*

EC, 2011a. *WHITE PAPER Roadmap to a Single European Transport Area—Towards a competitive and resource efficient transport system COM(2011)0144 final*

EC, 2011b. *Commission Communication. A Roadmap for moving to a competitive low carbon economy in 2050, COM (2011)112.*

FHWA, 2012. Climate change and extreme weather vulnerability assessment framework. *Federal Highway Administration of U.S. Department of Transportation*, FHWA Publication No: FHWA-HEP-13–005.

Filbrandt, U., 2008. Sustainable port development container terminals in Bremerhav. In Zanke, U; Roland, A; Saenger, N; Wiesemann, J.U; Dahlem, G (ed.), *Proceedings of the Chinese-German joint symposium on hydraulic and ocean engineering, 299–303, August 24–30, 200. Darmstadt.*

Geerlings, H. & van Duin, R. 2011. A new method for assessing CO2-emissions from container terminals: a promising approach applied in Rotterdam. *Journal of Cleaner Production* 19(6–7):657–666.

Gobierno de España, 2012. Estudio de costes de paso de contenedor por terminales. Memoria de comunicación. Ministerio de Fomento. Puertos del Estado.

Golbabaie, F., Seyedalizadeh Ganji, S.R., Arabshahi, N., 2012. Multi-criteria evaluation of stacking yard configuration. *Journal of King Saud University—Science*, 24(1), 39–46.

Guo, X. & Huang, S.Y., 2012. Dynamic Space and Time Partitioning for Yard Crane Workload Management in Container Terminals. *Transportation Science*, 46(1), 134–148.

He, J., Huang, Y., Yan, W., Wang, S., 2015. Integrated internal truck, yard crane and quay crane scheduling in a container terminal considering energy consumption. *Expert Systems with Applications* 42(5), 2464–2487.

Kaysi I.A. & Nehme, N., 2015. Optimal investment strategy in a container terminal: A game theoretic approach. *Maritime Economics and Logistics*, advance online publication.

Lee, B.K., & Kim, K.H., 2012. Optimizing the yard layout in container terminals. *OR Spectrum*, 35(2), 363–398.

Li, Y., Ping, H., Ma, Z.-H., Pan, L.-G., 2014. Statistical analysis of sudden chemical leakage accidents reported in China between 2006 and 2011. *Environmental Science and Pollution Research* 21 (8), 5547–5553.

Linstone, H.A. & Turoff M., 1975. *The Delphi method: techniques and applications.* Addison-Wesley Pub. Co., Advanced Book Program.

Lu, H.J., Zhen, H.Q., Chang, D.F., 2012. Visualization Method of Bucket Terminal Yard Working Data. *Applied Mechanics and Materials* 178, 2770–2774.

Molero, GD., 2016. *Análisis de criterios de diseño básico de una terminal de contenedores de sustancias químicas peligrosas aplicando el proceso analítico jerárquico (AHP).* Universitat Politècnica de València.

Monfort, A., Aguilar J., Vieira, P., Monterde, N., Obrer, R., Calduch, D., Martín, A.M., Sapiña, R., (2011) *Manual de capacidad portuaria: aplicación a terminales de contenedores*, (Fundación VALENCIAPORT, Valencia).

Monios, J. & Bergqvist, R., 2015. Intermodal terminal concessions: Lessons from the port sector. *Research in Transportation Business and Management*, 14, 90–96.

Monios, J. & Wilmsmeier, G., 2012. Giving a direction to port regionalisation. *Transportation Research Part A: Policy and Practice*, 46(10), 1551–1561.

Nogal M., O'Connor A., Caulfield B., Brazil W., 2016. A multidisciplinary approach for risk analysis of infrastructure networks in response to extreme weather. *Proceedings of 6th Transport Research Arena, April 18–21, 2016, Warsaw, Poland.*

Ocalir-Akunal, F.V., 2016. Using Decision Support Systems for Transportation Planning Efficiency (pp. 1–475). Hershey, PA: IGI Global.

Palacio A, Adenso Díaz, B., Lozano S., 2015. A decision-making model to design a sustainable container depot logistic network: the case of the Port of Valencia. *Transport*, Article in Press.

Peilin, Z., Jian, M., Long, Y., 2012. *Research on Layout Evaluation Indexes System of Dangerous Goods Logistics Port Based on AHP, 115(2), 943–949. Software Engineering and Knowledge Engineering: Theory and Practice. Advances in Intelligent and Soft Computing.* Springer.

Romero, N., Nozick, L.K., Xu, N., 2016. Hazmat facility location and routing analysis with explicit consideration of equity using the Gini coefficient. *Transportation Research Part E* 89, 165–181.

Saanen, Y., VanMeel, J., Verbraeck, A., 2003. *The design and assessment of next generation automated container terminals. Simulation In Industry Published: 15th European Simulation Symposium, Delft, The Netherlands,* 577–584.

Saaty, T.L., 1980. *The Analytic Hierarchy Process, Planning, Piority Setting, Resource Allocation.* McGraw-Hill, New York.

Saaty, T.L., 2016. *The analytic hierarchy and analytic network processes for the measurement of intangible criteria and for decision-making. International Series in Operations Research and Management Science,* 233, 363–419.

Tramarico, C.L., Salomon, V.A.P., Marins, F.A.S., 2015. Analytic Hierarchy Process and Supply Chain Management: A Bibliometric Study. *Procedia Computer Science* 55(Itqm), 441–450.

UNCTAD, 2014. *United Nations Conference on Trade and Development, Review of Maritime Transport 2014,* United Nations Publication.

WPCI, 2016. World Ports Climate Initiative, http://wpci. iaphworldports.org/project-in-progress/cargo-handling-equipment.html, August, 2016.

Yang, C.-C., Tai, H.-H. Chiu, W.-H., 2014. Factors influencing container carriers' use of coastal shipping. *Maritime Policy & Management* 41(2), 192–208.

Yang, Y.C., Lin, C.L., 2013. Performance analysis of cargo-handling equipment from a green container terminal perspective. *Transportation Research Part D: Transport and Environment* 23, 9–11.

Zhang, X., Miller-Hooks, E. Denny, K. 2015. Assessing the role of network topology in transportation network resilience. *Journal of Transport Geography* 46, 35–45.

Transport Infrastructure and Systems – Dell'Acqua & Wegman (Eds)
© 2017 Taylor & Francis Group, London, ISBN 978-1-138-03000 1

Pedestrian paths: Safety and accessibility for all users

E. Pagliari
Technical Department Coordinator, Automobile Club d'Italia, Rome, Italy

S. Balestrieri
Technical Department, Automobile Club d'Italia, Rome, Italy

ABSTRACT: The proposal includes an assessment of quality, under the point of view of accessibility and safety of pedestrian Paths. The aim of this project is to understand how safe and accessible our pedestrian Paths really are (especially for disabled people). In an urban context, pedestrians are the most vulnerable road users. We propose a new assessment methodology that aims at improving the quality and safety of pedestri-ans' accessibility to our streets: are our roads really safe and accessible for everyone? Which are the main ob-stacles/design mistakes in planning a pedestrian route accessible to everyone, in particular to disabled people? What are the principal suggestions to improve the quality of our streets? The project will provide objective evidence in response to all these questions through the comparison of different situations in order to highlight their strengths and weaknesses; such comparison will follow the criteria of efficiency, safety and accessibility; a 5 level rating is used (excellence in case of 5 stars, serious problems if 1 star) in order to inform on the present state of roads safety and accessibility, highlighting the lacks, deficiencies and shortcomings that re-strict a city actual practicability; proposals are made for the correct design of pedestrians safety; the study provides helpful suggestions to remove the obstacles that make it difficult, in particular for disabled people, to walk along our streets. The methodology illustration will include definition of the following: parameters and indicators for the assessment, checklists, test pro-tocol and evaluation grid. The evaluation will be conducted considering three test criteria (functionality, safety and accessibility) and three categories of disabled pedes-trians (blind, deaf and wheelchair people). The results of two pilot tests on the following types of pedestrian route will be presented: from house to school, from public transport stop to touristic place or commercial area.

1 INTRODUCTION

Pedestrians make up one among the most critical issue of urban mobility. Typical problems may affect both the safety degree and the accessibility level of dedicated paths.

Statistics on road accidents show a constant or increasing number/percentage of pedestrians involved, even when the number of total casualties is decreasing. In Italy, in 2015 there have been 601 deaths of pedestrians, showing an increase of 4,0% with respect to 2014, a year during which deaths were also increased by 4,3% with respect to 2013.

Accessibility for everyone and particularly for temporary or permanent disabled pedestrians still represents a critical issue. Street crossing time is adjusted to a "standard" pedestrian who walks at a speed of one meter per second, while there are various situations in which people may have to walk at a slower pace: elderly people, parents carrying their infants on a stroller, people carrying heavy shopping bags and those who are affected by a temporary or permanent disability. Accessibility is often compromised also as a result of bad habits and misbehaviors such as illegal parking on pedestrian reserved areas, footways occupied by commercial activities or illegal advertisements, etc.

The present study is meant to find a solution to the above mentioned critical aspects by offering a methodology for the assessment of pedestrian Paths' safety and accessibility. The methodology proposed involves the adoption of a 5 star rating method: with a 5 stars score, the pedestrian route is safe and accessible, with a 1 star score, it presents serious deficiencies. The final part of the analysis provides design solutions to be implemented in order to improve safety and accessibility on each route.

This document is made of four sections, the first of which is the present introduction. In the second section the assessment methodology devised by ACI for pedestrian Paths is outlined. In the third section two pilot projects are illustrated, which were conducted in order to gauge the assessment method in two different contexts. In the first case, a pedestrian route from a public transport stop to a place of interest for tourists was analyzed, while in the second case, the pedestrian route between a public transport stop and a school was taken into consideration. The fourth and last section is devoted to some final considerations and to an illustration of the future steps to be taken.

2 METHODOLOGY

The proposed evaluation method for pedestrian paths is in line with the one used for assessing safety and accessibility of pedestrian crossings by the FIA (Fédération Internationale de l'Automobile) European Pedestrin Crossing Assessments (EPCA), This project was promoted and led by the Automobile Club d'Italia (ACI).

The project—carried out and developed between 2008 and 2011—provides a survey of about 1,000 pedestrian crossings in 22 countries and 46 towns throughout Europe. The method was set up by ACI in cooperation with "La Sapienza" Rome University.

The weighting process is based on the technique of cross-comparison, is reviewed by a qualified focus group (analytic hierarchy process) and then validated through in-depth investigations on severe road crashes involving pedestrians.

Two different checklists have been developed: the first one for pedestrian crossings at intersections (see Figure 1) and the second one at road arcs.

Using the checklist, the evaluation process takes into account 27 safety indicators and/or factors grouped into 4 safety categories.

Crossing system (12 safety factors, weight: 23%)

- ○ Crossing distance (from sidewalk to sidewalk).
- ○ Pedestrian-vehicles conflict points.
- ○ Pedestrian refuge islands (crossing islands).
- ○ Width pedestrian refuge island in relation crossing distance.
- ○ Exclusive pedestrian signal phase.
- ○ Green and Transition phases efficiency.
- ○ Red phase efficiency.
- ○ Pedestrian countdown signal.
- ○ Pedestrian push—button.
- ○ Crossing road surface maintenance.
- ○ Crossing traffic signs maintenance.
- ○ Crossing road markings maintenance.

Daylight visibility (5 safety factors, weight: 26%)

- ○ Minimum approach sight distance (distance needed for a driver to recognize the presence of a pedestrian waiting to cross at the pedestrian crossing).

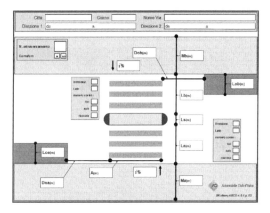

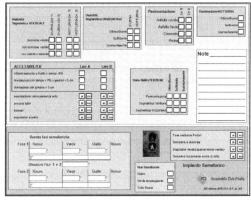

Figure 1. Sample of pedestrian crossing checklist.

- ○ Visibility of Pedestrian crossing signs (for drivers).
- ○ Visibility of road markings (for drivers).
- ○ Pedestrian crossing width.
- ○ Specific traffic direction markings (e.g. triangles/arrows or "Look left/Look right" road markings).

Nighttime visibility (4 safety factors, weight: 32%)

- ○ Lighting conditions.
- ○ Minimum approach sight distance in the night time (distance needed for a driver to recognize the presence of a pedestrian waiting to cross at the pedestrian crossing).
- ○ Visibility of Pedestrian crossing signs at night time(for drivers).
- ○ Visibility of road markings at night time (for drivers).

Accessibility (11 safety factors, weight: 19%)

- ○ Presence of dropped or ground level kerbs.
- ○ Ramp width.
- ○ Step height for the sidewalk.

○ Suitable angle between sidewalk and crossing (90°).
○ Presence of tactile paving (for visually impaired people).
○ Presence of acoustic devices (for blind or partially sighted pedestrians).
○ Presence of obstacles (parked vehicles, utility poles, signs, holes, etc.) that could be a hazard for approaching pedestrians or pushing them to cross outside the crossings.
○ Kerb width.
○ Specific traffic direction markings (e.g. triangles/arrows or "Look left/Look right" road markings).
○ Animated pictograms (colored men) for walk or don't walk phases (moving pedestrian).
○ Pedestrian countdown signal.

The differences between crossings with and crossing without traffic lights were considered in the evaluation process (different evaluation degrees).

Crossings were classified according to an overall assessment scale; moreover, they were also rated with reference to each safety indicator. The evaluation was carried out on a five point scale system, as follows: Very Good (++), Good (+), Sufficient (o), Unsatisfying (–), Poor (–). Furthermore, points of strength and weakness, as well as possible improvement actions have been highlighted for each pedestrian crossing.

The method proposed is not meant however to assess safety and accessibility of pedestrian crossings only: it aims at evaluating complete pedestrian Paths, including side-walks and pedestrian areas. For this purpose, the pedestrian route taken into consideration was first divided into homogeneous stretches, which were classified according to two categories:

1. homogeneous pedestrian stretch (on pedestrian reserved path, e.g. footpaths);
2. pedestrian crossing (on carriageways where both vehicles and pedestrians pass).

The method of the EPCA project (see above) was employed for category 2 (pedestrian crossings), while as for category 1 (pedestrian Paths) a new evaluation methodology has been developed by ACI Technical Department taking advantage of the same approach used for pedestrian crossings.

The technique of cross-comparisons was used, with revision by a qualified focus group (analytic hierarchy process) and subsequent validation through data on accidents involving pedestrians.

An ad hoc check list for homogeneous stretches has been developed (see Figure 2 below).

Using the checklist, the evaluation process takes into account 23 safety indicators and/or factors grouped into 3 safety categories.

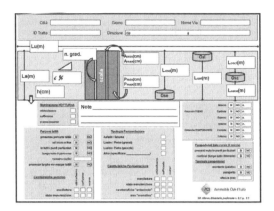

Figure 2. Sample of pedestrian route checklist.

General feature (22 safety factors, weight: 40%)

○ pedestrian path lenght
○ pedestrian path wide
○ footpath slope
○ height pedestrian path with respect to the roadway
○ fixed obstacle
○ temporary obstacle
○ number driveways
○ promiscuity pedestrians/cyclists
○ stairs presence and steps number
○ step raise (stairs)
○ step tread (stairs)
○ type of paving
○ manufacture paving
○ maintain paving
○ anti-skid features
○ color rendering paving road
○ presence and type tactile route
○ number codes tactile route
○ presence plates and/or tactile maps
○ presence barriers delimiting pedestrian paths
○ type of barriers delimiting pedestrian paths
○ height barriers delimiting pedestrian paths

Night visibility (2 safety factors, weight: 20%)

○ lighting conditions
○ color rendering paving road

Accessibility (16 safety factors, weight: 40%)

○ pedestrian path wide
○ footpath slope
○ fixed obstacle
○ temporary obstacle
○ number driveways
○ promiscuity pedestrians/cyclists

- maintain paving
- anti-skid features
- color rendering paving road
- presence and type tactile route
- number codes tactile route
- presence plates and/or tactile maps
- stairs presence (more than one step)
- step raise (stairs)
- step tread (stairs)
- number of steps (stairs)

Similarly to pedestrian crossings, homogeneous pedestrian stretches were classified according to an overall assessment scale; moreover, they were also rated with reference to each safety indicator. The evaluation was based on the same five level rating system as used for pedestrian crossings. Furthermore for each homogeneous pedestrian stretch, points of strength and weakness have been highlighted, as well as possible actions to improve its safety.

The evaluation of the entire pedestrian route is carried as follows:

- an overall evaluation of all pedestrian crossings included in the pedestrian route (5 evaluations, an overall one and one according to each of the 4 relevant safety categories);
- an overall evaluation of all homogeneous pedestrian stretches included in the pedestrian route (4 evaluations, an overall one and one for each of the 3 relevant safety categories);
- an overall evaluation of the whole pedestrian route (4 evaluations, an overall one and one for each of the 3 relevant safety categories).

Eg. assessment crossing road markings maintenance
- Safety factor: **crossing road markings maintenance** (weight **12,00%** if single assessment, **2,76%** if overall assessment)
- Assessment = **Acceptable**
- Rating (don't to weight value):
  √ Very Good = 1,00
  √ Good   = 0,75
  √ Acceptable = 0,50
  √ Weak   = 0,25
  √ Poor/Missing = 0,00
- Spreadsheets:

3 "PILOT" PROJECTS

3.1 *Pedestrian paths from a public transport stop to a place of interest for tourists or a commercial area*

The choice of the pedestrian route to be analyzed was inspired from the Extraordinary Jubilee declared by Pope Francis in 2015 and 2016, during which pilgrims are invited to walk along a route across Rome, before passing through the "holy door".

Following this suggestion, a pilgrim/pedestrian route was traced from Termini Railway Station in Rome, crucial point of the urban commuting network, to the Basilica of Santa Maria Maggiore (See Figure 3).

The assessed pedestrian route was subdivided into 17 homogeneous stretches and 14 crossings. The survey was conducted in the second half of October 2015.

On the overall, the pedestrian route under assessment achieved a pass mark; however, some deficiencies in infrastructure design and a number of clear shortcomings in management and maintenance were observed.

As regards infrastructure design inadequacies, the following ones were detected:

- incorrect placement of some fixed obstacles along the pedestrian route and/or on the footway (lamp posts too far from the kerbside, phone boots or advertising totems placed too much in the middle of the footway section, as well as parking meters, rubbish bins or boxes, etc.);
- deficient quality of paving manufacture (for example, newly maintained pavings are quickly deteriorated);
- poor quality of employed materials (for instance, tactile routes get dirty and in bad conditions very soon);
- grades for wastewaters drainage are sometimes not aligned to gully grid;
- no tactile routes for the blind and visually impaired, or incorrectly designed tactile routes;
- poor night lighting.

The following management and maintenance deficiencies are to be pointed out:

Table 1. Assessment of the safety factor and/or indicator.

Safety factor	Don't to weight value—rating/assessment—(A)	Weight for single assessment—(B)	Weight value for single assessment —[A×B]	Weight for overall assessment—(C)	Weight value for overall assessment—[A×C]
crossing road markings maintenance	0,50	12,00%	0,0600	2,76%	0,0138

Table 2. Assessment of the safety category and/or total assessment.

Total assessment rating	Sum of the weight value (for overall assessment) of safety factors $\Sigma [A_i \times C_i]_{i=1, n}$
5 (++) Very Good	>= 0,80
4 (+) Good	>= 0,50 and 0,80
3 (0) Acceptable	>= 0,35 and 0,50
2 (−) Poor	>= 0,15 and 0,35
1 (−) Very Poor	0,15

Figure 3. Evaluated pedestrian path.

○ the presence of many temporary obstacles to pedestrians' transit on both sides of the route, as well as in the middle; as a consequence, pedestrians are often forced to a sort of slalom, which represents a serious difficulty especially for the blind and visually impaired; among the temporary obstacles, the following were encountered: commercial activities' advertisements, legal or illegal commercial stalls, café and restaurant tables; illegally parked vehicles or bicycles;

○ often inadequate maintenance of paving surface; consequent risks for pedestrians;

○ no maintenance of traffic signs and road markings;

○ unclean and untidy pedestrian routes;

○ deficient or no check on compliance to the "Street regulations" included in the General Plan for Urban Traffic in force in the Rome Municipality.

Assessment results of pedestrian crossings were as follows:

⌣ among the 14 pedestrian crossing subjected to assessment, no one rated as excellent;

○ only 3 out of 14 pedestrian crossings scored "good";

○ 6 pedestrian crossings out of 14, i.e. more than 40% of the total crossings assessed, didn't pass the test and were judged insufficient;

○ on the overall, pedestrian crossings submitted to test were assessed as "insufficient";

The strengths—rather few—were as follows:

○ raised surface crossways (only two instances);

○ kerb extentions (only two instances);

○ tactile routes with correct signals (only four instances);

○ adequate street crossing time;

○ adequate night lighting (only three instances).

On the overall, the weaknesses can be listed as follows:

○ barriers delimitating pedestrian paths were scarcely noticeable and in bad conditions;

○ poor overall conditions of paving, traffic signs and road markings;

○ inadequate night lighting;

○ tactile routes with incorrect or deficient signals (for instance, unmarked traffic lights, traffic lights marked only on one side of the crossing, tactile routes unaligned with ramps);

○ presence of temporary obstacles on the road crossing or on the dedicated footpath leading to a pedestrian road crossing (illegal parked vehicles, rubbish bins, park meters, commercial stalls, etc.);

Suggested interventions were as follows:

○ replace and/or repair barriers delimitating pedestrians paths;

○ arrange dedicated lighting;

○ maintain paving, traffic signs and road markings;

○ improve the signals and the information conveyed on tactile routes;

○ use kerb extentions in parking areas and near public transport stops;

○ improve accessibility by removing all kind of obstacles.

Evaluation of the homogeneous pedestrian route stretches led to the following results:

○ none of the homogeneous pedestrian route stretches subjected to test was rated "excellent";

○ only 3 homogeneous pedestrian route stretches out of 17 scored "good";

○ up to 8 homogeneous pedestrian route stretches out of 17—i.e. more than 50% of the totality, did not pass the test, and were judged "insufficient";

○ the totality of the homogeneous pedestrian route stretches got an overall rate of "just above fail".

On the overall, the strengths—rather few and to be found mainly on the stretches making up the route which links Stazione Termini to Basilica di Santa Maria Maggiore—can be listed as follows:

○ uniform road paving, with adequate traction;
○ the presence of widenings, both at the start and at the end of the route;
○ large footway;
○ the presence of a tactile map;

When it comes to weaknesses, major shortcomings were detected on the following:

○ many permanent obstacles on the footways, by the roadway, reduce the space for pedestrian transit (lamp posts, road signs);
○ tactile routes are inadequate or absent; in particular, no directional indicators for walkways and no adequate division between walkways areas and roadways were found;
○ many movable obstacles on both sides and in the middle of walkways reduce the space for transit;
○ some areas with rough pavement were detected;
○ Barriers delimitating pedestrian paths are scarcely noticeable and in bad conditions (some were pulled away);
○ many dangerous footway gratings on walkways were detected;
○ frequent crowded times occurring during the day;
○ deficiencies in cleanliness and tidiness were found.

Suggested interventions are as follows:

○ removal and repositioning of permanent obstacles in order to make the walkway uniform;
○ removal of temporary obstacles to increase the space on walkway and make it uniform;
○ improve neatness and tidiness;
○ improve maintenance and general quality of paths;
○ improve accessibility for disabled people (tactile routes, ramps, directional indicators):
○ adjust the width of footways and footpaths.

The "pilot" pedestrian route, from Termini Central Railway station to Basilica of Santa Maria Maggiore and return, including both pedestrian crossings and homogeneous pedestrian route stretches, got an overall rate of "just above fail".

The route under analysis scored a "passing" mark on general safety parameters and nocturnal visibility, while considerable deficiencies were detected on accessibility, which was judged inadequate.

3.2 *Pedestrian path from home/public transport stop to school*

The pedestrian path that has been taken into account links a school, located in Via Stabilini, to the stops of the bus lines serving this southeastern area of Rome, which is next to the Great Ring Road (GRA, Grande Raccordo Anulare). Students usually follow this route to reach the school after stepping out of the bus, and on their way back (bus stops: Ciamarra/Rizzieri, Rizzieri/Oberto; bus lines: 046, 20, 500, 558, 559—see also Figure 4 below).

The path has been divided into 13 homogeneous stretches and 9 pedestrian crossings. The survey was carried out during the first half of July 2016.

The pedestrian path has been deemed overall acceptable, both as to the infrastructure design and the management and maintenance works.

Pedestrian crossings has scored the highest rates, both overall and for each safety indicators.

The pedestrian stretches has been deemed overall "sufficient", but the results has been different depending on the safety indicator taken into account. The stretches are "excellent" in terms of night visibility, "sufficient" as regards general features and "insufficient" for accessibility.

The following must be highlighted in terms of design deficiencies:

○ some fixed obstacles along pedestrian paths/footways are not very well located (light or road sign poles are too far from the edge of the sideway, dumpsters and/or bins etc.);
○ almost total absence of tactile paths for blind and visually impaired people;
○ cycling routes too close to footways;
○ the presence of driveways with interference between pedestrians and vehicles.

On the contrary, both road signs and the quality of surface maintenance have been deemed sufficient and/or good.

Figure 4. Evaluated pedestrian path.

The following must be highlighted as regards the 9 pedestrian crossings assessed:

○ no pedestrian crossing has scored excellent;
○ 5 pedestrian crossings out of 9 have scored "good";
○ 3 pedestrian crossings out of 9 have scored "insufficient";
○ 1 pedestrian crossing has not passed the test ("poor" result);
○ The overall result of all pedestrian crossings assessed has been "good".

The following points of strength have been highlighted:

○ footways at the grade of the street (in 2 cases only);
○ kerb extension;
○ correct and consistent information on tactile paths (in two cases only);
○ adequate night lighting.

The following points of weakness must be pointed out:

○ no tactile paths (in 7 out of 9 cases);
○ visibility between pedestrians and drivers (and vice-versa) strongly limited by the shape of the footways (in 1 case only);
○ cross-slope in some cases higher than 8%.

The following suggestions are made:

○ improving the information on tactile paths;
○ kerb extension in case of parked vehicles or public transport stops;
○ improving accessibility (reduction of cross-slope).

As for the assessment of the homogeneous pedestrian stretches the following must be highlighted:

○ none of the 13 pedestrian stretches has scored excellent;
○ 5 pedestrian stretches out of 13 have scored "good"—that is 40% approximately;
○ 8 pedestrian stretches out of 13 have scored "acceptable"—i.e. 60%;
○ all pedestrian stretches have passed the test;
○ the overall result of all pedestrian stretches has been acceptable.

On the whole, the following points of strength have been highlighted:

○ road surface provides adequate grip;
○ kerb extension at the beginning and end of the path;
○ wide footways;
○ "good" night lighting;
○ almost total absence of fixed or temporary obstacles.

As to the points of weakness, major shortcomings were detected as follows:

○ no tactile paths; in particular, no directional indicators to walkways and no adequate division between walkways areas and roadways;
○ presence of some dumpsters that narrow the pedestrian road;
○ no pedestrian barriers;
○ presence of some driveways;
○ simultaneous presence of bikers and pedestrians (in one case).

The following suggestions are made:

○ improving accessibility for disabled people (tactile routes, ramps, directional indicators, pedestrian barriers etc.);
○ repositioning of some permanent obstacles (drinking fountain, dumpsters, etc.);
○ separating cycling and pedestrian paths.

It must be pointed out that the assessment of this second type of pedestrian path, which links the bus stops to the school located in Via Stabilini, has allowed us to improve our methodology. Specifically, new indicators have been included in order to take into account also driveways and bike paths crossing the pedestrian route.

4 CONCLUSIONS

The project —"Pedestrian Safety and Accessibility for all users"—which was applied on some pedestrian routes in Rome and whose results were illustrated in the present study, represents only a first step.

Further development will involve testing of the same methodology on more pedestrian routes, with consequent re-adjustment of the evaluations parameters and of the weights underlying the system as well as of the assessment methodology itself.

Other important aspects requiring improvement and further analysis are those linked to the different forms of disability, such as, for example, blindness, deafness and mobility impairment.

Sharing this experience at European level will be crucial in order to extend, homogenize and complete the available range of project solutions. In this connection—as it was the case in the past for pedestrian crossings—ACI will rely on the FIA network, which brings together the Automobile Clubs across the world, starting from the European region.

The final objective of the ACI initiative is to draft a manual for good standards in design and planning of pedestrian routes in a urban context.

Therefore a further step will be the drawing up of guidelines for planning and designing of pedestrian routes, which should review, update and complete an earlier publication on pedestrian crossings realized by ACI in 2011 ("Guidelines for Planning and Design of Pedestrian Crossings").

REFERENCES

ACI (2011). *Guidelines for Planning and Design of Pedestrian Crossings*; <www.aci.it>.

AIIT (2001). *Fermate del trasporto pubblico extraurbano*, quaderno n° 4 di Quaderni di tecnica del traffico e dei Trasporti; Udine, Tipografica-Basaldella di Campoformido.

AIIT (2005). *Fermate del trasporto pubblico urbano*, quaderno n° 5 di Quaderni di tecnica del traffico e dei Trasporti; Udine, Tipografica-Basaldella di Campoformido.

AIIT (2010). *Piani di dettaglio del traffico urbano e piani di intervento per la sicurezza stradale urbana*, quaderno n° 7 di Quaderni di tecnica del traffico e dei Trasporti; Forlì, Egaf editrice srl.

Assemblea Federale della Confederazione Svizzera (2011). *Legge federale sulla circolazione stradale (LCStr)*; <www.admin.ch>.

Busi, R. (2001). *Tecniche per la sicurezza in ambito urbano*; Forlì, Egaf editrice srl.

Comitato economico e sociale europeo (2008). *Parere del Comitato economico e sociale europeo in merito alla Proposta di regolamento del Parlamento europeo e del Consiglio relativa alla protezione dei pedoni e degli altri utenti della strada vulnerabili*; Gazzetta Ufficiale dell'Unione europea.

Comitato economico e sociale europeo (2011). *Verso uno spazio europeo della sicurezza stradale: orientamenti strategici per la sicurezza stradale fino al 2020*; Gazzetta Ufficiale dell'Unione europea.

Consiglio Federale Svizzero (2010). *Ordinanza sulla segnaletica stradale (OSStr)*; <www.admin.ch>.

Consiglio Federale Svizzero (2011). *Ordinanza sulle norme della circolazione stradale (ONC)*; <www.admin.ch>.

Department of transportation (2007). *The official High Way Code*; UK, <www.direct.gov.uk>.

Department of transportation, traffic and road way section (2001). *Designing sidewalks and trail for access*; USA, <www.FHWA.dot.gov>.

Department of transportation, traffic and road way section (2007). *Traffic line manual*; Oregon <www.oregon.gov>.

Federal High Way Administration (2009). *Manual on uniform traffic control devices (MUTCD;)* <www.MUTCD.FHWA.dot.gov>.

Kane County Council of Mayors & Development Department & Division of Transportation & Forest Preserve District (2000). Pedestrian Design Guide, chapter 4 of *The Kane County 2020 Transportation Plan*; <www.co.kane.il.us>.

Land Transport NZ (2007). Crossings, chapter 15 of *Pedestrian Planning and Design Guide*; Wellington, Land Transport NZ.

Luci e Illuminazione (2001). *Illuminotecnica—Requisiti illuminotecnici delle strade con traffico motorizzato*; Milano, UNI.

Ministère de l'Equipement et du Transport (2009). *Code de la Route*, Paris; <www.legifrance.gouv.fr>.

Ministero delle infrastrutture e dei trasporti (1992). *Nuovo Codice della Strada*; Gazzetta Ufficiale.

Ministero delle infrastrutture e dei trasporti (2001). *Norme sulle caratteristiche funzionali e geometriche delle intersezioni stradali*; CNR.

Ministero delle infrastrutture e dei trasporti (2001). *Norme funzionali e geometriche per la costruzione delle strade*; Gazzetta Ufficiale.

Ministero delle infrastrutture e dei trasporti (2004). Modifica del decreto 5 novembre 2001, n. 6792, recante *«Norme funzionali e geometriche per la costruzione delle strade»*; Gazzetta Ufficiale.

Northern Ireland Assembly, United Kingdom (2006). *The Zebra, Pelican an Puffin Pedestrian Crossing Regulations*; UK, Stationery Office Limited.

Parlamento europeo e Consiglio (2003). *Direttiva 2003/102/CE relativa alla protezione dei pedoni e degli atri utenti della strada vulnerabili prima e in caso di urto con un veicolo a motor,*; Gazzetta Ufficiale dell'Unione europea.

Parlamento europeo e Consiglio (2005). *Direttiva 2005/66/CE relativa all'impiego dei sistemi di protezione frontale sui veicoli a motore*; Gazzetta Ufficiale dell'Unione europea.

Parlamento europeo e Consiglio (2008). *Direttiva 2008/96/CE sulla gestione della sicurezza delle infrastrutture stradali*; Gazzetta Ufficiale dell'Unione europea.

Parlamento europeo e Consiglio (2009). *Regolamento n.79/2008 concernente l'omologazione dei veicoli a motore in relazione alla protezione dei pedoni e degli altri utenti della strada vulnerabili,*; Gazzetta Ufficiale dell'Unione europea.

Portland State University (2005). *Establishing Pedestrian Walking Speed*; Portland State University.

Potter, S.M. (2004). *Pedestrian Slip Resistance testing to AS/NZS 3661:181993 and AS/NZS 4986:2004 for resene paints Ltd*; <www.resene.co.nz>.

Presidente delle Repubblica (2010). *Legge 120 del 29/07/2010*; Gazzetta Ufficiale.

Presidente della Repubblica (2011). *D. Lgs n.35 Attuazione della direttiva 2008/96/CE sulla gestione della sicurezza delle infrastrutture*; Gazzetta Ufficiale.

UNI (1998). *Prestazioni della segnaletica orizzontale per gli utenti della strada*; Milano, UNI.

UTTIPEC & Delhi Development Autority (2009). *Pedestrian Design guide Line*; New Delhi, <www.uttipec.nic.in>.

Zilm, D. (2005). *Pedestrian kerb ramps and footpaths construction check-list*; Australia, <www.prospect.sa.gov.au>.

The following websites, in addition to those mentioned above have also been consulted:
– <www.planningportal.gov.uk>
– <www.whellchair-ramps.co.uk>
– <www.bfu.ch>

Transport Infrastructure and Systems – Dell'Acqua & Wegman (Eds)
© 2017 Taylor & Francis Group, London, ISBN 978-1-138-03009-1

Designing principles of the efficient and safe urban cycling infrastructure

Y.V. Trofimenko & E.V. Shashina
Moscow Automobile and Road Construction State Technical University (MADI), Moscow, Russia

I.V. Markin
Bicycle Transport Research Institute, Moscow, Russia

ABSTRACT: The hierarchical structuring of requirements and approaches to the cycle network designing on the city and municipal levels, formation cycle corridors as well as cycle network linear sections and intersections are given. 9 typical in architectural and planning terms measures for the cycling infrastructure objects in Russian cities are identified: ensuring the sight distances; traffic flows calming; ensuring the prohibition of vehicular access to the cycling infrastructure; bicycle traffic organization on one-way streets; location of bike lanes and vehicles parking; mutual location of pedestrian and bike paths on the sidewalks; one-level crossings organization—intersections; one-level crossings organization—roundabouts; location of the bike paths and bus stops. Using expert-analytical techniques (GIS-technology), the safety and ecological performance (Copert-4 methodology) of abovementioned measures were evaluated on existing and future cycling routes in Kazan and Kaliningrad.

Currently in Russian cities bike paths are located in recreation areas, which offer sufficient comfort and traffic safety due to the lack of traffic, but on streets and city roads with combined traffic of cars and cyclists and intersections in one level, the safety problem are of particular relevance.

The main goal of cycling infrastructure development is to promote cycling as an alternative to car trips and to provide opportunities for safe and convenient travel by bicycle (Trofimenko, Zege 2013).

1 PRINCIPLES OF CYCLE INFRASTRUCTURE DESIGNING

The first principle of designing cycle infrastructure is the territorial planning—the identification of new opportunities for the use of the urban area for mobility. The second principle is functionary—improving the efficiency and comfort of cycling. The architectural-planning decisions of road network should reduce the risk of accidents with cyclists, the severity of their consequences and should take into account the needs and capabilities of different age groups and categories of cyclists (Trofimenko, Sova 2013).

According to the Bicycle Transport Research Institute there are five main requirements for cycling infrastructure:

1. Safety. Cyclists do not pose any significant danger, but they feel vulnerable when moving in the traffic flow. The main risk lies in the vehicles and bikes speed and weight difference. Safety can be provided in three ways reducing the number of collisions:
   - reducing the vehicle density,
   - reducing the speed to 30 km/h and below,
   - separating cyclists in space and in time from heavy traffic. Places where it is impossible to avoid a cyclist and a car contact (intersections and crossings) should be organized especially carefully to ensure the safety of all road users.
2. Rectifiable path. The cyclist should reach the destination more directly. Detours and, accordingly, the travel time must be minimized. This makes cycling very competitive against the car at short distances. Influence factors are:
   - the number and length of detours,
   - the number of stops at intersections,
   - traffic light control,
   - the road grades etc.
3. Connectivity or continuity. The cyclist should reach the destination freely and without interference. Cyclists will appreciate the advantages of bike paths, if they are interconnected into a single network, and allow to navigate without dismounting from the bike and not stopping for curbs, crossings, intersections, and other obstacles. Every home, every office and institution should be accessible by bicycle, i.e. they should have access ways for reaching cycle network. Connectivity means also the connection with

other transport networks, mainly bus stops and public transport hubs.

4. Route attractiveness. Cycling infrastructure needs to be integrated into a pleasant environment. Despite the fact that perception is very variable and individual, special attention should be paid to it while designing. Vibrant streets, urban gardens, the use of the surrounding area, distance to the road network, the number of crossings with other transport streams, road grades are of great importance.

5. Comfortable ride. The bike ride should be calm, with minimum physical and mental effort. Situations that require stopping, changing speed, excitement, risks should be avoided. Poor road surface, for example, can create excessive vibration, shock. Bottlenecks and high curbs can make hard overtaking and keeping the bike in balance. Other factors that affect comfort are:

- the band width for cyclists and the possibility to drive parallel to each other,
- the field of view parameters,
- the level of noise pollution,
- the level of air pollution,
- the wind direction and force,
- the number and angle of road grades,
- the nature of the lighting (brightness and direction),
- the distance from potentially dangerous objects.

1.1 Design of linear sections

To ensure comfortable and safe conditions for cycling a designer must know the size of the bike, its performance characteristics, and requirements for cycling infrastructure objects. These factors determine the permissible turning radius, road grade and visibility distance. In many cases, the designed elements of single cycling elements are controlled by the nearby roadway and by the design decision of the street or road.

The following factors should be taken into consideration while designing bike lanes:

- functional assignment (binder, distributing or providing direct access road),
- environment (buildings' type, inside or outside build-up environment),
- general transport situation (traffic density and speed),
- the geometric parameters of the bike lanes (available width, number of lanes etc.).

While integrating cycling infrastructure into existing urban transport network two principles are used: separation and overlapping of traffic flows.

Although the separation of different kinds of traffic flows appears to be the most effective measure for ensuring the safety of vulnerable road users, it is difficult to implement it in all the streets and cycling routes because of the limitations of urban space and budget. When choosing one or other option the governing requirement should be safety.

To determine the optimal split of vehicle and bicycle traffic in the city a scheme shown in Figure 1 can also be used.

It is not always possible to divide transport flows and to protect all the road users inside the build-up environment. Therefore, a choice of a bike lane type on a particular road or street depends entirely on local traffic conditions. Possible design solutions are presented in Table 1. There are three main things that are commonly used and can be arranged quickly, easily and not expensive:

1. bike lanes in both directions on streets with one-way traffic,
2. combined bus and bike lanes,
3. increased cycle stop area at intersections.

At permitted speeds of over 50 km/h the risk of fatality in a collision between a car and a cyclist becomes very large, so cyclists should be protected from passing parallel transport through physical separation of bike lanes with bollards, curbs, and other structure elements.

1.2 Design of intersections

About 70% of all accidents with serious consequences occur at intersections. Many of them occur as a result of collision of turning car with straight-moving

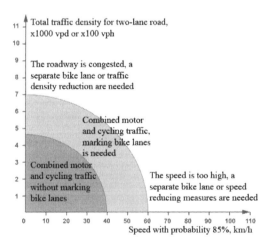

Figure 1. Recommended operating conditions for cycling separation (recommended by Transport Research Institute).

Table 1. Selection of cycle facilities depending on the density and speed of traffic flows*.

Type of street or road	Maximum (allowed) speed of traffic flow, km/h	Traffic density, adjusted vehicles per hour	Section of cycling transport network, cyclists density is equal or less than 250 cyclists per hour	Main cycling route, cyclists density is more than 250 cyclists per hour
Laneway, pedestrian street	–	–	Combined cycling and pedestrian movement	
Local street	30	1…400	Combined movement with vehicles or bike lane within roadway	Organization of cycle street, bike paths separated from the roadway
		400…1000	Bike lane, combined with the roadway	
		>1000	Bike path separated from the roadway with parking, landscaping area or structure elements	
Distributing/ main street	50/70	Doesn't matter	Bike path separated from the roadway with parking, structure elements, landscaping area, lateral dividing strip	

* These recommendations are incorporated in a new draft of Chapter 11 of SP 42.13330.2011. A set of rules. The urban development. The planning and construction of urban and rural settlements (approved by order of the Ministry of regional development of The Russian Federation from 28.12.2010 № 820).

cyclist. However, the presence of intersections and their number greatly affects the comfort and rectification of cycle routes. Therefore, a special attention should be given to the intersections of vehicle and cycling flows: the cyclists should be able to cross intersections, turn left and right safely, quickly and comfortably.

The choice of design solutions should provide the required level of safety, speed, comfort and it largely depends on the purpose of bike route, special conditions of arrangement (inside or outside build-up environment) and the traffic speed and density.

Safety is the main requirement while designing an intersection—prevention a dangerous situation with a simple, intuitive design. It is achieved by:

– providing the visibility that is crucial: cyclists need to be in sight of the motorist as long as possible. Therefore, the main recommendation in the case of separate cycling facilities is to bend bike paths closer to the roadway long before the intersection,
– minimizing the difference in the speed of the cyclist and the vehicle: the vehicle speed should be close to the bicycle speed of 20…30 km/h,
– using the additional elements of road facilities: traffic islands, rollup band, extended zone before the stop line, lanes for overtaking bicycles.

Speed is one of the key issues when crossing intersections on the bike. Delays caused by waiting at intersections, greatly increase the cycling trip duration. Design and regulation should be intended to minimize wait time. The following measures should be taken:

– providing the priority in traffic for cyclists,
– creation of traffic lights' short cycles,
– organization of "green wave",
– making possible overtaking on the right,
– making logical and direct crossing of intersections,
– avoiding phased crossing of intersections.

Comfort is achieved by turning radius maintenance which allows the cyclist to maneuver easily without significantly reducing the speed and without being displaced from his lane.

These requirements are implemented in the following main types of intersections:

– simple intersections is the basic variant of intersections on roads with mixed traffic and a speed limit of 30 km/h,
– single-lane roundabouts is a safe solution for active traffic, because the cyclists are positioned between the slower cars. Several lanes is much riskier and should be designed with a separate bike path around it,
– signalized intersections are relatively risky and involve some waiting time. Nevertheless, they are essential on the main roads with heavy traffic flows. Design solutions should provide clear visibility of cyclists, allow short and easy maneuvers of cyclists, reduce the waiting time,
– multilevel crossings (tunnels, bridges, overpasses) should be used to cross the busiest roads and bypass complex and dangerous intersections.

Table 2 contains summarized recommendations for the design of different types of intersections.

1.3 *The formation of cycle corridors*

The cycle network designing should start with the identification of the main places of attraction and communication directions from the point of view of potential cyclist.

At first the situation is considered citywide, then within each district and microdistrict. It is necessary to identify the main potential directions of transport corridors for cycling, to determine the direction of traffic flows and congestion on the streets and roads network. Within these corridors a search is made for possible alternative routes on parallel streets, landscaped areas to provide main cycle links between districts. When doing this several points should be considered:

1. cycling network doesn't have to repeat the streets and roads network scheme, as it can actively use off-street space,
2. bike paths should be as far as possible from the active traffic,
3. trip distance of more than 7 km are practiced mostly as recreation and tourism, and rarely are business or household on daily basis,
4. existing cycle routes in the city in the absence of bicycle infrastructure can help to identify priority goals, additional ways to overpass difficult sections, but are mostly of sporadic transitory nature and cannot serve as the basis for the formation of a network of bike paths.

The main cycle directions citywide are usually the following:

- residential zone—the historic centre,
- residential zone—a large city recreation area, a sports centre,
- residential zone—a shopping centre,
- residential zone—the suburbs.

These trips are focused on the furthest point of gravity that's why they require a high average speed and in relation to the territory of the host they are mainly of transit nature without interruption. Transit segments of cycle routes should be the most direct and shortest. For tracing, road-doubles on the perimeter of residential areas and districts, streets and roads with the inactive traffic, shortcuts through landscaped common areas should be chosen with cyclists priority identification.

The main elements of the cycle transit network are separate bike path or different variants of combined vehicle and cycling traffic, for example the cycling lane on the roadway or cycle street.

After determining communication destinations citywide each district is considered separately.

The main directions within districts and microdistricts are:

- home—transfer hub to rapid transit (subway, railway, bus station),
- home—school, kindergarten, hospital, civil service, domestic service, shop,
- home—regional recreation area,
- home—home.

Table 2. Designing guidelines for various types of intersections.

Recommended intersections	Field of use	Main criteria for design	Basic elements of design solutions
Simple intersec tion	Quiet roads with a speed limit of 30...50 km/h. All bicycle routes inside the settlements	Traffic density and number of intersections with bike paths. Equivalent roads or roads with different levels of priority according to signs and markings	The cycle route has the priority in traffic. Entry and exit from the intersection. Traffic island
Roundabout	Moderately busy roads with the permitted speed of 50 km/h or more. Moderately busy cycle routs of the city, district and local importance. Inside and outside the settlements	Road hierarchy, traffic density, bandwidth. Bike paths, bike lanes or combined traffic	One or two lanes. The size of the roundabout. A bypass lane for cyclists. A tunnel for cyclists
Adjustable crossroads	Busy roads with speed of above 50 km/h. Active bike paths of the city and district importance. Inside and outside the settlements	Amount of cyclists. Desired waiting time	Strong regulation. Detection of the cyclist. Increased area before the stop line

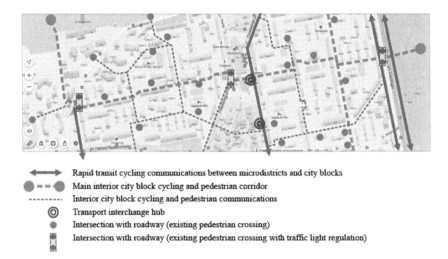

Rapid transit cycling communications between microdistricts and city blocks
Main interior city block cycling and pedestrian corridor
Interior city block cycling and pedestrian communications
Transport interchange hub
Intersection with roadway (existing pedestrian crossing)
Intersection with roadway (existing pedestrian crossing with traffic light regulation)

Figure 2. The formation of cycling and pedestrian corridors within the district based on the existing pedestrian flows (Northern Tushino, Moscow, Russia).

Table 3. Ranking of safety ensuring measures on cycling routes in the cities of Kazan and Kaliningrad (Russia) (fragment).

Route	The weight of measure, %								
	1	2	3	4	5	6	7	8	9
Kazan									
Azino-2	29	0	29	0	0	29	3,3	0	9,7
Bank of Kaban lake	40,8	0	40,8	0	0	0	18,4	0	0
Bank of river Kazanka	38,1	0	36,5	0	0	0	25,4	0	0
Total 15 routes									
Kaliningrad									
Ostrovskogo street	25	0	25	0	0	50	0	0	0
Gaydara street	41,2	0	35,3	0	0	0	23,5	0	0
Generala Chelnokova street	50	0	38,5	0	0	0	11,5	0	0
Total 44 routes									

It is important to mention that there are interior routes that determine the format of most cycle trips. According to the results of a survey conducted by the Bicycle Transport Research Institute in 2014 more than 2/3 of the cycle trips destinations in Moscow are located within a residential zone (within 5 km around the house).

Tracing of interior cycle routes should be approached very carefully from the point of view of safety, since such trips are made by cyclists of different ages, skills and behaviour on the road. The maximum protection of this level cycling infrastructure from traffic should be provided. The main cycle routes are recommended to be considered through pedestrian and green corridors within residential zones. Detailed tracing of one of the corridors in the Northern Tushino (Moscow) can be seen in Figure 2.

The main cycling and pedestrian corridor (orange) is about 2.4 km long and connects two districts, two recreation areas and passes near a metro station. At the intersections with the roadway there are already street-level pedestrian crossings with traffic light regulation. There are also bus stops that allow to leave the bike in the parking and to take the bus. The cyclist can also go to transit "speed" bike path (green arrows). Cycling and pedestrian corridor allows cyclists not to use streets to get to the subway, schools, clinics, recreation areas, bank, etc.

The formation of such cycling and pedestrian corridors allows the maximum removal of cyclists from traffic on the streets and roads network. Intersections with traffic occur in just a few points that represent the existing pedestrian crossings with traffic light regulation.

Vehicle laneways through such areas should be converted to bike streets with cyclists priority in traffic. Such laneways should provide access to the entrance of a house, but they should not be through for vehicles. Road markings and signs should clearly define the bicycle advantage and a speed reduction to 30 km/h there.

In fact, through traffic in microdistricts should be limited to vehicles and provided for cyclists. This will provide safe travel through all the area for all kinds of cyclists without entering the street or roadway. For fast transit traffic, the combined vehicle-cycling traffic on the main road doubles (with the cyclists priority and a mandatory speed reduction for cars) or bike lanes on the streets, if there are no the road-double, can be used. It is necessary to reduce the vehicles speed to 30 km/h when using roads-doubles along main district streets for cycling.

2 MEASURES INCREASING OBJECTS OF CYCLING INFRASTRUCTURE COMFORT AND SAFETY

On the basis of the analysis of foreign experience of cycling transport networks functioning, foreign and domestic regulatory guidance documents the following typical in architectural and planning terms measures to increase objects of cycling infrastructure comfort and safety are identified (CROW Record, 2007, FHWA Guidance, 2014, Green Book, 2004, London Cycling Design Standards, 2014, RASt 06, 2006, Trofimenko, Galyshev, 2015):

– ensuring the sight distances,
– traffic flows calming,
– ensuring the prohibition of vehicular access to the cycling infrastructure,
– bicycle traffic organization on one-way streets,
– location of bike lanes and vehicles parking,
– mutual location of pedestrian and bike paths on the sidewalks,
– one-level crossings organization—intersections,
– one-level crossings organization—roundabouts,
– one-level crossings organization—bike paths crossings with streets and roads,
– pedestrian crossings and bike paths in a one-level,
– location of the bike paths and bus stops,
– bicycle parking,
– cycle routes' grades,
– engineer protection methods of cyclists.

The analysis of existing and future cycling routes in the cities of Kazan and Kaliningrad (Russia) was made and the most relevant measures ensuring the cyclists safety for Russian cities were identified in order to assess the impact of these measures

on greenhouse emissions reduction (Trofimenko, Shashina 2016):

1. ensuring the sight distances,
2. traffic flows calming,
3. ensuring the prohibition of vehicular access to the cycling infrastructure,
4. bicycle traffic organization on one-way streets,
5. location of bike lanes and vehicles parking,
6. mutual location of pedestrian and bike paths on the sidewalks,
7. one-level crossings organization—intersections,
8. one-level crossings organization—roundabouts,
9. location of the bike paths and bus stops.

Next, the importance (frequency of occurrence) of each of the measures on each prospective route in Kazan and Kaliningrad (Trofimenko, Shashina 2016) was estimated using GIS "Yandex-maps". It was found out that the most important (most frequent) measures in both cities are (see Table 3):

– ensuring the sight distances (1),
– ensuring the prohibition of vehicular access to the cycling infrastructure (3),
– mutual location of pedestrian and bike paths on the sidewalks (6),
– one-level crossings organization—intersections (7).

3 ASSESSMENT OF THE LEVEL OF SAFETY OF CYCLING INFRASTRUCTURE

The methodological approach involves during the infrastructure designing the assessment of potential conflict situations which occur under the influence of various disturbances (traffic, speed) at the appropriate spatial conditions for parties involved (cyclists, pedestrians). Based on this assessment the analysed transport network check is made, which determines the road spatial conditions for its sections (segments between equivalent road intersections), roadway and sidewalk width, presence, type and efficiency of facilities, including visibility at the intersection, the presence, type and size of the cycle infrastructure objects and also for intersections of equivalent roads of the analysed road network in the form of the relevant requirements for each type of intersection.

Based on the above-described approach a methodology for the safety assessment of bicycle routes was developed in MADI (see its scheme in Figure 3).

The safety of cycling infrastructure is estimated by the severity of conflict situations σ, which represents the probability of occurrence of accidents involving cyclists (Shashina 2014). When $\sigma = 1$, an accident is unavoidable, when $\sigma = 0$ the probability

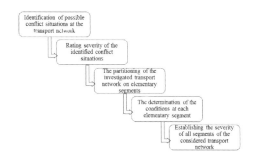

Figure 3. Scheme of safety assessment of a cycle route.

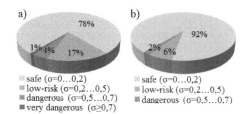

safe (σ=0...0,2)
low-risk (σ=0,2...0,5)
dangerous (σ=0,5...0,7)
very dangerous (σ≥0,7)

safe (σ=0...0,2)
low-risk (σ=0,2...0,5)
dangerous (σ=0,5...0,7)

Figure 4. Safety levels of potential cycling routes: a) Kazan, b) Kaliningrad.

of an accident is equal to zero. 4 ranges of danger degrees are set to assess the cycling routes danger: safe ($\sigma = 0...0,2$); low-risk ($\sigma = 0,2...0,5$); dangerous ($\sigma = 0,5...0,7$); very dangerous ($\sigma \geq 0,7$).

These qualitative levels of road safety make easier the comparison of results from different studies. Ongoing or future plans to achieve a certain level of safety are an indicator while creating technologies and roads facilities, cycle infrastructure objects with the required safety level.

It is found out that from 72.1 km potential routes of Kaliningrad 92% will be safe, 6%—low-risk and 2%—dangerous. From 116.73 km potential routes of Kazan 78% will be safe, 17%—low-risk, 4%—dangerous and 1%—very dangerous (Fig. 4).

4 THE REDUCTION OF AIR POLLUTION AND GREENHOUSE GAS EMISSIONS FROM THE CITY TRANSPORT SYSTEM THROUGH THE DEVELOPMENT OF CYCLE MOVEMENTS

It is assumed that the cycle routes safety assurance will increase the number of people who prefer cycling to driving, which will lead to the reduction of emissions of polluting substances from the exhaust gases, including Greenhouse Gases (GHGs).

Source data of the analysed cycle routes for the calculation of the gross GHG emissions were:

– the design speed of the vehicles;

– the traffic density in one direction;
– the cyclists density in one direction;
– the length of the route.

Gross GHG emissions (CO_2, CH_4, N_2O) (coefficients of reduction to CO_2-EQ CO_2 : CH_4: $N_2O = 1 : 25 : 298$ (GEF, UNFCCC)), and gross emissions of other polluting substances (CO, NOx, PM) from the vehicles were determined by the COPERT 4 (version 11.2) on each of the considered cycle routes in both cities (Kazan and Kaliningrad) for two scenarios:

1. the lack of bicycle infrastructure; people prefer to use a personal car;
2. the presence of bicycle infrastructure, ensuring the safety of cyclists on potential routes; some people use a bicycle instead of personal car for business, work cultural and general trips.

For calculations were used the results of previously executed research into the characteristics of traffic flows in the cities of Kazan and Kaliningrad.

In particular the calculations used the vehicle park structure distribution by types of passenger cars and environmental class, climatic characteristics of the cities. Vehicles mileage was taken equal to the length of settlement sections of the streets and roads with potential routes.

Traffic density for the first scenario was taken according to the results of previously performed studies. Traffic density for the second scenario was taken equal to the difference of traffic density according to the first scenario and the cyclists density, i.e. the decrease in the density that is achieved through the people redistribution.

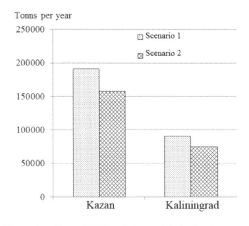

Figure 5. Gross GHG emissions with the implementation of measures to ensure the security of Cycling infrastructure for the two scenarios in the city of Kazan and Kaliningrad, tCO_2-EQ/year.

An assumption was also made that the density of movement of vehicles and cyclists is typical during daylight hours (about 10 hours a day) daily (365 days a year). It takes into account only the urban movement (Urban) with given design speeds.

The results of calculations of GHGs gross emissions is shown in Figure 5.

From Figure 5 it follows that the introduction of measures to ensure the cycling infrastructure safety can provide the total reduction in gross GHG emissions of the transport system of Kazan on 32842.2 tonnes CO_2-EQ/year (17.2%), transport system of the city of Kaliningrad—on 15446.0 tonnes CO_2-EQ /year (17%).

To quantify the impact of specific measures to ensure the safety of cycling infrastructure on reducing energy consumption, air pollution and greenhouse gas emissions let us consider the example of 6 cycle routes with different cycle paths types in the cities of Kazan and Kaliningrad (3 in each city). It appears that the implementation of these measures will enhance the safety and comfort of cycle routes, which should encourage people to use bicycle instead of a private car.

It is assumed that the selected routes will be equipped with bike lanes of various types. They are held in different parts of the urban areas and are typical for them.

To evaluate the contribution of different activities, ensuring safety, reducing gross emissions of GHGs the weight of each of them (see Table 3) is multiplied by the calculated decrease in gross GHG emissions on the route.

It is established that in the cities the contribution of different measures to ensure the safety of cycling infrastructure in gross GHG emissions from transport will be different.

In Kazan the greatest contribution to the reduction in gross GHG emissions gives activity 1 – ensuring the sight distances – 12166 tonnes of CO_2-EQ/year (37.1% of the total reduction from all activities, then activity 3 – ensuring the prohibition of vehicular access to the cycling infrastructure – 11357.1 tonnes of CO_2-EQ/year (34.6%), followed by the event 7 – one-level crossings organization—intersections – 5242.3 tonnes of CO_2-EQ/year (16%). The other 6 events from the point of view of reducing GHG emissions are less important because their contribution is only 12.3%.

Approximately the same situation is observed in Kaliningrad. But there are already 4 measures that are the most significant. The greatest contribution to the reduction in gross GHG emissions also gives the event 1 – 4666.8 tonnes of CO_2-EQ/year (30.2% of the total decrease from all the activities) activity 3–4415.6 tonnes of CO_2-EQ/year (28,6%), activities 6 and 7 – approximately 2930 tonnes of CO_2-EQ/year (19%). The contribution of other measures in the reduction of gross emissions of GHGs is not more than 3.3%.

5 CONCLUSIONS

The most effective measures to ensure the safety of cycling infrastructure in terms of gross GHG emissions reduction and energy efficiency in the analysed cities are:

- ensuring the sight distances,
- ensuring the prohibition of vehicular access to the cycling infrastructure,
- mutual location of pedestrian and bike paths on the sidewalks,
- one-level crossings organization—intersections.

So it is recommended to use these measures while designing potential cycling routes and cycle networks in the other cities of the Russian Federation, because they provide not only the implementation of the required level of safety of cycling infrastructure, but also the maximum decrease in gross greenhouse gas emissions of city transport system.

REFERENCES

CROW Record 85—Design Manual for bicycle traffic, 2007. Drukwerk.
FHWA Guidance: Bicycle and Pedestrian Provisions of Federal Transportation Legislation, 2014. Washington, D.C.: US Department of Transportation.
Green Book. Policy on Geometric Design of Highways and Streets, 2004. Washington, D.C.: AASHTO.
London Cycling Design Standards, 2014. London: Transport for London Consultation and Engagement.
Richtlinien für die Anlage von Stadtstraßen. RASt 06, 2006. Kologne: Forschungsgesellschaft für Straßen—und Verkehrswesen, Arbeitsgruppe Straßenentwurf.
Shashina, E.V. 2014. Development of scientific-methodical bases of estimation of reliability of a bus driver in conditions of conflict and emergency situations. Moscow: inaugural dissertation.
Trofimenko, Y.V., Galyshev A.B., 2015. Assessment of environmental and economic benefits from the development of cycling in large cities. *Vehicle Fleet Operator* No 4: 29–31.
Trofimenko, Y.V., Shashina, E.V., 2016. The methodology and results of the safety assessment of bicycle infrastructure in Russian cities. *Papers of The Twelfth International Conference. Organization and Traffic Safety Management in Large Cities*: 63–70.
Trofimenko, Y.V., Sova, A.N., Burenin, V.V., 2013. About the development of cycling transport and cycling infrastructure in Moscow. *Herald of MADI* No 4: 98–102.
Trofimenko, Y.V., Zege, S.O., Zege, O.S., 2013. Back to the future or the development of cycling in Moscow. *Integral* No 3 (71): 60–61.

Transport Infrastructure and Systems – Dell'Acqua & Wegman (Eds)
© 2017 Taylor & Francis Group, London, ISBN 978-1-138-03009 1

Analysis of the maritime traffic in the central part of the Adriatic

Zvonimir Lušić
Faculty of Maritime Studies Split, Split, Croatia

Danijel Pušić
Croatian Hydrographic Institute-Split, Split, Croatia

Dario Medić
Faculty of Maritime Studies Split, Split, Croatia

ABSTRACT: This paper analyzes maritime traffic in the central part of the Adriatic Sea as well as risk of collisions and groundings on the main sailing routes passing through this area. For the purpose of this study the main sailing routes will be those that pass in the vicinity of the island Palagruža, i.e. through the area bounded from north by line that connect islands Sušac and Jabuka and from the South by line connecting island Pianosa and cape Gargano. Although this part of Adriatic is almost completely open sea, large concentration of ship traffic cause increased probability of maritime accidents, especially collisions. Two main traffic flows, northwest and southeast, are concentrated in two narrow paths, mainly safely separated due to traffic separation schemes in vicinity. But these main traffic flows overlap with routes of other vessels; fishing vessels, yachts, transversal traffic, etc. accordingly causing increased probability of accident. Calculation of probability of collisions and groundings, in this paper, will be based on AIS traffic data and IWRAP software. Accordingly, some recommendations how to reduce risk of accidents will be given.

Keywords: Central part of Adriatic, Island Palagruža, main longitudinal sailing routes, AIS traffic, collisions and groundings.

1 INTRODUCTION

The biggest ports in the Adriatic are situated on its northwestern coasts. Accordingly, the greatest traffic can be found on sailing routes which connect these ports and the Strait of Otranto, i.e. the exit to the Adriatic. Regarding these sailing routes, the biggest concentration can be found in the main sailing route, which mainly goes through the open sea. This sailing route is defined by three key points: separation zones of the Northern Adriatic, separation zone near the island Palagruža and the Strait of Otranto. As a result, a massive concentration of traffic is grouped into two separate navigational directions; one northwestern, the other southeastern with the width of several naut. m. (Figure 1). The aforementioned separation zones separate opposite waterways hence reducing the possibility of collisions in opposite courses. If the traffic in the central part of the Adriatic is observed, i.e. in the larger area of Palagruža, at first glance it could be concluded that the marine traffic is regulated well, with a relatively small probability of collisions and groundings. This is mostly contributed by the established separation zones, VTS system and

Figure 1. Ais Traffic-2014.
Source: HHI-map 101Source: Marinetraffic.

a relatively large distance to the coast. However, the constant growth of the traffic concentration shows that the danger of accidents, especially collisions, is in fact not negligible. The main reasons are crossing of longitudinal sailing route with the transversal ones, branching of the main waterway in the area of the Northern Adriatic in two parts, overlapping of the main sailing routes with fishing zones, passenger ships traffic and nautical tourism ships, economical activity etc.

This paper will analyze the effect of the remaining ship traffic on the ships that move in central longitudinal sailing routes in the open sea of the Adriatic, i.e. that move in the larger area of the island Palagruža. For the collision and grounding assessment IWRAP software will be used, i.e. the model based on the quantitative determination of potential collisions and groundings (IWRAP). Input data for the analysis will be obtained by gathering AIS ship tracking in the summer months of the 2015. IWRAP is a modeling tool that can estimate the frequency of collisions and groundings in a given waterway based on information about traffic volume/composition and route geometry [www.IALA]. The assessment of the number of collisions and groundings based on the IWRAP methodology can be justified in the cases when the amount of traffic is big enough and when the traffic and ships characteristics are known.

2 GEOGRAPHIC AND NAVIGATION ANALYSIS OF THE AREA

Based on its geographic and sailing characteristics the Adriatic Sea is a unique sailing area [Lakoš, 1980]. Sailing in the Adriatic Sea as the specific and unique area as it is, in the sense of safe navigation requires good knowledge of hydro-navigational and hydro-meteorological conditions. In the summer period navigation in the Adriatic, particularly around the island area, requires indispensable knowledge and respect of the navigation conditions which derive from hydro-navigational and hydro-meteorological conditions in certain areas. Quick change of weather, especially the sudden appearance of gale in the entire and especially northern part of the Adriatic, a large number of navigational obstacles, numerous islands, islets, shallows, rocks, reefs, narrow passages and canals and a large number of non-merchant ships (fishing vessels, small vessels, yachts…) make navigation in the Adriatic Sea demanding, particularly in the coastal area.

The island Palagruža is the largest of a small group of islets lying southeast of Biševo island, nearly in the middle of the Adriatic, some 35 miles far from the Croatian coast. Two lights are exhibited from the summit: the main light is exhibited from the lighthouse—stone tower on dwelling, and a secondary light about 20 m east—white column on hut. The island is 91 m high, rocky and almost barren. The steep and rocky south coast is almost inaccessible. Off the west extremity of the island lie the above-water rocks Volići and Pupak. Small craft can find some shelter in the cove Stara Vlaka on northern side, and in the cove Žalo on southern side of the island. The rocky islet of Mala Palagruža (51 m) lies southeast of the island, and is bordered by above and below-water rocks and shoals. The islet Kamik Od Tramuntane stands to NE and the islet Kamik Od Oštra to SE. The islet of Galijula is situated about 3 miles southeast of Palagruža island. In the broader area of Palagruža there are no greater dangers, the distance from island line to nearest island on east coast is circa 23 naut. m., and to island Pianos 24 naut. m., to Gargano 28 naut. m. The greatest danger is represented by hydro-meteorological conditions. Prevailing wind is maestral and the greatest danger is represented by jugo which can develop waves up to 7 m high.

3 EXISTING SAFETY MEASURES

The area around the island Palagruža belongs to the outside island area with a very notable traffic density, right after the Strait of Otranto by the number of ships navigating the area. Due to convergence of the navigational direction, i.e. merging of two navigational directions in the narrow area around the island Palagruža, and the great traffic density, system-oriented navigation was established and all for the purpose of improving the safety of navigation. The traffic separation scheme in the area around the island Palagruža (Figure 2) was established in 1989 by the Command of sailing routes of foreign warships, foreign tankers, foreign nuclear ships and other foreign ships that transport nuclear or other dangerous and harmful

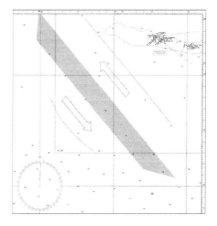

Figure 2. Sailing close to the island Palagruža. Source: HHI, 1995 map 100-23.

substances during harmless passage through the territorial sea of the Socialistic Federative Republic of Jugoslavia. The scheme itself consists of the traffic separation zone of two (2) naut. m. of total width. The area of traffic separation is determined by the central line which connects points with coordinates: a) 42° 04,7'N – 016° 24,6'E i b) 42° 23,0'N – 015° 56,2'E. Sailing paths continue laterally to the separation zone in three (3) naut. m. to the both sides, in which case in the sailing path closer to the Italian coast the navigational direction is 136°, and in the path close to the island Palagruža, it is 316°. Separation zones one naut. m. wide are situated next to both external edges of the traffic separation system's sailing paths, in purpose of separating the traffic in the system from the traffic in the area of the coastal navigation. The position of the traffic separation system is the most suitable because the passage through the island Palagruža and the coast, i.e. Pianos island on the approximately same distance (somewhat closer to Palagruža island, i.e. somewhat deeper in the territorial waters of the Republic of Croatia), ensures optimal usage of radar devices with the scope of determining the ship position.

The traffic separation scheme is not accepted by the IMO (International Maritime Organization), and its establishment occurred based on the IMO Resolution principle of the existing (regular) maritime traffic flow. Principle of the existing (regular) maritime traffic flow makes a general assumption according to which navigational direction measures should respect the usual maritime-traffic flows as much as possible. The basis of this principle is the need to introduce separation measures in a certain area to navigational practice as simply and in as short adjustment period of time as possible. Analyzing the so far accepted navigational direction measures, this principle proves fundamental. In practice, there are no examples of not being respected consistently. Considering that the traffic separation system near the island Palagruža is in the territorial waters of the Republic of Croatia the Croatian VTS (Vessel Traffic Service) is applied in the area. Vessel traffic service's main scope is the navigation safety and this system reduces the risk of collisions at sea, makes better control of the ships and ecological protection of the sea and the islands. Such systems are placed in the zones of dense traffic, where the freedom of ship movement is limited by the size of the maritime area, existence of obstacles in navigation, limited depths or unfavorable meteorological conditions. Traffic separation scheme close to the island Palagruža is situated in the Croatian VTS sector B. Sector B—comprises the parts of internal waters and the part of the territorial waters of the Republic of Croatia from the border (external border of the territorial waters)

to the VTS sector limits: Rijeka, Zadar, Šibenik, Split, Ploče, Dubrovnik and Pula—maneuver sector. VTS service gives Information Service (IS) significant to safe navigation of sea crafts in the VTS area on VHF channels Ch10 and Ch60. All sea crafts are obligated to communicate with the VTS service, Port Authority and port management and to vigil over VHF radio channel of the VTS sector in which they are located. Participation in the VTS services is obligatory for:

a. Ships of 150 gross tonnage,
b. Ships over 50 m and more in length,
c. Ships on international voyages,
d. Sea crafts with limited maneuver abilities,
e. Sea crafts that transport dangerous substances or pollutants,
f. All sea crafts regardless of the length, tonnage or purpose that represent or are in a situation of potential navigation risk or navigation safety, safety of people or environment protection,
g. Ships that tow or push other sea crafts, regardless of their length.

The vigil on Ch10 is obligatory for those sea crafts that are in accordance with the directives of the Resolution of the IMO's Maritime Safety Committee, MSC: 139(76) from December 5, 2002. which are obligated to participate in the Adriatic Reporting System (ADRIREP). The operational area of the mandatory ship reporting system covers the whole Adriatic Sea, north from the latitude 40° 25'.00 N. The area is divided into 5 (five) sectors, each of them assigned to a competent authority, operating on a designated VHF channel.

4 MAIN SAILING ROUTES AND STRUCTURE OF THE TRAFFIC

On the main longitudinal sailing route more longitudinal sailing routes can be formed. Generally, they can be divided into the main longitudinal sailing route in the central part of the Adriatic and those closer to eastern and western coasts (Figure 3). Considering the fact that the east coast is more indented than the west coast, the choice of longitudinal sailing routes along it is greater. Most frequently, they are divided into those out of and inside the outer line of islands of the east coast. The main longitudinal sailing route extends in the central Adriatic in the NW-SE direction following the direction of the main longitudinal axis of the Adriatic Sea and is 400 naut. m. long in total. In its central part, it goes between islands Palagruža and Pianos, i.e. the circle which is made of islands Sušac, Pianos and cape Gargano. The greater part of the longitudinal sailing route extends in the area of sufficient depth and width where there are no

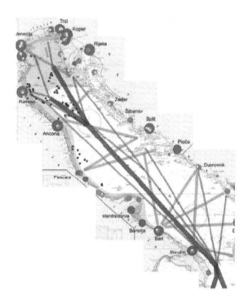

Figure 3. The main sailing routes in the Adriatic. Source: Komadina et al, 2013.

Table 1. Ship traffic in the international navigation according to the main northern Adriatic ports.

Rijeka	1.230*
Koper	2.032
Trieste	3.949*
Venezia	3.402
Monfalcone	768
Ravenna	3.122*
Ancona	4.482

Source: Port statistics 2014*/2015

significant dangers to navigation, with the exception of danger of collision with the opposite or transverse traffic, danger of grounding on the final parts of the path or in the broader area of Palagruža island, possible unfavorable hydro-meteorological conditions and similar.

The main longitudinal sailing routes in the Adriatic are formed between bigger ports on the eastern coast (Rijeka, Zadar, Šibenik, Split, Ploče, Dubrovnik, Bar, Durres) and the ports on the western coast of the Adriatic (Ravenna, Ancona, Pescara, Bari, Brindisi). These routes are formed according to the criterion of the shortest distance, if the hydro-meteorological conditions allow it (wind and waves above all) and are not moved more than 50 naut. m. from the coast. The ship traffic between the ports of the eastern and the western coast are multiple times smaller in comparison to the longitudinal traffic. The biggest parts of the mentioned traffic are passenger lines between bigger ports of the eastern and western coast [Lušić, Kos, 2006].

Overall traffic in the longitudinal sailing routes can be relatively well estimated based on the number of ships that sail into the ports of the Northern Adriatic. According to the available data approximately 19.000 ships annually sails into the most important ports of the north Adriatic in the international navigation (Table 1). The biggest part of the mentioned traffic is on the longitudinal sailing route, i.e. it can be estimated that its daily load is up to 50 ships in one direction daily.

Based on the AIS data it can also be concluded that daily, in every moment, 100–200 ships sail in the Adriatic [marinetraffic]. Circa 20% of all vessels are tankers and circa 10% of all vessels report dangerous cargo on board [Matika, 2013] [Adria VTS, 2009].

The traffic in the narrow area of Palagruža on the main sailing route is approximately circa 10–15 ships daily in one direction [marinetraffic]. According to the collected AIS data for June 2015 in the area of TSS Palagruža in every direction daily, yearly, there are circa 11 ships [AIS-PFST]. The greatest traffic density is in the Northern Adriatic area and in the directed navigation area SW of Palagruža. This increased density is a consequence of relatively small width of opposite sailing routes due to which the traffic is concentrated in the relatively small area, especially around Palagruža where TSS is positioned so that it does not cross the borders of the Croatian territorial waters. In Figure 4 there is a current view of navigation in the central part of the Adriatic.

Figure 5 shows annual traffic density according to the type of ship based on the AIS data for 2014. It is clear that only fishing boats and recreational crafts do not follow the standard longitudinal sailing route. Almost immediately when observing the state of navigation it can be concluded that, in the Adriatic, mainly cargo ships move on the longitudinal sailing routes, with a great number of tankers. Equally, it can be concluded that fishing boats are formed in the vicinity of this area very often.

For the purposes of this paper AIS data about the movement of ships in the area of directed navigation around Palagruža are analyzed, from June 22–29, 2015 and the results are shown in Figures 5 and 6 and the Table 2. The results of the analysis, reduced to an annual level, result in the ship traffic from 4200 for NW direction and 4241 for SE direction, i.e. that is a daily average of circa 11,5 ships. The collected data confirm the expected structure and traffic distribution and the data that tankers make 20% of the traffic on the main longitudinal sailing route.

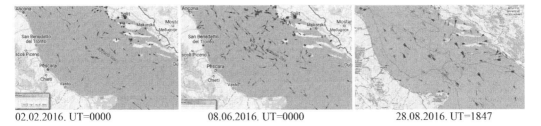

| 02.02.2016. UT=0000 | 08.06.2016. UT=0000 | 28.08.2016. UT=1847 |

Figure 4. State of ship traffic in a given moment (cargo-green, tanker-red; fish.-purple; passenger-blue, other-black; vectors-30 min).
Source: Marinetraffic-August, AIS Station PFST-February, June.

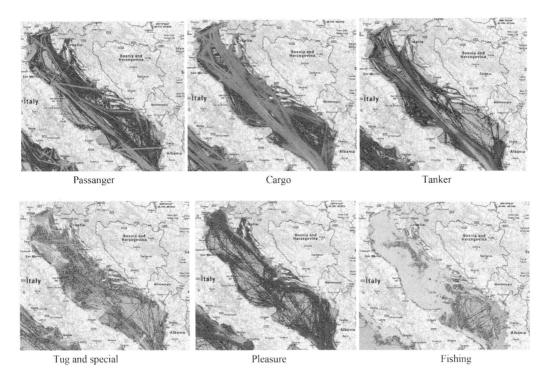

| Passanger | Cargo | Tanker |
| Tug and special | Pleasure | Fishing |

Figure 5. Traffic density according to the type of ship, year 2014.
Source: Marinetraffic.

Table 2. Average speed and draft.

	SE		NW	
	Average speed	Average draft	Average speed	Average draft
Oil produkt	12,6	6,7	12,5	6,7
General	14,3	5,4	13,6	5,4
Passenger	21,6	3,5	20,1	3,5
Pleasure	10,3	3,4	17,7	3,4
Other	13,8	3,4	12,4	3,4

Because of the interruption and mistakes in the AIS signal reception, it was expected that the traffic calculated in such manner should be greater than obtained, also for the same reason there is a relatively great number of ships of unknown type. Also, the data was collected in the part of year when the traffic in the Adriatic is significantly greater than average, but as the structure of the traffic is mainly made of cargo ships, seasonal changes in this sailing route are not so expressed as in the others (Figure 4).

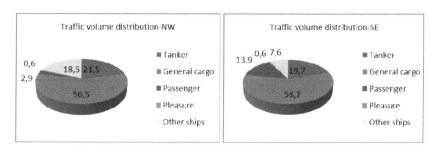

Figure 6. Traffic volume distribution 06/2016.

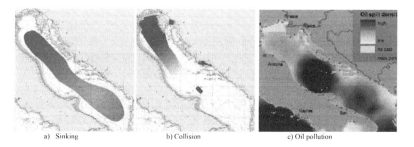

a) Sinking   b) Collision   c) Oil pollution

Figure 7. Areas of increased risks.
Source: Zec et al., 2009; Case study report-The Adriatic Sea, 2011.

5 ACCIDENT ANALYSIS

The number of accidents of big merchant ships in the Adriatic is relatively small. Generally speaking, total number of collisions and groundings is great; however this number is mainly relevant to smaller vessels that sail very close to the coast. For instance, in the area of Croatian responsibility, based on the interventions by Port Authority, in the year 2015 there were 11 collisions, 43 groundings, 48 flooding, 9 fires, and so on [www.mmpi.hr]. Examples of greater accidents: sinking of the dry cargo transportation ship "Cavtat" in Otranto in 1974 in Murter sea with 1300 t VCM; fire on ro-ro ship "UND Adriyatik" 13 naut. m. SW of Rovinj in 2008; collision of cargo ships off Ravenna in 2014 and so on. In the area of open sea in the Central Adriatic it is important to mention eventual sinking of fishing boats in 2014, after the collision with a merchant ship circa 30 naut. m. before TSS north Adriatic. Since there is no greater number of collisions and groundings in open parts of the Adriatic, these can only be estimated according to some other similar areas in the world. Potential places of groundings derive from positions of the most protruding parts of the coast towards the open sea, and as far as the collisions go, these are mainly places where crossings and narrowing of sailing routes occur. For collisions it will be the area of northern part of central Adriatic (Figure 7).

The main longitudinal sailing routes are especially dangerous because it is the main route for

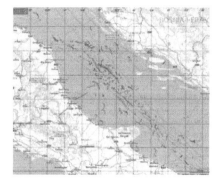

Figure 8. Oil slicks detected in the Adriatic.
Source: Morović et al, 2015.

tankers. For bigger ports of the Northern Adriatic (Trieste, Venice, Omišalj and Koper) it is annually transported circa 58 million T [Morović, Ivanov, 2011]. So far there were no greater tanker accidents. Smaller slicks, as a consequence to daily operations, were mostly widespread precisely along this route (Figure 8).

Additional potential danger for environment is found in gas fields of the Northern Adriatic, as well as the plan to spread the plan to exploit gas and oil along the entire Adriatic. Figure 9 shows the positions of the existing exploitation fields (red) and the plan of researching areas of gas and oil in the area of the Croatian part of the Adriatic [Strategic study, 2015].

Figure 9. Exploitation and researching areas. Source: Strategic study, 2015.

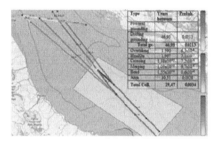

Figure 10. Results of the analysis of collisions and groundings—IWRAP.

6 RESULTS OF THE IWRAP ANALYSIS

As it was already mentioned, AIS ship traffic was analyzed in details in the central Adriatic area, from 12 to 29 June 2015. If the traffic is observed on the main longitudinal sailing route which goes through TSS Palagruža it can be concluded that the probability of collisions and groundings in this area is minimal or, in other words, the expected return period of collision is over 2.000 years, and for groundings 50 years. A small probability like this is expected considering that due to the existence of the system-oriented navigation the opposite waterways are almost completely separated and the transversal traffic is negligible. However, the situation changes considerably if taking into consideration that the main longitudinal sailing route crosses with many small ships' routes. For example, if in the larger area of the island of Palagruža (marked area in Figure 10) there are 50 ships in average, in 24 hours' time (30 m of average size) and that each one of them within the 24 hours spends 20 hours sailing and 4 hours drifting, the probability of collision increases significantly, with $4,17 \times 10^{4}$ to $1,36 \times 10^{2}$. There are no important changes for groundings. Furthermore, if the area of observation expands to TSS Northern

Adriatic (Figure 10) the probability of collisions and groundings increases minimally, under condition that a number of smaller vessels within the referential area stays the same. The results confirm that the critical points for collisions are somewhat more to the west than TSS Palagruža, points in which sailing routes branch towards TSS Northern Adriatic.

The results of the analysis also show that the probability of tanker participation in potential accidents is very high, especially when observing collisions of bigger ships in the longitudinal sailing route. The results of the analysis also show that the probability of tanker participation in potential accidents is very high, especially when observing collisions of bigger ships in the longitudinal sailing route.

7 RECOMMENDATIONS

Generally, it has to be emphasized that the Adriatic Sea is recorded with regard to accidents and casualties as a high risks area. According to the IMO frequency of accidents in the Adriatic Sea is recorded at a five times higher level than the world average, a fact mostly due to speed boats and yachts accidents [Kačić, 2011].

System-oriented navigation on the main longitudinal sailing route of the Adriatic exists and efficiently separates opposite waterways. Improvements can be made in the establishment of the traffic separation system in Otranto and in the improvement of the traffic separation system near Palagruža, first of all in mutual agreement of the interested countries and acceptance by the IMO. Additional traffic separation systems closer to eastern coasts are not to be expected for now, however suggestions and plans exist. The establishment of hydrocarbon exploitation areas in the Central and Southern Adriatic could consequently demand changes in navigation regulation in the Adriatic, i.e. changes of the existing and establishment of additional regulatory measures as well as navigation supervision. Wider area of Palagruža is an important fishing area as well and in that sense significant fishing vessels traffic can be expected in the future as well as the dangers of collisions. Corresponding movement supervision of these ships and officer training are objective measures which can have a preventive influence on navigation safety.

Navigation supervision service exists in the Adriatic, but the activity needs to be directed towards better coordination of coastal countries and the improvement of the system itself with focus on active tracking of ship traffic. Considering the consequences which can cause accidents of bigger ships, especially tankers, expansion and restructuring of power to act in situations of crisis seem justified, as well as the improvement of the coordination of coastal countries.

8 CONCLUSION

The open sea area of the Central Adriatic at first glance seems like an area in which maritime accidents of bigger ships rarely happen and that navigation in this part of the Adriatic is safe in comparison to other parts. The reason for this are established separation zones, established navigation supervision service, small intensity transversal traffic, moderate traffic on the main longitudinal route and lack of bigger accidents. Precisely the lack of bigger accidents is one of the main reasons of fake safety. In other words, the main longitudinal route is tankers' main path from Otranto to bigger ports in the NW Adriatic and any bigger accident of these ships would have catastrophic consequences to coastal countries, their economies, environment etc.

If AIS ship traffic is observed, it can be concluded that this traffic on the main longitudinal sailing route is of moderate intensity, mostly consistent throughout the entire year. The greatest danger is the narrowing and crossing of sailing routes in the part between TSS Northern Adriatic and TSS Palagruža. In the summer months this part of the Adriatic is additionally loaded by ships and yachts traffic, and somewhat increased transversal traffic. Also, a great danger is found in fishing vessels that frequently concentrate in this area crossing their routes with the main longitudinal sailing route. The results of the analysis show that the probability of bigger ships collisions with these vessels in the part of the mentioned TSS is the biggest.

Bearing in mind further increase in maritime traffic in the Adriatic, the same or bigger fishing or other vessels traffic, hydrocarbon exploitation expansion to the entire Adriatic and negative or better yet catastrophic effect of any bigger accident on the main longitudinal sailing route, the need to additionally improve the navigation safety system is unquestionable. There is room for that, especially in the establishment of more effective and modern ship supervision system, improvement of regulatory measures of navigation regulation (ships routing), improvement of system of action in situations of crisis and better coordination between countries, preventively as well as after the accidents.

REFERENCES

Adria VTS-služba nadzora i upravljanja pomorskom plovidbom na Jadranu, report v2.1, 2009.

IMO-a, MSC.139(76).

Kačić, H.: Traffic separation schemes in the Adriatic Sea, Round table "EU Maritime Policy and the (Northern) Adriatic" organized by the Maritime Law Association of Slovenia (MLAS), Portorose, 2011.

Komadina, P.; Brčić, D.; Frančić, V.: VTMIS služba u funkciji unaprjeđenja sigurnosti pomorskog prometa i zaštite okoliša na Jadranu, Pomorski zbornik 47–48 (2013), 27–40.

Lakoš, S.: Prilog istraživanju regulacije plovidbe Jadranom—Doktorat, Pomorski fakultet u Rijeci, 1980.

Lušić, Z.: Kos, S.: Glavni plovidbeni putovi na Jadranu, Naše more 53 (5–6), Dubrovnik, 2006, 198–205.

Morović. M.; Ivanov, A.: Oil Spill Monitoring in the Croatian Adriatic Waters:needs and possibilities, ACTA ADR IAT., 52(1): 2011, 45–56.

Morović. M.; Ivanov, A.; Oluić, M.: Oil spills distribution in the Middle and Southern Adriatic Sea as a result of intensive ship traffic, ACTA ADR IAT., 56(2): 2015, 145–156.

Nadzor i upravljanje pomorskim prometom u unutarnjim morskim vodama i teritorijalnom moru Republike Hrvatske, Hrvatski hidrografski institut, Split, 2015.

Peljar I, Jadransko more-istočna obala, Hrvatski hidrografski Institut, Split, 2012.

Pre-accesion maritime transport strategy of the republic of Croatia, The Ministry of the Sea, Tourism, Transport and Development, Zagreb, 2005.

Prometno-plovidbena studija, plovno području Split, Ploče i Dubrovnik, Pomorski fakultet u Rijeci, Rijeka, 2014.

Prometno-plovidbena studija, plovna područja: Primorsko-goranske, Ličko-senjske, Zadarske i Šibensko-kninske županije, Pomorski fakultet u Rijeci, Rijeka, 2015.

Službeni list SFRJ, Broj 21—str 555, od 17. Ožujka 1989.

Strateška studija o vjerojatno značajnom utjecaju na okoliš Okvirnog plana i programa istraživanja i eksploatacije ugljikovodika na Jadranu, Ires ekologija d.o.o. za zaštitu prirode i okoliša, Zagreb, 2015.

The potential of Maritime Spatial Planning in the Mediterranean Sea" Case study report: The Adriatic Sea, Policy Research Corporation, Brussels, 2011.

Zec, D.; Maglic, L.; Simic Hlaca, M.: "Maritime Transport and Possible Accidents in the Adriatic Sea", Presentation in Dubrovnik, 2009 (https://bib.irb.hr/datoteka/450379.Maritime_Transport_and_Possible_Accidents_in_the_Adriatic.pdf)

https://luka-kp.si/eng/311 (Port of Koper-statistics 2015, August 2016)

http://www.mppi.hr/UserDocsImages/SAR%20stat%20 2015%201.I.-31.XII.%2020–1_16.pdf (SAR akcije-RH, september 2016.)

http://www.iala-aism.org/products/technical/risk-management-tools.html (IWRAP MkII, August 2016)

http://www.autoritaportuale.ancona.it/images/RAPPORTO_STATISTICO_2015.pdf (Port of Ancona-statistics 2015, August 2016)

http://www.porto.monfalcone.gorizia.it/public/allegati/23-4-2016-11-59-56_354084.pdf (Porto Monfalcone-2015, August 2016)

http://www.port.ravenna.it/traffico-porto-di-ravenna-il-2014-si-chiude-con-lottimo-risultato-di-un-88/ (Traffico porto di Ravenna 2014-August 2015)

http://www.porto.trieste.it/wp-content/uploads/2015/04/ESPO-Statistics-2014-EN-2.pdf (Port of Trieste-statistics, August 2016)

https://www.port.venice.it/files/page/apvstatistiche2015_0.pdf (PORT OF VENICE, August 2016)

Transport Infrastructure and Systems – Dell'Acqua & Wegman (Eds)
© *2017 Taylor & Francis Group, London, ISBN 978-1-138-03009-1*

Study on distribution of bicycle sharing system facility based on Markov chain model

R. Liu, J.F. Dai, J.X. Lin, Q. Shi, Z. Sun, F. Yang & J.C. Huang
Beijing Urban Transportation Infrastructure Engineering Technology Research Center, Beijing University of Civil Engineering and Architecture, Beijing, China

ABSTRACT: unreasonable layout and scale of stations is one of key factor that often lead to unbalance borrowing and returning phenomenon in bicycle sharing system. This problem will reduce the operation efficiency of the whole system. In order to solve this problem, Markov Chain Model is applied to establish a borrowing and returning probability matrix, then the steady state probability matrix will be got, via the step transition, and the model will be tested by using a linear regression model. By analyzing the predicted borrowing and returning probability of stations with different land use, the bike use proportion and initial station scale of the system will be identified. In addition, we can evaluate scale allocation reasonability of the established stations by using the model in this paper, propose the proportion of bikes and lock piles at future, make the whole system reach the maximal natural balance, and attract more users to select public bikes.

1 INTRODUCTION

To promote green public travel and solve travel terminal coverage problem during recent years, more and more countries start to promote the Bicycle Sharing System (BSS). The public bike replaces partial public transportation and walking [1], and alleviates pressure on the buses, so the BSS is quickly constructed in different countries in the world and is extensively promoted in China. Till 2014, the quantity of public bikes in China reaches 79.3% of total public bikes in the world [2]. China becomes the typical country to use the BSS. Although the quantity of bikes is ranked as the first position, construction and operation of this system is still on exploration and continuous perfection phase. The specific problems are described as follows: the borrowing and returning bikes are not balanced at stations, so it reduces system's appeal to the users and increases the operation cost of operators. From the investigation data, 42% users from the central districts in Beijing think that unreasonable scale and layout of the stations and non-timely scheduling leads to unbalance in the system. The station A and B at a traffic travel cell have different land properties and borrowing and returning demands, so the station A only allows for rental and the station B only allows for returning, which frequently occurs. To effectively utilize bikes, realize relative balance of bike using among stations, and satisfy the borrowing and returning demand of the riders to most extent, researchers study this problem in different aspects.

T. Benarbia proposes a Petri net model, which divides the station stock into engorgement, penury and desired phase and selects different scheduling schemes by phase transition to rebalance borrowing and returning at stations (Benarbia et al. 2013). Rainer-Harbach identifies effective unloading quantity of scheduling bikes by using variant neighborhood search method and other methods (Rainer-Harbach et al. 2013). Kloimullner rebalances stations by using dynamic scheduling algorithm (Kloimuller et al. 2014). The above methods are based on the later scheduling operation and manual adjustment of bikes and free bike lock piles. Christine Fricker proposes to realize natural balance by guiding users to return bikes to stations with smaller saturation and considers rebalance problems in the early scheduling period (Fricker et al. 2012). Ahmadreza considers land utilization when analyzing flow characteristics of public bikes (Ahmadreza et al. 2014).

Mode choices of BSS users are highly associated with factors such as location and land property of stations. Borrowing and returning characteristics will satisfy certain regulation. With elapse of time, the system will be prone to stabilization. Such stability is suitable for Markov chain model. This model establishes the transition matrix by using quantity of the borrowing and returning bikes among stations. The matrix will gradually become stable via step transition, so the feature vectors of this matrix can be solved to represent quantity of the borrowing and returning bikes of stations under steady state. Land blocks of same property will have similar demands for borrowing and returning. This paper will classify stations according to land property in the same Traffic Analysis

Zone (TAZ) and analyze proportion of borrowing and returning in different land use. This proportion can directly reflect allocation of bikes and lock piles when the system is stable and reaches borrowing and returning balance state. Allocate or adjust the station scale with the proportion can reduce the difference between rental and return and alleviate difficulties in borrowing and returning.

The structure of this paper is as follows: the basic knowledge of Markov chain is introduced in section 2. Section 3 illustrates the Markov chain model of bicycle sharing system, and describes the settings, solving and verification method. Section 4 briefly presents a real-world case study of five central districts in Beijing, and obtained the allocation proportion of bikes and lock piles. Finally, conclusions and future work are summarized in section 5.

2 MARKOV CHAIN MODEL

For a random process $\{X(t), t \in T\}$, if the condition $t_0 < t_1 < \cdots < t_n < t \in T$ is satisfied, we can get the following equation:

$$p[X(t) \leq x \mid X(t_n) = x_n, X(t_{n-1}) = x_{n-1}, \cdots, X(t_0) = x_0] = P[X(t) \leq x \mid X(t_n) = x_m] \quad (1)$$

This random process is called as the Markov process. The Markov chain is one special case of the Markov process. Its time parameters and state space are dispersed. If $n > n_1 > n_2 > \ldots > n_k$, the Markov chain is represented as follows: Getting started

$$P(X_n = j \mid X_{n_1} = i_1, X_{n_2} = i_2, \cdots, X_{n_k} = i_k) = P(X_n = j \mid X_{n_1} = i_1) = p_{i_1,j}^{m_k,n} \quad (2)$$

If $p_{i,j} = P(X_i = j \mid X_0 = i), i, j \in I$, p_{ij} is called as the transition probability and indicates the probability of transition from the state i to the state j. The matrix based on p_{ij} which called transition probability matrix is constructed as follows:

$$P = \begin{bmatrix} p_{00} & p_{01} & p_{02} & \cdots \\ p_{10} & p_{11} & p_{12} & \cdots \\ p_{20} & p_{21} & p_{22} & \cdots \\ \vdots & \vdots & \vdots & \ddots \end{bmatrix} = (p_{ij}) \quad (3)$$

Elements in each row indicate probability of transition from one state to different states, so the sum of probabilities in one row is 1, namely:

$$p_{ij} \geq 0, \sum_{j=1}^{n} p_{ij} = 1 \quad (4)$$

When $\{X_m, m \in M\}$ is homogeneous Markov chain, for any $i, j \in E$ in the state space E, if no π_j constant depending on i makes $\lim_{n \to \infty} P_{ij}^{(m)} = \pi_j$, this Markov chain is ergodic. For an ergodic Markov chain, regardless of starting state in the system, when the transition steps m is enough, the probability of transition to the state j will approach to π_j, namely the system will become steady state after a period (Bao.2102).

At this time, the Markov chain should satisfy the following three prerequisites (Li.2010):

1. The transition matrix will not vary with time;
2. Total states keep constant in the prediction period;
3. The system change process has no aftereffect.

3 MARKOV CHAIN MODEL OF BICYCLE SHARING SYSTEM

3.1 Model assumption

1. Travel of BSS users is purposeful, so travel probabilities of the whole system will be distributed by certain law on a day. This law will be stable and will not significantly change with time, so it can satisfy the requirement of Markov chain model that transition matrix will not vary with time.
2. The distribution places, quantity of stations total bikes and total lock piles in the whole system are constant, so it is deemed that the states inside the system keep stable, which satisfies the requirements of constant states in the prediction period;
3. The bicycle sharing system is relatively independent and system operation suffers from smaller external interference. A user selects a bike from any stations based on the desire to reach the destination. Selection only depends on his own travel and is independent of travel by previous user of the same bike at this station, so the system satisfies the requirements that system change has no aftereffect.

The bicycle sharing system satisfies three prerequisites required by the Markov chain model, so it can be used to establish the state transition matrix.

3.2 State transition matrix

By sorting data of BSS, the OD pair of the borrowing and returning among stations can be obtained, so the system state transition matrix P is constructed as follows:

$$P = \begin{bmatrix} a_{11} & a_{12} & \cdots & a_{1n} \\ a_{21} & a_{22} & \cdots & a_{2n} \\ \vdots & \vdots & \ddots & \vdots \\ a_{n1} & a_{2} & \cdots & a_{nn} \end{bmatrix} \qquad (5)$$

When the borrowing matrix is P, a_{ij} represents borrowing probability of the user who borrows a bike at the station i and returns it at the station j. When the matrix P is a returning matrix, a_{ij} represents returning probability of the user who borrows a bike at the station j and returns it at the station i

3.3 Steady state probability vector

The state transition matrix satisfies three prerequisites of the Markov chain model, so the bike transition probability distribution of the whole system will reach a steady state with increase of transition steps, namely:

$$\omega P = \omega \qquad (6)$$

$$\sum_{i=1}^{n} \omega_i = 1 \qquad (7)$$

P is a steady state probability matrix. ω is called as the steady state probability vector, namely the feature vector of the transition probability matrix, and represents borrowing and returning probability at different stations when the system is under steady state. The probability sum of all stations is 1. It is analyzed and computed based on OD and different stations are considered in a same block, so the derived probability matrix will directly reflect mutual relations between borrowing and returning conditions at different stations instead of the independent characteristics.

3.4 Model verification

A linear regression method is used to verify reliability of Markov chain model. The computed bor-

rowing and returning probabilities of stations are regressed and fitted with the average daily proportions of actual stations. Reliability is verified by the goodness of fit with the regression equation.

As shown in Figure 1, the disperse points indicate the predicted average borrowing probability of stations. The fitting equation is $Y = ax$. If the predicted average borrowing probability is fully equal to the actual borrowing probability, all disperse points will fall onto the fitting curve, and at this time $a = 1$, the fitting equation is $Y = x$. The predicted value is more approaches to the actual value while a is prone to 1. If the fitting priority R^2 approaches to 1, it indicates that the equation has higher fitting degree for predicted values.

4 CASE STUDY

4.1 Case introduction

As shown in Figure 2, bikes can be parked and borrowed at any stations in five central districts of Beijing. The stations are plentiful, the land properties are diversified, and the borrowing and returning demand is huge in five districts, so it can provide enough data for statistics and analysis. In addition, the investigation data show that 15% users think stations insufficient for borrowing and returning, 13% users think lock piles insufficient for timely returning, and 14% users think bikes insufficient for timely borrowing. The Markov chain model is used to solve unbalanced allocation and difficulties in both borrowing and returning of the current stations in these areas.

4.2 Application

4.2.1 Create OD matrix

To sort data of bikes on borrowing and returning from stations in five districts of Beijing, the borrowing OD matrix in 24 hours can be derived (as shown in Table 1). The elements in the matrix

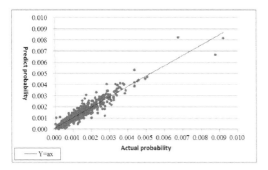

Figure 1. Comparison of predict probability and actual probability.

Figure 2. Station distribution map of BSS station in five districts of Beijing.

Table 1. Borrowing OD matrix in 24h.

Station number	1	2	3	4	5	...	702
1	6	0	0	1	0	...	0
2	1	1	3	3	3	...	0
3	0	2	7	1	4	...	0
4	1	11	4	6	3	...	0
5	1	2	5	1	11	...	0
6	1	1	0	1	0	...	0
7	5	3	1	1	0	...	0
8	1	2	1	0	2	...	0
9	1	0	0	0	0	...	0
10	0	0	0	0	0	...	0
⋮	⋮	⋮	⋮	⋮	⋮	⋮	0
702	0	0	0	0	0	...	0
Total	151	185	105	107	99	...	7

Table 2. Returning OD matrix in 24h.

Station number	1	2	3	4	5	...	702
1	6	1	0	1	1	...	0
2	0	1	2	11	2	...	0
3	0	3	7	4	5	...	0
4	1	3	1	6	1	...	0
5	0	3	4	3	11	...	0
6	0	2	1	2	1	...	0
7	1	5	1	0	0	...	0
8	0	2	0	0	0	...	0
9	0	0	0	0	0	...	0
10	0	0	0	0	0	...	0
⋮	⋮	⋮	⋮	⋮	⋮	⋮	⋮
702	0	0	0	0	0	...	0
Total	141	178	98	106	91	...	4

Table 3. Borrowing probability matrix in 24h.

Station number	1	2	3	...	702
1	0.0397	0.0000	0.0000	...	0.0000
2	0.0066	0.0054	0.0286	...	0.0000
3	0.0000	0.0108	0.0667	...	0.0000
4	0.0066	0.0595	0.0381	...	0.0000
5	0.0066	0.0108	0.0476	...	0.0000
6	0.0066	0.0054	0.0000	...	0.0000
7	0.0331	0.0162	0.0095	...	0.0000
8	0.0066	0.0108	0.0095	...	0.0000
9	0.0066	0.0000	0.0000	...	0.0000
10	0.0000	0.0000	0.0000	...	0.0000
⋮	⋮	⋮	⋮	⋮	0.0000
702	0.0000	0.0000	0.0000	...	0.0000

Table 4. Returning probability matrix in 24h.

Station number	1	2	3	...	702
1	0.0426	0.0056	0.0000	...	0.0000
2	0.0000	0.0056	0.0204	...	0.0000
3	0.0000	0.0169	0.0714	...	0.0000
4	0.0071	0.0169	0.0102	...	0.0000
5	0.0000	0.0169	0.0408	...	0.0000
6	0.0000	0.0112	0.0102	...	0.0000
7	0.0071	0.0281	0.0102	...	0.0000
8	0.0000	0.0112	0.0000	...	0.0000
9	0.0000	0.0000	0.0000	...	0.0000
10	0.0000	0.0000	0.0000	...	0.0000
⋮	⋮	⋮	⋮	⋮	0.0000
702	0.0000	0.0000	0.0000	...	0.0000

Table 5. Borrowing and returning steady state vectors of BSS.

Station number	1	2	3	...	702
Borrowing ω_p	0.00229	0.00280	0.00153	...	0.00004
Returning ω_d	0.00236	0.00295	0.00169	...	0.00016

represent the bikes borrowed from the station A (column title station) and returned to the station B (row title station). The returning OD matrix can be also derived (shown as the Table 2). The elements in the matrix represent bikes borrowed from the station B (row title station) and returned to the station A (column title station).

4.2.2 *Generate state transition matrix*
As shown in Tables 3 and 4, the matrix for ratios of borrowing and returning bikes at different stations to total borrowing and returning bikes at all stations can be derived, namely borrowing and returning probability matrix.

4.2.3 *Solve steady state probability vector*
The steady state probability vectors of the borrowing and returning bikes, as shown in Table 5, can be computed according to the state transition matrix which derived from the Table 3 and 4 by using the equation (6), namely ratio of borrowing or returning bikes at each station to total bikes of whole bicycle sharing system in five central districts.

4.2.4 *Result verification*
Compare the computed steady state probability vector value with borrowing and returning probability on a day and verify the probability of the stations by using the linear regression model.

In order to ensure the stability of the state transition matrix of the Markov chain model, three common business days in summer are selected to compute the average state transition matrix, then solve the steady state probability vectors, and compare the vectors value with the actual probability of the fourth business day.

As shown in 3 and 4, x axle indicates the actual borrowing and returning probability, y axle represents the average borrowing and returning probability, the trend line is $Y = ax$. If the predicted average probability is equal to the actual probability, all sparse probability points will be distributed on the trend line in the figure. The prediction probability data will be fitted by using the fitting equation $Y = ax$ in a one-variable linear manner to get the linear correlation coefficient, as shown in Table 6.

The fitting equation for predicting average borrowing and returning probability approaches to $Y = x$, the correlation coefficient is higher than 0.90. It can effectively describe proportion of borrowing and returning bikes at different stations to total bikes in the whole system.

4.3 Result analysis

The borrowing and returning probability of stations in five districts of Beijing can be obtained by

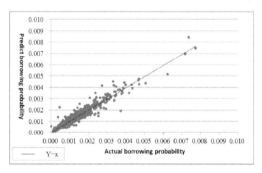

Figure 3. Comparison of predict borrowing probability and actual borrowing probability.

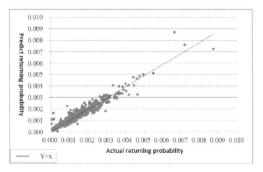

Figure 4. Comparison of predict returning probability and actual returning probability.

Markov chain model, namely the ratio of borrowing and returning bikes. The stations in the system are divided into the bus stops, administrative areas, railway station, public transit hub, park, residence, commercial district, scheduling station, school and hospital by land property. The stations of same properties are sorted. As shown in Figure 5, the borrowing proportion of stations in 24 hours is roughly equal to the returning proportion. The proportion of the borrowing and returning of the bus stop, administrative area, park, residence, school and hospital is within 4.5%–6.5%. The proportion of the railway station and commercial district is higher and is within 8%–10%. The proportion of the public transit hub is over 15%. The proportion of the scheduling station is significantly higher and is more than 30%.

The analysis shows that the morning and night peak period occurs at 07:00~09:00 and 17:00~19:00 on the business days (Pfrommer et al. 2013). To satisfy the requirements of the Markov chain model for matrix stability, the four-hour data in the morning and night peak hours are selected together for consolidation and analysis. The data is sorted to get the borrowing and returning proportion distribution diagram of the stations with different land properties, as shown in Figure 6.

The borrowing proportions and returning proportions are different in the peak period. The proportion of borrowing of commercial district, railway station and public transit hub is higher than returning. The proportion of returning of the scheduling station is higher than borrowing. Compared to the whole-day data, the returning of bikes

Table 6. Reliability verification of predict probability of borrowing and returning.

Probability property	Fitting equation	R^2
Borrowing	$y = 0.9878x$	0.9121
Returning	$y = 0.9748x$	0.9084

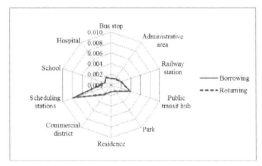

Figure 5. Predicted borrowing and returning proportion of stations with different land properties in 24h.

at the scheduling station is more centralized in the peak hour.

As shown in Table 7, the actual borrowing and returning bikes are similar to the predicted, which verifies reliability of the predicted value. In addition, the proportion of the borrowing and returning is similar, which indicates that total borrowing bikes and total returning bikes are balanced. So the predicted data can truly reflect the proportion when the system is stable.

Since the data includes peak value and flat peak value, it can reflect borrowing and returning bikes of the stations. Therefore, select it as the scale allocation proportion of the bikes and lock piles at stations with different land properties. The lock piles are fixed and the bikes are flowing during actual operation. Only the allocation proportion of the lock piles at stations can be obtained and the fixed bike allocation proportion cannot be obtained. As shown in Table 8, in order to compare the lock pile proportion of actual stations with allocated proportion derived from analysis, it shows that the truly configured scale of stations at the bus stop,

administrative area, railway station, park, residence, commercial district, school and hospital blocks is bigger, so their capacity should be properly reduced. The truly configured scale at the public transit hub and scheduling station is too small to satisfy borrowing and returning demands, so the stations should be expanded. In addition, after the station lock piles are allocated according to the data in the table 8, the total configured bikes in the system can be identified according to the ratio of bikes to lock piles (0.80) (Zhang. 2013) and the bikes required by each land property station is identified according to the bikes allocation proportion.

Based on the above data, the following suggestions are proposed for scale layout of stations with different land properties in one TAZ:

1. Strengthen the block division, divide a cell into several public bike blocks, strengthen travel function in the block (Li et al. 2009), and divide different stations by land properties;

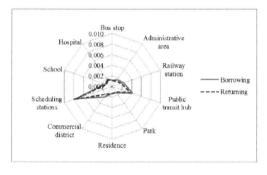

Figure 6. Borrowing and returning proportion of stations with different land properties in peak period.

Table 8. Summary of lock piles and bike allocation proportion of stations with land properties.

Land properties	Actual lock pile proportion	Allocated lock pile proportion	Allocated bike proportion
Bus stop	10%	5%	5%
Administrative area	10%	6%	6%
Railway station	10%	8%	8%
Public transit hub	8%	15%	16%
Park	8%	5%	5%
Residence	10%	6%	5%
Commercial district	12%	10%	9%
Scheduling station	12%	33%	33%
School	11%	6%	5%
Hospital	10%	7%	7%

Table 7. Summary of actual proportion and prediction proportion at stations with different land properties.

Land properties	Borrowing in 24h		Returning in 24h		Borrowing in peak period		Returning in peak period	
	Predicted value	Actual value	Predicted value	Actual value	Predicted value	Actual value	Predicted value	Actual value
Bus stop	5%	5%	5%	5%	5%	8%	5%	8%
Administrative area	6%	6%	5%	6%	6%	6%	5%	6%
Railway station	8%	8%	8%	8%	9%	10%	8%	10%
Public transit hub	16%	16%	16%	15%	17%	16%	16%	15%
Park	6%	5%	6%	5%	6%	6%	6%	6%
Residence	5%	5%	5%	6%	5%	8%	6%	7%
Commercial district	9%	9%	9%	10%	10%	8%	8%	8%
Scheduling station	31%	33%	32%	33%	30%	19%	35%	21%
School	6%	5%	5%	6%	6%	7%	5%	7%
Hospital	7%	7%	7%	7%	7%	11%	7%	11%

2. Choose the scheduling station as the center and the interchange of the public bike block (Li et al. 2009). The stations such as the public transit hub and railway station with higher borrowing and returning proportion can perform functions of a scheduling station. The scheduling bikes should be stored to satisfy the peak hour scheduling requirement of surrounding stations.
3. Perform scale allocation or scale verification for different stations in one TAZ with different land properties, reasonably allocate the bikes and lock piles quantity, or adjust the scale among adjacent stations.

5 CONCLUSIONS

The Markov chain model is used to solve the proportion of the borrowing and returning bikes at different stations under steady states during the whole system. It can correctly reflect the borrowing and returning conditions of a station. The station scales with different land properties can be identified by the proportion. In addition, the proportion also can verify the scale of established stations, adjust allocated facilities at stations which under highly unbalanced state, reduce the difference between borrowing and returning, improve operation efficiency and service level of the system, and attract more users to select the public bike for travel.

This paper does not fully consider complexity of the stations with different land properties. The multiple land properties of one station can be classified as primary and secondary. E.g. the primary property of the station A is commercial, but the bus stops are located around it, and the transfer demands will also affect the borrowing and returning actions, so the bus stop can be identified as secondary property. Multiple properties of stations should be classified in future research and be analyzed in detail to increase reliability of the stations allocation.

ACKNOWLEDGEMENT

This work has been carried out under the support of Beijing Talent Cultivation Foundation (2013D005017000003), Beijing philosophy and social science planning project (16GLC048) and Ministry of housing and the Ministry of science and technology project (2016-K2-034).

REFERENCES

Bao Na. (2012). Study on the location decision and scheduling of city Public Bicycle's rental station model, Chang'an University.

Benarbia, T., Labadi, K., Darcherif, A. M., Barbot, J. P., & Omari, A. (2013). Real-time inventory control and rebalancing in bike-sharing systems by using a stochastic Petri net model. International Conference on Systems and Control (pp.583-589).

Benarbia, T., Labadi, K., Omari, A., & Barbot, J.P. (2013). Balancing dynamic bike-sharing systems: A Petri nets with variable arc weights based approach.

Faghih Imani A, Eluru N, Elgeneidy A M, et al. (2014). How Does Land-Use and Urban Form Impact Bicycle Flows--Evidence from the Bicycle-Sharing System (BIXI) in Montreal,Transportation Research Board Annual Meeting.

Fricker, C., & Gast, N. (2012). Incentives and redistribution in homogeneous bike-sharing systems with stations of finite capacity. Euro Journal on Transportation & Logistics, 1–31.

Kloimüllner, C., Papazek, P., Hu, B., & Raidl, G. R. (2014). Balancing Bicycle Sharing Systems: An Approach for the Dynamic Case.Evolutionary Computation in Combinatorial Optimisation. Springer Berlin Heidelberg.

Li Lihui, Chen Hua, Sun Xiaoli. (2009). Bike Rental Station Deployment Planning in Wuhan, Urban Transport of China, Vol.7, No.4, July 2009.

Li Zhenghao. (2010). Analysis of urban public bicycle scale in future, Energy Conservation & Enviromental Protection in Transportation (2),44–46.

Meddin R, DeMaio P. The Bike—sharing World Map[EB/OL].2015[2014-01-26].http://www.bikesharing world.com.

Pfrommer, J., Warrington, J., Schildbach, G., & Morari, M. (2013). Dynamic vehicle redistribution and online price incentives in shared mobility systems. Intelligent Transportation Systems IEEE Transactions on, 15(4), 1567–1578.

Rainer-Harbach, M., Papazek, P., Hu, B., & Raidl, G. R. (2013). Balancing Bicycle Sharing Systems: A Variable Neighborhood Search Approach. Evolutionary Computation in Combinatorial Optimization. Springer Berlin Heidelberg.

Wei Zhu, Yuqi Pang, De Wang, & Harry Timmermans. (2013). Travel Behavior Change after the Introduction of Public Bicycle Systems: Case Study in Minhang District, Shanghai. Meeting of the Transportation Research Board (Vol.43, pp.76–81).

Zhang Jianguo. (2013). Research on urban public bicycle vehicle allocation problem. (Doctoral dissertation, Southwest Jiaotong University).

Transport Infrastructure and Systems – Dell'Acqua & Wegman (Eds)
© 2017 Taylor & Francis Group, London, ISBN 978-1-138-03009-1

A study on the influence of parking policy mix on travel model choice behavior within the scope of heterogeneous travel purposes

Hai Yan
Beijing Key Laboratory of Traffic Engineering, Beijing University of Technology, Beijing, China

Mowen Ran & Mengtian Li
Beijing University of Technology, Beijing, China

ABSTRACT: Parking policies have significant impacts to travel mode choice and traffic system operation. Due to heterogeneous travel purpose, people perceive the same parking policy mix with different extent. In order to understand the existing reason about different perception for the same policy, a multiple-factor behavior model has been developed, by considering the factors of parking price, public transit travel time, traffic conditions and parking supply etc. Three major findings obtained from this study are as following: the first finding is that travel mode choice has been significantly affected by dynamic traffic. Second, parking price and bus travel time have important effects on the mode choice of commuting drivers. Third, the non-commuting drivers are more sensitive to parking price than commuting drivers only when making mode choice. These findings provided a theoretical reference for zoning and interval management of the parking policy mix.

1 INTRODUCTION

There is a clear tendency that more families would like to have their own car(s) in China and a lot of Chinese cities are plagued by traffic congestion, air quality deterioration due to the boom in car traffic. The city management authorities have actively implemented the Traffic Demand Management (TDM) measures to optimize the structure of residents travel and guide the rational use of private vehicles, among which Parking Demand Management (PDM) is an important part of TDM. PDM mainly includes parking supply, parking price, parking taxes control, and law enforcement, etc.

When PDM policies are being designed, governments are usually expected to reduce vehicle use intensity and improve the proportion of public transit by hybrid parking policy, and achieve the goal of "balance the dynamic with the static".

Previous studies have disclosed that car travelers react differently to parking policy for different travel purposes (Shiftan, 1999; Shiftan, 2002), and the degree is also different between drivers. However, few studies have been dedicated to the influence of dynamic traffic on PDM effect. To this end, two issued have been addressed in the study: one is which kind of hybrid parking policy can work effectively under dynamic traffic and the other is how car travelers with different travel purposes will react to hybrid parking policy.

2 LITERATURE REVIEW

2.1 *Existing research works on parking price policy*

In past years, several studies have been conducted to the influence of parking price policy on travel mode choice behavior (Guan. H.Z. et al. 2006; Shi. F. et al. 2009; Mei. Z.Y. et al. 2010; Su. Q. & Zhou. L. 2012). Peng et al. (1996) conducted a survey on the impact of parking price for travel mode choice of commuters, and established a set of Nested Logit models including three factors: public transit accessibility, work location, and residence location. The study found that parking price has significant impact on commuting travel. That is when parking price rises and the convenience of public transit being improved, the proportion of driving alone will reduce and the ratio of utilizing public transit will increase correspondingly. Still and Simmonds (2000) found that parking price significantly affects the choice on travel mode and destination, and in the long term, it will affect economic activity and land utilization. D'Acierno et al. (2006) set up a parking pricing model in order to increase the proportion of travel mode of public transit, and provide a number of policy advices about optimizing parking meter standards, improving public transit service. Albert and Mahalel (2006) established a Multinomial Logit model to study the impact of parking price and congestion charge on travel mode choices. The results show that travelers are more sensitive to the congestion charge and how the congestion charge influences the choice of travel time; but travel model is mainly affected by parking price. Huanhuan Yin (2007) analyzed the accepting attitude of Beijing residents towards parking price, and the influence of parking price on travel choice behavior, which can be served as a basis for assessment parking price policy and management parking. Yu Ge (2010) established a Multinomial Logit model for the travel behavior during morning peak hours and the results show travel choice behavior vary with the parking price for different travel purpose. A large proportion of car trips will be transfered when improving public transit service and increasing the parking price. Danwen Bao (2010) conducted a research on resident travel mode choice sensitivity to bus fares and parking price changes based on Nanjing parking survey data, and this study could provide a theoretical basis for pricing parking and bus fares. Lv. G.L. & Sun. Z.A (2014) analyzed traffic congestion conditions and existing parking charge in Shenzhen, proposed a general plan for parking charge adjustment according to different zones, types of parking and parking time. Based on the characteristics of car commuting travel, such as long parking time and intensive arrival/departure at parking lots, presented two parking charge schemes. Also point out that parking adjustment fees should be charged for vehicles at off-street parking lots and corresponding profits should go to the municipal budget. Ommeren. & Russo. (2014) according to welfare-maximizing principles, proposed the price of parking must vary per day given shifts in daily demand. Study the economic consequences of not doing so by estimating the employees' parking demand at an organization that varies the price of parking by day of the week. Using a difference-in-differences methodology estimate the effect of the employees' parking price on demand. The deadweight loss of free parking is about 10% of the organization's parking costs.

2.2 Existing research works on travel purposes

There have been a number of studies focused on travel purposes (Habib. & K.M.N. 2012; Liang, X. et al. 2013; Fan. H.B. 2014; Cheng. A.W. 2015). Shiftan and Burd-Eden (2001) analyzed travel intention of the driver by building a Logit model under different parking policies. Their results show that the commuters are more likely to change their traffic mode or traveling time, but the non-commuters could have more choice for their changes and they are also more sensitive to any policy than commuters. Lanhui Liu (2003) established a Logit Model to analyze weekday and holiday travelers parking choice behavior based on the survey data in Xidan area, Beijing. The study found that the impact of

walking distance and parking price on the travel mode choice during the weekdays is less than that during holiday. Kelly and Clinch (2006) built a Probit model based on parking survey data in Dublin, Ireland, and analyzed the parking price effects on commuters and non-commuters, calculated the price sensitive interval, and provided a number of suggestions on the parking pricing. William H.K. Lam (2006) established a time-varying network equilibrium model which analyzed the influence of walking distance, parking capacity, parking fees on different travel purpose of parking demand. And could be utilized to evaluate the effects of improvements in parking facilities and different parking policies. Huanmei Qin (2008) established weekday and holiday travel choice behavior models using disaggregate theory, and found that the most sensitive range of the mode choice for parking rates change. Yaosheng Yong (2008) established weekday and holiday parking length choice models, which provide reference to regulation of parking time and improvement in turnover of parking facilities. Chi Pan (2012) compared commuter and non-commuter parking demand characteristics and parking behavior from multiple dimensions based on the investigation of parking behaviors in Dalian. The result shows that differentiated management measures should be corresponding to different purposes of parking in order to manage static traffic system efficiently. Lu. X.S. & Huang. H.J. et al (2013) based on the bottleneck theory, using a hierarchical Logit model to describe commuters' mode choice behaviors, and then mode choice equilibrium equations under elastic demand are constructed. It is shown that when transit and park-and-ride place are operated by government and the working area parking lot belongs to a private enterprise, lower fares and higher parking fees in working area can effectively encourage parking interchanging. Numerical results also support the current differentiation parking charge policy in Beijing

From the previous research, the role of a single policy function has been widely addressed, but few works had been concentrated to the effect of the hybrid parking policy on the travel mode choice behavior for different travel purpose.

This study aims to develop a disaggregate travel choice behavior model oriented to travel purpose under different traffic states estimated based on Beijing parking RP and SP survey data, and analyze the effects of the hybrid parking policy under different dynamic traffic states.

3 PARKING BEHAVIOR SURVEY

The survey was conducted at several urban parking lots in Beijing, among drivers who used those

Table 1. Variables and level design of SP survey.

Trip purpose	Parking price	Public transit travel time	Traffic states	Possibility of finding the parking space
commute	Unchanged			--
	increase 1 CNY/h*	same with car travel time	jam	
	increase 1.5 CNY/h			
	increase 2 CNY/h	decreased by 20%	free flow	--
	increase 5 CNY/h			
non-commute	Unchanged			
	increase 2 CNY/h	same with car travel time	jam	less possibility
	increase 4 CNY/h			
	increase 6 CNY/h	decreased by 20%	free flow	higher possibility
	increase 10 CNY/h			

* CNY (Chinese Yuan), CNY ¥ 1 = USD $ 0.15.

parking lots. 350 questionnaires were retrieved and the effective samples are 317. The overall proportion was 62.5% of drivers who in those parking lots. The survey contents combined Revealed Preference (RP) and Stated Preference (SP). The questionnaire included three parts: (1) the individual information; (2) parking behavior Revealed Preference; (3) parking behavior Stated Preference.

In view of parking price, public transit travel time and traffic congestion degree are more significant factor to travel mode choice (Daniel Baldwin Hess 2001), therefore, the above three variables are considered in the part of SP questionnaire survey. Furthermore, as non-commuting travel are more sensitive to parking information versus commuting(Lanhui Liu 2003), parking price increment level has increased, and possibility of finding the parking space was added as a variable. The respondents could choose one travel mode among private car, public transit and change travel (change time, change destination or cancel the travel) under 10 factor combinations. Variables and level design of Stated Preference survey are summarized in Table 1.

4 MODEL ESTIMATION AND ANALYSIS

In the analysis of travel behavior, multimode (two or more) competition should be taken into account, Multinomial Logit (MNL) model is a common modeling approach. To study the impact of hybrid parking policies on the parking choice behavior, this study used disaggregate theory to set up MNL model. The structure of MNL model is as follows:

$$P_{in} = \frac{e^{V_{in}}}{\sum_{j=1}^{J_n} e^{V_{jn}}}$$ (1)

where V_{in} is the deterministic component of utility for option i by individual n; P_{in} is the probability for travel mode i by individual n; where J_n is the set of available travel modes.

The deterministic component of the utility can be expressed as a linear function of different influencing factors.

$$V_{in} = \sum_{k=1}^{k} \theta_k X_{ink}$$ (2)

where k is the number of variables, θ_k is the corresponding coefficient to be estimated, and X_{ink} is the variable k when individual n chooses option i.

After computing the test of independence of variables and analyzing relationship between variances by Statistical Product and Service Solutions (SPSS), this study established travel choice model using the most significant influential variables. The commute travel choice model includes four factors: the added parking price, car travel time, public transit travel time, parking duration and so on. On the other hand, the Non-Commute model has one more variable—possibility of finding the parking space. These model was calibrated based on the survey data in Beijing city by TransCAD. The results are shown in Table 2.

From Table 2 all variables coefficients obtained theoretically signs, the parameters of the t test absolute value is greater than 1.65, which indicate that variables for different purpose of travel choice behavior have significant impacts within 90% confidence level.

The added parking price coefficient is negative, indicating that when parking price increased, travelers are less willing to drive. The car travel time coefficient is negative, showing that as driving time increases, the tendency to drive is decreased. However, the public transit travel time coefficient is negative, indicating that when

Table 2. Model calibration results.

Variable name	Commute				Non-Commute			
	free flow		jam		free flow		jam	
	coefficient	t test	coefficient	t test	coefficient	t test	coefficient	t test
car travel mode constant	2.234	13.277	3.370	14.183	4.017	11.076	2.600	11.818
public transit travel mode constant	1.343	7.502	3.024	11.625	1.198	8.465	2.253	5.584
Added parking price	−0.300	−10.378	−0.670	−8.151	−0.391	−11.559	−0.636	−1.913
car travel time	−0.059	−7.911	−0.103	−10.214	−0.047	−6.307	−0.063	−7.047
public transit travel time	−0.065	−8.024	−0.111	−10.421	−0.057	−7.013	−0.153	−7.589
parking duration	−0.005	−4.456	0.003	3.950	−0.003	−5.009	0.002	3.087
possibility of finding the parking space	--	--	--	--	−2.114	−8.482	−1.064	−7.197
L(0)	−2086.029		−1391.418		−2085.741		−1434.669	
L(θ)	−1569.791		−947.625		−1687.199		−1153.442	
−2(L(0)-L(θ))	1032.476		887.585		797.085		562.454	
ρ^2	0.247		0.319		0.191		0.196	
$\bar{\rho}^2$	0.245		0.315		0.188		0.191	

the longer travel time is, fewer travelers tend to take public transit. Parking duration coefficient in traffic congest model is positive, but in traffic clear model the coefficient is negative, indicating that drivers tend to change the travel time under the congest state for a long time parking, not vice versa. For non-commute travel, possibility of finding the parking space coefficient is negative, indicating that with the possibility decreased, the proportion of non-commute travelers who still drive is decreased.

ρ^2 and $\bar{\rho}^2$ are two indexes of goodness-of-fit for model evaluation. Generally, the model accuracy is high when ρ^2 and $\bar{\rho}^2$ are both between 0.2 and 0.4. Table.2 shows that ρ^2 and $\bar{\rho}^2$ for the models are between 0.19 and 0.32, which show that these models are reliable.

4.1 Model sensitivity analysis to added parking price

According to aggregate analysis theory, mean value method is used and variables mean value Xin is applied into the formula (3), to solve the probability of traveling choice:

$$S_i = P(i|\overline{X}, \theta) \qquad (3)$$

Based on the results of the established model, the proportional changes of car travel with the increase of parking price can be calculated. The scenarios are shown in Table 3 and the results are illustrated in Figure 1.

Table 3. Explanation of hypothesis.

Travel purposes	Dynamic traffic operating state	Symbol
Commute	free flow	P1
	jam	P2
Non-Commute	free flow	P3
	jam	P4

As shown in the Figure 1 dynamic traffic state has significant effects on the behavior of travel mode choice. The patterns of commuting travels and non-commuting travels are similar to each other. With the added parking price increasing, the proportion of car travelers is decreased. And, the change can significantly influence the travel mode choice when the value varies from 0~8 CNY/h. However, the effect is not obvious when it is more than 10 CNY/h.

Traffic state has significant effect on the behavior of travel mode choice. Under the same parking price, the proportion of give up driving under jam conditions is larger than that under free flow conditions. In other words, the role of parking prices to adjust the travel choice under jam conditions is greater than under free flow conditions, and drivers are more likely to change travel choice

To analyze the sensitiveness of the travel choice to added parking price, the proportion that people still use cars for travel is calculated for different variation of added parking price. The symbol is shown in Table 4 and the results are shown in the Figure 2.

1032

As shown in the Figure 2 both the distributions of commuting travel and non-commuting ones are similar, which means, the sensitivity of car travelers increases as the parking price rises before reaching the peak point, and then decreases. The non-commuters are more sensitive to parking price rises than commuters.

The price policy can work well in the corresponding ranges. For commuters, the sensitive ranges are from 0 to 8 CNY/h under the jam and from 6 to 10 CNY/h under free flow; for non-commuters, the ranges are: from 0 to 6 CNY/h and from 4 to 10 CNY/h under for the both cases, respectively.

4.2 Model sensitivity analysis to public transit travel time

Based on the results of the established model, the proportional changes of car travel with the public transit travel time influence can be calculated. The scenarios are listed in Table 5 and the corresponding results are shown in Figure 3.

As shown in the Figure 3 commuters are more sensitive to public transit travel time than the non-commuters, and sensitiveness gradually increases with decrease of public transit travel time. For the non-commuters, the decrease in public transit travel time has little influence. It is not obvious to change their mode choice when the public transit travel time less than 40% drop.

For the non-commuters, car travel drops by about 25% when added parking price is 2 CNY/h. Compared to no-growth, the sensitivity of traveling choice to public transit travel time changes little. The proportion of commuter car travel will decrease by about 15%, when added parking price is 2 CNY/h. Compare to remain unchanged, the sensitiveness to public transit travel time will increase.

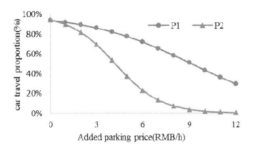 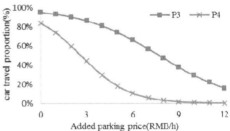

Figure 1. Proportion of choosing private car with change in added parking price.

Table 4. Explanation of sensitivity analysis.

Symbol	The change of added parking price (CNY/h)
1	0–2
2	2–4
3	4–6
4	6–8
5	8–10
6	10–12
7	12–14

Table 5. Explanation of hypothesis.

Travel purposes	Dynamic traffic operating state	Added parking price (CNY/h)	symbol
Commute	jam	0	P1
		2	P2
Non-Commute		0	P3
		2	P4

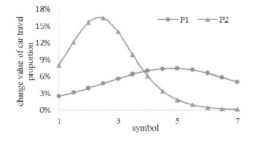 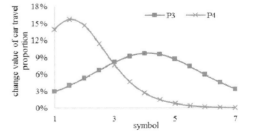

Figure 2. Sensitivity analysis of the model.

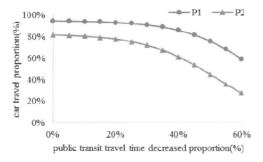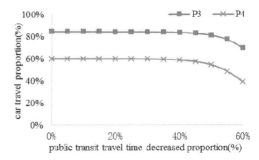

Figure 3. Proportion of choosing private car with change in public transit travel time decreased proportion.

5 CONCLUSIONS

In this paper, from the point of view of parking behavior Revealed Preference and Stated Preference, conducted the questionnaire survey at several urban parking lots in Beijing. Based on the survey data, select the variables which have significant effects on the travel choice, used disaggregate theory developed multiple-factor commute and non-commute travel choice behavior model, by considering the factors of parking price, public transit travel time, and parking supply etc. Calibrated these model by TransCAD, get all variables coefficients, according to aggregate analysis theory, using mean value method analyzed commuter and non-commuter mode sensitivity to parking price and public transit travel time, and discussed the effects of the hybrid policy under different traffic states. Based on the survey, the followings can be concluded.

There is difference of travel mode choice between commuters and non-commuters, it has not the same degree of sensitivity to parking price and public transit travel time etc. Overall, the non-commuters are more sensitive to parking price than commuters; commuters are more sensitive to public transit travel time than the non-commuters.

Traffic state has significant effect on the behavior of travel mode choice. Jam conditions compared to free flow conditions, drivers are more likely to change travel choice, the role of parking prices adjust travel choice enhance, the parking price policy will be more effective to play.

Therefore, it should be make differentiated parking policy mix to manage different travel purposes demand under different traffic state.

Parking price have a significant impact on effects on the non-commute travel mode choice. Improvement of public transit travel time played little effect on it. Therefore, the most effective PDM policy mix for business district is: control the supply of berths, parking price growth range is from 4 to 10 CNY/h in clear zone, and from 0 to 6 CNY/h in congestion zone.

The adoptability of PDM policies varies in place and time, the research findings will provide a reference for making zoning or timing policies more precisely, and achieving the parking management objective of "balance the dynamic with the static".

ACKNOWLEDGMENT

This thesis is supported by the National Natural Science Foundation of China (51308018).

REFERENCES

Albert & Mahalel. 2006. Congestion Tolls and Parking Fees. A Comparison of the Potential Effect on Travel Behavior. *Transport Policy*. 13: 496–502.

Bao. D.W. 2010. The impact analysis of parking fees on travel mode choice. *Transportation systems engineering and information*. (6):80–85.

Cheng. A.W. 2015. Analysis on the Parking Behavior of Central Business Districts-a Case Study of Chengdu. Southwest Jiaotong University.

D'Acierno et al. 2006. Optimization models for the urban parking pricing problem, *Transport Policy*, 13:34–48.

Daniel Baldwin Hess. 2001.The Effect of Free Parking on Commuter Mode Choice. Evidence from Travel Diary Data. *Transportation Research Record*. 35–42.

Fan. H.B. 2014. Research on Transfer Model of Car-commuter Trip Based on Nested Logit. Beijing Jiaotong University.

Ge. Y. 2010. The influence of the parking price on the travel behavior in city center area. *Mathematics in Practice and Theory*, (2):41–47.

Guan. H.Z, Li. Y, Qin. H.M. 2006. Analysis on the TDM-based Trip-mode Regulation Survey in Megapolis Downtown. *Journal of Beijing University of Technology*. 4: 338–342.

Habib & K.M.N. 2012. Modeling commuting mode choice jointly with work start time and work duration. *Transportation Research Part A:Policy and Practice*.46(1):33–47.

Kelly. & Clinch. 2006. Influence of varied pricing tariffs on parking occupancy levels by trip purpose. *Transport Policy*. 13 (6):487–495.

Liang. X et al. 2013. Influencing Factors of Parking Choice Behavior in Central Business Area. *Journal of Transport Information and Safety*.31:27–30.

Liu. L.H. 2003. Study on Parking Behavior in Big City's Shopping Center. Beijing University of Technology.

Lu. X.S. & Huang. H.J. et al. 2013. Mode choice equilibrium and pricing mechanisms considering peak trip chain. *Systems Engineering-Theory&Practice*.33: 167–174.

Lv. G.L. & Sun. Z.A. 2014. Parking Charge Adjustment in Shenzhen. *Urban Transport of China*.12: 12–17.

Mei. Z.Y, Xiang. Y.Q, Chen. J et al. 2010. Optimizing model of curb parking pricing based on parking choice behavior.*Journal of Transportation Systems Engineering and Information Technology*.10(1): 99–104.

Ommeren. & Russo. 2014. Time-varying parking prices. *Economics of Transportation* 3: 166–174.

Pan. C. 2012. Variation Analysis of Parking Behavior with Different Travel Purposes. *Traffic information and security*. (1):39–42.

Peng Z, Dueker K.J., Strathman J.G. 1996. Residential location, employment location, and commuter responses to parking charges. *Transportation Research Record*, 1556: 109–118.

Qin. H.M. 2008. A Study of the effect of parking price on the mode of inhabitant trip behavior-with the cars, public transit and taxi in Beijing as an example. *China Civil Engineering Journal*. (8):93–98.

Shi. F. et al. 2009. Class of Comprehensive Optimization of Congested Road-Use Pricing and Parking Pricing. *System Engineering and Information Technology*. 9(1): 74–79.

Shiftan. 1999. Responses to parking restrictions: lessons from a stated preference survey in Haifa and their policy implications. *World Transport Policy and Practice* 5(4), 30–35.

Shiftan. 2002. The effects of parking pricing and supply on travel patterns to a major business district. In: Stern, E., Salomon, I., Bovy, P.H.L. (Eds.), *Travel Behavior*. Edward Elgar, Cheltenham.

Shiftan. & Burd-Eden. 2001. Modeling response to parking policy. *Transportation Research Record*, (1765):27–34.

Still. & Simmonds. 2000. Parking Restraint Policy and Urban Vitality. *Transport Reviews*. 20(3):291–316.

Su. Q. & Zhou. L. 2012. Parking management, financial subsidies to alternatives to drive alone and commute mode choice in Seattle.*Regional Science and Urban Economics*. 42(1–2): 88–97.

William H.K. Lam. 2006. Modeling time-dependent travel choice problems in road networks with multiple user classes and multiple parking facilities. *Transportation Research Part B*. (40): 368–395.

Yao. S.Y. 2008. Research on Relationship Between Parking-charge and Parking Behavior in CBD. *Journal of Hebei University of Technology* (10):110–114.

Yin. H.H, Qin. H.M, Guan. H.Z. 2007. Beijing residential parking survey. *Urban Transport of China*. 5(6):49–53.

Transport Infrastructure and Systems – Dell'Acqua & Wegman (Eds)
© *2017 Taylor & Francis Group, London, ISBN 978-1-138-03009-1*

Classification and recognition model for the severity of road traffic accidents

Jianfeng Xi, Lincan Wang, Tongqiang Ding & Weifu Sun
College of Traffic, Jilin University, Changchun City, P.R. China

ABSTRACT: In order to improve the classification accuracy and classification speed on the severity of road traffic accidents, the work proposed the method of classification and identification based on the combination of rough set theory and Support Vector Machines (SVM). Rough set theory was used to calculate the significant degree of each attribute in "people, vehicles, roads and environment". Based on the sorting of significant degree, the influencing factors of accident severity were dealt with feature extraction. Then the influencing factors with high significant degree were inputted into SVM. Taking general accidents and severe accidents as two classification labels, the work established the classification and identification model of the severity of road traffic accidents. At last, the validity of the method was analyzed and verified in the case of statistical data of road traffic accidents in one region of China. The results showed that the classification and identification model of the severity of road traffic accidents, based on rough set theory and SVM, can improve the accuracy of recognition and reduce the calculation work load of pattern identification. It has good classification and recognition as well as good generalization.

1 INTRODUCTION

Recently, the number of traffic accident casualties remains high with the increase of road mileage, vehicle population and motorization. Therefore, it is necessary to focus on the influencing factors of the severity of road traffic accidents as well as the classification and recognition model of accident severity.

Researchers at home and abroad have tried different analysis models of traffic accident severity from different perspectives. The most widely used is discrete choice model based on Logit/Probit models (Savolainen et al. 2011, Chang & Yeh 2006): Bedard et al. (2002) used multiple Logit model to study the influence of drivers, collision types and vehicles on severity accidents. Yau (2004), with the use of Logit model, studied the influence of region characteristics, drivers, vehicles, environment, etc. on the severity of vehicle accidents. Yao et al. (2014) proposed a prediction method, consisted of support vector machine and tabu search, to detect freeway incidents. Ma et al. (2011) used Logit model to study the influencing factors of the severity of road tunnel accidents. Meanwhile, they studied road safety evaluation based on accident severity by using fuzzy Delphi method. It is noted that most of the empirical data sets of classical statistical models (Savolainen et al. 2011) cannot satisfy the strict assumed conditions of discrete choice model, thus resulting in biased estimation of parameters. Thus, many scholars have applied some intelligent classification models (e.g., neural network model (Delen et al. 2006), Bayesian model (Juan et al. 2011) and decision tree model) for analysis modeling of accident severity. The results showed that intelligent classification model has higher classification accuracy and generalization (Xie et al. 2007, Yu et al. 2013, Yao et al. 2015).

2 RESEARCH METHODS

2.1 *Rough set theory*

Rough set theory, proposed by Polish mathematician Z Pawlak (Pawlak & Skowron 2007, Pawlak 1982, Pawlak 1984) in 1982, is a new mathematical tool to analyze incomplete, uncertain data. It includes the theories for the calculation of the minimum invariant set (kernel) and the solution for the minimum rule set (reduction) in large amounts of data (Yu et al. 2013). Without the need of additional information and prior knowledge, the existing information can be used to recognize the classification and significant degree of objects in specific conditions. Using rough set theory for decision analysis is to obtain useful information from the given data sets and establish rough set model in the original data. A reasonable decision scheme can be made through data compression, reduction, importance of attribute, etc (Kwak & Choi 2002, Zhao & Wang 2002).

To analyze the influencing factors of the accident severity with the use of rough set algorithm, it mainly includes: extraction and conversion of traffic accident information, selection of subsample groups, calculation of correlation degree (significant degree) between four factors ("people, vehicles, roads and environment") and traffic accident consequence based on subsample groups, validation of calculation results using test, and calculation results of the given significant degree (Xi et al. 2009). The process of the algorithm is shown as Figure 1.

2.2 Support Vector Machine (SVM)

SVM, the supervised classification algorithm based on VC dimension theory of statistics and structural risk minimization theory, can solve the problems such as Support Vector Classification (SVC), *Support Vector Classification* (SVR) and distribution estimation. Among the classification models and methods, SVM, according to limited information of samples, can find the best compromise between the complexity of models and learning ability. It is a powerful tool in the field of classification, which can solve practical problems (e.g., small samples, nonlinearity and high dimensionality) and overcome the disadvantage that neural networks are easy to fall into local optimal solution (Yuan & Cheu 2003, Zhang 2008).

From the perspective of data mining, the severity of road traffic accidents can be regarded as the classification of SVM (Zhang & Wang 2008). The core idea is to map input vectors to a high-dimensional space where the optimal hyperplane is established through pre-selected non-linear mapping (kernel function) (Yao et al. 2014).

For the training set T of linear classification,

$$T = \{(x_1, y_1), (x_2, y_2), (x_3, y_3), \ldots, (x_m, y_m)\} \in (X \times Y)^m \tag{1}$$

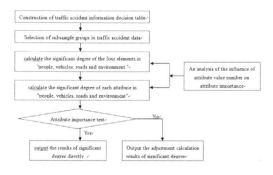

Figure 1. Process of mining algorithm of traffic accidents based on rough set theory.

where $x_i \in X = R^n, y_i \in Y = \{-1, +1\}, i = 1, 2, 3, \ldots, m$. There should be a hyperplane H in the linear classification. It satisfies $(w \cdot x_i) + b = 0$ where $w \in R^n, b \in R$.

Based on the specification in the form of hyperplane, its necessary and sufficient conditions and the maximum margin principle, the followed quadratic programming problem can be solved by solving the optimization of the sample set:

$$\min_{w,b} \quad \tau(w) = \frac{1}{2} \| w \|^2$$
$$s.t. \quad y_i((w \cdot x_i) + b) \geq 1, i = 1, 2, \ldots, m \tag{2}$$

Its Language function can be defined as:

$$L = \frac{1}{2} \| w \|^2 - \sum_i a_i \{ y_i [(w \cdot x_i) + b] - 1 \} \tag{3}$$

Its dual problem is (Sanchez A 2003)

$$\max d(a) = \sum_{i=1}^{m} a_i - \frac{1}{2} \sum_{i=1}^{m} \sum_{j=1}^{m} y_i y_j a_i a_j (x_i, x_j)$$
$$s.t. \quad a_i \geq 0 \forall i \tag{4}$$
$$\sum_{i=1}^{m} y_i a_i = 0$$

We solve the optimal a_i^* in the equation and classification threshold b^* which can be obtained by any support vector.

Thus, the optimal classification function is

$$f(x) = \text{sgn} \{(w, x) + b\}$$
$$= \text{sgn} \left\{ \sum_{i=1}^{n} a_i^* y_i (x_i \cdot x) + b^* \right\} \tag{5}$$

where n is the number of support vectors.

In the case of linear inseparable, we can introduce non-negative slack variable $\xi_i \geq 0 (i = 1, 2, \ldots, m)$ and penalty factor $C > 0$ to "soften" the constraints and penalize the case that ξ_i is with large value in the objective function. Then, the optimization problem becomes

$$\min_{w,b,\xi} \quad \frac{1}{2} \| w \|^2 + \sum_{i=1}^{m} \xi_i$$
$$s.t. \quad y_i((w \cdot x_i) + b) \geq 1 - \xi_i \quad i = 1, 2, \ldots, m \tag{6}$$

For separable nonlinear problem, the kernel function $K(x_i, x_j)$ realizing linear mapping should be introduced to Eq. (4). Different kernel function $K(x_i, x_j)$ can constitute different SVMs, and the general form of its decision function is

$$f(x) = \text{sgn}\left\{ \sum_{i=1}^{m} a_i^* y_i K(x_i \cdot x) + b^* \right\} \quad (7)$$

Gaussian Radial Basis Function (RBF) is selected as follows:

$$K(x_i, x_j) = \exp[-|x - x_i^2| / \sigma^2] \quad (8)$$

3 CLASSIFICATION AND RECOGNITION MODEL OF SEVERITY CLASSIFICATION OF ROAD TRAFFIC ACCIDENTS

Severity classification model of accidents, based on SVM, mainly includes two stages—1) Training stage: samples are established based on input and output for SVM training; 2) recognition stage: based on the output of unknown samples, the SVM can obtain recognition output by their input.

Given l groups of traffic samples, the influencing factor of the severity of road traffic accidents is X, and $X = \{x_1, x_2, x_3, \ldots, x_l\}$; the accident severity is Y. The work combines rough set theory and SVM to recognize severity classification of road traffic accidents, and the whole modeling process includes the following steps:

1. Selection of input and output variables
Analyze the characteristics of traffic accident data, and establish information decision table based on rough set theory.

Determination of input variables: Apply rough set theory to calculate the significant degree of four elements (people, vehicles, roads and environment) and their attributes. Select the important variables with the significant impact on the severity of road traffic accidents as eigenvectors.

Determination of output variables: The severity of road traffic accidents is selected as output variables of the model, and the classification labels of the accident severity are defined as two types—general accidents and severe accidents.

2. Selection of study samples
It is the basis for SVR classification model to select appropriate study samples which are divided into training and test ones. Training samples are used to study and train, and test samples to verify the validity of the model. The work selected the statistics data of road traffic accidents and related influencing factors as study samples.

3. Selection of kernel functions
Selection of kernel functions has big influence on the accuracy of the classification model of severity of SVR road traffic accidents. However, the theory of kernel function selection is not perfect. It is believed that radial basis function kernel is better than other when dealing with non-linear samples (Yu et al. 2015). So the work selected Radial Basis Function (RBF) kernel $K(x_i, x_j) = \exp[-|x - x_i^2| / \sigma^2]$ as the kernel function of the classification model.

4. Selection of structure parameters
RBF is an important kernel function, but the selections of kernel parameter σ and penalty parameter C have big influence on the regression accuracy of SVM. It is a heavy work to determine the parameters by trying the methods. As an intelligent optimization algorithm, genetic algorithm is independent of whether the search space is continuous, or differentiable restrictions, with a wide range of applications. Therefore, genetic algorithm was used in the work for optimization of kernel parameter σ and penalty parameter C.

5. Model Training
Through the data of training samples, take RBF as kernel function. According to the selected structure parameters in Step (4), apply model training. Then substitute training model into Step (6), and training results into Step (7).

6. Test prediction
Based on Step (5), use the trained SVR model for the prediction of test samples, and substitute predicted results into Step (7).

7. Training accuracy test
After modeling, select the composite classification accuracy based on Cross Validation (CV) to evaluate modeling accuracy.

8. Output training and test results.

4 EMPIRICAL STUDY

4.1 Characteristic analysis of dataset

The work selected accident data of one region in China from 2006–2013 as the object of empirical study. There were 4320 accident information from database system in all, including 2042 property damage accidents, 1922 injury accidents and 356 fatal accidents. Characteristic variables of influencing factors of road traffic accidents were from "Accident Basic Information" dataset.

4.1.1 Establishment of decision table of traffic accident information
Traffic accident data presents super-spatial structure which is radial, multi-dimensional and multi-level. According to rough set theory, its sample matrix can be regarded as a knowledge representation system—the knowledge representation system of traffic accidents. Therefore, in the sample matrix of traffic accident data, each column corresponds to one attribute of rough set, and each element to one

attribute value of corresponding attribute in its column. Based on different types of attributes, attributes can be divided into condition attributes and decision attributes, thus establishing the decision table of traffic accident information (Yu et al. 2015, Yao et al. 2015, Skowron & Rauszer 1992). The decision table of traffic accident information is shown in Table 1.

There are a total of 56 attributes in the accident database. From the influencing factors of the severity of traffic accidents, there are only 3 ones for decision attributes—the number of deaths, the number of injuries and property damages; the rest 53 belong to condition attributes. Due to its complexity, contribution attributes were combined and classified. The work used stratified analytic method for initial dimensionality reduction, focusing on the establishment of personnel information attribute sets, vehicle information attribute sets, road information attribute sets and environment information attribute sets.

4.1.2 Calculation of Significant degree of each attribute in "people, vehicle, roads and environment"

Four decision sub-tables successively corresponding to "people, vehicle, road and ring" were taken as new study objects. The sample size of each decision sub-table was equal to that of the decision table, but the number with condition attribute is less than that of the decision table. To obtain more stable and accurate calculations and t-test, 10 subsample groups were selected from each decision sub-table. Each subsample group was dealt with rough set algorithm to obtain the significant degree of each condition attribute, compared with decision attribute, in the decision sub-table corresponding to each subsample group. Eventually, 10 calculation results of

significant degree were obtained for any condition attributes. The mean value of 10 calculation results were regarded as the significant degree of the condition attribution compared with decision attribute. The calculation equation is as follows:

$$r_{a_{ij}} = \frac{\sum_{k=1}^{10} r_{c_i}^k - |pos^k_{c_i - a_{ij}}(D)|/|U|}{10} \quad (9)$$

where a_{ij} is the j-th attribute in attribute set c_i;
$r_{a_{ij}}$ the significant degree of the attribute a_{ij};
$pos^k_{c_i - a_{ij}}(D)$ the number of elements in $c_i - a_{ij}$ positive region of D;
$k = 1, 2 \ldots\ldots 10$, the ordinal number of subsample group.

Taking the traffic accident data of 4,320 roads in one region of China as analysis object, the work selected the severity of traffic accidents as decision attribute of all decision sub-tables. Based on the significant degree of each condition attribute compared with the severity of traffic accidents, 30 influencing factors were selected from 53 condition attributes as input variables of SVM classification model. The specific analysis results are shown in Tables 2–4.

4.2 SVM modeling and calculation

Based on the comparison of rough set theory to the significant degree of attributes, the work focused on the establishment of independent variable set (including driver attributes, vehicle attributes, driving environment attributes, road attributes and accident attributes), with 30 variables. The attribute variables definition is shown as Table 5.

Table 1. Decision table of traffic accident information.

	Condition attribute							Decision attribute
Sample number	Light condition	Visibility	Gender	Age	Personnel type	Traffic mode	...	Fatal
1	1	4	1	4	99	7	...	1
2	1	4	1	4	99	2	...	0
3	1	4	1	3	13	1	...	2
4	1	4	1	5	52	18	...	2
5	1	3	1	4	14	1	...	1
6	1	3	1	5	14	1	...	0
7	2	1	1	3	16	1	...	0
8	2	1	2	5	16	18	...	3
9	1	4	2	3	14	1	...	1
...	...	...	...	...	...	...	...	...

Table 2. Significant degree of human's attributes to traffic accident severity.

	Gender	age	Residence type	Personnel type	Traffic mode	...	Driving years	Fault behavior
Significant degree	0.0254	0.1435	0.0920	0.1429	0.0836	...	0.1323	0.1990

Table 3. Significant degree of vehicle's attributes to traffic accident severity.

	Plate number type	Load volume	Overload volume	Third party insurance	Vehicle legal condition	...	Vehicle safety condition	Vehicle using character
Significant degree	0.0424	0.0314	0.0176	0.0278	0. 0324	...	0.0300	0.0472

Table 4. Significant degree of road's attributes to traffic accident severity.

	Rodeside protection	Highway classification	Road separation	Road surface condition	Road surface type	Junction type	...	Road alignment	Road type
Significant degree	0.0722	0.0604	0.0505	0.0247	0.0532	0.0605	...	0.0568	0.0242

Table 5. Attribute variables definition.

Variables type	Variables	Variable declaration
Driver Attributes	Gender	0 (Male); 1 (Female)
	Age	0 (16–25); 1 (26–35); 2 (36–45); 3 (46+)
	Driving Years	0 (1–5); 1 (6–10); 2 (11–15); 3 (16–20); 4 (20+)
	Residence Type	0 (Agricultural Residence); 1 (Non Agricultural Residence)
	Accident Liability	0 (Secondary Liability); 1 (Equal Liability); 2 (Primary Liability); 3 (Full Liability)
	Traffic mode	0 (Non-Motorized Vehicle); 1 (Van); 2 (Coach); 3 (Others)
Vehicle Attributes	Driving License Type	0 (Motor Vehicle); 1 (No License); 2 (Others)
	Number Plate Type	0 (Motor Cycle); 1 (Cars); 2 (No Number Plate); 3 (Others)
	Vehicle Legal Condition	0 (Legal); 1 (Illegal)
	Vehicle Safety Condition	0 (Safe); 1 (Unsafe)
	Vehicle Application	0 (Passenger Transport); 1 (Freight Transport); 2 (Others)
	Overload	0 (Yes); 1 (No)
	Third Party Insurance	0 (Insurance); 1 (No Insurance)
	Driving Condition	0 (Going Straight); 1 (Turn); 2 (Turning Around); 3 (Others)
Driving Environment Attributes	Weather	0 (Clear day); 1 (Others such as rain, snow, fog, wind)
	Visibility	0 (200 Meters +); 1 (100–200 Meters); 2 (50–100 Meters); 3 (50 Meters & below)
	Light Condition	0 (Daylight); 1 (Dark with Streetlight); 2 (Dark without Streetlight)
	Traffic Signal	0 (Yes); 1 (No)
Road Attributes	Road Cross Section	0 (Motorway); 1 (Non-motorized transport); 2 (Mixed Road); 3 (Others)
	Road Type	0 (Urban Road); 1 (Highway); 2 (Others)
	Road Alignment	0 (Straightness); 1 (Others)
	Road Physical Separation	0 (Separation); 1 (No Separation)
	Road Condition	0 (In Good Condition); 1 (Broken)
	Road Surface Condition	0 (Dry); 1 (Slippery)
	Junction Type	0 (Usual one); 1 (Intersection); 2 (Others)
	Roadside Safeguard Type	0 (NO Safeguard); 1 (Safeguard)
	Road Surface Type	0 (Asphalt); 1 (Others)
Accident Attribute	Accident Type	0 (Frontal Impact); 1 (Side Impact); 2 (Rear-end Collision); 3 (Others)
	Accident Time	0 (6:00–18:00); 1 (18:00–6:00)
	Month	0 (Spring & Autumn); 1 (Summer & Winter)

Table 6. Optimum contrast.

	SVM	Rough Set+SVM
SVs	56	30
Accuracy (%)	88.312	94.403

With the Matlab software platform, the work used SVM package for SVM modeling of empirical data and kernel parameter optimization; the sequence minimization algorithm was also used for the solution. Considering the observed frequency of fatal accidents was only 356, the work defined two classification labels for accident severity—general accidents (2,042 events, accounting for 47.3%) and severe accidents (2,278 events, accounting for 52.7%) which included injuries (1,922 events) and fatalities (356 events).

10% of the total samples were selected as training set in stratified random sampling, and the rest 90% as test set. The bend number of Cross Validation (CV) is 5 in the training process. In the process of kernel parameter optimization, the genetic algorithm set evolutional generation as 100, and the population as 20. After the calculation with Matlab, it could be found that optimal combination of corresponding parameters was $C = 1$, and $\sigma = 0.7$.

Based on the parameters from optimization, the work established prediction model where test sample data was substituted for calculation. In addition, the work used SVM to test the classification model of accident severity instead of rough set theory. Table 6 shows different effects.

5 CONCLUSIONS

Taking the data of road traffic accidents in one region of China as the sample, the work proposed the classification and recognition model of the severity of road traffic accidents based on rough set theory and SVM. The main tasks were as follows: 1) After analyzing sample data, we established the decision table of traffic accident information. 2) Through rough set theory, we obtained the significant degree of each attribute in "people, vehicles, roads and environment". After sorting based on the significant degree, variables were restructured and simplified to improve the scientific city of input parameters in the classification model as well as the accuracy of the model. 3) According to two classification labels—general accidents and sever accidents, SVM was used to establish the classification model of the accident severity. The availability of the model was verified through examples.

The results showed that based on rough set theory and SVM, the combined classification compared with a single classification method has high fitting ability and prediction accuracy. Meanwhile, it can improve the calculation speed under the condition of large data quantity, thus better realizing the identification and prediction of the severity of traffic accidents in regions or road sections.

REFERENCES

Bédard, M., Guyatt, G.H., Stones, M.J. & Hirdes, J.P. 2002. The independent contribution of driver, crash, and vehicle characteristics to driver fatalities. *Accident Analysis & Prevention* 34(6): 717–727.

Chang, H.L. & Yeh, T.H. 2006. Risk factors to driver fatalities in single-vehicle crashes: comparisons between non-motorcycle drivers and motorcyclists. *Journal of Transportation Engineering*, 132(3): 227–236.

Delen, D., Sharda, R. & Bessonov, M. 2006. Identifying significant predictors of injury severity in traffic accidents using a series of artificial neural networks. *Accident Analysis & Prevention* 38(3): 434–444.

Juan, D.O., Randa Oqab, M. & Calvo, F.J. 2011. Analysis of traffic accident injury severity on spanish rural highways using bayesian networks. *Accident Analysis & Prevention* 43(1): 402–411.

Kwak, N. & Choi, C.H. 2002. Input feature selection by mutual information based on Parzen window. *IEEE transactions on pattern analysis and machine intelligence* 24(12): 1667–1671.

Ma, Z., Shao, C., Ma, S. & Ye, Z. 2011. Constructing road safety performance indicators using fuzzy Delphi method and grey Delphi method. *Expert Systems with Applications* 38(3): 1509–1514.

Pawlak, Z. 1982. Rough sets. *International Journal of Information and Computer Sciences* 11: 341–356.

Pawlak, Z. 1984. Rough classification. *International Journal of Man-Machine Studies* 20(84): 469–483.

Pawlak, Z. & Skowron, A. 2007. Rudiments of rough sets. *Information Sciences* 177(1): 3–27.

Sánchez A., V.D. 2003. Advanced support vector machines and kernel methods. *Neurocomputing* 55(03): 5–20.

Savolainen, P.T., Mannering, F.L., Lord, D. & Quddus, M.A. 2011. The statistical analysis of highway crash-injury severities: a review and assessment of methodological alternatives. *Accident Analysis & Prevention* 43(5): 1666–1676.

Skowron, A. & Rauszer, C. 1992. The discernibility matrices and functions in information systems. *Theory & Decision Library* 11: 331–362.

Xi, J.F., Wang, X.Y. & Wang, S.W. 2009. Analysis method based on rough set theory for road traffic accident. *Journal of Changchun University of Science and Technology (Science and Technology)* 32(2): 257–259.

Xie, Y.C., Lord, D. & Zhang, Y.L. 2007. Predicting motor vehicle collisions using bayesian neural network models: an empirical analysis. *Accident Analysis & Prevention* 39(5): 922–933.

Yao, B.Z., Hu, P., Zhang, M. & Jin, M. 2014. A support vector machine with the tabu search algorithm for freeway incident detection. *International Journal of Applied Mathematics & Computer Science* 24(2): 397–404.

Yao, B.Z., Yao, J.B., Zhang, M.H. & Yu, L. 2014. Improved support vector machine regression in multi-step-ahead prediction for rock displacement surrounding a tunnel. *Scientia iranica* 21(4):1309–1316.

Yao, B.Z., Wang, Z., Zhang, M.H., Hu, P. & Yan, X.X. 2015. Hybrid Model for Prediction of Real-Time Traffic Flow. *Proceedings of the Institution of Civil Engineers –Transport* 169(2):88–96.

Yao, B.Z., Yu, B., Hu, P., Gao, J.J. & Zhang, M.H. 2015. An improved particle swarm optimization for carton heterogeneous vehicle routing problem with a collection depot. *Annals of Operations Research* 242(2):1–18.

Yau, K.K.W. 2004. Risk factors affecting the severity of single vehicle traffic accidents in Hong Kong. *Accident Analysis & Prevention* 36(3): 333–340.

Yu, B., Li, T., Bao, H.L., Li, Y.H. & Yao, B.Z. 2013. Real-time stop-skipping strategy for bus operations at a terminal. *Road & Transport Research* 22(1): 26–38.

Yu, B., Kong, L., Sun, Y., Yao, B.Z. & Gao, Z.J. 2015. A bi-level programming for bus lane network design. *Transportation Research Part C Emerging Technologies* 55: 310–327.

Yu, B., Song, X.L., Yang, Z.M. & Yao, B.Z. 2015. k-Nearest Neighbor Model for Multiple-Time-Step Prediction of Short-Term Traffic Condition. *Journal of Transportation Engineering-ASCE 2015* 142(6): 04016018.

Yu, B., Zhu, H.B., Cai, W.J., Ma, N., Kuang, Q. & Yao, B.Z. 2013. Two-phase optimization approach to transit hub location—the case of Dalian. *Journal of Transport Geography* 33(4): 62–71.

Yuan, F. & Cheu, R.L. 2003. Incident detection using support vector machines. *Transportation Research Part C Emerging Technologies* 11(03): 309–328.

Zhang, X.G. 2000. Introduction to statistical learning theory and support vector machines. *Acta Automatica Sinica* 26(1): 32–41.

Zhang, J.H. & Wang, Y.Y. 2008. A rough margin based support vector machine. *Information Sciences* 178(9): 2204–2214.

Zhao, K. & Wang, J. 2002. A reduction algorithm meeting users' requirements. *Journal of Computer Science and Technology* 17(5): 578–593.

Transport Infrastructure and Systems – Dell'Acqua & Wegman (Eds)
© 2017 Taylor & Francis Group, London, ISBN 978-1-138-03009-1

Research on the method of multi-resource traffic data fusion for the calculation of urban road traffic operation index

Song Zeng
Shenzhen Qianhai Cloudist Technology and Services Co. Ltd., Shenzhen, Guangdong Province, China

Huaikun Xiang
Shenzhen Polytechnic, Shenzhen, Guangdong Province, China

ABSTRACT: At present, the computation of the urban road traffic operation index is mainly based on the GPS data of floating cars in China. Because the number of the floating cars is not enough to cover the whole road network efficiently and the GPS data is affected by the positioning error, the road traffic operation index based on the GPS data only of the floating cars is not sufficiently accurate, which inevitably influences the validity and authority of the index. The key aim of the multi-resource traffic data fusion is to provide online, dynamic and accurate data for the computation of the index. In this paper, a segment of an urban expressway equipped with the geomagnetic detectors and the floating cars in Shenzhen is selected as the scenario for the data collection. Three data fusion methods are proposed, respectively based on the confidence in traffic sensors' property, the artificial neural network and the least squared nonlinear support vector machine (abbreviated as LS-SVM). The method based on confidence is simple and straightforward and its computing speed is very fast, but the confidence level of the traffic sensor is required according to its operating performance. The accuracy and reliability of the method based on the artificial neural network relies on the structure and the size of the training sample for the neural network. The method based on LS-SVM shows the best data fusion accuracy.

1 INTRODUCTION

Urban road traffic operation indices used in many cities in China are a quantitative evaluation for road traffic congestion, which is to evaluate the overall operation condition of traffic on an urban road network. In order to quantify traffic congestion, the traffic management authorities and research institutes in many countries have developed a series of local traffic operation indices, such as V/C ratio, the ratio of congestion time to free-flow traffic time, the economic value of congestion time and so on (L.Y. Wang, L et al., 2016). Schrank and Lomax proposed an experience index, Road Congestion Index (RCI), which can be obtained by the weighted average of the ratio of the vehicle mileage to the lane mileage within the corresponding area of the highway and the length of all main roads. RCI is used for the macroscopic evaluation of regional traffic operation performance, representing the long cycle performance of the traffic situation (M. Boarne et al., 1998 & M.H. Liu, 2008). Some studies in Europe focus on the evaluation of traffic congestion from the perspective of economic value. They put forward 15 evaluation indexes, including traveler average cost, the satisfaction of transportation system, the ratio of predicted costs and actual costs, etc. Many

urban traffic management authorities and traveler information service providers in China have carried out a series of the investigations and application of various traffic index models proposed so far, such as Beijing, Guangzhou, Nanjing, Shenzhen, Shanghai, etc. It is found that the calculation method of exponential model can be done by means of the serious congestion mileage ratio, ratio of travel time and mixed evaluation.

To calculate the road traffic operation index from domestic first-tier cities, the single source of GPS data of floating car for traffic congestion modeling and operation index calculation are used. The GPS data of floating car obtained by the vehicle that installed positioning and wireless communication device, owning the information of the speed and position of the vehicle during the running process. The data were uploaded through the real-time communication network to the traffic management center of collection and processing (R.H. Liao, et al., 2016).

From the quality of the GPS data itself, there are many sources of error, such as SA interference error, multi-path effect error, the deviation of antenna phase center for GPS receiver. In addition, since most of the GPS data of floating car to calculate the road traffic operation index from the domestic first-tier cities come from the taxi, and the taxi as a characterization of the operating state of the urban road traffic exist obvious shortage and uneven distribution problems. Tale Shenzhen as an example, currently the number of running taxi is 16000, while the urban road mileage is 6900 kilometers. If all 16000 taxis run on the road every day and evenly distributed, the traffic density of taxi is 2.3 vehicle/km. Thus, only usage of the GPS data of taxi for urban road traffic operation index modeling and calculation, there will be a larger error, and the reliability be poor. Therefore, it is necessary to investigate the multi-resource traffic data fusion method for urban road traffic operation index (Z.C. He, et al., 2016). The accuracy and reliability of the calculation results of road traffic operation index can be improved by fusing the data that collected by the existing road traffic sensors.

Data fusion was first used in the military field, and it is a multi-level and multi-dimensional process of handling detection, interconnection, evaluation and combination of multi-resource information and data. It aims to obtain accurate target identification, complete and timely battlefield situation and threat assessment. The paper takes Shenzhen as an example that using the expressway geomagnetic sensor and floating car GPS sensor as the object of the research of data fusion method, the interval average speed of vehicles, collected by vehicle license plate recognition system on the expressway, is used as the true value of data fusion. Three practical data fusion methods are proposed, respectively based on the performance of confidence, neural networks and nonlinear regression support vector machine.

2 DESIGN OF MULTI-RESOURCE TRAFFIC DATA FUSION FRAMEWORK

While we have two or more senor data, we can consider data fusion based on multi-resource sensors to improve the quality and reliability of traffic parameters analysis. Fig. 1 shows the framework of data fusion based on multi-resource traffic sensors.

According to the different levels of data abstraction, the US department of defense JDL of DFS divided data fusion into three levels, namely data level, feature level and decision level data fusion. Data level fusion is a comprehensive analysis of untreated observation information on the original observation information layer that collected by various sensors. And it is generally only applied in the same sensors.

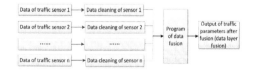

Figure 1.   The framework of data fusion based on multi-resource traffic sensors.

Data level fusion can be realized functions as follow: data validation, elimination and complement. Data association, namely the division of data sets. Data registration, such as unified temporal and spatial expression, GIS-T matching etc.. Feature level fusion is to extract the original information from the sensors to comprehensive analyze and process the feature information. It is able to realize traffic characteristic analysis, such as travel speed, travel distribution etc.. It can also be used for traffic incidents, such as traffic congestion status, traffic accidents and so on. Decision level fusion is to each information source provides information for analysis and progressing, a plurality of detection results, makes a higher fusion based on basic judgments and decision, and it can realize the traffic status evaluation, traffic management scheme decision etc.. Obviously, the urban road traffic operation index belong to traffic characteristic level, therefore, the article focuses on the characteristic of multi-resource traffic data level fusion method.

According to the ideas of feature level data fusion, it can be roughly classified into the following four steps:

1. Fusion data determination, focus on the existing multi-resource sensors analysis, clear fused the data sources;
2. Fusion method selection: according to the data, and combined with the road traffic condition, choose appropriate fusion methods and models;
3. Fusion experimental analysis: combined with the layout of the test facilities and road traffic conditions, choose the appropriate sections to do the fusion analysis, especially the feature level fusion and decision level fusion.
4. Fusion effect evaluation: develop the corresponding measurement scheme and collect data as the true value to evaluate the effect of data fusion.

3 MULTI-RESOURCE DATA FUSION METHOD BASED ON LS-SVM

Support Vector Machine (SVM) is developed on the basis of statistical learning theory of VC dimension theory and structural risk minimization principle. According to the limited sample information in the complexity of the model (i.e. learning accuracy of a particular training sample) and learning ability (i.e. ability of identify any sample without error), SVM can seek the best compromise in order to obtain the best generalization ability.

3.1   Multi-resource data fusion model based on LS-SVRM

LS-SVM uses the least squares linear system as the loss function. The advantages of LS-SVM method is that it can solve the problem of large scale, simplify the computation algorithm, and improve the study speed siginificantly. Figure 2 is the multi-resource traffic sensors data fusion model diagram of SVM.

3.2 The algorithm of multi-resource traffic data fusion based on LS-SVRM

Step 1. Preparation of training data and test data. First, select the training data to train the SVM model, and the function relationship between detection data and the real value of sensors is found by learning. Supposed that there are l group data x_{ip} from m detectors in a certain period, among them $i=1,2,...,l$, $p=1,2,...,m$. Then each set of data from m detectors could be referred to as $X_{io} \in R^m$, the input data for training can be referred to as $X=(x_{1o},x_{2o},...,x_{lo})$. x_{lo} represents the vector consisting of a set of i th data for each detector. Suppose that the measured value in the period is $y(y \in R^l)$, , and then the output data for training is $Y=y$.

Step 2. Choose appropriate generalization parameters ε (also known as the loss function), upper bound C and kernel function $K(x,x')$. The most commonly used generalization parameters ε in SVM is insensitivity loss function. The formula is $c(x,y,f(x)) = \tilde{c}(f(x)-y)$, and $\tilde{c}(\xi) = \max(|\xi|-\varepsilon,0)$. Upper bound C is also known as the penalty factor and refers to the upper bound of α in the judgment function. Currently, the commonly used method to select C is Cross Validation. The basic idea is to divide the training sample into n parts and only take one part to train each time, then use the model to test other training samples. After the whole process, we can calculate the percentage of correct identification of samples, then use others to repeat above steps. Each C will get a percentage, then select the percentage of maximum C as the final parameters of the model. The

most commonly kernel function are D order polynomial kernel function, Gaussian RBF kernel function, Sigmoid kernel function, Linear Spine kernel function, Exponential RBF kernel function etc.

Step 3. Constructing support vector machines, in order to solve the optimization problems.

$$\min_{\alpha^{(*)} \in R^{2l}} \frac{1}{2} \sum_{i=1}^{l} \sum_{j=1}^{l} (\alpha_i^* - \alpha_i)(\alpha_j^* - \alpha_j) K(x_{io}, x_{jo}) + \varepsilon \sum_{i=1}^{l} (\alpha_i^* + \alpha_i)$$
$$- \sum_{i=1}^{l} y_i (\alpha_i^* - \alpha_i) \quad (1)$$

$$s.t. \begin{cases} \sum_{i=1}^{l} (\alpha_i - \alpha_i^*) = 0 \\ 0 \le \alpha_i \\ \alpha_i^* \le \dfrac{C}{l} \end{cases} \quad (2)$$

Then the optimal solution is $\bar{\alpha} = (\bar{\alpha}_1, \bar{\alpha}_1^*, ..., \bar{\alpha}_l, \bar{\alpha}_l^*)^T$.

Step 4. Construct the decision function, that is, to establish the functional relationship between input data and the target values:

$$f(x) = \sum_{i=1}^{l} (\bar{\alpha}_i^* - \bar{\alpha}_i) K(x_i, x) + b \quad \text{and}$$

$$b = y_j - \sum_{i=1}^{l} (\bar{\alpha}_i^* - \bar{\alpha}_i)(x_i \circ x_j) + \varepsilon, j=1,2,...,l$$

Step 5. Input the training sample data to do the SVM training. When the error rate of support vector and the training time are satisfied, we can get the combination of kernel function and loss function of SVM and its relevant parameters. The error rate of support vector is equal to the ratio of number of support vectors and the number of samples. The smaller the error rate, the better the training results is, and the length of training is shorter, which indicates that the training efficiency is higher.

Step 6. Input the test data set and complete the data fusion test based on the training of the support vector machine.

4 METHOD VALIDATION

4.1 Location selection

The paper takes sections of North Central Avenue in Shenzhen, Guangdong province, as our research object to verify the method. After research and testing facilities layout situation, combined with the layout of facilities, we choose a representative section (including 2 sections) to analyze, the section is the North Central Avenue Qiaoxiang

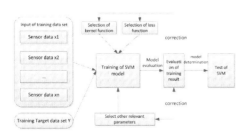

Figure 2. Multi-resource traffic data fusion model of SVRM.

village-North Central Avenue Xinzhou overpass. Through analysis of the data from GPS, geomagnetic and vehicle license plate recognition, we can calculate the standard deviation between geomagnetic sensors and GPS. Then confidence is obtained by normalizing treatment and we can do the data fusion on the basis (fusion of interval average speed characteristics obtained by the two kinds of sensor).

4.2 Data cleaning

For a single time slice of the missing data, we adopted the exponential smoothing repair method. For the recognition of the similarity, we can consider the model of K-mean clustering and cloud similarity flexibly. The GPS data repaitd algorithm, based on gradient direction process, as shown in Figure 4.

4.3 Data fusion performance evaluation index

The error evaluation of the data fusion results can be used in a variety of indicators, the general evaluation indicators include:

1. ME (Mean Error)
The mean error is the arithmetic mean of the random error of all measured values in the equal precision measurement. The calculation method is as follows, among them: is the fusion value, is the true value.

$$ME = \frac{\sum_{i=1}^{n}(Y_i - F_i)}{n} \quad (3)$$

Figure 3. The section of method validation and the schematic diagram of traffic sensor.

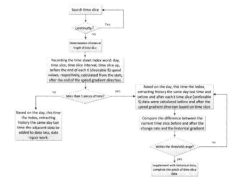

Figure 4. Data cleaning process based on gradient direction.

2. MAD (Mean Absolute Deviation)
MAD is the average of the absolute value of the deviation from the arithmetic mean and the error of the predicted value. The calculation method is as follows:

$$MAD = \frac{\sum_{i=1}^{n}|Y_i - F_i|}{n} \quad (4)$$

3. MPE (Mean Percentage Error)
MPE is the percentage of prediction error. The calculation method is as follows:

$$MPE = \frac{\sum_{i=1}^{n}\left(\frac{Y_i - F_i}{Y_i} \times 100\right)}{n} \quad (5)$$

4. MAPE (Mean Absolute Percentage Error)
MAPE is the deviation between the predicted and actual values divided by the absolute value of the ratio of the actual value calculated average and is expressed as percentage of forecast error indicators in the relative form. The calculation method is as follows:

$$MAPE = \frac{\sum_{i=1}^{n}\left(\frac{|Y_i - F_i|}{Y_i} \times 100\right)}{n} \quad (6)$$

5. MSE (mean square error)
MSE is the average of the mean value of the difference between the predicted and the actual value. The smaller the index value, the higher the prediction accuracy is. The calculation method is as follows:

$$MSE = \frac{\sum_{i=1}^{n}(Y_i - F_i)^2}{n} \quad (7)$$

6. SDE (Standard Deviation Error)
SDE is a measure of the degree of dispersion of the average value of a set of data. The larger the SDE, the larger of difference between the average and most representative numerical value is. If SDE is smaller, the data is closer to the average value of the numerical. Standard deviation is an important indicator of accuracy.

For finite samples, the standard deviation of freedom is, at this time the standard deviation is also known as the error, for:

$$SDE = \sqrt{\frac{\sum_{i=1}^{n}(Y_i - F_i)^2}{n-1}} \quad (8)$$

7. LSE (Least Squared Error Method)

If the LSE between the test data and the true value is less than the difference between the single source and the real value, then the data fusion model is valid. The calculation method is as follows:

$$LSE = \frac{\sqrt{\frac{\sum_{i=1}^{n}(Y_i - F_i)^2}{n}}}{\frac{\sum_{i=1}^{n}F_i}{n}} \qquad (9)$$

Comprehensive analysis, ME, MAD, MSE, SDE were affected by the level of time series data and measurement unit. MPE, MAPE, LSE eliminates the influence level of time series data and unit of measurement, reflects the relative value of error. SDE can better reflect the dispersion degree of error distribution, thus indicating the reliability of the data in a certain extent. MPE reflects the size of the error distribution and the calculation is simple, the least square error and LSE reflects fluctuations in the size of the data. Data fusion can be used for comparison with data fusion results and single sensor data quality. Therefore, the project focused on the choice of SDE index, MPE index and LSE index respectively to characterize the reliability of the data, the error and the validity of the model.

4.4 Data fusion performance evaluation

According to the working principle and procedure of the above SVM, the traffic sensor data can be obtained by this data fusion. The analysis of the multi-resource data fusion results in the morning rush hour (7:30–9:30).

Evaluation of the fusion effect, wherein SDE = 0.22, MPE = 0.01, LSE = 0.03. The comparison of detection accuracy of single detector and the precision of fusion results are shown in Fig. 5 to Fig. 7.

We can see that the fusion data is much closer to the true value than single data from the above results. From the index of LSE, we can see that the result of fusion is very good. So the fusion model that applied to the section of the period is appropriate.

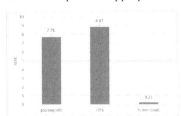

Figure 5. The SDE comparison of fusion result based on LS-SVM in the morning rush hour and the data of single detector.

1. The analysis of the multi-resource data fusion results in the evening rush hour (18:05–19:30).

Evaluation of the fusion effect, wherein SDE = 6.00, MPE = 3.12, LSE = 0.10. The comparison of

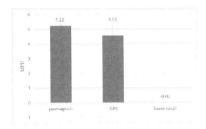

Figure 6. The MPE comparison of fusion result based on LS-SVM in the morning rush hour and the data of single detector.

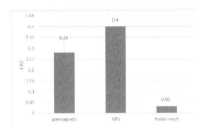

Figure 7. The LSE comparison of fusion result based on LS-SVM in the morning rush hour and the data of single detector.

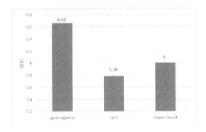

Figure 8. The SDE comparison of fusion result based on LS-SVM in the evening rush hour and the data of single detector.

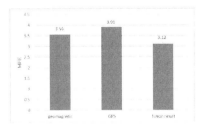

Figure 9. The MPE comparison of fusion result based on LS-SVM in the evening rush hour and the data of single detector.

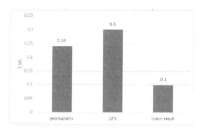

Figure 10. The LSE comparison of fusion result based on LS-SVM in the evening rush hour and the data of single detector.

Table 1. Evaluation of data fusion method.

Features	Period	Accuracy			SDE	
		Geomagnetic	GPS	Weight of sensor	SVM	NN
Morning rush hour	7:30–9:30	7.61	8.87	6.27	0.22	3.63
Evening rush hour	18:05–19:30	6.66	5.78	4.70	6.00	1.67
Ascent of rush hour	6:30–7:30	3.68	7.06	3.79	5.58	0.99
Decline of rush hour	9:30–10:30	4.11	3.53	2.77	2.79	0.77
Other	0:00–6:30	5.77	9.04	5.82	1.41	2.72

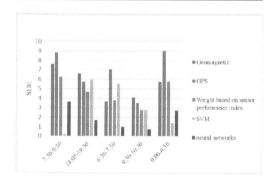

Figure 11. The SDE comparison of single detector and the method of multi-resource data fusion in different period.

detection accuracy of single detector and the precision of fusion results are shown in Fig. 8 to Fig. 10.

Since the reliability of the evening rush hour geomagnetic data distribution is poor from the above results, it makes that the SDE of data fusion result is lower than the GPS data. In other word, the reliability of GPS data in the period is higher than the geomagnetic data. But the detection results of data fusion of MPE and LSE index are better than the single sensor data, so the data fusion in the period is effective.

In order to compare to the method of LS-SVM data fusion, the paper also use the fusion method based on sensor performance index weight and fusion data based on artificial neural networks. Because of the space limit, given the following three kinds of direct comparison of different methods of data fusion in the SDE index on the summary results, as shown in Table 1 and Figure 11.

4.5 Conclusions

The magnetic detection accuracy is better than that of GPS from the detection effect of a single detector. For the comparison of multi-sensor fusion methods in different periods, the LS-SVM fusion method is better than other methods in terms of precision and reliability. Some insightful findings obtained in this investigation include:

1. In the morning rush hour, the effect of LS-SVM is optimal among the three proposed methods. The difference between any two of the three methods is not very large, though.
2. In the evening rush hour, the effect of the artificial neural network, is optimal and the effect of LS-SVM is second among the three methods.
3. For the start of the rush hour, the effect of the multi-resource data fusion method based on the sensor performance index is the worst. Actually, the difference between the two methods respectively based on LS-SVM and the artificial neural network is not very large. The previous numerical results show that the data fusion method based on the artificial neural network is relatively good.
4. For the end of the rush hour, the effect of the data fusion method based on the sensor performance index is the worst among the three proposed methods.
5. For the other period, the effect of the data fusion method based on the sensor performance index is the worst among the three.

The result in this paper has been applied in traffic surveillance data process in Shenzhen, and the data fusion methods will be improved with more and more field data.

REFERENCES

Boarnet, M., E. Kim & E. Parkany, 1998. Measuring traffic congestion [J]. Transportation Research Record, 1998(1512): 93–99.

He, Z.C., Y.Q. Zhou & Z. Yu, 2016. Regional traffic state evaluation method based on data visualization [J]. Journal of Traffic and Transportation Engineering, 2016(1): 133–140.

Liao, R.H., J. Zhou & H.L. Xu, 2016. Study on congestion index of urban traffic network considering link weights [J]. Chinese Journal of Systems Science, 2016(1): 84–88.

Liu, M.H. 2008. Evaluation model and algorithm of hierarchical traffic congestion in large cities [D]. Beijing: Beijing Jiaotong University, 2008.

Wang, L.Y., L. Yu, J.P. Sun & G.H. Song, 2016. Research and application of traffic operation index [J]. Traffic Information and Security, 2016(3):1–9, 26.

Transport Infrastructure and Systems – Dell'Acqua & Wegman (Eds)
© 2017 Taylor & Francis Group, London, ISBN 978-1-138-03009-1

The optimed project: A new Mediterranean hub-based Ro-Ro network

Paolo Fadda, Gianfranco Fancello, Claudia Pani & Patrizi Serra
CIREM—University of Cagliari, Italy

ABSTRACT: This study is part of the OPTIMED project funded under the European programme ENPI CBCMED—Cross Border Cooperation in the Mediterranean—aimed at optimising the trade network between the high Tyrrhenian arc and the south-eastern shore of the Mediterranean basin. The project focuses on the Mediterranean Ro-Ro Freight sector. The main objective of the study is to design a new optimised corridor between the two shores of the Mediterranean sea. The new corridor, founded on the hub-based concept, is composed by two hub ports: Porto Torres (Italy) and Beirut (Lebanon) serving the north-western area and the south-eastern area, respectively. The new network must be able to ensure scheduled services enabling shipping companies, and consequently all the operators using their services, in order to obtain more reliable delivery times, as opposed to the current system whereby delivery times cannot be guaranteed. The application-area includes 24 ports located within 8 Mediterranean countries. A data collection process has been performed in order to collect the information necessary to define the existing scenario and to characterise nodes (ports) and arcs (links) of the new network. The collected data are divided into demand and supply variables. In order to characterize the new designed transport network in terms of optimal services frequencies, capacities and schedules a two-step optimization approach has been defined: the first step defines the optimal frequencies and capacities of connection services along the various routes; the second step defines an optimal timetable for the organization of the services identified by the previous step. Origin/Destination matrices have been used to calibrate the network and alternative scenarios of future demand have been built-up, in order to determine how the new network structure can be able to match changing demand over the years. Environmental balances have been performed to assess the environmental efficiency of the new network with respect to the existing transport options.

1 INTRODUCTION

The Mediterranean sea is an attractive market for shipping operators, in particular considering its geographical location at the centre of the major international shipping routes. Anyway, the development of maritime transport and logistics sector in the Mediterranean area still needs to be improved in order to ensure more efficient and sustainable trade relations between the north-western and south-eastern shores of the Mediterranean basin. The flow of exported-imported goods between the two Mediterranean shores is estimated in around 38 million tons (2012), of which more than 90% is transported by maritime routes. Looking at the existing East-West Mediterranean services today there exists a limited number of regular and scheduled ro-ro services connecting the two Mediterranean shores. Besides, in the majority of cases such services appear fragmented, not integrated and characterised by high service's time and low frequencies. In addition to these services based on scheduled departures and established itineraries, it is possible to resort to on-demand services in which departures and itineraries depend on the specific transport requests of the period. Because of their nature, spot services are typically characterised by not scheduled boarding times, uncertainty and low reliability of delivery times with consequent high logistics costs. On the other side, land transport mode still appears attractive on several itineraries because of the reliability of its door to door services and the temporal continuity of its links. However, the dependence of several Mediterranean areas from road transport brings a number of well-known problems related to high carbon emissions, road congestion and safety issues. In order to overcome such issues the existing maritime transport system needs to be rethought and rationalized. The OPTIMED project proposes to transform the logic of distribution of goods by sea: from not integrated and fragmented spot connections to an innovative two-hub based network option characterised by scheduled, regular and reliable frequencies and more competitive delivery times.

This paper illustrates the main activities and outputs of the OPTIMED project. The structure of the paper is as follows: section 1 introduces the new proposed maritime network structure, section 2 briefly illustrates the data collected, section 3 describes the optimisation approach designed to characterise the new transport option in terms of optimal itineraries, frequencies, capacities and service schedules, section 4 illustrates the resulting network configuration and its parameters, section 5 illustrates the performed feasibility studies, finally section 6 concludes.

2 A NEW NETWORK LAYOUT

The OPTIMED project proposes to rationalise the transport of goods within the Mediterranean basin by proposing an innovative integrated network option based on the hub and spoke paradigm and characterised by scheduled, regular and reliable frequencies and more competitive delivery times. A new two-hub based network structure has been defined by identifying as hub ports the port of Porto Torres (Italy) for the western side of the Mediterranean basin and the port of Beirut (Lebanon) for the eastern side. Both ports benefit from a barycentric position within the basin of reference and from useful port and back-port areas to be used for cargo handling operations.

Each hub port serves a set of ports according to the *hub and spoke* distribution paradigm in which all traffic volumes move along spokes connected to the hub in a system of connections arranged like a wire wheel.

The main Ro-Ro ports located in the High-Tyrrhenian arc involved in trade with the countries located on the eastern shore of the Mediterranean basin and the main Ro-Ro ports located in the South-Eastern Mediterranean shore involved in trade with the countries located on the western shore have been selected. The port selection process has been performed in several stages according to the following three main criteria:

1. geographical location;
2. nature of the traffic served;
3. analysis of ongoing east-west commercial relationships among the countries involved in the project.

In a first stage, geographical criteria have been used to define the area of analysis. The selected Ro-Ro ports are distributed across the 8 countries involved in the project: France, Italy and Spain for the western side and Cyprus, Egypt, Lebanon, Syria and Turkey for the eastern one. According to the geographical location criteria, the main aim has been to ensure a consistent coverage of the area involved. The various ports have been selected taking into account the average distance in nautical miles between each port and the hub of reference.

In a second stage, traffic requirements have been introduced to reduce the number of ports involved.

Figure 1. New OPTIMED network layout.

Since the project concerns the optimisation of a new Mediterranean Ro-Ro trade network, only the ports serving a consistent share of Ro-Ro traffic have been included in the analysis. As a consequence other ports, despite their importance, have been excluded from the selection process.

In a third stage, commercial requirements have been introduced to select the set of ports to be involved in the project allowing to identify the main ports involved in goods exchange between the eastern and the western shores of the Mediterranean basin.

The selected ports are:

– Valencia, Sagunto, Tarragona, Barcelona, Marseille, Sète, Toulon, Savona, Genoa, La Spezia, Livorno, Civitavecchia, Naples and Salerno, for the western Mediterranean region;
– Mersin, Lattakia, Tartous, Tripoli, Damietta, Alexandria, Port Said and Limassol, for the south-eastern Mediterranean region.

Figure 1 shows the proposed OPTIMED network layout.

3 THE DATA COLLECTION PROCESS

An extensive data collection process has been performed by interviews and questionnaires in order to collect the data and information necessary to define the existing demand and supply scenario and to characterise nodes (ports) and arcs (links) of the new network. The collected data includes both demand and supply variables.

The analysis of the demand scenario has been necessary to improve the knowledge of the actual demand characteristics first of all at a general level between countries, then at a more specific level among ports. More specifically, the demand analysis has concerned:

– determination of the import/export volumes between the involved countries in terms of tons of freight and economic value (€). The collected

Table 1. Annual Origin—Destination matrix (1000 tons), year 2012, all modes.

O/D	France	Italy	Spain	Cyprus	Egypt	Lebanon	Syria	Turkey
France	/	/	/	509,79	1826,37	938,28	81,61	2614,58
Italy	/	/	/	307,95	2165,58	1124,42	128,93	5760,73
Spain	/	/	/	160,94	905,82	392,04	129,16	2995,21
Cyprus	55,34	23,87	17,93	/	/	/	/	/
Egypt	2419,67	3867,63	2122,33	/	/	/	/	/
Lebanon	12,52	56,39	72,68	/	/	/	/	/
Syria	4,95	47,98	7,80	/	/	/	/	/
Turkey	1295,52	4599,02	2097,31	/	/	/	/	/

demand data have been further classified by product category (TARIC Classification) and by transportation mode (sea, road, rail, air);
– determination of the trade goods volumes, both ro-ro and container traffic, between the selected ports.

Demand data have been collected for the five year period 2008–2012.

General trading volumes between the two areas under study are provided in Table 1 in order to offer a general view of the market in the involved shores.

The analysis of the supply system has been performed in order to portray the existing scenario of maritime services in the Mediterranean area and to collect the data and information necessary to characterize the elements, both arcs and nodes, that will compose the new OPTIMED network. The performed collection process has concerned:

– infrastructural, operating and cost-related information for each of the 24 ports taking part in the new network; information has been collected by means of a supply collection form which was sent to the 24 ports involved in the OPTIMED project;
– census of the existing Mediterranean Ro-Ro services in order to gain a general overview of the liner ro-ro services that currently operate in the Mediterranean sea. Collected variables include: name of the company that provides the service, business area, name of the service, frequency, port rotation, transit times and total service times.
– technical and operational characteristics of the Mediterranean Ro-Ro fleet currently in operation. Data collected for each vessel concern: name, year built, type, power engine, capacity, service speed).

4 THE OPTIMISATION APPROACH

In order to characterize the new designed transport network in terms of optimal services frequencies, capacities and schedules a two-step optimization approach has been developed:

1. the first step defines the optimal frequencies and capacities of connection services along the various routes;
2. the second step defines an optimal timetable for the organization of the transport service, i.e. it defines departure and arrival dates and times for the various linking services (both mother and feeder) identified by the previous step.

Step 1

A specially designed Mixed Integer Linear Programming Model (MIPM) for the Services Frequency Selection in a Hub and Spoke Maritime Network has been developed in order to define the optimal frequencies and capacities of the connection services along the various routes.

The Services Frequency Selection model has been designed in order to meet two conflicting goals:

– the maximisation of the service quality for transporters, i.e. the maximisation of services frequencies, in order to guarantee the availability of the transport service;
– the maximisation of the load coefficient for each service in order to ensure the economic convenience of the system for those companies that will operate the various services.

Step 2

In order to design a competitive new transport option it is very important to ensure that the various services are spread from alternative ports along the week so as to guarantee every day a good coverage of the whole area of interest. The goal of the second phase was to provide a feasible schedule able to maximise the weekly coverage of each port while minimising waiting times at the two hub ports. The "weekly coverage" of a port can be defined as the number of days, within a week, in which at least one feeder is leaving from that port or from a port in the nearby. For instance, if a feeder is leaving from Genoa on Monday and Wednesday, while a feeder is leaving from La Spezia on Tuesday and Thursday, both ports will have a weekly coverage equal to 4, while, if feeders from La Spezia leave on Monday and Thursday, both ports will have a weekly coverage equal to 3. In so doing, transporters can decide to use the most convenient port of departure for their goods depending on the day they are planning to send their goods.

The definition of the optimal services' timetable has been performed by means of a specially designed Mixed Integer Linear Programming model that tries to maximise the weekly coverage of each port while minimising waiting times for connection in the two hub ports.

The model takes as input the weekly frequencies and ships categories for each service defined by the frequency selection model (step 1) and provides as outputs the weekly schedule and the weekly coverage for each port.

5 TESTING OF THE OPTIMISATION APPROACH: FINAL NETWORK CONFIGURATION

The designed two-step optimisation approach has been tested on different scenarios of demand and different configurations of vessels capacity and speed. Among the various demand scenarios tested, the scenario corresponding to 100% of existing ro-ro volumes and 10% of container volumes has been selected as more representative of the demand of transport that will likely characterize the new OPTIMED network in the coming years. For the selected demand scenario, the following assumptions concerning the service's configuration have been taken:

– three classes of feeder vessels, characterised by different capacities (lm) have been considered:
  – small feeder vessels: 1350 lm;
  – medium feeder vessels: 2520 lm;
  – large feeder vessels: 3320 lm.

The average service speed of feeder vessels is 18 knots.

– three classes of mother vessels characterised by different capacities (lm) have been considered:
  – small mother vessels: 4600 lm;
  – medium mother vessels: 6350 lm;
  – large mother vessels: 7700 lm.
    The average service speed of mother vessels is 21 knots.
– The maximum occupation rate allowed for the vessels operating the new services is at the level of 90% for both categories of ships, mother and feeder.

In the testing-phase, the selected scenario has been first solved by using the developed Mixed Integer Programming Model for the Services Frequency Selection in a Hub and Spoke Maritime Network. The model returns as outputs the optimal frequencies along the various routes (both feeder and inter-hub routes) together with their used and residual capacities. For each service, the highest between the two frequency values in the two directions of the service has been selected as final frequency value of the service.

The following assumptions have been made:

– every vessel has to perform a complete roundtrip service (outward and return journey);
– empty load units (both containers or trailers) should always return to their origin port.

Tables 2, 3 and 4 show the detail of the resulting western feeder services, eastern feeder services and inter-hub mother services, respectively. The various services are detailed in terms of:

– number and optimal capacity of the vessels that operate the service, in terms of linear meters (lm);
– optimal weekly frequency of the service;
– total weekly capacity (lm) of the service.

In a second stage, starting from the optimal services frequencies identified by the previous step, the optimal weekly schedule for the transport service organisation has been defined in order to

Table 2. Western feeder services.

Feeder service name	Feeder vessel (lm)			Weekly frequency	Weekly capacity (lm)
	1350	2520	3320		
Barcelona-Porto Torres-Barcelona	0	1	0	1	2.520
Genoa–PortoTorres-Genoa	2	0	0	2	2.700
La Spezia–Porto Torres–La Spezia	1	0	0	1	1.350
Marseille–Porto Torres-Marseille	5	1	0	6	9.270
Naples–Porto Torres-Naples	1	0	0	1	1.350
Sète–Porto Torres-Sète	1	0	0	1	1.350
Valencia–Porto Torres-Valencia	1	0	1	2	4.670

Table 3. Eastern feeder services.

| Feeder service name | Feeder vessel (lm) | | | Weekly frequency | Weekly capacity (lm) |
	1350	2520	3320		
Mersin–Beirut-Mersin	1	4	1	6	14.750
Lattakia–Beirut-Lattakia	1	0	0	1	1.350
Damietta–Beirut-Damietta	0	1	0	1	2.520
Alexandria–Beirut-Alexandria	1	0	0	1	1.350
Port Said–Beirut–Port Said	1	0	0	1	1.350
Limassol–Beirut-Limassol	1	0	0	1	1.350

Table 4. Inter-hub mother services.

| Mother service name | Mother vessel (lm) | | | Weekly frequency | Weekly capacity (lm) |
	4600	6350	7700		
Porto Torres–Beirut – Porto Torres	5	0	0	5	23.000

maximise the weekly coverage of each port while minimising waiting times for connection in the two hubs. Such configuration of the service has been implemented into the new web-based logistics platform developed within the project in order to support the new optimised trade network and to enable users and shippers to plan shipping and identify the best transport option.

6 FEASIBILITY STUDIES

In order to characterize the new transport option in terms of produced air pollutant emissions and consumptions, several energy and environmental balances for a number of significant services taking part in the new network have been calculated. Balances have been performed assuming to perform the transport service once by using the maritime services offered by the new designed hub-based network, once by using the existing Mediterranean regular services, and finally by means of the road mode. Specifically:

– the calculation of emissions and consumptions for OPTIMED routes has been performed by using a well-established state-of-the art procedure (Trozzi, 2010) based on defined emission factors (Entec UK Limited, 2002) that considers separately the three fundamental navigation phases:
  – cruise (sailing time);
  – hoteling (time spent at the port);
  – manoeuvring.

Unit values of CO_2 emissions and fuel consumption (kg per tonne of transported goods) have been calculated by dividing total emissions and

consumptions that characterise a vessel trip by the number of tonnes transported in the same trip according to the proposed OPTIMED scheme;

– concerning the calculation of emissions and consumptions for existing regular services, since the actual payload of the analysed services was unknown, in order to estimate related unit values the maximum vessel load factor that characterizes the corresponding service in the Optimed route has been assumed as load factor for the analysed service;[1]

– the calculation of CO_2 emissions and fuel consumptions for a selection of road routes have been performed by using an *activity-based* calculation method (Cefic, 2011) based on the following formula:

CO2 emissions = Transport volume by transport mode × average transport distance by transport mode × average CO2-emission factor per tonne-km by transport mode

Emissions have been calculated using the average CO_2-emission factors recommended by McKinnon and Piecyk (2010) for road transport operations.

Average fuel consumption have been calculated by multiplying the average fuel consumption factor of a truck by the number of travelled kilometres. The assumed average truck consumption index is 0,35 l/km. Moreover, CO, NO_x, PM_{10}

---

[1]The maximum load factor is the maximum of the three load factors that characterises the three sea legs composing the corresponding OPTIMED service.

1055

Table 5. Unit values of CO_2 emissions (kgCO_2/tonne of transported goods) for a selection of west-east routes.

Route	Road route	OPTIMED service	Existing Med service
Barcelona—Mersin	221,7	143,7	226,5
Barcelona—Tartous	236,9	159,7	na
Barcelona—Damietta	306,7*	157,7	na
Marseille—Mersin	194,3	140,5	na
Marseille—Tartous	208,5	156,5	na
Marseille—Alexandria	235,3*	154,6	263,6
Genoa—Mersin	173,6	139,3	216,8
Genoa—Tartous	189,3	155,4	na
Genoa—Damietta	208,9*	153,4	na

Table 6. Unit values of fuel consumption (kg/tonne of transported goods) for a selection of west-east routes.

Route	Road route	OPTIMED service	Existing Med service
Barcelona—Mersin	68,3	45,2	71,2
Barcelona—Tartous	73,0	50,2	na
Barcelona—Damietta	94,5	49,6	na
Marseille—Mersin	59,9	44,2	na
Marseille—Tartous	64,2	49,2	na
Marseille—Alexandria	72,5	48,6	82,9
Genoa—Mersin	53,5	43,8	68,2
Genoa—Tartous	58,3	48,9	na
Genoa—Damietta	64,4	48,3	na

This road route includes a ferry.

and VOC emissions (kg per single travel) produced by a single truck along the selected routes of interest have been calculated by assuming the emission factors (g/km) provided by the National Atmospheric Emissions Inventory (2013).

Tables 5 and 6 summarise, respectively, the unit values (kg/tonne of transported goods) of CO_2 emissions and fuel consumptions for a selection of significant west–east (and vice versa) routes. Specifically, the second column refers to road option, the third column to OPTIMED service, while the fourth column to existing scheduled maritime service.

Looking at the data, unit values of emissions and consumptions appear significant lower for OPTIMED services. Moreover, while unit values of consumptions and emissions concerning the analysed maritime services have been calculated by taking into account the return flows of empty units, unit values concerning the road transport option have been estimated by assuming the case of trucks travelling at full load[2].

Estimates confirm the positive environmental and energy impact of the system proposed: the shift of the flows of goods to a more efficient and optimised Mediterranean transport option can imply lower environmental impacts and energy consumptions. This type of transport not only reduces traffic on roads and motorways, it also significantly reduces air pollution with positive effects on air quality and human health, and it saves on freight costs and on travel time thanks to optimised and reliable ro-ro services.

7 CONCLUSIONS

This study has proposed a new rationalised Mediterranean network between the eastern and the western shores of the Mediterranean basin. The new corridor, founded on the hub-based concept, is composed by two hub ports: Porto Torres (Italy) and Beirut (Lebanon) serving the north-western area and the south-eastern area, respectively.

New layouts of the shipping connection structure, characterised by frequencies of services and journey times more competitive than the current arrangement, have been proposed. Moreover, alternative scenarios of future demand had been built-up, in order to determine how the new network structure can be able to match changing demand over the coming years. Environmental balances have confirmed the environmental efficiency of the new network with respect to the existing transport options.

[2]The data concerning the actual percentage of empty running was not available.

The new hub-based network intends to create new opportunities finalised at the matching between supply and demand among the involved ports leading to positive economical and commercial impacts for transport and logistics operators and for territories in general.

The OPTIMED project contributes to make maritime transport connections between the north-western and the south-eastern shores of the Mediterranean basin more efficient and more reliable by creating a new optimised sea transport corridor. In the coming years, the project outcomes can provide an answer to the need for more efficient and sustainable trade relations within the Mediterranean transport system by promoting hub-based network structures, port cooperation policies, short sea shipping and Motorways of the Sea initiatives.

ACKNOWLEDGEMENTS

This research is part of the OPTIMED project funded by the ENPI CBC Mediterranean Sea Basin Programme. The project partnership includes: Autonomous Region of Sardinia (Italy), European School of Short Sea Shipping (Spain), CIREM —University of Cagliari (Italy), ASCAME— Association of Mediterranean Chambers of Commerce (Spain), North Sardinia Port Authority (Italy), Chamber of Commerce of Beirut and Mount Lebanon (Lebanon), Lebanese Ministry of Public Works and Transport (supporting partner).

The authors would like to thank all those who have collaborated on this work.

REFERENCES

Cefic, E.C.T.A. 2011. *Guidelines for Measuring and Managing CO2 Emission from Freight Transport Operations.*

Entec UK Limited, 2002. *Quantification of emissions from ships associated with ship movements between ports in the European Community*, European Commission Final Report, July 2002.

McKinnon, A. & Piecyk, M. 2010. *Measuring and Managing CO_2 emissions.* Edinburgh: European Chemical Industry Council.

National Atmospheric Emissions Inventory, 2013. Report.

Trozzi, C. 2010. *Emission estimate methodology for maritime navigation.* Conference Proceedings, US EPA 19th International Emissions Inventory Conference, San Antonio, Texas, September 27–30, 2010.

Transport Infrastructure and Systems – Dell'Acqua & Wegman (Eds)
© 2017 Taylor & Francis Group, London, ISBN 978-1-138-03009-1

Evaluating the capability of deterministic and stochastic simulation tools to model oversaturated freeway operations

D. Jolovic & P. Martin
New Mexico State University, Las Cruces, NM, USA

A. Stevanovic
Florida Atlantic University, Boca Raton, FL, USA

S. Sajjadi
PTV America, Portland, OR, USA

ABSTRACT: The HCM2010 freeway facilities methodology offers a supplemental computational engine FREEVAL, which is a macroscopic/mesoscopic tool. It enables users to implement HCM based freeway analyses quickly for both undersaturated and oversaturated conditions. Vissim is a microscopic traffic simulation tool that enables users to model real world traffic conditions with high levels of accuracy. There are few studies which assess the capability of FREEVAL to replicate oversaturated field conditions. Further, the FREEVAL tool was recently updated and coded in a new Java environment. This paper addresses a gap in the literature by contrasting the methodologies behind the two tools and by offering an explanation and discussion of their outputs in terms of density and space mean speed. The study covers three major HCM freeway segment types: basic, on-ramp, and weaving for oversaturated conditions. The assessment reveals that both tools are capable of replicating oversaturated conditions reliably after the calibration/validation process.

1 INTRODUCTION

The 2010 Highway Capacity Manual (HCM, 2010) suggests a methodology for analyzing oversaturated freeway conditions. Oversaturation occurs when vehicular demand becomes greater than the available capacity (i.e., demand-to-capacity ratio exceeds 1.0). Oversaturation accounts when the Level-of-Service (LOS) is F, and densities are greater than 45 passenger cars per mile per lane. To analyze oversaturation, the HCM2010 suggests the use of two types of tools: 1) macroscopic FREEway EVALuation (FREEVAL) simulation tool, which is the part of the current manual, or 2) microscopic traffic simulation software, commercially available (HCM, 2010).

The FREEVAL computational engine is a macroscopic tool designed to faithfully implement the operational analysis for both undersaturated and oversaturated directional freeway facilities (Trask et al. 2015). First developed in 2001, it went through several refinement iterations. Current FREEVAL, which is used in this research, is fully coded in a Java environment and named FREEVAL2015a.

There are several widely spread microsimulation tools on the US market—TSS Aimsun, PTV Vissim, Transmodeler, Corsim, etc. This paper applies PTV Vissim as an alternative tool to model and analyze the oversaturated freeway segment. Generally, microsimulation tools will produce more accurate results but require more input, and can be time-consuming. On the other hand, FREEVAL was created to be a simple tool for quick facility assessment. The results may not be as comprehensive as from the micro simulator, due to its macroscopic nature.

The goal of this paper is to report an evaluation of the two different tools for freeway assessment and to draw conclusions on their ability to properly replicate oversaturated conditions in terms of space mean speeds and density. Jolovic & Stevanovic, 2013 showed that microsimulation was successful in assessing heavy freeway congestion, while FREEVAL struggled to do so. The previously tested version was developed in MS Excel powered by Visual Basic for Applications.

Several previous studies have shown that microsimulation and HCM methodologies can differ in LOS ranges, comparative delays and densities (Hall, et al. 2000), (Bloomberg et al. 2003), (Milam et al. 2006), (Jolovic D. et al. 2016). However, the differences are not substantial.

A German version of FREEVAL is based on the US FREEVAL (Hartmann, Vortisch, & Schroeder, 2014), (Hartmann & Vortisch, A rationale for enhancing the German Highway Capacity Manual to incorporate oversaturated freeway facility analysis, 2016). It still lacks fully operational FREEVAL for oversaturated conditions.

The paper covers three major HCM freeway segment types (basic, on-ramp, weaving) for oversaturated conditions. Field data are collected from the Performance Measurement System (PeMS) online database in California, for the I-880 freeway. FREEVAL and Vissim models are calibrated and validated using the obtained data (e.g., vehicle speeds and traffic volumes per lane in 15-min intervals). The outputs of both tools are evaluated and compared against the field data.

2 METHODOLOGY

The test bed is a part of I-880 Nimitz freeway, located between Stevenson Blvd and Mowry Ave in Freemont, California. The test bed consists of two basic segments, one on-ramp and one weaving segment. The outline is shown in Figure 2. The weaving segment has an auxiliary lane between Stevenson and Mowry ramps. Field data were collected from California's PeMS online database for May 18th, 2016 from 3:15 PM to 7:15 PM.

The microsimulation model was developed, calibrated and validated for the 4-hour period, from 3:15 PM to 7:15 PM, with a 15-min warm up time. The Vissim model is calibrated using 15-min traffic volumes and vehicular speeds from May 18, 2016. Calibration is a process of matching specific field and modeled values (e.g., speeds, volumes). Validation is a process of testing whether the model is capable of replicating actual field data.

Vissim is stochastic in nature, so five simulation run outputs were averaged. Then we compared the results with the field values.

When assessing oversaturation, the modeler should be able to collect, or to estimate the demand (i.e., the total amount of vehicles willing to travel the facility). However, our data consisted of traffic flows only (i.e., the number of vehicles that traversed the facility). We did not have access to camera recordings which would allow us to observe the queueing and the total vehicular demand. This problem was approached by artificially increasing traffic inputs in the model, until the speeds from the model and from the field matched. This approach was not viable for all the time intervals. In those cases the authors posed desired speed decisions in specific spots in the model, to lower the speeds on the freeway mainline. The range of desired speeds is determined based on the speed data per lane. This approach is used before on freeways directional facilities.

The default driving behavior and lane change parameters were not changed.

Figure 2 and Figure 3 show the calibration results for Vissim. Figure 2A) shows a high coefficient of determination ($R^2 = 0.934$) which means that the simulation closely match field conditions. The trend line equation shows that for each vehicle in the field, there are 0.946 vehicles in the simulation. For the weaving area, the match is even closer (see Figure 2B).

Figure 3 shows that the heavy congestion starts around 4:30 PM and lasts until 6:30 PM. That is the oversaturated time frame of the analysis. The time

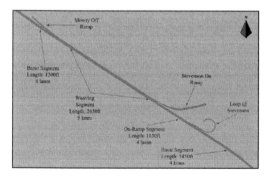

Figure 1.   Segment of I-880 Nimitz freeway in California.

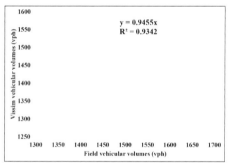

A) Volume match for I-880 before Stevenson on-ramp

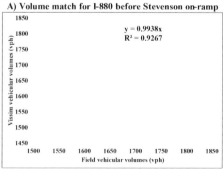

B) Volume match for I-880 weaving area

Figure 2.   Volume calibration for Vissim model.

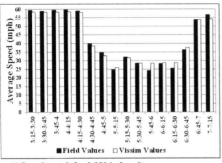

A) Speed match for I-880 before Stevenson on-ramp

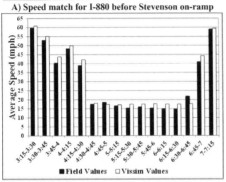

B) Speed match for I-880 weaving area

Figure 3. Speed calibration for Vissim model.

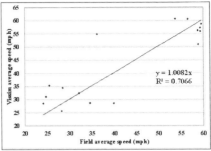

A) Speed validation for basic freeway segment

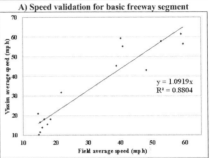

B) Speed validation for weaving freeway segment

Figure 4. Vissim speed validation.

periods before and after that (3:15–4:30 PM, and 6:45–7:15 PM) experienced no congestion. They were modeled because the analysis in FREEVAL tool has to start and end with undersaturated conditions, (i.e., all vehicles have to clear the model) and the authors wanted to model and contrast the same time frame in both tools. For heavy congested conditions, the speeds were as low as 15mph, as presented in Figure 3B).

Figure 4 presents validation results in terms of vehicle speeds for the same period (3:15–7:15 PM) but for a different date. It shows the model can be applied to a different day with high confidence in terms of vehicular speeds.

The FREEVAL model has four freeway segments: two basic segments, one on-ramp and one weaving segment. Segment lengths are coded according to the HCM supplemental material and FREEVAL user guide. We developed 16 time intervals, each one 15-min long, to cover 4 hours of analysis. When modeling FREEVAL, the user has to ensure that the first and the last time intervals are not oversaturated, as stated in the FREEVAL manual. Otherwise, the results will be biased. Figure 5 shows an example of FREEVAL input.

We coded FREEVAL with traffic input, free flow speed for mainline, on and off ramps, and heavy vehicles percentages. We plotted speed output against Vissim calibrated speeds for each 15-min interval. The initial outputs showed that

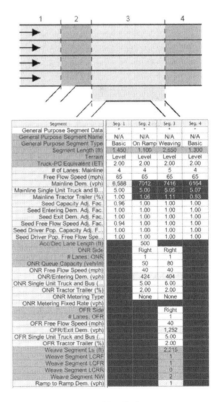

Figure 5. Input FREEVAL window.

1061

FREEVAL needed an adjustment to match speeds closely with the field data. We altered Capacity Adjustment Factors (CAF) for every segment and achieved a good match between field and modeled data, as recommended by the user guide. FREEVAL contains one more adjustment factor called Speed Adjustment Factor (SAF). However, the SAF is a static location factor and should be altered only to model weather or geometric effects on free flow speed (Zegeer, et al., 2014). Thus, this factor was not used for calibration purposes.

When a user changes CAF, the speed-flow relationship of the segment changes. In our model, CAF factors range from 0.65 to 1.00. CAF factors should not be used to increase capacity (i.e., CAF should not be larger then 1.00), as the FREEVAL guidelines recommend.

3 RESULTS AND CONCLUSIONS

Figure 6 shows the vehicle speeds comparison results between Vissim model output, field data and FREEVAL output. The results show that macroscopic FREEVAL model can be calibrated to match microsimulation Vissim's outputs and field data with a high degree of confidence even for oversaturated conditions. This clearly shows that FREEVAL2015a has improved when compared to previous versions which were built in MS Excel.

Figure 7 shows density comparisons for three different segments: basic, on-ramp, and weaving segments. The density is matched for the field values, Vissim output values and FREEVAL output. The field density is calculated by dividing traffic flows by speeds obtained in the field. Vissim's density is calculated according to the procedure which can be found elsewhere (Jolovic D., Stevanovic, Sajjadi, & Martin, 2016). The density from Vissim is calculated for the same segment lengths as coded in FREEVAL. Both Vissim and FREEVAL density outputs follow the field density trend. Table 1 shows the Pearson product moment correlation coefficient (R^2) between the field and Vissim densities and the field and FREEVAL densities. A close match is achieved for all of the segments, for both Vissim and FREEVAL tools.

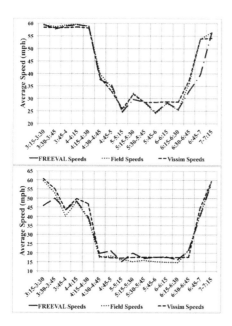

Figure 6. Speed match between field data, Vissim and FREEVAL output for basic (upper part) and weaving (lower part) segment.

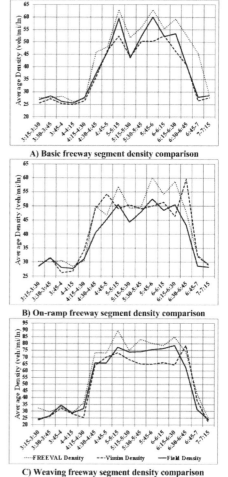

A) Basic freeway segment density comparison

B) On-ramp freeway segment density comparison

C) Weaving freeway segment density comparison

Figure 7. Density assessment for 3 freeway segment types.

Table 1. R^2 values depicting how reliably Vissim and FREEVAL replicate field density.

Segment	Vissim vs. field density R^2	FREEVAL vs. field density R^2
Basic	0.952	0.875
On Ramp	0.913	0.958
Weaving	0.803	0.971

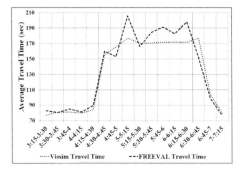

Figure 8. Total travel time output match between FREEVAL and Vissim.

However, the FREEVAL tends to overestimate density for all tested segments, which is also confirmed in previous study for undersaturated conditions. Higher density leads to higher total travel time on a modeled corridor as shown in Figure 8. The heavy congestion starts around 5 PM and lasts two hours and FREEVAL will reports higher travel times than Vissim. For all but two time periods, the discrepancy between travel times is less than 10%. Since FREEVAL is macroscopic in nature, these differences are in acceptable ranges.

Showing that FREEVAL is capable of modeling heavily congested conditions will allow practitioners to quickly assess problematic freeway segments. For example, by modeling current traffic conditions accurately, one can evaluate the impacts of proposed strategies (e.g., increased/decreased demand, additional capacity, managed lanes). Evaluation can also involve work zones, weather impacts and incident occurrence. FREEVAL is open source tool and it could be suitable for state and government agencies use. It can provide reliable and quick initial freeway facility estimates.

Modeling oversaturated road conditions is challenging because the demand exceeds capacity of the facility. The detectors will not record traffic demands, only traffic volumes. This paper assessed two conceptually different tools: Vissim microscopic traffic simulator and FREEVAL macroscopic tools, which is based on HCM2010 freeway methodology. The major conclusions are:

1. Both FREEVAL and Vissim are capable of assessing oversaturated freeway conditions in terms of speed and density.
2. Vissim is more time consuming when it comes to calibration and validation of the model.
3. The Capacity Adjustment Factor (CAF) should be altered to calibrate FREEVAL. The Speed Adjustment Factor (SAF) should not be applied.
4. Proper demand estimation will lead toward better FREEVAL outputs.
5. FREEVAL tends to overestimate density which confirms previous findings.
6. FREEVAL is a good option for state and government agencies to evaluate certain strategies and scenarios quickly.

This paper covers only three segments at one freeway facility. The field data collection was limited to traffic flows and spot vehicle speeds. Future research should test additional freeways facilities across the US, and have longer segments evaluated (e.g., model evaluate several weaving segments and on/off ramps). Also, it would be beneficial to have higher granularity of data from the field, such as vehicle headways, and traffic video recordings to observe the weaving maneuvers.

REFERENCES

Bloomberg, L. & Swenson, M. & Haldors, B. 2003. Comparison of Simulation Models and the HCM. *TRB 2003 Annual Meeting*, (p. 20p). Washington, D.C.
Hall, L.F. & Bloomberg, L. & Rouphail, N.M. & Eads, B. & May, A. 2000. Validation Results for Four Models of Oversaturated Freeways Facilities, *Transportation Research Record: Journal of the Transportation Research Board*: 161–170.
HCM 2010. *HCM 2010*. Washington, D.C.: TRB.
Jolovic, D. & Stevanovic, A. 2013. Evaluation of VISSIM and FREEVAL to Assess an Oversaturated Freeway Weaving Segment, *TRB Annual Meeting*, (p. 16p). Washington, D.C.
Jolovic, D. & Stevanovic, A. & Sajjadi, S., & Martin, P.T. 2016. Assessment of Level-Of-Service for Freeway Segments Using HCM and Microsimulation Methods. *7th International Symposium on Highway Capacity and Quality of Service* (pp. 403–416). Elsevier.
Milam, R.T. & Stanek, D. & Chris, B. 2006. The Secrets to HCM Consistency Using Simulation Models. *85th TRB Annual Meeting*, (p. 11p). Washington, D.C.
Trask, L. & Aghdashi, B. & Schroeder, B. & Rouphail, N. 2015. Freeway facilities and reliability analysis computational engine for the HCM 6th Edition: A guide for multimodal mobility analysis. Raleigh, NC: ITRE at North Carolina State University.
Zeeger, J. & Bonneson, J. & Dowling R, Ryus & P. & Vandehey, M. & Kittelson, W. (2014). *Incorporating Travel Time Reliability into the Highway Capacity Manual*. Washington DC: TRB.

Transport Infrastructure and Systems – Dell'Acqua & Wegman (Eds)
© 2017 Taylor & Francis Group, London, ISBN 978-1-138-03009-1

Evaluation method of urban intersection safety under mixed traffic

Tao Chen & Mengxue Li
Chang'an University, Xi'an, China

Lin Sun
Beijing Automotive Technology Center, Beijing, China

Bin Chen
Sichuan Vocational and Technical College of Communications, Sichuan, China

Yan Gao
The Institute for Traffic Management of Ministry of Public Security, Wuxi, China

ABSTRACT: The safety of intersections directly influences the traffic situation of the whole city. To improve intersection safety, this paper presents an evaluation method of urban intersection safety based on Traffic Conflict Technique (TCT) under mixed traffic involving vehicle, bicycle and pedestrian. The intersection risk degree is used as evaluation index, and the traffic conflict risk degree is calculated by using factors of Time-To-Collision(TTC), conflict speed and conflict angle. The conflict risk degrees include vehicle-vehicle conflict risk degree, vehicle-bicycle conflict risk degree and vehicle- pedestrian conflict risk degree. Then, the evaluation model of urban intersection safety is presented. Based on the model, the evaluation method is proposed using analysis of Vissim and SSAM (Surrogate Safety Assessment Model) Finally, series observation experiments are conducted to obtain the parameters of two urban intersections. The evaluation method is proved to be effective by the simulation.

1 INTRODUCTION

Compared with other urban road locations, urban intersections generate more traffic crashes because of considerable conflicts in motorized and non-motorized traffic, conflicts between vehicle traffic and bicycle traffic, vehicle traffic and pedestrian, bicycle traffic and pedestrian. Urban intersection congestion caused by mixed traffic has been a problem of city traffic management in China. About 30% of urban traffic crashes take place at intersections (Annual Bulletin, 2015). Many traffic accidents occur in urban intersection because of the chaotic traffic (Zhang et al. 1994). Hence, it is necessary to propose effective methods to estimate the dangerous intersections.

Some evaluation methods have been proposed for the problems of intersection safety. The traffic safety evaluation model currently is based mainly on historical traffic crash data or the Traffic Conflict Technique (TCT).

The safety evaluation model based on crash data is a postmortem analysis method intended for use after accidents. Obtaining the accident characteristics of small samples, a long collection cycle and stochastic processes necessitate a long period in determining safety improvement outputs; the length of time consumed translates to increased safety risks (Chin et. al 1997 & De et. al 2002). Chittoori et al. (2015) studied the skewed intersections using accident rate as evaluation index.

The safety evaluation model based on TCT focuses on conflicts arising in motorized traffic. Jarvis et al. (2012) developed an evaluation method based on traffic conflict technique, and the method is used to study the influence of vehicle-vehicle conflict to intersections. Sien Zhou et al. (2011) presented evaluation model to estimate intersections under mixed vehicle-bicycle traffic. Taking vehicle–vehicle conflict for research object, Wael K et al. (2014) analyzed intersection safety using PET (Post-encroachment time) as evaluation index. Some simple index is chosen to establish the evaluation model. Vedagiri et al. (2015) studied the evaluation method of uncontrolled intersections under mixed traffic conditions, and TTC (time-to-collision) is chosen as an evaluation index. Kiefer et al. (2005) developed an inverse TTC model to implement motorized traffic crash alerts when thresholds were surpassed. Previous work mainly focused on analysis of vehicle-vehicle conflict. And the selected evaluation index such as TTC and PET involved few parameters. For this reason, evaluation of intersection safety can not be accurate and comprehensive.

The structure of the paper is as follows. Section 2 introduces the evaluation model of urban intersection safety under mixed traffic and the proposed evaluation method. The simulation application using PTV VISSIM and SSAM are presented in Section 3. Sections 4 and 5 present discussions and conclusions drawn from the application, respectively.

2 MODELS AND METHOD

Using intersection risk degree as evaluation index, this paper calculated the traffic conflict risk degree, and established the intersection safety evaluation model. The parameters involved in the models included the conflict type, conflict number, TTC, conflict speed, and conflict angel, which were obtained by the simulation using PTV VISSIM and SSAM after field investigation.

2.1 Intersection traffic safety evaluation model

The traffic conflict risk degree reflects the potential risk of traffic conflict (Liu et al. 1997). In order to represent the risk degree accurately, typical parameters of traffic conflict risk model should be chosen firstly.

As the traffic conflict identification parameters, TTC is the time to the conflict of the vehicles (Jiang et al. 2015). It is an important index of the dangers of traffic conflict. As a result, TTC can be set as a parameter of the model.

Previous studies have shown that there is a certain relationship between the severity of the traffic and vehicle speed (Quaassdorff et al. 2016). Hence, in the process of traffic conflict, vehicle speed can represent the traffic conflict risk degree to a certain extent.

Because of the different traveling speed and direction of the conflict vehicles, the angle of the collision also has an influence on the severity of the crash when the conflict converts into collision (Alhajyaseen, 2015). In order to characterize the conflict risk degree more accurately, it is necessary to consider the size of the conflict angles. The conflict angle can be calculated by the speed angles between the conflict vehicles.

$$\alpha = |\alpha_1 - \alpha_2| \tag{1}$$

where: α is the conflict angle; α_1 is the speed angle of the front conflict vehicle; α_2 is the speed angle of the rear conflict vehicle.

In order to combine vehicle speed with conflict angle, the relative velocity of the conflict participants is used as the parameter of conflict risk degree:

$$V = \|v_1 - v_2\| = \sqrt{v_1^2 + v_2^2 - 2\cos\alpha} \tag{2}$$

where: V = relative speed of the conflict vehicles; v_1 = velocity vector of the front vehicle, it is the vehicle speed when one conflict party is pedestrian; v_2 = velocity vector of the rear vehicle, it is 0 when one conflict party is pedestrian.

Based on the previous analysis, TTC and relative speed can be used to represent the conflict risk degree, and the smaller the TTC and vehicle speed, the more serious the traffic conflict. The conflict risk degree is can be calculated as:

$$S = \frac{V}{TTC} = \frac{\|v_1 - v_2\|}{TTC} = \frac{\sqrt{v_1^2 + v_2^2 - 2\cos\alpha}}{TTC} \tag{3}$$

where: S = traffic conflict risk degree.

The value of TTC can be 0 sometimes, which lead to the useless of the formula (3). The correction conflict risk degree can be given as:

$$S = \frac{V}{e^{TTC}} = \frac{\|v_1 - v_2\|}{e^{TTC}} = \frac{\sqrt{v_1^2 + v_2^2 - 2\cos\alpha}}{e^{TTC}} \tag{4}$$

where: e is the constant.

Then, different type of conflict risk degree can be calculated. According to the different participants of traffic conflict, there are three kind of conflict risk degree: vehicle-vehicle conflict risk degree (R_v), vehicle-bicycle conflict risk degree (R_n) and vehicle-pedestrian conflict risk degree (R_p). The calculation is introduced as follow.

Vehicle-vehicle conflict risk degree:

$$R_v = \sum_{i=1}^{N_v} SV_i \tag{5}$$

where: i = the ith vehicle-vehicle conflict; N_v = the number of vehicle-vehicle conflict; SV_i = the ith vehicle-vehicle conflict risk degree.

Vehicle-bicycle conflict risk degree:

$$R_n = \sum_{i=1}^{N_n} SN_i \tag{6}$$

where: i = the ith vehicle-bicycle conflict; N_n = the number of vehicle-bicycle conflict; SN_i = the ith vehicle-bicycle conflict risk degree.

Vehicle-pedestrian conflict risk degree:

$$R_p = \sum_{i=1}^{N_p} SP_i \qquad (7)$$

where: i = the ith vehicle-pedestrian conflict; N_p = the number of vehicle-pedestrian conflict; SP_i = the ith vehicle-pedestrian conflict risk degree.

The intersection traffic safety evaluation model was composed by the vehicle-vehicle conflict risk, vehicle-bicycle conflict risk, and the vehicle-pedestrian conflict risk. The intersection traffic safety evaluation model was expressed as:

$$R = W_v \times R_v + W_n \times R_n + W_p \times R_p \qquad (8)$$

where: R = the intersection risk degree; W_v = the weights of vehicle-vehicle conflict influence on the safety of intersection; W_n = the weights of vehicle-bicycle conflict influence on the safety of intersection; W_p = the weights of vehicle-pedestrian conflict influence on the safety of intersection.

In this paper, we have analyzed the scoring method and adapted the way of scoring by the experts after a lot of investigations, and the scoring results were shown in Table 1 (Cheng et al. 2004).

The intersection risk degree was calculated and considered as the intersection traffic safety evaluation index. The risk degree of each intersection was calculated when evaluating the intersection traffic safety. The higher risk degree, the more dangerous the intersection, which was used to determine the relative risk intersection by comparing the risk values of the intersections.

2.2 Evaluation method

Based on above analysis, the evaluation method of urban intersection traffic safety is put forward. The implementation process is as follows:

1. Collect survey data. Field investigation is necessary to prepare the microscopic simulation parameters of the intersection, such as the intersection geometric parameters, signal control parameters, the traffic parameters and driver parameters.

Table 1. The weights of different traffic participants.

Traffic participants	Weight
Vehicle-Vehicle	0.25
Vehicle-Bicycle	0.33
Vehicle-Pedestrian	0.42

2. Vissim simulation. The intersection simulation model is set up via VISSM. Using the survey data, the intersection network, the signal control system, and the traffic flow were established in VISSIM and can be simulated virtually. Then, trajectory data can be obtained as output files.
3. Conflict analysis. Vehicle trajectory data is analyzed to identify conflict number, TTC, conflict speed, and conflict angle by using SSAM. And the conflict type can be judged according to the conflict angel: when conflict angel is less than 30°, the conflict is rear end; when conflict angel is larger than 85°, it is a crossing conflict; the others can be consider as lane change.
4. Calculate intersection risk degree. Calculate the intersection risk degree by the established evaluation models. Then, compare the intersection safety risk to determine the dangerous intersection.

3 APPLICATIONS

Field investigation was conducted at two intersections in Xi'an, China. The survey data involves traffic flow, traffic speed, traffic composition, traffic direction and traffic light time etc. Then, the data was used to simulate the traffic condition of the intersections by Vissim.

Then, trajectory data was for 60 minutes simulation for each intersection was used for SSAM analysis. The distribution of conflict types are shown in Figure 4 and Figure 5, and the number of conflicts are displayed in Table 2. Finally, the intersection risk degree can be obtained according to proposed evaluation model.

Based on Table 2, the total number conflicts of the intersection 2 was more than the intersection 1. The main conflicts in the intersection 1 were vehicle to vehicle conflicts, and the rear end conflicts made up most of the portion. While the main conflicts in the intersection 2 were vehicle to non-vehicle

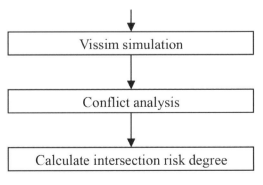

Figure 1. Flow chart of the evaluation method.

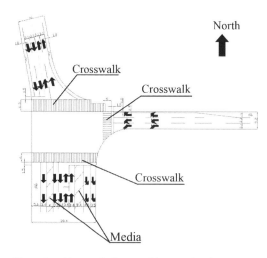

Figure 2.   Geometric figures of intersection 1.

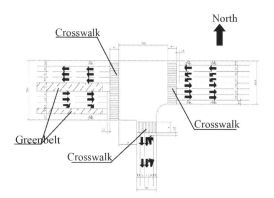

Figure 3.   Geometric figures of intersection 2.

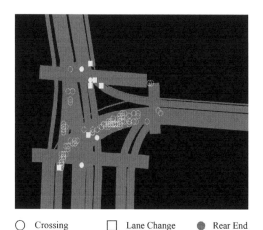

○ Crossing   □ Lane Change   ● Rear End

Figure 4.   Conflict types of intersection 1.

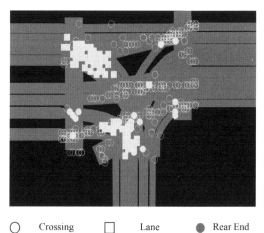

○ Crossing   □ Lane   ● Rear End

Figure 5.   Conflict types of intersection 2.

Table 2.   Number of conflicts.

	Intersection 1	Intersection 2
Total	110	399
Crossing	4	33
Rear End	100	231
Lane Change	6	135
Vehicle-Vehicle	100	229
Vehicle-Bicycle	10	161
Vehicle-Pedestrian	0	9

Table 3.   Intersection risk degree.

	Intersection 1	Intersection 2
R	100.3	527.1
R_v	346.4	1096.4
R_n	41.6	686.9
R_p	0	62.6

conflicts, vehicle to vehicle conflicts, and the rear end conflicts and the lane change conflicts made up most of the portion.

According to Table 3, the intersection 2 had a bigger risk degree than the intersection 1, that was 527.1. Compared with intersection 1, it was more dangerous, which catered to the real condition.

4   DISCUSSIONS

1. The total number of conflicts of intersection 2 is 399, which is far larger than intersection 1. At the intersection 1, the main component of the

traffic conflicts is vehicle-vehicle conflict, which is 91% of all. To conflict type, rear end accounts considerable proportion. At the intersection 2, besides vehicle-vehicle conflicts, there are 161 vehicle-bicycle conflicts. In addition to rear end, lane change accounts about one-third of whole conflicts.

2. The traffic conflicts are mainly distributed in the left-turn lane from east to south at intersection 1, and the major type is rear end. At intersection 2, many lane change conflicts occur at west exit and south exit, and rear end conflicts are mainly distributed in eastbound lane.

3. Evaluate results prove that intersection 2 is more dangerous, its intersection risk degree reaches up to 527.1, which is far above intersection 1. Traffic accident data from the traffic management department also shows that there are more accidents occur at intersection 2.

5 CONCLUSIONS

Studies were mainly focused on the macroscopic evaluation of the intersection safety, and few for microscopic analysis. Furthermore, the present intersection traffic safety evaluation method has difficulty to get the effective accident data. The scientific and the validity of the method can not be guaranteed. Considering the influence of bicycle and pedestrian to the traffic condition, this paper proposed a safety evaluation method to estimate urban intersections in China. This method can evaluate and predict the safety of urban intersections, and can also provide support for urban intersection design. Simulation using the data surveyed at two intersections shows the feasibility of the method, and traffic accident data given by traffic management department proves correctness of the method, which can be a reference for the future study of the intersection safety.

ACKNOWLEDGMENT

This work was supported by Science Foundation of Chinese Ministry of Transport (2015319812200); the Open Project Program of the key Laboratory of road traffic safety of Ministry of Public Security in China (2016ZDSYSKFKT05); Chinese Universities Scientific Fund (310822162018); and the Natural Science Foundation of Shaanxi Province of China (2016 JM5013).

REFERENCES

Annual Bulletin of Traffic Accident (2014). 2015. Traffic Administrative Bureau of Ministry of Public Security of People's Republic of China.

Alhajyaseen, WKM. 2014. The development of conflict index for the safety assessment of intersections considering crash probability and severity. *Procedia Computer Science*. 32: 364–371.

Alhajyaseen, WKM. 2015. The integration of conflict probability and severity for the safety assessment of intersections. *Arabian Journal for Science and Engineering*. 40 (2):421–430.

Chin, H.-C. & Quek, S.-T. 1997. Measurement of traffic conflicts. *Safety Science*. 26 (3): 169–185.

Chittoori Bhaskar, C.S.; Khanal, M.; Harelson, D. 2015. Safety evaluations for skewed intersections on low-volume roads case study. *Transportation Research Record*. 2472:236–242.

Cheng, W. & Li, J. 2004. The application of the fuzzy clustering method in evaluation of traffic safety based on traffic conflict at intersection. *Communication and Transportation Systems Engineering and Information*. 4 (2):48–55.

Chunping, Z. & Tianran J. 1994. Research on traffic safety evaluation at urban road intersection. *Journal of Tongji University*. 22 (1):47–52.

De Leur, P. & Sayed, T. 2002. Development of a road safety risk index. *Transportation Research Record*. 1784: 33–42.

Jarvis, A.; Tarek, S.; Mohamed, H.Z. 2012. Safety evaluation of right-turn smart channels using automated traffic conflict analysis. *Accident Analysis and Prevention*. 45: 120–130.

Jiang, X.; Wang, W.; Bengler, K. 2015. Intercultural analyses of Time-to-Collision in vehicle- pedestrian conflict on an urban midblock crosswalk. *IEEE Transactions on Intelligent Transportation System*. 16 (2): 1048–1053.

Kiefer, R.J.; LeBlanc, D.L.; Flannagan, C.A. 2005. Developing an inverse time-to-collision crash alert timing approach based on drivers' last-second braking and steering judgments. *Accident Analysis & Prevention*. 37(2): 295–303.

Liu, X. & Duan, H. 1997. Research on standard program of traffic conflict techniques at intersections. *Journal of Highway and Transportation Research and Development*. 14(3):29–34.

Quaassdorff, C.; Borge, R.; Perez, J.; Lumbreras, J.; de la Paz, D.; de Andres, J.M. 2016. Micro scale traffic simulation and emission estimation in a heavily trafficked roundabout in Madrid. *Science of the Total Environment*. 566: 416–427.

Sien, Z.; Jian, S.; Xiao, A.; Keping, L. 2011. Development of a conflict hazardous assessment model for evaluating urban intersection safety. *Transport*. 26(2):216–223.

Vedagiri, P. & Killi Deepak, V. 2015. Traffic safety evaluation of uncontrolled intersections using surrogate safety measures under mixed traffic conditions. *Transportation Research Record*. 2512:81–89.

Transport Infrastructure and Systems – Dell'Acqua & Wegman (Eds)
© *2017 Taylor & Francis Group, London, ISBN 978-1-138-03009-1*

Sustainable development of the transport system through rationalization of transport tasks using a specialised travel planner

I. Celiński, G. Sierpiński & M. Staniek
Faculty of Transport, Silesian University of Technology, Katowice, Poland

ABSTRACT: The article provides a discussion on the chosen approach to rationalization of the tasks performed by public transport operators. The operating methodology was founded on utilisation of the features offered by a tool known as Green Travelling Planner (GT Planner). It is a specialised travel planner created for purposes of multimodal and eco-friendly travelling. The planner is capable of setting very accurate parameters of demand in the road network on a level being far superior than questionnaire-based surveys of transport behaviour patterns. The precision with which the data concerning the demand parameters are established makes it possible to determine exact physical locations of their occurrence in a geocentric system of the transport network. The surveys conducted by the authors imply that, within the area of the Upper Silesian conurbation (Poland), such a planner will allow for a sample representative of transport behaviour patterns to be obtained on a level as high as even 6% of the population (once it becomes available to the general public). Exact knowledge of the structure of demand in the transport network enables undertaking substantiated efforts aimed at sustainable development of transport. The pursuit of this goal can be reinforced by rationalization of transport-related tasks based on the data acquired by means of GT Planner. There is a perfect correspondence between the methodology proposed and the extensive framework of actions undertaken by the EU (European Commission 2011).

1 INTRODUCTION

Contemporary transport systems are undergoing a significant transformation. Traffic flows close to the capacity are observed in most sections of the road network. At the same time, the potential of public transport has not been utilised to the fullest extent. Sustainable development of transport should lead to minimisation of negative environmental impact and satisfy people's needs (Beckerman W. 1994, Grizzle R.E. 1994). Public transport in most cities is based on the historically shaped transport networks, socio-economic and settlement structures. Such supply-related configuration of the transport system occurred in correlation with the limitations of the transport infrastructure network and technological constraints of public transport operators. One should also mention restrictions applicable to the possibility of a thorough analysis of processes that shape the demand for public transport (with dynamic distribution in time and space). First and foremost, in previous decades, no such dynamism was observed in terms of relocation of traffic generators and absorbents as it is today. The current relocation of objects characterised by high traffic-generating potential causes that traffic streams counted in thousands of road users move between different sections of the road network. In the

conurbation of Upper Silesia (Poland), for instance, forming a very compact grouping of 19 municipalities with a dense transport network and cluster development, examples of the above phenomenon can be observed in locations where a former mine having been closed down is replaced by a shopping centre (like in downtown Katowice, Poland). In such cases, in a vast territory of the given urban area, both spatial and temporal distribution of traffic changes within mere several months. These and similar examples of drastic changes to the traffic structure are also observed in other countries (Aguiléra A. et al. 2009). As a consequence of the foregoing, sustainable development of transport was and is hindered due to the processes of supply of and demand for public transport services which proceed differently. Within the last several decades, the dynamic changes being observed have applied to both supply of and demand for means of public transport. However, it is for the unbalanced demand, which, after all, is very difficult to balance as such, that transport has been developing unsustainably. It may even be claimed that due to the aforementioned issues, a fair share of demand for public transport services is now artificially suppressed. Unsatisfied demand in terms of public transport causes an increase in the individual traffic volume. The method presented in this article highlights well-grounded paths towards sustainability of transport and solutions to the related problems (Chakravarty A.K. & Sachdeva Y.P. 1998, Ionescu G. 2016, Kanister D. 2008, Okraszewska R. 2008, Monzon-de-Caceres A. & Di Ciommo F. 2016, Silva Cruz I. & Katz-Gerro T. 2015, Stanley J. & Lucas K. 2014).

In recent years, more and more of the parameters that determine the demand in the transport network have been available owing to Information Technologies (IT). Such information is available in relation to the location of demand generation and absorption, travel motivation and other aspects. Dynamic parameterisation of demand for public transport also presents an opportunities to accelerate the sustainable development of transport.

The article provides a discussion on the public transport demand parameterisation based on a specialised multimodal travel planner known as GT Planner. Accurate knowledge of the processes which shape/promote the public transport demand allows for rationalization of transport tasks. Demand measuring by means of GT Planner closely reflects customer expectations about the transport system, thus enabling the goal sustainable transport development (reasonable and community expected) to be pursued. Emphasis has been put on two levels of issues: the existing and the suppressed demand.

The matter of calibration of demand parameters in a relation between travel planner users and transport network users has also been addressed. The examples provided in the paper are based on partial results of the international project entitled "A Platform to analyse and foster the use of Green Travelling options" implemented under the ERA-NET programme and co-financed by the National Centre for Research and Development.

2 GT PLANNER CHARACTERISTICS

Planning is the very basis of most actions undertaken in the institutional as well as the personal dimension. With regard to travelling, planning has usually served the purposes of occasional travels made in longer distances. The development of technology has significantly increased the availability of IT tools capable of supporting or even replacing traditional travel planning in a simple manner. Nowadays, the foregoing also applies to obligatory travelling.

Travel planners should offer features which face up to the user's actual needs. The most popular planners are tools developed for specific urban areas. Their functionality is often limited to planning of public transport routes. Universal planners, on the other hand, namely those which are

not restricted to certain geographic limits, typically support planning of travels made by passenger car only. A disadvantage of such basic planners is usually the inability to fully analyse the available options for travelling between pre-defined points of start and destination.

Under the international project entitled "A platform to analyse and foster the use of Green Travelling options" (ERA-NET programme), a tool named Green Travelling Planner (GT Planner) was developed. It is not only a universal, but primarily a multimodal travel planner (Esztergár-Kiss D. & Csiszár Cs. 2015, Sierpiński G. et al. 2014, 2016, Sierpiński G. 2017). The algorithms implemented in the solution have enabled supporting the travelling person in the travel planning in several ways. Firstly, GT Planner is capable of indicating the optimum route for the specific (user-defined) travelling mode. It features eleven pre-defined travel types (including electric cars, which may require recharging at local charging stations, or urban bike and car rental services). Multimodal connections based on the Park & Ride and Bike & Ride systems have also be included. Another way to make use of the algorithms implemented is the option of comparing all the driving/walking options available within the given area in a route between points defined by the traveller. Not only does this solution provide the requested information, but also performs the educational function. Even for the travels made on a daily basis, travelling persons may not be aware of other options to cover the given route, and therefore their choice may not be reasonable or optimum. Moreover, four optimum route search algorithms have been implemented in the planner based on the following criteria: the quickest, the shortest, the cheapest and the greenest route.

The above features of GT Planner combined with access to information from multiple databases (defining characteristics of the transport infrastructure, public transport organisation, elevation traffic conditions etc.) provide a wide spectrum of possibilities and enable a comprehensive review (for all input data simultaneously) of travelling options available in the chosen area (the planner may be applied to any area). With regard to the problem discussed in the article, first and foremost, it is possible for the travelling person to acquire knowledge about more environment-friendly travelling opportunities along pre-defined routes. As for the change of transport behaviour patterns, it translates directly into the demand for eco-friendly travelling opportunities (public transport, bicycle, walking or motivation to buy an electric car). Secondly, since users frequently submit queries concerning specific modes of travelling between specific points, and these are archived in the planner, it supports the available organisational changes in public transport oriented towards identification of suppressed demand. This problem has been addressed more extensively further on in the article.

3 ROAD NETWORK DEMAND

The various ways in which demand is satisfied in the transport network are predominantly moulded/limited by three factors. The first of them is the existing transport infrastructure, the second one is the range of services offered by public transport operators, and the third is the spatial structure of the settlement and the socio-economic system (Handy S. et al. 2005., Lu X. & Pas, E.I. 1999). Bearing these limitations in mind, the travelling-related needs of inhabitants of the given area can be satisfied to a certain extent. Characteristics of the transport infrastructure constitute an exogenous variable for the demand structure, upon which public transport operators as well as people travelling by individual means of transport cannot exert any relevant impact (in a short time horizon). The portfolio of services rendered by public transport operators may be influenced to a far larger extent (for the sake of demand satisfaction). The problem can be brought down to the fact that the temporal and spatial structure of demand for individual transport network elements has been poorly recognised. Furthermore, the surveys currently conducted account for the existing demand only. Consequently, the general function of demand observed in the transport network may be noted as the following set of values:

$$d_t = \{G, A, T, TM, TP, \Delta t, R, MOT...\} \qquad (1)$$

where d_t = demand characteristic at time t; G = set of travel generation points; A = set of travel absorption points; T—set of travel start times; TM—set of transport modes; TP—set of transport routes, Δt—time window shift for travel start, R—set of space resistance functions; MOT– set of travel motivations.

The description of demand expressed as equation (1) disregards transport infrastructure parameters, including traffic organisation and control as well as congestion characteristics of the transport network which also affect the manner in which the transport network demand structure is moulded. They will be discussed with reference to rationalization of transport-related tasks in public transport.

In a broader perspective, the transport network demand function is composed of two disjoint sets described by equation (1):

$$d_t = d_o + d_s \qquad (2)$$

where d_t = overall demand in road network; d_o = observed (recordable, measurable) demand; d_s = suppressed demand (cannot be observed).

The set representing the demand being observed (recordable and measurable) pertains to transfers made in the transport network, namely those which can be described and parameterised in any manner. Regardless of the foregoing kind of demand, there is also supressed demand in the transport network (Szarata 2010, 2013). Supressed demand describes these transfers which cannot be realized (and consequently also observed and/or measured) for the following reasons:

- absence of suitable means of transport at the travel generation or absorption point, in space and/or time,
- long time to walk to the point of access to technical infrastructure of the given means of transport,
- low frequency at which transport fleet runs within the given transport network area,
- other, often subjective dependences of definitely local nature (and etiology).

The main distinction between the demand types introduced above, both observed and supressed, can be brought down to different values in sets TM and TP. Therefore, the key aspect of the analysis in question is the choice of means of transport (Buehler R. 2011). Certain insignificant differences may occur in the sets of shifts of the travel start time window and the function of space resistance. Parameter Δt defines the time by which the given travel may be accelerated/delayed. The parameter value affects the parameters of optimisation of the transfer route in travel planners. Time, in this respect, is generally the key factor in the demand analysis (Millward H. & Spinney J. 2011).

Therefore, demand in the transport network should be satisfied in three spheres:

- reflecting the observed demand,
- identifying the suppressed demand in the transport network and taking it into account in the transport services delivered,
- changing the culture of mobility (Okraszewska R. et al. 2014).

These three disparate, yet complementary trends in action should, on the other hand, reflect the existing transport infrastructure and its expansion capabilities, the existing traffic organisation and control as well as the possibilities to change them. Moreover, while adapting the infrastructure and the transport offer to the demand, one should take congestion data into account. Interesting insights can be reached through analysis of demand in individual age and profession groups (Limanond T. et al.

2011). However, it is beyond the thematic scope of this article to explain these phenomena.

Both the observed and the supressed demand may be established using the specialised travel planner, i.e. GT Planner. Actual (non-supressed) demand can be directly determined by means of the travel planner feature which consists in archiving the queries submitted at the planner server. Supressed (hidden) demand may partially be analysed using GT Planner as well, with reference to these queries submitted to the planner for which the GTAlg algorithm did not return any transfer route. Consequently, the planner user was forced to choose a different travelling mode to cover the distance between the start and the destination. What GT Planner also enables is the acquisition of indirect data used to establish the characteristics of supressed demand. In order to obtain specific values for this type of demand, one needs to apply extended computation methodology.

Figure 1 illustrates GT Planner's basic functionality in terms of parameterisation of the demand observed in the transport network. Queries concerning travel routes submitted to GT Planner may be saved against most parameters described in equation (1). The GT Planner user leaves data regarding their demand characteristics at the server in the form of a set (specific demand footprint on the server):

$$d_{usi(ti)} = \{G, A, T, TM, TP, R, ..., MOT\} \qquad (3)$$

where d_{usi} = demand characteristic of the ith network user at interval ti.

The travel time shift window, as accounted for in equation (1), has not been implemented in GT Planner yet. GT Planner users, while making use of its functions, form a representative sample with regard to the area serviced by the given travel planner. In highly developed countries, where nearly 100% of households are provided with Internet access, the foregoing theoretically makes it possible to obtain a 100% representative sample which nearly perfectly corresponds to the actual population. In 2014, the authors conducted extensive research of travel behaviour patterns displayed by inhabitants of the Upper Silesian conurbation in Poland. Results of the research, based on a sample of 14,000 inhabitants, imply that ca. 49% of them make use of navigation while travelling. Car navigation systems are capable of capturing nearly identical demand characteristics as the travel planner. In the same area, 41% of inhabitants plan their travels, which closely coincides with the studies performed in USA (Gretzel U. et al. 2007). Among those surveyed, 6% use travel planners on a daily basis. Popularity of travel planners grows year by year, which is particularly noticeable where there is a complicated system of public transport or a dense transport network. The foregoing data imply that it is possible to compare demand parameters using a sample of at least 6%, which is still 500% more than the typical sample size used in large agglomerations (Sierpiński G. 2016). Furthermore, such data are collected in a continuous manner (in GT Planner) which makes it possible to capture travel connections not always revealed in traditional surveys of transport behaviour patterns, which particularly applies to transfers made rarely and optionally. To recapitulate the foregoing, data concerning the observed demand may be collected across the aforementioned area with the accuracy of 6% up to 40% of the entire population number (the latter being a very optimistic threshold). These values show approximately to what degree traffic models may improve as a result of enhanced parameterisation of the demand structure in the given transport network.

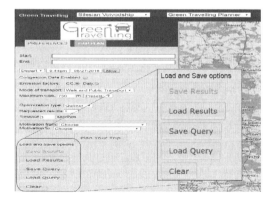

Figure 1. GT Planner—demand parameterisation feature.

4 SUPPRESSED DEMAND

Parameterisation of supressed demand may be conducted in a different manner (Szarata 2010, 2013). First and foremost, supressed travels cannot be physically observed as they are not being physically made at the current moment in the road network. Nevertheless, they are variants of the travels currently made for obligatory motivations and perhaps, to a certain extent, they do not exist under optional motivations at all. The notion of supressed demand may also apply to those streams whose observed volume is disproportionate to their traffic-generating potential. For instance, a person does not travel to a large shopping centre because of insufficiently good transport connection with the store location, at the same time, not being in possession of individual means of transport. Supressed demand may partially be analysed in GT Planner with reference to these queries submitted to the planner for which GTAlg did not return any response (transfer route parameters).

Equation (4) illustrates characteristics of supressed demand in the transport network.

$$d_t = \{G, A, T', TM', TP', \Delta t', R', MOT...\} \qquad (4)$$

where -'- corresponds to modified parameters from equation (1).

Therefore, this type of demand comprises the same traffic generation and absorption points and the same travelling motivations as those accounted for under the observed demand; what differs, however, is all the remaining parameters. As for the transfers connected with the supressed demand, the transport network user changes the travelling time (insignificantly—optional travels), means of transport, transfer routes in the transport network. What also changes is the set of preferences towards the time window shift for the travel start time as well as values of the space resistance function for the given transfer.

Figure 2 provides an example of the difference between travels included in the sets of observed and supressed demand. In this case, travelling by train is supressed (darker line in the figure). The lighter line in the figure marks the travel made to the location of supressed travel (this user prefers train). Such a transfer cannot be made due to unavailability of railway connection at the appropriate time of the day (however, there is an actual railway line between the travel start and destination points). As regards the given user of the transport network, walking to railway stations (grey cubes) may prove dangerous, since reaching the destination may cause exceedance of permissible space

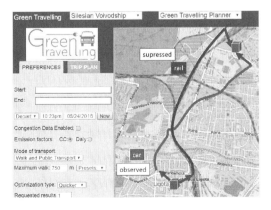

Figure 2. Observed vs. suppressed travel example.

resistance values for transfers made on foot etc. In such a case (see Figure 2), the train timetable should be adjusted in order to unlock the supressed demand. At the given travel generation/absorption point, one can establish a set of parameters affecting the suppression of travel by specific means of transport. Therefore, in the transport network, it should be delimited and divided into spatial territories, then to be characterised in terms of values of the demand suppression indices with regard to the given means of transport:

$$\sup rd_{TMx} = f(Y, A, T, P, s, sf...) \qquad (5)$$

where: sup rdTMx = suppression of means of transport x; x = {train, tram, bus}; Y—tram being present in the given spatial territory; respectively A,T—bus and train being present, P—share of walking paths in the transport network, s—mean slope, sf—safety parameters for pedestrian traffic etc.

Equation (5) expresses complex dependences: the presence of different means of transport within the same territory conditions the possibility to use them in a different manner. Availability of a bus supresses the possibility of using a train. Nevertheless, as in the example provided above, one can determine suppression indices for individual means of transport in any chosen spatial territory of the transport network. This, in turn, makes it possible to determine at least one alternative travel in the set of supressed demand for specific locations of traffic generation and absorption points (territory). Consequently, with reference to equation (2), for each travel from the set of those observed, it is possible to add one supressed travel with the highest value of the suppression index in the given spatial territory given by equation (5). Alternatively, it may be substituted by a travel for which GT Planner has not returned any transfer route. The very process of indexing spatial territories of the road network is performed using a GFTS (General Transit Feed Specification) standard database as well as smart parsers of data acquired from OSM (Open Street Maps). Based on the data stored according to the GTFS standard, values of availability are determined for the given means of transport in the given spatial territory at the given time (suppression indices are saved in dynamic arrays). Using the OSM data, one can read other parameters indicating the potential for suppression of the given means of transport in the chosen spatial territory.

The manner in which the spatial territory indexing procedure is conducted in the road network may be illustrated as shown in Figure 3. In this example (concerning the conurbation of Upper Silesia, Poland), it was assumed that fur-

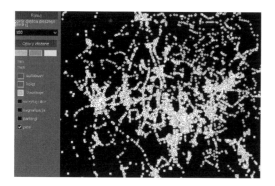

Figure 3. Indexing tool for transport network demand suppression.

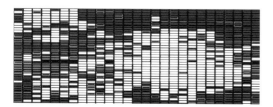

Figure 4. Index of demand suppression for PuT in the Upper Silesian conurbation.

ther transfers made on foot to distant points of access to public transport were supressed by the falling number of pedestrian crossings located in the roads along the transfer route. Consequently, the small number of pedestrian crossings in the given territory increases the time required to arrive at public transport stops. Besides technical aspects, there is also a psychological dimension in the travel suppression process (Choocharukul K. et al. 2008). Figure 4 provides a digital map for an index of this type. The lighter the colour in the given spatial territory, the smaller the demand suppression for public transport based on trams and/or buses. It corresponds to spatial territories where, in the demand model, each observed travel completed by individual means of transport can be substituted with one travel using public means of transport (the closer the colour is to white, the more the said assumption is legitimate).

If a travel is made in such a transport network based on the generator and the absorbent located in a spatial territory with the index being favourable to public transport, then the said action is fully justified. Using individual means of transport for purposes of this travel is due to other premises than purely transport-related ones (subjective perception of the transport network).

5 NETWORK VS. PLANNER CALIBRATION

Part of the travels planned (and identified) by means of GT Planner will never be actually realized The foregoing stems from various reasons: users testing the planner, abandoning a specific transport behaviour pattern, mistakes in using GT Planner, cases of the transport network being closed not taken into account by the planner, random reasons, no transfer route returned by the server etc. Therefore, in the total balance of travels made across the entire transport network, one can calculate a certain proportion:

$$dc = \left(j_p / j_r \right)_{\Delta t} \tag{6}$$

where: dc = planned demand accomplishment coefficient; j_p = total number of travels planned in the transport network measured and expressed as Δt using GT Planner (or/and car navigation tools); j_r = number of total travels observed in the physical network in a time interval;

Quantity $1/dc$ determines the accuracy of representation of the actual demand in the given transport network using GT Planner covering the respective area. The authors claim that this value may be approaching 1 as we witness further development of information technologies and car navigation solutions. It would mean attaining the capacity to represent nearly the entire demand in the transport network across the whole population.

The value of coefficient dc, when the exact demand structure is known (with the accuracy of individual transfer routes), can be calculated in cross-sections of transport network $(1/dc)^{int}$, where: int—road network cross-section.

Two notes should be made with regard to the value of coefficient (6):

– it is used to measure the reliability of tools similar to GT Planner as far as estimation of transport network demand is concerned (to a small extent, also with reference to hidden demand),
– the difference between the number of transfers established in the given measuring cross-section of the transport network and the number planned in the given planner covering the area in question may be used to calculate transit traffic in the given measuring cross-section. Furthermore, to some extent, the aforementioned difference indicates the quality of the services offered by public transport operators.

In transport networks, one can measure the actual number of transfers made in specific measuring cross-sections. This goal may be accomplished using traffic sensors, such as induction, video, radar and other types. On account of measurements conducted with reference to the entire multimodal cross-section (one which features more than a single transport mode), the measurement result expressed in the number of means of transport should be converted into the number of persons (or travels). It should be emphasised how dynamically real-time road traffic imaging methods evolve, this being a phenomenon favourable to the methodology discussed.

It is for the acquisition of data concerning the observed demand in the transport network cross-sections that traffic control (management) tools have been obtained. The value of coefficient dc^{int} may be noted as the following complex function:

$$dc^{int} = f(q_T, \overline{q}, C, d_{sup}, IND...) \qquad (7)$$

where: dc^{int} = coefficient of demand accomplishment in cross-section int; q_t = transit traffic intensity in the given cross-section; $\overline{q}$ = intensity of travels planned and not made; C—technical parameters of the transport network cross-section; d_{sup} = supressed demand in the given transport network cross-section; IND = index of suppression of the road network spatial territory where the cross-section is located. Addressing this problem is beyond the thematic scope of this article.

6 SUSTAINABLE DEVELOPMENT METHODOLOGY USING GT PLANNER

The authors claim that sustainable transport development by traffic modelling may proceed in two stages. It may be pursued as it has been done so far, namely by building well-grounded traffic models based on specific transport behaviour patterns being recorded (Janic M. 2016). In this article, the authors have indicated specific tools to build more reliable traffic models. Using such tools as GT Planner, one can theoretically build traffic models for samples that are representative at 6% and more. In practice, it translates into nearly costless augmentation of the sample size compared to those typically used in traffic model building by 500% with minimum financial outlays. Classical traffic structure models are incapable of representing supressed transport network demand sufficiently well. By indexing spatial (transport) territories of the transport network with the data acquired from OSM structures, one can determine suppression indices for individual means of transport within their areas. The data to be used for purposes of such an activity are available in data structures used in OSM maps. Based on an analysis of suppression

indices, in each spatial territory, one can identify an alternative means of transport with regard to transfers made by individual means of transport—one which ensures conditions favourable to its utilisation in the given territory. The foregoing offers a possibility to replace travels unfavourable from the sustainable transport development perspective in the traffic model with travels made by other means of transport. What becomes replaced is the least supressed travel by public transport. Figure 5 illustrates the working methodology to be assumed in this respect.

As shown in Figure 5, the actions undertaken in order to pursue the goal of sustainable transport development, assuming that one is familiar with the observed and the supressed demand (and accepting certain inaccuracies in this respect), are as follows:

- correction of public transport lines,
- correction of spatial arrangement of public transport access points (not necessarily connected with the line routing adjustment),
- changes to the traffic organisation which may unlock the supressed demand in the given spatial territory,
- changes in the scope of traffic control which may unlock the supressed demand in the given spatial territory.

However, a description of the foregoing is beyond the thematic scope of this article.

Most of the aforementioned actions allow for efficient rationalization of transport-related tasks performed by public transport operators using the resources at hand (they are minor organisational changes). Some changes, such as the relocation of network access points, require certain expenditures. Nevertheless, they are still far less costly than any physical interference in the network structure.

Much larger outlays are needed when introducing changes to the routing of the linear and nodal transport infrastructure as well as attempting to build new sections of the transport network.

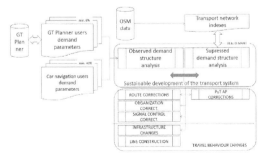

Figure 5. GT Planner based sustainable methodology.

Nevertheless, it is for the latter that one can near completely unlock the supressed demand. The authors claim that by undertaking efforts reaching beyond the infrastructure it is possible to unlock the supressed demand for public transport services to a considerable extent. In the course of the relevant studies, the authors developed simple tools used for supressed demand identification, which already allows for the development of transport to be more sustainable compared to what has been observed so far.

7 CONCLUSIONS

Using a travel planner offering features identical (or similar) to those provided by GT Planner as well as data delivered by car navigation systems makes it possible, as the authors claim, to succeed in parameterisation of the observed demand on the level of 6 up to 40 per cent. The foregoing conclusion relies on the analysis of statistical data from surveys of transport behaviour patterns conducted within the area of the Upper Silesian conurbation (with ca. 2.5 million inhabitants). There are interesting studies of the travel planning phenomenon pertaining to highly developed countries (Gretzel U. et al. 2007). They address a wider context of travel planning which also includes the support of guides. In these countries, 44.2% of those surveyed plan their travels 2 months in advance. In contrast, only 0.4 per cent of them are planning travels on an ongoing basis (on the go). Among those surveyed, 97% use Internet data sources. If one managed to persuade these people to plan their travels using travel planners (and not only guides) by way of targeted actions, it would translate into a 4,000 percent increase in the size of representative samples used in traffic modelling.

All things considered, demand suppression, although justified by financial reasons, technical and organisational parameters of the infrastructure, characteristics of traffic and, last but not least, lack of knowledge about of the latter, is invariably deleterious from the social perspective. The foregoing stems from the fact that the demand suppression process is iterative in nature. The historical maladjustment of the range of public transport services to the supply was duplicated throughout consecutive decades (generations) of changes occurring in the transport network structure as well as the contemporary mistakes. Further novel solutions are introduced every decade, yet they never account for the supressed demand. This issue essentially resembles erosion in the way it proceeds, for it will ultimately lead to a collapse of the transport system.

In developing countries, where the opportunities to shape the transport system on an ongoing basis are more widely open, the methodology proposed in this article should be obligatory (Verma A. Ramanayya T.V. 2014). For this purpose, in the said countries, one must also identify the supressed demand and extensively emphasise the role of the culture of mobility (Okraszewska R. et al. 2014).

It should be noted that the methodology discussed in this article highlights both the means and the actions intended for the sake of sustainable transport development which only require small financial expenditures (Stanley J. & Lucas K. 2014). Not only is the application of GT Planner for sustainable transport development purposes cheap, but it also requires uncomplicated computational procedures.

ACKNOWLEDGEMENTS

The present research has been financed from the means of the National Centre for Research and Development as a part of the international project within the scope of ERA-NET Transport III Future Travelling Programme "A platform to analyze and foster the use of Green Travelling options (GREEN_TRAVELLING)".

REFERENCES

Aguiléra, A., Wenglenski, S., Proulhac, L. 2009. Employment suburbanisation, reverse commuting and travel behaviour by residents of the central city in the Paris metropolitan area, *Transportation Research Part A* 43: 685–691.

Beckerman, W. 1994. *Sustainable Development: Is it a Useful Concept?* Environ. Values.

Buehler, R. 2011. Determinants of transport mode choice: a comparison of Germany and the USA, *Journal of Transport Geography* 19: 644–657.

Chakravarty A.K., Sachdeva Y.P. 1998. Sustainable transport policies for developing countries. In Freeman P., Jamet C. (eds.) *Urban Transport Policy: A Sustainable Development Tool.* Balkema, Rotterdam: 87–93.

Choocharukul K., Van, H.T., Fujii, S. 2008. Psychological effects of travel behavior on preference of residential location choice, *Transportation Research Part A* 42: 116–124.

Esztergár-Kiss D., Csiszár Cs. 2015. Evaluation of multimodal journey planners and definition of service levels. *International Journal of Intelligent Transportation Systems Research* 13: 154–165.

European Commission. 2011. *White Paper: Roadmap to a Single European Transport Area—Towards a competitive and resource efficient transport system.* COM(2011) 144.

Gretzel, U., Yoo, K.H., Purifoy, M. 2007. Online travel Reviever Study, Laboratory for Intelligent Systems in Tourism Texas A&M University, 4–7, College Station, Texas. http://www.tripadvisor.com/pdfs/OnlineTravel-ReviewReport.pdf

Grizzle R.E. 1994. Environmentalism Should Include Human Ecological Needs, *Bioscience 44*/4: 263–268.

Handy, S., Cao, X., Mokhtarian, P. 2005. Correlation or causality between the built environment and travel behavior? Evidence from Northern California, *Transportation Research Part D* 10: 427–444.

Ionescu G. 2016. *Transportation and the Environment: Assessments and Sustainability.* CRC Press Taylor & Francis Group.

Janic M. 2016. *Transport Systems: Modelling, Planning, and Evaluation.* CRC Press Taylor&Francis Group.

Kanister, D. 2008. The sustainable mobility paradigm. *Transport Policy* 15: 73–80.

Khattak, A.J., Rodriguez, D. 2005. Travel behavior in neo-traditional neighborhood developments: A case study in USA, *Transportation Research Part A* 39: 481–500.

Limanond T., Butsingkorn T., Chermkhunthod Ch. 2011. Travel behavior of university students who live on campus: A case study of a rural university in Asia, *Transport Policy* 18: 163–171.

Lu, X., Pas, E.I. 1999. Socio-demographics, activity participation and travel behavior, *Transport. Res. Part A* 33: 1–18.

Millward, H., Spinney, J. 2011. Time use, travel behavior, and the rural–urban continuum: Results from the Halifax STAR project, *Journal of Transport Geography* 19: 51–58.

Monzon-de-Caceres A., Di Ciommo F. 2016. *CITY-HUBs: Sustainable and Efficient Urban Transport Interchanges.* CRC Press Taylor & Francis Group.

Okraszewska R. 2008. *Przestrzenne sytuacje konfliktowe wywołane rozwojem systemu transportowego w warunkach równoważenia rozwoju,* Gdańsk University of Technology.

Okraszewska R., Nosal K., Sierpiński G. 2014. The Role Of The Polish Universities In Shaping A New Mobility Culture—Assumptions, Conditions, Experience. Case Study Of Gdansk University Of Technology, Cracow University Of Technology And Silesian University Of Technology. *Proceedings of ICERI2014 Conference*: 2971–2979.

Our Common Future. Report of the World Commission on Environment and Development, 1987. http://www.un-documents.net/wced-ocf.htm

Sáez A.E., Baygents J.C. 2014. *Environmental Transport Phenomena.* CRC Press Taylor & Francis Group.

Sierpiński G., Staniek M., Celiński I. 2014. Research And Shaping Transport Systems With Multimodal Travels—Methodological Remarks Under The Green Travelling Project. *Proceedings of ICERI2014 Conference*: 3101–3107.

Sierpiński G., Staniek M., Celiński I. 2016. Travel behavior profiling using a trip planner. *Transportatrion Research Procedia* 14C: 1743–1752.

Sierpiński G. 2017. Technologically advanced and responsible travel planning assisted by GT Planner. *Lecture Notes in Network and Systems* 2: 65–77.

Silva Cruz, I., Katz-Gerro, T. 2015. Urban public transport companies and strategies to promote sustainable consumption practices, *Journal of Cleaner Production,* http://dx.doi.org/10.1016/j.jclepro.2015.12.007

Stanley, J., Lucas, K. 2014. Workshop 6 Report: Delivering sustainable public transport, *Research in Transportation Economics* 48: 315–322.

Szarata A. 2010. Modelowanie symulacyjne ruchu wzbudzonego i tłumionego, *Transport Miejski i Regionalny* 3:14–17.

Szarata, A. 2013. The simulation analysis of suppressed traffic. *Advances in Transportation Studies*: 29: 35–44.

Verma A., Ramanayya T.V. 2014. *Public Transport Planning and Management in Developing Countries.* CRC Press Taylor & Francis Group.

Transport Infrastructure and Systems – Dell'Acqua & Wegman (Eds)
© 2017 Taylor & Francis Group, London, ISBN 978-1-138-03009-1

Tram and light rail transit in modern urban mobility systems

R. Emili
AIIT, Rome, Italy

ABSTRACT: Since 1960, in Rome as in all western world's cities, begins the exponential growth of individual transport by car. The consequence of this phenomenon was a gradual cities transformation: road saturation, increasingly intolerable levels of air pollution and noise, migration of inhabitants from downtown to suburbs and exasperated car commuting. These changes have led to sort out these issues is mandatory to rethink mobility systems traditionally based on highways and subway lines, much expensive in terms of *Life Cycle Cost* compared to cheaper and innovative *Light Rail* systems. The paper considers the comparison of *Life Cycle Cost* between different public transport modes, highlighting more favorable economic terms for *Light Rail* systems compared to other modes for a range between 5,000 and 12,000 phpdt. Combining tradition with innovation three innovative systems have set examples followed in many other cities around the world: Strasbourg *Euro Tram*, Dallas *Light Rail Transit*, Karlsruhe *Tram-Train* network.

1 INTRODUCTION

Since 1960, in all western world's cities, begins the exponential growth of individual transport by car.

As time passes, the increase in the individual motorization rate has caused a transformation of human habits and consequently changed the cities' shapes and layout (see Fig. 1).

These changes have also affected urban mobility systems using extensive tram networks.

Even in Rome, since 1960s, cars began to invade the streets, causing the need to have more city areas used by cars. Therefore, major public works were focused on continuous expansion of highway systems. The Rome tram system, at the time very large, was considered an obstacle to movement and parking of cars. For this reasons it was considered necessary to eliminate the trams to the city's streets. Throughout the years there has been a continuous global growth of cars. In fact, today, the rate of the growth cars has consistently outpaced that of human population.

In cities, this phenomenon has produced saturation of the available road surfaces, as well as increasingly intolerable levels of air pollution and noise (see Fig. 2).

In Rome, this process has produced the following paradoxical situation:

- Based on the 2014 census, residents in Rome are 2,872,021 people, with an average population density of 2,230.9 inhabitants/km^2;
- The cars owned by Rome's people are on average about 0.7 car/inhabitant, which corresponds to the presence in Rome of about 2 millions of cars;
- The area required for car parking is on average estimated at 12.35 m^2;

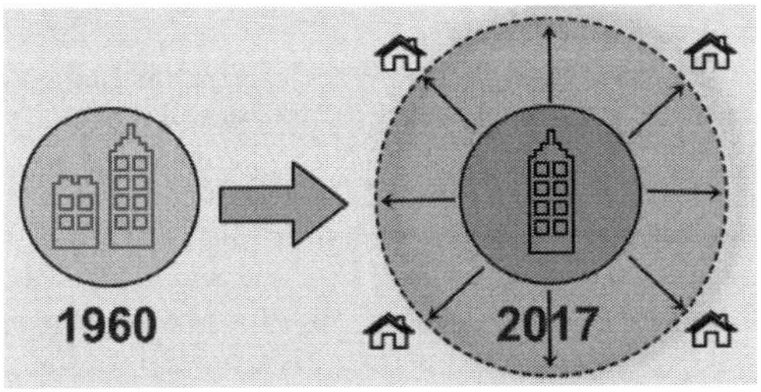

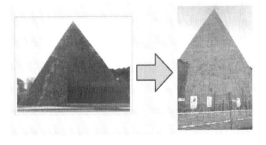

Figure 2. Rome, air pollution damage on the marble surface of the monument named *Pyramid of Caius Cestius*: to the left before the restoration works carried out in 2015, to the right after works.

Figure 1. The shape changes of the cities as a result of the mass motorization phenomenon.

- As a result, the area occupied by cars owned by the average resident population on 1 km² was approximately 0.01945 km², which is equivalent to about 1.9% of urban area;
- In Rome the artificial surface extension used for road networks amounted to about 1.47% of the entire municipal artificial surface (Italian National Conference ASITA, year 20011);
- The confrontation between areas occupied by municipal roads (1.47%) and those occupied by cars of resident people (1.9%), derive the following paradoxical situation (see Fig. 3): the area required for resident people's cars is 29.25% larger than the road network usable for their movement.

From statistical data regarding the movements of population living outside the municipality of Rome, carried out by the statistical research institute (ISTAT 2014), it follows that on weekdays traffic from areas outside the municipality of Rome would be an average magnitude of about 320,000 cars/day. Clearly this situation is further worsening compared to that shown in Fig. 3.

The paradox explains the following circumstances:

1. Expansion of motorcycle transportation mode;
2. Block of city traffic in rainy days that force a greater use of cars and the diminished use of motorcycles.

This urban disorder causes a migration process of the Italian resident population in Rome's central areas towards more peripheral areas (see Fig. 4). Several problems arise from this migration process, as the exasperated car commuting and social decay in suburbs due to the absence of a valid public transport linkage to the city downtown.

These changes concern all western world cities. In order to sort out these issues it is necessary to rethink mobility systems not only being based on the use of expensive highways and subways, but also on cheaper solutions in terms of both construction and operational *Life Cycle Costs(LCC)*.

For these reasons, considerations must be made on the reintroduction of modern trams and on the utilization of existing secondary railways, consti-

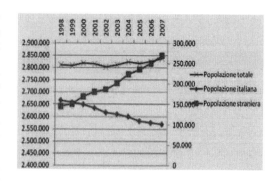

Figure 4. Rome, change resident population in the municipality area between 1998 and 2007.

tuting more economically sustainable mobility systems, especially for low-density residential areas.

In this regard it is especially significant all that was designed and built in the cities of Strasbourg *Euro Tram*, Dallas *Light Rail Transit*, Karlsruhe *Tram-Train* network.

2 LIFE CYCLE COST COMPARISON OF PUBLIC TRANSPORT SYSTEMS

An important component of a transport system design activity is to define the capital need and its use program, so that they can meet different needs according to priorities. The *Capital Project* includes *Life Cycle Cost* analysis.

In the 90 s, for the construction of a tram line in Rome (6 km long), to cross the city central areas, *Trastevere* district, a *Life Cycle Cost* analysis was made on a 30 years technical lifespan, to compare the construction costs and maintenance/operating costs of a LRT (Light Rail Transit) mode with respect to those required for an BRT (Bus Rapid Transit) mode.

The hypothetical calculations for an offer of a daily intake of passengers traveling in a single direction having a time pattern shown in Fig. 5, that during the daily service has little difference to the maximum peak (in this case between 7.30 am and 8.30 am).

The comparative Life Cycle Cost analysis on 30 years was repeated for different values of transportation capacity and for some possible economic cycles having different values of the inflation rate and interest rate on capital investment.

The Figure 6 shows the LCC graphs (value Euro 1998, inflation rate index 3%, interest rate index on capital investment 5%), obtained by adding up costs for Capital Expenditure (CAPEX) and for Operational Expenditure (OPEX) of the two systems in comparison, in function of: M*aximum*

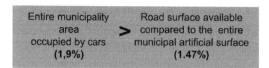

Figure 3. Rome, comparison between the area occupied by resident population's cars (on the left) and the area made available from the Municipal road network for their circulation (to the right).

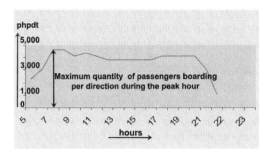

Figure 5. Daily offer diagram of the passengers boarding per direction on the central tram line n° 8 in Rome (through the *Trastevere* district).

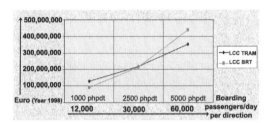

Figure 6. LCC comparison between a modern tram line (light rail) and a BRT line (Bus Rapid Transit).

PHPDT (stands for *Peak Hour Peak Direction Traffic:* quantity of passengers boarding per direction during the peak hour) and *Boarding passengers per direction/Day*.

Figure 6 shows that for more than 2,500 phpdt (or beyond a daily offer of 30,000 passengers/day) the *Rail Light* mode is cheaper than the *Bus Rapid Transit* mode.

For 5.000 phpdt, the different incidence for CAPEX and OPEX between the two projects in comparison is as follows (see Figs. 7, 8):

– CAPEX *Rail Light*: 41%;
– CAPEX *Bus Rapid*: 15%.
– OPEX *Rail Light*: 59%;
– OPEX *Bus Rapid*: 85%.

That comparison shows more initial expenditure required by the LRT system, that corresponds to a greater expense of a BRT system to be incurred during 30 years of operation. Comparative analysis can be extended, for higher transport performance, to other modes such as: Bus Rapid Transit, Light Rail Transit (at grade and elevated), Monorail and Metro.

For example, by applying the calculation method to a generic line having a 20 km length and a peak demand of 12,000 phpdt (levels of 12,000–15,000 phpdt are generally considered the feasibility limits

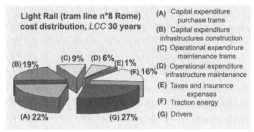

Figure 7. Light Rail: costs' incidence (5,000 phpdt).

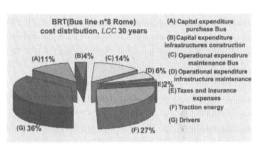

Figure 8. BRT: costs' incidence (5,000 phpdt).

for a LRT system), in the hypothesis of a economic cycle that approximates an yearly increment of 9% for workforce and 5% for other costs (see References: 8th Urban Mobility India Conference & Expo). In this case, setting equal to the number *1.0* the value of a *Light Rail Transit system (at grade)*, the calculation of each LCC per seat for every transport modes referred to above would lead to a costs ratio as shown in Fig. 9. In conclusion of the above, to the writer's opinion, it is possible draw the following considerations:

1. In high level demand corridors beyond 12,000 phpdt, the *Metro* mode is the only feasible choice;
2. In corridors of medium level demand, between 12,000 and 5,000 phpdt, the *Light Rail* modes are perhaps the best choice;
3. Between 5,000 and 2,500 phpdt, the choice lies in two modes: *Light Rail Transit* and *Bus Rapid Transit*. More detailed analysis is needed;
4. In low demand level corridors, under 2,500 phpdt, the choice lies on the BRT and Bus modes.

On average the *Index Construction Cost per km* (CAPEX/km), has the following quantity orders:

A. CAPEX *Metro* (underground) ≈ €120–200 millions/km;
B. CAPEX *Light Rail Transit* (at grade) ≈ €30–60 millions/km.

Metro rail	1.9
Monorail	1.8
Light Rail Transit (elevated)	1.5
Light Rail Transit (at grade)	1.0
Bus Rapid Transit	1.5

Figure 9. Offering of 12,000 phpdt, prices 2007. Results of the economic comparison of different transport modes in terms: LCC per seat.

These economic considerations have helped to guide the choice of important cities towards public transport in Light Rail modes, considered more effective technically and economically more efficient, especially for residential low-density areas.

3 STRASBOURG (EUROTRAM)

Strasbourg has about 270,000 inhabitants. Also in Strasbourg, as in many cities of western world, trams were removed in 1960 to leave urban space for car traffic and parking needs.

After 30 years, In 1990, to solve the problems created by phenomenon *Mass Motorization* (increasing traffic and pollution) it was decided to build a new *Public Rapid Transit* system.

The discussion on the system to choose was restricted between a system VAL (*Véhicule Automatique Léger*) and a modern tramway that could meet the *Total Quality* criteria, as summarized in Fig. 10.

At the end the decision was made for a modern Tram, for the following reasons:

1. A Capital Expenditure (CAPEX) about four times lower than VAL system;
2. Minimization of the Operational Expenditure (OPEX);
3. The tram line can be easily associated with redevelopment and pedestrianization of some valuable areas, such as city central area (In this case Place Kebler and adjacent streets);
4. The access to trams is possible with simple platforms with a decking situated just 30 cm from the surface road.
5. Minimizing construction time and inconveniences caused to population.

As already mentioned, the Strasbourg's tram project was developed according to the principles of the so-called *Total Quality* (see Fig. 10), which adheres to the following practical guidelines.

**Total Quality Criteria
used for the Tram designe**

1) Maximization of positive qualities
(merits exaltation)
2) Minimization of negative qualities
(defect reduction)
3) Maximization of safety levels

4) Minimization of construction and
operating costs

Figure 10. Four principles inherent the *Total Quality* policy applied to the project of a modern tramway system.

A. Maximizing of the positive qualities (exaltation of merits compared to competing modes)

A.1. Compared to Bus Rapid Transit mode, the bond constituted by the rails allows more length of the vehicles to increase transport capacity of each rolling stock. In fact, the original EUROTRAM had 33 m length and capacity 210 passengers (counted 4 pass/m^2). The successive model has a length of 43 m and the capacity of 288 passengers;

A.2. EUROTRAM has low floor and provides step-free boarding. It was designed to allow passengers in wheelchairs, as well as those with strollers and bicycles, to embark and disembark more quickly and safely (see Fig. 11);

A.3. Each tram consists of intercom coaches to promote a better passenger distribution inside.

A.4. A tram design adapted to the needs of urban insertion and characterization of every specific city.

B. Minimizing of the negative qualities (defects reduction compared to competing modes)

B.1. The main weakness of old traditional trams, compared to bus mode, was the operating rigidity due to the constraint of avoiding rails obstructions and, possibly, reverse travel direction to limit that service interruption only on the obstructed stretch. For these reasons EUROTRAM vehicles were equipped with driving cabs on both ends, so as to allow reverse travel on special railway switches arranged on line, avoiding the construction of bulky rings for march reversal of tram ride.

B.2. A weakness of the tramway mode is coming from the impossibility of overcoming

Figure 11. Eurotram vehicles at the *Homme de Fer* stop (photo in year 2012).

Figure 12. Dallas Central Business District and her very large residential area around.

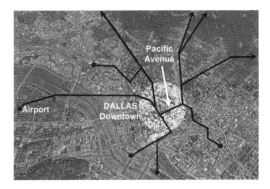

Figure 13. The radiating conformation of the LRT network deploys the transport service on most of the residential areas surrounding Dallas downtown.

any obstacle located on rails causing service interruption or delay of which users waiting at subsequent stops are unaware. To mitigate this situation on Strasbourg's tramway, all systems commonly in use on metropolitan lines were introduced: centralized control system, waiting times information, voice communication with users waiting at stops, etc.

B.3. To mitigate interference with road traffic and allow increasing commercial speeds, some streets have been devoted exclusively to trams. In correspondence of the road intersections have been installed special traffic lights for giving priority to the tramway passage. In addition, for elimination of any interference with existing structures on road surface, was built a line section in underground, with included a tram stop in correspondence with the main railway Station.

The first line section Strasbourg's tram was inaugurated in 1994 (9.8 km length).

The success of this innovative project in terms of urban qualification, transport effectiveness and economic efficiency has been so impressive that it reintroduced the tram mode all over the world.

The Strasbourg's tram system had been extended to a length of 42,7 km and today it is constituted by six lines (about 54 km in commercial operation).

For a network extension (11.8 km length) it was estimated an investment of €397.53 millions (year 2008, source: *Railway technology.com*), which corresponds to an *Index of the Cost Construction of Line* about €33.7 millions/km.

4 DALLAS *LIGHT RAIL TRANSIT* SYSTEM

Dallas is a typical American city (see Figs. 12, 13), with its central nucleus that spread over a vertical dimension (Downtown or Central Business District surrounded by a multitude of residential areas with very low population density (1000–1400 inhabitants/km^2) connected to downtown with a large highways network, always saturated by cars traffic at rush hour.

An exasperated car commuting has arisen with expansion of residential areas all around Dallas' central business district, which made it necessary to rethink the mobility system based on private cars.

These were motives why in the 90's works were initiated for the construction of a public transport system on rails called *LRT (Light Rail Transit)*.

The *LRT* system provides that rolling stock can travel on peripheral lines, or underground, with railway driving modes (driving assisted by an automatic signal system to allow speeds up to 110 km/h) and, instead, in residential areas they can travel on the street, at grade, with tram driving modes (travel to sight).

The Dallas' main road, Pacific Avenue, was closed to car traffic and devoted entirely to transit of the LRT vehicles (see Figs. 14, 15). Here the rolling stock proceeds at grade with tramway driving

Figure 14. Pacific Avenue. You can observe the abundant road signs prohibiting transit for cars (photo in year 2000).

Figure 15. Pacific Avenue entirely dedicated to the new public transport system on rails. On the left you can see the complicated system for lifting the disabled in wheelchairs to the level of the specific platform for their embarkation, located one meter above the rail (photo in year 2000).

mode. After Pacific Avenue the line through the rest of the city in tunnel, inside of which is also situated West End Station.

Originally every LRT rolling stock was composed by two articulated coaches, equipped with a driving cab at each end.

The length of one LRT rolling stock so composed was 28.24 m, with capacity of about 200 people.

The access to the vehicles load platform was located at a height of 100 cm from the rails, therefore passengers access was through means of steps (a realization that at first showed a substantial American mistrust of low floor tram solution, similarly to what was done in Strasbourg). To allow access to disabled persons in wheelchairs and baby strollers, on the station were installed ramps or lifting mechanisms on a platform situated at the height of 100 cm and conveniently positioned to allow driver's assistance (see Figures 15 and 16).

In 1996 the first stretch of 20 miles was inaugurated. Today, Dallas's LRT system extends approximately for 145 km and carries an average of 103,100 passengers per weekday. For the great success it was necessary to increase the vehicle's capacity, creating a new Super Light Rail Vehicle (SLRVs), by separating each LRT unit in two sections on their articulation joint and inserting an entirely new coach between them. Thereby rendering every LRT vehicle a three-coach operational unit, having a capacity of about 250 passengers. So was possible to constitute train of up four operational unit, having overall length 150.57 m and 1,000 passenger capacity. The middle coach of every *Super Light Rail Vehicle* has a low floor to allow a direct access (no steps) to the passenger in wheelchairs, as well as those with strollers and bicycles (each SLRV middle section was also equipped with a bicycle rack). By 2014 DART had converted all its No. 115 units LRVs in SLRVs. Today, with the purchase of No. 48 new units SRLT, is operating a total of No. 163 units SLRVs.

The Figure 13 shows the radial configuration that was given to the network for extending the service to the widest areas possible. In correspondence of each peripheral station large interchange parking were constructed. The project was carried out over a period of 20 years and it would seem to have required a financial investment very limited to

Figure 16. Underground stop *West End.* (photo in year 2000).

1086

that needed for an equivalent conventional metro system.

5 KARLSRUHE (TRAM-TRAIN NETWORK)

To meet the same requirements as described for Dallas, an original and innovative rail system called *TRAM-TRAIN* was adopted in Karlsruhe (Germany).

This system involves use of special rolling stock that can travel indifferently on railway lines and on tram lines (see Figures 17 and 18).

To create a rail system of this type it is necessary to solve regulatory, technical and organizational issues.

The German guidelines for heavy railway operation (EBO) are different from German guidelines for tramway (BOStrab).

The traditional trams need modifications, to be able to operate in a DC power environment, as well as with AC power. Consequently, a dual-mode light rail vehicle was developed.

Figure 17. Karlsruhe, Kaiserstraße. On the right is transiting a Tram operated by the municipal company VKB. On the left is transiting a vehicle TRAM-TRAIN operated by the specialist firm AVG (photo in year 2012).

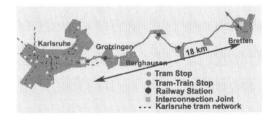

Figure 18. The first TRAM-TRAIN line from Karlsruhe to Bretten.

In 1992 the first *TRAM-TRAIN* line was inaugurated, from Karlsruhe to Bretten (see Fig. 18).

The Tram-Train vehicles operates between Karlsruhe (Marktplatz) and Grötzingen like a tram, respecting guideline BOStrab for tramway. In Grötzingen the Tram-Train vehicles change electric voltage (750VDC to 1,500VAC) and then operate as a heavy rail vehicle respecting EBO rail specifications on 18 km tracks towards Bretten. In Grötzingen the train's accountability is transferred from AVG tram driver to the operation manager of Deutsche Bahn AG.

This goal was possible only through the co-operation and co-ordination between cities political Administrators, as well as Local Transport Companies VKB (Karlsruhe tram operator), AVG (tram-train operator), Regional Rail Transport Companies (DB Regio AG-Rhein Neckar Region, DB Regio AG-Baden-Württemberg Regionen, and All of these operators are coordinated by KVV (Transport Association of Greater Karlsruhe).

Today, the Tram-Train network consists of more than 663 km (see Fig. 19) and, on basis the information provided by AVG, in the case of *Karlsruhe-Pfinzta* line, users grew from 4,000 passengers/day of the old traditional rail link, to 25,000 passengers/day for new Tram-Train link.

Today the Karlsruhe's Tram and Tram-Train networks carries 170 millions passengers per year.

In Karlsruhe's down town approximately 2 km of tramway line run through Kaiserstrasse, the main shopping street mostly pedestrianized (see Fig. 17). On that street, traffic of tram-train vehicles and trams has become so intense to make it necessary to built a tunnel on the southern branch line, from Marktplatz to Augartenstrasse, to carry underground very of this traffic. The *Stadtbahn* tunnel has been under construction since 2010.

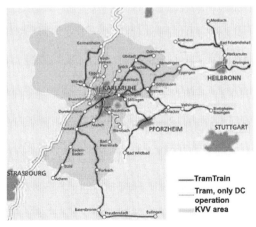

Figure 19. Karlsruhe, Tram and Tram-Train networks.

This transportation mode, using existing networks, requires a capital expenditure (CAPEX) virtually limited to purchasing only Tram-Train vehicles. Consequently, as Karlsruhe's experience shows, for the construction of Tram-Train systems, the main issue are not economical but rather of coordination and cooperation between Institutions and Operating Agencies at various levels, politicians, administrative, regulatory and operational.

6 CONCLUSIONS

At the conclusion of the above, it is the writer's opinion that combining tradition and innovation the modern Light Rail systems are technically and economically the most effective and efficient public transport systems for a range between 5,000 and 12,000 phpdt.

These systems are optimal for the connection between the ancient city with the extensive low density housing areas that in time have grown all around.

The Systems brought into service in Strasbourg (Eurotram), Dallas (Light Rail Transit) and Karlsruhe (Tram-Train network) are good examples that drive the relaunch of the local public transport by rail all over the world.

REFERENCES

Di Somma, A. 2011. Italian National Conference ASITA. In, *The development of the urban fabric of Rome since the war to date*. Reggia di Colorno: Parma.
Emili, R. 2008. 3th National Tramway Conference.
Singal, B.I. 2015. 8th Urban Mobility India Conference & Expo. In, *Life Cycle Cost Analysis of alternative Public Transport Modes*. New Delhi.
Utilization of the Life Cicle Cost analysis for evalutation of Tram mode compared with Bus mode. Rome.

All photos and graphics contained in the paper were performed by the author, with the exception of the following:

Fig. 4, documentation made public by Rome Municipality;

Fig. 12, tourist documentation (year, 2000)

Fig. 13, author's processing from the web program "Google Earth";

Figs. 18 e 19, documentation made public by KVV (Karlsruher Verkehrsverbund) and AVG (Albtal-Verkehrs-Gesellschaft).

Transport Infrastructure and Systems – Dell'Acqua & Wegman (Eds)
© 2017 Taylor & Francis Group, London, ISBN 978-1-138-03009-1

A methodology for evaluating the competitiveness of Ro-Ro services

V. Marzano & A. Papola
University of Naples Federico II, Naples, Italy

F. Simonelli
University of Sannio, Italy

ABSTRACT: This paper proposes a methodology for the appraisal of the effectiveness of Ro-Ro services in competition with the all-road alternative. The methodology is presented and applied with reference to the analysis of the intermodal corridor Italy-Spain. The relevant ports are firstly analyzed in terms of their maritime accessibility, that is measuring the factors usually taken into account by shipping companies in choosing new routes/services. Then, a potential basin and an effective basin are defined for each port, respectively the first encompassing the set of o-d pairs whose maritime shortest path passes through that port, the second containing the o-d pairs with an actual convenience in choosing Ro-Ro with respect to the road mode. The basins are evaluated with reference both to an ideal maritime supply, that is with all ports connected among themselves, and to the current maritime supply between Italy and Spain. This allows the identification of new potential attractive routes, whose effective basin are in turn determined by applying the same proposed methodology. Some aggregate impact measures (i.e. variation of vehicles · km and tons · km) useful for appraisal purposes are also calculated.

1 INTRODUCTION

The enhancement of the motorways of the sea between countries within the Mediterranean and the North Sea is a key topic in the transportation planning strategies of the European Union. Consistently, there is an increasing demand for studies and methodologies allowing for an effective appraisal of new maritime services, both in the private (e.g. maritime companies, freight shippers and forwarders) and in the public (e.g. regional agencies and administrations) sectors. In more detail, there is usually need for measuring the competitiveness of a port, or of a system of regional ports, in terms of both seaside and landside transportation accessibility, in order to evaluate the capability of attracting new maritime services with a significant demand basin. With reference to this issue, two main types of maritime connections can be identified in practice, depending on whether the sea mode is the only feasible transportation alternative (e.g. services between Northern Africa and European Mediterranean countries) or there is contemporary availability of the road mode (e.g. services between Italy and France/Spain).

This paper proposes an appraisal methodology, based on the implementation of a transportation supply model, allowing for a quantitative evaluation of the aforementioned transportation

accessibility and for the identification of new potentially attractive maritime services, in contexts with a remarkable competition with the all-road alternative. The methodology is presented and described through its application to the intermodal corridor between Italy and Spain, that is the study area is represented by the ports of western Italy and southern Spain. In more detail, the focus will be on the characteristics of the Italian ports. From a practical perspective, the methodology can be subdivided in two main phases. The first is focused on the preliminary identification of new maritime routes and is made up by three steps:

- firstly, maritime accessibility of ports under study is evaluated. Maritime accessibility is defined by measuring, for each pair of ports, some performance parameters usually taken into account by shipping companies in order to define the most convenient routes;
- then, the potential basin of each port is determined as the set of all the o-d pairs whose shortest path using sea mode passes through the considered port. This allows checking the overall accessibility of the port, both landside and seaside, as it comes from its geographical position, independently of the current level of transportation services;

- finally, the effective basin of each port is determined as the set of all the o-d pairs whose shortest path using sea mode passes through the considered port and is characterized by a generalized transportation cost convenient with respect to the competing all-road shortest path. This allows addressing properly the competition with the road alternative and also helps limiting the basin wherein transportation demand can be effectively captured by the Ro-Ro service.

The second phase, once potential new maritime services are identified accordingly with the outcomes of the preceding analyses, is based on their appraisal, that is on the identification of their effective basin and of their impacts on the overall freight system between the two countries.

The paper is organized as follows. Firstly, the freight supply model and the demand estimate adopted throughout the study are presented. Then, the two phases of the methodology are reported in two separate sections. Finally, conclusion and research developments are discussed.

2 MODEL IMPLEMENTATION

2.1 Supply model

The transportation supply model adopted for the study is part of a wider supply model, whose study area encompasses 49 countries, i.e. the EU27 countries plus Eastern Europe and all other Mediterranean coastal countries, developed as part of a transportation DSS for policy appraisal at a European level, see Cascetta et al. (2013) for details. The level of zonization is consistent with the NUTS2 geonomenclature, corresponding to the regional level for both Italy and Spain. As a result, 20 zones are identified in Italy and 15 in Spain. Three different networks are implemented for road, rail and maritime services respectively, with connections representing rail-road and sea-road intermodal modes. In more detail, rail mode takes into account both traditional and combined services, and sea mode considers both Ro-Ro and containerized services: for this aim, two different databases have been collected for containerized and not containerized services respectively, starting from web-based available information collected by ports, maritime companies and national authorities. With reference to the analytical model, level-of-service attributes are calculated using cost functions available in the literature, mainly taken from Russo and Modafferi (1999), Torrieri et al. (2001) and Vitillo (2011). Notably, with reference to the maritime mode, instead of applying a frequency based approach — see Cascetta et al. (2009) — a waiting time of 4 hours has been assigned to

each port. Finally, transportation generalized costs are calculated as linear combination of travel times and monetary costs, using the VTTS of 35€/h, as suggested for instance in Jeannesson-Mange (2006) and de Jong et al. (2014). The overall supply model has been then validated through comparison with aggregate measures of travel times and costs reported in several surveys carried out by national statistics bureaus across Europe.

2.2 Demand estimation

Freight flows between Italy and Spain should be estimated at the NUTS2 geographical level, consistent with the zonization level of the supply model. The starting point for the estimation is represented by the databases provided by the Italian (ISTAT) and Spanish (INE) national statistics bureaus. In more detail, ISTAT provides statistics on the external trade of Italy (COEWEB database). For each one of the ten commodity classes of the one digit NST/R goods nomenclature, the amount of trade between each Italian region and the Spain is available in value and quantity. Similar data are provided by INE about import/export between Spanish regions and Italy. Trade quantities are also available for four transport modes (road, sea, air, rail). In that respect, since Ro-Ro flows are recorded partly in road mode and partly in sea mode, together with container and bulk goods, data have been post-processed in order to extract only road and Ro-Ro flows, that is the target demand of the Ro-Ro service under study. In turn, the whole o-d matrix has been obtained through disaggregation of Italian regional flows towards Spain among Spanish regions and, similarly, of Spanish regional flows towards Italy among Italian regions, proportionally to the total amount of road trade generated and attracted by each region, available on EUROSTAT basis.

In order to account for the effects of the economic downturn, a comparison of trade data between 2007 and 2013 (latest available year at the time of this research) revealed a decrease in trade and then a slight recovery. Thus, in order to take into account a full recovery from the economic downturn, 2007 trade data have been assumed as reference. The total amount of trade between Italy and Spain by road and Ro-Ro was in 2007 greater than 7 millions of tons (7.143.891). The first 20 o-d regional pairs (out of 300) involve mostly Northern Italian regions from one side and Cataluña from the other side: for instance, trade between Lombardia and Cataluña has been 500.000 tons/year. It is worth mentioning that, apart from the mentioned regions, other significant flows are between Lazio and Cataluña (67.000 tons/year) and Campania and Andalucia (28.000 tons/year).

3 IDENTIFICATION OF POTENTIAL SERVICES

3.1 Definition of the maritime accessibility of ports

The first step for the identification of new potential maritime services between Italy and Spain is the evaluation of the maritime accessibility of ports under study, that is Savona, Genova, Livorno, Civitavecchia, Napoli, Salerno and Palermo from the Italian side and Barcelona, Valencia and Tarragona from the Spanish side. Maritime accessibility has been defined by calculating, for each o-d pair of ports, the following performance parameters usually taken into account for by shipping companies in order to define the most convenient routes:

- robustness of the schedule with reference to different speeds of the vessel: this allows absorbing any delays due, for instance, to unfavorable weather conditions. In the calculations vessel speeds from 16 to 24 knots have been considered;
- number of vessels needed for supplying a given weekly frequency;
- possibility of triangulation towards a third port, trying to minimize the inactivity of the vessel. In more detail, the distance of the furthest port to be reached with 1 and 2 departures/week respectively has been calculated;
- possibility of designing a "regular" schedule, i.e. with departures at the same time across days within the week.

Therefore, the analysis aims at selecting, within all the feasible pairs of ports, a subset of possible more attractive routes to be served. All calculations were carried out by means of the supply model described in the preceding section; results are reported in Table 1.

From this standpoint, all o-d pairs of ports within the study area can be regarded as potentially interesting, since there are not dominated ports both in the Italian and Spanish basins. Notably, the current supply pattern as per 2014 encompasses almost the most attractive pairs of ports. With reference to potential routes to be taken into consideration, the most promising are the Napoli/Salerno-Barcelona/Tarragona, Palermo-Barcelona/Tarragona and Savona-Barcelona/Valencia. Notably, the maritime services Napoli-Barcelona is preferable to Napoli-Tarragona, since the former allows 3 departures per week with constant schedule at a speed of 24 knots. Moreover, in the case of 2 departures per week, the vessel may serve with a triangulation a port within a radius of 206 nautical miles, therefore useful for a connection with Palermo (the distance between Napoli and Palermo is about 170 nautical miles).

Table 1. Maritime accessibility of ports under analysis.

origin	destination	distance [nautical miles]	vessel speed [knots]	overall transit time [h]	# departures/week	inactivity time [h]	furthest port for triangulation with 1 service/week [nautical miles]	furthest port for triangulation with 2 services/week [nautical miles]	# vessels needed for 3 departures/week	allows constant departure time
civitavecchia	barcellona	447	16	63.85	2	40.31	258	97	2	Y
			20	52.68	3	9.97	20	0	1	N
			24	45.23	3	32.31	292	98	1	Y
	tarragona	486	16	68.74	2	30.53	180	58	2	Y
			20	56.59	2	54.82	468	194	2	Y
			24	48.49	3	22.53	174	39	1	N
	valencia	585	16	81.11	2	5.78	0	0	2	N
			20	66.49	2	35.02	270	95	2	Y
			24	56.74	2	54.52	558	231	2	Y
genova	barcellona	372	16	54.47	3	4.59	0	0	1	N
			20	45.17	3	32.48	245	82	1	Y
			24	38.98	4	12.08	49	0	1	N
	tarragona	437	16	62.59	2	42.82	279	107	2	Y
			20	51.67	3	12.99	50	0	1	N
			24	44.39	3	34.82	322	113	1	Y
	valencia	547	16	76.34	2	15.32	59	0	2	Y
			20	62.67	2	42.65	347	133	2	Y
			24	53.56	3	7.32	0	0	1	N
livorno	barcellona	391	16	56.81	2	54.37	371	153	2	Y
			20	47.05	3	26.85	188	54	1	Y
			24	40.54	4	5.83	0	0	1	N
	tarragona	440	16	62.97	2	42.07	273	104	2	Y
			20	51.97	3	12.08	41	0	1	N
			24	44.64	3	34.07	313	108	1	Y
	valencia	543	16	75.82	2	16.36	67	1	2	Y
			20	62.26	2	43.48	355	137	2	Y
			24	53.22	3	8.35	4	0	1	N
napoli	barcellona	542	16	75.79	2	16.42	67	2	2	N
			20	62.23	2	43.54	355	138	2	Y
			24	53.19	3	8.42	5	0	1	N
	tarragona	613	16	84.58	1	83.42	603	270	2	Y
			20	69.26	2	29.48	215	67	2	Y
			24	59.05	2	49.90	503	203	2	Y
	valencia	679	16	92.82	1	75.18	537	237	2	Y
			20	75.86	2	16.29	83	1	2	N
			24	64.55	2	38.91	371	137	2	Y
palermo	barcellona	577	16	80.16	2	7.69	0	0	2	N
			20	65.73	2	36.55	286	103	2	Y
			24	56.10	2	55.79	574	239	2	Y
	tarragona	607	16	83.88	2	0.25	0	0	2	N
			20	68.70	2	30.60	226	73	2	Y
			24	58.59	2	50.83	514	209	2	Y
	valencia	689	16	94.11	1	73.89	527	232	2	Y
			20	76.89	2	14.22	62	0	2	N
			24	65.41	2	37.18	350	127	2	Y
savona	barcellona	351	16	51.90	3	12.30	34	0	1	N
			20	43.12	3	38.64	306	113	1	Y
			24	37.27	4	18.93	131	18	1	N
	tarragona	409	16	59.16	2	49.68	333	135	2	Y
			20	48.93	3	21.22	132	26	1	N
			24	42.11	3	41.68	404	154	1	Y
	valencia	534	16	74.77	2	18.46	84	10	2	Y
			20	61.42	2	45.17	372	146	2	Y
			24	52.51	3	10.46	30	0	1	N
salerno	barcellona	570	16	79.21	2	9.59	13	0	2	N
			20	64.97	2	38.07	301	110	2	Y
			24	55.47	3	1.59	0	0	1	N
	tarragona	632	16	86.97	1	81.03	584	260	2	Y
			20	71.17	2	25.65	177	48	2	Y
			24	60.65	2	46.71	465	184	2	Y
	valencia	739	16	100.33	1	67.67	477	207	2	Y
			20	81.87	2	4.27	0	0	2	N
			24	69.56	2	28.89	251	77	2	Y

3.2 Definition of the potential basin of ports

Once investigated the maritime accessibility of each pair of ports, with prior exclusion of unattractive pairs in the view of the shipping companies, the overall accessibility of a port, i.e. taking into account both its maritime and access/egress landside connections, should be measured. This accessibility is measured by means of the calculation of the potential basin of each port, defined as the subset of all the o-d pairs of regions whose shortest path using sea mode makes use of the considered port. The shortest path is calculated using transportation generalized cost as link impedance. Moreover, for a better representation of the phenomenon, each o-d pair has been weighted through economic attributes, that is GDP and number of firms, available on EUROSTAT basis. This allows calculation of the magnitude of each potential basin, in terms of percentage of these attributes on the overall Italy-Spain system. It should be stressed that such analysis does not consider the current supply pattern: in fact, all o-d pairs have been directly connected with ideal services running at a given vessel speed, variable between 16 and 24 knots. Otherwise, if only current services were taken into account, the actual accessibility of the port would be biased by the actual supply pattern (i.e. a port would present a reduced potential basin only because of a limited current number of weekly departures), and also possible threats deriving from competition between routes would not be correctly evidenced. On the contrary, the proposed approach allows defining the overall accessibility of the port as it comes from its geographical position, independently of the current level of transportation services and of the actual freight demand. Results are reported in Table 2 where, for practical purposes closer ports (e.g. Napoli and Salerno, Barcelona and Tarragona) are grouped together.

The widest potential basin is offered by the ports of Genova and Savona, encompassing 127 o-d regional pairs (out of 300) for Barcelona/Tarragona and Valencia routes—considering vessel speeds up to 20 knots—covering almost 45% of the GDP of the overall Italy-Spain system. However, the faster is vessel speed the lower is the basin: at 24 knots, the number of o-d pairs fails to 101 (35% of total GDP). This is mainly due to the competitiveness of the port of Livorno, whose services towards Barcelona/Tarragona and Valencia are profitable for vessel speed of 24 knots. In more detail, the potential basin of Livorno increases with the speed of the vessel, from a minimum of 53 o-d pairs (approximately 15% of overall GDP and firms) to a maximum of 79 pairs (24% GDP). The situation of Civitavecchia is completely different, with a potential basin reduced from 105 o-d pairs (30% GDP and firms) at 16 knots to 45 o-d pairs at 24 knots (15% GDP and firms). The potential basin of the ports of Napoli and Salerno is constant from 20 to 24 knots, with 60 o-d pairs (about 17% of GDP and firms), while for lower speeds is significantly smaller. Finally, the potential basin of the port of Palermo does not depend on the speed, since encompasses a limited number of captive o-d pairs. In conclusion, the Genova/Savona cluster has a wide but unstable potential basin, since there are some o-d pairs attracted by the competition of the port of Livorno. Analogously, Civitavecchia has a very attractive geographical position, but with a threat represented by competing services from Livorno and from the Napoli/Salerno cluster. Notably, Livorno is the only port able to consolidate its potential basin when vessel speed increases. Finally, Southern Italian ports are characterized by a captive potential basin, with a slight integration of o-d pairs captured from Napoli/Salerno to the port of Civitavecchia for increasing values of the vessel speed.

3.3 Definition of the effective basin of ports

Conclusions drawn in the previous section have been reached independently of the current maritime service supply and not taking into account the competition with the all-road alternative. Therefore, in order to measure explicitly the presence of such competing mode, the previous analysis has been particularized by introducing for each port an effective basin. This is defined as the set of all the o-d pairs of regions whose shortest path using sea mode passes through the considered port and is characterized by a generalized transportation cost lower than a prefixed percentage threshold with respect to the competing all-road shortest path. Therefore, the effective basin can be seen as the subset of the potential basin wherein transportation demand can be effectively captured by the Ro-Ro service. The shortest path is calculated, as described above, with reference to the transportation generalized cost; three different convenience thresholds have been taken into account, respectively 0%, 5% and 10%. The analysis has been carried out again under the hypothesis of all o-d pairs of ports connected by services running at the same speed, variable between 16 and 24 knots. This also allows checking the robustness and the opportunities of different o-d pairs through evaluation of the change of their effective basins for different vessel speeds. Results are reported in the following Table 3, replying the structure of Table 2.

Comparing Table 3 (threshold 0%) and Table 2, a significant reduction of the size of the basin of the Genova/Savona cluster, whose effective basin encompasses for 24 knots 44 o-d pairs (15% GDP and firms) in contrast with the 127 o-d pairs of the corresponding potential basin, can be observed. Livorno also falls from the 79 o-d pairs of the potential basin to the 59 (20% GDP and firms) of the effective basin. On the contrary, the effective basin of the port of Civitavecchia is unchanged, mainly with reference to the maritime connections

Table 2. Summary of the potential basins of the ports.

origin	destination	16 knots			18 knots			20 knots			22 knots			24 knots		
		o-d pairs	% GDP	% firms	o-d pairs	% GDP	% firms	o-d pairs	% GDP	% firms	o-d pairs	% GDP	% firms	o-d pairs	% GDP	% firms
civitavecchia	barcelona/tarragona	91	26.24%	26.46%	117	21.08%	21.09%	39	12.06%	11.04%	36	11.38%	10.42%	30	9.44%	8.49%
civitavecchia	valencia	14	4.65%	4.98%	6	2.12%	2.09%	6	2.12%	2.09%	9	2.80%	2.70%	15	4.75%	4.63%
livorno	barcelona/tarragona	39	10.82%	11.50%	39	10.82%	11.50%	39	10.82%	11.50%	52	16.94%	17.59%	65	19.93%	20.48%
livorno	valencia	14	5.18%	5.55%	14	5.18%	5.55%	14	5.18%	5.55%	14	5.18%	5.55%	14	5.18%	5.55%
napoli/salerno	barcelona/tarragona	-	-	-	26	5.16%	5.37%	52	14.18%	15.42%	48	13.47%	14.54%	40	11.12%	11.86%
napoli/salerno	valencia	-	-	-	8	2.53%	2.89%	8	2.53%	2.89%	12	3.30%	3.77%	20	5.59%	6.45%
palermo	barcelona/tarragona	11	4.00%	4.20%	10	3.56%	3.71%	10	3.56%	3.71%	10	3.56%	3.71%	10	3.56%	3.71%
palermo	valencia	4	1.35%	1.49%	5	1.79%	1.98%	5	1.79%	1.98%	5	1.79%	1.98%	5	1.79%	1.98%
genova/savona	barcelona/tarragona	117	43.53%	41.65%	117	43.53%	41.65%	117	43.53%	41.65%	104	37.41%	35.56%	91	34.42%	32.66%
genova/savona	valencia	10	4.23%	4.18%	10	4.23%	4.18%	10	4.23%	4.18%	10	4.23%	4.18%	10	4.23%	4.18%

Table 3. Summary of the effective basins of the ports under study for different convenience threshold with respect to the all road alternative.

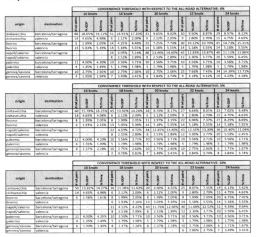

origin	destination	CONVENIENCE THRESHOLD WITH RESPECT TO THE ALL-ROAD ALTERNATIVE: 0%														
		16 knots			18 knots			20 knots			22 knots			24 knots		
		# od pairs	%GDP	%firms	# od pairs	%GDP	%firms	# od pairs	%GDP	%firms	# od pairs	%GDP	%firms	# od pairs	%GDP	%firms
civitavecchia	barcelona/tarragona	68	18.65%	19.11%	51	16.91%	17.20%	32	9.65%	9.02%	30	9.50%	8.87%	29	8.97%	8.12%
civitavecchia	valencia	14	4.65%	4.98%	6	2.12%	2.09%	6	2.12%	2.09%	9	2.80%	2.70%	15	4.75%	4.63%
livorno	barcelona/tarragona	8	1.99%	2.05%	14	4.25%	4.48%	23	7.25%	7.70%	30	10.22%	10.70%	45	14.73%	15.31%
livorno	valencia	13	5.03%	5.39%	14	5.18%	5.55%	14	5.18%	5.55%	14	5.18%	5.55%	14	5.18%	5.55%
napoli/salerno	barcelona/tarragona	-	-	-	24	4.95%	5.14%	48	13.46%	14.60%	45	12.83%	13.87%	40	11.12%	11.86%
napoli/salerno	valencia	-	-	-	8	2.53%	2.89%	8	2.53%	2.89%	12	3.30%	3.77%	20	5.59%	6.45%
palermo	barcelona/tarragona	11	4.00%	4.20%	10	3.56%	3.71%	10	3.56%	3.71%	10	3.56%	3.71%	10	3.56%	3.71%
palermo	valencia	4	1.35%	1.49%	5	1.79%	1.98%	5	1.79%	1.98%	5	1.79%	1.98%	5	1.79%	1.98%
genova/savona	barcelona/tarragona	10	2.75%	2.66%	10	2.75%	2.66%	10	2.75%	2.66%	25	7.68%	7.43%	34	14.39%	13.71%
genova/savona	valencia	4	1.95%	1.94%	7	3.49%	3.41%	8	3.84%	3.74%	9	4.18%	4.12%	10	4.29%	4.18%

origin	destination	CONVENIENCE THRESHOLD WITH RESPECT TO THE ALL-ROAD ALTERNATIVE: 5%														
		16 knots			18 knots			20 knots			22 knots			24 knots		
		# od pairs	%GDP	%firms	# od pairs	%GDP	%firms	# od pairs	%GDP	%firms	# od pairs	%GDP	%firms	# od pairs	%GDP	%firms
civitavecchia	barcelona/tarragona	60	15.78%	16.25%	45	13.92%	14.24%	28	8.76%	8.17%	27	8.64%	8.01%	23	7.02%	6.48%
civitavecchia	valencia	14	4.65%	4.98%	6	2.12%	2.09%	6	2.12%	2.09%	9	2.80%	2.70%	15	4.75%	4.63%
livorno	barcelona/tarragona	8	1.99%	2.05%	8	1.99%	2.05%	11	2.97%	3.15%	22	6.96%	7.37%	27	8.29%	8.69%
livorno	valencia	6	1.93%	2.16%	13	5.03%	5.39%	14	5.18%	5.55%	14	5.18%	5.55%	14	5.18%	5.55%
napoli/salerno	barcelona/tarragona	-	-	-	22	4.59%	4.72%	44	12.45%	13.40%	43	12.33%	13.30%	36	10.40%	11.04%
napoli/salerno	valencia	-	-	-	8	2.53%	2.89%	8	2.53%	2.89%	12	3.30%	3.77%	20	5.59%	6.45%
palermo	barcelona/tarragona	11	4.00%	4.20%	10	3.56%	3.71%	10	3.56%	3.71%	10	3.56%	3.71%	10	3.56%	3.71%
palermo	valencia	4	1.35%	1.49%	5	1.79%	1.98%	5	1.79%	1.98%	5	1.79%	1.98%	5	1.79%	1.98%
genova/savona	barcelona/tarragona	8	2.37%	2.28%	10	2.75%	2.66%	10	2.75%	2.66%	5	1.79%	2.60%	6	1.71%	1.67%
genova/savona	valencia	-	-	-	2	0.75%	0.82%	7	3.49%	3.41%	8	3.84%	3.74%	8	3.84%	3.74%

origin	destination	CONVENIENCE THRESHOLD WITH RESPECT TO THE ALL-ROAD ALTERNATIVE: 10%														
		16 knots			18 knots			20 knots			22 knots			24 knots		
		# od pairs	%GDP	%firms	# od pairs	%GDP	%firms	# od pairs	%GDP	%firms	# od pairs	%GDP	%firms	# od pairs	%GDP	%firms
civitavecchia	barcelona/tarragona	50	13.82%	14.27%	38	11.28%	11.63%	20	4.90%	4.55%	25	8.07%	7.55%	19	6.13%	5.62%
civitavecchia	valencia	14	4.65%	4.98%	6	2.12%	2.09%	6	2.12%	2.09%	9	2.80%	2.70%	15	4.75%	4.63%
livorno	barcelona/tarragona	6	1.58%	1.61%	8	1.99%	2.05%	8	1.99%	2.05%	14	5.09%	5.39%	20	6.75%	7.02%
livorno	valencia	-	-	-	6	1.93%	2.16%	13	5.03%	5.39%	14	5.18%	5.55%	14	5.18%	5.55%
napoli/salerno	barcelona/tarragona	-	-	-	21	4.11%	4.22%	43	11.76%	12.66%	40	11.68%	12.52%	32	9.39%	9.84%
napoli/salerno	valencia	-	-	-	8	2.53%	2.89%	8	2.53%	2.89%	12	3.30%	3.77%	20	5.59%	6.45%
palermo	barcelona/tarragona	11	4.00%	4.20%	10	3.56%	3.71%	10	3.56%	3.71%	10	3.56%	3.71%	10	3.56%	3.71%
palermo	valencia	4	1.35%	1.49%	5	1.79%	1.98%	5	1.79%	1.98%	5	1.79%	1.98%	5	1.79%	1.98%
genova/savona	barcelona/tarragona	6	1.40%	1.40%	8	2.37%	2.28%	8	2.37%	2.28%	-	-	-	-	-	-
genova/savona	valencia	-	-	-	-	-	-	-	-	-	6	2.86%	2.75%	7	3.49%	3.41%

towards Barcelona: at 16 knots there are 82 o-d pairs (24% of GDP and firms), 68 involving Barcelona and 16 Valencia. The same applies for Southern Italian ports, whose effective basin coincides with the potential basin. These considerations, together with the geographical structure of the study area, suggest the existence of an "arc effect", that is the competitiveness of the maritime service with respect to the corresponding all-road alternative increases as the ratio between the length of the road path and the length of the maritime path increases. The point of equilibrium of this effect can be located in proximity of the port of Civitavecchia: in fact, the competition of the road mode becomes significant for the Northern Italian ports and for Barcelona/Tarragona, while Southern Italian ports and Valencia serve mode-captive basins. Notably, for Northern ports, the competition within the maritime mode arises, with an interesting elasticity with respect to the vessel speed. For instance, as vessel speed increases, there is a reduction of competitiveness for the Civitavecchia-Barcelona route, leading to 45 o-d pairs for 24 knots (15% of GDP and firms), while the Civitavecchia-Valencia route firstly decreases (6 o-d pairs at 20 knots) and then increases up to 15 o-d pairs (5% of GDP and firms) for 24 knots. This is partly due to the increase of the Genova/Savona-Valencia route, whose effective basin firstly increases and then decreases (4 o-d pairs at 16 knots, 8 o-d pairs at 20 knots and again 4 o-d pairs at 24 knots). Therefore, Barcelona and Valencia are better connected with Civitavecchia for extreme values of vessel speed (16 or 24 knots), while for intermediate speeds (20 knots) there is a stronger competition with the Genova/Savona

cluster. A cross sectional comparison of the different thresholds in Table 3 allows drawing some conclusion relatively to the robustness of the effective basins. Indeed, a significant size reduction from the 0% to the 10% thresholds suggests a strong competition of the all-road alternative, therefore a slight worsening of the performances of the maritime service in that context would result in a considerable loss of competitiveness. In that respect, the aforementioned arc effect applies again: the effective basins of the Genova/Savona cluster are remarkably lower for 10% with respect to 0%. This means, as expected, that the convenience of the maritime route is very similar to that of the all-road route, with the same magnitude in terms of times and costs. Vice versa, southern ports and routes are characterized by a significant convenience of the maritime service, leading to more captive basins. In that respect, the effective basin of the Napoli/Salerno cluster for 24 knots is the largest. The same applies also to Spain: the Barcelona/Taragona cluster is more attractive but less stable than Valencia, whose effective basins are very similar in their structure to those of Southern Italian ports. Finally, the comparison between potential and effective basins allowed for a better understanding of the accessibility of the ports under study and of the opportunities and threats coming from the competition with the all-road alternative. Therefore, taking into account the current supply of maritime services between Italy and Spain, as per 2014, in the following attention will be focused on new maritime services between Napoli and Barcelona (2 departures/week) with triangulation to Palermo (1 departure/week), and between Savona and Valencia (up to 3 departures/week).

4 EVALUATION OF NEW MARITIME SERVICES

4.1 Introduction

The preliminary identification of potential maritime services carried out in the previous section aimed at identifying routes satisfying necessary conditions, both in the light of the shipping companies and of the wideness of the effective demand basin, to be attractive for new maritime services. Notably, these analyses have been carried out under the assumption of an ideal maritime supply pattern, with all ports connected among themselves through services running at the same speed. Obviously, a proper appraisal of the identified potential routes Napoli-Barcelona (with triangulation to Palermo) and Savona-Valencia would require relaxing this assumption, that is the approach of the previous section should be applied with reference to the current supply pattern—both in terms of routes, speeds and fares. As a result, Table 4 reports the result of the same calculation of Table 3 with reference to the current supply pattern; that is, the current real effective basins of the ports under study are analyzed.

Table 4 shows that the 52% of the overall freight flow between Italy and Spain (equal to 7.143.891 tons) are attracted by the Ro-Ro services for a 0% convenience threshold; the same percentage reduces to 21% for a 5% threshold and to the 13% for a 10% threshold. Notably, considering the current supply, a mean load factor of 0.70 and a cargo capacity of 2000 tons/departure, the overall yearly capacity of the intermodal corridor is approximately 2 million tons, between the total amount transported for convenience threshold of 0% and 5%: this confirms the reliability of the approach and the assumption of deterministic mode choice behavior. The significant variation of the effective basin with respect to the convenience threshold can be explained in the light of the arc effect described in the previous section. Indeed, Northern ports are characterized by a richer demand basin, but with a larger instability due to the significant competition of the all-road alternative, while Southern ports are characterized by a more limited but stable effective basin. In quantitative terms, moving from the convenience threshold of 0% to 10%, the percentage of freight demand captured by the Genova/Savona cluster falls from 61% to 26%, while that of the Napoli/Salerno cluster increases from 4% to 14%. Consequently, with reference to the competition with all-road alternative, the route Savona-Valencia is expected to capture more demand with respect to the Napoli-Barcelona, but with a more unstable market.

4.2 Napoli-Barcelona

The effective basin of Napoli-Barcelona route is determined by introducing the route in the current supply pattern, with a fare similar to the current services and a transit time coming from a mean vessel speed of 20 and 24 knots respectively. Results are reported in Table 5.

Consistently with the above, both at 20 and 24 knots there is no substantial increase (about 1%) in the total amount of demand attracted by the intermodal corridor Italy-Spain for any convenience threshold. Moreover, there is no remarkable advantage in running the service at 24 speed rather than 20, with a significant advantage for the shipping companies. The new service does not interfere with the services from Livorno and Northern ports, while subtracts an amount of freight demand to the port of Civitavecchia equal to the load of the new service is subtracted. There is no competition between the ports of Napoli and Salerno, since the flows of the latter to/from Valencia do not change between current and future scenarios. The overall demand flow is approximately 400.000 tons/year for 24 knots and 0% convenience threshold and about 230.000 tons/year for 20 knots and 10% convenience threshold. The robustness of the effective basin is actually an opportunity for the shipping companies, together with the possibilities of triangulation and of constant schedule pointed out in the maritime accessibility analysis of the port of Napoli. In more detail, taking into account a

Table 5. Analysis of the basin of the maritime service Napoli-Barcelona.

SPEED 20 KNOTS				
CONVENIENCE THRESHOLD WITH RESPECT TO THE ALL-ROAD ALTERNATIVE: 0%				
origin	destination	% GDP	% firms	freight demand [tons/year]
civitavecchia	barcelona	15.75%	15.42%	520,980
genova	barcelona	15.86%	15.10%	2,408,862
livorno	barcelona/tarragona	6.55%	6.75%	368,996
livorno	valencia	1.93%	2.16%	112,421
napoli	barcelona	14.16%	15.11%	261,341
salerno	valencia	3.72%	4.22%	71,574
CONVENIENCE THRESHOLD WITH RESPECT TO THE ALL-ROAD ALTERNATIVE: 5%				
origin	destination	% GDP	% firms	freight demand [tons/year]
civitavecchia	barcelona	14.03%	13.75%	479,224
genova	barcelona	5.53%	5.46%	586,560
livorno	barcelona/tarragona	3.90%	3.88%	170,883
napoli	barcelona	12.86%	13.61%	240,922
salerno	valencia	3.72%	4.22%	71,574
CONVENIENCE THRESHOLD WITH RESPECT TO THE ALL-ROAD ALTERNATIVE: 10%				
origin	destination	% GDP	% firms	freight demand [tons/year]
civitavecchia	barcelona	8.46%	8.50%	234,385
genova	barcelona	3.67%	3.59%	243,996
livorno	barcelona/tarragona	3.68%	3.67%	158,330
napoli	barcelona	12.38%	13.11%	232,321
salerno	valencia	3.72%	4.22%	71,574
SPEED 24 KNOTS				
CONVENIENCE THRESHOLD WITH RESPECT TO THE ALL-ROAD ALTERNATIVE: 0%				
origin	destination	% GDP	% firms	freight demand [tons/year]
civitavecchia	barcelona	9.90%	9.06%	390,933
genova	barcelona	15.86%	15.10%	2,408,862
livorno	barcelona/tarragona	6.55%	6.75%	368,996
livorno	valencia	1.93%	2.16%	112,421
napoli	barcelona	20.73%	22.29%	400,906
salerno	valencia	3.72%	4.22%	71,574
CONVENIENCE THRESHOLD WITH RESPECT TO THE ALL-ROAD ALTERNATIVE: 5%				
origin	destination	% GDP	% firms	freight demand [tons/year]
civitavecchia	barcelona	8.61%	7.91%	359,123
genova	barcelona	5.53%	5.46%	586,560
livorno	barcelona/tarragona	3.90%	3.88%	170,883
napoli	barcelona	19.62%	21.03%	389,594
salerno	valencia	3.72%	4.22%	71,574
CONVENIENCE THRESHOLD WITH RESPECT TO THE ALL-ROAD ALTERNATIVE: 10%				
origin	destination	% GDP	% firms	freight demand [tons/year]
civitavecchia	barcelona	3.51%	3.15%	116,267
genova	barcelona	3.67%	3.59%	243,996
livorno	barcelona/tarragona	3.68%	3.67%	158,330
napoli	barcelona	18.28%	19.45%	361,022
salerno	valencia	3.72%	4.22%	71,574

Table 4. Effective basins of the current maritime supply between Italy and Spain for different convenience threshold with respect to the all-road alternative.

CONVENIENCE THRESHOLD WITH RESPECT TO THE ALL-ROAD ALTERNATIVE: 0%				
origin	destination	% GDP	% firms	freight demand [tons/year]
civitavecchia	barcelona	27.36%	27.81%	756,709
genova	barcelona	15.86%	15.10%	2,408,862
livorno	barcelona/tarragona	6.55%	6.75%	368,996
livorno	valencia	1.93%	2.16%	112,421
salerno	valencia	6.02%	6.67%	95,720

CONVENIENCE THRESHOLD WITH RESPECT TO THE ALL-ROAD ALTERNATIVE: 5%				
origin	destination	% GDP	% firms	freight demand [tons/year]
civitavecchia	barcelona	24.33%	24.68%	694,095
genova	barcelona	5.53%	5.46%	586,560
livorno	barcelona/tarragona	3.90%	3.88%	170,883
salerno	valencia	5.69%	6.30%	92,570

CONVENIENCE THRESHOLD WITH RESPECT TO THE ALL-ROAD ALTERNATIVE: 10%				
origin	destination	% GDP	% firms	freight demand [tons/year]
civitavecchia	barcelona	16.87%	17.45%	423,355
genova	barcelona	3.67%	3.59%	243,996
livorno	barcelona/tarragona	3.68%	3.67%	158,330
salerno	valencia	5.35%	5.94%	88,619

triangulation with the port of Palermo, a further increase of about 30.000 tons/year is observed, mainly subtracted to Civitavecchia and, in smaller part, to Salerno. A further exploration of the effective basin of the Napoli-Barcelona route is reported in Figure 1, depicting the effective basins of the route for 20 and 24 knots speed respectively and for 10% convenience threshold; colors indicate how many times a region enters with o-d pairs within the effective basin. The stability of the basins is therefore confirmed.

4.3 Savona-Valencia appraisal

The same approach of the previous section has been also applied to the Savona-Valencia route. The effective basin is reported in Table 6 for the 20 and 24 knots speeds respectively.

The differences with the case of the Napoli-Barcelona are significant. Firstly, there is a remarkable increase in the level of total demand attracted by the intermodal corridor, equal to 9% for 20 knots and to 10.5% for 24 knots at a 0% convenience threshold. This obviously comes from the centrality of the port of Savona with respect to the regions generating the most of freight flows. Interestingly, apart from the expected natural reduction of traffic flows from Genova, there is also a significant quote of demand subtracted to Civitavecchia and Livorno. Moreover, there is a remarkable elasticity of the effective basin with respect to the vessel speed. Finally, there is no appreciable interaction between the new route Savona-Valencia and the Napoli/Salerno cluster.

Table 6. Analysis of the basin of the maritime service Savona-Valencia.

SPEED 20 KNOTS				
CONVENIENCE THRESHOLD WITH RESPECT TO THE ALL-ROAD ALTERNATIVE: 0%				
origin	destination	% GDP	% firms	freight demand [tons/year]
civitavecchia	barcelona	25.65%	26.18%	695,151
genova	barcelona	11.39%	10.74%	1,670,607
livorno	barcelona/tarragona	6.55%	6.75%	368,996
salerno	valencia	6.02%	6.67%	95,720
savona	valencia	11.96%	12.16%	1,200,289

CONVENIENCE THRESHOLD WITH RESPECT TO THE ALL-ROAD ALTERNATIVE: 5%				
origin	destination	% GDP	% firms	freight demand [tons/year]
civitavecchia	barcellona	22.62%	23.05%	632,537
genova	barcellona	4.94%	4.81%	560,366
livorno	barcellona	3.90%	3.88%	170,883
salerno	valencia	5.69%	6.30%	92,570
savona	valencia	10.67%	10.86%	980,818

CONVENIENCE THRESHOLD WITH RESPECT TO THE ALL-ROAD ALTERNATIVE: 10%				
origin	destination	% GDP	% firms	freight demand [tons/year]
civitavecchia	barcellona	16.87%	17.45%	423,355
genova	barcellona	3.67%	3.59%	243,996
livorno	barcellona	3.68%	3.67%	158,330
salerno	valencia	5.35%	5.94%	88,619
savona	valencia	3.52%	3.40%	534,206

SPEED 24 KNOTS				
CONVENIENCE THRESHOLD WITH RESPECT TO THE ALL-ROAD ALTERNATIVE: 0%				
origin	destination	% GDP	% firms	freight demand [tons/year]
civitavecchia	barcellona	25.65%	26.18%	695,151
genova	barcellona	5.03%	4.75%	817,247
livorno	barcellona	5.37%	5.58%	293,688
salerno	valencia	6.02%	6.67%	95,720
savona	valencia	20.63%	20.45%	2,229,764

CONVENIENCE THRESHOLD WITH RESPECT TO THE ALL-ROAD ALTERNATIVE: 5%				
origin	destination	% GDP	% firms	freight demand [tons/year]
civitavecchia	barcellona	22.62%	23.05%	632,537
genova	barcellona	1.15%	1.18%	57,347
livorno	barcellona	2.72%	2.72%	95,575
salerno	valencia	5.69%	6.30%	92,570
savona	valencia	18.06%	18.00%	1,888,233

CONVENIENCE THRESHOLD WITH RESPECT TO THE ALL-ROAD ALTERNATIVE: 10%				
origin	destination	% GDP	% firms	freight demand [tons/year]
civitavecchia	barcellona	16.87%	17.45%	423,355
genova	barcellona	1.08%	1.12%	55,513
livorno	barcellona	2.50%	2.50%	83,022
salerno	valencia	5.35%	5.94%	88,619
savona	valencia	15.64%	15.66%	1,559,144

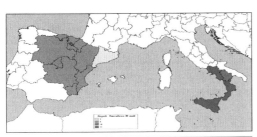

Figure 1. Effective basins of the Napoli-Barcelona route for two different vessel speeds (20 and 24 knots respectively).

The effective basins are depicted in Figure 2 with reference respectively to 20 and 24 knots speeds and for 10% convenience threshold; again, colors indicate how many times each region contributes with o-d pairs to the effective basin. Notably, there is a significant difference with the situation of the Napoli- Barcelona route. In more detail, for 20 knots the basin is narrow, encompassing only the regions more directly connected to the ports of Savona and Valencia. At 24 knots the effective basin is significantly wider, but with only few o-d pairs for each region.

4.4 Aggregate impacts

The adopted methodology allows also calculation of aggregate measures useful as input variables for cost-benefit analysis and for measuring other exogenous impacts (e.g. pollution). For this aim, Table 7 reports the absolute value in the current scenario and the variation of vehicles·km and tons·km due to the introduction of the new routes;

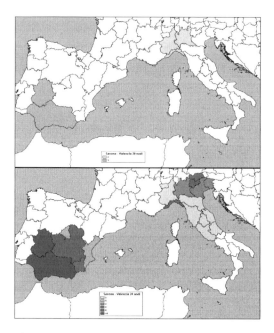

Figure 2. Effective basins of the Savona-Valencia route for two different vessel speeds (20 and 24 knots respectively).

Table 7. Aggregate impacts of the new routes Napoli-Barcelona and Savona-Valencia.

ABSOLUTE VARIATION OF TONNES KM ON ROAD			
scenario	0% threshold	5% threshold	10% threshold
current	7,545,856,423	9,618,535,693	10,313,240,452
napoli barcelona 20 knots	-29,993,629	-33,362,497	-68,370,428
napoli barcelona 24 knots	-60,671,909	-92,821,903	-101,383,183
savona valencia 20 knots	-534,294,219	-1,049,438,602	-603,899,488
savona valencia 24 knots	-915,788,933	-1,622,095,193	-1,626,350,592

ABSOLUTE VARIATION OF VEHICLE KM ON ROAD			
scenario	0% threshold	5% threshold	10% threshold
current	377,292,821	480,926,785	515,662,023
napoli barcelona 20 knots	-1,499,681	-1,668,125	-3,418,521
napoli barcelona 24 knots	-3,033,595	-4,641,095	-5,069,159
savona valencia 20 knots	-26,714,711	-52,471,930	-30,194,974
savona valencia 24 knots	-45,789,447	-81,104,760	-81,317,530

the conversion from vehicles to tons has been carried out assuming a mean load factor of 20 tons/vehicle. The variation of vehicles·km for each o-d pair has been calculated by considering the sum of the kilometers for the all-road alternative and the road access/egress kilometers for the maritime alternative.

A first comment is that in the current scenarios, in spite of the overall increase in tons transported by road from 48% to 90% for the 0% and the 10% convenience threshold respectively (see Table 3), the corresponding increase in tons·km is less than 30%. This is mainly due to the fact that the non-captive tons are those transported between closer o-d pairs, essentially between Northern

Italy and Eastern Spain. In accordance with the results of the previous section, the impact of the Savona-Valencia route is greater than the Napoli-Barcelona (10 times larger reduction of tons · km) as a consequence of the higher quantity of tons attracted in the effective basin. However, that impact is significantly reduced for increasing convenience thresholds with respect to the Napoli-Barcelona, because of the robustness of the effective basin of the latter.

5 CONCLUSIONS

This paper proposed an analysis of the Ro-Ro intermodal corridor Italy-Spain, through a methodology based on the availability of a transportation supply model. The ports within the study area have been firstly analyzed in terms of their maritime accessibility, that is measuring the factors usually taken into account by shipping companies for choosing new routes/services. Then, for each port a potential basin and an effective basin have been defined, respectively the first encompassing the set of o-d pairs whose transportation cost shortest path passes through that port, the second containing the o-d pairs with a convenience (expressed in percentage of the transportation cost) in using the maritime mode with respect to the road mode. These analyses have been carried out with reference both to an ideal maritime supply, with all ports connected among themselves, and to the current maritime supply between Italy and Spain.

As a result, the methodology identifies potential attractive routes in the light both of the maritime companies (i.e. making most effective use of the vessels in terms of weekly schedule and so on) and of the freight demand (i.e. convenience with respect to the all-road alternative). The analysis is also useful for port authorities, which can identify their target basin and potential competing ports in a simple manner, without cumbersome and expensive surveys. Obviously, those tenets should be regarded as necessary conditions for the effective establishment of new SSS routes, and further practical implementation issues should be taken into account for the effective implementation of the route. In other words, the methodology for the analysis of competitiveness proposed in the paper can be applied in a first step for identifying a set of potential routes, and then the actual choice of the most convenient service can be done in a second step considering other practical aspect. This helps in the transferability of the methodology to contexts different from the Mediterranean basin. Notably, in spite of its simplicity, the methodology proposed seems to be reliable and effective, even under the hypothesis of deterministic mode choice behavior.

REFERENCES

Cascetta E., Marzano V., Papola A. (2009). Schedule-based passenger and freight mode choice models for ex-urban trips. In: N.H.M. Wilson, A. Nuzzolo. Schedule-Based Modelling of Transportation Networks. Theory and Applications, Springer ed. (New York) vol. 46, pp. 241–250.

Cascetta E., Marzano V., Papola A., Vitillo R. (2013). A multimodal elastic trade coefficients MRIO model for freight demand in Europe. In: M. Ben-Akiva, H. Meersman and E. Van de Voorde (eds). Freight Transport Modelling, Emerald ed., pp. 45–68.

de Jong G., Kouwenhoven M., Bates J., Koster P., Verhoef E., Tavasszy L., Warffemius P. (2014). New SP-values of time and reliability for freight transport in the Netherlands. Transportation Research Part E: Logistics and Transportation Review, Volume 64, April 2014, Pages 71–87.

Jeannesson-Mange E. (2006). Short sea shipping cost benefit analysis. *Proceedings of ETC 2006 conference, Strasbourg, France.*

Marzano, V., Papola, A., and F. Simonelli F. (2008). A large scale analysis of the competitiveness of new short-sea shipping services in the Mediterranean. *Proceedings of the 2008 ETC Conference*, Leeuwenhorst, Netherlands.

Russo, F. and F. Modafferi. (1999). Modelli per la determinazione di tempi e prezzi del trasporto marittimo containerizzato nel bacino del Mediterraneo. *Working paper 2/99 DIMET*, Università di Reggio Calabria (in Italian).

Torrieri, V., Gattuso D. and G. Musolino (2001). Cost functions for freight transport in multimodal networks and scenario evaluations for Euro-Mediterranean container shipping. *Proceedings of 9th World Conference on Transportation Research*, Seoul.

Vitillo, R. (2011). Potenzialità e impatti dell'intermodalità nel bacino Euro-Mediterraneo: sviluppi teorici e prospettive operative. PhD Thesis (in Italian).

Transport Infrastructure and Systems – Dell'Acqua & Wegman (Eds)
© 2017 Taylor & Francis Group, London, ISBN 978-1-138-03009-1

Airport ground access logit choice models for fixed track systems

I. Politis, P. Papaioannou & G. Georgiadis
Department of Civil Engineering, Aristotle University of Thessaloniki, Thessaloniki, Greece

ABSTRACT: In this paper, binary logit models are developed in order to identify the mode and user specific parameters that affect airport ground access mode choice. The airport under examination, the International Airport "Macedonia" of Thessaloniki, is the second busiest airport in Greece however only buses connect the airport with the city center. A combined revealed and stated preference study examined the significant factors that influence airport visitors' choices against three future alternative fixed track public transport systems. Likelihood ratio tests examined whether user specific attributes are improving the overall fit of the models or not. The results indicate significant relation of mode specific variables with the final choice. User specific attributes of gender, age and trip frequency was found in the case of bus and taxi users to have an explanatory power on mode choice. Finally, travel and cost elasticities for the different examined cases are presented and commented.

Keywords: logit choice models, mode and user characteristics, ground access, travel and cost elasticities

1 INTRODUCTION

Policy makers and local/governmental authorities, in order to avoid airport capacity problems, normally invest on new runways and/or terminals or build a new airport when physical or space constraints hider expansion. In most cases increases in passenger demand lead to the expansion and redesign of the existing transport access infrastructure for accommodating the increased traffic demand to and from the airport. Given the highway congestion problems which inevitably affect the air-city links, the introduction of a high capacity, fast and convenient public transport mode in the form of a fixed track system (sub-urban rail, metro, or tram) is often proposed and chosen as the preferred alternative.

This paper examines the mode and user specific attributes (socioeconomic and trip characteristics) that may have an explanatory power on airport access mode choice in the case of fixed track systems. The airport under examination is the International Airport "Macedonia", located at the suburban area of the city of Thessaloniki in northern Greece. Three fixed track systems namely metro, tram and monorail, were examined as alternatives to the existing travel modes. Trip makers to the airport were classified into three segments: car, bus and taxi users.

2 UNDERTAKEN RESEARCH

Thessaloniki International Airport "Macedonia" (IATA code: SKG, ICAO code: LGTS), operates since 1930, and it is the second largest state owned airport in Greece. It is located approximately 16 km from the city center of Thessaloniki. The airport constitutes a key player for the development of Northern Greece, and is the main hub for nearby tourist destinations. The airport area is positioned by the sea, and as a result there is only one access point to the airport facilities. At the moment, it is accessible only by car, public transport (buses) and taxi. Public Transport connection to the city center of Thessaloniki is rather poor, since only one bus line runs every 30 min in winter time and every 15 min during summer months. The actual modal split of trips to and from the airport is estimated as follows: 54.5% by car, 18.5% by bus and 27% by taxi (THEPTA, 2011). These findings are in line with the work of Ashley & Merz (2002) who found that airports without rail-link connections could not achieve public transport shares greater than 30%.

The transport master plan of the city, anticipates a 5.1 km length connection with the terminal of the metro line which is under construction and is located at the eastern suburban area of the city. The connection is anticipated to be done by a fixed track system such as the metro (extension to the metro line under construction), a tram or a monorail system. A feasibility study, conducted by the Public Transport Authority (THEPTA) of the city within the framework of the CIVITAS CATALIST initiative, gave a comparative advantage to the monorail option in terms of cost and benefit benchmarking evaluation (THEPTA, 2011).

A combined Revealed (RP) and Stated Preference (SP) questionnaire survey was conducted at the users of the current available airport access modes, i.e. the car users, the bus users and the taxi users. The questionnaire was distributed only to people who were ready to depart as well as to employees and passenger accompanying persons. For the development of the choice models, 357 responses were finally considered as valid. The RP questionnaire consisted of 4 sections. The first section is about the trip characteristics of the respondent. At the second section of the questionnaire, a brief introduction of the three examined alternative modes was given to the respondents together with respective photos and maps, illustrating the future connection plans of the airport with the metro terminal. The third section pertains to the SP experiment of hypothetical scenarios regarding the choice of the fixed track systems. Finally, the last section consists of questions regarding interviewees' socio-economic characteristics. The SP experiment was a binary choice set and was designed as follows: the interviewer, urged the interviewee to recall in his mind his/her last trip to the airport. Afterwards, hypothetical scenarios with the three fixed track systems were given to him/her regarding this trip and the respondent stated his/her willingness to change travel mode in a 6-point Likert scale (from "definitely the alternative mode" to "definitely my current mode". For each of the alternative modes (called also brands in SP theory), it was decided to examine three attributes, namely the total travel time, the total travel cost (out of pocket cost) and the number of transfers. In order to keep the scenarios tested at a manageable number, it was decided that the levels of the attributes should be three: Low, Medium and High. Finally, nine scenarios were selected for each user segment, with respect to the dominance and transitivity of the choices. For each one of the current mode users (car, bus and taxi), different attribute levels were considered for the alternative modes so as to create plausible hypothetical scenarios. Table 1 presents the attribute levels for each mode of the SP survey. Travel time is expressed in minutes and cost in euros; both are expressed as absolute differences of the revealed values of time and cost for the current mode used. The transfers are absolute numbers and are independent from the number of transfers the interviewee does with his/her current mode.

Table 1. Attribute levels for the SP experiment.

	Metro	Tram	Monorail
Car Users			
Travel Time	(−10/0/+10)	(−10/0/+10)	(−10/0/+10)
Cost	(−5/−3/−1)	(−1/−5/−3)	(−3/−1/−5)
Transfers	(0/2/1)	(2/1/0)	(1/0/2)
Bus Users			
Travel Time	(0/−10/−20)	(0/−10/−20)	(0/−10/−20)
Cost	(+1/+2/+3)	(+3/+1/+2)	(+2/+3/+1)
Transfers	(0/2/1)	(2/1/0)	(1/0/2)
Taxi Users			
Travel Time	(−10/0/+10)	(−10/0/+10)	(−10/0/+10)
Cost	(−5/−15/−10)	(−10/−20/−15)	(−15/−10/−20)
Transfers	(0/2/1)	(2/1/0)	(1/0/2)

Table 2. Beta coding and parameters for model 1 & 2.

Model 1

Vmetro		Vcar	
beta	Parameter	beta	Parameter
β_1	metro$_{time}$	β_4	car$_{time}$
β_2	metro$_{cost}$	β_5	car$_{cost}$
β_3	metro$_{transfers}$	β_6	car$_{transfers}$
			ASC$_{car}$

Model 2

Vmetro		Vcar	
beta	Parameter	beta	Parameter
β_1	metro$_{time}$	β_4	car$_{time}$
β_2	metro$_{cost}$	β_5	car$_{cost}$
β_3	metro$_{transfers}$	β_6	car$_{transfers}$
		β_7	freq_everyday
		β_8	freq > 4times_month
		β_9	freq_4times_month
		β_{10}	freq_3times_month
		β_{11}	freq_2times_month
		β_{12}	freq_1times_month
		β_{13}	purp_work
		β_{14}	purp_accomp
		β_{15}	Age_26–35
		β_{16}	Age_36–45
		β_{17}	Age_46–55
		β_{18}	Age_56–65
		β_{19}	Age_plus
		β_{20}	gender
			ASC$_{car}$

3 MODEL SPECIFICATION

Two binary choice models were set up for each type of user; a simplified model including only mode specific variables like travel time, travel cost and number of transfers (model 1) and an extended model (model 2) including also *socio-economic* variables like the age, the gender, the frequency and the trip purpose. Age 18–25, Man, seldom frequency visit and fly purpose were set as reference categories for the dummy variables coded at the choice model experiments respectively. Table 2 presents, in its general form, the parameters of the utility functions for the two binary choice models (Model 1 and Model 2) for the current system (Vcurrent) against the new system (Vnew). In total 3 (user segments) × 3 (alternative fixed track systems examined) × 2 (simplified and extended models) = 18 binary choice models were developed. The models were calibrated through the freeware package BIOGEME (Bierlaire, 2003).

4 MODEL RESULTS AND INTEPRETATION

The 3 tables at the APPENDIX, illustrate the parameter estimates of the binary choice models for each type of user respectively; furthermore, they present the results of the Likelihood Ratio tests for the comparison of the simplified (Model 1) and the extended models (Model 2).

Apart from the unstandardized betas estimates (b), the tables also present the standardized betas (β) for comparative analysis of the strength of the prediction across the variables. The calculation of the standardized betas was done through the following formula (Kaufman, 1996):

$$SS^{\Delta P} = \left[\frac{1}{1+\exp^{-\left(\ln\frac{P_{Ref}}{1-P_{Ref}}\right)+\frac{1}{2}\hat{b}s}} - \frac{1}{1+\exp^{-\left(\ln\frac{P_{Ref}}{1-P_{Ref}}\right)-\frac{1}{2}\hat{b}s}} \right] \tag{1}$$

where: P_{Ref} = a probability value used as a reference point, $\hat{b}$ = the unstandardized logistic regression coefficient, and s = the sample standard deviation. In order to use the formula, the following steps were taken: a) calculation of the sample standard deviation for each variable, b) calculation of the logistic regression predicted probabilities, c) calculation of the mean of those values to obtain P_{Ref}, and d) substitute s, $\hat{b}$, and P_{Ref} into the function (King, 2007).

4.1 *Interpretation of the car users model*

The models developed for the car users, are the only ones that indicate a statistical significant *Alternative Specific Constant (ASC)*, i.e. an a priori preference for one mode when time, cost and number of transfers is equal for both modes. All

6 models, indicate a negative ASC for the car, meaning a relative preference of the car users for the alternative. A priori preferences to the transit alternatives have been observed in the cases of mid length congested corridors (Angel, 2011) and in the case of specific demand segments for airport ground access mode choice studies (Gosling, 2008). The comparison of the constants among the proposed alternatives, indicate a strong preference to the metro system, followed by the tram and the monorail respectively.

Travel time appears to have a statistical significant contribution to the mode choice only for the metro and monorail model. This outcome is plausible since car drivers consider the tram option as a surface mode which will occupy a part of the road infrastructure and will provide similar travel characteristics with the car option. Also, the number of transfers for the metro and the monorail system were found to have a statistical significant contribution to the mode choice. This finding shows that the transfers to a (underground or elevated rail link) mode of public transport cause significant disutility for the car drivers, in comparison to the tram system where surface transfer from/to the transit mode is assumed. Finally, for the tram model, alternative specific beta coefficients were used for the travel cost as explanatory variables.

Regarding the investigation of the user specific attributes, the analysis shows a limited contribution of these factors; the age was found to be statistically significant on the mode choice in the cases of metro and tram models and trip frequency was found to be an important factor for the monorail case.

In the cases of metro and monorail, the magnitude of the travel time standardized coefficients are almost double compared to those for the number of transfers. Additionally, the standardized estimates for the user specific variables have lower values both from the travel time and the number of transfers. The cost parameter for the tram models, i.e. travel cost for car and travel cost for tram, are of almost equal importance. Finally, the Likelihood Ratio tests revealed that the inclusion of the user specific variables have zero effect on mode choice interpretation, regarding the car users, since they do not improve the explanatory power of the choice models with the only exception of the tram model.

4.2 *Interpretation of the bus users model*

Alternative specific constant was not found to be statistically significant for any of the six models examined. This actually means that the bus users do not have an a priori preference for their current or for the alternative proposed mode of transport. This finding is in line with the findings of similar studies that argue about the relative preference of existing public transport users over the alternative proposed rail public transport mode (Axhausen et al, 2001, Ben-Akiva & Morikawa, 2002, Scherer, 2010).

The parameters of travel time and travel cost (fares) are statistically significant whereas the number of transfers is not. This could be considered as a plausible outcome since the bus users are used to travel through transfers and therefore they do not give any extra (dis)utility to transfers as it is the case for the car drivers. Both for cost (fare) and time attributes, the beta coefficients were set as generic.

In contrary to the findings of the car user models, it was found that transit user specific attributes play significant role in the choice of mode process. Gender differences were identified, with females being inclined towards the bus (see the odds ratio results given in Table 3). Additionally, trip frequency to the airport was found to be an important factor for mode selection as well as the age level for the case of the tram alternative.

The standardization of the parameters led to the conclusion that travel time is considered as the most important factor for the mode choice for the bus users as well. However, in this case, the standardization of the user specific variables indicates that they also have strong influence on the mode choice compared to the mode specific variables of time and cost as the Likelihood Ratio tests also confirm.

4.3 Interpretation of the taxi users model

The last set of models also indicate a non a priori preference among the current or future travel modes. In all the cases, the alternative specific constant was found non-significant, something that has been also confirmed by various similar studies (Castillo-Manzano, 2010, Jou et al, 2011, Roh, 2013).

The models indicate that the taxi users are highly discouraged to make a transfer within their trip. In all the examined models, the number of transfers was statistically significant, as well as the trip cost whereas travel time was not found to be significant for the monorail and tram cases.

Regarding the investigation of the user specific parameters it was found that the majority of them are of significant importance. Age, trip purpose and gender differentiations were identified for the majority of the examined alternatives.

The usage of the standardized beta shows that travel cost seems to be dominant factor for the mode choice and the transfers from/to the alternative transport mode were found to have importance.

Finally, the Likelihood Ratio diagnostic tests, indicate that user specific variables indeed improve the predicting capability of the models.

4.4 Travel cost and travel time elasticities

The general formula for the calculation of the (direct point) elasticities of each individual in a basic discrete choice scheme can be written as (Dunne, 1984, Louviere et al, 2000):

$$E_{X_{ikq}}^{P_{iq}} = \frac{\partial P_{iq}}{\partial P_{ikq}} \cdot \frac{X_{ikq}}{P_{iq}} \tag{2}$$

Table 3. Summary of odd ratio calculations.

Beta	Variables and reference variables	Car users			Bus users			Taxi users		
		M1	M2	T	M1	M2	T	M1	M2	T
Age: *Reference category 18–24 years old*										
b15	Aged 26 to 35 years old							4.6	2.8	9
b17	Aged 46 to 55 years old	1.7		1.8				5.4	4	4.7
b19	Aged over 66 years old						0.1		35.2	17.6
Frequency: *Reference category travel seldom*										
b7	every day				3					
b9	travel 4 times per month								35.2	17.6
b11	travel 2 times per month		3.5							2.3
b12	travel 1 time per month				0.2	0.4				
Gender: *Reference category man*										
b20	Woman				2.4	2.4	3.1	2.3		
Purpose: *Refernce category fly*										
b14	Escort								0.2	0.3

M1: Metro, M2: Monorail, T: Tram.

Or by using the quotient rule of the derivatives (i.e., $\partial e^{az}/\partial Z = ae^{az}$), equation (2) can be written as:

$$E_{X_{ikq}}^{P_{iq}} = \beta_{ik} X_{ikq} (1 - P_{iq}) \qquad (3)$$

From equation (3) it can be easily concluded that direct elasticity approaches zero as P_{iq} approaches unity and approaches $\beta_{ik} X_{ikq}$ as P_{iq} approaches to zero. In order to obtain aggregate elasticities, an approach known as "sample enumeration method" was used since the usage of sample average values for the X_{ikq} or the $\hat{P}_{iq}$ (average estimated P_j) is not correct for the non-linear logit models. In the sample enumeration approach, individual probability is first calculated and then aggregate by weighting each individual elasticity by the individuals' estimated probability of choice (Louviere et al, 2000).

The mathematical expression of the above is written as:

$$E_{X_{jkq}}^{\bar{P}_i} = \left(\sum_{q=1}^{Q} \hat{P}_{iq} E_{X_{jkq}}^{P_{iq}} \right) \bigg/ \sum_{q=1}^{Q} \hat{P}_{iq} \qquad (4)$$

where: $\hat{P}_{iq}$ is an estimated choice probability and P_i refers to the aggregate probability of choice probability i. Table 4 presents the travel cost and the travel time elasticities, both for the simplified (model 1) and the extended model (model 2), whether of course these calculations are feasible.

From Table 4, it can be concluded that both travel time and cost are inelastic in most of the cases, with the exception of cost elasticities for the taxi users. Travel time found to be more inelastic compared to the respective elasticities for travel cost (i.e. the travel time elasticities have lower values compared to the respective travel cost elasticities) with the only exception of the bus users and the tram alternative.

From the results it can be also concluded that bus users are more sensitive to travel time in comparison to car users for all the alternative modes examined, with the exception of the metro alternative. On the other hand, taxi users were found to be more sensitive compared to bus users regarding the travel cost.

Between the modes, the metro seems to have larger cost and travel elasticities compared to the other alternatives which means that the mode with the greater potential due to a percentage change on the current travel conditions of time and cost is the metro system.

Finally, the inclusion of the user specific attributes in the models, though it seems it does not have a great effect on the magnitude of the elasticities, it causes a relative increase in the travel time elasticities (i.e. the travel time elasticities are higher in most cases for model 2) and respectively a relative decrease in the cost elasticities (i.e. the travel cost elasticities are lower for model 2 with the exception of the tram alternative).

ACKNOWLEDGMENTS

The survey in which this paper refers to, is part of a project entitled: "Connection of *Thessaloniki International "Macedonia" Airport with the future fixed track rapid system network of Thessaloniki. Initiatives and experiences derived from Public Transport Development in European Cities".* The project was funded under the 6th European Union Framework Program, Grant agreement no: CATALIST/6FP/call4/010. The authors would like to express their sincere thanks to the Thessaloniki Public Transport Authority for the provision of the raw data.

Table 4. Travel time and cost elasticities.

	Metro		Monorail		Tram	
Current mode	Model 1	Model 2	Model 1	Model 2	Model 1	Model 2
Travel Time Elasticities						
Car	0.281	0.279	0.018	0.019	0.002	0.002
Bus	0.260	–	0.242	0.282	0.487	0.532
Taxi	–	–	–	–	–	–
Travel Cost Elasticities						
Car	–	–	–	–	–	–
Bus	0.449	–	0.374	0.320	0.422	0.312
Taxi	15.773	14.951	3.059	2.855	0.829	1.161

– : Not found statistically significant.

REFERENCES

Angel, I. (2011). A User Preferences Analysis of Light Rail Transit and Bus Public Transport Systems. In *Transportation Research Board 90th Annual Meeting* (No. 11–3187).

Ashley D. & Merz S.K., *"Airport Rail Links – A Post Audit"*, 5th Annual Queensland Transport Infrastructure Conference, 2002.

Axhausen KW, Haupt T., Fell B, Heidl U, (2001). Searching for the Rail Bonus-Results from a panel SP/RP study. *EJTIR*, *1*(4), 353–369.

Ben-Akiva, M., & Morikawa, T. (2002). Comparing ridership attraction of rail and bus. *Transport Policy*, *9*(2), 107–116.

Bierlaire, M., *"BIOGEME: a free package for the estimation of discrete choice models"*, Proceedings of the 3rd Swiss Transport Research Conference, Ascona, Switzerland, 2003.

Castillo-Manzano, J.I. (2010). The city-airport connection in the low-cost carrier era: Implications for urban transport planning. *Journal of Air Transport Management*, *16*(6), 295–298.

Dunne J.P. (1984), Elasticity measures and disaggregate choice models, *Journal of Transport Economics and Policy, p.p.* 189–197.

Gosling, G.D. (2008). *Airport ground access mode choice models* (Vol. 5). Transportation Research Board.

Jou R.C., Hensher D., Hsu T.L., "Airport Ground Access Mode Choice Behavior after the Introduction of a New Mode: A Case Study of Taoyuan International Airport in Taiwan." *Transportation Research Part E: Logistics and Transportation Review* 47(3) p.p. 371–81, 2011.

Kaufman, R.L., *"Comparing effects in dichotomous logistic regression: A variety of standardized coefficients"*, Social Science Quarterly, 77, 90–109, 1996.

King J., *"Standardized Coefficients in Logistic Regression"*, Annual meeting of the Southwest Educational Research Association, San Antonio, Texas, Feb. 7–10, 2007, on line at http://www.ccitonline.org/jking/homepage/

Louviere J.J., Hensher D.A., Swait J., *"Stated Choice Methods"*, Cambridge University Press, 2000.

Roh, H.J. (2013). Mode Choice Behavior of Various Airport User Groups for Ground Airport Access. *Open Transportation Journal*, *7*, 43–55.

Scherer, M. (2010). Is Light Rail More Attractive to Users Than Bus Transit? *Transportation Research Record: Journal of the Transportation Research Board*, *2144*(1), 11–19.

THEPTA (Thessaloniki Public Transport Authority), *"Connection of Thessaloniki International "Macedonia" Airport with the future fixed track rapid system network of Thessaloniki. Initiatives and experiences derived from Public Transport Development in European Cities"*, Feasibility Report, CIVITAS CATALIST initiative, Grant agreement no: CATALIST/6FP/call4/010, Thessaloniki, June 2011, Greece.

APPENDIX

Table A1. Parameters estimates of the binary choice models for the CAR USERS.

Variable	Alternative Mode: Metro						Alternative Mode: Monorail						Alternative Mode: Tram					
	Model 1			Model 2			Model 1			Model 2			Model 1			Model 2		
	b	β	p	b	β	p	b	β	p	b	β	p	b	β	p	b	β	p
ASCcar	−1.920 (−5.26)		0.00	−2.047 (−5.46)		0.00	−0.755 (−3.89)		0.00	−0.801 (−4.07)		0.00	−1.740 (−3.06)		0.00	−1.840 (−3.35)		0.00
Specific mode																		
β₁																		
β₂													−0.182 (−3.30)	−0.339	0.00	−0.192 (−3.45)	−0.356	0.00
β₃	−0.887 (−4.07)	−0.166	0.00	−0.895 (−4.09)	−0.168	0.00	−0.441 (−2.69)	−0.089	0.01	−0.444 (−2.69)	−0.089	0.01						
β₄													−0.048 (−3.39)	−0.353	0.00	−0.051 (−3.52)	−0.366	0.00
user																		
β₁₁										1.25 (1.85)	0.055	0.06						
β₁₇				0.541 (1.68)	0.050	0.09										0.566 (2.05)	0.056	0.04
Generic																		
βₜᵢₘₑ	−0.148 (−6.80)	−0.276	0.00	−0.149 (−6.82)	−0.278	0.00	−0.096 (−5.47)	−0.190	0.00	−0.097 (5.49)	−0.192	0.00						
βcost																		
N	367			367			364			364			361			361		
ρ²	0.284			0.290			0.174			0.181			0.024			0.032		
adj ρ²	0.272			0.274			0.162			0.165			0.012			0.016		
logL(0)	−254.385			−254.385			−252.306			−252.306			−250.226			−250.226		
logL(β)	−182.117			−180.685			−208.362			−206.583			−244.209			−242.100		
−2LL																		
LR test	2.86						3.56						4.22					

Standard errors of the parameters estimation are excluded from the table due to page length constraints/ t student values in parenthesis.

Table A2.   Parameters estimates of the binary choice models for the BUS USERS.

| Variable | Alternative Mode: Metro | | | | | | Alternative Mode: Monorail | | | | | | Alternative Mode: Tram | | | | | |
| | Model 1 | | | Model 2 | | | Model 1 | | | Model 2 | | | Model 1 | | | Model 2 | | |
	b	β	p	b	β	p	b	β	p	b	β	p	b	β	p	b	β	p
Specific user																		
β_7				1.103 (2.48)	0.101	0.01												
β_{12}				-1.830 (-3.80)	-0.127	0.00				-0.986 (-2.11)	-0.066	0.03						
β_{19}																-2.240 (2.55)	-0.108	0.01
β_{20}				0.887 (4.41)	0.103	0.00				0.858 (2.93)	0.094	0.00	1.130 (3.91)	0.137	0.00			
Generic																		
β_{time}	-0.067 (-2.07)	-0.132	0.04				-0.089 (-7.03)	-0.164	0.00	-0.106 (-7.09)	-0.193	0.00	-0.124 (-8.37)	-0.248	0.00	-0.152 (8.47)	-0.300	0.00
β_{cost}	-0.587 (-3.03)	-0.111	0.00				-0.846 (-9.59)	-0.154	0.00	-0.758 (-7.48)	-0.138	0.00	-0.750 (-6.77)	-0.153	0.00	-0.618 (-4.99)	-0.126	0.00
N	300			300			308			308			305			305		
ρ^2	0.045			0.159			0.294			0.326			0.253			0.306		
adj ρ^2	0.035			0.145			0.285			0.308			0.244			0.287		
logL(0)	-207.944			-207.944			-213.489			-213.489			-211.410			-211.410		
logL(β)	-198.570			-174.824			-150.658			-143.804			-157.914			-146.729		
-2LL																		
LR test				47.49						13.71						22.37		
	difference significant at 99% confidential level						difference significant at 99% confidential level						difference significant at 99% confidential level					

Standard errors of the parameters estimation are excluded from the table due to page length constraints/t student values in parenthesis.

Table A3. Parameters estimates of the binary choice models for the TAXI USERS.

	Alternative Mode: Metro						Alternative Mode: Monorail						Alternative Mode: Tram					
	Model 1			Model 2			Model 1			Model 2			Model 1			Model 2		
Variable	b	β	p	b	β	p	b	β	p	b	β	p	b	β	p	b	β	p
Specific mode																		
β_1	-0.065 (-2.99)	-0.104	0	-0.067 (-2.95)	-0.107	0												
β_2	-1.34 (-5.93)	-0.876	0	-1.384 (-5.67)	-0.887	0												
β_3	-7.84 (-6.33)	-0.878	0	-8.16 (-6.10)	-0.878	0	-3.03 (-8.69)	-0.448	0	-3.92 (-8.95)	-0.562	0	-0.47 (-4.11)	-0.081	0	-0.397 (-3.09)	-0.069	0
β_5	-1.21 (-6.62)	-0.565	0	-1.29 (-6.53)	-0.597	0												
user																		
β_9										3.56 (3.25)	0.084	0				2.87 (3.00)	0.077	0
β_{11}																0.851 (1.98)	0.051	0.05
β_{14}										-1.85 (-2.30)	-0.094	0.02				-1.18 (-1.97)	-0.068	0.05
β_{15}				1.53 (4.41)	0.116	0				1.02 (2.91)	0.084	0				2.2 (6.67)	0.203	0
β_{17}				1.69 (4.51)	0.1183	0				1.39 (3.75)	0.105	0				1.54 (4.49)	0.131	0
β_{19}										3.56 (2.37)	0.06	0.02				2.87 (2.16)	0.055	0.03
β_{20}										0.835 (2.81)	0.078	0						
Generic																		
β_{cost}							-0.299 (-9.42)	-0.228	0	-0.431 (-9.15)	-0.326	0	-0.092 (-8.43)	-0.079	0	-0.162 (-9.25)	-0.14	0
N	370			370			377			377			373			373		
ρ^2	0.373			0.433			0.342			0.423			0.183			0.321		
adj ρ^2	0.358			0.409			0.334			0.392			0.175			0.29		
logL(0)	-256.464			-256.464			-261.316			-261.316			-258.544			-258.544		
logL(β)	-160.722			-145.479			-172.017			-150.764			-211.256			-175.469		
-2LL LR test	30.49						42.51						71.57					
	difference significant at 99% confidential level						difference significant at 99% confidential level						difference significant at 99% confidential level					

Standard errors of the parameters estimation are excluded from the table due to page length constraints/t student values in parenthesis.

Transport Infrastructure and Systems – Dell'Acqua & Wegman (Eds)
© *2017 Taylor & Francis Group, London, ISBN 978-1-138-03009-1*

A simulation model for managing port operations

S. Gori & M. Petrelli
University of Roma Tre, Rome, Italy

ABSTRACT: The objective of this study is to create a simulation tool that involves the development and the management of the entire port infrastructure, using historical and real-time data, working with design and operating scenarios. Such infrastructures are characterized by several, different activities, often in overlap in space and time, involving critical issues about terminals capacity and reliability of ships scheduling. Such issues are growing for importance due to the combined effect of the increasing level of freight and the increasing difficulties in medium and long term investment projects. The simulation model has been realized with a discrete-event approach using the software ARENA and a tool for a graphic user interface, implemented in the C# language. The model has been implemented to simulate the activities in Civitavecchia, a multipurpose port, using real world data.

1 INTRODUCTION

The simulation of the activities of highly used environments such as seaports is of increasing importance. These infrastructures are characterized by numerous, different activities, often in overlap in space and time, while the increasing level of freight traffic and, in some cases, also passenger traffic involves issues of the terminals capacity, the congestion on the internal navigation system and the reliability of ships scheduling. Satisfactory performance is absolutely required by the different users of services and terminals. This is particularly true for a complex infrastructure, such as a port, in which several activities are still today not managed as a standard industrial process.

The objective of this study is to create an effective tool that, differently from many other models, takes into account and works with the development and the management of the whole port, using historical and real-time data and working with design or operating scenarios. In fact, the increasing availability of powerful computational apparatuses has allowed the use of sophisticated program to model the behavior of single components with an adequate level of detail and, if needed, in real time.

A simulation model is able to describe all processes that are necessary to define the services and the internal movements in a port, allowing the analysis and extraction of data, the simulation of different scenarios to study any kind of deficiencies and to identify potential solutions.

The simulation model includes the representation of both the sea and the land side of port and it make possible to write, study, and compare, over time, moment by moment, all relationships which interact with one another defining the best configuration of the port environment. Such system can be viewed as a complex system containing several entities with interfering attributes. In particular, micro simulation models could be applied to different scenarios involving complex interactions in space and in time to analyze the performances of single port components or, using forecasting traffic data, to help the design of infrastructural and management improvements of the port or to support the choice of navigation priorities and berth allocation for vessels in real-time processes. Micro simulation models use a discrete-event approach to represent each single service and each movement of different typology of vehicles over time. The activities of the port system are represented as a chronological sequence of events requiring a set of parameters, rules and policies that are used to represents a choice.

The simulation model described in this paper is an evolution of a previous model of Ancora et al. (2012). Starting from this promising model, the main differences with respect to the paper previously mentioned are represented by some important modification of the model (arrival time of the vessels, explicit representation of the berth capacity and correlated berth allocation problem). Instead, the main novelties of this paper with respect to the existing literature are represented by the simulation of port facilities in their entirety in combination with a very detailed level of representation of port activities, similarly to other simulation models of single specific aspect. In addition, differently from other models, the extensive phase of calibration and validation of the model is not only limited to running different scenarios observing differences in output, but it involves the comparison of outputs with specific real world data.

The model has been applied to simulate the operation of the Port of Civitavecchia using real world data. This has been chosen as it is one of the main Ro-Ro freight traffic and cruise services in the Mediterranean Sea, organized with many multipurpose berths and an important bottleneck represented by the internal navigation channel.

The paper is structured as follows: a state-of-the-art about the port simulation studies is presented in the second section; the third section describes the characteristics of the simulation model; the fourth section describes the results of the phases of calibration and validation of the model while the last section concerns some concluding remarks.

2 LITERATURE REVIEW

The simulation is a methodology that is performed to understand the behavior of a real system without disrupting its environment. According to this specific value, the simulation has been used in many, different systems such as urban, economic, production, transportation and maritime field. In such field, for example, the simulation methods have been carried out to analyze the impact of different terminal layouts and to determine the optimum level of equipment investment (Hayuth et al., 1994).

Several works for investigating, analyzing, evaluating and improving port activities are carried out in the literature. In some cases, each of them involves a specific area and/or activity by means of the simulation while the simulation of the port, as a unique and complex infrastructure, is not dealt with a large number of authors in literature. In fact, the works reported in literature range from studies of very general approach, usually about port performance analysis, to very specific aspects. These are generally related to the optimization of a single definite operation. The great evolution of operational research has produced large and renewed interest for this second family of problems, mainly related to the container terminals management. Voss & Stahlbock (2004) and Steenken & Voss (2008) represent the most complete reviews about the main operation problems as container pre-marshalling problem, landside transport, stowage planning problem, yard allocation problem.

A relatively restricted number of works has been published on the simulation of port facilities. El Sheik et al. (1987) propose a simulation model of a developing world port, focusing the attention in the analysis of the ship-to-berth allocation rules. Ramani (1996) has developed an interactive computer simulation model in order to support the logistic planning of container operations. The traffic stream of the Bosphorus is modeled in AWE-SIM by Köse et al. (2003) investigating the effects of the new pipe line to be built on the strait traffic. Guenther et al. (2006) represent the transportation activities in an automated container terminal analyzing especially the possible dispatching strategies and the impact of stochastic variations in handling and transportation times. Van Asperen et al. (2007) study a specific model for the analysis of the jetty capacity of ports involving special attention to capture the ship arrival process. Koch (2007) propose a simulation model to allow the assessment of the operational and the cost impact of introducing upstream monitoring of cargo containers within port facilities. Cortés et al. (2007) simulate the freight transport process in the inland port of Seville, considering all existing types of cargo and testing several development scenarios. Also Arango et al. (2011) choose the simulation of the Seville inland port to study the problems associated with allocating berths for containerships. This work produce an optimization model for the berth allocation problem integrated with simulating techniques involving some specific operations (truck and containerships arrivals, berth assignment, towing vessels and berths). Longo et al. (2013) study the performance of the seaports by means of a discrete-event simulation evaluating impacts of some critical factors as inter-arrival headways, handling times and volumes of cars and trucks to be loaded and unloaded. Uğurlu et al. (2014) have modeled BOTAS Ceyhan Marine Terminal by using Awesim simulation program. The resulting simulation model provides information about terminal capacity, terminal congestion, loading times, maneuvering time, types of ships arriving to port and transportation capacity.

3 THE SIMULATION MODEL

The model for the simulation of the port activities is an innovative model developed by means of a discrete-event simulation, adopting the SIMAN simulation language. Such language combines power, flexibility and ease of deployment by means of a system of modules that represent logical steps, animations and the collection of statistics necessary for the system description. This allows simulating real world data in more detail than a static model based on average values and deterministic processes. The simulation model has been realized using the software ARENA and an additional tool for supporting users with a Graphic User Interface (GUI), implemented in the C# language. This tool helps decision makers about the design and the management phase in the port infrastructure offering a very easy way to set the input data and to modify these to create different scenarios. The model output results permit to identify the main critical issues in the system analyzing the vessels arrivals, berths usage, waiting times, delays and queue values for different kind of vectors.

The model consists of a series of building blocks, organized for the reproduction of the different processes that take place within the port. All these blocks are a combination of elementary modules.

The model has been implemented to simulate all the port activities related with the arrival and the departure of any vessel. According to this approach, the model reproduces each single service and each movement of different typologies of vectors in the sea side and the land side of the port. The activities of the port system are represented as a chronological sequence of events based on a set of parameters, rules and policies.

In particular, the model of the port is composed by different module groups (Fig. 1) that represent the following operations: a) arrival of vessels; b) vessels navigation in the port channel; c) vessels evolution and mooring operation; d) truck and cars arrivals; e) handling operation at berths; f) departure of vessels. These modules are defined and calibrated specifically for each kind of ship and typology of freight transported. For each vessel are defined several relevant properties as size (tonnage), length (not all the berths are available for long vessels), product (each vessel handles just one specific type of product), expected date and time of arrival, assigned berth, shipping company, expected date and time of departure, quantity of freight transported disaggregated in loading or unloading, number of equipment and workers used for handling. These data about the arriving vessels are automatically downloaded, by means of a web service, from the official online platform used by the shipping agents for the request of port call and berthing.

The simulation of the port activities starts with the creation of the vessels with all related attributes and

their insertion in the simulation process. The vessels enter in the navigation channel to reach the basin of evolution, make the evolution and start the mooring operations. The simulation of the internal channel of navigation is performed taking into account the rules to manage the use of this obliged passage. In particular, the structure of the sub-model "channel navigation" permits to take into account the priority for the vessels departing from the port, considering a different capacity of the channel for the entrance (only 1 vessel) or for the exit movements (till 4 vessels). These rules ensure a greater capacity in terms of berths available. The system used the first-come-first-served allocation strategy. Hence when a vessel arrives at the port it has to wait in the queue until the navigation channel becomes free. From simulation are recorded, moment by moment, all data concerning the transition and use of the channel, the presence of a queue and his length, both for arriving and exiting vessels.

After this, each vessel, berth loaded attribute assigned, is simulated during the evolution and the mooring operation. Each berth has its own time of call, which depends on the available space inside the port and the type of call requested. The mooring operation, in the model, is simulated as one busy time slot using the berth. Even here, all types of vessels have a different average time of mooring, based on what the vessels handle and the berth configuration. This level of detail allows estimating an additional spent time, increasing the performance of the model, making it as close as possible to represent the real world data.

After the phase of mooring, the model starts the unloading and/or loading operation simulation. Everything is simulated with different resources that take into account the number of rosters and workers used and the handling equipment at disposal. Each handling operation is performed differently depending on the type of product and movement involved. For instance, the cruises ships perform unloading, loading and transit of passengers; the coal ships make unloading phase following a simulated chain of steps (cranes with crampons downloading from the hold into the hoppers from which the coal is discharged into the truck trailers); car carrier ships are simulated calculating the turnaround time that employees take to make a complete roster: starting from the internal space of the vessel, reaching the berth, travelling in the land side of port to find the parking and the car storage point. An example of an unloading and/or loading operation process is illustrated in Fig. 2.

The simulation of the ro-ro and ro-pax vessels handling operation is performed, for each vessel, considering the events of arrival in the port of the cars and of the trucks to be loaded. The arrival processes are defined using a specific probabilistic

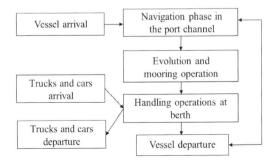

Figure 1. Overview of the simulation model.

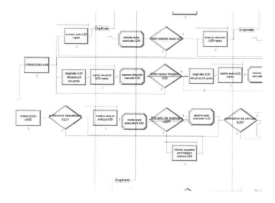

Figure 2. Modules blocks reproducing handling operation at Ro-Ro berth.

distribution, different for cars (a gamma distribution with different parameters depending on the different destinations of ferries) and trucks (a constant arrival process during the day).

4 CALIBRATION AND VALIDATION OF SIMULATION MODEL

The model has been implemented to simulate the activities within the Port of Civitavecchia. This port is organized with many multipurpose berths and an important bottleneck represented by the internal navigation channel. The largest traffic components are the Ro-Ro freight traffic and the cruise services. Civitavecchia is located on the Tyrrenian Sea, in the Central Italy, close to the city of Rome and it can be accessed by rail, road and motorway. It has always been one of the Italy's main ports for the Ro-Ro ships and ferries services directed to Sardinia, Sicily and other international connections with regions in the western and southern part of the Mediterranean Sea. Due to the closeness with the city of Rome, in the last years, Civitavecchia is increasing more and more as one of the main ports for the cruises traffic in Europe with more than 2 million of passengers every year. Data about arriving vessels are automatically downloaded, by means of a web service, from GIADA the official online platform for the request of port call and berthing in Civitavecchia.

After the phase of construction of the simulation model, described in the previous section, a rigorous phase of model calibration is performed adopting a trial-and-error technique based on engineering judgment and experience. The calibration involved checking the model results against observed field data. The collected data used for this phase included vessels traffic for the years 2011 and 2012, with details about date and time of arrival, assigned berth and quantity of freight handled. These data have been used in combination with the information obtained by an extensive series of interviews with different workers and managers of the port activities.

A large amount of replications are carried out to calibrate the simulation model. Data from the previous simulations have been extracted and compared with field data at disposal. According to the comparison of the model outputs, the values of the model parameters and, if eventual necessary, also the sequence of processes are refined in a iterative procedure with the aim to better fit with real world data at disposal.

After the calibration phase, the model validation phase is structured in two different sequential steps. The first step, as follows called the verification phase, has to ensure that the model works correctly by verifying that the simulation model properly translated the conceptual simulation model (flowcharts and assumptions) into a working program. The second step, as follows called the validation phase, is testing the model to evaluate its accuracy and applicability determining whether the conceptual simulation model is an accurate representation of the system.

The verification phase is performed using traffic data of 5 different weeks characterized by different level of vessels traffic and typologies (week of July, August, October and December of year 2014, week of January of year 2015). In other word, the model runs under a variety of settings of input parameters to check whether the model gives reasonable output. The outputs of these replications of the model are related to the occupancy rate of the berths, the occupancy rate of the navigation channel, the number of vessels in queue in the phase of arrival and departure (see Table 1). As expected, the outputs show different level of occupancy and vessels in queue according to the different simulation period, characterized by different number of calls and different typologies of vessels traffic.

Table 1. Simulation output of the verification phase.

Simulation period	July 2014	August 2014	October 2014	December 2014	January 2015
Call at port (#)	69	66	60	40	31
Vessels in queue	31%	28%	45%	18%	7%
Channel occupancy rate	14%	13%	14%	7%	6%
Ro-Ro berths occupancy rate	32%	26%	34%	39%	35%

The validation phase is performed using data of the final week of the month of July of year 2014 with the maximum level of traffic in terms of vessels and with very detailed data at disposal for 50 vessels (expected and effective arrival and departure times, assigned berths, etc.). Additional replications of the simulation model are performed to evaluate his capacity to reproduce real world data. This activity permits to identify two different critical areas. The first one is the arrival and the departure time of the vessels. In the previous version of the model, the arrival time is established according to the expected time of arrival (ETA) and the average delay observed for the Ro-Ro vessels producing, in some cases, especially for the other kind of vessels, unrealistic arrival times and consequently departure times. To better represent these data, the arrival time is adjusted taking into account also the observed distribution of time differences between effective times of arrival and ETA for all kind of vessels. The results of this model refinement are reported in Table 2 showing an important reduction of the total and the average time differences between effective and simulated times of arrival of vessels.

The second one is related with the berth allocation introducing an explicit consideration of the real capacity of the berths. In this way, the model explicitly involves the evaluation of the possibility to receive additional vessels and the possibility to readdress the vessels in other, feasible berths. The results of this refinement of the model are reported in Fig. 3 showing a different and more correct usage of the berths between the two versions of the simulation model.

Finally, it is important to remark that these refinement permits to better represent some specific major aspects of the port activities without any loss of capacity of representation of the previous model. The new version of the model simulates a number of calls equal to the previous one. This model consistency is also confirmed by the other outputs indicator used to evaluate port

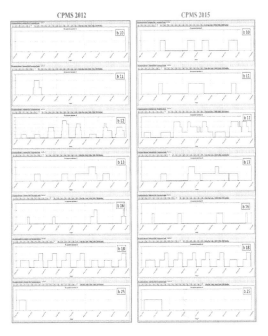

Figure 3. Comparison of berths utilization between old and new version of simulation model.

performance (vessels in queue, berth occupancy rate, navigation channel occupancy rate, etc.).

5 CONCLUSIONS

The simulation of a complex infrastructure as a port is studied by a limited number of authors in literature. Generally, the existing works reported in literature range from analysis of few and general elements, usually about port performance assessment, to very specific aspects. These are usually related to the optimization of a single specific operation.

Starting from these considerations, the implemented model has proved to be robust and effective in the representation of the port activities evolution, involving complex interactions respect to the space and the time. In addition, the goodness of fitting with real world data is due to the iterative activity of analysis of the results and the comparisons with detailed real world data modifying processes and parameters to upgrade the simulation model.

Table 2. Differences of arrival and departure times between old and new version of simulation model.

Differences between simulated and effective	Old version of model	New version of model
Total difference in arrival times (min)	383	345
Average difference in arrival times (min)	10	9
Total difference in departure times (min)	2,614	604
Average difference in departure times (min)	55	14

REFERENCES

Ancora, V. & Carbone, S. & Gori, S. & Petrelli, M. 2012. A model for the microsimulation of port activities.

In *Proceedings of International Conference on Traffic and Transport Engineering, Belgrade, 29–30 November 2012*.

Arango, C. & Cortes, P. & Mununzuri, J. & Onieva, L. 2011. Berth allocation planning in Seville inland port by simulation and optimization. *Advanced Engineering Informatics* 25: 452–461.

Cortes, P. & Mununzuri, J. & Ibanez, J.N. & Guadix, J. 2007. Simulation of freight traffic in the Seville inland port. *Simulation Modelling Practice and Theory* 15: 256–271.

El Sheik, A.A.R. & Paul, R.J. & Harding, A.S. & Balmer, D.W. 1987. A microcomputer-based simulation study of a port. *The Journal of Operational Research Society* 38(8): 673–681.

Guenther, H.O. & Grunow, M. & Lehmann, M. & Neuhaus, U. & Yilmaz, I.O. 2006. Simulation of transportation activities in automated seaport container terminals. In *Proceedings of the Second International Intelligent Logistics Systems Conference*.

Hayuth, Y. & Pollatschek, M.A. & Roll, Y. 1994. Building a port simulator. *Simulation* 63(3): 179–189.

Koch, D. 2007. PortSim—a port security simulation and visualization tool. In Proceedings of the 41st annual IEEE international Carnahan conference on security technology: 109–116.

Köse, E. & Başar, E. & Demirci, E. & Güneroğlu, A. & Erkebay, Ş. 2003. Simulation of marine traffic in Istanbul Strait. Simulation Modelling Practice and Theory 11: 597–608.

Longo, F. & Huerta, A. & Nicoletti, L. 2013. Perfomance analysis of a Southern Mediterranean Seaport via Discrete-Event simulation. *Journal of Mechanical Engineering* 59(9): 517–525.

Ramani, K.V. 1996. An Interactive Simulation Model for the Logistics Planning of Container Operations in Seaports. *Simulation* 66: 291–300.

Steenken, D. & Voss, S. 2008. Operations research at container terminals: a literature update. *OR Spectrum* 30: 1–52.

Uğurlu, Ö. & Yüksekyıldız, E. & Köse, E. 2014. Simulation Model on Determining of Port Capacity and Queue Size: A Case Study for BOTAS Ceyhan Marine Terminal. *The International Journal on Marine Navigation and Safety of Sea Transportation* 8(1): 143–150.

Van Asperen, E. & Dekker, R. & Polman, M. & De Swaan Arons, H. 2003. Allocation of ships in a port simulation. In *Proceedings of the 15th European Simulation Symposium*.

Voss, S. & Stahlbock, R. 2004. Container terminal operations and operations research: a classification and literature review. *OR Spectrum* 26: 3–49.

Transport Infrastructure and Systems – Dell'Acqua & Wegman (Eds)
© 2017 Taylor & Francis Group, London, ISBN 978-1-138-03009-1

Air transport hinterland in Adriatic and Ionian region: Equity and connectivity matters

O. Čokorilo
Faculty of Transport and Traffic Engineering, University of Belgrade, Belgrade, Serbia

ABSTRACT: The European Union Strategy for the Adriatic and Ionian Region (EUSAIR) as one of four EU macro—regional strategies provides a framework for contributing to the economic, social and territorial cohesion. The importance of air transport connections to the hinterland is particularly observed in the proposed research taking into account all relevant challenges of air transport system and infrastructure. The overall objective of the proposed research is to contribute to expanded, improved and safer air transport networks, which will attract new investments to the poorer regions, promote trade and contribute to the connectivity within EU transport corridors and the Adriatic and Ionian region. Analysis is focused on existing infrastructure, connections and available aircraft fleet capacities based on three possible scenarios which are identified on the recommendations of available databases (EASA, Eurocontrol, Boeing, Airbus, etc.). The purpose of the paper is to provide the beneficiaries with a comprehensive transport platform development in accordance with transport policy as a tool for strategic planning, development and design of air transport systems and infrastructure.

1 INTRODUCTION

Hinterland connections in Adriatic-Ionian macro-region that covers coastal areas in six different countries (Italy, Slovenia, Croatia, Montenegro, Albania and Greece) are a precondition of ports development in this region. The paper provides the evaluation of the integration among Adriatic ports and their hinterland enabled with air transport network that allows to identify needs and priorities, bottlenecks, potentials for passenger transport services/lines and their possible future integrations (Čokorilo, et al., 2015).

It is important to notice at this point, road transport which is of the utmost importance for the functioning of ferry and cruiser traffic flows in Adriatic-Ionian region. Expressed flexibility and ability to quickly respond to modern transport demands have enabled the largest share of road transport at the level of the whole transport market. Rapid road transport development can lead to congestions on the main routes in ports as well, and it may have a negative influence on the environment and the health of the population and decrease of traffic safety level. Therefore, it is necessary to make conditions for redirection of the demand to other transport modes with the aim of controlling the excessive development of the road transport.

Railways are important part of European transport system and significant contributor to achieving the sustainable transport in the future. Road and railway traffic networks are an indispensable prerequisite for the development and competitiveness of the ports in the Adriatic-Ionian region.

New air links within the Adriatic region could also considerably improve mobility and accelerate economic integration and cooperation processes. However, the problem of intraregional connectivity prevails, where majority of destinations from and to the Adriatic airports are in the Western Europe and minor of all air transport operations in the region are realised within the Adriatic network. Underdeveloped connections between the Adriatic ports and major cities represent a barrier for fast and convenient travel within the region. Equity issues are also important for future macro-region development since mobility is in relation with accessibility, not just for certain regions but to different users groups. Social aspect of travellers habits is important for building future trips demands. Therefore, air transport by its nature could provide excellent platform for the region connectivity (for example, new infrastructure investments are more cheaper comparing to road or railways networks).

2 LITERATURE REVIEW

Air transport development is a multi functional process which mostly depends on aviation market in terms of regional economic development and liberalization (Graham, 1997; Graham, 1998; Graham and Guyer, 1999; Thompson, 2002). Contemporary regional studies and research are focused mostly on economical aspects due to the touristic development within particular zones (Yamaguchi, 2007; Brooke, et al., 1994; Prideaux, 2000). Moreover, some authors are oriented on modelling environmental issues within the certain region (Abeyratne, 1999; Houyoux, et al., 2000) or climate change problems (Mickley, et al., 2004). Decision makers are more oriented in creating policies which will enable air transport market defragmentation and connectivity improvements (Button et al., 1998; Steiner et al., 2008; Šimecki et al., 2013; Čokorilo and Čavka, 2015).

3 LONG TERM AIRLINE PASSENGER FORECASTS—GLOBAL AIRLINE PASSENGER GROWTH

In this section it was examined the long term airline passenger forecasts published by aircraft manufacturers Boeing and Airbus. Both have produced a broad long term global market forecast for the period 2013 to 2032 using 2012 as the base year. Boeing and Airbus employ similar methodologies to form the forecast. At an aggregate level the two sets of predictions are largely comparable with each other. However, there are some key differences between the two manufacturers forecasts, which will be discussed whenever these influence the estimates results at a macro level.

In its 2011 market outlook, Boeing's forecast for 2030 was for 13.3 trillion RPK (revenue passenger kilometres) worldwide. The most up-to-date analysis produced by the American manufacturer predicts 14.7 trillion RPK by 2032. The average annual growth rate is similar but revised downward marginally (5.1% in 2011 versus 5.0% in 2013). Airbus points out in its forecast that historically (since the 1970s) air traffic has doubled every

Table 1. Boeing and Airbus forecast comparison.

Manufacturer	Boeing	Airbus
RPK (trillion) 2012	5.5	5.5
RPK (trillion) 2032	14.6	13.9
Total Growth 2012–2032	164%	151%
Average Annual Growth Rate	5.0%	4.7%

Source: Boeing, Airbus.

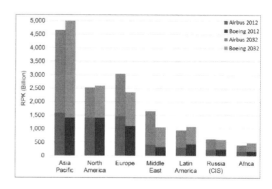

Figure 1. Boeing and Airbus Regional Forecast Comparison 2012–2032.
Source: Boeing, Airbus.

Figure 2. Intra and Inter-regional RPK Annual Average Growth Rates 2012–2032.
Source: Boeing.

fifteen years and will do so again by 2025. In its previous forecast, Airbus predicted average annual RPK growth of 4.8% between 2011 and 2031. This is in agreement with the most recent forecast by the European manufacturer.

3.1 Inter and intra-regional traffic flow growth

The Boeing Current Market Outlook provides a breakdown of inter- and intra-regional RPK forecast growth. In the figure above a diagram of the major flows is presented. Within the circles is the expected intra-regional RPK growth between 2012 and 2032. The arrows indicate the percentage growth on inter-regional traffic flows.

The forecast growth in RPK in the next twenty years is concentrated in traffic to, from or within the Asia Pacific region (including China). When China is included in growth rates for traffic within Asia Pacific, the aggregate growth rate is 6.4%. However when China is measured separately, it accounts for a growth rate of 6.9%. The lowest RPK growth is expected in the intra-North American market. The forecasted RPK growth is of 2.3%. The comparison of these figures with the previous Boeing market outlook indicates that most of the average annual growth rates are lower than those stated in the previous forecast. A relative growth in these rates is detected only for Latin America, Africa and Middle East regions. In the previous Boeing forecast the highest RPK growth for inter-regional traffic flows was attributed to the Europe-China market, followed by the Middle East-Asia Pacific segment. However, as this figure shows, the highest rate of forecast growth on inter-regional traffic flows is now predicted to be on Middle East-Asia Pacific routes (7.2% per year), reflecting the expected continued use of the Middle East for transfers between Europe/North America and Asia Pacific. Europe to China growth rates have fallen from 7.4% to 6.1% since the previous forecast.

3.2 Regional flows

In its latest Global Market Forecast for the period 2012 to 2032, Airbus has examined traffic flows

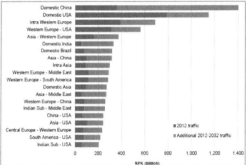

Figure 3. Largest 20 traffic flows in 2032.
Source: Airbus.

and provided data for traffic routes at a detailed level. From this data the largest overall flows by volume can be determined. In terms of the largest traffic flows in absolute volume, domestic China will overtake the domestic U.S. market. While experiencing growth rates below the world average over the forecast period, traffic flows within Western Europe and across the Atlantic remain the next two largest passenger markets (Figure 3).

4 ADRIATIC IONIAN AIR TRANSPORT IN FACTS

Main infrastructure and traffic indicators are shown in Table 2 and Table 3.

The analysis of future air transport developments is crucial for medium and long-term infrastructure capacity planning and for increasing

Table 2. Infrastructure indicators.

Airport	Number of runways/ passenger terminals	Main runway length (m)
Trieste	1/1	3000
Dubrovnik	1/1	3,300
Pula	1/1	2,946
Bari	2/1	2,820
Ancona	1/1	2,962
Rijeka	1/1	2,500
Venice	2/1	3,300
Pescara	1/1	2,419
Portorož	1/1	1,201
Tivat	1/1	3,252
Tirana	1/1	2,750
Split	1/1	2,550
Corfu	1/1	2,373

Table 3. Indicators of passenger and cargo traffic.

Airport	Passengers 2012	Cargo (t) 2012	Passengers 2013	Cargo (t) 2013	Passengers 2014	Cargo (t) 2014
Trieste	880,543	636	853,981	573	740,000	453
Dubrovnik	1,480,470	357	1,522,629	376	1,584,471	291
Pula	358,320	–	377,428	–	360,556	–
Bari	3,791,977	1,999	3,601,377	2,033	3,677,160	2,061
Ancona	564,476	6,864	501,689	6,680	480,673	6,990
Rijeka	72,762	–	140,776	–	101,939	–
Venice	8,192,296	33,112	10,579,186	45,662	10,723,442	44,426
Pescara	563,187	1,221	548,217	721,1	556,679	44
Portorož	n/a	n/a	n/a	n/a	n/a	n/a
Tivat	725,392	–	868,423	–	910,933	–
Tirana	1,665,331	1,875	1,757,342	2,164	1,810,305	2,324
Split	1,424,013	649	1,581,734	462	1,752,657	429
Corfu	n/a	n/a	n/a	n/a	2.383.353	n/a

the efficiency of the aviation system. Future demand and region opportunities could be clearly understand by the SWOT analysis which has been conducted in order to evaluate the strengths, weaknesses, opportunities and threats involved in air transport development within the Adriatic Ionian region (Table 4).

Table 4. SWOT analysis.

S
Excellent geographic position and level of air routes network development within and/or outside the region;
Available airports infrastructure resources;
Available resources for aviation sector connection to port, road and rail resources;
Strong membership in European aviation organizations of Adriatic Ionian region countries;
Available sources for potential regional airports network development and its connectivity with the rest of Europe.

W
Insufficiently usage of existing airport infrastructure (particularly for small size airports with seasonal character of aircraft operations);
Lack of modern air navigation system (in some airports);
Lack of financial resources for air mode infrastructure development;
Weak and insufficiently steady political position of some non EU countries;
National monopoly on several airports;
Lack of cargo resources;
Lack of airports interconnection by rail;
Insufficient airports utilization by passengers and aircraft due to their capacities;
Lack of certain airports interconnection with ports.

O
Interests of airports development as an open markets in European air transport network;
Future possibility to air cargo transport increase;
Possibility for establishing international and regional port connectivity by air;
Possibility for reducing travel time within the region by air connection.

T
Lack of aviation infrastructural development projects in relation to linking hinterland by air connectivity;
Manifestation of partial and local interests inside the countries;
Long lasting economic crisis aiming the passenger decreasing trend;
Airports seasonal character.

5 SCENARIO ANALYSIS FOR AIR TRANSPORT DEVELOPMENT

Due to the lack of intraregional connectivity within the Adriatic Ionian basin, certain parts of the region have limited access to regional, European and global markets. Mentioned imbalance of accessibility to services, markets and opportunities for further social and economic progress is an obstacle for overall development of the Adriatic Ionian region.

According to abovementioned statistics, it is projected that in Europe passenger traffic will rise at an annual rate of 3.8% to 2032, reaching 2.35 trillion RPK. This is a downward revision of the European market which, in the previous forecast, was estimated to grow to 2.88 trillion in 2030. This decrease is largely due to the performance of the aviation market in the base years; 1.22 trillion RPK in 2010 in contrast to 1.11 trillion RPK in 2012. The core reason for the under-performance is the effect of the economic downturn that has impacted the European region since 2008/2009. The IHS Global Insight GDP forecast for Europe for the years 2012 to 2032 estimate a 2% annual increase against 1.9% projected between 2010 and 2030, thus showing increasing confidence in improving economic conditions going forward.

The existing connectivity network and air transport frequencies within the Adriatic Ionian region are bellow growing needs and demands of the travelling public. While connections with main European destinations are dominant and all leading European air carriers already operate in the Adriatic region, presently less than 15% of all airlines commercial activities are realised within the Adriatic Ionian network. Therefore, an underdeveloped connection between individual capitals and major cities of the Adriatic Ionian region represents a barrier for fast and convenient travel within the region.

Key determinants of air transport connections development are based on contracts for the joint operations of air carriers and associated partners within airline alliances. Star Alliance, with Austrian Airlines, Lufthansa, Croatia Airlines and Adria Airways as members, dominates in this region, particularly within SEE area (Mantecchini et al., 2013). Accordingly, the highest frequency air transport connections in the Adriatic Ionian region are linking the Adriatic Ionian region and Western Europe, with the largest number of flights to European nodal airports (Rome, Frankfurt, Munich and Vienna).

Passengers travelling by air have to consider fixed time blocks, respecting their duration, mostly set by air transport. One of the major disadvantages of air transport to rail and to some extent to road is the check in time, asking the passengers to be at the airport much sooner before the actual flight. In general, for European airports it is two hours for economy travellers on international flights and about 90 minutes for domestic. Travel time used in analysis is calculated as a sum of waiting time at the airport before the flight, flight time and waiting time at the airport after the flight. Time spent at the seasonal airports in Adriatic Ionian basin before and after the flight (45 minutes) is lower in comparison to European airports due to its smaller capacity and volume of traffic.

5.1 State of art and scenario development

The analysis of the future air transport development of the Adriatic basin is neither possible from the regional, national or EU integrated view but by an assessment from an international perspective. The evaluation basis is the macroeconomic valence of the economic instrument: air transport as a mean of transport for persons and high-quality goods. The possible socio-economic effects at a national and regional level are only then assessable if the international possibilities for development are known and the prerequisites necessary for it were defined.

5.1.1 Transport modal cooperation

Air transport is not possible without the other means of transport. These serve as conveyer and distributor. The national air traffic services are in demand of the road and railway traffic ahead of all things. Because of the dimension of the catching and target areas of the air traffic passengers and the air freights, a narrow networking of the airports as an interface to the road and railway traffic is an indispensable prerequisite for the development and competitiveness of an airport. This is obliged to the interests of the respective country and the international interests as a national instrument tied regionally. With the integration of the surface-bound transport services in aeronautics these traffics also undergo their internationalisation and the operative and administrative necessities connected with the integration.

5.1.2 International airports and local influence

A usual analysis of the international airports as a municipal traffic location by the local government politics, because of the high political importance for dedicated politicians of the commune or region, makes these frequently forget that today certainly the needs of these locations cannot be satisfied with the assessment bounds of the municipal or regional possibilities. On the contrary necessities are given, which exceed the fortune of the municipal or regional finances. By then visible limitations

have been the reasons in many cases that within a limited possibility for development the existing chances were not taken. The egoisms of powerful decision makers are also the reason for undesirable developments in other cases. These are afterwards reversible only by the resulting replacement of main facilitating economic actions with an enormous effort only.

5.1.3 Intermodal cooperation

The great and effortful problems are the coordination of the interests and the obligations of the cooperating traffic carriers. The contractual specification of the cooperation is possible only when the basis of the cooperation is grounded on modular contracts, which contains the possibility of further developing from the regular experiences of the cooperation and facilitating the agreements further in the mutual interest. Narrow-minded attitude to be frequently stated is one of the risks often appearing and reason for the failure of developments to adhere to allegedly indispensable interests and to prevent a flexibility and optimisation of the cooperation. Therefore it is necessary to put a high independent authority in charge of the project, whose strategy is not determined by a tactical essential. The principle of the cooperation is the renunciation at the tactical level in favour of a higher profit at the strategic level. The power renunciation at the local and regional level makes the common gain possible at the national and international level.

5.1.4 Infrastructure requirements for flexibility

1. Not all observed airports are connected to the existing rail networks of the region.
2. All airports are capable of serving tourist flights as arriving tourists will be able to reach their final destinations by port, rail or bus services.
3. Short term capacity problems at certain airports are solvable, which excludes high infrastructure investments to serve possible spot problems, leaving finance for long term investments.

5.2 Scenarios

The overall goal of future transport development within the Adriatic Ionian region should be based on attracting international transport flows and increasing regional development. Air transport is recognized for adequate tool for connecting ports and hinterland due to the easiest of new routes establishment. The existing capacity of airports and aircraft is sufficient for the forthcoming period. As it is shown before, an important parameter of future air transport development is GDP which is evaluate for Europe as 2% of average increase until 2032. As a key driver in aviation development, this parameter is function of RPK which from the Boeing statistics could be measured round 5% of growth rate for intra and inter regional flights for the Adriatic Ionian region. Air transport has been in constantly growth rate in the last decade.

Some negative aspects within the region are related to the post war development of transport infrastructure in the SEE at the beginning of the last decade which was tightly connected to the national links without a comprehensive regional view. However, in the process of economic development of each country within the region, the need for intraregional accessibility in all aspects became even more important. On the other hand, constant pressure from the EU towards the integration of the region and the establishment of political, territorial and economic cohesion became the one of the preconditions for the EU accession process.

Three potential scenarios are related to air transport development within the Adriatic Ionian region:

5.2.1 International and Intercontinental Flights Development (optimistic scenario—Best)

It is not possible to expect that the future development of airports in the Adriatic Ionian region would be equal for all countries, but there is still great opportunity to develop new routes which will continue to expand the number of potential passengers. The possible way of future development should be based on hub connectivity (Athens, Roma, Milan, Belgrade, Zagreb, etc.) with the expected growth rate of 5% RPK.

5.2.2 Regional and International Flights Development (realistic scenario—Modest)

This scenario consider existing flights routes between Adriatic Ionian region and other European cities. The expected growth rate is evaluate as 2% of RPK growth, according to GDP and forecast until 2032. This scenario will cover minor growth of passengers, while the growth rate is perceived on the basis of tourist destinations within the region.

5.2.3 Municipal and Local Oriented Airports (pessimistic scenario—Worst)

This scenario is not expected to be held within the large number of airports within the Adriatic Ionian region, but still some future trends should bring reductions in RPK or number of operations for some seasonal airports. Above all, some global economy drivers, Ukrainian crisis, migrants flows, etc. could provide reductions in larger airports but not more than 5% in the total RPS within the region.

Air transportation has become the key infrastructure for global economic development. Despite uncertainties and potential challenges that

may slow down growth, air transport services must continue to expand in a safe, secure, and sustainable manner. This entails substantial investments for markets such as Adriatic Ionian region with expected passenger growth to average between 1–2% annually. However, even less rapidly growing countries must invest in modern airport and air traffic management and surveillance infrastructure. Operators around the globe can only secure sustainable and profitable air services when achieving high energy efficiency through modern aircraft and efficient operations.

6 CONCLUSIONS

It can be concluded that future measures in infrastructure investments could bring good basis for air transport development as well as touristic attractions. Flexibility in travel time, or new intra-regional routes could significantly bring much more passengers using air transport services. The existing potential is not reach its limits so attractiveness and solutions for modal share are the key drivers for increasing load factor on the existing routes.

Moreover, proposed best scenario would increase air transport services and would open Adriatic-Ionian market to be much more oriented and extended to other regions. All this improvements should accelerate economic integrations via open market issues, revitalization of aviation market dedicated to general aviation and similar, and make the aviation sector more competitive to other modes by opening new lines within the region and wider.

REFERENCES

Abeyratne, R.I. 1999. Management of the environmental impact of tourism and air transport on small island developing states. *Journal of Air Transport Management* 5(1): 31–37.

Brooke, A.S., Caves, R.E. & Pitfield, D.E. 1994. Methodology for predicting European short-haul air transport demand from regional airports: an application to East Midlands international airport. *Journal of Air Transport Management* 1(1): 37–46.

Button, K., Haynes, K. & Stough, R. 1998. *Flying into the future: air transport policy in the European Union.* Edward Elgar Publishing. 199 p.

Čokorilo, O.; Čavka, I. 2015. The role of air transport development in Adriatic-Ionian macroregion. *International Journal for Traffic and Transport Engineering* 5(1): 344–359.

Čokorilo, O., Ivković, I., Čavka, I., Twrdy, E., Zanne, M. & Ferizović, A. 2015. Hinterland Connections of Adriatic-Ionian Region. In *Proceedings of International Scientific Conference—Cooperation Model of the Scientific and Educational Institutions and the Economy*, Zagreb, Croatia. 61–67.

Graham, B. 1997. Regional airline services in the liberalized European Union single aviation market. *Journal of Air Transport Management* 3(4): 227–238.

Graham, B. 1998. Liberalization, regional economic development and the geography of demand for air transport in the European Union. *Journal of Transport Geography* 6(2): 87–104.

Graham, B. & Guyer, C. 1999. Environmental sustainability, airport capacity and European air transport liberalization: irreconcilable goals? *Journal of Transport Geography* 7(3): 165–180.

Houyoux, M.R., Vukovich, J.M., Coats, C.J., Wheeler, N.J. & Kasibhatla, P.S. 2000. Emission inventory development and processing for the Seasonal Model for Regional Air Quality (SMRAQ) project. *Journal of Geophysical Research: Atmospheres (1984–2012)*, 105(D7): 9079–9090.

Mickley, L.J., Jacob, D.J., Field, B.D. & Rind, D. 2004. Effects of future climate change on regional air pollution episodes in the United States. *Geophysical Research Letters* 31(24): 1–4.

Prideaux, B. 2000. The role of the transport system in destination development. *Tourism management* 21(1): 53–63.

Steiner, S., Šimecki, A. & Mihetec, T. 2008. Determinants of European air traffic development. *Transport Problems* 3(4): 73–84.

Šimecki, A., Steiner, S. & Čokorilo, O. 2013. The Accessibility assessment of regional transport network in the South East Europe. *International Journal for Traffic and Transport Engineering* 3(4): 351–364.

Thompson, I.B. 2002. Air transport liberalisation and the development of third level airports in France. *Journal of Transport Geography* 10(4): 273–285.

Yamaguchi, K. 2007. Inter-regional air transport accessibility and macro-economic performance in Japan. *Transportation Research Part E: Logistics and Transportation Review* 43(3): 247–258.

Transport Infrastructure and Systems – Dell'Acqua & Wegman (Eds)
© 2017 Taylor & Francis Group, London, ISBN 978-1-138-03009-1

Which role for tramways in the next years

G. Mantovani
Independent Consultant, Rome, Italy

ABSTRACT: In the first decades of last century, tramways were the main means of local public transport. Later different reasons caused a progressive decline, not with the same extent everywhere, but in the seventies a revival began and is still in progress, even though discussed. Reasons of decline and revival are outlined. Then various innovative characteristics of the modern tramways are examined, in order to highlight the offered advantages. However, pros and cons must be considered, to understand when the construction of new lines or systems is useful and sustainable. Thus, it is attempted to define the field of advisable use of tramway (rather than buses, BRT, metro), in terms of transport demand and other factors. In conclusion, the need to select proper tramway plans and to support them with a proper general policy for mobility is reminded and some advice is given to develop good new systems.

1 A VERY CONCISE HISTORY OF TRAMS

1.1 Origins (1830–1880)

To outline the future of trams it seems useful to briefly retrace their history. The starting point can be traced back to 1832, when the first street urban railway line was opened by John Stephenson in New York, or perhaps more accurately to 1853, when, still in New York, the Frenchman Alphonse Loubat introduced a kind of grooved rail, which avoided the normal standard railway track projections onto the road surface, therefore eliminating an obstacle to the circulation of other vehicles and pedestrians. A guided means of transportation was born, suitable for development in urban areas. In the following decade, Loubat promoted the realization of the "chemin de fer américain" in Paris and in 1855 the first European tramway service was introduced in that same city. The spread of the new system was rapid. The first British line dates back to 1860, in Birkenhead; the first German line to 1865, in Berlin. In Italy, in particular, the first city to introduce the circulation of trams was Turin in 1872, followed by Naples (1875), Milan (1876) and Rome (1877).

The motivation to develop railways and tramways in the nineteenth century was basically energy-related. The regularity of surfaces and the reduced friction, characteristic of the motion of metal wheels on metal rails, required significantly less traction force, the mass to be transported being equal, than what was necessary for vehicles with wheels moving on road surfaces, such as omnibuses. At the time when urban transport was still dominated by horse traction, the rail therefore made it possible to reduce the number of horses required to haul a given payload. This led to trams gradually replacing omnibuses.

The horse traction however posed significant problems: these concerned sheltering and care for the horses, and the amount of manure on the roads, particularly significant in larger networks requiring extremely high numbers of animals. To eliminate such problems, the focus became mechanical traction; various techniques were tested and more or less broadly adopted: mainly steam but also gas, compressed air (another form of energy storage on the vehicle) engines, etc.. The use of steam engines on vehicles obviously had a number of negative effects in urban areas and eventually led to the development of traction based on underground cable in continuous motion (due to the action of fixed machinery), gripped by vehicles through a special lever; a system that became widespread in many cities, in San Francisco for instance—where two lines still survive today as a living museum—and in Melbourne.

1.2 Wide expansion (1880–1930)

It was electricity that offered the solution. Fyodor Pirotsky's 1880 brief experiment in St. Petersburg was not pursued and the first electric tramway went into operation in Berlin (Figure 1) thanks to Werner von Siemens (with vehicles equipped with a 7.5 kW engine powered initially by the two rails, to which a voltage of 180 V d.c. was symmetrically applied). It was Frank Sprague however who first achieved in 1888 a truly significant accomplishment, in Richmond U.S.A. (the power was from an overhead line, through a pole equipped with a

Figure 1. Siemens's electric tram in Berlin (1881).

contact wheel, at a voltage of 550 V d.c., which in fact became a standardized value for urban tramway applications). The spread of electric traction was rapid and Sprague's system widely used, which determined a considerable development of tram systems. In 1890 the first electric tramways appeared in France (in Clermont-Ferrand) and in Italy (in Rome, using a special series feeding technique, soon discarded, and in Florence, on the suburban line to Fiesole).

Interestingly, feeding through overhead lines was met with opposition, for aesthetic or safety reasons. Even at the time, alternative systems were developed, not exclusively by means of batteries, which limited power and/or autonomy. Tram lines were also built (for instance in London and in Paris) with power transmission through conductors set in a conduit located under the road, at the center of the track. Shoes, supported by a rod mounted on the vehicle and passing through a slot in the flooring, slid on conductors. These solutions however were complex and in need of constant maintenance. Attempts were also made to collect the current from contacts placed on the road surface, with devices designed to turn them on only when a vehicle passed over them, for safety purposes; they were not successful but they somehow acted as precursors to modern ground-level power feeding systems in operation today (Bordeaux is an important case).

The tramway became for many decades—from the end of the nineteenth century to the 30s of the last century—the dominant means of urban public transport worldwide, also because demographic, social and urban developments determined a significant increase and modification of transportation in cities, especially in the new suburbs. The tram was no more for a select audience but became 'everyone's coach'; economic problems also ensued, which in many cases led to the municipalization of services.

In its heyday, the tramcar enjoyed significant developments. The need to enhance its capacity resulted initially in convoys comprising one or more trailers and in vehicles of ever-increasing size. To improve behavior at curves and to avoid excessive loads on each axle, the two-axle vehicles were gradually replaced by vehicles with two two-axle bogies, pivoting with respect to the car body. Subsequently articulated vehicles were introduced, obtained both by combining 2 preexisting two-axle vehicles through a short suspended car body (Bern, 1932; Rome, 1936) and by building new structures based on the Jacobs bogie (the first attempt of this type dates back to 1926, but its spread was made possible by the "carousel", an ingenious portal that allows a very functional articulation betwwen the car bodies, invented by Mario Urbinati and used for the first time on vehicles in Rome, 1938). At the time, forms of standardization of vehicles appear, due for instance to the success of Peter Witt's project for Cleveland in 1915 (trams of the Witt type are still in service in Milan) and later required by the President's Conference Committee (U.S.A.) in the 1930s. The PCC trams were built in very large numbers, also under license (the Czechoslovakian ČKD Tatra deserves mention, which produced over 10,000 trams just of the T3 type).

1.3 Decline (1930–1970)

The Thirties marked the beginning of a decline that in many countries became dramatic in postwar years. To give a sense of its magnitude one can mention, for instance, that in 1940, after many decommissions had already taken place, there were 45 Italian cities with an urban tramway system, while in 1973 only four were left (Turin, Milan, Rome and Naples, not considering Trieste's surviving line, atypical and mostly suburban). In 1966, only three cities in France (Lille, Saint Etienne and Marseille) had kept their tramways, and extensive decommissions had taken place in other European countries. Not everywhere though, since in Central Europe (Germany, Switzerland, Austria) and in the Eastern Europe trams continued to play an important role. Even outside Europe the decline was not generalized, one need only consider the extensive network in Melbourne.

The causes for decommissioning were manifold: the development of metros in major cities, the technical progress of buses and trolley buses—often in contrast with the obsolescence of tramways and tramcars -, the space required by private vehicles on the roads, the damages caused by the WWII. Lobbying by bus manufacturer and oil companies was sure a contributing factor, and widespread was the illusion that public transport could become a marginal service in cities where everyone would have a car or at least a motorbike.

Therefore, a decline due to reasons that were largely not intrinsic. Modernization of networks and far-sightedness in mobility policy would have avoided the decline.

1.4 *Renaissance (1970-present)*

In the Seventies the increase of congestion in the streets, with its many harmful effects, reestablished the necessity of efficient public transport as a key element of urban mobility. A new role for the tram emerged, with its specific function as a primary system in medium-sized cities and as a complementary system to metros in larger cities. Table A shows data regarding the spread of tram systems in the world. What emerges is the great development in France, where in 30 years as many as 26 cities have reintroduced trams on their roads (Laisney, 2011). The first was Nantes, in 1985. Particularly interesting is the case of Strasbourg, where the first line opened in 1994 (Figure 2). A strong political controversy opposed metro to trams, which eventually ended in favour of the latter, with the result that a very successful network was developed, which now includes six lines. It is also interesting that trams have been developed primarily in medium-sized cities (with 100,000 to 400,000 inhabitants), but also in major ones, already equipped with extensive underground networks (greater Paris, with 14 metro lines: 8 new tram lines, over 100,000 passengers per day on the T3 semicircular line. Lyon, with 4 metro lines: 5 new tram lines) and in smaller cities (for example Valenciennes, 50,000 inhabitants including the surrounding municipalities reached by one of two tram lines).

Figure 3 is a graphical representation of the extension of new tram systems in some European countries from 1985 to the present. In none of the other countries that had almost suppressed trams a development comparable to the one in France did occur. In Italy, the first new tramway system was built in Messina in 2003, which unfortunately has been inadequately managed. There are currently 8 new systems in operation, so far consisting of only one line, an exception being Venice/Mestre (2 lines)

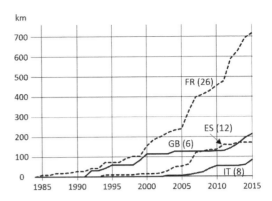

Figure 3.   Cumulative extension of new systems in some European countries (years 1985–2015).

and Palermo (4 lines, 3 of which branch out to the outskirts from a common section). Four old systems (Milan, Turin, Rome and Naples) were never discontinued, but to some extent reduced and expanded. (Mantovani G., 2014)

In countries where most of the networks have been preserved, a technological, functional as well as infrastructural modernization has been provided, gradually aligning them with newly built systems. In Germany there are now 52 tram networks in operation. The one in Berlin had previously been confined to the former eastern sector, but new routes have been created (recently, allowing access to the new Hauptbahnhof) and wide renovation of rolling stock is ongoing. In Munich, a city that also has extensive metro and S-Bahn networks, the intention of suppressing tramways has long since been abandoned and indeed new routes have been built. In several cities tramway subways have been constructed in central areas. The Hannover system, which combines a proper metro and a tram network, is particularly interesting: three underground lines cross the central area of the city, with functional interchange stations, and then disperse into 8 tram routes into the outskirts. Turning to Switzerland, the case of Zurich deserves mention, where a 1972 referendum rejected the plan for metro, and a 2 km tunnel, which had already been built, was inserted into the tram network. Two recent projects have brought the number of lines up to 15.

2   CHARACTERISTICS OF NEW TRAMWAYS

2.1   *Vehicles*

2.1.1   *Dimensions*
A progressive increase in the length of vehicles was determined by two major factors: the need to further increase capacity (to satisfy the demand

Figure 2.   Strasbourg, 26.11.1994, Opening of line A.

for transport and also to decrease operating costs, reducing the impact of the cost of the driver on the cost of ridership) and the elimination of trailers (to facilitate the distribution of passengers and for safety reasons). Today the typical length is around 30–32 m, but lengths of around 40–42 m are also widespread, and even longer trams are in circulation, reaching 56 m in the series that recently came into service in Budapest. The width depends on various constraints, but in the new systems it preferably measures 2.40 m—which facilitates the interior fitting as opposed to smaller ones—and, whenever possible, 2.65 m.

2.1.2 Low floor

To allow step-less boarding, floors typically have a height of circa 35 cm with respect to the top of rail (in new and unloaded vehicles), at least along most of the vehicle's length and always in correspondence to doors. With platforms 30 cm high there is a maximum height difference of 5 cm, which is considered acceptable. However, a smaller difference is preferable, and to this end a system of self-levelling of the vehicle is adopted. Thanks to the very small horizontal gap between the platform and the vehicle, a feature offered by guided paths, appropriate conditions are obtained for the passage of wheelchairs, with no need for retractable steps. A low floor is generally preferred along the entire length, if necessary with slightly sloping connecting parts, but there are those who consider it functional to also have a higher floor at both ends (connected with one or two steps), which allows for the installation of conventional motor bogies. Mention should finally be made of the 332 ultra-low-floor trams in Vienna, with thresholds 19 cm above the ToR, devised for boarding almost flush with ordinary curbs. However, they have required a unique vehicle architecture, expensive to manage, and their supply will be discontinued.

2.1.3 Vehicle architecture

The difficulty of placing traditional pivoting and Jacobs bogies under low-floor carbodies has resulted in the development of innovative vehicle architectures. For example, one might mention trams with non-integral low-floors (with traditional motor bogies, high floors at both ends of the vehicle and special solutions for articulations) and trams with integral low floors developed in Germany in the 80s (with two pairs of independent wheels under each carbody, operated through cardan transmissions). Today, widespread amongst all major manufacturers are solutions involving integral low floors based on suspended carbodies, supported at both ends by shorter carbodies mounted on non-pivoting bogies. Therefore there is minimal occurrence of relative motion on the horizontal plane, limited to what the suspensions

allow for; the suspended carbodies must also measure in length less than traditional architectures and as a result the number of joints is relatively high (e.g. 6 for a 32 m tram). These solutions, however, do not perform well when entering curves, especially when they are of a small radius: the wear of rails and wheels is greater than in traditional trams and passengers might experience annoying lateral accelerations. To overcome these drawbacks some manufacturers (Škoda, Alstom) have found ways to insert pivoting bogies under low floor carbodies, with the counterpart of some mechanical complexity and greater encumbrance on board in correspondence with wheels. Modular construction is generally used, to contain production costs and, at the operational stage, to facilitate maintenance, while the vehicles can easily be lengthened, if need be, with the inclusion of additional carbodies. Vehicle customization, often requested by clients, is achieved through specially designed front shells.

2.1.4 Traction equipment

Asynchronous motors have been used extensively for a long time. They are characterized by reduced size and simple maintenance, and are controlled by electronic equipment (traction inverters and braking choppers), which offer excellent speed regulation, with superior performance compared to traditional motors (especially in the acceleration and deceleration phases). A versatile recovery of braking energy is also obtained (used for on board auxiliary services or transmitted to the overhead line, if receptive, or accumulated in super-capacitor banks; therefore never dissipated in resistors).

2.1.5 Running without an overhead line

Overhead lines have a negative visual affect in urban areas with monuments or having any other historical value, and in other instances might be subject to potentially hazardous interference (for example trees). Various techniques have therefore been developed for powering without an overhead line, which fall into two categories: on-board energy storage (by means of batteries, rechargeable in sections with overhead lines, or by means of supercapacitors, charged very quickly during bus stops) and ground current collection (by means of a contact line whose segments, for obvious safety reasons, are powered only when covered by the vehicle). These solutions have been made possible by technological advancements, but cost and maintenance requirements suggest that doing away with overhead lines should only be limited to routes where it is truly necessary.

2.1.6 Monitoring, diagnostics, information

Modern trams have effective control and diagnostic systems, as well as numerous additional functions such as localization, data exchange with the central control stations, surveillance of driver attention levels, real time communication of both visual and audio information to passengers, etc.

2.1.7 *Some critical issues.*

Some of the consequences of technological evolution should be highlighted. The greater complexity and the many different kinds of electronic apparatuses raise fears that the life expectancy of vehicles, traditionally very high, will be reduced, and they will require at least one revamping of electronic equipment at mid—life. Maintenance as well now requires staff to be skilled in using various types of equipment and to receive specialized training; as a consequence, it is assigned increasingly to the manufacturer, at least for an initial period. To minimize such potential drawbacks, a good technical knowledge of the clients is required, which among other things allows for the drafting of careful specifications, aimed also to preserve to tramways the feature of open systems.

2.1.7 *Trams on tyre*

Rubber-tyred vehicles have been developed, in which the driving function is entrusted to a single central rail, where small guide wheels have the function of steering the rubber-tyred wheels. Following the failure of other vehicles, the market has narrowed down to the Translohr tram by NTL—a subsidiary of Alstom—currently in service in 7 cities, 2 of which are in Italy. The main claimed advantages of this system are the reduction of infrastructure costs, the possibility of very small radius curves and the ability to circulate on slope percentages of up to 13%. However, counterparts exist and one must consider that it is a closed system, that leads to establishing an exclusive bond with a supplier.

2.1.8 *Tram-Train*

Today the mythical system of Karlsruhe is operated on a network of over 500 km and is composed of 14 lines of various type (seven are T-T strictly speaking, that is combining circulation on railway lines, together with trains, and on tram lines). The huge success of this system gave rise to many feasibility studies in several countries, but just about ten true T-T systems have been or are being realized until now. That of Kassel deserves a special mention for its extension; others are in Germany, France, Spain and Great Britain. Different systems exist, without mixed circulation of trains and trams (e.g. Cagliari) or being just tram-derived vehicles circulating only on railways (e.g. Nantes).

A true T-T poses many technical problems (especially regarding safety) but generally they can be solved, as the systems in operation demonstrate.

It seems that the feasibility rather depends on the possibility to create combined routes being functional and serving an adequate demand, on the residual capacity of the railway section, perhaps on some open-mindedness too. It seems also that T-T fits better medium-sized towns, as the tram section of a T-T route can allow to reach several places of the town directly.

2.2 *Lines*

2.2.1 *Track construction*

Ballastless tracks are now used more and more frequently. Many different techniques exist, some using prefabrication methods. They offer better geometrical steadiness, simple maintenance and longer duration. An exhaustive comparison of the various techniques—with regard to performances, construction and maintenance costs—should be very useful.

2.2.2 *Measures to ensure speed and regularity*

Surface public transport often requires dedicated lanes, so that users are guaranteed satisfactory speed and, most importantly, regularity, but also so that operating costs are reduced. These objectives are particularly important for tramways, considering their function as primary networks and the high passenger flow. It is therefore appropriate trams be assigned their own dedicated lane (possibly shared only with well-regulated buses), with cutting-edge priority traffic lights at intersections, and underpasses used to cross very congested nodes. Lines with high speed performance and regularity can therefore be focused on, for which the term Light Rail Transit (LRT) has been coined (broadly equivalent to *Metrotranvia* in Italian and *Stadtbahn* in German). This should be the way to build new tramways.

2.2.3 *Alignment versatility*

A tramway does not necessarily have the same characteristics along its entire alignment. LRT sections in the outskirts can coexist with slow sections in central pedestrian zones, to combine the advantage of greater speed due to the predominance of LRT sections with a direct service to the center. It should be noted that the circulation of trams in pedestrian areas, for instance at a limited top speed of 15 km/h, grants pedestrians greater safety compared to that of other vehicles, as demonstrated in many circumstances.

2.2.4 *Proper environmental integration, urban regeneration*

The construction of a good tramway is not achieved simply by laying the rails on the roads, with no attention to urban context. The project must take into account the urban landscape it will go through and the activities linked to the various locations, with particular attention to areas of greater value. The construction of the tramway should be associated with accurate regeneration projects, extending to the entire road section, including also the suitable renewal of street furniture. To this end, the design team must possess multidisciplinary skills. Accomplished examples of how tramways can be well integrated into the urban context can be found in French projects.

2.2.5 Centralized control

The central control room of a modern tramway is equipped with systems that monitor in real time the regularity of circulation and the condition and status of vehicles, as well as remote controlling the equipment (both lines and power supplies). In the event of circulation disruptions, the control centre operator may arrange for appropriate corrective action, with the assistance of a computerized system.

3 TRAMWAYS: WHEN AND HOW THEY SHOULD BE DEVELOPED

3.1 Pros and cons

In trying to outline the future of tramways, we must evaluate their advantages and disadvantages compared to other urban transport systems. In Table 1, metros, tramways and bus lines are very roughly compared in many respects. Some aspects are easily quantifiable by nature—at least assuming ranges of values—some are not.

However, unquantifiable aspects, as comfort and system image, must be considered, because they contribute to the choice between public and private transport, thus affecting the modal split.

3.1.1 Line capacity

Obviously, it depends on the vehicles' capacity and the headway. Assuming a minimum range of 120 seconds for tramways and bus lines (to limit irregularity of runs and the resulting loss of capacity) and 90 seconds for the metros (assumed to be automatic) the maximum capacities are as shown in Fig. 4.

3.1.2 Operating speed

Typical values are also shown in Fig. 4. For trams and buses circulating in mixed-traffic, speed depends on traffic conditions and values can be very low, which is annoying to users and causes higher costs for the operator. Reserved lanes, efficient priority traffic lights and some infrastructure projects can gradually raise the operating speed, turning tram systems into LRTs and bus lines

Table 1. A comparison of transit systems.

	Bus	Bus Rapid Transit	Tram	Light Rail Transit	Subsurface Light Metro	Deep Metro
Max line capacity, kpax/h*direction	3	4	6	10	20	40
Typical operating speed, km/h	14	20	14	20	30	33
System accessibility [1]	5	4	5	4	2	1
Boarding easiness [1]	3	3	5	5	5	5
Kinematic comfort of passengers [1]	3	3	5	5	5	5
Visual comfort of passengers [1]	5	5	5	5	1	1
Environmental quality [1,2]	2	2	5	5	5	5
Protected lanes; required road cross section, m	7	7	6	6	-	-
Operation with snow and heavy rain [1]	3	3	4	4	5	5
Possibility of operation connected with railways	NO	NO	YES	YES	YES	YES
Typical realization cost M€/km	3	9	15	25	90	150
Best case operation cost €/place*km	0.06	0.07	0.04	0.04	0.03	0.03

[1] Indicative mark: 1 = worst. 5 = best.

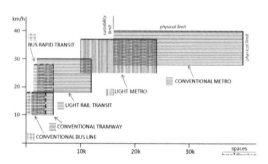

Figure 4. Capacity and operating speed ranges for different transit systems.

into BRTs (Bus Rapid Transit), however without matching metros, favored by having an entirely segregated route. It should be noted that users are interested in door-to-door travel time, and therefore, on limited distances, the metro's reduced accessibility compensates for their increased speed, as shown in Fig. 5.

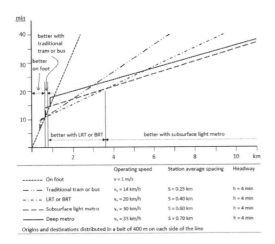

	Operating speed	Station average spacing	Headway
-------- On foot	v = 1 m/s		
—- - — Traditional tram or bus	v_o = 14 km/h	S = 0.25 km	h = 4 min
—- - - — LRT or BRT	v_o = 20 km/h	S = 0.40 km	h = 4 min
— — Subsurface light metro	v_o = 30 km/h	S = 0.60 km	h = 4 min
——— Deep metro	v_o = 33 km/h	S = 0.70 km	h = 4 min

Origins and destinations distributed in a belt of 400 m on each side of the line

Figure 5. Door-to-door travel time for different transit systems.

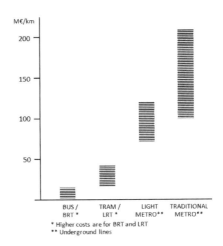

* Higher costs are for BRT and LRT
** Underground lines

Figure 6. Indicative comparison of construction costs for different transit systems.

3.1.3 Construction costs

The construction costs (Capex: right of way and vehicles) depend on many factors, but are placed in distinct brackets for bus lines, tramways and metros, as shown in Fig. 6, based on data coming from direct experience and found in the literature. These costs should be taken into account to quantify the debt service, to be added to operating costs.

3.1.4 Operating costs

The operating costs (Opex) per km and rolling stock unit depend on many factors: organizational scheme, number of people assigned to different duties, extent of maintenance work required on facilities and vehicles and on vehicle capacity, capacity of each rolling stock unit (which affects the incidence of drivers cost), etc,. It is very difficult to give common values, because one finds wide variations of factors from one system to another. Furthermore, data from transit operators are often not homogenous or unavailable. Very rough ranges are shown in Fig. 7.

3.1.5 Specific cost

It appears useful to refer to a specific indicator, i.e. the total cost per km and place offered, taking into account Opex and also Capex reimbursement. In calculating this indicator, other uncertainties apply, such as the interest on Capex and the life cycle of the various system components. However, it results that tram/LRT solution offers economic advantages in a very limited range of cases. With such a conclusion, opinions diverge significantly in the literature, both in favor and not of tram/LRT, confirming how difficult it is to arrive to a general consensus in the presence of multiple criteria to be applied, that can be considered of variable importance.

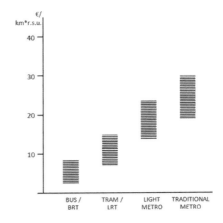

Figure 7. Indicative comparison of operating costs for different transit systems (referred to the rolling stock unit).

3.2 Mobility policies

Mobility policies must also be consistent. One cannot proclaim that public transport is a priority and then set aside good LRT projects, such as the Termini-Aurelio line in Rome (undisputedly a useful complement to the insufficient metro network) for fear of the biased dissent of motorists and shop owners. To oppose that scheme, ridiculous technical or environmental motivations were put forward.

In some cases, an LRT system can also be preferred to a metro, because of the possibility of reaching the required capacity through more LRT lines, which also ensure greater direct coverage of the territory, at a lower cost than metros. This is the case, for example, in Florence, where the first line of the planned LRT system (Fig. 8) has already

Figure 8.   Florence, T1 trams in viale Fratelli Rosselli.

Table 2.   Main data of T1, the first LRT line of Florence (Mantovani G., 2014).

Route: Central Railway Station to Scandicci		
Operator: Gest Spa (Ratp Dev Group)		
Strictly reserved right of way		
Vehicle type: multiarticulated, 5 carbodies, 3 trucks		
Line length	7.4	km
Stops, including terminuses	14	
Gauge	1435	mm
Line voltage	750	V
Number of vehicles	17	
Vehicle length	32.1	m
Vehicle width	2.4	m
Vehicle capacity (at 4 standing pass./m²)	202	pass.
Maximum allowed speed	50	km/h
Operating speed	1.5	km/h
Weekday patronage	44000	pass./d

been in service for 6 years with considerable success. About 24% of the patronage is composed of new users of public transport, pushed away from cars. Main data about this line are quoted in Table 2. But the development of the whole system is delayed (or perhaps prejudiced) owing to the decision to cancel the planned route section in city centre following a drastic and specious pedestrianization. In many European cities, trams and pedestrians coexist satisfactorily and safely in historical centres, but these examples are ignored. An underground central section (almost 4 km) is under evaluation, but considerable problems and high costs are expected, due to peculiar conditions of Florence centre (Fantechi & Mantovani 2015).

4   CONCLUSION

We have seen that tramway or, rather, LRT is the necessary solution in a range of line capacity (roughly, from 4,000 to 10,000 places/h), which cannot be effectively served with buses and does not justify the costs of a metro.

However, we must consider other factors (travel comfort first, but also environmental qualities,

system perception, etc.) difficult to monetize, which favor trams and certainly make them more appreciated by users. There are also useful indirect effects, of social and even economic nature. Among them, significant improvements of the modal split.

Therefore, there are reasons to prefer tramway or LRT also for lower capacities, even when the total costs (Capex reimbursement and Opex) are somewhat higher than in case of buses. It is necessary then to address the issue of mobility policies, in contrast to the vision currently dominant in certain countries, according to which reduction of costs is the overriding goal, which inevitably drives important groups of users away from public transport. It is intolerable to consider public transport just as a service for poor people, who must accept to waste their time, to travel in uncomfortable conditions, etc.. Quality public transport is a key factor for the livability of cities and therefore adequate financial resources must be made available to ensure its quality. Among other things, the increase of resources must also derive from reasonable increases of fares, when they are too low, otherwise subsidizing the neediest groups of users. Those who can are willing to pay for quality.

Let us conclude with some concise suggestions aimed at achieving valuable, successful tram or LRT systems:

– Operate within the framework of solid (and respected) town planning
– Avoid to plan tramways or LRT where the transport demand does not justify them
– Perform first of all exhaustive feasibility studies
– Include urban renewal in the project (keeping costs separately)
– Take into account the experience gained in comparable situations
– Ensure high quality design and specifications
– Include multidisciplinary skills in the design team
– Launch prompt and careful activities for public communication and consultation.

REFERENCES

Fantechi, A. & Mantovani, G. 2015. L'inserimento delle tranvie nei centri storici: una rassegna e il caso di Firenze. *Atti del 6° Convegno "Sistema Tram"*, Roma, 19–20 Marzo 2015. Roma: AIIT.
Laisney, F. 2011. *Atlas du tramway dans le villes françaises*. Paris: Editions Recherches.
Mantovani, G., 2014. Il ritorno del tram a Firenze, *La Tecnica Professionale* 2014(2): 32–41.
Mantovani, G.. 2014. Tramway and LRT in Italy: the case of Florence and general issues; *Proceedings of International Conference on Traffic and Transport Engineering, Beograd, 27–28 November 2014.* Belgrade: Scientific Research Center Ltd.

Author index